Jane's Electronic Solutions

Jane's provides the most comprehensive open-source intelligence resource on the Internet. It is your ultimate resource for subscription-based security, defence, aerospace, transport, and related business information, providing you easy access, extensive content and total control.

Jane's offers you the choice of how you want to receive our information. Whether you want to integrate data into your organisation's network or access the data online, Jane's can provide you with critical information quickly and easily, in a format to suit you.

www.janes.com

Intelligence and Insight You Can Trust

D1031859

10142020S

Jane's

World Railways
2005-2006

Edited by Ken Harris
Contributing Editor, Manufacturers: Jackie Clarke

Forty-seventh Edition

Total number of entries 3,043 New and updated entries 1,861
Total number of images 1,318 New images 386

Visit jwr.janes.com and view the list of latest updates that have been added to the online version of *Jane's World Railways* subsequent to this print edition.

Bookmark jwr.janes.com today!

Jane's World Railways online site gives you details of the additional information that is unique to online subscribers and the many benefits to upgrading to an online subscription. Don't delay, visit jwr.janes.com today and view the list of the latest updates to this online service.

ISBN 0 7106 2710 6
"Jane's" is a registered trademark

Copyright © 2005 by Jane's Information Group Limited, Sentinel House, 163 Brighton Road, Coulsdon, Surrey, CR5 2YH, UK

In the US and its dependencies
Jane's Information Group Inc, 110 N Royal Street, Suite 200, Alexandria, Virginia 22314, US

Printed and bound in Great Britain by Cambridge University Press, Cambridge

"For me, it's perfect to combine travel, conferencing and office work!"

Maria Ramirez Gonzales, independent PR consultant, Madrid

A19100-V900-F759 C1-X-7600 Theim 067-040796

It's all in a day's work as the world's fastest series-production train takes you to your destination – with top speeds of 350 km/h. The Velaro is a prime example of ongoing customized development of a proven high speed platform. It sets new standards in passenger comfort as well as providing impressive business and conference facilities. ETCS Trainguard ensures a high standard of safety. Specific after-sales service and maintenance concepts help maximize availability.

With trains like the Velaro or the magnetic-levitation Transrapid we're leaders in high speed technology. Our know-how also extends to the necessary infrastructure, such as electrification, signaling and turnkey systems. To find out more go to:
www.siemens.com/transportation

SIEMENS

efficient rail solutions

Contents

Contents continued on page [5]

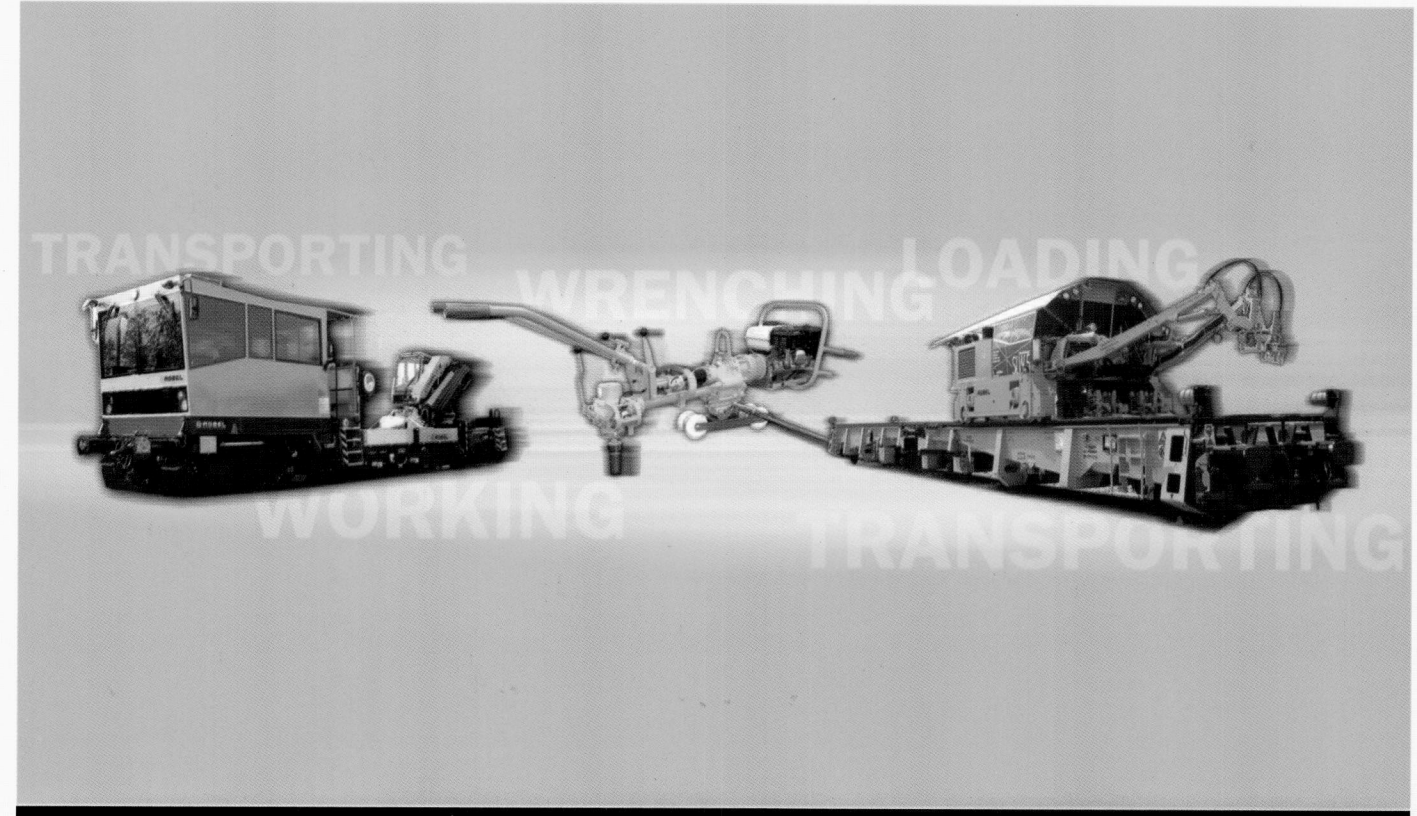

ROBEL Bahnbaumaschinen GmbH · Industriestraße 31 · 83395 Freilassing · Germany
Tel: +49 (0) 8654/609-0 · **Fax:** +49 (0) 8654/609-100
E-mail: info@robel.info · **Internet:** www.robel.info

Quality Policy

Jane's Information Group is the world's leading unclassified information integrator for military, government and commercial organisations worldwide. To maintain this position, the Company will strive to meet and exceed customers' expectations in the design, production and fulfilment of goods and services.

Information published by Jane's is renowned for its accuracy, authority and impartiality, and the Company is committed to seeking ongoing improvement in both products and processes.

Jane's will at all times endeavour to respond directly to market demands and will also ensure that customer satisfaction is measured and employees are encouraged to question and suggest improvements to working practices.

Jane's will continue to invest in its people through training and development to meet the Investor in People standards and changing customer requirements.

www.janes.com

Intelligence and Insight You Can Trust

AUTOMATION SOLUTIONS

TCMS: Train Communication and Management Systems
- Integrated automation: Controllers, central or distributed I/Os, application engineering.
- Specially adapted for use in retrofit projects and for new vehicles.

CAN: A standard drawn from automotive engineering put to good use in rail vehicles.

www.selectron.ch

a company of
Schneider
Electric

Jane's World Railways website: jwr.janes.com

Does your rolling stock suffer from dirty, oily & wet compressed air?

Get the worst out of your compressed air...
www.domnickhunter.com/railways

domnick hunter

How to use *Jane's World Railways*

Jane's World Railways is designed to give the reader data on railway systems and railway equipment manufacturers across the globe. A companion volume, *Jane's Urban Transport Systems*, gives details on metro, light rail and bus operations in the world's larger cities.

Jane's World Railways is divided into two main sections. The first half of the book is a country-by-country listing of the railway systems and operators of the world. For most countries, basic data is given on the transport ministry, including key personnel and contact details. Network maps are also provided. System operators then follow, with address, telephone and fax numbers, web site, key personnel, track gauge, route length and, where appropriate, length of track electrified and electrification system.

The text description of each system is broken down into a number of standard subheadings: Political background, Finance, Passenger operations, Freight operations, Intermodal operations, New lines, Improvements to existing lines, Traction and rolling stock, Signalling and telecommunications, Electrification and Track. Motive power details are given in tables. The smaller systems do not necessarily have entries for each standard subheading, while for the larger systems the standard subheadings are broken down with italicised subheads on individual subjects.

This section is followed by a list of operators of international rail services in Europe.

The second half of the book concentrates on the manufacturing industry; it is broken into a number of chapters, detailing firms active in different sectors such as locomotive manufacture, permanent way equipment and so on. At the front of each chapter are two company listings: an alphabetical listing and a listing by country. Each entry gives details of the manufacturer including address, telephone, fax, web site, key personnel, information on products or services provided, recent contracts and, where appropriate, background information and recent developments.

The manufacturers and services sections include coverage of rolling stock leasing companies, information technology systems and consultancy services. A separate section covers railway associations and agencies.

To help users evaluate the data of this edition, the following identifiers have been used:

- **VERIFIED** The editor has made a detailed examination of the entry's content and checked it's relevancy and accuracy for publication in the new edition to the best of his ability.

- **UPDATED** During the verification process, significant changes to content have been made to reflect the latest position known to Jane's at the time of publication.

- **NEW ENTRY** Information on new equipment and/or appearing for the first time in the title.

- **NEW** New images are identified as **NEW** and some are followed by a seven-digit number for ease of identification by our image library.

A full list of all entries indicating their current status is provided in the index.

Total number of entries 3,043 New and updated entries 1,861
Total number of images 1,318 New images 386

Visit jwr.janes.com and view the list of latest updates that have been added to the online version of *Jane's World Railways* subsequent to this print edition.

All rights reserved. No part of this publication may be reproduced, stored in retrieval systems or transmitted in any form or by any means, electronic, mechanical, photocopying, recording or otherwise, without the prior written permission of the Publishers. Licences, particularly for use of the data in databases or local area networks are available on application to the Publishers. Infringements of any of the above rights will be liable to prosecution under UK or US civil or criminal law.

Copyright enquiries
Contact: Keith Faulkner, email: copyright@janes.com

British Library Cataloguing-in-Publication Data.
A catalogue record for this book is available from the British Library.

DISCLAIMER This publication is based on research, knowledge and understanding, and to the best of the author's ability the material is current and valid. While the authors, editors, publishers and Jane's Information Group have made reasonable effort to ensure the accuracy of the information contained herein, they cannot be held responsible for any errors found in this publication. The authors, editors, publishers and Jane's Information Group do not bear any responsibility or liability for the information contained herein or for any uses to which it may be put.

This publication is provided for informational purposes only. Users may not use the information contained in this publication for any unlawful purpose. Without limiting the generality of the foregoing, users must comply with all applicable laws, rules and regulations with regard to the transmission of facsimilies.

While reasonable care has been taken in the compilation and editing of this publication, it should be recognised that the contents are for information purposes only and do not constitute any guidance to the use of the equipment described herein. Jane's Information Group cannot accept any responsibility for any accident, injury, loss or damage arising from the use of this information.

Jane's Electronic Solutions

Jane's online service

For sheer timeliness, accuracy and scope, nothing matches Jane's online service

www.janes.com is the most comprehensive open-source intelligence resource on the Internet. It is your ultimate online facility for security, defence, aerospace, transport, and related business information, providing you with easy access, extensive content and total control.

Jane's online service is subscription based and gives you instant access to Jane's information and expert analysis 24 hours a day, 7 days a week, 365 days a year, wherever you have access to the Internet.

To see what is available online in your specialist area simply go to **www.janes.com** and click on the **Intel Centres** tab

Once you have entered the **Intel Centres** page, choose the link that most suits your requirements from the following list:

- Defence Intelligence Centre
- Transport Intelligence Centre
- Aerospace Intelligence Centre
- Security Intelligence Centre
- Business Intelligence Centre

Jane's offers you information from over 200 sources covering areas such as:

- Market forecasts and trends
- Risk analysis
- Industry insight
- Worldwide news and features
- Country assessments
- Equipment specifications

As a Jane's Online subscriber you have instant access to:

- *Accurate and impartial* information
- *Archives* going back five years
- *Additional reference content*, source data, analysis and high-quality images
- *Multiple search tools* providing browsing by section, by country or by date, plus an optional word search to narrow results further
- *Jane's text and images* for use in internal presentations
- *Related information* using active interlinking

www.janes.com

Jane's

Intelligence and Insight You Can Trust

Jane's Libraries

To assist your information gathering and to save you money, Jane's has grouped some related subject matter together to form 'ready-made' libraries, which you can access in whichever way suits you best – online, on CD-ROM, via Jane's JIS or through Jane's Data Service.

The entire contents of each library can be cross-searched, to ensure you find every reference to the subjects you are looking for. All Jane's libraries are updated according to the delivery service you choose and can stand alone or be networked throughout your organisation.

www.janes.com

Jane's Defence Equipment Library

Aero-Engines
Air-Launched Weapons
Aircraft Upgrades
All the World's Aircraft
Ammunition Handbook
Armour and Artillery
Armour and Artillery Upgrades
Avionics
C4I Systems
Electro-Optic Systems
Explosive Ordnance Disposal
Fighting Ships
Infantry Weapons
Land-Based Air Defence
Military Communications
Military Vehicles and Logistics
Mines and Mine Clearance
Naval Weapon Systems
Nuclear, Biological and Chemical Defence
Radar and Electronic Warfare Systems
Strategic Weapon Systems
Underwater Warfare Systems
Unmanned Aerial Vehicles and Targets

Jane's Defence Magazines Library

Defence Industry
Defence Weekly
Foreign Report
Intelligence Digest
Intelligence Review
International Defence Review
Islamic Affairs Analyst
Missiles and Rockets
Navy International
Terrorism and Security Monitor

Jane's Market Intelligence Library

Aircraft Component Manufacturers
All the World's Aircraft
Defence Industry
Defence Weekly
Electronic Mission Aircraft
Fighting Ships
Helicopter Markets and Systems
International ABC Aerospace Directory
International Defence Directory
Marine Propulsion
Naval Construction and Retrofit Markets
Police and Security Equipment
Simulation and Training Systems
Space Directory
Underwater Technology
World Armies
World Defence Industry

Jane's Security Library

Amphibious and Special Forces
Chemical-Biological Defense Guidebook
Facility Security
Fighting Ships
Intelligence Digest
Intelligence Review
Intelligence Watch Report
Islamic Affairs Analyst
Police and Security Equipment
Police Review
Terrorism & Security Monitor
Terrorism Watch Report
World Air Forces
World Armies
World Insurgency and Terrorism

Jane's Sentinel Library

Central Africa
Central America and the Caribbean
Central Europe and the Baltic States
China and Northeast Asia
Eastern Mediterranean
North Africa
North America
Oceania
Russia and the CIS
South America
South Asia
Southeast Asia
Southern Africa
The Balkans
The Gulf States
West Africa
Western Europe

Jane's Transport Library

Aero-Engines
Air Traffic Control
Aircraft Component Manufacturers
Aircraft Upgrades
Airport Review
Airports and Handling Agents –
    Central and Latin America (inc. the Caribbean)
    Europe
    Far East, Asia and Australasia
    Middle East and Africa
    United States and Canada
Airports, Equipment and Services
All the World's Aircraft
Avionics
High-Speed Marine Transportation
Marine Propulsion
Merchant Ships
Naval Construction and Retrofit Markets
Simulation and Training Systems
Transport Finance
Urban Transport Systems
World Airlines
World Railways

Jane's
Intelligence and Insight You Can Trust

Glossary

AAR	Association of American Railroads
AC	Alternating current
ACI	Automatic car identification
AFC	Automatic fare collection
APTA	American Public Transit Association
ATC	Automatic train control
ATO	Automatic train operation
ATP	Automatic train protection. System which takes control of train in event of failure of driver to respond to adverse signals.
AVE	*Alta Velocidad Española*. Spanish name given to French design of high-speed train.
AVI	Automatic vehicle identification
BOT	Build, operate, transfer
CAD	Computer-aided design
CKD	Completely knocked down
CRT	Cathode ray tube
CTC	Centralised traffic control
cwr	Continuously welded rail
DBOM	Design, build, operate and maintain
DC	Direct current
demu	Diesel-electric multiple-unit: a dmu with electric transmission.
dmu	Diesel multiple-unit: a self-powered diesel train.
DPTAC	Disabled Persons Transport Advisory Committee
DVMD	Distance/velocity measurement device
EC	EuroCity. European international intercity passenger service conforming to predetermined levels of quality.
E & M	Electrical and mechanical
EMD	Electro Motive Division (former name of locomotive manufacturing division of General Motors).
emu	Electric multiple-unit: a self-powered electric train.
ERTMS/ETCS	The European Rail Traffic Management System/European Train Control System.
EuroCity	European international passenger service meeting predetermined standards for performance and accommodation.
GDP	Gross domestic product, a measure of national output.
GE	General Electric
GIS	Geographical Information System
GM	General Motors
GNP	Gross national product, a measure of national income, counting GDP plus income from foreign investments.
GPS	Global positioning system (satellite tracking technology).
GTO	Gate turn-off thyristor employed in traction control systems.
HVAC	Heating, ventilation and air-conditioning.
Hz	Hertz (unit of measurement of the electrical frequency of AC power systems).
IC	InterCity. European domestic intercity passenger service conforming to predetermined levels of quality.
ICE	InterCity Express. The German design of high-speed train.
ICN	InterCity Night. Brand name for German Rail's overnight services.
ICT	InterCity Tilt. The German design of intercity tilting train.
IGBT	Integrated gate bipolar transistor. Leading edge technology for control of electric traction systems.
Indusi	German driver vigilance system
kV	Kilovolts. 1 kV = 1,000 V
kW	Lilowatts. 1 kW = 1,000 W
LCD	Liquid crystal display
LED	Light-emitting diode
LGV	*Ligne à Grande Vitesse*. French high-speed line
LIM	Linear induction motor
LRT	Light rapid transit (modern tram system)
LRV	Light rail vehicle, for operations on urban systems (a modern tram).
LZB	*Linienzugbeeinflussung* (ATP)
Maglev	Magnetic levitation
MW	Megawatts. 1 MW = 1,000 kW
NBS	*Neubaustrecke*. Newly constructed line, usually for high-speed services, in Germany.
OEM	Original equipment manufacturer
ppa	Passengers per annum
Pendolino	Italian tilting train, designed by Fiat.
RoLa	*Rollende Landstrasse*. 'Rolling highway' piggyback system for conveying lorries.
S-Bahn	High-frequency heavy rail suburban passenger service in Austria, Germany and Switzerland.
Shinkansen	Japanese high-speed railway system
TAV	*Treno Alta Velocita*. The Italian design of high-speed train.
TEU	Foot equivalent units (standard method of measuring container capacity/traffic).
TGV	*Train à Grande Vitesse*. The French design of high-speed train.
TOC	Train Operating Company
TVM	French-designed cab-signalling system, as used on the TGV network.
UIC	Union Internationale de Chemin de Fer. International trade association and lobbying organisation for railway operating companies, based in Paris.
UIRR	International Union of Road-Rail Transport Companies. Formed to develop and operate combined transport in Europe.
UNIFE	International trade association for European railway equipment manufacturers, based in Brussels.
VVVF	Variable voltage, variable frequency. Leading edge technology for the control of electric traction systems.
X2000	Swedish tilting train, designed by Adtranz (now Bombardier Transportation).

Locomotive wheel arrangement

A classification system is used whereby the number of driven axles on a bogie or frame is denoted by a letter (A = 1, B = 2, C = 3) and the number of undriven axles is noted by a number. The letter 'o' after a letter indicates that each axle is powered individually.

The most common types are Co-Co, two three-axle bogies, all axles individually powered, and Bo-Bo, a similar machine but with two-axle bogies rather than three-axle ones. In the former Soviet bloc, many double and triple unit locomotives exist, such as Bo-Bo+Bo-Bo.

Jane's Market Intelligence

An authoritative resource providing you with the ability to keep track of vital industry personnel, evaluate competitors, identify potential buyers, partners and business opportunities and build your acquisition strategies. Jane's market intelligence will ensure you have the complete industry picture and save you time gathering information from a variety of sources.

www.janes.com

Jane's
Intelligence and Insight You Can Trust

101500205

Alphabetical list of advertisers

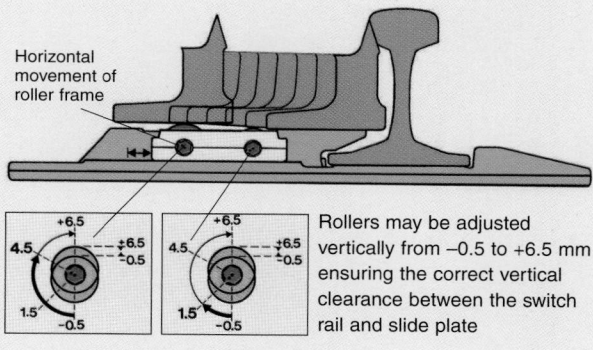

Schwihag

INTEGRATED ROLLER SLIDE PLATES:
lubrication free – simple to install
for use in new assemblies or existing layouts
with full or shallow depth switches

Horizontal movement of roller frame

Rollers may be adjusted vertically from –0.5 to +6.5 mm ensuring the correct vertical clearance between the switch rail and slide plate

The adjustment of rollers to a maximum height of 6.5 mm allows a greater distance to be spanned between roller plates.

No special precautions are required to allow mechanical tamping of the switch assembly.

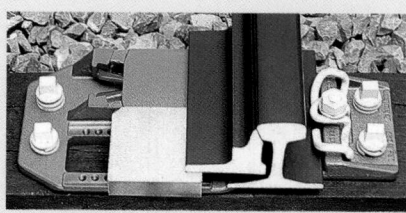

Metal cover as protection of the rollers from ballast during the installation of the switch assemblies

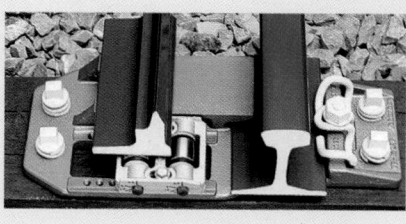

Major advantages:
- Roller slide plate is fully interchangeable with standard slide plate
- Mounted directly onto bearer
- Friction free movement of switch blade
- Minimal maintenance
- Environmentally friendly
- Cost effective

progress and improvement of Schwihag fastening system for switches and crossings

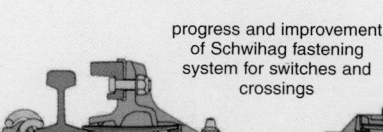

IFAV check rail plate with SSb 2 spring clip

IBAV slide plate with SSb 2 spring clip

Schwihag
Gesellschaft für Eisenbahnoberbau mbH

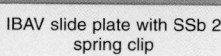

CH-8274 Tägerwilen, Lebernstrasse 3
Phone: ++41/71/666 88 00
Facs: ++41/71/666 88 01
E-mail: info@schwihag.com
www.schwihag.com

High impact solutions for targeting the railway market

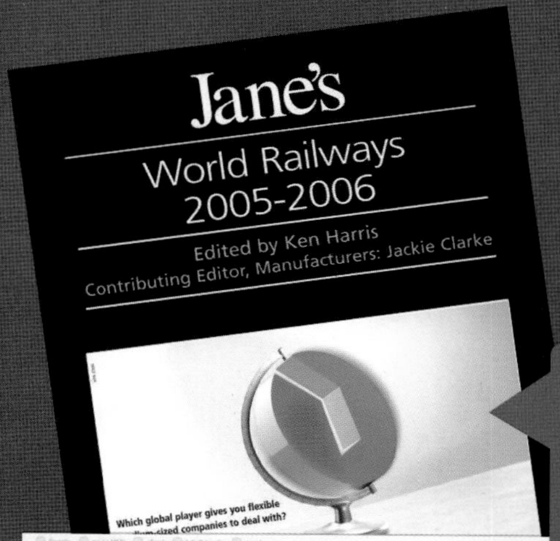

Jane's World Railways

This valuable yearbook is the foremost information source on the railway industry, giving you a truly global perspective on the development of railway systems in 140 countries and states worldwide.

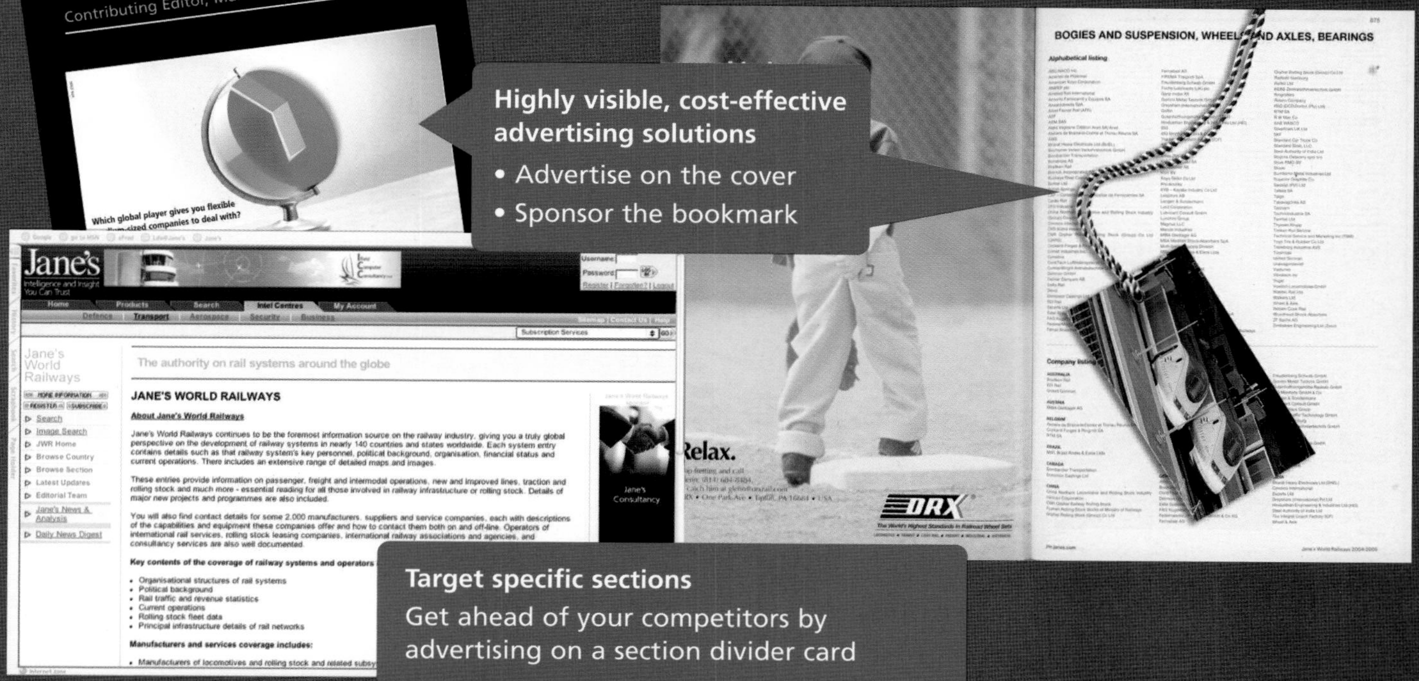

Jane's World Railways 2005-2006
Edited by Ken Harris
Contributing Editor, Manufacturers: Jackie Clarke

Highly visible, cost-effective advertising solutions
• Advertise on the cover
• Sponsor the bookmark

Target specific sections
Get ahead of your competitors by advertising on a section divider card

Why Advertise?

• Quality readership - every copy purchased
• Exceptional shelf life - minimum of one year
• High frequency of use
• Editorial excellence by world leading experts

How do you benefit?

• Qualify for a free copy of the yearbook (min half page, four colour ad).
• Book the front cover and sponsor jwr.janes.com for free!
• Increased focus with specific section targeting.
• Stand out above your competitors - book an ad next to their listing.
• 1000 additional bookmarks for your own promotions (Offer applies to bookmark position).
• All advertisers also appear in janes.com's "View Advertisers" section for up to a year.

To advertise in Jane's World Railways, please contact

UK Tel: +44 (0) 20 8700 3738
US Tel: +1 703 236 2410
Email: transadsales@janes.com

Jane's
Intelligence and Insight You Can Trust

EDITORIAL AND ADMINISTRATION

Director: Ian Kay, e-mail: Ian.Kay@janes.com

New Media Publishing Director: Sean Howe, e-mail: Sean.Howe@janes.com

Publisher: Jonathan Grevatt, e-mail: Jonathan.Grevatt@janes.com

Content Services Director: Anita Slade, e-mail: Anita.Slade@janes.com

Content Systems Manager: Jo Agius, e-mail: Jo.Agius@janes.com

Pre-Press Manager: Christopher Morris, e-mail: Christopher.Morris@janes.com

Team Leader: Neil Grace, e-mail: Neil.Grace@janes.com

Content Editor: Hannah Leech, e-mail: Hannah.Leech@janes.com

Production Controller: Laura-Jane Walker, e-mail: Laurajane.Walker@janes.com

Content Update: Jacqui Beard, Information Collection Team Leader
Tel: (+44 20) 87 00 38 08 Fax: (+44 20) 87 00 39 59
e-mail: yearbook@janes.com

Jane's Information Group Limited, Sentinel House, 163 Brighton Road, Coulsdon, Surrey CR5 2YH, UK
Tel: (+44 20) 87 00 37 00 Fax: (+44 20) 87 00 37 88

SALES OFFICES

Send Europe and Africa enquiries to: *Mike Gwynn – Head of Information Sales*
Jane's Information Group Limited, Sentinel House, 163 Brighton Road, Coulsdon, Surrey CR5 2YH, UK
Tel: +44 (0) 20 8700 3750   Fax: +44 (0) 20 8700 3751
e-mail: customerservices.uk@janes.com

Send US enquiries to: *Robert Loughman – Sales Director*
Jane's Information Group Inc, 110 N Royal Street, Suite 200, Alexandria 22314, US
Tel: (+1 703) 683 37 00   Fax: (+1 703) 836 02 97   Telex: 6819193
Tel: (+1 800) 824 07 68   Fax: (+1 800) 836 02 97
e-mail: customerservices.us@janes.com

Send Asia enquiries to: *David Fisher – Group Business Manager*
Jane's Information Group Asia, 78 Shenton Way, #10-02, Singapore 079120
Tel: +65 6325 0866   Fax: +65 6226 1185
e-mail: asiapacific@janes.com

Send Australia/New Zealand enquiries to: *Russell Smith – Business Manager*
Jane's Information Group, PO Box 3502, Rozelle Delivery Centre, NSW 2039, Australia
Tel: +61 (0)2 8587 7900   Fax: +61 (0)2 8587 7901
e-mail: oceania@janes.com

Send Middle East enquiries to: *Ali Abdellatif Siali – Regional Sales Manager*
Jane's Information Group, PO Box 502138, Dubai, United Arab Emirates
Tel: +971 4 390 2336   Fax: +971 4 390 8848
e-mail: mideast@janes.com

Send Japan enquiries to: *Norihisa Fukuyama – Information Consultant*
Jane's Information Group, Palaceside Building, 5F, 1-1-1, Hitotsubashi, Chiyoda-ku, Tokyo 100-0003, Japan
Tel: +81 (0)3 5218 7682   Fax: +81 (0)3 5222 1280
e-mail: japan@janes.com

Send India enquiries to: *T.C. Martin – Information Consultant*
Jane's Information Group, PO Box 3806, New Delhi 110049, India
Tel/Fax: +91 (0) 11 26 51 61 05
e-mail: india@janes.com

ADVERTISEMENT SALES OFFICES

Head Office
Jane's Information Group
Sentinel House, 163 Brighton Road, Coulsdon, Surrey CR5 2YH
Tel: (+44 20) 87 00 37 00   Fax: (+44 20) 87 00 38 59/37 44
e-mail: transadsales@janes.com

Tracy Attwooll – Advertisement Sales Manager, Transport
Tel: (+44 20) 87 00 37 41   Fax: (+44 20) 87 00 38 59/37 44
e-mail: tracy.attwooll@janes.com

Mat Stevens – Senior Sales Executive, Transport
Tel: (+44 20) 82 76 47 22   Fax: (+44 20) 72 87 77 65
e-mail: matt.stevens@janes.com

US/Canada
Jane's Information Group
110 N Royal St. Suite 200, Alexandria, Virginia 22314, US
Tel: (+1 703) 683 37 00   Fax: (+1 703) 836 55 37
e-mail: transadsales@janes.com

Katie Taplett – US Advertising Sales Director
Tel: (+1 703) 683 37 00   Fax (+1 703) 836 55 37
e-mail: katie.taplett@janes.com

Sean Fitzgerald – Account Executive
Tel: (+1 703) 683 37 00   Fax: (+1 703) 836 55 37
e-mail: sean.fitzgerald@janes.com

Northern US and Eastern Canada
Linda Hewish – Advertising Sales Executive
Tel: (+1 703) 683 37 00   Fax: (+1 703) 836 55 37
e-mail: linda.hewish@janes.com

South Eastern US
Kristin D Schulze – Advertising Sales Manager
PO Box 270190, Tampa, Florida 33688-0190
Tel: (+1 813) 961 81 32   Fax: (+1 813) 961 96 42
e-mail: kristin.schulze@janes.com

Australia: Richard West *(UK Head Office)*
Benelux: Mat Stevens *(UK Head Office)*
China and Hong Kong: Mat Stevens *(UK Head Office)*
France: Mat Stevens *(UK Head Office)*
Germany and Austria: *MCW (Media and Consulting Wehrstedt)*
Tel: (+49 34) 74 36 20 90   Fax: (+49 34) 74 36 20 91
e-mail: info@wehrsteddt.org

Iran: Eideh Info Company
Ali Jahangard
19 4th Street, Ghaem Magham Avenue, Tehran, Iran
Tel: (+98 21) 873 59 23
e-mail: eideh@mavara.com

Israel: *Oreet – International Media*
Tel: (+972 3) 570 65 27   Fax: (+972 3) 570 65 26
e-mail: liat_h@oreet-marcom.com

Italy and Switzerland: *Ediconsult Internazionale Srl*
Tel: (+39 010) 58 36 84   Fax: (+39 010) 56 65 78
e-mail: genova@ediconsult.com

Middle East: Mat Stevens (see UK Head Office)

Russian Federation and Ukraine: Vladimir N Usov
Tel/Fax: (+7 3435) 32 96 23
e-mail: uvn125@uraltelecom.ru

Scandinavia: *The Falsten Partnership*
Tel: (+44 1273) 77 10 20 Fax: (+ 44 1273) 77 00 70
e-mail: sales@falsten.com

Singapore: Mat Stevens (see UK Head Office)

South Africa: *Richard West* (see UK Head Office)

South Korea: *JES Media Inc*
Contact: Young-Seoh Chinn
Tel: (+82 2) 481 34 11/13   Fax: (+82 2) 481 34 14
e-mail: jesmedia@unitel.co.kr

Spain: Via Exclusivas SL
e-mail: viaexclusivas@viaexclusivas.com

For all other areas, contact Tracy Attwooll (UK Head Office)

ADVERTISING COPY
Linda Letori (Jane's UK Head Office)
Tel: (+44 20) 87 00 38 56   Fax: (+44 20) 87 00 38 59/37 44
e-mail: linda.letori@janes.com

For North America, South America and Caribbean only:
Lia Johns (Jane's US address)
Tel: (+1 703) 683 37 00   Fax: (+1 703) 836 55 37
e-mail: lia.johns@janes.com

Jane's Users' Charter

This publication is brought to you by Jane's Information Group, a global company with more than 100 years of innovation and an unrivalled reputation for impartiality, accuracy and authority.

Our collection and output of information and images is not dictated by any political or commercial affiliation. Our reportage is undertaken without fear of, or favour from, any government, alliance, state or corporation.

We publish information that is collected overtly from unclassified sources, although much could be regarded as extremely sensitive or not publicly accessible.

Our validation and analysis aims to eradicate misinformation or disinformation as well as factual errors; our objective is always to produce the most accurate and authoritative data.

In the event of any significant inaccuracies, we undertake to draw these to the readers' attention to preserve the highly valued relationship of trust and credibility with our customers worldwide.

If you believe that these policies have been breached by this title, you are invited to contact the editor.

A copy of Jane's Information Group's Code of Conduct for its editorial teams is available from the publisher.

www.janes.com

Intelligence and Insight You Can Trust

The world's first dual-gauge electric locomotive, completed in Spain in late 2004 by a consortium comprising Talgo and Team SA, which supplied the traction equipment. The 74 tonne four-axle L-9202 machine has been designed to work uninterrupted between Spain's classic 1,668 mm gauge system and the standard gauge adopted for the country's new high-speed lines. This interoperability also calls for a dual-voltage capability; the locomotive is therefore able to operate under both 3 kV DC and 25 kV AC power supplies. It is also equipped for the signalling and driver vigilance systems installed on each network and is prepared for ERTMS operation. One-hour output under either power supply is 3,600 kW, with a designed maximum speed of 260 km/h on high-speed lines and 220 km/h on classic routes. While this prototype is a dual-cab unit, the series locomotives that are expected to follow will be single-cab to act as power cars at each end of a rake of Talgo coaching stock and will feature different electrical equipment and a revised body profile (Bryan Philpott)
NEW / 1122866

Executive Overview

Middle East rail expansion responds to growing traffic demands

While the Middle East often occupies the media headlines, it is rarely to highlight rail developments in the region. However, it is here that many major development plans are being taken forward with the aim of boosting economic development and, in some instances, opening up the prospect of strengthening international ties with neighbours that continue to witness conflict.

The opening in May 2005 of the 800 km Mashhad–Bafq line in eastern Iran marked the latest stage in the rapid expansion of that country's rail network. Connecting at Bafq with the line from Tehran to the expanding Gulf port of Bandar-e Abbas, the new railway offers a direct link to the sea from Turkmenistan and landlocked central Asian countries via the Mashhad–Sarakhs line, commissioned in 2000. Built in just four years, the single-track line passes through largely mountainous terrain, necessitating the construction of 23 major bridges and some 5,800 km of tunnelling. As well as providing access to the east from Bandar-e Abbas, the Mashhad–Bafq line also provides an outlet for significant reserves in the region of coal and iron ore.

The Islamic Iranian Republic Railways (RAI) network is struggling to keep pace with traffic demand, both for passengers and freight. In 2003 some 14 million passengers and 28 million tonnes of freight were handled on a system totalling around 8,000 route-km. Plans unveiled by the Iranian government in 2005 foresee the present network expanding by half to accommodate a doubling of existing traffic levels.

The most significant new domestic lines will be a direct link between Qom and Esfahan, supplementing the existing route via Bäd, with an extension to the city of Shiraz in the south. This is already under construction. Also planned is a line from Bam in the southeast to the port of Chabahar, on the Gulf. It is, however, the prospect of new international links that provide the most intriguing possibilities, as these set the scene for Iran increasingly to serve as a key strategic rail crossroads for the entire region.

The first of these to be completed will be the long-planned line between Bam and Zahedan, to where an existing broad gauge (1,676 mm) line from Mirjaveh to provide a link with the Pakistan network. Much of the formation of this line has been completed. Also long in gestation has been a link from Qazvin, northwest of Tehran, to the border with Azerbaijan at Astara. This key strategic scheme took an important step forward in May 2005 when the railways of Iran, Azerbaijan and the Russian Federation finally signed an agreement to establish a joint company to oversee construction of the 511 km line. The need for this link has become more acute following suspension for political reasons of services between Iran and Azerbaijan via the border crossing at Jolfa and through Armenia. When completed, the railway will offer a north-south link between the Baltic and the Gulf.

RAI also views as vital the westward expansion of its network from Arak to Kermanshah, on which work has started, with studies under way to take this line forward to the Iraqi capital, Baghdad. Eventually this would create a link through Iraq to Syria. With construction proceeding on the Trans-Kazakhstan line, discussed in the Executive Overview of the last edition of Jane's World Railways, a link across western Iran could form a key component of a possible standard-gauge rail connection between China and the Mediterranean. The Iranian and Iraqi networks are also to be connected by a short Khorramhsahr–Basra link in the far south.

And finally RAI is also studying a line to the border with Afghanistan, from where further construction may bring the railway to Heart. Already a 148 km branch has been built from the new Mashhad–Bafq line eastwards to Sangan, the site of extensive iron ore deposits; only a short extension of this line is required to reach the border with Afghanistan.

While railway construction in Iran remains firmly in the state sector, in Saudi Arabia private sector investors are to play a major role in the Saudi Landbridge scheme. Outlined in last year's Executive Overview, this foresees construction of a 950 km line across the Arabian Desert to link Riyadh, the western limit of the existing Saudi Railways Organisation (SRO) line from the port of Dammam, with Jeddah, on the Red Sea. Also included in the project is a 115 km line northwest from Dammam to Jubail, a major industrial centre on the Gulf coast. In early 2005 pre-qualification started for bidders for what is claimed to be one of the Middle East's largest ever Build-Operate-Transfer (BOT) contracts, covering construction and operation of the line, and also including taking over the existing SRO system. Passenger services will

Alignment of A1 Express Line linking Tel Aviv and Jerusalem (Aharon Gazit)

NEW/ 1122867

become the subject of a separate franchise agreement on the line's completion. A contract award is expected in 2006.

Early indications were that the railway would be built to a high specification both for passenger and freight services. In the case of the former, these could operate at up to 220 km/h, while freight traffic would largely consist of double-stack container trains. Freight trains would complete the 1,400 km journey in around 18 hours. It is planned to construct the line for 25 tonne axleloads and to create an alignment that will allow for future doubling. All services will be diesel-operated.

A separate scheme in Saudi Arabia, the Western Railway, is to see construction of a 570 km north-south line linking Al Madinah, Jeddah and Mecca, providing access for pilgrims to the countries two holy cities. Studies have also been undertaken into a 1,100 km line running northwest from Riyadh to provide an outlet for reserves of bauxite and phosphates. Should all these projects reach fruition, the Saudi rail network will have increased from the present 1,400 route-km to over 4,100 route-km.

The remarkable growth of Israel's railway also continues, with more significant developments in the year under review here. In October 2004 a line was commissioned by Israel Railways (IR) serving Tel Aviv's Ben Gurion International Airport. As well as meeting the needs of airport users via direct access to the terminal building, the new station also acts as a commuter park-and-ride facility to provide an alternative to Tel Aviv's notorious traffic congestion.

The line to the airport is at present a single-track branch from the Tel Aviv–Lod main line. Eventually it will form part of the so-called A1 Express Line between Tel Aviv and Jerusalem. Due to be completed in 2009, the 40 km double-track line is to be engineered for 160 km/h running. However, it has not been necessary for passengers in Jerusalem to wait until 2009 for a rail service. In 2005 rehabilitation was completed of the original line from Tel Aviv via Bet Shemesh, returning to the city at the new Malkha station after a gap of seven years.

IR's transformation has been remarkable. In 1990 the railway was carrying just 2.5 million passengers. Current predictions are that by 2010 passenger traffic will reach 65 million. Between 2000 and 2004 patronage increased by more than 75 per cent from 13 million to 22.9 million as capacity was added to the system by new line construction, track doubling and the rehabilitation of sections of the railway. Attractive, modern stations have also played an important part in persuading car drivers to use rail. This provision of infrastructure has been matched by investments in rolling stock that have seen the progressive introduction of a large fleet of IC3 dmus and the recent procurement of sets of double-deck push-pull coaching stock.

Further expansion of the IR network is planned to meet continuing growth in passenger demand, with several new lines projected in the greater Tel Aviv area and a continuing programme of double-tracking. In 2005 work started in the north on the 'Jezreel Valley Line' from Haifa to Bet She'an, close to the border with Jordan. Eventually this is expected to provide a cross-border connection, fulfilling an accord reached between the governments of Israel and Jordan as long ago as 1995. Also under construction is a 70 km line from Ashqelon to Be'er Sheva, to be completed in 2008.

However, the most notable new line project is focused on freight, providing the Red Sea port of Eilat with a rail link. At some 300 km, this would enable maritime container traffic to run via Be'er Sheva to Ashdod, on the Mediterranean coast, offering a faster land-bridge alternative to the Suez Canal. Upgrading the Be'er Sheva–Har Zin section has already begun and preliminary design work on the alignment between Har Zin and Eilat has been completed.

Another capacity expansion strategy being pursued by IR is electrification. Having abandoned thoughts of electrifying its entire passenger network, in 2005 the railway invited new tenders for wiring a core network, expected to be around 300 route-km. Among routes covered would be those from Tel Aviv to Nahariya in the north, to Ashqelon in the southwest and the new A1 line to Jerusalem.

Rail seeks to improve its environmental credentials – and cut fuel costs

Increasingly strong legislation governing rail vehicle emissions has accelerated the study and implementation of alternative traction technologies and fuel sources. In North America, British Columbia-based RailPower Technologies Corporation has started to make a significant impact with its range of hybrid locomotives intended mainly for rail yard and short haul operations and designed to fall well within the requirements of the US Environmental Protection Agency's stringent 'Tier 2' locomotive emission regulations.

Given these advantages, it is hardly surprising that Green Goat has attracted the attention of North American operators, especially those with extensive marshalling yard activities. In the US state of California, BNSF and Union Pacific have signed agreements with the California Air Resources Board (CABD) to reduce emissions at their rail yards, with a commitment to cut exhaust particulates by 20 per cent by June

RailPower Technologies Corporation's GG20B Green Goat hybrid shunting locomotive (RailPower Technologies Corporation)   *NEW* / 0585013

2008. Both operators have now ordered batches of Green Goats for yard service in California and by August 2005 contracts for the supply of around 100 examples had been received by RailPower from various US and Canadian operators. Canadian Pacific Railway has ordered no fewer than 35 Green Goats. In the case of California, grants are made to assist with the purchase of the locomotives, with a similar arrangement applying under the Texas Emissions Reduction Program (TERP). Here four cabless Type GG20B Green Goats have been ordered by BNSF with the assistance of TERP grants.

Green Goat is not a completely new locomotive. It is remanufactured from an existing life-expired General Electric or General Motors/EMD machines, usually of Types B23-7 or GP-9 or larger respectively – the recycling of old equipment enhancing the environmental qualities of the rebuilt locomotive. Initial conversions were carried out in Canada but in April 2005 the first US-produced Green Goat was completed and despatched to a BASF chemicals plant in Texas. Further US production capacity was guaranteed in July 2005 with the signing of agreements with Super Steel to produce locomotives for the East Coast market at its Schenectady, New York, plant.

The success of the Green Goat concept has not gone unnoticed by other US traction manufacturers and in 2005 it was reported that general Electric was developing a 3,280 kW hybrid line-haul locomotive. This would divert dynamic braking energy to charge lead-free batteries that would store power to provide an 'on-demand' boost of up to 1,500 kW, cutting emissions and reducing fuel consumption.

Meanwhile, RailPower has moved to introduce its hybrid technology to a wider market. In 2005 licensing agreements were signed with Brush Traction, covering the UK and Ireland markets, and with Swedish Train Technology (STT) to serve Scandinavia. As this edition of Jane's World Railways moved towards publication, STT was preparing a Type T43H demonstration hybrid locomotive for completion by the year-end. Brush Traction too was planning to produce a demonstrator using the chassis and running gear of a retired Class 86 electric locomotive.

Elsewhere in the world, other operators are seeking ways of cutting their fuel costs and reducing harmful emissions. In Thailand, state rail operator SRT in August 2005 commenced a six-month trial using compressed natural gas (CNG) instead of diesel. If the programme is successful, SRT could extend the use of CNG to its whole fleet. A similar pilot programme was initiated in Brazil by the Ferrocarril Central Andino during 2005.

Another Brazilian operator, the Vitória a Minas Railway, in 2005 switched to using renewable biological fuel in place of diesel after successful trials with two locomotives. And in Sweden a Class Y1 railcar dubbed 'Amanda' has been converted to run on biogas – fuel produced from organic material and waste products. Partners in the project include rail maintenance firm EuroMaint, Svensk Biogas AB, a subsidiary of a publicly owned utilities group in Linköping, Banverket and various Swedish public transport and municipal agencies. The vehicle was initially set to work on the Linköping– Västervik line.

With increasingly frequent reminders that the supply of fossil fuels is not infinite and that prices globally can be affected by climate and conflict, as well as by the more usual economic forces of supply and demand, rail operators will be paying even closer attention to this major cost area. Factor in increasing political pressure for rail to play its part in reducing the causes of climate change and the adoption of alternative fuel sources and traction technologies looks more and more necessary.

Ken Harris, Editor
Brighton, October 2005

Acknowledgements

It is a pleasant task again to thank and acknowledge the contributors, correspondents and friends who have assisted in various ways during the preparation of this latest edition of *Jane's World Railways*. Their specialised knowledge, helpful submissions and photographic contributions are essential features of both hardcopy and online versions of this product.

While we trust it is not evident in its pages, this edition is the first to have been produced using a new content management software system. This technology is being rolled out across the entire Jane's product range and is intended to enable editors to respond more quickly to market developments that call for changes in content and to provide a better structure for the data for users of the Internet versions of each title. This has called for special efforts this year from all involved in the production of *Jane's World Railways* and this is gratefully acknowledged here.

Jackie Clarke, Contributing Editor, Manufacturers and Services, has again successfully processed the high volume of changes to our coverage of the industry's suppliers, introducing many improvements along the way while ensuring that changes are reflected accurately. This work continues to involve close co-ordination of content with that of our sister publication, *Jane's Urban Transport Systems*.

Kate Hainsworth, Editorial Assistant, continues to play an essential role in undertaking research and in maintaining effective mailing information for our data-gathering purposes. More recently, Kate has also taken on a share of the content updating – a reflection of the volume of change taking place in the rail supply industry.

We also thank Mary Webb, Editor of our sister publication, *Jane's Urban Transport Systems*. Continuing liaison ensures that the content of our two publications is harmonised to deliver optimum value to our customers.

Thanks and acknowledgements for information and pictures in this edition are due to many friends, contributors and correspondents around the world. They include Toma Bacic, John C Baker, Edward Barnes, Colin Boocock, Chris Bushell, Roger Carvell, Bruce Evans, Brian Garvin, Aharon Gazit, K K Gupta, Andrzej Harassek, David Haydock, Flávio Francesconi Lage, Michal Málek, Bryan Philpott, Milan Srámek, Marcel Vleugels, Quintus Vosman, David C Warner, Brian Webber, Rory Wilson and Philip Wormald. The contributions of many other photographic contributors are also acknowledged.

Many of the maps in this edition have been updated, serving as a reminder that rail network change more often than might at first be apparent. Thanks for interpreting those changes are due to our graphic artist, Barrie Compton, who continues to excel at the complex task of updating the maps.

In-house editorial support at our Coulsdon offices in the UK has been ably provided by Hannah Leech and her predecessor Lizzy May, with additional support by Diana Barrick. Thanks are due to all for their hard work, support and humour. The support of the editorial management team, Neil Grace and Jon Grevatt, is also appreciated, as is the work of the production team at Coulsdon.

Thanks are also again due to Jacqui Beard, who oversees our information collection activities and is for many companies in the rail industry often the first point of contact in Jane's.

Jane's World Railways is now also well established as a Jane's Online product, available via the Internet. As a consequence, updating is a year-round activity. Even before this printed version was published, substantial updates and additions had already been made to our data resource.

Updating *Jane's World Railways* continues to depend on the cooperation of the railway undertakings and companies featured in it. We are therefore pleased to acknowledge and thank the many organisations which have responded so generously to our requests for the data that enables *Jane's World Railways* to provide such extensive coverage of the railway industry worldwide. We also invite any users of *Jane's World Railways* to provide corrections, updates or any new information that they consider might enhance the product.

Ken Harris, Editor

As Contributing Editor to both *Jane's World Railways* and also our sister title, *Jane's Urban Transport Systems*, my role is to oversee the coverage of the companies that appear in the sections contained within the Manufacturers and Services umbrella. This year has included the additional challenges of learning the new system by which the Jane's titles are now being produced and this in itself has taken us a step further toward providing our customers with a product that will reflect improvements in structure, access to information and enhanced continuity.

The task of updating the high number of entries contained within *Jane's World Railways* is very much aided by the valuable contribution from our contacts at the companies featured who diligently supply us with their editorial amendments and new developments from around the world. I am always very grateful to the individuals who take time to collate, in many cases, large volumes of information, often from within different divisions of their company in order for us to reflect such timely and comprehensive content.

I also would like to thank the editorial team, Kate Hainsworth, for her hard work on our data-gathering efforts and research and most recently for her additional contribution in commencing work on editorial processing. My thanks also to Ken Harris for his specialist knowledge, continuous support, guidance and good humour.

Jackie Clarke, Contributing Editor, Manufacturers and Services

Ken Harris

Before assuming the editorship of *Jane's World Railways* in 1997, Ken Harris was publisher and editor of an international rail industry journal *Passenger Rail Management* and also acted as an organiser of conferences on topics relating to the rail industry, air-rail links and intermodal freight. His current Jane's responsibilities also include contributing to special studies and consultancy projects and working in close editorial collaboration with *Jane's World Railways* sister product, *Jane's Urban Transport Systems*. He also acts as a consultant to an international rail industry exhibitions company.

Earlier in his career, Ken served as publishing director for Jane's surface and air transport products, with responsibility for the general management and development of the list.

He is married with two daughters and lives in Brighton, East Sussex, UK.

Jackie Clarke

Jackie Clarke joined Jane's Information Group in 1989, working within the Transport portfolio of titles. Jackie has remained within this market area of the company as it evolves from a traditional publishing organisation to an e-business. During this time, Jackie has worked in the various capacities of Assistant Editor for the titles *International ABC Aerospace Directory*, *Jane's Airports and Handling Agents*, *Jane's Road Traffic Management*, *Jane's World Airlines* and most recently as Contributing Editor, (Manufacturers) for *Jane's World Railways*. She has also recently gained additional responsibility for the Manufacturers section of *Jane's Urban Transport Systems*, the sister title to *Jane's World Railways*.

Jane's Transport Intelligence

Jane's transport portfolio offers you access to the world's most accurate and authoritative reference and analysis information relating to the aviation industry and the railway and urban transport industries. Listings of manufacturers with products and contact details ensures you can locate suppliers and identify market opportunities.

www.janes.com

Jane's

Intelligence and Insight You Can Trust

101580205

FREE ENTRY/CONTENT IN THIS PUBLICATION

Having your products and services represented in our titles means that they are being seen by the professionals who matter – both by those involved in procurement and by those working for the companies that are likely to affect your business. We therefore feel that it is very much in the interests of your organisation, as well as Jane's, to ensure your data is current and accurate.

- **Don't forget** – You may be missing out on business if your entry in a Jane's product is incorrect because you have not supplied the latest information to us.

- **Ask yourself** – Can you afford not to be represented in Jane's printed and electronic products? And if you are listed, can you afford for your information to be out of date?

- **And most importantly** – The best part of all is that your entries in Jane's products are TOTALLY FREE OF CHARGE.

Please provide (using a photocopy of this form) the information on the following categories where appropriate:

1. Organisation name: _____

2. Division name: _____

3. Location address: _____

   _____

4. Mailing address if different: _____

   _____

5. Telephone (please include switchboard and main department contact numbers, for example Public Relations, Sales, and so on):

   _____

6. Facsimile: _____

7. e-mail: _____

8. Web sites: _____

9. Contact name and job title: _____

   _____

10. A brief description of your organisation's activities, products and services: _____

    _____

    _____

11. Jane's publications in which you would like to be included: _____

    _____

Please send this information to:
Jacqui Beard, Information Collection, Jane's Information Group,
Sentinel House, 163 Brighton Road, Coulsdon, Surrey, CR5 2YH, UK
Tel: (+44 20) 87 00 38 08
Fax: (+44 20) 87 00 39 59
e-mail: yearbook@janes.com

Copyright enquiries:
Contact: Keith Faulkner
e-mail: copyright@janes.com

Please tick this box if you do not wish your organisation's staff to be included in Jane's mailing lists ☐

JWR

RAILWAY SYSTEMS AND OPERATORS

RAILWAY SYSTEMS AND OPERATORS

RAILWAY SYSTEMS AND OPERATORS

Afghanistan

Ministry of Transport

PO Box 2509, Ansari Walt, Kabul
Tel: (+93) 210 15

Key personnel
Minister of Transport: Syed Anwari

Political background
In 1982 the first railway tracks appeared in Afghanistan with completion, after three years' work by Afghan and Soviet labour, of an 816 m combined rail and road bridge over the Abu Darja river, the border with the former USSR, now Uzbekistan, and the projection over it of a rail link from the Bukhara–Dushanbe line near Termez to Hairaton in Afghanistan. This penetration was to be continued into Afghanistan, beginning with a 200 km line to Pali-Khumri, some 160 km north of Kabul, but progress was blocked by the mountainous terrain, the long annual periods in which the area is blanketed by heavy snow and the unstable political situation within the country.

A new prospect that would place Afghanistan astride a Central Asian railway was opened up in 1992 when Pakistan's Economics Minister offered the Central Asian republics aid for the construction of a railway through Afghanistan to an emergent Arabian Sea port in Baluchistan at Pasni. Later, a revised plan emerged for an 800 km trans-Afghan line linking the existing Pakistan Railways route at Chaman with Kushka in Turkmenistan, via Herat and Kandahar. This has been accorded priority by the Pakistan government, and was the subject of an accord signed in March 1994 between the Pakistan, Afghan and Turkmenistan governments. A World Bank-funded US$1.5 million feasibility study into the scheme was published in 2000. Construction cost of the line is estimated at US$600 million.

Albania

Ministry of Transport and Telecommunications

Bulevardi Deshmoret e Kombit, Tirana
Tel: (+355 4) 22 57 75
e-mail: mtt@mtt.gov.al
Web: www.mtt.gov.al

Key personnel
Minister: Spartak Poçi

UPDATED

Albanian Railways (HSH)

Hekurudhat Shqiptare
Drejitoria e Pergjithshle e Hekurudhave Shqiptare, Rue Skenderbeg, Durrës
Tel: (+355 52) 223 11   Fax: (+355 52) 220 37

Key personnel
Director General: Leonard Jani
Directors
    Finance: Jani Kona
    Operations: Petraq Pando
    Technical: Kristaq Prifti
    Infrastructure: Kozma Plepi
    Head of International Relations: Miranda Jani

Gauge: 1,435 mm
Route length: 447 km

Political background
Following deposition of the hardline Communist government, consideration was given in the early 1990s to abandonment of the badly rundown state railway system. However, after a study on the future of the railway by CIE Consult of Ireland under World Bank auspices, it was decided that efforts would be made to revitalise HSH. Since then, funding for improvements has been provided the World Bank, the European Union and the governments of the United States and Italy.

The condition of the network remains poor and in 2004 UK-based consultancy Scott Wilson was appointed by the Albanian government to develop proposals to modernise and re-equip HSH. These proposals contributed to a five-year national transport plan that was under discussion at draft stage at the end of 2004.

Organisation
Following the CIE Consult report, HSH was restructured into two business units (passenger and freight) and two service units (infrastructure and rolling stock).

Passenger operations
Passenger services are operated on the Tirana–Durrës, Tirana–Vlorë, Tirana-Pogradec and Tirana–Shkodër (Durrës) routes. Only one class of accommodation is provided. A regular interval between Durrës and Tirana is proposed. Passenger services were withdrawn in 1996 from the Milot–Rrëshen line, which was closed in early 1998.

Consultants' proposals include a possible commuter service between Tirana and Durrës.

Freight operations
The principal freight flow is a daily international service linking Montenegro with Tirana. The opening in 2004 of a rail link to the port of Durrës was expected to boost traffic further, while further benefits will result from construction of a line to Macedonia (see *New lines*).

New lines
A link from the northern Albanian railhead at Shkodër into Montenegro was finished in 1986. This was Albania's first rail connection with a foreign railway, but it suffered track damage during civil unrest in 1997. Rehabilitation of this section was completed in March 2003.

An agreement is in place to extend the line from Pogradec via Korce to Flórina in Greece; design work was completed in 1995, but no physical construction has taken place.

In 2004 an agreement was signed with the Italian government to construct a further international link from Lin, on the Elbasan–Pogradec section, to the Macedonian border at Radozda. This will connect with a future 60 km line via Struga to Kicevo in Macedonia which would form part of European Transport Corridor VIII from the Adriatic to the Black Sea.

In October 2004 a 2 km branch serving the port of Durrës was opened.

Improvements to existing lines
In 2004 it was announced that Italian military engineers were to undertake rehabilitation of the Rrogozhinë–Elbasan–Pogradec line in connection with the scheme to build a new link from this line at Lin to Macedonian border (see *New lines*). Italian contractors were also to take part in upgrading the Shkodër–Durrës line.

Traction and rolling stock
The only serviceable locomotives in 2004 were 25 Czech-built 1,007 kW (1,350 hp) T669 Co-Co diesels, operational survivors of 61 delivered by

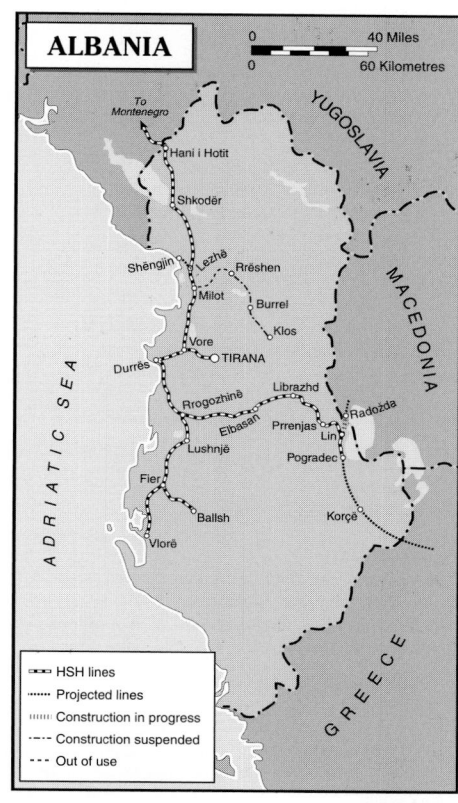

NEW/0585091

CKD Praha between 1968 and 1990. Coaching stock mostly comprised ex-Austrian or -Italian vehicles, many reportedly in poor condition.

Total freight vehicle stock is around 2,000, but no more than 400 wagons are serviceable.

Signalling and telecommunications
Although colourlight signalling was installed between Durrës and Tirana, Durrës and Elbasan, and Durrës and Laç, most of this was reported to be out of use in 2004, operations instead regulated by radio despatching.

Track
Rail: 38, 43, 48, 49 kg/m in 12 to 24 m lengths
Sleepers: Wood, duo-bloc concrete
Min curve radius: 300 m
Max axleload: 21–24 tonnes
Max speed: 60 km/h

UPDATED

Algeria

Ministry of Transport

Rue Didouche Mourad 119, Algiers
Tel: (+213 2) 74 06 99
Fax: (+213 2) 64 66 37

Key personnel
Minister: Selim Saâdi

Algerian National Railways (SNTF)

Société Nationale des Transports Ferroviaires
21 Boulevard Mohamed V, Algiers
Tel: (+213 2) 71 15 10
Fax: (+213 2) 63 32 98

Key personnel
Director General: Ing Abdelhamid Lalaimia
Secretary General: Ahmed Halfaoui
Directors
Operating: Ali Leulmi
Human Resources: Abderrahmane Belkadi
Infrastructure: Mourad Soliman Benameur
Rolling Stock: Lakhdar Saadi
Finances: Djamel Djenas
Planning: El Berkenou
Purchasing: Abdelhamid Moudjebeur
Studies: Tahar Bouifrou
Audit: Moukhtar Rahal
Regional Directors
Algiers: Kerdel Ramdane
Annaba: Abdelhamid Benboudjemaa
Constantine: Mohammed Chérif Handel
Oran: Yacine Bendjaballah
Working Directors
Algiers: Zaki Fouad Azzouz
Constantine: Mustapha Makhloufi
Mohammadia: Abderrahmane Belkacemi
Sidi-Bel-Abbès: Mohammed Boumaaza
Director of Freight Vehicle Operations:
Abdelmalek Hamzaoui
External Relations Manager: Ms Houriadib

Gauges: 1,432 mm; 1,055 mm
Route length: 2,888 km; 1,085 km
Electrification: 283 km at 3 kV DC

Organisation
The network consists primarily of two standard-gauge coastal lines running east and west from Algiers: about 550 km westward to the railhead at Akid Abbès (where a connection with Moroccan

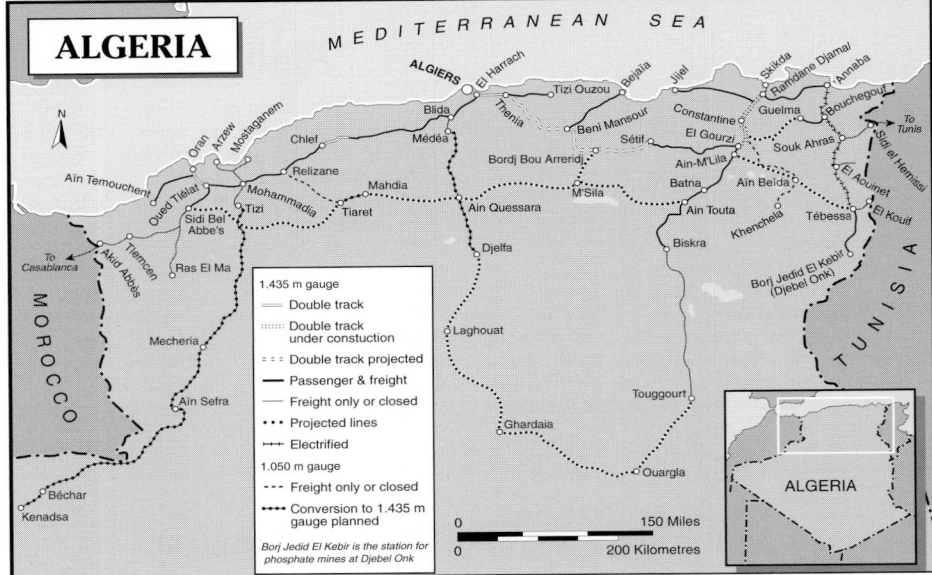

Railways, broken in 1976, was reactivated in 1989), and about 370 km eastwards to a connection with the 520 km north-south line at El Guerrah. In addition to standard-gauge spur lines, a 300 km (partly electrified) 1,435 mm gauge line runs parallel with the Tunisian border, providing international connecting services at Souk-Ahras with Tunisian National Railways (SNCFT) from the port of Annaba to Djebel Onk. Major narrow-gauge lines run from Tizi to Béchar and Blida to Djelfa; conversion to standard gauge is proposed.

In 1987 a number of subsidiaries were formed to free the railway's management for full attention to transport. Setirail and Estel are the infrastructure and signalling/telecommunications subsidiaries; the others include Infrafer and Infrarail, new construction and tracklaying; Restau-Rail, on-train catering; Rail-Express, small freight consignments door-to-door; and STIM, multimodal transport.

The railway is managed by nine central divisions and four regional administrations. In 2000 12,510 staff were employed.

Passenger operations
SNTF operates long-distance locomotive-hauled passenger services on nine main routes, the best served being Algiers–Chlef–Oran (422 km), with five daily services in each direction. Three

classes of accommodation are provided: luxury (air conditioned), first class and second class. A suburban network also serves Algiers.

In 1999 32 million passenger journeys were made, nearly 90 per cent of these on suburban and commuter services.

Freight operations
Petroleum products form SNTF's principal freight traffic, in 2000 generating over one quarter of the 8 million tonnes lifted. Volumes of phosphates from the mines at Djebel Onk are also significant, much of the output conveyed over the electrified line to Annaba for export. Also carried is traffic for the El Hadjar steel complex, which gets its ore in 1,500 tonne trains from mines at Quenza and Bou Khedra, about 190 km south of Annaba (whence the imported coal for its coking plant is also ferried by unit train).

SNTF employs bogie-changing to facilitate the transfer of rail vehicles from standard to 1,055 mm gauge tracks.

Traffic (million)	1998	1999	2000
Passenger journeys	34.1	31.4	28.3
Passenger-km	1,163	1,069	1,142
Freight tonnes	8.29	7.8	7.8
Freight tonne-km	2,174	2,057	2,029

New lines
A joint Algerian-Indian study group developed plans for a first section of the High Plateau route, the 146 km from Ain Touta, on the line south from Constantine to Biskra, to M'Sila, location of an aluminium plant. Construction of this line was in progress in 2000, although less than one quarter of the scheme was reported to be completed. The line is being engineered for 160 km/h with long-welded 54 kg/m rail on twin-block sleepers, the latter manufactured in a plant established at Ain Touta.

SNTF contemplates extension of its standard-gauge system across the heart of the Sahara, looping southward from Touggourt in the southeast, through Ouargla then northwest via Ghardaia and Laghouet to Ain Quessara, on the projected High Plateau route. The first section of this project, from Touggourt 210 km south to Hassi Messaoud features in SNTF's network development programme.

Improvements to existing lines
SNTF has an extensive programme of network development which includes track doubling and upgrading, renewals, and realignments and reconstruction. However, progress has been hampered by Algeria's economic and political difficulties.

Among large-scale projects completed is the double-tracking of the line between Ramdane Djamal, 67 km north of Constantine and El

Class 040 YDA General Motors 1,055 mm gauge diesel-electric locomotive (Marcel Vleugels)   0089081

Newly laid trackwork in the industrial suburbs of Algiers   0089080

Gourzi, 38 km south of Constantine. This vital link between Algiers and the petrochemical port of Skikda, as well as Annaba, has been doubled throughout the 67 km from Ramdane Djamal to Constantine and the 20.6 km south from El Gourzi to El Khroub, along with realignments.

In conjunction with installation of heavier UIC 54 welded rail on concrete sleepers of SL Type U (1,722 per km), this scheme has raised permissible freight speed from 60 to 90 km/h and wagon axleload from 18 to 28 tonnes. Following installation of modern automatic signalling, the line's train operating capacity has doubled and a throughput of 7 million tonnes a year is possible.

Also completed is doubling of the 43.5 km from El Harrach, on the outskirts of Algiers, to Thénia, a project funded by an Austrian loan.

Track renewals are in progress between Ramdane Djamal and Annaba. To create a relief route between Constantine and Annaba, resuscitation of the 95 km El Khroub–Guelma connection, abandoned in the 1950s, is under way. Upgrading and conversion to standard-gauge is planned for the 600 km metre-gauge line between Mohammadia and Bechar.

Traction and rolling stock

On 1,432 mm gauge SNTF operates 19 electric and 194 diesel locomotives, 59 locotractors and 12 twin-unit diesel railcars. Coaching stock totals 422, and there are 10,118 wagons. The narrow-gauge traction fleet comprises 30 diesel locomotives and there are 33 coaches.

A new fleet of electric locomotives came into service in 1996 on the Djebel Onk–Annaba ore haul. Alstom supplied 14 Co-Cos rated at 2,400 kW, designed to haul trains of up to 2,700 tonnes.

On the diesel front, General Motors has supplied 10 Co-Cos rated at 2,400 kW, the first of which was delivered in 1994. A further 50 units of various types were supplied by GM through to 1996, some as kits for assembly locally. Three diesel shunting locomotives rated at 448 kW were supplied in 1995 by local manufacturer Ferrovial. An order was also anticipated for a new fleet of 10 diesel multiple-units for Algiers suburban services.

Signalling and telecommunications

Automatic block signalling is installed on the Algiers–Thénin (50 km) and El Harrach–El Affroun (50 km) sections.

Resignalling and a complete renewal of the telecommunications network ranks high. Among other things, the railway aims to make track-to-train radio communication a standard feature on its principal routes. Electrically operated mechanical signals are gradually being replaced by colourlight displays throughout the system. On new lines and upgraded tracks automatic signalling is being installed. SNTF has signed an agreement in principle with Siemens for formation of a joint company to manufacture and install signalling equipment. As a result, a programme of installing modern interlocking systems has been undertaken at 33 stations between El Affroun and Oued Tlelat.

Diesel locomotives

Class	Builder's type	Wheel arrangement	Output kW	Speed km/h	Weight tonnes	No in service	First built	Mechanical	Builders Engine	Transmission
Standard gauge										
060 DD	GT 26 W	Co-Co	2,400	120	120	27	1971	GM	GM	E GM
060 DF	GT 26 W	Co-Co	2,400	120	120	25	1973	GM	GM	E GM
060 DG	GT 26 W	Co-Co	2,400	120	120	15	1976	GM	GM	E GM
060 DH	GT 22 W	Co-Co	1,600	120	120	24	1976	GM	GM	E GM
060 DL	GT 26 W	Co-Co	2,400	120	120	25	1982	GM	GM	E GM
060 WDK	GL 18 M	Co-Co	800	100	78	5	1977	GM	GM	E GM
060 DJ	U 18 C	Co-Co	1,400	100	96	25	1977	GE	GE	E GE
060 DM	GT 26 HCW-2A	Co-Co	2,400	125	125	10	1985	GM	GM	E GM
040 DH	GL 18 B	Bo-Bo	800	80	80	5	1990	GM	GM	E GM
040 DH	GL 18 B	Bo-Bo	800	80	80	8	1993	GM	GM	E GM
060 DP	GT 26 HCW-2A	Co-Co	2,400				1994	GM	GM	E GM
1,050 mm gauge										
040 YDA	GL 18 M	A1A-A1A	800	80	72	24	1977	GM	GM	E GM
060 YDD	GL 18 2C/2M	Co-Co	1,200	80	80	5	1989	GM	GM	E GM

Diesel railcars

Class	Builder's type	Cars per unit	Motor cars per unit	Power/car kW	Speed km/h	No in service	First built	Mechanical	Builders Engine	Transmission
ZZN 200	ALN 668	2	2	286	120	14	1972	Fiat (Iveco)	Fiat (Iveco)	Fiat (Iveco)

Electric locomotives

Class	Wheel arrangement	Power kW	Speed km/h	Weight tonnes	No in service	First built	Builders Mechanical	Electrical
6CE	Co-Co	1,492	80	130	17	1972	LEW	Škoda
6FE	Co-Co	2,400	–	132	14	1995	GEC Alsthom Transporte	ACEC

Track

Rail: UIC 54 has been adopted as the standard for main line renewals
Sleepers: Concrete twin-block installed on 1,732 km
Max gradient: 3.2%
Max axleload: 20 tonnes

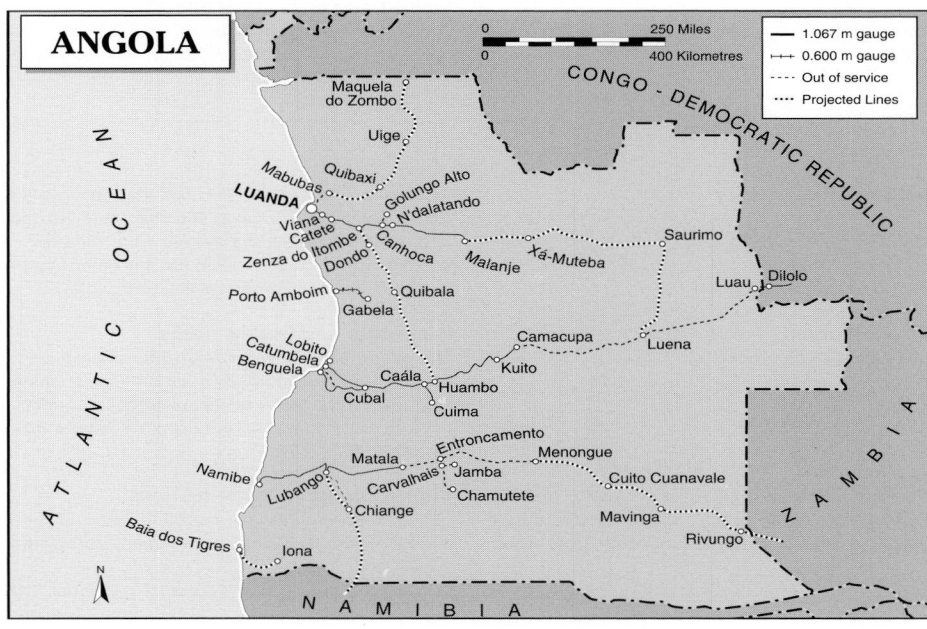

SNTF employs bogie changing facilities such as these lifting jacks at Relizane to enable wagons to move between standard and 1,055 mm gauge networks (Marcel Vleugels)   0524891

Angola

Ministry of Transport and Communications

PO Box 1250-C, Luanda
Tel: (+244 2) 33 77 93/33 77 44
Telex: 3108 Mitrans AN

Key personnel
Minister of Transport: André Luis Brandao

Amboim Railway

Caminho de ferro do Amboim
Estaçao Puerto Amboim

Key personnel
General Manager: A Guia

Gauge: 600 mm
Route length: 123 km

Angola, Congo
0058517

Organisation

The railway is a single line between the port of Amboim and the coffee-growing region at Gabela. The first 40 km out of Amboim is fully operational, the remainder intermittently so.

Passenger and freight operations

Steam-hauled mixed trains are run when circumstances permit. The main freight commodities carried are timber, agricultural products and coffee. A railcar-operated service also operates intermittently between Amboim and Gabela.

Improvements to existing lines

A partial rehabilitation has been accorded priority, and a new telecommunications system is urgently required.

Traction and rolling stock

There are six steam locomotives, 12 wooden-bodied coaches and 60 freight wagons, around 50 per cent of which are operational. A Wickham-built railcar is also in service.

Benguela Railway

Companhia do Caminho de ferro de Benguela
Praça 11 de Novembro no 3, Lobito
Tel: (+244 72) 226 45
Fax: (+244 72) 401 33
e-mail: CFB/LOBITO-rocha@ebonet.net

Key personnel

Chairman: Dr Guilherme Magaihães Pratas
Administrator and General Manager:
   Dr Daniel João Quipaxe
Administrator and Commercial Director:
   Eng Enesto Alpoim Ferreira da Rocha
Administrators:
   Eng José Manuel Miranda Vaz de Carvalho
   Eng Carlos dos Santos Braz
   Cristiano Reis d'Almeida

Gauge: 1,067 mm
Route length: 1,301 km

Organisation

The Benguela Railway should be a major traffic route to the sea for Zambian and Congolese copper, but the connection from the port of Lobito across Angola to the Congo Democratic Republic border at Dilolo, where it is connected with SNCC, has been disrupted by guerrilla action since 1975.

In late 1996, an Italian company, Tor di Valle, won a contract to reconstruct the out-of-use section of the railway. The company was to be paid by the exploitation of eucalyptus forests. A further

Benguela Railway diesel locomotives

Class	Wheel arrangement	Power kW	Speed km/h	Weight tonnes	No in service	First built	Mechanical	Builders Engine	Transmission
D.101-D.314	Co-Co	1,582	109	90	6	1973	GE USA/ GE Brazil	GE 7FDL12	E GE
D.20-D.25	Bo-Bo	493	46	60	–	1988	GE Brazil	Cummins NT 855L	E GE
D.11-D.12	C	313	40	42	–	1972	Barclay	Paxman	H Barclay
D.1-D.4	C	313	27	41	–	1962	North British	Paxman	H Voith/NBL
–	B	309	50	44	–	1960	Moyse	Poyaud	E Moyse

move to rehabilitate the line came in 1997, with the awarding of a contract to remove landmines from the Cubal-Kuito section. Work on this started in 1998.

At the end of 1998 the Benguela Railway employed 2,146 staff.

Passenger and freight operations

Regular passenger trains operate over the 33 km coastal line between Lobito and Benguela. In 1998 mixed train operation resumed on a usually weekly basis over the 957 km section between Benguela and Luena. If circumstances permitted, a mixed train was also running between Luena and Luau, near the border with Congo, Democratic Republic, but international services remained suspended.

Traffic	1997	1998
Passenger journeys (000)	5.8	2.8
Passenger-km (million)	141.9	;69.6
Freight tonnes (000)	101.5	57.4
Freight tonne-km (million)	0.69	0.42

Traction and rolling stock

In 1998 the active fleet comprised six main line diesel locomotives and two shunters, with the remainder of the tractive stock unserviceable. The line-haul diesels are GE Type U20C. The 10 passenger cars are survivors of 12 second-hand vehicles supplied from South Africa in 1992–93. The railway owns 750 wagons.

Type of coupler in standard use: AAR M201/10A
Type of braking in standard use: Vacuum
Track
Rail: BS 60A/N 30 kg/m, BS 80A 40 kg/m, BS 20A 45 kg/m, CFB 60 30 kg/m
Sleepers: Wood 2,000 × 250 × 140 mm, spaced 1,460/km in plain track, 1,491/km in curves; steel
Fastenings: Elastic spike
Min curve radius: 100 m
Max gradient: 2%
Max axleload: 15 tonnes

Direcçao Nacional dos Caminhos de Ferro

PO Box 1250-C, Luanda
Tel: (+244 2) 33 97 94/33 02 33
Fax: (+244 2) 33 99 76
Tx: 3108 Mitrans AN

Key personnel

Director: R M da C Junior

Gauge: 1,067 mm; 600 mm
Route length (four railways combined): 2,648 km; 123 km

Organisation

Three previously independent railways (the Amboim, Luanda and Namibe Railways) are now amalgamated in a national system, while the Benguela Railway is supervised by the Ministry of Transport but retains its own administration. Because of the country's continuing guerrilla warfare, the four railways have so far been unable to integrate operations fully or handle international traffic consistently.

Freight and passenger operations

Despite operating difficulties, in 1991 the four railways carried a total of 3.6 million tonnes of freight, 45.3 million tonne-km. Passenger journeys amounted to 5 million, for 246 million passenger-km. Passenger traffic held up well through to 1993, but freight was 20 per cent down at 2.8 million tonnes.

Traffic (All four railways)	1991	1993
Passenger journeys (million)	5.0	5.0
Passenger-km (million)	246.2	n/a
Freight tonnes (million)	3.6	2.8
Freight tonne-km (million)	45.3	n/a

Luanda Railway

Caminho de ferro de Luanda
PO Box 1250-C, Luanda
Tel: (+244 2) 700 61/732 70

Key personnel

Director: A Alvaro Agante

Gauge: 1,067 mm
Route length: 479 km

Organisation

The line runs from the port of Luanda to Malanje, serving an iron, cotton and sisal producing region. A branch runs 55 km south from Zenza to Itombe and Dondo. The 600 mm gauge line from Canhoca to Golungo Alto has been closed. Efforts to rehabilitate the system were reported in late 1997, with a US$15 million funding being sought for second-hand locomotives and track work.

Passenger and freight traffic

The main freight flow is of agricultural produce to Luanda, which amounted to 2.5 million tonnes in 1993. Passenger traffic is also of some importance, in particular a suburban service into Luanda from Viana and Baia (36 km).

Improvements to existing lines

Rehabilitation of the line started in 1989, but only a few short sections have been completed.

Poor track conditions and inoperable signalling render train working problematic to the east of N'dalatando.

Traction and rolling stock

Of the 35 steam locomotives, only one-third is available at any time; these work the Luanda suburban service. The diesel situation is worse, with no more than a few of the 50-strong fleet in operating condition. All four Fiat diesel railcars are out of service. In addition, there are 50 coaches and about 1,800 wagons.

Namibe Railway

Caminho de ferro do Namibe
Caixa Postal 130, Sá de Bandeira, Namibe

Key personnel
General Manager: J Salvador

Gauge: 1,067 mm
Route length: 858 km

Organisation
The Namibe Railway consists of a 756 km line running from Namibe inland to Menongue via Lubango, Matala and Entroncamento. A line of some 150 km linking Lubango with Chiange was destroyed in the war.

Passenger and freight operations
In normal times the Cassinga ore fields yielded traffic of some 6 million tonnes annually, but, at last report in late 1994, only the 424 km from Namibe to Matala was in operation and trains were still being affected by military action. Only some 500,000 passengers were carried in 1993, plus less than 100,000 tonnes of freight.

Traction and rolling stock
In 1993 only six or eight diesel locomotives were in operational condition. The programme announced in 1990, under which 11 locomotives, 59 passenger cars and 50 freight wagons were to be rehabilitated, had still not been completed. Currently about 20 coaches are in working order, along with 60 wagons.

Argentina

Ministry of Infrastructure

Secretariat of Transport
Hipólito Irigoyen 200, Piso 12, Oficina 1209, C1086 AAB, Buenos Aires
Tel: (+54 11) 43 81 89 11
Fax: (+54 11) 48 14 18 23

Key personnel
Minister: Nicolás Gallo
Secretary: Ignacio Alfonso
Under-Secretary: Alba Thomas Hatti

América Latina Logística (ALL-BAP)

Avenida Santa Fe 4636–piso 3°, C1425 BHV Buenos Aires
Tel: (+54 11) 47 78 24 00
Fax: (+54 11) 47 78 24 08
e-mail: eoliver@all-logistica.com.ar
Web: http://www.all-logistica.com.ar

Key personnel
Director General: Eduardo Oliver
Directors
   Commercial: Vlad Simian
   Administration and Finance: Roberto Monteiro
   Operations: Rubén Chaparro
   Human Resources: José María Ohrnialian
   Institutional Relations: Carlos Guaia

Gauge: 1,676 mm
Route length: 5,350 km

Organisation
On 5 June 1992 the Consorcio Ferrocarril Central group was awarded the 40-year concession to operate 5,400 km of the General San Martín railway and 706 km of the Sarmiento railway (both 1,676 mm gauge). This network, which became the Buenos Aires al Pacifico/San Martín (BAP) system, was the third freight concession to be granted, and was considered to be potentially very profitable.

The CFC consortium was headed by Industries Metalúrgicas Pescarmona (IMPSA), which held a 60 per cent stake and whose main manufacturing plant in Mendoza is connected to BAP lines. IMPSA was also the main partner in the CFM consortium later awarded the Urquiza concession. Other members of CFC were Román Maritima (25 per cent), Transapelt and Hugo G Bunge. Railroad Development Corporation (which controls the Iowa Interstate Railroad (qv)) was chosen as operator, with Conrail as technical consultant.

In 1997 BAP unified operations with Ferrocarril Mesopotámico General Urquiza (qv), where there are overlapping shareholding interests. In August 1998 the IMPSA-led consortium announced its intention to sell its share in BAP, along with its interests in the Ferrocarril Mesopotámico General Urquiza (qv), to a Brazilian consortium, América Latina Logística (now Garantia Partners), which operates that country's Ferrocarril Centro-Atlantico

ALL train of cereal hoppers loaded with soya beans for export at Ludueña Junction, Rosario, behind a former Spanish National Railways Class 321 Alco diesel locomotive (Angel Ferrer)   0089089

ALL freight at Rosario yard en route to Puerto San Martín, with a former Spanish National Railways Class 321 locomotive as motive power (Angel Ferrer)   0089090

and Ferrovia Sul-Atlantico networks. This resulted in the adoption of a new name for the system, ALL-BAP.

Freight operations
Operations as the Buenos Aires al Pacífico commenced on 26 August 1993, several months later than planned due to problems with the transfer of staff from FA. The lines covered by the concession link the provinces of Mendoza and San Juan in the Andean foothills to Buenos Aires, crossing the provinces of San Luís, Córdoba and Santa Fe in the process. In 1996 BAP began direct services to the Río de la Plata terminal in the port of Buenos Aires following completion of a new link.

The traffic base inherited from FA consisted mainly of trainload movements of petroleum, cereals, coke and limestone. The new operators have since developed wagonload, intermodal (both container and piggyback) and door-to-door services and to broaden the range of commodities carried.

Traffic flows, especially of containers, from Mendoza to the port of Buenos Aires have become especially significant, with trains assembled at Palmira yard, near Mendoza. The majority of BAP's traffic is eastward to Buenos Aires and the port and industrial centre of Rosario.

Trains are operated by a crew of two and have end-of-train telemetry devices.

Traffic (million)	1998–99	2000
Freight tonnes	3.42	2.9

Improvements to existing lines
At the commencement of BAP operations the maximum permitted speed on the San Martín east-west main line was 120 km/h, although many branch lines were only suitable for speeds of 12 km/h. Of the US$150 million BAP intended to invest in the first five years of the 30 year concession period, US$45 million was to be spent on track improvements and a continuous-rail welding programme. An infrastructure analysis matrix has been deployed to assess the commercial desirability of infrastructure improvements and to determine those which should take priority over others.

In 1997 the loss of oil traffic led to closure of the Malargue branch, although expanding cement traffic led to the reopening of the Justo Daract–San Luis branch.

Traction and rolling stock
At the end of 2000 the locomotive fleet totalled 102. Main line traction included 20 Alco-engined Class 321 Co-Co diesel-electric locomotives acquired from Spanish National Railways (RENFE).

At the end of 2000, ALL-BAP operated 3,910 freight wagons, of which 1,314 were unserviceable. Recent acquisitions include 46 high-capacity wagons for palletised goods, 130 container wagons and 148 general purpose freight wagons.

Signalling and telecommunications
The old British semaphore and staff block signalling system has been replaced by a North American radio-based track warrant system. In Mendoza province, a 185 km section of CTC dating from the 1960s and expanded in the 1980s was removed by BAP during the IMPSA tenure.

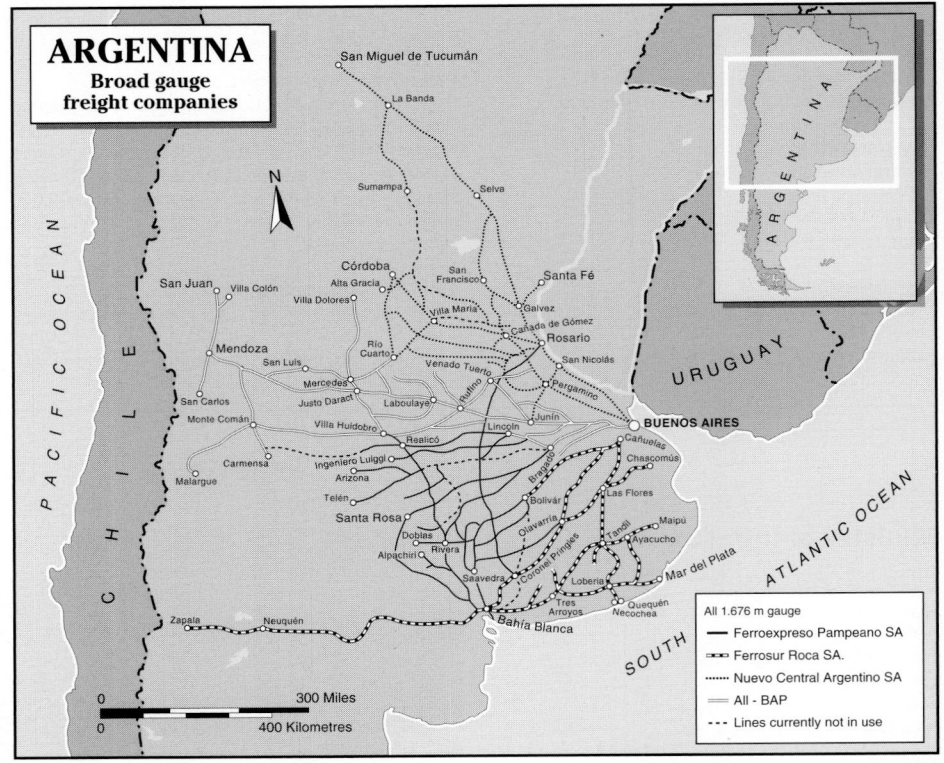

ARGENTINA
Broad gauge freight companies

All 1.676 m gauge
━━━ Ferroexpreso Pampeano SA
▭▭▭ Ferrosur Roca SA.
┄┄┄ Nuevo Central Argentino SA
═══ All - BAP
- - - Lines currently not in use

0533220

Diesel locomotives

Class	Builder's type	Wheel arrangement	Output kW	Speed km/h	Weight tonnes	No in service	First built	Builders Mechanical	Engine
Alco RSD 16	Alco USA	Co-Co	1,380	90	118	21	1968	Alco 321-B	E GE
Alco RSD 35	Alco Canada	Co-Co	930	90	98	50	1962	Alco 321	E GE
Alco 321	Alco Spain	Co-Co	1,570	120	119	20	1979	Alco 321-C	E GE
Gaia	Argentina	Co-Co	1,380	75	99	4	1968	Alco 321-B	E GE
Loco Tractor	Argentina		520	40	50	9		Detroit Diesel	M

América Latina Logística (ALL-FMGU)

Avenida Santa Fe 4636–piso 3°, C1425 BHV Buenos Aires
Tel: (+54 11) 47 78 24 00
Fax: (+54 11) 47 78 24 08
e-mail: eoliver@all-logistica.com.ar
Web: http://www.all-logistica.com.ar

Key personnel
See entry for *América Latina Logística (ALL-BAP)*.

Gauge: 1,435 mm
Route length: 2,739 km

Organisation
In January 1993 the Consorcio Ferrocarril Mesopotámico (CFM) consortium was awarded a 40-year concession to operate FA's Urquiza railway, taking over in October 1993. The majority partner in the CFM consortium with a 71 per cent holding was IMPSa, which also headed the CFC group (see entry for América Latina Logística (ALL-BAP)). Other companies involved in the CFM consortium were Pescarmona, Alesia, Olmatic SA, and Petersen Thieley Cruz. As with BAP, US Class I railroad Conrail was chosen as operator, assisted by Railroad Development Corporation. FMGU began operations on 12 October 1993.

In 1997 operations were unified with those of the Buenos Aires al Pacifico (see América Latina Logística (ALL-BAP)), in which FMGU shareholders have interests. In August 1998 the owning consortium announced its intention to sell its share in BAP, along with its interests in the Ferrocarril Mesopotámico General Urquiza, to a Brazilian consortium, América Latina Logística (now Garantia Partners), which operates that country's Ferrocarril Centro-Atlantico and Ferrovia Sul-Atlantico networks. The Argentine government authorised the transaction and ALL took over the system on 26 May 1999.

Combined staff for ALL-BAP and ALL-FMGU totalled 920 in 2000.

Freight operations
ALL Mesopotámico comprises the former FA 1,435 mm gauge network in the east of the country and by far the best maintained, still carrying acceptable levels of traffic at the time of privatisation. It extends from Buenos Aires north to Encarnación, where it connects with FCPCAL of Paraguay via the Roque González de Santa Cruz bridge over the River Paraná. ALL Mesopotámico also connects with AFE of Uruguay across the Salto dam.

The traffic base consists mainly of soya beans, cereals, aggregates, forestry products and containers, moved largely as wagonloads. Crews of two and end-of-train devices have replaced FA work practices and brake vans. In 2000, the network carried 1 million tonnes of freight.

Intermodel operations
The use of a 250 tonne breakdown crane has allowed the company to start handling containers at Mendoza, resulting in intermodal services now being offered to customers. Future strategy is to concentrate on more logistics-oriented services.

Improvements to existing lines
During the 40-year period of the concession, CFM undertook to invest US$166 million. Whilst ALL Mesopotámico infrastructure generally did not demand the attention given to that of ALL Central, the 130 km branch to Corrientes and the 110-year-old truss bridge at Agua Pey requires attention in the near future.

Traction and rolling stock
At the end of 2000, 36 locomotives were in operation. The locomotive fleet consists of some of the more recent General Motors deliveries to FA and some older General Electric locomotives. ALL Central's Mendoza facility was being used in preference to the Urquiza locomotive works at Paraná, with running repairs undertaken at Alianza. A consolidated wagon repair works has been established at Alianza, on the outskirts of Buenos Aires, to serve both the ALL fleets.

At the end of 2000, 2,139 wagons were in use. The fleet has been receiving airbrakes and knuckle couplers.

To improve its intermodal operations, ALL has acquired 20 'Roadrailers', which it introduced in Argentina.

Signalling and telecommunications
Computer-assisted radio train dispatching with track warrants has been introduced. A microwave system has replaced a semaphore signalling system and also provides business communications in the place of the unreliable Argentine telephone network.

Comisión Nacional de Regulación de Transporte (CNRT)

Avenida Maipú 88–piso 5°, 1084 Buenos Aires
Tel: (+54 11) 43 18 35 48
Fax: (+54 11) 43 18 36 62

Key personnel
President: Jose Emilio Bernasconi

Formerly the Unidad de Coordinación del Programa de Reestructuración Ferroviaria, and incorporating functions of the former Ferrocarriles Metropolitanos, this organisation now oversees all concessions for interurban and suburban passenger and freight networks in Argentina.

With the transfer in December 1998 of Ferrocarril Belgrano Cargas to the private sector, all of the former Argentine Railways network, which totalled nearly 34,000 km before the start of privatisation, passed out of central government ownership.

The number of passengers using metropolitan rail services grew from 212 million in 1993 to 456.1 million in 1997, while the figure for interurban passengers declined from 11 million in 1990 to just 2.6 million in 1997. Freight tonnage, which in 1993 stood at 9.5 million tonnes, doubled to 18.9 million tonnes in 1997. Tonne-km increased from 5 billion to 9.8 billion in the same period.

Cooperativa de Trabajo Ferroviario Urquiza de Paraná

Paraná, Entre Rios

Key personnel
President: Ramón Ismael Claria

Gauge: 1,435 mm
Route length: 120 km

Organisation
After a delay of eight months, Cooperativa de Trabajo Ferroviario Urquiza de Paraná introduced regular passenger services on the 120 km Paraná-Nogoyá line in Entre Rios province on 23 December 1998. The line was owned by freight operator Ferrocarril Mesopotámico General Urquiza (FMGU) (see América Latina Logística (ALL Mesopotámico)), which initially was not taking any track access fee for the use of its infrastructure. Following the takeover by ALL from IMPSA of the FMGU system, the new concessionaire stated that it wished to discuss track access fees with Cooperativa de Trabajo Ferroviario Urquiza for its passenger services. However, a derailment in July 1999 of a Paraná-Nogoyá passenger train due to poor track conditions brought services to a halt.

The co-operative consists of 120 staff, mostly former Argentine Railways employees who were not taken on when FMGU took up its concession in 1993.

Passenger operations
Until March 1999, the service was intended to run twice weekly in each direction, becoming daily therafter. Nine intermediate stations are served in an area of 300,000 inhabitants. Train accommodation includes first and tourist class coaches.

Under the Proyecto Ferroviario Provincial scheme, Entre Rios province was working with FMGU and the Cooperativa to explore reopening the 314 km Paraná–Concordia main line for passenger operations, and some track repairs may be carried out under a state-financed employment arrangement. The line has seen no regular passenger trains since 1992, but it is felt that restoration of services would provide a valuable east-west social and tourist link across Entre Rios province.

Traction and rolling stock
The Cooperativa owns three diesel locomotives (two GE U13s and one General Motors G22), and 15 passenger coaches, including Pullman, first and tourist class vehicles and luggage vans.

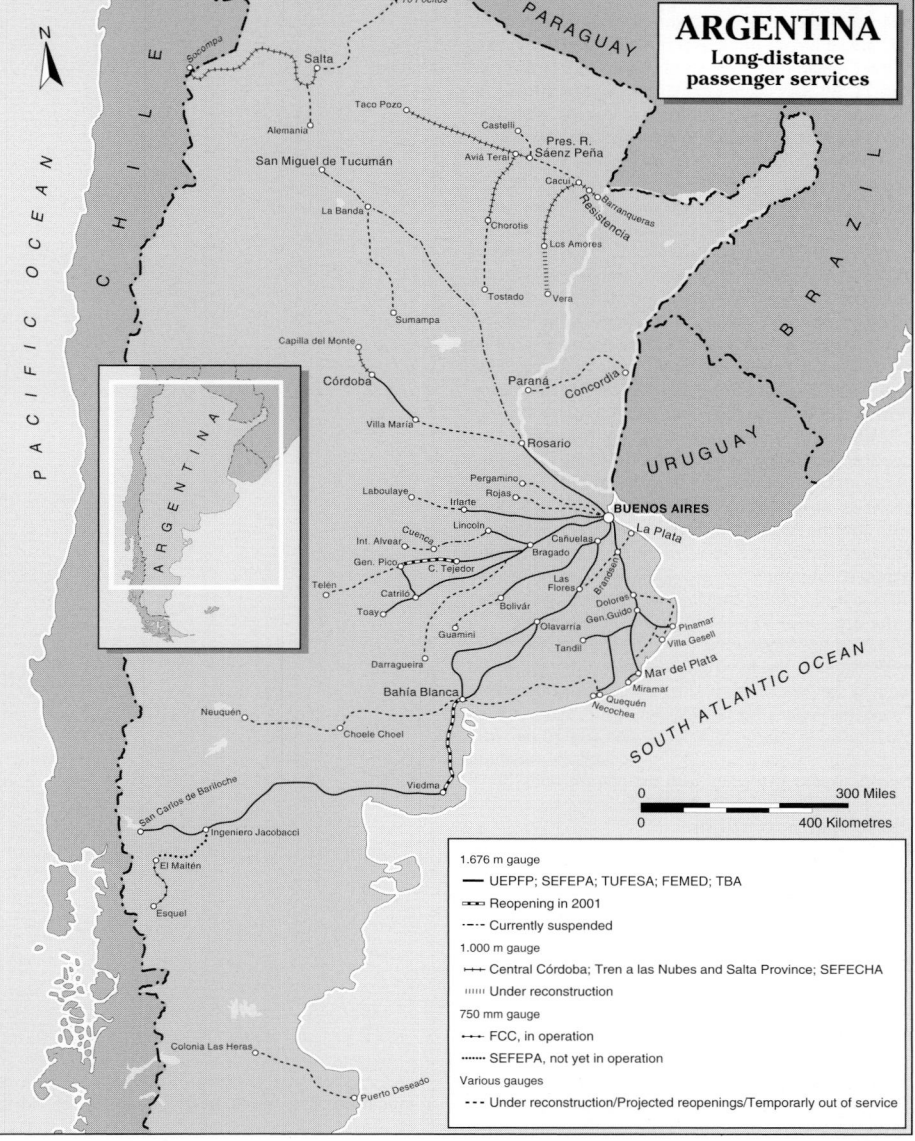

Ferrocarril Austral Fueguino (FAF) (Tranex Turismo SA)

Avenida Corrientes, 538–piso 5°, 1043 Buenos Aires
Tel: (+54 11) 43 25 06 81
Fax: (+54 11) 43 26 35 40

Key personnel
President: Enrique Díz

Passenger operations
FAF runs a 7.2 km 500 mm gauge tourist line linking Ushuaia and the Tierra del Fuego National Park. It is the world's most southerly railway. Passenger numbers have grown from 7,000 in its first season to 23,000 in the summer of 1996/97, with a 33 per cent increase reported in December 1997.

Improvements to existing lines
A further 700 m track extension is under construction, and the company has announced its intention eventually to serve Ushuaia.

Traction and rolling stock
FAF owns two steam locomotives, three diesel locomotives, 16 passenger coaches and 10 service wagons.

Ferrocarril Belgrano Cargas SA (FC BC)

Padre Mugica 426, 1104 Buenos Aires
Tel: (+54 11) 45 10 35 00

Key personnel

Chairman and General Manager:
  Dr Graciela Isabel Coria
Vice-President: Dr Angel Luis Stafforini
Head of Commercial Department: Martina Laplane

Gauge: 1,000 mm
Route length: 7,347 km

Political background

The metre-gauge General Belgrano system, now FC BC SA, serves the far northwest of Argentina with routes to the Chilean and Bolivian borders. On 6 November 1999 FC BC SA became the last portion of the former Argentine Railways system to be privatised, with a 30-year concession awarded to a consortium comprising the railway staff union, Unión Ferroviaria (99 per cent) and the Argentine government (1 per cent), which had been operating the network on a temporary basis since 1997. The privatisation mode applied was the same as that for the Metropolitanos lines, with some financial contribution from the national government. This provided annual state financial support for investments of 50 million pesos during the first five years, followed by a total of 142 million pesos during the following ten years. Transurb was to assist FBC technically, and was to be responsible for operations. However, the financial support did not materialise

Under the terms of a revised agreement, the governments of nine northern provinces obtained the right to operate freight and passenger services throughout their territory (3,313 km), with international connections at border stations. Tenders were called in April 2000, with successful bidders intended to provide their own motive power and rolling stock. The northern provinces generate about 80 per cent of FC BC SA traffic.

In 2004 the Argentinian government was assessing privatisation bids for the FC BC SA network, which was to be operated under a 50-year concession. Proposals were for a 75 per cent stake to be transferred to the private sector, leaving 20 per cent with Unión Ferroviaria, 4 per cent with the northern provinces and 1 per cent with the government.

Passenger operations

The Salta–Socompa line is the only section on which a private company operates a passenger service, run under concession awarded by Salta province. Other parts of the metre-gauge network have been transferred or leased to Chaco and Santa Fé provinces for the operation of passenger services (see entry for Servicios Ferroviarios Chaqueños (SEFECHA) in Argentine section of *Railway systems and operators*).

Freight operations

The main commodities handled are mining and ore products, grain and container traffic, the latter playing an important role in movements to and from Chile.

Improvements to existing lines

The 7,343 km network needs extensive repairs on many sections. Some track renewals have been carried out since 1994, including the section between Güemes and Pocitos in the north of the country, where the line has been partially rebuilt. Attention has also been given to the Salta–Socompa corridor, which is currently the only link with Chile, and where traffic has grown significantly. The aim is to allow train

speeds of 50 km/h over the entire network, which will call for investments of 100,000 pesos per km. It is estimated that around 20 per cent of all rail and 50 per cent of all sleepers may need to be replaced.

Traction and rolling stock

In 2004 FC BC SA operated around 120 diesel locomotives, many of them General Motors G22s.

Signalling and telecommunications

US-based Comsat Mobile Communications has supplied a satellite-based communications system to handle train movements from FC BC's control centre at Salta.

UPDATED

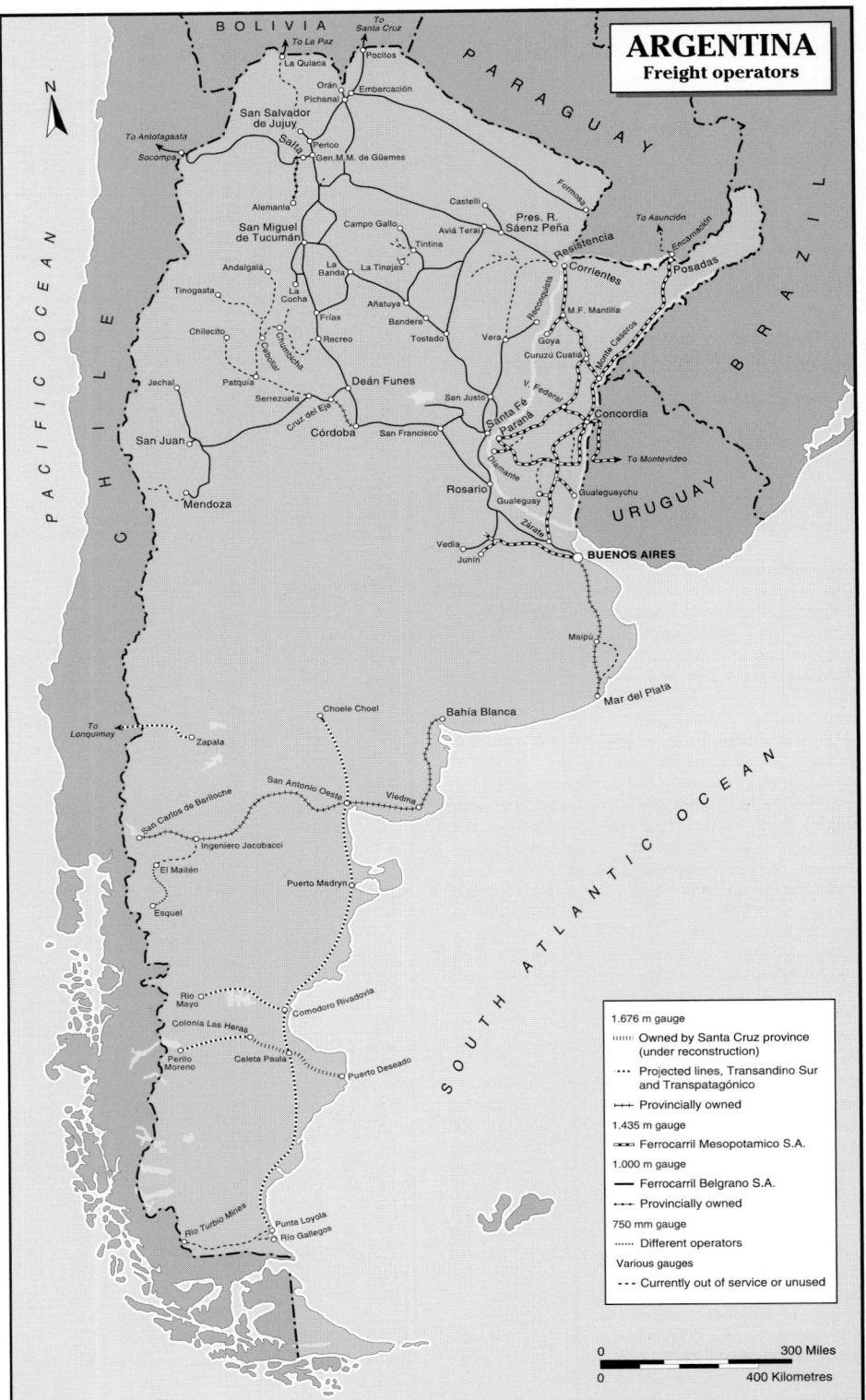

ARGENTINA
Freight operators

Map legend:

1.676 m gauge
||||||| Owned by Santa Cruz province (under reconstruction)
··· Projected lines, Transandino Sur and Transpatagónico
+++ Provincially owned
1.435 m gauge
▭▭▭ Ferrocarril Mesopotamico S.A.
1.000 m gauge
— Ferrocarril Belgrano S.A.
··· Provincially owned
750 mm gauge
····· Different operators
Various gauges
--- Currently out of service or unused

0   300 Miles
0   400 Kilometres

0058520

Ferrocarril Córdoba Central (FCC)

Estación R del Busto, 5009 Córdoba, CBA
Tel: (+54 351) 482 22 52
Fax: (+54 351) 482 22 52

Key personnel

Director: Horacio Cao

Gauge: 1,000 mm
Length: 158 km

Organisation

FCC operates freight and passenger (including the *Tren de las Sierras* tourist operation) services between the city of Córdoba and Cruz del Eje, on a route closed by FA in 1977. FCC came into being in response to the call for tenders issued by the province of Córdoba, and began operations on 5 December 1993. Services on the final 43 km stretch of line, from Capilla del Monte to Cruz del Eje, were reported as temporarily suspended in early 1998.

Passenger traffic	1997	1998	1999*
Passenger journeys	24,806	25,251	17,893
(*January-October)			

Passenger operations

Services are operated on Saturdays and Sundays all year round, with additional trains during the summer period and during school holidays in winter. In 1997 24,806 passengers were carried compared with 27,919 in 1996.

Delivery in 2000 of a Zanello-built railcar was reported as likely to be used for a daily return journey on the Córdoba–Capilla del Monte line.

Improvements to existing lines
Between July 1997 and June 1998 US$2.5 million were spent on infrastructure and rolling stock improvements. In 1999 a sleeper replacement programme was under way and in a subsequent phase of improvement work tunnels and bridges may undergo major repairs.

Traction and rolling stock
FCC owns three diesel locomotives (two unserviceable), three diesel railcars (two unserviceable) and 14 passenger coaches (nine unserviceable). In 1999 a refurbished Alco RSD-35 diesel locomotive was acquired from Córdoba workshops for use on the passenger service and a twin-unit Zanello railcar was delivered and due to enter revenue service in July 2000.

Elderly General Electric locomotive re-engined with an Alco power plant heading an FCC passenger train at Cosquin (Peter Lais)
0058586

Ferrocarriles Del Chubut (FdeC)

Carlos Pellegrini 841, 9210 El Maitén, Chubut
Tel: (+54 29 45) 49 51 90
Fax: (+54 29 45) 49 51 90

Passenger operations
This company operates one weekly regular service between El Maitén and Esquel on a 750 mm gauge branch line. Other steam-hauled services are run for tourists, including a daily Esquel–Nahuel Pan (20 km) train during summer months.

Traction and rolling stock
FdeC operates 6 steam locomotives and 10 passenger coaches.

Ferrocarriles Mediterráneos (FeMed)

Boulevard Perón 101, 5000 Córdoba
Tel: (+54 351) 42 82 14
Fax: (+54 351) 42 82 14

Key personnel
President: Julio Badra

Gauge: 1,676 mm

Organisation
This provincially supported regional passenger operator initiated services over Nuevo Central Argentina (qv) tracks between Córdoba and Villa María (140 km) in October 1997. Passenger services on this route had been withdrawn four years earlier.

Passenger operations
Two trains each day are provided between Córdoba and Villa Maria (142 km), in 1999 carrying more than 60,000 passengers. A proposal for a first class overnight service between these communities had still not been implemented in 2000, nor had a plan for a service class overnight train between Rosario and Buenos Aires. In 2000 a service between Córdoba and Alta Gracia (48 km) was expected to commence shortly, as were trains on other lines radiating from Córdoba.

Improvements to existing lines
FeMed has renovated Córdoba Mitre station, which is to form the hub of the city's rail and road public transport system. However, poor track conditions are reported to be hampering the development of FeMed services. Poor track, especially between Villa Maria and Rosario, has prevented the restoration of passenger services between Córdoba and Buenos Aires, with line speeds at some locations as low as 15 to 25 km/h. Córdoba province was expected to provide funding of peso1.2 million for track improvements in 2000 to raise line speeds to 60 to 70 km/h.

Traction and rolling stock
FeMed owns seven diesel locomotives (three unserviceable) and 60 passenger coaches, including six Pullman cars, sleeping and restaurant cars and some car-carrying wagons. Of these, 42 have been refurbished by FeMed and up to 12 more may be similarly treated.

Ferrocoop

Cooperativa de Trabajo de Olavarría Limitada
Pringles 3100, 7400 Olavarría
Tel: (+54 22 84) 44 05 15
Fax: (+54 22 84) 44 62 60

Gauge: 1,676 mm

Organisation
On 19 December 1996, Servicios Ferroviarios Patagónicos (SEFEPA) (qv) signed an agreement with the Ferrocoop co-operative to permit the latter to operate passenger services on the Bahía Blanca–Neuquén line for 30 years. Three services were envisaged: Cipolletti–Neuquén, Villa Regina–Plottier, and a daily train to Cinco Saltas. The co-op would assume responsibility for all costs and was expected initially to transport 4,800 passengers/day.

Traction and rolling stock
Initially services were to be provided by four three-car air conditioned dmus built by local supplier Zanello SA. Each is powered by two 360 kW Cummins engines. Ordered from Spanish National Railways in 1998 were four second-hand Class 593 dmus.

FerroExpreso Pampeano SA (FEPSA)

Avenida Córdoba 320, piso 4, C1054AAP Buenos Aires
Tel: (+54 11) 45 10 49 00
Fax: (+54 11) 45 10 49 30; 49 45

Key personnel
General Manager: Rodolfo Glattstein

Gauge: 1,676 mm
Route length: 4,953 km

Organisation
FEPSA was the first private enterprise to complete a major takeover of FA operations. FEPSA is 80 per cent owned by Coinfer, an Argentine/US investment consortium comprising four Argentine companies – Techint Compañía Técnica, Sociedad Comercial del Plata, Gesiemes and Riobank Internacional – and the Iowa Interstate Railroad. Iowa Interstate, which has a 2 per cent stockholding, is providing technical advice on operation, traction and rolling stock. The Union Pacific Railroad of the US is serving as technical consultant on telecommunications and computerisation. Argentine Railways retains a 16 per cent stockholding in FEPSA. The residual 4 per cent is held by FEPSA employees.

On 1 November 1991, FEPSA took over 5,094 route-km of FA's 1,676 mm gauge grain-carrying lines centred on the routes between Rosario (northwest of Buenos Aires) and Bahía Blanca and between Huinca Renanco and Bahía Blanca; the routes formed part of the Mitre, Sarmiento and Roca railways. FEPSA has an option to extend the 30-year concession by 10 years. In 1998, the company initiated renegotiation of its existing concession with the aim of restructuring its investment programme.

In 1997 FEPSA employed 994 staff.

Freight operations
FEPSA's principal cargoes include wheat, maize, soya beans, sorghum, sunflower seeds and the pellets and oils derived from them.

The US grain company Cargill has completed a rail-served grain storage facility at the port of Bahía Blanca with a storage capacity of 60,000 short tons. Improvement to warehousing facilities at Bahía

Blanca should allow FEPSA to begin importing fertiliser, thereby ensuring return loads for trains discharging export grain.

Experiments with double heading of 60-wagon block grain trains (formerly single-headed trains of 30 wagons were the norm) have shown that it is now possible for FEPSA to operate 4,700 tonne trains. Future traction policy will involve using small switching locomotives on branch lines and concentrating large wagon movements between a limited number of hub yards. In 1996, a co-operation agreement was signed with Ferrosur

Roca (qv) to integrate and co-ordinate freight services. Steel traffic is also handled jointly with Nuevo Central Argentino. Door-to-door services have also been developed.

Traffic (million)	1997	1998–99
Freight tonnes	3.428	2.606
Freight tonne-km	1,375,000	938,257

Traction and rolling stock
In 1997 FEPSA operated 47 diesel locomotives and around 1,600 wagons.

Signalling and telecommunications
Signalling is by cab radio and fax via a microwave network, permitting a track warrant system of train control and operation, replacing the former station-to-station staff, or rod, train movement authority. Communication between Bahía Blanca and Buenos Aires is by satellite.

Ferrosur Roca SA (FR)

Bouchard 680 – piso 8°, 1106 Buenos Aires
Tel: (+54 11) 43 19 39 00
Fax: (+54 11) 43 19 39 01
e-mail: ferrosur@impsat1.com.ar

Key personnel
General Manager: Sergio do Rego
Directors
    Commercial: Mario Casasco
    Administration and Finance: Ricardo Wagner
    Human Resources: Carlos Sánchez Obertello
    Operations: Gustavo Romera
    Operations Resources: Elbio Armanazqui

Gauge: 1,676 mm
Route length: 3,110 km

Organisation
In December 1992, the concession to operate the General Roca railway was awarded to the Ferrosur Roca consortium; it was the sole bidder. Cement manufacturer Loma Negra owns 65 per cent of the new company, with other members being Acindar, Petroquímica Comodoro Rivadavia, Decavial, Banco Francés and the Asociación de Cooperativas Agrarias. Operations began on 11 March 1993. Although initial management of the company was Canadian, local managers have since taken over. At the end of 1998 FR employed 720 staff.

The Roca railway extends due south from Buenos Aires and serves the country's main agricultural belt. Three deep water ports, Bahía Blanca, Quequén and San Antonio Oeste, are also served by the railway. There is an inland branch which runs due west of Bahía Blanca and crosses the important fruit-growing area located in the upper valley of the Río Negro, terminating at Zapala.

In 1996, FR and FEPSA decided to co-ordinate their activities, with FEPSA personnel moving into FR's modern headquarters building. FR closed its own workshops in Bahía Blanca and switched all activities to FEPSA's own facilities in the same city. Motive power now moves freely between the two networks, as do wagons, while radio frequencies have been harmonised.

Freight operations
FR traffic has maintained a steady climb since it took over the concession in 1993, and in August 1997 for the first time lifted over 400,000 tonnes of freight. The main traffic base consists of construction materials, which account for around 75 per cent of total tonnage, petroleum, petrochemicals and hazardous cargos, such as caustic soda, and grain. Polyethylene traffic between Bahía Blanca and Buenos Aires will reach 270,000 tonnes in 2000 and FR plans to increase this figure to 370,000 tonnes in two years. However, aggregates traffic has fallen from 400,000 to 255,000 tonnes per month. The company has a contract to move 120,000 tonnes of cement and 250,000 tonnes of clinker annually for Loma Negra, Argentina's main cement producer. Grain is conveyed to the port of Quequén and to flour mills.

In February 1996 FR initiated its Frigotren service, conveying fresh fruit from the Alto Valle del Río Negro to Buenos Aires for both domestic consumption and for export. The train formation includes a generator vehicle to provide refrigeration power during the train's 40 hour journey.

Ferrosur Roca's Frigotren service for fresh fruit includes a generator car in the train formation to provide power for refrigeration units
0010762

FR also carries intermodal traffic destined for Chile via the Pino Hachado Pass 'Corredor Bioceánico' route.

Two-person locomotive crews, consisting of a driver and conductor (the latter to handle all point work), operate all trains.

Traffic (million)	1997	1998	1999
Freight tonnes	4.500	4.125	4.364
Freight tonne-km	1,635	1,547	1,708

New lines
FR is providing technical project management services for the Argentine section of the Southern Trans-Andean Rail Link (STAR) scheme. STAR involves the construction of a new 220 km link between the 1,676 mm gauge railheads at Lonquimay in Chile and Zapala in Argentina.

Construction of the 160 km Argentine section is to be undertaken using public funds, while the Chilean government has attempted to attract private finance (principally in Europe) for the 60 km section of the link that would lie within its borders. In 1994 the Argentine province of Neuquén contracted consulting engineers Bechtel to produce a revised scheme for the link; Neuquén provided US$60,000 towards the study, with the US Trade Development Agency contributing US$380,000.

It is hoped that the completion of the Lonquimay–Zapala link would create a freight route linking Talcahuano on the Pacific with Bahía Blanca on the Atlantic. Annual freight traffic between Chile and Argentina across the southern Andes has been reckoned to be in the order of 1.7 million tonnes.

Improvements to existing lines
FR plans to invest US$173 million in the first 15 years of its concession, although it will have to pay a total of US$47 million to the government in fees during the full 30-year life of the concession. Eventual maximum speed on trunk routes has been pitched at 68 to 77 km/h. Investment in track improvements will total US$86 million.

Track improvements to the main line in the province of Buenos Aires were well advanced in

2000, with initial sections already completed. The line carries around 80 per cent of its traffic.

Traction and rolling stock
At the end of 1998 the FR locomotive fleet comprised 57 units, including 25 GM 1687 kW GT22CW machines built 1972–86. A further three refurbished locomotives have been procured, and in 2000 four additional 1,640 kW locomotives were delivered by Gevisa. These were to be deployed on cement and chemicals traffic.

At the end of 1998 FR operated 2,200 freight wagons. Recent fleet developments included the refurbishment of 84 tank wagons for bulk petroleum products, conversion of existing vehicles into 300 cement wagons and 120 container flat wagons, and the conversion of six covered wagons for the secure transport of hazardous materials. Refurbishment and conversion of an initial batch of 100 cement wagons was in progress in 2000.

Signalling and telecommunications
The previous station-to-station staff authority system of operation has been replaced by one of track warrants, developed by CN. To make this possible, much initial investment has gone into train radio, end-of-train telemetry, microwave communications and computer data processing to permit use of a Computer-Assisted Manual Block System (CAMBS). Dispatching takes place at Olavarría.

Track
The main trunk route between Buenos Aires and Bahía Blanca has stone-ballasted track with continuous welded rail. Much of the rest has been poorly maintained in recent years.

Rail: Some 2,700 km of track is laid with 49 and 50 kg/m rail, while 1,370 km is 42 kg/m. The remainder of the network uses 28–36 kg/m
Sleepers: Timber and steel
Spacing: 1,394–1,722/km on plain track; 1,474–1,722 on curves
Min curve radius: 160 m
Max gradient: 1.4%
Max axleload: 20 tonnes

Ferrovías

Avenida Ramos Mejia 1430 – piso 4°, 1104 Buenos
Aires
Tel: (+54 11) 45 11 88 33
Fax: (+54 11) 45 11 88 43

Key personnel
President: Dr Gabriel Romero
Vice-President: Osvaldo R Aldao
General Manager: Felix Imposti
Company Accountant: Hector S Cimo

Gauge: 1,000 mm
Route length: 132.6 km

Organisation
The concession for FEMESA's diesel-operated
Belgrano North line from Buenos Aires' Retiro
station to Villa Rosa was awarded to the Ferrovías
consortium, led by Leddevi Construcciones, which
had Portuguese Railways (CP) and Barcelona
Metro as operators. CP was subsequently forced
to withdraw, due to budgetary cutbacks at home;
a 1 per cent share in Ferrovías was acquired by
Transurb Consult of Belgium, the new technical
advisor to the group. Other members of the
consortium include Seminara Constructores and
the bus operator Cooperativa de Transportes
Automotores de Cuyo.

On 3 February 1994 a 10-year operating contract
was signed by government and Ferrovías for
the Belgrano North route. Operations began on
30 April 1994. In 1998–99 Ferrovías renegotiated
its concession with a view to obtain a 20-year
extension in return for extensive investment (see
'Improvements to existing lines').

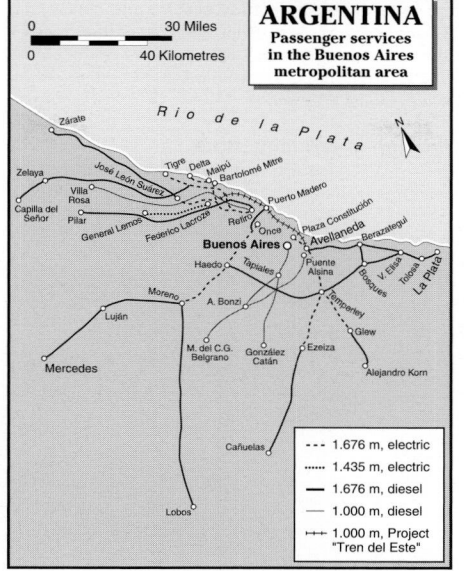

0058519

Passenger operations
Initial efforts to improve service quality focused
on reducing the number of cancellations,
improving passenger security and providing
the travelling public with cleaner and better-lit
trains. Ferrovías claims to have reduced fare
evasion by some 30 to 40 per cent by carrying out
more ticket inspections, and plans to introduce a
closed-station system. Ticketing was to switch to

preprinted tickets issued by PCs, with electronic/
magnetic tickets to work barriers planned for the
long term.

In 1997 Ferrovías carried 32.3 million fare-paying
passengers.

Improvements to existing lines
A subsidy of US$196 million was negotiated for
the concession period, the bulk of which was
earmarked for infrastructure improvements. Track,
level crossings, signalling and telecommunications
were felt to be areas of particular concern.

In 1995, a contract worth US$23 million was
signed with local companies for the renewal of
18 route-km of double track between Buenos Aires
and Carapachay. Ferrovías has also undertaken a
refurbishment programme covering the 22 stations
it serves, financed by its own resources to the tune
of US$1.5 million.

Under the terms of a renegotiated concession,
Ferrovías plans to electrify the 62 km Retiro–Villa
Rosa line, on the Belgrano North system,
reconstructing all 22 stations and providing a fleet
of new emus.

Traction and rolling stock
Ferrovías operates a fleet of 21 General Motors
G22 CU diesel-electric locomotives. At the end
of 1997, the Ferrovías passenger fleet comprised
101 coaches.

Signalling and telecommunications
The Ferrovías system is equipped with automatic
colour-light signalling.

Metropolitano

Avenida Santa Fe 4636 – piso 2, C1425BHV Buenos
Aires
Tel: (+54 11) 47 78 58 00
Fax: (+54 11) 47 78 58 78

Key personnel
Managing Director: Carlos R Beltrán Simó

Gauge: 1,676 mm, 1,000 mm
Route length: 316.3 km, 66.3 km; 119 stations
Electrification: 47.5 km at 25 kV 50 Hz AC

Organisation
The 10-year concessions to operate suburban
services on the Roca, San Martín and Belgrano
South routes were awarded in January 1993 to
the Trainmet consortium, consisting of Empresa
Argentina de Construcciones, Ormas SA, Román SA,
DGT Electronica and 63 independent bus operators.
US operator BART and JARTS of Japan were to
act as partners. Trainmet, which subsequently
changed its name to Transportes Metropolitanos,
began operating the three franchises as TMS, TMB
and TMR between April 1994 and January 1995.
Subsequently, the company has operated under

the name Metropolitano. Trainmet now owns
51 per cent of the capital stock but in mid-2000 a
legal case was in progress with Ormas SA.

In 1999 Metropolitano began renegotiating its
concession to obtain a 20-year extension to 2023 in
return for an investment programme amounting to
an estimated US$1.5 billion.

Passenger operations
TMS
Trainmet began operating San Martín suburban
services as Transportes Metropolitanos San Martín
(TMS) on 1 April 1994. The TMS system comprises a
diesel-worked 1,676 mm gauge route (56.3 route-km)
between Buenos Aires (Retiro) and Pilar.

TMB
Operation of Belgrano South suburban services
as Transportes Metropolitanos Belgrano (TMB)
began on 1 May 1994. The TMB system comprises
66.3 route-km of diesel-worked 1,000 mm gauge
lines.

TMR
Operation of Roca suburban services as Transportes
Metropolitanos Roca (TMR) began on 1 January
1995. The system comprises 259 route-km of

1,676 mm gauge lines, only 202 km of which carry
passenger services, radiating from Buenos Aires'
Plaza Constitución terminus. Electrified routes
serve Temperley, Glew and Ezeiza, and services to
Cañuelas, Alejandro Korn, Berazategui, La Plata
and Haedo are diesel-worked.

Combined ridership on the three systems rose
from 88 million passenger journeys in 1994 to
207 million in 1997.

Traction and rolling stock
In 1997 the Metropolitano fleet comprised 68
1,676 mm and 20 1,000 mm gauge diesel
locomotives mostly of General Electric and General
Motors designs, 295 1,676 mm gauge coaches,
plus 58 baggage cars, and a further 91 metre-
gauge coaches. The Roca system emu fleet totalled
187 units, and a dmu was also in operation on the
metre-gauge system.

In 1997, Metropolitano received 10 Canadian-built
General Motors diesel locomotives, of which five
were for broad-gauge operation and five for metre-
gauge. Other broad-gauge diesel locomotives were
procured by leasing. More recently, two new U20C
units for the metre-gauge Belgrano Sur section
were supplied by Gevisa.

Metrovías SA

Bartolomé Mitre 3342, 1201 Buenos Aires
Tel: (+54 11) 49 59 68 00
Fax: (+54 11) 48 66 30 37
e-mail: info@metrovias.com.ar
Web: http://www.metrovias.com.ar

Key personnel
General Manager: Roberto Macías
Planning Manager: José A Barbero

Gauge: 1,435 mm
Route length: 25.6 km
Electrification: 25.6 km at 600 V DC

Organisation
In 1993, the Metrovías consortium was awarded
a 20-year concession to run electric suburban
services on the Urquiza line as well as the 36.5 km

Buenos Aires metro system. Headed by Roggio
SA (a construction company based in Córdoba),
Metrovías originally included Cometrans (an
association of 30 independent bus companies
in Buenos Aires), with Burlington Northern and
Morrison Knudsen of the USA providing technical
assistance.

Metrovías, in the guise of Trenes de Buenos
Aires (TBA) (qv), was also awarded concessions
to operate electric suburban services on the Mitre
and Sarmiento lines; it had originally been hoped
that TBA operations would start in June 1994.
Following a change in TBA's corporate structure,
which involved Morrison Knudsen selling its
holding in Metrovías to Roggio SA (giving the latter
a 72 per cent stake), TBA operations on the Mitre
and Sarmiento lines began on 29 May 1995. With
Burlington Northern Santa Fe now a member of the
TBA consortium, Metrovías has received technical
assistance from Transurb Consult of Belgium.

In early 1999, Metrovías was renegotiating its
contract with a view to a 20-year extension.

Finance
The company is to receive US$130 million in annual
subsidies to operate both the Urquiza service and
that of the metro system. It was hoped that cost
reduction and control of rampant fare evasion
would permit the subsidy to be eventually used
to upgrade the service on offer. The concession
contract allows for fare rises of 3 per cent per year
once service and quality targets have been met.

Passenger operations
Having begun operations on the Urquiza route
on 1 January 1994, Metrovías recorded a total of
22.53 million paying passengers in that year as a
whole, with 9.23 million train-km operated. The
number of passengers carried was 35 per cent up
on 1993. In 1995, 23.26 million paying passengers

were recorded and 9.67 million train-km operated, with demand 39 per cent up on 1993. In 1997, 25.2 million passengers were carried.

Improvements to existing lines
Metrovías is to provide US$436 million of investment for the Urquiza route and the metro system over the lifetime of the concession. Major infrastructure work on the Urquiza route required by the concession contract includes track renewal, level crossing elimination and improvements to the third-rail electrification system and its cabling. By the start of 1996, Metrovías had begun track renewal, cabling and third-rail work costing US$8.2 million, US$1.9 million and US$0.4 million respectively.

Traction and rolling stock
Metrovías trains are usually formed of three two-car emu trainsets, with each trainset comprising one motor and one trailer car. As from November 1996, Spanish rolling stock builder CAF assumed all responsibility for the maintenance of emu cars, using workshops at Lynch and Rubén Dário.

Metrovías also operates three diesel locomotives.

Nuevo Central Argentino SA (NCA)

Avenida Alberdi 50, S2013EPN Rosario, Santa Fe
Tel: (+54 341) 437 65 61; 65 64
Fax: (+54 341) 439 23 77
e-mail: info@nca.com.ar
Web: www.nca.com.ar

Key personnel
President: Miguel Acevedo
General Manager: Horacio Díaz Hermelo
Operations Manager: Ernesto Gutierrez
Infrastructure and Track Maintenance:
    Daniel Zurbriggen
Logistics Manager: Hugo Zanelli

Gauge: 1,676 mm
Route length: 5,100 km

Organisation
On 2 July 1992 control of Argentine Railways' General Mitre railway passed to Nuevo Central Argentino for a period of 30 years; it was the second concession to be granted. The Mitre network connects six industrial cities in the North Central region (Tucumán, Córdoba, Rosario, San Nicolás, Campana and Río Tercero) with Buenos Aires. It also serves Argentina's most fertile grain-producing area, the Pampa Húmeda, providing access to several bulk-commodity ports on the Paraná river.

The concessionaire is a consortium of two Argentine companies headed by Aceitera General Deheza (AGD), an oil and seed producer which is also one of the NCA's main customers; and the Asociación de Cooperativas Agrarias Argentinas, a major grain co-operative.

Originally, the US-based regional railroad Montana Rail Link (a Burlington Northern Santa Fe subsidiary), the Washington company RBC Associates and the transport management company Anacostia & Pacific were associates.

In 2004 NCA employed 1,202 staff.

Passenger services
While NCA is a freight railway, passenger services provided by other companies operate over its

NCA train of hopper wagons   0122592

tracks. In 2004 Ferrocentral was running four trains each week over NCA tracks between Córdoba and and Buenos Aires via Rosario (700 km). Trenes de Buenos Aires (TBA) also runs two trains a week over NCA lines between Retiro (Buenos Aires) and Rosario, the Zárate–Retiro portion operating under TBA's own concession.

A service over NCA tracks linking Tucumán and Buenos Aires was active until 1996, discontinued until 1997, resuming again until March 2000 and finally stopped in 2004.

Freight operations
NCA ran its first train on 23 December 1992. In 1997 4.8 million tonnes were carried.

Traffic growth has continued, and by 2004 some 8.3 million tonnes annually were being carried. Grain products are the most important commodity handled: 2004 was the eighth consecutive year in which more than 1.2 million tonnes were transported, an achievement assisted by the rehabilitation of previously unserviceable wagons. Movements of copper concentrate for a major customer, Minera Alumbrera, totalled 800,000 tonnes during 2002, representing 11 per cent of NCA traffic. Other major commodities handled include aggregates, slag and clinker, which accounts for 20 per cent of traffic. NCA also provides a total logistics service that takes into account the needs of production processes, warehousing, port facilities and export requirements. Other recent traffic developments include the transport of imported car components to assembly plants in Córdoba and the movement of raw materials for the cement industry from Córdoba to the Campana plant in Buenos Aires. Possible future traffic flows were being studied in 2004, including cars, fertiliser and copper concentrate from other mines in northwest Argentina.

NCA has also been pursuing a policy of upgrading yards and improving access to its network via new branches and sidings, as well as focusing on the reliability of its services and equipment.

Traffic (million)	2002	2003	2004
Freight tonnes	7.273	8.082	8,326
Freight tonne-km	3,380	3,758	3,708

Improvements to existing lines
A 1,470 km core network consisting of main lines linking Buenos Aires to Rosario and Tucumán, and Rosario to Córdoba, now have stone ballast, while many wooden sleepers and sections of rail have been replaced. NCA has been pursuing a programme of renewing or rehabilitating branch lines and yards.

In 2004 NCA absorbed a previously closed branch of some 260 km from Sumampa to Villa del Rosario, linking Córdoba and Santiago del Estero provinces and increasing the total network length from 4,836 to 5,100 km.

NCA mixed freight service powered by an Alco locomotive   0122591

Traction and rolling stock

The fleet available for line service at December 2004 comprised 72 locomotives plus several GAIA 970 kW diesel-electric locomotives that have been refurbished for shunting operations at main yards and terminals.

The NCA wagon fleet amounted to 5,254 vehicles at the end of 2004, comprising grain hoppers (around half the fleet), tank wagons, bulk cement and flat wagons, and mineral hoppers, reflecting the system's diversified traffic base. The refurbishment of unserviceable wagons, inherited at the time the concession was won, has continued

with further projects and studies being undertaken in this area.

Heavy repair shops are located at Villa Diego near Rosario, where a further two smaller shops also handle minor repairs. Other smaller shops are also to be found at Córdoba and Tucumán.

Signalling and telecommunications

Before assuming operational control of the Mitre railway in 1992, NCA installed a VHF cab-signalling system on its main lines to Tucumán and Córdoba, boosted by repeater stations across the network.

Two-person locomotive crews handle 2,000-tonne trains consisting mostly of 40 wagons, and all movements are controlled from a computer-aided dispatching centre located in Rosario. An IBM AS/400-based wagon tracking system is also in operation.

NCA technicians have developed a system for the remote radio control of level crossing gates.

In 2005 a GPS satellite tracking system was under development to determine locomotive and train location.

UPDATED

Servicios Ferroviarios Chaqueños (SEFECHA)

García Meron 5800, 3514 Fontana, Chaco
Tel: (+54 37 22) 47 55 50
Fax: (+54 37 22) 47 55 50

Key personnel

President: Ing Manuel Emilio Vecchi
Vice-President: Sergio Gabriel Peyrano
Chief Engineer: Mario Acevedo

Organisation

SEFECHA was set up in 1997 to provide passenger services over some 500 km of metre-gauge routes in Chaco province, mostly operating over track owned by Ferrocarril Belgrano Cargas (qv).

Passenger operations

SEFECHA's initial daily service between Cacuí and La Sabana (117 km) proved successful and in 1998 new services were added from Presidencia

Roque Saenz Peña to Chorotis (188 km) and to Taco Pozo (311 km). The latter is operated three times weekly in each direction, serving communities in the extreme northeast of Santiago del Estero province. In 1999, the La Sabana service was extended to Los Amores, in the extreme north of Santa Fe province and an extension south to Gobernador Vera was likely in 2000. A suburban service from Puerto Barranqueras via the Chaco provincial capital, Resistencia, to Cacuí, Puerto Tiros (16.5 km) was also introduced in 1999 with 11 trains daily in each direction. SEFECHA expected to increase this to 16 in each direction to create virtually a metropolitan service. The largest obstacle to the reintroduction of services is reported to be numerous unmanned level crossings on the track, which is owned by Ferrocarril Belgrano Cargas (qv).

Improvements to existing lines

SEFECHA has funded the installation of automatic barriers at two level crossings in Resistencia

and in 1999 ordered similar equipment for an additional 26 sites within the city's boundaries. Plans exist to upgrade the 10.5 km section of line between Resistencia and Puerto Barranqueras in collaboration with FBC, and additional passing loops are planned at two intermediate stations.

Two stations and 15 stops have been created for the Puerto Barranqueras–Cacuí service.

Traction and rolling stock

Four two-car MAN dmus and six Ferrostaal railcar-trailer sets have been acquired second-hand from Spanish local operator SFM in Mallorca and refurbished by Igarreta in Buenos Aires. The Ferrostaal units have been equipped with air conditioning and audio passenger information systems.

Servicios Ferroviarios Patagónicos (SeFePa)

Laprida 240, 8500 Viedma, Rio Negro
Tel: (+54 29 20) 42 21 30
Fax: (+54 29 20) 42 74 13

Key personnel

Managing Director: German Jalabert

Gauge: 1,676 mm, 750 mm
Route length: 817 km, 237 km

Political background

SeFePa was established by the province of Rio Negro to operate passenger services within its boundaries upon the withdrawal of federal support for intercity passenger trains.

Freight services are also operated over SeFePa's southern region, particulary to San Carlos de Bariloche. In late 1997 it was announced that through freight services were to be reintroduced between Buenos Aires and Bariloche. These are operated by Ferrocargas del Sur.

During 1999 SeFePa passenger operations were to be taken over for a 10-year period by Ferrotransportes Patagónicos SA (FTP), a subsidiary of Mar del Plata-based bus operator Platamar. Initially SeFePa's ex-FA rolling stock will be used, but this is expected to be replaced by Zanello-built railcars and dmus based at Córdoba. Service frequencies are to be increased and line speeds will be raised after infrastructure improvements have been carried out. The journey time for the 819 km San Carlos de Bariloche–Viedma service is expected to be reduced from 17 to 12 hours after infrastructure rehabilitation. At a later date FTP is also expected to take over SeFePa freight services, leaving SeFePa responsible only for infrastructure maintenance.

Passenger operations

There is no longer a through service between Plaza Constitución (Buenos Aires) and San Carlos de Bariloche. UEPFP trains run as far as Carmen de Patagones and SeFePa provides services from this latter point to San Carlos de Bariloche with its own equipment.

Former Spanish National Railways Class 321 diesel-electric locomotive in service with SeFePa at San Carlos de Bariloche (Bryan Philpott)　　　0023712

SeFePa has taken over responsibility for the Ingeniero Jacobacci-El Maiten portion of the 750 mm gauge route from Ingeniero Jacobacci to Esquel, although the 'Old Patagonian Train' remains out of service in Rio Negro province. There are plans to transfer provision of tourist train services to a private-sector operator but no potential candidates have expressed an interest. Ingeniero Jacobacci is also on SeFePa's Viedma–San Carlos de Bariloche 1,676 mm gauge route. Operated between El Maiten and Esquel by the province of Chubut, the 402 km steam-operated Ingeniero Jacobacci-Esquel line both attracts tourists and provides essential transport services in a very remote part of Argentina.

SeFePa was reported to be enjoying healthy traffic in 1996, with ridership up 50 per cent on Ingeniero Jacobacci–Bariloche trains, and 100 per cent between Bariloche and Viedma.

In December 1999, SeFePa obtained a national government subsidy of US$280,000 for upgrading permanent way and rolling stock refurbishment. In 2000, the company was operating a twice-weekly passenger train between Viedma, on the Atlantic coast, and San Carlos de Bariloche, receiving an annual subsidy of US$100,000 from the provincial Finance Ministry to maintain services.

Traction and rolling stock

In 2000, SeFePa was still seeking motive power and rolling stock from former state-owned railways. Nine diesel locomotives and some 200 freight wagons were due to be transferred to provincial systems.

In 1994 Rio Negro purchased for Pta312 million a batch of used rolling stock from RENFE. Three Class 321 Alco diesel-electric locomotives, four sleeping cars, six first-class coaches, three luggage/generator vans and two diesel multiple-units were purchased.

Trenes de Buenos Aires (TBA)

Avenida Ramos Mejía 1358 – piso 2°, 1104 Buenos
Aires
Tel: (+54 11) 43 17 44 00
Fax: (+54 11) 43 17 44 16

Key personnel
President: S C Cirigliano
Vice-President: J Crawford
General Manager: Roberto Agosta
Director, Operations: H Payne

Gauge: 1,676 mm
Route length: 419 km
Electrification: 93.9 km at 800 V DC

Organisation
Trenes de Buenos Aires (TBA) was formed by the
members of the Metrovías consortium (qv), which
had been awarded the concessions for suburban
services on the Mitre and Sarmiento systems in
1993. TBA's shareholders are Morrison Knudsen
(41.65 per cent), Cometrans (41.65 per cent) and
Burlington Northern Santa Fe (16.7 per cent).

In 1998 TBA was granted a 20-year extension
to its current concession, which runs until 2005.
The new agreement provided for investments of
US$2.2 billion in the Mitre and Sarmiento lines and
an elimination of government subsidy by 2000.

Passenger operations
TBA began operating the Mitre and Sarmiento
systems on 29 May 1995. FEMESA had carried
37.8 million fare-paying passengers on the Mitre
system and 59.9 million on the Sarmiento system
in 1994. However, by 1996, TBA had boosted
patronage on the Mitre line to 69.8 million
passengers and on the Sarmiento route to
99.3 million. In 1997, combined ridership on the two
systems had risen to 192.3 million.

In 1997 a US$25 million magnetic ticketing
system was introduced.

Also in 1997 TBA launched a three-class daily
service between Retiro (Buenos Aires) and Rosario
Sur (294 km). Since February 2000, these trains
have been running into Rosario Norte station
and lightweight dmus have been substituted for
locomotive-hauled stock during weekdays.

*Lightweight diesel trainset comprising two power cars and a trailer at Rosario Norte with the daily
return working to Retiro (Buenos Aires)* (Angel Ferrer)   0089092

Improvements to existing lines
Under its original 10-year concession, TBA was
to invest US$405 million. Rolling stock was the
immediate priority, but TBA also made provision
for track renewal, an extensive programme
to eliminate level crossings, and station
modernisation. World Bank funding was secured
by the government in 1995 for the rebuilding of the
Caballito–Liniers section of the Sarmiento system
to eliminate 18 level crossings.

Under the concession-extension agreement
finalised in 1998, TBA was to electrify the
Bancalari–General Rodriguez and Mariano Acosta
branches on the Sarmiento system and extend
Mitre line electrification to Maquinista Savio.
Track improvements were also planned between
Buenos Aires and Rosario, and a major station
refurbishment programme was to be undertaken.

TBA signed a US$95 million contract with
Siemens Argentina in 1995 for the renewal and
maintenance of traction power supply equipment,
including substations at Floresta and Morón on the
Sarmiento network.

In October 1997 TBA reopened a 26 km line
linking the city terminal at Castelar and Puerto
Madero, including a restored 5 km underground
link with two intermediate stations. The service
is dmu-operated, and is part of a TBA strategy to
develop a regional express network in the Buenos
Aires area by linking the Mitre and Sarmiento
networks via a cross-city tunnel link.

Traction and rolling stock
TBA inherited 367 electric multiple-unit cars,
43 hauled passenger coaches and 96 diesel
multiple-units from FEMESA. Refurbishment of
362 emu cars and 43 coaches had been made
an investment priority, with 130 emu cars to be
refurbished by the end of 1997. These include
Toshiba-built emus used on the Mitre electrified
network. Work on these commenced at TBA's
Castelar workshops in 1997. Now known as UMAP
(Unidad Múltiple Argentina de Pasajeros) units,
each three-car set has been fitted with streamlined
cab-ends, new seating and air conditioning.

In 1997 the TBA fleet comprised 29 diesel
locomotives (25 serviceable), 78 coaches, 418 emu
cars 374 serviceable), 11 dmus. Recent locomotive
acquisitions include 15 Class 313 Co-Co diesel-
electric locomotives purchased from RENFE of
Spain in 1995.

Future rolling stock procurement planned under
TBA's 1998 concession extension agreement
included 492 emu cars and new vehicle for diesel-
operated routes.

Second-hand dmus have been ordered from
Spanish National Railways and some of these were
thought likely to work on the Retiro–Rosario Norte
route.

Signalling and telecommunications
TBA drivers have been issued with radios to
communicate with dispatchers, and on lightly
used diesel-operated routes track warrant control
has replaced manual block signalling. From
1997, TBA hoped to replace semaphore signals
on 5 km of the Mitre main line from Retiro to
Empalme Maldonado and carry out signalling
improvements at Once, Caballito, Flores,
Floresta, Liniers and Moreno on the Sarmiento
system.

Refurbished Toshiba-built TBA Mitre line emu (Peter Lais)   0058588

Trenes & Turismo SA

Caseros 441, 4000 Salta, STA
Tel: (+54 387) 421 63 94
Fax: (+54 387) 431 12 64
Buenos Aires office:
Esmerelda 1008, 1007 Buenos Aires (CF)
Tel: (+54 11) 431 88 71

Key personnel
Chairman: Julio Ruiz de los Llanos
Vice-President: Corina Lewin
Directors: Miguel Desimone,
   Eduardo Lewin

Gauge: 1,000 mm
Route length: 218 km

Organisation
Trenes & Turismo SA is a consortium comprising
La Veloz del Norte and Dinar, both of which
are involved in tourism and road transport. It
previously traded as Movitren SA. The company
operates the famous 'Tren a las Nubes' (Train to
the Clouds) tourist service between Salta and the
Viaducto La Polvorilla, 4,197 m above sea level,

on the Salta-Socompa line owned by Ferrocarril Belgrano Cargas (FBC) (qv). FBC operates freight trains and one weekly passenger service over the entire 570 route-km to and from Socompa. As well as the Tren a las Nubes, which has been operating throughout the year since the summer 1998/99 timetable, Trenes & Turismo also runs the 'Tren a la Quebrada del Inca' (Train to the Valley of the Incas) several times during winter months between Salta and Gobernador M Sola (92 km), on the second zigzag of the line.

As Movitren, the company received its first concession to run the Tren a las Nubes service from Argentine Railways in 1992. In mid-1997 a renewal of the concession until 2026 was granted.

The company employs 60 staff.

Passenger operations

In 1999, 27,890 passengers were carried, compared with 24,000 in 1996. Both Tren a las Nubes and Tren a la Quebrada del Inca services convey first class coaches and a restaurant car and are staffed with hostesses and medical personnel, the latter necessary to deal with altitude sickness among passengers.

A third tourist train, the 'Tren al Portal Andino' (Train to the Gateway to the Andes), is planned over part of the line to Viaducto la Polvorilla between Salta and Campo Quijano (41 km), but in early 1999 a date for its introduction had yet to be announced. Also on Trenes & Turismo's agenda is the introduction of the long-planned international service, the 'Tren del Sol' (Train to the Sun), from

Salta to Santa Cruz de la Sierra, in eastern Bolivia, but Ferrocarril Oriental of Bolivia is reportedly also keen to start a similar service. Trenes & Turismo is understood also to be planning to bid for a suburban service from Cerillos through Salta to a station to the south of the provincial capital.

Traction and rolling stock

Trenes & Turismo owns 14 coaches, including two baggage cars, an air conditioned restaurant car and a Pullman bar coach. The locomotive usually employed is a GM GT-22CU owned and maintained by FBC and leased by Trenes & Turismo. The procurement of new locomotives and coaches has been considered, as well as railcars or railbuses for local services.

Tucumán Ferrocarriles SA (TuFeSA)

Corrientes 1075, 4000 San Miguel de Tucumán, Tucumán
Tel: (+54 381) 430 38 95
Fax: (+54 381) 430 19 09

Key personnel

General Manager: Viviana Totongi

Political background

Owned and funded by regional government, TuFeSA in October 1997 reinstated passenger services between Buenos Aires and Tucumán (1,170 km) after a one-year suspension. Trains ran

over 1,676 mm gauge tracks owned by Nuevo Central Argentino (qv), and were to operate without subsidy.

However, following the death in November 1998 of its President, David Giménez, TuFeSA ran into serious financial difficulties and in March 2000 the provincial authority withdrew the concession. Services ceased and their resumption was expected to be delayed. By April 2000, four consortia had expressed an interest in taking over the concession but no agreement had been concluded.

Improvements to existing lines

TuFeSA planned to take over the long-disused Tucumán–Juan Bautista Alberdi line (100 km) to rehabilitate it for local services.

Traction and rolling stock

TuFeSA invested in the refurbishment of a fleet of 36 coaches taken over from Tucumán province. Some of these were destroyed in an accident in January 1998, leaving TuFeSA with 24 serviceable vehicles.

TuFeSA also owns six diesel locomotives (four unserviceable) and planned to lease two GT-22 machines from Ferrocarril Belgrano Cargas and equip them with broad-gauge bogies.

Unidad Ejecutora del Programa Ferroviario Provincial (UEPFP)

General Hornos 11 – piso 4°, 1154 Buenos Aires
Tel: (+54 11) 43 05 51 74
Fax: (+54 11) 43 05 59 33

Key personnel

General Manager: Guillermo Crespo
General Co-ordinator: Dr Alberto Trezza

Gauge: 1,676 mm
Route length: 793 km

Political background

UEPFP was established by Eduardo Duhalde, governor of the province of Buenos Aires, after the federal government ceased to support long-distance passenger services in March 1993. The company initially contracted Ferrocarriles Argentinos to run services, and on 27 August 1993 began operations itself with a fleet of ex-FA locomotives and coaches.

UEPFP now runs services supported by Buenos Aires and La Pampa provinces. It owns the Buenos Aires–Mar del Plata and General Guido–Pinamar lines, totalling 412 km.

Traffic (million)	1997	1998	1999*
Passenger journeys	2.4	1.87	1.7

* 1 Jan–31 Oct

Passenger operations

UEPFP passenger operations are marketed as 'Ferrobaires' and comprise the Atlantic and Pampas zones. The Atlantic zone comprises services from Buenos Aires to Mar del Plata, Pinamar, Tandil, Quequén, Necochea, Bahía Blanca, Carmen de Patagones and Bolivar; the Pampas zone comprises services from Buenos Aires to Darregueira, Santa Rosa-Toay, General Pico, Pasteur, Cuenca, Iriarte and Rojas.

The Atlantic zone is considered to be potentially profitable, serving coastal resorts south of Buenos Aires and attracting much holiday traffic. To serve the principal resort of Mar del Plata, UEPFP operates 'El Marplatense' featuring refurbished air conditioned rolling stock. By late 1994, eight trains were in operation daily (with an extra return service at weekends) in each

UEPFP's 'El Bahiense' overnight service at Bahía Blanca Sud station behind a General Motors locomotive (Peter Lais)   0058590

direction between Buenos Aires and Mar del Plata, covering the 400 km in 4 hours 50 minutes. In 1999 it was reported that UEPFP had ordered Talgo coaches for service on this route. After completion of track improvements, these vehicles would allow speeds to be raised to 160–170 km/h, cutting the end-to-end journey time to under three hours.

On 21 May 1994 UEPFP reopened the 72 km General Guido–General Madariaga route, closed in 1978. A new branch to Pinamar opened in 1996 and the company also reintroduced services on the General Madariaga–Vivoratá line using 80 km/h dmus. Ferrobaires declare the Pinamar service a great success, exceeding all expectations, with 27,000 tickets sold in January 1997 alone. Buenos Aires–Pinamar intercity services resumed in November 1997 following a programme to reinstate track on the final 21 km of the 101 km General Guido–Pinamar line. In 1997/98 a service of two return trains a day was scheduled to complete the 346 km journey in 4 hours 35 minutes.

New lines

In 1996, UEPFP and Spanish National Railways began conversations regarding the creation of a high-speed rail link between Rosario and

Mar del Plata using the 'ring' line around western and southwestern Buenos Aires. Similar plans have also been advanced by Japanese, Italian and German interests. In a first phase of this project UEPFP plans to introduce tilting rolling stock to cut journey times.

A new line from Dolores towards Bahía Blanca featured in 1998 proposals to upgrade the Buenos Aires–Mar del Plata line (see below).

A branch line from Pinamar to the coastal resort of Villa Gesell is under construction and further branches from the Buenos Aires-Mar del Plata main line are under consideration.

Improvements to existing lines

In 1998 and 1999, plans were being developed to upgrade for 160 km/h running the line from Buenos Aires to Mar del Plata at an estimated cost of US$600 million. The project would include resignalling, some double-tracking, and the elimination of level crossings.

Traction and rolling stock

In 1998 the UEPFP fleet comprised 70 diesel locomotives and 302 passenger coaches. Most of the latter have been refurbished in UEPFP workshops, although in 2000 work commenced on the modernisation of 34 vehicles in the Chascomús workshops of Emepa. In addition, seven diesel multiple-units (each seating 170 passengers) have been purchased from RENFE of Spain for US$1.3 million, for use on routes with low traffic levels.

In July 1995 an agreement was signed with GEC Alsthom Transporte and RENFE of Spain to form a

company to maintain traction and rolling stock at La Plata.

In 1999 UEPFP was reported to have ordered around 60 Talgo coaches under a leasing arrangement for 160 km/h services on the Buenos Aires–Mar del Plata line. The vehicles were to be built locally.

Yacimientos Carboníferos Río Turbio SA (RFIRT)

Gdor Lista 790, 9400 Río Gallegos, Santa Cruz
Tel: (+54 29 66) 42 08 74
Fax: (+54 29 66) 42 08 74

Passenger operations

Plans for the operation of a tourist train have been announced by the mayor of Río Turbino. This would

be steam-hauled using wooden-bodied coaches currently in store. The scheme has provincial government support.

Freight operations

RFIRT is a privately owned company running coal trains on 750 mm track between the Andes Australes mines and the ports of Río Gallegos and Loyola. Two return services a day are operated using four Faur diesel-hydraulics acquired second-hand from

Signalling and telecommunications

The agreement signed with RENFE in 1995 also made provision for Spanish assistance with the resignalling of the Buenos Aires–Mar del Plata route.

Bulgaria in 1996, replacing former steam traction, which had held sway for more than 40 years. Two of the 1,000 hp locomotives have been remotored using Caterpillar D379 traction motors, while the other two retain their Romanian originals. A fifth Faur has been cannabilised.

The service is operated by fewer than 100 workers. Attempts to introduce radio signalling were defeated by the topography; and crews are now contacted by cell phone.

Armenia

Ministry of Transport and Communications

10 ul Zakian, 375015 Yerevan
Tel: (+374 2) 52 88 10
Fax: (+374 2) 56 05 28

Key personnel
Minister: Andranik Manoukyan

0088711

Armenian Railways

Tigran Metsa ul 50, 375005 Yerevan
Tel: (+374 1) 52 04 28; 57 36 20
Fax: (+374 1) 57 36 30
e-mail: mail@arway.am; frdrail@yahoo.com

Key personnel
Director-General: A Khrimyan
First Deputy Director-General: V Karagezyan
Deputy Directors-General: A Danielyan,
    R Stephanyan, E Tarposhyan
Head of Staff: V Arakelyan
Department Heads
    Finance and Economics: S Marutyan
    Investment and Construction: Sh Barsekyan
    Infrastructure: L Badalyan
    Rolling Stock: A Edigaryan
    Technical Policies: S Harutynyan

Supply: S Sargsyan
Operations: H Khrimyan
Traffic Safety: L Levin
Financial Monitoring: S Ordukhanyan
Accounting: V Avetisyan
External Relations: N Bdoyan
Automotive Transport: B Babakhanyan
IT: S Saratikyan
Personnel: A Martirosyan

Gauge: 1,520 mm
Route length: 845 km
Electrification: 3.3 kV DC

Political background
Armenian Railways became a state-owned joint stock holding company in 2002, with three separate subsidiary entities responsible for infrastructure, rolling stock and operations.

In recent years traffic has been effectively paralysed due to a suspension for political reasons of services at three of the country's four border crossings–into Azerbaijan via Nakhichevan and via Idzhevan, and into Turkey via the border crossing at Akuryan (closed in 1993). Akuryan is ready to accept traffic should the border point be reopened and is equipped to exchange traffic between 1,520 mm and standard gauge. The link with Georgia at Ayrum is fully operational and capable of handling 30 pairs of trains a day. With no seaboard of its own, Armenia moves all its rail import and export traffic by this route to and from the Georgian ports of Batumi and Poti. These provide links with Bulgaria and Ukraine. Due to civil unrest in the Abkhazia region of northwest Georgia there has been no rail connection with the Russian network since 1992, further isolating the Armenian system. However, there were efforts brokered by Russia in 2004 to bring about a restoration of rail communications. Pending that development, the railways of Armenia and Russia agreed in November 2004 to set up a joint venture company to develop rail freight links between the two countries, focusing initially on the establishment of a train ferry service.

Recent moves to divest non-core activities have included the transfer in 2004 to Elster-Metronika of the railway's traction power supply and control system.

Organisation
The rail network has existed as a separate entity since 1992. It comprises a southern portion of the former SZhD's Trans-Caucasus Railway and is fully electrified. Backbone of the system is the 295 km single-track line from the capital, Yerevan, through Gyumri (formerly Leninakan) and Vanadzor (formerly Kirovakan) to the border with Georgia at Ayrum. This currently forms the only international connection, with a Yerevan–Tbilisi through passenger service reintroduced in 1997. There are 75 stations.

A few sections of the Yerevan–Ayrum line were doubled in response to a heavy increase in traffic following the disastrous 1988 earthquake which struck the country, but the second track has since been removed following a substantial fall in demand for capacity resulting from the break-up of the Soviet Union. Currently only 0.8 per cent of the network is double-tracked.

The line south of Erazhkh leading the Azerbaijani republic of Nakhichevan has been closed since 1990. In February 1996 passenger services ceased on the 84 km electrified route between Yerevan and Sevan. Also out of use are the electrified lines from Sevan eastwards along Lake Sevan to Zod (121 km, closed 1995) and northwards to Idzhevan (48 km, closed 1990). The latter line includes the 8 km Megradzhorsk tunnel.

In July 2004 Armenian Railways employed 4,673 staff.

Passenger and freight traffic
Political unrest has decimated traffic on this predominantly freight network. In 1991, the railway carried 29.1 million tonnes of freight for 4,200 million tonne-km, while passenger traffic amounted to 2.8 million journeys and 300 million passenger-km. By 2000 this had declined to a mere 1.4 million tonnes (354 million tonne-km), while passenger traffic had halved to 1.1 million journeys. However, following the split of the railway into three business units and despite difficult international links, freight operation were reported to have generated a profit in 2002.

Improvements to existing lines
In 2003 a Siemens digital communications system using fibre optic cables was commissioned on the Yerevan–Ayrum line.

Traction and rolling stock
In 2004 the traction fleet totalled 110 diesel and 90 electric locomotives. Electric locomotive classes include the VL8, VL10, and derivatives, (some of which were out of service), while main line diesels include the M62 and TEM3 types (many out of service). There are 28 emus, 220 coaches and 3,000 wagons, many stored unserviceable. In 2004 Armenian Railways was granted World Bank credits covering the rehabilitation of rolling stock following studies by Canarail.

Electrification
The Armenian Railways network is 98 per cent electrified using the 3.3 kV DC system; the exception is the short cross-border section between Karzhyvan and Nyuvedi, which provides a link with the country's isolated republic of Nakhichevan but remained out of use in 2004.

UPDATED

Australia

Department of Transport and Regional Development

PO Box 594, Canberra, ACT, 2601
Tel: (+61 2) 62 74 71 11
Fax: (+61 2) 62 57 25 05
Web: www.dotars.gov.au

Key personnel
Minister: Hon John Anderson, MP

Political background
Railways in Australia were built initially by state governments (to four different gauges) to serve local needs. When it was necessary to connect Western Australia with South Australia by rail, the Commonwealth government built the line across the Nullarbor Plain (as standard-gauge but isolated from that gauge elsewhere). As the tracks extended to state borders, there arose the problem of the break of gauge. Although interstate trade grew, each state had a vested interest in retaining as much industry and trade as possible within its state boundaries.

The Second World War showed the folly of a change of gauge each time a border was reached. Post war, efforts were slowly made to convert track (or provide new track as between Albury and Melbourne) to standard gauge. It was only in 1995 that all state capitals were directly connected by standard gauge, although they had been connected via Broken Hill since 1970.

While much of the inconvenience of not having one standard gauge has been overcome, the disadvantage of having geographically based railways all with different standards is now beginning to be addressed. In 1991 the former National Rail Corporation Ltd was established to provide interstate (only) freight services over the tracks of the state systems. All governments have since agreed (to a greater or lesser degree) to introduce competition to their rail activities.

By 1997 a few private-sector organisations were operating services over the tracks of others. Each state has worked towards providing third-party access to its tracks while the Commonwealth government and State governments have established a body to regulate interstate standard-gauge third-party access, Australian Rail Track Corporation (ARTC) (see entry in Australia section of *Railway systems and operators*). So it is now theoretically possible for a company (which meets acceptable standards) to run train services on any government organisation's tracks.

Each state is dividing its passenger, freight and track businesses into discrete operations, though the problem of freight trains getting equal rail access at commuting times is starting to be addressed by funding to create dedicated trackage. There has been movement towards the standardisation of safe-working practices and radio systems. Fortunately, the physical train aspects (coupler height, braking systems, locomotive multiple-unit compatibility) have, by chance, been uniform. There are 22 different signalling regimes and 18 different radio systems, with some drivers having to be conversant with 11 different safe-working systems. It remains to be seen whether future signalling on the standard gauge network will follow existing local practice or will be designed to national standards. Given the time it is taking to eliminate semaphore signalling in New South Wales and Victoria, where considerable numbers of local mechanical interlockings remain, it may be many years before uniform standards are a reality.

In 2004 there were the following permutations of railways:

1. Railways that provided everything including track, locomotive, crew and wagons (QR and private mine-to-port railways)
2. Railways that provided locomotive, crew and wagons but ran on another's track
3. Operators that provided wagons but hired locomotives and crews and ran on another's tracks (for example Specialized Container Transport)
4. Operators that provided locomotives and crews but hauled others' wagons over another's tracks (operators providing terminal shunting only).

The rail industry has long been stressing the imbalance between government support for road and rail transport. It is claimed that rail access costs five times more than road for a similar transport task.

State and Commonwealth Transport Ministers have stated that the interstate network should provide the following levels of service:

1. Less than 2 per cent of track to be subject to speed restriction.
2. At axleloads to 21 tonnes, maximum speed to be 115 km/h, with average speed of 80 km/h.
3. At axleloads between 21 and 25 tonnes, maximum speed to be 80 km/h, with average speed of 60 km/h.
4. Crossing loops to accommodate train lengths of 1,800 m on most routes and 1,500 m on the Melbourne–Brisbane route.

They also wish to establish a protocol whereby on-time operations are rewarded.

Important developments in early 2004 included the agreement by the New South Wales government to lease non-urban sections of that state's rail network to the federal government agency, Australian Rail Track Corporation, and the opening of the Alice Springs-Darwin line.

The former agreement followed years of stalemate between the two levels of government which saw neither prepared to invest in the physical improvements which all agreed were overdue. The agreement will now see large sums invested in sensible upgrading which will assist rail to become competitive with road on the east coast route where road has a natural advantage of a more direct route and the financial advantage of years of government investment.

While the Alice Springs–Darwin railway fills a void on the rail map of Australia, in the national context it is largely irrelevant as initially only one freight a day has been run each way. Many believe government money would have been better invested elsewhere where more would have benefited. However, now that it is open, it is the responsibility of the Northern Territory government and the citizens and businesses of the territory to deliver on their claim that they will be able to develop links with Asia which will see Darwin become the major port of entry/exit for imports/exports.

A parliamentary inquiry is to be conducted to review the connections between road and rail networks and ports. Because of Australia's federal structure, state boundaries and the resulting routes of railways, many industries cannot use rail to reach the nearest port.

Depending on which contractors are counted, it is estimated the rail industry employs directly about 41,000 people. An industry problem is a nationwide shortage of train drivers and the ageing of the workforce. Despite opportunities to earn a good income, young people are apparently reluctant to join industries with 24-hour/7-day operations or where employment is offered in locations which have an adverse impact on families.

Apart from the government public transport rail segment mentioned above there is a thriving private rail freight sector in Australia, mainly serving the Queensland sugar industry and in the Pilbara region of Western Australia.

Organisation
Today there are about 4,000 km of 1,600 mm gauge (mainly in Victoria), 19,000 km of 1,067 mm gauge (mainly in Queensland, Western Australia and Tasmania) and 18,000 km of standard gauge (1,435 mm).

A feature of recent years has been the move away from operators staying within their home state's boundaries. To give a continental view of the following operator listings, we summarise the national and state situations (clockwise).

Australia: Pacific National, the previously government-owned freight operator provides services in New South Wales, Victoria, South Australia and Western Australia. Non-government operators include Australian Railroad Group and Specialized Container Transport. The national passenger service is provided by Great Southern Railway.

Queensland: The state government-owned QR is vertically integrated, and there are no plans to sell. Pacific National Queensland started freight operations on QR track in 2005. Airtrain Citylink operates the airport railway branch in Brisbane.

New South Wales: The state government provides urban passenger (CityRail) and rural passenger (Countrylink) services and owns track (Rail Infrastructure Corporation). There are also non-government freight haulers (for example Pacific National and Silverton). A major development in 2004 was the leasing of most non-urban track to the federal government agency, Australian Rail Track Corporation.

Victoria: The sale of rail operations to private companies has seen Connex eventually operating all the Melbourne suburban passenger network. Freight services were provided by Freight Australia until its 2004 acquisition by Pacific National, operating exclusively on broad gauge but also with competition on the standard gauge network. The government retains ownership of the track and right of way, though responsibility for access rests with the private operators (without their having the power to reject competition).

South Australia: The state government owns the urban passenger operator (TransAdelaide). All freight is privately operated and track privately owned (Australia Railroad Group). Government attempts to revive services in the southeast had by 2004 failed to attract an operator despite assurances that broad gauge routes would be converted to standard.

Western Australia: The state government owns the passenger operator, the Western Australian Public Transport Authority. Interstate freight operations have been sold to Australian Railroad Group. Pacific National also operates on the standard gauge. Track is leased to WestNet Rail.

Northern Territory: A new line from Adelaide to Darwin has been built for AustralAsia Railway Corporation, the government agency representing the three governments that contributed funds for its construction and the ultimate owner of the track. A BOOT contract has been signed with Asia Pacific Transport, which is operating freight services via its FreightLink Pty Ltd subsidiary. Passenger services are operated by Great Southern Railway.

Tasmania: The privately owned freight operator and track owner is Pacific National Tasmania.

The Australian Rail Operations Unit is a non-statutory body established within the department on 1 January 2000 under an intergovernmental agreement on rail operational uniformity.

Passenger operations
Nationwide urban passenger carryings in 2003 were 466 million journeys in urban areas, mostly in the two largest cities, Sydney and Melbourne, and about 9 million journeys in rural areas. Carryings are static or rising with population at best in most states due to government investment in roads and private investment in cars. For many people, trains do not travel directly between their houses and work locations. Rural and tourist operations are static in the absence of investment and due to airline competition. Passenger train-km are estimated at 46 million for New South Wales, 33 million for Victoria, 18.5 million in Queensland, 8.5 million in South Australia and 7.5 million in Western Australia. Each of the eastern state capitals is to introduce smartcards for payment of fares. Unfortunately there has been no move to

ensure the same standards, so with three different suppliers it is likely that cards from one state will not be valid in another.

Freight operations

Statistics show that interstate rail freight volumes rose from 8.4 million tonnes in 1991 to over 12 million tonnes in 2001. The total tonnage of rail freight in Australia, including that of mineral lines, increased from 340 million tonnes in 1991 to over 544 million tonnes in 2002–03. Nationally, minerals account for 80 per cent of rail freight traffic, this tonnage shared roughly equally by coal and iron ore. It should be remembered that it is geology that dictates where these minerals are mined and that there is no viable alternative to rail transport. Freight train-km are rising on the standard gauge network and in Queensland and Western Australia. Net tonne-km for freight was estimated at 158 billion in 2002–03.

New lines

Schemes for new lines are regularly promoted. These currently include a continuation of the line from Victoria through western New South Wales to Brisbane, Gladstone and Tennant Creek (see entry for Australian Transport & Energy Corridor Ltd in Australia section of *Railway systems and operators*).

A new passenger line is being built in Perth, capital of Western Australia, to serve the region to the south of the city. This follows the impressive increase in passenger carryings on the existing network since electrification.

Improvements to existing lines

The Victoria government is funding an upgrade to existing tracks for fast train operations (up to 160 km/h) from Melbourne to Ballarat (119 km), Bendigo (162 km), Traralgon (158 km) and Geelong (73 km). Under the Fast Train project new trains would run about 20 km/h faster than at present, reducing travelling times by one third. An economic feasibility study has been in progress since 2001 to investigate the benefits of converting most of the state's rural broad gauge track to standard gauge. At present, only 29 per cent of the network is standard gauge. Enthusiasm within government waned following a change of administration and with the realisation that few benefits would flow to state taxpayers. The government has seen mostly public disinterest in its Fast Train project, with public perception that the large sums needed to save a few minutes in journey times would be largely irrelevant to most travellers on relatively few trains. The public has been alerted by the media to the reality that a more frequent slower service can be better than a faster infrequent service. Moreover, the conversion of track to standard gauge is unlikely to bring electoral gains as it is unlikely to stimulate employment.

The Commonwealth Government, through ARTC, in 2004 assigned a A$450 million funding package to upgrade the Brisbane–Sydney east coast interstate route. This followed a 2003 announcement that A$872 million was to be invested in upgrading other lines in New South Wales as a result of the state government granting a 60-year lease to ARTC

to operate its rail infrastructure. Currently the east coast route carries only around 15 per cent of freight on this corridor.

The mining companies of Western Australia (iron ore) and Queensland (coal) continue to experience increasing demand for their minerals and are keen to exploit their resources while market conditions are favourable. These exports can only reach port by rail, so there is likely to be further and continuing expansion of track and capacity in those areas. Several extensions are planned in Western Australia and a 100 km coal line is being constructed to the Rolleston coal mine in central Queensland.

Signalling and telecommunications

Reflecting the history of the rail network, signalling systems across the nation vary greatly. Most of the heavily trafficked sections have appropriate colour light signalling, with lighter traffic sections having some form of Train Order. Electric staff and staff and ticket can still be found where few trains operate. Of the major cities, only Sydney does not have a centralised signalling centre controlling most of its suburban area. Crossing loop lengths are a constraint on some lines but much of this problem is likely to be solved with the takeover of much of the New South Wales track by ARTC and a subsequent injection of new funding. ARTC is investigating the implementation of a GPS-based, cab signalling system which would see the elimination of wayside signals.

UPDATED

Airtrain Citylink Limited

PO Box 66, Pinkenba, Queensland 4008
Tel: (+61 7) 32 16 33 08
Fax: (+61 7) 32 16 33 61
e-mail: info@airtrain.com.au
Web: www.airtrain.com.au

Key personnel

General Manager: Martin Earp
Sales Manager: Greg Hand

Gauge: 1,067 mm
Route length: 8.5 km
Electrification: 8.5 km at 25 kV AC 50 Hz

Political background

In May 1996, the Queensland government granted the company authority to build and operate a passenger railway between Toombul and Brisbane's airport for 35 years. Between Toombul and Brisbane City trains use existing QR tracks. Contracts were signed in February 1999.

Organisation

The company has the support of Transfield, a major construction company, Macquarie Bank, and other appropriate organisations.

Finance

The construction of the line cost A$190 million. The company based its strategy on obtaining about 15 per cent of the existing airport traffic. This was the first project in Australia where the developer accepted the patronage risk.

Passenger operations

Services commenced on schedule on 7 May 2001, with QR operating trains under contract. Patronage was forecast to start at about 2.5 million per annum rising to 5 million per annum after 5 years. Brisbane airport has experienced continuing rises in the volume of operations in recent years, with all forecasts showing this trend continuing for many years to come. Initial patronage was disappointing

QR SMU emu forming an Airtrain service at Toombul shopping town (Brian Webber)   **NEW**/1114968

but long-term forecasts remain as projected. As a consequence off-peak services were reduced and evening services cancelled in 2005 in step with demand.

Two innovative products have been introduced by the company to attract business for tourists visiting Queensland's Gold Coast. AirtrainConnect offers a door-to-door service with porter assistance to the Airtrain platform at Brisbane Airport and limousine service from Nerang station to the passengers' Gold Coast accommodation. Airtrain Smartpass provides transfers to local theme parks and unlimited use of local bus services, as well as use of Airtrain Citylink.

New lines

The 8.5 km elevated branch line required about two years to construct. Design work was complete in mid-1999 and construction started in July of that year. QR was contracted to provide track, overhead catenary and signalling for A$11.7 million.

The line runs on the longest bridge construction in Australia, with stations serving Domestic and International air terminals.

Traction and rolling stock

QR operates services, mainly with its IMU fleet, co-ordinating these with services to Robina on the Gold Coast.

Signalling and telecommunications

These are to the same standards as those for a QR line. QR provides train control services.

Track

The line was built to current Queensland Rail standards though with fairly steep grades for grade separation with roads and a future airport runway. While it is expected only passenger emus will use the line, it could be used by freight trains in the future if that became desirable. Cwr and concrete sleepers are used. The 8.5 route-km of track is supported on a viaduct, Australia's longest rail structure, comprising 2,300 piled foundations with 258 concrete columns and headstocks spaced between 30 and 45 m apart. The initial 500 m are double track, as is the section between the two stations.

UPDATED

Asia Pacific Transport Pty Ltd (APT)

Freight Link Pty Ltd
GPO Box 2750, Adelaide, South Australia
Tel: (+61 3) 83 01 12 07
Fax: (+61 3) 83 01 13 77
Web: www.asiapactrans.com.au

Key personnel

Chief Executive Officer: Bruce McGowan
General Manager, Operations: Tony Aldridge

Gauge: 1,435 mm
Track length: 1,415 km

Organisation

APT has supervised the building of and will operate the 1,415 km Alice Springs–Darwin railway until 2051. In addition, APT assumed a lease on and is responsible for maintenance of the 830 km line between Tarcoola and Alice Springs. It won these rights from the government organisation AustralAsia Railway Corporation (see entry in Australia section of *Railway systems and operators*), to which the railway will revert after 50 years. APT is a consortium of the following companies: Halliburton KBR (consortium leader); Barclay Mowlem Holdings (see entry in *Permanent way components, equipment and services* section); John Holland Group (see entry in *Permanent way components, equipment and services* section); Macmahon Holdings; Australian Railroad Group (see entry in Australia section of *Railway systems and operators*); and SANT Holding.

Construction was undertaken by ADrail, a joint venture of Halliburton KBR, Barclay Mowlem, John Holland and Macmahon. An operating company within APT, FreightLink, has been established to promote the railway and a new intermodal terminal at Darwin's East Arm Port. Australian Railroad Group is providing the rail operation for FreightLink.

Freight operations commenced in January 2004, followed in February by 'The Ghan' passenger service.

Passenger operations

Great Southern Railway, operator of 'The Ghan' service, runs its train from Adelaide to Darwin weekly. Additional services have been programmed by the company for its winter 2005 programme. Tourist activities are provided at the intermediate stops of Tennant Creek and Katherine.

Freight operations

APT's operating subsidiary, FreightLink, is promoting services over the line linking the Northern Territory and other locations within Australia and international services using the new line as a land bridge between southern Australian ports and Darwin, providing access to Asia.

Initially five trains per week were operated in each direction. Initial loadings exceeded targets though with little utilisation by shippers moving container traffic to/from Asia. Revenue has been below forecast due to the lack of high-yield freight, such as fuel. The journey time for the 2,955 km Adelaide–Darwin route is 43 hours. The line is capable of handling trains of up to 1.8 km and can sustain double-stack container operations. These were expected to start in 2005.

FreightLink container train at Katherine headed by an FQ Class locomotive (FreightLink)   0585028

Improvements to existing lines

Additional sidings are to be provided 160 km north of Tennant Creek for the loading of 600,000 tonnes of manganese ore annually from the Bootu Creek mine for haulage to Darwin.

Traction and rolling stock

In 2003 EDI Rail delivered four FQ Class (Clyde-GM model GT46C) 2,860 kW diesel-electric locomotives and 30 'five-pack' (five-section) container wagons. The contract covers provision of maintenance services for this equipment for 10 years. Two FJ Class shunting locomotives have been procured for use at Alice Springs and Darwin. Australian Railroad Group assists with other traction and rolling stock as required.

Signalling and telecommunications

The new single-track line has four intermediate crossing loops, each of 1,850 m, with motorised, self-restoring points operated locally by push-button or by radio remote control by the drivers of approaching trains. There are many block points which allow monitoring of train progress and enable protection of track maintenance activities. Additional sidings are provided at the two significant intermediate towns, Tennant Creek and Katherine. South of Alice Springs, there are 13 loops in the 830 km from the junction with the transcontinental line, Tarcoola.

Train control is provided by Australian Railroad Group from Adelaide using its train order system using satellite telephone technology. Crossing loop spacing confirms the distances covered by the line: Alice Springs–Illoquara (228 km); Illoquara–Tennant Creek (238 km); Tennant Creek–Newcastle Waters (292 km); Newcastle Waters–Katherine (351 km); Katherine–Berrimah (307 km).

Track

It is estimated that temperature variations of between −10° and 65°C during the coldest night and hottest day could result in 1.2 km expansion in track.

Rail: 60 kg/m
Sleepers: Prestressed concrete sleepers weighing 280 kg
Spacing: 1,370 per km.
Fastenings: Pandrol clips
Maximum gradient: 1.2 per cent, north of Alice Springs
Bridges: There are six major bridges, 97 minor ones and 1,220 culverts. The largest is the 510 m Elizabeth River concrete bridge, 14 km south of Berrimah, the northern terminus. The Ferguson River bridge, unused for 27 years since the closure of an earlier narrow gauge line, was upgraded and re-used.

UPDATED

For details of the latest updates to *Jane's World Railways* online and to discover the additional information available exclusively to online subscribers please visit
jwr.janes.com

AustralAsia Railway Corporation

GPO Box 4796, Darwin, Northern Territory 0801
Tel: (+61 8) 89 46 95 95   Fax: (+61 8) 89 46 95 78
e-mail: rail@aarc.com.au
Web: www.aarc.com.au
www.nt.gov.au/railway

Key personnel

Chairman: Paul Tyrrell
Chief Executive Officer: Brendan Lawson

Political background

The AustralAsia Railway Corporation (AARC) is a statutory corporation established in 1997 by the governments of Northern Territory and South Australia to manage the concessioning of the construction and operation of the railway linking Alice Springs and Darwin.

In 1911 the federal government promised to construct a railway linking South Australia with Darwin as part of their agreement to take over responsibility of the Northern Territory from that state. However, no timescale was set. The Tarcoola–Alice Springs line was built between 1975 and 1980 and following completion the federal government pledged A$10 million for a route survey between Alice Springs and Darwin.

By 1995 the governments of Australia, Northern Territory and South Australia agreed to contribute large sums if private enterprise provided the majority of the funds, eventually contributing A$165 million each, together with A$80 million of stand-by funding. The federal government has leased the 830 km Tarcoola–Alice Springs railway for a nominal rental fee, while a consortium of private interests, Asia Pacific Transport Consortium Pty Ltd (APT) (see entry in Australia section of *Railway systems and operators*), contributed the remainder (about A$850 million) as a build/own/operate/transfer back (BOOT) venture covering the new line between Alice Springs and Darwin. AARC has granted APT a 50-year concession, at the end of which ownership of the railway will revert to it. The concession also covers operation of the Tarcoola–Alice Springs line.

Operations on the line commenced in January 2004.

UPDATED

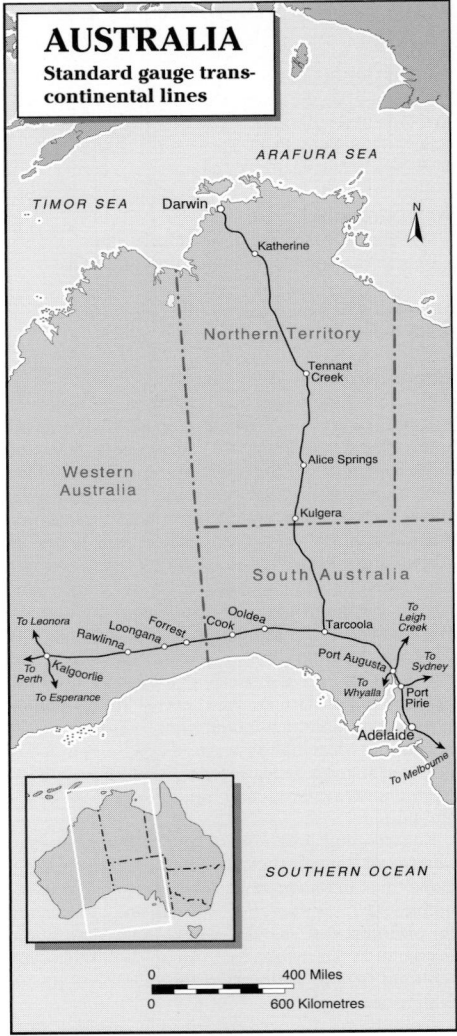

AUSTRALIA
Standard gauge trans-continental lines

0572295

Australian Railroad Group Pty Ltd (ARG)

2-10 Adams Drive, Welshpool, Western Australia 6106
GPO Box S1422, Perth, Western Australia 6845
Tel: (+61 8) 92 12 25 00
Fax: (+61 8) 92 12 27 30
Web: www.arg.net.au

Key personnel

Chief Executive Officer: Mike Mohan
Deputy Chief Executive Officer: Murray Vitlich
General Manager, Finance: John Cleland
General Manager, Corporate Development: Ian Thomson
Chief Safety Officer: Mike Lundell
Chief Accounting Officer: Stuart Gale
Manager, Risk Assessment: David Best
National Manager, Human Resources: Caroline Hudson
Manager, Information Technology: Brad Fisher
General Manager, Operations: Mike Irvine
National Equipment Manager: Bruce Carswell
National Marketing Manager: Ken Potts
Transport Manager, Eastern Region: Bert Easthope
Transport Manager, Western Region: Peter Satie

Gauge: 1,600 mm; 1,435 mm; 1,067 mm
Route length: approximately 10,000 km

Political background

The group was established following the purchase of several government-owned railway operations in South Australia and Western Australia.

Organisation

Australian Railroad Group Pty Ltd is jointly and equally owned by Wesfarmers, a public corporation, and Genesee & Wyoming Inc (GWI) of US. It acquired Westrail Freight in December 2000 from the state government of Western Australia for A$323 million. To complete the transaction GWI contributed Australia Southern Railroad, which it purchased in August 1997, and its interest in the Asia Pacific Transport Consortium, which was selected to operate the Alice Springs–Darwin line.

The Group is the largest private rail owner/operator in Australia, although its major activities are in the states of South Australia and Western Australia. In 2003 the group won a significant contract in New South Wales and it operates occasionally to Melbourne. It also provides locomotives, wagons, crews and management services to the operator of the Darwin railway. The company has about 1,000 employees.

Finance

In 2001–02 revenue was A$330 million. Expenses were A$242 million, yielding a profit of A$88 million.

Freight operations

ARG hauls about 50 million tonnes of freight annually. Operations are complicated by the historical dilemma of three gauges. Broad-gauge operations are limited to a small section of track in the Adelaide area while narrow-gauge operations are concentrated on the Eyre Peninsula in South Australia and the former Westrail network in southwest Western Australia. On the standard-gauge, ARG trains can be seen on main lines between Perth and Sydney or

Standard gauge L Class diesel locomotive at Forrestfield Grain Handling Terminal, Western Australia (ARG)
***NEW**/1115033*

1,067 mm gauge S Class diesel locomotive with a train of bauxite at Kwinana, Western Australia (John Bollans)   ***NEW**/1115035*

Standard gauge L Class diesel locomotive leaving Toll's container depot at North Fremantle, Western Australia (ARG)   ***NEW**/1115034*

Melbourne as well as assisting the new freight operation to Darwin. In 2002 the group hauled 866,000 carloads, with grain and iron ore each contributing 20 per cent of this figure and alumina and bauxite a little less.

Broad gauge lines (288 km)
Traffic comprises limestone and grain. Limestone traffic totals some 500,000 tonnes annually, transported seven days a week from Penrice Quarry to suburban Adelaide in a 22-wagon consist. Grain is hauled from the Balaklava, Burra and Kapunda branches. On this section 145 wagons are available.

Standard gauge lines
In 2001–02 33 million tonnes of freight was hauled on this gauge. ARG trains can be seen between Perth and Sydney or Melbourne (via Adelaide) as well as in rural New South Wales and South Australia.

Narrow gauge lines
The isolated Eyre Peninsula network (748 km) transports about 1.3 million tonnes of gypsum from Kevin to the port at Thevenard (60 km) and up to a million tonnes of grain annually to Port Lincoln, following a good season. Some 240 wagons are available on this section. An independent study has indicated that A$20 million is required to be invested in this network to secure its future. This funding would need to come from government. The Western Australian operation (4,049 km) primarily hauls bulk export commodities (grain, minerals, woodchips) to six ports in the southwest.

Traction and rolling stock
ARG operates around 200 locomotives and 4,000 wagons. Of the locomotive fleet in 2002, 13 were on broad gauge track in South Australia, 55 were narrow gauge units in Western Australia and 24 were based at Port Lincoln on the Eyre Peninsula. Locomotives are moved between states and even-

ARG diesel locomotives

Class	Wheel arrangement	Power kW	Speed km/h	Weight tonnes	No in service	First built	Mechanical	Builders Engine	Transmission
A*	Co-Co	1,063	100	89	3	1960	Clyde	GM 12-567C	E GM
AB*	Co-Co	1,230	100	96	4	1969	Clyde	GM 12-645E	E GM
ALF/32	Co-Co	2,460		130	8	1994**	Clyde	GM 16-645E3	E GM
CK/10	Bo-Bo	708	100	70	6	1960	Clyde	GM 8-567/8-645E	E GM
CLF/30	Co-Co	2,460	130	128.5	7	1993**	Clyde	GM 16-645E3	E GM
CLP/30	Co-Co	2,460	140	131	10	1993**	Clyde	GM 16-645E3	E GM
DA/900	Co-Co	1,640	90	97	7	1972	Clyde	GM 16-645E	E GMD
DB*	Co-Co	1,640	90	110	13	1982	Clyde	GM 16-645E	E GM
GM	A1A-A1A	1,390		110	9	1967	Clyde	GM 16-567B	E GM
K	Co-Co	1,454	100	114	6	1967	EE	EE 12CSVT	E EE
L	Co-Co	2,386	100	134	10	1967	Clyde	GM 16-645E	E GM
P*	Co-Co	2,000	90	101	17	1989	Goninan	GE 7 FDL 12	E GE
Q	Co-Co	3,095	115	134	19	1997	Clyde	GM 710G3B-ES	E GM
S*	Co-Co	2,424	90	118.5	11	1998	Clyde	GM 710G3B-ES	E GM
22	Co-Co	1,490	124	110	16†	1969	Clyde	GM 16-645E	E GM
500	Bo-Bo	373	64	57	4	1964	SAR	EE 4SRKT	E
700	Co-Co	1,490		115	5	1970	A E Goodwin	Alco 251C	E
830/DA	Co-Co	671		76		1960	AE Goodwin	Alco 251C	E

* 1,067 mm gauge
** Rebuilt
† Includes locomotives requiring overhaul before entering service

gauges to suit operational and maintenance needs. The locomotive fleet comprises small numbers of many types, reflecting under-investment by previous owners.

Signalling and telecommunications
ARG is responsible for some train control and safe working in South Australia. Elsewhere train control is provided by track owners.

Track
ARG is responsible for broad and narrow-gauge track in South Australia. Its subsidiary, WestNet Rail, is responsible for track in Western Australia.

UPDATED

Australian Rail Track Corporation Limited (ARTC)

PO Box 10343, Gouger Street, Adelaide, South Australia 5031
Tel: (+61 8) 82 17 43 66   Fax: (+61 8) 82 17 45 78
e-mail: track@artcom.com.au
Web: www.artc.com.au

Key personnel
Managing Director: David Marchant
Directors: Dale Budd; Robert Maher; Martine Pop
General Managers:
 Engineering and Infrastructure: Malcolm Owens
 Finance and Administration: Geoff Atkinson
 Operations and Customer Service: Denise McMillan

Gauge: 1,435 mm
Route length: 7,026 km

Political background
The federal government-owned Corporation was established in February 1998 following

inter-governmental agreement to establish an organisation solely to manage access to infrastructure development of the interstate standard-gauge network. It had become apparent to the various governments that the continuation of state control of the network was untenable and only an organisation with overall responsibility could enable the rail industry to compete with road transport, which deals with only one government agency, and to a lesser degree, the shipping industry.

The Corporation is negotiating to establish equitable and transparent access pricing arrangements on the standard-gauge network and is aiming to introduce common standards, including those for safe working, throughout Australia.

Organisation
In November 1997 Australia's Transport Ministers signed an agreement to establish a single company, Australian Rail Track Corporation (ARTC), to manage the interstate standard-gauge rail network. It owns and manages the track between

Kalgoorlie (Western Australia) and Broken Hill (New South Wales) and Wolseley (South Australia) and some branches; manages, under a 15-year lease, the interstate standard-gauge track within Victoria; has rights to sell access between Kalgoorlie and Kwinana (Western Australia) under a wholesale access agreement with WestNet Rail (see entry in Australia section of *Railway systems and operators*) and has a working relationship with QR about the use of the 135 km of standard gauge track in south Queensland. The Tarcoola–Alice Springs branch, which it owns, is leased to AustralAsia Railway Corporation (see entry in Australia section of *Railway systems and operators*).

ARTC has negotiated with the New South Wales government to take over responsibility for the main line tracks within that state; an arrangement which commenced on 1 July 2004 for a projected 60 years. This agreement includes the Hunter Valley coal lines, which handle around 55 trains per day and have an annual capacity of 102 million tonnes, and the dedicated freight route to Sydney's port. Ownership remains with the state of New South

Wales but ARTC will be responsible for investment decisions.

In 2002–03, there were 11,813 services on the (then) network, an increase of 5 per cent over the previous year. The percentage of trains entering the ARTC network behind schedule, is a concern, although most of these suffered no further delay while on ARTC tracks and many regained time. In 2004 operators with access agreements with ARTC included: Australia Railroad Group; Countrylink; Freight Australia (pending acquisition by Pacific National); Great Southern Railway; Pacific National; Patrick Rail; QRNational; Silverton Tramway; and Specialized Container Transport.

When ARTC was formed in 1997, rail's share of land transport on the east-west corridor was 69 per cent. By 2004 this had grown to 81 per cent, giving cause for optimism that the 20 per cent share handled by rail on the east coast north-south corridor can be greatly increased. The release of funding for the New South Wales network, delayed by political considerations for several years, should give marketers of rail transport a better opportunity to gain business.

Improvements to existing lines

An audit of track condition in 2001 revealed upgrading estimated at A$507 million to be necessary. This figure was twice the government's funding commitment but well short of the A$1 billion identified by a government committee in 1995. Most of the work required is in New South Wales but on interstate routes on which that state's government was unwilling to invest. The audit suggested A$398 million was needed on the Melbourne–Sydney–Brisbane corridor, of which ARTC controlled only the Melbourne–Albury section. A figure of A$146 million was identified as necessary to start on the Sydney Freight Priority Project. The Sydney commuter rail network is a critical bottleneck which causes many hours of delay to vital freight movements, with a block on freight traffic in effect at peak hours.

As part of the New South Wales lease, ARTC is required to invest A$872 million in the first five years. A$180 million will be spent to construct additional track for freight traffic running south within the Sydney metropolitan area, while A$145 million will be invested in upgrading Hunter Valley lines. A further A$175 million is earmarked to improve signalling, lengthen crossing loops between Sydney and Albury and replace the elderly bridge over the Murrumbidgee River at Wagga Wagga. A$123 million will be spent on the North Coast Line towards Brisbane, including

replacing electric staff working north of Casino with CTC. Elsewhere, other improvements include replacing timber bridges between Cootamundra and Werris Creek to eliminate speed restrictions and allow heavier axle-loads and eliminating height restrictions between Parkes and Broken Hill to allow double-stacking of containers on that section and west to Perth. This expenditure should bring track standards towards the situation where rail transport can be competitive with road transport, which in Australia often has the benefit of more direct routes.

The 2004–05 Federal budget also included a one-off grant of A$450 million towards upgrading the Sydney-Brisbane east coast interstate main line to eliminate operating constraints which include short passing loops, sharp curves and outdated train control equipment.

Track improvements have enabled line speed forecasts in 1997 to be achieved. These were that track should allow 115 km/h running with 20 tonne axleloads, 110 km/h with 21 tonne axleloads and 80 km/h with 23 tonne axleloads.

Wheel Condition Monitors have been installed at Lara in Victoria, Port Germein in South Australia and Parkeston in Western Australia to detect potentially defective wheels before the fail. A Rail Bearing Acoustic Monitor near Port Pirie analysed more than 700,000 bearings in its first year, identifying a significant number of defects.

The A$3.5 million Port Augusta track upgrading project has been completed. An additional track bypassing the existing station provides a faster route and a 1.8 km crossing loop. The area has been provided with improved signalling, controlled by ARTC.

A 1,800 m crossing loop has been provided at Dry Creek, Adelaide, to improve the operational interface with TransAdelaide and ARG trackage.

A$2 million has been budgeted to upgrade the iconic Murray River bridge on the Adelaide–Melbourne line in eastern South Australia. The 1926 structure will be strengthened to 30 tonne axleload standard with 60 kg/m rail and a ballast deck.

ARTC has implemented a national Code of Practice for operations and safe working on its track and it is expected that the code will later be extended to the tracks of other owners to standardise everyday activities associated with train running.

Traction and rolling stock

ARTC owns a number of track maintenance wagons and vehicles, although much equipment is provided by contractors and locomotives are hired when required.

Signalling and telecommunications

Train control previously in Melbourne has been centralised in Adelaide. The aim is for 96 per cent of services to arrive on time. Safe-working systems include CTC in the Melbourne and Adelaide metropolitan areas and between Adelaide and the Victorian border, train order between Adelaide and Kalgoorlie, and Section Working Authority (similar to train order) in Victoria country areas. The takeover by ARTC of some New South Wales trackage saw the organisation assume responsibility for its train control. There is much scope to upgrade signalling in New South Wales by replacing semaphore signalling with colour light signalling. However, the government has announced a A$20.3 million allocation to ARTC to develop an in-cab signalling system based on GPS technology, which should lead to the removal of the present signals. The Advanced Train Management System is to be developed in association with Lockheed Martin. The government is financing an extension of mobile phone coverage into areas of rail significance where otherwise it may not be economic for phone companies to provide the technology. There is now the real prospect of the eventual standardisation of driver instructions and signal displays across the network.

ARTC is providing train control facilities for the private Leigh Creek–Port Augusta operation.

Track

Standard rail: Flat bottom throughout, weighing 60, 53, 47, and on some branch lines 40 kg/m
Joints: Fishplates, bolts: but all main lines are cwr
Rail fastening: Dog and screw spikes, Pandrol and McKay Safelok elastic rail spikes and clips, T-headed bolts and nuts
Crossties (sleepers): untreated hardwood 2,500 × 230 × 115 mm; CR2 prestressed concrete 2,514 × 264 × 211 mm; AN3/AN4 prestressed concrete 2,500 × 264 × 211 mm; AN6 prestressed concrete 2,500 × 264 × 240 mm
Spacing: 1,600 to 1,300 per km
Filling: Crushed stone and gravel ballast
Min curve radius: 14.5°
Max gradient: 2.5%; between Mitcham (Adelaide) and Mount Lofty (22.4 km) the grade averages 1 in 53
Longest straight: 477 km–Nullarbor Plain
Max axleload: 23 tonnes (permissible axleloads in Victoria have been raised from 20 to 21 tonnes as a result of track upgrading)
Highest station: Peterborough, South Australia (532 m)

UPDATED

Australian Transport & Energy Corridor Limited (ATEC)

14 Argyle Place, Argyle Street, Albion, Queensland 4010
Tel: (+61 7) 32 62 81 77   Fax: (+61 7) 32 62 81 99
e-mail: atecrail@aire.com.au
Web: www.aire.com.au

Key personnel

Chairman: Everald Compton, AM
General Manager: John Balassis

Gauge: 1,435 mm
Route length: approximately 4,500 km

Background

This organisation is the proponent of a railway through country areas between Melbourne and Darwin, known as the Australian Inland Rail Expressway (AIRE). It is a private company with 15 shareholders, all involved in the rail or construction industries. A separate company, AIRE Pty Ltd, has been established as a joint venture equally owned by ATEC and Australian Rail Track Corporation Ltd (ARTC) (see entry in Australia section of *Railway systems and operators*).

The project has received support from those who will benefit and has had some encouragement from governments. There was initially an insistence that the project should and could be built without

government financial commitment but it is not clear how profitable operations could be achieved other than in the long term.

Since the idea was first floated, the Alice Springs–Darwin railway has been constructed with the result that what may have been a monopoly will now have a competitive route to southern markets. A problem for the AIRE concept is that there needs to be construction in several locations before a through route between Melbourne and any interim terminus could be established. Transport between any interim termini and ultimate destinations would need to be by road and Australian experience is that once sizeable volumes are loaded on truck they generally continue by truck. At March 2005 there had been no letting of contracts for construction. The reason offered was that it was necessary for the state of New South Wales to first come to an agreement with ARTC over track responsibilities, since concluded, before the proponents of this project could negotiate with the New South Wales track-owner.

The project is being promoted in a climate where there is little enthusiasm for investment in rail infrastructure. The Alice Springs–Darwin line is notable for the involvement of governments who, it can be argued, built the line for political and "national development" reasons rather than the idea that the project made undeniable economic sense and would be a good investment. It may be the case that other more worthwhile

projects may have received the green light if that project did not proceed. Similarly the AIRE is competing with other important rail projects and may not be seen as worthy of government and industry support, in a political climate where all governments are experiencing difficulties raising additional revenue to meet ever increasing community expectations for basic services. ATEC is currently concentrating its efforts on plans to provide a route from Werris Creek to Gladstone, via existing track to Moree and North Star and by new standard-gauge construction via a largely existing route in Queensland. This has been costed at A$2 billion. It is notable that this route does not serve Brisbane and attempts to gain coal traffic from existing and future mines west of Toowoomba. This is a revision of the original proposal of a Melbourne–Darwin line, which now seems unlikely to be considered viable. However ATEC remains totally optimistic that the project is viable and will proceed during the next decade.

Hopes were raised in April 2005 when the national government announced it was funding a feasibility study into the building of the Melbourne–Brisbane section, although this could be yet another delay to actual construction. The government stresses that private industry needs to take the risks involved should the study suggest that this alternative to the present route should be built. It is politically attractive for governments to support schemes involving development in rural and regional areas.

New lines
Planning is for a new standard gauge track to join North Star, northern New South Wales with Tennant Creek, Northern Territory, with branch lines to the Queensland coastal ports of Brisbane, Gladstone and Townsville. The branch lines are required to serve locations which would be generators of traffic. The recommended route mainly uses existing track between Melbourne and North Star, the current terminus of a line in New South Wales which is the nearest railhead to Queensland. Between North Star and Tennant Creek new construction is required, although it is predicted that some existing rail corridor, currently with 1,067 mm gauge track, will be used. Existing track between Melbourne and North Star would need to

be upgraded in many locations to raise existing speed limits and to enable double stacking of containers, a provision seen as essential to provide a definite advantage over the existing coastal route. Plans call for the line to handle 1,800 m trains running at a maximum speed of 115 km/h.

Although not included in the current proposal, it seems that if this project were to proceed, QR's existing South Western Line to Goondiwindi and the Moura Line would need to be standard-gauged, or less likely, dual-gauged. It is not clear how the Queensland Government might respond to this although, interestingly, QR has initiated a standard-gauge operation.

In 2005 the proposal had evolved into two aspects: a bypass of Sydney for through traffic; and a route for coal from southwest Queensland to Gladstone. Politicians were starting to acknowledge that the large sums to be spent to provide a freight

network through Sydney may be better spent on this alternative. There are several coal discoveries that could be developed now that prices and demand have risen. This would make a link to Gladstone viable, although it is not clear how long environmental concerns will allow continuing increases in coal consumption throughout the world.

Freight operations
This company is intended to facilitate construction of the railway. Once a section of the route is complete, operators will be encouraged to locate loadings for their trains, perhaps by diverting existing traffic from the present coastal route, which has the impediment of Sydney for much traffic.

UPDATED

---

bhpbilliton Iron Ore Railroad

PO Box 231, Nelson Point, Port Hedland, Western Australia 6721
Tel: (+61 8) 91 73 67 13   Fax: (+61 8) 91 73 67 89
Web: www.bhpbilliton.com

Key personnel
Vice-President, Railways: Mike Darby
Managers
    Operations: Lindsay Morrison
    Track and Signal: Eugenio Alvarez
    Rolling Stock: Russell Donnelly

Gauge: 1,435 mm
Route length: 729 km

See map in entry for WestNet Rail in the *Railway systems and operators* section.

Organisation
The railway has two operations: the Mount Newman Joint Venture Railway and the Mount Goldsworthy Joint Venture Railway. These two operations are managed from one central location at Nelson Point Railroad Administration, Port Hedland.

Freight operations
The Mount Newman line runs from Newman north to Port Hedland (Nelson Point) (489 km), with three branches: at 393.4 km (Jimblejar Junction, or JBJ), a 32 km spur to Jimblejar mine; at 402 km, a 3 km spur to Orebody 23/25; and at 281 km, a 30 km spur to Yandi.

The Mount Goldsworthy line runs from Yarrie west to Port Hedland (Finucane Island) (208 km). It has a 5 km spur line to Nimingarra at 165 km.

Both lines are single-track. The Mount Newman line has 21 passing loops, mostly of 3.5 km although the latest to be provided are 4.5 km to allow for possible longer train lengths in the future. The Mount Goldsworthy line has five 1.2 km loops. On the Yandi branch, two locomotives hauling formations of 208 wagons operate from Port Hedland to Yandi (315.8 km); two banking locomotives assist during loading and then

Diesel locomotives

Class	Wheel arrangement	Power kW	Speed km/h	Weight tonnes	No in service	First built	Mechanical	Builders Engine	Transmission
SD40-2	Co-Co	2,240	112	167	8	2003*	GM EMD	GM 16-645E3	E GM
AC6000CW	Co-Co	4,660	120	198	8	1999	GE	GE 7HDL-16A	E GE
CM40-8	Co-Co	3,130	112	195	36	1991	Goninan	GE 7FDL-16	E GE
CM39-8	Co-Co	3,060	112	195	4	1988	Goninan	GE 8FDL-16 or 7FDL-16	E GE
SD70ACe	Co-Co	3,207	–	–	13	–	EMD Canada	EMD 710	E EMD

*Date of acquisition
*On order 2005

provide banking to Shaw (216 km from Hedland), before detaching and returning to Yandi. The loaded train then goes forward to Hedland behind two locomotives.

Operations on the Mount Newman line normally see eight trains in service, a mix of two- and three-rake formations, the latter comprising 312 wagons. The two-rake trains are hauled either by three 4,476 kW GE AC6000CW locomotives or by four 2,984 kW GE Dash 8 machines. On the Mount Goldsworthy line, four trains are operated daily, formed of one Dash 8 locomotive and 90 hopper wagons. Trains are driver-only-operated and are claimed to be the longest and heaviest in the world.

In 2004 about 90 million tonnes were hauled to port, with additional contracts likely to increase this figure to 100 million tonnes. In 2005 the iron ore mining industry was experiencing heavy demand, with China as a major consumer.

Improvements to existing lines
An impressive marshalling yard has been built at Jimblebar Junction to enable trains to be uncoupled and combined. The yard is 8.5 km long and is believed to be the longest in the southern hemisphere.

A new line of some 30 km was under construction in 2003 to extend the line from Yandi to Mining Area 'C'.

Traction and rolling stock
The combined traction fleet for the two lines totals 56 diesel locomotives, including eight

GE AC6000 and 40 Dash 8. In 2003 these were joined by eight secondhand General Motors SD40-2 locomotives procured as a stopgap measure pending new locomotive orders. New traction was to take the form of 13 SD70ACe 3,207kW/4,300 hp locomotives, ordered from EDI Rail but to be built by EMD Canada.

The wagon fleet consists of over 2,000 rotary dump wagons used on the Mount Newman line and 283 bottom dump wagons on the Mount Goldsworthy line. The latest addition to the wagon fleet is the Type Golynx vehicle, 720 of which were being supplied by United Goninan in 2004. Of lighter, stainless steel construction, aerodynamic design and larger capacity, these have dramatically reduced fuel consumption and increased tonnages carried.

Signalling and telecommunications
The railway is controlled by CTC, supplemented by track-to-train radio, from a control centre at Port Hedland. Interlockings can function automatically in the event of any failure in the CTC telemetry. Train operation is protected by: 12 hot box/hot wheel detectors, one cold wheel detector, one wheel impact monitor and three in-motion weighbridges.

Automatic Equipment Identification (AEI) transponders are fitted to each item of rolling stock and readers are located at Headland and at each mine to define the source of the ore being carried. All locomotives are equipped with Harmon ATP equipment, Epic electronic air brakes and Locotrol 3 equipment. ATP wayside equipment is located on the main line and is used to send in-cab messages to drivers when entering or departing from speed-restricted stretches of line.

Track
On the Mount Newman main line, 89 per cent of track is on concrete sleepers. A programme in place in 2003 was to see remaining timber and steel sleepers replaced by concrete over the following three years. Track inspection is undertaken on a daily and weekly basis. The following equipment is used: an instrumented ore wagon; an ultrasonic rail testing car; a Plasser EM80 recorder car; and special-purpose Hi-Rail vehicles.

Track maintenance crews are based at Hedland and Newman, as well as at the Redmont Track Centre, some 200 km south of Port Hedland. Mobile flash butt welding equipment has been acquired to improve in-track weld quality.

Two GE AC6000CW locomotives and a GM EMD SD40-2 (centre) pass Goldsworthy Junction with a 312-wagon train of empties, assisted by two sets of mid-train 'helper' locomotives (Ian Francis)

NEW/1115027

Rail: 66 and 68 kg/m continuous welded rail in 400 m lengths (standard carbon on tangent and head-hardened on curves and some tangents); some 47 kg/m rail on the Yarrie line
Sleepers: Concrete: 300 wide × 240 deep × 2,600 mm long (89 per cent of main line)
Timber: 200 wide × 150 deep × 2,400 mm long (9.5 per cent)

Steel: 120 mm deep × 9 or 10 mm thick (1.5 per cent)
Spacing: Steel/concrete 600 mm; timber 533 mm
Fastenings: Concrete: Pandrol; Timber: Pandrol, dog spikes; Steel: Traklok
Min curvature radius: 528 m
Max gradient: Newman line: 1.5% (empty); 0.55% (loaded); Yarrie line: 1.04%

Max axleload: Newman line–38 tonnes; Yarrie line–26 tonnes

Type of brake: Air
Type of coupling: Alliance automatic

UPDATED

CityRail

PO Box K348, Haymarket, New South Wales 1238
Tel: (+61 2) 82 02 20 00   Fax: (+61 2) 82 02 21 11
Web: www.cityrail.nsw.gov.au

Key personnel
Chief Executive Officer: Howard Lacy
General Manager, CityRail Stations: Bob Irving

Route length: 900 km (approx)
Gauge: 1,435 mm
Electrification: 1.5 kV DC

Organisation
CityRail is a business group of Rail Corporation New South Wales following the most recent restructuring in January 2004. It operates an extensive rail network and provides suburban and regional passenger services from Sydney to Lithgow in the Blue Mountains (156 km), Goulburn (225 km) in the Southern Highlands, Nowra (153 km) in the Illawarra, and South Coast areas, and to Scone (315 km) and Dungog (245 km) in the north. It provides around 3,000 services each weekday, carrying approximately 930,000 passengers. These services operate between 306 stations over some 2,080 km of track, serving a population of 4.9 million. CityRail operates over tracks owned and maintained by Rail Corporation New South Wales (see entry in Australia section of *Railway systems and operators*).

The government introduced the restructuring implemented in 2004 following several disastrous events (a collision, a bridge closure and a derailment following the death of a driver) which brought the government and its operators into public disrepute. Despite evidence at public inquiries that many of the problems were caused by too many restructurings in the recent past, it was felt that the splitting of above-rail and below-rail activities had resulted in organisations working in competition with each other rather than aiming to achieve common goals. There has been an acknowledgment by government and operators that the system and timetabling had been too complex and interdependent, with numerous potential pathing and connection weak spots.

The government, suffering daily embarrassment in the media due to constant train late running and cancellations, has decided to concentrate effort and finances in the near future to creating so-called 'Clearways'. The system is being split into routes which will remain independent with their own trains, crews, tracks and platforms. This is seen as the only way to restore CityRail

Double-deck electric multiple-units

Class	Cars per unit	Motor cars per unit	Motors per car	Power/ motor kW	Speed km/h	Cars in service	First built	Builders Mechanical	Electrical
V**	–*	–	–	–	115	242 (M–123; T–119)	1970	Comeng	–
S/R	–*	–	–	–	115	359 (M–207; T–152)	1972	Comeng	–
S/R	–*	–	–	–	115	150 (M–80; T–70)	1978	Goninan	–
K	–*	–	–	–	115	160 (M–80; T–80)	1981	Goninan	–
C	–*	–	–	–	115	56 (M–28; T–28)	1985	Goninan	–
T (Tangara)	–*	–	–	–	115	138	1987	Comeng	–
G (Tangara)**	–*	–	–	–	115	80 (M–40; T–40)	1994	Goninan	–
M (Millennium)	4	2	–	–	130	141 (M–71; T–70)	2002	EDI Rail	–

*Trains may run as 2-, 4-, 6-, 8- or 10-car formations of the same type
**Interurban units, equipped with toilets and more comfortable seating

Diesel railcars or multiple-units

Class	Cars per unit	Motor cars per unit	Motored axles/car	Power/ motor kW	Speed km/h	Cars in service	First built	Builders Mechanical	Engine	Transmission
620 Class										
NPF	2	1	2	227 × 2	122	6	1961	NSWGR	Cummins NTA855 R4	H Voith
NTC	2	1	–	–	122	6	1961	NSWGR		
Endeavour										
TE	2	2	2	353	145	14	1994	ABB	Cummins KTA 19R	H Voith
LE	2	2	2	353	145	15	1994	ABB	Cummins KTA 19R	H Voith

as a credible transport option. CBD underground tracks have little spare capacity, platforms can be overcrowded, and there is no potential for more services.

CityRail operates over the tracks of RailCorp.

Finance
In 2002–03 farebox recovery was about 24 per cent of overall operating costs. The balance was provided by government to cover non-commercial services, concession fares and capital works projects.

The State Budget continues to provide funds at traditional levels for the gradual improvement of services and infrastructure. Priority for funding has moved to the purchase of new rolling stock and the completion of construction of the Chatswood–Epping branch.

Passenger operations
Services are provided on nine routes from the city and on two branch lines. Services through the CBD cross the Sydney Harbour Bridge, run through the City Underground Loop or continue to the Eastern Suburbs Line. There is only one class of fare for

travel and the busier stations have ticket-activated barriers to control entry/exit.

Patronage declined slightly in 2003–04 to 273.3 million journeys, compared to 277.6 million journeys in 2002–03. Two security guards now travel on all services after 19.00.

Weekend services were significantly reduced in July 2004 to enable train crews to be relieved from constant overtime until more drivers can be trained. A new timetable sees a slowing of services with longer dwell times at stations to achieve on-time running.

The government has announced the closure of the 5 km branch between Broadmeadow and Newcastle and its three stations. Newcastle is currently the CBD terminus and focus station for this important city. Passengers will transfer to/from buses at Broadmeadow, which will have additional track provided to cater for terminating services. It will become the terminal and change point for electric services from the south and railcar services to the north.

New lines
Construction work has started on a new line linking Chatswood on the North Shore line, and Epping on the Northern line. Three new stations will be provided on this 12 km underground line, which will offer an alternative route between the city and some northern suburbs. It is expected that a rearrangement of services will enable frequencies more accurately to match demand when services commence in 2008. The government has reviewed the original proposal to extend from Epping to Parramatta and has announced that potential patronage is insufficient to justify construction of that section.

Improvements to existing lines
Under CityRail's 'Easy Access' programme 21 stations will be upgraded at a cost of A$37.8 million.

There have been reports that track and signalling infrastructure improvements have not kept pace with demand for increased service capacity with the result that it is now difficult to timetable sufficient trains to meet present or anticipated future demand. Routine track maintenance also

Millennium emus at Lidcombe, in the Sydney suburbs (Ian Francis)   **NEW**/1115001

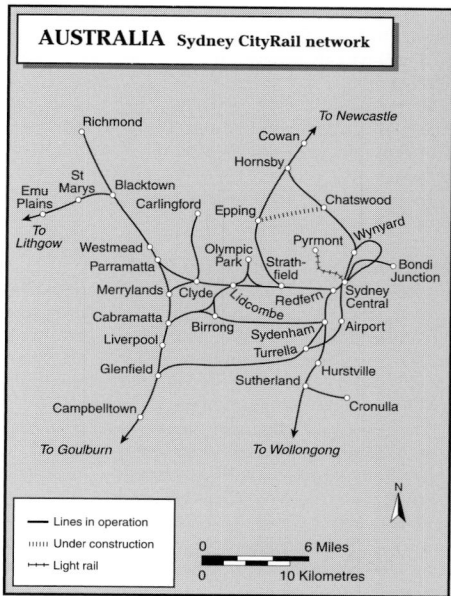

AUSTRALIA Sydney CityRail network

0573895

Tangara 'T' Series emu formation at Sydney's Central station (Brian Webber)   **NEW**/1115002

causes frequent service disruption, with a lack of diversionary tracks around work sites. In addition, the absence of a central signalling control centre makes the overall control of operations more difficult during periods of service disruption.

Additional turnback facilities are being provided at Bondi Junction and Edgecliff at a cost of A$54 million to give flexibility to operations on that line. The major junction and service terminus at Hornsby is to be provided with another platform and additional tracks to provide more reliable paths for passenger and freight services. Three turnback sidings were opened at Macdonaldtown in inner Sydney in September 2004 at a cost of A$17 million to tackle an imbalance of services. Double track is to be provided on sections of the Cronulla and Richmond lines at costs of A$145 million and A$80 million to assist punctuality, currently affected by single-line sections.

Traction and rolling stock
CityRail operates 1,457 double-deck emu cars (Tangara suburban 367; other suburban 770; outer

suburban 80; interurban 240). Currently 1,312 are rostered in peak-hour service. Additionally 28 Endeavour and six 620 Class two-car diesel railcars are used.

The system is receiving its first new trains in a decade with the delivery of 35 four-car double-deck 'Millennium Trains'. The contract was let in October 1998 to Evans Deakin Industries, whose Sydney subsidiary EDI Rail is building the trains. The first was delivered in 2002 with the remainder due in 2005. The contractor is to be responsible for maintenance until 2016. The four-car trains will have driving trailer cars at each end, with the motor cars in the centre. Each set seats 464 in 'walkover' seats, with standing space for a further 620 passengers.

United Goninan has been awarded a contract to build 81 Outer Suburban carriages (20 four-car trains and spare car) at a cost of A$262 million. Each set will seat 432 passengers and will feature selective door controls to enable guards to open only appropriate doors at short platforms.

One third of CityRail's fleet is not air-conditioned, now a passenger expectation. Those cars are expected to be replaced by carriages for which tenders are being considered. It is anticpated that contracts will be awarded for 290 double-deck

and 208 single-deck cars. The decision to return to single-deck construction is based on the need for smaller trains to serve shorter runs and on their quicker loading times.

The six two-car 620/720 railcars have been refurbished with improved cab facilities and repaints. They run services west from Newcastle. These vehicles will be replaced in 2006 by 14 two-car dmus being built by United Goninan under a A$90 million contract let in December 2002.

A driver vigilance monitoring system has been fitted to all suburban trains.

Electrification
Electrification was initially introduced in 1926 and has been extended to most passenger lines in the Sydney residential area. Most services are provided by electric trains. Diesel trains operate beyond the end of electrification at Kiama, Macarthur and Newcastle.

Type of coupler in standard use: Passenger, automatic; Endeavour, Scharfenberg
Type of braking in standard use: Westinghouse air; Endeavour air and hydrodynamic

UPDATED

Comalco Railway

Post Office, Weipa, North Queensland 4874
Tel: (+61 7) 40 69 89 69   Fax: (+61 7) 40 69 83 68
Web: www.comalco.com

Key personnel
Railway Superintendent: John Bartholomew

Gauge: 1,435 mm
Route length: 19 km

Freight operations
Owned by Comalco Minerals and Alumina, the railway connects Andoom (mine) with Weipa (port). Operations commenced in 1972 and the line carries some 10 million tonnes of bauxite ore annually.

Traction and rolling stock
The railway is operated with one 1972-built Clyde-EMD GT26C and one 1990-built Clyde-EMD JT42C 3,000 hp/2,240 kW diesel-electric locomotives, 120

bottom-discharge ore (124 tonne gross) wagons and nine freight wagons.

Track
Rails are 67.5 kg/m on hardwood sleepers at 2,020/km spacing. The line includes a 1 km long concrete bridge.

UPDATED

Connex Melbourne

PO Box 5092BB, Melbourne, Victoria 3001
1 Spring Street, Melbourne, Victoria 3000
Tel: (+61 3) 96 10 24 00   Fax: (+61 3) 96 10 26 00
Web: www.connexmelbourne.com.au

Key personnel
Chief Executive Officer: Bruce Hughes

Gauge: 1,600 mm
Route length: 372 km
Electrification: 1.5 kV DC overhead

Political background
The Melbourne passenger train network was government-owned and -operated until 1999. In 1997 it was split into two systems, Bayside Trains and Hillside Trains prior to privatising by the novel method of granting two franchises. The government awarded the Hillside Trains contract to Connex, part

of the Veolia Environnement (formerly Vivendi) group, while National Express was awarded the Bayside franchise. In July 2000 Hillside Trains was renamed Connex Trains Melbourne and Bayside Trains subsequently became M>Train.

Both operators experienced contractual and financial difficulties due to over-optimistic forecasts and a deficient ticketing system. As a result, in December 2002 Connex signed an Interim Operating Agreement with the Victoria government reducing the term of the franchise from 2014 to the end of 2003. Negotiations between the government and National Express failed and the company decided to terminate its franchise.

In February 2004 the government awarded Connex the franchise to operate the entire Melbourne train network for at least the following five years and the company became sole operator of the network on 18 April 2004. The contract provides for a government subsidy.

Organisation
Objectives set by the government as part of the new franchise are primarily aimed at enhancing staff presence on the network, particularly after dark.

Of Melbourne's 209 stations, 67 are designated Premium Stations and are staffed for all train services. Under new franchise initiatives another 31 stations will be staffed during the morning peak and a minimum of 20 staffed in the afternoon. There will also be an increased staff presence on late evening trains.

While it took more than two years to divide the network into Bayside and Hillside franchises, Connex has set a time limit of 18 months to re-integrate operations of the two former systems.

Asset management of Connex Melbourne's infrastructure is contracted to Mainco, which is 30 per cent owned by Connex and 70 per cent by ALSTOM Australia. The network includes 714 km of track, 66 substations, 122 km of transmission lines,

Three-car emu supplied by Siemens (Siemens TS)   ***NEW***/0585294

2,438 signals, 390 bridges and various communication systems. There are also 210 level crossings.

The combined operation employs some 2,600 staff, including those working for maintenance partners.

Traffic
There were 133 million passenger journeys on the combined Melbourne suburban network in 2003–04 compared with 131 million in 2002–03.

A challenge for the network in 2006 will be meeting the demand generated by the staging of the Commonwealth Games in Melbourne.

Passenger operations
Connex Melbourne operates services on 15 electrified lines. These are radial in nature and extend from the city centre to Melbourne's eastern, northeastern, southeastern, bayside, northern and western suburbs. Trains also operate to Melbourne's Showgrounds and Flemington racecourse when required. Connex also operates a diesel-hauled service between Frankston and Stony Point.

Since 18 April 2004 Connex has been responsible for 1,867 scheduled Monday-Thursday services, 1,869 on Fridays, 1,472 on Saturdays and 1,172 on Sundays.

More than 25,000 free car parking spaces are provided at stations.

Improvements to existing lines
In 2004 the Victoria government was undertaking preparatory work for a 9 km extension of the Broadmeadows line to Craigieburn, to where train operations are scheduled to begin in 2006. A new station will be provided at Roxburgh Park and train stabling will be necessary at the terminus.

The important Spencer Street station in the CBD has undergone redevelopment since 2004.

The government plans to provide increased track capacity between Caulfield and Dandenong, and between Jolimont and Clifton Hill as well as improving junction trackwork at those two stations. There are also plans to provide duplicated track between Greensborough and Eltham on the Hurstbridge line. All these enhancements will improve train punctuality and provide scope for additional services. A proposed extension of the Epping line to Morang South is to be the subject of an inquiry.

Traction and rolling stock
The former Hillside fleet inherited by Connex Trains comprised 75 six-car sets, 45 built by Comeng and 29 by Hitachi. All 58 three-car air conditioned X'Trapolis emus supplied by ALSTOM were delivered on budget and on time. They incorporate contemporary facilities such as CCTV surveillance cameras, pushbutton-operated doors, GPS-driven passenger information displays and improved wheelchair access. In addition, since 2000 ALSTOM has undertaken the refurbishment of the Comeng-built units to bring their interiors to a standard comparable with the new X'Trapolis trains.

The company has also taken over the former M>Train fleet, comprising 98 three-car trains built by Comeng and recently refurbished to include provision of improved lighting and information displays, more comfortable seats and CCTV in each car. In mid-2000 Siemens, Austria, was awarded a contract to supply and maintain 62 new three-car emus partly to replace 56 existing Hitachi-built three-car vehicles dating from the 1970s. The first examples of the new fleet, the design of which is an example of Siemens 'MoMo' vehicle concept, entered service in March 2003 and by the end of 2004 17 were in service.

All trains have been equipped with an automatic park brake after a runaway while a driver was changing ends.

By the end of 2005 the entire fleet will be either new or refurbished.

Electrification
The Melbourne suburban network is entirely electrified. In recent years the overhead power supply has been extended to enable trains to serve locations further from the centre.

UPDATED

ALSTOM X'Trapolis emu at Werribee (ALSTOM)
0576722

CountryLink

Level 1, Sydney Central Station, Sydney, New
South Wales 2000
PO Box K349, Haymarket, New South Wales 1238
Tel: (+61 2) 82 02 20 00
Web: www.countrylink.info

Key personnel

General Manager: Greg McLoed
Product and Planning Manager: Luke O'Dwyer
Sales Manager: Nadine Clench

Gauge: 1,435 mm
Track length operated: 4,338 km

Political background

Further restructuring of railways in New South
Wales is in progress in 2005. Involved in this,
CountryLink, which is a business group of RailCorp,
combines the functions of the StateRail authority
with the greater metropolitan functions of the Rail
Infrastructure Corporation. The RailCorp structure
also includes CityRail (see entry in Australia section
of *Railway systems and operators*).

Organisation

CountryLink operates long-distance passenger
services (train and connecting coach) to 360
destinations, primarily in New South Wales but
also across borders into Queensland, Victoria and
the Australian Capital Territory. The New South
Wales services are provided under a community
service obligations contract arrangement with the
state government.

CountryLink operates over the tracks of Australian
Rail Track Corporation Limited in New South Wales
and Victoria and of other infrastructure authorities
in Queensland.

Finance

In 2002–03 costs were A$217 million while farebox
revenue was A$43 million.

Passenger operations

CountryLink services carried 2.13 million
passengers in 2004, operating a daily average of
more than 14,000 km. Return train services operate
between Sydney and 10 destinations:

Daily services to Brisbane (990 km), Casino
(806 km), Grafton (700 km), Armidale (579 km),
Moree (666 km) and Dubbo (462 km)

Xplorer (left) and XPT trains at Sydney Terminal station (Brian Webber) **NEW**/1115039

Diesel railcars or multiple-units

Class	Cars per unit	Motor cars per unit	Motored axles/car	Power/ motor kW	Speed km/h	Units in service	First Built	Builders Mechanical	Builders Engine	Builders Transmission
XPT	7–9	2	4	1,500	160	*	1981	Comeng/ ABB	Paxman VP185	E Brush
Xplorer	2–6	2–6	2	353	145	23	1994	ABB	Cummins KTA-19-R	H Voith

*19 power cars, 60 trailers

Twice daily to Canberra (330 km)
Twice daily (daylight and overnight) to
Melbourne (950 km)
Weekly services to Broken Hill (1,125 km) and
Griffith (640 km).

Patronage has been showing a slow decline,
dropping below 1.9 million journeys in 2003–04.

The weekly weekend Sydney–Griffith return
services previously run by a locomotive-hauled
train were replaced by an Xplorer set in May 2000.
The weekly service to Broken Hill was reinstated
in July 2002. In 2004 the Murwillumbah service
was cancelled beyond Casino, enabling the 129 km
branch to be mothballed. Buses now provide
connections to the many small towns in the district.

Traction and rolling stock

CountryLink's all-diesel fleet consists of:
XPT trains with a power car at each end of a set
of trailers.

Xplorer railcars which can be coupled and
uncoupled en route to allow a train to separate and
serve two destinations.

CountryLink operates 19 XPT power cars, 52 XPT
cars for seated passengers, eight first class XPT
sleeping cars and 23 Xplorer railcars.

The XPT power cars have been re-engined with
new VP185 Paxman engines. By 2005 tenders
had closed for the second refurbishment of the
XPT fleet, which is intended to upgrade them for
another ten years of service.

The New South Wales government has
announced that it intends to call tenders for a
three-car diesel passenger set for a Sydney–Broken
Hill service, to be delivered early 2005 and in
service mid-2005. This set will include some
first class seating, buffet facilities and onboard
entertainment.

Type of coupler: XPT, automatic; Xplorer,
Scharfenberg
Type of braking: XPT, Westinghouse air; Xplorer,
air and hydrodynamic

UPDATED

XPT set forming a Sydney–Grafton service at Thornton, near Newcastle, New South Wales
(Ian Francis) **NEW**/1115040

CRT Group Pty Ltd

PO Box 334, Altona North, Victoria
Tel: (+61 3) 92 90 17 00 Fax: (+61 3) 93 91 71 50
e-mail: email@crtgroup.com.au
Web: www.crtgroup.com.au

Key personnel

Executive Directors: Colin Rees; Phillip Rees
General Manager, Intermodal: John McNamara

Organisation

CRT is a privately owned logistics group with
activities that combine rail, road and shipping
activities. In the rail sector it specialises in
intermodal operations.

Freight operations

Daily intermodal services are operated between
Altona North (Melbourne) and Yennora (Sydney)
and from both terminals to Albury/Wodonga.
Services are also operated to Port Botany and
Sydney Haulage Container Park. There is also a Port
Rail Shuttle service linking the Port of Melbourne
with Altona North.

In May 2004 a three times-weekly service was
initiated between Altona North and Brisbane
in collaboration with QR and an Altona North–
Adelaide service is planned.

Traction and rolling stock

CRT Group does not operate a main line fleet.
However, in 2002 the company procured two

CargoSprinter freight dmu power units from
Windhoff AG in Germany. The design of the units
was adapted to suit Australian conditions and to
enable them to convey 9 ft 6 in containers. Each
power unit is equipped with two Volvo 530 kW
truck engines and hydraulic transmission. Up to
eight wagons can be marshalled between a pair
of power units. The CargoSprinters have been used
for demonstration purposes and on the Port Rail
Shuttle service in Melbourne.

CRT Group also owns three 485 kW Caterpillar-
engined Class 73 B-B diesel-hydraulic locomotives,
two of which are used for yard shunting.

UPDATED

Freight Australia

PO Box 1646N, Melbourne, Victoria 3001
Tel: (+61 3) 96 19 13 11   Fax: (+61 3) 96 19 45 55
e-mail: inquiries@freightaustralia.com.au
Web: http://www.freightaustralia.com.au

Key personnel

Chief Executive Officer: Marinus van Onselen
General Manager, Commercial: Garry Molloy
General Manager, Access: Helen Franklin

Gauge: 1,600 mm; 1,435 mm
Track access (in Victoria): 3,458 km

Political background

With the reorganisation of Victoria's government rail operators, V/Line Freight was corporatised on 1 July 1997 and in February 1999 the government announced the privatisation of the business via a consortium wholly owned by RailAmerica Inc of the USA. The consortium bid A$163 million for the purchase of V/Line Freight's assets and the 45-year lease of access to the track involved. Operations under the new ownership commenced in May 1999.

In 2003 RailAmerica initiated a policy of selling its overseas assets. This resulted in the company reaching agreement in March 2004 to sell Freight Australia to Pacific National for A$285 million, subject to regulatory approval.

Organisation

Staff numbers grew from 570 to nearly 700 in the first two years of privatisation, reflecting business growth. The company has adopted a decentralised management structure to assist the marketing effort.

Finance

In 2001–02 revenue was A$174 million against expenses of A$137 million. The severe drought in Victoria reduced loadings of grain. One customer, a grain marketer, contributed a quarter of revenue, while access fees from V/Line, the passenger operator, represented another 15 per cent. Productivity has been improved by a policy of ignoring state borders to capture profitable business wherever it can be won.

Freight operations

Freight Australia provides freight services on two gauges and resources for services run by other operators. Reference to the map accompanying Victoria Rail Track Corporation shows that effectively there are two independent systems, broad gauge and standard gauge, and while based in Victoria Freight Australia has spread its operations into New South Wales.

In 2001–02 8 million tonnes of freight were carried. About 4 million tonnes of grain, 370,000 tonnes of cement, 550,000 tonnes of minerals/quarry and a million tonnes of containers are hauled annually. The Australian Wheat Board has signed a five-year contract with Freight Australia to haul most of the state's grain to Melbourne for domestic use or to Geelong or Portland for export. At the peak of the 2001 export season, FA was hauling 17,000 tonnes of grain daily to ports. There has been strong growth in container traffic at several provincial centres.

The two lines into southern New South Wales carry a considerable number of wheat or rice containers. Locomotives have been hired from other operators to handle additional traffic when necessary.

Contracts secured in 2002 included one from Nestlé Purina, covering the transportation of cereals for five years, and a three-year deal with Shell Oil company of Australia to convey bulk fuel in New South Wales.

In May 2003 FA commenced regular operations between Melbourne and Sydney, two of Australia's largest generators of freight traffic.

FA won a five-year contract to provide hook-and-pull services for SCT between Melbourne and Adelaide and across the Nullarbor Plain to Perth. To avoid refuelling stops, an innovative in-line fuelling system has been introduced, enabling locomotives to receive fuel from a tank car as the journey progresses.

Freight Australia provides safe working, crews or locomotives for other operators, including West Coast Railway.

Improvements to existing lines

FA has reopened a 74 km freight-only line between Sale and Bairnsdale to capture timber products traffic.

Traction and rolling stock

The V/Line Freight traction fleet at sale comprised 86 diesel-electric locomotives, of which 15 are standard gauge (11 G Class, three H Class, and one X Class). While most date from the 1960s and 1980s, the three S Class units are over 40 years old.

In 2001 Freight Australia rebuilt six G Class locomotives with 2,600 kW GM 645F3B engines and repowered six X Class to 2,235 kW. Seven A Class units are to be equipped with Super Series

control equipment. EDI/Clyde has delivered one new V Class unit.

Some locomotives have been stored due to a downturn in traffic after drought conditions affected grain loadings.

Freight Australia owns 1,189 grain hopper wagons, 275 box cars, 902 container flats and 79 tank wagons.

In 2001 700 grain wagons were being rebuilt to increase their capacity from 76 to 100 tonnes, install improved loading and discharge systems and equip them with bogies allowing their operating speed to be increased to 115 km/h.

Track

Freight Australia holds a sub-lease of most country track in the state of Victoria. It must grant access to any other operator of both passenger and freight services.

The broad gauge lines west from Ouyen and Heywood into South Australia see no trains. The new standard gauge line to Wolseley provides an alternative for any traffic. Freight Australia has surrendered two lines to the government: Toolamba-Echuca and Maryborough-Castlemaine, both of which duplicated other lines.

Gauge conversion

The Victoria state government has been investigating the desirability of converting some or all broad-gauge track to standard gauge. In 2004 the latest statements suggested that there was little enthusiasm for conversion as the financial benefits were difficult to quantify. A figure of A$96 million over five years has been allocated in state budgets but this would provide for conversion of only a small distance. It is estimated that if 27 per cent of broad gauge track (2,000 km on 13 lines) were converted then 70 per cent of traffic would be able to be part of the national standard gauge network.

There are three track standards, all with welded rail: 47 kg/m rail on wooden sleepers for operations at up to 115 km/h; 47 or 53 kg/m rail on wooden sleepers for 130 km/h; and 60 kg/m rail on concrete sleepers for 160 km/h.

UPDATED

Diesel locomotives: 1,600 mm or 1,435 mm gauge

Class	Builder's type	Wheel arrangement	Power kW	Speed km/h	Weight tonnes	No in service	First built	Mechanical	Builders Engine	Transmission
A	AAT22 C-2R	Co-Co	1,840	115	121	7	1983*	Clyde	GM 12-645 E3B	E GM
G	JT26 C-255	Co-Co	2,460	115	127	31	1984	Clyde	GM 16-645 E3B	E GM
H	G188	Bo-Bo	820	100	81	5	1968	Clyde	GM 8-645E	E GM
P	G18HB-R	Bo-Bo	826	100	77	5	1984*	Clyde	GM 8-645E	E GM
S	–	Co-Co	1,450/ 1,340	133	123	4	1957	Clyde	GM 16-567C	E GM
T	G88	Bo-Bo	710/ 826	100	69	13	1959	Clyde	GM 8-567 CR or 8-645E	E GM
V	GF46C	Co-Co	3,095	115	134	1	2002	EDI Rail	GM 710 G3B-ES	E GM
X	–	Co-Co	1,450/ 1,640	115	118	24	1966/ 1974	Clyde	GM 16-567 E or 16-645E	E GM
Y	–	Bo-Bo	480	64	68	14	1963	Clyde	GM 6-567C or 6-645E	E GM

* Rebuilt

Three G Class diesel-electric locomotives leave Port August with an Adelaide-bound freight, having crossed the Nullarbor Plain (Brian Webber)   0547786

H Class locomotive shunting containers near Portland depot, Victoria (Brian Webber)
0547787

Great Southern Railway (GSR)

PO Box 445, Marleston Business Centre, Marleston, South Australia 5033
Tel: (+61 8) 82 13 10 44   Fax: (+61 8) 82 13 44 80
e-mail: stationmaster@gsr.com.au
Web: www.trainways.com.au

Key personnel
Chief Executive Officer: Tony Braxton-Smith
Director, Finance: (vacant)
Director, Marketing and Sales: Alan Stuart
Director, Operations: Brian Duffy
Director, Guest Services: Bronwyn Schoen
General Manager, Safety: Keith Hunt

Gauge: 1,435 mm

Background
Following its decision to sell off the assets of the Australian National Railway Commission, the Commonwealth Government of Australia executed a contract with the Great Southern Railway Consortium for the acquisition of the passenger rail business in November 1997. The company became a wholly owned subsidiary of Serco Group Pty Ltd in October 1999.

Passenger operations
Carrying some 250,000 passengers annually, Great Southern Railway is a leading provider of long distance rail services in Australia and the owner of the country's three celebrated tourist trains, The Indian Pacific, The Ghan and The Overland. The company also operates its 'Trainways' programme of holiday packages that include rail travel, accommodation, tours and one-way air tickets.

With completion of the new transcontinental link to the north in early 2004, The Ghan runs as a weekly return service between Adelaide and Darwin, crossing almost 3,000 km through

The Ghan passing through a typical central Australian landscape (GSR)   ***NEW**/1067947*

Australia's 'red centre'. A second weekly return service operates between Adelaide and Alice Springs, the heart of the Australian outback. In 2005 the peak tourism period (May to July) was to see two weekly return Adelaide–Darwin services and three between Adelaide and Alice Springs.

Three classes of accommodation are provided: Gold Kangaroo, with à la carte dining, private sleeping cabins and en-suite facilities; Red Kangaroo Sleeper, with shared twin sleeping accommodation and private washing facilities; and Red Kangaroo Daynighter, providing reclining seat accommodation. Dining and lounge cars are provided for Gold and Red Kangaroo services.

The twice-weekly Indian Pacific runs 4,352 km from Perth to Sydney via the Nullarbor Plain and Adelaide, with accommodation provided in three classes, Gold Kangaroo, Red Kangaroo and Red Kangaroo Daynighter, similar to the Ghan.

The Overland is operated four times a week in each direction between Adelaide and Melbourne, running during daytime from Adelaide and overnight from Melbourne. Gold and Red Kangaroo Daynighter levels of service are provided. The service is financially supported by the Victoria and South Australia state governments.

Traction and rolling stock
GSR is the owner of 111 passenger cars and 14 motorail wagons, which are maintained under contract by United Goninan.

The company does not own any locomotives: traction and crews are hired from Pacific National and primarily consists of NR Class locomotives, some painted in Ghan colours. The contract was renewed for five years from July 2002.

UPDATED

Pacific National Pty Ltd (PN)

Locked Bag 90, Parramatta, New South Wales 2124
Tel: (+61 2) 98 93 25 00   Fax: (+61 2) 98 93 25 01
Web: www.pacificnational.com.au

Key personnel
Chief Executive Officer: Stephen O'Donnell
Chief Financial Officer: Mal Grimmond
General Managers
  Commercial: Robert Jeremy
  Operation Services: John Fullerton
  Coal: Peter Winder
  Intermodal: Graham Lyon
  Rural and Bulk: Mike Bandinette

Gauge: 1,435 mm; 1,600 mm
Route access: RailCorp (New South Wales)–2,865 km; ARTC (Victoria, South Australia, Western Australia)–4,490 km; QR (Queensland)–131 km; Westnet (West of Kalgoorlie, Western Australia)–710 km

Political background
The former National Rail Corporation Ltd (NR) was formed in September 1991 to create a competitive and commercially viable rail freight business. Its shareholders were the Commonwealth Government of Australia and the states of New South Wales and Victoria. The company commenced commercial operations in April 1993. It took over all of the interstate rail freight business conducted by five separate state-based rail authorities, including the assets predominantly used in that business.

At the end of January 2002, the Australian government announced that NR was to be sold to National Rail Consortium Pty Ltd (NRC Pty) in a combined sale with the former FreightCorp. The value of the transaction, which was completed at the end of February 2002, was A\$1.172 billion. Trading as Pacific National (PN), the company is owned jointly by Patrick Corporation and Toll Holdings, two of Australia's largest public companies.

In March 2004 Pacific National reached agreement with Rail America to purchase Freight Australia for

A\$285 million. Freight Australia operated mainly on the broad gauge network in Victoria although it also ran standard gauge services in New South Wales. The company purchased the former Tasrail operation in the island state of Tasmania in late 2004 (see entry for Pacific National Tasmania in the Australia section of *Railway systems and operators*).

Pacific National Queensland commenced operations in March 2005, running daily services on the coastal 1,600 km 1,067mm gauge route in Queensland.

Organisation
PN uses its own employees to operate its own intermodal terminals and trains, using wagons and locomotives it owns, leases or hires. Access to track is obtained through contracts with authorities owned by state and commonwealth governments, which also provide all train control. Having begun operations in 1993 as NR, PN is now the country's largest carrier of long-distance freight by rail.

Passenger operations
While PN does not operate passenger services, its locomotives are seen at the head of Great Southern Railways' (see entry in Australia section of *Railway systems and operators*) interstate 'Indian Pacific', 'Ghan' and 'Overland' services as PN provides 'hook and pull' services for GSR.

Freight operations
Under the new combined operation, PN provides interstate transport on the standard gauge network and intrastate services in New South Wales and Victoria. Its market share ranges between 15 per

Two NR Class diesel locomotives at Kagaru, Queensland, with a container train (Brian Webber)
***NEW**/1115036*

cent on the Sydney–Melbourne route to 40 per cent on the long–haul east coast-Perth corridor (across the Nullarbor Plain). It is estimated that in 2005 PN will haul over 115 million tonnes of freight.

In late 2003 PN signalled its intention to compete with QR on the Queensland network when it ordered a fleet of 1,067 mm gauge diesel locomotives (see Traction and rolling stock). Operations in the state commenced in March 2005.

The former NR signed a contract with BHP worth A$1.5 billion over eight and a half years to provide transport between manufacturing and distribution centres. A daily train of blue metal is provided from Dunmore Quarry, south of Sydney, to Cooks River, Sydney.

Following improvements to track, speeds on the Melbourne–Adelaide link have increased and 21 tonne axleloads are now allowed. Maximum train length on that route is now 1,500 m with loads to 5,000 tonnes.

Between Adelaide and Perth, clearances have been improved to 6.7 m to enable the regular double-stacking of containers on well wagons. On this route, NR now has competition from Patrick Stevedores, Specialized Container Transport and Toll Rail.

NR trains up to 1,800 m long operate in the east–west corridor and trains up to 1,300 m in the north–south corridor. Infrastructure improvements have been made to permit 1,500 m operations.

Services have been withdrawn on the Maroona-Portland line in western Victoria.

Intermodal operations

In 2001, the former NR was carrying some 600,000 containers annually.

The Internet-based FreightWeb e-commerce system, on line since August 1997, is a first for rail in Australia. FreightWeb provides customers with secure access to book and trace containers across the network, and has received strong market acceptance.

SeaTrain

Four weekly SeaTrain container services run between Brisbane and Sydney port areas. Mayne Logistics (Boxcar) now gives all its business to PN.

Trailerail

Trailerail has been operated by NR and now PN since 1 July 1996. This unit's freight is carried in road/rail trailers of three types: dry or refrigerated pantechnicons or curtain-sided trailers. Two trains are operated weekly between Melbourne, Adelaide and Perth and return, offering services which compete directly with road transport. NR introduced a Sydney–Perth service in February 1999.

Improvements to existing lines

Although PN is not responsible for track conditions, the operator has benefited by the continued reduction in the number of speed restrictions across the standard-gauge network. A new Y connection at Parkes, New South Wales, which cost only A$2 million, has resulted in substantial time savings, sometimes of up to three hours. The introduction of train order working between Parkes and Broken Hill has also substantially reduced transit times.

A new operating yard to accommodate 1,500 m trains was opened adjacent to Sydney freight terminal in time for the 2000 Olympic Games. The A$15 million project provided stabling for three trains.

Traction and rolling stock

Standard gauge

The NR Class locomotives, used principally on interstate hauls, have a fuel capacity of 12,500 litres, allowing them to run through between Brisbane and Melbourne at 115 km/h where track permits. One locomotive can haul 1,780 tonnes in Victoria. The type features a variable horsepower system which allows drivers to conserve fuel when the locomotives are running on sections not requiring full power or with light loads.

Supplementing the NR Class now are some 81, AN, BL, and DL Class locomotives used as trailing

One of 13 PN Class locomotives ordered for operations on the 1,067 mm gauge network in Queensland, seen at the Maryborough plant of its builder, EDI Rail (Brian Webber)   **NEW**/1115037

Two 90 Class locomotives at East Maitland with an empty coal train (Ian Francis)   **NEW**/1115038

Diesel locomotives (ex-NRC and Freight Corp – 1,600 mm or 1,435 mm gauge)

Class	Builder's type	Wheel arrangement	Power kW	Speed km/h	Weight tonnes	No in service	First built	Mechanical	Builders Engine	Transmission
BL	JT26C-2SS	Co-Co	2,240	150	128	10	1983	Clyde	GM 16-645E3B	E GM
48/PL	DL531G	Co-Co	710	120	75	114	1959	Goodwin	Alco 6-251B	E GE/AEI
80	CE615A	Co-Co	1,492	130	121	10	1978	Comeng	Alco 12-251CE	E Mitsubishi
81	JT26C-2SS	Co-Co	2,240	133	129	97	1982	Clyde	GM 16-645E3B	E GM
82	JT42C	Co-Co	2,259	115	132	55	1994	EMD	GM 710G3A	E EMD
C	GT26C	Co-Co	2,240	132	132	2	1977	Clyde	GM 16-645E3	E GM
DL	AT42C	Co-Co	2,240	150	122	14	1988	Clyde	GM 12-710 G3A	E GM
90	GT46CW-M	Co-Co	2,836	115	165	31	1994	EMD	GM 719G3A	E EMD
AN	JT46C	Co-Co	2,860	150	130	9	1992	Clyde	GM 16-710 G3A	E GM
NR	CV40-9i	Co-Co	3,000	115	132	120	1996	Goninan	GE 7FDL-16	E GE

Diesel locomotives (ex- Freight Australia – 1,600 mm or 1,435 mm gauge)

Class	Builder's type	Wheel arrangement	Power kW	Speed km/h	Weight tonnes	No in service	First built	Mechanical	Builders Engine	Transmission
A	AAT22C-2R	Co-Co	1,840	115	121	7	1983*	Clyde	GM 12-645E3B	E GM
G	JT26C-255	Co-Co	2,460	115	127	31	1984	Clyde	GM 16-645E3B	E GM
H	G188	Bo-Bo	820	100	81	5	1968	Clyde	GM 8-645E	E GM
P	G18HB-R	Bo-Bo	826	100	77	5	1984*	Clyde	GM 8-645E	E GM
S	–	Co-Co	1,450	133	123	4	1957	Clyde	GM 16-567C	E GM
T	G88	Bo-Bo	710	100	69	13	1959	Clyde	GM 8-567cR or 8-645E	E GM
V	GF4^C	Co-Co	3,095	115	134	1	2002	EDI Rail	GM 12-710 G3B-ES	E GM
X	–	Co-Co	1,450/ 1,640	115	118	24	1966/ 1974	Clyde	GM 16-567E or 16-645E	E EMD
Y	–	Bo-Bo	480	64	68	14	1963	Clyde	GM 6-567C or 6-645E	E GM

* Date rebuilt

or shunting units. The use of electric locomotives in the Sydney area ended in June 2002.

The former NR purchased 32 120 tonne coal wagons from Adtranz (now Bombardier Transportation) for a Macquarie Generation contract in the Hunter Valley of New South Wales. The operator has also spent A$10 million for 73 wagons of three permanently coupled skeletal platforms to haul steel products, from ANI Engineering, Mittagong, New South Wales.

NR Class maintenance is contracted to Goninan, which provides all services except refuelling throughout Australia. Each locomotive visits the Spotswood, Melbourne maintenance depot every 122 days, and continuous computer links to the locomotives provide diagnostic analysis.

The Australian Wheat Board purchased 51 NGXH 100 tonne gross grain wagons from China and leased them to PN to run with the rail company's 54 similar wagons.

Broad gauge (1,600 mm)
A fleet of 107 locomotives and 2600 wagons was purchased with Freight Australia in 2004. Of those about 75 locomotives and 1,600 wagons run on this gauge.

Narrow gauge (1,067 mm)
In late 2003 PN ordered 13 PN Class diesel-electric locomotives for operations over the Queensland network. Deliveries commenced in February 2005 from EDI Rail's Maryborough, Queensland plant. The locomotives are similar to QR's 4000 Class from the same builder. In addition, 178 three-segment container wagons have been purchased to

Pacific National 81 Class locomotives operating stone traffic in connection with extensions to the port of Brisbane (Brian Webber)   0583007

run between Brisbane's Moolabin Yard, Townsville and Cairns. Three MKA Class locomotives perform shunting duties.

Signalling and telecommunications
Train control and signalling functions are provided by the owners of track on which PN trains operate.

PN's need to monitor its train operations cross all mainland Australian states has seen the use of mobile telephones pioneered for communication between locomotives and the company's control centre. PN also has a truly national train radio system, which provides transparent and seamless communication across all states and track owners.

A computer system is being installed to enhance further PN's ability to track its trains and customers' freight and to manage freight resources. Supported by AEI readers in and near terminals, the system will handle operational train and marshalling yard management, as well as locomotive and wagon deployment and management.

UPDATED

Pacific National Tasmania

PO Box 140, Newstead, Launceston, Tasmania 7250
Tel: (+61 3) 63 37 22 11   Fax: (+61 3) 63 37 22 19
Web: www.pacificnational.com.au

Key personnel
Chief Executive Officer, Pacific National:
 Stephen O'Donnell

Gauge: 1,067 mm
Track length: 805 km (630 km in use)

Political background
Originally a state government enterprise, Tasrail was a component of the Commonwealth government-owned Australian National which was auctioned in August 1997.

Tasrail was purchased by a consortium led by Wisconsin Central Transportation Corporation, Berkshire Partners, and Fay Richwhite and Company (of New Zealand), which has also bought into rail operations resulting from privatisation in New Zealand and the United Kingdom. The group paid A$22 million for the Tasmanian rail freight business including track and infrastructure, 30 locomotives, rolling stock and modern maintenance workshops. Ownership of Tasrail land passed to state government control to be leased to ATN. The new owners took over at midnight on 14 November 1997, with a 50-year lease on the track.

In April 1998 ATN acquired the Pasminco Emu Bay Railway (the Melba Line) for a purchase price of A$7.8 million.

Following the acquisition of Wisconsin Central by Canadian National, Australian Transport Network, including Tasrail was sold in February 2004 to Pacific National Pty Ltd, the major freight operator on the mainland network. The trading name Pacific National Tasmania was adopted.

Subsequent to the acquisition of the network by Pacific National, the Commonwealth Government designated the main lines from Hobart to Burnie and from Western Junction to Bell Bay as part of the National Rail Network.

Organisation
As well as operating services, Pacific National Tasmania also undertakes maintenance and upkeep of track and infrastructure.

The company employs 170 staff.

Freight operations
The island of Tasmania does not represent an ideal market for rail, as distances are short and there is little bulk traffic available. The distance between the two major cities, Hobart and Launceston, is only about 150 km. Additionally, the railway now competes with B-double trucks, which have been allowed to operate on the state's major roads. However, Pacific National see growth opportunities in several areas, including forest products and container movements.

Most trains require two or three locomotives, with most track seeing two or three trains on weekdays. About a third of the traffic is bulk cement from Railton to Devonport, a short haul.

Services on the North East line from Launceston were suspended in October 2004.

New workshop facilities for wagon and locomotive maintenance at East Tamar Junction, Launceston, were completed in September 1993, resulting in improvements in productivity and maintenance efficiency, and reducing delays in returning locomotives and rolling stock to traffic.

Traction and rolling stock
The purchase in 2003 by Pacific National included the acquisition of 39 ageing locomotives and 668 wagons in Tasmania and a number of standard gauge wagons in Victoria. Three 2140 Class locomotives have been stored while 3 similar MKA Class have been obtained.

Tasrail (now Pacific National Tasmania) freight powered by a 2020 (D) Class locomotive and two other diesels crossing the Firth River near Devonport (Brian Webber)   0573218

Signalling and communications
A communications upgrade has been completed, with cellular telephones and UHF radios now installed in all locomotives and track maintenance vehicles. Track Warrant Control system is in use.

Track
The track west of Burnie sees no traffic following damage, while the North East line has seen no traffic since October 2004.
Tunnels: 3

UPDATED

Diesel locomotives

Class	Wheel arrangement	Power kW	Weight tonnes	No in service	First built	Mechanical	Builders Engine	Transmission
2020	Co-Co	1,491	110	2	1971	Clyde	GM 16-645E	E GM
2150	Bo-Bo	597	58.4	2	1961	TGR[1]	EE 6SRKT	E EE
2140	Co-Co	1,339	89.5	6	1967	EE, Rocklea	EE 12CSVT	E EE[3]
2110	Co-Co	1,380	97.5	4	1972	EE, Rocklea	EE 12CSVT 11	E EE
2114	Co-Co	1,752	97.5	5	1973	EE, Rocklea	EE 12CSVT 111	E EE
2120	Co-Co	1,752	92	6	1973	EE, Rocklea	EE 12CSVT 111	E EE[5]
2100	Co-Co	1,339		1	1996	EE, Rocklea	EE 12CSVT 111	E EE[2]
2101	Co-Co	1,339		1	1997	EE, Rocklea	EE 12CSVT 111	E EE[2]
2000	Co-Co	1,119	91	5	1964	Clyde	GM 12-645E	E EMD[4]
QR	Co-Co	1,119	91	3	1964	Clyde	GM 12-645E	E EMD[4]
DC	A1A-A1A	1,230	82.75	1	1978	Clyde	GM 12-645E	E EMD[4]

[1] Tasmanian Government Railways, Launceston.
[2] Rebuilt from 2120 Class.
[3] Acquired from Queensland Rail 1988.
[4] From Tranz Rail (New Zealand).
[5] Acquired from Queensland Rail 1987.

Pilbara Rail Company

PO Box 21, Dampier, Western Australia 6713
Tel: (+61 8) 91 43 63 30   Fax: (+61 8) 91 43 63 45

Key personnel
Chief Operating Officer: Geoff Neil

Gauge: 1,435 mm
Route length: 899 km (HI–638 km; RR–261 km)

Organisation
From April 2002 Pilbara Rail Company took over the rail operations, though not the ownership, of Hamersley Iron Ore and Robe River Iron Associates, following these two mining organisations coming under the common majority ownership of Rio Tinto. The move also facilitated the operation of trains for both companies on a single section of track and avoided building a separate parallel route to the new mine at West Angelas.

The railway operates solely to convey iron ore from the mines to two ports at Dampier, in the remote Pilbara area of Western Australia. It has the capacity to handle approximately 110 million tonnes per annum. In recent years there has been strong demand for iron ore, particularly from China, with the Pilbara mines securing a substantial market share. There was an increase in exports of 3 million tonnes in 2002.

Hamersley Iron line
The original 280 km section to Mount Tom Price opened in 1966, with the line extended a further 100 km to Paraburdoo in 1972. Since then 56 km have been doubled and a 40 km spur was constructed to serve Brockman Mine in 1991. In 1994 another spur line was constructed to a new mine at Marandoo, since extended to West Angelas (68.5 km) in 2001–02.

Robe River line
This railway enabled the first haul of iron ore in August 1972 from Pannawonica to Cape Lambert port (192 km).

Freight operations
Hamersley Iron line
This line has several branches to serve mine loading points. The main line is from Dampier to Rosella (251 km), from where there is a branch to the west to Brockman (294 km) and another east to Yandicoogina (440 km), which has a 68.5 km branch from Juna Downs to West Angelas, a mine operated by Robe River. The main line continues to Wombat Junction (280 km), where there is a branch to the west to Tom Price (295 km) with the line terminating at Paraburdoo (385 km).

Trains from the mines at Tom Price, Brockman and Marandoo consist of two diesel-electric locomotives hauling 226 wagons each of 105 tonnes nominal capacity. This results in gross train weights of 29,000 tonnes. Wagons are coupled in pairs by a solid drawbar with rotary couplings connecting each pair. A train is approximately 2.3 km in length and is the heaviest and longest employing head—end locomotive power anywhere in the world. The main line configuration permits following train movements at 15 minute headways. There are about nine services each way per day.

Two Pilbara Rail GE Dash 9-44CW locomotives with a loaded iron ore train for Dampier (Ian Francis)
NEW/1115030

Diesel locomotives

Class	Wheel arrangement	Power kW	Speed km/h	Weight tonnes	No in service	First built	Mechanical	Builders Engine	Transmission
Dash 9-44CW/ CM44-9CW	Co-Co	3,430	105	197	60[1]	1995	GE	GE 7FDL16	*E* GE
CM40-8M[2]	Co-Co	3,130	110	195	12	1989	GE	GE FDL16	*E* GE

[1] Includes six locomotives on order in 2005
[2] Robe River locomotives, rebuilt from C636/M636

The maximum opposing grade to loaded trains on the Tom Price–Dampier section is 0.33 per cent, while empty trains returning to the mine negotiate a maximum adverse grade of 2 per cent. These grades and the gross loads of trains permit an exact balance of locomotive power. On the Tom Price–Paraburdoo section there is a constant compensated grade of 0.42 per cent against loaded trains. Three banking locomotives are required by loaded trains for the 100 km journey between Paraburdoo and Tom Price to overcome this adverse grade. At Dampier trains are unloaded in rotary dumpers at either the Parker Point or the EII terminal. At Parker Point pairs of wagons are uncoupled from the train prior to dumping.

When the tare weights of wagons and work trains are added, the single-line railway sees about 85 million tonnes of traffic annually, making it one of the heaviest tonnage single-line railways in the world.

Robe River line
The railway provides transport of ore from Pannawonica, now Deepdale, to Cape Lambert port. Ore tonnage totalled a record 31.2 million tonnes in 1999–2000. Single-operator trains with four locomotives and about 200 wagons travel between the mine and port in under four hours. There are three consists in use making about five return runs daily. The railway has the capacity to carry well over 32 million tonnes annually.

Traction and rolling stock
Hamersley Iron line
A fleet comprising 29 main line diesel locomotives, 2,430 ore wagons and 126 maintenance vehicles is available. The General Electric Dash 9 locomotives, the most powerful in Australia, have microwave ovens and CD players for crew comfort. In 2005 the company had on order a further six Dash 9 locomotives from GE/United Goninan.

Robe River line
The line received 11 new General Electric C44-9W locomotives in 2004. These enable two sets to be run with these locomotives and another to be run with existing CM40-8M machines.
Couplers in standard use: Fixed and rotary
Braking in standard use: Westinghouse air

Signalling and telecommunications
Hamersley Iron line
A CTC system, using motorised point operations and block signalling, controls traffic. Radio communication is maintained with train crews and track personnel and provides an emergency back-up service should the CTC fail. The communication system consists of a microwave bearer and UHF mobiles. This is the first entirely cab-signalled system in Australia, controlling trains on about 600 km of track.

Robe River line
Train movements are authorised by train order from Cape Lambert control centre using VHF radio.

Track
The two lines were originally separate but now have a connection known as Weston Creek between 71 km on the Robe River line and 77 km on the Hamersley line. Intensively used high-axleload lines require considerable continuous maintenance. The railway has an impressive selection of maintenance machinery including tamping machines, ballast regulators, a track recording vehicle, rail grinder, P811 resleepering machine and 33 ballast hoppers with two plough wagons.

Hamersley Iron line
The main line is predominantly single-track with passing sidings at approximately 20 km intervals. Heavy-duty 68 kg/m rail, continuously welded, is used throughout, with alloy and head-hardened steels utilised for high-wearing curve sections.
Rail: 68 kg/m
Crossties (sleepers): Concrete, 2,590 × 265 × 211 mm
Spacing: 1,640/km + 1,539/km
Fastenings: Pandrol clips
Max speeds: Loaded train 80 km/h, empty train 70 km/h, freight train 80 km/h, light engines 100 km/h
Max gradients: Against empty 2.03%, against loaded ex-Tom Price 0.33%, against loaded ex-Paraburdoo 0.42%
Max altitude: About 750 m near Tom Price

Robe River line
Three crossing loops are provided in the 200 km between Deepdale load point and Cape Lambert port unloading centre.
Rail: 68 kg/m
Crossties (sleepers): Concrete 2,600 × 280 × 235 mm

Four Dash 8 locomotives lead a Robe River loaded iron ore train across the Hamersley Iron line on the descent to Port Lambert (Wolfram Veith)
0121174

Spacing: Concrete, 1,538/km
Fastenings: Concrete, Mackay Safelock
Min curvature radius: 3°
Max gradient: 1.29% (empty), 0.5% (loaded)
Max axleload: 36 tonnes

UPDATED

QR

PO Box 1429, 305 Edward St, Brisbane, Queensland
4001
Tel: (+61 7) 32 35 22 22   Fax: (+61 7) 32 35 17 99
e-mail: qrati.bit.net.au
Web: www.qr.com.au

Key personnel
Chairman, Railways Board: Bronwyn Morris
Chief Executive: Bob Scheuber
Group General Managers:
    Chief Operating Officer: Tony Drake
    Passenger Services: Michael Scanlan
    Shared Services: Brian Bock
    Rolling Stock and Component Services: Andy Taylor
    Infrastructure Services: Glen Mullins
    Network Access: Stephen Cantwell

Gauge: 1,435 mm (interstate line); 1,067 mm; dual-gauge
Route length (owned): 99 km; 9,099 km; 36 km. An additional 293 km of 1,067 mm gauge route see no services
Route length (not owned): 35 km of 1,067 mm gauge
Electrification: 1,877 route-km; 2,453 track-km of 1,067 mm gauge at 25 kV 50 Hz AC

Political background
QR is a government-owned corporation established under the state Queensland Transport Infrastructure Act. All shares are held on behalf of the state by the state Treasurer and the Minister for Transport and Main Roads. A nine-member, government appointed board guides the management team.

Organisation
QR is Australia's largest rail network and one of Australia's major transport businesses. The change of name from Queensland Rail has been introduced to reflect the company's future direction in not restricting its activities either to its home state or to Australia. Operations currently consist of the following groups: QRNational; Passenger Services; Infrastructure Services; Network Access; Workshops; Shared Services. This latest structure has separated commercial operations from those dependent on government Community Service Obligation grants whilst remaining as one integrated organisation. It has also placed Network Access at arm's length from QR operator groups to enable equal opportunity for third party operators. In 2005, a major competitor, Pacific National Queensland, commenced running on QR tracks. QR will be able to invite the private sector to participate in joint venture operations, particularly where there are opportunities for expansion. QR intends to remain an integrated organisation keeping train operations and infrastructure together. Unlike other Australian railways, QR believes that, worldwide, this is the structure of the most successful operators. QR has expanded activities to interstate and overseas as contractors under the name iQR.

QR has purchased an interstate operation, now named QRNational (see entry in Australia section of *Railway systems and operators*), to run standard-gauge freight services into other states.

In June 2004 QR employed 13,658 staff.

Finance (A$ million)	2001–02	2002–03	2003–04
Sales	2,058	2,112	2,178
Profit	252	170	191

In 2003–04 sales revenue improved for the sixth consecutive year.

QR has a new financial structure and a requirement to pay income tax equivalent to the Queensland government.

QR emu activity in Brisbane (Brian Webber) 0554904

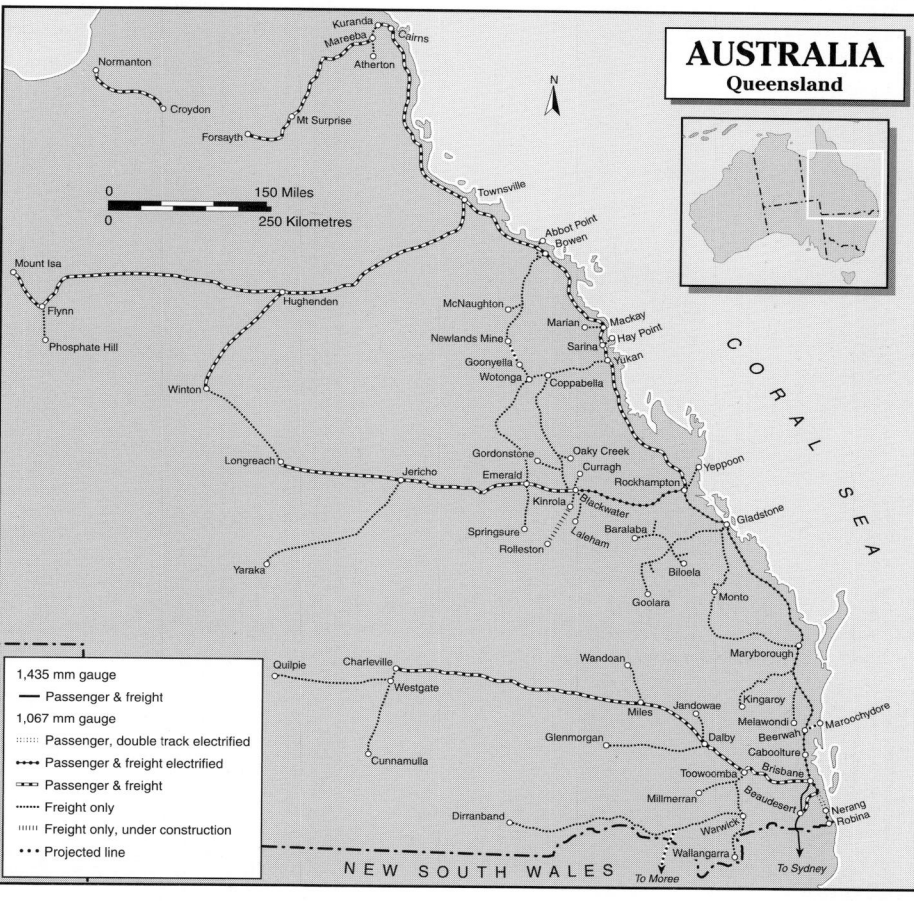

AUSTRALIA
Queensland

1,435 mm gauge
— Passenger & freight
1,067 mm gauge
∷∷∷∷ Passenger, double track electrified
••••• Passenger & freight electrified
▭▭▭ Passenger & freight
····· Freight only
ıııııı Freight only, under construction
••• Projected line

NEW/1115026

Northbound QR diesel Tilt Train near Narangba (Brian Webber) 0554900

Passenger operations
Citytrain (urban)
QR operates eight suburban and three interurban routes in Brisbane and adjoining cities with most routes provided with half-hourly services. Patronage in 2003–04 was 48.1 million journeys, continuing the increases of recent years. About 158,000 people are carried each weekday with 4,848 services provided weekly. A new, more

robust timetable has provided many stations with a 15 minute off-peak weekday service. Punctuality improved to 97 per cent on-time to within three minutes. QR also provides trains for Airtrain (see entry in Australia section of *Railway systems and operators*) services.

QR has provided all stations with mini-computers and ticket vending machines, contributing to a 15 per cent increase in revenue. The introduction

2300 Class diesel electric locomotive leaving Winton for Longreach and Brisbane with a livestock train (Brian Webber)   0554901

QR Speno rail grinder working between Bundaberg and Isis (Brian Webber)   0554903

QR Diesel Tilt Train at Mayne (Brian Webber)   *NEW*/1115022

Cairns (1,654 km; 'Sunlander' and 'Tilt Train' three times and twice weekly respectively)
Townsville (1,319 km; 'Sunlander' twice weekly)
Longreach (1,325 km; 'Spirit of the Outback' twice weekly)
Charleville (777 km; 'Westlander' twice weekly)
and between Townsville and
Mount Isa (977 km; 'Inlander' twice weekly).

Unlike most other Australian rail systems, which have scaled down long-distance passenger services because of increasing road and air competition, Traveltrain has upgraded rolling stock, focused on travel through scenic areas and actively marketed its products through its travel centres, travel agents and wholesalers. However, patronage is slowly declining.

A high-profile QR activity is the running of 15-carriage Kuranda Scenic Railway tourist trains from Cairns, ascending a range with spectacular views, 15 tunnels and passing two waterfalls. The two daily services carry about 450,000 passengers each year.

The unique rail motor (built in 1950) called the 'Gulflander' continues to run its weekly 153 km service on the isolated Normanton–Croydon line, attracting tourists to the remote Gulf of Carpentaria region.

The two electric Tilt Trains commenced running the daylight services between Brisbane and Rockhampton from November 1998. Additional Tilt Train services between Brisbane and Bundaberg commenced in July 1999. The services operated by the Tilt Trains have proved very popular and even resulted in competing airlines reducing their fares. A Tilt Train servicing depot has been established at Gracemere, near Rockhampton. Two diesel Tilt Trains have provided Brisbane-Cairns services since June 2003, replacing some 'Sunlander' services and allowing withdrawal of older coaches.

Passenger traffic	2001–02	2002–03	2003–04
Journeys (million)			
Traveltrain	1.08	1.05	1.04
Citytrain	45.4	46.2	48.0

Freight operations

QR now brands its freight operations as QRNational, complete with a new locomotive colour scheme, only the third in 50 years.

In 2001–02 QR was the first Australian transport company to haul more than 150 million tonnes of freight in a single year. Coal continues to make up the majority of freight business, with QR carrying a record 143 million tonnes in 2003–04. Most coal is hauled to the Central Queensland ports of Hay Point and Gladstone, with smaller quantities to Abbot Point and Fisherman Islands and to domestic power stations. The state government is planning for the industry's future infrastructure needs in the expectation that railed coal production will rise from 143 million tonnes in the year to 30 June 2004 to 202 million tonnes in the year to 30 June 2010. The company's other freight business includes minerals, containers, industrial products, grain and general merchandise.

The general traffic wagon fleet has the capacity to haul three million tonnes of grain, 1.7 million tonnes of raw sugar and 800,000 head of cattle annually. QR is the only Australasian system now hauling cattle; 600,000 head are carried in most years.

With the recommencement of the Phosphate Hill project, QR is hauling a million tonnes of fertiliser to the coast with a similar volume of acid being hauled west from Townsville.

New container terminals have been provided at Cloncurry and Ayr.

Pacific National has has started to compete with QR on North Coast Line services. This presents a major challenge for QR, as Pacific National will gain from its owners, Patrick and Toll, the significant tonnages currently conveyed on QR services.

Freight traffic	2001–02	2002–03	2003–04
(millions)			
1,067 mm gauge			
Tonnes carried			
Coal	135.0	138.6	143.5
Minerals and other freight	18.4	17.9	18.4

of integrated ticketing with bus operators in the southeast has resulted in more convenient and cheaper travel with significant increases in patronage.

Most stations have been equipped with closed circuit television and many trains have been equipped with video surveillance cameras in an effort to combat anti-social behaviour. Each evening two or three services on each line are advertised as 'Guardian trains' and carry two security guards to improve passenger security and confidence.

Traveltrain (country)

QR provides air-conditioned passenger services between Brisbane and:
Bundaberg (351 km; 'Tilt Train' weekdays)
Rockhampton (622 km; 'Tilt Train' daily)

Two 3200 Class electric locomotives operated via the Locotrol system marshalled mid-train in an empty coal consist at Yukan (Brian Webber)   0576710

Total	153.4	156.5	161.9
Net tonne-km	40,503	n/a	n/a

New lines
Freight lines
Further development of Queensland coal and mineral deposits has prompted expansion of the existing network to service new mines and associated processing industries.

A new coal mine at Hail Creek required a 52 km connection to QR's Goonyella line, which opened in late 2003. The mine should be producing 5.5 million tonnes annually after four years.

The international mining company, Xstrata, has committed to opening a new coal mine at Rolleston, for which QR is building a 110 km, A\$240 million extension to its Kinrola branch in central Queensland. Production is expected to start in late 2005, with an output of 3.2 million tonnes, increasing to 8 million tonnes by 2008. Grain may also be hauled over the line.

A new 6.1 km electrified branch with a balloon loop has beenprovided to serve Moorvale mine, which is on the Peak Downs line south of Coppabella. In the same area a new crossing loop has been provided at Winchester.

The government has announced funding of A\$25 million to prepare for construction of a 78 km connecting line between the Goonyella system and Newlands to enable coal to be hauled to the government-owned Abbot Point port. The present coal loading facilities of Hay Point and Dalrymple Bay are privately owned and are unlikely to be able to handle additional export coal, for which miners are keen to win contracts at this time of high demand.

Passenger lines
An environmental impact assessment study has been released for public comment for an extension of the Gold Coast line beyond its present terminus, Robina, to Reedy Creek, Andrews and Elanora. The government has announced that the line will be extended 4 km to a new terminus at Reedy Creek by 2007.

On the Sunshine Coast north of Brisbane, the government has announced that the 20 km section from a junction with the North Coast Line at Beerwah to Caloundra will be built by 2015 and extended to Maroochydore (another 20 km) by 2020.

Improvements to existing lines
Citytrain network
A fourth track has been provided between Mayne and Northgate and a third track has opened between Northgate and Lawnton which includes one 18-span and one six-span bridge.

Additional platforms and track have been provided at Park Road on Brisbane's Citytrain network at a cost of A\$13 million. With bidirectional signalling from there to the central business district stations, it is now possible to run more peak hour services.

Country network
A major regrading of the track at The Leap, north of Mackay, has been completed to enable a larger through load to be hauled.

Most coal lines have now been approved for 80 km/h operation by block trains and for 100 km/h when other trains run.

The 977 km line between Townsville and Mount Isa is being upgraded to allow increased axleloads (20 tonnes) and higher operating speeds (80 km/h

for mineral trains). Works completed include the replacement or upgrading of bridges, steel resleepering of the 590 km between Hughenden and Mount Isa and commencement of heavy-duty turnouts to all main line loops. A programme to replace all timber sleepers between Townsville and Hughenden with prestressed concrete sleepers and to install heavier rail cascaded from other lines was completed by late 1998.

A new passenger station has been built at Townsville to serve Tilt Train services, which commenced in June 2003.

Other improvements
The A\$370 million project to upgrade the Rockhampton–Cairns line has been completed

IMU emu crossing the Nerang River bridge with a service to Brisbane's airport (Brian Webber)
NEW/1115023

allowing 100 km/h running south of Townsville. The Rosewood–Helidon track relaying project will be completed thanks to a further budget allocation of A\$13.3 million. A similar sum is being invested between Miles and Muckadilla employing part-used rails from elsewhere. Large sums are being spent on timber bridge and low-speed turnout replacement.

The state government is to fund a third track between Salisbury and Kuraby (9.5 km) and to double Ormeau-Coomera (7 km) and Helensvale– Robina on the Gold Coast line and Mitchelton-Keperra on the Ferny Grove line. Costed at A\$342

Electric railcars

Class	Cars per unit	Motor cars per unit	Motored axles/car	Power/kW motor	Speed km/h	Units in service	First built	Builders Mechanical	Electrical
1979 emu	3	2	4	135	100	67	1979	Walkers	ASEA
1982 emu	3	1.5	4	135	100	20	1983	Walkers	ASEA
1988 ICE	2	2	4	135	120	8*	1988	Walkers	ASEA
1994 SMU	3	2	4	180	100**	42	1994	Walkers	ABB
1995 IMU	3	2	4	180	140	14	1995	Walkers	ABB
1998 Tilt	3	2	2		160	4	1998	Walkers	ABB

* plus 4 trailers
** some 115 km/h

Diesel railcars

Class	Cars per unit	Motor cars per unit	Motored axles/car	Power/kW motor	Speed km/h	Units in service	First built	Builders Mechanical	Engine	Transmission
2003 Tilt	9	2	4	2 × 1,350	160	2	2003	EDI Rail	MTU	H Voith

Electric locomotives

Class	Wheel arrangement	Output kW	Speed km/h	Weight tonnes	No in service	First built	Builders Mechanical	Electrical
3100/3200	Bo-Bo-Bo	2,900	80	110	82	1986	Comeng	Hitachi/GEC
3300/3400	Bo-Bo-Bo	2,900	80	113	22	1994	Clyde	Hitachi/GEC
3500/3600	Bo-Bo-Bo	2,900	80	110	50	1986	Walkers/ASEA	ASEA/Clyde
3550[1]	Bo-Bo-Bo	2,900	80	110	29	1988	Walkers/ASEA	ASEA/Clyde

[1] Ex-3900 Class upgraded for coal traffic

Diesel locomotives

Class	Builder's type	Wheel arrangement	Power kW	Speed km/h	Weight tonnes	No in service	First built	Mechanical	Builders Engine	Transmission
1550[3]	GL 22C	Co-Co	1,119	100	91	6	1975	Comeng/Clyde	EMD 12-645 E	E EMD/Clyde
2400	GL 22C	Co-Co	1,119	100	91	14	1977	Comeng/Clyde	EMD 12-645 E	E EMD/Clyde
2450	GL 22C	Co-Co	1,119	100	91	9	1979	Comeng/Clyde	EMD 12-645 E	E EMD/Clyde
2470	GL 22C-2	Co-Co	1,119	100	91	34	1980	Comeng/Clyde	EMD 12-645 E	E EMD/Clyde
1720	GL 18C	Co-Co	746	100	62.5	44	1966	Comeng/Clyde	EMD 8-645E	E Clyde
2100[3]	GL 26C	Co-Co	1,492	100	93–97	10	1970	Comeng/Clyde	EMD 16-645E	E EMD/Clyde
2130[3]	GL 26C	Co-Co	1,492	100	93–97	9	1974	Comeng/Clyde	EMD 16-645E	E EMD/Clyde
2141[3]	GL 26C-AC	Co-Co	1,492	100	93–97	5	1973	Comeng/Clyde	EMD 16-645E	E EMD/Clyde
2150[3]	GL 26C-2	Co-Co	1,492	100	93–97	5	1978	Comeng/Clyde	EMD 16-645E	E EMD/Clyde
2170[3]	GL 26C-2	Co-Co	1,492	100	93–97	40	1982	Comeng/Clyde	EMD 16-645E	E EMD/Clyde
2300[2]	GL 22C	Co-Co	1,329	100	93	60	1998[2]	Clyde/QR	EMD 12-645E	E EMD
2600	U 22C	Co-Co	1,640	80	109	13	1983	Goninan, Qld	GE FDL12	E GE
2800	CM 30-8	Co-Co	2,240	110	117	50	1995	Goninan, Qld	GE 7CDL12	E GE
4000	GT 42CU	Co-Co	2,260	100	120	49	2000	EDI Rail (Walkers/Clyde)	EMD12N710G	E EMD/ Siemens

[1] stored
[2] rebuilt from 1550/2400/2450/2470 Classes
[3] Many being upgraded with 12-cylinder turbocharged engines, ballasted to 110 tonnes and re-designated 2250 Class

Newly built 4000 Class diesel-electric locomotive at EDI Rail's Maryborough facility (Brian Webber)
NEW/1115025

2300 Class diesel locomotive in the latest QRNational livery on road transport for its return to service after overhaul (Brian Webber)
NEW/1115024

million, these projects are expected to take three years to complete and will allow the running of much needed additional services.

QR has called tenders for 8.7 km of track duplication on the Central Line between Windah and Grantleigh to include an additional bridge at Gogango Creek. A previous tender was for Wallaroo–Dingo (22 km).

Traction and rolling stock

QR's locomotive fleet at June 2004 comprised 330 diesel locomotives and 183 electric locomotives, while 10,304 (mainly coal and container) wagons and 985 service vehicles were available.

The passenger fleet comprised 449 urban emu cars, 12 electric and 18 diesel Tilt Train cars and 184 Traveltrain locomotive-hauled coaches. Delivered in 2003 were five power cars for the air-conditioned Traveltrain fleet to replace older vehicles.

The Maxi overhaul programme to upgrade and renumber locomotives as they pass through the workshops has ended with 60 2300 Class diesel-electric locomotives in traffic.

The additional eleven 4000 Class diesel-electric locomotives have entered service costing A$69 million. A number of older locomotives are being upgraded with turbochargers and weighted to provide greater haulage capacity, re-entering service as the 2250 class. The 3900 class electric locomotives are being upgraded and transferred from passenger and general freight work to coal services.

A contract has been signed with United Goninan to rebuild three electric locomotives as prototypes for the remaining 1986-built units. It is anticipated that rebuilt locomotives incorporating modern equipment will reduce the number of traction units required on coal hauls, allowing for the operation

of more consists using the same number of locomotives.

Thirty sets of improved SMU units have entered service on Brisbane Region passenger services, replacing the remaining locomotive-hauled sets. Another 24 sets have been ordered from and EDI Rail and Bombardier joint venture at cost of A$212 million. Fifteen of the three-car trains will be to IMU

interurban configuration with toilets, while the remainder will be SMU types for suburban work. They are to be delivered by late 2008.

Two nine-car diesel Tilt Trains entered service in June 2003 between Brisbane and Cairns. Like the two existing electric Tilt Trains, they are able to run at 160 km/h where track conditions permit.

Recently 400 new container flat wagons have been delivered and a further 319 existing wagons have been upgraded for heavier loads and 100 km/h running. In 2005 250 coal wagons were being delivered for Rolleston services.

Signalling and telecommunications

Safe-working systems in use are:
Remote-controlled signalling (CTC): 3,119 km
Train order: 6,382 km
Ordinary staff: 643 km (low traffic branches)
Electric staff: 75 km (1,435 mm line)

Radio is used extensively, with all trains having direct access to train controllers while en route.

Centralised Traffic Control is in use between Brisbane and Townsville and between Ipswich and Toowoomba, allowing improved running times and withdrawal of employees from remote crossing loops. Direct Traffic Control, a train order type system involving computers on locomotives, has been extended to lightly used lines. The impressive Universal Traffic Control, a PC-based signalling system for the Brisbane suburban area has been commissioned. It has allowed fewer staff to control 268 km of track, allowing for 1,916 routes with 1,069 signals and 669 points.

A$7 million is to be spent providing CTC to replace electric staff on the standard-gauge line between Acacia Ridge (Brisbane) and the state border, where the QR network connects with ARTC.

Electrification

All Brisbane suburban services (over six million kilometres annually) are run by emus, while Traveltrain passenger and freight services are hauled by diesel-electric locomotives. Central Queensland coal services from the Blackwater and Goonyella areas are hauled by up to five electric locomotives, including two or three mid-train, often 3900 Class, controlled by the driver with Locotrol or Distributed Power remote equipment. Many Blackwater area coal trains are now hauled by 4000 Class diesel-electric locomotives equipped for Distributed Power. They are also used on the Moura, Newlands and Mount Isa lines.

Several short sections have been electrified in connection with the opening of additional coal mine trackage and the McArthur branch.

Track

Double track is provided in the Brisbane area between Caboolture and Helidon, Sandgate, Mitchelton, Ormeau and Manly, with further additional tracks between Lawnton, Lindum, Salisbury and Corinda. In Central Queensland, double track extends from Gladstone

2800 Class locomotive on a general freight mostly formed of intermodal traffic near Theebine (Brian Webber)
0576707

Two 4000 Class locomotives on a loaded train of 75 VALQ coal hoppers from Newlands, leaving the North Coast line at Kaili en route for Abbot Point (Brian Webber)   0576708

to Rocklands (112 km), with a further 50 km west of Rocklands (Westwood-Windah, Tunnel-Wallaroo), and from Jilalan to Broadlea (130 km), on the Goonyella coal route. In North Queensland, Nome to Townsville (20 km) is double track.

A A$42 million re-equipping of the on-track maintenance fleet is at an advanced stage with deliveries of three new high production continuous action tampers, ballast regulators, sleeper inserting machines and on-track welders. A new A$8 million 72 tonne rail grinder was supplied by Speno Australia in August 1997. QR estimates the Speno Star 40 M1 self-propelled rail grinder will save A$50 million in five years.

A new A$14 million Plasser ballast cleaning machine has been acquired. It weighs 245 tonnes empty and is 88.5 m long. The impressive machine is similar to units in use in Europe and the USA but is the first in the world on 1,067 mm gauge. With three ballast storage wagons, it represents the largest single contract for track maintenance machinery ever in Australia. It will be used on the Goonyella Line.

A major A$374 million programme to relay the Rockhampton–Townsville (702 km) section of the North Coast line with 60 kg/m rail on concrete sleepers has been completed. Some 1.5 million sleepers and 310 turnouts have been installed. This has allowed the line's maximum speed to be raised to 100 km/h.

Standard rail: Flat bottom 60, 53, 50, 47, 41, 40, 31, 30 and 20 kg/m rail has been used throughout the state, dependent on line class. New construction has been standardised to 60 kg/m rail for heavy-haul lines, 50 kg/m as the normal main line standard and 41 kg/m for lighter trafficked lines.
Joints: 6-hole bar fishplates.
Welded rail: Rails are purchased in 27.4 m lengths and flashbutt welded at depot into lengths up to 110 m. Long-welded rails are laid in lengths up to 220 m on unplated track and to unrestricted lengths on plated track. Heavy-haul lines and other lines with prestressed concrete sleepers are continuously welded. Site welding is generally by the thermite process though extensive work using a mobile flashbutt welding machine has been undertaken.
Tracklaying: Relaying of track is predominantly carried out by tracklaying machine. In 1997/98 170 km was relaid, well over budget, with the machine laying a record 12.63 km in one week. Some 350,000 prestressed concrete sleepers worth A$22.1 million are being inserted in the Townsville to Hughenden section over two years.
Sleepers: Mostly unimpregnated local hardwood timber 2,150 × 230 × 115 mm or 150 mm thick on the older heavy-haul lines. Prestressed concrete sleepers are used extensively for new construction including heavy-haul lines. Steel sleepers are

being installed on a continuous face for over 450 km between Hughenden and Mount Isa and on shorter sections of other lines. They are also replacing timber sleepers on a 1 in 3 or 1 in 4 pattern for almost 1,000 km between Rockhampton and Cairns. Extensive installation of treated timber sleepers has also been undertaken in recent years, although current policy is no longer to purchase timber sleepers.
Sleeper spacing: Normally 610 mm in main line or heavy-haul tracks for timber, 685 mm for concrete and steel.
Fastenings: Normal standard 16 mm square dogspikes and springspikes with 115 mm thick timber sleepers; 19 mm square dogspikes used with 150 mm thick timber sleepers and indirect fasteners on curves used on older heavy-haul lines. The use of elastic rail spikes has now been discontinued. Indirect fastenings are used with concrete and steel sleepers.
Ballast: Mainly crushed rock in new work but river gravel used on some branch lines.
Max curvature: Generally minimum radius of 100 m though new construction to 300 m radius at least.
Max gradient: Generally not exceeding 1 in 50, recently 1 in 90.
Max altitude: 925 m at The Summit, Southern line
Max permitted speed: Freight trains–100 km/h; long-distance passenger trains–100 km/h for the 'Lander' services and 80 km/h for other services; suburban emus–100 km/h; interurban ICE units–120 km/h; interurban IMUs–140 km/h. Tilt trains operate at 160 km/h where allowed on the North Coast line
Max axleload: 26 tonnes on some mineral lines, 20 tonnes on North Coast line and electrified tracks, 15.75 or 10.6 tonnes elsewhere
Bridge loading: All bridges on important lines can carry loading equivalent to Coopers E25-E30. Many equivalent to Coopers E35 and most new construction to this standard. Heavy-haul mineral lines have bridges built to carry Coopers E50 loading. QR's longest pre-stressed concrete bridge is the 856 m Coomera River bridge on the Gold Coast line, opened in 1996.
Tunnels: 51

UPDATED

QRNational

CasinoSuite 1, 91 Frederick Street, Merewether, New South Wales 2291
Tel: (+61 2) 49 63 07 00   Fax: (+61 2) 49 63 51 41
e-mail: info@interail.com.au
Web: www.freight.qr.com.au

Key personnel
General Manager: Catherine Baxter
Business Development Managers: Neill Bencke, John Quilter

Political background
The Queensland State Government-owned QR (see entry for QR in Australia section of *Railway systems and operators*) had been keen to enter the interstate standard gauge market for some time. This was a strategy to expand interstate business opportunities to provide a more national focus. An opportunity arose to purchase an existing operation, Northern Rivers Railroad (NRR), based in Casino, New South Wales. It had already obtained approvals for track access. To fulfil this strategy, QR formed a subsidiary, QRNational, trading as Interail Australia, to purchase the rolling stock and operations of NRR and to provide services on the interstate standard gauge network. The takeover was finalised on 31 May 2002. In 2005 the Interail name was replaced in most publicity by QRNational, in line with the rebranding of its parent.

The company operates over lines owned by track authorities, mainly Australian Rail Track Corporation.

Freight operations
The company initially concentrated on providing locomotives and crews for ballast and maintenance trains to the Rail Infrastructure Corporation (see entry in Australia section of *Railway systems and operators*). It also offers train maintenance and first

QRNational 423 Class (ex-QR 1,067 mm gauge 1502 Class) and 421 Class locomotives with a Casino–Glenapp welded rail train (Brian Webber)   *NEW*/1115000

response services such as breakdown and recovery under contract. In March 2003 it commenced a contract to haul coal between Duralie mine and Stratford washing plant in the Hunter Valley region of New South Wales. It is anticipated that about one million tonnes of coal a year will be hauled for at least seven years.

A second coal contract was won in September 2003 to move over 800,000 tonnes of coal per year between Newstan Mine and Vales Point Power Station, south of Newcastle. A further more significant coal haul contract in the Hunter Valley has been secured to carry 10 million tonnes annually from Mount Arthur Mine to port.

General freight services, mainly of containers, were commenced in April 2004 between Melbourne (Dynon and Altona), Sydney (Yennora) and Brisbane (Fisherman Islands) using leased locomotives and container wagons. Initially three services a week were run but this was quickly doubled.

Traction and rolling stock
In 2005 the company owned 10 locomotives of various classes, and also leases locomotives to meet operational requirements. QR has gauge-converted six of its previously withdrawn 1502 Class locomotives to become the 423 Class units in the standard gauge fleet. Four locomotives from the Freight Australia (Victoria) operation have been obtained for the Melbourne–Brisbane service. Nine 5000 Class locomotives and 310 Type VEA 106 tonne wagons have been purchased from United Goninan for the Mount Arthur coal haul contract. These locomotives are equipped with a 3,000 kW GE 7FDL16 engine, Distributed Power systems and they and the wagons have a through-wired electronic control pneumatic braking system.

UPDATED

Rail Corporation New South Wales (RailCorp)

Level 2, Sydney Central Station, Sydney, New South Wales 2000
PO Box K349, Haymarket, New South Wales 1238
Tel: (+61 2) 93 79 30 00   Fax: (+61 2) 93 79 53 10
Web: www.staterail.nsw.gov.au

Key personnel
Chief Executive: Vince Graham

Gauge: 1,435 mm
Electrification: 1.5 kV DC

Political background
On 1 July 1996 the New South Wales government initiated a reform programme which at the time claimed it would improve public transport services by separating the infrastructure and train management functions of the former State Rail Authority and establishing four separate entities.

From that date, SRA became responsible for the operation of all passenger services of the CityRail and Countrylink business units (see entries in Australia section of *Railway systems and operators*) and retained those assets essential to its business, including rolling stock and stations.

All rail freight services were established under the separate FreightCorp organisation (since sold); the Railway Services Authority (subsequently Rail Services Australia) was established to supply goods and services to the rail industry including infrastructure, while the Rail Access Corporation became responsible for ownership and management of the New South Wales rail permanent way. In January 2001 these two organisations merged to become Rail Infrastructure Corporation.

On 1 January 2004 Rail Corporation New South Wales (RailCorp) was formed to merge the State Rail Authority of New South Wales and the metropolitan functions of the Rail Infrastructure Corporation. This further change reversed the previous decision to separate above- and below-rail activities, which has been found to result in a loss of accountability and direction. The new corporation is only responsible for passenger operations as government freight activities were sold to Pacific National Pty Ltd. Country maintenance functions will continue either under the geographically reduced Rail Infrastructure Corporation or Australian Rail Track Corporation Limited.

Finance
The former State Rail Authority had revenue in 2002–03 of A$2,099 million and expenditure of A$2,018 million. Revenue came from fares and government subsidy while expenditure was primarily staff costs and rail access fees.

Passenger operations
Details of operations and traction and rolling stock appear under headings CityRail and Countrylink.

Freight operations
In 2002–03 the state's rail network carried 41.6 billion tonne-km gross of freight. Nine commercial operators have access to the network, with Pacific National the dominant operator. Non-grain freight continues to rise, with a notable increase on the routes to/from Port Botany, Sydney, which has seen a significant reduction in truck movements in the congested area.

New lines
Parramatta Rail Link
As originally planned the Parramatta Rail Link involved construction of a new 28 km line between Westmead/Parramatta on the Western line (via Epping on the Main North line) and Chatswood on the North Shore line. The aim of this project was to provide a passenger bypass of the lines through Sydney's central business district to relieve capacity constraints on the inner Western line. The new line, which was to have been mainly underground, would provide commuters with a direct link between Parramatta, located in the west of Sydney and the city's second major business

1115029

district, the Macquarie Park educational end employment centre and the Lower North Shore. The line would add five new stations to the Sydney suburban network, while seven existing stations would be rebuilt or upgraded.

In 2001 the state government announced a four-year commitment to fund A$1 billion of rail improvements including a start of planning and pre-construction on the 15 km Chatswood-Epping section of the link to Parramatta. Planning approval was given in February 2002. An initial phase, including a new Parramatta Transport Interchange, is to be completed by 2006; the Chatswood-Epping section is scheduled for completion by 2008.

However, in August 2003 the New South Wales government announced that construction of the 13 km Epping-Parramatta section of the scheme was to be deferred indefinitely, arguing that the line did not meet cost-benefit requirements.

Other lines
A further new line has been proposed in Sydney's northwest from Beecroft on the Parramatta-Hornsby line to an initial terminus at Mungerie Park (Rouse Hill). It is predicted that this area will see rapid residential growth during the next 10 years. Of the line's 19 km the initial 11 km will be underground. Six stations are proposed though no dates for construction have been set. Trains could continue to the city by the future Chatswood-Epping line. The line could be further extended to join with the Blacktown-Richmond line at either Riverstone or Vineyard.

Construction of new lines is no longer a priority, attention instead focused on resolving problems including the provision and financing of reliable services on the existing network.

Improvements to existing lines
CityRail network
In early 2004 the CityRail service suffered from chronic late running and cancellation problems, which embarrassed the government and led to a change of focus from investing in more infrastructure, to ensuring the reliability of current operations. A six-year A$1 billion investment programme is intended to implement projects to simplify operations on the CityRail network in a bid to improve service reliability. Plans include designating five routes as 'Rail Clearways' and separating operations on 14 routes. It is accepted that the network is too complex, with a problem on one service affecting those on other lines hours later. Additional track is to be provided to duplicate the Cronulla line and similar provision will increase capacity on several trunk routes. There are also

to be additional platforms and turnback tracks at Berowra, Homebush, Hornsby, Lidcombe, Liverpool, Macarthur and Revesby. Three turnback tracks have been provided at Macdonaldtown (CBD area). A crossing loop is to be installed on the Carlingford line to enable half-hourly frequency services to be operated.

A A$30 million project to provide a second track between Marayong and Quakers Hill on the outer suburban Richmond line, has been completed, improving reliability of timekeeping on the branch. Two new crossovers, resignalling and overhead wiring replacement between Quakers Hill and Schofield as well as an upgrade of facilities at Bankstown Stabling sidings added a further A$6 million to the project.

A turnback facility has been provided at Central to enable trains from the Airport line to terminate when necessary. A$10.2 million has been spent to signal this improvement. A turnback track is being provided at Bondi Junction (a terminus and not a junction) at a cost of A$54 million to enable more trains to run on the line to relieve overcrowding.

Other lines
In an unexpected move, the government announced that the Broadmeadow–Newcastle branch will be closed. This will take the railway away from the Newcastle central business district and require train users to change to buses for the final few kilometres. Justification for this proposal is that the ground-level railway divides the area with several level crossings. Currently electrified services from Sydney and diesel railcar services from the Hunter Valley provide a frequent service.

Signalling and telecommunications
CityRail network
A new Sydney Rail Management Centre has been commissioned to centralise the functions of train control, maintenance, management and passenger information. A shortcoming of the system in the recent past has been the absence of a Central Signalling Centre with most signalling controlled locally. A Metropolitan Signal Control System project has been commenced to provide more accurate timetable information and automated reporting of delays. In 2003 the first of the new Control Centres was opened at Sydenham, where a quick response team of technical staff is based. The A$31 million Sefton Park Area resignalling project has commenced. Sefton Park is one of the locations where the network is vulnerable due to many conflicting movements at a triangular junction.

Electrification

Overall, 1,760 km of track is electrified. Overhead electrification extends from Newcastle (168 km from Sydney) in the north to Bowenfels in the west (160 km), Glenlee in the south (60 km) and to Kiama, in the Illawarra (119 km). All other passenger routes in Sydney are all electrified. It is planned to extend electrification by 34 km to the end of the Illawarra line at Nowra.

Track

The state's rail network of 8,700 km of track includes more than 5,000 bridges.

Double track extends from near Muswellbrook (289 km) in the north to Junee (486 km) in the south, Wallerawang (171 km) in the west, and Unanderra (88 km) in the Illawarra. Quadruple (or more) track is provided between Newcastle and Maitland in the north, between Sydney and St Marys in the west and between Sydney and Hurstville (Illawarra line). Responsibility for freight-carrying sections of this part of the network now rests with the Australian Rail Track Corporation.

A third track has been provided between Turella and Kingsgrove on the East Hills line in Sydney.

The Sydney Harbour bridge, though not owned by a rail company, has the longest span (503 m) of any railway bridge in Australia and carries the tracks 60 m above the water.

RailCorp has obtained a Plasser RM900-HD ballast cleaner, claimed to be the world's largest. It is 135 metres long, weighs 430 tonnes and can run at 80 km/h.

Rail: Type to AS1085: 30, 36, 40, 47, 53 and 60 kg/m.
Crossties (sleepers): Timber or concrete, 230 × 130 × 2,440 mm, spaced 1,666/km in plain track, 1,818/km in curves.
Fastenings: Resilient
Min curvature: 160 m (10.9°)
Max axleload: 25 tonnes
Max gradient: 1 in 30 (3.3%)
Longest double-track tunnel: Woy Woy, 1.8 km
Longest bridge: Hawkesbury River, 843 m
Recent bridge standard: Coopers E60

UPDATED

Rail Infrastructure Corporation (RIC)

GPO Box 47, Sydney, New South Wales 2001
Level 15, 55 Market Street, Sydney, New South Wales 2000
Tel: (+61 2) 92 24 30 00   Fax: (+61 2) 92 24 39 00
e-mail: info@ric.nsw.gov.au
Web: www.ric.nsw.gov.au

Key personnel

Chief Executive Officer: John Cowling

Route length: 2,448 km
Gauge: 1,435 mm

Political background

RIC was formed as a state-owned corporation in January 2001 under the Transport Administration Amendment (Rail Management) Bill 2000. It is a merger of Rail Access Corporation (RAC), previously owner of the New South Wales rail network, and Rail Services Australia (RSA), which was responsible for its maintenance. These two organisations had been set up in 1996 as part of the reform of the New South Wales railway system. Accordingly, RIC inherited RAC's functions of overseeing safety and maintenance of the infrastructure and of selling access rights to train operators and those of RSA in the delivery and management of infrastructure assets.

Following further amendments by government to the structure and responsibility for Sydney area rail passenger services and the lease of many New South Wales main lines to the Australian Rail Track Corporation Ltd, from January 2004 RIC was responsible only for track maintenance and train control of most country branch lines, other functions passing to Rail Corporation New South Wales.

Improvements to existing lines

The state government has announced A$21 million funding for continued maintenance of nine country grain-only branch lines totalling 610 km and suspension of operations on four others totalling 354 km. Those to receive funding are lines which although in recent years have only seen seasonal or occasional grain trains, are seen to carry sufficient traffic to be a viable and necessary alternative to road haulage. The lines are: Camurra–North Star (84 km); Burren Junction–Merrywinebone (52 km); Bogan Gate–Tottenham (115 km); West Wyalong–Lake Cargellico (112 km); Ungarie–Naradhan (60 km), Griffith–Hilston (108 km), The Rock–Boree Creek (57 km); and Koorawatha–Greenthorpe (22 km).

Those lines on which operations will be suspended are: Barmedman–Rankins Springs (115 km); West Wyalong–Burcher (54 km); Yanco–Willbriggie (40 km); and Binnaway–Gwabegar (145 km). The state of the track, low potential traffic volumes and an alternative road network have influenced the closure decision. The announcement was a politically difficult one for government as any decision to remove facilities from rural areas is always very unpopular. The Casino–Murwillumbah branch (129 km) has seen its only services withdrawn.

UPDATED

Silverton Rail

PO Box 415, Parramatta, New South Wales 2124
Level 5, 20 Charles Street, Parramatta, New South Wales 2150
Tel: (+61 2) 96 33 57 77   Fax: (+61 2) 96 33 57 66
e-mail: info@silverton.net.au
Web: www.silverton.net.au

Key personnel

Managing Director: Graham Clements
NSW State Manager: John McArthur
Regional Manager, Broken Hill: Doug Aikins

Gauge: 1,435 mm

Organisation

Now a subsidiary of Transcorp Pty Ltd, the company originated as the connecting railways between Broken Hill (New South Wales) and Silverton (South Australia), hauling minerals and supplies for the mines of Broken Hill. With the conversion of the railway through Broken Hill to standard-gauge, the company remained as a provider of shunting services in Broken Hill. The governments of Australia allow third-party operators on to the tracks of the government systems so Silverton has expanded its activities into the niche market of providing crews and short-term leasing of locomotives and wagons to other operators.

Silverton Rail employs around 100 staff.

Freight operations

A major Silverton Rail customer is Namoi Cotton. The company is also responsible for a Cobar–Newcastle ore haul and containerised grain movements

Class 442 and 48 locomotives during ballasting operations at Tragalbar, New South Wales (Brian Webber)
NEW/1115028

from Nyngan. It has been successful in obtaining maintenance train contracts with Rail Infrastructure Corporation in Western New South Wales.

Traction and rolling stock

The company owns 30 operational locomotives, including six 442 Class, three 44 Class, one 45 Class, two C Class and ten 48 Class, as well as around 100 wagons. It also owns most of the surplus locomotives that Pacific National was required to sell as a condition of its privatisation. They include 25 80 Class diesel and 58 electric locomotives, mostly inoperable. It is unclear what the future holds for these purchases but they could prove valuable as traffic circumstances change into the future. The electric locomotives are the only ones available for the Sydney area electrification, should electric locomotive haulage again be desired.

UPDATED

For details of the latest updates to *Jane's World Railways* online and to discover the additional information available exclusively to online subscribers please visit
jwr.janes.com

Specialized Container Transport (SCT)

7 Westlink Court, Altona, Victoria 3018
Tel: (+61 3) 99 31 53 33   Fax: (+61 3) 93 69 97 47
e-mail: information@sct.net.au
Web: www.sct.net.au

Key personnel

Chairman: Peter J Smith
Chief Executive Officer: Martin A Svikis

Gauge: 1,435 mm

Political background

Specialized Container Transport decided that it should operate its own trains after experiencing difficulties dealing with existing operators. It introduced a private, weekly freight service between Melbourne, Adelaide and Perth in July 1995, breaking 100 years of government monopoly of rail services. Today it operates three weekly services between those three cities.

Freight operations

Patronage has necessitated that trains be run three times each week. SCT built an A$9 million rail freight terminal in Perth in 1998; in 1999 a new terminal in Melbourne was commissioned at a cost of A$20 million and a A$7 million facility was opened in Adelaide in 2001.

SCT claims to have continually broken records, notably introducing Australia's first refrigerated rail service and achieving trailing loads of 4,400 tonnes in 53-wagon formations through the Adelaide Hills and 6,300 tonne 74-wagon trains elsewhere. These operate at speeds of up to 110 km/h.

Traction and rolling stock

Main line locomotives and crews are provided by other operators. SCT has purchased eight locomotives for shunting at Adelaide and Perth. The company has invested considerably in rolling stock, with a fleet of over 250 vehicles. Additional box wagons have been delivered from QR's Queensland workshops. SCT also has a substantial fleet of road vehicles in most states of Australia to complement its rail operations.

Track

The company's trains operate by agreement over track managed by the Australian Rail Track Corporation.

UPDATED

TransAdelaide

136 North Terrace, Adelaide, South Australia 5000
Postal: GPO   Box   2351,   Adelaide,   South Australia 5001
Tel: (+61 8) 82 18 22 00   Fax: (+61 8) 82 18 22 06
e-mail: info@transadelaide.sa.gov.au
Web: www.transadelaide.sa.gov.au

Key personnel

General Manager: Bill Watson
Manager, Rail Systems: Valdis Evele
Manager, Fleet Services: Jim Sandford
Manager, Business Services: Dennis Huxley
Manager, Corridor Services: Randall Barry

Gauge: 1,600 mm
Length: 120 route-km

Political background

Following the dissolution of the State Transport Authority, TransAdelaide was launched on 4 July 1994. It operates train and tram services under a contract until 2010 to the Office of Public Transport, which has responsibility for policy and planning aspects of public transport.

The current government has released its State Strategic Plan, which sets as an objective the doubling of public transport use to 10 per cent of all weekday travel by 2018. The current five-year plan includes consideration of electrification of the network rather than ultimate replacement of the ageing diesel railcar fleet. Adelaide is the only mainland state capital city in Australia not to have an electric urban train service. It is possible/likely that the opportunity would be taken to standardise the rail gauge at the same time.

Organisation

The Authority controls the metropolitan passenger railway system of Adelaide. Some track is shared with freight services to and from the country system. In 2004 TransAdelaide employed 545 staff, 362 of which were involved with train and tram operations.

Passenger operations

The Authority operates five suburban rail routes serving 84 stations to termini at Outer Harbor (22 km), Gawler (41 km), Noarlunga (30 km), Grange (13 km) and Belair (22 km), providing 30-minute off-peak/20-minute peak frequency railcar services. The years of steady decline in public transport patronage in Adelaide, a city of about one million residents, has been arrested, with average growth in patronage of 4.46 per annum experienced since 2001.

In 2003–04 11.19 million train passengers were carried, continuing the increase in patronage that followed a decline until 1998–99, when 7.4 million passenger journeys were made. A survey has revealed that school and tertiary level students form the largest group of travellers, with 60 per cent using rail services five days per week.

The Authority continues to encourage passengers to purchase tickets 'off-board' from licensed ticket

TransAdelaide 3100 Class dmu passing Dry Creek yard, Adelaide (Ian Francis)   **NEW**/1115003

Diesel railcars or multiple-units

Class	Cars per unit	Motor cars per unit	Power/ car kW	Speed km/h	Vehicles in service	First built	Mechanical	Builders Engine	Transmission
2000	2/3	1	395 × 2	130	M12	1980	Comeng	Cummins KTA 19R	H Voith
2100	2/3	1	–	–	T18	1980	Comeng	–	
3000	1	1	354	100	M30	1988	Comeng/Clyde	Mercedes-Benz	E ABB
3100	2	2	354	100	M40	1988	Comeng/Clyde	Mercedes-Benz	E ABB

vendors, though onboard ticket vending machines are available. Only seven stations are staffed.

Improvements to existing lines

TransAdelaide has accelerated work across its rail system to comply with the Disabled Discrimination Act. Work has focused on wheelchair access, ramps, platform condition, access to pedestrian mazeways, and new signage.

To enhance customer confidence in onboard safety, TransAdelaide has engaged a private security company to provide guards on all services after 19.00. Three-quarters of the railcar fleet is now equipped with CCTV, with similar equipment also provided at many stations and stabling areas.

The 10.8 km Glenelg to City tram service is being upgraded for use by new light rail vehicles, with associated track and platform upgrades. Concrete sleepers and power augmentation are also being provided. The new service will be fully operational by mid-2006.

A programme was underway in 2004 to upgrade all points and crossings and to carry out bank stabilisation. TransAdelaide was also proceeding with the staged concrete re-sleepering of all lines. By late 2004 the Outer Harbour Line had been completed.

Traction and rolling stock

In 2004 the fleet comprised 100 railcars: 70 3000/3100 Class and 30 2000/2100 Class, six of which were out of commission. These are equipped with ticket vending machines and security cameras.

The 3000/3100 Class have closed circuit TV to allow drivers to observe passengers boarding and alighting and have been equipped with electromagnetic emergency brakes. United Gonna maintains the fleet under contract.

The Glenelg to City tram service will be served by nine Bombardier Flexity light rail vehicles.

Signalling and telecommunications

The principal signalling system is CTC. In 2004–05 this was upgraded.

Coupler in standard use: Scharfenberg fully automatic.

Track

Rail: Australian Standard 47 kg/m (177 km); 53 kg/m (45 km)
Crossties (sleepers): Hardwood timber 2,800 × 260 × 130 mm; steel (BHP M7-5 section) 2,595 × 260 × 127/146 mm
Spacing: 1,315/km plain track and curves for timber and steel; 1,492/km for concrete
Fastenings: 19 mm² dogspike with sleeper plates. Elastic fastenings on steel sleepers, points and crossings
Min curvature radius: 200 m
Max gradient: 1 in 45
Max axleload: 21 tonnes
Max line speed: 90 km/h
Diesel railcars or multiple-units

UPDATED

V/Line

Level 23, 570 Bourke Street, Melbourne, Victoria 3000
PO Box 5343, Melbourne, Victoria 3001
Tel: (+61 3) 96 19 59 00   Fax: (+61 3) 96 19 50 00
Web: www.vline.com.au

Key personnel
Managing Director: Andrew Neal

Political background
When the government privatised V/Line Passenger, it was bought by National Express Group Australia, which also won the Bayside Trains suburban passenger contract. National Express withdrew from its contracts in Victoria in December 2002 after being unable to achieve its financial objectives. A state government-owned company, V/Line Passenger Pty Ltd, took over operations in October 2003. It was expected that tenders will again be called during 2005 to re-privatise this service.

Organisation
V/Line employs 630 staff, including about 200 drivers.

Finance
In the nine months between October 2003 and June 2004, farebox revenue was A\$35 million and government subsidies were A\$84.4 million. Total revenue was A\$125.6 million.

Passenger operations
V/Line operates country passenger services between Melbourne and South Geelong, Ballarat, Ararat, Bendigo, Echuca, Bairnsdale, Albury, Shepparton, Warrnambool and Swan Hill. Sprinter railcars work many services. Some Bendigo services have been extended to Eaglehawk on a trial basis. As a result of election promises, services are again being provided to Ararat and Bairnsdale, the latter after a gap of 10 years. Trains run over tracks owned by ConnexTrains Melbourne or Pacific National.

In 2004 1,154 services serving 68 stations carried about 130,000 passengers a week or 6.9 million a year.

The state government has been negotiating with V/Line and other possible operators to reintroduce services on the Leongatha and Mildura lines. Both had passenger services in the past and the government is budgeting A\$32.7 million to upgrade the track and A\$12 million a year as a subsidy.

Traction and rolling stock
The locomotive fleet comprises 41 units, all of Clyde/General Motors origin. The hauled passenger coach fleet totals 139 vehicles and, together with the Sprinter railcars, these provide V/Line with 62 trains for traffic. Vehicle maintenance is contracted to EDI Rail.

In November 2001, Bombardier Transportation signed a contract to build 29 two-car V'locity 160 dmus, with an option for an additional 10 similar units. Against this option, nine more units were subsequently ordered. Deliveries are scheduled to take place between October 2004 and late 2006. The contract includes maintenance provision for 15 years, with its total value set at A\$535 million.

UPDATED

V/Line Sprinter railcars leaving Ballarat for Melbourne (Brian Webber)   0547034

V/Line N Class locomotive with a South Geelong–Melbourne passenger service at Geelong (Brian Webber)   0547035

Diesel locomotives: 1,600 mm gauge

Class	Builder's type	Wheel arrangement	Power kW	Speed km/h	Weight tonnes	No in service	First built	Mechanical	Builders Engine	Transmission
A	AAT22C-2R	Co-Co	1,840	115	121	4	1983*	Clyde	GM 12-645E3B	E GM
N	JT22C HC-2	Co-Co	1,840	115	124	25	1985	Clyde	GM 12-645E3B	E GM
P	G18HB-R	Bo-Bo	826	100	77	8	1984*	Clyde	GM 8-645E	E GM
Y	G-6B	Bo-Bo	480	64	68	4	1963	Clyde	GM 6-567C or 6-645E	E GM

* RebuiltYard shunters

Diesel railcars: 1,600 mm gauge

Class	Cars per unit	Motored axles	Power/ motor kW	Speed km/h	No in service	First built	Mechanical	Builders Engine	Transmission
Sprinter	1	4	235	130	21	1993	Goninan	Deutz	H Voith
V'locity	2	2	559	160	38*	2005	Bombardier	Cummins QSK19R	

* On order

VicTrack Access

Victoria Rail Track Corporation
GPO Box 1681P, Melbourne, Victoria 3001
Tel: (+61 3) 96 19 88 50   Fax: (+61 3) 96 19 88 51
e-mail: victrack@victrack.com.au
Web: www.victrack.com.au

Key personnel
Chairman: Tom Quirk
Chief Executive: John Sutton

Gauge: 1,600 mm; 1,435 mm
Route length: 3,235 km; 451 km; 22 km dual-gauge

Political background
In the reorganisation of Victoria's rail system, the VRT (also known as VicTrack Access) was created to provide arm's-length control of the state's track infrastructure excluding the metropolitan system.

Organisation
VRT commenced operations on 1 July 1997 as an independent business unit responsible for the provision of access to and asset management of the non-electrified rail infrastructure. It is intended that VRT remain in government ownership.

VRT's remit is that train control should become 'commercial and efficient', ultimately not requiring government funding. Ten operators had network access approval at the commencement of operations. The two-gauge issue is one for the government to contemplate and fund. As a first step, the Victoria and Commonwealth governments agreed to bring their Wodonga–Kalgoorlie and Broken Hill–Kalgoorlie tracks under single management of the Australian Rail Track Corporation Ltd (see entry in Australia section of *Railway systems and operators*) from mid-1998.

With the sale of the government freight business, country track was leased first to Freight Australia. With its purchase by Pacific National, transfer of the lease to that organisation has been negotiated under conditions designed to guarantee access by other operators and to allow improvements thought desirable and funded by the state government.

Improvements to existing lines

Former 1,600 mm gauge lines in the west of the state (Murtoa–Hopetoun; Dimboola–Yaapeet; Ararat–Portland), isolated by the standard-gauging of the Geelong–Adelaide route, have been converted to standard gauge at a cost of A\$20.4 million to enable continued haulage of grain. Ararat–Maryborough (88 km) was converted to standard gauge in April 1996 (but has been out of use since November 2004), while Maryborough–Dunolly (22 km) is to be dual-gauged. The former broad-gauge main line between Ballarat and Ararat (92 km), which lost services in 1995, has been restored to carry a passenger service, operations re-starting from July 2004.

The major Melbourne station, Spencer Street, is being rebuilt to provide modern, attractive facilities to encourage increased patronage on suburban and country services.

The Victoria government has announced a A\$550 million regional fast rail project to upgrade lines and reduce travel times between Melbourne and the provincial centres of Ballarat (118 km), Bendigo (162 km), Geelong (73 km) and Traralgon (158 km). The upgrade will include modernised signalling to allow 160 km/h running, rail replacement and level crossing improvements. It is expected that the improvements will reduce the fastest journey times by about 20 minutes, except to Geelong, where due to the shorter distance from Melbourne, the journey time will reduce by 6 minutes. First contracts were let in mid-2002 and the whole programme is due to be completed by mid-2005. However, some disappointment was expressed in the industry at the decision not to relay with gauge-convertible sleepers.

A tentative start has been made on providing standard gauge track to supplement or replace broad gauge track. The government has allocated A\$96 million to re-arrange track in some freight yards but justification for converting main or branch lines to standard gauge is regarded as more difficult. Whilst having one gauge in the state seems to be sensible, the benefits and savings are harder to quantify and are likely to pass to freight train operators rather than to the government, taxpayers or passengers.

The government has indicated its support for three extensions of existing electrification with overhead being provided above existing track between Broadmeadows and Craigieburn while the Epping branch will be extended to South Morang and the Cranbourne line extended to East Cranbourne, both relaying of previously closed track.

Signalling and communications

From August 1997 VRT took over responsibility for train control of services using its track.

Most of the VRT system is controlled by a simple train order working system. Other systems in use are CTC (577 km), staff and ticket (336 km), automatic block signalling (322 km), electric staff

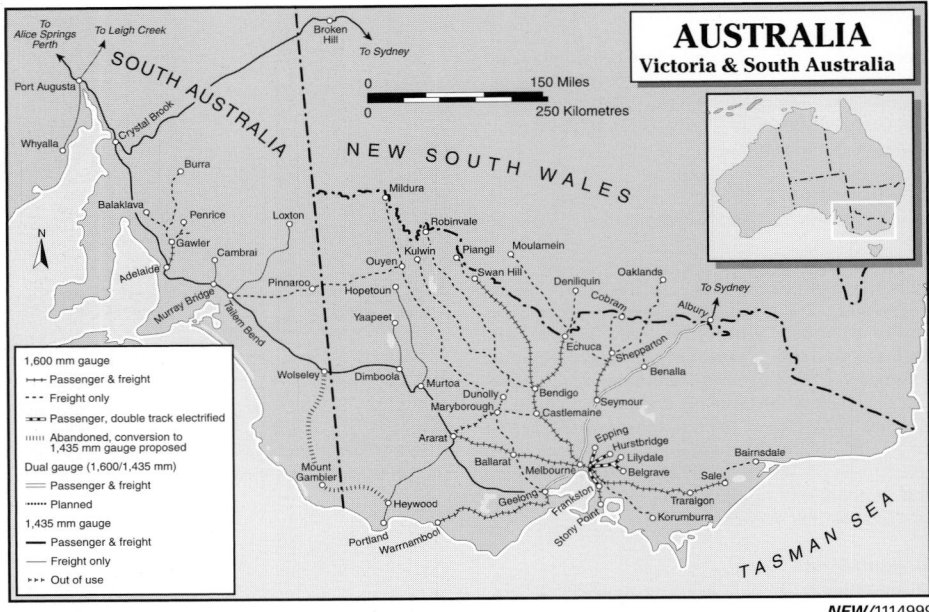

AUSTRALIA
Victoria & South Australia

NEW/1114999

(318 km), double line block (240 km) and automatic and track control (194 km).

Track

The 1,600 mm gauge country lines from Melbourne to Bendigo and Geelong and the suburban network are mainly double-track. In connection with the Fast Passenger Train project, Kyneton-Bendigo will become single-track with three long crossing loops. This is due to clearances in tunnels and historic bridge standards not being adequate for higher speed trains passing on them.

Geelong line track circuits were altered in 1997 to allow 130 km/h running by Sprinters and 120 km/h operation by passenger trains (up from 115 km/h).

Welding of the 47 kg rail on the 200 km Geelong–Warrnambool section has been completed.

Standard rail: Flat-bottomed 47, 53 and 60 kg/m rail rolled in 13.72 m lengths (speeds of up to 115 or 130 km/h respectively are allowed on 47 or 53 kg/m rail, while 160 km/h will be allowed for new trains on 60 kg/m rail on concrete sleepers)

Crossties (sleepers)

Timber: Non-treated Australian hardwoods (Red Gum, Ironbark, Box Stringbark and Messmate)
Dimensions: 1,600 mm gauge 2,705 × 250 × 125 mm; 1,435 mm gauge 2,590 × 250 × 125 mm
Spacing: 685 mm centres
Concrete: Prestressed concrete with cast-iron shoulders to take Pandrol rail clips
Dimensions: 2,670 × 275 × 145 mm at midspan (208 mm deep at ends). Rail seat canted at 1 in 20
Spacing: 670 mm centres

Fastenings

Timber: Most track fastened with dogspikes. Sleeper plates used on all tracks except 60 lb/yd branch lines, double-shouldered and canted at 1

in 20. 'Fair' deep bow one-piece rail anchors used instead of pads. Approximately 150 km of track relaid in 60 kg/m rail on rolled double-shoulder sleeper plates with Pandrol clips and three lock spikes per rail foot
Concrete: Pandrol rail clips, rail pads and insulators used on 53 and 60 kg/m rail laid on concrete sleepers
Ballast: Generally broken stone, usually basalt, but granite, rhyodicite and diabase also used. For rail lengths up to 27 m, 250 mm bearing depth with 50 mm shoulder width. For long or continuously welded rail, 300 mm deep with 405 mm shoulder width
Trackwork design standards: Curves of less than 2,400 m radius transitioned. Main line curves for 100 km/h traffic to be 830 m radius minimum, while for 50 km/h main line traffic minimum radius should be 400 m
Welded rail: Standard 13.72 m rail lengths welded into 27.5 to 82 m lengths at the central flashbutt welding depot, Spotswood. Once laid, rails Thermit-welded into 328 m lengths or continuously welded rail. Stress-control measures taken during field welding to ensure the continuously welded rail is in an unstressed condition within the temperature range of 33 to 38°C

Maximum gradient

Main line: 2.08% = 1 in 48
Branch line: 3.33% = 1 in 30
Max speeds: 160 km/h Vlocity railcars, 130 km/h–Sprinter railcars, 115 km/h–locomotives
Max axleload: 22.6 tonnes
Max altitude: 591.3 m near Wallace, Melbourne–Ballarat line
Longest straight: 38.3 km between Glenorchy and Murtoa, Western line
Tunnels: 3

UPDATED

Western Australia Public Transport Authority (PTA)

Public Transport Centre, West Parade, Perth, Western Australia 6000
Tel: (+61 9) 83 26 20 00 Fax: (+61 9) 83 26 26 59
e-mail: enquire@pta.wa.gov.au
Web: www.pta.wa.gov.au;
www.transwa.wa.gov.au;
www.newmetrorail.wa.gov.au

Key personnel

Chief Executive Officer: Reece Waldock
Executive Director, Finance and Contracts:
John Leaf
Director, People and Organisational Development:
Cliff Gillam
Director, Policy Unit: Sue McCarrey
Director, Transperth: Mark Burgess

General Managers
Transwa: Kim Stone
Transperth Train Operations: Pat Italiano
Network and Infrastructure: Hugh Smith
Project Director, NewMetroRail: Garry Willox
Director, City Project, NewMetroRail:
Richard Mann
Legal Officer: Michelle D'Adamo
Manager, Corporate Communications: David Leith

Political background

The Western Australian Government Railways Commission was established under the (State) Government Railways Act 1904. Westrail, the previously state government-owned freight operator in Western Australia, was privatised in December 2000 and its haulage task is now performed by the Australian Railroad Group Pty Ltd. The overview of track access is now performed

by WestNet Rail (see entry in Australia section of *Railway systems and operators*). This left WAGR's remaining role as that of operator of passenger services in the state. From July 2003 its residual responsibilities were incorporated in the PTA by enactment of the Public Transport Authority Bill 2003. This gave the authority responsibility for all rail, bus and ferry transport in the state, including those previously operated by Transperth, as well as rail and road infrastructure. Country passenger rail services are branded Transwa, while urban services continue under the Transperth name.

Passenger operations

Under the Transperth brand, the PTA provides suburban bus, train and ferry passenger services. The electrified rail system runs on 97 route-km of 1,067 mm gauge double-track covering four lines and serving 58 stations. Some 90 per cent of the

4,600-plus services run each week arrived within three minutes of their scheduled time in 2003–04. This was down from a fairly consistent 98 per cent, reflecting prolonged industrial action and some civil works through the year, but was still better than most other Australian metropolitan rail systems. In 2003–04 Transperth Trains recorded total boardings of 31.1 million.

Transperth offers secure parking (either patrolled or locked-compound) at 17 major suburban stations. Other safety and security initiatives include emergency buttons/phones on trains and platforms, improved lighting, and video surveillance on trains and platforms. It has employed and trained a force of more than 200 Transit Guards (including roving patrols) and by May 2005 will have installed a world class digital surveillance and 24-hour monitoring/alarm/communication system which includes ID-quality cameras at all platform entry points. A fully integrated contactless smart card ticketing system will be in operation across the whole system by August 2005.

The PTA operates three country passenger rail services under the Transwa brand name. Its flagship is the standard-gauge 'Prospector', which runs a daily service (twice on Mondays and Fridays) between Perth and Kalgoorlie (653 km). New high-speed Prospector railcars were introduced in June 2004, reducing the journey time to about six hours. A contractor is responsible for onboard catering. The narrow-gauge (1,067 mm) 'Australind' runs twice daily in each direction between Perth and Bunbury (183 km). The 'AvonLink' provides a weekday commuter service between Northam and Midland (109 km), where passengers can transfer to and from the urban network. This is supplemented by a service which runs between East Perth and Merredin on Monday, Wednesday and Friday.

Country rail patronage was down fractionally at 255,803 in 2003–04. About half of all passengers travel on concession fares. Transwa also operates a fleet of modern road coaches, completely replaced in 2003, which, supplemented by some contract coaches, serves 275 locations through the southwest corner of the state.

New lines
The NewMetroRail project will almost double the existing network by 2007, primarily by building a new Southern Suburbs Railway line with 10 stations (including two underground) to the coastal city of Mandurah, about 75 km south of Perth. The project also includes a 2 km spur line to Thornlie, a southeastern suburb; a tunnel through the CBD to link up with the northern line at Perth station; the construction of a depot at Nowergup (opened June 2004), 2 km north of the new (and northernmost) station at Clarkson, opened together with 5 km of track in October 2004 at a cost of A$8.7 million. By early 2007 the network will comprise 69 stations on five lines.

Traction and rolling stock
In 2004 the PTA operated, on 1,067 mm, five 'Australind' diesel railcars and 48 two-car emus on the Perth suburban network. On standard gauge it operated eight 'Prospector' diesel railcars on Transwa services. These were supplemented in 2003–04 by four additional dmus–two two-car and one three-car Prospector sets and a two-car AvonLink set.

A consortium of EDI Rail and Bombardier Transportation was awarded a A$437 million contract to supply 31 three-car stainless steel-bodied emus to serve extensions to the electrified Perth suburban network. Deliveries of the units, being built at Maryborough, Queensland, began in mid-2004. By late 2004 six of these had been introduced on the Northern Suburbs Railway (NSR). The contract includes vehicle maintenance for a 15-year period at a new depot at Nowergup and also provides an option for ten more trains of the same type. The introduction of three- and six-car trains has required the extension of platforms on the NSR and at other selected stations.

NEW ENTRY

Transperth Class BEA/BET/BEB emu running in the motorway meridian at Glenalough with an evening peak Perth–Whitfords service (Ian Francis)   ***NEW**/1115004*

Transwa Class WDA/WDB/WDC dmu forming a morning Kalgoorlie–East Perth service at Ashfield (Ian Francis)   ***NEW**/1115005*

Diesel railcars

Class	Cars per unit	Motor cars per unit	Motored axles/car	Output/ motor kW	Speed km/h	Units in service	First built	Mechanical	Builders Engine	Transmission
WCA	5	5	2	335	144	5	1972	Comeng	Cummins K19	H Voith T113r
WCE	3	–	–	–	144	3	1972	Comeng	–	–
ADP	3	3	2	406	110	3	1987	Comeng	Cummins K19	H Voith T311r
ADQ	3	2	2	406	110	2	1987	Comeng	Cummins K19	H Voith T311r
WDA	1	2	2	2 × 386	200	3	2003	United Goninan	Cummins N14E-R3	H Voith T212 bre
WDB	1	2	2	2 × 386	200	3	2003	United Goninan	Cummins N14E-R3	H Voith T212 bre
WDC	1	2	2	2 × 386	200	1	2003	United Goninan	Cummins N14E-R3	H Voith T212 bre
WEA	1	2	2	2 × 386	160	1	2003	United Goninan	Cummins N14E-R3	H Voith T212 bre
WEB	1	2	2	2 × 386	160	1	2003	United Goninan	Cummins N14E-R3	H Voith T212 bre

Electric railcars or multiple-units

Class	Cars per unit	Motor cars per unit	Motored axles/car	Output/ motor kW	Speed km/h	Units in service	First built	Builders Mechanical	Electrical
AEA	1	1	4	195	110	48	1990	Walkers	ABB
AEB	1	1	2	195	110	48	1990	Walkers	ABB
BEA	1	1	4	216	130	31*	2004	EDI Rail	Bombardier
BEB	1	1	4	216	130	31*	2004	EDI Rail	Bombardier
BET	1	–	–	–	130	31*	2004	EDI Rail	Bombardier

* Under delivery from 2004

WestNet Rail

GPO Box S1422, Perth, Western Australia 6845
Tel: (+61 8) 92 12 28 00   Fax: (+61 8) 92 12 29 21
Web: www.arg.net.au

Key personnel

General Manager: Tim Ryan
Civil Infrastructure Manager: Jelle Sibma
Commercial Manager: Paul Larsen
Control and Communications Systems Manager:
 John Ursic
Customer Services: Ian Burnett

Gauge: 1,067 mm; 1,453 mm; dual-gauge
Route length: 3,877 km; 1,212 km; 172 km

Political background

With the privatisation of Westrail, a 50-year lease to manage the Western Australian intrastate rail network was granted to WestNet Rail, a subsidiary of Australian Railroad Group Pty Ltd. WestNet Rail is responsible for track maintenance, train control and setting track access charges. Network ownership remains with the state. A regulatory overview and arbitration mechanism is provided by the Office of the Western Australian Independent Rail Access Regulator, established under the Railways (Access) Act 1998 to ensure access to the network on fair commercial terms.

Improvements to existing lines

A road/rail deviation, the Southern Transport Corridor, is being built at Geraldton to provide a more direct route and to remove the rail line from the foreshore. The project, costing A$72.2 million, is being funded by the state government.

A$9 million is being spent to rehabilitate the Picton–Manjimup section to allow continued operation of woodchip trains to Bunbury. The government is assisting with funding a 900 metre dual-gauge loop and terminal at the port of Fremantle as throughput is expected to triple by 2012.

A major upgrade has been undertaken on the Toodyay–Miling line.

Signalling and telecommunications

A A$52 million signalling upgrade has enabled all signals and points between Koolyanobbing and Kalgoorlie to be remotely controlled from Merredin.

Track

On the Perth–Kalgoorlie transcontinental route, 555 km of track has concrete sleepers and a maximum speed for passenger services of 160 km/h and 100 km has timber sleepers, with a maximum speed of 130 km/h. Freight trains do not exceed 115 km/h.

UPDATED

Austria

Ministry of Transport

Radetzkystrasse 2, A-1031 Vienna
Tel: (+43 1) 711 62   Fax: (+43 1) 713 78 76
Web: http://www.bmvit.gv.at

Key personnel
Minister: Hubert Gorbach
Head of Railways: Dr K Bauer

UPDATED

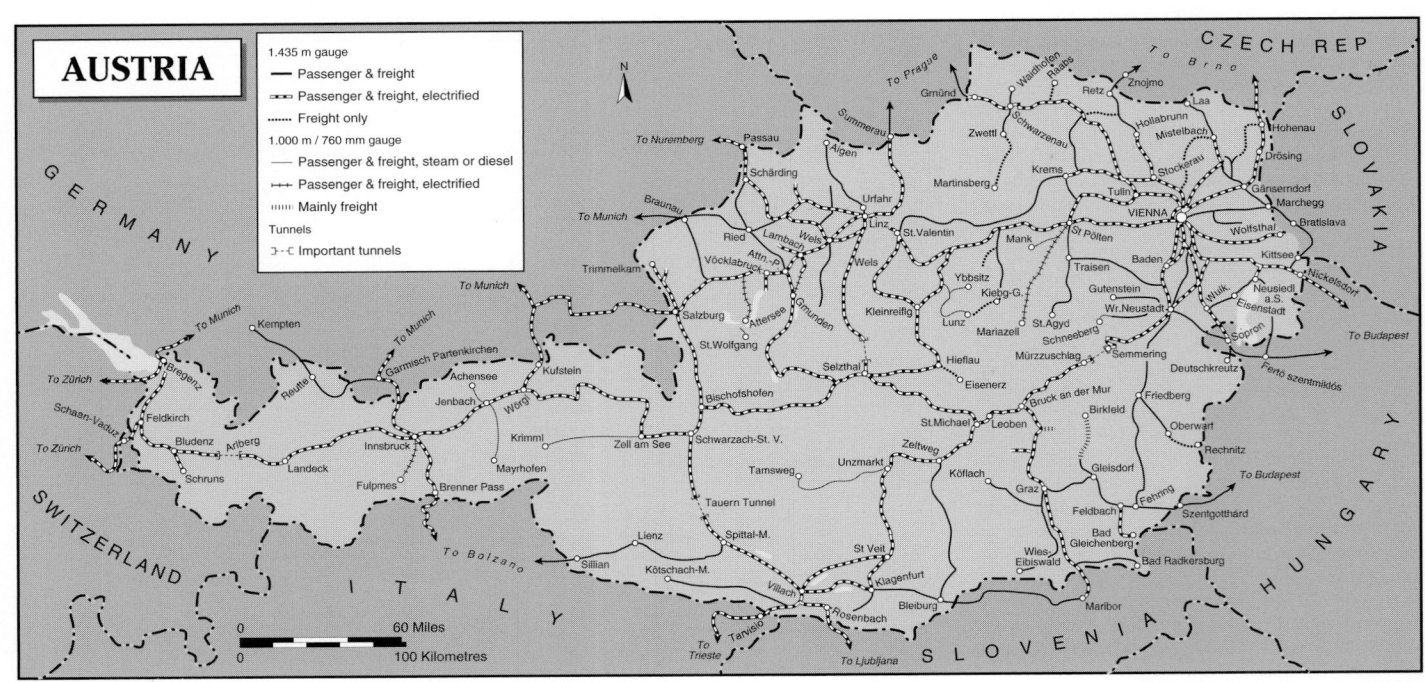

Austrian Federal Railways (ÖBB)

Österreichische Bundesbahnen
Elisabethstrasse 9, A-1010 Vienna
Tel: (+43 1) 580 00   Fax: (+43 1) 580 02 50 01
Web: http://www.oebb.au

Key personnel
Chairman of the Supervisory Board:
  KR Franz R Rottmeyer
Members of the Board of Management
  Director General: Rüdiger vorm Walde
  Deputy Director General: Dipl-Ing Helmut Hainitz
  Director: Ferdinand Schmidt
Directors of Business Units
  Freight Traffic: Mag Anton Hoser
  Passenger Traffic: Dr Gerhard Stindl
  Traction: Dipl-Ing Herwig Wiltberger
  Technical Services: Dipl-Ing Dr Alfred Zimmermann
  Operations and Capacity Management:
    Dipl-Ing Peter Klugar
  Planning and Engineering: Dipl-Ing Thomas
    Türinger
  Infrastructure: Dipl-Ing Michael Zuzic
  Building Works and Maintenance:
    Dipl-Ing Christian Fuchs
  Signalling and Systems Engineering:
    Dipl-Ing Helmut Steindl
  Telecommunications: Dipl-Ing Ewald Hladky
  Energy: Dipl-Ing Manfred Irsigler
Directors of Central Services
  Planning, Control and Accounting:
    Mag Rudolf Wotruba
  Personnel: Dr Wolfgang Moldaschl
  Purchasing and Materials Management:
    Mag-Ing Johann Göbel
  Informatics: Walter Linn
  Finance: Mag Gerhard Leitner
  Estate Management: Ing Mag Heinz Redl
Directors of Staff Units
  Audit Office: Mag Norbert Wagner
  Communications: Dr Viktoria Kickinger
  Equity Holdings and International Relations:
    Mag Edmund Hauswirth
  Change Management: Mag Erich Pirkl
  Legal Affairs: Mag Andrea Ergert

Gauge: 1,435 mm; 1,000 mm; 760 mm
Route length: 5,294 km standard gauge; 5.8 km
1,000 mm gauge; 334.6 km 760 mm gauge
Electrification: 1,435 mm: 3,263 km at 15 kV 16²⁄₃ Hz
AC, 2.2 km at 3 kV DC, 760 mm: 84 km at 6.5 kV
25 Hz AC

Political background
A new Bundesbahngesetz (Railways Act/BBG)
reorganised ÖBB from 1 January 1993. The railway
became a company (combining elements of public
(AG) and private (GmbH) status) with a distinct legal
personality and subject to normal company law.

The state pays for new infrastructure and
contributes five-eighths of the cost of track
maintenance, as agreed in 1987, but it neither covers
deficits nor makes any statutory grant to the railway.
ÖBB as operator pays a track-access charge. Loss-
making but socially necessary local services are
grant-supported at a level set by the Ministers of
Transport and Finance jointly. Any services over and
above the basic specification must be 'purchased' by
the *Länder* governments, which received significant
extra federal funding out of oil tax revenues to
this end. The *Länder* were slow to conclude their
contracts with the railway, being unwilling to pay
the state-owned carrier for its services and desirous
of using the tax-money for other transport services
they had long bought in, thus freeing their own
money for use elsewhere. After the Minister of
Transport intervened in 1994, cuts were postponed,
thus causing a requirement for additional state
support. The contracts were all eventually signed,
that with Vienna being the last, at the end of 1998. In
2001 these contracts were under review.

Organisation
ÖBB meets European Union Directive 91/440 in
separating for accountancy purposes infrastructure
from transport services, thus creating a framework
for third-party access. An infrastructure sector
and a commercial sector are each divided

Three-car emu on the 760 mm gauge Mariazellerbahn (D Trevor Rowe)   0089079

into a number of business-units, each with its
own bottom-line responsibility. There are also
central services and central staffs. The Board of
Management, increased from three members to
five in August 1997, was reduced again to three in
May 2001. Dipl-Ing Hainitz, the only member of the
old board to have his contract renewed, continues
to be responsible for infrastructure, and Ferdinand
Schmidt, who joined the railway from Lauda-Air,
took on responsibility for passenger traffic. There is
no longer a board member for freight traffic.

The number of employees has continued to fall
to approximately 49,000 at the end of 2000. Since
1993 productivity had risen to 66 per cent by the
end of 2000.

Finance

**Table 1: Transport Operations Sector: Profit & Loss
Account (Sch billion)**

	1997	1998	1999
Sales	26.6	27.3	27.6
Other income	5.2	6.3	6.0
Total income	31.8	33.6	33.6
Personnel costs	14.5	14.9	15.2
Depreciation	3.9	4.0	3.3
Expenses on property, plant & equipment	9.0	9.6	9.8
Infrastructure user charges	3.4	3.5	3.7
Total expenditure	30.8	32.0	32.0
Operating profit	1.0	1.6	1.6
Financial result	−0.7	–	–
Profit	0.3	–	–

Table 2: Rail Cargo Austria: Sales (Sch million)

	1997	1998	1999
Express-Cargo	1,573	–	–
Complet-Cargo	8,316	–	–
Combi-Cargo	1,084	–	–
Public Benefit Service Income	2,049	2,062	2,084
Total	13,022	12,800	14,100

Table 3: Passenger Business (Sch million)

	1997	1998	1999
Rail business	5,742	5,416	5,525
Bus operations	1,318	1,330	1,354
Navigation	39	39	39
Public Benefit Service income	6,100	6,458	6,292
Total	13,199	13,243	13,210

Passenger operations
Long-distance
Following the introduction of the first stage of
a national fixed-interval passenger timetable,
Neuer Austrotakt 1991 (NAT91), a second stage
was planned between 1995 and 1997, and a third
and final stage for 2000. NAT91 focused on ÖBB's
seven InterCity (IC) routes, offering an hourly
service pattern between Vienna and Graz, Villach,
Selzthal, and Innsbruck via Salzburg and Zell am
See, and 2-hourly services over other routes.
Additionally, peak-hour InterCity services on the
Vienna–Salzburg–Bregenz, Vienna–Innsbruck, and
Vienna–Villach–Salzburg routes featured

'SuperCity' train-pairs, timed to facilitate day-
return business trips and offering enhanced
facilities, and subject, like EuroCity (EC) services
running through Austria, to a supplementary fare.

Although the new fixed-interval service increased
train-km by 30 per cent and costs by 20.4 per cent,
passenger receipts rose only by 6.8 per cent. Despite
a reduction from 94 million to 92 million train-km in
1995, services in that year covered only 73 per cent
of their costs, making a loss of Sch1,537 million.
In 1996 train-km were further cut, to 85 million. The
fixed-interval principle stayed, but the 'SuperCity'
trains and a number of the more lightly loaded
early and late services were withdrawn, and on
some routes through trains were replaced by
connections.

The 1996 timetable also brought a major change
in on-board service provision. Wagons-Lits, the
long-standing provider, lost the on-train contract
(though retaining station restaurants and sleeper
services) to the Austrian company Trainristo,
and the number of trains with full restaurant-car
service was reduced from 152 to 60 (with another
76 running with foreign restaurant cars). At-seat
service, delivered by ÖBB staff, was extended.

Both the number of passengers and the load-
factor rose steadily in 1999 and 2000, with the latter
reaching 2.02 passengers per train-km operated.
In 2000 ÖBB joined its German and Swiss
counterparts DB and SBB in establishing the Trans
Europe Excellence Rail Alliance (TEE), which aims
to offer a unified cross-border rail service in Central
Europe. It was originally planned that 116 tilting
trains would be ordered by the alliance, of which
32 would belong to ÖBB. However, in late 2001, DB
withdrew from the planned joint procurement of
rolling stock, although it was to retain its marketing
role in the alliance. Meanwhile, 2001 saw major
accelerations on the routes between Graz and Linz,
Salzburg, and Innsbruck.

Since 1 January 2001 it has been possible to
buy tickets for domestic journeys online, paying
for them by credit card and printing them on a
domestic printer. This represents a world first for
Austria. The VorteilsCard, which can already be
used for payment at stations, was due to be made
usable for payment on the Internet in mid-2001.

ÖBB now accepts electronic payment at its
principal stations, and the equipment in use is
capable of working with all current cards and with
the future *Quick* electronic purse. The *VorteilsCard*,
which replaced the old Half Price Pass in 1996, was
made available with a *Eurocard* function in mid-
1997, becoming the first Austrian credit card to
carry the holder's photograph.

Suburban and urban
At the end of 1995 a 10-year Sch30,000 million
investment programme in S-Bahn and U-Bahn
facilities in the Vienna area began. The S-Bahn
works include greatly increased park-and-ride

provision at principal stations, a better link to Vienna International Airport, and extension of the line to Heiligenstadt to run southeastwards along the Handelskai on the Danube west bank.

The airport link is being provided by development of the Pressburger Bahn as S-Bahn line S7. In late 1997 the annual investment in this project was doubled to Sch840 million to allow completion for 2002 instead of for 2006. The work is now expected to be finished in 2003, and the city section of the route was closed in mid-2001 to allow major engineering works to take place.

The S70 is also being planned. This involves a link from a rebuilt Südtiroler Platz station past the Schweizer Garten and the Ostbahn and then curving to the Donauländebahn and running on to a junction with the S7 at Klein Schwechat.

Salzburg and Graz are also to have S-Bahns. Salzburg's plans involve four new city stations and a 15-minute service on the core (common) route section between Salzburg and Freilassing, where a third line is also to be provided. Cost of the stations and this widening will be Sch38.55 million, split between city, *Land,* and federal government. Completion is envisaged for about 2005.

The developments being considered by the Styrian *Länder* and the city of Graz for the Graz area, which envisage the provision of S-Bahn services on existing lines, follow the cancellation of plans for a Karlsruhe-type network. In addition to ÖBB, the city transport services and the various minor railways operating in the area are involved. Where ÖBB cannot provide an economic rail service, it will, in consultation with the states, propose a bus or group taxi service as an alternative and will also support the states if they wish to find an alternative rail operator. Seven lines were closed to passenger traffic on 10 June 2001 and five others were being assessed.

Traffic (million)	1998	1999	2000
Passenger journeys	179.1	181.7	182.7
Passenger-km	7,971	–	–
Freight tonnes	76.5	78.0	84.7
Freight tonne-km	15,348	–	–

Freight operations

Since 1993 the freight business has traded as Rail Cargo Austria, in three divisions: Complet-Cargo (wagonload traffic), Combi-Cargo (intermodal traffic), and Express-Cargo (less-than-wagonload business).

Freight operations have seen considerable growth in recent years, though pressure on rates has meant that the increase in traffic has not been matched by a similar growth in profits. The 1997 freight record was exceeded in 1998, again in 1999, and rose further to 84.7 million tonnes in 2000.

Rail Cargo Austria is investing heavily in the expansion of its Logistik Centres in the period up to 2004. The investment will see expansion of 19 centres, construction of a new freight centre in Vienna, and an enlargement of the Linz goods station. The three new Logistik Centres, at Bischofshofen, Wels, and Wörgl, awarded ISO 9001 quality management certification at the beginning of 1997 at the end of a one-year assessment period, are all flourishing. Austria's transit significance is likely to grow, and substantial improvements to cross-border routes are planned to meet developing needs.

To improve freight operation in the Vienna area and to free up line capacity for additional S-Bahn services, the Floridsdorfer Hochbahn, a 4 km line built in 1916 and abandoned after the Second World War, was restored to provide a new freight route between the northern main lines and the Kledering marshalling yard. Diversion of goods traffic, primarily from Retz, Gmünd, and Krems, to this electrified single line linking Jedlersdorf with Leopoldau has allowed the very busy line through Vienna Mitte to be freed almost entirely of freight trains and has also relieved the heavily used section of the Franzjosefbahn between Tulln and Vienna of some of its traffic.

In 1998, the Wieselburg–Gresten line was converted from narrow gauge to standard gauge in order to facilitate the working of goods traffic.

ÖBB has joined with DB in establishing a locomotive pool, EuroTraction, to which each railway is contributing 25 new high-performance

Class 1014 dual-voltage electric locomotive in the Vienna suburbs with empty coal wagons returning to the Czech Republic (Ken Harris)   0109841

locomotives in a first step. The aim is to reach a figure of 1,500 machines by 2005, thus making modern freight locomotives available in the free market for operators requiring haulage.

Intermodal operations

Much of the recent growth in intermodal traffic stems from political moves to restrict heavy lorries. In 1989 the Transport Minister barred all freight vehicles above 7.5 tonnes, domestic and foreign, from night-time use of Austrian roads and the government underwrote a six-month cut in ÖBB piggyback rates over the Brenner Pass, on which route intermodal services were vigorously expanded.

A by-product of the European Union Schengen open frontiers policy has been a switch back to road by significant numbers of ÖKOMBI customers on the Brenner route (of whom 50 per cent come from third countries), and, in an attempt to counter this, two additional trains were put on each way from 1 February 1998 to give a two-hourly interval service throughout the 24 hours and prices were significantly reduced for a trial period, by up to 41 per cent for day-transits and up to 24 per cent for night-transits. New Class 1012 locomotives have been brought into use on the Brenner route.

New lines

ÖBB is pursuing a major capital programme to increase capacity and raise speeds on its principal network. In 1987 government approved the *Neue Bahn* ('New Railway') plan in principle, and execution and funding of the first phase were agreed in 1989, when the High Performance Lines Act was passed. The high-performance network is essentially five major axes: the Donauachse (Regensburg–Passau/ Salzburg–Vienna–Nickelsdorf(–Budapest)); the Pyhrn–Schoberachse (Regensburg–Passau–Wels– Pyhrn–Schoberpass–Graz–Spielfeld(–Marburg)); the Brennerachse ((Munich–) Kufstein–Brenner(–Verona)); the Tauernachse ((Munich–)Salzburg–Rosenbach (–Laibach)); and the Südbahnachse (or Pontebbana) ((Warsaw–)Hohenau–Vienna–Tarvis(–Trieste)) – broadly the network of lines making up the main trunk routes for international traffic.

The first phase envisaged commitment of Sch44 billion up to 1998, on a mixed-funding basis as used in Austria for motorway construction. A state-owned company, Eisenbahn-Hochleistungsstrecken-AG (HL-AG), was created on 3 April 1989 to manage the planning and execution of new infrastructure and upgrading projects. HL-AG reports directly to the Ministry of Economics and Transport, and the scope of its work is determined exclusively by orders issued by that ministry in collaboration with the Ministry of Finance. Legislation, in the form of a Treaty of Co-operation with ÖBB, sets out the framework for collaboration with the federal railways. While HL-AG deals with most of the new construction, ÖBB is handling major route improvement schemes and planning the Brenner corridor improvements either side of the new Innsbruck avoiding line. Between 1988 and 1994 some Sch22 billion

were spent on the elimination of a number of significant bottlenecks, a prerequisite for achieving the high freight traffic growth rates of recent years.

The Rail Infrastructure Financing Act of July 1996 allocated Sch60 billion to be spent at a rate of Sch12 billion per year over the following five years. This was the first time ever that investment funds have been allocated on a long-term basis, and the figure represents a continuation of the level of spending of 1995.

The 2000 expenditure on the principal routes was: Donauachse Sch1,600 million, Tauernschse Sch700 million, Brenner Sch200 million, Arlberg Sch1,400 million, Pontebbana Sch1,200 million, Pyhrn-Schober Sch1,200 million, Vienna area Sch900 million, other lines Sch3,900 million.

The Brenner line

The Brenner route, linking Munich with Innsbruck and Verona, now ranks first of the 14 principal trans-European infrastructure projects supported by the European Union. The principal attraction for Austria of the planned base tunnel is the elimination of a substantial amount of lorry traffic known to be causing significant environmental damage. The existing line has already been extensively upgraded. Although clearance work on the Austrian section to accept RoLa piggyback road vehicles of up to 4.05 m in height was finished in 1989, enlargement of tunnel clearances on the Italian side is not yet complete, so RoLa trains terminate at Brenner.

ÖBB opened the 15 km Innsbruck avoiding line (12.7 km in Austria's longest tunnel) in 1994. In 1995, work was begun on quadrupling between the German frontier at Kufstein and the start of the Innsbruck avoiding line at Baumkirchen. Also begun was installation of automatic block signalling and reversible working on both tracks of the northern ramp of the Brenner.

No firm date for a start on the 65 km base tunnel, which will reach deep into Italy, has been reached. In late 1994 Austria, Germany and Italy agreed that an optimal scheme for the Brenner route would see new infrastructure on the whole 400 km between Munich and Verona at a total cost of €12.5 billion, the tunnel taking €4.5 billion, works north of it €2.8 billion, and works south of it €5.2 billion. Up to 400 trains daily, double the current figure, could be run.

Südbahn and Semmering base tunnel

Work is proceeding to improve the capacity of the Südbahn; this is the Semmering Pass route from Vienna to Bruck an der Mur, Graz, and Villach. To permit more local services between Vienna and Wiener Neustadt (about 50 km out), Südbahn main line traffic will eventually be rerouted over the Ostbahn main line to Grammatneusiedl, where a new spur will take trains on through Wampersdorf to Wiener Neustadt over a doubled and upgraded branch line. The rerouteing will also improve Südbahn access to the Kledering marshalling yard.

ÖBB's scheme to bypass the steep and curvaceous Semmering Pass section of the Südbahn, which it describes as the sole section of the main European network with branch line characteristics, received the go-ahead in 1993. Gradients of 1:40 and curves down to a radius of only 170 m, a distance twice that of a direct line, tight tunnel clearances, problems in the use of banking locomotives, and high maintenance costs all conspire to make the historic Ghega line unsuitable for today's traffic and unavailable for RoLa operations.

The Sch6.1 billion solution will provide by 2002 a new double-track railway between Gloggnitz and Mürzzuschlag with a summit tunnel more than 12 km long and several smaller tunnels. The ruling grade will be 1 in 91, distance will be reduced by some 19 km, capacity on this very busy route will be greatly enhanced, and passenger trains will save 30 minutes.

However, in April 1998 a provisional stop to the plans was announced. At the end of September a commission of experts appointed to review the Semmering Base Tunnel Project reported, recommending not only a delay of at least two years for the tunnel scheme (to await a constitutional court decision) but also bringing forward plans for a Vienna–Schwechat–Eisenstadt–Sopron line. The commission held that the basic case for a high-capacity north-south link remained convincing, but it felt that as a consequence of the 1989 opening up of Eastern Europe, the traffic projections that had been made were too high and that sufficient capacity could be found on the Semmering line till 2015 and on the Aspang line till 2025. It also noted that the base tunnel project would bring with it a need for further additional and expensive projects, and it suggested other measures for consideration in the meantime. In particular, it noted that Hungary's possible accession to the European Union could mean that an alternative and very attractive 'plains' solution would become possible, also taking the place of the *Süd-Ost Spange* long-term proposals. The Semmering line itself was declared a UNESCO World Heritage Site in December 1998.

Another complication was that the *Land* of Niederösterreich argued before the Austrian Constitutional Court that the Semmering base tunnel plans were unconstitutional because they contravened its environmental protection law, but on 25 June 1999 the court dismissed this claim and gave the *Land* until the end of the year to lodge a new complaint (which cannot prevent the building of the line but can only ask for conditions to be set, such as an application to be considered in the light of public interest).

Mid-April 2000 saw the go-ahead for the first stage of the Koralmbahn, between Klagenfurt and Grafenstein. The last consent required was that the Defence Minister, needed because the line will cross land belonging to the Gradnitz ammunition store. The 14 km of new railway will cost Sch2,000 million and will be financed by the rail infrastructure company. The Hochleistungsstrecken AG is responsible for the planning and the execution of the project and was expected to begin the tendering process in mid-2000.

Improvements to existing lines
Environmental measures
The first agreement on railway noise-protection measures for an entire *Land* was signed in August 1997 in Salzburg by the federal transport minister, those responsible for transport and environment in the *Land* Salzburg, and ÖBB. This formally regulates the planning and realisation of noise-protection measures for the whole of the *Land,* the cost of projects being met half from government railway funds, a quarter by the *Land,* and a quarter by the local authority affected.

Westbahn
The Westbahn is the main east-west transversal connecting Vienna with St Pölten, Linz, and Salzburg. Westbahn modernisation for 200 km/h speeds began before the opening up of the east and Austria's accession to the European Union. The changed situation requires a capacity increase that can be obtained only by full quadrupling between Vienna and Wels (and possibly later on to Salzburg).

Two twin-track, mixed-traffic lines are envisaged throughout, linked every 25 to 30 km, one of them a high-speed railway. The new concept builds on the old. Thus, for example, where major realignments are planned, four-tracking will be obtained by retaining the (modernised) original route. Work so far has raised capacity by 10 per cent; the first accelerations came in 1996.

Working westwards from the Austrian capital, between Vienna and St Pölten there is to be a new high-speed line on the north side of the Danube, and there will also be a 25.7 km St Pölten freight avoiding line (with a 3.9 km tunnel). Much of the work between St Pölten and Attnang-Puchheim is now complete, with a mixture of upgrading of the historic line and new construction. The existing Attnang-Puchheim to Salzburg line will for the time being be maintained as it is but HL-AG was commissioned in early 1990 to plan a new line at a possible cost of Sch12 billion.

According to the Hochleistungsstrecken AG, progress on the upgrading of the Westbahn at the end of 1998 was as follows:

Work completed (35.2 km)	*km*
Melk (in part) (May 1997)	7.0
Krummnussbaum–Säusenstein (March 1994)	7.0
Pyhrnbahn connection (Marchtrenk–Traun) (May 1994)	13.2
Lambach (January 1995)	3.9
Breitenschützing–Schwanenstadt (June 1997)	4.1
St Peter-Seitenstetten	6.2
Haag-St Valentin	13.3
Work in progress (49.2 km)	
St Pölten–Prinzersdorf	7.5
Loosdorf	4.7
Melk	12.1
Sarling–Ybbs	4.3
Kottingburgstall	1.1
Work at the planning stage (203.1 km)	
Vienna–St Pölten	49.1
St Pölten goods avoiding line	24.8
Prinzersdorf–Gross Sierning	4.5
Rohr node	2.1
Pöchlarn–Krummnussbaum Junction	4.7
Ybbs/D–Hubertendorf	4.5
Hubertendorf–Blindenmarkt	4.2
Blindenmarkt–Amstetten	6.3
Aschbach–Krenstetten	8.2
St Valentin station reconstruction	2.5
Enns avoiding line	10.8
Asten–Linz Hauptbahnhof	11.4
Attnang/Puchheim-Salzburg	70.0

Early in 2000 it was announced that the Austrian and German governments and the two national railways had signed agreements for the development of the main railway lines linking Austria with Bavaria. There is to be a step-by-step improvement of the Munich–Mühldorf–Freilassing–Salzburg–Wels–Linz line and of the line from Nuremberg through Passau to Wels. On the Mühldorf–Simbach–Neumarkt–Kallham line there will be capacity guarantees.

Ostbahn
The Ostbahn is the main line into Hungary, crossing the border at Nickelsdorf/Hegyeshalom.

It was upgraded between 1990 and 1996 for 140/160 km/h operation, with major renewals, alterations to stations, and modernisation of signalling and safety equipment, and there were significant accelerations to the Vienna–Budapest services in the 1997 timetable.

Together, the Westbahn and Ostbahn form the 'Donauachse'. These two routes are being linked by a new connecting line in Vienna through the Lainzer tunnel and the new Vienna station (see 'Major new stations').

Schober Pass
The north-south Schober Pass line between Selzthal and St Michael is the central link in the (Germany–)Linz–Graz(–Slovenia) route. Until recently largely single track, the route would have been unable to provide sufficient operating capacity for the regular-interval passenger timetable and the planned increase from 90 to 150 freight trains daily. The case for this and other north-south increased operating capacity projects was further strengthened by events in eastern Europe.

There has been an extensive double-tracking, realignment, and resignalling programme, which has allowed a major acceleration of passenger services in the 2001–02 timetable. A new curve at Selzthal to save Graz-Salzburg trains reversal there will contribute to faster timings. Total cost of all the works, including elimination of level crossings, is put at Sch4.7 billion.

South of the Schober Pass, the 28 km section between St Michael and Bruck an der Mur, over which north-south Schober Pass trains share the route with east-west Südbahn trains, is being quadrupled. Boring of a 5.4 km Galgenberg Tunnel for a Leoben avoiding line is the first step. North of the pass, the Pyhrnbahn from Selzthal north to Linz is due to be doubled and resignalled before the end of the century.

A 13-km single-track Linz avoiding line, linking the Pyhrnbahn at Traun with the Westbahn at Marchtrenk, at the approach to Wels, was opened in 1994.

Tauern line
The trans-Alpine north-south line from Salzburg through the Tauern Tunnel to Rosenbach on the Slovenian border, where trains can go on further south via the Karawanken Tunnel route, is being doubled at a cost of Sch600 million.

Enlargement of the 280 m Untersberg Tunnel near Schwarzach St Veit in a Sch46 million scheme has permitted Tauern route piggybacking of 4.05 m high road vehicles.

Long-term plans cover improvements to the north of the Tauern route between Salzburg and Schwarzach St Veit (already double-track, but where curves are being eased) and to the south of it between Spittal-Millstättersee and Rosenbach. Realignments where feasible will lift maximum speed to 130 km/h, 140 km/h over parts of the southern section. The aim is to raise freight train capacity from 110 to 150 a day.

Arlberg line
The Arlberg is the key east-west trans-Alpine route in the west of Austria, linking Innsbruck and points

Class 1014 electric locomotive leaving the Hungarian border station at Hegyeshalom with a Vienna–Budapest express (Eddie Barnes)

0089082

to the east with Switzerland. Progressive double-tracking continues, with attention now focused on the 25 km segment between Ötztal and Landeck on the eastern approaches to the Arlberg Tunnel. Associated work aims to raise line-speed to 140 km/h, provide extra capacity for an S-Bahn service, and eliminate the railway's division of some communities by rerouteing, principally in the Ötztal to Kronburg section. Given the steep sides of the valley, this necessitates tunnelling, and eventually 12 km of the 25 km section will be underground. Preliminary work on the Schnann to St Jakob section has begun and plans are under way for the Langen to Klösterle section.

The two-track Arlberg Tunnel has had its clearances enlarged to allow RoLa piggybacking with 4.05 m high road-vehicles.

Links with Eastern Europe
In 1995 the governments of Austria, the Czech Republic and Germany agreed a programme to upgrade the Vienna–Prague–Berlin corridor, and a further agreement was signed between ÖBB and ČD in October 1997.

Studies have been commissioned by ÖBB and its Hungarian counterpart MÁV into electrification of the line between Graz, Szentgotthard and Szombathely.

Three lines will improve links with the Slovak capital, Bratislava. The Pressburger Bahn is being developed as far as Schwechat as the S7 link to Vienna International Airport (see above). Beyond the airport, more double-tracking and some new construction will extend the line beyond its present terminus at Wolfsthal to Kittsee and a junction with the line from Parndorf, extension of which to Bratislava began in late 1994.

The Parndorf route uses the Budapest main line as far as Parndorf, whence the existing single-track branch to Kittsee has been rebuilt for 160 km/h speeds and later doubling. A 2.5 km extension suitable for 140 km/h leads on to the border, and on the Slovak side the line is linked to Petržalka, across the Danube from Bratislava. The full service, of both D-trains and local trains, began running on 1 August 1999 and loadings soon rose above expectations. Plans to work the route with ÖBB Class 1014 bi-current locomotives had to be dropped because there was no opportunity for Slovakian Railways to provide the necessary kilometre exchange. On the Austrian side Class 1046s are being used, and through trains change to a ŽSR Class 240 at Petržalka (where three tracks are divided in the middle by isolating sections).

The third route into Slovakia runs from Vienna via Marchegg to Bratislava. This has now lost its fast trains to the new route via Parndorf.

Major new stations
In Vienna, the Südbahn at Meidling (where there will also be a junction with the Donauländebahn for freight traffic use) is being linked by a 14 km new line, with all but 2 km in tunnel, under the Lainzer Tiergarten and the city's southwestern residential area, with the Westbahn and with the new high-speed line at Purkersdorf Sanatorium. A new Vienna station on the site of the present Südbahnhof is planned, with a major urban redevelopment as well as the station works.

In 1997 ÖBB began work on a Sch6.2 billion programme for the modernisation and redevelopment of its principal stations, the aim being to create a more attractive environment for the customer, with many more service and trading outlets. Private capital will be involved wherever possible. This follows on from the *Neue Bahn* plan to spend Sch650 million on 32 stations to create public transport interchanges, provide parking space and station garages, and offer a wide range of other passenger services and amenities. The stations involved are: Vienna West, South, North, Rennweg, Floridsdorf, Hütteldorf, and Heiligenstadt; Baden, Wiener Neustadt, Bruck an der Mur, Graz; Leoben, Klagenfurt; Krems; St Polten, Linz, Wels, Attnang-Puchheim, Salzburg, Bischofshofen, Schwarzach St Veit, Innsbruck, and Feldkirch. This last will be the first station in western Austria to be totally renovated. The aim was to complete the project in 2000 at a total cost of Sch62.6 million (in collaboration with the local

authority). For Linz a planning company (formed by ÖBB, the city of Linz, and the Raiffeisenlandesbank Oberösterreich) has been set up to manage a project with a total value of Sch3–4 billion for station improvement, Sch1 billion for development of the station as a local traffic hub for rail, and Sch1.6 billion for development of offices for the

Land (federal state) government of Oberösterreich, along with flats and shops.

Traction and rolling stock
At the end of 1998 ÖBB had 715 electric locomotives, 465 diesel locomotives and 17 steam locomotives. There were also 226 emus and 129 dmus.

Siemens-built Class 1016 Taurus electric locomotive leaving Vienna Kledering marshalling yard with a wagonload freight (Ken Harris)   0109842

Diesel locomotives

Class	Wheel Arrangement	Power KW	Speed km/h	Weight tonnes	No in service	First built	Mechanical	Builders Engine	Transmission
1,435 mm gauge									
2043	B-B	1,035	110	70	72	1964	JW	JW 400 (01–4) LM 1500 (5+)	H Voith
2143	B-B	1,035	110	68	75	1965	SGP	SGPT 12c	H Voith
2048	B-B	808	100/65	64	34	1991	MAK	CAT3512DI	H Voith
2050	Bo-Bo	1,140	100	75	12	1958	Henschel	GM 12-567c	E GM
2016	Bo-Bo	2,150	140	80	70*	2002	Siemens	MTU 16V 4000 R41	E Siemens
Shunting locomotives									
2060	B	129	30/60	27	16	1954	JW	JW 200	H Voith
2062	B	250	40/60	32	48	1958	JW	JW 400	H Voith
2067	C	398	65	49	111	1959	SGP	S 12a/S12na	H Voith
2068	B-B	820	50/100	68	60	1989	JW	JW 480D	H Voith
2070	B-B	740	100	72	60*	2001	Vossloh	–	H Voith
760 mm gauge									
2090	Bo	72	40	13	1	1930	SGP	Saurer BXD	E Syst Gebus
2190	Bo	86	45	13	1	1934	SGP	SGP SU8	E Syst Gebus
2091	1-Bo-1	114	50	22	4	1936	SGP	SGP R 8	E Syst Gebus
2092	C	88	20	17	3	1943		Deutz ABM 517	H Voith
2095	B-B	405	60	32	15	1958	SGP	S 12a	H Voith

*Includes locomotives on order

Electric locomotives

Class	Wheel arrangement	Output kW Continuous/ One-hour	Speed km/h	Weight tonnes	No in service	First built	Builders Mechanical	Electrical
1010	Co-Co	3,260/3,990	130	110	18	1955	SGP	ABES
1110	Co-Co	3,260/3,990	110	110	28	1956	SGP	ABES
1014/1114	Bo-Bo	3,000/3,400	160	74	18	1993	SGP	BES
1016/1116*	Bo-Bo	6,400	230	86	400**	1999	Siemens/ÖBB	Siemens
1040	Bo-Bo	1,980/2,020	80	80	12	1950	Lofag	ABES
1041	Bo-Bo	1,980/2,020	80	83	20	1952	SGP	ABES
1141	Bo-Bo	2,100/2,400	110	80	28	1955	SGP	ABES
1042	Bo-Bo	3,336/3,600	130	84	198	1963	SGP	BES
1042.5	Bo-Bo	3,808/4,000	150	84		1966	SGP	BES
1142	Bo-Bo	3,336/3,600	130	84	54	1963	SGP	BES
1044	Bo-Bo	5,000/5,310	160	84	122	1974	SGP	BES
1044.2	Bo-Bo	5,000/5,310	160	84	88	1989	SGP	BES
1245	Bo-Bo	1,504/1,780	80	83	2	1934	Lofag	ABES
1046	Bo-Bo	1,360/1,550	125	67	13	1956	Lofag	ABES
1146	Bo-Bo	2,400	140	73	2	1987	Lofag	Elin
1163	Bo-Bo	–/1,600	120	80	20	1994	Graz	Adtranz
1822	Bo-Bo	4,400/-	140	83	5	1991	SGP	ABB
Shunting locomotives								
1063	Bo-Bo	1,520/2,000	100	82	50	1983	SGP	BES
1064	Co-Co	1,520	100	113	10	1985	SGP	BES
760 mm gauge, 6.5 kV 50 Hz								
1099	C-C	310/405	45	50	15	1911	Krauss	Siemens

*Class 116 is dual-voltage (15 kV AC 16 2/3 Hz/25 kV AC 50 Hz)
**Includes units on order

Bilevel City Shuttle push-pull set powered by a Class 1142 electric locomotive forming a Vienna S-Bahn service at Simmering Ostbahn (Ken Harris) 0109843

Czech Republic, Slovakia, France, and Belgium. Rated at 6.4 MW, they have a maximum tractive effort of 300 kN and a maximum speed of 230 km/h. They weigh 86 tonnes, and their length is 19.28 m. In freight traffic they are able to work trains of between 1,600 and 2,000 tonnes at speeds of up to 120 km/h. A major share of the construction work is being carried out in Austria–some components are bought from ÖBB workshops and final assembly is in Linz Works.

In 2001, ÖBB sold its nine surviving ASEA-built Class 1043 electric locomotives to Swedish operator Tågab (qv).

In 2001, Elin EBG was awarded a contract to equip 10 Class 1044 electric locomotives with radio remote-control equipment for mid-train working.

As part of it plans to build up a new fleet of standard locomotives, in 1998 ÖBB ordered 40 new main line diesel locomotives and 60 diesel shunters, with options on a further 200 locomotives.

The main line locomotives are being supplied by Siemens Austria, with first deliveries due in early 2002. Designated Class 2016 and also known as the 'Hercules' type, they are diesel-electric units with three-phase traction motors and are intended for both freight and passenger services. An option has been confirmed on 30 additional locomotives to the 40 originally ordered, and further options exist on 80 more.

The Class 2070 'Hector' B-B diesel-hydraulic shunting locomotives are being supplied by Vossloh Schienenfahrzeugtechnik, forming the company's G800 model. Deliveries commenced in 2001 and options exist on 90 additional units.

Pending the availability of the new main line diesel locomotives, ÖBB is leasing a small number of Soviet-built Class 232 high-powered diesel locomotives from DB AG to work freight services between the Slovak border and Vienna.

Also on order from Bombardier Transportation/Elin in 2001 were 11 three-car Talent emus for the Salzburg area and 40 four-car Talent emus for the Vienna S-Bahn. ÖBB is also to obtain a number of new dmus.

Passenger vehicles
The ÖBB stock of passenger coaches fell from 3,436 at the end of 1995 to 3,287 at the end of 1996, and then rose to 3,583 at the end of 1998.

The first phase of *Neue Bahn* investment earmarks Sch1.6 billion for new traction and rolling stock (though over the whole *Neue Bahn* programme forecast expenditure in this area totals Sch12.5 billion). Most of the first-phase money is going on development and evaluation of new 200 km/h passenger car prototypes with sophisticated amenities. ÖBB aims eventually to have trains in international service capable of using the 300 km/h new lines in neighbouring states. SGP has developed an SGP-300 range of guided-wheelset bogie designs including a model with 300 km/h capability.

In September 1995 the railway works at St Pölten rolled out ÖBB's first push-pull driving trailer, designated Class 80–75. After the initial production run of 14 vehicles, 75 were built by 1999 by conversion of 26.4 m Jenbacher vehicles built between 1982 and 1987 at a total cost in the order of Sch615 million.

Tenders were sought in 1994 for 50 bilevel coaches and 10 bilevel driving coaches. The contract went to ARGE Doppelstockwagen (a consortium of Siemens Verkehrstechnik/SGP and Jenbacher) in March 1995 for a sum of roughly Sch700 million. Add-on orders took the total number of vehicles on order to 240, to be formed as 40 six-car push-pull sets and branded 'City Shuttles'. The bilevel cars began to enter service in May 1998 on the Südbahn and then came into use on the Ostbahn in October 1998.

ÖBB's latest SGP Class Rh 4090 narrow-gauge electric (6.5 kV AC 25 Hz) multiple-units can be used in three- or four-car formations, two powered vehicles and two trailers, or a power-car, a trailer, and a driving trailer. Two sets can be coupled. Maximum speed is 70 km/h. They have modern seating, enclosed gangways, and closed-system toilets.

ÖBB is also spending Sch2,800 million on 660 refurbished City Shuttle cars. This stock is for use in

In early 1997 ÖBB opened a new depot at Villach, where 110 staff look after 160 locomotives, 250 passenger coaches and 4,000 freight wagons. This is a third major modern installation, alongside Knittelfield and Linz. The Linz works are being extensively modernised and will, in future, look after all ÖBB's high-performance passenger equipment. The railway aims in the future to exploit its modern facilities more extensively by tendering for international work.

Locomotives
The last Class 1044 locomotive was delivered in 1995. The first three locomotives of Class 1012, the 82 tonne, 6,000 kW, Bo-Bo successor to Class 1044, were due to be delivered in June 1996 but they were not actually acquired until December 1996, after considerable negotiation about costs, the price paid being Sch70 million per machine instead of the Sch90 million originally asked. They are used in place of Class 1044s on domestic services, so they have only been authorised for 160 km/h operation instead of the design speed of 230 km/h.

For through freight working between Germany and Italy via the Brenner, an Adtranz-Siemens-SGP consortium has delivered five prototypes of a dual-voltage (15 kV AC/3 kV DC) 82 tonne Class 1822 Bo-Bo rated at 4,300 kW with a top speed of 140 km/h. Since June 1997 these locomotives have been deployed on passenger workings over the Brenner in an integrated ÖBB/FS service running in timings laid down for FS electric traction which the previous Austrian diesels could not maintain.

In 1997 ÖBB sought tenders for 200 high-performance locomotives, with installed power of 6.5 MW and a top speed of 230 km/h, to be used on both fast passenger and heavy freight services. In view of the fall in prices of recent years, it hoped to be able to obtain a purchase price of only Sch40 million per locomotive. By the end of 2000 orders for 400 of the type had been placed with Siemens AG Österreich. Deliveries of these 'Taurus' locomotives began early in 2000. The locomotives, designated Classes 1116 (dual-voltage) and 1016, will be able to work not only in Germany and Austria but also, in the case of Class 1116, in Hungary, the

Diesel railcars or multiple-units

Class	Cars per unit	Motor cars per unit	Motored axles/car	Power/ motor KW	Speed km/h	Units in service	First Built	Builders Mechanical	Engine	Transmission
1,435 mm gauge										
5047	1	1	2	419	120	100	1987	JW	OM444LA	H Voith
5147	2	1	4	838	120	10	1992	JW64	OM444LA	H Voith
8081	–	–	–	–	90	3	1964	–	–	–
760 mm gauge										
5090	1	1	4	235	70	17	1986	Knotz	MAN-D 2,866 LUE	E BBC

Electric railcars or multiple-units

Class	Cars per unit	Motor cars per unit	Motored Axles/car	Output/motor KW	Speed km/h	Units in Service	First built	Builders Mechanical	Electrical
1,435 mm gauge									
4010	6	1	4	620	150	29	1964	SGP	BBC
4020	3	1	4	300	120	120	1978	SGP	BES
4030.1/2	3	1	4	250	100	50	1956	SGP	BES
4030.2	3	1	4	315	120	22	1962	SGP	BES
4130	3	1		315	120	2	1958	SGP	Siemens
4855*	1	1		480	120	2	1989		Elin
760 mm gauge									
4090	3/4	1/2				2	1994	SGP	Elin

*15 kV/800 V DC51.

Carinthia, Styria, Upper Austria, Tirol, and Salzburg. The work of refurbishment includes making the vehicles suitable for mobile telephones, providing power supplies for PCs, installing new general purpose power supply equipment and fitting modular air conditioning systems. The first vehicles entered service in 2001 and the work is scheduled for completion in 2004.

ÖBB is also in the course of renewing its BahnBus fleet, with the purchase of 440 new buses at a cost of Sch1,100 million.

Twenty new couchette coaches, costing Sch300 million, were brought into service in 2001 on trains to Berlin, Hamburg and Paris, thus completing the modernisation of ÖBB's night stock.

In 2001 tenders were invited for 116 dual-voltage 200–230 km/h tilting trainsets to be ordered jointly by ÖBB, German Rail and Swiss Federal Railways to equip the three companies' Trans Europe Excellence Rail Alliance (TEE) (see Passenger operations, Long-distance). Of these, 32 were to be procured by ÖBB, with the first examples entering service in 2004.

Signalling and telecommunications
Electronic signalling installations are being developed by two companies, Alcatel Austria and Siemens AG. Each is pursuing its own software technology.

With the commissioning of new signalling and LZB (the German form of automatic train protection) in March 1993 over the 25 km between Linz and Wels, 200 km/h running became possible for the first time in Austria. Later that year LZB was commissioned from Wels to Lambach and Attnang-Puchheim. The new Salzburg electronic signalling control centre was brought into use in May 1996. This is an Alcatel installation replacing five previous signalboxes.

Between mid-October and mid-November 1999 the first trials of European Train Control System (ETCS) took place on the Parndorf-Hungarian border section of the Vienna-Budapest main line (which will be ÖBB's first main line to be equipped with the system). The line was fitted with transponders, and two locomotives were appropriately adapted (1014.015 from ÖBB and V63.156 from MÁV).

Track
Standard rail
Standard-gauge: 60.34, 53.81, 49.43 kg/m
Narrow-gauge: 35.65 kg/m
Length
Standard-gauge: 30 and 60 m
Narrow-gauge: 20 m
Crossties (sleepers)
Standard-gauge: impregnated wood 2,600 × 260 × 160 mm; concrete 2,600 × 300 max × 200 mm max; also some steel
Narrow-gauge: impregnated wood 1,600 × 200 × 130 mm; concrete 1,500 × 200 max × 160 mm max

Crossties spacing
Standard-gauge: 600–700 mm (1,667–1,429 per km)
Narrow-gauge: 700–810 mm (1,429–1,235 per km)
Rail fastening
Standard-gauge: resilient fastening, ribbed slabs, clips and bolts
Narrow-gauge: ribbed plates and elastic clips
Filling
Standard-gauge: broken stone ballast 30–65 mm
Narrow-gauge: broken stone ballast 25–35 mm
Thickness under sleepers
Standard-gauge: 200–300 mm
Narrow-gauge: 150 mm
Min or sharpest curvature
Standard-gauge: 9.7° = min radius of 180 m
Narrow-gauge: 29.1° = min radius of 60 m
Max gradient compensated
Standard-gauge: 4.6 per cent
Narrow-gauge: 2.5 per cent
Gauge width with max curvature
Standard-gauge: 20 mm
Narrow-gauge: 20 mm
Max super elevation
Standard-gauge: 160 mm
Narrow-gauge: 60 mm
Max axleload
Standard-gauge: 22.5 tonnes
Narrow-gauge: 12 tonnes

Graz–Köflach Railway (GKB)

Graz–Köflacher Bahn und Busbetrieb GmbH (GKB)
Köflacher Gasse 41, A-8020 Graz
Tel: (+43 316) 598 70   Fax: (+43 316) 59 87 16
e-mail: sales@gkb.at
Web: http://www.gkb.at

Key personnel
Managing Director: Franz Weintögl

Gauge: 1,435 mm
Route length: 96.5 km

Organisation
The railway, which is operated as an autonomous entity, heads south from its own station at Graz to Lieboch, where it branches northwest to Köflach and south to Wies-Eibiswald. From the Wies-Eibiswald branch a further short freight-only branch runs east to Gleinstätten. In addition to its rail services, the company operates 27 bus routes in West Steiermark. There are 450 staff.

A subsidiary company, LTE Logistik Transport GmbH, was established by GKB as an open access freight operator following the award of a licence in 1999 to run passenger and freight trains over the Austrian network. Operations commenced

Class 1500 Jenbacher/Henschel diesel-hydraulic locomotive with double-deck stock forming a Köflach-Graz service approaching Graz Köflacherbahnhof (John C Baker)   0583350

in August 2001, hauling cement traffic between Marchegg and Vienna Liesing. Initially, locomotives from the Vossloh hire pool were used, pending delivery from the same company of two Type G1209 diesel-hydraulics in late 2001.

There are two other subsidiaries: LBB Lavamuend Bahn Betriebs GmbH, established in May 2002, and Graz-Köflacher Bahn- und Busbetrieb Deutschland, created in October 2003.

Passenger operations
The GKB operates diesel railcar and local and commuter push-pull services, carrying around 10 million passengers annually. Trains run roughly hourly to/from both Köflach and Weis-Eibiswald.

Freight operations
Freight trains serve both Köflach and Weis-Eibiswald. Tonnage totals around 500,000 tonnes annually.

Traction and rolling stock
The company operates two steam locomotives, 13 diesel locomotives, 13 VT70 diesel trainsets, two VT10 diesel railcars and three trailers, 30 passenger coaches, including 15 bilevel passenger cars, and 98 freight wagons. The VT70 dmus are articulated twin-units with MTU engines and ABB electric transmissions, built by Simmering-Graz-Pauker to Linke-Hofmann-Busch design under licence.

Signalling and telecommunications
Track-to-train radio communication, supplied by AEG-Westinghouse, became operational in 1992.
Type of coupling: UIC-coupler, railcars and railbuses excepted
Type of braking: Compressed air

Diesel line-haul locomotives

Class	Wheel arrangement	Power kW	Speed km/h	Weight tonnes	No in service	First built	Builders Mechanical	Builders Engine	Builders Transmission
1500.1–6	B-B	1,103	100	64–72	6	1975	Jenbacher/ Henschel	Jenbacher LM1500	H Voith L720rU2
700	C	515	48	6	1	1977	MaK	MaK	H Voith L4r4U2
600	C	441	60	48	3	1973	Jenbacher	Jenbacher JW600	H Voith L26StV
1700	B-B	1,700	100	80	1	2003	Vossloh	MTU 12V 396TC14	H Voith
1100	B-B	808	100	64	2	1961	Henschel	Caterpillar 3512DI-TA	H Voith HDHL 216rs

Diesel relicars

Class	Cars per unit	Motor cars per unit	Motored axles/car	Power/ motor kW	Speed km/h	Units in service	First built	Builders Mechanical	Builders Engine	Builders Transmission
VT70	2	2	2	228	90	5	1980	SGP/LHB	Büssing BTYUE	E BBC AC-DC
VT70	2	2	2	237	90	8	1983	SGP/LHB	Büssing D2866 LUE	E BBC AC-DC
VT10	1	1	2	2 × 110	90	2	1953	Uerdingen	Büssing U10	M ZF-Gmeinder

Track
Rail: (B) (S49) 49.43 kg/m
Crossties (sleepers): Wood, thickness 160 mm
Concrete, thickness 200 mm
Spacing: 1,538/km plain track and curves
Fastening: Rippenplatte and Pandrol
Min curvature radius: 181.25 m
Max gradient: 0.015%
Max axleload: 20 tonnes

*Class 1100 Henschel diesel-hydraulic locomotive
with a freight service at Graz* (John C Baker)
0583351

LTE Logistik- und Transport-GmbH

Reininghausstrasse 3, A-8020 Graz
Tel: (+43 316) 598 72 35   Fax: (+43 316) 598 72 39
Web: www.lte.at

Vienna office
Absberggasse 47, A-1103 Vienna

Key personnel
Managing Director: Dr Georg Pammer
Managing Director and Head of Operations:
   Gerhard Eibinger
Manager, Operations: Hugo Koroschetz

Subsidiaries
LTE Logistik a Transport Slovakia sro
LTE Logistik a Transport Czech Republik sro
Lombardiniho 22b, SK-83103 Bratislava, Slovakia
Tel: (+421 2) 50 58 28 13   Fax: (+421 2) 50 58 32 89
e-mail: ladislav.patz@lte.sk
Managing Director: Ladislav Patz

Background
LTE is an Austrian open access freight operator
established as a joint venture by the Porr group
of infrastructure companies and Graz–Köflacher
Eisenbahn GmbH (GKB). Freight operations
commenced in 2001.

Subsidiaries have been established in the Czech
Republic and Slovakia.

In January 2005 LTE became a founding member
of the European Bulls railfreight alliance (see entry
in International section of *Railway systems and
operators*), which provides cross-border services
linking several countries.

Freight operations
Traffic gained by LTE has included movements
of cement for Holcim, spoil trains in connection
with construction of the Lainzer tunnel and
the delivery of Desiro UK emu vehicles from
Siemens' Vienna works to Wildenrath, Germany,
in collaboration with the German operator NIAG.
Other rolling stock delivery contracts have also
been secured.

LTE also co-operates with European Bulls
alliance partner rail4chem to move chemicals
between Wittenberg, Germany, and Linz, Austria,
and in conveying petroleum from Bitterfeld to
St Valentin.

In September 2003, in collaboration with
German operator TXLogistik, LTE introduced a
bimodal service between Brennero, Italy, and
Flensburg, Germany, using Dispolok ES 64 U2
electric locomotives. An intermodal service is also
operated between Graz and Duisburg, Germany,
using LTE's Class 185 electric locomotive.

Traction and rolling stock
LTE owns two Vossloh G 1206 and one G 1700
diesel-hydraulic locomotives and one Bombardier
Class 185 electric locomotive. A Class 2016 diesel-
electric locomotive is also hired from Siemens
Dispolok.

UPDATED

Stern & Hafferl Light Railways

Stern & Hafferl Verkehrs Gesellschaft mbH
PO Box 122, A-4810 Gmunden, Austria
Tel: (+43 7612) 79 52 07   Fax: (+43 7612) 79 52 02
e-mail: sekretariat@stern-verkehr.at
Web: www.stern-verkehr.at

Key personnel
President: Dipl-Ing Jochen Döderlein

Organisation
The group operates the following railways:

Gauge: 1,435 mm
Linz–Eferding–Waizenkirchen–Neumarkt–Kallham/
Peuerbach, 58.9 km, electrified at 750 V DC (Linzer
Lokalbahn)
Lambach–Vorchdorf, 15.5 km, electrified at 800 V
DC (Vorchdorferbahn)
Lambach–Haag am Hausruck, 26.3 km, electrified at
800 V DC and 15 kV 16²/₃ Hz (Haager Lies)

Gauge: 1,000 mm
Gmunden–Vorchdorf, 14.7 km, electrified at 800 V
DC (Traunseebahn)

*Type GTW 2/6 electric railcars used on Linzer Lokalbahn services, which are operated by Stern &
Hafferl Light Railways* (Milan Šrámek)
0109844

Vöcklamarkt–Attersee, 13.4 km, electrified at 800 V
DC (Attergaubahn)

Traction and rolling stock
The group owns seven electric locomotives, two
diesel locomotives, 29 passenger cars, 40 light

rail vehicles and 53 freight wagons. Also operated
are eight 14 GTW 2/6 lightweight electric railcars,
all equipped for dual-voltage (750 V DC/15 kV AC
16²/₃ Hz) operation. These are used on the Linzer
Lokalbahn.

UPDATED

Styrian Provincial Railways (StLB)

Steiermärkische Landesbahnen (StLB)
PO Box 893, Radeyzkystrasse 31, A-8011 Graz
Tel: (+43 316) 812 58 10   Fax: (+43 316) 81 25 81 25
e-mail: office@stlb.at
Web: http://www.stlb.at

Key personnel
General Manager: F Brünner
Managers
   Finance: H Wittmann
   Traffic: A Pint

Chief Engineers
   Mechanical and Electrical: R Zeller
   Track: F Brünner

Organisation
The group operates the following railways:

Gauge: 1,435 mm
Feldbach-Bad Gleichenberg, 21 km, electrified at
1.8 kV DC
Gleisdorf-Weiz, 15 km
Peggau-Übelbach, 10 km, electrified at 15 kV AC

Gauge: 760 mm
Weiz-Oberfeistritz (Feistritztalbahn), 11.8 km
Unzmarkt-Tamsweg (Murtalbahn), 65.5 km
Mixnitz–St Erhard, 10.4 km, electrified at 800 V DC.

In addition, StLB has an operating concession
to provide local services on the line from
Graz to Mogersdorf, close to the border with
Hungary.

As well as operating freight services over its
own system, StLB was also expected to use its
Gmeinder-built diesel locomotives on ÖBB tracks
under open access rights.

Traction and rolling stock

The group owns eight narrow gauge steam, one standard gauge and four narrow gauge electric locomotives, seven standard gauge and nine narrow gauge diesel locomotives, four standard gauge electric railcars, two standard gauge and five narrow gauge diesel railcars, six trailers, 34 passenger cars, five baggage and postal cars, and 194 freight wagons. The most recent traction acquisitions were three 1,100 kW B-B diesel-hydraulic locomotives from Gmeinder.

StLB Gmeinder-built 1,100 kW diesel-hydraulic locomotive on display at InnoTrans 2002 before delivery (Ken Harris)
0528627

Wiener Lokalbahn AG (WLB)

Eichenstrasse 1, A-1120 Vienna
Tel: (+43 1) 90 44 40   Fax: (+43 1) 90 44 43 50
e-mail: office@wlb.at
Web: www.wlb.at

Key personnel

General Managers: Josef Pfelz, Dipl-Ing Günther Zimmerl

Gauge: 1,435 mm
Length: 30.4 km
Electrification: 30.4 km at 850 V DC

Organisation

Majority owned (96 per cent) by the city of Vienna, WLB operates an interurban light rail line south from Vienna, as well as several bus routes. The line has also traditionally carried freight, and since March 2001 WLB has been an open access operator.

Passenger operations

WLB operates a high-frequency service over the Badner Bahn, between Vienna Opera and Baden Josefsplatz, partly operating on the city's tram infrastructure and partly on its own double-track line. In 2002 rail passenger journeys totalled 7.53 million compared with 7.06 million the previous year.

Freight operations

On its own lines WLB generates around 180,000 tonnes of freight annually. Since 2001, when the company gained an open access licence to operate in Austria and Germany, it has progressively gained both intermodal and bulk products traffic.

Two WLB T 2500 low-floor trams forming a Vienna Opera–Baden Josefsplatz service at Schöpfwerk (John C Baker)
***NEW**/0585072*

In 2005 container, swapbody and unaccompanied trailer services included trains linking Vienna and Duisburg Intermodal Terminal, operated in collaboration with Duisport rail, and Bremerhaven–Austria workings in co-operation with Rurtalbahn in Germany. In 2004 WLB secured a contract to move kerosene from Ingolstadt, Germany, to Kleinschwechat, Austria.

Traction and rolling stock

For its passenger services WLB operates 26 Düwag trams dating from 1979 to 1993 and six Type T 2500 three-section low-floor vehicles delivered by a Bombardier-led consortium in 2000. Three more of these were on order in 2005. Freight operations are handled by five diesel locomotives, including three ex-DB refurbished V100.4 machines, and three leased Siemens Type ES64 U2 electric locomotives. Additional diesel and electric locomotives were to be leased to cater for growing traffic.

UPDATED

WLB Type ES64 U2 electric locomotive at Wien Heidestrasse with a container service from Dortmund (John C Baker)
***NEW**/0585071*

Azerbaijan

Azerbaijani Railways (AZR)

Azerbaycan Dövlet Demir Yolu
ul Mustofaeva 230, 370010 Baku
Tel: (+994 12) 98 44 67   Fax: (+994 12) 98 85 47

Key personnel
President: V M Nadirli
Director General: Z A Mamedov
Deputy Director: M S Panakhov
Manager, International Affairs: RT Zeinalov

Gauge: 1,520 mm
Route length: 2,957 km (of which 270 km
   non-operational)
Electrification: 1,278 km at 3 kV DC

Political background
Having borders with Russia, Georgia, Armenia and
Iran, Azerbaijan gained independence from the
USSR in 1991. The dispute with Armenia concerning
the territory of Nagorno-Karabakh, and the cutting
of both links with Russia in Abkhazia and Chechnya,
had a serious effect on traffic levels.

In 1999 228 route-km in the country's
southeastern area were reported to be controlled
by Armenian military forces, preventing through
operations via Nakhichevan to Iran.

Privatisation of AZR was reported to be a
possibility following the introduction of enabling
legislation in 1999–2000.

Organisation
The current AZR organisation was established in
August 1995. There are three operational divisions
based in Baku, Gyandzha and Nakhichevan. These
are overseen by AZR's general management, based
in Baku.

The network comprises the whole of the former
SZhD's Azerbaijani Railway. Two main lines extend
from the capital, Baku: the northern runs along the
Caspian coast to Makhachkala in Russia, the other
heads south to Alyat before turning inland to serve
Kyurdamir, Yevlakh and Akstafa before reaching
Tbilisi in Georgia. Both lines are double-track and
electrified.

A third main line, only partially electrified,
follows the Iranian border to Nakhichevan, from
where there is an electrified link to Iran at Jolfa.

A total of 825 km of AZR's route length is double-
track.

AZR operates a cross-Caspian ferry from Baku to
Turkmenbashi in Turkmenistan.

In April 1997 agreement was reached between
the governments of Azerbaijan, Georgia, and
Ukraine to establish a joint international company
to run daily rail and train ferry services linking the
three countries.

Passenger operations
In 1989, traffic stood at 2,020 million passenger-km. By
1999, this had declined to 422 million passenger-km.

Passenger trains are now operated on three key
routes. The most important connects Baku with
Yalama, on the border with Russia, and in 2002
was served by through international trains on the
following routes: Baku–Moscow (three train pairs
per week); Baku–Kiev (two pairs); Baku–Kharkov
(two pairs); Baku–Rostov na Donu (four pairs);

and Baku–Astrakhan (four pairs). In addition, there
were two daily Baku–Yalama internal stopping
services. International operations over the Baku-
Yalama route only became possible after an easing
of tension in the Russian republic of Chechnya
in 1995. These services were subsequently
strengthened after the opening by the Russians in
1998 of the new Kizlyar–Kizil'yurt line, which avoids
Chechnya. This is now used by all international rail
traffic between the two countries.

The east-west Baku–Yevlakh–Beyuk–Kyasyk–
Georgia (Tbilisi) route, with branches, is operated
by five pairs of commuter trains running mostly in
morning and evening peaks ands is also served by
an overnight Baku–Tbilisi international train.

The southernmost line is served by a pair
of overnight trains each day linking Baku with
Goradiz, at the frontier with the Armenian-held
territory of Nagorno-Karabakh, and by a pair of
Baku–Astara trains. Separated from the rest of
the AZR network, a twice-weekly service between
Nakhichevan and Tabriz was restored in February
1997, linking Azerbaijan and Iran after a nine-year
break. Through coaches to/from Tehran were
introduced in July 2001.

A relatively dense commuter service using emus
is operated around Baku and in Gyandzha.

Freight operations
The country is rich in oil, ore and other resources,
and agrarian produce and chemicals have also been
important in freight shipments. The collapse of the
Soviet Union hit traffic levels hard: down from
92 million tonnes (44,900 million tonne-km) in 1989
to 7 million tonnes (2,100 million tonne-km) in 1996.
Traffic then rose sharply in 1997, thanks to which the
railway was able to contribute 12 billion manat to
the state budget. Freight shipments were 9.1 million
tonnes in 1997, 10.6 million in 1998, and 11.7 million
in 1999, 5.1 billion tonne-km being recorded in 1999.
Oil and petroleum products traffic originating on
the Apsheron peninsula east of Baku accounted for
85.9 per cent of this tonnage, with most of the
remainder consisting of building materials.

As part of the Traceca agreement, a Baku –Tbilisi-
Poti unit container train linking Azerbaijan and
Georgia was launched in November 1996.

New lines
The last new line to have been completed was that
running 158 km from Yevlakh, on the Baku–Tbilisi
main line, northwards to Belokany. Inauguration
took place in 1987, during the period of Soviet
rule. Also started in 1987 was a line from Padar to
Shemakha, of which only 24.5 km of track as far as
Geylyarchyol had been laid up to 2002. Completion
of the line looked unlikely in view of the country's
poor economic situation.

A second route into Iran has been long planned
from the Astara port terminus of the line south from
Baku. Its immediate goal is a mere 7 km across the
border, to a site where customs and warehousing
facilities are to be provided for Iranian imports
which currently languish awaiting clearance at
Astara. Preliminary works started in early 1993, and
some construction was undertaken in 1994–95.

Traction and rolling stock
In 1999 the locomotive fleet comprised 190 Class
VL8 and 62 Class VL11 3 kV DC electric locomotives
and diesel-electric locomotives of classes TEM2
(128 examples), 2TE10M (22), 3TE10M (24), 2M62
(12) and ChME3 (50). Large amounts of these
were, however, stored as unserviceable, and it was
estimated that more than half of the locomotive
fleet was in need of replacement. Non-powered
rolling stock consisted of 875 passenger coaches,
of which 100 were unserviceable, and around
25,400 freight wagons.

In late 1997 the railway owned 25,000 freight
wagons. New and refurbished oil tank wagons were
also required, both for export of Azerbaijani oil and
for transit oil from Turkmenistan and Kazakhstan.

To restore the Baku–Rostov service, passenger
stock had to be leased from Russia's North
Caucasus Railway.

Signalling and telecommunications
A total of 1,126 route-km of the AZR network
is equipped with automatic block signalling.
The introduction in 1996 of container services
between Azerbaijan and Georgia necessitated
laying optical fibre cable and modernising
signalling between Baku and the Georgian
border at Beyuk-Kyasyk.

Electrification
Azerbaijan was one of the pioneers of Soviet
electrification, with the 33 km from Baku to
Sabunchi and Surachami energised at 1.2 kV DC
in 1926. Conversion to 1.5 kV followed in 1940,
and to 3 kV in the 1960s. The most recent project
was the Jolfa–Nopachen line, energised in 1989.
Electric traction is responsible for 80 per cent of
all traffic.

AZR yard scene, with TEM2 and VL11 locomotives in evidence (Norman Griffiths)   0088190

Bangladesh

Ministry of Communications

Bhavan 7, 8th Floor Secretariat, 1000 Dhaka
Tel: (+880 2) 83 16 65   Fax: (+880 2) 86 43 70
Web: www.moc.gov.bd

Key personnel
Minister: Nazmul Huda
State Minister: Salahuddin Ahmed
Secretary: Md Shafiqul Islam
Joint Secretary, Rail: Md Abu Jafar

UPDATED

Bangladesh Railway (BR)

Rail Headquarters, Chittagong
Tel: (+880 31) 50 01 20/50 01 39

Key personnel
Director-General: Muzatfizur Rahman

Gauge: 1,676 mm; 1,000 mm
Total route length: 946 km; 1,822 km

Organisation
The system is under the overall control of a Director General who heads the Railway Division of the Ministry of Communications. For day-to-day operational management the railway is partitioned into two administrative zones. The West Zone is mainly broad-gauge, the East metre-gauge. The two zones were unconnected except by a metre-gauge wagon ferry, and were operated separately, until opening of the dual-gauge line over the Bagabandhu bridge over the Jamuna River in 2001.

At the end of FY2002 BR employed 35,540 staff.

Finance
In 2002 operating revenues amounted to BDT3,884 million; addition of public service obligation and welfare grants brought total revenue to BDT4,901 million. Operating expenditure was BDT5,355 million, producing a net loss of BDT454 million.

Traffic (million)	2002
Passenger journeys	38.7
Passenger-km	3,970
Freight tonnes	0.9
Freight tonne-km	908

Passenger operations
Passenger traffic remained buoyant as freight tonnage declined to below 1 million tonnes in 2002. This was stimulated by the new journey opportunities opened up by the cross-Jamuna route and completion of dual-gauging to allow through running of broad-gauge trains from the West Zone to Dhaka.

Current works are aimed at raising the top speed of existing metre-gauge services from 75 to 100 km/h. Purchase of 250 new coaches was authorised in 2003, among them a batch of 50 vehicles being delivered during 2005–06 from Indonesia's PT Inka at a cost of BDT869 million. These are air conditioned sleeping and seating cars for broad-gauge services.

Early in 1997, India and Bangladesh reached agreement in principle on the restoration of passenger services between the two countries, which have been suspended since the war of 1965. First priority will be a Calcutta-Jessore route.

Freight operations
BR's principal freight task is now hauling containers between the port of Chittagong and Dhaka, using box wagons converted into flats. Distribution of petroleum products is also seen as a strategic priority, but movements have been hampered by the poor condition of the tank wagon fleet; some tankers are over 60 years old. Relief will be provided by a new batch of 100 tank wagons ordered in early 2005 from China's Liaoning Mechanical Col.

In this sector, too, the emphasis is on long-haul traffic, particularly to the remote northern regions of the country. General freight haulage has continued to decline on account of poor quality service, with many non-containerised imported goods arriving in Chittagong now being distributed by road.

New lines
The long-planned USD696 million, 4.8 km road-rail bridge over the Jamuna River, was finally opened in 2001 after several years' delay. Further work involved dual-gauging the route from the bridge to Dhaka, completed in 2004. Associated track improvements and resignalling work have created a route over which broad-gauge trains now run at up to 120 km/h. Work continues on dual-gauging the broad-gauge main line northwards from Ishurdi to Santahar and Parbatipur (475 km).

Feasibility studies for three new lines and a Dhaka–Laksam direct line have been completed in recent years, but any new construction is problematic on cost grounds and the latter project was initially shelved despite the obvious benefit to Dhaka–Chittagong trains. However in 2005, six groups were shortlisted by the Ministry of Communications after submitting bids covering electrification of the Dhaka–Chittagong line, plus construction of a cut-off that would reduce the distance between the two cities from 346 km to 267 km.

Traction and rolling stock
In 1999 BR was operating 272 diesel-electric (68 broad-gauge and 204 metre-gauge) and 266 broad-gauge and 1,011 metre-gauge coaches. Freight stock comprised 3,134 broad-gauge and 10,600 metre-gauge freight wagons. Considerable numbers of all types of vehicle were stored out of service or were otherwise unsuitable for use. A long-term aim has been a complete renewal of the mixed motive power, which includes units of British, Canadian, Hungarian, Japanese and US manufacture. However reliance on foreign aid militates against this. The latest batch of locomotives being delivered in 2005, 11 metre-gauge units, comes from South Korean builder Rotem and is funded by loans from the South Korean Economic Development Corporation Fund.

Signalling and telecommunications
A 1,700 km fibre optic network was constructed in the 1990s along routes connecting the East and West zones of the railway, with a radio link across the division that then existed at the

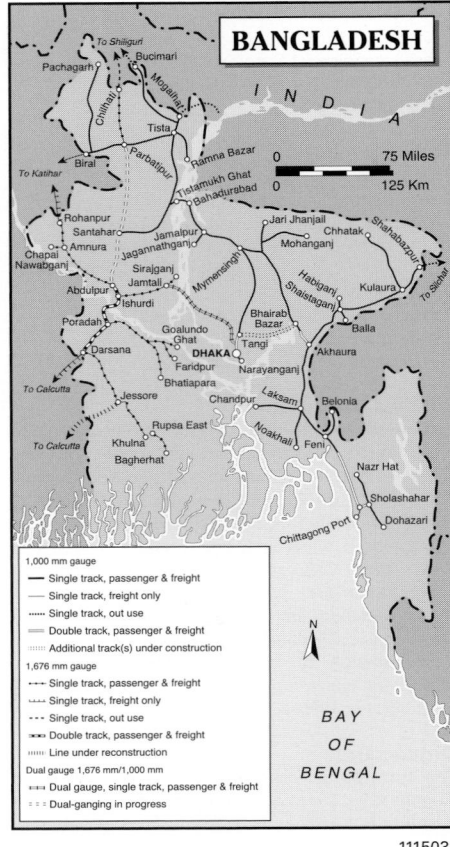

1115031

Jamuna River. The work included the supply and installation of digital automatic telephone exchanges and telephone instruments. In addition, the contract included a train control system for telecommunication between trains and railway stations, civil engineering work, installation, training and maintenance.

In 2005 LG Industrial Systems was working on a USD11.7 million contract to provide modern electronic interlocking at 10 stations on the line between Akhaura and Sylhet.

Track
Rail types and weight: 75 lb 'A', 90 lb 'A' FF BSS rails, 50 lb 'R', 60 lb 'R', 90 lb 'R' FF BSS rails, 50 NS, 50 ISR and 80 lb
Crossties (sleepers)
Thickness: Wooden sleeper BG 5 in; steel through sleeper BG ½ in, cast-iron CST/9 block; wooden sleeper MG 4½ in; steel through sleeper MG 11–32 in; cast-iron sleeper CST/9 (block)
Spacing: N + 1, N + 2, N + 3, N + 4, N + 5
Rail fastenings: Fishplates, fish boxes, dogspikes, bearing plates, anchor bearing plates, round spikes, steel keys and steel jaws and rail anchors of different sizes
Signal and train control installations: Relay and mechanical interlocking

UPDATED

Belarus

Ministry of Transport and Communications

ul Lenina 17, 220030 Minsk
Tel: (+375 172) 68 74 07   Fax: (+375 172) 27 56 48

Key personnel
Minister: A V Lukashov
Head of Presidential Cabinet: V I Rakhmanko

Political background
In mid-1996, a governmental decree transferred control of the national railway from the Ministry of Transport and Communications to direct oversight by the Presidential Cabinet.

Belarussian Railways (BCh)

ul Lenina 17, 220030 Minsk
Tel: (+375 172) 25 48 60   Fax: (+375 172) 27 56 48

Key personnel
Director General: V I Gapeev
First Deputy Director General: V I Jerelo

Deputy Director General, Way and Works:
A A Sivak
Deputy Director General, Personnel: V F Abramov
Chief Engineer: V N Shubaderov
Deputy Director General, Traction and Rolling
Stock: V V Balakhonov
Manager, External Relations: A P Sergiyenko

Gauge: 1,520 mm; 1,435 mm
Route length: 5,497 km; 15 km
Electrification: 874 km at 25 kV 50 Hz

Political background

Belarus signed wide-ranging Union Treaties with
Russia in 1996 and 1997. The country straddles the
Berlin–Moscow main line and, in conjunction with
Poland, potentially provides, from a Russian point of
view, a more reliable form of access than Lithuania to
the Russian enclave of Kaliningrad on the Baltic coast.
In March 1998 a permanent Belarussian/Russian
working group for railway co-operation was set up.

In 1998, a Railway Law, four years in the making,
was passed. Previously, Soviet-era procedures had
been preserved, although many of these had no
legal foundation. The new law, among other things,
defines railway property and thereby makes more
difficult piecemeal appropriation under the guise of
privatisation. The wearing of railway uniforms and
the name of the railway itself are now legalised.

Unlike in some other eastern European countries,
privatisation has not become an issue for BCh.

Organisation

BCh was formed after the break-up of the former
Soviet Railways (SZhD) in 1992 and comprises SZhD's
Belarussian Railway. It is mainly a transit railway. In
2002 the number of employees totalled 103,259.

Finance

Despite a precipitous collapse, freight traffic
remains profitable, especially transit traffic, for
which the tariffs are substantially higher than
domestic rates. Any surplus are used to offset
losses on passenger services. The government
no longer provides financial support, so the
subsequent recovery depended on cost-cutting and
the marketing of auxiliary services.

Traffic	1999	2000	2002
Passenger journeys (million)	165	168	152.9
Passenger-km (billion)	16.9	17.7	14.3
Freight tonnes (million)	51.6	88	93.6
Freight tonne-km (billion)	30.5	31.4	34.2

Passenger operations

At the time of the break up of SZhD, Belarus
was recording over 18 billion rail passenger- km
annually. After a decline in the mid-1990s, traffic has
subsequently recovered, albeit not to former levels.

Freight operations

Traffic in 1991 was recorded at 65.6 billion tonne-
km, but by 1995 this had dropped to 25.5 billion.
A subsequent recovery has seen volumes grow
again but not to the levels experienced during the
Soviet era. Transit traffic provides the main source
of the railway's freight revenue.

Improvements to existing lines

Electrification began in the 1960s on the Minsk-
Molodechno and Minsk–Pukhovchi lines. Now two
double–track routes form the core of the system:

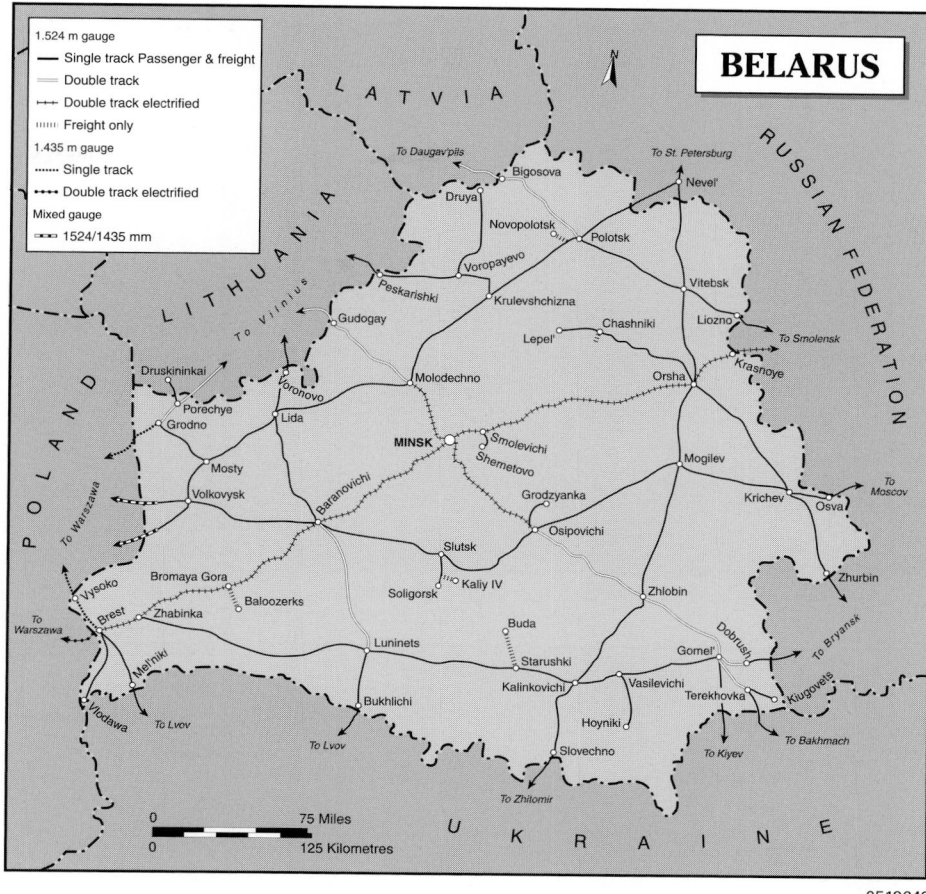

0519649

Belarus Railways Class TEP70 diesel-electric locomotive (Edward Barnes)

0558626

a line from Minsk to Vilnius, Lithuania, which is
electrified as far as Molodechno; and a northeast
to southwest line crossing the country from the
Russian Federation to Poland, passing through
Orsha, Minsk, Baranovichi and Brest, which is
electrified throughout.

Traction and rolling stock

The locomotive stock totals some 500 units and
includes ChS4t electrics plus ChME3, M62 family,

TEP60 and TEP70 diesels, together with DR1P, DR1A
and ER9 family multiple-units. In 1999, BCh took
delivery of a prototype diesel Class DDB1 trainset
from the Demikhovo dmu plant in Russia. The train
is formed of two M62 diesel locomotives and 10
refurbished trailers in a push-pull configuration.

Belgium

Federal Mobility & Transport Department

Rue Brederode 9, B-1004 Brussels
Tel: (+32 2) 237 67 11   Fax: (+32 2) 230 18 14
Web: www.mobilit.fgov.be

Key personnel

Minister: Renaat Landuyt
Secretary-General: Pascal Hertsens

UPDATED

Dillen & Le Jeune Cargo (DLC)

Airport Business Center, Luchthavenlei 7a, B2100
Deurne
Tel: (+32 3) 844 97 02   Fax: (+32 3) 844 97 03
e-mail: info@dlcargo.com
Web: www.dlcargo.com

German office
Am Victoriaturm 2, D-68159 Mannheim
Tel: (+49 621) 122 23 38   Fax: (+49 621) 122 22 37

Key personnel

Chairman and Chief Operating Officer:
Ronny Dillen
Chief Executive Officer: Jeroen Le Jeune

Organisation

DLC was the first open access operator to run
freight trains in Belgium, commencing services
on 3 April 2002 with a three times-weekly train of
BMW car components for export from Regensburg,
Germany, to Antwerp. DLC now operates trains

DLC JT42CWR locomotives stabled at Antwerp (Quintus Vosman)   ***NEW**/1115032*

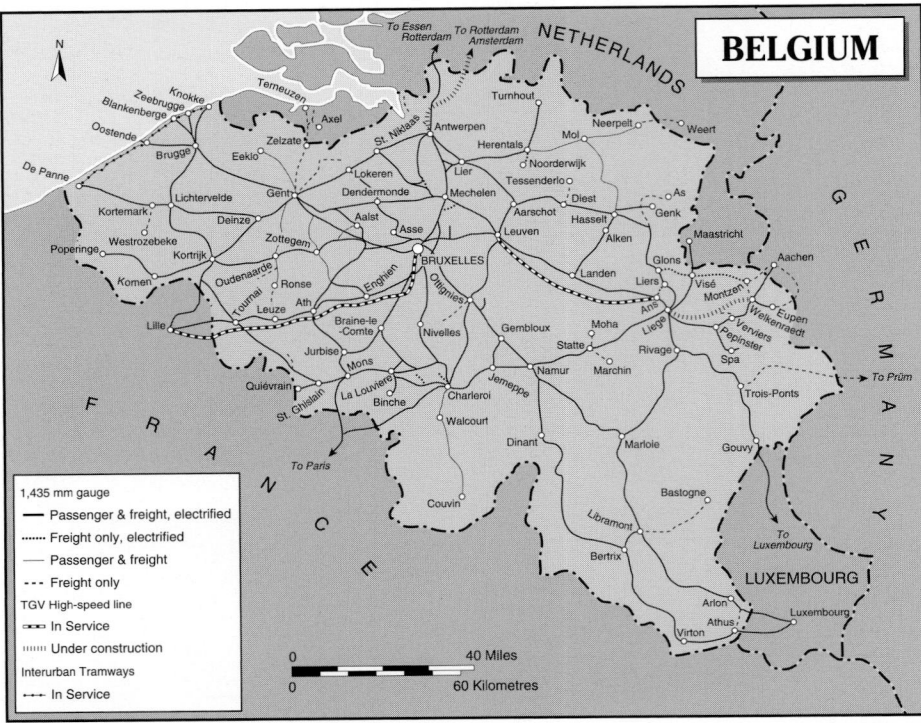

***NEW**/1115042*

from several points in Belgium to Germany and the Netherlands and has offices in Germany. The company is 40 per cent owned by Hupac (see entry in the Freight section of *International operators of rail services in Europe*).

Traction and rolling stock

DLC uses 10 hired General Motors JT42CWR diesel-electric locomotives ('Class 66') plus one hired Siemens Taurus electric locomotive.

UPDATED

Société Nationale des Chemins de fer Belges (SNCB) Nationale Maatschappij der Belgische Spoorwegen (NMBS)

85 Rue de France, B-1060 Brussels
Tel: (+32 2) 525 21 11   Fax: (+32 2) 525 40 45
Web: www.sncb.be
    www.nmbs.be

Key personnel

Chairman of the Board of Administration:
  Alain Deneef
Chief Executive: Karel Vinck

Directors
  Traction and Rolling Stock: Jean Denayer
  Passenger Services: Leo Pardon
  B-Cargo: Marc Descheemaecker
  Finance: Luc Lallemand
  Operations: (vacant)
  Network: Alex Migom
  Infrastructure: Jean-Marie Raviart
Public and Press Relations: France Nivelle

Subsidiary

B-Cargo
Rue des Deux Gares 80, B-1070 Brussels
Tel: (+32 2) 525 86 62   Fax (+32 2) 525 87 09
e-mail: cargo@b-rail.be
Web: www.b-rail.be

Gauge: 1,435 mm
Route length: 3,521 km
Electrification: 2,624 km at 3 kV DC, 303 km at 25 kV 50 Hz AC

Political background

In October 1992 SNCB was statutorily reconstituted as a public limited company with the state holding all shares. New financial arrangements were concluded at this time, under which international passenger traffic and all wagon and part-load freight operations were excluded from state

support; in these sectors SNCB is required to be fully commercial.

The early 1990s economic recession hit SNCB hard and prompted a series of economy measures in 1994–95. This resulted in a severe bout of labour unrest in late 1995, which led to a tripartite agreement between SNCB, government and unions to put the corporation on a fresh financial footing. The agreement, reached in January 1996, was to see €9.2 billion invested in the railway over the period 1996–2005. About a third of this would go to high-speed line construction, the rest to upgrading the historic network.

The state undertook to provide €396–570 million annually towards the investment plan. In addition, the state reimburses the railway for the cost of meeting its social obligations.

The Belgian government has created a special fund for financing new infrastructure, Financière TGV, consisting of both public and private contributions; interest will be paid on investments with the fund. The fund will invest €3.1 billion in the high-speed line project.

SNCB's Board of Directors approved an action plan, 'Target 2005', based on the tripartite agreement, in February 1996. On 1 January 1998, the enterprise was restructured as 10 'activity centres', five 'service centres' and seven 'central co-ordination units'. This will include the separation of infrastructure management according to EU rules.

In May 2001, a new 12-year investment plan was agreed. Over this period €16 billion will be invested to cope with forecast traffic growth of 50 per cent. The main priorities will be:
– purchase of new stock (€3.4 billion);
– completion of the TGV network (€2.38 billion);
– modernisation of main lines;
– creation of a Brussels RER system;
– development of rail links to Brussels airport;
– development of rail links to ports.

The new plan includes a call for investment from the private sector, the most likely source being the port authorities. Investment in 2002 was €1,381 million, up 11 per cent on 2001. Of this, €565.7 million went on the classic network and €494.2 million on the future TGV network to give a total of 77 per cent on infrastructure. €233.4 million was invested in traction and rolling stock.

At the same time, the board of SNCB was reduced from 18 to 10 members. The trade unions lose representation but are part of a new 'strategic council', together with regional transport operators.

The appointment of Karel Vinck as chief executive and the review of SNCB's situation which followed led to the action plan 'MOVE', which was presented in June 2003 and is aimed at restoring the company's viability. In passenger traffic the accent will be on improving service quality, while by 2007 an increase of 12 per cent in traffic is expected on a network essentially unchanged from the present. International services are likely to be cut back, leaving mainly Thalys and Eurostar high-speed operations. Traffic on these is expected to rise by 60 per cent to 12.8 million passengers annually by 2007. In the freight sector a reduction of 40 per cent in costs is necessary and SNCB is likely to abandon unprofitable services. In all sectors productivity will be raised by 25 to 30 per cent, thus allowing a reduction in staff numbers of 10,000. In order to achieve the savings and complete new infrastructure, SNCB asked the Belgian government to increase investment in the period up to 2007 as well as writing off the company's historic debt.

In March 2004 the Belgian Council of Ministers approved the reorganisation of Belgian Railways. From 1 January 2005, SNCB will be divided into three autonomous state-owned companies– infrastructure manager Infrabel and a train operator which will retain the name SNCB/NMBS, the two subsidiaries coming under a holding company, SNCB/NMBS Holding. The holding will only have a 50 per cent stake in Infrabel, the state retaining 50 per cent. Staff will be employed by SNCB Holding but will be seconded to the two subsidiaries. Work on separating activities will take place during 2005. As part of this process, the state will take over SNCB's historic debt. This rose by 23.4 per cent in 2003 to over €6 billion.

In 2003 SNCB's total staff was 40,096, down 2.6 per cent on 2002. Under the reorganisation plan this will fall to 37,000 by the end of 2005, thanks to voluntary early retirement and non-replacement of staff. About 21,000 will be seconded to the operating company and 15,000 to Infrabel.

Finance (million €)

Revenue	2000	2001	2002
Passenger	887.3	917.7	967.0
Freight	344.5	323.0	332.8
ABX	133.1	114.5	115.2
Infrastructure Management	628.6	642.9	663.2
Other	192.5	173.2	171.3
Total operating Revenue	2,186.0	2,171.3	2,249.5
Other revenue	1,450.7	1,456.5	1,544.9
Total revenue	3,636.7	3,634.8	3,794.4

Expenditure	2000	2001	2002
Materials	211.9	223.8	230.7
Services	672.2	641.8	722.2
Personnel	2,306.8	2,392.0	2,536.0
Depreciation	454.6	474.1	498.4
Provision for risk	−147.7	−160.8	−150.8
Other	11.2	25.8	89.2
Total expenditure	3,509.0	3,596.7	3,925.7

SNCB reported a loss of €131 million in 2002 compared with a surplus of €38 million in 2001. Turnover was up 4.3 per cent to €3,794.4 million but operating costs rose 9.1 per cent to €3,925.7 million. Of the 169.3 million passenger journeys recorded in 2003, 154.9 million were on domestic services, the remainder on international trains.

Traffic (million)

	2001	2002	2003
Passenger journeys	160.3	164.9	169.3
Passenger-km	8,117	8,260	8,276
Freight tonnes	57.1	57.2	55.7
Freight tonne-km	7,080	7,297	7,293

Passenger operations
Passenger revenue overall rose by 6.9 per cent in 2002, to reach €967 million. Domestic traffic was up 2.9 per cent and revenue by 4.9 per cent, while international traffic grew by 2.2 per cent and revenue by 7.2 per cent.

Eurostar
Eurostar trains to London via the Channel Tunnel introduced in November 1994 (for details see United Kingdom entry) and TGV services to Paris in January 1995 (for details see France entry), have dramatically improved international rail services to and from Brussels.

Eurostar has become market leader on this route. In March 1999, a centralised management structure—Eurostar Group—was implemented to determine commercial policy and the development of all Eurostar activities.

Eurostar traffic has been slower to develop than on the Paris–London route and the basic service remains every two hours. The opening of the Belgian high-speed line on 14 December 1997 cut journey times to London by 40 minutes to 2 hours 40 minutes. This was reduced further to 2 hours 15 minutes non-stop in September 2003 with the opening in the UK of the first section of the Channel Tunnel Rail Link. Eurostar traffic fell by 5.9 per cent to 1.5 million passengers in 2002, revenue falling by 9 per cent.

Thalys
In June 1996 the Paris TGVs, marketed as Thalys, began running the first section of the Belgian high-speed line. Frequencies were improved to hourly, with the journey taking 2 hours between the French and Belgian capitals. The opening of the complete line cut the journey time to 85 minutes and was accompanied by an increase in frequency. Success has been such that the service frequency was increased to half-hourly in September 2002.

Five to eight weekday Thalys services from Paris run beyond Brussels to Amsterdam. Seven of the Paris trains run beyond Brussels to Liège and six to Aachen and Cologne. One pair of return Thalys services per day run from Ostend to Paris via

Thalys high-speed trainsets on Belgium's high-speed line at Wasmes (QuintusVosman) 0580494

Bruges and Ghent and one pair between Namur and Paris via Charleroi and Mons.

A Thalys service from Amsterdam to Disneyland, Paris, operates during summer months as do five pairs of trains between Brussels and Disneyland, calling at Charles de Gaulle/Roissy airport. In 2001 Air France withdrew its Brussels to Roissy service, passengers now using Thalys. Thalys also operates to the French Alps in winter and Amsterdam–Marseille in summer. In 2002 Thalys traffic grew by 3.5 per cent and revenue by 8.1 per cent, with 6 million passengers carried.

ICE
From December 2002 Germany Rail (DB) introduced a three times-daily ICE service from Brussels to Cologne and Frankfurt, taking 3 hours 45 minutes. This was reduced by 12 minutes when ICE3 sets started to use the Leuven-Liège high-speed line on 1 July 2004.

TGV
A Brussels–Nice through TGV introduced in 1995 proved successful: this prompted the introduction of additional through TGVs to Lyon, Marseille, Nice, Perpignan and Grenoble. Almost all of these trains run via Lille, where they attach a second unit. In winter, direct TGV services to the Alps have been introduced. These services continue to attract passengers: in 2002 revenue was up 13.1 per cent to €6.9 million from 6.1 million in 2001.

Other international services
A review of international services in 2002 led to the withdrawal of trains from Brussels to Milan, Rome, Lourdes and Biarritz. SNCB has also withdrawn from joint operation of Brussels–Vienna, Paris–Amsterdam–Berlin/Hamburg, Paris–Amsterdam, Brussels–Chur and Hamburg–Bordeaux routes. SNCB operates a range of motorail trains which registered 99 million passenger–km in 2002, up 25.6 per cent on 2001.

Domestic services
Domestic passenger traffic rose by 2.9 per cent in 2002, while revenue increased by 4.9 per cent. Traffic is dominated by travel to and from the Brussels area. A regular-interval domestic timetable, carefully designed to facilitate interchange, provides connections between internal InterCity (IC), InterRegional (IR) and Local (L) services. The services are among the most intensive in Europe. International expresses fit into this pattern at less-frequent intervals and the basic service is supplemented by large numbers of direct trains, mainly to/from Brussels during Monday-Friday peak periods. At weekends, additional direct trains run to/from tourist areas, particularly coastal resorts. In July 2003 trains to the coast carried a record 760,000 passengers.

The IC/IR system was completely revised in May 1998. The main features were more through services to Brussels, Brussels Airport and Lille, and improved co-ordination with the TGV system. Minor reductions in poorly used services were made in 2002, while Mol–Hasselt and Chareleroi–Couvin services were improved to IR standard thanks to the introduction of new Class 41 dmus and increased train frequencies.

In March 2004 SNCB launched online sale of tickets on the Internet. The service was the first in Europe to cover the majority of existing tickets, without including reservations in most cases.

Brussels RER
The 1980s STAR 21 plan for Belgian Railways foresaw the creation of a Brussels RER (Regional Express) network of nine routes. Most will converge on the cross-city Brussels Junction railway, while the others will provide important orbital connections. Services would be more frequent than at present and fed by express buses. In 1991 141,000 passengers daily travelled by train from within a 30 km radius of Brussels to the city centre between seven and nine o'clock in the morning, a market share of only 30 per cent. The object of the RER is to bring this share close to 50 per cent.

A report in 1998 set the completion of an RER network at 2008. In 2004 the cost of the project was put at €1.720 billion plus €790 million (2000 prices) for rolling stock.

Work on the RER started in April 2004 with quadrupling from Watermael to Schuman and on a new 1,250 m tunnel from Schuman to Josaphat, the most important element of the plan. Other parts of the plan include quadrupling from Brussels to Halle (complete), Leuven (in progress), Ottignies and Nivelles.

It was expected that 50 double-deck emus would be ordered in late 2004 for delivery from 2008, with an option for a further 107 units. In the period leading up to opening of the RER, SNCB is reorganising Brussels suburban services, with increased frequencies, under the branding City Rail.

In May 2000 the government asked SNCB to add a third track to the Brussels–Denderleeuw line.

Brussels airport
In May 1998 a larger station opened at Brussels airport on a new alignment of the branch line from Zaventem. This was part of the work on expanding the airport passenger terminal. The new alignment will eventually allow a spur off the Brussels–Antwerp line and a further spur to the Brussels–Leuven–Liège line. Total cost of this work, known as the Diablo project, is €74 million. The second of these was under construction in 2001.

Also planned is a tunnel connection between the Brussels–Ottignies line at Schuman and Schaerbeek–Josaphat. This will create a second north-south RER route, possibly for direct services between the European Union headquarters complex at Schumann and the airport station.

Freight operations
In 2003 freight volumes were stable and revenue increased by 8.8 per cent. The activity remains loss-making but should break even in 2007.

SNCB's freight services are now marketed under the 'B-Cargo' logo, using the stylised 'B' that is the railway's trademark. B-Cargo's strategy is based on three activities—regional transporter serving ports from Dunkirk to Vlissingen, corridor manager and specialist in certain key sectors. B-Cargo aims to create group synergies with SNCB's Inter Ferry Boats (IFB) intermodal, ABX parcels, Rheinkraft steel logistics (formerly part of ABX) and Haeger & Schmidt inland navigation subsidiaries.

Restructuring of IFB converted a €110 million loss in 2002 into a small profit in 2003. As part of the reorganisation of the SNCB Holding, resources will be specifically dedicated to B-Cargo during 2004–05.

Rail freight traffic is dominated by Antwerp, which generates around half of all traffic, either in the port or its hinterland. The ports of Gent and Zeebrugge also generate large volumes. The other main traffic generator is heavy industry, particularly steel production, concentrated around Liège and Charleroi. This traffic is declining and B-Cargo is therefore seeking to diversify its activities. In 2000 B-Cargo lost all coal traffic to power stations after the generating company Electrabel switched to barge transport. Transit traffic between the Netherlands and southern Europe via France is important.

SNCB now reviews private sidings served on a biannual basis. The number of marshalling yards has been reduced to four: Gent Zeehaven, Antwerp Nord, Kinkempois (Liège) and Monceau (Charleroi). Minor yards still open include Schaerbeek (Brussels), Stockem and Montzen.

International services, particularly 'Eurailcargo' wagonload services, are increasingly important to SNCB, Antwerp–Munich traffic growing by 29 per cent in 1997 and Antwerp to Malaczewicze (Poland) by 60 per cent.

Under the 'New Cargo' part of the 'MOVE' plan, a Cargo Operating Centre for all freight services will be created in 2004 together with an automatic wagon tracing system. Costs will be pushed down thanks to higher productivity and cuts in loss-making services.

In December 2004 B-Cargo was to recast the freight timetable, introducing regular interval timings and shuttle-style services on the main Antwerp–Basle, Rotterdam–Antwerp–Paris (the latter in concert with Fret SNCF) and Antwerp–Duisburg routes with the aim of increasing average speeds and optimising use of traction.

In March 2004 B-Cargo signed an agreement with the trade unions under which ancillary activities will be transferred in-house. Part of the agreement will involve staff being increasingly multi-skilled.

SNCB's ABX parcels and logistics subsidiary expanded rapidly from 1998 to 2001 through acquisitions in 35 countries, employing 16,000 staff and with an annual turnover of almost €3 billion. However, this has resulted in a massive debt which is now weighing down the parent's accounts. In October 2003 SNCB started to restructure ABX, transferring profitable activities to a new holding, ABX Logistic Worldwide.

In early 2004 SNCB applied for a safety certificate for freight operations on the French network. Until now, SNCB has always worked in co-operation with SNCF on cross-border services.

B-Cargo employs approximately 8,000 staff.

Intermodal operations

SNCB sees intermodal freight as a growth area, both for maritime containers and for inland swapbodies. This traffic in 2002 grew by 6 per cent to form 31 per cent of the railway's traffic in 2002 compared with 12 per cent in 1990.

A domestic shuttle 'Railbarge' service for containers operates between Antwerp and Zeebrugge, the 'Port Express' service links Antwerp with Rotterdam on a daily basis and a Zeebrugge–Dunkirk service is also provided.

A new intermodal terminal, known as 'Main Hub,' near Antwerp Noord marshalling yard, opened in late 2000. Capacity is 600,000–700,000 TEU annually and this can eventually be doubled.

In 1998, SNCB created a subsidiary, InterFerryBoats (IFB), through the merger of Interferry and Ferryboats. IFB specialises in operating terminals for intermodal traffic and logistics. It operates five terminals in Belgium and a facility in Dunkerque, France. It also organises the operation of intermodal shuttles via its hub at Muizen. IFB has less serious problems than ABX but is being restructured and integrated within B-Cargo.

New lines

Paris-Brussels high-speed line

The complete Belgian section of the Brussels–Paris high-speed line, from Fretin near Lille (France)

Class 77 diesel locomotive at Lier with an Antwerp trip freight (John C Baker)   0580490

Class 13 electric locomotive at Rémilly, France, with a Belgium-bound intermodal service from Basle (John C Baker)   0580491

to Lembeek, opened on 14 December 1997. The new line cut Brussels–Paris and Brussels–London journey times by around 40 minutes. Upgrading work continues on the classic line between Lembeek and Brussels Midi station. A six-track cut-and-cover tunnel through Halle was completed in 1999 and 220 km/h operation will soon be possible most of the way from Lembeek to Brussels. A flyover at the mouth of Midi station is to be completed by 2004.

Brussels-Antwerp upgrading

At an estimated cost of €116 million, two tracks of the existing quadruple line from Brussels to Antwerp are being upgraded for 160 km/h by 2005. The work includes: complete track renewal, including increase of inter-track space to 2.25 m to simplify mechanised maintenance; resignalling for reversible working and accompanying installation of crossovers; and renewal of the electric traction current supply system.

From Berchem, on the southern outskirts of Antwerp, construction of a through line in tunnel under the city began in 1998. The new line will serve an underground station beneath the Antwerp Central terminus before joining the Roosendaal line near Antwerp Dam. Thus, services to and from the Netherlands will avoid their present reversal in Antwerp Central, saving more than 10 minutes. Completion is scheduled for 2006.

Construction of the new HSL-Zuid high-speed line from Antwerp to Schiphol, near Amsterdam, started in 2000; completion date is estimated at the end of 2006. One station in Belgium will be built on the new line–at Noorderkempen near the Netherlands border. When the line is open, best Brussels–Amsterdam timings will be reduced from 2 hours 41 minutes to 1 hour 43 minutes. Cost of the Belgian section of line is €622 million, of which €367 million will be paid by the Netherlands government.

Brussels–Liège–Aachen–Cologne

The new 300 km/h line from Leuven to Liège opened in December 2002. Thalys and Ostend-Eupen

InterCity services (the latter limited to 200 km/h) were diverted on to the new line. Initially reductions in journey times were minor due to continuing work on the rest of this route. Liège Guillemins station is being completely rebuilt (see below). The line from Brussels to Leuven is being widened and upgraded for 220 km/h running by 2005. From Brussels Midi to Liège, 105 km, the transit time will be 39 minutes when the project is completed. From Liège to Aachen, 41.5 km, new line and upgrading will bring the travel time down from 42 to 28 minutes.

SNCB has created three subsidiary companies to manage construction of the Belgian elements of the high-speed lines: TUC Rail SA for the infrastructure; Eurostation SA for the works at Brussels Midi and Antwerp Central stations; and Euro-Liège TGV, for the new station in Liège.

Liège Guillemins station, with evidence of reconstruction to provide a modern facility for high-speed services (Marcel Vleugels)   0580493

Improvements to existing lines

In 2002 140 km were approved for 300 km/h running, 116 km for 160 km/h, 621 km for 140 km/h, 239 km for 130 km/h, 1,259 km for 120 km/h and 1,105 km for less than 120 km/h.

The lines from Brussels to Antwerp, Halle and Leuven are being upgraded as part of the high-speed lines programme (see section above).

Other parts of the network are being upgraded for 160 km/h. Routes being tackled include Charleroi to Antoing; Brussels to Namur; and Brussels to Ostend. Parts of the last named route will be quadrupled and improved for speeds of up to 200 km/h. In most cases, the spacing between tracks has to be widened from 140 to 160 cm during track renewal work.

Improving the Brussels-Luxembourg line, on which trains are limited to 130 km/h, has been made an EU priority project. Present plans for the upgrade include quadrupling the Brussels-Ottignies line (also part of the Brussels RER plan) and modernising the remainder of the route with a small number of new high-speed cut-offs. Total cost is €860 million. This project will replace plans for a putative high-speed line to Rhisnes, near Namur, and Gosselies, near Charleroi.

The 1996–2005 plan for freight concentrates investment on raising limits to 22.5 tonnes per axle and speeds to 120 km/h, plus standardisation on a large loading gauge for intermodal traffic. In addition, there will be investment in extra tracks on the approaches to Antwerp and Zeebrugge. Track has been upgraded on the 'Athus–Meuse' line at the same time as electrification (see below).

Agreement was reached in 2000 on the reopening of the 'Iron Rhine' route between Antwerp and the Ruhr region of Germany. However, in 2003 work had stopped due to disagreement between Belgium and the Netherlands over the cost of modernising the route.

It has also been agreed that a new freight line east of Antwerp should be built to link Antwerp Noord yard and Lier to bypass the congested city area. Another new line from the port to the Roosendaal–Vlissingen line in the Netherlands is at the planning stage. A line from the port to the Gent line awaits finance. Doubling of track and electrification is in progress on the line to the west bank port in Antwerp.

The SNCB infrastructure department is to reorganise its maintenance activities around 22 'infrastructure logistics centres' (CLIs) by 2006.

Major new stations

A mammoth reconstruction of Brussels Midi (South) station is under way. To be executed in four phases, the project had to be carried out in stages because of the need to keep the busy 22-platform station fully operational throughout the rebuilding. When complete the station will form part of a major urban redevelopment project embracing 120,000 m² of offices and conference facilities, and 20,000 m² of retail space. A new underground parking area for 2,500 cars was completed in 2001.

The first three phases of the scheme are concerned with completion of a terminal and platforms for high-speed train services on the west side of the station. As all passengers for Eurostar services to London must first pass through security checks, an airport-style enclosed departure area for terminal platforms 1 and 2 has been created, with its own bar and toilet facilities.

Through platforms are used mainly by Thalys (Paris–Brussels–Cologne/Amsterdam) TGV trains. The main concourse, a wide gallery with the usual offices and facilities along each flank, including a travel centre with 32 open counters, has been rebuilt to more spacious proportions. Overall roofing has been provided for all platforms.

Liège Guillemins station is being completely rebuilt as part of SNCB's high-speed plans. There will be a major remodelling of the track layout and a new building designed by the Spanish architect, Santiago Calatrava.

Antwerp Centraal station is being rebuilt. The former terminal with 10 platforms is being expanded to provide 14 platforms on three levels. Capacity will be doubled by 2007 by the provision

Diesel locomotives

Class	Wheel arrangement	Power hp	Speed km/h	Weight tonnes	No in service (January 2004)	First built	Mechanical	Builders Engine	Transmission
52/53/54	Co-Co	1,720	120	108	18	1955	AFB	GM	E GM/Smit
55	Co-Co	1,950	120	110	34	1961	BN	GM	E ACEC/SEM (licence GM)
62	Bo-Bo	1,425	120	80	87	1961	BN	GM	E GM
91	B	335	35 40	36	26	1961	Cockerill/BN ABC	GM	H Twin-Disc Q Cockerill
84	C	550	30	55.8	3	1958	ABR/B&M	ABC 6 DUS	H Voith L37U
82	C	650	60	57	73	1965–66	BN/ABR	ABC 6 DXS	H Voith L217U
73	C	750	60	56	74	1965–68	BN	Cockerill-Ougrée 6TH 695 SA	H Voith L217U
74	C	750	60	59	10	1977	BN	ABC 6 DXS	H Voith L217U
77	B-B	1,540	60/100	90	123*	1999	Vossloh	ABC 60 ZC-1000	H Voith L4

B&M = Baume & Marpent   * of 180 on order.

Electric locomotives

Class	Wheel arrangement	Output kW continuous/ one-hour	Speed km/h	Weight tonnes	No in service (January 2004)	First built	Builders Mechanical	Electrical
22	Bo-Bo	1,740/1,880	130	87	15	1954	BN	ACEC/SEM
23	Bo-Bo	1,740/1,880	130	93.3	81	1955	AM	ACEC/SEM
25	Bo-Bo	1,740/1,880	130	83.9	7	1960	BN	ACEC/SEM
25.5[3]	Bo-Bo	1,740/1,880	130	85	8	modified 1973	BN	ACEC/SEM
26	B-B	2,240/2,355[1] 2,470/2,590	130	82.4	34	1964	BN	ACEC
16[2]	Bo-Bo	2,620/2,780	160	82.6	7	1966	BN	ACEC
20	Co-Co	5,130/5,150	160	110	24	1975	BN	ACEC
27	Bo-Bo	4,150/4,250	160	85	60	1981	BN	ACEC
21	Bo-Bo	3,130/3,310	160	84	60	1984	BN	ACEC
11[3]	Bo-Bo	3,130/3,310	160	84	12	1985	BN	ACEC
12[4]	Bo-Bo	3,130/3,310	160	84	12	1986	BN	ACEC
13	Bo-Bo	5,000	200	90	60	1998	ALSTOM	ALSTOM
15[5]	Bo-Bo	2,780	160	78	3	1962	BN	ACEC

[1] First five locomotives only.
[2] Dual-voltage 1.5 kV/3 kV DC.
[3] Tri-voltage 1.5 kV/3 kV DC/25 kV AC.
[4] Quadri-voltage 1.5 kV/3 kV DC/15 kV 16⅔ Hz/25 kV 50 Hz.
[5] Dual-voltage 3 kV DC/25 kV AC 50 Hz.

Diesel multiple-units

Class	Cars per unit	Motored axles per car	Power/ car kW	Speed km/h	Units in service (January 2004)	First built	Builders Mechanical	Engine
41	2	2	485	120	96	2000	ALSTOM	Cummins

Electric multiple-units

Class	Cars per unit	Motor cars per unit	Motored axles per car	Power/ car KW	Speed km/h	Units in service (January 2004)	First built	Builders Mechanical	Electrical
AM62/AM63/AM65	2	2	2	310	130	114	1962	BN/Ragheno	ACEC
AM66	2	2	2	340	140	187	1970	BN/SNCB	ACEC
AM75	4	2	4	1,360	140	44	1975	BN	ACEC
AM80	3	1	4	1,240	160	138	1981	BN	ACEC
AM86	2	1	4	680	120	51	1988	BN	ACEC
AM96	3	1	4	1,400	160	120*	1996	Bombardier	ALSTOM

* Includes 50 dual-voltage 3 kV DC/25 kV.

Class 11 electric locomotive at Kalmthout with an Amsterdam-Brussels express (John C Baker) 0580492

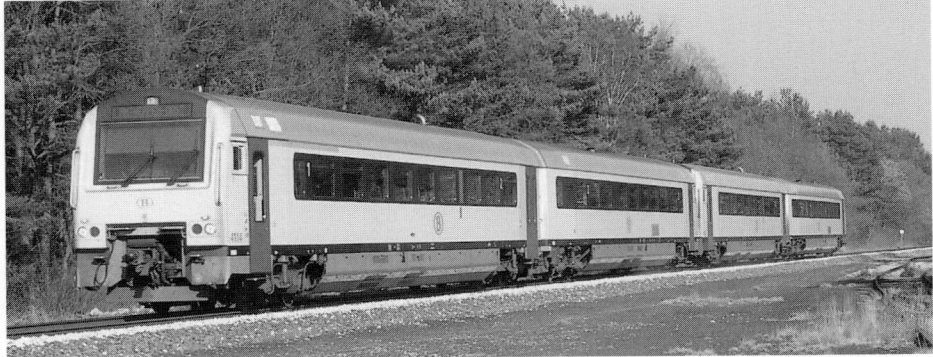

Two ALSTOM-built Class 41 dmus forming a Neerpelt-Antwerp service at Balen (Quintus Vosman)
0554367

Class AM96 dual-voltage emu, together with an AM65 unit, receiving attention at SNCB's main workshops at Mechelen (Marcel Vleugels)
0554369

of through tracks below street level. In December 2003 the upper level was reopened to traffic.

Traction and rolling stock

In April 2004 SNCB traction and rolling stock comprised 397 electric locomotives, 477 diesel locomotives, 1,684 emu vehicles, four Eurostar trainsets, seven Thalys trainsets, 192 dmu vehicles, 1,395 passenger coaches and 15,629 freight wagons.

SNCB is investing heavily in both new rolling stock and refurbishment at present. In December 1995 SNCB and CFL (Luxembourg Railways) ordered a total of 80 3 kV DC/25 kV AC locomotives from Alstom. Designated Class 13 by SNCB, the locomotives cost €3.48 million each and deliveries were completed in late 2001. The locos are managed in a pool with the 19 CFL machines. Main duties are freight on the Antwerp–Luxembourg corridor plus Ostend–Eupen (at speeds up to 200 km/h) and Liège–Luxembourg passenger trains. In 2002 these locomotives started to operate through to St Louis, in France.

In the short term, SNCB foresees the need for around 140 new electric locomotives to replace older types and is consulting neighbouring countries over the need for multi-voltage designs. A specification for Class 14, a four-voltage locomotive design, has been drawn up. A tender was launched in early 2004 for 40 locomotives plus 20 options. These will be dedicated to freight traffic and will replace Class 13 on freight work, allowing the latter to power Type M6 double-deck stock. They will also replace Class 25.5 on services to Rotterdam.

In June 1997 SNCB ordered 90 Class 77 B-B 1150 kW diesel-hydraulic locomotives from Siemens (now Vossloh) for €1.5 million each. These will have ABC power units manufactured in Belgium. The locomotives will be used mainly to replace a heterogeneous fleet used for shunting and short-distance freight duties. Delivery started in late 1999. In 2001 SNCB ordered 90 additional locomotives and was considering an order for 90 more of the class to replace its entire main line diesel fleet.

At the same time SNCB ordered 80 Class 41 two-car dmus seating 150 passengers from Alstom Transporte of Spain at a cost of €1.76 million each.

In 2001, SNCB ordered 16 additional units of this type. The units were delivered from late 2000 to 2003, replacing all remaining diesel locomotive-hauled passenger trains. In 2002 SNCB modified six Class 55 Co-Co diesel-electric locomotives for operation on the 'Iron Rhine' route from Antwerp to Mönchengladbach, Germany. This involved adding Dutch and German signalling equipment.

In June 2004 SNCB and Netherlands Railways placed a joint order worth €260 million with Ansaldobreda for 12 high-speed trainsets, with an option for 14 more, to operate the Brussels–Amsterdam service at 250 km/h from 2007 when the HSL-Zuid opens. The trains will be able to operate off three different voltages and will have signalling equipment for the two countries.

In 1999 an order was placed with Bombardier Transportation for 210 Type M6 double-deck coaches for commuter services. They are air-conditioned and equipped to a much higher standard than previous stock. Delivery was due to be completed in 2004. In 2003 SNCB ordered a further 112 cars, 52 of which will be driving trailers. The coaches will be used in push-pull mode with Class 13 locomotives.

SNCB is carrying out refurbishment work on its fleet of 578 Type M4 hauled coaches and 181 two-car AM66-79 emus.

In the freight sector, B-Cargo has recently introduced 200 type Shimns bogie covered wagons for carrying steel coil, built by Costamasnaga

in Italy, and has unveiled the prototype of the future type Rils covered wagon for the carriage of pallets, steel coils or rolls of paper. This has a new fast load blocking system known as 'Easy Clamp'. In 2000 SNCB started to receive 700 Lgnss 500 Sgnss wagons for container transport from Astra, Romania.

High-speed rolling stock

SNCB owns four of the 38 Eurostar trainsets used jointly with SNCF and Eurostar UK Ltd for the London–Paris/Brussels high-speed operation (for details see the United Kingdom entry).

Seventeen four-voltage Thalys sets are jointly operated for the Paris–Brussels–Cologne service by SNCB, NS, SNCF, and DB. A maintenance depot for Eurostar and Thalys sets has been built at Brussels-Forest. Two NS and two DB sets as well as the seven Belgian sets are maintained there by SNCB.

Signalling and telecommunications

SNCB is progressively installing a track-to-train radio system, employing equipment supplied by Alcatel-Bell. Because of the country's dual language and the need for drivers to use the system with facility in both French- and Flemish-speaking territory, the system uses illuminated cab displays of pictogram codes rather than telephonic communication between control centre and train crew.

SNCB is in the course of replacing its traditional Automatic Train Control (ATC) system, based on contact between brushes mounted beneath traction units and track-mounted 'crocodiles', with an inductive transponder system known as Train-Balise-Locomotive (TBL). The TBL system–being installed first on the new high-speed lines–brakes the train automatically if the authorised speed is exceeded. The existing Brussels–Lille high-speed line is equipped with TVM 430 cab-signalling but Leuven–Liège is equipped with TBL2, the upgraded version of TBL, as it is used by both TGVs at 300 km/h and locomotive-hauled trains at 200 km/h.

SNCB promised to accelerate both programmes after a head-on crash near Leuven in March 2001.

Signalling and track circuits are gradually being modified on main routes to allow trains with three-phase drive to operate over these lines.

Modernisation of switching technology for Antwerp marshalling yard has been carried out by Siemens and the Duisburg-Wanheim unit of Thyssen AG. The work included delivery of the MSR32 guidance system (a radio-controlled system for shunting locomotives) as well as systems installation.

Electrification

83 per cent of the SNCB network is electrified, and 93 per cent of passenger trains and 72 per cent of freight trains are electrically hauled.

While the Belgian network has thus far used the 3 kV DC system, new high-speed lines are being electrified on the 25 kV AC system. The 142 km Dinant–Athus 'Athus-Meuse' line–an important freight artery between Antwerp and Luxembourg–has also been electrified at 25 kV. This project was finished in December 2002.

SNCB is to electrify the remaining 7 km from Montzen to the German border by 2005, completing the Antwerp/Liège–Aachen corridor.

Track

77 per cent of the SNCB network has been laid with continuously welded rail (cwr). Some 182 km of cwr were installed in 1997.

Two Class 13 electric locomotives at Livange with a heavy B-Cargo cross-border freight consisting of new cars and tank wagons from France to Belgium (John C Baker)
0554366

Standard rail: Flat bottom, 50 and 60 kg/m main track, 50 kg/m secondary track
Length: Main track: 243 m rails long-welded. Secondary track: jointed 28 m rails
Joints: 4-hole fishplates
Rail fastenings: Sole plates and screws, mostly K-fastenings on wood sleepers. New track Pandrol fastenings. Pads are inserted under the rail when concrete sleepers are used.
Crossties (sleepers): Existing track: generally oak, 2,600 × 280 × 140 mm. Sections of welded-rail track have been laid with three types of concrete sleeper: Type RS (two blocks joined by a steel bar) with Type RN flexible rail fastenings; Type VDH (two blocks joined by a steel bar) with Pandrol fastenings; and Type DMD (monobloc prestressed) with Pandrol fastenings.
Spacing: 1,667/km on main line track; 1,370–1,590/km secondary routes
Filling: Broken stone or slag
Min curvature radius
Main line: 2.18° = 800 m
Secondary line: 3.5° = 500 m

Running lines: 8.75° = 200 m
Sidings: 11.7° = 150 m
Max gradient: 2.5%
Max altitude: 536 m at Hockai on Pepinster-Trois Ponts line
Max axleload: Certain locomotives have axleloads of 24 tonnes. Except for certain bridges they can operate anywhere on the system, subject to speed restriction.

UPDATED

Benin

Ministry of Public Works and Transport

PO Box 16, Cotonou
Tel: (+229) 31 33 80

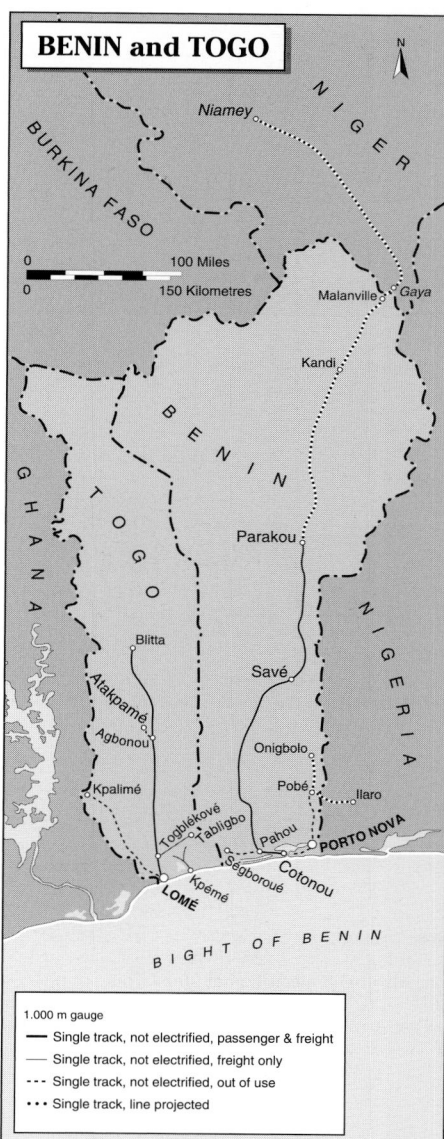

BENIN and TOGO

1.000 m gauge
— Single track, not electrified, passenger & freight
— Single track, not electrified, freight only
--- Single track, not electrified, out of use
••• Single track, line projected

0058497

Key personnel
Minister: Joseph Sourou Attin
Director General: A Glele

NEW ENTRY

Organisation Commune Benin-Niger des Chemins de Fer et des Transports (OCBN)

PO Box 16, Cotonou
Tel: (+229) 31 33 80   Fax: (+229) 31 41 50
Tx: 5210

Key personnel
Director General: Raphael Bonou
Director of Motive Power and Rolling Stock: Edmond Agbla
Director of Way and Works: Alzouma Younsa
Director of Operations: Gabriel Alaye
Director of Supplies: J Hinson
Director of Finance: M Ousseini
Director of Personnel: A Sonon

Gauge: 1,000 mm
Route length: 578 km

Organisation
OCBN operates, on behalf of Niger and Benin, a single-track metre-gauge railway consisting of the Northern line from Cotonou to Parakou via Pahou (438 km), the Eastern line from Cotonou to Pobé (107 km), and the Western line linking Pahou and Ségboroué (33 km). From Parakou freight traffic is transported by road to the Niger capital of Niamey.

Passenger operations
In 1994 the railway carried 600,000 passengers (107 million passenger-km).

Freight operations
In 1994 OCBN carried 250 million tonne-km of freight.

New lines
A cherished project is the extension of the Northern line from Parakou to Niamey, Niger's capital, a distance of 650 km. At present, Niger traffic has to be road-hauled to Parakou. An agreement was signed in 1976 between Niger and Benin for construction of a rail link, since three-quarters of Niger's exports are channelled through Cotonou, and work was started in 1978 but made scant progress. Neither the World Bank nor any other aid agency has been prepared to help finance the scheme.

Improvements to existing lines
A successful application was made in 1991 to France's Fund for International Co-operation (CCCE) for a loan worth US$8.6 million. It was applied to rehabilitation of the Benin segment of the Cotonou-Parakou Northern line, a project which has since been completed.

Further rehabilitation work was undertaken using fresh supplies of new rail delivered from British Steel in 1994.

OCBN has so far been unsuccessful in gaining external finance for modernisation of the Cotonou-Pobé Eastern line and its extension to a cement factory at Onigbulo with a projected output of 500,000 tonnes a year. Half the output would be for Nigeria, and Belgian finance has been offered for construction of a railway from the cement plant to Ilaro, northwest of Lagos, in Nigeria.

Traction and rolling stock
At last report, locomotives in operation totalled eight Alsthom BB500 and 12 Alsthom BB600 diesel-electrics, plus six shunting tractors. Soulé railcars totalled seven and other stock consisted of 31 Soulé passenger coaches and trailers and 296 freight wagons.

In 1999, it was reported that a credit from the West African Development Bank was being sought to purchase several new locomotives and freight wagons.

Bolivia

Ministry of Economic Development

Palacio de Comunicaciones, piso 18, Avenida Mariscal Santa Cruz, La Paz
Tel: (+591 2) 37 73 20   Fax: (+591 2) 37 13 47

Key personnel
Secretary of Transport: A Revollo

Bolivian National Railways

Empresa Nacional de Ferrocarriles (ENFE)
Empresa Ferroviaria Andina (FCA) (former ENFE Andean network)
PO Box 428, Estación Central, Plaza Zalles, La Paz
Tel: (+591 2) 35 12 03   Fax: (+591 2) 39 21 06

Key personnel
President: Abraham Monasterios Castro
General Manager: José Taborga

Empresa Ferroviaria Oriental (FCO) (former ENFE Eastern network)
PO Box 108, Santa Cruz
Tel: (+591 33) 34 84 67   Fax: (+591 33) 32 75 07

Key personnel
General Manager: Ing Fernando Marticorena Zillerulo

Gauge: 1,000 mm
Route length: (FCA) 2,275 km; (FCO) 1,244 km

Political background

All rail services in Bolivia are now run by the private sector. In 1995 the Ministry of Capitalisation invited bids for a 50 per cent shareholding in a new public/private corporation holding ENFE's assets. Of the seven bidders which came forward, Cruz Blanca–now CB Transportes–was successful, paying US$39 million for a 40-year operating concession. The remaining 50 per cent of ENFE has been retained by the government to form the basis of private pension funds for the country's citizens. The money paid by Cruz Blanca is being invested in the rail network.

Under the terms of the concession, the company was required to invest US$25 million in the network over seven years. However, Santa Cruz has decided to spend US$32 million in just five years, much of which will finance infrastructure upgrades to increase line speeds.

In 1996 Antofagasta and Bolivia Holdings, parent of Chile's Antofagasta, Chili and Bolivia Railway (FCAB), purchased a 73 per cent stake in Cruz Blanca's share of ENFE's Andean network and 23 per cent of its share of the Eastern network. FCAB connects with the Andean network at Ollagüe.

Organisation

Bolivia is a landlocked country and lack of communications has made virtually impossible the sort of economic development which the country needs. The country's two railways, the Andean (2,275 km) and Eastern networks (1,244 km) connect only via Argentina. They are of major importance as a means of access to ports on the Pacific and Atlantic oceans via neighbouring countries. These international railway connections, some of which have fallen into disrepair, are as follows: with Chile to the Pacific ports of Arica and Antofagasta; with Argentina to the Atlantic ports of Rosario and Buenos Aires; with Brazil to the Atlantic port of Santos; with Peru (by ship across Lake Titicaca to Puno) to the Pacific port of Matarani.

Traffic
(EFO only)

	1996	1997
Passenger journeys (000)	303.9	324.2
Passenger-km (million)	132.9	136.7
Freight tonnes (000)	869.8	894.7
Freight tonne-km (million)	513.2	524.2

Passenger operations

Passenger operations were progressively run down during the final years of state ownership of the ENFE network and now represent around 6 to 7 per cent of total revenue. Although the Eastern network concession requires FCO to provide minimal passenger services, CB Transportes has invested in this area. All existing passenger rolling stock has been refurbished and 11 second-hand coaches have been bought from the Ferrocarril Arica-La Paz (FCALP). Services are operated between Santa Cruz de la Sierra and Corumbá and between Santa Cruz de la Sierra and Yacuiba.

In 1999 FCA was only operating services between Oruro and Villazón.

Freight operations

In 1998 FCO was reported to have carried over 1 million tonnes of freight. The company planned to develop domestic intermodal traffic to link FCO and FCA networks and was seeking to increase cross-border traffic with Argentina and Brazil.

New lines

A technical assistance agreement with the US Federal Railroad Administration was signed in December 1998 aimed at reopening studies into a rail link between the FCA and FCO networks. The new 388 km line would connect Aiquile and Santa Cruz de la Sierra.

Traction and rolling stock

In 1997 FCO operated 22 diesel locomotives, two diesel railbuses, 51 passenger coaches and 727 freight wagons. At last report, the Andean network operated 27 locomotives, seven railbuses, 53 passenger coaches and 727 freight wagons.

In 1997 FCO invited tenders for 200 box wagons, 75 flat wagons and 45 high-sided wagons.

Signalling and telecommunications

Recent investments have concentrated on communications and control systems, including GPS train-positioning technology.

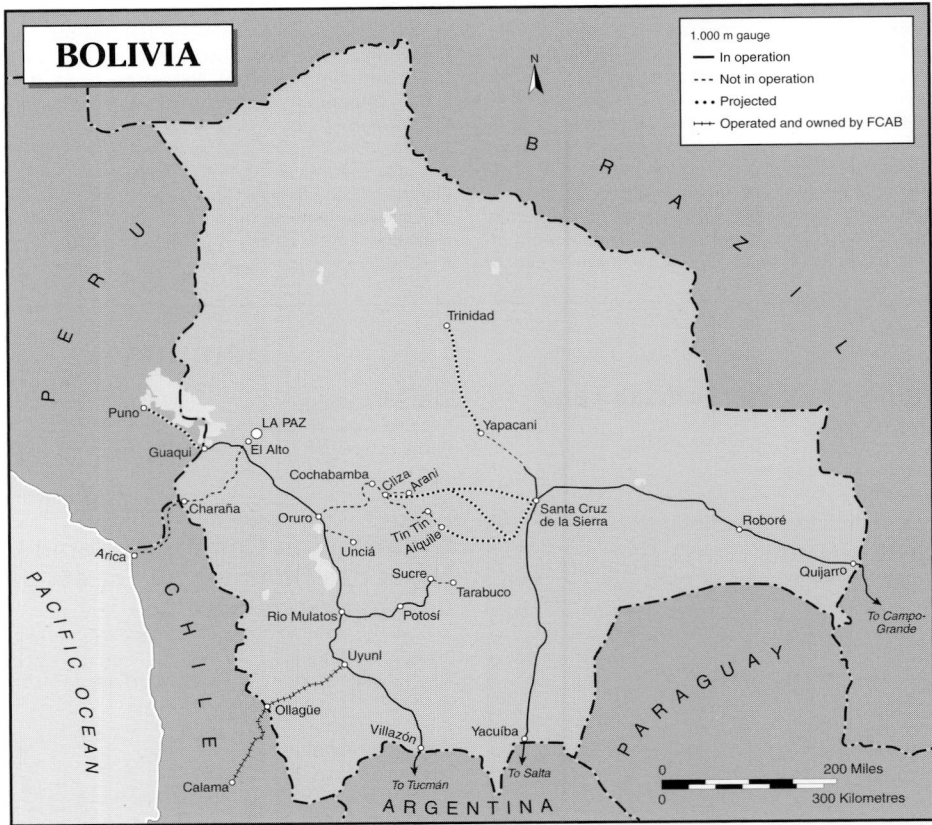

0114280

Santa Cruz Estacion on Bolivia's Eastern Network   0001854

Diesel locomotives

Class	Wheel arrangement	Power kW	Speed km/h	Weight tonnes	No in service	First built	Mechanical	Builders Engine	Transmission
950	Bo-Bo-Bo	970	70	81.6	9	1968	Hitachi/ Mitsubishi	MAN VGV 22/30 ATL	E Hitachi/ Mitsubishi
1000	Bo-Bo-Bo	1,550	100	90	7	1978	Hitachi/ Mitsubishi	MTU 12 V 956TB 11	E Hitachi/ Mitsubishi
521	Bo-Bo	280	25	30	1	1968	Hitachi	Hitachi MAN RGV 18/12TL	H Hitachi
841	Bo-Bo	395	40	55	1	1980	Hitachi/ Mitsubishi	MTU V396 TC 12	H Hitachi
U 20C	Co-Co	1,550	103	89.9	6	1977	GE	FDL 12 GE	E GE
U 10B	Bo-Bo	590	103	50.8	7	1977	GE	Caterpillar D 398	E GE
980	Bo-Bo	551	60	66.5	1	1987	Sulzer	MTU	E Sulzer
846	Bo-Bo	480	60	66.4	2	1950	Sulzer	Sulzer	EM Sulzer

Diesel railbuses

Cars per unit	Motor cars per unit	Motored axles/car	Power hp per motor	Speed km/h	No in service	First built	Mechanical	Builders Engine	Transmission
2	1	2	340/240	80	8	1967/78	Ferrostaal	Cummins NHHRTO-6	M Zahnfabrik
2	1	2	335/240	90	3	1978	Ferrostaal	Cummins	H Voith T 211 R

Track

Rail: ASCE 29.76, 37.2 and 39 kg/m
BSS 32.24 and 37.2 kg/m
Crossties (sleepers): Wood and steel
Spacing: 1,400–1,640/km
Min curvature radius: 15°
Max gradient: 3% compensated on curves
Max axleload: 18 tonnes

Ferroviaria Oriental SA (FCOSA)

PO Box 3569, Avenida Montes Final s/n, Santa Cruz de la Sierra
Tel: (+591 3) 46 39 00   Fax: (+591 3) 46 39 20
e-mail: fcosa@fcosa.com
Web: http://www.fcosa.com

Key personnel

General Manager: Jaime Valencia Valencia
Chief Commercial Officer: Edgar Dick Noya
Chief Administration and Finance Officer: Ricardo Fernández Durán
Chief Operating Officer: Ricardo Forli
Chief Mechanical Officer: Franklin Majie Ríos
Infrastructure and Environment Manager: Fernando Osorio Rodriguez

Gauge: 1,000 mm
Route length: 1,187 km

Political background

A 40-year concession to operate the geographically separate eastern portion of the network of Bolivian National Railways (ENFE) was awarded in 1996 to a consortium in which Genesee & Wyoming (for further details see entry in the USA portion of the *Railway systems and operators* section) is the major shareholder and strategic partner. G&W holds a stake of 48.61 per cent in Trenes Continentales, which owns a shareholding of 50 per cent in the operating concession. Other leading members of Trenes Continentales are investment and funding institutions, while some neighbouring railways also hold minor stakes. The remaining shares in the concession-holding consortium are Bolivian pension funds.

Organisation

The FCOSA system comprises two lines radiating from its hub at Santa Cruz de la Sierra: to the east, a 643 km line runs to Quijarro, a port town on the Paraguay river with a rail link into the Brazilian Ferrovia Novoeste system at Corumbá; and a 534 km line southwards to Yacuiba, where a cross-border connection into Argentina is made with the Ferrocarril Belgrano Cargas network.

Passenger operations

On the line to Quijarro, FCOSA operates a daily express service six days a week in each direction, a railcar service three days a week and a mixed train twice a week. By 2002, some 500,000 passengers annually were being carried, representing an increase of around 25 per cent since 1997.

Freight operations

In 2002, FCOSA carried around 1 million tonnes of freight, an increase of around 90 per cent on the figure achieved in 1997. Products handled include export minerals, soya, cereals and oils and imported manufactured goods. The east line to Brazil is the more important of the railway's two routes.

Traction and rolling stock

At the beginning of 2003, FCOSA operated 24 diesel locomotives: eight GE U20C, nine GE U10B, four Hitachi/Mitsubishi Bo-Bo-Bo machines and three GM GR12U Co-Cos. All were inherited from ENFE except the three GM machines, which were acquired from STF, Colombia, and regauged from 914 to 1,000 mm. Also in the fleet were two Ferrostaal railcars, 54 coaches and 698 freight wagons.

Bosnia-Herzegovina

Bosnia-Herzegovina Railways Public Corporation

Bosnia-Herzegovina Railways Public Corporation (BHŽJK)
Željeznice Bosne i Hercegovine (ŽBH)
Musala 1, 71000 Sarajevo
Tel: (+387 71) 61 84 48
Fax: (+387 71) 144 62 55

Key personnel

Director of Traction and Maintenance:
 Miroslav Mehmedbasič
Željeznice Republike Srpske (ŽRS)
Svetog Save 71, 74000 Doboj, Republika Srpska
Tel: (+381 53) 22 40 50
Fax: (+381 53) 22 40 50

Key personnel

Director General: Dipl Ing Marinko Biljanovič

Gauge: 1,435 mm
Route length (1991): 1,021 km
Electrification (1991): 795 km at 25 kV 50 Hz AC

0087933

Political background

Under the terms of the 1995 Dayton Peace Agreement an accord to form the Bosnia-Herzogovina Railways Public Corporation (BHŽJK) was signed in April 1998. The role of the corporation is to co-ordinate the activities of the railway companies which emerged as a result of the ethnic divisions which formed the basis of the Dayton Agreement. These were ŽBH, responsible for lines and services in the Federation (Bosnian Croats and Bosnian Muslims) and ŽRS, which operates lines in Republika Srpska (Bosnian Serbs). Subsequently, a third organisation ŽHB emerged to manage Federation routes where there was Bosnian Croat dominance, although this body has not received official recognition. In late 1998 it appeared probable that ŽBH and ŽHB would merge to form a new Federation Railway (ŽFBH) organisation.

Passenger operations

By 1998, ŽBH commenced running local services from Sarajevo to Konjic and Zenica and some freight trains, while ŽHB was operating local trains south from Mostar. The revival of passenger services continued in 1999–2000. By early 2000, ŽBH was running services on nine routes. The Sarajevo–Mostar–Capljina saw the highest frequency, with four trains daily each way. The international 'Bosna Ekspres' was running once daily in each direction between Sarajevo, Mostar and Ploče, Croatia. ŽRS was providing a higher level of service, with trains operating on 13 routes. In general, trains do not cross inter-entity borders; passengers are required to change trains at these points.

It was expected that the following services would be reintroduced during 2000: Sarajevo–Bihać Sarajevo–Doboj–Zagreb; Doboj–Zagreb–Ljubljana; and Banja Luka–Doboj–Tuzla–Žvornik–Belgrade. In addition, ŽBH planned to introduce a Tuzla–Brcko service.

Freight operations

During the NATO-Yugoslav war, ŽRS freight traffic fell drastically to about 10 per cent of pre-war levels, as most movements had previous been to and from the Federal Republic of Yugoslavia. During 2000, there was a revival of this traffic. ŽBH freight loadings were less affected by the war. Freight trains regularly cross all border points, both inter-entity and international. A large volume of ŽBH traffic and a fair part of that of ŽRS is SFOR-related.

Improvements to existing lines

Restoration of heavily damaged infrastructure, including the reconstruction of many damaged bridges, began in 1995/96. Italian and British military railway engineers have played a prominent role in this process, while funding has been provided by the United States Agency for International Development and the Swedish International Development Co-operation Agency, among others.

Traction and rolling stock

ŽBH traction in early 2000 comprised: 29 Class 441 and two ex-Croatian Railways Class 1141.1 25 kV electric locomotives; eight Class 661, 28 Class 642, one Class 643 and five ex-German Rail Class

---

For details of the latest updates to *Jane's World Railways* online and to discover the additional information available exclusively to online subscribers please visit
jwr.janes.com

232 diesel-electrics; four ex-German Rail Class 212 diesel-hydraulics; and eight Class 732 and two Class 733 shunters. The Class 1141.1 machines were acquired to work 'Bosna Ekspres' services between Sarajevo and Ploče, Croatia.

Also owned were six Class 411 emus (four serviceable) and 31 steam locomotives.

ŽRS traction in regular use in early 2000 comprised: 11 Class 441 electric locomotives; seven Class 661, two Class 642 and two Class 643 diesel-electrics; and five ex-German Rail Class 212 diesel-hydraulics. One Class 734 diesel-hydraulic shunter was also reported to be active.

Signalling and telecommunications

In 1999, signalling on most ŽRS lines was reported to be inoperable and trains were being controlled by radio despatching.

ŽBH also uses radio despatching on all lines except Sarajevo-Konjič, where colourlight signalling is in use.

ŽBH Class 1141.1 electric locomotive on the operator's only international service, the Sarajevo-Ploče 'Bosna Ekspres', at Zenica (Frank Valoczy)
0099113

Botswana

Ministry of Works and Transport

Private Bag 007, Gaborone
Tel: (+267) 395 85 00
Fax: (+267) 391 33 03

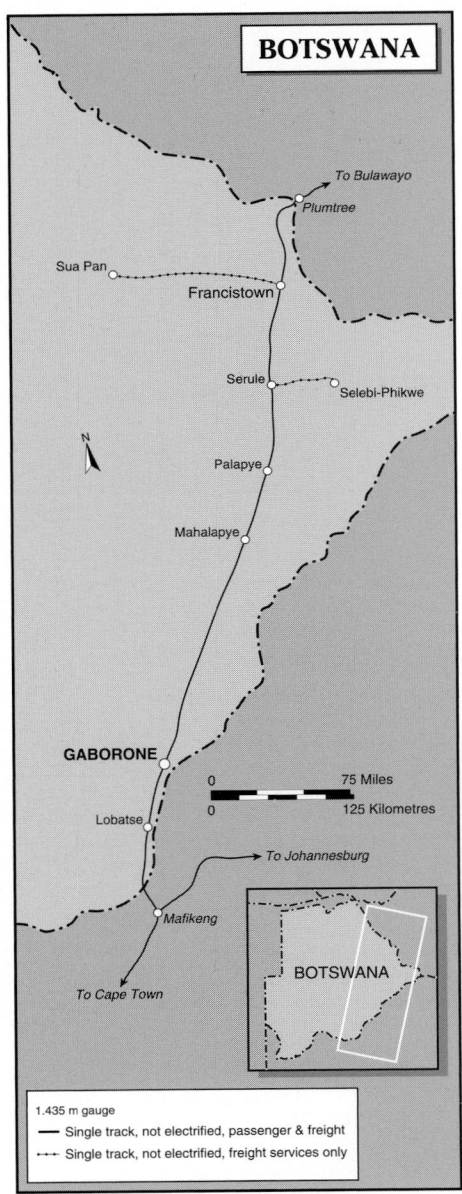

BOTSWANA

To Bulawayo
Plumtree
Sua Pan
Francistown
Serule
Selebi-Phikwe
Palapye
Mahalapye
N
GABORONE
0   75 Miles
0   125 Kilometres
Lobatse
To Johannesburg
Mafikeng
BOTSWANA
To Cape Town

1.435 m gauge
— Single track, not electrified, passenger & freight
•·· Single track, not electrified, freight services only

0058498

Key personnel

Minister: Tebelelo Seretse
Permanent Secretary: G N Thipe
Deputy Permanent Secretaries: T M Lekankan;
  G L Tlogelang

Botswana Railways (BR)

Private Bag 0052, Mahalapye
Tel: (+267) 41 13 75
Fax: (+267) 41 13 85
e-mail: botrail@info.bw

Key personnel

General Manager: A Ramji
Assistant General Manager (Finance):
  P K Sengupta
Assistant General Manager (Human Resources):
  Batlhatswi S Tsayang
Assistant General Manager (Infrastructure):
  S S Kapoor
Assistant General Manager (Rolling Stock): Bennet
  M Katai
Assistant General Manager (Business
  Management): C B Botana
Number of staff employed (1999): 1,168

Gauge: 1,067 mm
Route length: 888 km

Political background

The country is traversed by 640 km of main line between Ramatlabama (north of Mafikeng, South Africa) and Bakaranga (south of Plumtree, Zimbabwe) with three branch lines, formerly managed by National Railways of Zimbabwe. Botswana Railways took over operation of the railway in 1986 and the following year took control of the assets and administrative services. The 175 km branch from Francistown to serve soda ash deposits at Sua Pan was opened in 1991.

In 1999, serious concern was expressed in Botswana regarding the impact on BR traffic, and therefore financial viability, of the Bulawayo–Beitbridge line in Zimbabwe, which opened during the year.

Finance (million pula)

Revenue	1996–97	1997–98	1998–99
Passengers	10.0	10.5	10.3
Freight	73.4	97.8	109.4
Other income	7.1	10.2	13.2
Total	90.5	118.5	132.9
Expenditure			
Total	74.8	104.8	115.0
Traffic	1996–97	1997–98	1998–99
Passenger journeys (000)	574	496	360
Freight tonnes (million)	1.97	2.57	2.81
Freight tonne-km (million)	795	1,111	1,282

Class BD 2 GM-built diesel locomotives at Lobatse with a train of chemical salt destined for Durban and shipment to Argentina (Marcel Vleugels)
0089794

Diesel locomotives

Class	Builder's type	Wheel arrange-ment	Power kW	Speed km/h	Weight tonnes	No in service	First built	Mechanical	Builders Engine	Transmission
BD 1	UM 20C	Co-Co	1,500	103	96	10	1982	Krupp	GE 7FDL 12	E GE
BD 2	GT22LC-2	Co-Co	1,700	107	96.6	20	1986	GM	GM 645E3B-12	E GM
BD 3	U15C	Co-Co	1,340	107	97.9	10	1991	GE	GE FDL8	E GE

Passenger operations

Passenger numbers recorded a significant fall in 1998–99 to 360,000 compared with 496,000 the previous year. This was mainly a result of fierce competition from road transport. Revenue held up due to increases in promotional fares.

Through-running to South Africa was resumed in early 1995 when the Bulawayo–Lobatse train was extended to Mafikeng.

Freight operations

Freight traffic achieved modest growth to 2.81 million tonnes in 1998–99, compared with 2.57 million tonnes in the previous year. Domestic traffic increased by 12 per cent, mainly due to a growth in coal hauled for Soda Ash Plant. The development of intermodal facilities at Gabcon, an inland port and container terminal in Gabarone, and a similar facility, Francon, serving the north of Botswana, also contributed to domestic traffic growth. Gabcon handled 2,894 TEU in 1997–98, a figure higher than the terminal's design capacity.

Import traffic fell by 2 per cent to 0.79 million tonnes in 1998–99. Transit traffic increased by 22 per cent to 1.15 million tonnes, mainly as a result of a routing policy and favourable haulage rates agreed with South African operator, Spoornet, which was intended to provide a sound long-term basis for stable cross-border traffic.

New lines

In a bid to regain traffic levels lost following opening of the Bulawayo–Beitbridge line in Zimbabwe, BR has proposed the construction of a direct line to Zambia to provide an alternative corridor for freight between South Africa and Zambia and Congo, Democratic Republic.

Improvements to existing lines

Track rehabilitation between Mahalapye and Radisele has been completed. This is the first such project undertaken by Botswana Railways using its own planning, design and engineering resources. In 1999 a similar scheme was in progress between Serule and Linchwe. Scheduled for completion by June 2000, this will complete the rehabilitation of BR's main line.

Traction and rolling stock

Botswana Railways operates 10 Krupp diesel-electrics, 20 Type GT22LC-2 diesel-electrics and 10 short-haul/shunting diesel locomotives. Freight wagon stock totals 1,025; there are 52 passenger coaches and one Cummins-powered railcar supplied by Union Carriage in 1993.

The GE-built Class BD 3 locomotives have been uprated from 1,120 kW to 1,340 kW under a re-engineering programme.

Recent deliveries include: 330 wagons for soda ash traffic over the Sua Pan line and coal haulage from Morupule; and 41 air conditioned passenger cars, including five buffet and six sleeping cars. The wagon contract was awarded to China National Machinery Import & Export Corporation and the passenger car contract to Mitsui (but the vehicles were built in South Africa). Generator cars have been subject to a re-engining programme.

BR plans to install solid state event recorders in its locomotives.

Signalling and telecommunications

The Radio ElectronicToken Block system now controls 698 km of line. A modification and improvement plan was implemented in 1993 and work continues to improve reliability. A back-up VHF radio system has been installed to help keep trains moving when there is a communications breakdown. An automatic block signalling scheme has been installed between Rakhuna and Mafikeng, a distance of 10 km.

A network of microcomputers has been installed using Novell Netware. A vehicle control system has been written 'in house', primarily to calculate vehicle hire charges and produce operational statistics, and the first phase of a personnel management system has been implemented. Systems to monitor vehicle utilisation, schedule preventive maintenance and automate the daily operating diary have been developed by the railway as an adjunct to the vehicle control system. A new stock control system was implemented in 1994.

BR has suffered from theft of copper wire from communications lines, necessitating the provision of back-up facilities from the national telecommunications company.

Type of coupler in standard use

Passenger cars: Alliance 8X6 vacuum
Freight cars: Alliance 8X6

Type of braking in standard use, locomotive-hauled stock: Vacuum

Track

Rail type and weight: Flat-bottom 40 kg/m (378 km), 50 kg/m (510 km)
Sleepers: Concrete, spaced 1,430/km in 50 kg/m cwr plain track, 1,500/km in curves; steel, spaced 1,430/km in 40 kg/m plain jointed track, 1,500/km in curves
Fastenings: Fist on concrete sleepers, clip bolt on steel
Min curvature radius: 200 m
Max gradient: 1.6%
Max permissible axleload: 17.2 tonnes

Brazil

Ministry of Transport

Esplanada dos Ministérios, Bloco R, 6° Andar, 70044-900 Brasilia DF
Tel: (+55 61) 224 01 85
Fax: (+55 61) 225 09 15
Web: http://www.transportes.gov.br

Key personnel

Minister of Transport: Anderson Adauto

Carajás Railway (EFC)

Estrada de Ferro Carajás
Avenida dos Portugueses s/n, Praia do Boqueirão, São Luís, Maranhão, CEP 65085-850
Tel: (+55 98) 218 44 05
Fax: (+55 98) 218 45 20

Key personnel

Director, Northern System: Jayme Nicolato Corrêa
General Managers
  Railway Operation: Carlos Eduardo Fontenelle Carriero
  Mechanical and Electrical Maintenance: Ronaldo José Costa

Gauge: 1,600 mm
Route length: 892 km

Organisation

EFC is owned by the mining group Companhia Vale do Rio Doce (CVRD), which also owns the 1,000 mm gauge Vitória a Minas Railway (EFVM) (qv). Brazil's largest holder of mineral rights, CVRD

EFC diesel locomotives

Builder's type	Wheel arrange-ment	Power kW	Speed km/h	Weight tonnes	No in service	First built	Mechanical	Builders Engine	Transmission
C30-7B	Co-Co	2,465	80	180	6	1984	GE	GE 7FDL16	E GE
C40-8	Co-Co	3,018	80	180	4	1989	GE	GE FDL16	E GE
SD40-2	Co-Co	2,429	80	180	27	1985	GM	EMD 645E3C	E EMD
SD60M	Co-Co	3,056	80	180	2	1991	GM	EMD 16-710G3A	E EMD
GT26-CU2	Co-Co	2,208	80	180	2	1984	GM	EMD-16-645E3B	E EMD
C36-7B	Co-Co	2,720	80	180	33	1984	GE	GE 7FDL	E GE
C44-9	Co-Co	3,350	80	180	7	1997	GE	GE 7FDL	E GE
SL80	Bo-Bo	898	42	65	2	1993	GE	GE	E GE

is the world's largest producer of iron ore and pellets. The company also mines maganese, kaolin and potash, and has a significant presence in the aluminium and woodpulp markets. EFC and EFVM are operated as the company's Northern System and Southern System respectively. As well as its railway operations, the CVRD portfolio also includes ports.

CVRD's Northern System has implemented a policy of forming partnerships to create multimodal transport links covering rail, road, river and ports, including collaborations with other rail operators.

Finance (US$ million)

Revenue	1998	1999
Passengers	3.56	1.99
Freight*	43.89	31.87
Total	47.40	33.86

*General cargo, excluding iron ore

Expenditure

Staff/personnel	39.33	23.83
Materials and services	78.54	50.62
Total	117.87	74.45

Traffic (million)	1998	1999
Passenger journeys	0.538	0.459
Passenger-km	191	163
Freight tonnes	50.13	47.74
Freight tonne-km	42,117	40,090

Passenger operations

Principally a freight railway, EFC operates three trains (of around 16 coaches including executive and standard classes) a week in each direction between São Luís and Parauapebas (861 km).

Freight operations

EFC's principal traffic is iron ore, moved from the mine at Carajás to Ponta da Madeira in trains comprising 204 ore wagons with a total capacity of 20,910 tonnes, hauled by two locomotives. A return trip, including crew changes, refuelling, loading and unloading, usually takes 58 hours and on average six loaded and six empty trains are in operation each day.

Other freight traffic in 1999 amounted to around 4.6 million tonnes. Soya products, fertiliser, pig

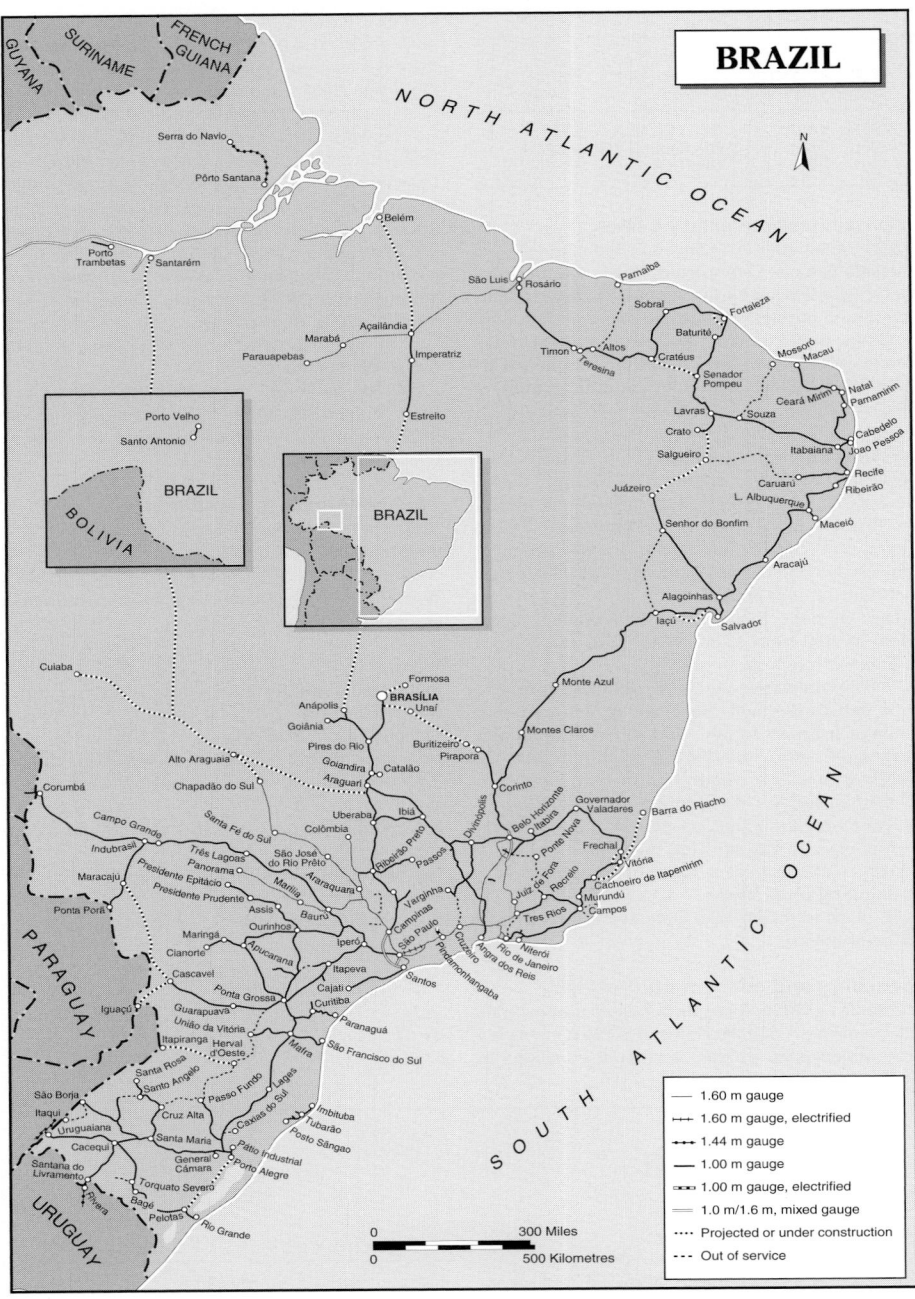

BRAZIL

Map legend:
- 1.60 m gauge
- 1.60 m gauge, electrified
- 1.44 m gauge
- 1.00 m gauge
- 1.00 m gauge, electrified
- 1.0 m/1.6 m, mixed gauge
- Projected or under construction
- Out of service

0   300 Miles
0   500 Kilometres

0114281

iron, petroleum products, cement, road vehicles and other commodities are carried on freight trains composed of around 80 wagons.

Traction and rolling stock
At the end of 1999 the rolling stock fleet comprised 83 diesel-electric locomotives, 37 passenger coaches (including two buffet cars and three luggage vans), 3,528 ore wagons and 728 other freight wagons.

Signalling and telecommunications
Since 1989 EFC's main line operation has been supported by a comprehensive CTC system, including wayside equipment such as power point motors, track circuits, cab signalling and hotbox detectors, onboard equipment including ATC and event recorders, and a control centre.

To increase capacity, EFC has installed solid state interlocking, constant warning time grade crossings, new hotbox hot wheel detectors and electrocode equipment. The analogue communications network is being replaced by a fibre optic system.

EFC is studying the development and implementation of a CTC system which will support GPS for train-position and locomotive health monitoring systems and an integrated maintenance centre, and an auxiliary traffic control system for train working optimisation and meet/pass planning which will be introduced to increase capacity.

Track
Rails: 68 kg/m long-welded
Fastenings: Pandrol or Denik
Crossties (sleepers): Timber, creosote-treated, thickness 170 mm, spaced 540 mm between centres
Min curve radius: 860 m
Max gradient: 0.4% loaded trains, 1% empty trains
Max axleload: 31.5 tonnes

Central

Compania Estadual de Engenharia de Transporte e Logistica
Edificio Barão de Maua 318, Avenida Francisco Bicalho, CEP 20220-310 Rio de Janeiro
Tel: (+55 21) 22 33 40 90   Fax: (+55 21) 25 02 51 77
Web: www.central.rj.gov.br

Key personnel
President: Albuino Cunha de Azevedo
Operations Director: Roberto Ramos dos Santos

Gauge: 1,000 mm
Route length: 73 km

Political background
Central is the successor to Flumitrens, which operated the whole Rio suburban network after the state of Rio de Janeiro took over control from Companhia Brasiliera de Trens Urbanos in December 1994. Most of the network was privatised in 1998 and is now operated on a 25-year concession by SuperVia (see entry in Brazil section of *Railway systems and operators*). Flumitrens continued to run the two metre-gauge lines which were not part of the concession. Central, which is the holding company for the state-owned assets operated by SuperVia, and which also runs Rio's

GE U12B diesel-electric locomotive operated by Central (Joao Bosco Setti)   ***NEW**/1114950*

Santa Teresa tourist tramway, assumed control of the two lines when Flumitrens went into liquidation in 2001.

Passenger operations
Central operates two unconnected lines with sparse services but huge potential for growth. A 42 km line from Guapimirim and Magé provides interconnection with SuperVia's trains at

Saracuruna, while on the other side of Guanabara Bay an isolated route links Visconde de Itaborai with Niteroi (33 km). In 2004 neither was carrying more than about 1,000 passengers daily. Growth is limited by the poor state of the infrastructure and the fact that only two trains are available to serve the Guapimirim line and just one for the Niteroi line.

The routes serve heavily populated corridors and Central has proposed development of both lines to support a frequent service. Central's plans for the Guapimirim line have already been submitted to Rio's Secretariat of Transport; they cover upgrading of the infrastructure, station improvements and provision of new rolling stock. The Niteroi line is estimated to have the potential to attract some 300,000 passengers daily. Under the Rio Mass Transit Project (PET), bids were sought at the end of 2004 for modernisation studies that would be financed by the Brazilian development bank BNDES. No capital funding has yet been forthcoming for either project.

Traction and rolling stock
The sparse fleet comprises three GE U12B diesel-electrics, eight Pidner coaches and four Vickers, all of 1950s vintage.

UPDATED

Companhía Brasilera de Trens Urbanos (CBTU)

Estrada Velha da Tijuca 77, CEP 20531-080 Rio de Janeiro, RJ
Tel: (+55 21) 288 19 92
Fax: (+55 21) 571 61 49

Key personnel
President: Luiz Otavia Mota Valadares
Finance and Administration: Jorge Miguel Felippe

Organisation
For many years, CBTU was the operator of nearly all suburban passenger rail networks in major Brazilian cities. Progressively, these systems have been transferred to control by the states: São Paulo took over CBTU lines serving the city and integrated them with CPTM; Rio de Janeiro's Flumitrens has been transferred to SuperVia; the systems of Fortaleza and Salvador were transferred to state ownership in 1998 and 1999 respectively; and Recife's suburban systems were taken over by Metrorec and Companhía Pernambucana de Trans. This has left CBTU with the Demetrô system in Belo Horizonte and networks serving João Pessoa, Maceió, Natal and Teresina. It was expected that Demetrô would be transferred to Minas Gerais state during 2000, and studies were in progress to explore how the remaining systems could be similarly handed over to their states.

Passenger operations
Belo Horizonte (Minas Gerais)
The 21.3 km Demetrô 1,600 mm gauge suburban system is double-track and electrified at 3 kV DC overhead. It runs from Eldorado via Calafate, Central and Santa Terza to Minas Shopping, with 14 stations open to passenger traffic. Around 25 million passengers were carried in 1999.

In April 1999, prequalification bids were invited for a 6.5 km extension from São Gabriel to Vía Norte. Estimated to cost US$99 million, this will feature two tunnels and two of the five stations will provide bus transfer facilities. Also under construction in 2000 was a 10 km branch from Entroncamento/Rodoviário to Barreiro.

Later extensions planned include: Eldorado-Betim (29 km); Barreiro–Ibirité (12 km); Vía Norte–Santa Luzia (13 km); and Minas Shopping to Venda Nova and Ribeirao das Neves. When completed, the entire network is expected to carry around 125 million passengers annually.

Of the original series of 15 emus, five were refurbished in 1999 and 10 more were ordered for delivery in 2000.

In August 1999, a new maintenance centre at São Gabriel, north of Minas Shopping, was commissioned.

Joao Pessoa (Paraíba)
A 32 km line running from Santa Rita via João Pessoa to Cabedelo is operated by CBTU. The single-track line has eight stations and is operated by two Alco RS-8 diesel locomotives and 17 coaches (14 serviceable). Around 1.5 million passengers are carried annually.

Maceió (Alagoas)
A 32 km single-track line runs from Maceió to Utinga, Rio Largo and Lourenço de Albuquerque, with services provided by two diesel locomotives and 21 coaches. In 1999, 2.3 million passengers were carried, excluding 'account' passengers travelling on free passes.

Natal (Rio Grande do Norte)
CBTU operates an 18-station 56 km network radiating from Natal to Parnamirim and Ceará Mirim. In 1998, 1.4 million passengers were carried.

Rolling stock comprises three Alco RS-8 diesel locomotives and 20 coaches.

It is expected that responsibility for the system will be transferred to Rio Grande do Norte within the next few years. Parts of the Northern Line, between Ribiera and Soledade (11.3 km) may be doubled and electrified to convert it to light rail.

Teresina (Piauí)
A short line traversing Teresina city centre to reach Timon, Maranhao (8 km) is served by suburban trains and marketed as 'Sistema de Trens Urbanos (STU) Teresina'. In 1998, 1.8 million passengers were carried. The present system was operated by Brazilian Federal Railways (RFFSA) as 'Metro Teresina' from 1990 until 1996, when CBTU took it over. Rolling stock comprises five Ganz-Mavag diesel trainsets transferred from Rio Grande do Sul. Three of these units are serviceable but only one is operating at any time.

Projects have been developed to construct a 9 km branch from Centro to Bela Vista, in the south of the city, and another branch would run from Frei Serafim to Bela Vista (7 km). These two lines would generate an additional 30,000 passengers a day, five times the current level of traffic. In the east of the city, completion of a 1.2 km branch to Bandeirante still awaited completion in 2000, construction having commenced in 1992. Proposals have been made to convert the system to light rail operation but financial problems faced by Piauí state have so far prevented these being implemented.

Traction and rolling stock
In 2000, CBTU was refurbishing 20 'Pidner' coaches for use on the Joao Pessoa, Maceió and Natal systems.

Companhia Ferroviária do Nordeste (CFN)

Avenida Francisco Sá 4829, CEP 70 310-002, Fortaleza, Ceará
Tel: (+55 85) 286 25 25
Fax: (+55 85) 286 61 56

Key personnel
President: Wagner Bittencourt de Oliveira
Directors
  Operations: Lauro Fassarella
  Commercial: Carlos Kopiptke
  Finance: Martiniano Dias

Gauge: 1,000 mm
Route length: 4,633 km

Political background
The CFN network was the poorest of the former Brazilian Federal Railways (RFFSA) regions. It combines the former subdivisions of Fortaleza, Recife and São Luis and connects with the Carajás Railway at São Luis and with Ferrovia Centro Atlántico at Proporía, south of Maceió. Of the 4,633 km network, only 557 km are in operable condition; the remainder needs varying amounts of rehabilitation work. A deficit of Cr50 million was recorded in 1994, when the workforce totalled more than 5,000. By January 1997, this figure had been reduced to 2,027 in preparation for privatisation. This occurred in January 1998, when the system was taken over by a consortium led by Taquari, part of the Vicunha group, with 40 per cent of the equity. Other consortium members are CSN, CVRD and Bradesco, each with a 20 per cent stake. In spite of modest freight traffic prospects, early indications were that privatisation had been successful.

Passenger operations
In 1999, the following passenger services continued over CFN tracks: suburban services around Fortaleza, João Pessoa, Maceió, Natal, Recife and on a short stretch of line through Terezina. All these services were run by CBTU or by local state-owned companies. Only the Recife-Cabo-Ribeirão service could be considered interurban. Services between Recife and Caruarú are still operated for special events.

Freight operations
During the first year under private ownership, CFN was expected to achieve 1.2 billion tonne-km thanks to a new commercial strategy. Through daily freight services branded 'Expressos' and running to fixed schedules have been introduced on the Fortaleza-São Luis, Fortaleza-Recife and Recife-Maceió corridors and a 30 per cent discount has been offered to new customers. Main commodities carried include alcohol, sugar, cement, scrap metal, steel products and aluminium. In 1998 CFN launched a bid to win oil products traffic back to rail, potentially generating 70 million tonne-km annually, and it was also expecting to introduce container services.

In 1999, traffic increased to around 1 billion tonne-km, compared with 640 million in 1998, and revenue increased by 49 per cent to Cr26.2 million. Manufactured products have been the main contributor to this increase, which was the highest of all Brazilian railways. Tonnages of cement, beer and coke increased especially significantly, with rises of 317, 164 and 113 per cent respectively in the amounts loaded in 1999 compared with the previous year. Loadings of cereals are also increasing significantly. Because of the poor condition of many freight wagons, which are unable to accept palletised goods, much freight is loaded manually.

New lines
A new line, projected in the 1980s, was the Transnordestina Railway. As its name suggests, this would traverse the country's far northeast inland, linking up to three metre-gauge lines that head inland from the Atlantic coast between Salvador and Fortaleza. The Transnordestina scheme is divided into three segments: the first, on which some construction work has been carried out during the 1990s, would connect Petrolina, in southwest Pernambuco, with Salgueiro (231 km); the second, for which detailed planning has been completed, will provide a 113 km link between Salgueiro and Missão Velha; the third line will connect Piquet Carneiro, on the Fortaleza-Missão Velha-Crato line, with Crateús, on the Fortaleza-Terezina line (178 km). In conjunction with this project, several existing lines are to be upgraded or rehabilitated. These include: Salgueiro-Caruarú (468 km); Missão Velha-Fortaleza (580 km); and Crateús-Terezina (290 km). Also likely is a new bridge from Petrolina to Juázeiro, including a 6 km link to provide a connection with the Ferrovia Centro Atlántico (qv). Connecting lines south of Juázeiro could also be upgraded.

The entire Transnordestina system would be single-track and has been estimated to cost US$815 million. Between 1991 and 1994, US$7.17 million was spent on the Petrolina-Salgueiro section and CFN has reached agreement with the federal government to raise US$322 million for this stretch of line, with this funding being found by CFN and the Northeast Development Agency (SUDENE). Work started in early 1999 and this first phase of the project was expected to be completed by 2001.

CFN also plans to build a new line diverging from the existing system and the planned Transnordestina at Salgueiro, running 120 km west to Araripina, a town at the centre of a gypsum-producing region in the extreme west of Pernambuco state. At Araripina, between 1.5 and 2 million tonnes of gypsum are produced annually and a potential export contract with a US-based company could add an additional 1.5 million tonnes of production. Construction will be simplified by the flat and dry terrain, with the project costed at an estimated Cr78 million. Construction was expected to start in 2000 or 2001 and take less than one year.

At Uniao dos Palmares, in Alagoas state, a short branch line is to be built, at an estimated cost of Cr300,000, to provide a connection to Usina Santa Clotilde, which serves several sugar cane mills producing over 200 million litres of fuel-alcohol annually. CFN aims to capture 10 to 20 per cent of this traffic.

Improvements to existing lines
In February 2000, CFN received a 15-year loan of Cr62 million from the National Bank for Social and Economic Development (BNDES) to help modernise its network. This forms part of a Cr569 million programme to modernise the entire CFN system, and was to be complemented by a

contribution of Cr39 million from the railway's own resources. The money was intended to be spent during 2000 to eliminate critical bottlenecks, especially the repair of several bridges. The largest rehabilitation projects include the 453 km Terezina-São Luis and 468 km Caruarú-Salgueiro lines, and it has also been decided to reconstruct the 243 km Souza-Mossoró line.

In May 1999, construction was started of a 16 km bypass line southwest of Fortaleza to divert freight trains from the city centre, reducing traffic on lines used by Metrofor suburban services. The line runs from Aracapé to Caucaia. At a later stage, a new freight terminal and marshalling yard may be built at Aracapé and the freight line from the port of Mucuripe to Aracapé and Vila des Flores could be separated from the existing suburban line. The bypass could also be used in future by diesel-hauled suburban passenger services (see entry for Metrofor). Completion was expected in December 2001.

On the outskirts of São Luis, a new container terminal is to be built, possibly also with 1,600 mm gauge tracks to enable intermodal traffic to be exchanged with CVRD's Carajás Railway.

Traction and rolling stock
With the help of 20 General Motors G12 diesel-electrics procured form other parts of Brazil, CFN commenced operations with 71 locomotives, of which 55 were serviceable. The average age of the locomotive fleet in 1998 was 33 years, and failures were frequent. The freight wagon fleet comprised 2,302 vehicles, including 441 transferred by RFFSA from other regions. In 1998, 1,861 wagons were available for traffic. Any passenger vehicles have been transferred to regional operators.

Rolling stock maintenance workshops are located at Fortaleza and Recife.

Signalling and telecommunications
Train despatching is generally by radio. As a result, the number of staffed stations has been reduced from 108 to 34 and the number of train control centres from six to two.

Track
Most parts of the network are laid with 25 or 30 kg/m rail. Sleepers are generally timber, spaced 1,500 to 1,666 per km.

Companhia Paulista de Trens Metropolitanos (CPTM)

Rua Boa Vista 185, 9° andar, 01014-001 São Paulo, SP
Tel: (+55 11) 32 93 44 45   Fax: (+55 11) 32 93 45 46
e-mail:mbandeira@cptm.sp.gov.br
Web: www.cptm.sp.gov.br

Key personnel
President: Mário Manuel Seabra Rodrigues Bandeira
Directors
Operations and Maintenance: José Luiz Lavorente
Administration and Finance: Antônio Kanji Hoshikawa
Planning: Mário Manuel Seabra Rodrigues Bandeira
Engineering: Stanislav Feriancic

Gauge: 1,600 mm; 1,000 mm; dual 1,600 mm/1,000 mm
Route length: 192 km; 18 km; 60 km
Electrification: 270 km at 3 kV DC

Political background
CPTM was created by the São Paulo state government on 28 May 1992 to run the city of São Paulo's suburban passenger services, with a view to better integration with the city's metro system. CPTM began operation of the former FEPASA's (for further details see Ferrovías Bandeirantes SA in the *Railway systems and operators* section) suburban routes in August 1993 and took over the former Companhia Brasiliera de Trens Urbanos (CBTU) system on 27 May 1994. At the end of 2004 CPTM employed 5,573 staff.

Traffic (million)	2002	2003	2004
Passenger journeys	345.4	353.8	368.8
Passenger-km	6,413	6,591	6,850

Passenger operations
Initially, the former FEPASA and CBTU systems were known as the West System and the East System respectively. Operations are now organised as six routes, Lines A to F.

Lines currently operated are:
Line A (Luz-Jundiaí, 60.5 km), comprising 17 stations with metro and Line B interchange at Barra Funda.
Line B (Júlio Prestes-Amador Bueno, 42 km), comprising 24 stations with Line C interchange at Osasco and Presidente Altino.
Line C (Osasco-Jurubatuba, 24.3 km), comprising 15 stations.
Line D (Luz-Rio Grande da Sierra, 37.2 km), comprising 14 stations.
Line E (Brás-Estudantes, 50 km), comprising 15 stations with metro interchanges at three.

Ex-RENFE Class 2100 emu at Pinheiros Station (Line C) (Flávio Francesconi Lage)   0536192

Line F (Brás-Calmon Viana, 38.8 km), comprising 10 stations with Line E interchanges at Brás, Tatuapé and Calmon Viana.

Improvements to existing lines
East Project (Expresso Leste)
The East region is one of the most populous in the São Paulo metropolitan area, generating demand for 2.5 million journeys daily, mainly to the centre of the city. To meet this demand a limited stop 'Expresso Leste' service between Guaianazes and Luz (24 km) was introduced on Line E, calling at six stations only and completing the journey in 33 minutes.

South Project
This proposed scheme aims to modernise the section of Line C between Pinheiros, where connection will be made with the future Line 4 of the metro system, and Santa Amaro, on metro Line 5, providing it with the characteristics of a surface metro and improving connections between

BRAZIL São Paulo Suburban network (CPTM)

— 1,600 mm gauge
═ 1,000/1,600 mm mixed gauge
••• 1,600 mm gauge projected

0580980

CPTM electric multiple-units

Class	Cars per unit	Motor cars per unit	Motored axles/car	Output/ motor kW	Speed km/h	Units in Service	First built	Builders Mechanical	Builders Electrical
1100	3	1	4	344	100	16	1956	Budd/Mafersa/ Cobrasma	Toshiba/ Hitachi/Sepsa
1400	3	1	4	344	90	13	1976	Mafersa	GE
1600	3	1	4	344	90	13	1978	Mafersa	GE
1700	4	2	4	315	90	19	1987	Mafersa	Hitachi/ Toshiba
2000	4	1	4	300	90	29	1999	CAF/Alstom	Adtranz
2100	3	1	4	290	140	42	1974	CAF	Toshiba/ Sepsa
3000	4	2	4	374	90	10	2000	Siemens	Siemens
4400	3	1	4	315	90	18	1965	FNV/Cobrasma	Hitachi/Toshiba
4800	3	1	4	168	90	2	1958	Nippon/ Kawasaki/Kinki	Toshiba
5000	3	1	4	223	90	70	1978	Cobrasma	Brown Boveri/ MTE
5500	2	1	4	250	90	24	1979	Mafersa/ Sorefame	Villares/ ACEC

the southern and western parts of the São Paulo conurbation and the city centre.

Centre Line Integration project

In 2005 work continued on the US$95.1 million 'Integração Centro' (Centre Line Integration) project, which interlinks CPTM's six lines at Barra Funda, Luz and Brás and also provides interchange with the metro system. Work commenced in 2001 and covers rehabilitation and construction work to provide four electrified surface tracks on a 7 km central section of the São Paulo suburban network, with a unified signalling and control system that will enable operation at 3-minute headways. Also included in the scheme is the modernisation of Brás and Luz stations, providing escalators and lifts and access for passengers with impaired mobility.

New lines

Airport line

In 2005 CPTM plans to launch its proposals for a rail link from Barra Funda to Gov André Franco Montoro airport at Guarulhos. The 33 km line would comprise 17 km of surface alignment, 8 km in tunnel and 8 km elevated. CPTM plans to award a DBOT concession for the project. The same line will be used to offer a high-quality service to be named 'Trem de Garulhos' between the two largest cities in the São Paulo metropolitan area, São Paulo and Garulhos. At Garulhos it is planned to establish a rail-bus interchange terminal and a station will be provided at the proposed future location of São Paulo's university.

Traction and rolling stock

In 2005 CPTM operated a fleet of 795 emu cars.

In 2000 CPTM received the last of 30 Class 2000 four-car emus for Line 6 built by a Franco-Spanish consortium (including CAF, GEC Alsthom and Adtranz) known as COFESBRA, to a design similar to Class 447 for RENFE of Spain.

In 1998 CPTM received first deliveries of 48 refurbished ex-Spanish National Railways (RENFE) Class 440 emus (Class 2100) and from 2000, 10 Class 3000 emus supplied by Siemens for Line C modernisation.

A fleet modernisation programme covering 26 older trainsets, including Class 1700 emus, was initiated in 2001. Completion was expected in 2005. Other types have also been the subject of refurbishment programmes, including Class 1700.

Budd-type emu at São Paulo (Flávio Francesconi Lage)   0536193

CPTM diesel locomotives

Class	Wheel arrangement	Power kW	Speed km/h	Weight tonnes	No in service	First built	Mechanical	Engine	Builders Transmission
3700	Bo-Bo	716	90	74	8	1968	LEW	MGO V12 BSHR	E GMF
3000	Bo-Bo	477	65	57	1	1967	GE	Caterpillar D-379BV8	E GE
6000	Bo-Bo	1,342	100	109	4	1952	GE	GE 7FDL	E GE
6000	Bo-Bo	1,193	100	109	4	1952	Alco/GE	Alco 244	E GE
7000	Bo-Bo	970	90	80	1	–	GM	GM645	E GM

In 2005 CPTM also operated 18 diesel locomotives.

Signalling and telecommunications

ATC is in use on Lines A, B, C and D (162 km) and partially on the Line E extension (18 km); on the remainder of Line E and on Line F a partial ATS system is used (70.8 km).

Track

Rail

Type: TR 57 (57 kg/m) (201 km); TR 57 (67 kg/m) (20 km); TR 68 (50 kg/m) (20 km)

Sleepers

Wood: 2,800 × 240 × 170 mm; 2,000 × 220 × 160 mm
Spacing: 1,833/km; 1,660/km
Concrete: 2,800 × 300 × 250
Spacing: 1,667/km (concrete)
Fastenings: Pandrol or Denik elastic
Min curve radius: 200 m
Max gradient: 2.43%
Max axleload: 36 tonnes

UPDATED

Estrada de Ferro Do Amapá (EFA)

Macapá, Território do Amapá

Gauge: 1,440 mm
Route length: 194 km

Organisation

The EFA is the only standard-gauge railway in Brazil. It extends over 194 km from the River Amazon port of Pôrto Santana to Serra do Navio, in central Amapá. The northernmost railway in Brazil, the line was originally built to carry manganese deposits mined at Serra do Navio and was inaugurated in January 1955 with the assistance of the US-based Bethlehem Steel and Eximbank of Washington. Passenger operations began in 1957. The line is privately owned and operated by Indústria e Comércio de Minérios SA (Icomi), which holds the mining concession until 2003, when the railway will be transferred to the state government of Amapá. After that date, the line is to continue to operate under state government administration,

assisting the development of agriculture along its route. Passenger services are also likely to continue both for tourism and social purposes. The financial position of the railway is reported as sound and it operates without deficit.

Passenger operations

Services are provided three times a week in each direction, carrying about 150,000 passengers annually. The line is served by 15 stations and halts.

Freight operations

The main traffic is manganese, of which more than 600,000 tonnes are carried annually. Over 5,000 tonnes of general freight, mostly agricultural products including tapioca, are also conveyed, with rail providing an essential means of transport in a region poorly served by roads. An expansion of agricultural traffic is foreseen.

Traction and rolling stock

The locomotive fleet comprises five GM SW-1200 diesel-electrics, three acquired in 1955 and

two in 1966, and an SW-1500 from the same manufacturer, delivered in 1971. There are 130 freight wagons, including 108 open wagons for manganese transport. Passenger services employ one small railcar and five coaches, plus a composite passenger/baggage car and two brakes. Other vehicles are used as mobile accommodation for track maintenance crews and the railway boasts an ambulance car for use in emergencies. Maintenance is undertaken at EFA's workshops at Pôrto Santana.

Track

Rail type and weight: 90-AS of 44.64 kg/m
Sleepers: wood
Ballast: crushed stone
Max gradient: 1.5%
Min curve radius: 305.6 m

Estrada de Ferro Trombetas (EFT/MRN)

Mineração Rio do Norte (MRN), Porto Trombetas, Municipality of Oriximiná, Pará

Key personnel

Manager: Reginaldo Pedriera Lapa

Gauge: 1,000 mm
Route length: 30 km

Organisation

The line commences at Porto Trombetas on the river Trombetas in the northwest of Pará state, some 80 km from Oriximiná and more than 300 km northwest of Santarém, and extends to the Rio do Norte bauxite mines. The railway employs 65 people for train operation and maintenance in a community that exists purely to exploit bauxite reserves and is considered to be highly profitable. Major shareholders in MRN are

Companhía do Rio Doce (CVRD) (46 per cent), Grupo Votorantim (10 per cent) and Alcan (24 per cent).

Freight operations

Bauxite is the railway's sole traffic, amounting to 9 to 11 million tonnes annually. Up to 15 22-wagon trains are run daily in each direction. Unloading is undertaken by a Viardor wagon-tippler and the wagons are equipped with automatic centre couplers.

Traction and rolling stock
The locomotive fleet comprises five GM G-12 970 kW diesel-electrics. There are 90 freight wagons, all high-sided four-axle vehicles with a payload capacity of 70 tonnes.

Track
Rail type and weight: TR-68 welded
Sleepers: CD 50 concrete
Fastenings: Pandrol E 209
Ballast: crushed stone

Max gradient: 1%
Min axleload: 32 tonnes

Ferrovia Centro-Atlântica (FCA)

Rua Sapucai 383, CEP 30150-904, Floresta, Belo Horizonte-MG
Tel: (+55 31) 32 79 55 00   Fax: (+55 31) 32 79 55 81
e-mail: comunicacaofca@centro-atlantica
Web: www.centro-atlantica.com.br

Gauge: 1,000 mm
Route length: 7,080 km

Key personnel
President: Thiers Manzano Barsotti

Political background
In June 1996 the second concession under the government's privatisation programme for RFFSA, for the Centre-East Network, was sold for the minimum price of Cr317 million. The so-called Tacumã Consortium, which bought the concession, consisted of CVRD subsidiary Tacumã Mining, Valia (the CVRD pension fund), intermodal operator Interférrea, local steel and cement producers, plus Judori Administração Empreendimentos e Participaçães. From the US came shortline operator Railtex, the Bank of Boston and investment group Ralph Partners. The buyer was the only bidder at the auction. The new company took over operations on 1 September 1996.

In February 2000 CVRD acquired the Railtex shareholding for US$6.4 million to become FCA's major shareholder, with a 20 per cent stake. This step was undertaken because FCA largely depends on CVRD's EFVM line for traffic.

Organisation
FCA operates all former RFFSA metre-gauge lines in the states of Espírito Santo, Minas Gerais, Goiás, Bahía, Sergipe, Rio de Janeiro and the south of Alagoas, serving such major centres as Belo Horizonte, Brasília, Rio de Janeiro and Salvador. It connects with the EFVM system at Belo Horizonte and Vitória, with Ferroban at Uberaba and Araguari and with CFN at Maceió, all of which are also 1,000 mm gauge. At Belo Horizonte, Rio de Janeiro, Tres Rios, Barra Mansa and Cuzeiro, connections are made with the broad gauge lines of MRS Logística. At Senador Canedo, southwest of Brasilia, FCA will connect with the Ferrovia Norte-Sul, under construction in 2004.

Traffic (million)	2002
Freight tonnes	23.7
Freight tonne-km	8,252

Bombardier MX 620 diesel locomotive with an FCA container train (Flávio Francesconi Lage)
***NEW**/0585052*

FCA General Motors DDM 45 diesel locomotive, formerly operated by EFVM
(Flávio Francesconi Lage)
***NEW**/0585053*

Freight operations
The main commodities carried by FCA are cement from various producers in northern Minas Gerais and Bahía, chemicals from Bahía and agricultural products from various locations in Minas Gerais and Goiás. On the Belo Horizonte—Uberaba line, fertiliser from Araferíl is also transported, while in Minas Gerais steel and iron ore are also important commodities. Petroleum products are also a major source of traffic, as are containers. Increasing volumes of transit freight from CFN are handled, especially on manufactured goods such as beer, sugar and cement to centres of consumption in central southern Brazil.

Passenger operations
FCA does not operate passenger services but the company shares tracks with SuperVía (formerly Flumitrens) in the Rio de Janeiro and Niterói regions. Similar arrangements are in place around Salvador, while between Lourenço de Albuquerque and Maceió, FCA shares tracks with neighbouring freight operator CFN and with CBTU, which runs some suburban passenger services.

Traction and rolling stock
In 2004 FCA's fleet comprised some 400 diesel locomotives and around 10,000 wagons. All locomotives are equipped with satellite tracking equipment.

UPDATED

Ferroeste

Estrada de Ferro Paraná Oeste SA
Avenida Iguaçu 420, 7° Andar, Curitiba PR, CEP 80230-902
Tel: (+55 41) 321 31 51
Fax: (+55 41) 233 21 47
e-mail: ferroest@pr.gov.pr
Web: www.pr.gov.br/ferroeste

Key personnel
President Director: Martin Roeder
Technical Director: Leopoldo de Castro Campos
Finance, Administration and Juridical Director:
  Samuel Gomes dos Santos

Gauge: 1,000 mm
Route length: 248 km

Political background
Ferroeste is owned by the government of the state of Paraná, one of the most prosperous states in Brazil due to its highly developed agriculture and growing agro-industrial activity. The railway runs through an area with substantial agricultural production requiring efficient transport of both produce and materials such as fertiliser.

The first phase of the line, the Guarapuava–Cascavel section, required an investment of US$350 million. This enabled commercial operations to begin in the first quarter of 1996. However, at the end of 1996, a 30-year concession (with a possible 30-year extension) was awarded to the Ferropar (Ferrovia Paraná SA) consortium comprising Brazilian companies Gemon, FAO and Pound SA. Ferropar began operations on 1 March 1997. America Latina Logistica do Brasil S/A (ALL) is also a shareholder in Ferropar.

Freight traffic
In 2004 Ferropar carried 1.3 million tonnes of freight. Grain predominates, with the state of Paraná responsible for some 25 per cent of Brazil's production, but soya beans, bran, wheat, corn, cement and fertilisers are also carried in significant quantities, with much of the agricultural produce destined for the port of Paranaguá.

New lines
As well as controlling the performance of its subcontracted concessionaire, Ferroeste is conducting studies and undertaking projects related to the railway's extension to Foz do Iguaçu and Guaíra, a development considered necessary to integrate Brazil into the continent's rail transport market and one greatly encouraged by Mercosul.

The 171 km link from Cascavel to Foz do Iguaçu will be started as soon as funding of US$130 million can be raised from private sector sources, feasibility studies having determined the level of state grants necessary to render the scheme attractive to investors. Additional investment will be need to fund the fleet of 70 locomotives and 1,052 wagons needed to operate the line. Once operational, the Cascavel-Foz do Iguaçu line will have capacity to handle 4.52 million tonnes annually.

Expansion of the network to Guaíra (169 km), close to the border with Paraguay, is seen as a way of funnelling the agricultural output for export from the neighbouring parts of Paraguay and the Brazilian states of Mato Grosso and Mato Grosso Sul via Ferroeste to the port of Paranaguá.

Traction and rolling stock
In 2005 operations on the Ferroeste system were handled by four General Motors GT22CUM-2

diesel-electric locomotives and 20 wagons owned by Ferropar and six similar locomotives and 120 wagons owned by ALL.

Ferronorte

Gauge: 1,600 mm
Route length: 430 km; additional 608 km under construction or planned

Organisation

A soya farming magnate in the states of Mato Grosso and Goiás who also fronts one of the country's major banks, Olacyr Francisco de Moraes, formed a company, Ferronorte, to build a line from the Ferronorte system at Santa Fé do Sul, northwest of São Paulo, into Mato Grosso. The new railway would eventually be 4,000 km long and two-pronged, forking at Cuiabá (Mato Grosso) into lines heading northwest to Porto Velho and north to Santarém, on the Amazon. In early 1995 Ferronorte's founder announced further plans to construct a railway from Santarém to the western extremity of the Carajás Railway (EFC). The distance from Santa Fé do Sul to Cuiabá will be 1,038 km. En route, at Alto Araguaia, a 550 km branch will run to Uberlandia, where there are soya processing plants.

Track

The Ferroeste system employs 45 kg/m rail on a mixture of concrete (main line) and wooden (yards) sleepers. Minimum curve radius is 250 m.

Services on an initial 310 km from Santa Fé do Sul to Aparecida do Taboado commenced in May 1999, followed by the opening in August 1999 of the 120 km Chapado do Sul–Alto Taquari. During 1999, work commenced on the 390 km section from Alto Taquari to Rodonópolis. It was anticipated that this would be completed in 2001, before that year's harvest. By this time, Ferronorte expects to be carrying between 10 and 12 million tonnes annually.

Since 1998, ownership of Ferronorte has been held by the Ferronorte Participações SA (Ferropasa) consortium, in which the Itamarati group, whose president is Mr de Moraes, owns 16.05 per cent. Ferropasa also holds a 36 per cent share of Ferroban (qv). In 1999, Ferropasa was awarded a grant of Cr200 million towards the Cr850 million construction cost of the Alto Taquari–Rodonópolis section by the Amazonia Investment Fund (Finam), a sum that by law had to be matched by Ferroban.

Ferronorte's purpose is the movement of rice, soya and grain for export to the Atlantic ports of Santos (near São Paulo) and Rio de Janeiro. Soya production in particular is expanding fast in Mato

Maximum gradient is 1.5 per cent for export traffic and 1.8 per cent for import flows.

UPDATED

Grosso and at present its only outlet is by road transport, which is markedly more expensive than rail for bulk movement.

Ferronorte is one of four railways which in 2000 jointly took over the 200 km network in the port of Santos from port authority Codesp. Ferroban, Ferrovia Novoeste and MRS Logística are also participants in a company established to replace Codesp's own rail operations.

Traction and rolling stock

In 1999, leasing arrangements were announced covering the acquisition of 50 GE Dash 9–44CW diesel-electric locomotives.

In December 1998, Trinity Industries Inc, based in Dallas, announced that its Brazilian subsidiary, Trinity Rail do Brasil, had entered a consortium with Companhia Comércio e Construcoes to build 300 grain hopper wagons for Ferronorte. Production commenced early in 1999. Additional wagons were ordered from Johnstown to provide a total initial fleet of 600.

Ferrovías Bandeirantes SA (Ferroban)

Praça Marechal Floriano Peixoto s/n°, Centro, CEP 13013-120 Campinas, São Paulo

Key personnel

President: José Carlos Nunes Marreco
Director, Administration & Finance: Sérigo Suney Gabizo
Director, Operations and Infrastructure: Joao Gouveia Ferrao Neto
Managers
  Permanent Way: Alvaro Delmont
  Mechanical Department: Joao Carlos Novaes
  Commercial: Pedro A Cutini
  Control: Pedro F Theberge
  Finance: Floriano P da Costa Neto
  Human Resources: José Homero B Elias
  Information Technology: Joarez Casagrande
  Contracts: Fernando Soria Henriquez

Gauge: 1,600 mm; 1,000 mm; dual 1,600/1,000 mm
Route length: 1,491 km; 2,517 km; 336 km (total route length 4,855 km)
Electrification: 463 km of 1,600 mm gauge, 581 km of 1,000 mm gauge and 78 km of dual gauge, all at 3 kV DC

Political background

After several attempts to privatise the former São Paulo State Railways (FEPASA) network, the federal government decided early in 1998 to incorporate it into the residual Brazilian Federal Railways (RFFSA) to speed up its transfer to the private sector. This move allowed São Paulo state to reduce its debt to the federal government.

RFFSA managed and operated FEPASA's freight and passenger operations for nearly 12 months under the name Malha Paulista. On 10 November 1998, Malha Paulista was sold at the Rio de Janeiro stock exchange to the Ferrovías consortium, led by Ferropasa Participações, formed by the neighbouring Ferrovia Novoeste (qv) and Ferronorte (qv) railways, and including Companhia do Rio Doce and several banks and investment companies. Ferrovías took over operations under the name Ferrovías Bandeirantes SA on 1 January 1999, having paid Cr245 million for the network.

Ferrovia Sul Atlántico/América Latina Logística has acquired track access on most metre-gauge lines in the São Paulo–Presidente Epitácio corridor. Ferronorte had trackage rights to run over the entire line from Santa Fé to the port of Santos. Similar trackage rights are expected to be granted

FEPASA freight at Campinas headed by four General Electric U20C locomotives (F F Lage)   0010790

Diesel-electric locomotives

Class	Wheel arrangement	Power kW	Speed km/h	Weight tonnes	No in service	First built	Builders Mechanical	Builders Engine	Builders Transmission
3100	Co+Co	447	80	64	8	1948	GE	CB	GE
3200	Bo-Bo	894	138	71.2	13	1957	GE	CB	GE
3500	Co-Co	671	95	68.1	7	1957	GE	Alco	GE
3600	Bo-Bo	652	100	56.7	13	1961	GM	GM	GM
3600	Bo-Bo	652	100	60.5	14	1960	GM	GM	GM
3650	Bo-Bo	976	100	74.9	21	1957	GM	GM	GM
3700	Bo-Bo	574	90	70	16	1969	LEW	SACM	LEW
3750	Bo-Bo	835	100	74	9	1968	LEW	SACM	LEW
3800	Co-Co	1,491	103	108	106	1974	GE	GE	GE
7000[1]	Bo-Bo	1,304	105	110.6	16	1958	GM	GM	GM
7050[1]	Bo-Bo	976	124	80	17	1958	GM	GM	GM
7760[1]	Bo-Bo	574	90	74	25	1967	LEW	SACM	LEW
7800[1]	Co-Co	1,491	103	108	26	1977	GE	GE	GE

[1] Broad gauge.

Electric locomotives

Class	Wheel arrangement	Output kW	Weight tonnes	No in service	First built	Builders Mechanical	Builders Electrical
2000	1-Co+Co-1	1,729	130	22	1943	GE	GE
2050	1-Co+Co-1	1,729	108	18	1943	Westinghouse	Westinghouse
2100	Bo-Bo	1,371	72.7	30	1968	GE	GE
6100[1]	2-Co+Co-2	2,846	122	19	1982	Westinghouse	rebuilt FEPASA
6150[1]	2-Co+Co-2	2,846	122	8	1982	Westinghouse	rebuilt FEPASA
6350	Co-Co	3,269	144	10	1967	GE Brasil	GE Brasil
6370[1]	2-Co+Co-2	2,846	165	3	1940	GE	GE
6450[1]	2-Do+Do-2	3,470	184	5	1951	GE	GE
EC362	Bo-Bo	2,480	98	2	1984	50 c/s Gp	50 c/s Gp

[1] Broad gauge.

to Novoeste as soon as the Baurú-Santos metre-gauge corridor is upgraded.

Ferroban is one of four railways which in 2000 jointly took over the 200 km network in the port of Santos from port authority Codesp. Ferronorte, Ferrovia Novoeste and MRS Logística are also participants in a company established to replace Codesp's own rail operations.

Finance

In 1999, Ferroban invested Cr75 million in an emergency programme to rehabilitate infrastructure and repair rolling stock, including some expenditure on improvements to workshops. In 2000, Cr156 million is to be spent on the purchase of new locomotives, the refurbishment of existing machines and the overhaul of freight wagons.

In the same year, an additional Cr65 million was to be spent on rehabilitation of two lines: Sante Fé do Sul–São Paulo and Campinas-Santos. In its first five years, Ferroban plans to invest Cr500 million in improvements to infrastructure and rolling stock.

Passenger operations

Ferroban took over some long-distance services from RFFSA but all were suspended in February 1999 after safety concerns were voiced by the São Paulo state authorities. Until then, these services were self–supporting. Some services were reinstated in August 1999 but only as a short-term step to fulfil contractual obligations. These included services three times weekly on Campinas-Panorama, Itirapina–São José do Rio Preto and Sorocaba–Apiaí routes, all without first class accommodation or catering vehicles. Four months later, these services were reduced to twice-weekly and further slowed, with very poor punctuality. With rarely more than 50 per cent occupancy during the peak travel period of January and February, these services look increasingly unlikely to undergo projected privatisation.

Freight operations

In 1998, Malha Paulista, then under RFFSA control, carried 9.7 million tonnes of freight, well under the figure of 13.8 million tonnes initially claimed early in 1999. In 1999, a slight upturn was achieved with 10.28 million tonnes carried, but an additional 12.8 million tonnes offered to Ferroban could not be carried due to a lack of locomotives and rolling stock. Loadings of around 15 million tonnes were thought likely for 2000.

Intermodal operations

Despite several past studies having been undertaken into possible container services to and from the port of Santos, intermodal operations have not been developed on the former FEPASA Malha Paulista system. Cnaga, a private-sector transport operator, has been studying the possibility of using the Ferroban network to move road-rail trailers from a terminal at Boa Vista, near Campinas, to Pelotas, in Rio Grande do Sul state. Paper would be conveyed northbound and rice southbound.

FSA/ALL has reinstated a weekly São Paulo–Uruguayana–Buenos Aires container service, which was first run by RFFSA prior to privatisation, and the same company is also undertaking trials with road-railers in this corridor. Neighbouring Ferronorte and Novoeste also plan to convey containers over Ferroban tracks.

New lines

Late in 1998, the connection between Santa Fé do Sul and the new Ferronorte railway was put into operation, completing a project involving a new 2,600 m double-deck road bridge over the River Paraná between Rubinéia and Santa Fé do Sul.

Improvements to existing lines

A priority for Ferroban is to improve access to the port of Santos for the neighbouring systems Ferronorte and Novoeste, both of which hold shares in Ferroban. Consequently, almost all infrastructure investment is made on the broad–gauge Campinas–Itirapina–Santa Fé do Sul and metre-gauge Mairink–Botucatú–Baurú lines, including the Campinas–Mairink–Santos line.

Three GE-built metre-gauge electric locomotives on FEPASA freight service at Mayrink (Günter Sieg)   0023713

On the broad-gauge corridor, some 200 locations requiring urgent repair or improvements have been identified. Axleloads are to be increased to 30 tonnes and crossing points will be lengthened to 1,500 m. In 1999, 200,000 sleepers were replaced on the Itirapina–Santa Fé do Sul line. On the Mairink-Santos section, additional tracks are being provided at the station yards at Embú Guaçú, Aldeinha and Caucaia and the marshalling yards at Paratinga and Perequê were to be remodelled in 2000. When completed in 2002, these improvements should allow Ferroban to run trains of 6,000 tonnes gross.

After these improvements, a key remaining problem will be the over-utilisation of the port of Santos, although privatisation of this facility was underway in 2000. Also to be resolved is the high rate of accidents on Ferroban lines, with an increase in trespass and the violation of level crossing regulations since the virtual disappearance of fast passenger trains. Some local authorities have started to play a role in attempting to cut the number of such incidents.

An additional problem facing Ferroban is the poor condition of some branch lines, resulting in a reduction or even suspension of services. Such measures have resulted in pressure from customers to make the necessary investments to improve or restore such services.

Traction and rolling stock

At the beginning of 2000, Ferroban owned 84 diesel and 45 electric locomotives for its broad-gauge network and 207 diesel and 72 electric locomotives for metre gauge. The fleet also contained four broad-gauge diesel railcars and two demotorised units on metre gauge, and 7,551 broad-gauge and 3,317 metre-gauge wagons. Of the combined wagon fleet of 10,868, some 7,900 were serviceable. The combined passenger vehicle fleet for both gauges comprised 252 coaches, including 46 recently refurbished. Around 150 were reported to need urgent repairs.

The average age of the diesel locomotive fleet was 39 years, while that of Ferroban electric locomotives was over 40 years. Some electric locomotives have been withdrawn because of their age and the planned de-electrification of the

network. Locomotive availability at the end of 1999 stood at 52 per cent.

In 2000, 60 GE Class 3800 U20 locomotives were due to be overhauled and re-engined. It was also reported that 20 new metre-gauge locomotives could be acquired, possibly from China. Also planned was the rebuilding of 550 unserviceable freight wagons by private contractors and the overhaul of 400 to 500 operational vehicles in Ferroban's own workshops.

Signalling and telecommunications

A modern telecommunications system has been installed across much of the network in conjunction with Embratel.

A new operations control centre at Campinas was to be commissioned in 2000, with a capability to track every train and freight wagon from loading to unloading points. Train despatching was being upgraded by Engesis to enable GPS tracking via Omnisat to be implemented. New end-of-train devices have been installed by Linksat on 130 locomotives.

Track

Rail
Type: TR 37, TR 45, TR 50, TR 55, TR 57, TR 68
Weight: 37, 45, 50, 55, 57, 68 kg/m
Crossties (sleepers)
Wood: 1,000 mm gauge 2,000 × 220 × 160 mm; 1,600 mm gauge 2,800 × 240 × 170 mm
Spacing: 1,000 mm gauge 1,600/km; 1,600 mm gauge 1,667/km
Rail fastenings: GEO or K; ML
Concrete block: 1,000 mm gauge 680 × 290 × 211 mm; 1,600 mm gauge 680 × 290 × 239 mm
Spacing: 1,500/km
Fastenings: FN
Concrete (monobloc): (1,000 mm gauge only) 2,000 × 220 × 210 mm to 2,000 × 320 × 242 mm
Spacing: 1,500/km
Fastenings: RN
Min curve radius: Main lines 150 m; branches 90 m
Max gradients: Main lines 2%; branches 3%
Max axleload: 1,000 mm gauge 20 tonnes; 1,600 mm gauge 25 tonnes

---

Ferrovía Norte-Sul (North-South Railway)

Operated by Companhia Vale do Rio Doce (CVRD)

Gauge: 1,000 mm
Route length: 226 km (additional 1,974 km planned)

Organisation

The Brazilian government projected this line during the 1980s to create a new transport corridor from central Brazil to the north coast at São Luiz and to Belém on the mouth of the Amazon. When complete, the line will follow an existing highway and the Tocantins river to run from Goiânia, in

Goiás state, via Miracema do Norte and Estreito, in Tocantins state, Porto Franco and Imperatriz to Açailândia (Maranhão state), and then north to Belém.

The project is divided into three phases. The first brings the line from an interchange with the Carajás Railway (EFC) at Açailândia into Tocantins state; the second foresees the construction of a line from Senador Canedo, near Goiânia; and the third covers the 600 km line from Açailândia to Belém.

In 1989, the first 106 km from Açailândia to Imperatriz were put into service, and EFC operates and maintains this section. Work on the section south of Imperatriz began in 1989

using labour from the Brazilian army, but this ended after a few months for financial reasons. In 1994 and again in 1996 some earthworks were undertaken by the army, but a major obstacle has been the need for a 1,300 m bridge over the Tocantins river at the border between the states of Maranhão and Tocantins. In 1999, Valec, the state concessionaire for the line, recommenced work between Imperatriz and Estreito using Cr100.4 million of public funds. Opening of this section took place on 29 October 1999. Completion of the entire 2,200 km line, which will traverse some 1.8 million km² of fertile agricultural land, could require an investment estimated at US$1.6 billion and take five years. In 2000,

Cr49 million was budgeted for infrastructure work on the scheme.

Privatisation of the Norte-Sul project was under discussion in 2000 because the Federal government foresees little prospect of funding completion of the project with public funds.

Passenger operations
A passenger service is run three times weekly between Açailândia and Estreito, providing a connection to and from São Luiz via EFC's broad-gauge service.

Freight operations
Freight consists mainly of agricultural products grown in the Imperatriz area, including soya beans and cereals. Fertiliser, general freight and fuel are also carried. In 1999, around 2.2 million tonnes were carried, 18 per cent more than the previous year. From 2000, a new fertiliser plant at Imperatriz was expected to generate an additional 300,000 tonnes of traffic annually. Once the entire line is completed, it is expected to handle around 14 million tonnes of freight annually, including soya, cereals, cellulose and timber. It will also improve the movement of freight to Belém and other Amazonian destinations.

All freight has to be transhipped to EFC's broad-gauge trains at Açailândia.

Traction and rolling stock
All rolling stock is owned by CVRD, which transferred equipment from its Vitória a Minas Railway.

Track
Sleepers: wood (Açailândia-Imperatriz); concrete (Imperatriz-Estreito)
Ballast: crushed stone
Max gradient: 1.5%
Min curve radius: 385 m

Ferrovia Novoeste (New West Railway)

Key personnel
Director: Glen Michael
Transport Director: Sergio Julian Cardoso
Administration and Finance: Homero Boretti Elias
Logistics and Permanant Way: Edmundo Dias do Amaral
Marketing: Ricardo Lopes
Engineering: Melvin Jones

Gauge: 1,000 mm
Route length: 1,600 km

Political background
The privatisation of the RFFSA network as seven 30-year regional freight operating concessions (including FEPASA) began in earnest in 1996. The concession for the route from Bauru to Corumbá and its branch to Ponta Porã (known as the Western Network) was auctioned at the Rio de Janeiro stock exchange on 5 March. The concession was acquired for Cr62.36 million (US$63.4 million) by a consortium of Brazilian and US investors, led by the Noel Group (owners of Illinois Central) and including Chemical Bank, Bank of America, Brazil Railway Partners and Western Rail Investors. Operations under the new ownership began in June 1996.

The concession for the Western Network is renewable for a further 30 years, with the concessionaire leasing infrastructure and rolling stock from RFFSA. As of March 1996, the Western Network's rolling stock fleet comprised 88 diesel-electric locomotives and 2,600 freight wagons, with half of the locomotives and some 7 per cent of the wagons out of service for want of maintenance and spare parts. Under the terms of the concession, the government requires some Cr359 million to be invested in the Western Network, with rolling stock, track maintenance and the upgrading of structures and communications regarded as priorities.

Ferrovia Novoeste is one of four railways which in 2000 jointly took over the 200 km network in the port of Santos from port authority Codesp. Ferroban, Ferronorte, and MRS Logística are also participants in a company established to replace Codesp's own rail operations.

Traffic (million)	1997	1998
Freight tonnes	2.54	2.88
Freight tonne-km	1,490	1,579

Freight operations
In 1998 Ferrovia Novoeste carried 2.88 million tonnes for 1.58 billion tonne-km, the principal commodities carried include petroleum products, ores, grain and fertiliser and manufactured products bound for Bolivia via Corumbá.

Ferrovia Sul-Atlantico (South Atlantic Railway)

Ferrovia Sul-Atlantico (FSA)

Key personnel
President: José Paulo Oliveira Alves

Gauge: 1,000 mm
Route length: 6,349 km

Political background
In December 1996, RFFSA's (qv) Southern Network was sold for US$208 million, 37 per cent higher than the reserve price, to a consortium of Railtex and Ralph Partners of the US and Banco Garantia, Judore and Interferrea of Brazil. The consortium took over operations on 1 March 1997. Cr300 million was to be invested within the first two years of the concession period, aimed at boosting revenue by 50 per cent. Some 100 locomotives were to be rebuilt in this period. Oil traffic would account for 45 per cent of the total hauled.

The new railway operates in the states of Paraná, Santa Catarina and Rio Grande do Sul and carries 50 per cent of all RFFSA traffic.

Traffic (million)	1997	1998
Freight tonnes	11.7	15.58
Freight tonne-km	6,250	8,534

Ferrovia Tereza Cristina (FTC)

Ferrovia Tereza Cristina SA
Rua dos Ferroviários 100, Oficinas, Tubarão, SC
Tel: (+55 48) 626 47 77
Fax: (+55 48) 626 43 25
e-mail: ftc@matrix.com.br

Key personnel
General Manager: Benony Schmitz Filho

Gauge: 1,000 mm
Route length: 164 km

Political background
A concession to operate RFFSA's former Tuberão division, in Santa Catarina province in the extreme south of Brazil, was awarded in January 1997 to a consortium of Gemon, Interfinance SA Participações and Santa Lúcia Agro-Indústria e Comércio. The isolated metre-gauge network serves a coal-producing region, with access to the port of Imbituba.

Freight operations
FTC mainly conveys coal from deposits near Siderópolis. Most is supplied to Gerasul, an energy producer. Agricultural products and containerised traffic are also handled. Tonnages carried in 1999 increased by nearly 30 per cent compared with the previous year.

Traffic (million)	1997	1998	1999
Freight tonnes	2,070	2,255	2,900
Freight tonne-km	149	166	230

Improvements to existing lines
In 1999, FTC invested Cr4 million in infrastructure and rolling stock. Work included completion of resleepering 82 km of track under a programme started in 1998 and the construction of a 2 km branch from Siderópolis to coal mines at Rio Deserto and Beluno. The Cr700,000 cost of this project is being borne jointly by FTC and the coal producers.

Traction and rolling stock
FTC owns 10 diesel locomotives, all of which were operational in 2000. Freight wagons consist mainly of coal hoppers, together with general freight vehicles and flatcars for container transport. The railway owns some steam locomotives and passenger coaches, but none of these were in service in 2000. In April 2000, Santa Catarina state began to develop plans for tourist services using this stock.

Metrofor

Autarquia de Região Metropolitana de Fortaleza
Rua Jose Laurenço, Aldeota, Fortaleza 60000, Ceará
Tel: (+55 85) 212 40 34

Gauge: 1,000 mm
Route length: 45 km

Political background
In 1998, Ceará state took over responsibility for the former CBTU suburban network in Fortaleza. This development unlocked investments of US$511 million, including a US$268 million loan from Japan's Eximbank which had been frozen since 1992. The state of Ceará is also making a contribution of US$58 million.

Organisation
The network consists of two separate metre-gauge lines. The southern line runs from the city's João Felipe main station via Paragaba, Mondubim, Aracapé and Maracanaú to Vila das Flores (25 km). The western line runs from João Felipe via Alvaro Wayne and Antônio Bezerra to Caucáia (20 km).

Passenger operations
Both lines are served by diesel locomotive-hauled trains operating at regular intervals on the southern line and irregularly on the western line. In 2000, Metrofor was carrying some 30,000 passengers daily. Fares covered only 18 per cent of operating costs.

Improvements to existing lines
The US$511 million mentioned above is to be used to boost the capacity of the Metrofor system to 185,000 passengers daily. In a first phase, the

southern line is to be remodelled and electrified for operation by emus. Existing stations are to be rebuilt and new ones created. A new maintenance facility is to be built at Vila das Flores. Rail access to João Felipe station will be taken underground, eliminating the use of busy level crossings in the city centre. Also in the first phase, a new line will be built from Aracapé to Caucáia to divert freight trains from the city centre (see entry for Companhia Ferroviária do Nordeste).

The western line will be remodelled but not electrified. Services to Caucáia will be provided by refurbished diesel trains operating at a 20 min frequency. Existing stations will receive a facelift and some new ones are to be built.

In a later phase, freight lines from Vila das Flores to Parangaba, junction of the line from the port of Mucuripé, will be separated from Metrofor's suburban line. This phase also foresees the construction of a new 6 km branch from Jereissati

to Maranguape, replacing a long-disused line from Maracanaú.

Traction and rolling stock
Services are provided by a fleet of six diesel locomotives and 45 coaches. Under plans to electrify the southern line, 10 four-car emus are to be acquired.

Metrorec

Trem Metropolitano de Recife
Rua José Natario 478, 50900-000 Recife
Tel: (+55 81) 251 09 33   Fax: (+55 81) 251 48 44

Gauge: 1,600 mm; 1,000 mm
Route length: 22 km; 32 km
Electrification: 22 km at 3 kV DC

Organisation
Formerly operated by CBTU (qv), Metrorec operates a suburban passenger network in and around Recife. The majority shareholding is by Pernambuco state.

Passenger operations
Metrorec operates an electrified 1,600 mm gauge double-track line from Recife to Ipiranga, Coqueiral and Jabatao (17 km) and from Coqueiral to Terminal

Intermodal de Passageiros (TIP) (5 km). There are 17 stations. Also operated is a 32 km diesel-operated metre-gauge line from Recife south to Cajueiro Seco and Cabo. Together, the two systems carry around 120,000 passengers daily, the broad-gauge lines accounting for around three-quarters of these.

A second ex-CBTU state-owned operator, Companhía Pernumbaca de Trens (CPT), provides one weekday return service on the metre-gauge line between Recife and Ribeirao via Cabo (88 km) as well as occasional long-distance trains between Recife and Caruarú (146 km) via CFN's Salgueiro line. It was expected that CBT would take over diesel-operated services on the line to Cabo once electrification of the Recife–Cajueiro Seco section was complete (see below).

Improvements to existing lines
Early in 1998, upgrading commenced of the metre-gauge line, including regauging and electrifying

the 21 km Recife-Cajueiro Seco section. Eleven new stations are to be provided. The scheme includes provision of a third metre-gauge track alongside the newly converted section to enable diesel-hauled passenger and freight trains to continue to reach Recife Central station. On the original electrified network, a 4.5 km extension is planned from TIP to Timbi. Funded jointly by the World Bank and the Brazilian federal government, the US$204 million project was scheduled for completion by the end of 2001.

Traction and rolling stock
On broad-gauge electrified lines, services are provided by 25 four-car emus supplied in 1984 by Brazilian manufacturer San Matilde. Metre-gauge rolling stock comprises seven diesel locomotives and 41 coaches.

MRS Logística

MRS Logística S/A

Headquarters
Praia de Botafogo, 228/sala 1201-E, Botafogo 22359-900, Rio de Janeiro–RJ
Tel: (+55 21) 551 14 50   Fax: (+55 21) 552 26 35
e-mail: cgi@mrs.com.br
Web: http://www.mrs.com.br

Operations and administration
Avenida Brasil, 2001, Centro 36060-010, Juiz de Fora–MG
Tel: (+55 32) 239 26 00   Fax: (+55 32) 239 36 09

Key personnel
Chief Executive Officer: Mauro R F Knudsen
Chief Financial Officer: Alberto Régis Távora
Development Director: Henrique Aché Pilar
Production Director: Rinaldo Bastos Vieira Filho

Gauge: 1,600 mm
Route length: 1,674 km
Electrification: 8 km at 3 kV DC

Political background
MRS was established in August 1996 by the MRS Consortium, which was formed with the aim of acquiring from the Brazilian government a concession to operate RFFSA's (qv) 1,674.1 km South-Eastern Network (the former Regions 3 and 4). MRS' main shareholders are its main customers: MBR and Ferteco Mineração, and four steel-producing companies, Companhia Siderúrgica Nacional (CSN), Usinas Siderúrgicas de Minas Gerais (Usiminas), Companhia Siderúrgica Paulista (Cosipa) and Gerdau S/A Siderúrgica. Other shareholders include Ultrafértil, ABS Empreendimentos Imobiliáros, Participações e Serviços, Celto Intergração Multimodal. In addition, former RFFSA employees were granted rights to subscribe up to ten per ecnt of the company's shares.

In November 1996 MRS was granted a 30-year concession by the federal government to be the exclusive provider of freight transport on the network and was also granted a lease by RFFSA for the same term to use its operational assets. The consortium paid the auction price of US$870 million for the concession and on 1 December 1996, now in the form of a corporation named MRS Logística S/A, took over operations.

Diesel locomotives

Class	Wheel arrangement	Power kW	Speed km/h	Weight tonnes	No in service	First built	Builders Mechanical	Builders Engine	Transmission
SD18	Co-Co	1,340	104	163	16	1961	GM	GM 567	E GM
SD38	Co-Co	1,493	97	163	34	1967	GM	GM 645E	E GM
SD40-2M	Co-Co	2,238	104	180	38	1980	GM	GM 645E3	E GM
U20C	Co-Co	1,493	104	120	23	1981	GE	GE 7FDL 12	E GE
U23C	Co-Co	1,680	112	180	76	1975	GE	GE 7FDL 12	E GE
U23C1	Co-Co	1,680	112	165	13	1975	GE	GE 7FDL 12	E GE
U23CA	Co-Co	1,940	112	180	27	1987	GE	GE 7FDL 12	E GE
U23CE	Co-Co	1,680	112	180	16	1995	GE	GE 7FDL 12	E GE
720	Bo-Bo	537	90	80	17	1956	GE	Alco 251 6K	E GE
RSD12	Co-Co	1,493	104	163	2	1986	Alco	Alco 251	E GE
U5B	Bo-Bo	448	64	51	9	1961	GE	Caterpillar D379	E GE
U6B	Bo-Bo	522	64	53	7	1967	GE	Caterpillar D379	E GE
EFCB	Bo	201	30	20	3	1969	EFCB	Detroit Diesel 401	E Leece Leville
RS3	Bo-Bo	1,194	104	109	3	1952	Alco	Alco 251	E GE
Hitachi	Bo-Bo	746	25	115	2	1980	Hitachi	Alco 251B	E Hitachi

Electric locomotives

Class	Wheel arrangement	Power kW continuous	Speed km/h	Weight tonnes	No built	First built	Builders Mechanical	Builders Electrical
Hitachi	B-B	2,460	45	118	9	1980	Hitachi	Hitachi

At the end of 1998, the company employed 3,299 staff.

Finance
Revenue (Cr million)	1997	1998
Freight	397.8	444.2
Total	397.8	444.2

Expenditure	1997	1998
Staff/personnel	147.0	116.8
Materials and services	132.0	140.3
Depreciation	3.9	13.1
Asset/leasing	54.9	63.0
Total	337.8	333.2

Traffic (million)	1997	1998
Freight tonnes	51.0	52.8
Freight tonne-km	20,400	21,400

Freight operations
The MRS network is located in the most developed region of Brazil, which accounts for some 65 percent of the country's GDP. Its lines provide the most direct transport link between the iron ore producing region in Minas Gerais state and the Atlantic ports of Guaíba, Rio de Janeiro, Santos and Sepetiba.

In 1998 iron ore both for export and for the domestic market was responsible for

approximately 70 per cent of the railway's traffic, with 36.6 million tons carried, while other steel-related products accounted for another 13 per cent. Total haulage was 52.8 million tonnes for 21.4 billion tonne/km, resulting in a gross revenue of Cr444.2 million.

The MRS network includes an 8 km electrified section which incorporates the rack-operated Old Serra Incline (10.7%), for which a small fleet of specialised Hitachi-built 3 kV DC electric locomotives is retained.

New lines
Two of MRS Logística's customers have built branch lines to connect their facilities to the network. BASF has built a 3 km spur to the MRS Logística line at Guaratinguetá (São Paulo state) and fertiliser and a chemicals producer is investing in similar facilities at its Santos terminal.

Traction and rolling stock
At the end of 1998 the MRS Logística traction fleet comprised 286 diesel locomotives and nine electric locomotives. The freight wagon fleet consisted of 9,305 vehicles; a further 1,699 wagons owned by the railway's clients were in use on the network. In 1999, 450 wagons for mineral traffic were ordered from Maxion (270) and T'Trans (180). Since it took

over the network in 1996, MRS Logística has given investment priority to rehabilitation and modernisation of the motive power and rolling stock fleets.

Signalling and telecommunications
CTC covers 1,290 route-km of the MRS network. Since 1996 MRS Logística has invested in improvements to its signalling and telecommunications systems and has also introduced a satellite communications system. The company has also procured a train operations simulator.

Track
Weight: 68 kg/m (1,072/km); 57 kg/m (602 km)
Sleepers
Wood: 280 × 24 × 17 cm
Spacing: 1,850/km
Concrete: twin-block
Spacing: 1,493/1,667/km
Fastenings: Pandrol clips, coach screws and RN System
Min curve radius: 300 m
Max gradient: 2%
Max axleload: 30 tonnes

MRS Logística GE-built diesel-electric locomotive on a container train at Piaçaguera 0058627

MRS Logística handles large volumes of iron ore traffic
0087936

Rede Ferroviaria Federal SA (RFFSA)

Praça Procópio Ferreira 86, 20224-900 Rio de Janeiro
Tel: (+55 21) 233 57 95   Fax: (+55 21) 263 31 28

Key personnel
President: I Popoutchi

Political background
The state-owned operator RFFSA has been privatised in the form of six 30-year freight operating concessions. By the end of 1997, all six RFFSA regions had been transferred to the private sector. Also in 1997, Ferrovia Paulista SA (FEPASA) was transferred to RFFSA allowing it to be privatised (see Ferrovías Bandeirantes SA). A concession was awarded for this system in November. RFFSA retains ownership and responsibility for the infrastructure of the railways which have been the subject of concessions. The new operators therefore have to enter into contracts for infrastructure use and maintenance with RFFSA.

Finance
Privatisation of RFFSA was prompted by the withdrawal of most former subsidies and compensations as desperate efforts were made in the early 1990s to stabilise the national economy, driving the railway into technical bankruptcy in the process. By 1991 the daily loss had soared to US$1 million and accumulated debt stood at US$1.2 billion. The latter had reached US$1.84 billion in 1994 and was expected to rise to some US$2.56 billion in 1995, when the intention was that funds generated by the auction of operating concessions would go towards settling RFFSA's accumulated debt. The enabling law was passed in 1995.

SuperVia

Praça Cristiano Ottoni, Sala 445, Rio de Janeiro RJ, CEP 20221

Gauge: 1,600 mm; 1,000 mm
Route length: 198 km; 22 km (total route length 220 km)
Electrification: All 1,600 mm gauge lines at 3 kV DC

Political background
The suburban network serving the city of Rio de Janeiro passed from the control of Companhia Brasiliera de Trens Urbanos to the state of Rio de Janeiro in December 1994. Services were subsequently operated by Flumitrens. In 1996 it was decided to initiate privatisation and in November 1998 the Supervia consortium, comprising Bolsa 2000, CAF and RENFE, took over operations. The 25-year concession covers operation of the entire 1,600 mm gauge network and a short section of metre-gauge. Flumitrems (now Central) continues to operate two other metre-gauge lines that were not part of the concession.

A three-year programme to upgrade the badly rundown network commenced in 1999 with a loan from the Inter-American Development Bank. A total of Cr324 million has been invested in rehabilitation of the infrastructure and rolling stock. By late 2004 121 of the 144 trainsets had been modernised, 31 km of track rebuilt, 274,000 new sleepers laid and 52 stations refurbished. Some 80 per cent of the 992 track-km of overhead line has been replaced or retensioned, and the signalling rehabilitated where corrosion and vandalism had taken their toll. Poor drainage meant that train operation often had to be suspended during periods of heavy rain. This has been rectified by extensive excavations, clearance of drains and construction of new soakaway channels.

In 2004 SuperVia employed 2,267 staff.

Passenger operations
SuperVia operates the largest suburban system in Brazil, with electrified 1,600 mm gauge routes running from Central terminal (formerly

Flumitrens electric multiple-units

Class	Cars per unit	Motor cars per unit	Motored axles/car	Output/motor kW	Speed km/h	No in service	First built	Builders Mechanical	Electrical
400M	3	1	4	315	90	–	1964	FNV/Cobrasma	Hitachi/Toshiba
400	3	1	4	255	90	–	1964	FNV/Cobrasma	GE
500	4	2	4	315	90	–	1977	Nippon Sharyo	Hitachi/Toshiba
700	4	2	4	315	90	–	1980	Mafersa	Hitachi/Toshiba
800	4	2	4	280	90	–	1980	Santa Matilde	GE
900	4	2	4	279	90	–	1980	Cobrasma	MTE
1000	3	1	4	315	90	–	1954	Metro-Vick	Hitachi/Toshiba/Villares

Dom Pedro II) to Santa Cruz, Paracambi, Belford Roxo and Saracurana. The 1,000 mm portion links Saracurana with Guapimirim. There are 89 stations.

In 1997 Flumitrens had recorded 71 million passenger journeys but by 2003 SuperVia had managed to raise the figure to 114 million. Average daily ridership rose from 150,000 to 380,000 during the same period. Most routes have a 30-minute interval service off-peak.

Traction and rolling stock
At the end of 2004 the 1,600 mm gauge fleet comprised 540 emu cars formed into 144 sets of three or four cars. Out of this fleet 135 trains are required for daily service. The 1,000 mm gauge fleet comprised 20 diesel-electric locomotives and 53 passenger coaches.

Twelve Class 900 emus were refurbished, with IGBT chopper equipment supplied by Adtranz Switzerland replacing the original camshaft control

system. Further modernisation and refurbishment has been carried out by SuperVia, with only 23 sets still to be tackled at the start of 2005. With the newest cars nearly 25 years old, a start has been made on renewing the fleet. In March 2004 an order was placed with a Mitsui/Rotem consortium for 20 four-car air-conditioned emus to be delivered starting in November 2005.

Trensurb

Empresa de Trens Urbanos de Porto Alegre SA
Rua Ernesto Neugebauer 1985, Bairro Humaitá, Porto Alegre RS, CEP 90250-140
Tel: (+55 51) 33 71 50 00   Fax: (+55 51) 33 71 51 66
e-mail: seapobib@trensurb.com.br
Web: http://www.trensurb.com.br

Key personnel
Director-President: Marco Arildo Prates da Cunha
Director, Administration and Finance:
    Paulo Roberto Cardoso Thimóteo
Director, Operations: Luís Carlos De Cesaro
Expansion and Development Superintendent:
    Humberto Kasper

Gauge: 1,600 mm
Route length: 34.5 km
Electrification: 34.5 km at 3 kV DC

Political background
Trensurb was founded in 1980 to develop and operate a high-capacity regional suburban rail system northwards from the existing RFFSA line in central Porto Alegre to Novo Hamburgo, an important development corridor. Its principal shareholder is the Brazilian federal government (96 per cent), with remaining shares held by the Rio Grande do Sul state government and Porto Alegre city council. Transfer to local state control has been discussed. This would require the federal government to assume responsibility for the line's accumulated debt.

In December 2003 the three levels of government, federal, state and municipal, agreed to study an integrated network of medium- to high-capacity public transport in the Porto Alegre Metropolitan Region in which Trensurb would be a participant.

Organisation
Trensurb is managed on behalf of the federal government by an executive directorate comprising: the Presidency; a Directorate of Administration and Finance; and a Directorate of Operations. Staff levels stood at 1,123 in December 2003.

Trensurb train on Low Vibration Track laid on the line's elevated section   0593310

Japanese-built emu at one of Trensurb's stations   1029002

Trensurb electric multiple-units

Class	Cars per unit	Motor cars per unit	Motored axles/car	Output/ motor kW	Speed km/h	Units in service	First built	Builders Mechanical	Builders Electrical
100	4	2	4	315	90	25	1984	Nippon Sharyo/ Hitachi-Kawasaki	Japanese consortium

Passenger operations
Trensurb operates a 17-station electrified route linking São Leopoldo in the northern suburbs of Porto Alegre with the city centre at Mercado and also serving the city's airport. The line was opened from Mercado to Sapucaia do Sul in 1985, extended by 3.9 km to Unisinos in São Leopoldo in 1997 and by a further 2.4 km elevated section to São Leopoldo in 2001.

At peak periods, with 19 trains in service, services operate at 5-minute intervals under ATC control with ATS and cab signalling. Off-peak service frequency is every 10 minutes.

Traffic (million)	2001	2002	2003
Passenger journeys	39.6	41.29	44.68

New lines
In 2004 the favoured option for a second line was for a 20-station 21 km route, mostly underground, linking Azenha in the south and Fiergs (Sarandi). Also planned is a 3.1 km branch to link Aeropuerto station on Line 1 with the planned Cairú station on

Line 2 to provide access to the existing rolling stock maintenance depot.

In 2004 bids to build a 9.3 km extension to Line 1 northwards from São Leopoldo to Novo Hamburgo were being evaluated.

Traction and rolling stock
Trensurb operates a fleet of 25 four-car air-conditioned emus built by a Japanese consortium of Hitachi, Nippon Sharyo and Kawasaki.

Signalling and telecommunications
Movements are regulated using CTC from a Centralized Operational Center, commissioned in 2002.

Track
Weight: 57 kg/m welded
Sleepers: Concrete, twin-block, conventionally ballasted; elevated section from Unisinos to São Leopoldo employs LVT (Low Vibration Track)

Vitória a Minas Railway (EFVM)

Vitória-Minas Railway (EFVM)
Avenida Dante Michelini 5500, Ponta de Tubarão
PO Box 8001, Vitória 29090-900, Espírito Santo
Tel: (+55 27) 335 34 20   Fax: (+55 27) 335 33 50
Web: http://www.cvrd.com.br

Key personnel
Southern System Director: Juarez Saliva de Avelar
General Managers
    Commercial: Mauro Oliviera Dias
    Control: Sílvio Renato Ribiero Louro
    Permanent Way: Antonio José Cuzzol
    Logistics: Elías David Nigri
    Sales: Cleber Cordeiro Lucas
    Marketing: Flávio Barbosa Montenegro
    Programming: Paulo Afonso Polese
    Railway Operations: Jayme Nicolatto
    Traffic Control and Train Operations:
        Arnaldo Soares Silva
    Rolling Stock Maintenance:
        Lidemberg José Rosa Cesário

Gauge: 1,000 mm
Route length: 898 km

Organisation
EFVM is owned by the mining group Companhia Vale do Rio Doce (CVRD), the world's largest producer of iron ore, which also owns the Carajás Railway (EFC) (qv).

At the end of 1998 EFVM employed 2,846 staff.

Passenger operations
EFVM operates daily trains between Vitória and Belo Horizonte and on its Desembargador Drumond–Itabira branch line. First, second and a

superior air conditioned class are provided and main line trains also include a restaurant car. It is reported that there are plans to introduce a peak-season Vitória–Belo Horizonte overnight service.

Freight operations
The railway's principal role is to transport iron ore from mines at Itibara, east of Belo Horizonte, to Port Tubarão, Vitória. This accounts for around 78 per cent of freight traffic. Steel, steel products, coal, chalk, soya products, cellulose and cereals are

Four Macosa-built General Motors GT26CU2s at General Carneiro (F F Lage)   0010791

also carried, as are cars, containers, manufactured goods and just-in-time goods. General freight traffic amounts to around 5.3 million tonnes annually.

Traffic (million)	1997	1998	1999
Passenger journeys	1.488	1.175	1.255
Passenger-km	358.6	287.1	n/a
Freight tonnes	104.7	104.8	101.2
Freight tonne-km	56,599	55,443	52,691

New lines

Jointly with the government of Espírito Santo, EFVM plans to build a new 17.5 km line from Frechal, the first station out of Vitória on the Belo Horizonte line, to Viana, on the FCA line to Rio de Janeiro. This would carry freight trains around Vitória, easing access to its port, and could also serve as a feeder to the planned new Vitória–Cachoeira do Itapemirim line, which will follow the coast to avoid the mountainous region south of Viana, easing the flow of freight traffic from southern Espírito Santo to Rio de Janeiro. The first 10 km of the new line may be electrified to be used as the first section of a 45 km light rail system in and around Vitória.

Improvements to existing lines

Double-tracking of the 62 km Desembargador Drumond–Costa Lacerda line was expected to be completed in July 2000 at a cost of Cr115 million. Construction of the second part of the 28 km Capitao Eduardo–Costa Lacerda deviation was also nearing completion in 2000. This scheme has cost Cr143.5 million. With these improvements, most of the 705 km Vitória–Belo Horizonte line will be double-track.

Traction and rolling stock

In 2000, EFVM operated 218 diesel locomotives, the most recent additions being 10 GE Dash-9 WC units with Bo-Bo-Bo-Bo axle arrangement delivered in 1997. In 2000, an additional batch

EFVM diesel locomotives

Class	Wheel arrangement	Power kW	Speed km/h	Weight tonnes	No in service	First built	Builders Mechanical	Builders Engine	Builders Transmission
G12	Bo-Bo	1,063	60	76	7	1956	GM	GM 12-567C	E GM
G12	Bo-Bo	1,063	60	76	18	1956	GM	GM 12-645E3	E GM
G16	Co-Co	1,455	60	101	38	1962	GM	GM 16-645E3	E GM
GT-GM	Co-Co	2,240	60	121	4	1973	GM	GM 16-645E3	E GM
GT-MAC	Co-Co	2,240	60	121	28	1978	GM/Macosa	GM 16-645E3	E GM
U26C	Co-Co	2,126	60	120	1	1981	GE Brasil	GE 7FDL12	E GE Brasil
GT-VIL	Co-Co	2,462	60	138	6	1982	GM/Villares	GM 16-645E3B	E GM
MATE*	Bo-Bo+Bo-Bo	1,063	28	125	1	1987	GM	GM 12-567C	E GM
DDM-45	Do-Do	2,910	60	162	72	1989	GM	GM 20-645E3	E GM
GT-MP	Co-Co	2,462	60	138	6	1991	GM/Villares	GM 16-645E3B	E GM
Dash-8	Bo-Bo-Bo-Bo	3,085	60	160	6	1991	GE Brasil	GE 7FDL16	E GE Brasil
DDM-MP**	Do-Do	2,910	60	162	6	1993	GM	GM 20-645E3B	E GM
Dash-9	Bo-Bo-Bo-Bo	3,085	60	160	12	1996	GE	GE 7FDL16 efi	E GE
Dash-9 WC	Bo-Bo-Bo-Bo	3,085	60	160	10	1997	GE	GE 7FDL16 efi	E GE

* EFVM conversion based on two G12s.
** Rebuilt from DDM-45s with new engines, generators, traction motors and computerised instrumentation.

of 15 similar Dash-9 locomotives was on order from GE's Brazilian licensee, GEVISA. The railway's rolling stock fleet comprised 64 passenger coaches, including six restaurant cars, and 13,117 wagons.

During 2000, it was planned to equip 100 locomotives with GE Harris Locotrol remote control equipment at a cost of US$5.5 million, enabling EFVM to increase train lengths from 160 to 258 cars. This was intended to alleviate pressures on line capacity, which can average a train every 72 mins during a 24 hour period.

Signalling and telecommunications

CTC is used on 540 km of double-track main line and 185 km of branches. Train despatching is employed on a non-signalled 134 km branch line. Signal aspects are reproduced in locomotive cabs. Each signal block has the added protection of derailment detectors, supplemented by hot box and broken wheel detectors.

Braking in standard use: Locomotives, 26L (AAR); hauled stock, ABD mechanical empty/load

Track

Rail type: 136 RE 68 kg/m (2,494 km)
Sleepers
Wood: 2,300 × 240 × 170 mm
Spacing: 1,852/km in plain track and in curves
Fastenings: Denick
Steel: 2,200 × 260 × 200 mm
Spacing: 1,667/km in plain track and in curves
Fastenings: Denick
Concrete: 2,300 × 260 × 200 mm
Spacing: 1,667/km in plain track and in curves
Minimum curve radius: 12° 40' (90.47 m)
Max gradient: 2.7%
Max axleload: 25 tonnes

Other Brazilian railway projects

Many projects for new lines in Brazil await realisation. In many instances, these were developed during state ownership of Brazilian Federal Railways (RFFSA) but not taken forward due to a lack of finance. Some of the more significant projects are listed below.

Rio de Janeiro–São Paulo–Campinas high-speed line

Under a technical co-operation agreement between Brazil and Germany, a plan to build a high-speed passenger line in the Rio de Janeiro–Campinas corridor has been drawn up. The 500 km railway would serve an area which accounts for some 20 per cent of the population of Brazil. A decision on whether to build the line was expected in 2000.

Ferrovía Norte do Espírito Santo

This proposed 336 km line would run from Barra do Riacho at Aracruz, near Vitória (Espírito Santo state) to Taquarí (southern Bahía state). It would be used for transporting timber from Bahía to Aracruz for cellulose and paper production and paper and cellulose from a plant near Taquarí for export via the port of Vitória. Some 8.6 million tonnes of freight annually have been forecast. Construction cost is estimated at around US$300 million. Companhia Vale do Rio Doce (CVRD) would be the line's operator, but private investors are sought as no public funds are available for its construction.

Ferrovía Litorânea Sul

A 147 km railway has been proposed to connect Vitória and Cachoeiro do Itapemirim, both in Espírito Santo state, to create a coastal route which would replace the existing inland line. The latter is in poor condition and is unable to carry more than 400,000 tonnes of freight annually. Traffic demand between the two locations could be as high as 6 million tonnes. The cost of the line is estimated at US$113 million.

Another significant proposal is for an 18 km line around Cachoeiro do Itapemirim from Cobiça via Moro Grande to Monte Cristo, taking rail traffic out of the city centre, where street running is currently necessary. RFFSA and the city council commenced construction of this line in 1989 and two years later the central 8 km had been completed. Completion of the remaining 10 km is still required.

Variante do Paraguaçu

The existing railway from Iaçu to Salvador passes through the streets of the historic towns of São Felix and Cachoeira, with curve radii down to 60 m. A combined rail and road bridge between the two communities is in poor condition, preventing the operation of heavy trains, and steep gradients are encountered to reach towns on the river Paraguaçu. A proposed new line is intended to eliminate this bottleneck, beginning at Salvador Pinto and running south of the existing line for 76 km, compared with the current 123 km, to Candeias, also replacing the existing line between Santo Amaro and Candeias. Work commenced in 1990 and 7 km of earthworks from Salvador Pinto to Baragogipe were completed,

with 2.5 km of track laid. Some US$12 million of an estimated total cost of US$120 million have already been spent but work has been suspended until additional funds can be secured.

Ramal Pirapora–Unaí

The state government of Minas Gerais and Companhia Vale do Rio Doce (CVRD) have undertaken studies into this 285 km line, which would open up a fertile agricultural region in the northwest of the state. Traffic forecasts suggested that some 3 million tonnes of agricultural products annually could be carried by the line. State funding of US$200 million has been proposed for the US$432 million project, the remainder to come from CVRD, which would also operate the line. However, Federal government pressure in 1996 prevented Minas Gerais state from making its contribution to the scheme and construction was postponed. CVRD has also drawn up plans for connecting lines from Unaí to Luziânia, where a connection with the former RFFSA main line to Brasilia would be made. From Brasilia, a 65 km branch northeast to Formosa has also been proposed.

Ferrofrango

The government of Santa Catarina state has developed proposals to build a 170 km branch line from Herval d'Oeste through Concórdia, Seara and Chapecó to Itapiranga, a chicken-rearing region. As well as meat products in refrigerated containers, other agricultural products could be carried. Private investors are being sought to participate in this scheme.

Bulgaria

Ministry of Transport and Communications

9 Diakon Ignatiy Street, Sofia 1000
Tel: (+359 2) 940 95 00; 94 34
Fax: (+359 2) 987 18 05
e-mail: mail@mtc.government.bg
Web: www.mtc.government.bg

Key personnel
Minister: Nikolay Vassilev
Secretary General: Rositza Vladimirova

Executive Agency, Railway Administration
Tel: (+359 2) 940 94 27   Fax: (+359 2) 987 67 69
Executive Director: Georgi Nikolov

The Agency was established in 2001 as an organisation within the Ministry of Transport and Communications to regulate the rail sector.

UPDATED

Bulgarian State Railways (BDŽ EAD)

Ivan Vazov Street 3, 1080 Sofia
Tel: (+359 2) 981 11 10   Fax: (+359 2) 987 71 51
Web: www.bdz.bg

Key personnel
Chair: Anelia Krushkova
Executive Director: Nasko Tsanev
Directors
   Freight: Dimitar Kupenov
   Passenger: Aleksander Aleksandrov
   Operations: Danail Vanchev
   Technical Support: Ivan Donchev
   Traction: Slavi Angelov
   Finance: Minail Dragiev
   Administration: Ivan Valchev
   Safety: Ivan Lalov
   Moveable Asset Procurement: Tatyana Vasileva
   International Relations and Protocol:
      Kunka Kirkova

Political background
The new Railway Transport Act came into force on 1 January 2002. In compliance with its provisions, the former Bulgarian State Railways was split into two independent entities: a railway operator (BDŽ EAD) and an infrastructure manager, the National Railway Infrastructure Company (HK). Within the Ministry of Transport and Communications the Executive Agency, Railway Administration was established to provide regulation. The Act introduced European standards to the railway sector in Bulgaria. The regulatory environment has been completely revised and adapted to comply with EU requirements for transport market liberalisation and equal infrastructure access rights, clear rules for transport safety and for the issue of licenses for emerging railway operators. BDŽ EAD has started the separation of financial accounting of passenger and freight services. The company has also adopted a large-scale modernisation programme of wagon and coaching stock fleet, implementation of which was under way in 2005.

Traffic (million)	2001	2002	2003
Passenger journeys	41.8	33.7	35.2
Passenger-km	2,990	2,598	2,517
Freight tonnes	19.3	18.5	20.1
Freight tonne-km	4,904	4,627	5,274

Organisation
In 2004 BDŽ EAD employed 17,600 staff.

Passenger operations
Passenger traffic in 2003 showed slight growth in comparison with 2002 but in general has declined compared with levels of previous years. Loss-making, lightly used services on some secondary lines have been withdrawn. The average journey distance for 2003 was 71.5 km. Passenger fares are kept at low levels in compliance with PSO (Public Service Obligation) requirements. Under way in 2004 was the procurement of new dmus for domestic operations, as well as of high-quality sleeping cars, which from the beginning of 2005 are to operate within the composition of existing trains from Sofia to Vienna and Ljubljana. This will be followed by the introduction of Euronight trains along the same two routes by the end of 2005. In 2002 BDŽ EAD introduced two passenger coaches designed for the needs of people with impaired mobility, and a further delivery of such coaches was arranged for the end of 2004.

Catering for the attraction of journeys with historic rolling stock is another aspect of the passenger service development. In 2004 BDŽ EAD was operating two standard-gauge steam locomotives gauge and a narrow-gauge locomotive was being restored.

Freight operations
Freight volumes in 2003 increased in comparison with 2002 and 2001, the result of a general revitalisation of the national economy. This growth trend is further confirmed by the comparative analyses of freight volumes for the first quarters of 2004 and of 2003. During this period in 2004 there was growth of 4 per cent

Siemens-built Class 10 Desiro dmu at Levski (Philip Wormald)   **NEW**/1122855

Škoda-built Class 44 electric locomotive on the Mezdra–Sofia line (Milan Šrámek)   0569058

Class 55 diesel-hydraulic locomotive with a local service from Levski at Svishtov (Philp Wormald)
NEW/1122856

Class 43.500 electric locomotive (left) and Class 07 diesel on passenger services at Samuil (Philp Wormald)   **NEW**/1122857

760 mm gauge diesel-hydraulic locomotive at Razlog with a service to Dobrinishte (Edward Barnes)

0569062

in transported volumes and of 11 per cent in net revenue. In 2003 ores and scrap metal accounted for 23 per cent of all traffic, solid mineral fuels for 22 per cent, processed and unprocessed non-ore raw materials for 14 per cent, and chemicals and chemical products for 7 per cent.

Most rail freight to and from CIS countries is carried via the Varna-Ilichevsk-Poti/Batumi ferry link. The difference in track gauges between Bulgaria and CIS countries requires bogie-changing at the ferry complex in Varna.

BDŽ EAD considers that its main priority is the development of combined transport projects for accompanied, unaccompanied and container consignments, characterised by high service quality, fixed delivery terms, attractive prices and protection of cargo. It envisages services along the east-west transport axis, from Turkey to central and western Europe. In October 2003 a container block train was commissioned for operation between Sofia and the Greek port of Thessaloniki.

Traction and rolling stock

In 2004 BDŽ EAD was operating 269 electric and 333 diesel locomotives, 1,433 carriages, 74 electric multiple-units, four dmus and 17,508 wagons. Main line diesels were from Romania and the former Soviet Union with Hungarian and East German industry supplying shunters. Electric locomotives are mostly from Škoda (Czech Republic), although some have also been procured from Romania. The age of the fleet is a matter of concern to BDŽ and in the early 2000s some steps were being taken to address this.

In December 2003 BDŽ EAD initiated tendering processes for the supply of new dmus and two high-speed trainsets. The first result of this process occurred in January 2005 when BDŽ EAD signed a contract with Siemens TS covering the supply of 25 two-car Desiro dmus. The first example was handed over in the same month. Final assembly of most and maintenance of the fleet will be undertaken in Bulgaria at workshops in Varna by a BDŽ EAD-Siemens joint venture. An option in the contract, covering the supply of 25 Desiro Classic emus (15 three-car and 10 four-car) for main line services, was taken up in June 2005. As with the earlier dmu order, final assembly of the emus is to be undertaken at BDŽ's Varna workshops with delivery scheduled for 2007–08.

In 2000 the first of two programmes to modernise the BDŽ EAD electric locomotive fleet was initiated. At least six Class 46 Co-Cos have been modernised by Koncar, Zagreb, and re-geared for 150 km/h running, becoming Class 46.2. The same company has also undertaken the refurbishment of two Škoda-built Class 44 locomotives, including the fitting of thyristor control equipment. They are designated Class 46.3.

In 2003 BDŽ EAD acquired some secondhand coaches from DB Regio in Germany.

UPDATED

Electric locomotives

Class	Wheel arrangement	Output kW	Speed km/h	Weight tonnes	No in service	First built	Builders
42	Bo-Bo	2,800	110	88	2	1965	Škoda
42.100[1]	Bo-Bo	2,800	110	88	17	1965	Škoda
43	Bo-Bo	3,020	130	84	1	1971	Škoda
43.300[2]	Bo-Bo	3,020	110	84	6	1992	Škoda
43.500[3]	Bo-Bo	3,020	110	84	42	1971	Škoda
44	Bo-Bo	3,040	130	84	76	1975	Škoda
44.300[4]	Bo-Bo	3,040	130	81	2	1975	Škoda
45	Bo-Bo	3,040	110	84	60	1982	Škoda
46	Co-Co	5,100	130	126	34	1986	Electroputere
46.100[5]	Co-Co	5,100	160	126	3	1986	Electroputere
46.200[6]	Co-Co	5,100	130 or 150	123	6	1986	Electroputere
61	Bo-Bo	960	80	74	20	1991	Škoda

[1] Refurbished in Bulgaria, 1984–94.
[2] Class 43, 44 and 45 locomotives refurbished in Bulgaria 1981–84 following accident damage.
[3] Re-geared in Bulgaria for 110 km/h (former Class 43).
[4] Refurbished by Koncar, Croatia, 2004.
[5] Re-geared in Bulgaria for 160 km/h (former Class 46).
[6] Refurbished by Koncar, Croatia from 1999 (two for 130 km/h, four for 150 km/h).

Diesel locomotives

Class	Wheel arrangement	Power kW	Speed km/h	Weight tonnes	No in service	First built	Builders
04	B-B	1,620	120	1	10	1963	H SGP
06	Co-Co	1,540	100	117	49	1966	E Electroputere
07	Co-Co	2,205	100	119	74	1972	E Lugansk
07.100*	Co-Co	2,205	100	124	3	1972	E Lugansk
51	Bo-Bo	442	80	62	11	1960	E VEB Hennigsdorf
52	D	442	30/60	60	65	1965	H Ganz-Mávag
55	B-B	920	60/100	68	109	1969	H Faur
760 mm gauge							
75	B-B	810	70	48	9	1966	H Henschel
76	B-B	810	70	52	4	1975	H 23 August
77	B-B	810	70	52	5	1988	H 23 August
80	C	283	30	36	1	1967	H Henschel
81	B-B	294	50	24	3	1982	H Kambarsg

*Inverter installed for train heating from 2001.

Electric railcars or multiple-units

Class	Builder's type	Cars per unit	Motor cars per unit	Motored axles/car	Output/ motor kW	Speed km/h	Units in service	First built	Builders Mechanical	Electrical
32	(CE)P-25	4	2	4	165/210	130	68	1970	RVZ	RVZ
33	(CE)P-33	4	2	4	170/210	120	6	1990	RVZ	RVZ
–	Desiro	3	–	–	–	160	15	2007*	Siemens/BDŽ	Siemens
–	Desiro	4	–	–	–	160	10	2007*	Siemens/BDŽ	Siemens

* On order for delivery 2007–08.

Diesel railcars or multiple-units

Class	Cars per unit	Motor cars per unit	Motored axles/car	Power/ motor kW	Speed km/h	Units in service	First built	Mechanical	Builders Engine	Transmission
10	2	2	1	275	–	25*	2005	Siemens	–	H
18	3	2	2	600	100	3	1967	Ganz-Mávag	Ganz-Mávag	M
19	1	1	4	2×195	125	1	1960	SGP	SGP	H

* Delivery in progress in 2005.

National Railway Infrastructure Company (NRIC)

3 Ivan Vazov Street, Sofia 1080
Tel: (+359 2) 988 33 81   Fax: (+359 2) 932 46 66
Web: www.rail-infra.bg

Key personnel
Chairman: Georgi Nikolov
Director General: Dimitar Gaidarov
Deputy Director General: Veselin Kojuharov
Head of Legal Affairs: Asia Roshkova
General Manager, Traffic Safety: Kiril Angelov
Head of Accounts: Tania Jekova
Director, Movements and Capacity Management:
   Dipl-Ing Angel Angelov
Director, Infrastructure Repair and Maintenance:
   Nisim Primo
Director, Strategy Development and Investment
   Policy: Maria Chakarova
Director, Property Management and Land Planning:
   Antonii Lolov
Director, Finance: Anton Ginev

Gauge: 1,435 mm; 760 mm
Route length: 4,049 km; 245 km
Electrification: 2,710 km at 25 kV 50 Hz AC

Political background
In preparation for its eventual accession to the EU, Bulgaria separated responsibility for its national railway infrastructure from Bulgarian State Railways in 2002. NRIC is now responsible for operating, maintaining and developing the system, subject to controls by the state regulatory body, the Executive Agency, Railway Administration.

Organisation
The Bulgarian railway network plays a pivotal role in providing a link between Western Europe, various Balkan countries and the Middle East via Turkey. It also offers connections between CIS countries and republics that once formed part of the former Yugoslavia. Connections are made with the Greek system at Kulata/Promahon, with the Romanian network at Ruse/Giurgiu, Kardam/Negru Voda and at Vidin/Kalafat by river ferry, with Serbia at Dragoman/Dimitrovgrad and with Turkey at Svilengrad/Kapi Kule. Train ferry facilities at Varna also link Bulgaria with Ilichovsk, Ukraine. However, network capacity is limited by infrastructure constraints. Although some trunk routes have seen some modernisation, much of the system remains limited to speeds no higher than 100 km/h.

In a bid to improve its financial position HK in 2002 implemented a programme of closures of loss-making lines, including most of the 760 mm gauge routes.

NRIC is organised as directorates responsible for: movement and capacity management; infrastructure repair and maintenance; strategic development and investment policy; property management and land planning; finance; and international project management. These report via the Deputy Director General. Safety and security functions report directly to the Director General.

In 2004 NRIC employed 15,800 staff.

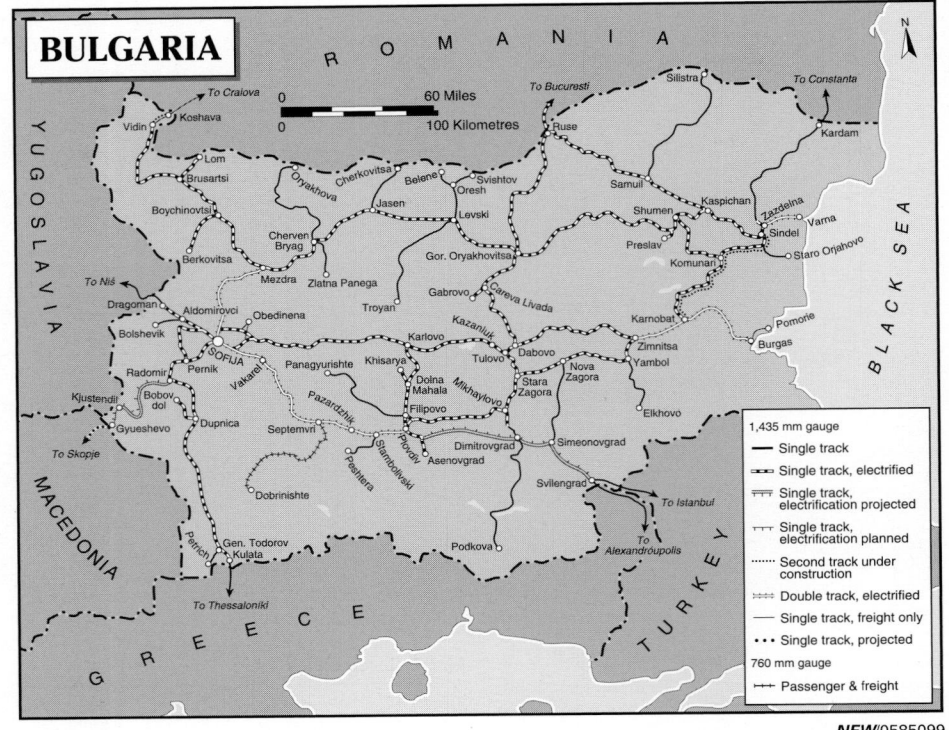

NEW/0585099

New lines
Link with Macedonia
Plans have been developed to establish a cross-border electrified rail link between Bulgaria, at Gyueshevo, and the Former Yugoslav Republic of Macedonia. Most of the work on the Bulgarian side of the border had been completed in 2005 but the Macedonian portion of the link remained incomplete.

Improvements to existing lines
Serbia–Bulgaria–Turkey
The main focus of investment in the Bulgarian network is on improvements to the trunk route from Serbia via Sofia and Plovdiv to the Turkish border at Svilengrad, which forms part of Pan-European Transport Corridors IV and IX. The focus will increase in importance when the planned Bosphorus tunnel linking European and Asian Turkey is completed. Central to this project is upgrading and electrification of the Plovdiv–Svilengrad section (153 km), for which international funding has been secured. The existing single-track line is to be electrified, part-doubled and upgraded for 160 km/h running and 22.5 tonne axleloads. The design of upgraded sections and of the electrification system will allow for possible future running at 200 km/h. Completion of upgrading the Sofia–Plovdiv section (156 km) is scheduled for 2008, but the Plovdiv–Svilengrad section is expected to be finished by the end of 2005.

Improved links with Romania
International funding is also being provided to help finance the construction of a second, double-track bridge across the Danube at Vidin, increasing capacity for traffic to and from the Romanian network (Corridor IV). Freight facilities at Vidin are also being improved. Work is scheduled for completion by 2007.

Macedonia–Black Sea
In 2005 upgrading was also well advanced of the east-west route from the Black Sea port of Varna via Gorna Oryakhovitsa, Pleven and Sofia to the Macedonian border and on to Albania (Corridor VIII). This route includes the Kjustendil–Gyueshevo section (see New lines). Completion had been scheduled for 2006 but is expected to take place later than this.

Signalling and telecommunications
Lines equipped with CTC include Sofia–Plovdiv and Sofia–Karlovo. ATP is also provided on the former route. Some fibre-optic communications cable has been installed.

Track
While 60 kg/m rail is used on high-density lines, 49 kg/m rail predominates. Rail is laid on a mixture of timber or prestressed concrete sleepers with K-type fastenings.

UPDATED

Burkina Faso

Ministry of Infrastructure, Transport & Housing

PO Box 7011, Ouagadougou 03
Tel: (+226) 30 62 11
Web: http://www.mith.gov.bf

Key personnel
Minister: Hyppolite Lingani

SOPAFER-B

Société de gestion du patrimoine ferroviaire du Burkina
PO Box 192, Ouagadougou 01

Tel: (+226) 31 35 99
Fax: (+226) 31 35 94

Key personnel
Director General: Oumar Zongo

Gauge: 1,000 mm
Route length: 622 km

Political background
The country was long-served by the Abidjan-Niger Railway under a joint arrangement with neighbouring Ivory Coast (qv). In 1986, following a disagreement over the wisdom of embarking on the 375 km Tambao extension in Burkina Faso, the railway was split. This move did little to help performance in either country and much freight traffic deserted to the roads.

By 1992, both railways were in a poor physical state and carrying little traffic. After considering total abandonment, the two countries decided that the railways should be managed as a single entity once again and bids were sought from private operators to form a joint holding company, acting as concessionaire. In 1993, a contract was awarded to the Sitarail group, a consortium comprising private investors (holding 67 per cent of the capital), the Burkina and Ivorian governments (15 per cent each) and Sitarail staff (3 per cent). The Sofrerail arm of French consultant Systra is the operating partner of the consortium. Private sector operation of the railway started in August 1995.

All fixed installations, infrastructure and rolling stock remain state-owned in the hands of two companies, SIPF and Sopafer, but Sitarail will maintain the infrastructure and rolling stock for the 15-year duration of the contract. For full details of the Sitarail operation in both countries, see the Ivory Coast entry.

Passenger operations

Operations in January 1998 consisted of: a daily passenger service between Ouagadougou and Abidjan, in Ivory Coast; a Saturday-only round-trip offered on the Ouagadougou and Kaya extension; and a daily 'luxury' (first class) railcar introduced between Ouagadougou and Bobo Dioulasso. However, due to a lack of spare parts to maintain locomotives and rolling stock in serviceable condition, by June 1998 the frequency of the Ouagadougou-Abidjan train had again been reduced to three times a week. A few weeks later the first class railcar on the Ouagadougou-Bobo Dioulasso route was also suspended.

Freight operations

There was some growth in imported oil traffic in the early 1990s, accounting for much of the meagre freight business, and trial hauls were made of manganese ore brought by road from Tambao to the Kaya railhead for forwarding to Abidjan. Freight carried in Burkina in 1995 amounted to about 200,000 tonnes, about 92 million tonne-km. Early indications were that tonnage was on the increase as a result of the concessioning agreement.

Sitarail's freight plans are based on development of the export manganese traffic and exploitation of zinc mined at Perkoa, some 30 km from the railway at Koudougou.

New lines

Construction of the Tambao extension has begun, with the aim of tapping rich manganese deposits in the region. The government bought 6,000 tonnes of used rail from Canadian National and in 1989 funds were obtained from the UN Development Programme for engagement of consultants to manage work on the initial 105 km from Ouagadougou to Kaya, which was opened in 1993.

The Burkina Faso government then advanced CFAFr6.2 billion for construction of the remaining 271 km from Kaya to Tambao. In 1990 Canac International of Canada was contracted by the UN Development Programme to oversee the first stage and assist in planning the second.

Earlier, in 1988, Canac International had secured a C$2.3 million turnkey contract to supply and install

a telecommunications network over the new line. The work included installation of five microwave sites between Ouagadougou and Kaya. In addition, a VHF/FM communications network would be installed to link train stations and train crews.

These works are currently in abeyance on account of the need to concentrate on rehabilitation of the existing line. Finance totalling some US$5 million for this was agreed by the West African Development Bank in 1996 (see Ivory Coast entry).

Improvements to existing lines

In late 1995, Sitarail invited tenders for supply of rail and sleepers to begin a programme of upgrading throughout the railway. Spare parts to refurbish GM locomotives were also sought.

Traction and rolling stock

At the start of 1996, the Sitarail combined operation had available 20 diesel locomotives, 17 shunters, 40 coaches and 600 wagons. Burkina's 1994–96 plan provided for purchase of three diesel-electric locomotives and 30 mineral wagons, while 60 tank wagons were acquired through the national oil company Sonabhy.

A deal has been agreed with Projects & Equipment Corporation of India for launch of local freight wagon production. The Indians are to set up a manufacturing plant at Bobo Dioulasso, get it going with a supply of wagon kits for assembly, then oversee transition to local manufacture from scratch.

Map caption: BURKINA FASO and COTE D'IVOIRE

1.000 m gauge
— Single track, not electrified, passenger & freight
••• Single track, line projected
-·-·- Old project, never realised

0058499

Cambodia

Ministry of Public Works and Transport

PO Box 65, Phnom Penh
Tel: (+855 23) 72 36 15

Key personnel
Minister: Khy Teng Lim

Royal Railway of Cambodia

Chemin de fer du Cambodge
Central Railway Station, Sangkat Srach Chak, Khan Daun Penh, Phnom Penh
Tel: (+855 23) 72 41 43
Fax: (+855 23) 72 58 97

Key personnel
Director: Pich Kim Sreang

Gauge: 1,000 mm
Route length: 602 km

Organisation
Because of the country's internal unrest, most of the rail network was out of use during the 1970s and 1980s, but in 1993 trains were operating on all sections except for the 48 km between Sisophon and the Thai border at Poipet. There are two main lines. The Old line runs westwards from Phnom Penh to Sisophon (339 km), whence the route to Poipet was removed in the 1970s. The New line, opened in 1969, links Phnom Penh with the country's only deep water port at Kompong Som, 263 km distant.

At the end of 1998 1,732 staff were employed.

Traffic (million)	1997	1998
Passenger journeys	0.530	0.438
Passenger-km	49.3	43.9
Freight tonnes	0.170	0.294
Freight tonne-km	36.1	75.7

Passenger and freight operations
In 1999 passenger services were operated on the Phnom Penh-Sisophon and Phnom Penh-Sianoukville lines, with passenger coaches also being added to some freight services. Freight traffic has shown growth following some rehabilitation of the railway's infrastructure and an improved economic environment. However, a shortage of freight wagons and the considerable age of existing vehicles was claimed by the railway in 1997 to be hampering traffic development.

For details of the latest updates to *Jane's World Railways* online and to discover the additional information available exclusively to online subscribers please visit
jwr.janes.com

New lines

A 450 km line linking Phnom Penh and Ho Chi Minh City (formerly Saigon) has been projected for a number of years but consistently cancelled owing to hostilities. The line would fill one of the missing links in the Trans Asian Railway Project. In 1996, Malaysia offered to finance and carry out a feasibility study for this route and reopening from Sisophon to Poipet; the two projects would close the missing links in a 5,500 km Singapore-Beijing dual-gauge route.

Improvements to existing lines

In 1999 government plans were announced to restore the line between Sisophon and Aranyaprathet, on the Thai border.

Traction and rolling stock

In 1998 the railway's serviceable fleet amounted to 18 diesel locomotives. Backbone of the diesel-electric fleet were nine Alsthom-built B-B 900 kW Class BB 1050 units fitted with MGO-V12 BZSHR engines, delivered in 1967–69. Between 1990 and 1994, ČKD Praha of the Czech Republic supplied four Class BB 1010 736 kW Bo-Bo diesel-electrics. The shunting locomotive fleet comprises two French-built machines delivered in 1957 and two ČKD-built 310 kW two-axle units with MTU engines, supplied in 1994.

In 1998 the passenger fleet comprised 12 coaches and a German-built railbus dating from 1969; there were 167 wagons.

Signalling and telecommunications

Both the former telegraph and radio signalling systems were destroyed and the current service is manually signalled.

Track

Rail of 30 kg/m (329 km) and 43 kg/m (264 km) on a mixture of steel and wooden sleepers is in use. Maximum permissible axleload is 20 tonnes for upgraded lines and 13 tonnes for older lines. Minimum curve radius is 300 m.

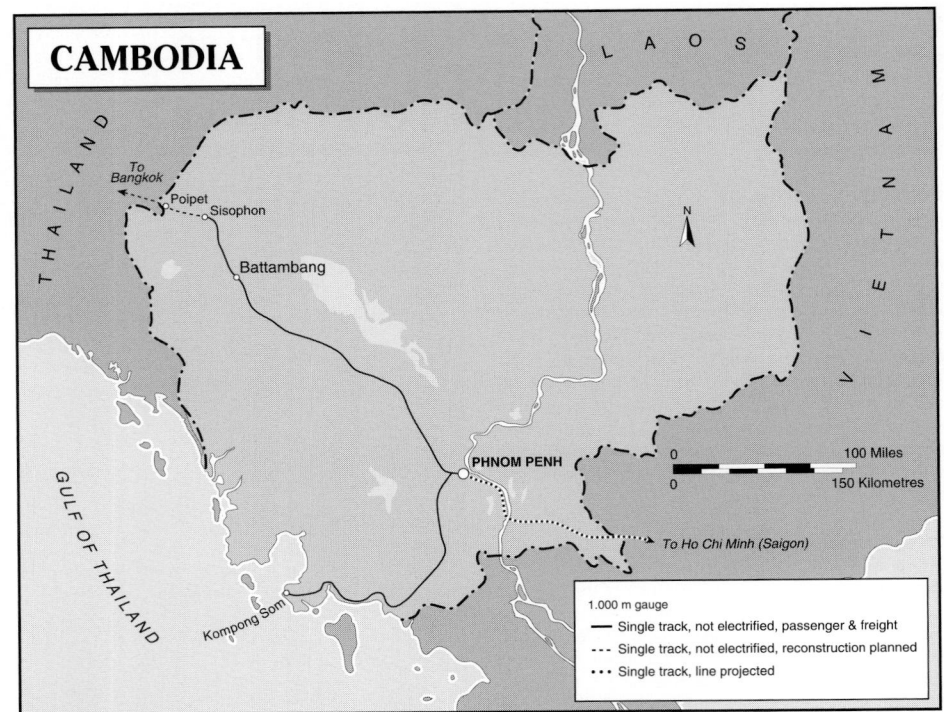

0089960

Cameroon

Ministry of Transport

PO Box 8043, Yaoundé
Tel: (+237) 23 22 36

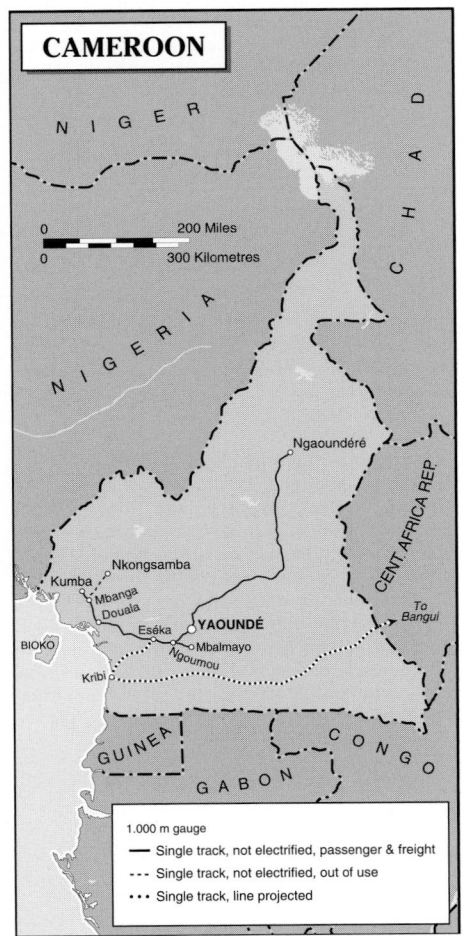

0058501

Key personnel

Minister: John Begheni-Ndeh

Cameroon Railways (Camrail)

Gare Centrale de Bessengue, PO Box 766, Douala
Tel: (+237) 340 60 45; 340 82 51
Fax: (+237) 340 82 52
e-mail: patrick.claes@camrail.cm

Key personnel

Administrator and General Manager: Patrick Claes
Directors
  Finance: Jean Philippe Barruel
  Operations: Claude Kontcho
  Planning and Procurement: François Bosco
  Rolling Stock: Georges Lomre
  Human Resources: Michel Ossock
  Fixed Installations: Vincent Tshiong Ngalula
Senior Manager, Passenger Service:
  Daniel Ebele Mbedi
Chief of Division for Management Control:
  Ebénéser Nzonlia
Chief of Key Management Indicators Office:
  Rose Asongwe

Gauge: 1,000 mm
Route length: 987 km

Political background

Moves towards privatisation of the former Regifercam railway system started in 1994. In 1998 it was announced that a 20-year concession to operate the railway was to be split between two bidders, Saga, a subsidiary of the French Bollore group, which now formally holds the concession, and Comazar, which operates the system. Foreign shareholders own a 77.4 per cent stake in Camrail, with local investors holding 9.07 per cent, employees holding 3.53 per cent and the government retaining 10 per cent of the capital. Operations under the concession started in April 1999.

Organisation

The Cameroon system consists of two single-track lines: the West line running from Bonaberi to Kumba and Nkongsamba; and the much longer Transcam line, opened in 1974, between Douala and Ngaoundéré.

In December 2004 Camrail employed 2,391 staff compared with 2,542 at the end of 2003.

Traffic (000s)	2002	2003	2004
Passenger journeys	1,111	1,108.8	1,003.2
Passenger-km	308,500	318,800	297,300
Freight tonnes	1,943.7	1,830.6	1,835.2
Freight tonne-km	1,179,000	1,089,100	1,114,300

Diesel locomotives

Class	Wheel arrange-ment	Power kW	Speed km/h	Weight tonnes	No in service	First built	Mechanical	Builders Engine	Transmission
CC2200	Co-Co	1,540	107	101	29	1980	Bombardier	Alco V12 251 C4	E CGE
BB1200	Bo-Bo	700	70	56	8	1969	Alsthom	SACM MGO V16 ASHR	E Alsthom/MTE
BB1100	Bo-Bo	600	60	68	20	1981	Alsthom	SACM MGO V12 ASHR	E Alsthom/MTE
BB1000	Bo-Bo	600	60	68	4	1978	Moyse	SACM MGO V12 ASHR	E Moyse
CC2600	Co-Co	1,940	100	111	6	1974	General Motors	EMD 16-645E3	E General Motors

Passenger operations

Passenger services include a daily intercity train in each direction on the Transcam line between Douala and Yaoundé (264 km), which is accomplished in 3 hours 50 minutes. Overnight services are also operated on a daily basis between Yaoundé and Ngaoundéré in addition to daytime trains on selected journeys. On the Western line, trains operate between Mbanga and Kumba and between Otele and Mbalmayo.

Freight operations

Camrail runs 13 freight services daily, as well as some additional services subject to demand. Principal commodities carried include: bulk petroleum products; logs and timber products; cotton; fertiliser; bulk alumina; sugar; cereals; and bagged cement. The railway also carries container traffic.

Improvements to existing lines

In 2002, the French government granted a loan to Camrail to cover upgrading of the Ka'a–Belabo–Pangar section (263 km) of the Yaoundé–Ngaoundéré line.

Traction and rolling stock

In December 2004 Camrail operated 67 diesel locomotives, 1,410 freight wagons, including 145 privately owned and 121 service wagons, and 71 passenger coaches (including eight luggage vans, six service vehicles, four buffet/dining cars and eight sleeper/couchette cars).

Signalling and telecommunications

The Yaoundé–Ngaoundéré section (521.3 km) is operated using track-to-train radio. An additional 263 km are controlled by solar-powered colourlight signalling with interlocking at intermediate stations. A microwave telecommunications network is in use.

Track

Rail: Vignole 30 kg/m (637 km) and 36 kg/m (275 km); 54 kg/m (45 km)
Crossties (sleepers): Timber and steel, thickness 130 mm
Spacing: 1,500/km plain track; 1,714/km curves
Fastenings: Sleeper screw (stiff); sleeper screw and Nabla (elastic)
Min curvature radius: 120 m
Max gradient: 1.6%
Max axleload: 20 tonnes

UPDATED

Canada

Transport Canada

Rail Policy Directorate, Surface Policy, Place de Ville, Tower C, Ottawa, Ontario K1A 0N5
Tel: (+1 613) 998 19 18
Fax: (+1 613) 998 26 86

Rail Safety Directorate
Tel: (+1 613) 998 29 84
Fax: (+1 613) 990 77 67
Web: www.tc.gc.ca

Key personnel

Minister: Jean-C Lapierre
Deputy Minister: Louis Ranger
Assistant Deputy Minister, Policy: Kristine Burr
Director General, Surface Policy: Guylaine Roy
Assistant Deputy Minister, Safety and Security: Marc Grégoire
Director General, Rail Safety and Security: Luc Bourdon

Political background

Transport Canada is the federal department responsible for transport. The department works to help ensure that Canadians have the best transportation system by developing and administering policies, regulations and programmes for a safe, efficient and environmentally friendly transport system; contributing to Canada's economic growth and social development; and protecting the environment.

Rail Policy Directorate

The Rail Policy Directorate is responsible for monitoring Canada's rail industry to ensure that it meets the varied requirements of the Canadian economy and Canadian shippers. In the federal context, this is done mainly through the Canada Transportation Act (CTA), which was enacted by Parliament in 1996. In addition to providing ongoing policy advice to the Minister of Transport on the broad range of factors that pertain to Canada's railway industry, the Directorate is also responsible for administering the subsidy to VIA Rail, a crown corporation providing national passenger services (see entry in Canada section of *Railway systems and operators*). It is also responsible for the federal government's fleet of approximately 12,400 hopper cars used in the transportation of western grain.

Canada Transportation Act (CTA)

The emphasis of the CTA is on enhancing trade and the viability and competitiveness of the Canadian transport system, reducing regulatory intervention and encouraging more innovative services. Further stimulus to rail carrier efficiency was provided by easing the process of plant rationalisation and encouraging the creation of shortline railways without federal investment or subsidies. In addition to commercial sales, the process has allowed provinces and communities to intervene in the public interest to acquire uneconomic rail lines important to local economies as an alternative to road and highway investment. This new policy has proved a major success, with 37 new shortline companies created between 1996 and 2000, compared to only 11 new ones in the previous half-decade.

CTA Review

Section 53 of the CTA requires that the Act be reviewed every five years. In response to this statutory requirement, the Minister of Transport created a Canada Transportation Act Review Panel in 2000 to assess whether the existing legislation provides Canadians with an efficient, effective, flexible and affordable transport system; and, where necessary or desirable, recommend amendments to the national transport policy as established in the legislation, and to any statutes covered by the review.

The Panel concluded that the Canadian rail system is fundamentally competitive and efficient. More specifically, it found, among other things, that the rail system works well for most users, most of the time; that there is not evidence that railways are earning excessive profits, and that market abuse is not systemic or widespread.

Structure of Canada's rail system

Canadian National Railway (CN) and Canadian Pacific Railway (CPR) are the dominant carriers in Canada, accounting for about 90 per cent of the rail industry activity and revenues. However, they operate about 70 per cent of the total domestic rail network, with shortline and regional railways operating the remaining 30 per cent.

Canada's rail system remained relatively stable in 2004. The only major development was a large transfer of track, approximately 2,300 km, when CN completed the takeover of BC Rail, a regional railway owned by the province of British Columbia.

Hopper cars

Transport Minister Jean-C Lapierre and Finance Minister Ralph Goodale announced in March 2005 that the Government of Canada would open negotiations with the Farmer Rail Car Coalition concerning the possible transfer of the federal railway hopper car fleet to the coalition.

In mid-2005 there were approximately 12,400 railway hopper cars in the government-owned fleet and they form the core of the rolling stock used by Canadian National Railway and Canadian Pacific Railway to move grain. These cars are provided at no cost to the railways for the transportation of grain from the Prairies for export through the ports of Vancouver, Prince Rupert, Thunder Bay and Churchill.

Intercity passenger rail services

The federal government provides most of Canada's intercity passenger rail services through VIA Rail Canada. VIA's financial performance has deteriorated since 2003, when a series of events beyond its control resulted in a significant decrease of its revenue-cost ratio from an all-time high of 64.5 per cent in 2002 down to 58.9 per cent in 2004. VIA Rail carried close to 3.9 million passengers in 2004. VIA Rail's federal government operating subsidy is fixed at CAD169 million.

In April 2000 the Government of Canada announced an infusion of slightly more than CAD400 million over five years for capital expenditures to address safety, health and capacity issues that were facing VIA in the late 1990s. VIA exhausted this special capital funding in 2004.

Rail Safety Directorate

The Rail Safety Directorate's mission is to advance the safety and sustainability of rail transportation. The Rail Safety Programme develops, implements and promotes safety policy, regulations, standards and research, and in the case of rail/road grade crossings, funds safety improvements. Regional inspectors monitor for compliance with the approved regulatory regime requirements and if necessary, investigate and enforce wherever non-compliance is discovered.

The Rail Safety Programme works toward the protection of life, health, the environment and property of the Canadian public; and towards a rail transportation system that is harmonised both domestically and internationally. This work is done under the authority of the Railway Safety Act (RSA), 1989. Amendments were made to the RSA in 1999 specifically to modernise and enhance the legislative and regulatory framework of Canada's rail transportation system. These amendments provide the railways with more responsibility in managing operational safety, while the public and other interested parties have a greater say on issues related to rail safety.

A major and key initiative is the adoption of Railway Safety Management Systems (SMS) – a formal framework for integrating safety into day-to-day railway operations. Railway SMS Regulations, which came into effect on 31 March 2001, require all federally regulated railways to implement and maintain a SMS. The Rail Safety Directorate developed an ongoing industry education and awareness programme and a SMS audit programme to assess company SMS documentation (pre-audit), and evaluate the implementation and effectiveness of a company's SMS (verification audit). The Rail Safety compliance programme is evolving towards integrating the Railway Safety Management System approach with the functional discipline compliance programmes – Equipment, Operations and Engineering.

UPDATED

Canadian Transportation Agency (CTA)

Ottawa, Ontario K1A 0N9
Tel: (+1 888) 222 25 92
Fax: (+1 819) 953 83 53
Web: http://www.cta-otc.gc.ca

Key personnel

Chairman: Marian Robson
Vice-Chair: Gilles Dufault
Members: M Baljinder, Mary-Jane Bennett, Guy DeLisle, S Gill, George Proud, Beaton Tulk, Liette Lacroix Kenniff
Director-General, Air and Accessible Transportation: Gavin Currie
Director-General, Rail and Marine Transportation: Seymour Isenberg

Director-General, Corporate Management: Joan MacDonald
General Counsel and Secretary: Claude Jacques
Director, Communications: Craig M Lee

Political background

The CTA was created in 1996 with the issuance of the Canada Transportation Act, succeeding the National Transportation Agency. The Agency is a quasi-judicial tribunal of the government of Canada that provides economic regulation over rail (as well as certain air and maritime) operations. The Agency helps to ensure that all modes of federally regulated transport are accessible to persons with disabilities.

With respect to railway matters, federal regulation normally applies to railways that operate across national or provincial boundaries. The Rail

and Marine Branch of the CTA acts as a resolution mechanism for shipper-carrier and carrier-carrier disputes involving service and competitive access, as well as disputes involving railway interaction with highways, utility companies, municipalities and landowners.

Other regulatory activities include railway line transfer and discontinuance, railway line construction, and monitoring the insurance adequacy of federally regulated railways. The Branch also undertakes statutory costing activities, including preparation of an annual (maximum) rate scale for the rail movement of western Canadian grain, and maintains an audit and compliance function.

L'Agence Métropolitain de Transport (AMT)

500 Place d'Armes, Suite 2525, Montreal, Quebec H2Y 2W2
Tel: (+1 514) 287 24 64
Fax: (+1 514) 287 24 60
Web: www.amt.qc.ca

Key personnel

Chief Executive: Joël Gauthier
Vice Presidents
    Administration and Finance: Céline Desmarteaux
    Planning: James Byrns
    Commuter Rail: Raynald Bélanger
Judicial Section and Corporate Executive Secretary: Denise Gosselin
Director, Commuter Trains: Paul Dorval
Director, Communications and Marketing: Manon Goudreault

Gauge: 1,435 mm
Route length: 197 km
Electrification: 30 km at 25 kV AC

Political background

Since 1996, the former CN commuter line connecting Central station, Montreal with Deux-Montagnes and the former CP line from Lucien-L'Allier (formerly Windsor) station to Dorion and Rigaud were taken over by L' Agence métropolitain de transport (AMT), an organisation created by the government of Quebec to co-ordinate and promote public transport in the greater Montreal area. As well as being charged with developing the commuter rail network, AMT is the planning authority for regional public transport services like express buses, reserved lanes and the metropolitan terminus.

Under the stewardship of AMT, the Montreal commuter rail network has grown from two to five lines, with additions and extensions planned. By 2005 ridership had grown to 14 million passengers annually compared to 7 million in 1996 with two commuter lines.

Passenger operations

The electrified former CN route from Montreal Central to Deux-Montagnes (30 km, 12 stations) was re-commissioned in 1995 after a three-year modernisation project.

Diesel push-pull services are operated on four routes: Montreal (Lucien-L'Allier)–Dorion/Rigaud (64 km, 19 stations); Montreal (Lucien-L'Allier)–Blainville (45 km, 10 stations); Montreal (Central)–Mont-Saint-Hilaire (35 km, seven stations); and Montreal (Lucien-L'Allier)–Delson (23 km, eight stations).

New lines

AMT plans to add two further lines to its network. The first, a 43 km line linking Montreal (Central) and the northeast of the metropolitan area of Repentigny-L'Assomption, would serve eight new stations, with trains using the Deux-Montagnes line as far as Mont-Royal. A second line, of 52 km, would link Montreal (Lucien-L'Allier) and Mascouche and would serve three new stations, with trains using the Blainville line as far as Saint-Martin.

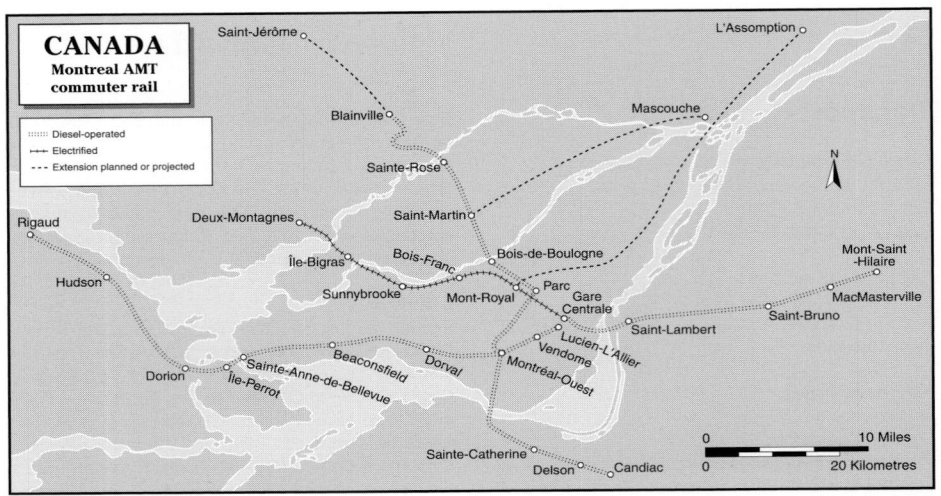

NEW/1114976

Bombardier-built MR-90 emu vehicles at Deux-Montagnes station   0544576

Diesel locomotives

Class	Wheel arrangement	Power kW	Speed km/h	Weight Tonnes	No in service	First built	Builders
F40PH	Bo-Bo	2,240	–	130	5	1974	GM
GP9-GC418	Bo-Bo	1,800	–	112	4	1959	GM/AMF
F59PH-I	Bo-Bo	2,400	–	–	11	1999	GM

Electric multiple-units

Class	Cars per unit	Motor cars per unit	Motored axles/car	Output/motor kW	Speed km/h	No in service	First built	Builders Mechanical	Electrical
MR-90	2	1	4	284	120	29	1994	Bombardier	GE

Also planned is a 16 km extension of the Montreal Lucien-L'Allier–Blainville line to Saint-Jérôme, with two additional stations.

Improvements to existing lines
In 2005 a station at Candiac was opened on the Montreal–Delson line. Other recent initiatives by AMT include improvements to signalling and parking facilities.

Traction and rolling stock
The Montreal–Deux-Montagnes line is served by 29 two-car MR-90 emus built by Bombardier from 1994. Services on the four diesel-operated lines are handled as push-pull formations by 21 diesel-electric locomotives of GM origin. The passenger coach fleet comprises 135 vehicles plus six generator cars for head-end power. In 2004–05 the coaching stock fleet was being supplemented by 22 bi-level cars built by Bombardier. Initial deployment of these was on the Montreal–Dorion/Rigaud line.

UPDATED

New bi-level coaches supplied to AMT by Bombardier in 2004–05 (AMT)
NEW/1141811

Alberta RailNet Inc

9808 96 Street, Grand Prairie, Alberta T8V 7T9
Tel: (+1 780) 831 04 07
Fax: (+1 780) 539 17 11
e-mail: railnet1@telusplanet.net
Web: http://www.albertarailnet.com

Key personnel
Vice-President and General Manager: Greg Pichette
Directors of Operations: Joe Gizanich,
Bruce Whiteman

Chief Mechanical Officer: Dan Sparks
Market Development Officer: Wendy Kemper
Administration Manager: Char Cote

Gauge: 1,435 mm
Route length: 588 km

Organisation
ARN is a subsidiary of North American RailNet (qv in USA section) which in June 1999 took over the operation of three former CN subdivisions in northwestern Alberta. Trackage begins 58 km northeast of Jasper and runs north to Grande Prairie, from where lines go northwest to Hythe and northeast to Tangent. The railway handles approximately 40,000 carloads annually, including coal, forest and industrial products and grain.

Algoma Central Railway

Algoma Central Railway Inc
PO Box 9500, Sault Ste Marie, Ontario P6A 6Y1
Tel: (+1 705) 541 28 50

Key personnel
President: Edward A Burkhardt
Vice-President, Marketing: William R Schauer
Marketing Director, ACRI: Domenic A Palumbo

Gauge: 1,435 mm
Route length: 518 km

Organisation
Following approval in December 1994 from Canada's National Transportation Agency, the rail assets of the Algoma Central Corporation were purchased by Wisconsin Central Transportation Company (WCTC) (qv in the US section) through a wholly owned subsidiary WC Canada Holdings Inc. Algoma Central Railway Inc has been the operating subsidiary since February 1995.

Finance
In acquiring the Algoma Central, WCTC paid US$16.1 million for property and rolling stock, including 23 locomotives and 879 passenger coaches and freight wagons. Additionally, WCTC paid US$8.2 million for the rights of way in a partnership transaction involving the province of Ontario. Algoma Central's financial results are now consolidated with those of parent WCTC.

Passenger operations
ACRI provides the thrice-weekly passenger service each way between Sault Ste Marie and Hearst, complemented by seasonal tourist trains (of which the best known is the Agawa Canyon service).

Freight operations
Under WCTC management, daily freight service operates between Sault Ste Marie and Oba, and on the Michipicoten branch. Service north of Oba to Hearst is Monday-Friday only. Marketing and customer service functions are now integrated with WCTC. Iron ore, steel and forest products are the dominant commodities.

Traction and rolling stock
The 23 diesel locomotives absorbed into WCTC ownership comprised a mix of EMD 2,200 kW Type SD40, EMD 1,500 kW Type GP38-2, EMD 1,100 kW Type GP7 and one switcher. Several of the higher-power locomotives have been redeployed; 11 ex-VIA steam-generator-equipped EMD FP9 units have been acquired for the passenger trains and seven are in service.

Coupler in standard use
Passenger cars: AAR Type E, F, H
Freight cars: AAR Type E, F, H
Braking in standard use, locomotive-hauled stock: Air Type 26

Track
Rail type: 115 RE, 100 RE, 85 CPR, 80 ASCE
Sleepers: Wood
Spacing: 1,822/km
Fastenings: Splice bars
Min curve radius: 1°
Max gradient: 1.7%
Max axleload: 30.4 tonnes

BC Rail

BC Rail Ltd
PO Box 8770, Vancouver, British Columbia V6B 4X6
Tel: (+1 604) 986 20 12
Fax: (+1 604) 984 52 01
Web: http://www.bcrail.com

Gauge: 1,435 mm
Route length: 2,314 km

Organisation
BC Rail Ltd was the railway operating unit of the BCR Group of companies, a Crown corporation, wholly owned by the province of British Columbia. In November 2003 it was announced that the British Columbia government had agreed to sell the business to CN, together with a long-term operating lease to operate over and maintain its infrastructure, which would remain in the ownership of the province. The acquisition was valued at C$1 billion in cash and was subject to approval by the Canadian competition authorities. CN also announced that it was seeking proposals from interested parties to operate tourist passenger services in British Columbia.

BC Rail freight at Squamish led by a Class GEF 40 (General Electric Dash 8-40CM) locomotive (Quintus Vosman)
0528637

The railway's principal yard facilities are at North Vancouver, Squamish, Lillooet, Williams Lake, Quesnel, Prince George, Fort St James, Mackenzie and Fort St John. Rail-served industrial parks are situated at Williams Lake, Prince George, Mackenzie, Fort St John, Dawson Creek, Fort St James and Fort Nelson.

At the end of 2001, BC Rail employed 1,953 staff, compared with 1,999 in 2000.

Passenger operations

BC Rail's scheduled passenger services between North Vancouver and Prince George were discontinued at the end of October 2002. From November 2002, a daily service between D'Arcy and Seton Portage and on demand from Seton Portage to Lillooet was introduced to serve local communities. The service employs two specially constructed 20-seat two-axle lightweight railcars designated Rail Shuttle Vehicles (RSVs). Ticket sales are handled on an advance booking basis by the Seton Lake Indian Band, while the vehicles are operated by BC Rail.

Freight operations

BC Rail serves some 800 shippers (150 on BCR Properties sites) in the province of British Columbia. Main commodities carried are forest products, coal and agricultural products. A 'trailer on flat car' intermodal service was discontinued in February 2002. The railway's traffic volume for 2001 totalled 13.99 million gross revenue tonnes compared with 15.27 million in 2000.

Traction and rolling stock

At the end of 2001, BC Rail rostered 125 diesel locomotives; nine Budd diesel railcars; 23 passenger coaches; and 9,146 freight wagons.

The GEF40 Dash 8 locomotives are being overhauled and upgraded to 3,300 kW with Dash 9 electronics. A seven-year remanufacturing programme of 27 Alco-engined yard/transfer locomotives has been completed.

The seven GF6C 50 kV AC electric locomotives formerly used on the Tumbler Ridge branch were withdrawn in October 2000.

Major wagon repairs and rebuilding are undertaken by BC Rail's Squamish workshop facility, which performs similar work on locomotives. Lighter running repairs are undertaken by workshops at Prince George.

Electrification

The use of electric traction on the 129 km Tumbler Ridge branch line ceased in 2000.

Track

Rail: 50, 60 and 68 kg/m
Crossties (sleepers): Softwood, 178 × 228 × 2,400 mm; steel, 130 × 300 × 2,500 mm
Spacing: Wood 510 mm; steel 610 mm

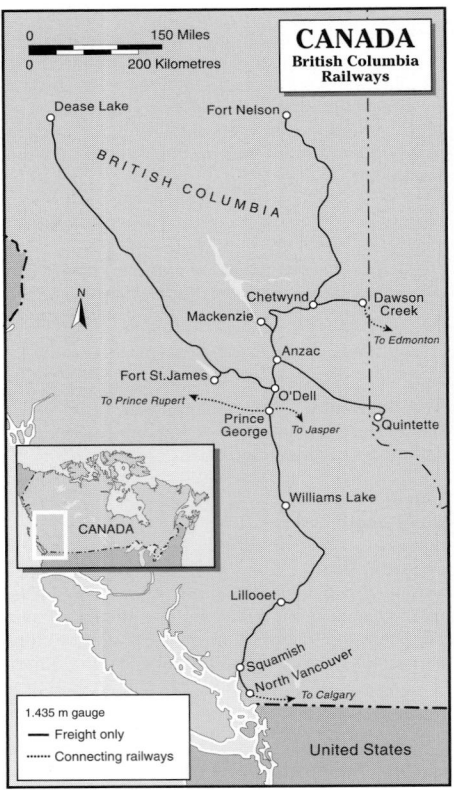

BC Rail is upgrading its GE Dash 8 locomotives to 3,300 kW, with Dash 9 electronics   0019553

0528691

Diesel locomotives

Class	Wheel arrangement	Power kW	Speed km/h	Weight tonnes	No in service	First built	Mechanical	Builders Engine	Transmission
M420	Bo-Bo	1,500	105	112.5	4	1973	MLW	MLW Alco 251-12	E GE
C420	Co-Co	1,500	105	119.3	2	1966	Alco	Alco 251–16	E GE
SD40	Co-Co	2,200	105	192.5	23	1978	GM	EMD 645-16	E EMD
Slug	Bo-Bo	1,200	105	109	10	1981	MLW	–	E GE
CRS20	Bo-Bo	1,500	105	113.6	27	1990 R	MLW/BCR	Caterpillar 3516	E KATO/GE
GEF36	Bo-Bo	2,700	105	n/a	15	1980	GE	GE 7FDL16	E GE
GEF36	Co-Co	2,700	105	192.5	6	1981	GE	GE 7FDL16	E GE
GEF39	Bo-Bo	2,900	105	130	3	1988	GE	GE 7FDL16	E GE
GEF40	Co-Co	3,000	105	192.5	26	1990	GE	GE 7FDL16	E GE
GEF44	Co Co	3,000	105	192.5	4	1995	GE	GE 7FDL16	E GE

Diesel railcars

Class	Cars per unit	Power kW	Speed km/h	Weight tonnes	No in service	First built	Mechanical	Builders Engine	Transmission
RDC1	1	520	129	53.5	6	1956	Budd	Cummins	Twin Disc
RDC3	1	520	129	53.5	3	1956	Budd	Cummins	Twin Disc

Fastenings: Cut spikes and anchors on timber sleepers, elastic clips on steel sleepers
Min curve radius: 110 m

Max gradient: 2.2%
Max axleload: 29.9 tonnes

Canadian National

Canadian National Railway Company
935 de la Gauchetière Street Ouest, Montréal, Quebec H3B 2M9
PO Box 8100, Montréal, Quebec H3C 3N4
Tel: (+1 514) 399 54 30
Web: http://www.cn.ca

Key personnel

President and Chief Executive Officer:
  E Hunter Harrison
Executive Vice-President and Chief Financial
  Officer: Claude Mongeau
Executive Vice-President, Sales and Marketing:
  James M Foote
Senior Vice-President, Law, Public Affairs and Chief
  Legal Officer: Sean Finn
Senior Vice-President, People: Les Dakens
Senior Vice-President, Eastern Canada Division:
  Keith Heller
Senior Vice-President, Operations: Jack T McBain
Vice-Presidents
  Financial Planning: Ami Haasz
  Intermodal: William K Berry
  Commercial Development: Cliff L Carson

Risk Management: John Dalzell
Labour Relations, North America: Kim A Madigan
Pacific Division: David P Edison
Grains and Fertilisers (Canada):
  S Ross Goldsworthy
Forest Products: Stan Jablonski
Midwest Division: Gordon Trafton
Chief Transportation Officer: Ed Harris
Gulf Division: Peter Marshall
Prairie Division: Keith Creel

Corporate Development: François Hebert
Engineering, Mechanical and Supply Management:
  Sameh Fahmy
Investor Relations: Robert E Noorigian
e-Business, Sales and Marketing: Anita Ernesaks
Sales and Market Development: Janice Murray
Automotive: Andy Gonta
Sales, Industrial Products: Jean-Jacques Ruest
US Government Affairs: Karen B Phillips
Chief Information Officer: Fred Grigsby

General Motors SD75I and SD40-2 locomotives leading a CN intermodal service at Spence's Bridge, British Columbia (Quintus Vosman)   0547793

Corporate Comptroller: Serge Pharand
Treasurer: Ghislain Houle

Gauge: 1,435 mm
Route length: 28,682 km in Canada and the USA

Background

Incorporated in 1919 as a state-owned corporation, CN became a publicly traded company on 18 November 1995 when the government of Canada sold all its shares to private investors in Canada, USA and overseas. The privatisation, the largest in Canadian history, involved the sale of 83.8 million common shares, generating gross proceeds to the government of C$2.28 billion. It followed management's implementation of an aggressive business plan three years before, which had produced a dramatic improvement in CN's financial performance and operational efficiency. While exposing the company to greater market discipline, privatisation unleashed a dynamic, entrepreneurial spirit throughout the organisation, prompting a steadily improving performance as CN focused on business growth, cost containment, competitive advantage for customers, and shareholder value.

Organisation

CN operates a network of nearly 29,000 route-km of track in Canada and the United States, generating revenues of C$6.110 billion in 2002 from the movement of a diversified and balanced traffic mix. It is the only railroad in North America with a network spanning both Canada and mid-America and connecting three coasts, the Atlantic, the Pacific and the Gulf of Mexico. CN serves the ports of Vancouver, Prince Rupert, Montreal, Halifax, New Orleans, and Mobile, Alabama, and the key cities of Toronto, Buffalo, Chicago, Detroit, Duluth/ Superior, Green Bay, Minneapolis/St Paul, Memphis, St Louis, and Jackson, Mississippi, with connections to all points in North America, including Mexico.

In 1999 CN acquired the Illinois Central Corporation (IC), which operated a 5,400-km network in the US Midwest. The acquisition fulfilled CN's 'three-coast' strategy, adding to its Pacific-to-Atlantic network a north-south corridor to the Gulf of Mexico. CN finalised the integration of IC operations with its own in 2001, a process that involved the co-ordination of network operations, traction distribution and rolling stock allocation, and information technology systems. In particular, CN modified and upgraded its Service Reliability Strategy (SRS) information system for tracking shipments and enabling scheduled service, extending SRS to the entire former IC territory in October 2000. The transfer to a single system proceeded smoothly.

Targeting growth through further strategic acquisitions, CN purchased the Wisconsin Central Transportation Corporation (WC) for C$1.301 billion in 2001, adding 4,500 km and 2,200 employees to the CN system. After the US Surface Transportation Board unanimously approved the transaction on 7 September, formal integration of WC into CN began on 9 October. The acquisition represented a logical move for CN: WC's trackage in Wisconsin and Michigan lay at the very heart of CN's three-coast network, and for a number of years the two railroads had been co-operating under a long-term haulage agreement. With ownership of WC's network, CN was able to secure the link between its eastern and western routes. The acquisition reinforced CN's position as North America's leading carrier of forest products, the dominant element in WC's traffic mix. Through its acquisition of WC, CN acquired 40.9 per cent of English Welsh and Scottish Railway in the UK. CN's interests in Australian Transport Network Ltd in Tasmania, Australia and in Tranz Rail Holdings Ltd, also acquired through the acquisition of WC, were sold in 2002 for aggregate net proceeds of C$69 million.

The acquisition of IC and WC prompted modifications to CN's corporate structure, beginning with the establishment of five new geographic divisions in mid-1999: Eastern Canada; Prairie; and Pacific in Canada; and Midwest and Gulf in the US. Aimed at bringing CN closer to customers and promoting quicker response and stronger relationships, the new structure strikes a balance between activities better managed centrally with

CN coal train at Windy Point, Alberta. The company handles coal for export, primarily for steelmaking in Japan, but also for utilities in North America and abroad   0134253

Diesel locomotives: Canadian operations

Class	Builder	Model	No in service	First built	Modified	Rating kW	Speed km/h
Road							
GR-12w	GM	GMD-1	1	1959	1983–87*	895	105
GR-12z	GM	GMD-1	1	1960	1987–90*	895	105
GR-12zc	GM	GMD-1	1	1959	1986–90*	895	105
GR-12u	GM	SW-1200RS	1	1959		895	105
GR-12y	GM	SW-1200RS	5	1960		895	105
GR-412a	GM	GMD-1B	23	1989 (Rebuilds)		895	105
GR-412b	GM	GMD-1B	15	1989 (Rebuilds)		895	105
GR-418a	GM	GP-9RM	7	1981–82 (Remanufactured)	1987†	1,343	105
GR-418b	GM	GP-9RM	9	1982–83 (Remanufactured)	1987†	1,343	105
GR-418c	GM	GP-9RM	12	1984 (Remanufactured)	1987†	1.343	105
GR-418d	GM	GP-9RM	12	1984 (Remanufactured)		1,343	105
GR-418e	GM	GP-9RM	7	1989–90 (Remanufactured)		1,343	105
GR-418f	GM	GP-9RM	12	1991 (Remanufactured)		1,343	105
GR-420b	GM	GP-38-2	34	1972–73	1982–83†	1,492	105
GR-420c	GM	GP-38-2	45	1973–74	1982–83†	1,492	105
Road Freight							
GF-620a	GM	SD38-2	4	1975		1,492	105
EF-640a	GE	8–40CM	30	1990		2,984	105
EF-640b	GE	8–40CM	25	1992		2,984	105
EF-644a	GE	9–44CWL	23	1994		3,282	105
EF-644b	GE	9–44CWL	40	1997		3,282	105
EF-644c	GE	9–44CWL	40	1997–98		3,282	105
EF-644d	GE	9–44CWL	40	2000		3,282	105
EF-644e	GE	9–44CWL	40	2002		3,282	105
GF-30c	GM	SD-40	1	1967		2,238	105
GF-30d	GM	SD-40	3	1967–68		2,238	105
GF-30e	GM	SD-40	2	1969		2,238	105
GF-30k	GM	SD-40	2	1971		2,238	105
GF-30m	GM	SD-40	4	1971		2,238	105
GF-30n	GM	SD-40-2	20	1975		2,238	105
GF-30p	GM	SD-40-2	16	1975		2,238	105
GF-30q	GM	SD-40-2	15	1976		2,238	105
GF-30r	GM	SD-40-2	19	1978		2,238	105
GF-30s	GM	SD-40-2	10	1979		2,238	105
GF-30t	GM	SD-40-2	30	1980		2,238	105
GU-30u-x	GM	SD-40-2	35	1980		2,238	105
GF-636a	GM	SD-50F	40	1985–86		2,685	105
GF-636b	GM	SD-50F	20	1987		2,685	105
GF-638a	GM	SD-50DAF	4	1986		2,835	105
GF-638b	GM	SD-60F	59	1989		2,835	105
GF-640a	GM	SD-70I	26	1995		3,000	70
GF-643a	GM	SD75I	139	1996		3,208	112
GF-643b	GM	SD75I	34	1997		3,208	105
GF-643c	GM	SD75I	104	1999		3,208	105
GF-630a-c	GM/CN	SD-40Q	29	1992–95 (Remanufactured)		2,238	105
GF-430a	GM	GP-40-2L	24	1974		2,238	105
GF-430b	GM	GP-40-2L	9	1974		2,238	105
GF-430c	GM	GP-40-2L	27	1975		2,238	105
GF-430d	GM	GP-40-2	2	1977		2,238	105
GF-430e	GM	GP-40-2	2	1974**	1991–2	2,238	130
GF-430f	GM	GP-40-2	1	1974**	1991–2	2,238	130
GF-430g	GM	GP-40-2	3	1975**	1991–2	2,238	130
Switchers							
GS-418a	GM/CN	GP-9RM	8	1985 (Remanufactured)		1,343	105
GS-418b/c	GM/CN	GP-9RM	58	1990–93 (Remanufactured)		1,343	105
GS-418d	GM/CN	GP-9RM	5	1993 (Remanufactured)		1,343	105
GY-418a	GM/CN	GP-9RM	13	1985–86 (Remanufactured)		1,343	105
GY-418b	GM/CN	GP-9RM	18	1986 (Remanufactured)		1,343	105
GY-418c	GM/CN	GP-9RM	7	1987/90 (Remanufactured)		1,343	105
GY-418d-f	GM/CN	GP-9RM	39	1988–90 (Remanufactured)		1,343	105
GS-412a	GM/CN	SW-200RB	10	1987 (Rebuild)		895	105
GH-20b	GM	GP-38-2	27	1973	1977–85	1,492	105
Boosters							
GY-00b	GM	YBU-4M	8	1980	1985–86	–	105
GY-00c	GM/CN	YBU	4	1986 (Remanufactured)	–	105	
GY-00d	GM/CN	YBU	18	1986 (Remanufactured)	–	105	
GY-00e/f/g	GM/CN	YBU	41	1987–90 (Remanufactured)	–	105	
GY-00m	MLW/CN	MS-7	7	1991 (Remanufactured)	1964–96	–	105
GH-00a	GM	HBU-4	19	1978	–	105	
GH-00b	GM	HBU-4	4	1980	–	105	
GH-00c	GM	HBU-4M	4	1980	1986	–	105

* Rebogied from A1A-A1A to B-B.   † Weight reduced.   ** Former GO Transit locos converted for freight haulage.

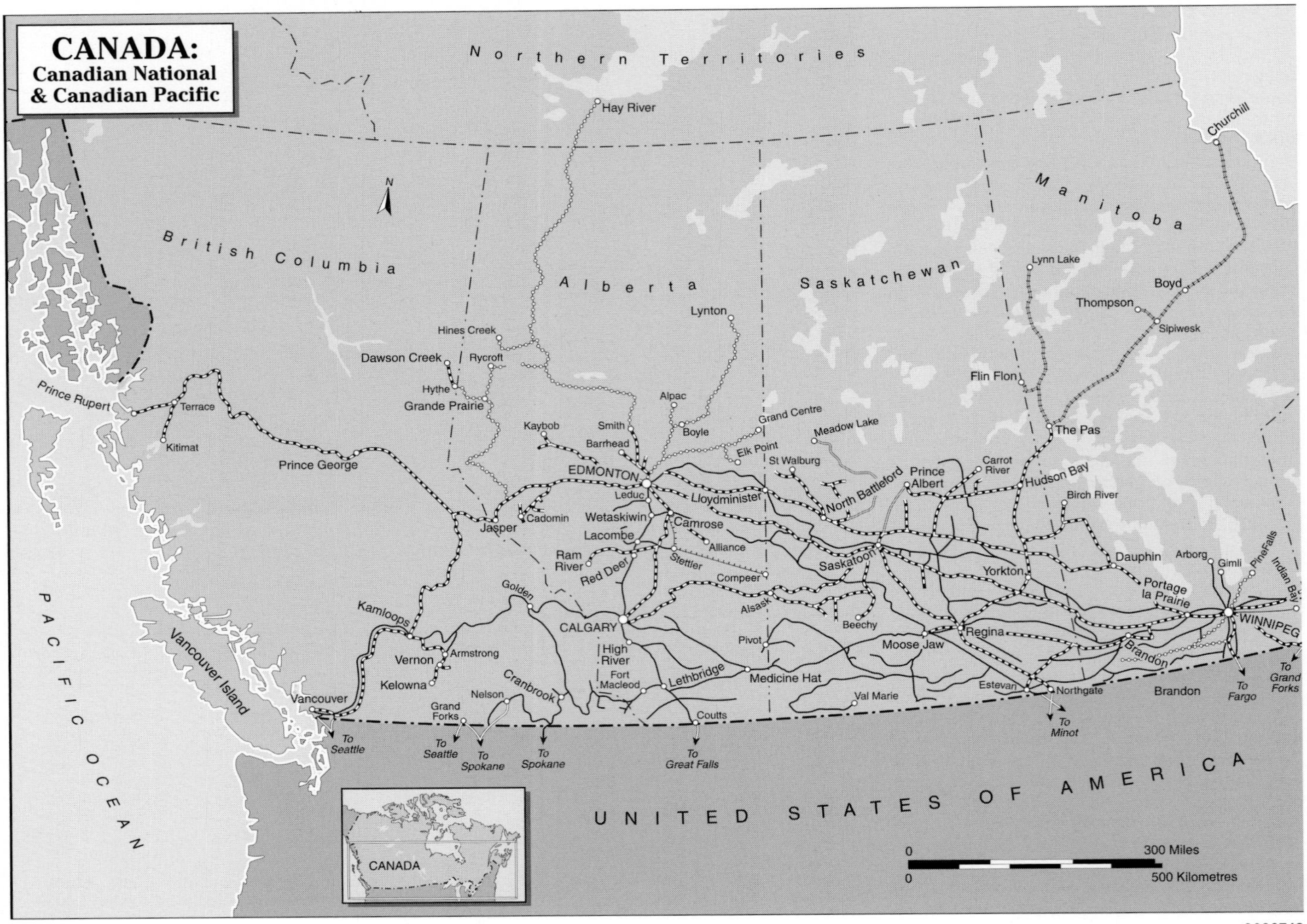

CANADA:
Canadian National
& Canadian Pacific

0092748

those best managed at a more local level. Network operations, marketing, and management of the largest customer accounts remains centralised, but each division has full responsibility for local operations, sales, and financial performance. When WC joined the CN system, its territory became a sixth division, named Wisconsin Central. At the beginning of 2003 CN realigned its divisions and merged the WC Division into the Midwest Division, thereby reducing its operating divisions to five.

In October 2003 CN announced that it was to acquire the railroad and related holdings of the US-based Great Lakes Transportation LLC for US$380 million. The acquisition covered the Duluth, Missabe and Iron Range Railway Co (DN&IR), the Bessemer and Lake Erie Railroad Co, the Pittsburgh & Conneaut Dock Co, a small harbour switching company, and Great Lakes Fleet Inc, owners of eight ships operating on the Great Lakes.

In November 2003 CN announced that it was to acquire from the British Columbia government the shares of BC Rail, along with the right to operate over its trackbed under a long-term lease. CN would be responsible for infrastructure maintenance. The acquisition for a cash price of C$1 billion was subject to approval by Canada's competition authorities. CN subsequently invited proposals from third party organisations to run tourist passenger services in British Columbia.

At the end of 2002, CN employed approximately 22,114 people.

Finance

CN revenues totalled C$6.110 billion in 2002, an increase of 8 per cent over C$5.652 billion the year before and 58 per cent higher than revenues in 1995, the year CN was privatised. Operating expenses increased by 17 per cent, from C$3.97 billion in 2001 to C$4.641 billion in 2002. CN's operating ratio (expenses as a proportion of revenues) increased from 70.2 per cent in 2001 to 76 per cent in 2002. Net income in 2002 was C$800 million, compared with C$1,040 million in 2001.

In 2002 five of CN's seven business units registered revenue gains: forest products (22 per cent); petroleum and chemicals (19 per cent); automotive (14 per cent); metals and minerals (14 per cent); and intermodal (9 per cent). Grain and fertilisers revenues declined 15 per cent and coal by 4 per cent. Carloads increased from 3,821,000 in 2001 to 4,164,000 in 2002.

The 2002 and 2001 financial information detailed above is based on US GAAP and includes data relating to WC from 9 October 2001.

Freight operations

CN markets its freight transportation services through seven business units based on major commodity groups. In 2002 the petroleum and chemicals unit generated 19 per cent of CN's freight revenue, metals and minerals 9 per cent, forest products 22 per cent, coal 5 per cent, grain and fertilisers 17 per cent, intermodal 18 per cent, and automotive 10 per cent. Fifty-seven per cent of this revenue comes from US domestic and transborder traffic, 19 per cent from overseas traffic, and 24 per cent from Canadian domestic traffic.

CN pursues business growth for the company and competitive advantage for customers by offering high-quality service and single-line access to a greater range of markets. With its industry-leading innovative service plan, CN operates a truly 'scheduled railroad', with precise scheduling and a trip plan for each car or container that directs the shipment from origin to destination. The service plan and scheduled operations enable CN to meet customers' demands for faster transit times and reliable on-time performance. At the same time, CN can exercise tight cost control through more efficient use of its assets. CN continues to make progress towards its target of 95 per cent trip compliance for its merchandise and intermodal traffic.

Marketing alliances and operational agreements with other carriers enable CN to offer customers greater reach as well as shorter transit times to markets. In 2000 CN and Canadian Pacific (CP) initiated an agreement on train operations along a 262 km stretch of the Fraser Canyon in British Columbia. With westbound trains using CN trackage and eastbound trains running along CP's, the arrangement has speeded service on both railroads. A haulage agreement with Burlington Northern Santa Fe (BNSF) for Illinois and Iowa agricultural products has resulted in improved transit times since it was implemented in October 2000, while an interline service agreement with BNSF gave shippers of new carload traffic extended access to markets in western Canada and the US when it went into effect in November the same year.

General Electric Dash 9-44CWL locomotive at Jasper, Alberta (Quintus Vosman)   0547794

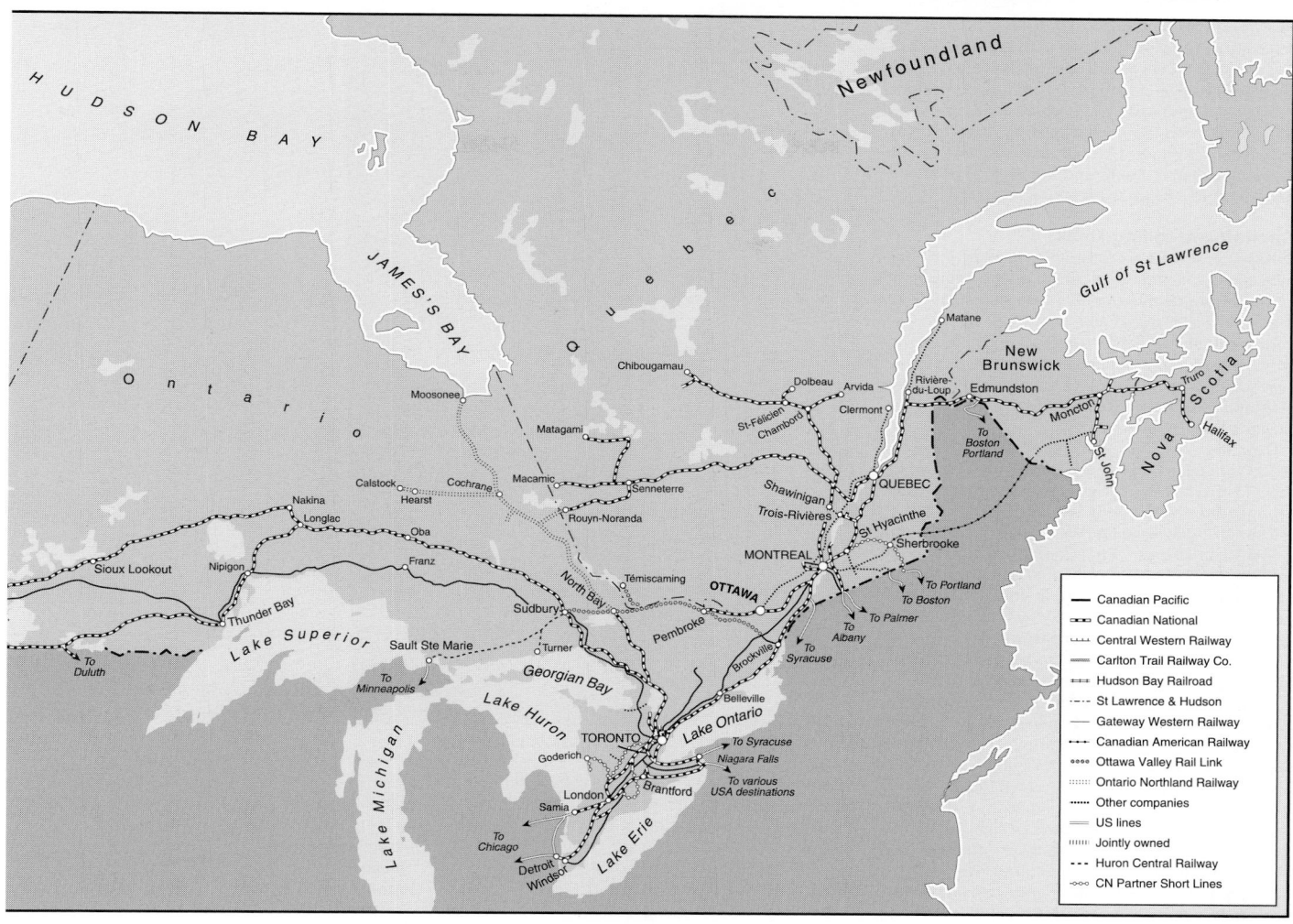

0100061

As a result of better asset utilisation and as part of a drive to achieve even greater efficiency, CN introduced a Guaranteed Car Supply programme in 2000, initially for customers who ordered centerbeam, centerstakes or double-door boxcars. The Guaranteed Car Order programme was expanded to include all eligible railcar fleets (covered hoppers, gondolas, boxcars, and so on) by November 2002.

Intermodal operations

Intermodal business accounted for about 18 per cent of CN's freight revenues in 2002. The intermodal business unit includes two market segments, domestic and international. The domestic segment handles consumer products and manufactured goods moving within Canada and the United States and generated 57 per cent of intermodal revenues in 2002. It offers two sales channels: retail, which supplies full door-to-door service through CN's trucking group, and wholesale, which provides service to trucking companies, carload freight forwarders, and intermodal marketing companies supplying their own transportation service packages to shippers. The international segment transports import-export containers for ocean shipping companies and contributed 43 per cent of the business unit's revenues in 2002.

The intermodal group has a strong base of strategic alliances with numerous trucking companies and other transportation service providers to supply seamless service throughout North America. An agreement with CSX Intermodal, initiated in 2001, led to new truck-competitive services between Vancouver and New York, Toronto and markets in Florida, and Toronto and New York. Also in 2001, CN and Union Pacific implemented an agreement covering NAFTA traffic moving between Canada, New England, and the Detroit area and south Texas, with connections to Mexico via Grupo Transportación Ferroviaria Mexicana.

CN operates 20 strategically located intermodal hubs and satellites in Canada and the United States,

Diesel locomotives: US operations

Class	Builder	Model	No in service	First built	Rating kW	Speed km/h
EF-640c	GE	D8-40C	12	1994	2,984	105
GF-630d	EMD	SD40-3	50	1967–71 (Remanufactured)	2,238	105
GF-630e	EMD	SD40	6	1967 (Remanufactured)	2,238	105
GF-630f	EMD	SD40	12	1969	2,238	105
GF-630g	EMD	SD40	1	1966 (Remanufactured)	2,238	105
GF-630h	EMD	SD40	16	1966 (Remanufactured)	2,238	105
GF-630i	EMD	SD40X	1	1966 (Remanufactured)	2,238	105
GF-630j	EMD	SD40-2	2	1975	2,238	105
GF-630k	EMD	SD40-2	1	1979	2,238	105
GF-630l	EMD	SD40-2	39	1975	2,238	105
GF-630m	EMD	SD40-3	21	1975–76 (Remanufactured)	2,238	105
GF30f	EMD	SD40	6	1969	2,238	105
GF30g	EMD	SD40	5	1970	2,238	105
GF30j	EMD	SD40	3	1970	2,238	105
GF30y	EMD	SD40-2	8	1975	2,238	105
GF-430h	EMD	GP40	1	1968	2,238	105
GF-30z	GM	SD40	1	1970	2,238	105
GF-30za	GM	SD40-2	5	1973	2,238	105
GF-430k	EMD	GP40-2	1	1973	2,238	105
GF-430l	EMD	GP40-2	1	1979	2,238	105
GF-430m	GM	GP40	17	1987–91 (Remanufactured)	2,238	105
GF-430n	EMD	GP40	2	1967	2,238	105
GF-430p	EMD	GP40	19	1970	2,238	105
GF-636c	EMD	SD45	42	1967–71	2,685	105
GF-636d	EMD	SP45	32	1967–71	2,685	105
GR-415a	GM	GP7	7	1952 (Remanufactured)	1,302	105
GR-417a	EMD/GTW	GP9	28	1955–58 (Remanufactured)	1,343	105
GR-20a	EMD	GP38	3	1966–69	1,492	105
GR-20d	EMD	GP38AC	2	1966–71	1,492	105
GR-420g	EMD	GP38-2	32	1972	1,492	105
GR-420h	EMD	GP38-2	8	1975	1,492	105
GR-420j	EMD	GP38-2	6	1977	1,492	105
GR-420k	EMD	GP38-2	12	1978	1,492	105
GR-420l	EMD	GP38-2	20	1978–79	1,492	105
GR-420m	EMD	GP38-2	5	1980	1,492	105
GR-420n	EMD	GP38-2	14	1972	1,492	105
GR-420p	EMD	GP38-2	38	1974	1,492	105
GR-420q	GM	GP38-2	6	1981	1,492	105
GR-625a	EMD	SD35	1	1965	2,100	105
GS-12a	EMD/IC	SW7/SW9	27	1949–56 (Remanufactured)	895	105
GS-415a	GM	SW1500	15	1969–71	1,305	105

including the 52-hectare Gateway Terminal near Chicago. The company is also a partner in the Deltaport Terminal at Vancouver. In 2001 CN invested C$50 million in new terminals at Edmonton (completed in 2001) and Montreal (completed in 2002). The company is currently implementing a new gate technology system, Speed Gate, which provides truckers with self-serve entry to terminals,

thus increasing throughput while virtually eliminating queuing at entry points.

RoadRailer service, introduced on the busy Montreal-Toronto corridor in 1999, continues to make inroads in this competitive marketplace, attracting a number of major accounts. The service uses unique dual-mode technology whereby highway trailers can be easily transferred to specially designed rail bogies and hauled as a train. RoadRailer combines the flexibility of highway transport with the efficiency of rail, enhancing CN's ability to compete with trucking for time-sensitive shipments in short-haul markets. CN has also initiated new expedited train services with redesigned schedules, cutting transit times between Toronto or Chicago and western points by as much as 24 hours and between Toronto and eastern centres by up to six hours.

Traction and rolling stock

CN's locomotive fleet objectives aim to increase utilisation and efficiency through the retirement of less productive units, the acquisition of more efficient ones and a range of process improvements. Fleet productivity continues to move ahead rapidly, thanks in part to improved asset utilisation as a result of the scheduled service plan and to efficiencies created by a major thrust towards longer trains.

Over the past four years, CN has been able to reduce its locomotive fleet significantly even while moving more freight. The active locomotive fleet dropped from 2,300 units in early 1998 to 1,450 at the end of 2002, excluding WC. Acquisitions of high-power units also serve to upgrade the efficiency of the fleet. In 2000, CN acquired 40 new GE 4,400 hp locomotives, 30 additional new units in 2002 and ordered a further 30 for delivery by the end of 2004. Motive power efficiency, measured in terms of gross ton-miles per available horsepower (gtm/hp), increased by 45 per cent from 1998 to 2002.

In 2002, as part of the ongoing motive power renewal programme, CN awarded a six-year contract to General Motors Electro-Motive Division to remanufacture the engines of 300 high-horsepower locomotives. Ranging from 2,835 to 2,984 kW (3,800 to 4,000 hp), the units represent almost 30 per cent of the 1,100-unit main line fleet with horsepower of 3,000 or more. The remanufactured engines will add to the locomotives' efficiency and reliability and contribute significantly to CN's goal of optimum asset utilisation. The company is also working to achieve further efficiency gains by combining smaller trains into longer ones hauled by the new, more powerful locomotives.

Operational efficiencies have also enabled CN to make significant reductions to its wagon fleet in recent years. The active railcar fleet dropped from

CN locomotive against the Chicago skyline   0134254

82,900 units in early 1998 to 61,500 at the end of 2002, excluding WC. Fleet productivity in terms of railcar velocity rose by 25 per cent during the same period.

Track

Rail: Currently, sections are being bought in four categories: 136 CN (a special section for head-hardened rail used only in curves), 136 RE, 132 RE (limited amounts for maintenance) and 115 RE.

Welded rail

Sections are electric pressure flash-butt welded into strands of about 1,480 ft (451 m) in central plants. After unloading at a laying site, rail may be electric pressure flash-butt field welded by portable plants into longer lengths before laying, or welded by aluminothermic process after laying.

Crossties (sleepers)

Thickness

Wood: Main lines: 7 or 6 in (180 or 150 mm)
  Branch lines: 6 in (150 mm)

Concrete (CN 60B): 8 in (200 mm) at rail seat

Spacing

Wood: Main lines: 1,932/km

Concrete: 1,640/km

Rail fastenings

Wood: 6 or 5½ × ⅝ in (150 or 140 × 16 mm)

Concrete: Pandrol

Canadian Pacific

Canadian Pacific Railway Company
Suite 500, Gulf Canada Square, 401 9th Avenue SW, Calgary, Alberta T2P 4Z4
Tel: (+1 403) 218 70 00   Fax: (+1 403) 205 90 00
Web: http://www.cprailway.com

Key personnel

President and Chief Executive Officer:
  R J Ritchie
Executive Vice-Presidents
  Operations: E V Dodge
  Commercial: J H McDiarmid
  Finance: G C Halatsis
Vice-Presidents
  Legal Services: Ms M M Szel
  Mechanical Services: N R Foot
  Information Services: A H Foster
  Intermodal and Auto: L M Allen
  Resource Products: F J Green
  Agricultural Products & Coal: R A Sallee
  St Lawrence & Hudson: J Coté
  Engineering Services: E J Rewucki
  Government and Public Affairs: D W Flicker
  Human Resources: R A Shields
  Field Operations: P A Pender
  Transport Services: W P Bell
Director, Commuter Rail: R O'Meara

Gauge: 1,435 mm
Route length: 25,000 km

Soo Line Railroad Company

Soo Line Building, PO Box 530, Minneapolis, Minnesota 55440, USA
Tel: (+1 612) 347 80 00   Fax: (+1 612) 347 80 59

CP grain train near Lake Louise, Alberta   0058675

St Lawrence & Hudson Railway Company Limited

PO Box 6042, Station Centre-Ville, Montréal, Québec H3C 3E4
Tel: (+1 514) 395 51 51   Fax: (+1 514) 395 77 54

Key personnel

President: Jacques Coté
Commercial Director: Mary McCarthy
Chief Operating Officer: Paul Gilmore

Organisation

Canadian Pacific Railway Company is currently the fifth largest railway system in North America based on route-kilometres. The railway is a wholly owned subsidiary of Canadian Pacific Limited, which also owns CP Hotels, Fording Coal, CP Ships, and is a majority shareholder in PanCanadian Petroleum.

In 1996 CPRC renamed its former eastern operating unit the St Lawrence & Hudson Railway to include the Delaware & Hudson Railway in the USA. At the same time it effectively scrapped its western operating unit. CPRC also introduced a new concept, the internal short line, on StL&H territory in Ontario, with separate labour agreements and management techniques adopted to operate a 140 km branch line northeast of Toronto. The relationship with CP Limited has also changed in that CPRC is now a stand-alone corporation rather than an operating division of the parent.

CPRC is undergoing a major restructuring which has included, *inter alia*, the relocation of its head office from Montréal to Calgary during 1996, bringing the administrative functions closer to the operations heart of the business, that is into the western provinces, and involving the relocation of 700+ employees. By eliminating a further 1,750 administrative positions the company has generated an annual saving of C$100 million from the reorganisation. Overall, 11 management levels have been compressed into six since 1995.

The declared design for CPRC is a 'wishbone' network consisting of Vancouver-Winnipeg plus Winnipeg-Toronto and Winnipeg-Chicago.

At the end of 1997 the company employed 19,776 staff, a fall of 6.6 per cent over the previous year.

The three core management functions within CPRC are now:
1. Commercial: to market and sell the railway
2. Operations: to run trains
3. Support services: to maximise company effectiveness and performance.

Soo Line Railroad

The Soo Line company, based in Minneapolis, Minnesota, continues to operate as a separate entity because of different laws and labour contracts which apply in the USA. It operates in 11 US states: Illinois, Indiana, Iowa, Kansas, Kentucky, Michigan, Minnesota, Missouri, North Dakota, South Dakota and Wisconsin. The Shoreham workshops in the Twin Cities support a fleet of 350 locomotives (252 road units and 98 yard/transfer units).

In 1996, CPRC entertained bids from 15 potential purchasers of the Chicago-Kansas City main line, as well as various grain lines in northern Iowa and southern Minnesota. In April 1997, it was announced that a new company, I&M Rail Link (qv in USA section), with a controlling interest held by the Washington Companies (see Montana Rail Link in USA) would start up operation on these lines with 25 locomotives and 1,000 cars from CP as part of the undisclosed sale price. CPRC retains a minority equity position.

Delaware and Hudson Railway

This northeastern US rail operation was purchased out of bankruptcy proceedings in 1991. Since then, CP has invested more than C$80 million to upgrade the railway. It is now part of the CP-owned St Lawrence & Hudson Railway. In addition to its intermodal business, traffic on the D&H includes fertiliser, clay, grain, food products, machinery, coal and forest products.

St Lawrence & Hudson Railway

The St Lawrence & Hudson Railway was formerly CP's Eastern Operating Unit, a Montréal-based management structure responsible for all lines and operations in the Montréal-Toronto-Chicago traffic lane, plus all CP lines in the northeastern USA. It competes in the area historically challenged by topographic constraints on rail operations, a dense highway network and abundant truck competition. In part this new creation was the response to the privatisation of CN and came into effect in December 1995; industry observers see a stand-alone condition in the near future. Presently, the operating unit has two management layers: Operations (to address service reliability and cost-effectiveness of running trains) and Commercial (to focus on customer service, accounts management and traffic growth). CPRC provides the third management area, namely Support Services. The 434 power units that CPRC identifies as property of the St Lawrence & Hudson are maintained at St Luc (Montréal), Toronto and Brighamton, New York.

E&N Railfreight

CPRC created a separate business unit, called E&N Railfreight, to serve 275 km of line on Vancouver Island. E&N is a freight-only operator but offers leases to passenger (VIA) or tourist interests.

Line sales

CPRC has pressed ahead with its rationalisation plan, filing its first three-year plan in August 1996

and its first update of same in April 1997. The St Lawrence & Hudson also has a three-year plan. By mid-1995 CP had disposed of all routes in the provinces of New Brunswick and Nova Scotia and the USA state of Maine.

In January 1995, CP sold the 670 km Sherbrooke-St John route to three newly established short line operators. The portion between Sherbrooke, Québec, and Brownville Junction, Maine, passed to the Canadian American Railroad Company (qv); the remainder was made into two contiguous railways (Eastern Maine in the US; New Brunswick Southern in Canada) formed by the Irving Group of New Brunswick. Earlier, in 1994, CP had abandoned 381 km of lines in eastern Québec and sold 92 km in Nova Scotia, now operated as the Windsor & Hantsport. During 1996 CP sold a further 384 km in southern Québec and the US state of Vermont to Iron Road Railways (qv in USA).

In November 1996, 615 km were leased to the newly formed Northern Plains Railroad (qv in the USA section). Called 'the Wheat Lines', the many branches fan out across North Dakota. Connection with CP is made at Thief River Falls, Minnesota, at the east end, and Kenmare, North Dakota, at the west end.

In the April 1997 three-year plan, 1,655 km of branch lines in the prairie provinces have been marked for rationalisation. CP anticipates 595 km are prospective short lines and the balance may be abandoned after proceedings are exhausted. In total, 7,000 km (including the St Lawrence & Hudson) were identified as underperforming. Some 3,240 km in 31 segments across Canada, ranging from 5 to 291 km in length, have been advertised for expressions of interest.

Finance

In 1996, CPRC had a fourth consecutive profitable year before restructuring charges, reporting net operating income of C$405.4 million and an operating ratio (expenses as a percentage of income) of 85.9, a four point improvement over a year earlier. Gross revenues were C$3,772.1 billion, essentially unchanged from 1995.

Freight operations

Salient statistics on the scale of freight operations on CPRC in 1997 are: 300 billion gross tonne-km, up 1.3 per cent over 1996; revenue tonne-km at 178.8 billion, up 2.4 per cent; gross tonne-km per employee at 14,892, an improvement of 9.3 per cent.

Overall, traffic rose slightly in 1997 from 1996, with grain and coal, sulphur and fertiliser traffic all showing revenue-tonne-km growth of over 10 per cent.

Intermodal operations

CPRC offers both domestic and import/export intermodal services, although without a port in the Maritimes (far eastern provinces) following recent line disposals, CPRC appears now to be excluded from a major traffic lane. CPRC has an interline service agreement with Conrail to expedite movements of containers between Montréal/Toronto and the ports of New York and New Jersey. The company operates 24 intermodal terminals in Canada and the USA.

In Vancouver CPRC is building a C$37 million replacement terminal, while a new C$27 million facility is being constructed in Calgary: both were scheduled to open in 1999. In Toronto, the Vaughan terminal is the subject of a C$17.8 million expansion scheme.

Finances (C$ million)	1994	1995	1996
Revenues	3,665.1	3,779.4	3,772.1
Expenses	3,621.8	4,540.1	3,241.3
Net income (loss)	43.3	(760.7)	530.8
Net income from real estate/other	21.0	15.9	(125.4)[1]
Net income after taxes, from railway	64.3	(744.8)[2]	405.4

[1] Includes interest expense.
[2] In 1995, a C$1.1 billion restructuring charge was booked.

Unaudited figures for 1997 indicated revenue of C$3,718.7 million, up 4.4 per cent on a restated 1996 figure of C$3,559.4 million. Freight revenue in 1997 accounted for C$3,428.7 million, the greatest growth being in grain, which produced a 13.2 per cent increase to C$844.8 million. Operating income was up 32.7 per cent to C$802.1 million. Asset rationalisation, including disposals of uneconomic lines, generated proceeds of C$450 million. Capital expenditure totalled some C$700 million, a new record for CPRC.

Diesel locomotives before assignment to operating units

Class	Power kW	First built	No in service	Builder (Rebuilder)	Engine
Road Freight					
SD40; 40A; 40B	2,238	1966–71	112	EMD	GM16-645E3
SD40M-2	2,238	1995*	10	EMD-MK	GM16-645E3
SD40-2, [2F]	2,238	1972–89	589	EMD [AMF]	GM16-645E3
GP40	2,238	1966–67	28	EMD	GM16-645E3
SD60	2,835	1987–89	58	EMD	GM16-710G3
SD60M	2,835	1989	5	EMD	GM16-710G3
SD80MAC	3,730	1998	10	EMD	GM20-710G3
M630	2,238	1969–70	1	MLW	MLW16-251E
M636	2,238**	1970–71	4	MLW	MLW16-251F
AC4400CW	3,283	1995	83	GE	GE 16V 7FDL16
AC4400CW	3,283	1997–98	91	GE	GE 16V 7FDL16
Road Switchers					
GP9	1,305	1981–88*	49	EMD/CP	GM16-645C
GP30; 35	1,680	1963–66	25	EMD	GM16-645D3
GP30C	1,492	1963	3	EMD	GM16-645D3
GP38	1,492	1970–71	21	EMD	GM16-645E
GP38-2	1,492	1983–86	185	EMD	GM16-645E3
GP39-2	1,715	1978	2	EMD	GM16-645
SD39	1,715	1968	2	EMD	GM16-645
RSD17	1,790	1959	1	MLW	MLW16-251B
RS18R	1,343	1989–90*	67	MLW/CP	MLW12-251C
RS23	746	1959–60	20	MLW	MLW6-251B
C-424	1,790	1963–66	50	MLW	MLW16-251B
Yard Switchers					
SW8	596	1984*	2	EMD/CP	GM8-567B
SW9	671	1982–83*	12	EMD/CP	GM8-567B
SW900	671	1984–85*	3	EMD/CP	GM8-567C
SW1200RS	895	1981–85*	90	EMD/CP	GM12-567B, C
SW1500	1,120	1966	2	EMD	GM12-645E
GP7	1,120	1987–88*	24	EMD/CP	GM16-645B, C
GP9	1,305	1980–87*	184	EMD/CP	GM16-645C
GP9M	1,305	1954	5	EMD	GM16-645C
GP15C	1,120	1990–91	7	EMD	CAT3512
MP15AC	1,120	1975–76	32	EMD	GM12-645E
SD10	1,343	1952–54	3	EMD	GM16-567C, E
F7B	1,120	1983*	1	EMD	GM16-567B
Slugs	n/a		7	EMD	n/a

* Date of rebuild.
** 1 rated at 2,312 kW (Caterpillar 3,608 engine).

TOFC

To meet fierce competition on the Montreal-Toronto corridor, StL&H has introduced its drive-on/drive-off 'Iron Highway' service, which eliminates the need to lift trailers onto rail vehicles, and operates on a slot reservation system accessible by telephone or the Internet.

RoadRailer

In addition to trailer and container on flat car (TOFC/COFC) traffic, CP hauls RoadRailer bimodal units. Triple Crown Service Inc, the RoadRailer operating subsidiary of the US railway Norfolk Southern, has extended its network into Canada with a Detroit-Toronto service which is operated by CPRC.

Triple-stack potatoes

In a joint venture involving Soo Line Railroad, container manufacturer Stoughton Composites, Thermo King and logistics company C H Robinson, the last tested a triple-stack container configuration in 1995. Three stacked 6 ft 4 in boxes (equal to the usual pair of 9 ft 6 in boxes) have been used to transport potatoes from Thief River Falls, Minnesota, to a processing plant at Frankfort, Indiana. The 6 ft 4 in containers are designed to 'weigh out' and 'cube out' simultaneously and avoid what would be dead space in a 9ft 6in container.

Improvements to existing lines

In recent years, CP has spent in the range of C$600-885 million annually to maintain its roadbed and equipment. The three-year projection for 1997–1999 is C$1.8 billion. Included are C$14.2 million for a new locomotive shop in St Paul; C$50 million for a rebuilt Bensenville (Chicago) Yard; and C$40 million for a new intermodal yard at Pitt Meadows (Vancouver). Among the improvements that have been undertaken recently include C$15 million spent on enlarging 47 tunnels, which has allowed CP to introduce double-stack container services between Vancouver and Chicago and on its principal route across Canada. The last job completed was the 13 km Connaught Tunnel in Rogers Pass, in 1995.

The company spent C$27.5 million to enlarge the Detroit River Tunnel connecting Windsor, Ontario, with Detroit, Michigan (USA). Work has also been undertaken on rehabilitation of the Niagara River Bridge, linking Ontario with the US state of New York. Enlargement work on the Detroit River Tunnel was completed in April 1994, with the first double-stack container train between Montréal and Chicago running on 6 May 1994 via the newly enlarged tunnel. By April 1995 CP recorded a 30 per cent increase in traffic through the tunnel following completion of the project. Another project in 1995 involved the installation of 5 km of double-track east out of Moose Jaw, Saskatchewan, to relieve queueing.

Traction and rolling stock

At the end of 1996 the CPRC locomotive fleet (that is with Soo Line and St Lawrence & Hudson Railway not included, and after dispositions) comprised

Transcontinental double-stack container train leaving Vancouver for Toronto and Montreal   0001839

821 diesel locomotives. To alleviate a major motive power shortage in 1995, units were leased from EMD, Helm, Conrail, GATX and rebuilders such as Precision National and National Railway Engineering. The 'loaner' fleet peaked at 240, was reduced to 135 by the first quarter of 1996 and then in early 1997 CP experienced a renewed shortage and leased 140 more units; 60 came from Norfolk Southern and 11 from Montana Rail Link.

Over 600 units (close to 40 per cent) are SD40, SD40-2 and SD40-2F models, delivered between 1970 and 1989 and these are shared by all the CP operating units. CP has retired all its high-horsepower MLW locomotives with the exception of one unit re-engined with a Caterpillar diesel; CP has assigned most of its lower-horsepower MLW units to the St Lawrence & Hudson Railway and this fleet is also in decline due to its age and maintenance costs.

By using a new asset management programme featuring enhanced preventive maintenance and quick response repairs, CP has been able to add 60 locomotives to its available fleet, the equivalent of power for 15–20 trains. CP's capital investment programme for 1995 included major expansion of its Moose Jaw, Saskatchewan, locomotive workshops with the doubling of the number of staff employed there.

Recent locomotive purchases

Having decided to invest C$200 million in its locomotive fleet in 1994, CPRC ordered 40 AC4400CW diesel-electric locomotives from GE in November 1994, the railway's first units with AC drives. The order was increased to 83 in early 1995, and deliveries continued through the end of 1995 and early 1996. The AC4400CW units are all assigned to Calgary, Alberta, for western coal and other bulk commodity service; three new locomotives are replacing five existing DC drive units.

CP subsequently ordered a further 181 of the AC4400CW class for delivery 1997–98, and 60 SD90/43MACs from EMD to be supplied during 1998.

Freight wagons

At the beginning of 1996, CPRC's revenue freight wagon fleet exceeded 50,000 owned or leased, 21,000 of which were covered hoppers for the transport of bulk commodities such as grain, potash and fertiliser. Recent orders have concentrated on this type of wagon, which has proved to be in short supply at times of peak grain traffic. In common with other Canadian carriers, CPRC has the potential to acquire a share of the fleet of 10,000 hoppers which have yet to be disposed of by the Canadian government. During 1996, 1,400 covered hopper cars were added at a cost of C$105 million. Also, the last tranche of steel coal cars (397 cars in 1996) were remodelled to increase payload by eight tonnes per car.

Cando Contracting Ltd

830 Douglas Street, Brandon, Manitoba R7A 7B2
Tel: (+1 204) 725 26 27   Fax: (+1 204) 725 41 00
e-mail: info@candoltd.com
Web: http://www.candoltd.com

Key personnel

President: Gordon Peters
Controller: Colleen MacCarl
Asset Manager: Doug Phillips
Sales Manager: Brent Montague
Abandonment Manager: Alex Burr
Installation Manager: Jerry Lovas

Alberta office

35D Rayborn Crescent, St Albert, Alberta T8N 5B6
Tel: (+1 780) 418 23 53
Fax: (+1 780) 418 23 65

e-mail: don.barr@candoltd.com
General Manager: Don Barr

Ontario office

160 Edward Street, St Thomas, Ontario N5P 1Z3
Tel: (+1 519) 637 87 56   Fax: (+1 519) 637 13 62
e-mail: doug.peters@candoltd.com
General Manager: Doug Peters

Cando operates four shortlines totalling some 560 km in Canada:

Athabasca Northern Railway Ltd (323 km), established in October 2000 to operate the former CN line between Boyle and Linton, Alberta, with three locomotives;
Barrie/Collingwood Railway (101 km), municipally owned by the Ontario communities it serves and operated by Cando since January 1998;

Central Manitoba Railway Inc (81 km), former CN lines in Manitoba linking Winnipeg–Pine Falls and Winnipeg–Graysville, operated with a fleet of four locomotives;
Orangeville Brampton Railway (55 km), municipally owned via the Orangeville Railway Development Corporation and operated by Cando with a twice-weekly service.

In addition, Cando provides rail switching and freight handling services for industrial concerns.

Cape Breton & Central Nova Scotia Railway

121 King Street, PO Box 2240, Stellarton, Nova Scotia B0K 1S0
Tel: (+1 902) 752 33 57
Fax: (+1 902) 752 66 65
Web: http://www.railamerica.com

Key personnel

General Manager: Peter Touesnard

Gauge: 1,435 mm
Route length: 394 km

Organisation

The Cape Breton & Central Nova Scotia Railway (CBNS) was acquired by RailTex in October 1993. When RailTex was taken over by RailAmerica, the railway became part of that company's North American Rail Group in February 2000. It operates 394 km of ex-Canadian National trackage east from its CN connection at Truro, across Nova Scotia to Sydney, where there is a steel works.

Freight operations

Carloadings in 2001 were approximately 26,000. Principal commodities carried are scrap iron, limestone, forest products and grain.

CANADA
Short lines in the Maritime Provinces

1.435 m gauge
- ⋯ Quebec, North Shore & Labrador Railway Passenger & freight
- ⊢⊣ Cartier Railway, Freight only
- ‐ ‐ Arnaud Railway, Freight only
- ▬ Cape Breton & Central Nova Scotia Railway Freight only
- ⊔⊔⊔ Canadian-American Railway, Freight only
- ○○○○ New Brunswick Southern Railway Freight only
- ▭▭▭ New Brunswick East Coast Railway Freight & VIA Rail Passenger
- ‐·‐ Chemins de fer Baie des Chaleurs Freight & VIA Rail Passenger
- ⋈⋈ Matapédia Railway Freight & VIA Rail Passenger
- ▭▭ Canadian National
- ⋯⋯ VIA Rail, Passenger only

0058503

Carlton Trail Railway (CTRW)

77-155th Street East, Prince Albert, Saskatchewan S6V 1E9
Tel: (+1 306) 763 94 70   Fax: (+1 306) 763 94 71
Web: www.omnitrax.com

Key personnel

Vice-President, Marketing (OmniTRAX Canada): Iris Thornton

Gauge: 1,435 mm
Route length: 523 km

Organisation

A subsidiary of OmniTRAX Inc (see US section of *Railway systems and operators*), CTRW operates unconnected segments of former CN lines in northwest and northeast Saskatchewan, comprising Saskatoon via trackage rights to Warman and on to Paddockwood and from North

Battleford via trackage rights to Speers and via Shellbrook to Meadow Lake. Connection is made with CN at North Battleford and Saskatoon. Principal commodities handled are forest products, grain and grain products.

UPDATED

Cartier Railway

Cie Minière Québec Cartier
Port Cartier, Duplessis County, Québec G5B 2H3
Tel: (+1 418) 766 23 21   Fax: (+1 418) 768 24 28
Web: www.qcmines.com

Key personnel

President and Chief Executive Officer: Guy Dufresne
Vice-Presidents
    Operations Technology: Gaston Morin
    Operations Management: François Pelletier
    Finance: J Roy
General Manager, Railway: Serge A Michaud
Divisional Manager, Mechanical Maintenance: Gérard Sirois

Gauge: 1,435 mm
Route length: 416 km

Organisation

Cartier Railway is a wholly owned subsidiary of Québec Cartier Mining.

Freight operations

The 307 km railroad built to convey iron ore concentrate from Lac Jeannine (mine and

concentrator site) to Port Cartier (harbour site) was completed in 1960. In 1972 a 138 km extension was built to Mont Wright where a second concentrator was built for another iron ore mine.

In 1975–76 a 4.8 km bypass was constructed to transport crude ore from Fire Lake (which was opened to compensate for the closing of the Lac Jeannine mine). However, Fire Lake mining ended in 1985 and today only Mont Wright workings are exploited.

The ore is hauled in unit trains of 160 wagons operated with two GE 4400 AC locomotives on the head end, a two-person crew and no caboose. The railroad normally operates five such trains daily throughout the year to match the concentrator production.

Train size best matches the cycle time of the fixed installations, which provide continuous loading and discharge at the port, the latter by a double-car Strachan & Henshaw rotary dumper which works at the rate of 3,800 tonnes per hour. In recent years the annual volume of ore transported and dumped has been 12 to 14 million tonnes.

Traction and rolling stock

The fleet consists of 20 diesel-electric line-haul locomotives (including 12 GE 4400 AC units) and 950 ore wagons and 380 miscellaneous vehicles.

Signalling and telecommunications

Computerised control has been superimposed on the Centralised Traffic Control (CTC) system, which is supplemented by centrally controlled hot box detectors and switch-point heaters. Radio communication with crews is employed for dispatch of train orders.

Track

Main line track is entirely cwr, employing 132 lb rail until 1988, but since then 136 lb has been adopted as the standard weight.
Rail: Standard carbon in tangents, low alloy in curves
Crossties (sleepers)
Type: Hardwood 2,590 × 228 × 177 mm
Spacing: 1,851/km
Fastenings: Cut spikes
Min curvature radius: 250 m
Max gradient: 1.35% against empty trains; 0.4% compensated against loaded trains
Max axleload: 31 tonnes
Type of braking: Wabco 26L
Type of coupler: CF70HT

UPDATED

Central Western Railway

Suite 306, Associated Centre, 13220 St Albert Trail, Alberta T5L 4W1
Tel: (+1 403) 742 25 03   Fax: (+1 403) 742 14 77
Web: www.railamerica.com

Key personnel

General Manager: Shawn Smith
Business Development Manager: Daniel Fletcher

Gauge: 1,435 mm
Route length: 166 km

Organisation

A subsidiary of RailAmerica, the CWR comprises former CN and CP lines in southern Alberta between Stettler and Consort and Morrin and Munson. Approximately 2,800 carloads are handled each year, with grain the main commodity carried.

UPDATED

E&N Railway Company (1998) Ltd

PO Box 581, 23 Esplanade, Nanaimo, British Columbia V9R 5L3
Tel: (+1 250) 754 92 22   Fax: (+1 250) 754 53 18
Web: http://www.railamerica.com

Key personnel
General Manager: Anne Venema

Gauge: 1,435 mm
Route length: 291 km

Organisation
Operated by RailAmerica, the ENR serves Vancouver Island, linking Victoria in the south with Courtenay via a line running along the island's east coast. The railway handles approximately 8,500 carloads annually, mostly of forest and paper products, minerals and chemicals.

Interchange is made with CP via a Wellcox–Vancouver train ferry.

Essex Terminal Railway Company

1601 Lincoln Road, PO Box 24025, Windsor, Ontario N8Y 4Y9
Tel: (+1 519) 973 82 22
Fax: (+1 519) 973 72 34
e-mail: bmckeown@etr.ca
Web: http://www.etr.ca

Key personnel
President: Brian G McKeown
Vice-President: Terry J Berthiaume

Controller: Teresa Boutet
Operations Superintendent: Edward G Clough
Road Master: Bill Comboye
Car Foreman: Robert Bulmer
Locomotive Foreman: Robert Woods

Gauge: 1,435 mm
Route length: 35 km

Organisation
The ETR provides industrial switching services between the east side of Windsor and Amhertsburg, Ontario, and is strategically located close to one of the key border crossings into the USA. Commodities carried include forest products, agricultural and chemical products, machinery, steel, grain, salt and scrap. Interchange is made with CN, CP, CSXT and NS.

Traction and rolling stock
The ETR operates five diesel-electric switching locomotives.

Genesee-Rail-One

6650 rue Durocher, Building 1, Outremon, Quebec H2V 3Z3
Tel: (+1 514) 273 57 39   Fax: (+1 514) 273 99 38
Web: http://www.gwrr.com

Key personnel
President: Mario Brault
General Manager, HCR: Garth Rushton
Manager, Marketing, HCR/QGR: Bill Sclater

Organisation
A subsidiary of Genesee & Wyoming Inc (qv in USA section), Genesee-Rail-One operates two former Canadian Pacific Rail lines in eastern Canada: the Huron Central Railway (HCR) resulted from the acquisition from CP Rail in July 1997 of the line between Sudbury and Sault Ste Marie, in northern Ontario; the Québec–Gatineau Railway (QGR) began operating in November 1997 and links Quebec, Montreal, Gatineau and Hull, Quebec. The QGR handles some 49,000 carloads annually.

Traction and rolling stock
In 2002, the HCR was operating 13 locomotives, all of General Motors design and comprising GP9, SD45 and GP40 types. The QGR fleet totalled 23, a mix of General Motors GP35, GP38, GP40, SD45 and SW1500 types.

Goderich-Exeter Railway

126 Weber Street West, Building #2, Kitchener, Ontario N2H 3Z9
Tel: (+1 519) 749 80 00   Fax: (+1 519) 749 80 88
Web: www.railamerica.com

Key personnel
General Manager: Bob Decicco
Business Development Manager: Cheryl Ford

Gauge: 1,435 mm
Route length: 256 km

Organisation
Based on the former CN Goderich and Exeter subdivisions, the GEXR was initially acquired by RailTex, and was that company's first acquisition outside the USA. RailAmerica took over operations in February 2000 following its acquisition of the interests of RailTex.

The railway links Goderich with Silver, with branches to Centralia and London. Interchange with CN is made at London and McMillan Yard, Toronto, and with CP at Kitchener and Guelph. Approximately 19,000 carloads annually are handled. Principal commodities carried are limestone, food and farm products, plastics, forest products and automotive components.

UPDATED

GO Transit

20 Bay Street, Suite 600, Toronto, Ontario M5J 2W3
Tel: (+1 416) 869 36 00   Fax: (+1 416) 869 35 25
Web: www.gotransit.com

Key personnel
Chairman: Peter Smith
Vice-Chairman: Dr Gordon Chong
Vice-Chair: Hazel McCallion
Managing Director and Chief Executive Officer: Gary W McNeil
Director, Corporate Services and Secretary: Jean M Norman
Director, Financial Services: Frances Chung

Gauge: 1,435 mm
Route length: 361 km

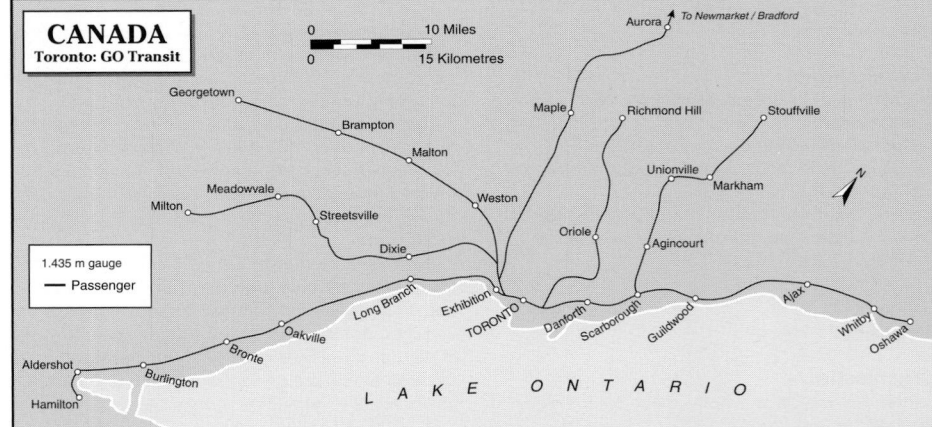

0003048

Political background
GO Transit is the provincially funded, inter-regional public transit service for the Greater Toronto Area (GTA) and Hamilton. Created in 1967 by the Province of Ontario, GO has grown from a single rail line along Lake Ontario's shoreline into an integrated network of trains and buses connecting downtown Toronto with the surrounding communities. In a transfer of responsibilities, the Province handed over the funding of GO to the GTA municipalities at the beginning of 1998. In January 2002 the Province took back responsibility for GO Transit. Legally known as the Greater Toronto Transit Authority, GO is once again a Crown agency of the Province of Ontario.

GO (originally an abbreviation of Government of Ontario) Transit serves a territory of over 8,000 km² with a population of 4.9 million, which continues to increase. It runs an integrated bus and rail passenger network with a total annual ridership of more than 43 million. Over the seven years to 2003, ridership grew by 40 per cent.

In May 2004 a new stream of funding for improvements to GO services was announced jointly by the Ontario and Federal governments.

C$1.05 billion is to be made available to fund a 10-year expansion programme, comprising 12 major projects. These include improvements to Toronto Union station; 36 km of three-tracking on the Lakeshore East and West lines to eliminate bottlenecks and raise capacity for expansion of GO Train services; 33 km of new track on the Georgetown route; improvements to the Milton, Bradford and Stouffville lines; and new buses and park-and-ride stations to expand the GO Bus network to serve Peterborough, Cambridge, Niagara Falls, and the Guelph-Kitchener-Waterloo area.

Passenger operations

Starting with a single rail route along Lake Ontario, GO now operates seven lines (taking into account that the Lakeshore line has been split into East and West lines) and 55 stations, with Toronto's Union station as the system's hub. In addition to 179 trains run daily, in 2005 GO Transit operated 1,503 bus trips. In 2004 train and directly related bus services carried an average of 150,000 passengers each weekday.

Train services are run under contract over Canadian National tracks (in six corridors) and St Lawrence & Hudson Railway – Canadian Pacific's eastern operating division (one corridor) – by CN and StL&H crews to GO Transit specification, but GO Transit owns the infrastructure of the Lakeshore East line between Pickering and Whitby.

In July 2000 GO Transit took ownership of the platforms at Toronto's Union Station. A C$100 million 10-year investment programme is planned to increase capacity and improve facilities at the terminal. A new Union Station Bus Terminal, adjacent to the rail facility, was completed in 2002.

Improvements to existing lines

In March 2003 the Canadian federal government agreed to contribute C$435 million to a C$1.2 billion programme of capacity and other improvements to the GO Transit network. The rest of the funding was to be provided by communities served by GO.

The programme includes: provision of a third track from Port Credit to Oakville and from Burlington to Aldershot; a new storage yard and improvements to signalling on the Lakeshore and Union corridors (Scarborough–Union Station–Hamilton); grade separations and improvements to track and stations on the Georgetown and Milton corridors, enabling a doubling of train frequencies; upgrading stations on the Milton line and at some Lakeshore locations to accommodate 12-car trains; grade separations, track improvements and new stations on the Bradford and Stouffville corridors; track improvements and a new service to Barrie; and the introduction of a bus-rail service to Peterborough, Cambridge, Niagara Falls and Guelph/Kitchener/Waterloo. Initial contracts for improvements to the Lakeshore West line from Toronto to Hamilton were let at the end of 2003. The Barrie service is due to commence in mid-2007. Provision of a third track between Don River and Scarborough, on the Lakeshore East line, is scheduled for completion by 2009, as are the improvements to the Bradford, Georgetown and Stouffville corridors. Station upgrades on the Milton and Lakeshore lines to accommodate longer trains is scheduled for completion by 2008.

Traction and rolling stock

GO Transit operates a standardised fleet consisting of 45 F59PH diesel-electrics from GM-Canada (jointly designed by GM and GO Transit) and a 375-strong passenger coach fleet exclusively composed of double-deck vehicles, built by UTDC and Bombardier. Bombardier

GO Transit Type F59PH diesel locomotive arrives at Toronto with a morning peak commuter service (Colin Boocock)
NEW/1114954

GO Transit ordered more Bombardier BiLevel cars in 2004 to bring its fleet up to 371 vehicles 0580215

also holds a contract to maintain GO's rolling stock. Mechanical problems with the F59PHs mean that the fleet may be retired at half-life; funding was granted in 2005 for the replacement of the fleet, expected to lead to the purchase of 27 replacement locomotives for delivery in 2006–08, with an option to procure a further 26.

Coupler in standard use (passenger cars): H-tightlock

Braking in standard use (locomotive-hauled stock): 26L, 26LUM and 26C

UPDATED

Great Western Railway (GWR)

PO Box 99, Shaunavon, Saskatchewan S0N 2M0
Tel: (+1 306) 297 27 77   Fax: (+1 306) 297 25 08
Web: www.greatwesternrail.com

Key personnel
General Manager: Stacey Wallis

Gauge: 1,435 mm
Route length: 515 km

Organisation
GWR started operations in September 2000 when CP's Shaunavon, Vanguard, Altawan and Noteku subdivisions in Saskatchewan were acquired by track materials supplier Westcan Rail. In November 2004 GWR was bought by a grouping of farmers and other local interests. Lines run southwest from an interchange with CP at Swift Current and Assiniboia, Saskatchewan, to Bracken, serving 27 grain elevators and private sidings. Some 2,600 cars are handled annually.

Traction and rolling stock
At the beginning of 2005 GWR operated two MLW 240 diesel-electric locomotives of 1974 vintage.

UPDATED

Greater Vancouver Transportation Authority (GVTA) – TransLink

West Coast Express
Suite 295, 601 West Cordova Street, Vancouver, British Columbia V6B 1G1
Tel: (+1 604) 689 36 41   Fax: (+1 604) 689 38 96
Web: www.westcoastexpress.com

Key personnel
President and Chief Executive Officer: Doug Kelsey

Gauge: 1,435 mm
Route length: 65 km

Political background
As the regional transport authority for the Vancouver area, TransLink is responsible for

transit projects, commuting options, major roads, air quality and intelligent transport systems. Its Transit Programme, which includes the West Coast Express operating subsidiary, accounts for some 70 per cent of the authority's budget.

Passenger operations
In May 1994 BC Transit, the responsibilities of which were taken over by GVTA in April 1999,

entered into an agreement with CP Rail to undertake development work for the introduction of a commuter rail service serving Vancouver and the northeast sector of the Greater Vancouver Regional District (GVRD). Marketed as West Coast Express (WCE), services began on 1 November 1995 over a 65 km CP route which is all double-track and equipped with CTC (centralised traffic control) between Mission and central Vancouver. Journey time (with six intermediate stops) is 73 minutes, comparing favourably with two hours by bus. WCE research indicates that 75 per cent of riders were previously motorists. The service contract was awarded to CP Rail and the maintenance contract to VIA Rail.

Traffic	2002	2003	2004
Passenger journeys (million)	1.9	1.9	2.0

Service frequency is five trains in the direction of peak flow only. Opening daily ridership was 5,000. WCE carried its 10 millionth passenger in October 2001. Daily ridership for 2004 averaged 8,250 during peak months.

In 2003 WCE extended its operating hours with a service called TrainBus, operated under contract by Gray Line. A coach-bus departs from Waterfront station each evening, calling at every WCE station. An additional TrainBus, launched in September 2003, starts after the last train leaves Mission in the morning. Future plans include the provision of additional coach services and the commissioning by 2008 of two additional stations.

West Coast Express trains at Waterfront Yard, Vancouver   0593969

All fare/ticket types have provision for passengers to transfer to other transit modes (bus; SeaBus; SkyTrain); 60 per cent of train riders make a transfer. WCE issues a reusable smartcard, Xpress, for some ticket purchases.

In 2004 WCE employed 13 staff.

Traction and rolling stock
WCE is operating with five 2,200 kW GM model F59PHI locomotives in push-pull mode with 37 Bombardier-built bilevel cars. In addition, one locomotive is on loan from Sounder Transit in Seattle. In 1998 a global positioning satellite (GPS) system was installed, enhancing the ability to monitor train location and on-time performance.

UPDATED

Hudson Bay Railway (HBRY)

PO Box 2129, 728 Bignell Avenue, The Pas, Manitoba R9A 1L8
Tel: (+1 204) 627 20 07   Fax: (+1 204) 623 30 95
Web: www.omnitrax.com

Key personnel
Vice-President, Marketing (OmniTRAX Canada):
    Iris Thornton

Gauge: 1,435 mm
Route length: 1,303 km

Organisation
A subsidiary of OmniTRAX Inc (see US section of *Railway systems and operators*), HBRY operates former CN lines in northern Manitoba. These comprise lines from The Pas northeast to Churchill, on Hudson Bay, and from The Pas north to Lynn Lake. Connection is made with CN at The Pas.

Principal commodities carried include ores and concentrates, copper and zinc, wheat and barley, forest products and petroleum products. HBRY also operates an intermodal service.

A passenger service is operated under contract with VIA Rail Canada.

UPDATED

Huron Central Railway

See entry for Genesee-Rail-One.

Lakeland & Waterways Railway

Suite 306, Associated Centre, 13220 St Albert Road, Edmonton, Alberta T5L 4W1
Tel: (+1 780) 448 58 55
Fax: (+1 780) 439 56 58
Web: www.railamerica.com

Key personnel
General Manager: Tim Husel
Business Development Manager:
    Angela Bourbonnais

Gauge: 1,435 mm
Route length: 164 km

Organisation
Operated by RailAmerica, the LWR comprises a network of four shortlines running north and east from Edmonton. The railway handles approximately 10,500 carloads annually, commodities carried including petroleum coke, grain, forest products and petroleum products. Interchange is made with CN at Edmonton and with the ANY at Boyle.

UPDATED

Kelowna Pacific Railway

2806 27th Avenue, Vernon, British Columbia V1T 9K4
Tel: (+1 250) 549 13 18   Fax: (+1 250) 549 15 89
Web: www.knighthawk.ca

Key personnel
President: Tom Rothfels
General Manager: Dave Hanratty

Gauge: 1,435 mm
Route length: 175 km

Organisation
Owned by KnightHawk Rail Ltd, a division of the Knighthawk contract air freight handling operation, KPR started operations in December 1999. It comprises former CN trackage in the Okanagan Valley, running from a connection with CN at its Kamloops yard and Campbell Creek (over 41 km of CP trackage rights) to Armstrong and Kelowna. Parts of the Okanagan valley route are operated through trackage rights over the Okanagan Valley Railway (see entry in Canada section of *Railway systems and operators*). About 16,000 carloads are handled annually, mainly forest products, grain and industrial products.

Traction and rolling stock
At the beginning of 2005 KPR operated 10 diesel-electric locomotives.

UPDATED

Ex-Central Kansas Railway GM locomotives in Kelowna Pacific service at Vernon, British Columbia
(Quintus Vosman)
NEW/0585081

Mackenzie Northern Railway

Roma Junction Yard Office, PO Box 7648, Peace River, Alberta T8S 1T2
Tel: (+1 780) 332 64 00   Fax: (+1 780) 332 29 04
Web: www.railamerica.com

Key personnel
General Manager: Grant Beattie
Business Development Manager: Daniel Fletcher

Gauge: 1,435 mm
Route length: 1,046 km

Organisation
Operated by RailAmerica since July 1999, the MNR is a former CN line which runs north from Smith, Alberta, to Hay River, Northwest Territories. The railway handles approximately 26,500 carloads annually, mainly of fuel, grain pulp, forest products and chemicals. Interchange with CN is made at Smith.

UPDATED

New Brunswick Southern

New Brunswick Southern Railway Co Ltd
11 Gifford Road, PO Box 5666, Saint John, New Brunswick E2L 5B6
Tel: (+1 506) 632 47 12
Fax: (+1 506) 632 58 18
Web: http://www.nbsouthern.com

Key personnel
General Manager: B L Bourgeois

Gauge: 1,435 mm
Route length: 190.5 km

Organisation
As part of the January 1995 disposal of the Canadian Atlantic Railroad by Canadian National, 190.5 km located in New Brunswick were acquired by the J D Irving Company and are now operated as the NBSR using six GP9 units overhauled either by OmniTrax (qv in the USA section) or by AMF Technotransport. The property consists of a 136 km line from the border at McAdam to Saint John, plus a 55 km branch from McAdam to Saint Stephen. The line continues as the Eastern Maine Railroad (also owned by Irving but served by the Candian American Railway (qv)) for 168 km into the US from McAdam to Brownville Junction, where it connects with the Bangor and Aroostook Railway.

NB Southern Railway and road operator Sunbury Transport collaborate with CP Rail and Canadian American Railroad to provide intermodal services.

Okanagan Valley Railway (OKAN)

4710 31st Street, Suite 104, Vernon, British Columbia V1T 5J9
Tel: (+1 250) 503 24 43   Fax: (+1 250) 503 24 47
Web: www.omnitrax.com

Key personnel
Vice-President, Marketing (OmniTRAX Canada): Iris Thornton

Gauge: 1,435 mm
Route length: 151 km

Organisation
A subsidiary of OmniTRAX (see US section of Railway systems and operators), OKAN serves the Okanagan Valley in British Columbia, running south from Sicamouse, where connection is made with CP, via Armstrong to Kelowna. Connection with CN is made at Armstrong. Principal commodities carried include sand, forest products, grain, chemicals and fertilisers, cement and soda ash.

UPDATED

Ontario Midwestern Railway Company

Organisation
This company was formed in the first half of 1997 to buy or lease 175 km of CP track within Ontario, from Mississauga to Owen Sound.

Ontario Northland (ONRail)

555 Oak Street E, North Bay, Ontario P1B 8L3
Tel: (+1 705) 472 45 00
Fax: (+1 705) 476 55 98
e-mail: info@ontc.on.ca
Web: http://www.ontc.on.ca

Key personnel
President and Chief Executive Officer (of the ONTC): K J Wallace
Vice-Presidents
Finance and Administration: S G Carmichael
Telecommunications: R S Hutton
Transportation Services: R G Leach
Superintendent, Train Operations: J Thib

Gauge: 1,435 mm
Route length: 1,211 km

Political background
Ontario Northland (ONRail) is a component of the multimodal Ontario Northland Transport Commission's operations. In October 2002 the Commission entered exclusive negotiations with Canadian National (CN) to acquire ONRail in pursuit of its policy of privatisation. These negotiations were terminated by CN in June 2003. Subsequently the Ontario government decided not to proceed with privatisation of the company.

Organisation
The system lies at the eastern rim of the province. It runs from North Bay, where it connects with Canadian National and CPRC routes westward from Ottawa, to Moosonee on James Bay, the southward-probing neck of Hudson Bay.

Trading as Rail Contract Shop, ONRail's repair and maintenance facility at North Bay offers its services to industrial customers in North America.

In addition to rail services, ONTC offers bus services, ferry services on the Great Lakes, and telecommunications. It left the air passenger business in March 1996.

In 1991 agreement was reached for the transfer to Ontario Northland of 240 km of CN line between Cochrane and Calstock. This section connects with Ontario Northland at Cochrane. The main line has four additional short branches.

Passenger operations
ONRail runs the Northlander passenger service from Cochrane to Toronto Union station daily except Saturdays, travelling over 367 km of CN track between Toronto and North Bay. North of Cochrane, to Moosonee, a mixed train, the Little Bear, runs three times a week September-June and twice-weekly in July and August when it is augmented by a tourist train, the Polar Bear Express.

In February 1999, ONRail's seasonal 'Snow Train' service commenced operations, carrying passengers and their snowmobiles between Toronto and Cochrane.

Freight operations
Freight, chiefly lumber, pulp, newsprint, chemicals, petroleum products and ores, is the backbone of Ontario Northland's business. Express freight services are provided between Cochrane and Moosonee.

Traction and rolling stock
The railway rosters 30 line-haul and six switching diesel-electric locomotives, 46 passenger coaches and 810 freight wagons. The passenger fleet includes 26 single-level commuter coaches acquired from GO Transit (the last six were purchased in 1994) and comprehensively renovated as first-class long-haul coaches in Ontario Northland workshops.

Three FP7A locomotives have been rebuilt with Caterpillar 3516 engines rated at 1,529 kW. Seven diesel locomotives without cabs have been acquired for conversion to provide head-end power for ONRail's refurbished passenger coaches.

Signalling and telecommunications
ONRail employs radio despatching.

Track
Rail: 125.77 kg/m
Crossties (sleepers)
Wood: Thickness: 180 mm
Spacing: 1,886/km
Fastenings: 4 spikes per sleeper on tangent, 8 spikes per sleeper on curves
Min curvature radius: 300 m
Max gradient: 1.5%
Max axleload: 29.5 tonnes (65,000 lb)

Diesel locomotives

Class	Wheel arrangement	Power kW	Speed km/h	Weight Tonnes	No in service	First built	Mechanical	Builders Engine	Transmission
SD40-2	C-C	2,237.1	104.6	170.7	8	1973	GM	645E3	E GM Main Gen
GP38-2	B-B	1,492.5	104.6	115.67	10	1974	GM	645E	E GM Main Gen
FP7A	B-B	1,529	113.0	117.11	3	1994*	GM	Cat 3516	E GM Main Gen
GP-9	B-B	1,304.97	104.6	117.34	6	1956	GM	567C	E GM Main Gen

*Repowered.

O-Train

OC Transpo
City of Ottawa Transit Services
1500 St Laurent Boulevard, Ottawa, Ontario K1G 0Z8
Tel: (+1 613) 741 64 40   Fax: (+1 613) 741 73 59
Web: www.octranspo.com

Key personnel

Chair, OC Transpo: Janet Stavinga
Transit Services Director: Gordon Diamond
Communications and Marketing: Joan Weinman

Organisation

Established in 1972, OC Transpo has exclusive rights to operate within the area of the 775,000 population Regional Municipality of Ottawa-Carleton, operating extensive bus services in the region. Since 2001 these have been operated by the City of Ottawa.

Passenger operations

In October 2001 OC Transpo commissioned its O-Train service, running at a 20 min frequency between Bayview and Greenboro (8 km). Serving five stations, at which services are closely integrated with bus operations, the system uses European-style lightweight diesel railcars running over an existing CP freight line. This is single-track with a passing facility at Carleton. A park and ride facility is provided at Greenboro. Following track improvements, the service frequency was increased to 15 min from September 2003.

New lines

A rapid transit expansion plan for Ottawa, which includes extension of the O-Train service, has been approved by the city's council. The plan covers the period from 2004 to 2021, when the population of Ottawa is forecast to be 1.2 million. As a first step,

in 2005 studies were in progress into extending the existing line by around 5 km from Greenboro to Leitrim Road. Opening could be achieved by 2006. This is seen as part of a longer-term plan to create an electrified 30 km light rail system linking the Rideau Centre with Barrhaven, with a possible spur to the city's airport. A completion date of 2009 has been cited for this scheme. Future plans also include the development of a 47 km east-west line across the city from Kanata to Orléans.

Traction and rolling stock

For the Bayview–Greenboro line, OC Transpo acquired three three-car Talent lightweight diesel railcars built in Germany by Bombardier Transportation. Bombardier also undertakes maintenance.

In the longer term, low-floor electrically powered vehicles are to be procured.

UPDATED

Ottawa Valley Railway

445 Oak Street East, North Bay, Ontario P1B 1A3
Tel: (+1 705) 472 62 00   Fax: (+1 705) 472 25 27
Web: www.railamerica.com

Key personnel

General Manager: Grant Bailey
Business Development Manager: John Winkler

Gauge: 1,435 mm
Route length: 626 km

Organisation

Operated by RailAmerica, the OVR provides services over CN lines between Smiths Falls and Coniston, Ontario, and on a branch from Mattawa, Ontario, to Temiscaming, Quebec. The railway handles approximately 56,700 carloads annually.

Principal commodities carried are chemicals, ores, metal products and wood pulp. Express intermodal services are also operated. Interchange is made with CP at Cartier, Smiths Falls and Sudbury, with Ontario Northland at North Bay and with CN at Pembroke and North Bay.

UPDATED

Quebec Central Railway

Vallée Junction, Quebec G0S 3J0
Tel: (+1 418) 253 62 92
Fax: (+1 418) 253 62 93

Key personnel

General Manager: Danny Giguere

Organisation

Originally extending to some 600 route-km, this railway was built to serve asbestos mines, and closed with the collapse of that industry in the 1980s. Although the line was largely abandoned, it was still owned leasehold by CP, which in 2000

sold out to trucking company Express Marco. Reconstruction with provincial and federal government funding saw train operations resume in mid-2001, hauling principally forest products and grain. The reopened portion extends to 323 route-km, comprising a main line between Charny, Quebec, where there is interchange with CN, and Sherbrooke, where traffic is exchanged with Canadian American; a branch runs some 125 km from Vallée Junction to Lac Frontière. Parts of the railway are used by two tour train operators.

Traction and rolling stock

QCR operates a mixed fleet of nine locomotives, including two GP11 and three FP7A.

Québec–Gatineau Railway

See entry for Genesee-Rail-One.

Québec North Shore & Labrador Railway

Québec North Shore & Labrador Railway Co
100 rue Retty, Sept-Iles, Québec G4R 3E1
Tel: (+1 418) 968 74 95   Fax: (+1 418) 968 74 98

Key personnel

President and Chief Executive Officer: M D Walker
General Manager: Marc Duclos
Manager, Materials and Services: John Turnbull
Superintendents
   Transport and Traffic: Michel Lamontagne
   Equipment Maintenance: Gilbert Sarazin
   Maintenance of Way, Signals and Communications:
     Louis Gravel

Gauge: 1,435 mm
Route length: 639 km

Organisation

Begun in 1950 by its then newly formed owners, the Iron Ore Company of Canada (IOCC), with the shareholding support of several US steelmakers, the 573 km main line of the Québec North Shore & Labrador (QNS&L) runs from Schefferville south to Sept-Iles on the St Lawrence River. Schefferville, the railhead for the Ungava ore tract in the Labrador peninsula, is just inside the Québec border, but otherwise the northern half of the route is enclosed by Newfoundland. Within this section a 58 km branch was run from Ross Bay westward to Labrador City in 1960.

Passenger operations

Using ex-VIA Rail Budd RDC diesel railcars, QNSL operates passenger services weekly from Sept-Iles

to Schefferville and twice-weekly from Sept-Iles to Labrador City. Additional services are operated according to demand during summer months. Car-carrier wagons are attached on most services to convey passengers' cars to locations lacking road access.

Freight operations

IOCC stopped mining in the Schefferville area in 1982, but still serves the railhead as there are no roads north of Sept-Iles. Mining is now concentrated in the Carol Lake area, at the extremity of the Labrador City branch.

Despite the savage winters in the region, QNS&L functions all year round. In winter, however, it only moves processed (beneficiated) ore in pellets, because of raw ore's propensity to freeze.

QNS&L runs loaded ore trains varying in length from 117 to 265 wagons; the latter trail about 3.3 km behind their lead locomotives and gross over 33,700 short tons, but a 117-wagon train weighs at least 14,000 short tons. Normal power is two 2,200 kW GM-EMD SD40 locomotives at the front end, but when a train is made up to 165 wagons or more, mid-train helper units, radio-controlled by the Locotrol system from the lead locomotive, are added. Twenty-two of the type SD40-2 units were Locotrol-equipped by AMF Technotransport in 1994 and redesignated

SD40-2CLCs. In 1997, IOCC announced a C$14 million contract with GE/Harris Railway Electronics for a train control system aimed at enhancing performance of a train powered by multiple locomotives.

Traction and rolling stock

Resources consist of 55 diesel-electric locomotives, 2,400 freight wagons, 11 passenger coaches and six Budd RDC diesel railcars bought used from VIA in 1994. In 1994 the QNS&L acquired its first GE locomotives, three Dash 8-40C units bought new. In 1998, the railway ordered 11 General Electric C44-9W locomotives.

Signalling and telecommunications

Centralised Traffic Control (CTC) is in operation on 416 km of QNS&L's 573 km main line. The railway is controlled from Sept-Iles.

Track

Rail: 65.5 kg/m
Crossties (sleepers)
Treated hardwood: 177.8 × 228.6 × 2,743.2 mm
Spacing: 2,080/km
Fastenings: Standard track, 165 mm (5½ in)
Min curvature radius: 220 m
Max gradient: 1.32%
Max axleload: 32.5 tonnes

Diesel locomotives

Class	Wheel arrangement	Power kW	Speed km/h	Weight tonnes	No in service	First built	Mechanical	Builders Engine	Transmission
SD40/40-2	C-C	2,238	114	174	48	1968	GM	GM	*E* GM
GP-9	B-B	1,305	114	109	4	1954	GM	GM	*E* GM

Québec Railway Corporation

1130 Sherbrooke Street West, Suite 310, Montréal, Québec, H3A 2M8
Tel: (+1 514) 982 99 44   Fax: (+1 514) 849 23 19
Web: http://www.domino-hq01.cn.ca

Key personnel
President: Serge Belzile
Vice President, Operations: Gilles Richard
Vice President, Finance: Alain Tessier

Gauge: 1,435 mm
Route length: 1,064 km

Freight operations
Through acquisitions from CN, QRC has established a network of lines on the north and south shores

of the St Lawrence River in eastern Québec and New Brunswick. Wagons are transferred between the two shores and into the North American rail network via the Cogéma train ferry.

The QRC bought 144 km of line linking Limoilou and Clermont from CN in 1994. It is operated as the Chemin de Fer Charlevoix Inc.

In December 1996, the company bought CN's line between Matapédia and Chandler (235 km) which it operates as the Chemin de Fer Baie des Chaleurs.

In December 1997, QRC acquired from CN the 168 km Matapedia Railway, linking Mont-Joli, Quebec and Campbellton, New Brunswick. In March 1999, it acquired from CN the Rivière-du-Loup–Mont-Joli and Mont-Joli–Matane lines (360 km). Together these are operated as the Matapedia and Gulf Railway (CFMG). At Matapedia,

connection is made with the Chemin de Fer Baie des Chaleurs. VIA Rail uses the Rivière-du-Loup line at Campbellton for its Maritime service.

In October 1998, QRC purchased the 157 km Pembroke, Ontario-Coteau, Québec line and gained operating rights over VIA Rail trackage between Coteau and Ottawa. It operates this line as the Ottawa Central Railway.

RailAmerica, Inc

RailAmerica operates several regional railroads and shortlines in Canada. See entry in United States of America section

RaiLink Ltd

In 1999 RaiLink was purchased by RailAmerica Inc (qv in USA section) for C$73.2 million.

RailTex Canada Inc

This company was a wholly owned subsidiary of RailTex Inc, which was acquired in February 2000 by RailAmerica Inc (qv in the USA section).

Rocky Mountaineer Vacations

1150 Station Street, Suite 100, Vancouver, British Columbia V6A 2X7
Tel: (+1 604) 606 72 00   Fax: (+1 604) 606 72 95
Web: www.rockymountaineer.com;
   www.whistlermountaineer.com

Key personnel
President and Chief Executive Officer:
   Peter Armstrong
Executive Vice President and Chief Operating Officer:
   James E Terry
Executive Vice President, Marketing and
   Communications: Graham Gilley

Passenger operations
Formerly Rocky Mountaineer Railtours, Rocky Mountain Vacations commenced operations in 1990 as the Great Canadian Railtour Company, following the privatisation of VIA Rail's daylight service through the Canadian Rockies, initiated in 1988. The company operates a range of tourist passenger services over CN and CP lines, offering a total of 146 departures in 2003. Services are provided in both directions between Vancouver, British Columbia and Jasper, Banff and Calgary, Alberta, running from mid-April to mid-October, as well as in December.

In 2004 additional rolling stock was ordered for two new services to commence in 2006, the

Two General Motors GP40-2 locomotives with a westbound Rocky Mountaineer service at Rupert, in suburban Vancouver (Brian Webber)
NEW/0585080

Fraser Discovery Route, between Whistler, Prince George, Alberta, and Jasper, Alberta, and the Whistler Mountaineer, between North Vancouver and Whistler, British Columbia.

Traction and rolling stock
Rocky Mountaineer Vacations operates five General Motors GP40-2 locomotives. Rolling stock totals more than 80 vehicles and includes 12 GoldLeaf Service bilevel dome cars built by Colorado Railcar,

with three more on order for the new services to be introduced in 2006.

The company's operations and maintenance facility is located at Kamloops, British Columbia.

UPDATED

Southern Ontario Railway

241 Stuart Street, Hamilton, Ontario L8R 3H2
Tel: (+1 905) 777 12 34   Fax: (+1 905) 777 01 85
Web: www.railamerica.com

Key personnel
General Manager: Stuart Thomas
Business Development Manager: John Winkler

Gauge: 1,435 mm
Route length: 87 km

Organisation
Operated by RailAmerica, the SOR provides freight services over CN lines linking Nanticoke, Caledonia, and Brantford, Ontario.

The railway handles approximately 45,000 carloads annually, mainly of fuel, grain and steel. Interchange with CN is made at Brantford and Hamilton, Ontario.

UPDATED

Southern Railway of British Columbia (SRY)

Southern Railway of British Columbia Ltd
2102 River Drive, New Westminster, British Columbia V3M 6S3
Tel: (+1 604) 521 19 66   Fax: (+1 604) 526 09 14
e-mail: kdoiron@sryraillink.com
Web: www.sryraillink.com

Key personnel
President: John F van der Burch
Vice President, Marketing and Sales:
   Ken W Doiron
Chief Mechanical Officer: Mike Moy

Manager Maintenance of Way: Bob Ewanchuck
Controller: Paul Tompkins

Gauge: 1,435 mm
Route length: 200 km

Organisation
Formerly the property of the British Columbia Hydro & Power Authority, the Southern Railway of British Columbia (SRY) assumed its current name in September 1988 when it was sold to the Itel Rail Corporation. In October 1994 it was acquired by Washington Companies, a natural resources and transport conglomerate that owns the Montana Rail Link system.

SRY's main line connects Vancouver (New Westminster) and Chilliwack, a distance of 130 km. There are several branches, including one to the industrial centre of Annacis Island. More than half of SRY's current 75,000 annual wagonloads of traffic is made up of cars imported through Annacis. Interchange is made with CN, CP, the Burlington Northern, Santa Fe and Union Pacific.

Traction and rolling stock
SRY has 23 locomotives, which are mostly switchers and over 2,000 freight wagons.

SRY also undertakes rolling stock engineering projects for other customers.

UPDATED

St Lawrence & Atlantic Railroad (Quebec) Inc

Chemin de Fer St-Laurent Atlantique (Quebec) Inc
415 Rodman Road, Auburn, Maine 04210, USA
Tel: (+1 207) 782 56 80   Fax: (+1 207) 782 58 57
Web: http://www.gwrr.com

Gauge: 1,435 mm
Route length: (with SLR) 416 km

Organisation

Together with the St Lawrence & Atlantic Railroad Company (SLR) in Maine, the St Lawrence & Atlantic Railroad (Quebec) Inc (SLQ) forms a contiguous 416 km route linking Ste Rosalie, Quebec and Portland, Maine, crossing the border at Norton, Vermont. The two companies formed part of the US-based Emons Transportation Group, which in February 2002 was acquired by Genesse & Wyoming Inc (qv in the USA section). SLQ connects directly with the CN system and indirectly with CP Rail via the Iron Road System.

Intermodal services are offered via a subsidiary company, Maine Intermodal Transportation. Together, SLQ and SLR claim to have the only route in Northern New England cleared for hi-cube, double-stack intermodal operations.

Toronto Terminals Railway Company Ltd

Suite 402, Union Station, Toronto, Ontario M5J 1E6
Tel: (+1 416) 864 34 40   Fax: (+1 416) 864 34 87
Web: http://www.ttrly.com

Key personnel

Director of Operations: Sam Spares
Supervisor, Rail Operations: Joe Fenech
Manager, Real Estate: Pio Mammone

Organisation

A wholly owned subsidiary of Canadian National and Canadian Pacific, TTR operates Toronto's municipally owned Union Station under a management contract with the city. The company's operations cover all tracks within a 5.8 km section in the vicinity of the station, handling some 40 intercity and 152 GO Transit commuter trains daily. Associated operations and through freight services result in some 350 train movements daily.

In 1983, a 600 m dive-under was constructed to reduce conflicting movements by freight services.

In February 2000, TTR announced that by 2003 it intended to transfer its lines and operating interest by sale, lease or transfer.

VIA Rail Canada

VIA Rail Canada Inc
3 Place Ville-Marie, Suite 500, Montréal, Québec H3B 2C9, Canada
Tel: (+1 514) 871 60 00   Fax: (+1 514) 871 62 27
Web: www.viarail.ca

Key personnel

President and Chief Executive Officer: Paul Côté
Chief Strategy Officer: Christena Keon Sirsly
Chief Operating Officer: John Marginson
Chief Financial Officer: J Roger Paquette
Vice-Presidents
   Marketing: Steve Del Bosco
   Procurement, Real Estate and Environment:
      Mike Greenberg
General Counsel and Corporate Counsel:
   Carole Mackay
Director, Public Affairs: Paul Raynor

Gauge: 1,435 mm
Route length operated: 6,524 km

Political background

VIA Rail Canada Inc came into being as a Crown Corporation in January 1977 as a creation of the then Trudeau administration. At that time it took over management of all rail passenger services previously operated by CN and CP Rail, except commuter services. Since then its fortunes have been inextricably tied to the cabinet of the government in office because the law requires only an order in council rather than a vote in parliament to alter VIA's future. The current mandate is clearly to apply private industry principles to administration of the corporation with the intent of determining how close it can come to self-sufficiency. For all its performance gains in the last few years, VIA is still considered expendable in some quarters of Canada's political spectrum.

VIA contracts with the Canadian government for the provision of those rail passenger services specified by the Minister of Transport. In turn, VIA contracts with railway companies for the operation of these services and with non-railway companies for the provision of incidental goods and services; 92 per cent of track used is contracted from CN.

In April 2000 the government announced a C$402 million five-year capital investment programme in VIA Rail. Projects covered by this funding include: signalling improvements, level crossing protection and track upgrades for the Montreal–Ottawa route; a major upgrade of London, Ontario station and other stations; and the development of new waste management procedures, initially on the Quebec City–Windsor corridor and later on other parts of the VIA Rail Canada network.

Finance

At the end of 2003, the company's workforce stood at 3,051, compared with 3,054 in 2001.

Revenues in 2003 were C$250.3 million compared with C$270.8 million in 2002 (down 7.6 per cent), while

VIA Rail refurbished Budd RDC railcar on Vancouver Island at Victoria forming a service to Courtenay (Colin Boocock)   ***NEW**/1114964*

operating expenses totalled C$431.4 million, a rise of 1.7 per cent. Capital expenditure stood at C$77.4 million, down 21 per cent on the previous year. The Canadian government made a total contribution of C$263.5 million for operating deficits, capital improvements and reorganisation charges. The revenue/cost ratio in 2002 was 58.5 per cent, down from 64.5 per cent a year earlier.

Traffic	2001	2002	2003
Passenger journeys (000)	3,865	3,981	3,789
Passenger-km (million)	1,482	1,526	1,379

Passenger operations

The VIA Rail network comprises four main groups of services totalling some 480 trains per week: the Quebec City-Windsor corridor, which accounts for 85 per cent of the corporation's ridership and 70 per cent of its revenue; Western services between Toronto and Vancouver, which primarily serve the tourism market; Eastern services, linking the Atlantic regions with central Canada via the Montreal–Halifax and Montreal–Gaspé routes; and Northern services in British Columbia, Ontario, Quebec and Saskatchewan, providing rail transport in regions where alternative modes are very limited or non-existent.

VIA Rail attributed the decline in patronage in 2003 to the combined effects of the SARS virus outbreak in the Toronto area and other health concerns, anxiety about the impending US war with Iraq and unpredictable phenomena such as forest fires in western Canada and a hurricane in the Maritimes region.

Improvements to existing lines

In 2002 VIA Rail completed a C$28 million upgrade of its Alexandria subdivision between Montreal and Ottawa. Work included track and level crossing improvements to allow 160 km/h operations and provision of a CTC system. This allowed VIA Rail to accelerate services between these cities by 20 minutes from 2003.

A new station was opened at Fallowfield (Barrhaven) in western Ottawa in 2002, providing passengers with improved access to trains to

Class EPA-42 (GE Type P42DC) diesel locomotive passing Paris, Ontario, with a Windsor–Toronto service (Colin J Marsden)   ***NEW**/1115013*

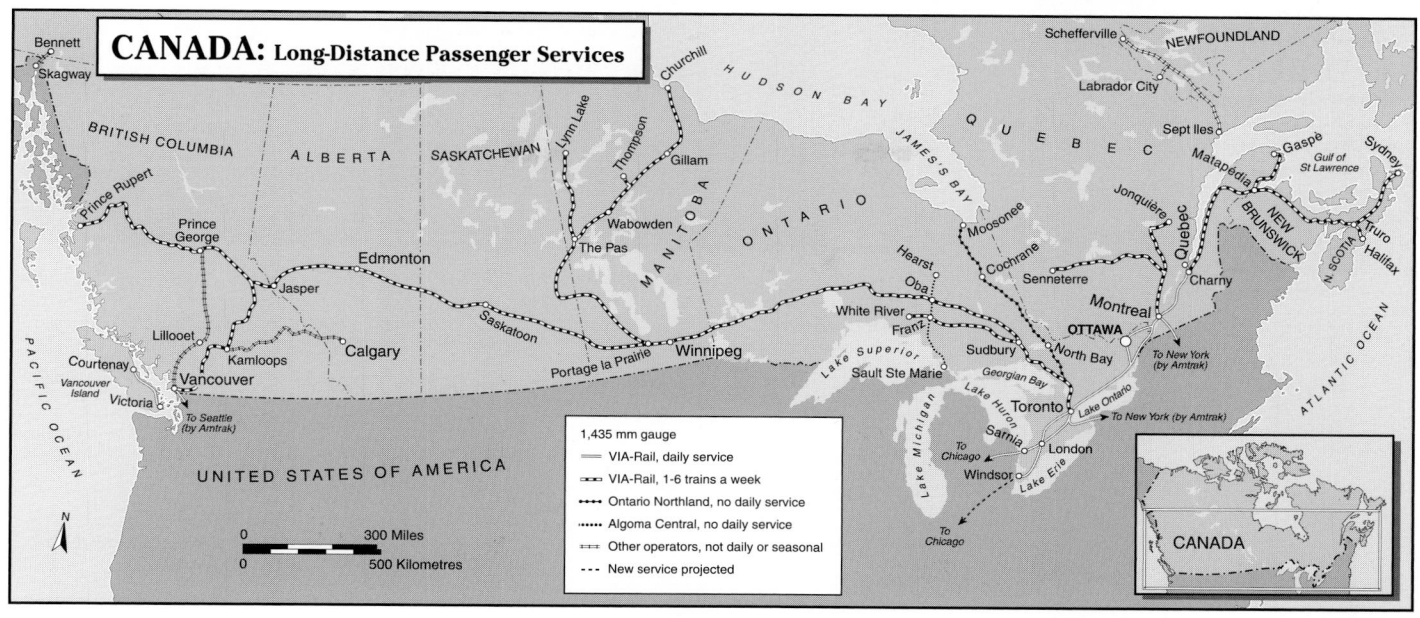

CANADA: Long-Distance Passenger Services

1,435 mm gauge
— VIA-Rail, daily service
▭▭▭ VIA-Rail, 1-6 trains a week
••••• Ontario Northland, no daily service
•••••• Algoma Central, no daily service
▭▭▭ Other operators, not daily or seasonal
– – – New service projected

NEW/1114966

and from Toronto. Major improvement schemes have been completed at Halifax station and at Vancouver's Pacific Central Station.

Traction and rolling stock
In 2004 VIA Rail's traction fleet consisted of 75 main line diesel-electric locomotives and two switchers.

The latest additions to the fleet are 21 Type P42 'Genesis' diesel locomotives ordered from General Electric in 2001 to replace LRC-2 and LRC-3 power cars.

The traditional passenger coach fleet consisted of 314 vehicles, including 131 LRC-2 and LRC-3 coaches delivered between 1978 and 1983, some of which are in store, and five Budd RDC diesel railcars. These have been the subject of a refurbishment programme by Industrial Rail Services, Moncton. Four of the RDCs are in use on the 'Malahat' service on Vancouver Island; the fifth is used between Sudbury and White River, Ontario. The hauled fleet has been supplemented by the purchase from ALSTOM of 139 UK-built passenger

Diesel locomotives

Class	Builder's type	Wheel arrangement	Power kW	Speed km/h	Weight tonnes	No in service	First built	Mechanical	Builders Engine	Transmission
GPA30a	F40PH2	Bo-Bo	2,238	153	129	20	1986	GMD	16-645E3C	E GMD
GPA30b	F40PH2	Bo-Bo	2,238	145	129	8	1987	GMD	16-645E3C	E GMD
GPA30c	F40PH2	Bo-Bo	2,238	145	129	26	1989	GMD	16-645E3C	E GMD
EPA-42	P42DC	Bo-Bo	2,985	175	112	21	2001	GE	16-7FDL16	E GE
GS-10	SW1000	Bo-Bo	746	100	–	2	1966	GMD	8-645E	E GMD

GMD General Motors Diesel (Canada)
CGE Canadian General Electric
B/MLW Bombardier/Montréal Locomotive Works
GE General Electric

coaches originally ordered by a consortium of operators for subsequently abandoned international overnight services between Great Britain and the European mainland. VIA Rail Canada has branded these vehicles Renaissance cars, introducing them initially on the Montreal–Toronto 'Enterprise' overnight service and day trains linking Quebec City, Montreal and Ottawa in

2002 and on the Montreal–Halifax 'Ocean' service in 2003.

Traction and rolling stock maintenance is undertaken at depots at Montreal, Vancouver and Winnipeg.

UPDATED

Windsor & Hantsport Railway

Windsor & Hantsport Railway Co Ltd
PO Box 578, Windsor, Nova Scotia B0N 2T0
Tel: (+1 902) 798 07 98   Fax: (+1 902) 798 08 16

Key personnel
General Manager: James Taylor

Gauge: 1,435 mm
Route length: 906 km

Organisation
The Windsor & Hantsport (WHRC) began operations in August 1994 after CP Rail sold 90 km of line, eight RS23 diesel locomotives and 76 80-ton hopper wagons to Iron Road Railways (see entry

in USA section of *Railway systems and operators*). Connection is made with the CN system at Windsor Junction, Nova Scotia, from where the line runs to Kentville. Principal commodities carried are cereals, gypsum, forest products and vegetable oil. WHRC also operates a seasonal tourist train, the 'Evangeline Express'.

UPDATED

Chile

Ministry of Transport & Communications

Amunategui 139, Santiago
Tel: (+56 2) 672 65 03   Fax: (+56 2) 699 51 38

Key personnel
Minister of Transport: N Irueta

Antofagasta (Chili) and Bolivia Railway plc (FCAB)

Ferrocarril de Antofagasta a Bolivia
Bolívar 255, Casilla ST, Antofagasta

Tel: (+56 55) 20 67 00   Fax: (+56 55) 20 62 20
e-mail: webmaster@fcab.cl
Web: http://www.fcab.cl

Key personnel
General Manager: Miguel V Sepúlveda
Planning and Development Manager:
   Marcelo F Contreras
Services (Traffic) Manager: L Bernardo Schmidt
Sales Manager: Carlos Yanine
Commercial Manager: Carlos E Acuña
Financial Manager: Pablo E Ribbeck
Human Resources Manager:
   Victor F Maldonado
Manager, Infrastructure: Juan Pavez

Manager, Engineering Maintenance: José Brown
Manager, Workshop: Carlos Bastías

Gauge: 1,000 mm
Route length: 911.33 km

Organisation
FCAB is entirely self-supporting financially. Founded in 1888, the company is listed on the London Stock Exchange. The railway runs from the Pacific port of Antofagasta to the Argentine border at Socompa on one route (over Ferronor track between Augusta Victoria and Socompa), and to the Bolivian border at Ollagüe on the second. FCAB has become the principal transport company

For details of the latest updates to *Jane's World Railways* online and to discover the additional information available exclusively to online subscribers please visit
jwr.janes.com

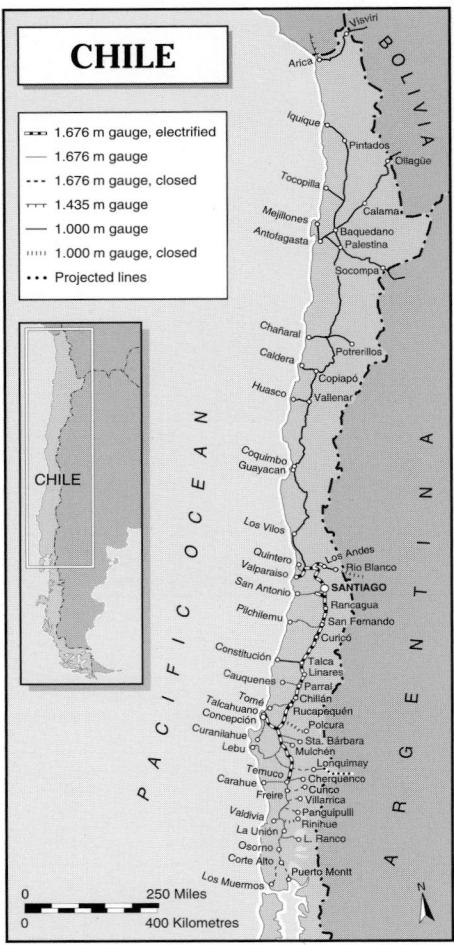

CHILE

- ☒☒☒ 1.676 m gauge, electrified
- —— 1.676 m gauge
- - - - 1.676 m gauge, closed
- ⊤⊤⊤ 1.435 m gauge
- —— 1.000 m gauge
- ⅲⅲⅲ 1.000 m gauge, closed
- ••• Projected lines

0058527

Former QR 2100 Class diesel-electric acquired for use on the FCAB network   1048552

Diesel locomotives

Class	Builder's type	Wheel arrangement	Power kW	Speed km/h	Weight tonnes	No in service	First built	Mechanical	Builders Engine	Transmission
600		Bo-Bo	373	55	45	2	1958	GM	Cummins NT.855	E Westinghouse
900	GA 8	Bo-Bo	708	55	52	3	1965	GM	EMD 8-567C	E EMD
900	GA 18	Bo-Bo	820	57	54	1	1969	GM	EMD 8-645-E	E EMD
900	G 18U	Bo-Bo*	820	96	72	1	1977	GM	EMD 8-645-E	E EMD
1400	GR 12	Co-Co	902	95	90	13	1961	GM	EMD 12-567C	E EMD
1400	NF210	Co-Co	940	96	103	18	1957	GM	EMD 12-567C	E EMD
1430**	GL 26C	Co-Co	940	80	97	10	2001	Comeng/ Clyde	EMD 16-645E	E EMD/Clyde
2000	GL26	Co-Co	1,492	96	–	11	–	GM	EMD 16-645E	E EMD

* Formerly A1A-A1A; centre axles removed
** Acquired from QR, Australia, built 1970

in northern Chile, responding to customer needs with three types of service: a door-to-door bimodal service that complements the company's rail transport activities with those of its own truck services; the design and construction of wagons to meet customers' specific requirements; and the provision of complete storage and shipment services.

In May 2004 FCAB employed 602 staff. Personnel development has been achieved through the establishment of 'quality circles' and an internal intranet has been created to enhance employee communications, with each part of the company provided with its own website.

Passenger operations

While primarily a freight operator, FCAB also operates passenger trains. In 2003 a service was operating weekly in both directions between Calama, Chile, and Uyuni in Bolivia via Ollagüe. In total 730,000 passengers were recorded in 2003.

Freight operations

In conjunction with FCAB's bulk trainload and wagonload business, door-to-door service has been offered since 1988 in conjunction with the company's road haulage subsidiary, TRAIN Company Ltd. TRAIN is a major truck operator in northern Chile, with more than 100 vehicles active in the mining, industrial and service markets.

In 2003 FCAB transported 3.89 million tonnes, an increase of 4 per cent on the figure achieved in the previous year. As well as being Chile's leading company in the transport of sulphuric acid and other hazardous materials, FCAB is also the main carrier of copper and mining products in the north of the country.

Anticipated new contracts have led FCAB to project an increase in traffic of 2 million tonnes in the period 2004–12.

Traffic	2001	2002	2003
Freight tonnes (million)	3.53	3.74	3.89
Freight tonne-km (million)	926.2	962.5	948.4

Traction and rolling stock

In May 2004 the FCAB fleet comprised 50 diesel-electric locomotives. The freight wagon fleet totalled 1,451. The fleet has been progressively reduced in size since 1998 as older vehicles are retired and replaced with new equipment offering increased capacity.

FCAB uses the Q Tron QES 1000 computerised control system for its locomotives, achieving increases in haulage capacity and improved maintenance planning.

Coupler in standard use, freight and passenger cars: AAR automatic
Braking in standard use, locomotive-hauled stock: 26L and 65L with straight control

Signalling and telecommunications

Train control is by VHF radio throughout the system. FCAB employs a computerised communications (Intranet) and control system linked to customers, enabling them to access details of progress of their shipments and to receive periodic reports and packing lists. In 2002, studies were in progress by FCAB into the possible use of a VHF radio positioning for its locomotives.

In the field of information technology, FCAB has two interconnected Local Area Networks (LANs) in Antofagasta and another in Calama, together forming a Wide Area Network (WAN). Some 250 PCs are connected to the WAN.

Track

Rail: 24.83 kg/m, 177.77 km; 32.24 kg/m, 306.19 km; 37.2 kg/m, 204.17 km; 37.22 kg/m, 3.02 km; 42.16 kg/m, 6.42 km; 42.21 kg/m, 188.99 km
Crossties (sleepers): Wood (Coigüe) 188 × 254 × 1,027 mm
Spacing: In plain track: 1,422/km; in curves: 1,422/km, 1,490/km
Fastenings: Screw spikes, Pandrol clips
Min curvature radius: Main line 10°, branches 15°
Max gradient: 3.4%
Max permissible axleload: 15 tonnes

Chilean State Railways (EFE)

Empresa de los Ferrocarriles del Estado
Avenida Libertador Bernardo O'Higgins 3322,
Santiago
Tel: (+56 2) 779 07 07
Fax: (+56 2) 776 26 09

Key personnel

President: Hugo Trevelli
Vice President: Guilleromo Artria
General Manager: Jaime Moncada
Managers
   Human Resources: Patricio Corvalan
   Legal Affairs: Patricio Morales
   Operations and Passenger: Jaime Contreras
   Merval: Andres Link
   Rehabilitiation Plan: Fernando Verbal
Comptroller: Ernesto Opazo

Gauge: 1,676 mm; 1,000 mm
Route length: 2,831 km; 402 km
Electrification: 1,317 km at 3 kV DC

Political background

Under legislation passed in 1992 which established the company in its present form, EFE was granted some measure of autonomy and allowed to meet its objectives by means of contracts, concessions and joint ventures with third parties. Under this legislation, EFE's objectives were to be set out in three-year development plans agreed with the Chilean government, the first of which was approved in August 1993.

As required by the 1993–96 plan, a controlling 51 per cent interest in EFE's freight subsidiary FEPASA (qv) was sold to the Transportes del Pacífico consortium in January 1995. As well as disposing of its non-core activities, EFE has been charged with increasing the involvement of the private sector in the railway system it currently operates. This was to take the form of concessions to operate passenger services, and concessions to upgrade and maintain sections of track to specified standards. It was expected that concessions for passenger services and track maintenance on a particular route could, if advantageous, be awarded to the same bidder.

EFE invited bids for a track maintenance concession on the Santiago–Chillán (to allow speeds of 140 km/h) and Chillán–Temuco/ Concepción–Valdivia sections of its Santiago–Puerto Montt main line in 1996, so that work could begin in 1997. FEPASA was to undertake maintenance on those lines and branches where there are no passenger services.

The long-planned privatisation of EFE passenger services was deferred due to global economic conditions, especially in Asia. However, in May 1997 the Ministry of Transport and Telecommunications announced plans to invest US$300 million in EFE. Half of this was to be directed at the Santiago–Puerto Montt line, for which an operating concession had been due to be awarded in early 1998, although no contract has been reported. It was hoped that a state investment in infrastructure upgrades would provide the impetus for a private-sector operator to emerge. Track conditions south of Temuco, some 700 km south of Santiago, have led to a temporary withdrawal of services, with a lack of rolling stock also a factor.

EFE metre-gauge diesel railcar at Talca (Marcel Vleugels)　　　　0058676

Passenger operations

After years of declining traffic, EFE is now handling growing numbers of passengers. A downturn in traffic on the Merval system, which provides suburban services between Puerto station, Valparaíso, and Limache (43 km), was attributed to modernisation work which led to train cancellations. Since completion of most of this work in November 1999, journey times have been reduced and service frequencies increased. Services continued to be provided by Class AEL emus, but replacements in the form of refurbished ex-RENFE Class 444 units from Spain were expected in 2001.

Since 1998, the number of services on the Metrotren system between Santiago and Rancagua (85 km) has increased steadily to 15 each way on weekdays and 17 on Sundays. All are provided by refurbished ex-RENFE Class 444 emus.

Metrotren services are now administered jointly by EFE and Metro de Santiago, with the possibility of a private-sector partner being sought in the future.

Servicios Regionales comprises Renaico– and Laja–Concepción–Talcahuano services and a metre-gauge Talco–Constitución railbus operation, while Largo Recorrido covers Santiago–Temuco overnight trains (one all-year and a second economy-class service in summer months) and three daytime trains between Santiago and Chillán.

Since December 1999, Biotren has provided suburban services on the Chiguayante-Concepción-Talcahuano route, with 14 services in each direction on weekdays and six on Sundays, all provided by refurbished Class AEL emus. The introduction of this service and the extension of regional services from Renaico and Laja to Talcahuano necessitated reopening the 5 km Concepción–Talcahuano branch to passenger traffic.

Despite enthusiasm from some politicians, there appears little prospect of reinstating passenger services on lines south of Temuco, due to poor track conditions. However, there are plans to revive services on the Santiago-Melipilla line to become part of the Metrotren system. Electrification and modernisation of this section could be carried out by an eventual concessionaire for the service.

Freight operations

Freight operations on the main EFE network passed out of state ownership in January 1995 with the privatisation of FEPASA, while those undertaken by its Arica–La Paz Railway subsidiary passed to private-sector operator Ferrocarril de Bolivia al Pacífico (qv) in November 1997.

New lines

In April 1999, the Chilean and Argentine governments agreed each to contribute funding of US$140 million towards a proposed 220 km rail link from Lonquimay to Zapala in Argentina, which would complete the so-called Transandino del Sur route. The Ferrocarril del Pacífico (qv) would be responsible for upgrading the Talcahuano–Lonquimay line feeding the new route, while Ferrosur Roca would undertake similar improvements on the Argentine side.

Improvements to existing lines

At the start of 1996, branches from the Valparaíso–Santiago–Puerto Montt trunk route, totalling 418 route-km of 1,676 mm gauge and 73 route-km of 1,000 mm gauge, were out of use. Reports from late 1997 say that on some double-track lines only one is in use due to track condition.

EFE has received US$120 million from the government for a three-year investment programme. Most of this is being spent on the Merval and Metrotren corridors. In 2000, modernisation work between Santiago and Rancagua was in progress and was expected to be completed early in 2001, enabling trains to run at speeds of up to 140 km/h.

Traction and rolling stock

In 1997 the EFE broad-gauge fleet comprised: 22 steam locomotives (2 serviceable); 39 diesel locomotives; 45 electric locomotives; 5 diesel railcars; 29 emus; 131 passenger coaches (plus 96 in private ownership); and 347 freight wagons

Diesel locomotives

Class	Builder's type	Wheel arrangement	Power kW	Speed km/h	Weight tonnes	No in service	First built	Builders Mechanical	Engine	Electrical
D-16 00	253/253	Co-Co	1,305	120	114.5	5	1954	GE	Alco 244	*E* GE USA
Dt-13 100	U-13-C	Co-Co	1,063	95	85	5	1967	GE	GE FDL-8	*E* GE USA
D-7100	040 DE	Bo-Bo	615	90	72	5	1963	B&L	SACM-MGO 12V-175-A5	Various European builders
Dt-6000	–	Co-Co	492	80	64	3	1954	GE	CB FWL-6T	*E* GE USA
D-5100	U-5-B	Bo-Bo	447	70	50	2	1963	GE	Caterpillar D-379	*E* GE USA
Dt-3000	–	Bo-Bo	223	48	40	1	1953	GE	Cummins HBI-600	–

Electric railcars or multiple-units

Class	Cars per unit	Motor cars per unit	Motored axles/car	Output/motor kW	Speed km/h	No in service	First built	Builders Mechanical	Electrical
AEZ	4	2	4	225	160	4	1973	Kawasaki	Toshiba
AEL	4	2	4	310	130	3	1973	Kawasaki	Toshiba/Hitachi
AES	2	1	4	190	130	12	1977	Fiat-Concord	SEL/Siam di Tella

Electric locomotives

Class	Wheel arrangement	Output kW	Speed km/h	Weight tonnes	No in service	First built	Builders Mechanical	Electrical
E-32	Co-Co	3,400	130	136	4	1962	Breda	Marelli
E-30	Bo-Bo	2,265	130	98	5	1962	Breda	Ansaldo-Marelli
E-17	Bo-Bo	1,950	90	76	2	1973	Breda	Ansaldo-Marelli

Maintenance concessionaires were to receive subsidy where the track standard required by EFE was above that which could be funded through access charges. It was reported that EFE was to retain responsibility for electrification, major bridges, signalling and dispatching.

Overseeing the project is the state-appointed Transport Investment Infrastructure Planning Commission (Sectra), which sets standards for track maintenance on all routes. Sectra will also determine minimum passenger services, covering such aspects as speed, frequency, capacity and standards of accommodation.

Organisation

On 29 September 1995 EFE reorganised its activities as six subsidiary limited companies; a property management company was later established on 3 October 1995. With the exception of Ferrocarril de Arica-La Paz SA, all are concerned with the trunk 1,676 mm gauge Valparaíso–Santiago–Puerto Montt route and its 1,000 mm and 1,676 mm gauge branches. Most of the trunk route is electrified, as far south as Temuco.

The separate 1,000 mm gauge system extending from La Calera, near Valparaíso, north to Iquique has been operated by a separate state-owned company, Ferronor (qv), since 1990.

With the exception of some 900 route-km placed under the responsibility of FEPASA, infrastructure maintenance and management on the core EFE system has been undertaken since September 1995 by Infraestructuray Tráfico Ferroviario SA (Infrastructure and Railway Traffic Ltd). Suburban passenger services are operated in the Valparaíso area by Merval (Metro Regional de Valparaíso SA – Valaparaíso Regional Metro Ltd) and in the Santiago area by Ferrocarriles Suburbanos SA (Suburban Railways Ltd) and marketed as 'Metrotren'. Long-distance passenger services between Santiago and Puerto Montt are operated by Ferrocarril de Pasajeros SA (Passenger Railway Ltd), which is marketed as 'Via Sur'. In turn, it was split into separate units in 1999, including Largo Recorrido (long-distance) and Servicios Regionales (regional). On 2 December 1999, Tren Regional del Bio-Bio (Biotren) came into existence, providing regional services around Concepción.

Traffic	1997	1998	1999
Passenger journeys (million)			
Merval	5.434	6.352	5.268
Metrotren	1.847	2.667	3.343
Biotren	–	–	0.028*
Other EFE	0.981	0.949	1.341
Total	8.262	9.968	9.980

* Services introduced 2 December 1999

(including metre-gauge). The metre-gauge fleet comprised: 2 steam locomotives; 21 diesel locomotives; 5 electric locomotives; 9 diesel railcars; and 18 passenger coaches. Large sections of each fleet were unserviceable.

An initial batch of five Class 440 emus procured from Spanish National Railways (RENFE) was joined by a further five examples in late 1998. Further rolling stock procurement from Spain was announced in April 1999. By 2002 EFE will acquire from RENFE 22 Class 440R and five Class 444 emus, one Class 308 diesel-electric locomotive, and two sets of coaching stock. The Class 444 units were expected to be used to improve journey standards on Santiago–Chillán long-distance services.

Coupler: Automatic
Braking: Air

Signalling and telecommunications
At the start of 1996, the following signalling systems were in use on the EFE network: Alameda–Puerto (187 km), mechanical signalling with electromechanical interlocking and track circuits; Alameda–Talca (249 km), mechanical signalling and interlocking, track circuits; Talca–Cabrero (208 km), mechanical signalling with electrical interlocking, track circuits; Cabrero–Temuco (233 km), electric signalling and interlocking, train staff working; San Rosendo–Talcahuano (83 km), electrical interlocking and train staff working; Temuco–Puerto Montt (389 km) and various branches (971 km), train staff working.

Under EFE's 1994–97 rehabilitation plan, signalling was to receive investment totalling US$3,52 million and telecommunications US$5.62 million. EFE and Chilesat (a private communications company) were to install an optical fibre network between Santiago and Temuco and San Rosendo and Concepción. The scheme was to receive US$1.75 million from EFE's rehabilitation budget and was to be used for ground-to-ground and later ground-to-train communication.

Class AEL Japanese-built four-car Metrotren emu at Rancagua (Marcel Vleugels)   0058677

Electrification
A total of US$5.27 million was to be spent on EFE's 3 kV DC electrification system under the company's 1994–97 rehabilitation plan. This investment was intended to compensate for several years of deferred maintenance, rather than fund any further electrification. Reports from late 1997 suggest that electrically powered operation north of Santiago is confined to the Merval commuter service. Some removal of overhead power supply equipment north of the capital has been undertaken by EFE.

Overhead power supply equipment on the Santiago–Nós section of the Metrotren system has been modernised and in 2000 similar work was in progress on the Paine–Talagante line.

Track

Rails	Lengths laid 1995
(kg/m)	*(km)*
60	563.3
50	972.9
40	1,453.7
30	457.1

Crossties (sleepers): Wood
Cross-section: 250 × 150 mm
Spacing: 1,800/km
Min curvature radius, main lines: 1,000 mm gauge, 80 m; 1,676 mm gauge, 180 m
Max gradient: Adhesion, 6%; rack, 8%
Max permissible axleloading: 18 tonnes (1,000 mm gauge), 25 tonnes (1,676 mm gauge)

FEPASA

Ferrocarril del Pacífico SA
La Concepción 331, Providencia, Santiago
Tel: (+56 2) 235 16 86   Fax: (+56 2) 235 09 20

Key personnel
General Manager: E Valdatta
Chief Operating Officer: Paul Victor
Chief Mechanical Officer: David L Powell
Director, Operations: Robert G Muilenberg
Transportation Superintendent, Southern Division: J David Wallace

Gauge: 1,676 mm; 1,000 mm
Route length: 2,085 km; 37 km
Electrification: 841 km

Political background
FEPASA was created in 1993 to operate freight services on EFE routes, excluding the Arica-La Paz Railway. It was initially a 99 per cent-owned EFE subsidiary, with the Chilean government holding the remaining 1 per cent. In January 1994 a 51 per cent stake in FEPASA (including the government's 1 per cent) was sold to the Transportes del Pacífico consortium for US$30 million. The consortium formally took control of FEPASA on 23 January 1995 under a 20-year concession.

Transportes del Pacífico comprised the Chilean holding company Cruz Blanca, Estrella Americana (a Chilean pension fund company) and San Pablo Bay Railway Company. The last is an affiliate of the US Anacostia and Pacific Company, which was to provide financial, technical and operating expertise.

Organisation
FEPASA operates freight services over EFE infrastructure, excluding the Arica-La Paz Railway. Responsibility for scheduling and track maintenance remains with EFE for the core Valparaíso-Puerto Montt route, its branch to Concepción and between San Rosendo and Talcahuano, Paine and Talagante and Santiago and San Antonio. On these routes, FEPASA pays track access fees to EFE. On some 900 km of routes where no EFE passenger trains operate, FEPASA is responsible for scheduling and track maintenance.

Finance
Bidders for the 51 per cent stake in FEPASA were required to provide details of the development strategies, operational and investment plans as far as 2000, commercial and financial policy and equipment maintenance programmes they intended to implement. Transportes del Pacífico has undertaken to invest US$88 million in FEPASA by 2000, with US$28 million earmarked for freight wagons, US$20 million for locomotives and US$28 million for track improvements.

Of the US$30 million Transportes del Pacífico offered to pay for its stake in FEPASA, 40 per cent was to take the form of a down payment, with the balance to be paid over a period of five years. At the time of the sale, FEPASA's annual revenue was reported to be around US$40 million. The company's new owners expected to make a loss of some US$1 million in their first year of operations (1995), and move into profit thereafter.

Breda-built Class E-32 3 kV DC electric locomotive on a northbound grain train at Talca (Marcel Vleugels)   0010795

However, the company admits that neither traffic nor profits have matched expectations, due in part to a policy of concentrating on reducing costs rather than gaining traffic. Business is also reported to have suffered from traffic losses following the closure of coal mines and the withdrawal of rail access to a cellulose plant.

In 1997 FEPASA announced a US$47.5 million programme of investment in infrastructure upgrades, locomotive acquisitions, and wagon refurbishment.

Freight operations
In 1994, prior to privatisation, FEPASA carried 5.132 million tonnes of freight on the EFE system and recorded 1.104 billion tonne-km. In total 4.2 million tonnes were carried in 1998, when 920 million tonne-km were recorded.

Bulk products, such as minerals (copper) and forest products (cellulose), form the principal component of FEPASA traffic base, but major growth in intermodal traffic is also foreseen. The new owning consortium aimed in the short term to cut operating costs and raise productivity, and train weights were to rise from 630 tonnes to around 1,100 tonnes, with a maximum of 3,000 tonnes. This was to require the use of locomotives operating in multiple.

FEPASA's new owners hope to increase traffic to over 2 billion tonne-km a year by 2000. Staff numbers were to be reduced through the introduction of new operating methods, equipment and technology (including an information system to track wagon movements), as developed in the USA and subsequently introduced to Argentina. EFE freight operations had required a staff of 2,600, which fell to 1,900 upon the creation of FEPASA in 1993. At the start of 1996, FEPASA had a staff of 700, including 400 formerly employed by EFE.

Improvements to existing lines
The raising of train weights in the short term was to require the lengthening of some passing loops. In the long term, conversion of the 1,000 mm gauge Los Andes-Río Blanco branch (the remaining 34 km of the former Transandine Railway to Argentina via Juncal) to 1,676 mm gauge was projected.

GE-built Alco-engined FEPASA Class D-1600 dating from the mid-1950s on southbound freight at Chillán (Marcel Vleugels)
0010796

This would eliminate the present transhipment of copper concentrate containers at Los Andes.

Traction and rolling stock
FEPASA passed into private ownership with 19 electric locomotives, 90 diesel locomotives (of 13 different models) and 4,800 wagons. Due to financial constraints, EFE had been deferring maintenance for some years and FEPASA's new owners intended to rationalise and modernise the locomotive fleet with a view to obtaining 100,000 km a year from each unit.

Abandonment of electric operation has been suggested to lower track-access charges paid to EFE, which in 1995 began trials of a microprocessor-based onboard system to measure and record traction current consumption by electric locomotives and multiple-units.

In January 1996 FEPASA signed a contract with National Railway Equipment Company, USA, for the supply of eight remanufactured General Motors SD39-2M Co-Co diesel-electric locomotives. In 1997 the company announced plans to acquire 15 rebuilt General Motors SD40 locomotives and to refurbish 1,600 wagons as part of a US$47.5 million investment programme.

Signalling and telecommunications
Responsible for regulating train movements on routes with no EFE passenger service, FEPASA's new owners intend to convert to a train warrant system, install train radio and replace brake vans with end-of-train devices.

Ferrocarril de Bolivia al Pacífico

Key personnel
President: José Saavedí Banzer

Gauge: 1,000 mm
Route length: 206 km

Political background
A concession to operate the Chilean section of the former 209 km Ferrocarril de Arica-La Paz SA was awarded to a new company on 1 August 1997 in return for half-year royalties of 46.5 million pesos. The successful bid for the 25-year concession came from C B Transportes, part of the Cruz Blanca group, whose subsidiary Ferrocarril Oriental de Bolivia

operates the 248 km section of the line in Bolivia. The consortium also includes Metropolitana Bolivia Ltda, Panamerican Securities SA, and Bolivian businessman José Saavedí Banzer. Operations under new ownership began in November 1997.

A 1904 treaty requires the Bolivian government to retain ownership of the infrastructure of its portion of the line between Charaña and Alto de La Paz. However, the government is reported to have suggested the FBP should pay US$300,000 per year to lease the Bolivian section for 10 years, the lease including the use of 80 wagons, as well as container-handling facilities.

Organisation
The new company operates with around 20 commercial and traffic management staff. Locomotive and wagon

maintenance has been outsourced from a separate company, Maestranza AG, which was reported to be creating a subsidiary to provide train crew.

Freight operations
Since being concessioned, the railway has witnessed a significant increase in traffic, with a 45 per cent rise to 183,000 tonnes recorded in the first year of operation under new ownership.

Traction and rolling stock
The operational fleet available in November 1997 comprised eight diesel locomotives and 350 wagons. In 1998 a rehabilitation programme for traction and rolling stock was started.

Ferronor

Empresa de Transporte Ferroviario SA
Avenida Alessandri 042, Casilla 62, Coquimbo
Tel: (+56 51) 33 25 00   Fax: (+56 51) 33 25 03
e-mail: ferronor@ferrornor.cl
Web: http://www.ferronor.cl

Key personnel
General Manager: Angel Gajardo
Commercial Manager: Pablo Arranz
Finance and Administration Manager:
  Eduardo Mendez

Gauge: 1,000 mm
Route length: 2,235 km

Political background
Privatisation of Ferronor was undertaken in 1997 when the Andrés Pirazzoli construction

and transport company bought Ferronor for US$12 million, double the bid of its nearest rival. The Chilean company subsequently sold a 55 per cent equity stake to US short line operator RailAmerica, which assumed operational control in February 1997.

Organisation
Ferronor's La Calera–Iquique main line crosses the Antofagasta (Chili) and Bolivia Railway (FCAB) at Augusto Victoria and Baquedano. Ferronor's route from Augusta Victoria to Socompa and Belgrano Cargas of Argentina is connected to its main line at Palestina via FCAB.

In January 2002, Ferronor took over the operation of the Ferrocarril Portrerillos, linking Portrerillos with Diego Almagro and Barquitos (91 km), under a 15-year agreement with copper producer Codelco.

Freight operations
Ferronor carries approximately 110,700 carloads annually, including iron ore, potassium chloride, copper ore, limestone, LPG and food products.

A new policy implemented by RailAmerica concentrates on Ferronor's core copper and iron ore traffic, as well as on international business with Argentina.

Copper concentrates and iron ore are moved from mines at the northern end of the system to the Pacific ports of Chañaral, Caldera, Huasco and Coquimbo. Traffic levels are subject to fluctuation in line with the market price of these commodities, but the temporary closure of the Cerro Colorado iron ore mine was offset by a contract to carry 2,000 tonnes/month of El Melón cement from La Calera to Copiapó and limestone in the reverse direction. A subsequent rise in iron ore prices and the reopening of the Cerro Colorado and Copiapó mines have recently boosted Ferronor traffic levels, with a 20-year contract signed with Compañia Minera Huasco (CMH) in 1996 to transport ore from Los Colorades mine to a pellet plant at Huasco. Shipments commenced in the third quarter of 1998.

In March 1998 a US$68 million contract was won from SQM Nitratos SA to transport some 500,000 tonnes annually of potassium chloride and potassium sulphite from the Minsal Salt Flats mine to Coya Sur and Tocopilla. In November 1998, the scope of this contract was expanded, increasing its value to US$80 million over 11 years.

In September 2002, Ferronor concluded a five-year contract with Norgas SA to convey compressed natural gas from northwest Argentina to locations in the northern region of Chile.

Traction and rolling stock

In 2002, Ferronor's fleet comprised 32 diesel locomotives (including nine for yard service) and around 700 wagons. Ferronor's workshops at Coquimbo carry out repairs for other railways as well as non-railway work. In 1998 the company

General Electric diesel locomotive on Ferronor iron ore train at Vallenar (Marcel Vleugels)   0010797

announced that project financing associated with its 20-year CMH contract (see 'Freight operations') would fund the acqusition of six locomotives, 66 freight wagons and the construction of a new maintenance facility.

Ferronor's takeover of operations on the Ferrocarril Portrerillos in January 2002 led to the transfer of nine diesel locomotives and 186 wagons.

China

Chinese Railways (CR)

Ministry of Railways, 10 Fuxing Men, Beijing 100844
Tel: (+86 10) 63 24 69 15   Fax: (+86 10) 63 98 10 65

Key personnel
Minister of Railways: Liu Zhijun
Vice-Ministers of Railways: Sun Yong Fu,
    Cai Qinghua, Hu Yadong, Lu Dongfu
Chairman: Hua Maokun
General Manager: Zhang Zhengqing
Chief Economist: Wang Zhaocheng
Technical Director: Zhou Yumin
Planning Director: Cao Qing
Operations Director: Chang Guozhi
Workshops Director: Tan Datong
Manager, International Co-operation:
    Mrs Tang Wensheng

Gauge: Almost entirely 1,435 mm, some 1,000 mm and 750 mm
Route length:
    National railways: 60,446 km (end 2003)
    Joint-venture railways: 7,738 km (end 2003)
    Local railways: 4,818 km (end 2003)
Electrification: 18,758 km at 25 kV 50 Hz AC (end 2003)
Multiple track: 24,650 route-km (end 2003)

Political background
The Ministry of Railways controls 14 geographically based railway administrations centred on Beijing, Chengdu, Guangzhou, Harbin, Hohhot, Lanzhou, Liuzhou, Jinan, Kunming, Nanchang, Shanghai, Shenyang, Urumqi and Zhengzhou. Until recently it also controlled most of the country's rail manufacturing and supply industries via five corporations: China National Railway Locomotive & Rolling Stock Industry Corporation (LORIC); the Railway Engineering Corporation; the Railway Construction Corporation; the Railway Materials Corporation; and the China National Railway Signal & Communications Corporation.

At the end of 1997 CR employed some 3.4 million staff. However, in 1998 the government stated its intention to separate the five ministry-run corporations and implement other structural change, with the aim of cutting core rail network staff numbers. Behind this move was a bid to stem heavy losses sustained by the network in recent years.

This decision was implemented by the Ministry of Railways in late 2000 when the five corporations and ten universities and colleges were officially placed under the administration of other non-railway government departments. This structural change reduced the number of Ministry of Railways workers to 2.4 million of which 1.5 million were directly engaged in railway transportation.

As part of this restructuring LORIC was split into two autonomous organisations, the China Northern Locomotive and Rolling Stock Industry (Group) Corporation (CNR) and the China Southern Locomotive and Rolling Stock Industry (Group) Corporation (CSR). These two organisations are large state-run key enterprises and each serves as an umbrella for the ex-LORIC rolling stock plants located in its area.

The Ministry of Railways also indicated its intention to gradually undertake more reforms from 2001 by splitting freight business, passenger business and network management into independent divisions. The Ministry will concentrate more on safety and regulation aspects.

While no timetable has been publicly set the MoR has taken a decision to move towards an overall structure for the national railway system whereby infrastructure is separated from operations. A unified national railway network company, a corporation funded by the state, is to be established for managing the railway infrastructure. A number of passenger and freight transport companies, in the form of corporations with modern organisational structures, are being created to engage in railway passenger and freight transport on a competitive basis.

Three specialist freight companies have been established dealing respectively with container, express cargo and special cargo. As part of ongoing development these companies are to be further reformed to improve corporate governance.

Current policy is for progressive devolution of authority to individual railway regions, and joint-stock ownership of non-trunk routes is being encouraged. In 1993 the Guangzhou area administration was reconstituted as the first autonomous railway organisation – the Guangzhou Railway Corporation – with almost 4,000 route-km and 172,000 staff.

In a further move in 1993, it was announced that the state railway monopoly would be ended, and

new operating standards and pricing structures introduced to enable the railway to gear itself more closely to the needs of a fast growing economy. Five regions were selected in 1996 as the first batch of autonomous railways.

A separate administration, the Guangshen Railway Company, runs the Chinese section of the Kowloon-Canton Railway, the 147 km line from Guangzhou (Canton) to the Hong Kong border's end-on junction with the former British section. This railway was allowed to produce its own timetables and set its fares independently from 1996, prior to the offer of shares on the Hong Kong stock market and in the USA. Funds raised by the flotation financed purchase of high-speed trains for the Guangzhou-Kowloon route; an X2000 tilting train from Adtranz arrived in early 1998 and continues in service on this important route.

In a move aimed at developing the corporate structure and improving the efficiency of the national railways, the Ministry of Railways announced in March 2005 that the number of levels of railway management would be reduced from four to three. The level of railway sub-administration has been eliminated by absorbing these organisations into the higher railway administration level. Efficiencies are expected as overlapping functions will be reduced.

The demand for new secondary lines in China, often for coal-hauling purposes, has stimulated the development of the joint-venture method of financing. Joint-venture partners are often provinces and autonomous regions, coal mining enterprises and the Ministry of Railways. By the end of 2003 some 7,700 km of such railways were in operation, including some electrified lines.

The government has also encouraged local authorities to build and operate their own railways, where such investment would stimulate regional economic development. At the end of 2003 there were approximately 75 local railways with a total route length of 4,800 km. Additional lines and

Hanlu dmu on the Jitong joint venture railway (Bruce Evans)   0554909

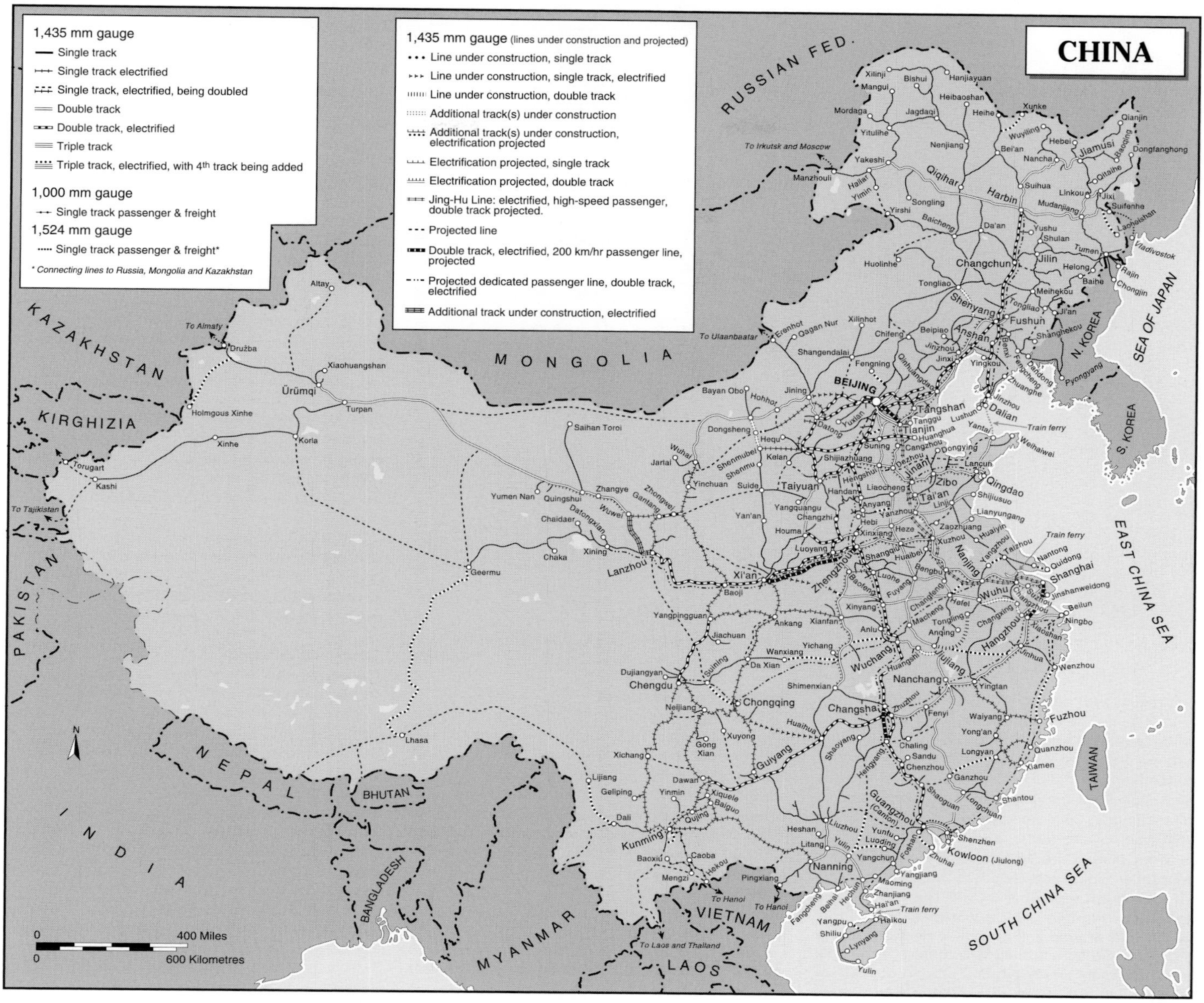

CHINA

1,435 mm gauge
- Single track
- Single track electrified
- Single track, electrified, being doubled
- Double track
- Double track, electrified
- Triple track
- Triple track, electrified, with 4th track being added

1,000 mm gauge
- Single track passenger & freight

1,524 mm gauge
- Single track passenger & freight*

*Connecting lines to Russia, Mongolia and Kazakhstan

1,435 mm gauge (lines under construction and projected)
- Line under construction, single track
- Line under construction, single track, electrified
- Line under construction, double track
- Additional track(s) under construction
- Additional track(s) under construction, electrification projected
- Electrification projected, single track
- Electrification projected, double track
- Jing-Hu Line: electrified, high-speed passenger, double track projected.
- Projected line
- Double track, electrified, 200 km/hr passenger line, projected
- Projected dedicated passenger line, double track, electrified
- Additional track under construction, electrified

1114969

extensions are under construction. Narrow gauge railways of 762 mm gauge made up 30 per cent of the total route length at the end of 1999 but many of these have been closed or in a few cases rebuilt to standard gauge.

In January 1999 the government indicated that foreign investment in the Chinese rail network, hitherto not permitted, would be encouraged on a trial basis from 2000. The present policy in China is to open up the rail freight market in three stages following formal membership of the WTO. In the first stage foreign investors may purchase up to 49 per cent of shares in new Sino-foreign rail freight joint ventures. In the second stage, after

2006, foreign investors will be allowed to hold the majority of shares in the joint venture three years after China's entry into the WTO. In the third stage foreign investors will be allowed to establish their own rail freight transport companies six years after China's entry into the WTO.

In another initiative aimed at attracting foreign capital into the railway market in China, railway enterprises are being encouraged to issue stocks on overseas stock markets. The Ministry of Railways announced in 2004 that it was planning to allow domestic and foreign investors to establish shareholding companies to participate in projects related to major passenger lines. By doing this it

is hoped that public funds would be attracted for such projects.

Foreign investment is being encouraged for new line construction and a list of rail projects open to foreign investment has been published.

The Ministry of Railways considers the attraction of domestic and foreign capital as a priority task to ease the severe shortage of funding for railway construction. Annual investment in railway construction has averaged around US$7 billion in recent years whereas around five times this figure has been available for road construction. At present capital for railway construction comes largely from government sources, such as central and local government funds and loans from the State Development Bank. With the severe bottlenecks that exist at present in the railway network, reform of the financing structures is seen as an urgent need. As part of the strategy to attract funds railway departments are considering establishing shareholding companies to attract private capital.

A pilot railway project involving significant private financing through bank loans is due to open in 2006. Ownership of the 45 km Quzhou–Changshan joint venture railway will be shared between the private Changshan Cement Co (32.5 per cent), the Ministry of Railways (35 per cent) and local government (32.5 per cent).

Recent years have seen a massive surge in demand for railway freight capacity in China (see statistics below), particularly for the movement of bulk commodities such as coal, oil and ores. The lack of capacity is considered to be a key constraint on economic development. To help address the capacity shortfall, the rate of construction of new lines has been increased. A major new initiative

Later Class SS9 electric locomotives feature as restyled exterior design, seen here with an example of the type at Beijing station with a service for Qingdao (Marcel Vleugels)   0580483

Class SS8 electric locomotive at Beijing West with a sleeper service to Xian (Marcel Vleugels)
0580484

has been the decision to build thousands of km of new dedicated passenger lines which when commissioned will allow many existing key main lines to become mainly freight routes, eliminating the inefficiencies inherent with mixed high-capacity freight and passenger operations.

In January 2004 the State Council of China approved a medium- and long-term railway network plan which includes the new dedicated passenger lines and sets the ambitious target of adding an additional 28,000 route-km to the national network by 2020. It is also envisaged that by this date around half the public railway network will be electrified and half will consist of multiple tracks. To meet these targets, around 30,000 km of route will need to be electrified and 25,000 km double-tracked over the 16 years from 2004.

In January 2005 the Minister of Railways, Liu Zhijun, announced that China would invest approximately US$12 billion in railway construction in 2005. This amount is approximately double the amounts invested annually in recent years.

Passenger operations

Demand for passenger and freight transport considerably outstrips both infrastructure and rolling stock capacity. Between 1968 and 1989 passenger traffic quadrupled, whereas the number of trains run rose by only 1.7 per cent. Consequently, severe overcrowding was common. From 1995 to 1997 passenger journeys declined but then reverted to an increasing trend. The maximum annual number of passenger journeys that had been recorded by the end of 2003 was in 2002, when 1,056 million passenger journeys were made. The annual number dropped to 973 million passenger journeys in 2003, the fall attributed to the SARS outbreak that occurred early in the year and the subsequent restrictions placed on travel in China. For 2004 it has been reported that the total number of passenger journeys was 1,120 million, resuming the rapid upward trend.

Over the five years from 1997 to 2002 passenger journeys increased at an average rate of around 3 per cent per year. The demand for additional

trains, though, is increasing more rapidly as the average travel distance is increasing and there is a demand for higher quality services resulting from increased passenger affluence. This is leading to a need for greater numbers of coaches with increased space per passenger. Between 1990 and 2000 the total number of passenger coaches in service increased by 32 per cent from 27,300 to 36,000. The proportion of hard- and soft-berth coaches, the types offering improved comfort and increased passenger space, increased from 20 to 34 per cent of the total fleet over this period.

The decline in passenger traffic experienced in the mid-1990s was attributed primarily to the growth in competition from road-based transport. China's network of intercity highways is expanding rapidly with over 34,000 km of dual-carriageway highway now in use. This infrastructure is allowing the development of greatly improved intercity bus and coach services that present significant competition to railway passenger services. The National Expressway Network Plan recently approved by China's State Council envisages that within 30 years a further 85,000 km of expressway will be added. By then the expressway network will connect all cities with a population of 200,000 or more.

Due to China's size, competition from air transport also represents a major challenge for the railway system. Within the next 20 years China is projected to become the world's second largest commercial aviation market after the United States.

Accelerated services

To counter the threat posed by expanding road and air passenger transport, the national railways embarked on a massive improvement strategy from the mid-1990s. A key development in passenger operations has been the introduction of 160 km/h and other accelerated services. Initially attention was focused on four key routes: Beijing–Shanghai; Beijing–Guangzhou; Beijing–Shenyang and Harbin–Dalian. In October 2000 160 km/h services were also introduced on the Xuzhou–Zhengzhou–Lanzhou–Urumqi line and this was followed by similar accelerations on the Beijing–Shenzhen, Shanghai–Nanchang, Wuhan–Xiangfan, Luoyang–Xiangfan and Xiangfan–Chongqing lines. Associated with this have been significant improvements to many adjoining sections of the network and the introduction of more powerful locomotives to allow higher average speeds on these sections where a limit of 120 km/h still applies. By the end of 2004 around 1,450 new locomotives capable of hauling the 160 km/h trains and around 750 locomotives for other accelerated services were operating.

The fifth major stage in the speeding up programme for passenger services was initiated on 18 April 2004. Notable features of this included the introduction of 19 pairs of new 'Z' class trains on important routes, the introduction of new operating procedures such as through running of locomotives and the provision of special waiting rooms and ticket purchasing facilities for these trains at major stations. The trains are formed of new rolling stock designed for 180 km/h operation, hauled for the most part by new SS7E and SS9(G) type electric locomotives and twin-unit DF11G type diesel-electric locomotives delivered from 2003.

Together with improvements to fixed infrastructure the new measures have allowed journey times on routes such as Beijing–Shanghai to be reduced by up to 2 hours. The fastest trains on the Beijing–Shanghai route now take 12 hours for the journey compared to 14 hours previously.

A notable feature of the new services is that the accommodation provided is almost all of soft-seat, hard-berth and soft-berth types and these trains thus represent a major improvement in quality of service. With their introduction the number of high quality seats and sleeping berths available on the various routes has been significantly increased.

The expansion of the improved passenger services is a major continuing programme. A sixth major stage in the overall programme was planned for 2005 but has been postponed for technical reasons. The sixth stage when introduced will see China with a network of around 22,000 km on

Traffic (million)	1999	2000	2001	2002	2003
Passenger journeys	976	1,019	1,052	1,056	973
Passenger-km	405,000	442,000	477,000	497,000	479,000
Freight tonnes	1,569	1,655	1,926	2,042	2,212
Freight tonne-km	1,258,000	1,334,000	1,457,000	1,552,000	1,724,000

Diesel-powered intercity trainset on a Shenmu Bei–Baotou service at Singing Sands
(Colin Boocock)
0554910

Class DF4D diesel locomotive with a passenger service approaching Weihe (Ian Francis)   0554911

which speeds above 120 km/h are permitted. On the key trunk routes the numbers of accelerated trains are being increased and efforts are being made to remove the remaining obstacles to 160 km/h operation on sections where this speed is presently not permitted. The infrastructure is being upgraded on some sections of line to permit operation at up to 200 km/h. On main lines radiating from major cities along the trunk routes accelerated services are also being introduced, in many cases in the form of short-distance intercity services.

Separate series of locomotive-hauled passenger rolling stock are being delivered for 120, 160 and now 180 km/h service and around 25 per cent of new stock is of double-deck layout, with versions for long-distance overnight travel and for shorter distance intercity services. Several types of diesel push-pull trains, of single- and double-deck layout, and electric push-pull trains are either under development or in trial operation on intercity services. In late 2000 production double-deck diesel-electric NZJ2 class push-pull trains capable of 170 km/h were introduced on several Beijing–Tianjin services, augmenting existing locomotive-hauled trains.

In October 2001, as part of efforts to improve the standard of passenger services on key routes where intense competition from airline services exist, a new overnight daily train was introduced between Beijing and Shanghai. In a departure from traditional make-up each train includes four coaches with luxury compartments. Types of accommodation provided include two-bed compartments and luxury compartments with separate bar. All compartments have their own washrooms.

Dedicated passenger lines
Separation of passenger and freight traffic is now put forward as the solution to congestion on the busiest axes. Consequently, studies began of a pilot scheme for a 200 km/h line reserved for passenger trains. The two routes reviewed for the experiment were Beijing-Tianjin (137 km) and Guangzhou to the Hong Kong border at Shenzhen

(143 km). Upgrading of the latter route for 160 km/h running was completed in 1994, and public service at that speed was inaugurated in March 1995.

An intensive passenger train service with many trains travelling at up to 160 km/h is now in operation between Guangzhou and Shenzhen. A fleet of Chinese-built 200 km/h electric push-pull trains are now in service on this route. High-speed services between Hong Kong and Guangzhou also follow this route and are provided by the X2000 tilting trains and the KCRC-owned KTT trains.

Construction of China's first dedicated passenger main line railway between Qinhuangdao and Shenyang is now complete. This 400 km double-track electrified line is designed for 200 km/h operation and will help to eliminate the freight bottleneck on the existing line, one of China's busiest. Construction and operation of the new line, which will initially only have six stations along its length, is providing opportunities for technical advances in many aspects of railway engineering in China.

Plans outlined under China's 10th Five-Year Plan (2001–2005) envisaged the route-length of line on which 160 km/h trains can be operated steadily increasing to a projected total of around 14,000 km by 2005. The network will comprise the Beijing–Shenyang, Beijing–Shanghai, Beijing–Nanchang–Shenzhen, Xuzhou–Lanzhou–Urumuqi, Shanghai–Chongqing–Chengdu and Guangzhou–Shenzhen lines. 200 km/h services will be operated between Guangzhou and Shenzhen in the south and between Beijing and Shenyang in the north. For the latter operation the new dedicated passenger line between Qinhuangdao and Shenyang will be used, while the existing Beijing–Qinhuangdao line has been upgraded to permit the higher speed.

Longer term planning will see the network of passenger only lines growing to a length of around 12,000 km by 2020. Dedicated passenger lines are soon to be constructed between Beijing and Tianjin, Shanghai and Hangzhou, Wuhan and Guangzhou, Zhengzhou and Xi'an, Jinan and Qingdao and Guangzhou and Zhuhai, with several more envisaged. Several of these new dedicated

passenger lines are being designed for eventual operation at speeds of up to 300 km/h.

Station facilities
Passenger traffic growth on China's railways has required expansion of station facilities at many centres. One of the major projects now at the construction stage is for a second main station in Shanghai to be known as Shanghai South Station. The present Shanghai station handles around 150,000 arriving and departing passengers each day and is considered to be close to its maximum capacity. The new station is to be located in the southwestern part of the city and will mainly handle trains on the Hangzhou route and those between Shanghai and other southern destinations.

Work has also recently commenced on a massive new main railway station for the major southern city of Guangzhou. This station, estimated to cost around US$1.8 billion, is being designed to handle 80 million people per year. It is expected to be completed in 2008 and will provide direct interchange with metro and light rail lines serving the Guangzhou area.

Freight operations
China continues to experience rapid economic growth. Official figures indicate that in 2003 the economy grew by around 9 per cent. For 2004 the government set an official growth rate target of 7 per cent in an attempt to slow growth and reduce the risks of overheating in key economic sectors.

The rapid economic growth has ensured that the railway system continues to strain as demand for bulk resources such as coal, iron ore and oil increases. Published estimates indicate that in 2003 China consumed 27 per cent of world steel, 31 per cent of world coal and 40 per cent of the world's cement production. Many of China's harbours are experiencing bottlenecks and long delays and it is estimated that up to 35 per cent of requests for rail wagons cannot be met at present. Many main lines are reported to be operating at capacity and in some cases well over capacity, such as on sections of the Beijing-Shanghai line and the Datong-Qinhuangdao coal line.

Economic growth in China resulted in freight transported on the national railway system rising from 1,405 million tonnes and 986,019 million tonne-km in 1988 to 1,618 million tonnes and 1,304,000 million tonne-km by 1997. Between 1997 and 1999, freight traffic growth levelled off, but from 2000 resumed, with 2,212 million tonnes and 1,724,000 million tonne-km being recorded for 2003. To this must be added the freight traffic carried by the new joint-venture railways which by 2002 was of the order of 60 million tonnes and 70,000 million tonne-km. Approximately 40 per cent of the freight traffic volume carried on the national railway system is coal, this traffic accounting for 881 million tonnes in 2003. Much of this originates in coal-rich Shanxi Province and in Inner Mongolia.

Despite the significant rise in rail freight traffic, China's economic growth is such that rail's share of the total freight market fell from around 70 per cent in 1990 to around 50 per cent in 2000. Much of the shift has been to road transport, where a massive expansion of the country's highway and expressway network has greatly increased medium- to long-distance capacity.

At the end of 2003 around 510,000 freight wagons were in service on the national railways in China, up from 365,000 wagons in 1990. The total in 2000 included around 253,000 open wagons, 93,000 covered wagons, 25,000 flat and container wagons and 38,000 tank wagons.

Development of railway freight operations at present are based on the following basic parameters for freight trains set by the Ministry of Railways:
- Ordinary freight trains of up to 4,000 tonnes capable of running at up to 90–100 km/h;
- High-capacity freight trains of 5,000 tonnes capable of running at up to 90–100 km/h;
- Express freight trains of maximum 1,500 tonnes capable of running at up to 120 km/h;
- High-capacity coal trains on dedicated lines of up to 10,000 tonnes.

Class NZJ2 diesel-powered double-deck intercity trainset (Bruce Evans)   **NEW**/1114981

The Ministry of Railways has been raising individual train weights by up to a third with 3,000 to 4,000 tonnes now the norm on many main lines. On the key trunk lines, on which 160 km/h passenger services are being operated, many freight trains of 5,000 tonnes are now being operated to maximise capacity on these lines. These developments have been fraught with several problems, such as restricted yard siding and loop capacities.

Freight train speeds

The maximum speed of ordinary freight trains in China at present is restricted to 75 to 85 km/h due to limitations presented by factors such as bogie, coupler and brake system design. Intensive research and development efforts are being made to find ways to eliminate these obstacles to increased freight train speeds. Among the projects being undertaken are several aimed at developing high-stability freight bogies. Specialist engineering organisations from several foreign countries including the USA and South Africa have been assisting in these projects.

The introduction into regular service of wagons fitted with modified standard freight bogies allowing an increase in maximum permitted speed to 90 km/h commenced in earnest in 2001. Most new general freight wagons leaving the production lines at China's wagon works are now fitted with the Type Z8A swing-motion bogie and a major retrofitting programme to equip existing wagons with similar refurbished bogies is now underway.

The drive to increase freight train speeds in China is in part related to the introduction of passenger services operating at speeds of up to 160 or 200 km/h on many key main lines. China's unique situation, where intense passenger and freight services are operated on the same lines, requires freight trains to be accelerated to improve operational compatibility and to achieve required line capacities.

Average point-to-point freight train speeds on important lines are being increased through electrification and the use of more powerful locomotives allowing higher speeds on adverse grades. On key non-electrified lines Class DF8B 3,700 kW diesel-electric locomotives are replacing the ubiquitous, less powerful DF4. Diesel freight locomotives with more powerful engines for this application are under development at several of China's locomotive works.

Plans for increasing express freight trains speeds and for an express freight service network are initially focused on parcels and container traffic on the busy trunk lines. Two types of wagon with a maximum designed speed of 120 km/h are now in production. These are the P65 boxcar for parcels traffic, equipped with type 2D cross-braced bogies, and the X1k container flat wagon, with type Zk3 bogies with independently sprung axles.

China's first production freight locomotive type designed specifically for 120 km/h operation, the SSJ3 Co-Co electric fitted with AC traction motors, is soon to enter service with an initial batch of 60 on order. It is envisaged that most new freight locomotive types will in future be designed for 120 km/h maximum operating speed as against the present standard of 100 km/h.

For express parcel services between Beijing and Guangzhou, Beijing and Shanghai and Beijing and Harbin, 72 special purpose coaches forming six trainsets were put into service in May 2004. These trains are designed for operation at speeds of up to 160 km/h.

Another significant technical advance relating to freight train operation now being incorporated in series production of freight rolling stock is the increase of maximum axle-load from 21 to 25 tonnes. As could be expected, initial production is of open coal wagons for unit-train operation on the special purpose Datong Qinhuangdao coal line. Prototypes of a 25-tonne axleload development of the standard general purpose open wagon for coal, coke and ore transport and of a 25-tonne axleload covered wagon have also been built.

For container traffic trials are being undertaken with new 25-tonne axleload rolling stock suitable for double-stack operation. China's first double-stack container train of 38 such wagons, with a container capacity of 160 TEUs, ran between

Class DF4 diesel locomotive with a coal train from Fuxin to Minzhu (Colin Boocock)   ***NEW***/1114983

Class SS4(G) twin-section electric locomotive with a freight service between Zhengzhou and Luoyang (Bruce Evans)   ***NEW***/1114980

Class G70T tank wagon (Bruce Evans)   0580488

Beijing and Shanghai in April 2004 and has since commenced regular operation.

Coal traffic

In both new line construction and electrification, expansion of coal-carrying capacity is of paramount importance. China's fast-growing domestic energy needs are 70 per cent met by coal and at the same time exports are rising steadily. Between 1999 and 2003 coal production in China increased by 54 per cent. China is now the world's largest coal producer, with around 1,700 million tonnes of coal mined in 2003 and 1,960 million tonnes in 2004, an increase in one year of 17 per cent. Despite this achievement, high internal transport costs and transport bottlenecks have led to China also importing coal from a number of countries, primarily for power stations in the economically vibrant southeastern coastal provinces.

China's principal coalfields are found in Shaanxi, Shanxi, Inner Mongolia, Henan, Shandong, Ningxia, Guizhou, Anhui and Heliongjiang provinces. Coal flows are predominantly north-to-south and west-to-east, and form over half the traffic on some main lines.

A 10-year programme, aimed to raise rail coal-carrying capacity to over 600 million tonnes a year, involved the upgrading of 12 existing coal routes and construction of eight more, plus a lift of maximum trainloads on key routes from 3,500 to 5,000 tonnes. Coal traffic requirements dominated much of the railway expansion and electrification programmes executed since the 1980s.

On the special 650 km Datong–Qinhuangdao coal line, trains of up to 10,000 tonnes are now being operated. This line was initially designed to transport up to 100 million tonnes of coal per year. This volume is now being exceeded with 120 and 150 million tonnes carried in 2003 and 2004 respectively and a further increase expected in 2005. The first trials of trains of 20,000 tonnes have been successfully undertaken on the line and regular operation of such formations together with other technical improvements is expected to allow capacity to be increased soon to 200 million tonnes per year. In 2004 the Ministry of Railways placed two significant orders with Siemens and ALSTOM and their respective domestic partners for a total of 360 twin Bo-Bo electric locomotives with three-phase AC propulsion, many of these intended for use on the Datong–Qinhuangdao line.

China's second major coal line, forming part of the Shenhua Project for developing the huge 225 billion tonne Shenfu Dongsheng coalfield in Shaanxi and Inner Mongolia, is continuing to develop. The full line from Shenmubei via Shuozhou to the new harbour at Huanghua, south of Tianjin, is now complete. The third phase of the overall project, which also includes further development of the mines, a coal terminal at the port, a large power station and provision of a shipping fleet for transporting coal, is due for completion in 2005, when production and shipping capacity is planned to reach 60 million tonnes per year.

In 2004 the Ministry of Railways announced that double-tracking of the Shenhua Baotou–Shenmu line would be underway in 2005. This joint venture railway serves mainly to bring coal from the Dongsheng and Shenmu areas to the national railways at Baotou for onward distribution to China's west and northeast. The Ministry also announced that three new special-purpose coal railways are being planned for Shanxi Province, where existing lines can no longer cope with demand.

Private sector participation
In late 2001 the MoR made public its three-stage plan for allowing foreign participation in railway freight transport. In the first stage, from 2002 to 2004, foreign investors will be permitted to hold minority shares in railway freight companies. In the second stage, from 2004 to 2006, majority shareholdings by foreign investors will be allowed and from 2006 it is planned that the railway freight sector will be completely open.

Intermodal operations
Following the launch of through service on the Alatau Pass route to Russia in 1992, nine container terminals for landbridge traffic have been established by the Chinese. These are at the port of Lianyungang, and at Tanggu, Hohhot, Erlianhot, Zhengzhou, Xi'an, Lanzhou, Urumqi and Druzhba. Using this route, transits to European destinations from Japan are 2,700 km shorter than via the Trans-Siberian. Japanese industry has access to the new international route via the ports at Lianyungang and Shanghai. Lianyungang port is being extended to raise capacity from 14 to 18.5 million tonnes a year, financed by a loan from the Japanese government.

Collaboration with Japan intensified during 1997 with the establishment of International Freight Railway Systems Co Ltd, a joint venture agreement between China Railway Foreign Service Corporation and JR Freight. This foresaw the establishment of maritime links from Japan feeding intermodal traffic (and general freight) into the CR network for onward rail distribution both in China and to neighbouring countries. Intermodal traffic for Europe was to be handled on new services using the China Land Bridge in a Japanese bid to establish a rail alternative to maritime links.

Intermodal traffic is also the focus of a feasibility study, started in 1997 and funded by the United Nations and the Asian Development Bank, into the Euro-Asia Continental Bridge project, linking east coast ports in China with northwest Europe via Xinjiang province in western China and Kirghizia.

There has been rapid growth in domestic container transport. In recent years for the first time block loads of domestic containers have become a more familiar sight, boosted by production of a number of types of wagon to cater for this traffic. A design of container wagon now being produced is the Type X1k designed to operate at speeds up to 120 km/h. A double-stack container wagon design has also now been developed and the first service using a train formed of 38 such wagons commenced between Beijing and Shanghai in April 2004.

While container transport on China's railways is still geared primarily towards internal movement of freight using domestic-sized containers, the MoR intends actively to expand the use of international standard (ISO) containers particularly for land-bridge and other international transport. All container wagons now being supplied to China's railways are equipped to carry domestic or international standard containers.

Class K13NK hopper wagon (Bruce Evans)

0580489

In an important step towards encouraging international freight traffic to switch from sea to rail for journeys between east Asia and Europe, March 2005 saw the running of a trial container train from Hohhot in Inner Mongolia to Frankfurt, a distance of 9,800 km. The train carried 100 loaded ISO containers and took 17 days for the journey. The service has been set up by Railion and will operate under a multinational agreement between the railways of Germany, Belarus, Russia, Mongolia and China, providing for a twice-monthly service.

The double-tracking of the 800 km Urumqi–Alashankou line is part of an international initiative to boost cross-border capacity with Kazakhstan and to eventually create a standard-gauge link to Europe. Associated with this project is one to build a 320 km standard-gauge line paralleling the existing 1,520 mm gauge line between Dostyk on the Chinese border and Altagai in Kazakhstan, where a new transshipment hub is under construction. This new line is due for completion in 2005 and will initially be worked by Chinese locomotives. At a later date the broad-gauge line will be regauged to create a double-track standard-gauge line from the Chinese border to Altagai.

New lines
New lines are coming on stream all the time, partly to relieve pressure on the heavily occupied trunk routes in the east of the country, where the bulk of the network is concentrated, and partly to extend railways into the western and southern provinces, which are poorly provided with rail transport.

The Ninth Five-Year Plan, covering the period 1996–2000, was to see 8,100 km of new lines completed at a cost of Y330 billion. The total included several routes to improve access to the Shanxi and Inner Mongolia coalfields, and to the Tarim Basin oil deposits. The greater portion of new construction (6,100 km) was to be funded by the state, while 2,000 km was to be paid for by regional authorities.

The Tenth Five-Year Plan, for the period 2001 to 2005, confirmed that the expansion of China's railway system was set to continue at a rapid pace. Under the new plan China intended to invest US$42 billion in railway construction over the five-year period to 2005. Construction would include around 6,000 km of new lines, 3,000 km of double-tracking, 5,000 km of electrification and improvements to many of the key main lines to permit 160 km/h passenger train operation. By 2005 the national railway network, including joint-venture lines, is expected to total 75,000 km, of which 20,000 km would be electrified and 25,000 km would be double-track.

In 2004 announcements were made indicating that the annual rate of investment in railway construction was to be doubled for the next four years to a figure of around Y100 billion.

Long-term plans approved by China's State Council in 2004 are targeted at expanding the public rail network from about 72,000 km of route at present to around 100,000 km by 2020. Two of the main objectives will be to create around 12,000 km of dedicated passenger lines including express passenger networks centred on Beijing, Shanghai and Guangzhou and to add around 16,000 km of new line in western China. The aim of this is to promote development of this region and to increase capacity of existing lines. In addition to new lines, the long-term plan includes about 13,000 km of double tracking and 16,000 km of electrification.

For 2005 the Ministry of Railways has indicated that 58 new railway projects are due to be started and that around 700 km of new line, 500 km of double-tracking and 900 km of electrification are due for completion.

Major new lines opened to traffic since 2002 include:
- Xi'an–Ankang (Shaanxi Province) (245 km)
- Xinyi–Changxing (Jiangsu) (640 km)
- Shenmu–Yan'an (Shaanxi) (390 km)
- Daxian–Wanxian (Sichuan) (160 km)
- Anbian–Meihuashan (Sichuan and Yunnan) (370 km)
- Meizhou–Kanshi (Guangdong and Fujian) (150 km)
- Yangquan–Shexian (Shanxi and Hebei) (140 km)
- Zhanjiang–Hai'an (Guangdong) (140 km)
- Xilinhot–Sangendalai (Inner Mongolia) (150 km)
- Suning–Huanghuagang (Hebei) (230 km)
- Liupanshui–Baiguo (Guizhou) (120 km)
- Qinhuangdao–Shenyang (Hebei and Liaoning) (420 km)
- Sections of Xi'an–Hefei (Hefei, Shaanxi, Henan, Hubei and Anhui)
- Shaoyang–Yongzhou (Junan) (120 km)
- Jiaozhou–Xinyi (Shandong) (300 km)

Major new line construction projects in progress in 2004 included:
- Xi'an–Hefei (Shaanxi, Henan, Hubei and Anhui) (955 km)
- Chongqing–Huaihua (Chongqing, Sichuan, Guizhou, Hunan) (625 km)
- Jixian–Tianjin (Tianjin) (120 km)
- Changjiabu–Jingmen (Hubei) (180 km)
- Golmud–Lhasa (Qinghai and Tibet) (1,120 km)
- Wenzhou–Fuzhou (Zhejiang and Fujian) (260 km)
- Ganzhou–Longyan (Fujian and Jiangxi) (280 km)
- Suiyang–Dongning (Heilongjiang) (100 km)
- Chongqing–Suining (Chongqing and Sichuan) (150 km)
- Nanjing–Hai'an (Jiangsu) (210 km)
- Wanxian–Yichang (Chongqing, Sichuan and Hubei) (420 km)
- Dali–Lijiang (Yunnan) (170 km)
- Liuyuan–Dunhuang (Gansu) (170 km)
- Jinghe–Korgas (Xinjiang) (300 km)

New passenger lines
With the very heavy passenger and freight rail traffic levels experienced on many of China's key double track main lines, and with the need to increase passenger train speeds to maintain rail's competitiveness, the Ministry of Railways' policy now is to construct dedicated passenger lines in the busiest corridors.

The 420 km Qinhuangdao–Shenyang line (the Qin-Shen line), now open but in early 2004 still not in full operation, has been designed as an electrified double-track line, initially for 160 km/h operation. It is planned that the maximum permissible speed on this new line will be increased in 2005 to 200 km/h and eventually to 250 km/h. Work on further upgrading of the existing Beijing–Qinhuangdao line to increase the maximum permissible speed to above 160 km/h is continuing. Completion of these two projects will allow a significant speeding up of passenger services on the busy Beijing–Shenyang axis.

The new 1,300 km Beijing–Shanghai line has been designated China's first true high-speed line, intended initially for 300 km/h but eventually for an ultimate maximum speed of 350 km/h. After lengthy deliberation, the Chinese government announced in 2003 in favour of adopting conventional high-speed rail instead of maglev technology for this line but more recently there have been indications that a final decision may not have been made. Construction of the initial section of the line, linking Shanghai and Nanjing, was due to start before the end of the 2001–2005 five-year plan. This section will be around 300 km in length, with around 240 km on viaduct. Nine stations including the terminal facilities at Nanjing and Shanghai are envisaged.

New dedicated 200 km/h passenger lines were announced for the Shanghai–Hangzhou and Changsha–Hengyang corridors while the Ministry of Railways indicated that the Beijing–Guangzhou corridor would see China's second 350 km/h line. Maglev and 350 km/h conventional high-speed rail technology are also being considered for the Shanghai–Hangzhou line.

In 2004 the Ministry of Railways announced that by 2020 it was planned that China would have around 12,000 km of dedicated double-track passenger railways. Work on over 3,000 km of this total would commence soon, with new lines in the Beijing–Tianjin, Wuhan–Guangzhou, Zhengzhou–Xi'an, Jinan–Qingdao and Lanzhou–Xining corridors the top priority. These lines are being designed for 200 km/h and in some cases 300 km/h operation.

The dedicated passenger line between Guangzhou and Wuhan includes a second major rail/road bridge over the Yangtze River at Wuhan. Construction of this major structure will start in 2005. The double-deck bridge is almost 5 km long with a main span of 500 m. As with most of the other major rail/road bridges in China, trains will run on the lower deck while the upper deck will carry a multi-lane road. Completion of the bridge is expected in 2008.

Plans for the development of a network of regional express passenger lines are evolving for the Guangzhou and Pearl River Delta area. The first new line in this network will connect Guangzhou and Zhuhai, bordering Macao. This will be a 140 km double-track electrified line with 14 stations and designed for a maximum speed of 200 km/h. The line, projected to cost around US$2.5 billion, will be a joint venture between Guangdong Province and the Ministry of Railways. Construction was expected to start in 2004. Eight further dedicated express passenger lines are planned for the region, with a network totalling 930 km envisaged by 2020.

Plans for a 200 km/h electrified express passenger railway in Fujian Province have also recently been announced. This line would extend from Fuzhou along the coast to Xiamen, a distance of about 275 km. It would eventually form part of an express passenger line extending from Shanghai, via Hangzhou, Ningbo, Wenzhou, Fuzhou, Xiamen and Shantou to Shenzhen.

Corridor capacity expansion

A second group of new line construction projects in China is aimed at the expansion of the grid of high-capacity corridors. Many of these will focus on the following three corridors: Dalian–Changxing (west of Shanghai); Shimenxian (Hunan)–Zhanjiang (Guangdong); and Nanjing–Xi'an (Jiangsu, Anhui, Henan, Shaanxi and Hubei).

A key proposal of the planning for the new corridor between Dalian and Changxing, which will ease the flow of rail traffic between the southern tip of Liaoning Province and the southern and central parts of China, is the introduction of train ferries between Dalian and Yantai, eliminating a long detour via Tianjin. Progress to date on the overall project includes completion of the line from Xinyi to Changxing, opened to traffic at the end of 2002, the start of construction of the Jiaozhou–Linyi line, completion of double-tracking between Lancun and Taocun and the ordering of three train ferries for the 150 km Dalian to Yantai sea crossing and

Class SS7D 160 km/h 'tri-bo' electric locomotive with a passenger service near Zhengzhou (Bruce Evans)   **NEW**/1114982

start of construction of the 12 km railway and ferry dock at Yantai. The overall project is due to start operation in mid 2006.

The Shimenxian–Zhanjiang scheme, the southern section of the existing north-south corridor linking Luoyang, Liuzhou and Zhanjiang, includes some new construction and doubling of existing tracks.

Improvements to the east-west corridor between Nanjing and Xi'an are required to augment the existing busy route via Xuzhou and to provide a more direct route for coal traffic between the Ningxia and Shaanxi coal-mining areas and the industrialised and developed eastern coastal areas.

Other new lines

On the basis of China's Tenth Five-Year Plan (2001–2005), the national railway system, including joint-venture public railways, was projected to total 75,000 km by 2005. In the longer term, the public railway network is projected to exceed 100,000 route-km by 2020 and 120,000 route-km by 2050. Some of the major projected lines include:

A line linking Lhasa, the capital of Tibet (Xizang), with the national railway network. Tibet remains the last province of China to be connected to the national system. Two routes were proposed, one an extension of the new Guangtong–Dali line in Yunnan Province and the other an extension of the existing 815 km Xining–Golmud line in Qinghai. In October 2000 the Chinese government formally announced the decision to proceed with the 1,100 km line from Golmud to Lhasa. Preliminary work including the construction of the first 100 km section of the line between Golmud and Wangkun took place in 2001 and major construction work on the full line started towards the end of 2001 and is now well advanced.

A line in Xinjiang linking Kashi, at the end of the recently completed Korla–Kashi line, with the Kyrgyzstan border town of Torugart. New lines would be constructed within Kyrgyzstan to provide international links to Kazakhstan and Uzbekistan and western Europe, forming a southern Trans-Asian route between China and Europe that is significantly shorter than the Alatau Pass route through Kazakhstan. Construction of this line was expected to start in 2004.

A line south from Kunming to the borders with Thailand, Laos and Myanmar. This would form part of the eventual Pan-Asia Railway Network linking China's Yunnan Province with these countries as well as with Cambodia, Vietnam and Malaysia.

A line from Zhongwei (Ningxia Autonomous Region) eastwards towards Taiyuan.

A line from Shizuishan eastwards to Dongsheng. A new east-west corridor would be created as a new line is presently under construction between Dongsheng and Jungar Qi, the present terminal of an electrified line from Datong.

A new 360 km line along the eastern coast between the cities of Wenzhou in Zhejiang and

Fuzhou in Fujian. This line will form the central section of an eventual 700 km line along the east coast, linking the cities of Ningbo and Shenzhen.

A new line from Chongqing to Nanjing generally following the course of the Yangtze River.

The construction of the Golmud to Lhasa railway is one of the greatest railway construction challenges engineers have faced. Much of the 1,100 km line is at altitudes of around 4,000 m, with the highest point at 5,070 m, and it traverses extremely rugged and inhospitable terrain with many areas of permafrost and seasonally frozen ground. Tunnels and bridges make up around 7 per cent of the length of the new line and the longest tunnel is around 1.7 km. Construction is now well in progress along the route and officials have indicated that they hope to complete this massive engineering task by 2007, with test running due to start in mid-2006.

In recent years China has greatly increased the rate of development of railway infrastructure in the poorer and more remote western areas of the country. The Tenth Five-Year Plan covering the period from 2001 to 2005 will see about 40 per cent of the total planned investment for large and medium railway projects directed to projects in the west involving new lines and improvements to existing lines.

Construction of a further line to tap into the coalfield in the Zhungeer region east of Dongsheng is expected to begin soon. This is a 115 km line that would extend southwards from Hohhot to link up to the existing Shenhua Datong–Zhungeer line. It was recently announced that the Kuwait Fund was to provide a US$3.5 million loan to finance the construction.

A 150 km single-track link between Meizhou in Guangdong and Kanshi in Fujian was recently opened to traffic. This is the first direct rail connection between these two provinces and is a valuable link in the network gradually forming along the coast south of Hangzhou.

Completion was achieved in 2000 of the 10.6 km combined road/rail bridge across the Yangtze River to replace the Wuhu train ferry. Construction of this bridge began in 1997. The bridge's completion has helped to ease traffic on the heavily utilised Nanjing–Shanghai line, as some trains for Hangzhou and points further south now use the route via Wuhu.

Major new general-purpose lines on which construction was expected to commence in 2005 include an electrified single-track line of 180 km linking Zhangjiakou in Hebei with Jining in Inner Mongolia and the first section of the line that will eventually connect Linhe in Inner Mongolia with Hami in Xinjiang. The first line will shorten the distance between Beijing and the major Inner Mongolian cities of Hohhot and Baotou by 130 km. The second line when completed will be around 1,000 km long and will significantly reduce the rail distance between China's far western regions and Beijing.

Joint-venture railways

Since the 1980s and 1990s there has been a steady increase in the proportion of new lines constructed as joint-venture railways where the Ministry of Railways, local governments and authorities share the investment costs.

By the end of 2003 more than 20 joint-venture railways, totalling about 7,700 km, were operating with many more under construction or in planning. In addition, on a number of the operating systems extensions, branches and capacity increasing works were under construction. During the Ninth Five-Year Plan for the national railways (1996–2000) joint-venture railways were expected to account for over 45 per cent of the total new route-length constructed. It is expected that during the period of the Tenth Five-Year Plan, from 2001 to 2005, development of joint-venture railways will account for a similar proportion of total route-length constructed. Approximately 40 per cent of the capital investment in the joint-venture lines has come from provincial authorities, major cities and other organisations.

Several of the joint-venture railways have now developed into major railway operations with locomotive fleets of between 50 and 100 units. The Jining-Tongliao, Sanshui-Maoming, Xi'an-Yan'an and Shenmu-Huanghuagang railways are examples. The Ministry of Railways' mid-term planning for the period to 2020 shows that several of the now established joint-venture railways are due to be double-tracked.

The joint-venture model for railway development and operation was introduced to speed up the expansion of the railway system, to reform the investment system and to open up new funding channels. Prior to the introduction of this model, funding for public railway development had come almost entirely from the Ministry of Railways.

Joint-venture railways are run as corporations using market-orientated systems for construction and management. Many joint-venture railways are set up as limited liability or joint stock companies. The standardised management system developed for the joint-venture railways includes requirements for establishing boards of directors and supervisors for the company. The companies are required to be managed independently and to have full responsibility for their own financial affairs.

Joint-venture railways are required to use competitive bidding for activities such as railway construction and for acquisition of rolling stock. This is to control costs through market competition. The contract systems for this are gradually being adapted to conform to international practice.

Foreign investment in joint venture railways is now permitted in China. One of the first cases where this has occurred is on the Pingyan railway linking Pinghu on the Shenzhen-Guangzhou line with the harbour of Yantian north of Shenzhen. A 65 per cent share in this railway was reported to have been bought in late 2001 by a foreign company with interests in the Yantian port development.

A pilot railway project involving significant private financing through bank loans is due to open in 2006. Ownership of the 45 km Quzhou-Changshan joint venture railway will be shared between the private Changshan Cement Co (32.5 per cent), the Ministry of Railways (35 per cent) and local government (32.5 per cent).

New lines being constructed as joint-venture railways include the following:
- Xinhe–Changxing (Jiangsu Province)
- Tongling–Jiujiang (Anhui and Jiangxi Provinces)
- Jixian–Tianjin (Tianjin City)
- Wenzhou–Fuzhou (Zhejiang and Fujian Provinces)
- Changjiabu–Jingmen (Hubei Province)
- Qingding–Luling (Anhui Province)
- Gangzhou–Longyan (Fujian and Jiangxi Provinces).
- Dali–Lijiang (Yunnan Province)
- Quzhou–Changshan

Recently completed projects using investment systems similar to the joint-venture railway model include the new Yangtze River bridge at Wuhu and the doubling of CR's Xiaoshan–Ningbo railway.

The first railway project in China involving a sea crossing, between Guangdong and Hainan Provinces, was being opened in stages between 2003 and 2004 using the joint-venture model. This project involves three components, the 135 km Zhanjiang-Hai'an railway on the mainland, two train ferries for the 23 km sea crossing and the 157 km Haikou-Chahexi railway on Hainan island. The joint-venture partners include the Ministry of Railways and the two provinces.

The rate of construction of local railways, public railways in which there is no investment by the Ministry of Railways, has increased in recent years as local authorities, provincial governments and other bodies such as mining concerns have sought to encourage economic development. At the end of 2003 there were about 75 local railways totalling 4,800 km in operation with around 20 more totalling 1,800 km under construction. About 30 per cent of the local railway route-length at the end of 1999 was narrow-gauge (762 mm) but many of these lines have been closed.

Improvements to existing lines

Current improvements to existing lines in China focus on three main areas: track-doubling to improve line capacity; electrification (see 'Electrification' section); and improvements to raise line speeds. In 1999 major track-doubling schemes for completion by 2002 covered over 3,000 route-km. In some cases, where difficult terrain is encountered, such as Chengdu-Yangpingguan (400 km), track-doubling involves extensive engineering such as new tunnels and major viaducts. In all 660 km and 1,296 km of double-tracking were opened to traffic in 2000 and 2001 respectively. China expects to complete about 3,000 km of double-tracking during the course of the Tenth Five-Year Plan extending to the end of 2005. Lines on which track-doubling has been recently completed include the following:
- Xiangtan-Liupanshui on the Zhuzhou–Kunming line (Hunan and Guizhou Provinces)
- Tianshui–Lanzhou (Gansu Provinces)
- Yanzhou–Rizhao (Shandong Province)
- Longchuan–Changping (Guangdong Province)
- Taocun–Yantai (Shandong Province)
- Xiaoshan–Ningbo (Zhejiang Province)

Lines in the process of being doubled in 2002 included the following:
- Shimenxian–Huaihua (Hunan Province)
- Yakeshi–Hailaer (Inner Mongolia)
- Litang–Nanning (Guangxi Province)
- Lanzhou–Wuwei (Gansu Province)

Other lines due to be double-tracked during the Tenth Five-Year Plan include the following:
- Hangzhou–Xuancheng (Zhejiang and Anhui Provinces)
- Nanjing–Wuhu (Jiangsu and Anhui Provinces)
- Taiyuan–Yuanping (Shanxi Province)
- Kunming–Zhanyi (Yunnan Province)
- Wuhan–Jiujiang (Hubei and Jiangxi Provinces)
- Ankang–Chongqing (Chongqing City and Sichuan Province)

Mid-term planning for the period through to 2020 for China's national railways includes considerable amounts of double-tracking on the programme. Lines shown to be double-tracked by 2020 include the following:
- Lanzhou–Golmud (Gansu, Qinghai)
- Lanzhou–Shizuishan (Gansu, Ningxia)
- Urumqi–Alashankou (Xinjiang)
- Turpan–Korla (Xinjiang)
- Chengdu–Daxian (Sichuan)
- Chengdu–Nanning (Yunnan, Guangxi)
- Guiyang–Liuzhou Kunming (Sichuan, Yunnan)
- Kunming–(Guangxi, Guizhou)
- Liuzhou–Henyang (Guangxi, Hunan)
- Guangzhou–Hechun (Guangdong)
- Yingtan–Wuhu (Jiangxi, Anhui)
- Xi'an–Ankang (Shaanxi)
- Chongqing–Xiangfan (Chongqing, Sichuan, Shaanxi, Hubei)
- Xiangfan–Wuhan (Hubei)
- Baotou–Xi'an (Inner Mongolia, Shaanxi)
- Goubangzi–Qiqihaer (Inner Mongolia, Jilin, Liaoning)

China has comparatively few sections of line with more than double tracks. One of these is the very busy Guangzhou–Shenzhen line in the south of the country which at present is triple-track. To increase capacity further and allow separation of passenger and other services, work on adding a fourth track was started in 2001. China's other main triple-track line, between Beijing and Tianjin, is also due to be quadrupled soon.

Under the ninth Five-Year Plan, the Ministry of Railways has been implementing improvements to raise line speeds for passenger services on main trunk routes (see 'Passenger operations' section). Work has included: bridge strengthening; replacement of level crossings with grade-separated structures; easing of curves; installation of high-speed turnouts; and some replacement of rail and sleepers. Similar work is also being done on lines feeding the trunk routes.

Traction and rolling stock

By the end of 2003 the combined locomotive total for the national railway system and the joint venture railways was over 16,000, of which approximately 300 were steam, 11,400 diesel and 4,600 electric. At end of 2001 the locomotive total for the various local railways was about 350, of which approximately 220 were steam and 130 diesel. Around 160 of these were for the 762 mm narrow-gauge lines.

About 500 new diesel locomotives and 350 electric locomotives are delivered to the national railways and other public systems annually. Steam traction has ended on the national railways but continues to play a small but rapidly diminishing role on some joint-venture and local railways. In addition to the new units supplied to the public railways, approximately 200 new diesel and electric locomotives are delivered to mining and industrial concerns each year. Export orders are also steadily increasing.

New locomotive construction for the six years to 2003 by the CNR and CRS (ex-LORIC) factories until recently administered by the Ministry of Railways was as follows:

	1999	2000	2001	2002	2003
Diesel	671	624	622	737	771
Electric	295	295	391	417	349

The Ministry of Railways expected that from 2001 to 2005 more than 50 per cent of the locomotives constructed for the national railways will be electric. To meet the targeted requirement for electric locomotives new production lines were established at the hitherto diesel-only factories at Dalian and Ziyang, and these now augment production at Zhuzhou and Datong, the two plants that have supplied most of China's electric locomotives. The first locomotives from the new production lines were delivered in late 2000. By 2003 roughly equal numbers of diesel and electric main line locomotives were being delivered annually to the national and joint-venture railways. For secondary and shunting duties at present only diesel types are being supplied as China has placed no priority on introducing electric designs for such duties.

With the almost complete elimination of steam traction and the greater pace of electrification, deliveries of diesel locomotives to the national and joint-venture railways have dropped from a peak of around 800 per year in the mid-1990s to about 500 per year. This drop has however been matched by an increase in production of diesel locomotives for local railways, mines, industry and power stations, allowing the rapid elimination of steam traction on these systems as well.

The year 2004 was significant in that it saw a massive surge in orders for locomotives and multiple-units placed with foreign suppliers in Europe, Japan and the United States. These included three orders for 200 km/h passenger emus, two orders for AC-motored twin-unit electric locomotives and an order for high-altitude diesel locomotives for the new Golmud–Lhasa line. The orders were the largest placed for imported designs and equipment since the mid-1980s.

Diesel locomotives

Around 5,000 of the locally produced 2,430 kW DF4/4A/4B (Dong Feng 4) Co-Co diesel-electrics in both freight and passenger versions are now in service. Although limited production of the DF4B is continuing for joint venture, local railways and industry, the principal present production DF4 type is the 2,940 kW DF4D. This has been produced in at least five versions, 80 km/h shunter, 100 km/h general-purpose, 120 km/h passenger and 160 km/h passenger. With production of diesel locomotives for passenger service greatly reduced, DF4D deliveries to the national railways now consist mainly of the 100 km/h mixed-traffic version. In 2003 delivery was made of the first batch of general-purpose DF4Ds fitted with radial-motion bogies, China's first production locomotives to be so equipped.

In late 2003 series production began at both the Beijing and Sifang works of the DF7G a medium-powered diesel-electric locomotive for shunting and secondary duties. This type has supplanted the DF5 and DF7C as the standard production secondary duty locomotive for the national railways.

The high-powered DF8B freight locomotive produced by the Qishuyan and Ziyang works is the most powerful freight type in domestic production, at 3,680 kW. This is China's first main line diesel designed for a 25-tonne axleload as against the 23-tonne standard and is being introduced on the key non-electrified main lines where heavier and faster freight haulage is required in terms of national railway policy. Around 700 DF8B locomotives are now in service, with production continuing at around 200 units per year. Several experimental units with radial motion bogies have been constructed and are now on trial.

Qishuyan's main production locomotive type until late 2003 was the DF11 for 160 km/h passenger services. By the end of 2003 about 450 of these units were in traffic.

The most notable new diesel locomotive design introduced in China recently has been the twin-unit double Co-Co DF11G. This type has been developed to haul the new 'Z' class trains introduced in April 2004 on the Beijing-Shanghai route. The schedules for these trains incorporate a two-hour reduction in overall journey time between the two major centres. The DF11G is the first production diesel type for the national railways to incorporate head-end-power for train power requirements.

Further planned development of the DF11 family includes versions with AC traction motors and models equipped with more powerful diesel engines.

The main line diesel locomotives (DF4D, DF8B and DF11G) now in production for the national railways are considered to be third-generation types. Development of fourth-generation diesel locomotives is underway and technological advances include new high-power, low-emission, diesel engines, AC traction motors, greater use of electronic control and fault-diagnosis equipment and radial and other more sophisticated bogies.

Both the Dalian and Qishuyan works have developed new high-power engines with the assistance of US and European specialists. In 2000 Dalian began testing prototypes of DF8B (freight) and DF11 (160 km/h passenger) type diesels fitted with a new 12-cylinder 3,820 kW engine, while at Qishuyan a new 16-cylinder 4,260 kW engine was successfully factory-tested in 2001 and is being service-tested in the prototype of a new freight locomotive design unveiled in 2002. Series production of locomotives equipped with these more powerful engines has not yet occurred.

China's first AC-motored diesel locomotives were completed in 1999 and 2000 and are now undergoing service testing. These prototypes included Dalian-built DF4Ds with Siemens traction motors, a Qishuyan-built DF8CJ, Zifang-built DF8BJ, Beijing-built DF7Js and Sifang-designed NJ1 shunting locomotives.

In late 2004 the Ministry of Railways announced that it had placed an order with General Electric of the US for 78 high-powered diesel-electric locomotives for use on the new line between Golmud and Lhasa (Xizang/Tibet) to be completed in 2007. These custom-designed locomotives are to be built in the US. They will be powered by a GE 16-cylinder engine and will have GE AC traction motors. Deliveries will begin from the end of 2005.

The main operational goals shaping main line locomotive development are the needs for operating 5,000-tonne general freight trains at up to 100 km/h and at high average speeds, express

Sifang-built Class DF7C for shunting and local freight and passenger work (Bruce Evans)   0580485

Class DF8B diesel locomotive with a heavy freight train at Zhengzhou (Bruce Evans)   0554913

freights at 120–140 km/h and express passenger trains at 140–200 km/h.

The recent commissioning of the Harbin–Dalian and Guangzhou–Wuhan main line electrification and the elimination of steam traction have allowed the national railways to accelerate the rate of withdrawal of older diesel locomotives constructed in the 1970s and 1980s. Types being withdrawn include early examples of the DF4 family, DFH3 and BJ passenger diesel-hydraulics and the Romanian-built ND2 and ND3 types.

Diesel and electric multiple-units

In recent years several of the Chinese locomotive and rolling stock works have designed and built diesel multiple-unit/push-pull passenger train sets as part of the massive efforts to improve intercity passenger services on key main lines. Several of these designs are now on test with prototypes and, in some cases, production batches in service. The Beijing–Tianjin express intercity services are now worked by a fleet of double-deck NZJ2 push-pull dmus. A second fleet of NZJ2 dmus has been introduced on Lanzhou–Xining services in western China. Most of the new sets are designed for 160 km/h operation. However, less sophisticated 120 km/h diesel push-pull sets, some with sleeper accommodation, are also being marketed to the joint venture railways to provide high-quality general and tourist passenger services.

As part of a strategy to develop high-speed passenger services on selected routes, China has been developing its emu designs for operation at speeds of up to 270 km/h. A 220 km/h prototype unit designed by Changchun Railway Car Works, formed of two power cars, eight single-deck trailers and two double-deck trailers, was completed during 1998. The prototype unit was a joint project of the Zhuzhou Electric Locomotive Works, the Changchun Car Company, the Puzhen Locomotive and Rolling Stock Works and the Tangshan Rolling Stock Works.

In September 2000 a prototype 25 kV AC 200 km/h trainset was rolled out at the Zhuzhou Electric Locomotive Works. The 'Blue Arrow' unit comprises a 4,800 kW power car, five trailers and a driving trailer, although it was expected that longer versions with two power cars would be produced. A fleet of these trains is now in service on the Guangzhou–Shenzhen route.

China's first 270 km/h emu, the 'China Star' type, is being produced initially for service on the new and upgraded Beijing–Shenyang route. This design set a Chinese record of 321.5 km/h during tests in 2002 on the new Qinhuangdao–Shenyang line.

In addition to the 200 and 270 km/h designs, several other types of emu are in use or on trial on various lines. China's first domestically developed modern emu was handed over for revenue-earning services in April 1999. The 'Spring City' trainset was produced jointly by the Changchun Car Plant, The Zhuzhou Electric Locomotive Research Institute and the Kunming Railway Administration. It is in use for tourist services between Kunming and the nearby 'Stone Forest'.

For accelerated intercity passenger services in mountainous parts of the country China's railway industry is developing tilting emu and dmu designs. The first of these emerged in 2003 and is formed of power cars supplied by the Dalian works and coaches produced at the Tangshan works.

The most high-profile agreement with a western company in the 1990s was a deal signed in November 1996 with Adtranz, whereby China was to rent a 25 kV AC version of the Swedish X2000 tilting high-speed train for two years, possibly purchasing it thereafter. This arrived in China in March 1998, and has been running between Guangzhou, Shenzhen and Kowloon since 2000.

A significant development in October 2004 saw the Ministry of Railways announce three large orders with Japanese, French and Canadian companies and their respective Chinese partners for fleets of emus for prime passenger services on key electrified main lines.

A Japanese consortium headed up by Kawasaki together with their Chinese partner Nanche Sifang Locomotive received an order for 60 200 km/h eight-coach emus. These are to be based on the Japanese E2-1000 series Shinkansen trains. The first three sets will be built in Japan and will be followed by six sets to be sent to China in knocked down state. The remaining 51 sets are to be built by Nanche Sifang in China. Delivery of the first trains is expected in 2006.

A similar order for the same number of emus was placed with ALSTOM and Changchun Railway Company, its Chinese partner. These trains will be non-tilting Pendolino derivatives. The first set is due to enter commercial service in 2007. As in the case of the Japanese order, the first three sets are to be built overseas, the next six sets supplied as knocked down sets and the remainder built in China.

The third order was for 20 eight-coach emus to be supplied by Bombardier with its joint venture partner Bombardier Sifang Power, based in Qingdao. The trains will be assembled at Qingdao with bogies and other equipment being supplied from various Bombardier plants in Europe.

Electric locomotives

To cater for the increased rate of electrification the Zhuzhou Electric Locomotive Works and the Datong Locomotive Works expanded output and two new electric locomotive production lines were established at the Dalian Locomotive and Rolling Stock Works and Ziyang Diesel Locomotive Works in the late 1990s. Between 350 and 450 electric locomotives are now delivered to the national railways and joint-venture railways annually, many of these high-powered twin-unit freight locomotives. Main production types in 2004 were the SS4(G) and SS3B(G) twin-unit freight locomotives and the SS7E and SS9(G) 160 km/h passenger locomotives. China also builds main line electric locomotives for export.

A Bo-Bo+Bo-Bo development of the SS3, the SS4(G), now China's most numerous electric locomotive type with almost 1,000 in service, is in batch production at the Zhuzhou, Datong and Dalian plants. It weighs 184 tonnes, is rated at 6,400 kW and has a top speed of 100 km/h. Small numbers of the technically improved SS4B have been supplied to the joint-venture railways forming part of the Shenhua coal project. A further version, the SS4C, with axleload increased from 23 to 25 tonnes has also been developed but had not entered batch production in 2004. The 4,800 kW SS6 and SS6B Co-Co emerged in 1992 and 1993 and feature traction equipment developed in association with Hitachi. The Bo-Bo-Bo SS7, also rated at 4,800 kW, is designed for heavy haulage on steeply graded and sharply curved routes. The SS7B is a freight version with 25 tonne axleload, while the SS7C is in production for 120 km/h passenger service.

In 2003 production of the SS3B at Zhuzhou, Ziyang and Datong was superseded by production of the first batches of China's most powerful locomotive, the SS3B(G), a 8,700 kW Co-Co+Co-Co design. The first 80 SS3B(G)s were put into service at Lanzhou, Guiyang and Ankang during the year. The new design is based on the earlier SS3B and incorporates new technology to improve reliability and to facilitate maintenance.

For 160 km/h passenger services five electric locomotive types have so far reached series production stage. The SS8 Bo-Bo, rated at 3,600 kW, was the first standard type for 160 km/h services and over 240 are now operating. This type set a Chinese rail speed record on 24 June 1998, when one attained a maximum speed in excess of 240 km/h on a section of the Beijing–Guangzhou main line south of Zhengzhou. Batch production of China's second electric type for 160 km/h services, the 4,800 kW SS9 Co-Co, began in 2001 with units being delivered to the northeastern railway administrations for use on premium passenger services on the recently electrified Harbin–Dalian main line. An improved version, the SS9(G), followed off the production line at Zhuzhou from late 2002. Around 120 were in service by the end of

Class SS9 Co-Co passenger electric locomotive (Bruce Evans)　　　0554906

Class SS4 twin-section electric locomotives at SuJiaTung (Colin Boocock)　　　0580487

Newly built Class YW25G hard-berth air-conditioned sleeping car (Bruce Evans)　　　0580486

2004 on the Beijing–Guangzhou and Beijing–Harbin/Dalian routes. Production of a fourth 160 km/h electric locomotive type, the 4,800 kW SS7D Tri-Bo, has also begun, at the Datong works. These are operating on the Zhengzhou–Baoji route, sections of which pass through mountain areas and have numerous curves. Datong is also now producing the SS7E Co-Co type and around 60 of these were in service by the end of 2004 with batch production continuing.

China's first domestically produced electric locomotive featuring three-phase AC traction motors, the 4,000 kW AC4000 produced by the Zhuzhou plant, commenced trials in 1997. Since then prototypes of three further 4,800 kW AC-motored designs, all of Bo-Bo layout for 200 km/h passenger haulage, have been placed into service. Two of these designs are classified as DJ and DJ2, the first having European-supplied motors and electrical equipment and the second domestically produced motors and equipment. The DJ2 was completed in 2001 and has been given the name 'Ao Xing' or 'Olympic Stars'.

In 1999 Siemens Transportation Systems established STEZ, a joint-venture company with the Zhuzhou Electric Locomotive Works and the Zhuzhou Electric Locomotive Research Institute to produce locomotives and traction components in China. The first products of the new company, Siemens Traction Equipment Ltd, Zhuzhou, were 20 Class DJ1 three-phase 6,400 kW eight-axle AC electric freight locomotives. The first three were built at Siemens' Graz plant in Austria, with deliveries commencing in September 2001. The remaining 17 locomotives were constructed at Zhuzhou in 2002–03. These locomotives are now in use on heavy-haul coal traffic on the Datong-Qinhuangdao line. Production by the new company is expected to grow to 70 to 100 locomotives per year with export production also planned for the future.

In late 2004 Siemens and Zhuzhou in partnership were awarded the first large-scale order for AC-motored electric locomotives. This order, worth around US$470 million, is for 180 9,600 kW eight-axle units for delivery from mid-2006 to the end of 2007. The locomotives are to be built at Zhuzhou with development and engineering work to be undertaken at Siemens facilities in Europe. These locomotives, which will have a top speed of 120 km/h, are intended for heavy-haul operations such as those on the special-purpose coal line linking Datong and Qinhuangdao.

In June 2004 an agreement was signed between ALSTOM and the China Northern Locomotive and Rolling Stock Industry (Group) covering co-operation in the manufacture at Datong of electric locomotives based on the Prima design. Initially the partnership will build heavy freight machines, later producing conventional freight and high-speed passenger variants. In October 2004 it was announced that the Ministry of Railways had placed an initial order with ALSTOM and Datong for 180 high-powered twin unit AC-motored electric locomotives based on ALSTOM's Prima design. The first 12 locomotives are to be built in France, with the remainder to be built at Datong. The units are intended mainly for heavy-haul operations such as those on the Datong–Qinhuangdao coal line.

Future electric locomotive development in China will be focused on producing a series of AC-motored designs for freight workings up to 140 km/h and for passenger workings up to 200 km/h. Power per axle will be increased to 1,600 kW, more advanced bogies will be developed and greater use will be made of electronic control and fault detection and diagnosis.

The Datong works has produced China's first design for 120 km/h freight workings. This is classified as the SSJ3. It is an AC-motored Co-Co fitted with radial motion bogies. This new type has been approved for batch production by the Ministry of Railways and initial orders placed for locomotives to be built at the Dalian works. Equipment for these locomotives includes Toshiba traction motors and Voith Turbo final drives. It is expected that over 400 of these locomotives will be built in the next few years.

Passenger coaches

Total production at the country's four coach manufacturing plants in the five years to the end of 2003 is given in the table below.

	1999	2000	2001	2002	2003
Coaches	1,780	3,244	3,291	2,728	1,539

Most of the passenger vehicles now being supplied to the national railways by the four coach manufacturing plants in China are based on the standard Type 25 coach initially developed at Changchun with Brel assistance. A wide selection of variants, of both single-deck and double-deck format, is now being produced for 120 and 160 km/h running. These coaches are produced in soft- and hard-sleeper and soft- and hard-seat configurations with hard-seat being the predominant form. Baggage vans, power cars and dining cars, the last in single- and double-deck format, are also produced.

The latest series of Type 25 coaches is designated as 25T. These are being supplied for the 2004 phase of the broad programme to accelerate main line passenger

Type C64K open wagon (Bruce Evans)   **_NEW_**/1114984

Class SS8 Bo-Bo electric locomotive with a Wulumuqi-Beijing service at Handan (Colin Boocock)   **_NEW_**/1114985

services in China. These vehicles have a design speed of 180 km/h. Soft- and hard-berth sleeping car and dining car versions are being delivered.

The joint-venture Changchun Adtranz Railway Co Ltd was established in January 1997, and planned an annual output of 160 vehicles. In 1996 South Korea's Hanjin Heavy Industries Co Ltd supplied 30 coaches suitable for 200 km/h running.

Coach production at a new works in Qingdao is set to expand following establishment of a joint venture between the China National Railway Locomotive and Rolling Stock Industry Corporation (LORIC), Canadian manufacturer Bombardier Transportation and Power Corp. From early 2003 an important new coach type began entering service when the first of an initial order for 300 coaches from this new plant established by the Bombardier Sifang Power Transportation joint venture was delivered to the national railways. The new company received a follow-up order for 17 dining cars and 21 sleeping cars in 2003.

A consortium of Bombardier Sifang Power Transportation and the Sifang Locomotive and Rolling Stock Co received an order worth about US$280 million from the Ministry of Railways in early 2005 to supply 361 coaches for services on the new railway to Lhasa in Tibet (Xizang). In addition to standard coaches, the order includes 53 special tourist coaches. These vehicles will be delivered between December 2005 and May 2006 and will be formed into sets to operate services between Lhasa and Beijing, and Shanghai and Guangzhou. The coaches will have state-of-the-art technology for high-altitude service (up to 5,000 m), while the special tourist coaches will include luxury compartment coaches with individual showers and coaches with panoramic viewing facilities.

Freight wagons

Total freight wagon production at the various MoR-associated CNR and CSR works for the five-year period from 1999 to 2003 is given in the table below.

	1999	2000	2001	2002	2003
Wagons	18,054	25,550	30,040	28,412	27,432

Most of these wagons have been supplied to the national railways, with a small proportion going to industry or for export. Additional production of freight wagons for industry and the public railways is undertaken at several heavy engineering works not linked to the MoR.

The main production types at present are the C64K general purpose open gondola (mainly for coal traffic) and the P64A covered wagon. A wide variety of other types is also produced including various tank wagons, hopper wagons and car transporters. Special purpose open coal wagons designed for a 25 tonne axleload are now being built for use in unit-trains on the Datong–Qinhuangdao coal line. An enlarged 25 tonne axleload version of the C64 general purpose gondola has also been developed and should soon enter production for the transport of ores and coal. A 25-tonne axleload container wagon design for double-stack services has also been produced.

In line with MoR policy to increase freight train speeds, from 2001 most new freight wagons, apart from those types already equipped with more sophisticated bogies, have been fitted with Type Z8A swing-motion bogies to permit operation at increased speeds of up to 90–100 km/h.

At the end of 2003, rolling stock comprised 40,487 passenger cars and 510,327 wagons (up from 34,346 passenger cars and 437,686 wagons in 1997).

On the approximately 75 local railways, rolling stock in 1998 amounted to 231 steam and 155 diesel locomotives, 200 passenger cars and 3,317 wagons.

Signalling and telecommunications

To meet line capacity requirements China's busy mixed-traffic main lines are equipped with automatic block signalling that on the busiest lines allows minimum intervals between trains of 5 to 7 minutes. Signalling system development in China is now focused on increased use of digital and computer-aided equipment and on providing locomotive-based signalling and speed-limiting systems on lines where quasi-high speed passenger services are operated.

Overall control of all signalling and telecommunications systems and projects is in the hands of the China National Railway Signal & Communications Corporation, with a staff of 23,000. Nevertheless the scale of work has led to joint ventures and contracts with foreign suppliers. Elin of Austria has supplied a track-to-train radio system embracing some 600 route-km south of Beijing. The country's first solid-state interlocking is being supplied by GEC-General Signal; it will be installed at Xiao Li Zhang, on the Zhengzhou–Wuchang line.

Ansaldo and its US subsidiary Union Switch & Signal are installing computerised interlockings for 13 stations on the new Beijing–Guangzhou line, along with a hump computer process control system for a marshalling yard on the route.

In 1997, the China Academy of Railway Sciences signed a co-operation agreement with Japan's Railway Technical Research Institute covering the joint development of automatic train control for a possible Chinese high-speed line. Japanese suppliers Nippon Sharyo, Keisan and Daido will equip a test track near Beijing.

During 1998–1999 Siemens commissioned China's first electronic interlocking at Fuyang marshalling yard, Anhui Province.

In October 1998 Shanghai GPT, a joint venture between GEC Marconi and Shanghai Railway Communications, won a US$10 million contract from the Ministry of Railways to upgrade Chinese Railways' communications system.

Recently contracts for the provision of GSM-R systems (radio communications systems tailored for railways and based on GSM – Global System for Mobile Communications) have been awarded for the upgraded Datong–Qinhuangdao coal line and the new Golmud–Lhasa line. These systems will be used for a number of functions including data communication, train control, train dispatch and train management. The contract for the Datong–Qinhuangdao line was awarded to Huawei Technologies while that for the Lhasa line was awarded to Nortel.

Electrification

The Ministry of Railways is planning to maintain the rate of electrification during the period of the tenth Five-Year Plan extending to 2005 at about 1,000 km per year.

Electrification at 25 kV AC 50 Hz has been a high priority since the early 1980s. Total route-km under wires amounted to 18,758 km by the end of 2003. In 2002 517 km of electrified line was put into operation. In 2001 the length of electrified line put into operation was significantly higher, in excess of 2,500 km, as the year saw energisation of the Harbin–Dalian, Guangzhou–Shaoguan, Chenzhou–Wuhan and Xi'an–Ankang lines. In 2003 a considerable length of double-tracking of electrified sections was put into operation.

One of China's major recent electrification projects, the electrification throughout of the key north-south Beijing–Guangzhou–Shenzhen mainline, was completed in 2001. Full electric services on the last two non-electrified sections of this line, between Guangzhou and Shaoguan and Chenzhou and Wuhan, were introduced during the course of the year. The overall scheme embodies CTC, fibre optic cabling, jointless track circuits and computerised interlockings. Wiring began at the northern end of the route, between Beijing and Zhengzhou (697 km) in 1991, and was completed in 1998. During the same period electrification of the Zhengzhou–Wuhan section was also completed.

Further south, the transversal from Kunming in the far southwest to Hangzhou, south of Shanghai, is in course of electrification. The 1,538 km from Kunming to Guiyang was energised throughout in 1993, along with the route northwards through Guiyang to Chongqing. Electric services over the 821 km from Guiding, just outside Guiyang, to Zhuzhou were inaugurated in April 1997. World Bank credits are supporting the double-tracking and wiring of the 892 km onwards from Zhuzhou to Hangzhou. Double-tracking has been completed between Hangzhou and Shanghai.

In recent years more attention has been given to electrification of the heavily utilised double-track main lines in eastern China. At present, work is underway on the Shijiazhuang–Dezhou, Jinan–Qingdao, Zhengzhou–Xuzhou and Shanghai–Hangzhou lines.

In 1999 electrification projects in progress for completion by 2002 covered nearly 5,300 route-km, a mixture of new and existing lines or works associated with track-doubling. Electrification projects in progress at the end of 2004 included:
- Xi'an–Nanyang (Shaanxi and Henan) (407 km new single line)
- Chongqing–Huaihua (Chongqing, Sichuan, Guizhou, Hunan) (625 km new single line)
- Zhengzhou–Xuzhou (Henan, Anhui) (340 km existing double-track line)

Diesel locomotives (National railways, including joint-venture railways)

Class	Wheel arrangement	Power kW	Speed km/h	Weight tonnes	No in service Dec 2004	First built	Mechanical Builders	Transmission
DF	Co-Co	1,320	100	126	10	1964	Dalian, Qishuyan, Chengdu, Datong	E
DF4	Co-Co	2,430	100/120	138	300	1969	Dalian, Ziyang	E
DF4A	Co-Co	2,430	100	138	360	1976	Dalian	E
DF4B	Co-Co	2,430	100/120	138	4,300	1984	Dalian, Datong, Sifang, Ziyang	E
DF4C	Co-Co	2,650	100	138	920	1985	Dalian, Datong, Sifang, Ziyang	E
DF4CK	A1A-A1A	2,650	160	126	2	1999	Ziyang	
DF4D/ DF4DF	Co-Co	2,940	100/120/ 140/160	138	1,150	1996	Dalian	E
DF4DD	Co-Co	2,940	80	138	20	1999	Dalian	E
DF5[1]	Co-Co	1,210	80/100	138	1,000	1984	Dalian, Sifang	E
DF5D	Co-Co	1,210	100	138	10	1999	Dalian	E
DF7	Co-Co	1,470	100	135	250	1982	Beijing 7 Feb	E
DF7B	Co-Co	1,840	100	135	190	1990	Beijing 7 Feb	E
DF7C	Co-Co	1,470/ 1,840	100	135	580	1991	Beijing 7 Feb	E
DF7D	Co-Co	1,840	100	138	220	1995	Beijing 7 Feb	E
DF7G	Co-Co	1,840	100	138	150	2003	Sifang	E
DF8	Co-Co	3,310	100	138	140	1984	Qishuyan	E
DF8B	Co-Co	3,680	100	138/150	650	1997	Qishuyan, Ziyang	E
DF9	Co-Co	3,610	140	138	2	1990	Qishuyan	E
DF10D	Co-Co	2,200	100	138	20	1994	Dalian	E
DF10F	Co-Co+Co-Co	4,400	160	240	8	1996	Dalian	E
DF11	Co-Co	3,610	160	138	450	1992	Qishuyan	E
DF11Z	Co-Co+Co-Co	7,220	160	276	4	2002	Qishuyan	E
DF11G	Co-Co+Co-Co	7,220	160	276	70	2003	Qishuyan	E
DF12[1]	Co-Co	2,430	100	138/150	40	1997	Ziyang	E
ND2	Co-Co	1,540	120	120	50	1974	Electroputere	E
ND3	Co-Co	1,540	100	126	50	1985	Electroputere	E
ND5	Co-Co	2,940	120	138	400	1984	GE	E
DF21[2]	Co-Co	–	–	–	6	–	Sifang	E
BJ[1]	B-B	1,985	90/120	92	50	1970	Beijing 7 Feb	H
DFH2	B-B	920	60	60	10	1973	Ziyang	H
DFH3	B-B	1,980	120	92	50	1972	Sifang	H
DFH5	B-B	790	40/80	86	200	1976	Ziyang	H
DFH21[2]	B-B	810	50	60	100	1977	Sifang	H
NY6	C-C	3,160	105	138	5	1972	Henschel	H
NY7	C-C	3,680	110	138	10	1972	Henschel	H

[1] Some built for 1,520 mm gauge.
[2] 1,000 mm gauge.

Diesel locomotive prototypes

Class	Wheel arrangement	Power kW	Speed km/h	Weight tonnes	No in service Dec 2004	First built	Mechanical Builders	Transmission
DF7E	Co-Co	1,840	100	150	2	1999	Beijing 7 Feb	E*
DF7F	Co-Co	2,650	100	138	1	2000	Beijing 7 Feb	E*
DF8B	Co-Co	3,820	100	150	2	2000	Dalian	E*
DF11	Co-Co	3,820	160	138	1	2000	Dalian	E*
NJ1	Co-Co	1,050	80	138	6	1999	Sifang	E†
DF4DJ	Co-Co	2,940	145	138	2	2000	Dalian	E†
DF7J	Co-Co	?	100	138	3	2002	Beijing 7 Feb	E†
DF8BJ	Co-Co	4,000	100	138/150	1	2001	Ziyang	E†
DF8BJ	Co-Co	?	100	138	1	2002	Qishuyan	E†
DF8CJ	Co-Co	4,260	100	138/150	1	2001	Qishuyan	E†

* DC traction motors
† AC traction motors

Electric locomotives (National railways, including joint-venture railways)

Class	Wheel arrangement	Power kW	Speed km/h	Weight tonnes	No in service Dec 2004	First built	Builders
SS1	Co-Co	3,780	93	138	720	1968	Zhuzhou
SS3	Co-Co	4,350	100	138	725	1978	Zhuzhou, Datong
SS3B	Co-Co	4,350	100	138	850	1992	Zhuzhou, Datong, Ziyang
SS3B(G)	Co-Co+Co-Co	8,700	100	276	140	2003	Zhuzhou, Datong, Ziyang
SS4/SS4(G)	Bo-Bo+Bo-Bo	6,400	100	184	1,150	1985	Zhuzhou, Datong, Dalian, Ziyang
SS4B	Bo-Bo+Bo-Bo	6,400	100	184	60	1999	Zhuzhou (Hitachi)
SS6	Co-Co	4,800	100	138	50	1991	Zhuzhou (Hitachi)
SS6B	Co-Co	4,800	100	138	140	1992	Zhuzhou (Hitachi), Datong
SS7	Bo-Bo-Bo	4,800	100	138	100	1992	Datong
SS7C	Bo-Bo-Bo	4,800	120	132	130	1998	Datong
SS7D	Bo-Bo-Bo	4,800	160	126	50	1999	Datong
SS7E	Bo-Bo-Bo	4,800	160	126	60	2002	Datong
SS8	Bo-Bo	3,600	160	88	240	1996	Zhuzhou
SS9	Co-Co	4,800	160	126	40	1998	Zhuzhou
SS9(G)	Co-Co	4,800	160	126	120	2002	Zhuzhou
8G	Bo-Bo+Bo-Bo	6,400	100	184	90	1987	Novocherkassk
6K	Bo-Bo-Bo	4,800	100	132	80	1987	Kawasaki, Mitsubishi
8K	Bo-Bo+Bo-Bo	6,400	100	184	145	1986	50 Hz Group
DJ1	Bo-Bo+Bo-Bo	6,400	120	184	20	2001	Siemens, Zhuzhou

- Shijiazhuang–Dezhou (Hebei) (180 km existing double-track line)
- Jinang–Qingdao (Shandong) (370 km existing double-track line)
- Wuwei–Zhangye (Gansu) (250 km existing double-track line)
- Goubangzi–Haicheng (Liaoning Province) (90 km existing single-track line)

Other electrification projects programmed to start during the course of the Tenth Five-Year Plan included the following:

- Beijing–Shanghai (Beijing, Tianjin and Shanghai cities, Hebei, Shandong, Anhui, Zhejiang) (1,460 km existing double- and triple-track line)
- Shimen–Huaihua (Hunan) (350 km existing single-track line)
- Tianjin–Shenyang (Tianjin City, Hebei, Liaoning) (710 km existing double-track line)
- Shanghai–Zhuzhou (Shanghai City, Zhejiang, Jiangxi, Hunan) 1,150 km existing double-track line)
- Luoyang–Xiangfan (Henan, Hubei) (370 km existing double-track line)

Electrification projects that are planned to start after the end of 2005 include the Zhangye–Jiayuguan line in the west of the country and the Xiangfan–Huaihua line in central China.

Track

The standard rail section now in use in China is 60 kg/m and by 2000 this accounted for around 58 per cent of the rail in use on national railway lines. 50 kg/m rail is also in use while the old standard of 43 kg/m is rapidly being eliminated. 75 kg/m rail is also in use on a few sections of line. The use of long-welded track has also rapidly increased with around 38,000 km in place by 2003.

The use of prestressed concrete sleepers with elastic fastenings for main lines is now standard. Concrete sleepers with bolted fastenings are used for less important tracks. Timber sleepers are used for most turnouts, at locations in main lines where the use of concrete sleepers is not appropriate, and on many less important tracks.

The massive efforts made in recent years to speed up and otherwise improve passenger services on many main lines has required a huge programme of turnout replacement and other track upgrading work. To suit the higher speeds more sophisticated turnouts, many with movable frogs and supported on concrete bearers, have been installed. Over the five years from 1995 to 2000 almost 7,500 new turnouts were installed as part

Class DJ1 two-section electric locomotive built by STEZ, a joint venture between Siemens Transportation Systems, Zhuzhou Electric Locomotive Works and Zhuzhou Electric Locomotive Research Institute   0116570

Henschel-built Class NY7 diesel-hydraulic locomotive with the China Orient Express at Kangzhuang (Ian Francis)   0554914

of the massive programme to accelerate passenger services. This programme is continuing.

China is steadily mechanizing permanent way work and around 40 new large track maintenance machines, some domestically manufactured and others imported, are put into operation each year.

UPDATED

Kowloon–Canton Railway Corporation

KCRC House, 9 Lok King Street, Fo Tan, Sha Tin, New Territories, Hong Kong
Tel: (+852) 26 88 13 33   Fax: (+852) 26 88 09 83
Web: www.kcrc.com

Key personnel

Chairman: Michael Tien Puk-sun
Directors
Chief Executive (Acting): Samuel Lai
Senior Director, Capital Projects: K K Lee
Senior Director, Transport: Y T Li
Director, Property: Daniel Lam
Director, Finance: Lawrence Li
Director, Human Resources: Mimi Cunningham

East Rail
Gauge: 1,435 mm
Route length: 34 km
Electrification: 34 km at 25 kV 50 Hz AC

Light Rail
Gauge: 1, 435 mm
Route length: 36.15 km
Electrification: 36.15 km at 750 V DC overhead

Organisation

KCRC is wholly owned by the Hong Kong Special Administrative Region Government. It currently provides three domestic passenger rail services: East Rail, a 34 km suburban mass transit railway

operating between Kowloon and Shenzhen, with a branch to Ma On Shan at Tai Wai; West Rail, a mass transit service running between the North West New Territories and urban Kowloon; and Light Rail, formerly a stand-alone transit service and now also a feeder service for West Rail. All of these networks are supported by feeder buses.

In addition to its domestic service, East Rail operates intercity passenger services to Guangzhou with its own rolling stock and provides access for intercity trains running to and from cities on the mainland, including Guangzhou, Shanghai and Beijing. East Rail also carries freight, mainly to and from the interior of the mainland. Furthermore, the corporation develops property projects along its railway networks with joint-venture partners.

During 2004 two new extensions were added to the East Rail line, expanding its reach to the Ma On Sha area and returning the KCR to Tsim Sha Tsui for the first time since the 1970s. The 1.1 km Tsim Sha Tsui Extension runs from Hung Hom to East Tsim Sha Tsui and Ma On Shan Rail extends East Rail 11.4 km from Tai Wai to Wu Kai Sha. The East Rail network now has a total track length of 46.5 km and has 22 stations. It provides an electrified mass transit rail service connecting urban Kowloon with both the North East New Territories and the mainland via Lo Wu.

At the end of 2004 KCRC employed a total of 5,874 staff, of which 4,227 were with the Transport Division, 867 with Capital Projects, 109 with Property and 671 with corporate and other services.

Passenger operations

East Rail provides both domestic and cross-boundary services, provided by emus in 12-car formations. The East Rail fleet comprises 444 cars.

East Rail currently operates more than 500 train trips daily, with headways of about 3 minutes during peak hours.

Since electrification in 1983, East Rail's patronage has increased from about 40,000 to 798,800 passenger trips per day in 2004.

East Rail operates intercity passenger services to 13 cities in mainland China, including Dongguan, Foshan, Zhaoqing, Guangzhou, Shanghai and Beijing, services to the last two having been introduced in May 1997. In August 1998, the first double-deck 'Ktt' through train was launched between Kowloon and Guangzhou (181 km). Operated push-pull style by Adtranz-built electric locomotives at each end of a 12-car double-deck two-class trailer formation, this train runs at up to 160 km/h to complete the journey in 1 hour 40 minutes. In June 2003 the frequency of this service was increased from seven to eight round trips a day, increasing again to 10 round trips in October 2003. In June 2001 a direct through train service was introduced between Donggang (Changping) and Kowloon, running on Fridays, Saturdays, Sundays and public holidays.

Phase 1 of the Light Rail system was opened in September 1988 and has subsequently been extended to create the present 36.15 km 68-stop network. Eleven routes are operated. At peak hours, services run as frequently as every

78 seconds at stops in the busiest section. Daily patronage in 2002 was 313,600.

Traffic (million)	2002	2003	2004
East Rail			
Passenger journeys	296	278	292
Passenger-km	4,540	4,183	4,385
Light Rail			
Passenger journeys	114	106	132

Freight operations

The Freight Department transports containers, general cargo and livestock between Hong Kong and the mainland and provides freight transport, intermodal and freight forwarding services in co-operation with the Ministry of Railways and its subsidiaries. Services are provided to some 60 cities in the mainland and more than 15 other cities worldwide.

About 16 freight trains operate daily, including eight inbound and eight outbound. In 2004 inbound freight volume was 0.5 million tonnes. Outbound traffic amounted to 0.3 million tonnes plus 0.4 million head of livestock.

Freight traffic is exclusively conveyed in wagons belonging to mainland railway authorities. The Freight Department operates five goods yards at Hung Hom, Mong Kok, Sha Tin, Sheung Shui and Fo Tan, plus a marshalling yard at Lo Wu. Together KCRC freight yards can accommodate up to 700 TEUs, 500 wagons and 1,000 lorries. In addition, the department maintains seven freight representative offices in mainland China.

Container shuttle services are available to various cities in mainland China, including Changsha, Chengdu, Chongqing, Dongguan, Kunming, Lanzhou, Nanjing, Shanghai, Shenzhen, Shijiazhuang, Ulaanbaatar, Urumqi, Wuhan, Xi'an and Zhengzhou.

New lines

West Rail, Phase 1

West Rail, Phase 1, is a 30.5 km double-track line from Nam Cheong in west Kowloon to Tuen Mun in the North West New Territories. The journey time along the entire alignment takes just 30 mins, a saving of 30 mins compared with road.

West Rail is integrated with KCRC's Light Rail system in the North West New Territories and provides interchanges with the MTRC system in urban Kowloon. It also offers an interchange at Nam Cheong with MTRC's Tung Chung Line and a second interchange at Mei Foo with MTRC's Tsuen Wan Line. Public transport interchanges are provided to allow passengers to use feeder services such as buses, minibuses and taxis.

The West Rail system comprises nine stations, a depot and a headquarters building housing a central operations centre.

East Rail extensions

The 7.4 km Lok Ma Chau Spur Line will provide a second passenger rail crossing to the mainland to relieve congestion at the Lo Wu boundary crossing. The finalised scheme is based on a tunnel-cum-viaduct approach. It will branch off the existing East Rail north of Sheung Shui and then go underground until it reaches Chau Tau. From here the railway rises gradually on viaduct and ends at Lok Ma Chau station. This new facility will incorporate customs and immigration clearance facilities similar to those at Lo Wu station. Major construction contracts, including those for the tunnels, viaduct and the terminus at Lok Ma Chau, were awarded by the end of 2002. The line's target opening date is mid-2007.

Other lines

Lok Ma Chau Spur Line

The Government of the Hong Kong Special Administrative Region endorsed the Corporation's construction of the Lok Ma Chau Spur Line in June 2002. Construction commenced in January 2003 and the project is targeted for completion in 2007.

Sha Tin to Central Link

KCRC won the bid to plan, build and operate the Sha Tin to Central Link in 2002. In 2004 the corporation completed the scheme's design and in September submitted the final proposal to the

Refurbished KCRC Metro-Cammell emu entering University station (T V Runnacles)   0104696

Examples of a 250-car fleet supplied by the Japanese IKK consortium for both East Rail and West Rail services   0122539

KCRC electric multiple-units

Cars per unit	Motor cars per unit	Motored axles/car	Output/ motor kW	Speed km/h	Units in service	First built	Builders Mechanical	Electrical
3	1	4	228	120	117	1981–91	Met-Cam	GEC Traction
4/2	2/0	4	240	120	32	2001	IKK	IKK
7				130	22	2002	IKK	IKK

KCRC diesel locomotives

Builder's type	Wheel arrangement	Power kW	Speed km/h	Weight tonnes	No in service	First built	Mechanical	Builders Engine	Transmission
G12	Bo-Bo	840	80	66	1	1954	Clyde	GM 567	E GM
G12	Bo-Bo	977	80	67	3	1957	Clyde	GM 567	E GM
G16	Co-Co	1,342	80	94	3	1961	GM	GM 567	E GM
G16	Co-Co	1,342	80	94	1	1965	Clyde	GM 567C	E GM
G26-CU	Co-Co	1,492	80	94	1	1973	Clyde, GM	GM 645E	E GM
G26-CU	Co-Co	1,492	80	92	2	1976	GM	GM 645E	E GM
8000	Bo-Bo	2,000	90	80.5	5	2003	Siemens	MTU	

KCRC electric locomotives

Class	Wheel arrangement	Power kW continuous	Speed km/h	Weight tonnes	No built	First built	Builders Mechanical	Electrical
TLN001/TLS002	Bo-Bo	5,000	160	84	2	1997	SLM	Adtranz

government. The scheme and its implementation programme are subject to the government's final decision.

Kowloon Southern Link

The 3.8 km Kowloon Southern Link (KSL) will extend West Rail from Nam Cheong station at Sham Shui Po to East Rail's East Tsim Sha Tsui station. Joining East Rail with West Rail will provide passengers with a convenient and direct transfer between these two rail corridors, forming a unified network.

In September 2002 the government invited the Corporation to proceed with the detailed planning and design of the KSL project. It was initially gazetted in March 2004, while proposed amendments to minimise disruption during construction were gazetted in January 2005. Construction was expected to start in mid-2005 with a forecast completion date of 2009. Completion of the project will provide a 30 minute journey time from Tin Shui Wai to Tsim Sha Tsui East. This will extend West Rail from Nam Cheong station to East Tsim Sha Tsui station on East Rail, joining the two major KCRC railway corridors at the southern section. The new line will have two underground stations.

Northern Link

The Northern Link will join East Rail and West Rail at the northern section. It will provide a much-needed corridor between the northeast and northwest districts of the New Territories, as well as linking West Rail to Lok Ma Chau for cross-boundary passengers.

Traction and rolling stock

A HK$1.3 billion three-year mid-life rolling stock refurbishment programme was completed in 1999, enabling the whole fleet of 348 cars to carry 15 per cent more passengers.

In March 1999 KCRC awarded a HK$3.1 billion contract for the design, supply, testing and commissioning of 250 emu cars to a Japanese consortium of Itochu-Kinki-Kawasaki (IKK). Of these, 154 cars are configured as 22 seven-car trains for West Rail, Phase 1. The remaining 96 cars are used on the existing East Rail line. All 96 cars for East Rail had been commissioned by May 2002. Delivery of the 154 West Rail cars, which started in April 2002, was completed in May 2003.

East Rail also has 12 diesel locomotives for freight operations.

The Light Rail system fleet totals 119 vehicles.

Five Eurorunner diesel-electric locomotives ordered from Siemens Transportation Systems in 2001 and similar to the Class 2016 'Hercules' machines supplied to Austrian Federal Railways entered service in 2004.

Signalling and telecommunications

East Rail

East Rail operates an integrated computerised signalling and telecommunication system designed to comply with the Automatic Train Protection (ATP) and Automatic Train Operation (ATO) systems. It comprises a Train Control System (TCS), a Centralised Solid State Interlocking System (CSSI), a Power Control System (PCS) and an Integrated Control and Communications System (ICCS).

Operated on screen-based workstations, the TCS provides a complete range of train control and monitoring functions. The signalling system is centralised at the East Rail Control Centre, which provides manual and automatic route setting and allows bidirectional working in both up and down tracks. The PCS system controls and monitors the 25 kV power supply to the trains and the ICCS allows effective station control and voice

communications between operating parties by employing Pulse Code Modulation (PCM) and Fibre Optic Distribution.

In general, the ICCS provides control and monitoring of the railway systems, traction power control, long line public address, passenger information, miscellaneous equipment, voice communications and integrated radio.

Coupler in standard use

Passenger cars: AAR coupler between units; bar coupler between cars
Freight wagons: AAR coupler
Braking in standard use, locomotive-hauled stock: Air

Track

Rail: UIC 54 (54 kg/m)
Crossties (sleepers): Prestressed concrete, 203 mm thick, spaced 700 mm centre to centre
Rail fastenings: Pandrol rail clip, Type PR429A with Pandrol glass-reinforced nylon insulators
Min curvature radius: Main line 270 m; sidings 150 m
Max gradient: 1 in 100
Max speed: 120 km/h throughout
Max axleload: 25 tonnes

UPDATED

MTR Corporation

Airport Express
GPO Box 9916, MTR Tower, Telford Plaza, Kowloon Bay, Hong Kong
Tel: (+852) 29 93 21 11 Fax: (+852) 27 98 88 22
Web: www.mtr.com.hk

Key personnel

Directors
    Chairman: Dr Raymond K F Ch'ien
    Chief Executive Officer: C K Chow
    Managing Director, Operations and Business Development: Phil Gaffney
    Project Director: Russell Black
    Property Director: Thomas Ho Hang-kwong
    Finance Director: Lincoln Leong Kwok-kuen
    Legal Director and Secretary: Leonard Turk
    Human Resources Director:
      William Chan Fu-keung
    Deputy Operations Director: Andrew McCusker
    Head of Operations: Wilfred Lau

Gauge: 1,435 mm
Route length: 33.3 km
Electrification: 34.8 km at 1.5 kV DC, overhead

Political background

In October 2000 partial privatisation of MTR Corporation was achieved when trading in the corporation's shares commenced on the Stock Exchange of Hong Kong. This reduced the government's shareholding to 77 per cent.

Organisation

The Airport Railway (AR) was developed and built by the MTR Corporation, Hong Kong's metro operator, to provide a direct rail link between Chek Lap Kok airport on Lantau Island via Kowloon to Central, on Hong Kong island. The line was also designed to provide a fast Tung Chung Line suburban service to North Lantau. Tung Chung Line operations started on 22 June 1998, while Airport Express services followed on 6 July, coinciding with the commissioning of the new airport.

Details of MTR's metro network may be found in *Jane's Urban Transport Systems*.

Passenger operations

Airport Express services are scheduled to run every 12 minutes, completing the 34.8 km journey in 23 minutes. In-town check-in facilities managed by MTR Corporation are provided at Hong Kong Station in Central District and Kowloon. Airport Express patronage was adversely affected by the SARS outbreak in

MTR Corporation Tung Chung (third from left) and Airport Express (right) emus alongside examples of the operator's metro stock   *NEW*/1110221

2003, resulting in passenger journeys for the year totalling 6.85 million, compared with 9 million in 2001. In 2004 traffic levels recovered, with 8 million passengers carried on Airport Express services.

A 10-minute service frequency is provided on the Tung Chung Line. During the morning and evening peak periods, frequency is increased to an average of eight minutes, with alternate trains turning back at an intermediate station at Tsing Yi, where the average headway is to be enhanced to four to five minutes.

New lines

In 2002 MTR Corporation was authorised to build a 3.5 km spur from Yam O, on the Tung Chung line, to a planned Disney theme park at Penny's Bay on Lantau Island, which is due to open to the public in the second half of 2005. By mid-2003 all major contracts had been let. The cost of the scheme, which includes the construction of two new stations, is estimated at HK$2 billion.

Improvements to existing lines

In December 2003 a new station was opened at Nam Cheong on the Tung Chung Line. It was built as an interchange between the Tung Chung Line and the Kowloon-Canton Railway Corporation's new West Rail, connecting the northwest New Territories with urban Kowloon.

MTR Corporation is building the Asia World-Expo station to enable the Airport Express to serve Asia World-Expo, an international exhibition centre under construction at Hong Kong International Airport. This will provide a rail service to the Central Business District and the MTR network as a whole. Completion of the station is anticipated by the end of 2005 to tie in with opening of Asia World-Expo.

A second platform for Airport Express services is planned at Hong Kong station to provide anticipated capacity for airport traffic and additional Tung Chung Line capacity.

Traction and rolling stock

Emus of a common basic design but with different access and interior layouts were supplied by an Adtranz/CAF joint venture. Initially 11 seven-car Airport Express and 12 seven-car Tung Chung Line trains were ordered, both types with a 135 km/h capability. Airport Express units incorporate one baggage car to convey containerised checked-in baggage. Train design and infrastructure provide for the addition of extra cars up to ten- and eight-car formations respectively as traffic grows.

In June 2004 MTRC placed an order with Rotem, South Korea, for four eight-car emus to handle additional traffic generated by the Yam O–Penny's Bay line. Delivery is scheduled for 2006–07.

UPDATED

Colombia

Ministry of Transport

Avenida Eldorado, CAN Edificio Minobras, Santa Fe de Bogotá
Tel: (+57 1) 222 44 11/75 77
Fax: (+57 1) 222 16 47/11 21

Key personnel
Minister: J Bendex Olivella
Deputy Minister: J A Latorre Uriza

El Cerrejon Coal Railway

Cra 54 No 72-80, AA52499, Barranquilla
Tel: (+57 5) 877 78 98   Fax: (+57 5) 877 78 98

Key personnel
Superintendent: M Mendoza
Senior Operations Supervisor: R Stand
Senior CTC Supervisor: J Gonzalez
Senior Maintenance of Way Supervisor: F Acuña

Gauge: 1,435 mm
Route length: 150 km

Organisation

Constructed by Morrison Knudsen in several months less than the originally scheduled timescale, this railway was built at a cost of US$300 million for Carbocol, the national coal-mining corporation, and Intercor, an Exxon affiliate. These two organisations have jointly developed South America's biggest coal-producing project in an opencast operation in the northeastern province of Guajira. The railway employs 117 people.

Freight operations

The railway moves some 15 million tonnes of coal annually (12.7 million tonnes/88,400 million tonne-km in 1992; 13.5 million tonnes in 1993) from the El Cerrejon complex to a shipment port close to the Venezuelan border at Bahia de Portete. The line was laid with 61.8 kg/m rail, continuously welded, on timber sleepers. The route is through scrubland, which helped to restrict ruling gradients to 0.3 per cent against loaded trains, 1 per cent in the reverse direction.

At the El Cerrejon mine complex, where trains are overhead-loaded from a pair of 10,000 tonne-capacity silos, and at the port, where the 91 tonne coal wagons are bottom door-discharged, track layouts are in loop form to allow merry-go-round operation of the trains. Thus, each 93-wagon train is planned to achieve three return trips within the 24 hours.

In 1999 coal traffic was promised a boost of up to 6 million tonnes a year by an agreement with other mining companies to move their increased production to Bahia de Portete.

Traction and rolling stock

The railway operates with 244 hopper wagons supplied by Ortner and 164 Autoflood II automatic discharge wagons manufactured by Johnstown America Corporation. The locomotives are eight General Electric Type B36-7 2,685 kW Bo-Bos.

Coal hoppers built by Johnstown America Corporation for the El Cerrejon system employ 3CR12 stainless steel from Cromweld
0021530

Map legend:
- ┼┼┼ 1.435 m gauge
- ─── 0.914 m gauge
- - - - Out of service
- ····· Projected lines

0519650

Ferrovias

Empresa Colombiana De Vias Ferreas
Calle 31 No 6-41, Piso 20, Santa Fe de Bogotá
Tel: (+57 1) 287 98 88   Fax: (+57 1) 287 25 15

Key personnel

President: Luis Diego Monsalve Hoyos
General Secretary: Nelson Morse Santos
Vice-Presidents
  Finance and Administration:
    Clara Inés Renjifo Saavedra
  Operations: Eric Morris
  Infrastructure: Jorge Enrique Hincapie

Gauge: 914 mm
Route length: 3,154 km

Political background

The national railway company FNC was liquidated in 1992, with a wholly state-owned company, Ferrovias, taking over responsibility for the rehabilitation, maintenance and development of the remunerative parts of the FNC infrastructure.

The government intended trains to be operated by a mix of private and joint public-private companies. Of the latter type, the Sociedad Colombiana de Transporte Ferroviario SA (STF) was formally established in 1991 to run a public rail freight service over the core main line from Lenguazaque and Bogotá to Santa Marta, and from Chiriguana to Santa Marta.

Other operators included STFO, offering public freight service on the Pacific line between Buenaventura and Cali; and Tren Metropolitano de Cali, providing passenger service around Cali. Tourist trains were established between Bogotá and Nemocón and around Medellín.

It was also intended to privatise workshop and repair facilities.

A Spanish consortium of RENFE, engineering consultancy Ineco and investment bank Socimer, was awarded a 12 month contract in September 1995 to develop and implement a programme for the private sector to upgrade, maintain and operate key routes via a system of concessions. Tenders were called in 1997 for a 30-year concession to rehabilitate and also operate the so-called Atlantic Network. This consists of the Bogotá-Santa Marta trunk route (969 km) and a further 400 km of branches: Bogotá-Belencito; Espinal-Neiva; and Puerto Berrío-Medellín. In February 1998 the concession was awarded to the 13-member Fepaz consortium, headed by the Colombian company Emcarbón and Dragados FCC International of Spain. However, the award was subsequently withdrawn from Fepaz for contractual reasons.

In mid-1999 it was announced that a consortium led by Dragados and including Spanish National Railways (RENFE) had been selected as preferred bidder. After restoring the main line, the concessionaires expect to handle some 20 million tonnes of freight annually.

Bids for a second concession were invited in May 1998 for the 650 km Pacific network, the main line of which connects Buenaventura, Calí and La Felisa, with a branch from Zarzal to Tebaida and Armenia. The concession will be for 30 years, and will cover renovation and operation of the line. Investments of up to US$190 million were reported to be needed in infrastructure upgrades, with the government expected to contribute US$120 million.

A sole bid was received from a consortium led by a Spanish company, Actividades de Construcción y Servicios, and in late 1998 this was being reviewed by the government.

Finance

Ferrovias is financed by 10 per cent of national petrol tax revenue and by tolls from rail infrastructure users. The investment budget for 1996 totalled 124.56 billion Colombian pesos, with 110.01 billion pesos allocated to major track improvements, 13.55 billion pesos for signalling and 1 billion for telecommunications.

New lines

There are two long-planned new line schemes which have remained on the back burner as attention has focused on the rehabilitation of the trunk Bogotá-Santa Marta and Medellín-Buenaventura routes and the recovery of traffic volume. One is a cut-off between Ibague and Armenia that would obviate the circuitous journey from Bogotá to the Pacific port of Buenaventura via Puerto Berrio. The proposed new 85 km cut-off would entail driving a 22 km-long tunnel. Achievement would reduce the rail distance from Bogotá to Buenaventura by no less than 500 km.

The other project is a 180 km link between Saboya and Puerto Carare, known as the Carare Railway. This line would cut the transport distance between the Caribbean port of Santa Marta and Bogotá by 350 km. Ruling gradient would be 2.4 per cent, with the route climbing from a point 100 m above sea level to an altitude of 2,560 m at Saboya. Minimum curve radius would be 200 m, with 56 kg/m rails to sustain 20 tonne axleloads in moving coal traffic from the Checua and Lenguazaque deposits. Construction would entail some 90 tunnels aggregating over 30 km in length and about 50 significant bridges or viaducts.

A new 93 km alignment of the Cali-Medellin line's section between La Pintada and Caldas, derelict since flooding and landslides in 1976, is also being reconsidered. Another proposal involves the construction of 240 km of new railway to connect the northern ports of Cartagena and Barranquilla with the Santa Marta line at Fundación.

Proposals for a new 1,435 mm gauge route for export coal from Cúcuta, west of Gamarra, to La Concha and La Ceiba on Lake Maracaibo in Venezuela were announced by Ferrovias in May 1996. The 16 km Colombian section of the 126 km route was costed at US$33 million, and it was hoped to put a construction concession out to tender by mid-1997.

Improvements to existing lines

Ferrovias launched an emergency programme of rehabilitation in 1990. An immediate aim was to reduce the frequency of derailments caused by poor track, which had prompted many customers to switch to road transport, and raise average operating speeds from the nadir of 15 km/h on key routes. The programme of track work, well under way by 1991, covered 1,100 track-km, and was raising maximum permissible speed on level track to 40 km/h, and in mountain territory to 25 km/h.

Under a long-term plan costed at US$357.2 million and scheduled for completion by 2000, track was to be thoroughly renewed, using 45 kg/m rail and concrete sleepers except in the mountain territory, in the following route-by-route order of priority: Bogotá-Santa Marta (965 km); Medellín-Grecia/Puerto Berrio (187 km); La Caro-Belencito (181 km); Bogotá-La Caro-Lenguazaque (110 km); and Yumbo-Buenaventura (158 km).

However, by 1995 it was apparent that Ferrovias would, henceforth, be looking to the private sector to undertake comprehensive route upgrades. In May 1996 it was reported that upgrading was under way on the La Loma-Santa Marta and Grecia-San Rafael sections of the network, with completion expected in 1997. The operation of unit coal trains between La Loma and Santa Marta has been proposed by Drummond Mining (see below).

Oil and coal traffic

According to forecasts, oil industry development was likely to generate up to 11,000 tonnes of freight a day in the northeast of the country by the end of the 1990s and local state governors have joined with commercial interests in pressing for rehabilitation of the 258 km Bogotá-Belencito route. The oil industry has also pressed for US$2 million to be spent on the renovation of the 618 km south-north line from workings at Neiva to Barrancabermeja, whence oil could be piped to a Caribbean port.

Colombia has developed a major coal export trade, and route upgrades have been mooted to cater for such traffic. Drummond Mining Corporation, USA, has stated its intention to develop mines around La Loma and has proposed operating unit coal trains to the Caribbean port of Ciénaga, near Santa Marta. Track access fees paid by Drummond trains could generate some US$12 million for Ferrovias; initial movements of some 3 million tonnes of coal a year have been forecast, rising to 13 million tonnes by 2000.

Track

Rail: ASCE 37 kg/m
Crossties (sleepers): Wood, 1.82 × 0.15 × 0.20 m; concrete, 1.90 m x 0.12 m x 0.22 m
Fastenings: Track spikes, drive spikes Deenik and Pandrol
Spacing: In plain track 1,666/km in curves 2,000/km
Min curvature radius: 80 m
Max gradient: 4.3%
Max speed: 62 km/h
Max axleload: 16 tonnes
Max altitude: 2,900 m

STF

Sociedad Colombiana de Transporte Ferroviario SA
Calle 72 No 13-23, Piso 2, Santa Fe de Bogotá
Tel: (+57 1) 255 86 84
Fax: (+57 1) 255 87 04

Key personnel

President: Dr Luis Fernando Zea Llano
Vice-Presidents:
  J G Arango Arango, J E Rojas, L F Vélez

STFO

Sociedad de Transporte Férreo de Occidente SA
23-47, Estación Ferrrocarril 2° Piso, Cali
Tel: (+57 23) 660 33 14   Fax: (+57 23) 660 33 20
e-mail: trapacif@cali.cetcol.net.co

Organisation

STF took over rail operations from the liquidated FNC in 1992, providing freight service over the core main line from Lenguazaque and Bogotá to Santa Marta, and from Chiriguana to Santa Marta. The Antofagasta (Chili) and Bolivia Railway provided assistance in determining methods of operation. The state owns 51 per cent of STF's capital. The intention was that the company's private partners would help fund renovation and new purchases, as well as provide STF working capital.

Key personnel

General Manager: Robert Antonio Guzman

Freight operations

STFO provides freight service between Buenaventura on the Pacific coast to Cali and Cartago, operating as Transpacífico. Five diesel locomotives carried 233,000 tonnes for 30.5 million freight tonne-km in 1996.

Freight operations

In 1997, 0.7 million tonnes of freight were carried, generating 346 million tonne-km. STF employed 236 staff at the end of that year.

Traction and rolling stock

In 1997 STF operated 43 diesel-electric locomotives and 1,079 wagons.

Transferreos

Transferreos Ltda
Terminal de Transporte, Medellín
Tel: (+57 4) 267 11 57/70 78

Key personnel
General Manager: John Jairo Castañeda

Passenger operations
Transferreos commenced passenger operations between Medellín and Barrancabermeja in 1992.

In 1993, a tourist train was introduced between Medellín and Cisneros.

Congo-Brazzaville

Ministry of Transport & Civil Aviation

PO Box 2148, Brazzaville
Tel: (+242) 81 43 34   Fax: (+242) 83 09 16

Key personnel
Minister: Isidore Mvouda

Chemin de Fer Congo-Océan (CFCO)

PO Box 651, Pointe Noire
Tel: (+242) 94 05 61   Fax: (+242) 94 12 30

Key personnel
General Manager: J Kidzouani
Technical Manager: J Koutoundou
Directors
  Operations: G Goma-Boukoulou
  Traction and Rolling Stock: B Nykoulou
  Track and Buildings: J Kimbatsa-Koudimba
  Supplies: J-P Niazaire
  Personnel: Joachim Ndebeka
  Finance and Accounting: Jean Claude Tchibassa
  Information Systems: Pierre Boussi
  Research and Development:
    Joseph Yongolo-Tchizinga
  Legal: Ted Nguimbi-Manitou

Gauge: 1,067 mm
Route length: 609 km

Political background
The Congo-Océan Railway extends from Pointe Noire to Brazzaville. At Mont Belo it connects with the Comilog Railway (qv), over which it also operates services. CFCO is a department within the Agence Transcongolaise des Communications (ATC) which controls intermodal transport by sea, river and rail.

In 1997, the Ministry of Economics, Finance and Planning retained Canadian consultants CPCS to advise on privatisation of the railway by way of a 20-year concession. The Comilog Railway (qv) would be included in this process. The successful bidder would be required to undertake infrastructure improvements and renew the traction and rolling stock fleet. Moves towards privatisation were subsequently suspended due to political instability, but the process was to be restarted with a new call for expressions of interest.

Passenger operations
In 1991 passenger-km totalled 547 million. By 1995, however, passenger-km had declined to 302 million.

Freight operations
Freight tonne-km stood at around 400 million in the early 1990s, but by 1995 had declined to 267 million, with 713,000 tonnes carried. Main commodities handled are timber, petroleum products and cement.

New lines
In the long term it is hoped to bridge the Congo river and connect CFCO with the Kinshasa–Matadi line in the Democratic Republic of Congo. Tecsult International of Canada has studied a combined road and double-track rail bridge. The plan includes double-tracking of 33 km between Loubomo and Mont Belo, where the 285 km Comilog line

diverges to the border of Gabon; and between M'Filou and Brazzaville.

Improvements to existing lines
The first 91 km realignment of the Pointe Noire–Brazzaville line, between Loubomo and Bilingua, was completed in 1985. As a result, speed ceilings were raised to 60 km/h for freight and 80 km/h for passenger trains and transit times were reduced by an hour for passenger trains, two hours for Comilog manganese trains and three hours for other freight. This scheme, which was funded by the World Bank and over a dozen other agencies, raised the line's freight-operating capacity beyond its previous 21 trains daily to 31.

Following studies by Tecsult, further realignment and track with welded 46 kg instead of 30 kg rail has progressed between Bilingua and Tahitondi.

Restoration work by SNCF International following the civil war in 1998–99 enabled services to be resumed between Brazzaville and Pointe Noire in August 2000.

Traction and rolling stock
At the start of 1998 the fleet comprised 40 locomotives, one diesel railcar, 71 passenger cars and 1,118 freight wagons.

A new fleet of eight dual-purpose diesel-hydraulics went into service in 1994. These are Type DGH 1000 B-B units built by ABB Henschel and powered by 800 kW SACM engines. Top speed is 90 km/h in line-haul mode and 45 km/h when shunting.

In 2001, CFCO received six GM Type GT26MC diesel-electric locomotives second-hand from Spoornet, South Africa.

Comilog Railway

Gauge: 1,067 mm
Route length: 285 km

Organisation
Built by the Compagnie Minière de l'Ogooue (Comilog), this railway connects the Congo-Océan Railway's Mont Belo station (200 km from Pointe

Noire) with M'Binda, in Gabon. Public service over this line is now provided by the Congo-Océan Railway (qv). Privatisation plans initiated in 1997 for the Congo-Océan Railway would also affect the Comilog system.

Freight operations
Until the late 1980s the Comilog Railway was moving some two million tonnes of manganese

ore per annum, but since the 1987 completion of the Transgabon Railway most of the traffic has been switched to that route, for shipment through the Gabonese port of Owendo.

Democratic Republic of Congo

National Office of Transport & Communications (ONATRA)

BP 98, Kinshasa 1
Tel: (+243 12) 247 61/224 21   Fax: (+243 12) 248 92

Gauge: 1,067 mm
Route length: 366 km

Organisation
Under the Mobutu regime which was deposed in May 1997, ONATRA had been operating the

Kinshasa-Matadi line on a lease from the national railway company SNCZ. Little is known about its operations under the new Kabila regime.

The Kinshasa-Matadi line is a vital part of the so-called Voie Nationale, an export-import route wholly within the former Zaïre which avoided the political problems of reaching the sea by way of railway lines running through the country's frequently unstable neighbours. The Voie Nationale consists of a rail segment from Sakania to Ilebo,

then river transport to Kinshasa, and finally use of the isolated rail segment from Kinshasa to the port of Matadi. In 1991, it carried 0.5 million tonnes of freight, for 158 million tonne-km, and 0.6 million passengers.

For details of the latest updates to *Jane's World Railways* online and to discover the additional information available exclusively to online subscribers please visit

jwr.janes.com

Société Nationale des Chemins de Fer Congolais (SNCC)

PO Box 297, Place de la Gare, Lubumbashi, Shaba Region
Tel: (+243 2) 22 34 30   Fax: (+243 2) 70 34 42 33

Key personnel

President and Chief Executive Officer (acting):
Muzinga Ilunga

Gauge: 1,067 mm; 1,000 mm; 600 mm
Route length: 3,621 km; 125 km; 1,026 km
Electrification: 858 km of 1,067 mm gauge at 25 kV 50 Hz AC

Political background

The National Railway Company is a state-owned organisation set up in May 1997 when forces loyal to new president Laurent-Desire Kabila, who had just deposed the former president Mobutu, seized the assets of Sizarail, a mixed public/private organisation owned by South African, Belgian and local interests. The South Africans had a 51 per cent stake in Sizarail, held by the Comazar joint venture company; the South African railway company Spoornet was a majority shareholder in Comazar, with Belgian National Railways also holding a stake. At least 14 locomotives and some wagons belonging to Spoornet were taken by the new regime, and the total value of assets involved was put at around US$70 million. The seizure led to the cessation of international traffic, cutting the southern and eastern sides of the country off from the outside world; however, later in 1997, there were signs that SNCC was prepared to reach accommodations with its neighbours, and was seeking co-operation with Sizarail. At the same time it was reported that government approval had been given for a US$300 million upgrading programme.

Sizarail had been created in 1995 to take over operation of the former state-owned railway Société Nationale des Chemins de Fer Zaïrois (SNCZ) and had been revitalising the rundown system.

Organisation

Sizarail's predecessor organisation, SNCZ, had been created in 1974 by the merger of five railways: La Compagnie des Chemins de Fer Kinshasa–Dilolo–Lubumbashi (KDL); Les Chemins de Fer des Grands Lacs (CFL); Matadi–Kinshasa (CFMK); Mayumbe (CFM); and Chemins de Fer Vicinaux Zaïrois (CVZ).

The former KDL railway serves the important mining centres of Shaba province (now renamed Katanga), Likasi, Kolwezi and Mososhi, and other mining and industrial areas such as the manganese mine at Kisenge, cement works at Lubudi, collieries at Leuna, and diamond mines at Mbuji-Mayi. Expanding agricultural and forest product industries have developed along the line of its route. The electrified territory is the 606 km of ex-KDL line from Lubumbashi to Kamina, and part of the branch from the line's mid-point at Tenke to the Angolan border at Dilolo, 252 km.

The isolated Ubundu–Kisangani line is metre-gauge. The 600 mm gauge, also isolated, is found on the Bumba-Mungbere line and its Bondo branch. Both these sections are in the east of the country.

Internationally, the system connects at Dilolo with the CF de Benguela (CFB) in Angola for access to the Atlantic port of Lobito; at Sakania with Zambia Railways and further on, Zimbabwe Railways and South Africa's Spoornet; and via Zambia with the TAZARA railway for access to Dar es Salaam.

In 1999, SNCC and National Railways of Zimbabwe signed agreements covering technical co-operation on the rehabilitation of track, infrastructure and rolling stock. This was particularly aimed at providing a more effective rail freight corridor from Congo to Zimbabwe via Zambia.

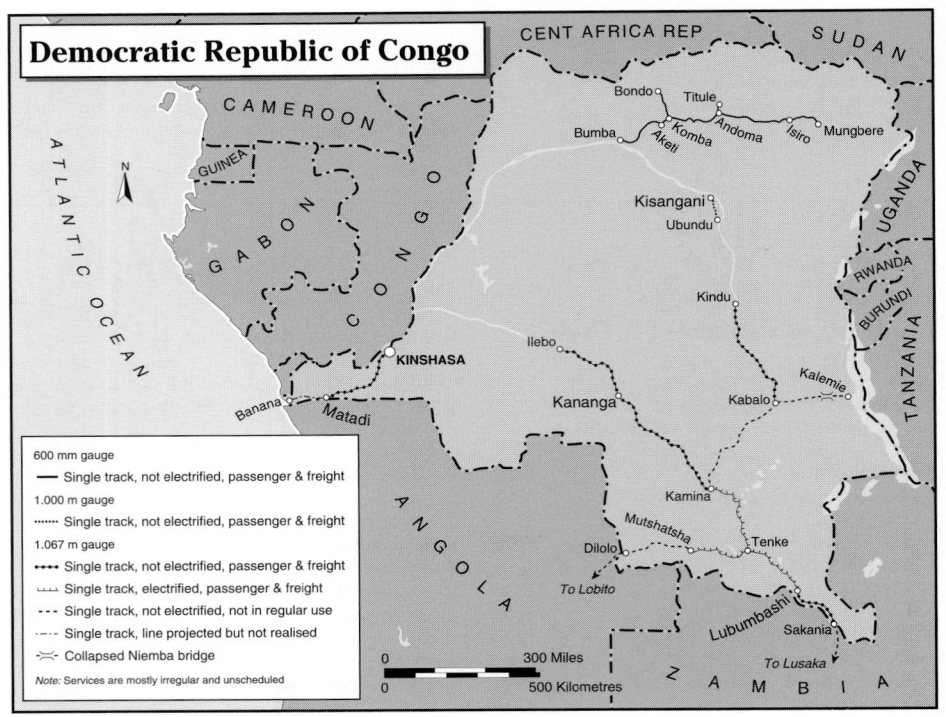

0058506

Diesel locomotives

Class	Wheel arrangement	Power hp	Weight tonnes	First Built	Builders Mechanical	Transmission
Line-haul						
1200	A1A-A1A	1,500	80	1968	Hitachi	E
1300	Co-Co	1,650	87.3	1969	General Electric	E
1400	Co-Co	1,650	87.3	1969	General Electric	E
1500	Bo-Bo	1,310	56.5	1967	Krupp Krauss-Maffei	E
61/63	0-6-0	250	13.5	1955	Atelier Metallurgique Tubize	H
81–85	B-B	510	31	1969	Nippon Sharyo	H
86–92	B-B	510	32	1974	Nippon Sharyo	H
Shunters						
71/73	0-4-0	335	16	1958	Cockerill	H
1161–1173	0-4-0	320	30	1965–68	Cockerill	H
1201–1241	B-B	510	45	1968–69–72	Hitachi Kawasaki	H
501–510	B-B	660	40	1973	Thyssen-Henschel	H
1011	0-8-0	1,000	58	1959	ARB FUF/HSP	–
21/22/25	0-6-0	110	16	1958	SA Moteur Moes	–
506–508	Bo-Bo	60	40	1980	Waremme	–

Electric locomotives

Class	Wheel arrangement	Output kW	Speed km/h	Weight tonnes	First built	Builders
2200	Bo-Bo	1,620	65	76	1956	ACEC
2300	Bo-Bo	1,505	65	73.7	1958	ACEC
2400	Bo-Bo	1,620	70	60	1960	ACEC
2450	Bo-Bo	1,620	75	60	1964	ACEC
2500	Bo-Bo	1,600	70	62	1969	Hitachi
2600	Bo-Bo	2,400	60	93	1976	Hitachi

Traction and rolling stock

At last report, for 1,067 mm gauge operation, SNCZ owned 86 line-haul and 59 shunting diesel locomotives, 51 electric locomotives, 282 passenger cars, 30 railcars and 4,793 freight wagons. Metre-gauge stock comprised three diesel locomotives, 9 passenger cars and 78 freight wagons; and 600 mm gauge stock, 14 diesel locomotives, five diesel railcars, 16 passenger cars and 329 freight wagons.

In 1999, SNCC reached agreement with National Railways of Zimbabwe (NRZ) to lease 20 refurbished passenger coaches for two years, with an option to extend the contract.

Track

Standard rail: 29.3 and 40 kg/m on KDL; 24.4 and 29.3 kg/m on CFL; 33.4 and 40 kg/m on ONATRA CFMK; 18 kg/m on ONATRA CFM; 9 to 33.4 kg/m elsewhere

Joints: Fishplates and bolts

Crossties (sleepers): Chiefly steel, also wood and concrete

Spacing: 1,250/km plain track, 1,500 km in curves

Rail fastenings: By clips and bolts to steel sleepers. RN flexible fastenings to concrete sleepers

Min curvature: 100 m

Max gradient: 15%

Max altitude: 1,614 ft at Dilongo-Yulu near Tenke on Bukama line

Max permitted speed
Electrified lines: 52 km/h
All other lines: 45 km/h

Max axleload: 15 tons nominal; 20 tons in special cases

Costa Rica

GM diesel loco and passenger service 0580999

Puente Atenas bridge and tourist train 0581000

Ministry of Public Works and Transport

San José
Tel: (+506) 222 86 81   Fax: (+506) 255 02 42

Key personnel
Minister: Rodolfo Silva
Director, Transport Division: H Blanco

Costa Rica Railways (Incofer)

Instituto Costarricense de Ferrocarriles
Calle Central, avenida 20 y 21, San José
Tel: (+506) 221 07 77   Fax: (+506) 222 34 58

Key personnel
President: Juan Ramón Rivera Rodriguez

Gauge: 1,067 mm
Route length: 278 km

Political background
Incofer was created in 1985 to undertake the modernisation of the system created by the merger of the National Atlantic and Pacific railways (Incofer's Atlantic and Pacific divisions) in 1977. In 1987 it took over the 250 km Ferrocarril del Sur network formerly operated by the Compañía Bananera de Costa Rica. Passenger services on the Ferrocarril del Sur ceased about 1990 and freight services followed around one year later, with the whole network dismantled between 1992 and 1994.

Due to its worsening finances, the government ordered Incofer to cease operations in June 1995 and put the railway into a care and maintenance regime while private-sector participation was sought.

In November 1996, the government invited bids to operate and maintain parts of the national rail network, with concessions to be granted based on promised levels of private-sector investment. However, no award was made. A second call for bids with broadly similar conditions was made in August 1997, and an award had been expected during the first half of 1998. However, none was made due to a lack of potential investors.

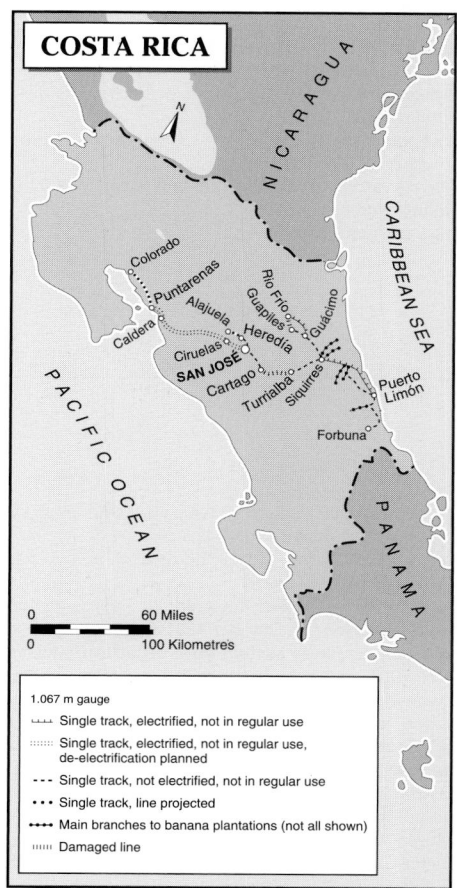

COSTA RICA

0   60 Miles
0   100 Kilometres

1.067 m gauge
⊢⊣⊢⊣ Single track, electrified, not in regular use
╍╍╍ Single track, electrified, not in regular use,
       de-electrification planned
--- Single track, not electrified, not in regular use
••• Single track, line projected
•••• Main branches to banana plantations (not all shown)
⊪⊪⊪ Damaged line

Costa Rica 0058507

In July 2000, the Costa Rican National Council for Concessions (CNC) announced its intention during the following month to initiate a new round of bidding for concessions to operate the system by publishing a request for qualifications. Prequalified firms or consortia would then be invited to respond to a request for proposal prior to submitting a bid. However, in mid 2004 Incofer was still a government company.

On environmental grounds the government is being pressurised to implement suburban services in four cities, including San José, where the TREM project envisages a local service to Heredia. The World Bank has made funding available to future concession holders to buy new equipment or improve infrastructure. Incofer's property may be ceded to concessionaires for use as security on loans.

Passenger operations
At the beginning of 2000, Incofer reinstated Saturday and Sunday round-trips on the San José–Puntarenas line for a trial period of three months. The trial proved highly popular and often failed to offer sufficient capacity for passenger demand. In 2004 the service was running in the summer season and between San José and Puerto Caldera only, with passenger-journeys amounting to some 5,000 annually.

Freight operations
In 1994 Incofer carried a total of 661,349 tonnes of freight, having carried 739,000 tonnes in 1993. A decade later, after resumption of limited freight operations, the railway's principal source of income was transport of bananas to Atlantic ports for export, with some 5 million boxes handled in 2003. Other freight included 14,000 tonnes of steel. This latter traffic looked set to expand in 2004, with plans to haul imported raw materials from the coast to the Grupo Pujol plant at Guápiles and move larger quantities of finished products.

New lines
A major outstanding project, discussed for some time past, is the construction of an effective link between the Atlantic and Pacific divisions, latterly connected only by a steeply graded line through the streets of San José. At Alajuela, close to the country's international airport, the two divisions are only 3 km apart and an elevated single-track

connection has been proposed at an estimated cost of US$70 million, including rolling stock for suburban passenger services.

Traction and rolling stock
When Incofer operations ceased in June 1995, the traction and rolling stock fleet included three diesel-electric locomotives formerly employed on Canadian National's Newfoundland system; two 14 tonne railbuses with 200 hp Cummins engines assembled by Incofer in its own workshops; a Romanian-built diesel-hydraulic B-B shunter converted to a 15 kV 20 Hz electrohydraulic unit by Incofer; and a pair of two-car diesel multiple-units acquired from FEVE of Spain in 1993. In 2004, the fleet comprised 9 diesel locomotives, three two-car dmus, two passenger coaches and 535 wagons.

The initial 15 kV 20 Hz AC electric system was operated by AEG locomotives of 1929 and Siemens locomotives of 1956. In 2004 11 electric locomotives remained on the fleet list despite the lack of infrastructure.

Type of coupler: Standard
Type of braking in standard use: Westinghouse air
Axleload: 16 tonnes

Electrification
The former Pacific Railway running 128 km from San José to Puntarenas was electrified at 15 kV 20 Hz AC in 1929–30. Between 1977 and 1982 modernisation and electrification of the Atlantic Railway was put in hand; the 132 km Limón-Río Frío main line was completely relaid with 43 kg/m long-welded rail on concrete sleepers, new yards were installed at both ends, bridges strengthened for 16 tonne axleloads, and electrification at 25 kV 60 Hz AC executed. The entire system was dismantled after the 1995 closure.

A fleet of 12 dual-voltage (25 kV 60 Hz/15 kV 20 Hz AC), 62 tonne 1,200 kW Bo-Bo electric locomotives was supplied by electrification contractors 50 c/s Group.

Track
Rail: ASCE 42.5
Sleepers: Wood or concrete spaced 1,600/km
Fastenings: Pandrol, Nabla RN, spikes
Min curve radius: 80 m
Max gradient: 4.25%
Max permissible axleload: 18 tonnes

Côte d'Ivoire

Ministry of Public Works Transport and Communications

PO Box V6, Abidjan 01, Ivory Coast
Tel: (+225) 34 73 15   Fax: (+225) 211 73 29

Key personnel
Minister: Ezan Akele
Director of Land Transport: Y B Quattara

For map see entry for **Burkina Faso**.

Ivory Coast Railway – SIPF

Société Ivoirienne de gestion du Patrimoine Ferroviaire
1 rue du chemin de fer, PO Box 1415, Abidjan 16
Tel: (+225) 21 96 24   Fax: (+225) 21 39 62

Key personnel
President: Youssouf Bakayoko
Director General: Bernard Tra Bi
Technical Director: Joseph N'da Ezoa

Director, Finance and Administration:
Edouard Bahi Kore

Political background
SIPF is the shell corporation administering railway assets, both infrastructure and rolling stock. Operations are now run by Sitarail (qv), which leases the assets from SIPF.

Sitarail

Transport Ferroviaire de Personnel et de Marchandises
PO Box 1216, Abidjan 16, Côte d'Ivoire
Tel: (+225) 21 62 93   Fax: (+225) 21 28 58
e-mail: sitarail@africaonline.co.ci

Key personnel
Chief Executive: A A Thiam

Gauge: 1,000 mm
Route length: 660 km

Political background
The railway is the Côte d'Ivoire portion of the former Abidjan-Niger Railway that was jointly owned and managed by Côte d'Ivoire and Burkina Faso (formerly Upper Volta). In 1986 the partnership broke up. In 1988 the countries seemed to settle their differences, but this rapprochement was short-lived, and each country managed its part of the railway separately until 1993.

The former operator SICF, created in 1989, was immediately in serious financial difficulty, unable to meet its costs from revenue. Local analysis estimated that the railway was in urgent need of CFAFr1.2 billion to keep going, and of at least CFAFr14 billion for rehabilitation. Services to Burkina Faso continued to run, but operating difficulties on both sides of the border persuaded the two governments that a common management was perhaps the better option after all. At the end of 1992 it was decided to call tenders for private-sector operation of the railway.

Two groups put forward proposals, and those of the Sitarail joint venture consortium were preferred. Sitarail comprises various local industrial groups including Saga, SDV and Société Ivoirienne de Café et Cacao working with French consultants Systra (Sofrerail), Transurb Consult and others. These hold 67 per cent of the capital, the remainder being held by the two countries' governments (15 per cent each) and Sitarail staff (3 per cent). Two state-owned companies, SIPF and Sopafer, retain ownership of railway assets, including infrastructure, property and rolling stock.

After a long period of negotiation, Sitarail took over operation of the railway as a single entity in August 1995 with some 1,800 of the original staff complement of 3,600. The World Bank and Caisse Française de Développement provided funding to pay off the redundant workforce.

The concession was let for 15 years, but after seven years Sitarail must be prepared to grant track access to third party operators.

In November 2001, the governments of Burkina Faso and Côte d'Ivoire signed an agreement to establish a joint railway infrastructure investment fund. The fund was to be administered by representatives of each government and of Sitarail.

Passenger operations
Sitarail abandoned unprofitable local services when it took over the concession, but retained a vestigial long-distance passenger service. The company carried some 0.4 million passengers in 1996. Service provision was severely limited by the amount of rolling stock available, with no more than a thrice-weekly Abidjan-Ouagadougou express being operated in early 1996.

Freight operations
Despite a good network of permanent roads the railway is a vital link with Burkina Faso, which particularly relies on it for transport of freight to and from the coast.

Some 85 per cent of traffic to Burkina Faso consists of petroleum products, containers, fertiliser and clinker. In the reverse direction, cotton, agricultural produce and livestock are the principal commodities carried.

Under the nationalised operator, the railway declined in importance, reaching a low point of 250 million tonne-km in 1994, down from over 600 million in the early 1980s. Prospects for recapturing freight lost during the years of decline are thought to be favourable: in 1996 492 million tonnes were carried for 428 million tonne-km. Two potential traffic flows are manganese and zinc from mines in Burkina Faso for export from Abidjan.

Improvements to existing lines
The new operator immediately sought bids for rail and other track components to inaugurate an upgrading programme over the entire route in both countries. CFAFr17.6 billion has been allocated to infrastructure improvements in the period 1996–2000. In 1999, funding to upgrade the Abidjan-Ouagadougou line was obtained from the International Development Association.

Traction and rolling stock
At the start of 1997 the Sitarail fleet comprised 13 diesel locomotives, 17 shunting locotractors, 2 railcars, 17 coaches and 443 wagons. A further 114 private-owner wagons were in service. In late 1995, priority was being given to obtaining spare parts for rehabilitation of the remaining five GM diesel-electrics out of the original fleet of 25. In the company's five-year plan, some CFAFr13.7 billion will be spent on overhauling rolling stock; in late 1996, Sitarail went out to tender for 20 bogie hopper wagons.

Signalling and telecommunications
A total of 21 stations have been equipped with colourlight signalling and power point operation based on SNCF's NSI relay system. Installation was by a French subsidiary of ABB. Sitarail plans to spend CFAFr1 billion on upgrading the telecommunications system.

In 1999 Sitarail invited tenders for the supply of a fibre optic telecommunications system to replace existing equipment between Abidjan and Ouagadougou.

Track
Rail: 30 and 36 kg/m
Ballast: Granite, 800–1,200 litres/m, hard sandstone, 700 litres/m
Sleepers: Metal, 1,550/km; concrete monobloc (Blochet), 1,357/km
Min curvature radius: 500 m, being raised to 800 m
Ruling gradient: 10%
Max axleload: 17 tonnes

Croatia

Ministry of the Sea, Tourism, Transport and Development

Vladimira Nazora, 10000 Zagreb
Tel: (+385 1) 378 45 00   Fax: (+385 1) 378 45 20
Web: www.vlada.hr

Key personnel
Minister: Bozidar Kalmeta

UPDATED

Croatian Railways Ltd

Hrvatske Željeznice doo (HŽ)
Mihanovićeva 12, 10000 Zagreb
Tel: (+385 1) 378 33 00   Fax: (+385 1) 378 33 26

e-mail: hrvatske.zeljeznice@hznet.hr
Web: www.hznet.hr

Key personnel
President and General Manager:
   Josip Tomislav Mlinarić
Deputy President: Ivica Klarić Kukuz
Director, Finance: Ante Dedić
Executive Director, HŽ Cargo: Davorin Kobak
Director, Traction and Rolling Stock: Mladen Nikšić
Director, Operations: Ivica Jurić
Director, Infrastructure: Zoran Tomšić
Director, Human Resources: Josip Stanković

Gauge: 1,435 mm
Length: 2,726 km

Electrification: 1,199 km, of which 1,066 km at 25 kV AC, and 133 km at 3 kV DC

Political background
A week after Croatia declared its independence from Yugoslavia in October 1991 the new republic severed its railway system from Yugoslav Railways (JŽ). HŽ is now a limited liability company, wholly owned by the state.

The railway sustained serious damage in the war in the region in the early 1990s. However, HŽ benefits from its advantageous position between central and western Europe and the Balkan region. It is intersected by three important transit corridors: Munich–Zagreb–Belgrade–Athens (Corridor X); Rijeka–Zagreb–Budapest–Lvov (Corridor Vb);

and Budapest-Beli Manastir-Sarajevo-Ploče (Corridor Vc).

In response to financial pressures, HŽ has continued to shed staff: at the end of 2003, the railway employed 14,905, down from 16,077 at the end of the previous year.

As a candidate country for membership of the European Union, the Croatian parliament in 2003 approved legislation separating accounting for infrastructure and transport operations from January 2006.

Organisation

Operationally, HŽ is organised in three sectors: a commercial sector, which manages freight and passenger services; an infrastructure sector that covers traffic management, civil engineering and signalling/electrotechnical activities; and a traction and rolling stock sector responsible for the provision, maintenance and inspection of locomotives, coaches and wagons. Fully owned subsidiaries include companies responsible for: rail vehicle manufacturing; rail vehicle maintenance and overhaul; passenger vehicle cleaning; overhaul and maintenance of track, structures and buildings; the design and production of signalling/electrotechnical equipment; security/fire fighting; and rail catering.

Traffic (million)	2001	2003
Passenger journeys	36.7	35.98
Passenger-km	1,240	1,163
Freight tonnes	12.0	13.3
Freight tonne-km	2,249	2,744

Passenger operations

Delivery in 2004 of Class 7123 tilting dmus enabled HŽ to reduce journey times on the Zagreb–Split route and to increase the number of daily return trips on this route from one to three.

Freight operations

Main commodities carried are iron ore and other minerals, timber and wood products, fertilizers, oil and petroleum products and cereals.

Intermodal operations

HŽ operates five container terminals in Zagreb, Našice, Osijek, Rijeka and Split, and owns a fleet of around 400 container flat wagons. HŽ is a shareholder member of the Intercontainer-Interfrigo international intermodal group. Hungary and Italy are the two most important countries with which container traffic is exchanged.

New lines

A new 23 km route has been studied from Jurdani, west of Rijeka, to Lupoglav Novi, on the Istrian peninsula. At present HŽ's Istrian network is isolated from the core system and traffic has to pass through Slovenian territory, increasing the Croatian network's costs. The main construction challenge would be a tunnel of either 12.5 km or 14.5 km under the Učka range at Čičarija. Several attempts have been made to start this project but it remains unfulfilled.

In 2003 work commenced on a new chord from Žabno, on the Klostar–Križevci line, to Gradec, on the Križevci–Zagreb line. To be completed by 2005, the single-track line will enable trains to run directly from Klostar to Zagreb without the need for reversal at Križevci.

Improvements to existing lines

In 2003 HŽ announced infrastructure investments of €1.3 billion, of which over €600 million was to be spent on the three European Corridors detailed above in *Political background*.

The Lika line from Ogulin to Knin and Split was reopened in 2004 after closure or modernisation.

In 2005 248 route-km were double-track. HŽ plans to double-track three further sections of line: Dugo Selo–Koprivnica–Hungarian border; Zagreb–Sisak–Novska; and Zagreb–Krapina.

Traction and rolling stock

At the beginning of 2005 HŽ was operating 125 diesel and 107 electric locomotives, 66 diesel shunters, 72 dmus, and 25 emus. The hauled coaching stock fleet stood at 471 vehicles and there were 7,371 freight wagons.

0121612

Refurbished Class 2062-100 General Motors diesel locomotive at Perkovic with a passenger service from Split (Philip Wormald)
NEW/0585089

Modernised Class 1061 electric locomotive at Rijeka with a train from Moravice (Philip Wormald)
NEW/0585089

Deliveries commenced from Bombardier in 2004 of an order for eight Class 7123 two-car diesel tilting trainsets similar to the Class 612 units operated by DB in Germany. HŽ planned to deploy the vehicles on Zagreb–Split, Zagreb–Sibenik and Zagreb–Zadar services.

In 2004 HŽ completed a two-year programme to refurbish 20 examples of its fleet of GM-built Class 2062 diesel-electric locomotives, rebuilt machines becoming sub-class 2062.1. The modernisation was undertaken by TŽV Gredelj, with UK-based Turner Rail Services acting as main contractor and participation by General Motors.

Another modernisation programme completed in 2004 covered 15 Class 1141 electric locomotives, including retrofitting thyristor control equipment. Refurbished by Koncar, the upgraded machines carry a Class 1141.3 designation.

In 2000 HŽ commenced a programme to refurbish its Ansaldo-built Class 1061 electric locomotives. Class 7121 dmus and Class 6011 and Class 6111 emus have also been the subject of refurbishment programmes, as have some hauled passenger coaches and sleeping cars.

Electrification

The single-track line from Moravice via Rijeka to Šapjane on the Slovenian border is energised at 3 kV DC, while the remainder of HŽ electrification is at 25 kV AC 650 Hz. Conversion of this pocket of DC electrification to the standard 25 kV AC is HŽ's current main electrification project.

In 2001, plans were announced to electrify the line between Strizivojna and Osijek (48 km), in eastern Croatia, to enable IC services to run from Zagreb without a traction change.

Electrification of the Lika line from Ogulin to Knin and Split is planned, with completion expected in 2006.

Track

Rail: 60 UIC, S 49, and some 35–45 kg/m
Sleepers: Beechwood, prestressed concrete
Fastenings: Elastic SKL1 and SKL2, Fixed-K
Minimum curve radius: 150 m
Max gradient: 2.7%
Max permissible axleload: 22.5 tonnes

UPDATED

Diesel locomotives

Class	Wheel arrangement	Power kW	Speed km/h	Weight tonnes	No in service	First built	Mechanical	Builders Engine	Transmission
2041	Bo-Bo	606	80	64	18	1962	B&L/Dakovič	SACM	E B&L
2042	Bo-Bo	680	80	64	3	1966	B&L/Dakovič	SACM	E B&L
2043	A1A-A1A	1,454	124	100	6	1960	General Motors	General Motors	E General Motors
2044	A1A-A1A	1,820	124	103	31	1981	GM/Dakovič	General Motors	E General Motors
2061	Co-Co	1,454	124	114	24	1960	General Motors	General Motors	E General Motors
2062	Co-Co	1,640	124	103	54	1973	General Motors	General Motors	E General Motors
2062-100	Co-Co	1,492	124	103	20	1973*	General Motors	General Motors	E General Motors
2063	Co-Co	2,461	124	120.1	13	1973	General Motors	General Motors	E General Motors
2131	C	294	60	42.5	–	1963	Jenbacher Werke	Jenbacher Werke	H
2132	C	441	60	44	–	1986	Jenbacher Werke	Jenbacher Werke	H
2133	C	511	60	54	–	1957	MaK	MaK	H
2141	B-B	1,176	120	68.2	1	1977	Duro Daković	SACM	H

B&L = Brissoneau & Lotz
* Modernised by TŽV Gredelj 2002–04

Electric locomotives

Class	Wheel arrangement	Output kW	Speed km/h	Weight tonnes	No in service	First built	Builders Mechanical	Electrical
1061[1]	Bo-Bo-Bo	2,640/3,150	120	108–112	23	1960	Ansaldo	Ansaldo
1141–000[2]	Bo-Bo	4,080	120	78	46	1970	SGP	ASEA/Rade Končar
1141–100[2]	Bo-Bo	4,080	120	82	8	1987	Gredelj/Dakovič	Rade Končar
1141–200[2]	Bo-Bo	4,080	140	82	12	1981	Gredelj/Thyssen	Rade Končar
1141-300[3]	Bo-Bo	3,840	140	82	15	1981	Gredelj/Dakovič	Rade Končar
1142[2]	Bo-Bo	4,400	160	82	13	1981	Gredelj/Dakovič	Rade Končar
1143	Bo-Bo	4,080	120	–	1	–	SGP	Rade Končar
1161[2]	Bo-Bo-Bo-	4,080	120	129	2	1988	Ansaldo/Gredelj	Ansaldo/Uljanik

[1] 3 kV
[2] 25 kV
[3] Modernised by Koncar 2002–04

Diesel railcars or multiple-units

Class	Cars per unit	Motor cars per unit	Motored axles/car	Power/motor kW	Speed km/h	No in service	First built	Mechanical	Builders Engine	Transmission
7021	5	2	4	386	120	1	1973	B&L	SACM	E
7121	2	1	4	106	120	37	1981	Macosa	MAN	H
7221	2	1	1	110	90	21	1959	Uerdingen	MAN	M
7122*	1	1	2	147	130	20	1979	Fiat/Kalmar	Fiat	H
7123	2	2	2	560	160	8	2004	Bombardier	Cummins QSK 19	H Voith

* Purchased second-hand from Swedish State Railways in 1995

Electric railcars or multiple-units

Class	Cars per unit	Motor cars per unit	Motored axles/car	Power/motor kW	Speed km/h	No in service	First built	Builders Mechanical	Electrical
6011*	4	2	4	145	110	9	1964	Pafawag	Pafawag
6111**	3	1	4	300	120	23	1977	Ganz-Mavag	Ganz-Mavag

*3 kV
**25 kV

Class 7121 dmu at Dalj forming a service to Osijek
(Philip Wormald)
***NEW*/0585090**

Cuba

Ministry of Transport

Avenida Rancho Boyeras y Tulipán, Havana
Tel: (+53 7) 81 45 05/81 47 80   Fax: (+53 7) 33 51 18

Key personnel
Minister of Transport: Álvaro Pérez

Cuban Railways (UFC)

Unión de Ferrocarriles de Cuba
Edificio Estación Central, Egido y Arsenal, Havana
Tel: (+53 7) 62 15 30   Fax: (+53 7) 33 86 28

Key personnel
Director General: Ing Pastor Pérez Fleites
Directors
   Locomotives and Rolling Stock: J Noya
   Finance: N Marrero
   Commercial: Dr C M Caballer
   Traffic: R Boffil
   Permanent Way: J C Miranda
   Signalling: R Morales
   Personnel: L Pereda

Gauge: 1,435 mm
Route length: 4,226 km
Electrification: 140 km at 1.2 kV DC

Organisation
The UFC network is operated as a fully integrated state enterprise by the Ministry of Transportation. It adopted its current name in 1998 after a restructuring that created business units handling various commercial and central functions. A divisional operational structure is in place, covering nine divisions.

At the end of 2001 UFC employed 16,454 staff.

Traffic (million)	2001
Passenger journeys	12.5
Passenger-km	1,766.6
Freight tonnes	3.45
Freight tonne-km	806.9

Passenger operations
Since 1992, passenger traffic is reckoned to have increased by some 20 per cent due to the rationing of petrol for road vehicles, a measure resulting from the end of Comecon trading arrangements and the US blockade. However, services suffer from outdated equipment and poor reliability.

Services on the electrified Havana–Matanzas line, the former Hershey Railway, are marketed under the Felcuba brand name.

Freight operations
Freight traffic comprises mostly sugar and its by-products of rum and molasses, tobacco and citrus fruit. Other commodities carried include cereals, cement, foodstuffs, fuel, minerals and salt. Traffic levels have fallen considerably since the 1980s, when some 13 million tonnes were carried annually. However, there has recently been a revival in sugar traffic, which has recovered from low production in the mid-1990s.

Improvements to existing lines
UFC's key project in 2003 was modernisation of the Havana–Santiago de Cuba line (830 km), the most important on the network. A key objective is to increase capacity for passenger and freight traffic and to raise line speeds from 100 to 140 km/h. This will necessitate the introduction of new rail welding techniques and improved track maintenance practices. Also required is modernisation of the semi-automatic signalling in place on around half of the route and extending its coverage to control the entire line. Initial tranches of funding for this project have been made available by the Cuban government.

Improvements have been made to track and bridges on the electrified Havana–Matanzas line to coincide with the introduction of emus procured

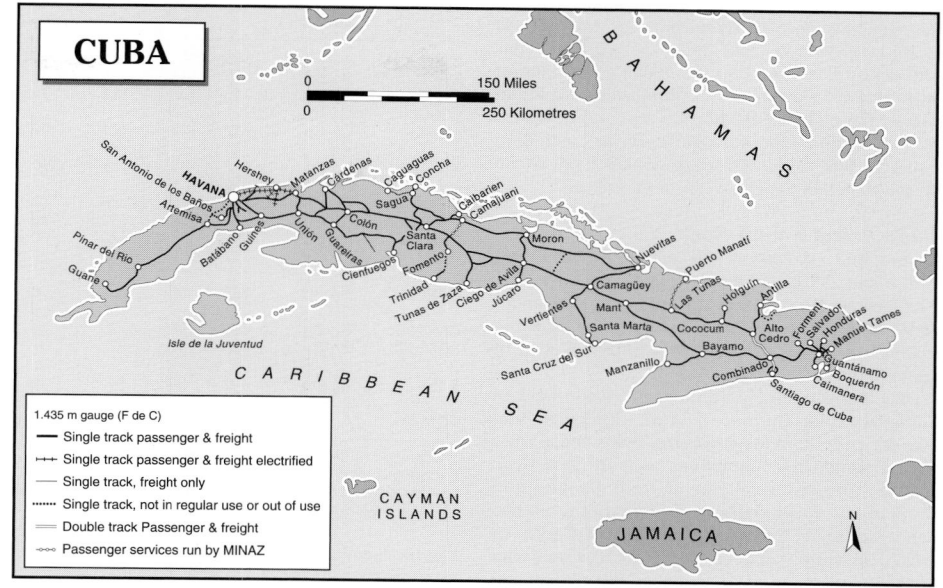

C36-7 diesel locomotive procured from Mexico and converted to A1A-A1A axle arrangement, seen at Cienfuegos port (Wilfrid F Simms)
0552774

Former Catalan Railways emu in service on the Havana–Matanzas 1.2 kV DC line (Lennox MacEwan)
0058683

second-hand from Spain in 1998 (see 'Traction and rolling stock' section).

In 1997 a major new station at Santiago de Cuba was brought into use, and in 1998 improved rail facilities were commissioned at the port of Matanzas, from which much of Cuba's sugar production is shipped. In the 1990s a new gound level station, Havana La Coubre, was developed east of the elevated Havana Central and now acts as the terminal for all local services, including trains to and from Matanzas, which have to be diesel-hauled between La Coubre and Elisa.

Traction and rolling stock

At the end of 2001 UFC's active traction fleet comprised 225 diesel and 12 electric locomotives, 42 diesel and 14 electric railcars, 673 passenger cars and 4,548 freight wagons. Large parts of the rolling stock fleet were out of service for want of spare parts.

UFC diesel traction is an eclectic assortment of Canadian, East German, French, Hungarian and Soviet types. From the end of the 1990s, these were supplemented by locomotives on North American outline procured secondhand from Canada and Mexico. For many years the dominant model has been the Soviet 1,912 kW Type TE-114K Co-Co, of which 96 were supplied, although the majority of these are reported to be unserviceable despite a re-engining and refurbishment programme initiated in 1998 at UFC's Camagüey workshops. Other recent locomotive acquisitions are 16 T-458 Bo-Bo machines obtained second-hand from the Czech Republic. Also reported is the acquisition secondhand from France of a BB63000 locomotive and some Hungarian DVM-9 machines.

UFC's 12 vintage electric locomotives are used on the 1.2 kV DC Havana–Matanzas line, the former Hershey Railway.

The most recent additions to the diesel railcar fleet are five Budd RDC vehicles obtained in 1998–99 from CN and CP in Canada and two former German Class 771/772 railbuses, although the latter were reported to be stored at UFC's Havana workshops in 2002. Also in the fleet are around 20 Pionero and 10 Ligero railbuses built locally by the former Ferrocarriles de Cuba and Taino respectively, an additional Budd railcar, survivors of a fleet of 15 Fiat Concord dmus converted to operate as single units and five Brill railcars that operate in the Guantanamo area.

UFC is now required by national policy (and economic and political circumstance) to seek local rolling stock manufacture and a local company, Empresa Productora de Equipos Ferroviarios of Cardenas, began supply of diesel railbuses based on the Ganz-Mávag design. The plant's first 85 passenger cars have been delivered to UFC. The Ministry of Steel, Iron & Machine Industry has produced two 'Taino' trains at its Cardenas works; these have run in Las Tunas province and can carry 300 passengers. In 1998 a major purchase of second-hand rolling stock was supplied from France in the form of 44 former Trans-Europe-Express (TEE) stainless steel-bodied coaches. All were overhauled before shipment and the contract also covered supplies of spare parts and training for UFC staff. In 2001, UFC also acquired some 150 locomotive-hauled coaches from DB in Germany.

In 1998 Catalan Railways (FGC) in Spain sold to UFC five three-car and three four-car standard-gauge 1950s-built emus for the electrified Havana–Matanzas line. The Pta315 million acquisition was financed by the International Commerce Bank of Cuba and Banco Bilbao Vizcaya in Spain. The Brill railcars detailed in the accompanying table are retained for tourist services.

Type of coupler in standard use: Semi-automatic
Type of brake in standard use: Air (50% Matrosov, 30% Westinghouse and 20% DAKO)

Signalling and telecommunications

On the main Havana–Santiago da Cuba line, replacement has proceeded of its almost exclusively manual point and block-telephone operation. Hitherto only nine of some 100 stations have had central point working, but track-to-train and train-to-train radio has been provided. Installation of a semi-automatic block system and relay interlockings has been completed on the 16-station section between Havana and Santa Clara (207 km) and in 1991 was continuing to Santiago de Cuba.

Electrification

The 147 km 1.2 kV DC electrified section on UFC links Havana with Matanzas. In 1999, agreement was reached with Catalan Railways in Spain to assist planning of a 12 km extension of this electrification to allow operations into Havana Central station.

Track

Rail: P50 51 kg/m
Sleepers: Prestressed concrete: 2,460 mm, spaced 1,520–1,840/km; creosoted pine: 2,750 mm, spaced 1,520–1,840/km
Fastenings: Elastic and rigid with track rails and screw bolts

Min curvature radius: 150 m
Max gradient: 3.0%
Max permissible axleload: 22 tonnes over 230 km

Sugar railways

The numerous railways linking the sugar plantations and factories are in sum of greater extent than UFC and at latest report totalled 7,742 km. Of the total, some 65 per cent is standard-gauge. Serving over 100 of the island's 154 sugar plants, these railways are mostly operated by the Ministry for Sugar (MINAZ) and employ around 900 locomotives, 380 of them steam, and over 30,000 wagons. The diesel fleet mainly comprises locomotives of USSR origin, with over 440 supplied between 1977 and 1990.

Diesel locomotives

Class	Wheel arrangement	Power kW	Speed km/h	Weight Tonnes	No Delivered	First Built	Builders Mechanical	Builders Engine	Builders Transmission
GMD 1	A1A-A1A	895	105	108	20	1958[1]	GM Canada	GM 12-567C	E GM
SW 1200RS	Bo-Bo	895	105	101	1	1956[2]	GM Canada	GM 12-567C	E GM
C30-7/C36-7/U36C	A1A-A1A	2,240	105	182	20	1973[3]	GE	GE FDL-16	E GE
E114K	Co-Co	1,912	120	121	102	1978	Voroshilovgrad	Kolomna 1A-5D49	E Jaricov
MX-624	Co-Co	1,912	135	112	50	1975	Bombardier	Alco 251E	E GE Canada
M62-K	Co-Co	1,234	100	120	32	1974	Voroshilovgrad	14D4DT2	E Kharkov
TEM-4K	Co-Co	735	100	120	40	1964	Bryansk	PDIT	E
TEM-2K	Co-Co	757	100	120	60	1974	Bryansk	PDITM 6-cyl	E Kharkov
TEM-15K	Co-Co	757	100	108	25	1988	Bryansk	Kolomna 2-6D49 8-cyl	E -
DVM-9	Bo-Bo	735	90	76	70	1969	Ganz-Mávag	16VCE17/24	E Ganz Electric
GM-900	Bo-Bo	662	90	72	58	1955	GM	8-567C	E GM
BB63000	Bo-Bo	606	90	72	32	1965	B&L[4]	MG0V12BSH 22	E B&L
T-458	Bo-Bo	492	80	74	16	1964	ČKD	ČKD 6S 310 DR	E ČKD
TGM-25	C	294	50	46	38 (?)	1970	Voroshilovgrad	–	H -

[1] Acquired 1999–2001 from Canadian National
[2] Acquired 1999 from Canadian National
[3] Ex-Mexican National Railways, built 1973–86, overhauled by Alstom 2001, rebogied from Co-Co
[4] B&L = Brissoneau & Lotz

Electric locomotives

Class	Wheel Arrangement	Output kW	Speed km/h	Weight tonnes	No in service	First built	Builders
GE 7230B	Bo-Bo	588/882	60	55	12	1920	GE

Electric railcars or multiple-units

Class	Cars per unit	Motor cars per unit	Output/motor KW	Speed km/h	No in service	First built	Builders
Brill	1	1	4 × 55	70	2	1923	Brill
M400	4	3	4 × 110	80	3	1943*	FGC/Cenemesa
M500	3	2	4 × 110	80	5	1943*	FGC/Cenemesa

*Ex-Catalan Railways, modernised 1985, delivered as 19 motor cars and eight trailers

Cuban-built diesel-mechanical railbus at Trinidad (Wilfrid F Simms)   0552775

Czech Republic

Ministry of Transport

PO Box 9, Nábřeží Ludvíka Svobody 1222/12, CZ-110 15 Prague 1 – Nové Město
Tel: (+420 972) 21 11 11   Fax: (+420 972) 23 11 84
e-mail: posta@mdcr.cz
Web: www.mdcr.cz

Key personnel

Minister: Milan Šimonovský
Under-Secretary, Railways: Vojtěch Kocourek
Director, Railways: Jaroslav Soušek

UPDATED

Czech Railways (ČD)

České dráhy a.s.
Nábřeží Ludvíka Svobody 1222/12, CZ-110 15 Prague 1 – Nové Město
Tel: (+420 2) 51 43 11 11; 51 43 22 99
Fax: (+420 2) 51 43 20 81; 51 43 24 00
e-mail: press@cd.cz
Web: www.cd.cz

Key personnel

Chairman, Supervisory Board: Vojtěch Kocourek
Board of Directors
  Chairman of Executive Committee and
    Director-General: Petr Kousal
  First Under-Secretary and Under-Secretary for
    Personnel and International Affairs: Josef Bazala
  Under-Secretary, Economy: Ivan Foltýn
  Under-Secretary, Commercial and Operations:
    Jiří Kloutvor
  Under-Secretary, Infrastructure: Petr David
Management Board
  Director-General's Office: Oldřich Zatloukal
  Strategy: Miloslav Jakeš
  Personnel: Mojmír Bakalář
  Economy Management: Stanislav Brablík
  Finance: Vladimir Filip
  Accounting and Taxation: Josef Šedivý
  Legal: Jan Blecha
  Property Undertaking: Pavel Krejci
  Supply and Sales: Tomáš Nachtman
  Investment: Jiří Bureš
  Planning: Milan Růžička
  Defence, Protection and Ecology: Jiří Fiala
  Safety Inspection: František Raška
  Infrastructure Operations: Antonín Krejčí
  Passenger Traffic: Jiří Kafka
  Freight Transport: Oldřich Mazánek
  Traction and Rolling Stock: Rostislav Novák
  Technical Development: Jan Bartek
  Technical Inspection: Jaroslav Dvořák
  Civil Engineering: Josef Koudelka
  Buildings: Pavel Novák
  Electrical and Power: Jan Matějka
  Automation and Electrotechnics: Zdeněk Thun
  Signalling and Telecommunications:
    Karel Plachetka
  Safety: Miroslav Kochař
  Internal Audit and Control: Vladimír Honzák
  International Affairs: Jiří Havlíček
  UIC: Ivo Malina
  Public Relations: Zdeňka Celá
Regional Managers
  Prague: Miroslav Jasenčák
  Plzeň: Josef Chaloupek
  Ústí nad Labem: Milan Fryč
  Brno: Pavel Surý
  Ostrava: Uršula Broschová

Political background

On 26 February 2001 the Czech government ratified a new reform programme for ČD. After being referred back for revision several times during 2001, the programme was finally approved by parliament on 5 February 2002. Signed by President Václav Havel on 19 February, the legislation proposed splitting ČD's infrastructure management from operations and the creation of two separate operating businesses from 1 January 2003. From that date ČD became a state-owned joint-stock company engaged only in the management and operation of

rail transport and shedding all responsibility for the management, maintenance and development of the railway's infrastructure. These functions were taken over by a new state–owned organisation, Railway Infrastructure Authority (Správa železniční dopravní cesty (SŽDC) – see entry in Czech Republic section of *Railway systems and operators*). ČD as was formally founded on 11 July 2002 with assets totalling Kcs39.5 billion, including the value of motive power, rolling stock and buildings. With six regional directorates (OPŘ, its structure now consists of four independent groups based on DB AG's model, with businesses responsible for passenger traffic, freight, operations management and ancillary activities, such as the railway's travel agency and its research institute. Ten years since foundation of the original ČD, the company has been cleared of all debt. ČD as is controlled by the government through a steering committee and a supervisory board.

With the radical change of separating infrastructure and operations, the economy of the railway is relatively independent of the state for the first time in its history. ČD plans to concentrate on building its revenue freight operations in an environment in which as much as 9 per cent of rail freight on the Czech network is carried by private operators. ČD's long-serving General Manager, Dalibor Zelený, was recalled by the transport minister in March 2003 and a new management board appointed.

ČD's transformation programme has been backed by an EU grant of €1.97 million, 57.5 per cent of which has been spent on a 'twinning project'. This was officially launched in July 2002 in partnership with Spanish National Railways (RENFE). From March 2004 RENFE helped ČD with the application of EU legislation with a particular view to harmonising charges for the use of rail infrastructure, defining the relationship between train operators and infrastructure owners and with assisting in the process of transforming the human resources sector.

In September 2003 ČD revealed a new organisational structure as part of its extensive four-stage programme to transform the joint-stock company into a holding company. The aim of the first stage was to establish single management control in the areas of finance, economy, accounting, legal affairs and personnel in a bid to improve the railway's business performance, particularly in the freight and passenger sectors. ČD's Commercial and Operations (DOP) and Infrastructure (DDC) divisions merged to establish central management control. This allowed the elimination of 150 senior management positions from the railway's restricted payroll. Six regional management centres were cut to five and ancillary activities are being consolidated into a group of subsidiary companies in which ČD will be the majority shareholder.

Hiving off these ancillary activities in this way as part of ČD's development strategy was approved in May 2004. The pilot project, Traťová Strojní Společnost (TSS) based in Pardubice, was established in January 2005 and involves

Class 714 diesel-electric locomotive, rebuilt from a Class 735 machine for secondary and shunting duties (Milan Šrámek)
0062501

Class 471 double-deck emus for Prague suburban services (Milan Šrámek)
0552777

Class 912 part-low-floor driving trailer for use with Class 812 railcars (Michal Málek)   0552778

Class 340 electric locomotive, converted to a dual-voltage (25 kV/15 kV) machine for cross-border services between the Czech Republic and Austria (Milan Srámek)   **NEW**/1114970

the operation of track maintenance machines. Initially wholly owned by ČD, it is hoped that TSS's business portfolio will extend to other undertakings, in particular infrastructure operator SŽDC but also into foreign markets as well, and will lead to major improvements and modernisation of track maintenance.

Hiving off three other non-core businesses was to follow in 2005. These are involved in telecommunications (ČD-Telekomunikace), the railway's research institute (VÚŽ) and a travel agency (ČD Travel). Six more subsidiaries are to be set up in 2006 covering railway staff health services, its training institute, holiday and housing agencies, traction power supply network and materials supply centres. The final phase of restructuring should see the separation of long-distance and regional passenger operations and of freight traffic. However, this step faces huge political opposition.

In September 2004 a pilot project was established, setting up a legally independent company in Frankfurt/Main, Germany, to handle Czech Railways foreign representation. CD Generalvertretung GmbH is wholly owned by ČD and acts in accordance with German law. Its aim is to enable the railway to compete in the European market and to increase its freight traffic revenues. The establishment of further representative companies was expected to follow.

In 2004 ČD carried 179 million passengers, up by 3.5 per cent on the figure for 2003. Freight traffic totalled 79.64 million tonnes, down 6.7 per cent on 2003. The fall in freight traffic was caused by strong competition from cross-border road transport after

the Czech Republic's entry to the EU and a decline in the movement of traditional commodities such as coal and timber. Revenues from passenger and freight traffic totalled Kcs5.28 billion and Kcs18.2 billion respectively.

According to the railway's business plan for 2005, revenues should reach Kcs46.4 billion (of which Kcs5.5 billion from passenger traffic and Kcs17.8 billion from freight), with expenses totalling Kcs47.1 billion. The overall loss will total Kcs690 million, compared to Kcs998 million in 2004. The railway expects to carry 181 million passengers.

Continuing restructuring and the need to increase productivity will entail a further fall in the number of employees. In 2004 6,800 people left the railway; ČD planned to eliminate a further 4,568 posts from the company's 70,500-strong payroll in 2005.

In March 2002 ČD became a full member of the Community of European Railways.

Privatisation

The proposed break-up and sale of parts of the Czech Railways network has been cancelled following the election in July 1998 of a Social Democratic government. While some branch lines may be leased out to private operators, the new government favours retaining the whole rail network under unified state ownership.

The previous Czech government had continued to examine ways of introducing private capital to the running of the railways, with the aim of rectifying an investment backlog, rendering ČD more competitive against other modes and reducing the level of state support. ČD's electrification (Elektrizace Železnice Praha a s), intermodal (ČSKD-Intrans) and

dining and sleeping car (JLV a s) divisions have already been privatised, in addition to nine major workshops and fringe activities such as restaurants and the railway health service. During the first phase, privatisation earned Kcs3.6 billion.

Under legislation passed in 1994 ending the monopoly of the state railways, private companies holding a licence from the Railway Authority were able to bid to operate closed or unprofitable sections of the ČD network. A new organisation, the Railway Authority (DU – Drážni Úřad), was established by the Ministry of Transport in April 1994, charged with awarding operating licences to private sector companies and harmonising their operations with those of ČD.

An historic agreement was signed by the National Property Fund and private operator Jindřichův Hradec Local Railways (JHMD) in February 1998, when the latter took over operation of two 760 mm local lines radiating from Jindřichův Hradec in southern Bohemia. JHMD had leased the lines from ČD since March 1997 after their abandonment by the national operator two months earlier, and rapidly recorded substantial increases in freight and passenger traffic.

Also initiated in 1998 was the process of leasing ČD lines to private operators. By February 1999 five rural lines covering 100 route-km had been leased: Trutnov–Svoboda nad Úpou (10 km), Nové Sedlo u Lokte–Krásný Jez (24 km), and Sokolov-Kraslice (23 km), all to private operator Viamont Ústí nad Labem; Milotice–Vrbno pod Pradědem to OKD Doprava; and Šumperk–Petrov nad Desno–Sobotín (29 km) to Železnice Desná.

An Association of Railway Companies (SŽS) was founded in October 1995 representing organisations wishing to purchase or operate over ČD lines or to obtain access to the state network.

Traffic (million)	2001	2002	2003
Passenger journeys	188.3	175.0	172.0
Passenger-km	7,262	6,562	6,483
Freight tonnes	88.0	82.6	85.3
Freight tonne-km	17,366	16,130	16,396

Passenger operations

Passenger traffic has suffered both from economic recession and competition from private bus companies offering cheaper fares and faster timings over routes such as Prague–Brno and Prague–Ústí nad Labem. ČD now accounts for only 7.4 per cent of passenger journeys in the Czech Republic. In response to this, the new management board elected in March 2003 decided to reduce costs for domestic journeys and extend the choice of discounts, with effect from 1 July 2003. Significant savings have been introduced on the longest and most frequent connections, such as Prague–Brno (255 km), where the cost of a single ticket fell by 23 per cent, making the journey cheaper (but not faster) than road coach services on the same route. Supplementary fares for EC/IC trains have also been reduced, while fares on ČD's premier service, 'Manažer' SC train pair 502/503, have been cut by 26 per cent. The number of trains operating to a clockface timetable has also been increased.

In the period of the 2003–04 timetable ČD was operating 7,360 passenger trains daily, of which 6,492 were local stopping services (down by 0.9 per cent on the previous year). This decrease in stopping services was a result of ČD being forced to discontinue to cross-subsidise the operation of passenger traffic from freight revenues. In the 2004–05 timetable, however, the figure rose again slightly to 6,795. Overall punctuality rose to 91.7 per cent (90.4 per cent in 2003).

With the introduction of the 2000–01 timetable, ČD accelerated express services on the Prague–Brno main line, authorising for the first time 160 km/h running over upgraded sections of Corridor 1 (see 'Improvements to existing lines' section). Diverted via Pardubice and Česka Třebová instead of the traditional route via Havlíčkův Brod, EuroCity trains cover the 257 km line in 2 hours 42 minutes at an average speed of 94.4 km/h, the fastest-ever rail connection between the two cities. Further journey-time reductions will follow completion of Corridor 1 upgrading in 2003 and the introduction of new Class 680 tilting trainsets (see 'Traction and rolling stock' section).

Class 754 diesel-electric locomotive with a Brno-bound passenger service at Znojmo (Michal Málek)
***NEW**/1114971*

Through daily EuroCity and InterCity services now connect Prague with Aarhus, Berlin, Bratislava, Budapest, Dresden, Hamburg, Košice, Nuremberg, Sylt, Vienna, Warsaw, Westerland and Zvolen. Two Vienna–Warsaw and Warsaw–Budapest EuroCity services serve the east of the Czech Republic.

Internally, service on the Prague–Bohumín route has been improved with introduction of InterCity services Nos 500/501 'Ostravan' and Nos 504/505 'Jan Perner', using refurbished coaches. IC 520/521 connects Prague with Zlín in central Moravia. ČD is attempting to attract more passengers to these routes, which as yet have no firm highway competition.

In January 1997, ČD introduced a high-quality service, the 'Manažer', on the route between Prague and Bohumín (366 km). Hauled by Class 162 3 kV DC electric locomotives, SuperCity (SC) train 502/503 consists of three first-class coaches and a restaurant car. Due to speed restrictions on the route caused by extensive works on ČD's Corridor 1 upgrading, the train now completes the journey in 4 hours 9 minutes, a commercial speed of 88.2 km/h. Free at-seat refreshment, newspapers and free parking at Ostrava's main station area are provided. By May 2000, 'Manažer' had carried 120,000 passengers. Another 1997/98 timetable innovation was the provision to carry accompanied private cars on the 'Hornád' (now 'Laborec') international overnight service between Prague and Poprad, in Slovakia's High Tatras region. By the end of 2004 the now daily service had carried 5,731 accompanied cars, of which 1,563 were in 2004. In February 2004 a similar service was launched on the 'Cassovoa' international overnight service between Prague and Košice (708 km), in the far east of Slovakia. Operated three times a week, the train carried 310 cars during its first year of operation.

In 2003 ČD introduced a seasonal weekly motorail train the 'Jadran Expres', between Prague and Split on Croatia's Adriatic coast. In its first year, this successful service carried 5,778 passengers and 613 cars, with an overall occupancy of almost 75 per cent. These figures increased to 7,450 and 715 respectively in 2004.

Losses resulting from passenger traffic are mainly generated by regional operations. To decrease these, ČD is striving to cut the cost of providing these services by using less operationally demanding rolling stock, reducing the number of employees used and using smartcards for ticketing. An extensive system of fare discounts has proved very successful too.

Another solution for rural lines mainly near the German border could be provided by establishing partnerships with foreign or private operators such as DB Erzgebirgsbahn, Connex or Keolis. With the December 2003 timetable change, ČD and DB subsidiary Erzgebirgsbahn launched a new cross-border regional service from Karlovy Vary to Zwickau, Germany, operated by DB Class 642 Desiro railcars. Two pairs of weekend-only trains are operated via Potůčky and Johanngeorgenstadt. In April 2004 a similar service, two train pairs at weekends only, began operating on the Chomutov–Chemnitz route via Vejprty and Cranzahl. In May 2004 a third ČD/DB weekend service operated by Desiro railcars was added between Děčín and

Dresden, using the main line in the River Labe valley via Dolní Žleb, Bad Schandau and Pirna. These services are mainly targeted at weekend leisure traffic, using the German 'Schönes Wochenende' ticket, which is valid in ČD lines in northern and western Bohemia.

In June 2004 Connex extended its existing Desiro-operated InterConnex Binz–Stralsund–Berlin–Zittau service to Liberec via Hrádek nad Nisou and Chrastava. Since the same month, Liberec has been served by RegionalExpress trains from Dresden. The latter is operated four times a day on Saturdays and Sundays only.

Class 471 3 kV DC double-deck suburban emu (CKD Vagonka)   ***NEW**/1066406*

A proven strategy to attract passengers to suburban services around major cities is developing a common tariff policy with regional authorities and other operators, enabling a single ticket to be used on all modes.

In November 2002 Grandi Stazioni SpA of Italy won a 40-year contract for the refurbishment and management of commercial and retail spaces at ČD's three major stations at Praha-Hlavní nádraží, Karlovy Vary and Mariánské Lázně. Grandi Stazioni will transform the stations into modern terminals with extensive commercial facilities, investing Kcs37.9 million and Kcs53.1 million at Mariánské Lázně and Karlovy Vary respectively by 2005 and Kcs658 million at Praha-Hlavní nádraží by 2008. The agreement was signed in December 2003, with work due to start in mid-2006.

In November 2004 ČD signed an accord with a joint-stock company, Masaryk Station Investment (MSI), paving the way for revitalisation of the listed Masarykovo nádraží (Masaryk station) in Prague. This will involve major refurbishment of all passenger and commercial areas including a station hotel at a total cost of Kcs9 billion. The railway has a 34 per cent share in the project. ČD wants to undertake further similar projects in the near future to upgrade major stations throughout the network. These would be managed by its new subsidiary, ČD-Reality.

Freight operations

ČD forms the Czech Republic's principal means of transporting bulk commodities, mainly coal (32.2 per cent of ČD's total freight volume), steel products (9.23 per cent), ores (8.96 per cent), raw

Class 812 Liaz-engined diesel railcar, converted from a Class 810 vehicle (Michal Málek)   0132660

materials (9.22 per cent), forest products (5.83 per cent), cereals and other foodstuffs (2.78 per cent) and chemical products (3.55 per cent). Given ČD's advantageous geographical location at the heart of Europe, international traffic is vital to the railway, accounting for around 61.4 per cent of freight volume. Transit traffic accounted for 9.8 per cent of freight revenues in 2003.

There was a serious loss of traffic after the demise of communism – in 1989 railways now in the Czech Republic carried 226 million tonnes – and the decline continued after Czechoslovakia was split in two, but with a 25 per cent market share of freight traffic (40 per cent in 1996) ČD still performs well in comparison with other EU countries, maintaining its fourth position among the largest freight hauliers. In 2002 it operated 2,358 regular daily freight services, 112 of these block trains. Electric traction accounted for 90 per cent of freight tonnage moved.

Measured in tonne-km, ČD hauls 98 per cent of all rail freight, the rest handled by a handful of open access companies. In 2003 open access companies operating under licences granted by the Railway Authority (DÚ), principally OKD-Doprava and Viamont, carried 7.94 million tonnes of freight, mainly coal and oil products, accounting for 8.5 per cent of the country's total rail freight volume.

In terms of quality, ČD is attempting to make itself more competitive by maximising opportunities for black train operation and upgrading the existing freight fleet or purchasing new wagons that meet customers' changing needs. Equipment is also leased from neighbouring railways.

An agreement has been reached between ČD and DB covering the operation of cross-border freight services. With a pilot project tested on Lovosice–Dresden piggyback trains (see below) from February 2002, locomotive drivers of both railways have since the 2004/05 timetable change, been operating deep into the other country's network, eliminating a crew change at the Děčín/Bad Schandau border crossing. Using ČD/Railion Class 372/180 dual-voltage electric locomotives throughout, ČD drivers operate to Dresden while Railion drivers may reach Lovosice, Mělník, Všetaty, Nymburk, Kutná Hora or Praha-Uhříněves.

On 17 March 1999, ČD signed an accord with Austrian Federal Railways to create a common freight rates policy. A similar agreement was signed in early March with Slovenian Railways and Polish State Railways. Through this harmonisation of tariffs, the four railways hoped to capture much lucrative north-south freight traffic from Railion, the joint freight company formed by Germany's and the Netherlands' national operators.

ČD's domestic express parcels service, 'ČD-Kurýr', was available at 181 stations in 2005.

Intermodal operations
Intermodal traffic in the Czech Republic is increasing sharply, accounting for 8.5 per cent of ČD's freight volume in 2003. In 2003 a total of 7.3 million tonnes were recorded (up 19.9 per cent on 2002), compared with just 1.6 million tonnes handled in 1994.

There are three principal companies operating container trains from their own terminals. The operation of container shuttles to and from northern German ports is mostly in the hands of Metrans Praha, established in 1991, which operates 23 and 27 train pairs per week between its terminal at Praha-Uhříněves and Hamburg Waltershof/Süd and Bremerhaven Seehafen respectively. Metrans also operates 16 pairs of inland container shuttles per week between Praha-Uhříněves and Želechovice nad Dřevnicí, Dunajská Streda, Slovakia, and Györ, Hungary, the last-mentioned introduced in June 2002. Hamburg port operator HHLA is the majority shareholder in Metrans (50.1 per cent), while 34.2 per cent of the shares have been held by Railion since 1999. Praha-Uhříněves is the Czech Republic's largest container terminal; the company also operates facilities at Želechovice nad Dřevnicí in Moravia and at Dunajská Streda in Slovakia, and partly owns terminals at Galanta, Slovakia, and Györ, Hungary, through its subsidiary Metrans Danubia. To boost its train capacity, Metrans ordered 400 Type Sggrss six-axle container wagons from Tartavagónka Poprad for delivery in 2004–05.

A second private intermodal company, ČSKD-Intrans, formerly a subsidiary of Czechoslovak State Railways, now operates four pairs of 'Hansa-Prague Shuttle' container shuttle services between its major terminal at Praha-Žižkov and Hamburg Waltershof/Süd and Bremerhaven. Since January 2003 80 per cent of ČSKD-Intrans has been owned by Intercontainer-Interfrigo (ICF) or Intercontainer Austria. ČSKD's once extensive network of inland feeder terminals has been reduced to just one at Přerov, plus two in Slovakia: Bratislava-ÚNS and Košice. The company operates its trains in collaboration with ČD and ITL Eisenbahn.

The third intermodal operator, the Rotterdam-based company, European Rail Shuttle (ERS), incorporating the major shipping companies Maersk-Sealand and P+O Nedlloyd, now runs its extensive container shuttles from a new terminal at Mělník-Labe. Until January 2003 the company collaborated with ČSKD-Intrans at Praha-Žižkov. ERS now operates five train pairs per week from Mělník-Labe to Rotterdam Maasvlakte Delta/Waalhaven RSC ('Bohemia Express'), four pairs to Bremerhaven-Nordhafen and three pairs to Hamburg. Between July 1999 and January 2002, ERS launched three continental container feeder services to and from Praha-Žižkov (now Mělník-Labe), serving: Bratislava-UNS (three times weekly from July 1999); Budapest-Józsefváros (twice weekly from November 2001); and Bucuresti-Sud (weekly from January 2002). In March 2004 ERS introduced a new container inland shuttle between Mělník-Labe and Sládkovičovo, Slovakia. The company also uses private container terminals at Uherský Brod, Kopřivnice and Plzeň.

In a bid to boost traffic, the Czech government in 2000 provided a grant of Kcs443.4 million (Kcs250 million in the previous year) to support the development of intermodal transport. The grant was mainly used to fund the purchase of type Sgnss container wagons (see *Traction and rolling stock* section), of which around 500 have been introduced since 1996. Part of the grant was also used to support *Rollende Ländstrasse* piggyback services.

An agreement was signed on 12 June 2001 by Czech transport minister Jaromír Schling and Russia's minister of railways, Nikolai Aksyonenko, covering the construction of a multi-modal international logistics and trans-shipment terminal at Bohumín-Vrbice in northern Moravia. Valued at US$200 million, the project also involved the construction of a new 70 km 1,520 mm gauge extension to the existing Polish State Railways LHS broad-gauge line built in the 1970s and running 394 km from the Ukrainian border at Hrubieszów to Slawków, near Katowice.

The Israel-based company, Shiran General Trade AG, which was to provide 30 per cent of the total cost of the scheme, expected to reduce journey times by two weeks between China, Japan and other countries in Asia and Central Europe by using the Trans-Siberian rail land-bridge and to cut transit costs by 20 per cent. The lowest traffic forecasts for the Bohumín terminal were for 100,000 containers annually and approximately 15 million tonnes of

raw materials, mostly iron ore. The reluctance of Poland to support a new multimodal terminal on the Czech side, however, led to a re-examination of these plans and in September 2002 it was agreed that the main terminal would be built in Slawków, Poland, with Bohumín now featuring as an ancillary standard gauge terminal. It was also agreed that the project would be guaranteed exclusively by the Czech and Polish governments.

As a result of the Czech Republic's entry to the EU in May 2004, eliminating border delays and generally simplifying road movements, the country's only piggyback service between Lovosice and Dresden, Germany, came to an end in June 2004 due to poor loadings.

Traction and rolling stock
At the start of 2004 ČD operated 938 electric locomotives (564 DC; 247 AC; 145 dual-voltage), 1,373 diesel locomotives, 791 diesel railcars, 95 emus, 4,675 passenger cars and 35,615 freight wagons.

In September 2002 ČD officially joined the Eurofima rolling stock leasing agency, when it bought a stake of 0.5 per cent for €15 million. In December 2003 the railway's stake in the agency was increased to 1 per cent. This would allow access in 2005 to €45 million, €16.12 million of which would assist the purchase of Class Ampz/Bmz 200 km/h coaches, €13.22 million for the procurement of Class 471 suburban emus and €15.66 million for Class 380 three-voltage electric locomotives (see below).

Pendolinos ordered
In August 1995, ČD ordered 10 seven-car Class 680 tilting trains for use on the Berlin-Prague-Vienna route over ČD's upgraded 160 km/h Corridor 1; with the new trains, the German and Austrian capitals will be 6 hours 35 minutes apart, compared to today's best timing of 9 hours 24 minutes covered by EC 172/173 with nine intermediate stops. The Kcs4.89 billion financial package for the acquisition of the trains was put together by a consortium comprising ČSOB of the Czech Republic, Creditanstalt Finanziaria Milano of Italy, Kreditanstalt für Wiederaufbau Frankfurt of Germany, Česká spořitelna, Creditanstalt Bankverein Wien, and Creditanstalt Praha.

In 1998, ČD cut the order from 10 sets to seven as a result of increased manufacturing costs attributed to unpredicted construction problems, inflation and devaluation of the Czech koruna.

The trains are based on the ETR470 Pendolino trains in use in Italy. The Czech trains are the world's first three-voltage tilting trains, capable of operating on 25 kV 50 Hz AC, 15 kV 16⅔ Hz AC and 3 kV DC. Total output is 3,920 kW. The trains are fitted with both radio communications equipment and the GSM-R digital communications system. They are also equipped with ETCS Level 2 signalling and train control equipment, together with interface modules to work with signalling on ČD, DB and ÖBB networks. The active tilt system will tilt the cars at up to eight degrees.

Initially, construction was to be undertaken by a consortium led by ČKD Dopravni systémy, with final

Class 371 160 km/h dual-voltage electric locomotive for use on Prague-Dresden international passenger services, stabled at Dresden (Colin Boocock)   *NEW*/1114972

Principal diesel locomotives

Class	Wheel arrangement	Power kW	Speed km/h	Weight tonnes	No in service (2005)	First built	Mechanical	Builders Engine	Transmission
701	B	147	40	22	6	1957	ČKD	Tatra 930–51	M ČKD
702	B	147	40	24	10	1968	ZTS Martin	Tatra 930–51	M ČKD
703	B	170	40	24	24	1969	ZTS Martin	Tatra 930–51	HM ČKD
704	B	250	65	28	20	1988	ČKD	Liaz M2-650	E ČKD
710	C	302	60/30	40	2	1961	ČKD	ČKD 12V 170 DR	HD ČKD
714/714.2[1]	Bo-Bo	520/600	80	64/60	56	1992	ČKD	Liaz 6Z 135 M1.2CT	E ČKD
720	Bo-Bo	551	60	61	11	1958	ČKD	ČKD 6S 310 DR	E ČKD
726	B-B	625	70/35	56.6	1	1963	ZTS Martin	ČKD K12 170 DR	HD ČKD
730	Bo-Bo	600	80	69.5	17	1978	ČKD	ČKD K6S 230 DR	E ČKD
721	Bo-Bo	551	80	74	13	1963	ČKD	ČKD K6S 310 DR	E ČKD
735	Bo-Bo	926	90	64	46	1973	ZTS Martin	Pielstick 12PA 4185	E ČKD
742	Bo-Bo	883	90	64	352	1977	ČKD	ČKD K6S 230 DR	E ČKD
751	Bo-Bo	1,102	100	75	84	1964	ČKD	ČKD K6S 310 DR	E ČKD
751.3[4]	Bo-Bo	1,102	100	74	15	1969	ČKD	ČKD K6S 310 DR	E ČKD
752[5]	Bo-Bo	1,213	100	71	1	1996	ČKD	ČKD K6S 310 DR	E ČKD
753	Bo-Bo	1,325	100	73.2	26	1968	ČKD	ČKD K12V 230 DR	E ČKD
750[2]	Bo-Bo	1,325	100	72	81	1991	ČKD	ČKD K12V 230 DR	E ČKD
754	Bo-Bo	1,460	100	74.4	58	1975	ČKD	ČKD K12V 230 DR	E ČKD
770	Co-Co	993	90	114.6	19	1963	ČKD, SMZ	ČKD K6S 310 DR	E ČKD
771	Co-Co	993	90	115.8	43	1968	SMZ Dubnica	ČKD K6S 310 DR	E ČKD
781	Co-Co	1,472	100	116	18	1966	KMZ Lugansk	VSZ 14 D 40	E CHZE
731	Bo-Bo	880	70	72	51	1988	ČKD	ČKD K6S 230 DR	E ČKD
749[3]	Bo-Bo	1,102	100	75	58	1992	ČKD	ČKD K6S 310 DR	E ČKD
743	Bo-Bo	800	90	66	10	1987	ČKD	K6S 230 DR	E ČKD
708	B	300	80	34	13	1995	ČKD	Liaz M1.2C M640D	E ČKD
799[6]	B	37	10	22	41	1992	ČKD	Zetor 5301	Battery/DE ČKD

[1] Rebuilt from Class 735.   [2] Rebuilt from Class 753.   [3] Rebuilt from Class 751 and 752.   [4] Ex-Class 752.
[5] Rebuilt from Class 751, with electronic control.   [6] Rebuilt from Class 700–703.

Diesel railcars or multiple-units

Class	Motored axles/car	Power/motor kW	Speed km/h	No in service (2005)	First built	Mechanical	Builders Engine	Transmission
809/810/811	1	156	80	516	1975	Studénka	Liaz ML634	HM/Praga
812*	1	242	80	1	2001	Pars Sumperk	Liaz M640SE	HM Voith DIWA
820	2	206	70	7	1963	Studénka	Tatra T 930–4	HD ČKD
830	2	301	90	27	1949	Studénka	ČKD 12V 170 DR	E ČKD
831**	2	308	90	35	1952	Studénka	6L 150 PV-3	E ČKD
842	2	408	100	37	1988	Studénka	Liaz ML640F	HM Allison 4TB 741R
843	2	600	110	31	1995	Studénka	Liaz M1.2C-ML640D	E ČKD
850	2	515	110	15	1962	Studénka	ČKD 12V 170 DR	HD ČKD
851	2	588	110	16	1967	Studénka	ČKD 12V 170 DR	HD ČKD
852	2	588	120	9	1968	Studénka	ČKD 12V 170 DR	HD ČKD
853	2	588	120	9	1969	Studénka	ČKD 12V 170 DR	HD ČKD
854†	2	596	120	32	1997	Studénka	Caterpillar 3412	HD ČKD
860	4	442	100	2	1975	Studénka	6PA 4 H 185	E ČKD

* Refurbished Class 810.   ** Re-engined Class 830 (1981–91).   † Rebuilt Class 852 and 853.

assembly taking place at ČKD's plant at Prague-Zličín for delivery in 1997. In September 1999, however, ČD expelled ČKD for financial reasons. In January 2000 it was finally agreed that construction, including final assembly, would be taken over by Alstom (formerly Fiat Ferroviaria) at its Savigliano plant in Italy, with traction equipment supplied by Siemens. In October 2000 the contract was ratified by the consortium of banks funding the trains' purchase. A Czech local content proportion of 47 per cent in the project has been achieved, with ČKD Vagónka, Lekov Blovice, Škoda Trakční motory and VÚKV Praha participating. Seats are also supplied by a domestic manufacturer, Borcad.

The prototype unit was rolled out of the ALSTOM Ferroviaria plant in Savigliano in March 2003. An official presentation took place in Prague in June of the same year. However, continuing problems with international type approvals and other difficulties forced ČD to suspend further deliveries in September 2003. This resulted in a delay to commissioning trials, with acceptance trials on ÖBB and DB Netz still awaited in early 2005. A resumption of deliveries took place in early 2004, with completion expected during 2005. ČD plans to deploy the seven-strong fleet in full international commercial service from the December 2005 timetable change, some 10 years after the initial order for the type was placed. Trial revenue operations with one set commenced in December 2004 between Prague and Děčín, but continuing problems with electromagnetic compatibility still prevented Class 680 from operating over the part of ČD's Corridor 1 south of Svitavy electrified at 25 kV AC 50 Hz. Technical modifications to the trains' electrical systems were expected to resolve this by mid-2005.

On 18 November 2004 unit 680.001 established a Czech national speed record of 237 km/h over a 25 kV section of the Brno–Břeclav route between Zaječí and Rakvice.

Locomotives
In March 2004 ČD awarded Škoda Dopravní Technika (ŠDT) a contract to supply 20 Class 380 three-voltage (25 kV 50 Hz AC, 15 kV 16⅔ Hz AC and 3 kV DC) electric locomotives. The 6,050 kW 86 tonne Bo-Bo machines with a maximum speed of 200 km/h are intended for heavy freight and main line passenger services on both the domestic network and into neighbouring Slovakia, Austria, Germany, Hungary and Poland. Traction equipment features asynchronous three-phase motors, IGBT modules and electronic control. The first example is expected to emerge in 2007, with the whole fleet available by 2009. The value of the contract is Kcs2.7 billion, which will be partly funded by Eurofima.

ČD has identified a need to refurbish a large number of electric locomotives to adapt them for higher speeds, resulting in a programme to upgrade some principal types for 160 km/h operation, the planned maximum speed in main corridors.

In 1992 a Class 150 3 kV DC 4,000 kW locomotive was refurbished for 160 km/h operation and later redesignated Class 151. Series modernisation started in 1995 and by January 2003, 13 locomotives of the 26-strong fleet had been treated. Modifications include upgraded bogies, spring suspension and train protection equipment.

A similar programme is being implemented for the dual-voltage (3 kV DC/15 kV 16⅔ Hz) 3,060 kW Class 372 machines, which will become Class 371. These share Prague–Dresden services with Deutsche Bahn's similar Class 180 locomotives. A prototype conversion in 1996 was followed by two more in 1997. An additional three locomotives had been modified by June 2001 at a cost of Kcs53 million, with a seventh conversion scheduled for 2002. From the 2000–01 timetable, Class 371 locomotives took over haulage of all EuroCity trains on the Prague-Dresden route. Reconstruction is carried out by Škoda Plzeň together with ČMŽO Přerov. In mid-2003 ČD took delivery of one Class 180 electric locomotive from Railion and designated it Class 371.2. With large-volume deliveries of modern electric locomotives under way to Railion, it is believed the latter's entire fleet of Class 180 machines will be sold to ČD.

Rolled out during 2003 was the first of three former Class 240 25 kV AC electric locomotives converted also to operate under the 15 kV 16⅔ ½Hz AC system as dual-voltage machines to handle cross-border services on the newly electrified line between Horní Dvořiště and Summerau, Austria. Designated Class 340, conversion was taking place at ČD's České Budějovice depot in close collaboration with Škoda Plzeň. The two remaining locomotives followed during 2004. They share cross-border duties with ÖBB Class 1116 dual-voltage machines that have been working regularly to Horní Dvořiště on fast services since the 2002–03 timetable.

ČD has embarked upon a programme to rationalise and modernise its diesel fleet, disposing of obsolete designs and locomotives made redundant by falling freight traffic. Older shunting types have been replaced by new Classes 704, 708, 714/714.2 and 731 locomotives. The once-large fleet Class 781 Soviet-built diesel-electric locomotives were finally withdrawn in 2002 and Class 753 was due to follow in 2003.

Recent modernisation programmes have included the refurbishment in the period 1991–94

Prototype Russian-built two-car dmu for regional lines, seen at Kúty, Slovakia, while on delivery to ČD in 2003 (Quintus Vosman)

0569063

of 119 Class 753 diesel locomotives to create Class 750, and a similar modernisation of 60 Class 751 locomotives into a new Class 749 between 1992 and 1995. Steam heat equipment in these machines has been replaced by electric heating.

Refurbishment of Slovak-built Class 735 locomotives into a new Class 714 began in 1992, with 29 Class 714 and 23 Class 714.2 examples returned to service by the end of 1997. A notable feature of these machines is dynamic braking rated at 1,020 kW. In late 2004/early 2005 ČD took delivery of the remaining eight Class 714.2 machines of an order placed in 1995 for 40 refurbishments. Although a few locomotives had been almost completed in 1997, the contract was later terminated due to the bankruptcy of the builder, ČKD DS. Final assembly of the incomplete locomotives is being undertaken by ŽOS Česká Třebová.

Two prototypes of a new two-axle 300 kW diesel-electric locomotive design for local services, the Class 708, emerged in the second half of 1995. They are fitted with a Type M1.2C engine from Liaz and a 472 kW electric brake. Ten were manufactured in 1996, entering ČD service in February 1997.

Multiple-units and railcars

Electric multiple-units form the most obsolete part of ČD's traction pool, with 83 per cent of the fleet considered life-expired in 2001.

In September 1995, ČD placed a Kcs2.3 billion order for 10 3 kV DC emus for Prague suburban services to replace Class 451 and 452 units. Designated Class 471, these double-deck units with aluminium bodyshells are manufactured by ČKD Vagónka. Power bogies and electrical equipment, including IGBT control systems, are produced by Škoda Plzeň.

Four two-car (Class 471 power car plus Class 971 driving trailer) and six three-car (Class 471 plus Class 071 trailer plus Class 971) units have been ordered, although this was later changed to all three-car sets, and these can be coupled into nine- or 12-car formations. Both versions are rated at 2,000 kW, with an electric brake capacity of 1,700 kW.

The first power car was completed in 1997, and by the end of 1998, one two-car unit and one three-car set were completed. They were put into revenue operation on the Prague–Kolín–Pardubice line in late July 2000. Credit secured in August 2000 enabled the purchase of three more three-car units by the end of 2000 for Kcs519 million. Some slippage in delivery of the vehicles occured due to the transfer of ČKD Vagónka's plant to Vítkovice. On 20 August 2001, the first three-car unit was rolled out of the new workshops and by the end of 2001 two additional vehicles had followed. Five three-car units were delivered in 2002–03 under a Kcs930 million state-guaranteed credit secured in December 2001. In 2004 two further emus were delivered and a sixteenth vehicle was handed over in January 2005. ČD is highly satisfied with Class 471, although it can only purchase a few each year due to limited funds. Another 14 emus of this type are on order for delivery by 2010 to bring the total fleet to 30.

In May 2002 ČD placed a Kcs2.5 billion contract for five six-car dual-voltage (3 kV DC/25 kV AC 50 Hz) double-deck emus designated Class 675. With a top speed of 160 km/h, they will be employed on Prague–České Budějovice and Brno–Ostrava intercity routes. With 84 first class and 466 second class seats, a prototype should emerge in 2006. Škoda Plzeň is responsible for bogies and electrical equipment.

In 1997 the private sector maintenance company Pars Nova Šumperk completed the refurbishment of a Class 853 diesel-hydraulic railcar, which became the new Class 854. Modernisation work featured the installation of a new Caterpillar diesel engine, a new Intelo Lokel control system, new seats, thermal windows and upgraded toilets. Revenue operation with the prototype began in January 1998. In March 1999, a second railcar was similarly treated. The excellent operational results led ČD in late 2000 to order serial refurbishment of its Class 852/853 vehicles, starting in 2000. By the end of 2004 a total of 32 railcars had been modernised in three batches, becoming Class 854, with a further 10 modernisations ordered for 2005.

Principal electric locomotives

Class	Wheel arrangement	Line voltage	Output kW continuous/ one hour	Speed km/h	Weight tonnes	No in service (2005)	First built	Builders Mechanical	Electrical
100	Bo-Bo	1.5 kV	360/440	50	48	2	1956	Škoda	Škoda
113	Bo-Bo	1.5 kV	400/960	50	64	4	1973	Škoda	Škoda
110	Bo-Bo	3 kV	800/960	80	72	28	1971	Škoda	Škoda
111	Bo-Bo	3 kV	800/880	80	72	34	1981	Škoda	Škoda/ČKD
121	Bo-Bo	3 kV	2,032/2,344	90	88	28	1960	Škoda	Škoda
122	Bo-Bo	3 kV	1,990/2,340	90	85	49	1967	Škoda	Škoda
123	Bo-Bo	3 kV	1,990/2,340	90	85	29	1971	Škoda	Škoda
130	Bo-Bo	3 kV	2,040/2,340	100	86.8	41	1977	Škoda	Škoda
140	Bo-Bo	3 kV	2,032/2,344	120	82	5	1953	Škoda	Škoda-Sécheron
141	Bo-Bo	3 kV	2,032/2,344	120	84	22	1957	Škoda	Škoda
150	Bo-Bo	3 kV	4,000/4,200	140	82	13	1978	Škoda	Škoda
151[1]	Bo-Bo	3 kV	4,000/4,200	160	82	13	1978	Škoda	Škoda
163	Bo-Bo	3 kV	3,060/3,400	120	85	88	1984	Škoda	Škoda/ČKD
163.2[2]	Bo-Bo	3 kV	3,060/3,480	120	85	14	1991	Škoda	Škoda/ČKD
162	Bo-Bo	3 kV	3,060/3,480	140	85	24	1991	Škoda	Škoda/ČKD
181	Co-Co	3 kV	2,790/2,890	90	120	69	1961	Škoda	Škoda
182	Co-Co	3 kV	2,790/2,890	90	120	36	1963	Škoda	Škoda
210	Bo-Bo	25 kV	880/984	80	72	37	1973	Škoda	Škoda
230	Bo-Bo	25 kV	3,080/3,200	110	88	84	1966	Škoda	Škoda
242.2	Bo-Bo	25 kV	3,080/3,200	120	84	82	1975	Škoda	Škoda
240	Bo-Bo	25 kV	3,080/3,200	120	85	30	1968	Škoda	Škoda
263	Bo-Bo	25 kV	2,930/3,060	120	85	2	1984	Škoda	Škoda/ČKD
340[5]	Bo-Bo	25 kV/15 kV	3,080/3,200	120	86	3	2003	Škoda	Škoda
362[3]	Bo-Bo	3 kV/25 kV	3,060/3,400	140	87	14	1980	Škoda	Škoda/ČKD
363	Bo-Bo	3 kV/25 kV	3,060/3,400	120	87	116	1980	Škoda	Škoda/ČKD
371[4]	Bo-Bo	3 kV/15 kV	3,060/3,400	160	84	7	1997	Škoda	Škoda/ČKD
372	Bo-Bo	3 kV/15 kV	3,060/3,400	120	84	9	1991	Škoda	Škoda/ČKD

[1] Rebuilt 1998 from Class 110. [2] Reconstructed from Class 150 for 160 km/h operation 1994–2002.
[3] Bogies exchanged 1993–2000. [4] Class 372 rebuilt for 160 km/h operation (includes 371.201 ex-Railion 180.001).
[5] Class 240 rebuilt 2003–04 to dual-voltage for operation into Austria.

Electric multiple-units

Class	Cars per unit	Line voltage	Motor cars per unit	Motored axles/car	Output/ motor kW	Speed km/h	No in service (2005)	First built	Builders Mechanical	Electrical
451	4	3 kV	2	4	165/190	100	36	1961	Studénka	MEZ
452	4	3 kV	2	4	165/190	100	9	1974	Studénka	MEZ
460	5	3 kV	2	4	250/270	110	25	1974	Studénka	MEZ
470	5	3 kV	2	4	240/260	120	2	1990	Studénka	MEZ
560	5	25 kV	2	4	420/465	110	9	1966	Studénka	MEZ
471	3	3 kV	1	4	500	140	15	1997	Studénka	Škoda
680	7	3 kV/25 kV/ 15 kV	4	2	490	230	3	2003	ALSTOM	ALSTOM, Siemens

ČD foresees similar refurbishment of the entire fleet of Class 853 railcars.

In September 2001, Pars Nova Šumperk completed the refurbishment of a Class 810 two-axle diesel-mechanical railcar to become new Class 812. The vehicle is powered by a new Liaz M640SE engine rated at 240 kW with Voith DIWA hydro-mechanical transmission. Main structural features are new front-ends of aluminium and a strengthened bodyshell. Other new features include Intelo automatic train control equipment, electro-pneumatic plug doors, driver's cab air conditioning and new seats and windows. Multiple-unit operation is possible with railcars of Classes 811, 843 and 854.

In September 2002 the same manufacturer rolled out a prototype of the Class 912 part-low-floor driving car, which is to form train compositions with Class 812 railcars. Rebuilt from a Class 010 (Baafx) non-powered trailer, the 21 tonne 80 km/h vehicle has laminated front-ends, electro-pneumatic plug doors, dethermal windows, wheelchair-accessible toilets, new seats and driver's cab air conditioning. Intelo-Lokel microprocessor control is installed. Regular operation with the prototype began in February 2002.

Class 812/912 will become the basis for new Class 814/914 permanently coupled two-car units rebuilt from Class 810 railcars and Class 010 trailers. A total of 100 such vehicles should be rebuilt in 2006–09 for Kcs2 billion, providing more comfortable and efficient rolling stock for ČD's numerous regional lines. The part low-floor dmus will be equipped with TEDOM (Liaz) Type ML640SE-Euro II diesel engines rated at 242 kW, new interiors and upgraded heating. There will be 85 seats in each vehicle and its maximum speed will be 90 km/h. A contract for a prototype unit was placed with Pars Nova Šumperk in January 2005 for delivery in November of the same year.

In August 2003 the Russian rolling stock builder Metrovagonmash Mytishchi delivered a prototype two-car diesel-hydraulic railcar designated Type RA-731. Based on the Class 6341 dmus supplied to MÁV, Hungary, and built in close collaboration with Czech subcontractors, the stainless steel-bodied railcar was procured by a private company, Elektromechanika Úvaly, with the aim of selling it to ČD. It is powered by a pair of MTU Type 6R183 TD13H engines rated at 315 kW each, through a hydrodynamic Voith transmission. The 80 tonne vehicle has 120 seats and a maximum speed of 120 km/h. Designated Class 835, the prototype commenced commissioning trials in December 2003 and in February 2005 trial revenue operation commenced on lines in southern Moravia.

ČD had initially considered procuring up to 40 of these vehicles as a method of clearing Russian debt to the country but this caused a dispute with the government regarding the selling price. ČD now inclines towards the widespread use of more contemporary low-floor designs such as Siemens' Desiro.

Passenger coaches

At the end of 2004 ČD's passenger vehicle fleet stood at 4,447, excluding non-powered railcar trailers. Of these, only 461 were capable of 160 km/h operation and 45 for 200 km/h.

In 1995, ČD ordered 45 new 200 km/h coaches from a consortium of MSV Studénka and Siemens Austrian subsidiary SGP for EC and IC services, mainly over the Corridor 1 Berlin-Prague-Vienna artery. The Kcs2 billion order comprised 26 Type Bmz second-class and nine Type Ampz first-class coaches, along with 10 Type WRRmz (now Type WRmz) dining cars. MSV was responsible for manufacture of bodyshells, underframes and seats. Final assembly took place at the Siemens SGP plant in Vienna. The first Type WRmz car was rolled out in September 1997, with all completed by the year-end. The first two Type Ampz cars were completed in December 1997 and the remainder were operational by October 1999. Production of Type Bmz second class coaches began in 1998. The whole fleet of 26 coaches had been handed over to

ČD by the end of 2000. Financing was arranged by a consortium of domestic banks, led by Konsolidační Banka.

As a follow-on contract, in 2003 ČD ordered a further 26 200 km/h coaches for delivery in 2005–06. Built by a consortium of Siemens and ČKD Vagónka, the contract is for 11 Type Ampz first class saloon-type and 15 Type Bmz second class compartment-type, all with pressure-tight bodyshells, full air-conditioning, tinted windows, automatic plug doors and retention toilets. Final assembly will take place at the Siemens SKV plant at Prague-Zličin (formerly ČKD Transportation Systems), with 35 per cent of the work undertaken by domestic suppliers. The value of the contract is €43 million, funded through Eurofima loans. The first four Ampz and seven Bmz coaches are due for delivery in 2005.

In March 2005 ČD ordered 12 Type WLABmz 200 km/h couchette cars for international overnight services. Apart from ordinary first and second class compartments, the air-conditioned coaches will also house high-standard compartments with showers and a closed WC circuit. Delivery is scheduled for 2006–07.

During 2000, nine original Type Bmee second class coaches built in East Germany in 1986–7 were rebuilt as Type WLAB couchette cars at a cost of Kcs147 million. They are equipped with disc brakes, air conditioning, vacuum toilets and automatic plug doors to meet RIV standards for international operation on fast overnight trains from Prague to Bucharest, Kosice, Stuttgart and Warsaw.

Freight wagons

In response to changes in the rail freight market, especially a continuing decline in volumes carried of bulk products such as coal and iron ore and an increase in the transport of more sophisticated cargo, ČD has been obliged to restructure its freight wagon fleet to cope with customer needs and to operate to international (RIV) standards at speeds of up to 100 km/h. The most urgent need has been procurement of Types Sgnss, Sggmrss, Tdns, Tadns, Hbills, Hbbillnss, Habbillns and Kns.

In November 2004 ČD placed orders for 380 new freight wagons of four types worth Kcs1.2 billion. Lostr Louny was to supply 50 Type Sggmrss six-axle 120 km/h twin-unit container flats with a loading capacity of 106 tonnes and 100 Type Tadnss four-axle hoppers with sliding roof for the transport of dry bulk materials. OOS Ostrava will be responsible for 200 Type Habbillnss covered four-axle wagons with sliding side-walls and 30 Type

Zacna 95 m³ tank wagons will be supplied by Astra Vagoane Arad of Romania. Delivery of all types is due by June 2006.

Modernisation of older types is also high on ČD's agenda. By the end of 2005 OOS Ostrava will have refurbished 950 wagons of three types under a Kcs1.003 billion contract awarded in mid-2004. The order includes Type Tams four-axle wagons with tarpaulin roof for metal products, rebuilt from Type Eas-u. ČD will refurbish an additional 500 wagons of this type, following 153 treated in the first half of 2002. A second contract will see a further 150 Type Rils four-axle wagons with tarpaulin cover rebuilt from Type Res 51 flats, following an initial 83 completed in 2002. The third order involves the modernisation of 300 Type Gbgkks wagons with sectional sliding roof. Upgrading Type Falls 11 coal hoppers to become Type Falls 54 with modified Y25Rs bogies to allow 100 km/h international operation is also to proceed, following an initial batch of 200 converted in 2002.

ČD is also hiring 50 Type Shimmns four-axle steel coil wagons (built by Tatravagónka Poprad) from AAE under a six-year lease.

UPDATED

Railway Infrastructure Authority (SŽDC)

Správa železniční dopravní cesty, so
Ulice Prvního pluku 367/5, CZ-186 00 Prague 8 – Karlin
e-mail: szdc@szdc.cz
Web: www.szdc.cz

Key personnel

General Manager: Jan Komárek
Director, General Manager's Office:
  Anna Nováková
First Under-Secretary: Bohuslav Navrátil
Under-Secretary for Economy: Zita Karasová
Under-Secretary for Economy: Miroslav Konečný

Gauge: 1,435 mm; 760 mm
Route length: 9,421 km; 20.2

Electrification: 1,633 km at 3 kV DC; 24 km at 1.5 kV DC; 1,263 km at 25 kV 50 Hz AC

Political background

SŽDC was officially formed by the Ministry of Transport in September 2002 to own and manage the national network's entire infrastructure portfolio and to sell the railway's surplus, non-core property to repay ČD's accumulated debt. The property portfolio includes some potentially lucrative sites in Prague, plus track and machinery centres, repair shops, staff training centres, holiday and convalescent resorts, hospital facilities and staff quarters. At the start of 2003 SŽDC's total assets reached Kcs84 billion. SŽDC is controlled by a board of directors under the auspices of the Ministry of Transport.

Due to SŽDC's limited staff, operation, maintenance and development of the infrastructure

was initially contracted for three years to the national train operator, Czech Railways (ČD) (see entry in Czech Republic section of *Railway systems and operators*). In line with EU legislation, both ČD and private operators pay SŽDC access fees to use its infrastructure, with ČD paying Kcs1.5 billion for passenger traffic and Kcs4.6 billion for freight traffic per year.

In December 2003 SŽDC became a full member of the Community of European Railways.

New lines

A re-examination has taken place of plans formulated in 1988 for a high-speed (VRT) network. The VRT network would be some 700 km long, engineered for 250–300 km/h operation and electrified at 25 kV AC. Total cost is estimated at up to Kcs180 billion.

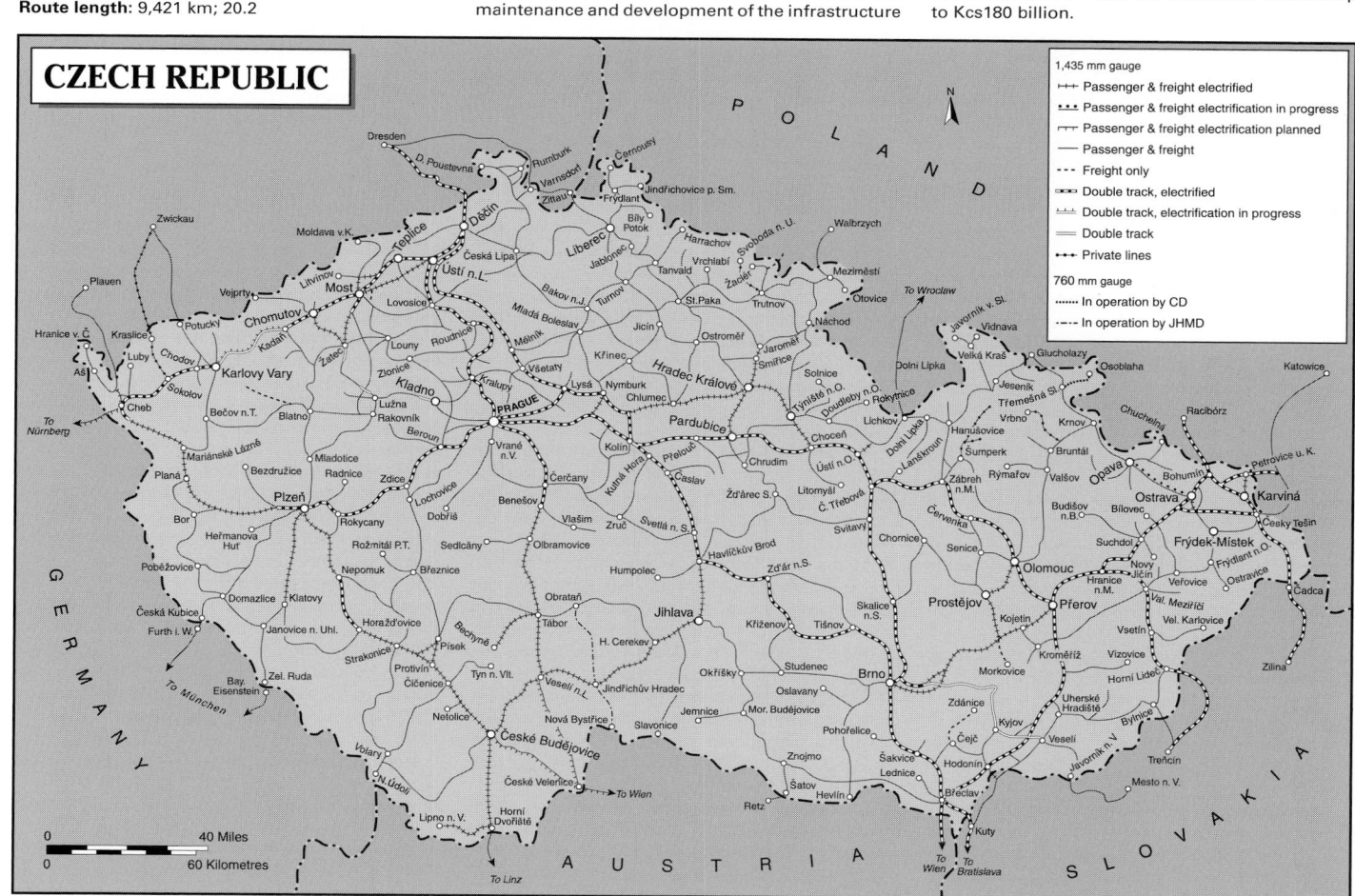

NEW/1114974

New tunnel at Veprelc on the Kralupy nad Vltavou–Vraňany section of the modernised Corridor 1
(Michal Málek)   0552779

The first section of the VRT network would be the Czech part of the Berlin–Prague–Vienna route. The Prague–Vienna via Gmünd route was one of three that was the subject of a trilateral agreement between the Czech minister of transport and his German and Austrian counterparts in July 1995, providing for construction of high-speed lines in future. The other routes in the agreement were the Prague–Nuremberg and Prague–České Budějovice–Linz lines.

At the start of 1994 a private company, PRaK, was formed to construct a rail connection between Prague and Kladno, a distance of 35.7 km. Two route models for the line were being assessed in 2002: one favouring a classic double-track electrified railway roughly following the existing route, the other a newly built S-Bahn-type light railway. Despite the project having suffered several postponements and the absence of a final decision on the nature of the line, it is assumed that construction will take place in conjunction with extensive remodelling of the capital's rail junction in 2003–08. The line would both provide commuter services and serve the international airport at Ruzyně, where a further expansion programme is under way. The government is striving hard to involve the private sector in building and financing the rail link, but no progress had been made by early 2005. Another option considered is the extension to the airport of Line A of the capital's metro, a proposal backed by the Prague municipal authorities. The cost of the project has so far soared to Kcs20-25 billion.

In March 2000 construction began on a new alignment between Březno and Chomutov on the Prague–Chomutov main line to replace existing infrastructure affected by mining subsidence. Built by Metrostav Praha, the new single-track alignment will be 7.1 km with a new station at Droužkovice. Construction will entail boring a 1.758 km tunnel, which will be the longest on the ČD network. The alignment is designed for a maximum speed of 100 km/h and it is proposed to electrify the section at 3 kV DC along with electrification of the Kadaň–Karlovy Vary main line in 2006. Since May 2003, however, construction has been delayed owing to the tunnel's sagging. The alignment is now due for completion in June 2006, with operation starting in September of the same year. The Kcs1.5 billion cost of the project is being met by the mining company active in the area.

On 15 December 2003 completion was achieved of a new alignment into Mladá Boleslav-město station, which has undergone major refurbishment. The aim of the reconstruction was to locate the existing track in a 8.5 m trench to eliminate several busy level crossings on a 1.6 km section of the Mladá Boleslav-město-Stará Paka line, which also sees freight trains serving the nearby Škoda car plant. The new alignment, signalled by Type ESA 11 remote control, necessitated a new below-ground station with two island platforms. The Kcs483 million project was funded by the Mladá Boleslav city authority, the state budget and Škoda.

Improvements to existing lines

Despite upgrading of two main international corridors having been in full swing for almost a decade (see below), rail infrastructure is still in an unsatisfactory condition and is severely under-funded. To restore track to an effective condition would require a total investment of Kcs132 billion, with a further Kcs20.5 billion needed for rehabilitation of signalling equipment and track structures. Modernisation works accounted for Kcs6.2 billion in 2003, with Kcs10.6 billion spent on track maintenance.

Corridor 1

In February 1994 the Czech government announced the formal inauguration of track renewal on Corridor 1, the country's key rail route. This forms the largest investment of the century on the national rail network, and is being carried out without interrupting regular services.

The 454 km Corridor 1 is double-track throughout and fully electrified, part at 3 kV DC and part at 25 kV AC 50 Hz. It runs from the German border at Děčín through Prague, Kolín, Česká Třebová and Brno to the Austrian border at Břeclav. It forms part of a proposed central European 160 km/h rail link between Berlin and Vienna, which after completion of upgrading should reduce the journey time between both capitals from the best in 1998 of 10 hours 16 minutes to 6 hours 30 minutes. The domestic journey time between the Czech Republic's two largest cities, Prague and Brno, will be cut from 3 hours 10 minutes to 2 hours 30 minutes. Work on the scheme started in 1993 and was completed in 2004. The project was divided into 28 sections, and incorporates

697 bridges and 14 tunnels, giving a total of 388.6 route-km renewed.

The Czech government approved a funding scheme for Corridor 1 in November 1994, with Kcs9 billion coming directly from the budget and Kcs15.43 billion to be paid from credits guaranteed by the government. Finance for the project came in the form of credits and loans from the European Bank for Reconstruction and Development, the European Investment Bank, Japan's EXIM Bank, the Kreditanstalt für Wiederaufbau in Germany, and non-guaranteed loans from the domestic Česká spořitelna and Komerční banka institutions. However, it subsequently become clear that the cost of upgrading would rise by 48 per cent to Kcs36.5 billion due to environmental demands and inflation, and an additional Kcs1.57 billion of funding was approved in December 2001.

A Czech-German memorandum was signed on 1 March 1996 to promote modernisation of the Prague–Berlin section of Corridor 1. A similar Czech-Slovak memorandum covering the important Prague–Bratislava axis was signed on 28 October 2000. By 2006 the entire route should be upgraded to 160 km/h for passenger services and 120 km/h for freight operations, with the best journey time cut from the current 4 hours 19 minutes to 3 hours 48 minutes.

The final two sections of Corridor 1, Ústí nad Orlici–Česká Třebová (6.3 km) and Záboří nad Labem–Přelouč (18.3 km) were completed in March and October 2004 respectively. These were partly funded through the EU's ISPA programme, which provided €14.3 million and €45 million respectively. Formal completion of these works took place during a ceremony at Přelouč on 12 October 2004.

In fact, this did not signal final completion of Corridor 1, as modernisation of tunnels at Děčín and Nelahozeves (totalling 4.1 route km) and of the Brandýs–Ústí nad Orlici section (10.3 km) has been postponed for financial reasons. Also to be upgraded is track between Prague Bubeneč and Úvaly (28.5 km), which forms part of an extensive 'Nové Spojení' remodelling programme for the capital's rail system (see below).

Outside Corridor 1 proper, modernisation of eight major stations along the route is scheduled for completion in 2015–29. The stations are those at Břeclav, Česká Třebová, Chocheň, Děčín, Kolín, Pardubice, Ústí nad Labem and Ústí nad Orlici. The border station at Děčín was treated first between 2001 and 2003 at a cost of Kcs1.45 billion. Work at Chocheň began in 2003 for completion in May 2005 at a cost of Kcs950 million.

Corridor 2

In August 1995 the government announced the go-ahead for a similar upgrade of the then ČD's Corridor 2, which forms part of European rail corridor E65 from Gdynia to Rijeka. In the Czech Republic, this links Břeclav–Přerov–Ostrava–Petrovice u Karviné with a branch from Přerov to Česká Třebova, a total of 323 km. Top speed will be 160 km/h.

Of ŠZDC's four corridors, this north-south link is regarded as the most lucrative, accounting for some 40 per cent of freight transit traffic across the country and 22 per cent of the railway's total revenue, but with competition from Slovak Republic Railways' parallel Bratislava–Trnava–Žilina–Čadca route. The route received European Union support at the Pan-European Transport Conference in Helsinki in June 1997, and the same month also saw the signing of an agreement between the Czech Republic and Poland covering mutual assistance in developing the corridor.

By 2004 the core Corridor 2 route of 215 km, which forms part of the international Vienna-Warsaw route, will have been upgraded for 160 km/h running, while upgrading the 108 km branch between Přerov and Česká Třebová, which provides a link with Corridor 1, will be completed by 2007. The total length of route-km renewed will be 299.1 km.

The government approved financing of the corridor in March 1996. The cost of the project is now expected to be Kcs36.9 billion, compared with an original estimate of Kcs25 billion. Funding arrangements will be similar to those adopted for Corridor 1, with Kcs18.9 billion coming directly from the government. A further Kcs11.5 billion will be provided by government-guaranteed credits (from the European Investment Bank and Germany's Kreditanstalt für Wiederaufbau) and Kcs6.5 billion by non-guaranteed credits (from Česká spořitelna, Deutsche Bank, the European Bank for Reconstruction and Development, and the Kuwait Fund for Arab Economic Development).

The project is divided into 14 sections. Work on the first section, the 20.4 km between Hodonín and Moravský Písek, began in September 1997. Completion was achieved in December 1999. Work on a second section, Břeclav–Hodonín (20 km), was completed in November 2000. During 1999 work started on three sections: Otrokovice–Huštěnovice (11.8 km, completed January 2001); Moravský Písek–Huštěnovice (18.5 km, completed November 2001); and Přerov–Hranice na Moravě (28.2 km, completed in October 2002).

In June 2000 work started on the 24.1 km Otrokovice–Přerov section. This was completed in December 2002. In 2001 work on the remaining three sections to the Polish border was started: Ostrava–Petrovice u Karviné (23.8 km), completed in November 2002; Studénka–Ostrava (23.4 km), completed in December 2003; and Hranice nad Morave–Studénka (32.5 km), completed in June 2004.

Modernisation of the 108 km branch between Přerov and Česká Třebová is divided into five sections, with work on the first, Olomouc–Červenka (18.3 km) starting in May 2002 and completed in September 2003. In August 2002 work began on a second section between Česká Třebová and Krasíkov (22 km), which involves a new 1.789 km alignment with three tunnels, costed at Kcs4 billion. Completion was in 2004. Work on another expensive section, between Krasíkov and Zábřeh na Moravě (14.6 km), began in April 2004. This incorporates two new cut-offs of 1.8 km and 2.6 km and three tunnels. Completion is due in 2006 at a cost of Kcs3.6 billion, of which €72.8 million comes from the EU's ISPA programme. This is the Czech Republic's largest infrastructure project to be supported by EU funding.

In May 1998 ČD, Austrian Federal Railways and Polish State Railways signed a Memorandum of Understanding on modernisation of the Vienna–Břeclav–Ostrava–Warsaw route. By 2005, a total of €1.58 billion is to be invested in upgrading the 677 km route, cutting 2 hours from the present best Vienna–Warsaw journey time of 7 hours 48 minutes. The current 2 hours 30 minutes domestic journey time between Ostrava and Břeclav (184 km) will be cut by 45–60 minutes.

Corridors 3 and 4

It is estimated that work on the Cheb–Prague (220 km) and Dětmarovice–Mosty u Jablunkova–Slovak border (56 km) sections of the west-east Corridor 3 could start in mid-2004, with completion expected in 2010–14. The government approved modernisation of Corridor 3 in June 2002, at an estimated cost of Kcs56 billion. The section between Plzeň and Cheb in western Bohemia is divided into eight sub-sections, with work on the first partly double-track section, Plzeň–Stříbro (30.1 km), due to start in 2005. This scheme involves a 7 km alignment with two tunnels. Upgrading the entire Plzeň–Cheb section has been costed at Kcs7 billion.

Work on Corridor 4, linking Děčín–Prague–České Budějovice–Horní Dvořiště/České Velenice, was stated as a priority by the government in July 1999 and finally approved in December 2001. Work is to begin in 2005, with completion expected around 2010 at a cost of Kcs38.6 billion, generated from the state budget, EU funds and private loans. Modernisation will reduce by 30 minutes the present best timings for conventional trains of 2 hours 20 minutes between Prague and České Budějovice.

After completion of the Ceské Budejovice–Horní Dvoriste electrification in June 2001 (see 'Electrification'), Corridor 4 between Prague and Ceské Budejovice is completely under the wires but only 26 per cent of the route is double-track. With feasibility studies completed, modernisation is divided into five sections: Horní Dvoriste–Ceské Budejovice (58 km); Ceské Budejovice–Veselí nad Luznicí (39 km); Veselí nad Luznicí–Tábor (27 km); Tábor–Benesov (54 km); and Benesov–Praha–Hostivar (49 km). Unlike Corridors 1 and 2, the project will involve both upgrading of the existing line and the construction of seven new cut-offs (Nemanice–Sevetín; Horusice–Veselí nad Luznicí; Soběslav–Doubí; Sudomerice–Hermanicky; Chotoviny–Sudomerive; Votoce–Olbramovice; and Olbramovice–Tomice). Commencement of work on the first subsection, Prague Hostivař–Strančice, was due in early 2005, followed by the Strančice–Benešov section in November 2005.

Prague 'Nové Spojení' (New Connection)

In December 2004 reconstruction of the Seifert bridges at Prague main station was completed. This formed the first stage of the modernisation of the capitals rail junction lines between the major stations at Hlavní nádraží (main station), Masrykovo nádraží (Masaryk station), Libeň Vysoč any and Holešovice. Entitled 'Nové Spojení (New Connection), the project will have a fundamental influence on improving rail transport in the Czech capital and will establish a common connection between Corridors 1, 3 and 4, with similar technical and safety standards.

Construction will entail laying 28.5 km of new, partly covered track in difficult terrain, 43 km of optical cable, 29 km of track cable and 16 km of 3 kV DC overhead power supply. The main feature of the project will be two double-track tunnels under Vítkov Hill totalling 2.68 km and the construction of four flyovers. Total cost of this long-term project is Kcs7.85 billion, which will be met by state funds. Undertaken by a consortium of Skanska ŽS, SSŽ, Metrostav and Subterra, work formally began in August 2004 with completion due in December 2010.

Brno station modernisation

An ambitious scheme valued at Kcs29 billion has been drawn up for the relocation of the main station in Brno. Due for completion in 2014–17, this would see the station relocated 800 metres from the city centre to a new site at Zvonařka, where a park-and-ride area would be provided. Dress & Sommer AG of Stuttgart was appointed in September 2003 to co-ordinate the preparatory stage of this scheme.

Signalling and telecommunications

By the end of 2003 automatic block signalling had been installed on 2,789 route-km of the SŽDC system. Between 1995 and 2002, 3,463 route-km were equipped with train radio at a cost of Kcs950 million.

During 2004 the stations at Poděbrady and Ražice were equipped with Type ESA 11 modern electronic interlocking equipment supplied by AŽD Praha at a combined cost of Kcs221 million. This equipment is now installed at 61 stations throughout the network.

After disputes with a rival supplier that delayed the project for three years, Kapsch CarrierCom AG was finally awarded a contract for a pilot installation of GSM-R wireless communications technology over the 201 km section of Corridor 1 between Děčín and Kolín via Orague in May 2004. The project is worth Kcs255 million but the cost of installation over the entire corridor (see above) could reach up to Kcs6 billion. Installation started in December 2004 and trial operation of the firs section, Prague–Kolín, was due in June 2005. The Czech section will connect with the route from Děčín into Germany which went live in January 2005. The Czech supplier AŽD Praha is collaborating on the project.

Electrification

Electric traction is available over 32.5 per cent of SŽDC route-mileage and hauls 90 per cent of all traffic (2003). The north operates at 3 kV DC, the south at 25 kV AC. There are six junctions of the electrification systems: at Kutná Hora, 73 km east of Prague; Králův Dvůr on the Prague–Plzeň route; Nedakonice on the Přerov–Břeclav route; Benešov on the Prague–Tabor route; Nezamyslice on the Přerov–Brno route; and Svitavy on the Česká Třebová–Brno route. Local routes between Tábor and Bechyňe (24 km) and Rybník and Lipno nad Vltavou (22 km) are energised at 1.5 kV DC.

The largest single electrification project since the 1980s was the wiring of the 91 km route between Brno and Česká Trebová via Svitavy and Blansko, a project started in April 1996. Commissioned on 21 January 1999 at a cost of Kcs1.319 billion, this double-track route is electrified partly at 25 kV AC 50 Hz and partly at 3kV DC, with new substations at Blansko, Svitavy and Opatov, and completes the wiring of Corridor 1 (see 'Improvements to existing lines').

On 7 June 2001 the first electric train ran over the 58.6 km single-track route from Ceské Budejovice to the Austrian border at Horní Dvořiště. Electrification of this route at 25 kV AC 50 Hz began in December 1998 and included extensive track upgrading, installation of new signalling equipment, the construction of two substations and the modernisation of eight stations, including that at Horní Dvořiště. Some funding for the Kcs2.8 billion project was provided by the European Union's Phare programme. On 10 December 2001, the entire corridor between the Adriatic and the Baltic became electrified on completion of a scheme by Austrian Federal Railways to install overhead power supply equipment on its 5.1 km section between the border with the Czech Republic near Horní Dvořiště and Summerau. As part of this scheme, in October 2003 ŠZDC began conversion from 1.5 kV DC to 25 kV AV 50 Hz of the 22 km regional line between Rybník and Lipno nad Vlatavou in southern Bohemia. Originally electrified at 1.2 kV DC in 1913 and converted to 1.5 kV DC in 1956, this was one of the pioneering electrification projects on the network. Costing Kcs240 million, conversion is divided into two stages, with final completion due in June 2005. The line will be fed by a new substation at Lipno nad Vltavou.

In January 2004 electrification work began on one of the last diesel-worked double-track sections on the ŠZDC network, the 46.7 km line between Kadaň and Karlovy Vary in northwest Bohemia. Undertaken by Skanska and EŽ Praha, the project involves installation of a 25 kV AC 50 Hz overhead power supply, new signalling equipment and renovation of seven bridges. The line will be fed by a new substation at Kadaň. With a maximum line speed of 100 km/h, journey times will be cut by 10–12 minutes. The total cost of the scheme is estimated at Kcs2.7 billion, with completion due in June 2006.

In November 2004 pre-electrification works were launched on the 27.2 km section of single-track line between Ostrava-Svinov and Opava-Východ in northern Moravia. By October 2006 the line will be electrified at 3 kV DC at a cost of Kcs1.73 billion. Work also includes complete renewal of track and signalling, bridge maintenance, the modernisation of six stations and the construction of a new substation at Opava.

A few short sections have been proposed for electrification at 3 kV DC during the period 2005–10: Letohrad–Lichkov-Polish Border (23 km); Kunčice–Ostrava (8 km); Lysá nad Laben–Milovice (5 km); Kutná Hora–Kutná Hora Město (2 km); and Otrokovice–Zlín–Vizovice (25 km). Also under consideration is energising at the Austrian standard of 15 kV AC 16.7 Hz the Znojmo–Šatov–Retz ÖBB route (11 km), which could be partly funded by the EU, and electrification at 25 kV AC 50 Hz of the 101 km line between Brno and Jihlava via Střelice and Okříšky.

Track

Main lines are generally laid with 49 kg/m rail, secondary lines with 30 to 40 kg/m. However, almost 10 per cent of all route-km has been relaid with 65 kg/m rail, since the lines concerned, carrying freight trains of increasing weight, are recording 60 to 80 million tonne-km of traffic a year. Most of the system allows maximum axleloads of 20 tonnes.

Rail is welded in long sections, fastened to wood or concrete sleepers spaced 1,450 to 1,500 per km. New Type B-91 200 km/h sleepers have been laid on upgraded sections since 1993. On main lines, minimum curve radius is 300 m and maximum speed 160 km/h.

The Corridor 1 upgrade is being laid with UIC 60 rail supplied by Moravia Steel Třinec and Voest-Alpine on concrete sleepers from ŽPSV Uherský Ostroh, with Vossloh fastenings. UIC GC clearances are being adopted, with a maximum axleload of 22.5 tonnes.

Type of rails: S49, R65, UIC60
Sleepers: SB8, U94, B91, B915, B90
Fastenings: Vossloh, Pandrol

UPDATED

Desná Railway (ŽD)

Železniční Desná
Connex Morava as
Vítkovická 3056/2, Ostrava-Moravská Ostrava
Tel: (+420 583) 24 22 42; 21 95 39
Fax: (+420 583) 24 22 42
e-mail: desna@connexmorava.cz
Web: www.zeleznicedesna.cz

Key personnel
Chairman: Ondřej Kopp
Director, Operations: Karel Mičunek

Organisation
Železnice Desná, a federation of River Desná valley municipalities, began operation of the 9 km Šumperk–Petrov nad Desnou–Sobotín local railway in northwest Moravia on 1 May 1998.

The line forms a section of the Czech Railways (ČD) route from Šumperk to Kouty nad Desnou, where operations were interrupted in July 1997 due to extensive flood damage. Restoration of the line, which ČD had refused to operate, was undertaken by Bohumín-based private construction company Stavební obnova železnic (SOŽ) using funds from the EU's Phare programme. SOŽ was also appointed to operate the line.

The missing 13 km section between Petrov nad Desnou and Kouty nad Desnou was reopened in May 1999 at a cost of Kcs68.5 million, with the 5 km leg to Velké Losiny opened in December 1998.

From October 2002 operations were taken over by Connex Moravia. This was the first instance of a foreign-owned operator participating in the Czech rail market.

Since October 2003 ŽD has been fully integrated into the complex transport system of the Olomouc region.

Passenger and freight operations
ŽD operates 19 daily pairs of passenger services serving 14 stations and carries some 5,000 tonnes of freight each month.

Traction and rolling stock
Rolling stock comprises four Class 810 diesel-hydraulic railcars and three Class 010 trailers, leased initially from ČD. These were transferred into ŽD ownership in December 2002. Freight services are handled by SOŽ's own fleet of one Class 730 and two Class 742.5 diesel-electric locomotives.

UPDATED

Jindřichův Hradec Local Railways (JHMD)

Jindřichohradecké místní dráhy as
Nádražní 203/II, CZ-377 01 Jindřichův Hradec
Tel/Fax: (+420 384) 36 11 65
e-mail: hruska@jhmd.cz
Web: www.jhmd.cz

Key personnel
Chairman of Administrative Council and General
   Manager: Jan Šatava
Deputy General Manager and Head of Operations:
   Jan Hruška
Director, Accounting: Marta Zrzavá
Director, Traffic: Jiří Kolář
Director, Technical: Václav Hodinář

Gauge: 760 mm
Route length: 79 km

Organisation
Founded in 1994, JHMD is a private-sector rail company providing passenger and freight services on the former Czech Railways (ČD) Jindřichův Hradec–Nová Bystřice and Jindřichův Hradec–Obratář narrow-gauge lines in southern Bohemia. These were bought from ČD for a symbolic fee of Kc1 after the national operator ceased operation in January 1997 following heavy losses (see 'Political background – Privatisation' under Czech Railways).

Passenger operations
In 2001 JHMD carried 331,000 passengers for 6.23 million passenger-km, compared with 317,000 (5.87 passenger-km) in 2000. Additional journeys were recorded on steam-hauled excursions.

JHMD Class 705.9 760 mm gauge diesel-electric on display after refurbishment and re-engining (Michal Málek)   0573982

Freight operations
Small volumes of freight are carried, cotton, steel plate, scrap and timber featuring among commodities handled.

Traction and rolling stock
The JHMD fleet comprises seven ex-ČD Class 705.9 diesel-electric locomotives built by ČKD, one steam locomotive (a Class U47.0 owned by the National Technical Museum), 11 passenger coaches and

169 support bogies for carrying ČD standard-gauge wagons.

In 2003 two Class 705.9 diesel locomotives underwent refurbishment at ŽOS Česká Třebová workshops, involving the installation of a Liaz M1.2C 640S 242 kW engine and a Siemens alternator.

UPDATED

OKD Doprava

OKD Doprava, as
Nádražní 93/2967, CZ-702 62 Ostrava 1-Moravská Ostrava
Tel: (+420 59) 616 61 11; 616 62 35
Tel (Milotice–Vrbno pod Pradědem):
   (+420 554) 75 17 66
Fax: (+420 59) 611 67 48
e-mail: obchod@okd-doprava.cz; vrbno@okd-doprava.cz
Web: www.okd-doprava.cz

Key personnel
General Director: Oldřich Faiman
Head of Supervisory Board: Pavel Sokol
Operations Manager: Bohumil Bonczek
Commercial and Finance Manager:
   Miroslav Langer
Investment Manager: Otto Roháč
Economy Manager: Miroslava Vitečková
Director, Railway Transport: Luvík Simerák

Organisation
Founded in 1994, OKD Doprava is the transport arm of Northern Moravia-based mining company

OKD, and is the Czech Republic's largest private rail operator, running freight trains throughout the ČD network. It also owns more than 400 route-km of lines in the Ostrava-Karviná coal basin. In addition, the company is active in the overhaul and refurbishment of freight wagons and is certified to ISO 9002 in this field.

In February 2004 OKD Doprava set up a wholly owned subsidiary in Slovakia, ŽDD Bratislava,

with 41 employees and four diesel electric locomotives.

Passenger operations
In January 1998, OKD restored traffic on the 20.5 km rural line from Milotice to Vrbno pod Pradědem, which it leases from ČD. The company invested over Kcs30 million in rehabilitation of the line after it sustained heavy flood damage during 1997.

OKD Doprava Class 753.7 diesel-electric locomotive, re-engined with a Caterpillar power unit (Quintus Vosman)   *NEW*/1114973

Freight operations

Coal is the principal commodity carried. In 2003 the company carried 10 million tonnes of freight.

In June 2004, in collaboration with PTKiGK of Poland, OKD Doprava began operating its first international service, carrying gravel between Drahotuše in the Czech Republic and Rybnik, Poland, over the Bohumín/Chalupki border crossing.

The first international train involving ŽDD Bratislava ran between Križoviany nad Váhom and Týniště nad Orlicí (via Kúty/Lanžhot) in October 2004.

Traction and rolling stock

Passenger traffic on the Milotice–Vrbno pod Pradědem is handled by one Class 810 railcar and a Class 010 trailer leased from ČD. Freight services are operated by a fleet of 12 Class 770.5 and three Class 771 six-axle diesel-electric locomotives and a large fleet of Class 740 Bo-Bo industrial locomotives (two of which have been refurbished as Class 740.4). There are also two Class 181 3 kV DC electric locomotives (ex-ČD) and some 2,000 freight wagons, mainly Type Wap coal hoppers.

The latest acquisitions are 12 Class 753.7 Bo-Bo 100 km/h diesel-electric locomotives purchased from ČD in 2002–04 and upgraded with a 1,455 kW Caterpillar engine by ČMKS at its Česká Třebová workshops. In 2004 OKD Doprava took delivery from ČMKS of four Class 752.6 (ex-ČD Class 753) four-axle locomotives fitted with a 990 kW ČKD Type K6S 310DR engine. Two more Class 753 locomotives are leased from a private company, Šauer.

The fleet of the company's Slovakian subsidiary, ŽDD Bratislava, comprises four diesel-electric locomotives of Classes T448.0, T669.05 and T669.15.

UPDATED

SD – Kolejová dopravá as

Tusimice 7, CZ-432 01 Kadan
e-mail: info@sd-kd.cz
Web: www.sd-kd.cz

Key Personnel

President: Ing František Maroušek

Gauge: 1,435 mm
Route length: 51.5 km
Electrification: 30.5 km at 3 kV DC

Organisation

SD – Kolejová dopravá as has, since 2002, been a wholly owned subsidiary of North Bohemian Mines plc, Chomutov (Severočeské doly as Chomutov). It manages brown coal transport on two standard-gauge lines from a loading point at Tusimice in northwest Bohemia (Doly Nástup Tusimice – DNT). A partly double-track 12.5 km line, commissioned in 1965, runs northeast to Kadaň the location of the Czech Republic's largest brown coal-fired power station, the 1,490 MW Prunéřov Power Plant (EPR); a second 10 km single-track line, operated since 1978, runs east from Tušimice to Březno u Chomutova, where trains are handed over to ČD motive power. Both lines were electrified in 1979–82 and are leased from their owner, Severočeské doly, as.

SD – Kolejová dopravá as also operates the 29 km line to the Doly Bílina (DB) loading point at Ledvice on behalf of Severočeské doly, as.

Freight operations

The line to the Prunéřov Power Plant carries up to 25 pairs of trains daily, each formed of 18 four-axle Type Falls hopper wagons. On this line there is a maximum rising gradient of 1.9 per cent. Some 7.5 million tonnes were transported in 2004. On the line to Březno u Chomutova two trains per day are operated, each formed of 32 Type Eas wagons. In 2004 1 million tonnes of coal were carried.

On the Ledvice line SD – Kolejová doprava undertakes coal loading and train formation for haulage by other operators, in 2004 handling 5.7 million tonnes (215,000 wagons).

Traction and rolling stock

The electric traction stock comprises: four 5,220 kW Class 184.5 Bo-Bo-Bo GTO chopper-controlled 122-tonne locomotives (Škoda Type 93E, 1994–99); and nine 2,040 kW Class 130 Bo-Bo 86-tonne locomotives (Škoda, 1977). The diesel fleet comprises: two 993 kW Class 770.5 Co-Co; one 970 kW Class 744.7 Bo-Bo; one 883 kW Class 740.4 Bo-Bo; one 627 kW Class 724.7 Bo-Bo; one 552 kW Class 721.5 Bo-Bo; and three 300 kW Class 704.5 two-axle shunters. The Class 744.7, 724.7 and 704.5 machines are equipped with Caterpillar engines. There are also 210 modern Type Falls hopper wagons.

UPDATED

SD – Kolejová dopravá Class 184.5 electric locomotive (SD – Kolejová dopravá)   ***NEW**/1144124*

SD – Kolejová dopravá Caterpillar-powered diesel traction line-up, with (left to right) Classes 724.7, 744.7 and 704.5 (SD – Kolejová dopravá)   *1144126*

Type Falls four-axle coal hopper wagon
(SD – Kolejová dopravá)
***NEW**/1144127*

For details of the latest updates to *Jane's World Railways* online and to discover the additional information available exclusively to online subscribers please visit
jwr.janes.com

Viamont as

Železničářská 1385/29, CZ-400 03 Ústí nad Labem-Střekov
Tel: (+420 47) 530 01 11   Fax: (+420 47) 530 01 00
Web: www.viamont.cz

Key personnel

Chairman of Administrative Board: Miroslav Plíhal
Director, Passenger Traffic: František Kozel
Director, Freight Traffic: Pavel Škarabela
Director, Infrastructure: Přemysl Vacek
Director, Commercial: Jiří Namyslov
Director, Finance: Michal Libánský
Director, Control and Inspection: Václav Poduška
Director, Strategy: Vladimír Trtík
Director, Marketing: Jan Klobouček
Director, Legal Affairs: Vladimír Mašek
Director, Personnel: Marie Danihlíková

Organisation

Viamont is a private-sector company founded in August 1992, with headquarters at Ústí nad Labem, in northern Bohemia. Its main activities include railway operations, rolling stock overhaul and refurbishment, and track modernisation.

In January 2005 Viamont became a founding member of the European Bull railfreight alliance (see entry in International section of *Railway systems and operators*) that provides cross-border services.

Passenger operations

In December 1997 Viamont restored passenger services on the 10 km local line between Trutnov and Svoboda nad Úpou. The line had been closed by Czech Railways (ČD) in the previous September due to heavy losses, and is now leased to Viamont. The company now operates 22 pairs of passenger trains on the line.

In February 1998, Viamont was also selected to operate two more ČD rural lines in western Bohemia, Nové Sedlo u Lokte–Krásný Jez (24 km) and Sokolov–Kraslice (23 km). Viamont operations on the Sokolov–Kraslice route commenced in May 1998, with 12 pairs of trains daily. Services were extended up to Hraničná on the German border in July 1998. The cross-border line to Klingenthal in Germany was reopened in early June 2000 at a cost of Kcs18 million and DM7.5 million respectively, with the support of the European Union. In October 1999, work started on restoration of the 52 m border bridge which was removed during the partition of Germany. The new bridge was installed in September 2000. The route now forms part of the Egronet Czech-German frontier regional railway system.

In 2002 Viamont carried 738,000 passengers on the Sokolov–Kraslice line, 13 per cent of which was cross-border traffic, and 20,170 tonnes of freight. To increase passenger service quality, Viamont in

Under a collaborative arrangement with Viamont, German operator Vogtlandbahn operates Siemens Desiro railcars, such as this example at Cheb, on the Sokolov–Zwickau route (Edward Barnes)   0573981

close collaboration with Vogtlandbahn in Germany began trial operations with Siemens-built Class 642 Desiro railcars on the Sokolov–Zwickau route in July 2003. The weekend-only service (eight train pairs) was subsequently made regular and extended to Karlovy Vary (two train pairs) in October 2003. A further extension to Mariánské Lázně/Marktredwitz via Cheb followed two months later.

Plans to resume operations on the Nové Sedlo u Lokte–Krásný Jez line, closed to traffic in 1997, are not to be implemented due to high costs.

Freight operations

Since November 1995 the company has been operating one or two pairs of coal trains daily over 211 km of the Czech main line network from a loading point at Březno u Chomutova to a power station at Dolní Beřkovice, carrying some 130,000 tonnes each month. Other commodities, such as construction materials, are also carried. In 1998 Viamont carried 2.8 million tonnes of freight, accounting for 3 per cent of all rail freight in the Czech Republic. However, this had decreased to 1.5 million tonnes by 2003 due to a decline in the consumption of coal, which accounts for 90 per cent of Viamont's traffic, with revenues from it totalling Kcs117 million in 2002. Viamont is also heavily involved as a subcontractor in programmes to

upgrade international railway corridors passing through the Czech Republic.

Freight operations on the Trutnov–Svoboda nad Úpou line began on 1 April 1999.

Together with rail4chem (Germany), LTE (Austria), FNM Cargo (Italy) and Comsa Rail Transport (Spain), Viamont in January 2005 became a founding member of European Bulls, an association of open access private rail freight operators (see entry in International section of *Railway systems and operators*).

Traction and rolling stock

The Viamont traction fleet comprises seven diesel-electric locomotives of Classes 720, 740 and 742.5 and three Class 710 diesel-hydraulics. A fleet of 220 Type Wap coal hoppers is hired from ČD.

For local passenger services on the Trutnov–Svoboda nad Úpou line, Viamont has leased from a private company, Lokotrans, two Class 830 and one Class 810 railcars and one Class 010 trailer. Operations on the Sokolov–Kraslice–Hraničná route employ two Class 714.2 and one Class 720 diesel-electric locomotives and eight Class 020 trailers, all leased out from ČD.

UPDATED

Denmark

Ministry of Transport and Power

Frederiksholms Kanal 27, DK-1220 Copenhagen K
Tel: (+45) 33 92 33 55   Fax: (+45) 33 12 38 93
e-mail: trm@trm.dk
Web: www.trm.dk

Key personnel

Minister: Flemming Hansen
Permanent Secretary: Thomas Egebo

UPDATED

National Rail and Ferry Authority

Trafikstyrelsen for jernbane og færger
Adelgade 13, DK-1304 Copenhagen K
Tel: (+45) 72 26 70 00   Fax: (+45) 72 26 70 70
e-mail: info@trafikstyrelsen.dk
Web: www.trafikstyrelsen.dk

Key personnel

Director General: Jens Andersen
Assistant Director General: Ulrik Winge
Operations Director: Anna Katrine Barslund
Projects Director: Martin Munk Hansen
Safety Director: Peter Sloth

Political background

The National Rail and Ferry Authority was formed on 1 July 2003 and is subordinate to the Ministry of Transport. It carries out a number of functions that had previously been split between different authorities: transport planning and coordination, regulating the operation of the network, granting and supervising franchises, safety policy and supervision, and advising the Ministry of Transport and the Danish Parliament on policy and strategy. These functions were previously carried out by Banestyrelsen, DSB, the Ministry of Transport and the railway safety authority Jernbanetilsynet; the last

has been abolished and its accident investigation role passed to a new organisation. The authority also has a similar role for ferry services.

One major reason for the creation of the authority was to avoid conflicts of interests in the franchising process. In 2001 Arriva Tog was awarded two franchises to operate local passenger services in north and west Jutland, a lower bid from DSB having been rejected as non-viable. Problems arose because the Ministry was at one and the same time DSB's owner, the franchising authority and the ministry responsible for the appeals board.

Between 2005 and 2014 the Authority will be responsible for awarding franchises for over one third of the local and regional services currently operated by Danish State Railways. The Helsingør–Copenhagen–Malmo group of services is the first franchise to be offered for tender: a contract for seven to 10 years starting in June 2006 is intended.

Organisation

There are no departments or sections within the authority: working groups are established to cover each task as required.

UPDATED

Banedanmark

Banedanmark – Danish National Railway Agency
Amerika Plads 15, DK-2100 Copenhagen Ø, Denmark
Tel: (+45) 82 34 00 00   Fax: (+45) 82 34 45 72
e-mail: bane@bane.dk
Web: www.bane.dk

Key personnel

Director General and Service Director:
 Jasper Rasmussen
Operations Director: Henrik Jørgensen
Infrastructure Management Director: Eigil Sabroe
Finance Director: Karsten Bjerregaard
Managers
  Service: Sonny Nielsen
  Safety: Thomas Böhme
Management Secretariat: Mette Ehlers Mikkelsen
Human Resources: Charlotte Smidt
Equipment: Erik Meier Girke
Internal Auditor: Henning H Larsen
Information Technology Manager:
 Kenneth Lau Rentius
Public Relations: Inger Petersen Thalund
Safety Director: Peter Sloth

Gauge: 1,435 mm
Route length: 2,157 km
Electrification: 171 route-km at 1.5 kV DC (Copenhagen suburban system), 430 route-km at 25 kV 50 Hz AC

Political background

As a result of legislation in the Danish parliament, DSB (Danish State Railways) was split in two on 1 January 1997: DSB continued to operate trains, while the newly formed Banestyrelsen took over responsibility for the infrastructure, traffic control, capacity management and planning. From 1 January 1999 freight operators have been allowed 'open access' to Banestyrelsen's (now Banedanmark's) network and similar rights have been available to passenger operators to compete with DSB since 1 January 2000.

Banedanmark is required to seek tenders for work on the infrastructure throughout the network from outside companies. The service division is at 'arms' length' from the main company in order for it to be able to compete for contracts.

In 2003 Banestyrelsen's Strategic Planning Unit was transferred to the National Rail and Ferry Authority and on 1 March 2004 Banestyrelsen was renamed Banedanmark.

Organisation

Banedanmark is divided into four independent divisions: Operations, Service, Infrastructure Management and Finance. There are Public Relations and Safety sections responsible to the Director General, as are the Management Secretariat and the Internal Auditor.

Operations is divided into Traffic Management, Planning and Information Technology sections; the first is responsible for traffic control and the daily management of operations. Infrastructure Management is the owner of the rail system and is responsible for its upkeep and development. Finance includes the Human Resources section.

The Executive Board consists of the Managing Director and the other division directors, but for day-to-day management the Board also includes the Human Resources, Safety and Service managers. The Supervisory Board has six members.

In 2005 Banedanmark employed 2,514 staff.

Finance

Banedanmark receives support from the Danish Ministry of Transport for its management and modernisation of the network: the current agreement covers the years 2005 and 2006, when Banedanmark

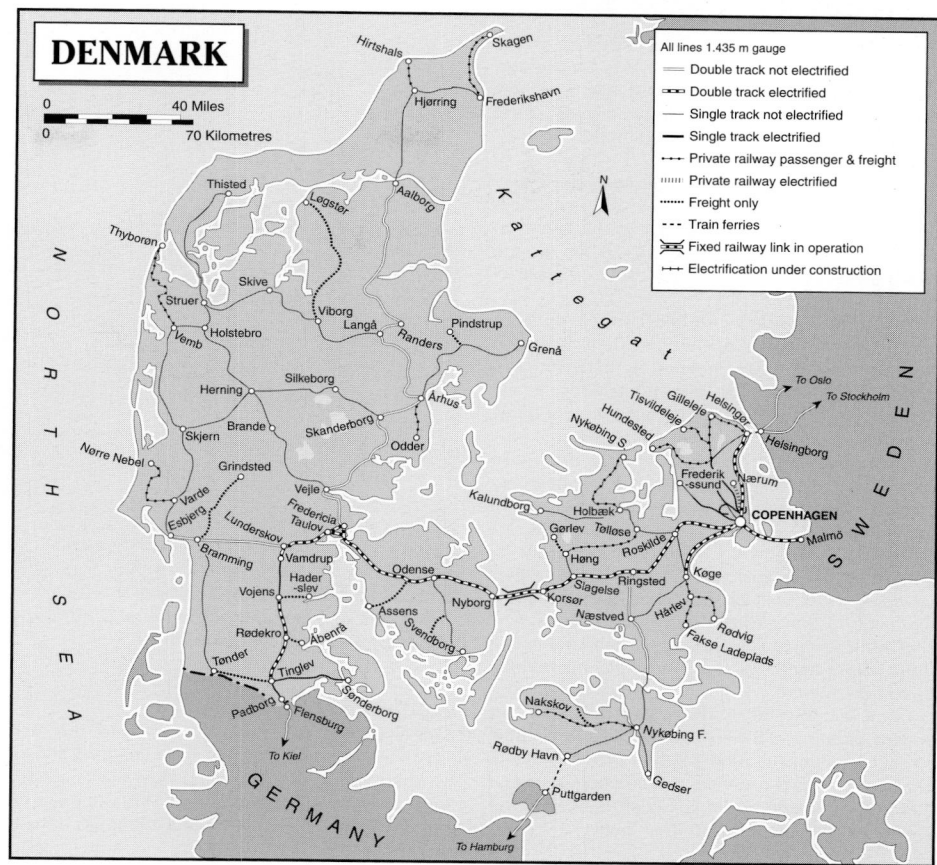

0137838

was to receive approximately DKr2.2 billion each year. The outline agreement for 2007–14 will see Banedanmark an average of around DKr2.4 billion a year over this period. Banedanmark has entered into contracts with all the operators on the network.

New lines
Great Belt fixed link

The long-cherished scheme for a fixed crossing of the 18 km Storebælt (Great Belt) waterway between Copenhagen on the island of Zealand and the rest of the country, opened to trains in June 1997. The 6.6 km low rail and road West Bridge runs from Funen to the island of Sprogø, whence there is the 8 km rail-only twin-bore East Tunnel to Zealand; road traffic uses the 6.8 km elevated East Bridge, opened in mid-1998.

The former Banestyrelsen was not directly involved in the construction of the link, but as its administrator collects the access charges paid by train operating companies and pays an annual fee which, from 1999 onwards, amounts to approximately DKr500 million.

Høje Taastrup station, on the busy section of line between Copenhagen and Roskilde, which has been the subject of capacity enhancement studies (John C Baker)   0137756

Fixed link with Sweden

An agreement to build a fixed crossing of the Øresund waterway was signed by the Danish and Swedish Transport Ministers in 1991. The 16 km link was completed by the Danish-Swedish Øresund Consortium in 2000 at an estimated cost of DKr18 billion.

From the Danish side the double-track railway and four-lane motorway enter a 3.5 km tunnel before rising to the 4 km long artificial island of Peberholm. Both then cross the 7.8 km bridge which gives a 57 m clear height above water level in its 1.1 km centre section.

On the Danish side a new line, partly in cut-and-cover tunnels, from Copenhagen Central station to the city's airport was opened by Banestyrelsen in 1998. The station at the airport lies under a new airport terminal building. Loops allow freight trains to avoid the platform lines and there is also a new servicing depot just to the east of the station. A 4 km cut-off allows freight trains from Sweden to Jutland and the rest of Europe to avoid central Copenhagen.

The link is electrified at Banestyrelsen's 25 kV AC 50 Hz, with the change to the Swedish 15 kV AC 16 Hz taking place at Lernacken, at the Swedish end of the link. The change from Danish to Swedish ATC and traffic control takes place on Peberholm, but Danish right-hand running is normally maintained through to Malmö.

The link is owned by the Øresund Consortium and the Danish and Swedish railways pay an annual rental for its use. In 2004 this was DKr400 million. This sum does not cover the construction costs and the motorway side of the link is cross-subsidising the railway as part of government policy.

Rødby–Puttgarden fixed link

Since 1992 the Danish government, in cooperation with the German and Swedish governments, has been investigating ways of improving traffic links between Scandinavia and the Continent. A new link across the Fehmarn Belt could create a 345 km rail route between Copenhagen and Hamburg, compared with over 500 km via the present border crossing at Padborg. The distance would be further reduced if a high-speed line was built on either side of the Fehmarn Belt and a Copenhagen–Hamburg journey time of 2 hours would be possible. Most freight traffic would continue to be routed via the Great Belt and Padborg.

Studies of different types of link have been completed and progress was made in 2004 in exploring the method of financing the link.

Copenhagen–Ringsted capacity expansion

In May 1997 the Danish parliament decided to investigate ways of increasing capacity between Copenhagen and Ringsted, the busiest section of line in Denmark. Banestyrelsen produced a report evaluating three options: expanding the existing line to four main line tracks throughout (only Høje Taastrup to Roskilde currently has four main line tracks); constructing a new double-track line between Copenhagen and Ringsted via Køge; or a combination of these two options, with the new line starting near Høje Taastrup. None of these options has yet been adopted but planning restrictions on the route of the new line from Copenhagen to Ringsted via Køge have been retained. A plan to extend the S-Train system to Roskilde using dual-voltage emus was adopted in January 2001 but postponed indefinitely in May 2002. A further evaluation of the options for the corridor was under way in 2005.

This Class MZ diesel locomotive shunting at Odense is part of a small traction fleet retained by Banedanmark (Philip Wormald)   *NEW*/1114998

Improvements to existing lines

Banedanmark still intends to double the two remaining sections of single track, totalling 35 km, between Vamdrup and Padborg on the line from Fredericia to the German border. This will improve the competitiveness of freight traffic as it will complete the double-track electrified route through Denmark from Sweden to Germany via the Great Belt. It is also considering doubling the lst section of single track between Copenhagen and Holbæk.

As part of DSB's 'Good Stations for All' project Banestyrelsen and DSB opened five stations for the new suburban service around Aalborg in December 2002 and December 2003. Two stations for the new suburban service to the southwest of Århus opened in December 2003 and October 2004. Upgrading of the Odense–Svendborg line was completed in late 2003 and work on the Århus–Hornslet–Grenaa line was finished in January 2005.

In 2004 Banedanmark completed a project to raise capacity through the centre of Copenhagen: improvements facilities for terminating trains at Østerport station will be completed in 2005. On the S-Train network doubling of the Frederikssund line was completed in 2002 and a new interchange opened at Flintholm in January 2004. The Copenhagen Ring Line was extended south from Flintholm to a temporary terminus in January 2005 and will reach its eventual terminus at Ny Ellebjerg in 2007. Work to increase train speeds throughout the S-Train network is also in progress. Construction of a fourth S-Train line westwards out of Copenhagen Central station is expected to start in 2007.

Signalling and traffic control

Automatic traffic control has been commissioned on over 600 km of Banestyrelsen main lines and is overseen by three control centres, at Copenhagen Central, Roskilde, and Fredericia. In the event of failure, supervision can also be taken over by the Network Control Centre in Copenhagen. The 'silent' element of passenger information is also handled by the automatic control system. A new automatic control system for the Copenhagen S-Train system,

based on the Siemens VICOS OC 501 technology, is to be installed in 2005–06.

In 1994 DSB started commissioning an Automatic Train Protection (ATP) system that permits 180 km/h running on its network and this has now been extended to over 800 km of main lines. It covers all main lines in Denmark except the 49 km section between Hobro and Aalborg, but installation on this section is now under consideration. The installation is based on the Siemens ZUB123 system and mobile units are now required on all motive power running on main lines. A simpler system allowing 120 km/h operation was installed on nearly 400 km of secondary lines on Funen and in Jutland in late 2004 and early 2005.

Electrification

The 171 route-km Copenhagen suburban S-Train system operates on 1.5 kV DC overhead supply.

DSB started main line electrification at 25 kV AC 50 Hz east of the Great Belt in the early 1980s. It was extended to the German border at Padborg to coincide with the opening of the Great Belt link. The branch line from Tinglev to Sønderborg, near the German border, and the lines into Fredericia were electrified at the same time. The line from Copenhagen to Copenhagen Airport was electrified from its opening, as was the Øresund link. No further electrification is planned.

Track

Rail: Flat-bottom, 60 kg/m (1,641 km), 45 kg/m (1,199 km), 37 kg/m (52 km)

Crossties (sleepers): twin-block concrete 2,328 × 209 × 290 mm with Sonneville fastenings. Monobloc concrete sleepers 2,500 × 206 × 280 mm with Vossloh fastenings have been used as standard since 1989

Spacing: 1,600/km

Min curvature radius: 300 m

Max gradient: 1.25% (1.56% in the Great Belt tunnels)

Max axleload: 22.5 tonnes

UPDATED

Danish State Railways (DSB)

Danske Statsbaner
Sølvgade 40, DK-1349 Copenhagen K
Tel: (+45) 33 14 04 00   Fax: (+45) 33 14 04 40
e-mail: dsbkomm@dsbkomm.dsb.dk
Web: www.dsb.dk

Key personnel

President and Chief Executive Officer: Keld Sengeløv
Executive Vice-President and Chief Financial
   Officer: Søren Eriksen

Division Directors
   Operations: Jørn Webler
   Sales: Bjørn Wahlsten
   S-Train: Benny Würtz
Deputy Operations Director: Kaj S Lund
International Manager: Jens Otto Daugaard
DSB Detail Manager: Klaus Lorenzen
Corporate and Service Functions
   Safety Manager: Thomas A Olsen
   Corporate Planning: Ove Dahl Kristensen
Deputy Operations Director: Kaj S Lund

Corporate and service functions:
   Purchasing Manager: Lars N Pedersen
   Management Secretariat: Lone Lindsby
   Public Relations Management: Anna Vinding
   Design Manager: Pia Bech Mathiesen
   Financial Manager: Bartal Kass
   Human Resources Manager: Hans Munck
   Information Technology Manager:
      Peter Lundsteen
   Chief Lawyer: Peder Nedergaard Nielsen
   Chief Auditor: Leif Frandsen

Political background

On 1 January 1997 DSB was split in two: infrastructure was allocated to a new company, Banestyrelsen (see entry for Banedanmark in the Denmark section of *Railway systems and operators*), while DSB continues to operate trains. Since the sale of its freight division to the Railion Group (see entry for Railion in the Denmark section of *Railway systems and operators*) in 2001, DSB has been entirely a passenger operator and accounts for 80 per cent of total Danish passenger traffic; the remainder is carried by smaller 'local' railway companies and by Arriva Tog (see entry for Arriva Tog in the Denmark section of *Railway systems and operators*).

Since January 2000 private companies have been able to provide passenger services in competition with DSB, subject to capacity. There are two types of passenger transport carried out as a public service: 'negotiated' transport and 'tendered' transport. All services not offered for tender are handled as negotiated transport.

Since 1 January 1999 DSB has been an independent publicly owned corporation which operates services under contract to the Ministry of Transport. DSB S-Train (see entry for DSB S-Train in the Denmark section of *Railway systems and operators*) has been a wholly owned subsidiary of DSB since 1 January 1999 and its services are covered by a separate contract. DSB receives no other public funds: costs and investment are financed from operating income and loans.

A contract covering a 10-year period starting in 2005 was agreed between DSB and the Ministry of Transport in late 2003. DSB will receive DKr2.4 billion a year compared with DKr2.9 billion in 2004. There will be a rise in train-km of 26 per cent but provision has been made to allow franchises to be sought for up to one-third of regional services.

Organisation

DSB has a board of directors, five line units and 10 corporate and service functions; there is also a nine-member Supervisory Board. The Board of Directors consists of two members: a managing director and a business director. These two, together with the directors of S-Train, Sales and Operations line units, form the Executive Board. The line units are Sales, Operations, DSB International, DSB Detail and S-Train, the last two are wholly owned subsidiaries.

DSB International, Financial Services, Human Resources, Information Technology, Purchasing and the Legal Department are responsible to the Business Director. DSB Detail, the Management Secretariat, Public Relations, Design and Traffic Planning Departments are responsible to the Managing Director. The Deputy Operations Director has responsibility for safety and is also responsible to the Managing Director. The Chief Auditor is responsible to the Board of Directors and the Supervisory Board.

DSB International pursues business opportunities outside Denmark; it is responsible for the Roslagståg operation in Stockholm and DSB Tågvärdsbolag in Skåne (Scania), both in Sweden.

DSB Detail is the former independent DSB Restaurant and Kiosk operation, taken over in 2002. Over 100 kiosks have been transferred to either DSB Sales or DSB S-Train and their staff trained to sell tickets, allowing ticket offices to be closed at all but around three dozen major stations. It continues to act in an advisory role for the commercial management of kiosks and from 1 January 2004 took over the operation of onboard catering.

At the end of 2004 DSB (excluding S-Train and DSB International's operations) had 6,700 employees, of which 245 were employed by DSB Detail.

Finance

DSB income comes largely from two sources: operating revenue and contracts, but it may also raise loans. The total financial support received in 2004 was DKr2,573 million, with S-Train and other subsidiary operations receiving a further DKr1,395 million.

In 2004 DKr879 million was invested in new IC4 and IR2 trains.

In 2004 DSB made a pre-tax profit of DKr947 million, a fall of just under 5 per cent on the figure for 2003.

Demonstrating the flexibility of DSB's multiple-unit operations, at Odense two IR4 emus bring up the rear of a northbound train headed by an IC3 dmu (Philip Wormald) ***NEW**/1114995*

IR4 emu at Høje Taastrup with a Roskilde–Copenhagen Østerport local service (John C Baker)

0576991

Class MQ Siemens Desiro diesel railcar approaching Odense with a service from Svendborg (Philip Wormald) ***NEW**/1114996*

Diesel locomotives

Class	Wheel arrangement	Power kW	Speed km/h	Weight tonnes	No in service	First built	Mechanical	Builders Engine	Transmission
ME	Co-Co	2,270	175	115	37	1981	Thyssen-Henschel/ Scandia	GM 16-645-E3B	E BBC H Voith
Tractor	B	94	45	17	12	1966	Frichs	Leyland UE 680	H Voith
MK	B	390	60	40	1	1996	Siemens	MTU 8V183TD13	H Voith

Finances (DKr million)

Revenue	2002	2003	2004
Operating income	3,895	3,888	4,052
Operating subsidy	3,642	4,147	3,968
Other	1,283	1,877	1,895
Total	8,820	9,912	9,915

Operating expenditure	2002	2003	2004
Staff/personnel	2,751	3,089	3,207
Other	3,611	4,273	3,975
Total	6,362	7,362	7,182
Depreciation	1,184	1,298	1,483
Financial costs	−181	−255	−303
Pre-tax profits	1,093	997	947
Tax	319	300	283
Net profit	774	697	664

Traffic (million)	2002	2003	2004
Passenger journeys	150	148	151
Passenger-km	5,541	5,397	5,509

Passenger operations

DSB's passenger operations are based on a regular interval timetable, with most of its stations served by at least one train an hour in each direction. Main line InterCity services consist of a core route from Copenhagen via Odense to Århus and Aalborg; most other major towns have a through train from Copenhagen every two hours. An overnight service operated by IC3 trainsets runs between Copenhagen and the main towns in the west of the country. East of the Great Belt most stations have a regular interval local service to Copenhagen; the majority of these trains run through the centre of Copenhagen to terminate at Østerport station on the north side of the city centre. The Copenhagen–Helsingør line sees six trains an hour in the peaks; the basic service is now part of the Øresund service and is worked by Øresund trainsets. West of the Great Belt local services run between regional centres, although Arriva Tog has taken over some of these. There are limited suburban services around Aalborg, Odense and Århus; the latter, along with Esbjerg, also has local services provided by Arriva Tog.

A major reorganisation of services involving the introduction of limited stop 'Lyntog' services and the introduction of regular 180 km/h running took place when the Great Belt Link opened in June 1997. On 27 September 1998, the Copenhagen–Kastrup Airport line was opened, served by frequent local trains to Copenhagen and Lyntogs to destinations west of the Great Belt. Local trains were extended over the fixed link across the Øresund when it opened on 1 July 2000 to form a 20 minute frequency Copenhagen-Malmö shuttle. The link is also used by long distance services to destinations in Sweden and DSB trains to Ystad, where connections are made with a fast ferry to the island of Bornholm. From June 2001 the Øresund shuttle was extended northwards from Copenhagen to Helsingør. Trains have also been extended beyond Malmö on the Swedish side.

Since the mid-1990s, DSB has been working on the 'Good Trains for All' (GTA) project. The main elements of this are: an increase in the number of trains operated (by 2006 an increase of 25 per cent compared with 2000 service levels is planned); faster services; extension of the regular interval timetable; and operation of all except a handful of services by new or recently built rolling stock. There will also be new or improved services around Aalborg, Århus, Esbjerg and Odense, as well as improvements to local services, with better connections and more frequent services. Outer suburban services around Copenhagen should also run more frequently. The plan should have been complete by 2006, but since 2001 a new hourly limited stop service between Copenhagen and Århus has been introduced and this was extended to Aalborg at the beginning of 2005. Services to southern Jutland and the German border at Padborg have been improved. A new suburban service around Aalborg serving new stations started in December 2003. The Odense–Svendborg service was also increased and accelerated from December 2003, as was the Århus–Hornslet–Grenaa service from January 2005.

Two IC3 dmus at Langa en route from Aalborg to Copenhagen (Colin Boocock)   0576993

Class ET/FT dual-voltage emu at Ørestad, forming a Copenhagen Østerport–Malmö service via the Øresund link (John C Baker)   0137759

Class MG (IC4) high-speed dmu supplied by Ansaldobreda   0580477

Electric locomotives

Class	Wheel arrangement	Line voltage	Output kW continuous	Speed km/h	Weight tonnes	No in service	First built	Builders Mechanical	Electrical
EA	Bo-Bo	25 kV	4,000	175	80	12	1984	Thyssen-Henschel/ Scandia	BBC

Sleeping car services are now reduced to one international train between Copenhagen and destinations in Germany, although DSB maintains a small fleet of couchettes for use on charters and for trains to ski resorts.

Traction and rolling stock

DSB operates 12 Class EA line-haul electric locomotives; 37 Class ME line-haul diesel-electric locomotives; 13 Class MK diesel shunting locomotives and tractors; 96 Class MFA/FF/MFB diesel-hydraulic trainsets (IC3); 81 Class MR/MRD diesel-hydraulic trainsets (another 15 are on hire to Arriva); 12 Class MQ diesel-mechanical trainsets; 44 Class ER/FR 25 kV four-car emus (IR4); and 24 Class ET/FT 'Øresund' 25 kV/15 kV three-car emus; and 239 passenger coaches.

The IC3 diesel trainsets are used for most InterCity services and a handful are fitted with either German or Swedish ATC equipment: those with the former work the Copenhagen–Hamburg service. In mid-2003 DSB acquired four secondhand trainsets of IC3 design from Sweden. The IC3 trainsets are being fitted with low-emission engines. The IR4 emus are used for local services between Copenhagen and Fredericia and on the InterCity service to Odense and Sønderborg; IC3 and IR4 trainsets are capable of being worked in multiple. The line-haul diesel and electric locomotives are mostly used on local and suburban services east of the Great belt and on weekend reliefs. The MR/MRD dmus are mostly used on local services in Jutland.

DSB's 'Good Trains for All' programme requires the present fleet of dmus and emus to be enlarged with vehicles of modern design. Between 2000 and 2002 DSB took delivery of 24 three-car Øresund emus from Adtranz (now Bombardier Transportation) for use on the Øresund link. These are interworked with 25 identical sets received by Swedish State Railways between 2000 and 2003.

The first of 83 Class MG (IC4) four-car high-speed diesel-mechanical trainsets arrived from the Italian manufacturer AnsaldoBreda in August 2003. The trainsets have 208 seats, including 21 tip-up places, and one car has a low-floor to improve access for passengers with special needs. Each articulated trainset is powered by four Iveco 560 kW Euro III engines driving through ZF 16 AS 2603 gearboxes. Maximum service speed is 180 km/h, although the trains' design allows for upgrading for 200 km/h operation. The first sets are expected to enter service in early 2006, allowing completion of the 'Good Trains for All' project in 2007. There is an option for 67 similar units but only the IC2 units mentioned below are likely to be acquired.

In late 2000 DSB hired three two-car Siemens Desiro dmus from Angel Trains for use on the Odense–Svendborg line, designating them Class MQ. In mid-2002, these units were returned as part of a deal that brought to Denmark 12 units of improved specification. In November 2002 DSB signed

Class ME diesel-electric locomotive at Holbæk with a regional express service for Kalundborg formed of double-deck stock (Philip Wormald) **NEW**/1114994

Diesel railcars or multiple-units

Class	Cars per unit	Motor cars per unit	Motored axles/car	Power/ motor kW	Speed km/h	No in service	First built	Mechanical	Builders Engine	Transmission
MR/MRD	2	2	2	191	130	96[1]	1979	Duewag/ Scandia	KHD F12L413F	H Voith
MFA/FF/ MFB (IC3)	3	2	2	250	180	96	1989	Ascan-Scandia/ Duewag	KHD BFBL-513-CP	H ZF 5HP600
MQ	2	2	1	315	120	12	2000	Siemens	–	H Voith
MG/FG/ FH/MG	4	2	2	560	180	83[2]	2003	Ansaldobreda	Iveco Euro III	M ZF 16 AS 2603
'IC2'	2	–	–	–	180	23	–[3]	Ansaldobreda	Iveco Euro III	

[1] 39 Class MR/MRD dmus are leased to Arriva Tog, leaving 57 in DSB service.
[2] Under delivery in 2005.
[3] On order for delivery in 2006.

Electric multiple-units (main lines) 25 kV

Class	Cars per unit	Motor cars per unit	Motored axles/car	Output/ motor kW	Speed km/h	No in service	First built	Builders Mechanical	Electrical
ER/FR FR/ER	4	2	4	480	180	44	1992	ABB Scandia	ABB Traction
ET/ FT/ET*	3	2	4	290	180	24	1999	Bombardier	Bombardier

* dual-voltage (25 kv AC 50 Hz/15 kv AC 16^2/$_3$ Hz).

a contract with AnsaldoBreda for 23 180 km/h two-car IC2 diesel trains for local and regional services. Delivery is expected during 2006.

Following the experimental hire of a rake of double-deck coaches from Swiss Federal Railways from September 1999 to April 2001, DSB is hiring 67 double-deck coaches, including 14 driving trailers, from Porterbrook Leasing

from late 2002 to the end of 2006. The coaches have been built by Bombardier Transportation to the standard design used in Germany. They are mainly used east of the Great Belt and are powered by 20 Class ME diesel locomotives that have had their push-pull equipment modified to suit these vehicles.

UPDATED

DSB S-Train

DSB S-Tog A/S
Sølvgade 40, DK-1349 Copenhagen K
Tel: (+45) 33 14 04 00   Fax: (+45) 33 14 04 40
e-mail: s-tog@s-tog.dsb.dk
Web: www.dsb.dk/s-tog

Key personnel

Managing Director: Benny Würtz

Gauge 1,435 mm
Route length: 171 km

Organisation

S-Train is a wholly owned subsidiary of DSB that provides suburban passenger services over 171 km of independent 1.5 kV DC lines in the Greater Copenhagen area. Operation of the Hillerød–Helsingør 'Lille Nord' line passed to The Local Railway at the beginning of 2005. Fare levels are set by the Greater Copenhagen Transport Authority (HT).

A 10-year contract for provision of S-Train services for the period 2005–14 contains targets for service levels and punctuality; it will also see

an increase in train-km of 13 per cent. The contract specifies the amount of support to be paid by the Ministry of Transport and Power. In 2004 this was DKr1,249 million, but it will fall to around DKr800 million by 2014. A new contract for 2005–14 will see an increase in train-km of 13 per cent.

At the end of 2004 S-Train employed 1,837 staff.

Passenger operations

In 2004 the S-Train system carried 89.6 million passengers for 1,161 million passenger-km, compared with 87.9 million (1,136 million passenger-km) in 2003. The timetable is based on a 20-minute interval service on each of 11 routes,

with all but seven of the 85 stations served by more than one route for much of the day.

In January 2004 the S-Train division and Banestyrelsen opened the new Flintholm station. This provides an interchange between the Copenhagen Metro and S-Train's Frederikssund and Ring lines. The Ring Line project started in 1998 and has seen the closure of the Frederiksberg–Vanløse line, with its formation being used by the Metro, and the diversion of the Hellerup–Vanløse shuttle to Flintholm. In January 2005 the shuttle was extended south from Flintholm, via a new station at Danshøj (on the S-Train line to Høje Taastrup), to a temporary terminus near Ellebjerg.

DSB S-Train electric multiple-units (1.5 kV)

Class	Cars per unit	Motor cars per unit	Motored axles/car	Output/ motor kW	Speed km/h	No in service	First built	Builders Mechanical	Electrical
MM/FU/MU/FS	4	2	4	147	100	43	1975	Scandia	GEC
FC/MC/MC/FC	4	2	4	150	100	8	1986	Scandia	ASEA
SA/SB/SC/SD	8	6	8	180	120	92	1995	ALSTOM (LHB)	Siemens
SE/SF/SG/SH	4	3	4	180	120	31*	2004	ALSTOM (LHB)	Siemens

* On order.

S-Train emus at Ishoj (Colin Boocock)   0576995

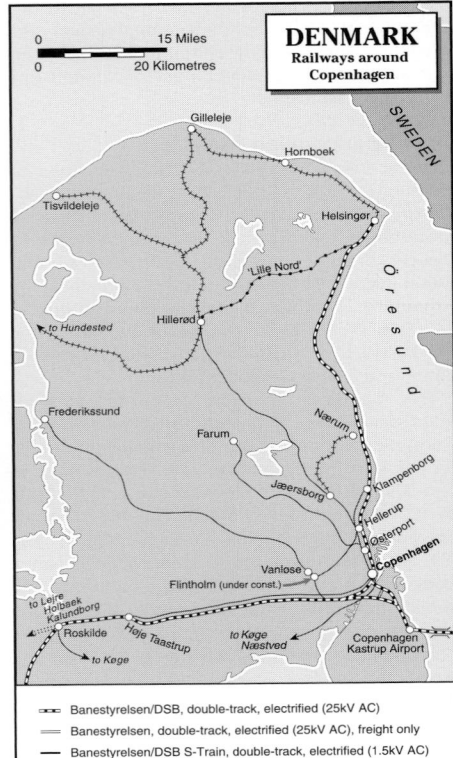

The Ring Line will be completed when the eventual terminus at Ny Ellebjerg in the south of the city is reached in January 2007.

Traction and rolling stock
In 1999 delivery commenced of 112 eight-car single-axle articulated emus ordered in 1997 from Linke-Hoffmann-Busch (now ALSTOM) and Siemens. These are similar to eight units delivered in 1995–97 and will replace the two- and four-car emus built by Scandia and Frichs between 1966 and 1978; all of the two-car sets have been withdrawn and withdrawal of the four-car sets has started, but five four-car sets are expected to be retained. Fifteen of the eight-car sets will now be delivered as 30 four-car sets and an additional four-car set has been ordered; one eight-car set has been withdrawn. Eight four-car sets built by Scandia in 1986 are to remain in service.

DKr1,400 million was invested in new trains in 2004.

UPDATED

0552974

Arriva Tog

Drewsensvej 1, DK-8600 Silkeborg
Tel: (+45) 87 23 51 51
Fax: (+45) 87 23 51 23
Web: www.arrivatog.dk

Key personnel
Managing Director: Karsten Røn Andersen
Traffic Manager: Martin Enevoldsen
Customer Services Manager: Jens Thousgard
Engineering Manager: Ivan Skødt Andersen
Safety Manager: Jens Arne Jensen

Organisation
In December 2001 Arriva was awarded two franchises to operate local services over 583 km of lines in west and central Jutland for eight years from January 2003. The operator receives an average annual subsidy of DKr156 million, a saving of 30 per cent on the previous amount of subsidy given to DSB. In May 2002 an additional contract was agreed with the Ministry of Transport to provide more peak hour services than Arriva's draft timetable had originally proposed.

Arriva Danmark, Arriva Tog's parent company, took over operation of the Western Railway (see entry for Western Railway in Denmark section of *Railway systems and operators*) on 1 June 2002. Management of the Western Railway was moved to Arriva Tog's office in Silkeborg in 2003.

Traction and rolling stock
During 2004 29 two-car Coradia LINT regional diesel trainsets were delivered from ALSTOM's Salzgitter, Germany, plant. They are owned by the leasing company Angel Trains Ltd. Of the 39 Class MR/MRD diesel railcars initially hired from DSB by Arriva Tog, 15 will be retained for additional peak hour services.

UPDATED

Class MR/MRD dmu at Tønder forming a service to Esbjerg, with Connex-operated NordOstseeBahn railcar from Niebüll at an adjacent platform (Colin Boocock)   0576994

Århus-Odder Railway (HHJ)

A/S Hads-Ning Herreders Jernbanen
Banegårdsgade 3, DK-8300 Odder
Tel: (+45) 86 54 09 44   Fax: (+45) 86 54 41 70
e-mail: hhj@hhj.dk
Web: www.odderbanen.dk

Key personnel
Chairman: Poul B Heise
Managing Director: Mikael Tittel Langager
Assistant Manager: Henning Møller

Gauge: 1,435 mm
Route length: 26.5 km

HHJ Duewag railcar and trailer at Odder
(Edward Barnes)
0576998

Organisation

HHJ runs from the DSB station in Denmark's second largest city, Århus, south to the town of Odder. Passenger services only are provided.

Passenger operations

In 2004 HHJ carried 1.046 million passengers.

During the period 2005–07 it is intended to spend approximately DKr72 million on raising speed limits on the line and installing automatic train protection so that it will be possible to combine HHJ with DSB's Århus–Hornslet–Grenå line.

Traction and rolling stock

HHJ owns one diesel-hydraulic locomotive, six Duewag-Scandia two-car diesel trainsets.

UPDATED

The Danish Railway (DJ)

Dansk Jernbane ApS
Brorsonsvej 21, DK-6270 Tøller
Tel: (+45) 74 67 24 00 Fax: (+45) 74 67 46 61
e-mail: info@dansk-jernbane.dk
Web: www.dansk-jernbane.dk

Key personnel

Managing Director: Ulrich Krey
Assistant Manager: Pascal Koch

Organisation

DJ is a subsidiary of the German Norddeutsche Eisenbahn Gesellschaft (see entry for EuroLuxCargo SA in the Luxembourg section of *Railway systems and operators*) and was formed to operate freight

trains within Denmark. In late 2004 it took over the freight business of the Western Railway.

Traction and rolling stock

DJ operates two former DSB Class MY diesel-electric locomotives acquired from the Western Railway.

NEW ENTRY

Greater Copenhagen Local Railways (HL)

Hovedstadens Lokalbaner A/S
Gammel Køge Landevej 3, DK-2500 Valby
Tel: (+45) 36 13 20 61 Fax: (+45) 36 13 20 96
e-mail: nkv@hl-as.dk

Key personnel

Managing Director: Niels Munch Christensen

Gauge: 1,435 mm
Route length: 184 km

Organisation

On 1 January 2001 the Greater Copenhagen Development Authority (HUR) took over responsibility for the local railways within its area: the jointly managed Gribskov Railway (42 km) and the Hillerød–Hundested Railway (39 km) in north Zealand; the Helsingør–Hornbæk–Gilleleje Railway (24.5 km) along the north Zealand coast; the Lyngby–Nærum Railway (8 km) in Copenhagen's northern suburbs; and the Eastern Railway (50 km) to the south of Køge. The last-named line is geographically only partially within HUR's area of responsibility but agreement has been reached for it to fall within the organisation's authority. On 17 May 2002, HUR set up two companies: Greater Copenhagen Local Railways (HL), which owns the infrastructure and rolling stock; and the Local Railway (Lokalbanen A/S), which operates the lines. The infrastructure of the 21 km of the Hillerød–Helsingør line not shared with Banedanmarks's Copenhagen–Helsingør line was also taken over, but the rolling stock used on this line did not pass into HL's ownership until the beginning of 2005.

Traction and rolling stock

HL owns seven former DSB diesel electric locomotives, one tractor, 11 two-car diesel trainsets (including six Adtranz Flexliners and one Siemens Desiro), 35 diesel motor coaches and 40 trailer cars.

Siemens Desiro railcar operating on the Helsingør-Hornbæk-Gilleleje line at Helsingør (John C Baker)
0137761

Gribskov line Flexliner railcar at Slotspavillonen (John C Baker)
0137762

All were taken over from the absorbed lines and are operated by the Local Railway.

At the beginning of 2005 HL confirmed an order for 27 ALSTOM Coradia LINT 41 two-car diesel

trainsets to be delivered in 2006–07 for use in north Zealand.

UPDATED

The Local Railway

Lokalbanen A/S
Nordre Jernbanevej 36, DK-3400 Hillerød
Tel: (+45) 48 29 88 00 Fax: (+45) 48 29 88 30
e-mail: post@lokalbanen.dk
Web: www.lokalbanen.dk

Key personnel

Managing Director: Rösli Gisselmann

Organisation

Having taken over the local railways lying within its area at the beginning of 2001, the Greater Copenhagen Development Authority (HUR) set up two new companies on 17 May 2002: Greater Copenhagen Local Railways (HL) (see entry for Greater Copenhagen Local Railways (HL) in Denmark section of *Railway systems and operators*), which owns the infrastructure and rolling stock of five absorbed railways; and the Local Railway, which was to operate the five lines. The Local Railway is wholly owned by HUR.

RegioShuttle railcar at Jaegerborg on the Lyngby–Nærum line (John C Baker)
0576989

Passenger and freight operations

Commuter traffic linking the communities served with Copenhagen forms the main activity on all five lines. In 2004 passenger traffic on the individual lines was:

Eastern: 0.89 million passengers (17.57 million passenger-km)

Gribskov: 1.64 million passengers (32.35 million passenger-km)

Hillerød–Hundested: 1.6 million passengers (35.67 million passenger-km)

Helsingør–Hornbæk-Gilleleje: 0.86   million passengers (9.75 million passenger-km)

Lyngby–Nærum: 0.95   million   passengers (5.73 million passenger-km)

There is no freight traffic on four of the former railways but in 2004 the Hillerød–Hundested branch carried 113,300 tonnes (2.9 million tonne-km) from the steelworks at Frederiksværk.

The Local Railway took over operation of the Hillerød–Helsingør line from DSB S-Train in January 2005.

UPDATED

Flexliner railcar at Kagerup on the Gribskov line (John C Baker)   0576990

Lolland Railway (LJ)

A/S Lollandsbanen
Banegårdspladsen 5, DK-4930 Maribo
Tel: (+45) 54 79 17 00   Fax: (+45) 54 79 17 61
e-mail: lj@lollandsbanen.dk
Web: www.lollandsbanen.dk

Key personnel

Managing Director: Benny Esmann Jensen
Manager: Claus Pedersen

Gauge: 1,435 mm
Route length: 50.2 km

Organisation

The Lolland Railway runs from a junction with the Banedanmark Copenhagen–Rødby line just west of Nykøbing Falster westwards across the island of Lolland to the town of Nakskov. From the intermediate station of Maribo there is a goods line to the small port of Bandholm; this line is now normally only used by a railway museum.

Passenger and freight operations

In addition to operating passenger and freight services on its own line, the railway operates all freight services south of Næstved in association with Railion. In 2004 the railway carried 1.05 million

passengers for 26.67 million passenger-km and 25,064 tonnes of freight for 445,872 tonne-km.

Traction and rolling stock

LJ operates six locomotives (including five ex-DSB diesel-electrics), five tractors, four Adtranz Flexliner two-car dmus, six Duewag/Scandia or Uerdingen motor coaches and eight trailers.

UPDATED

North Jutland Railways (NJ)

Nordjske Jernbaner A/S
Skydebanevej 1, DK-9800 Hjørring
Tel: (+45) 96 24 22 20   Fax: (+45) 96 24 22 21
e-mail: info@njba.dk
Web: www.njba.dk

Key personnel

Managing Director: Preben Vestergaard

Gauge: 1,435 mm
Route length: 57.6 km

Organisation

NJ was formed in January 2001 by the merger of the 17.9 km Hjørring-Hirtshals Railway (HP) and the 39.7 km Skagen Railway (SB); both lines link the main Banedanmark line in north Jutland with the ports of Hirtshals and Skagen respectively.

Passenger and freight operations

As well as operating passenger and freight services over its two lines, NJ also handles all freight traffic north of Aalborg and operates a combi-terminal at Hirtshals.

The Sgane line improvements detailed below enabled NJ to increase the number of daily trains in each direction from 12 to 21.

Siemens Desiro railcar (left) and one of the Duewag units that have largely been replaced (NJ)
*NEW/*1143608

Improvements to existing lines

In late 2003 relaying of the Skagen line was completed and this has produced reductions in journey times as the new Siemens Desiro vehicles can operate at 100 km/h between Frederikshavn and Hulsig and at up to 120 km/h on the Hulsig–Skagen section.

In 2004 the signalling system on the Hirtshals line was replaced, with interlockings provided between stations and the installation of a CTC system that was integrated with the existing system on the Skagen line. Both lines are now regulated from a CTC facility at Skagen.

Traction and rolling stock

The NJ fleet consists of four ex-DSB diesel locomotives, two shunting tractors, five Duewag/Scandia and Uerdingen motor coaches and five trailer coaches. In late 2004 and early 2005 the line received seven Siemens Desiro diesel trainsets and these have replaced most of the older Duewag/Scandia and Uerdingen vehicles.

UPDATED

Railion Denmark

Railion Denmark A/S
Sydvestvej 21-23, DK-2600 Glostrup
Tel: (+45) 33 54 18 00   Fax: (+45) 33 54 18 54
e-mail: railion@railion.dk
Web: www.railion.dk

Key personnel

Managing and Sales Director: Christian Thing
Operations Director: Bo Eklund
Finance Director: Andres Schneider
Human Resources and Administration Manager:
    Hans Schlichter
Safety Manager: Benny Spangsborg

Political background

The change of DSB's status on 1 January 1999 and competition for freight traffic on the Danish network under open access rules from the same date resulted in rationalisation within DSB's freight

division. After a failed attempt to set up a joint freight company with the Norwegian and Swedish state railway companies, DSB finally agreed to sell the division to Railion, a subsidiary of German Rail (DB AG). The sale was finalised on 27 June

Diesel locomotives

Class	Wheel arrangement	Power kW	Speed km/h	Weight tonnes	No in service	First built	Mechanical	Builders Engine	Transmission
MZ	Co-Co	2,270	143	116.5	7	1967	N&H, Frichs	GM 16-645E3	E Thrige
MZ	Co-Co	2,865	165	123	13	1977	N&H, Frichs	GM 20-645E3	E GM
MK	B	390	60	40	41	1996	Siemens	MTU 8V 183TD13	H Voith
Tractor	B	94	45	17	20	1966	Frichs	Leyland 680	H Voith

2001, although it was backdated to 1 January 2001. Railion paid DKr170 million, which DSB then used to purchase 2 per cent of Railion. Only wagonload, intermodal and international traffic passed to Railion, as the sundries traffic had been sold to a consortium of road hauliers in May 2000 and most of this business has since been lost to rail.

Organisation

Railion Denmark is divided into three divisions: Finance; Operations; and Sales. Operations is responsible for all aspects of the operation of trains and terminals, including train planning. Information technology is responsible to Finance.

Freight operations

There are intermodal terminals at Århus, Copenhagen, Esbjerg, Høje Taastrup, and Taulov. Overnight services operate between Høje Taastrup and Århus, Esbjerg and Taulov. A weekly service between Høje Taastrup and Esbjerg runs in cooperation with DFDS-Tor Line. International services include trains to France and Italy, services from Sweden conveying timber and paper products and an overnight intermodal service from Taulov and Århus to Helsingborg in Sweden. The Scandinavian Maritime Express runs from Copenhagen and Århus to Bremerhaven conveying containers destined for overseas shipment, and a similar service between Taulov and Hamburg is run in cooperation with P&O Nedlloyd started in 2003. An intermodal service from Taulov and Padborg to Hamburg is run in cooperation with Kombi Dan and German Kombiverkehr. Wagonload traffic was further reduced during 2004 and now only consists of a core network serving the most important terminals, but some private lines provide feeder services and this is expected to increase in the future.

Traction and rolling stock

Railion Denmark operates 13 dual-voltage and 10 AC-only line-haul electric locomotives of Classes EG and EA; 20 Class MZ line-haul diesel-electric locomotives; 41 Class MK diesel shunting locomotives and tractors; and 603 wagons. The Class EG locomotives are capable of operating into both Germany and Sweden.

UPDATED

Class EA electric locomotive passing Hedehusene with an intermodal service bound for Germany (Philip Wormald) **NEW**/1115011

Railion Class EA (left) and EG electric locomotives at Padborg (Colin Boocock) 0576997

Electric locomotives

Class	Wheel arrangement	Power kW continuous	Speed km/h	Weight tonnes	No in service	First built	Builders Mechanical	Electrical
EA	Bo-Bo	4,000	175	80	10	1984	Thyssen-Henschel/Scandia	BBC
EG*	Co-Co	6,500	140	129	13	1999	Siemens	Siemens

*Dual-voltage (25 kV AC 50 Hz/15 kV AC 16$^{2/3}$ Hz)

Vemb–Thyborøn Railway (VLTJ)

Lemvigbanen A/S
Banegårdsvej 2, DK-7620 Lemvig
Tel: (+45) 97 82 32 22   Fax: (+45) 97 81 08 10
e-mail: vltj@lemvigbanen.dk
Web: www.lemvigbanen.dk

Key personnel
Managing Director: Knud Vigsø

Gauge: 1,435 mm
Route length: 59.5 km

Organisation

The line runs from the Banedanmark station at Vemb in northwest Jutland via the town of Lemvig, where the railway's depot is located, to the small port of Thyborøn.

Passenger and freight operations

In 2004 the VLTJ carried 259,890 passengers for 5.39 million passenger-km and 36,133 tonnes of freight for 1.87 million tonne-km. Most of the railway's freight traffic is generated by a chemical works near Thyborøn.

Traction and rolling stock

The VLTJ owns two ex-DSB diesel locomotives, two tractors, four Duewag/Scandia two-car diesel trainsets, one Uerdingen single unit and one former Swedish State Railways single unit.

UPDATED

Western Railway (VNJ)

Vestbanen A/S
Svinget 11, DK-6800 Varde
Tel: (+45) 76 95 21 00   Fax: (+45) 76 95 21 10
e-mail: vestbanen@vestbanen.dk
Web: www.vestbanen.dk

Gauge: 1,435 mm
Route length: 38 km

Organisation

VNJ runs from a junction with the Banedanmark network at Varde on the west Jutland coast line now operated by Arriva Tog (see entry for Arriva Tog in the Denmark section of *Railway systems and operators*) to Nørre Nebel.

On 1 June 2002 operation of VNJ was taken over by Arriva Danmark, but ownership remains with the local authority.

In late 2004 VNJ's freight business and locomotives passed to the newly formed Danish Railway (see entry in the Denmark section of *Railway systems and operators*).

Traction and rolling stock

VNJ operates three Duewag/Scandia and Uerdingen two-car diesel trainsets.

UPDATED

West Zealand Local Railways (VL)

Vestsjællands Lokalbaner A/S
Jernbaneplads 6, DK-4300 Holbæk
Tel: (+45) 59 48 50 00   Fax: (+45) 59 44 23 40
e-mail: post@vlb.dk
Web: www.vlb.dk

Key personnel
Managing Director (Acting): Niels Larsen

Gauge: 1,435 mm
Route length: 100 km

Organisation

VL was formed on 21 May 2003 by the amalgamation of the Høng-Tølløse Railway (HTJ) and the Odsherreds Railway (OHJ); the railways were both 50 km long and were jointly managed. The HTJ ran between Slagelse, on the Banedanmark main line from Copenhagen to Jutland, and Tølløse on the Copenhagen–Kalundborg line. The OHJ ran from Holbæk to the coastal town of Nykøping (Zealand).

The railway and the local authority are considering two 5 km extensions over disused lines: one north from Høng to Gørlev, the other south from Slagelse to Antvorskov.

Passenger and freight operations

An hourly service is in operation on both lines, with some extra afternoon trains between Slagelse and Høng. Some trains run over the Banedanmark network between Tølløse and Holbæk. In 2003 the lines carried 1.7 million passengers, giving 33.9 million passenger-km, and 11,525 tonnes of freight, giving 439,000 tonne-km.

Traction and rolling stock

The company operates a fleet of four former DSB diesel-electric locomotives, 13 two- or three-car diesel-trainsets (including three Adtranz-built Flexliners), one tractor and six coaches.

UPDATED

Dominican Republic

Ministry of Works and Communications

Avenida San Cristobal, Santo Domingo

Key personnel
State Secretary: E Williams

Central Romana Railroad

Central Romana, La Romana

Key personnel
President: C Morales T
Vice-President and General Superintendent:
    R J Rivera
Director, Purchases: B R Grullon

Gauge: 1,435 mm
Route length: 375 km

Freight operations
The railway operates 13 locomotives and 950 freight wagons for the transport of sugar cane.

A further 240 km of 558 mm, 762 mm and 1,067 mm gauge track is operated by the private Angelina, CAEI and Cristobal Colon sugar cane systems.

Dominica Government Railway

Santo Domingo

Gauge: 762 mm
Route length: 142 km

Freight operations
The railway operates four locomotives and 72 freight wagons. The main freight traffic comprises bananas from Guayubin moved to the port of Pepillo for export.

A total of eight sugar cane systems are operated by the semi-nationalised CEA group, totalling 986 km on 762 mm, 889 mm and 1,067 mm gauges.

Ecuador

Ministry of Public Works & Communications

1184 Avenida 6 de Diciembre y Wilson, Quito
Tel: (+593 2) 24 26 66

Key personnel
Minister: P J López T
Under-Secretary: G Uzcátegui P

State Railways of Ecuador (ENFE)

Empresa Nacional de los Ferrocarriles del Estado
PO Box 159, Calle Bolivar 443, Quito
Tel: (+593 2) 21 61 80

Key personnel
Director General: Mario Arias Salazar
Traffic Manager: Vicente Cevallos Cazar
Directors
    Administration: S Gudino C
    Finance: G Vintimilla
    Technical: M Herrera R
Motive Power Superintendent: E Benavides
Transport and Telecommunications Engineer:
    W Idrovo
Permanent Way Engineer: Marco Redrobán A
Managers
    Quito-San Lorenzo Division: G Gallo
    Sibambe-Cuenca Division: M Montalvo

Gauge: 1,067 mm
Route length: 965.5 km

Political background
In September 1999, Ecuador's National Modernisation Council (CONAM) presented proposals for privatising the national rail network. CONAM expected to sell three lines for the following prices: Quito–Guayaquil for US$97 million; Ibarra–San Lorenzo for US$10 million; and Ibarra–Cajas for US$1.5 million. A technical and economic study was to have been drawn up by CONAM but was still not available in April 2000. The change from the sucre to the US dollar as national currency is reported to have caused economic difficulties which would have an impact on the proposed privatisation.

Organisation
ENFE is composed of three divisions. The main line (the 446.7 km Guayaquil–Quito Division) connects Durán, located on the opposite bank of the river from Ecuador's main port of Guayaquil, with Quito, which lies at some 2,800 m altitude in the the the Andes. From Durán the line runs across low-lying plains for 87 km to Bucay, at the foot of the western slopes of the Andes. Over the next 79 km the line climbs 2,940 m at an average grade over the whole section of 3.7 per cent (1 in 27). The line strikes many sharp curves, and several stretches are laid on a grade of 5.5 per cent (1 in 18), including a double zigzag which was required to negotiate a particularly awkward mountain outcrop known as the *Nariz del Diablo* (Devil's Nose). Once the summit of this section is reached at Palmira, 3,238 m in altitude and 166 km from Durán, the line remains in the high Sierra, never falling below 2,500 m, and rising to 3,609 m at the overall summit of Urbina, 264 km from Durán.

After severe floods in 1992 and 1994 the line was completely relocated and rebuilt near Tixan (from Alausí down to the *Nariz del Diablo* to avoid future breaches). In 1992 floods and high waters of the Rio Chanchán destroyed many bridges on the Sibambe–Huigra–Bucay section, and they have mostly been replaced by new steel structures. Floods during the El Niño weather phenomenon washed away several parts of the lowland line between Durán and Bucay and also some stretches between Bucay and Huigra, making any use of the lower sections of this line impossible. Some repairs have been carried out by ENFE near Huigra. In 1999, ENFE estimated the cost of rebuilding the Durán–Bucay–Sibambe line at US$1 million, with work expected to be completed by October 2000.

In late 1999, a weekly round trip by a mixed train between Quito and Riobamba was introduced, in addition to the three times-weekly railbus between Riobamba/and Sibambe and the Sunday Quito–Cotopaxi excursion train.

The 373.4 km Quito–San Lorenzo Division runs northwest from Quito to the coastal town of San Lorenzo, near the border with Colombia. The Quito–Ibarra section of this Division has been out of regular service since the mid-1980s, and subsequently a short section was made inoperable when a parallel road was widened. In the early 1990s a railcar service was introduced between Ibarra and Otavalo to convey tourists to the Otavalo Indian market, but rolling stock problems

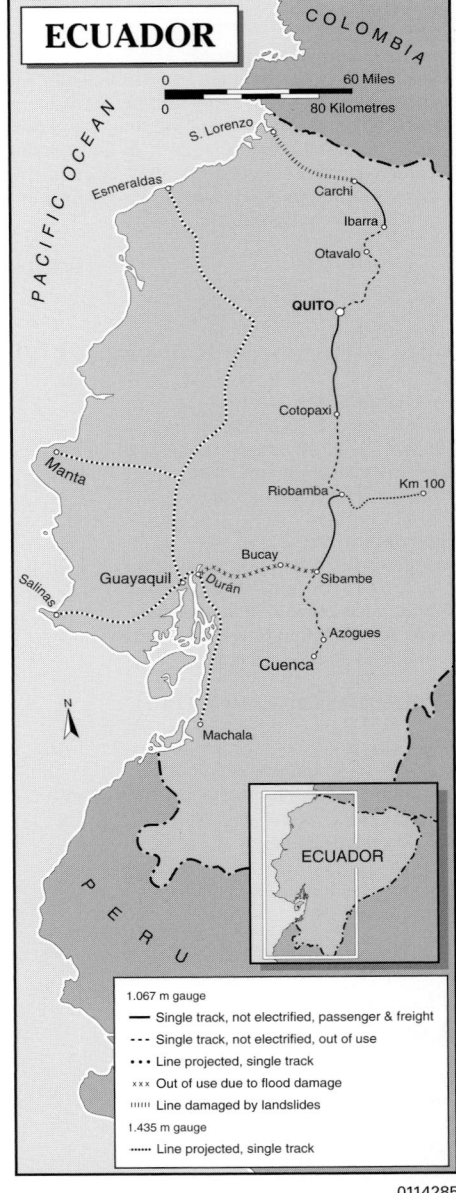

ECUADOR

0	60 Miles
0	80 Kilometres

1.067 m gauge
—— Single track, not electrified, passenger & freight
--- Single track, not electrified, out of use
••• Line projected, single track
××× Out of use due to flood damage
⁞⁞⁞⁞ Line damaged by landslides
1.435 m gauge
······ Line projected, single track

0114285

have since led to its withdrawal. The Ibarra–San Lorenzo section has also suffered from landslides after heavy rainfall and has been repaired on several occasions. The most vulnerable section is around Carchi.

Until 1997 the railway had an effective monopoly in both passenger and freight transport in this part of the country, but the construction of a road on a generally parallel route has had a severe impact on traffic. Heavy rains in early 1999 again disrupted traffic and by May 1999 no signs of reconstruction had been reported.

ENFE's Sibambe–Cuenca Division links Sibambe, 131 km from Durán on the main line, with Cuenca, an important provincial capital in the southern part of the country. On its northernmost section, near Sibambe, the line has been blocked on several occasions due to landslides and in the mid-1990s only sporadic traffic was operated on the southernmost part of the line between Azogues and Cuenca. However, in 1996 the line was closed completely near Cuenca when the track was covered in asphalt as part of a road improvement scheme.

In December 1999, 517 km of line were reported to be open for services.

Late in 1999, staff members were reduced to fewer than 400.

Passenger and freight operations

Passenger traffic has fallen from some 1.5 million passenger journeys in 1993 to around 110,000

in 1998. Freight traffic fell from 37,000 tonnes to 5,000 tonnes over the same period.

New lines

The principal ambition of ENFE has been to build a new north-south axis connecting the oil port of Esmeraldas, in the north of the country on the Pacific coast, with Machala, the inland centre of banana plantations south of Guayaquil. It would not entail difficult civil engineering. There would be a case for up to three branches to the line, one to serve a major economic development area based on prawn farms at Manta.

ENFE has also undertaken studies to improve rail access to Guayaquil. Both transhipment by either lorry or river barge and a new rail barge were considered. In 1996 US oil company Occidental Exploration & Production commissioned a feasibility study for a new 100 km 1,435 mm gauge route to bring freight and passengers to oilfields under development in the Amazon basin. This would have 300 tonne freight trains running at 50 km/h.

Improvements to existing lines

In April 1994, the ENFE board approved a Su212 billion investment programme which included track improvements and resleepering on the San Lorenzo–Ibarra and Riobamba–Durán sections of the system, as well as station modernisation.

Traction and rolling stock

The nine new Bo-Bo-Bo locomotives ordered from GEC Alsthom in 1991 with French aid began trials in Ecuador in August 1992. Designated Class 2400, each machine cost US$3.3 million; they are powered by SEMT Pielstick Type 12 PA4 V200 VG engines with a UIC rating of 1,780 kW.

ENFE's other motive power comprises a mix of Alco DL535B Co-Cos and GEC Alsthom 960 hp B-B-Bs, steam locomotives and a handful of diesel railcars. Much of the fleet is unserviceable and many items of equipment are marooned by severed rail connections.

In December 1999 the fleet stood at:

Locomotives and rolling stock: 13 diesel-electric (4 reported serviceable), 9 steam (none serviceable); 15 railcars (4 serviceable); 50 passenger coaches (less than half serviceable); 120 freight wagons (fewer than 100 serviceable).

Type of coupler in standard use: Automatic
Type of brake in standard use: Air

Track
Rails: 35, 30, 27.5 and 22.5 kg/m
Sleepers: Wood, 2,000 × 200 × 180 mm
Spacing: 1,700/km
Max curve radius: 20°
Max gradient: 5.5%
Max axleload: 15 tonnes

Egypt

Ministry of Transport and Communications

Cairo
Tel: (+20 2) 355 55 66   Fax: (+20 2) 355 55 64

Key personnel
Minister: Hamdi Abdel Salam

Egyptian National Railways (ENR)

Station Building, Ramses Square, Cairo
Tel: (+20 2) 575 10 00; 577 13 88
Fax: (+20 2) 574 00 00

Key personnel
Chairman: Eng Mohamed Arafa Al-Newaem
Deputy Chairmen
  Regions: Eng Harrafee Mahmoud Abdelqaoui
  Permanent Way and Signalling:
    Eng Aschraf Salamee Abu Zeid
  Technical, Purchasing and Stores:
    Magdee El-Ezb Amine Shalabee
  Construction: Eng Nawal Taha Mahmoud
  Financial and Administration Affairs:
    Rushdy Mahmoud El-Khatib
  Operations and Commercial Affairs:
    Eng Eid Abd El Kader Metwalli Awad

Gauge: 1,435 mm
Route length: 5,063 km
Electrification: 62 km at 1.5 kV DC

Organisation
Egyptian National Railways extends from the Mediterranean up the Nile Valley, serving the Nile Delta, Cairo, Alexandria, Port Said, Ismailia, Suez and connecting at Sadd el Ali, its southernmost point, with the river steamers of Sudan Railways. From El Quantara, on the Port Said–Ismailia line, a branch runs east following the coast and connects with Israel Railways; it has been disused for many years. A significant recent expansion of the network was the opening in 1996 of an west-east line in Upper Egypt, from Abu Tartour to the Red Sea port of Safaga.

In 2003 ENR employed some 73,000 staff.

Traffic (million)	1999	2000
Passenger journeys	842	842
Passenger-km	60,040	60,052
Freight tonnes	10.84	12.5
Freight tonne-km	–	–

Passenger operations
In modern times ENR has been primarily a passenger railway. Passenger traffic grew to some 462 million journeys by 1990 (compared with 313 million journeys in 1980) and jumped substantially thereafter to reach 775 million

journeys in 1997 (60.62 billion passenger-km). This growth continued into 2000, when 842 million passenger journeys were recorded. This was due largely to the huge success of the cross-city tunnel linking Cairo's two busiest commuter lines. In 2003 ENR was operating 1,315 passenger trains daily.

Three 10-car turbotrain sets built by ANF Industrie provide prime service over the 208 km between Cairo and Alexandria. It is hoped eventually to exploit this equipment's 160 km/h capability to achieve a Cairo–Alexandria transit in 1 hour 30 minutes, with each of the trains

0552563

completing three round trips daily. At present, however, they are confined to 140 km/h, and that only over certain sections. In the Cairo area speed cannot exceed 60 km/h for some 25 km distance.

Luxury air conditioned overnight trains using refurbished stock operate on the Cairo–Luxor–Aswan route. Infrastructure upgrading on this popular tourist route has included double-tracking and resignalling. Daytime services have benefited considerably from the track upgrading, permitting operation at 140 km/h of four trainsets supplied by ABB Henschel (now Adtranz) and Hyundai.

A computerised seat reservation system is used for services equipped with air-conditioned stock. ENR plans to extend this system to 32 principal stations by the end of 2004.

Freight operations
Freight too has advanced, though ENR has a market share of only around 12 per cent. In 1990, 8.6 million tonnes were hauled for 2,827 million tonne-km, but this had increased to 12.5 million tonnes by 2000. The main constituent is ore carried on the Baharia–Helwan line, with wheat and oil next in importance. Container traffic has also experienced substantial growth, sufficient to encourage ENR to commission studies into the establishment of intermodal terminals, or 'dry ports', in major towns and cities.

New lines
Freight carryings are expected to receive a further boost of some 7 million tonnes annually with the opening of the long-planned line to tap phosphate deposits at Abu Tartour. The 235 km initial section of this export corridor, linking the Cairo–Aswan main line at Qena with the Red Sea port of Safaga, was opened in 1984, but construction of the western portion up to Abu Tartour was not started until much later. Work on this difficult project, which includes a crossing of the Nile at Nag'Hammadi, was completed in October 1996. A 46 km branch south from El Kharga to Baris is planned.

The infrastructure of a 70 km line from Port Said to the Nile Delta has been completed.

Another longstanding project is that for construction of a line into neighbouring Libya from ENR's northwestern terminal at Salûm, now revived as a joint venture between Egyptian and Libyan interests. Feasibility studies of the 130 km alignment have been made by ENR and in 1996 it was announced that Libya would build the line to Salûm from Tobruk with technical assistance from Egypt. This would connect the ENR system to a proposed 3,170 km network in Libya.

The long-closed link between Ismailia and the Sinai peninsula was re-established in November 2001 following the opening of a new rail/road swing bridge across the Suez Canal at El Ferdan. At 640 m, this is the world's longest swing bridge. It was built by Consortium El Ferdan Bridge, led by Krupp Stahlbau of Hanover. This is part of a larger project to provide a 225 km rail link from Ismailia to Rafah, close to the Gaza border. The first phase of the scheme, which has been completed, is a 100 km line from El Ferdan to Beer Al-Abd with six stations. The line to Rafah is being built for 160 km/h running, and will be CTC-controlled.

In 1998 ENR undertook preliminary studies into a 500 km line to Sudan, linking Aswan with the northern Sudanese town of Wadi Halfa. Other projects studied or developed include: a

ENR local service headed by a Henschel/GM AA22T locomotive at Luxor (Colin Boocock)   0059393

Diesel locomotives

Class	Wheel arrangement	Power kW	Speed km/h	Weight tonnes	No in service	First built	Mechanical	Builders Engine	Transmission
3222/3445 (JT22MC)	Co-Co	1,845	140	111	45	1979	EMD, GM Canada	645E3-12	*E* EMD
3016 (AA22T)	Co-Co	1,845	120	122	95	1976	Henschel	645E3-12	*E* EMD
3085 (AA22T-2)	Co-Co	1,845	120	122	123	1981	Henschel	645E3-12	*E* EMD
3001/3189/3271 (AA22T/DB)	Co-Co	1,845	120	122	56	1979	Henschel	645E3-12	*E* EMD
3222/3445 (JT22MC)	Co-Co	1,845	–	–	45	1979	EMD, GM Canada	645E3-12	*E* EMD
34xx (G12)	A1A-A1A	1,230	–	–	4	–	EMD	645E-12	*E* EMD
3601/3801 (G22W/G22W AC)	Bo-Bo	1,230	105	80	299	1977	EMD, GM Canada	645E-12	*E* EMD
2001 (DE2550)	Co-Co	1,845	80	132	45	1995	Adtranz	645E3-12	*E* EMD
2101 (DE2550)	Co-Co	1,845	–	–	23	1998	Adtranz	645E3-12	*E* EMD
2301 (CP18-61)	A1A-A1A	1,380	–	–	30	1997	General Electric	FDL-8	*E* GE
GA-900*	Bo-Bo	900	–	–	30	2004	ALSTOM	MTU 8V-396	*E* ALSTOM

*On order for delivery from 2004

new branch line to serve iron ore deposits near Aswan; a branch from the Cairo–Suez line to serve a proposed new port on the Gulf of Suez; and a branch from the planned Ismailia–Rafah line to serve a new port east of Port Said.

In 2000, a contract was signed with Spanish consultants to conduct a study into a high-speed line between Alexandria and Cairo, including a possible extension to Aswan. This project now features in ENR's long-term ambitions.

Improvements to existing lines
A recent priority has been renovation of the 345 km line built shortly after the Second World War with Soviet aid from Helwan to Baharia, which is primarily an ore and coal carrier. Improvements included complete track renewal and protective measures to prevent service interruptions caused by drifting sand. CTC is managed from a control centre in Cairo.

Much of ENR's track maintenance and renewal is undertaken by private-sector contractors, with both Franco-Egyptian and German-Egyptian joint venture companies active.

In the longer term ENR foresees electrification of its Cairo–Alexandria main line.

Traction and rolling stock
In 2003, rolling stock totalled 671 main line diesel locomotives, three 10-car turbotrains, 50 multiple-units, 3,160 passenger cars, 38 diesel railcars and 10,598 freight wagons.

A fleet of 700 bottom-discharge wagons for the Abu Tartour phosphates traffic was supplied by local manufacturer Semaf. They are hauled by 45 DE2250 single-cab GM-powered 1,845 kW diesel-electrics delivered during 1995–96 by Adtranz Germany. A further 23 similar locomotives, but in a dual-cab configuration, have been delivered for more general duties in the Nile Delta area.

In 2000 ENR ordered 30 MTU-powered Type GA900 900 kW diesel-electric shunting locomotives from ALSTOM in Valencia. Deliveries were due to commence in 2004.

Electrification
The double-track suburban main line from Cairo to Helwan (42 km) is electrified at 1.5 kV DC, and forms Line 1 of the Cairo metro system.

El Salvador

Ministry of Public Works

1a Avenida Sur 630, San Salvador
Fax: (+503) 271 01 63

Key personnel
Minister: Jorge Alberto Sansivirini
Deputy Minister: Roberto Bara Osegueda

El Salvador National Railways (FENADESAL)

Ferrocarriles Nacionales de El Salvador
PO Box 2292, Avenida Peralta 903, San Salvador
Tel: (+503) 71 56 32   Fax: (+503) 71 56 50

Key personnel
President (CEPA): C H Figueroa
Managers
  General (CEPA): A German Martinez
  Operations (CEPA): J A Nunez
    Jose Eriberto Erquicia

General (FENADESAL): Tulio Omar Vergara
Purchasing: Luis Antonio Guzmán
Finance: Fredy Antonio Mayora
Traffic: R Marín
Personnel: Wilfredo Ciudad Real
Transportation: L A Carballo
Rolling Stock and Equipment:
  Julio Fernando Pienda
Maintenance of Way: Andres Abelino Cruz

Gauge: 914 mm
Route length (operational): 283 km

Organisation

FENADESAL was formed from two railways which were formerly the property of overseas companies: the Salvador Railway, which passed to the state in 1965 under the name of Ferrocarril de El Salvador (FES); and the International Railways of Central America (IRCA), a railway undertaking that included the railway system and port at Cútuco, which was nationalised in 1974 under the name of Ferrocarril Nacional de El Salvador (FENASAL).

The two undertakings were merged under state control in May 1975 (together with the port of Cútuco) and became FENADESAL, which is administered by the port authority CEPA.

The railway is divided into three districts: District No 1 which comprises San Salvador (the capital) to the port of Cutúco in the east of the country (252 km); District No 2 which runs from San Salvador to the frontier of El Salvador with Guatemala (146 km), and a branch to Santa Ana, in the west of the country (20 km); District No 3 which runs from San Salvador to the port of Acajutla, on the Pacific Ocean (104 km), and includes a branch from Sitio del Niño to Santa Ana in the west (40 km).

Finance

The railway is heavily subsidised. In 1997 its revenues from passenger and freight traffic and other sources totalled c11.9 million, against total expenditure of c48.1 million.

Investment totalling c3.07 million was planned for 1996, with c1.25 million to be spent on major track improvements, c0.28 million on telecommunications, c0.24 million on computer systems and c1.3 million on a locomotive fuelling facility.

Traffic (million)	1996	1997
Passenger journeys	0.313	0.460
Passenger-km	4.8	7.1
Freight tonnes	0.161	0.313
Freight tonne-km	17.3	–

Passenger operations

By the start of 1996 FENADESAL had ceased to operate passenger services, with the exception of over a short stretch of line between Armenia and Sonsonate, which received some attention by introducing a refurbished railbus which operates four times daily in each direction. On market days in Armenia the railbus is replaced by a mixed train. Occasional services hauled by either steam or

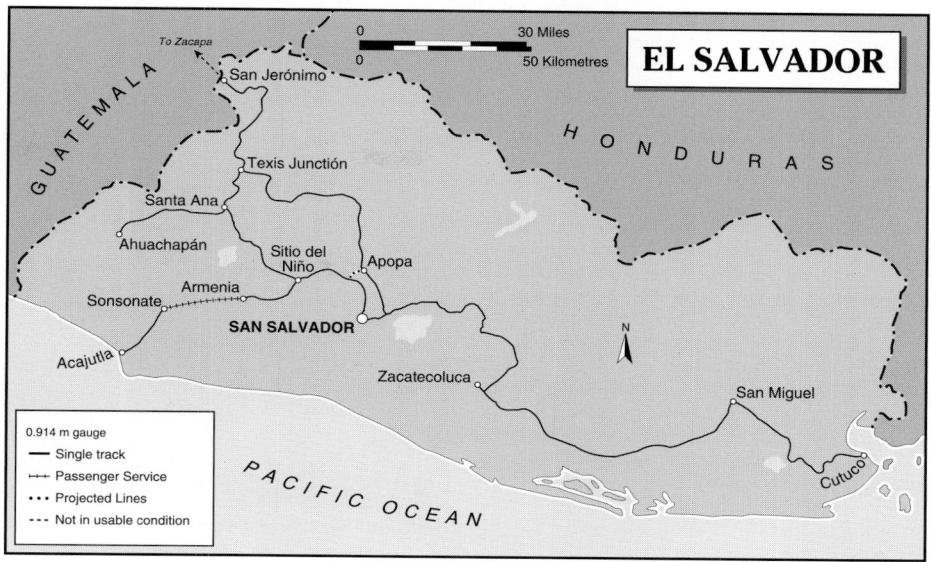

0058560

diesel traction, mostly for tourist groups, operate between Sonsonate, Armenia and Sitio del Niño.

Improvements to existing lines

A group from Taiwan has expressed interest in upgrading existing track to allow international trains to run through to the port of Cortés in neighbouring Honduras. Ferrovías Guatemala is studying the possibility of reconstructing the international link from San Jerónimo to Zacapa in Guatemala to provide a corridor for international traffic to Mexico, the USA and the port of Puerto Barrios in Guatemala.

Traction and rolling stock

Operational resources at the end of 1997 totalled: 10 diesel-electric locomotives (plus 2 unserviceable); 1 steam locomotives (plus 1 awaiting restoration);

2 railcars; 16 passenger coaches; 370 freight wagons.

Track

Rail: ASCE 54, 60, 70 and 75 lb/yd
Sleepers: Hardwood 7 ft x 6 in by 8 in
Fastenings: Standard, spike and angle bar
Spacing: 1,725/km in plain track 1,850/km in curves
Min curvature radius: 328 ft
Max gradient: 3%
Max axleloading: 12.5 tonnes

Diesel locomotives

Class	Builder's type	Wheel arrangement	Power kW	Speed km/h	Weight tonnes	No in service	First built	Mechanical	Builders Engine	Transmission
850	GA-8	Bo-Bo	595	57	61	10	1965	GM	GM 567 CR	E EMD
700	U6B	B-B	280	60	47	2	1956	GE	Caterpillar D-397	E GE

Eritrea

Ministry of Transport

PO Box 204, Asmera
Tel: (+291 1) 11 43 07
Fax: (+291 1) 12 70 48

Key personnel
Minister: Dr Giorgis Teklemikael

Eritrean Railways

PO Box 569, Asmera
Tel: (+291 1) 11 43 07   Fax: (+291 1) 12 70 48

Key personnel
Director: Amanuel Ghébrésélassié

Gauge: 950 mm
Route length: 306 km

Political background

Following achievement of independence from Ethiopia in 1993, the government moved to revive parts of the former Northern Ethiopia Railway. This 306 km line, linking Ak'ordat with the capital Asmera and the Red Sea port of Mits'iwa had been out of use since 1978. Consultants determined that the original alignment could be rehabilitated, and a small fleet of rolling stock rebuilt using existing spare parts.

0547488

By the end of 2000, the Mits'iwa–Ghinda section had been reopened and reopening of the remaining section to Asmera, 118 km from Mits'iwa, was achieved in 2003.

Plans exist to continue reconstruction west of Asmera to Bisha and to create a new international link to Kassala, Sudan.

Passenger and freight operations
A railcar-operated passenger service and tourist trains operate between Mits'iwa and Ghinda. A daily freight service is provided between Damas and Ghinda. Services were to be extended to Asmera on rehabilitation of the line from Ghinda.

Traction and rolling stock
Serviceable traction in 2003 comprised two diesel locomotives, including one Krupp-built diesel-hydraulic machine, three steam locomotives of Italian origin and two Fiat railcars. Two Soviet-built road-rail vehicles were also in use.

ER Krupp-built B-B diesel-hydraulic locomotive at Mai Atal (Graham Vincent)   0536195

Estonia

Ministry of Economic Affairs and Communications

Harju 11, EE-15072 Tallinn
Tel: (+372 6) 25 63 42   Fax: (+372 6) 31 36 00
e-mail: info@mkm.ee
Web: www.mkm.ee

Key personnel
Minister: Edgar Savisaar
Secretary-General: Marika Priske
Deputy Secretary-General, Maintenance of Transport: Andres Tint
Executive Officer, Roads and Railway Department: Eda Rembel
UPDATED

Estonian Railway Inspectorate

Lastekodu Street 31, EE-10113 Tallinn
Tel: (+372 6) 05 74 01   Fax: (+372 6) 05 74 11
e-mail: raudteeinspektsioon@risnp.ee
Web: www.rinsp.ee

Key personnel
Director General: Jüri-Karl Seim

Organisation
The Estonian Railway Inspectorate is a government agency under the supervision of the Ministry of Economic Affairs and Communications. Its Surveillance Department is responsible for the safe and efficient operation of railways in Estonia, covering industry monitoring, technical supervision, establishing standards and technical norms, granting operating licences and certificates. The Administration Department maintains registers of rail traffic and the Development Department manages and co-ordinates investment funding and organises the work of the Capacity Allocation Commission.
NEW ENTRY

Edelaraudtee AS

Kaare 25, EE-72213 Türi
Tel: (+372 38) 571 23   Fax: (+372 38) 571 21
e-mail: edel@edel.ee
Web: www.edel.ee

Key personnel
Director: Kulvi Pukka

Gauge: 1,520 mm
Route length: 265 km

Political background
Edelaraudtee (South West Railway) commenced operations as an independent business in January

Map of Estonia showing the rail network, with legend:
1,524 mm gauge
— In operation
+++ Electrified
∘-∘-∘ Independently owned (out of use)
- - - Out of use
-·-·- Passenger services suspended

1114993

Edelaraudtee Class DR1A diesel trainset at Lehtse forming a Tapa–Tallinn service (Philip Wormald)
*NEW/*1114991

1997 in anticipation of rail privatisation legislation approved by parliament during the same year. In 2000 Edelaraudtee, which also owns the South Western narrow-gauge lines and runs passenger trains elsewhere, was sold to the British company GB Railways via its subsidiary GB Railways Eesti AS.

Passenger operations
Edelaraudtee provides diesel-operated domestic passenger services on the Tallinn–Lelle–Pärnu/Viljandi, Tallinn–Narva, Tallinn–Tartu, Tartu–Valga and Tartu–Orava routes.

Freight operations
Freight services are provided on the Tallinn–Lelle–Pärnu/Viljandi routes.

Traction and rolling stock
Edelaraudtee operates a fleet of 11 diesel-electric locomotives of ChME3 and M62 types. DR1A diesel trainsets are used for passenger services.
UPDATED

Elektriraudtee AS

Vabadse pst 176, EE-10917 Tallinn
Tel: (+372 6) 73 74 06   Fax: (+372 6) 73 74 00
e-mail: info@elektriraudtee.ee
Web: http://www.elektriraudtee.ee

Key personnel

Chairman: Ardo Ojasalu
Managing Director: Tarmo Olgo
Management Board Members:
   Alar Kaup, Rein Riisalu

Political background

Initially a wholly owned subsidiary of Estonian Railways, Elektriraudtee commenced operations as an independent business on 1 January 1999 in response to rail privatisation legislation approved by parliament in 1997. In November 2000 ownership of the company passed from Estonian Railways to the state, which subsidises services under an agreement dating from April 1998. Full or partial privatisation of the company is planned.

Passenger operations

Elektriraudtee provides suburban commuter services on a six-line 132 km electrified network around Tallinn and in Harju county. Trains serve 36 stations.

In 2000–01 passenger journeys totalled 3.6 million.

Traction and rolling stock

Services are provided by a fleet of 69 emu cars built by Riga, Latvia. Some of these have been re-formed as two- and three-car units for use on less heavily used services.

Rolling stock maintenance is undertaken at a depot at Pääsküla.

Elektriraudtee Type ER2S four-car emu at Tallinn Balti Jaam (Eddie Barnes)   0059395

Electric railcars or multiple-units

Class	Builder's type	Cars per unit	Line voltage	Motor cars per unit	Motored axles/car	Output/ motor kW	Speed km/h	Units in service	Builders Mechanical	Electrical
ER1		4/6	3.3 kV	2/3	Bo-Bo	200	130	7	M: Riga (RVR) T, P: Kalinin	Riga (RER)
ER2	62–61	4/6	3.3 kV	2/3	Bo-Bo	200	130	4	Riga (RVR)	Riga (RER)
ER12	62–251	4/6	3.3 kV	2/3	Bo-Bo	200	130	3	Riga (RVR)	Riga (RER)

Estonian Railways (EVR)

AS Eesti Raudtee
Pikk Str 36, EE-15073 Tallinn
Tel: (+372 6) 15 86 10   Fax: (+372 6) 15 87 10
e-mail: raudtee@evr.ee
Web: www.evr.ee

Key personnel

Chairman, Supervisory Board: Edward A Burkhardt
Vice-Chairman, Supervisory Board:
   Guido Sammelselg
Supervisory Board Members: Jüri Käo;
   Henry Posner III; Marek Uusküla
Chairman of the Management Board and Chief
   Executive Officer: Christopher Aadnesen
Deputy Chief Executive Officer: Riivo Sinijärv
Finance Director: Stephen Archer
Infrastructure Director: Kaido Simmermann
Marketing Director: Rene Varek
Director: Mark Rosner

Gauge: 1,520 mm
Route length: 695 km

Political background

Following the declaration of independence by the three Baltic states of the former USSR in 1991, each country set up its own railway organisation. Estonian Railways became functional at the start of 1992. In 1997 the government approved plans to restructure EVR in preparation for privatisation. This followed studies by Belgian, Finnish and Irish consultancies. As part of this process three additional companies were established covering parts of the former state system: Edelaraudtee; Elektriraudtee and EVR Ekspress (see entries in Estonia section of *Railway systems and operators*). This enabled EVR to focus on its core business, primarily freight, as privatisation of it was pursued. A shareholding of 66 per cent of EVR was sold, making it the first Eastern European state railway company to be privatised, and the first to maintain vertical integration.

The successful bidder for EVR was Baltic Rail Services (BRS), a consortium of UK-based Jarvis International (25.5 per cent), US-based Railroad Development Corporation (5 per cent), Rail World

(25.5 per cent) and Estonian investment company Ganiger OU (44 per cent). The transaction was completed in August 2001. At the end of 2001 the AIG Emerging Europe Infrastructure Fund joined the BRS consortium.

With Estonia now a member of the EU, open access conditions prevail, with capacity allocation functions performed by a government agency, the Railway Inspectorate via a Committee for Capacity Allocation.

Traffic (million)	2001	2002	2003
Freight tonnes	38.47	42.13	42.09
Freight tonne-km	–	–	–

Organisation

EVR is organised as four principal divisions: Transportation; Infrastructure; Rolling Stock; and Tapa Depots. In January 2004 the company employed 2,674 staff.

Freight operations

Freight traffic is mainly concentrated on through routes to/from Moscow and St Petersburg. Principal commodities carried by international services include oil products, accounting for some 70 per cent of traffic, cereals, chemicals and fertilisers. Development of this transit traffic is regarded as a key priority. The main commodities carried by domestic services are oil shale and oil products from the Kohtla-Jarva and Johvi mines eastwards to Narva, and forest products.

Traffic between the border with Russia and Estonia's ports increased by 20 per cent in 2002 but has levelled out the following year. EVR

Two EVR GE Type C36-7i diesel-electric locomotives at Lehtse on the Tapa–Tallinn line with a heavy oil train from Russia (Philip Wormald)   **NEW**/1114992

attributed this to icing conditions in Estonian ports, leading to traffic being diverted elsewhere. This underlines the competitive environment in which EVR operates, with Russian shippers able to route export traffic via ports in other Baltic countries as well as those in Russia.

Modernisation of the traction fleet has enabled EVR to increase the maximum trailing load of its services from 4,000 to 5,500 tonnes.

New lines
Estonia is a participant in the proposed Rail Baltica project to create a new standard gauge high-speed rail link between the Baltic states and Poland. While this would be primarily a passenger line, it potentially could carry freight, with an impact on EVR business. In mid-2005 a clear strategy for the project remained to be established.

Improvements to existing lines
A priority for EVR has been the reconstruction and upgrading of the crucial Tallinn–Narva line, which handles transit traffic to/from the Russian Federation. This was started before privatisation in 1997 and continued under the network's new ownership. Ongoing work has included doubling single-line sections and renewing track with long welded UIC60 rail, concrete sleepers and modern

Diesel locomotives

Class	Wheel arrangement	Power kW	Speed km/h	Weight tonnes	No in Service	Mechanical	Builders Engine	Transmission
ČME3	Co-Co	1,000	95	123	12	ČKD-Sokolovo	ČKD K65310DR	E ČKD-Trakce
C36-7i	Co-Co	2,640	–	160	58	GE	GE 7FDL16	E GE
C30-7i	Co-Co	2,240	–	160	19	GE	GE 7DFL12	E GE

fastenings or in other instances welding existing jointed R65 rail.

To accommodate heavier US-built diesel locomotives, the EVR main line network was upgraded in 2002–03 to accept an axleload of 31 tonnes. Passing loops on single-line sections, especially linking Russia's October Railway and Tapa, were being lengthened to enable longer trains to be run, and widespread improvements to the railway's signalling system were implemented.

A turntable has been installed at Muuga, Tallinn, to turn the single-cab GE locomotives that EVR acquired. At the same site a new locomotive and fuelling facility has been established.

Traction and rolling stock
As part of its upgrading strategy, during 2002 EVR acquired 77 Type C36-7i and C30-7i ex-UP and

Conrail diesel-electric locomotives refurbished and upgraded at GE plants in Mexico to replace most of the Russian- and Ukrainian-built machines that it inherited. Only 12 Class ChME3 heavy shunting locomotives were retained from the Soviet era.

EVR has strengthened its wagon fleet by acquiring 500 tank wagons, 300 covered hopper wagons and 50 container flats from Russian builders.

Track
Rail type: R43, R59, R65, UIC 60
Sleepers: wood; concrete (1,840/km)
Min curvature radius: 600 m
Max gradient: 1.6%
Max permissible axleload: 31 tonnes on main lines; 22.5 tonnes elsewhere

UPDATED

EVR Ekspress AS

Telliskivi 62, Tallinn
Tel: (+372 6) 615 67 22   Fax: (+372 6) 631 12 30

Key personnel
Chairman: Vahur Karniol
Management Board: Tarmo Jürisson; Anti Selge

Political background
EVR Ekspress commenced operations as an independent business on 1 April 1999 in response to rail privatisation legislation approved by parliament in 1997. Following an international bidding competition, a majority shareholding of 51 per cent was acquired by a private company, the Fraser Group.

The remaining share is held by Estonian Railways.

Passenger operations
EVR Ekspress took over the international passenger services of Estonian Railways. The main routes are to Moscow, St Petersburg and Warsaw.

Ethiopia

Ministry of Transport and Communications

PO Box 1238, Addis Ababa
Tel: (+251 1) 51 61 66   Fax: (+251 1) 15 80 45

Key personnel
Minister: Dr A Hussein

For map see entry for **Eritrea**.

Chemins de Fer Djibouti-Ethiopien (CDE)

PO Box 1051, Addis Ababa
Tel: (+251 1) 51 72 50   Fax: (+251 1) 51 39 97

Key personnel
General Manager (Acting): Abraha Habte-Egzi

Gauge: 1,000 mm
Route length: 781 km

Political background
In 2004 there were moves to part-privatise the railway on the basis of a 51 per cent concession. A shortlist of six bidders was approved by the railway's board in October 2004. It was expected that the successful candidate would be named in June 2005.

Organisation
The railway runs from the port of Djibouti to Addis Ababa, a route length of 781 km, of which 100 km are in Djibouti. Since 1982 the railway has been under joint control of the republics of Djibouti and Ethiopia, with headquarters in Addis Ababa. The transport ministers of the two countries occupy the positions of president and vice-president.

Traffic
Both rail and highway traffic and development have been severely affected by guerrilla activity since the 1970s. Rail traffic volumes have also been depressed by economic factors and poor availability of traction. Privatisation proposals

foresee investments that would raise the railway's freight capacity from around 240,000 tonnes annually to 1.5 million tonnes.

Traction and rolling stock
Line-haul diesel locomotive stock consists of Alsthom units delivered between 1955 and 1985. In 1997 these were supplemented by four second-hand Alsthom locomotives procured from FEVE in Spain. There are also seven locotractors and six 400–700 kW diesel railcars, all supplied from France, two by Soulé in 1984–85. Around 30 passenger coaches and some 500 wagons are believed to feature in the railway's stock.

Track
Rail type: 20 kg, 25 kg, 30 kg
Sleepers: Metalbloc
Fastenings: Clips and bolts
Max axleloading: 13.7 tonnes

UPDATED

Finland

Ministry of Transport & Communications

PO Box 31, FIN-00023 Government
Eteläesplanadi 16-18, Helsinki
Tel: (+358 9) 160 02   Fax: (+358 9) 16 02 85 96
e-mail: info@mintc.fi
Web: www.mintc.fi

Key personnel
Minister: Leena Luhtanen
Permanent Secretary: Juhani Korpela
Director, Railways and Aviation Unit:
  Lampinen Reino

UPDATED

Finnish Rail Administration (RHK)

Ratahallintokeskus
Kaivokatu 6, PO Box 185, FIN-00101 Helsinki
Tel: (+358 9) 58 40 51 11   Fax: (+358 9) 58 40 51 00
e-mail: info@rhk.fi
Web: www.rhk.fi/english/index.html

Key personnel
Director General: Ossi Niemimuukko
Directors
  Traffic System: Anne Herneoja
  Safety: Kari Alppivuori
  Maintenance: Markku Nummelin

Project Management: Kari Ruohonen
Administrative: Hannu Mäkikangas
Senior Legal Counsel:
  Rami Metsäpelto
Head of International Affairs:
  Kari Konsin
Communications Manager:
  Timo Saarinen
Financial Manager: Lisbeth Laine
Environmental Manager: Arto Hovi
IT Manager: Teuvo Eronen

Gauge: 1,524 mm
Route length: 5,741 km
Electrification: 2,619 km at 25 kV 50 Hz AC

Political background

Under the Rail Network Act 1994, ownership of Finland's railway infrastructure rests with the Finnish Rail Administration, which started operations on 1 July 1995 as a government agency reporting to the Ministry of Transport and Communications.

The Finnish Rail Administration is a small supervisory body, with a headquarters staff of about 100. It has overall responsibility for network development, railway safety, operator licensing and type approvals, and awards contracts (annual value about €400 million) for track maintenance, construction work and traffic control. The main supplier is VR-Track Ltd (a division of the national railway company, Finnish Railways), but all infrastructure work became subject to competitive tendering in 1997.

In 2004 RHK initiated a process of inviting tenders for track maintenance in northern Finland with the intention of awarding contracts to take effect from July 2005. Previously this work has been undertaken against a fixed price annual agreement with VR-Track Ltd. By 2010 similar tendering arrangements will have been put in place for the whole network.

The Administration is funded by an annual budget allocation and by charges levied for use of the infrastructure. The track fees paid by VR Ltd (see entry for Finnish Railways) equate to about €50 million annually. VR Ltd remains the sole train operator at present, but the Rail Network Act includes provision for other operators to apply to run services.

New lines

Kerava–Lahti direct line

Work is in progress on a 74 km direct line (63 km of new track) between Kerava (north of Helsinki) and Lahti, to supplement the present route via Riihimäki. The twin aims of the project are to improve rail connections to eastern Finland and to cater for projected traffic growth on the Helsinki–St Petersburg corridor. The new line will cut the Helsinki–Lahti journey time, currently 1 hour 15 minutes via Riihimäki, by 30 minutes. Around 80 per cent of the new line follows the alignment of the Helsinki–Lahti motorway. Construction commenced in 2004 and the line is scheduled for commissioning in 2006. The project is costed at €331 million.

Helsinki airport line

A master plan for a new line to serve Helsinki's international airport was completed at the end of 2003. The 18 km double-track electrified line will run between the Martinlaakso line and the main line from the capital to the north, forming a loop with the Kerava and Leppävaara suburban lines. The plan foresees the construction of seven stations, including one to serve the airport terminal, and will entail some 8 km of tunnelling in the airport area. Construction is expected to start in 2008. Estimated cost of the project is €300 million.

Improvements to existing lines

Main line upgrades

In 2004 track superstructures were being renewed on the Tampere–Jyväskylä, Luumäki–Säkäniemi,

Kokemäki–Rauma and Seinäjoki–Oulu lines. Plans for the last-mentioned include raising the line speed to a maximum of 200 km/h for passenger traffic and increasing the axleload limit to 25 tonnes for 100–120 km/h freight services.

The Seinäjoki–Oulu line features in a list of priority projects prepared by a ministerial working group. Other lines identified for development include the Tampere–Seinäjoki and Lahti–Luumäki sections. Progressing these will depend on funding being included in the state budget.

Helsinki suburban capacity enhancements

A fourth track for suburban services between Helsinki and Tikkurila (on the Tampere line) opened in 1996 and an extension to Kerava was completed in August 2004. Third and fourth tracks between Helsinki and Leppävaara (on the Turku line) opened in August 2001. These new tracks allow local traffic to be separated from long-distance traffic, permitting a significant increase in local train capacity.

Yard strategy

The ministerial working group endorsed RHK's view that Ilmala yard, Helsinki, which maintains and services the network's passenger stock fleet, is in need of renewal and this features in the list of priority projects.

A study by RHK into freight yard requirements up to 2025 identified the development of marshalling yards at Kouvola and Tampere as a priority. It also affirms the need for rail freight users, such as ports and industrial undertakings, to share the cost of investing in terminal facilities.

Signalling and telecommunications

At the end of 2004 a total of 2,568 route-km was equipped with automatic block systems, and 2,448 km with Centralised Traffic Control (CTC).

Automatic Train Protection (ATP) is to be introduced throughout the network by 2006, with the exception of a few lightly used freight-only lines. Lines with ATP totalled 3,479 km at the end of 2004.

The provision of a GSM-R train radio and operational communications system to replace the existing analogue network was underway in 2004. To be completed by the end of 2006, the system will cover some 5,000 km of track and yards. It employs equipment being supplied by Siemens and will be operated by Corenet Oy, with VR as the main user.

Electrification

At the end of 2004 a total of 2,619 route-km was under catenary, approximately 45 per cent of the network. About 75 per cent of all traffic is now hauled by electric traction, and the long-term aim is to raise this to over 80 per cent.

The first electrification scheme in northern Finland, between Tuomioja and Raahe (28 km), was completed in 2001. Electrification of the Oulu–Rovaniemi line (208 km) was completed in December 2004, and the Oulu–Kontiomäki and Iisalmi – Kontiomäki – Vartius lines will follow by 2006.

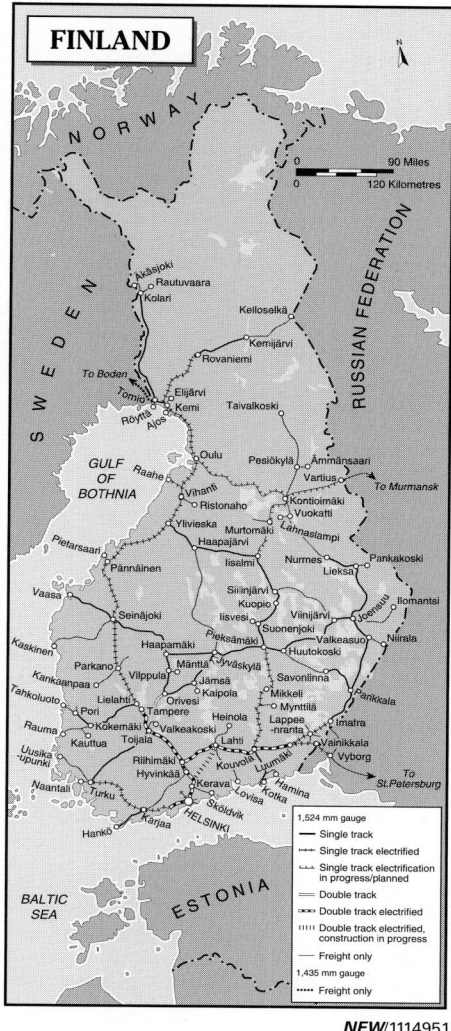

NEW/1114951

In 2004 RHK was undertaking a study into the electrification of a further 600 km of line. Its report was due in 2005.

Track

Rail type: K 30 (30 kg/m), K 43 (43.567 kg/m), K 60 (59.74 kg/m), 54E1 (54.43 kg/m), 60E1 (60.34 kg/m)

Sleepers: Concrete 2,600 × 300 × 220 mm (3,533 km); wood 2,700 × 240 × 160 mm

Spacing: 1,640/km

Rail fastenings: Pandrol for concrete sleepers; Hey-back for wooden sleepers

Min curve radius: 300 m

Max gradient: Normally 1.25% (9 route-km with gradients up to 2%)

Max permissible axleload: 24.5 tonnes Russian wagons; 22.5 tonnes (25 tonnes on selected routes) Finnish wagons

UPDATED

Finnish Railways – VR-Group Ltd

VR-Yhtymä Oy
Vilhonkatu 13, PO Box 488, FIN-00101 Helsinki
Tel: (+358) 307 10   Fax: (+358) 30 72 17 00
Web: www.vr.fi

Key personnel

VR-Group Ltd
 Chairman: Martin Granholm
 Chief Executive: Henri Kuitunen
 Group Directors
  Finance: Veikko Vaikkinen
  Administration: Pertti Saarela
  Development: Mirja Mutikainen
  Corporate Communications: Martti Mäkinen
VR Ltd
 Managing Director: Tapio Simos
 Freight Services Director: Ilkka Seppänen

Passenger Services Director: Antti Jaatinen
Technical Director: Markku Pesonen
Operations Director: Tapio Myllymäki
VR-Track Ltd
 Managing Director: Teuvo Sivunen
 Director, Business Development: Seppo Karanko
 Head of Railway Consulting: Harry Harjula
 Director: Arto Kukkola
 Chief Electrical Engineer: Lassi Matikainen
 Project Manager, Export: Pasi Aarnio

Gauge: 1,524 mm (see entry for RHK)

Organisation

Since 1 July 1995 the Finnish State Railways (Valtionrautatiet, VR) has been a limited-liability company, with all shares owned by the state. On the same date ownership of rail infrastructure was transferred to the Finnish Rail Administration.

The parent company, VR-Group Ltd (VR-Yhtymä Oy), has about 300 staff providing financial, administrative, information technology and property management services. Its two principal subsidiaries are transport operator VR Ltd (VR Osakeyhtiö), with 8,380 staff, and VR-Track Ltd (Oy VR-Rata Ab), with 2,460 employees engaged in infrastructure construction and renewal work.

Finance

In 2003 the VR Group recorded a net profit of €41 million on a turnover of €1.16 billion. The train-operating subsidiary, VR Ltd, had a turnover of €649 million, of which freight accounted for €352 million and passenger traffic €297 million.

Socially necessary passenger services receive an annual state subsidy of €37 million. The Finnish Rail Administration purchases infrastructure maintenance from VR-Track Ltd and traffic control services from VR Ltd.

Finances (€1,000)

Profit and loss account	2001	2002	2003
Net turnover	1,150,873	1,139,936	1,160,494
Changes in stocks and work in progress	–1,351	–2,016	213
Production for own use	28,738	33,084	26,777
Profits from associated companies	223	184	115
Other operating income	11,783	8,814	11,638
Materials and services	314,223	310,481	313,491
Personnel expenses	569,833	581,801	558,336
Depreciation	109,526	114,419	115,896
Other operating expenses	147,636	146,520	158,479
Total expenses	1,141,219	1,153,221	1,146,202
Financial income and expenses	12,258	10,602	8,309
Profit before extraordinary items and taxes	61,306	37,383	61,344
Extraordinary expenses	0	–2,510	0
Income taxes	–17,685	–10,994	–17,600
Minority interest	–1,647	–2,223	–2,433
Profit for year	41,974	21,656	41,310

Traffic (million)	2001	2002	2003
Passenger journeys	55.0	57.7	59.9
Passenger-km	3,282	3,305	3,338
Freight tonnes	41.7	41.7	43.5
Freight tonne-km	9,857	9,664	10,047

Passenger operations

VR Ltd operates long-distance and local rail services throughout Finland. In total some 60 million journeys were made in 2003, 48 million commuter journeys and 11.9 million long-distance journeys. Commuter journeys increased by 4.8 per cent while the number of long-distance journeys rose by 2.3 per cent. Passenger-km totalled 3.3 billion.

VR plans to introduce tilting high-speed trains on all main intercity routes by 2010, under the brand name Pendolino. Two prototype sets entered trial service on the 194 km Helsinki–Turku route in late 1995 (see 'Traction and rolling stock'), and full commercial operation at up to 200 km/h began in mid-1997, cutting journey time from 2 hours to 1 hour 45 minutes. Sixteen further trains have been ordered, seven of which were in service in 2003. New Pendolino services were introduced from Helsinki to Jyväskylä in October 2001 and to Oulu and Kuopio in June 2002.

The next major step in the programme will see services intensified on the new line between Kerava and Lahti, 100 km north of Helsinki, in 2006. The new double-track line is designed for 220 km/h running, with no level crossings, and will shorten journey times between the capital and Lahti and eastern Finland by almost 30 minutes.

The introduction of 220 km/h links between major centres forms part of a wider plan by VR and the government for a nationwide integrated transport system, Vali 2012. The other elements of the plan are conventional locomotive-hauled InterCity services, operating at 160–200 km/h with stops every 50–100 km, and local services operated by multiple-units or railbuses, with a maximum speed of 120–160 km/h and stops every 5–30 km. These rail services will be complemented by bus and shared taxi connections to stations. A regular-interval service with clock face departures and improved connections was introduced on the busiest intercity routes in southern Finland in June 2002.

VR has purchased 92 double-deck intercity coaches, all of which were in service by mid-2002. The 200 km/h air conditioned vehicles provide facilities which include provision for passengers with impaired mobility, bicycle stowage, a family compartment and children's play area.

In the long term VR also aims to run international Pendolino services between Helsinki and St Petersburg (443 km) once the necessary infrastructure improvements are in place. The present journey time of 6 hours 15 minutes could be cut to as little as 3 hours when the infrastructure is fully upgraded on both sides of the border.

Three daily through trains currently link Helsinki with Russia. The 'Repin' day train to St Petersburg and the 'Tolstoi' night train to Moscow are formed of

Class Sm3 Pendolino tilting trainset at Helsinki station, where a new roof has been added as part of an upgrading programme (Ken Harris)   0143316

Class Sr2 electric locomotive at Tampere with an InterCity service (Edward Barnes)   0567276

Diesel locomotives

Class	Wheel arrangement	Power kW	Speed km/h	Weight tonnes	No in service	First built	Mechanical	Builders Engine	Transmission
Dv 12	B-B	1,000	125	64	192	1964	Lokomo/ Valmet	Tampella-MGO	H Voith L 216 rs V 16 BSHR
Dv 12	B-B	1,000	125	68		1974	Valmet	Tampella-MGO	H Voith L 216 rs V 16 BSHR
Dv 12	B-B	1,000	125	66		1965	Lokomo/ Valmet	Tampella-MGO	H Voith L 216 rs V 16 BSHR
Dv 15	D	620	75	60	7	1958	Lokomo/ Valmet	Tampella-MAN W8V 22/30 AmA	H Voith L 217 U
Dv 16	D	700	85	60	28	1962	Lokomo/ Valmet	Tampella-MAN W8V 22/30 AmAuL	H Voith L 217 U
Dr 14	B-B	875	75	86	24	1969	Lokomo	Tampella-MAN R8V 22/30 ATL	H Voith L 206 rsb
Dr 16	Bo-Bo	1,500	140	84	2	1987	Valmet	Wärtsilä Vasa 8 V22	E Strömberg
Dr 16	Bo-Bo	1,677	140	84	20	1987	Valmet	Pielstick 12 PA4-V-200 VG	E Strömberg

Electric locomotives

Class	Wheel arrangement	Power kW continuous/ one-hour	Speed km/h	Weight tonnes	No in service	First built	Builders Mechanical	Electrical
Sr 1	Bo-Bo	3,100/3,280	140	84	103	1973	Novocherkassky	Strömberg
Sr 1	Bo-Bo	3,100/3,280	160	84	7	1973	Hyvinkää	Strömberg
Sr 2	Bo-Bo	5,000/6,000	210	82	46	1996	SLM/Transtech (Talgo)	Adtranz (Bombardier)

Electric railcars or multiple-units

Class	Cars per unit	Motor cars per unit	Motored axles/car	Output/ motor kW	Speed km/h	Units in service	First built	Builders Mechanical	Electrical
Sm 1	2	1	4	215	120	50	1968	Valmet	Strömberg
Sm 2	2	1	4	155	120	50	1975	Valmet	Strömberg
Sm 3	6	4	2	500	220	9	1994	Fiat/Transtech/Alstom	Parizzi
Sm 4	2	2	–	–	160	10	1999	Fiat	Parizzi

Russian rolling stock, while the 'Sibelius' day train to St Petersburg employs VR stock. Some 260,000 journeys annually are made using these services.

Within Finland a comprehensive network of overnight services, conveying sleeping cars and car-carriers, links the main cities of Helsinki, Turku and Tampere with the north and east of the country. In 2003 20 new double-deck sleeping cars were ordered for these services.

Since 1996 the entire train fleet has been equipped with a passenger announcement system, which uses satellite positioning to determine where each train is at any time and to synchronise multilingual station and connection announcements accordingly.

VR operates a comprehensive website for its customers, enabling the most popular tickets for daytime travel to be bought online. This highly popular facility is highly rated by the tourism industry.

Freight operations

Marketed under the brand name VR Cargo, freight traffic has boomed in recent years. In 2003 the volume of rail freight totalled 43.5 million tonnes, an increase of 4 per cent on the previous year. Of this total, 25.0 million tonnes came from domestic traffic and 18.5 million tonnes from international services. Domestic traffic increased by 1 per cent and international volumes by 9 per cent.

Most international traffic covers movements between Finland and Russia or transit traffic to third countries. At 14.4 million tonnes, traffic between Finland and Russia in 2003 increased by 14 per cent on 2002. More than half of the total comprises raw timber imports into Finland.

Transit traffic via Finland to third countries totalled 3.2 million tonnes. Container traffic to and from east Asia grew strongly, up 93 per cent compared with 2002. Some 100,300 TEU were carried between Finland and the Asia-Pacific region, an increase of 6 per cent on the previous year's figure. With a 12-day transit time, the route between Vainikkala through Siberia to Nakhodka, on the Pacific, offers a viable alternative to sea transport. Consumer goods from the Asia-Pacific region are imported into Finland and stored in transit warehouses, while exports from Finland are also carried on this route.

Traction and rolling stock

At the end of 2003 VR deployed in commercial traffic 156 electric locomotives, 273 line-haul diesel locomotives, 140 diesel shunters, nine six-car Pendolino trainsets, 110 two-car suburban emus, 742 locomotive-hauled passenger coaches (including 111 sleeper and 58 catering cars), 24 car-carriers and 11,324 freight wagons.

By the end of 2003, VR had taken delivery of all of its order for 46 Class Sr2, a 6,000 kW electric locomotive based on the Swiss Federal Railways Class 460. The 210 km/h locomotives were built by the Adtranz (later Bombardier Transportation)/SLM consortium, with Finnish manufacturer Rautaruukki Oy (later Transtech Oy and subsequently acquired by Patentes Talgo of Spain) producing the bodies and undertaking final assembly.

Finnish manufacturer Talgo has also supplied 92 double-deck low-floor intercity coaches to VR, all of which are in service by the end of 2003. The fleet comprises 64 second-class cars and 28 low-floor service cars with space for prams, wheelchairs and cycles. The aluminium vehicles are built for 200 km/h operation.

In 2003 VR placed an order with Talgo for 20 aluminium-bodied double-deck sleeping cars for overnight services to Lapland. Their introduction in 2006 will enable VR to retire 36 older vehicles. VR also plans to refurbish 76 older sleeping cars.

In 2003 VR placed an order with Talgo for 15 double-deck car-carriers, delivery of which is scheduled for 2005–06.

Fiat Ferroviaria (now ALSTOM) has supplied 10 two-car Class Sm4 emus for Helsinki suburban

Class Sm4 emu supplied by Fiat (now ALSTOM Transport) for Helsinki suburban services, seen at Helsinki with an earlier Class Sm2 unit (Ken Harris)   0567278

Class Dr16 diesel-electric locomotive at Rovaniemi with the 'Santa Claus Express' to Helsinki (Edward Barnes)   0567277

Class Sr1 Soviet-built electric locomotive at Tampere (Ken Harris)   0567279

services, all of which had been delivered by the end of 1999. The units are of aluminium low-floor construction, giving platform-level access throughout. Each air-conditioned unit seats 180 and stands 100, with wheelchair, pram and cycle space. Maximum speed is 160 km/h. The order includes an option on up to 40 more units. A follow-on order for 20 of these was placed in June 2002 for delivery in 2004–05.

In August 2001 VR placed an order with ČKD Vagónka to supply 16 railbuses for regional lines. The 63-seat vehicles will be equipped for driver-only operation and up to three units can be operated in multiple. Delivery is due to start in 2005. The contract includes an option for an additional 20 similar vehicles.

A programme to refurbish the 50 Class Sm1 emus used on Helsinki suburban services was completed in 2001. Technical upgrading and complete interior modernisation will extend

the life of the units, built in 1968–73, by at least 15 years. Next in line for refurbishment are the 50 newer aluminium-bodied Class Sm2 emus. The first refurbished example of this type entered trial service in June 2002, with series refurbishment scheduled for 2004–20.

Pendolino S220 units – Class Sm3

In 1995 VR took delivery of two prototype six-car electric trainsets to a 220 km/h design based on the Fiat Pendolino tilt-body units of Italian Railways (FS). Assembled in Finland by Rautaruukki Oy, designated Class Sm3 and branded Pendolino S220, they entered full commercial service in mid-1997.

The Sm3 features Fiat's modified bogie design used in the ETR460 second-generation tilt-body trainsets for FS, in which all tilt and lateral active suspension apparatus is contained within the bogie. Unlike the Italian units, the Sm3 has

For details of the latest updates to *Jane's World Railways* online and to discover the additional information available exclusively to online subscribers please visit
jwr.janes.com

electromagnetic brakes on each bogie and is equipped for multiple working.

Following experience with the two prototypes, VR placed an order for eight production units with Fiat Ferroviaria in December 1997. The production trainsets have a higher seating capacity (309) and an improved air conditioning system, and were manufactured entirely at Fiat's (now ALSTOM's) Savigliano plant. Seven of the eight sets were in service by the end of 2003. Eight additional trainsets were ordered in 2002 for delivery in 2004–06. In 2003 the two prototype sets were modified to conform to the production Class Sm3 trainsets in terms of seating arrangements and other equipment.

Type of coupler in standard use: Screw on locomotive-hauled stock; automatic coupler on emus, Pendolino trainsets and some locomotive types
Type of braking in standard use, locomotive-hauled stock: KE-GP; KE-GP-A; KE-GP + Mg; KE-PR; KE-PR + Mg

UPDATED   *63-seat railbus supplied to VR by ČKD Vagónka* (CKD Vagónka)   *NEW*/1066405

France

Ministry of Transport, Public Works, Tourism and Maritime Affairs

Ministère des Transports, de l'Equipement, du Tourisme et de la Mer
Grande Arche Sud, F-92055 Paris La Défense Cedex 04
Tel: (+33 1) 40 81 21 22   Fax: (+33 1) 40 81 39 97
Web: www.equipement.gouv.fr

Key personnel
Minister: Dominique Perben
Press Relations Officer: Corinne Meutey
  Tel: (+331) 40 81 31 59   Fax: (+33 1) 40 81 31 64

UPDATED

French Railways Infrastructure Authority (RFF)

Réseau Ferré de France
92 avenue de France, F-75648 Paris Cedex 13
Tel: (+33 1) 53 94 30 00   Fax: (+33 1) 53 94 38 00
Web: www.rff.fr

Key personnel
President: Jean-Pierre Duport
Director General: Jean-Marie Bertrand
Director, Development: Hervé de Tréglodé
Director, Finance: Patrick Persuy
Director, Network: Jean-Michel Richard
Director, Investment: J A Schneck
Director, Personnel: C Parent
Director, Communications: Luc Roger

Gauge: 1,435 mm; 1,000 mm
Route length: 28,918 km (in operation, of which 1,547 km high-speed lines); 167 km
Electrification: 5,859 km at 1.5 kV DC; 8,256 km at 25 kV 50 Hz AC; 122 km at other voltages

Organisation
RFF was established at the beginning of 1997 as a 'Public Establishment of Industrial and Commercial Character' (EPIC) to assume the management and development of, and investment in, the national rail infrastructure. It took over from SNCF control of all track, bridges and other structures, station platforms, intermodal freight terminals, marshalling yards, electrification infrastructure, signalling and telecommunications.

These assets are valued at EUR30.7 billion. The new EPIC has to balance its accounts from year to year, receiving revenue from SNCF of EUR2, 240 billion in 2004 – and other operators for access to the network plus an infrastructure grant from central government and subsidies from national and local government for individual projects.

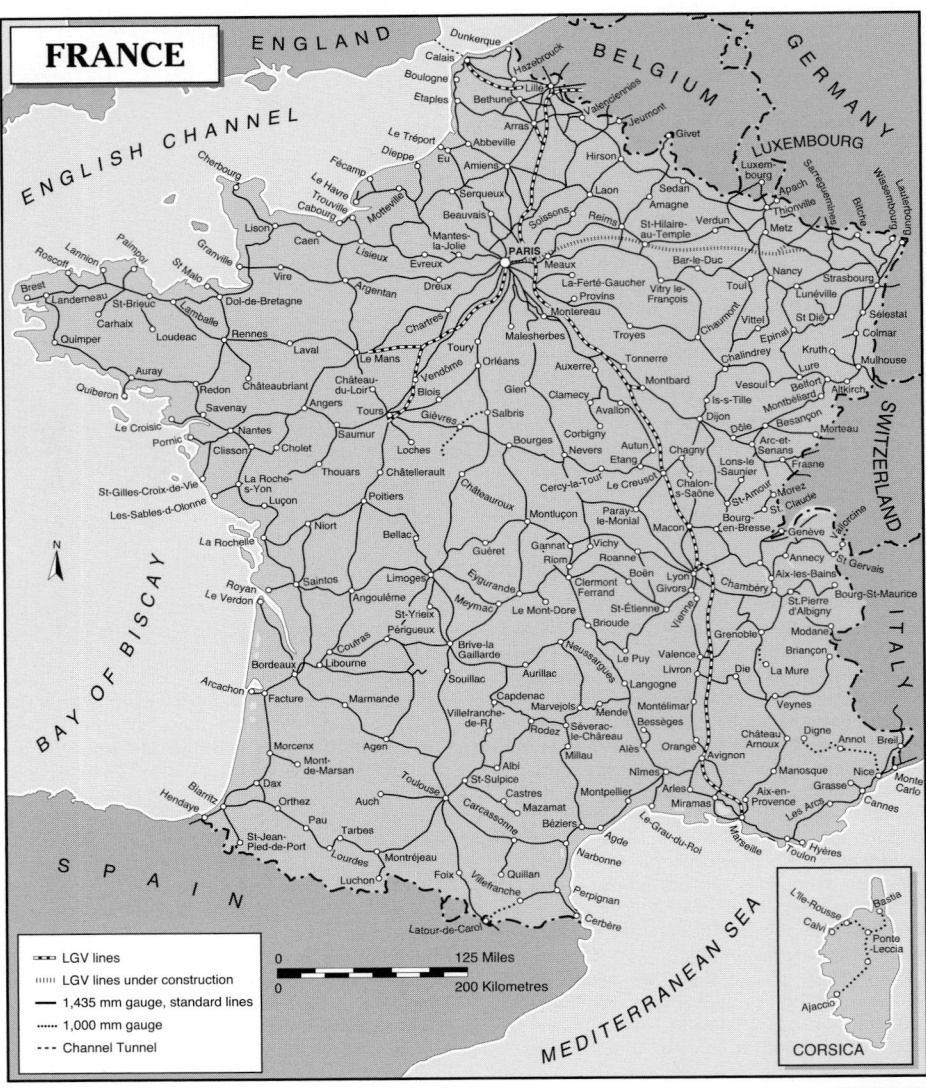

NEW/0585079

In 2004 RFF's investment totalled EUR2,432 million (11 per cent up on 2003), of which EUR1,014 million was committed to extensions to the TGV network, EUR444 million to modernisation of the classic network and EUR728 million on maintenance of the system. SNCF was paid EUR2.535 billion in management charges.

RFF is pressing the French government to write off the majority of its debt, which grew from EUR20.5 billion in 1997 to EUR26.5 billion in 2004. Annual debt servicing costs are around EUR2,500 million. Of this, the majority is supposed to come from central government. However, in early 2004 RFF was owed EUR3,000 million and thus the debt is increasing.

Maintenance and operation is subcontracted to SNCF which prepares the timetable and manages traffic movements as well as monitoring, maintaining and repairing installations. In this aspect, the situation in France differs from that in most other European countries where infrastructure has been completely separated from operations.

Following worries, particularly from trade unions, that the split of RFF/SNCF was a prelude to privatisation, the French government created the *Conseil Supérieur du Service Public Ferroviaire* (CSSPF), a consultative body which will give its opinion on major changes in rail policy and co-ordinate RFF and SNCF.

SNCF has created a new department to manage stations, improve the passenger environment and enhance interchanges with other transport modes. Investment was to be doubled to EUR305 million by 2003.

On coming to power in 2002 the new government ordered an audit of all rail infrastructure investment and concluded that within the existing budget all projects could not be carried out within the timescale originally planned. Many projects agreed for the 2000–06 period have been delayed by central government budget restrictions. These have also led to reduced maintenance and an increase in temporary speed restrictions. In mid-2003 the government was looking for new sources of finance, including a tax on lorries and public-private partnerships, in order to complete the programme.

State-Region contracts
All major rail construction and upgrading projects concerning the regions are governed by State-Region contracts which are renewed every seven years. Most projects have elements of finance from both central and regional government, some also involving the *départements* and cities.

Freight line closures
The radical restructuring of Fret SNCF (see entry for French National Railways in France section of *Railway systems and operators*) is leading to the withdrawal of freight services on many marginal rural routes.

Access charges
RFF introduced a provisional track access charging system in 1997 which has since been revised with higher prices. Charges vary according to the route taken and time of day on busy lines, but are highest in peak periods (06.30–09.00 and 17.00–20.00). The network is divided into four categories of line with different charges. Congested lines around large cities have the highest charges, with prices falling on high-speed lines, less again on interurban lines, and lowest on the rest of the network, which charge EUR0, 23 per km. This system favours freight traffic. The charge is split into a right of access, reservation of a path and use of the path. In 2001 RFF was proposing to reduce the right of access charge in order to favour new operators while increasing charges for use of a path. There are other charges for using marshalling yards and 'stabling' wagons. RFF raised charges to SNCF in 1999 from EUR930 million to EUR2,240 million in 2004. In order not to endanger SNCF's recovery, the difference between charges and costs is initially financed by the state, the payment reaching EUR1,385 million in 2003.

At the end of 2003 RFF employed 534 staff.

A guarantee of open access to outside rail companies operating international services was made law in 2003, with the French state providing licences. However, this is being implemented very slowly. In June 2005 the first open access operation started in France, with Connex hauling quicklime from Dugny, near Verdun, to the Saarbrücken area of Germany.

In 2004 Eurotunnel and German operator rail4chem obtained operating licences for France but neither was expected to start services before 2005. UK-based railfreight operator EWS has applied for a licence and hopes to start services in 2005.

Finances
RFF reported a loss of EUR650 million in 2004, a figure which has declined each year from €1,700 million in 2000. In 2004 the improved result was mainly thanks to a government grant of EUR800 million to reduce RFF's debt.

Revenue (EUR million)	2003	2004
Track access charges	1,953.1	2,239.7
State contribution (CCI)	1,395.1	1,110.4
Other	965.5	1,147.9
Total	4,303.7	4,498.0

Expenditure		
Management payment to SNCF	2,630.3	2,639.5
Studies and work	862.8	1,031.0
Other operating expenses	128.7	152.5
Depreciation	971.0	971.0
Other	+55.9	+85.7
Total	4,542.9	4,714.3

New lines
France's first high-speed line (Ligne ä Grande Vitesse or LGV), LGV-Paris Sud-Est (LGV-PSE), from the suburbs of Paris to Lyon, opened in two stages in 1981 and 1983 and was extended to Valence in 1994. It was followed by the Atlantique line from Paris to Le Mans and Tours in 1989–90, the Nord Europe from Paris to Lille and Calais in 1993, and the Jonction (Interconnexion) linking lines avoiding Paris in 1994–96. The LGV-Méditerranée opened from Valence to Marseille and Nîmes in 2001. The LGV-Est Européen was approved in 1999 and will open from Paris to Baudrecourt, near Metz, in 2007.

In December 2003, the French government gave the go-ahead for:
• Perpignan-Figueras (work started in 2004) and the Nîmes–Montpellier bypass (start in 2006);
• the eastern branch of the LGV Rhin-Rhône (work to begin in 2006, inauguration in 2010);
• LGV Sud Europe-Atlantique (Angoulême–Bordeaux starting in 2008, Tours–Angoulême in 2010, completion in 2012–14);
• LGV Bretagne–Pays de la Loire (start in 2009);
• LGV Est Baudrecourt–Strasbourg (start in 2010).

The financing of these lines in not confirmed and dates may slip.

The 'LGV Routes' map shows the current extent of existing routes and of plans approved or at an advanced stage. See earlier editions of *JWR* for full details of the background to existing lines.

LGV-Méditerranée
The LGV-Med, which extends southward from Valence to L'Estaque (218.7 km) on the approach to Marseille (3 hours from Paris), and from Avignon to Manduel-Redessan near Nîmes (32.7 km) opened on 10 June 2001. The line has three new stations, at Valence, Avignon and Aix-en-Provence. The route was revised several times and cut back from Lunel near Montpellier, but costs increased – from EUR2.71 billion to EUR3.69 billion (this sum includes the upgrade of the Paris Sud-Est line to 300 km/h standards, see below) – reducing the project's estimated rate of return from 12 to 6.4 per cent, below the 8 per cent limit that SNCF put on financing projects entirely from its own funds. The state therefore contributed EUR366 million towards the project. The European Investment Bank has made loans totalling EUR396 million for the line.

There are now more than 1,050 km of uninterrupted high-speed lines from near Calais to the outskirts of Marseille. On 27 May 2001, SNCF carried out an 'endurance run' with a standard TGV Réseau set and completed this journey in exactly 3 hours 30 minutes, achieving an average speed of almost 305 km/h.

The LGV Méditerranée is now to be extended as the LGV Provence-Alpes-Côte d'Azur, taking high-speed running from Aix-en-Provence to Toulon and the French Riviera, as a high priority. In 2005 public debate was under way on possible routes. The best option would put Nice at 3 hours 40 minutes from Paris. According to the route chosen, costs would vary from EUR4.9 to EUR7.5 billion.

LGV-Est Européen
A new 406 km line from Paris to Strasbourg received its 'Declaration of Public Utility' in 1996 but the stumbling block for this project was its low estimated rate of return. Savings have been achieved by limiting the first stage of construction to the 300 km Vaires–Baudrecourt (Paris–Metz) section, plus 44 km of links with existing lines, at a cost of EUR3.2 billion (1997 prices). This will

mean projected travel times will be achieved from Paris to Reims (45 minutes), Metz and Nancy (both 1½ hours) but that Strasbourg will be 2 hours 20 minutes from Paris rather than 1 hour 50 minutes as planned, although still an improvement on the present best of 4 hours. Construction is being supported by the EU, Luxembourg and the regions, *départements* and cities served en route.

Construction started in early 2002. Completion will be in 2007.

There will be three new stations on the line: Champagne–Ardenne, near Reims; Meuse, near Verdun; and Lorraine, between Nancy and Metz. The Lorraine station will be built near the local airport and motorway. A second station, 15 km to the east at Vandières, where the LGV crosses the existing Metz–Nancy line, has been approved to provide interchange with regional trains.

The line is to be designed for operation at 350 km/h although, initially, TGV sets will be limited to 320 km/h. Trains employed will be refurbished TGV Réseau sets and German ICEs.

Frankfurt (Main) will be 3 hours 45 minutes from Paris (6 hours at present), Basle 3 hours 30 minutes (4 hours 45 minutes) and Munich 5 hours 30 minutes (8 hours 20 minutes).

Work on the second stage, from Baudrecourt to Strasbourg, with a link to the German high-speed network, is now expected to begin in 2010.

LGV Languedoc-Rousillon
The LGV-Languedoc-Roussillon is a 325 km extension of the Nîmes branch of the LGV-Méditerranée to Perpignan, linking into new lines to Barcelona and Madrid. This line will now be built in stages.

A new 60.5 km line from Manduel to Lattès, avoiding Nîmes and Montpellier, with new stations serving each city, is to be built by 2010. There will also be 20 km of links to existing lines. This will be a mixed TGV and freight line and will cost EUR1.05 billion, including minor modernisation of the Montpellier–Perpignan line. The Déclaration d'Utilité Publique for the line was published in May 2005.

The other section to be built quickly will be from Perpignan to Le Perthus, (24.6 km) continuing to Figueras in Spain (19.9 km), where it will connect to the Spanish high-speed line to Barcelona and Madrid. This will also be a mixed passenger/freight line and will include a 8.2 km tunnel under the Pyrenees of which 7.3 km will be in France. Cost of the latter will be EUR549 million plus EUR122 million for installations in Perpignan. Work started in 2004 for completion in 2008. Perpignan–Barcelona timings will be cut from 2 hours 45 minutes to 50 minutes.

The central section between Montpellier and Perpignan has yet to find full financing and is forecast to cost EUR2.5 billion. RFF is carrying out feasibility studies of the line but completion is not expected before 2015–20.

LGV Rhin-Rhône
This project consists of 425 km of new line – a west-east line linking the LGV Sud-Est to Dijon and Mulhouse via Belfort and a north-south section from this line to Lyon.

The first section of the eastern branch will run 189 km from Auxonne, east of Dijon to Petit Croix, near Belfort, with new stations at Auxon, near Besançon, and at Meroux, near Belfort. Cost is put at EUR1.35 billion. The route of the line was approved in 2001 and declared of public utility in 2002. Construction of this section is expected to start in 2006 for completion in 2010. At a later date, a western branch will link Dijon to the Paris–Lyon LGV.

A southern branch, from a point near Dole to the outskirts of Lyon, with a new station in the Bresse area, will allow much faster services between Germany/Strasbourg and Lyon/south of France. As the line will not be heavily used by passenger trains, it is SNCF's intention to use it to divert freight traffic from congested lines in the Dijon area. This has complicated the exact choice of route, with five variations under study. Completion is not expected before 2015.

LGV Bretagne-Pays de la Loire

The route for the 200 km LGV BPL from Le Mans to Rennes was chosen in late 2000. A new line will avoid Le Mans to the north then continue south of the existing line to the outskirts of Rennes. A branch to Sablé-sur-Sarthe will link to the existing Le Mans–Nantes line, already largely upgraded for 220 km/h, and another will serve Laval. On completion Paris–Rennes times will be cut from 2 hours to 1 hour 20 minutes. The line is estimated to cost EUR2 billion. Construction is expected to begin in 2009 and may be staged. Tilting TGV sets are to be used to reduce journey times to the west of Rennes, allowing Paris to Quimper and Brest journey times of 3 hours.

LGV Sud Europe Atlantique

Work is advancing on the LGV-SE scheme, extending high-speed running over the 343 km from Tours to Bordeaux at a cost of EUR3.7 billion. This will cut the Paris–Bordeaux journey time from 3 hours to 2 hours 12 minutes. A route close to the existing line via Poitiers and Angoulême has been confirmed. It is intended to phase work, completing construction of an Angoulême–Bordeaux section by 2013 at a cost of EUR1.7 billion, followed by Tours–Angoulême. The public inquiry into the 121 km Angoulême–Bordeaux section began in early 2005.

Studies are taking place into the upgrading of the Bordeaux-Daxline for 220 km/h and into upgrading the high-speed line from Bordeaux to Toulouse. A public debate of the latter started in June 2005. Opening would not be before 2016.

Most of the remainder of the 1991 TGV Master Plan for 16 projects totalling 3,442 km of new lines was abandoned in early 1999. Instead, some classic routes will be upgraded to carry tilting TGVs.

Liaison Transalpine Lyon–Turin

This project is based on the construction of a 52.7 km base tunnel from St Jean-de-Maurienne, France, to Susa, Italy, on the Lyon–Turin route, with completion planned by 2015 at a cost of EUR10 billion. The tunnel is to have bores sufficiently large to allow its use by shuttle trains carrying complete lorries. Full details appear in the entry for Italy. The French will also build a new high-speed line from Lyon to Chambéry, a new freight line from Ambérieu to Chambéry and a mixed-use line from Chambéry to the base tunnel. The base tunnel is estimated to cost EUR7 billion, which would be financed 25 per cent each by the French and Italian states, 20 per cent by the EU and 30 per cent by private sources.

Work is also taking place on access routes: the Grenoble–Montmélian line will be electrified while the Valence–Grenoble line is to be doubled and electrified.

Prior to opening of the Alpine base tunnel, the line from Ambérieu to Modane via Chambéry and the Fréjus tunnel to Italy will be upgraded, with the tunnel enlarged to B+ loading gauge by 2006. A shuttle service carrying complete lorries was introduced between France and Italy in November 2003.

EOLE

A new underground route for Paris with main line clearances, known as EOLE (East-West Express Link) or RER line E, opened in July 1999 and now routes some Gare de l'Est local trains via new subterranean stations at Magenta (between Gares de l'Est and Nord) to Haussmann, near the St Lazare terminus.

The line will later be extended to Pont Cardinet, where it will connect with suburban lines in the west of Paris. This project has now been integrated within Liaison Normandie-Vallée de Seine.

Liaison Normandie-Vallée de Seine

In order to link Normandy with Roissy-Charles de Gaulle airport and the TGV network, studies were taking place in 2001 into the use of the EOLE line by TGVs. A link with EOLE and capacity improvements between Epône and Achères and in Mantes-la-Jolie are being studied.

Improvements to existing lines

Between 1996 and 2001 the LGV Paris Sud-Est, opened in 1981, was completely upgraded. Renewal work involved the upgrading of the line's signalling system to TVM 300 standards and provision of new track and ballast. This allowed train headways to be reduced from 5 to 4 minutes (later to 3 minutes) and maximum speed raised from 270 to 300 km/h for the opening of the LGV-Méditerranée.

SNCF and Swiss Federal Railways have agreed to share the cost of reopening and upgrading the La Cluse to Bellegarde 'Haut Bugey' line which will reduce Paris–Geneva TGV journey times by 20 minutes from 3 hours 30 minutes. Cost is estimated at up to EUR275 million, with work due to start in 2006 for completion in 2007. The Dole–Pontarlier line will also be upgraded at a cost of EUR61 million, cutting 25 minutes from Paris–Lausanne timings.

Paris region master plan

A 2000–2006 plan places the emphasis on development of orbital routes. An increase in population of 15 per cent, mostly in the outer suburbs, will heighten demand for transport by 60 per cent. Journeys between outer suburbs are expected to grow by 70 per cent.

To respond to demand for inter-suburb travel, existing lines will be adapted and linked by new sections of line. The Grande Ceinture (GC) freight-only line from Sartrouville to Noisy-le-Sec (first phase of the 'Tangentielle Nord') is to reopen by 2012 at a cost of EUR335 million, with a branch from Pontoise to Epinay-sur-Seine at a later stage. This will involve widening from two to four tracks over some sections. 'Tangentielle Ouest-Sud' will create a line from Achères in the northwest to Melun in the southeast. This will use existing lines from Cergy to Achères, the reopened and electrified western GC to Versailles, existing lines to Epinay-sur-Orge, a new line from Epinay to Evry, then existing lines to Melun. This will cost EUR381 million. No date has yet been set for this project. An initial section of the GC from St Germain to Noisy-le-Roi reopened to passengers in December 2004.

Removing freight congestion

A special effort will be made during the period 2000–10 to eliminate bottlenecks, in particular to enable freight traffic to grow. Specific projects include: quadrupling the river Garonne bridge in Bordeaux by 2008 at a cost of EUR190 million; adding a fourth track between Toury and Cercottes (Paris–Orléans route) by 2009 at a cost of EUR40 million; reopening and electrification of the Rouen avoiding line (2006, EUR45 million); widening Dunkerque–Hazebrouck–Douai (2008, EUR85 million); providing a third track on the Sucy–Bonneuil–Valenton line (Paris Orbital) (2007, EUR56 million); and laying a fourth track between Metz and Woippy, (EUR27 million), the last-mentioned completed in 2004.

An eastern bypass line for Lyon is being considered.

The French and Spanish governments have agreed to the reopening of the Pau–Canfranc line and to conversion of the Canfranc–Zaragoza line to 1,435 mm gauge to create a relief route for freight.

Regional upgrading
Paris–Limoges–Toulouse

RFF, SNCF and the regions concerned are to finance an upgrade of this line but plans to use tilting TGVs were dropped in 2003. Cost is EUR242 million, of which EUR96 million is for infrastructure. Work will include eliminating level crossings and reinforcing the 1,500 V DC supply. Speeds of up to 200 km/h will be possible north of Vierzon, 170 to 200 km/h from Vierzon to Limoges and 160 km/h south of Limoges.

The Centre region is to finance reopening of the 76 km Chartres–Orléans line by 2008 at a cost of EUR120 million. See also *Electrification*.

RFF, SNCF and the Basse Normandie region are financing a EUR109 million upgrade of the cross-country Caen–Rennes line, including electrification at 25 kV AC from Lison to Saint Lô and construction of a new 8 km branch to the Mont St Michel,

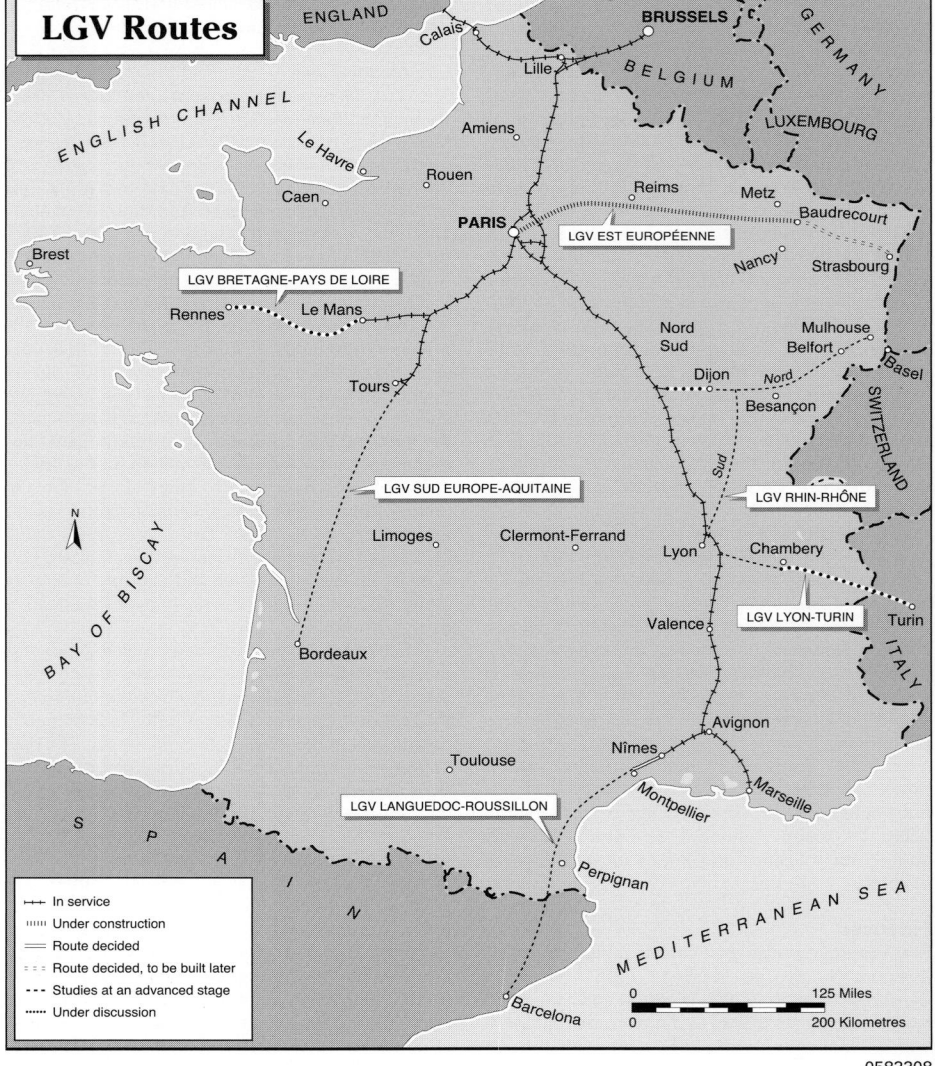

LGV Routes

Legend:
- ┥┝ In service
- ⁞⁞⁞⁞ Under construction
- ═══ Route decided
- ⁞⁝⁞ Route decided, to be built later
- - - - Studies at an advanced stage
- ······ Under discussion

0583308

France's premier tourist site outside Paris, to be completed by 2008.

The Paris–Clermont-Ferrand lines is to be upgraded with the aim of bringing the end-to-end journey time to below three hours.

On the Mediterranean coast, the 17 km Cannes–Grasse line has been rebuilt and electrified for passenger trains at a cost of EUR40 million as part of a plan for a Cannes–Nice RER-type system. The line reopened in March 2005.

Signalling and telecommunications

SNCF has installed an automatic train protection system, a version of the L M Ericsson automatic train and speed control system developed jointly with MTE-Alsthom. SNCF designates the system as KVB (*Contrôle de Vitesse par Balises*). The system, employing track-mounted transponders, identifies temporary speed restrictions as well as signal aspects. All 5,000 main line locomotives and around three-quarters of the network's 16,000 signals have been equipped. A new version, designated KVBP, is being developed for suburban lines with heavy traffic where the existing version reduces capacity.

ETCS

SNCF is participating in the validation of the EIRENE digital radio standard for European railways which is part of the European Train Control System (ETCS). EIRENE will be tested until 1999 on a 32 km section of the Jonction high-speed line and 55 km of the Nord Europe line.

Electrification

All main passenger and freight lines in France, except for Paris–Troyes–Belfort, are now electrified. In general, lines north and east of Paris are electrified at 25 kV AC and those south of Paris at 1.5 V DC. Almost all recent projects, including fill-ins in the 1.5 V DC area, have been at 25 kV AC necessitating the use of dual-voltage traction and rolling stock.

New schemes to go ahead in the period 2000–06 include: Gretz–Troyes (completion 2008); Tours–Vierzon (2008); Lison–St Lô (2006); Nantes–Les Sables d'Olonne (2008); Rennes–St Malo (December 2005); and St-Etienne–Firminy (2005). All these schemes also include upgrading for higher speeds. Electrification of branches from Nancy to Remiremont (78 km) and

St Dié (51 km) plus Reims–Chalons-en-Champagne is included in the LGV-Est Européen project. The Remiremont and St Dié projects were inaugurated in May 2005.

Other electrification schemes under study include: Bourges–Saincaize (2010); Marseilles–Aix-en-Provence; Quimper-Landerneau; Lutterbach-Thann; and Gières–Montmélian (2010).

Track

Rails: 60 kg/m where traffic exceeds 30,000 tonnes/day; 50 kg/m elsewhere; UIC 60 on LGVs

Sleepers: Wood (oak or tropical wood) 150 mm thick; concrete (mono- or duo-block) 220 mm thick

Spacing: 1,600–1,722/km

Fastenings: Wooden sleepers: rigid or screw or elastic (Types NR and NABLA); concrete sleepers: elastic or NR or NABLA type

Min curve radius: 150 m (in depots)

Max gradient: 4%

Max axleloading: (on selected routes): 22.5 tonnes

UPDATED

French National Railways (SNCF)

Société Nationale des Chemins de Fer Français
34 rue du Commandant Mouchotte, F-75699, Paris Cedex 14
Tel: (+33 1) 53 25 60 00
Web: www.sncf.com

Key personnel

President: Louis Gallois
Director (President's Cabinet): Patrick Ropert
Executive Director: Guillaume Pepy
Director (Strategy): Elisabeth Borne
Director (European Affairs): Luc Alliadière
Director (Research and Technology):
    Philippe Renard
Secretary, Council of Administration:
    Michèle Audibet
Directors
    Finance, Purchasing and Data Processing:
        Jean-Pierre Menanteau
    Human Resources: Pierre Izard
    Operating: Jacques Couvert
    Production: (vacant)
    Station Developments: Pascal Lupo
    Traction & Rolling Stock: Roland Bonnepart
    International Development: Jean-Pierre Loubinaux
Secretary General: Paul Mingasson
Business Directors
    Long-distance Passenger (Voyageurs
        France-Europe): Mireille Faugère
    Press Service: Pierre Youang
        Tel: (+33 1) 53 25 73 88
    Transport Public: Bernard Sinou
    Press Service: Joëlle Tournebize
        Tel: (+33 1) 53 25 86 09
    Ile de France Passenger: Thierry Mignaux
    Press Service: Virginie Abadie-Dalle
        Tel: (+33 1) 53 25 70 60
    Freight: Marc Veron
    Press Service: Françoise Ragot
        Tel: (+33 1) 53 25 78 93
Director of Communications: Bernard Emsellem
Head of Press Service: Jean-Paul Boulet

Political background

SNCF is a 'Public Establishment of an Industrial and Commercial Character' or EPIC, being entirely state-owned but with a legally autonomous status. It was originally created on nationalisation in 1938. The EPIC is now part of a larger group. The EPIC SNCF itself made a surplus of EUR128 million in 2004 compared with a loss of EUR204 million in 2003. Revenue was up by 5.7 per cent overall, long-distance passenger traffic by 7.5 per cent, regional passenger by 4.6 per cent and Paris region by 4.6 per cent.

Under arrangements to separate SNCF's operations from infrastructure management, the government accepted future responsibility for infrastructure finance and EUR25.6 billion

of SNCF's debt, the part linked to infrastructure investment and past infrastructure losses. The infrastructure debt was transferred to a separate EPIC, known as *Réseau Ferré de France* (RFF) (see entry under France section of *Railway systems and operators*) in 1997. RFF now owns all SNCF rail infrastructure as defined by European Union rules. SNCF remains sole manager of the network, having responsibility for maintaining track and signalling, train control, allocating train paths and approving new operators. SNCF's own trains receive first priority for train paths. SNCF's infrastructure branch produced a surplus of EUR85 million in 2002 compared with EUR94 million in 2002.

Following a period of trials, all regional councils took over full responsibility for planning and finance of local rail transport from 2002. This leaves SNCF merely as a supplier of trains and staff, for which it retains its monopoly. Most regions are investing in new and refurbished rolling stock and boosting service frequencies.

As a consequence of the passage of a law on air quality, all local authorities have produced a study of local transport needs (PDU). Most have opted to improve public transport and some will invest in freight infrastructure as part of state/region contracts in the period 2000–2006.

SNCF put together a new strategy in 1996 under a new president, Louis Gallois. His appointment was confirmed for a further five years in 2003.

Following a major public opinion survey, a series of 47 action programmes was drawn up. In general, Louis Gallois has adopted a strategy of consolidation without the sort of massive investment seen in recent years and reducing

losses by increasing business volumes within existing resources. This is being done by improving passenger services by increasing frequencies, operating trains with greater capacity, improving stations and information, simplifying and cutting fares. Customers were surveyed again in late 1997 with results showing SNCF had improved its image.

Although still state-owned, SNCF now bids for contracts to manage rail services within and outside France, particularly in Europe through its subsidiary Keolis (see below). Through Keolis (formerly VIA-GTI) SNCF jointly manages some passenger operations abroad.

Organisation

SNCF is divided into four businesses, each of which has its own budget and bottom-line responsibility: Long-Distance Passenger (Voyageurs France-Europe); Urban and Regional Passenger (Transport Public – TP); Paris Region Passenger (Ile de France) (which now comes under TP); and Freight (Fret SNCF). Each Business Director contracts with the other departments for the means of production and back-up services. Each of the 20 geographical regions has its own director, who has control of local passenger services and reports directly to the president. Regional and Paris Region passenger services are heavily subsidised while other activities are required to break even.

SNCF has moved slowly since 2000 to assign dedicated resources to each division. A plan to allocate staff more closely to activity sector, known as Cap Clients, was shelved in April 2001 after a two-week strike.

TGV Duplex bilevel high-speed trainset at Lyon Part Dieu forming a service to Paris (John C Baker)

0554630

Apart from the rail core in France, SNCF has full or part ownership of over 550 other companies connected with transport throughout the world, including road hauliers, hotels, travel agencies and ferry operators. Employing over 33,000 staff, these are managed through the company's subsidiary, SNCF Participations. Indeed, through its subsidiaries, SNCF is the main road haulier in France. In 2000 68 per cent of turnover came from, and 84 per cent of staff were employed in, the core rail business.

SNCF employed 174,755 staff in 2003, 1.6 per cent fewer than in 2002, with a total of 238,360 in the group as a whole (up 1.1 per cent).

Traffic (million)	2001	2002	2003
Passenger journeys	875	–	–
of which Paris suburban	560	–	–
Passenger-km	52,840	54,260	53,080
of which Paris suburban	8,810	10,110	9,990
Freight tonnes	126.3	–	–
Freight tonne-km	50,400	50,040	46,840

Finance

The SNCF group recorded revenue of EUR22.1 billion (EUR15.5 billion for the rail activity) in 2004, compared with EUR20.8 billion (EUR15.4 billion) in 2003. A surplus of EUR323 million was recorded in 2004 compared with a surplus of EUR11 million in 2003.

In 2003 SNCF invested EUR2.14 billion, of which EUR1.17 billion went on traction and rolling stock.

The French state contributes very heavily each year to SNCF's accounts. In 2003 it contributed EUR5,370 million, of which EUR2,376 million went to the pension fund, allowing staff to retire at 50 (drivers) or 55. EUR1,546 million was contributed to regional services, EUR677 million for debt servicing, EUR618 million in investment grants and EUR479 million for fares reductions for certain categories.

Passenger operations

SNCF's passenger operations are divided into three sectors:

Voyageurs France-Europe (VFE), formerly known as Grandes Lignes, for long-distance services including TGV;

Transport Public (TP) for some intercity services, local semi-fast and stopping trains;

Ile de France (IDF) for Paris region suburban trains, now within the remit of TP but with its own director.

At the beginning of 2005 SNCF separated a number of less important long-distance passenger services from VFE and rebranded them Trains Inter Régionaux (TIR). They are now managed by Transport Public. Apart from the Paris–Le Havre and Paris–Cherbourg routes, all lose money. SNCF is seeking to withdraw some services and recast others in concert with regional councils.

VFE and IDF have their own directors but responsibility for regional services is delegated to local managers who negotiate directly with regional councils, albeit with co-ordination by the TP support service. SNCF has reorganised the boundaries of several of its regions to correspond more closely to the territory of regional government.

Long-distance services

VFE is SNCF's main activity, in terms of turnover and the contribution made to infrastructure from the surplus of revenue over direct costs. This surplus was EUR523 million in 2004 compared with EUR257 million in 2003. In 2001 VFE traffic was 52.8 billion passenger-km. TGV services, which extend well beyond the limits of the new infrastructure, produce a growing proportion of GL revenue – reaching 58 per cent in 2001. Passenger-km on TGV services rose from 10.49 billion in 1988 to 37.4 billion in 2001 but fell on 'classic' services from 37 billion to 15.4 billion.

SNCF is carrying out improvements line-by-line to long-distance services with the aim of improving market share and countering air competition and new motorways. Services have been recast on the following routes, mainly producing frequency increases within existing resources: Paris–Limoges, where competition is coming from a new motorway; Paris–Strasbourg and Paris–Basle, where air competition is increasing.

Class CC72000 diesel-electric locomotive in service with SNCF's VFE passenger business (Edward Barnes)   ***NEW*/**1122833

In 1998 SNCF introduced its first-ever strict clock-face interval timetable on a major route from Paris to Lyon. This was followed by Paris-Lille. Success was such that further lines are benefiting, including Paris-Nantes in May 1999 and Paris-Rennes in December 2000.

The LGV Méditerranée high-speed line opened on 10 June 2001, allowing a three-hour timing for the 250 km from Paris to France's second city, Marseille. Following the success of regular interval timetables elsewhere, the service is exactly hourly off-peak and half-hourly in the peaks. Services are operated mainly by TGV Duplex sets. Most other cities south of Lyon saw major cuts in journey times. In its first year, rail travel on the corridor rose by 38 per cent and in its second year by 8.5 per cent.

A major part of plans on non-TGV routes has been the modernisation of 710 of SNCF's 3,300 Corail day coaches, which on average are 20 years old. Series refurbishment, at a cost of EUR50,000 per coach, started in 1995 and modernised vehicles were introduced on services from Paris to Cherbourg, Limoges, Metz, Strasbourg, Basle, Le Havre and Clermont-Ferrand. Since 2003 SNCF has been undertaking a more profound refurbishment of Corail stock. The newly branded Corail Téoz stock were introduced on Paris–Clermont Ferrand services in 2003, Paris–Strasbourg in 2004 and Paris–Limoges in 2005.

SNCF relaunched overnight services in May 2000 with air conditioned stock, fixed train formations, a 'reception area', improved facilities in couchette coaches and much-increased security. In 2004 trains were rebranded Lunéa. Scheduled stops between midnight and 05.30 have been abolished.

International services

To improve management of international passenger services and to stem the general decline in their use, SNCF has moved to create joint ventures (*Groupement d'Intérêt Economique* or GIE) with other national operators.

Apart from Eurostar services, the first joint venture, concerning Paris–Lausanne/Bern TGV services, was created in 1993 with Swiss Federal Railways. The first benefit from this agreement was the extension from December 1995 of one Paris–Lausanne service to Brig in order to tap the winter sports market. The GIE has extended the Bern service to Zürich and refurbished the nine three-voltage TGV sets capable of operation into Switzerland. In January 2000, the status of the GIE was changed and the organisation renamed Rail France Suisse SAS (see entry in Operators of international rail services in Europe section) and in 2002, the service was renamed Lyria. In 2001, these services carried 900,000 passengers and turnover rose by 5.5 per cent. In 2005 the Paris–Geneva TGV service will be transferred to Lyria and in 2007 Paris–Zurich services via the LGV Est will also join Lyria.

In 1994 SNCF and Italian Railways (FS) created a GIE known as 'Alpetunnel', a company which will study the future TGV Lyon-Turin. At the same time, SNCF and FS agreed on measures designed to boost passenger figures on France-Italy services by 50 per cent. The result was a direct daytime TGV service from September 1996 between Paris and Milan via Turin. At the same time FS ETR.460 'Pendolino' tilting trainsets took over the Milan–Turin–Lyon service. In September 1997 SNCF and FS relaunched overnight services between Paris and Italy under the new 'Artesia' branding. In 2001 Artesia night services carried 932,000 passengers. Turnover rose by 7 per cent. However, low-cost airlines and high costs led SNCF to cut back overnight services and withdraw the Lyon–Milan service in 2003–04. On the other hand, Paris–Milan TGV services are to be expanded.

Class Z 22500 double-deck emu for Ile de France services (ALSTOM)   0554548

In 1994 SNCF and Belgian Railways signed a similar agreement with the aim of developing Paris–Brussels services, as well as extensions to Antwerp and Liège. The company Westrail International was formed to market Thalys TGV services to Amsterdam and Cologne (see entry for Belgium). In May 1999 this company was renamed Thalys International.

On 14 December 1997 the new Belgian high-speed line was opened in its entirety (see Belgium entry for details), and four-voltage PBKA Thalys TGVs were introduced on a new two-hourly Paris–Brussels–Cologne service.

In 1995, SNCF created two GIEs in conjunction with Spanish Railways (RENFE). The first is charged with studying the proposed TGV line between Narbonne and Barcelona. The second aims to develop night services between Paris and Madrid/Barcelona. The latter became a limited company named Elipsos in 2001. From 1996 to 2001, traffic grew by 52 per cent and revenue by 87 per cent on these routes.

In 2000 in partnership with German Rail (DB), Luxembourg Railways and Swiss Federal Railways, SNCF established a joint venture limited company, Rhealys SA, to develop the market for services linking the four countries. However, in 2005 SNCF has formed a joint venture with DB to operate Paris–Germany services via the LGV Est, the Swiss services will now be operated by Lyria and Luxembourg services will be operated by SNCF.

In order to reinforce its presence abroad, SNCF in 1997 acquired British Rail International, a subsidiary of British Rail specialising in selling rail travel abroad and rail holidays in Britain. This is now managed as Rail Europe.

Ticketing
SNCF introduced the Socrate electronic reservation system, based on American Airlines' Sabre system, in 1992–93 (see earlier editions of *Jane's World Railways* for details). Socrate has allowed SNCF to introduce a flexible market pricing system, with fares no longer linked to the number of kilometres travelled. Socrate manages the allocation of reduced price fares in order to maximise revenue per train, basing decisions on historical data on loadings and the progression of reservations. All TGVs have compulsory reservations, as have upgraded Corail Téoz services.

SNCF has the most successful Internet ticket sales site in Europe. Tickets can be ordered by credit card and are posted free or can be printed by the user's computer. SNCF sold 35 per cent of long-distance tickets by Internet in 2003. In 2004 the site was achieving annual sales of 16 million tickets for EUR648 million. From 2003 SNCF used these facilities to offer cheaper advanced sale and last-minute ticket offers.

In December 2004 SNCF launched the iDTGV service, for which all tickets are sold by the Internet. The initial daily train pair between Paris and Toulon was successful, with 77 per cent occupation, and was followed by a Paris–Montpellier daily train pair in 2005.

Paris region (Ile de France) suburban services
SNCF services in the Paris region are now marketed under the name Transilien. Services have achieved about 10 billion passenger-km since 1991, despite extensions to the RER system. Fraud is a major problem, causing revenue losses of up to 20 per cent on some lines.

Transport provision in the Paris region is governed by five-year contracts (*contrats de plan*) between the Ile de France region and the state. Specific projects are outlined in the entry for French Railways Infrastructure Authority (RFF) in the France section of *Railway systems and operators*.

In the same way that responsibility for rail transport was transferred to regional councils in 2002, a similar transfer is to be made in July 2005 to the Syndicat des Transports d'Île-de-France (STIF).

Major developments of the Paris network centre on the RER (Regional Express) network, started in 1969, which links suburban lines across the centre of the city. The network consists of five lines, A, B, C, D and E. The RER is operated by the SNCF, in the case of Lines A and B jointly with the Paris transport

Class Z 23500 two-car regional emu forming a Nord-Pas-de-Calais service at Hazebrouck (John C Baker)   0583010

Class X73500 railcar at Reims (Milan Šrámek)   **NEW**/1122832

authority RATP; SNCF now has 1,282 route-km of suburban operation in the region. All are electrified.

Three-quarters of Paris region traffic is with discounted fares. Operating costs are met 40 per cent by fares (but employers pay half the cost of their employees' annual tickets or Orange Card monthly/weekly tickets); 40 per cent by the *Versement Transport*, the payroll tax levied from employers; and 20 per cent by public authorities (the state, 70 per cent; departmental authorities, 30 per cent).

Regional services
Regional services (*Train Express Régional* – TER) achieved 9.2 billion passenger-km in 2003 and turned in a surplus of EUR237 million compared with EUR213 million in 2002.

Since 2002 regions have established public transport plans, then negotiate five-year contracts for train services with SNCF, which retains the monopoly over operations. Regions have complete responsibility for financing all train services.

To finance regional service deficits, the state transfers EUR1.5 billion per year to the 20 regions, approximately double the previous subsidy. Expansion and improvement in the quality of regional services has been rapid in the past decade and passenger use has risen steadily. At present most improvements are being made in increasing frequencies, introducing new trains and opening new stations in the suburbs of major cities.

Rolling stock for TER services has been developed in consultation with the regions (see 'Traction and rolling stock' section).

In 2000 SNCF took a 43 per cent share in French urban transport operator VIA-GTI. The company was then merged with SNCF bus and coach operator Cariane and the resulting company was renamed Keolis in 2001. Through Keolis SNCF is now seeking contracts for joint urban and regional transport operations both in France and abroad. In May 2004 private equity group 3i took a controlling interest of 53.3 per cent in Keolis, although SNCF still effectively runs the company.

SNCF has appointed a manager charged with developing short-distance 'peri-urban' services around large towns. The first fruit of this will be the conversion of the Paris region Aulnay-Bondy line to light rail standards in 2006. Fifteen light rail vehicles were ordered from Siemens for this line in 2002, with an option for 64 more. Karlsruhe-style projects to run LRVs on tramway and RFF systems are under development for Mulhouse and Strasbourg.

The Nord-Pas de Calais region introduced limited TGV services in 2000, taking advantage of local high-speed lines and TGV sets in marginal time.

Freight operations
SNCF's rail freight subsidiary, Fret SNCF, had a turnover of EUR6,393 million in 2004, up 1.7 per cent on 2003. The activity lost EUR382 million compared with EUR447 million in 2003. Traffic was stable.

Rail's share of the French freight market has fallen from 46 per cent to around 25 per cent since 1974, but this figure is still the highest in the EU, where the average is 14 per cent. However, revenue per tonne-km is falling regularly — by 2.5 per cent in 1997. Rail accounts for only 29 per cent of SNCF's freight business, which includes road haulage companies.

In commercial terms, SNCF's freight activity is divided into business sectors: coal and steel; oil, chemicals and metals; manufactured goods; agricultural products; automobile, exceptional loads and military; and wood, paper and building materials.

SNCF published separate figures for freight for the first time in 1999, recording a loss of EUR137 million, compared with EUR85 million in 1998. The losses subsequently rose from EUR69 million in 2000 to EUR447 million in 2003.

Fret SNCF registered 46.8 billion tonne-km in 2003. Despite hopes of doubling traffic to 100 billion tonne-km by 2010, traffic has fallen regularly from 55.4 billion tonne-km in 2000.

Faced with increasing losses, SNCF appointed Marc Véron head of Fret SNCF in 2003 and announced the 'Plan Fret' which is designed to return Fret SNCF's conventional traffic to break-even by 2006. The plan is based on the simplification of train operations, concentrating trunk hauls on five main corridors known as Grands Axes and 12 local distribution areas known as Zones Locales. Most locomotive classes are now restricted to specific routes. The aim is to increase locomotive and staff productivity. At the same time, uneconomic traffic is being streamlined or dropped and rates increased where necessary.

This will be accompanied by investment in information technology, in particular a system to reserve space in a particular train. Investment in a limited number of new locomotive types will increase reliability, reduce maintenance costs and fleet size. The Plan Fret will require an injection of some EUR800 million from the French government. This received EU approval in 2005 on condition that France opens up domestic freight traffic to competition by 31 March 2006.

In 2004-05 four main marshalling yards, 16 freight nodes and 100 loading points were closed and 2,500 jobs shed by natural wastage. Initial results in 2004 showed improved timekeeping, better loaded trains (7 per cent better from 2003 to 2004) and higher locomotive and wagon utilisation (13 per cent better). However, traffic was down by 15 per cent in the first quarter of 2005, mainly due to frequent strikes.

Cross-border traffic is seen as a growth area. International traffic represents 50 per cent of tonne-km, although cross-border market shares are well below domestic market shares in adjoining countries. Particular points of action are the establishment of SNCF delegations abroad, the elimination of locomotive and staff changes at border points, concentration of traffic for certain countries on 'hub' freight yards and 'one-stop shops' for the sale and tracking of capacity on cross-border services. In 2001 SNCF and Trenitalia set up common subsidiary, Sideuropa, to develop steel traffic between France and Italy. In the same year, SNCF and DB agreed procedures for the cross-border operation of locomotives and crews.

Fret SNCF employs approximately 25,000 staff.

Wagonload traffic
The wagonload network was rationalised considerably in the 1980s. Although its traffic declined from 31.8 billion tonne-km in 1980 to 15.9 billion in 1993, it accounts for 28 per cent of all SNCF freight and almost half total freight revenue. However, the number of terminals served has been halved to 1,250, and service to some 2,800 private sidings was withdrawn between 1989 and 1993. In addition, since 1993, the number of major marshalling yards was cut from 27 to 19. Only some 239 principal terminals (*Gare Principal Fret*, or GPF), generating at least 10 wagonloads daily, now have direct connection with a marshalling yard. Some major GPFs are linked by through trains avoiding the main marshalling yards. The plan has obtained Day-A-Day-B transits for 70 per cent of wagonload traffic. As part of the Plan Fret, marshalling yards in Achères, Lille and Toulouse have closed while others such as Hausbergen (Strasbourg) have reduced their activity.

Southbound intermodal service passing Gemeaux headed by a Class BB36000 intermodal service (John C Baker)   ***NEW*/1122831**

Eurostar high-speed trainset employed on LGV Nord domestic services, seen at Paris Gare du Nord (Milan Šrámek)   0583009

Class BB27000 dual-voltage electric locomotive with a Germany-bound freight service at Sierck les Bains (John C Baker)   ***NEW*/1122830**

As part of the Plan Fret, a large number of stations handling timber are to close where trainloads are not generated.

All freight trains now run at a top speed of at least 100 km/h where track alignment permits. All SNCF main lines are passed for axleloads of 22.5 tonnes, so that four-axle wagons can gross up to 90 tonnes, which means payloads can reach at least 60 tonnes.

Block trains
This traffic is relatively stable. Despite a general downward trend in movement of heavy products, SNCF is aiming for a 5 to 10 per cent increase in traffic, mainly through better service.

Fret SNCF is hoping to lengthen freight trains which are normally limited by traction and infrastructure to 750 m and 1,800 tonnes. Together with DB Cargo, Fret SNCF has selected ALSTOM and SAB WABCO to develop a future European Freight Intelligent System (EFIS), a standard electronic brake control and communication system for freight trains. This may allow trains to be extended to 2,250 m. Trials with pairs of coupled freight trains, with the centre locomotive controlled from the leading one, were due to begin in late 2002.

In 2001 SNCF started operating 900 m container trains from the port of Fos to Lyon and 800 m trains of empty ore wagons from Lorraine to Dunkerque.

Automatic Vehicle Identification

Fret SNCF is now using an AVI system on some 2,200 wagons carrying loads such as bottled beverages or household appliances which may suffer damage during transit. The system employs Amtech vehicle-mounted transponders which incorporate shock detectors and trackside interrogators.

Sernam

Sernam (*Service Nationale de Messageries*) is SNCF's 100 per cent-owned parcels carrier. Sernam lost EUR110 million on a turnover of EUR518 million in 2000 despite continuous reorganisations in recent years. Sernam was absorbed by SNCF's road haulage subsidiary Geodis in 2001 and now uses only one rail service. SNCF is now to sell the subsidiary after the EU ruled out further subsidies.

In October 1998 Sernam freight services started operating at 200 km/h, the first trains of this type in the world to do so. Trains between Paris and Orange operate partially over the Paris Sud-Est high-speed line, using Class BB22200 electric locomotives with TVM 430 cab signalling. These haul covered bogie vans which have been tested at up to 279 km/h.

Intermodal operations

Intermodal traffic now accounts for 25 per cent of SNCF freight. Traffic is stable at around 13.8 billion tonne-km. Domestic traffic is managed and marketed by two companies, Novatrans (the French UIRR company) and CNC, with the swapbody figuring in the businesses of both. Novatrans is 60 per cent owned by road haulage interests, 40 per cent by SNCF. CNC is 71 per cent owned by SNCF Participations. CNC is open to all users, while Novatrans deals exclusively with road hauliers. Novatrans operates 23 terminals and 1,600 wagons and runs 100 trains daily. CNC runs 100 trains daily, operates some 17 terminals in France, 5,500 flat wagons, and 9,000 containers and swapbodies. It is associated with 150 hauliers deploying some 2,000 road vehicles.

Most intermodal traffic is by direct overnight trains running at 120 to 160 km/h, but CNC also operates a service known as 'Combi 24' — a network of evening trains converging on a hub near Paris then, after shunting, departing and giving morning arrivals. Novatrans also has a hub operation centred on Noisy-le-Sec near Paris. The government is paying SNCF a yearly grant of around EUR76 million towards new terminals. Investment includes EUR457 million in terminals and enlarging loading gauges on the Paris-Le Havre and Kehl-Dijon-Modane routes. During the period 2001–03, 17 new or extended terminals will be brought into operation. Bordeaux Hourcade came into service in December 2001. The largest of the terminals is at Dourges, near Lille, which opened in December 2003.

Fret SNCF has developed some intermodal enterprises independent of Novatrans and CNC. One is Chronofroid, with which SNCF aims to recover some of the two million tonnes of refrigerated perishables traffic it has lost to road since 1980. In 1990 a similar operation was launched with tank swapbodies for chemical products, in this case with international service over distances of more than 500 km as an objective. Hence its title of TransEuroChem.

As a trial, SNCF and RFF are to give higher priority to freight services on the Metz-Dijon-Modane route through reorganisation of pathing and track maintenance. This route carries many intermodal services which are particularly time-sensitive.

TGV trainsets

	TGV-PSE	TGV-PSE	TGV-A	TGV-R	TGV-R	TGV-Duplex	Thalys (TGV-PBKA)
	dual-voltage dual-class	tri-voltage dual-class	dual-voltage dual-class	dual-voltage dual-class	tri-voltage dual-class	dual-voltage double-deck	four-voltage dual-class
Length over couplers	200.19 m	200.19 m	237.6 m	200.19 m	200.19 m	200.19 m	200.19 m
Tare weight in working order	386 tonnes	386 tonnes	490 tonnes	383 tonnes	385 tonnes	380 tonnes	380 tonnes
Weight available for adhesion	194 tonnes	194 tonnes	136 tonnes	136 tonnes	136 tonnes	137 tonnes	135 tonnes
Continuous power rating							
At 25 kV 50 Hz	6,450 kW	6,450 kW	8,800 kW	8,800 kW	8,800 kW	8,800 kW	8,800 kW
At 1.5 kV DC	3,100 kW	3,100 kW	3,880 kW	3,880 kW	3,880 kW	3,880 kW	3,680 kW
At 15 kV 16²/₃ Hz	–	2,800 kW	–	–	–	–	–
Max speed	270 km/h	270 km/h	300 km/h	300 km/h	300 km/h	300 km/h	300 km/h
Seating capacity							
1st class	69/110	110	116	120	120	197	120
2nd class	276/240	240	369	257	257	348	257

SNCF Fret Vossloh-built Class BB61000 diesel-hydraulic locomotive near Rémilly with a Saarbrücken–Metz wagonload freight (John C Baker)   **NEW**/1122829

Bombardier-built Class Z 27500 (AGC) regional emu (Bombardier)   0567266

Class X73500 railcar arriving at Lyon Part-Dieu with a local service from Roanne (John C Baker)   0554629

In early 2005 Fret SNCF announced a rescue plan for intermodal services which lost EUR116 million in 2004, either by Fret SNCF, CNC or Novatrans. Five CNC terminals will close, as will the Paris hub operation. Operation of other terminals will be opened to competitive tendering. CNC will in future concentrate on maritime traffic and Novatrans on domestic traffic.

Piggyback services

Following a fire in the Mont Blanc road tunnel in which 39 people died, and increasing concern about the congestion and pollution of Alpine valleys by growing lorry traffic, the French authorities have moved to promote the carriage of complete lorries by train. The types of wagon used elsewhere in Europe have been rejected in favour of a new design, Modalohr, which does not require small wheels. An initial service using two trains, each carrying 18 complete trucks or 27 unaccompanied trailers and operating four train pairs per day, was introduced in November 2003 between Aiton, France, and Turin, Italy. Usage has been low due to limits on loading gauge in advance of tunnel enlargement between France and Italy.

In 2004 SNCF announced a project known as 'Route Roulante 2006' to operate similar services on long-distance north-south corridors such as Luxembourg–Spain or Lille–Marseille. A feasibility study is in progress with the aim of launching a service in 2006.

Traction and rolling stock

At the beginning of 2004 SNCF owned 379 TGV trainsets, 1,764 electric, 1,679 diesel and 1,197 tractor locomotives, 1,042 emus (3,725 cars, TGV sets excluded), 1,388 diesel railcars/trailers and 7,656 hauled passenger cars. Freight stock totalled 47,680 SNCF-owned and 64,258 privately owned.

Current TGV orders

In mid-2005 the TGV fleet consisted of 107 eight-car TGV Sud-Est, 105 10-car TGV Atlantique, 79 eight-car TGV Réseau and 75 eight-car TGV Duplex. In addition, there are three postal TGVs.

In 2000–01 SNCF ordered an additional 52 TGV Duplex sets for use on services between Paris and the south of France, freeing PSE and Réseau sets for other services. In November 2003 SNCF ordered a further seven TGV Duplex, taking the total to 89 sets. In addition, SNCF ordered 15 sets of Duplex trailers. These will be mated with 15 pairs of TGV Réseau power cars, bringing the total of double-deck sets to 104. The 15 sets of TGV Réseau trailer cars thus freed will be refurbished and combined with 30 new three-voltage power cars known as TGV POS (Paris Ostfrankreich Süddeutschland), which will operate Paris–Germany services over the LGV Est Européen from 2007. A pair of TGV POS power cars were delivered by ALSTOM in September 2004 and were tested throughout 2005.

TGV Est services from 2007 will be launched with the 38 refurbished TGV Réseau sets capable of 320 km/h, the 15 POS sets mentioned above and five or six German ICE sets.

Other TGV derivatives which are jointly owned by SNCF are 27 eight-car Thalys sets (10 tri-voltage, 17 quadri-voltage) and 38 Eurostar sets (31 18-car and 7 14-car).

Eurostar

Details of the 38 SNCF trains in the jointly owned Eurostar fleet for operation between London and Paris/Brussels via the Channel Tunnel and TGV-NE will be found in the United Kingdom entry.

Of the Eurostar fleet, 29 sets are three-voltage but nine sets belonging to SNCF have been additionally modified for operation under 1.5 kV DC in order to allow operation from London to the south of France over the Paris Sud-Est LGV. They are used for a winter sports service linking London with Bourg St Maurice in about 8 hours during the skiing season and for a London–Avignon service in summer. Lower than expected traffic levels on Eurostar have led SNCF to use three sets between Paris and Lille, thus freeing six TGV Réseau sets.

Future TGV development

A major project to develop a new generation of TGV has been abandoned. Instead, a variety of tests are

being carried out in order to improve the TGV design. These concern eddy current brakes to allow higher speeds and measures such as adding valances to bogies to reduce noise. In 2001 and 2002, ALSTOM carried out tests with an 'AGV' part rake.

ALSTOM and SNCF have also tested a TGV prototype with automatic body tilting supplied

by Fiat. The train, converted from a TGV-PSE set, ran trials from April 1998 to April 2000. Of the development budget, EUR4.7 million was financed by the French government PREDIT project, EUR12.8 million by ALSTOM and EUR8.4 million by SNCF. Plans to run tilting TGVs from Paris to Limoges were dropped in 2004.

Electric locomotives: principal classes

Class	Wheel arrangement	Line voltage	Output kW	Speed km/h	Weight tonnes	No in service	First built	Builders Mechanical	Electrical
BB8100/8000	Bo-Bo	1.5 kV	2,100	105	92	3	1949	Alsthom	Alsthom
BB8500/8700[3]	B-B	1.5 kV	2,940	100/140	78	78	1963	Alsthom	Alsthom
BB7200	B-B	1.5 kV	4,360	160[1]	84	237	1977	Alsthom	MTE
BB9200	Bo-Bo	1.5 kV	3,850	160	82	86[2]	1957	MTE	MTE
BB9300	Bo-Bo	1.5 kV	3,850	160	82			MTE	MTE
BB9700	Bo-Bo	1.5 kV	3,850	160	82			MTE	MTE
BB9600	B-B	1.5 kV	2,210	140	59	28	1959	Fives-Lille	CEM
CC6500[3]	C-C	1.5 kV	5,900	100/220	115	16	1970	Alsthom	MTE
BB16000	Bo-Bo	25 kV	4,130	160	84	57[5]	1958	MTE	MTE
BB16100	Bo-Bo	25 kV	4,130	160	84			MTE	MTE
BB15000	B-B	25 kV	4,360	160	88	61	1971	Alsthom	MTE
BB16500[3]	B-B	25 kV	2,580	100/140	74	145	1958	Alsthom	Alsthom
BB17000[3]	B-B	25 kV	2,940	90/140	78	105	1964	Alsthom	Alsthom
BB20200[3]	B-B	25/15 kV	2,940	100/140	80	6	1969	Alsthom	Alsthom
BB22200 (2-current)	B-B	25/1.5 kV	4,360	160[4]	89	202	1977	Francorail	Francorail
BB25100 (2-current)	Bo-Bo	25/1.5 kV	4,130 3,400	130	84	53	1963	MTE	MTE
BB25200 (2-current)	Bo-Bo	25/1.5 kV	4,130 3,400	160[6]	84	46	1964	MTE	MTE
BB25500[3] (2-current)	B-B	25/1.5 kV	2,940	100/140	78	137	1963	Alsthom	Alsthom
BB26000 (2-current)	B-B	25/1.5 kV	5,600	200	90	233	1988	GEC Alsthom	GEC Alsthom
BB27000 (2-current)	Bo-Bo	25/1.5 kV	4,200	140	–	152[7]	2000	ALSTOM	ALSTOM
BB36000 (3-current)	B-B	25/3/1.5 kV	5,600	200	90	60	1997	GEC Alsthom	GEC Alsthom
BB37000/37500 (2-current)	Bo-Bo	25/3/1.5 kV	4,200	140	–	29[8]	–	ALSTOM	ALSTOM
Class 92	Co-Co	25 kV/750V	5,000	140	126	9	1994	Brush	ABB

[1] 102 units geared for higher tractive effort and 100 km/h maximum; 3 units geared for 200 km/h maximum.
[2] Total for BB9200, BB9300 and BB9700.
[3] Monomotor bogies – 2 gear ratios.
[4] 8 units geared for 200 km/h maximum.
[5] Total for BB16000 and BB16100.
[6] 33 locomotives reduced to 130 km/h.
[7] Total in service at end of 2004 of an order for 180.
[8] Locomotives on order: 29 Class BB 37000; 1 Class BB 37500.

Electric multiple-unit power cars: principal classes

Class	Cars per unit	Line voltage	Motor cars per unit	Motored axles/car	Output/ unit kW	Speed km/h	Units in service	First built	Builders Mechanical	Electrical
Z5300	3/4	1.5 kV	1	4	1,180	130	94	1965	Fives-Lille/ CFL-CIMT/ De Dietrich	Jeumont
Z5600	4/6	1.5 kV	2	4	2,700	140	52	1982	ANF-CIMT	TCO
Z7100	2/4	1.5 kV	1	2	940	130	18	1960	Decauville/ De Dietrich	Jeumont/ Oerlikon
Z7300 Z7500	2	1.5 kV	1	4	1,275	160	90	1980	Francorail	MTE
Z6100	3	25 kV	1	2	615	120	59	1964	SFAC-CFL/ De Dietrich	CEM-SW/ Alsthom
Z6300	3	25 kV	1	2	615	120	20	1965	CFL-Fives Lille/ De Dietrich	CEM-SW/ Alsthom
Z6400	4	25 kV	2	4	2,350	120	75	1976	CFL	Alsthom/ TCO
Z8100	4	1.5/25 kV	2	4	2,500	140	51	1979	SFB-ANF	TCO
Z8800	4	25 kV	2	4	2,800	140	58	1985	Alsthom/ ANF-CIMT	TCO
Z9500 Z9600	2	1.5/25 kV	1	4	1,275	160	56	1982	Francorail	MTE
Z11500	2	25 kV	1	4	1,275	160	22	1987	Alsthom	Alsthom
Z20500	4/5	25/1.5 kV	2	4	2,800	140	194	1988	GEC Alsthom/ ANF	GEC Alsthom
Z20900	4	25/1.5 kV	2	4	2,800	140	54*	1988	ALSTOM	ALSTOM
Z92050	4	25/1.5 kV	2	4	2,800	140	6	1988	GEC Alsthom/ ANF	GEC Alsthom
Z21500	3	25/1.5 kV	2	4	880	200	57	2003	ALSTOM/ Bombardier	ALSTOM
Z22500	5	25/1.5 kV	2	4	3,000	140	53	1995	GEC Alsthom	GEC Alsthom
Z23500	2	25/1.5 kV	1	4	1,700	140	80	1998	ALSTOM/ Bombardier	ALSTOM
TGV-PSE 23000	10	25/1.5 kV	2	4 + 2	6,450 3,100	270	98	1978	Alsthom/ Francorail	Alsthom/ Francorail
TGV-PSE 33000	10	25/15/ 1.5 kV	2	4 + 2	6,450 3,100 2,800	270	9	1981	Alsthom/ Francorail	Alsthom/ Francorail
TGV-A	12	25/1.5 kV	2	4	8,800 3,880	300	105	1989	GEC Alsthom/ ANF/ De Dietrich	GEC Alsthom
TGV-R	10	25/1.5 kV	4	4	8,800 3,800	300	49	1992	GEC Alsthom/ De Dietrich	GEC Alsthom
TGV-R	10	25/1.5 kV + 3 kV	4	4	8,800 3,800	300	40	1994	GEC Alsthom/ De Dietrich	GEC Alsthom
TGV-Duplex	10	25/1.5 kV	4	4	8,800 3,800	300	75	1996	GEC Alsthom/ De Dietrich	GEC Alsthom
Thalys PBKA	10	25/1.5 kV 3/15 kV	4	4	8,800 3,800	300	17	1996	GEC Alsthom/ De Dietrich	GEC Alsthom

*18 more on order.

Diesel locomotives: principal classes

Class	Wheel arrangement	Power kW	Speed km/h	Weight tonnes	No in service	First built	Mechanical	Builders Engine	Transmission
68000	A1A-A1A	1,660	130	106	7	1963	CAFL	Sulzer 12LVA 24	E CEM
68500	A1A-A1A	1,645	130	105	19	1963	CAFL	SACM-AGO V12 DSHR	E CEM
72000[4]	C-C	2,250	160	110	85	1967	Alsthom	SACM-AGO V16 ESHR	E Alsthom
61000	B-B	1,000	100	70	23[2]	2002	Vossloh	Caterpillar	H Voith
63000	Bo-Bo	355/435	80	68	31	1953	B&L	Sulzer 6LDA	E B&L
63500	Bo-Bo	450	80	68	489[2]	1956	B&L	SACM-MGO V12 SH	E B&L
66000	Bo-Bo	830	120	70	375[3]	1959	Alsthom	SACM-MGO V16 BSHR	E CEM
66400	Bo-Bo	830	120	70			Alsthom	SACM-MGO V16 BSHR	E CEM
66700	Bo-Bo	830	120	70			Alsthom	SACM-MGO V16 BSHR	E CEM
67000[4] 67200[1]	B-B	1,440	90	80	80	1963	B&L MTE	SEMT-Pielstick 16PA4	E MTE
67300[4]	B-B	1,440	90/140	80	87	1967	B&L MTE	SEMT-Pielstick 16PA4	E MTE
67400	B-B	1,525	140	83	228	1969	B&L	SEMT-Pielstick 16PA4	E MTE
Y7100	B	130	54	32	207	1958	Billiard/ Decauville	Poyaud 6PYT	H Voith
Y7400	B	130	60	32	465	1963	Decauville/ De Dietrich/ Moyse	Poyaud 6 PYT	M BV Asynchro
Y8000 Y8400[5] (2 gears)	B	215	30/60	36	525	1977	Arbel-Fauvet Rail	Poyaud V12-520NS	H Voith

[1] 67200 as 67000 but with TVM cab signal for LGV.
[2] Leased from Angel Trains Cargo.
[3] Total for 66000, 66400 and 66700.
[4] Monomotor bogies (2 gears).
[5] As Y8000 but with remote control.

Diesel and gas-turbine multiple-units

Class	Cars per unit	Motor cars per unit	Motored axles/car	Power kW	Speed km/h	No in service	First built	Mechanical	Builders Engine	Transmission
X2100 X2200	1	1	2	440	140	52 58	1980 1985	ANF	SFAC/Saurer SJS-S 1DHR	H Voith
X2800	1	1	2	426	120	42	1957	Decauville/ RNUR	SACM-MGO V12 SH	HM Maybach
X2720	2	2	2	605	140	7	1955	De Dietrich	SACM-MGO V12 SH	H Maybach
X4300 X4500	2	1	2	295	120	169	1963	ANF	Poyaud/SFAC Saurer SDHR	M De Dietrich
X4630	2	1	2	295	120	106	1971	ANF	SFAC/Saurer SDHR	H Voith
X4750	2	1	2	440	140	53	1975	ANF	SFAC/Saurer S1DHR	H Voith
X4900	3	2	2	590	140	13	1975	ANF	SFAC/Saurer SDHR	H Voith
X72500	2/3	2	2	1,200	160	117	1997	ALSTOM	MAN	H Voith
X73500/ 73900	1	1	2	456	140	331	1999	ALSTOM	MAN	H Voith

Modernised Class X4750 dmu at Saint-Dié (Marcel Vleugels)   **NEW**/1122834

In 2000 SNCF, FS and DB announced that they would work together on future high-speed train design, while ALSTOM and Siemens created a consortium to work together on this project.

Locomotives
Following difficulties in supplying enough locomotives for the freight activity, SNCF divided the main line locomotive fleet by activity sector

at the beginning of 1999. Locomotives were allocated to: VFE (352 electric, 77 diesel); Fret SNCF (1,310 electric, 578 diesel); Action Régionale (261 electric, 222 diesel); Ile de France (141 electric); and Infrastructure (five electric, 172 diesel). Fret SNCF is to reorganise operations to increase locomotive productivity to allow withdrawal of the oldest units. Locomotives are increasingly organised in small, dedicated fleets to

increase reliability for targeted customers. In 2001, 40 locomotives were dedicated to CNC intermodal services from the Villeneuve hub. Transport Public locomotives and rolling stock are now nominally allocated region by region and are increasingly based closer to their area of operation.

SNCF has placed orders with ALSTOM for 210 new 'Prima' electric locomotives for freight services. Fret SNCF specified a relatively simple 4,200 kW design, 180 to be designated Class BB27000 with dual-voltage capability for use throughout France, with asynchronous motors, IGBT technology and a maximum speed of 140 km/h. In tendering, the accent has been placed on low cost and reliability. The remainder of the order is for 29 Class BB 37000, with additional 15 kV AC capability for operation in Germany and Switzerland plus one BB 37500 with a 3,000 V DC capability for Belgium or Italy. The latter may now be supplied as a Prima 6000 four-voltage locomotive by ALSTOM. Deliveries of BB 37000 were completed in 2004 while delivery of BB 27000 will be completed in 2005.

SNCF already has 60 Class BB 36000 tri-voltage electric locomotives for operation in Belgium and Italy. Regular operation into Italy on the Modane route started in late 2003. Operation into Belgium is also increasing.

Fret SNCF suffered a severe locomotive shortage in 2000, with many trains stopped for lack of traction. To help solve the shortage of diesel traction, SNCF hired 23 Vossloh G1206 locomotives from mid-2001. They are designated Class BB 61000.

In early 2004 Fret SNCF ordered 400 Bo-Bo diesel locomotives from ALSTOM's Prima range with the 2,000 kW MTU engine and transmission supplied by Siemens. These will be designated Class BB 75000 and will be delivered from 2006 to 2015. They will replace most of the present Fret SNCF line haul fleet, including pairs of smaller locomotives at present employed in multiple. Later in 2004, Fret SNCF ordered 160 lower-powered diesel locomotives designated Class BB 60000 from ALSTOM. These will be built at the Valencia plant in Spain which has now been taken over by Vossloh. Prior to delivery of the new locomotives, Fret SNCF is equipping 50 Class BB 66000 and 81 BB 66400 locomotives with new MTU 12V 4000 R41 engines of 1,040 kW.

In order to reduce emissions and fuel consumption, SNCF is to re-engine all 525 Class Y8000 and Y8400 shunters with Renault engines by 2007. In addition, around 25 Class CC72000 locomotives will be re-engined. In 2003 SNCF also announced the refurbishment and re-engining of 200 Class Y7100 shunters.

Paris suburban stock
SNCF Paris suburban rolling stock at the start of 2003 totalled 2,992 emu cars (751 sets) and 1,125 push-pull trainset cars. Of the combined fleet, 2,513 cars were double-deck.

SNCF has ordered 60 ALSTOM Prima electric locomotives for Paris suburban services. These are designated Class BB 27300 and will be identical to Fret SNCF's BB 27000 except for being equipped to operate passenger trains. They will replace older classes powering double-deck stock in push-pull mode.

In mid-2005 SNCF was expected shortly to place a large order for 330 four- or five-car single-deck suburban emus, known as Nouvelle Automotrice Transilien (NAT), which will replace the last remaining stainless-steel emus dating from the 1960s.

Trains for the regions
The majority of contracts signed by the SNCF and the regions have featured funding from the latter for the purchase of new rolling stock.

The regions participated in design of a new two-car diesel trainset from ALSTOM, classified X72500 and capable of 160 km/h. Features include the use of two 'disposable' 300 kW MAN lorry engines per power car, with Voith hydraulic transmission, low-floor entrances, air conditioning and a streamlined nose.

The cost per two-car unit is EUR4.12 million. In total 117 units, fully financed by the regions, were delivered from 1997 to 2003. Units are used mainly on medium-distance limited-stop services, many of them inter-regional. Of the total, 73 units are two-car, both of them powered, while 44 units are three-car with a non-powered centre car.

In order to provide economical operation on minor lines, SNCF and DB placed a joint order (40 each) for 80 light single railcars with De Dietrich and LHB (now ALSTOM) in October 1996. The French railcars, costing EUR1.45 million each, are classified X73500, and are powered by two engines totalling 456 kW. They have a 140 km/h maximum speed, seating for 61 passengers, with a low-floor centre section, and are built for one-person operation. The initial order had grown to 331 units by 2004 as the regions added their own requirements. A version of this design with signalling equipment for operation into Germany is designated Class X73900. Fifteen of these units now operate Metz/Strasbourg–Saarbrücken and Strasbourg–Offenburg. Delivery was completed in 2004.

In September 2001 SNCF named Bombardier Transportation preferred bidder to supply a new fleet of up to 500 160 km/h multiple-units for services which it operates on behalf of regional authorities. Designated Autorail à Grande Capacité (AGC), the new trains are intended to replace Class X2720, X4300, X4500, X4630 and X4750 dmus. Contracts for 330 units were placed by the end of 2003, leading to first deliveries in January 2004. Further contracts in 2005 took the total on order to 427 units.

The AGC is articulated, with a partial low floor, and will be built in two-, three- and four-car versions. Transmission will be electric. Both Class X 76500 dmu and Class Z 27500 emu versions will be supplied and some will be supplied as Class B 81500 dual-mode vehicles capable of operating either as dmus or as emus on electrified lines. Most will be a version with diesel plus 1.5 kV DC but eight units will be diesel plus 1.5 kV DC/25 kV AC. Construction will be undertaken at Bombardier's Crespin plant, near Valenciennes.

SNCF received five new metre-gauge dmus from CFD for its Blanc–Argent line in 2002–03. It also received two panoramic emus from Stadler for the metre-gauge Villefranche–La Tour de Carol line in January 2004 and has ordered a further three similar trains for the metre-gauge St Gervais–Vallorcine line.

SNCF owns a subsidiary, Chemins de Fer de Corses (CFC), to operate the metre gauge network in Corsica. The French government has sanctioned significant modernisation of the CFC network and 15 two-car dmus are on order from CFD.

Deliveries of the three-car dual-voltage Class Z21500 emu capable of 200 km/h commenced in mid-2002. The design has been ordered by the Centre (15 sets), Bretagne (17), Pays-de-la-Loire (14) and Poitou-Charentes (5) regions for express long-distance services.

SNCF ordered 98 examples (327 cars in sets of two to five cars) of a new generation of double-deck emu (TER2N NG) from ALSTOM. The units have a top speed of 160 km/h and have all cars powered. The first units were delivered in 2004. A further 22 three-car sets were ordered in June 2005.

TGV Réseau high-speed trainset leaving Paris Gare du Nord (Milan Šrámek)   0583008

Ile de France Transilien-branded Class Z 6400 four-car emu in the Paris suburbs at Puteaux (Milan Šrámek)   0583011

The regional authorities are increasingly financing refurbishment of existing trains to render them more attractive and increase their useful life. A further 71 Z2 emus are now to be refurbished and equipped with air-conditioning.

In early 2005 SNCF awarded ALSTOM a EUR37 million contract to convert 48 Corail coaches into driving trailers for use at up to 200 km/h.

France Wagons

In 1993 SNCF created a subsidiary, France Wagons (FW), charged with management of its fleet of 59,000 wagons. The wagon fleet, worth EUR762 million, plus the accompanying EUR213 million debt, was transferred to the wholly owned subsidiary which rents back the wagons to SNCF. SNCF hopes to improve its wagon productivity, which is much lower than that of French private wagons. SNCF wagons made an average of only 15 loaded journeys carrying 501 tonnes over a total of 5,500 km in 1993. Wagon numbers were cut from 76,000 in 1993 to 54,000 in 2000, while traffic rose. There is an increasing trend towards the dedication of wagons to specific traffic flows. One such case concerns wagons conveying tobacco, increasing productivity by 50 per cent.

UPDATED

CDG Express

40 rue d'Alsace, F-75475 Paris Cedex 10
Tel: (+33 1) 40 18 83 80 Fax: (+33 1) 40 18 60 52
e-mail: contact@cdgexpress.org
Web: www.cdgexpress.org

Key personnel
President: Armand Toubol
Project Manager: Jean-Michel Belhomme
Director, External Relations: Pierre-Henri Gronier

Background
CDG Express is a European economic interest group (GIE) formed jointly by the Paris airports authority, Aéroports de Paris (ADP), Réseau Ferré de France (RFF) and French National Railways (SNCF) to develop plans for a fast rail link between Roissy-Charles de Gaulle airport and central Paris. At present only 19 per cent of Roissy's 48 million annual passengers reach the airport by rail.

As originally conceived, the project consisted of a 25 km electrified line from the city's Gare de l'Est to the airport, including 14 km of new line. This would have included a section following the TGV Jonction line, 2.5 km on the surface and a 10.8 km tunnel to Noisy-le-Sec, from where the CDG Express line would follow the existing RFF lines for the 9 km to Paris Est.

Following public consultation, alternative routes were studied, leading to the adoption of one known as 'La Virgule' which would share with freight traffic and regional express services two tracks of the existing line from the approaches to Gare du Nord via Aulnay-sous-Bois to Mitry before taking a new line following the alignment of the LGV Jonction line to the airport. Compared with the scheme originally proposed, this route, while slightly longer at 32 km, reduces the amount of tunnelling costly needed to just 1.2 km and employs 23 km of existing infrastructure. The project also includes changes to the service pattern of the existing RER Line B link to Roissy. This will now be restricted to one pair of tracks, while CDG Express services will use the fast lines, shared with regional services to Crépy-en-Valois and Laon.

At Gare de l'Est, complete redevelopment is planned to coincide with the commissioning in 2007 of the TGV Est service. This will include improvements intended to treat Gare de l'Est, Gare du Nord and Magenta (RER 'EOLE' Line E) stations, together with access to five Métro lines, as a single transport hub. A city air terminal at platform level with 28 desks for in-town check-in is to be provided for CDG Express services, which will use two dedicated tracks. At Roissy-Charles de Gaulle, CDG Express services will terminate at the existing TGV station, where track layout modifications will be carried out and passenger and baggage handling facilities provided.

A service frequency of four trains an hour is foreseen with a non-stop journey time of

20 minutes. It is projected that by 2015 the service will be used by 7.3 million passengers annually.

Projected rolling stock requirements are seven 160 km/h six-car emus, each providing capacity for around 400 passengers. Initial traffic projections suggest at least 6 million passengers a year. Costs are estimated at €630 million for infrastructure and €110 million for rolling stock. Final detailed design is due to be completed in 2006, leading to a government Declaration of Public Utility (DUP) in 2007. A 42-month construction period is expected to start in 2008, with completion due in April 2012.

UPDATED

CFD

Cie des Chemins de Fer Départementaux
9-11 rue Benoît-Malon, F-92150 Suresnes
Tel: (+33 1) 45 06 44 00   Fax: (+33 1) 47 28 48 84
e-mail: fdc@cfd.fr
Web: www.cfd.fr

Key personnel
President, CFD Group: François de Coincy
President, CFD Bagnères: Patrick Esnault

Organisation
CFD has built trains and operated local lines in France since 1881. It manufactures small diesel locomotives and railcars at Bagnères-de-Bigorre (the former Soulé plant), specialising in narrow gauge trains. The workshop also carries out refurbishment work and has contracts with RATP to upgrade metro stock.

In early 2000, SNCF, through its subsidiary VFLI (65 per cent) and CFD formed joint-ventures, CFD Industrie, to run a workshop at Montmirail, and Voies Ferrées du Morvan, which operates the 87 km Autun–Avallon line in central France.

UPDATED

CFTA

Société Générale de Chemin de Fer et de Transports Automobiles
Parc des Fontaines, 169 avenue Georges Clemenceau, F-92735 Nanterre Cedex
Tel: (+33 1) 46 69 30 00   Fax: (+33 1) 46 69 30 01
Web: www.connex.net;
   www.cfta.fr;
   www.connextradition.com

Key personnel
Chairman: François Peter
Director General: Jean Pierre Fremont
Director, Freight: Yves Cautin

Organisation
CFTA is a railway operating subsidiary of Connex, the transport business of the French conglomerate Veolia Environnement. CFTA has existed in one form or another for 120 years, with particular experience of operating rural branch lines. In France, CFTA operates several lines totalling 696 km on behalf of SNCF, providing station, traction and track maintenance staff. CFTA staff drive both SNCF trains and CFTA's fleet of diesel locomotives and multiple-units.

CFTA is bidding for further operating contracts as these come up for renewal but has also lost historic contracts to other operators. The company has recently won the contract to operate Trains de Mouettes from Saujon to La Tremblade, the Chemin de Fer de la Mure from Saint Georges-de-Commiers to La Mure (25 km), near Grenoble, and the Chemin de Fer de la Rhune, a rack line in the Pyrenees mountains. CFTA also manages the metre-gauge Chemin de Fer de La Provence line from Nice to Digne (151 km). It also operates steam trains in summer on the Paimpol–Guingamp line. These are operated by the subsidiary Connex Tradition.

Passenger operations
CFTA operates passenger services over the Carhaix–Guingamp–Paimpol lines in Brittany plus Longueville-Provins east of Paris. On the Provins line, SNCF passenger stock is used whilst CFTA supplies a locomotive for freight. In the case of the Brittany lines, CFTA uses two-axle, driver-only-operated railbuses in conjunction with the local regional council.

Freight operations
Freight subsidiary CFTA Cargo operates freight trains on behalf of SNCF over 11 lines, mainly in central and eastern France. In 2000, CFTA lost the contract for 100 km of lines in the Nevers area to SNCF.

CFTA also manages private sidings. In 1994, the company took over Socorail, a company specialising in sidings for the oil industry, acquiring 80 shunting locomotives. Both CFTA Cargo and Socorail are part of the Connex Industries subsidiary.

In June 200, CFTA Cargo started open access freight services (the first in France), carrying quicklime from Dugny-sur-Meuse to Völkingen in Germany. The company uses Vossloh G1206 diesel locomotives hired from Angel Trains Cargo.

Traction and rolling stock
CFTA operates 25 main line diesel locomotives, 100 shunting locomotives and 10 diesel railcars. The company has workshops at Gray which tender for outside overhaul work.

UPDATED

T3M

Transport du 3me Millénaire
5 route de Stains, F-94380 Bonneuil-sur-Marne
Tel: (+33 1) 41 94 16 50   Fax: (+33 1) 41 94 16 51
e-mail: fa.t3m@wanadoo.fr

Key personnel
President: Jean Mourot

Organisation
In October 2000 T3M was founded as a wholly owned subsidiary of Connex and launched an intermodal rail service between the river port of Bonneuil in the Paris suburbs and Lungavilla in northern Italy. In February 2004 the company was sold to three entrepreneurs. In September 2004 T3M launched a service between Bonneuil and Marseille.

Traction and rolling stock
TAB hires double wagons from AAE for the service. Traction is provided by the domestic national operators, SNCF and Trenitalia.

UPDATED

Voies Ferrées Locales et Industrielles (VFLI)

6 rue d'Amsterdam, F-75009 Paris
Tel: (+33 1) 55 07 81 00   Fax: (+33 1) 55 07 82 68
e-mail: info@vfli.fr
Web: www.groupevfli.com

Key personnel
President: Gérard Gibot

Organisation
VFLI is a subsidiary of SNCF holding SNCF Participations, which was set up in 1997 to manage local freight lines and industrial branches on a low-cost basis. VFLI operates some 40 industrial branches and has several bigger subsidiaries, including Voies Ferrées des Landes, which works certain lines around Mont-de-Marsan, and the MDPA potash mine network near Mulhouse.

In 2000 VFLI formed a joint venture with CFD (see entry in France section of *Railway systems and operators*), Voies Ferrées du Morvan (VFM) to operate the 87 km Avallon–Autun line and also took over former CFD workshops at Noyon, plus Locorem, near Liège, Belgium. VFLI also part-owns the CFD workshops at Montmirail, E&T in Lyon, Locotract in Arles and Serma in Lançon de Provence. All of these carry out locomotive refurbishment work and VFLI hires out these locomotives to industrial users in France and Belgium.

In November 2001 VFLI took over operation of the rail system of Houllières du Bassin de Lorraine (HBL), a 210 km network serving coal mines in northeast France, the last of which closed in May 2004. VFLI is using HBL's workshops as a base for locomotive refurbishment and hire. A new subsidiary, VFLI Cargo, was set up to operate the former HBL network, which carried 4 million tonnes in 2002.

From January 2003 all of VFLI's locomotives were transferred to a new subsidiary, Gemafer.

In mid-2003 VFLI won a contract to manage two industrial branches in Romania for Lafarge, a major cement manufacturer.

In 2003 VFLI presented Evolis, a Bo-Bo diesel locomotive type rebuilt with two 330 kW Deutz engines from SNCF's withdrawn Class BB 63000 or similar locos supplied to industry. VFLI was supplying 10 similar rebuilds to RATP and is increasingly taking over management of SNCF's withdrawn diesel locomotives.

In 2004 VFLI was due to supply over 50 diesel locomotives (including 40 hired from EWS in the UK) plus drivers for tracklaying on the LGV Est Européen high-speed line. A new subsidiary, Fertis, was set up for this purpose.

In 2003 VFLI had 636 staff, was operating over 500 km and recorded a turnover of €48.6 million.

Traction and rolling stock
In 2003 VFLI owned 88 shunters and 75 main line diesel locomotives plus 800 freight wagons.

UPDATED

Gabon

Ministry of Tourism, National Parks and Transport

PO Box 3974, Libreville
Tel: (+241) 70 11 62/76 32 40

Key personnel
Minister: A M Miyakou

Transgabon Railway (Transgabonais)

Compagnie d'Exploitation du Chemin de Fer Transgabonais
PO Box 2198, Libreville
Tel: (+241) 70 24 78; 70 07 24; 70 05 74
Fax: (+241) 70 20 38

Key personnel
President and Director-General: Christian Bongo
Deputy Director-General: Bernard Maisonnier
Adviser to the President: Joël Engone Ndong
Directors:
  Operations: Christan Renamy
  Human Resources: Timothée Mihindou
  Finance: Jean Luca

Gauge: 1,435 mm
Route length: 814 km

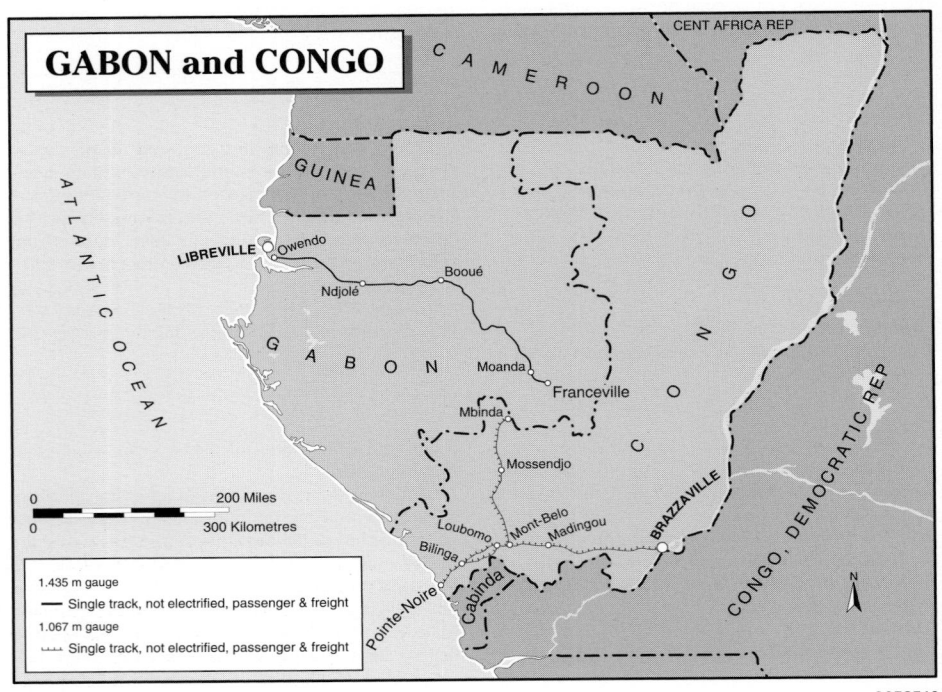

0058510

Political background
Construction of the former Gabon State Railways (OCTRA) from Libreville/Owendo to Booué started in 1974 and the first section between Owendo and N'Djolé (183 km) opened in 1979. Plans originally called for construction first of the main Owendo–Booué section, and later extensions south to Franceville and north to iron ore fields at Belinga. However, the economic case for continuous construction of the Booué–Franceville line was found to be overwhelming, as there are large deposits of manganese ore at Moanda in the Haut–Ogooué as well as extensive reserves of timber.

Services were inaugurated over the 162 km from N'Djolé to Booué in 1983, and the Booué–Franceville section was completed in 1986.

Privatisation occurred in 1999, when the Transgabonais consortium commenced a 20-year concession to operate the line. Since then, personnel numbers have been reduced from 1,867 to 1,363 in 2003.

Passenger operations
Transgabonais serves 20 stations. In 2002, passenger traffic amounted to 221,000 journeys for 97.5 million passenger-km.

Freight operations
In 2002, Transgabonais carried 2.952 million tonnes for 1,553 million tonne-km. Manganese is the principal commodity hauled, generating around 60 per cent of the railway's tonnage, with forest products accounting for a further 30 per cent.

Improvements to existing lines
A priority project is the creation of an intermodal terminal at Franceville to capture traffic to and from neighbouring countries in central Africa, including the Central African Republic, Congo and the Democratic Republic of Congo. Transgabonais foresees an eventual capacity at this facility of 20 million tonnes annually, compared with 4 million tonnes carried to Franceville in 2002.

Other projects undertaken since privatisation include: renovation of station buildings; the lengthening of passing loops to permit the operation of longer trains; provision of a satellite-based communications system; and construction of a new station and provision of supplementary storage facilities for forest products at Owendo.

Traction and rolling stock
In 2003, Transgabonais operated 18 main line diesel locomotives, including some Bombardier 2,685 kW MX636 machines, designated locally CC200. There were also seven shunting locomotives.

Passenger cars total 40, of which 10 have been refurbished. The wagon fleet totals 800, including 400 for forest products. Some refurbishment has been undertaken and new vehicles were acquired in 2002 from Maxion, Brazil.

Track
Rail: 50 kg/m Vignoles-type
Crossties (sleepers): Wood, 1,670/km
Max gradient: 0.8%
Max axleload: 25 tonnes

Georgia

Ministry of Transport

Kazbegi ul 12, 380012 Tbilisi
Tel: (+995 32) 93 25 46   Fax: (+995 32) 93 91 45

Key personnel
Minister: Merab Adeishvili

Georgian Railway (GRZhD)

Gruzinskaya Zhelezneya Doroga
Avenue Carica Tamara 15, 380012 Tbilisi
Tel: (+995 32) 95 25 27   Fax: (+995 32) 95 25 27
e-mail: webmaster@georail.org.ge
Web: http://www.railway.ge

Key personnel
Chairman and Managing Director:
  David Onoprishvili
Deputy Directors General
  Freight and Passenger: T Ghvaberidze
  Rolling Stock: O Tukhareli

0058561

Construction: I Khucishvili
Technical: T Donadze
Finance: R Mitaishvili
Foreign Affairs: I Melkadze
Legal: R Giorgadze

Gauge: 1,520 mm, 912 mm
Route length: 1,575 km, 37 km
Electrification: 3.3 kV DC; 37 km at 1.5 kV DC

Political background

A strategic development plan under preparation in 2004 foresees restructuring of GRZhD as a state-owned joint stock company, with privatisation an eventual possibility.

Organisation

The network comprises the greater part of the Caucasian Railway of the former SZhD. Its principal route is the electrified double-track running from close to the border with Russia on the Black Sea coast at Sochi/Veseloe, via Sukhumi, Ochamchire, Samtredia, Zestafoni, Khashuri and Gori to Tbilisi, where it divides. One line runs to Baku in Azerbaijan, the other to Yerevan in Armenia. On the main line between Zestafoni and the Azerbaijan border, 90 per cent of freight is in transit between Black Sea ports, Armenia and Azerbaijan. The connection with the Russian network in the northwest has been closed since the early 1990s due to political unrest in the Abkhazia region. Plans were being developed in 2004 to reinstate this link when political conditions permit.

Much of the terrain is inhospitable. There are severe gradients on the Zestafoni–Khashuri route, which rises 2,000 m in 60 km (2.8 per cent ruling gradient); there are radii as tight as 160 m. Train weights are restricted to 2,500–3,000 tonnes with three locomotives.

Workers employed in railway trades totalled 15,662 in 2001. Operating staff totalled 12,404.

Passenger and freight operations

Despite civil unrest since Georgia's independence in 1991, Georgian Railway began a perceptible recovery in 1996, with restoration of old schedules and a reduction of freight tariffs. Freight trains now take one day to cross from Batumi to Baku, compared to 13 in 1994. Five million tonnes of freight were carried in 1996 and by 2003 the figure had risen to 16 million tonnes (for comparison, 36.2 million tonnes were carried in 1989). From 1996 a container service has been provided between the Black Sea port of Poti and Baku, Azerbaijan. In 2003 some 80 per cent of freight tonnage was transit traffic, while the railway was handling 90 per cent of domestic freight traffic.

Passenger traffic stood at 2.1 million journeys in 2003, down slightly from 2.3 million in 2000.

In 1997 the Tbilisi–Moscow passenger service was restored after four years' cessation; it operates via Baku.

The narrow-gauge Borzhomi–Bakuriani line serves tourism and skiing traffic.

New lines

The long-projected Arkhot Railway connection across the Caucasus mountains from Tbilisi to

Class VL10 3.3 kV DC twin-section electric locomotive at Khashuri (Norman Griffiths)   0547796

Ordzhonkidze in Russia came back on to the agenda when the country gained independence. Three routes have been canvassed for the tortuous ascent from the Tbilisi area to the Arkhot pass, that favoured being a direct line from Gori, some 80 km west of Tbilisi, up the valley of the Liakhvi river to Tskhinvali. The alignment would then strike northeast through a summit tunnel in the Kazbegi area to reach Tarskoye in the Terek valley, thence following the river to Ordzhonkidze.

The proposed double-track and electrified route would extend to some 180 km and would require substantial lengths of bridging and tunnelling. The summit bore alone would exceed 20 km in length. Six intermediate stations are planned.

In 1995 Georgia and Turkey agreed to build a new line between Kars and Tbilisi via Akhalkalaki to replace the existing line via Armenia. This would involve an upgrade of the existing 190 km from Tbilisi to Akhalkalaki and a new 32 km link to the Turkish border. The scheme remained a possibility in 2004, when an alternative line south from Batume, on the Black Sea coast, to Turkey was also being studied. However, while more lightly engineered and requiring only 15 km of new line in Georgia, this would entail construction of some 250 km in Turkey.

Improvements to existing lines

Completion of work on the river bridge at Natanebi has reconnected the Black Sea port of Batumi with

the railway network. The Marelisi–Zestafoni section of the main line is being slowly reconstructed and the cross-border link to Armenia at Sadakhlo was being upgraded in 2002 under the Traceca programme.

Traction and rolling stock

In 2001 the electric locomotive stud of Classes VL8, VL10, VL11 and VL22, numbered 226, including examples stored out of use. The diesel locomotive fleet totalled 154, including examples stored out of use.

The passenger coach fleet numbered 590 vehicles and the freight fleet 16,000 wagons, of which 5,000 were in service.

An initial batch of four DR1A dmus has been modernised with MTU engines in place of the original Russian units.

Electrification

The network is fully electrified at 3.3 kV DC, the 37 km stretch at 1.5 kV DC being the steeply graded 912 mm gauge branch from Borzhomi to Bakuriani (possibly the only remaining line at this gauge in the former USSR).

Germany

Ministry of Transport

Krausenstrasse 17-20, D-10117 Berlin

Key personnel

Minister: Kurt Bodewig
Secretary of State: M Henke
Parliamentary Secretary: Manfred Carstens
Railways Department Manager: H G Gern

Federal Railway Administration (EBA)

Eisenbahn-Bundesamt
Vorgebirgsstrasse 49, D-53119 Bonn
Tel: (+49 228) 98 26; 0   Fax: (+49 228) 98 26; 199
Web: http://www.bund.de

Key personnel

President: Horst Stuchly
Vice-President and Head of Division 1:
   Ralf Schweinsberg
Head of Division 2: Dr-Ing Jens Böhlke
Head of Division 3: (vacant)
Head of Division 4: Peter Schäfer

The Restructuring of the Railways Act 1993 (*Eisenbahnneuordungsgesetz (ENeuOG)*), provided for the creation of the Federal Railway Administration (*Eisenbahn-Bundesamt (EBA)*) as an independent body under the Federal Ministry of Transport to exercise the rights and duties of the state other than those specifically within the sphere of the Federal Ministry with respect to the restructured railway industry. The EBA came into existence on 1 January 1994. It is based in Bonn, and there are 15 branch offices, one in each of the former DB/DR regional divisions.

The EBA is made up of a President's office and four divisions. Each of the 15 branch offices mirrors the organisation of the headquarters.

At presidential level there is also a Preliminary Investigation Bureau, upon the work of which the Federal Audit Office bases its instructions.

The divisions are:

Division 1: Legal Matters, Approval of Plans, Central Services (four departments)

Division 2: Infrastructure, Operations, MAGLEV (four departments)

Division 3: Rolling stock, plants (five departments)

Division 4: Infrastructure investment (four departments)

Division 1 deals with legal questions relating to the supervision of the railways and with the central services of the EBA. The railway supervisory authorities in the federal states base their regulations for regional railways on those applied at federal level by this division of the EBA. Division 2 deals with all matters of safety connected with infrastructure and operation and with all matters of approval and exemption under the EBO and the signalling regulations, and Division 3 deals with similar questions affecting rolling stock. Division 4 deals with the rationing of the funds assigned by the federal government to the EBA for investment in infrastructure (some DM10 billion a year).

Loans are normally provided interest-free, repayable on a depreciation basis within 40 years. Non-repayable grants are occasionally given, however, on the basis of economic considerations verified by this division. Each project involving federal funds is the subject of a specific agreement covering full details of the scheme, funding methods, and the amount to be spent, and each agreement is based on a specified set of rules.

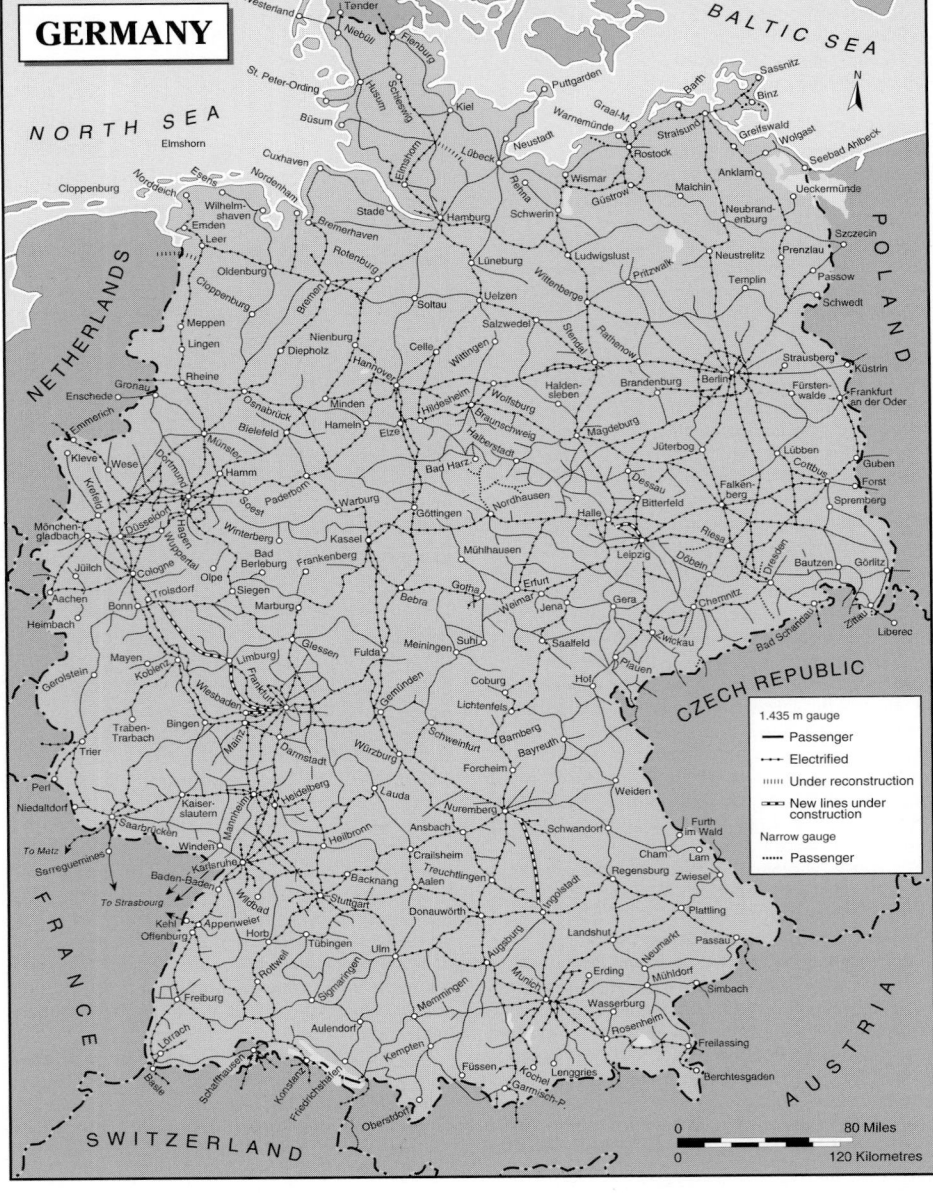

0114289

Federal Railways Fund (BEV)

Bundeseisenbahnvermögen
Kurt-Georg-Kiesinger Allee 2, D-53175 Bonn
Tel: (+49 228) 307 70   Fax: (+49 228) 307 71 60
e-mail: info@bev-bahnimmobilien.de
Web: http://www.bev-bahnimmobilien.de

Key personnel

President: Herr Heine
Head, Division 1: Herr Schilling
Head, Division 2: Herr von Niebelschütz
Head, Division 3: Herr Linder
Head of Public Affairs: Herr Scheuber

The *Bundeseisenbahnvermögen* (BEV) (Federal Railways Fund) is the third of the trio of organisations established by the *Eisenbahnne uordnungsgesetz (EneuOG)* 1993. Its functions are the administration and management of non-operational railway estate (currently valued at over DM11.3 billion), co-operation with the Federal Debt Administration in the management of transferred accumulated debt of the former state railways, management of some 70,000 personnel in civil service grades assigned to DB AG, responsibility for the pensions of railway staff in such grades and operation of staff welfare schemes (principally the provision of railway housing).

The BEV has no independent legal capacity – it is technically a *Sondervermögen* (special asset) of the Federal Republic, as DB and DR were previously – although it can act, sue and be sued under its own name. It has its own economic management and accounting functions, and is authorised to raise funds by the issue of debt securities and treasury bills or by taking up loans.

Any proper expenditure not covered through income is borne by the federal budget. The total financing requirement is up to DM15 billion a year.

The headquarters of the BEV are in Bonn. Organisationally, there is a two-tier structure: three divisions (and a number of special departments) report to the president, and between five and seven *Referate* report to the head of each division. The First Division deals with personnel matters, education, organisation and central services; the Second Division handles all legal affairs, medical matters, property and housing; and the Third Division is responsible for finances, information technology and telecommunications. There are regional offices in Berlin, Cologne, Essen, Frankfurt am Main (with an outpost in Saarbrücken), Hannover (with an outpost in Hamburg), Karlsruhe (with an outpost in Stuttgart), Munich and Nuremberg.

AKN Railway

AKN Eisenbahn AG
Rudolf-Diesel Strasse 2, D-24568 Kaltenkirchen
Tel: (+49 4191) 93 30
Fax: (+49 4191) 93 33 09
e-mail: info@akn.de
Web: www.akn.de

Key personnel

Managing Director: Johannes Krusznski
Operations Manager: Dr Ing Klaus Franke

Gauge: 1,435 mm
Route length: 254.9 km (117.4 km owned, 137.5 km DB Netz)

Organisation

The present company was created by the merger of several local government-owned lines in and around Hamburg.

In 2004 AKN employed 327 staff.

Passenger operations

The main line from Hamburg to Eidelstedt via Kaltenkirchen to Neumünster runs in its southern section effectively as a diesel-operated S-Bahn. It connects with the electrified S-Bahn at Hamburg-Eidelstedt. Through services via DB tracks to Hamburg Hbf are planned. The line from Norderstedt to Ulzburg is operated on behalf of Verkehrsgesellschaft Norderstedt and connects with the Hamburg metro at Norderstedt. The Neumünster–Heide line (61 km) is operated under contract for DB.

In 2000 AKN secured a contract to operate services on the Heide–Büsum line (20 km), initially for three years from November of that year.

In December 2002 AKN took over operation of the Neumünster–Bad Segeberg–Bad Oldesloe route in

a joint venture with Hamburger Hochbahn, NBE nordbahn Eisenbahngesellscahft mbH (nordbahn).
   Traffic statistics for 2003 were 12.3 million passengers and 118 million passenger-km.

Albtalbahn (AVG)

Albtal-Verkehrs-Gesellschaft mbH (AVG)
Tullastrasse 71, D-76131 Karlsruhe
Tel: (+49 721) 610 70   Fax: (+49 721) 61 07 50 09
e-mail: avg@karlsruhe.de
Web: www.kvv.de

Key personnel

Managing Director and Operations Manager:
   Dr-Ing E h Dipl-Ing Dieter Ludwig
Assistant Operating Manager:
   Dipl-Ing Siegfried Lorenz

Gauge: 1,435 mm
Route length: 468 km (includes 261 km of own routes and 207 km owned mostly by DB Netz)
Electrification: 750 V DC and 15 kV 16⅔ Hz AC

Allgäu-Express GmbH (ALEX)

Obere Eicher Strasse 20, D-87435 Kempten
Tel: (+49 831) 512 86 10   Fax: (+49 831) 512 86 19
e-mail: servus@alexpress.de
Web: www.alexpress.de

Freight operations
Traffic statistics for 2003 were 43,000 tonnes and 1.7 million tonne-km.

Organisation
AVG is affiliated with the municipal transport company of Karlsruhe.

Passenger operations
The original AVG line ran from Karlsruhe via Ettlingen to Bad Herrenalb/Ittersbach. Since 1966 operations have been progressively extended over DB lines which have been either bought or leased or are operated under contract by AVG. AVG services now form an S-Bahn network extending from Karlsruhe to destinations that include Bad Herrenalb, Bad Wildbad, Baden-Baden, Bietigheim-Bissingen, Bruchsal-Menzingen/Odenheim, Bretten–Eppingen, Forbach im Schwarzwald, Freudenstadt, Heilbronn, Hochstetten, Ittersbach, Mühlacker, Pforzheim and

Key personnel
Managing Directors: Wolf-Dieter Deuschle, Christian Saxer

Gauge: 1,435 mm
Route length: 173 km (operated)

Traction and rolling stock
The AKN fleet comprises 39 two-car dmus and eight diesel locomotives.

UPDATED

Wörth. Achern is to be added to the network in 2004–05. AVG regional routes penetrate the centre of Karlsruhe via tramway tracks.

Freight operations
Traffic statistics for 2001 were 249,000 tonnes.

Traction and rolling stock
A fleet of 211 electric light rail vehicles, owned by AVG, Verkehrsbetriebe Karlsruhe and DB, are operated in a pool. Out of these 117 are equipped for dual-voltage operation. Also in the fleet in 2004 were eight diesel locomotives.

UPDATED

Organisation
Allgäu-Express GmbH is a joint venture between EuroThurbo GmbH, a subsidiary of the Swiss-based operator Thurbo AG (see entry in Switzerland section of *Railway systems and operators*), and Regentalbahn AG (see entry in Germany section of *Railway systems and operators*), established to take over regional express services on the Munich–Kempten–Oberstdorf line (173 km).

Passenger operations
ALEX operates a two-hourly two-class service on the Munich–Kempten–Oberstdorf route, alternating with DB trains. A bistro-style catering service is provided.

Traction and rolling stock
Services are operated by five Siemens ER20 EuroRunner 2,015 kW diesel-electric locomotives leased from the Dispolok pool and 22 ex-DB coaches refurbished by PFA.

NEW ENTRY

ER20 EuroRunner diesel-electric locomotive with a Munich–Oberstdorf ALEX service at Sonthofen (Philip Wormald)
***NEW**/0585077*

Augsburger Localbahn (AL)

Augsburger Localbahn GmbH
Friedberger Strasse 43, D-86161 Augsburg
Tel: (+49 821) 56 09 70   Fax: (+49 821) 560 97 45
e-mail: stefan.blaas@augsburger-localbahn.de
Web: www.augsburger-localbahn.de

Key personnel
Managing Director: Udo Schambeck
Operations Manager: Jürgen Privinsky
Business Development Manager: Stefan Blaas

Gauge: 1,435 mm
Route length: 40 km

Organisation
Public freight line serving the city of Augsburg and its environs. The company is mainly owned by its clients, with the city of Augsburg holding a minority share. In 2003 AL employed 48 staff.

Freight operations
Connects with DB at Augsburg-Ring and serves many private sidings. Now also operates freight

services over DB lines. In 2004 AL carried 900,000 tonnes, 430,000 tonnes on it own network and 470,000 tonnes on DB lines.

Traction and rolling stock
Eight diesel locomotives, including four Class V100.4 rebuilt ex-Deutsche Reichsbahn Class 201.

UPDATED

BASF

BASF AG
Gebäude B818, Carl-Bosch-Strasse 38, D-67056, Ludwigshafen, Germany
Tel: (+49 621) 605 52 96

Key personnel
Logistics Manager: Bernd Flickinger

Gauge: 1,435 mm
Route length: 209 km

Organisation
Internal railway operations undertaken by the chemicals company BASF. Traffic over the main line network is handled by rail4chem, in which BASF is a shareholder. However, BASF has retained one local flow, moving limestone from Stromberg to its Ludwigshafen plant. It also moves containers from Ludwigshafen to Fermershein.

Traction and rolling stock
For its internal operations BASF has 20 diesel locomotives. The fleet includes three refurbished

Class V100.4 machines used for the Stromberg limestone traffic and Fermershein container movements. The company also owns some 900 wagons.

UPDATED

Bayerische CargoBahn GmbH (BCB)

Bahnhofplatz 1, D-83607 Holzkirchen
Tel: (+49 8024) 99 71 62
Web: www.connex-gruppe.de

Key personnel
Managing Director: Heino Seeger
General Manager: Stephan Schreier

Organisation
Established in February 2002 as an open access freight operator, BCB is a wholly owned subsidiary of Connex Cargo Logistics.

Freight operations
Regional open access freight services, since July 2003 including cross-border operations into Austria. BCB also provides traction for infrastructure maintenance work. The company collaborates both with other open access operators. These include assisting Hafen und Güterverkehr Köln (HGK) with deliveries of aviation spirit to Munich Airport and cooperating with Württembergische Eisenbahn Gesellschaft (WEG) in the movement of sand to a glass-making plant in Neuburg. BCB also participates in the NeCoSS container shuttle service that links Nuremberg and Schweinfurt with Bremen, Bremerhaven and Hamburg.

Traction and rolling stock
BCB operates two locomotives, oneType G1206 and one G2000, both built by Vossloh Locomotives.

NEW ENTRY

Bayerische Oberlandbahn (BOB)

Bayerische Oberlandbahn GmbH
Bahnhofplatz 1, D-83607 Holzkirchen
Tel: (+49 8024) 99 71 71   Fax: (+49 8024) 99 71 11
e-mail: auskunft@bayerischeoberlandbahn.de
Web: www.bayerischeoberlandbahn.de;
    www.connex-gruppe.de

Key personnel
Managing Director: Heino Seeger
Commercial Director: Christoph Kraller

Gauge: 1,435 mm
Route length: 120 km

Organisation
Formerly a joint venture of Connex (formerly DEG) (75 per cent) and Bayerische Zugspitzbahn (25 per cent), BOB is now a wholly owned subsidiary of Connex Verkehr GmbH. A 15-year contract was awarded to BOB in 1998 by Bayerische Eisenbahngesellschaft, an executive agency of the state of Bavaria, to run passenger services over the Munich–Holzkirchen–Bayrischzell,

Munich–Schaftlach–Tegernsee and Munich–Lenggries lines, a network of 120 km. Well over 4 million passengers are carried annually.

Bayerische CargoBahn GmbH (BCB), a wholly owned freight-operating subsidiary of Connex Cargo Logistics GmbH, is a sister company. Another Connex-owned sister company, Bayerische Instandhaltungsgesellscahft für Schienenfahrzeuge mbH (BIGS), maintains the BOB fleet at workshops in Lenggries.

Traction and rolling stock
The fleet initially comprised 17 five-section Integral dmus built by IVT. These were augmented in July 2004 by three-car Talent dmus supplied by Bombardier Transportation.

NEW ENTRY

Bayerische Oberlandbahn IVT-built ID5 'Integral' dmu
(Ken Harris)
0533632

Behala

Berliner Hafen- und Lagerhausgesellschaft mbH
Westhafenstrasse 1, D-13353 Berlin
Tel: (+49 30) 39 09 50   Fax: (+49 30) 39 90 51 39
e-mail: info@behala.de
Web: www.behala.de

Key personnel
Managing Directors: Horst Schuberth;
    Peter Staeblein

Gauge: 1,435 mm
Track length: 27 km

Organisation
Founded in 1923, the Berlin port railway serves the Hafen Neukölln, Osthafen, Südhafen and Westhafen.

Freight operations
Traffic statistics for 1999 were 1.1 million tonnes.

Traction and rolling stock
Three diesel locomotives.

UPDATED

Bentheimer Eisenbahn (BE)

Bentheimer Eisenbahn AG (BE)
Industriegebiet Nordhorn-Süd, Otto Hahn Strasse 1,
D-48529 Nordhorn
Tel: (+49 5921) 803 30
Fax: (+49 5921) 80 33 55
e-mail: info@bentheimer-eisenbahn.de
Web: www.bentheimer-eisenbahn.de

Key personnel
Managing Director: Dipl.-Betr wirt Peter Hoffmann
Operations Manager: Dipl-Ing Ulrich Walther

Gauge: 1,435 mm
Route length: 76 km

Freight operations
The only German regional railway of which the service extends across the border to Coevorden, the Netherlands. BE operates container terminals at Nordhorn and Coevorden and connects with DB at Bad Bentheim and with NS at Coevorden. Also runs freight trains over DB Netz tracks to destinations that include Münster, Rheine and the Hanover area.

Traffic statistics for 1999 were 581,000 tonnes and 31.7 million tonne-km.

Traction and rolling stock
In 2004 the fleet comprised 11 diesel locomotives and some 70 freight wagons.

UPDATED

Bodensee–Oberschwaben–Bahn GmbH (BOB)

Kornblumenstrasse 7/1, D-88046 Friedrichshafen
Tel: (+49 7541) 50 50   Fax: (+49 7541) 50 52 21
e-mail: info@bob-fn.de
Web: www.bob-fn.de

Key personnel
Managing Director: Dipl-Betr.wirt Manfred Foss

Gauge: 1,435 mm
Route length: 42 km (entirely over DB Netz tracks)

Organisation
Owned by city and local authorities, the BOB operates local trains north from Friedrichshafen via Ravensburg to Aulendorf. Services started in 1993. Operated by Hohenzollerische Landesbahn (HzL).

Passenger operations
Traffic statistics for 2001 were 845,000 passengers and 12.9 million passenger-km.

Traction and rolling stock
Seven diesel railcars.

UPDATED

boxXpress.de GmbH

Köhlfleetdamm 5, D-21129 Hamburg
Tel: (+49 40) 74 05 19 31   Fax: (+49 40) 74 05 19 33
e-mail: boxxinfo@boxxpress.de
Web: www.boxxpress.de

Key personnel
Managing Director: Joachim Kregel

Organisation
boxXpress.de was formed in June 2000 as an open access operator, handling maritime container traffic between north German ports and inland terminals. Its shareholders are: ERS (European Rail Shuttle) Holding BV, a subsidiary of Maersk Sealand and P&O Nedlloyd; Eurogate Intermodel GmbH; and TX Logistik AG.

Freight operations
boxXpress.de operates container services between the ports of Bremerhaven and Hamburg and terminals at Konwestheim (Augsburg), Munich and Nuremberg. Gemünden serves as an intermediate hub. Capacity in 2004 was approximately 230,000 TEUs.

Traction and rolling stock
In 2004 the traction fleet took the form of 10 6,400 kW 15 kV electric locomotives leased from Siemens Dispolok: two Type ES 64 F and eight Type ES 64 U2. Some 280 Type Sgmmrss 90' container wagons were leased from AAE.

NEW ENTRY

BSL – Olefinverbund GmbH

PO Box I163, D-06217 Merseburg
Tel: (+49 3461) 49 21 92   Fax: (+49 3461) 49 80 18

Key personnel
Managing Directors: Bart Groot; Heino Zell
Operating Managers
   Schkopau plant: Dipl-Ing Jürgen Jahnke
Böhlen plant: Dipl-Ing Michael Landgraf
Logistics, rail traffic: Dipl-Wirtsch-Ing
   Peter Ludwig

Gauge: 1,435 mm
Track length: 94.5 km

Organisation
Industrial railway.

Freight operations
Traffic statistics for 1997 were 6.1 million tonnes and 20.6 million tonne-km, compared with 5.6 million tonnes and 16.2 million tonne-km in 1996.

Traction and rolling stock
15 locomotives, 86 freight wagons.

Butzbach–Licher Eisenbahn AG (BLE)

Himmrichsweg 3, D-35510 Butzbach
Tel: (+49 6033) 961 50   Fax: (+49 6033) 96 15 15
e-mail: mail@ble-online.de
Web: www.ble-online.de

Key personnel
Chief Executive: Dipl-Ing Peter Berking
Operating Manager, Rail: Carsten Hessler

Gauge: 1,435 mm
Route length: 24 km (owned, freight only); 172 operated (DB Netz)

Organisation
Owned and operated by Hessische Landesbahn.

Passenger operations
BLE operates regional services on a group of lines northeast of Frankfurt. In 1998 the company took over operation of the Friedberg–Friedrichsdorf line (16 km) from DB and in 1999 operations on the lines from Frankfurt to Friedberg, Hungen and Nidda (70 km) from DB.
   From January 2001 BLE took over services on the Gelnhausen–Nidda–Hungen–Giessen (70 km) and Friedberg–Hanau (32 km) lines.

Freight operations
Limited services continue over two unconnected sections of the original network, connecting with DB at Butzbach and Bad Nauheim respectively. Traffic statistics for 2003 were 38,000 tonnes compared with 67,000 tonnes in 1999.

Traction and rolling stock
Passenger traffic is handled by 23 of a fleet of 70 GTW 2/6 railcars ordered by Hessische Landesbahn and delivered by Stadler in 2000–01. Two small diesel locomotives are used for freight services.

UPDATED

Chemion Logistik GmbH

Bayer-Chemiepark, Gebäude X6, D-51368 Leverkusen
Tel: (+49 214) 303 39 00   Fax: (+49 214) 303 39 01
e-mail: chemion@chemion.de
Web: www.chemion.de

Key personnel
Managing Director: Jürgen Sommer

Organisation
Chemion was established by Bayer AG in 2001 as a logistics company to take over its railway operations and those of the former Eisenbahn Köln-Mulheim-Leverkusen. Traffic is moved between Cologne and the Bayer plants at Dormagen, Leverkusen and Uerdingen.

Freight operations
Traffic statistics for Chemion's constituents for 2000 were 631,000 tonnes.

Traction and rolling stock
Thirteen diesel locomotives.

UPDATED

Connex Verkehr GmbH

Postfach 94 02 58, D-60460 Frankfurt am Main
Tel: (+49 69) 789 00 01
Fax: (+49 69) 78 90 01 91
e-mail: post@connex-gruppe.de
Web: www.connex-gruppe.de

Background
A subsidiary of the diversified French-based Veolia Environnement group, Connex Verkehr GmbH is a holding company for extensive rail and bus operations in Germany. Passenger rail and activities are grouped under the German parent company, while rail freight businesses are subsidiaries of Connex Cargo Logistics GmbH.

Berlin office
Georgenstrasse 22, D-10117 Berlin
Tel: (+49 30) 20 07 33 43   Fax: (+49 30) 20 07 32 00

Key personnel
Chairman: Ragnar Nordström
Managing Directors: Jérôme Jeauffroy, Hans Leister

Organisation
Connex Verkehr GmbH is a holding company for the following railway operations:
Bayerische Oberlandbahn GmbH (Holzkirchen)
Bayerische Instandhaltungsgesellschaft für
   Schienenfahrzeuge mbH (Holzkirchen)
LausitzBahn GmbH (Görlitz)
Nord-Ostsee-Bahn GmbH (Kiel)
NordWestBahn GmbH (Osnabrück)
Ostmecklenburgische Eisenbahngesellschaft mbH
   (Neubrandenburg)
Ostseebahn GmbH (Rostock)
Rheinische-Bergische Eisenbahn GmbH (Mettmann)
Württembergische Eisenbahn-Gesellschaft mbH
   (Waiblingen).

In March 2002, Connex Regiobahn launched its first long-distance open access main line passenger service. Using Talent dmus, the InterConnex service replaced an Interregio route abandoned by DB AG and links Rostock, Berlin, Leipzig and Gera. It operates without subsidy. In December 2002 a second route, Zitta–Stralsund, was added using stock from the Lausitzbahn. In 2002 the service was altered to start from Liberec, Czech Republic.

Connex Cargo Logistics GmbH

Georgenstrasse 22, D-10117 Berlin
Tel: (+49 30) 39 60 11 90   Fax: (+49 30) 39 60 11 70
e-mail: cargo@connex-gruppe.de

Organisation

Connex Cargo Logistics GmbH is a holding company for the following railway operations:
Bayerische CargoBahn GmbH (Holzkirchen)

Farge-Vegesacker Eisenbahn-Gesellschaft mbH (Bremen)
Hörseltalbahn GmbH (Eisenach)
Industriebahn-Gesellschaft Berlin mbH (Berlin)
NeCoSS GmbH (Bremen)
Niederbarnimer Eisenbahn AG (Berlin)
NordWestCargo GmbH (Osnabrück)
Regiobahn Bitterfeld GmbH (Bitterfeld)
RCB Rail Cargo Berlin GmbH (Berlin)

Schöneicher-Rüdersdorfer Strassenbahn GmbH (Schöneiche)
Teutoburger Wald Eisenbahn AG (Gütersloh)

UPDATED

Dortmunder Eisenbahn (DE)

Dortmunder Eisenbahn GmbH (DE)
Speicherstrasse 23, D-44147 Dortmund
Tel: (+49 231) 983 95   Fax: (+49 231) 983 96 06
e-mail: service@dortmunder-eisenbahn.de
Web: www.dortmunder-eisenbahn.de

Key personnel

Managing Director, Technical and Operating
   Manager: Herr Kerkeling
Managing Director, Commercial:
   Herr Hutschenreiter

Gauge: 1,435 mm
Route length: 50 km (excluding industrial lines and sidings)

Organisation

Public freight line serving the port of Dortmund. Also acts as industrial railway for the Hoesch and Krupp steelworks in Bochum. DE is jointly owned by Dortmunder Hafen AG and Thyssen Krupp Stahl. Provides third-party maintenance services.

Freight operations

DE freight trains also run over DB lines. Traffic statistics for 2003 were 15.98 million tonnes. These figures include both public and non-public traffic.

Traction and rolling stock

34 diesel locomotives, 696 freight wagons.

UPDATED

Duisport Rail GmbH

Bliersheimer Strasse 42-52, D-47229 Duisburg
Tel: (+49 203) 80 31   Fax: (+49 203) 80 32 32
Web: www.duisport.de

Key personnel

Managing Directors: Thomas Schlipköther,
   Markus Teuber
Operations Manager: Uwe Hinz
Marketing: Michael Lirschner, Volker Schmitz

Gauge: 1,435 mm
Track length: 120 km, including sidings

Organisation

Founded in April 2001 as a wholly owned subsidiary of Duisburger Hafen AG, Duisport Rail GmbH provides rail services in the port area. It is also involved with partners in linking factories such as the Bayer chemicals plant at Dormagen with the port and in operating long-distance freight services from Duisburg Intermodal Terminal.

Traction and rolling stock

In 2004 the locomotive fleet comprised seven shunters (including six ex-DB Class 360s) and five main line machines (two Class 202/203, one Class 207 Vossloh G1206 plus hired traction).

NEW ENTRY

Duisport Rail Class 207 (Vossloh G1206) diesel-hydraulic locomotive (Duisport, Köppen)
***NEW**/0585031*

Eisenbahn und Häfen GmbH (EH)

Franz-Lenze-Strasse 15, D-47166 Duisburg
Tel: (+49 203) 521   Fax: (+49 203) 522 54 52
Web: www.eisenbahn-und-haefen-gmbh.de

Key personnel

Managing Directors: Dieter Biersching,
   Dipl-Ing Manfred Redeker
Head of Rail Operations: Dipl-Ing Gerd Kullik

Gauge: 1,435 mm
Track length: 505 km
Electrification: 62 km at 1.5 kV DC

Organisation

EH is the industrial railway system of Thyssen Krupp Stahl AG and is a wholly owned subsidiary of the company. It serves several other industrial clients, but is not a common carrier. The company operates two private ports on the river Rhine at Schwelgern and Walsum.

In 1998 a subsidiary company, EH Güterverkehr GmbH, was formed to provide commercial rail freight and related services. Some 10 million tonnes are handled annually in co-operation with Railion Deutschland and private operators.

Freight operations

Originally operating in the Duisburg, Dinslaken and Oberhausen areas, EH is now also responsible for the

operation of industrial lines at Mülheim and Krefeld. About one-third of traffic is exchanged with DB.

EH locomotives run freight trains over DB lines, while DB locomotives penetrate into the EH network.

Traffic statistics for 2004 were 73.4 million tonnes and 398.9 million tonne-km.

Traction and rolling stock

EH operates 95 diesel locomotives (17 of these are dual-mode electric locomotives and have a diesel generator). All locomotives are equipped for remote control. Electric traction to be abandoned gradually by 2007. Also in the fleet are 1,966 freight wagons of different types.

UPDATED

EKO Trans

EKO Transportgesellschaft mbH
Werkstrasse 1, D-15890 Eisenhüttenstadt
Tel: (+49 3364) 37 50 60   Fax: (+49 3364) 37 22 30
e-mail: eko_trans@eko.arcelor.com
Web: www.eko-trans.de

Key personnel

Managing Director: Dipl-Ing (FH) Udo Thiede
Operations Manager: Dipl-Ing (FH)
   Gerhard Timpel

Gauge: 1,435 mm
Track length: 155 km

Organisation

EKO Trans is a wholly owned subsidiary of EKO Stahl GmbH, part of the Arcelor Group. Originally established to operate the industrial system of the company's steelworks, EKO Trans now also runs freight trains over DB lines as an open access operator.

Freight operations

EKO Trans' centre of operations is a marshalling yard at Ziltendorf, on the Frankfurt (Oder)–Cottbus line, where traffic is exchanged with the DB Netz system. Open access operations are run on an 'own account' basis and in collaboration with DB Cargo and other operators. Traffic statistics for 2001 were 6.7 million tonnes and 100.2 million tonne-km.

Traction and rolling stock

EKOTrans operates 18 diesel locomotives, including nine modernised Class 260 for internal operations, two Class 293 and one 2,240 kW Class 242, one Class 143 electric locomotive and 245 wagons.

UPDATED

Elbe-Weser Railways (EVB)

Eisenbahnen und Verkehrsbetriebe Elbe-Weser GmbH
Bahnhofstrasse 67, D-27404 Zeven
Tel: (+49 42 81) 94 40   Fax: (+49 42 81) 944 30
e-mail: bahn@evb-elbe-weser.de
Web: www.evb-elbe-weser.de

Key personnel

Managing Director: Dipl-Vw Ulrich Koch
Operating Manager: Heinz Schulze

Gauge: 1,435 mm
Route length: 285 km

Organisation

EVB was formed in 1981 by the merger of the Bremervörde–Osterholz and the Wilstedt–Zeven–Tostedt railways, which only carried freight traffic. In 1991 the Bremerhaven–Bremervörde–Stade, Bremervörde–Rotenburg–Brockel and Hesedorf–Hollenstedt lines, totalling 156 km, were taken over from DB. In 1993 the Buxtehude–Harsefeld Railway was taken over. EVB also operates bus services.

EVB is a shareholder in Metronom, formed in 2002 to operate main line regional services on Hamburg–Bremen and Hamburg–Uekzen routes from December 2003. Other members of the consortium are Bremer Strassenbahn AG, Hamburger Hochbahn AG and Osthannoversche Eisenbahn AG.

EVB has also established a joint venture with DB Regio to operate NordseeBahn services between Bremerhaven and Cuxhaven since December 2003.

Passenger operations

Passenger services are operated from Bremerhaven via Bremervörde-Harsefeld to Buxtehude and

EVB two-car Class 628-type dmu (Marcel Vleugels)   0531916

onwards over DB tracks to Hamburg-Neugraben, where a connection is available to the Hamburg S-Bahn. EVB also operates the Moorexpress railcar service between Stade and Osterholz via Bremervörde.

Traffic statistics for 2001 were 980,000 passengers and approximately 14.7 million passenger-km, compared with 910,000 passengers and some 13.8 million passenger-km in 1999.

Freight operations

EVB handles container traffic between the deep-sea ports of Hamburg, Bremen and Bremerhaven. EVB trains also run over DB lines, including container trains between Bremerhaven/Bremen/Hamburg and Stuttgart/Nuremburg. Traffic in 2001 amounted to 890,000 tonnes.

EVB is also a member of the Neutral Container Shuttle System (NeCoSS), a co-operation agreement that also includes Bremer Spedition Acos Transport GmbH and Connex Cargo Logistics. Together the companies operate open access container services in Germany.

Traction and rolling stock

In 2004 the fleet comprised five dmus (similar to DB Class 628), five diesel railcars and 14 diesel locomotives (including six ex-DB Class 211, one Vossloh G2000, one ex-DB Class 219, one ex-DB Class 220 and one ex-DB Class 232. There were also two ES64 U2 6,400 kW Taurus electric locomotives on hire from Siemens Dispolok.

For NordseeBahn services to be introduced between Bremerhaven and Cuxhaven in December 2003, EVB operates and maintains nine Alstom Coradia LINT 41 dmus acquired by the Lower Saxony regional transportation authority.

UPDATED

Erfurter Industriebahn (EIB)

Am Rasenrain 16, D-99086 Erfurt
Tel: (+49 361) 74 20 70   Fax: (+49 361) 742 07 27
e-mail: info@erfurter-bahn.de
Web: www.erfurter-bahn.de

Key personnel

Managing Director:
   Dipl-Betr wirt Heidemarie Mähler
Operating Manager: Erwin Nönig

Gauge: 1,435 mm
Route length: 15 km (owned)

Organisation

Originally an industrial freight line owned by the city of Erfurt. Since 1991 EIB has provided passenger services under contract to Thüringen state.

EIB is also a joint venture partner with Hessische Landesbahn GmbH (HLB) in the SüdThüringenBahn (STB) (see entry in the Germany section of *Railway systems and operators*), founded in 2000.

Passenger operations

Railcars are operated under contract to Thüringen state over the Erfurt–Döllstadt–Leinefelde line, alternating with DB RE (regional) trains. In May 1999, services were extended to Kassel Wilhelmshöhe and in May 2000 EIB commenced

EIB Adtranz RegioShuttle lightweight diesel railcar (Quintus Vosman)   0063197

operations on the Gotha-Bad Langensalza line and services on the Apolda–Ilmenau line were also subsequently added to the network.

In 2003 EIB was awarded a contract to operate services on the Schweinfurt–Gemünden and Schweinfurt–Meiningen routes from 2005.

Freight operations

The main commodity carried is coal for a power station. EIB also runs an industrial line in a chemical plant at Rudolstadt-Schwarza.

Traction and rolling stock

Twenty three Adtranz/Stadler RegioShuttle RS 1 diesel railcars, one Bombardier Itino railcar and four diesel locomotives, including three Type V100s.

UPDATED

For details of the latest updates to *Jane's World Railways* online and to discover the additional information available exclusively to online subscribers please visit
jwr.janes.com

Farge–Vegesacker Eisenbahn (FVE)

Farger Strasse 128, D-28777 Bremen
Tel: (+49 421) 686 46   Fax: (+49 421) 68 35 60

Key personnel
Managing Directors: Dipl-Ing Heinz Thomas Schare,
   Heinz Wolfgramm
Operations Manager: Martin Mertens

Gauge: 1,435 mm
Route length: 10 km

Organisation
Associated with and operated by Connex Cargo
Logistics GmbH.

Passenger operations
In 2004 plans were being discussed at local
government level to introduce a passenger service
between Farge and Bremen Hbf.

Freight operations
FVE connects with DB at Bremen-Vegesack.
Traffic is mainly coal for a power station. Traffic
statistics for 1997 were 490,000 tonnes and
5 million tonne-km.

Traction and rolling stock
Traction is provided by Nord West Cargo (Connex
Cargo Logistics).

UPDATED

FLEX

Background
FLEX Verkehrs AG was a subsidiary of
Norddeutsche Nahverkehrsgesellschaft (NNVG)
and was established in 2002 to operate regional
services under a 12-year contract with the
state of Schleswig-Holstein. In August 2003
the company initiated bankruptcy proceedings
after reportedly encountering difficulties
in the apportionment and collection of revenue.
   The state of Schleswig-Holstein subsequently
appointed NordOstBahn (NOB) (see entry
in Germany section of *Railway systems and
operators*), a Connex Gruppe subsidiary, to
take over the contract for two years from
November 2003.
   A new contract was to see operation of the
service revert to DB in 2005–06.

UPDATED

Frankfurt-Konigsteiner Eisenbahn AG (FKE)

Bahnstrasse 13, D-61462 Königstein
Tel: (+49 6174) 290 10   Fax: (+49 6174) 29 01 15
e-mail: mail@fke-online.de
Web: www.fke-online.de

Key personnel
Chief Executive: Dipl-Ing Peter Berking
General Manager, Rail: Wolfgang Köhler

Gauge: 1,435 mm
Route length: 16 km (owned)

Organisation
Controlled by Hessische Landesbahn, which has
a 51 per cent stake, FKE's own line extends from
Frankfurt-Höchst to Königstein (16 km), but services
over this line now start at Frankfurt Hbf. Operations
on the Brandoberndorf–Bad Homburg (42 km) and
Bad Homburg–Frankfurt Hbf (19 km) lines started in
1993 and since 1997 FKE has worked the Frankfurt-
Höchst–Bad Soden line (7 km). FKE also operates
bus services.

Passenger operations
A busy suburban carrier, FKE connects with
the Frankfurt S-Bahn at Frankfurt-Höchst and
Friedrichsdorf. Some peak-hour trains run through
over DB tracks to Frankfurt (Main) Hbf.

Traction and rolling stock
FKE operates 24 dmus (20 VT 2E built by LHB in
1987–92, one VT 628 unit (Duewag 1994–95), two
VT 629 (Duewag 1995) and one Stadler GTW2/6).

UPDATED

Freiberger Eisenbahngesellschaft mbH (FEG)

Carl Schiffner Strasse 26, D-09599 Freiberg
Tel: (+49 3731) 30 07 77   Fax: (+49 3731) 300 77 22
e-mail: info@freiberger-eisenbahn.de
Web: www.freiberger-eisenbahn.de

Gauge: 1,435 mm
Route length: 31 km

Key personnel
Managing Directors: Henrik Behrens,
   Michael Engelhardt

Background
FEG is a subsidiary of Rhenus Keolis GmbH & Co
KG, a joint venture formed by Rhenus and Keolis.

Passenger operations
In November 2000 FE took over the operation
of the Freiberg–Mulda–Holzhau line in Saxony
until 2019. Around 225,000 passengers are carried
annually.

Traction and rolling stock
Three Adtranz RegioShuttle RS1 single-unit
railcars.

UPDATED

Georgsmarienhütte Eisenbahn (GET)

Georgsmarienhütte   Eisenbahn   und   Transport
GmbH
Klöckner Strasse 1, D-49110 Georgsmarienhütte
Tel: (+49 5401) 39 43 60   Fax: (+49 5401) 39 43 73

Key personnel
Managing Directors: Wilfried Hülsmann,
   Dipl-Ing Hubert Unland (also Operations Manager)

Gauge: 1,435 mm
Route length: 7.3 km

Organisation
Industrial steelworks line.

Freight operations
GET connects with DB at Hasbergen. Line haul
traffic is handled by DB with through trains
from Osnabrück, while GET handles shunting
movements and internal works traffic. Traffic
statistics for 2001 were 837,000 tonnes and
6.7 million tonne-km.

Traction and rolling stock
Five diesel shunting locomotives.

UPDATED

German Railways Group (DB)

Deutsche Bahn Gruppe
Potsdamer Platz 2, D-10785 Berlin
Tel: (+49 30) 29 70   Fax: (+49 30) 29 72 61 30
Stephensonstrasse 1, D-60326 Frankfurt am Main
Tel: (+49 69) 973 30   Fax: (+49 69) 97 33 75 00
Web: http://www.bahn.de

Key personnel
Chairman of the Supervisory Board:
   Dr Michael Frenzel
Directors
   Chairman and Managing Director:
      Hartmut Mehdorn
   Finances and Controlling:
      Dipl-Betriebsw Diethelm Sack
   Marketing: Klaus Daubertshaüser
   Personnel and Social Affairs: Stefan Garber

Property, Law and Passenger Stations:
   Dieter Ullsperger
Research and Technology:
   Dr Karl-Friedrich Rausch
Passenger Traffic: Dr Phil Christoph Franz
Freight Traffic: Dr Bernd Malmström
Infrastructure: Dipl-Ing Roland Heinisch
Infrastructure Business and Network Planning:
   Dr Thilo Sarrazin
Purchasing and Materials: Klaus-Bend Bapp
Department of International Affairs:
   Klaus Ebeling
Public Affairs: Dieter Hünerkoch

Gauge: 1,435 mm; narrow gauges
Route length: 38,384 km; 66 km
Electrification: 18,652 km (15 kV 16$\frac{2}{3}$ Hz AC except
for Berlin and Hamburg S-Bahn systems)

Political background
Under the Restructuring of the Railways Act
1993 (Eisenbahnneuordungsgesetz (ENeuOG)),
which came into force on 1 January 1994, the
former West German Deutsche Bundesbahn,
East German Deutsche Reichsbahn, and the
Railway Property in West Berlin were fused as
the Federal Railway Assets. These were then
divided into a public section and a commercial
section. The public section is divided into the
Federal Railway Office (Eisenbahn-Bundesamt
(EBA)), and the Office for Federal Railway Assets
(Bundeseisenbahnvermögen (BEV)), which deals
with the non-operational railway estate, the civil
service personnel of the former state railways
(who could not because of their status pass
under the control of a joint-stock company), and
inherited debts. The commercial section of the
railway became Deutsche Bahn AG, charged with
managing the railway industry according to good

Class 605 ICE TD tilting diesel trainset for Dresden-Nuremberg services (David Haydock)   0114245

DB's ICE Fleet

Type	No	Introduced	Seating capacity	Max speed km/h
ICE 1	59	1991	700 approx	280
ICE 2	44	1996	368	280
ICE 3	50	2000	391	330
ICE T	43	1999	398	230
ICE TD	20	2000	196	200

business principles in line with German company law. To do this, business sectors, each with its own bottom-line responsibility, were established.

During the 1999 financial year DB AG was converted into a holding company, the Deutsche Bahn Gruppe (German Railways Group), under which come five subsidiaries, each with its own bottom-line responsibility. These are:

DB Netz, which operates, maintains, and markets the 64,821 km of track belonging to DB and provides infrastructure capacity and services in return for track access charges to some 100 railway traffic operators;

DB Reise & Touristik, which runs more than 1,400 long-distance passenger trains a day carrying some 400,000 passengers and, through 13 subsidiaries ranging from the AMEROPA travel agency to the MITROPA AG catering undertaking, provides tourism-related services at home and abroad;

DB Regio, which carries some 4 million passengers a day in 29,000 trains linking principal centres and in the regions;

DB Cargo, which runs some 6,200 goods trains a day to transport some 300 million tonnes of freight a year and owns some 131,000 goods wagons with a capacity of some 6 million tonnes, as well as more than 4,300 locomotives;

DB Station & Service, which operates 5,794 passenger stations and stopping-places throughout Germany.

The changes were envisaged from the start of the restructuring programme and are intended to pave the way for private-sector participation in the national railway.

Further changes were agreed early in 2000. Whilst not affecting the devolution of bottom-line responsibility brought in with the second stage of the railway reforms at the beginning of 1999, these are designed to match responsibilities to customer expectations and to remove divisions of answerability, experience having shown that the structure developed since the 1994 reforms is too complex and not sufficiently customer-focused. It has also become clear that the multitude of individual businesses adds unnecessary costs. The main changes involve the creation of a Marketing Division at Group Management level, bringing the whole of the passenger business together

into a single Business Sector, concentration of purchasing into the Chairman's Division from 1 January 2001, creation of a special Property Sector, and rearrangement of groups and service areas like Communications or Legal Services whose activities cross Divisional boundaries. As far as possible, the Sectors will have largely parallel structures, with central functions for finance and audit, marketing and commerce, personnel, and either production and technology or operations.

As a result of these alterations there are now three organisational levels in DB. At the top of the pyramid comes the Group Management. The middle level is made up of five Business Sectors: Infrastructure, Passenger Traffic, Goods Traffic, Passenger Stations and Property. At the base level are the Business Units that have the task of delivering the products for the Business Sectors. These are essentially the existing companies and their structure is not changed – they continue to have full responsibility for production and bottom line. They are: Long Distance Passenger Traffic, Regional Passenger Traffic, Urban Passenger Traffic, European Freight, Regional Freight, Logistics, Stations and Construction & Repair.

In late 2000 the federal government announced an additional DM2,000 million a year of state support for the railway.

Performance Indicators (€ million)

	1999	2000
Turnover	15,630	15,465
Pre-tax profit	91	37
Employees at 31 December	241,638	222,656
Fixed assets	34,071	32,815
Balance sheet total (equity & liabilities)	37,198	39,467
Equity capital	8,701	8,788
Gross investment	8,372	6,892
Net investment*	3,229	3,250
Depreciation	1,966	2,052
Pre-tax cash-flow**	2,107	2,113

* Gross investment less grants towards construction costs from the federal, Land, and/or local authorities.

** Pre-tax profit plus depreciation of fixed assets and changes in pension reserves.

Traffic (million)

	1998	1999	2000
DB Reise & Touristik			
Passenger journeys	149	146.5	144.8
Passenger-km	34,275	34,897	26,226
DB Regio			
Passenger journeys	1,450	1,534	1,568
Passenger-km	31,324	37,949	38,162
DB Cargo			
Freight tonnes	289	279	301
Freight tonne-km	73,273	71,494	80,634

Traction and rolling stock

All the Deutsche Bahn Gruppe's traction units and rolling stock is assigned to the separate businesses. Overall, at the end of 2000, the total stock stood at 12 steam locomotives, 3,514 electric locomotives, 2,248 diesel locomotives, and 1,262 shunting locomotives. There were 2,076 electric multiple-units (excluding the DC stock of the Berlin and Hamburg S-Bahn systems), 2,001 diesel multiple-units, and 216 railbuses. At the end of 2000 the ICE fleet included 120 ICE 1 power cars and 46 ICE 2 power cars.

DB Regio had 1,637 bilevel passenger coaches. DB Reise & Touristik had 705 ICE 1 trailer cars and 309 ICE 2 trailer cars (45 of them being driving trailers). There were 264 ICE 3 vehicles and 42 ICE T trainsets. In addition to these, there were 1,847 IC/EC coaches and 1,518 IR/D coaches. The night stock comprised 631 vehicles.

DB Cargo had 131,178 freight wagons, and there were 60,611 privately owned wagons and 5,510 hired wagons on the system. Of the owned freight wagons, 33,169 were covered, 40,406 open, 55,895 flats, and 1,708 silo wagons.

The InterCity Express (ICE) family

The ICE 1 and ICE 2 fleets are complete.

One ICE 1 was destroyed in the Eschede accident in June 1998, the cause of which appears to have been a broken tyre. All except two of the fleet of 60 ICE 1s had been equipped with Class 064 resilient wheels, authorised for use in 1992, in order to try to eliminate vibration problems that had been encountered. At the instance of the EBA these have now been exchanged for monobloc wheels of the type used in the ICE 2s, at a total cost of some DM30 million. The change brought a recurrence of the vibration problems that led to the introduction of the tyred wheels in the first place, and as a provisional solution DB has specified turning the wheels every 240,000 km. In parallel, investigations are going on into the possibility of inserting a 10 mm damper layer between rail and sleeper on the high-speed lines to reduce the vertical stiffness of the track (which is a contributing factor to the problem). In the longer term DB hopes to be able to return to the use of tyred wheels with ultrasound monitoring. The railway is also examining a suggestion for the acoustic monitoring of bogies during the journey.

In August 1994 DB AG placed a 'run-on' order for 50 eight-car ICE 2-2s from Siemens/AEG. In fact, these trains, of which the first entered service in 2000, are very different from their predecessors and are now designated ICE 3. The design reduces weight still further, cuts fuel consumption and costs, and produces a train suitable for pan-European operation. The ICE 3 is a genuine unit-train: traction is distributed throughout the entire train, with roughly every second axle powered. Advantages include better transmission of tractive effort to rail, reduction of slipping, more seats for the same train length, and a more favourable weight distribution. Axleload is under 17 tonnes. Maximum speed is 330 km/h and the train can work on gradients of up to 40 pro mille. Four trains are to be three-voltage (15 kV 16²/₃ Hz AC, 1.5 kV DC, and 25 kV 50 Hz AC) and nine are to be four-voltage (adding the Belgian 3 kV DC) for cross-border workings; the remaining 37 trains are to be equipped only for the German/Austrian/Swiss 15 kV 16²/₃ Hz AC (though trains working into Switzerland, of course, require an additional pantograph). Four similar trainsets are being supplied to Netherlands Railways (qv) for services into Germany. Studies are being undertaken into the possibility of equipping a second series of DB AG ICE 3s with tilt.

In 2001 an additional 13 ICE 3s and 28 ICE Ts were ordered. In late 2001, DB withdrew from

ICE T set forming a Singen-Stuttgart service at Horb (Colin Boocock)   0089563

Class 611 RegioSwinger tilting dmus forming a Cologne-Saarbrücken service at Saarburg (John C Baker)   0114246

the planned joint procurement with its Austrian and Swiss counterparts ÖBB and SBB of a fleet of 116 tilting trainsets for TEE Alliance cross-border services. DB had intended to acquire 50 units.

DB has indicated that locomotive haulage of long-distance passenger services will end by 2010, by when trainsets of the ICE family will be used for all traffic of this type.

Tilting trains
DB AG's experience of tilting trains dates back to the highly successful introduction of the Fiat-system Class VT610 trains in North Bavaria in the early 1990s. The success of this operation led to a general reappraisal of future policy (also connected with tighter financial conditions), as a result of which three new types of tilting train were ordered.

For InterCity services on electrified lines 40 tilting unit-trains of Classes 411 and 415 are being delivered. The first units came into service in 1999 on Stuttgart-Zurich services. Externally, these trains resemble the ICE family, and they share internal design and equipment with the ICE 3 units. The ICE Ts use the latest, all-under-floor, Fiat tilt technology. With up to 8° body-tilt and a maximum uncompensated lateral acceleration of 2.0 m/s², the ICE Ts can run through curves up to 30 per cent faster than classic stock and can reduce journey times by 15 to 20 per cent. Top speed is 230 km/h and the trains are formed of five or seven vehicles, the majority being seven. Because the ICE T has a slightly smaller profile than the ICE, on account of the need to allow for the tilt, there is a marginal reduction in comfort. The motors are positioned lengthways under the floors and deliver a performance similar to that of the ICE 3 (though with a lower maximum speed). The seven-car sets include a restaurant, the shorter sets have a bistro.

For InterCity services on non-electrified lines, 20 diesel-powered four-car 200 km/h tilting trains (Class 605) are being built in a DM280 million contract with Siemens (with its Duewag AG subsidiary) and Deutsche Waggonbau AG. The first vehicles were in production at the start of 1998 and the trains came into use in April 2001 on Dresden–Nuremberg services. Each train offers 195 seats (41 first, 148 second, and six in a mother-and-child compartment) and there is an all-standing bar-area. Addition of a fifth car for strengthening is possible, and the sets can run in multiple. Motors and tilt-mechanism (the Siemens electromechanical system) are bogie-mounted.

The third new class of tilting train is designed for local passenger traffic and is the Class 611. Adtranz has supplied 50 diesel-powered two-car units with a top speed of 160 km/h which use a tilt technology with a military origin, developed from that in the Leopard tank, and a Voith hydraulic transmission. Each train offers 141 seats, and a closed toilet system is used. There have been many problems with these trains. A further 197 generally similar Class 612 two-car units were ordered by DB Regio from Adtranz (now Bombardier Transportation), while 21 were ordered by DB Reise&Touristik. Deliveries were in progress in 2001.

New electric locomotives
In 1994 DB AG ordered equipment to a value of DM4 billion. In addition to the ICE 3 and the three classes of tilting trains mentioned above, 420 locomotives (of the three types described below) and 339 S-Bahn units were ordered, with options for a further 500 locomotives and 200 S-Bahn units to a value of DM3.5 billion. Deliveries began before the end of 1996, and the arrival of the new stock is allowing a large part of the obsolete equipment from the 1960s and 1970s, now unreliable and expensive to maintain, to be scrapped. The urgent need for the replacement of locomotives is proved by the fact that 20 per cent of all daily perturbations in traffic are currently caused by locomotive faults and failures.

Class 103 has been replaced by the new Adtranz Class 101, of which 145 have been built. These three-phase, AC drive, 6,000 kW locomotives have a top speed of 220 km/h. They are being used initially on the InterCity network, but as more ICEs and ICTs come on stream they will gravitate towards freight traffic.

There will be 170 representatives of the new Class 152 from Krauss-Maffei/Siemens. This class, based on the EuroSprinter supplied to Spain, Portugal, and most recently Greece, is a heavy freight locomotive, also with a 6,000 kW rating and a 140 km/h maximum speed. Deliveries began in 1997. Along with the 101s, these machines will replace the 30-year-old Class 150s.

Over the next few years Classes 110, 139, 140, 141 and 143, some of them more than 30 years old, will begin to be replaced by a first delivery

Electric locomotives

Class	Origin	Line voltage	Wheel arrangement	Power kW	Speed km/h	Weight tonnes	No in service	First built	Builders Mechanical	Electrical
101	–	15 kV	Bo-Bo	6,000	220	–	145	1997	Adtranz	Adtranz
103	DB	15 kV	Co-Co	7,440/ 7,780	200	114	129	1970	Henschel	Siemens
109	DR	15 kV	Bo-Bo	4,200	120	82	4	1962	LEW	LEW
110	DB	15 kV	Bo-Bo	–	140	86.4	374	1956	Krauss-Maffei	Siemens
	DB	15 kV	Bo-Bo	3,620/ 3,700	140	86[1]		1963	Henschel/Krupp	AEG/BBC
111	DB	15 kV	Bo-Bo	3,620/ 3,700	160	83	226	1974	Krauss-Maffei	Siemens
112	DR	15 kV	Bo-Bo	4,200	160	83	128	1991	LEW	LEW
113	DB	15 kV	Bo-Bo	3,620/ 3,700	120	86	11	1962	Krauss-Maffei/Henschel	Siemens
120	DB	15 kV	Bo-Bo	4,400/ 5,600	160/ 200[2]	84	62	1979	Krauss-Maffei/ Krupp/Henschel/ Thyssen	BBC
127	–	15 kV	Bo-Bo	6,400	230	–	1	1993	–	Siemens
128	–	15 kV	Bo-Bo	6,400	220/ 250	–	1	1994	–	AEG
139[3]	DB	15 kV	Bo-Bo	3,620/ 3,700	110	86	47	1957	Krauss-Maffei/Krupp/	AEG, BBC
140	DB	15 kV	Bo-Bo	3,620/ 3,700	110	83	792	1957	Thyssen/ Henschel	Siemens
141	DB	15 kV	Bo-Bo	2,310/ 2,400	120	67	379	1956	Henschel	BBC
142	DR	15 kV	Bo-Bo	2,920	100	82	54	1962	LEW	LEW
143	DR	15 kV	Bo-Bo	3,720	120	82.8	634	1984	LEW	LEW
145	–	15 kV	Bo-Bo	4,200	140	–	80	1998	Adtranz	Adtranz
146	–	15 kV	Bo-Bo	4,200	160	–	31	2001	Bombardier	Bombardier
150	DB	15 kV	Co-Co	4,410/ 4,500	100	126/ 128	168		Krupp/Henschel/ Krauss-Maffei	AEG, BBC, Siemens
151	DB	15 kV	Co-Co	5,982/ 6,288	120	118	169	1973	Krupp/ Krauss-Maffei	AEG/BBC/ Siemens
152	–	15 kV	Bo-Bo	6,000	140	–	170	1997	Krauss-Maffei	Siemens
155	DR	15 kV	Co-Co	5,400	120	123	267	1974	LEW	LEW
156	DR	15 kV	Co-Co	5,880	125	–	4	1991	–	–
171	DR	25 kV	Co-Co	3,660	80	–	11	1965	–	–
180	DR	15/3 kV	Bo-Bo	3,080	120	84	20	1988	Skoda	Skoda
181	DB	15/25 kV	Bo-Bo	3,000/ 3,240	150	84	1	1968	Krupp	AEG
181.2	DB	15/25 kV	Bo-Bo	3,200/ 3,300	160	83	25	1975	Krupp	AEG
182	–	15/25 kV	Bo-Bo	6,400	140	–	25[4]	2001	Siemens	Siemens
184	DB	1.5/3/ 15/25 kV	Bo-Bo	3,240	150	–	1	1967	–	–
185	–	15/25 kV	Bo-Bo	4,200	140	–	400[4]	2001	Bombardier	Bombardier

[1] Nos 110.288 onwards, which are Class 110.3; remainder are Class 110.1.
[2] Nos 120.001–4 only.
[3] The principal difference between Classes 139 and 140 is that the former has rheostatic braking for heavily graded routes.
[4] Number of locomotives ordered.

DB AG lightweight diesels unit on order/delivered

Class	Type
640 001 to 640 030	Alstom single-unit trains
641 001 to 641 040	De Dietrich DB/SNCF joint local units similar to Class 640
642 001 to 642 150	Siemens-Duewag two-unit trains
643 001 to 643 075	Bombardier-Talbot diesel-hydraulic dmus
644 001 to 644 058	Bombardier-Talbot diesel-electric dmus
646 001 to 646 030	Adtranz GTW2/6
648 001 to 648 ???	Alstom Coradia LINT two-car units
650 101 to 650 127	Adtranz single-unit RegioShuttle RS1
670 101 to 670 128	Double-deck railbuses of the second series

Diesel locomotives

Class	Origin	Wheel arrange-ment	Power kW	Speed km/h	Weight tonnes	No in service	First built	Mechanical	Builders Engine	Transmission
202	DR	B-B	900	100	65	380	1969	LEW	12 KVD 21 AL-3	H Pirna
204	DR	B-B	1,100	100	65	65	1969	LEW	12 KVD 21 AL-4	H Pirna
211	DB	B-B	760	90/100	62	34	1961	–	MTU MD 12 V 538 TZ MTU MB 12 V 493 TZ	H
212	DB	B-B	930	100	63	241	1962	–	MTU MB 12 V 652 TA	H
213	DB	B-B	930	100	63	10	1962	–	MTU MB 12 V 652 TA	H
215	DB	B-B	1,430/ 1,849	140	77.5	133	1968	–	MTU MB 16 V 652 TB MTU 12 V 956 TB 10	H
216	DB	B-B	1,300	120	77.5–77	138	1964	–	MTU MB 16 V 652 TB	H
217	DB	B-B	1,430	140	77	13	1965	–	MTU MD 16V 652 TB	H
218	DB	B-B	1,840/ 2,060	140	80		1968	–	MTU MA 12 V 956 TB MTU 12 V 956 TB 11 Pielstick 16 PA 4 V 200	H
218.2–4	DR	C-C	1,470	120	95	404	1966	Babelsberg	2 × 12 KVD 18/21 A II	H Pirna
218.9	DR	B-B	1,840	90/140			1979	–	–	H
219	DR	C-C	1,980/ 2,200	120	99	167	1962	Bucharest	2 × M820 SR	H
229	DR	C-C	2,760	140	–	20	1992	–	–	H
232	DR	Co-Co	2,200	120	112.4	482	1973	Voroshil-ovgrad	5 D 49	E Voroshil-ovgrad
234	DR	Co-Co	2,200	140	–	64	1992	–	–	E
259		Co-Co	2,460	96	121	1	1990	General Motors	General Motors	E General Motors
290	DB	B-B	810	70/80*	77/77.8	216	1964	–	MTU MU 12 V 652 TA	H
291	DB	B-B	810	80	76–90	46	1965	–	Mak 8 M 282 AK	H
293	DR	B-B	735	65	64	2	1981	LEW	12 KVD 18/21 A3	H Pirna
294	–	B-B	1,995	40/80	–	193	–	–	–	H
295	–	B-B	1,995	40/80	–	57	–	–	–	H
298	DR	B-B	750	80		80				

*Nos 290.001–20 only.

Electric railcars or multiple-units (excluding Berlin and Hamburg S-Bahn stock)

Class	Cars per unit	Line voltage	Motor cars per unit	Motored axles/car	Power kW	Speed km/h	Weight tonnes	No in service	First built	Builders Mechanical	Electrical
401[1]	–	15 kV	–	–	4,800	280	80	120	1989	KM, Krupp, Henschel	ABB, AEG, Siemens
402[2]	–	15 kV	–	–	5,000	280	80	44	1996	–	–
403	8	15 kV	–	–	8,000	330	–	37[4]	1999	–	–
406	8	15 kV	–	–	8,000	330	–	13[4]	1999	–	–
411	7	15 kV	–	–	4,000	230	–	32[4]	1999	DWA/Duewag/ Fiat	Siemens
415	5	15 kV	–	–	3,000	230	–	11[4]	1999	DWA/Duewag/ Fiat	Siemens
420/ 421	3	15 kV	–	–	2,400	120	138	480	1971	MBB/MAN	AEG/BBC/ SSW
423	4	15 kV	–	–	2,350	140	–	190[4]	1999	Adtranz	Alstom
424	4	15 kV	–	–	2,350	140	108	40[4]	1999	Adtranz/DWA/ Siemens	–
425	4	15 kV	–	–	2,350	160	108	156[4]	1999	Adtranz	Alstom
426	2	15 kV	–	–	1,175	160	60.9	43[4]	1999	Adtranz/DWA/ Siemens	–
445	3	15 kV	–	–	3,600	140	179	1[4]	1999	Adtranz/DWA/ Siemens	–
450	3	15 kV/ 750 V	–	–	560	100	–	4	1994	Duewag/ABB/ WV	–
'Thalys'[3]	–	4-cur	–	–	8,800	300	–	4	1997	Alstom	Alstom

[1] ICE 1 power car.
[2] ICE 2 power car.
[3] Thalys power car.
[4] Number ordered.

of 80 of Adtranz's new Hennigsdorf-built Class 145, of 4,200 kW continuous power and with a 140 km/h maximum speed. Delivery began in early 1998. Like the new Class 101, Class 145 belongs to the Adtranz modular family of locomotives, so major components have all already been fully tested in Class 101. In 2001 deliveries commenced from Bombardier of 31 Class 146 locomotives, a 160 km/h passenger version of the Class 145. An option for a further 400 similar machines was taken up in July 1998; these will be designated Class 185 and will be dual-frequency, to allow

their use in cross-border freight traffic as well as in domestic traffic. Deliveries commenced in April 2001. They are being delivered at 50 a year from the beginning of 2001, flowing on from completion of the Class 145 order at the end of 2000. Three pre-serial production locomotives will be built. Other multisystem locomotives on order in 2001 were 25 Class 182 dual-voltage (25 kV AC/15 kV AC) Siemens 'Taurus' units and 100 Class 189 machines. The first of the former type was to be handed over to DB Cargo in August 2001.

Diesel locomotives
In 2001, orders were placed by DB Cargo for replacement engines to modernise 140 Class 232 Soviet-built diesel-electric locomotives. The existing Kolomna 5D49 engine is being replaced by a 2,235 kW 12D49 unit from the same manufacturer. The move followed the earlier trial installation of similar engines in two locomotives of the same class.

Multiple-unit fleet renewal
DB Regio is spending almost DM6 billion in the greatest ever investment programme in new trains for local traffic, designed to fit services for the new century. Included in the orders are 440 S-Bahn sets and electric units for regional traffic as well as the Class 605 diesel sets discussed in the 'Tilting trains' section above. Additionally, more lightweight and double-deck units are being bought. A further DM1 billion will go on the modernisation of existing stock.

The Munich, Stuttgart, and Frankfurt/Main S-Bahn systems are receiving 100 new Class 423 units, built by Adtranz with the participation of Alstom and having a floor height of 998 mm. These replace the Class 420/421 equipment dating from 1972. There is an option for a further 200 sets.

An order for 60 units of a new Class 424 was placed with a consortium of Adtranz, Siemens, and DWA (Bombardier Transportation). These were to have a 760 mm floor height, and the electronics were to be largely underfloor. Of the 60, 45 were for the new Hannover Expo 2000 S-Bahn, Dresden, and Leipzig; the remainder were to be deployed in the Mannheim area. The same consortium was given an order for a further 136 four-car units designated Class 425, a variant of the Class 424 design with more seats, fewer doors, and a 160 km/h maximum speed, for Express S-Bahn work on longer routes, together with 43 two-car units designated Class 426 and essentially shortened Class 425s. Classes 425 and 426 are provided with entry steps to allow them to work on lines without high station platforms. All three classes were to be articulated. It was then announced in early 1996 that the Class 424 design would not be built. Instead, an additional 60 Class 425s are being constructed, provided that automatic entry steps can be provided to deal with platform heights of 380, 560, and 760 mm. Examples of both Class 425 and 426 entered service with DB Regio in 2001.

The new Class 474 units for Hamburg's S-Bahn began to enter service in August 1997. All were in traffic by July 2001. Floor level in the new units is 100 mm lower than in their predecessors, to help all passengers but especially the disabled. Glass dividing walls allow a good view through the whole unit, and each vestibule has facilities for voice connection to the driver, thus contributing to greater passenger safety. DB's 2005 energy-saving programme is taken into account, and the braking systems return energy to the overhead line. The total DM800 million order was for 103 trains.

Lightweight multiple-units
To revitalise its regional services, DB AG has ordered large numbers of new lightweight diesel units mainly from the 'RegioSprinter' and 'Talent' families.

Coaching stock
Bilevel cars now feature significantly in local operations all over Germany. A series of such vehicles, and matching driving trailers, have been produced with a 160 km/h capability to work on RE services in the Berlin area. DB Regio has ordered a further 98 bilevel cars from Deutsche Waggonbau GmbH. These include 23 driving trailers and 75 trailers. This latest order is part of a contract dating from 1996 which provided for the delivery of 250 vehicles and an option for 350 more. The 98 are

Class 101 electric locomotive at Singen with an InterRegio service (David Haydock)   0114247

Diesel railcars or multiple-units

Class	Cars per unit	Motor cars per unit	Motored axles/car	Power kW	Speed km/h	Weight tonnes	No in service	First built	Mechanical	Builders Engine	Transmission
605	4	–	–	1,700	200	–	20	–	–	–	E
610	2	–	–	970	160	–	20	1992	Duewag/ MAN	MTU	E
611	2	–	–	1,118	160	116	50	1997	Adtranz	–	H
612	2	–	–	1,118	160	116	200	1999	Adtranz	–	H
614	3	2	2	370	140	124.5	42	1972	O & K/ Uerdingen	MAN	H
624/634[1]	3	2	2	330	120	115.5– 118.1	44	1961	MAN/ Uerdingen	MAN	H
627	1	1	2	285/287	120	36	8	1974	–	–	H
628	2	1	2	2 × 210/ 1 × 375	120	64	–	1974	Duewag/ KGB MBB	–	H
628.2	2	1	2	410	120		466	1987	Duewag/ KGB/MBB	–	H
628.4	2	1	2	485	120		–	1992	–	–	–
640	1	–	–	315	120	38.9	30[3]	1999	LHB 'LINT'	–	M
641	1	–	–	514	140	47	40[3]	1999	De Dietrich	–	H
642	2	–	–	514	120	66.1	150[3]	1999	Siemens/ Duewag 'Regio- Sprinter'	–	M
643	2	–	–	5–630	120	67	75[3]	1999	Bombardier- Talbot 'Talent'	–	H
644	3	–	–	1,000	120	84.3	59[3]	1999	Bombardier- Talbot 'Talent'	–	E
646	2	–	–	550	120	–	42[3]	1999	GTW/2/6	–	E
650	1	–	–	514	120	37.7	27[3]	1998	Adtranz/ DWA 'Regio- Shuttle'	–	H
670	1	–	–	250	100	–	5	1997	DWA	–	M
771/172	Max 2	2	1	132	90	22	65	1960	Bautzen	6 kVD 18	M Bautzen
772/172	Max 2	2	1	132	90	22	89	1964	Bautzen	6 kVD 18	M Bautzen
796[2]	1	1	2	110	90	20.9	329	1953	MAN/ Uerdingen/ WMD	–	M

[1] Type 624 rebuilt with air suspension bogies.
[2] Two-axle railbus: works as required with non-powered trailers, but not in permanent set formations.
[3] Number of units ordered.

ordered under that option. The 160 km/h cars are being built in Görlitz at a total cost of DM185 million. There will be first-class-only vehicles in which the seats will have fold-down tables and in which the saloons will be closed off by glass doors to try and prevent the draughts that passengers have complained of, and six vehicles will have a Train Caf catering area, with wardrobe. There will be a satellite information system to show the next stop as well as the normal p/a installation. There will be provision for fitting ticket-machines at a future date.

In 2001 a major programme of withdrawal of old restaurant cars began.

All DB AG passenger vehicles are to be fitted with retention toilets by 2002 as another step in the railway's general environmental-awareness programme.

Freight vehicles

DB Cargo followed up its summer 1998 order for 1,570 new wagons and other transport-units with a further order for 1,600 goods wagons and swapbodies. Additionally, 1,550 wagons were to be converted in railway workshops, thus bringing its investment to more than DM500 million. The new order included 550 17.25 m slide-wall wagons for the transport of high-value industrial and commercial goods (like washing-machines, dishwashers, or coffee-makers) and palletised goods from the chemical industry, and there is an option on a further 300 of the same sort. Of this fleet, 300 wagons were to conform to European gauge, whilst the other 250 will be to the larger German gauge. The balance of the second order comprised a further 200 covered four-axle self-unloading wagons for potash and other fertiliser; 350 bilevel car-transporters (150 of them with modern gauge-change equipment); 100 bogie-wagons with rolling roof for water-sensitive clay transport from the Westerwald to Italy and France, some of them with the new automatic coupler; 100 covered bogie wagons with heavy-duty unloading equipment and radio local control for plaster (Rea-Gips); 200 heavy-duty swap-bodies for roll-paper transport; another 100 special wagons with sliding doors for car-part movement and the conversion of 1,850 carriers for combined transport, 50 of them with a 22.5-tonne axleload.

DB Cargo is currently investing a substantial amount in a large number of new and refurbished freight vehicles. On top of the summer 1998 order for 1,570 new wagons and a further order for 1,600 new vehicles and 1,550 conversions, the company announced a 1999–2000 shopping list including 2,450 units (new builds, rebuilds, and rentals). Further large orders for new freight rolling stock have been placed as part of the ongoing modernisation of the fleet, including a DM21 million order in the spring of 2001, split as DM8 million for 50 new bulk lime wagons and DM13 million for rebuilding 200 more wagons for timber transport.

The CargoSprinter, a freight dmu which appeared in October 1996 and which was described in earlier editions, saw regular commercial service on trains between Frankfurt airport and the north but is no longer in regular use.

In conjunction with the railways of Austria and Italy, DB AG has been developing a 'quiet' freight train for the Brenner route. An experimental train of 30 wagons features disc brakes, bogie shrouds and special wheels to minimise transmissions of vibrations.

Signalling and telecommunications

DB AG holds 50.2 per cent of the telecommunications company Mannesmann Arcor AG & Co, based in Frankfurt am Main. Arcor has at its disposal a national ISDN network of 40,000 km, with which it aims to be Germany's second national telecommunications operator as well as to provide for all the telecommunications requirements of a modern railway. In 2000, the company was replacing its existing analogue radio network with a digital GSM-R network.

In a rationalisation move at the beginning of 1998, DB AG took over from Arcor all non-telecommunications services, bringing them under its own Anlagen und Haus (AHS) service-organisation and taking on 1,379 Arcor staff. AHS handles railway-specific services like the installation and servicing of ticket machines, safety systems, office and communications technology, information systems like destination indicators, platform clocks, and p/a systems.

A Central Control Office for the whole of DB's network, located in Frankfurt am Main and replacing previous control offices in Berlin and Mainz, came into use in 1997. This is the Rechnerunterstützte Zentrale Betriebsleitung (RZBL), where 40 staff directly supervise all Long Distance Passenger Business services trains as well as principal goods trains and special workings. Some 1,800 trains a day are watched from seven workstations from start to finish of their journeys in real time - and connections, substitute trains, special stops, and diversions can all be decided. Some 4,000 connections a day are supervised and each month the staff deal with between 12,000 and 15,000 cases of conflict. Additionally, the 15 Regional Traffic Offices are co-ordinated from here and contact is maintained with foreign administrations. DB Netz aims eventually to have the 17,000 km network on which 90 per cent of rail traffic flows controlled from just seven Traffic Centres.

At the end of 2000 there were 122 electronic signalboxes (ESTw) in use, an increase of 18 during the year. Some are very large installations, covering cities like Dresden, Frankfurt am Main and Hannover.

A serious accident to the Gläserner Zug in late 1995 dramatically illustrated the risks when a train leaves a platform against a red signal. To prevent such accidents in future, the Indusi (PZB) driver vigilance system is to be modified, the new form being designated PZB 90. Additional monitoring beyond the signal and additional stop monitoring are provided, so that if any train does start against the signal it will be braked at a speed as low as 25 km/h.

Computer-aided control

Nuremberg was the first hub on the historic system to commission a computer-aided traffic control centre (RZu). Similar installations are being completed at Frankfurt and Karlsruhe by SEL, which supplied the Nuremberg apparatus, and at Cologne and Hannover by Deutsche Philips. Total cost of these projects is DM126 million.

In an RZu, train movements are monitored by train describer apparatus, and visual display units (VDUs) in the control centre depict graphically train performance against schedule. In the light of real-time progress the computer will propose individual train priorities for optimal adherence to the timetable.

The CIR-ELKE (Computer Integrated Railroading – Erhöhung der Leistungsfähigkeit im Kernnetz) was fully commissioned in June 2001 on the 197 km Karlsruhe-Basle route (chosen because it can be worked on a self-contained basis in terms of motive power). It is now believed that the capacity-enhancement benefits expected from this system can be more than matched by the benefits that will come from more recent signalling developments, and it is unlikely that CIR-ELKE will be extended further. Nevertheless, expenditure on the system is not regarded as wasted, as much has been learnt and some features can be transferred.

Track

Standard rail: Type S49, weighing 49.5 kg/m; Type S54, 54.5 kg/m; Type S64, 64.9 kg/m. Lengths generally 30–120 m
Type of rail joints: 4- and 6-hole fishplates
Sleepers (crossties): Wood; steel; reinforced concrete.

Wood sleepers impregnated beech, fir or oak, 2,600 × 260 × 160 mm

Steel, 2,600 × 9 mm weighing 86.3 kg

The latest type of RC sleeper (Spannbetonschwelle B58) weighs 235 kg, is 2,400 mm long, 190 mm thick under rails, 280 mm wide at bottom and 136 mm at top
Spacing: 650–800 mm

Class 423 articulated emu built by Adtranz for Cologne services (Colin Boocock)   0114209

Rail fastenings: Baseplates and bolts, clips and spring washers with thin rubber or wood (poplar) pad between rail and plate; resilient rail spikes with wood and concrete sleepers and resilient rail clips with steel sleepers.
Max gradient: Main lines: 2.5%
   Secondary lines: 6.6%
Max curvature: Main lines: 9.7° = min radius 180 m
   Secondary lines: 17.5° = min radius 100 m
Maximum superelevation: 150 mm on curves of 300 m radius and above

Rate of slope of superelevation: Generally 1:10 V (V = speed in mph). On occasion this may be increased to 1:8 V up to 1 in 400. On reverse curves the permissible limit is 1:4 V up to 1 in 400
Max altitude: Main line: 967 m between Klais and Mittenwald. Highest station Klais, 933 m
   Secondary line: 969 m between Bärenthal and Aha on the Titisee-Seebrugg line
Max axleloading: 22.5 tonnes

DB Netz

DB Netz AG
Theodor-Heuss-Allee 7, D-60486 Frankfurt am Main
Tel: (+49 69) 26 53 20 02   Fax: (+49 69) 26 53 20 07

Key personnel
Board Members
Chairman: Dipl-Ing Roland Heinisch
   Infrastructure: Prof Dr rer pol Ulf Häsler
   Marketing and Commercial Affairs:
      Dagmar Haase
   Operations: Dipl-Ing Gerhard Schinner
   Personnel: Leuthold Lewin
   Technology and Development:
      Dipl-Ing Holger Schulze-Halberg

In 2000 DB Netz had a turnover of €3,525 million and spent €3,896 billion on its capital account. The business employed 53,554 people. The 1997 figures were DM156 million, DM8.7 billion, and 78,370 employees.

To conform to EU directive 91/440, DB AG issued a price-list for the use of the railway network by all customers, internal and external in mid-1994. This was modified from 1 January 1995 after protests from the states about the impact of certain of its provisions on the regular interval services they wished to purchase for local routes. There were further substantial modifications to the system in 1998 and a completely new tariff, conforming to all national and EU requirements, was introduced in April 2001. The aim remains to cover the full costs for the use of the infrastructure through the access charges.

The Order for the Use of the Railway Infrastructure came into effect in December 1997. The infrastructure operator must seek consensus agreement between the parties where there are competing claims for paths but, if this cannot be achieved, the path is given to the party willing to pay most for it. The fact that one of the parties may be a more valuable customer by reason of his interest in other lines must play no role - this being to prevent the incumbent businesses of DB AG from overwhelming smaller undertakings. Under normal circumstances the infrastructure operator must not treat different applicants for the same line differently. To achieve pricing transparency the framework permits rebates for bulk capacity purchase within certain limits.

The Deutsche Institut für Wirtschaftsforschung (DIW) calculated that in 1996 only 52 per cent of full track costs were actually being met through access charges, as required by the 1994 reforms, with a payment of DM7.1 billion and full costs of DM13.5 billion. Only the Local Passenger Traffic

Business was paying sufficient to meet fully its share of track costs. The DIW asserted both that a full coverage of costs would not be possible and that under existing conditions the lower access charges demanded by the states could not be obtained. It therefore urged abandonment of this aim and its replacement by a requirement to cover all recurrent costs in full but only a clearly defined portion of the capital costs.

After significant cost rises in major projects, the DB AG supervisory board decided to introduce common standards throughout the country for such schemes and to establish an investment committee to keep a close central view on progress. All major projects now require the approval of the chairman of the board of management and the finance director. Significant changes are also being made in the way in which routine engineering works are planned, by 'bundling' tasks for 22 designated corridors, to reduce costs and improve the use of time and facilities.

New lines (Neubaustrecken – NBS)
After the creation of DB AG at the start of 1994 all outstanding new-line projects were reappraised in the light of both possible cuts in federal funding and an intention to pursue a wide-scale deployment of tilting stock. Although the aim remains to maintain investment levels, a significant portion of the money now comes from the sale of surplus lands rather than from the state.

The Hannover-Würzburg and Mannheim-Stuttgart new lines came into use in 1991 (although the Nantenbach Curve, the final section of the Hannover-Würzburg project, was not opened until three years later), and the Spandau-Oebisfelde line was brought into service in 1998. These lines are engineered for mixed traffic working, and the first two pass through geographically difficult and quite densely populated countryside. Construction was very expensive, with a substantial proportion of the cost attributable to environmentally necessitated measures. At an average of DM50 million per kilometre the Hannover–Würzburg line is the most expensive railway ever built (the French LGV Sud-Est cost only a third as much).

Cologne-Frankfurt NBS
Fixed-price contracts to a value of DM4,000 million were let during 1996 for the construction of this 300 km/h line and work is now well advanced over the whole length. The section serving the new Frankfurt Airport InterCity station was brought into use in May 1999. Completion of the whole project is expected in 2002, two years later than originally planned. There are problems of cost over-runs.

From Cologne two additional tracks are being built beside the present Right Rhine main line as far as Siegburg, where the new line turns away and follows the direct course of the A3 motorway through the hills. A third of the length is in tunnel. Because this is to be a purely passenger railway and because the terrain traversed is very difficult, the design parameters have been relaxed to allow a ruling gradient of 40 pro mille and a minimum curve radius of 3,500 m (use of slab track throughout allowing cant of up to 180 mm).

Cologne-Bonn airport will be served by a loop, the cost of which is being shared between the federal treasury, the Nordrhein-Westfalen Land government, and the airport authority.

Nuremberg-Ingolstadt NBS
The only other new line on which substantial progress is at present being made is that between Nuremberg and Ingolstadt. This 80 km new line (31 km in tunnel) is also due for completion in 2002. This is so far the only major rail construction project to be privately financed, the federal government having taken the necessary powers in the 1996 Finance Act. Optimisation in planning and construction and exploitation of all possibilities in awarding contracts made it possible to reduce the originally envisaged investment costs so that the total burden on the state sinks to DM622 million a year over 15 years, or roughly DM9 billion – well beneath the permitted level of DM622 million a year for 25 years (DM15.6 billion).

Leipzig-Dresden NBS/ABS
Work began in late 1993 on a scheme to create 106 km of new high-speed infrastructure between Leipzig and Dresden, partly by rebuilding the existing line and partly by long sections of new construction such as a Riesa avoiding line. Shortage of money forced a simplification of the plans and the Riesa avoiding line no longer figures. A significant portion of the work was completed in early 1999 (although no progress has been made so far on the final approach to Dresden, where decisions remain to be taken on the upgrading of the Berlin-Dresden main line).

Rastatt-Offenburg NBS
The 193 km Karlsruhe-Offenburg-Basle main line is being upgraded, part NBS and part ABS. One of Europe's principal international arteries, the route unites the main lines from Mannheim and Heidelberg at Rastatt and channels traffic from the whole of northwest Europe into Switzerland.

An NBS section over the 49 km between Rastatt and Offenburg, generally parallel to the existing

Driving trailers working with refurbished double-deck coaches on local services at Kaiserslautern
(Colin Boocock)   0114207

line, was nearing completion in 2000. Between Offenburg and Basle the original plans envisaged reconstruction of the existing railway for 200 km/h operation and the provision of a third track, but an extra pair of tracks is now to be provided throughout because it is expected that Swiss schemes to increase trans-Alpine rail freight capacity will generate considerable additional traffic.

Other schemes
There is little progress to report on the NBS schemes for lines between Erfurt and Ebensfeld, Halle/Leipzig and Erfurt, and Stuttgart and Günzburg. A short section of the Leipzig-Erfurt line, at the Leipzig end, is under construction.

Improvements to existing lines (Ausbaustrecken – ABS)
Ausbaustrecken are lines that have undergone major upgrading, including, where necessary, realignments and the provision of additional tracks, to maximise the scope for 200 km/h running. All German railway lines on which speeds of over 160 km/h are operated must be equipped with LZB continuous cab signalling, and no level crossings are permitted.

The 1990 reunification brought a shift in priorities: during the 1980s emphasis had been on a strengthening of north-south links, but more recently the focus has been on east-west routes. In 1991 a programme of 17 major transport works, nine of them rail, was agreed under the title of the Deutsche Einheit programme.

The period 1991–97 saw DM42,000 million spent in the 'Five New Federal States', but even so plans for most of the schemes were simplified to save money. In 2001 work was still in progress on the Lübeck-Stralsund route.

As a consequence of the abandonment of the Berlin Hamburg 'Transrapid' project the already-modernised main line between the two cities is now being further upgraded to allow speeds of up to 230 km/h. The work being done includes the long-delayed total modernisation of the railway through Wittenberge.

Infrastructure schemes in progress in the 'new federal states' that are not elements of the Deutsche Einheit programme include:

A DM 900 million investment-programme on the so-called Sachsenmagistrale between Dresden and Hof being carried out in agreement with the states of Saxony and Bavaria. This forms part of the ABS across South Germany from Karlsruhe through Stuttgart and Nuremberg to Leipzig and Dresden. New ICE TD trains will in due course reduce the Dresden-Hof timing from 3 hours 21 minutes to 2½ hours and Dresden-Nuremberg from 5 hours to about 4 hours. Bayreuth will acquire a much-improved link with Hof by the construction of the Schlomener Kurve, commissioned on 1 September 2000, which will cut at least 10 minutes off local train timings between the two cities and also allow Bayreuth to be linked into DB Reise&Touristik's long-distance network.

There is a DM177 million programme for the doubling, upgrading, and resignalling of the Weimar-Gera line.

Modernisation of the 84 km main line east from Berlin to Frankfurt an der Oder and Poland should be completed in 2002. The line is being upgraded for 160 km/h operation and resignalled to reduce times by quarter of an hour to 36 minutes.

Upgrading between Berlin and Dresden: DB AG and ČD have agreed to accelerate services between the two cities by short-term renovation and modernisation of the existing infrastructure coupled with the introduction of tilting trains. There is, however, a serious delay in the delivery of the new rolling stock. The Berlin-Prague-Vienna/Bratislava-Budapest corridor is to be integrated into the European high-speed network, and the German share involves upgrading the 160 km/h Dresden line for 200 km/h operation.

Berlin's railway system is being restored and modernised. Work began in late 1995 on a DM10 billion north-south tunnel through the heart of the city with a new central station at the Lehrter Stadtbahnhof, where the north-south route is crossed by the Stadtbahn west-east main line and S-Bahn route. The old direct lines from the south, from Halle/Leipzig and Dresden, are to be restored and re-routed at Papestrasse to the new alignment. At its northern end the new line will connect both ways with the (Inner) Ring, currently under restoration, enabling trains from the west and north to reach the new low-level Lehrter Bahnhof platforms. On the Ring and a little to the east, work on the Nordkreuz junctions for main line and S-Bahn trains is now well advanced. Although the plan to run a new S-Bahn line parallel to the north-south main lines has been partially put on hold to save money, the alignment is being protected. Eastwards from Zoologischer Garten the 114-year-old Berlin Stadtbahn has been rebuilt. The new Berlin-Spandau station was completed at the end of 1998. Reconstruction of Gesundbrunnen station is in progress. By 2000, major concerns had arisen about cost over-runs and plans for savings were being made.

Cologne-Aachen ABS
The line from Cologne to the Belgian frontier beyond Aachen is the link between the Cologne-Frankfurt NBS and the West European high-speed network. Between Cologne and Düren 250 km/h main lines are being provided, and from Düren to Aachen the existing line is to be upgraded for higher speeds. The full scheme will not be complete until 2005.

Dortmund-Kassel ABS
This ABS scheme covers the 215 km line from Dortmund via Hamm and Paderborn to a junction with the Hannover-Würzburg NBS at Kassel. Financial constraints have brought about a simplification of the plans, and after 10 years the work has still not been completed.

Ebensfeld-Nuremberg ABS
Instead of continuing the future NBS from Erfurt through to Nuremberg, an ABS is proposed from Ebensfeld southwards, involving the laying of two additional tracks alongside the well-aligned existing railway at a cost of DM3.3 billion.

Günzburg-Augsburg(-Munich) ABS
This is the continuation of the proposed new line between Stuttgart and Günzburg to form a high-speed line between Stuttgart and Munich. Two additional tracks, suitable for 230 km/h working, are being built over the 43 km between Olching (near Munich, where the separate S-Bahn tracks end) and Augsburg in a 5-year project costing DM1.1 billion.

Ingolstadt-Munich ABS
This is the ABS continuation from the outskirts of Ingolstadt to Munich of the NBS south from Nuremberg to Ingolstadt. Work should be complete in 2003, when timings between Nuremberg and Munich will be cut to an hour.

Karlsruhe-Basle
As stated above, much more of the work on this 197 km line will now be NBS rather than ABS.

Fulda-Frankfurt am Main-Mannheim ABS
This is the connection between the Hannover-Würzburg NBS at Fulda and the Rhine Valley route. The line passes through difficult country to gain the Rhine-Main plain, and over the first 25 km or so speeds are limited to 110 km/h. The ABS works will make 200 km/h possible over 55 km of the remainder of the section through to Frankfurt. Work on the 79 km Riedbahn line thence to Mannheim is now essentially complete. It has included the laying of a third track to allow overtaking at eight locations, the elimination of 27 level crossings, and realignments at 11 points to permit 200 km/h over 62 km of its length.

Munich-Markt Schwaben-Mühldorf-Freilassing ABS
Prospective EU single-market growth of north-south freight traffic, Austria's accession to the EU, and that country's restrictions on lorry transit-traffic demand an increase in cross-border capacity. An ABS scheme, including electrification, has been developed for the 120 km route from Munich to Freilassing through Markt Schwaben, to create a high-quality relief route for the heavily taxed main line from Munich to Freilassing via Rosenheim, at a cost of some DM1 billion. Mühldorf's considerable commuter traffic to Munich will also benefit from the electrification.

Strasbourg-Appenweier and Saarbrücken-Mannheim ABS
To connect the German ICE network with France's TGV-Est Européen from Paris to Strasbourg, there are ABS schemes for the lines between Strasbourg and Appenweier and between Saarbrücken and Mannheim. There is no progress to date.

Other improvements
The single-track line between Gross Gleidingen and Hildesheim is to be doubled over most of its length at a cost of DM175 million.

Prior to reunification, the Hannover-Hildesheim-Goslar-Bad Harzburg line was due for singling. Instead, the Hildesheim-Goslar section is being upgraded for use by NeiTec trains and DM100 million is to be spent further east on the Halberstadt-Halle line to bring it up to 160 km/h standards where possible, with the aim of an RE service with Class 612 tilting trains between Hannover and Halle taking about 2½ hours.

Work began in March 1998 on a second 1,300 m double-track tunnel to sort out the bottleneck between Mainz Süd and Mainz Hbf.

Level crossings
DB Netz, working together with federal, state, and local authorities, is engaged in a major programme to improve safety at the 26,000 level crossings remaining on the system. New and upgraded lines are free of crossings, since it is a requirement of the Eisenbahn Bau- und Betriebsordnung that there should be no lines on lines where trains run at speeds above 160 km/h, and the current aim is to eliminate all existing crossings on heavily trafficked routes by the end of the decade. Others will be provided with up-to-date safety technology. These measures are expected to cost roughly DM630 million a year.

DB Reise&Touristik

DB Reise&Touristik AG
Stephensonstrasse 1, D-60326 Frankfurt am Main
Tel: (+49 69) 265 71 30   Fax: (+49 69) 26 51 42 88

Key personnel
Board Members
Chairman: Dr phil Christoph Franz
  Marketing and Commercial Affairs:
    Dr rer pol Ingo Bretthauer
  Production and Technology:
    Dipl-Ing Karl-D Reemtsema
  Finance and Audit: Dr rer pol Rolf Kranüchel
  Personnel: Jens-Uwe Bruysten

In 2000 the Long Distance Passenger Business had a turnover of €3,463 million (1999: €3,257 million) and capital expenditure of €499 million. The business employed 30,293 staff.

InterCity Express
ICE 1s first entered service in 1992 on a route from Hamburg to Munich using both the Hannover-Würzburg and Mannheim-Stuttgart high-speed lines (Neubaustrecken (NBS)). All 60 first-generation trains were in service for the start of the 1993–94 timetable, running on four routes including a new one between Berlin and Munich via Frankfurt am Main. In the first serious incident with an ICE, one train was lost in the Eschede accident in June 1998.

The first ICE 2s were introduced in June 1996, as long trains without either restaurant cars or driving trailers. By June 1997 the catering vehicles were available and the trains took over the two-hourly Berlin-Cologne service, still as long trains. With the delivery of the driving trailers, the fleet was available for use as designed for the May 1998 timetable.

The ICE 3 came into service on special trains to the Hannover EXPO 2000 in the 2000–2001 timetable, both with DB and with NS. The ICE T seven-car units first came into revenue-earning service on the Stuttgart-Zürich service in the 1999–2000 timetable; they were also used during that timetable-year on a number of other services. The five-car units came in with the 2000–2001 timetable on the Dresden-Frankfurt am Main service. The ICE TD entered commercial service in 2001 on trains between Dresden and Hof and between Munich and Zurich.

Two foreign designs of high-speed train now work into Germany. In December 1997 'Thalys' services worked by multivoltage TGV units began to run between Paris and Cologne, with one pair of trains being extended in 1998 to run to and from Düsseldorf, and in March 1998 Cisalpino services using Fiat Pendolino trainsets began working into Stuttgart from Zurich.

Other long-distance services
In addition to its ICE services, DB Reise&Touristik operates locomotive-hauled domestic IC (InterCity) and international EC (EuroCity), IR (InterRegio) services, and, through a subsidiary, various overnight services.

Most IC locomotive-hauled trains now run as fixed train-sets with driving trailers and are maintained in fixed formation. In 2001 the 117 trainsets were being repainted into the ICE livery and were also being internally refurbished. The aim is to offer a single image and a common quality standard for long-distance services.

InterRegio trains were planned to connect regional centres. In practice, the role of the IR services was never entirely clear and the numbers of such services have been cut back, most significantly in the June 2001 timetable. Replacing them are more stops by IC trains, extra regional services and some trains of a new category, the InterRegioExpress (IRE).

TEE Alliance
DB AG has established with ÖBB and SBB in Austria and Switzerland respectively the TEE Alliance, with the aim of offering customers common levels of service across the borders of the three countries. In 2001 the three railways were seeking tenders for the supply of 116 tilting trainsets, to enter service in 2003.

Berlin-bound ICE 2 trainset at Duisburg-Kaiserberg (Wolfram Veith)   0089562

Hamburg-Cologne 'Metropolitan' service passing the new airport station at Düsseldorf (Wolfram Veith)   0089557

Sleeper services (DB Nachtzug)
In 1999 responsibility for all sleeper services was passed to the DB Reise&Touristik subsidiary DB AutoZug GmbH and plans were announced for the development by 2001 of a network of more than 20 routes, 10 of them appearing in the May 1999 timetable. Other rolling stock is being modernised to offer a consistent standard of comfort and service (including conversion of the four-bed compartments in the new bilevel sleeping-cars to two-bed compartments) and a clearer and simpler pricing system is to be offered. The Talgo ICN (InterCityNight) services has been integrated into the new concept.

Since September 2000 the night train between Berlin and Malmö has been operated by the Georg Verkehrsorganisation of Frankfurt am Main, using leased rolling stock. This is the first regular long-distance service to be taken over by a private operator.

Metropolitan GmbH
A DB Reise&Touristik subsidiary, Metropolitan GmbH, was formed to offer direct, fast connections between major centres. The first such service began running between Cologne and Hamburg in August 1999. Trains call only at Düsseldorf and Essen and take about 3 hours 20 minutes for the journey (significantly slower than was aspired to at the planning stage). The rolling stock used offered at first three different types of accommodation in one class, 'Silence' for those who wished to rest, 'Club' for those wanting to talk, and 'Office' for those wanting to get on with their work whilst on the train. There is now an additional type of accommodation so that second class can be catered for. Two Class 101 locomotives are dedicated to the service (and liveried accordingly). The first year's

operations were viewed as successful though load-levels remained relatively low.

Customer services
In an attempt to accelerate and simplify ticket purchase DB Reise&Touristik has introduced large numbers of vending machines to sell long-distance tickets.

DB Reise&Touristik launched its 'Surf+Rail' scheme in late 1999. Fixed-price second-class tickets, inclusive of reservations, are available at four price-levels for a limited number of through connections between major cities only by Internet credit card purchase. The tickets are valid only in the designated train and only for the named passenger. The tickets must be booked at least three days before travel, must include a Saturday night away, cannot be exchanged, and have no validity on DB Regio. The customer receives a confirmation of the booking by e-mail and prints his/her own ticket on a sheet of A4 paper. This shows the booked connection, with times and seat numbers and the passenger's name. On-train staff have a list of Surf+Rail tickets for their train, which allows proper checking. The scheme was an immediate success and its scope has been significantly extended.

Since early 1999 it has been possible to reserve 'Mobile-telephone Seats' in non-smoking accommodation in all ICEs. The vehicles are marked with pictograms and the word 'Handy' is printed on the reservation-ticket. Repeaters in the trains ensure good reception in these vehicles. 'Quiet Areas' have also been introduced in the ICE fleet.

On the motive power side, DB Reise&Touristik aims to concentrate all its motive power maintenance in just four places: Berlin, Hamburg, Munich, and Saarbrücken.

DB Cargo

DB Cargo AG
Rheinstrasse 2, D-55116 Mainz
Tel: (+49 6131) 15-9   Fax: (+49 6131) 156 02 19
Web: http://www.db-cargo.de

Key personnel

Board Members
Chairman and Commercial Director:
Dr Bernd Malmström
Operations and Technology:
Dr-Ing Eh Karl-Heinz Jesberg
Finance and Audit: Wilhelm Wegscheider
Personnel and Legal Affairs:
Birgit Gantz-Rathmann

DB Cargo's turnover of €3,831 million in 2000 showed a rise of €290 million on the 1999 figure, and capital expenditure was €405 million, down on the €501 million of the previous year. There are 38,555 employees, against 40,995 a year earlier.

DB Cargo operates through five profit centres: Coal & Steel, Building Materials & Waste Disposal, Oil & Chemical Products, Industrial Goods, and Commercial Goods, Agriculture & Forestry.

In late 2000 DB AG announced plans for DB Cargo to reposition itself in all three of its product areas: trainload and mass goods traffic is increasingly to be worked using modern rolling stock, with prices adjusted to meet competitive needs; the railway will aim to double its share of the container traffic market; and because the wagonload traffic service no longer meets the needs of the market, it will be improved by concentration on fewer collection points and increased speeds. The company believes that it has to offer a wider range of container services to try to render redundant the present time-consuming marshalling, collecting and transhipment arrangements, and it aims to draw lorries and inland shipping into the railway's chain so as to be able to offer transport from a single supplier with improved punctuality. The plan is known as MORA C (Marktorientiertes Angebot Cargo). At the end of 2000 the wagonload business made up 40 per cent of the total freight business of 75,000 million tonne-km. Of this, 85 per cent came from 320 major customers, where there was a growth potential of at least 5 per cent, and the other 15 per cent came from a spread of some 7,000 customers whose needs amounted to only a few wagons each per week or month, with growth potential assessed at only 1.9 per cent. This part of the business was making a loss of DM168 per wagon moved. DB Cargo was to assess all its 2,100 goods locations and to undertake discussions with

customers about how best their needs might be met. The range of products on offer would include containers worked in Combined Traffic service, a diversion of the traffic to other goods locations, and co-operation with other railways or road hauliers.

Under MORA C, DB Cargo aims to improve its wagonload traffic out-turn by at least DM500 million a year. The reforms are to be introduced in stages for completion in 2004, by which time DM5,000 million will have been invested, not just in modern rolling stock and locomotives, but also in GPS systems, general growth, development of a sharper customer focus and the raising of quality, all in order to improve rail's competitiveness. The company's long-term aim is to increase rail freight by 42 per cent over a 15-year period, raising performance from 75,000 million tonne-km in 2000 to 120,000 million in 2020.

The premium level of service is provided by the InterCargo trains, which guarantee overnight siding-to-siding connection with departures between 15.00 and 17.00 and arrivals between 07.00 and 09.00.

The international counterpart of InterCargo-Express is the EurailCargo network; this targets second-morning delivery over distances up to 1,500 km.

In June 1998 DB Cargo and NS Cargo declared their intention to fuse their two undertakings and establish a new European rail transport company (open to further partners), with a working name of Rail Cargo Europe. The new company, now known as Railion, is established under German law with headquarters in Mainz and it started operations

at the end of 1999. DB Cargo is also entering into close working relationships with other railways.

In collaboration with SNCB/NMBS (Belgium) and NS (the Netherlands), DB AG is involved in the revivification of the so-called 'Iron Rhine' route between the Belgian port of Antwerp and the German river port of Duisburg. On 28 March 2000, the three countries signed an agreement providing for a resumption of a trail service of 15 trains a day, to start by the end of 2000 and to continue till 2008. On the German side, DM50 million is already earmarked for the work from the government's Anti-Stau-Programm funds.

Intermodal operations

DB AG's KLV (Kombinierter Ladungsverkehr) intermodal unit operates trains carrying containers, swapbodies, and piggyback unaccompanied road trailers. The KLV portfolio also includes both domestic and international (Austria, the Czech Republic and North Italy) Rollende Landstrasse (RoLa), or 'Rolling Highway' – trains using ultra-low-floor well-wagons on small-wheeled bogies to form a continuous roadway for the drive-on/drive-off conveyance of accompanied truck and truck-and-trailer rigs.

The RoLa service between Germany and Lovosice in the Czech Republic, up the Elbe Valley from Dresden, introduced in 1994 in an area where the roads are far from suitable for heavy traffic, has been very successful and Saxon and Czech negotiators have agreed both on improvements and on the maintenance of the service through to the start of the next century.

A RoLa piggyback service heads south through Königstein towards the Czech Republic (Wolfram Veith)   0089560

Gauge-changing system being developed by DB Cargo and Knorr Bremse to allow the exchange of traffic with broad-gauge networks of former Soviet countries (David Haydock)   0114244

Class 152 electric freight locomotive on an intermodal service at Ulm (David Haydock)   0114248

DB Regio

DB Regio AG
Stephensonstrasse 1, D-60326 Frankfurt am Main
Tel: (+49 69) 26 56 10 70   Fax: (+49 69) 26 51 42 88
Web: http://www.dbregio.de

Key personnel

Board Members
Chairman: Dr Phil Christoph Franz
Production and Technology: Dr-Ing Joachim Trettin
Finance and Audit: Dr rer pol Armin F Schwolgin
Personnel and Legal Affairs: Heinrich Brüggemann

Subsidiary companies

S-Bahn Berlin GmbH
Invalidenstrausse 19, D-10115 Berlin
Tel: (+49 30) 29 74 38 16

S-Bahn Hamburg GmbH
Steinstrasse 12, D-20095 Hamburg
Tel: (+49 40) 39 18 39 04

DB Regio, the local passenger traffic business, had a total turnover in 2000 of €7,517 million (1998: €7,328 million), of which orders from the federal states accounted for €4,330.9 million (1999: €4,283.4 million). Capital expenditure was €1,305 million for the year, against €1,311 million in the previous year. There were 52,769 employees at the end of 2000 (1999: 55,605). There are 14 regions and eight workshops. The company owns 17 regional bus companies, four DB ZugBus companies, the S-Bahn companies for Berlin and Hannover, and the Usedomer Bäderbahn.

From 1 January 1996 responsibility for the support of local rail transport passed from the federal government, which had been spending about DM6 billion a year, to the states. These received additional money to buy in from DB AG those loss-making services they or local authorities deemed to be socially necessary, the support being based on the 1993–94 timetable's service pattern. They decide which services are to be retained and how they are to be provided. They can contract with DB AG to run them; they can negotiate with an existing independent railway to provide a service, using the general right of access to the national infrastructure; they can even buy the infrastructure involved for themselves; or they can buy a replacement road service.

It is to be noted that a 1997 DB AG report suggested that more than 11,000 km of the 40,000 km network was either completely unprofitable or offered no possibility of commercial exploitation in the longer term. New and more effective ways of managing local networks are currently being explored.

Local services are branded as Regional Express (RE), RegionalBahn (RB), and S-Bahn. A new category, the InterRegioExpress (IRE) was introduced in 2001 to indicate especially fast RE services or RE trains running to replace IR services.

Class 650 Adtranz RegioShuttle railcars at Ulm (David Haydock)   0114243

DB Regio, like DB Reise&Touristik, is investing heavily in new rolling stock and has a 10-year programme to eliminate hauled-stock working.

S-Bahn systems

As long as no increase in subsidised operating costs is involved, the federal government has an obligation to grant-aid the capital cost of S-Bahn work. Under the 1967 Municipal Transport Finance Act (Gemeinde-verkehrsfinanzierunggesetz (GVFG)) the state dedicates part – currently DM0.054 per litre – of its oil tax revenues to meeting 60 per cent of the cost of urban transport infrastructure improvements approved and financially supported locally.

Two of the S-Bahn systems are now operated by separate companies set up as subsidiaries of the Deutsche Bahn Gruppe. These are S-Bahn Berlin GmbH, established in January 1995, and S-Bahn Hamburg GmbH, established two years later. In all other areas with S-Bahn operations, services are provided by DB Regio directly.

In Berlin, restoration of the S-Bahn network to its 1930s extent continues under a specially financed programme. Rolling stock is being replaced. Significant numbers of the new Class 481 trains are now running, but it will still be some time before all pre-war units have been replaced.

In Hamburg more than DM900 million is being invested in the system over the next few years. A link to join Fulsbüttel Airport to the S-Bahn at Hamburg-Ohlsdorf is being planned. New trains of Class 474 are coming into service. An Adtranz/Alstom consortium is supplying 45 of these three-car emus, in which the end-cars in each set are powered by Adtranz equipment, with four water-cooled 125 kW asynchronous AC motors under GTO inverter control. An option to buy a further 58 units at a cost of some DM350 million has been exercised, with delivery starting in 1999.

The expansion of other S-Bahn systems continues apace. In Cologne, work is well under way on the provision of segregated S-Bahn tracks between Cologne and Düren as part of the Cologne-Aachen ABS project, and an airport link, is being built. In Dresden a DM326 million project sees the restoration of four tracks between Dresden Hbf and Pirna to allow the segregation of S-Bahn services from main line traffic on the international route to the Czech Republic. Other work involves station renewals and an airport link which was opened in early 2001, and further measures planned include widening of the Arnsdorf and Meissen lines – work on the latter being dependent on progress on the upgrading of the Dresden-Leipzig/Berlin main lines. In the Halle/Leipzig area the DM450 million Halle-Leipzig S-Bahn line is now under construction. At 32 km it will be 6 km shorter than the existing route, mainly because of a different approach to Leipzig. It will require some 26 km of new construction and will have 10 intermediate stations.

A substantial S-Bahn network was developed in Hannover in time for EXPO 2000. Lines are:
S1 Stadthagen-Wunstorf-Hannover Hbf-Weetzen-Haste (that is, forming a loop)
S2 Nienburg-Wunstorf-Hannover Hbf-Weetzen-Haste
S3 Hannover Hbf-Lehrte-Celle
S4 Bennemühlen-Hannover Hbf-Hameln
S5 Flughafen Langenhagen-Hannover Hbf-Hameln.

S1 and S2 running hourly gives a 30-minute service between Wunstorf and Haste via Hannover Hbf, as do S4 and S5 between Langenhagen and Weetzen. Total cost of the north-south S-Bahn is put at DM690 million.

The Rhine-Main S-Bahn system, centred on Frankfurt am Main, continues to grow. A 'Messe' S-Bahn station has been built in Frankfurt. In the Rhine-Neckar area an agreement was signed in 1996 for the construction and financing of the new DM340 million S-Bahn. MVV (Mannheimer Versorgungs- und Verkehrgesellschaft) and DB were originally competitors as potential operators, but in late 1996 they came together and put forward a joint proposal for working both the proposed Neustadt-Speyer-Ludwigshafen-Mannheim-Heidelberg-Biberach-Bruchsal regional line and all local traffic between the Palatinate and Odenwald. The Neustadt-Speyer-Ludwigshafen-Mannheim-Heidelberg-Bruchsal/Eberbach route is being modernised throughout, quadrupled between Ludwigshafen and Mannheim (with a new

Class 644 Talent dmus forming a Gummersbach service at Cologne Hbf (Colin Boocock)   0089564

Rhine bridge, on which work began in 1997), and electrified between Speyer and Schifferstadt. There will be a half-hourly service, with trains dividing in Schifferstadt for Neustadt and Speyer and in Heidelberg for Bruchsal and Eberbach. Extensions to Mosbach (Baden), Karlsruhe, and Kaiserslautern are envisaged later, and further lines are proposed to follow the initial east-west route. A cost-benefit analysis has given positive forecasts for a Worms-Frankenthal-Ludwigshafen-Mannheim route with a connection to the BASF works station.

In 2001, DB Regio won a 12-year contract to operate a new Rhein-Neckar S-Bahn network of some 240 route-km around Mannheim and Heidelberg. Due to commence in December 2003, services will be provided by 40 Class 425 emus. Also under discussion is a route from the South Hessen towns of Biblis, Bürstadt and Lampertheim through Mannheim and on over the Rheintalbahn via Schwetzingen/Hockenheim to Karlsruhe. To the north, the Rhine-Ruhr-Wupper S-Bahn is to be extended from Dortmund to Hagen, and other extensions in the Dortmund area are being planned, and a new S9 line, between Haltern, Essen, and Wuppertal-Vohwinkel, is due for completion by the turn of the century.

In Bavaria, the Munich S-Bahn received its second connection to the new Munich Airport with the opening in 1998 of the so-called Neufahrner Spange, 6.7 km in length, to complete a 19 km route from Ismanning coming in from the west. The total airport service is now six trains an hour. A second tunnel is to be built to solve the problems of congestion on the Munich city tunnel line. Separation of local traffic is required on the future ABS Munich-Ingolstadt route, and a DM200 million quadrupling between Obermenzing and Dachau is planned to allow an expansion of the Munich-Dachau S-Bahn service. During 2001–02, 144 new Class 423 emus are being brought into service to provide a DM900 million upgrade for Munich's S-Bahn. In Nuremberg construction work on the 26 km section from Nuremberg to Roth currently being modernised should be complete by the turn of the century and planning for the future is now focused on the Nuremberg-Fürth-Forchheim axis.

Class 474 Hamburg S-Bahn third-rail emu at Hamburg Hbf (Marcel Vleugels)   0089561

In June 2001, work commenced on infrastructure work for the creation of a cross-city S-Bahn network in Leipzig. The project includes a double-track tunnel under the city to link the Bayerischer Bahnhof terminus, in the south, with Leipzig main station, which lies north of the centre.

Mixed running

Heavy rail shares tracks with light rail (the trams of the Albtal Railway (qv) in the Karlsruhe area. The aim was to establish a large regional network at low cost by connecting light rail routes and DB AG, allowing trams to travel out into the surrounding areas on DB AG tracks shared with ordinary trains. By the end of 1994 a fleet of 36 dual-system light rail cars was available, able to operate on both 750 V DC and 15 kV AC. The initial route was opened in 1992 and other routes have been added. The operation is proving enormously successful and further extensions are planned.

Services began in the autumn of 1997 on a similar mixed system in Saarbrücken, on a first section between Saarbrücken and Saargemünd (France). This now extends northwards via Riegelsberg and Heusweiler to Lebach, giving a total length of 44 km of which 19 km is new construction, the remainder existing DB AG or SNCF tracks. Future connections to the University of the Saarland and to Forbach are planned. Total cost will be some DM540 million, with the federal government providing DM214 million through the GVFG and the Land Saarland DM224 million. The first 15 trains were built by Bombardier and in July 1998 the same manufacturer received an order for a further eight vehicles of the same type.

There is a development that sees vehicles built to heavy rail standards running on to a tramway in Zwickau, where the Regental Bahnbetriebs GmbH ordered 10 'RegioSprinters' equipped to meet tramway requirements so that they can run over the 3 km of tramway (which will be converted from metre-gauge to mixed-gauge) between Zwickau Hbf and the city's Neumarkt (new market place).

DB Station & Service

DB Station & Service AG, D-60326 Frankfurt am Main
Tel: (+49 69) 26 50   Fax: (+49 69) 26 52 45 69

Key personnel
Board Members
  Chairman: Dieter Ullsperger
  Station Development: Martin Lepper
  Finance and Audit: Dipl-Ing Maximilian Kittner
  Personnel: Alfred Possin

In 2000 the passenger stations business spent €552 million (1999: €554 million) on the maintenance of its 5,794 stations (1999: 5,876). There were 5,015 employees, against 5,593 the previous year.

Major new stations
'Station 21' projects have been announced for Dortmund, Frankfurt am Main, Lindau, Magdeburg, Mannheim, Munich, Saarbrücken, Stuttgart, and Ulm. The first and most advanced is 'Stuttgart 21', for which four schemes (from 27 submitted) were awarded prizes in the architectural competition. These are now being worked over prior to a final decision and funding questions are being carefully examined.

Following on from the successful completion of the DM400 million modernisation of Leipzig station, work on a DM130 million project at Cologne has been completed, providing a new travel and service centre, a high-quality waiting area, about 65 shop units, and a new second frontage, as well as a total restoration of the historic parts of the building. DM110 million were spent on a thorough modernisation of Hannover Hbf before Expo 2000. There is also to be a DM850 million rebuilding of Dortmund station. Major works programmes have been completed at Rostock, Stralsund, Berlin Lichtenberg and Berlin Ostbahnhof.

Modernisation continues at Frankfurt am Main. The first stage of major improvements to Schwerin

station has also been completed. Dresden Hbf is to be restored and modernised according to the plans of the British architect Sir Norman Foster (who was responsible for the rebuilding of Berlin's Reichstag building as the home of the German parliament).

Alongside the projects already mentioned, DB Station & Service is engaged in major modernisation of other stations. These include Aachen, Bielefeld, Bochum, Bremen, Gelsenkirchen, Hamburg-Dammtor, Koblenz, Mannheim, Mülheim, Münster, Oberstorf, Oranienburg, Siegburg, and Weimar. In many cases the structures are listed. The longer-term aim is to modernise all 6,000 stations to make them customer-friendly transport centres, retaining wherever possible the old structures whilst equipping them to meet the needs of today's

travellers. Some 70 per cent of all stations have so far seen improvement works.

Bookholzberg (between Bremen and Oldenburg) was the first station to receive the new *Plus-Punkt* treatment for upgrading small stations where facilities at present are either non-existent or very much run down. A basic element of a square, red module with glazed sides provides waiting space and houses information and service elements: ticket-machine, ticket-canceller, timetable-display, fares, route-diagram, area map, card-telephone with emergency button, and waste-bins. Additional 'packages' can then be added to provide features like video surveillance, a telephone with service-buttons (taxi-call, bus services), and drinks and snack machines, or even more elaborate facilities

Station improvement work in progress at Rostock (Marcel Vleugels)   0114208

like toilets, cycle-stand, kiosk, complaints box, money-changer, newspaper sales, and so on. More than 1,000 stations will be provided with *Plus-Punkt* furniture.

Furthermore, DB Station & Service recently spent DM5 million on construction of its first modular station, at Memmingen (on the line from Ulm to Kempten). The prototype uses elements which can be adapted to suit any ground plan. In this instance 1,000 m² are covered by a single-storey building. The company is planning 20 such stations nationwide.

DB Station & Service now offers a 'KonferenzService' of meeting rooms at more than 40 of Germany's largest stations, with the advertising slogan 'Meeting rooms with ICE connections'.

Rail-air links
Lufthansa and 22 foreign airlines offer air tickets including a coupon for internal German journeys to or from their international departure or arrival airport which can be exchanged for a rail ticket without further charge at stations or on a train. Deutsche Bahn offers all comers a Rail-&-Fly ticket for travel to and from airports on two zonally based scales with substantial discounts for family or group travel. Furthermore, deals are now in place

with two major holiday firms for rail travel between home and departure airport to be included in the price of the holiday.

Reserved accommodation serviced by Lufthansa staff is provided in selected IC trains between Düsseldorf and Frankfurt Airport. Stuttgart passengers using Frankfurt Airport can check in or reclaim their baggage at Stuttgart Hbf and travel in reserved accommodation between there and the airport.

A frequent fast service between the centre of Berlin and Schönefeld Airport is made available by the provision of dedicated accommodation in certain RE services. The trains run between 04.30 and 23.00 and the fare is the standard flat fare of the Berlin-area transport system.

DB Reise & Touristik hopes to develop an InterCity connection to the new Munich Airport. Now the second S-Bahn link has been completed, construction of a short spur between the main line west of Munich Hbf at Pasing and S-Bahn Line S1 at Moosach would enable IC trains from the Augsburg direction to reach the airport station and, if required, to serve Munich Hbf as well by completing the S-Bahn circle to arrive in the city from the east.

In May 1999 the new Frankfurt Airport station was opened. Situated on the first section of the Cologne–Rhine/Main NBS to be brought into service, this new station has long-distance services not only to and from Frankfurt am Main Hbf but also direct to Mannheim and the south.

Main line and S-Bahn connections are under construction to serve Cologne/Bonn Airport, the costs being shared by the railway, the federal government, the *Land*, and the airport company. Rail links have been enhanced at Düsseldorf Airport by the construction, at a cost of DM87 million, of a new station on the Düsseldorf–Duisburg main line, to be linked to the airport by a people mover and served by some 88 long-distance trains a day, as well as S-Bahn services.

The Hannover S-Bahn system, completed in time for the opening of Expo 2000, includes a double-track branch to the airport at Langenhagen from the northern line of the S-Bahn system.

A new station serves the airport at Dresden, and one is planned for Hamburg. These will be for S-Bahn services.

GVG Verkehrsorganisation GmbH (GVG)

Savigny Strasse 80, D-60325 Frankfurt am Main
Tel: (+49 69) 74 95 74
Fax: (+49 69) 74 99 16
Web: www.berlin.night-express.com

Organisation
GVG is an independent travel company which provides an overnight service between Berlin Ostbahnhof and Malmö, Sweden, via Sassnitz. Operated in collaboration with SJ in Sweden, the service's frequency varies from three times a week to daily in each direction, according to the season.

Traction and rolling stock
The GVG locomotive fleet is formed of four Class 109 LEW-built electric locomotives (former DR Class E11) and one shunter.

NEW ENTRY

Hafen und Güterverkehr Köln (HGK)

Hafen und Güterverkehr Köln AG
Bayenstrasse 2, D-50678 Cologne
Tel: (+49 221) 39 00   Fax: (+49 221) 390 13 43
Web: www.hgk.de

Key personnel
Chairman of Supervisory Board:
   Franz-Josef Knieps
Chief Operating Manager, Railway Operations:
   P Schumacher
Chief Operating Manager, Railway Infrastructure:
   Manfred Eising

Gauge: 1,435 mm
Route length: 98.2 km (owned)

Organisation
Controlled by the city of Cologne (Stadtwerke Köln GmbH), HGK was formed in 1992 by merging the freight operations of the Köln–Bonner Eisenbahn (KBE), the Köln–Frechen–Benzelrather Eisenbahn (KFBE) and Cologne's public ports and port railways.

In 1998 HGK took a shareholding of 18 per cent in a joint venture with Zanders (a paper producer), P&O and the community of Bergisch Gladbach, creating BGE Eisenbahn Güterverkehr GmbH. The company operates an intermodal terminal in Bergisch Gladbach in October 1999. A second logistics centre was opened in Düren in 2002.

In April 2002 HGK became a member of Swiss Cargo Deutschland GmbH, with a shareholding of 44 per cent. Other members of the venture are Swiss SBB Cargo AG (51 per cent) and Swiss intermodal operator, Hupac Intermodal SA. The company also has shareholdings in: CTS Container-Terminal GmbH (15 per cent), which operates a terminal at Cologne-Niehl port; DKS (24.5 per cent), a communications company in the public transport sector; Knapsack Cargo (26 per cent), operator of a rail-served chemicals terminal in Hürth; and Rail Consult Gesellschaft für Verkehrsberatung mbH (25 per cent). HGK also held a 25.1 per cent shareholding in Netherlands-based open access rail operator Short Lines BV, but this company withdrew from freight haulage operations in 2004.

In September 2004 HGK was granted a licence to operate over the Netherlands railway network.

Passenger operations
Interurban light rail services of Cologne and Bonn city transport authorities use HGK lines.

Freight operations
HGK operates intermodal and international trains between Cologne and the North Sea ports of Antwerp and Rotterdam in association with other operators and between Cologne and Brescia, Milan and Oleggio (Italy) via Basel through its stake in SBB Cargo Deutschland. Some 20 trains a day are operated over this north-south axis. There is also a

service between Ingolstadt and Vienna, and Poland was added top the network in 2004.

Domestic services cover the whole of Germany. In 2003 HGK won some 4 million tonnes of new business annually through a strategic partnership with Transpetrol.

In 2003 HGK carried 8.5 million tonnes for 1.145 million tonne-km.

Traction and rolling stock
In 2004 HGK operated 39 diesel locomotives, 10 electric locomotives and some 465 freight wagons.

HGK diesel locomotives (principal types)

Class	Wheel arrangement	Power kW	Speed km/h	Weight tonnes	No in service	First built	Mechanical	Engine	Builders Transmission	
DE 11	Co-Co	2,650	160	126	3	1989	Krupp-MaK	MaK 12M282	E	–
DH 31	B-B	808	70	80	8	1961	KHD	Caterpillar 3508 DI-TA	H	Voith
DE 61	Co-Co	2,460	120	126	12	1999	EMD	EMD 12N-710G3B-EC	E	EMD
1001	B-B	1,500	120	–	3	2002	Vossloh	GM	H	Voith

HGK electric locomotives

Class	Wheel arrangement	Power kW continuous	Speed km/h	Weight tonnes	No in service	First built	Mechanical	Builders Electrical	
2000/4000	Bo-Bo	4,200	140	80	10	2000	Adtranz/Bombardier	Adtranz/Bombardier	

HGK Class DE 61 (EMD Type JT42CWR) with a Germany-bound intermodal service from Rotterdam at Dordrecht, Netherlands (Quintus Vosman)   ***NEW*/**0585042

The locomotive fleet includes three Type DE 1024 MaK-built 2,650 kW Co-Co demonstrators dating from 1989 and 16 Type DE 1002 1,320 kW Bo-Bos built in 1987 by the same manufacturer. The Class DH 31 machines were rebuilt in 1994–95 with Caterpillar engines. Since 1999 the fleet has been supplemented by 12 General Motors JT42CWR Co-Co locomotives similar to the Class 66 machines supplied by the same manufacturer to EWS in the UK.

The fleet was supplemented from 2000–01 by 10 Type 145 electric locomotives supplied by Bombardier Transportation and leased from Porterbrook Leasing and by Vossloh G1206 diesel-hydraulic locomotives.

UPDATED

Hamburg Port Railway

Hafenbahn Hamburg
Freie und Hansestadt Hamburg, Wirtschaftsbehörde, Strom- und Hafenbau, Abteilung Hafenbahn
Dalmannstrasse 1, D-20457 Hamburg
Tel: (+49 40) 28 47 25 41

Key personnel
General Manager: Dipl-Ing Reinhard Höfer
Assistant Manager: Dipl-Ing H-Michael Röfer

Gauge: 1,435 mm
Route length: 403 km

Organisation
Hamburg port railway is not a public carrier. It is operated by DB, but carries out its own track maintenance.

Freight operations
Traffic statistics for 1997 were 23.7 million tonnes and 106.8 million tonne-km, compared with 21.9 million tonnes and 98.6 million tonne-km in 1996.

Traction and rolling stock
Two diesel locomotives.

Harz Narrow Gauge Railways

Harzer Schmalspurbahnen GmbH (HSB)
Friedrichstrasse 151, D-38855 Wernigerode
Tel: (+49 3943) 55 80   Fax: (+49 3943) 55 81 12
e-mail: info@hsb-wr.de
Web: www.hsb-wr.de

Key personnel
Managing Director: Matthias Wagener

Gauge: 1,000 mm
Route length: 132 km

Organisation
HSB is owned by local authorities in the region within which it operates. The lines of the former Nordhausen–Wernigeroder Eisenbahn and Gernrode–Harzgeroder Eisenbahn were taken over from the former German State Railways (DR) in 1993. They are now operated as the Selketalbahn line.

Passenger operations
HSB serves the Brocken mountain, a popular tourist destination not accessible by road. Suburban services between Nordhausen and Ilfeld, operated by diesel railcars, have been improved by introducing a fixed interval timetable and opening additional halts. Traffic statistics for 2001 were 1.2 million passengers and 33 million passenger-km.

Freight operations
Transporter bogies replaced transporter wagons in 1996. Traffic statistics for 2001 were 7,000 tonnes and 129,000 tonne-km.

New lines
Conversion to narrow gauge is planned of the 8.5 km line between Gernrode and the town of Quedlinburg, a world heritage site, under a €6 million scheme funded by Sachsen-Anhalt Land. To be completed in 2006, the extension will increase the length of the HSB to 140 km.

Traction and rolling stock
HSB traction and rolling stock comprises 25 steam locomotives, 10 diesel railcars, six line-haul diesel locomotives, six diesel shunting locomotives, 77 coaches, 55 freight wagons, 79 transporter wagons and 40 transporter bogies.

UPDATED

HellertalBahn (HTB)

HellertalBahn GmbH
Bahnhofstrasse 1, D-57518 Betzdorf/Sieg
Tel: (+49 2741) 97 35 75
e-mail: info@hellertalbahn.de
Web: www.hellertalbahn.de

Gauge: 1,435 mm
Route length: 43 km (operated)

Background
Formed in 1999, HellertalBahn GmbH is a joint venture between the Hessische Landesbahn (HLB), the Siegener Kreisbahn and the Westerwaldbahn.

Passenger operations
Since September 1999, the HellertalBahn has provided regional services on the Betzdorf–Neunkirchen–Haiger–Dillenburg line, which runs through the states of Rheinland-Pfalz, Nordrhein-Westfalen and Hessen.

Traction and rolling stock
Three Stadler GTW 2/6 dmus procured by the HLB.

UPDATED

Hessische Landesbahn GmbH (HLB)

Mannheimer Strasse 15, D-60329 Frankfurt
Tel: (+49 69) 242 52 40
Fax: (+49 69) 24 25 24 60
e-mail: mail@hlb-online.de
Web: www.hessenbahn.de
      www.hlb-online.de

Key personnel
Managing Director: Dipl-Ing Peter Berking

Organisation
HLB was founded in 1955 as holding company for the state-owned local railways in Hesse.

At present it owns/part-owns and operates the following railways:
Butzbach–Licher Eisenbahn AG (BLE) (94.4 percent)
Frankfurt–Königsteiner Eisenbahn AG (FKE) (51 per cent)
Kassel–Naumburger Eisenbahn AG (KNE) (51 per cent)
In addition, HLB participates in three joint venture rail operations:
Hellertalbahn GmbH (HTB) (33.3 per cent), jointly with the Siegener Kreisbahn and the Westerwaldbahn
SüdThüringenBahn GmbH (STB) (50 per cent), jointly with the Erfurter Industriebahn
Vectus Verkehresgesellschaft mbH (74.9 per cent), jointly with Westerwaldbahn GmbH

For more details of each of these see individual entries in Germany section of *Railway systems and operators*.

In 2004 a consortium of HLB and Hamburger Hochbahn (HHA) won a 10-year contract to operate services on Göttingen–Kassel, Kassel–Fulda, Göttingen–Bebra and Bebra–Eisenach lines from December 2006.

Traction and rolling stock
Rolling stock operated by HLB subsidiaries in 2003 totalled 50 diesel railcars. In addition there were 17 trams operated by KNE and five diesel locomotives. Vehicles owned by HLB comprised 30 DWA-built GTW 2/6 diesel railcars and one MaK DE 1002 diesel-hydraulic locomotive.

UPDATED

HzL Hohenzollerische Landesbahn AG

Hohenzollerische Landesbahn AG
Bahnhofstrasse 21, D-72379 Hechingen
Tel: (+49 7471) 180 60   Fax: (+49 7471) 18 06 12
Web: http://www.hzl-online.de

Key personnel
Chief Executive: Hans-Joachim Disch
Rail Operations Manager: Dipl-Ing Bernhard Strobel

Gauge: 1,435 mm
Route length: 107 km (owned); 350 km (operated)

Organisation
HzL is local government-owned. Freight and passenger trains also operate over DB lines.

Passenger operations
HzL passenger services operate between Hechingen and Sigmaringen. The Sigmaringen-Tübingen line (42 km) is operated under contract to Baden-Württemberg state. Other services are operated for DB. Traffic statistics for 1999 were 5.9 million passengers and 112.9 million passenger-km, these figures include bus operations (55 buses).

In 2000 HzL won a contract from the Baden-Württemberg regional authorities to operate a group of services in the Black Forest area. The company took over the so-called 'Ringzug' services in September 2002.

HzL also manages services on the Bodensee-Oberschwaben Bahn (BOB) (qv).

Freight operations

Traffic statistics for 1999 were 328,000 tonnes and 27.2 million tonne-km. Principal traffic is industrial salt bound for Burghausen, Bavaria, which is hauled by HzL between Stetten and Ulm.

Traction and rolling stock

Twelve diesel locomotives, 33 diesel railcars (including 22 Adtranz RegioShuttle units), five driving trailers and three trailers.

The procurement of additional diesel railcars for the 'Ringzug' services (see Passenger operations) was anticipated.

Two HzL Adtranz RegioShuttle railcars at Tübingen with an Albstadt-Ebingen service (David Haydock) 0109847

Hörseltalbahn (HTB)

Hörseltalbahn GmbH
Adam-Opel-Strasse 100, D-99817 Eisenach
Tel: (+49 3691) 66 31 60   Fax: (+49 3691) 66 31 62
e-mail: info@hoerseltalbahn.de
Web: www.hoerseltalbahn.de

Key personnel

Managing Directors:
Hans-Dieter Stützer, Heinz Wolfgramm

Gauge: 1,435 mm
Route length: 9 km

Organisation

HTB is a wholly owned subsidiary of Connex Cargo Logistics GmbH.

Freight traffic

HTB primarily serves an Opel car manufacturing plant at Eisenach. It also collaborates with DB Cargo and Intercontainer Interfrigo.

Traction and rolling stock

Three diesel locomotives.

UPDATED

InfraLeuna

InfraLeuna Infrastruktur und Service GmbH
Postfach 11 11, D-06234 Leuna
Tel: (+49 3461) 43 30 01   Fax: (+49 3461) 43 42 90
Web: www.infraleuna.de

Key personnel

Managing Director: Andreas Hiltermann
Head of Logistics: Wolfgang Pautsch

Gauge: 1,435 mm
Track length: 65 km

Organisation

InfraLeuna Infrastruktur und Service GmbH is a subsidiary of InfraLeuna GmbH, which manages and develops an extensive industrial site in Leuna that accommodates chemicals facilities, refineries and power generating facilities, as well as utilities. Its extensive logistics services and facilities include an industrial railway, that provides shunting services within the site and connection with the DB network. Ancillary services include the rental and servicing of freight wagons, including tank wagon cleaning.

InfraLeuna intends to undertake open access operations on the DB network.

Freight operations

Traffic statistics for 2001 were 6.1 million tonnes and 43.2 million tonne-km.

Traction and rolling stock

The locomotive fleet totals 17, including one Vossloh G2000, five V100.4 and one V100.1, two ex-DR V180 C-C units and eight shunters. There are also 16 freight wagons.

UPDATED

Karsdorfer Eisenbahn (KEG)

Karsdorfer Eisenbahngesellschaft mbH
Westendamm, D-58239 Schwerte
Tel: (+49 2304) 98 28 30   Fax: (+49 2304) 982 83 13
e-mail: info@privatbahn.de
Web: http://www.privatbahn.de

Key personnel

Managing Director: Bernhard van Engelen
Operations Manager: Dipl-Ing Konrad Höft
General Manager, Burgenlandbahn: Uwe Rückreim
General Manager, Freight: Ralf Jentges
Director, Research and Testing Centre:
Dan M Costescu

Gauge: 1,435 mm
Track length: 30 km

Organisation

KEG was founded in 1993 to take over the industrial railway of a cement works and subsequently entered the market as an open access freight operator. Subsequently, the company diversified its operations. As well as freight and passenger operations, these now include the provision of traction and rolling stock overhaul and maintenance and vehicle testing and measurement services via the company's Research and Testing Centre.

In September 2001, KEG acquired Waggonbau Brüninghaus GmbH (qv in the *Freight vehicles and components* section), which manufactures and repairs freight wagons and other rail vehicles.

KEG Class 2100 diesel-electric locomotives at Rheine depot (David Haydock) 0109848

Passenger operations

Since 1995 KEG has run passenger services for DB. In 1996 the company took over management of the Rügensche Kleinbahn (qv). In a joint venture with DB in 1999, KEG founded Burgenlandbahn GmbH to operate the national operator's former routes in the Naumburg/Zeitz/Merseburg region of Sachsen-Anhalt.

KEG also operates the Citybahn Chemnitz system in Stollberg and holds a contract to run the Rügensche Kleinbahn tourist railway (qv).

Freight operations

In 1997 KEG carried 43,000 tonnes for its own operations. In 1999, the company inaugurated a service conveying aviation fuel from Lingen (Ems) to Munich airport, a distance of some 840 km over DB tracks. Subsequently, additional contracts to carry petroleum products have been secured.

The company also operates permanent way maintenance trains under contract to DB.

Traction and rolling stock

KEG operates more than 40 diesel locomotives, nine diesel railcars and 20 freight wagons. Delivered by Bombardier DWA in 1999 were 18 LVT/S lightweight railcars which are leased to Burgenlandbahn.

In 1999, KEG acquired from Romanian State Railways four Class 60 Sulzer-engined diesel-electric locomotives for use on the aviation fuel traffic detailed above. The 1,546 kW machines were overhauled before delivery from Romania. By 2002 additional acquisitions had brought to 20 the number of this type (KEG Class 2100) in the fleet. In the same year KEG introduced to traffic its first Type 7000 5,100 kW electric locomotive, a refurbished Electroputere/ASEA-built machine also acquired from Romanian State Railways.

In 2002, the fleet was augmented by an ES 64 U2 'Taurus' electric locomotive from the Siemens Dispolok hire pool.

Kassel–Naumburger Eisenbahn (KNE)

Kassel–Naumburger Eisenbahn AG
Wilhelmshöher Allee 252, D-34119 Kassel
Tel: (+49 561) 93 07 40   Fax: (+49 561) 930 74 21
e-mail: mail@kne-online.de
Web: www.kne-online.de

Key personnel
Chief Executive: Dipl-Ing Peter Berking
General Manager: Dipl-Ing Veit Salzmann

Gauge: 1,435 mm
Route length: 51 km

Organisation
Majority-owned (51 per cent) and operated by Hessische Landesbahn (see entry in Germany section of *Railway systems and operators*).

Passenger operations
An extension of the Kassel tramways operates over 2.9 km of the KNE line in Baunatal, which has been electrified for the purpose. Passenger services over the former DB Wabern–Bad Wildungen line (17 km) are operated by KNE under contract.

Regionalbahn Kassel, a joint venture between KNE and the Kassel tram operator KVG, took over the Kassel–Kaufungen–Hessisch Lichtenau line, the 24 km Lossetalbahn, from DB in 1998 for conversion into an electrified light rail line.

Freight operations
Freight traffic is mainly for the Volkswagen car plant at Baunatal. Traffic statistics for 2001 were 1.3 million tonnes and 10.6 million tonne-km.

Traction and rolling stock
Two diesel locomotives, three Stadler GTW 2/6 diesel railcars for Kassel–Wabern–Bad Wildungen services and 17 electric tramcars operated in a pool with Kassel tramways.

UPDATED

Laubag

Lausitzer Braunkohle AG – Zentraler Eisenbahnbetrieb
Knappenstrasse 1, D-01968 Senftenberg
Tel: (+49 3573) 780   Fax: (+49 3573) 78 24 24

Key personnel
Chief Executive: Prof Dr Ing Kurt Häge
Director, Railway: Dipl-Geol Jürgen Kobus

Gauge: 1,435 mm
Route length: 376 km
Electrification: 325 km at 2.4 kV DC overhead

Organisation
Laubag is an industrial railway system operated by a lignite mining company in the Cottbus area, serving five opencast mines and three power stations. Connections with DB are made at Peitz and Spreewitz. The railway also serves other industrial enterprises.

Freight operations
Traffic statistics for 1999 were 52.7 million tonnes and 698.8 million tonne-km. Through lignite trains for Berlin and elsewhere are transferred to DB.

Traction and rolling stock
The Laubag fleet comprises 72 Bo+Bo 1,380 kW Type EL2 electric locomotives built by LEW Hennigsdorf, 25 diesel locomotives and 476 freight wagons.

Signalling and telecommunications
A Siemens SICAS centralised traffic control system was completed in 2000.

LausitzBahn GmbH

Zittauer Strasse 71/73, D-02826 Görlitz
Tel: (+49 3581) 33 99 00   Fax: (+49 3581) 33 95 05
e-mail: kundenservice@lausitzbahn.de
Web: www.lausitzbahn.de

Key personnel
Managing Director: Dipl Ing (FH) Andreas Trillmich
Head of Sales and Marketing: Christof Schulze

Gauge: 1,435 mm
Route length: 127 km

Organisation
Lausitzbahn GmbH is a wholly owned subsidiary of Connex Regiobahn GmbH. The company was set up after Connex won a contract in June 2002 to provide regional rail services between Cottbus and Zittau, initially for a three-year period commencing 15 December 2002.

From December 2004 LausitzBahn took over the management of passenger operations on the Leipzig–Bad Lausick–Geithain line (44 km) under a three-year contract.

Passenger operations
In December 2002 LausitzBahn commenced services on the 127 km Cottbus–Görlitz–Zittau line. Trains call at 17 stations or halts. From the same date, part of the route, between Görlitz and Cottbus, has been used by InterConnex long-

Siemens Desiro railcar operated by LausitzBahn   0533633

distance regional services. Operated jointly by LausitzBahn GmbH and another Connex subsidiary, Ostmecklenburgische Eisenbahngesellschaft mbH (see entry in Germany section of *Railway systems and operators*), the Friday-to-Monday services link Görlitz and Stralsund (492 km).

In December 2003 LausitzBahn commenced one daily return service between Görlitz and Dresden-Neustadt.

Traction and rolling stock
Services are provided by nine two-car part-low-floor Siemens Desiro diesel railcars leased from Angel Trains International.

UPDATED

MecklenburgBahn GmbH (MEBA)

Ludwigsluster Chaussee 72, D-19061 Schwerin
Tel: (+49 385) 399 00   Fax: (+49 385) 397 61 53
e-mail: info@mecklenburgbahn.de
Web: www.mecklenburgbahn.de

Key personnel
Managing Director: Dipl-Ing Norbert Klatt

Gauge: 1,435 mm
Route length: 79 km

Organisation
MEBA operates local passenger services in the *Land* of Mecklenburg-Pommern under a seven-year contract that commenced in 2001. There are 20 staff.

Passenger operations
MEBA operates regional services over the Rehna–Schwerin–Parchim line (79 km). Passenger numbers total around 400,000 annually.

Traction and rolling stock
Services are operated with six ALSTOM Coradia LINT 41 two-car diesel railcars.

NEW ENTRY

metronom Eisenbahngesellschaft mbH

St Viti Strasse 15, D-29525 Uelzen
Tel: (+49 581) 97 16 40   Fax: (+49 581) 971 64 19
e-mail: info@der-metronom.de
Web: www.der-metronom.de

Key personnel
Managing Director (Technical): Dr-Ing Carsten Hein
Managing Director (Commercial): Henning Weize
Company Secretary: Karl Schwinke

Organisation
MetroRail GmbH was established in 2002 by a consortium of German urban and regional operators to fulfil a contract placed by the Lower Saxony regional transport authority (LNVG) to operate regional services on the Hamburg–Bremen and Hamburg–Uelzen routes from December 2003 until December 2010. The members of the consortium are: Bremer Strassenbahn AG (BSAG) (5 per cent); NiedersachsenBahn (NB) (69.9 per cent); and Hamburger Hochbahn AG (25.1 per cent). NB is a joint venture company established by Verkehrsbetriebe Elbe-Weser GmbH (EVB) (40 per cent) and Osthannoversche Eisenbahn AG (OHE) (60 per cent).

Passenger operations
The contract placed with metronom Eisenbahngesellschaft mbH provides for 17 daily

metronom Class 146.1 electric locomotive and double-deck stock at Uelzen after arrival with a service from Hamburg (Philip Wormald)
NEW/0585049

train pairs between Hamburg and Bremen and 19 between Hamburg and Uelzen, with additional services at peak periods.

Traction and rolling stock
Services are provided by 10 Class 146.1 5,600 kW electric locomotives and 66 double-deck coaches (including 10 driving trailers) supplied by Bombardier Transportation. The coaching stock is formed as three eight-car and seven six-car sets. The Class 146.1 locomotive is a 160 km/h passenger derivative of the Class 185 freight machine. All rolling stock has been procured by LNVG. Maintenance is undertaken at a facility at Uelzen.

UPDATED

Mitteldeutsche Eisenbahn GmbH (MEG)

Postfach 1461, D-06204 Merseburg
Tel: (+49 3461) 49 22 49   Fax: (+49 3461) 49 63 90
Web: www.meg-bahn.de

Key personnel
Managing Directors: Dipl-Ing Koch, Dr-Ing Sonntag

Background
Formerly the industrial railway system of chemical and petrochemical plants in the Merseburg area of Sachsen-Anhalt, MEG was established as an open access operator in 1999. It is owned jointly by DB Cargo (80 per cent) and Transpetrol GmbH (20 per cent). The company also has facilities in Böhlen, Rüdersdorf and Schkopau.

Freight operations
These include open access movements of chemicals and petrochemicals, cement and coal. MEG also provides traction for infrastructure trains and undertakes services such as vehicle repair and tank wagon cleaning and maintenance.

Traction and rolling stock
In 2005 the diesel locomotive fleet totalled 44, including one Class 204, seven Class 228, two Class 229, three Class 232, 22 Class 346 and nine Vossloh Type G1206. In 2003 MEG also acquired from DB four Class 156 5,600 kW electric locomotives.

NEW ENTRY

Molli

Mecklenburgische Bäderbahn Molli GmbH & Co KG
Am Bahnhof, D-18209 Bad Doberan
Tel: (+49 38203) 41 50   Fax: (+49 38203) 415 12
e-mail: molli-bahn@t-online.de
Web: www.molli-bahn.de

Key personnel
Managing Director:
  Dipl-Wirtschefterin Angelika Münchow
Operating Manager: Dipl-Ing Jan Methling

Gauge: 900 mm
Route length: 15 km

Organisation
The Bad Doberan–Kühlungsborn narrow-gauge line was taken over from DB in 1995. Freight traffic was discontinued.

Passenger operations
Molli connects Kühlungsborn, a popular resort on the Baltic coast, with DB at Bad Doberan. The steam-operated line is also a tourist attraction.

Traction and rolling stock
Five steam locomotives and 37 coaches.

UPDATED

NBE nordbahn Eisenbahngesellschaft mbH & Co KG

Rudolf-Diesel-Strasse 2, D-24568 Kaltenkirchen
Tel: (+49 4191) 93 33   Fax: (+49 4191) 93 38 40
e-mail: info@nordbahn.info
Web: www.nordbahn.info

Key personnel
Managing Directors: Wolfgang Dirksen;
  Dipl-Kfm Nis Nissen

Gauge: 1,435 mm
Route length: 45 km

Organisation
NBE nordbahn Eisenbahngesellschaft mbH (nordbahn) is a joint venture company established by AKN Eisenbahn AG and Hamburger Hochbahn AG to operate services over the 45 km Neumünster–Bad Segeberg–Bad Oldesloe regional line in Schleswig Holstein. Services commenced on 15 December 2002.

Traction and rolling stock
Services are operated with five ALSTOM Coradia LINT 41 diesel railcars.

UPDATED

NEG Niebüll GmbH

Norddeutsche Eisenbahngesellschaft GmbH
Bahnhofstrasse 6, D-25899 Niebüll
Tel: (+49 4661) 98 08 80   Fax: (+49 4661) 980 88 19
e-mail: info@neg-niebuell.de
Web: www.neg-niebuell.de

Key personnel
Managing Directors: M Calmes, A Kremer, A Krey

Operations Manager: W Mohrbach

Gauge: 1,435 mm
Route length: 18 km

Organisation
NEG Niebüll links the DB station at Niebüll with Dagebüll, railhead for the islands of Föhr and Amrum. The present company took over the business in 2003 following the bankruptcy of the previous operator, Nordfriesische Verkehrsbetriebe AG (NVAG).

Passenger operations
As well as providing a year-round link between Niebüll and Dagebüll, NEG Niebüll also exchanges through coaches with DB from Berlin, Cologne, Frankfurt am Main, Dresden and Prague between April and October. Its rail services carry some 300,000 passengers annually.

Freight operations

NEG Niebüll runs freight trains northwards from Hamburg, on various lines radiating from Neumünster and over the DB/Banestyrlesen line from Niebüll to Tønder (Denmark).

Traction and rolling stock

NVAG operates one diesel railcar, two trailer cars (including a driving trailer) and five diesel locomotives, including a G2000BB and a G1206, both supplied by Vossloh.

UPDATED

Neukölln–Mittenwalder Eisenbahn (NME)

Neukölln–Mittenwalder Eisenbahn AG
Gottlieb-Dunkel-Str 47, D-12099 Berlin-Tempelhof
Tel: (+49 30) 70 09 03 50   Fax: (+49 30) 703 30 78
e-mail: info@neukoelln-mittenwalder-
eisenbahn.de
Web: www.neukoelln-mittenwalder-
eisenbahn.de

Key personnel

Directors: Eberhard Conrad; Sven Tombrink
Operations Manager: Dipl-Ing Dieter Radzuweit

Gauge: 1,435 mm
Route length: 9 km (owned); 25.6 km (operated)

Organisation

NME is a freight line in southern Berlin, connecting with DB at Hermannstrasse (Berlin-Neukölln).

Freight operations

Principal commodities carried are household waste, coal and fuel oil. Traffic statistics for 2001 were 1 million tonnes and 6.1 million tonne-km.

Traction and rolling stock

Six diesel locomotives, eight freight wagons.

UPDATED

Neusser Eisenbahn (NE)

Neuss-Düsseldorfer Häfen GmbH & Co KG, Neuss-Neusser Eisenbahn
Hammer Landstrasse 3, D-41460 Neuss
Tel: (+49 2131) 532 30   Fax: (+49 2131) 532 31 05
Web: www.nd-haefen.de

Key personnel

Principal: Dipl-Ing Ludwig von Hartz
Railway Operations Manager: Frank Türger

Gauge: 1,435 mm
Track length: 53 km

Organisation

The railway is a division of the Neuss municipal port authority.

Freight operations

The Neusser Eisenbahn serves 75 private sidings. It also operates freight services over DB lines. Traffic statistics for 2001 were 2.8 million tonnes and 11.2 million tonne-km.

Traction and rolling stock

Ten diesel locomotives, including two G2000BB, one G1700BB, one G1300 and two G1205BB, and 78 freight wagons.

UPDATED

Niederbarnimer Eisenbahn AG (NEB)

Georgenstrasse 22, D-10117 Berlin
Tel: (+49 30) 396 01 10   Fax: (+49 30) 39 60 11 70
e-mail: home@neb.de
Web: www.neb.de

Key personnel

Managing Directors: Dipl-Volkswirt Detlef Bröcker,
   Dr-Ing Christian Kuhn

Background

A subsidiary of Connex Regiobahn GmbH, NEB has a 15-year contract commencing in December 2005 to take over from DB services on the 34 km Heidekrautbahn from Karow, in the northern suburbs of Berlin, to Gross Schönebeck, and on the 10 km branch from the intermediate station at Basdorf to Schmachtenhagen. Between 1998 and 2002 extensive modernisation took place of the line and its passenger facilities.

Traction and rolling stock

NEB ordered five three-car Bombardier Talent dmus to provide services on the line.

UPDATED

Niederrheinische Verkehrsbetriebe AG (NIAG)

Homberger Strasse 113, D-47441 Moers
Tel: (+49 2841) 20 50   Fax: (+49 2841) 20 56 70
e-mail: info@niag-online.de
Web: www.niag-online.de

Key personnel

Chief Executives: Dipl-Ing Otfried Kinzel (chair)
   Dr rer pol Dipl-Kfm Gerhard Brückner
Rail Operations Manager: Dipl-Ing Manfred Diehl

Gauge: 1,435 mm
Route length: 36 km

Organisation

NIAG is a county council-owned transport company which provides road passenger services and rail freight services.

Freight operations

Lines owned by NIAG extend from Moers to Rheinberg and to Hoerstgen/Sevelen. Ore and coal are the main commodities carried. The railway also operates its own port on the river Rhine at Orsoy. Freight trains are also operated over DB lines from Moers to Trompet and Homberg. In 2003 2.5 million tonnes were carried, compared with 2.7 million tonnes in 2002.

Traction and rolling stock

10 diesel locomotives, including five MaK Type G1202BB/G1204B/G1205BB, two ex-DB Class 216 and a rebuilt ex-DB Class 211. There are also some 40 freight wagons.

UPDATED

Nord-Ostsee-Bahn (NOB)

Nord-Ostsee-Bahn GmbH
Raiffeisenstrasse 1, D-24103 Kiel
Tel: (+49 431) 73 03 60   Fax: (+49 431) 730 36 50
e-mail: post@nord-ostsee-bahn.de
Web: www.nord-ostsee-bahn.de

Gauge: 1,435 mm
Route length: 384 km

Key personnel

Managing Director: Karl-Heinz Fischer
Fiance Manager: Dipl-Betriebswirt Christian Kemp
Operations Manager: Rainer Blüm
Commercial Manager:
   Dipl-Kaufm Carsten Carstensen
Marketing and Communications Manager:
   Suzanne Thomas

Background

NOB is a wholly owned subsidiary of Connex Verkehr GmbH.

Passenger operations

In November 2000 NOB took over operations on three routes in Schleswig-Holstein: Kiel–Rendsburg–

Class 185 electric locomotive on a regional express service the NOB-operated Hamburg–Padborg (Denmark) route (Philip Wormald)
NEW/0585068

Schleswig–Husum (101 km); Husum–Bad St Peter–Ording (43 km); and 50 per cent of services on the Kiel–Neumünster line (31 km). The Niebull–Tønder (Denmark) (18 km) route was added to the network in 2003.

In November 2003 NOB took over operation of two-hourly regional express services the 191 km Hamburg–Neumünster–Flesburg–Padborg (Denmark) route following the insolvency of the previous operator, FLEX.

In 2003 NOB was awarded a 10-year contract, effective from December 2005, by the Schleswig-Holstein regional transport authority to operate long-distance regional passenger services the 'Marschbahn' between Hamburg and Westerland (243 km), on the island of Sylt. The company captured the traffic after a tendering process in which DB was the losing bidder.

Traction and rolling stock

For local services radiating from Kiel NOB uses nine Alstom LINT 41 dmus. An LEW-built NE 81 railcar is used on Niebull–Tønder services and the fleet also includes two Siemens Desiro twin-articulated units and an older railcar on loan to other Connex group companies.

Hamburg–Neumünster–Flesburg–Padborg services are handled by two Bombardier-built Class 185 and two Siemens Type ES64 U2 electric locomotives hired from Angel Trains and Siemens Dispolok respectively.

For its Hamburg–Westerland services, NOB is to procure 14 3,000 kW Type G3000BB diesel-hydraulic locomotives of a new design from Vossloh for the non-electrified section north of Itzehoe and four Bombardier Class 146.1 5,600 kW electric locomotives for the southern section to Hamburg

and possibly on to Berlin or Cologne. Coaching stock is also to be supplied by Bombardier in the form of 90 single-deck coaches to be formed into four-or six-car push-pull sets. Deliveries of locomotives and rolling stock from Bombardier were due to begin in 2005 but it was expected that 12 Type ME 26 diesel-electric locomotives would be hired from Siemens Dispolok for the December 2005 commencement of services pending delivery of the Vossloh machines.

UPDATED

NordWestBahn (NWB)

NordWestBahn GmbH
Alte Poststrasse 9, D-49074 Osnabrück
Tel: (+ 49 1805) 60 01 61   Fax: (+49 541) 34 47 93
e-mail: dialog@nordwestbahn.de
Web: www.nordwestbahn.de

Gauge: 1,435 mm
Route length: 720 km

Key personnel
Managing Directors: Dipl-Ing Hansrüdiger Fritz;
Dipl-Sozialwirt Martin Meyer

Background
NWB is a joint venture between Connex Regiobahn GmbH (64 per cent), Stadtwerke Osnabrück AG (26 per cent) and Verkehr und Wasser GmbH Oldenburg (10 per cent), operating services on behalf of the regional transport authorities in Niedersachsen and Nordrhein-Westfalen.

NWB employs 300 staff.

Passenger operations
In November 2000 NWB took over from DB Regio operation of a 310 km network covering Bremen–Delmenhorst–Osnabrück and Osnabrück–Oldenburg–Esens/Wilhelmshaven lines in the Weser-Ems region. In December 2003, jointly with Teutoburger Wald Eisenbahn (TWE), NWB took over further services from DB Regio. These were Münster–Bielefeld, Bielefeld–Paderborn, Bielefeld–Lage–Altenbeken, Paderborn–Holzminden and the

NWB ALSTOM Coradia LINT 41 dmu at Osnabrück (Quintus Vosman)   **NEW**/0585070

Bielefeld–Osnabrück line in Nordrhein–Westfalen. Around 8 million passengers annually are carried.

Freight operations
In 2001 Connex-Gruppe and Stadtwerke Osnabrück AG established NordWestCargo GmbH with the aim of expanding its operations into the freight market. See entry in Germany section of *Railway systems and operators.*

Traction and rolling stock
Services are provided using a fleet of 29 ALSTOM Coradia LINT 41 two-car dmus, six Siemens Desiro two-car dmus and 19 three-car and six two-car Bombardier Talent dmus.

UPDATED

NordWestCargo GmbH (NWC)

Hafenstrasse 5, D-49090 Osnabrück
Tel: (+49 541) 34 49 00   Fax: (+49 541) 34 49 46
e-mail: hafen@stw-os.de
Web: www.nordwestcargo.de

Key personnel
Contact: Dr Heino Schulz

Background
Established in 2001, NWC is an open access freight operator jointly owned by Connex Logistics

GmbH (51 per cent) and Stadtwerke Osnabrück AG. Its commercial activities are managed by Connex. In 2003 NWC took over the freight operations of the Bremen-based Farge-Vegesacker Eisenbahn (FVE), mainly coal traffic to Farge power station.

Freight operations
Operations centred on Osnabrück were supplemented in 2003 by the movement of trains of domestic waste from Groningen, Netherlands, to a power station near Bremen,

operating these in conjunction with the Dutch operator, ACTS.

Traction and rolling stock
Five diesel locomotives, including three MaK G1300BBs and one Vossloh G2000BB. Traction from other Connex group companies is also used.

NEW ENTRY

MVV OEG AG

Luisenring 49, D-68159 Mannheim
Tel: (+49 621) 29 00   Fax: (+49 621) 290 23 24
Web: www.mvv.de; www.rnv-online.de

Key personnel
Chief Executive Officer: Dr Rudolf Schulter
Directors: Dr Werner Dub, Hans-Jürgen Farrenkopf,
Karl-Heinz Trautmann

Gauge: 1,000 mm
Route length: 59 km
Electrification: 750 V DC overhead

Background
Formerly Oberrheinische Eisenbahn Gesellschaft AG (OEG), the company's management and operations were merged in 2000 with those of the Mannheim municipal operator, MVV Verkehr AG, adopting its current name. This move created a combined urban tram and interurban light rail network of 195 km and a bus network of 217 km.

In October 2004 MVV Verkehr/MVV OEG, together with urban and interurban rail and bus operators serving Heidelberg and Ludwigshafen, formed Rhein-Neckar-Verkehr GmbH (RNV) as a holding company to co-ordinate and oversee their activities in the region.

Passenger operations
MVV OEG is an interurban electric line with through running with Mannheim and Heidelberg tramways. Traffic figures for 2001 were 15.9 million passengers and 172 million passenger-km. Figures include traffic gained using a small bus fleet.

Traction and rolling stock
The fleet comprises 42 multisection electric light rail vehicles.

UPDATED

Ostdeutsche Eisenbahn GmbH (ODEG)

Bahnhof 1, D-19370 Parchim
Tel: (+49 3871) 609 93 15
e-mail: info@odeg.info
Web: www.odeg.info

Key personnel
Managing Directors: Dr Ralf Böhme; Dietmar Knerr

Gauge: 1,435 mm
Route length: 175 km (in Mecklenberg-Vorpommern)

Organisation
ODEG is a joint venture initially established by Hamburger Hochbahn AG (HHA) and Prignitzer Eisenbahn GmbH (PEG) to run regional services on a 175 km network in Mecklenburg-Vorpommern. In 2004 ODEG commenced a second contract to run regional services on a group of lines to the east of Berlin and in Brandenburg.

Passenger operations
In December 2002 the company commenced a six-year contract to run services on Hagenow Land–Ludwiglust–Parchim–Karow–Waren–Neustrelitz and Neustrelitz–Mirow routes. In December 2004 ODEG took over from DB Regio the operation of services over four routes in Berlin–Brandenburg: Berlin Lichtenberg–Beeskow–Frankfurt (Oder); Berlin Lichtenberg–Tiefensee; Berlin Lichtenberg–Eberswalde–Seelow–Frankfurt (Oder); and Eberswalde–Templin. The contract is for 10 years.

ODEG Stadler-built RS1 railcar (Quintus Vosman)   *NEW*/0585065

Traction and rolling stock
Services in Mecklenberg-Vorpommern are operated using seven Stadler RegioShuttle RS1 railcars. For Berlin-Brandenburg lines 25 additional RS1 railcars were ordered from Stadler. Ownership of these latter vehicles is divided between HHA (12) and PEG (13).

UPDATED

Osthannoversche Eisenbahnen (OHE)

Osthannoversche Eisenbahnen AG
Biermannstrasse 33, D-29221 Celle
Postfach 1663, D-29206 Celle
Tel: (+49 5141) 27 60   Fax: (+49 5141) 27 62 58
e-mail: info@ohe-transport.de
Web: www.ohe-transport.de

Key personnel
Chief Executive: Dr rer pol Jens Jahnke
Directors: Dipl-Ing Claus Luessmann;
  Heinrich Lindhorst;
  Dipl-Ing Olaf Ernst; Dipl-Kfm Udo Gantzke;
  Dipl-Verw Betr wirt Sliwinski

Gauge: 1,435 mm
Route length: 321 km (owned), 106 km (DB Netz operated)

Background
The Land of Lower Saxony is the majority shareholder in OHE, with 40.2 per cent of the capital, while the German state owns 33.8 per cent and DB Regio AG 8.9 per cent.

Organisation
OHE Group subsidiary companies and shareholdings also include various bus, road transport and river port activities, as well as the 20 km Rinteln-Stadthagener Verkehrs GmbH (RStV) freight railway, west of Hanover.

Passenger operations
In 2001 a consortium led by OHE and including Bremer Strassenbahn (BSAG), Elbe-Weser Railways (EVB) and Hamburger Hochbahn (HHA) secured a contract to operate regional express services on Hamburg–Bremen and Hamburg–Uelzen routes from 2003 (see entry for 'metronom' in Germany section of *Railway systems and operators*).

Freight operations
Formed in 1944 by the merger of five local railways, OHE now operates freight services over the following lines:
  Celle–Wittingen (51 km)
  Wittingen–Rühen (35 km)
  Beedenbostel–Mariaglück (6 km)
  Celle-Soltau (59 km)
  Beckedorf–Munster (24 km)
  Soltau–Lüneburg (57 km)
  Winsen (Luhe)–Hützel (41 km)
  Winsen (Luhe)–Niedermarschacht (18 km)
  Lüneburg–Bleckede (24 km)
  Wunstorf West–Mesmerode (6 km)
  Since 1990 OHE has also operated freight trains over DB Netz lines and provides locomotives for DB permanent way maintenance trains. Traffic statistics for 2004 were 1,506,000 tonnes and 104.9 million tonne-km (1,350,000 tonnes for 61.9 million tonne-km in 2003). A further 87,000 tonnes (1.8 million tonne-km) were carried by RStV (65,000 tonnes and 1.4 million tonne-km in 2003).

Traction and rolling stock
The fleet comprises 25 diesel locomotives, one Class 185 electric locomotive, two diesel railcars for excursion traffic and three freight wagons. Diesel locomotives include four Bombardier/General Electric 2,460 kW Blue Tiger machines, three ex-DB Class 216s and 11 MaK-built diesel-hydraulic B-Bs in the 1,200–1,600 kW power range.

UPDATED

Bombardier/General Electric Blue Tiger diesel-electric locomotive operated by OHE (OHE)   *NEW*/1110276

OME

Ostmecklenburgische Eisenbahngesellschaft mbH
Warliner Strasse 25, D-17034 Neubrandenburg
Tel: (+49 395) 430 84 10   Fax: (+49 395) 430 84 99
e-mail: service@omebahn.de
Web: www.omebahn.de

Key personnel
Managing Director: Steffen Höppner
Operations Manager: Ulrike Wildt

Gauge: 1,435 mm
Route length: 356 km

Background
OME is a wholly owned subsidiary of Connex Regiobahn GmbH.

Passenger operations
OME provides services over a network of DB lines in the Mecklenburg-Vorpommern region of northeast Germany based on Neubrandenburg. Lines

operated are: Bützow–Neubrandenburg–Stettin (Poland); Neustrelitz–Neubrandenburg–Stralsund; Gustrow–Rostok; and Bergen–Putbus–Lauterbach Mole. Services are co-ordinated with InterConnex regional express services on the Rostock–Gera–Berlin and Stralsund–Berlin–Dresden routes.

Freight operations
OME provides shunting services for industrial customers in Neubrandenburg.

Traction and rolling stock
Passenger services are operated by 10 three-section Talent dmus supplied by Bombardier Transportation and four Siemens Desiro two-car dmus, all leased from Angel Trains Europa. Shunting services for freight customers are handled by a LEW-built V60D diesel-hydraulic locomotive.

UPDATED

*OME Bombardier Talent dmu
at Neubrandenburg
(David Haydock)
NEW/0585064*

PCK Raffinerie GmbH

Passower Chaussee 111, D-16303 Schwedt
Tel: (+49 3332) 460   Fax: (+49 3332) 46 54 80
e-mail: info@pck.de
Web: www.pck.de

Key personnel
Managing Directors: Dr Hans-Otto Gerlach
   Hans-Joachim Knust
Director, Logistics: Ralph Fenselau
Rail Operations Manager: Dipl-Ing Jochen Bismark

Gauge: 1,435 mm
Track length: 120 km

Organisation
Industrial railway of a petrochemical plant at Schwedt on the river Oder.

Freight operations
Around 4.8 million tonnes are handled by rail annually, accounting for around 50 per cent of the plant's output.

Traction and rolling stock
Six shunting and seven main line diesel locomotives.

UPDATED

PE Cargo GmbH

Märkisches Ufer 34, D-10179 Berlin
Tel: (+49 33968) 50 70   Fax: (+49 33968) 507 22
e-mail: knoblauch@pe-cargo.de
Web: www.pe-cargo.de

Key personnel
Managing Directors: Mathias Tenisson,
   Dr Uwe Knoblauch
Sales Manager: Herr Pohle

Organisation
PE Cargo GmbH is a freight subsidiary of PE Arriva AG (formerly Prignitzer Eisenbahn Holding AG) and a sister company of the regional passenger rail operator, Prignitzer Eisenbahn GmbH, handling commercial freight services using open access rights, providing traction for works trains and supplying traction to other operators.

The formation of PE Arriva AG was a result of the acquisition in 2004 of Prignitzer Eisenbahn Holding AG by the UK-based transport services company Arriva plc.

Freight operations
In 2001 PE Cargo handled 660,000 tonnes of freight, including 300,000 tonnes of coal and 170,000 tonnes of cement. Traffic also includes construction materials which are hauled from Sachsendorf to the Berlin area using open access rights.

Traction and rolling stock
In 2005 the traction fleet comprised 10 refurbished V200 main line diesel-electric locomotives, 12 V60 and 2 V22 shunters, one V270 machine and one Class E94 electric locomotive. PE Cargo also operates a small wagon fleet.

UPDATED

Prignitzer Eisenbahn GmbH (PEG)

Pritzwalker Strasse 8, D-16949 Putlitz
Tel: (+49 30) 684 08 43 30
Fax: (+49 30) 684 08 43 40
e-mail: berlin@prignitzer-eisenbahn.de
Web: www.prignitzer-eisenbahn.de

Key personnel
Managing Director: Dr Ralf Böhme

Organisation
PEG is a subsidiary of PE Arriva AG, formerly Prignitzer Eisenbahn Holding AG, following the acquisition of the latter by the UK-based transport services company Arriva plc in 2004. As well as operating regional rail services on its own account on three groups of routes (see below), PEG is also a joint venture partner with Hamburger Hochbahn in Ostdeutsche Eisenbahn GmbH (ODEG) (see entry in Germany section of *Railway systems and operators*), which from December 2002 took over services on a group of lines in Mecklenburg-Vorpommern.

Other subsidiaries of PE Arriva AG are PE Cargo GmbH (see entry in Germany section of *Railway*

PEG Type VT 643 Bombardier Talent dmu at Enschede, Netherlands, forming a Dortmund-bound service (Quintus Vosman)

NEW/0585093

systems and operators), Prignitzer Lokomotiv und Waggonbau GmbH (PLW) (traction and rolling stock overhaul and maintenance) and ImoTrans GmbH (traction and rolling stock hire). The company is also a joint venture partner with Hugo Stinnes AG in Ostmecklenburgische Bahnwerk GmbH (OMB), which undertakes rolling stock maintenance and overhaul at the former DB works at Neustrelitz.

Passenger operations

Since 1996 PEG has operated regional lines in Brandenburg around the Pritzwalk area, northwest of Berlin. In 2001, in partnership with Hamburger Hochbahn AG, PEG secured a contract to operate the 175 km Ostdeutsche Eisenbahn GmbH (ODEG) network in Mecklenburg–Vorpommern covering Hagenow Land–Ludwigslust–Parchim–Karow–Waren–Neustrelitz and Neustrelitz–Mirow. The contract took effect in December 2002.

In December 2002 PEG commenced services in Nordrhein–Westfalen on the Duisburg-Ruhrort–Oberhausen–Dorsten route (36 km) and in December 2004 took over regional services on the Dortmund–Coesfeld–Gronau–Enschede (Netherlands) route (105 km).

Traction and rolling stock

Pritzwalk area services employ eight Stadler Type VT 650 RegioShuttle railcars. For the Ruhr contract, the company procured six Bombardier Talent Type VT 643 dmus, supplementing these with 11 similar vehicles for the Dortmund-Enschede line. In 2005 the fleet also included three DWA Type VT 670 double-deck railcars and five Type VT 798 railcars.

UPDATED

rail4chem

rail4chem Eisenbahnverkehrsgesellschaft mbH
Schützenbahn 60, D-45127 Essen
Tel: (+49 201) 430 40   Fax: (+49 201) 430 41 99
e-mail: info@rail4chem.com
Web: www.rail4chem.com

Key personnel

General Manager: Matthias Raith
Assistant General Manager: Hartmut Gasser
Head of Finance: Christoph P König
Head of Planning and Sales: Michael Roggenkamp
Co-ordinator, Key Account Management:
  Ralf Bickert
Head of Operations and Technology: Dirk Munder

Organisation

Rail4chem is an open access freight operator formed in 2001 by the German-based chemicals group BASF and three forwarding companies, Bertschi, Hoyer and VTG-Lehnkering. Each company holds 25 per cent of the company's shares. In October 2002 Hoyer's rail operations subsidiary, Hoyer Railserv, transferred its activities and equipment to rail4chem.

Wholly owned subsidiary companies have been established in the Netherlands (rail4chem Benelux BV) and in Switzerland (rail4chem Transalpin AG). In the case of the former, assets and traffic were taken over in late 2004 from Netherlands-based open access operator Short Lines.

In January 2005 rail4chem became a founding member of the 'European Bulls' alliance (www.european-bulls.com) of open access freight companies with partners in Austria, the Czech Republic, Italy and Spain. The alliance was formed to provide seamless cross-border services operated to common standards.

Rail4chem General Motors Class 66 locomotive at Rotterdam Europort (Quintus Vosman)   *NEW*/0585092

Freight operations

Rail4chem specialises in the transport of chemicals and petroleum products and operates both domestic and international services, including a traffic flow between the Ruhr and Geleen, Netherlands, and from Grosskorbertha, Germany, to Antwerp, Belgium. Major customers include BASF and Shell. Intermodal services are also operated. At the end of 2004 a container service was launched from Rotterdam in the Netherlands to Brescia, Italy, running five times a week in each direction. Operations in Italy are handled by Ferrovie Nord Cargo, one of rail4chem's European Bulls partners.

A train control centre is located at the company's Essen headquarters and a cargo service centre is sited in Nordhausen.

With its partner companies, rail4chem recorded around 2,000 million tonne-km in 2004.

Traction and rolling stock

In 2005 the rail4chem electric locomotive fleet totalled 13: five Bombardier Class 145; seven Bombardier Class 185; and one Siemens ES64U. The diesel locomotive fleet consisted of 10 locomotives: three Vossloh G1206; two Vossloh G2000; and five General Motors Class 66. The diesel fleet includes machines taken over from Short Lines in 2004.

UPDATED

Regentalbahn AG

Die Länderbahn
Bahnhofsplatz 1, D-94234 Viechtach
Tel: (+49 9942) 94 65 10   Fax: (+49 9942) 94 65 28
e-mail: info@laenderbahn.com
Web: www.laenderbahn.com

Subsidiaries

Regentalbahn AG is a holding company for or has a shareholding in the following companies or operations:
Regental Bahnbetriebs GmbH (RBG)
Bahnhofsplatz 1, D-94234 Viechtach
Tel: (+49 9942) 94 65 30   Fax: (+49 9942) 94 65 38
e-mail: regentalbahn@laenderbahn.com
  Responsible for Oberpfalzbahn (joint venture with DB Regio), Waldbahn and Regental Cargo activities.

Vogtlandbahn GmbH (VBG) (see entry in Germany section of *Railway systems and operators*)

Regental Fahrzeugwerkstätten GmbH
A rolling stock maintenance and repair facility established in 1989, handling work for Regental AG and other operators.

Granitwerk Prünst
A producer of granite track ballast in which Regentalbahn AG holds a 50 per cent stake.

Allgäu Express GmbH (ALEX)
A joint venture with EuroThurbo GmbH, a subsidiary of the Swiss-based Thurbo AG. See entry for Allgäu Express GmbH (ALEX) in Germany section of *Railway systems and operators* and entry for Thurbo AG in Switzerland section of *Railway systems and operators*.

Two RBG RS1 RegioShuttle railcars at Deggendorf, operating a Waldbahn service (Quintus Vosman)   *NEW*/0585075

Key personnel

Managing Directors: Dr Johann Niggl, Tobias Richter

Gauge: 1,435 mm
Route length: 47 km (owned), 273 km (RBG-operated)

Organisation

Regentalbahn AG is a holding company for several rail-related companies in Germany (see above).

In October 2004 the UK transport services group Arriva became its major shareholder when it acquired the state of Bavaria's stake of 76.9 per cent in the company.

RBG's original line, from Gotteszell to Viechtach, is now freight only. The Lam–Kötzting line, taken over in 1973, is passenger only (see below), with through running over the DB line to Cham. The Plattling–Bayerisch Eisenstein, Zwiesel–Grafenau and Zwiesel–Bodenmais lines are operated for DB using the 'Waldbahn' brand name.

In June 2001 Oberpfalzbahn, a joint company established by RBG and DB Regio, took over the operation of the Schwandorf–Cham–Waldmünchen, Cham–Furthinwaldand Cham–Kötzting–Lam lines. Connection is made with the Vogtlandbahn at Schwandorf.

In May 1998 a subsidiary company, Vogtlandbahn GmbH, was established to operate services over 263 km of DB lines in the Zwickau/Plauen area of Saxony under a 15-year contract. It has since further expanded its network (see entry in Germany section of *Railway systems and operators*).

In December 2003 Allgäu Express GmbH (ALEX), a joint venture between Regentalbahn AG and EuroThurbo GmbH, took over operation of most regional express services between Munich and Oberstdorf, Bavaria (see entry in Germany section of *Railway systems and operators*).

Passenger operations
The two RBG passenger operations, Oberpfalzbahn and Waldbahn, cover networks totalling 273 km and serve 52 stations.

Freight operations
The freight business of RBG, Regental Cargo, is an open access operator based in Neuenmarkt, Oberfranken, handling own-account traffic, providing traction for works trains and working with partners to run container services between Hof and the ports of Bremerhaven and Hamburg.

Traction and rolling stock
Regentalbahn/Regental Cargo: six diesel locomotives (including two MaK G1202BB, one Vossloh G1206BB and two G2000BB), six diesel railcars (including two former DB battery railcars rebuilt in own workshops for diesel operation), two driving trailers, three coaches.
Waldbahn: 14 Adtranz/Stadler RS1 RegioShuttle diesel railcars.
Oberpfalzbahn: 11 Adtranz/Stadler RS1 RegioShuttle railcars.

UPDATED

Regiobahn Bitterfeld (RBB)

Regiobahn Bitterfeld GmbH (RBB)
Chemiepark Areal C, Strasse am Landgraben 5, D-06749 Bitterfeld
Tel: (+49 3493) 784 00   Fax: (+49 3493) 784 01
Web: www.connex-gruppe.de

Key personnel
Managing Director: Dipl-Ing Meinhardt

Gauge: 1,435 mm
Route length: 70 km

Organisation
This railway was created by the merger of three industrial lines, was taken over by DEG in late 1995 and now forms part of Connex Cargo Logistics GmbH. Thirty-five private sidings are served.

Freight operations
In 2004 RBB carried some 990,000 tonnes.

Traction and rolling stock
The RBB fleet comprises five V60 and five V100.4 diesel locomotives, one V22 shunter and one Vossloh G2000 diesel locomotives. A Bombardier Class 185 electric locomotive from the Connex Cargo Logistics pool is also used.

UPDATED

Rheinbraun

Rheinbraun Aktiengesellschaft – Gruben- und Grubenanschlussbahnen
Stüttgenweg 2, D-50935 Cologne
Tel: (+49 221) 48 00   Fax: (+49 221) 480 13 51
Web: http://www.rheinbraun.de

Key personnel
Chief Executive: Dr Dieter Henning
Operations Manager: Dipl-Ing Werner Koenigs

Gauge: 1,435 mm
Route length: 318 km
Eectrification: 6.25 kV AC 50 Hz overhead

Organisation
Industrial line of lignite mining company.

Freight operations
The railway carries rubble and lignite, connecting with DB and HGK lines. Axleload of 35 tonnes and loading gauge exceed those of DB. Traffic statistics for 1999 were 112.4 million tonnes and 2.1 billion tonne-km.

Traction and rolling stock
The fleet comprises 51 electric locomotives, 21 diesel locomotives and 818 freight wagons. Most of the electric locomotives are due for replacement, dating from the 1950s: in 1999, delivery commenced of 10 type EL 2000 machines ordered from Adtranz. These have been developed from the design of DB Class 101 and 145 locomotives.

Rhein-Haardtbahn GmbH (RHB)

Industriestrasse 3-5, D-67063 Ludwigshafen
Postfach 21 12 23, D-67063 Ludwigshafen
Tel: (+49 621) 50 50   Fax: (+49 621) 505 22 20
e-mail: info@rhein-haardtbahn.de
Web: www.rhein-haardtbahn.de

Key personnel
Managing Directors: Dipl-Vw Giso Rocker
   Dr-Ing Gerhard Weismüller
Operations Manager: Dipl-Ing Günter Quass

Gauge; 1,000 mm
Route length: 16 km
Electrification: 750 V DC

Organisation
RHB is an electrified interurban railway, associated with Ludwigshafen municipal tramways.

In October 2004 RHB, together with urban and interurban rail and bus operators serving Heidelberg and Mannheim, formed Rhein-Neckar-Verkehr GmbH (RNV) as a holding company to co-ordinate and oversee their activities in the region.

Passenger operations
RHB runs from Mannheim via Ludwigshafen to Bad Dürkheim with through running over Mannheim and Ludwigshafen tramways. Traffic statistics for 2003 were 1.9 million passengers and 16.3 million passenger-km.

Traction and rolling stock
Six Düwag electric railcars and two trailers, plus one shunting locomotive. Ludwigshafen (VBL) trams are also used over the line.

UPDATED

Rheinische-Bergische Eisenbahn GmbH (RBE)

Bergstrasse 1B, D-40822 Mettmann
Tel: (+49 2104) 30 52 00   Fax: (+49 2104) 30 52 15
e-mail: rbe@connex-gruppe.de
Web: http://www.connex-gruppe.de

Key personnel
Managing Director: Udo Winkens

Gauge: 1,435 mm
Route length: 34 km

Organisation
RBE is a subsidiary of Connex Regiobahn GmbH (qv in Connex Verkehr GmbH entry).

Passenger operations
In 1999 RBE took over operations on Route S28 of the Düsseldorf S-Bahn, covering the Kaarst–Neuss–Düsseldorf (17 km) and Mettmann-Düsseldorf (16 km) branches, operating them under the Regio-Bahn brand name.

New lines
Plans exist to extend the Mettmann branch to Wuppertal by 2003.

Traction and rolling stock
The fleet comprises eight Bombardier Talent two-car dmus. In 2001 four additional units of the same type were ordered.

For details of the latest updates to *Jane's World Railways* online and to discover the additional information available exclusively to online subscribers please visit
jwr.janes.com

Rhenus Keolis

NL Bielefeld, Meisenstrasse 65, D-33607 Bielefeld
Tel: (+49 521) 927 37 12   Fax: (+49 521) 927 37 22
Web: http://www.eurobahn.de

Key personnel
Managing Director: Jörg Kiehn

Background
Formerly Eurobahn, Rhenus Keolis is a joint venture owned by Keolis and Rhenus to operate regional rail services in Germany. The company also operates the Freiberger Eisenbahngesellschaft (qv).

Passenger operations
In May 2000, the company took over passenger operations on the Bielefeld–Herford–Bünde–Rahden line (the 'Ravensberger Bahn') and the Bielefeld–Lemgo line, both in Nordrhein–Westfalen. During the previous year, the company took over operations on the Alzey–Kirchheimbolanden line in Rheinhessen.

In December 2000, the company commenced an open access service over DB tracks between Bielefeld and Cologne.

Traction and rolling stock
Services are provided with Adtranz RS1 RegioShuttle single-unit railcars: three are employed on Ravensberger Bahn services, two on the line in Rheinhessen.

Rügensche Kleinbahn (RüKB)

Rügensche Kleinbahn GmbH & Co
Binzer Strasse 12, D-18581 Putbus
Tel: (+49 38301) 80 10   Fax: (+49 38301) 801 15
e-mail: info@rasender-roland.de
Web: http://www.rasender-roland.de

Key personnel
Managing Director: Bernhard van Engelen
Operations Manager: Dipl-Ing Konrad Höft
Local Operations Manager: Jochen Warsow

Gauge: 750 mm
Route length: 24 km

Organisation
The Putbus–Göhren line on the island of Rügen was taken over from DB in 1996, and is operated by Karsdorfer Eisenbahn (qv).

Passenger operations
In 1997 131,400 train-km were operated.

New lines
The railway is to be extended from Putbus to Lauterbach (2 km) over DB tracks. A third rail needs to be laid for this purpose.

Improvements to existing lines
The line is to be completely rehabilitated at a cost of DM24 million. Maximum speed is to be increased from 30 to 50 km/h.

Traction and rolling stock
Six steam locomotives, one diesel locomotive, 29 coaches, 6 baggage cars. Three diesel railcars are on order for delivery in 2001.

Ruhrkohle

Ruhrkohle Bergbau AG – RAG-Bahn und Hafenbetriebe
Talstrasse 7, D-45986 Gladbeck
Tel: (+49 2043) 50 13 20   Fax: (+49 2043) 50 15 60

Key personnel
Managing Directors: Dr-Ing Gerhard Hartfeld
   Dipl-Ing Norbert Rüsel

Gauge: 1,435 mm
Track length; 479 km
Electrification: 242 km at 15 kV AC 16²/₃Hz

Organisation
The industrial railway of the Ruhrkohle AG coal mining company operates an extensive network of lines in the Ruhr industrial area, although the system is shrinking with the decline of coal mining. The railway connects with DB at 29 locations. Freight trains are also operated over DB lines, and Ruhrkohle owns and operates 12 ports on inland waterways. The Werne-Bockum-Hövel railway (12 km) is the only line operated as a public carrier.

Freight operations
Traffic statistics for 1997 were 63.7 million tonnes and 265.1 million tonne-km, compared with 67.4 million tonnes and 288.2 million tonne-km in 1996.

Traction and rolling stock
Ruhrkohle operates a fleet of 106 diesel locomotives, 28 electric locomotives and 3,464 freight wagons. In 1999 11 1,500 kW B-B diesel-hydraulic locomotives were supplied by Vossloh.

In June 2001, the first of six Bombardier Class 145 electric locomotives was delivered. Initial duties included coal traffic from the Oberhausen area to a power station at Lahde.

Ruhrkohle diesel locomotive with empty coal hoppers for the Fürst Leopold mine at Dorsten (Wolfram Veith)   0063200

Ruhr-Lippe Eisenbahn

Regionalverkehr Ruhr-Lippe GmbH (RLG)
Krögerweg 11, D-48155 Münster
Tel: (+49 251) 627 00   Fax: (+49 251) 627 02 22

Key personnel
Managing Director: Dr Ing Eberhard Christ
Rail Operations Manager:
   Dipl-Ing Hansrüdiger Fritz

Gauge: 1,435 mm
Route length: 42 km

Organisation
The local bus company, owned by WVG (qv), has responsibility for three separate railway lines connected with DB at Hamm, Soest and Neheim-Hüsten respectively. Hamm–Vellinghausen is the busiest line, with heavy coal traffic for a power station. Rail services are operated by WLE (qv).

Freight operations
Traffic statistics for 1997 were 823,000 tonnes and 8.4 million tonne-km, compared with 643,000 tonnes and 5.7 million tonne-km in 1996.

Traction and rolling stock
Three diesel locomotives.

Rurtalbahn GmbH (RTB)

Kölner Landstrasse 271, D-52351 Düren
Tel: (+49 2421) 39 01 42   Fax: (+49 2421) 39 01 35
e-mail: info@rurtalbahn.de
Web: www.rurtalbahn.de

Key personnel
Managing Directors: Guido Emunds,
  Hans-Pieter Niessen, Achim Schmitz

Gauge: 1,435 mm
Route length: 90 km, of which 75 km with passenger service

Organisation
A small local freight operator until 1993, Dürener Kreisbahn GmbH (DKB) took over the former DB branches from Düren south to Heimbach and north to Jülich and Linnich. These lines were rehabilitated and a frequent passenger service with Siemens Duewag lightweight railcars was established.

In 2003 DKB joined with RATH GmbH to form RTB with shareholdings of 25.1 per cent and 74.9 per cent respectively.

RTB's Jülich–Puffendorf line (15 km) was out of use in 2004.

Passenger operations
Passenger services are operated from Düren north to Linnich (26 km), south to Heimbach (30 km) and southeast to Zülpich (19 km).

Freight operations
Traffic statistics for 2001 were 209,000 tonnes and 870,000 tonne-km.

Rurtalbahn Siemens-built RegioSprinter diesel railcar (Quintus Vosman)   **NEW**/0585032

In July 2004 RGB in collaboration with Wiener Lokalbahnen launched an open access container service linking Passau and Bremerhaven.

Traction and rolling stock
Passenger services are operated with 17 Siemens Duewag RegioSprinter lightweight railcars. There are seven diesel locomotives, including a Vossloh G1206 B-B diesel-hydraulic acquired in 2004.

A Taurus electric locomotive leased from Siemens Dispolok is used for the Passau–Bremerhaven intermodal freight service.

UPDATED

SüdThüringenBahn (STB)

Background
STB is a joint venture established by Erfurter Industriebahn (EIB) (qv) and Hessische Landesbahn (HLB) (qv) to operate regional services in Thüringen.

Passenger operations
In January 2001 STB commenced services on the Wernshausen-Zella-Mehlis line and in March took over some services on the Eisfeld-Eisenach route. Local trains between Erfurt and Meiningen were added to the network in June 2001.

Traction and rolling stock
Services are provided using Adtranz RS1 RegioShuttle railcars, including seven units ordered in October 2001 to serve lines in northern Thüringen.

SWEG

Südwestdeutsche Verkehrs AG (SWEG)
Friedrichstrasse 59, D-77933 Lahr
Tel: (+49 7821) 270 20   Fax: (+49 7821) 27 02 35
e-mail: info@sweg.de
Web: http://www.sweg.de

Key personnel
Chairman: Hans Joachim Disch
Directors: Johannes Müller; Hans-Peter Schiff;
  Bernd Strobel

Gauge: 1,435 mm
Route length: 110 km

Organisation
The present company was formed in 1962 to consolidate various regional railways in the Baden part of Baden-Württemberg state, some of which were abandoned by their previous owners. Currently the company owns and operates the following lines:
  Schwarzach–Achern–Ottenhöfen (10 km)
  Biberach–Oberharmersbach (11 km)
  Bad Krozigen–Münstertal (Münstertalbahn) (13 km)

Riegel–Endingen–Breisach   (Kaiserstuhlbahn) (40 km)
  Meckesheim–Waibstadt–Aglasterhausen–Hüffenhardt (36 km, of which 19 km leased from DB)
  Bühl–Schwarzach–Söllingen (15 km, freight only)
  SWEG also runs extensive bus operations with a fleet of 280 vehicles.

Subsidiary companies
Ortenau S-Bahn GmbH
Hauptstrasse 26, D-77652 Offenburg
Tel: (+49 781) 239 30   Fax: (+49 781) 923 93 10

A wholly owned subsidiary of SWEG, the Ortenau S-Bahn is a 104 km network radiating from Offenburg to Achern, Appenweier and Bad Greisbach, Hausach, and Kehl. Services are provided by Adtranz RS1 RegioShuttle railcars.

Breisgau S-Bahn GmbH (BSB)
Besanconstrasse 99, D-79111 Freiburg (Breisgau)
Fax: (+49 7821) 27 02 45

BSB is a joint venture between SWEG and Freiburger Verkehrs AG (VAG) providing a high-frequency service between Freiburg and Breisach using Adtranz RS1 RegioShuttle railcars.

Passenger operations
Traffic statistics for 1997 were 2 million passengers and 17.6 million passenger-km, compared with 2.1 million passengers and 16.6 million passenger-km in 1996.

Freight operations
Traffic statistics for 1999 were 203,000 tonnes and 2.9 million tonne-km.

Traction and rolling stock
Six diesel locomotives, 55 diesel railcars, 24 coaches and two freight wagons. The fleet includes 26 Adtranz RS1 RegioShuttle railcars. Nine vehicles of the same type are used on Breisgau S-Bahn services.

Teutoburger Wald Eisenbahn (TWE)

Teutoburger-Wald-Eisenbahn AG (TWE)
Am Grubenhof 2, D-33330 Gütersloh
Tel: (+49 5241) 160 67   Fax: (+49 5241) 252 45

Key personnel
Managing Directors: Dipl-Ing Thomas Schare
  Dipl-Ing Thomas Schulte
  Heinz Wolfgramm
Operations Manager: Karl Gottwald

Gauge; 1,435 mm
Route length: 104 km

Organisation
Associated with and operated by Connex Cargo Logistics GmbH. The TWE workshops at Lengerich are a central facility for all Connex Gruppe railways and also carry out work for third parties.

In 2000 it was reported that local authority funding was to finance the reopening to passenger traffic of the Gütersloh–Harsewinkel and Gütersloh–Verl lines by 2005.

Freight operations
The line extends from Ibbenbüren via Lengerich and Gütersloh to Hövelhof and connects with DB at each of these four locations. TWE freight trains also run over DB lines.

Traffic statistics for 1997 were 391,000 tonnes and 25.6 million tonne-km, compared with 296,000 tonnes and 16.7 million tonne-km in 1996.

Traction and rolling stock
Five diesel locomotives and nine freight wagons.

Trans Regio

Trans Regio Deutsch Regionalbahn GmbH
Schönbornstrasse 7, D-54295 Trier
Tel: (+49 651) 991 26 76
e-mail: info@trans-regio.de
Web: http://www.trans-regio.de

Gauge: 1,435 mm
Route length: 101 km

TX Logistik AG

Rhöndorfer Strasse 85, D-50364 Bad Honnef
Tel: (+49 2224) 77 90   Fax: (+49 2224) 77 92 09
e-mail: info@txlogistik.de
Web: www.txlogistik

Key personnel
Chairman: Raimund Stüer

Organisation
Established in 1999, TX Logistik is predominantly owned by the company's management and private shareholders but the Italian national rail company, Trenitalia SpA, holds a stake of 15 per cent. It was among the first of Germany's open access freight operators, initially running services for boxXpress.de (see Germany section of *Railway systems and operators*), in which TX Logistik has a 15 per cent shareholding. In 2003 subsidiaries were established in Austria (TX Logistik GmbH Österreich) and Switzerland (TX Logistik GmbH Schweiz), and in 2004 a Swedish subsidiary, TX Logistik AB Schweden, was set up. There are two other wholly owned subsidiary companies: TX Consulting GmbH and TX Service Management GmbH. TX Logistik holds licences to operate in Austria and Switzerland. In Austria the company collaborates with LTE Logistik- und Transport GmbH (see entry in Austria section of *Railway systems and operators*).

In 2004 the company employed 175 staff.

Freight operations
TX Logistik operates whole and part train-load conventional freight and intermodal services, as well as offering terminal operations and locomotive hire. Cross-border operations into Italy and the Netherlands commenced in 2003 and the company's network now includes Antwerp and Rotterdam and Italian destinations served by both

Background
Trans Regio is a joint venture between Trier-based bus operator Moselbahn Gesellschaft mbH and Rheinische Bahngesellschaft AG, Düsseldorf's public transport operator.

Passenger operations
In May 2000 Trans Regio took over passenger operations on two lines in the Rheinland-Pfalz region: Andernach–Mayen (subsequently extended to Kaisersesch (43 km)); and Kaiserslautern–Kusel (45 km). In June 2001 the company took over services on the Bullay–Traben-Trarbach 'Moselwein-Bahn' line (13 km).

Traction and rolling stock
Nineteen Adtranz RS1 RegioShuttle railcars.

Type ES 64 U2 electric locomotive leased from Siemens Dispolok heading a TX Logistik train of new BMW cars (TX Logistik)   *NEW*/0585050

the Gotthard and Brenner routes. Among early contracts secured was the movement of new cars from plants in southern Germany to northern ports. Other commodities handled include chemical products and steel. TX Logistik is also certified in Germany in the management of the disposal of domestic waste.

Traction and rolling stock
In 2004 the fleet operated by TX Logistik and its subsidiaries totalled 31 electric locomotives and 11 diesel locomotives, including: 13 Type ES 64 US 6,400 kW electric locomotives leased from Siemens Dispolok; 11 Bombardier Class 185 5,600 kW electric locomotives leased from Angel Trains Cargo; one Bombardier Blue Tiger 2,460 kW diesel locomotive; two Vossloh G800BB and one G1206 diesel-hydraulics leased from LSG and two Type V100.1 shunters from GATX Leasing. The wagon fleet totalled some 450.

NEW ENTRY

Usedomer Bäderbahn (UBB)

Usedomer Bäderbahn GmbH (UBB)
Am Bahnhof 1, D-17424 Heringsdorf
Tel: (+49 38378) 27 10   Fax: (+49 38378) 271 14

Key personnel
Managing Directors: Dipl-Ing Jörgen Bosse
   Dipl-Kfm Corinna Hinkel
Marketing and Technical Director:
   Dipl-Ing Hans-Joachim Kohl
Operations Manager: Dipl-Ing Andreas Pinske

Gauge: 1,435 mm
Route length: 79 km

Organisation
UBB is a wholly owned subsidiary of DB Travel & Tourism, serving the island of Usedom, a popular seaside resort. Lines were taken over from the parent company in 1995.

Passenger operations
Traffic statistics for 1999 were 1.5 million passengers, compared with 1.2 million in 1998.

New lines
A fixed link with the German mainland at Wolgast across the river Peene was completed in 2000. At the other end of the line an extension across the Polish border to Swinoujscie is planned, with a further extension to the border town of Garz, but

UBB GTW 2/6 railcar at Peenemünde from Zinnowitz (David Haydock)   0109852

there will be no physical connection with Polish State Railways.

Traction and rolling stock
Most of the fleet of 14 diesel railbuses and 11 trailers has been replaced by 14 Type GTW 2/6 Class 646 diesel railcars from Stadler/Adtranz/Bombardier. Two diesel locomotives are also operated.

vectus Verkehrsgesellschaft mbH

Bahnhofsplatz 2, D-65549 Limburg
Tel: (+49 6431) 584 50 Fax: (+49 6431) 58 45 21
e-mail: info@vectus-online.de
Web: www.vectus-online.de

Key personnel
Managing Directors: Dipl-Ing Veit Salzmann;
Dipl-Ing Horst Klein

Gauge: 1,435 mm
Route length: 222 km

Background
vectus Verkehrsgesellschaft mbH is a joint venture company set up in 2003 by Hessische Landesbahnen GmbH (74.9 per cent) and Westerwaldbahn GmbH (25.1 per cent) to operate regional services on a four-line network around the city of Limburg. Awarded by the Rheinland-Pfalz Nord and Rhein-Main-Verkehrsverbund public transport authorities, the 10-year contract commenced in December 2004.

Passenger operations
vectus serves four lines radiating from Limburg forming the Westerwald-Taunus network:

ALSTOM Coradia LINT 41 diesel railcar for vectus on display at InnoTrans 2004 (Ken Harris)
NEW/0585017

Limburg–Wiesbaden; Limburg–Koblenz; Limburg–Siershahn; and Limburg–Au (Sieg). There are 46 stations and halts.

Traction and rolling stock
Services are operated with a fleet of 10 single-car Coradia LINT 27 and 18 two-car Coradia LINT 41 diesel railcars supplied by ALSTOM from its Salzgitter plant in 2004.

NEW ENTRY

Verkehrsbetriebe Kreis Plön (VKP)

Verkehrsbetriebe Kreis Plön GmbH
Diedrichstrasse 5, D-24143 Kiel
Tel: (+49 431) 70 58 11 Fax: (+49 431) 70 58 60

Key personnel
Managing Director: Dipl-Betr wrt Günter Gloe
Rail Operations Manager: Harald Hansen

Gauge: 1,435 mm
Route length: 24 km

Organisation
VKP is a county council-owned transport company providing passenger services by road and freight services by rail.

Passenger operations
Jointly with DEG (qv) VKP was the successful bidder for the operation of passenger services over DB lines from Kiel to surrounding locations in Schleswig-Holstein. A new company, Nord-Ostsee-Bahn (NOB) (qv) manages these operations.

Freight operations
VKP owns and operates the Kiel–Schönberger Eisenbahn and also operates a short line serving a power station in Kiel. The main commodity carried is coal. Traffic statistics for 1999 were 761,000 tonnes and 6.1 million tonne-km.

Traction and rolling stock
One diesel locomotive.

Verkehrsbetriebe Peine-Salzgitter (VPS)

Verkehrsbetriebe Peine-Salzgitter GmbH (VPS)
Am Hillenholz 28, D-38229 Salzgitter
Tel: (+49 5341) 21 35 41 Fax: (+49 5341) 21 45 76

Key personnel
Managing Directors: Dipl-Ing Arndt Frielinghaus
(also Operations Manager)
Dipl-Kfm Detlef Mehl

Gauge: 1,435 mm
Route length: 69 km

Organisation
UPS is a steelworks industrial railway which also serves other clients. Operations are split into three divisions: Peine; Salzgitter (connecting with each other); and Ilsenburg. VPS also runs freight trains over DB lines. VPS locomotives are available for lease and its workshops also carry out work for third parties. The company owns and operates two ports on inland waterways.

Freight operations
Traffic statistics for 1999 were 34.5 million tonnes and 208 million tonne-km. About one third of all traffic is exchanged with DB.

Traction and rolling stock
The VPS fleet comprises 53 diesel locomotives and 775 freight wagons. The company also develops special freight wagons for the needs of the steel industry.

Vogtlandbahn GmbH (VBG)

Ohmstrasse 2, D-08496 Neumark
Tel: (+49 37600) 77 71 01 Fax: (+49 37600) 77 72 51
email: info@vogtlandbahn.de
Web: www.vogtlandbahn.de

Gauge: 1,435 mm
Route length: 595 km

Organisation
Established in January 1998, Vogtlandbahn GmbH is a wholly owned subsidiary of Regentalbahn AG (see entry in Germany section of *Railway systems and operators*), initially to take over from DB the operation of regional services in the Zwickau/Plauen areas of Saxony under a 15-year contract. Since then the company has extended its network to include routes in Bavaria and Thüringen and to run into the Czech Republic.

Passenger operations
The VBG network extends from Gera, Schleiz and Zwickau Neumark in the north via Hof, Marktredwitz, Weiden and Schwandorf to Regensburg in the south. In 2000 cross-border services to Cheb, Czech Republic, were reintroduced after a gap

Vogtlandbahn Siemens Desiro railcars at Weiden forming services to Hof and Schwandorf (Quintus Vosman)
NEW/0585076

of 55 years and these have since been extended further into the Czech Republic to Marianske Lazne. At Schwandorf a connection is made with the Oberpfalzbahn, jointly owned by Regentalbahn (via its RBG subsidiary) and DB Regio.

VBG serves 118 stations and halts, 12 of these in the Czech Republic; it operates over 56 route-km of the Czech network.

Traction and rolling stock
In 2005 VBG operated 18 Siemens RegioSprinters and 24 two-car Siemens Desiro railcars. Some of the former are adapted for street-running over the tram network in Zwickau.

NEW ENTRY

Volkswagen

Volkswagen Transport GmbH & Co OHG (VAG)
Porschestr 102, Südkopf-Center, D-38436 Wolfsburg
Tel: (+49 5361) 26 30   Fax: (+49 5361) 26 34 10

Key personnel
Managing Director: Dipl-Kfm Johannes M Fritzen
Operations Manager: Dipl-Ing Gerd Schönitz

Gauge: 1,435 mm
Track length: 174 km

Organisation
Industrial railway of the Volkswagen automobile plant.

Freight operations
In 1999 5.7 million tonnes were carried.

Traction and rolling stock
The fleet comprises 12 diesel locomotives and 295 freight wagons.

Wanne–Herner Eisenbahn und Hafen (WHE)

Wanne–Herner Eisenbahn und Hafen GmbH (WHE)
Am Westhafen 27, D-44653 Herne
Tel: (+49 2325) 78 80   Fax: (+49 2325) 78 84 30
e-mail: info@whe.de
Web: http://www.whe.de

Key personnel
Managing Director: Betr wirt Karl-Heinz Wick
Manager: Friedhelm Unger
Operations Manager: Ludwig Funke

Gauge: 1,435 mm
Route length: 13.7 km

Organisation
WHE connects with DB at Wanne Übergabebahnhof and also connects with Ruhrkohle AG. It owns and operates a canal port for coal traffic.

Freight operations
The main commodity carried is coal. WHE serves several collieries and a power station. Traffic statistics for 2003 were 5.8 million tonnes and 21.6 million tonne-km.

Traction and rolling stock
The WHE fleet comprises eight diesel locomotives and 131 freight wagons.

UPDATED

Westerwaldbahn (WEBA)

Westerwaldbahn GmbH
Bindweide, D-57520 Steinebach
Tel: (+49 2747) 922 10   Fax (+49 2747) 92 21 20
e-mail: info@westerwaldbahn.de
Web: http://www.westerwaldbahn.de

Key personnel
Operations Manager: Dipl-Ing Horst Klein

Gauge: 1,435 mm
Route length: 31 km

Organisation
WEBA is operated directly by the county council running freight services on the Scheuerfeld–Bindweide–Oberdreisbach (17 km) and Bindweide–Rosenheim (3 km) lines. Since May 1998 the Westerwaldbahn has also operated freight services on behalf of DB Cargo on the Altenkirchen–Raubach line and is expected to assume ownership of this line in the future. The Betzdorf–Daaden line (10 km) was taken over from DB in 1994 and is operated for passenger traffic.

Passenger operations
Traffic statistics for 1999 were 1.5 million passengers and 15.1 million passenger-km. This includes bus operations with a fleet of 20 buses.

Freight operations
82,000 tonnes were carried in 1997.

Traction and rolling stock
The WEBA fleet comprises one VT 628 dmu and one VT23/24 railcar for schools traffic, five diesel locomotives, nine freight wagons and one driving trailer.

Westfälische Landes-Eisenbahn (WLE)

Westfälische Landes-Eisenbahn GmbH
Krögerweg 11, D-48155 Münster
Tel: (+49 251) 627 00   Fax: (+49 251) 627 02 22

Key personnel
Managing Director: Dr Ing Eberhard Christ
Rail Operations Manager: Dipl-Ing Hansüdiger Fritz
Local Operations Manager: Reiner Maier

Gauge: 1,435 mm
Route length: 120 km

Organisation
WLE is a subsidiary of Westfälische Verkehrsgesellschaft mbH (WVG) (qv). Its main line runs from Münster via Neubeckum and Lippstadt to Warstein (102 km, including 6 km leased from DB). Freight services over DB lines are also operated in the Lippstadt area. Main traffic is stone from quarries at Warstein to cement works in the Beckum area. WLE is also responsible for operating the railway divisions of Regionalverkehr Münsterland and Ruhr-Lippe Eisenbahn (qv).

Freight operations
Traffic statistics for 1997 were 1.3 million tonnes and 78.1 million tonne-km, compared with 78.4 million tonne-km in 1996.

Traction and rolling stock
The WLE fleet comprises 17 diesel locomotives and 83 freight wagons.

Westfälische Verkehrsgesellschaft mbH (WVG)

Westfälische Verkehrsgesellschaft mbH
Krögerweg 11, D-46155 Münster
Tel: (+49 251) 627 00   Fax: (+49 251) 627 02 22

Key personnel
Managing Director: Dr Ing Eberhard Christ
Deputy Managing Director:
   Dipl-Kfm Dieter Eichner

Organisation
WVG is a holding company for several bus companies and local railways owned by the Westfalen provincial administration. At present it owns and operates the following railways:
   Regionalverkehr Münsterland GmbH
   Ruhr-Lippe Eisenbahn GmbH (qv)
   Westfälische Landes-Eisenbahn GmbH (qv).

Wismut

Wismut GmbH – Anschlussbahn
Jagdschänkenstr 29, D-09117 Chemnitz
Tel: (+49 371) 81 20

Gauge: 1,435 mm
Track length: 57 km

Organisation
Wismut is the industrial railway of the former Soviet-German uranium mining company. Mining operations have ceased. The railway is still used in connection with site restoration work.

Freight operations
Traffic statistics for 1997 were 625,000 tonnes and 14.6 million tonne-km.

Traction and rolling stock
The fleet comprises nine diesel locomotives and 70 freight wagons.

Württembergische Eisenbahn-Gesellschaft (WEG)

Württembergische Eisenbahn-Gesellschaft mbH
Seewiesenstrasse 21-23, D-71334 Waiblingen
Tel: (+49 7151) 303 80 11   Fax: (+49 7151) 303 80 19
Web: http://www.connex.gruppe.de

Key personnel

Managing Director: Wolf-Dieter Deuschle

Gauge: 1,435 mm
Route length: 86 km

Organisation

WEG is a subsidiary of Connex Regiobahn GmbH. It owns and operates the following local railways in Württemberg:

Nebenbahn Ebingen–Onstmettingen (Talgangbahn)
Nebenbahn Gaildorf–Untergröningen (Obere–Kochertalbahn)
Nebenbahn Korntal–Weissach (Strohgäubahn)
Nebenbahn Nürtingen–Neuffen (Tälesbahn)
Nebenbahn Vaihingen–Enzweihingen (Stadtbahn).

WEG also operates the Schönbuchbahn (Böblingen–Dettenhausen) and the Wieslauftalbahn (Schorndorf–Rudersberg) for the respective local authorities. Passenger trains of WEG also operate over DB lines.

Passenger operations

Certain trains of the Strohgäubahn run through to Stuttgart-Feuerbach over DB tracks to provide better connections with the Stuttgart S-Bahn. WEG carries around 4.3 million passengers annually.

Traction and rolling stock

The fleet comprises 22 diesel railcars, including eight RegioShuttle units and 11 trailers.

Ghana

Ministry of Transport & Communications

PO Box M38, Ministry Branch Post Office, Accra
Tel: (+233 21) 66 64 65   Fax: (+233 21) 66 71 14

Key personnel

Minister: E K Salia
Deputy Ministers: P M G Griffiths, A K Peasah
Chief Director: Dr W A Adote
Director, Planning: E A Kwakye
Director, Administrative: G P Ansah
Planning Officer, Railways: P Azumah

Ghana Railway Corporation (GRC)

PO Box 251, Takoradi
Tel: (+233 31) 21 81; 25 05   Fax: (+233 31) 237 97

Key personnel

Managing Director: M K Arthur
Deputy Managing Directors
  Engineering: Raymond Afeke
  Administration and Operations: S K Agboletey
Financial Controller: T Cofie
Personnel and Administrative Manager:
  J Y Domson-Appiah
Chief Civil Engineer: S Barnes
Chief Mechanical and Electrical Engineer:
  A A Edoh
Chief Signalling and Telecommunications
  Engineer: J Obeng
Traffic Manager: E R O Quaye
Controller of Supplies: G Nuamah
Chief Internal Auditor: R K Johnson
Principal, Central Training Institute: K Owusu-Adjei
Public Relations Officer: J Abaka-Amuah

Gauge: 1,067 mm
Route length: 953 km

Political background

In recent years traffic on GRC has declined as heavy road transport has taken much of its traffic. A problem of overstaffing was addressed by a series of redundancies announced in 1993, the staff complement being reduced to around 4,500 by 1995.

A major policy review, the Railway Policy Reform and Restructuring Study, was undertaken by foreign consultants in 1995. This study proposed a range of substantial changes in GRC's organisational structure, financial restructuring, staff reduction to 1,500 and privatisation of non-core business activities.

Organisation

The railway network is in the form of a letter 'A', the apex being at Kumasi and the two feet at the port of Takoradi and at Accra; the chord connects Huni Valley and Kotoku on the two legs. Branch lines run to Sekondi, Prestea, Kade, Awaso and Tema.

Passenger operations

GRC operates daily passenger trains between Accra and Kumasi, in the north, and three services daily from Kumasi to Takoradi, in the southwest. The Accra–Takoradi direct service was suspended in 1998 due to heavy road competition. With the exception of two Kumasi–Takoradi return trains, most services are overnight trains conveying both seating and sleeping accommodation. Local services are operated between Accra and Koforidua and on the Dunkwa–Awaso branch. The long-awaited reopening of the Accra–Tema suburban line had not occured by mid-1999.

Freight operations

Freight traffic is mainly export-oriented and its main commodities are cocoa, timber, bauxite and manganese; the last two commodities together account for more than 90 per cent of the total freight tonnage. GRC is also engaged in haulage of cement and other imports from the coast to the interior, as well as agricultural produce from sources to marketing centres.

Some 715,000 tonnes were carried in 1994 for 125.7 million tonne-km. The German consultancy DE-Consult has identified considerable potential traffic which would be available with a more reliable operation.

All operations are hampered by the malfunctioning signalling system, poor quality infrastructure and badly maintained rolling stock. The relatively high derailment frequency (one every five days on average) is a natural outcome of some of these problems.

Intermodal operations

A potential intermodal terminal site has been identified at Fumisua, near Kumasi, for movement of container traffic from and to Takoradi and Tema ports. At the present time, however, there are no container handling facilities on GRC and the rail connnection to Tema was severed in 1995 as a result of road construction.

New lines

In 1990 Japanese aid was secured for construction of a 96.5 km line from the port of Tema to Akosombo, on Lake Volta, a valuable link in an intermodal transport chain to the north of the country and to Burkina Faso. However, when the link between Accra and Tema was severed in 1995 as a result of road construction, the future of the project was put in doubt.

In the longer term GRC is hopeful of building two other lines. One would run from Awaso to Sunyani, and the other from Bosuso to Kibi. Both would have exploitation of bauxite deposits as their objective.

Improvements to existing lines

In the late 1980s, the railway completed a rehabilitation project on the Kumasi–Takoradi Western line (including the Awaso branch), which is the main export trunk route. Foreign exchange inputs for the project, totalling US$45 million, came mainly from the World Bank and African Development Bank, but also from donor countries (France, Switzerland and the UK). The project covered track rehabilitation, replacement of signalling (using solar power) and telecommunication equipment, rehabilitation of locomotives and rolling stock, equipment of wagons with roller bearings, workshop modernisation, setting up of a Central Training Institute and staff training. Unfortunately, by 1995 much of the rehabilitated equipment had fallen into disrepair, particularly the solar-powered signalling, leading to serious operational problems on the route.

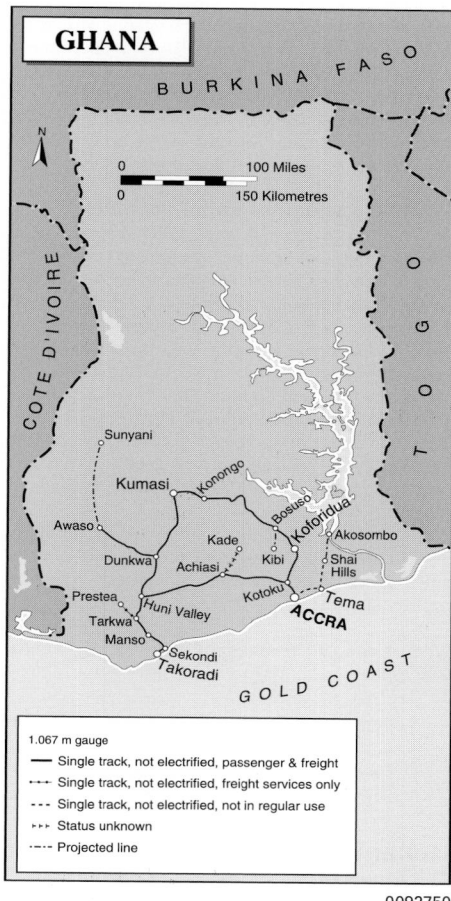

0092750

In 1988 the Italian government granted US$30 million towards similar rehabilitation of GRC's Central and Eastern lines, and also towards a microwave radio network. With further World Bank support of US$9 million, work on the 306 km Eastern line from Accra to Kumasi was started. The available funds also provided for attention to the 182 km Kotoku–Huni Valley section of the Central line from Accra to Takoradi.

Ghana's railway network has excessive curvature in several sections, and this tends to reduce line capacity. For example, between Kumasi and Takoradi 32 per cent of the 270 km Western line consists of curved track. Replacement of existing 39.7 kg/m rail by 44.7 kg/m rail has been carried out on the Manso–Huni Valley section; despite possessing suitable equipment, only a very small section of GRC track has been continuously welded. In 1995 German consultants DE-Consult carried out an in-depth technical and economic survey of the Western line, which identified considerable traffic potential and recommended a programme of investment in new air-braked freight wagons and in technical rehabilitation measures, though falling short of proposing extensive realignment.

Traction and rolling stock

In 1995 GRC operated 38 diesel locomotives and 10 diesel-hydraulic shunters, 139 passenger coaches and 810 wagons.

All of the passenger coaches currently in service were built in 1986 by the Görlitz and Bautzen factories in East Germany. These include buffet, first and second class day and sleeping cars and five tourist saloons.

With the aid of finance from the French Central Bank for Economic Co-operation, GRC took delivery of nine Type AD26C 1,850 kW diesel locomotives of the 2601 class from GEC Alsthom in 1993. The units are mainly employed on the Western line for manganese and bauxite haulage. The contract provided for GEC Alsthom to maintain the locomotives for four years and also to re-equip the railway's locomotive depot at Takoradi. In early 1998 French funding agency CFD approved a FFr 12.4 million grant to extend this maintenance contract until 1999.

In 1991 Ghana secured a US$96 million credit to assist the second Transport Rehabilitation Project. Besides infrastructure work, purchase and repair of locomotives and rolling stock, and procurement of technical and consultancy assistance would be undertaken with the money. During 1995 three Henschel locomotives of the 1661 class were delivered, together with the first tranche of 100 covered freight wagons equipped with Scheffel self-steering bogies. A total of 14 new General Motors locomotives were delivered in 1996.

Signalling and telecommunications

Under a World Bank project, GRC changed from the previous semaphore system to colourlight signals powered by solar panels in all 42 stations on the Western line. Regrettably, by 1995, lack of maintenance and suitable spare parts meant that the only functioning section was the short stretch between Takoradi and Harbour Junction and between Angu and Tarkwa. On other sections of the Western line, operation is by a ticket system based on the stationmaster obtaining telephonic permission to proceed into a section from the next station. This system is unreliable and leads to excessive delays.

The telecommunications network is similarly only partly functioning, with the Takoradi control office being regularly out of contact with up-country stations.

Type of coupler in standard use: No 2 Alliance/ABC type

Diesel locomotives

Class	Wheel arrangement	Power kW	Speed km/h	Weight tonnes	No in service	First built	Mechanical	Builders Engine	Transmission
2601	Co-Co	1,700	80	96.5	8	1993	GEC Alsthom	CAT 3606L	*E* GEC Alsthom
1851	Co-Co	1,380	96	84	2	1969	English Electric	EE 12CSVT	*E* English Electric
1661	Co-Co	1,119	80	86	3	1995	ABB Henschel	GM 12-645E	*E* ABB Henschel
1651	Co-Co	1,119	96	85	10	1978	Thyssen-Henschel	GM 645E	*E* Thyssen-Henschel
1401	Co-Co	966	100	83	3	1959	Henschel	GM 567C	*E* Henschel
751	Bo-Bo	508	88	54	2	1954	English Electric	EE 6 SRKT	*E* English Electric
721	Bo-Bo	501	80	57	4	1986	Daewoo	MTU 8V396TC12	*E* Toshiba SE-223
701	Bo-Bo	441	64	60	6	1982	Brush	Rolls-Royce DV8TCE	*E* Brush
541	C (0-6-0)	447	45	43	10	1975	Thyssen-Henschel	Henschel 12V 1516A	*H* Voith L4, 4V2

Type of brake in standard use: Locomotives, air; carriages and wagons, vacuum

Track
Standard rail, type and weight

Western line
Takoradi–Manso: RBS 39.7 kg/m
Manso–Huni Valley: BSA 44.7 kg/m
Huni Valley–Kumasi: RBS 39.7 kg/m

Eastern line
Kumasi–Konongo: BS 29.8 kg/m
Konongo–Juaso: BA 'A' 39.7 kg/m
Juaso–Osino: RBS 39.7 kg/m
Osino–Tafo: BS 29.8 kg/m
Tafo–Accra: RBS 29.8 kg/m

Central line
Huni Valley–Nyinase: RBS 39.7 kg/m
Nyinase–Twifu: BS 'A' 39.7 kg/m
Twifu–Akenkausu: BS 29.8 kg/m
Akenkausu–Achiasi Junction: BSA 39.7 kg/m
Achiasi Junction–Kotoku Junction: RBS 39.7 kg/m
Prestea branch line: RBS 29.8 kg/m
Awaso branch line: ASCE 37.2. kg/m
Joints: 4-hole fishplates
Crossties (sleepers): Standard steel; and wood 127 × 254 × 1,981 mm
Spacing: 1,365/km steel, 1,602/km wood

Rail fastenings
Wood sleepers: Dogspikes, Macbeth spike anchors, Type T3 (UK) elastic rail spikes. Elastic rail spikes Type ES18 and DS18 (Germany)
Steel sleepers: Keys, ABK clips
Filling: Mainly crushed granite and manganese ore, some gravel

Max curvature
Takoradi-Kumasi 8° 40' = radius of 202 m
Kumasi-Accra 8° 40' = radius of 202 m
Central line 8° 40' = radius of 202 m
Prestea branch 17° = radius of 103 m
Awaso branch 6° = radius of 291 m
Max gradient: 1.25% = 1 in 80; except Prestea branch 2.5% = 1 in 40
Longest continuous gradient: 10 km with ruling grade of 1.25% and max curves of 8° 40'
Max altitude: 286 m near Kumasi

Max axleloading
Takoradi-Kumasi-Accra 16 tonnes
Central line 16 tonnes
Prestea branch 12½ tonnes
Awaso branch 16 tonnes
Gauge widening on sharpest curve: 13 mm (theoretical)
Superelevation on sharpest curve: 89 mm (theoretical)
Rate of slope of superelevation: 13 mm per rail length

Greece

Hellenic Ministry of Transportation & Communications

Xenofontos Street 13, GR-10191 Athens
Tel: (+30 1) 325 12 11-19   Fax: (+30 1) 323 90 39
Web: http://www.yme.gr

Key personnel
Minister: Michalis Liapis

Hellenic Railways Organisation (OSE)

Organismos Sidirodromon Ellados
1-3 Karolou Street, GR-10437 Athens
Tel: (+30 1) 524 83 95
Fax: (+30 1) 524 32 90; 524 62 39
Web: http://www.ose.gr

Key personnel
President and Managing Director:
Konstantinos Giannakos

Gauge: 1,435 mm; 1,000 mm; dual-gauge, 1,435 mm and 1,000 mm; 750 mm
Route length: 1,565 km; 961 km; 23 km; 22 km
Electrification: 764 km at 25 kV AC 50 Hz (in progress)
The railway from Athens to the Peloponnese, serving Patras and southern Greece, is metre-gauge.

0525465

GREECE

1.435 m gauge
— Single track
+++ Single track, electrified
---- Single track, projected
++++ Single track, being upgraded and electrified
= Double track
== Double track, being upgraded and electrified
++++ Double track, electrified under construction
--- Not in use
1.000 m gauge
○ Reconstruction stopped
— Single track
∷∷∷ Dual gauge (1.435/1.000 m)
►►► Not in use
750 mm gauge
∷∷∷ Single track
600 mm gauge
····· Single track partly in use

Subsidiaries

Wholly-owned OSE subsidiary companies include:

Ergose

Ergose was established by OSE to manage and implement its investment programme, covering projects such as upgrading and electrification of the Athens–Thessaloníki main line and other major construction projects. Its constitution allows it to provide these services to third parties.

Gaiaose

Gaiaose manages OSE's property interests and activities.

Proastiakos SA

Proastiakos was established to manage train operations in large urban areas. These include the new suburban rail network serving Athens.

Thriassio SA

Thriassio's main responsibilty is operation of the freight centre being developed at Thriassio Pedio to serve the Athens region, covering operation of rail services to and from the site, provision of new rail infrastructure or development of the existing network to serve it and the provision of consultancy services to the freight sector.

Political background

In 2004 Greece was yet to implement EU legislation regarding the separation of train operations and infrastructure. Therefore OSE remains a vertically integrated network.

Organisation

Pending eventual adoption of EU legislation regarding the separation of train operations and infrastructure, in 2003 OSE established an organisational structure based on independent service units. This is headed by six general directorates responsible for: development and planning; infrastructure; administration; technical support; freight services; and passenger services.

Traffic (million)	2000
Passenger journeys	12.3
Passenger-km	1,583
Freight-tonnes	2.2
Freight tonne-km	350

Passenger operations

OSE operates passenger services on its standard gauge main trunk line north from the capital, Athens, to Thessaloníki. From here, an electrified line runs northwest to Idomeni and the border with Macedonia. Services also run east from Thessaloníki via Alexandroúpolis to Pythion, from where there is a link to Turkey, and to Ormenion and on to Bulgaria. To the southwest of Athens, a metre-gauge system runs to the Peloponnese, with lines serving Patras and Kalamata.

Freight operations

One of the most important lines for freight is that from Thessaloníki to Idomeni, on the border with Macedonia.

In collaboration with private sector partners, a new freight facilities complex and intermodal terminal for the Athens region was under construction at Thriassio Pedio in 2004.

New lines

Athens suburban lines

Major work is in progress to create an electrified standard gauge suburban network in Athens. This includes a line from the port of Piraeus via Athens Larissa station and an interchange known as SKA (Acharnes Transportation Centre), in the north of the city, to Eleftherios Venizelos (EV) airport in the east. This section of the project, scheduled for completion in time for the Olympic Games in August 2004, involves construction of 20 km of new line and the upgrading of an existing 10 km. Between Plakentia and the airport, the line will be shared by OSE and metro trains. The line provides a 30-minute journey time between Larissa station and EV.

Subsequent stages of the project provide for a new double-track line from SKA to west to Corinth and on to Kiato, including a new bridge over the

Class H560 electric locomotive preparing to leave Thessaloníki with the 'Hellas Express' service to Belgrade (Colin Boocock)
0583113

Class A470 diesel-electric locomotive leaving Drama with an Athens–Ormenion passenger service (Colin Boocock)
0583114

Hellenic Shipyards dmu at Corinth, on the metre gauge Peloponnese system (Edward Barnes)
0583115

Diesel locomotives

Class	Wheel arrangement	Power kW	Speed km/h	Weight tonnes	No in service	First built	Mechanical	Builders Engine	Transmission
1,000 mm gauge									
A9100	Co-Co	993	96	80.3	11	1965	Alco	Alco 6-251D	E GE 761 A3
A9200	Co-Co	1,200	90	80	13	1967	Alsthom	P PA4-185/V12	E GE Canada
A9400	B-B	2 × 240	90	48	17	1967	Mitsubishi	G-M V8-71N	H Niigata
1,435 mm gauge									
A100	C	485	60	51	27	1962	Krupp	M GT06A-V12	H Voith L27 zub
A150	B-B	520	70	48	11	1973	Faur	M-MB820Bb	H Voith
A170	B-B	560	70	48	6	1978	Faur	M V12MB820Bb	H Voith
A200	Bo-Bo	785	105	64.6	7	1962	Alco	Alco 251 B	E GE
A220	Bo-Bo	795	109	63.5	3	1973	GE	C D398B-V12	E GE
A450	Co-Co	2,015	149	120	20	1973	MLW	Alco V12-251 F	E GE Canada
A470	Bo-Bo	2,100	160	90	35	1997	Adtranz/ Bombardier	MTU	Adtranz
A500	Co-Co	2,650	149	124	8	1974	MLW	Alco V16-251 F	E GE Canada

Abbreviations: C: Caterpillar; M: Maybach; Mek: Mekydro; P: Pielstick

Corinth Canal. North of SKA suburban trains will use the existing line to Inoi, Thiva and Khalkis. The entire project is scheduled for completion in 2008.

Western Railway Corridor

In 2003 OSE launched plans to build an electrified standard gauge network of lines in the west of Greece, including a link to the Peloponnese. Totalling 741 km, the Western Railway Corridor

comprises four main sections: Kalampaka–Ioannina–Igoumenitsa (153 km); Kalampaka–Kozani (113 km); Ioannina–Andirrio (with a line to the port of Platiliyiali, Greece's deepest) (210 km); and a fourth line via a 9.5 km tunnel from Andirrio to Rio, on the Pelopennese, and on to Patras and Kalamata (265 km).

Among the aims of the project are provision of improved transport infrastructure in western Greece and better links via its ports, notably at Igoumenitsa and Platiliyiali, with other EU countries: at present rail connections with the rest of Western Europe are via non-EU countries and suffer either from quality constraints or from the legacy of conflict. These factors have contributed to the Western Railway Corridor being identified as part of the EU's Trans-European Network, with funding earmarked for 2006–14.

It is intended that the new lines will be single-track and generally engineered for 120–160 km/h. In all, there will be 126.9 km of tunnels and 30.1 km of bridges. Tenders for engineering design for the first phase were invited in 2004. Completion of the project is planned for 2014.

Improvements to existing lines
Athens–Thessaloníki upgrading
The principal project, in hand since 1978, has been electrification, doubling and realignment of sections of the Athens–Thessaloníki main line (517 route-km), together with track renewal employing UIC54 continuously welded rail on two-block concrete sleepers, with minimum curve radius of 2,000 m. Some parts of the line have been engineered for 200 km/h running, while on some sections provision is made for future operations at 250 km/h. This axis carries more than half the railway's total traffic.

By mid-2004 some 80 per cent of electrification work had been completed, with the entire project expected to be finished by the end of 2005. Among major works has been the creation of a new 38.8 km double-track alignment to replace the difficult single-track section between Evangelismos and Leptocarya. This was commissioned in January 2004. Thanks to improvements already made, in February 2004 OSE was able to introduce an IC service with a 4 hour 30 minute journey time between Athens and Thessaloníki, compared with the 6 hours it took previously.

Except between Larisa and Plati (134 km), electric signalling will be operative and a modern telecommunications system will cover the whole axis from Athens to Idomeni. There will be electrified double track over the whole route between Athens and Thessaloníki except for Lianokladi–Domokós (65 km) and the project includes driving two new tunnels through the foothills of Mount Olympus, in northern Greece.

The investment programme also includes upgrading the line from Thessaloníki to the Bulgarian border to provide an improved alternative link to Central Europe.

A subsidiary project has seen the 80 km metre-gauge Paleofarsalos–Kalambaka line converted to standard-gauge. The line connects with the Athens–Thessaloníki route.

Thessaloníki suburban/regional lines
In 2004 OSE was implementing improvements to lines radiating from Thessaloníki with the aim of improving suburban and regional services. A major element of this project is full refurbishment of the line from Plati, on the northern end of the Athens–Thessaloníki route, to Kozáni and Florina, covering 235 km in total. The eventual network will also cover lines north to Idomeni, east to Serres and south on the main line to Athens as far as Platamonas.

Peloponnese improvements
An upgrade of the metre gauge Athens–Tripoli–Kalamata line was in progress in 2004 with the aim of cutting the overall journey time from 5 hours 35 min to 4 hours.

Improved port access
OSE plans to improve rail access to ports at Alexandroúpolis, Patras and Thessaloníki.

Traction and rolling stock
At the beginning of 2002 OSE operated 108 standard-gauge and 31 metre-gauge diesel locomotives, 130 standard-gauge and 83 metre-gauge dmu vehicles; 400 standard-gauge and 140 metre-gauge coaches.

For the Athens–Thessaloníki and Thessaloníki–Alexandroúpolis routes, German industry manufactured 12 four-car 160 km/h intercity dmus in 1989. In 1994, a further eight five-car

Class H560 electric locomotives handle freight traffic on the line to Macedonia (Artemis Klonos)
0524927

LEW-built diesel trainsets, such as this example at Volos, are the mainstay of OSE intercity services (Edward Barnes)
0524923

Diesel railcars or multiple-units

Class (running numbers)	Car per unit	Motor cars per unit	Motored axles/car	Power/car kW	Speed km/h	No in service	First built	Mechanical	Builders Engine	Transmission
1,435 mm gauge										
560	2	1*	2	—	115	17	2003	Stadler/ Hellenic Shipyards	MTU	E
601	4	2	4	1,180	160	12	1989	LEW	MTU 396TC13	E S
651	5	2	4	1,180	160	8	1995	Adtranz	MTU 396TC13	E S
660	2	2	2	275	120	8	2004	Siemens	–	M
91	3	2	2	895	140	6	1976	G-M	G-M-SEMT	HD Voith L520-RO
701	2	2	4	305	120	12	1990	MAN	MAN D2842/ME	H Voith T320RZ
1,000 mm gauge										
–	2	1*	2	–	100	12	2004	Stadler/ Hellenic Shipyards	MTU	E
6461	3	1	4	850	100	11	1983	G-M	P 8PA4-185V	HD Voith KB380/1
6501	3	2	4	398	–	10	1991	HS	–	–
6521	2	1	2	305	–	10	1991	HS	–	–
750 mm gauge										
3001–3003 2	1	2	260	40	3	1959	Billard	MerY15-536	E	
3004–3006 2	1	2	375		3	1967	Decauville	Mer MB836-B/L	E	

Abbreviations: G-M: Ganz-Mávag; HS: Hellenic Shipyards; Mek: Mekydro; Mer: Mercedes; P: Pielstick
* powered centre module.

Electric locomotives

Class	Wheel arrangement	Power kW continuous/ one-hour	Speed km/h	Weight tonnes	No in service	First built	Builders Mechanical	Electrical
H560	Bo-Bo	5,000	200	80	30*	1997	Krauss-Maffei	Siemens

* Includes 24 on order in 2004.

trains were ordered from AEG Bahnsysteme (now Adtranz) for these lines. Assembly took place at the former AEG Hennigsdorf works. In 1995, eight extra trailer cars were ordered from DWA's Bautzen factory to make the second series into five-car sets.

Adtranz built 26 Class A470 (now Class 220) 2,100 kW three-phase diesel-electrics, delivery commencing in 1998. The design of these locomotives allows for eventual conversion to electric traction. Ten additional locomotives of the same design were delivered by Bombardier in 2003–04.

In late 1997 Siemens/Krauss-Maffei commenced delivery of six Class H560 EuroSprinter-based 5,000 kW electrics for hauling express passenger and freight trains over the newly electrified section from Thessaloníki to the Macedonian border. In March 1998 a further 24 of these locomotives were ordered for services on the Athens–Thessaloníki main line.

For local services, OSE ordered 29 GTW 2/6 Class 560 low-floor diesel railcars from a consortium of Stadler and Hellenic Shipyards: 17 for the standard gauge system and 12 for the metre gauge network. Delivery of the standard gauge units commenced in 2003, while the metre gauge vehicles were to follow in 2004.

The dmu fleet was further supplemented in 2004 by the arrival of eight Siemens Desiro two-car units, designated Class 660 by OSE.

OSE's first emus were also ordered in March 1998, with a contract for 20 five-car units for Athens suburban services to be supplied by Siemens Transportation Systems in co-operation with Hellenic Shipyards. The 160 km/h aluminium-bodied articulated vehicles will be based on Siemens' Desiro design.

For the metre-gauge system, a total of 10 Class 6501 air conditioned three-car trainsets for intercity operation have been built by Hellenic Shipyards. Two of the cars in each unit have each bogie powered by a 398 kW 2,100 rpm engine. Ten two-car units were subsequently supplied from the same source and are designated Class 6521.

Recent coaching stock acquisitions include 185 intercity coaches ordered in 1998 from Siemens and constructed by local builders and 79 second class intercity vehicles supplied by Bombardier in Hungary from 2003. Refurbishment of an additional 107 existing coaches and of vehicles acquired secondhand from DB AG, Germany has also been undertaken.

In 2004 refurbishment was in progress of the eight MLW-built Class A500 diesel-electrics, with improvements to both engine and electronics.

Electrification

Priority has been given to the 25 kV 50 Hz AC electrification of the 587 km Athens–Thessaloníki–Idomeni main line, which has the backing of the European Union. First section to be tackled was the 76 km of single line from Thessaloníki to the Macedonian border at Idomeni, where track was renewed and curvature eased. This route carries three times the freight of the Athens–Thessaloníki line. Commissioning took place in 1997. In 1996 a consortium led by ABB and including Adtranz (now Balfour Beatty Rail) won a US$145 million contract to electrify the Athens–Thessaloníki route. This is scheduled for completion in 2005.

Also included in the project, and forming part of a plan to create an Athens electrified suburban network, is upgrading and electrification of the branch from Inoi, on the main Athens–Thessaloníki line, to Khalkis, due to be completed in 2006.

Coupler in standard use: 1,435 mm gauge, UIC 520–521

Brake in standard use: Air, mostly Knorr

Track

Rail: 1,435 mm gauge, UIC 50, 54; narrow-gauge, 31.6 kg/m

Sleepers: 1,435 mm gauge, reinforced concrete twin-block (Vagneaux type) 680 × 290 mm; steel 2,550 × 260 mm; timber 2,600 × 250 mm. Narrow-gauge, steel and timber

Fastenings: 1,435 mm gauge, RN and Nabla for concrete sleepers; K direct fastenings for wood or steel

Min curve radius: 1,435 mm gauge, 300 m; narrow-gauge, 110 m

Max gradient: 1,435 mm gauge, 2.8%; narrow-gauge, 2.5%

Max permissible axleload: 1,435 mm gauge, 20 tonnes; narrow-gauge, 14 tonnes

Max permissible speed: 90–100 km/h; 120 km/h parts of Athens–Thessaloníki main line

UPDATED

Guatemala

Ministry of Communications, Infrastructure and Housing

8th Avenue and 15th Street, Zona 13, Guatemala City
Tel: (+502 2362) 605 15

Key personnel

Minister: Arquitecto Eduardo Castillo Arroyo
Vice-Ministers: Ing Federico Moreno;
  Ing Roberto Díaz; Lic Lilia del Valle;
  Arquitecto José Luis Gándar

UPDATED

Bandegua Railway

Cia de Desarrollo Bananero de Guatemala Ltd
Edificio La Galeria 5° Nivel, 7 Avenue 1444, Zone 9, Guatemala City
Tel: (+502 2) 34 03 78   Fax: (+502 2) 32 21 52

Key personnel

General Manager: G K Brunelle
Director of Engineering: E Casado
Railroad Superintendent: G Aguirre
Mechanical Superintendent: M Pérez

Gauge: 914 mm
Route length: 102 km

Traction and rolling stock

The company operates 11 diesel locomotives, 17 diesel railcars, seven passenger coaches and 101 freight wagons.

Ferrovías Guatemala (FEGUA)

Compañia Desarrolladora Ferroviaria SA
24 Avenida 35-91, zona 12, Guatemala City
Tel: (+502 2) 485 01 30   Fax: (+502 2) 485 01 35

Key personnel

General Manager: Renato Fernández Ravelo
Vice-President, Operations: William J Duggan

Gauge: 914 mm
Route length: 784 km

Political background

Guatemala's railway was nationalised in 1968. At this time, the railway had significant flows of passengers, coffee and bananas, although a severe lack of investment in the ensuing 20 years effectively eroded this traffic base. From a 1980s yearly average in excess of 500,000 tonnes of freight carried, FEGUA was down to little more than 100,000 tonnes per year by the mid-1990s. All passenger traffic ceased in 1994, and the entire network closed in March 1996, just as the government had begun privatisation.

In 1997, US-based Railroad Development Corporation beat one other firm to win a 50-year concession to run the system, paying the government US$10 million, plus 5 per cent of revenues for the first two years of operation and 10 per cent annually thereafter. A further US$10 million will need to be spent on upgrading the main Guatemala City–Puerto Barrios line. RDC established a Guatemalan affiliate, Compañia Desarrolladora Ferroviaria SA (CODEFE) to prepare for a resumption of operations. The concession was signed in October 1997 and endorsed by Guatemala's congress in April 1998. Operations under new ownership commenced in 1999. The company trades as Ferrovías Guatemala.

The relocation of facilities in Guatemala City will involve RDC abandoning the present Central station in favour of a new container terminal in the industrial zone. The company also expects to generate extra income by selling right-of-way to pipeline, electricity, and fibre optic communications concerns.

Passenger operations

No scheduled passenger services are operated and Ferrovías Guatemala has no plans to introduce such trains. However, charter trains for group travel are run and the first of these was operated in March 2000 as part of a five-day round-trip between Guatemala City and Puerto Barrios. Traction was provided by a Baldwin steam locomotive.

Freight operations

Through freight operations resumed on the 318 km line between Guatemala City and Puerto Barrios in November 1999. Two or three trains daily run over the refurbished line, journeys taking about 15 hours in each direction. Cement and other construction materials are the main commodities carried.

New lines

The governments of Guatemala, Canada, El Salvador, Honduras, Mexico and the USA, together

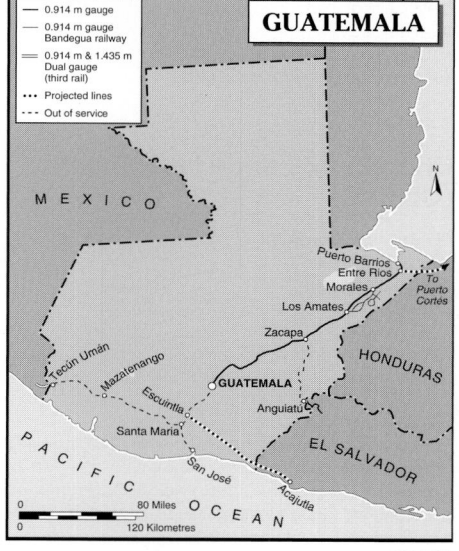

0114290

with the World Bank and the Interamerican Bank for Development have undertaken studies into a new international rail link to connect Guatemala, El Salvador and Honduras with the NAFTA states. In Guatemala, the scheme would necessitate rebuilding the existing line from Zacapa to Anguiatú and San Salvador, which is currently out of use, providing a link from El Salvador to Honduras.

Improvements to existing lines

Despite difficulties such as stolen rail and washouts due to Hurricane Mitch, Ferrovías Guatemala has rehabilitated the Atlantic line between Puerto Barrios and Guatemala City. In a second phase, rehabilitation is planned of the Pacific route from Esquintla to Retalhuleu and Tecún on the Mexican border. This will serve as a feeder route into Mexico for containers. It is intended to establish a transhipment point for transferring this traffic from the 914 mm gauge of Guatemala to the Mexican standard-gauge network at the Mexican border station of Ciudad Hidalgo. Containers for Guatemala City will be unloaded at Esquintla. In a later phase, the Esquintla–Guatemala City line may be restored to service.

Traction and rolling stock

Prior to the 1996 cessation of operations the FEGUA fleet totalled 28 General Electric diesel and two steam locomotives, 68 passenger coaches and 1,523 wagons. Recent additions are 10 Type MX620 1,492 kW locomotives from Bombardier; serviceable examples of this type and of 15 General Electric U10s locomotives will form the traction for RDC's revival of operations.

One of several Baldwin steam locomotives has been returned to traffic for passenger charter services and the restoration of two more was in hand for similar work in 2000.

Track

Rail: 30 kg/m
Sleepers: Timber, spaced 1,800/km in plain track, 1,000/km in curves
Max curvature: 6°
Max gradient: 3.3%
Max permissible axleload: 20 tonnes

Prior to the granting of an operating concession to the Railroad Development Corporation, an MLW diesel locomotive heads an eastbound freight train near Agua Caliente   0016434

Guinea

Ministry of Trade, Transport & Tourism

PO Box 715, Conakry

Key personnel
Minister: I Sylla

For map see entry for **Senegal**.

Chemin de Fer de la Guinée (ONCFG)

PO Box 581, Conakry

Key personnel
Director: M K Fofana

Gauge: 1,000 mm; 1,435 mm
Route length: 662 km; 175 km

Organisation

The railway links Kankan and the limit of navigability of the Upper Niger at Kouroussa with the port of Conakry. Crossing the Fouta Djalon mountains on gradients as steep as 2.9 per cent, and with curves of 100 to 150 m radius, it was lightly laid, without ballast, and has consequently deteriorated badly with rising axleloads. Some renovation of track and rolling stock has been carried out under foreign aid programmes.

The government's aim is to reconstruct the entire railway, regauging it to standard gauge. As part of this scheme, a 134 km standard-gauge line was built in 1974 with technical and financial aid from the USSR. Running from Conakry to Kindia, the line parallels the metre-gauge line for almost its entire length and is used exclusively for the transport of bauxite from mines at Kindia. The line is operated jointly by ONCFG and the Société de Bauxite de Kindia (SBK), its sole user. From Dabola, at Km 442 on the metre-gauge line, a branch line of some 140 km runs north to bauxite mines near Tougué. This line is jointly operated by ONCFG and the Société de Bauxite Dabola-Tougué (SBDT).

Passenger operations

Passenger operations using railcars between Conakry and Kankan ceased towards the end of the 1980s due to a lack of spare parts for rolling stock and poor track conditions. A limited passenger service has been run between Conakry and Kindia by attaching a passenger coach to a freight train twice-weekly but in many cases passengers are simply carried in a freight wagon. The total number of passengers carried in 1998 totalled around 50,000.

Freight operations

Freight operations are confined to the 442 km Conakry–Dabola section of the main line; east of Dabola there was in 2000 no traffic due to the poor state of the line's infrastructure. One daily train conveys bauxite from the mines at Tougué to Dabola, where the ore is transferred to metre-gauge wagons for movement to the port at Conakry. While bauxite accounts for around 80 per cent of traffic, iron ore, uranium, diamonds, graphite, chalk and gold are also transported. From Kindia to Conakry only bauxite is carried. Trains on this line are more frequent, and loading also takes place at Débélé, about midway on the route. Loadings from Tougué and Kindia total some 4 million tonnes annually.

New lines

With assistance from Slovakian Railways, the 140 km Dabola–Tougué line was completed in 1998, using second-hand rails from the Slovakian network. Additional projects include rebuilding the entire Kindia–Dabola–Kankan main line as a standard-gauge system. New lines from Kankan towards the border with Burkina Faso in the north and Ivory Coast in the south are also projected, although no steps have been taken towards their construction. Most of these projects date back to 1958 to 1984. During this period, in 1974, a new station was built at Simbaya, on the outskirts of Conakry, but never used for passengers. Adjacent USSR-built workshops are responsible for the maintenance of standard-gauge rolling stock.

Traction and rolling stock

The metre-gauge traction fleet consists of 30 main line diesel locomotives and several small shunters, although in 1999 only two Alsthom locomotives built in 1982 and two General Motors machines dating from 1986 were serviceable, together with a pair of two-axle French-built shunters. Two former Czech Railways Class 710 locomotives were acquired in 1998 and converted to metre gauge at Nymburk workshops in the Czech Republic. These are exclusively used on services to Dabola.

Standard-gauge locomotives for the line to Kindia consist of 13 USSR-built Type TEM 2T six-axle machines acquired in 1998. In July 1996, the Czech company Prime International obtained 18 former Czech Railways Class 770 and 771 diesel locomotives and 35 Falls-type freight wagons from Slovakian Railways originally purchased for a subsequently cancelled contract with Islamic Iranian Republic Railways. One of the locomotives and several of the wagons were lost in a storm during shipment: those vehicles which reached Guinea were in 2000 stored out of use at Simbaya. Three of the locomotives eventually entered service on the Dabola–Tougué line.

None of ONCFG's 16 metre-gauge railcars was serviceable at the end of 1999 and scrapping of most was reported to be imminent. Of 20 bogie passenger coaches, no more than five remained usable. Some 200 of the fleet of 500 freight wagons were serviceable, some of which were tank wagons that included eight-axle examples imported from Congo during the 1980s.

On the standard-gauge line to Kindia, most of the freight wagons are four-axle vehicles built in the USSR. Wagons regauged for metre-gauge operation are used on the Dabola–Tougué line.

Industrial railways

From Conakry, the metre-gauge Chemin de fer Conakry–Fria (CFCF) line runs parallel to the ONCFG main line for 30 km before diverging to the north to Fria, 142 km from Conakry. Owned by SBK and opened in 1960, the line serves Fria bauxite mines and its aluminium plant. It is single-track, with 46 kg/m rails on steel sleepers. Rolling stock comprises three Alsthom 820 kW diesel-electric locomotives and 61 50 tonne wagons. The line carries some 900,000 tonnes of bauxite annually.

Also operated by SBK is the Chemin de fer de Boké (CFB), located around 250 km northwest of Conakry. This standard-gauge line runs 136 km from the port of Kamsar via Boké to Sangaredi. Moving more than 12 million tonnes of bauxite annually from the mines at Sangaredi, at the foot of the Peul hills, it is the country's most important carrier of this commodity. Built by a consortium of European contractors and opened in 1973, the line has UIC 60 kg/m continuously welded rail laid on steel sleepers. As well as bauxite, around 50,000 tonnes of other merchandise and 150,000 passengers are carried annually in five trains a week in each direction. The railway owns 17 US-built diesel locomotives, 460 ore wagons, 39 other freight wagons and three passenger coaches.

Honduras

Ministry of Communications, Public Works and Transport

Barrio La Bolsa, Comayaguela
Tel: (+504) 33 76 90   Fax: (+504) 33 92 27

Key personnel
Minister: G Aparicio

Honduras National Railway (FNH)

Ferrocarril Nacional de Honduras
PO Box 496, San Pedro Sula
Tel: (+504) 53 40 80; 53 18 79
Fax: (+504) 52 80 01

Key personnel
General Manager: Diego Zúniga Muñoz
Directors
   Administration and Finance: J O Garcia
   Commercial and Marketing Services:
      A Ramos Carrazco
   Mechanical: F Maury Martell
   Operations: J Manuel Benegas

Gauge: 1,067 mm; 914 mm
Length: 279 km; 420 km

Organisation
The existing FNH system includes 420 km of 914 mm gauge track which was formerly part of the Vaccaro and Standard Fruit system, which was nationalised in 1976. All lines are out of use but not dismantled. The former Tela Railroad lines, 164 km of 1,067 mm gauge, formerly leased to and operated by United Brands Company, were put under FNH control in 1983, but operated as a separate entity until the mid-1990s. About 90 km of this system remain serviceable. The remaining 115 km of 1,067 mm gauge track belongs to FNH and consists mainly of the Puerto Cortés–San Pedro Sula–Potrerillos main line and many smaller branches, of which that from Chamelecón to La Lima is the most important. Most of these lines are in operation or at least extant.

Damage to the railway by Hurricane Mitch principally affected the lines east of Tela and around La Ceiba, where all infrastructure was destroyed. Lines south of Puerto Cortés towards San Pedro Sula were affected to a lesser extent. After some minor repairs, freight services resumed during 1999 between Puerto Cortés and San Pedro Sula and other inland points.

Passenger operations
After a complete suspension of passenger services in mid-1998, no trains ran for almost 18 months. Early in 2000, FNH reinstated regular scheduled passenger services on the 76 km Puerto Cortés–Tela line with a round-trip every Friday and Sunday.

Freight operations
Most traffic runs between Puerto Cortés and San Pedro Sula and consists principally of fruit and other agricultural products. Some fertilisers and construction materials are also transported.

New lines
Nothing has resulted from a project unveiled in 1983 to construct up to 350 route-km of new lines. Work on the planned San Pedro Sula–Yoro line was initiated but halted, but this and other schemes, such as the Puerto Castillo–Sonaguera and Arenal–Rio Bonito lines, remain on the national agenda awaiting an improvement in economic conditions or private investment funds. Some of the projects form part of a scheme for a pan-Central American railway (see entry for Guatemala).

Traction and rolling stock
The 1,057 mm system is operated with two steam and 34 diesel locomotives, 37 passenger coaches, 22 diesel railcars and 1,960 wagons. The 914 mm fleet comprises eight diesel-electric and two diesel-mechanical locomotives, 17 diesel railcars, 16 passenger coaches and 530 wagons. Operations on the 1,067 mm Tela system are conducted with 28 diesel locomotives, 18 railcars, 70 passenger coaches and 1,324 freight wagons.

Hungary

Ministry of Economy and Transport

Pf 111, H-1880 Budapest Honvéd utca 13-15, H-1055 Budapest V
Tel: (+36 1) 336 78 40   Fax: (+36 1) 336 78 38
e-mail: kovacs@gkm.hu
Web: http://www.gkm.hu

Key personnel
Minister: István Csillag
Deputy State Secretary for Transport:
   Ferenc Kovács
Railways Department: Lajos Horváth

Györ-Sopron-Ebenfurt Railway (GySEV/RoeEE)

Györ-Sopron-Ebenfurti-Vasút Rt
Raab-Oedenburg-Ebenfurter Eisenbahn AG
Szilágyi Dezsö-tër 1, H-1011 1 Budapest
Tel: (+36 1) 224 58 99   Fax: (+36 1) 224 58 08

Key personnel
General Manager: Dr János Berényi
Deputy General Managers:
   Dr László Fehérvári, Csaba Siklos
Director – Sopron: Dr Tibor Jozan
Director – Wulkaprodersdorf: Dr Csaba Székely
Personnel Director: György Tabori
Commercial Director: Dr Tibor Varga
Finance Director: Eszter Menich

Gauge: 1,435 mm
Route length: 163 km (Hungary), 65 km (Austria)
Electrification: 218 km at 25 kV 50 Hz AC

Organisation
A majority shareholding in GySEV is owned by the Hungarian government (61 per cent), with the Austrian government holding 33.3 per cent. The Port of Hamburg owns 4.3 per cent of the equity.

The 62 km Sopron–Szombathely line was taken over by GySEV from Hungarian State Railways (MÁV) in 2001. In 2003 negotiations were in progress concerning the takeover by GySEV of

the Szombathely–Szentgotthárd–Austrian border line (54 km).

Passenger operations
In 2000, GySEV carried 1.35 million passengers. Fast services are operated between Sopron and Budapest. Sopron–Vienna services are operated by Austrian Federal Railways.

In December 2001, GySEV took over services on the Sopron–Szombathely route. These were previously operated by MÁV.

Freight operations
In 2000, GySEV handled 6.2 million tonnes of freight.

Intermodal operations
In March 1996, Intercontainer-Interfrigo launched a new intermodal service between the Dutch port of Rotterdam and Sopron entitled 'Hansa-Hungaria-Container-Express'. This operates daily. At GySEV's Sopron terminal, which was expanded in 1999, containers are either transhipped to feeder trains for Hungary, the Balkan countries and Turkey or to road vehicles. Piggyback traffic is also significant, accounting for around one-fifth of freight tonnage.

Improvements to existing lines
Following its acquisition by GySEV, the Sopron–Szombathely line is the subject of a major

upgrading programme to be completed by 2005. Electrification was commissioned in 2002; further modernisation will see new signalling equipment installed and track upgraded for 160 km/h operations.

Plans have been developed to realign a 1,350 m section of the Sopron–Ebenfurth line in tunnel in Sopron to minimise the environmental impact of increased freight traffic.

Traction and rolling stock

In early 2003 the railway owned 15 Class V43 electric, 14 Class M44, one Class M42 and two Class M40 diesel locomotives, two single-car and one two-car Austrian-built dmus, 42 passenger cars and 241 wagons. The coaching stock fleet was supplemented in late 2003 by the acquisition of 10 secondhand vehicles from DB AG, enabling some older ex-MÁV coaches to be retired.

In addition, in 2002 GySEV introduced five dual-voltage (15 kV AC/25 kV AC) Class 1047.5 'Taurus' 6,400 kW electric locomotives supplied by Siemens. They are used mainly on freight traffic between Hungary and Austria.

Electrification

Electrification of the Sopron–Szombathely line was commissioned in December 2002.

GySEV Class 1047.5 Siemens-built 'Taurus' electric locomotive with a piggyback service at Gramatneusiedl, Austria (Quintus Vosman)   0547797

Hungarian State Railways Co Ltd (MÁV)

Magyar Államvasutak Rt
Andrássy út 73-75, H-1940 Budapest
Tel: (+36 1) 322 06 60   Fax: (+36 1) 342 85 96
e-mail: mail@mav.hu
Web: http://www.mav.hu

Key personnel

Chairman: Dr László Udvari
General Manager: Zoltán Mándoki
Special Directors
   Passenger Services: Ferenc Vizsy
   Freight Services: Imre Kovács
   Infrastructure: Dr Tibor Zsákai
Assistant General Managers
   Economics: Annamaria Benczédi
   Personnel: Flórián Kugler
   Strategy: Dr István Tömpe
Chief of International Affairs Department:
   György Tábori

Gauge: 1,524 mm; 1,435 mm; 760 mm
Route length: 36 km; 7,519 km; 219 km
Electrification: 2,628 km at 25 kV 50 Hz AC

Political background

Like all railways of the former Communist East European countries, MÁV steadily lost traffic after 1989 in the face of political and industrial change. Freight traffic more than halved in the 1989–95 period, from 104 to 46.4 million tonnes. Passenger traffic, stagnant since the mid-1980s, dropped 30 per cent as unemployment rose and public and private road transport competition intensified. Traffic in both sectors has since stabilised and in 2003 43.7 million tonnes and 158.2 million passengers were carried.

At the end of 2003 the number of staff employed was 52,100. Ancillary businesses such as engineering workshops were transferred to separate companies in the early 1990s.

From 1 January 1994 new legislation distinguished between the passenger and freight operations of MÁV and its infrastructure, as well as its engineering and property management, each accounting separately as business units within the organisation from 2003. Main lines will continue to be owned by the central government, while branch lines or operations over them may be handed over to local authorities at a later stage. Private-sector firms will be able to apply to run railway services, subject to government licence. Moreover, freight operations covering 20 per cent of main line capacity will be opened up to international competition by the end of 2006 and completely in 2007 in accordance with EU rules. This follows Hungary's accession to the Union in May 2004.

In addition, under the Railways Act of 1993, since 1 February 1995 a contract has existed between the state and MÁV Ltd. It defines which infrastructure elements remain exclusively in state (treasury) ownership, the fee to be paid for their use and public service obligations whose deficits will be covered by the state in accordance with EU legislation. This contract will contribute to the consolidation of MÁV's finances.

The government also repeatedly exempted MÁV from substantial historic debts, guaranteed its long- and short-term loans and increased its grant for passenger services, although this still lagged behind operating costs.

Keeping abreast of EU legislation the government in 2001 established a separate, independent body to manage the allocation of network capacity and to set fees for its utilisation.

At the beginning of 2004 the government determined the restructuring of MÁV Ltd and the implementation of the reforms necessary on the country's accession to the EU. This included a reduction of 20 per cent in staff numbers by 2006, the initiation of a process to develop a structure of regional railway bodies and the establishment of mechanisms for the refund of public service obligations and for measuring infrastructure performance.

Traffic (million)	2001	2002	2003*
Passenger journeys	159.6	161.9	158.2
Passenger-km	9,902	10,408	10,224
Freight tonnes	43.8	43.0	43.7
Freight tonne-km	7,426	7,244	7,693

* Provisional figures.

Passenger operations

MÁV joined the western European EuroCity system with the 1988 conferring of EC status on the Vienna–Budapest 'Lehar'. In 2004 there were six EC links and four EuroNight services from/to Budapest. After four years of planning a direct Budapest–Sarajevo night service was introduced in March 2003. Altogether, 40 international passenger services operate over the MÁV network.

Domestically, MÁV introduced supplementary-fare intercity services radiating from Budapest in 1990. These now cover 13 routes ranging from 180 to 330 km.

From 1997, new InterPici services began, running over distances of 30 to 60 km to provide connections with smaller communities and offer a standard of travel close to that of intercity services.

From June 2001 MÁV rebranded certain intercity services 'InterCity Rapid'. Using 200 km/h CAF-built coaches and attracting higher-than-normal seat reservation fees, these limited stop services were introduced on the Budapest–Debrecen and Budapest–Miskolc–Nyíregyháza routes. To strengthen cross-border traffic towards the EU via Austria, four pairs of InterRegio trains between Györ and Vienna and one pair each between Budapest and Wiener Neustadt and Budapest and Graz were introduced from the end of 2002.

In the restructuring on Hungarian railways, the government and MÁV will focus on the development of intercity services and suburban transport. The aim of the company is to provide hourly services on routes within a 60–80 km radius of Budapest and two services per hour within a 30–40 km radius of the capital.

Class 1047 and V43 electric locomotives at Budapest (Quintus Vosman)   0583263

Freight operations

Freight traffic has been badly affected by the political and economic changes in Hungary since 1989, with some plants which generated heavy traffic having closed and new private-sector road operators having taken other freight. As Hungary is a landlocked country with its economy based on foreign trade, international traffic is very important, with over half of MÁV's freight tonnage crossing the border. The co-operation of multinational companies in planning their plants for international traffic offers new opportunities for rail freight, while in the domestic market the delivery of stone and aggregates for extending Hungary's motorway network promises gains.

In preparation for the liberalisation of transport in line with EU requirements, MÁV's freight business has signed medium-term contracts with its domestic key account clients, thus securing nearly half of its core market.

MÁV's wagon fleet has been drastically reduced to reflect the reduced tonnage being carried, from a fleet of around 70,000 vehicles in the 1980s to 16,000 in 2003. Inadequacies of the wagon fleet in terms of obsolescence and unsuitability of vehicle types continue to hamper the development of freight traffic.

The freight business unit has developed a concept for wagon management aimed at using 11,000 modern vehicles of appropriate types through purchase, leasing and hiring as well as through selling and liquidation.

Intermodal operations

To offset the drop in heavy industrial traffic such as coal and steel, MÁV is promoting container and piggyback transport.

The railway-owned international container company Intercontainer-Interfrigo (ICF), along with several business partners, established Pannoncont, a Budapest-based company, in mid-1992. Intercontainer services run between Budapest and Trieste in northern Italy, to Rotterdam and also to Thessaloniki and Istanbul.

In conjunction with MÁV, Hungarokombi, the Hungarian member of UIRR (the international piggyback organisation), operates RoLa ('rolling motorway') services for trucks accompanied by their drivers. RoLa services were first introduced between Wels (Austria) and Budapest in 1992. Since 1993 services have operated between Wels and Kiskundorozsma, on the southeastern border of Hungary, and from 1997 from Kiskundorozsma to Sežana, Slovenia. In 1996 Ganz-Hunslet built 50 RoLa wagons for MÁV. In 2003 102,000 lorries and 303,000 TEUs were carried, contributing 12 per cent to MÁV's traffic volume.

The Budapest Intermodal and Logistic Centre is being developed with MÁV's participation on the southern perimeter of the capital, funded in part

Class V63 electric locomotive at Nyíregyháza with a Chop–Budapest service (Colin Boocock)

0143313

Push-pull trainset forming a local service at Nyíregyháza, powered by a Class MDmot lightweight diesel-electric locomotive (Colin Boocock)

0143314

Diesel locomotives

Class	Builder's type	Wheel arrangement	Power kW	Speed km/h	Weight tonnes	No in service (2004)	First built	Mechanical	Builders Engine	Transmission
M28.1/M28.2	Rába M033	B	100	30	20	21	1955	Rába	Ganz	*M* Rába
M32	DHM6	C	260	60	35	9	1973	Mávag	Ganz	*H* Ganz
M40.0/M40.1/M40.2/ M40.9/M40.203 & 214	DVM6	Bo-Bo	740	100	76	26	1966	Mávag or MTU	Ganz or Caterpillar	*E* Ganz Electric
M41.2/M41.23	DHM7	B-B	1,320 or 1,500	100	66	106	1973	Mávag	Pielstick/Ganz	*H* Voith/Ganz
M43.1/M43.12	LDH45	B-B	330	60	48	75	1974	Aug 23	Aug 23	*H* Brasso Hydrom
M44.0/M44.4/M44.5	DVM2	Bo-Bo	440	80	62	109	1956	Mávag	Ganz/Caterpillar	*E* Ganz Electric
M47.1/M47.12/M47.13/M47.2	LDH70	B-B	630	70	48	55	1974	Aug 23	Aug 23	*H* Brasso Hydrom
M61	M60	Co-Co	1,430	100	108	2	1963	Nohab	GM	*E* GM
M62.0/M62.3/M62.5/M62.6	M62	Co-Co	1,470	100	120	83*	1965	Lugansk	Kolomna	*E* Charkov

* Includes 9 1,524 mm gauge.

Diesel railcars or multiple-units

Class	Builder's type	Cars per unit	Motor cars per unit	Motored axles/car	Power/ motor kW	Speed km/h	Units in service (2004)	First built	Mechanical	Builders Engine	Transmission
MDmot		5	1	4	590	100	37	1970	Mávag	Mávag	–
Bzmot 300	Bzmot	1	1	1	206	80	77	1975	Studenka	Rába D10	*HM* Voith
Bzmot 400	Bzmot	1	1	1	210	80	23	1997	Studenka	Volvo	*HM* Voith
Bzmot 200	Bzmot	1	1	1	228	80	158	1978	Studenka	MAN D2866	*HM* Voith
6341	RA-V	2	2	2	2 × 315	100	6	2002	Metrovagonmash	MTU	*H* Voith
6342	Desiro	2	2	2	2 × 375	120	2	2003	Siemens	MTU	–

by contributions from the Hungarian government and the European Union. These facilities will be completed by 2006 at a cost of some US$75 million. The first phase, a terminal with an initial annual capacity of 75,000 TEUs, was put in operation in 2003.

New lines

A 44 km direct rail link between Hungary and Slovenia was officially opened in May 2001. The 19 km Hungarian section has been built for operations at up to 120 km/h and the alignment of the route has been designed to allow eventual upgrading for 160 km/h. The line includes a 1.4 km viaduct, claimed to be the longest in Central and Eastern Europe.

Plans to develop a 23 km rail link between Budapest Nyugati station and the city's Ferihegy airport have been developed. If adopted, the scheme would be implemented by a joint venture company in which MÁV would hold a share. A study commissioned by MÁV and property developerTriGranit, which reported in early 2001, found that the link could be completed at a cost of US$80 million. Some private sector funding would be required to complete the scheme. Around one-third of the line's route would entail new construction, including a 650 m tunnel to provide access to the airport terminal. A journey time of 19 minutes is foreseen, with a service frequency of three trains an hour.

Improvements to existing lines

Major MÁV improvements planned with the help of European Union funding include: Zalalövö–Zalaegerszeg–Boba (83 km), which forms part of the Pan-European Transport Corridor 5, covering track upgrading and electrification; Budapest–Györ–Hegyeshalom (191 km, Corridor 4), track and signalling improvements by 2006; and Lökösháza (114 km, Corridor 4), covering upgrading and track doubling.

EU funding is also helping to finance upgrading the Budapest–Ujszász–Szolnok route, 100 km on Corridor 4, by the end of 2003 and the Cegléd–Kiskunfélegyháza line (58 km) by 2005.

As a result, five sections of pan-European Corridors 4 and 5 in Hungary, accounting for nearly 600 km of track, will have been rehabilitated by 2006 with co-financing by the EU and by EIB loans.

Traction and rolling stock

At the end of 2003 MÁV was operating 460 electric locomotives, 484 diesel locomotives, 24 emus, 303 motor coaches and dmus, 2,876 hauled passenger cars (including 79 buffet/dining and 88 sleeper/couchette cars) and 17,542 wagons.

In 1998 a programme was launched to upgrade locomotive-hauled stock for suburban services. With the help of a loan from the European Bank for Reconstruction and Development, 200 vehicles have been refurbished and modernised. In October 2001 MÁV awarded Bombardier Transportation a contract to refurbish 136 suburban coaches at its Dunakeszi facility. Deliveries were completed in July 2003.

In 2000 work started on refurbishment and re-engineering with Caterpillar or MTU power units of Classes M40, M41, M47 and M62 diesel locomotives. Of these two manufacturers' engines, Caterpillar units were selected for locomotives refurbished from 2003.

Examples of Class V43 electric locomotives have been refurbished and modified for operation with modernised Type Bhv push-pull coaching stock sets, forming the V43.2000 sub-series.

In 2002 MÁV and GySEV (see entry in Hungary section of *Railway systems and operators*) jointly purchased respectively 10 and five Class 1047 dual-voltage electric locomotives with an output of 6.4 MW and a maximum service speed of 230 km/h. Built by Siemens Transportation Systems, these are of the 'Taurus' design, similar to the Austrian Federal Railways Class 1116.

In 2001 Siemens won a contract to supply 13 articulated two-car Desiro dmus for regional services for delivery by the end of 2003. Between 2002 and 2004 MÁV is also acquiring 40 Type RE-1 two-car dmus from Metrovagonmash in Russia under an arrangement intended to repay Russian debt to Hungary.

MÁV Class M47.13 diesel-hydraulic shunting locomotive, originally built by FAUR, Romania, and subsequently rebuilt with MTU engine, at Miskolc (Edward Barnes) 0554371

Electric locomotives

Class	Builder's type	Wheel arrangement	Speed km/h	Weight tonnes	No in service	First built (2004)	Builders Mechanical	Engine
V43.1/V43.2	VM14	B-B	130	80	338	1964	Mávag	Ganz Electric
V46	VM16	Bo-Bo	80	80	60	1983	Mávag	Ganz Electric
V63.0/V63.1	VM15	Co-Co	120/160	116	52	1975	Mávag	Ganz Electric
1047	ES64U2	Bo-Bo	160/230	86	10	2002	Siemens (Krauss Maffei)	Siemens

Electric railcars or multiple-units

Class	Cars per unit	Motor cars per unit	Motored axles/ car	Output kW	Speed km/h	No in service (2003)	First built	Builders Mechanical	Electrical
BDVmot	4	1	4	1,444	120	19	1988	Ganz-Hunslet	Ganz Electric
BVhmot	4	1	4	1,755	160	2	1995	Ganz-Hunslet	ABB/Ganz Ansaldo
BVmot	4	1	4	1,755	120	3	1996	Ganz-Hunslet	ABB/Ganz Ansaldo

As part of the restructuring mentioned above, in 2004 the Hungarian government authorised MÁV to acquire 30 emus for suburban services with an option for 30 more. Tender invitations were to be issued in mid-2004 for delivery in 2005.

Signalling and telecommunications

Nearly half of the system is equipped with automatic colourlight signalling and modern integral signalling centres have been installed at two-thirds of all stations: automatic block signalling is installed over 2,356 route-km, while a further 129 km are controlled by CTC; 749 km are controlled by centralised traffic supervision, including the 180 kmroute between Budapest and Hegyeshalom.

Track-to-train radio is operational on 1,400 km of the radial routes from Budapest.

In consortium with Austria's Alcatel, MÁV has developed and installed a train control system on the Vienna–Budapest main line as a pilot project for Level 1 of the European Train Control System currently under development in other parts of Europe. A similar system was installed in 2002 on the new link to Slovenia and is to be provided for the entire Budapest–Vienna line by 2004.

Some World Bank aid has contributed to development of a computerised Traffic Management Information System (TMIS). This provides real-time data on traction and vehicle movements, marshalling yard operation and wagon distribution as well as fulfilling client information functions. Some 800 terminals throughout the railway will be linked to it; it was brought into operation in 1996, and the first clients went online during 1997. Since 2001 the system has also been integrated into the western European Hitrail organisation for wagon tracing.

TMIS is the first phase of a major development programme in MÁV informatics, valued at around US$150 million. Subsequent phases cover ticketing, seat reservations, and passenger information systems: these elements are being delivered by IBM, while information technology systems to manage MÁV finances were supplied by ICL in 2000.To serve these systems, and to create capacity

for a competitive public telecommunications service, the programme included the construction by Siemens of a fully digital data transmission network of 2,600 km of fibre optic cables during the three-year period 1999 to 2001.

Electrification

In 1997, Siemens was awarded a DM280 million contract to electrify the following routes:

Székesfehérvár–Szombathely (170 km); Újpest–Vácrátót (30 km); and Balatonszentgyörgy–Murakeresztúr (45 km). Electric traction was introduced in 1998 on the last two routes, while the first-named was energised at the end of 2000.

The next electrification projects to be undertaken are: the new line from Boba to Hodoš, Slovenia (100 km) by 2006, with EU funding; and Györ–Celldömölk (70 km) by 2007, with an EIB loan.

Track

Rail: 60, 54 or 48 kg/m, in some places lighter
Sleepers: Concrete, steel and timber. Length, 2,420 mm; height, 190 mm; width, base, 280 mm; width, top, 200 mm
Fastenings: GEO 'K' type, SKL, Pandrol Fastclip, Nabla
Min curve radius: 150 m
Max gradient: 2%
Max permissible axleload: 22.5 tonnes

All new rail is 54.3 kg/m laid on prestressed concrete sleepers with a ballast depth of 500 mm. Minimum curve radius is 1,300 m and normal top speed is 140 km/h.

Under the Communist regime, five-year plans saw programmed upgrading of track for heavier axleloads and higher speeds. Due to lack of funding, this upgrading fell into abeyance for many years.The minimum aim is to eliminate most speed limitations related to the condition of track on the international main lines that are sections of the EU's Trans-European Rail Freight Network, totalling 2,300 km. In 1998 a rehabilitation programme was adopted to complete this task by 2010.

India

Indian Railway Board (IR)

Rail Bhavan, Raisina Road, New Delhi 110 001
Tel: (+91 11) 23 38 89 31 41   Fax: (+91 11) 23 38 44 81
e-mail: secyrb@rb.railnet.gov.in
Web: http://www.indianrailways.com

Key personnel

Minister for Railways: Laloo Prashad Yadav
Railway Board Members
 Chairman: R K Singh
 Financial Commissioner: V Viswanathan
 Electrical: S C Gupta
 Engineering: S P S Jain
 Mechanical: P N Garg
 Staff: S M Singla
 Traffic: R N Aga
 Secretary: V N Mathur
Additional Members/Advisers
 Budget: N Krishnamurthy
 Adviser (IR): P K Sharma
 Adviser (L&A): Uttam Chand
Civil Engineering: Budh Prakash
 Commercial: DVR Sharma
 Economic Adviser: K V Krishnan
Electrical: B M Lal
 Finance: A S Tiwari
 Adviser (Finance): S C Gupta
 Information Technology: Arun Dubey
 Legal Adviser: P K Malhotra
 Mechanical: T K Biswas
 Mechanical, Production Units: V K Bhargava
 Planning: Sumant Chak
 Safety and Coaching: (vacant)
 Adviser (Signalling): K K Bajpayee
 Staff: U V Acharya
 Stores: R C Saxena
 Telecommunications: R C Sharma
 Tourism and Catering: P C Joshi
 Traffic: L R Thapar
 Works: O P Agarwal
 Vigilance Adviser: Shiv Kumar
Director General, Railway Protection Force:
 A K Pandey
Director General, Health Services: Dr D K Das
Executive Directors
 Coaching: V N Mathur
 Corporate Co-ordinater: S C Agnihotri
 Safety: M G Arova
 Statistics and Economics: V Sahai
 Accounts: S Balachandran
 Efficiency and Research: R K Tandon
 Telecom Development: V G Rameshkumar
 Signalling: A K Saxema
 Track: Madan Lal
General Managers, Manufacturing and other units
 Chittaranjan Locomotive Works: S Jain
 Varanasi Diesel Locomotive Works: J K Kohli
 Integral Coach Factory: S Kumar
 Rail Wheel Factory: J N Pant
 Rail Coach Factory: M Sirajuddin
 Railway Electrification, CORE: C R Kalsi
 Metro Railway, Calcutta: Vinod Kumar
 NF Railway Construction: N Ramasubramanyan

IR liquefied petroleum gas wagon (K K Gupta)   0547042

Research, Design and Standards Organisation:
 G K Wadhwa
Principal, Railway Staff College: G K Garg
Director, IR Institute of Civil Engineering: A K Goel
Director, IR Institute of Signalling and
 Telecommunications: P S Vankatraman
Director, IR Centre for Advanced Management and
 Technology: C B Midha
Chief Administrative Officers
 Diesel Component Works: A K Gupta
 Central Organisation for Modernisation of
 Workshops: Ashok Baijal
Chairman, Konkan Railway Corporation (ex officio):
 S P S Jain

Gauge: 1,676 mm; 1,000 mm; 762 mm and 610 mm
Route length: 45,718 km; 14,406 km; 3,106 km
Electrification: on 1,676 mm gauge 16,095 km at
25 kV 50 Hz AC, 433 km at 1.5 kV DC; 165 km of
1,000 mm gauge at 25 kV 50 Hz AC

Organisation

Indian Railways, organised as a central government undertaking, is Asia's largest and the world's second largest state-owned railway system under unitary management. The Ministry of Railways functions under the guidance of the Minister of Railways. Day-to-day affairs and formulation of policy are managed by the Railway Board, comprising the Chairman, Financial Commissioner and functional members. Until 2003, the railway was made up of nine zonal systems: Central, Eastern, Northern, North Eastern, Northeast Frontier, Southern, South Central, South Eastern and Western. Due to the changing traffic patterns caused by large-scale gauge conversion and opening of the Konkan Railway, IR decided in 1996 to create seven more zones. These are the East Coast, East Central, North Central, North Western, South Western and West Central railways and the South East Central Railway, with headquarters at Bhubaneswar, Hajipur, Allahabad, Jaipur, Hubli, Jabalpur and Bilaspur. East Central Railway and North Western

Railway became operational on 1 October 2002. The remaining five new zonal railways commenced operations on 1 April 2003.

IR is a monopoly organisation and enjoys considerable economies of scale in its operations. Though the time is not yet ripe for its corporatisation, several changes in its conceptual framework are necessary in order to enable it to cope with the challenges of the twenty-first century.

For IR the major challenge in investment planning is to strike a proper balance between its dual role as a public utility on the one hand and a commercial enterprise to be run on sound business principles on the other.

Rail transport is likely to become a bottleneck in the growth of the economy unless adequate investments are planned in basic infrastructure in the tenth five-year plan (2002–07). Estimated IR freight traffic projections in the terminal year (2006–07) of this plan are 624 million tonnes and 369 billion tonne-km, against 2001–02 (terminal year of the ninth five-year plan) figures of 489 million tonnes and 323 billion tonne-km. Estimated passenger traffic projections for 2006–07 are 5.885 billion passengers and 625 billion passenger-km, compared with 2001–02 figures of 5.000 billion passengers and 473 billion passenger-km, including suburban passengers.

With these estimates in mind, the government announced a Rs150 billion five-year rail development plan including Rs80 billion to remove capacity bottlenecks in critical sections of line, including the so-called 'Gold Quadrilateral', to improve rail connectivity to ports. The programme calls for the construction of four major bridges. The government has set up a Special Purpose Vehicle (SPV), National Rail Vikas Nigam, to plan and manage the programme, undertake project development, mobilisation of financial resources and to implement these projects.

Safety

In recent years IR has experienced several major accidents that have led to heavy loss of life. These have been attributed to human factors, to equipment or infrastructure failure and to sabotage.

In June 2001 a Mangalore–Chennai mail train derailed while passing over a bridge near Kozhikode, Kerala, killing 50 people. This prompted the government to create a guaranteed Special Railway Safety Fund (SRSF) of Rs170 billion to be utilised over six years. The fund is being used to replace life-expired track (12,260 km), repair or replace 262 bridges, items of rolling stock and signalling systems at 1,560 stations.

In January 2003 a rear-end accident occurred between a Kochiguda–Manmad express and a freight train on the South Central Railway. This incident resulted in the deaths of more than 20 passengers and was attributed to failures by railway staff. It was subsequently decided that IR would recruit about 20,000 'Group D' staff and to improve safety-related training.

A 10-year Corporate Safety Plan (2003–13) was formulated by IR and presented to parliament in August 2003. Safety targets of IR as a whole have

Non-air conditioned 'Chair Class' passenger coach produced for IR by ICF (K K Gupta)   0547039

been laid down in this plan, the broad objective of which is to reduce the rate of accidents per million train-km from the present level of 0.44 to 0.17 by 2013.

IR is also installing an Anti-Collision Device (ACD) developed by Konkan Railway Corporation on 1,736 route-km under a programme due for completion by December 2004. Extensive installation of track circuiting on running lines is being implemented and design features are also being introduced to improve the crashworthiness of passenger coaches.

Finance

In the railway budget for FY2003–04, IR's annual plan, including expenditure of Rs23.11 billion on safety-related works through the SRSF, was kept at Rs129.18 billion. This was Rs6.03 billion higher than the revised estimate of the previous year (FY2002–03). At the Interim budget stage in January 2004 this was revised to Rs139.18 billion. This outlay is to be financed by budgetary support from the government of Rs65.77 billion and balance resources were to be generated internally or through market borrowings by railways, except that Rs5.0 billion for the new Udhampur–Baramulla line will have separate allotment. As a populist measure, in the budget for 2003–04, there was no increase in the freight rates or in passenger fares.

IR's budget estimate for FY2003–04 is based on a target of 540 million tonnes of freight carried against a target of 510 million for FY2002–03. This was revised to 550 million tonnes at the interim budget stage in January 2004. For passenger traffic, a 2.8 per cent increase was recorded. Gross revised traffic receipts as well as total revised working expenses for FY2003–04 are both lower at Rs425.55 billion and Rs393.27 billion respectively and net revenue is Rs41.48 billion.

The Railway Fare & Freight Committee set up to examine the whole structure of fares and freight rates has made several recommendations for attracting more traffic, raising additional revenue and reducing costs. Many of its recommendations have been accepted by IR, but implementation will be in stages. However, the government has yet to solve overmanning in the public sector, and the tradition that a public-sector post is effectively a job for life. IR's staff, for example, totalled 1.511 million as at 31 March 2002 for a wage bill of Rs190.37 billion, an increase of Rs1.96 billion over 2000–01.

IR has initiated several steps to augment earnings and curtail expenditure. The railway also plans to raise revenue from other non-traditional sources. These steps are expected to be pursued vigorously to ensure adequate funds are available to invest in the replacement of life-expired or obsolete assets and modernisation of the system.

Revenue (Rs million)	1999–2000	2000–01	2001–02
Passengers	104,060	112,792	120,687
Freight	220,610	233,051	248,454
Other income	6,573	7,032	9,445
Total	331,243	352,875	378,586

Expenditure		2001–02	2001–02
Working expenditure	256,449	275,344	287,028
Reserve funds	51,990	71,329	75,904
Total	308,439	346,673	362,932

Traffic (million)		2000–01	2001–02
Passenger journeys	4,585	4,833	5,093
Passenger-km	430,666	457,022	493,488
Freight tonnes	478.2	504	522.2
Freight tonne-km	308,039	315,516	336,445

Gauge conversion

The thrust of gauge converstion projects in the tenth plan (2002–07) is on completing the works that provide connectivity to ports and industry and those which enhance the capacity of a saturated system or remove traffic bottlenecks. A total of 2,365 km of gauge conversion is planned during the plan period compared with 2,103 km in the ninth plan. An outlay of Rs7.33 billion was provided in IR's budget for 2003–04 for gauge conversion work and a physical target of converting 775 km was set. IR signed a MoU with the government of Gujrat and other participants in January 2004 for gauge

Class WCAU3 dual-voltage broad gauge emu supplied by Alstom, BHEL and ICF (K K Gupta)

0547041

Plasser & Theurer RM80 ballast cleaner in operation (K K Gupta)

0547040

Computerised Passenger Reservation System office at Secunderabad, South Central Railway

0583116

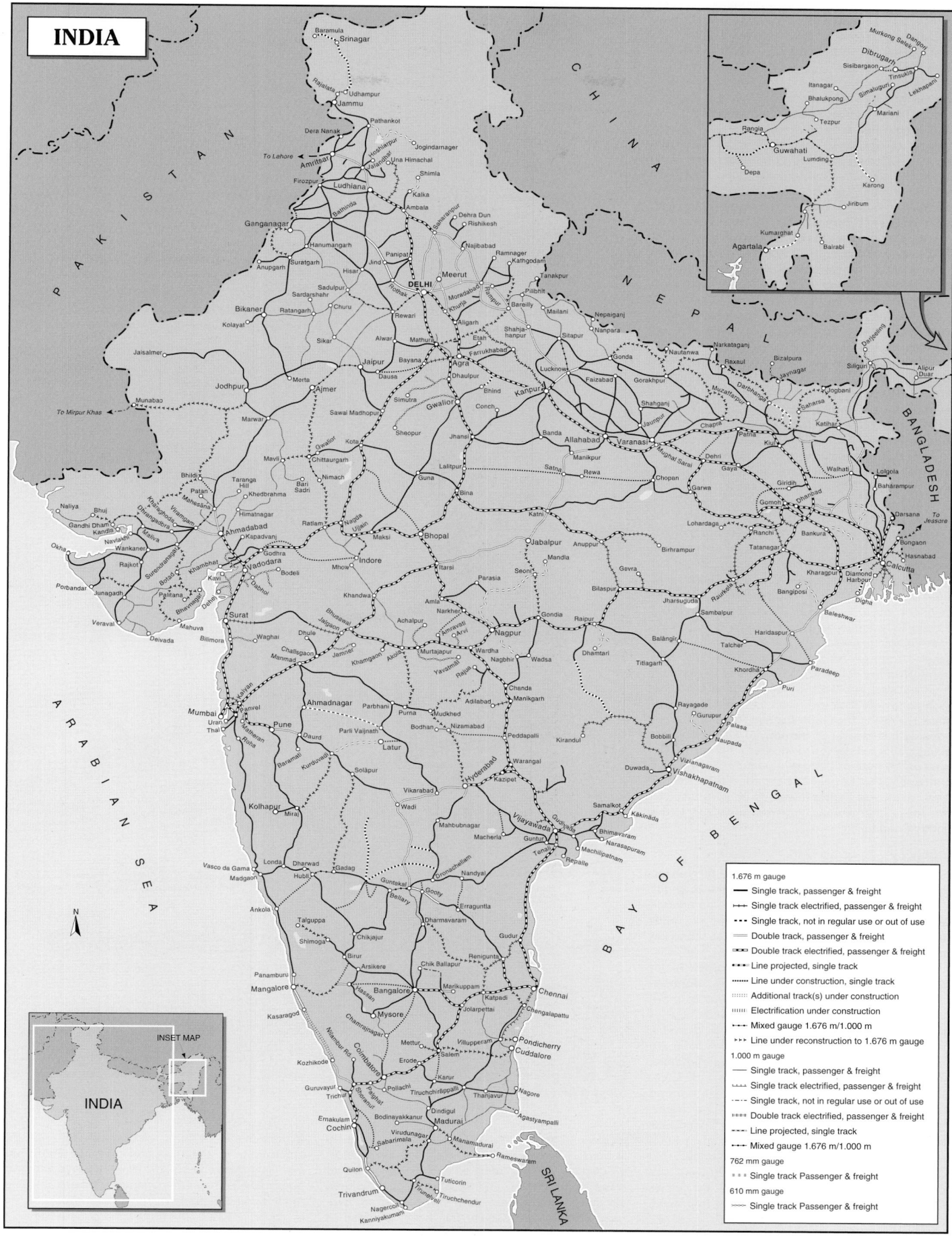

INDIA

1.676 m gauge
— Single track, passenger & freight
⊢⊢ Single track electrified, passenger & freight
--- Single track, not in regular use or out of use
═ Double track, passenger & freight
▭▭▭ Double track electrified, passenger & freight
••• Line projected, single track
····· Line under construction, single track
⋮⋮⋮ Additional track(s) under construction
⊪⊪⊪ Electrification under construction
⊢•⊢ Mixed gauge 1.676 m/1.000 m
▸▸▸ Line under reconstruction to 1.676 m gauge

1.000 m gauge
— Single track, passenger & freight
⊥⊥⊥ Single track electrified, passenger & freight
-·-· Single track, not in regular use or out of use
⊞⊞⊞ Double track electrified, passenger & freight
••• Line projected, single track
⊢•⊢ Mixed gauge 1.676 m/1.000 m

762 mm gauge
═ ═ ═ Single track Passenger & freight

610 mm gauge
∼∼∼ Single track Passenger & freight

0547605

For details of the latest updates to *Jane's World Railways* online and to discover the additional
information available exclusively to online subscribers please visit
jwr.janes.com

conversion of the 313 km Gandhidham–Palanpur section at an estimated cost of Rs4.53 billion. During the interim budget review in January 2004 two more gauge conversion projects totalling 569 km were approved.

Passenger operations

There is continuous demand for introduction of new trains and increased frequency of existing service, especially 'Rajdhani' and 'Shatabdi' expresses. During 2002–03, 13 pairs of trains were added to the timetable and 16 pairs extended. Additional main line emu and dmu services were introduced over several sections to segregate short- and long-distance traffic. Railbus service was introduced on some branch lines with low traffic densities. In addition, every year a large number of trains are run and extra coaches provided to clear extra demand during summer, winter and Puja holiday periods.

Despite operational, resources and track capacity constraints, the introduction of 50 pairs of train services, increases in the frequency of 13 pairs of trains and the extension or augmentation of several other services were proposed in the FY2003–04 budget to meet demand. Most of these have since been introduced.

IR is now providing further improved facilities for passengers such as the installation of CCTV at important reservation offices to monitor the activities of touts and anti-social people; the provision of Rajdhani Express-type features on long-distance services; the installation of Interactive Voice Response System (IVRS), a computerised passenger enquiry system providing information on reservations, and train arrivals and departures; and the provision of instant confirmed reservations to passengers required to travel at short notice in over 100 long-distance trains on payment of a surcharge. With the networking project fully implemented, it is now possible for passengers to book outward and return journey tickets from or to any station from any PRS terminal in the country. Computerised reservation facilities were available at 702 locations including several remote locations such as Leh, Manali and state capitals in the North-Eastern Region. Internet booking has also been introduced and tickets are being delivered to the customer's home. At the time of the interim budget in January 2004 it was announced that 17 pairs of new trains to be named 'Sampark Kranti Express' were to be introduced to provide quicker connectivity from the states to the national capital.

Freight operations

In FY2002–03 freight grossed 518.74 million tonnes, growth of 5.3 per cent over the previous year. During FY2001–02, indices for IR broad-gauge wagon utilisation were wagon turn-round – 7.2 days; wagon-km per day – 191.5 km; and net tonne-km per wagon day – 2,223.

The chief measures taken to enlarge freight operating capacity (and also to enhance productivity and efficiency) are recourse to more unit train working, block train segregation of high-capacity, roller bearing-equipped wagons, pursuit of as many 4,500 tonne trainloads as are feasible within existing passing loop parameters, and raising train speeds to a maximum of 80 km/h. Between 90 and 100 per cent of all coal, ore and petroleum product traffic moves in unit trains. Point-to-point fast freights known as 'Speed Link Express', introduced for movement of general goods between Delhi, Calcutta, Mumbai and Chennai, brought encouraging results and are to be extended to other corridors. IR also views upgrading of terminals as a means of improving freight traffic performance.

As a result of overall economic recovery, IR was set to achieve its freight budget loading target of 550 million tonnes during 2003–04, and a target of 570 million tonnes has been set for FY2004–05. However, IR is greatly concerned at slippage in its market share of freight traffic to 40 per cent. Its objective is to raise this share to 50 per cent during the first decade of the new millennium. In IR's budget for 2003–04 no increase was made to freight rates for any commodity. On the other hand, IR is making innovative marketing efforts to

Class WDM2 diesel locomotive heading a Jan Shatabdi express passenger service
 0583117

Reach stackers are used to handle some of IR's growing container traffic
0583118

Class WAP4 25 kV AC electric locomotive and IR's latest ALSTOM-LHB air-conditioned coaches
0583119

increase traffic in other sectors than core products like coal, iron and steel and raw materials for steel production. IR has reduced the classification of some commodities and is also giving 'volume discounts' to become more competitive. A two-point rake loading facility has been provided at some stations and wagon rakes are being made available in less than 48 hours. IR is planning to facilitate warehousing at existing railway terminals as well as at privately operated sites as a marketing tool to attract traffic.

IR is developing a computerised freight operations information and control system (FOIS) to provide real-time information to customers in regard to bookings, movements and delivery of freight consignments through customer service cells to be set up in all zonal railways. The Rake

Management System, developed for specific application to unit train and rake operation, has been commissioned on the zonal railways. The next phase, the Terminal Management System, is being implemented to improve wagon turn-round.

Intermodal operations

The Container Corporation of India (Concor) was set up in 1988 under the administrative jurisdiction of the Ministry of Railways (see Indian Railway Board). Its remit is to market freight movement in containers and to develop the necessary infrastructure for multimodal operations. At present, Concor operates 51 domestic and international container-handling terminals. Inland Container Terminal at Dadri, 45 km from Delhi, is the latest. Traffic has risen from 66,187 TEU in FY90–91 to 1,380,000 in 2002–03. Package deals have been agreed with several major industries for through carriage of raw materials to production sites plus movement of finished products to destination.

Scheduled 'Contrac' trains with specified frequencies and guaranteed transit times run between key centres. New container terminals (ICDs) were opened in 2001–02 at Balasore, Bhusawal, Jaipur, Jamshedpur, Jodhpur, Khodiyar, Mairaj, Pondicherry, Rajkot, Salem Market and Vaizag. Through the operations of Concor, IR is now firmly involved in multi-modal transport. Concor handles international traffic at the ports of Mumbai, Nhava Sheva, Chennai, Calcutta, Haldia, Kandla, Tuticorin, Cochin and Vizag. Concor procured 735 high-speed wagons during 2002–03 and at the end of March 2004 had 3,997 such wagons.

Traction and rolling stock

During 1996–97 Adtranz completed delivery of 30 three-phase 4,475 kW locomotives under an order placed in 1993 for 20 six-axle freight locomotives with top speed of 100 km/h, and 10 for 160 km/h passenger service. The contract also involved transfer of technology to Chittaranjan Locomotive Works (CLW). Under this agreement CLW started the manufacture of three-phase freight locomotives (Class WAG9) in 1998 and passenger locomotives (Class WAP5) in 1999. The Class WAP7 passenger variant followed in 2001. The WAP7 version is capable of hauling prestigious Shatabdi Express services formed of 24 coaches at speeds of up to 140 km/h. By 31 January 2004 CLW had manufactured 38 examples of Class WAG9 and 17 WAP7 locomotives.

On 1 March 2004 IR had a total of 3,011 electric locomotives: 2,787 broad gauge AC; 137 broad gauge AC/DC; 58 broad gauge DC; and 19 metre gauge AC.

On the diesel side, in 1996 IR ordered 20 2,980 kW diesel-electrics with AC traction motors from General Motors USA, and undertook production of these locomotives, which are designated Class WDG4, from 1998–99 at its Diesel Locomotive Works (DLW) in collaboration with GM. The first of these was rolled out in August 1999. Production of the Class WDP4 passenger version has also been developed at DLW.

With a view to developing export business in complete locomotives, a design and development centre has been created at CLW. During 2001, IR production units manufactured 102 broad-gauge diesel and 82 electric locomotives, and 2,384 coaches including 71 suburban emu cars and 62 dmu cars.

The Railways Ministry's 2003–04 budget provided for supply to IR of 85 diesel and 69 electric locomotives, 147 suburban emus, 80 main line emus and 47 dmus, 1,760 coaches and 25,500 wagons.

At 31 March 2003 IR's fleet comprised 34 MG and 19 NG steam, 4,616 diesel (BG and MG), 5,136 motor and trailer emu/memu cars (BG and MG), 22 diesel railcars, 38,979 coaches and 213,919 wagons (BG and MG).

Type of coupler in standard use: Passenger cars, screw and automatic buffer coupling; freight wagons, Alliance II and centre buffer.

Type of brake in standard use: Most broad-gauge freight stock built after 1982 is airbraked; others vacuum, as is all passenger stock except that for the 'Rajdhani' and 'Shatabdi' expresses which are airbraked. Phased programme under way for conversion of broad-gauge passenger stock to airbrakes.

Signalling and telecommunications

Following an accident at Firozabad in 1995, IR has been urgently pressing ahead with installation of track circuiting at block posts on all trunk routes and important main lines. Furthermore, the main emphasis is on replacement of mechanical interlockings at main line wayside stations by panel interlockings on an age/condition basis, and in busy yards by route-relay interlockings. At the end of FY2001–02, IR had 2,224 panel and 229 route-relay systems and 14 solid-state interlockings were in operation. New technologies are digital axle-counters, audio frequency track circuits, failsafe multiplexers and LED signal units.

Electric multiple-units

Class	Cars per unit	Motor cars per unit	Motored axles/car	Output/ motor kW	Speed km/h	First built	Builders Mechanical	Electrical
1.5 kV DC BG								
WCU14	3	1	4	139/187	105/80	1969	Jessop/BEML	TDK/BHEL/Japan
WCU15	3	1	4	139/187	105/80	1970	ICF/BEML/Jessop	BHEL
25 kV AC BG								
WAU-4	3/4	1	4	167.3	80/90	1967	ICF	BHEL/Hitachi
25 kV AC/1.5 kV DC BG								
WCAU	3	1	4	187	100	2001	ICF	Alstom/BHEL
25 kV AC MG								
YAU-	4	1	4	126.5	65	1966	ICF	Hitachi/Fuji
YAU- (Thyristor)	4	1	4	90	65	1991	ICF	GEC Alsthom/ UK BHEL
MEMU	4	1	4	163.3	100	1994	ICF	BHEL

Electric locomotives

Class	Wheel arrangement	Output kW continuous/ one hour	Speed km/h	Weight tonnes	No in service	First built	Builders Mechanical	Electrical
Broad-gauge 1.5 kV DC								
WCM6	Co-Co	3,430	105	120	2	1996	CLW	CLW
WCG2	Co-Co	3,130/1,220	90	132	57	1971	CLW	CLW
Broad-gauge 25 kV AC								
WAM4	Co-Co	2,715/2,870	120	112.8	471	1971	CLW	CLW
WAP1/3	Co-Co	2,800/2,910	130/140	107	64	1980	CLW	CLW
WAP4/6	Co-Co	3,730/3,990	140	107	294	1995	CLW	CLW
WAP5	Bo-Bo	2,985/4,000	160	78	14	1996	ABB	ABB (3 phase loco)
WAP7	Co-Co	4,565	140	123	17	2001	CLW	Bombardier/CLW (3 phase loco)
WAG5	Co-Co	2,870/3,250	80	118.8	1,171	1978	CLW	CLW/BHEL
WAG7	Co-Co	3,730/3,990	100	123	673	1992	CLW	CLW
WAG6A	Bo-Bo-Bo	4,475/4,560	100	123	6	1988	ASEA	ASEA
WAG6B	Bo-Bo-Bo	4,475/4,560	100	123	6	1988	Hitachi	Hitachi
WAG6C	Co-Co	4,475/4,560	100	123	6	1988	Hitachi	Hitachi
WAG9/ WAG9H	Co-Co	4,565	100	123	61	1996	Adtranz	Adtranz/CLW 3 phase loco
Broad-gauge dual-voltage 25 kV AC/1.5 kV DC								
WCAM1	Co-Co	2,715/2,870* 2,185	120/80	113	52	1975	CLW	CLW
WCAM2	Co-Co	3,505* 2,160	120/80	113	20	1995	BHEL	BHEL
WCAM3	Co-Co	3,730* 3,432	105	121	53	1996	BHEL	BHEL
WCAG1	Co-Co	3,730* 3,432	100	128	12	1998	BHEL	BHEL
Metre-gauge 25 kV AC								
YAM1	Co-Co	1,215/1,300	80	52	18	1965	Mitsubishi	Mitsubishi

* First figures apply to AC operation, subsequent figures to DC

Diesel locomotives

Class	Wheel arrangement	Power kW	Speed km/h	Weight tonnes	First built	Mechanical	Builders Engine	Transmission
Broad-gauge								
WDM2	Co-Co	1,790	120	112.8	1962	DLW	Alco/V251B	E BHEL
WDM3A	Co-Co	2,310	120	112.8	1994	DLW	Alco/V-251B/upgraded	E BHEL
WDG3A	Co-Co	2,310	100	123	1996	DLW	Alco	E BHEL
WDP1	Bo-Bo	1,715	120	80	1995	DLW	Alco	E BHEL
WDM4	Co-Co	1,965	120	113	1962	GM	GM567D3	E GM
WDM6	Bo-Bo	895	75	70	1981	DLW	DLW/251D	E BHEL
WDM7	Co-Co	1,475	105	96	1987	DLW	DLW-Alco/251-B12 Eye	E BHEL
WDS4	C	520	65/27	60	1969	MaK/CLW	MaK/CLW/6M282A(k)	H/HM KPC Voith L4v2U2
WDS4A	C	490						H
WDS4B		520						H
WDS4D		450	65	60	1968	CLW	MaK/6M282A(k)	H Voith L4r2u
WDP3A	Co-Co	2,310	160	117	1998	DLW	DLW	
WDG4	Co-Co	2,984	100	126	2000	GM/DLW	GM/DLW	
WDP4	Co-Co	2,984	140	126	2001	GM/DLW	GM/DLW	
WDS5	Co-Co	795	109	126	1967		Alco/251-B	E GE
WDS6	Co-Co	1,045	62.5	126	1977	DLW	Alco/DLW/251D	E BHEL
Metre-gauge								
YDM1-IR	B-B	472	80	46.6	1955	MaK/CLW	MaK/CLW/6M282A	H Voith
YDM-2	B-B	520	75	48	1986	CLW	CLW/MaK/6M282A(k)	H Voith
YDM3	B-B	1,055	80	58.5	1961	GM	GM 12 567c	E GM
YDM4	Co-Co	1,045	96	72	1961	Alco/DLW	Alco 251-D	E BHEL
YDM4A	Co-Co	1,045	96	67	1964	MLW Canada	MLW/251-D	E GE Canada
YDM5	C-C	1,035	80	69	1964	GM	GM 12 567c	E GM
Narrow-gauge								
NDM1	B'-B'	110	33	29	1955	Arn Jung	MWM TRHS/518 S	H Voith L33 U
NDM5	B-B	365	50	22	1987	CLW	Cummins/KTA 1150L	H Voith L2r2zu2
ZDM2R	B'-B'	520	50	32	1964	MaK	Maybach/MD 435	HM MaK
ZDM3	B'-B'	520	50	35	1971	CLW	CLW/MaK/6M282A(k)	HM Kirloskar
ZDM4A	1B-B1	520	50	38.5	1984	CLW	CLW/MaK/6M282A(k)	HM Kirloskar
ZDM5	B-B	365	50	22	1990	CLW	CLW/MaK/6M282A(k)	H Voith L4r 22

Other areas of investment are increasing line capacity and improved safety, by provision of tokenless block, track circuiting, block-proving by axle counters, and interlocking of level crossings. During 2001–02 98 level crossings were interlocked.

After successful trials, the Anti-Collision Device developed by Konkan Railway Corporation is being installed on a 1,700 km broad gauge section of the Northeast Frontier Railway. Similar work has been sanctioned on a total of 1,700 km on the Northern, South Western and South Central Railways.

On the telecommunications side, IR is replacing analogue microwave by digital systems, installing electronic telephone exchanges in place of electromechanical, and providing UHF radio links over busy sections. During 2001–02 668 route-km of long-haul digital microwave was installed, bringing the total lines equipped to 6,352 route-km. At the end of FY2001-02 IR had fibre-optic communication on 5,782 route-km. It has been decided to provide such communications on all newly electrified sections of line, and for replacement of existing copper cables according to age and condition.

In 2000, IR decided to commercially exploit its right of way by laying and leasing optical fibre cables along over 62,000 km and to modernise the rail communications system to improve operation and safety. Rail Tel Corporation was set up to implement this plan. Some 150 major cities and towns and 1,500 stations are expected to be connected to the fibre-optic network during FY2003-04.

To provide a quick communications link from a site in event of an accident or emergency, satellite phones are being provided at each zone and divisional headquarters.

In IR's budget for FY2003-04, an outlay of Rs6,890 million was provided for signalling and telecommunications work.

Electrification

In 1925 1.5 kV DC traction was introduced on IR and at present this is confined to the 433 km suburban network on the Central and Western railways in the Mumbai area. The 25 kV 50 Hz system was adopted as standard in 1957 and has ben used for all subsequent electrification schemes. During recent years, the focus has been on electrifying the trunk routes linking Mumbai, Delhi, Calcutta and Chennai; and five of the seven major routes between these cities are energised. Electrification of the Chennai–Calcutta and Chennai–Mumbai routes is in progress, with energisation also planned for the high-density routes in and around the mineral-rich areas of Bihar, West Bengal and Orissa.

At 31 December 2003 electrification extended to 16,614 route-km. This was 26 per cent of total route-km, but during electric traction was handling 50 per cent of passenger train-km and 61.2 per cent of gross tonne-km on broad-gauge.

In 1991 the Asian Development Bank approved a loan worth US$225 million towards the US$617 million expansion of operating capacity between New Delhi and Calcutta. This focuses on creation of a second electrified route with modern signalling between the two cities.

Electrification of the 218 km Patratu–Sonnagar section and Sitarampur–Danapur–Mughalsarai (562 km) has been completed.

Freight train hauled by a Class WDM2 diesel-electric locomotive (K K Gupta)   0114649

IR assigns high priority to electrification of the remaining high-density routes operated by diesel traction in a move to reduce dependence on imported petroleum products. During the ninth plan (1997–2002) 2,484 route-km were electrified. During 2001–02, electrification of 603 route-km was completed. This comprises: ER 165 km; NR and NER 35 km; SER 210 km; SR 101; and WR 92 km). In FY2003-04 an outlay of Rs1.23 billion has been assigned and a target of electrifying 350 route-km has been set. Now IR's aim is to complete electrification of the entire Kolkatta–Chennai route and provide an alternative electrified route on the Asansol–Mughalsarai section.

A 2 × 25 kV autotransformer system is now in operation on the heavy haul route Bina–Katni–Anuppur–Bishrampur/Chirimiri sections.

Track

With the progressive introduction by IR of long-welded rail and concrete sleepers, the extension of mechanised track maintenance is a priority. The World Bank has financed substantial purchases of equipment. Besides using its own resources, IR is also resorting to funding from the Asian Development Bank and recourse to leasing and contracting of mechanised maintenance. During 2001–02, 63,719 km of mechanised tamping was completed. High-output tampers are being used for straight track, and Unimat machines for turnouts. In 2001–02 40,177 turnouts were tamped using these machines. Dynamic track stabilisers and ballast regulators are also being used to improve retention of packing by tampers. The procurement of timber sleepers for main lines has been terminated and

concrete sleepers are now installed. At the end of FY2001-02 IR had a total of 44,737 track-km laid on concrete sleepers.

In FY2002-03 renewal was completed of 4,776 track-km. With rapid progress being made on gauge conversion, the aim is to liquidate all arrears of track renewal on the busiest broad-gauge sections. Ultrasonic rail testing has been introduced on all major routes. An outlay of Rs26.05 billion has been provided in IR's budget for FY2003-04 for track renewal (3,300 track-km of primary renewal and 550 track-km of secondary renewal). Of this amount Rs18.29 billion was from the Special Railway Safety Fund.

Research and development

Projects undertaken by IR's Research Standards & Development Organisation (RDSO) fall into four categories: improvements in traction and rolling stock; improving infrastructure; improving safety; and improving reliability. Some of the important projects being implemented are: the use of aluminium alloys in wagon design (to improve payload-to-tare ratio); lightweight BOXN wagons; development of goods brake vans for operation at 100 km/h; the design of crashworthy passenger coaches; an LHB variant of the ACON coach; development of a train-actuated axle detector type warning device and of an indigenous digital axle counter.

To give a greater focus to research and development activities the status of RDSO has been elevated from that of an 'Attached Office' to the Railway Board to that of a zonal railway from 1 January 2003.

Central Railway

400001 Chatrapati Shivaji Terminus, Mumbai
Tel: (+91 22) 22 62 15 51   Fax: (+91 22) 22 62 45 55
Web: http://www.centralrailway.online.com

Key personnel

General Manager: S B Ghosh Dastidar
Additional General Manager: S P Chaudhary
Senior Deputy General Manager: A K Upadhyay
Chief Administrative Officer (Construction):
   S P Vatsa
Chief Electrical Engineer: R K Sareen
Chief Commercial Manager: S K Nanda
Chief Operating Manager: R N Verma
Principal Chief Engineer: C K Narsimhan
Chief Mechanical Engineer: A K Rao
Chief Medical Director: Dr H L Parmar
Financial Adviser and Chief Accounts Officer:
   Sudha Chobe
Chief Signal and Telecommunication Engineer:
   R C Tripathi
Controller of Stores: G B Bhatnagar
Chief Personnel Officer: Mohd Irshad

Data logger at the route relay interlocking facility at Mumbai's central control cabin   0089095

Chief Security Officer: R K Sharma
Chief Public Relations Officer: Sumil Jain

Gauge: 1,676 mm; 762 mm
Route length: 3,193 km; 573 km
Electrification: 1,339 km at 25 kV 50 Hz AC; 369 km at 1.5 kV DC

Organisation

CR formerly had divisional headquarters at Mumbai, Bhusawal, Nagpur, Jabalpur, Solapur, Jhansi and Bhopal. However, the creation of new zonal railways saw the Bhopal, Jabalpur and Jhansi Divisions, together with a new Agra Division, transferred away. Now it mainly serves the state of Maharashtra.

By virtue of its central location, CR is of pivotal importance to the IR network, connecting north and south India and east and west India. Besides carrying heavy transit traffic, it has a high loading potential of its own.

Revenue (Rs million)	2000–01	2001–02
Passengers	21,294	22,766
Freight	33,235	39,171
Other income	2,844	3,424
Total	57,373	65,361

Expenditure (Rs million)	2000–01	2001–02
Staff/personnel	n/a	n/a
Materials and services	n/a	n/a
Depreciation	n/a	n/a
Financial charges*	n/a	n/a
Total	52,694	54,811

* Pension fund and payment to general revenue.

Traffic (million)	2000–01	2001–02
Passenger journeys	1,446	1,416
Passenger-km	111,112	115,013
Freight tonnes*	117.8	115.0
Freight tonne-km	48,748	57,658

* includes non-revenue traffic.

Passenger operations

CR transports nearly 3.5 million passengers on about 1,600 passenger trains every day. In 2001–02, passenger-km constituted 23.8 per cent of the IR total and are the highest amongst the zonal railways.

The Mumbai area suburban services account for 80 per cent of the total originating passengers and 28 per cent of total passenger-km, with an average trip length in 2001–02 of 28.9 km. Suburban traffic has grown from 452 million originating passengers to 1,014 million since 1970. CR runs 1,160 suburban trains a day in the Mumbai area, while the number of 12-car trains has increased from 182 to 188 per day. The network has also been extended by a double-track line across Thane Creek to a satellite city, New Mumbai on the mainland, to absorb population from the overcrowded Mumbai island.

In 2000–01 originating passenger journeys totalled 1,265 million, compared to the previous year's total of 1,235 million, of which suburban traffic accounted for 1,014 million and non-suburban for 247 million on the broad-gauge system. The balance of 3.8 million passengers was carried on narrow-gauge lines. Total passenger-km was 115,013 million.

Future planning is based on an annual growth rate of 4 per cent for suburban and 3 per cent non-suburban traffic. To cater for the rise in suburban traffic, the network in the Mumbai area is being expanded and strengthened. Track-doubling on the Belapur–Panvel line was completed in 2000. Construction of fifth and six lines between Kurla and Thane has also been undertaken. Additional emu rakes are being introduced and signalling works are under way. It is also planned to reduce progressively the headway of suburban services during peak hours initially from 5 to 4 minutes, and later 3 minutes. The pace of track renewal is being stepped up, and passenger information systems are being made more efficient. The World Bank has approved a loan of Rs16.13 billion for the Mumbai suburban sections of the Central Railway and the Western Railway.

Intercity emu of Central Railway   0001709

In 1995 CR commissioned one of the two new terminal stations in Mumbai, at Kurla on the city's outskirts. Its five platforms can accommodate the 26-car trains which IR plans to introduce. This terminal is already being expanded to cater for three additional pairs of trains. A new station has also been built at Habibganj (Bhopal).

Freight operations

Freight consists chiefly of bulk commodities such as coal, cement, fertiliser, petroleum products, food grains and raw materials to steel plants. Total originating revenue traffic up to the end of February 2004 for FY2003–04 was 37.2 million tonnes and the operating ratio was 84.8.

Until 1992–93, annual growth of originating and total traffic had been almost 4 per cent since the mid-1980s. To meet the increased demand, extensive electrification, doubling of tracks, track renewals and improved signalling are in various stages of completion. Higher capacity electric locomotives have also been introduced. Freight trains with trailing loads of 4,700 tonnes are already running and their number is being increased.

A rapid increase in ISO container traffic followed the 1989 opening of Jawaharlal Nehru Port at Nhava Sheva in New Mumbai. This is served by an offshoot of the new suburban line across Thane Creek.

New lines

Among major new line construction projects in progress are Panvel–Karjat (28 km), Amravati–Narkher (138 km) and Baramati-Lonand (54 km). Construction of the 250 km Ahmednagar–Parli Vaijnath line in the underdeveloped area of Marathwada is also in progress. In the budget for 2003–2004 Rs 593 million was allocated for these projects.

Improvements to existing lines

During 2001–02 double-tracking of 45 km was completed. In the budget for FY2003–04, Rs600 million was provided for new double-tracking works and those in progress. Gauge conversion of 52 km of narrow-gauge between Pandhar and Kurduwadi was completed in 2000. Phases III and IV of the conversion of the Miraj-Latur narrow-gauge line (359 km) were in progress in 2004 and Rs250 million was provided in the FY2003–04 budget for this work. In addition Rs3,550 million was provided for track renewals, of which Rs1,290 was from the Special Railway Safety Fund.

Traction and rolling stock

In March 2004 the railway was operating on broad gauge 268 AC and 124 DC or AC/DC electric locomotives. At 31 March 2003 it had 713 broad gauge diesel locomotives, 372 emu motor cars and 740 trailers, 3,654 passenger coaches (including 1,500 sleeping and 52 dining cars), and 61,448 freight wagons (four-wheel equivalent). On narrow gauge the railway was operating three steam and 41 diesel locomotives, four railcars, 163 passenger coaches and 238 freight wagons (four-wheel equivalent).

Signalling and telecommunications

The present emphasis is on rehabilitation of ageing assets, along with enhancement of safety, reliability and application of modern technology. Safety aids are being installed like block-proving by axle counter, track circuiting of whole station areas, provision of two distant signals and introduction of route-relay and solid-state interlocking. Approval has been given for the provision of optical fibre cables on several new sections of the CR network, including: Delhi–Agracant; Bhusawal–Nagpur; Wardha–Ballarshah; and Pune–Wadi. During FY2003–04, Rs 610 million was allocated for these works. A total of 128 data loggers and 877 audio frequency rack circuits have been provided so far.

Electrification

A major project in progress is conversion of the Mumbai Division's traction power supply system from 1.5 kV DC to 25 kV AC by Mumbai Rail Vikas Corporation. The Vasai Road–Panvel–Jassai section has been completed, resulting in a reduction in running time of 1 hour to 1 hour 39 minutes. In 2004 work on the Igatpuri–Kasara and Karjat-Lonavla–Pune sections was also in progress for completion in March 2008.

Coupler in standard use
Passenger cars: Screw, Schaltbau (emus)
Wagons: CBC AAR-type screw; CBC Alliance II

Braking in standard use
Locomotive-hauled stock: Vacuum and air
Emus: Air

Track (BG)
Rail: 60, 52 and 44 kg/m FF
Crossties (sleepers): Monobloc concrete, 2,750 × 70 × 230 or 151 mm; wooden, 2,750 × 250 × 130 mm; steel, of various sizes
Sleeper spacing: 1,660/km on A routes; 1,540/km on B, C and D routes; 1,310/km on E routes
Fastenings: Keys, elastic fastenings such as Pandrol clip
Max gradient: 2.9% (1 in 34)
Max permissible axleload: 22.5 tonnes

Container Corporation of India Ltd (Concor)

2nd Floor, Le Meridian Commercial Tower, Raisina Road, New Delhi 110001
Tel: (+91 11) 23 75 31 64; 65; 67; 68; 69
Fax: (+91 11) 23 36 84 24
e-mail: concor.co@sprintrpg.ems.vsnl.net.in

Key personnel

Managing Director: A K Kohl
Director, International Marketing and Operations: P G Thyagarajan
Director, Domestic Division: S C Misra
Director, Projects and Services: Rakesh Mehrotra
Director, Finance: Arun N Pai
Executive Director, Finance: Runa Mukherjee
Group General Managers
    Strategic Planning: K Sathianathan
    Customs: R Pant
    Technical: A K Khosla
    Marketing and Commercial: A K Gupta
    Planning and Development: P S Nerwal

Organisation

Container Corporation (Concor) activities cover three distinct areas: carrier, terminal operator, and containerised freight service (CFS) operator. As a carrier, Concor conveys containerised international cargo by scheduled trains between ports and inland terminals. Terminal operations are carried out at 51

Reach stacker loading containers at a Concor terminal (K K Gupta)   0547043

facilities throughout India. As a CFS operator, Concor provides value-added services by offering transport, warehousing and import/export cargo handling.

See also 'Intermodal operations' under the entry for 'Indian Railway Board'.

Developments

In late 2004 Concor announced that it planned to create 20 to 25 additional inland container depots over the following four years. Nine of these were to be commissioned in FY2004-05.

Traction and rolling stock

By the end of FY2004-05 Concor expected its fleet of container-carrying wagons to total around 7,600.

UPDATED

East Central Railway

844101 Hajipur, Bihar
Tel: (+91 6224) 747 28   Fax: (+91 6224) 747 38

Key personnel

General Manager: R S Varshney
Secretary to the General Manager: J S P Singh
Additional General Manager: A K Jhingran
Chief Engineer: A K Gupta
Financial Adviser and Chief Accounts Officer: Ram Prakash
Chief Commercial Manager: G R Vij
Chief Electrical Engineer: S P Singh
Chief Mechanical Engineer: P Agarwal
Chief Personnel Officer: C L Bharti
Chief Medical Director: Dr M M Mohanty
Chief Signal and Telecommunications Engineer: Puran Singh
Controller of Stores: B K Sinha
Chief Operating Manager: A K Jhingran

Chief Administrative Officer (Construction): Shyam Kumar
Chief Security Commissioner: L Kumar

Gauge: 1,676 mm
Route length: 2,681 km
Electrification: 1,314 km at 25 kV AC 50 Hz

Organisation

East Central Railway is one of several new zonal railways that the Ministry of Railways decided to create in 1996. It became operational on 1 October 2002 and comprises the former Danapur, Dhanbad and Mughalsarai Divisions of the Eastern Railway and the former Samastipur and Sonepur Divisions of the North Eastern Railway. It serves mainly Bihar state and parts of Jharkhand and Chattisgarh states.

New lines

There are a number of sanctioned new line projects, including Koderma–Tillaiya and Koderma–Ranchi.

Rs2,220 million was provided in the budget for 2003–04 for these works.

Improvements to existing lines

Two major gauge conversion projects that have been approved are of the Jaynaar–Darbhanga–Narkatiaganj (268 km) and Samastipur–Khagaria sections (86 km). Of the four major bridges that IR has decided to construct, three will be on the East Central Railway (at Patna and Monghyr over the River Ganges and a bridge over the River Kosi).

Traction and rolling stock

In 2004 the East Central Railway fleet comprised 258 electric and 154 diesel locomotives on broad gauge and 28 metre gauge diesel locomotives. It also operated 2,482 broad gauge and 380 metre gauge passenger coaches and 39,274 broad gauge and 975 metre gauge freight wagons (four-wheel equivalent).

East Coast Railway

751023 Chandrashekarpur, Bhubaneswar, Orissa
Tel: (+91 674) 230 00 29; 230 07 73
Fax: (+91 674) 230 01 96

Key personnel

General Manager: S R Chaudhuri
Secretary to the General Manager: G C Ra
Financial Advisor and Chief Accounts Officer: S Mookerjee
Chief Commercial Manager: L N Sadhangi
Chief Engineer: K Gangopadhya
Chief Mechanical Engineer: K K Rao
Chief Medical Director: Dr G Nagabhuhan
Chief Signal and Telecommunications Engineer: P Samal
Controller of Stores: S K Prasad
Chief Security Officer: S C Sahoo

Gauge: 1,676 mm; 762 mm
Route length: 2,421 km; 91 km
Electrification: 1,087 km at 25 kV AC 50 Hz

Organisation

East Coast Railway is one of several new zonal railways that the Ministry of Railways decided to create in 1996. It became operational on 1 April 2003 and comprises the former Khurda Road, Sambalpur and Waltair Divisions of the South Eastern Railway. It serves the state of Orissa and parts of Andhra

Class WAG5 electric locomotive powering a passenger service on the East Coast Railway's Kottavalasa–Kiruldul line (K K Gupta)   0547044

Pradesh and Chattisgarh. The region is very rich in mineral and natural resources.

Passenger operations
From April to December 2003 the East Coast Railway carried 30.97 million originating passengers. In 2004 39 Passenger Reservation Systems were in operation on the network.

Eastern Railway

17 Netaji Subhash Road, Calcutta 700 001
Tel: (+91 33) 222 71 20   Fax: (+91 33) 248 03 70

Key personnel
General Manager: Shyam Kumar
Additional General Manager: Ramesh Chand
Senior Deputy General Manager: (vacant)
Financial Adviser and Chief Accounts Officer:
  Ranjan Tewary
Chief Operating Manager: S R Thakur
Chief Commercial Manager: J K Mitra
Chief Electrical Engineer: H B Singh
Chief Personnel Officer: C K Sharma
Principal Chief Engineer: S K Vij
Chief Mechanical Engineer: D N Mathur
Chief Medical Director: Dr S B Sarcar
Chief Signal and Telecommunications Engineer:
  A K Kapoor
Controller of Stores: S Paul
Chief Security Commissioner: B I Passah
Chief Public Relations Officer: S Majundar

Gauge: 1,676 mm; 762 mm
Route length: 2,250 km; 132 km
Electrification: 1,218 km at 25 kV 50 Hz AC

Organisation
ER used to be described as the 'Black Diamond' railway on account of its prime task of supplying coal to power stations.

However, the Danapur, Mughalsarai and Dhanbad Divisions of the Eastern Railway were transferred to the new East Central Railway (see entry in India section of *Railway systems and operators*), which commenced operations on 1 October 2002. It now mainly serves West Bengal.

Passenger operations
Originating passenger journeys amounted to 564.5 million between April 2003 and January 2004. Some 1.56 million daily journeys were made on the Calcutta suburban network. Almost all main line passenger services are electrically operated. The compositions of emu trains running on the Sealdah and Howrah Divisions are mostly of nine or 10 coaches, but in December 2002, ER for the first time introduced three pairs of 12-coach trains on the Sealdah Division. Greater use of 12-coach formation is foreseen in these two divisions.

Each day ER runs 177 mail/express services, 869 emu services on the suburban sections of the Howrah and Sealdah lines, 96 non-suburban dmus and 59 main line emus (MEMUs). MEMU services were introduced on Barddhaman–Purulia and Asansol–Jhaja routes in August and September 1998 respectively.

As well as extending platorms to accommodate 24-coach trains, ER is taking additional steps to improve its services to passengers, including: provision of additional ticket counters; construction of platform shelters; the supply of drinking water; and the establishment of a system to communicate train running information. Computerised reservation is now available at 296 locations. Information regarding the availability of reservations on 21 important long-distance trains over Eastern and South Eastern networks is distributed via a cable TV service in Calcutta and surrounding areas.

Freight operations
In the period April 2003 to January 2004 ER achieved a total loading of 31 million tonnes of freight. After restructuring, the operating ratio dropped to 341.9 but has since improved to 188.1.

Improvements to existing lines
Several development works are in progress or planned to improve the infrastructure. These

Freight operations
The railway loaded 52.01 million tonnes of freight up to January 2004, the third highest on the Indian railway network.

New lines
Construction of over 850 km of new lines has been approved. The longest is Khurda Road–Bolangir

Eastern Railway emu specially prepared for an inaugural run on the newly electrified Katwa–Bandel section (K K Gupta)   0547045

Coal train during loading on the Eastern Railway (K K Gupta)   0114653

include: construction of new lines; restoration of dismantled routes; track-doubling of busy single-track lines and construction of new road overbridges and provision of improved traffic facilities. ER's budget for 2003–04 provided a sum of Rs954 million for these works. In addition Rs2,100 million was provided for track renewal, of which Rs1,000 million was from the Special Railway Safety Fund.

Traction and rolling stock
In March 2004, after restructuring, the railway operated 197 electric locomotives on broad gauge. The previous year its fleet consisted of 521 diesel locomotives, 420 emu motor cars and 892 trailers, 4,314 passenger coaches, and 72,470 freight wagons (four-wheel equivalent). On narrow gauge the fleet comprised five diesel locomotives, five railcars and 40 passenger coaches.

Signalling and telecommunications
Apart from signalling works to improve line capacity on busy sections, ER is at present concentrating on replacement of old electromechanical installations by route-relay and panel interlockings. Some of these will be of the solid state (SSI) type. Track circuiting or provision of axle-counters has been sanctioned on all major ER routes and a number of level crossings are to be interlocked to improve operational safety.

(289 km). In the interim budget approved by the Indian parliament in early 2004 construction of another two new lines totalling 154 km was authorised.

Optical fibre cable is being provided to replace copper cable in several sections on an 'age-cum-condition' basis. The old relay interlocking system at Howrah was replaced in September 2003. During FY2003-04 an outlay of Rs546 million has been provided for signalling and telecommunications work.

Electrification
Electrification of Krishnanagar–Lalgola (128 km) was in progress in 2004. Rs154 million was provided in 2003–04 for this project.

Coupler in standard use
Passenger cars: Drawbar with screw coupling
Freight cars: Centre buffer coupler; CBC with transition coupling; drawbar with screw coupling

Braking in standard use
Locomotive-hauled stock: Air and vacuum brake

Track (BG)
Rail: 52 and 60 kg/m in long welded rail panels
Crossties (sleepers): prestressed concrete – spacing 1,680–1,540 km CST-9 – spacing 1,562/km; wooden
Fastenings: Elastic rail clips and malleable or spheroidal graphite cast-iron inserts
Min curvature: 8°
Max gradient: 2.25%
Max permissible axleload: 22.5 tonnes

Konkan Railway Corporation Ltd (KRC)

Belapur Bhavan, Sector 11, CBD Belapur, Navi Mumbai (New Bombay) 400 614
Tel: (+91 22) 27 57 20 15; 20 16; 20 17; 20 18
Fax: (+91 22) 27 57 24 20
e-mail: general@konkanrailway.com

Key personnel
Chairman: S P S Jain
Managing Director: B Rajaram
Director, Operating and Commercial:
Dr K K Gokhale
Director, Way and Works: D G Diwate
Finance Director: R K Sinha
Financial Adviser and Chief Accounts Officer:
R Ravikumar
Chief Track Engineer: D S Chauhan
Chief Mechanical Engineer: Deepak Gupta
Chief Electrical Engineer: A K Bharadwaj
Chief Signal and Telecommunications Engineer:
A A Bhatt
Chief Operating Manager: Vivek Sahai
Deputy General Manager: Ashish Apte
Chief Manager, Information Technology:
Sharad Saxena
Controller of Stores: R K Ahirwar
General Manager, Projects: A F Sheware

Gauge: 1,676 mm
Route length: 760 km

Organisation
This major railway opened fully in January 1998. It extends from Roha, south of Mumbai, down the coast to Mangalore. It provides a much shorter route than the historic Mumbai–Mangalore line, which also entailed a break of gauge. The new line, in addition, opens up remote areas of Maharashtra, Goa and Karnataka states.

In 1990 the Konkan Railway Corporation was formed to take over construction of the 760 km

KR Ro-Ro piggyback train   0594524

route and operate it on completion. Initially the line is a single 1,676 mm gauge track with 58 intermediate stations, built through difficult country even by Indian standards. There are 171 major bridges, the longest of which is the Sharawati at 2,065 m, and the highest, Panvel Viaduct, 64 m, and a further 1,670 minor structures. The 92 tunnels total 83.6 km, with the longest, Karbute Tunnel, near Ratnagiri, extending to 6.5 km and using incrementally launched concrete box girder construction for the first time in India. Engineered for 160 km/h top speed, ruling gradient is 0.67 per cent and minimum curve radius 1,250 m. Track is 52 kg/m rail on prestressed concrete sleepers. The entire route is equipped with specially designed points and crossings with thick-web switches and laid on concrete sleepers. Stations have panel interlockings and colour light signals; communication is by fibre optic link. All routine activities of the railway have been computerised, including ticketing, parcels booking and commercial coaching and train operations.

In addition to its train operations, KRC has also introduced several new technological concepts to the railway market. These include: a GPS-based Anti-Collision Device (ACD) system; a Self-Stabilising Track System (SST); a Railway Applications Package (RAP) suite of software modules covering many fields of railway operations; and a computer-based Rolling Stock Health Analyst (ROSHAN) system. A programme

KRC passenger train   0594523

to install 1,770 ACDs on 1,736 route-km of the Northeast Frontier Railway was underway for completion in December 2004.

The company has also undertaken construction and consultancy projects. It has been awarded the work of constructing the most critical Katra-Laoli (90 km) section of a new railway in Jammu and Kashmir.

Traffic (million)	2000–01	2001–02	2002–03
Passenger journeys	23.54	22.65	26
Freight tonnes (originating)	1.29	1.19	1.35
Freight tonne-km	81.98	55.62	86.08

Passenger operations
KRC in 2003 operated 13 pairs of Express and five pairs of passenger trains, including the Delhi–Trivandrum Rajdhani Express. In 2004 KRC proposed adding one more pair of trains. The company operates a Town Booking Agency (TBA), a computer-based system that enables passengers living in remote villages to have tickets delivered to their homes.

Freight operations
The company is endeavouring to increase its market share of freight traffic. A major initiative was the introduction in 1999 of a Roll-on/Roll-off (Ro-Ro) piggyback service for lorries, believed to be the first of its kind in India.

North Central Railway

211001 Allahabad, Uttar Pradesh
Tel: (+91 532) 256 10 07   Fax: (+91 532) 256 10 08

Key personnel
General Manager: I P S Anand
Secretary to General Manager: S S Negi
Financial Adviser and Chief Accounts Officer:
Suman Kumar
Chief Commercial Manager: A K Shrivastav
Chief Engineer: A N Verma
Chief Electrical Engineer: Amar Nath
Chief Mechanical Engineer: Dev Raj
Chief Medical Director: Dr K SudhaRao
Chief Operating Manager: Girish Chandra
Chief Personnel Officer: Om Prakash
Chief Signal and Telecommunications Engineer:
P K Shrivastava
Controller of Stores: Mahesh Chandra
Chief Security Officer: Haranda

Gauge: 1,676 mm
Route length: 3,062 km
Electrification: 25 kV AC 50 Hz

Organisation
North Central Railway is one of several new zonal railways that the Ministry of Railways decided to create in 1996. It became operational on 1 April 2003 and comprises the former Allahabad Division of the Northern Railway, the Jhansi Division of the Central Railway and the newly created Agra Division. It mainly serves the state of Uttar Pradesh and parts of Madhya Pradesh, Rajasthan and Bihar.

Passenger operations
North Central Railway carries over 338,000 originating passengers every day. In 2003–04 it introduced 32 new trains.

Freight operations
NCR constitutes less than 5 per cent of the IR network but it carries 16 per cent of its total freight

traffic. Commodities handled include coal to power stations, steel, cement, fertilisers, food products, cereals and petroleum products.

New lines
There are two new lines, Etaah–Mainpuri (59.5 km) and Etaah–Agra (110 km) at different stages of construction. Rs300 million was provided in FY2003-04 for these lines.

Improvements to existing lines
Gauge conversion of Agra Fort–Bandikui was in progress in 2004. Provision of third and fourth lines was in progress on the Kanpur–Panki and Tundla–Yamuna Nagar sections.

A major project costing over Rs5,000 million and funded by KFW of Germany will provide a state-of-the-art signalling and telecommunication system on Ghaziabad section.

In 2003–04 Rs1,860 million for track renewals and Rs590 million for signalling were provided. Of these totals, Rs980 million and Rs380 million respectively were from Special Railway Safety Fund.

North Eastern Railway

Gorakhpur 273012, Uttar Pradesh
Tel: (+91 551) 220 10 41   Fax: (+91 551) 220 12 99

Key personnel
General Manager: J P Batra
Senior Deputy General Manager: R Sinha
Chief Personnel Officer: A K Tiwari
Financial Adviser and Chief Accounts Officer:
R Ashok
Chief Commercial Manager: V K Jaiswal
Chief Engineer: Onkar Singh
Chief Administration Officer, Construction:
Vijay Kumar
Chief Operations Manager: K Chaturvedi

Chief Electrical Engineer: R K Sapre
Chief Signals and Telecommunications Engineer:
G D Chandolia
Chief Mechanical Engineer: P K Gupta
Chief Medical Director: Dr S K Mehta
Chief Public Relations Officer: M K Singh
Controller of Stores: A K Mukhopadhya
Chief Security Officer: Jaya Singh

Gauge: 1,676 mm; 1,000 mm
Route length: 1,520 km; 1,983 km
Electrification: 20 km at 25 kV AC 50 Hz

Organisation
Increases of population and economic development now exert extreme pressure on the

still predominantly metre-gauge network of the NER. It covers the state of Uttar Pradesh and parts of Bihar and Uttaranchal, and moves traffic for Nepal. It provides a vital link to the northeastern states of the country.

The Samastipur and Sanpur Divisions have been transferred to the East Central Railway, which commenced operations in October 2002.

Passenger operations
In 2002–03 originating passenger journeys for the railway including the now-separate East Central Railway totalled 202 million, of which 67.5 per cent were on broad gauge, for 41,174 million passenger-km.

New lines

In the interim budget for 2004 construction of two new broad gauge lines (of 58 and 35 km) was approved.

Improvements to existing lines

Conversion to 1,676 mm gauge of NER's entire trunk route from Lucknow to Katihar was completed before the current programme. This had greatly improved rail access to Assam, while regauging of the Bhatni–Varanasi line simplified interchange of freight with the Central, Eastern and Northern railways.

Under the current programme, conversion of Burhwal-Sitapur (98 km), Varanasi–Allahabad (125 km) and Mankapur–Katra (90 km) has been completed. In addition, the Rampur–Kathgodam conversion/construction project enabled introduction of direct broad-gauge trains between Kathgodam and Delhi/Calcutta. In 2004 gauge conversion works were in progress on several other sections and a sum of Rs582million was earmarked for these works in the 2003–04 budget. Due to the broad-gauge conversion of important rail sections, many metre-gauge sections have been isolated. Rail bus services have been provided in such sections. In the interim budget for 2004–05, gauge conversion of one such section, Aunihar-Jaunpur (59 km), was approved.

Construction of a new 8.6 km line between Faizabad and Katra, which includes a rail bridge over the River Saryu, was completed and inaugurated by the Indian Prime Minister in February 2004.

The circular railway at Lucknow, which includes nearly 20 km of the NER, was electrified in 2002 and main line emu services are now running over it. This is the first section of the NER to be electrified.

Traction and rolling stock

In March 2003 (before restructuring) NER operated, on metre gauge, 124 diesel locomotives, 1,061 passenger coaches and 3,417 wagons (6,352 four-wheel equivalent); and on broad gauge, 195 diesel locomotives, 2,560 passenger cars and 3,419 wagons (7,488 four-wheel equivalent).

Rail bridge over the River Saryu, opened in 2004 0583120

Gauge conversion work in progress on the Kanpur–Kasganj section 0583121

Northeast Frontier Railway (NFR)

Maligaon, Guwahati 781 011, Assam
Tel: (+91 361) 257 04 22   Fax: (+91 361) 257 11 24

Key personnel

General Manager (acting): Shyam Kumar
General Manager, Construction:
  N Ramasubramanian
Additional General Manager: R G Nair
Financial Adviser and Chief Accounts Officer:
  Mrs Sabita Gopal
Senior Deputy General Manager: C I Langovin
Chief Commercial Manager: Z A Siddique
Chief Operating Manager: A Lal
Chief Engineer, Co-ordination: Parthasary
Chief Electrical Engineer: U C D Shreni
Chief Mechanical Engineer: S K Suri
Chief Signal and Telecommunications Engineer:
  P Swarup
Controller of Stores: S C Anihotri
Chief Personnel Officer: A K Nigam
Chief Medical Director: Dr K Ghosh
Chief Public Relations Officer: L Sarma
Chief Security Commissioner: S P Ram

Gauge: 1,676 mm; 1,000 mm; 610 mm
Route length: 1,758 km; 2,106 km; 87 km

Organisation

NFR serves the whole of Assam and North Bengal, parts of North Bihar, the states of Arunachal, Manipur, Meghalaya, Mizoram, Nagaland, Sikkim and Tripura. Its 610 mm gauge component is the world-famous Darjeeling–Himalaya Railway. In December 1999, this mountain railway became the

second line to be accorded 'World Heritage Site' status by UNESCO.

Besides playing a vital role in the transport of people and essential commodities, NFR is also of strategic importance, since the region is practically enveloped by international borders. NFR covers one of the most picturesque regions, overlooked by the Himalayas. Apart from serving well-known tourist centres like Darjeeling, Shillong and the wildlife sanctuaries of Kaziranga, Manas and Jaldapara, it also covers the vast tea-garden belts

of North Bengal and Assam. From Darjeeling to Dibrugarh, several stretches of NFR are prone to damage by floods or landslides during monsoons. To maintain this vital communication link with the least interruption has always been a challenging task.

Passenger operations

Direct mail/express trains link Guwahati with Delhi, Mumbai, Bangalore, Calcutta, Chennai, Patna, Lucknow, Kanpur, Varanasi, Allahabad, Cochin and Trivandrum.

Northeast Frontier Railway Class YDM4 metre gauge diesel-electric locomotive at Siliguri
(Colin Boocock) 0547046

In January 1998, the Delhi–Guwahati Rajdhani Express was extended once a week to Dibrugarh. In FY2002–03 originating passenger traffic totalled 29.6 million journeys. In 2004 there were 46 Passenger Reservation Systems in use on the network.

Freight operations
Inward freight traffic consists mainly of essential commodities such as food grains, salt, sugar, cement and steel. Outward traffic, which is smaller in volume, comprises petroleum products, coal, bamboo, timber, jute, tea, and dolomite. In FY2002–03 revenue-earning originating traffic totalled 7.08 million tonnes.

New lines
Since 1979 NFR has had an independent construction organisation which has given a boost to the development of the railway network in the region. Present policy is to build rail links between the capitals of the region, and accordingly construction of a 119 km line from Kumarghat to Agartala was included in the 1996–97 budget and construction work on this project started in November 1998. The 21 km Kumarghat–Manu section of this line was commissioned in December 2002.

An important project was construction of the second rail/road bridge across the Brahmaputra river at Jogighopa, along with 142 km of broad-gauge line from Jogighopa and Goalpara and thence to Guwahati. The 2.3 km bridge and line from Jogighopa to Goalpara (20 km) was inaugurated in April 1998.

Another current major investment is the Rs17.67 billion Bogipeel bridge over the River Brahmaputra, together with connecting lines. It will link Dibrugarh and Lakhimpur districts of Assam and it will play a key role in providing infrastructure to boost development of the region.

Improvements to existing lines
In 1998–99, conversion to broad gauge of the Mariani–Jorhat (18 km) and Sibsagar–Moranhat (39 km) sections was completed. Conversion of New Jalpaiguri–Siliguri–New Bongaigaon (280 km) and Lumding–Silchar sections is in progress. An outlay of Rs1,630 million has been provided in the FY2003-04 budget.

In the interim budget for 2004–05 gauge conversion of Rangiya-Murkangsselek (510 km) was approved.

Traction and rolling stock
In March 2004 NFR was operating, on 1,676 mm gauge, 136 diesel locomotives, 1,296 passenger cars and 14,109 wagons (four-wheel equivalent);

Northeast Frontier Railway 0-4-0ST steam locomotive at Darjeeling with a school train from Kurseong (Colin Boocock)   0547047

on 1,000 mm gauge, 125 diesel locomotives, 537 passenger cars and 6,297 wagons (four-wheel equivalent); and on 610 mm gauge, 14 steam and two diesel locomotives, 36 passenger cars and four wagons.

Type of coupler in standard use: Passenger cars, screw; freight cars, CBC/screw type
Type of braking in standard use: Vacuum

Signalling and telecommunications
Signalling and telecommunications works are in progress throughout the region. The work of laying optical fibre cable on the Furkating–Tinsukia–Dibrugarh route is in progress at a cost of Rs187 million. Intra Voice Response System (IVRS) equipment has been installed at Guwahati and New Jalpaiguri stations, providing reservation status and availability information. IVRS was also being commissioned at other major stations. An outlay of Rs340 million for signalling and telecommunications works has been provided in the 2003–04 budget, of which Rs90 million was from the Special Railway Safety Fund.

Track
Rail: Flat-bottom 60, 52, 44.61, 37.13, 29.76, 24.8 kg/m
Crossties (sleepers): Wood, steel trough, concrete
Thickness: Wood: BG, 127 mm; concrete, 210 mm
MG, 114 mm
NG, 114 mm
Spacing
Main lines: BG, 1,540/km
MG, 1,596/km
NG, 1,230/km
Branch lines: BG, 1,309/km
MG, 1,344/km
Fastenings:
Wooden sleepers: CI/MS bearing plates and rail screws
Steel trough sleepers: loose jaws with keys
Cast-iron sleepers: keys
Concrete sleepers: elastic rail clips
Max gradient: BG 0.64%; MG 2.7%; NG 4.344%
Max permissible axleload: BG 22.9 tonnes; MG 12.7 tonnes; NG 7.6 tonnes

Northern Railway

Baroda House, Kasturba Ghandi Marg,
New Delhi 110 001
Tel: (+91 11) 23 38 72 27   Fax: (+91 11) 23 38 45 03

Key personnel
General Manager: R R Jaruhar
Additional General Manager: V K Kaul
Senior Deputy General Manager:
S K Budhalalatati
Financial Adviser and Chief Accounts Officer:
G Suman
Chief Commercial Manager: D R Sharma
Principal Chief Engineer: J P Shukla
Chief Mechanical Engineer: M K M Agarwal
Chief Medical Director: Dr D Sharma
Chief Electrical Engineer: R S Grover
Chief Operations Manager: Mathew John
Chief Signal and Telecommunications Engineer:
N K Goel
Chief Administrative Officers, Construction:
A P Mishra; Rakesh Chopra
Controller of Stores: S Kumar
Chief Personnel Officer: A Swami
Chief Security Commandant: S Awasthy
Chief Public Relations Officer: D P S Sandhu

Gauge: 1,676 mm; 1,000 mm; 762 mm
Route length: 6,459 km; 88 km; 261 km
Electrification: 668 km of 1,676 mm gauge at 25 kV 50 Hz AC

Organisation
NR's territory extends from Delhi through Punjab, Haryana, Himachal Pradesh, parts of Jammu and Kashmir, Uttar Pradesh and Uttaranchal states.

The former NR Bikaner and Jodhpur Divisions have been transferred to the newly created North Western Railway and the former NR Allhabad Division to the North Central Railway.

Passenger and freight operations
In 2002–03 the railway recorded 365.6 million originating passenger journeys and 39.56 million tonnes of originating revenue freight.

Eleven out of 28 pairs of Shatabdi/Jamshatabdi Express trains of the entire IR system run over NR and all 16 pairs of Rajdhani Express services originate or terminate in the Delhi area, which handles nearly 0.65 million passengers daily. NR operates over 450 special trains to different parts of the country during the summer season and over 80 special services during the Durga Puja holiday period. NR is now operating several pairs of 24-coach trains to increase passenger-carrying capacity. In October 2002 emu services were inaugurated on the Lucknow circular railway. NR has introduced the facility of booking train tickets via the telephone. A Rail Credit Card system, launched in association with a multinational bank, has been introduced for this purpose. While presenting the interim budget in parliament in January 2004, the Railway Minister announced

that it was proposed to introduce 17 'Sampark Kranti' express trains from Delhi to major cities in the country.

NR carries food grains from the Punjab and Haryana to almost all parts of the country, and loads 1,400 to 1,500 wagons daily during harvest.

New lines
Work on the new broad-gauge line from Jammu to Udhampur (53 km) was nearing completion in 2004. The project features high embankments, deep cuttings and a 2.5 km tunnel. It has been decided to extend this line from Udhampur to Srinagar and Baramula via Katra (290 km) as a national project. An outlay of Rs5,000 million has been provided in the FY2003-04 budget for this and Rs608 million for other new line projects. The Udhampur–Baramula project is due to be completed by August 2007. Work is also in progress on 83.74 km on the Nangal–Talwara broad-gauge project, part of which has been commissioned. In the interim budget of January 2004 two more new line works, Rewari–Rohtak (81 km) and Jind–Sonepat were approved.

Improvements to existing lines
Double-tracking works on the Muradnagar–Meerut city (29.5 km) and Gaziabad–Hapur sections was completed during FY2000-01 and work on several other sections is in progress. An outlay of Rs481 million for track-doubling projects has been provided in the FY2003-04 budget. Electrification

Passenger service hauled by a Class WAP3 electric locomotive (K K Gupta)   0114655

Northern Railway dmu (K K Gupta)   0114656

of the Ludhiana–Amritsar has been completed and work on the Ambala–Moradabad section was in progress in 2004. During 2002 electrification of 35 km of the circular railway around Lucknow area under Northern Railway and North Eastern Railway jurisdiction was completed. Rs3,690 million has been provided for track renewals during FY2003–04.

A major passenger terminal is being constructed at Anand Vihar in east Delhi from where all east- and northeast-bound traffic will be handled.

Traction and rolling stock
In 2004 there were 212 electric locomotives, 72 emu motor coaches, 145 emu trailers, 48 memu motor coaches and 141 memu trailers in service. Electric traction hauls 62.2 per cent of freight and 42.8 per cent of passenger traffic. On 31 March 2003 NR had 463 main line diesel broad gauge locomotives, 29 narrow gauge and 98 shunting diesel locomotives and 4,533 broad gauge passenger cars and 21,739 freight wagons (49,875 four-wheel equivalent).

Signalling and telecommunications
NR has installed the world's largest route relay interlocking (RRI) at Delhi main railway yard, for which it has been awarded a certificate by Guinness World Records Ltd.

Eleven stations on the Delhi–Ambala section have been equipped with networked dataloggers.

These send data to a central computer where the data is processed and analysed to generate online reports which assist in monitoring the health of the signalling system.

Besides fitting of track circuits and axle counters to enhance operational safety, major works in progress on NR include: replacement of interlocking equipment at intermediate stations; installation of train-to-control mobile communications on the Delhi–Ludhiana section; replacement of analogue by digital microwave; and provision of optical fibre cables.

In the 2003–04 budget, an outlay of Rs680 million was provided for NR signalling and telecommunications works. This included Rs500 million from the Special Railway Safety Fund.

North Western Railway

302006 Jaipur, Rajasthan
Tel: (+91 141) 222 26 95   Fax: (+91 141) 222 29 36

Key personnel
General Manager: R M Agarwal
Secretary to the General Manager: Ani Lal
Chief Control Manager: K K Singh
Chief Engineer: P S Baghel
Financial Adviser and Chief Accounts Officer:
  R C Punia
Chief Electrical Engineer: R K Gupta
Chief Mechanical Engineer: Nikhilesh Jain
Chief Signal and Telecommunications Engineer:
  S K Vashistha
Controller of Stores: M K Surana
Chief Administration Officer (Construction):
  A Singh

Chief Personnel Officer: G L Meena
Chief Medical Director: Dr S Ganguli
Chief Operating Manager: K K Gupta
Chief Security Officer: C Thamodaran
Chief Public Relations Officer: S B Gandhi

Gauge: 1,676 mm; 1,000 mm
Route length: 2,578 km; 2,875 km

Organisation
North Western Railway is one of several new zonal railways that the Ministry of Railways decided to create in 1996. It became operational on 1 October 2002 and comprises the Jodhpur and reorganised Bikaner Divisions of the Northern Railway and the reorganised Ajmer and Jaipur Divisions of the Western Railway. It mainly covers areas of Rajasthan and parts of Haryana and Punjab.

New lines
Three new lines totalling 296.7 km are under construction, including Ajmer–Pushkar (93 km) A sum of Rs1,150 million was included in the budget for FY 2003–04.

Improvements to existing lines
Gauge conversion work on several sections, including Udaipur–Chittorgarh (114 km), were in progress in 2004. Sums of Rs1,510 million for track renewals and Rs160 million for signalling and telecommunication work were sanctioned for FY2003–04, of which Rs1,250 million and Rs100 million respectively were from the Special Railway Safety Fund.

South Central Railway

Rail Nilayam, Secunderabad 500 371, Andhra Pradesh
Tel: (+91 40) 27 83 30 20   Fax: (+91 40) 27 83 32 03

Key personnel
General Manager: T Stanley Babu
Additional General Manager: A K Jain
Senior Deputy General Manager: Ramesh Chandra
Financial Adviser and Chief Accounts Officer:
  Geetha Thopal
Chief Administrative Officer (Construction):
  Amar Singh
Chief Commercial Manager: K Bhasker
Principal Chief Engineer: R Bhargava
Chief Electrical Engineer: Satish Kumar
Chief Mechanical Engineer: M L Gupta
Chief Medical Director: Dr Z Hussain
Chief Signal and Telecommunications Engineer:
  P R Gaundon
Chief Operating Manager: H G Sharma
Controller of Stores: Mangat Rai
Chief Personnel Officer: V K Manglik
Chief Security Commissioner: B Kamal Kumar
Chief Public Relations Officer: N V Ramana Reddy

Class WDP4 DLW-built GM-powered diesel-electric locomotive heading a South Central Railway express passenger service
    0583123

Gauge: 1,676 mm, 1,000 mm
Route length: 4,582 km, 1,171 km
Electrification: 1,604 km of 1,676 mm gauge at 25 kV 50 Hz AC

Organisation
SCR was set up in 1966 from portions of the Southern and Central railways. As part of the restructuring of the Indian railway network in April 2003 the Hubli Division of South Central Railway was transferred to South Western Railway.

Passenger and freight operations
Prestigious Rajdhani Express services were introduced from Delhi to Secunderabad in January 1998. In 1999, the reservation system at Secunderabad was networked with those at New Delhi, Calcutta, Chennai and Mumbai, enabling reservations to be made to any station on the IR network.

On the Ligampalli–Hyderabad–Secunderabad–Falaknuma section (43.4 km) SCR moved 162 million passengers during April 2003–January 2004. The issue of reserved tickets using computerised systems accounts for 94 per cent of sales of this type, with 80 passenger reservation centres in operation. During 2003–04 nine such centres were commissioned. Facilities for purchase of rail tickets using credit cards are available at 36 passenger reservation centres.

In February 2004 SCR implemented the second phase of the Multimode Transport System (MMTS), a joint venture between IR (SCR) and the government of Andhra Pradesh to integrate rail and road transport on the Secunderabad–Hyderabad–Ligampalli and Secunderabad–Falaknuma sections. These serve some of the most densely populated areas of the metropolitan region. MMTS forms a vital part of the railway's plan to strengthen its services here, with train services supported by a strong network of bus links.

During FY 2003–04, SCR had moved 35.7 million tonnes of freight by January 2004. The major commodities are coal, cement, food, cereals, fertilisers, petroleum products and general merchandise. The Freight Operating Information System (FOIS) has been commissioned at 21 locations on SCR to ensure effective monitoring of freight movements and giving real-time information on wagon locations to customers.

New lines
The new 35 km Karimnagar–Peddapali line opened to goods traffic in April 2000 and to passenger traffic in February 2001.

Construction of over 900 km of new lines in different sections is in various stages of progress. Of these, the largest project is a 246 km new line between Mahbubnagar and Munirabad, estimated to cost Rs4.2 billion.

Improvements to existing lines
Double-tracking of several sections is in progress. During 2001–02, track-doubling of the Gudar–Renigunta section (25 km) was completed. Patch doubling of the Gooty–Renigunta section has been approved in the 2001–02 works programme at a cost of Rs3.04 billion. Gauge conversion of the Purna–Akola section (269 km) was approved in the 2000–01 works programme at an estimated cost of Rs 2.28 billion.

In 2003–04 an outlay of Rs787 million has been provided for gauge conversion works.

Traction and rolling stock
At the start of 2004 the railway operated on 285 broad electric and 406 broad and metre gauge diesel locomotives, 6,290 passenger coaches (including 570 air-conditioned) and 34,569 wagons (four-wheel equivalent).

Electrification
Electrification of the Renigunta–Guntkal section on the Chennai–Mumbai main line is to be done through Rail Vikas Nigam Ltd.

Class WAG1 electric locomotive undergoing overhaul at South Central Railway's Lallagudam, Secunderabad, depot
0583122

Class WDM2A diesel-electric locomotive at the 1.535 km Vagoda Tunnel on the Nandyal–Giddalur section
0583124

SCR emu at Secunderabad forming the inaugural Multimode Transport System service in February 2004
0583152

South East Central Railway

RE Office Complex, 495004 Bilaspur, Chattis garh
Tel: (+91 7752) 23 20 04; 23 71 50
Fax: (+91 7752) 26 85 50
Web: http://www.secrailway.com

Key personnel

General Manager: Pramod Kumar
Secretary to General Manager: S K Pankaj
Senior Deputy General Manager: O P Agarwal
Financial Advisor and Chief Accounts Officer:
  R C Sethi
Chief Commercial Manager: A Gupta
Chief Engineer: D D Dewangan
Chief Electrical Engineer: V P Rao
Chief Mechanical Engineer: M Kapur
Chief Medical Director: Dr D Dhanabalu
Chief Operations Manager: M S Jayanth

Chief Personnel Officer: S Natrajan
Chief Signal and Telecommunications Engineer:
  G R Mali
Controller of Stores: A Kishore
Chief Security Commissioner: S K Mishra

Gauge: 1,676 mm; 762 mm
Route length: 1,602 km; 799 km
Electrification: 1,256 km at 25 kV AC 50 Hz

Organisation

South East Central Railway is one of several new zonal railways that the Ministry of Railways decided to create in 1996. It became operational on 1 April 2003 and comprises the former Bilaspur and Nagpur Divisions of the South Eastern Railway and the newly created Raipur Division formed by dividing the Bilaspur Division.

Passenger operations

In the period April 2003–January 2004 51.94 million passenger journeys were made.

Freight operations

In the period April 2003–January 2004 67.89 million tonnes (51,509 million tonne-km) were handled.

Traction and rolling stock

In 2004 the broad gauge traction fleet stood at 142 electric and 84 diesel locomotives; there were also 46 narrow gauge diesel locomotives.

South Eastern Railway

11 Garden Reach Road, Kidderpore, Calcutta 700 043
Tel: (+91 33) 24 39 12 81   Fax: (+91 33) 24 39 78 26

Key personnel

General Manager: R R Bhandari
Additional General Manager: Y P Gupta
Senior Deputy General Manager: Asot Chaturvedi
Financial Adviser and Chief Accounts Officer:
  Sushma Pandey
Chief Operating Manager: H K Padhee
Chief Commercial Manager: S B Bhattacharya
Chief Mechanical Engineer: D Ray
Principal Chief Engineer: Rajat Mitra
Chief Administrative Officer (Construction):
  A K Ganguly
Chief Electrical Engineer: R P Rehan
Chief Medical Director: Dr Y C Bhushanan
Chief Personnel Officer: H S Pannu
Chief Signals and Telecommunications Engineer:
  G G Biswas
Chief Security Commissioner: B Raj
Controller of Stores: S K Sen

Gauge: 1,676 mm; 762 mm
Route length: 2,273 km; 158 km
Electrification: 2,079 km at 25 kV 50 Hz AC

Organisation

SER was created in 1955 and was the successor to the Bengal Nagpur Railway. In April 2003 the railway was split into three parts. The former Khurda Road, Sambalpur and Waltair Divisions of the SER were transferred to the newly created East Coast Railway; and the former Bilaspur and Nagpur Divisions were transferred to the South East Central Railway. This left the SER with the Adra, Chakradharpur, Kharagpur Divisions and the newly constituted Ranchi Division.

With a total route km reduced to 2,431 km, SER now serves the state of West Bengal and parts of Jharkhand and Orissa.

Passenger and freight operations

Before restructuring the South Eastern Railway achieved freight loadings of 201.6 million tonnes during 2002–03, the best-ever achieved by any zonal railway in India. With greatly reduced route length, SER achieved 56.7 million tonnes during the first nine months of 2003–04. After restructuring, SER is continuously searching for new traffic, improving the infrastructure, optimising use of rolling stock and better marketing strategies.

New lines

The 57 km Tamluk–Kanth section of a new line project has been completed and dmu services have started over it. Work on the remaining 32 km portion to Digha was in progress in 2004 for completion by the end of the year.

Improvements to existing lines

Gauge conversion of the Ranchi–Lohardaga line was due to be completed by June 2004 and its

SER freight terminal at Shalimar 0583125

Construction work in progress on the new Tamluk–Digha line in West Bengal 0583126

extension to Tori by December 2004. Work to provide a third track between Kharagpur and Panskura on the Howrah–Mumbai main line is to be done through Rail Vikas Nigam Ltd.

Traction and rolling stock

In March 2004 SER was operating on 1,676 mm gauge 307 electric locomotives. In April 2003 it operated 128 emu/memu power cars and 288 emu/memu trailers, 558 diesel locomotives, 4,194 coaches and 120,907 wagons (four-wheel equivalent); and on its 762 mm gauge lines 65 diesel locomotives.

Electrification

During 1999–2000 electrification of 126 route-km was completed. It is proposed to undertake

electrification of the 89 km Tamluk-Digha line. When this is done, emu services will be introduced between Howrah and Digha.

Track (BG)

Rail: 60, 52, 44.6 kg/m
Crossties (sleepers): Steel 106 mm thick; CST/9 122 mm thick; concrete 210 mm thick; timber 125 mm thick
Spacing
On straight track: 1,660/km, 1,540/km
On curves (438 m): 1,660/km, 1,540/km
Fastening: Elastic on cwr, conventional elsewhere
Max gradient: 2%
Max axleload: 22.86 tonnes

Southern Railway

Park Town, Chennai 600 003
Tel: (+91 44) 25 35 34 55   Fax: (+91 44) 25 35 14 39

Key personnel
General Manager: V Anand
Additional General Manager: K K Pande
Senior Deputy General Manager: (vacant)
Financial Adviser and Chief Accounts Officer:
   S Parthasarthy
Chief Operating Manager: P Sudhakar
Principal Chief Engineer: D C Mitra
Chief Mechanical Engineer: J Ghosh
Chief Commericial Manager: T Verghese
Chief Medical Director: Dr G C Raju
Chief Personnel Officer: P Kumar
Chief Electrical Engineer: R Mohan Dass
Chief Signals and Telecommunications Engineer:
   V Shankar
Controller of Stores: R Ramesh
Chief Security Commissioner: K S Balasubramanian
Chief Public Relations Officer: S Sridhar
Chief Administrative Officers, Construction:
   N Aravindam, Uday Shankar

Gauge: 1,676 mm; 1,000 mm
Length: 2,990 km; 2,241 km

Organisation
At the time the IR network was restructured in April
2003 the former Bangalore and Mysore Divisions
were transferred to create the South Western
Railway. SR now comprises five divisions, Chennai,
Madurai, Pal ghat, Trichi and Trivandrum. It serves
the states of Tamilnadu and Kerala.

Passenger and freight operations
Originating passenger traffic in FY2002–03 grossed
465.11 million journeys. To meet increasing
passenger demand, many important trains are now
operated with 24 coaches. During summer 2002
SR ran over 978 special trains to handle holiday
traffic.

SR also operates the Chennai MRTS commuter
rail system, Phase I of which (between Chennai
Beach and Thirumalli) was commissioned in
October 1997. Phase II from Thirumalli to Vellachery
(10.3 km) was inaugurated in January 2004. MRTS
is a broad gauge double-track part-elevated/part-
surface railway electrified at 25 kV AC.

Revenue freight traffic in FY2002–03 amounted
to 24.07 million originating tonnes.

As part of control office computerisation, SR has
developed a Live Train Position Display System
(LTPDS) and installed it at Chennai Central.

Emu on the Southern Railway's recently completed Chennai Beach–Thirumalai rapid transit line
0019830

New lines
A major SR project is the new Karur–Salem line. Work
on linking the twin terminals of Chennai Central and
Egmore by an elevated railway at an estimated cost
of Rs930 million has also been started.

Improvements to existing lines
As part of the gauge conversion project between
Chennai Beach and Tiruchirappalli, a parallel
broad-gauge line has been opened between Beach
and Tambaram (29 km). In 2004 SR was preparing
to complete the final phase of gauge conversion
between Tambaram and Egmore. There are
several other sections on which conversion work
is in progress. In the 2000–01 works programme,
conversion of two more sections, Tiruchy–Mana
Madurai (150 km) and Villupuram–Kapadi (161 km),
was approved, at an estimated cost of Rs1.75 billion
each. In the 2003–04 budget an outlay of Rs902
million was provided for conversion projects.

Track-doubling of several busy sections is in
progress. An outlay of Rs814 billion was provided
for doubling projects in 2003–04.

SR is making rapid progress in track
modernisation, with some 3,344 km already laid
with concrete sleepers. A computerised track
management system is being introduced on
high-speed routes. During 1996–97 a total length
of 6,344 km of track, including welds, was tested
ultrasonically. In 2003–04 budget provision was

made for Rs2,250 million for track renewal work,
of which Rs1,550 was from the Special Railway
Safety Fund.

Electrification
During 2001–02 electrification of 101 route-km was
completed. Electrification of the 324 km Erode–
Palghat–Ernakulam section has been completed.
Electrification of 320 km Ernakulam-Trivandrum
section has been in progress and was nearing
completion in 2004.

Traction and rolling stock
At March 2004 the railway operated on 225 electric
locomotives on 1,676 mm gauge and 18 on metre
gauge. In March 2003, before restructuring, SR
operated on broad gauge 390 diesel locomotives,
134 emu power cars and 319 trailers, 5,084 coaches
and 26,539 wagons (four-wheel equivalent); and on
metre-gauge 10 steam, 101 diesel and 18 electric
locomotives, 45 emu power cars and 105 trailers,
two diesel railcars, 907 coaches and 2,612 wagons
(four-wheel equivalent).

Diesel push-pull trains provide suburban-
style services in non-electrified sections on the
Guruvayur–Ernakulam–Alleppey–Kottayam–Quilon
route. During 2000–01 main line emu services
between Arakkonam and Jolarpettai were
introduced.

South Western Railway

580023 Hubli, Karnataka
Tel: (+91 836) 236 08 86   Fax: (+91 836) 236 52 09

Key personnel
General Manager: R N Aga
Secretary to the General Manager: Prabhat
Financial Adviser and Chief Accounts Officer:
   M Ramachandran
Chief Commercial Manager: K V N Das
Principal Chief Engineer: P Sriram
Chief Electrical Engineer: P S Bajwa
Chief Mechanical Engineer: R Sharma
Chief Medical Director: Dr H Dhambolu

Chief Operating Manager: A Jacob
Chief Personnel Officer: N Swaminathan
Chief Signal and Telecommunications Engineer:
   S Manohar
Controller of Stores: H J Alva
Chief Security Commissioner: B Mohan

Gauge: 1,676 mm; 1,000 mm
Route length: 2,587 km; 451 km
Electrification: 25 kV AC 50 Hz

Organisation
South Western Railway is one of several new zonal
railways that the Ministry of Railways decided
to create in 1996. It became operational on 1

April 2003 and comprises the former Bangalore
and Mysore Divisions of the Southern Railway
and the Hubli Division of the South Central
Railway.

Passenger and freight operations
During 2003 19 new pairs of passenger trains were
introduced on the railway. Freight loadings were
26 million tonnes.

Traction and rolling stock
At the end of 2003 the railway was operating
168 broad gauge and 12 metre gauge locomotives,
90 box wagons, 39 BoBx 'N' wagons and 1,572
other wagons.

West Central Railway

482001 Jabalpur, Madhya Pradesh
Tel: (+91 761) 262 74 44   Fax: (+91 761) 262 81 33

Key personnel
General Manager: D K Gupta
Secretary to the General Manager: A K Kankane
Senior Deputy General Manager: A Dhemre
Chief Commercial Manager: Radha Charan
Financial Adviser and Chief Accounts Officer:
   S Ananthnarayanan

Principal Chief Engineer: K K Sharma
Chief Electrical Engineer: P C Sehgal
Chief Mechanical Engineer: B P Singh
Chief Medical Director: Dr P S Saluja
Chief Operating Manager: V N Tripathy
Chief Signal and Telecommunications Engineer:
   Krishna Kumar
Chief Personnel Officer: Manoj Pande
Controller of Stores: V P Raheja
Chief Security Commissioner: Anil Sharma
Deputy General Manager and Chief Public Relations
   Officer: R Mishra

Gauge: 1,676 mm
Route length: 2,925 km
Electrification: 25 kV AC 50 Hz

Organisation
West Central Railway is one of seven new zonal
railways that the Ministry of Railways decided
to create in 1996. Operations started in 2003.
It comprises the former Bhopal and Jabalpur
Divisions of the Central Railway and the Kotah
Division of the Western Railway, serving Madhya
Pradesh and parts of Rajasthan and Uttar Pradesh.

New lines

There are two important new line works approved on the railway: Ramganj Mndi-Bhopal (260 km) and Lalitpur–Satna/Mahoba–Khajuraho/Rewa–Singrauli. In the FY2003-04 budget Rs524.5 million was provided for these projects.

Western Railway (WR)

Churchgate, Mumbai 400 020
Tel: (+91 22) 22 00 56 70   Fax: (+91 22) 22 01 76 31
Web: http://www.westernrailwayindia.com

Key personnel

General Manager (Interim): Y Z Ansari
Additional General Manager: S S Godbole
Senior Deputy General Manager: Amitabha Dutta
Principal Chief Engineer: B B Sharan
Chief Administrative Officer, Survey and
   Construction: Vid Prakash
Chief Commercial Manager: M G Arora
Chief Electrical Engineer: Sukhbir Singh
Chief Mechanical Engineer: T N Perti
Chief Signal and Telecommunications Engineer:
   S Ramasubramanian
Chief Personnel Officer: B B Modgil
Financial Adviser and Chief Accounts Officer:
   Sudha Choubey
Controller of Stores: S Mandhani
Chief Operating Manager: K C Jena
Chief Public Relations Officer: Shailendra Kumar

Gauge: 1,676 mm; 1,000 mm; 762 mm
Route length: 3,315 km; 2,284 km; 876 km
Electrification: 1,715 km at 25 kV 50 Hz AC; 63 km at 1.5 kV DC

Organisation

Formed in 1951, WR is India's second-largest railway. It now serves the whole of Gujarat state, parts of Rajasthan, Maharashtra and Madhya Pradesh. At the time of restructuring of IR in 2002–03 the former Ajmer and Jaipur Divisions of the WR were transferred to the North Western Railway, which commenced operations in October 2002, and the former Kotah Division was transferred to the West Central Railway. WR now consists of Mumbai Central, Vadodra, Ratlam, Rajkot Bhavnagar and newly formed Ahmedabad Divisions.

Passenger operations

The railway is IR's busiest passenger carrier and its most profitable component, with an operating ratio of 87.71 per cent in FY2001–02. WR covers the commuter network of Mumbai, running 1,007 emu commuter trains daily and bringing a morning peak flow of almost 450,000 passengers into the city's Churchgate terminal, where trains arrive at 1.8 minute intervals.

Since 1960 greater Mumbai's population has trebled and WR's suburban passenger journeys soared from 236 million to 1,075 million in 2001–02. But it has not been possible to raise the number of trains operated in proportion. Consequently, some peak-hour trains, with a seating capacity of only 900 or so, carry 3,000 to 3,500 commuters each. Some relief has been provided by raising the standard train formation from 9 to 12 cars, and 357 such services are now in operation.

During 2000–01 WR introduced new long-distance trains between Okha and Dehradun and Ahmedabad and Nagpur, in addition to several other express trains, and increased the frequency of its two Rajdhani express trains to Delhi. During 2003 17 new mail/express and passenger trains were introduced and during 2004 coaching stock of the Mumbai-New Delhi Rajdhani Express sevices have been replaced by state-of-the-art ALSTOM LHB coaches.

In 2001–02 originating passenger journeys totalled 1,401 million, for 77,747 million passenger-km. Computerised reservation has now been extended to almost all major stations.

Freight operations

WR was the first of the Indian railways to cater for container traffic in 1966, and current services

Improvements to existing lines

A programme covering 92 km of double-tracking has been authorised, for which Rs170 million was approved in the FY2003–04 budget. Rs2,368 million were approved in the same budget for track renewals and Rs241 million for signalling and telecommunications works. Of these totals

include dedicated trains from Mumbai to the New Delhi dry port at Tughlakabad. WR also operates a guaranteed transit Mumbai–New Delhi wagonload service under the title 'Speed-Link Express'. Most of WR's bulk commodity unit trains, like three of the country's other major bulk-hauling railways, are now made up to 4,500-tonne formations of BOX-N wagons with double-headed locomotive power. In 2001–02, originating freight traffic amounted to 37.1 million tonnes, for 43,927 million tonne-km.

New lines

A major work in progress is a new 316 km line between Godhra and Indore. The Dewas–Maksi portion of this project was completed in January 2003. Total expenditure of Rs351.6 million was provided for new line projects in the FY2003–04 budget for WR.

Improvements to existing lines

Extension of a fifth track from Santacruz to Borivali has been completed, and quadrupling and electrification between Borivali and Virar is in progress under the same scheme. Provision of Rs3.04 billion was made in the budget for FY2003–04

Rs600 million and Rs191 million respectively were from the Special Railway Safety Fund.

Traction and rolling stock

In March 2004 the traction fleet included 368 electric locomotives.

Commuter activity at a WR station (K K Gupta)   0547049

Train management system at WR's Mumbai control centre (K K Gupta)   0547050

for track renewals. Of this Rs1,670 million was from the Special Railway Safety Fund.

Major gauge conversion projects include Gandhidham–Palanpur, Junagarh–Veraval and Mehsana—Viramgam.

Traction and rolling stock

In March 2004 WR had 210 broad gauge electric locomotives (72 AC/DC, the rest AC). On 1 April 2003 WR was operating 903 emu/memu coaches (out of which 293 were motor coaches, the rest trailers), 295 broad gauge diesel locomotives, 3,556 broad gauge coaches and 41,617 wagons (four-wheel equivalent). On metre gauge there were 21 steam and 236 diesel locomotives, 1,276 coaches and 5,800 wagons (four-wheel equivalent). On narrow gauge WR had 22 diesel locomotives.

Signalling and telecommunications

Various signalling and telecommunications works in progress include: provision of track circuits in station yards; block proving by axle-counters; replacement of existing electromechanical interlockings by panel interlocking, some of which will be SSI; provision of 4/6 quad cable and optical fibre cable; replacement of route relay interlockings at Bandra and Mahim; replacement of analogue microwave by digital microwave. An outlay of Rs458 million was provided for signalling and telecommunications in the FY2003–04 budget.

Electrification

Work to convert the Mumbai suburban area 1.5 kV DC traction power supply system to 25 kV AC has been undertaken and the same was expected to be completed by December 2004. The work of converting DC emu rakes into AC/DC was also in progress in 2004 and 13 rakes have been converted by mid-year.

Coupler in standard use

Passenger cars: Screw
Freight cars: Centre buffer couplers and screw coupling

Braking in standard use

Loco-hauled stock: Vacuum and air; emu: air

Track

Rail, types and weight

Broad-gauge	Type/specification
Flat bottom	
65 kg steel rails	Gost 8160 & 8161.56
Wear-resistant rails	
60 kg/m	UIC 860/0 grade 'C'
Wear-resistant rails	
52 kg/m	UIC 860/0 grade 'B'
Medium manganese	
flat bottom 60, 52	IRS speen T12
and 44.6 kg/m	
Metre-gauge	
Medium manganese	
44.6, 37.1 and	IRS speen T12
29.7 kg/m	

Sleepers: Wooden, cast-iron, concrete and steel sleepers. The standard now adopted for high-speed trunk routes is concrete
Concrete: Monobloc, thickness at centre 180 mm, thickness at rail level 210 mm, length 2,750 mm. With elastic rail clips
Wooden
1,676 mm gauge 2,750 × 250 × 130 mm
1,000 mm gauge 1,800 × 200 × 115 mm
762 mm gauge 1,500 × 180 × 115 mm
Steel/cast-iron: CST-9 for 1,676 and 1,000 mm gauge
Spacing: 1,660/km high-speed routes; 1,310 to 1,540/km other routes
Max curvature: 10° (175 m radius) on 1,676 mm gauge (BG)
Max gradient: 2.9% on BG
Max axleload: 22.9 tonnes on 1,676 mm gauge

Indonesia

Department of Transport

8 Medan Merdeka Barat, 10110 Jakarta Pusat
Tel: (+62 21) 35 15 96; 36 13 08   Fax: (+62 21) 36 13 05

Key personnel
Minister: Agum Gumelar

Indonesian Railway Public Corporation (Perumka)

Perusahaan Umum Kerata Api
Jalan Perintis Kemerdekaan, 1, Bandung 40113, Java
Tel: (+62 22) 43 00 31/430 39/430 54
Fax: (+62 22) 43 00 62/503 42

Key personnel
Chief Director: Dr Anwar Suprijadi
Corporate Secretary: S Sonny
Director, Personnel: A S Harsono
Director of Finance: E Haryoto
Director of Marketing: T B Padmadiwirja
Director of Operations: Adi Witjaksono
Director, Technical: S E Saputro
Chief of Planning and Development: S Siregar
  Accounting: Hadijono
  Development: Soegeng G
  Way and Works: P J Sujatno
  Signalling and Telecommunications: M Iyad
  Workshops: Soeparman
  Traction: Soekamto
  Rolling Stock: N Sumarna

Gauge: 1,067 mm; 750 mm
Route length: 5,961 km; 497 km (partly undergoing gauge conversion)
Electrification: 125 km at 1.5 kV DC

Organisation

The railway network in Indonesia is confined to two islands: Java, where the main system of some 4,967 km is located, and Sumatra.

Political background

Perumka's current concern is to overtake past shortfalls in investment so as to satisfy rising demand for its services. It has been helped by a government decision in 1992 to accept responsibility for maintenance of the railway infrastructure.

This followed liberalisation in 1991, when Perumka was released from full government control to become a public corporation, a move that was completed in 1992. It has commercial freedom in all but fixing third-class passenger fares; those remained under government control and at levels unvaried since 1984. Perumka also gained freedom to borrow in the domestic banking market.

Starting in 1992, Perumka planned to make some of its more intensively used lines in Java accessible to private passenger and freight service operators. The railway would continue to run its own services in competition with any private-sector entrepreneurs. The primary aim of the move was to satisfy rising demand for transport without straining Perumka's own resources. The first such private service started in late 1995.

Passenger operations

Perumka draws almost two-thirds of its income from a passenger business that has steadily increased since the 1970s. In 1991 this sector recorded 66 million journeys and 10,417 million passenger-km, rising spectacularly to 95.4 million journeys and 12,224 million passenger-km in 1993. But with more than one-third of total volume represented by third-class travel at government-controlled fares in the Jakarta area, the growth has not been matched by a commensurate increase in revenue.

A through-ticketing deal agreed in 1992 with the Indonesian national airline Garuda is seen as one way of helping Perumka regain the business travellers lost to internal air services. Some upgrading of 'Eksekutif' class coaches has been undertaken and a few new vehicles ordered, but major purchases must await the fruits of Perumka's new flexibility to charge market fares for better quality accommodation.

A start was made during 1995 with introduction of a new faster service between Jakarta and Surabaya, christened JS950 in commemoration of the fiftieth anniversary of the national railway. Cutting four hours off current timings, the trains are formed of GE locomotives and a new fleet of coaches built locally by PT Inka.

Freight operations

Annual freight carryings have soared even more strikingly, from 800 million tonne-km in 1982–83 to almost 4 billion in 1993, when freight tonnage reached 15.7 million. This reflects the rapid advance of the country's liberalised economy, which has generated a 26 per cent rise of non-oil exports and a 12 per cent growth of non-oil manufacturing since 1983. Long-term estimates show an even more substantial rise than that envisaged for passenger traffic, with 35 million tonnes forecast by 2005.

Intermodal operations

A recent feature in the freight sector has been intermodal development, which has increased ninefold since 1987 to over 24,000 TEU a year. The government plans creation of five rail-connected inland container terminals to serve maritime container traffic. To maximise potential, however,

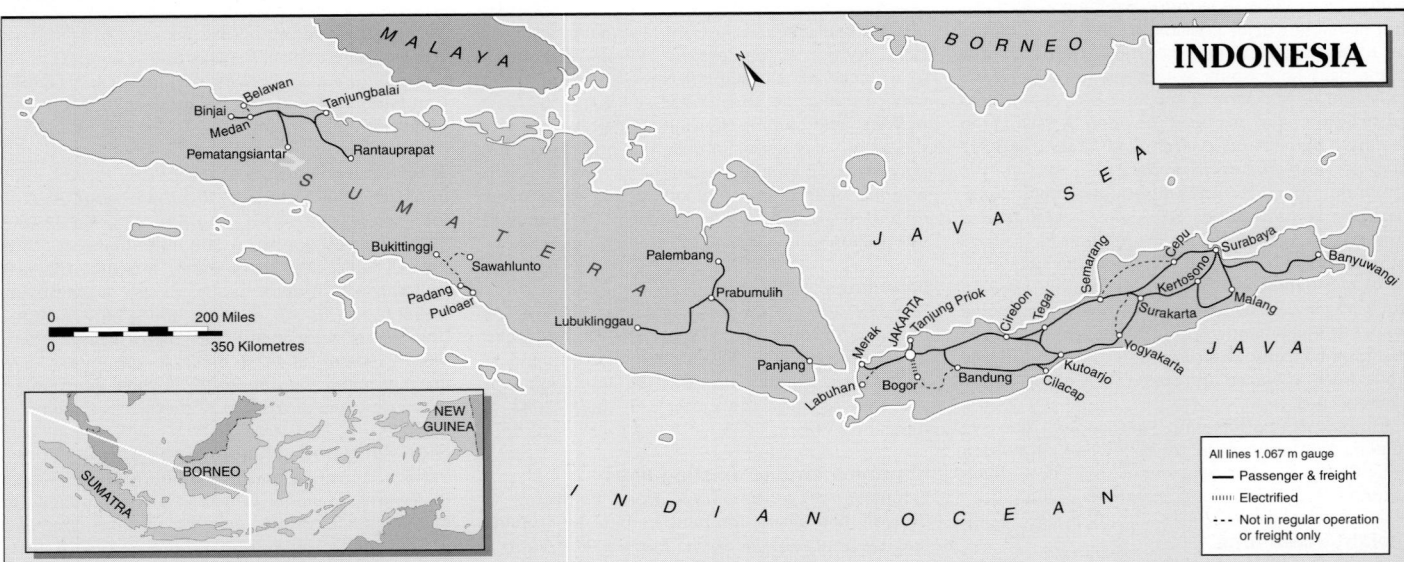

0058533

more track improvement is required and also some work on clearances, which in at least one area restrict the railway to movement of standard ISO container sizes.

Over its 180 km main line between Bandung and Jakarta Perumka runs a weekly unit train for maritime containers to American President Lines' charter. For domestic traffic Perumka's four 17-wagon container trains run daily each way between the port of Tangjungpriok, near Jakarta, and the Gedebage terminal near Bandung. Another container train plies between Cigading port and Bekasi, on the outskirts of Jakarta. A 17 km line has been built from Tangjungpriok to Cakung for the benefit of container traffic.

Improvements to existing lines

The country has substantial coal reserves, which the government sees as a major source of exports, as well as fuel for much-needed additions to electricity-generating capacity and for cement manufacture.

The first major investment in Perumka's bulk freight capability was a US$1.3 billion project to upgrade the 410 km 1,067 mm gauge route in Sumatra for 40-wagon unit train movement of 2.5 to 3 million tonnes of coal a year from the Bukit Asam field to a new south coast port of Tarahan, for shipment thence to an electricity-generating station near Merak in West Java. Parts of the route are also used by other bulk freight flows. The works raised maximum permissible axleloadings from 13 to 18 tonnes and line speed to 60 km/h for freight and 90 km/h for passenger trains. Mechanical signalling was retained, but traffic control by single-line token has been superseded by radio control through a new UHF/VHF radio network.

The Bukit Asam-Tarahan port coal haul could eventually gross 12 million tonnes a year. The investment needed and the operational economics of quadrupling the scheme's present design capacity have been studied by a Canadian consultancy.

The Sumatran coal railway from Padang to Solok has been upgraded, as has the 120 km line from the port of Merak, west of Jakarta, to Serpong (needed for coal destined for a cement works at Cibinong in West Java).

In 1999 it was announced that the Asian Development Bank was to assist in financing a feasibility study into the reopening of the 486 km line from Langsa to Banda Aceh, on the northern tip of Sumatra. Total cost of rehabilitating the line, which runs through a region which has witnessed fighting between separatists and government security forces, is estimated at US$540 million.

Jakarta-Merak upgrade

In late 1996, Davy British Rail International Ltd (DBRI) completed the rehabilitation of the 120 km rail link between Jakarta and Merak, in a £45 million contract for Perumka.

The project involved the design of a new 30 km spur linking with the Jakarta-Merak line, and complete rerailing with 140 km of new R42 continuously welded rail, to give a traffic speed

Diesel locomotives

Class	Wheel arrangement	Power kW	Speed km/h	Weight tons	First built	Engine	Builders Transmission	Builders Repowering
CC 200	Co-Co	1,200	90	96	1951	GE 12V 244E	E	Alco 12V 250 (4 locos)
CC 201	Co-Co	1,450	100	82	1976	GE 7 FDL 8	E	
BB 200	A1A-A1A	650	120	74.8	1956	GM 8567CR	E	
BB 201	A1A-A1A	1,000	120	78	1964	GM G12 567C	E	
BB 202	A1A-A1A	745	100	65	1968	GM GL8 645E	E	
BB 203	A1A-A1A	1,290	100	78	1978	GE 7 FDL 8	E	
BB 204	B-2-B	745	60	55	1982	MTU 12V 396 TC 12	H Voith hydrostatic	
BB 300	B-B	500	75	36	1956	MB 820 B	H	
BB 301	B-B	1,120	120	52	1962	MTU 12V 652 TB11	H L630 r U2	MB 12V 652TB11 (23 locos)
BB 302	B-B	820	80	44	1969	MTU 12V 493	H L520 r U2	
BB 303	B-B	860	90	44	1971	MTU MB12V 493	H L52 r U2	
BB 304	B-B	1,120	120	52	1974	MTU 12V 652	H L720 r U2	
BB 306	B-B	640	75	40	1983	MTU 8V 396TC 12	H L4 r 42 U2	
C 300	C	260	30	30	1964	MB 836 B	H L203 U	
D 300	D	250	50	34	1956	MB 836 B	H 2WIL1.15	GM 8V92 (40 locos)
D 301	D	250	50	28	1960	MB 836 B/2	H 2WIL1.5	GM 8V71 (6 locos) GM 12V 71 (6 locos) MWMTD 232V12 (13 locos)

Electric railcars or multiple-units

Class	Cars per unit	Motor cars per unit	Motored axles/car	Output/motor kW	Speed km/h	First built
MCW 5	4	2	4	230	120	1976
VCW 8		2	4	230	120	1978
BN-Holec	4	2	4	180		1993
Hyundai-ABB	4	2	4	180		1993

Diesel railcars or multiple-units

Class	Cars per unit	Motor cars per unit	Motored axles/car	Power/motor kW	Speed km/h	First built	Builders Engine	Builders Transmission
MCDW 300/ MCW 300	1	1	4	160	90	1964	8V 71	H Diwabus
MCW 301	2	2	2	135	90	1976	DMH17H	H TC 2A
MCW 302	2	2	2	215	90	1978	DMH17SA	H TCR 2–5

of 80 km/h. Other tasks included rehabilitating 14 existing stations and creating three new ones, with associated trackwork, and uprating 70 bridges – including masonry substructures and steel superstructures – to take increased axleloads.

In addition, the project team installed a new colourlight signalling system (based on centralised traffic control – CTC), a VHF utility radio system and a lineside telephone system; and provided a signalling and telecommunications training centre. Civil works comprised improvements to the track alignment, drainage and earthworks; and reballasting.

Traction and rolling stock

In 1992 Perumka was operating 563 diesel locomotives, 30 four-car emus, 147 diesel railcars, 1,262 passenger cars and 12,683 freight wagons.

PT Inka, the BN Division of Bombardier Eurorail of Belgium and Holec of the Netherlands agreed joint production of seven four-car emus for the Jakarta suburban network, delivered in 1993. The first three trains were assembled by BN (now Bombardier Transportation) in Belgium, the remainder locally by PT Inka. A further 25 sets have been built by BN.

Meanwhile, Hyundai of Korea supplied two prototype four-car emus, of which the asychronous motors, GTO thyristors and Micas S power control system were provided by ABB.

Local builder Inka, in collaboration with Hitachi, in 1997 commenced delivery of 24 four-car emus for Jabotek services. The 1.6 kV DC stainless steel-bodied units feature AC traction motors and VVVF inverter control.

Passenger car needs, to cope with severe overcrowding in third class and satisfy government requests for more business travel capacity, comprised 420 economy class, 55 executive class and 245 business class coaches. Inka continues to meet this demand, in 1997 supplying 10 economy class and 18 executive class coaches. Bogies for 30 coaches for the Jakarta-Surabaya run were supplied by GEC Alsthom.

A further 24 diesel railcars were ordered from Mitsubishi/Hitachi.

In 1997 Inka was supplying Perumka with 118 coal hopper wagons.

Signalling and telecommunications

Over 2,580 route-km is equipped with Siemens & Halske mechanical tokenless block, located at some 300 centres in Java and 45 in Sumatra. All-relay interlockings are operational at numerous traffic centres.

Under an A$115 million contract, Westinghouse Brake & Signal of Australia has installed new colourlight signalling controlled by the Westrace system on the 300 km route between Cirebon on the north coast and Yogyakarta in the southeast. Computerised signalling master control and

New station on the Jakarta-Merak line, on which upgrading was completed in 1996   0003057

communication centres are being established at Cirebon, Yogyakarta and Purwokerto; all 43 stations on the route have local control centres. One problem has been what to do about the many unofficial crossings on the line; under the old mechanical system, cans of stones suspended from the signal wires rattled when the signal changed, warning crossing keepers to drop bamboo barriers over crossings. This system will not be feasible with colourlights and how these crossings will be controlled, or if they will be allowed to continue in existence, has yet to be established.

Westrace signalling is also being installed on 130 km between Tasikmalaya and Kroya in central Java.

The railway's first solid-state interlocking has been installed by GEC Alsthom, along with Sigview ATC, to control a critical 20 km section of the Tanggerang line in suburban Jakarta.

Electrification
The 1.5kV DC electrified system in Java covers the line between Jakarta and Bogor (55 km) and some sections around the city of Jakarta. It is planned to expand electrification under the Jabotabek scheme into a regional network of nine lines and some 220 km embracing the satellite cities of Bekasi and Tanggerang.

So far work has chiefly concerned track renewal, double-tracking and installation of some new halts and signalling improvements. For example, GEC Alsthom Signalling was contracted in 1992 to modernise signalling on the 25 km Tanah Abang- Serpong line. Electrification proceeds – but slowly – on the Tanggerang line and also on the western line from Jakarta to Manggarai, where it is now operational over the new elevated alignment opened in 1992. Rolling stock shortages prevented immediate utilisation of this route for a 12 minute interval service between Jakarta and Gongondia.

Two further phases of the scheme were authorised in 1995, including rebuilding more track on elevated alignment.

In 1998, the Indonesian government announced proposals to electrify main lines on Java, with Japanese companies invited to finance and equip the scheme.

Track
Standard rail
R54 54–43 kg/m; R50 50.4 kg/m; R14A 42.59 kg/m; R14 41.52 kg/m; R3 33.4 kg/m; R2 25.75 kg/m
750 mm gauge: R10 16.4 kg/m
600 mm gauge: ID 12.38 kg/m
Crossties (sleepers): Wood 130 mm thick; concrete 195 mm thick; steel 100 mm thick
Spacings: 1,666/km plain track, 1,700/km curves
Rail fastening: Rigid: dog or screw spike for R2 rail; Klem plate KI/KK for R42 rail; Klem plate KE/KF for R2 rail; Dorken spike; double elastic spring clip F type with rubber pad; double elastic Pandrol clip and rubber pad; single elastic Pandrol clip
Max gradient: 2.5%; 7% (rack sections)
Min curvature radius
Main line: 300 m
Max altitude: 1,246 m near Garut, Java
Max axleload
Main line: 14 tons

International

BTZ

Bayerische Trailerzug Gesellschaft für bimodalen Güterverkehr mbH
Poccistrasse 7, D-83306 München, Germany
Tel: (+49 89) 747 14 80   Fax: (+49 89) 74 71 48 22
e-mail: info@btz-bimodal.de
Web: http: www.btz-bimodal.de

Key personnel
Managing Director: Kurt Pelster

Organisation
Founded in 1991, BTZ was until January 2002 a subsidiary of Europäische Trailerzug Beteiligungsgesellschaft mbH (ETZ). In 1996, an Italian subsidiary, BTZ Italia srl, was established. Following the acquisition in November 2001 of a majority shareholding in ETZ by Wabash National Corporation, ownership of BTZ passed in January 2002 to Bimodal Verwaltungsgesellschaft (BVG) (51 per cent) and Brenner Schienentransport AG (49 per cent), a subsidiary of Brenner Autobahn AG.

In October 2003 it started insolvency proceedings and services were suspended.

Services
In June 1995, BTZ began operating RoadRailer bimodal vehicles on a trans-Alpine route from Munich, Germany, to Verona, Italy. In 2002 routes served were Cologne–Verona and Hamburg–Verona.

BTZ operated 460 trailers and 272 bogies.

Cisalpino AG

Postfach 5757, Parkterrasse 10, CH-3001 Bern, Switzerland
Tel: (+41 31) 329 09 09   Fax: (+41 31) 329 09 19
e-mail: info@cisalpino.com
Web: www.cisalpino.com

Key personnel
President: Dr Massimo Ghenzer
Managing Director: Ing Lucio Gastaldi

Organisation
Cisalpino AG was set up jointly in November 1993 by the Swiss and Italian railways to operate an international train service using ETR470 tilting trainsets. Trenitalia (Italian Railways) originally had a 50 per cent holding; SBB (Swiss Federal Railways) 40 per cent, and BLS Lötschbergbahn 6 per cent; the remaining 4 per cent was held jointly by certain Swiss cantons. In 2003 Trenitalia and SBB became the only shareholders, each holding 50 per cent in the company.

Services
Services are operated on the following routes:
  Zurich–Milan
  Stuttgart–Zurich–Milan
  Zurich–Milan–Florence
  Geneva–Milan–Venice
  Basle–Milan
  Four train pairs operate via the Gotthard route, two pairs use the Simplon route and one pair runs via the

Cisalpino ETR470 tilting trainset at Thun, Switzerland (Edward Barnes)   ***NEW**/1114955*

Lötschberg line. For the route between Zurich and Stuttgart Cisalpino AG acts as a leasing company.

Traction and rolling stock
Cisalpino's rolling stock fleet comprises nine tilting ETR470 trains built by Fiat Ferroviaria. The nine-car trains each comprise a restaurant/bar car, three first- and five second-class cars.

The ETR470s have dual-voltage capability, allowing them to cope with the Italian (3 kV DC) and Swiss and German (15 kV 16²/₃ Hz AC) electrical systems. Apart from this, they are much the same technically as the ETR460 Pendolino trains operating Italian domestic services. However, for Alpine conditions, the ETR470s are differently geared from an ETR460 so as to obtain higher tractive effort but with a reduced maximum speed of 200 km/h.

In 2003 tenders were invited for 14 additional trains, with an option for 20 more, capable of 250 km/h to meet anticipated increased passenger demand generated by the commissioning of new high-speed lines in Germany and Italy. This resulted in ALSTOM being awarded a contract in March 2004 for 14 seven-car Pendolino trainsets to be delivered in 2007–08.

UPDATED

CityNightLine

CityNightLine CNL AG
Postfach 7377, Bahnhofplatz 15, CH-8023 Zurich, Switzerland
Tel: (+41 1) 247 75 75   Fax: (+41 1) 247 75 76
e-mail: direktion@citynightline.ch
Web: http://www.citynightline.ch

Key personnel
Director: Christian Zogg

Organisation
CityNightLine CNL AG was set up by the national railways of Germany, Austria and Switzerland to operate luxury 'CityNightLine' sleeper services on international routes between the three countries. First trains ran in May 1995. In the face of poor financial results, ÖBB pulled out of the consortium in 1996, leaving just DB AG of Germany (60 per cent) and SBB of Switzerland (40 per cent) as shareholders. Subsequently, DB AG became the sole shareholder.

Services
CityNightLine services initially operated nightly on three routes: Hamburg–Zurich (the 'Komet'), Berlin–Zurich (the 'Berliner') and Vienna–Dortmund (the 'Donau Kurier'). A fourth route, Vienna–Zurich (the 'Wiener Walzer'), was taken over by ÖBB when it left the consortium and is now run as a conventional EuroNight service. Subsequently, a portion of the Zurich–Berlin service was extended to Leipzig and Dresden and branded the 'Semper', and in 2001 the summer-only 'Sirius' weekend

service between Zurich and Dortmund was added. A planned service expansion from December 2002 was to see Amsterdam served by the 'Pegasus' from Zurich and the 'Pollux' from Munich.

CityNightLine services comprise three levels of accommodation. At the top end are 'A' class deluxe sleeping compartments, with toilet and shower in the compartment. 'B' class is similar to conventional sleepers, with beds and wash basins in the compartments. 'C' class features reclining seats. Continental breakfast is served to A and B class passengers.

In 2001, CityNightLine services carried 493,000 passengers, compared with 459,000 in 2000.

Traction and rolling stock
The double-deck sleeping cars accommodating A and B class passengers were built new for CityNightLine services. Schindler of Switzerland, SGP of Austria and Bombardier Talbot of Germany built 18 cars each.

European Bulls

European Bulls Railfreight Alliance
Moezelweg 136a, Portnumber 5601, NL-3198 LS
Europoort, Rotterdam, The Netherlands
Tel: (+31 181) 25 10 47   Fax: (+31 181) 25 10 31
e-mail: info@european-bulls.com
Web: www.european-bulls.com

Key personnel
Managing Director: Robert Spierings

Organisation
European Bulls is an alliance of European open access rail freight companies established in January 2005 to provide seamless cross-border services delivered to a common set of standards. Its founding members were:
COMSA Rail Transport (Spain) (associate member)
Ferrovie Nord Cargo (Italy)
LTE Logistik- und Transport (Austria)
rail4chem (Benelux, Germany, Switzerland)
Viamont (Czech Republic)

In February 2005 the first locomotive to bear European Bulls branding, a Class 185 electric locomotive operated by rail4chem, was handed over by builders Bombardier.

NEW ENTRY

European Rail Shuttle BV (ERS)

PO Box 59018, NL-3008 PA Rotterdam, Netherlands
A Plesmanweg 61 K-L, NL-3088 GB Rotterdam, Netherlands
Tel: (+31 10) 428 52 22   Fax: (+31 10) 428 52 12
e-mail: info@ersrail.com
Web: http://www.ersrail.com

Key personnel
Managing Director: Frans Zoetmulder

Background
Established in 1994, ERS is a joint venture involving shipping companies that include Maersk-Sealand and Nedlloyd/P&O to provide container rail shuttle services between the port of Rotterdam and destinations in Europe.

Corporate developments
ERS is a major shareholder (47 per cent) in boxXpress.de, established in 2000 to provide container rail transport services between German North Sea ports and the country's hinterland. In 2003, services were provided from Hamburg and Bremerhaven to Augsburg, Munich, Nuremberg and Stuttgart.

In October 2002, ERS established ERS Railways (for further details see ERS Railways in the *Railway systems and operators* section, Netherlands)

as a rail traction provider with its own fleet of locomotives and wagons.

Services
Operates intermodal shuttle services between Rotterdam and Belgium, Bulgaria, the Czech Republic, Denmark, Germany, Greece, Hungary, Italy, Poland and Slovakia. In 2003 the company was operating 250 trains each week and handled 410,000 TEUs.

Eurostar Group

Eurostar House, Waterloo Station, London SE1 8SE, UK
Tel: (+44 20) 79 22 61 80   Fax: (+44 20) 79 22 44 24
Web: www.eurostar.com

Key personnel
Chairman: Guillaume Pepy
Chief Executive: Richard Brown
Chief Operating Officer: Jacques Damas
Directors:
Finance: Ian Nunn
Commercial: Nick Mercer
Customer Service: Nicolas Petrovic
Communications: Paul Charles
Human Resources: Marc Noaro
Corporate and Legal Services: Victoria Wilson

Organisation
The Eurostar Group board includes representatives of the three companies which jointly operate the service, Belgian National Railways (for further details, see Société Nationale des Chemins de fer Belges in *Railway systems and operators*, Belgium), Eurostar (UK) Ltd (for further details, see Eurostar (UK) Ltd in *Railway systems and operators*, UK) and French National Railways (for further details, see French National Railways in *Railway systems and operators*, France).

Passenger operations
Since 1994 Eurostar has established itself as the leading carrier from London to Brussels and Paris and has the largest share of the air/rail market on its core routes.

A London–Paris Eurostar set photographed soon after leaving the Channel Tunnel, with Calais Fréthun station in the background (David Haydock)
0110188

From June 2000 Eurostar's catering contract was taken over by the Momentum consortium, a joint venture between Granada Food Services of the UK and Italian caterer Cremonini SpA.

As well as providing passenger travel, Eurostar offers a range of services for the carriage of urgent packages. Esprit Europe provides a fast door-to-door delivery service from London and Ashford to Brussels and Paris, while Esprit Global and Esprit UK respectively offer similar worldwide and domestic UK services.

Traction and rolling stock
The Eurostar fleet of high-speed trainsets is shared among the three railways which form the Eurostar Group. Details are provided within their individual entries.

Interior refurbishment of the fleet began in 2004.

UPDATED

Eurotunnel

Operating headquarters: *France*
Siège d'Exploitation, BP69, F-62904 Coquelles, Cedex
Tel: (+33 3) 21 00 60 00   Fax: (+33 3) 21 00 60 01
Web: www.eurotunnel.com

United Kingdom
Cheriton Parc, Cheriton High Street, Folkestone CT19 4QS
Tel: (+44 1303) 28 22 22   Fax: (+44 1303) 85 03 60
London office
Golden Cross House, 8 Duncannon Street, London WC2N 4JF

Tel: (+44 20) 74 84 50 16   Fax: (+44 20) 74 84 51 55
Paris office
19 blvd Malesherbes, F-75008 Paris
Tel: (+33 1) 55 27 39 59   Fax: (+33 1) 55 27 37 75

Key personnel
Chairman (Non-executive): Jacques Gounon

Chief Executive: Jean-Louis Raymond
Deputy Chief Executive: Hervé Huas
Directors
   Shuttle Services: Pascal Sainson
   Commercial Passenger Division: Jo Willacy
   Commercial Freight Division: Dirk Broek
   Strategy and Development: Dave Pointon
   Infrastructure: Alain Bertrand
   Business Services: Dominique Bon
   Operational Finance: Claude Lienard

Gauge: 1,435 mm
Route length: 60 km
Electrification: 60 km at 25 kV, 50 Hz

Cars leaving a Eurotunnel shuttle service   0525665

Political background
Anglo-French government agreement to construction of a rail-only tunnel under the English Channel was announced on 20 January 1986. The scheme chosen by the British and French governments was that proposed by the consortium of the Channel Tunnel Group and their French partners, France Manche (CTG-FM), which is now known as Eurotunnel.

Agreement on payments to be made by British and French Railways for use of the Channel Tunnel was reached in July 1987 after protracted negotiation. Minimum usage charges will be paid each month, even if the traffic in any one month does not reach the forecast level. The railways agreed to pay these usage charges for the first 12 years of the Tunnel operation. In return, the railways won entitlement to use 50 per cent of the capacity of the Tunnel, as varied from time to time through the concession period. Protection clauses allowing for reduction in minimum usage charges if the Tunnel is not available for use were also written into the agreement.

The Tunnel was handed over by the builders, Trans Manche Link, to the operators, Eurotunnel, in December 1993. The Tunnel complex comprises a service tunnel 4.8 m in diameter, connected every 375 m by cross-passages on either side to two railway tunnels 7.5 m in diameter. These tunnels run for 50 km, with 38 km under the sea, connecting terminals at Cheriton near Folkestone, Kent, and Coquelles, near Calais in northern France.

The first shuttle trains carrying lorries ran on 19 May, while the first conventional freight train (carrying new Rover cars bound for Italy) ran on 1 June. Through London-Paris/Brussels Eurostar passenger trains and car-carrying shuttles began in late 1994.

Finance
Construction of the Tunnel system cost some £10 billion. This was well in excess of preconstruction estimates, and the cost over-run put some strain on the funding of this wholly privately financed project. Over 60 per cent of the revenue was expected to be derived through tourist traffic in the summer months: Eurotunnel was hoping that its first summer of full operations, in 1995, would be sufficiently fruitful to see the company through to a steady income stream. However, fierce competition from ferries constrained revenues and in September 1995 Eurotunnel suspended interest payments on its £8 billion debts in order to force its creditor banks into a financial restructuring of the company's debt load.

An £8.5 billion restructuring of the company's finances, which involved the banks swapping debt for equity and the two governments agreeing an extension of the company's concession beyond 2052 until 2086, was approved by the company's shareholders in July 1997. The concession extension also received the approval of the two governments in return for payment of 59 per cent of Eurotunnel's pre-tax profits between 2052 and 2086. Further refinancing took place in 2002 and in 2004 shareholders replaced the board of directors at the company's annual general meeting.

Passenger operations
Passenger operations through the Channel Tunnel are of two types. Cars, caravans, camper-vans and trailers, motorcycles and road coaches are transported through the Tunnel on Eurotunnel's shuttle trains, while London-Paris/Brussels foot passengers are conveyed on Eurostar trains.

Eurotunnel trains operate between elaborate terminals just beyond the tunnel portals, at Cheriton in England and Coquelles in France. The main running lines at each end are in a loop, with a flyover to avoid conflict between incoming and outgoing trains, so that Eurotunnel trains are in continuous Anglo-French circuit without reversal. Terminal-to-terminal journey time is 35 minutes.

To use Eurotunnel shuttle trains, car and coach drivers simply drive off the motorway into the Eurotunnel terminal, negotiate customs and immigration posts and then drive on to the train. Unlike the 'Eurotunnel Freight' services for lorries, the car and coach users of the service stay with their vehicles for the transit of the Tunnel, and no railway coaches with seating are provided for them.

In 2004 the number of cars conveyed by Eurotunnel shuttles totalled 2,101,323, down 8 per cent on the figure for 2003 and continuing a gradual decline since 1999, when 3,260,166 cares were carried. At 63,467 vehicles in 2004, coach numbers fell by 12 per cent compared with 2003.

The number of passenger using Eurostar services to transit the Channel Tunnel increased by 15 per cent to 7,276,675 in 2004, continuing a recovery from the 6,602,817 recorded in 2002.

Freight operations
Freight operations through the Channel Tunnel are of two types. Lorries are transported through the Tunnel on Eurotunnel's 'Eurotunnel Truck Shuttles', while intermodal and conventional rail freight is conveyed in trains operated by the national rail operators, EWS and SNCF.

To use the Eurotunnel Truck Shuttles, lorry drivers simply drive off the motorway into the Eurotunnel terminal, negotiate customs and immigration posts and then drive on to the train. A minibus driving along the platform picks the drivers up and delivers them to a club car for transit of the Tunnel. Another minibus on the other side of the Channel takes the drivers back to their vehicles. Journey time is 35 minutes start to stop.

The number of lorries carried in 2004 was 1,281,207, slightly fewer than the 1,284,875 conveyed in 2003. Conventional rail freight tonnage in 2004 recovered slightly to 1,889,175, 8 per cent up on the previous year but far from the 2000 peak of 2,947,388 tonnes.

Traction and rolling stock
Locomotives
Eurotunnel initially bought 38 5.6 MW Bo-Bo-Bo electric locomotives with GTO converter three-phase AC drive for shuttle operations. They were built by Brush of Loughborough, with electrical equipment from ABB (now Adtranz) of Switzerland. There is a locomotive at each end of each shuttle

train, with either machine capable of powering the shuttle should the other one fail. Subsequent orders brought the total fleet of this type to 40 mixed-use machines (Class 9/0) and 13 freight-only units (Class 9/1). One Class 9/0 locomotive has been withdrawn. Upgrading of these locomotives was in progress in 2005, with modified machines designated Class 9/8.

In 1999 Eurotunnel ordered an additional seven 7 MW locomotives to handle increased freight vehicle traffic. Delivered in 2001–02, these became Class 9/7.

Eurotunnel also operates five Krupp/MaK Type DE1004 1,180 kW diesel-electric locomotives and 12 Hunslet/Schoma 270 kW diesel-hydraulic locomotives for yard and maintenance work.

Passenger shuttles
The Euroshuttle Consortium Wagon Group, comprising chiefly the Canadian-owned multinational Bombardier, was contracted to build 108 covered single-deck transporters for coaches; 108 double-deck covered passenger transporters for private cars; and 18 single-deck and 18 double-deck loading cars.

The passenger shuttle cars are formed into nine single-deck and nine double-deck sets, each of 12 transporters. Each set is flanked by loader vehicles. The normal passenger train shuttle format combines a bilevel and single-level set, each with its two loader cars, and has a locomotive at each end. It thus totals 30 vehicles and is 792 m long.

Truck shuttles
The shuttles for lorries were built by an Italian consortium comprising Breda and Fiat Ferroviaria. The initial order was for 228 single-deck transporters for lorries, 33 loading wagons for same, and 19 club cars in which the lorry drivers ride through the Tunnel. A standard 'Eurotunnel Truck Shuttle' consist is: Bo-Bo-Bo locomotive, club car, loading/unloading wagon, 14 lorry transporter wagons loading/unloading wagon, Bo-Bo-Bo locomotive. Subsequently orders were placed for 216 additional shuttle wagons and three club cars.

Signalling and telecommunications
The initial operating plan provides for a 3-minute headway in each direction. Half the 20 hourly paths in each direction can be occupied by Eurotunnel's shuttle trains. The Tunnel is equipped with the TVM430 signalling and Automatic Train Protection (ATP) system adopted for French Railways' TGV Nord. The maximum speed possible is 200 km/h, but in practice the Eurotunnel shuttle trains are limited to 140 km/h, Eurostar trains to 160 km/h and freight trains to 100 km/h.

Track
Rail: UIC 60 kg/m
Sleepers: Twin block concrete encased in rubber boot and cast into floor slab with no tie bars
Spacing: 600 mm
Fastenings: Sonneville S75
Max gradient: 1.1%

UPDATED

Hupac SA

Viale R Manzoni 6, CH-6830 Chiasso, Switzerland
Tel: (+41 91) 695 28 00   Fax: (+41 91) 695 28 01
Web: www.hupac-intermodal.ch

Key personnel
Chairman: Hans-Jörg Bertschi
Deputy Chairman: Daniel Nordmann
Secretary: Peter Hafner
Managing Director: Bernhard Kunz

Background
Founded in 1967 and owned by road hauliers (72 per cent) and railway companies (28 per cent), Hupac SA is a holding company for a group of firms providing intermodal transport and logistics services.

In 1998, Hupac SA acquired a majority shareholding in the Netherlands-based intermodal transport company, Trailstar NV.

In April 2001, Hupac SA, BLS Lötschbergbahn (BLS) and Swiss Federal Railways (SBB) established a joint venture company RAlpin AG, to run a 'rolling highway' piggyback service between Freiburg in Breisgau, southwest Germany, to Novara, Italy. This followed a tender invitation issued by the Swiss Federal Office of Transport to provide piggyback services via the upgraded Simplon route through the Alps. Each company has a shareholding of 30 per cent in the company; the remaining 10 per cent was reserved for Italian State Railways.

Subsidiaries

Fidia SpA
Viale Rimembranza 10/12, I-28047 Oleggio, Italy
Tel: (+39 0321) 99 81 55   Fax: (+39 0321) 913 42
Manager: Paolo Paracchini

Hupac Intermodal NV
A Plesmanweg 151, NL-3088 GC Rotterdam, The Netherlands
Tel: (+31 10) 495 25 22   Fax: (+31 10) 428 05 98
President: Bernhard Kunz
Director of Operations: M Jansen

Hupac Intermodal SA
Viale R Manzoni 6, CH-6830 Chiasso, Switzerland
Tel: (+41 91) 695 28 00   Fax: (+41 91) 695 28 01
Web: http://www.hupac-intermodal.ch
Chief Executive: Bernhard Kunz

Hupac GmbH
Zum Umschlagbahnhof 2, D-78224 Singen, Germany
Tel: (+49 7731) 87 90 60   Fax: (+49 7731) 87 90 65
Manager: Rudi Mager

Hupac SpA
Via Dogana 8/10, I-21052 Busto Arsizio, Italy
Tel: (+39 0331) 60 85 11   Fax: (+39 0331) 38 28 80
Manager: Francesco Crivelli

Termi SA
Viale R Manzoni 6, CH-6830 Chiasso, Switzerland
Tel: (+41 91) 695 29 55   Fax: (+41 91) 695 28 05

Terminal Singen TSG GmbH
Zum Umschlagbahnhof 2, D-78224 Singen, Germany
Tel: (+49 7731) 879 00   Fax: (+49 7731) 87 90 16
Directors: Hans-Joachim Güntner, Rudi Mager

Services

Hupac Intermodal SA provides intermodal container, swapbody, unaccompanied semi-trailer and accompanied piggyback services linking Italy and northern Europe, with some connecting services also running west to Perpignan, France, to destinations in Austria and to the Nordic countries. In Germany and Switzerland, Hupac has a train operating licence and provides traction for some of its services.

Hupac SpA operates the company's Busto Arsizio terminal near Milan; Hupac GmbH holds the company's operating licence in Germany; and Terminal Singen TSG GmbH operates the Singen terminal in southern Germany. Termi SA is responsible for the planning and development of intermodal terminals and transhipment facilities.

Traction and rolling stock

The Hupac fleet includes three Siemens ES 64 US EuroSprinters leased from Dispolok, four Class D753.7 refurbished diesel-electric locomotives acquired from the Czech Republic for use in Italy and two diesel-hydraulic B-B machines. The company also owns around 2,900 wagons for its intermodal traffic. These include 60 new-generation piggyback wagons delivered in 2004 by Bombardier Transportation for services linking Germany, Switzerland and Italy.

UPDATED

Intercontainer-Interfrigo

Intercontainer-Interfrigo (ICF) SA
Margarethenstrasse 38, PO Box, CH-4002 Basel, Switzerland
Tel: (+41 61) 278 25 25   Fax: (+41 61) 278 24 45
e-mail: icf@icfonline.com
Web: http://www.icfonline.com

Key personnel

Managing Director, Finance and Administration: Franz Böni
Director, Sales and Operations: Patrice Pinoli
Director, Human Resources: Angela Kopp
Managers, Information Technology:
  Jean-Luc Helmer, Jean-Marie Portha

Organisation

Intercontainer-Interfrigo is owned by 27 European railway companies. Its customers include freight forwarders, transport companies, shipping agents and shipping lines. Joint-venture companies with ICF participation include:

  Allied Continental Intermodal Services, Reading, UK
  Bahnhof-Kühlhaus AG, Basel, Switzerland
  CLB, Bettembourg, Luxembourg
  Intercontainer Austria GesmbH, Vienna, Austria
  Intercontainer Scandinavia AB, Gothenburg, Sweden
  Intercontainer Ibérica, Madrid, Spain
  NV Interferry, Antwerp, Belgium
  Italcontainer SpA, Milan, Italy
  Optimodal Nederland BV, Rotterdam, Netherlands
  Pannoncont Kft, Budapest, Hungary
  Polcont, Spólka zoo, Warsaw, Poland
  Transthermos GmbH, Munich, Germany.

Services

One of Europe's leading international combined transport operators, Intercontainer-Interfrigo runs a network of customised intermodal services, including shuttles, block trains and hubs with connections to other terminals. These span western Europe and provide links with the Russian Federation and associated states. The company also provides temperature-controlled transport services using a fleet of specialised rail wagons.

UPDATED

Linx AB

Box 249, SE-401 24 Gothenburg
Tel: (+46 31) 10 49 51   Fax: (+46 31) 10 49 93
e-mail: webmaster@linx.se
Web: http://www.linx.se

Key personnel

Managing Director: Reidar Jignéus
Directors
  Operations: Carl-Johan Lonntrop
  Marketing: Lars-Olle Ericsson
  Human Resources: Bjørn Bjørnsson
  Finance: Ulla Grath

Organisation

Linx AB is a joint venture, equally owned by Norwegian State Railways (NSB) and Swedish State Railways (SJ), established in May 2000 to develop, operate and market intercity services on the Oslo–Karlstad–Stockholm and Oslo–Gothenburg–Malmo–Copenhagen corridors. However, in mid-2004 its was announced by the two partners that Linx was to be liquidated by the end of that year in the face of mounting losses.

Passenger operations

Linx took over services on the Copenhagen–Gothenburg–Oslo and Oslo–Karlstad–Stockholm routes in June 2001.

Traction and rolling stock

Linx is initially leasing seven X2 high-speed tilting trainsets from SJ, equipped for dual-voltage operation to enable them to use the Øresund Link to Denmark. The trains have been refurbished internally and branded in Linx livery by TGOJ at its Tillberga works.

Linx X2 tilting trainset at Gothenburg
(Edward Barnes)
0536200

For details of the latest updates to *Jane's World Railways* online and to discover the additional information available exclusively to online subscribers please visit
jwr.janes.com

Lyria

56 rue de Londres, F-75008 Paris, France
Tel: (+33 1) 42 93 38 74   Fax: (+33 1) 42 93 39 02

Organisation

Under its original name, Rail France Suisse SAS, Lyria was formed in January 2000 by French National Railways (SNCF) and Swiss Federal Railways (CFF/SBB) as a Société par Actions Simplifiée (SAS), a company with a simplified corporate structure, to develop high-speed train services on the Paris–Lausanne and Paris–Neuchâtel–Bern–Zurich routes. It succeeded the TGV France-Suisse joint venture formed by the two railways in 1993. The name Lyria was adopted in March 2002. The company is responsible for service quality, development and delivery; pricing policy; marketing; promotion; and contracts for provision of on-board services. Since March 2002, the last has been undertaken by Chef-Express, part of the Italian Agape group.

Services

Services are operated under the branding 'Ligne de Coeur', using TGV trainsets. Paris–Lausanne is served by up to five trains daily in each direction, while two daily return trains (one on Saturdays)

TGV service from Paris arriving at Zurich Hbf (John C Baker)   0524952

link Paris with Zurich. During the ski season, some Lausanne services are extended to Brig at weekends.

Traction and rolling stock

TGV France-Suisse services are provided by nine three-current TGV-PSE trainsets, one of which was

formally acquired by SBB in 1993. A two-year fleet refurbishment programme was completed in 1999.

UPDATED

STVA

Société de Transport de Véhicules Automobiles Immeuble Le Cardinet, PO Box 826, F-75828 Paris Cedex 17, France
Tel: (+33 1) 44 85 56 78   Fax: (+33 1) 44 85 57 00
e-mail: stva@stva.com
Web: http://www.stva.com

Key personnel

President of the Executive Board: J P Bernadet
Managing Director: Y Fargues
Deputy Managing Director, Commercial and
   Operations: J Henry
Deputy Managing Director, Finance and
   Administration: J J Pronzac

Services

Operates a fleet of automobile transporters and offers full service (predelivery operations) throughout Europe.

Thalys International SC

20 Place Stéphanie, B-1050 Brussels, Belgium
Tel: (+32 2) 548 06 00   Fax: (+32 2) 511 29 44
Web: www.thalys.com

Key personnel

Managing Director: Jean-Michel Dancoisne
Deputy Managing Director: Nathalie Dereume

Organisation

Originally named Westrail International, Thalys International was established as a joint-venture company by Belgian National Railways (SNCB) and French National Railways (SNCF) to develop and manage high-speed rail services on the Paris–Brussels–Antwerp–Amsterdam and Paris–Brussels–Liège–Cologne routes. German Rail (DB AG) and Netherlands Railways (NS) participate in the operation by commercial agreement and retain the option to become full joint-venture partners as services develop to and from their domestic networks. The company is responsible for service quality, development and delivery; communications; marketing; and contracts for provision of onboard services.

Services

Services are operated with high-speed trainsets under the 'Thalys' branding. Routes operated are: Paris–Brussels; Paris–Brussels–Cologne; Paris–Brussels–Antwerp–Rotterdam–Schiphol Airport–Amsterdam; and Brussels–Roissy–Charles de Gaulle-Marne la Vallée. Some Paris–Brussels services operate to/from Belgian domestic destinations, including Bruges,

Thalys PBKA trainset forming an Ostend – Paris at Oostkamp (John C Baker)   0533631

Ghent and Ostend. Two levels of service, Comfort 1 and Comfort 2, are available, the former offering access to a personal steward and increased onboard and off-train facilities.

From March 2001, five daily Thalys services in each direction between Brussels Midi and Roissy-Charles de Gaulle replaced all Air France flights between the Belgian capital and Paris' principal airport following a partnership agreement between the two carriers.

From September 2002, the frequency of Paris–Brussels services was increased to half-hourly. At the same time, dedicated platforms at Paris Gare du Nord were assigned to Thalys services.

Journey time reductions were to result from the opening in December 2002 of the high-speed line between Louvain and Liège, Belgium.

In December 2003 Thalys inaugurated a daily service between Paris and Brussels National Airport.

Traction and rolling stock

Thalys services are provided by 17 four-voltage PBKA (Paris–Brussels–Köln–Amsterdam) high-speed trainsets and 10 three-voltage PBA sets bearing Thalys livery. Nine of the PBKA sets are owned by SNCB, of which two are leased to DB AG, six belong to SNCF, and two to NS. The PBA units are owned by SNCF. The trains are used in a common pool. Technical details can be found in the entry for France in the Railway systems section.

UPDATED

Transfesa

Transportes Ferroviarios Especiales SA
Musgo 1, Urb La Florida, E-28023 Madrid, Spain
Tel: (+34 91) 387 99 00   Fax: (+34 91) 372 90 59
e-mail: transfesa@transfesa.com
Web: www.transfesa.com

Key personnel

Chairman and Executive Director: Emilio Fernández Fernández

General Manager: Luis Del Campo Villaplana
Division Managers
   Industrial Logistics: Arturo Boix Faubell
   Motor Vehicles: Abraham Peralta Arroyo
   Chemicals and Bulk Products:
      Juan Diego Pedrero Sancho
Road Transport: Jaime González López
General Cargo: Juan Diego Pedrero Sancho
Finance: José González Rodríguez
Organisation and Resources:
   Pablo Rodríguez Mosquera

Technical: José L Sánchez Humanes
Controller: Carmen Romero de la Calle
Communications and Corporate Image:
   Julián Gacimartín Quiñones

Services

Transfesa is a pan-European rail-based operator capable of delivering the entire range of logistics services within the supply chain, including warehousing, consolidation and distribution. The company owns a fleet of 8,000 interchangeable-axle wagons, 2,500 swapbodies and 341 vehicle-carrying and general cargo wagons for its international operations. It also owns two axle-changing facilities on the Franco-Spanish border at Cerbère and Hendaye, and warehouses and terminals at different locations in Europe.

In 2004 the company recorded 235,599 (234,816 in 2003) rail consignments (3,239 million tonne-km compared with 3,158 the previous year), of which 80 per cent were in cross-border traffic and the remainder within the Iberian Peninsula. While the automotive market is the key sector (80 per cent

of tonne-km in 2004), Transfesa also transports chemicals, granular products, perishables, electrical appliances, paper and steel products. Road consignments increased to 72,486 in 2004.

UPDATED

UIRR SC

International Union of Combined Road-Rail Transport Companies
31 rue Montoyer, Bte 11 B-1000 Brussels, Belgium
Tel: (+32 2) 548 78 90   Fax: (+32 2) 512 63 93
e-mail: headoffice.brussels@uirr.com
Web: http://www.uirr.com

Key personnel
Chairman: Werner Külper
Director General: Rudy Colle

Full members
UIRR SC

Austria
Ökombi
Taborstrasse 95, A-1200 Vienna
Tel: (+43 1) 331 56 0   Fax: (+43 1) 331 56 300

Belgium
TRW
100 Avenue du Port, Bte 1, B-1000 Brussels
Tel: (+32 2) 421 12 11   Fax: (+32 2) 425 59 59

Czech Republic
Bohemiakombi
Opletalova 6, CZ-11376 Prague 1
Tel: (+420 2) 24 24 15 76
Fax: (+420 2) 24 24 15 80

Denmark
Kombi-Dan
Thorsvej 8, DK-6330 Padborg
Tel: (+45 74) 67 41 81   Fax: (+45 74) 67 08 98

France
Novatrans
21 rue du Rocher, F-75008 Paris Cédex 08
Tel: (+33 1) 53 42 54 54   Fax: (+33 1) 43 87 27 98

Germany
Kombiverkehr
PO Box 940153, D-60459 Frankfurt/Main
Tel: (+49 69) 79 50 50   Fax: (+49 69) 79 50 51 19

Hungary
Hungarokombi
Szilagyi Dezso tér 1, H-1011, Budapest
Tel: (+36 1) 224 05 50   Fax: (+36 1) 224 05 55

Italy
Cemat
Via Valtellina 5-7, I-20159 Milan
Tel: (+39 02) 66 89 51   Fax: (+39 02) 66 80 07 55

Netherlands
Trailstar
Albert Plesmanweg 151, NL-3088 GC Rotterdam
Tel: (+31 10) 495 25 22   Fax: (+31 10) 428 05 98

Norway
Kombi-Nor
c/o Kombi-Dan, Thorsvej 8, DK-6330 Padborg, Denmark
Tel: (+45 74) 67 41 81   Fax: (+45 74) 67 08 98

Poland
Polkombi
ul Targowa 74, PL-03-734 Warsaw
Tel: (+48 22) 619 79 14   Fax: (+48 22) 619 00 00

Portugal
Portif
Avenue Sidono Pais, 4-4-P.3, P-1000 Lisbon
Tel: (+351 1) 52 35 77   Fax: (+351 1) 315 36 13

Slovakia
Eurotrans
Kuzmanyho 22, PO Box B-2, SK-01092 Zilina
Tel: (+421 41) 62 24 47   Fax: (+421 41) 62 56 28

Slovenia
Adria Kombi
Tivolska 50, SLO-1000 Ljubljana
Tel: (+386 61) 131 01 57   Fax: (+386 61) 131 01 54

Spain
Combiberia
Rafael Herrera 11, 3° Pta 308, E-28036 Madrid
Tel: (+34 91) 314 98 99   Fax: (+34 91) 314 93 47

Sweden
Swe-Kombi
Hamntorget 3, S-252 21 Helsingborg
Tel: (+46 42) 12 65 65   Fax: (+46 42) 13 88 46

Switzerland
Hupac
Viale R Manzoni 6, CH-6830 Chiasso
Tel: (+41 91) 695 29 00   Fax: (+41 91) 683 26 61

UK
CTL
179/180 Piccadilly, London W1V 9DB
Tel: (+44 171) 355 46 56   Fax: (+44 171) 629 57 14

Associate member
CNC
8, avenue des Minimes, F-94302 Vincennes
Tel: (331) 43 98 70 00   Fax: (+33 1) 73 74 18 12

Services
The UIRR was founded in 1970 and its central objective is to ensure a more sustained development of rail transport of swapbodies and containers as well as of semi- trailers and lorries by private transport hauliers.

UIRR members operate cross-border intermodal services throughout western Europe and control over 55 per cent of all European combined transport. Local road hauliers are major shareholders in the individual national member companies.

UPDATED

Unilog NV

Leuvensesteenweg 443, B-2812 Muizen, Belgium
Tel: (+32 15) 42 20 11   Fax: (+32 15) 42 38 29
e-mail: tony.davis@unilog.be

Key personnel
Commercial Manager: Tony Davis

Services
Operates intermodal services between Belgium (Muizen), Germany and the UK (Daventry and Manchester) via the Channel Tunnel.

Unilog carries all types of containers and swapbodies, with a maximum height of 9 ft 1 in and a maximum width of 2,550 mm.

The headquarters for Unilog is at Dry Port Muizen near Mechelen, halfway between Antwerp and

Brussels. The terminal is owned and operated by IFB (Inter Ferry Boats), majority owned by Belgian National Railways).

Onward connecting rail services to and from Muizen are operated by TRW, providing links with Italy, Spain and other countries.

UPDATED

VSOE

Venice Simplon-Orient-Express Ltd
A division of Orient-Express Hotels Inc
Sea Containers House, 20 Upper Ground, London SE19PF, UK
Tel: (+44 171) 805 50 60   Fax: (+44 171) 805 59 08

Key personnel
President: Simon M C Sherwood
Chief Executive: Nicholas R Varian
Vice-President and Treasurer: Peter Parrott
Director of Public Relations: Pippa Isbell

Services
VSOE operates the following services:
A luxury train between London, Paris, Düsseldorf and Venice, Florence, Rome and Prague.
A vintage Pullman train on day and weekend excursions within the UK.
A luxury train in Asia between Singapore, Kuala Lumpur, Bangkok and Chiang Mai.
A deluxe river cruiser on the Ayerarwady river in Myanmar.

In December 1998, VSOE, in partnership with Queensland Rail, was to launch the luxury 'Great South Pacific Express' service from Brisbane to Cairns and Sydney.

Iran

Ministry of Roads & Transportation

PO Box 15185-1498, Shahid Kalantary Building, Tehran
Tel: (+98 21) 646 41 57   Fax: (+98 21) 564 70 86
Web: www.mrt.ir

Key personnel
Minister: Ahmad Khorram
Vice-Minister: Eng Mohammad Saeid Nejad

UPDATED

Islamic Iranian Republic Railways (RAI)

Rahahane Djjomhouriye Eslami Iran
Shahid Kalantari Building, Rah Ahan Square, Tehran 13185-1498
Tel: (+98 21) 564 16 00; 564 39 46
Fax: (+98 21) 51 24 60 23; 565 03 53
e-mail: rail-rai@neda.net
Web: www.irirw.com

Key personnel
Vice Minister, Roads and Transportation, Chairman and President, RAI: Eng Mohammad Saeidnejad
Board Member, Substitute President:
    Mohammad Moslehi
Board Member, Vice-President, Technical and Infrastructure: Eng Abdolmajid Shahidi
Board Member, Vice-President, Finance and Administration: S Hassan Mousavi Nejad
Vice-President, Operations and Movement:
    Eng Noureddin Aliabadi
Vice-President, Fleet Affairs:
    Eng Mohammad Talafi Noqani
Director General, Public Relations:
    Hamid Sediqpour

Gauge: 1,435 mm; 1,676 mm
Route length: 7,172 km; 94 km
Electrification: 146 km at 25 kV 50 Hz AC

Political background
The structure of the railway was changed in 1990 from that of a state-owned entity to a limited company affiliated to the Ministry of Roads and Transport.

Restructuring has been based on decentralisation to 13 regional managements, with a split also effected between infrastructure and operations. Eventually, it is aimed to allow access to RAI tracks for private operators.

A continuing major expansion of the rail network in Iran is in progress to enable the country to exploit its mineral resources and develop its rail freight transit role. This is expected to create a network of around 10,600 route-km by 2014.

RAI has indicated a willingness to undertake co-operation with the private sector in fields that include operations, track maintenance and rolling stock procurement, although expressly not for infrastructure development and ownership. RAI cites provision of rolling stock for the projected high-speed line between Qom and Esfahan as a possible example of this type of partnership.

Organisation
RAI's key routes run from the ports of Bandar Khomayni and Khorramshar on the Persian Gulf to Tehran; from the capital northwest to Razi on the Turkish border and Djolfa on the Azerbaijan border; and east from Tehran to the border with Turmenistan at Sarakhs. Tehran also radiates lines north-eastward to Bandar Turkhman the Caspian sea port of Amirabad, Mashhad and Sarakhs on the Turkmenistan border, and southeast to Bafq, Kerman and the port of Bandar-e Abbas. Further to the southeast is a still isolated 94 km 1,676 mm gauge line from Zahedan to the Pakistan border at Mirjaveh, although RAI is constructing a link between this section and its main network.

The electrified part is the final 146 km of the route to Azerbaijan, from Tabriz to Djolfa. The electrification was undertaken by Technoexport

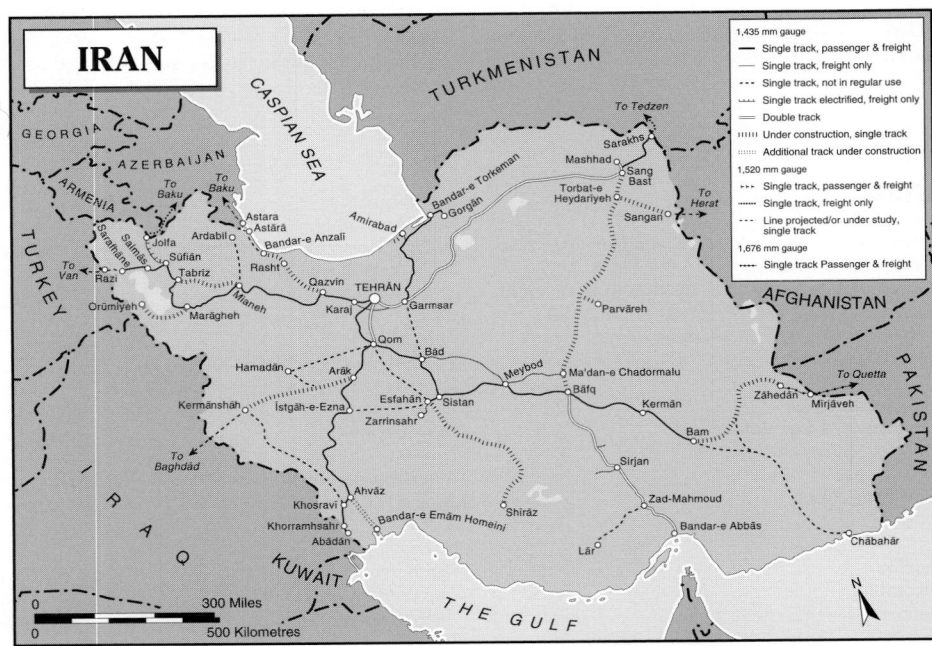

NEW/0585094

and the traction is furnished by eight 3,600 kW locomotives based on Swedish Railways' Rc4. Built by SGP of Austria, the units have ABB electrical equipment.

In 1996 a government-owned affiliate company, Raja Passenger Trains (see entry in Iran section of *Railway systems and operators*), was established to manage and operate passenger rail services on the RAI network. RAI-owned subsidiaries also include the Railway Personnel Saving Fund Institure, Railway Consulting Engineers of Iran (Metra) and companies responsible for producing the railway's ballast and sleeper requirements.

Traffic (million)	1999	2002
Passenger journeys	10.7	14.3
Passenger-km	6,451	8,640
Freight tonnes	23.0	26.5
Freight tonne-km	14,082	15,840

Passenger operations
Passenger services on the RAI network are managed and operated by an affiliate company, Raja Passenger Trains (see entry for Raja Passenger Trains in the Iran section of *Railway systems and operators*).

Freight operations
Minerals generate 50 per cent of RAI's freight tonnage, oil products 16 per cent. Transit traffic to and from the Central Asian republics is of rapidly increasing importance, and the line from Mashhad

into Turkmenistan places Iran astride major routes to Europe and the Persian Gulf.

The 2,013 km transit route from Turkey to Sarakhs has capacity for around 2 million tonnes annually; similar capacity is offered by the 1,937 km link from Azerbaijan via Jolfa to Sarakhs.

RAI anticipated that the opening of the Turkmenistan route and the commissioning of the Bafq–Bandar-e Abbas line would eventually raise freight tonnage to an annual 33 million tonnes.

Both Iran's rail-connected Gulf ports are equipped for container handling: Bandar-e Emäm Homeini has an annual handling capacity of 450,000 TEU, while at Shahid Rajai (Bandar-e Abbas) there is nominal capacity for 320,000 TEU annually.

For cross-border operations connecting with the 1,520 mm gauge former Soviet system, bogie-changing facilities era located at Jolfa and Sarakhs, the latter with a daily capacity to change 200 bogies.

New lines
In 2005 RAI was constructing over 2,800 route-km of new lines, with a further 2,445 km under study.

Bandar Abbas railway
Construction of a 700 km line from Bafq to Bandar-e Abbas, to connect the port of Shahid Rajai, iron ore mines of Golegohar, and copper reserves at Sarchesmeh to the existing railway network, was completed in 1995. This strategically important route also provides better access to the Persian Gulf than the line to Korramshahr.

RAI's latest traction acquisitions are Alstom Prima Type AD43 C 2,880 kW diesel-electric locomotives produced in passenger (seen here) and freight versions   0554607

A line running west from the Bafq–Bandar-e Abbas line at Zad Mahmoud to Lar was under study in 2005.

Silk Road railway
In 1996 RAI commissioned the 180 km connection from Mashhad to Sarakhs on the Turkmen frontier, where it meets the 130 km line built simultaneously from the Central Asian main line at Tajan. Together with the new line to Bandar Abbas, this connection will give the Central Asian republics an important new outlet to the Persian Gulf and Europe. Bogie-changing facilities are provided at Sarakhs, where a free trade zone has been established.

Other projects
Construction continued in 2005 of a 506 km line from Esfahän to Shiraz, Iran's only major city not served by rail. The line has to surmount very difficult terrain in its last 150 km through the Zagros Mountains, necessitating construction of 19 significant tunnels. Improved access from Qom to Esfahän would be achieved by a new direct 280 km link which was the subject of studies by RAI in 2005. This line is to be designed for 350 km/h operation.

Provision of a still shorter route to the Persian Gulf is the aim of a direct Mashhad–Bafq line (805 km). This will reduce the distance from the Central Asian republics to Bandar Abbas by some 1,900 km. Work on the Mashhad–Chadormalou section was in progress in 2005, when the line was expected to be opened. A 148 km branch under construction east from this line to Sangan is intended to serve significant mineral deposits and studies were under way in 2005 to extend this line eastwards across the Afghanistan border to Herat. Further south on the same line, a minerals branch to Parvareh was being built in 2005.

Construction continued in 2005 of a 539 km line from Kerman to Zahedan, which is at present the terminal railhead of the 1,676 mm gauge cross-border line from Pakistan. While a change of gauge will be necessary at Zahedan, this line will complete a through rail connection between Europe and the Indian subcontinent. Services over the section between Kerman and Bam New Citadel commenced in 2003. The Bam–Zahedan section (304 km) is expected to be completed in 2006.

Also the subject of studies in 2005 was a 595 km line southeast from Bam to the Gulf port of Chabahar, where development is planned.

A 556 km line under construction from Arak, on the Qom–Khorramhsahr line, west to Kermanshah is due to be completed by 2007. In the longer term this is expected to provide an important link to Iraq.

In northwest Iran a new more direct line is under construction between Mianeh and Tabriz, shortening the distance for traffic to and from Azerbaijan and Turkey. RAI is also studying a line from Mianeh to Ardabil, which could eventually continue to the Azeri border at Parsabadimishli. A 183 km branch from the existing Mianeh–Tabriz line is under construction in 2005 to serve Orümiyeh.

Work was also in progress in 2005 on the initial section of a second route from Iran into Azerbaijan, connecting with that country's line south from Baku to the Caspian Sea port of Astara. Leaving the Tehran–Mianeh line at Quazvin and running via Rasht to Bandar-e-Anzali, this will eventually avoid the need to pass through Armenia using the present route into Azerbaijan via Djolfa.

A by-pass line south of Tehran has been completed, enabling freight traffic to bypass congested routes in the capital.

Other new lines that were the subject of studies by RAI in 2005 included:
Garmsar–Bäd (192 km)
Esfahän–Azna (320 km)
267 km of lines to connect Hamadan to the rail network
A short link connecting Khorramhsahr with Basra, Iraq.

Improvements to existing lines
Doubling of track continues on lines where limits of capacity have been reached. One key route that has been upgraded in this way is the Tehran–Mashhad

Trains carrying petroleum products cross on the Tehran–Mashhad line　　　0554608

General Motors GT26-CW diesel-electric locomotive undergoing refurbishment at RAI's Karaj workshops　　　0554609

RAI employs mechanised aids such as this Plasser & Theurer SVM-1000 high-speed tracklayer for its rail expansion schemes　　　0554610

line (926 km), where continuous welded rail has been provided and the route upgraded for 160 km/h running with the potential to increase this to 200 km/h at a future date. This project was completed in March 2003. A feasibility study into electrification of this line was under way in 2005.

Double-tracking of the Tehran–Ghom line (151 km) was completed in March 1999 and is to be extended to Esfahän. A second track is also planned for the Ahwaz–Bandar Khomayni (120 km) line.

Improvements to passenger facilities include provision of a new station at Bandar Abbas, construction of a third platform at Mashhad and the provision of additional stations on the Tehran–Qom and Bafq–Bandar Abbas routes. In 2002 an international terminal for travellers from and to Turkey was commissioned at Tehran station.

With the increase of the length of trains and of their tonnage, the loops of all stations on the Tehran–Ghom, Tehran–Mashhad, Andimeshk–Ahwaz and Tehran–Tabriz lines have been lengthened.

Traction and rolling stock
At the end of 2004 RAI's serviceable motive power fleet consisted of around 550 diesel-electric and five electric locomotives (similar to the Swedish Rc4 design), plus three diesel-hydraulic shunters, one dmu and the three Turbotrains bought from ANF-Frangeco in 1982. The wagon fleet totalled 16,549, and there were 1,261 passenger cars. The passenger rolling stock fleet is assigned to RAI's affiliate, Raja Passenger Trains.

At the end of 1997 a major contract was signed with GEC Alsthom (now ALSTOM Transport) for the supply of 100 Type AD43 C 2,880 kW diesel-electric locomotives with MAN B&W engines. The contract provided for the first 20 locomotives and subassemblies for the next five manufactured in France. These latter machines and the remaining 75 units are being built by Wagon Pars at its plant at Arak, under a technology transfer arrangement which will also see some component manufacture in Iran. Thirty are being supplied with modified suspension for passenger use (Type AD43 C1; the remaining 70 (Type AD43 C2) are ballasted to a 23-tonne axleload for freight traffic. At the end of 2004, 23 of these locomotives were in traffic. A dedicated maintenance depot for these locomotives is planned at Mashhad.

In 2004 RAI stated that it intended to purchase 50 Bo-Bo diesel electric locomotives from Siemens for passenger work.

By the end of 2004 RAI had completed the refurbishment of 30 General Motors and 12 General Electric diesel locomotives at its Karaj workshops as part of a six-year programme to upgrade some 220 machines.

There is also a project to upgrade 37 General Motors GT26-CW locomotives for 160 km/h operation of passenger services. This includes fitting ZTR microprocessor control equipment and renewing electrical and control circuits.

In 2000 an order was placed with Siemens Transportation Systems to supply 20 Class DH4-1 four-car intercity dmus for use by Raja Passenger Trains on the Tehran–Mashhad route. Five are being built by Siemens at the TVT Nova facility in Maribor, Slovenia; the remainder will be supplied in kit form for assembly in Iran by Wagon Pars. The first examples appeared in 2004.

Two five-coach prototype Intercity diesel trains have been bought by Iran from DSB of Denmark; built in 1982, they were the forerunners of the successful Danish IC/3 Class. The two prototypes were refurbished at DSB's workshops in Århus before delivery to Iran in 1995; they have been put into service on the Tehran–Mashhad line. In 2004 RAI also procured 52 ex-DSB passenger coaches.

In November 2004 South Korean manufacturer Rotem announced that in a technology transfer agreement with Iran Khodro Rail Transport Industries (IRICO) it was to supply 120 suburban dmu cars. The first 24 vehicles are to be built in Korea, the remainder in Iran. Deliveries are due to be completed by 2008.

In March 1998 the first batch of 135 refurbished ex-RENFE passenger coaches was shipped from Spain to Iran. Outstanding orders for new coaches

CTC control centre governing movements on the Tehran–Mashhad route   0554612

Diesel locomotives

Builder's type	Wheel Arrangement	Power kW	Speed km/h	Weight tonnes	No in Service 2005	First built	Mechanical	Engine	Transmission
AD43C1	Co-Co	2,880	150	123	9	2002	Alstom	MAN B&W	E
AD43C2	Co-Co	2,880	110	123	14	2002	Alstom	MAN B&W	E
G8	Bo-Bo	715	84	68.4	4	1958	GM-EMD	567C	E
G12	Bo-Bo	980	105	72.7	70	1958	GM-EMD	567C	E
G16	Co-Co	1,450	105	103	1	–	GM-EMD	567C	E
G18	Bo-Bo	746	105	64.8	2	1968	GM-EMD	567C	E
G22W	Bo-Bo	1,118	105	74.2	24	1975	GM-EMD	645E	E
GT26-CW	Co-Co	2,235	120	119.6	141	1971	GM-EMD/ GM Canada	645E3	E
HD10C	Bo-Bo	798	100	69	22	1971	Hitachi	0398TA	E
U30C	Co-Co	2,250	105	119.7	27	1992	GE Canada	7FDL	E
C307i	Co-Co	2,250	110	132.1	21	1993	GE Canada	7FDL	E
2M62	2 × Co-Co	2,530	100	252	5	–	Lugansk	14D40	E
060-DA	Co-Co	1,544	120	120	4	–	Electroputere	Sulzer	E
Secmafer 16	D	550	60	45	1	1970	Secmafer	Detroit	H
Secmafer 8	B	250	60	25	2	1970	Secmafer	Poyaud	H

Electric locomotives

	Wheel arrangement	Output kW	Speed km/h	Weight tonnes	No in service	First built	Mechanical	Electrical
40-700RCH	Bo-Bo	3,480	135	78	5	1982	Asea	Asea

in 2005 included 150 160 km/h vehicles from China and a similar number from domestic builder Wagon Pars. RAI had also invited bids for 150 double-deck coaches.

Growing numbers of privately owned freight wagon are in service on the RAI network. In 2004–05 1,300 were introduced and RAI was to provide arrangements for the private purchase of 5,000 more.

Signalling and telecommunications
Major signalling projects in progress or planned for implementation in 2005 include: provision of CTC on the Sharud–Mashhad section of the Tehran–Mashhad line and installation of electric signalling at four stations; installation of electric signalling at 13 stations on the Bafq–Bandar-e Abbas line and provision of CTC at Sirjan; provision of eight block section on the Tehran–Garmsar line; provision of electric signalling at five stations on the Shourab–Maybod line; and continuing renewal of the signalling system on the Qom–Ahwaz line.

Projects under study or to be implemented in 2003 were: provision of ATC and a block system on the Tehran–Mashhad route; provision of an ATC system on the Tehran–Tabriz route; provision of electric signals and CTC on the Badroud–Meybod route; the second phase of renewal of the signalling system on the Qom–Ahwaz line; Under implementation or planned in 2005 were: the provision of fibre-optic communications systems and infrastructure on several key routes totalling 1,143 km; and the installation of a radio communications network covering around 1,000 km of the route to Azerbaijan.

Track
Standard rail
Type U33 46.3 kg/m (2,200 km installed)
Type 2A, 38.4 kg/m in 12.5 m lengths (149 km)
Type 3A, 33.5 kg/m in 12 m lengths (63 km)
UIC 50 kg/m in 12.5 m lengths (31 km)
UIC 60 kg/m in 18 m lengths (Zahedan–Mirjaveh) and continuously welded (Mashhad–Sarakhs; Bafq–Bandar Abbas; and Tehran–Garmsar) (4,730 km)
Rail joints: 4- and 6-hole fishplates; and welding
Crossties (sleepers): Creosote-impregnated hard wood, steel and mono- or twin-block concrete. Wood 2,600 × 250 × 150 mm. Steel 2,550 × 260 × 60 mm Concrete 2,500 × 290 × 24 mm. A mono-block concrete sleeper production plant at Shahroud was commissioned at the end of 1999.
Spacing: 1,680/km (Steel 1,600/km)
Rail fastenings
Wood sleepers: sole plates, screws and bolts
Steel sleepers: clips and bolts
Filling: Part broken stone, and part river ballast; minimum 200 mm under sleepers
Max curvature: 7.9° = min radius 220 m
Longest continuous gradient: 16 km of 2.8% (1 in 36) grade between Firouzkouh and Gadouk
Max altitude: 2,177 m near Nourabad station
Max axleloading: 20 tonnes (22.5 tonnes Bafq–Bandar-e Abbas)
Max permitted speed
Freight trains: 55 km/h
Passenger trains: 80 km/h
170 km/h on parts of Tehran-Mashhad line, and 160 km/h on other new or upgraded lines

UPDATED

Raja Passenger Trains Co

1 Sanaee Street, Karimkhan Zand Avenue, Tehran
PO Box 15875/1363
Tel: (+98 21) 883 51 59   Fax: (+98 21) 883 43 40
e-mail: info@rajatrains.com
Web: www.rajatrains.com

Key personnel

Chairman and Managing Director:
  Dr Mohsen Seyyed Aghaee
Directors: Eng Djavad Mosadeghi;
  Eng Mojtaba Shivapour; Eng Muhammad
  Moezziddin
Deputy Executive: Eng Abbas Ghorban Ali Beyg
Administration and Finance Deputy: Eng Seyyed
Akhavan
Vice-President: Eng Seyyed Mustafa Kheradmand
Technical Deputy: Eng Mojtaba Shivapour

Organisation

Established in October 1996, Raja Passenger Trains
is a wholly owned subsidiary of the government-
owned Islamic Iranian Republic Railways (RAI) and
charged with the operation, marketing and ticketing
of all passenger services on the Iranian network.
Its wide-ranging brief includes the development
and upgrading of the passenger coaching fleet,
including participating in the procurement of new
locomotives and rolling stock, the development
of domestic passenger services and facilities,
managing the concessioning of train operations
to private-sector companies, encouraging joint-
venture projects to develop passenger rail transport
in Iran and the development of new routes within
Iran and passenger rail links with neighbouring
countries. The company is also responsible for
recruitment and training.

The company has three main departments,
responsible for: operations; technical issues, mainly
relating to rolling stock; and administration and
finance. In 2004 some 2,900 staff were employed.

Passenger operations

Services are operated on five main domestic routes
linking major cities in Iran and on four international
routes. In 2003–04, 82 per cent of domestic services
were operated under concessions let by Raja
Passenger Trains. Much ticket retailing is also
undertaken by private sector companies.

Passenger numbers have continued to grow.
In 2000–01 11.7 million passengers were carried.
In 2003–04 numbers had risen to 16.1 million.
Forecasts by the company see this figure more
than doubling by 2009 as new lines are opened and
an expansion of services is implemented.

On the domestic network recent developments
include the introduction of Tehran–Kerman–Bam
services following completion of the Kerman–Bam
section of the new line to Zahedan, an Isfahan–
Bandar-e Abbas service and trains between Yazd
and Mashhad.

Cross-border services are operated from Tehran
to Nakhichevan, Azerbaijan, and in co-operation
with Syrian Railways and Turkish State Railways
respectively, services have been introduced on
Tehran–Damascus and Tehran–Istanbul routes.
There is a Zahedan–Quetta service linking Iran and
Pakistan. A service is also planned linking Tehran,
Tashkent (Uzbekistan) and Almaty (Kazakhstan).

The development of passenger services is
hampered both by train speeds and by the
preponderance of single track on the RAI network:
only the Tehran–Mashhad and Tehran–Qom lines
are double-track. The issue of train speeds has been
partially addressed by the arrival of Alstom-built
AD43 C 160 km/h diesel locomotives and by the
adaptation of some of RAI's older GM locomotives
for this higher speed.

Future plans foresee developments in regional
and commuter services, with moves made in 2004
to procure rolling stock suitable for such traffic.

Traffic (million)	2002-03	2003-04
Passenger journeys	14.3	16.1
Passenger-km	8,600	9,300

New lines

Studies were under way by RAI in 2005 into an
electrified high-speed line linking Tehran and
Isfahan. Capable of sustaining operations at up
to 350 km/h, this line would possibly involve a
private sector operator. Raja Passenger Trains will
also be active in providing services over new lines
constructed under Iran's extensive rail network
expansion programme.

Traction and rolling stock

In 2003–04 Raja Passenger Trains had at its disposal
1,012 coaches and 249 service vehicles such as
catering cars, parcels/baggage vans and generator
cars. Traction is provided by RAI.

To upgrade and expand its rolling stock fleet,
Raja Passenger Trains has been placing significant
orders. In 2000 20 four-car intercity dmus were
ordered from Siemens. Designated Class DH 4-1
and intended initially for service on the Tehran–
Mashhad route, these 160 km/h steel-bodied air-
conditioned units are designed specially to operate
in the harsh climatic conditions of northeast Iran.
The first five are being built by Siemens at plants
in Europe, while the remaining 15 will be supplied
to Iran in kit form for assembly at the Wagon Pars
plant. First examples were completed by Siemens
in 2004.

For suburban services 40 three-car dmus were
ordered in 2004 from a consortium comprising Iran
Khodro Industrial Group (principal shareholder
in Wagon Pars), South Korean rolling stock
manufacturer Rotem and Marubeni Corporation.
Eight trains are to be manufactured in Korea,
the remainder in Iran by Wagon Pars. Delivery is
scheduled for 2006.

Also in 2004 RAI/Raja Passenger Trains stated
that contracts had been signed with Siemens to
supply 80 Bo-Bo diesel-electric locomotives for
passenger services.

In 2001 nine ex-Class 141 two-car two-axle railbuses
were acquired from the UK for commuter services.

In 2002 the company ordered 150 coaches (138
first class coaches and 12 buffet cars) for intercity
services from Sifang, China. In addition, 336 bogies
were ordered to re-equip existing vehicles in the
fleet. Domestic manufacturer Wagon Pars has also
received orders to supply 200 passenger coaches.
Transporter wagons have also be procured to carry
accompanied cars. In 2004 Raja Passenger Trains
also stated its intention to procure 150 double-deck
coaches.

The private sector participation of the Kerman
Development Organisation has resulted in an
agreement to fund the acquisition of two sets of
dedicated vehicles for the Tehran–Kerman route.

UPDATED

Tehran Urban & Suburban Railway Company (TUSRC)

PO Box 15878 13113-4661, 37 Miremad Avenue,
Tehran 15878
Tel: (+98 21) 874 01 11; 01 12
Fax: (+98 21) 874 01 14; 01 15
e-mail: info@tehranmetro.com
Web: www.tehranmetro.com

Key personnel

Chairman and Managing Director:
  Mohsen Hashemi
Vice-Chairman: Masood Ahmadi
Deputy Managing Director, Finance/Administration:
  Jafar Rabiee
Deputy Managing Director, Construction:
  M Majdardakani
Deputy Managing Director, Operations: B Ababafi
Deputy Managing Director, Equipment: F Zarrini
Deputy Managing Director, Human Resources:
  Mohammad Montazeri

Gauge: 1,435 mm
Route length: 40.1 km
Electrification: 40.1 km at 25 kV 50 Hz AC

Organisation

TUSRC was established to develop and construct
an urban rail network serving the Iranian capital,
including a 41.5 km electrified suburban heavy
rail system that forms Line 5. Running west
from the centre of Tehran at Sadeghieh, where
interchange is made with Line 2 of the Tehran
metro system, the line provides connections to
the satellite cities of Karaj and Mahr Shahr. In
2005 six stations were operational; eventually the
line will have 12 stations including Sadeghieh. It
is being constructed and equipped by the China

*Type TM1 electric locomotive built by Zhuzhou Electric Locomotive Works for the Tehran Urban &
Suburban Railway Company*
0063748

International Trust, Investment and International
Co-operation Corporation, the China National
Technical Import and Export Corporation and the
China North Industry Corporation. Services on the
initial section to Karaj began in 1999.

TUSRC is now controlled by Tehran's municipality
and also operates and is developing the city's
metro system.

New lines

The first section of the line, between Tehran
and Karaj (31.5 km), has four stations and was
commissioned in February 1999; the remaining
10 km two-station section to Golshar was
inaugurated in March 2005. In 2005 construction
was in progress of a 1.4 km extension to serve
depot facilities at the line's western end.

A 25 km extension westwards to Hashtgerd is planned. Three other lines have also been studied: south from the end of metro Line 1 at Haram-e-Motahhar via Imam Khomeini Airport to Vavan; southeast from the end of projected metro Lines 8 and 9 at Moshiyeh to Varamin; and east from a proposed extension to existing metro Line 2 at Pardis to Damavand.

Traction and rolling stock
Services are provided by 20 3,200 kW SS5 Bo-Bo electric locomotives built by Zhuzhou Electric Locomotive Works and 58 double-deck coaches supplied by Changchun Railway Car Works, these operating as eight-car push-pull sets with a locomotive at each end. The 12 locomotives supplied for the line's first phase are designated Class TM1; eight additional machines, delivered in 2005, feature restyled cab exteriors and are designated Class TM2.

Signalling and telecommunications
The line is equipped with TVM300 automatic train control equipment supplied by Ansaldo subsidiary CS Transport under a subcontract placed by China North Industry Corporation. ATP is also provided.

UPDATED

Iraq

Ministry of Transport & Communications

Kanat Street, Baghdad

Key personnel
Minister: Luai Hatim

Iraqi Republic Railways (IRR)

Damascus Square, Baghdad

Key personnel
Director General: Mr Salam

Gauge: 1,435 mm
Route length: approximately 2,200 km

Organisation
Since closure of the metre-gauge network in 1988, IRR's network consists principally of a 1,435 mm gauge main line linking the Persian Gulf port of Um Qasr and Basra with Baghdad, there splitting to form routes to the Syrian border at Husaiba (opened in 1987) and El Yaroubieh. Branches to Kirkuk and Akashat to exploit mineral deposits are also of recent construction.

The ravages of three wars have left the system in very poor condition and in 2004 US-led efforts were in hand to rehabilitate it. Train speeds were limited to 80 km/h and were often restricted to 40 km/h.

In 2004 IRR employed around 12,000 staff.

Passenger operations
Before the 2003 war IRR was handling around 1 million passengers annually for some 380 million passenger-km. In 2004 daily services were operated between Baghdad and Al Mawsil, Al Qaim and Basra. Since July 2003 there has also been an overnight sleeper service between Al Mawsil and Aleppo, Syria.

Journey times were extended due to low speeds imposed by track conditions and because of security.

Freight operations
Pre-war carryings were of over 3 million tonnes annually (872 tonne-km). In 2004 the bulk of traffic was military and humanitarian, mostly from the port of Umm Qasr to the country's interior. However, efforts were being made at reconstruction to assist the revival of freight traffic.

New lines
The first of several new line projects, completed in the 1980s, was a 404 km line from Baghdad via Radi and Haditha to the Syrian border at Husaiba, with a 115 km branch from Al Qaim to the phosphate mines at Akashat in the west, so as to link the latter with a fertiliser plant at Al Qaim, in the Euphrates valley. Like all new IRR lines, the Baghdad–Husaiba was engineered for 250 km/h, partly for ease of maintenance in the foreseeable future, when speeds would be limited to 140 km/h for passenger and 100 km/h for freight.

From Haditha on this route a 252 km transversal, built initially as single track but with provision for later doubling, was completed via Baiji to Kirkuk in 1987. This project involved bridging the Euphrates, Tigris and Therthar rivers. The line was built with CTC, track-to-train radio and hot box detectors. During the 2003 war this line was severed east of Baiji by the bombing of the Euphrates bridge.

In the longer term, IRR could revive elements of the extensive new line construction programme in place for many years. This includes a line northeast from Basra via Al Kut to Baghdad, running east of the existing route between those two cities. A further line would then run from Baghdad to Kirkuk and Al Mawsil, again running to the east of the existing main line. IRR had also already started building a railway leaving the existing Basra–Baghdad line at Ad Diwaniyah to serve An Najaf and Karbala before rejoining it south of the capital.

A link from Al Mawsil to the Turkish border has also been proposed.

Improvements to existing lines
In 2004 the US Agency for International Development (USAID) was leading the rehabilitation of the IRR system. Its main focus was to improve access to rail facilities at the port of Umm Qasr under a project due to be completed at the end of that year. A more ambitious scheme covers construction of a new track on the trunk route between Umm Qasr via Baghdad to the Syrian border at Rabiya (around 1,100 km), parallel to the existing alignment.

Traction and rolling stock
Military action and attacks by insurgents, coupled with backlogs in overhaul and maintenance, had by 2004 reduced a locomotive fleet that once numbered around 200 to some 70 active main line units and around 30 shunters. In 2004 delivery was in progress of 30 Type TE114 (Class DEM 2800) main line units from Lugansk, Ukraine, and six Type TEM18 (Class DES 3300) heavy shunters from Russian builder Bryansk, both supplied under the UN's 'oil-for-food' programme. Pending delivery of the DEM 2800 units, most main line duties were handled by survivors of a batch of 50 Chinese-built Class DEM 2700 locomotives and the Henschel-GM DEM 2500s.

Other recent deliveries are eight Class DHS 130 diesel-hydraulic shunters from Tülomsas, Turkey. They are based on the builders' Class DH 9500 for Turkish State Railways.

All 72 of the Class DEM 4000/DEM 4100 2,685 kW diesel-electrics supplied by French industry in the 1980s were out of service, their future uncertain.

Deliveries were due to begin in 2005 of 240 container flats procured by the Iraqi Coalition Provisional Authority from WagonySwidnica in Poland.

Signalling and communications
IRR's signalling and train control system was inoperable in 2004. Most movements were controlled by train order. Provision of an effective signalling and communications system has been identified as a priority by IRR and agencies working to restore the network.

UPDATED

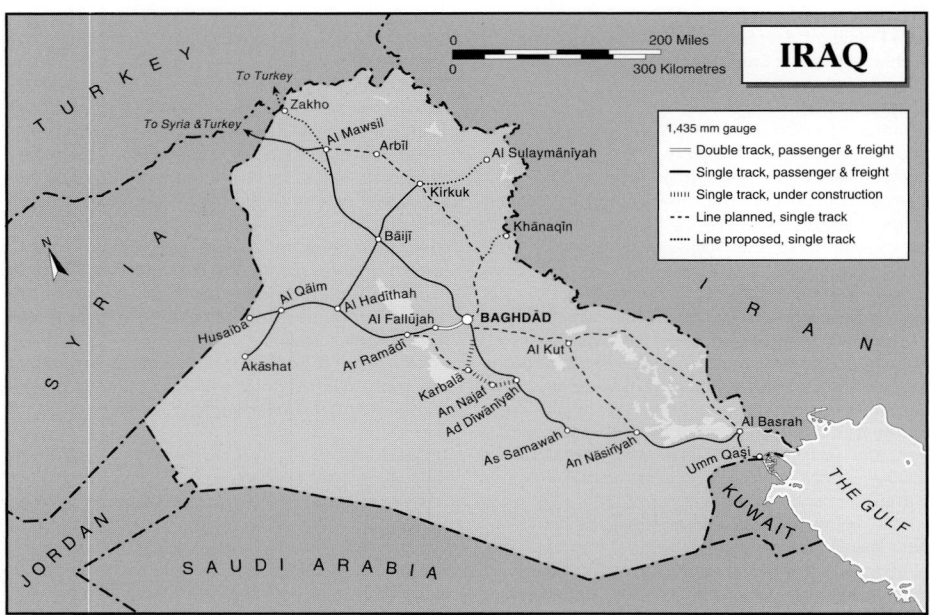

0583354

IRR Diesel locomotives

Class	Wheel arrangement	Power kW	Speed km/h	Weight tonnes	No in service (2004)	First built	Mechanical	Builders Engine	Transmission
DEM 2300	Co-Co	1,492	–	–	3	1975	MLW	Alco	E GE
DEM 2400	Co-Co	1,640	–	–	8	1981	Macosa	GM 645E	E GM
DEM 2500	Co-Co	1,680	–	–	27	1982	Henschel	GM 645E	E GM
DEM 2700	Co-Co	1,492	–	–	31	2001	Dalian	12V240ZJ	E Dalian
DEM 2800	Co-Co	1,960	–	–	30*	2004	Lugansk	–	E Lugansk
DES 3100	Co-Co	820	90	115	17	1979	ČKD	K6S 310DR	E ČKD
DES 3300		895			6*	2004	Bryansk	–	E Lugansk
DHS 100	B-B	450			5	1973	Nippon Sharyo	–	H
DHS 130	B-B	745	80	68	8	2002	Tülomsas	Cummins	H Voith

* delivery in progress in 2004.

Ireland

Department of Transport

Transport House, 44 Kildare Street, Dublin 2
Tel: (+353 1) 670 74 44   Fax: (+353 1) 670 96 33
Web: www.transport.ie

Key personnel

Minister: Martin Cullen
Secretary General: Julie O'Neill

IRELAND

1.600 m gauge
— Passenger & freight
‣‣‣ Passenger only
--- Freight only
⊐⊏⊐ Electrified

0 80 Miles
0 120 Kilometres

0089966

Iarnród Éireann (IE)

Connolly Station, Amiens Street, Dublin 1
Tel: (+353 1) 836 33 33   Fax: (+353 1) 836 47 60
Web: http://www.irishrail.ie

Key personnel

Chairman: Dr John Lynch
Directors: Paul Cullen, Gerry Duggan, Plev Ellis,
   Tras Honan, Patrick Lynch, Anne Marie Mannix,
   Bill McCamley, Paul Prescott, Joe Meagher
Managing Director: Joe Meagher
General Manager, Infastructure Division:
   Joe Leahy
Manager, IE Freight: S Aherne
Manager, New Business (Freight): Donie Horan
Manager, InterCity: J P Walsh
Manager, Safety: Ted Corcoran
Manager, Human Resources: J Keenan
Manager, Finance and Administration and
   Company Secretary: Richard O'Farrell
Chief Engineer, Infrastructure: Brian Garvey
Chief Mechanical Engineer: J McCarthy
Manager, Strategic Planning: Tom Finn
Manager, Procurement and Materials
   Management: Pat Mullan
Manager, Resources and Central Traffic Control:
   O Doyle
Manager, Media and PR: Barry Kenny
Manager, Network Catering: Tom Mythen
Manager, Suburban Rail: Michael Murphy
Manager, Marketing: Ray Kelly
Manager, Passenger Services (South/West):
   Willie O'Connor
Manager, Passenger Services (North/East):
   Bertie Corbet
Manager, Navigator Freight Agency: Shay Hart
Manager, Bulk Freight: Patsy Mee
District Manager, Heuston: Noel McKenna
District Manager, Connolly: Tom Devoy
District Manager, Galway: Gerry Glynn
District Manager, Cork: Tim Sheehan
District Manager, Limerick: Brian Kelly

District Manager, Waterford: Paul Cheevers
District Manager, Suburban: Christy Stapleton
Port Manager, Rosslare Harbour:
   Walter Morrissey
Freight Manager, East (Dublin):
   Frank Spellman
Freight Manager, West (Galway): Gerry Mongan
District Freight Manager, Waterford: Peter Roche
District Freight Manager, Cork: Oliver O'Donovan
District Freight Manager, Limerick: Joe Whelan

Gauge: 1,602 mm on cwr track; 1,600 mm on jointed track
Route length (open to traffic): 1,947 km
Electrification: 45.5 km at 1.5 kV DC

Political background

In late 2000 the government published proposals to restructure public transport in Ireland, with significant likely implications for IE. The proposed reforms foresee the division of IE into two independent companies responsible for infrastructure and operations, with the future possibility of either transferring the operations company to the private sector or of franchising all or some services. Also anticipated is the creation of an independent body to develop major projects in partnership with the private sector, establishing an independent public transport regulator and

setting up a safety authority to cover all public rail operations. It was expected that legislation covering these reforms would be introduced in the Irish parliament during 2001. A second reform phase is expected to provide for the establishment as separate companies of the subsidiaries of Córas Iompair Éireann, the state transport holding company and parent of IE.

Organisation

Iarnród Éireann is responsible for all rail services in the state including the Dublin Area Rapid Transit system. It is also responsible for rail freight and the company's own road freight services, for catering services, and for the operation of Rosslare Europort, which caters for sailings to the UK and the Continent.
   At the end of 2001 IE employed 5,897 staff compared with 5,358 in 2000.

Passenger operations

IE offers a nationwide service of diesel-hauled passenger trains, including a recently upgraded Enterprise service between Dublin and Belfast, while in the Dublin area it operates an electrified suburban system, known as Dublin Area Rapid Transit (DART). The DART system operates over 45.5 route-km electrified at 1.5 kV DC from Bray, south of Dublin Connolly Station, to Howth, in the north, with a fleet of 40 two-car emus. The

Class 8200/8400 1.5 kV emu supplied by Alstom for DART suburban services undergoing commissioning at Dublin Fairview depot (Colin Boocock)   0109857

Class 201 diesel locomotive at Malahide with a Dublin Pearse–Drogheda service formed with a Mark 3 push-pull set (Colin Boocock)   0524935

Diesel locomotives

Class	Wheel arrangement	Power KW	Speed km/h	Weight tonnes	No in Service	First built	Mechanical	Builders Engine	Transmission
071	Co-Co	1,800	145	102	18	1976	GM	GM12-645 E3C	E
121	Bo-Bo	700	120	65	11	1961	GM	GM 8-645CR	E
141	Bo-Bo	700	120	68	35	1962	GM	GM 8-645 CR	E
181	Bo-Bo	820	120	68	11	1966	GM	GM 8-645 E	E
201	Co-Co	2,219	165	112	32	1994	GM	GM 12-710G3 B	E

Electric multiple-units

Class	Cars per unit	Motor cars per unit	Motored axles/car	Output/motor KW	Speed km/h	Cars in service	First built	Builders Mechanical	Electrical
8100/8300	2	1	4	130	100	76	1983	LHB	GEC Traction
8200/8400	2	1	–	–	110	10	2000	Alstom	Alstom
8500/8600	2	1	–	–	–	16	2000	Mitsui	–

Diesel multiple-units

Class	Cars per unit	Motor cars per unit	Motored axles/car	Speed km/h	Cars in Service	First built	Mechanical	Builders Electrical	Transmission
2600	2	2	2	110	17	1993	Tokyu Car	Cummins NTA 855-R1	H Niigata DW14G
2700	2	2	2	120	25	1998	Alstom	Cummins NTA-855	H Niigata DW14G
2800	2	2	2	–	20	2000	Mitsui	Cummins	Niigata
2900	4	–	–	–	80	*	CAF	–	–

* On order in 2001.

Mitsui-built Class 2800 dmus at Dublin Connolly (Colin Boocock)   0524933

Mitsui-built Class 8500/8600 DART emu at Dublin Pearse forming a Howth-Bray service (Colin Boocock)   0524934

DART is currently being extended southwards to Greystones and northwards to Malahide. Services over these extensions, which add 7.5 route-km to the network, commenced in April 2000 and June 2000 respectively.

Traffic	1999	2000	2001
Passenger journeys (000)	32,765	31,721	34,206
Passenger-km (million)	1,457.6	1,389.1	1,515.3
Freight tonnes (000)	2,901	2,707	2,612
Freight tonne-km (million)	526	490.8	515.7

Freight operations
IE's major freight customers are the brewing, construction and fertiliser industries. The railway also hauls maritime containers and offers a parcels service known as 'Fastrack'.

Revenue from freight in 2001 increased by 3.0 per cent on the figure for 2000. Traffic increases were achieved in kegs, cement, mineral ores and ammonia distribution.

Intermodal operations
Containers form an important part of IE's freight traffic, with around 130,000 TEU carried each year. IE provides daily (Monday to Friday) domestic common-user services between Dublin and Ballina, Dundalk, Galway, Limerick, Sligo, Waterford and Westport. Three common-user services run each day (Mondays to Fridays) between Dublin and Cork.

Improvements to existing lines
As part of a €546 million safety investment programme, IE is upgrading almost 640 km of track, replacing jointed rail with cwr on concrete sleepers. This will result in improvements to all radial routes from Dublin on the Intercity network. Branch lines are also receiving investment.

The Safety Investment Programme, covering the period 1999–2003, was approved in 1999 by the Minister for Public Enterprise and also covers improvements to bridges and fencing, upgrading of signalling systems and improvements to level crossings.

Traction and rolling stock
IE's motive power fleet at the start of 2002 consisted of 110 diesel locomotives, 64 dmu vehicles and DART emu. Freight wagons totalled some 1,830, and there were 181 passenger coaches.

In February 1998 IE placed an order with Alstom for five Class 8200/8400 two-car 1.5kV DC emus for DART suburban services. Assembly of the trains, which comprise a motor car plus trailer, was also undertaken in Spain, with delivery completed in 2000. The new vehicles are able to operate in multiple with existing DART stock.

In February 1999 IE placed an order with Mitsui, Japan, for 10 Class 2800 two-car diesel railcars and 16 Class 8500/8600 emu cars for DART services. Delivery of both types was completed in 2000. The Class 8500/8600 vehicles are formed as two two-car half-sets to create a four-car unit.

In 2000 IE ordered 15 Class 2900 four-car dmu vehicles from CAF in Spain. An additional five four-car units were added to the order in 2001.

IE expected to place contracts for the supply of new intercity coaches in 2002.

Talgo-Transtech in Finland received an order to supply 24 container flats by the end of 2001. These were due to enter service on the Dublin-Cork line in late 2002.

Signalling and telecommunications
The Dublin suburban line is controlled from a computer-based VDU console at Connolly station. The system utilises automatic route-setting and optimisation of public road level crossing closures, the latter being supervised by closed-circuit television. The CTC building also houses a similar console for control of 458 route-km of main line railway. The system permits operation of cab signalling equipment fitted to all locomotives, as well as train radio, and is being extended to cover further lines. In 1997 coverage was extended on the Belfast route to reach the border with Northern Ireland.

DART electric trains in the Dublin suburban area are fitted with Automatic Train Protection (ATP) equipment, activated by coded currents passed through the running rails, detected by coils on the emus and processed on board. Lineside signalling has been retained for diesel-hauled trains on the route.

Locomotives are fitted with a Continuous Automatic Warning System (cab signalling). Both emus and locomotives are fitted with AEG-Telefunken train radio allowing two-way transmission of fixed-message telegrams and voice communication with the controlling signalman at the appropriate CTC centre (suburban or main line). Since 1996 this system has been extended beyond CTC-controlled areas, allowing drivers to maintain communications with local signal cabins at all times. The equipment also permits voice communication between train drivers and on-train staff using hand-held portable radios.

Type of coupler in standard use: Passenger cars, buckeye or screw; emus, Scharfenberg and Dellner automatic; freight cars, screw or Instanter
Type of braking: Locomotive-hauled stock, vacuum or air

Track
Standard rail: Flat bottom 113, 95, 92 and 85 lb/yd, 54 kg/m, 50 kg/m; bullhead 95, 90, 87 and 85 lb/yd
Crossties (sleepers): Timber 2,590 × 255 × 125 mm; concrete 2,475 × 220 × 180 mm
Spacing: Concrete 1,144/km straight, 1,556/km curved track; timber, 1,313/km straight, 1,422/km curved track
Rail fastenings: Timber sleepers: CI chairs and sole plates; concrete sleepers: H-M (Vossloh) fastenings, Pandrol and H-M (Vossloh) on crossings

Min curvature radius
Running lines: 115 m
Sidings: 80 m
Max gradient: 1 in 40
Longest continuous gradient: 8.45 km, with 1% ruling gradient
Max altitude: 165 m at Stagmount, Co Kerry
Max axleloads: 18.8 tonnes for locomotives; 15.75 tonnes for wagons; 18.8 tonnes for specific traffic on bogie wagons

---

Israel

Ministry of Transport

97 Jaffa Road, Jerusalem 94342
Tel: (+972 2) 622 86 94
Web: www.mot.gov.il

Key personnel
Minister: Meir Shitrit

UPDATED

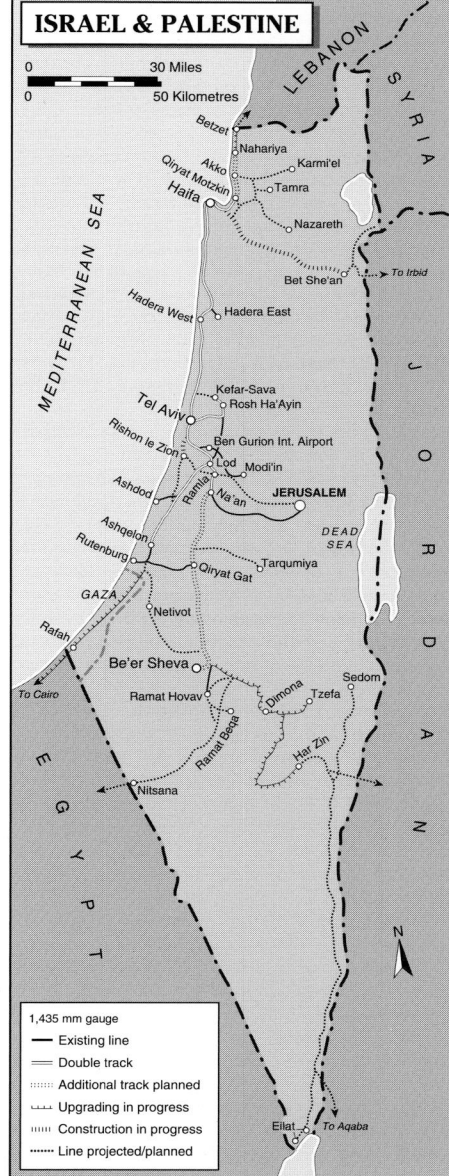

ISRAEL & PALESTINE

1,435 mm gauge
— Existing line
═ Double track
⋯ Additional track planned
⊥⊥ Upgrading in progress
⊓⊓ Construction in progress
⋯ Line projected/planned

NEW/1122851

Israel Railways (IR)

Central Station, PO Box 18085, Tel Aviv 61180
Tel: (+972 3) 693 74 01   Fax: (+972 3) 693 74 80
Web: www.israrail.org.il

Key personnel
Director General: (vacant)
Deputy General Directors
    Rolling Stock: Dudu Gabai
Operation and Maintenance: Harel Even
Traffic: Tzwika Silberschmidt
Development and Design: Illya Volkov
Marketing: Avi Hefetz
Finance & Economics: Ilan Tzurshdai
Personnel & Human Resources: Isaac Seri
Passenger Traffic: Michael Reiff
Cargo Traffic: Yossi Carmeli
Directors
    Rolling Stock Maintenance: Sami Cohen
    Rolling Stock Development: Zeev Gordon
    Signalling and Telecommunications: Dubov
    Purchase and Supply: Shai Eisenberg
    Economics, Budget & Trade: Eli Dalizky
    Finance: Roni David
    Public Relations and Spokesman: Benny Naor

Gauge: 1,435 mm
Route length: 853 km

Political background
IR was long operated as a separate economic enterprise by the Ministry of Transport, but in 1988 the system became part of a new public enterprise; the Ports & Railways Authority. One

objective was to reallocate some of the former Ports Authority's accumulated funds to railway investment; government policy for the past few years has been to renovate the railway, principally its passenger operations, as an attractive alternative to the rising volume of motor transport. Another aim of the change was to allow the railway more commercial freedom to react to changing market conditions.

In March 1997 IR was separated from the Ports & Railway Authority while remaining under its control. In March 1998 the Israeli government introduced to the Knesset legislation to establish IR as an independent fully state-owned company, with a nine-member board which would include appointees from National Infrastructure, Transport, and Finance ministries.

Traffic (million)	2002	2003	2004
Passenger journeys	17.5	19.8	22.9
Passenger-km	1,116	1,278	1,423
Freight tonnes	7.9	7.7	7.9
Freight tonne-km	1,100	1,278	1,423

Passenger operations
In 2004 IR was operating 288 trains daily, serving 42 stations. Traffic continues to grow rapidly as IR opens new lines, increases line capacity and adds new rolling stock to its fleet.

Freight operations
In terms of tonne-km, freight traffic grew by 6 per cent in 2004.

Dominant traffic is potash from the Dead Sea and phosphates from Oron and Har Zin, which are conveyed to the port of Ashdod in unit trains of up

The station at Ben Gurion International Airport, served by a new line which opened in October 2004 (Aharon Gazit)

NEW/1122849

Jerusalem's Malkha station, terminus of the rehabilitated link with Tel Aviv that was commissioned in 2005 (Aharon Gazit)   ***NEW**/1122850*

ALSTOM 'Mega' diesel locomotive and Bombardier-built push-pull driving/generator cars stabled at Lod depot (Aharon Gazit)   ***NEW**/1122847*

to 4,000 tonnes (west of Dimona) hauled by two GM Type G26CW 1,640 kW Co-Co diesel-electrics in multiple. Some goes to chemical plants on Haifa Bay, 310 km from Har Tzin. For the rest, IR's freight is chiefly bulk movement of containers, grain and oil. Container traffic is also carried, mainly to and from the ports of Ashdod and Haifa, and to a lesser extent, Eilat.

Israeli government proposals, to switch from road to rail the haulage of domestic refuse from a concentration site at Hiriya, near Tel Aviv, (some 150 km south to the Rotem Plain, southeast of Be'er Sheva), remained unfulfilled in 2005.

New lines

A 9 km single-track branch from the Tel Aviv-Lod line to Ben Gurion International Airport opened in October 2004. As well as serving the airport's international terminal via a dedicated station, the line also provides a park-and-ride commuter facility for travellers to Tel Aviv, Haifa and Nahariya. In 2005, construction of a 20 km extension from Ben Gurion to Mod'in, was progressing. It will be served by two stations. In the longer term, part of this line will form part of the so-called Express Line (A1) linking Tel Aviv and Jerusalem. Expected to open in 2009 at an estimated cost of ILS3.5 billion,

the 40 km double-track line will include a 20 km tunnel, the first on the IR system. It will be engineered for 160 km/h running.

In the Tel Aviv area IR plans several lines, as part of a developing suburban network, and in the longer term electrification of some routes is proposed. Projects include a new 16 km double-track line from Tel Aviv Hahagana to Rishon Le-Zion West, due for completion in late 2008 and eventually to be extended to Ashdod (19 km). Creating a direct link between Tel Aviv and Ashdod will eliminate the need for trains to use the circuitous route via Lod.

Other new lines planned include: Ashqelon via Sederot and Netivot to Be'er Sheva, with completion projected for 2008; a line from Akko, on the Haifa–Nahariya coastal line, inland to Karmiel, also for completion in 2008; a link from Ra'anana (west of Kefar Sheva) to the Tel Aviv–Haifa line (2008).

The long-awaited project for a railway to the Red Sea port of Eilat continues to receive Israeli government support and has been the subject of detailed design work by IR. Construction of a 180 km link from Be'er Sheva to Eilat would create a land-bridge alternative to the Suez Canal. It is thought that initially the line would only carry freight traffic, enabling the 300 km between the ports of Ashdod and Eilat to be covered in three and a half hours. An economic viability study of the project was completed in 2004 by US consultancy Mercator Transport Group. Preliminary design has previously been completed for the section from Har Tzin to Eilat, with a branch to the Jordanian port of Aqaba. A 32 km branch from this line would serve Israel's Sedom potash works at the southern end of the Dead Sea.

In early 1995 an accord was signed between the Israeli and Jordanian governments approving, in principle, construction of a line extending some 90 km from the port of Haifa to Irbid in northern Jordan, designed to provide access to the Mediterranean for Jordanian import/export traffic. The first section of the route, so-called the 'Jezreel Valley Railway', would utilise the derelict alignment of the former Hedjaz Railway branch from Deraa to Haifa, which was closed when Israel was created in 1948. This project was revived in October 2002, when the Israeli government announced that it had started detailed design on its section of this line. A double-track line is envisaged eventually. Work on the core section of the line, the 60 km from Haifa to Bet She'an, commenced in 2005, with completion planned for 2008. For their part, the Jordanian authorities would complete a long-proposed link from Irbid to Amman.

In late 1996 Israel reached agreement with the autonomous Palestinian authority to restore the long-disused 18 km rail link between Ashkelon and Gaza, although no progress has been made due to the political situation in the country.

In 2004 a 12.5 km branch was commissioned from the line between Be'er Sheva's North and Central stations to serve a terminal handling hazardous materials at Ramat-Hovav. An eventual extension of this line would complete a projected link to the Egyptian border at Nitsana.

Improvements to existing lines

Restoration and upgrading of the 35 km Bet Shemesh–Jerusalem section of the original line from Tel Aviv was completed in April 2005 at a cost of ILS125 million, enabling passenger services to be reintroduced between Israel's two principal cities after a gap of seven years. The new Malkha station serves as the line's terminus in Jerusalem.

Double-tracking of the Tel Aviv–Haifa line was completed in 1999, cutting the journey time between the two cities to 37 minutes. IR has undertaken design work on quadrupling the entire Tel Aviv–Haifa route to cope with anticipated future traffic levels. Electrification is foreseen, enabling speeds of up to 200 km/h to be achieved eventually.

Double-tracking schemes planned include: Qiryat Motzkin–Nahariya (2005); Tel Aviv–Kefar Sava (2006); Na'an-Be'er Sheva (2007).

A third track has been provided on sections of the Ayalon Railway, IR's key route through central Tel Aviv.

Diesel locomotives

Class	Wheel arrangement	Power kW	Speed km/h	Weight tonnes	No in service	First built	Mechanical	Engine	Builders	Transmission
G12	Bo-Bo	1,230	105	76	13	1954–66	GM	GM 12-645E		E GM
G16	Co-Co	1,450	107	124	3	1960–61	GM 567C	GM 16-567C		
G26CW	Co-Co	1,640	124	99	9	1971–79	GM	GM 16-645E		E GM
G26CW-2	Co-Co	1,640	124	116	6	1982–86	GM	GM 16-645E		E GM
GT26CW-2	Co-Co	2,200	124	119	1	1989	GM	GM 16-645E3		E GM
Mega	Bo-Bo	2,460	160	90	35	1998	ALSTOM	GM 12-710G3B		E GM
Semi-Mega	Co-Co	2,460	120	114	8	1998	ALSTOM	GM 12-710G3B		E GM
GA900	Bo-Bo	900	80	14	3	1998	GEC Alsthom	MTU 8V 396		E GEC Alsthom

Diesel railcars or multiple-units

Class	Cars per unit	Motor cars per unit	Motored axles/car	Rated power kW	Speed km/h	Units in service	First built	Mechanical	Builders Engine	Transmission
IC3	3	3	4	300	160	48	1992	ABB-Scania/ Adtranz	Deutz	H ZF

In 2004–05 rehabilitation was in progress of the Be'er Sheva–Dimona section (36 km) of the line into the Negev Desert in preparation for the introduction of passenger services. The project includes realignment of several sections totalling 3 km.

Traction and rolling stock

Diesel locomotive stock at the end of 2004 comprised 75 main line units (35 assigned to passenger work, 40 to freight) and three shunters. A fleet of 48 three-car IC3 dmus was also in operation. Coaching stock totalled 188, of which 90 were double-deck and 37 were modern ALSTOM-designed vehicles. Railway-owned freight stock totalled 615 wagons, of which 550 were assigned to revenue traffic; some 500 privately owned vehicles are also in use.

In 2004–05 IR supplemented its then 40-strong fleet of IC3 dmus with eight similar vehicles procured from Sweden, where they had been replaced by electric traction. Refurbishment and overhaul of the trains before they entered IR service included replacement of the existing Deutz engines with more powerful units from the same manufacturer and the installation of air-conditioning equipment with increased capacity.

In 2001 IR received its first double-deck four-car push-pull coaching stock sets from Bombardier Transportation's Görlitz plant in Germany. By 2005 the number of these vehicles had increased to 90. The driving car of each set includes a generator to provide power for train services. In 2004 IR placed a follow-on contract with Bombardier for 54 double-deck coaches to create six new five-car push-pull sets and to strengthen existing four-car formations to five-car sets. Delivery was due to commence at the end of 2005, with assembly and the production of some major sub-assemblies undertaken in Israel by IAI's Ramta Division at its plant in Be'er Sheva.

In May 2005 a cooperation agreement was signed between Bombardier and IAI's Ramta Division as part of a joint bid to supply to IR 86

An IR IC3 dmu in typical terrain on the rebuilt line to Jerusalem (Aharon Gazit)　**NEW**/1122848

single-deck coaches, with options on up to 500 more.

An additional eight three-car IC3 dmus were due to enter service from September 2001 and orders for seven more were expected. Under licence from GEC Alsthom Transporte, local builder Haargaz assembled 37 loco-hauled coaches, including five with driving cabs for push-pull operation and auxiliary diesel engines providing a train power supply. Older passenger coaches have been refurbished, and most of the fleet is now air conditioned.

Signalling and telecommunications

In August 2002 SEL-Alcatel was awarded a contract for the supply of a network management and traffic control system for the entire railway. Located at Hof-Ha-Carmel station, Haifa, the system's control centre became operational in 2004.

Electrification

A new call for tenders to electrify key sections of the IR network was made in May 2005. It is expected that initially around 300 route-km would be electrified, including the lines from Tel Aviv to Nahariya in the north, to Jerusalem via the new A1 line in the east and to Ashqelon in the south. This follows the earlier pricing of proposals to electrify the entire IR passenger network, rejected on grounds of cost.

Type of coupler in standard use: Screw
Type of brake in standard use: Air

Track

Rail: 60 kg/m ; 54 kg/m (351 km); 50 kg/m (226 km; 49 kg/m; 46 kg/m; 37 kg/m
Crossties (sleepers): Concrete monobloc 300 kg, timber, and steel
Spacing: 1,720/km, concrete; 1,666/km, timber and steel
Fastenings: Vossloh
Min curvature radius: 140 m; 600 m min, new projects
Max gradient: 2.68% (Bet Shemesh-Jerusalem)
Max permissible axleload: 22.5 tonnes

UPDATED

Italy

Ministry of Infrastructure and Transport

Piazzale Porta Pia 1, I-00161 Rome
Tel: (+39 06) 064 41 21
Web: www.infrastrutturetrasporti.it

Key personnel
Minister: Pietro Lunardi

The issuing office for open access licences is:
Ministry of Infrastructure and Transport
Railway Supervision Structure VIG4

Via Caraci 36, Rome
Tel: (+39 06) 41 58 34 51　Fax: (+39 06) 41 58 34 10
e-mail: vig4@trasportinavigazione.it
Contact: Mario Vivaldi

UPDATED

Appulo–Lucane Railways (FAL)

Ferrovie Appulo Lucane Srl
Corso Italia 6, I-70123 Bari, Italy
Tel: (+39 080) 572 52 33
Fax: (+39 080) 524 50 17
e-mail: info@fal-srl.it
Web: www.fal-srl.it

Key personnel
Chairman: Alessandro Tamburrino
Administrative Director: Domenico Magrone

Operating Director: Alessandro di Macco
Company Secretary: Clorinda Drago

Gauge: 950 mm
Route length: 183 km

Organisation
FAL was one of two railways formed in 1991 by splitting the former Ferrovie Calabro–Lucane into separate operations based on the Bari and Cosenza networks, the latter becoming Calabria Railways (qv). It became an independent company in January 2001. The FAL network connects Bari with Avigliano and Matera in Basilicata and Puglia regions. In 2003 FAL was building a new 11 km line from Bari to Bitritto.

In 2004 FAL employed 714 staff.

Traction and rolling stock
Traction and rolling stock comprises 20 railcars and 22 trailers, six locomotives and 95 wagons. The company called for tenders worth €26.4 million for new or refurbished dmus in 2005.

UPDATED

Arezzo Railways (LFI)

La Ferroviaria Italiana
Via Guido Monaco 37, I-52100 Arezzo
Tel: (+39 0575) 398 81; 32 42 94
Fax: (+39 0575) 284 14; 32 48 01
e-mail: info@lfi.it
Web: www.lfi.it

Key personnel
Director General: Vincenzo Balzini
Administrative Director and Deputy Director General: Giulio Bigozzi
Operations Director: Armando Selmi

Gauge: 1,435 mm
Route length: 85 km
Electrification: 85 km at 3 kV DC

Organisation
In 2002 Arezzo and Siena provinces sold their 30 per cent share in FFI to a group of companies including the Central Umbria Railway (FCU) (see entry in Italy section of *Railway systems and operators*) and Paris urban transport operator RATP. LFI then joined Consorzio Rasena, which operates local transport in Toscana province.

LFI operates frequent passenger services from Arezzo to Sinalunga (40 km) and Pratovecchia Stia (45 km), as well as buses. The company is rebuilding a 5 km section of the Stia line with the intention of increasing speeds, avoiding risk of landslips and generating new freight traffic including block coal trains for a cement plant in Rassina.

In 2001 LFI obtained an operating licence for RFI tracks and was negotiating the takeover of both regional passenger and freight services over FS lines around Arezzo, including the Arezzo–San Giovanni Valdarno–Florence-Pisa route.

*Ex-SNCB Class 56 stainless steel-bodied emu
in service with LFI* (Marcel Vleugels)
0552785

In January 2005 LFI formed two new joint-stock subsidiary companies, splitting train operations from infrastructure. These are:
* Trasporto Ferroviario Toscano (TFT SpA) (operations)
* Rete Ferroviaria Toscana (RFT SpA) (infrastructure)

Traction and rolling stock
The railway operates nine electric and five diesel locomotives, nine electric railcars, 18 trailers and 60 wagons. In 1999 LFI bought three Type AM56 two-car emus from Belgian National Railways and these entered service after refurbishment by Metalmeccanica Milanesio in Moretta.

In 2005 TFT was awaiting delivery from ALSTOM of four three-car Minuetto emus. These were to be branded 'ETT' (Electro Triple Train).

UPDATED

Bari-Nord Railway (FT)

Ferrotramviaria SpA-Ferrovie del Nord Barese
Piazza Moro 50B, I-70122 Bari
Tel: (+39 080) 578 95 11
Fax: (+39 080) 523 54 80
e-mail: info@ferrovienordbarese.it
Web: www.ferrovienordbarese.it

Key personnel
President: Enrico Maria Pasquini
Managing Director: Dr Ing Nicola Nitti
Operations Manager: Dr Ing Cesare Soria
Administration Manager: Dr Giambattista Angiuli

Gauge: 1,435 mm
Route length: 70 km
Electrification: 70 km at 3 kV DC

Organisation
FT operates the Bari-Barletta line via Bitonto (70 km). The company is doubling much of this line and is building a new 9.3 km line from Bari Lamasinata to Ospedale San Paulo which will later be extended to Regiani. This line will open in 2006 and is to be followed by a new 7 km underground loop off the Bari-Bitonto line to serve Bari airport.

In early 2000 FT created a joint venture, known as Rail Traction Company (RTC), with three other

partners, which now operates freight services from Verona to Munich, Germany. FT received its own operating licence in 2004.

Traction and rolling stock
The railway operates two electric locomotives, 15 electric railcars and 11 control trailers, four coaches and 16 wagons. In 2005 the first of six new three-car emus were received from ALSTOM.

UPDATED

Calabria Railways (FC)

Ferrovie della Calabria Srl
Via Milano 28, I-88100 Catanzaro
Tel: (+39 0961) 89 61 11
Fax: (+39 0961) 74 70 07
e-mail: mail@ferroviedellacalabria.it
Web: www.ferroviedellacalabria.it

Key personnel
Administrator: Francesco Alberto Covello
President: Giuseppe Mario Scali
Director General: Giuseppe Lo Feudo
Railways Operating Manager: Aldo Ghionna
Traction Manager: Giuseppe Volpi

Gauge: 950 mm
Route length: 231 km

Organisation
FC was one of two railways formed in 1991 by splitting the former Ferrovie Calabro–Lucane into separate operations based on the Cosenza and Bari networks, the latter becoming Ferrovie Appulo–Lucane (see entry in Italy section of *Railway systems and operators*). It became an independent company in January 2001. The FC network connects Cosenza with San Giovanni in Fiore and Catanzaro, the last-mentioned with a section of Strub rack, and also includes two branches from Gioia Tauro totalling 41 km. In 2002 the railway achieved

1.4 million passengers (24.4 million passenger-km), up from 23.9 million in the previous year. Part of the network is now used only by steam-hauled tourist services.

In December 2003 FC introduced a new shuttle service between Cosenza and Casali.

Traction and rolling stock
Traction and rolling stock comprises 41 railcars and seven trailers, 12 locomotives, four coaches and 56 wagons. In late 2003 FC called for tenders to supply four adhesion railcars, four rack-fitted railcars and five trailer cars.

UPDATED

Casalecchio–Bologna–Vignola Railway (ATC)

Azienda Trasporti Consorziali, Bologna
Via Saliceto 3, I-40128 Bologna
Tel: (+39 051) 35 01 11   Fax: (+39 051) 35 01 77
e-mail: atc.dg@atc.bo.it;
Web: www.atc.bo.it

Subsidiary
Società Suburbana FBV
Via S Donata 25, Bologna
Tel: (+39 051) 421 79 11   Fax: (+39 051) 421 79 30
e-mail: info@suburbanafbv.it
Web: www.suburbanafbv.it

Key personnel
General Manager, Suburbana FBV:
   Claudio A Claroni
Railway Manager: Fabio Formentin

Gauge: 1,435 mm
Route length: 24 km
Electrification: 3 kV DC

Organisation
ATC operates public transport in the city of Bologna as well as serving industrial sidings in and around the city. In 1996, the 24 km Casalecchio–Vignola line and the 3 km branch Vignola–Confine were closed for rebuilding with the aim of introducing a frequent Bologna-Vignola service. In July 2002, a new joint venture company, Suburbana FBV, was established by ATC and the Emilia Romagna Railway (FER) (see entry in Italy section of *Railway systems and operators*) to operate the Casalecchio–Vignola line. Both companies have a shareholding of 50 per cent in the new venture.

The line was reopened from Casalecchio to Bazzano in September 2003, then from Bazzano to Vignola in September 2004. A new chord at

Casalecchio allows a direct hourly service from Bologna without reversal. This was initially operated by FER dmus. It is expected that the replacement of these, by modernised ex-Belgian emus in late 2005, will allow services to run on a more frequent half-hourly basis, compared to the current hourly service.

Traction and rolling stock
The railway owns five electric and two diesel locomotives, and seven wagons. ATC has acquired two new Type E122 two-car emus from Ansaldobreda and four two-car 1950s emus from Belgian National Railways, modernised for the Suburbana FBV service.

UPDATED

Central Umbria Railway (FCU)

Ferrovia Centrale Umbra Srl
Largo Cacciatori delle Alpi 8, I-06121 Perugia
Tel: (+39 075) 57 54 01   Fax: (+39 075) 573 52 57
e-mail: fcu@fcu.it
Web: www.fcu.it

Key personnel
Administrator: Vannio Brozzi
Administrative Manager: Fausto Pucci
Operating Manager: Mauro Fagioli

Gauge: 1,435 mm
Route length: 152 km
Electrification: 152 km at 3 kV DC

Organisation
In 2001 FCU passed under the control of the Umbria region and let the contract to re-electrify its network to ALSTOM, ABB, Bonciane and Sirti. Work will be completed by 2005. The Perugia Ponte San Giovanni–Perugia Santa Anna section was doubled in 2002–03. FCU has an open access licence and in December 2002 extended certain services over RFI metals to Orte and Terontola.

Traction and rolling stock
The railway owns two electric and five diesel locomotives, 42 diesel railcars and 60 wagons. The diesel locomotives mainly haul coal trains from Perugia to Marsciano. In 2004 FCU called for tenders worth €12 million for four emus with an option for two more. Delivery will take place in 2006–07.

UPDATED

Circumetnea Railway (FCE)

Managed by Government Commission, Rome
Ferrovia Circumetnea
Via Caronda 352A, I-95128 Catania
Tel: (+39 095) 541 11 11   Fax: (+39 095) 43 10 22
e-mail: info@circumetnea.it
Web: www.circumetnea.it

Key personnel
Director General: Carlo Pino

Gauge: 950 mm; 1,435 mm
Route length: 110 km; 4 km
Electrification: 4 km at 3 kV DC

Organisation
This narrow-gauge railway forms an incomplete circle around Mount Etna in Sicily. In 1998, 31 million passenger-km were achieved, down from 36.7 million the previous year.

In June 1999 the on-street section of the railway in central Catania was replaced by a 3.8 km underground, standard-gauge electrified line from Porto to Borgo, where the narrow-gauge line now terminates. There are three intermediate stations. The 'metro' service is provided by four two-car emus purchased from Firema in 2001.

FCE intends to convert a further 20 km of line, to Paterno, with at least six new stations. There are also plans for a new branch from central Catania to Fontanarossa, 10 km to the north.

Traction and rolling stock
The railway operates 27 diesel railcars, nine trailer coaches and four two-car emus. A further four emus were ordered from Firema in early 2005.

UPDATED

Circumvesuviana Railway

Circumvesuviana Srl
Corso Garibaldi 387, I-80142 Naples
Tel: (+39 081) 772 21 11   Fax: (+39 081) 772 24 50
e-mail: circum@vesuviana.it
Web: www.vesuviana.it

Key personnel
Government Commissioner:
  Dr Ing Gaetano Danese
General Manager: Dr Fernando Rigo
General Secretary: Dr Domenico Sica
Rail Operations Director: Dr Ing Michele di Matteo

Gauge: 950 mm
Route length: 142 km
Electrification: 142 km at 1.5 kV DC

Organisation
The Circumvesuviana operates services under contract to the Campania region over an extensive territory around the volcano, Vesuvius, to the east of Naples stretching to Baiano, in the province of Avellino, and to Sarno in the province of Salerno, and also skirts the Bay of Naples to a terminus in Sorrento. The Sorrento branch is largely in tunnel, one of which under Monte Faito is 4.8 km long with a station inside the bore.

In 2003 38.7 million passengers were carried for 524 million passenger-km.

The railway has a continuing programme of heavy engineering works to increase operating capacity and improve operation. In March 2003 a new double-track line opened from Napoli Collegamento to Casalnuovo, on the Baiano line. A new branch from Volla, on this new line,

Mock-up of new articulated emu for the Circumvesuviana Railway by Ansaldobreda and Firema, displayed at the 2005 UITP Congress in Rome (Ken Harris)   *NEW*/1122805

via Madonelle to San Giorgio a Cremano on the Sorrento line enables a service to be run from San Giorgio to Casalnuovo. This line will later be extended to the new Naples high-speed station at Afragola. A new chord opened in June 2004, which allows a direct Naples-Madonelle-San Giorgio a Cremona servce.

Other plans include double-tracking of the Sorrento branch between Morgine and Castellammare di Stabia; double-tracking and realignment of the Sarno branch between Torre Annunziata and Pompei, a total of 14.5 km; and construction of a new 30 km double-track branch from the Baiano line at Nola to Avellino. The line from Pomigliano to Alfa Sud was extended 2.9 km to Acerra in February 2005. The line will eventually be extended to a new interchange at Acerra with the RFI Naples–Cancello line.

Traction and rolling stock
The company operates 118 articulated three-car emus, two three-car 'heritage' emus, three diesel shunting locomotives, four Bo-Bo electric locomotives, two coaches and 26 wagons. A further 23 three-car articulated emus were ordered from Ansaldobreda and Firema in 2004 for delivery from 2006.

UPDATED

Cumana & Circumflegrea Railways (SEPSA)

Società per l'Esercizio di Pubblici Servizi Anonima (SEPSA SpA)
Via Cisternia dell'Olio 44, I-80134 Naples
Tel: (+39 081) 735 41 11   Fax: (+39 081) 735 42 92
e-mail: direzione.ferro@sepsa.it
Web: www.sepsa.it

Key personnel
President: Nicola Martino
General Manager: R Bianco

Gauge: 1,435 mm
Route length: 47 km
Electrification: 47 km at 3 kV DC

SEPSA emus at Torregaveta (David Haydock)   0580498

Organisation

The Cumana (20 km) and Circumflegrea (27 km) railways (also known as Ferrovia Cumana-Circumflegrea e Autolinee) connect Naples with Torregaveta, serving a densely populated area with an all-day 20-minute service on each branch, doubling to 10 minutes all day within the boundaries of Naples. In 2001 SEPSA's rail services carried 18.5 million passengers. SEPSA also operates bus services in the Naples area.

Doubling is in progress on both lines. In 1996 the government allocated SEPSA €75 million towards the €108 million cost of a new line to serve a university at Monte Angelo. Work started in 2001 on a 5.5 km connection with three new stations between Kennedy on the Cumana line and Soccavo on the Circumflegrea line. A new station is to be built at Cilea in order to provide interchange with the Naples metro system. Plans also exist for a new line from Licola, on the Circumflegrea, to Mondragone,

and for a line linking the Circumflegrea and FS lines north of Naples with a new airport at Grazzanise.

SEPSA employs 799 staff on rail services.

Traction and rolling stock

The railway operates 30 two-car emus, one diesel locomotive and 10 wagons. In 2004 €55.8 million was allocated by the state and Campania, for the purchase of 15 new emus.

UPDATED

Del Fungo Giera Servizi Ferroviari SpA (DFG)

Via Borra 35, I-57123 Livorno
Tel: (+39 0586) 82 96 05   Fax: (+39 0586) 21 96 08
e-mail: info@delfungogiera.com
www.delfungogiera.com

Key personnel

President: Dominico Libro

Organisation

A rail operating subsidiary set up in 2000 by a well-established transport company, which handles 2.5 million tonnes of freight per year, including 700,000 tonnes by rail, particularly chemicals. A daily return service carrying chemicals between Alessandria and Livorno was started in April 2003. DFG intends to expand into full operation of rail freight services.

Traction and rolling stock

In 2005 DFG was using 26 Class E 636 electric locomotives hired from Trenitalia Cargo. The company has signed a contract with ALSTOM to purchase 50 tri-voltage electric locomotives similar to the SNCF Class BB 36000 and 30 Prima diesel-electric locomotives. Tests with an SNCF BB 36000 took place in 2002.

UPDATED

Domodossola–Locarno Railway (SSIF)

Società Subalpina di Imprese Ferroviarie SpA
PO Box 60, Via Mizzoccola 9, I-28845 Domodossola
Tel: (+39 0324) 24 20 55   Fax: (+39 0324) 452 42
e-mail: vigeinfo@tin.it
Web: www.cobatool.com/vigezzina/

Key personnel

General Manager: Ing Daniele Corti

Gauge: 1,000 mm
Route length: 32 km (and 20 km in Switzerland)
Electrification: 32 km at 1.35 kV DC

Organisation

SSIF operates through services to Locarno jointly with the Swiss railway FART. The very scenic line is commonly known as the Centovalli.

Traction and rolling stock

The railway operates 14 electric railcars, 11 coaches and 27 wagons.

In 2004 SSIF ordered three 3-car emus from Officine Ferroviarie Veronesi.

UPDATED

Emilia Romagna Railways (FER)

Ferrovie Emilia Romagna Srl
Via Zandonai 4, I-44100 Ferrara
Tel: (+39 053) 297 93 11   Fax: (+39 053) 297 93 14
e-mail: info@fer-online.it
www.fer-online.it

Gauge: 1,435 mm
Route length: 279 km (including Bologna–Vignola (33 km)

Organisation

On 1 January 2001 four government-controlled independent operators in the province of Emilia Romagna merged to form FER. Participating railways are the Bologna–Portomaggiore (FBP), Padane (FP), Parma–Suzzara (FPS) (formerly part of Venete Railways) and Suzzara–Ferrara (FSF) systems. FER is owned by the Emilia-Romagna region (59.44 per cent) plus eight provinces within the region.

FER has an open access operating licence and in 2003 launched a new freight service carrying clay and containers from the port of Ravenna to Dinzzano (see entry for the Casalecchio–Bologna–Vignola Railway (ATC) in the Italy section of *Railway systems and operators*).

In July 2002 a new joint venture company, Suburbana Bologna Vignola (FBV), was established by FER and the Casalecchio–Bologna–Vignola Railway (ATC) (see entry in Italy section of *Railway systems and operators*) to operate the Casalecchio–Vignola line (see entry for ATC). Both companies have a shareholding of 50 per cent in the new venture.

In 2004 FER employed some 450 staff.

Freight operations

The company carried 1.3 million tonnes or 149.4 million tonne-km in 2004.

Improvements to existing lines

In 2003 FER was electrifying the Suzzara–Ferrara line and was to build a bypass avoiding Ferrara

FER Fiat-built dmu on the Bologna–Vignola line (Quintus Vosman)   *NEW*/1122806

station for direct freight trains. A 12 km link from Portamaggiore to Dogato is under construction and will link FBP and FP lines. In April 2005 FER called for tenders to electrify the Bologna–Portomaggiore line.

The province also plans to reopen three lines to traffic by 2010: Modena–Vignola (20 km); Ferrara–Copparo (20 km); and Ostellato–Porto Garibaldi (30 km). It is reported that at a later date, Reggiane Railways (ACT), ATC Bologna and the Modena Sassuola Railway (ATCM) could also join FER.

Traction and rolling stock

The merged company owns 16 diesel locomotives, 51 diesel railcars and 34 coaches. It plans to use the recently rebuilt FSF workshops at Sermide for traction and rolling stock maintenance. FER's

locomotive fleet mainly consists of 10 former DB Class 220 B-B diesel-hydraulics. In 2003 the company received two former FS Class 668.1600 diesel railcars and five trailers, all refurbished. In 2005 the company received three Class E 464 electric locomotives from Bombardier. There is an option on three more. The locomotives will be used with 18 ex-SBB coaches, modernised by a local consortium.

In 2004 FER purchased six Class 342 Bo-Bo electric locomotives from Slovenian Railways (SŽ). They will be classified E 640 in Italy. The company also purchased three former Romanian Railways Class 60 Co-Co diesel locomotives and 10 former SŽ Class 311/315 emus.

UPDATED

Ferrovie Udine-Cividale Srl (FUC)

Via Pesciera 30, I-33100 Udine
Tel: (+39 0432) 58 18 44   Fax: (+39 0432) 58 18 83
e-mail: info@ferrovieudinecividale.it
Web: www.ferrovieudinecividale.it

Gauge: 1,435 mm
Route length: 15 km

Organisation

FUC was separated from Sistemi Territoriale SpA in January 2005, having once been one of three lines forming Ferrovie Venete. The railway is owned by the Friuli-Venezia-Giulia region.

Passenger operations

Passenger services are operated between Udine and Cividale.

Traction and rolling stock

In 2005 rolling stock totalled six diesel railcars, two driving trailers and one diesel locomotive. In late 2004 FUC ordered three two-car (plus two options) and two four-car (plus one option) GTW dmus from Stadler. They will be delivered from autumn 2006.

NEW ENTRY

FS Holding SpA (FS)

Piazza della Croce Rossa 1, I-00161 Rome
Tel: (+39 06) 441 01   Fax: (+39 06) 44 10 51 86
Web: www.ferroviedellostato.it

Key personnel

President and Managing Director, FS: Elio Catania
Managing Director, Trenitalia: Roberto Testore
Directors
  Finance and Administration: Maurizio Basile
  Human Resources: Francisco Forlenza
  External Relations: Silvio Sircana
  Passenger: Massimo Ghenzer
  Cargo: Giuseppe Smeriglio
  Logistics: Giuseppe Smeriglio
  Engineering: Emilion Maestrini
  Strategy: Vincenzo Soprano
  Legal: Maurizio Marchetti
  International Affairs: Alfredo Macchiatti
  Technical: Rocco Segreti
  Rolling Stock: Roberto Testore

Subsidiaries

Trenitalia SpA
Piazza della Croce Rossa, I-10061 Rome
Web: www.trenitalia.com;
  www.cargo.trenitalia.it;
  www.regionale.trenitalia.it

In 2000 FS split off its operating division as Trenitalia SpA. Initially there were three train-operating divisions: Passeggeri (long-distance passenger); Regionale (local passenger services including city suburban); and Cargo. Train fleets were divided between the three (see 'Traction and rolling stock'). Trenitalia's Rolling Stock Technology Unit (formerly Unitá Tecnologie Materiale Rotabile – UTMR), based in Florence, was responsible for the acquisition of new rolling stock and the technical specifications for modifications to or refurbishment of existing equipment.

At the start of 2005, Trenitalia was reorganised into two operating divisions:
Direzione Generale Operativa Passeggeri, divided into two business units:
– Passeggeri Nazionale e Internazionale
– Passeggeri Locale
Direzione Generale Operativa Logistica to run Trenitalia Cargo, which changed its name to Trenitalia Global Logistic in late 2004

Class E464-powered push-pull set operated by the TILO Italian-Swiss joint service in the Lombardia and Ticino regions (Quintus Vosman)   ***NEW**/1122816*

UTMR was renamed Direzione Manutenzione e Logistica at the same time.

Grandi Stazioni SpA

(see entry in Italy section of *Railway systems and operators*).

Italferr SpA

(see entry in *Consultancy services* section)
Italferr is the engineering consultancy division of FS.

Centostazioni SpA

In the mould of Grandi Stazioni, Centostazioni SpA (formerly known as Medie Stazioni) was created by FS (60 per cent shareholding) and private companies to refurbish, restyle and manage over a five-year period 103 medium-sized FS stations with an annual traffic of 457 million passengers. The company allocated €230 million for this work.

Ferrovie Real Estate

Via Arno 64, I-00198 Rome
Tel: (+39 06) 85 27 96 04   Fax: (+39 06) 85 27 95 77
Web: www.metropolis-spa.it

Ferrovie Real Estate (formerly known as Metropolis SpA) was established by FS in 1991 to manage and realise the value of the railway's real estate assets, including disused land.

Political background

In 2000 FS reorganised according to EU rules, initially becoming the holding company for infrastructure authority RFI before it became an independent state-owned company in 2001, the train operating arm, Trenitalia, Grandi Stazioni, formed to upgrade major stations, and TAV, responsible for building high-speed lines. The last-named is now a subsidiary of RFI. This followed approval by the Italian parliament in July 1998 of the adoption of European Union directives covering separate accounting of train operations and infrastructure functions.

As a result of this change, from 1 June 2000 Italian railways were liberalised on EU lines and any licensed operator can now operate over RFI tracks and use its passenger and freight facilities. FS licensed its own operating arm under the name Trenitalia. It operates long-distance passenger, regional passenger and freight services. The FS group has an annual turnover of some €7.6 billion, €3 billion of which is in passenger receipts, while most of the remainder comes from the state.

In 2001 the company moved into profit after a long period during which losses were incurred. This was mainly possible thanks to savings in labour costs. Due to the separation of infrastructure provision from operations, Trenitalia will have to pay track charges, as will other operators. FS retains exclusive rights to infrastructure management.

Italy has more than 20 independent local railways, many of which now have open access licences and are keen to start such services. Apart from these, a small number of new operators have started freight train operation, the most significant being RTC (see entry).

In early 2005 the first franchise for passenger train operation was granted to a joint venture formed by Trenitalia and Sistemi Territoriali (see entry in Italy section of *Railway systems and operators*). The €70.3 million, six-year contract from December 2006 is for operation of local passenger trains on 10 lines in the Veneto region, around

Class E464 electric locomotive with a Passeggeri Locale service at Rome Termini (Ken Harris)   ***NEW**/1122814*

ALSTOM-built Minuetto emu near Bologna (Quintus Vosman)   *NEW*/1122815

Venezia and Verona. Services in the Liguria region were out to tender in mid-2005.

In early 2004 Trenitalia's head announced that the Global Logistics division would be split off and possibly privatised when profitable. Trenitalia Global Logistics made a loss of some €250 million in 2003. The subsidiary is said to have 14,200 staff, 2,500 of which would move to RFI, which would take over shunting operations.

Traffic (million)	2003	2004
Passenger journeys	497.9	491.8
Passenger-km	45,221	45,818
Freight tonnes	82.1	83.6
Freight tonne-km	22,500	23,400

Finance

FS announced its first-ever surplus in 2001 of €29 million, compared with a loss of €683 million in 2000 and €3,889 million in 1997. The surplus rose to €78 million in 2002 but dropped to €31 million in 2003. Traffic receipts in 2003 were stable at €3,006 million. Total turnover fell from €7,731 to €7,601 million, while the cost of production rose from €6,898 million to €6,959 million.

To help FS towards financial self-sufficiency, the state took over its accumulated debt and met the costs of a programme begun in 1990 to reduce staff numbers through voluntary retirement. Between 1989 and 1997 the workforce was cut from 209,000 to 120,000 and productivity (measured in traffic units per employee) rose by 55 per cent. By 1999, staff numbers had been further reduced to 114,500.

In 2003 the FS group had 101,946 staff, 57,096 of which were with Trenitalia and 36,820 with RFI.

Loss-making routes

In the 1993–95 period a core network of some 5,200 route-km was defined, where FS would have freedom of commercial judgement and operate without subsidy. Over the rest of the system it was initially anticipated that some 2,000 route-km of heavily loss-making regional lines would close in order to eliminate up to 80 per cent of historic government cash support of the railway's revenue. However, these heavy cuts did not take place, and decisions on the future of unremunerative regional lines has been transferred to the regions.

Investment

In 2003 the FS group invested €7,208 million, 31 per cent more than in 2002. Of this, €2,042 million went on the existing network, €3,839 million on the new high-speed network and €1,143 million on new and refurbished trains.

Independent railways

There are more than 20 secondary railways in Italy. From 1936 to 2000, 13 of these were directly controlled by the Ministry of Transport through government commissioners. Control of almost all of these companies has since been transferred to local government and several mergers have taken place. These railways are increasingly under the control of regional councils.

Passenger operations

Long-distance

These are operated by Trenitalia's Passeggieri division. In 2003 long- and medium-distance passenger traffic was 67.6 million passengers (up 12.5 per cent on 2002), accounting for 24.93 billion passenger-km (down 4 per cent).

In 1997 FS implemented its plans for a three-tier network of long-distance regular-interval services. Services operated between Italy's principal cities and to France and Switzerland with tilting and non-tilting high-speed rolling stock are marketed as Eurostar Italia, connecting with second-tier InterCity trains serving regional centres and provincial capitals. The third tier is formed by 'Interregionali' regional trains radiating from principal nodes. These are operated by the Regionali activity. By 2001 Eurostar had captured a 17 per cent share of Italy's long-distance passenger market. In 2003 Eurostar traffic rose by 5 per cent but international traffic fell by 19.4 per cent.

In December 2004 Trenitalia launched a low-cost train branded Treno OK. The train is second class only and operates one return service per day between Rome and Milan with single fares as low as €9. The service is operated by ETR 450 tilting sets. A second service between Rome and Bari was launched in March 2005.

International

FS and SNCF of France have formed 50/50 joint ventures to market and develop high-speed and overnight services between Italy and France. Handling some 1 million passengers a year by the start of 1995, overnight services linking Paris with Milan, Florence, Venice and Rome were restructured in 1995 with improved timetables and new fares, and are now marketed under the Artesia brand-name. New or refurbished rolling stock was provided from 1996.

September 1996 saw introduction of two daily return services worked by SNCF TGV Réseau high-speed trainsets on the Paris–Turin–Milan route, with a journey time of 6 hours 35 minutes for Paris–Milan. At the same time, FS ETR 460 trainsets began working one return Milan–Lyon and two return Turin–Lyon services. The latter were withdrawn in 2003 at the same time as a third Paris–Milan TGV was introduced. This service will be further increased in 2006 by the addition of trains operated by ETR.500 sets.

Trenitalia also owns 50 per cent of Cisalpino (see entry in *Operators of international rail services in Europe* section), the company operating tilting trains from Italy to Zurich and Geneva in Switzerland. Cisalpino has subsequently taking over responsibility for all passenger trains between Italy and Switzerland, including those using hauled stock.

In December 2004 Trenitalia and Austrian Railways (ÖBB) started to jointly promote cross-border intercity trains under the Allegro brand, with global pricing for tickets.

Regional services

Regional passenger services are operated by Trenitalia's Passeggeri Locale division. These are divided into longer distance Interregionale, Diretto, Passeggeri Locale and Metropolitano services, the last for suburban areas (see below). These will be rebranded Cityexpress, Regioexpress, Regionali and Suburbani in December 2004.

Trasporto Regionale carried 430.32 million passengers in 2003 (up 1.5 per cent on 2002) for 20.291 billion passenger-km (also up 1.5 per cent).

In 2004 Trenitalia and Swiss Federal Railways (SBB) formed a joint company, TILO, with the aim of integrating cross-border services in the Lombardia and Ticino regions. SBB has ordered new dual-voltage FLIRT emus from Stadler for these services.

The provinces are increasingly taking over responsibility for investment in local rail networks and rolling stock. Toscana is to finance a variety of local passenger and freight schemes worth €83 million. The Veneto region is investing in the creation of a local network known as Sistema Ferroviario Metropolitano Regionale. The estimated cost of the first phase is €350 million, while a €300 million second phase will include new rail links to Venice and Verona airports.

Class E636 electric locomotive with a tank train at Milan Lambrate (Marcel Vleugels)   0583017

Two Class E633 electric locomotives with a piggyback service near Novara (David Haydock) 0583016

In early 2001 Passeggeri Locale announced that it would re-staff many of its 1,500 unstaffed stations, reopen waiting rooms, install ticket and vending machines and security video cameras. Voluntary organisations were to be invited to set up in some stations.

Suburban

Suburban operations are managed by 21 local business units which are structured to facilitate their eventual privatisation or separation from FS proper. Funding is now sourced locally, with FS receiving five-year contracts to operate services. However, initially optional competitive tendering will become mandatory when contracts fall due for renewal. In 1995 Rome suburban services came under a common fares scheme also adopted by the city's bus, tram and metro networks.

On 1 February 2001 Trenitalia, together with Naples city council and municipal transport operator ANM, founded Metronapoli, a common subsidiary to operate local transport. Trenitalia's holding is 38 per cent. The company is licensed to run over RFI lines and includes city transport plus the 17 km RFI Napoli Gianturco–Pozzuoli Solfatara line, which is largely underground and known as the Passante. Metronapoli has taken over 20 FS Class ALe 724 four-car emus, which operate an all-stations service over the Passante, running every six minutes in the peaks.

Freight operations

In the freight sector, Trenitalia Cargo was mandated to lift its market share from 12.5 per cent in 1997 to 14.5 per cent by 2000, when a rise to 28.9 billion train-km was targeted as part of the company's strategic plan. In the event, traffic in 2004 totalled 83.6 million tonnes for 23.4 billion tonne-km, a rise of 3.7 per cent. Domestic traffic rose by 9 per cent, with the main rise from container traffic. International traffic fell slightly.

In late 2004 Cargo was renamed Global Logistic, due to the strategy of introducing new door-to-door logistics services in concert with road hauliers.

Controllo Centralizzato Rotabili (CCR), a central real-time and computer-based data transmission system, now monitors freight operation through input via 400 terminals of activity at all 2,800 freight-generating and reception points.

Approximately half of FS freight is international. Rail carries 22 per cent of traffic through the country's ports – 54 per cent through Trieste, 34 per cent at La Spezia, 30 per cent at Livorno and 29 per cent through Genoa, Italy's busiest port.

Over 50 per cent of FS Global Logistics' traffic is concentrated in the northern part of the country and only 10 per cent in the far south.

A strategic alliance between FS and SBB in Switzerland was signed in March 1998 to improve the competitiveness of international rail freight between the two countries. However, this failed due to unreliability on the part of FS and SBB has set up a competing company in Italy, SBB Cargo Italy, to haul trains between the border at Chiasso and the Milan area. BLS is co-operating with FNME for similar reasons.

FS has collaborated with Swiss (BLS and SBB), German, and Dutch rail networks to establish the North-South Freightway, which became operational in February 1998. Linking Italian ports to northern Europe, this allocates and manages train path through participating countries via a single control centre.

In December 2002 FS, ÖBB and DB created the Brenner Rail joint-venture with a control room in Innsbruck, Austria, to co-ordinate timetables, motive power and information systems. In 2003 Trenitalia and RENFE launched a joint-venture, Logistica Mediterranea Cargo, with the aim of developing freight traffic between Italy and Spain.

In 2004 Trenitalia acquired a 15 per cent shareholding in the German open access rail freight company, TX Logistics (see entry in Germany section of *Railway systems and operators*), and is developing traffic via Switzerland with the latter. However, Trenitalia has lost large quantities of traffic in the Milan area to domestic operator FN Cargo, the Swiss company SBB Cargo and German operator Railion Italia.

In 2003 Trenitalia Cargo carried out trials with trains of up to 3,200 tonnes with the aim of raising train weights, particularly for steel traffic.

Trenitalia Global Logistics has a wholly owned subsidiary, SerFer (Servizi Ferroviari srl) which manages private sidings and freight terminals including the port networks of Genoa and Naples. SerFer moves 600,000 wagons annually, has 400 staff and a fleet of over 170 locomotives. In early 2002, SerFer gained an open access freight licence and is now operating main line freight services over short distances, hiring electric locomotives from Global Logistics.

Intermodal operations

Intermodal freight is managed by Italcontainer, of which 71 per cent share is owned by Trenitalia, 25 per cent by Intercontainer-Interfrigo and 10 per cent by CEMAT, the Italian piggyback company. CEMAT is 41.25 per cent owned by FS. In intermodal business FS itself is essentially a wholesaler of track space and train operation.

Intermodal traffic now constitutes 40 per cent of FS freight. Trenitalia Global Logistics is Europe's leading intermodal carrier, with 33.8 million tonnes handled in 1999. Recent improvements have included a move from 100 to 120 km/h operation of intermodal trains on the north-south trunk route from Milan and a 3 per cent advance in the commercial speed of such services.

Grants for equipment and terminals

Legislation of 1990 provided financial inducements for companies to invest in rail-based intermodal transport. A 20 per cent contribution to capital cost is on offer for purchase of containers or swapbodies and other apparatus needed for intermodal operation, and a 10 per cent rebate of the cost of rail movement.

Legislation of 1990 put up almost €500 million towards the creation in key commercial areas of multifunction 'freight villages', or Interporti, equipped to deal in both conventional rail and intermodal freight, and where the rail installations are surrounded by warehousing and other value-adding activities that combine to offer a full logistics service. FS is supported in this programme by CEMAT, and further finance for most of the Interporti has come from local authorities, banks and other private-sector bodies. The nine 'first level' Interporti are members of Assointerporti, a European Economic Interest Group set up to associate similar enterprises throughout the EU.

Diesel locomotives

Class	Wheel arrangement	Power kW	Speed km/h	Weight tonnes	No in service	First built	Mechanical	Builders Engine	Transmission
D343*	B-B	995/1,015	130	60	44	1967	Fiat/OM/Sofer Omeca/Breda	Fiat 218 SSF/ Breda-Paxman 12YJCL	E TIBB/OCREN
D443	B-B	1,400	130	72	49	1966	Fiat/OM/Sofer/ IMAM/Reggiane	Fiat 2312 SSF/ Breda-Paxman 12YLCL	E ASG/OCREN
D345	B-B	995	130	61	145	1974	Breda/Sofer/ Savigliano	Fiat 218SSF	E TIBB/Marelli/ Italtrofo
D445	B-B	1,560	130	72	149	1974	Fiat/Omeca	Fiat 2112SSF	E Ansaldo
D141	Bo-Bo	515	80	64	29	1962	TIBB/Reggiane	Fiat MB 820B	E TIBB
D143	Bo-Bo	420	70	65	49	1942	TIBB/OM	OM	E TIBB
D145	Bo-Bo	850	100	72	100	1982	Fiat/TIBB	BRIF ID 36 SS12V Fiat 8,297.22 × 2	E TIBB/Parizzi
D146	B-B	900	100	–	32	2003	Firema	–	H
D147	B-B	900	100	–	1	2005	Firema	Caterpillar 3512	H
235	C	160	50	34	17	1957	Badoni	BRIF 1D36 N8V	H Voith
225	B	129	50	32	106	1955	Breda/Jenbach/ Greco/Sofer/ IMAM	Breda/ Jenbach/Deutz	H Breda/Voith
245	C	258	65	48	407	1962	Reggiane/OM/ Breda/ IMAM/ Ferraro/Greco	MB820-Fiat D26N12V BRIF JW 600 CNTR OM-SEV	H BRIF-Voith Llt24
214	B	95	35	22	494	1964	Badoni/Greco	Fiat 8217-02,001	H BRIF-Voith Llt33
216	B	118	30	21	45	1965	Badoni/Simm	OM DGL	H Von Roll
255	C	500	53.5	53	30	1991	Badoni/Greco	BRIF ID 3658V	H Voith

* Refurbishment/re-engining in progress in 2005.

Diesel railcars

Class	Motored axles/car	Power/ motor kW	Speed km/h	No in service	First built	Mechanical	Builders Engine	Transmission
ALn668	2	110–170	110–130	693	1956	Omeca/Breda/ Fiat Savigliano	Fiat	M
ALn663	2	170 × 2	120/130	120	1983	Fiat Savigliano	Fiat 8217.32	M
ALn501	4	560 × 2	130	95	2004*	ALSTOM	Iveco	H Voith

* Includes units on order in 2005.

Electric trainsets

Class	Cars per unit	Line voltage	Motor cars per unit	Motored axles/car	Output/ motor kW	Speed km/h	No in service	First built	Builders Mechanical	Electrical
ETR 450	9	3 kV	9	2	315	250	15	1987	Fiat	Marelli
ETR 460	9	3 kV	6	4	500	250	29	1988	Fiat	Parizzi
ETR 470	9	3 kV/15 kV	6	4	500	200	9	1996	Fiat	Parizzi
ETR 480	9	3 kV[1]	6	4	500	250	15	1996	Fiat	Parizzi
ETR 500	14	3 kV	2	4	1,100	300	60	1992	Trevi	Trevi
–	7	3/25 kV	–	–	–	250	12	–[3]	ALSTOM	ALSTOM

[1] Conversion in progress for additionally operating from 25 kV.
[2] Five also equipped for 1.5 kV DC and 25 kV AC; all power cars to be replaced by dual-voltage (3/25 kV) vehicles on order 2005.
[3] On order for delivery 2006-07.

Electric locomotives

Class	Wheel arrangement	Line voltage	Output kW Continuous/ one hour	Speed km/h	Weight tonnes	No in service	First built	Builders Mechanical	Electrical
E402A	Bo-Bo	3 kV	5,600	250	84	47	1988	Reggiane/ Fiat/Breda	Ansaldo
E402B	Bo-Bo	3/15 kV	5,600	250	84	80	1996	Ansaldo/ Fiat/Breda	Ansaldo
E402C	Bo-Bo	3/25 kV	6,000	–	–	24	–[2]	Ansaldobreda	Ansaldobreda
E405	Bo-Bo	3 kV	6,000	–	–	42	2003	Bombardier	Bombardier
E412	Bo-Bo	3/15 kV	6,000	200	87	20	1997	Adtranz	Adtranz
E424	Bo-Bo	3 kV	1,500/1,660	100/120	73	105	1943	Breda/Savigliano/ Ansaldo/ Reggiane/ Brown Boveri/OM	Breda/ Savigliano/ Ansaldo/ Marelli/ Brown Boveri/CGE
E444	Bo-Bo	3 kV	4,000/4,440	200	83	113	1967	Savigliano/ Breda/Casaralta/ Fiat	OCREN/ Asgen/ Savigliano
E464	Bo-Bo	3 kV	3,000/3,500	160	72	388[1]	1999	Bombardier	Bombardier
E636	Bo-Bo-Bo	3 kV	1,890/2,100	110	101	242	1941	Breda/ Brown Boveri/ Savigliano/OM/ Reggiane/Pistoiesi	Breda/Brown Boveri/ Savigliano/ CGE/ Marelli/ Ansaldo
E645	Bo-Bo-Bo	3 kV	3,780/4,320	120	110	93	1958	Breda/ Brown Boveri/ Savigliano/OM/ Reggiane/Pistoiesi/ IMAM	Breda/Brown Boveri/ Savigliano/ CGE/ Ansaldo/ OCREN
E646	Bo-Bo-Bo	3 kV	3,780/4,320	140	110	198	1961	Breda/ Brown Boveri/ Savigliano/OM/ Reggiane/Pistoiesi/ IMAM	Breda/Brown Boveri/ Savigliano/ CGE/Marelli/ Ansaldo/ OCREN
E652	B-B-B	3 kV	4,950	160	106	174	1989	ABB/Sofer/ Casertane	ABB/Ansaldo/ Marelli
E656	Bo-Bo-Bo	3 kV	4,200/4,800	150	120	451	1975	TIBB/Sofer/ Casaralta Reggiane/ Casertane	TIBB/Italfrafo/ Asgen Marelli/ Ansaldo/ Retam
E632	B-B-B	3 kV	4,350/4,900	160	103	65	1982	Fiat/TIBB/Sofer	Ansaldo/ Marelli/TIBB
E633	B-B-B	3 kV	4,350/4,900	130	103	143	1979	Fiat/TIBB/Sofer	Ansaldo/ Marelli/TIBB
E322	C	3 kV	190	50	36	2	1961	FS	TIBB
E323	C	3 kV	190	32/60	46	6	1966	TIBB	TIBB
E324	C	3 kV	190	32/60	45	1	1966	TIBB	TIBB

[1] Including examples on order in 2005.
[2] On order 2005.

Electric multiple-units: power cars

Class	Cars per unit	Line voltage	Motor cars per unit	Motored axles/car	Output/ motor kW	Speed km/h	No in service	First built	Builders Mechanical	Electrical
ALe 582	2/4	3 kV	1/2	4	280	140	90	1987	Breda	Marelli/ Ansaldo
ALe 601	1	3 kV	1	4	218	200	40	1961	Casaralta	Casaralta/ OCREN
ALe 803	3[1]	3 kV	1	4	218	130	47	1961	Stanga/ Savigliano/ IMAM	Savigliano/ Sofer
ALe 801 ALe 940	4[2]	3 kV	2	4	218	140	62	1976	Stanga/ Fiore/Aetal/ Lucana/Sofer	Marelli/ Stanga/ Fiore/ Lucana/ Sofer
ALe 644	4	3 kV	1	4	415	140	6	1980	Breda	Breda
ALe 724	4	3 kV	1	4	305	140	89	1982	Breda	Marelli/ Ansaldo
ALe 841	4	3 kV	2	4	218	200	18	1994	Converted from ALe 601	
ALe 642	2–3	3 kV	1/2	4	305	140	60	1991	Breda/Fiore	Ansaldo/ Lucana
ALe 426	4[3]	3 kV	2	2	318	140	188	1997	Breda/Firema	Adtranz/ Ansaldo
ALe501	3	3 kV	2	2	312	160	96[4]	2004	ALSTOM	ALSTOM

[1] Including two Type Le 803 trailers.
[2] Including two Type Le 108 trailers.
[3] Including two Type Le 736 trailers . ALe 506 power car.
[4] Includes units on order in 2006.

Traction and rolling stock

Total FS rolling stock resources at the beginning of 2003 comprised 1,953 electric locomotives, 405 diesel locomotives, 128 electric trainsets, 1,690 emu cars, 813 diesel railcars, 1,099 shunting locomotives, 10,951 hauled passenger coaches and 49,600 freight wagons.

During 1999 FS divided its locomotive fleet by activity: Passeggeri (long-distance passenger) has 301 electric and 35 diesel locomotives; Passeggeri Locale (regional passenger) has 654 electric and 139 diesel locomotives; FS Global Logistics has 870 electric and 217 diesel locomotives. All high-speed trainsets have been allocated to Passeggeri and all dmus, emus and railcars are assigned to Passeggeri Locale. This split has led to modifications such as the upgrading of Passeggeri Locale's Class E 633 electric locomotives from 140 to 160 km/h operation and the downgrading of Global Logistics' Class E 656 to lower speeds but with higher tractive effort. In both cases renumbering has taken place.

Pendolino trainsets

Trenitalia has a fleet of 29 ETR 460 tilt-body trainsets from Fiat, capable of 250 km/h. These are nine-car, like the earlier ETR 450, but with only six cars motored, due to the use of more powerful three-phase asynchronous motors with GTO inverter control. Body width is 2.8 m, and improved sound insulation and full pressure sealing have been applied to the aluminium alloy bodies. The first ETR 460s entered service in 1995.

Production of a batch of 15 dual-voltage ETR 480 trainsets prepared for subsequent fitting of 25 kV 50 Hz AC equipment for eventual operation on Italian high-speed routes was completed in 1999. Modification for 25 kV AC operation, necessary on the Roma–Napoli line from 2006, started in 2004.

Nine dual-voltage ETR 470 electric trainsets with the Fiat active body-tilt system operate on the Geneva–Milan–Venice, Basle/Berne–Milan and Stuttgart–Zurich–Milan routes. A joint venture company, Cisalpino AG (see entry in *Operators of international rail services in Europe* section) has been established to manage these services.

In 2004 Trenitalia ordered 12 new generation Pendolino trainsets, and Cisalpino ordered 14 sets. The new trains will all be seven-car formations capable of operating in multiple at 250 km/h. Delivery will start in 2006.

ETR 500

Track parameters for Italy's planned high-speed (AV) network, dictated by need to make the new lines usable by fast freight as well as passenger trains, obviated need of body-tilt in the Type ETR 500 train for dedicated AV operation at 300 km/h. Design of a prototype was entrusted to a Breda-led consortium known as TREVI and background to its development is covered in past editions of *Jane's World Railways*.

The decision taken in 1993 to electrify the new AV routes at 25 kV 50 Hz AC has resulted in design changes to the ETR 500. The initial batch of 30 ETR 500 trainsets was ordered before this decision was taken, and thus has been delivered configured for DC only. Deliveries of the second batch of 30 ETR 500 trainsets equipped with both AC and DC traction equipment was completed in 2001. Both batches will be able to operate from a 1.5 kV DC supply (the norm in southern France) at reduced power. Other modifications in comparison with the prototypes include reduction of maximum axleloading from 19 to 17 tonnes, which would permit operation over the French high-speed network.

In 2003 Trenitalia ordered 60 dual-voltage Class E 404.600 power cars from Trevi for the earlier ETR 500 sets. The displaced E 404.100 3,000 V DC power cars will be redeployed with classic hauled intercity coaches. Bombardier won the contract to convert the E.404.100 power cars in June 2005. All ETR.500 sets are now formed of 12 coaches.

Electric locomotives

FS has a long history of ordering Bo-Bo-Bo designs due to the number of mountainous routes on the Italian network. Of the present fleet of 1,953 electric locomotives, 1,349 are of Bo-Bo-Bo designs. Current deliveries are all of the Bo-Bo axle configuration.

Five Class E402 prototypes were built by a consortium including Ansaldo, Fiat and Breda. Equipped with a three-phase drive, the E402 is an 84-tonne, 5,600-kW, Bo-Bo locomotive designed for 250 km/h, with a tractive effort of 180 kN at 102 km/h and of 92 kN at 200 km/h.

Following delivery of 45 E402A (3 kV DC only), FS has received 80 E402B (3 kV DC/25 kV AC 50 Hz) electric locomotives, which operate both passenger and freight services. Twenty of the latter are also equipped to operate freight traffic to and from France under the 1.5 kV DC system.

From mid-1997 FS took delivery of 20 Class E412 locomotives from Adtranz. Designed for international services to Austria and Germany, this 200 km/h design can run under both 3 kV DC and 15 kV AC 16⅔ Hz catenary, and at reduced performance under 1.5 kV DC. However, in 2003 the class had still not been approved outside Italy.

In 2002 Trenitalia Cargo agreed to buy from Bombardier 42 Class E405 electric locomotives originally destined for Poland. These are 3 kV DC-only versions of the E412 three-voltage machines. All had been delivered by early 2004. The locomotives work in multiple on the Verona–Tarvisio route. Trenitalia Global Logistics has said it will buy 70 new electric locomotives in the period 2003–07. Trenitalia Global Logistics has ordered 24 Class E402C locomotives which will be able to operate off 3 kV DC or 25 kV AC supplies.

In 1996 FS placed an order valued at €88 million with Adtranz (now Bombardier) for an initial batch of 50 lightweight E464 electric locomotives. The 3,000 kW (continuous) locomotives, which feature a driving cab at one end only, are used on regional and suburban push-pull passenger services. Further contracts have brought the total on order to 388, of which 240 had been delivered by the first quarter of 2005. As these locomotives are delivered, Passeggeri Locale is transferring the older Class E646 and E656 to the Global Logistics division. Trentalia has said that its intention is to eventually transfer all six-axle electric locomotives to freight work.

In the absence of large numbers of new locomotives, Trenitalia Global Logistics is refurbishing Class E636 and E645 Bo-Bo-Bo electric locomotives to improve conditions for traincrew.

TAF emus

From 1997 to 2001 FS and FNME took delivery of 72 four-car aluminium-bodied double-deck electric multiple-units. Construction was undertaken by a consortium of Ansaldo, Adtranz, Breda and Firema. Designated TAF (Treno ad Alta Frequentazione), each trainset comprises two motor and two trailer cars, with a maximum speed of 140 km/h and accommodation for 475 seated passengers. Duties for early FS deliveries included services linking Rome city centre with Fiumicino airport. In 2001 FS ordered an additional 27 units and FNME ordered five more.

New double-deck stock

Trenitalia has ordered 300 new double-deck coaches, with an option for 150 more. Known as 'Vivalto' and intended for suburban services, the vehicles are to be supplied by the CO.RI.FER consortium formed of Fervet SpA, Officine Magliola SpA, Officine Ferroviarie Veronesi SpA and Rail Service International SpA (RSI). The first 20 five-car sets of coaches were expected to be delivered in 2005.

Minuetto emus and dmus

In early 2001 Trenitalia ordered 200 three-car 'Minuetto' articulated trainsets from ALSTOM, broken down as 110 160 km/h emus and 90 130 km/h dmus for regional services. Some of these have been financed by the provinces. In mid-2005 firm orders had been placed for 96 emus and 95 dmus. Deliveries began in 2004 and the first trains entered service in early 2005. All feature partial low floors and are air-conditioned. Dmus are powered by Iveco engines with hydraulic transmission.

In addition, the Passeggeri Locale division has embarked on a policy of adding air conditioning equipment to all rolling stock with a medium-term projected service life. This includes the majority of

ETR480 Pendolino tilting trainset at Bologna (Quintus Vosman)   **NEW**/1122817

ETR 500 high-speed trainset at Bologna Centrale (Quintus Vosman)   0583015

TAF double-deck emu at Villastallone (Quintus Vosman)   **NEW**/1122818

Class E412 electric locomotives on a test train at Arezzo (Marcel Vleugels)   0554633

diesel railcars and MDVC and Piano Ribassato low-floor locomotive-hauled stock.

As in France and Germany, regional authorities in Italy are beginning to finance and own rolling stock dedicated to services for which they are responsible.

New shunting locomotives

Trenitalia has ordered 33 Class D146 900 kW B-B diesel-hydraulic locomotives from Firema to replace Classes D141 and D143, the latter dating from 1942. The first of the new machines entered service in early 2003. In early 2005 the last of the 33 locos was outshopped as Class D147, with a Caterpillar 3512 engine.

Diesel locomotive refurbishment

Trenitalia received a refurbished Class D343 locomotive with a new Caterpillar 3512 DITA engine of 1,480 kW from Gredelj of Croatia in early 2005. The company is now to refurbish a further 40 locos, to be renumbered Class D347, by 2007.

Trenitalia is also re-engining 134 Class 245 diesel shunters with Fraschini engines.

UPDATED

Class D146 Firema-built diesel-hydraulic locomotive (Ken Harris)   0554632

Gargano Railway (FG)

Ferrovie del Gargano Srl
Strada Communale esterna 82 S Ricciardi, I-71016 San Severo
Tel: (+39 0882) 22 14 14   Fax: (+39 0882) 24 76 45
e-mail: fergagano@fg.nettuno.it
Web: http://www.ferroviedelgargano.com

Key personnel

Director General: V Scarcia
Operating Manager: A Oliva

Gauge: 1,435 mm
Route length: 56 km
Electrification: 56 km at 3 kV DC

Organisation

FG operates the Peschici–San Severo line on the Gargano peninsula, with some services running over FS tracks to Foggia and Bari under an open access licence, gained in 2002. The company also operates buses.

In 2003 tenders were called for construction of a new cut-off line between San Severo and Sannicandro–Garganico, serving the town of Apricena. This will cut journey times by 15 minutes. Work is in progress on reopening the former FS Foggia–Lucera line (20 km), on which FG is to take over operations.

FG electric locomotive on a passenger service at Rodi Garganico (Quintus Vosman)   0552783

Traction and rolling stock

The railway operates four electric locomotives, 10 electric railcars, nine passenger cars and 28 freight wagons. In 2003 FG called for tenders to refurbish four electric railcars and two driving trailers.

VERIFIED

Genoa–Casella Railway (FGC)

Managed by Government Commission, Rome
Ferrovia Genova-Casella
Via alla Stazione per Casella 15, I-16122 Genoa
Tel: (+39 010) 83 73 21
Fax: (+39 010) 837 32 48
e-mail: fgc@ferroviagenovacasella.it
Web: www.ferroviagenovacasella.it

Key personnel

Managing Director: Paolo Gassani

Gauge: 1,000 mm
Route length: 25 km
Electrification: 24 km at 3 kV DC

Organisation

A scenic line climbing the hills north of Genoa. FGC carries around 250,000 passengers annually.

Future plans include the construction of a new cut-off 4.5 km in length and of a new terminus in Genoa to connect with the upper station of the Granarolo funicular. €2 million was allocated to modernisation work during 2005.

Traction and rolling stock

The railway operates 12 electric railcars, three electric and one diesel locomotives, 16 coaches and 24 wagons. MOre modern fleet acquisitions include two Adtranz-built Type E46A electric railcars.

UPDATED

Grandi Stazioni

Via G Giolitti 34, I-10185 Rome
Tel: (+39 06) 47 84 11   Fax: (+39 06) 482 39 15
e-mail: info@grandistazioni.it
Web: www.grandistazioni.it

Background

With 60 per cent of its capital, FS founded the company Grandi Stazioni in 1998 with the aim of remodelling 14 of the biggest stations in Italy, which are used by 600 million people a year and generate a turnover of €350 million. Roma Termini served as a model and was completed in January 2000. Rebuilding of the remaining stations started in 2003, the work expected to be completed over a period of three years. Cost will be some €557 million, of which €261 million will come from the state.

In 2004 Grandi Stazioni had a turnover of €152.4 million and generated profits of €13.7 million.

Grandi Stazioni has won contracts for similar station development work with Czech Railways.

UPDATED

MetroCampania Nord Est

Via Don Bosco, I-80141 Naples
Tel: (+39 081) 599 32 54   Fax: (+39 081) 599 32 53
e-mail: resp.locale@alifana.it
www.alifana.it

Key personnel
Director General: G Racioppi
Director, Alifana Railway: A Marescotti
Director, Benevento-Naples Railway:
  G de Iudicibus

Gauge: 1,435 mm
Route length: FA – 41 km, FBN – 49 km
Electrification: 49 km at 3 kV DC (FBN)

Organisation
MetroCampania Nord Est, formerly Ferrovie Alifana e Benevento-Napoli (FABN), comprises the former Alifana Railway (FA) Piedmonte Matese–Santa Maria Capua Vetere line (41 km) and the Benevento–Naples Railway (FBN) (49 km), trains on both lines running through to Naples over FS tracks. MetroCampania Nord Est also operates buses.

The ex-FA line is now being modernised and electrified at 3 kV DC. FA also owns a 35 km direct line from Santa Maria Capua Vetere to Naples via Aversa but this has not operated since 1976. In 1996 the government agreed to fund 50 per cent of the cost of converting this line from 950 mm to standard gauge and taking it from Piscinola to Capodichino airport, with the municipality providing an additional €155 million to extend tracks to Garibaldi/Centrale in central Naples. The Piscinola–Garibaldi line, which will be electrified at 1.5 kV DC, will be shared with the Naples metro. Work began in 1997. Tenders were called in 2004 to build the 3.3 km Piscinola–Capodichino Aeroporto section. The 12 km section

FABN Class 668 dmu at Caserta (David Haydock) 0580501

from Aversa to Piscinola has already been rebuilt underground but the tunnel, ready since 1990, still awaits track and overhead power supply equipment. Conversion from Santa Maria Capua Vetere to Aversa is still under study. The Piscinola–Mugnano section was due to open in 2005 and Mugnano–Aversa in 2006, electrified at 1.5 kV DC and operated by Naples metro trains.

The ex-FBN line is already fully electrified. Plans exist to build a cut-off to reduce the line's length by 7 km, eliminate level crossings and raise speeds from 70 km/h to between 100 and 140 km/h.

FABN obtained an open access operating licence in early 2003.

Passenger operations
In 2002 FABN carried 1.6 million passengers.

Traction and rolling stock
FA operates 16 diesel railcars, three trailers, two diesel locomotives and nine freight wagons. FBN operates three electric locomotives, five electric railcars, one two-car emu, three three-car emus and eight wagons. In 2003–04 the FA line received three refurbished railcars, former FS Class 668.1400 units. In 2004 the Campania region allocated €31.7 million for eight emus for the FA line.

UPDATED

Metropolitana di Roma SpA (Met.Ro)

Via Volturno 65, I-00185 Rome
Tel: (+39 06) 46 95 20 80   Fax: (+39 06) 46 95 22 84
e-mail: direzione@atac.roma.it
Web: www.metroroma.it

Key personnel
Chairman: S Bianchi
Managing Director: Roberto Cavalieri
Operating Manager: A Maranzano
Manager, Roma Lido: D Ceci
Manager, Roma Pantano: U Zumbo
Manager: Roma Viterbo: U Montanari

Gauge: 1,435 mm, 950 mm
Route length: 131 km, 18.5 km
Electrification: 102 km of 1,435 mm gauge at 3 kV DC, 29 km of 1,435 mm gauge at 1.5 kV DC, 18.5 km of 950 mm gauge at 1.5 kV DC

Political background
In 1995 Rome bus and tram operator ATAC merged with COTRAL, which operated three suburban rail routes, the Rome metro and regional bus services and in 2000 was brought under the same management as Rome's metro network as Metropolitana di Roma SpA (Met.Ro). ATAC now owns all infrastructure and rolling stock while Met.Ro manages Metro lines A and B plus three suburban lines. An integrated fares structure including FS suburban services was introduced following the merger. The company has a licence to operate over RFI tracks and has discussed new services with ATM, Milan.

Passenger operations
The Met.Ro suburban passenger network comprises three electrified routes: Rome–Viterbo (102 km, 1,435 mm gauge); Rome–Lido di Ostia (29 km, 1,435 mm gauge); and Rome–Pantano Borghese (17.8 km, 950 mm gauge). Passenger journeys for each line in 2004 were 24 million, 29 million and 11 million respectively.

Improvements to existing lines
The Rome Piazzale Flaminio–Prima Porta section of the Viterbo route (Line F) has been upgraded and doubled. Doubling is in progress on the line to Montebello. The rest of the Viterbo line is to be

modernised at a cost of €103 million. The Pantano Borghese route is to be rebuilt to standard gauge on its inner section to become semi-metro Line G, connecting with metro Lines A and B at Termini via a new city-centre tunnel. From Torre Angela to Pantano the line is being doubled and rebuilt to metro standards with segregated tracks to become part of metro Line C in 2011. The line is currently closed from Grotte Celoni (12.6 km from Rome) to Pantano.

On the Lido di Ostia route (Line E) a tender for construction of an 18 km single-track extension

Met.Ro three-car emu at Acqua Acetosa (David Haydock) 0580497

For details of the latest updates to *Jane's World Railways* online and to discover the additional information available exclusively to online subscribers please visit
jwr.janes.com

from Lido di Ostia to Tor Vaianica was launched in mid-2005. A further line could be built from Tor Vaiancica to Pomezia on the FS Roma–Formia line.

Traction and rolling stock
Met.Ro's electric multiple-unit fleet comprises 74 cars for the Viterbo route, 34 vehicles for the Pantano Borghese route and 162 cars for the Lido di Ostia route.

In 2002 delivery started of 10 Type MRP 236 three-car emus from ALSTOM and Costaferroviaria (now CostaRail). These are to operate on the Rome–Viterbo line, where trains dated back to 1932. The order included an option on a further six units. Six double-articulated trainsets were ordered from the same manufacturers for the Pantano route. The Ostia route is to receive trains cascaded from the metro system.

UPDATED

North Milan Railway Group (FNM)

Gruppo Ferrovie Nord Milano
Piazza Cadorna 14, I-20123 Milan
Tel: (+39 02) 851 11   Fax: (+39 02) 851 17 08
e-mail: infocare@ferrovienord.it
Web: www.ferrovienord.it

Key personnel
President: Norberto Achille
Director General: Marco Piuri
Production and Operations Manager, North Milan:
  Ing Luigi Legnani
Operations Manager, Brescia-Edolo:
  Ing Federico Bonafini
Commercial Director: Gianni Scarfone

Gauge: 1,435 mm
Route length: 326.8 km, of which 308.6 km in use
Electrification: 200.3 km at 3 kV DC

Background
FNM is by far the largest train operator in Italy after Trenitalia and has traditionally been owned by the Lombardia region, which still holds 58 per cent of its capital. FS now holds 15 per cent of its capital.

Having long been mainly a suburban passenger operator through subsidiary Ferrovie Nord Milano Esercizio (FNME), with little freight traffic, FNM is now expanding quickly into this latter sector under EU open access rules, using the Ferrovie Nord Cargo (FN Cargo) brand name, and is currently acquiring more than 20 locomotives to double its traction fleet. Revenue from freight operations rose from €6.8 million in 2002 to €19 million in 2003. In January 2005 FN Cargo became a founding member of the European Bulls railfreight alliance that provides cross-border services. FN Cargo also

ES64F4 four-voltage electric locomotive hired from Siemens Dispolok by FN Cargo
(Quintus Vosman)
NEW/1122807

Class 520 diesel-electric locomotive (ex-Czech Railways Class D752) at FMN's Iseo depot
(David Haydock)
0583012

operates international services partnered with BLS Cargo and SBB Cargo.

In 2001 FNM created a new intermodal freight subsidiary, Eurocombi SpA, in conjunction with freight forwarder Merzario. Most traffic so far has been between the Milan area and Domodossola or Chiasso, where trains are handed over to BLS or SBB, although the company was due to start freight operations in the Naples area during 2004.

In March 2003 FNM acquired 70 per cent of Cargo Clay, a company managing the transport of clay from Germany to Italy and of ceramics in the opposite direction. FNME also operates buses.

FNM is now working with SNCF subsidiary Keolis with the aim of winning operating contracts in Italy, outside its traditional operating area.

Organisation
FNM serves the northern suburbs of Milan with a main four-track route to Saronno, where it forks to Como and to Laveno on Lake Maggiore. The latter is single track beyond Malnate. Also from Saronno a single-track branch heads west to Novara, and a freight-only branch east to Seregno. A further double-track route runs from Bovisa to Seveso, beyond which single track extends to Canzo-Asso. In 1993 FNM took over operation of the 108 km non-electrified Brescia North Railway, linking Edolo with Rovato and Brescia.

In 2002 FNM employed 2,754 staff.

Passenger operations
In 2002 Milan suburban services carried 48.1 million passengers (1,109 million passenger-km), while the Brescia–Edolo service handled 1.2 million passengers (36.8 million passenger-km).

As well as operating conventional suburban services, FNME provides a dedicated link to Milan's Malpensa airport. The company launched

Electric locomotives

Class	Wheel arrangement	Line voltage	Output KW	Speed km/h	Weight tonnes	No in service	First built	Builders Mechanical	Electrical
600	Bo + Bo	3 kV	1,030	75	63	2	1928	OM	CGE
610	Bo + Bo	3 kV	1,030	80	61	4	1949	Breda	CGE
620	Bo + Bo	3 kV	2,250	130	72	6	1985	TIBB	Ansaldo
630	Bo + Bo	3 kV	3,650	120	80	9	1991	Škoda	Škoda
640	Bo-Bo	3 kV	1,880	120	81	8*	1968	Ansaldo	Ansaldo
660	Bo-Bo-Bo	3 kV	2,905	120	110	3**	1962	Ansaldo	Ansaldo

* includes examples on order in mid-2005.
** on order in mid-2005.

Electric multiple-unit power cars

Class	Line voltage	Motored axles/car	Output/ motor kW	Speed km/h	No in service	First built	Builders Mechanical	Electrical
700	3 kV	4	183	80	5	1929	OM	TIBB
730	3 kV	4	272	80	3	1932	Tallero	TIBB
740	3 kV	4	272	80	7	1929	Tallero	CGE
740	3 kV	4	272	90	9	1953	Breda	CGE
740	3 kV	4	272	90	8	1957	Breda	CGE
750	3 kV	4	280	130	24	1982	Breda	Ansaldo/Marelli
760	3 kV	4	910	140	54	1998	Breda/Firema	Adtranz/Ansaldo

Diesel locomotives

Class	Wheel arrangement	Power kW	Speed km/h	Weight tonnes	No in service	First built	Builders Mechanical	Engine	Transmission
145	Bo-Bo	950	100	72	3	1995	Fiat	Fiat	E
500	Bo-Bo	383	75	47	5	1971	TIBB	Fiat	E
510	C	400	80	42	4	1979	Breda	Breda	E
520	Bo-Bo	1,455	100	77	18	1968	ČKD	Caterpillar	E

its 'Malpensa Express' from Cadorna terminus in Milan in May 1999, shortly after the airport began to handle scheduled international flights. In 2002 Malpensa Express services carried 1.6 million passengers (74.9 million passenger-km).

Freight operations
In 2002 Milan area freight movements totalled 213,000 tonnes. Traffic on the Brescia–Edolo line amounted to 133,000 tonnes.

Improvements to existing lines
In 1984 the government approved execution of a €134 million modernisation plan. The main item was 17 km of four-tracking on FNME's Como main line between Bovisa and Saronno, accompanied by level crossing elimination and station reconstruction, plus two new stations. This project was finished in 1999.

FNME's Milano Nord Bovisa station was linked to the Passante line (see entry for Rete Ferroviaria Italiana in the Italy section of *Railway systems and operators*) in December 1997.

The 4 km from FNME's present Milan terminus to Bovisa carries over 300 trains a day, two-thirds of which continue to Saronno. This section is being widened from two to four tracks. Quadrupling beyond Bovisa to Cadorna will enable separation of fast and slow services. This is scheduled for completion in 2006. In May 2003 FNME opened a new station at Milano Nord Domodossola on this section, replacing the facility at Milano Nord Bullona.

The first phase of a €237 million scheme to serve Milan's Malpensa airport, by doubling 20.5 km of the Saronno–Novara branch as far as Bivio Sacconago and constructing from there a 13 km airport link, was completed in 1999. Track doubling from Bivio Sacconago to Vanzaghello went out to tender in early 2004.

In 2004 the Italian government granted €74.5 millionfor upgrading and electrification of the 15 km Saronno–Seregno line, including doubling 10 route-km and construction of seven stations. The line is part of a putative Bergamo–Malpensa Airport line.

Completion of these projects, including the Passante, is forecast to boost FNME daily train working to almost 500 services.

TAF double-deck emu forming a Malpensa Express service at Saronno (David Haydock)   0583013

Traction and rolling stock
Rolling stock in service at the start of 2004 comprised 23 electric locomotives, 24 diesel locomotives, 110 emu power cars, 15 diesel railcars, 270 trailer vehicles and passenger coaches and 96 freight wagons. The fleet includes many double-deck coaches, which often operate with Class E750 single-deck railcars.

FNME has a fleet of 21 four-car double-deck TAF *(Treni ad Alta Frequentazione)* electric multiple-units for Passante services (see FS entry), plus six more which operate Malpensa Express airport services. The type is designated Class EA 760 for the driving motor cars and Class EB 990 for the intermediate trailers. FNME also purchased nine 3,290 kW Bo-Bo electric locomotives built (but never delivered) by Škoda for the former Czechoslovakian State Railways.

In 2003 the company ordered 18 three-car and nine five-car double-deck emus from the Associazione Temporanea d'Imprese consortium formed of Ansaldobreda, Firema and Keller, with options for 12 and six more units respectively. This was followed by an order for six single-deck articulated emus capable of 160 km/h which will be used on Malpensa Express services, replacing TAF emus.

In 2003 and 2004 FNM acquired 18 Class D752 diesel-electric locomotives formerly operated by Czech Railways (ČD) for freight services, designating them Class DE 520. The fleet is supplemented by eight Class 342 3 kV DC electric locomotives acquired from Slovenian Railways designated Class E 640 and three Class 1061 Bo-Bo-Bos, to become FNM Class E 660, from Croatian Railways, the latter to be delivered in 2005.

In 2005 FN Cargo hired five ES64F4 four-voltage electric locomotives from Siemens Dispolok. They are used to haul a European Bulls service from Brescia to Rotterdam as far as Emmerich on the Dutch border, as well as for services within Italy.

Czech-built Class E 630 electric locomotive with a passenger service at Milan Nord Cadorna (Marcel Vleugels)   0583014

Signalling and telecommunications
Automatic block controls 89.8 route-km, of which 63.8 km is double-track and 17 km quadruple-track. FS-type semi-automatic block controls 4 km. There are 287 signal-protected level crossings on the system.

Track
Rail (km installed): RA 36 kg/m (42.4); FS 46 kg/m (4.9); UNI 50 kg/m (427.5); UNI 60 kg/m (2.1)
Sleepers (km installed): Timber (124.5), FS concrete monobloc (94.4) and bibloc (254.3)
Size: Timber 2,600 × 260 × 150 mm
Monobloc 2,300 × 300 × 190 mm
Bibloc 2,300 × 263 × 217 mm
Spacing: 1,500 per km
Fastenings: Direct for RA 36; Type K for FS 46; Direct UNI 50 for monobloc sleepers; Type RN and Nabla for bibloc sleepers
Min curve radius: 250 m
Max gradient: 3%
Max permissible axleload: 20 tonnes

TAF four-car double-deck emu forming a Passante line service at Milano Dateo (Quintus Vosman)
0530200

UPDATED

Rail Traction Company SpA (RTC)

Via Sicilia 66, I-00187 Rome
Tel: (+39 06) 42 04 67 01   Fax: (+39 06) 42 01 13 79
e-mail: direzione@railtraction.it
Web: www.railtraction.it

Key personnel

President of the Administrative Board:
  Ferdinand Willeit
Managing Director and Chief Executive Officer:
  Giuseppe Sciarrone
Technical Director: Claudio Ubertini
Marketing and Sales Director: Francesco Grotti
Finance and Administration Director:
  Alessandro Di Nallo
Human Resources Director: Adriao Tomaro

Organisation

Rail Traction Company was formed in February 2000 as an open access freight operator, initially to serve the Verona–Brennero–Munich Brenner route. The company started operations on the Trieste–Tarvisio–Munich route in mid-2005 and intends to extend its operating area southwards in future.

Shareholders in the company are: Società di Trasporti su Rotaia (STR), owned by the Autostrada del Brennero motorway operating company, Ferrotramviaria SpA, a private construction company and transport operator; SAE SpA, an intermodal operator based in Trento; Reset 2000 Srl, an investment and management company; and Fercam SpA, an intermodal operator based in Bolzano. In 2004 Railion took a 30 per cent stake on RTC and is to work closely with the company in developing traffic on the Brenner route. In mid-2005 capital shares were held as follows: STR – 33.97 per cent; Railion Deutschland – 30.07 per cent; Ferrotramviaria – 16.98 per cent; SAE – 8.49 per cent; Reset 2000 – 6.99 per cent; and Fercam – 5.50 per cent.

To enable its trains to operate on the German rail network, RTC has signed an agreement with Lokomotion SpA, which holds an operating licence to run services in Germany. Ownership of Lokomotion is divided between

A Class 189 electric locomotive hired from Siemens Dispolok leads a Class EU43 machine on an RTC Brenner line intermodal service (RTC)   **NEW**/1146662

Kombiverkehr (50 per cent), RTC (30 per cent) and STR (20 per cent).

Day-to-day aspects of RTC services are managed from an operations centre in Verona. In 2005 RTC had 1003 staff.

Freight operations

Operations commenced in October 2001 with two pairs of intermodal trains between Verona and Munich (Reim) on behalf of CEMAT and Kombiverkehr, initially providing haulage as far as Brennero. Since then RTC has expanded into operating trains carrying BMW cars to Verona and has extended services to the Brescia area.

In 2005 RTC was operating 12 daily trains each way between Italy and Germany. In 2004 the company operated 4,120 trains and transported 2,145 million tonnes, 80 per cent of this in combined transport. This constituted around 45 per cent of combined transport on the Verona–Munich route.

In May 2005 RTC activated a new combined service between Cervignano del Friuli and Munich via the Tauern line. This diversification of activity,

which includes services via the Tarvisio pass is regarded by the company as a first step towards gaining access to the emerging markets of the new European Union countries.

Traction and rolling stock

The company acquired from Bombardier Transportation eight Class EU43 electric locomotives originally built for Polish State Railways (PKP). These are dual-voltage (3 kV DC/15 kV AC) machines. In 2004 RTC acquired from Siemens Dispolok five four-voltage electric locomotives of the same design as the Railion Class 189 machines. The use of these machines has enabled RTC to operate cross-border services over the Tauern line without frontier traction changes. From July 2005, interoperable services were also extended to Brenner route trains.

In April 2004 RTC bought two former Czech Railways Class D.753 Bo-Bo diesel locomotives for operations on less demanding lines.

UPDATED

Railion Italia Srl

Via Umberto Giordano, I-15100 Alessandria
Tel: (+39 0131) 21 87 88   Fax: (+39 0131) 24 07 00
Web: www.railion.it

Sales office

Railion Italia Services Srl
Via Vittor Pisani 8, I-20124 Milan
Tel: (+39 02) 67 10 07 48   Fax: (+39 02) 66 71 38 55

Key personnel

President of the Management Board:
  Dr Christian Heidersdorf
Chief Operating Officer: Giuseppe Arena

Chief Financial Officer: Olaf Müller
Sales (Milan): Marcus Ringeisen

Organisation

Railion Italia is the Italian subsidiary of Railion, formed in April 2004 when Railion bought a 95 per cent stake in Strade Ferrate del Mediterraneo Srl (SFM), a private freight operator established in 2003. The company was formally launched in May 2005. It has some 60 staff.

Freight operations

Railion Italia operates freight trains between the Swiss-Italian frontier stations of Domodossola,

Luino and Chiasso and the north of Italy, handing traffic over at those locations to Railion Deutschland or to its partner, BLS Cargo. At start-up the company was operating 50 international trains each week.

Traction and rolling stock

In mid-2005 Railion Italia had a fleet of seven Vossloh Type G2000 diesel locomotives on hire from Angel Trains Cargo.

NEW ENTRY

Reggiane Railways (ACT)

Azienda Consorziale Trasporti Reggio Emilia
Viale Trento Trieste 11, I-42100 Reggio Emilia
Tel: (+39 0522) 92 76 11   Fax: (+39 0522) 92 76 74
e-mail: actre@actre.it
Web: www.actre.it

Key personnel

General Manager: S Cavaliere

Gauge: 1,435 mm
Route length: 77 km

Organisation

ACT is owned by local towns in Reggio Emilia province, operating passenger services on lines from Reggio Emilia to Guastalla (28 km), Sassuolo (22 km) and Ciano d'Enza (26 km). In 2002 the company carried 841,000 passengers.

The Sassuolo line serves Europe's main ceramics industry and ACT hauled 1.5 million tonnes

Ex-Swiss Federal Railways coaches configured in a diesel-operated push-pull formation with an ACT passenger service at Sassuolo (Quintus Vosman)   **NEW**/1122808

of incoming clay and outgoing tiles in 2002, connecting with FS at Reggio Emilia. The freight yard at Dinazzano is to be expanded in the next few years to cope with growing traffic. ACT created a joint venture with FS to manage this facility, taking a shareholding of 51 per cent. The company has also built a new yard at S Giacomo di Guastalla designed to handle 350,000 tonnes of steel annually.

ACT has obtained an open access operating licence and is planning to operate freight services to Livorno.

ACT has also revealed a plan to reopen the 17 km Carpi–Bagnolo in Piano line for passenger and freight traffic. The project would cost €142 million. ACT also operates local buses.

Traction and rolling stock
The railway operates 13 main line and six shunting locomotives plus 22 railcars. The railway acquired seven new Vossloh G2000 diesel-hydraulic locomotives in 2003–04. ACT's Reggio Santa Croce depot has been licensed by Vossloh for the whole of Italy. The depot is also carrying out refurbishment work on behalf of Vossloh.

UPDATED

Vossloh G2000 diesel-hydraulic locomotive with an ACT freight service at Bosco (Quintus Vosman)
NEW/1122809

Rete Ferroviaria Italiana (RFI)

Piazza della Croce Rossa 1, I-00161 Rome
Tel: (+39 06) 441 01
Web: www.rfi.it

Key personnel
Chief Executive: Mauro Moretti

Gauge: 1,435 mm
Route length: 16,030 km
Electrification: 10,358 km at 3 kV DC

Subsidiary
TAV
Treno Alta Velocita SpA
Via Mantova, I-00198 Rome
Tel: (+39 06) 47 84 11   Fax: (+39 06) 482 39 15
Web: www.tav.it
Chief Executive: Umberto Bertele
Managing Director: Antonio Nicci

Political background
RFI was formed as an independent state-owned company on 1 July 2001, taking in Italian Railways' former infrastructure division (Divisione Infrastruttura). It remains wholly owned by FS. RFI owns and manages the Italian rail network, initially taking over those lines on which Trenitalia trains operate but later adding lines over which independent companies operate.

RFI has control over TAV, the subsidiary responsible for designing and building new high-speed lines. Formed in 1991, TAV was initially owned jointly by FS (45 per cent) and by a dozen banks, nine of them Italian, two French and one German, which together put up 60 per cent of the company's starting capital. In 1998 the company became 100 per cent FS-owned. TAV was granted a 50-year concession to design, construct and market the high-speed network in 1991. Train operation and rolling stock and infrastructure maintenance will be the responsibility of FS.

RFI's roles include the issue of safety certificates to licensed train operating companies operating over its network and charging such operators tolls for access to the network.

New lines
Rome–Florence Direttissima
Italy's first high-speed line, the 254 km Rome–Florence Direttissima, became fully operational in 1992 with the opening of its final 44 km section from Arezzo South to Figline. It is electrified on the 3 kV DC system. The final stretch was made traversable at 300 km/h to allow for testing of new ETR 500 rolling stock. Currently, all trains are limited to 250 km/h in commercial service over the Direttissima.

RFI is studying how to upgrade the Direttissima for 300 km/h, which will involve conversion to 25 kV AC electrification. At its northern end, the Direttissima is to be linked to the Florence–Bologna high-speed line by an orbital line around Florence running mostly on the surface.

High-speed routes: network plan
The TAV network was approved in 1986 as T-shaped, running south-north from Battipaglia (south of Salerno) to Milan via the Rome-Florence Direttissima, and west-east from Turin to Venice via Milan.

In 1993 the government made available the funds for a start of the high-speed project, having agreed to provide 40 per cent of the total cost of the infrastructure. The components authorised were Rome–Naples, Florence–Bologna, Bologna–Milan and Milan–Turin.

The total cost of building the approved high-speed infrastructure (964.7 km) was originally put at €14.9 billion. This had risen to almost €29 billion by 2002, €18 billion of which was still to be found. It has been proposed that TAV's concession should be extended from 2041 to 2061 to take this into account.

Delays to the high-speed programme have been caused by the vetting of construction contracts following allegations of corruption and by opposition on environmental grounds, although the new routes have been designed to follow existing motorway corridors where possible. In 2003 634 km of line were under construction, plus 58 km of connections with the existing network and four new stations.

The new high-speed routes are to be electrified at 25 kV 50 Hz AC rather than 3 kV DC, and have a minimum curve radius of 5,450 m and a maximum gradient of 1.8 per cent. Maximum axleload is to be 18 tonnes. Signalling and train control will be derived from the French TVM 430 system, supplied by Ansaldo Trasporti and its French subsidiary CS Transport. Although the new lines are to be engineered for a maximum speed of 300 km/h, in 1995 it was announced that high-speed services would not operate above 250 km/h for an indeterminate period, in response to environmental concerns. Passenger services over the new routes will be provided by high-speed trainsets at 300 km/h and locomotive-hauled trains operating at up to 220 km/h. The high-speed network will also be used by freight trains (principally intermodal services) operating at up to 160 km/h. There will be major new stations at Bologna, Florence, Naples and Turin. Rome Tiburtina station will be completely redeveloped.

High-speed routes: Rome–Naples
Work began on the €5.2 billion Rome–Naples high-speed line at 11 sites in 1994. Its opening is expected in December 2005. Initially the line

will only be served by a two-hourly shuttle as access lines into Rome and Naples have not been completed. Existing services operated by ETR sets will be diverted via the new line in December 2006 when this work is finished. The route follows the A1 motorway for most of its length and comprises 204.5 km of new construction and 15.5 km of upgraded infrastructure to gain access to Rome and Naples. Other connections with existing routes are to be provided at Frosinone, Caserta and Cassino. Electric trains ran for the first time in December 2002 over a part-completed section of the line.

High-speed routes: Florence–Bologna
The final alignment of the 78.3 km Florence–Bologna high-speed line, costing €4.7 billion, was approved in 1995. Work started in 1996, with a consortium led by Fiat and construction company Impregio to build the major part (71.5 km) of the route to the outskirts of Florence. It is expected that construction of much of the line will be completed by 2005 but opening has been delayed until July 2008. Crossing the Apennine range, 66.8 km of the principal 71.5 km section will be in tunnel. A new station, to be served only by high-speed services, is to be built in Florence.

High-speed routes: Bologna–Milan
The existing route between Bologna and Milan is one of the most congested on the FS network. The cost of the new route has risen to €6.2 billion. There are to be eight connections with existing lines and an intermediate station at Reggio Emilio. Work started in November 2000. Completion is expected in April 2007. The start of construction was delayed by a redesign to allow freight trains to use the line more easily.

The high-speed route comprises 182 km of new infrastructure starting at Melegnano, 21 km south of Milan Central. The subject of controversy during the early planning stages, much of it runs parallel to the A1 motorway as far as Bologna's outskirts. Some 38 km of the alignment is carried on viaducts and bridges. The line is to pass below Bologna in tunnel, with a new station below the existing FS facility. Completion of the new station has been delayed to 2010. The line will cut the Milan–Bologna journey time from 1 hour 45 minutes to 1 hour.

With future high-speed services in mind, two extra tracks were commissioned in 1996 on the Milan–Melegnano route between Milan Rogoredo and San Giuliano Milanese (7 km).

High-speed routes: Milan–Turin
The alignment of this 125 km route has caused controversy, particularly in the Novara area, and is planned to follow the A2 motorway. New infrastructure is to begin at Certosa, 9 km from Milan Central, and run to Settimo Torinese, 8 km from Torino Porta Nuova. Connections with existing

lines are to be provided at Santhia and Novara. The cost of the project is estimated at €6.9 billion.

Construction started in February 2002 and completion of the Turin–Novara section (86 km) is planned for February 2006 with Novara–Milan (39 km) following in 2009. When complete, the line will cut the Turin–Milan journey time from 1 hour 40 minutes to 50 minutes.

High-speed routes: Milan–Venice

Not included in the budget detailed above, construction of this line was originally costed at €2.79 billion, excluding connections to existing lines. In 2002 the cost of the Milan–Padua section had risen to €7.65 billion. The route comprises 212 km of new line, starting 20 km from Milan Central at Melzo. It follows the SS1 motorway to Brescia (where connections with the existing network are to be provided), and then the A4 motorway to south of Verona where it rejoins the existing FS system. The remainder of the high-speed route links Verona and Padua, with a connection to the existing network at Vicenza. High-speed services will reach Venice over the existing 23 km Padua–Mestre route, upgraded for 220 km/h and quadrupled. Construction of the rest of the route started in 2004 for completion by 2008.

High-speed routes: Milan–Genoa

A new line from Tortona, on the route from Milan, to Genoa was given the go-ahead in 2003. Construction cost is estimated at €4.7 billion. Two-thirds of the new line will be underground, almost the whole way from Novi Ligure to the outskirts of Genoa, including a 37 km tunnel. Opening would be in 2011 at the earliest.

The costs detailed above do not include those of associated remodelling of rail networks in urban areas. Costs for this will be: Bologna €1,105 million; Florence €1,211 million; Milan €73 million; Naples €374 million; Rome €568 million; and Turin €129 million.

Norms for TAV high-speed lines
Maximum speed: passenger trains – 300 km/h; freight trains – 120 km/h
Train headway: 5 minutes
Max curve radius: 5,450 m
Max gradient: 1.8 per cent (exceptionally 2.5 per cent)
Distance between track centres: 5 m
Max axleloads: locomotives – 25 tonnes; coaches – 22.5 tonnes
Power supply: 25 kV AC 50 Hz
Signalling: ERTMS Level 2

High-speed routes: Lyon–Turin

In September 2000, the French and Italian governments agreed to build a base tunnel of 52.7 km between St Jean-de-Maurienne and Susa, on the Lyon-Turin route. Initial test boring work started in early 2002. Completion is scheduled for 2015 and the project will cost an estimated €10 billion. The main aim of the scheme is to divert freight traffic via the tunnel, which would be built to allow 'rolling highway' services (see entry for French Railways Infrastructure Authority, RFF). Both French and Italian authorities will link the tunnel to their respective high-speed networks. By 2006/07, the two railways will enlarge the existing loading gauge to UIC 'B' standards. In 1994 FS and SNCF of France formed the Alpetunnel joint venture to undertake technical and economic studies for the base tunnel (see entry for France).

Cross-city routes: Milan

In December 2004 the 18.4 km Passante cross-city line in Milan was completed, construction having started in 1982. The double-track line starts in the west of the city at Certosa on the lines from Turin and Domodossola, with a branch from North Milan Railway (FNME) lines at Lancetti, then passes under the city centre, serving five stations which give interchange with Milan's three metro lines, then emerges onto the surface at Porta Vittoria, linking to the line to Verona at Pioltello. A link to the Bologna line at Rogoredo will open subsequently. The first section of the line, from FNME's Bovisa station to Porta Venezia, opened in December 1997. From December 2004 Trenitalia services consisted

of an quarter-hourly service eastwards from Certosa, trains running half-hourly from Novara and Gallarate, with one train an hour originating from Varese. The same service operates in the peak and off-peak times. Trenitalia trains are operated by TAF double-deck emus and double-deck stock powered by Class E 464 electric locos. The FNME service consists of half-hourly services from Saronno, and a quarter-hourly service from Bovisa, every other train starting from Affori. In the peak, trains run through from Affori to Seveso, Meda and Mariano Comense. FNME trains are a mixture of single- and double-deck emus. All trains terminate at Porta Vittoria except for those from Gallarate/Varese, which extend to Pioltello.

Services will eventually operate every 20 minutes at peak periods and hourly at other times on five routes: Novara–Lodi, Gallarate–Codogno, Malpensa–Brescia, Saronno–Treviglio and Seveso–Pavia.

Further FS investment associated with the scheme has included four-tracking from Pioltello to Treviglio (22 km), approved in 1995 for completion around 2000 at an estimated cost of €362 million. Two extra tracks were commissioned in 1996 between Rogoredo and San Giuliano Milanese (7 km), on the route to Lodi and Codogno. The upgrading of the Milan orbital route connecting Certosa, Lambrate, Porta Vittoria, Rogoredo and Porta Romana has also been proposed; four-tracking between Pioltello and Lambrate has also been undertaken.

Cross-city routes: Turin

In 1983 FS signed an agreement with the Turin city authorities aiming to establish a regional rail system similar to Milan's. A minimal through

service started operation in September 1997 and a 3.3 km route from Lingotto to Porta Susa via Zappata opened in September 1999. SATTI has built a new station on its Dora to Ceres line to serve Turin's Caselle airport.

The second phase will involve quadrupling the line from Torino Stura to Torino Porta Nuova and a construction of a new underground station at Porta Susa. This is expected to take at least 10 years.

Bridge to Sicily

A long-standing project to build a road/rail bridge over the Straits of Messina to Sicily was approved in 2001. The cost of the 3.3 km span is put at €5.7 billion and about 40 per cent this is sought from private sources. The bridge would allow up to 200 trains per day to cross between Reggio di Calabria and Messina. At present passenger and freight trains cross on ferries.

Improvements to existing lines

Only 6,400 of the 15,000 route-km of the RFI network is double track. Many lines are being doubled at present. In 2002 RFI doubled 31 km of line and opened 9 km of completely new line plus 31 km of cut-offs; 71 km of line were electrified and 288 km put under the Sistema di Comando e Controllo; 135 level crossings were eliminated during the year.

Pontebbana route

The Pontebbana route between Villach, Austria, and Trieste/Venice via the Tarvisio Pass and Udine already carries 11 per cent of FS international traffic, almost the same as the Brenner, and is heading for major growth of its freight traffic. Its single line and outdated electrification have consequently

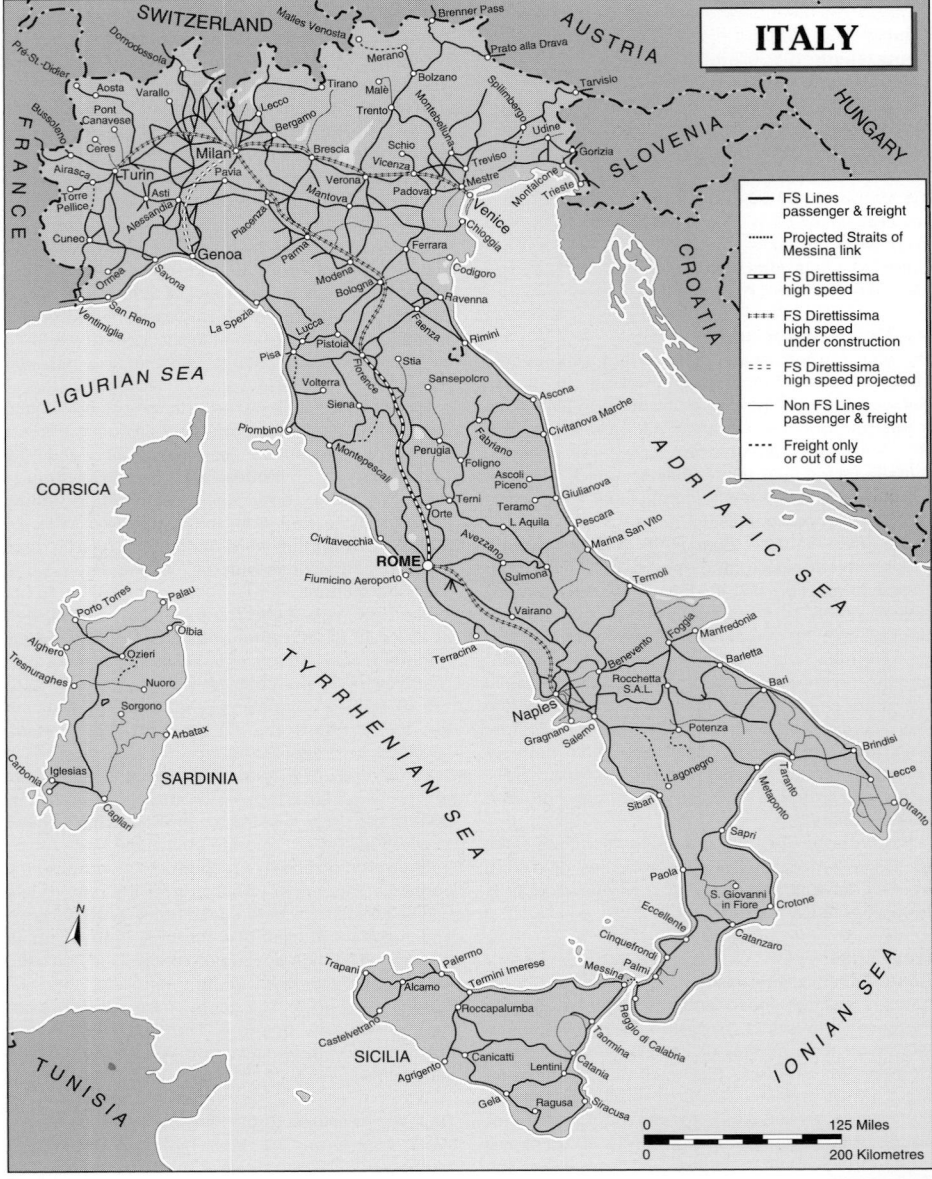

Italy

become more of a burden year by year, prompting FS to launch a programme of double-tracking, track rebuilding for heavier axleloads, realignment and grade easement, re-electrification and resignalling with automatic block for reversible working. New tunnelling is involved in obtaining an alignment for the second track over 85 per cent of the route. Work was completed in 2001.

Rome
Major projects in the Rome area have included a new orbital link from St Peters via Vigna Clara to the Rome–Florence route north of Tiburtina. The link includes a 4.6 km tunnel under Rome's northern suburbs. Other upgrading projects include the double-tracking of the Tiburtina–Guidonia route and quadrupling between Casilina and Ciampino.

Work has already been completed on doubling the San Pietro–Cesaro line and on complete electrification of the route between San Pietro and Viterbo (82 km). Thirteen stations have been rebuilt.

These projects form the basis of a scheme to dramatically increase suburban rail provision in Rome by 2000. Cross-city services would be provided between Fiumicino and Fara Sabina and between La Storta and Ciampino, sharing the southern Trastevere–Ostiense–Tuscolana orbital route and operating in conjunction with St Peters–Vigna Clara–Tiburtina and Tiburtina–Guidonia services. Investment totalling €2.07 billion would be required for the scheme, which also includes refurbishment work at Rome's Termini, Tiburtina, St Peters, Ostiense and Trastevere stations.

Other schemes
RFI is gradually upgrading the Verona–Bologna and Bologna–Ancona–Foggia–Bari Adriatic route. Work includes track-doubling and elimination of level crossings. In 2002 the 12.6 km section from Vasto

San Salvo to Porto di Vasto Adriatico was doubled. RFI reopened and electrified the Merano–Malles line in May 2005.

A new 12.6 km double-track tunnel was opened in late 2001 between Messina and Villafranca Tirrena in Sicily, replacing a 5.5 km single-track structure with adverse gradients. The new tunnel cuts Messina–Palermo journey times by 20 minutes and the use of banking locomotives for freight trains has been eliminated.

Swiss Federal Railways and Italian Railways are to reopen/build a new 18 km line from Mendrisio, Switzerland, to Arcisate, Italy, to create a more direct route from the Gotthard line to Varese, Milan and Malpensa airports. The cost of the project, due to open in 2010, is estimated at €293 million.

Additional central government funding has been secured to fund upgrading of existing routes: Padua–Mestre (€176 million); Orte–Falconara (€206.6 million); Caserta-Foggia (€279 million); the Sardinian network (€103.3 million); and Palermo–Messina–Catania (€372 million).

In April 2004 the 45 km Pisa–Valda route reopened to freight traffic after refurbishment and electrification. The route avoids tunnels south of Livorno which are too small for high-cube containers.

The go-ahead was received in early 2004 for a new 4.5 km loop serving Verona airport.

Signalling and telecommunications
For speeds above 150 km/h, automatic block with coded current cab repetition is being adopted. Cab repetition is integrated with automatic speed control. For speeds up to 250 km/h on the Rome–Florence Direttissima the following speeds are encoded (km/h): 250-230-200-150-100-60-30.

In 1996 Ansaldo completed the first phase of the €98 million Naples–Reggio di Calabria route modernisation project, comprising the 200 km

between Battipaglia and Paola. This section, controlled by a new CTC installation with 22 interlockings at Sapri, received automatic block signalling with coded current cab repetition, a centralised hotbox detection system and improved public address and wayside telephone systems.

In 1997 Ansaldo Trasporti was awarded a €129 million contract to build operations control centres at Genoa, Naples and Venice covering traffic on Genoa-Rome, Bologna–Brennero, and Bologna–Bari routes. Implementation of the entire programme is scheduled for the end of 2000.

In 2002, Siemens received a €168 million contract from RFI to develop a GSM-R mobile radio communication system to be introduced in 2005, replacing existing equipment.

Electrification
The historic network in Italy, and the Rome–Florence high-speed line, is electrified at 3 kV DC; however, future high-speed lines will be energised at 25 kV 50 Hz AC.

In 2003 new electrification included Giulianova–Teramo and Bassano del Grappa–Castelfranco Veneto.

Track
Monoblock prestressed concrete sleepers are being used almost exclusively in current track upgrading and doubling projects (though a test installation of slab track was recently completed). Length of sleeper is 2.3 m but for new track, where speeds may exceed 160 km/h, a new design with a length of 2.6 m is being adopted. For rail renewals, and all new lines, UIC 60 (60 kg/m) rails are being used, fastened with K-type clips. RFI replaces about 850 km of track each year.

UPDATED

Sangritana Railway (FAS)

Ferrovia Adriatico-Sangritana Srl
Via Dalmazia 9, Piazzale della Stazione, I-66034 Lanciano
Tel: (+39 0872) 708 14   Fax: (+39 0872) 70 85 00
e-mail: sangritana@sangritana.it
Web: www.sangritana.it

Key personnel
Managing Director and Operating Manager: Dr Ing Antonio Bianco

Gauge: 1,435 mm
Route length: 126 km
Electrification: 103 km at 3 kV DC

Organisation
In January 2001 FAS became a limited liability company, wholly owned by the Abruzzo region. FAS operates the following lines: Marina San Vito (FS)–Lanciano–Crocetta–Archi–Castel di Sangro (104 km, out of use in 2004); Ortona Marina–Crocetta (38 km), of which only Ortona Marina–Crocetta (10 km, freight only) is open at present; Torino di Sangro (FS)–Piazzano di Atessa (9 km, freight only).

A new line, Lanciano–San Vito, is under construction to eliminate level crossings. Through Lanciano–Pescara journey times will be cut from 60 to 25 minutes. Another new line, from Torino di Sangro to Archi, is under construction. The company is also to modernise the Bomba Torricella–Quadri Borello section. FAS plans to extend passenger services over RFI lines in the near future.

The Caldari line may be reopened to Guardiagrele if freight traffic warrants. The Piazzano line was built in the 1980s to serve a car production plant and has been managed by FAS since 1991. The railway has received an open access operating licence and has ambitions to take over the management of other local industrial branches and, possibly, local freight workings from FS.

In 2002 FAS employed 364 staff.

Traction and rolling stock
The railway operates four electric and 12 diesel locomotives, 11 electric railcars, six coaches and 28 wagons. The most recent locomotive acquisitions, in 2002–03, were six Class D752 Bo-Bo diesel-electrics formerly operated by Czech Railways and refurbished by ČMKS Holding and the Inekon Group. They are used on the 9 km Torino di Sangro (FS)–Piazzano di Atessa line. In 2001 FAS acquired a MaK G1100 diesel-hydraulic from the Vossloh hire fleet. In 1999 FAS bought 10 Type AM56 two-car emus from Belgian National Railways. Five of these were refurbished by Ansaldo Trasporti in 2004–05.

The company has ordered four ALSTOM Minuetto three-car dmus for delivery from 2005.

In 2004 FAS was tendering for four new or secondhand electric locomotives rated at 4,500 kW.

UPDATED

FAS refurbished ex-Czech Railways Class D752 diesel-electric locomotive at Lanciano
(Quintus Vosman)
0552784

Sardinian Railways (FdS)

Managed by Government Commission, Rome
Ferrovie della Sardegna
Via Cugia 1, I-09129 Cagliari
Tel: (+39 070) 34 23 41   Fax: (+39 070) 34 07 80
e-mail: fdsdc@tin.it
Web: www.ferroviesardegna.it

Key personnel
Director General: E Porceddu

Gauge: 950 mm
Route length: 614 km

Organisation
FdS was formed in 1989 with the merger of Strade Ferrate Sarde and Ferrovie Complementari della Sardegna. Routes operated are Cagliari–Sorgono/Arbatax, Tresnuraghes–Macomer–Nuoro and Alghero–Sassari–Palau Marina. Regular passenger services operate over 219 km of the network on the following routes: Sassari–Alghero (30 km); Sassari–Nulvi (35 km); Sassari-Sorso (11 km); Macomer–Nuoro (61 km); and Cagliari-Isili (82 km). The remainder of the system was used by freight and tourist services only. The company also operates bus services in Sardinia.

A 950 mm gauge tramway system is under construction in Sassari. The eventual aim is to take over FdS lines to Sorso and Alghero, with a 10 km extension to Fertilia airport in prospect from the latter location. The urban Cagliari–Monserrato section of the line to Isili is currently being doubled and electrified for a light metro service. A 1.7 km branch from Monserrato to Policlino was approved in mid-2005.

Traction and rolling stock
The railway operates four steam locomotives, 16 diesel locomotives, 39 diesel railcars, 53 coaches and 19 wagons. In 2003 FdS called for tenders for nine new two-car dmus, with an option of a further 15.

UPDATED

Sassuolo–Modena Railway (ATCM)

Azienda Trasporti Collettivi e Mobilità
Piazza Manzoni 21, I-41100 Modena
Tel: (+39 059) 30 80 62   Fax: (+39 059) 41 68 50
e-mail: info@atcm.mo.it
Web: www.atcm.mo.it

Key personnel
President: Laura Tosi
General Manager: Silvano Cavaliere

Gauge: 1,435 mm
Route length: 20 km
Electrification: 20 km at 3 kV DC

Organisation
ATCM was renamed in 2001, when it became a public company, albeit with all shares held by 47 towns in the Modena province. A new, partially underground 3 km connection from Modena ATCM station to the more central FS station opened in January 2004, allowing interchange with other passenger and freight services. The eventual intention is that ATCM should operate services to Bologna, Carpi and Rolo over FS lines.

In 2004 ATCM carried 244,000 passengers.

Ex-SNCB Class 54 emu at Modena (Quintus Vosman)   **NEW**/1122810

ATCM gained an open access freight licence in mid-2002.

ATCM had 606 staff in 2004.

Traction and rolling stock
The railway operates two electric locomotives, five emus, six coaches and 10 wagons. The emu fleet includes three two-car units acquired second-hand from Belgian National Railways.

UPDATED

Sistemi Territoriali SpA (ST)

Piazza Giacomo Zanellatos 5, I-35131 Padua
Tel: (+39 049) 74 49 99   Fax: (+39 049) 77 43 99
e-mail: info@sistemiterritorialispa.it
Web: www.sistemiterritorialispa.it

Key personnel
President: Dr Gian Michele Gambato
Director: Sergio Bertonasco

Gauge: 1,435 mm
Route length: 57 km

Organisation
Sistemi Territoriali, formerly Ferrovie Venete, originally comprised two geographically separate routes, the Ferrovia Adria–Mestre (FAM) (57 km) and the Ferrovia Udine–Cividale (FUC) (15 km). In January 2005 the latter line was placed under separate management, becoming Ferrovie Udine–Cividale Srl (see entry in Italy section of *Railway systems and operators*).

ST falls within the authority of the Veneto region, which has allocated €49 million for upgrading the Adria-Mestre line.

ST holds an open access licence and in April 2005 took over freight services on the Vicenza–Schio line from Trenitalia. In 2005 the company won a joint contract with Trenitalia (see entry) to operate 10 lines in the Veneto region. This will be effective for six years from December 2006.

Ex-Czech Railways ST Class D 753 diesel locomotive at Rovigo (Quintus Vosman)   **NEW**/1122811

Traction and rolling stock
In 2005 rolling stock totalled 10 diesel railcars, three driving trailers and six diesel locomotives. In 2003 the company received two former Czech Railways Class 753.7 diesel-electric locomotives and in 2004 bought three more. In the meantime ST hired four Class 668.1700 railcars from Trenitalia in early 2004.

In 2004 ST ordered two GTW 2-4/12 and one GTW 2-2/6 diesel-electric railcars from Stadler for delivery in 2006, to operate Adria–Mestre line services.

UPDATED

Società Automobilistica Dolomiti (SAD)

Web: www.sad.it

Organisation

SAD operates train services on the 60 km Merano–Málles line, which reopened in May 2005. The line's infrastructure is managed by Struttura Trasport Alto Adige (STA) on behalf of Bolzano province, which financed its refurbishment. A connection is made at Merano with Trenitalia services to Bolzano.

SAD also operates trains on the Ferrovia del Renon/Rittnerbahn, a short tourist line in the mountains above Bolzano, as well as bus and cable-car services in the region.

Traction and rolling stock

Eight Stadler GTW 2/6 two-car dmus operate the service. The first was rolled out in October 2003.

UPDATED　　*SAD Stadler GTW 2/6 dmu at Merano* (Quintus Vosman)　　*NEW*/1122819

South Eastern Railway (FSE)

Ferrovie del Sud-Est e Servizi Automobilistici Srl
Via G Amendola 106/D, I-70126 Bari
Tel: (+39 080) 546 21 11　Fax: (+39 080) 546 22 11
e-mail: fsudest@fseonline.it
Web: www.fseonline.it

Key personnel

Director General: N Aversano

Gauge: 1,435 mm
Route length: 474 km

Organisation

This is geographically the largest private railway in Italy, operating most lines in the Otranto peninsula in the Puglia region: Bari–Mungivacca–Conversano–Putignano–Martina Franca–Taranto (113 km); Mungivacca–Casamissima–Putignano (43 km); Martina Franca–Francavilla Fontana–Lecce (103 km); Lecce–Zollino–Nardo–Gallipoli–Casarano (75 km); Zollino–Maglie–Gagliano Léuca (47 km); Novoli–Nardo–Casarano–Gagliano–Léuca (75 km); and Maglie–Otranto (18 km). FSE also operates buses.

In 2003 FSE was granted an open access freight licence.

FSE diesel railcar at Nardo (Quintus Vosman)　　0580496

Traction and rolling stock

The railway operates 28 diesel locomotives, 52 diesel railcars and 31 trailers, 50 coaches and 181 wagons. The railcar fleet includes 10 ex-FS ALn 668.1400 Fiat single-unit railcars, which have been modernised and converted into two-car sets at Alitransport's Leon d'Oro works.

UPDATED

Trentino Trasporti SpA

Via Innsbruck 65, I-38014 Gardolo
Tel: (+39 0461) 82 10 00　Fax: (+39 0461) 03 14 07
e-mail: info@ttspa.it
Web: www.ttspa.it

Key personnel

President: Vanni Ceola
Director, Operations: Giancarlo Crepaldi
Director, Infrastructure: Mauro Dorigoni

Gauge: 1,000 mm
Route length: 66 km
Electrification: 66 km at 3 kV DC

Organisation

FTM merged with bus operator Atensina in November 2002 to become Trentino Trasporti SpA, but is still usually known by its old name. Trentino Trasporti is mainly owned by Trento province (73.75 per cent) and the town of Trento (18.75 per cent).

Principally a passenger operator (2 million journeys in 2004), the railway also moves 1,435 mm gauge freight wagons using a fleet of 12 transporter bogies and four transporter wagons. There is a 2.3 km dual-gauge section connecting the FS network at Trento with a factory at Gardolo.

Construction of a 10 km, four-station extension from Malé to the winter sports resort of Marilleva was completed in May 2003 and studies have been conducted into a further 7 km extension to Fucine. The company also runs buses.

There were 142 rail staff in 2004.

FTM emu at Tassulo (Quintus Vosman)　　0580495

Traction and rolling stock

The rolling stock fleet comprises one electric locomotive, five electric railcars and two trailers, and nine three-car articulated emus. The first of 14 two-car part-low-floor Coradia emus from ALSTOM was delivered in December 2004 to enable service frequencies to be increased on Trento–Mezzo Lombardo and Trento–Malé routes.

UPDATED

Turin Transport Group (GTT)

Gruppo Torinese Trasporti SpA
Corso Turati 19/6, I-10128 Turin
Tel: (+39 011) 576 41
Fax: (+39 011) 576 43 30
e-mail: gtt@gtt.to.it
Web: www.gtt.to.it

Key personnel
President: Giancarlo Guiati
Managing Director: Davide Gariglio

Gauge: 1,435 mm
Route length: 81 km
Electrification: 55 km at 3 kV DC

Organisation
At the beginning of 2003 Turin's transport operators, ATM and SATTI, merged as Gruppo Torinese Trasporti (GTT). Within GTT the railway activity encompasses the former SATTI's two passenger lines, metro operations, freight services and rolling stock maintenance. GTT was awarded the contract to manage construction and operation of Turin's first metro line, which will open in 2005.

The company gained an open access licence in September 2001.

Passenger operations
GTT operates local services over two routes, Turin–Ceres (43 km, electrified as far as Germagnano) and Turin–Rivarolo–Pont Canavese (38 km). Around 3.8 million passenger journeys are recorded annually.

Co-ordination of FS and former SATTI services started on 1 September 1997 with a dmu service linking FS at Chieri via the city centre to Rivarola Canavese. A second through service is planned, linking Germagnano station on the Ceres line via the city centre to a new destination which remains to be finalised.

Operations for both lines are to be controlled from a single new centre at Ciriè.

Improvements to existing lines
The Turin–Germagnano line reopened in April 2001 after construction of a new deviation to

GTT Fiat Type Y low-floor emu at Settimo Torinese (David Haydock)   0580500

serve Turin airport and an underground deviation serving Caselle on the Turin–Ciriè section. Service frequencies have since been more than doubled. Rebuilding after a landslip plus re-electrification on the Germagnano–Ceres section is continuing. Electrification of the 22 km Settimo Torinese–Rivarolo Canevese line was completed in March 2002 at a cost of €5.2 million.

In the long term, the Turin–Ceres route is to be upgraded to regional metro standards and connected to the cross-city route currently under construction for FS (see entry for Italian Railways in *Railway systems and operators*).

Traction and rolling stock
At the beginning of 2004 GTT operated four diesel and four electric locomotives, 20 diesel railcars and 12 trailers, seven Fiat Type Y two-car low-floor emus, eight ex-Belgian National Railways emus, 15 coaches and 15 wagons.

In 2004 GTT ordered 10 ALSTOM Minuetto three-car emus, with an option for nine additional units. They are scheduled for delivery in 2005–06. In 2004 GTT called for tenders for refurbishment and re-engining of its dmu fleet.

UPDATED

Japan

Ministry of Land, Infrastructure and Transport

2-1-3, Kasumigaseki, Chiyoda-ku,
Tokyo 100-8918
Tel: (+81 3) 52 53 81 11
Web: http://www.mlit.go.jp

Key personnel
Minister: Chikage Oogi
Senior Vice-Minister: Gotaro Yoshimura
Parliamentary Secretary: Yosuke Takagi

Japan Railway Construction Public Corporation (JRCC)

Sanno Grand Building, 2-14-2, Nagata-cho,
Chiyoda-ku, Tokyo
Tel: (+81 3) 35 06 18 94
Fax: (+81 3) 35 06 18 90

Key personnel
President: Mitsuhiko Matsuo

Organisation
The Corporation was set up in 1964 to construct railways on behalf of the government for subsequent leasing or transfer to railway operating companies. It is responsible for the construction of all new Shinkansen lines and has taken over much other new construction work from the JR Group and other railways.

Japan Railways Group (JR)

Political background
Privatisation
Japanese National Railways (JNR) was statutorily disbanded in 1987 and its assets, operations and liabilities were distributed among a number of new companies, known as the Japan Railways Group. The dismemberment legislation provided that JNR's passenger business, its infrastructure and its assets, on JNR's 1,067 mm gauge network be distributed geographically between six companies, three on Honshu island and one each on Hokkaido, Shikoku and Kyushu.

Initially, all remained in the public domain. Only the Hokkaido, Shikoku and Kyushu companies started free of any inherited debt liabilities, but all three required subsidy for their current operations, which was provided through government-established Management Stabilising Funds.

Privatisation of the new companies was the ultimate objective of JNR's dismemberment. At the start of 1991 the Transport Ministry announced that two million shares in each of the three biggest companies, JR East, JR Central and JR West, would be put on the market in 1992, but, with the subsequent serious downturn of the Tokyo stock market, the placing was postponed. The sale of half the JNR Settlement Corporation's holding of four million JR East shares eventually took place in October 1993 and was heavily oversubscribed. JR Central and JR West flotations were scheduled for 1994–95 but the adverse effects of the Kobe earthquake on revenue and profitability led to postponement; sales of JR West shares took place in 1996, and JR Central followed in October 1997. The

proceeds from the sales are intended to be used to reduce the ¥26,200 billion debt inherited from JNR.

Since 1984 over 35 local companies have been established to take over loss-making JNR/JR rural lines. The new operators are known as third-sector companies, because they are a hybrid of private and local community finance.

Previously, Japanese railway business was governed by two sets of statutes, one to regulate JNR and one covering other railways. This has been superseded by new legislation covering all railway business. It has reduced the degree of regulation, with provisos that railway safety and customer services are not impaired. A licence is required to run a railway business and railway facilities are subject to inspection. Furthermore, fares and charges must be approved in advance by the Minister of Transport, although written notice is considered adequate for discounted fares and charges. Finally, train schedules must be submitted to the Ministry in advance of implementation.

Shinkansen network
The 1,435 mm gauge Shinkansen network was at first vested in a Shinkansen Property Corporation, which leased the infrastructure to the new companies for train operation. The companies were responsible for upkeep of the infrastructure. After protracted negotiations, terms were agreed in 1991 for sale of the network later in the year to the three companies operating it.

The proceeds were applied in part to the financing of Shinkansen network extensions begun in 1992, and also to clearing by the JNR Accounts Settlement Corporation of more of the accumulated debt left from the JNR regime.

East Japan Railway Co (JR East)

2-2-2 Yoyogi, Shibya-ku, Tokyo 152-1578
Tel: (+81 3) 53 34 11 51   Fax: (+81 3) 53 34 11 10
Web: http://www.jreast.co.jp

Key personnel

Chairman: Masatake Matsuda
President and Chief Executive Officer:
   Mutsutake Otsuka
Executive Vice-Presidents
   Life-style Business Development Headquarters:
      Eiji Hosoya
   Railway Operations Headquarters: Yoshio Ishida
   Corporate Planning Headquarters: Satoshi Seino
Executive Directors
   Railway Operations Headquarters, Facilities and
      Construction Departments: Nobuyuki Hashiguchi
   Finance, Personnel and Health and Welfare
      Departments: Makoto Natsusme
   Railway Operations Headquarters, Marketing,
      Transport and Rolling Stock and Credit Card
      Development Departments and IT Business
      Project: Yasutomo Shirakawa
   Technology Planning Department, Corporate
      Planning Headquarters, JR East Research and
      Development Center, Transport Safety
      Department, Railway Operations Headquarters:
      Yukio Arimori
   Tokyo Branch Office: Hiroshi Okawa
   Administration, Inquiry and Audit, Public Relations
      and Legal Departments: Tetsujiro Tani
   Life-style Business Development Headquarters:
      Yoshiaki Arai

Gauge: 1,067 mm; 1,435 mm (Shinkansen)
Route length: 6,573.1 km; 956.3 km
Electrification: 2,671.6 km at 1.5 kV DC (1,067 mm);
1,887.4 km (1,067 mm) at 20 kV 50 Hz AC and
956.3 km (1,435 mm) at 20 kV 50 Hz AC

Organisation

The railway runs rail passenger transport and
related activities in the Tohoku region and the
Tokyo metropolitan region, including the Tohoku
and Joetsu Shinkansen. Some 304 Shinkansen
and 12,192 conventional trains were operated daily
from December 2001. In April 2002, 74,050 staff
were employed.

Finance (million yen) Consolidated

	2000–01	2001–02
Operating revenue	2,546,041	2,543,378
Operating expenditure	2,222,289	2,227,038
Transportation, other services and cost of sales	1,722,744	1,712,324
Sales, general and administrative expenses	499,546	514,714
Operating income	323,751	316,339

Operating revenues and expenses

In FY2001 (2001–02), operating revenue decreased
by 0.1 per cent from ¥2,546 billion in FY2000
(2000–01) to ¥2,543.4 billion (US$19,123 million).
This was due to a decrease in revenues from
passenger transport, partially offset by an increase
in revenues from the station space utilisation
business.

Revenues from railway operations were ¥1,667.6
billion (US$12,538 million), a decrease of 0.8 per
cent. Revenues from Shinkansen passes increased
but revenues from non-pass travellers decreased
on both Shinkansen and conventional lines.

Shinkansen network revenues decreased
1.0 per cent to ¥458.4 billion (US$3,447 billion)
as the proportion of passengers using bargain-
priced travel products increased despite an
expansion of Shinkansen passenger-km. Revenue
from Shinkansen passes was up 5.1 per cent to
¥21.3 billion (US$161 million) and revenue other
than that from passes was down 1.3 per cent to
¥437.1 billion (US$3,286 million).

Tokyo metropolitan area network revenues
decreased 0.3 per cent to ¥841.5 billion (US$6,327
million). Revenue from commuter passes was
down 0.7 per cent to ¥346.1 billion (US$2,602
million). Non-commuter revenues remained at the

same level as the previous year at ¥495.5 billion
(US$3,725 million).

Intercity and regional network revenues declined
1.5 per cent to ¥367.6 billion (US$2,764 million).
Commuter pass revenues decreased 0.2 per cent to
¥119.9 billion (US$901 million) and non-commuter
revenues were down 2.0 per cent to ¥247.7 billion
(US$1,863 million).

Revenues from businesses other than railway
activities increased 1.1 per cent to ¥875.8 billion
(US$6,585 million). This was mainly attributable
to increases in revenues in retail businesses and
restaurants within stations.

Operating expenses increases 0.2 per cent
(¥4.7 billion) from ¥2,223.3 billion in FY 2000–01
to ¥2,227.0 billion (US$16,745 million). Operating
expenses accounted for 87.6 per cent of operating
income in FY2001 and 87.3 per cent in FY2000.

Traffic (million)	1999–2000	2000–01	2001–02
Passenger journeys	5,893	5,862	5,846
Passenger-km	125,998	125,344	124,916

Passenger operations

Tokyo metropolitan area network

The Tokyo metropolitan area network consists
of 1,117.4 km of lines operating within a radius
of approximately 100 km from Tokyo station,
an area with a population of over 32 million.
This network accounts for 61.01 per cent of
JR East's total passenger-km and 44.3 per cent of
its total operating revenues. Demand for commuter
rail services is immense and the metropolitan
area is the focus of considerable investment,
aimed at reducing the serious overcrowding
and lack of capacity. The company's Tokyo New
Network 21 project involves lengthening trains,
increasing train services and running faster trains
on existing lines, together with infrastructure
improvements – such as the electrification of the
outer suburban Hachiko line which was completed
in 1996.

On the Yamanote line orbiting central Tokyo,
where the peak service operates at 2½ minute
headways, standard train formations have been
extended by one car to 11. The line's latest emu
cars have six pairs of double doors in each
bodyside to accelerate passenger loading and
detraining, and foldaway seats to increase peak
standing space to ease congestion. Similar cars
were introduced on the Yokohama line in 1994,
allowing an increase in train length from seven
to eight cars, and on the Keihin–Tohoku line
in May 1995 and subsequently on the Saikyo
and Chuo–Sobu lines. Extra capacity is also
provided by bilevel cars inserted in Series 211

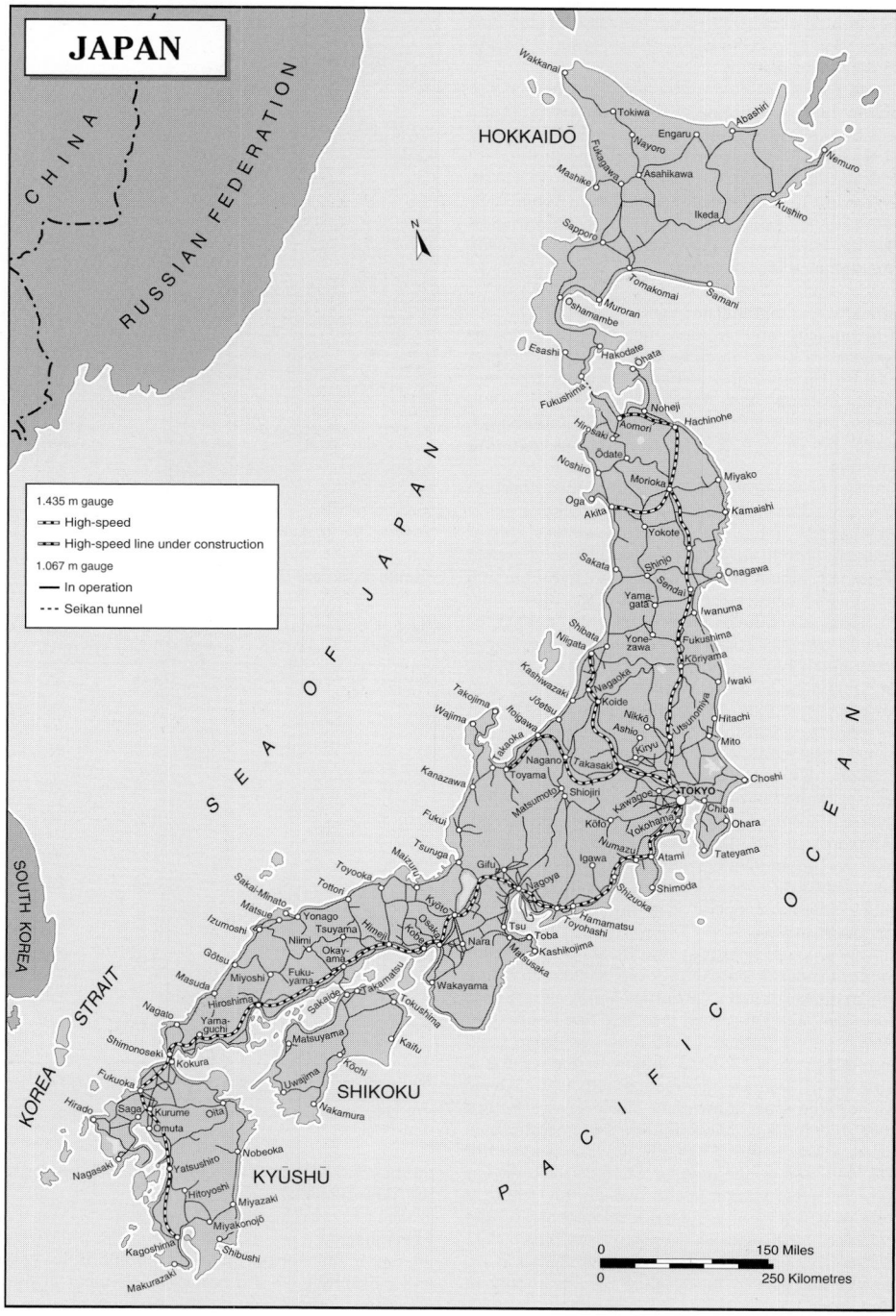

0554471

electric trains operating south of Tokyo on the Tokaido line. In 1992 a new Series 215 design of 10-car emu formed exclusively of bilevel cars was introduced between Tokyo and Odawara. A further capacity increase has been provided by opening up freight tracks for new passenger services, notably the extension of the Saikyo Line to Ebisu and the Shonan–Shinjuku line, which commenced operations in 2001.

Introduction of new cars in the Tokyo metropolitan area

Series 209 10-car trainsets were introduced on the Keihin–Tohuku line in April 1993. Including the six-car sets operating on the Nambu line from Kawasaki from September 1996, the total in 2002 stood at 624 cars in service. These lightweight trains cost only two-thirds of that of a conventional emu. An extended maintenance cycle produces a further reduction in costs. JR East replaces some 400 cars each year to replace existing commuter stock, around 200 of which have been manufactured at the company's own Niitsu Rolling Stock Plant, supplemented by deliveries from traditional suppliers.

The low cost concept has also been employed in the development of the 15-car Series E217 outer suburban emus which now operate many services on the Sobu–Yokosuka lines. These units include two double-deck 'Green Class' cars and 13 standard class cars, 10 of which have only longitudinal seats to maximise capacity.

The Series E501 emus built for services on the Joban line incorporate converter-inverters made in Germany by Siemens. The Series E501 features 50 Hz AC/DC dual-mode and inverter control.

Series E231 cars were introduced in 2000. These are built to a wider profile as a congestion reducing measure, feature a Train Information System (TIMS) and are of a design that minimises maintenance requirements.

Intercity and regional networks

The intercity and regional networks operate on 5,464.4 km of rail lines serving areas throughout eastern Honshu, and include all non-Shinkansen lines outside the Tokyo metropolitan area, generating 24.8 per cent of the company's passenger-km and 19.3 per cent of total operating revenue.

Intercity services are being upgraded through the provision of improved connections with Shinkansen lines, increased frequencies, faster speeds, and through the introduction of newly designed trains tailored to the requirements of specific services. The Series 255 'Boso View Express' emus which were introduced in 1993 for limited express 'Super View Sazanami' and 'Super View Wakashio' services between Tokyo and the Boso Peninsula are designed for business and leisure travel, featuring both individual and group seating arrangements and compartments for stacking surfboards and golf bags.

Series E351 emus began running on 'Super Azusa' limited express services on the Chuo line between Tokyo (Shinjuku) and Nagano prefecture (Matsumoto) in December 1993. This series features an innovative suspension system for a smoother ride, longer trainsets giving 30 per cent more seating capacity and large windows for panoramic views.

New types of energy-saving Series E257 cars for express services were introduced on the Joban line 'Fresh Hitachi' train in 1997, on the Tohoku line 'Super Hatsukari' in 2000 and on the Chuo line 'Azusa' and 'Kaiji' in 2001. Series 701 AC emu cars have been introduced in the Tohoku region to reduce travel times and increase operational efficiency. These trains incorporate lightweight, stainless steel bodies that offer energy consumption and maintenance savings. The Series 701 has a maximum speed of 110 km/h and is operated as a two-car or three-car set. The latest variant is the two-car 701–5000, 10 sets of which have provided local services on the Tazawako Line (Akita Shinkansen) since March 1997. Series E127 emu cars have been introduced in DC electric sections in the Nagano district in Niigata prefecture.

In April 1999 JR East relaunched overnight services between Sapporo and Ueno using 12-car

Series 215 bilevel 10-car suburban emu at Odawara (Wolfram Veith)   0023015

Series 209 'disposable' emu   0001710

Principal JR East diesel railcars

Class	Cars per unit	Motor cars per unit	Motored axles/car	Power/ motor kW	Speed km/h	Weight tonnes	Cars in service	First built	Mechanical	Builders Engine	Transmission
28	1	1	1	135	95	34.1	11	1961	F/NT	DMH17H D/S/NT	TC2A S/NC
58	1	1	2	135 × 2	95	39.4	32	1961	F/NT	DMH17H D/S/NT	TC2A S/NC
30	1	1	1	135	95	32.4	3	1962	F/NT/N	DMH17H D/S/NC	TC2A S/NC
52	1	1	2	135 × 2	95	36.6	24	1962	F/NT/N	DMH17H D/S/NC	TZ2A S/NC
40	1	1	1	165	95	37.3	110	1977	F/NT	DMF15HSA D/S/NT	DW10 S/NC
47	1	1	1	165	95	35.9	28	1977	F/NT	DMF15HSA D/S/NT	TC2A S/NC
48	1	1	1	165	95	36.2	71	1979	F/NT	DMF15HSA D/S/NT	TC2A S/NC
100/ 101	1	1	1	245	100	24.9	64	1990	F/NT	DMF11HZ DMF14HZT	DW14B NC
110	1	1	1	313	100	29.4/ 29.9	89	1990	F/NT	DMF13HZA DMF14HZA	DW14A Voith
111	1	1	1	313	100	28.9/ 29.4	47	1991	F/NT	DMF13HZA DMF14HZA	DW14A-B NC
112	1	1	1	313	100	28.4/ 28.9	47	1991	F/NT	DMF13HZA DMF14HZA	DW14A-B NC

Abbreviations: D: Daihatsu Motor; F: Fuji Heavy Industries Ltd; Fe: Fuji Electric; H: Hitachi; K: Kawasaki Heavy Industries; Kn: Kinki Sharyo; Me: Mitsubishi Electric; Mh: Mitsubishi Heavy Industries; N: Nippon Sharyo Seizo; NC: Niigata Converter; NT: Niigata Engineering; S: Shinko Engineering; Se: Shinko Electric; T: Toshiba; To: Toyo Denki Seizo

double-deck emus. The trains are operated under the 'Cassiopeia' brand-name.

Shinkansen

JR East's Shinkansen started operations in 1982 and currently covers a network extending in five directions: Tohoku; Yamagata; Akita; Joetsu; and Nagano. The number of trains currently operating each day is 304, three times the figure at the start

of operations. Some 250,000 passengers use these services daily.

Tohuku Shinkansen (Tokyo-Morioka – 535.3 km)

The Tohoku Shinkansen commenced operations in June 1982 between Omiya and Morioka. Subsequently the section between Omiya and Ueno opened in March 1985 and that between Ueno and Tokyo in June 1991. From a figure of 94 daily services to and from Tokyo in 1987, JR East in December 2001 was running 153 trains each day. Initially trains ran at a top speed of 210 km/h but this was increased to 275 km/h in March 1997 with the introduction of new rolling stock. As a consequence, travel times between cities served has been reduced dramatically to 2 hours 21 minutes between Tokyo and Morioka (a reduction of 48 minutes compared with the 1987 journey time). In December 2002, the section between Morioka and Hachinohe (96.6 km) was scheduled to open, reducing the fastest journey between Tokyo and Hachinohe by 40 minutes to 2 hours 50 minutes. The number of passengers commuting to work or school by Shinkansen is also growing rapidly within 100 km of Tokyo as the commuting area expands and capacity during morning and evening peaks is being enhanced by increasing the number of trains, introducing double-deck trains that provide seating for 1,634 passengers and other measures.

Joetsu Shinkansen (Tokyo–Niigata – 333.9 km)
The Joetsu Shinkansen commenced operations in November 1982, six months after the commissioning of the Tohoku Shinkansen, and was completed between Tokyo and Niigata in June 1991. One feature of its operations is a large number of passengers during winter months travelling to and from a ski resort near Echigo-Yuzawa station, creating seasonal variations in demand. The number of daily trains has increased from 65 in 1987 to 95 in 2002. Maximum speed permitted in 275 km/h and the fastest trains connect Tokyo and Niigata in 1 hour 37 minutes. As with the Tohoku Shinkansen, the number of commuters using Joetsu Shinkansen services is increasing rapidly within 100 km of Tokyo and capacity is being increased by the deployment of double-deck eight- and 12-car trainsets, the latter providing seats for 1,229 passengers.

Yamagata and Akita Shinkansen (Tokyo–Yamagata – 359.9 km; Tokyo–Akita – 662.6 km)
Yamagata services between Tokyo and Yamagata were inaugurated in July 1992 and extended from Yamagata to Shinjo in December 1999. Thirty trains are operated daily between Tokyo and Yamagata, 16 of these extended to Shinjo. Akita Shinkansen services were inaugurated in March 1997. A total of 28 daily services operate between Tokyo and Akita, with two additional trains running between Sendai and Akita.

Yamagata Shinkansen trains between Tokyo and Fukushima and Akita Shinkansen services between Tokyo and Morioka run over the Tohoku Shinkansen line and are coupled to that route's trains. Through operation over Shinkansen and converted conventional lines eliminates the need for passengers to change trains at intermediate stations. The maximum speed on converted lines, which necessitated a large-scale gauge conversion programme, was raised to 130 km/h, greatly reducing travelling times: the fastest journey time between Tokyo and Yamagata is now 2 hours 29 minutes (a reduction of 53 minutes) and that between Tokyo and Akita 3 hours 49 minutes (a reduction of 77 minutes). This through operation does not add to train density between Tokyo and Omiya, where each Shinkansen competes for capacity, or to demand for the use of platforms at Tokyo station.

Nagano Shinkansen (Tokyo–Nagano – 222.4 km)
Nagano Shinkansen services were inaugurated in October 1997, just before the Nagano Olympics. This was the first new 'full standard' Shinkansen line in Japan for 15 years. Initially, 24 return services were provided between Tokyo and Nagano and four between Tokyo and Karuizawa. This has since been amended to provide 27 return services on the Tokyo–Nagano route and one between the capital and Karuizawa.

Features of the Nagano Shinkansen include a continuous gradient with a maximum of 3 per cent between Annaka–Haruna and Karuizawa and an alignment between Tokyo and Nagano for which 54 per cent is in tunnel. Maximum line speed is 260 km/h and the fastest Tokyo–Nagano journey time is 1 hour 19 minutes. Future plans foresee extension of the line to Kanazawa, in the Hokuriku area.

Series 215 intercity emu forming 'Super View Odoriko No 1' service from Tokyo to Matsumoto (Wolfram Veith)   0023016

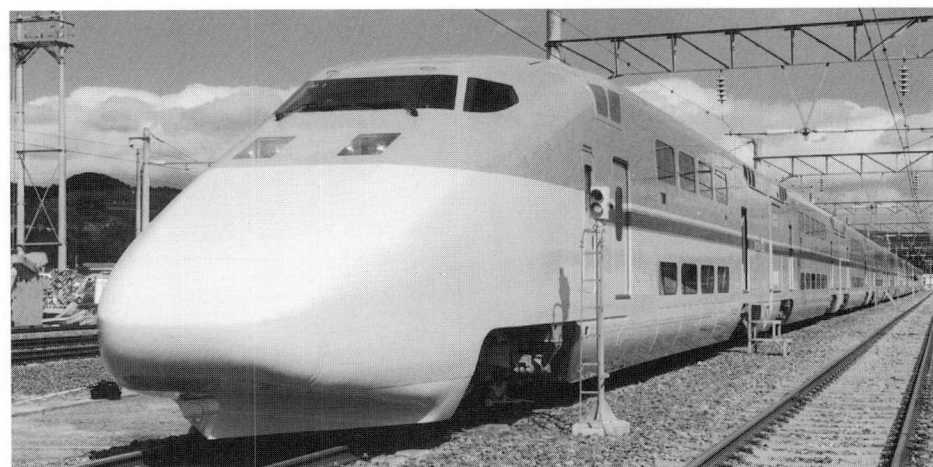

JR East Series E1 double-deck Shinkansen trainset   0518977

Series E2 Shinkansen trainset   0500749

Principal JR East diesel locomotives

Class	Wheel arrangement	Power kW	Speed km/h	Weight tonnes	No in service	First built	Mechanical	Builders Engine	Transmission
DD51	B-2-B	2,200	95	84	6	1966	H/Me/K	2 × 745 kW DML61Z	H 2 × DWZA
DD14	B-B	1,000	70	58	12	1966	K	2 × 373 kW DMF31SB-R	H 2 × DS1.2/1.35
DD15	B-B	1,000	70	55	4	1961	N	2 × 373 kW DMF31SB	H 2 × DS1.2/1.35
DE10	AAA-B	1,350	85	65	46	1969	K/N/H	DML61ZB	H DW6
DE11	AAA-B	1,350	85	70	9	1970	K/N	DML61ZA, B	H DW6
DE15	AAA-B	1,350	85	65	27	1970	N	DML61ZB	H DW6

Traction and rolling stock

As of April 2002, JR East owned two steam, 109 electric and 113 diesel locomotives, 11,699 Shinkansen and emu cars, 546 dmu cars, and 324 hauled passenger coaches.

In parallel with the increase in transport volume, the number of Shinkansen cars has also been increased. With the planned extension of the Tohoku Shinkansen to Hachinohe in December 2002, the E2 trainset fleet is being expanded. The existing 15 eight-car Series E2 trainsets are being increased to 10-car sets and 14 additional units will be progressively added to the fleet. By 2005, 120 Series 200 emu trainsets will be refurbished and 240 Series 200 emu cars will be renewed.

JR East has converted its maintenance works at Niitsu, Niigata prefecture, into a full-scale manufacturing plant for the production of a proportion of its own rolling stock. Production began in 1993, and the annual output between 1994 and 2001 was 1,319 cars. Vehicles produced comprise Series 209, E217 and E231 cars.

Signalling and telecommunications

The principal signalling systems comprise:
ATS-P (Automatic Train Stop-Pattern): 1,422 km
ATC (Automatic Train Control): 1,022 km
CTC (Centralised Traffic Control): 6,206 km
ATOS (Autonomous Decentralized Transport Operation Control System): 570 km
Fibre optic cables: 2,906 km

ATS-P is an automatic brake control and signalling system which is being installed on busy sections in the Tokyo metropolitan area. Its computerised system checks intervals between trains and train speed using data sent from a signal to the train.

ATOS employs the latest computer and information technology and is also used on busy sections of the Tokyo metropolitan area. It provides the control centre with detailed information on train position and delay status and incorporates functions automatically to control train routes according to the schedule database and their routeing within stations. This enhances provision of passenger information in event of service disruption. In addition, safety during track maintenance is enhanced by the ability of staff to use a computer terminal to prevent trains entering a work site.

Fibreoptic cables have been laid in the Tokyo metropolitan area to provide a self-supporting communications network which will ensure the maintenance of safe and reliable train operations and also act as a network for various kinds of information.

Track

Rail type and weight: 50 kg N rail; 60 kg N rail
Sleepers
PC (concrete) sleepers:
Shinkansen: 2,400 × 254.6 × 330 mm
Conventional lines: 2,000 × 174 × 240 mm
Wood sleepers:
Conventional lines: 2,100 × 140 × 200 mm
Fastenings
Shinkansen: direct fastening 8-type for slab track; 102-type and Pandrol-type for PC sleeper ballasted track
Conventional lines: 5-type; Pandrol-type
Min curve radius: Shinkansen: 4,000 m
Min curvature radius on trunk lines: 300–800 m, depending on train speed and tonnage carried, as well as on the number of sleepers used
Max gradient: Tohoku and Joetsu Shinkansen 3.5%, Yamagata and Akita Shinkansen 3.8%; conventional lines 3.5%
Max permissible axleload: 17 tonnes

Sleeper spacing per km
1st grade line, PC (prestressed concrete): 1,760; wood: 1,920
2nd grade line, PC: 1,560; wood: 1,640
3rd grade line, PC: 1,560; wood: 1,560
4th grade line, PC: 1,480; wood: 1,480
On sharp curves and sharp gradient sections, the number of wood sleepers shown is increased by 2/25 m.
Shinkansen lines: 1,720

Series E3 train for the Akita Mini-Shinkansen coupled to an older Series 200 unit at Omiya (Wolfram Veith)   0023017

Principal JR East electric locomotives

Class	Wheel arrangement	Line voltage	Output kW	Speed km/h	Weight tonnes	No in service	First built	Builders
ED75	B-B	20 kV	1,900	100	67.2	12	1968	Me/H/T
EF64	B-B-B	1.5 kV	2,550	115	96	14	1964	T/K/To
EF65	B-B-B	1.5 kV	2,550	115	96	19	1965	T/To/N
EF81	B-B-B	1.5 kV 20 kV	2,550 2,370	115	100.8	61	1968	H/Me

Principal JR East electric railcars or multiple-units

Class	Cars per unit	Motor cars per unit	Motored axles/car	Output/motor kW	Speed km/h	Cars in service	First built	Builders Mechanical	Electrical
1.5 kV DC									
103	10	6	4	110	100	935	1964	N/T/H/K/Kn	H/T/Me/Fe/To
107	2	1	–	–	100	54	1988		
201	10	6	4	150	100	781	1979	N/T/H/K/Kn	H/T/Me/Fe/To
203	10	6	4	150	100	170	1982	N/T/H/K/Kn	H/T/Me/Fe/To
205	10	6	4	150	100	1,413	1984	N/T/H/K/Kn	H/T/Me/Fe/To
207	10	6	4	150	100	10	1986		
209	10	4	4	95	110	1,028	1993	R/T	M/T/To
113	4	2	4	120	100	766	1963	N/T/H/K/Kn	H/T/Me/Te/To
115	4	2	4	120	100	718	1966	N/T/H/K/Kn	H/T/Me/Fe/To
211	10	4	4	120	110	575	1985	N/T/H/K/Kn	H/T/Me/Fe/To
215	10	4	4	120	120	40	1992	N/N	To/Se
165	3	2	4	120	110	55	1962	N/T/H/K/N	H/T/Me/Fe/To
167	4	2	4	120	110	35	1965		
169	3	2	4	120	110	15	1968		
251	10	6	4	120	120	40	1990	R/S	To/Se/M
253	3	2	4	120	130	99	1991	T/S	H/M/Se/To
301	10	8	4	110	100	50	1966		
E127	2	1	4	120	110	50	1994	K	To/Me
E217	11	4	4	95	120	745	1994	K/Tk	Me/To
255	9	4	4	95	130	45	1993	Kn/Tk	T/To
E351	8	4	4	150	130	60	1993	H/N	H/To
E231	10	4	4	95	120	953	2000	NI/K/Tk	Me/H/T/Fe
Dual-voltage 1.5 kV/20 kV									
417	3	2	4	120	100	15	1978		
403	4	2	4	120	100	40	1966		
415	4	2	4	120	100	303	1971	N/K/Kn	H/T/Me/Te/To
455	3	2	4	120	110	105	1962	N/T/H/K/Kn	H/T/Me/Fe/To
183	4	2	4	120	120	195	1972	N/T/K/Kn	H/T/Me/Fe/To
185	4	2	4	120	110	227	1980	N/T/K/H/Kn	H/T/Me/Fe/To
189	9	3	4	120	120	118	1974		
485	4	2	4	120	120	252	1968	N/T/H/K	H/T/Me/Fe/To
489	9	6	4	120	120	2	1971		
583	4	2	4	120	120	24	1968	N/T/H/K/Kn	
651	17	4	4	120	130	99	1988	K	
E501	10	4	4	120	120	60	1994	Tk	Si/H/T
20 kV AC									
717	3	2	–	–	110	30	1985		
719	3	1	–	–	110	108	1990		
200*	12	12	4	230	240	384	1980	N/T/H/K	H/T/Me/Fe/To/Se
400*	6	6	4	210	240	84	1990	T/H/K	H/T/M/Fe To
701	2	1	4	125	110	266	1992	K	Me/T/Fe
25 kV AC									
E1*	12	6	4	410	240	72	1993	H/K	H/T/Me/To
E2*	8	6	4	300	275	232	1995	H/K/N	H/Me/T/To
E3*	5	4	4	300	275	116	1995	K/TK	Me/To/H/T
E4*	8	4	4	420	240	192	1997	H/K	Me/To/H/T

* Shinkansen

Abbreviations: D: Daihatsu Motor; F: Fuji Heavy Industries Ltd; Fe: Fuji Electric; H: Hitachi; K: Kawasaki Heavy Industries; Kn: Kinki Sharyo; Me: Mitsubishi Electric; Mh: Mitsubishi Heavy Industries; N: Nippon Sharyo Seizo; NC: Niigata Converter; NI: JR East Niitsu Sharyo Seisakusho; NT: Niigata Engineering; S: Shinko Engineering; Se: Shinko Electric; T: Toshiba; To: Toyo Denki Seizo; TK: Tokyu Sharyo; Si: Siemens
Note: JR East leases Series 400 and Series E3 trains from Yamagata JR through Superexpress Holding Co and Akita Shinkansen Trains Holding Co respectively.

Central Japan Railway Co (JR Central)

Head office: JR Central Towers, 1-1-4 Meieki, Nakamura-ku, Nagoya, Aichi 450-0003
Tel: (+81 52) 564 23 17 Fax: (+81 52) 587 13 00
Tokyo Head office, International Department: Yaesu Center Building, 1-6-6 Yaesu, Chuo-ku, Tokyo 103-8288
Tel: (+81 3) 32 74 97 27 Fax: (+81 3) 52 55 67 80
Web: jr-central.co.jp

Key personnel

Chairman: Yoshiyuki Kasai
President: Masayuki Matsumoto
Executive Vice-Presidents: Toshiaki Araya,
  Tsuneo Hara, Masataka Ishizuka, Yoshiomi Yamada
Senior Executive Directors
  Secretariat and Administration: Akihiro Amaya
  Director General, Technology: Akio Seki
Executive Directors
  Conventional Lines Operations: Koshi Akutsu
  Shinkansen Operations: Takeshi Tategami
  Finance: Takao Innami
  Business Promotion: Toyonori Noda
Directors
  Corporate Planning: Mitsuru Nakamura
  Personnel: Kouei Tsuge
  Marketing: Masayuki Kono
  Transportation Safety: Naotoshi Yoshikawa
  Administration: Shin Kaneko
  Construction: Haruo Goto, Fujio Cho
Corporate Executive Officers
  Maglev System Development: Akira Nakagawa
  Technology Planning: Toshiaki Doi
Corporate Officers
  Technology Research and Development:
    Tsutomu Morimura, Masaki Seki
  Audit: Takashi Ono
  Washington Office: Osamu Nakayama
  Business Promotion: Junichi Hirasawa
  Shinkansen Operations: Kazumasa Ishizu
  Employee Training: Kazuhiro Yoshikawa
  Legal Affairs: Teruo Kachi
  Maglev System Development:
    Hiroshi Nakashima
  Shizuoka Branch Office: Kaoru Umemoto
  Kansai Branch Office: Akira Sugimoto

Gauge: 1,067 mm; 1,435 mm (Tokaido Shinkansen)
Route length: 1,425 km; 553 km
Electrification: 944 km at 1.5 kV DC (1,067 mm); 553 km at 25 kV 60 Hz (1,435 mm)

Organisation

Central Japan Railway Company (JR Central, also known as JR Tokai) commenced operations in April 1987 following the privatisation and break-up of Japanese National Railways (JNR). The core of JR Central's operations is the Tokaido Shinkansen, linking Tokyo, Nagoya and Osaka. The company also operates a network of conventional lines in central Japan, in the Nagoya and Shizuoka area. JR Central and its consolidated subsidiaries are strengthening affiliated businesses by making full use of the company's stations and trains. JR Central Towers, the core project of the group's affiliated business diversification plan, houses rental office units, a department store, hotel and other facilities.

In June 2002 JR Central reorganised its technological developments and established a new General Technology Division that gives the company an integrated internal technological structure. In July 2003 the company opened its own research and development facility in Komaki, Aichi Prefecture, further to reinforce its technological development organisation.

On 31 March 2004 JR Central employed 20,187 staff.

Finance

The Japanese economy, although beginning to show some signs of brightening, continue to provide an unclear situation in 2003-04. Given these economic conditions, JR Central prioritised the fundamentals of railway operation: safe and reliable transportation. The company strove to enhance its competitive edge and improve service, while also continuing to improve the operational skills of its personnel and upgrade various equipment and facilities.

Driving cab of a Series 700 trainset   0113581

Series 313 suburban emu with VVVF control   0113592

JR Central Series 300 Shinkansen trainset forming a southbound service on a concrete-stabilised embankment near Fujikawa (Wolfram Veith)   **NEW**/0585084

Various medium- and long-term projects were in progress during the year. For example, JR Central dramatically improved its Tokaido Shinkansen service in October 2004 by opening the new Shinagawa station, a project that the company has been engaged in since its launch, and by drastically revising the timetable focusing on the replacement of the Shinkansen fleet with new rolling stock capable of 270km/h operation.

In addition, with the aim of achieving further improvements to the Tokaido Shinkansen, above-ground facilities have been improved and conversion work has been undertaken in preparation for the start of a new ATC system at the end of fiscal 2006. JR Central has also designed and manufactured a pre-series trainset of the next generation N700, successor to the Series 700 rolling stock.

During 2003-04 JR Central continued to conduct geological, topographical and other surveys covering the entire route of the planned Chuo Shinkansen. Progress was also made in developing maglev technologies, through conducting running tests aimed at confirming long-term reliability and durability and reducing costs. As a result, consolidated operating revenues totalled ¥1,384 billion, up 1.5 per cent from the previous period. Factors contributing to this rise included an increase in transportation volume on the preceding term. Due to decreasing interest costs resulting from the reduction of total long-term debt, consolidated net income increased by 47.3% to ¥72 billion, and net income per share was ¥32,173.

Consolidated total long-term debt at the fiscal year-end in March 2004 totalled ¥4,050 billion, down ¥196 billion from a year earlier.

Finance (million yen)

(million yen)	2001–02	2002–03	2003–04
Operating revenues (non-consolidated)			
Railway	1,120,218	1,100,920	1,118,660
Other	8,312	8,425	9,124
Total	1,128,530	1,109,345	1,127,784
Operating expenses (non-consolidated)			
Railway	765,830	778,907	795,111
Other	5,791	5,712	5,353
Total	771,621	784,619	800,464

Traffic (million)	2001–02	2002–03	2003–04
Passenger journeys	498.4	492.4	495.8
Passenger-km	49,533	48,468	49,577

Passenger operations
Tokaido Shinkansen
As the principal transport artery linking Tokyo, Nagoya and Osaka, Japan's major metropolitan hubs, the Tokaido Shinkansen has seen its passenger volume increase in step with Japan's GDP growth. In 2004-05 the Tokaido Shinkansen offered up to 12 train departures per hour from Tokyo and conveys an average of 361,000 passengers per day. From its founding in April 1987 until March 1992, JR Central responded to growing demand from business travellers and long-distance commuters by increasing the frequency of Hikari and Kodama train departures respectively. In March 1992, when the timetable was revised, JR Central introduced Series 300 rolling stock for Nozomi trains with operating speeds up to 270 km/h. This cut journey times drastically. In March 1993 hourly frequency was introduced for fully fledged Nozomi services. In March 1999 the new Series 700 rolling stock (jointly developed with JR West) was introduced, providing increased comfort and noise protection. In October 2000 the number of Nozomi trains serving Shin-Yokohama station was doubled. In October 2001, more Nozomi trains were added to the fleet: one Nozomi departs every 30 minutes, with a total of 75 trains a day, an increase from 53 trains previously. In October 2003 the timetable was drastically revised in line with the opening of the new Shinagawa station.

Conventional lines network
JR Central operates a network of 12 conventional lines, which form a common network with the Tokaido Shinkansen. These lines have contributed substantially to the development of communities and the regional economy around Nagoya and Shizuoka. The annual number of passengers

Diesel locomotives

Class	Wheel arrangement	Power kW	Speed km/h	Weight tonnes	No in service	First built	Builders Engine	Transmission
DE10	AAA-B	1,350	85	65	6	1970	DML61ZB	H DW6
DE15	AAA-B	1,350	85	65	2	1978	DML61ZB	H DW6
DD51	B-2-B	2,200	95	76	2	1971	2 × DML61Z	H 2 × DW2A

Electric locomotives

Class	Wheel arrangement	Output kW	Speed km/h	Weight tonnes	No in service	First built
1.5 kV DC						
EF58	2C + 2C	1,900	120	113–115	2	1957
EF64	B-B-B	2,550	115	96	3	1964
EF65	B-B-B	2,550	115	96	2	1969
ED18	A1A-A1A	915	70	66	1	1954

Diesel railcars or multiple-units

Class	Cars per unit	Motor cars per unit	Motored axles/car	Power/motor kW	Speed km/h	No in service	First built	Bodies	Builders Engine	Transmission
28	2	2	1	180	95	2	1962		DMH17H	H TC2A or DF115A
58	2	2	2	180 × 2	95	3	1962		DMH17H	H TC2A or DF115A
58–5000	2	2	2	350	110	2	1965		DMF14HZB	H C-DW14A
65–3000	2	2	2	500	95	1	1970		DML30HSD	H DW4F
65–5000	2	2	2	500	110	1	1969		DML30HSD	H DW4F
11-0, –100, –200	1	1	2	330	95	35	1988		C-DMF14HZA	H C-DW15
11–300	1	1	2	350	95	6	1999		C-DMF14HZB	H C-DW15
40	1	1	1	350	95	14	1979		DMF14HZ	H DW10
47	2	2	1	220 350	95	5	1976		DMF15HSA DMF14HZB	H DW10 H DW10
48	2	2	1	350 220	95	40	1979		DMF14HZ BMF15HSA	H DW10 H DW10
85	3/4/5	3/4/5	2	350 × 2	120	80	1988	N/F/Nt	DMF14HZ	H DW14A
75	2	2	2	350 × 2	120	40	1993	N	DMF14HZB	H DW14A

Abbreviations: N: Nippon Sharyo Seizo Ltd; F: Fuji Juko; Nt: Niigata Tekko.

Electric railcars or multiple-units

Class	Cars per unit	Motor cars per unit	Motored axles/car	Output/motor kW	Speed km/h	Cars in service	First built
1.5 kV DC							
103	3, 7	2, 4	4	100	100	10	1965
113	3, 4, 6	2, 4	4	120	100	218	1963
115	3	2	4	120	100	63	1966
117	4	2	4	120	110	72	1982
119	1, 2	1	4	110	100	57	1982
123	1	1	4	120	100	7	1988
165	3	2, 4	4	120	110	12	1963
211-0	4	2	4	120	110	8	1986
211–5000	3, 4	2	4	120	110	242	1988
213	2	1	4	120	110	28	1989
285	7	2	4	220	130	14	1998
311	4	2	4	120	120	60	1989
311	2, 3, 4	1, 1.5, 2	2, 4	185	120	77	1999
313	2, 3, 4	1, 2, 2	4, 2	185	120	187	1999
371	7	5	4	120	120	7	1990
373	3	1	4	185	120	42	1995
381	4, 6	2, 4	4	120	120	16	1973
383	2, 4, 6	1, 2, 3	4	155	130	76	1994

Shinkansen trainsets (1,435 mm gauge)

Class	Cars per unit	Motor cars per unit	Motored axles/car	Output/motor kW	Speed km/h	Weight tonnes	Cars in service	First built
25 kV 60 Hz								
300	16	10	4	300	270	44.4	976	1990
700	16	12	4	275	285	44.3	768	1997

travelling on this network has increased by more than 20 per cent since JR Central was established in 1987. This reflects the success of various measures, including the introduction of faster and more modern rolling stock, increasing service frequencies and installing air-conditioning on all trains. The introduction of 'Wide View' rolling stock on limited express trains has also been popular. The company has synchronised Tokaido Shinkansen and conventional line limited express train timetables to create an integrated network of Shinkansen and Wide View trains. Commuters have benefited from the increased frequency of local trains during peak-demand morning and evening periods, as well as from the introduction of expanded rapid-train services that reduce travel times. Moreover, train intervals have been adjusted to provide timetables that better serve passenger needs.

In December 1999 JR Central conducted a comprehensive timetable revision for conventional lines to strengthen its competitiveness against other modes, improve profitability and enhance efficiency. With the resulting timetable revision, the company increased the frequency of rapid trains during peak hours and introduced the 'Center Liner' reserved seat-only rapid service on the Chuo Line. In October 2001 the frequency of rapid trains during morning peak hours was increased and the schedule of the Home Liner all-reserved seat rapid service for homeward-bound evening commuters was improved. In July 2002 JR Central introduced special reduced fare short-distance express train tickets (non-reserved seat-only) valid within 30 km to facilitate usage of express trains. In October 2003 the conventional line train schedule was changed in line with the Tokaido Shinkansen timetable revision to improve network integration.

New lines
The Chuo Shinkansen Maglev project
The Chuo Shinkansen is a national project planned in response to the perceived need for a service

complementing that now offered by the Tokaido Shinkansen. It is one of the basic projected routes and is identified as a 'Shinkansen line that merits construction' according to Article 4 of the Nationwide Shinkansen Railway Development Law. The Chuo Shinkansen will provide an alternate route connecting Tokyo and Osaka via Kofu, Nagoya and Nara.

In its capacity as the future management entity of the Chuo Shinkansen, JR Central considers the Superconducting Maglev suitable for adoption when the Chuo Shinkansen is realised because of its advanced characteristics and high-speed performance. With this in mind, the company has pursued the practical application of Superconducting Maglev technology since April 1997 by conducting trial runs on the Yamanashi Maglev Test Line in co-operation with the Railway Technical Research Institute.

On 14 April 1999 a trainset reached a world speed record (for manned maglev vehicles) of 552 km/h, enabling JR Central to verify the operability of maglev technologies at an intended maximum operating speed of 500 km/h. On 16 November 1999 passing tests were conducted using two trainsets, which passed at a closing speed of 1,003 km/h.

In March 2000 the practicability of a high-speed mass transportation system using Superconducting Maglev technology was verified by a Ministry of Transport (now Ministry of Land, Infrastructure and Transport) committee. JR Central is to continue to verify further the system's reliability, durability and cost-effectiveness, as well as improving its aerodynamic performance.

This continuing development has included the introduction in July 2002 of a new test vehicle to the Yamanashi Maglev Test Line, followed by new facilities such as a highly efficient power transformer, a new type of guideway as well as ground coils later in Autumn 2002. In April 2003 the committee conducted a mid-term evaluation and gave a high assessment that fundamental technology developments are making steady progress in achieving the initial objectives of research and development and in putting the Superconducting Maglev to practical use by the end of FY 2005. On 2 December 2003 the superconducting Maglev achieved a manned world-record speed of 581 km/h, breaking its 1999 record.

Improvements to existing lines

In September 2001 JR Central introduced a new multiple inspection train, claimed to be the first in the world to measure track, catenary, signalling and telecommunications facilities while travelling at speeds of up to 270 km/h. The train is intended for use on the Tokaido and Sanyo Shinkansen lines.

Traction and rolling stock

As at 31 March 2001 Shinkansen rolling stock (1,435 mm gauge) comprised 122 16-car trainsets of three types (see table). In March 1999, timetables were revised and Series 700 trainsets commenced commercial Nozomi services. Developed jointly by JR Central and JR West as a next-generation Shinkansen design for both railways, the 16-car 270 km/h unit features an innovative aerodynamic nose shape, aluminium alloy bodyshells and VVVF control.

On 1,067 mm gauge JR Central operated 1,119 emu cars and 229 dmu cars. Also included in the fleet were eight electric and 10 diesel locomotives and 26 hauled passenger cars.

Signalling and telecommunications

The Tokaido Shinkansen's ATC (Automatic Train Control) system features innovative two-frequency equipment which permitted the introduction of 270 km/h operation in early 1992. The Tokaido Shinkansen facilities support one of the safest and most reliable high-speed railway services in the world. JR Central is currently developing a new Automatic Train Control (ATC) system to replace the existing equipment. The new system will control train operating speeds using a design that will allow for higher levels of safety, reliability and maintainability, as well as smoother train control.

JR Central has made steady investments in centralised traffic control for its conventional

Maglev vehicle on the Yamanashi test track (Wolfram Veith)   **NEW**/0585085

Class 285 'Sunrise Express' emus operated jointly by JR Central and JR West at Tokyo station (Wolfram Veith)   **NEW**/0585086

Two JR Central Class 112 emus forming a Nagoya-bound service at Bentenjima (Wolfram Veith)   **NEW**/0585087

lines, and by the end of FY2000–01 the system was controlling 96 per cent of the 1,067 mm gauge network. A new automatic train stop system known as ATS-ST is now standardised throughout the network. A modification to the system activates an emergency brake application if a driver ignores a signal at danger.

Track
(1,067 mm gauge lines)
Rail: 30.1 to 60.8 kg/m (on main routes 60.8 or 50.47 kg/m)
Sleepers: Prestressed concrete or timber 2,000 × 240× 174 mm, spaced 1,480 to 1,920/km according to grade of route
Fastenings: elastic or rigid

Min curve radius: 300 to 800 m, according to maximum speed
Max gradient: 1 to 3.5%, according to maximum speed
Max axleloading: 16 tonnes

Track
(Tokaido Shinkansen)
Rail type: 60.8 kg/m
Sleepers: Prestressed concrete 2,400×330×255 mm, or 2,400 × 300 × 219 mm, spaced 1,720/km
Fastenings: Elastic
Min curve radius: 2,500 m
Max gradient: 2%
Max axleloading: 18 tonnes

UPDATED

West Japan Railway Co (JR West)

4-24, Shibata 2-chome, Kita-ku, Osaka, 530-8341
Tel: (+81 6) 375 89 81   Fax: (+81 6) 375 89 19
e-mail: ia@westjr.co.jp
Web: www.westjr.co.jp

Key personnel

Chairman: Shojiro Nan-ya
President: Takeshi Kakiuchi
Senior Managing Directors:
  Masayuki Sakata, Kenzo Tokuoka
Directors: Masataka Ide, Yasutada Ikeda,
  Kazuaki Maruo, Akio Nomura, Yoshio Tateishi

Gauge: 1,067 mm; 1,435 mm (Sanyo Shinkansen)
Route length: 4,388 km; 644 km
Electrification: 3,328 km at 1.5 kV DC (1,067 mm);
644 km (1,435 mm) and 322 km (1,067 mm) at
20 kV AC 60 Hz

Organisation

The railway runs passenger transport and related activities in the Hokuriku region and western Honshu, an area of 43 million inhabitants – 34 per cent of Japan's population. The network totals 50 lines, with 289 Shinkansen and 7,988 conventional trains operated daily in 2001. The company has steadily diversified into activities such as the travel trade and affiliated business, including restaurants and retail shops.

A total of 31.72 per cent of the shareholding in the company is owned by the state-owned Japan Railway Construction Public Corporation (qv), which was established in 1987 upon the reform of the former Japanese National Railways and partial privatisation of the network. Legislation introduced in 2001 created conditions which could lead to the eventual sell-off of this residual state shareholding.

Finance

Finance (JPY million)	2002–03	2003–04
Operating revenue	849,090	845,892
Operating expenses	745,796	740,416
Operating income	103,293	105,475
Net income (loss)	33,490	37,174

(Figures relate to JR West rail business only)

Traffic (million)	2002–03	2003–04
Passenger journeys		
(rail only)	1,772	1,789
Passenger-km	51,674	52,142

Passenger operations

Services on 1,067 mm gauge

JR West's conventional lines comprise the Urban Network of 14 commuter lines serving the metropolitan areas of Kyoto, Osaka and Kobe; major intercity and regional lines; and local lines. On the urban network, ridership fell in 2000–01 by 0.2 per cent to 1,431 million passenger journeys but volume increased by 0.1 per cent to 28,331 passenger-km. Passenger numbers on other conventional lines fell by 2 per cent to 400 million for 10,414 passenger-km, a decrease of 2.3 per cent.

Recent improvements to the urban network include: the introduction in March 2000 of two new services, JR Kobe (Kobe–Osaka) and JR Kyoto (Osaka–Kyoto), on the Biwako line between Kyoto and Nagahama, both using Series 223 Special Rapid 130 km/h emus; double-tracking between Hanazono and Nijo on the Sagano line; increased use of Rapid trains on the Nara line following provision of double-track between JR Fujinomori and Kyoto stations; the opening in March 2001 of Universal-city station to serve the Universal Studios Japan theme park.

On other conventional lines, the introduction of Series 683 emus has enabled JR West to increase the frequency and shorten the journey times of its Thunderbird interurban limited express services.

In 1998 JR West introduced a new luxury sleeping car train developed in collaboration with JR Central, the 'Sunrise Express'. This complements the existing Osaka–Sapporo 'Twilight Express' sleeping car service.

Local lines account for about 30 per cent of the total network but only 3 per cent of passenger-km.

Series 281 emu on 'Haruka' service at Nishioji connecting Kyoto and Osaka with Kansai International Airport (Wolfram Veith)   0099125

'Sunrise Express' bilevel 1,067 mm gauge emu for overnight service   0099128

Class 201 1.5 kV DC emu forming a local service at Kyoto (Marcel Vleugels)   0132664

JR West Series 700 'Rail Starâ' Shinkansen trainset forming a Tokyo–Hakata service at Hakata   0132665

Some lines have been transferred to third-sector companies, and JR West has established 27 regional operating units aimed at upgrading service and enhancing the profitability of the remaining lines on an individual basis. Operating economies introduced include staff undertaking a wider range of duties.

Shinkansen Services

In 2000–01, Sanyo Shinkansen services carried 58 million passengers, a decline of 0.2 per cent on the previous year, although volume increased by 1.3 per cent to 13,805 million passenger-km. Series 500 high-speed trainsets operate Nozomi services between Shin-Osaka and Hakata (Fukuoka) in 2 hours 17 minutes, while since March 2000, Series 700 trains have provided Hikari Rail Star services over this route. During the period 2000–01, these latter services achieved an average passenger load of 87 per cent of capacity.

Since October 2001 Nozomi services have additionally called at Shin-Kobe station to improve their competitiveness with airlines and three Hikari Rail Star services on the Shin-Osaka–Hiroshima route have been extended to Hakata.

New lines

JR West is a partner with Osaka city and prefecture in a third-sector venture to rebuild and introduce passenger services on the 20.4 km orbital Joto freight line running from Shin-Osaka around the eastern suburbs of the city.

The proposed Naniwasuji line would form a north-south route through the centre of Osaka from Shin-Osaka via Umeda to the JR Namba station, providing improved access to Kansai International Airport from northern Osaka and the Kyoto and Hyogo prefectures.

Traction and rolling stock

In 2003 the company operated on 1,067 mm gauge five steam, 47 electric and 61 diesel locomotives, 4,661 emu cars, 534 dmu cars, 253 hauled passenger coaches and 258 wagons. Shinkansen rolling stock comprised 804 emu cars.

JR West was able to deploy nine Series 500 sets by 1999. The first new-generation Series 700 Shinkansen trainsets, developed jointly with JR Central, commenced service in March 1999. Branded 'Rail Star', each set comprises eight cars, of which six are motored. Total seating capacity is for 816 passengers.

Signalling and telecommunications

ATS-P (Automatic Train Stop-Pattern) train protection equipment has been installed on the Hanwa line, including the Kansai Airport branch, the Osaka loop line, and the Tozai line. Future planned installations will cover the Osaka–Maibara section of the Kyoto line and Osaka–Aboshi on the Kobe line.

Track
(1,067 mm gauge lines)

Rail: 30 to 60.8 kg/m
Sleepers: Wood 140 × 200 × 2,100 mm; and concrete 174 × 156–240 × 2,000 mm
Sleeper spacing: Wood 1,480–1,920/km depending on class of route; concrete 1,480–1,760/km. On Class 2, 3 and 4 track with wooden sleepers, increased by 80 through curves of 600 m radius or less
Fastenings: Wooden sleepers, spike or spring clip with plate and pad; concrete sleepers, spring clip with pad
Min curve radius: 800, 600, 400 and 300 m for maximum speeds respectively of 110 km/h or over, 90–110 km/h, 70–90 km/h and 70 km/h
Max permissible axleload: 19 tonnes

JR West diesel locomotives

Class	Wheel arrangement	Output kW	Speed km/h	Weight tonnes	No in service	First built	Mechanical	Builders Engine	Transmission
DD15	B-B	745	70	55	8	1961	N D	DMF31SB S	H DS1.1/1.35
DD16	B-B	600	75	48	1	1971	K S/D	DML61Z H/K	H DW2A
DD51	B-2-B	1,640	95	84	13	1966	H/Mh/K S/D	DML61Z H/K	H DW2A
DE10	AAA-B	1,000	85	65	28	1969	K/N/H S/D	DML61ZB H/K	H DW6
DE15	AAA-B	1,000	85	65	11	1970	N/K S/D	DML61DZB H/K	H DW6

JR West diesel railcars

Class	Cars per unit	Motor cars per unit	Motored axles/car	Output/ motor kW	Speed km/h	Weight tonnes	Cars in service	First built	Mechanical	Builders Engine	Transmission
181	3	3	2	370	120	44.6	40	1968	F/NT/N D/S/NC	DML30HSE	H DW4E S/NC
28	2	2	1	135	95	34.3	29	1961	F/NT/N D/S/NC	DMH17H	H TC2A/DF115A S/NC
23	1	1	1	135	95	34.2	1	1966	F/NT/N D/S/NC	DMH17H	H TC2A/DF115A S/NC
33	1	1	1	185	95	34	2	1988	GW D/S/NC	DMF13HS	H DF115A NT/S
35	2	2	1	135	95	32	1	1961	F/NT/N D/S/NC	DMH17H	H TC2A/DF115A S/NC
37	2	2	1	155	95	31.6	2	1982	NT D/S/NC	DMF13S	H TC2A/DF115A S/NC
40	1	1	1	160	95	36.4	63	1979	F/NT D/S/NC	DMF15HSA	H DW10
47	2	2	1	160	95	35.5	184	1976	F/NT D/S/NC	DMF15HSA	H DW10 H S/NC
48	2	2	1	160	95	35.9	5	1971	F/NT D/S/NC	DMF15HSA	H DW10 S/NC
52	1	1	2	135 × 2	95	36	3	1962	F/NT/N D/S/NC	DMH17H	H TC2A/DF115A S/NC
53	1	1	2	135 × 2	95	39.7	3	1966	F/NT/N D/S/NC	DMH17H	H TC2A/DF115A S/NC
58	2	2	2	135 × 2	95	39.4	31	1961	F/NT/N D/S/NC	DMH17H	H TC2A/DF115 S/NC
65	2	2	2	375	95	42.9	19	1969	F/NT/N D/S/NC	DML30HSD	H DW4F S/NC
120	1	1	1	225	95	26.7	89	1992	NT D/S/NC	SA6D125H-1	H TACN-22–1605 S/NC
121	1	1	–	–	100	35.9	9	2003	–	–	–
126	2	2	–	–	100	34.6	20	2000	–	–	–
187	2	2	–	–	120	43.4	26	2000	–	–	–

JR West electric locomotives

Class	Wheel arrangement	Output kW	Speed km/h	Weight tonnes	No in service	First built	Builders Mechanical	Electrical
1.5 kV DC								
EF15	1-C-C-1	1,900	75	102	1	1947	H	H/T/Mh/F/To
EF58	2-C-C-2	1,900	100	115	1	1946	T	H/T/Mh/F/To
EF59	2-C-C-2	1,350	90	106.6	1	1963	H	H/T/Mh/F/To
EF60	B-B-B	2,550	100	96	1	1962	To	H/T/Mh/F/To
EF64	B-B-B	2,550	100	96	2	1964	T/K/To	H/T/Mh/F/To
EF65	B-B-B	2,550	110	96	17	1969	K/To/T/N	H/T/Mh/F/To
EF66	B-B-B	3,900	110	100.8	10	1973	K/To	H/T/Mh/F/To
EF81*	B-B-B	2,550/2,370	110	100.8	14	1968	H/Mh	H/T/Mh/F/To

* Dual-voltage 1.5 kV/20 Kv

Sanyo Shinkansen electric multiple-unit cars

Class	Cars per unit	Motor cars per unit	Motored axles/car	Output/motor kW	Speed km/h	Cars in service	First built	Builders Mechanical	Electrical
25 kV 60 Hz									
100	4/6	4/6	4	230	119	256	1987	K/Ki/H/N/Tk	F/T/M/H/Ty
300	16	10	4	300	270	144	1992	K/Ki/N/Tk	T/Ty/N
500	16	16	4	300	300	144	1995	K	
700	8/16	8/16	4	–	285	312	1999	–	–

Abbreviations: D: Daihatsu; F: Fuji Heavy Industries; GW: Goto Workshop; H: Hitachi; K: Kawasaki Heavy Industries; Ki: Kinki Sharyo; Mh: Mitsubishi Heavy Industries; Me: Mitsubishi Electric; MW: Matto Workshop; N: Nippon Sharyo Seizo; NC: Niigata Converter; NT: Niigata Engineering; S: Shinko Engineering; T: Toshiba; Tk: Tokyu Haryo; To: Toyo Denki Seizo; Ty: Toyo Electric

For details of the latest updates to *Jane's World Railways* online and to discover the additional information available exclusively to online subscribers please visit
jwr.janes.com

Principal JR West electric railcars or multiple-units (1,067 mm gauge)

Class	Cars per unit	Motor cars per unit	Motored axles/car	Output/motor kW	Speed km/h	Cars in service	First built	Builders Mechanical	Electrical
1.5 kV DC									
103	7	4	4	110	100	734	1964	K/Ki/H/N/Tk	H/T/Me/F/Ty
105	2	1	4	110	100	121	1980	Ki/H/Tk	H/T/Me/F/Ty
113	4	2/4	4	120	110	400	1963	K/Ki/H/N/Tk	H/T/Me/F/Ty
115	4	2/4	4	120	110	462	1962	K/Ki/H/N/Tk	H/T/Me/F/Ty
117	6	2/4/6	4	120	115	116	1979	K/Ki/N/Tk	H/T/Me/F/Ty
125	1	1	–	–	120	8	2002	–	H
183	4	2	4	120	120	90	1990	K/Ki/H/N	H/T/Me/F/Ty
201	7	4	4	120	100	224	1981	K/Ki/H/N/Tk	H/T/Me/F/Ty
205	7	4	4	120	110	48	1986	Ki/N	H/T/Me/F/Ty
207	3/4/7	1/2/3	4	155	120	484	1991	K/Ki/H	H/T/Me/F/Ty
211	2	1	4	120	120	2	1988	Ki	H/T/Me/Ty
213	3	1	4	120	110	37	1986	K/Ki/H/N/Tk	H/T/Me/F/Ty
221	2/4/6/8	1/2/3	4	120	120	474	1988	K/Ki/H/N	H/Ts/Me/Ty
223 Ty	2/6	1/3	4	180	120	500	1994	K/Ki	H/Mh/Mh/Me/T/
281	5	2	4	180	130	63	1994	K/Ki	H/Mh/Me/T/Ty
381	7	4	4	120	120	189	1978	K/Ki/H/N	H/T/Me/F/Ty
283	3/6	1/2	4	220	130	18	1996	K/Ki/H	
1.5 kV DC/20 kV AC									
413	3	2	4	120	110	31	1975	MW	H/T/Me/F/Ty
415	3	2	4	120	100	33	1990	K/Ki/H/N/Tk	H/T/N/F/Ty
419	3	2	4	120	100	45	1975	MW	H/T/Me/F/Ty
457	3	2	4	120	110	80	1969	H	H/T/Me/F/Ty
485	10	6	4	120	130	120	1968	K/Ki/H/N/Tk	H/T/Me/F/Ty
489	9	6	4	120	130	43	1971		H/T/Me/F/Ty
583	10	6	4	120	120	60	1968	K/Ki/H/N	H/T/Me/F/Ty
681	9	3	4	190	160	84	1992	K/Ki/H	T/M/Ty
683	3/6	–	–	–	–	141	2000	–	

Abbreviations: D: Daihatsu; F: Fuji Heavy Industries; GW: Goto Workshop; H: Hitachi; K: Kawasaki Heavy Industries; Ki: Kinki Sharyo; Mh: Mitsubishi Heavy Industries; Me: Mitsubishi Electric; MW: Matto Workshop; N: Nippon Sharyo Seizo; NC: Niigata Converter; NT: Niigata Engineering; S: Shinko Engineering; T: Toshiba; Tk: Tokyu Haryo; To: Toyo Denki Seizo; Ty: Toyo Electric

Track (Shinkansen lines)
Rail: 60.8 kg/m
Sleepers: Concrete 255 × 172–300 × 2,400 mm
Sleeper spacing: 1,720/km in plain track
Fastenings: Spring clip with pad
Min curve radius (main line): 4,000 m
Max permissible axleload: 17 tonnes

UPDATED

One of JR West's few locomotive-hauled services is the 'Twilight Express' sleeping car train linking Osaka and Sapporo. It is seen at Shin-Osaka behind a Class EF81 electric locomotive (Marcel Vleugels)
0132666

Hokkaido Railway Co (JR Hokkaido)

1-1 Nishi 15-chome, Kita 11-jo, Chuo-ku, Sapporo 060-8644
Tel: (+81 11) 700 57 31   Fax: (+81 11) 700 57 19
Web: http://www.jrhokkaido.co.jp

Key personnel

Chairman: Shin-ichi Sakamoto
President: Akio Koike
Executive Managing Director, General Manager of Railway Operations: Hirohiko Kakinuma
Managing Director, Planning: Kazuhiro Saku
Managing Director, Finance: Niichi Muto

Gauge: 1,067 mm
Route length: 2,500 km
Electrification: 434 km at 20 kV 50 Hz AC

Organisation

The company is responsible for rail passenger transport and related activities in Hokkaido, serving 473 stations on the northernmost Japanese island, and also including operations through the Seikan Tunnel. It operated 1,297 trains daily in March 2003. Despite large-scale closure of loss-making rural lines by JNR before the formation of JR Hokkaido in 1987, some very unremunerative lines are still operated in this area of low population density. Other activities include a bus operation, hotels, retail, property and construction.

Finance

In 2002 consolidated operating revenues were ¥150,900 million against expenditure of ¥178,700 million, producing an operating loss of ¥27,800 million.

Passenger operations

In 2002–03 JR Hokkaido recorded 121 million passenger journeys.

Since it took over the Hokkaido system, the company has developed more attractive services, both interurban for the island's own population and to its tourist areas for visitors. This has involved introduction of new types of diesel train that are more competitive against road transport, in terms both of passenger amenities and speed.

Seven-car Series 281 tilting diesel trainsets, derived from the 'HEAT 281' experimental train completed in 1992, entered service in 1994, cutting the fastest journey time between Sapporo and Hakodate to 3 hours.

Following tests with a prototype Series 283, developed from the earlier Series 281, tilting trains went into service on the Sapporo–Kushiro route in 1997. The Series 283 'Furico' sets operating the 'Super Ozora' limited expresses have independent wheel bogies and a four-speed gear system. In spite of difficult operating conditions, including

JR Hokkaido diesel locomotives

Class	Wheel arrangement	Power kW	Speed km/h	Weight tonnes	No in service	First built	Mechanical	Builders Engine	Transmission
DD51	B-2-B	1,618	95	84	–	1967	H/M/K	DML61Z S/NT/D	H DW2A H/K
DE10	AAA-B	993	85	65	–	1973	K/N	DML61ZB S/NT/D	H DW6 H/K
DE15	AAA-B	993	85	65	–	1969	N	DML61ZB S/NT/D	H DW6 H/K

Electric locomotives

Class	Wheel arrangement	Output kW	Speed km/h	Weight tonnes	No in service	First built	Builders Mechanical	Electrical
ED79	B-B	1,900	100	68	28	1971	H/M/T	H/M/T

sharp curves and heavy winter snowfall, the best timing was cut to 3 hours 34 minutes in 2001, when the service was increased from three to six return trips daily.

With extension of the Tohoku shinkansen to Hachinohe opened in December 2002, a new service of 'Super Hakucho' trains began running between Hakodate and Hachinohe using five-car 'HEAT 789' emus. The four daily trains, which cover the 272 km in 2 hours 52 minutes, connect with 'Hayate' shinkansen services to Tokyo. Planning

continues on the proposed 360 km shinkansen line from Aomori to Sapporo.

In contrast to the rest of Hokkaido, suburban services in the Sapporo area have seen growth of 3 to 10 per cent a year. Particularly busy are the routes to Teine and Ainosato-kyoikudai, on which the number of daily trains were increased in 2003 by 25 and 14 respectively.

A new station building was completed in Hakodate in June 2003, and in March 2003 the JR Tower hotel and shopping complex was opened in Sapporo.

Traction and rolling stock

In 2003 the railway operated 28 electric and 48 diesel locomotives, 534 diesel railcars and 351 emu cars.

Loco-hauled passenger trains are no longer operated, apart from through sleeping car trains from Honshu via the Seikan tunnel.

Kyushu Railway Co (JR Kyushu)

3-25-21 Hakataekimae, Hakata-ku, Fukuoka 812-8566
Tel: (+81 92) 474 25 01
Fax: (+81 92) 474 97 45
e-mail: service@jrkyushu.co.jp
Web: http://www.jrkyushu.co.jp

Key personnel

Chairman: Koji Tanaka
President: Susumi Ishihara
Director General, Business Development/Urban Development: Akimasa Hayami
Managing Directors
Director of Travel Services/Marketing and Sales/ Railway Operations/Ferry Operations: Yasuharu Maruyama
Director General of Corporate Planning/ Administration/Tokyo Branch Office/Okinawa Branch Office/JR Kyushu Hospital: Suichi Honda
Director General of Railway Operations: Toshiro Kameyama
Directors
Nishi-Kagoshima Station Development: Yoshihisa Kamino
General Manager, Northern Kyushu Regional Head Office: Takao Nishimura
Deputy Director, General Business Development/ General Manager, Administration: Hisaji Akizuki
General Manager, Finance and Accounting/ Auditing: Keisuke Saeki
General Manager, Oita Regional Office: Keio Kawano

Gauge: 1,067 mm
Route length: 2,101.1 km
Electrification: 1,177 km at 20 kV 60 Hz AC; 51.1 km at 1.5 kV DC

Organisation

The company runs rail passenger transport and related activities in the Kyushu region. It operated 2,925 trains daily from April 2002. Since formation of JR Kyushu in 1987, several local lines have been transferred to third-sector undertakings. The company also runs a hydrofoil service between Hakata and Pusan, South Korea. The company's bus services subsidiary was divested in April 2001.

Other parts of the JR Kyushu Group cover the development and exploitation of station buildings, leisure and retail activities, construction and food service businesses.

In 2003 JR Kyushu employed 10,140 staff.

Finance (100 million yen)	2000	2001	2002
Passenger revenue	1,175	1,142	1,136
Other revenue	430	387	361
Total revenue	1,605	1,529	1,497
Operating expenses	1,729	1,652	1,531
Pre-tax profit/loss	−123	−122	−34

Traffic (million)	2000	2001	2002
Passenger journeys	3,831	3,819	3,780
Passenger-km	196	194	189

Passenger operations

Operations cover 21 routes and serve 574 stations.

JR Kyushu has introduced several new types of rolling stock over recent years to upgrade its services. Refurbished Series 485 emus predominate on intercity services, but these are now supplemented by Series 783 'Hyper Saloon' emus introduced since 1990 and by nine-car Series 787 emus. The latter were introduced on 'Tsubame' services between Hakata and Kagoshima in 1992 to counter competition from parallel expressway bus services and short-distance air routes. Since

March 1994, Series 787 units have also operated daily 'Kamome' return services between Hakata and Nagasaki.

March 1995 saw the introduction of a new 'Wonderland Express' service between Hakata and Oita operated by four seven-car Sonic 883 VVVF emus with active body-tilt. These units feature a distinctive exterior design and a colourful interior based on an amusement park theme; they have a variety of seating patterns and a panoramic section.

In March 2000, Series 885 'Kamome' six-car tilting emus were introduced on Hakata-Nagasaki limited-stop express services, providing 16 daily return trips in each direction. The introduction of these trains marked the completion of a programme of enhancements to JR Kyushu's network of limited express services, 308 of which were being operated daily by the company in 2000.

Series 811 commuter emus have been introduced on the Town Shuttle network of routes linking the main urban areas in northern Kyushu.

On non-electrified lines, new Kiha 200- and 220-type dmus have been introduced on local services. Since 1992, Kiha 185 express dmus have been running on the Trans-Kyushu Highland Express routes linking Hakata with the Mount Yofu area and Kumamoto with the Mount Aso region.

In the 16 years up to 2002, JR Kyushu had opened 47 new stations on its network.

New lines

Construction began in 1991 of the 127 km standard gauge Nishi-Kagoshima–Shin-Yatsushiro section of JR Kyushu's first Shinkansen line, the Kagoshima route. This section is due to open in the early part of 2004 and will include five stations: Nishi-Kagoshima; Sendai; Izumi, Shin-Minamata (provisional name); and Shin-Yatsushiro (provisional name). Some 70 per cent of the line is in tunnel and the steepest gradient is 3.5 per cent.

The northern section of the Kyushu Shinkansen, called the Nagasaki route is also under construction. The two sections will create a 257 km line north to Hakata, where connection will be made with the Sanyo Shinkansen, and Nishi-Kagoshima. In March 1998, work started on the Shin-Yatsushiro–Funagoya section, mostly covering land acquisition and tunnel excavation, and in April 2001 construction started on the 40 km

northern section between Funagoya and Hakata. A 120 km branch to Nagasaki will leave the main line at Shin-Tosu; the alignment for this line was announced in February 1998.

Construction of both sections is being undertaken by Japan Railway Construction Corporation (JRCC), with central and local governments bearing the cost. Services will be operated by JR Kyushu, which will lease the track from JRCC.

The Kyushu Shinkansen will be worked by Series 800 six-car trainsets operating at a maximum speed of 260 km/h. When the entire line is commissioned the journey time between Fukuoka and Kagoshima will be reduced from 3 hours 49 minutes to 2 hours 20 minutes or less. In the longer term there is a government-sponsored project to develop a design of train capable of operating on different track gauges. This would allow through working between the Kyushu Shinkansen and conventional 1,067 mm gauge lines.

Improvements to existing lines

At Kagoshima, future terminus of the Kyushu Shinkansen, a major development is planned in the station area. As well as serving as a major transport hub linking rail, bus, tram and taxi services, the 'Alternative City' project will incorporate major business, retail and leisure facilities, as well as extensive parking.

A recent electrification project covered the Sasaguri Line, betwen Yoshizuka and Keisen, the Chikuho Main Line, between Orio and Keisen, and part of the Kagoshima main line, between Kurosaki and Orio. Together, the scheme covered 64.8 route-km and was undertaken to stimulate economic and social regeneration following a rundown in coal mining in the Chikuho area. The cost of the ¥14 billion scheme was shared by JR Kyushu (¥5.6 billion), Fukuoka Prefecture and communities along the line (¥6.4 billion), and government funding under legislation introduced in 1961 to assist the regeneration of former coal mining areas (¥2 billion). The JR Kyushu funding also includes a ¥4 billion interest-free loan provided under similar regeneration legislation.

Included in the project were: trackside power supply systems; additional crossing places to increase line capacity; and a new depot for the Class 817 emus introduced to serve the line. Five

Hitachi-built Series 883 Kamome tilting emus entered service between Hakata and Nagasaki in March 2000
0098050

stations were renovated and one, at Kotake, is to serve as a future 'park and ride' facility.

Traction and rolling stock

In 2002 the company operated 13 electric and nine diesel locomotives, 1,178 emu cars and 347 dmu cars and 60 freight wagons.

Between 1987 and 2001, JR Kyushu invested ¥131 billion in new passenger rolling stock, acquiring 841 new trains. Of these, 752 were emus, including 413 for commuter services, and 89 were dmus.

For the electrification of the Sasaguri and Chikuho lines (see 'Improvements to existing lines'), new Class 817 emus were introduced. Construction of these includes the use of materials considered friendly to the environment, such as aluminium, wood, glass and leather. Interior features include reversible seats made of wood and leather.

Signalling and telecommunications

CTC (Centralised Traffic Control) controls 80 per cent of the system; programmable route control covers 81 per cent of the system. The first phase of JR Hakata traffic control centre scheme was completed in 1997 as part of a strategy to integrate dispersed control functions.

Track
Rail: 50.4 and 60 kg/m
Sleepers: Prestressed concrete or timber 2,100 × 200 × 140 mm spaced 1,480, 1,560 or 1,760/km according to class of route
Fastenings: Elastic
Max axleload: 18 tonnes

Shikoku Railway Co (JR Shikoku)

8-33, Hamano-cho, Takamatsu 760-8580
Tel: (+81 87) 825 16 15   Fax: (+81 87) 825 16 19
e-mail: keiekikaku@jr-shikoku.co.jp
Web: www.jr-shikoku.co.jp

Key personnel
Chairman: Toshiyuki Umehara
President: Kiyohiro Matsuda
Senior Directors:
   Railway Operations: Tadashi Sano
   General Affairs: Atsushi Yamamoto
Marketing and Sales: Masafumi Izumi

Gauge: 1,067 mm
Route length: 855.2 km
Electrification: 235.4 km at 1.5 kV DC

Organisation
The company runs passenger transport and related activities in the Shikoku region. It operated 1,022 trains daily from October 2003.

Passenger operations
Traffic has risen significantly since the introduction of through trains from Japan's principal island, Honshu, over the Seto bridges opened in 1988. In FY2003 the railway recorded 50.8 million passenger journeys and 1,563 million passenger-km.

The service of 36-38 rapid trains each way daily to Shikoku connects Shikoku with the Sanyo Shinkansen at Okayama. There is also an overnight Tokyo–Shikoku sleeper service.

Traffic (million)	2002	2003
Passenger journeys	51.8	50.8
Passenger-km	1,591	1,563

Traction and rolling stock
At the beginning of FY2004 the railway operated seven diesel locomotives, 180 dmu cars, 100 single diesel railcars, 122 emu cars and 36 single electric railcars.

Equipment includes three Type TSE-2000 dmu cars with a top speed of 120 km/h. Employed on cross-island services between Okayama/Takamatsu and Kochi, they are the world's first diesel-powered vehicles to be equipped with an active body-tilt system, which is designed to raise acceptable curving speed by 20 to 30 per cent depending on curve radius. An upgraded version, the N2000 Special Express (Series 2400/2450/2500), with a top speed of 130 km/h, was introduced in 1995. This two-car unit features 260 kW engines to improve acceleration, calliper disk brakes and automatic wheelslip prevention.

The Series 8000 emus used on the electrified Okayama/Takamatsu–Matsuyama routes also feature active body-tilting. These units have VVVF three-phase drive motors, a maximum speed of 160 km/h and carbodies carried on pendulum roller bolsters. An onboard memory checks the train location through the ATS system and issues tilting signals. The pantographs are mounted on an independent carriage on guiderails, with linkages to the bogies to ensure correct positioning of the pantographs when body-tilt is activated.

In 2003 JR Shikoku introduced Series 5000 emus featuring one double-deck trailer for the Shikoku–Sanyo Shinkansen 'Marine Liner' connecting service shared with JR West, which acquired similar units. The vehicles won that year's Blue Ribbon design prize.

Class 5000 'Marine Liner' emu with two-class double-deck driving trailer (JR Shikoku)   **NEW**/1135455

JR Shikoku diesel locomotives

Class	Wheel arrangement	Power kW	Speed km/h	Weight tonnes	No in service	First built	Mechanical	Builders Engine	Transmission
DE10	AAA-A	1,250	85	65	7	1966	K/N/H	DML61ZB NT/S/D	H DW6 KH

JR Shikoku diesel railcars or multiple-units

Class	Motored axles/car	Power/ motor kW	Speed km/h	No in service	First built	Mechanical	Builders Engine	Transmission
1000	1	300	110	56	1989	NT	SA6D125-HD-1	NT DW14
TSE-2000	2	245 × 2	120	3	1989	F	SA6D125-H	NT TACN 22–1600
2000	2	245 × 2	120	10	1989	F	SA6D125-H	NT TACN 22–1600
2100	2	245 × 2	120	22	1989	F	SA6D125-H	NT TACN 22–1600
2150	2	245 × 2	120	7	1989	F	SA6D125-H	NT TACN 22–1600
2200	2	245 × 2	120	18	1989	F	SA6D125-H	NT TACN 22–1600
2400	2	260 × 2	130	6	1995	F	SA6D125-H	NT TACN 22–1600
2450	2	260 × 2	130	6	1995	F	SA6D125-H	NT TACN 22–1600
2500	2	260 × 2	130	4	1997	F	SA6D125-H	NT TACN 22–1600
185	1	185 × 2	110	32	1986	NT/F/N	DMF13HS NT	S/NC TC2A DF115A
65	2	375	95	6	1969	NT/F/N	DML30HS NT/S/D	S/NC DW4F
58	1	135 × 2	95	15	1962	NT/F/N	DMH17H NT/S/D	S/NC TC2A DF115A
54	1	185 × 2	95	12	1986	NT/F	DMF13HS NT	S/NC TC2A DF115A
47	1	165	95	42	1980	NT/F	DMF15HSA NT/S/D	S/NC DW10
40	1	165	95	11	1979	NT/F	DMF15HSA NT/S/D	S/NC DW10
32	1	165	95	23	1986	NT/F	DMF13HS NT DMH17H	S/NC TC2A DF115A
28	1	135	95	7	1961	NT/F/N	DMH17H NT/S/D	S/NC TC2A DF115A

Abbreviations: D: Daihatsu; F: Fuji Juko; K: Komatsu Seisakusyo; N: Nippon Sharyo Seizo Ltd; NC: Niigata Converter; NT: Niigata Tekko; S: Shinko Zouki

JR Shikoku electric railcars or multiple-units

Class	Cars per unit	Motor cars per unit	Motored axles/car	Output/ motor kW	Speed km/h	No in service	First built	Builders Mechanical	Electrical
111	4	2	1	100	100	3	1962	N/K	H/T/M/To/Se
113	4	2	4	120	100	4	1963	NT/H/K/Kn	H/T/M/Se/To
121	2	1	1	110	100	19	1986	H/K/Kn/Tk	H/T/M/To
5000	3	1	4	–	130	6	2003	Tokyu Car	–
6000	3	1*	4	160	110	2	1996	N	T
7000	1	1*	4	120	110	36	1990	K	Fuji
8000	3/4/5	2*	4	185	140	12	1992	N/H	To/T

* VVVF Inverter control
Abbreviations: H: Hitachi Ltd; K: Kawasaki Heavy Industries; Kn: Kinki Sharyo; M: Mitsubishi; N: Nippon Sharyo; Se: Shinko Electric; T: Toshiba; To: Toyo Denki Seizo

Signalling and telecommunications

A centralised traffic control centre and headquarters in Takamatsu opened in 1997, controlling the entire network.

Electrification

Electrification was begun to tie in with completion of the Seto bridge complex. Catenary between Takamatsu and Iyo on the Yosan line (206.3 km) was completed in early 1993, with Okayama-Matsuyama services carried out by Series 8000 tilting emus from March of that year.

Track

Rail: 60 kg/m (33 km), 50 kg/m (740 km) and 40 kg/m
Sleepers: Prestressed concrete or wood spaced 1,560/km in plain track, 1,640/km in curves
Fastenings: Spike (wood sleeper) or double elastic
Min curve radius: 200 m
Max gradient: 3.3%
Max axleload: 17 tonnes

UPDATED

Japan Freight Railway Co

3-13-1, Iidabashi, Chiyoda-ku, Tokyo 102-0072
Tel: (+81 3) 239 91 11   Fax: (+81 3) 239 91 30
Web: http://www.jrfreight.co.jp

Key personnel

Chairman: Yoshio Kaneda
President: Naohiko Itoh
Senior Managing Directors: Katsuji Iwasa, Ryoichi Yonemoto
General Managers: Masaaki Kobayashi, Haruki Kouno, Shuji Tamura

Organisation

Freight service is managed and marketed by JR Freight nationwide on the 1,067 mm gauge network. This concern owns its locomotives, wagons and terminals, but hires its track space from the six passenger railway companies. It has no marshalling yards and dedicates itself to bulk commodity and container traffic in trains using rolling stock modified to permit operation at up to 130 km/h. Some maintenance is subcontracted to the passenger railways.

Finance

JR Freight is a drastically slimmed-down enterprise compared with the freight activity of JNR at its end, when wage costs alone exceeded freight revenue by 25 per cent. The company has withdrawn from a third of the route-km served by JNR and is running 40 per cent fewer daily trains. It has also discarded most individual wagonload traffic in favour of complete trainloads from one terminal to another and increased labour productivity.

Operating revenues fell slightly to ¥157.4 billion in 2002–03. However, a pre-tax profit of ¥1.8 billion was achieved, mainly through the sale of surplus land.

To counter deteriorating performance the company adopted 'New Challenge 21', a three-year (2002–04) management plan, in 2001 with the aim of enhancing marketing, improving cost competitiveness, promoting planned capital investment and expanding non-railway business. In the longer term, the company also intends to increase the number of wagons per train on the Tokaido line, to introduce new locomotives to speed up services and to reduce railway employee numbers from 6,200 to 5,000 by 2004.

Finance (million yen)	2000–01	2001–02	2002–03
Operating revenues			
Railway	136,900	131,500	130,500
Other	27,500	28,700	30,100
Total	164,400	160,200	160,700
Operating expenses			
Total	164,700	159,800	158,100
Pretax profit/loss	−3,700	−5,600	3,800
Profit/loss after tax	−2,700	−3,200	1,600

Traffic (million)	2000–01	2001–02	2002–03
Freight tonnes	40.1	40.0	38.7
Freight tonne-km	21,900	22,000	21,900

Freight operations

A further decline in traffic levels occurred in 2002–03, with total freight tonnes down 0.2 per cent to 38.7 million, and tonne-km decreased 0.1 per cent to 21.9 billion. Bulk commodities fell by 4.6 per cent to 17.81 million tonnes while container traffic was unchanged at 20.85 million tonnes.

The most significant bulk commodities by weight are petroleum (53 per cent of total carryings), cement (11 per cent), limestone (4 per cent), paper and pulp (4 per cent), chemicals (3 per cent) and coal (2 per cent). However, containers are the biggest single traffic component, travelling 900 km on

Class EH500 dual-voltage electric locomotive at Tokyo with an intermodal service   0561558

Type M250 Super Rail Cargo 1.5 kV DC freight emu undergoing trials on the Tokaido line   0573219

Class EF210 electric locomotive at Shin Osaka (Marcel Vleugels)   0561559

average compared to 200 km or so for other traffic, and yielding 66 per cent of JR Freight's total revenue.

The company's total number of daily scheduled trains stood at 710 in 2003.

Intermodal operations

In 2002 425 container trains were running daily over trunk routes extending from Asahikawa and Kushiro in Hokkaido down to Fukuoka, Nagasaki and Kagoshima in Kyushu. One service runs throughout from Fukuoka to Sapporo, a distance of 2,130 km. 'Superliner' container trains connecting the main cities such as Tokyo, Osaka, Hiroshima and Fukuoka are permitted a top speed of 110 km/h. In all, 138 terminals are served by the container train system.

Width is the difficult dimension rather than height; since 1987 JR Freight has raised the latter from 8 ft (2.4 m) to 2.5 m in its own stock. Length was at first a standard 20 ft, but now JR Freight has available numerous 30 ft containers with 47 m³ capacity; that approximates to the cube available in the Japanese trucking industry's most widely used vehicles. JR Freight also deploys refrigerated and ventilated containers, as well as boxes, tanks and hoppers.

JR Freight has now accumulated a stock of new low-floor flatcars that can carry 40 ft, 8 ft 6 in high ISO containers of 30 tonnes gross weight, or two 20 ft boxes with a total gross weight of 40 tonnes.

Since 1995 JR Freight has been operating a daily car transporter service between Nagoya and Niigata using fully enclosed 'Car Rack' wagons. The sides and roof of the rail wagons lift upwards to accommodate a total of eight cars carried on two levels. These wagons can accommodate ordinary containers as return cargo.

Traction and rolling stock

In April 2003 JR Freight operated 582 electric and 254 diesel locomotives and 9,304 wagons. Recent deliveries include Class EF510 dual-voltage (1.5 kV DC/20 kV AC) Bo-Bo-Bo type with inverter-controlled motors and Class EH200 1.5 kV DC (Bo-Bo)-(Bo-Bo) machines.

In 2003 trials commenced with 'Super Rail Cargo' 1.5 kV DC freight emus designed for 130 km/h container services on the Tokaido trunk line.

Designated the M250, the train is formed of a pair of modules incorporating driving and propulsion functions at either end of a variable number of container-carrying wagons. Control is by VVVF inverter with three-phase asynchronous traction motors. Total output is 3,250 kW. Axle-load is 12.5 tonnes.

JR Freight is equipping the Joetsu, Sanyo, Tohoku amd Tokaido lines with a GPS-based train-tracking system.

Class DD51 diesel-hydraulic locomotive at Takinawa (Marcel Vleugels)   0561560

JR Freight diesel locomotives

Class	Wheel arrangement	Power KW	Speed km/h	Weight tonnes	No in service	First built	Mechanical	Builders Engine	Transmission
DE10	AAA-B	1,000	85	65	122	1967	K/N/H	DML61ZB	H DW6 (2000 series)
DE11 (2000 series)	AAA-B	1,000	85	70	4	1979	K/N	DML61ZA/B	H DW6
DD51	B-2-B	1,650	95	84	104	1966	H/M/K	2 × DML61Z or 2 × SA12V170	H 2 × DW2A
DF200	Bo-Bo-Bo	2,500	110	96	13	1992	K/T	MTU 12V396TE14	E
DF200 (50 series)	Bo-Bo-Bo	2,650	110	96	11	1999	K/T	Komatsu SDA12V170	E

JR Freight electric locomotives

Class	Wheel arrangement	Output KW	Speed km/h	Weight Tonnes	No in service	First built	Builders
1.5 kV DC							
EF64	Bo-Bo-Bo	2,550	100	96	113	1964	T/K/To
EF65	Bo-Bo-Bo	2,550	100	96	140	1964	K/To/T/N
EF66	Bo-Bo-Bo	3,900	110	100.8	75	1968	K/To
EF67	Bo-Bo-Bo	2,850	100	99.6	8	1981	To
EF200	Bo-Bo-Bo	6,000	120	100.8	21	1990	H
ET210	Bo-Bo-Bo	3,390	110	100.8	30	1995	M/K
EH200	(Bo-Bo)-(Bo-Bo)	4,520	110	134.4	2	2001	T
20 kV AC							
ED75	Bo-Bo	1,900	100	67.2	72	1963	M/H/T
ED76	Bo-2-Bo	1,900	100	87	25	1967	H/M/T
ED79	Bo-Bo	1,900	110	68	9	1989	T
Dual-voltage 1.5 kV/20 kV							
EF81	Bo-Bo-Bo	2,550/2,370	110	100.8	64	1968	H/M
EH500	(Bo-Bo)-(Bo-Bo)	4,000	110	134.4	22	1997	T
EF510	Bo-Bo-Bo	3,390	110	100.8	1	2001	M/K

Abbreviations: H: Hitachi; K: Kawasaki; M: Mitsubishi; N: Nippon Sharyo Seizo; T: Toshiba; To: Toyo Denki Seizo

Shinkansen network development

The network in operation at the start of 1999 and its owning and operating railways comprised:

JR East:

Tohoku Shinkansen	Tokyo to Morioka
Joetsu Shinkansen	Omiya to Niigata
Nagano Shinkansen	Takasaki to Nagano
Yamagata Mini-Shinkansen	Fukushima to Yamagata
Akita/Ou Mini-Shinkansen	Morioka to Omagari/Ou

JR Central:

Tokaido Shinkansen	Tokyo to Shin-Osaka

JR West:

Sanyo Shinkansen	Shin-Osaka to Hakata

The Shinkansen programme agreed in 1988 specified three types of extension:

Type 1. To full Shinkansen 1,435 mm gauge standard, engineered for 260 km/h by present equipment, 300 km/h by the next generation equipment in design.

Type 2. Addition of a third rail to existing 1,067 mm gauge, possibly with some realignment, with the use of small-profile trainsets restricted to 130 km/h on mixed-gauge.

Type 3. Infrastructure to full Shinkansen 1,435 mm gauge standard, but initially laid with 1,067 mm gauge track, engineered for 160 to 200 km/h.

Application of these concepts to the extensions approved in 1990 was as follows:

Hokuriku (later Nagano) Shinkansen
Type 1: Takasaki-Karuizawa (41 km)
Type 1: Karuizawa-Nagano (75 km)
Type 3: Nagano-Kanazawa (89 km)

Tohoku Shinkansen
Type 2: Morioka-Aomori (125 km)
Type 1: Numakunai-Hachinohe (but dual-gauged for use also by freight trains)

Kyushu Shinkansen
Type 3: Yatushiro-Kagoshima (128 km)

Construction work began on the Karuizawa-Nagano section of the Hokuriku, the Morioka-Aomori extension of the Tohoku, and the Yatushiro-Kagoshima section of JR Kyushu's first Shinkansen in September 1991. Construction commenced in October 1993 on the Itoigawa Uozu section of the Hokuriku Shinkansen. All were being built by the Japan Railway Construction Corporation (see above). Construction costs were to be shared between central government (40 per cent) and the JR companies, which will pay their share out of receipts from the Shinkansen lines they now own.

The outcome of a government review of the Shinkansen programme was announced at the beginning of 1994. The previously authorised sections of the Hokuriku, Tohoku and Kyushu lines retained priority. Environmental assessment studies of the Takefu-Osaka section of the Hokuriku line, the Nagasaki and Hokkaido branches of the Kyushu line and the Hokkaido line from Aomori to Sapporo commenced in 1994–95, although work on these sections was not to start before the next programme review in 1997.

A change in government policy saw a 25 per cent increase in funding for Shinkansen construction, which has enabled work to progress on some additional sections of the network. About 90 per cent of the ¥228 billion budget was allocated to the Takasaki-Nagano section of the Hokuriku Shinkansen, but work has also progressed on the Kumamoto section of the Kyushu line and the Morioka-Numakunai section of the Tohoku line, which will now be built to Type 1 rather than Type 2 standard.

The Takasaki-Nagano section was opened for public services in October 1997, but no date has been announced for completion of the section beyond Nagano to Toyama and Kanazawa.

Shinkansen rolling stock

Series 0
During the first 20 years after the opening of the Tokaido-Sanyo Shinkansen, rolling stock of essentially the same performance and accommodation was employed. The original cars of this Series 0 have now all been replaced.

Series 100
Series 0 has been complemented on the Tokaido-Sanyo Shinkansen by the Series 100, which combines pursuit of better aerodynamics, more effective noise control and economy in energy consumption with improvement of passenger comfort and amenities. Earlier Shinkansen types had all cars powered, but in the Series 100 four of a set's 16 cars are trailers. The powered cars, which are paired, each have a 230 kW motor of lighter (828 kg) and more compact design on each axle. A Series 100 set is lighter than a Series 0 set, even though two (or in some sets all) of a Series 100's four trailers are bilevels (the other two are the end cars with driving cabs). The comparison is 922 tonnes as against 967 tonnes for 16 cars with a full complement of 1,277 passengers. Maximum axleloading, at 15 tonnes, is one tonne below that of a Series 0.

Aerodynamic improvements in the Series 100 include a longer and reshaped nose and closer attention to smooth exterior surfaces, in particular by avoiding recessed windows. The effect has been to reduce drag coefficient by 20 per cent, compared with Series 0, and also noise emission. Moreover, even though the total installed power of a Series 100 (11,040 kW) is less than that of a Series 0 16-car set, the Series 100 has proved 17 per cent more economical in power consumption on a Tokyo-Sanyo 'Hikari' schedule. The 1.6 km/h/s acceleration rate of a Series 100 compares with the 1.0 km/h/s of a Series 0. Nose shaping is a factor in noise control as well as reduction of drag, since the noise emanating from an accelerating train's front end rises by a factor of between the fifth and sixth power of its speed.

With eddy current brakes on its non-motored trailers to complement the regenerative braking of the powered cars, the Series 100 has a braking performance superior to its predecessors.

Series 200
Advances in Shinkansen speed resulted from an exhaustive programme of research and high-speed tests begun in the late 1970s. It proved possible to limit lineside noise at 240 km/h to 79 dB(A), the figure previously achieved at 210 km/h to respect the statutory limit of 80 dB(A), largely by increasing the frequency of rail grinding to eliminate surface defects, and by pantograph alterations.

As built for JR East's Joetsu and Tohoku lines, the Series 200 trainsets were formed of two-car units, each of which had to run with a pantograph operative. The normal 12-car formation now runs with only three pantographs raised and a 25 kV bus-line laid along the tops of the cars distributes the current. Noise has been further curbed by adoption of three-stage pantograph springing to sustain contact with the overhead wire and to minimise the intermittent arcing that has been a perennial cause for environmentalist complaints; and by surrounding the pantograph with shielding, which limits transmission of noise to the lineside.

A programme of replacement of Series 200 units was expected to commence after the opening of the Nagano and Akita Shinkansen projects in 1997. However, in 1998 JR East commenced a refurbishment programme covering at least 100 cars.

Series 300
Test programmes demonstrated that speeds up to 300 km/h were feasible on the present track of all Shinkansen, but not with existing sets, which at that speed would break both the 80 dB(A) noise limit and also generate unacceptable vibration. Also, aerodynamic drag in tunnels would be too high. Consequently a new Series 300 train was designed and, as described above, was put into public service by JR Central on the Tokaido Shinkansen in 1992.

JR Central Series 300, Series 100 and Series 0 trainsets at Osaka depot (Wolfram Veith)　0023031

The Series 300 is distinguished externally by a dramatically reshaped nose-end and low-slung body. Floor level is 1.15 m and roof crown 3.6 m above rail, compared with 1.3 m and 4 m respectively for the Series 100. A considerable reduction of weight results from adoption of aluminium alloy body construction and more powerful traction motors. With six of a set's 16 cars non-motored, total weight is 691 tonnes, compared with the 922 tonnes of a Series 100. Maximum axleloading in the bolsterless bogies is 14 tonnes. Like the Series 100, the 300 adds eddy current to its braking systems; a change is that the 300 has regenerative instead of dynamic braking.

Traction is provided by three-phase AC 300 kW motors, four per motor car, supplied by a 3,000 kW pulse-width modulation converter with 4,500 V, 2,000 A GTO thyristors feeding 1,760 kVA VVVF inverters. Power-to-weight ratio is 17.37 kW/tonne. Designed top speed is 300 km/h.

Series 300X
In 1994 JR Central took delivery of a six-car series 300X experimental trainset, designed for commercial speeds of 350 km/h, and is undertaking a two-year programme of high-speed trials on an upgraded section of track between Kyoto and Maibara. The 300X achieved a record speed for Japan of 443 km/h in July 1996. The train features two different nose configurations (cusp shape and rounded-wedge shape) to test air resistance and noise, reduced cross-section smooth-sided aluminium-bodied cars with coupling bellows and undercarriage skirts, pantograph covers, dampers between cars, tilting system and active control system. Individual cars have been constructed by four different manufacturers using different body fabrication techniques in order to assess the scope for further weight reductions.

The Series 300X forms the basis for trains that may be required for the Chuo Shinkansen between Tokyo and Osaka via Kofu if the proposed maglev line does not proceed.

Series 400
For its Yamagata Mini-Shinkansen, as described above, JR East acquired 12 small-profile six-car Series 400 sets. All axles are powered by thyristor-controlled 220 kW motors. The cars are of the same height as those of the Series 200, to which the sets are coupled when running over the Tohoku Shinkansen and run at up to 240 km/h between Tokyo and the divergence of the Mini-Shinkansen at Fukushima. But the bodies are significantly narrower, so that each seating bay has one seat less per row than in a Series 200. A seventh car was subsequently added to each set.

In September 1991 a Series 400 was run at up to 345 km/h on the Tohoku Shinkansen.

Series E1
To meet rising demand for commuter services on its Tohoku and Joetsu lines, JR East launched the all double-deck 12-car Series E1 'MAX' (Multi Amenity Express) in March 1994, and commercial services commenced in July 1994. Because the car height has been increased, the front cars have been given an ultra-streamlined shape to improve their appearance. Doors have been widened to prevent congestion around the entrances and wheelchair lifts are installed in two cars. Cars are equipped with electronic information displays and FM radio receivers. Vending machines are available in some cars in place of restaurant or buffet facilities. Half the cars are motored, with 24 traction motors providing a total continuous rating of 9,840 kW. Maximum speed is 240 km/h.

Series E2
Services on the Nagano (formerly Hokuriku) Shinkansen are operated by eight-car Series E2 trainsets designed to achieve 260 km/h and climb 3.5 per cent gradients. Performance trials began following the delivery of a prototype set in May 1995, and revenue services began in October 1997. The train features a low-profile cross-section as on the Series 300 and a streamlined nose to reduce pressure when entering tunnels. To reduce aerodynamic noise, underfloor equipment is enclosed, plug doors are fitted flush with body sides and pantographs are surrounded by shields. VVVF-controlled 300 kW motors are fitted to six power cars and accept power supply at 50 or 60 Hz. The Series E2 and E3 units are designed so that noise and vibration levels at 275 km/h do not exceed those of existing 240 km/h trains.

An initial series of 19 trainsets was to be expanded by an additional nine units by March 2000.

Series E3
A prototype five-car Series E3 unit was delivered in April 1995 for test running prior to commercial operation on the Akita Shinkansen from 1997. Series E2 and E3 share main electrical components and bogies to reduce construction and maintenance costs but the E3 has a smaller cross-section to meet clearances on the Akita line. The unit features asynchronous motors and a maximum speed of 270 km/h, though speed is limited to 130 km/h on the regauged line between Morioka and Akita. Power is supplied at 25 kV 50 Hz AC on the Shinkansen and 20 kV 50 Hz AC on the converted line. Due to limited underfloor space, some equipment is roof-mounted under an aerodynamically designed cover. Series E3 sets are coupled to Series 200 or E2 trains for the run between Tokyo and Morioka. Five five-car units were initially constructed. In 1998 it was announced that these would be increased to six-car sets and that a further train, also of six cars, would be built.

Series E4

JR East introduced the new double-deck Series E4 from December 1997 on the Tohoku Shinkansen to alleviate rush-hour congestion, particularly on the Omiya to Oyama section. The newly designed 16-car trainset, consisting of two eight-car units, has total seating capacity of 1,634 and runs at 240 km/h. This will make it the highest capacity high-speed train in the world. Bodyshells are of aluminium alloy, contributing to an axleload of only 16 tonnes when fully laden. Four cars are motored and VVVF traction equipment is used. One-hour rating is 6,720 kW. Initially three sets were ordered by JR East and delivered in 1997. A further seven have been ordered for entry into service by March 2000.

STAR 21

A prototype nine-car train was completed for JR East by Kawasaki in 1992. Tagged STAR 21 (Superior Train for the Advanced 21st Century Railway), it is designed to prove various components and technologies with a view to future operation at 300 to 400 km/h. Consequently, it features some alternative concepts, a peculiarity most apparent in the differing outlines of its two end cars. Of these, one is designated Series 952, the other Series 953.

The unit features three types of body in which aerospace techniques and materials have been employed for weight saving, which has been so effective that the maximum axleloading is 10.5 tonnes. For the first time in Shinkansen rolling stock development, five of the cars are articulated. Eight different designs of bolsterless bogie

have been applied to the unit. Hollow axles and aluminium axleboxes are features.

Four of the cars are motored for a continuous rating of 2,640 kW to serve a total train weight of 256 tonnes. The motors are three-phase AC under VVVF control, and draw their current from pantographs on the third and seventh cars.

In November 1992 STAR 21 set a Japanese speed record of 358 km/h and in December 1993 reached 425 km/h, at that time second only to the French TGV-A's 515 km/h. Further work is now in progress to reduce noise levels associated with high-speed running.

Series 500

In 1992 JR West unveiled a six-car prototype variously designated Series 500 and WIN 350, the latter to mark its objective of commercial operating speed. As with JR East's STAR 21, the unit featured differently styled end cars for investigation of aerodynamics and noise suppression at high speed. Test running of the WIN 350 train enabled the development of measures to respond to environmental concerns and to ensure passenger comfort at operational speeds of 300 km/h.

In January 1996 JR West took delivery of a 16-car Series 500 preproduction train, based on the results of the WIN 350 research. The 500 entered commercial service at a maximum speed of 300 km/h between Shin-Osaka and Hakata in March 1997 and was scheduled to operate through to Tokyo in late 1997. This new train features a reduced cross-section lightweight aluminium body, 'smooth' surfacing technology with plug doors

and enclosed underfloor equipment modules, a distinctive 15 m long aerodynamic pointed nose, a wing-shaped pantograph to reduce aerodynamic and wire contact noise and active suspension to allow curves to be negotiated up to 20 km/h faster than other trains. All axles are motored and bolsterless bogies reduce vibration. The passenger capacity of 1,323 is the same as the 'Nozomi' trainsets. Series 500 trainsets entered service with JR West in March 1997.

Series 700

Test running of a prototype Series 700 Shinkansen trainset developed by rolling stock engineers from JR Central and JR West began in October 1997. The 16-car train has been designed as a low-noise, low-maintenance replacement for the Tokaido and Sanyo Shinkansen routes. Hollow extruded aluminium sections filled with sound insulation has been used for the bodyshell, while computer-controlled secondary suspension and inter-car dampers are provided to improve ride quality and reduce vibration. Twelve cars are motored, giving a total power rating of 13,200 kW. Seating capacity is for 1,323 passengers. Top speed is 270 km/h, although 285 km/h was expected to be the maximum on the Sanyo Shinkansen.

ATLAS

Japan's Railway Technical Research Institute is also working up designs for a 350 km/h Shinkansen train, in a project dubbed ATLAS (Advanced Technology for Low Noise and Attractive Shinkansen).

The rump of JNR

What remains of JNR was reorganised as the Japanese National Railways Settlement Corporation, which retains all the assets and liabilities that are not transferred to successor companies (including the Japan Railway Construction Corporation). Its tasks are:
Reimbursement of long-term liabilities and payment of interest;
Disposal of real estate and other assets in order to raise the necessary money;
Execution of necessary business activities to utilise the rights and meet the obligations transferred to the company from JNR;
Action to achieve re-employment of personnel made surplus by the Reform. (Some 93,000 JNR personnel were surplus to the needs of the new companies at the latter's formation.)

JNR Settlement Corporation
6-5, Marunouchi 1-chome, Chiyoda-ku, Tokyo 100
Tel: (+81 3) 32 14 79 59   Fax: (+81 3) 32 40 55 86
Chairman: Yasuo Nishimura
The corporation's function is disposal of the assets and long-term liabilities not transferred to new companies during the restructuring of 1987, and promotion of the re-employment of surplus employees. JNRSC took on most of the JNR debt in order to facilitate the privatisation of the ex-JNR companies, but 10 years after the JNR break-up the continued existence of ¥28,100 billion of debt (equivalent to nearly 6 per cent of gross domestic product) was a political embarrassment. Reduction of the inherited JNR debt is proving difficult because of lower than expected proceeds from railway privatisation and land sales due to the sluggish economy and collapse in property values. A proposal put forward in early 1997 was that the government should assume ¥13,800 billion of the outstanding debt; land sales might cover a further ¥8,000 billion. The balance of ¥6,300 billion required to enable the winding up of JNRSC might be financed by a surcharge on rail fares, an idea which is being strongly resisted by the ex-JNR companies. A subsequent government plan provided for outstanding long-term debt being paid off over a 50-year period and JNRSC being wound up. In October 1998 legislation was passed by the Japanese parliament to wind up the JNR Settlement Corporation. The government took over debt of ¥23,500 billion, to be written down over 60 years, while an outstanding sum of ¥4,300 billion

was passed back to railway undertakings in Japan, which would be expected to recover a comparable figure from the disposal of property.

Japan Telecom Co
1-7-4 Kudan-kita, Chiyoda-ku, Tokyo 102
Tel: (+81 3) 32 22 66 51   Fax: (+81 3) 32 22 66 60
Chairman: Kazumasa Mawatari
President: Koichi Sakata
Maintenance of railway telecommunication equipment and provision of general telecommunication services.

Railway Information Systems Co Ltd
6-5, Marunouchi 1-chome, Chiyoda-ku, Tokyo 100
Tel: (+81 3) 32 14 46 95   Fax: (+81 3) 32 40 55 93
Chairman: Yoshisuke Mutoh
President: Hiroyuki Hayashi
Information processing for railway companies, and related computerised information services.

Railway Technical Research Institute
2-8-38, Hikari-cho, Kokubunji-shi, Tokyo 185
Tel: (+81 425) 73 72 19   Fax: (+81 425) 73 72 55
Tokyo Office
Tel: (+81 3) 32 40 96 72

Miyazaki Maglev Centre
Tel: (+81 982) 58 13 03
Chairman: Yoshinosuke Yasoshima
President: Hiromi Soejima
Executive Directors: Nobuhisa Izumi, Toshiaki Sasaki

Organisation
Research and development activities to meet the requirements of the railway companies.

High-speed 1,067 mm gauge design
The Institute has developed a preliminary design for a 12-car lightweight train with active tilt-body apparatus for high-speed operation on the country's 1,067 mm gauge routes.

ATLAS Shinkansen project
As noted above, the Institute is developing a 350 km/h Shinkansen train design under the title of ATLAS.

Maglev system
Maglev is being considered as an option for the second Tokyo-Osaka high-speed route (see JR Central entry above). The objective is to operate at

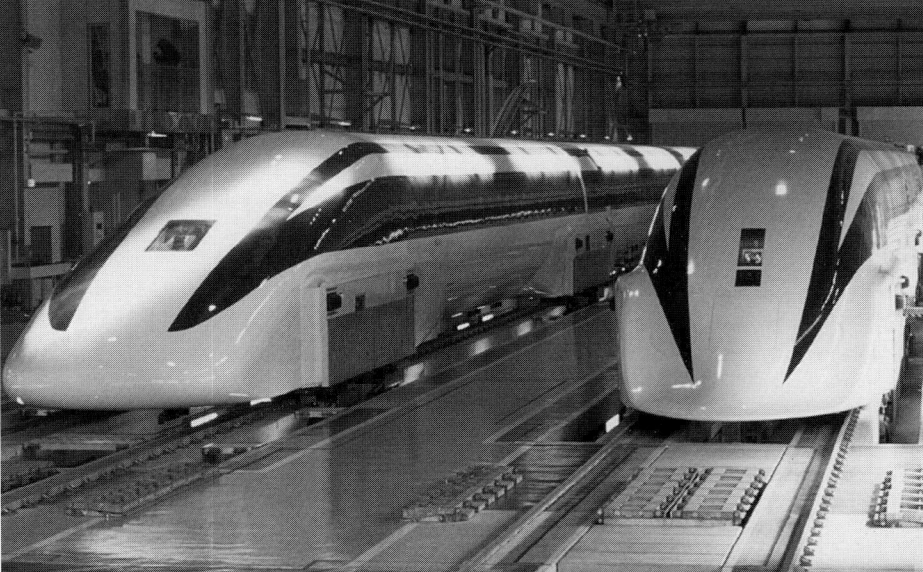
The two MLX-01 maglev vehicles at the Yamanashi test centre   0023032

500 km/h so as to cover the Tokyo-Osaka distance in 1 hour. To further this aim, the government authorised construction as a national project of a 43 km Yamanshi test track, roughly 100 km from Tokyo. The test track is being used to study vehicle behaviour in tunnels and when passing another car running in the opposite direction; operation and control of a full train service; permissible gradient and curvature parameters; and turnout performance. Progress with construction of the test track has been held up due to lack of government finance and land acquisition difficulties, but the Japan Railway Construction Corporation and JR Central have now completed an 18.4 km priority section (16 km in tunnel) between Sakaigawa and Akiyama.

A three-car test train known as MLX-01 was delivered in July 1995. This first MLX-01 comprises a 'double cusp' end car built by Mitsubishi, an 'aero wedge' end car by Kawasaki and a Nippon Sharyo middle car. The cars are linked by articulated levitation bogies which incorporate superconducting magnets and retractable take-off and landing wheels for use at low speeds. Aircraft technology has been used for the ultra-lightweight bodyshells which are constructed in a semi-monocoque style to a slightly smaller cross-section than Series 300 Shinkansen trains. A second five-car MLX-01 train was delivered in October 1997, permitting trials with two trains in operation at the same time.

Trials in superconducting mode, but without levitation, began in February 1997. Test running in levitation mode followed in May 1997. A manned record speed of 531 km/h was achieved in December 1997, and later the same month an unmanned speed of 550 km/h was reached. Tests continued in 1999, with the aim of proving the practicality of maglev in regular service by 2000.

Chichibu Railway

Chichibu Tetsudo
1-1, Akebono-cho, Kumagaya-shi 360
Tel: (+81 485) 23 33 11   Fax: (+81 485) 26 05 51

Key personnel
President: K Kakihara

Gauge: 1,067 mm
Route length: 79.3 km
Electrification: 79.3 km at 1.5 kV DC

Organisation
The railway operates passenger services on two lines which connect at Kumagaya in Saitama prefecture; the 14.9 km Hanyu line and 56.8 km Chichibu line. The latter serves a mountainous area northwest of Tokyo which attracts sightseers and provides limestone for cement, an important source of freight traffic for the railway. Interchange is available with JR, Tobu and Seibu lines and some Seibu services run through between Tokyo and Chichibu line destinations, mainly on Sundays and holidays; a steam-hauled tourist service also operates on the Chichibu line. A 7.6 km freight connection provides for the transfer of freight traffic to and from the JR network. Rolling stock comprises 22 electric locomotives, 67 emu cars and 168 wagons.

Hankyu Corporation

Hankyu Dentetsu
16-1, Shibata 1-chome, Kita-ku, Osaka 530
Tel: (+81 6) 373 50 92   Fax: (+81 6) 373 56 70

Key personnel
Chairman: Kohei Kobayashi
President: Motohiro Sugai
Senior Managing Directors: Yoshihito Utahashi, Norikazu Matsubara

Gauge: 1,435 mm
Route length: 146.8 km
Electrification: 146.8 km at 1.5 kV DC

Organisation
The Corporation was set up in 1907 to construct an interurban railway to develop suburban Osaka, and is now a diversified enterprise. The railway remains the cornerstone of the group, but other activities range from bus and taxi companies to real estate development, department stores, recreation centres and audio-visual entertainment.

Finance
Revenue from railway operations constitutes approximately 54 per cent of total operating revenue.

Passenger operations
The railway serves nine lines with 84 stations and runs over 1,000 eight-car trains daily on its Kobe line and to Kyoto, and over 700 a day to Takarazuka. The terminal in Osaka, built in the 1970s, has 10 platforms and is located in a complex including a 17-storey office building and an underground shopping mall. The rail exit from the terminal is six-track. The company is currently implementing a programme to replace existing Automatic Train Stop (ATS), Total Train Control (TTC) and other safety equipment with more advanced systems.

Hankyu electric multiple-units

Class	Cars per unit	Motor cars per unit	Motored axles/car	Output/motor kW	Speed km/h	Cars in service	First built	Builders Mechanical	Electrical
2000					110	38	1960	Alna Koki	
2300	4/7	2/4	4	150	110	78	1960	Alna Koki	Toshiba/Toyo Denki
2800	4/7	2/4	4	150	110	28	1964	Alna Koki	Toyo Denki
3000	3/4/6/8	2/4	4	170	110	114	1964	Alna Koki	Toshiba
3100	3/4/8	2/4	4	120	110	40	1964	Alna Koki	Toshiba
3300	6/8	4/6	4	130	110	126	1967	Alna Koki	Toshiba/Toyo Denki
5000	8	4	4	170	110	47	1968	Alna Koki	Toshiba
5100	8/10	4/6	4	140	110	90	1971	Alna Koki	Toshiba
5200	4/6	2/4	4	170	110	12	1970	Alna Koki	Toshiba
5300	6/7/8	4/6	4	140	110	105	1972	Alna Koki	Toshiba/Toyo Denki
2200	4	2	4	150*	110	10	1975	Alna Koki	Toshiba
6000	6/8/10	4/6/8	4	140	110	130	1976	Alna Koki	Toshiba
6300	8	4	4	140/150	110	72	1975	Alna Koki	Toshiba/Toyo Denki
7000	6/8/10	4/5/6	4	150	110	210	1980	Alna Koki	Toshiba
7300	6/8/10	3/4/5/6	4	150/180*	110	83	1982	Alna Koki	Toshiba/Toyo Denki
8000	6/8/10	3/4/5	4	170*	110	90	1988	Alna Koki	Toshiba
8300	6/7/8	3/4	4	170*	110	61	1989	Alna Koki	Toshiba/Toyo Denki
8200	2	1	4	200	110	4	1995	Alna Koki	

* VVVF inverter control.

Elevated tracks are under construction at six locations to eliminate level crossings.

Passenger journeys declined from 789 million in 1993–94 to 758 million in 1994–95 due largely to Japan's economic downturn and the effects of the Hanshin earthquake of January 1995. The railway experienced severe damage in the earthquake with services suspended on the main line between Nishinomiya and Kobe and on the Itami and Koyoen branch lines. Reconstruction works costed at ¥85.2 billion were completed by June 1995.

Rolling stock
The railway owns 1,343 emu cars, all air conditioned. The most recent motored cars have VVVF inverter control and the new Series 8200 units feature seats which fold away during the rush hour to increase passenger capacity.

Track
Rail: 50.4 or 60.8 kg/m
Sleepers: Wood: 230 × 2,400 mm; prestressed concrete: 300 × 2,400 × 170 mm
Fastenings: Double elastic
Spacing: 1,760/km
Min curvature radius: 100 m
Max gradient: 3.5%
Max permissible axleload: 17.78 tonnes

Hanshin Electric Railway

Hanshin Denki Tetsudo
1-24 Ebie 1-chome, Fukushima-ku, Osaka 533
Tel: (+81 6) 457 21 23

Key personnel
Chairman: Shunjiro Kuma
President: Masatoshi Tezuka

Gauge: 1,435 mm
Route length: 40 km
Electrification: 40 km at 1.5 kV DC

Organisation
The Hanshin Electric Railway is at the centre of the Hanshin Group of approximately 60 affiliated companies extending into transport, retailing, real estate, sports, leisure, construction and computer-related business. Rail operations account for 41 per cent of group turnover.

Passenger operations
The railway operates a main line between Osaka and Kobe with two branches. In 1993–94 the railway recorded 240 million passenger journeys and 2,249 million passenger-km. The Hanshin earthquake of January 1995 affected performance in 1994–95 with passenger journeys down to 229 million and passenger-km down to 2,187 million.

Services extend via the underground Kobe Rapid Railway on to the Sanyo Electric Railway to the west of Kobe; reciprocal through running services are operated, providing a metro-style service across central Kobe.

All cars and underground stations are equipped with air conditioning. Programmed Traffic Control (PTC) and Automatic Train Stop (ATS) systems have been introduced. The proportion of main line in tunnel or elevated is due to rise to 87 per cent within 10 years through elimination of level crossings.

The railway was severely damaged in the January 1995 earthquake, with buses replacing trains between central Kobe and the eastern suburbs and all through-running services via the Kobe Rapid Railway suspended. Full service was restored, however, by June 1995. From the total fleet of 314 cars, 41 were written off as a result of earthquake damage. New replacement vehicles comprised two 4-car Series 5500 units, three Series 8000 cars and five 6-car Series 9000 units delivered during 1995–96.

Iyo Railway

Iyo Tetsudo
4-1, Minatomachi 4-chome, Matsuyama-shi 790
Tel: (+81 899) 48 33 21

Key personnel
President: H Nagano

Gauge: 1,067 mm
Route length: 33.9 km
Electrification: 24.5 km at 750 V DC and 9.4 km at 600 V DC

Passenger operations
The railway operates 53 electric multiple-unit cars on two lines serving Matsuyama, on the island of Shikoku. Apart from its 33.9 km railway, it operates a 6.9 km urban tramway system in Matsuyama with 36 cars. There were 24 million passenger journeys in 1994–95.

Izu Hakone Railway

Izu Hakone Tetsudo
300, Daiba, Mishima-shi 411
Tel: (+81 559) 77 12 00

Key personnel
Chairman: Yoshiaki Tsutsumi
President: Kakuro Kato

Gauge: 1,067 mm
Route length: 29.4 km
Electrification: 29.4 km at 1.5 kV DC

Passenger operations
This Seibu Railway subsidiary operates two electric locomotives and 60 emu cars on the 9.6 km Daiyousan line in Kanagawa prefecture and the 19.8 km Sunzu line in Shizuoka prefecture. JR East limited express 'Odori-Ko' trains run through from Tokyo to the Sunzu line. The company also operates two funiculars, buses and a cruise ship. Passengers made 12.4 million journeys in 2000, accounting for 15 per cent of the company's revenue.

Izukyu Corporation

Izu Kyuko (Izukyu)
1-21-6 Dogenzaka, Shibuya-ku, Tokyo 150-0043
Tel: (+81 3) 34 96 71 55

Key personnel
President: Shun-ichi Oki

Gauge: 1,067 mm
Route length: 45.7 km
Electrification: 45.7 km at 1.5 kV DC

Passenger operations
The railway, a Tokyu Corporation affiliate, opened in 1961, serves the eastern coastline of the Izu peninsula from Ito to Izukyu-Shimoda. For the most part this is territory of the Fuji Hakone Izu national park, not far from Tokyo, and rolling stock is geared to tourist travel. At Ito the railway connects with the JR East branch from Atami. Most local services operate through to Atami and JR 'Odori-Ko' limited expresses run through from Tokyo to Izukyu destinations. The railway operates 68 emu cars and carried 6.3 million passengers in 2000; its operations account for 37 per cent of the company's revenue.

Keifuku Electric Railway

Keifuku Denki Tetsudo
3-20, Mibu-kayo-goshomachi, Nakagyo-ku, Kyoto 604
Tel: (+81 75) 841 93 81

Key personnel
Chairman: Minoru Miyashita
President: Seiya Yamakami

Gauge: 1,067 mm
Route length: 59.2 km
Electrification: 59.2 km at 600 V DC

Passenger operations
The railway operates 30 emu cars and two electric locomotives on three 1,067 mm gauge lines totalling 59.2 km at Fukui. It also operates a separate 11 km 1,435 mm gauge tramway system in Kyoto with a fleet of 28 cars. Railway operations account for 27 per cent of the company's revenue.

Keihan Electric Railway Co Ltd

Keihan Denki Tetsudo
2-27, Shiromi 1-chome, Chuo-ku, Osaka 540
Tel: (+81 6) 944 25 21   Fax: (+81 6) 944 25 01

Key personnel
Chairman: Minoru Miyashita
President: Akio Kinba
Executive Vice-President, Accounts, Finance, Purchases, Subsidiaries: Yutaka Ogura
Senior Managing Director, Railway Operation, Electrical Engineering, Rolling Stock: Rikuro Kimura
Chief Manager, New Business Planning and Development: Hiroshi Yoshida
Managing Director, Personnel, Audit, Information Systems: Junzo Takai
Deputy Chief Manager, Business Planning and Developments: Toru Nakanishi
Otsu Branch: Kazuyuki Hasegawa
Director, New Line Construction: Kihachiro Nakaichi
Railway Operation: Kimio Nishimura
Accounting and Finance: Hiroichi Shimizu

Gauge: 1,435 mm
Route length: 91.5 km
Electrification: 66.3 km at 1.5 kV (Keihan line); 25.2 km at 600 V DC (Otsu line)

Passenger operations
The railway's main Keihan line runs 51.6 km from its Yodoyabashi station in Osaka to the underground Demachi Yanagi station in Kyoto City, with two branches, the Uji and Katano lines. The 11.5 km of continuously elevated quadruple-track between Temmabashi and Kayashima stations is the longest on any Japanese private railway. The Otsu line is light rail extending 25.2 km from Kyoto Sanjo to Hamaotsu, Ishiyama-dera and Sakamoto.

Serving 89 stations, the railway recorded 401 million passenger journeys and 5,319 million passenger-km in 1994–95.

The railway generates 58 per cent of the company's income. The rest is derived from 65 subsidiary and affiliated companies covering activities ranging from bus, taxi and leisure boat services to hotels, retail stores and construction.

Keihan electric railcars or multiple-units

Class	Cars per unit	Motor cars per unit	Motored axles/car	Output/motor kW	Speed km/h	Cars in service	First built	Builders Mechanical	Builders Electrical
Keihan line 1.5 kV DC									
9000	8						1995		
8000	7	4	4	175	110	70	1989	K/S/N	To/T/M
7000	7	3	4	200	110	28	1989	K/S/N	To/T/M
6000	8	4	4	155	110	112	1983	K/S/N	To/T/M
5000	7	4	4	155	110	49	1970	K/S/N	To/T/M
3000	7	4	4	175	110	9	1971	K/S/N	To/M
2600	4/5/6	2/3	4	155	110	131	1978	K/S/N	To/T/M
	7	4							
2200	7/8	4	4	155	110	100	1964	K/S/N	To/T/M
2400	7	4	4	155	110	42	1969	K/S/N	To/M
1000	7	4	4	155	110	42	1977	K/S/N	T/M
1900	5	4	4	108	110	45	1956	S/N	T/M
7200	1/3/4	1/2	4	200	110	16	1995	K/S	
Otsu line 600 V DC									
700	2	2	2	70	70	10	1992	Ke/S/N	To/T
600	2	2	4	53	70	20	1984	Ke/S/N	To/T
80	2	2	4	45	65	16	1961	Ki/N	To/T
260	2	2	4	45	65	8	1957	Ki/N	To
350	2	2	2	85	65	7	1966	Ki/N	To

Abbreviations: K: Kawasaki; Ke: Keihan; Ki: Kinki; M: Mitsubishi; N: Nabco; S: Sumitomo; T: Toshiba; To: Toyo Denki Seizo

Future developments include through operation from Otsu to Kyoto City's new Tozai underground line with new Series 8000 four-car emus delivered in readiness for this and construction of a Nakanoshima line westward through Osaka City to connect with a new line serving the Kansai International Airport.

Traction and rolling stock
Series 8000 stock is for limited express services and features onboard television and a telephone booth. Five double-deck cars were ordered for delivery in 1997 and another five in 1998. They are based on the double-deck car incorporated into an older Series 3000 unit in December 1995 and will eventually see all limited express units including a double-deck car. The latest emu type is the eight-car Series 9000.

Signalling and telecommunications
Autonomous decentralised traffic control system (ADEC) is installed throughout the Keihan line. It links central control's mainframe computer and local processors by fibre optic cable, and allows overall traffic control to proceed simultaneously with independent control of an area's operation where that has become irregular.

Coupler in standard use: Tightlock automatic couplers; rotary key-block Tightlock automatic couplers; rod couplers
Braking in standard use: Electric command digital and electric pneumatic brake

Double-deck car added to existing Series 3000 set (Anthony Robins)   0001715

Track
Rail: Keihan line: 50 kg/m (144.2 km) and 60 kg/m; Otsu line 40 kg/m (47.8 km) and 50 kg/m
Crossties (sleepers): Keihan: concrete, thickness 170 mm; spacing 1,760/km; Otsu: wood, thickness 150–170 mm; spacing 1,520/km
Fastenings: Elastic (spring clip or F type)

Min curvature radius: Keihan line: 200 m; Otsu line: 45 m
Max gradient: Keihan: 3.3%; Otsu: 6.7%
Max permissible axleload: Keihan: 15 tonnes; Otsu: 8–15 tonnes

Keihin Electric Express Railway

Keihin Kyuko Dentetsu
2-20-20, Takanawa, Minato-ku, Tokyo 108-8625
Tel: (+81 3) 32 80 91 23   Fax: (+81 3) 32 80 91 93
Web: www.keikyu.co.jp

Key personnel
Chairman: Ichiro Hiramatsu
President and Representative Director:
  Masaru Kotani
Senior Managing and Representative Directors:
  Tsuneo Ishiwata, Yukimasa Saitoh
Managing Directors: Takayuki Goseki,
  Kiyoshi Ishikawa, Masahiro Satoh
Senior Managing Director: Masaru Okino

Gauge: 1,435 mm
Route length: 87 km
Electrification: 1.5 kV DC

Background
Keihin Electric Express Railway is part of the Keikyu Group, which comprises 78 companies covering diverse sectors including property, tourism and leisure and distribution, as well as transport. The Transportation Group consists of 19 companies, including scheduled bus services on 576 routes totalling 5,100 km and trucking businesses. However, in 2003 the company initiated the sale of its bus operations following intensified competition resulting from the deregulation of scheduled services.

Finance
In FY2004-05 the company posted transport operating revenues of JPY117 billion, a rise of 0.2 per cent from the previous year. Operating income was JPY16.3 billion.

Passenger operations
The railway extends from Shinagawa in Tokyo, southward to Yokohama and the Miura peninsula, and exercises through-running over the Asakusa line of the Tokyo metro. The railway recorded some 424 million passenger journeys in 2003–04.

In 1993 through services commenced between Keihin Kyuko's Haneda branch line, serving Haneda airport, and central Tokyo via the Asakusa line metro. Initially, passengers were required to transfer to the newly extended Tokyo Monorail for final access to the airport, but in November 1998 an underground extension of the Haneda line to the new airport terminal building was commissioned. Through services between Haneda and Narita airports, in collaboration with other operators, were implemented in November 1998. By mid-2004 the number of passengers using the service to Haneda had exceeded 100 million. Further growth was expected following the opening of a new terminal at the airport in 2004.

Improvements to existing lines
In late 2003 a new Shinkansen station opened at the railway's main terminal at Shinagawa.

Traction and rolling stock
The railway owns approximately 758 emu cars (606 motor cars and 152 trailers).

Track
Laid with 50 kg/m rail on 165 mm-thick prestressed concrete sleepers with elastic fastenings, with a minimum curvature of 60 m radius. Maximum axleloading is 13.7 tonnes.

UPDATED

Japan
Tokyo locality

0	10 Miles
0	15 Kilometres

Narita Airport

Tokyo Metropolis

Sengakuji
Shinagawa

Keikyu Kamata
Haneda Airport
Keikyu Kawasaki

Yokohama

Kamiooka

Keihin Express Electric Railway
— In operation
═ Operating rights

Yokosuka-chuo
Shinzushi   Uraga

Misakiguchi

0132277

Keihin electric railcars or multiple-units

Class	Cars per unit	Motor cars per unit	Motored axles/car	Output/ motor kW	Speed km/h	Cars in service	First built	Builders Mechanical	Builders Electrical
1000	2	2	4	75/90	110	170	1958	K/Tk/S	M/To/Ko
	4	4	4	75/90					
	6	6	4	75/90					
	8	8	4	90					
800	6	6	4	100	100	132	1978	K/Tk/S	M/To/Ko
2000	8	6	4	120	120	72	1982	K/Tk/S	M/To/Ko
	4	3	4	120					
700	4	2	4	150	110	44	1967	K/Tk/S	M/To/Ko
1500	6–8	6	4	100/120	120	166	1988	K/Tk/S	M/To/Ko
	4	4	4	100	120				
600	8	6	4	120	120	88	1993	K/Tk/S	M/To/Ko
	8	4	4	180					
	4	2	4	180					
2100	8	4	4	190	120	80	1998	K/Tk/S	Siemens/M/To/Ko
N1000	8	4	4	190	120	8	2002	K/Tk/S	Siemens/M/To/Ko

Abbreviations: K: Kawasaki; Ko: Koito; M: Mitsubishi; S: Sumitomo; Tk: Tokyu; To: Toyo Denki Seizo

Keio Teito Electric Railway

Keio Teito Dentetsu
9-1, Sekido 1-chome, Tama City, Tokyo 206
Tel: (+81 423) 37 31 41

Key personnel
Chairman: K Kuwayama
President: H Nishiyama

Gauge: 1,372 mm; 1,067 mm
Route length: 72 km; 12.8 km
Electrification: 84.8 km at 1.5 kV DC

Organisation
The railway runs northwest from Shibuya in
Tokyo and southwest from Shinjuku, exercising
through running over the Shinjuku line of the
Tokyo metro. The railway, one of 46 companies in
the group, generates 64 per cent of the group's
income. It operates 848 emu cars, and in 1996–97
recorded some 587 million passenger journeys and
6,938.1 million passenger-km, compared with
588 million and 6,954.7 million respectively in
1995. New Series 1000 emus with 20 m long cars
have been introduced to increase capacity on the
narrow-gauge Inokashira line.
    The company employs 4,303 staff.

VVVF inverter-controlled Series 1000 narrow-gauge emu on Keio's Inokashira Line (Takashi Kasai)
0023034

Keisei Electric Railway

Keisei Dentetsu
1-10-3 Oshiage 1-chome, Sumida-ku, Tokyo 131
Tel: (+81 3) 36 21 22 31

Key personnel
President: Hiroto Senoo

Gauge: 1,435 mm
Route length: 91.6 km
Electrification: 1.5 kV DC

Organisation
The railway stretches eastward from Ueno in Tokyo
to Chiba and to the station in the international
airport of Narita, to which it operates its 'Skyliner'
expresses at half-hour frequency. It has a reciprocal
through service with the Asakusa line of the Tokyo
metro and with the Chiba Express Electric Railway.
Passenger journeys in 1994–95 totalled some
280.4 million, passenger-km 3.860 billion. Rail
revenues form 62 per cent of the company's
gross, the remainder coming from buses and real

estate. The company has a half-share in Tokyo's
Disneyland, but has no rail connection with the
site. The railway operates 496 emu cars.

Track
Rail type and weight: 50N 50 kg/m
Crossties (sleepers): Type K2
Thickness: 172 mm

Spacing: In plain track 1,560/km; in curves 1,680/km
Fastenings: Dogspike (double elastic fastening)
Min curvature radius: 160 m
Max gradient: 3.5%
Max axleload: 14.75 tonnes

Keisei electric railcars

Class	Cars per unit	Motored axles/car	Output/motor kW	Speed km/h	First built	Builders Mechanical	Electrical
AE 100	6 Mo 2 Tr				1990	Tk/N	
AE	6 Mo 2 Tr	4	140	120	1972	Tk/N	T/To/M
3600	4 Mo 2 Tr	4	140	120	1981	Tk/N	T/To/M
3500	4 Mo	3	100	120	1972	Tk/N/K	T/To/M
3050	4 Mo	4	75	120	1959	N/Ki/Te	T/To/M
3100	4 Mo	4	75	120	1960	N/Ki/Te	T/To/M
3150	4 Mo	4	75	120	1963	N/Ki/Te	T/To/M
3200	4 Mo/6 Mo	3	100	120	1964	N/Ki/Tk	T/To/M
3300	4 Mo/6 Mo	3	100	120	1968	N/Ki/Te	T/To/M
3700	6 Mo 2 Tr				1991	N	
3400	6 Mo 2 Tr				1993		

Abbreviations: Ki: Kisha; M: Mitsubishi; N: Nippon; T: Toshiba; Te: Teikoku; Tk: Tokyu; To: Toyo; Mo: motor; Tr: trailer

Kinki Nippon Railway

Kinki Nippon Tetsudo (Kintetsu)
6-1-55, Uehommachi, Tennoji-ku 6-chome, Osaka 543
Tel: (+81 6) 775 34 44   Fax: (+81 6) 775 34 68
Web: http://www.kintetsu.co.jp

Key personnel
President: Akio Tsujii
Chairman: WA Tashiro

Gauge: 1,435 mm; 1,067 mm; 762 mm
Route length: 404.4 km; 162.3 km; 7 km
Electrification: 1.5 kV DC: (394.2 km of 1,435 mm
gauge; 159 km of 1,067 mm gauge)
    750 V DC: (10.2 km of 1,435 mm gauge; 7 km of
762 mm gauge)

Organisation
The Kinki Nippon Railway is part of the Kintetsu
group of about 250 associated companies,
ranging from hotels and construction concerns
to bus and taxi companies, and including Japan's
largest railway rolling stock manufacturer, Kinki
Sharyo.
    The Kinki Nippon has the most extensive
route-km of any of Japan's private railways and
lies third in the table of passenger movement,
carrying about two million daily. Extending
eastward from Namba and Abenobashi termini

Series 21020 'Urban Liner' emu
0595238

in Osaka to Kyoto, Nara, Nagoya and Ise Bay, the
railway runs limited expresses throughout the
Kinki and Tokai areas. Its main line is intercity,
running 190 km from Osaka to Nagoya, and
it also serves Nara, Kyoto and the Ise-Shima
National Park (136 km from Osaka). A 10.2 km
extension, 4.7 km of it in tunnel, was opened
from Ikoma to Nagata in 1986; at Nagata it makes
a junction with the Osaka metro, over which
through service is run to central Osaka.
    The railway has its own research laboratory and
generates 79 per cent of the income of its parent
company.

Passenger operations
In 2001–02 the Kinki Nippon recorded 664 million
passenger journeys and 12,874 million passenger-km.

Traction and rolling stock
Rolling stock comprises 1,999 emu cars, including
bilevel and vista-dome vehicles.
    In 2002 the company introduced its Series
21020 'Urban Liner' emus on its non-stop service
between Nagoya and Osaka. The Kinki Sharyo-built
units previously used on this service were due to
undergo refurbishment.

Kobe Electric Railway

Kobe Dentetsu
1-1, Daikai-dori 1-chome, Hyogo-ku, Kobe 652
Tel: (+81 78) 575 22 36   Fax: (+81 78) 577 24 67

Key personnel

President: Yasuo Ipponmatsu

Gauge: 1,067 mm
Route length: 64.4 km
Electrification: 1.5 kV DC

Konan Railway

Konan Tetsudo
23-5, Kita-Yanagida, Hommachi, Hiraka-machi, Minami-Tsugaru-gun, Aomori 036-01
Tel: (+81 172) 44 31 36

Key personnel

President: T Tarusawa

Nagoya Railroad

Nagoya Tetsudo (Meitetsu)
1-2-4, Meikei, Nakamura-ku, Nagoya-shi 450-8501
Tel: (+81 52) 571 21 11   Fax: (+81 52) 581 60 60
Web: www.meitetsu.co.jp

Key personnel

Chairman: Sokichi Minora
President: M Kimura

Gauge: 1,067 mm
Route length: 503 km
Electrification: 479 km at 1.5 kV DC; 59 km at 600 V DC

Organisation

Between 1941 and 1944 the private urban railways of the Nagoya region were knitted into a coherent regional network by the conversion of city-centre tram tracks into an inter-system connection focused on a new underground Shin-Nagoya station alongside the Japan Rail station. Since then the system has been rationalised by some closures, but new lines have been laid to cater for fresh suburban development, such as the Chita line in 1980. The network includes 36.6 km of 600 V DC tramway and light rail in Gifu prefecture. Besides the Nagoya Railroad, the diversified Meitetsu Corporation also runs bus, taxi, road freight, sea ferry and air-taxi services, hotels, restaurants and travel agencies, but the railway produces 61 per cent of total revenue.

Passenger operations

Shin-Nagoya station handles over 800 trains a day, with 25 trains hourly each way on the main route between Shin Gifu/Inuyama and Toyohashi/Toko-name/Kowa. Reciprocal through-running services operate between the Toyota and Inuyama lines and the Tsurumai line of the Nagoya metro. Airport

Organisation

This Hankyu subsidiary operates to Ao, Sanda and the hot-spring resort of Arima-Onsen via a steeply graded route through the Rokko mountain range north of Kobe. Within Kobe trains operate over 0.4 km of Kobe Rapid Railway tracks to an underground terminus at Shin-Kaichi where interchange is available with Hankyu, Hanshin and Sanyo services. A 5.5 km branch to serve a new town at Kobe-Sanda Garden City was completed in March 1996. Rolling stock comprises one electric locomotive and 167 emu cars. Railway

Gauge: 1,067 mm
Route length: 30.7 km
Electrification: 30.7 km at 1.5 kV DC

Organisation

The railway operates two unconnected electrified lines in and around Hirosaki, Aomori prefecture, from Hirosaki to Kuroishi and from Chuo Hirosaki to Owani Onsen. They carried 3.5 million passengers

operations represent 68 per cent of the company's revenue.

The Kobe earthquake of January 1995 caused damage estimated at ¥14.9 billion and resulted in the suspension of services between central Kobe and Suzurandai, a distance of 7.5 km, for approximately six months.

in 2000. Rolling stock comprises three electric locomotives, three diesel railcars and 43 emu cars.

Principal Nagoya Railroad electric railcars or multiple-units

Class	Cars per unit	Motor cars per unit	Motored axles/car	Output/motor kW	Speed km/h	Cars in service	First built	Builders Mechanical	Electrical
1000	4	2	4	150	120	92	1988	N	T
1200	6	4	4	150	120	64	1991	N	T
7000	4	4	4	75	110	106	1961	N	T
7500	6	6	4	75	110	42	1963	N	To
6500	4	2	4	150	110	96	1984	N	T
6800	2	1	4	150	110	78	1987	N	M
3500	4	2	4	170	120	136	1993	N	To

Abbreviations: M: Mitsubishi; N: Nippon; T: Toshiba; To: Toyo

Principal Nagoya Railroad diesel railcars or multiple-units

Class	Cars per unit	Motor cars per unit	Motored axles/car	Power/motor kW	Speed km/h	Cars in service	First built	Builders Mechanical	Engine	Transmission
20	1	1	2	187.5	80	5	1987	Fuji	Nissan	Shinko
30	1	1	2	187.5	80	4	1995	Fuji	Nissan	Niigata
8500	2/3	2/3	2	262.5	120	5	1991	Nippon	NTA-855-R	Niigata

Express services link Nagoya airport with Meitetsu and JR stations. Electrification was progressively standardised at 1.5 kV DC after the unification of the system, but 600 V DC survives on tramway and light rail lines in Gifu and to the north of that city. Passenger traffic in 2003–04 via all modes amounted to 331 million journeys.

New lines

An extension of the Toko-name Line was under construction in 2004 to provide a link from Shin-Nagoya station to Central Japan International Airport (35 km). To be commissioned in 2005, the airport is sited on reclaimed land offshore.

Traction and rolling stock

Rolling stock includes 1,097 emu cars and 41 tramcars.

Coupler in standard use: Passenger cars: Tightlock automatic; freight cars: automatic
Braking in standard use: Electromagnetic

Track

Rail: 50 kg/m; 37 kg/m
Crossties (sleepers): Prestressed concrete: 200 × 160 × 2,000 mm; wood: 200 × 140 × 2,100 mm
Spacing: 1,640/km
Fastenings: Tie plate, dogspike
Min curvature radius: 160 m
Max gradient: 3.5%

UPDATED

Nankai Electric Railway

Nankai Denki Tetsudo
1-60 Namba 5-chome, Minami-ku, Osaka 542
Tel: (+81 6) 644 71 20

Key personnel

Chairman: Shigeo Yoshimura
President: Taiji Kawakatsu

Gauge: 1,067 mm
Route length: 171.7 km
Electrification: 157.4 km at 1.5 kV DC; 14.3 km at 600 V DC

Passenger operations

The railway operates 724 emu cars, and in 1995–96 recorded 302.6 million passenger journeys. The main line runs from Namba in Osaka to Wakayama (64.2 km), with five branches, and the Koya

Series 31000 three-car emu

0063750

line links Osaka with the pilgrimage and resort destination of Koya-San. Reciprocal through-running services operate over the Semboku Rapid Railway, built to serve new town development to the south of Osaka. The company also operates a 14.3 km 600 V DC line at Wakayama and a 0.8 km funicular. Railway operations represent 63 per cent of the company's revenue.

New lines

A new 8.8 km spur connecting Nankai's main line with the new offshore Kansai International Airport opened in September 1994, with Nankai's 'Rapi:t' airport service operated by new Series 50000 stock. These distinctively styled five-car units have porthole-shaped side windows and an unusual external cab design. To compete with JR for airport traffic, Nankai proposes to lay mixed-gauge track along its 1,067 mm gauge main line to enable airport services to operate through central Osaka via the 1,435 mm gauge Sakaisuji line metro and on to Kyoto and Nara via the Hankyu, Keihan and Kintetsu systems.

Nishi Nippon Railroad

Nishi Nippon Tetsudo
11-17 Tenjin 1-chome, Chuo-ku, Fukuoka 810
Tel: (+81 92) 761 66 31
Fax: (+81 92) 722 14 05

Key personnel
Chairman: Reinosuke Ohya
President: Hisayuki Hashimoto

Gauge: 1,435 mm; 1,067 mm
Route length: 100 km; 21 km
Electrification: 1.5 kV DC; 600 V DC

Organisation
The railway, which owns 355 emu cars, operates a 75 km main line from Fukuoka to Omuta, with branches, and a separate 21 km line from Kaizuka in the eastern suburbs of Fukuoka to Tsuyazaki. It also operates a 5 km 600 V DC line with 19 cars.

Track is laid with 37, 40 and 50 kg/m rail on timber and prestressed concrete sleepers spaced at 1,560/km and allowing 16 tonnes maximum axleloading; minimum curvature is 160 m. Traffic totalled some 154.3 million passenger journeys and 2.105 billion passenger-km in 1994–95. Railway operations represent 17 per cent of the company's revenue.

Odakyu Electric Railway Co Ltd

Odakyu Dentetsu
8-3, 1-chome, Nishi Shinjuku, Shinjuku-ku,
Tokyo 160-8309
Tel: (+81 3) 33 49 21 51; 33 49 25 26
Fax: (+81 3) 33 46 18 99; 33 49 24 47
e-mail: ir@odakyu-dentetsu.co.jp
Web: www.odakyu-group.co.jp

Key personnel
President: T Matsuda

Gauge: 1,067 mm
Route length: 120 km owned: also operates over 56 km of other operators' tracks
Electrification: 1.5 kV DC

Organisation
The railway runs three routes southwest from Shinjuku, Tokyo, to Odawara, Hakone, Tama New Town, the Mount Fuji region and the coast west of Yokohama. It has through service arrangements with 6 km Hakone Tozan Railway (a subsidiary company) and from Matsuda to Nomazu over JR's Gotemba Line (50 km). A 10 km light rail system, the Enoshima Electric Railway, is also operated by the group, which has more than 100 individual companies and obtains 63 per cent of its income from rail transport. Other group transport activities include long-distance and charter bus services, and taxi and hire businesses.

In 2005 the company employed 3,577 staff.

Passenger operations
In 2004–05 the railway recorded 670 million passenger journeys. At peak periods the railway's core handles 29 trains each way per hour. A limited stop express service is operated between Shinjuku and Hakone. Named the 'Odakyu Romancecar', the service gained new rolling stock in March 2005.

The railway has adopted ATP (Automatic Train Protection) based on coded frequency circuitry and a total traffic control system with both relay and electronic interlockings.

Improvements to existing lines
Odakyu Electric Railway has been implementing a major capacity enhancement scheme in the Tokyo metropolitan area, expanding double-track sections of line to four tracks. A first 2.4 km section between Kitami and Izumi-Tamagawa was completed in 1997, followed by a second

Odakyu commuter trains on recently quadrupled track in the Tokyo suburbs
(Odakyu Electric Railway Co Ltd)   **NEW**/1143557

Series 50000 VSE emu for Odakyu Romancecar limited express service
(Odakyu Elctric Railway Co Ltd)   **NEW**/1143558

section, Setagaya-Daita–Kitami (6.4 km) in November 2004. In 2005 work was in progress on the final section of the scheme, Higashi-Kitazawa–Setagaya-Daita (1.6 km), for completion by 2014. As well as providing quadruple track, the project also entails elimination of level crossings.

Traction and rolling stock
The railway owns 1,065 emu cars. The latest vehicles to be procured include Model 50000 VSE emus for the Odakyu Romancecar service.

UPDATED

Ohmi Railway

Ohmi Tetsudo
3-1 Daito-cho, Hikone-shi 522
Tel: (+81 749) 22 33 01

Key personnel
President: T Kaida

Gauge: 1,067 mm
Route length: 59.5 km
Electrification: 59.5 km at 1.5 kV DC

Organisation
This company is a Seibu Railway subsidiary. It operates 33 emu cars, 10 electric and one diesel locomotive, five diesel railbuses and 20 wagons.

Emus operate between Maibara, an interchange with the Tokaido Shinkansen, and Ohmi-Hachiman, and also on the Toga branch line, but diesel railbuses have been introduced between Yokaichi and Kibukawa to reduce costs.

Sagami Railway

Sagami Tetsudo
2-9-14 Kitasaiwai, Nishi-ku, Yokohama 220
Tel: (+81 45) 319 21 11

Key personnel
Chairman: Masahiro Hoshino
President: Kojiro Tsushima

Gauge: 1,067 mm
Route length: 38 km
Electrification: 1.5 kV DC

Sanyo Electric Railway

Sanyo Denki Tetsudo
1-1, Oyashiki-dori 3-chome, Nagata-ku, Kobe 653
Tel: (+81 78) 611 22 11

Key personnel
Chairman: Tadashi Suzuki
President: Masami Ohkuni

Seibu Railway

Seibu Tetsudo
1-11-1 Kusunokidai, Tokorozawa-shi, Saitama 359
Tel: (+81 429) 26 20 35   Fax: (+81 429) 26 22 37

Key personnel
Chairman: Yoshiaki Tsutsumi
President: Hiroyuki Toda

Gauge: 1,067 mm; 762 mm
Route length: 176 km; 3.6 km (rubber-tyred)
Electrification: 1.5 kV DC

Organisation
Part of a multifaceted corporation that includes hotels, Japan's busiest department store, housing and road transport among its businesses, the railway serves the western suburbs of Tokyo with two main routes radiating from terminals on or near JR East's city-centre Yamanote loop, the 43.8 km Ikebukuro and 22.6 km Shinjuku lines. These lines throw off and are in some cases interconnected in the suburbs by 10 branches. The Ikebukuro terminus deals with an average of 700 train workings daily, with departures at 1½ minute headways in the evening peak. There is a reciprocal through service with the Yurakucho line of the Tokyo metro. The Ohmi Railway is a Seibu subsidiary.

Passenger operations
Traffic ran at some 661.8 million passenger journeys (9,489 million passenger-km) in 1994–95,

Passenger operations
This Yokohama suburban system comprises a 24.6 km main line from Yokohama to Ebina with a branch to Izumi-chuo. A 3.1 km extension from Izumi-chuo to connect with the Odakyu Enoshima line and extended Yokohama metro at Shonan-dai opened in 1999.

Rolling stock comprises 442 emu cars of seven types built by Hitachi and Tokyu Sharyo and four electric locomotives. Passenger journeys totalled 233 million in 2000; patronage has gradually declined from a record 251 million in 1995. Railway

Gauge: 1,435 mm
Route length: 63.3 km
Electrification: 63.3 km at 1.5 kV DC

Passenger operations
The railway operates between Himeji and Kobe with a fleet of 199 emu cars.

Within Kobe, trains run via the underground Kobe Rapid Railway and on to the Hankyu and

Series 2000 emu formation (Hiroshi Naito)   0023037

67 per cent of which was made on season tickets which generate 46 per cent of total receipts. The railway's operating ratio is approximately 87 per cent. Railway operations represent 46 per cent of the company's revenue.

Traction and rolling stock
The fleet of 1,133 emu cars comprises 12 types. Further Series 10000 seven-car units built by

Shimabara Railway

Shimabara Tetsudo
7385-1 Bentencho 2-chome, Shimabara-shi 855
Tel: (+81 95) 762 22 31

Key personnel
President: D Shirakuta

Gauge: 1,067 mm
Route length: 78.5 km

Passenger operations
The railway operates one diesel locomotive and 25 diesel railcars on its line serving the Shimabara peninsula in Nagasaki prefecture. Complete service over the line was restored in April 1997 following

Shin Keisei Electric Railway

Shin Keisei Dentetsu
16-16 Hatsutomi, Kamagaya-shi 273-01
Tel: (+81 473) 84 31 51

Key personnel
Chairman: Haruo Hosokawa
President: Naoyuki Takeuchi

Gauge: 1,435 mm
Route length: 26.5 km
Electrification: 1.5 kV DC

Passenger operations
The railway was constructed to serve suburban housing development in Chiba prefecture to the east of Tokyo, and connects with its parent Keisei Railway at Tsudanuma and with JR East's Joban

operations represent 29 per cent of the company's revenue.

Hanshin systems under a reciprocal through-service agreement. Through-running services were suspended, however, between January and June 1995 following the Hanshin earthquake which severely damaged the Kobe Rapid Railway. Railway operations represent 74 per cent of the company's revenue.

Hitachi have been introduced to allow an increase in limited express services on the Ikebukuro and Shinjuku lines.

closure of part of the middle section of the line due to volcanic activity.

line at Matsudo. Rolling stock comprises 194 emu cars operated as six- and eight-car trains. Railway operations represent 56 per cent of the company's revenue.

Tobu Railway

Tobu Tetsudo
1-2, 1-chome, Oshiage, Sumida-ku, Tokyo 131
Tel: (+81 3) 36 21 50 57

Key personnel
Chairman: Kaichiro Nezu
President: Takashige Uchida

Gauge: 1,067 mm
Route length: 464 km
Electrification: 464 km at 1.5 kV DC

Organisation
The railway's main line runs 135 km northward from Asakusa in Tokyo to Shimoimaichi and Nikko, with branches to Utsunomiya and Isezaki and through-service arrangements with the Hibiya line of the Tokyo metro. The Tojo line runs 75 km from Ikebukuro in Tokyo to Yorii, with through running from the Yurakucho line of the Eidan metro. There are 88 companies in the group, which earns 66 per cent of its revenue from the railway.

Passenger operations
Its daily average of almost 2.5 million passengers ranks second in the table of the busiest private railways in Japan; the total in 1994–95 was 945.3 million passenger journeys and 14,366 million passenger-km.

Series 100 'Spacia' limited express emu (Anthony Robins)   0001717

Improvements to existing lines
The railway is currently undertaking ¥84 billion worth of infrastructure improvements, including work completed at the important interchange of Kita-Senju. A further ¥196.1 billion will be spent on works including connection of the Isezaki line to the new Number 11 underground line. These projects are scheduled to be completed in 2004.

Traction and rolling stock
Rolling stock comprises 10 electric locomotives, 1,848 emu cars and 13 wagons.

Tokyo Express Electric Railway

Tokyo Kyuko Dentetsu (Tokyu)
5-6 Nanpeidai-cho, Shibuya-ku, Tokyo 150
Tel: (+81 3) 34 77 61 11   Fax: (+81 3) 34 96 29 65

Key personnel
President: Shinobu Shimizu
Executive Vice-President: Ototaro Horie
Senior Managing Directors: K Endoh, T Endoh,
   K Ihara, H Nagatoshi

Gauge: 1,372 mm; 1,067 mm
Route length: 5 km; 96 km
Electrification: 5 km (1,372 mm gauge) at 600 V DC; 96 km at 1.5 kV DC

Organisation
The railway is part of the Tokyu Corporation, the flagship of the massive Tokyu Group, which embraces 400 companies and nine non-profit institutions covering real estate enterprises, bus and taxi companies, department stores, supermarkets, construction companies, road freight, railcar building, advertising agencies, construction companies, shipping, airline (Japan Air System, the country's third largest carrier) and air freight activity. Railway operations represent 44 per cent of the company's revenue, in 2000 generating turnover of ¥128 billion.

The rail network is located in the southwest of the Tokyo metropolitan area and runs from Shibuya, Meguro and Gotanda on JR's Yamanote line to Kanagawa prefecture and Yokohama. The 9.4 km Shin-Tamagawa line out of Shibuya is effectively a metro, and the railway also has through-service arrangements with the Tokyo metro's Hibiya and Hanzomon lines.

Passenger operations
With a total of 947 million passenger journeys in 2000, the Tokyo Express Electric Railway is the busiest railway in the capital's suburban area, and carries more than any other private railway in the country. Services are provided on eight lines. Although it claims to be the cheapest Japanese railway to ride, it also claims to generate the biggest revenue per track-km. Average length of passenger journey is only 9.1 km. Annual passenger-km totals 8,680 million.

Improvements to existing lines
On its hugely congested Toyoko line the railway is undertaking a track-quadruling programme costing ¥210.8 billion. It is also elevating more of its infrastructure to alleviate road congestion. Construction started in 1995 on a 3.3 km underground section replacing part of the existing Mekama line and providing a connection with TRTA's Namboku and TMG's Mita metro lines to allow for through services; this was completed in 2002. Reconstruction and extension of the Oimachi line was also being undertaken to allow central Tokyo passengers to bypass the Den-en Toshi and Shin-Tmagawa lines.

Traction and rolling stock
In 1996 Tokyu operated 1,054 emus, most on 1,067 mm gauge, but including 18 two-car units on 1,372 mm gauge. Most of the vehicles have been built in the Corporation's own workshops. Tokyu Car Corporation is one of the group's subsidiaries.

Signalling and telecommunications
Automatic Train Control (ATC) is in use on 61.1 km of the system.

Track
Tokyu employs 50 kg/m rail on prestressed concrete sleepers of 150 to 160 mm thickness, with double elastic fastenings. Minimum curve radius is 160 m, maximum gradient 4 per cent, and maximum permissible axleload 15.5 tonnes.

Electric railcars or multiple-units

Class	Cars per unit	Motor cars per unit	Motored axles/car	Output/motor kW	Speed km/h	No in service	First built	Builders Mechanical	Builders Electrical
1.5 kV DC									
7000	2	2	4	70	100	2	1962	Tokyu	Hitachi
7200	3	2	4	110	100	32	1967	Tokyu	Hitachi/Toyo
7600[1,2]	3	2	4	110	110	9	1986	Tokyu	Toyo
7700[1,2]	4	2	4	170	120	56	1987	Tokyu	Toyo
8000	5, 8	3, 6	4	130	120	187	1969	Tokyu	Hitachi
8090	5, 8	3, 6	4	130	120	90	1980	Tokyu	Hitachi
8500	10	8	4	130	120	400	1975	Tokyu	Hitachi/Toyo
9000	5, 8	3, 4	4	170	120	117	1986	Tokyu	Hitachi/Toyo
1000	4, 8	2	2	130	120	113	1988	Tokyu	Hitachi/Toyo
2000	10	6		170	120	30	1991	Tokyu	Hitachi
600 V DC									
70	2	2	2	48.5	40	8	1942	Kawasaki	GE
80	2	2	2	74.6	40	6	1950	Hitachi/Toyo	Hitachi
150	2	2	2	60	40	4	1964	Tokyu	Toyo

[1] VVVF inverter control
[2] 7600 rebuilt from 7200; 7700 rebuilt from 7000

Toyama Chiho Railway

Toyama Chiho Tetsudo
1-36 Sakuramachi 1-chome, Toyama-shi 930
Tel: (+81 764) 32 51 11

Key personnel
President: H Ogata

Gauge: 1,067 mm
Route length: 93.2 km
Electrification: 93.2 km at 1.5 kV DC; 6.4 km at 600 V DC

Passenger operations
The Toyama Chiho Railway operates three routes from Toyama serving the coastal area and providing access to the northern Japan Alps. 'Alpen Express' trains connect the tourist destinations of Tateyama and Unazuki Onsen, with onward travel to Mount Tateyama and the Kurobe Dam being available via the Kurobe Gorge Railway. Rolling stock comprises 51 emu cars, two electric and six diesel locomotives, and three wagons. In 1995 the company acquired a three-car set of ex-Seibu 'Red Arrow' stock which has been modified for use on express services. The company also operates an urban tramway in Toyama with 19 cars.

Jordan

Ministry of Transport

PO Box 35214, Amman 11180
Tel: (+962 6) 551 81 11   Fax: (+962 6) 552 72 33
e-mail: info@mot.gov.jo
Web: www.mot.gov.jo

Key personnel
Minister: Saoud Nsairat

Developments
In early 2005 the Ministry of Transport selected five shortlisted consulting forms/consortia to study a railway development and implementation strategy for Jordan. The government foresees construction of a core standard gauge network comprising a North-South Railway (NSR) and an East-West Railway (EWR). This core network might be supplemented by other lines.

The study was to have four main objectives:

An assessment of the financial and economic viability of the development of a network from both a national and a regional perspective;

Establishing an implementation programme for the development of the network;

Finalising the alignment of the NSR and the EWR;

Provide the government with a means of stimulating interest in the international community in railway development in the country.

UPDATED

Aqaba Railway Corporation (ARC)

PO Box 50, Ma'an
Tel: (+962 3) 213 21 14   Fax: (+962 3) 213 18 61

Key personnel
Director General: Eng M M Krishnan
Assistant Director General, Technical: A Malkawi
Assistant Director General, Administrative:
  A Maitah
Directors
  Planning and Training: N Al-Sa'ud
  Mechanical: M Shafa Amri
  Traffic and Communications: I Al-Shorbaji
  Train Control: Subhi Rasheed
  Permanent Way: Y Al-Trawneh
  Finance: A Abu Tata
  Internal Control: Nabel A Al-Rawad
  Material and Stocks: I Abu-Darweesh
Director General's Office Manager: Odeh Zayadnah

Gauge: 1,050 mm
Route length: 292 km

Organisation
ARC is a public corporation, the sole function of which is haulage of phosphates from mines at El-Abyad, El-Hasa and Shediya to the Red Sea port of Aqaba over a network of new and reconstructed lines, which are passed for 16 tonne axleloads. Part of the route, from El-Hasa to Hettiya, is the original Hedjaz Railway rehabilitated; the remainder is new. ARC also connects with HJR at Batn-el-Ghul.

In the mid-1990s a study carried out by Transurb recommended that ARC be converted into a government-owned company to be sold to the private sector. Following a further consultancy exercise to provide technical, financial and legal advice, four groups submitted financial offers for a 25-year operating concession, with the Jordanian government retaining ownership of the infrastructure. An agreement was finally signed in August 1999 with a Wisconsin Central-led consortium to set up Jordan Rail. However, in March 2001 amendments were made to the concession deal that presented obstacles. One of these concerned levels of employment in Ma'an, with the King of Jordan giving an undertaking that no jobs or housing for railway staff would be lost in Ma'an. Another difficulty was that a plan to construct a new line to serve mines at Wadi II was not mature in terms of cost. These changes led to the withdrawal of strategic shareholders in the Jordan Rail consortium, while in the meantime Wisconsin Central was sold to Canadian National, which had no desire to expand outside North America. All these factors led to the cancellation of the agreement with the Jordan Rail consortium.

In 2004 the Jordanian government announced that it was to revive privatisation plans, with establishment of ARC as a government-owned company a first step. Subsequently the government would sell shares in the company to a qualified strategic partner selected after an international tendering process. The government expected this transaction to be completed by the end of 2005. Projects to attract investors include a long-standing proposal to build a line to the phosphate deposits at Shediya, using infrastructure suitable for the eventual laying of standard gauge tracks.

ARC traffic (million)	2002	2003
Freight tonnes	2.52	2.48
Freight tonne-km	520	490

Freight operations
In 2003 ARC carried 2.48 million tonnes of phosphates traffic and around two-thirds of the total output of the country's major producer, the Jordan Phosphate Mining Corporation. It was expected that in 2004 2.9 million tonnes would be carried. Despite the collapse of initial privatisation proposals, construction of a branch to a mine at Shediya is still seen as a key to capturing a greater share of remaining traffic, which is currently handled by road transport. The competitiveness of rail has been increased by weight restrictions imposed on trucks.

Future freight operations could extend beyond the movement of phosphates: there is a possibility of using the line to transport cereals from silos at Aqaba to Amman.

Passenger operations
Studies have been undertaken into the operation of a steam-hauled tourist service between Aqaba and Wadi Rum.

New lines
Long-planned proposals exist for a 22.5 km route linking phosphate mines at Shediya with ARC at Batn-el-Ghul, and a 16 km link from Aqaba port to a new fertiliser plant in an industrial area known as Wadi II. Implementation of these projects was to be driven by traffic requirements and in 2004 expected demand prompted the Ministry of Transport to place a contract with the French consultancy, Systra, to produce detailed designs for the Shediya line.

Improvements to existing lines
The originally designed capacity of the line was 1.5 million tonnes annually, but it has carried two-and-a-half times as much. A prime concern has therefore been the limitations of the existing infrastructure, with its curves of as little as 125 m radius and a 30 km gradient in places as steep as

NEW/1114959

1 in 37 near Aqaba. That incline restricts train length to 35 wagons of 42 tonnes payload and 64 tonnes gross each. The curvature dictated adoption of Scheffel cross-braced bogies in wagon deliveries from Gregg of Belgium and Samsung of South Korea.

The block signalling between El-Hasa and Umran is controlled by UHF radio, employing fail-safe frequency-division multiplex apparatus. Westinghouse Westbloc coded block colourlight signalling controls operation at five stations. A VHF radio system, using mobile sets, connects train and station staff with the Ma'an control centre.

Traction and rolling stock
ARC's traction fleet comprises 22 diesel-electric locomotives and the wagon fleet totals 260. Fleet availability had fallen during the stalled privatisation process but was revived from 2003 after government investment in the rehabilitation of 11 locomotives and 45 wagons.

Type of coupler in standard ARC use: AAR Alliance 2
Type of braking in standard ARC use: Air 26 Lavi

Track
Rail: S-49 (49 kg/m); BSS 70A (34.8 kg/m); and S-30 (30 kg/m)
Sleepers: Wood and monobloc concrete
Spacing: 1,666/km, 1,755/km in curves less than 300 m radius
Fastenings: DS 18 and Pandrol E2000
Min curvature radius: 125 m, 13.9°
Max gradient: 2.7%
Max permissible axleload: 16 tonnes

UPDATED

Diesel locomotives

Class	Wheel arrangement	Power kW	Speed km/h	Weight tonnes	No in service	First built	Mechanical	Builders Engine	Transmission
UL17C	Co-Co	1,305	120	94.3	8	1975	GE	7FDL8	*E* GE
U18C	Co-Co	1,380	120	94.3	3	1978	GE	7FDL8	*E* GE
U20C	Co-Co	1,600	120	108	11	1981	GE	7FDL12	*E* GE

Hedjaz Jordan Railway (HJR)

PO Box 4448, Amman
Tel: (+962 6) 489 54 13   Fax: (+962 6) 489 41 17

Key personnel

Director General: Eng Abdel Razzak Abu Al Filat
Deputy Director General:
   Eng Oussama Suleyman Kreychee
Administration: Zuhair Momany
Infrastructure: Eng Maarouf Al-Soudi
Building: Eng Khalif Mehawesh
Mechanical: Eng Rayed El-Helo
Procurement: Issa Mahmud
Personnel: Ahmed Al Nueymat
Traffic: Eng Ali Hassan Djad Allah
Traction and Rolling Stock: Eng Hani Nesour
Finance (acting): Nada Al Saadi

Gauge: 1,050 mm
Route length: 212.5 km

Political background

Even before the country's problems occasioned by the 1991 Gulf War, HJR was carrying little traffic. It is of marginal importance in the country's economy. However, in December 2001, the governments of Jordan, Saudi Arabia and Syria agreed to reconstruct the railway, including the cross-border section from Batn-el-Ghul to Al Madinah in Saudi Arabia.

Passenger operations

There is a service each way three days a week from Amman across the border to Damascus in Syria. In 2003 9,198 passengers were carried (return trips) on this service. School trains are operated between Al Mafraq and Al Qatranah, carrying 8,618 passengers in 2003. Diesel-operated tourist trains are also run between Al Qatranah and Dar'a, Syria. Current plans in 2005 were for the development of tourist traffic, with the planned rehabilitation of steam locomotives and rolling stock and improvements to passenger facilities.

Total passenger-km for 2003 was 1.54 million, up 13 per cent on the figure for 2002.

Freight operations

In 2003 freight traffic amounted to 1.78 million tonne-km, up 21 per cent on the figure for 2002. There is service across the border to Damascus in Syria.

Possible future traffic developments include the movement of fuel to phosphate company facilities.

New lines

Plans exist to build a branch to serve grain silos at Juwaida.

Improvements to existing lines

There have been long-standing plans to improve the railway between Amman and Zarqa. Some realignments as well as track renewals were envisaged. A strengthened passenger service was in mind, but its provision would involve procurement of additional rolling stock. The upgrading might also enable HJR to launch an oil feed service to the phosphate mines from Zarqa refinery.

Traction and rolling stock

HJR owns five GE Type U10 670 kW Caterpillar-engined diesel-electric locomotives and six oil-fired steam locomotives. Passenger rolling stock comprises nine coaches and around 100 wagons.

The overhaul of five steam locomotives and of some passenger coaches was planned in 2005 as part of a scheme to boost tourist traffic.

Type of coupler in standard use: Hook
Type of braking in standard use: Vacuum (locomotives air and vacuum)

Track

Rail: 21.5 kg/m flat-bottom
Sleepers: Steel fish-type, 1,830 mm long × 20 mm thick
Spacing: 1,450/km plain line, 1,600/km curves
Fastenings: Angle-type four bolt
Min curvature radius: Main line 100 m, turnouts 91.5 m
Max gradient: 2%
Max axleload: 10.5 tonnes

UPDATED

Kazakhstan

Ministry of Transport and Communications

Abay Prospekt 49, Astana
Tel: (+7 3172) 24 10 04; 75 77 01
Fax: (+7 3172) 32 10 58
Web: www.mtk.gov.kz

Key personnel

Minister of Transport: Kazmurat Nagmanov
Head of Railways Department: N Baydauletov

UPDATED

Kazakhstan State Railways (KTZ)

Pobedy Prospekt 98, 473000 Astana
Tel: (+7 3172) 77 07 11   Fax: (+7 3172) 75 38 91
Web: www.railways.kz

Key personnel

Managing Director: B A Baimukhanov
Executive Director: Kanat Tulemetov
Head of Administrative Board: RT Kuzembaev

Gauge: 1,520 mm
Route length: 13,700 km
Electrification: 3,700 km mostly at 25 kV AC but some 3 kV DC

NEW/0585062

Political background

Kazakhstan is large and sparsely populated. It is the ninth largest country in the world, extending from the Caspian Sea to northern China, and yet has a population which is the same size as Belgium's. With a history as the southern side of the Russian empire, its cities were built as fortresses on the frontier and communications links are chiefly long lines towards the heart of Russia, with few routes to the rest of Asia. Kazakhstan appeared as an independent entity upon the collapse of the USSR in 1991; the capital city has been moved from the frontier city of Almaty to Astana, in the centre of the country. Kazakhstan has maintained amicable and increasingly close relations with Russia.

In early 2001 it was announced that the railways were to be restructured. A first step towards this was the conversion of KTZ into a closed joint-stock company, ZAO KTZ, in March 2002, a move intended to improve management and accounting methods. It is intended that the state will retain ownership of the railway's infrastructure and rolling stock. A passenger operating company is to be formed, its service levels and fares determined by the government in return for subsidy. Competition is foreseen in the freight sector, including provision for major industries and shippers to run their own trains. A separate joint-stock company providing locomotive maintenance services is to be established by 2006.

Organisation

The network comprises most of the former Almaty, Tselinnaya and West Kazakhstan railways of the former SZhD. In 1996–97 there was a series of reorganisations, with these three constituent railways shedding divisions in order to create three new railways, making six, with limited independence. One of them was subsequently re-absorbed, leaving five operating regions, that are no longer subdivided into divisions, and have the status of state enterprises under the close supervision of Kazakhstan State Railways.

There are several lengthy main lines traversing regions of low population. The principal route is the 1,507 km Trans-Kazakhstan Railway running from Petropavlovsk on the Trans-Siberian Railway through Kokchetav, Astana and Solonichki to the Karaganda coalfield. This was later extended to Chu, on the Turkestan–Siberian route, and Lugovoy where it connects with lines into Kyrgyzia and Uzbekistan. The Turkestan–Siberian route runs 1,445 km from Semipalatinsk via Aktogay and Zhangiz-Tobe to Almaty and Chu. From Aktogay the line to the Chinese border at Druzhba now forms part of a through route from the Chinese capital Beijing to Russia. This, the Trans-Asian route, provides a Japan–Western Europe link that is claimed to be 2,500 km shorter than the Trans-Siberian route.

A third main line in the west of the country links Tashkent, in Uzbekistan, with Orenburg in Russia, via Aralsk, Kandagach and Aktyubinsk, a distance of 1,854 km.

In 2002 it was agreed that ownership of three lines totalling some 100 km within Kazakhstan, but operated by Uzbekistan Railways for reasons dating back to the Soviet era, would be transferred to KTZ.

The Kazakhstan system has many long stretches of single track; over one third of the network is double track.

In 2003 the railway employed around 150,000 staff.

Passenger operations
Following independence, passenger traffic was at first affected by rolling stock shortages, leading to a big reduction in through services to Russia, but business has recovered following delivery of new coaches and improved maintenance arrangements. In 1993, the railway recorded 20 billion passenger-km. By 1999 traffic had fallen to 8.89 billion passenger-km (14.3 million passenger journeys) but a rise to 10.45 billion passenger-km in 2002 hinted at signs of a revival.

Freight operations
In 2002 KTZ accounted for more than 70 per cent of Kazakhstan's freight transport. The railway's principal traffic is coal from the Karaganda and Ekibastuz fields. Most of this is moved over the busiest of the routes, the east-west Pavlodar–Tobol line, once part of a Soviet trunk route. Since independence, coal movements from Ekibastuz to various parts of the former Soviet Union have held up well, while other freight traffic has declined substantially. Other key commodities carried include iron ore and other ores, including bauxite and copper, and construction materials. The commissioning of new pipelines has led to a decline in movements of petroleum products. In 2001 a total of 133 billion tonne-km was achieved, with forecasts of a continuing upward trend. Transit traffic between China and Russia totalled 16.8 million tonnes in 2002.

New lines
With the aim of creating a more 'national' network and reducing the need for domestic traffic to pass through neighbouring countries, a number of new cut-off lines have been built, are under construction or are planned. The 184 km Konechnaya–Aksu line opened in June 2001. This, and the planned Charskaya–Ust Kamenogorsk line, will avoid Russian territory, while the new Arkalik–Zhezqazghan–Kzyil Orda line will provide a shorter

and all-Kazakh route between the Caspian Sea and China.

In the northwest of the country, construction of the 404 km Krasno-Oktyabrskiy–Donskoye line was completed in November 2004, avoiding the need for domestic mineral traffic to run via Russia.

An agreement was signed in 1997 between the governments of Kazakhstan and Turkmenistan to promote the building of a north-south corridor. This would link Yerelyevo on the Caspian Sea with Turkmenbashi in Turkmenistan, forming a direct rail route avoiding Uzbekistan. Major traffic flows between Russia and Iran were expected for this route.

In 2004 KTZ confirmed that it planned to construct a 3,038 km standard-gauge railway to form a key part of a link between China and western Europe. The line would provide a freight land-bridge between the Yellow Sea port of Lianyungang and Iran, gaining access there to Turkey. That country's cross-Bosphorus tunnel, now under construction, will provide a connection to the European network. The use of standard gauge would eliminate trans-shipment to 1,520 mm gauge vehicles at the China-Kazakhstan border. As well as necessitating major new line construction in Kazakhstan, the link would also require a 770 km link in Turkmenistan and 70 km in Iran. The Kazakh element that forms the bulk of the project would be achieved by a combination of completely new railway and provision of standard-gauge tracks alongside existing 1,520 mm lines, in some instances following new alignments.

Improvements to existing lines
Efforts have for some time been concentrated on easing and smoothing the well-paying international transit traffic. Developing the approaches to Druzhba has priority, especially work on the Aktogay–Druzhba line, for which credit agreements with Japan's OECF fund were reported to have been arranged in 2003. Gauge-changing equipment was installed in 1998 at Druzhba, where KTZ exchanges traffic with the standard-gauge Chinese Railways system.

Traction and rolling stock
In 2002 the locomotive fleet totalled some 1,800, a mix of Class VL80C electric machines and various diesel types, although only two-thirds of these were reported to be serviceable.

Diesel fleet developments
In a tie-up with GE of the USA, 16 Ukrainian-built 2TE10M twin diesel locomotives have been rebuilt with 7FDL-12 engines. One unit from the West Kazakhstan Railway was shipped to the USA for rebuilding, and GE supplied kits for the other 15 to be rebuilt locally.

In 2002 KTZ concluded a contract with China Northern Locomotive and Rolling Stock Industry (Group) Corporation (CNR) to supply 152 high-powered Type DF4D diesel-electric locomotives with GE traction equipment. The first two machines were to be built in China by CNR; the remaining 150 are to be assembled in KTZ's workshops at Chu, in the south of Kazakhstan, with final deliveries expected in 2007.

Talgo equipment
In 2000 KTZ agreed to purchase a six-car gauge-convertible Talgo Pendular trainset to be powered by its own traction. In 2003 additional cars were procured. One set was deployed on a non-daily overnight service between Almaty and Astana (1,350 km). Another set was used on the Almaty–Chimkent route. In July 2003 two additional 22-car Talgo Pendular Series 7 formations were handed over to KTZ, enabling both services to be operated daily. Unlike the earlier vehicles, these later cars were designed to operate in the extremes of temperature encountered in Kazakhstan. From 2004 haulage of Talgo services was taken over by three Class KZ4A 4,200 kW Bo-Bo 82 tonne electric locomotives built at the Zhuzhou plant in China. Prior to their arrival, services had been handled by Class TEP70 diesel-electric locomotives: despite the routes over which Talgo-equipped services are operated being electrified, existing KTZ electric locomotive types proved unsuitable. The 210 km/h Class KZ4A machines feature Siemens traction equipment and asynchronous traction motors.

In 2003 it was reported that KTZ was to procure passenger and freight rolling stock from Russia, financed by the latter country's debts to Kazakhstan.

The locomotive depots at Atbasar and Chu now undertake capital repairs, making Kazakhstan State Railways independent in this respect.

UPDATED

Kenya

Ministry of Transport & Communications

PO Box 52692, Ngong Road, Nairobi
Tel: (+254 2) 72 92 00   Fax: (+254 2) 72 63 62
Web: www.kenyaweb.com/government

Key personnel
Minister: John Michuki

UPDATED

Kenya Railways Corporation (KR)

PO Box 30121, Nairobi
Tel: (+254 2) 22 12 11   Fax: (+254 2) 34 00 49; 22 41 56

Key personnel
Chairman: J Mturi
Managing Director: Vitalis Ong'on'go
General Managers
   Commercial: F N Thuranira
   Technical: P E Okiring
   Finance: C Nyamai
Chief Mechanical and Electrical Engineer:
   A O Okuku
Chief Civil Engineer: M J Katsivo
Chief Signal and Telecommunications Engineer:
   V Hagono
Chief Traffic Manager: I K Ranote
Chief Accountant: P G Muguti
Chief Internal Auditor: R Otieno

Kenya

NEW/0585067

Supplies and Procurement Manager: D Bosire
Corporate Planning Manager: Dr S N Oresi
Human Resources Manager: T M Nagira
Information Systems Manager: O M Kilonzo
Business Development Manager: D N Muluka
Central Workshops Manager: J M Thuo
Passenger Services Manager: T C Masha
Quality and Safety Manager: B M Kimau
Public Relations Manager: J S Odhiambo
Director, Railway Training Institute: J L Muchera
Corporate Secretary: A K Maina
Estates Manager: E M Jenkins
Security Services Manager: P K Eggesa

Gauge: 1,000 mm
Route length: 2,778 km

Henschel-built Class 62 diesel-hydraulic locomotive with a mixed train at Kisumu (Edward Barnes)
***NEW**/0585066*

Political background

Long-discussed plans to privatise the KR network were moving forward in 2004–05 following agreement between the governments of Kenya and Uganda to offer a joint concession for most of their combined systems. Canarail and International Finance Corporation acted as advisers on the transaction. In early 2005 evaluation was in progress of shortlisted bidders for a 25 year concession to run freight services in both countries and a 7 year concession to operate KR's passenger services. It was expected that the successful bidder would take a 60 per cent stake in the new company, with the remaining shareholding divided between local and private interests. The handover to a private sector operator was expected in July 2005. The privatisation will exclude the Konza–Magadi line, which since 1995 has been operated for the movement of soda ash by the Magadi Railway Company under a leasing agreement with the government.

In connection with the privatisation, a World Bank feasibility study into the rehabilitation of the network, was in progress in 2005. Meanwhile, KR has been freed by the government from tariff controls and a system of subsidy has been established for the provision of uneconomic services. The railway has been given managerial freedom by the government and in 2005 was implementing a programme of staff reduction. However, economic liberalisation has exposed KR to increased road competition, prompting the company to restructure its marketing strategy.

It was expected that privatisation would see the establishment of three organisations: the railway company; Kenya Railways Management Authority; and Kenya Railways Safety Authority.

Traffic

Traffic (million)	2003–04
Passenger journeys	5.657
Freight tonnes	2.0

Passenger operations

KR handles substantial passenger traffic volumes on the Nairobi–Mombasa route, which is served by the daily 'Jambo Kenya Deluxe' overnight service. Other routes include Voi–Taveta, Nairobi–Kisumu–Butere and Nairobi–Nanyuki, the last two of these having services restored in 2003. In 2003–04 these main line services carried 527,783

passengers. However, KR's largest traffic volumes are carried by commuter services in Nairobi and its surrounding area, in 2002–03 handling 5.13 million passengers. Main line and commuter passenger journeys were up on KR forecasts by 36 and 13 per cent respectively, improvements attributed to a government restructuring of the road transport industry.

Freight traffic

In 2003–04 KR carried 2.0 million tonnes of freight compared with 2.3 million in 2002–03. It handles some 80 per cent of Kenya's freight and is the source of a similar percentage of the railway's revenue. The use of container services between Mombasa and Nairobi's Inland Container Depot has grown in importance. Already served by four daily shuttle services in each direction, the corridor was expected to benefit further from 2005 from improvements in handling facilities at the port and the acquisition of additional container-carrying wagons, as well as from commercial initiatives intended to increase the attraction of the service to shippers. These were expected to include more agressive sales and marketing strategies, coupled with competitive tariffs, the operation of block trains, and the provision of seamless services to and from Kampala, Uganda.

At Kisumu rail traffic is transferred to train ferries for movement across Lake Victoria to Tanzania and Uganda. Services to the latter have increased in importance in recent years.

Improvements to existing lines

The line west from the capital, Nairobi, to the border with Uganda at Malaba was the subject of upgrading in 2004. The line connecting Nairobi with the Indian Ocean port of Mombasa has also been recently upgraded. However, much of the network remained in need of upgrading and rehabilitation.

Radio links on the Nairobi–Kisumu line have been enhanced.

Traction and rolling stock

Fleet numbers in 2005 were 156 diesel locomotives, 588 passenger coaches, and 6,827 freight wagons.

Refurbishment of locomotives of Classes 93 and 94 by General Electric has been undertaken. Other locomotives have been overhauled by Bombardier.

Type of coupler in standard use: PH/DA
Brake system: Graduated automatic air

Track

Rail: 50, 60, 80 and 95 lb FB
Crossties (sleepers): Steel
Spacing: 1,485/km (1,568/km in curves)
Fastenings: 'K' type (Pandrol)
Max gradient: 3.5%
Max axleload: 18 tonnes

UPDATED

Diesel locomotives

Class	Wheel arrangement	Power kW	Speed km/h	Weight Tonnes	No in service	First built	Mechanical	Builders Engine	Transmission
94	Co-Co	2,172	72	101.8	10	1978	GE (US)	GE 7F DL12	E GE (US)
93	Co-Co	1,947	72	101.8	25	1978	GE (US)	GE 7F DL12	E GE (US)
92	1 Co-Co 1	1,901	72	118	12	1971	MLW	Alco 251F	E GE (Canada)
87	1 Co-Co 1	1,372	72	104.7	22	1960	EE	EE12CSVT	E EE
72	1 Bo-Bo 1	925	72	73	3	1972	GEC	EE8CSVT	E GEC
71	1 Bo-Bo 1	925	72	71	1	1967	EE	EE8CSVT	E EE
62	B-B	552	72	38	32	1977	Henschel	MTU493TZ MTU396TC	H Voith L520rU2
47	D	391	28	53	27	1977	Hunslet	RRDV8TCE	H Voith L2r3ZU
46	D	2 × 257	32	49	16	1967	Barclay	Cummins NT380	H Twin Disc CF 11500

Not all locomotives serviceable

Magadi Rail

PO Box 1, 00205 Magadi
Tel: (+254 20) 699 90 00   Fax: (+254 20) 699 93 58
e-mail: info@magadisoda.co.ke
Web: www.magadisoda.co.ke

Key personnel

Managing Director: James Mathenge

Gauge: 1,000 mm
Route length: 146 km

Background

Magadi Rail is the railway operating division of Magadi Soda Company, a wholly owned subsidiary of Brunner Mond that produces and exports

sodium carbonate (soda ash) at Lake Magadi, some 120 southwest of Nairobi. Rail services from the company's facility at Magadi to Konza, on the Nairobi–Mombasa main line, were formerly provided by Kenya Railways but, in a bid to improve the efficiency of its rail transport service, operations were taken over by Magadi Soda Company in the mid-1990s under a leasing arrangement with the Kenyan government. Magadi Rail undertook rehabilitation of the line, including provision of a radio communications and train control system that regulates train operations through to Mombasa.

Freight operations

Magadi Rail handles over 300,000 tonnes of sodium carbonate annually, moving this for export to the port of Mombasa in hopper wagons.

Traction and rolling stock

Magadi Rail leases five main line diesel locomotives and one diesel shunter from Kenya Railways. The main line units were refurbished and returned to service in 1998. There are also 150 wagons in the fleet.

NEW ENTRY

Korea, North

Korean State Railway (ZČi)

Zosun Tchul Zosun Mindzuzui Inminhoagug
Pyongyang
Fax: (+8502) 381 45 27

Key personnel
Minister of Railways: Kim En Sam

Gauge: Almost entirely 1,435 mm, but some narrow-gauge
Route length: 5,214 km
Electrification: approximately 3,500 km at 3 kV DC

Organisation
ZČi is organised into five regional operating divisions, each responsible for the Ministry of Railways. The Ministry is also responsible for the country's railway manufacturing industry.

Passenger operations
At last report, annual passenger journeys totalled some 35 million, passenger-km 3,400 million.

Freight operations
At last report, freight carryings totalled some 38.5 million tonnes and 9,100 million tonne-km. Around 70 per cent of the country's freight is reported to be carried by rail.

New lines
Recent completions include an 80 km railway from Jukchon to Onchon, for more direct access to the port of Nampo, and a 252 km line from Hyesan to Kanggye via Manpojin. In 1997 a 100 km electrified line between Wonsan and Mount Kumgang was inaugurated.

A new double-track route is planned for the Russian Federation border crossing, between Tumangang and Hassan near Najin.

Improvements to existing lines
The year 1992 was marked by launching of a programme of modernisation of the country's rail infrastructure and industry. The first sign of action was production of a plan for reconstructing the strategic northern main line parallel to the Chinese border between Hyesan and Musan, on which capacity was to be raised to 50 million tonnes a year. Double-tracking of the 337 km route started in 1993 and was carried out in two stages.

Studies published in South Korea in 1998 indicate that major investments are required on many major routes, including Pyongyang-Kaesong and Pyongyang-Sinuiji. On both of these lines, demand is reported to exceed capacity.

In 2000, reduced political tension between North and South Korea led to an agreement to reinstate the railway linking the two countries. This will necessitate reconstruction of a 20 km section of line from Pongdong-ni, in North Korea, across the border and the demilitarised zone to Munsan. Work on the South Korea portion of the link started late in 2000.

Improvements are also expected to be made to the rail link between North Korea and Russia following a meeting between the leaders of the two countries in 2000. Following an evaluation of the condition of the North Korean network, the Russian government indicated its willingness to assist in upgrading the line from Pyongyang to the border crossing at Hongui as part of a broader scheme to link the South Korean port of Pusan with the Trans-Siberian route at Khazan, Russia.

Traction and rolling stock
No details are known of the numbers of locomotives, coaches and freight wagons, but a fleet of more than 300 electrics is believed to be available. With just over 80 per cent of the network now under wires, 90 per cent of all freight is reported as hauled by electric traction. The latter began with importation of 10 2,030 kW Škoda

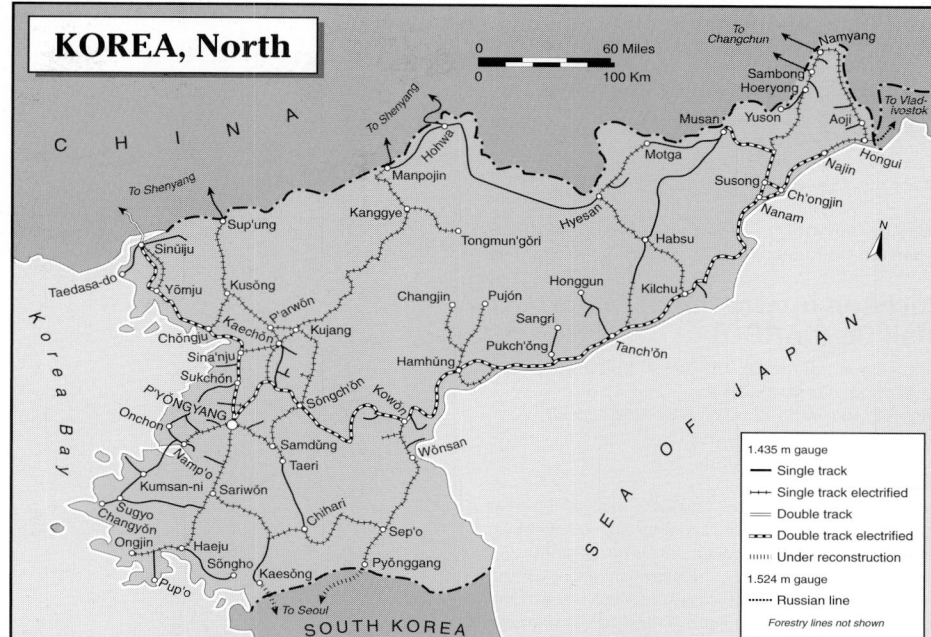

Domestically manufactured 'Red Flag' twin section 3 kV DC electric locomotive at Pyongyang (Thomas E Fischer)

0063330

Bo-Bos similar to the Type E449.0 of CSD, and of some USSR-built locomotives. Since the early 1980s locomotives as well as rolling stock have been produced by the country's Kim Jong Tae factory at Pyongyang. Production has focused on 8-axle twin section 'Red Flag' locomotives, which form the core of the electric traction fleet. In 1992 this plant was reported to be building bilevel passenger cars. Chinese-built passenger coaches are also in service.

Local production of rolling stock has been supplemented by delivery of seven French-built 3,600 hp diesel-electric Co-Cos similar to those delivered to Iraq. A first batch of seven was supplemented in 1987 by five more, built by De Dietrich with Alsthom electrical equipment.

In December 1996, six ex-DR Class 220 Type M62 diesel locomotives built at the Voroshilovgrad locomotive works arrived in North Korea from Germany, supplementing large numbers of similar machines presumed to have been procured from the former Soviet Union. In May 2000, 10 similar machines were delivered: six from Poland (former PKP Class ST44); three from Slovakia (former

ŽSR Class 781); and one from Germany (former DB Class 220).

Electrification
In 1992 electrification was reported complete on the Singhung-Hamjiwan stretch of the Lake Pujon line in the northeast; this line is to be wired throughout. By the close of 1992 electrification was completed from Haeju to Ongjin, completing the wiring from Pyongyang to the southwest extremity of the country.

Electrification is continuous throughout the 780 km from Pyongyang northeast to Vladivostok, and through passenger service (involving a bogie change at Tumangang, near Najin in Korea) between Pyongyang and Moscow was launched in 1987.

Electrification of the link to China via Hoeryong was completed in 1995, and upgrading to carry heavier traffic is in progress. Another partially built cross-border route in the region is being completed to cope with the increasing traffic with China.

Korea, South

Ministry of Construction & Transportation

1 Joongang-Dong, Kwacheon-shi, Kyungki-Do, 427712 Seoul
Tel: (+82 2) 504 90 24
Fax: (+82 2) 504 91 99
Web: http://www.moct.go.kr

Key personnel
Minister: Dong-Suk Kang

Incheon International Airport Rail Co (IIARCO)

8th Floor, Hyundai Annexe Building, 140-2 Gye-dong, Jongo-gu, Seoul
Tel: (+82 2) 520 56 10   Fax: (+82 2) 520 59 88
Web: http://www.iiarc.com

Background

Incorporated in 2001, IIARCO has a 30-year concession to operate an express rail service that will link Seoul with its international and domestic airports at Incheon and Gimpo respectively, both of which are located west of the capital. The line will be Korea's first to be operated as a concession and will also carry commuter traffic.

IIARCO's shareholders include: Hyundai Engineering and Construction (27 per cent); Posco Engineering and Construction (11.9 per cent); Daelim Industrial (10 per cent); Dongbu Corp (10 per cent); and the Korea Rail Network Authority (9.9 per cent).

IIARCO has awarded a turnkey contract to the Incheon Korean French Consortium (IKFC) to undertake project management of the line's construction and to provide its power supply, train control and communications systems. IKFC comprises ALSTOM, Eukorail and Rotem.

Passenger operations

IIARC foresees a half-hourly service frequency to the two airports, with a 45-minute journey time. A more intensive commuter service will be operated, serving intermediate stations. It is expected that services will be operated under the 'IREX' brand name.

New lines

The project involves construction of a 61.7 km standard gauge railway in two phases. The first phase (41 km), scheduled for commissioning in March 2007, will provide a link between the two airports. A second phase (20.7 km), due to be completed in 2009–10, will enable services to run into Seoul's main station. Around 60 per cent of the line will be in tunnel.

The line will be electrified at 25 kV AC 60 Hz. Train control will be provided by ALSTOM, using its Urbalis 200 system.

Connections will be provided with both Incheon and Seoul metro systems and by serving Seoul's main station, a link will also be provided with KTX high-speed services south from the capital, as well as with conventional main line services.

Traction and rolling stock

IIARC has ordered 24 six-car aluminium-bodied 120 km/h emus from Rotem to cover its requirements for both airport express and commuter services. Eighteen of these are intended for commuter services and feature four pairs of doors per side of each car and high-density interior; the remaining six trains for airport services will have only two pairs of doors per car side and will be provided with onboard luggage storage facilities.

Korail

920, Tunsan-dong, So-ku, Taejon 302-701
Tel: (+82 42) 472 30 14   Fax: (+82 42) 472 30 10
e-mail: cph-icd@mail.korail.go.kr
Web: http://www.korail.go.kr

Key personnel
Administrator: Kim, Se-ho
Deputy Administrator: Lee, Geun-guk
Directors General
  Corporate Planning: Shin, Kwang-soon
  Administration: Lee, Guen-Guk
  Supply: Kim, Cho, Jin-kyu
  Safety and Environment: Kim, Hae-soo
  Transportation Business: Yoon, In-kyum
  Business Development: Jung, Hyun-chul
  Civil Engineering: Kim, Sang-gyun
  Electrical Engineering: Soh, Jong-suk
  Rolling Stock: Kim, Sun-ho
  Construction: Im, Yong-Hyeon
  High-Speed Rail and Research: Kim, Chun-hwan
Director, International Cooperation Division:
  Yoo Hee-Bok

Political background

Restructuring of the South Korean rail industry in January 2004 following government legislation enacted in July 2003 saw the former Korean National Railroad split into two separate government agencies: Korail, which undertakes passenger and freight train operations; and the Korea Rail Network Authority (KR), which oversees new line construction and improvements to the existing network. As part

Inauguration of KTX high-speed services at Seoul station on 1 April 2004 (ALSTOM)   0583037

The map is titled **KOREA, South** with legend:

1,435 mm gauge
— Single track
⊢⊣⊢ Single track, electrified
— Single track, freight only
····· Single track, electrification planned
═ Double track
⊨⊨ Double track electrified
▪▪▪ High-speed line
••• High-speed line projected or under construction

0576728

of these changes Korail assumed responsibility for operation of KTX high-speed rail services from their inauguration in April 2004.

Korail funding comes from fares and station revenues, public service obligation grants and from other sources of government support.

The government intends that Korail will become a state-owned but independent corporation from January 2005. At that stage, the company would commence making payment to KR for use of its infrastructure, with fees initially linked to the cost of maintenance. A later stage of reform could see privatisation of Korail, either as a single entity, as separate freight and passenger companies, or as a series of regional operators.

Traffic (million)	1999	2000	2001
Passenger journeys	823	816	911
Passenger-km	28,356	28,097	29,228
Freight tonnes	42.1	45.2	45.1
Freight tonne-km	10,071	10,803	10,492

Passenger operations

Passenger traffic, which has continued to rise steadily, has been KNR's and now Korail's main source of income. The backbone of the system has been the 444 km double-track Kyongbu line, running between the nation's two principal cities, Pusan on the southeast coast and the capital city of Seoul in the northwest, via Taegu and Taejon. Principal intercity services in recent years on this and other key lines have been formed of 150 km/h Saemaul Express diesel-hydraulic trainsets comprising six trailers and two power cars equipped with MTU engines and Voith transmission. While it constituted less than 15 per cent of total route-km, the Kyongbu line accounted for nearly half of the system's operating revenues.

KTX high-speed services

The Seoul–Pusan corridor was selected in the 1990s as the route for a new high-speed line. The first section, between Seoul and Taegu, was formally inaugurated in April 2004, with 300 km/h KTX high-speed trainsets continuing to south Pusan via the classic line, which has been electrified pending construction of the new line from Taegu by 2008–10. KTX trainsets also run to Mokpo, leaving the high-speed line at Taejon. At launch there were up to 92 KTX services in each direction between Seoul and Taejon, with around two-thirds of these going forward to Pusan and the remainder to Mokpo. The Seoul–Pusan journey time has been cut from 4 hours 10 minutes by Saemaul Express using the classic line to 2 hours 40 minutes; on the Seoul–Mokpo route the previous 4 hour 42 minute timing has been reduced to 3 hours 15 minutes.

Seoul suburban services

Korail operates a 178.6 km electrified suburban network serving Seoul and its outlying metropolitan area. This is operated as seven lines serving 106 stations and carries nearly 900 million passengers annually. In 2004 several extensions to this network were under construction and others were at the planning stage.

Freight operations

Freight traffic has been in decline in recent years, falling from 60 million tonnes in 1993 to 45 million in 2001. Freight revenue is around a quarter of that derived by Korail from passengers.

Traction and rolling stock

As at June 2002 the fleet comprised 493 diesel and 96 electric locomotives, 1,672 emu cars, 610 dmu cars and three railcars. Other passenger stock in operation totalled 1,639 locomotive-hauled cars and there were 13,494 wagons. Some 5,000 privately owned freight wagons were also in use on the network.

Diesel locomotives

Class	Wheel arrangement	Power kW	Speed km/h	Weight tonnes	No in service	First built	Mechanical	Builders Engine	Transmission
2000	Bo-Bo	596	105	95.5	6	1955	GMC	GMC	E GMC D27
2100	Bo-Bo	746	105	87	28	1961	GMC	GMC	E GMC D75D
3000	Bo-Bo	653	105	75	39	1956	GMC	GMC	E GMC D47B
3200	Bo-Bo	653	105	73	26	1966	Alco	GMC	E GMC GE761
4000	Bo-Bo	977	105	78.5	5	1963	GMC	GMC	E GMC D57B1
4100	Bo-Bo	977	105	88.5	7	1966	GMC	GMC	E GMC D77B
4200	Bo-Bo	977	105	88	20	1967	GMC	GMC	E GMC D75B
7000	Co-Co	2,238	150	113	15	1986	Hyundai	GMC	E Hyundai D77B
7100	Co-Co	2,238	150	132	86	1975	GMC/Hyundai	GMC	E GMC/Hyundai D77B
7200	Co-Co	2,238	150	132	39	1971	GMC/Hyundai	GMC	E GMC/Hyundai D77B
7300	Co-Co	2,238	150	124	83	1989	Hyundai	GMC	E Hyundai D77B
7400	Co-Co	2,238	150	126	59	1997	Hyundai	GMC	E Hyundai D77B
7500	Co-Co	2,238	105	132	54	1971	GMC/Hyundai	GMC	E GMC/Hyundai D77B
7500	Co-Co	2,238	150	124	20	1996	Hyundai	GMC	E Hyundai D77B

Electric locomotives

KNR Series	Wheel arrangement	Output kW	Speed km/h	Weight tonnes	No in service	First built	Builders Mechanical	Electrical
8000	Bo-Bo-Bo	3,900	85	132	90	1972	BN/Alsthom	Alsthom/ACEC AEG/ABB
8000	Bo-Bo-Bo	3,900	85	132	4	1986	Daewoo	Daewoo/Woojin
8100	Bo-Bo	5,200	150	88	2	2000	Korea Rolling Stock Corporation	Siemens

Diesel railcars or multiple-units

Class	Cars per unit	Motor cars per unit	Motored axles/car	Power/car kW	Speed km/h	No in service	First built	Mechanical	Builders Engine	Transmission
DHC	6	2	4	1,469	150	4	1987	Daewoo/Hyundai	MTU 12V396TC13	H Voith L520RU
	8	2	4	1,469	150	51	1988	Daewoo/Hyundai/Hanjin	MTU 16V396TC13	H Voith L520 RZU2
DEC	5	2	4	723	110	2	1980	Daewoo	Cummins KTA 2 300L	
NDC	4	2	4	231	120	9	1984	Daewoo	Cummins NT855R1	H NT855R4T211R
CDC	5	3–5	2	231	120	36	1996	Daewoo	Cummins NT855R1	H Voith T211YZ

High-speed trainsets

Class	Cars per unit	Motor cars per unit	Motored axles/car	Power/motor kW	Speed km/h	Units in service	First built	Builders Mechanical	Electrical
KTX	20	2	6*	1,100	300	46	1997	ALSTOM/Rotem	ALSTOM/Rotem

* Includes power bogie in each trailer car adjoining power car

Electric railcars or multiple-units

Class	Cars per unit	Motor cars per unit	Motored axles/car	Power/motor kW	Speed km/h	Units in service	First built	Builders Mechanical	Electrical
Dual-voltage 25 kV/1.5 kV									
1000	10	6	4	120	110	66	1974	Hy/D/H	D/Hy/W/C/T/M
2000	10	5	4	200	110	30	1994	Hy/D/H	Hy/W/C/T/M
5000	10	5	4	200	110	42	1997	Hy	Hy/W/T/M
25 kV									
2000	6	3	4	200	110	22	1994	Hy/D/H	Hy/W/C/T/M
9000	10	6	4	120	110	2	1980	D	D/T
1.5 kV									
3000	10	5	4	200	110	16	1995	Hy	Hy/M

Abbreviations: C: Chung Gye, D: Daewoo, H: Hanjin, Hy: Hyundai, M: Mitsubishi, T: Toshiba, W: Woojin

Korail KTX high-speed trainsets (ALSTOM)   0583038

For details of the latest updates to *Jane's World Railways* online and to discover the additional information available exclusively to online subscribers please visit

jwr.janes.com

KTX high-speed trainsets

Rolling stock for Seoul–Pusan/Mokpo high-speed services comprises 46 935-seat trainsets based on TGV/Eurostar technology, comprising two power cars and 18 trailers, with the outer bogie of each end 'trailer' motored. Under a technology transfer contract with the ALSTOM-led Korea TGV Consortium, two preseries and 10 production trainsets were constructed by ALSTOM in France, with the remining 34 built in Korea by local rolling stock partners that became Rotem in January 2002. The first French-built set was completed in May 1997 and tested on SNCF high-speed lines before shipment to Korea.

In 2004 Korail was reported to be preparing proposals to procure 20 10- or 12-car trainsets for Mokpo line services for delivery in 2006–07.

New electric locomotives

In 2000 the former KNR took delivery of the first of two prototype Class 8100 electric locomotives. Built by Daewoo Heavy Industries (now Rotem) and Siemens and based on German Rail's Class 152 machines, these were the first three-phase asynchronous-drive electric locomotives to be used in Korea and were expected to form the basis of a future standard multipurpose design for the KNR network. The 88-tonne locomotives have a rated output of 5,200 kW and a top speed of 150 km/h. A technology transfer agreement covers series production of the type in Korea.

Class 8100 electric locomotive   0089056

Type of coupler in standard use: Passenger cars – tight lock type; freight cars – Shibada (E-Type)

Type of braking in standard use, locomotive-hauled stock: Air brake

---

Korea High Speed Rail Construction Authority

Political background

KHRC was the government body set up to plan and build South Korea's first high-speed railway, from Seoul to Pusan. Its assets and functions passed to the Korea Rail Network Authority (see entry in Korea, South section of *Railway systems and operators*) in January 2004.

Korea Rail Network Authority

Daelim Building, 452-3 Daeheung-dong, Jung-gu, Taejon
Tel: (+82 42) 481 32 64   Fax: (+82 42) 472 06 36
Web: http://www.korail.go.kr

Key personnel

Chairman and Chief Executive Officer:
   Chung Jong-Hwan

Gauge: 1,435 mm
Route length: 3,472 km
Electrification: 1,342 km at 25 kV AC 60 Hz; 19 km at 1.5 kV DC

Political background

The Korea Rail Network Authority (KR) came into existence in January 2004 following legislation enacted in July 2003 to split the Korean National railroad into two government agencies, one responsible for infrastructure management, the construction of new lines and upgrading of the existing network, the other (Korail) for train operation on both conventional lines and on the new high-speed railway.

A further step in South Korean railway reform is due to occur in January 2005, when Korail will be charged by KR for use of its infrastructure, a move seen as a prelude to the eventual privatisation of train operations.

Organisation

KR is organised as six divisions and at its establishment in 2004 employed some 1,500 staff.

New lines

Seoul–Taegu–Pusan high-speed line

Ownership of the assets of the Seoul–Taegu section of South Korea's first high-speed line passed from the Korea High Speed Rail Construction Authority (KHRC) to KR on its establishment in January 2004. Branded KTX, the line was formally inaugurated in April 2004, with TGV-derived trainsets operating at 300 km/h.

Contracts to construct the entire Seoul–Pusan line (412 km) were placed by KHRC with the Korea TGV Consortium, led by ALSTOM subsidiary Eukorail, in 1994. Subsequently, in 1998, it was decided that initially only the section from Seoul to Taegu (223.6 km of new line) would be constructed. The existing Taegu–Pusan line (118 km) has been electrified to allow KTX high-speed trainsets to work throughout. The line from Taejon to Mokpo (256 km), in the southwest of the country, has also been upgraded and electrified to provide a connection to the KTX service network.

The TVM430 cab signalling system is employed for train control, while the design of the overhead power supply system is derived from that of the LGV Nord in France.

In late 2001 the government authorised KHRC to build the second phase of the scheme between Taejon and Pusan via Kyongju to complete the high-speed alignment between Seoul and Pusan. The target date for completion of this second section is 2010.

The total cost of the project is estimated at US$17 billion.

Honam High-Speed Rail Project

This project aims to cut journey times for KTX services to Mokpo, which is currently reached by the classic Honam Line. This leaves the Seoul–Pusan high-speed route at Taejon and was upgraded and electrified to coincide with the introduction of high-speed services. Design work has since started on a new high-speed line between Suseo and Hyangnam, covering the northern section of the route, to be followed by a new alignment for the southern section from Iksan to Mokpo. The project is expected to be completed by 2015.

Other new lines

South Korea's National Basic Transportation Network Plan, a 20-year government transport strategy published in 2000, foresees an increase in the route-km of the country's rail network from 3,472 km in 2004 to 4,908 km by 2019. Plans include several

Viaduct on KR's Seoul–Taegu high-speed line (ALSTOM)   0583036

east-west lines as well as a railway running north from P'ohang to serve the country's east coast.

Seoul airport railway project
KR has a shareholding of 9.9 per cent in Incheon International Airport Railroad Co (IIARCO) (see entry in Korea, South section of *Railway systems and operators*), the private section consortium created to operate a dedicated express rail link serving Seoul's Gimpo (domestic) and Incheon (international) airports. The first phase of the scheme, linking the two airports, is due to open in 2007.

Improvements to existing lines
The National Basic Transportation Network Plan foresees an expansion of double-track from 38.1 per cent of the rail network in 2004 to 80 per cent by 2019. Over the same period the plan's objectives include an increase in the extent of the network electrified from 39.2 to 82 per cent.

Priority schemes in 2004 included electrification of the Iksan–Sunchon and Iksan–Chonan lines. Other projects were intended to ease congestion in metropolitan areas.

Restoration of links with North Korea
Restoration of the abandoned 12 km South Korean section north of Munsan of the former Seoul–Kaesong western rail link with North Korea has been completed and work north of the Demilitarized Zone (DMZ) was expected to be finished in 2004. In the east, work was in progress in 2004 to restore the 7 km section between Kojin and the DMZ. The utilisation of these restored lines will be governed by political conditions between the two countries but in the longer term South Korea aspires to sending freight traffic though North Korea and China to gain access to the Trans-Siberian Railway.

Signalling and telecommunications
CTC (Centralised Traffic Control) is installed in the Seoul suburban area and on the Kyongbu, Changhang, Taebaek and Yongdong lines. In total, 1,117.4 route-km are so controlled. Automatic Train Stop (ATS) equipment is installed over 3,119.9 route-km, with Automatic Train Control (ATC) over 52.1 km.

TVM430 cab signalling is employed on the Seoul–Taegu high-speed line.

KTX high-speed trainset on the portion of the Chonan–Taejon test section of track inaugurated in December 1999
0088213

Electrification
Electrification in place in 2004 amounted to 1,342 km at 25 kV AC 60 Hz and 19 km at 1.5 kV DC. See *Improvements to existing lines*.

Track
Main conventional lines are mostly laid with 50 kg/m rail, but since 1981 the former Korean National Railroad laid continuously welded 60 kg/m rail to heavily trafficked sections. Some secondary lines also have 50 kg/m rail, others 37 kg/m. Sleepers on main lines are now mostly of locally manufactured concrete, with Pandrol fastenings securing long-welded rail. On branch lines, timber sleepers prevail.
Rail: flat-bottom 37, 50 and 60 kg/m
Crossties/sleepers
Material: Wooden or prestressed concrete ties
Dimensions
Common tie: 15 × 24 × 250 cm

Switch tie: 15 × 25 × 280–460 cm or 23 × 23 × 250–300 cm
Total number installed
In plain track: 1,600 per km
In curves: 1,700 or a few more per km
Fastenings type
Wooden tie: elastic fastening type
PC tie: Pandrol fastening type
Min curvature radius: 200 m
Max gradient: 3.5% (3°)
Max permissible axleload: 25 tonnes
The Seoul–Taegu high-speed line employs 60 kg/m rail mostly laid on concrete sleepers with Pandrol fastenings but some station sections employ slab track.

Kyrgyzstan

Ministry of Transport and Telecommunications

ul Isanov 42, 720017 Bishkek
Fax: (+7 3312) 21 36 67

Key personnel
Minister of Transport and Telecommunications:
Kubanyeh Bekzhumaliyev

Kyrgyzstan Railways

Kyrgyzia Zhelezni Darogy (KZD)
ul L'va Tolstogo 83, 720009 Bishkek
Tel: (+7 3312) 25 30 54

Key personnel
General Director: Isa Ormukulov

Gauge: 1,520 mm
Length: 470 km

Political background
This state system emerged from the collapse of the USSR in 1991.

Organisation
The main railway, formerly part of the USSR's Alma-Atinskaya Railway, is a 323 km branch from the Trans-Kazakhstan line to the capital, Bishkek (formerly Frunze), and its terminus at Issyk-Kul

(Balykchi) on the shores of Lake Issyk-Kul. There are also several short branches in the south linking such cities and towns as Tashkumyr, Dzhalal-Abad, Osh and Kyzyl-Kia with the eastern Uzbekistan system.

Passenger operations
In 1999, there were 31.3 million passenger-km travelled.

In 2001 it was reported that passenger services (all originating at Bishkek) consisted of a daily train to Moscow, a mixed train into Kazakhstan, and a seasonal service to Dzhalal-Abad over the railways of Kazakhstan, Uzbekistan and Tajikistan. Domestic services were limited to a daily train to Balykchi (Issyk-Kul) and four suburban services.

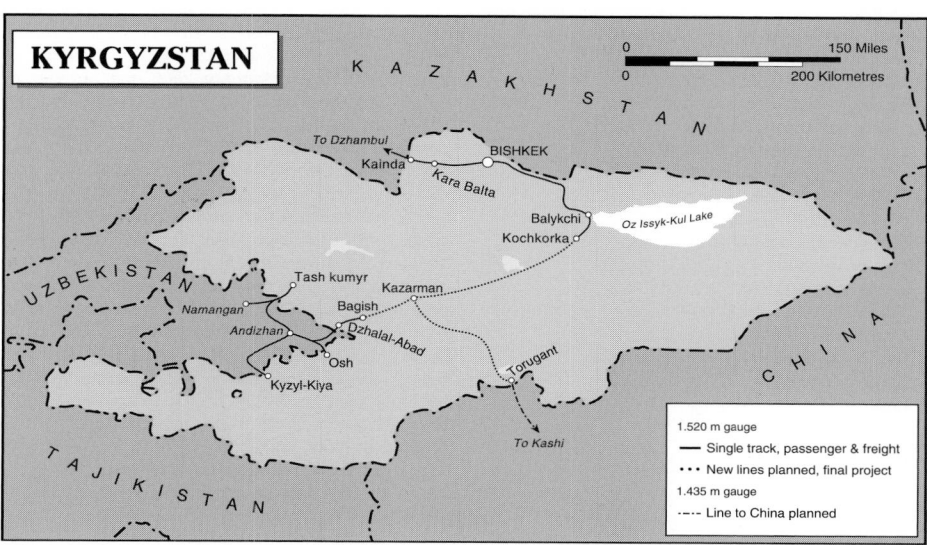

0114294

Freight operations

In 1999, there were 2.9 million tonnes travelling for 350 million tonne-km; virtually all freight crosses the border. Major commodities are coal (932,000 tonnes) and petroleum products (862,000 tonnes).

Most of the coal is imported: several lines built to serve domestic collieries are now closed. Domestic freight traffic is very sparse except between Bishkek and Tokmak.

New lines

The 60 km line from Balykchi (Issyk-Kul) to Kochkorka was opened in 2000. This is the first phase of a Trans-Kyrgyzstan line across the seismic Pamir range to link the northern railways with the southern railways at Dzhalal-Abad. According to a law passed in late 2000, part of this line will be incorporated in an international project providing a new link between China and Europe; a line will be built south from the Trans-Kyrgyzstan line, probably from Kazarman, to connect with Chinese Railways at Kashgar (Kashi). An accord agreeing to construction of the 256 km line was signed in June 2001 by the presidents of Kyrgyzstan and China and tendering was in progress to select consultants to undertake design of the project. The line is expected to open in 2005–06.

Traction and rolling stock

The railway is entirely diesel worked. There are 34 main line locomotives of classes 3TE10 and 2TE10, and 23 shunting locomotives of classes TEM2 and ChME3.

The rolling stock fleet comprises 520 coaches (of which 79 were out of use in 1996) and 2,616 wagons (727 not in use).

To satisfy complaints by other railways, a wagon repair workshop is to be established on the frontier at Belovskaya to ensure that freight vehicles requiring attention are rectified before leaving Kyrgyzstan territory.

Laos

Ministry of Transport

Vientiane

Key personnel

Spokesman: Songkane Luangmuninthorne

New lines

Laos has been without railway lines until now, but a 1,000 mm gauge rail link has been planned across the recently completed Friendship Bridge across the Mekong river, which is the border with Thailand (see Thailand entry for details). However, work on the 20 km line linking Vientiane, the Laotian capital, with the bridge was suspended in 1998 as a result of regional economic difficulties. The line was being funded by a joint venture of the Shaviriya Group of Thailand (75 per cent), through its subsidiary Pacific Transportation, and the Laotian government (25 per cent). The company was to have a 60-year operating concession, property development and telecommunications rights along the route and tax breaks on the first 18 years of operating revenues.

Under discussion by the joint venture partners are construction of a west-east new line from Vientiane to the Vietnamese capital Hanoi, and a line striking northwards from Vientiane to the Chinese border, and a line south-east from Vientiane to Champasak.

No work had started on the initial phase of the project by the end of 1999, and the Laotian government was reported to be seeking possible new partners to take it forward.

Latvia

Ministry of Transport and Communications

Gogola iela 3, Riga LV-1743
Tel: (+371 7) 02 82 22   Fax: (+371 7) 21 71 80
e-mail: satmin@sam.gov.lv
Web: www.sam.gov.lv

Key personnel

Minister: Ainars Slesers

UPDATED

Latvian Railway (LDZ)

Valsts akciju sabiedrība Latvijas Dzelzceļš
Gogoļa iela 3, Riga LV-1547
Tel: (+371 7) 23 49 40   Fax: (+371 7) 82 02 31
e-mail: info@ldz.lv
Web: www.ldz.lv

Key personnel

Chairman of the Board: Andris Zorgevics
Senior Vice-Chairman of the Board: Staņislav Baiko
Vice-Chairman of the Board, Finance Director:
  Rihards Peders
Directors
  Legal and Administrative: Uldis Pētersons
  Freight: Ēriks Šmuksts
  Finance: Igors Nikolajevs
  Strategic Development: Jānis Vēvers
  Personnel: Brigita Abike
  Economic: Vladimirs Grjaznovs
  Infrastructure: Mihails Jagodkins
  Rolling Stock: Jānis Pētersons
  Real Estate: Andris Burtnieks
  Technical Inspection: Arijs Sināts
Heads of key units
  Legal: Vēsma Upite
  Stores: Nikolajs Vasiļjevs
  External Relations: Valērijs Turko
  Computer Centre: Jānis Gulbis

Gauge: 1,520 mm; 750 mm
Route length: 2,270 km; 33.4 km
Electrification: 257.4 km of 1,520 mm gauge at 3.3 kV DC

Political background

The rail system of this Baltic state became legally established in August 1991 and started independent

0023696

operations in January 1992 with a workforce of around 24,000; by January 2005, the number of employees was down to 12,976 (excluding staff employed by the JSC Pasažieru Vilciens domestic passenger operator, JSC VRC Zasulauks rolling stock repair company and the LLC Dzelzceļa Apsardze railway security agency).

LDZ is the largest transport company in the Baltic states, its most significant function to provide services on the east-west and north-south transit corridors and to the Main Latvian ports of Ventspils, Riga and Liepāja. In 2004 the railway handled 51.1 million tonnes of freight, 5.6 per cent up on the figure achieved in 2002 and 40.2 per cent more than in 2000. Transit traffic in total accounted for 86.3 per cent of freight carried but transit traffic through Latvian ports represented 77.8 per cent of total shipments. Increases in both overland and maritime transit traffic is attributed to successful co-operation with operators in Russia, Belarus, Estonia, Lithuania, Kazakhstan and Ukraine among others.

LDZ is in the process of restructuring. This was initiated to adjust the railway to the requirements of a market economy. As a result, the formerly centralised undertaking has been transformed into a vertically integrated railway company with functional business units operating in different market areas.

On 5 May 2000, an LDZ shareholders' meeting approved the Restructuring Action Programme for 2000–03, anticipating the establishment as a holding company. Within this, member companies would operate in their respective market segments as independent legal entities. The restructuring programme had the support of the Latvian Cabinet of Ministers.

At the end of 2001, the first subsidiary company was established. Named Pasažieru Vilciens ('Passenger Train'), it is responsible for domestic passenger services, this first subsidiary company combined the activities of the former Elektrovilciens ('Electric Train') and Dīzeļvilciens ('Diesel Train') divisions. During 2003 three further subsidiaries were established: VRC Zasulauks, a joint stock company responsible for rolling stock overhaul and repair; Starptautiskie Pasažieru Pārvadājumi, a joint stock company providing international passenger services; and SIA Dzelzcela Apsardze, a limited liability company covering railway security.

Finance

In 2003 LDZ's audited profit was 5.7 million lats. This successful result was attributed to market-oriented management, a flexible tariff policy and an efficient marketing strategy, providing for a responsive freight rates structure, reduced costs and improved use of financial resources. Moreover, in 2003 LDZ received a subsidy from the Latvian government of 0.5 million lats to cover passenger rolling stock renovation.

With Pasažieru Vilciens established as a separate company, the following figures for passenger revenues in 2002 and 2003 only represent income from international services.

Finance (000 lats)			
Revenue	*2001*	*2002*	*2003*
Passengers	9,143	990	762
Freight	81,477	92,834	109,683
Parcels and mail	1	0	0
Other income	12,247	18,224	20,571
Total	102,868	112,048	131,016
Expenditure	*2001*	*2002*	*2003*
Staff/personnel/ materials/ services	90,543	90,641	110,186
Depreciation	8,687	9,422	15,584
Total	99,330	100,063	125,770
Traffic	*2002*	*2003*	*2004*
Passenger journeys (000)*	21,960	22,961	23,856
Passenger-km (million)	744	762	810
Freight tonnes (000)	40,100	48,355	51,058
Freight tonne-km (million)	15,020	17,604	16,877

* Passenger journey figures are derived from LDV's average passenger journey length indicator, which in 2003 was 80.7 km for general domestic traffic and 23.6 km for suburban traffic

Passenger operations

Passenger rail services in Latvia are provided by three companies: the joint stock company Starptautiskie Pasažieru Pārvadājumi, which runs international trains; the joint stock company Pasažieru Vilciens, which operates domestic services; and Gulbenes–Alūksnes Bānitis Ltd, which provides passenger services over the 750 mm gauge Gulbene–Alūksne line in the northeast of the country.

International services comprise the following: the 'Latvijas Ekspresis' (Moscow–Riga, daily); the 'Jūrmala' (Moscow–Riga, daily); the 'Baltija' (St Petersburg–Riga, daily), Riga–Gomel (every other day) and Riga–Simferopol (seasonally, every other day).

The main domestic lines link the capital Riga with Daugavpils, 218 km away in the southeast of the country, Rèzekne, in the east (224 km), Jelgava, in the south (43 km) and Lugaži, in the northeast (164 km).

In August 2001, services on the Riga–Ventspils and Riga–Liepāja routes were suspended due to poor patronage and rolling stock shortages. Services were being maintained on the busiest sections of these routes, between Riga and Tukums and Riga and Jelgava respectively.

Most routes are operated with Class DR1A and DR1P dmus. Suburban lines around Riga are operated by ER2 and ER2T 3 kV emus.

Freight operations

The decline of heavy industry since the break-up of the Soviet Union and the rise of road-based freight transport have seen freight tonnages on LDZ decline from around 50 million tonnes annually two decades ago to 28.8 million tonnes in 1995. Following a further considerable decline in traffic in 1998 and 1999, volumes started to grow in 2000 and in 2004 more than 51 million tonnes were carried. As in previous years, international transit traffic accounts for the largest share – 86.3 per cent in 2004, including oil and oil products, fertilisers, chemical products, ferrous metals, coal, foodstuffs, grain and timber.

Major investments are being made to develop container terminals at Latvian ports. At Ventspils

Class 2TE10M twin-unit diesel-electric freight locomotive at Jelgava (Norman Griffiths) 0569066

Modernised Class DR1A dmu 0569065

LDZ Class ER2 emu at Riga (Quintus Vosman) **NEW**/0585034

LDZ Class 2M62U twin-section diesel locomotive with a heavy coal train at Riga (Quintus Vosman) **NEW**/0585033

the Nord Natie Ventspils Terminàls facility has been commissioned, while wagon handling capacity at the port has also been increased.

Improvements to existing lines

An Infrastructure Improvement Study was commissioned by the European Union towards the end of 1997, aimed at identifying bottlenecks in the existing rail network, and in particular those parts affecting the east-west transit corridor. Following this study, in late 1998 LDZ was lent €34 million by the European Investment Bank to upgrade some 300 km of single track between the Russian border and Latvian ports to provide additional capacity. A second loan of US$20 million from the European Bank for Reconstruction and Development was to finance LDZ's new marshalling yard at Jùras Parks (Sea Park) and a northern rail bypass for the port of Ventspils.

During the period 2000–02 LDZ modernisation projects were granted €113 million of ISPA funding. This was matched by €37.6 million from LDZ and state budget resources. Four projects relating to the development of the east-west railway corridor financed by ISPA funding were approved: modernisation of the signalling system; replacement of 780 turnouts; modernisation of the rolling stock hotbox detection system; and construction of the Rēzekne II marshalling yard.

In 2003 LDZ was preparing four further projects financed by the EU's Cohesion Fund for completion by 2008: renovation of Riga Railway Junction; track renewal on the east-west corridor; construction of a second track on the Riga–Krustpils section; and introduction of the GSM-R radio-based communications system.

A US$6 million project to refurbish Riga's main station was started in 1998 by a Latvian-Norwegian joint venture company, Linstow Varner. Renovation of the existing building and creation of a new station complex were completed in September 2003.

Traction and rolling stock

Outside the Riga suburban system, trains are diesel-hauled. The decline in traffic has prompted withdrawal of older locomotives, with all remaining steam locomotives scrapped or sent to museums and many diesel locomotives taken out of service.

On 1 April 2005 the broad-gauge fleet comprised 196 diesel locomotives, 131 dmu cars and 221 emu cars. There were 24 passenger coaches (including four dining, 16 sleeper and four technical/service vehicles) and 5,333 freight wagons.

In 2003 LDZ received the first of its emu fleet to undergo refurbishment. Improvements include upgraded seating, installation of passenger information displays, modernised lighting and

Diesel locomotives

Class	Wheel arrangement	Power kW	Speed km/h	Weight tonnes	No in service	First built	Mechanical	Builders Engine	Transmission
1,520 mm gauge									
TEP70	Co-Co	2,900	160	129	15	1981	Kolomna	2A-5D49	E Kharkov
2TE10M	2 × Co-Co	2 × 2,200	100	2 × 138	10	1981	Lugansk	10D100	E Kharkov
2TE10U	2 × Co-Co	2 × 2,200	100	2 × 138	14	1990	Lugansk	14D40	E Kharkov
2M62	2 × Co-Co	2 × 1,270	100	2 × 120	40	1976	Lugansk	14D40	E Kharkov
2M62U	2 × Co-Co	2 × 1,270	100	2 × 126	30	1988	Lugansk	14D40	E Kharkov
M62	Co-Co	1,270	100	119	32	1965	Lugansk	14D40	E Kharkov
ChME3	Co-Co	993	95	121	55	1964	ČKD	K6S310DR	E ČKD
750 mm gauge									
TU2	Bo-Bo	240	50	32		1958	Kaluga	1D12	E Dinamo
TU7	B-B	294	50	24	2	1988	Kambarka	1D12	H Kaluga

Diesel railcars or multiple-units

Class	Cars per unit	Motor cars per unit	Motored axles/car	Power/ motor kW	Speed km/h	Cars in service	First built	Mechanical	Builders Engine	Transmission
DR1A	2–6	1–2	2	735	120	87	1973	Riga	M756	H Kaluga
DR1P	2–6	1–2	2	735	120	13	1973	Riga	M746	H Kaluga
DR1Am	2–3	1	2	745	120	30	1997*	Riga	MTU	H Voith
AR2	1	1	2	320	120	1	1997	Riga	NTA855R4	H Voith

* Rebuilt from Class DR1A.

Electric railcars or multiple-units

Class	Cars per unit	Motor cars per unit	Motored axles/car	Output/ motor kW	Speed km/h	Cars in service	First built	Builders Mechanical	Electrical
ER2	4–8	2–4	4	200	1420	162	1962	Riga	Riga
ER2T	4–8	2–4	4	235	130	64	1987	Riga	Riga
ER2M	4–8	2–4	4	200	130	4	1966	Riga	Riga
ER2I	4–8	2–4	4	200	130	11	1964	Riga	Riga

heating systems and the provision of retention toilets. The project was intended to extend the vehicles' working life by at least five years, after which a fleet replacement programme is foreseen.

Signalling and telecommunications

Automatic block signalling systems cover 1,064 km of LDZ's network, including 698 km on lines equipped with Centralised Traffic Control (CTC) systems; the 'Minsk' CTC system covers 81 km, the 'Neva' 431 km and the 'PChDC' 186 km. Semi-automatic block systems cover 925 km.

Various types of relay-based interlockings are installed at 164 of LDZ's 178 stations, operating 2,580 electrically locked sets of points. In June 2001, LDZ commissioned its first electronic interlocking, of the Ebilock-950 type. One of the largest of its type in Europe, this was supplied by Adtranz Signal (now Bombardier Transportation) and controls movements at Riga passenger station and at Tornakalns.

LDZ's telecommunications network consists of 561 km of fibre-optic cables with Gigabit Ethernet STM-1 and STM-4 SDH equipment, as well as 50 km of fobre-optic cables with LAN (100 points) equipment, 500 km of microwave radio lines with ATM equipment and 300 km of HDSL 2 Mb lines using copper cables. The telephone network has 110 exchanges supplied by Nortel.

Track

Rails: 75, 65, 50, 43 kg/m; 65 and kg/m is the most common, covering 1,572 km
Sleepers: Wood (2,750 × 180 × 250 mm), concrete (2,700 × 300 × 219 mm)
Spacing: 1,840/km in plain track, 2,000/km on curves
Fastenings: screw types: KB, W-14, SB-3, KD
Min curve radius: 200 m
Max gradient: 1.4%
Max axleloading: 23.5 tonnes

UPDATED

A/S Pasažieru Vilciens (PV)

Turgeòeve iela 14, Riga 50, LV-1547 Latvia
Tel: (+371 7) 583 40 09   Fax: (+371 7) 583 30 49
Web: http://www.pv.lv

Key personnel
Director: Vasilijs Hristins

Background

PV was created as a state-owned joint stock company to take over the non-suburban domestic passenger activities of Latvian Railway (LDZ). It commenced operations in January 2002. Previously there had been substantial cuts in intercity passenger service over the network.

Services were initially being operated using LDZ traction and rolling stock.

International passenger services are operated by Starptautiskie Pasažieru Pārvadājumi, an LDZ subsidiary, while the parent company also operates Riga suburban services.

Lebanon

Office des Chemins de Fer et du Transport en Commun (CEL)

PO Box 11-0109, 1107-2010 Beirut
Tel: (+961 1) 44 21 98
Fax: (+961 1) 44 70 07

Key personnel
President and Director General:
  Radwan Bou Nasreddine
Chief of Traffic and Operation: Faouzi Kharboutly

Chief of Traction and Rolling Stock:
  Ing Ahmad Adbed Aziz
Chief of Track and Structures: Ing Sayed Aouad
Chief of Finance: Gabriel Menassa
Head of Accounts: Carlos Maamari
Head of Administration: Fahed Choueiri
Chief of Stores: Fadi Baini
Chief of Personnel: R Chedid

Gauge: 1,435 mm; 1,050 mm
Route length: 319 km; 82 km

Organisation

The hostilities that ravaged the country during the 1980s affected the railway very severely and major parts of the system became unusable. The only services to have operated in recent years are short suburban service in Beirut operated with railbuses, and oil traffic south from Beirut to Saida. By 1999, these had been suspended.

Proposals developed during the early 1990s by French consultants Soferail to upgrade and electrify the Beirut–Tripoli line and to restore services on the line to Akkari in Syria have not been pursued.

In April 2001, the Czech consultancy firm Sudop Praha completed a study into rehabilitation of the section of line between Tripoli and Jôunié, north of Beirut.

In 2001 it was reported that tenders had been invited for the supply of rails for the reconstruction of the line from Beirut to the Syrian border.

0058564

Liberia

Introduction

There are three railways in Liberia, all originally constructed for iron ore transport. After several years of civil war, there is no report that any of the three are in operation.

Lamco Railroad

Roberts International Airport

Gauge: 1,435 mm
Route length: 267 km

National Iron Ore Company Ltd

PO Box 548, Monrovia

Gauge: 1,067 mm
Length: 145 km

Bong Mining Company

PO Box 538, Monrovia

Gauge: 1,435 mm
Length: 78 km

Organisation
Lamco, the Liberian American-Swedish Minerals Company, is an iron ore mining company which mined at Nimba and Tokadeh. The railway served principally to move ore, latterly amounting to some 6.5 million tonnes annually, from these mines

Organisation
The railway was opened in 1951. It operated 12 diesel locomotives, 253 ore and 28 other wagons. Annual traffic before the civil war totalled approximately 1 million tonnes.

At the start of the 1990s, before the onset of civil war, annual freight traffic grossed 6.7 million tonnes and 860 million tonne-km.

to the deep water port in Buchanan. There were 24 diesel-electric locomotives and 545 wagons. Operations ceased in 1989.

Libya

Department of Communications

PO Box 14527, Bab Ben Ghashir, Tripoli
Tel: (+218 21) 499 32   Fax: (+218 21) 401 06

Key personnel
Director General: Izz Al-din Al-Hinshiri
Director, Railways: Eng Alaeddin Al Weyfati

Railways Executive Board

PO Box 41748, Al Khames
Tel: (+218 21) 361 37 97; 360 94 85
Fax: (+218 21) 360 94 85
e-mail: genmana@libyanrailways.com
Web: http://www.libyanrailways.com

Key personnel
Chairman: Eng Mohammed Abdulssamed Ali
Director, General Affairs Office: Ali Mohammed Gholeh

General Managers
  Engineering Administration: Eng Abu Bakr Saeed
  General Administration of Materials:
    Eng Youssel Abdalla Al-Kabir
  Administrative and Financial Affairs:
    Mohammed Abdalla Al-Fitouri
  General Administration and Planning:
    Mohammed Bachir Al-Sayeh
  General Administration, Training and
    Operations: Eng Zayed Ali Al-Ghzeioui

Political background
With the dismantling of the British projection of 1,435 mm gauge from Egypt to Tobruk laid in the Second World War, no railways have run in Libya since 1965. Also discarded is the 950 mm system built around Tripoli and Benghazi on the eve of the First World War.

The present government, however, has revived plans to build a new 1,435 mm gauge system comprising a line along the Mediterranean coast from the Tunisian frontier via Tripoli to the Egyptian

frontier and a line south from Surt to Sebha, in the heart of a mineral resource area.

In 1992 the General Projects Office was established to take forward plans to develop the system and in 2000 this organisation became the Railways Executive Board. In March 2000 the Board signed a contract with the China Civil Engineering Construction Corporation (CCECC) to build the first phase of the network, from Ras Ajdir to Tripoli. This followed the establishment in 1998 of a joint Libyan-Chinese committee to oversee initial stages of the project.

Arrangements for operating the new railway remain to be decided, with concessioning stated to be one possible option.

In late 2000 the Board employed some 750 staff.

New lines
Current plans foresee an eventual network of 3,170 km comprising: a 2,170 km line from Ras Ajdir, at the Tunisian frontier, eastwards via Tripoli, Benghazi and Tobruk to make a connection with

the Egyptian system at Musa'id; and a line of some 1,000 km south from Surt via Waddan to Sebha, with a branch westwards to Tarot. A total of 96 stations will be provided.

The coast line is expected to carry both passenger and freight traffic, especially agricultural and petroleum products, while the line to the south will primarily carry iron ore from the Sebha area to a steelworks at Misratah, east of Tripoli.

The contract signed in 2000 with CCECC, valued at US$477 million, covers construction of the first section of the network, a 163 km line from Ras Ajdir to Tripoli and a 28 km link to the capital's port area. The contract also covers maintenance of the line for two years after its completion. The line is to be

provided with 16 stations and will be laid with UIC 54 continuous welded rail on concrete sleepers. It will feature 45 bridges but no tunnels.

Construction of the network is being divided into four sectors, with the Ras Ajdir–Tripoli line forming part of the first sector, which covers the section from the Tunisian frontier to Surt. In 2001 site preparation work had been completed and construction of this sector was in progress, with completion expected by 2003 at an estimated cost of US$10 billion. The remaining sectors cover: Surt–Benghazi; Benghazi–Musa'id; and Surt–Sebha. On the Benghazi–Musa'id section preparatory work was in progress between Tobruk and Musa'id in 2000.

Traction and rolling stock
The Railways Executive Board has estimated that it will require 244 diesel locomotives and some 8,600 passenger and freight vehicles to equip the complete network.

Signalling and telecommunications
Tenders for equipping the system call for CTC, colourlight signalling and automatic train control. Radio communication between train and control centre is also called for, together with a fibre optic telecommunications system.

Lithuania

Ministry of Transport and Communications

Gedlimino 17, LT-01505 Vilnius
Tel: (+370 5) 239 39 99   Fax: (+370 5) 212 43 35
e-mail: transp@transp.lt
Web: www.transp.lt

Key personnel
Minister: Zigmantas Balčytis
State Secretary: Dr Alminas Mačiulis
Director, Transit and Railway Transport:
    Česlovas Šikšnelis

Organisation
The ministry's Transit and Railway Transport Department is responsible for policy formulation, drafting legislation, international co-operation and promotion of the country's rail transport facilities. A State Railway Inspectorate under the ministry regulates the activities of railway undertakings in fields such as licencing, safety, rolling stock registration and arbitration of disputes between the infrastructure operator and network users.

UPDATED

Lithuanian Railways (LG)

AB Lietuvos Geležinkeliai
Mindaugo 12-14, LT-03603 Vilnius
Tel: (+370 5) 269 20 38   Fax: (+370 5) 269 20 28
e-mail: lgkanc@litrail.lt
Web: www.litrail.lt

Key personnel
Director General: Jonas Biržiškis
Deputy Director General, Freight and Passenger
    Transport: Juozas Senuta
Development Director: Albertas Šimenas
Infrastructure Director: Adomas Kazbaras
Internal Control Director: Gintautas Mikšta
Managers
    Passenger: Stasys Dailydka
    Freight Transport: Rimvydas Valys
    Rolling Stock: Virgilijus Jastremskas
    Infrastructure: Algirdas Panavas
    Investments: Irmantas Kandratavičius
    Technical Department: Alfredus Zubkevičius
    Economics: AušraŠpuraite
    Procurement: Saulius Braziulis
    Communications and Power Supply:
        Romanas Račas
    IT: Romualdas Kondratas
    Chief Accountant: Stasys Šimkunas
    Internal Audit: Irena Ungailiene
    Real Estate: Česlovas Norkus
    Personnel: Leonis Apulskis
    Legal: Liudmila Sarkisianc
    International Relations and Integration:
        Helma Jankauskaite
    Internal Administration: Liudmila Kirkiliene
    Chief Safety Manager: Petras Skripskis

Passenger business unit
Pelesos Street 10, LT-2030 Vilnius
Tel: (+370 5) 269 20 54   Fax: (+370 5) 212 09 26
e-mail: passenger@litrail.lt

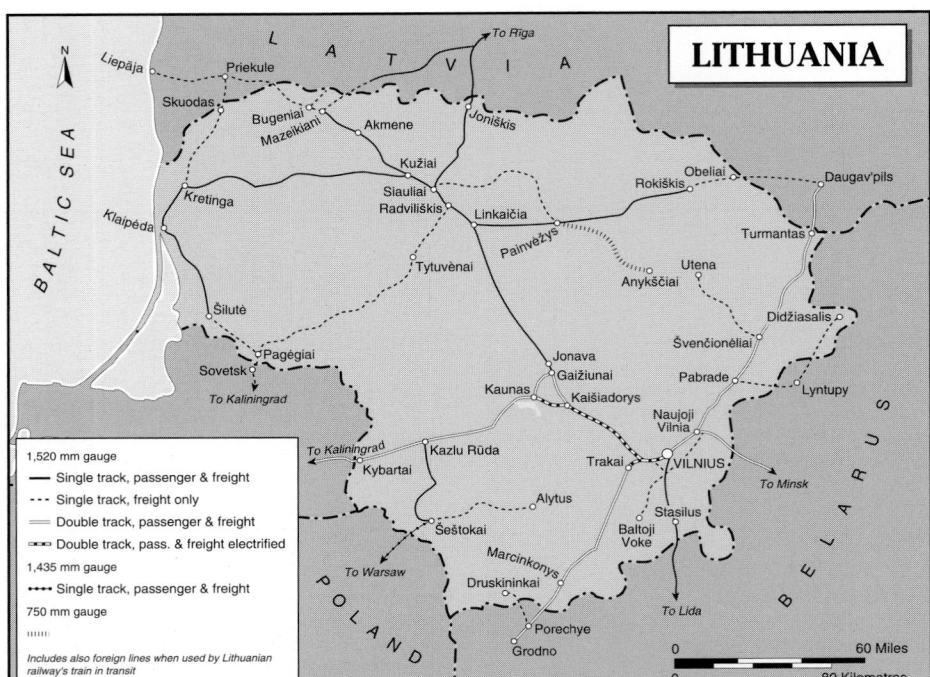

NEW/0585097

Gauge: 1,520 mm; 1,435 mm
Route length: 1,775 km; 22 km
Electrification: 122 km at 25 kV 50 Hz AC

Political background
The independent transport system of Lithuania was re-established at the break-up of the USSR in 1991, when sea, air and railway transport were transferred to the Ministry of Transport of the Republic of Lithuania. Since 1996 LG has been a state-owned joint-stock company.

In early 2005, following Lithuania's accession to the EU the previous year, the government announced its intention to split LG into three independent state-owned companies responsible for infrastructure, freight traffic and passenger services. Apart from complying with EU directives covering the structure of national rail networks, the move was also intended to eliminate cross-subsidy of passenger services by freight. The government

anticipated that restructuring would be completed by the end of 2005. This followed the approval of enabling legislation in December 2003 and will lead to the liberalisation of the country's rail market.

Relations with neighbouring countries, particularly Russia, are of paramount importance since domestic traffic accounts for less than 20 per cent of the total freight tonnage.

Responsibility for a surviving 67 km section of the Lithuanian 750 mm system linking Panevežys with Anykščiai was transferred in 2001 from LG to a separate public body, Aukštaitijos Siaurasis Geležinkelis.

At the end of 2004 LG employed 11,500 staff.

Traffic (million)	2001	2002	2003
Passenger journeys	7.7	7.2	–
Passenger-km	532.8	498.1	–
Freight tonnes	29.2	36.65	43.5
Freight tonne-km	7,741	9,767	

Diesel locomotives

Class	Wheel Arrangement	Power kW	Speed km/h	Weight tonnes	No in service	First built	Mechanical	Builders Engine	Transmission
1,520 mm gauge									
M62	Co-Co	1,470	100	119	42	1970	Lugansk	14D40	E Kharkov
2M62/ 2M62U	2 × Co-Co	2,940	100	238	91	1980	Lugansk	14D40	E Kharkov
TEP60	Co-Co	2,206	160	127	12	1970	Kolomna	11D45	E Kharkov
TEP70	Co-Co	2,942	160	129	4	1993	Kolomna	2A-5D49	E Kharkov
TEM2	Co-Co	883	100	120	48	1965	Bryansk	PD1M	E ETM
ChME3	Co-Co	994	95	121	41	1980	ČKD	K6S310DR	E ČKD

Totals of each type in service are approximate

Passenger operations

A generally efficient network of express buses, using an upgraded road infrastructure, has reduced rail passenger traffic significantly since independence. Other than the small electrified suburban network around the capital Vilnius and the link to Kaunas, all passenger services are diesel-worked, the majority in a fleet of ageing dmus built by Ganz of Budapest and RVR of Riga.

After a sharp decline in the post-Soviet era through the 1990s, passenger traffic stabilised in the early years of the present decade.

However, international passenger traffic has declined significantly, the former overnight through Tallinn–Warsaw train now operating on alternate days only, with a change of train necessary at the gauge-change station Sestokai. Only one through train, operating overnight, remains to connect Vilnius with the Latvian capital Riga. Other cross-border services link Vilnius with Moscow, St Petersburg and Minsk.

Freight operations

LG's freight operations have experienced strong growth in the first years of the present decade, in 2003 recording an increase in tonnage carried of 18.6 per cent compared with the previous year. The figure of 43.5 million tonnes was the highest achieved since LG was established as an independent company.

In 2002 LG carried 36.65 million tonnes. Of this, some 57 per cent was transit traffic, 26 per cent import/export and the remainder domestic freight. The most significant commodity sector was oil and petroleum products, accounting for 42 per cent of tonnage, followed by chemical and mineral fertiliser (14 per cent) and metals traffic (10 per cent).

Most transit traffic is handled through the ice-free port of Klaipeda, although some is also destined for the Russian enclave of Kaliningrad. Principal marshalling yards are located at Radviliškis and Vaidotai.

Two significant international intermodal service initiatives were launched in 2003. The 'Viking' provides a connection for containers, semi-trailers and piggyback vehicles between the Baltic at Klaipeda and the Black Sea at Odesa, exchanging traffic at Vilnius, Minsk and Kiev; the 'Baltic-Transit' operates weekly between Baltic ports in Estonia, Latvia, Lithuania amd Kaliningrad to Kazakhstan and Uzbekistan, with traffic from Lithuania and Kaliningrad tripped to Latvia to be formed into the trunk service.

The operations at Klaipeda have seen significant changes since the rail ferry service to Mukran in the then German Democratic Republic were inaugurated in 1986. The dedicated rail ferry service to Mukran has been replaced by a combined ro-ro road/rail operation using one Lithuanian and one German ferry, and additional ro-ro road-only services also operate to other ports in Germany, Denmark and Sweden.

Most of the non-domestic rail traffic is interchanged with other former USSR railways, particularly operating in the east-west direction. North-south international traffic is much less important, the major interchange point being at Sestokai with the Polish (PKP) system, where a change of gauge takes place.

New lines
Standard gauge link with Poland

Plans were originally unveiled in 1997 for the construction of a standard-gauge (1,435 mm) line from the Polish border, for a distance of 86.5 km via Mariajampole to Kaunas, Lithuania's second city. Here, it is planned to establish a common border station, a freight distribution centre and automatic gauge-changing facilities at a road/rail interchange at Rokai on the southern outskirts. Pre-investment and feasibility studies have since been carried out and further preparations for the scheme were in hand in 2004.

Rail Baltica

Along with the governments of Estonia and Latvia, the Lithuanian administration was a signatory on 2001 to an agreement to develop a high-speed rail link for passenger and freight traffic connecting Warsaw with Tallinn via Kaunas and Riga. Pre-design studies commenced in 2004 and the project would be undertaken in three phases: Warsaw–Kaunas

Class TEP70 diesel locomotive leaving Radviliškis with a Vilnius–Klaipeda passenger service (Philip Wormald)
***NEW**/0585100*

Class 2M62U twin-section diesel locomotive at Kena with a container train bound for Belarus (Philip Wormald)
***NEW**/1114947*

LG Class ER9M emu arriving at Riga, Latvia, forming a service from Lithuania (Philip Wormald)
***NEW**/1114948*

(projected completion 2010); Kaunas–Riga (2014); and Riga–Tallinn (2016). In October 2003 the European Commission added the scheme to its list of priority international transport links.

Improvements to existing lines

The LG network in recent years has benefited from extensive programmes of track renewal, infrastructure improvements and modernisation of signalling and communications systems. This has been principally concentrated on European Corridors I (Helsinki–Tallinn–Riga–Kaunas–Warsaw) and IX (B and D) (Kiev–Minsk–Vilnius-Klaipeda and Kaišiadorys–Kaunas–Kaliningrad) and has been partly funded by loans and subsidies from European financing institutions. However, the Lithuanian government estimates that it will take until 2015 before the network meets the highest standards established elsewhere in the EU.

A high priority has been given to modernising traffic management and telecommunications systems. Communications systems on the Kaišiadorys–Radviliškis line have been modernised with the help of a Danish government loan; a credit from the European Investment Bank has funded modernisation of signalling on the same section. Telecommunications renewal projects cover the Vilnius–Kaišiadorys, Kaišiadorys–Kybartai and Klaipeda–Šiauliai-Radviliškis lines.

Improvements to border stations aimed at speeding up transit traffic were also being implemented: reconstruction and development of Kybartai stations was started in 2004, funded by the EU's Phare programme.

Developments at Šeštokai (Mockava), terminal point for the standard-gauge line from Poland, include provision in 2000 of automatic gauge-changing equipment, installation of a 40-tonne gantry crane and commissioning of a liquid gas transfer facility.

Also planned for implementation in 2004 were reconstruction of facilities at Pauostis to serve the Klaipeda oil terminal and other developments at Klaipeda.

A major ongoing project to reconstruct and upgrade Vilnius station continued in 2005, with completion expected in 2006.

Traction and rolling stock

At the beginning of 2005, LG's fleet comprised 139 main line diesel locomotives, 91 diesel shunters, 44 dmus (mostly Class DR1A) and 16 emus (Class ER9M). In addition, one prototype RVR (Riga) single-unit AR2-type railcar was in service. Main line locomotives are mainly of Classes M62 and 2M62. Small numbers of Class TEP60 and TEP70 diesel locomotives are used for passenger traffic. The dmu fleet comprised Class D1 and DR1 types, while the emus are of the ER9M type.

In 2004 LG commenced a process of modernising its traction fleet, inviting tenders for the supply of 34 main line diesel locomotives. In the same year the company also commenced a programme to modernise 30 of its twin-section 2M62 locomotives.

Also in 2004 RVR in Latvia commenced a programme to upgrade 20 LG dmus.

At the beginning of 2005 the passenger coach fleet totalled 214 vehicles. The freight wagon fleet totalled 9,336 vehicles.

Signalling and telecommunications

Centralised traffic control covers 169 km, automatic block 445 km, and semi-automatic block 1,323 km.

Electrification

25 kV 50 Hz electric multiple-units operate between Vilnius and Kaunas, as well as providing commuter service for Vilnius.

Track

Rails: UIC 60
Crossties (sleepers): Concrete
Sleeper spacing: 1,640/km in plain track, 1,840/km in curves
Fastenings: Screws/bolts – DO type; in concrete sleepers – KB type
Min curve radius: 497 m
Max gradient: 1.5%
Max axleload: 25 tonnes

UPDATED

Luxembourg

Ministry of Transport

11 rue Notre Dame, L-2938 Luxembourg
Tel: (+352) 478 44 01   Fax: (+352) 24 18 17
Web: www.gouvernement.lu

Key personnel

Minister: Lucien Lux
Government Commissioner, Rail: Paul Schmit

UPDATED

EuroLuxCargo SA

P O Box 1803, L-1018 Luxembourg
Tel: (+352) 49 90 47 00   Fax: (+352) 49 90 47 09
e-mail: elc@cfl.lu
Web: www.cfl.lu

Key personnel

President: Alex Kremer
Director: Marc Calmes

Organisation

EuroLuxCargo is a wholly owned subsidiary of Luxembourg Railways (CFL) formed in 1997 to market the operator's freight services.

In 2001 EuroLuxCargo acquired a minor German railway, the Norddeutsche Eisenbahn Gesellschaft (NEG) of Uetersen, which holds a licence to operate throughout the country's main line network. The company has taken over from DB Cargo shunting activities around Trier and feeds scrap metal trains into the CFL network at Wasserbillig. NEG uses three diesel locomotives in Trier. In 2004 NEG took over the bankrupt NVAG, based in Niebüll, Germany.

UPDATED

Luxembourg Railways (CFL)

Société Nationale des Chemins de Fer Luxembourgeois
9 place de la Gare, L-1616 Luxembourg
Tel: (+352) 499 00   Fax: (+352) 49 90 44 70
e-mail: info@cfl.lu
Web: www.cfl.lu

Key personnel

President: Jeannot Waringo
Vice-President: Marc Glodt
Director General: Alex Kremer
Directors: François Jaeger, Nicolas Welsch, Marc Wengler
Company Secretary: Edouard Schwinninger
Managers
  Infrastructure: François Jaeger
  Network: Claude Mersch
  Finance and Data Systems: Marc Wengler
  Operations: Paul Lorang
  Traction and Rolling Stock: Marcel Barthel
  Fixed Plant and Signalling: Jean-Marie Franziskus
  Personnel: Nico Bollendorff
  Passenger Business: Nicolas Welsch
  Freight Business: Alex Kremer

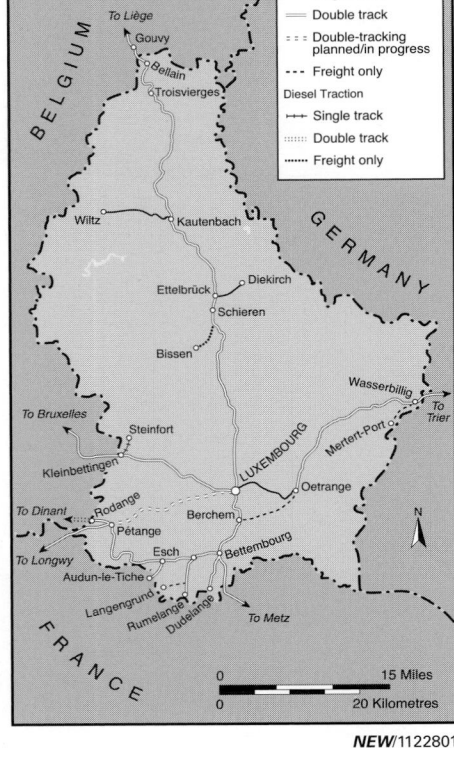

NEW/1122801

Gauge: 1,435 mm
Route length: 275 km
Electrification: 243 km at 25 kV 50 Hz AC; 19 km at 3 kV DC

Political background

Since its creation, shares in CFL have been owned by the Luxembourg, Belgian and French governments. In 1996, the Luxembourg government increased its stake in CFL from 63.25 to 94 per cent and took over responsibility for the railway's debt as part of a reorganisation plan. The state became sole owner of CFL's infrastructure and the government became responsible for infrastructure maintenance and modernisation. Belgium's share of CFL has fallen from 24.5 to 4 per cent and France's from 12.25 to 2 per cent.

CFL operates bus services which complement the rail network.

Organisation

CFL is directed by a joint board of 15 which includes members from the Luxembourg government, CFL, and one member each from the French and Belgian governments.

CFL is managed in four 'activity units' covering operations, technical, commercial and central services. Apart from these, there are eight subsidiaries, including businesses responsible for tourism, freight, intermodal, wagons, site development and insurance.

In 2004 CFL employed 3,100 staff, of which 104 in its subsidiaries.

Finance

In 2004 CFL's turnover rose by 6.84 per cent but the company lost €280,000. It invested €112 million in 2004, of which €85.7 million came from the Fonds du Rail government infrastructure fund.

CFL now gives details of performance by activity. While passenger activities were held to be 'in surplus', freight made a loss of €30 million in 2004.

Finances (€ million)	2002	2003	2004
Revenue			
Passenger traffic	32.06	32.65	33.61
Freight traffic	88.65	80.38	94.20
Other*	182.85	201.84	204.12
Compensation for public service obligations	76.17	83.12	93.27
Other	48.99	48.28	45.84
Total	431.28	446.17	471.04

* Remuneration for management of infrastructure ('Infragrant')

Expenditure	2002	2003	2004
Linked to passenger services	11.84	11.23	11.91
Linked to freight services	52.09	46.09	56.16
Materials	23.87	23.25	20.46
Personnel	191.17	196.69	202.60
Other	154.99	171.76	180.19
Total expenditure	433.96	449.02	471.32
Profit	−2.68	−2.75	−0.28

Traffic (million)	2002	2003	2004
Freight tonnes	17.9	16.2	17.1
Freight tonne-km	617	568	600

Passenger operations

In 2004 passenger numbers on domestic services rose by 2.4 per cent and by 5.1 per cent on international services.

In 2002 the Luxembourg government launched the project 'mobiliteit.lu', aimed at lifting public transport's market share from 14 to 25 per cent by 2020. It is to finance extensions to the rail network to achieve this. The first extension will be a branch off the Luxembourg–Ettelbruck line to the city's airport. This will be served by tram-train vehicles.

CFL is carrying out a major renewal of its passenger stock which will increase capacity by over half and bring down average age of stock from around 30 years to five years by early 2006. All new stock has been financed by government.

Commuter traffic from countries bordering Luxembourg has grown very strongly in recent years thanks to new season tickets and additional services. In 2001 for example, traffic from Metz grew by 21 per cent and from Longwy by 346 per cent. In early 2004 CFL revealed plans to extend several domestic services into neighbouring countries to attract even more cross-border commuters.

There are five main international services serving the country: Brussels–Luxembourg–Strasbourg–Basle; Luxembourg–Metz–Paris; Luxembourg–Metz–Nancy; Luxembourg–Trier; and Luxembourg–Liège.

Cross-border improvements

Electrification of the 'Athus–Meuse' line in December 2002 (see 'Electrification' below) has allowed most freight services to be diverted away from the direct Namur–Luxembourg line via Libramont. Following this, CFL and SNCB wish to invest in upgrading the line to bring down the Luxembourg–Brussels journey time.

CFL has ordered 12 ALSTOM double-deck Class 2200 three-car emus, three of which are to be used in common with SNCF for services to Metz and Nancy.

In May 2000, SNCB Type AM96 air conditioned emus were introduced on the Brussels–Luxembourg service. In June 2003 SNCB Type M6 double-deck stock was introduced on the busiest peak services on this route.

The Luxembourg–Paris service will be significantly accelerated in 2007 thanks to the new LGV Est Européen line (see section on France).

Freight operations

CFL freight traffic rose by 5.6 per cent in 2004, while revenue rose by 17.2 per cent. Domestic traffic fell by 17 per cent but international traffic rose by 7.5 per cent. The Luxembourg steel industry generates 36 per cent of all CFL freight traffic. The nature of this traffic has changed due to conversion to electric furnaces. Transport of traditional, heavy raw materials including iron ore, coke, limestone and fuel oil in block trains has to some extent been replaced by the movement of the new raw material, scrap metal, in wagonload services and traffic in semi-finished products. Transit traffic is now the most important sector, generating 43 per cent of traffic.

On 14 January 1998, Luxembourg became the control centre for the newly created international 'freight corridor' from Muizen in Belgium via Namur, Luxembourg, Metz and Dijon to Milan and Gioia Tauro, Italy and via Marseille, France to Barcelona, Spain. The corridor was used by an average of 95 trains per month in 2002. In 2004 CFL, together with SNCF and SNCB, created a joint subsidiary, SIBELIT, with the aim of improving services on this route.

EuroLuxCargo (see Organisation) has used its German subsidiary NEG to feed traffic into the CFL network at Wasserbillig.

Intermodal operations

Transiting intermodal traffic has increased considerably in the last decade, Luxembourg being situated on the main Rotterdam/Antwerp–Switzerland–Italy route.

The Bettembourg terminal is being developed as an international intermodal traffic hub and offers connections to Germany, France, Switzerland, Italy and Spain. Most of the intermodal traffic is generated by the ports of Rotterdam and Antwerp, and Milan.

New lines

In September 1992 the Transport Ministers of France and the Grand Duchy signed an agreement guaranteeing that Luxembourg would be served by TGV Est Européen trains when that French high-speed line becomes operational. Luxembourg will contribute €141 million to the cost of a spur from TGV Est Européen at Baudrecourt. Luxembourg will get four daily TGV services from Paris and – when TGV Est Européen is extended that far – one from Strasbourg. From 2007 Luxembourg will be 2 hours from Paris compared with 3 hours 30 minutes in 2001.

The Transport Minister has suggested that over the 16 km between the French frontier and Luxembourg City, where the existing double track is already heavily occupied, a third track should be added for TGVs. It could be arranged for a higher speed than the 140 km/h maximum of the present double track without incurring insupportable civil engineering costs.

Newly delivered Bombardier-built Class 4000 electric locomotives (Edward Barnes)   ***NEW**/1122802*

Newly built Class 2200 three-car double-deck emu at Raismes, France during trials (David Haydock)   0580502

Class 2100 diesel railcar at Kleinbettingen (David Haydock)   0580503

Class 3000 electric locomotive with a CFL freight from France at Livange (John C Baker)

0580504

In November 2003 Luxembourg's government approved the tram-train scheme to build a 15 km line from the Luxembourg–Ettelbruck line to Kirchberg and Findel. The line will cost €390 million. Work started in 2005 with opening expected in 2007.

In December 2003 CFL extended the Bettembourg–Dudelange Usines line to Volmerange-les-Mines in France with the aim of attracting commuters working in Luxembourg.

Improvements to existing lines

A new chord at Aubange opened in December 2004, allowing direct operation from the Athus–Meuse line to Longwy and southwards,. This enabled freight services to avoid the congested nodes of Luxembourg, Thionville and Metz and this has reduced transit traffic via the CFL network.

In 2004 CFL was due to begin work on doubling the direct Luxembourg–Pétange line. Work is due to be completed in 2007.

Traction and rolling stock

In mid-2005 CFL operated 39 electric (plus six on hire), 16 line-haul (plus eight on hire) and 22 shunting diesel-electric locomotives, 11 diesel tractors, six diesel railcars, 34 emu sets (75 cars) plus two Belgian three-car emus used on the Brussels service and two two-car dmus maintained by DB, 76 passenger coaches (including 16 Corail coaches of the SNCF-CFL pool) and 3,100 freight wagons.

In December 1995 CFL and SNCB ordered 80 5,000 kW dual-voltage locomotives with three-phase asynchronous AC motor drives from ALSTOM. This pool of locomotives is jointly operated, mainly on freight over the Antwerp–Luxembourg–Metz (France) route (see 'Electrification') via the Athus–Meuse line and passenger/freight on the Luxembourg–Liège line. CFL's share of the fleet is 19 Class 3000 locomotives. All CFL and SNCB locomotives were in service by early 2002.

CFL owns two SNCB Type AM80 three-car emus as its contribution to Brussels–Luxembourg services. The stock is maintained in Belgium. CFL also owns two DB Class 628.4 dmus, used in a pool with DB sets on the Luxembourg–Trier service, and maintained by DB.

At the end of 2000 CFL ordered six Class 2100 single-unit diesel railcars from ALSTOM, which were based on the SNCF Class X73500. They are used on branch lines, releasing emus for busier routes.

In 2001 CFL ordered 12 three-car Class 2200 double-deck emus from a consortium led by ALSTOM and with significant participation by Bombardier. They are based on the TER2N NG

design ordered by SNCF. The first three were delivered in 2004, to be followed by six in 2005 and three in 2006. Total cost is €101 million.

With an orer for new diesel locomotives in mind, CFL has hired two Vossloh Type G1206 and six Vossloh Type G1000 BB diesel-hydraulic locomotives, all with radio control, from Angel Trains Cargo. CFL's infrastructure department had four Cockerill and two Vossloh shunting locomotives on hire in 2005.

A new maintenance and servicing facility on the site of CFL's Luxembourg depot is under construction and will eventually replace existing workshops.

In early 2003 CFL ordered 85 double-deck coaches from Bombardier for €127 million and 20 Class 4000 dual-voltage electric locomotives (based on the DB Class 185) to power them and costed at €61.2 million. The locomotives and 67 intermediate coachs were delivered in 2004–05 and have replaced Class 3600. The coaches will supplant all existing hauled stock, providing a considerable increase in peak capacity. Driving trailers were being delivered in 2005. Pending deliveries of its Class 4000 locomotives, which commenced in September 2004, CFL hired Class 185 machines from Angel Trains Cargo. These will be returned in 2006.

CFL intends to purchase 15 tram-train vehicles, with an option for 29 more, for the new line to Findel.

In 2004 CFL had a fleet of 3,100 wagons, plus 490 wagons on hire.

Signalling and telecommunications

CFL is pursuing a plan to concentrate signalling of each route on a single centre, supported by the new installation at Luxembourg. Most of

the network is currently controlled from five signalling centres.

Following two head-on crashes in 1997, CFL has introduced a simple safety system known as MEMOR II+ to prevent a train passing a red signal. This was completed in late 2003. In the longer term, CFL is to be one of the first railways to introduce ERTMS/ETCS. The first section to be equipped will be Cruchten–Ettelbruck–Diekirch in 2004.

The SNCF KVB system is to be installed on the Luxembourg–Bettembourg frontier line.

Electrification

Electrification of Luxembourg's main routes was completed in 1993. Ninety-five per cent of Luxembourg's rail lines are now electrified, all on the 25 kV 50 Hz AC system, except the cross-border line from Brussels and Namur in Belgium which is electrified at 3 kV DC. International expresses travelling between Belgium and France via Luxembourg change locomotives in Luxembourg station. Most transiting freight trains are now hauled by SNCB Class 13/CFL Class 3000 dual-voltage electric locomotives from Belgium to France via Luxembourg.

SNCB completed electrification of the Luxembourg–Liège route in 2000 and of the Athus–Meuse route from Rodange in Luxembourg via Virton and Bertrix to Dinant in December 2002.

Track

Rails: UIC 60, 54 and U33
Crossties (sleepers): Wood
Thickness: 150 mm
Spacing: 1,435/km
Rail fastenings: 'K' fastenings

UPDATED

Diesel locomotives

Class	Wheel arrangement	Power kW	Speed km/h	Weight tonnes	No in service	First built	Builders Mechanical	Engine	Transmission
800	Bo-Bo	600	80	74	3	1954	AFB	GM 8–567B	E GM
850	Bo-Bo	615	105	72	*	1956	B & L	SACM MGO V-12 SH	E B & L
900	Bo-Bo	690	105	72		1958	B & L	SACM MGO V-12 SHR	E B & L
1800	Co-Co	1,435	120	110/114	16	1963	BN	GM 16–567C	E ACEC/SEM

* In 2005 combined fleet of Classes 850 and 900 totalled 13.

Electric locomotives

Class	Wheel arrangement	Output kW	Speed km/h	Weight tonnes	No in service	First built	Builders Mechanical	Electrical
3000	Bo-Bo	5,000	200	90	19	1998	ALSTOM	ALSTOM
4000	Bo-Bo	5,600	140	84	20	2004	Bombardier	Bombardier

Electric railcars and multiple-units

Class	Cars per unit	Motor cars per unit	Motored axles/car	Output/ motor kW	Speed km/h	Units in service	First built	Builders Mechanical	Transmission
250	2	1	2	308	120	5	1975	Carel et Fouché	MTE
260	3	1	2	308	120	2	1970	Carel et Fouché	CEM
2000	2	1	4	308	160	22	1990	De Dietrich	GEC Alsthom
2200	3	3	2	800	160	5*	2004	ALSTOM/Bombardier	ALSTOM

* On order for delivery 2004–06.

Diesel railcars

Class	Cars per unit	Motor cars per unit	Motored axles/car	Power/ motor kW	Speed km/h	Units in service	First built	Mechanical	Builders Engine	Transmission
2100	1	1	2	258	140	6	2000	ALSTOM/DDF	–	Voith

Macedonia, Former Yugoslav Republic of

Ministry of Transport and Communications

4 Crvena Skopska Opstina Square, 1000 Skopje
Tel/fax: (+389 91) 312 32 92; 312 62 28; 314 54 97
e-mail: contact@dtk.gov.mk
Web: www.dtk.gov.mk

Key personnel
Minister: Xhemail Mehazi

UPDATED

Macedonian Railways (MŽ)

Makedonski Železnici
PO Box 543, Železnička 50b, 1000 Skopje
Tel: (+389 92) 322 79 03   Fax: (+389 92) 46 23 30
e-mail: mz65dir@mt.net.mk
Web: www.mt.net.mk

Key personnel
Director General: Stojan Naumov
Deputy Director General: Dragi Apostolovski
Assistant Directors General
  Marketing: Todor Hadzi-Boskov
  Finance: Milco Smilevski
  Infrastructure: Ljubo Tanelovski
  Legal Affairs: Kosta Cikos (Acting)
  Traffic: Mirko Spirovski
  International Affairs: Ratko Stefanovski

Gauge: 1,435 mm
Route length: 699 km
Electrification: 223 km at 25 kV 50 Hz

Political background
MŽ came into being as the result of Macedonia declaring its independence from the former Yugoslavia in 1991. It is a member of the UIC (International Union of Railways) and employed 3,609 staff at the end of 2004, down from 3,656 in 2003.

Passenger operations
The most important routes run south from the capital Skopje to Gevgelija (166 km) on the Belgrade–Thessaloniki–Athens route (two pairs of international trains and two pairs of local trains a day) and from Skopje to Bitola (184 km, four pairs of local trains a day).

Traffic (million)	2002	2003	2004
Passenger journeys	0.930	0.901	0.917
Passenger-km	98	92	94
Freight tonnes	2.208	2.390	2.629
Freight tonne-km	334	373	424

New lines
The main line through Macedonia runs from the Greek border in the south to the Yugoslav border in the north. A new east-west line is under construction in the north of the country to complement this north-south line. This route, 56 km in length, will run from Kumanovo to the Bulgarian border. It is eventually intended to form part of an Adriatic to Black Sea through link, from Durrës in Albania to Burgas in Bulgaria; in Macedonia, this will require construction of another new line from Kičevo to the Albanian border.

Traction and rolling stock
At the end of 2004 MŽ was operating 16 electric and 35 diesel locomotives, four emus and 12 dmus, and one diesel railcar. Coaching stock amounted to 135 vehicles, including 42 coaches, 17 sleeping cars and 21 couchette cars; there were 1,651 freight wagons. In 2001–02 the traction fleet was supplemented by the acquisition of three Class 442 electric locomotives, converted by Končar from former Croatian Railways (HŽ) Class 1141 machines.

Signalling and telecommunications
Automatic block signalling is being installed on 39 km from Klisura to Gevgelija; it is already in

NEW/1114988

Former Croatian Railways Class 1141 electric locomotive modernised by Končar and now designated MŽ Class 442
0122618

Diesel locomotives

Class	Wheel arrangement	Power kW	Speed km/h	Weight tonnes	No in service	First built	Mechanical	Builders Engine	Transmission
661	Co-Co	1,434	124	108	21	1961	GM	GM	E EMD
643	Bo-Bo	680	80	65	3	1967	B & L	MGO	E B & L
642	Bo-Bo	606	80	63	5	1961	Duro Daković	MGO	E Končar
734	C	440	60	48	2	1960	MAK	Maybach GT06	H Voith
732	C	397	60	43.5	2	1965	Duro Daković	Jenbacher 600	H Voith
667	Co-Co	882	100	120	2	1981	BMZ SSSR	BMZ	E BMZ SSSR

Electric locomotives

Class	Wheel arrangement	Output kW continuous/ one hour	Speed km/h	Weight tonnes	No in service	First built	Builders Mechanical	Electrical
441	Bo-Bo	3,860/4,080	140	82	8	1973	ASEA/Končar	ASEA/Končar
442	Bo-Bo	3,860/4,080	120	82	3	2001*	Končar	Končar
461	Co-Co	5,100/5,400	120	120	5	1978	Electroputere	Electroputere

* Modernised ex-Croatian Railways locomotives.

place on a further 172 km between Tabanovci and Gevgelija.

Type of coupler in standard use: U-85t

Track
Rail: 30–54 kg/m; 49 kg/m rail is the most common, being installed on 483 km
Sleepers: Wooden (1,324,661 installed), concrete (19,312), metal (16,526)
Sleeper spacing: 1,660 per km, in plain and curved track
Fastening type: K system
Minimum curve radius: 250 m
Max gradient: 2.6%
Max permissible axleload: 22.5 tonnes

Diesel railcars and multiple-units

Class	Cars per unit	Motor cars per unit	Motored axles/car	Power/ motor kW	Speed km/h	Units in service	First built	Mechanical	Builders Engine	Transmission
712	3	2	4	206	120	12	1976	MACOSA/ Duro Daković	MAN	*H* Voith

Electric railcars and multiple-units

Class	Cars per unit	Motor cars per unit	Motored axles/car	Output/ motor kW	Speed km/h	Units in service	First built	Builders Mechanical	Electrical
412	4	2	4	170	120	4	1985	Riga	Riga

UPDATED

Madagascar

Ministry of Transport & Meteorology

PO Box 4139, Anosy, Antananarivo
Tel: (+261 2) 246 04
Fax: (+261 2) 240 01

Key personnel
Minister: Aimé Rakotondrainibe

MADAGASCAR

0 — 75 Miles
0 — 125 Kilometres

Imerimandroso
Ambatosoratra
Morarano-Chrome
Vohidiala
Toamasina
ANTANANARIVO
Lohariandava
Moramanga
Antsirabe
Vinaninkaréna
Fianarantsoa
Manakara
Ihosy
Toliara

1.000 m gauge
--- Single track, not electrified, passenger & freight
— Single track, not electrified, freight services only
••• Not completed line
-··- Future projects

0058512

Fianarantsoa-Côte Est Railway (FCE)

Gare FCE, Fianarantsoa
Tel: (+261 20) 755 13 54
e-mail: fce@blueline.mg
Web: www.fce-madagascar.com

Gauge: 1,000 mm
Route length: 163 km

Political background
FCE is the isolated southern line of the former state railway RNCFM which was not privatised with the main network in 2003 (see Madarail entry in Madagascar section of *Railway systems and operators*). Opened throughout in 1936, the line runs from the port of Manakara to Fianarantsoa in the southern highlands. Threatened with closure several times, the FCE line suffered from the same neglect that brought the main network to its knees at the end of the 1990s. Subsequently it almost succumbed to severe damage inflicted by cyclones Eline and Gloria in February and March 2000. While the Toamasina–Antananarivo main line was rebuilt immediately, no government financial assistance was provided for rebuilding the southern line. Here, restoration was undertaken by volunteers with the help of Swiss Federal Railways and some Swiss regional railway companies, plus World Bank finance, enabling RNCFM to provide at least a minimum level of service. A five-year rehabilitation plan saw the line largely rebuilt by 2004.

The plans also envisaged eventual concessioning of the line to the private sector, and in October 2004 the government sought bids for joint operation of the railway and the port of Manakara.

Operations
In 2004 FCE was being operated largely for tourists, the attraction being the magnificent scenery en route from the coast to the high plateau, with steep gradients, tortuous curves and 67 bridges and 48 tunnels. Nevertheless, some local freight and passenger traffic is being carried. Mixed trains run five days per week, with journey times of between eight and 12 hours, according to the quantity of freight on offer.

Tiraction and rolling stock
Three diesel locomotives from of a fleet of five were available for traffic at the end of 2004, plus 100 freight wagons, about 10 coaches and a 'Micheline' railcar dating from the French colonial era.

NEW ENTRY

Madarail

PO Box 1175, Antananarivo 101
Tel: (+261 20) 223 45 99   Fax: (+261 20) 222 18 83
Web: www.comazar.com/madarail

Key personnel
Managing Director General: François Lisambert
General Manager: Patrick Stevenaert

Gauge: 1,000 mm
Route length: 691 km

Political background
In 1982 the national railway RNCFM was established as a state-owned entity operating under the country's laws governing limited companies, which allowed some independence in commercial policy-making, but neglect of the infrastructure has left the railway poorly equipped to compete with an improving road network. Election of a new government on a liberalisation ticket in 1993 led to moves to open railway operations to private contractors, and in 1998 Canadian consultant CPCS Transcom assisted in developing concessions to operate the northern network.

The privatisation process stalled at the end of the 1990s, and during the country's political crisis of 2001 it was suggested that the rail network might be dispensed with altogether. But eventually the Comazar group subsidiary Madarail SA was awarded a 25-year concession, taking over operations in July 2003. The infrastructure remains the property of the state, while Madarail owns the rolling stock.

Organisation
The system as concessioned to Madarail comprises three main routes in the central-eastern region of the country. These are: the TCE (371 km), connecting Antananarivo inland with the port of Toamasina (formerly Tamatave); the MLA (Moramanga-Lac Alaotra, 167 km); and the TA (Antananarivo-Antsirabé, 153 km) serving the high plateau south of Antananarivo. The 163 km southern system, now known as the Fianarantsoa-Côte Est Railway (see entry in Madagascar section of *Railway systems and operators*), is not part of the concessioned network.

A five-year investment programme was started as soon as the concession was agreed. Costed at €37.5 million, rehabilitation of the network is being funded mainly by loans from the World Bank (€21 million) and the European Investment Bank (€11 million). In 2004 the concessionaire had 878 employees.

Operations
Madarail took over a railway that was practically defunct; only one locomotive remained in working order, and passenger service had been abandoned. Little freight was being moved on account of the poor state of the traction and rolling stock fleet. The first priority was rehabilitation of sufficient rolling stock to enable resumption of a reliable freight service between the port of Toamasina and the capital. Initially, the emphasis was on moving import/export traffic, including petroleum products, containers, cement, chrome ore and fertilisers. From a slow start, Madarail hoped to carry more than 300,000 tonnes in 2004.

By mid-2004 11 locomotives were operational, comprising seven newly acquired and four of the 1986 batch from Alstom. One of the latter was the first loco to undergo total refurbishment in Antananarivo workshops for a decade. Around 100 wagons were rehabilitated, including 20 container flats for the traffic between Toamasina port and Antananarivo. It was hoped to resume passenger operations during 2004, initially between Moramanga and the coastal town of Ambila Lemaitso.

A new bonded warehouse at Antananarivo was expected to open in mid-2005; this would be served by block container trains from Toamasina.

New lines
In the south of the country the mountainous territory has extraordinarily rich and very diverse mineral deposits. RNCFM had built a 27 km extension of the TA line to a new cement works 1,800 m above sea level in the Ibity mountain massif. Several other schemes totalling some 900 km were proposed in the 1980s and 1990s, but none materialised. Madarail hopes that the newly-stablised railway will revive the interest of international mining companies and could lead to further new construction.

Traction and rolling stock
In 2004 the operational fleet comprised 11 main line diesel locomotives, two road/rail shunting tractors, about 250 wagons and 10 coaches.

Signalling and telecommunications
With World Bank aid, the MLA line was equipped with a radio telecommunications system.

Type of coupler in standard use: Freight cars, Willison automatic, Madagascar type; passenger cars, De Dietrich, Soulé

Type of brake in standard use, locomotive-hauled stock: Automatic air; direct air; and vacuum

Track
Rails: S25, 26, 30, 36, 30 US, 30 East, 37 English
Sleepers: Wood 1,920 × 220 × 150 mm; steel 1,900 × 294 × 147 mm
Spacing: 1,666/km wood; 1,500/km steel
Fastenings: Screw (wood sleepers); frog (steel sleepers)
Min curvature radius: 80 m
Max gradient: 3.5%
Max axleload: 16 tonnes

UPDATED

Malawi

Ministry of Transport and Civil Aviation

Private Bag 322, Capital City, Lilongwe 3
Tel: (+265) 73 01 22   Fax: (+265) 73 38 26

Key personnel
Minister: Harry Thomson
Principal Secretary: J L Kalemera

Malawi Railways (1994) Ltd

PO Box 5144, Limbe
Tel: (+265) 64 08 44   Fax: (+265) 64 06 83

Key personnel
Chairman: Dean Lungu
General Manager: E R Limbe
Assistant General Managers: H T Thindwa,
    K S J Chenjerani, J A Kazembe
Chief Accountant: M Ndenya
Company Secretary: Vacant
Chief Marketing Manager: T Nnensa
Chief Traffic Manager: M F Mlenga
Chief Mechanical Engineer: H Chimwaza
Chief Engineer (Telecoms and Electrical):
    M F Kuntiya
Supplies Manager: H P Nyasalu

Gauge: 1,067 mm
Route length: 797 km

Political background
A major restructuring of the railway was carried out in 1995 under the auspices of the World Bank, and in 1996 the Malawi government invited consultants to advise on strategies for possible privatisation, with the aims of stimulating economic growth, increasing transport sector competition, and reducing government funding pressures.

During 1997, the Canadian-based Hickling Transcom (now CPCS Transcom) consultancy completed a study of the feasibility of privatisation; this led to a call for prequalification bids by early 1998 from organisations interested in taking over an operating concession and in 1999 Mozambique Ports & Railways (CFM) was selected to run the system. Under privatisation proposals, infrastructure would remain government-owned but would be maintained by the concessionaire. Passenger services would be supported by a public service obligation subsidy. Other details of the concession structure remained to be finalised. The confidence of prospective bidders was being raised by indications that the government of neighbouring Mozambique, through whose rail network Malawi traffic must pass for the transport of external trade, was planning to offer concessions to operate its CFM(N) rail network (qv) and the port of Nacala. The Malawi government has also announced that the winning bidder for the Malawi concession would be free to bid for the Mozambique operations, raising the prospects for unified management of the so-called 'Nacala rail corridor'. However, in 1999 the Malawi government was reported to be dissatisfied with the 16 per cent minority shareholding being offered in such a concession by the Mozambique government.

Organisation
A single-track line runs from Mchinji near the Zambian border through Lilongwe and Blantyre to the southern border with Mozambique. This line connects with the Mozambique port of Beira. A line from Nkaya to Nayuchi on the eastern border with Mozambique connects with the port of Nacala.

Following restructuring, passenger and cargo services on Lake Malawi are now run by a separate company, Malawi Lake Services. These connect with the rail system at Chipoka at the south end of the water.

Traffic (000)	1994–95	1995–96
Passenger journeys	537	339
Passenger-km	21,272	18,048
Freight tonnes	473	254
Freight tonne-km	78,340	43,431

Finance (MK 000)		
Revenue	*1994–95*	*1995–96*
Passengers	4,209	4,424
Freight	50,362	43,851
Miscellaneous	78,283	10,856
Total	132,854	59,131

Expenditure	1994–95	1995–96
Staff/personnel	78,939	19,791
Materials and services	100,199	15,671
Depreciation	6,199	17,422
Financial charges	7,912	3,378
Total	193,249	56,262

Passenger operations
In the early 1990s the railway was handling over 1 million passenger journeys annually. However, this has fallen sharply to around 30 per cent of this level. In the second half of the 1990s just a handful of passenger services remained, with no trains at weekends. Five times a week there is a mixed train between Balaka and Limbe, supplemented by an additional Blantyre-Limbe service on the north-south main line. On the eastern line a twice-weekly mixed train runs between Nkaya and Nayuchi, providing connections with international services to and from Cuamba in Mozambique.

The fall in traffic is attributed to slow journey times due to track conditions, leaving rail vulnerable to strong competition from fast and frequent bus services. Residual train services are mainly patronised by passengers travelling to markets with goods not permitted on buses.

Freight operations
Traffic has declined since the 1980s, when around 1 million tonnes annually were transported, to a current level of some 300,000 tonnes. Approximately half of this is international traffic. Main commodities carried are cement, grain, fuel products, fertilisers, and tobacco. Consultancy studies suggest that international traffic offers the best growth potential, especially via the Nacala corridor.

New lines
Proposals exist to develop a line from the Moatize coalfield in Mozambique through Malawi to connect with the existing line to Nacala. This would reach the Malawi system from the west at a point between Blantyre and Nkaya. The still

Examples of MR's two locomotive types are seen at Limbe: a recently refurbished MLW (Bombardier) Co-Co dating from 1980 and one of five CMI shunters built in 1993 (Eddie Barnes)   0023147

Diesel locomotives

Wheel arrangement	Power kW	Speed km/h	Weight tonnes	No in service	First built	Mechanical	Builders Engine	Transmission
Co-Co	1,120	116	86	12	1980	Bombardier	Alco 8-251-E	E Canada GE
C	380	56	43.5	5	1993	CMI	Cummins KTA 19L	H Twin Disc

unfinished line from Lilongwe to Mchinji would also be completed and a connecting line would be built westwards from Mchinji to Chipata (27 km) into eastern Zambia. Some earthworks for the Chipata extension were undertaken several years ago but no progress was made after a decision by the Zambian government to transfer track materials to Zambia Railways. However, in 1999 Malawi Railways announced that it intended to resume work on the project.

Improvements to existing lines
Rehabilitation of the Nkaya-Nayuchi line to the border of Mozambique (44 km) has been undertaken following agreement in mid-1995 of World Bank and USAID credits totalling US$28.6 million. Of the total, US$9.53 million is being spent on track rehabilitation, which includes bridge and structure strengthening, points and crossing work, production of 17,000 sleepers, rental of a track tamper and purchase of gang and inspection trolleys.

Traction and rolling stock
The locomotive fleet at the start of 1998 comprised 12 main line diesel-electrics (all now refurbished under a USAID programme), and five diesel-hydraulic shunters. The latter machines were supplied in 1993 by Cockerill Mechanical Industries of Belgium. The rolling stock fleet consisted of 389 freight wagons, 47 engineering service wagons, 29 passenger coaches and one special coach.

Type of coupler in standard use: AAR 10 Automatic profile
Type of brake in standard use: Vacuum

Track
Standard rail: BSR 30 kg/m, length 12.2 m; BSA 30 kg/m, length 48.8 m; BSA 40 kg/m, length 48.8 m and cwr
Sleepers: Timber, steel and concrete

	Spacing in plain track	Spacing in curved track
Timber	1,310/km	1,476/km
Steel (30 kg)	1,310/km	1,476/km
Concrete	1,460/km	1,640/km

Fastenings: Pandrol clips on concrete sleepers, clip and steel key, elastic rail spikes
Min curve radius: 111 m
Max gradient: 2.27%
Max axleload: 15 tonnes (old track) 18 tonnes (new track)

Malaysia

Ministry of Transport

Levels 5, 6 and 7, Block D5, Parcel D
Federal Government Administrative Centre, 62502 Putrajaya
Tel: (+60 3) 88 86 60 00   Fax: (+60 3) 88 89 25 37
Web: http://www.mot.gov.my

Key personnel
Minister: Dato'Sri Chan Kong Choy
Secretary General: Zaharah Shaari

Express Rail Link

Express Rail Link Sdn Bhd
25th Floor, Wisma UOAII, 21 Jalan Pinang, 50450 Kuala Lumpur
Tel: (+60 3) 21 64 22 77   Fax: (+60 3) 21 66 70 69
Web: http://www.myERL.com

Key personnel
Executive Chairman:
 Mohamed Nadzmi Mohamed Salleh
Chief Executive Officer: Dr Adnan Aminuddin
Vice-President, Marketing and Sales:
 Woo Yew Seong

Gauge: 1,435 mm
Route length: 57 km
Electrification: 57 km at 25 kV AC 50 Hz

Political background
In August 1997 Express Rail Link (ERL) was granted a 30-year government concession, with a 30-year extension option, to design, finance, construct, manage, operate and maintain an express rail system linking Kuala Lumpur Sentral at Brickfields and the city's new international airport (KLIA) at Sepang, south of the capital. Land acquisition for the scheme was undertaken by the Malaysian government. After some delay as a result of regional economic conditions, in October 1998 a DM1.3 billion turnkey contract to build the system was placed with SYZ, a consortium led by the Transportation Systems Group of Siemens AG, which owns a shareholding of 59 per cent, after a financing package was secured from Germany. Construction work is being undertaken by local consortium partner Yeoh Tiong Lay Sdn Bhd (YTL). Operations commenced on 19 April 2002.

Organisation
Shareholding in ERL, which was formally incorporated on 29 January 1996, is by Tabung Haji Technologies (60 per cent) and YTL Corporation (40 per cent).
 Under a contract signed in November 1999, a subsidiary company, ERL Maintenance Support

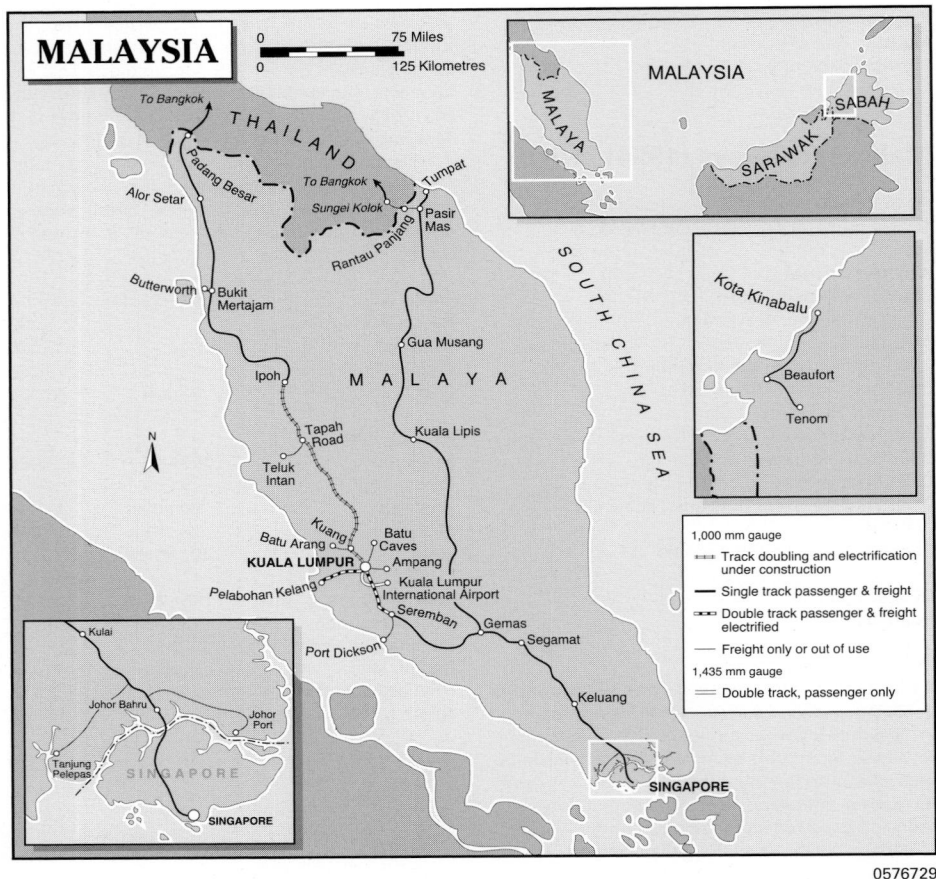

Sdn Bhd (E-MAS), has been established to operate and maintain the system for three years. Ownership of E-MAS is divided between Siemens (51 per cent) and ERL (49 per cent). It is intended that ERL will assume Siemens share of E-MAS after three years of operation of the system. At the start of operation in April 2002, E-MAS employed 327 staff.

Passenger operations
ERL plans to operate two services: KLIA Express, a non-stop service linking KL Sentral and KLIA, with a journey time of 28 minutes; and KLIA Transit, a commuter service linking the same two points but with additional stops at the three stations listed in 'New lines'. KL CAT, which is located in the KL Sentral development, offers city centre check-in and baggage transfer for departing passengers and a baggage check-out facility for arriving passengers. A service frequency of 15 minutes is provided for 21 hours each day. Revenue operations began in April 2002.

For KLIA Transit, a half-hour frequency is planned, with a 36 minute journey time. Interchange will be made at Bandar Tasik Selatan with KTM Komuter main line services and with the STAR light rail system and with the Putrajaya Monorail at Putrajaya's Western Transport Terminal. At KL Sentral, interchange will be made with KTM Komuter and main line services, the Putra rapid transit system and KL Sentral Titiwangsa services. Commencement of KLIA Transit revenue operations is scheduled for June 2002.

New lines
The project entailed construction of a 57 km double-track standard-gauge line, electrified at 25 kV AC 50 Hz. The alignment starts at a new Kuala Lumpur City Air Terminal (KL CAT), around 1 km south of the existing main KTM station, and parallels the existing line for some 17 km before diverging south to Sepang. Civil works started in June 1997, and include 39 bridges and three short

Siemens-built four-car Desiro ET articulated emu for the KLIA Express airport link 0110185

Traction and rolling stock

As part of the turnkey contract, Siemens Transportation Systems has supplied 12 four-car articulated emus based on the design of the Class ET 425 vehicles supplied to Deutsche Bahn AG and designated Type Desiro ET. Eight of the units are for KLIA Express services, and feature higher levels of comfort and seating for 156 passengers, as well as a stowage area for baggage containers; the remaining four units are high-density vehicles with seating for 144 but standing capacity for 396 passengers. Body construction is of extruded aluminium sections and all vehicles are air conditioned.

Assembly of the ERL fleet has been undertaken at the Siemens SGP Verkehrstechnik facility in Vienna, with bodyshell components produced at Uerdingen, Germany and bogies coming from Siemens' Graz, Austria plant.

Signalling and telecommunications

All train operations are controlled by a Traffic Management System (TMS), which has input/output devices in the Operations Control Centre (OCC), located at Salak Tinggi. Trains are driver-controlled under the guidance and control of an Automatic Train Protection (ATP) system.

An optical fibre bearer network centred on the OCC is employed to carry communications functions, including public address, clock and CCTV services. A SCADA system located at the OCC manages the traction power supply system and station infrastructure facilities for the entire network.

ERL electric multiple-units

Class	Cars per unit	Motor cars per unit	Motored axles/car	Output/motor kW	Speed km/h	Units in service	First built	Builders Mechanical	Electrical
ERL	4	2	4*	225	160	8	2001	Siemens	Siemens
CRS	4	2	4*	225	160	4	2001	Siemens	Siemens

* Units are articulated.

Track

Rail: UIC 54, continuously welded
Sleepers: prestressed concrete monobloc; slab track used in KLIA ERL station

cut-and-cover tunnels. Three intermediate stations are located at Bandar Tasik Selatan, Putrajaya, and Salak Tinggi. Depot and workshop facilities are located north of the last station. Track and signalling systems have been designed for train operations at speeds up to 160 km/h. Siemens has supplied system engineering services, including power supply, signalling, telecommunications and SCADA systems, as well as providing overall project management.

Malayan Railway (KTM)

Keretapi Tanah Melayu Berhad
Jalan Sultan Hishamuddin, 50621 Kuala Lumpur
Tel: (+60 3) 22 63 11 11
Web: http://www.ktmb.com.my

Key personnel

Executive Chairman: Tan Sri Dato' Thong Yaw Hong
Managing Director: Encik Mohd Salleh bin Adbullah
Directors
　Operations and Customer Service:
　　Mohd Zin Yusop
　Special Projects: P Satyamoorthy
　Finance: K Sinnappu
　Corporate Services: Dr Ismail Rejab
　Human Resources: Md Fauzi Hj Said
　Engineering: Haji Mazlan Waad
General Managers
　Permanent Way: Zainal Abidin Salleh
　Fleet Engineering: Sarbibi Tijan
　Signalling and Communications:
　　Lee Heng Cheong
　Human Resources: Abdul Mokti Zakaria
　Safety Operations: S Apputhurai
　Corporate Communications: Hamdan Ahammu
　Training and Development: K Sudharmin
　Traffic Operations: Samat Mahat
　Electrification: Dzulkifli Mohd Ali
　Freight Services: Abd Radzak
　Passenger Services: Hilmi Mohamad
　Property Management: Mohd Nasir
　Corporate Planning: Hilmi Mohamad
　Information Technology: Shafin Yunus
　Finance: Azman Ahmad Shaharbi
　Company Secretary: Nor Aida Othman

Gauge: 1,000 mm
Route length: 1,699 km
Electrification: 150 km at 25 kV 50 Hz AC

Political background

The government has long been seeking to privatise KTM. In early 1998 discussions regarding full privatisation continued between the government and the Marak Unggul consortium, comprising Renong, DRB-Hicom, and Bolton Properties, which had already taken over management control of the railway. Marak Unggul was reported to have commissioned consultants to review all aspects of KTM operations and development projects.

Rolling stock maintenance is now in the hands of a private-sector company, RailTech Industries, in which KTM has a 26 per cent holding.

Organisation

The railway's prime route is the 787 km main line from Singapore north through the capital, Kuala Lumpur (KL), to Butterworth, one of Malaysia's principal sea ports on the west coast of the peninsula. Short branches reach sea ports at Port Klang, Pasir Gudang and Tanjung Pelepas. The other major route is the 528 km East Coast line running northwards from a junction with the Singapore-Butterworth line at Gemas to Kota Bharu and Tumpat. Both lines link with the State Railway of Thailand.

At the beginning of 2003 KTM employed 5,024 staff.

Passenger operations

KTM operates both electric suburban services (branded 'Komuter') around the capital, Kuala Lumpur, and long-distance trains.

Suburban services operate on two electrified routes: Rawang–Seremban and Sentul–Port Klang. These interchange with each other and with the Putra metro and KLA Ekspres airport express service at Bandar Tasik Selatan.

Long-distance travel soared spectacularly in the 20 years to 1991, lifting the total of annual passenger-km from 620 million to 1,850 million;

Toshiba-built Class 24 diesel-electric locomotive heading a KTM passenger service at Taiping (Edward Barnes) 0554374

KTM diesel locomotives

Class	Wheel arrangement	Power kW	Speed km/h	Weight tonnes	No in service	First built	Mechanical	Builders Engine		Transmission
18	0-6-0	450	24	46.25	8	1979	Brush	MTU 6V 396 TC 12	E	Brush
19	Bo-Bo	480	78	58.4	10	1983	Hitachi	MTU 6V 396 TC 12	E	Hitachi
22	Co-Co	1,275	96.5	84	10	1979	EE-AEI	EE8CSVT-MK 111	E	EE
23	Co-Co	1,610	110	90	11	1983	Hitachi	SEMT/ SP12 PA4V	E	Hitachi
24	Co-Co	1,790	120	90	26	1987	Toshiba	SEMT/16V-PA4	E	Toshiba
25	C-C	1,120	107	89.6	17	1990	GM	EMD 8-645E 3C	E	GM
26	Co-Co	2,460	120	116.4	20	2003	Bombardier	GE 7FDL12	E	GE
'CKD8C'	Co-Co	2,460	110	120	2	2003	Dalian	Dalian 12V240ZJD-2	E	Dalian

Electric multiple-units

Class	Cars per unit	Motor cars per unit	Motored axles/car	Output/ motor kW	Speed km/h	Units in service	First built	Builders Mechanical	Electrical
81	3	2	4	190	120	14	1994	Jenbacher	Holec
82	3	2	4	–	120	21	1997	Hyundai	Mitsubishi
83	3	2	4	–	120	20	1997	Union Carriage	GEC Alsthom

thereafter patronage began to fall. Following opening of the North-South Expressway (NSE) road in 1994, passenger-km declined to 1,348 million; in 1995, there was a further fall, to 1,270 million. An encouraging revival to 1,396 million passenger-km (5.9 million passenger journeys) was recorded in 1996.

The stiff competition focused attention on the long-term need to reduce journey times between Kuala Lumpur and Singapore. This will be achieved by doubling and electrifying the whole line for 160 km/h operation by 2008.

In addition to ordinary train services, KTM operates day and night Singapore-Kuala Lumpur and Kuala Lumpur-Butterworth express trains, and a single daily express between Gemas and Tumpat on the East Coast line. In conjunction with the State Railway of Thailand, KTM runs a daily International Express between Butterworth and Bangkok. All express and night trains on the north-south main line are air conditioned. Intercity trains carry over 10,000 people daily, the KL–Singapore route being the busiest.

Orient-Express Hotels of the UK, operators of the Venice Simplon-Orient Express in Europe, runs a weekly luxury cruise train service, the Eastern & Oriental Express, between Singapore, Kuala Lumpur and Bangkok.

Traffic	1999	2000
Passenger journeys (000)	4,344	3,825
Passenger-km (million)	1,316	1,220
Freight tonnes (million)	2.8	5.5
Freight tonne-km (million)	907	916

Freight operations

Traffic has fallen slightly in recent years. Cement is the most important commodity carried, but fuel, gypsum and dolomite are among other products handled. International traffic is exchanged with the Thai system.

Intermodal operations

Container traffic is a fast-growing business, with 16 daily trains operated in 2003. Begun in 1974, this has become KTM's biggest freight earner. A significant component of the growth has been containers from southern Thailand, which have a quicker haul across the border at Padang Besar to a Malaysian port than to Bangkok. Around a quarter of southern Thailand's rubber exports are shipped through Butterworth. In April 1999, KTM launched a twice-weekly container landbridge service between Bangkok and Port Klang, providing a journey time of 60 hours.

The government has been promoting intermodal transport, and KTM formed a wholly owned subsidiary in 1988, the Multimodal Freight Company, which has acquired 225 highway tractors and 1,300 trailers so as to offer door-to-door service. This operates from all major ports and the inland container depots at Kuala Lumpur, Ipoh, Prai, Nilai and Padang Besar.

Improvement and expansion of rolling stock has been a recent KTM priority. Flatcar capacity has been added regularly, bringing the total to 3,500 TEU in 2003. Structures have been modified to gain clearance for 9 ft 6 in cube containers on certain routes.

New lines

In 2002 a 31.5 km line was opened linking Pelabauhan with the new port of Tanjung Pelebas, north of Johor Bahru. Built by the Indian company, Ircon, the line mainly carries container traffic.

For Komuter services the line from Sentul is to be extended to Batu Caves by 2006.

Design and supervision contracts for the construction of a 3 km branch from Butterworth to the port's north terminal were awarded in late 1997.

In 1998, Canadian consultancy Canarail was commissioned by Malaysian oil company Petronas to design an 80 km industrial line from its facilities around Kerteh to the port of Kuantan. Canarail had earlier completed a feasibility study into the project.

Improvements to existing lines

Doubling, resignalling and electrification of the line from (Kuala Lumpur–) Rawang to Ipoh is under way and scheduled for completion in late 2005. All level crossings will be eliminated to allow 160 km/h operation. The aim is to extend the Seremban–Rawang Komuter service to Tg Malim and introduce rapid intercity services between KL and Ipoh. In 2003 the government announced it was to award contracts to a joint venture of Malaysia Mining Corporation and Gamuda for upgrading, double-tracking and electrifying the two single track sections of the KTM north-south main line: from Ipoh to the border with Thailand at Padang Besar (339 km); and south from Seremban to Johor Bahru (297 km), giving better access to Singapore. However, at the end of the year the government decided to postpone the scheme.

A new main workshop and emu depot is planned at Batu Gajah by 2006. This would replace the existing works at Sentul.

The KTM network forms an element of a regional project under discussion by seven ASEAN countries to create a Trans-Asia Rail Link, with an upgrade of existing infrastructure favoured to improve connections with neighbouring Thailand.

Traction and rolling stock

At the beginning of 1997 KTM's traction fleet comprised 101 diesel locomotives. The passenger stock totalled 359, plus 18 three-car emu trains. Freight stock totalled 3,206 vehicles.

The KL Komuter fleet consists of 18 three-car emus supplied by Hunslet Transportation Projects/ Jenbacher, plus two batches each of 22 three-car sets from Mitsubishi/Hyundai and GEC Alsthom/ Union Carriage, all supplied in the 1990s.

Motive power shortages have been alleviated by delivery of 30 Indian Railways' Class YDM4 Co-Cos, hired on a power-by-the-hour basis from Ircon, which is also maintaining the locos with its own staff. There are also four MKA-2000 Co-Cos remanufactured and regauged by Morrison Knudsen Australia; they are English Electric units surplus to Australian National's requirements in Tasmania, which in 1995 were leased to the Malaysian firm Rail Tech Services for use by KTM as required. MKA retains another three such locomotives which are available to be sent to Malaysia if needed.

Ircon has also undertaken refurbishment of 12 Class 22 diesel-electric locomotives, and in 1997 was reported to be negotiating a contract to refurbish and maintain a further 15.

In 2003 KTM received 20 Class 26 2,460 kW Bombardier/General Electric Blue Tiger diesel locomotives. In addition, in 2003 two 2,460 kW CKD8C Co-Co diesel-electric locomotives were procured from CNR, China, and 20 additional machines rated at 2,610 kW were ordered from the same source for delivery from 2005.

Coupler in standard use: Hook and knuckle. Hook couplers are being progressively replaced by the automatic knuckle-type on all wagons; this should permit increases of up to 50 per cent in gross freight train weights.
Braking in standard use: Air and vacuum

Signalling and telecommunications

KTM is modernising signalling and communications on the south main line between Seremban and Johor Bahru. Route relay interlockings at stations and automatic block signalling, when completed, will have the entire distance from Singapore to Ipoh controlled by colourlight signals.

Electrification

Operation of the country's first electric trains started in 1995 with commissioning of the Kuala Lumpur suburban network. With 25,000 passengers daily being recorded within six months, capacity of the 18 three-car emus very quickly became strained, and a further 44 sets were ordered. By 2003 traffic had grown to over 60,000 passengers daily. Electrification is now in progress from Rawang to Ipoh.

Completion of the scheme saw the launch of an air conditioned commuter emu service on the routes from Kuala Lumpur to Sentul, Port Klang, Seremban and Rawang.

Track

Standard rail: Flat bottom in 12.2 m (40 ft) lengths
Main line: 40 and 60 kg/m
Rail fastening: Elastic spikes
Crossties (sleepers): Malayan hardwoods 242 × 127 × 2,000 mm (1,359 km); concrete (903 km)
Spacing: 1,666/km
Filling: 6 cm (2½ in) limestone ballast to a depth of 15 cm (6 in) under sleepers
Max curvature: 12.25° = radius of 142 m
Ruling gradient: 1%; except Taiping Pass 1.25%
Longest continuous gradient: 8.2 km on Prai-Singapore main line, with 1.25% grade, sharpest curve 142 m radius for a length of 320 m
Max altitude: 137 m near Taiping
Max axleload: 16 tons

Sabah Ministry of Transport and Communications

88999 Kota Kinabalu, Sabah
Fax: (+60 88) 23 98 52

Key personnel
Minister: W M Bumburing
Assistant Secretary, Railways: H Gunggut

Sabah State Railways

Jabatan Keretapi Negeri Sabah
Karung Berkunci 2047, 88200 Kota Kinabalu, Sabah
Tel: (+60 88) 546 11   Fax: (+60 88) 23 63 95

Key personnel
General Manager: M T Jaafar

Gauge: 1,000 mm
Route length: 134 km

Organisation
The railway links Tanjong Aru with Beaufort, running along the coastal strip before climbing inland along the Padas river to Tenom.

Freight and passenger operations
In 1992 the railway carried 0.4 million passengers and some 2.2 million tonnes of freight, mainly rice, rubber and timber products.

There is a daily return mixed train and several short workings by ancient diesel railcars.

Traction and rolling stock
In 1992 the fleet comprised 15 diesel locomotives, 21 coaches, 3 diesel railcars and 215 wagons.

Hitachi-built Bo-Bo diesel electric at Tanjong Aru (Eddie Barnes)   0023148

Mali

Ministry of Public Works and Transport

PO Box 260, rue Baba Diarra, Bamako
Tel: (+223) 22 59 67/8   Fax: (+223) 22 83 88

Key personnel
Minister: Mohammed Ag Erlaf
Director of Cabinet: M Sidibe
Technical Adviser, Railways: M Traore

For map see entry for **Senegal**.

Chemins de Fer du Mali (RCFM)

PO Box 260, rue Baba Diarra, Bamako
Tel: (+223) 22 59 68   Fax: (+223) 22 59 67

Key personnel
Director General: Lassana Kone
Deputy Director General:
   Hamadoun Assouman Cisse
Directors
   Traffic: Daouda Diane
   Finance: Djibril Nama Keita

Technical: Tounko Danioko
Planning: Mady Konate
Purchasing: Mohamed Traore
Personnel: Fodé Traore

Gauge: 1,000 mm
Route length: 729 km

Political background
The governments of Mali and Senegal have retained Canadian consultants CPCS Transcom to assist in the preparation and allocation of a concession to handle international traffic operations on the Bamako-Dakar line. Expressions of interest were being reviewed in early 1998.

Organisation
The former Dakar-Niger Railway starts at Dakar in Senegal and runs inland via Kayes to the River Niger. The present CF du Mali is that portion of the line inside its territory, the remainder being the CF du Senégal. To give Mali an alternative outlet to the Atlantic, a new line linking Bamako, capital of Mali, with Conakry, capital of Guinea, is planned with a route length of 800 km, of which 600 km will be in Guinea.

At the beginning of 1997 RCFM employed 1,686 staff.

Passenger operations
In 1996 RCFM recorded 189 million passenger-km and carried 763,000 passengers.

Freight operations
In 1996 RCFM recorded 256 million tonne-km of freight and carried 574,000 tonnes.

Recovery of a major international traffic role is a prime management objective and this has been recognised in the railway's contract with the government. The latter has proclaimed RCFM to be its main means for spurring development in the west of the country.

Traction and rolling stock
At the beginning of 1997 RCFM owned 25 line-haul diesel locomotives, 16 diesel-electric railcars and trailers, 41 passenger coaches and 382 freight wagons. The locomotive fleet includes Alsthom Type BB1100, GM Canada CC2200 and Alsthom CC2400 machines.

In common with RCFS of Senegal (qv), RCFM has recently acquired and adapted for metre-gauge 20 redundant French Railways Type B10t passenger cars of the 'Bruhat' type.

Mauritania

Ministry of Transport

Centre Administratif, Nouakchott

Key personnel
Minister: S M Deina

For map see entry for **Senegal**.

Mauritanian National Railways (SNIM)

PO Box 42, Nouadhibou
Tel: (+222) 574 51 74   Fax: (+222) 574 53 96
e-mail: snim@snim.com
Web: www.snim.com

Key personnel
Director General: Mohamed Saleck Ould Heyine
Director of Railway and Port:
   Mohamed Khalifa Ould Beyah

Head of Rolling Stock:
   Mohamed El Moctar Ould Taleb
Head of Permanent Way: Sidi Ould Sid'Ahmed
Head of Railway Studies: Cheikhna Ould Lamine

Gauge: 1,435 mm
Route length: 717 km

Organisation
The line, completed in 1963, runs from port facilities at Nouadhibou to Tazadit for the transport of iron ore from opencast mines at Kedia d'Ijdill, Guelb El

---

For details of the latest updates to *Jane's World Railways* online and to discover the additional information available exclusively to online subscribers please visit

jwr.janes.com

Diesel line-haul locomotives

Wheel arrangement	Power kW	Speed km/h	Weight tonnes	No built	First built	Mechanical	Builders Engine	Transmission
Co-Co	1,865	60	138	10	1961	Alsthom	SACM MGO V16 BSHR	E Alsthom
Co-Co	2,460	60	145	21	1982	EMD	EMD 645 E3	E EMD
Bo-Bo	630	70	73	8	1961	B&L	SACM MGO V12 ASHR	E B&L
Co-Co	2,240	60	145	6	1994	EMD	EMD	E EMD
Co-Co	2,240	60	145	5	1997	EMD	EMD	E EMD

Rhein and M'Haoudat. Built and originally operated by Miferma, the line was nationalised in 1974 and is now operated by Société Nationale Industrielle et Minière (SNIM).

Passenger operations

A daily passenger service is provided by incorporating two coaches in the formation of an ore train between Nouadhibou and Zouerate. An intermediate stop at Choum provides access to leading tourist sites in Mauritania. Flat wagons are also provided for the carriage of road vehicles. These services are managed by ATTM SA, an SNIM subsidiary.

Freight operations

Principal traffic is iron ore, carried in trainloads of 210 wagons grossing around 22,000 tonnes and hauled by three or four locomotives. Three return trips are made daily. Unloading at Nouadhibou is by rotary tippler.

Traffic (million)	2001	2002	2003
Freight tonnes	10.3	9.6	10.2
Freight tonne-km	6,955	6,448	6,852

Traction and rolling stock

Equipment at the beginning of 2005 consisted of 30 main line diesel-electric locomotives, eight shunting diesel-electrics of 630 kW, 1,200 revenue freight wagons and six passenger coaches.

Signalling and telecommunications

Operations are controlled by manual dispatching with HF and VHF radio.

Track

Standard rail: 54 kg/m and 68 kg/m UIC

Welded joints: Practically the whole line was laid with long-welded rails; 8 × 18 m railbars were flash-butt welded at the depot into 144 m lengths, which after laying were Thermit welded into continuous rail. Longest individual length of welded rail is 80 km

Crossties (sleepers): Type U28 steel and timber, weight 75 kg

Spacing: 600 mm

Rail fastening: Clips and bolts to metal sleepers, Nabla ties to timber sleepers

Max curvature: 1.75° = min radius of 1,000 m

Max gradient: 0.63% against loaded trains 1.3% against empty trains

Max altitude: 400 m

Max axleload: 26 tonnes

Max speed: Loaded trains 50 km/h; empty 60 km/h

Type of signalling: Radio control

UPDATED

Mexico

Secretariat of Communications & Transport

Avenida Universidad y Xola, Col Narvarte, 03028 Mexico City 12, DF
Tel: (+52 5) 519 74 56/92 03   Fax: (+52 5) 519 06 92

Key personnel
Secretary: Carlos Ruiz Sacristán
Under-Secretary: Aaron Dychter

Organisation
Overseeing the privatisation of the state railway industry is the Railway System Restructuring Committee, which is chaired by Carlos Ruiz Sacristán. The committee forms part of the overall Inter-Ministerial Divestiture Commission, tasked with disposing of state assets. Representatives from both Mexican railways and the government sit on the committee, with CS First Boston and Banca Serfin acting as financial advisors and Mercer Management Consulting co-opted as primary consultant. These private sector organisations have been valuing assets, overseeing the bidding process and facilitating due diligence financial investigations.

0058536

Ferrocarril Coahuila a Durango (LCD)

Durango, Dgo

Gauge: 1,435 mm
Route length: 978 km, plus trackage rights over 285 km

Organisation
Commonly known as the Línea Coahuila a Durango, Ferrocarril Coahuila a Durango SA operates several branch and secondary lines in the states of Chíhuahua, Coahuila, Durango and Zacatecas that were not taken over by the neighbouring Ferromex and TFM systems. Both of these larger railways provide trackage rights for LCD. The lines forming the system were transferred from FNM in April 1998.

LCD connects with Ferromex at Barroterán, Ciudad Frontera, Escalón, Felipe Pescador and Torreón and with TFM at Monclova, near Ciudad Frontera. The network comprises the Felipe Pescador–Durango–Torreón line, including parts of the former branches around Durango, and the Escalón–El Rey–Ciudad Frontera line, including the Sierra Mojada branch and several short branches around Barroterrán, in the northeast of Coahuila state. Several smaller branches and parts of secondary lines have been taken out of service due to a lack of traffic.

The company is owned and operated by Indústrias Peñoles, a Mexican mining and natural resources company, and Acerero del Norte, a steel producer. US short line operator Genesee & Wyoming acts as technical and operational advisor.

Passenger operations
The central part of the Ciudad Frontera–Escalón line is served by a mixed passenger and freight train running three times a week in each direction between Cuatro Cienegas and Sierra Mojada. The service is operated for social reasons because the region has poor road access. There are plans to extend the service to Ciudad Frontera.

Freight operations
Coal, iron ore, limestone and timber generate about 70 per cent of freight traffic, with the remainder consisting mainly of petroleum products and paper. Indústrias Peñoles and AHMSA (Altos Hornos de México SA) are by far the most significant customers. Almost all freight is exchanged with neighbouring railways.

Traction and rolling stock
LCD owns 27 GE U23B diesel locomotives acquired second-hand from Norfolk Southern in the USA. There are also six passenger coaches and an unknown quantity of freight wagons, mostly ex-FNM.

Ferrocarril del Istmo de Tehuantepec (FIT)

Key personnel
Director General: Gustavo Baca

Gauge: 1,435 mm
Route length: 207 km

Political background
Administered by the Secretariat of Communications and Transport (qv), FIT was established in 1999 as the infrastructure operator of a single-track line from Medias Aguas, 93 km inland from the Atlantic coast, to the port of Salina Cruz on the Pacific coast, connecting at Medias Aguas with Ferrosur and at Ixtepec with Ferrocarriles Chiapas-Mayab (FCCM).

The line was retained by the government mainly for strategic reasons, providing the key element of a coast-to-coast route. FIT does not run its own trains: Ferrocarriles Chiapas-Mayab has trackage rights over the entire line.

FIT has received funding of peso123 million from the proceeds of the privatisation of FNM. Of this, peso53 million is to be spent on track upgrading and maintenance, civil engineering work and acquisitions, while an additional peso12 million will be invested in improvements at the port of Salina Cruz. Remaining funds will be spent on engineering facilities and planning for the new Medias Aguas–Coatzacoalcos line (see below).

Passenger operations
See entry for Ferrocarriles Chiapas-Mayab (FCCM).

Freight operations
All freight operations are run by FCCM, which pays for trackage rights. FIT expects eventual annual traffic volumes to and from Salina Cruz to total 2.4 million tonnes, mostly in containers.

New lines
In April 1999, a 3 km line was opened at Salina Cruz to provide access to a new container handling terminal.

FIT plans a new 96 km line paralleling the existing Medias Aguas–Coatzacoalcos route owned by Ferrosur, thus completing a nationally owned coast-to-coast corridor.

Ferrocarriles Chiapas-Mayab SA de CV (FCCM)

Calle 43, 429C, Entre 44 y 46, Colonia Industrial, Mérida, Yucatán 97000
Tel: (+52 99) 930 25 00   Fax: (+52 99) 930 25 39
Web: http://www.gwrr.com

Key personnel
President and General Manager: Paul M Victor
Chief Financial Officer: Carlos Pereyra
Superintendent, Transportation:
  Sergio Tonioni Vega
Chief Mechanical Officer: Homero Walss
Chief Engineer: Fernando Osorio Rodriguez
Marketing Manager: Alejandro Garcia

Gauge: 1,435 mm
Route length: 1,805 km, plus trackage rights over 321 km

Organisation
FCCM is a wholly owned subsidiary of US short line operator Genesee & Wyoming (qv in the United States of America section) which took over former FNM lines in the states of Campeche, Chiapas, Oaxaca, Tabasco, Veracruz and Yucatán in September 1999. Its lines run from Coatzacoalcos to Mérida, with some branches in the Yucatán peninsula and from Ixtepec south to Tapachula and the border with Guatemala, including a short stretch into Guatemala to Tecún Umán station. This last facility is laid with mixed – gauge track to allow the entry of standard-gauge trains. FCCM also has trackage rights over Ferrocarril del Istmo de Tehuantepec (FIT) and Ferrosur lines (qv). Connection is made with the Ferrosur network at Coatzacoalcos and Medias Aquas.

The unused branches to Peto and Sotuta, totalling 181 km south and east of Acanceh, were to be dismantled by FCCM.

Passenger operations
FCCM runs a few government-subsidised socially desirable passenger services. These consist of trains three times a week in each direction between Coatzacoalcos and Campeche and between Coatzacoalcos and Tapachula. On behalf of the Yucatán state government, a weekend tourist train also runs between Mérida and Izamal.

Freight operations
Principal commodities carried include cement, cereals, chemicals, fertilisers, petroleum products, propane and sugar. FCCM also carries containers over FIT and Ferrosur tracks between the ports of Coatzacoalcos and Salina Cruz. Container movements on behalf of FIT were expected to reach 2.4 million tonnes from 2000. Apart from coast-to-coast traffic, most freight is carried on the Coatzacoalcos–Mérida line, with Ixtepec–Tapachula and Tapachula–Tecún Umán lines also seeing significant tonnages. The branches from Mérida to Acanceh, Valladolid and Tizimin see very little traffic and the Mérida–Progreso branch remained out of use in 2000, despite the presence of port facilities at the latter location.

The interchange of traffic with Guatemala consists of a train running most days conveying cement, chemicals and fertilisers from Mexico and agricultural products in the opposite direction. Road-rail trans-shipment is undertaken at Tecún Umán. Traffic may increase sharply when the Guatemalan line from Tecún Umán to Ciudad de Guatemala is restored to use, with the possible introduction of cross-border container services via a connection at Ciudad Hidalgo.

New lines
In collaboration with the states of Yucatán and Quintana Roo, FCCM is planning a 100 km line from Valladolid to Cozumel, connecting Quintana Roo to the Mexican railway network for the first time.

Ferromex

Ferrocarril Mexicano, SA de CV
Bosque de Ciruelos 99, 3 Piso, Col Bosques de Las Lomas, DF 11700 Mexico City
Tel: (+52 55) 52 46 37 00
Fax: (+52 55) 52 46 39 07
Web: www.ferromex.com.mx

Key personnel
Director General: A Cesar Pérez
Directors
  Marketing: Rogello Velez López De La Cuerda
  Commercial: Juan Manuel Correa Cuellar
  Operations: Ing Lorenzo Reyes Retana
  Administration and Finance:
    Enrique Nava Escobedo
  Purchasing:
    Ing Florentino Matadamas Hernández
  Corporate Services: Pedro R Dupeyron Vásquez
  Industrial Relations: Héctor Ojeda Milanés

Gauge: 1,435 mm
Route length: 8,460 km

Background
In June 1997 the former Grupo Ferroviario Mexicano was awarded the second 50-year concession to be offered by the government in its railway privatisation programme for the Pacific North network. Formal transfer took place in February 1998, establishing the present company. Ownership is shared by mining concern Grupo México SA de CV, with a 74 per cent stake, and Union Pacific of the USA, with 26 per cent. A 13 per cent shareholding held by Constructoras ICA, part of Mexico's biggest construction group, was sold to Union Pacific in 1999.

TFM holds a shareholding of 25 per cent in the Ferrocarril y Terminal del Valle de México (TVM), which operates marshalling yard and terminal facilities in and around the capital. Ferromex, Ferrosur and the Mexican government also each hold 25 per cent of TVM.

In 2002 a planned merger of Ferromex and Ferrosur was blocked by the Mexican government.

Organisation
The Ferromex network is territorially the largest in Mexico, comprising routes from Mexico City to Nogales and Mexicali via Guadalajara, to Ciudad Juárez via Torreón, and from Torreón to Piedras Negras, Monterrey and Tampico. Connection is made with UP at Mexicali/Calexico, Nogales, Ciudad Juárez/El Paso and Piedras Negras/Eagle Pass, with BNSF at Ciudad Juárez/El Paso and Piedras Negras/Eagle Pass and with Texas Pacific Transportation at Ojinaga/Presidio.

Ferromex also manages the 320 km Nacozari short line system east and southeast of Nogales on behalf of the concession holder, Son Grupo México.

In 2004 Ferromex employed some 6,200 staff.

Passenger operations
Ferromex operates the daily 'Chepe' tourist train on the Chihuahua–Topolobampo section of the former Ferrocarril Chihuahua al Pacífico. The Chepe is a fully air-conditioned train conveying a restaurant car. Another tourist train, the 'Tequila Express', runs on Saturdays between Guadalajara and Tequila and includes first class and buffet cars. Rolling stock for both trains has been completely refurbished. Passenger facilities at Chihuahua have been transferred from the old FdeC station to the former

Diesel locomotives

Class	Wheel Arrangement	Power KW	Speed km/h	Weight tonnes	No in service	First built	Mechanical	Builders Engine	Transmission
SW1000	Bo-Bo	8,956	–	–	14	–	GM-EMD	GM-EMD	E GM-EMD
G12	Bo-Bo	980	–	–	5	–	GM-EMD	GM-EMD 12-645	E GM-EMD
SW1504	Bo-Bo	1,120	–	–	18	–	GM-EMD	GM-EMD	E GM-EMD
GP38/GP38-2	Bo-Bo	1,492	–	–	52	–	GM-EMD	GM-EMD 16-567	E GM-EMD
GP40-2/SD40-2	Bo-Bo/Co-Co	2,240	–	–	162	–	GM-EMD	GM-EMD 16-645	E GM-EMD
RSD12	Co-Co	1,340	–	–	1	–	Alco	Alco 251	E GE
B23-7	Bo-Bo	1,680	–	–	3	–	GE	GE FDL12	E GE
C30-7/C36-7	Co-Co	2,240	–	–	186	–	GE	GE FDL16	E GE
AC4400CW	Co-Co	3,280	–	–	50	1999	GE	GE 7FDL16	E GE

FNM facility. Chartered tourist trains have also run over various Ferromex lines, including the 'Sierra Madre Express', operated with its own vehicles by Rail Passenger Services Inc of Tucson, Arizona.

In January 2000 Ferromex ceased to run state-subsidised passenger trains on the Torreón–Aguascalientes, Ciudad Victoria–Tampico and Piedras Negras–Saltillo–San Luís Potosí lines, but continued to run between Ojinaga and Topolobampo, and Guadalajara and Tequila Jalisco, as long as subsidy payments were maintained.

Freight operations

Main commodities carried include cereals (around 30 per cent of revenue), minerals, industrial products and cement. Some intermodal traffic is handled, especially from the port of Manzanillo, but Ferromex's terminal facilities for containers are limited. In 2003 an intermodal service between Mexico and the USA was launched in conjunction with BNSF, crossing the border at Ciudad Juárez/

El Paso. Intermodal is regarded by Ferromex as one of the markets favoured for future traffic growth, together with automotive distribution. Between 2001 and 2001 automotive revenues rose by 66 per cent, and intermodal by 10 per cent.

In 2003 a total of 36.8 million tonnes was handled, for 28,184 million tonne-km, generating peso6.466 billion. Tonne-km rose 2.2 per cent, largely accounted for by improvements in the agricultural, minerals and petroleum sectors.

Improvements to existing lines

Major investments have been made in improving the railway's infrastructure since concessioning in 1998. The Mexico City–Querétaro line has been the subject of rehabilitation, as have other parts of the network. Some 640 km has been relaid with new rail and a further 363 km with recovered rail. Significant improvements have also been made to CTC and communications systems, enabling Ferromex to handle increasing traffic.

In 2003 the company planned to invest US$45 million in operational infrastructure, much of this on the rehabilitation of 138 km of track and heavy maintenance of a further 252 km. In addition, US$25 million was earmarked for work on commercial infrastructure, including a new intermodal and automotive terminal at Guadalajara.

Traction and rolling stock

In 2004 the Ferromex locomotive fleet totalled 491 machines, of which 10 were dedicated to passenger service and 60 to shunting and yard tripping. The most modern of these are 50 General Electric AC4400CW machines delivered from 1999. Rolling stock comprised some 11,100 freight wagons, plus more than 2,800 leased and some owned vehicles not in traffic. There were also 53 passenger coaches.

UPDATED

Ferrosur

Ferrosur sel Sureste, SA de CV
Plaza Polanco Jaime Balmes 11 4/C, Col Los Morales Planco, 11510 Mexico City DF
Tel: (+52 55) 53 87 65 00   Fax: (+52 55) 53 87 65 95
Web: http://www.ferrosur.com.mx

Key personnel

Chief Executive: Daniel Torres
Directors
　Marketing: Luis Olivera
　Customer Services: Ricardo Trejo
　Chief Information Officer: Enrique Chavez

Gauge: 1,435 mm
Route length: 1,564 km

Background

FNM's former Sureste (Southeastern) railway was the third and smallest division of the former national railway to be privatised when Ferrosur, a consortium of aggregates and construction company Tribasa (66.6 per cent) and the Inbursa financing group, took over operations in

December 1998 for a 50-year concession. The Tribasa stake was subsequently sold to Frisco, a mining company which is part of the Grupo Carso conglomerate.

Ferrosur holds a shareholding of 25 per cent in the Ferrocarril y Terminal del Valle de México (TVM), which operates marshalling yard and terminal facilities in and around the capital. Ferromex, TFM and the Mexican government also each hold 25 per cent of TVM.

An attempted merger between Ferrosur and Ferromex was blocked in 2002 by the Mexican government.

Organisation

The Ferrosur network extends southeast from Mexico City to serve the port of Veracruz via Orizaba, to Coatzacoalcos, the location of an extensive petrochemical complex, and to Puebla. It is also responsible for operations on the 366 km Puebla–Oaxaca line, which carries very little freight traffic. At Medias Aguas and at Coatzacoalcos connection is made with the Ferrocarril Coahuila a Durango short line system. There is no connection with the US network, although a freight wagon

ferry, which commenced operations in 2000, links Coatzacoalcos and Mobile, Alabama.

In 2003 Ferrosur employed 2,100 staff.

Freight operations

Cereals, cement, chemicals, petroleum products and minerals are the main commodities carried, as well as container and automotive traffic passing through the port of Veracruz.

Improvements to existing lines

Track improvements made since privatisation include renewal of 58 per cent of main lines with continuously welded rail on concrete sleepers. A further 100 km were being rehabilitated or improved in 2003.

Traction and rolling stock

In 2003 the Ferrosur fleet comprised 170 diesel locomotives, all of GE manufacture and mainly C30-7, B23-7 and U36C types. Procurement was imminent of 15 GE 3,280 kW locomotives. The wagon fleet totalled 2,888.

Terminal Ferroviaria del Valle de México (TFVM)

Gauge: 1,435 mm
Route length: 297 km

Political background

The Mexico City Terminal Railway (TFVM) was privatised on 30 April 1998, with the three main

freight railway concessionaires, Ferromex, Ferrosur and TFM, each taking a 25 per cent stake.

Organisation

TFVM operates marshalling yard and terminal facilities in the Mexico City area. These include the Valley of Mexico marshalling yard, which has recently undergone a modernisation programme to increase its capacity from 800 to 2,000 wagons

a day. The capital's only intermodal facility, at Pantaco, is also run by TFVM.

Traction and rolling stock

TFVM took over 46 diesel locomotives and 192 freight wagons from FNM.

Transportación Ferroviaria Mexicana (TFM)

Transportación Ferroviaria Mexicana SA de CV
Periférico Sur No 4829, Col Parques del Pedregal, CP 14010 Mexico DF
Tel: (+52 55) 54 47 58 36   Fax: (+52 55) 54 47 58 30
e-mail: tfm@tfm.com.mx
Web: www.tfm.com.mx

Key personnel

Chief Executive Officer: Francisco Javier Rión

Gauge: 1,435 mm
Route length: 4,251 km

Background

TFM has operated the former North East network of Mexican National Railways (FNM) since June 1997, having won a 50 year concession with a 50 year extension option as part of the country's rail privatisation process. Ownership of TFM was initially shared between the Mexican government (20 per cent) and Grupo TFM (80 per cent). In turn, Grupo TFM's shareholders were the Mexican multimodal transport company, Transportaciónes

Marítimas Mexicanas (TMM)/Grupo Servia (48.5 per cent), the US firm Kansas City Southern Industries (KCS) (46.6 per cent) and the Mexican government (4.9 per cent).

In December 2004 Grupo TMM shareholders approved the sale of its voting interest, by then 51 per cent, to KCS. The purchase, which was subject to KCS shareholder approval, followed an earlier unsuccessful acquisition bid by the US company in 2003. This was rejected by TMM shareholders.

TFM holds a shareholding of 25 per cent in the Ferrocarril y Terminal del Valle de México (TVM), which operates marshalling yard and terminal facilities in and around the capital. Ferromex, Ferrosur and the Mexican government also each hold 25 per cent of TVM.

Organisation

The TFM network runs north of Mexico City via San Luis-Potosí to serve the US border crossings at Nuevo Laredo/Laredo and at Matamoros/Brownsville, making connections at both with UP and also with Texas Mexican at Laredo. Other lines from Mexico City serve the ports of Lázaro Cárdenas in the west, Tampico/Puerto Altamira in the east and Veracruz, to the southeast. TFM also

serves Guadalajara via trackage rights. The port of Veracruz is an extremely important traffic generator in its own right, handling around a quarter of all of Mexico's maritime cargo.

In 2003 TFM employed some 3,700 staff.

Freight operations

Principal markets served include the metals, cement, agriculture, automotive and chemicals and petroleum products industries.

The first phase of the Toluca Automotive Terminal was completed in April 2003 at a cost of USD26 million.

Intermodal operations

In 2003 some 8.5 per cent by revenue of TFM's business was intermodal, with double-stack services operated on Nuevo Laredo–Mexico City and Veracruz–Mexico City routes. RoadRailer traffic is also handled. Since it took over the system, TFM has increased the number of its intermodal terminals from two to seven. Among these is the USD11 million Salinas Victoria Intermodal Terminal in Nuevo León, completed in 2002, which replaced a temporary facility serving the Monterrey region.

Improvements to existing lines

A major priority for TFM has been upgrading of the Mexico City–Nuevo Laredo trunk line (1,283 km), including track upgrading with heavier and continuously welded rail, the wider use of concrete sleepers, upgrading and extension of the CTC system and the provision of additional hot axlebox detectors.

Improvements have been made at Sanchez Yard, Nuevo Laredo, resulting in improvements in the handling of cross-border traffic.

Traction and rolling stock

In 2003 the TFM locomotive fleet totalled 477 units, of which 150 were leased. These leased machines,

75 General Electric AC4400 diesels and 75 SD70-MACs from General Motors, together with 105 owned GE Super 7s, have enabled TFM to claim to operate the youngest fleet in the North American railfreight system.

Since it took over the business, TFM has through leasing acquired a modern wagon fleet that in 2003 totalled some 6,400 vehicles.

Signalling and telecommunications

Train movements are controlled by CTC over 1,500 km of the network and by Track Commanding Control (CMV) track warrants for 2,700 km. Freight tracking systems are in place to enable

customers to monitor their shipments via the Internet.

Electrification

The 25 kV AC 60 Hz electrified section of line between Mexico City and Querétaro remains energised for possible future use as part of a proposed suburban network for the capital but the use of electric traction was not adopted by TFM.

UPDATED

Moldova

Ministry of Transport and Communications

Boulevard Stefan cel Mare 134, MD-2012 Kishinev
Tel: (+373 2) 22 10 01   Fax: (+373 2) 54 65 64
Web: www.mci.gov.md

Key personnel
Minister: Vasile Zgardan

UPDATED

MOLDOVA

NEW/0585095

Moldovan Railways (CFM)

Căile Ferate Moldova
Vlaiku Pirkelab Str 48, MD-2012 Kishinev
Tel: (+373 2) 25 44 08   Fax: (+373 2) 22 13 80
e-mail: cfm@railway.md
Web: www.railway.md

Key personnel
Director General: Miron Gagaouz
First Assistant and Operations and Commercial
  Director: Vladimir Bogoev
Director, Technical and Development:
  Miron Gachkevich
Director, Social and Infrastructure:
  Vyacheslav Jordan
Departmental Heads
  International Relations: Valery Constantynov
  Freight: Victor Garkoucha
  Passengers: Vasily Volchok
  Finance: Lydia Gutjum

Operations: George Efrim
Rolling Stock: Leonid Gusakov
Traction: Ivan Vasilaki
Security: Vladimir Dolgopolov
Statistics: Marina Kniazeva
Legal: Oleg Petrovich

Gauge: 1,520 mm; 1,435 mm
Route length: 1,124 km; 14 km

Political background

Moldovan Railways was formed in 1992 in the break-up of SZhD (Soviet Railways) and comprises the latter's Moldovan section. It is controlled by the Ministry of Transport and Railways. Two lines traverse the rebellious Dnestr region in the east of the country. Civil war disrupted the fledgling independent railway, meaning that much of it was out of action until 1995, when most lines were back in traffic, including those in the Dnestr region. However, in 2004 there was a revival of separatist activity in the region, culminating in a blockade of railway facilities at Tighina (Bendery).

Bisecting the country from east to west, CFM's principal main line joins the Ukrainian port of Odessa with Romania, and forms part of trans-European Corridor IX. Traffic prospects are closely linked to the Ukrainian and Russian economies.

Passenger operations

International year-round or seasonal passenger trains to CIS countries operate between the capital, Kishinev, and Adler, Minsk, Moscow, Odessa, St Petersburg and Simferopol. In the north of the country, four pairs of daily international trains run between Kishinev via Ocniţa to Ivano-Frankovsk, southwest Ukraine, crossing the border between the two countries several times. The Mamalika–Larga section of this route is also served by frequent international stopping trains operating to Chernovtsky from Kiev and Ocniţa. In the south, a pair of passenger services is operated over both Kishinev–Reni and Basarabeasca–Etulia routes.

The main crossing point for services to and from neighbouring Balkan states is at Ungheni/Iasi, which sees a pair of daily express trains linking Kishinev and Bucharest and seasonal fast trains on Kishinev–Constanta, Sofia–Saratov, Varna–Minsk and Burgas–Minsk routes, as well two daily pairs of Ungheni–Iasa trains. Bogie-changing takes place at Ungheni.

Most internal day trains are operated by Hungarian-built Type D1 diesel multiple-units, at speeds up to 100 km/h.

Traffic (million)	2001	2002	2003
Passenger journeys	4.8	5.1	–
Passenger-km	324.8	354.9	–
Freight tonnes	10.6	12.6	14.8
Freight tonne-km	1,980.0	2,751.4	3,035.4

Freight operations

Important flows of freight traffic from and to CIS countries use the Kishinev–Tighina–Kuchurgan route to Ukraine, with coal, forest products, petroleum products and construction materials among the main commodities carried. Iron ore for Romania is carried via the Kuchurgan–

Tighina–Basarabeasca–Reni route. However, since 2004 freight movements via these routes has been disrupted by the separatist occupation of railway facilities at Tighina. Also important is the line from Balti Slobozia via Rybnitsa to Ukraine, especially for the movement of scrap metal, steel and cement.

The main border crossing into Romania is at Ungheni/Iasi. The crossing at Prut/Falciu has been closed for freight traffic since 1991.

New lines

The Romanian and Moldovan governments have discussed construction of a new line from Nicolina in Romania to the Moldovan capital, Kishinev. The new line, to be built at 1,435 mm gauge, would run parallel to the CFM's existing east-west 1,520 mm gauge line, a few kilometres to the south. In 2002, it was reported that plans for this line were to be finalised shortly.

Rehabilitation is planned of the 44 km line connecting Căinar and Revaca, in the southeast of Moldova. Out of operation since 1944, the line was to be restored to avoid the need to traverse the rebellious territory of Dnestr. Some progress on this scheme was made in 1993–94 but work was subsequently abandoned due to a shortage of funds. The scheme was revived in 2004 following the separatist blockade of Tighina, with reopening projected for the end of 2005.

Improvements to existing lines

In the early part of the current millennium, studies were undertaken into creating a standard gauge rail link between the Romanian border at Ungheni and the capital, Kishinev. No further progress has been reported.

Traction and rolling stock

In 2003 the CFM fleet totalled 242 diesel locomotives (61 2TE10M, 60 3TE10M, seven M62 and 114 ChME3 Classes, including many unserviceable). For suburban passenger operations, CFM operates 41 Class D1 Hungarian-built dmus. No new rolling stock had been obtained since 1984.

Signalling and telecommunications

A total of 221 route-km is double-track. Of this, 192 km together with 385 km of single-track line are equipped with automatic block signalling.

Electrification

In 1991–93, preparatory work connected with electrification at 25 kV AC 50 Hz of the Razdelnaya (Ukraine)–Tighina–Kishinev main line was undertaken. This section forms part of European Corridor IX, linking Iasi/Ungheni, Kishinev and Razdelnaya. However, by 1996, when work stopped due to a lack of funds, only the Razdelnaya–Kuchurgan section in Ukraine had been completed.

Track

Rail: 65 kg/m on main lines, 45–55 kg/m on secondary lines
Sleepers: Mostly wooden, spaced at 1,840 per km
Max axleload: 24 tonnes

UPDATED

Mongolia

Ministry of Roads, Transport and Tourism

Government Building 2, United Nations Street 5/2, Ulaanbaatar 210646, Mongolia
Tel: (+976 11) 31 55 33; 32 08 20; 33 09 71
Fax: (+976 11) 31 06 12
e-mail: info@mrtt.pmis.gov.mn
Web: www.mrtt.pmis.gov.mn

Key personnel
Minister: Gavaagiyn Batchuu

UPDATED

Mongolian Railway (MTZ)

Mongolin Tömör Zam
PO Box 376, Ulaanbaatar, Mongolia
Tel: (+976 11) 32 21 17   Fax: (+976 11) 32 83 60
Web: www.mtz.mn

Key personnel
President: V V Magdei
Chief Engineer: L Bold
Deputy Directors General: Yo Batsaikhan,
    V Otgondemberel, Ch Tsogtbayar,
    G Vandandagva

Gauge: 1,520 mm
Route length: 1,810 km

Organisation
MTZ is a state-owned autonomous company. Its core main line, the Transmongolian Railway, extends 1,110 km from the Russian Federation frontier at Sühbaatar to the Chinese border at Dzamïn-Uüd, with branches to mineral deposits at locations which include Baganuur and Shivee Ovoo/Sharingol (coal), Erdenet (copper) and Bor-Öndör (fluorspar). There is an isolated 349 km line, the Bayantumen section, in the east of the country, which links Mongolia with the Russian Federation at Solovyevsk with via a border station at Choybalsan, and throws off a branch to Erdes.

Access to the systems of neighbouring China and the Russian Federation is facilitated by agreements concluded in 1991 and 1992 respectively. These enable Mongolia free access to sea ports and the rights to move commodities through its neighbours' territories. The nearest Chinese port to Mongolia is Tianjin, some 1,700 km from Ulaanbaatar, while Vostochnyy, Russia, is 4,400 km from the capital. Rail is favoured over road transport, which is handicapped by a requirement to reload freight at border crossings.

Organisationally, most of MTZ's activities are managed by the Freight Traffic Department. As well as running both freight and passenger services, it is responsible for traffic control, provision of traction and rolling stock, technical matters, finance, safety and inspections, and personnel.

Passenger operations
Passenger traffic declined less dramatically than freight through to 1993. An increase was recorded in 1994, and there was a slight fall again in 1995, when 2.8 million passengers were carried. Since then, traffic has increased continuously and, in 2001, 4.1 million passengers were carried. Of these, just 4 per cent were international travellers. Rail faces strengthening competition from road transport. Since 2000, paved roads have been available from Ulaanbaatar to Sühbaatar, Darhan, Erdenet and Bagaanur, with a consequent adverse impact on rail, with its slower journey times and more frequent stops. Despite this, MTZ handles some 30 per cent of the country's passenger traffic.

In 1999 MTZ initiated its Tour Bureau, which has progressively introduced several regular special services for the tourism market.

Freight operations
From a peak of 17.8 million tonnes hauled in 1988, freight traffic declined to just over 7 million tonnes

annually in the period 1994–97. During this period, the country was undergoing a transformation from a planned production system to a market-oriented economy. Since then, transit traffic handled by MTZ has increased substantially, with tonnages more than doubled between 1999 and 2001. In 2001, at 2.16 million tonnes, transit traffic accounted for 21 per cent of freight tonnage handled by MTZ, and has since risen further with the growing use of the network to move crude oil from the Russian Federation to China. A break of gauge at the Mongolian-Chinese border means that all transit freight between the two countries has to be trans-shipped. Delays to traffic are often caused by shortages of wagons at the Chinese border station, Erenhot, and at its Mongolian counterpart, Dzamin-Uüd. Further restrictions have been caused by poor locomotive fleet availability.

Domestic traffic has also grown steadily. The main commodities carried are coal and copper concentrate, as well as appreciable volumes of construction materials, petroleum products, forest products, leather products and foodstuffs.

MTZ carries more than 95 per cent of Mongolia's freight traffic.

Traffic (million)	2000	2001	2002
Passenger journeys	4.291	4.100	3.981
Freight tonnes	9.181	10.148	11.637

Improvements to existing lines
Recent improvements have included the provision of eight additional stations or intermediate crossing points on the main line to assist the movement of transit traffic. In 2002 future plans included track improvements to enable line speeds to be increased, flood prevention measures and upgrades to principal stations to meet passengers' contemporary needs.

Japanese funding was also involved in a programme of works at the Chinese border station of Dzamïn-Uüd, where Mongolian 1,520 mm gauge tracks meet those of 1,435 mm gauge in the People's Republic. Additional and improved transhipment facilities have been provided for transit traffic to and from China, including a new container loading platform.

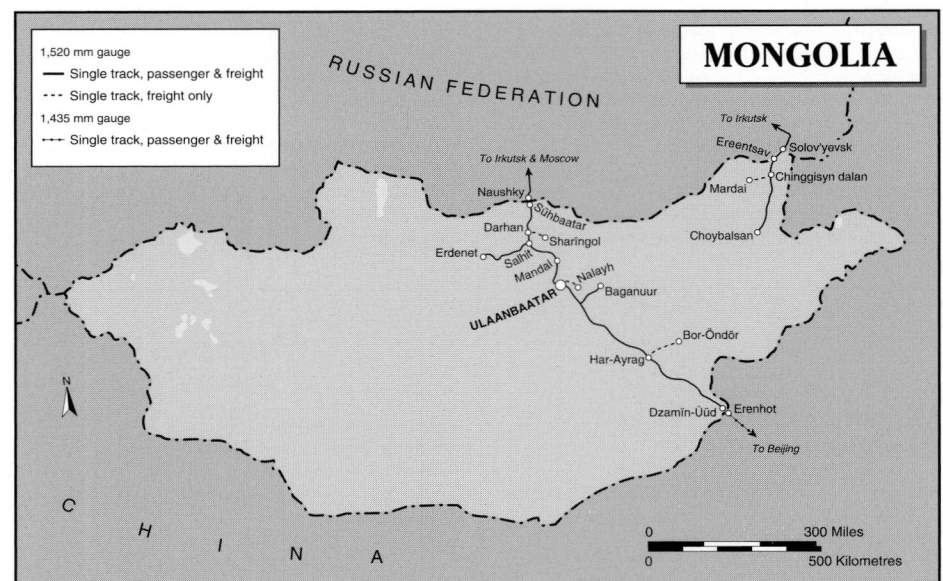

NEW/1115015

Diesel locomotives

Class	Wheel arrangement	Power kW	Speed km/h	Weight tonnes	No in service	First built	Mechanical	Builders Engine	Transmission
2M62	2 × (Co-Co)	2 × 1,472	100	2 × 120	64	1981	Lugansk	Lugansk	E Lugansk
M62	Co-Co	1,472	100	120	13	1987	Lugansk	Lugansk	E Lugansk
TEM2	Co-Co	883	100	120	28	1972	Briansk	Penza	E Briansk

Class 2M62M diesel locomotive with a passenger service at Sayinshand (Colin Boocock)

NEW/1115014

Possible future plans include the automation of train control and main line electrification.

Traction and rolling stock

In 2002 the stock of 107 diesel-electric locomotives is formed chiefly of USSR-built Type 2M62 4,000 hp 2 × Co-Cos and TEM2 1,200 hp Co-Cos. There are also 13 M62 units of 2,000 hp. Passenger cars total 280 and freight wagons 2,500. In 2002, re-engining of elements of the diesel locomotive fleet was being planned.

Signalling and telecommunications

Semi-automatic block is in operation between Ulaanbaatar and Dzamïn-Uüd and on several branches, totalling 1,577 km, while the Choybalsan-Ereentsav line (237 km) is worked by electric token. Radio communications are in place on the main line network and 90 per cent of MTZ's locomotives are equipped with driver vigilence devices.

Recent improvements have included the installation of an optical fibre communications network. By 2000, 1,405 km of MTZ's lines had been equipped with this system, which is also connected to similar systems serving the neighbouring rail networks of China and Russia. The system was also to be used for future steps to improve train operations and management control.

Track

Rail: Type R-43, 44.65 kg/m; Type R-50, 51.5 kg/m; Type R-65, 64.64 kg/m
Crossties (sleepers): Wood, 200 × 250 × 2750 mm
Fastenings: Conventional, Russian Federation standard
Min curvation radius: 296 m
Max gradient: 1.8%
Max axleload: 23.5 tonnes

UPDATED

Morocco

Ministry of Transport and Merchant Marine

PO Box 759, Rabat
Tel: (+212 37) 77 42 66   Fax: (+212 37) 77 45 78

Key personnel
Minister: Karim Ghellab

Moroccan Railways (ONCFM)

Office National des Chemins de Fer du Maroc
8 bis rue Abderrahmane Al Ghafiki, Rabat-Agdal
Tel: (+212 7) 77 47 47
Fax: (+212 7) 77 44 80
Web: http://www.oncf.org.ma

Key personnel
Director General: Ing Karim Ghellab
Director, Finance and Management:
   Mohamed Lahbib El-Gueddari
Traffic Manager: Abdesselam El Ghissassi
Transport Manager: Mohamed Smouni
Commercial Manager: Mohamed Rabie Khlie
Traction and Rolling Stock Manager (Central):
   Mohamed Soufi
Traction and Rolling Stock Manager:
   El Hassane Leqsiouer
Track Manager: Moha Khaddour
Electrification, Signalling and Telecommunications
   Manager: Rachid Bouslama
Human Resources Manager: Ali El Karram
Management and Finance Manager:
   Mohamed El Gueddari
Inspector General: Mustapha Benmoussa
Supply Manager: Bouchaib Belhaj
Data Processing Manager: Rokia Belkebir
International Relations: Larbi Aidi

Gauge: 1,432 mm
Route length: 1,907 km
Electrification: 1,003 km at 3 kV DC

Political background

ONCFM is a state-owned enterprise with its own legal entity and financial autonomy, working under the administrative umbrella of the Ministry of Transport. The Moroccan transport market was deregulated in 1993.

Relationships with government are currently managed by a Contract Plan signed in September 1996 and covering the five-year period up to 31 December 2000. This plan provides a budget of Dh2.2 billion for capital investment purposes. To coincide with the signature of this plan and to secure investment funding from international lending institutions, the government took over ONCFM's long-term debts at the end of 1996 and implemented a commercially based restructuring programme.

In early 1997, the World Bank and European Investment Bank approved loans of US$195 million towards the US$600 million cost of this restructuring and investment programme.

Organisation

The railway runs about 220 km south from Tangier to the Sidi Kacem junction with the northwest coastal line to Rabat. The latter continues to Marrakech via Sidi el Aïdi and has a spur to Oued Zem and east to Oujda, via Fès, to link up with Algerian Railways at the frontier. A line running south from Oujda skirts the Moroccan-Algerian frontier as far as the southeast railhead at Bouârfa.

At the beginning of 2001 ONCFM employed 10,308 staff.

Traffic (million)	1999	2000
Passenger journeys	12.2	13.1
Passenger-km	1,880	1,956
Freight tonnes	28.13	–
Freight tonne-km	4,794	–

Passenger operations

The 90 km main line between Rabat and Casablanca has been upgraded for 140 to 160 km/h over 76 km and is exploited by an hourly TNR

(*Train Navette Rapide*) service provided by emus built in Belgium by BN (now Bombardier Transportation) and ACEC (now Alstom) to a design based on SNCB's Type AM80. But, whereas the SNCB AM80 has regenerative braking, the Moroccan units have rheostatic. The Moroccan three-car sets can operate in multipled pairs. Completion of upgrading of the Kénitra–Meknès corridor in 2002 will see TNR services extended to Sidi Kacem.

UIC Type X air conditioned cars (see below), designed for 160 km/h operation, equip named trains between Rabat and Marrakech; Casablanca, Tangier, Fès and Oujda; Tangier and Marrakech; and Oujda and Marrakech.

A 13 km electrified branch connects the main line out of Casablanca to the south with the King Mohamed V International Airport. Belgian-built emus are also used on this service.

Freight operations

Phosphates, which account for around one-third of freight tonnage, are moved for export shipment in 78 three-axle wagon trains of 3,900 tonnes payload, 4,680 tonnes gross, over the Beni Idir–Khouribga/ Sidi Daoui-Casablanca and Sidi Azouz/Youssoufia– Safi electrified routes.

Class E 1300 electric locomotive with a mixed freight train at Rabat (Quintus Vosman)   0059397

Electric locomotives

Class	Wheel arrangement	Power kW continuous/ one-hour	Speed km/h	Weight tonnes	No built	First built	Builders Mechanical	Electrical
E 700	Bo-Bo	1,220	110	88	14	1948	Alsthom	Alsthom
E 800	Co-Co	2,295	80	132	7	1959	Alsthom	Alsthom, CEM
E 900	Co-Co	2,425	125	114	7	1969	Alsthom	Alsthom, MTE
E 1000	Co-Co	3,000	125	120	23	1975	Pafawag	Dolmel
E 1100	Co-Co	2,850	100	120	22	1977	Hitachi	Hitachi
E 1200	Co-Co	2,850	100	132	8	1982	Hitachi	Hitachi
E 1250	Co-Co	3,900	160	120	12	1984	Hitachi	Hitachi
E 1300	B-B	4,000	160	85.5	27	1989	GEC Alsthom	GEC Alsthom

Diesel locomotives

Class	Wheel arrangement	Power kW	Speed km/h	Weight tonnes	No built	First built	Builders Mechanical	Engine	Transmission
DF	C-C	2,250	135	108	14	1968	Alsthom	SACM-AGO V16 ESHR	E Alsthom
DG	Bo-Bo	588	90	72	75	1973	Fablok	HCP 8 VCD22T	E Dolmel
DH	Co-Co	2,427	105	117	18	1974	GM	EMD 16-645E3	E GM
DK	Co-Co	1,470	105	125	11	1983	GM	EMD 12-645E3	E GM
DI	Bo-Bo	735	85	88	18	1983	GM	EMD	E GM
DJ	Bo-Bo	820			19	1992	Brush		E Brush

New lines

A long-projected rail connection from Taourirt to the Mediterranean port of Nador appeared more likely in 1999 when a ministerial statement backed the scheme.

In 1999, government proposals were revealed to construct a new line eastwards from Tangier to Tetouan and the Spanish territory of Cueta.

Improvements to existing lines

In March 1997, ONCFM commenced a US$233 million upgrading of the 140 km main line between Kénitra (40 km north of Rabat) to the city of Meknès. An initial phase covers the 85 km section between Kénitra and Sidi Kacem, and involves double-tracking with new 54 or 60 kg/m continuously welded rail, realignment for 160 km/h running, elimination of level crossings, and renewal of overhead power supply equipment. Tenders were invited in September 1997 for the Sidi Kacem–Meknès section. The complete scheme is due to be completed by the end of 2002.

The line between Fès and Oujda has been relaid with 60 kg/m rail.

In 1999, catenary renewal was carried out on the Sidi el Aïdi–Oued Zem line and in the Casablanca area.

Traction and rolling stock

The fleet in 2000 included 131 diesel locomotives, 80 electric locomotives and 14 three-car emus. The passenger coach fleet totalled 414.

Under an agreement with De Dietrich, UIC Type X air conditioned cars of side-corridor layout have been built locally for ONCFM by Société Chérifienne du Matériel Industriel et Ferroviaire (SCIF), though some parts were supplied from France. In addition, 52 cars for local service have been acquired second-hand from Belgian Railways (SNCB).

Three Class DF French-built main line diesel locomotives have been refurbished.

In December 1996, seven 4,400 kW electric locomotives were ordered from GEC Alsthom, with an option on two more; ONCFM has now ordered 27 electric locomotives of this type. Alsthom-built Class E 900 electric locomotives were refurbished from 1999.

At the beginning of 2001, the ONCFM freight wagon fleet stood at 6,894.

Signalling and telecommunications

In 1999, ONCFM placed a Euro53 million contract with Alstom to renew signalling between Sidi el Aïdi and Marrakech and Kenitra and Oujda. The scheme covers 700 route-km and involves modernisation of 55 stations, including Casablanca, where a new electronically controlled signalling centre will be provided. First phases of the project are due to be commissioned early in 2001.

Manual block has been installed under the 1994–98 plan on the principal non-electrified lines from the Algerian border at Oujda to Fès and Tangier to Sidi Kacem. Other recent works include equipping the main line between Fès and Sidi el Aïdi with automatic speed control.

Electrification

The 584 km Fès–Marrakech line was electrified in the 1930s by the Paris–Orleans Railway. Renewal of catenary features in the upgrading and double-tracking of the Kénitra–Meknès line (see 'Improvements to existing lines').

ONCFM aspires to electrify its complete network, allowing it to make use of indigenously produced hydro-electric power rather than imported oil.

Track

Rail: Less than 45 kg/m (657 km); 45–54 kg/m (1,154 km); 54 kg/m over (1,253 km)

Crossties (sleepers) material: Concrete, 1,464 km; steel, 1,541 km; timber, 59 km

Number per km: 1,666 km

Fastenings type: Nabla and rigid fastenings

Min curvation radius: 300 m

Max gradient: 2.5%

Max permissible axleload: 22 tonnes

Class E 1250 electric locomotive at Sidi Bou Othmane with a Rabat–Marrakech train formed of Corail coaches built in Morocco under licence (Quintus Vosman) 0059398

Class DF French-built diesel-electric locomotives at Sidi Kacem (Quintus Vosman) 0059399

Mozambique

Ministry of Transport & Communications

PO Box 2158, Maputo
Tel: (+258 1) 42 71 73   Fax: (+258 1) 42 77 46

Key personnel
Minister: Tomas Salomão

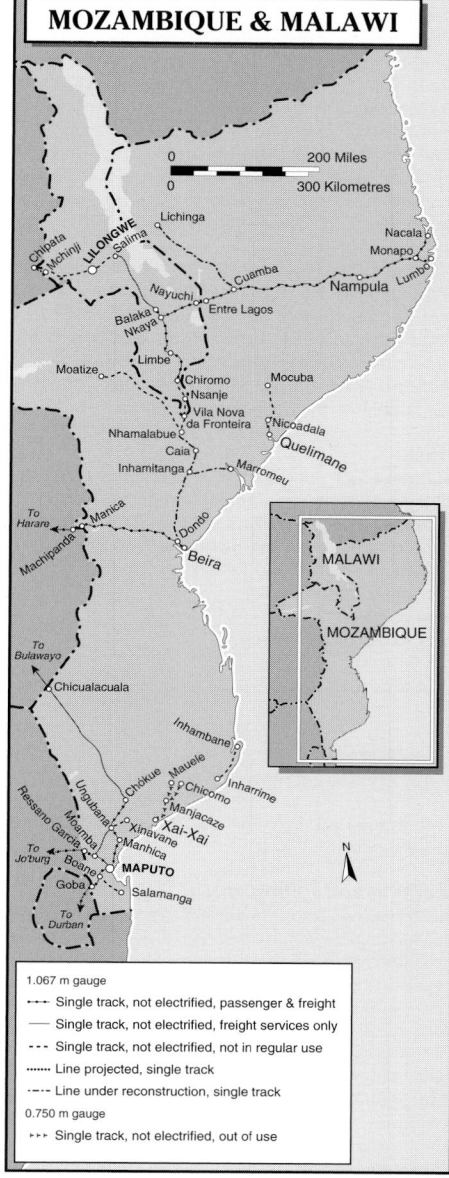

MOZAMBIQUE & MALAWI

0114300

Mozambique Ports & Railways (CFM)

PO Box 2158, Maputo
Tel: (+258 1) 42 71 73   Fax: (+258 1) 42 93 57

Key personnel
Chairman: Rui Fonseca
Director General: Mário A Dimande
Directors
  Engineering: Eng Anibal Laice
  Finance: B M Cherinda
  Commercial: D Gomes
  Computer Services: Avito Jequicene
  Purchases and Stores: Sancho Quipiço
  Personnel: I R Júnior
  Southern Railway: A Manave
  Central Railway: J A Felipe
  North Railway: F J Nhussi
  Zambésia Railway: O J Jaime

Gauge: 1,067 mm; 762 mm
Route length: 2,983 km; 140 km

Political background
In 1990 the Mozambique Railways (CFM) were joined with the country's ports of Maputo, Beira and Nacala in a new state corporation, Mozambique Ports & Railways. This changed the status of both activities from government agency to a financially accountable corporation.

In the following year a new law was enacted to give impetus to plans for revitalisation of CFM, envisaging for the first time the possibility of private-sector participation.

Freight traffic performance has been at the mercy of low motive power availability and poor security, both at Maputo port and en route. In an attempt to address the endemic congestion at Maputo in particular, several of the port's facilities have been franchised to private operators. Another long-standing problem demanding attention is the need for dredging at Maputo and Beira to permit larger vessels to call at the ports.

Because of CFM's poor performance, donor governments have been pressing for outright privatisation of the railway as a means of strengthening operational control. World Bank assistance was obtained in 1993 for the Maputo Corridor Revitalisation Technical Assistance Project, designed to review the options for private-sector involvement and negotiate its implementation.

In the second half of 1997, five operating concessions in the Maputo Corridor were put up for sale. The five concessions cover the 93 km from Komatipoort (South Africa) to Moamba and Maputo, the 528 km Limpopo line from Chicualacuala to Maputo, the 69 km route from Goba in Swaziland, the harbour at Maputo and CFM's motive power workshops. In 1998, Spoornet (qv) was named preferred bidder to operate freight services between Maputo and South Africa, while the Consortia 2000 group was selected as the successful bidder for two other freight concessions into the port, providing links to Chicualauala (and on to Zimbabwe) and to Goba (and on to Swaziland). In mid-1999, negotiations continued on all three concessions amid reports of difficulties reaching contract terms.

In February 1999, the Sociedade de Desenvolvimento do Corredor de Nacala (SDCN) consortium was named preferred bidder for an operating concession for the CFM-North line between the port of Nacala and the Malawi border. SDCN members include Railroad Development Corporation and Edlow Resources, both based in the USA, and Mozambique and Portuguese investors. A stake of 16 per cent was also to be offered to interests in Malawi, although reports in 1999 suggested dissatisfaction in Malawi at the size of this proposed shareholding. Transfer to SDCN was expected to occur on 1 October 1999.

Organisation
The railway is made up of five distinct systems linking the coastal ports to the hinterland, managed from four regional headquarters. These are:

CFM-North
Gauge: 1,067 mm
Route length: 919 km
This line runs from the port of Nacala, with a branch (at present closed) to Lumbo westward to Cuamba and Lichinga. A recently built line from Cuamba connects at Entre Lagos, on the border, with Malawi Railways and affords Malawi rail access to the port of Nacala.

CFM-Centre
Gauge: 1,067 mm
Route length: 994 km
From the port of Beira the line runs westwards to connect with Zimbabwe Railways at Machipanda. From Dondo Junction, 29 km from Beira, a line runs northward to connect with Malawi Railways with an extension from Dona Ana to Moatize. A branch line of 83 km links Inhamitanga with Marromeu.

CFM-Zambésia
Gauge: 1,067 mm
Route length: 145 km
An isolated line running from the coastal town of Quelimane to Mocuba.

CFM-South
Gauge: 1,067 mm; 762 mm
Route length: 930 km; 143 km
Three major international routes run from the port of Maputo to the Swaziland border at Goba (64 km); to the South African border at Ressano Garcia (88 km); and the Limpopo line to the Zimbabwe border at Chicualacuala (528 km).

The first joins up at the border of Swaziland with the Swaziland Railway, which connects the Umbovu Ridge iron ore complex at Kadake with the port of Maputo. The second continues into South Africa. The third line goes through Zimbabwe to Zambia, Botswana and the southeast of the Democratic Republic of Congo.

CFM-South also operates the two isolated lines linking Xai-Xai with Mauele and Chicomo (762 mm gauge), and Inhambane with Inharrime (1,067 mm gauge, 90 km).

At the beginning of 1997 CFM employed 10,558 staff.

Passenger operations
Improved passenger services depend on rolling stock availability; currently there are insufficient coaches to run through trains to Swaziland and Zimbabwe, though the Johannesburg-Maputo train resumed thrice-weekly operation in 1994 using Spoornet stock. A local service introduced between Maputo and Manhica proved popular immediately. Some relief was provided by delivery of 25 coaches second-hand from South Africa in 1994, and a batch of new vehicles was ordered from Zimbabwe builders.

Traffic figures for 1996 were 5.7 million journeys (5.5 million in 1995), 358 million passenger-km (251 million in 1995).

Freight operations
In the 1980s operation was severely disrupted by hostilities, with serious consequences for neighbouring railways for which Mozambique's ports provide a shipment outlet. Now that relations are improving, the prospects for rail are similarly reviving. In the short term, close attention is being given to retaining existing freight business, while new export traffic is being sought on the rehabilitated corridors. Closer co-operation with neighbouring railways should see more through running, joint cross-border operations, and streamlined customs and accounting practices.

Traffic figures for 1996 were 4.1 million tonnes (3.1 million in 1995), 983 million tonne-km (886 million in 1995).

New lines
Under study in 1998 was a 180 km line to convey iron ore from central Zimbabwe to Manica, on CFM's Beira line. Here a smelting facility established in a special trade zone would produce iron briquettes for export via Beira.

Improvements to existing lines
Reconstruction of CFM's deteriorated trunk route tracks has been a top priority, first in view of the demands for access to the ports for traffic to and from the country's landlocked neighbours, and latterly in the light of the new political situation following normalisation of relations with South Africa.

Loans from Canadian, French and Portuguese sources helped to fund a US$195 million rebuilding of the 538 km Northern line from Nacala to Cuamba with 40 kg/m long-welded rail on concrete sleepers for 20 tonne axleloadings. Work on the remaining 77 km section from Cuamba to the Malawi border at Nayuchi was also being undertaken in 1999 with the help of a grant from the Portuguese government. There is also a plan to boost passenger service on the route with EU assistance.

Diesel locomotives

Class	Power kW	Speed km/h	Weight tonnes	No in service	First built	Mechanical	Builders Engine	Transmission
1a	2,150	103	96	1	1966	GE	7FDL-12B3	E
2a	2,150	103	96	5	1968	GE	7FDL-12B7	E
3a	2,150	103	120	17	1973	GE	7FDL-12D10	E
4a	2,150	108	96	11	1979	GE do Brasil	7FDL-12D29	E
5a	2,150	108	96	17	1980	GE do Brasil	7FDL-12D29	E
6a	2,150	108	96	6	1984	GE	7FDL-12	E
7a	2,150	103	96	5	1990	GE	7FDL-12	E
8a	1,850	103	108	15	1991	GEC Alsthom	3606	E
DH-125	1,250	80	68	28	1980	Faur	6LDA 28B	H Brason
10a	1,100	70	64	6	1991	GE	3512	E
200	1,200	80	82.5	1	1963	AEI	Sulzer-6LDA-28B	E

In early 1998 CFM was also working on a two-phase US$ 26 million renovation of the branch north from Cuamba to Lichinga.

Export/import traffic is moving again in the Limpopo corridor northwards from Maputo, over which a partial renovation was completed in 1993. Nevertheless, a further programme of works is necessary before the line can be considered fully operational, and British, Canadian, German, Portuguese and Kuwaiti donors have proffered a further US$65 million towards the US$200 million cost.

Rehabilitation of the Goba line is also being undertaken, with 45 kg/m rail on concrete sleepers and resignalling. Assistance provided by the Italian government funded work on the initial 20 km from Goba to Boane, while a start on the remaining 63 km awaits funding from other sources.

A grant from the UK of US$20 million supported rehabilitation of the Beira corridor line west of Dondo Junction, starting with the severely graded 100 km near the Zimbabwe border, where 30 kg/m rail was replaced by 40 kg/m, curves eased to a minimum radius of 500 m and ruling gradient from 1.2 to 2.4 per cent. The objective is to double-track and resignal the 27 km between Beira and Dondo. Further work planned for this route will benefit from technical assistance provided by the Australian government. All relaying is being carried out with 54 kg/m rail on concrete rather than timber sleepers. The upgrading has led to a rapid growth in transit traffic between Zimbabwe and the container terminal at Beira port.

Rehabilitation of the Beira line to Malawi remains to be funded, after international assistance provided under the Beira Corridor Authority programme ended in 1996. A combined project, involving dredging and other works at Beira with upgrading of the railway, is costed at US$381 million.

Funding for a R65 million rehabilitation of the Maputo-Ressano Garcia line is being provided jointly by Spoornet of South Africa, CFM, and the private sector. Early completion of the route's rehabilitation, along with dredging of the harbour at Maputo, will allow this port once again to fulfil its role as the nearest to Pretoria and Johannesburg. However, the scheme also includes construction with private finance of a toll road to Maputo from Witbank in South Africa.

Traction and rolling stock

At the beginning of 1997 CFM's fleet comprised 83 diesel and 18 steam locomotives, 17 diesel railcars, 152 passenger cars and 7,186 wagons.

The most recent traction acquisitions have been 15 GEC Alsthom Type AD26C 1,850 kW diesel locomotives powered by Caterpillar Series 3606 engines, for use in the Limpopo corridor, and six World Bank-financed General Electric Caterpillar Series 3512-engined Type U10B 1,100 kW diesel locomotives delivered to the Beira Corridor Authority. Both series were supplied in 1991–2.

Signalling and telecommunications

Much of the existing signalling is deficient and the long-term aim is to resignal throughout the network. Rehabilitation of the Goba line includes installation of a new traffic control system, while resignalling and some doubling of the Ressano Garcia line is under study as part of the rehabilitation project mentioned above.

One of the most striking advances has been in the application of computers. A second phase of computerised traffic control was implemented in 1991, bringing on line a second computer to improve data capture. The system now encompasses wages, traffic and routeing statistics, and movement of ships at the port of Maputo. It will be extended to cover all financial, personnel and commercial matters and stores control.

In an associated project, US Aid has financed computerisation of all clerical activities at the port. The microwave link between Maputo and the South African border at Ressano Garcia is being supplemented by another between Maputo and the Zimbabwe border at Chicualacuala, bringing this route into the data capture system.

Track

Standard rail: 54 kg/m is replacing 30, 40 and 45 kg/m section when rehabilitation takes place
Crossties (sleepers): Timber, 2 × 0.24 × 0.13 m Concrete and twin-block concrete, 2 × 0.24 × 0.13 m
Spacing: In plain track, 1,500/km
Min curve radius: 100 m
Max gradient: 2.7%
Max axleload: 20 tonnes

Myanmar

Ministry of Rail Transport

PO Box 118, Yangon

Key personnel
Minister: U Win Sein

Myanmar Railways (MR)

PO Box 118, Bogyoke Aung San Street, Yangon
Tel: (+95 1) 28 44 55

Key personnel
Managing Director: U Thaung Lwin
General Manager: U Aye Mu
Departmental Heads
  Operating: U Tin Shwe
  Mechanical and Electrical: U Tun Aye
  Civil Engineering: U Kyaw San
  Finance: U Nyan Win
  Commercial: U Myint Wai

Gauge: 1,000 mm
Route length: 3,955 km

Organisation

The most important line connects the two principal cities, Yangon (Rangoon) the capital, and Mandalay 619 km to the north. MR has no connections with neighbouring railways. Extension of the Bago-Moatama line over the Salween river and into Thailand at Phisantouk has been studied, but is not seen as a priority.

Passenger operations

Just over five billion passenger-km were travelled in FY95-96.

Principal passenger services are operated on Yangon-Mandalay, Mandalay-Lashio, Mandalay-Myitkyina, Yangon-Ye, Thazi-Myingyan and Yangon-Pyay routes. Some overnight services on longer routes, such as Yangon-Mandalay, offer sleeping accommodation.

Freight operations

In FY95-96 traffic totalled 926 million tonne-km, almost double what it had been five years before. Principal freight commodities are timber, rice, sugar cane and aggregates.

New lines

A 102 km line south from Ye to Tavoy was opened in March 1999, having taken nearly five years to construct. The line was built by defence services personnel and members of the local population.

In 1999, defence personnel were working on the final stages of a new line from the regional capital of Pakokku to Gangaw and Kalay. New railway (and road) construction is carried out under the auspices of the State Peace and Development Council, which since 1988 has constructed nearly 1,400 km of new lines.

Rebuilding of the so-called 'Death Railway' link to Thailand was discussed by the two governments in 1996 but the project has not been taken forward.

Improvements to existing lines

Foreign aid has been sought to finance several major projects. One is conversion of Yangon's orbital commuter line to an electrified system operated by 19 locomotives and 105 passenger cars, the total cost of which is put at US$86.7 million. In 1992 bids were invited for resignalling of this line, but satisfactory tenders were not forthcoming. The work was readvertised at the start of 1996.

A second aspiration is the first stage of relaying the Yangon-Mandalay main line and equipping it with a VHF communications system.

Traction and rolling stock

At the beginning of 1996 MR's resources included 16 serviceable steam and 205 operational diesel locomotives. Additional locomotives were under repair or out of use. Some 600 passenger coaches and 3,800 freight wagons were also owned by the railway.

The greater part of MR's diesel fleet is French-built, with Alsthom Bo-Bo-Bos of 900 kW, 1,200 kW and 1,500 kW prominent. 1980s acquisitions included seven 375 kW diesel-hydraulic locomotives from Kawasaki and Sumitomo of Japan and 19 820 kW diesel-hydraulic locomotives from Krupp of Germany.

In 1993 MR took delivery of six Class DF 2016 1,500 kW diesel-electric locomotives from the Dalian rolling stock plant in northeast China's Liaoning province. A further six such locomotives were ordered later, along with four Class DF1264 900 kW units. In 1995 further orders valued at US$40 million were placed with Chinese builders for locomotives, rolling stock and spare parts, including nine MTU-powered diesel-electrics from Quingdao works.

Under a technology transfer arrangement in the 1980s, Daewoo of South Korea set up a factory at Mandalay capable of producing 60 passenger cars and 120 wagons a year.

In 1997, MR awarded a US$11 million order for 400 wagons to Fabrika Vagona Kraljevo of Yugoslavia.

Signalling and telecommunications

Most of the network remains under the control of semaphore signals and wire-based communications. Bids were sought in early 1996 for resignalling of the Yangon suburban network and three other locations.

Track
Standard rail: Flat bottom BS
Main line: 75 and 60 lb/yd (37.2 and 29.8 kg/m) in 39 ft lengths
Main branches: 60 lb/yd (29.8 kg/m)
Other branches and sidings: 50 lb/yd (24.9 kg/m)
Joints: Suspended; joint sleepers 14 in centres. Rails joined by fishplates and bolts
Welded track: 117 ft (35.7 m) lengths. Thermit welded *in situ*

Crossties (sleepers): Hardwood (Xylia Dolabriformis) and creosoted soft wood, 8 in × 4½ in × 6 ft (203 × 115 × 1,829 mm)
Spacing
  Main line: N × 3
  Branch line: N × 2 (N = length of rail in linear yards)
Rail fastening: Dog spikes, elastic rail spikes
Filling (ballast): Broken stone, 70 mm, shingle on branch lines
Thickness under sleeper: 150 mm
Max curvature
  Main line: 6° = radius of 955 ft (291 m)
  Branch line: 17° = radius of 338 ft (103 m)

Max gradient
  Main line: 0.5% = 1 in 200 compensated
  Branch line: 4.0% = 1 in 25 compensated
Max permitted speed
  Main line: 48 km/h
  Branch line: 32 km/h
Max axleload: 12 tons on 75 and 60 lb/yd rail
Bridge loading: Indian Railway Standard ML

Namibia

Ministry of Works, Transport and Communication

Ministry of Works, Transport and Communication
Private Bag 13341, 9000 Windhoek
Tel: (+264 61) 208 91 11   Fax: (+264 61) 22 85 60
Web: www.op.gov.na

Key personnel
Minister: Hon Moses Amweelo
Permanent Secretary: B Kathindi

UPDATED

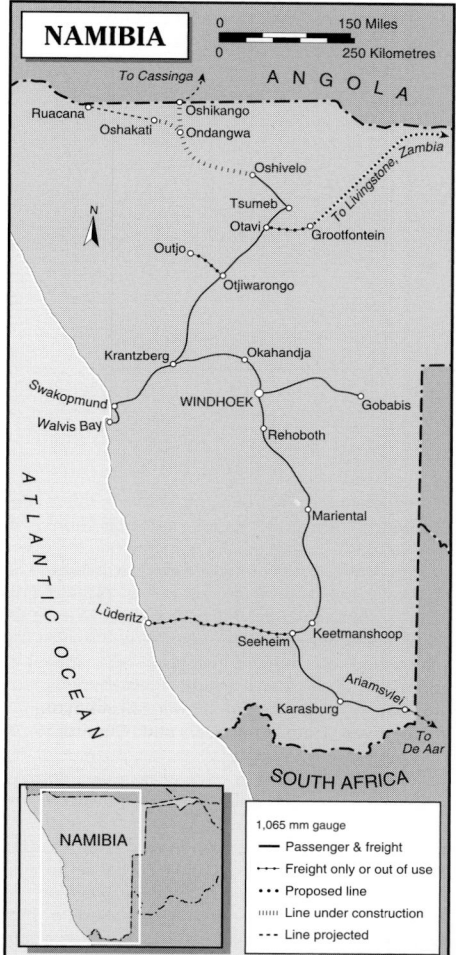

NAMIBIA

0 — 150 Miles
0 — 250 Kilometres

To Cassinga · ANGOLA
Ruacana · Oshikango · Oshakati · Ondangwa · Oshivelo
To Livingstone, Zambia
Tsumeb · Otavi · Grootfontein
Outjo
Otjiwarongo
Krantzberg · Okahandja
Swakopmund · WINDHOEK · Gobabis
Walvis Bay · Rehoboth
Mariental
ATLANTIC OCEAN
Lüderitz · Keetmanshoop
Seeheim · Ariamsvlei
Karasburg
To De Aar
SOUTH AFRICA

NAMIBIA

1,065 mm gauge
—— Passenger & freight
---- Freight only or out of use
••• Proposed line
‖‖‖ Line under construction
--- Line projected

NEW/1114977

TransNamib Holdings Ltd

Private Bag 13204, Windhoek 9000
Tel: (+264 61) 298 11 11   Fax: (+264 61) 22 79 84
e-mail: rail@transnamib.com.na

Key personnel
Chairman: Dr Klaus Dierks
Chief Executive Officer: John Shaetonhodi
Head of Train Operations: Reggie du Toit
Chief Mechanical Engineer: K V Asokan

Chief Engineer, Research and Development:
  Sakkie Engelbrecht
General Manager, Engineering: Matty Hauuanga
Marketing and Sales Manager: Brian Black
Chief Public Relations Officer:
  Olivia Kanyemba-Usiku

Gauge: 1,065 mm
Route length: 2,382 km

Political background
In 1988 the Namibian rail, road and air services, previously worked by South African Transport Services under contract, were formally handed over to the Namibian body that was renamed TransNamib in 1989. This followed a *de facto* transfer in 1985. The government is the principal shareholder. The state owns the railway infrastructure, with the Ministry of Finance the shareholding ministry and the Ministry of Works, Transport and Communication the regulatory ministry. The rail, road and air services are run as separate business units.

TransNamib has operated without outside funding or subsidy since its inception. After a gradual decline in performance, the company's fortunes have been turned round by a new management team which took over from 2002 and converted TransNamib from a loss-making enterprise to a profitable one. This has lead to some investment in modernising resources such as traction and rolling stock.

Passenger operations
In the early 1990s, business suffered from the expansion of minibus operations, some actually poaching passengers at stations. TransNamib responded by offering improved service at lower prices. In order to stem losses, the organisation revamped the whole passenger service by introducing coaches refurbished with airline-type seating and air conditioning attached to timetabled freight trains; these are known as 'StarLine' services and have proved very popular.

The luxury 'Desert Express' for tourists operates between Windhoek and Swakopmund/Walvis Bay. Sleeping car and restaurant facilities are included in the train formation. In 2005 a new Chinese-built dmu (see below) entered service on the same route.

Freight operations
The main constituents of the freight business are mining products, petroleum products, bulk liquids, building materials and containers, with a particular focus on traffic through the port of Walvis Bay. The rail link with South Africa has declined significantly as a result of severe road competition and shorter road distances from the Transvaal area.

New lines
Construction of a 247 km northern extension of the railway from Tsumeb to Ondangwa, close to the border with Angola continued in 2005. By February the line had been completed as far as Oshivelo. The Northern Railway Extension project will eventually see the line taken beyond Ondangwa to Oshikango and will also entail construction of a 35 km branch from Ondangwa to Oshakati that could be extended to Ruacana. The project is seen as giving Namibian companies better access to the market in Angola.

A rail link between Namibia and Zambia appeared to have moved closer to being realised in 1999, when the governments of the two countries agreed to construct a bridge over the Zambesi at Sesheke. This would be reached by extending the existing NamRail Walvis Bay–Grootfontein line to met a new Zambia Railways line diverging from its Livingstone–Mulobezi line. However, despite studies funded by the EU in 2001, no work on the project had been reported by 2005.

Improvements to existing lines
In 2005 rehabilitation was under way of the 139 km Aus–Lüderitz section of the Seeheim–Lüderitz line (318 km), on which the rails had deteriorated to such an extent that passenger services were suspended.

General Electric U20C diesel-electric locomotive with a freight train at Windhoek (Marcel Vleugels)
NEW/1114978

A container terminal adjacent to Van Eck power station in Windhoek was completed in 2004.

Traction and rolling stock

In 2005 the railway operated 45 survivors of a fleet of six GE U18C1 1,475/1,340 kW 1-Co-Co-1 and 65 GE Type U20C 1,605/1,490 kW Co-Co diesel-electric locomotives. TransNamib also has two steam locomotives for special trains. Other rolling stock comprises around 130 passenger cars and some 1,500 freight wagons. In 2004 a start was made on modernising the traction fleet with the delivery of four 1,640 Co-Co diesel-electric locomotives built in China at the Ziyang plant and supplied by China South Locomotive and Rolling Stock Industry Corporation (CSR). These are powered by Cummins QSK60 16-cylinder engines.

Chinese industry supplied a 120 km/h Cummins-powered dmu for luxury passenger services in 2005. More vehicles of this type were expected to be procured. The wagon fleet was supplemented the previous year by 30 tank wagons also supplied by Chinese manufacturers for the movement of petroleum products.

Signalling and telecommunications

Radio control now covers operation throughout the railway.

Coupler in standard use: SASKOP Type M on passenger cars, and Types M and S on wagons

Braking in standard use: Vacuum; two daily fast trains are air-braked

Track

Rail type and weight: HCOB 22 kg/m (185.4 km), 30 kg/m (1,405.8 km), 48 kg/m (916.5 km), 57 kg/m (78 km)

Crossties (sleepers): Concrete, steel, timber (on bridges and 1:12 pointwork)
Concrete, 2,057 × 230 × 250 mm
Steel, 1,917.7 × 305 × 106 mm

Spacing: In plain track 1,429/km, in curves 1,538/km

Rail fastening: Fist and Pandrol on concrete sleepers, bolt and nut in steel, screw on timber

Min curve radius: 200 m

Max gradient: 1.5%

Max axleload: 16.5 tonnes on 48 and 57 kg/m rail, 15 tonnes on 30 kg/m rail, 11.5 tonnes on 22 kg/m rail

UPDATED

Nepal

Ministry of Labour and Transport Management

Singhadurbar, Kathmandu
Tel: (+977 1) 424 78 42; 424 19 63
Fax: (+977 1) 425 68 77
Web: www.moltm.gov.np

Key personnel

Minister: Raghuji Panta
General Director, Transport Management:
Shanker Prasad Dhungana

UPDATED

Janakpur Railway (JR)

PO Box 309, Teku, Kathmandu
Tel: (+977 1) 23 33 30

Key personnel

Chief Executive: Kiran KIshor Ghimire
General Manager: Dirga Narayan Regmi

Gauge: 762 mm
Route length: 53 km

Organisation

The Janakpur Railway (JR) runs from Jaynagar in Bihar State, India, across the Nepal border north and west to Janakpur (32 km) and on to Bizulpura (21 km). However, in 2004 the Janakpur–Bizalpura was reported to be out of use. The railway is effectively operated by the Ministry of Labour and Transport Management, although there have been moves to establish it as a separate state-owned company.

The railway was originally built as a timber line designed to open the virgin jungle to the north of Janakpurdam. As the forest has long since been cut, the railway now operates primarily to provide access in an area with few roads. Passengers are the main source of revenue, with pilgrims to the temples of Janakpurdam forming the bulk of traffic. At latest report the railway recorded 1.6 million passenger journeys and 22,000 tonnes of freight.

Passenger operations

In 2004 a daily service of three trains in each direction was operated. Freight is conveyed in wagons attached to passenger services.

Traction and rolling stock

The traction fleet consists of four Indian-built Type ZDM5 diesel-hydraulic B-B locomotives, two of which were reported to be out of use in 2004. There are also several stored steam locomotives, some of which are used for special services.

UPDATED

Netherlands

Ministry of Transport & Water Management

Plesmanweg 1-6, PO Box 20901, NL-2500 EX Den Haag
Tel: (+31 70) 351 61 71   Fax: (+31 70) 351 78 95
Web: http://www.minvenw.nl

Key personnel

Minister: Mrs T Netelenbos
Secretary of State: Mrs M de Vries
Secretary-General: R Pans
Director-General, Passenger Transport:
M van Eeghen
Director, Mobility and Market: P Boot
Director, Infrastructure and Policy: M Olman
Head, Rail Infrastructure Investment and
Maintenance: P. de Booij
Directors General, Freight Transport:
B Westerduin, F J P Heuer
Head, Freight Transport Sectors: K Van Hout
Head, Railfreight: Mrs T Zwartepoorte
Head, Repositioning Government Agencies:
K van Hout

Within the framework of the railway reforms which are taking place in the Netherlands, the Minister of Transport has launched a policy document called *'De derde eeuw spoor'* (The Third Age of Rail). Its aim is to strengthen the position of rail transport. In three main areas it calls for action to be taken to improve the role played by rail.

The first is restructuring the managerial responsibilities for rail traffic. Progressively, responsibility for regional train services will be decentralised to regional authorities (provinces) and the services tendered. The first to be tendered will be the so-called 'contract sector services', which the national authorities pay NS Reizigers to run. This process will take into account the size of prospective operators' market shares: those with a share considered too large will be excluded.

In 1999, for some 30 services contracts have been made between the government and NS at a cost to the state of NS G155 million a year. The state will continue to be responsible for national train services. NS Reizigers will have the exclusive concession for running the national core network, which is currently served by most IC services and some fast trains. NS Reizigers will also have the exclusive right to operate domestic high-speed services on the HSL-Zuid line. For both exclusive concessions a performance contract will be made between NS Reizigers and the state in which obligations for NS are based on maintaining specified levels of service.

The second area addressed by the policy document concerns the balance between control by the authorities and the market. It is supposed that demands will be made on performance by the operators. Important here are growth of the numbers of peak-hour passengers and punctuality, but also some other points. It is probable that a bonus or penalty regime will be linked to performance.

The third area concerns the publicly owned, government-financed agencies within NS Holding: Railned, NS Railinfrabeheer BV and NS Verkeersleiding BV (Traffic Control). These all undertake tasks for which the government is responsible and will remain so after the separation of infrastructure and operations, as well as the opening up of the network for other or new operators. The government's intention is to take these three agencies out of NS Holding from 1 January 2000. In 1999 it was likely that an independent public institution would be established to handle capacity management, operations control and access. Possibilities about the way to organise infrastructure maintenance and construction projects were still being studied.

A track access charge is to be introduced on 1 January 2000. In fact, track access fees already notionally exist but the charge is rated at G0 per km. European directives prescribe the existence of a rating system for track access charges and the present Dutch one is in accordance to this legislation. The proposed change planned in January 2000 is in fact a change in rating. In mid-1999 a final decision remained to be taken regarding the method by which operators would pay these charges. The Minister of Transport has proposed that a total charge must generate G400 million annually across the network. This amount does not cover the costs incurred by the government for track maintenance. NS Reizigers, which operates a majority of trains on the network, is to pay G250 million and decentralised, regional and tendered train service operators are to pay G100 million. Freight should pay between zero and G50 million. The charge is to be introduced in stages, eventually with different tariffs for freight and passenger transport. The proposal is not definitive and is still under discussion. All train operators oppose the track access charge.

To implement the policy described above, the Minister of Transport is planning new legislation for public transport, including public train operations, and this was expected to be in parliament during 1999. The new law will also define rules governing competition and tendering for regional train services. Another move by the Minister of Transport aims to renew existing railway legislation to meet contemporary requirements. The new legislation will be in accordance with European Union directives, and will define the tasks and competences of the various authorities involved with rail traffic. It will also define the position of the three publicly financed government agencies mentioned above. The Minister aims to make the new law effective by 2001.

Freight traffic and limited passenger services have been liberalised. In theory, if an operator fulfils the conditions to operate services, there are no legal or policy obstacles to entering the market. In practice, conflicts can occur scheduling freight between other trains. Much of the capacity of the Netherlands network is used by the high-density passenger services operated by NS Reizigers, which runs trains to a rigid pattern and with high frequencies, especially on the main corridors. This makes it difficult to run freight services during daytime. Rules for managing the line capacity were being developed in 1999.

Open access

Recently the Dutch railway network has been opened up to new operators, who must fulfil certain conditions. Open access and other rail reforms on the Dutch railway network are currently founded on a temporary, non-legislative basis, but are to be covered by the new legislation detailed above. Until then the procedure for setting up a train operating company focuses on three main points: access to the profession; access to the market; and access to the traffic. The Ministry of Transport and the three publicly financed government agencies, Railned, NS Railinfrabeheer and NS Verkeersleiding, all part of the NS Holding, are involved in this process. Railned, which is responsible for track capacity management, timetables and safety, and is considered to act independently, serves as the interface between the operator and the other two agencies.

First, prospective entrants to the market must be recognised by the Minister of Transport as being a train operating company. For access purposes, rail traffic is split up into three categories: freight, limited/charter train services and public services. For the last, public transport legislation approved by the Minister of Transport is needed. For freight or restricted passenger services, recognition by the Minister is sufficient. An access contract comes into force if it is clear that the new operator is capable of running his intended services, staff have been examined satisfactorily, rolling stock has been approved and if the new operator accepts specific legal responsibilities. The operator must also have a certificate of safety, issued by Railned.

In 1996 Lovers Rail was the first company to operate public services after withdrawal of the

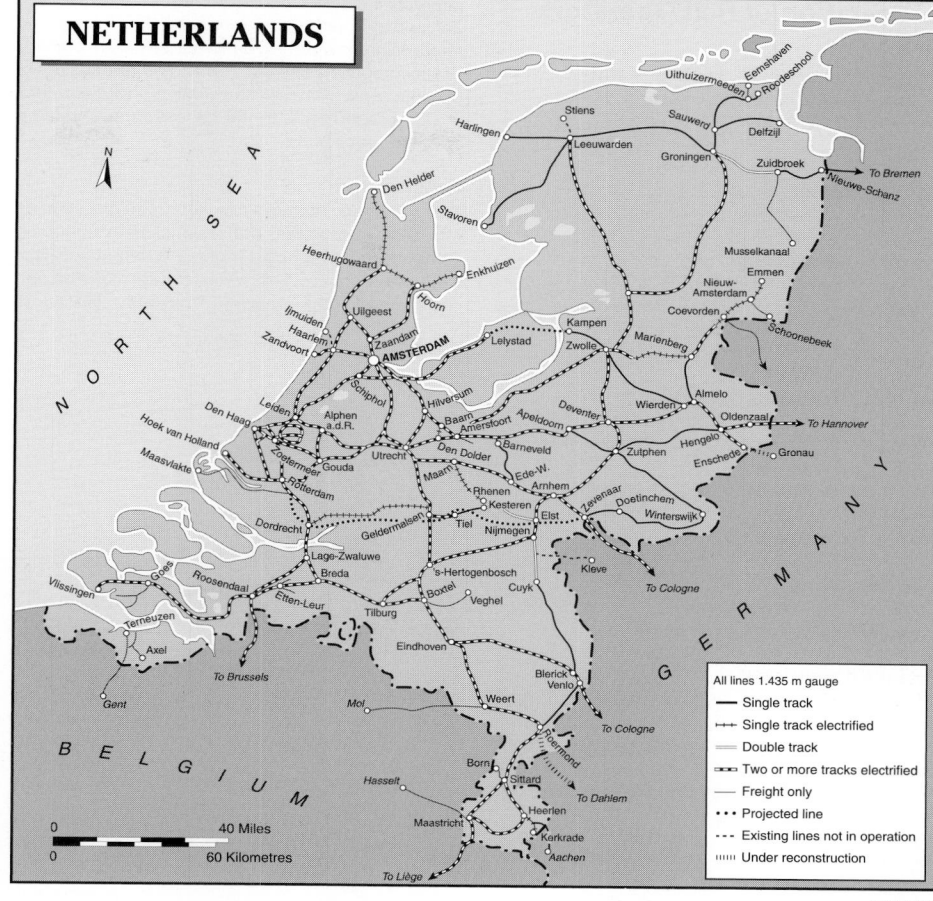

0088729

NS monopoly of passenger traffic, introducing trains between Amsterdam and the coastal town of IJmuiden. Unclear what kind of competition in public transport might emerge, the Minister of Transport decided that no new operators could enter this market of public transport, fearing the prospect of 'cherry-picking', companies bidding only to operate profitable routes, while NS Reizigers continued to run a nationwide network which included loss-making services. The minister will continue to refuse to grant licences for public services until the new legislation is in place.

In May 1998 another public passenger train service was started by a new train operator, Oostnet. In this case there was no cherry-picking; NS was unwilling to continue services on the Almelo-Mariënberg regional line for the subsidy offered by the government. In a pilot project for the decentralisation and tendering of regional train services, Oostnet, the regional bus operator in the eastern Netherlands, took over services, which are partially supported financially by regional

and local authorities. It is thought that regional authorities will be responsible for tendering for local transport services under the provisions of the new legislation.

By 1 April 1999, 22 operators had been recognised by the ministry. Twelve of these had received safety certificates and 10 had not. The safety certificate allows an operator to run independently on the Dutch rail network. The operators listed are involved in the operation of public and limited passenger transport, freight and historic train services.

Operators with a safety certificate: ACTS; Bentheimer Eisenbahn; Lovers Rail; NBM Rail; NS Cargo; NS Materieel; NS Reizigers; Oostnet; Short Lines; Structon Railinfra Materieel; Volker Stevin Rail & Traffic; and Zuidlimburgse Stoomtrein mij.

Operators without a safety certificate: Chem Trans Logistic; Nederland ERS Railways BV; Holland Spoor; Hoogovens Staal; HST/VEM; NDX Intermodal; NS Ultrasoonbedrijf; Rasrail; Stichting Museum Buurtspoorweg; and Stichting Stadskanaal Rail.

State Transport Inspectorate

Rijksverkeersinspectie
Johanna   Westendijkplein   115,   NL-2521   EN Den Haag
PO Box 10700, 2501 HS Den Haag
Tel: (+31 70) 333 70 00   Fax: (+31 70) 305 27 99

The rail division of the State Transport Inspectorate has been wound up. Its duties have been transferred to the rail safety division of Railned, leaving just one individual responsible for monitoring Railned's safety operations.

Transport Safety Board

Raad voor de Transport veiligheid
Prins Clauslaan 18, P.O.Box 95404, NL-2509 CK Den Haag
Tel: (+31 70) 333 70 00

Key personnel
President: HRH Prince Pieter van Vollenhoven
President, Rail Bureau: Mrs Schmitz

The Transport Safety Board assumed its responsibilities from 30 June 1999. The Board investigates all major transport accidents, including those on the rail network. The Board can also make recommendations regarding transport safety and the environment. Administratively, it is formed of four bureaux, of which one is responsible for rail.

Association of Rail Operators (BVS)

Branche Vereniging van Spoorvervoerders
PO Box 19040, NL-3501 DA Utrecht
Tel: (+31 30) 236 88 44
Fax: (+31 30) 236 83 38

Key personnel
President: J. Hoekwater

Political background
BVS was established to take care of the interests of the new rail operators in the Netherlands. These interests concern subjects such as procedures and difficulties in setting up train operating companies, relationships with government and government agencies such as Railned, RIB and other parts of NS Holding, and technical harmonisation issues. BVS is also concerned with the education and training of personnel.

Organisation
The members of the association can be split up into three categories: professional train operating companies, both of passenger and freight services; track maintenance and infrastructure construction companies; and railway museum operators who want to run trains on the Dutch network. In early 1999 there were nine members: ACTS Nederland; Lovers Rail; NBM Rail NV-materieel; Oostnet Groep; Short Lines; Stichting Holland Spoor; Structon Railinfra Materieel; Vereniging Historisch Railvervoer Nederland; and Volker Stevin Rail and Traffic.

ACTS Nederland BV

PO Box 19040, NL-3501 DA Utrecht, Kromme Nieuwegracht 58
Tel: (+31 30) 236 88 44   Fax: (+31 30) 236 83 38
e-mail: directie@acts-nl.com
Web: www.acts-nl.com

Key personnel
Director: R P T van Gansewinkel
Finance Manager: P H A van der Schoot

Freight operations
ACTS specialised at first in the transport of household waste. Initially, ACTS did not operate its own trains, instead hiring its container-carrying wagons to other operators, with haulage provided by NS Cargo. In early 1998 the company started to run its own trains.

ACTS subsequently expanded its activities, in March 1999 taking over from NS Cargo the domestic container service between the port of Rotterdam (Maasvlakte) and Veendam, in the far northeast of the Netherlands, and Leeuwarden, to the north.

In June 2002 the company added two international services to Germany to the five already operated domestically. By December 2004 about 50 trains per week were operated. In addition, ACTS occasionally provides traction for track maintenance work and special passenger trains.

Traction and rolling stock
At the end of 2004 the traction fleet owned by ACTS comprised four ex-NS Class 1200 electric locomotives, three Class 6700 diesel locomotives, former Class 62/63 machines bought by ACTS from Belgian National Railways, and three former German Type V60 locomotives for engineering trains and shunting duties. The company also leases one Class 60 and two Class 58 diesel locomotives (see below), two HE 1206s and one Class 66.

In 2002 ACTS concluded a 10-year contract to lease up to five Class 58 diesel-electric locomotives from the UK rail freight operator, English Welsh & Scottish Railway. The first two of these locomotives arrived in the Netherlands in 2003, while a third example of the type was due to arrive in early 2005.

The wagon fleet comprises 206 ACTS and 76 Sgns vehicles, together with 28 additional wagons leased from AAE and others.

UPDATED

Refurbished Class 58 locomotive, hired by ACTS from UK freight operator EWS (Quintus Vosman)
0583352

ERS Railways BV

PO Box 59018, NL-3008 PA Rotterdam, Netherlands
A Plesmanweg 61 K-L, NL-3088 GB Rotterdam, Netherlands
Tel: (+31 10) 428 52 22   Fax: (+31 10) 428 52 12
e-mail: info@ersrail.com
Web: http://www.ersrail.com

Organisation
ERS Railways was established in October 2002 as a subsidiary of European Rail Shuttle (ERS) (for further details see European Rail Shuttle BV in the *Operators of international rail services in Europe* section). The company provides open access rail freight services using its own fleet of locomotives and wagons, serving both ERS and other customers.

Traction and rolling stock
ERS Railways hires 10 General Motors Type JT42CWR ('Class 66') locomotives from Porterbrook and HSBC, and two Vossloh G 1206 locomotives from Angel Trains. Rolling stock consists of more than 800 container wagons leased from AAE and KombiWaggon.

JT42CWR locomotives at NedTrain's Tilburg workshops prior to their deployment with ERS Railways (Quintus Vosman)
0543493

Netherlands Railways (NS)

NV Nederlandse Spoorwegen
PO Box 2025, NL-3500 HA Utrecht
Moreelsepark 1, NL – 3511 EP Utrecht
Tel: (+31 30) 235 91 11   Fax: (+31 30) 233 24 58
Web: http://www.ns.nl

Key personnel

President and Chief Executive, NS Holding:
  Rob den Besten
Corporate Staff
  Communications: Mrs J C B Straatman
NS Reizigers BV (passenger business unit)
  Managing Director: H Huizinga
  Director, Marketing & Sales: M M D van Eeghen
  Operations Director: W A G Doebken
  Director, Staff & Organisation: R Lantain
NS Cargo BV (freight business unit)
Web: http://www.nscargo.nl; cargoweb.nl/NScargo
  Managing Director: E Smulders
  Commercial Director: Z van Asch van Wijck
  Finance Director: R van Haaren
NS Materieel BV (rolling stock engineering)
  Managing Director: Tj Stelwagen
  Commercial Director: R Knipping
  Finance Director: A Valk
NS Stations BV (station management)
  Directors: P Stulp; A van Engelen
NS Vastgoed BV (Property)
  Director: H Portheine
NS Opleidingen (training)
  Director: D Kruijd
NS Verkeersleiding BV (traffic control)
  Director: Mrs J Arts
NS Railinfrabeheer (rail infrastructure)
  Director: R Oliemans
Railned (capacity management)
  Director: P Ranke
NS Railway Police
  Directors: M van Asch van Wijck, G van Beek

Gauge: 1,435 mm
Route length: 2,808 km
Electrification: 2,061 km at 1.5 kV DC

Political background

NS has been transformed into an independent, commercially oriented business following the 1992 report of the Wijffels select committee, appointed by the Dutch minister of transport. Changes centred on separation of rail operations and infrastructure management – completed in 1994 – and the progressive change of NS from a block-subsidised organisation to one of separately accountable business units. In June 1998, it was announced that NS Cargo was to merge with its German counterpart, DB Cargo, to form a new joint company, Rail Cargo Europe. The fusion of the two organisations was expected to be complete by late 1999, when operations would be managed from new headquarters in Mainz, Germany.

Further changes now being implemented see the advent of limited on-rail competition in a transitional period which will stretch up to 2000. Lovers Rail (qv), a private company, began passenger operations on the line between Amsterdam and IJmuiden in August 1996. Several other new operators plan to introduce freight services.

In early 1998, the competitive model for rail transport after 2000 had not been decided upon, but the transport ministry was moving to prevent any one operator, including NS, from having more than 50 per cent of the country's rail market. NS was pressing the government to grant it exclusive rights to the intercity network and to end uncertainty as its proposals for rolling stock acquisition and upgrading had to be put on hold.

In 1996, NS drew state financial support of G66 million for operations in respect of unremunerative services and fare reductions deemed socially necessary. An agreement has been reached with the government over the continuation in service until at least mid-1999 of 29 unprofitable lines and the operation of seven new stations. For the period mid-1998 to mid-1999, the government has agreed to pay G155 million for this. A list of around 30 loss-making lines was drawn up in the mid-1990s and the Dutch government has moved to open up their operation to international tender. From May 1998 the company Oostnet is to take over the Almelo-Mariënberg line, initially using NS rolling stock. There are now moves to open up complete local stopping train networks to tender rather than concentrating on individual lines.

Organisation

In recent years NV Nederlandse Spoorwegen has been transformed into an independent commercially oriented business. The restructuring followed the method advised by the Wijffels select committee, appointed by the Dutch minister of transport. The first phase of this process has been completed and in mid-1999 the second stage was due to start soon. Its main feature will be the removal from railway control of the state-financed government-commissioned agencies, Railned, NS Railinfrabeheer and NS Verkeersleiding (see below). Although involved in performing functions in which neutrality and objectivity are required, these agencies currently remain part of NS Holding. This was a result of implementing European Union Directive 91/440, separating infrastructure and operations. The next phase will see a split between organisations with public functions and commercially oriented business units.

NS Holding is divided into two groups of companies: one comprises the three government-commissioned agencies mentioned above; the other is formed by NS Groep NV and includes the railway's commercial business units, each of which has independent company status. The business units are: NS Reizigers, NS Cargo, NS Stations, NS Vastgoed NS Materieel, NS Beveiliging Services, NS Facilitaire Bedrijven and NS Opleidingen. Some of these business units have shares or a participation in other companies, such as NS Reizigers, which is one of the three parties involved in the Syntus regional integrated transport company (qv). In mid-1998 it was announced that NS Cargo was to merge with DB Cargo, the freight business of German Rail. NS Holding will have a minority shareholding in the resulting new company, Rail Cargo Europe. NS Cargo was still recording losses during 1998, but the deficit had declined to G2 million. Break even was forecast to occur in 1999, just before the merger was due to take place.

State subsidies have been withdrawn in steps. NS received its last subsidy for train operations in 1997, marking the end of a five-year reduction process. On the contrary, NS chairman Rob den Besten proposed paying the shareholder, in the form the state, a dividend worth some G9 million from the positive results achieved during the year. Capital contributions from the state to help strengthen the commercial position of NS in its market were due to finish in 1999.

A new financial relationship was made between NS Reizigers and the state for running loss-making services. In addition, some services were 'bought' by the state to provide a public transport alternative to heavily congested roads. For 29 loss-making lines NS Reizigers receives G112 million annually, these payments continuing until the adoption of a tendering procedure for local or regional services. In the first years of the new millennium it is expected that tendering for providing service on regional networks – both rail and integrated rail and bus networks – will have become widespread. In future NS is expected to be given a 10-year concession to operate the national IC network, with a contract that incorporates service and performance obligations.

NS Reizigers is acting strategically so far as regional tendering is concerned, joining with other companies to form joint ventures. The government requires that no participant in such a joint company operating in the Dutch public transport market has a share larger than 50 per cent. For example, NS Reizgers has one third of the shares in the Syntus joint venture company.

Business units

The passenger (NS Reizigers) and freight (NS Cargo) businesses are described under 'Passenger operations' and 'Freight operations'.

NS Materieel manages major rolling stock workshops at Tilburg, Amersfoort and Haarlem. Smaller workshops are located at Maastricht, Rotterdam Fijenoord, Amsterdam Zaanstraat, Zwolle and Onnen (Groningen). NS Materieel has suffered from overcapacity and the company has been restructured with a reduced workforce and modernised processes at its facilities. Work is sought from domestic and foreign clients as well as from NS companies.

NS Stations develops commercial activities at stations under exclusive rights granted to it by government. The total number of retail outlets rose from 1,231 in 1997 to 1,268 in 1998, when revenues of G1.05 billion and profits of G139 million were generated. At some stations where a ticket office is considered too expensive to operate, tickets are sold at a Wizzl, a small supermarket selling products to passengers. Provision of new retail space is regarded as an essential element of station improvement schemes.

NS Vastgoed handles the group's property portfolio, developing commercial property in and around stations, including the provision of accommodation at new or rebuilt stations.

Service units

Other business units, which mostly sell services to other NS Holding companies, are: NS Beveiliging Services (Security); NS Facilitaire Bedrijven, responsible for administration, documentation and research; and NS Opleidingen (training). Some parts of NS Beveiliging Services have been sold. The railway police service, managed on a non-commercial basis by NS Beveiliging Services, is to be removed from NS from 1 January 2000 to become part of the national police force.

Two NS Cargo Class 6400 diesel locomotives head grain hoppers near Gilze-Rijen
(Alex Dasi-Sutton)

0023254

Class DM 90 two-car dmus at Zwolle (Quintus Vosman)   0058794

Joint ventures

NS and British Telecom have established a joint telecommunications company, Telforty, which operates a mobile telephone network in the Netherlands. Over G1 billion is being invested annually in the company and losses resulting from these start-up costs have had a negative effect on the consolidated results of NS Groep.

NS also has a stake in a rail infrastructure construction company, Strukton Groep BV and in engineering consultancy Holland Railconsult. Both were formerly part of the old NS organisation.

Government agencies

Outside the NS Groep but in 1999 still within NS Holding are three government-commissioned agencies: Traffic Control (NS Verkeersleiding) which regulates the allocation of train paths on a day-to-day basis between NS businesses and possible private-sector operators under the European Union's open access provisions; Railned, a capacity management agency which conducts forward planning of train paths; Rail Infrastructure (NS Railinfrabeheer) which is responsible for the management, upkeep, expansion and development of the system including the commissioning of contractors for the execution of work. These tasks are financed from the government infrastructure fund, Infrafonds. NS Railinfrabeheer contracts outside firms for maintenance, renewal and repair work; these include Strukton Railinfra bv (qv in Permanent Way Equipment and Services section), a joint venture which has taken over the assets of NS Infra Services Materieel.

Finance

NS finished 1997 with a positive result of G164 million, up from G105 million in 1996. Overall, revenue rose by 5 per cent.

The average number of full-time employees in the company during 1997 was 25,938, against 28,191 in 1996.

Traffic (million)	1996	1997	1998
Passenger journeys	306	316	321
Passenger-km	14,091	14,485	14,879
Freight tonnes	20.8	22.9	24.7
Freight tonne-km	3,123	3,406	3,778

Passenger operations

The performance the NS Reizigers passenger business unit continues to improve. In 1998 profits rose to G362 million compared with G158 million the previous year. Revenue rose by 9 per cent and much progress was made on efficiency and cost reductions. Passenger volume rose by 2.9 per cent in 1998 to 14.9 billion passenger-km. Punctuality rose slightly during 1998, from 82.9 per cent in 1997 to 83.0 per cent. NS Reizigers' target is 87 per cent. Preventive measures against poor punctuality are to include avoiding vulnerable scheduling in timetable planning and a reduction in the coupling and splitting of portion trains, which will

sometimes result in longer than necessary trains being run at certain times.

Between 1992 and 1998 state subsidies for passenger services have declined from almost G300 million to zero in 1998. In 1997 NS received G66 million for operations.

Passenger operations in the Netherlands are characterised by high-frequency, regular-interval services. For domestic services the lowest frequency is hourly. Most lines see two trains an hour, while on main lines the frequency is generally much higher, and routes like Amsterdam-Utrecht are served by four IC services an hour.

Although the three passenger service types introduced by NS in the early 1990s, Aglo/Regio, Interregio and Intercity, were perpetuated by NS Reizigers, these names have been abandoned recently. Nevertheless, the split of services into the three categories is in effect taking place. NS is aiming to run faster IC services by omitting calls at some stations, while adding fast service stops at other stations where traffic levels justify this. Other stations are served by stopping services which NS Reizigers can withdraw if loss-making. If the government wishes to retain loss-making services, it can sign a contract with NS Reizigers to run them. In 1999 the two parties signed a G155 million contract covering the operation of such services. A government programme to combat road congestion, called 'Samen Werken Aan Bereikbaarheid' (Working Together on Sustainability) partly covers the addition of further train services to NS operations.

In the 1998–99 timetable NS added 5 per cent more passenger services – 250 trains – compared with the previous year. NS claims that the new services generated some 10,000 additional passengers daily.

NS Reizigers is making preparations to run domestic high-speed IC services on the HSL-Zuid line. New rolling stock is to be ordered, but in 1999 its design had not been determined, although double-deck equipment seemed most likely.

In 1998 NS withdrew from operating the Almelo-Mariënberg regional non-electrified line, considered the heaviest loss-making of all Dutch train services. Services were taken over by Oostnet, a regional bus operator and part of the VSN-1 group, after a tendering procedure. More regional services are to be abandoned by NS Reizigers, although it will take shareholdings in joint venture companies/tendering to operate such services.

Transvision is a successful co-operative venture between taxi companies and NS Reizigers, carrying passengers to and from stations for a G6 ticket. Another alliance has been formed covering international passenger services: NS, KLM Royal Dutch Airlines, Schiphol airport, and Belgian and French National Railways have signed an agreement to develop a strategic co-operative alliance to operate trains using the HSL-Zuid to link the Netherlands, Brussels and Paris.

Freight operations

During 1998 NS Cargo freight traffic increased by 8 per cent up to 24.7 million tonnes. Much of this rise was attributable to a doubling of coal movements from the ports of Amsterdam and Rotterdam compared with 1997. The trailing load of such coal trains has been increased from 2,400 tonnes to 3,800 tonnes.

Container transport also rose by 10 per cent to a total of 490,000 units, although in a move to rationalise activities, NS Cargo withdrew most of its domestic container services. In 1999 NS Cargo lost its largest domestic container service, between Rotterdam Maasvlaket and Veendam container terminal, to ACTS (qv). In 1998 three new container services have been started: Rotterdam-Warsaw; Rotterdam-Mainz/Mannheim; and the Amsterdam-Frankfurt airfreight shuttle. Almelo-Poznan and Rotterdam-Barcelona, on behalf on Transfracht, were withdrawn. Traffic between Rotterdam and Eastern Europe, including Russian Federation countries, is growing rapidly. Nevertheless, rail's share in the market port's freight transport market was only 16 per cent in 1997.

Other freight sectors performed well in 1998: lime and cement traffic was up by 25 per cent to 300,000 tonnes and grain doubled up to 550,000 tonnes. Chemical transport was up by just 1 per cent.

An improving financial performance by NS Cargo saw losses decline from G6 million in 1997 to G2 million in 1998. NS Cargo was expected to break even in 1999, the year in which the merger with DB Cargo was due to take place. Revenue rose by 4 per cent to G371 million. The workforce was reduced from 1,837 in 1997 to 1,723. This includes approximately 300 drivers (240 in 1997) who were on loan from NS Reizigers. During 1999 drivers were assigned permanently either to NS Cargo or NS Reizigers.

International developments include fitting five Class 6400 diesel locomotives Class 6400 with Belgian automatic train protection and communications systems to enable them to operate intermodal services between the ports of Rotterdam and Antwerp, avoiding traction and crew changes at the border. Three more locomotives are to be similarly equipped. Four Belgian Class 25.5 dual-voltage electric locomotives, already equipped with the Dutch ATB train protection system, are being employed in through services between the two ports.

International through working of Dutch-German freight traffic was also to be introduced following the fitting of five Class 6400s with the German Indusi train protection system. These locomotives are to operate trains between the Ruhr and the port of Rotterdam, as well as regional freight services. DB Cargo is to equip five Soviet-built Class 232s with Dutch ATB equipment to haul heavy coal trains between Germany and Rotterdam and Amsterdam.

Freight services in the Arnhem area will be served from Oberhausen, Germany, instead of NS Cargo's principal yard at Kijfhoek, Rotterdam, from which freight is distributed throughout the Netherlands. Although the merger of NS Cargo and DB Cargo had not been completed when these changes were initiated, they can be seen as a consequence of the intention to combine.

New lines

There are two high-speed line projects in the Netherlands: the HSL-Zuid (high-speed south) line, linking Amsterdam with Belgium and France, and the HSL-Oost (high-speed east) line to Germany.

HSL-Zuid

Most of the legal procedures on the HSL-Zuid project have been completed and work was due to start in the second half of 1999, with completion planned for 2005. The 100 km project consists of two portions: from Hoofddorp (just south of Schiphol airport) to Rotterdam, and from the south of Rotterdam to the Dutch/Belgian border near Breda, mostly following the alignment of the E19 motorway. The line connects with the Belgian network at Antwerp. A link is to be provided to the existing Dutch network at Breda, enabling domestic services to use the high-speed line. The line will be engineered for 300 km/h and will employ a 25 kV

AC 50 Hz power supply. For domestic services NS Reizigers is to order trains for speeds of 220 km/h or possibly higher. It is anticipated that running domestic services on the high-speed line will change the IC network dramatically, yielding significant time savings over the relative short-distance journeys which characterise the Dutch rail system. An Amsterdam–Rotterdam journey using the high-speed line will take 28 minutes instead of the present 1 hour 5 minutes and Amsterdam–Breda will take 48 minutes compared with 1 hour 40 minutes. International services will also benefit: the Amsterdam-Paris journey time will be reduced from present 4 hours 44 minutes to 3 hours 3 minutes.

To minimise the environment impact of the new line, its southern part will mostly run parallel to the E19 motorway or the existing Rotterdam-Breda line. On the section between Rotterdam and Hoofddorp a 7 km twin-bore tunnel is planned to avoid disturbing of the 'Green Heart of Holland'.

The government has planned the HSL-Zuid project as a partnership between public and private sectors. At 1998 prices, the project will cost an estimated G8.9 billion, of which an Investment contribution of G1–1.5 billion is expected from the private sector. The line will be operated under a 25-year infrastructure management concession to be awarded by the government.

HSL-Oost

The HSL-Oost project is to be achieved by a major upgrading and extension of the present line from Utrecht to Arnhem and on into Germany. Options for raising line speeds were still being studied in 1999. However, for the heavily used Arnhem-Utrecht line partial quadrupling is considered unavoidable. The HSL-Oost project focuses on the line from Utrecht to the German border, but the Utrecht-Amsterdam line can also be considered part of this project. Plans for improving this section are running ahead of the main HSL-Oost scheme and are expected to receive ministerial approval in 1999.

Betuweroute

In 1994 the Dutch parliament approved construction of an east-west freight-only rail corridor, the so-called Betuwe line. Total cost is estimated at G9 billion at 1997 price levels. To distinguish the new freight-only line from the existing Dordrecht-Geldermalsen-Tiel-Elst line, the new corridor is now known as the Betuweroute. Its length will be 112 km, from Kijfhoek yard, Rotterdam, to the Dutch/German border near Emmerich, Germany. Most of the line will run parallel to the A15 motorway. At Valburg, between Arnhem and Nijmegen, a container terminal is planned. At Zevenaar, the Betuweroute will connect with the existing line from Arnhem to Germany to reach Emmerich. Construction work has started and most legal procedures had been completed by 1999. The Betuweroute is to be completed in 2005 and by 2010 is expected to carry some 43 million tonnes of freight. This projected figure compares with 24 million tonnes carried by NS Cargo in 1998.

Also part of the Betuweroute is the 48 km Havenspoorlijn (harbour line) which will act as a feeder for the port of Rotterdam. This will run west from Kijfhoek yard to the Maasvlakte. Work on the project has started and is due to be completed by 2003. A bridge at the Dintelhaven was completed in 1998 and at the beginning of 1999 work started on boring the Botlek tunnel in the port of Rotterdam. The Havenspoorlijn is to be electrified using the 25 kV AC 50 Hz power supply system. This will form a pilot project for this system, which will also be used throughout on the new sections of the Betuweroute.

Extensive measures have been necessary to protect the environment from the impact of the new line. There are five major civil engineering works between Kijfhoek and the German border: four tunnels and one bridge. The bridge will cross the Amsterdam-Rhine Canal at Tiel. The tunnels will also serve as crossings of rivers and canals. Although the operation of double-stack container services has so far not been approved, tunnels and bridges are being constructed with clearances adequate to accommodate them at a later stage.

The German and Dutch governments have reached agreement on investments in German network infrastructure to handle traffic generated by the new line. About EUR1 billion is to be spent upgrading German routes to be used by freight services destined for the Netherlands. Improvements will focus on the Emmerich-Oberhausen line and bottlenecks in the Ruhr. Despite the agreement, there were reported to be concerns among Dutch politicians about the implementation of these improvements.

Studies are being carried out into extensions of the Betuweroute northeast to the German border crossing at Bentheim and to the south from Nijmegen towards Venlo.

Line 11

Proposals are being developed for another international freight-only line, between the Netherlands and Belgium. Called 'Line 11', this would give direct access from the port of Antwerp to the Dutch network, connecting with the Vlissingen-Roosendaal line just south of Bergen op Zoom. Several options are being studied. The new line is needed because, in spite of the construction of the high-speed line between Rotterdam and Antwerp, the classic Roosendaal-Antwerp line is still projected to reach maximum capacity. There are also concerns that capacity will be exhausted on the Rotterdam-Roosendaal-Breda triangle. A decision about Line 11, or at least a solution to capacity problems between Roosendaal and Antwerp, is envisaged by 2003.

The 'Iron Rhine'

Besides Line 11, the Belgian government has been pressing for a direct, fast freight-only line between the port of Antwerp and the Ruhr, in Germany. Talks have taken place between Dutch, Belgian and German authorities for the reopening of the so-called 'Iron Rhine' from Antwerp to Mönchengladbach via the Dutch towns of Weert and Roermond. The infrastructure for this route exists but the Dutch-German border link east of Roermond has been closed, though not officially. A high-frequency service will require significant upgrading of the Dutch portion of the route and some new infrastructure may be required.

The Hanzelijn

The 'Hanzelijn' or Hanseatic line, is planned between Zwolle and Lelystad, although a definitive route has not yet been set. If built for standards suitable for 200 km/h running, journey time savings for IC services on the Zwolle-Amsterdam route could be over 20 minutes. Upgrading the existing Zwolle-Amersfoort-Amsterdam route would be very expensive, due to the high number of level crossings to be eliminated. Currently the project is scheduled for completion by 2007 at a cost of G2 billion.

Light rail

In 1997 the Minister of Transport selected six light rail projects for development, considering these

a valuable contribution to solving traffic and transport problems. Most of the plans remain the subject of studies. The six projects are:

- IGO+, which was to start in May 1999. This project aims to provide good-quality public transport based on integrating bus and rail services in a region in the east of the Netherlands. Rail routes involved are Doetinchem-Winterswijk and Winterswijk-Zutphen. Initially rail services are provided by hired NS conventional dmus but lightweight diesel railcars have been ordered. Decisions about infrastructure modifications are still to be made. (See entry for 'Syntus' later in the Netherlands section.)
- The introduction of light rail is being studied in south Limburg (Maastricht-Heerlen-Kerkrade). Tendering for the provision of regional services is envisaged. Infrastructure implications are not yet clear: some signalling and safety problems have occurred with the use of lightweight rail vehicles on the Dutch network and infrastructure modifications would be necessary. However, it is expected that the number of passengers could double as a result of the initiative.
- RandStadRail concentrates on the Rotterdam/ Den Haag/Zoetermeer triangle and has been proposed by NS, regional bus operator ZWN and the city transport authorities of Rotterdam (RET) and Den Haag (HTM) to relieve road congestion by linking main line and urban rail systems and adapting the existing Den Haag-Zoetermeer and Den Haag-Rotterdam Hofplein lines for light rail vehicles. Studies continue into a pilot project which could be operational by 2003.
- Randstadspoor Utrecht is a proposal to develop a regional public transport network in the Utrecht area and is based on the existing rail network around the city. The scheme is closely linked to urban planning in the region. In anticipation of the Randstadspoor Utrecht project, the minister has approved the construction of a new station to serve a new housing development at Houten.
- The Rijn-Gouwe light rail project covers the Gouda-Alphen and Rijn-Leiden lines. Train services are considered to be loss-making while road traffic is congested in the area. Options differ from running regional services with light rail vehicles on existing lines to creating a new route, based on the Karlsruhe model, for light rail vehicles through the centre of Leiden to the coastal communities of Katwijk aan Zee and Noordwijk. Plans are being studied.
- RegioNet is a plan for an extending and integrating public transport around Amsterdam by making comparatively small investments. The proposal calls for more studies in possible connections between different kinds of rail systems.

Improvements to existing lines

As track capacity limits on much of the Dutch rail system were reached at the end of the 1980s, the

Class 1600 electric locomotive on an IC service at Den Haag CS (Quintus Vosman)   0058798

DDAR double-deck stock propelled by an mDDM power car (Quintus Vosman)   0058790

government decided to execute a fast investment plan to solve the worst problems in the network. This investment programme was called Prorail. Major projects included: extending Utrecht CS station; quadrupling the Leiden-Den Haag-Rijswijkline; quadrupling the Schiphol line between Hoofddorp and Amsterdam; and providing better access to Amsterdam CS from the west by increasing the number of tracks from four to six. The Prorail projects have all been completed except the Schiphol quadrupling and station expansion from three to six platform tracks: this is scheduled for completion in 2001.

Much of the Prorail investment programme contained investments proposed by NS's own Rail 21 plan. However, one major difference between Rail 21 and Prorail was that NS proposed quality improvements while Prorail's aim was to raise rail traffic volumes.

A second wave of investments is now being implemented under the TTP (Second Tactical Package) programme, with a target completion date of 2005. Most TTP projects are in the Randstad and the main corridors to and from it. Most attention is paid to access to Amsterdam Schiphol airport and the port of Rotterdam.

Just one large investment was announced for a provincial line in the Netherlands: a partial doubling of the Groningen-Sauwerd line, where the line diverges to Delfzijl, Roodeschool and Eemshaven. This is the single-track line with the highest traffic density in the Netherlands.

There is one new line to be constructed under the TTP programme: the Hanzelijn, connecting Zwolle and Lelystad (see 'New Lines' section). Some of the projects have reached the construction phase, while for others plans are still being developed or have been delayed for financial reasons.

Many improvement schemes aimed at increasing track capacity have been planned or implemented, especially in the dense network in the west of the Netherlands. Capacity on the Amsterdam-Rotterdam-Dordrecht route has been raised by quadrupling the Leiden-Den Haag-Rijswijk section. Quadrupling of the Schiphol line, which serves the international airport, is in progress for completion by 2001. The underground Schiphol station is being enlarged from three to six tracks. South of Schiphol, at Hoofddorp, the layout of the station has been expanded from two to four tracks and a flyover has been constructed to avoid conflicting movements. The line between Dordrecht and Rotterdam has been quadrupled, except one small section at Barendrecht. A flyover at Rotterdam Lombardijen has been built to ease freight movements, and preparations have been made to connect the HSL-Zuid high-speed line and the new Havenlijn (harbour line), which will be part of the freight-only Betuweroute.

Other major infrastructure projects recently completed include: station and the junction improvements at Amersfoort; extension of the

western access to Amsterdam CS; quadrupling the Utrecht CS-Utrecht Overvecht line and construction of junction flyovers; doubling the single-track line between Heerhugowaard and Schagen and partially doubling the non-electrified Groningen-Leeuwarden line, where the maximum speed has been raised to 140 km/h.

Still in progress in 1999 were quadrupling of the Boxtel-Eindhoven section which is shared by Amsterdam-Maastricht and Rotterdam-Venlo services, with a flyover and extended station layout at Boxtel and a new layout at Arnhem station, including a flyover. The Woerden bottleneck has also been eased and recently ministerial approval was given for the quadrupling of the line between the Woerden station and the junction east of the station, where lines to Utrecht and Amsterdam diverge. Between Houten and Houten Castellum, on the Utrecht-Den Bosch line, a third track will be constructed. More capacity enhancements on this line are expected once plans for a suburban heavy rail operation in addition to existing main line traffic are approved.

Major improvements are also planned for the Amsterdam-Utrecht line: between Utrecht and Duivendrecht the line is to be quadrupled to separate fast and slow trains and to raise line speeds and a bilevel junction is to be constructed at Breukelen. This project was expected to be approved by the Minister of Transport during 1999.

Upgrading of the non-electrified Zwolle-Wierden line for 140 km/h running has been completed.

In 1999 NS Reizigers introduced a DM 90-operated fast service on this route with an extension to Enschede.

While most investment has been focused on the western Netherlands, the Minister of Transport has approved the partial doubling of the non-electrified Groningen-Sauwerd line.

At some locations new links are to be provided, making new services possible. Government approval has been given for the Gooiboog, a link by which trains can run directly from Utrecht to Almere via Hilversum. A new link near Amsterdam Sloterdijk is being studied so that trains can run directly from Zaandam to Schiphol airport. There are also plans to extend track capacity at some stations. Other investments at stations concern improved parking provision for cycles.

An investment programme of G300–350 million is being implemented to raise freight train axleloads to 22.5 tonnes. The programme is split up into three parts, the first of which started in 1997. The second was due to start during 1999 and the third in 2001.

Investments have also been made necessary to ensure environmental protection, and programmes to reduce noise on running lines and in yards are being developed. Some yards have posed particular problems: for example, the Amsterdam Watergraafmeer yard suffers from a permanent legal threat of closure due to violations of noise control legislation.

In the longer term other improvements to the existing rail network are being studied. Two of these are aimed at raising capacity without adding to existing infrastructure or constructing new lines. One possibility explored is overhead power supply conversion from the present 1.5 kV DC to 25 kV AC 50 Hz. So far it has been decided only to introduce 25 kV on two new lines: the HSL-Zuid southern high-speed line and the Betuwe line. Also being studied is the introduction of new train control and capacity management system, BB 21. This would be a moving block system based on Level 3 of the ERTMS (European Rail Traffic Management System). No decision to introduce either system has yet been made.

Traction and rolling stock
During the 1980s and 1990s NS replaced a major part of its first generation of post-war traction and rolling stock. The Class 1200 electric locomotives were withdrawn in 1998, apart from six bought by ACTS (qv), and remaining Class 1100 units were expected to be withdrawn during 1999. Older types of emus and dmus have also been withdrawn. The Mat' 54 two-car and four-car emus were withdrawn in 1997 and the Plan X two-car dmus were retired in 1998, finally replaced by DM 90 dmus on the international link between Heerlen and Aachen, Germany.

DD-IRM emu on an Amsterdam-Utrecht service (Quintus Vosman)   0058796

Diesel locomotives

Class	Wheel arrangement	Power kW	Speed km/h	Weight tonnes	No in service	First built	Builders Mechanical	Engine	Transmission
200/300	Bo	64	60	21	82	1934	Schneider	Stork	E Heemaf/ETI
600	C	300	30	41	45	1949	EE	EEC 6KT	E EEC
2200	Bo-Bo	650	100	74	8	1955	Schneider	Stork/Schneider	E Westinghouse
6400	Bo-Bo	1,180	120	80	120	1988	MaK	MaK	E ABB

Diesel railcars or multiple-units

Class	Cars per unit	Motor cars per unit	Motored axles/car	Power/motor kW	Speed km/h	No in service	First built	Builders Mechanical	Engine	Transmission
DE-II	2	2	2	193	110	3	1953	Allan	Cummins	E Smit
DE-III	3	1	4	182	130	41	1960	Werkspoor	SACM	E Smit
DH-I	1	1	2	212	100	19	1983	Uerdingen	Cummins	H Voith
DH-II	2	2	2	212	110	30	1981	Uerdingen	Cummins	H Voith
DM 90	2	2	2	320	140	53	1995	Duewag	Cummins	H Holec

Electric locomotives

Class	Wheel arrangement	Output kW	Speed km/h	Weight tonnes	No in service	First built	Builders Mechanical	Electrical
1300	Co-Co	2,885	135	111	15	1952	Alsthom	Alsthom
1200	Co-Co	2,235	130	108	9	1951	Werkspoor-Baldwin	Heemaf-Westinghouse
1100	Bo-Bo	1,925	135	83	13	1950	Alsthom	Alsthom
1600	B-B	4,540	160	83	58	1981	Alsthom	Jeumont-Schneider
1700	B-B	4,540	160	86	81	1991	GEC Alsthom	Schneider

Electric railcars or multiple-units

Class	Cars per unit	Motor cars per unit	Motored axles/car	Output/motor kW	Speed km/h	No in service	First built	Builders Mechanical	Electrical
mP	1	1	4	145	140	7	1965	Werkspoor	Smit
64-II	2	2	2	246	140	242	1964	Werkspoor	Heemaf/Smit
64-IV	4	2	4	246	140	31	1964	Werkspoor	Heemaf/Smit
SGM-II	2	2	4	330	125	30	1975	Talbot/SIG	Oerlikon/Holec
SGM-III	3	2	4	330	125	60	1980	Talbot/SIG	Oerlikon/Holec
ICM-III	3	1	4	312	160	94	1977	Talbot/Wegmann	Heemaf/Smit
ICM-IV	4	2	4	312	160	50	1991	Talbot/Wegmann	TCO/Holec
SM 90	2	2	4	300	160	9	1993	Talbot	Holec
DD-IRM-III	3	2	2	200	160	34	1994	Talbot	Holec
DD-IRM-IV	4	2	2	200	160	47	1994	Talbot/De Dietrich	Holec

Replacement of ageing stock has been made possible by recent deliveries of new equipment, most significantly 290 Class DD-IRM inter-regional emu cars, 53 Class DM 90 two-car dmus and 50 mDDM power cars to operate three- and four-car DD-AR push-pull sets. With the arrival of the last of the mDDM vehicles at the end of 1998, there were no outstanding items of traction or rolling stock on order by any of the NS train operating businesses.

The rapid delivery of the mDDM motorcoaches freed about half of the Class 1700 electric locomotives, which had been used exclusively with double-deck push-pull sets. The displaced locomotives have had their automatic couplers removed and are now used with conventional hauled stock. Modifications to their braking system are required before they can haul freight trains.

In 1997 DM 90 'Buffalo' dmus entered service on the Arnhem-Winterswijk and Nijmegen-Venlo-Roermond routes. Due to the relatively light weight of these units, it has been necessary to install axle-counter equipment on the lines served by them, as well as the latest ATB-NG (new generation) automatic train protection system. DM 90s are also used on the Leeuwarden-Groningen, Zwolle-Enschede and Heerlen-Aachen (Germany) routes, each of which have been provided with axle-counting equipment. For the last-named service three sets have been equipped with the German Indusi train protection system. It took some two years for approval to operate the DM 90 on the German network.

Due to the rapid growth in its passenger numbers, NS Reizigers decided in 1998 to refurbish the four-car Plan T (Mat' 64) emus. Originally these 31 units were to be withdrawn from 1999. A new interior is being provided and the luggage compartment has been converted into a second class non-smoking saloon, adding 31 seats per set. In all, around 1,000 seats will have been added to the NS Reizigers fleet and the trains are to run for a further eight years.

In 1999 NS was waiting to assess political attitudes towards the future of rail transport. The Minister of Transport's 'Third Age of Rail' policy statement, published in 1998, indicated a key role for NS Reizigers, which promptly announced its intention to order 128 coaches to lengthen its DD-IRM emus. Four-car units are to be extended to six cars and three-car sets to four. For the domestic IC services on the HSL-Zuid line, NS Reizigers plans to order high-speed double-deck trains with capability at least of 220 km/h and possibly higher.

Due to continuing growth in passenger volumes, NS Reizigers has been forced to take short-term steps to avoid a lack of seating capacity. Several of the oldest double-deck coaches have been modified for normal locomotive-hauled services, in combination with ICR coaches. In 1998 NS decided to return several Plan W coaches to service; they had been withdrawn in 1996. A further 25 of these vehicles were to be recommissioned in 1999. In addition, NS and Belgian National Railways (SNCB) have reached agreement to hire 80 Type K4 coaches, which had been sold to SNCB by French National Railways. They were due to start running with existing locomotive-hauled stock on Den Haag-Heerlen and Den Haag-Venlo IC services in late 1999. This would add a further 6,000 seats each day. The hire contract was to last for two years with an option for a further two years. In addition, a mid-life overhaul of the ICR fleet – including those used on 'Benelux' services – has been started.

In January 1999 the NS rolling stock fleet was split up, mainly between NS Reizigers and NS Cargo. All emus, dmus and coaches have been assigned to NS Reizigers, together with 81 Class 1700 and 20 Class 1600 (1638–1658) locomotives. During 1999 the latter were to be renumbered into Class 1800. All remaining Class 1100 and Class 1300 locomotives are now owned by NS Cargo.

Most of the diesel locomotive fleet was assigned to NS Cargo in the form of 120 Class 6400s and six Class 2200s. NS Reizigers has become the owner of some Class 600 diesel shunters, while other shunting locomotives have been acquired by other NS Holding business units, such as NS Materieel and Railpro. Others considered to be a surplus were sold to third parties like ACTS.

High-speed trainsets

NS owns two TGV trainsets for Amsterdam-Paris services. At the end of 1999 delivery from Siemens of the first of four ICE-M six-car four-voltage high-speed trainsets for Amsterdam-Cologne-Frankfurt services was due to take place. During 1998 and 1999 the type has been tested in Germany and was expected for trials on the Dutch network during the second half of 1999. Services using the new trains are due to begin in 2000. NS also had an option on another two ICE-M sets for services between Amsterdam and Berlin but by mid-1999 this had been converted into a firm order. Until introduction of the ICE-M sets on the international services between the Netherlands and Germany, DB IC coaches will be used for Amsterdam-Cologne trains, while Swiss stock will be employed on direct services between Amsterdam and Switzerland until ICE-M sets take over these services in 2002.

Signalling and telecommunications

NS is participating in European studies into the possibilities for capacity expansion by introducing new control and safety systems. Part of the research is concentrated on a system for the improved use of infrastructure capacity by means of automatic regulators and recommended speeds.

In September 1994, NS opened a new Siemens SIMIS type microcomputer interlocking signalling centre in Rotterdam, the largest of its type in Europe, to replace the existing relay-based centre. In 1995 a new traffic control centre was opened near Amsterdam Centraal station. Further centres are under construction in Amersfoort and Arnhem.

Electrification

A total of 75 per cent of the NS system is electrified at 1.5 kV DC, including all main routes. The only extension to this in progress at present is the electrification of the Rotterdam dock line. After previously rejecting the idea, NS decided in 1994 to reconsider the gradual conversion of its existing electrification from 1.5 kV DC to 25 kV 50 Hz AC. This merits consideration in the context of connections with the emergent continental high-speed network and because intensification of passenger traffic will demand traction equipment with enhanced acceleration and braking characteristics. The first applications of 25 kV 50 Hz AC will be on the Amsterdam–Belgium and Amsterdam–Germany high-speed lines and the Betuwe freight route. Apart from new traction orders, it is possible that Class 1600 and 1700 locomotives will be converted to dual-voltage operation. The decision on conversion to 25 kV 50 Hz AC will be made by the Dutch government.

Track

Standard rail, weights
Main lines: UIC 54 kg/m (3,025 m installed)
Branch lines: 46 kg/m (3,436 m installed)
Crossties (sleepers): Wood, 250 × 150 × 2,600 mm; twin-block concrete 230 × 300 × 2,250 mm; monobloc concrete 230 × 300 × 2,500 mm
Spacing: 1,667/km; 1,333/km
Fastenings: Wood, bolt or DE-clip; twin-block concrete, DE-clip; monobloc concrete, Vossloh clip
Min curvature radius: 300 m; 500 m on track with monobloc sleepers
Max axleload: 22.5 tonnes (20 tonnes on twin-block concrete-sleeper track)

NoordNed Personenvervoer BV

PO Box 452, NL-8901 BG Leeuwarden, Netherlands
Tel: (+31 58) 233 56 38   Fax: (+31 58) 233 56 36
Web: http://www.noordned_ov.nl

Key personnel
Director: H Donker

Political background
In 1999, NoordNed took over regional train services on two lines in the province of Friesland, in the far north of the Netherlands, after a tendering procedure. The two lines, Leeuwarden–Stavoren (51 km) and Leeuwarden–Harlingen Haven (26 km), both of which are not electrified, were among 33 routes on which NS Reizigers had indicated no continuing interest. Subsequently NS Reizigers transferred to NoordNed services between Leeuwarden and Groningen (54 km).

In May 2000, NoordNed extended its operations to cover all non-electrified routes in the north of the Netherlands. This followed a second tendering procedure covering the non-electrified Groningen–Sauwerd–Roodeschool, Groningen–Sauwerd–Delfzijl and Groningen–Nieuweschans regional lines in the province of Groningen. The last of these is partly double-track, the others single-track. Arriva was already active in the province of Groningen, operating both regional and city bus services.

The Groningen Noord–Sauwerd section, which is shared by trains to Delfzijl and Nieuweschans as well as by freight traffic to Eemshaven, is being partially doubled with government funding to increase capacity, the work including provision of a double-track bridge across the Van Starkenborch Canal. Completion is for 2003.

Still unclear in late 2000 was the situation regarding international service between Groningen and Leer in Germany, which crosses the border at Nieuweschans. Initially, NoordNed was also to take over this service, but German local authorities pulled out of the negotiations and NS Reizigers insisted that it would no longer operate the service, terminating the hire of the German Rail (DB AG) Class 624 diesel trainsets used for it. NoordNed has no rolling stock approved for operation on the German network. Meanwhile, upgrading of the link between the two communities for freight traffic

NoordNet Class DH-2 dmu at Franeker, on the Leeuwarden–Harlingen line (Quintus Vosman)
0103633

was under way for completion in 2000, funded by the Dutch and German governments and the European Union.

Organisation
NoordNed shares are held by NS Reizigers, British bus operator Arriva and banking interests. None of the parties has a share of more than 49 per cent. As a condition of tendering procedures, the Netherlands Minister of Transport demands that no party is allowed to have over 50 per cent ownership of an operating company.

Passenger operations
NoordNed is an integrated regional passenger transport operator, providing both train and bus services in Friesland. There is one train per hour between Leeuwarden and Stavoren, the frequency doubled to two trains per hour on the Leeuwarden–Sneek section. Three services per hour in each direction are provided between Leeuwarden and

Groningen following infrastructure upgrading and a partial track-doubling completed in 1999.

Traction and rolling stock
NoordNed has leased DM 90 and DH 'Wadloper' two-car and single-unit dmus from NS Financial Services, a newly created NS business unit based in Dublin, Ireland. The DM 90s, the newest type of dmu in the Netherlands, have undergone modifications to their control equipment to allow them to operate with the 100 km/h DH trainsets. The Leeuwarden–Groningen service is operated with DM 90s running at 140 km/h.

For the contract secured in May 2000, NoordNet was expected to continue to use the stock operated by NS Reizigers, the remaining Class DH single-unit railcars and Class 3100 and 3200 two-car dmus, again leased from NS Financial Services. It was also likely that some ageing Class Plan U diesel-electric three-car dmus from the early 1960s would be added to the fleet.

Short Lines BV

Zaltbommelstraat 10, NL-2089 JK, Rotterdam
Tel: (+31 10) 428 31 20   Fax: (+31 10) 429 49 84
Web: http://www.shortlines.nl

Key personnel
Directors: R Spierings, J Herijgers

Organisation
Short Lines was founded in 1997 to run open access freight services, the first operator to compete on the Dutch network with NS Cargo. The company has concluded an agreement with HGK, the Cologne port operator, for the provision of traction for intermodal services both between Rotterdam and Germany and within the Netherlands. HGK subsequently acquired a 25.1 per cent shareholding in Short Lines.

Freight operations
In mid-1998 a daily intermodal service was launched between Rotterdam and Born, some 20 km north of Maastricht, on behalf of a terminal operator at the latter location. In March 1999 a Rotterdam-Cologne intermodal service for Transfracht was started jointly by Short Lines and HGK. This runs six days a week and also conveys wagons for Munich and Austria. Trains run through without a locomotive change, although HGK and Short Line crews change at the border.

In 2000, after competitive tendering, Short Lines took over from Railion the operation of a major daily flow for Dutch chemicals group DSM. This train runs between the port of Rotterdam and a major chemical plant at Geleen, some 15 km north of Maastricht. DSM ranks among Railion's

five biggest customer. Also in 2000, Short Lines ordered two Cargo Sprinter freight dmus for just-in-time services between the Eindhoven and Rotterdam areas. The company was understood to be considering possible international services with such equipment.

Traction and rolling stock
Short Lines has no fleet of its own at present. Trains are hauled by three heavy MaK DE-1024 diesel locomotives, prototype machines which have been proven by German Rail (DB) as Class 240. These are the most powerful diesels to run on the

Dutch network and have been equipped with three automatic train protection systems: Indusi (DB), an ATP-system for running on the HGK docks network and ATB, the Dutch ATP system. MaK DE-1002 locomotives are also now approved for access on the Dutch network, subject to the provision of ATP equipment, and it was expected that some HGK machines of this type would be adapted for this system, possibly to work domestic services.

The Cargo Sprinters mentioned above were due to be delivered at the beginning of 2001.

In 2000 Short Lines was using some 140 container wagons.

HGK-owned MaK Type DE-1002 diesel locomotive operating a Short Lines container service between Rotterdam and Born (Quintus Vosman)
0103634

Syntus

PO Box 17, NL-7000 AA Doetinchem
Tel: (+31 314) 35 01 60
Fax: (+31 314) 33 26 51
Web: http://www.syntus.nl

Key personnel

Managing Director: Ruth Prins-Maatman

Organisation

Syntus was founded in December 1998. In May 1999 it took over regional passenger services on the Winterswijk–Doetinchem and Winterswijk–Zutphen lines from NS Reizigers. Shares in Syntus are divided equally between three companies: NS Reizigers, VSN-1 and Cariane, a French regional bus operator. As well as the rail routes mentioned above, the company operates the regional bus network, which is closely integrated with its train services. The project is seen as a first step towards creating a light rail-based regional integrated transport system.

With the approval of the Netherlands Minister of Transport, Syntus and NS Reizigers reached agreement that Syntus would take over the operation services on the Arnhem–Doetinchem line from 10 June 2001. Through services between Arnhem and Winterswijk have also been restored. When Syntus commenced operations, this service was split into two portions, with NS operating between Arnhem and Doetinchem and Syntus between Doetinchem and Winterswijk.

Syntus Alstom-built Coradia LINT 41 diesel railcar at Arnhem (Quintus Vosman)   0536251

Passenger operations

Syntus has doubled all frequencies during the day, offering throughout its network a bus or train every 30 minutes rather than at best hourly, as before. Remarkably, subsidies have remained unchanged. Efficiencies achieved include flexible staff deployment, with personnel equally likely to drive a train or bus or handle revenue protection tasks.

Traction and rolling stock

In April 1999 Syntus ordered 11 two-car Coradia LINT 41 lightweight diesel railcars from Alstom.

These were delivered during the first half of 2001. Subsequently, Syntus ordered nine additional Coradia LINT 41 railcars, delivery of which was completed in October 2002.

For through Arnhem–Winterswijk services, Syntus leases four DM 90 dmus from NS.

New Zealand

Ministry of Transport

PO Box 3175, Wellington, New Zealand
Tel: (+64 4) 472 12 53   Fax: (+64 4) 473 36 97
e-mail: info@transport.govt.nz
Web: http://www.transport.govt.nz

Key personnel

Minister: Pete Hodgson
Secretary for Transport: Robin Dunlop
Deputy Secretary: Roger Toleman

VERIFIED

New Zealand Railways Corporation (Ontrack)

Level 4, Wellington Railway Station, Bunny Street, Wellington
PO Box 593 Wellington
Tel: (+64 4) 495 30 00   Fax: (+64 4) 495 90 45
Web: www.nzrailcorp.nz

Key personnel

Chairman: Cameron Moore
Chief Executive: David George

Gauge: 1,067 mm
Route length: 4,128 km
Electrification: 411 km at 25 kV AC 50 Hz; 95 km at 1.5 kV DC

Political background

As part of the sale process of TranzRail to the Toll group, the government purchased back the track and infrastructure of the national rail network as at 30 June 2004. Responsibility for it, track maintenance, train control and track access and charges have been vested in this new body. The government paid a nominal NZ$1 plus an estimated NZ$50 million to compensate the Toll group for subleases and structures on the land. The Toll group was granted

exclusive access rights until 2070 for freight and the Wellington Metro services. The rights are subject to a "use it or lose it" provision if Toll Rail's traffic should fall below 70 per cent of the average 2003–04 levels on any line segment. Should this lead to another operator establishing services, Toll Rail is required to cooperate with any new operator granted access. Toll Rail may nominate one director to New Zealand Railways Corporation.

The Crown also made a commitment to spend NZ$100 million on upgrading the network and NZ$100 million on asset replacement, for example of line and sleepers, which will not be recovered from Toll Rail in access charges. Any amount over NZ$25 million spent on the South Island Midland line will be funded through track access charges which would doubtless be passed on to the major customer, the government-owned Solid Energy coal operation.

It is estimated that about 14 per cent of all domestic freight is carried by rail. It is a government policy that this should be increased by transfer from road. While the network is regarded as generally fit for purpose, there is a need to address the serious backlog of deferred maintenance and to provide for future improvements. These include upgrading and strengthening track and bridges for heavier axle loads and faster speeds; the lowering of tunnel floors to allow 9 ft 6 in containers to be carried and to improve signalling and increase the length of some crossing loops. In 2002–03 net traffic was 14.8 million tonnes or 3,853 million tonne-km.

Organisation

In addition to its headquarters and Central Regional Office in Wellington, Ontrack has its Northern Regional Office in Auckland and a Southern Regional Office in Christchurch.

In 2005 Ontrack employed 124 staff.

New lines

Ontrack, the Ministry of Transport and Toll Rail are to investigate the potential for building short

branches to the port at Marsden Point, near Whangerai, and to dairy factories at Edendale and Clandeboye.

Improvements to existing lines

A high priority has been the provision of a new bridge on the Rapahoe branch near Greymouth to ensure continuation of coal services.

A second track of 2.2 km has been provided on the Western line in Auckland between Mount Eden and Morningside to facilitate the running of additional peak hour services between New Lynn and Britomart. Planning is to be funded for duplication between New Lynn and Henderson on the same route.

On the South Island coal route, the Midland Line, the first crossing loop to be extended for 32-wagon trains has been completed at Mawherati.

Signalling and telecommunications

Most trains operate under either CTC (Centralised Traffic Control) or TWC (Track Warrant Control). Tranz Rail has centralised its train control operations in Wellington.

Electrification

The North Island Main Trunk line between Palmerston North and Hamilton has been electrified at 25 kV 50 Hz AC since 1988.

The 95 km Wellington suburban passenger system is electrified at 1.5 kV DC.

Track

During the period of previous ownership of the rail network it is acknowledged that there was underinvestment in maintenance. This is being addressed as part of the Crown takeover. It has been estimated that there is a national backlog of 560,000 sleepers needing replacement.

In a mountainous country like New Zealand, it is not surprising that the railway includes 150 tunnels (87 km) and 2,178 bridges (74 km). The three longest

---

For details of the latest updates to *Jane's World Railways* online and to discover the additional information available exclusively to online subscribers please visit

jwr.janes.com

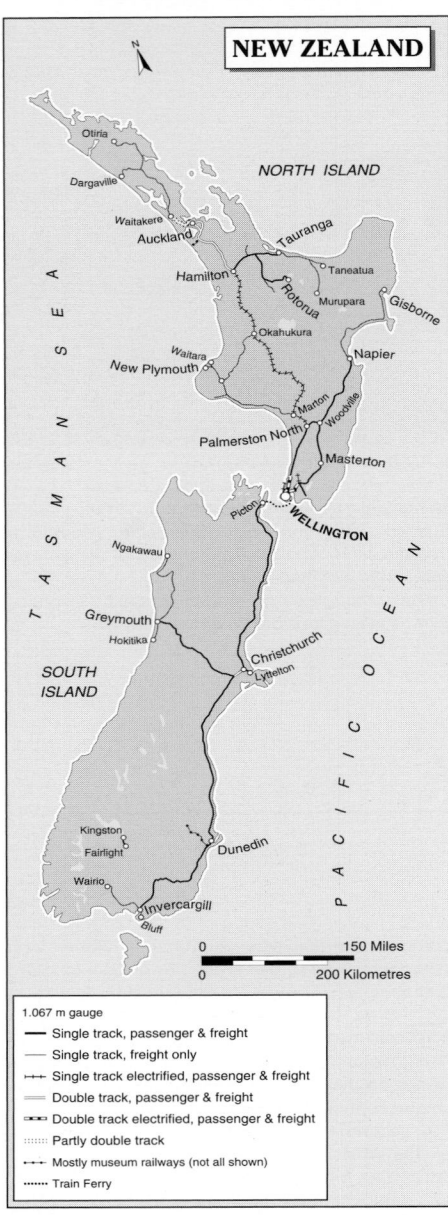

NEW ZEALAND

1.067 m gauge
— Single track, passenger & freight
— Single track, freight only
⊢⊣⊣ Single track electrified, passenger & freight
═══ Double track, passenger & freight
ᴄᴏᴄ Double track electrified, passenger & freight
····· Partly double track
⊢·⊢· Mostly museum railways (not all shown)
······ Train Ferry

NEW/0023773

tunnels are Kaimai (8.9 km) between Tauranga and Morrinsville; Rimutaka (8.5 km) between Upper Hutt and Featherston; and Otira (8.6 km) between Otira and Arthur's Pass. Tunnel clearances are being increased to cater for continuing increases in the height of 'standard' containers but some tunnels still currently restrict routes on which containers can be hauled. The longest bridge is over the Rakaia River (Christchurch-Picton) spanning 1,743 m, while the highest is over the Mohaka River (Hawkes Bay) standing 97 m over water height.

An EM80 track evaluation car regularly checks track standards.

Rail
Main line: 50 kg/m; 91 lb/yd; 85 lb/yd
Provincial lines: 91 lb/yd; 85 lb/yd; 75 lb/yd; 70 lb/yd
Branch lines: 70 lb/yd; 55 lb/yd
Welding method: Flash-butt in depot, Thermit in field. New rails flash-butt welded in depots into 76.8 m lengths and transported to site for laying. Short rail in track may be Thermit-welded into similar lengths. Continuous welded rail is formed by Thermit process with termination joints at extremities and epoxy glue joints, or Benkler encapsulated joints, and anchored. Over half of the track (2,050 km or 70 per cent of main and provincial lines) comprises continuous welded rail.

Crossties (sleepers)
NZ Pinus radiata (all lines) 67 per cent
Concrete (main lines only) 19 per cent
Hardwood remainder

Spacing
Main line: Timber 600 mm; concrete 700 mm

Fastenings
Main lines: Timber: Pandrol spring fastenings on bedplates; clips, crew spikes, spring washers on double-shoed bedplates. Spring clips and screw spikes without bedplates
Concrete: Pandrol spring fastenings with rubber or plastic pads and nylon insulators
Branch lines: Timber: screw spikes and bedplates and cascaded from higher ranking lines
Laying method: Concrete: by Tranz Rail designed and built sleeper-laying machine or by spot resleepering machinery
Timber: Laid manually either in face or by spot resleepering machinery

Dimensions
Concrete: 254 × 190 × 2,134 mm
Timber: 200 × 150 × 2,134 mm
Max gradient: South Island – 1 in 33 (in Otira Tunnel); North Island – 1 in 35 (Westmere Bank, New Plymouth line)
Max altitude: North Island – 814 m at Waiouru, 290 km north of Wellington on the North Island Main Trunk; South Island – 737 m at Arthur's Pass on the Midland line.
Max permitted speed: 100 km/h – passenger trains; 80 km/h – freight trains

UPDATED

Connex Auckland Ltd

Address
PO Box 105-355, Auckland
Level 7, Citibank Centre, 23 Customs Street, Auckland
Tel: (+64 9) 969 77 77   Fax: (+64 9) 969 77 00
e-mail: info@connexauckland.co.nz
Web: www.connexauckland.co.nz

Key personnel
General Manager: Chris White
Marketing and Communications Manager:
  Penny Hartill

Organisation
A subsidiary of Connex Group Australia, Connex Auckland Ltd was established to run suburban passenger trains in Auckland under a four-year contract which took effect in August 2004. The agreement is with Auckland Regional Council (ARC) and includes three one-year renewal options to reach a possible maximum of seven years. Services were formerly provided by Tranz Metro, a subsidiary of Tranz Rail (now Toll Rail). A new brand name, MaXX, has been adopted.

Services are operated over the tracks of the national infrastructure owner, New Zealand Railways Corporation, with the freight operator Toll Rail retaining freight access.

Passenger operations
Connex Auckland operates suburban services over a three-line network totalling 79 route-km. Around 1.1 million passengers are carried annually.

Former Perth diesel railcar, refurbished for use on Auckland area suburban services, at Newmarket (Brian Webber) 0573223

Diesel railcars

Class	Power kW	Weight tonnes	No in service	First built	Mechanical	Builders Engine	Transmission
ADK/ADB	276	33/16	9/9	1969	Comeng/WAGR	Cummins	*HM*
ADL/ADC	424	43/36	10/10	1981	Goninan	Cummins	*H*

Connex Auckland MaXX formation of SA coaches, generator car and two Class DC diesel locomotives stabled at Papakura (Brian Webber) **NEW**/0585037

The system received a boost in July 2003 with the commissioning of an extension from Auckland's main station to serve an underground five-platform terminus at Britomart, providing better access to the central business district. Further benefits will result from a planned NZ$100 million programme funded by ARC to upgrade 36 stations. The provision of a second track on a 2.2 km section of the Western line between Mount Eden and Morningside has allowed the provision of additional services from/to New Lynn. Capacity enhancements will also be achieved by further doubling of sections of the Western line.

Traction and rolling stock

Connex Auckland inherited a fleet of 38 dmu cars, including 10 two-car Goninan-built units displaced by electrification of the network in Perth, Western Australia. From 2004 these were being supplemented by two SX type sets of carriages originally from Brisbane, Australia and three Type SA push-pull sets formed of modernised and re-gauged ex-British Rail Mark 2 coaches. The SX trains are usually hauled by DBR class locomotives while the SA cars are pulled/pushed by refurbished General Motors Class DC diesel-electric locomotives. Each three-car SA set of coaching stock is supplemented by an SD Class driving trailer that also incorporates a generator for onboard services. Connex has leased two DBR Class and eight DC Class locomotives from Toll Rail to power these formations. Introduction of these new trains, to be completed by 2006, is part of an ARC-funded NZ$61 million rolling stock upgrading package that also covers improvements to the dmu fleet.

UPDATED

Toll Rail Ltd

Smales Farm, Northcote Road and Taharoto Drive, Takapuna, Auckland
Tel: (+64 4) 498 30 00
Fax: (+64 4) 498 32 59
e-mail: info@tollnz.co.nz
Web: www.tollrail.co.nz

Key personnel

Chief Executive Officer: Gary Taylor

Subsidiaries

Tranz Metro
Level 1, Wellington Railway Station, Bunny Street, Wellington
Tel: (Wellington) (+64 4) 489 30 00;
   (Auckland) (+64 9) 270 51 40
Web: /www.tranzmetro.co.nz

Hillside Engineering Group
PO Box 2146, Dunedin
Tel: (+64 3) 479 34 70   Fax: (+64 3) 479 32 09
e-mail: sales@hillsidenz.com
Web: www.hillsidenz.com
Business Manager: Kevin Kearney

Political background

For more than a century, railways in New Zealand were run as a government department. In 1982, New Zealand Railways Corporation was established as a statutory corporation with a commercial mandate. In 1986, it became a state-owned enterprise and in October 1990 the New Zealand government established New Zealand Rail Ltd as a limited liability company.

The deregulation of the transport industry and major restructuring were the hallmarks of these years. The restructuring culminated in the sale in 1993 of New Zealand Rail to a consortium comprising local banking interests, with rail expertise coming from the US railroad Wisconsin Central Transportation Corporation. The purchase included the tracks but not the land occupied by the railway, which remains in government ownership, the company maintaining a lease from the Crown to occupy land for its railway operations until 2030. In 1995 the company changed its name to Tranz Rail Ltd.

In October 2003 the Australian transport company, Toll Holdings Ltd, acquired the majority of shares via its New Zealand subsidiary Toll NZ for NZ$186 million and held 84 per cent by the end of the year. Another 8.3 per cent is held by a New York, US, fund manager. Even before the takeover bid, the New Zealand government had agreed to repurchase for a nominal sum the track infrastructure, to be operated and maintained through a government agency established in July 2004. Train operators will pay access fees.

Tranz Rail sold its 24 per cent holding in Australian Transport Network, the consortium that has taken over the railway system on the island of Tasmania and has a standard-gauge operation in Australia, to Pacific National Pty in February 2004.

In May 2004 Tranz Rail was renamed Toll Rail. Responsibility for freight forwarding and retailing freight services passed to another Toll NZ subsidiary, Toll Tranz Link.

Organisation

Toll Rail operates a nationwide rail network and is New Zealand's largest freight carrier. Following its 2004 acquisition of a majority shareholding in Tranz Scenic, Toll is also the operator of New Zealand's long-distance passenger rail business. In the same year the company's former Auckland commuter passenger operations passed to Connex but Toll retained responsibility for the Wellington Tranz Metro business, indicating that it was keen to remain in this sector. The Toll Rail group also includes Hillside Engineering Group, based in Dunedin, which undertakes general railway engineering, refurbishment and the supply of spares and consumables.

In 2003 the former Tranz Rail had 2,960 employees, compared with 3,757 the pervious year.

Finance

For the six months to December 2004, revenue was NZ$305 million. Margins were stated to show a solid underlying improvement. The share price has risen from about NZ$0.40 in 2003 to NZ$3.32 by March 2005.

Passenger operations

Long-distance services
In December 2001 Tranz Rail sold 50 per cent of its long-distance passenger operations to Tranz Scenic

A Class DFT locomotive leaves Otiria for Whangerei with a train of logs (Brian Webber) 0573221

DC Class locomotives in Toll Rail colours head a northbound freight through Bunnythorpe Loop (Brian Webber) **NEW**/1115043

2001 Ltd as part of its restructuring programme. However, in 2004 the new management re-purchased this stake following a change of policy regarding passenger operations. Services operated are: the Auckland–Wellington `Overlander'; the Picton–Christchurch `TranzCoastal'; the Christchurch–Greymouth `TranzAlpine', which runs over the Southern Alps; and the Palmerston North–Wellington `Capital Connection' service. The overnight 'Northerner' sitting car services between Auckland and Wellington were withdrawn in November 2004. The most popular service is the TranzAlpine with a record 200,000 passengers in 2004–05. Almost 700 passengers travelled on the busiest day of the summer.

Urban commuter services

Tranz Metro operates urban commuter services in Wellington (electric multiple-units). Locomotive-hauled services are also operated in the Wellington district, with the 'Wairarapa Connection' services linking Masterton (91 km) with the capital. Some 800,000 passengers journeys are made monthly. On weekday mornings about 10,000 passengers arrive at Wellington terminus on trains that arrive and depart every two or three minutes. Responsibility for operating the Auckland commuter services passed to Connex in July 2004.

Freight traffic	2000–01	2001–02	2002–03
Freight tonnes (million)	14.4	14.3	14.8

Freight operations

Since privatisation in 1993 Tranz Rail and now Toll Rail has continued to move record amounts of freight. Each week about 530 trains run on the North Island and 220 on the South Island. While tonnages continue to rise, much of this increase is carried over shorter distances. Major traffics are: agricultural and food products (35 per cent of total tonnage); forest products (17 per cent); manufactured goods (18 per cent); and coal (6 per cent). The Auckland–Wellington (North Island Main Trunk) section carries about 36 per cent of the total volume.

The company has also expanded into bulk milk transport with four consists running twice daily seven days each week at the peak of the season between Oringi, near Woodville, on the Napier branch, and Hawera, on the Taranaki branch. The customer is one of New Zealand's largest dairy companies. Finished dairy products from the same customer are also hauled by train to the port of New Plymouth.

A 10.4 km section of closed line has been reinstated between Morrinsville and Waitoa for haulage of dairy products.

Toll Rail's Tranz Link sister company provides freight transport by rail, road (emphasised by the development of a network of more than 300 owner/drivers) or sea. It also offers warehousing, distribution and freight management services, including a freight forwarding division based in Australia.

Toll Rail has pioneered the use of remotely controlled shunting locomotives. Forty locomotives of three classes have been equipped with the system, which enables safer and more precise shunting to take place. The system has been extensively tested and is fail-safe, as the locomotive stops if it does not receive a continuous signal from the shunter's equipment. In the future the equipment may be fitted to certain main line locomotives to enable a single-crewed locomotive to be used for wayside shunting.

Tranz Link has an Amicus 11 computer system connected to all freight terminals to monitor consignments. This has improved revenue collection and billing accuracy and has also allowed paperless documentation. The company has also introduced Ontrac, a state-of-the-art freight tracking system, using barcode technology applied to individual freight items rather than tracking the paper trail that goes with the freight, as with conventional systems.

An extension of Ontrac, Ontrac Direct, allows customers to track the progress of their freight in close to real time. Using the Internet, customers can follow their own consignments.

An EF Class electric locomotive together with a dead sister machine head a northbound freight towards Waioru on the North Island Main Trunk line (Brian Webber) **NEW**/1115044

Diesel locomotives

Class	Wheel arrangement	Power kW	Weight tonnes	No in service	First built	Mechanical	Builders Engine	Transmission
DBR	A1A-A1A	709	68	4	1980	GM	8-645C	E
DC	A1A-A1A	1,100	82.75	41	1978	GM	12-645E	E
DFT	Co-Co	1,845	87.6	26	1992	GM	12-645E3C	E
DX[1]	Co-Co	2,050	99	42	1972	GE	7 FDL-12	E
DH	Bo-Bo	678	54	6	1978	GE	CAT D398 B	E
DSC	Bo-Bo	2 × 175	41	27	1962	AEI/NZR	2 × Cummins NT855	E
DSJ	Bo-Bo	354	54	5	1987	Toshiba KTA-1150-L	Cummins	E
DSG	Bo-Bo	2 × 354	56	24	1981	Toshiba KTA-1150-L	2 × Cummins	E
DQ[2]	Co-Co	1,063	90	4	1964	Comeng/GM	12-567C	E
TR	C (0-6-0)	110–135	20	24	1936–77	Price/Hitachi/ Drewry/Bagnall	Cummins NH/ Gardner GL	H

[1] Three uprated to 2,240 kW; one uprated to 2,425 kW; one modernised and reclassified DXR.
[2] Purchased from QR, Australia; out of service.

Electric locomotives (25 kV AC 50 Hz)

Class	Wheel arrangement	Power kW	Weight tonnes	No in service	First built	Builders Mechanical	Electrical
EF	Bo-Bo-Bo	3,000	106.5	17	1988	Brush	Brush

Electric multiple-units (1.5 kV DC)

Class	Power kW	Weight tonnes	No in service	First built	Builders Mechanical	Electrical
DM/D	450	42.4	33 cars	1936–47	EE	EE
EM/ET	400	72.1	44/44	1982	Ganz-Mávag	–

DC Class locomotive at Westfield, Auckland, with the southbound 'Overlander' service to Wellington (Brian Webber) **NEW**/1115045

Ganz-Mávag two-car emu forming a Wellington Tranz Metro service at Petone (Brian Webber)
***NEW**/1115046*

A secure access procedure protects customers' commercially sensitive information.

The Toll Rail Service Centre represents another step in the company's philosophy of providing the best possible customer service. The facility provides a single point of customer contact around the clock seven days a week.

Toll also operates three roll-on/roll-off train ferries, and a seasonal fast ferry, across Cook Strait between Wellington and Picton. The ferries now operate 24-hour sailings, with over 5,000 sailings a year. They connect the 2,500 km of North Island track with the 1,500 km of South Island track. About 80,000 wagons are carried between the islands annually. A freight train from Auckland to Christchurch is run, providing a 24-hour transit time.

A programme to upgrade the ferry fleet resulted in the building in Spain of the 'Aratere', the first new ferry bought by the company in 15 years. One of the most modern vessels in New Zealand and specifically designed for travel across Cook Strait, she has a crossing time of 3 hours and helps reduce overall transit times for freight operators.

The former Tranz Rail won initial resource consent to build a new ferry terminal at Clifford Bay (38 km south of Blenheim) to replace Picton, which is reached through a sound. The cost of building the port would be offset by savings in the rail haul. The sea trip would be reduced by 30 minutes and the land trip to Christchurch by a similar time. Land was purchased in the 1990s and there is now optimism that the project will proceed.

The coal producer Solid Energy has renewed a contract for the carriage of coal across the South Island. World over-supply of the type of coal mined has resulted in reduced tonnages. The 13-year contract provides for rail loadings to increase to 3.8 million tonnes annually from 2007–08, requiring eight trains of 30 wagons daily, one additional. Solid Energy New Zealand Ltd has conducted trials in shipment by barge, reminding Toll Rail it is not entirely dependent on the rail link. Hillside

Engineering Group have delivered of 22 new coal hoppers for Solid Energy. They are used between the west coast and Lyttelton.

A new coal haul commenced in December 2004 between the port of Mount Maunganui and Huntley, conveying imported Indonesian coal to a power station. A fleet of 33 CE wagons has been built for this traffic.

The port of Napier has been chosen to handle new 4,100-TEU vessels. Toll Rail is hauling much of the resulting container traffic to Wellington or Palmerston North. There are also two trains each week to Gisborne.

Intermodal operations
Toll Rail has sought to increase carryings of containers and swapbodies, introducing 'Spaceliner' swapbodies in a bid to compete with road transport. Another new design of swapbody, the 'Iceliner', is a refrigerated swapbody that has been specially designed for the company's temperature-controlled meat and fish transport market.

A new traffic being developed in conjunction with the road industry is the transport of domestic waste to country areas for disposal. Six wagons have been adapted to carry the containers and to enable them to be transferred to road vehicles without cranes or forklifts.

Traction and rolling stock
In 2003 the former Tranz Rail operated 131 diesel main line, 113 diesel shunting, and 23 electric locomotives, 124 electric units, 38 diesel railcars and 159 main line passenger cars. The wagon fleet of 4,321 vehicles included 2,870 container flat wagons. Some locomotives have been leased to the operator of Auckland's passenger services, Connex Auckland.

From 1995 Tranz Rail purchased 25 second-hand Clyde-GM locomotives (now designated Class DQ) from Queensland Rail in Australia. In 1998 Tranz Rail acquired further second-hand locomotives in the

form of nine Class A and one Class AB machines from Westrail, Australia. These GM-powered locomotives, which are generally similar to the DQ units, await refurbishment and modifications at Hutt workshops.

Other locomotives are being equipped with a new wheel-slip control device, ditch lights and larger fuel tanks as they pass through workshops. A feature to improve employee safety is the provision of shunter's steps behind the locomotive headstock. The final eight DF class locomotives have been fitted with turbochargers.

The former Tranz Rail purchased a locomotive driving simulator to train drivers and encourage economical train running.

In December 2001 ALSTOM Transport won a seven-year contract, effective from April 2002, to take over the maintenance of all Tranz Rail's locomotives. The contract covers both preventative and corrective maintenance and overhaul and entailed ALSTOM taking over the railway's Hutt workshops in Wellington.

Upgrading work continues on a fleet of insulated wagons for the meat and dairy sectors. A new class of CC coal wagon, with a tare of 20 tonnes and designed to carry 70 tonnes, has been introduced. Upgrading of track will be required before a 22.5 tonne axleload can be accommodated. Twenty used coal hoppers (class CW) were purchased from Australia's Westrail in 1996 for Midland line traffic.

A new class of box wagons equipped with plug-type glass fibre doors has been introduced. Some 75 of the class, ZH, have being built at Hillside Engineering Group. The wagon has two doors on each side which move to open fully their half of the wagon side, making loading by forklift simple.

In 2002 Hillside commenced delivery of 22 new coal hoppers for Solid Energy New Zealand Ltd for traffic between the west coast and Lyttelton. Export coal traffic totalled 1.7 million tonnes in 2001 and Tranz Rail expects to double that figure by 2006.

Overhaul of Wellington's Ganz-Mávag multiple-units has been completed. The 44 two-car units date from 1982 and work most services. Other services are run by elderly English Electric units last overhauled in the 1980s. Greater Wellington Regional Council, Toll Rail and the government are to spend NZ$5.4 million to upgrade 36 of these carriages. Johnsonville line services had been under threat of being replaced by buses due to the condition of track and trains. The population of the capital is stagnant with most growth happening and forecast in or around Auckland.

Carriages used on the diesel locomotive-hauled Wairarapa trains are to be replaced by 18 rebuilt carriages previously purchased from Britain which will be transferred to the ownership of the Wellington Regional Council. The national government is providing about half of the NZ$20 million cost.

Coupler in standard use: Passenger cars and unit trains: AAR 'E'; freight wagons: `Norwegian' hook and pin; emus and railcars: Sumitomo tightlock
Type of braking in standard use: Air
UPDATED

Nicaragua

Ministry of Construction and Transport

PO Box 5, Managua

Key personnel
Minister: P Virgil

New lines
The former Ferrocarril de Nicaragua closed down in 1994, but in 1996 plans emerged for an ambitious

new railway linking the Atlantic ocean with the Pacific. The idea was that the 370 km line would provide a landbridge alternative to the Panama Canal, shifting containers and other freight from a new port at Monkey Point on the Atlantic seaboard to one of three sites being investigated on the Pacific coast.

In 1998 the Nicaraguan government named the international CINN consortium as preferred bidder to prepare a feasibility study into the scheme, the cost of which has been estimated at

US$1.5 billion. A less costly but longer alternative has been proposed by a rival consortium, SIT Global.

Ferrocarril Ingenio San Antonio

Gauge: 1,067 mm
Route length: 6 km

This line runs from Chichigalpa to Ingenio San Antonio. Initially built as a sugar cane railway, it has subsequently become a common-carrier, operated by Nicaragua Sugarestados. After an interruption of around a year caused by traction failures, operations resumed in 1999 following the repair of one of the fleet of three Davenport locomotives. Several converted freight wagons serve as passenger vehicles, carrying 150 to 250 passengers daily in three pairs of scheduled services. Travel is free of charge. Freight traffic is very limited, although molasses are carried occasionally.

Nigeria

Federal Ministry of Transport

Annex 3, New Federal Secretariat Complex, Shehu Shagari Way, Abuja
Tel: (+234 9) 52 37 05 13
Web: www.nopa.net/transport

Key personnel

Minister: Dr Abiye Sekibo
Minister of State: Eng Mohammed Musa

UPDATED

Nigerian Railways Corporation (NRC)

PMB 1037, Ebute Metta, Lagos
Tel: (+234 1) 774 73 20; 545 74 60
Fax: (+234 1) 583 13 67

Key personnel

Managing Director and Chief Executive:
A A Abubakar
Secretary: P I Onyeabor
Directors
Administration J B Dannana
Civil Engineering: Eng G F C Ezeani
Operations: J C Nwankwo
Engineering: Nkpubre O Nkpubre
Mechanical and Electrical: Eng Kabir Zayyana
New Lines: Eng E I Oradiegwu
Internal Audit: Chike Etiobi
Finance: A A Adamu
Managers
Signalling and Telecommunications:
Eng C M W Nwosu
Materials Management: Osmond Iroham
Assistant Director, Public Relations:
S D N Ndakotsu

Gauge: 1,067 mm
Route length: 3,505 km

Political background

In April 2005 the Nigerian government's Bureau of Public Enterprises invited expressions of interest from potential advisers in the concessioning of operations of the state-owned network. Privatisation, unlikely before 2006–07, is viewed as the only viable means of arresting the decline of a system which despite some injection of government funding in recent years remains in poor condition. It was expected that the first five years of a 25-year concession would see the successful bidder and the government working together on rehabilitation of the system.

Organisation

NRC is organised operationally as seven regional districts.
In January 2005 NRC employed 13,000 staff.

Passenger operations

In an attempt to bring some relief to the appalling traffic conditions in Lagos, NRC has been running limited peak-hour passenger services between Ifo and Lagos (48 km). In 2005 16 daily local services in operation. With the acquisition of five light axleload locomotives for passenger traffic, these services have improved considerably.

Freight operations

The poor state of the infrastructure and rolling stock contributed to a dramatic decline in traffic after 1989; in 1992 fewer than 200,000 tonnes were hauled (50 million tonne-km). With the adoption of a new marketing strategy and the acquisition from China of 45 main line locomotives, freight traffic saw increase. A significant success for NRC was the capture in 2004 of petroleum products traffic between Lagos and the north of the country for Oando plc. Rail's competitiveness has been assisted by damage and vandalism to the pipelines that led to the loss of much of this traffic in the first place, and by some funding from Oando to cover track and rolling stock rehabilitation.

New lines

Construction of a 277 km standard gauge link from the steelworks at Ajaokuta to the sea at Warri continued in 2005.
In 2005 construction was also still in progress of a new 19 km 1,067 mm gauge line to link Port Harcourt with the Federal Ocean Terminal oil and gas terminal at Onne.

Improvements to existing lines

In 1997 China's Civil Engineering Construction Corporation embarked on a large-scale rehabilitation of the network. The project was aimed at doubling line speeds to 80 km/h. Further contracts awarded to the same company in 2002 covered the Jebba–Minna and Minna–Baro lines. In the same year a programme was implemented to automate 133 level crossings. Other recent improvements were upgrading parts of the signalling system and installation of a digital microwave telecommunications system.

Traction and rolling stock

At last report, NRC owned 200 diesel locomotives, 287 passenger cars and some 4,000 freight wagons. The diesel locomotive fleet was reported to suffer from very low availability. A programme to rehabilitate 20 Class 1901 locomotives was initiated in 2004.
Between 1997 and 1999 Dalian Locomotive Works of China delivered 50 1,800 kW Class 2101 CKD8A diesel locomotives.
In mid-1997 NRC began taking delivery of new air-conditioned passenger coaches from the Sifang plant in China.

Track

As much as 80% of NRC track is reported in need of overhaul
Rail: BS60R 29.8 kg/m; BS70A 34.7 kg/m; BS80R, 80A 39.7 kg/m
Crossties (sleepers): Steel, 130 × 7.5 mm
Fastenings: Pandrol: K Type

UPDATED

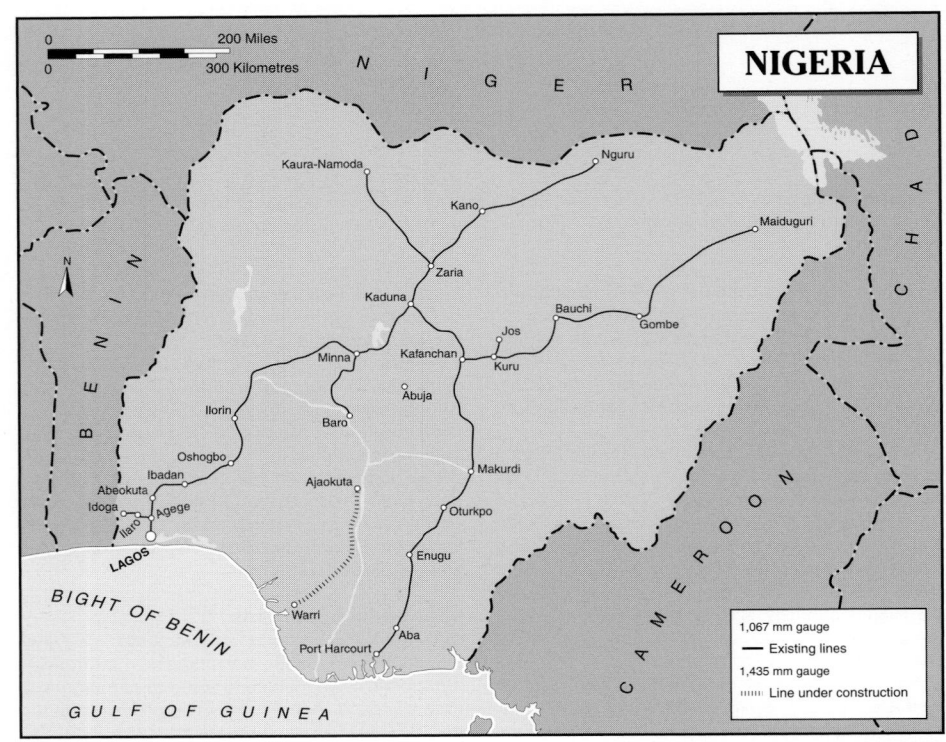

NEW/1114967

Norway

Ministry of Transport & Communications

Akersgata 59, PO Box 8010 Dep, N-0030 Oslo
Tel: (+47) 22 24 90 90
Fax: (+47) 22 24 95 70

e-mail: postmottak@sd.dep.telemax.no
Web: http://odin.dep.no/sd

Key personnel

Minister: Torild Skogsholm
Under-Secretary of State: Arnfinn Ellingsen

Secretary General: Per Sanderud
Director General, Rail: Pål Tore Berg

Norwegian National Rail Administration (JBV)

Jernbaneverket
PO Box 1162 Sentrum, 0107 Oslo, Sweden
Tel: (+47) 22 45 51 00   Fax: (+47) 22 45 54 99
Web: www.jernbaneverket.no

Key personnel

Director General: Steinar Killi
Executive Directors
   Infrastructure Management: Jon Frøisland
   Traffic Management: Arne Habberstad
   Market Relations and Communication:
      Svein Horrisland
   Strategic Planning: Anita Skauge
   Management Control: Stein O Nes
   Safety and Management Development:
      Ove Skovdahl
   Information: Jan Erik Kregnes
   International and Administrative Affairs:
      Ole M Drangsholt

Gauge: 1,435 mm
Route length: 4,077 km
Electrification: 2,680 km at 15 kV 16²/₃ Hz AC

Organisation

Jernbaneverket (Norwegian National Rail Administration) was established on 1 December 1996 under legislation passed by the Norwegian parliament, splitting the former Norwegian State Railways into NSB BA, a limited liability company, and Jernbanverket, a public body.

Jernbaneverket's Directorate in Oslo is responsible for regulatory functions such as scheduling and allocation of track capacity. The Directorate comprises the following departments: Market Relations and Communication; Strategic Planning; Management Control; Safety and Management Development; and Administrative and International Affairs. It also serves as the headquarters for Infrastructure and Traffic Management.

Infrastructure Management is responsible for the construction, management and maintenance of Norway's national rail infrastructure and for managing property, stations and terminals. It comprises three regional units, each with overall responsibility with its territory. A separate unit is responsible for constructing new infrastructure.

Traffic Management is responsible for operational management and provision of passenger information services. It comprises three traffic regions, which are subdivided into a total of eight train control areas. A separate unit, BaneEnergi, operates the traction power supply system, selling this to train companies.

In January 2005 Jernbaneverket had a total of 3,152 staff.

Finance

Parliament fixes annual appropriations for infrastructure maintenance and investment. The charges paid by train operators for use of the infrastructure are also set by parliament. Track charges reflect long-term marginal costs and may therefore differentiate between different types of rail traffic. They also take account of environmental factors and parity with other modes.

Jernbaneverket's expenditure is governed by a 10-year plan drawn up by the government and subject to annual review by parliament. After the 2004 review, the National Transport Plan for the period 2006–15 recommends capital expenditure on new infrastructure of NKr26.4 billion, of which more than half will go towards improvements in the Greater Oslo area. Over the same period, the plan recommends allocating a total of NKr30.6 billion to operation and maintenance of the rail network.

New lines

The line most recently opened is the Airport Line (Oslo–Gardermoen–Eidsvoll) serving Oslo's new international airport at Gardermoen, northeast of the city. Opened in August 1999, it is a 65 km mainly double-track line, engineered for train speeds of up to 210 km/h.

Improvements to existing lines

At the end of 2004 the Norwegian rail network had a total of 214 route-km with double track. In recent years priority has been given to extending double-tracking in the Greater Oslo area and on the Østfold (Oslo–Moss) and Vestfold (Drammen–Skien) lines, a policy which will be continued under the National Transport Plan 2006–15.

On the Østfold line, double-tracking of the 7 km Såstad–Haug section (between Moss and Halden) was completed in 2000, enhancing opportunities for trains to pass on this busy section of the Oslo–Gothenburg route.

On the Vestfold line, two new double-track sections of 6.9 km and 5.8 km near Åshaugen opened in 2001. A 500 m passing loop at Nykirke opened in 2002.

A similar programme of modifications is under way on the Dovre (Eidsvoll–Hamar) and Østfold (Oslo–Kornsjø) lines. The latter will also benefit cross-border services from Oslo to Stockholm and Gothenburg.

Meanwhile, work continues to improve track alignments and reduce the risk of snow blockage on the mountain section of the Oslo–Bergen line. In addition, projects have been undertaken to provide additional passing loops both in Flå and Gulsvik (between Hønefoss and Geilo).

To increase capacity on the busy Oslo–Drammen line, construction work started in January 2001 on Sandvika–Asker section (11.6 km) of new double- or multiple-track through a built-up area, with bridges over two rivers and a motorway and the remodelling of Asker station. The project is scheduled to be completed in August 2005.

Another planned improvement is the station development at Lysaker station, a bottleneck among interchange stations on the Oslo S–Asker line. The project involves integration with a new bus station and a terminal for a people mover serving Fornebu, as well as 0.9 km of new double-track line including a new bridge over the Lysaker river. Construction will be undertaken between 2005 and 2008.

Construction work on a new 5.6 km double-track line running direct from Lysaker to Sandvika is to start in 2007.

On the Jæren line planning is in progress for upgrading to double-track the Sandnes–Stavanger section (14.5 km).

Signalling and telecommunications

At 31 December 2004 about 60 per cent of the network was equipped with Centralised Traffic Control (CTC) and Automatic Train Control (ATC). In 2004 interlockings with ATC and CTC were installed on the Spikkestad line. On the Nordland line work to install CTC, ATC and new signalling is

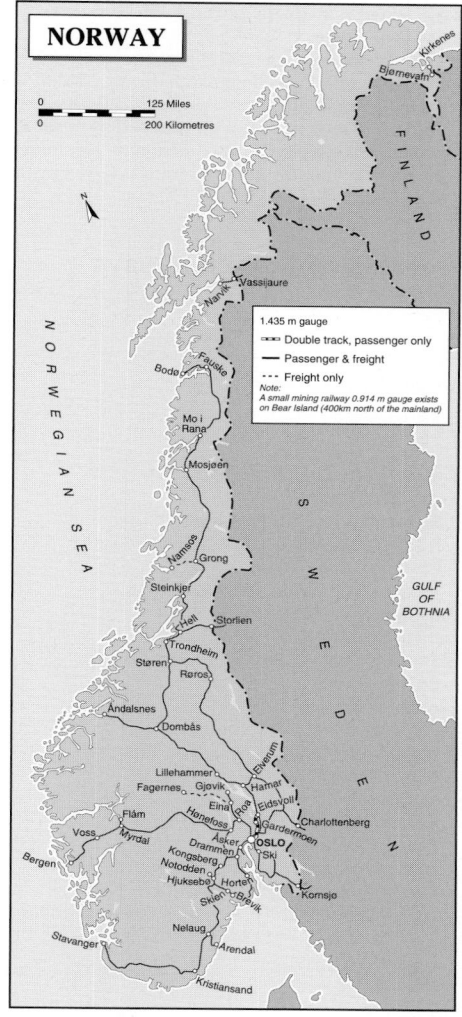

0023774

under way on the Grong–Mosjøen section, with the Mosjøen–Bodo section set to follow.

GSM-R

Norwegian GSM-R construction commenced in September 2003 and by May 2004 the first 80 km of line covered by the system was in operation. During the course of 2004 GSM-R was commissioned along 1,500 km of track; approximately 40 per cent of the network. It is intended that there will be 220

Jernbaneverket 'Malevogn' diesel-hydraulic track inspection vehicle built by Mer Mec and photographed at Hönefoss (Edward Barnes)
0552787

base stations in operation and 200 tunnels with GSM-R coverage.

Lines without train radio (40 per cent) have been given priority to be equipped with GSM-R. It will be the first mobile communications network to provide full coverage on all tracks and in tunnels, eventually replacing five analogue systems. Jernebaneverket will own, operate and maintain the GSM-R system, which will eventually cover 4,000 route-km.

Track
Rail type: S 49 (49.43 kg/m), S 54 (54.54 kg/m), UIC-54 (53.90 kg/m), UIC-60 (60.34 kg/m)
Sleepers: Mainly concrete, length 2,400–2,600 mm
Spacing: 1,700/km

Rail fastenings: Pandrol (PR341A, e-clip, Fastclip)
Max curvature: 8°
Max gradient: 2.5% (5.5% on Myrdal-Flåm line)
Max permissible axleload: 22.5 tonnes (25 tonnes on Narvik-Riksgränsen ore line)

UPDATED

Norwegian State Railways (NSB AS)

Prinsensgate 7-9, N-0048 Oslo
Tel: (+47) 23 15 00 00   Fax: (+47) 23 15 31 46
Web: http://www.nsb.no

Key personnel
Chairman: Olav Fjell
President and Chief Executive: Einar Enger
Executive Directors
  Economics and Finance: Kjell Haukeli
  Traffic Safety and Licensing: Tom Ingulstad
  Communications: Arne Wam
  Head of Secretariat: Steinar Norli
Rolling Stock: Jan Runesson
Human Resources: Roger Keiseraas
Managing Director, Passenger Business:
  Rolv Roverud
Directors
  Planning: Margareth Nordby
  Marketing: Marianne Ødegaard Ribe
  Sales: Karen K Hancke
  Operation: Øysten Risan
  Strategy: Arne Fosen
  Economics: Irene Katrin Thunselle
Managing Directors, subsidiaries
  CargoNet AS (rail freight): Kjell Froyslid
  Linx AB (joint with SJ, Sweden): Reidar Jignèus

Mantena AS (maintenance): Ole S Edvardsen
MiTrans AS (engineering): Tor Henriksen
Trafikkservice AS (cleaning): Knud Thiessen
ROM Eiendomsutvikling AS (property): Pål Berger
Arrive AS (IT): Eyvind Ranvig
Nettbus AS (bus operations): Arne Veggeland

Political background
Under legislation that came into force on 1 December 1996, Norwegian State Railways was reconstituted as a special limited company (NSB BA) with all shares held by the state. On the same date, the operations of the former NSB infrastructure sector were transferred to the Norwegian National Rail Administration (Jernebaneverket). With effect from 1 July 2002, the Norwegian government changed the status of NSB to that of a normal limited company, and NSB BA became NSB AS.

Although NSB has greater commercial freedom than it previously enjoyed, the structure established by the legislation specifically provides for a degree of continuing political control, in the form of public purchasing of services and parliamentary approval for investment plans. The minister of transport acts as the sole shareholder, appointing the company's non-executive directors and approving major decisions.

Competition to provide services on some routes hitherto operated by NSB was due to result from the government's decision to invite tenders for two pilot projects in 2004. These cover the Oslo–Jaren–Gjøvik line and the Bergen–Myrdal Vossbanen local service. If franchising these services proves successful, the government is expected to invite tenders for other state-supported routes from 2006.

Organisation
A major restructuring of the group from 1 January 2002 saw NSB AS becoming primarily a passenger train operator. Initially divided into two units, Passenger Train Eastern Norway and Passenger Train Regions, these were combined as a single business in December 2002 as part of a cost-saving strategy. The freight business was hived off into a separate company, CargoNet AS, owned 55 per cent by NSB and 45 per cent by Green Cargo, the Swedish rail freight company. The parcels business likewise became a separate company, Ekspressgods AS, but wholly owned by NSB. Other non-core activities, such as maintenance (Mantena AS), engineering (MiTrans AS), cleaning (Trafikkservice AS), IT (Arrive AS) and property (ROM Eiendomsutvikling AS), were also incorporated as separate subsidiary companies. The NSB group also includes Nettbuss, Norway's largest bus company and also a significant operator in Sweden.

Commercial and marketing responsibility for the Flåm Railway, which leaves the Oslo–Bergen line at Myrdal, was passed to local authority interests in 1998. Subsequently locomotives and coaches for this service were transferred to the same agency by NSB.

In January 2003, the former NSB subsidiary Flytoget AS, which runs trains on the high-speed line serving Oslo international airport, was separated from NSB to become a free-standing publicly owned company.

At 31 December 2002, the NSB Group had 9,138 employees (7,818 full-time equivalents).

Finance
In 2002 the NSB Group converted the previous year's loss of NKr33 million into a profit of NKr6 million. The parent company, NSB AS, recorded a profit of NKr44 million compared with NKr26 million in 2001, although these figures are not directly comparable due to corporate structural changes.

As a limited company, NSB AS is responsible for raising its own investment capital on the private loan market and is no longer subject to government borrowing consents. In 1999 NSB established a European long-term borrowing programme with a ceiling of €750 million and at 31 December 2002 loans of NKr3,151 million were outstanding. NSB BA has established a financial reserve of NKr2 billion via a syndicated credit facility which expires in July 2004. A government loan of NKr2.1 billion was converted to equity in 2001.

Bombardier-built Class BM93 tilting 'Talent' dmu operating a regional service at Hamar (Edward Barnes) 0524937

Finance (NKr million)	2000	2001	2002
Sales and other operating Income	5,622	5,595	6,073
Public purchase of services	1,303	1,530	1,637
Operating profit/loss	–304	108	108
Net /profit loss after tax	–325	–33	6
Assets	12,968	13,101	13,769
Interest, bearing liabilities	6,584	4,197	4,447
Equity	3,978	6,079	6,045

Traffic (million)	2000	2001	2002
Passenger journeys	54.0	53.5	49.8
Passenger-km	2,895	2,811	2,774

Passenger operations
NSB runs commuter services in the Oslo, Trondheim, Bergen, Stavanger and Kristiansand areas and long-distance trains throughout the country.

Class BM70 emu at Lillehammer (William A Willard) 0524939

Class EI18 electric locomotive with an intercity service at Ål (Edward Barnes) 0524941

Ansaldobreda Class BM72 articulated emu 0530158

Electric locomotives

Class	Wheel arrangement	Output kW	Speed km/h	No in service	First built	Builders Mechanical	Electrical
EI16	Bo-Bo	4,440	140	9	1977	Strømmen/Nohab/Hamjern	ASEA
EI18	Bo-Bo	5,880	200	22	1997	Adtranz	Adtranz

Electric railcars or multiple-units

Class	Cars per unit	Motor cars per unit	Motored axles/car	Output/ motor kW	Speed km/h	Units in service	First built	Builders Mechanical	Electrical
BM68A	3	1	4	160	100	21	1956	Skabo	NEBB
BM68B	3	1	4	160	100	9	1960	Skabo	NEBB
BM69A	2	1	4	297	130	13	1970	Strømmen	NEBB
BM69B	2–3	1	4	297	130	19	1973	Strømmen	NEBB
BM69C	3	1	4	297	130	14	1975	Strømmen	NEBB
BM69D	3	1	4	297	130	35	1983	Strømmen	NEBB
BM69E[1]	3	1	4	297	130	2	1983	Strømmen	NEBB
BM70	4	1	4	430	160	16	1992	ABB	ABB
BM72	4	2	2	637	160	36	2000	Ansaldobreda	Ansaldobreda
BM73/ BM73B	3	3	2	325	210	22	1999	Adtranz	Adtranz

[1] Rebuilt from BM69D in 1994 NEBB = Norsk Elektrisk-Brown Boveri.

Diesel railcars or multiple-units

Class	Cars per unit	Motor cars per unit	Motored axles/car	Power/ motor kW	Speed km/h	No in service	First built	Mechanical	Builders Engine	Transmission
BM92	2	1	2	2 × 360	140	15	1984	Duewag	2 × Daimler- Benz OM 424A	E BBC
BM93	2	2	2	2 × 360	140	11	2000	Bombardier	Cummins N14E-R	M

In 2002 passenger journeys fell by 6 per cent compared with the previous year. NSB attributed this decline to growing competition from car travel, newly deregulated long-distance coach operations and low-cost airlines, as well as some of its own operating difficulties. A marketing structure of three brands for passenger services, Puls (commuter), Agenda (regional) and Signatur (long-distance), was abandoned in 2003 in the face of poor customer acceptance and operational constraints.

NSB has introduced Class BM73 and BM73B tilt-body electric trainsets on many of its main long-distance routes (see 'Traction and rolling stock' section). With technical problems largely resolved, the new stock is able to operate at full speed. Journey times from Oslo are now as follows (previous best times in brackets): Kristiansand 3 hours 55 minutes (4 hours 45 minutes); Stavanger 6 hours 25 minutes (7 hours 30 minutes); Bergen 5 hours 30 minutes (6 hours 30 minutes); Trondheim 5 hours 30 minutes (6 hours 40 minutes). Other recent improvements have resulted from the introduction of Class BM93 Bombardier Talent tilting dmus on regional services and the gradual

commissioning of Ansaldobreda Class BM72 commuter emus.

Long-distance services from Oslo to Stavanger, Bergen and Trondheim are operated with modern locomotive-hauled stock, built in the 1980s and refurbished in 1994–96. As far as possible, day coaches and sleeping cars on these services now operate in fixed-formation rakes.

In May 2000 NSB and its Swedish counterpart, SJ, formed a joint venture company, Linx AB, to market and operate passenger services on the Oslo–Stockholm and Oslo–Gothenburg–Copenhagen routes. Operations began in June 2001. From February 2003 all Oslo–Stockholm services were formed of Linx X2 tilting trainsets.

Traction and rolling stock

At 31 December 2002, NSB's passenger motive power fleet included 22 Class EI18 and nine Class EI16 electric locomotives, 126 electric multiple-units, 32 diesel multiple-units and 297 passenger coaches.

After successful trials with a Swedish X 2000 train on the Oslo–Kristiansand route in 1996, NSB placed a follow-on order with Adtranz in March 1997 for 16 tilt-body trainsets based on the design of the Class BM71 Flytoget airport service emus. Designated Class BM73, the four-car units feature a revised interior for long-distance services with accommodation for 207 passengers. The trains operate the Oslo–Kristiansand–Stavanger, Oslo–Bergen and Oslo–Trondheim routes. Although the first sets entered service on the Kristiansand route in November 1999 and the Trondheim route in January 2000, they had to be withdrawn temporarily at the end of June 2000 pending attention to an axle problem which caused a set to derail on the Kristiansand line. A second follow-on order placed in early 2000 called for a further six sets, but with a revised interior layout (seating 243) for medium-distance services (Class BM73B).

In September 1997 NSB placed an order for 36 Class BM72 new-generation commuter emus with Italian manufacturer Ansaldo Trasporti (now Ansaldobreda). The four-car low-floor units seat 300 passengers and were delivered at the rate of three sets a month from July 2000, operating from mid-2002 on local services in the Oslo and Stavanger areas. NSB has an option on two further batches, each of 20 sets. However, running gear corrosion problems led NSB in early 2004 to suspend acceptance of the final 18 units of the order pending discussions with the manufacturer's subcontractors.

For services on the non-electrified Dombås–Åndalsnes, Hamar–Røros and Trondheim–Bodø lines, NSB uses 11 lightweight tilt-body Class BM93 Talent dmus from Bombardier Transportation.

The Class BM70 emus dating from 1992 have been refurbished.

CargoNet AS

Platousgt. 14-16, N-0048 Oslo
Tel: (+47) 23 15 45 45
e-mail: post@cargonet.no
Web: www.cargonet.no; www.cargonet.se

Key personnel
Chief Executive Officer: Kjell Frøyslid
Vice-Chief Executive Officer: Arild Drageset

Directors
Finance: Ivar Otterlei
Sales and Marketing: Knut Brunstad
Strategy: Bjarne Whist
Personnel: Kristin Moe
Traffic Safety: Liv Bjørnå

Background
Formerly NSB Gods, the freight operations business of Norwegian State Railways (NSB),

CargoNet AS was formed in January 2002. It is jointly owned by NSB AS (55 per cent) and the Swedish rail freight company, Green Cargo AB (45 per cent). As part of the arrangements to establish the company, Green Cargo's intermodal subsidiary in Sweden, formerly RailCombi AB, became a wholly owned subsidiary of CargoNet and is now CargoNet AS. Services are operated over tracks owned and managed by the Norwegian National Rail Administration.

Electric locomotives

Class	Wheel arrangement	Output kW	Speed km/h	No in service	First built	Builders Mechanical	Electrical
EI14	Co-Co	5,076	100	31	1968	Thunes	NEBB
EI16	Bo-Bo	4,440	140	11	1977	Nohab/Hamjern	ASEA

Diesel locomotives

Class	Wheel arrangement	Power kW	Speed km/h	Weight tonnes	No in service	First built	Mechanical	Builders Engine	Transmission
Di8	Bo-Bo	1,570	120	80	18	1997	Siemens	Caterpillar	E Siemens
CD66	Co-Co	2,385	120	126	6	2003	General Motors	GM12N-710G3B-EC	E General Motors
Skd226	B		218	70	34	12	1971	Kalmar	KHD F12 M 716	H Voith

Class CD66 diesel-electric locomotive (CargoNet) **NEW**/1122861

CargoNet Class EI14 electric locomotive with a container train at Hönefoss (Edward Barnes)

0585008

Organisation

The CargoNet group has an annual turnover of around SEK1.4 billion and in 2004 employed some 935 staff. These figures include the activities of Railcombi AB in Sweden.

Since January 2005 CargoNet's operations in Norway and Sweden have been fully integrated.

CargoNet AS's System train block train business is organised as a separate unit.

To provide access to the wider European rail network, partnership agreements have been signed with Hupac in Swtizerland and KombiVerkehr in Germany.

Freight operations

The main focus of CargoNet's activities is intermodal transport, although the company also handles block train (System train) services. In 2004 800,000 TEUs were handled at CargoNet's terminals in Norway and Sweden.

In Norway and Sweden intermodal services convey containers, ro-ro flats and semi-trailers. They cover a network centred on Alnabru terminal in Oslo and Malmö terminal in Sweden. CargoNet group trains serve the major communities of Åndalsnes, Bergen, Bodø, Narvik, Stavanger and Trondheim, as well as Drammen, Fauske, Kristiansand, Mo i Rana and Mosjøen in Norway, and Älmult, Borlänge, Gävle, Göteborg, Hallsberg, Helsingborg, Jönköping, Luleå, Norrköping, Stockholm, Sundsvall, Trelleborg and Umeå in Sweden. On some routes up to five daily trains are operated, running at an average speed of 72 km/h.

Specific CargoNet services giving onward international connections include Scandinavian Rail Express (SRE), linking Oslo and the port of Trelleborg for ferry services to northern Europe, Gothenburg Oslo Rail Express (GORE), serving Gothenburg for destinations that include the UK. Baltic Rail Express (BRE) operates between Oslo and Stockholm for onward links to terminals including Åbo and Helsinki in Finland, and on to Tallinn, Estonia. The Scandinavian Ruhr Shuttle (SRS), an intermodal train recently introduced by CargoNet, takes just 13.5 hours to cover the 918 km between Duisburg and Malmö, an average speed of 68 km/h. From Duisburg there are excellent onward connections to Ludwigshafen, Lyon, Munich, Rotterdam and Vienna. CargoNet services also feed container, swapbody and semi-trailer traffic into the Arctic Rail Express (ARE) flow that links Oslo with northern communities such as Gällivare in Sweden and Narvik and eastern Finnmark in Norway.

Traction and rolling stock

The main elements of the CargoNet traction fleet are all 31 ex-NSB Class EI14 electric locomotives, 11 Class EI16 machines and the fleet of 18 Class Di8 single-cab diesel-electrics acquired by NSB in 1997. The fleet was supplemented in 2003 by delivery of six Class CD66 diesel-electric locomotives built by General Motors (now EMD) and leased from HSBC Rail. They were acquired primarily to operate freight services on the Trondheim–Bodø route.

UPDATED

Class Skd 226 diesel-hydraulic shunting locomotive (right) alongside an older Class Di2 machine (CargoNet) **NEW**/1122862

Flåm Railway

Flåmsbana
PO Box 75, N-5742 Flåm
Tel: (+47) 57 63 21 00   Fax: (+47) 57 63 23 50
Web: www.flaamsbana.no

Key personnel

Managing Director, Flåm Utvikling: Olav Lühr
Operations Manager: Sivert Bakk
Sales Manager: Bengt Hammer

Gauge: 1,435 mm
Route length: 20 km
Electrification: 20 km at 15 kV AC 16²/₃ Hz

Background

Formerly part of the Norwegian State Railways (NSB) network, the Flåm Railway is a major tourist attraction. Descending from Myrdal, on the Olso–Bergen line, to the Sognefjord at Flåm, the line drops from 865.5 m above sea level to 2 m in just 20 km, offering passengers outstanding scenic views. In 1998 responsibility for promoting and marketing the line was transferred to Flåm Utvikling AS (Flåm Development Ltd), a publicly owned company. Subsequently locomotives and rolling stock were transferred to Flåmsbana by NSB, which retains responsibility for day-to-day train running.

Passenger operations

Five journeys in each direction are operated during the winter timetable, rising to 10 in high summer.

Class El 17 electric locomotives on the Flåm Railway (Flåm Utvikling AS)   ***NEW**/0585083*

Journey time is around one hour. There are 11 stations. In 2004 a record 459, 144 passengers were carried, compared with 417,540 in the previous year.

Traction and rolling stock

Services are operated using five ex-NSB Class El 17 Henschel-built 3,400 kW electric locomotives and 12 coaches. All are finished in the line's dedicated green livery.

UPDATED

Flytoget AS

PO Box 19 Sentrum, N-0101 Oslo
Tel: (+47) 23 15 90 00   Fax: (+47) 23 15 90 01
e-mail: flytoget@flytoget.no
Web: www.flytoget.no

Key personnel

Chief Executive Officer: Thomas Havnegjerde
Commercial Director: Sverre Hoeven

Background

Flytoget was originally established as a division of Norwegian State Railways (NSB) to operate dedicated services to Oslo's new Gardermoen Airport, which opened in October 1998. Trains initially ran over mostly classic lines until a new 48 km direct link from Oslo was commissioned in August 1999. In 2003 Flytoget was separated from NSB, forming a separate state-owned company reporting to the Ministry of Transport & Communications.

Traffic (million)	2002	2003	2004
Passenger journeys	3.9	3.9	4.2
Passenger-km	213	215	218

Passenger operations

The core weekday service provides six services between Oslo S and the airport every hour in each direction, half of these additionally calling at Lillestrøm and running west of Oslo to Asker. Flytoget trains compete with some NSB services.

Flytoget electric multiple-units

Cars per unit	Motor cars per unit	Motored axles/car	Output/ motor kW	Speed km/h	Units in service	First built	Builders Mechanical	Electrical
3	3	2	325	210	16	1997	Adtranz	Adtranz

Traction and rolling stock

Flytoget operates 16 three-car Class BM71 high-speed (200 km/h) emus that were procured for this service.

UPDATED

Flytoget Class BM71 emu at Lillestrøm
(Edward Barnes)
0524938

Ofotbanen AS (OBAS)

Postboks 333, N-8505 Narvik
Stasjonsveien 1, N-8515 Narvik
Tel: (+47) 76 92 32 50
Fax: (+47) 76 92 32 51
e-mail: post@ofotbanen.no
Web: www.ofotbanen.no

Key personnel

Managing Director: Terje Østensen

Background

OBAS is a private sector company formed in 2001 to operate passenger services over the Ofotbanen iron ore line between Narvik and Kiruna, Sweden.

Passenger operations

From June 2003, acting under a three-year contract with NSB, OBAS has been assisting Connex Tåg AB in the operation of trains on the Kiruna–Narvik leg of through services from southern Sweden. There were three daily services in each direction in the summer timetable and two in winter. OBAS also operates tourist charter services.

Traction and rolling stock

The OBAS fleet comprises five Class El 13 electric locomotives acquired from the Norwegian national freight operator, CargoNet, five ex-NSB Class Di 3 diesel-electric locomotives (of which two are unserviceable), three dmus and 20 passenger coaches.

UPDATED

Pakistan

Ministry of Railways

Islamabad
Tel: (+92 51) 82 52 47   Fax: (+92 51) 82 88 46

Key personnel
Federal Railways Minister: Javed Ashraf

Pakistan Railways (PR)

31 Sheikh Abdul Hameed Bin Badees, Lahore
Tel: (+92 42) 636 19 00   Fax: (+92 42) 30 61 93

Key personnel
Chairman, Executive Board: Muhammad Nasir
Managing Directors
  Freight: Muhammad Aslam
  Passenger: Abdul Quayyum
  Infrastructure: Nasir Amin
  Railway Resettlement Authority: Bashir Ahmad

Gauge: 1,676 mm; 1,000 mm
Route length: 7,718 km; 445 km
Electrification: 293 km of 1,676 mm gauge at 25 kV
50 Hz AC

Political background
Some faltering steps were made towards privatisation in the early 1990s. In 1992 it was announced that as a first step the government proposed to franchise out the lines from Lahore to Narowal and Faisalabad, and the Lodhran-Pakpattan route. Initial reports suggested that the franchisees, which took over at the beginning of 1993, had raised income on the routes, but this was due more to tighter security and control of ticketless travel than any immediate improvement to the train services. Both original franchises were later terminated and readvertised, but government approval for new leases was not immediately forthcoming.

In July 1997, disappointed by the lack of progress and the chaotic state of the railways, the government instituted a radical shake-up. World Bank-inspired reforms would see the railways and communications ministries merged into a single transport ministry, a rail regulatory authority set up, and PR divided into three bodies responsible for freight, passengers and infrastructure. A fourth business would manage railway-owned facilities such as schools and hospitals and would be responsible for the disposal of non-core businesses. Within three years they would be privatised. A task force, headed by economic affairs secretary Javed Burki, has been created to implement the restructuring plan, which was to include replacement of the existing board of management. The plan also foresaw massive cuts in PR's 110,000-strong workforce and the transfer of schools and hospitals for railway staff and their families to local authorities.

The division of PR into four businesses took place in September 1998, when rolling stock assets were divided between freight and passenger businesses. Management of the network, including signalling and train control functions, is being handled by the infrastructure unit. Coinciding with this restructuring was the creation of a new senior management board. Through the Privatisation Commission, the government plans to offer for sale or by concession its three core businesses, freight, passenger and infrastructure, as well as non-core activities such as rolling stock plants and sleeper manufacturing facilities and railway land and other property.

Organisation
PR comprises the whole of the North-Western system of the former British India rail network with the exception of lines in the southwestern Punjab. The main routes connect Karachi with Hyderabad, Multan, Lahore, Rawalpindi, Peshawar, Quetta and Zahedan. It was known as Pakistan Western Railway from 1961 to 1974. The 445 km metre-gauge network is in the southeast of the country.

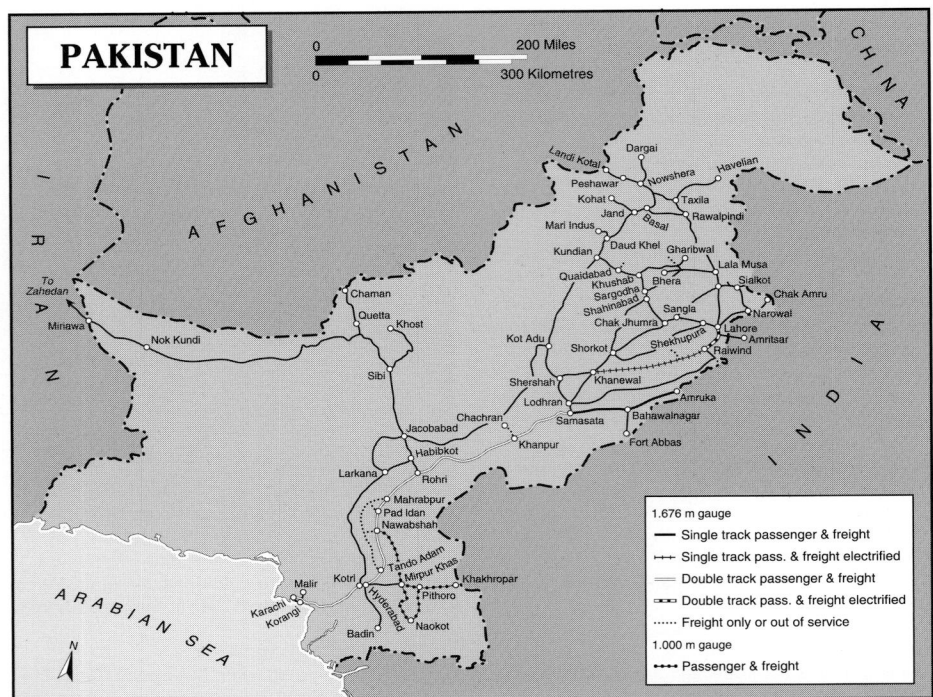

0021392

Following a railways ministry order that railway operation be divorced from non-railway activity, PR is now structured in several management units. These are: Railway Operations; Production Unit I, which includes the Islamabad coach works and the Moghalpura steel works; Production Unit 2, which includes the new Risalpur locomotive works; Production Unit 3, covering concrete sleeper manufacture and rail welding; Property Management & Development; and PR's Consultancy Service (PRACS).

Finance
PR's financial situation deteriorated rapidly during 1993–94 and, in mid-1994, the outgoing General Manager, Syed Zahoor Ahmad, warned of impending crisis if the bias in favour of road transport was not reversed and rehabilitation of the rail network begun. Matters declined further during 1995, and it was forecast that PR's deficit would rise from Rs1.3 billion in 1994–95 to Rs2.3 billion in 1995–96. In fact, the deficit for FY95-96 was Rs8 billion, an outcome which brought the railway close to bankruptcy.

Government's response in the eighth five-year plan (1993–98) was to allocate PR a share of some 30 per cent of investment planned for the transport sector, whilst road projects were to receive 57 per cent. But, due to poor economic conditions nationally, little of this money was forthcoming; by mid-1996 only Rs7.7 billion had been made available for the entire plan period, and this figure subsequently fell victim to the restructuring process detailed above.

Investment
Concerned that almost a third of its diesel locomotives are life-expired, PR developed plans to remanufacture 101 Alco units, but foreign exchange shortages meant that only six units had been re-engined by the start of 1996. The project was revived after agreement of a further loan of Rs5 billion from Japan in late 1995, but was subsequently cancelled under 1997 restructuring plans. However, procurement of 30 Blue Tiger diesel locomotives from Adtranz/General Electric went ahead.

Passenger operations
Passenger journeys reached a record 85 million in 1988–89. By 1994–95, this had fallen to 73.7 million journeys, for 18,904 million passenger-km. PR has since maintained this level of traffic.

General view at Rawalpindi station, with improvements in progress (John Bamforth)   0058811

Prospects were not improved by the poor condition of many coaches, with some services being withdrawn on account of rolling stock shortage. Multan works started a crash programme of light refurbishment to raise comfort standards. More coaches have been air conditioned, and upholstery has replaced hard seating in many second- and third-class cars. Some trains now have video entertainment.

In 1995 there was a move to reduce operating costs by substituting railcars for locomotive-hauled trains, but suitable cars were not available and the cost of construction or conversion was thought too high for the plan to succeed.

At various times during the 1990s plans were canvassed for introduction of high-speed trains, including construction of a new alignment between Karachi and Hyderabad. While the high cost militated against such schemes, PR nevertheless aims to equip all expresses with air conditioned coaches in an attempt to improve business prospects. By 1998, Islamabad works had built 250 air conditioned cars, including 150 economy-class saloons, 50 sleeping cars and 50 generator vans. Of these, 175 have been financed under an agreement with German donors which also covered technology transfer. The first 15 coaches were built in Germany.

Freight operations
In 1991–92 freight tonnage reached a low point of 7.6 million tonnes, for 5,962 million tonne-km. In the period 1993–98 tonnage was on average 5 to 6 million tonnes annually, with 5.1 billion tonne-km generated in 1995–96. PR has faced strong competition from road hauliers.

The first private-sector freight operators to run trains over PR tracks are likely to be involved in a scheme to raise three-fold, to 1.8 million tonnes, the amount of oil moved by rail to up-country power stations. Private operators would be permitted to run their own rolling stock over designated routes, and investment of some US$200 million will be required for 48 locomotives and 1,300 tank wagons. Several Canadian, US and South African parties expressed interest in running freight trains, but later complained of government lassitude in progressing the scheme. The franchise was readvertised by the government in June 1996, and by mid-1998 expressions of interest were reported to have been submitted by 14 prospective bidders.

In 1999, the private operation of container trains between Port Qasim, Lahore and Faislabad was the subject of tender invitations from PR.

New lines
Surveying has started on the first new railway to be built with private-sector funding. This is a 150 km line into the Thar desert from the Badin terminus of PR's branch from Hyderabad. Designed to tap reserves of coal and other minerals, the line will also facilitate construction of a 1,300 MW power station in the midst of the vast coalfield at Islamkot. The concession to build both railway and power station has been granted to a consortium of Hong-Pak United Power Generation, Tanson Development and American United Corporation.

In 1991 PR and Islamic Iranian Republic Railways signed an agreement to co-operate in development of the two systems. It provided for construction of the long-discussed 375 km connection between PR at Zahedan and IIRR at Kerman, although funding constraints have led to little progress with the scheme. A 40 km branch is to be built from the Zahedan line at Taftan to a copper mining development at Saindak.

Construction of several other lines has been canvassed in recent years, most recently a 650 km direct line from Quetta to Karachi, but the pressure on the investment budget makes it unlikely that a start will be made on any of them.

Traction and rolling stock
Under the 1998 restructuring, the passenger unit took over 268 locomotives and 2,239 coaches; the freight unit inherited 267 locomotives and 26,755 wagons.

Electric traction is used between Lahore and Khanewal using a fleet of 29 British-built Bo-Bo locomotives.

PR Adtranz-built 'Blue Tiger' 2,500 kW diesel-electric locomotive 0058810

Diesel locomotives

Class	Builder's type	Wheel arrangement	Power kW	Speed km/h	Weight tonnes	No in service	First built	Mechanical	Builders Engine	Transmission
HGMU-30	TV6125A2	Co-Co	2,462	120	120	30	1985	Henschel	GM USA 16-645E3C	E GM USA
GMU-30	GTCW-2	Co-Co	2,238	122	114.95	36	1975	GM	GM USA 16-645E3	E GM USA
GMU-15	GL-220	Co-Co	1,119	122	85.44	32	1975	GM	GM USA 12-645E	E GM USA
GMCU-15	G22CU	Co-Co	1,119	122	86.90	30	1979	GM Canada	GM USA 12-645E	E GM Canada
GEU-61	–	Bo-Bo	455	80	67.84	1	1954	GE	Cooper Bessemer USA-FWL-67	E GE USA
GEU-15	U-15-C	Co-Co	1,492	122	83.00	23	1970	GE	GE USA 7FDL-B4	E GE USA
GEU-20	U-20-C	Co-Co	1,119	122	96.00	40	1971	GE	GE USA 7FDL-B11	E GE USA
HAU-10	HFA-10A	Co-Co	746	72	120	4	1980	Hitachi	Alco USA 6-251E	E Hitachi
HAU-20	HFA-22A	Co-Co	1,492	120	102.6	28	1982	Hitachi	Alco USA 12-251GE	E Hitachi
HBU 20	HFA-22B	Co-Co	1,492	125	105	60	1986 1987	Hitachi/ Bombardier	Bombardier 12-251C4	E Hitachi
HPU-20	HFA-24P	Co-Co	1,492	120	101.3	10	1982	Hitachi	Pielstick 12PA4200VG	E Hitachi
ALU-95	DL-531	Co-Co	709	104	73.98	25	1958	Alco	Alco USA 6-251B	E GE
ALU-12	DL-535	Co-Co	895	96	75.00	49	1962	Alco	Alco USA 6-251B	E GE
ALU-18	DL-541	Co-Co	1,343	120	96.00	24	1961	Alco	Alco USA 12-251B	E GE
ALU-20	DL-543	Co-Co	1,492	120	102	52	1962	Alco	Alco USA 12-251C	E GE
ALU-24	DL-560	Co-Co	1,790	120	112.44	21	1967	Alco	Alco USA 16-251B	E GE
ALU-20R	DL-543	Co-Co	1,492	120	102	6	1986*	Alco	Bombardier 12-251G4	E GE
ARP-20	DL-212	AIA-AIA	1,492	120	109.06	23	1977*	Alco	Bombardier 12-251C4	E GE
ARU-20	E-1662	AIA-AIA	1,492	120	111.9	26	1976*	Alco	Bombardier 12-251C4	E GE
ARPW-20	DL-500C	Co-Co	1,492	120	102	42	1982*	Alco	Bombardier 12-251C4	E GE USA/ Canada
FRAU-75	–	Bo-Bo	560	69	68	2	1980*	Alsthom	Pielstick SEMT	E Alsthom PA-4

* Date of rebuilding.

Electric locomotives

Class	Wheel arrangement	Output kW	Speed km/h	Weight tonnes	No in service	First built	Builders Mechanical	Electrical
BCU-3DE	Bo-Bo	2,230	120	81.3	29	1966	AEI/Met Cam	English Electric

In 1996, PR ordered 30 'Blue Tiger' diesel locomotives from a consortium of Adtranz, Germany, and General Electric, USA. The new machines were delivered in 1998. With a rated output of 2,500 kW, and weighing 132.6 tonnes in service, these will comprise 20 110 km/h freight units and 10 150 km/h machines for passenger service.

In accord with a policy of self-sufficiency in rolling stock manufacture, a diesel-electric locomotive manufacturing works has been constructed near Risalpur at a total cost of Rs1,993 million. The foreign exchange component of that sum was Rs1,237 million. The project has the support of the Japanese government and a technology transfer agreement has been concluded with Hitachi. The first locomotive, a 150 kW shunter, was rolled out at the end of 1993, but the target of 18 locomotives set for 1995 had still not been achieved in 1998. Projected output is 25 locomotives per annum.

In 1998/1999 the Pakistan Railways Carriage Factory in Islamabad was undertaking mid-life refurbishment of PR passenger coaches, and in 1999 PR was given authority to refurbish 102 diesel locomotives.

Signalling and telecommunications
A major signalling project has been implemented comprising both conventional and modern

signalling works, such as provision of colourlight signals on the double-track main line between Lahore and Raiwind and of tokenless block on other important lines, starting with Lodhran-Khanewal-Faisalbad and Sangla Hill-Wazirabad. Following a call for international tenders, orders for design, supply and installation of the equipment were placed with Siemens.

Signalling over some 2,000 route-km of main lines is to be upgraded in a programme agreed in 1993. Japan's Marubeni Corp will carry out the work, with much equipment to be supplied by Siemens' local subsidiary.

A modern train and traffic control system has been installed on the Rawalpindi-Peshawar Cantt section of main line over a length of 174 km. The equipment was supplied by Aydin Monitor System of the USA.

Further projects, which had been scheduled for 1994–95 completion, provided for UHF communication over 598 km covering the Rawalpindi-Peshawar Cantt and Kot Adu-Attock City sections; and track circuiting of storage sidings at 94 stations on the Hyderabad-Peshawar line.

Type of coupler in standard use: Screw
Type of braking in standard use: Passenger cars (at 2/92): air, 872 cars; vacuum, 1,467 cars.
Freight cars: vacuum all

Electrification

In order to get the full benefits of electric traction and to remove operational bottlenecks, extension of electrification from Khanewal to Sama Satta is considered essential. Though planned for more than 20 years, and included in the eighth plan, the project has never been accorded any funds. Estimated cost is Rs1,174 million. Tenders to execute the electrification were called in 1989, but no contracts were let.

Track

Rail: 50 kg RE, 45 kg R BSS, 37.5 kg R BSS

Crossties (sleepers)

Type	Thickness	Spacing
PSC Monobloc	234 mm	1,640/km
RCC twin-block	231.77 mm	1,562/km
Wooden	125.152 mm	1,562/km
Steel trough	106.36 mm	1,562/km
CST 9 (cast-iron plates joined with tie bar)	133.35 mm	1,562/km

Fastenings
PSC/RCC sleepers: RM Type
Wooden sleepers: WI bearing plates with dog spikes; CI bearing plates with round spikes and keys
Steel trough sleepers: Mills spring loose jaws with keys
CST/9 CI plate sleepers: Keys
Min curvature radius: 10°
Max gradient: 4%
Max permissible axleload: 22.86 tonne

Panama

Ministry of Public Works

PO Box 1632, Panama 1
Tel: (+507) 32 55 05/32 55 72   Fax: (+507) 32 57 58

Key personnel
Minister: J A Dominguez
Secretary: E Perez Y

Chiriquí Land Company Railways

Chiriquí Land Company, Puerto Armuelles, Chiriquí, Apartado 6-2637 or 6-2638, Estafeta El Dorado, Panama City
Tel: (+507) 70 76 41   Fax: (+507) 70 80 64

Chiriquí National Railroad

Ferrocarril Nacional de Chiriquí
PO Box 12B, David City, Chiriquí

Key personnel
General Manager: M Alvarenga

Key personnel
General Manager: Cameron Forsyth
Assistant General Manager: Ricardo Flores
Technical Services Manager: Victor Mirones

Gauge: 914 mm
Route length: 152.5 km

Organisation
The railway, which was formerly divided into the Armuelles (133 km) and Bocas (243 km) Divisions, is dedicated to the transport of bananas. In 1992, the network was reduced to 152 km in length. Track is formed of 30 kg/m rails spiked to wooden sleepers spaced 1,600/km in plain track, 1,700/km in curves. Maximum permissible axleload is 20 tonnes and the maximum gradient 2 per cent.

Gauge: 914 mm
Route length: 126 km

Organisation
The railway, operated by the government of Panama, consists of a single line linking David City with Puerto Armuelles, in the region of the Costa Rican border.

Freight operations
Freight traffic in 1993 amounted to 410,670 tonnes.

Traction and rolling stock
The railway operates 17 diesel locomotives, seven diesel railcars and 380 freight wagons. Most powerful traction units are five 700 hp Caterpillar-engined locomotives, one a Whitcomb unit of 1948, the remainder GE of 1959 and 1970. Standard coupler is knuckle-type and braking of vehicles mechanical.

Traction and rolling stock
The fleet comprises five diesel locomotives, five diesel railcars, five passenger coaches and 24 freight wagons.

Panama Canal Railroad (PCRC)

Panama Canal Railway Company
Corozal West T-376, Panama
Tel: (+507) 317 60 70   Fax: (+507) 317 60 61
e-mail: info@panarail.com
Web: www.kcsi.com/pcrc

Key personnel
President: David L Starling

Gauge: 1,435 mm
Length: 77 km

Political background
National austerity kept the former Ferrocarril de Panamá (FCP) short of investment funds, with the result that both infrastructure and rolling stock became badly rundown, and traffic declined sharply. In 1996, a concession to operate the line was granted to a joint venture consisting of the US firms Kansas City Southern Industries, which owns Kansas City Southern Railway, and Mi-Jack Products (equipment supplier and terminal operator). A final contract was approved by the Panama government in 1998. The 25-year concession, which called for no

initial purchase cost, required PCRC to rehabilitate, operate and develop the line, together with its intermodal terminals, infrastructure and equipment, and to pay the government 5 per cent of gross revenue until its initial investment has been recovered and 10 per cent thereafter.

At the end of 2004 ownership of PCRC was divided between KCS (42 percent), Panama Holdings LLC (42 per cent) and International Finance Corporation (16 per cent).

PCRC offers an alternative to the parallel Panama Canal between the coastal ports of Manzanillo (Cristobál) and Balboa. The company eventually expects to run 32 trains a day, giving a capacity of 500,000 containers annually.

Passenger operations
Passenger operations, handled by PCRC subsidiary Panarail Tourism, resumed between Panama City and Balboa in 2001. The daily commuter service is supplemented by a luxury tour train marketed to cruise liner passengers.

Freight operations
Freight operations resumed in October 2001, with up to 10 container trains per day provided.

Some 17,500 containers were handled in 2002. The completion in December 2003 of Phase 1 of a planned expansion in port capacity at Balboa is expected to boost traffic.

Improvements to existing lines
Reconstruction and the construction and equipping of terminals took place over a two-year period at a cost of US$75 million. The entire trackbed was renewed by Kansas City Southern, and 1,524 mm gauge track replaced with standard-gauge laid with continuous welded rail. Some realignment was undertaken to raise speeds to 65–95 km/h. Work was completed in 2001.

Traction and rolling stock
In 2004 the motive power fleet comprised six diesel locomotives – five ex-Amtrak FP40s and a GP10 (rebuilt GP9). There were also 12 double-stack container wagons and six ex-Amtrak passenger coaches.

UPDATED

Paraguay

Ministry of Public Works & Communications

Ministerio de Obras Públicas y Comunicaciones
Cod Postal 1221, Oliva y Alberdi, Asunción
Tel: (+595 21) 414 90 00
Web: www.mopc.gov.py

Key personnel
Minister: Dr José Alberto Alderete

UPDATED

Ferrocarriles del Paraguay SA (FEPASA)

PO Box 453, Calle México 145 e/Eligio Ayala, Asunción
Tel: (+595 21) 44 32 73/44 78 48
Fax: (+595 21) 44 32 73
Web: www.ferrocarriles.com.py

Key personnel
President: Abg Lauro M Ramírez López
Secretary General: Adriana V Leguizamón Künzle
Managers
    Technical: Lauro Bernal
    Cultural: Lic Maribel Omar de Zárate
    Financial: Eduardo Benítez

Gauge: 1,435 mm
Route length: 36 km

Political background
Various efforts since the 1990s to privatise and rehabilitate Paraguay's railway system, the former Ferrocarril Presidente Carlos Antonio López (FPCAL), have met with no significant success. By 2005, the network (renamed Ferrocarriles del Paraguay SA (FEPASA), had been established as a semi-autonomous company in readiness for privatisation. However by this time it had been reduced from its former 441 km to a 30 km section southeast from Asunción, used for a tourist service and a 6 km line from Encarnación providing a connection with the Argentinian network. Materials from other parts of the main line linking the capital with Encarnación had been auctioned for scrap.
    In 2005 FEPASA employed 110 staff.

Passenger operations
A weekend tourist passenger service, the 'Tren del Lago' (The Lake Train) operates between Jardim Botánico, in the east of Asunción, to Luque.

Freight operations
Cross-border freight traffic is operated over the 6 km section between Encarnación and Posadas in Argentina, mainly exporting Paraguayan soya beans and imported goods moving in containers. This enables Paraguayan exports to reach Brazilian ports via the ALL Mesopotámico network

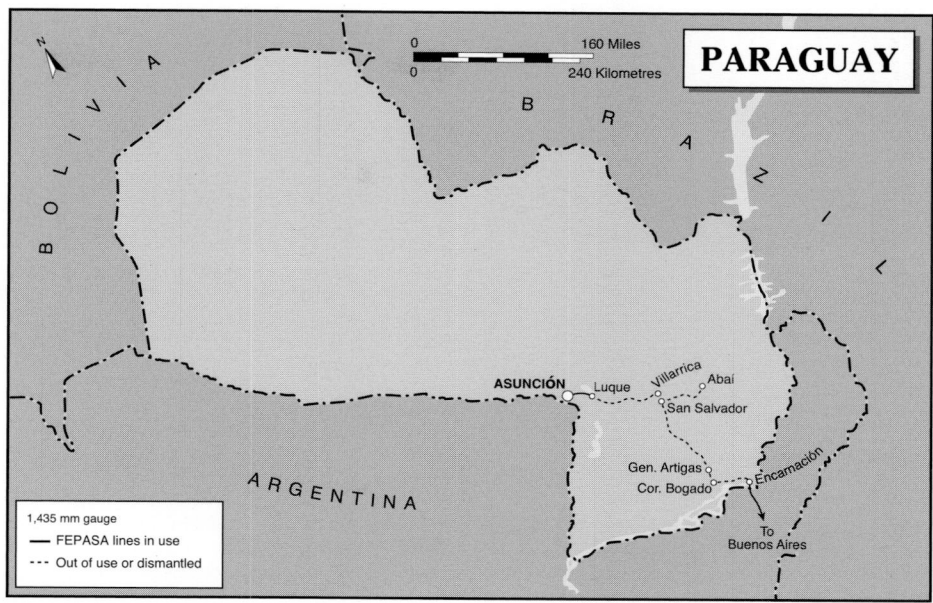

NEW/1115017

British-built wood-burning steam locomotive at the since-abandoned San Salvador station (Graham Vincent)

NEW/1115016

in Argentina. Some 210,000 tonnes were moved in 2003.

Improvements to existing lines
Plans exist to carry out infrastructure works that would allow extension of the Asunción–Luque tourist service to Sapucai (84 km).

Traction and rolling stock
In 2005 the FEPASA fleet comprised eight British-built wood-burning steam locomotives and two passenger coaches.

UPDATED

Peru

Ministry of Transport & Communications

Avenida 28 de Julio 800, Lima
Tel: (+51 1) 433 12 12   Fax: (+51 1) 433 04 27

Key personnel
Minister: Juan Castilla Meza
Secretary General: Dr J J Quelopana Rázuri

ENAFER-Peru

Empresa Nacional de Ferrocarriles del Perú
PO Box 1379, Jr Ancash 207, Lima 01

Tel: (+51 1) 428 94 40   Fax: (+51 1) 428 09 05
e-mail: comercial@mail.enafer.com
Web: http://www.enafer.com/

Key personnel
Chairman: Davis San Roman Benvante
Managers
    General: Raúl Rosales Trelles
    Operations: Alberto Mori Ito
    Financial and Administration: Carlos Reyes Huerta
    Commercial: Raúl Salas Cornejo
    Privatisation: Terry Medina Llerena
    Public Relations: José Luis Sánchez León Mantilla

Gauge: 1,435 mm; 914 mm
Route length: 1,296 km; 314 km

Political background
In 1990 the Peruvian government announced its intention to privatise ENAFER. At the same time it issued a decree permitting the construction of new railways competitive with the state system by private enterprise. In November 1991 President Fujimori confirmed that he intended to end the state regulation of the transport market and abolish the state railway's monopoly. It was proposed that management of infrastructure would be split from day-to-day operation of trains, with private companies invited to run services and paying rent for the use of the track.
    In 1997, the Commission for the Promotion of Private Industry was preparing proposals for privatisation; Mercer Consulting of the USA was assisting in the process.

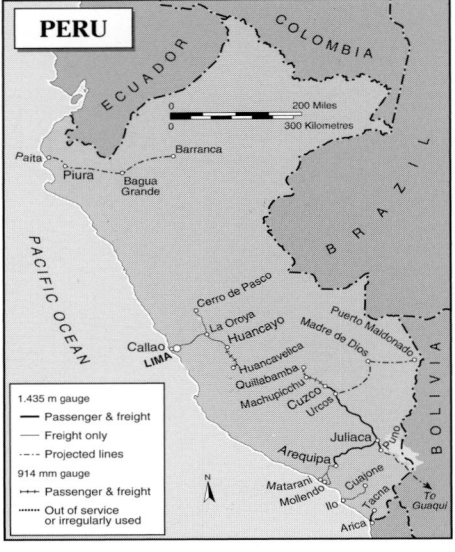

PERU

200 Miles
300 Kilometres

1.435 m gauge
— Passenger & freight
— Freight only
--- Projected lines
914 mm gauge
+++ Passenger & freight
+++ Freight only
····· Out of service
or irregularly used

0114303

Diesel locomotives

Class	Wheel arrangement	Power kW	Speed km/h	Weight tonnes	First built	Mechanical	Builders Engine	Transmission
	0-6-0	256	30	42	1964	Yorkshire Engine	Rolls-Royce C8TEL	*H* Rolls-Royce
	0-6-0	140	20	32.6	1950	Hunslet	Gardner 863	*M* Hunslet
DL-531	Co-Co	671	80	71.6	1958	Alco	Alco 6251B	*E* GE
DL-532-B	Bo-Bo	708	80	69.4	1974	MLW	Alco 6251B	*E* GE
DL-535-A	Co-Co	895	80	69.7	1967	Alco	Alco 6251C	*E* GE
DL-535-B	Co-Co	895	80	81.4	1963	Alco	Alco 6251B	*E* GE
DL-535B	Co-Co	895	80	80.7	1976	MLW	Alco 6251B	*E* GE
DL-535D	Co-Co	895	80	80.7	1964	MLW	Alco 6251B	*E* GE
DL-500-C	Co-Co	1,342	110	104	1956	Alco	Alco 12251C	*E* GE
DL-543	Co-Co	1,491	110	110	1962–63	Alco	Alco 12551B	*E* GE
DL-560-D	Co-Co	1,789	105	110	1964–66	Alco	Alco 16251B	*E* GE
DL-560-D	Co-Co	1,789	105	110	1974	MLW	Alco 16251E	*E* GE
GT-26CW-2	Co-Co	2,237	105	116	1982	GM	GM 16-645E3B	*E* GM
GT-26C2-2	Co-Co	2,237	105	116	1983	GM	GM 16-645E3B	*E* GM

In July 1999 a group of investors led by Sea Containers Limited, which owns the UK's Great North Eastern Railway franchise, was announced as the winning bidder for a 30-year concession, to operate ENAFER, extendable by an additional 30 years. The bidding vehicle was a divisible consortium, with Peruvian investors led by Oleacha & Co taking over the Central Line from the port of Callao to Puno and Cusco and Sea Containers and Peruval Corporation managing the Southern and Machu Picchu lines through the PeruRail SA consortium. The two companies each have a share of 50 per cent in PeruRail, but Sea Containers retain the right to make decisions on capital investments. A subsidiary of PeruRail will be the principal operator of the Southern and Machu Picchu lines which will have to pay access charges to its parent. In turn, PeruRail will also charge other operators for using its networks and for provision of rolling stock. PeruRail is required to pay the Peruvian government 33.375 per cent of such access charges, although during the first 5 years of the concession these payments may be offset by the cost of infrastructure improvements.

The concession includes all track, signalling, stations, workshops, ships and rolling stock. There is no purchase price for the concession other than US$5 million for spare parts and office equipment. All three networks are to be taken over by the respective operator from 21 September 1999. Sea Containers will manage the Southern and Machu Picchu lines for a fee and lease new rolling stock to both railways.

Organisation
ENAFER was formed in 1972 with the nationalisation of The Peruvian Corporation railways, a private company which ran most of Peru's railways and the Lake Titicaca ferry services. The system now comprises the Central, Southern and South Eastern Railways (the Machu Picchu line) with headquarters in Lima, Arequipa and Cusco respectively.

Passenger operations
Systemwide, ENAFER achieved 1.8 million passenger journeys (216 million passenger-km) in 1995. By mid-1995, passenger services had been suspended on the Central Railway (except for some weekend excursion services to San Bartolomé) but were in operation on the South Eastern Railway between Cusco and Quillabamba, and on the Cusco-Puno and Puno-Arequipa sections of the Southern Railway.

Freight operations
ENAFER carried 1.7 million tonnes of freight and generated 484 million tonne-km in 1994. In 1995, these figures declined to 1.5 million tonnes, 407 million tonne-km.

New lines
A US$1 billion, 1,800 km railway between the port of Bayover and Acre in Brazil has been proposed.

In 1997 the Peruvian government commissioned consultants to undertake a technical and economic feasibility study for a 150 km mixed-gauge line between Puno and Desaguadco. This would replace the existing Lake Titicaca ferry service, which imposes limitations on international services to Bolivia, with disruptions and reliability problems caused by low water levels in the lake. In recent years Bolivian transit traffic has been lost to the line serving the Chilean port of Arica, where road and rail connections are better.

In 1998 the Peruvian government announced it was to study three new non-connected lines: from the port of Paita, in the north of the country, to Barranca (650 km); from Tambo del Sol, on the existing Cerro de Pasco line, to Pucallpa (350 km); and Urcos, on the Southern Railway, to Puerto Maldonado, near the border with Bolivia (300 km).

Track
Standard rail: 34.7 and 39.7 kg/m
Crossties (sleepers): Peruvian hardwood
Made-up sleepers consisting of two blocks of reinforced concrete joined by a piece of used rail have been used in sidings and on straight stretches of main line
Spacing
Main line: 1,600–1,720/km
Branch line: 1,365–1,700/km
Rail fastenings
Soleplates and $^7/_8$ in coachscrews
Soleplates and $^5/_8$ in dog spikes
Pandrol fastenings are being fitted where new 35 kg/m rail is being laid
Max curvature: 17.5° = min radius 100 m
Max gradient: 4.7% (Central Railway), 4% (Southern Railway)
Max altitude: 4,839 m on Central Railway at La Cima siding on the Ticlio-Morococha branch, 173 km from Callao. On main line 4,782 m inside Galera Tunnel, 172 km from Callao
Max axleloading
Central Railway: 1,435 mm, 18 t; 914 mm, 14 t
Southern Railway: 1,435 mm, 17 t; 914 mm, 14 t
Tacna-Arica: 19.5 t
Bridge loading: Cooper E-40

Central Railway

ENAFER-Ferrocarril del Centro (FCC)
PO Box 1379, Ancash 207, Lima
Tel: (+51 1) 427 66 20   Fax: (+51 1) 428 10 75

Key personnel
Manager: Ing Terry Medina Llerena
Chief of Operations: Raul Liao Rengifo
  Infrastructure: Jorge Vigil Rojas
  Mechanical: Manuel Pinto Podesta
  Commercial: Fernando Tovar Madueño

Gauge: 1,435 mm
Route length: 380 km

Political background
The 332 km Lima-Huancayo line, plus the Cerro de Pasco line of the Empresa Minera del Centro del Perú (132 km) and the 127 km 914 mm Huancayo-Huancavelica line have been bundled together as a single concession for privatisation purposes.

Organisation
The standard-gauge main line runs from Callao to Huancayo where it connects with the 914 mm gauge line to Huancavelica, which was privatised in 1996. There are 66 tunnels with aggregate length of 8.9 km, 59 bridges and nine zig-zags (reversing stations).

The main line climbs from sea level to its highest point of 4,782 m in the Galera Tunnel in 171 km from Callao on an average gradient of 4 per cent. The highest point on the system is 4,818 m at a siding at La Cima on the Ticlio-Morococha branch. This makes it the highest standard-gauge line in the world. The steepest gradients occur in the first 222 km from Callao, at sea level, to La Oroya at 3,726 m above sea level.

Passenger operations
In 1999 an occasional tourist passenger service was being operated between Lima and Huancayo.

Traction and rolling stock
The Central Railway operates 25 diesel locomotives, 41 passenger coaches and 972 freight wagons.

Empresa Minera del Centro del Perú (Centromin Perú) SA

Railway Division
PO Box 2412, Lima 100
Tel: (+51 1) 476 10 10   Fax: (+51 1) 476 97 56

Diesel locomotives

Class	Wheel arrangement	Power kW	Speed km/h	Weight tonnes	No in service	First built	Mechanical	Builders Engine	Transmission
GR-12	Co-Co	980	105	174	5	1964	GM	GM 12/567C	*E* GM
GA-8	Bo-Bo	635	70	173	3	1964	GM	GM 8/567C	*E* GM
G22CW	Co-Co	1,120	105	107	2	1976	GM	GM 12/645E	*E* GM
G18W	Bo-Bo	745	115	66	1	1976	GM	GM 8/645E	*E* GM

Key personnel

Chairman: J C Barcellos
General Manager: J Merino
Central Manager of Operations: J C Huyhua
Services Manager: L Pérez
Superintendent of Railways: J Chávez
Chief Operations Officer: V Zúñiga
Chief Mechanical Officer: C A Hoyos
Supervising Engineer, Way and Structures: A Chang Way
Accountant, Railways: J Gutiérrez

Gauge: 1,435 mm
Route length: 212.2 km

Political background

In 1997, the government announced that the company would be merged with ENAFER (qv) prior to letting concessions for the operation of that railway. The Cerro de Pasco line has been bundled with the Central Railway for privatisation purposes. The transfer will be overseen by the Commission for the Promotion of Private Industry.

Organisation

The railway division of the Centromin mining company comprises two lines lying east of Lima branching from ENAFER's Central Railway. These are La Oroya-Cerro de Pasco (132.2 km) and Pachacayo-Chaucha (80 km), operated for the transport of concentrates, ores, raw materials and spare parts. Centromin's railway operations employed a staff of 282 at the end of 1996.

Finance (US$ million)

Revenue	1994	1995	1996
Freight	3.176	3.069	5.812
Other income	1.684	1.242	0.077
Total	4.860	4.311	5.889

Expenditure	1994	1995	1996
Staff/personnel	2.801	3.272	3.197
Materials and services	0.655	0.493	0.552
Depreciation	1.059	0.540	0.482
Total	4.515	4.305	4.231

Traffic (million)	1994	1995	1996
Freight tonnes	0.884	0.885	0.888
Freight tonne-km	103.497	101.347	105.866

Traction and rolling stock

The fleet in 1996 consisted of 10 diesel locomotives and 627 freight wagons.

Coupler in standard use: Sharon 10A
Braking in standard use: Air (valves K, AB, ABD Wabco Westinghouse)

Signalling and telecommunications

The railway uses a combination of mechanical, hand, telegraph, telephone and radio signalling.

Track

Rail type and weight: 9,020 45 kg/m (158 km); 7,040 35 kg/m (112 km)
Crossties (sleepers): Wood; 500,000 installed, spacing 1,850/km
Fastenings: Cut track spikes
Min curvature radius: 15° (La Oroya-Cerro de Pasco); 16° (Pachacayo-Chaucha)
Max gradient: 2.44% (La Oroya-Cerro de Pasco); 4.15% (Pachacayo-Chaucha)
Max permissible axleload: 19.8 tonnes

Ilo-Toquepala Railway

Ferrocarril Ilo-Toquepala
PO Box 2640, Lima
Tel: (+51 1) 436 15 65   Fax: (+51 5) 472 63 44

Key personnel

Manager: R D Alley
Resident Engineer: G Pasut
General Foreman: T L Chapman
Operations: D M Krinovich
Maintenance of Way Foreman: R Lungstrom

Gauge: 1,435 mm
Route length: 219 km

Organisation

The railway connects the port of Ilo with copper mines at Toquepala, with a branch serving deposits at Cuajone.

Traction and rolling stock

The railway operates 41 diesel locomotives and 629 freight wagons.

South Eastern Railway

ENAFER-Ferrocarril Sur-Oriente (FCSO)
Estación Marko Jara Schenone, Cuzco
Tel: (+51 84) 22 19 31   Fax: (+51 84) 23 35 51

Key personnel

Manager: Alberto Cruzado

Gauge: 914 mm
Route length: 186 km

Political background

Plans did exist to convert this to a standard-gauge line but not privatise it. However, both decisions have since been reversed, and the line is now to be sold off as a separate concession.

Organisation

Formerly managed as part of the Southern Railway, the 914 mm gauge route from Cuzco to Quillabamba via Machu Picchu emerged as a separate division in 1993.

Passenger and freight operations

The primary traffic is tourism, from Cuzco to Puente Ruinas, the station nearest to the Inca remains of Machu Picchu, for which there is no road access from Cuzco. This lack of a parallel road also dictates that most of the available freight traffic travels by rail, with transhipment at Cuzco. In 1997 there were two daily trains from Cuzco to Puente Ruinas and one mixed train from Cuzco to Quillabamba. Trailing load is severely restricted by the zigzags and gradients from Cuzco San Pedro terminus up

MLW Type DL535A diesel locomotive on passenger service at Cuzco San Pedro (Norman Griffiths)   0023264

to El Arco. In addition to drinks and snack, onboard services on passenger trains include first aid.

Traction and rolling stock

The South Eastern Railway operates seven 900 kW diesel locomotives, one MLW Type DL535 and six MLW Type DL535A/B, eight diesel railcars (2 Ferrostaal, 6 Macosa), 36 passenger coaches and 88 freight wagons.

Southern Railway

ENAFER-Ferrocarril del Sur (FCS)
PO Box 194, Avenida Tacna y Arica 201, Arequipa
Tel: (+51 54) 24 90 03
Fax: (+51 54) 23 16 03

Key personnel

Manager: Ing Terry Medina Llerena
Chief of Operations: Nicólas Rodríguez Nieto
  Mechanical: Víctor Franco Velarde
  Infrastructure: Ernesto Medina Barzola
  Commercial: Alejandro Torres Delgado

Gauge: 1,435 mm
Route length: 855 km

Organisation

The 1,435 mm gauge main line of the Southern Railway runs from the ports of Matarani and Mollendo on the Pacific coast to Juliaca, 462 km, where the line divides to Puno, 47 km, on Lake Titicaca; and to Cuzco, 338 km, where it connects with the 914 mm gauge South Oriental Railway to Quillabamba. The main line climbs from sea level to its highest point at Crucero Alto, 4,477 m, in 359 km from Matarani on an average gradient of 3 per cent.

ENAFER's Southern Railway operates shipping services on Lake Titicaca at an altitude of 3,818 m. The highest ferry service in the world runs on a 204 km route from Puno in Peru to Guaqui in Bolivia. The fleet includes one train ferry, one mixed passenger/freight vessel and one dredger.

Finance

Investment totalling US$22.15 million was planned for the Southern Railway in 1996, with US$12.6 million to be spent on major track improvements. Other major investment items included US$5.43 million to be spent on refurbishing diesel-electric locomotives, US$1.63 million on freight wagons and US$1.22 million on track maintenance machinery.

Passenger operations

In 1996 the Southern Railway recorded 267,943 passenger-km, down from the previous year's 302,447.

Traffic (million)	1995	1996
Freight tonnes	0.5	0.5
Freight tonne-km	216	259

Traction and rolling stock

The Southern Railway's serviceable fleet at the beginning of 1998 comprised 26 diesel locomotives, 33 diesel railcars, 81 passenger coaches and 917 freight wagons. Diesel locomotives types operated are Alco-built DL-535 and DL-543 models, Bombardier DL-560s, and General Motors GT26CW-2s.

Track

Standard rail: 39.7 kg/m (517 km) and 36.7 kg/m (315 km)

Crossties (sleepers): Hardwood (576 km) and concrete twin-block (217 km)
Spacing
Plain track: 1,700/km
Curves: 1,760/km
Rail fastenings: Tirafondos and Pandrol
Min curve radius: 102 m
Max gradient: 4.2%

Philippines

Department of Transport & Communications

Philcomcen Building, Ortigas Avenue, Pasig, Metro Manila
Tel: (+63 2) 631 87 61   Fax: (+63 2) 639 99 85

Key personnel
Secretary: Vincente Riveira

Philippine National Railways (PNR)

Torres Bugallon Street, Kalookan City, Metro Manila
Tel: (+63 2) 362 24 06   Fax: (+63 2) 362 08 24

Key personnel
General Manager: Jose Dado
Assistant General Management: Rafael Mosura, Jnr
Acting Assistant General Managers
   Maintenance: Erasto Laiz
   Operations: Ramon Jimenez
   Finance and Administration: Francisco Aure
Managers
   Legal Department: Antonio Holgado
   Public Affairs Department: Hilario Rojo
   Performance and Efficiency Board: Lynna Goyma
   OIC, Security and Investigation Services:
     Alexander Josol
   Training Operations Department: Ramon Jimenez
   Station Operations Department: Bonaparte Roque
   Rolling Stock Maintenance Department: Erasto Laiz
   Permanent Way Maintenance Department:
     Antonio Garcia Jnr
   Materials Management Department: Francisco Aure
   Controllership Department: Edilberto Manalo
   Treasury Department: Stalin Landas
   Personnel Services Department: Salvacion Bundoc
   Hospital Services: Armando Fuentes
   Real Estate Development: Edna Ramos

Gauge: 1,067 mm
Route length: 897 km, of which 492 km in operation

Political background

Plans to privatise PNR had been expected to be put before Congress in 1998 as a way of arresting the deterioration of the network. Ownership of infrastructure was expected to remain with the state, with private-sector concessionaires taking over operation, maintenance and upgrading of the system. No progress with these plans had been reported by early 1999.

Organisation

Much of the northern part of the network remains closed due to the poor condition of the infrastructure. Operations are currently concentrated on the 478 km Southern line from Manila to Legaspi City, on which major rehabilitation work was completed in 1995. Only the final 41 km from Polangui to Legaspi remain closed. Suburban trains run south from Manila on this line as far as Carmona.

Passenger operations

Passenger traffic has suffered from the poor state of the system; the number of journeys declined by more than half between 1990 and 1993.

Reconstruction of the line to Legaspi has seen the previous formation rebuilt with concrete sleepers, fresh ballast and part-worn rail
0003028

Long-distance passenger figures have continued to languish, but Manila suburban services bounced back sharply in 1993 following commissioning of 10 new diesel locomotives. Passenger journeys doubled to 4.6 million, holding steady for the following two years. Reopening throughout of Main Line South (see 'Improvements to existing lines' section below) is expected to improve carryings.

Freight operations

With nearly half the system closed, traffic has been in decline for several years. Tonnage amounted to only 5,000 in 1992 for 0.9 million tonne-km. By 1995 there had been a mild recovery, with 14,000 tonnes moving 3.9 million tonne-km, although the transport of construction materials for the Main Line South reconstruction artificially improved the figures.

Since 1997 container services operated by the Manila port authority have been run to a terminal near Santa Rosa (45 km).

New lines

There are long-term plans for construction of two new lines, once rehabilitation of the existing network is complete. A 153 km line to Sorsogon from Main Line South is planned, along with a 281 km route to open up the Cagayan Valley in the north of the country. Reinstatement of the 92 km Balagtas-Cabanatuan branch is also planned.

In 1997 it was reported that international funding was being sought for construction of a 134 km line from PNR's Main Line South at Comun to Matnog, the capital of Albay Province.

MCX

In March 1997, a 25-year concession to build a commuter line south from Manila was let by the government to a consortium led by property developer Ayala Land Inc. The US$600 million plan is to reconstruct and electrify the line from Manila to Calamba along the shore of Laguna de Bay. This would form part of a more ambitious scheme, the so-called MCX project, to link the capital with the port of Batangas, 100 km to the south. Rolling stock planned is 22 locomotives and 112 coaches. The consortium will own the trains and operate services, but the government has insisted that the track be handed to it upon completion so that other operators can use it.

Manila-Clark Rapid Railway System

In August 1997 government go-ahead was given to build a 130 km/h 102 km link, the Manila-Clark

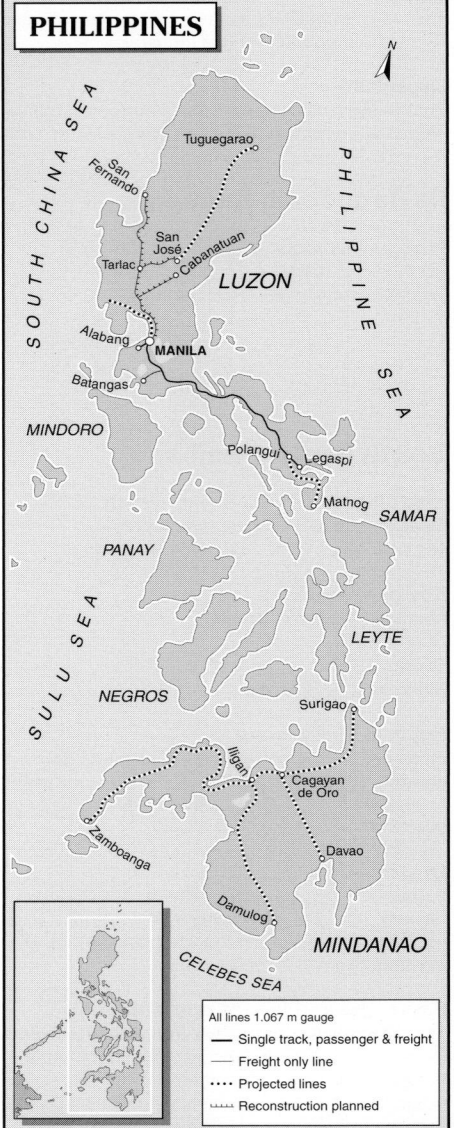

PHILIPPINES

All lines 1.067 m gauge
— Single track, passenger & freight
— Freight only line
··· Projected lines
⌐ Reconstruction planned

0114304

Rapid Railway System (MCRRS), from Manila to the development area on the former US Clark airbase. This involves reconstruction and electrification of the 25 kV Main Line North from Manila to 1,435 mm gauge, with construction of a short branch from Dan to Clark. Construction had been expected to get under way in 1998 following granting of a concession to North Luzon Railways (Northrail), a consortium in which the government-owned Bases Conversion Development Authority is the principal shareholder and in which PNR and RENFE of Spain also feature. However, Asian regional economic problems delayed the start of work, although the scheme's prospects looked brighter in 1999 with the acquisition of a 25 per cent stake in construction of the first phase by a Filipino-Japanese consortium, Fil-Estate Management, which is backed by Mitsui and Nishimatsu Construction, both of Japan. The government also took the opportunity to restructure the scheme, with construction entrusted to state-owned firms and private-sector companies required to invest in rolling stock.

The first 40.6 km phase, estimated to cost US$650 million, will run from Monumento, in Caloocan City, to Calumpit, in Bulacan province, ending at Clark airport. Under the second phase, the line will be extended to Subic, then to San Fernando La Union and San Jose.

Mindanao
In 1995 a plan to build a 2,000 km railway system on the Philippines' second largest island, Mindanao, was put forward by a consortium led by Malaysian engineering company Promet. Members include Spanish companies CAF and Cobra. A joint public/private-sector task force has been formed to take forward the proposed scheme, which would be constructed under a 12-year build-operate-transfer (BOT) scheme. In mid-1999 the government was close to inviting tenders for the project, which is expected to cost US$541 million. A high priority was being given to construction of the network, half of which the government expected to be operational by 2002.

One other important rail-related project is for construction of toll roads above PNR rights-of-way, allied to development of housing for up to 12,000 families. This is designed to remove the large number of squatters who currently live in shacks alongside the tracks.

Improvements to existing lines
A major project for the total rehabilitation of PNR's 437 km Main Line South, from Manila to Polangui

(the remaining 42 km to Legaspi have been closed since landslip damage in 1976) was largely completed in 1995. In 1996, rehabilitation of the sections from Naga to Legaspi (96 km) was carried out by the Australian contractor John Holland, with consultants TMG International undertaking supervision. PNR's goal was to have 10 trains a day running at speeds of up to 75km/h; the route is expected to carry 2.5 million passengers annually.

Except for a commuter train service from Manila to Malolos (37 km from Manila), the whole of the 266 km Main Line North from Manila to San Fernando La Union and the 55 km branch line from Tarlac to San Jose Nueva Ecija are closed to train operations due to heavy damage caused by storms. Local interests are co-operating with international investors in proposing a US$100 million package of improvements to the Main Line North, to serve K-Line's new container port at Subic.

Rehabilitation of the 56 km Calamba-Batangas branch is planned.

Traction and rolling stock
Chronic availability problems have plagued every PNR effort to improve services. In 1995 PNR operated 25 diesel locomotives, including 16 Japanese-built 1,120 kW machines supplied in 1992, 21 diesel railcars, 30 passenger cars and 266 wagons.

Acquisition of six more diesel-electric locomotives, repair of five others and refurbishment of 20 passenger coaches was going ahead under the Main Line South improvement project.

Signalling and telecommunications
In the Manila terminal area 13.6 km of double-track line with semaphore signals is controlled from interlocked cabins. On single-track lines elsewhere trains operate on the English 'staff' system or by telegraph or telephone communication and VHF radio from station to station.

Track
Standard rail
  Main line: 32.2 kg/m in 30 and 33 ft lengths
    37.2 kg/m in 60 and 25 ft lengths
  Branch lines: 32.2 kg/m in 30 and 33 ft lengths
Rail joints: Angle bars with slots for spikes
Crossties (sleepers)
  Main line: 'Molave' wood, 127 × 203 × 2,133 mm, spaced at 610 mm
  Branch line: 'Molave' wood, 127 × 203 × 2,133 mm, spaced at 610 mm
  Bridge ties: 'Yacal' wood, 203 × 203 × 2,438 mm, spaced at 406 mm
  A limited number of steel ties are also used
Rail fastenings: Track spikes; bolts with square nuts; 'Hipower' nutlock washer, elastic rail spikes
Max curvature
  Main line: 9.2° = min radius 190 m
  Branch line: 11½° = min radius 150 m
Max gradient: 1.2%
Max axleload: 16 tonnes
Max permitted speed: 60 km/h

Poland

Ministry of Infrastructure

Ministerstwo Infrastruktury
ul Chałubińskiego 4/6, PL-00-928 Warsaw
Tel: (+48 22) 630 10 00
Web: www.mi.gov.pl

Key personnel
Minister: Krzysztof Opawski
Deputy Minister (responsible for rail transport): Grzegorz Mędza
Director of Railway Department: Jakub Mieńkowski

Background
In March 2003 an entirely revised parliamentary act relating to railway transport came into force, meeting the requirements of EU directives. Under this new law the previous concessions, which were in many cases limited to a specific geographical area and/or to the movement of specific kinds of goods, were replaced by licences. A licence relates to the transport of persons or freight or to the lease (or hire) of traction units. One company may obtain all three types of licence. In accordance with the act on 1 June 2003, a new regulatory body; the Railway Transport Office (Urząd Transportu Kolejowego or UTK), was created. The UTK replaced the former Chief Railway Inspectorate (Główny Inspektorat Kolejnictwa or GIK) and among other duties is

responsible for granting licences. UTK also issues certificates of approval for the use in Poland all types of railway technical equipment that is safety-related, including all types of rail vehicles.

This new act and other legal regulations in Poland related to rail transport are considered to be liberal in comparison with other European countries.

Apart from the PKP Group, by mid-2005, 23 private operators had received a licence for the transport of goods and in one case an additional license was granted for the provision of passenger services. They were also issued for leasing (or hiring) traction units to other companies. 15 operators had received a licence for the transport of freight only (excluding traction leasing or hire), one company received one for the transport of passengers and another for passengers and freight (excluding traction leasing or hire). One company had been granted a licence for leasing (or hiring) traction units only. Six companies of PKP Group, as well as Koleje Mazowieckie Sp z o.o. (a joint venture of PKP Regional Services Ltd and the Mazowieckie regional authority), have licences for the transport of passengers and/or goods, some of these also hold them for leasing (or hiring) traction units. One company, SKM Warszawa, has a licence for passenger transport and for leasing/hiring traction units. Altogether 59 companies had been granted licences for various services. These includes 10 companies operating narrow-gauge

railways, as well as museum and tourist trains over PLK standard-gauge tracks.

Despite holding a license, not all companies have taken advantage of theirs to undertake rail transport services. Even those active in the market tend to operate on a very limited scale.

The most important non-PKP companies are:
- PCC Rail Szczakowa SA (formerly Kopalnia Piasku Szczakowa SA), in Jaworzno;
- Przedsiębiorstwo Transportu Kolejowego i Gospodarki Kamieniem Sp z o.o., in Zabrze;
- Przedsiębiorstwo Transportu Kolejowego i Gospodarki Kamieniem SA, in Rybnik;
- CTL Logistic SA (formerly Chem Trans Logistic Holding Polska SA), in Warsaw;
- Kopalnia Piasku Kotlarnia SA
- Kopalnia Piasku Kuźnica Wareżyńska SA
- DEC Sp zoo, in Warsaw;
- Rail Polska (Rail World Inc group).

The more important smaller companies include, among others: Kolej Bałtycka (Baltic Railway, Heavy Haul International group) and operators which are part of large industrial groups such as Pol-Miedz´ Trans (KGMH Polska Miedz´ SA group), Orlen Kol-Trans (PKN Orlen group), Kolej Baltycka (Baltic Railway, Heavy Haul International group) and Lotos Kolej (Rafineria Gdańska group).

UPDATED

CTL Logistic SA

Al Armii Ludowej 26, PL-00-609 Warsaw
Tel: (+48 22) 549 32 00 Fax: (+48 22) 549 32 03
e-mail: info@ctlhp.com.pl
Web: www.ctl.pl

Key personnel
Chairman: Jarosław Pawluk
Board Members
  Vice Chairman and Development Director: Janusz Glowacki
  Marketing: Katarzyna Izdebska
  Operations: Krzysztof Niemiec

Subsidiary
CTL Rail Sp z o.o.
ul Przemysłowa 10, PL-40-020 Katowice
Tel: (+48 32) 781 92 62 Fax: (+48 32) 781 92 63
e-mail: ctlrail@ctl.pl

Gauge: 1,435 mm
Route length: 130 km plus over 600 km of industrial branches

Organisation
Formed by 16 companies comprising forwarders and logistics firms, rolling stock companies and one of the former 'sand railway' companies, CTL

is Poland's largest freight forwarder and a growing carrier of chemicals, as well as undertaking traditional sand railway tasks. The company was formerly known as Chem Trans Logistic Holding Polska SA (CTL Polska).

The company employs around 2,000 staff.

Freight operations
CTL Logistic performs rail transportation activity through its subsidiary, CTL Rail Sp z o.o. The company operates block trains with its own locomotives on its private network as well as over PKP PLK tracks. CTL Polska forwards about 28 million tonnes of freight annually, of which

in 2004 8.2 million tonnes (1 billion tonne-km), was carried by CTL as a private railway. Most of the commodities carried are chemicals and petroleum products, but coal and stone are also handled.

In February 2003 CTL Polska and the German open access operator rail4chem Eisenbahnver-kehrsgesellschaft mbH (see entry in Germany section of *Railway systems and operators*) started operation of the first-ever private freight train service across the Polish-German border. A block train of tank wagons carrying petroleum products from an oil refinery at Leuna, Germany, hauled by CTL Polska diesel locomotive R4C-001 (similar to PKP's Class ST44), crossed the border in Guben/Gubin. Its cargo was destined for customers in the Katowice area of the Upper Silesia industrial region. An expansion of this service is planned, with rail4chem block trains operating from Zeebruge via Antwerp, Duisburg, Leuna and Schwarzheide into Poland. CTL and rail4chem also intend to create a joint company for the movement by rail of chemical products between Germany and Poland. However, cooperation between CTL and rail4chem subsequently ceased.

CTL has six trans-shipment terminals on the eastern border of Poland for transferring freight between standard and broad (1,520 mm) gauge systems.

Traction and rolling stock
CTL's rolling stock fleet consists of 145 electric and diesel locomotives, including about 30 main line locomotives such as Class ET21 and ET22 electrics

CTL Romanian-built Class ST43 diesel-electric locomotive at Dąbrowa Górnicza-Ząbkowice (Silesia) (Andrzej Harassek)   ***NEW****/1122835*

(according to PKP classification) and Class ST43 (Romanian-built) and ST44 (Soviet-built) diesels. Other locomotives are are diesel shunters including SM42- and SM48-type locomotives. There are also

over 4,000 wagons, including 1,500 tank wagons for chemicals and 1,500 open wagons of different types for coal transport.

NEW ENTRY

DEC Sp zoo

ul Twarda 30, PL-00-831 Warsaw
Tel: (+48 22) 697 91 00   Fax: (+48 22) 697 92 00
e-mail: dec@dec.pl
Web: www.dec.pl

Key personnel
Chairman of the Board: Bogdan Janicki

Organisation
Wholly owned by the American GATX Corporation, the company is Poland's largest owner and operator of tank wagons, mainly for petroleum

products. It operates block trains with PKP Cargo locomotives and traincrew, as well as groups of wagons within PKP freight services.

As well as acting as the main Polish carrier of liquid products, DEC also leases wagons to other operators, with around 2,000 wagons hired every day. Altogether DEC's tank wagons carry about 8.5 million tonnes annually, accounting for approximately 90 per cent of the Polish market.

In 2000 DEC had around 1,300 employees.

Traction and rolling stock
DEC owns about 11,000 tank wagons, some of them adapted for bogie changing to enable them

also to operate over 1,520 mm gauge tracks. The company has eight maintenance and wagon-washing centres and one overhaul facility. In addition to its wagon fleet, DEC also operates some 30 diesel shunting locomotives.

UPDATED

Kolej Bałtycka SA

Baltic Rail Joint Stock Company
ul Gerarda Merkatora 11, PL-70-767 Szczecin
Tel: (+48 91) 485 11 07   Fax: (+48 91) 485 11 08
e-mail: biuro@kolejbaltycka.pl
Web: www.kolejbaltycka.pl

Key personnel
Chairman of the Board: Richard M. Painter
Manager, Poland: Rafał Milczarski

Organisation
Kolej Bałtycka SA is a member of international Heavy Haul Power group. In Poland the company operates regular international freight trains of oil products and, sporadically, of bulk materials.

Traction and rolling stock
The company uses Canadian-built EMD diesel Class 29 locomotives (known elsewhere as 'Class 66'). It also has a number of tank wagons as well as a specialised set of 24 open wagons with a self-discharging system that enables the unloading of 1,000 tonnes of bulk material within one hour without the need of external equipment.

NEW ENTRY

Heavy Haul Power EMD JT42CWR locomotive in service with Kolej Bałtycka SA hauling an international oil products train at the border crossing at Tantow (Germany)/Szczecin Gumieńc (Poland) (Andrzej Harassek)   ***NEW****/1122836*

Koleje Mazowieckie Sp z o.o. (KM)

Mazovia (Mazowsze) Railways
ul Lubelska 1, PL-03-802 Warsaw
Tel: (+48 22) 47 38 716   Fax: (+48 22) 47 38 814
e-mail: sekretariat@mazowieckie.com.pl
Web: www.mazowieckie.com.pl

Key personnel
Chairman: Grzegorz Kucinski
Management Board
  Operations and Promotion: Grzegorz Kucinski
  Technology and Development:
    Czesław Sulima
Public Relations
Tel: (+48 22) 473 76 67   Fax: (+48 22) 473 88 14
e-mail: d.nowakowska@mazowieckie.com.pl

Organisation
KM is a joint venture train operating company established by the Mazowsze region (voievodship) authority (51 per cent) and PKP Przewozy Regionalne Sp z o.o. (PKP Regional Services) (49 per cent). From the beginning of 2005 KM took over from PKP Przewozy Regionalne all local services within the Greater Warsaw area and the Mazowsze region. All the trains are being operated over PKP PLK tracks. KM employs about 1,800 personnel.

Rolling stock
At present KM uses 184 Class EN57 emus hired from PKP Przewozy Regionalne. A modernisation programme has been started and by the end of 2005 40 trains will have been modernised. KM is gradually applying its new white-green livery to its trains.

Class EN57 emu in KM livery at Warsaw Zachodnia (West) station (Andrzej Harassek)   **NEW**/1122837

In June 2005 a letter of intent was signed between KM, the Mazowsze regional authority and the European Bank for Reconstruction and Development covering a loan for the purchase of new trains. Talks have also been conducted with the European Investment Bank. Altogether KM intends to spend about Zl 1 billion on new emus and push-pull trains. KM will also purchase nine second-hand light dmus from Deutsche Bahn (four Class VT 627 single-unit and five Class VT 628 two-car units). The vehicles will be used on the lightly used lines in the Mazowsze region. Four additional dmus might be purchased later.

NEW ENTRY

Kopalnia Piasku Kotlarnia SA

Kotlarnia Sand Mine Joint Stock Company
ul Dębowa 3, PL-47-246 Kotlarnia
Tel: (+48 77) 484 88 01   Fax: (+48 77) 484 88 00
e-mail: market@kotlarnia.com.pl
Web: www.kotlarnia.com.pl

Key personnel
Chairman of the Board: Krzysztof Kolczyk
Member: Urszula Wojkowska

Organisation
The company operates freight transport, mainly sand and coal, over its own tracks and over PKP PLK lines. Its rail infrastructure is managed by Kopalnia Piasku Kotlarnia – Linie Kolejowe Sp z o.o. (Kotlarnia Sand Mine – Railway Lines Ltd), also located at the Kotlarnia address. In 2004 the company carried over 3,800 tonnes (110,7 million tonne-km).

Traction and rolling stock
The company has five electric locomotives (Class ET21) and 18 diesels (Classes SM48 and SM42), as well as nearly 400 wagons.

NEW ENTRY

Kopalnia Piasku Kuznica Warezynska SA

Kuźnica Warężyńska Sand Mine Joint Stock Company
ul Letnia 1, PL-41-300 Dąbrowa Górnicza
Tel: (+48 32) 262 50 21   Fax: (+48 32) 262 09 53
e-mail: kuznica@kpkuznica.com.pl
Web: www.kpkuznica.com.pl

Key personnel
Chairman of the Board: Zbigniew Pucek

Organisation
The company operates freight transport, mainly sand and coal, over its own tracks and over PKP PLK lines. In 2004 the company carried over 2,100 tonnes (67.9 million tonne-km). The majority of shares in the company are held by PTKiGK Zabrze (see entry in Poland section of *Railway systems and operators*).

Traction and rolling stock
The company operates five electric locomotives (Class ET21) and seven diesels (Class SM48 and Czech-built Type T448).

NEW ENTRY

Lotos Kolej Sp z o.o.

Lotos Rail Ltd
ul Elbląska 135, PL-80-718 Gdańsk
Tel: (+48 58) 308 76 55   Fax: (+48 58) 308 76 78
e-mail: magadalena.cieszynska@lotoskolej.pl
Web: www.lotos.pl /lotoskolej/

Key personnel
Chairman of the Board: Henryk Gruca

Organisation
Lotos Kolej is a member of Lotos SA group, active in the Polish oil industry. Lotos Kolej is a railway freight operator, carrying mostly oil and liquid gas products. The company also hires tank wagons to other companies. In 2004 Lotos Kolej carried 0.7 million tonnes (191 million tonne-km). It employs around 230 staff.

Traction and rolling stock
The company has number of Class ST44, ST43, SM42 and SM48 diesel locomotives 40 tank wagons for liquid gas transport and a number of tank wagons for oil products.

NEW ENTRY

Lotos Kolej Class M62 (ST44) diesel locomotive with a train of tank wagons at Dąbrowa Górnicza-Ząbkowice (Silesia) (Andrzej Harassek)   **NEW**/1122839

PCC Rail Szczakowa SA

ul Bukowska 12, PL-43-602 Jaworzno
Tel: (+48 32) 753 77 11   Fax: (+48 32) 617 74 70
e-mail: sekret@pcc.railszczakowa.com.pl
Web: www.pcc.railszczakowa.com.pl

Key personnel

Chairman of the Board: Mieczysław Olender
Vice-Chairman: Józef Wdaniec

Organisation

The company operates over about 175 km of lines and branches, in 2004 carrying about 8.4 million tonnes (834 million tonne-km) annually, mainly of coal and construction materials. It is now the largest private operator over the PKP PLK network. In 2003 the company commenced the regular transport of coal and other bulk materials in block trains over PKP PLK lines using its own locomotives.

The rail division of Kopalnia Piasku Szczakowa SA (Szczakowa Sand Mine) has around 440 employees.

Traction and rolling stock

The company operates nearly 50 locomotives, including 19 Class 3E electric locomotives (PKP Class ET21), diesels of Type M62 (PKP Class ST44), TEM2 (SM48) and LDA2100 (ST43), and around 900 wagons.

UPDATED

PCC Rail Szczakowa Type TEM2 diesel locomotive on the company's tracks near Mysłowice, Silesia (Andrzej Harassek)
NEW/1122838

PKP Cargo SA

PKP Cargo Spółka Akcyjna
ul Grójecka 17a, PL-02-021 Warsaw
Tel: (+48 22) 474 43 20   Fax: (+48 22) 474 14 43
Web: www.pkp-cargo.pl

Key personnel

Chairman: Józef Marek Kowalczyk
Management Board
  Economics and Finance: Stanisław Cichoń
  Personnel and Administration: Władysław Krasuski
  Operations: Andrzej Waszczuk
  Marketing and Commercial Affairs:
    Zygmunt Siarkiewicz

Organisation

PKP Cargo employs about 47,500 staff.

Freight operations

PKP Cargo SA is the largest and principal Polish rail freight operator, with a share of about 91 per cent of the Polish domestic market in terms of tonne-km and 54 per cent in tonnes carried. The company also provides locomotives and drivers for PKP Intercity and PKP Regional Services.

The company operates large trans-shipment facilities between standard and Russian broad-gauge (1,520 mm), located near the border with Belarus (at Małaszewicze) and with Ukraine (at żórawica-Medyka). PKP Cargo operates 10 marshalling yards, 121 shunting stations, as well as three large multimodal terminals. About 1,500 freight terminals and loading points are also operated by PKP Cargo.

In 2004 PKP Cargo carried about 156 million tonnes (45 billion tonne-km) of freight. Domestic traffic accounted for 55 per cent of this, international traffic the remaining 45 per cent (including transit movements). The average transport distance is about 290 km. Coal still constitutes a major proportion of loadings – about 45 per cent.

Traction and rolling stock

The PKP Cargo locomotive fleet in 2004 totalled approximately: 1,250 electric freight locomotives (including two shunters); 1,600 diesel freight and shunting locomotives; 500 electric passenger locomotives; and 430 diesel passenger locomotives. The wagon fleet stood at around 82,000, 19 per cent of these specialised vehicles.

Everyday maintenance is performed in PKP Cargo's own depots, with major overhauls outsourced to ZNTK works or to manufacturers.

PKP has also about 25 steam locomotives. Some are not in working order, but about 15 can still be

Class ET22 electric locomotive with a PKP Cargo mixed freight service at Opole (Andrzej Harassek)
NEW/1122840

Diesel locomotives

Class	Wheel arrangement	Power kW	Speed km/h	Weight tonnes	First built	Mechanical	Builders Engine	Transmission
SM03	B	111	45	24	1959	Fablok	Nowotko	*M* Zastal
SM30	Bo-Bo	257	58	36	1959	Fablok	Nowotko DVSa-350	*E* Dolmel
SM42	Bo-Bo	588	90	72	1963	Fablok	HCP 8 VCD22T	*E* Dolmel
SP42	Bo-Bo	588	90	70	1966	Fablok	HCP 8 VCD22T	*E* Dolmel
ST43	Co-Co	1,544	100	116	1965	Craiova	Sulzer 12 LDA 28	*E* BBC
ST44	Co-Co	1,471	100	116	1966	Voroshilovgrad	Kolomna 14D20	*E* Charkow
SU45*	Co-Co	1,287	100	96	1967	HCP	HCP Fiat 2112SFF	*E* Dolmel
SU46	Co-Co	1,650	120	102	1974	HCP	HCP Fiat 2112SSF	*E* Dolmel
SM31	Co-Co	882	80	120	1976	Fablok	HCP	*E* Dolmel
SM48	Co-Co	882	100	116	1976	PZM-Lugansk	PDG-YM	*E* Charkow
SP32	Bo-Bo	1,300	100	74	1985	23 August	23 August M820SR	*E* Craiova

* rebuilt from SP45 – steam heating replaced with electric heating equipment (all former SP45 locomotives have been rebuilt or scrapped).

Electric locomotives

Class	Wheel arrangement	Output kW continuous/ one hour	Speed km/h	Weight tonnes	First built	Builders Mechanical	Electrical
ET 21	Co-Co	1,860/2,400	100	112	1957	Pafawag	Dolmel
EU06	Bo-Bo	2,000/2,080	125	80	1961	Metropolitan Vickers	English Electric Co.
EU07/EP07*	Bo-Bo	2,000/2,080	125	80	1963	Pafawag, HCP	Dolmel
ET22	Co-Co	3,000/3,120	125	120	1971	Pafawag	Dolmel
EP05**	Bo-Bo	2,032/2,344	160	80	1973	Škoda	Škoda
EP08	Bo-Bo	2,000/3,000	140	80	1973	Pafawag	Dolmel
ET40	Bo-Bo + Bo-Bo	4,080/4,680	100	164	1976	Škoda	Škoda
ET41†	Bo-Bo + Bo-Bo	4,000/4,160	125	167	1978	Cegielski	Dolmel
ET42	Bo-Bo + Bo-Bo	4,480/4,840	100	164	1978	Novocherkassk	Novocherkassk
EP09	Bo-Bo	2,920/3,230	160	84	1986	Pafawag	Dolmel
EM10††	Bo-Bo	960	80	72	1990	Ciegelski	Dolmel/Elta/IEL

* Class EP07 is a modernised version of Class EU07.
** Former Class EU05 125 km/h locomotives purchased in 1961 as continuation of series built for former Czechoslovak State Railways and subsequently rebuilt by PKP for 160 km/h running as Class EP05; only two units remained in service in 2005.
† Semi-permanently coupled twin-unit version of Class EU07.
†† All four locomotives rebuilt and modernised in 2004.

used. These are located at Wolsztyn, 80 km west of Poznań, and in Chabówka museum, 100 km south of Kraków. Two or three locomotives are used in daily passenger service on Wolsztyn-Poznań, Wolsztyn-Zbąszynek and Wolsztyn-Leszno routes. In Chabówka museum one serviceable Class SN61 diesel railcar (Ganz-Mávag, 1970s) and one Class EP03 Bo-Bo electric locomotive (ASEA, 1940s), as well as another SN61 railcar at Szczecin depot, are retained for use on special trains. For this activity PKP Cargo has a passenger operating licence.

UPDATED

Modernised Class ET22 electric locomotive in the latest PKP Cargo corporate colour scheme displayed at Warsaw Zachodnia station (Andrzej Harassek)
NEW/1122841

PKP Intercity Ltd

PKP Intercity Sp z oo
ul. Grójecka 17, PL-02-021 Warsaw
Tel: (+48 22) 474 55 00; 524 55 10
Fax: (+48 22) 474 43 51
Web: www.intercity.pl

Key personnel
Chairman: Jacek Przesluga
Management Board Members: Lucyna Krawczyk; Tadeusz Matyla
Public Relations: Anna Rosiek
Tel: (+48 22) 474 26 36
e-mail: a.rosiek@intercity.pl

Organisation
At the end of 2004 the company had about 2,000 employees.

Passenger operations
PKP Intercity operates all high-standard (so-called 'qualified') passenger trains. These include six pairs of Eurocity trains per day: three pairs between Warsaw and Berlin and one pair between Poznań and Berlin (all three under the brand name 'Berlin-Warszawa-Express'); one pair between Warsaw/Kraków and Vienna (two portions of the train are coupled/split at the Polish/Czech border station of Petrovice u Karvine); one pair linking Warsaw, Budapest and Vienna (the train splitting/combining at the Czech/Austrian border station at Břeclav); one pair between Warsaw and Prague; a pair of international Intercity trains (Kraków–Berlin–Hamburg); and 14 pairs of domestic Intercity and 29 pairs of Express trains.

Night services, consisting exclusively of sleeping and couchette cars, in early 2004 included five pairs of international 'hotel trains' (from Warsaw to Cologne, Vienna/Prague, Moscow and Vilnius, and from Kraków to Kiev). After Poland's accession to the European Union, the Warsaw–Cologne prestigious train was upgraded to EuroNight standard and extended to Brussels. Running at up to 200 km/h on some sections in Germany, the train consists of a 'sleeperette' car (with reclining seats), couchette cars and high-quality sleeping cars with eight standard three-berth compartments which can be arranged as single, double or tourist class and two two-berth 'business' class compartments equipped with private toilet and shower. A restaurant car is also provided. The train also conveys through sleepers from/to Moscow and Minsk, Belarus.

Class SU46 diesel-electric locomotive heading a PKP Intercity service on the non-electrified Reda–Hel line in northern Poland (Andrzej Harassek)   **NEW**/1122842

Class EP09 electric locomotive heading a PKP Intercity IC service formed of stock in the latest livery (Andrzej Harassek)   0585002

For details of the latest updates to *Jane's World Railways* online and to discover the additional information available exclusively to online subscribers please visit
jwr.janes.com

In 2005, apart from EN to Brussels, the PKP Intercity overnight international trains connected Warsaw with Vienna/Prague and Moscow, and Kraków with Kiev. The latter employs SUW 2000 gauge-adjustable wheelset technology that makes transfer from standard to broad-gauge track and vice versa faster and safer, with no need for bogie changes.

In April 2004 PKP Intercity introduced a new category of trains under brand name 'Tanie Linie Kolejowe' (low-cost or cheap railway lines) covering eight pairs of day trains and five pairs of overnight services. The ticket price is calculated according to the 'fast trains' tariff (cheaper than the 'express' tariff, which is applicable to other express and intercity trains). Obligatory reservation costs ZL3 instead of Zl9-20 in express and intercity trains. The limited number of second class tickets in each train is available at the promotional global price, including reservation, of Zl27 regardless of the distance of travel, under the condition that the ticket is bought at the ticket office (no phone or on-line reservation is possible) at least seven days before the journey. The price of a couchette is Zl20 instead of Zl50 in other trains and a sleeping car ticket costs Zl10 less than the normal price. There are neither complimentary drinks nor snacks, but the buffet cars are available. Also there are no blankets in the couchette (passengers can obtain these at an additional charge). The overnight trains are guarded by a special security service to prevent thefts.

In 2004 the company carried some 8 million passengers (2.8 billion passenger-km), 6.5 per cent down on the figure for the previous year.

Traction and rolling stock

PKP Intercity owns about 720 compartment and open saloon coaches, including some 158 modern 26.4 m UIC-Z1 and Z2 coaches, many of them air-conditioned, about 260 sleeping and 200 couchette cars, 16 dining (restaurant) cars and

Interior of new restaurant car forming part of the 'Jan Kiepura' EuroNight service
(Andrzej Harassek)   0585003

80 buffet cars. However, the company has no locomotives. These are hired from PKP Cargo SA, as are train drivers. The company has its own rolling stock depots where everyday maintenance is carried out. Major overhauls are performed by Rolling Stock Overhaul Works (Zakłady Naprawcze Taboru Kolejowego or ZNTK), independent companies which were split off from PKP in the early 1990s.

In 2003 PKP Intercity received from the Pojazdy Szynowe (Rail Vehicles) PESA Bydgoszcz SA its first modern sleeping car. The car was rebuilt from a Type Z2 standard compartment coach. The car is designed for 200 km/h operation. However, the

25AN/s bogies, equipped with SUW 2000 gauge-adjustable wheelsets, allow for 160 km/h operation (for 200 km/h different bogies would be required). The air-conditioned car has eight standard, three-berth compartments, which can be arranged as single, double or tourist class, and two two-berth 'business' class compartments. Further UIC Z1-type cars of similar standard, newly built at the H Cegielski works in Poznań, were delivered in 2004 for the 'Jan Kiepura' EuroNight service. The older type of German-built sleeping cars of 1970s and 1980s are gradually being modernised.

UPDATED

PKP Metallurgical Broad Gauge Railway Ltd (LHS)

PKP Linia Hutnicza Szerokotorowa Sp z oo
ul Szczebrzeska 11, PL-22-400 Zamość
Tel: (+48 84) 638 62 23; 677 77 00
Fax: (+48 84) 638 52 36
e-mail: lhs@pkp.com.pl
Web: www.pkp-lhs.pl

Key personnel

Chairman: Piotr Juś
Management Board Members
  Economics and Finance: Jolanta Piłat
  Commercial and Operations: Eugeniusz Tarka
  Technology and Development: Mieczyslaw Czuk

Gauge: 1,520 mm
Route length: 395 km

Organisation

The company manages infrastructure and operates trains over a 395 km broad (1,520 mm) gauge line from the Ukrainian border at Hrubieszów to Sławków, in the Upper Silesian industrial region. The line was built in the 1970s to transport iron ore from the former Soviet Union to Katowice steel works and sulphur from the Tarnobrzeg area to the Soviet Union. The line was then named LHS (Linia Hutniczo-Siarkowa, or the Metallurgical-Sulphur Line). Broad gauge was chosen to avoid difficult and costly trans-shipment at the border.

As well as its own line, LHS's main assets comprise four trans-shipment facilities, an installation for changing bogies and wheelsets from standard to broad gauge and vice versa, and a track with special equipment for adjusting the variable gauge wheelsets of the Polish-designed SUW-2000 system.

The company employs around 1,140 staff.

Freight operations

LHS holds a licence for goods traffic, therefore only freight trains currently operate on the line. In 2004 the company carried about 7.3 million tonnes

LHS Class ST44 diesel-electric locomotive with a trial piggyback service from Kiev, Ukraine, to Sławków, Poland (Stanisław J Szewczak)   1020699

LHS locomotives

Class	Wheel arrangement	Power kW	Speed km/h	Weight tonnes	First built	Builders Mechanical	Engine	Transmission
ST44	Co-Co	1,471	100	116	1966	Voroshilovgrad	Kolomna 14D20	E Charkow
SM48	Co-Co	882	100	116	1976	PZM-Lugansk	PDG-YM	E Charkow

of freight (2.6 billion tonne-km), mainly of iron ore (5.7 million tonnes) imported into Poland in block trains.

In March 2003 the first piggyback train carrying lorries arrived at Sławków terminal. It had conveyed 31 heavy road vehicles from the Ukrainian capital, Kiev, in 36 hours, including five hours of custom formalities at the border. The lorry drivers travelled in two sleeping cars attached to the train. Thanks to the larger clearances of broad gauge lines, there was no need to use the special, low-floor wagons with small-diameter wheels which are employed on standard gauge European lines. This pilot project was an element in the creation of a logistics terminal at Sławków, at the western end of the LHS line. It is estimated that

around 10 million containers are transported each year from the Far East to Europe. Some of these could be transported by rail over the Trans-Siberian Railway and onwards via Ukraine to Poland, to be re-loaded at Sławków onto standard gauge wagons or road vehicles.

Traction and rolling stock

LHS operates 50 Class ST44 diesel-electric freight locomotives and eight Class SM48 diesel shunters. The company does not own any commercial wagons. Most freight is transported in Russian or Ukrainian wagons, or in broad gauge wagons of other railways or PKP Cargo wagons suitable for wheelset- or bogie-changing.

UPDATED

PKP Rapid Commuter Railway in Tri-City Ltd (SKM)

PKP Szybka Kolej Miejska w Trójmieście Sp z oo
ul Morska 350 A, PL-81-002 Gdynia
Tel: (+48 58) 721 27 50   Fax: (+48 58) 721 29 91
Web: www.skm.pkp.pl

Key personnel
Chairman: Mikołaj Segien
Management Board Member: Piotr Małolepszy

Gauge: 1,435 mm
Route length: 32 km
Electrification: 3 kV DC

Organisation
PKP SKM operates most of the commuter services in the 'tri-city' area (Gdańsk/Sopot/Gdynia). It runs over its own 32 km double-track, electrified line linking Gdańsk, Sopot, Gdynia and Rumia, using separate tracks alongside the main line, and beyond to Wejherowo over the PKP PLK main line. SKM is also developing regional services far beyond the tri-city area. Although the dedicated line is managed and owned by SKM, many elements of infrastructure are shared with PLK, and signalling systems (station interlockings) cover both SKM and PLK tracks within the station areas, as do traction power supply substations.

At the end of 2003 SKM employed some 740 staff.

Passenger operations
In 2004 SKM carried 35.5 million passengers, running about 230 trains per day.

Traction and rolling stock
SKM operates 73 emus, which are serviced at one of the biggest depots in Europe for vehicles of this type. These comprise Classes EN57 and EN71, as well as Classes EW58 and EW60. The last-mentioned two types are being gradually withdrawn, while the other classes are being repainted in the new company's livery.

UPDATED

SKM Class EN57 emu in 'Tri-City' livery at Malbork (Andrzej Harassek)   ***NEW**/1122843*

SKM emus (all 3 kV DC)

Class	Cars per unit	Motor cars per unit	Motored axles/car	Output/ motor kW	Speed km/h	First built	Builders Mechanical	Electrical
EN57	3	1	4	152	110	1962	Pafawag	Dolmel
EN71	4	2	4	152	110	1976	Pafawag	Dolmel
EW58*	3	2	4	206	120	1975	Pafawag	Dolmel
EW60*	3	1	4	206	100	1992	Pafawag	Dolmel

* built exclusively for use on the SKM system.

PKP Regional Services Ltd

PKP Przewozy Regionalne Sp z oo
ul Grójecka 17, PL-02-021 Warsaw
Tel: (+48 22) 659 36 03; 474 25 95; 474 41 05; 830 09 49
Fax: (+48 22) 474 40 39
e-mail: p@pkp.com.pl
Web: www.pr.pkp.pl

Key personnel
Chairman: Leszek Ruta
Management Board
  Finance: Małgorzata Kuczewska-Łaska
  Commercial Affairs: Marek Nitkowski
  Technology and Operations: Henryk Szklarski
  Personnel: Andrzej Wciórka

Organisation
At the end of 2004 PKP Regional Services staff totalled around 22,000.

Passenger operations
The company operates the vast majority of passenger local, regional, and many long-distance fast trains (except IC Ex and TLK services) in Poland, as well as a number of night trains and international services (except Night Express, EC and EuroNight and some other services operated by PKP Intercity). Twenty-four Regional Services long-distance trains (including three international services) convey also PKP Intercity couchette and sleeping cars.

In 2005, after several years' break, the company reintroduced services on two secondary lines: Ełk–Olecko (northeast Poland) and Sławno–Darłowo (a seaside resort).

In 2004 the company carried 221 million passengers (14.8 billion passenger-km), 5 per cent down on the figure for the previous year. Each day the company operates around 3,400 trains, including 237 long-distance services.

In 2004 PKP Regional Services together with the Warsaw regional authority (Mazowieckie voievodship) created a new company, Koleje Mazowieckie (KM Sp z o.o.) – Mazowsze Railways Ltd (see entry in Poland section of *Railway systems and operators*). The new company has taken over local services in the Mazowieckie region.

Class SA106 railbus manufactured by PESA with a conventional trailer coach in matching livery at Mikołajki, northeast Poland (Andrzej Harassek)   ***NEW**/1122844*

Class SA109 twin-unit railbus manufactured by KOLZAM Racibórz, operating between Kołobrzeg and Goleniów in northwest Poland (Andrzej Harassek)   ***NEW**/1122845*

Traction and rolling stock

PKP Regional Services has about 3,000 compartment and open saloon coaches (24.5 m UIC-Y type), mostly built in the 1970s and 1980s, 760 four-car, articulated double-deck coach sets, 150 double-deck coaches, and 260 luggage, buffet and other special coaches. The company also owns 1,076 emus and 12 lightweight diesel cars (railbuses). A further 37 railbuses (including seven secondhand three-car Class 624/924 dmus purchased by Szczecin region from DB AG) operated by PKP Regional Services are owned by local authorities (voievodships). As with PKP Intercity, PKP Regional Services has no locomotives for hauled trains. These are hired from PKP Cargo. Train drivers, including those for emus and railbuses, are also hired from PKP Cargo. The company performs only day-to-day maintenance of coaches and emus in its own depots. Major overhauls are performed by Rolling Stock Overhaul Works.

In 2005 PKP Regional Services received €51 million from European aid programmes (structural funds) for the modernisation of passenger rolling stock and purchase of new vehicles. Using EU funding and a Polish contribution, the company is to acquire 11 160 km/h emus for services between Warsaw and Łódź, the second largest city in Poland, over the line being modernised also under structural funds programme (see entry for PKP PLK SA in the Poland section of *Railway systems and operators*). Further 75 existing emus will be modernised, as well as 81 passenger coaches, 24 of them to be adapted for passengers using wheelchairs.

UPDATED

Electric multiple units owned by PKP Regional Services (all 3 kV DC)

Class	Cars per unit	Axles/car	Motor cars per unit	Motored axles/car	Output/ motor kW	Speed km/h	First built	Builders Mechanical	Electrical
EN57	3	4	1	4	152 or 175*	110	1962	Pafawag	Dolmel
EN71	4	4	2	4	152	110	1976	Pafawag	Dolmel
ED72	4	4	2	4	152/195**	110	1993	Pafawag	Dolmel
ED73	4	4	2	4	195	120	1996	Pafawag	Dolmel

* motors replaced during modernisation.
** second batch, with higher output and top speed.

Lightweight diesel cars (railbuses), including those owned by local authorities and operated by PKP Regional Services

Class	Cars per unit	Axles/car	Motor cars per unit	Motored axles/car	Transmission	Output/ motor kW	Speed km/h	First built	Builders
SN81	2	2	1	1	mechanical	110	90	1988	Kolzam
SA101	2	2	1	1	hydraulic	200	90	1990	HCP
SA102	3	2	2	1	hydraulic	400	90	1993	HCP
SA104	2	2	1	2	hydraulic	157	90	1995	Kolzam
SA105 (213M)*	1	2	1	1	hydraulic-kinetic	250	100	2002	ZNTK Poznań
SA108 (215M)*	2	2	2	1	hydraulic-kinetic	2 × 250	100	2003	ZNTK Poznań
SA106 (214M)*	1	4	1	2	hydraulic	500	120	2001	PESA Bydgoszcz
SA107 (211M)*	1	2	1	1	hydraulic	190	100	2003	Kolzam
SA109 (212M)	2**	2	2	1	hydraulic	2 × 190	100	2004	Kolzam
SA110/ SA112***	¾	4	2	2	hydraulic	2 × 247	120	1964	MAN, Waggonfabrik Uerdingen

* owned by local authorities.
** three-car units are also planned.
*** Former DB Class 624/924 dmus, three- or four-car units consisting of two motor cars (Class 624) and one or two intermediate trailers (Class 924). In 2005 only two-car (motor + motor) formations were in use, with trailers kept as reserve.
Manufacturers' type designations are shown in parentheses.

Refurbished Class EN57 emu (Andrzej Harassek)   0585007

Class SA108 (Type 215M) two-car diesel railbus built by ZNTK for local authority-sponsored regional services (Andrzej Harassek)   0585006

PKP Warsaw Suburban Railway Ltd (WKD)

PKP Warszawska Kolej Dojazdowa Sp z oo
ul Batorego 23, PL-05-825 Grodzisk Mazowiecki
Tel: (+48 22) 755 55 64   Fax: (+48 22) 755 20 85
e-mail: wkd@wkd.com.pl
Web: www.wkd.com.pl

Key personnel

Chairman: Grzegorz Dymecki
Management Board
    Finance and Commercial Affairs: Jolanta Dałek

Gauge: 1,435 mm
Route length: 36 km
Electrification: 600 V DC overhead

Organisation

Inaugurated in December 1927, the present WKD system was one the earliest electrified railways in Poland. Then named Elektryczna Kolej Dojazdowa (EKD or Electric Suburban Railway), the line was privately owned. After the Second World War the railway was nationalised and absorbed within PKP. Its name was changed to Warszawska Kolej Dojazdowa (WKD), and it formed part of the PKP Warszawa region.

Prototype Class EN95 part-low-floor emu for the WKD Warsaw–Grodzisk Mazowiecki suburban line (Piotr Rydzyński)   0585004

WKD 600 V DC emus

Class	Cars per unit	Motor cars per unit	Motored axles/car	Output/ motor kW	Speed km/h	First built	Builders Mechanical	Electrical
EN94	2	2	2	56.5	80	1969	Pafawag	Dolmel
EN95	4*	2	2	250	90	2004	PESA	Medcom, IEL

* Articulated unit (Bo-2-2-2-Bo).

Interior of Class EN95 emu (Piotr Rydzyński)

0585005

Two Class EN94 LRV-type emus on the WKD Warsaw–Grodzisk Mazowiecki suburban line (Andrzej Harassek)

1020697

In 2005 privatisation of PKP Warszawska Kolej Dojazdowa Sp z o.o. was well advanced and it was likely this would be completed by the end of the year. All its shares will be sold to the consortium of local authorities of six counties served by the line, the Warsaw city authority and the Mazowieckie regional authority.

The line is separate from the main PKP network. The only connection between the two is provided over a non-electrified link between Pruszków main line station and Komorów WKD station. This track is used for supplies to WKD.

The double-track line is electrified using the 600 V DC overhead system and, due to short braking distances, is equipped with a two-aspect automatic line block system between Warszawa śródmieście WKD terminus and Podkowa Leśna Główna station. A further single-track section goes to Grodzisk Mazowiecki terminus, where the line's depot and the company's headquarter are located. From Podkowa a 3 km single-track branch to Milanówek is in operation.

WKD employs 210 staff.

Passenger operations
In 2004 WKD carried some 6.2 million passengers.

Traction and rolling stock
WKD services are provided by 35 purpose-built two-section articulated emus, which operate in pairs.

In June 2004 WKD received a prototype of a new emu design from Pojazdy Szynowe (Rail Vehicles) PESA Bydgoszcz SA. The Class EN95 unit is a four-car (Bo-2-2-2-Bo) articulated emu, specially designed for the WKD line. The unit has four 250 kW asynchronous traction motors and can accommodate 150 seated and 350 standing passengers. The floor in the sections between bogies is 600 mm above rail. Commercial service with the vehicle commenced in September 2004.

WKD also retains one serviceable vintage four-axle Class EN80 motor car built by English Electric in 1927. This is used for special trips.

UPDATED

Polish Railway Lines Joint Stock Company (PKP PLK)

PKP Polskie Linie Kolejowe Spółka Akcyjna (PKP PLK SA)
ul Targowa 74, PL-03-734 Warsaw
Tel: (+48 22) 619 98 83   Fax: (+48 22) 473 39 43
Web: www.plk-sa.pl

Key personnel
Chairman: Tadeusz Augustowski
Board Members
  Technology and Development:
    Krzysztof Groblewski
  Product Quality and Sales: Krzysztof Szwed
  Finance and Economics: Mirosław Pawłowski
  Personnel and Administration: Andrzej Krawczyk
  EU Co-operation: Grażyna Liberadzka
Public Relations: Krzysztof Łańcucki
  Tel: (+48 22) 473 22 00   Fax: (+48 22) 473 30 02
  e-mail: k.lancucki@pkp.com.pl

Gauge: 1,435 mm; 1,520 mm
Route length: 21,645 km; 234 km. 19,111 km are in operation (4,241 km of trunk lines, 10,173 km of primary (main) lines, 3,380 km of secondary lines and 1,318 km of local lines). 11,945 km of lines are considered of national importance. Operations on remaining lines suspended.
Electrification: 11,842 route-km at 3 kV DC

Political background
In accordance with the provisions of the parliamentary act on railway transport and the act of 8 September 2000 on the commercialisation, restructuring and privatisation of PKP, PKP PLK SA was established as infrastructure manager for the Polish national rail network (except the railway lines operated by WKD Ltd, SKM in Tri-City Ltd and LHS Ltd). PLK is responsible for its operation and maintenance and for investment in it. The

PLK computerised CTC signalbox (Andrzej Harassek)

0585001

infrastructure is owned by PKP SA and hired to PLK. These assets will be gradually taken over by PLK as progress is made in clarifying the legal status of the land occupied by the railway. New assets created as a result of investments will be owned by PLK.

Initially PKP SA was the sole owner of all PLK shares, which would be gradually taken over by the state treasury against value of government subsidies. The state finances investments on the main railway lines (lines of national importance). Investments on other lines and maintenance of the entire network is financed from PLK's own funds, which are generated by access charges from train operators, bond issues, loans and other sources. According to legislation, PLK shares held by PKP SA and the State Treasury are non-disposable. This means the PLK SA cannot be sold to the private sector.

Organisation
PLK's primary activity is to assign access to the network for train operators and to develop schedules based on their requirements, as well as the development of a regulatory regime and pricing policies. In 2004 four PKP group companies operated trains over PLK lines: Intercity, Cargo, Regional Services and SKM (the last-mentioned

also has its own commuter line between Gdańsk and Gdynia). The new company, Koleje Mazowieckie Sp z o.o. (Mazowieckie Railways Ltd. – a joint venture of PKP Regional Services Ltd and Mazowieckie regional authority operating regional passenger services in Mazowieckie region), also operates on PLK lines. In addition, there were 27 non-PKP operators, which include three Czech companies; České Drahy as, Connex Česka Železnični sro, and Railtrans sro, and German operator Lausitz Bahn GmbH. The Czech companies operate freight and local passenger trains on the so-called 'privileged transit' lines near Głuchołazy and between the Czech town of Liberec and Zittau, Germany (the line runs 3 km through Polish territory). Lausitz Bahn operates the Zittau–Görlitz passenger service via Polish tracks, based on the similar privileged transit principle.

In 2004 the network handled 235.87 million train-km, broken down as: PKP Regional Services 50.8 per cent; PKP Cargo 30.5 per cent; PKP Intercity 7.1 per cent; SKM in Tri-City 0.3 per cent; and others (non-PKP) 2.2 per cent. A total of 9.1 per cent of train-km constitutes movements for the internal needs of PKP Group companies, mainly works trains and so-called 'service trains' (for example, empty passenger trains), as well as light engine movements.

PLK employs about 44,700 staff.

Improvements to existing lines

PLK investments are concentrated on the main international lines which form pan-European transport corridors. In 2004 the total of PLN993.1 million of investments were financed by the state budget (28 per cent), the EU's Phare and ISPA/Cohesion funds (27.5 per cent), EIB loans (26 per cent), PLK's own resources (18 per cent) and other sources (0.5 per cent). In 2002 the first two ISPA projects started. These cover the modernisation of the E 20 line between Mińsk Mazowiecki and Siedlce (east of Warsaw, towards Belarussian border at Brest). In 2003 further ISPA projects include modernisation of the Poznań node (along the E 20 line), modernisation of the eastern section of the E 20 line, from Siedlce to Terespol (the border crossing to Belarus) and modernisation of the E 30 line between Legnica and the German border in Görlitz (together with freight branch to Horka). From 2004 all former ISPA projects have been continued under the Cohesion Fund.

Further projects, to be co-financed from the Cohesion Fund's 2004–06 programme, include the first phase of modernisation of the Warsaw–Gdynia section of the E 65 north-south link and the first phase of the Wrocław–Poznań E 59 line modernisation (the Wrocław–Rawicz section), while structural funds will contribute to improvements to services between Warsaw and Łódź, Poland's second largest city. The EIB loan continues to contribute mainly to modernisation of the Opole–Wrocław–Legnica section of the E 30 line. On this section Phare funding was also used in 2003, the last investment using finance from this source. Other lines being modernised from the state budget and PLK sources include the Grodzisk Mazowiecki–Zawiercie E 65 line (the Central Trunk Line or Centralna Magistrala Kolejowa (CMK)), and the Łowicz–Kutno section of line E 20.

Proposals in the National Development Plan for 2007–13, to be co-financed from the Cohesion Fund, include completion of the E 20 line modernisation, the second phase of modernisation of the Warsaw–Gdynia and Wroclaw–Poznan lines, modernisation of further sections of the E 65 line south of Warsaw, the start of modernisation of the E 59 Poznan–Szczecin land CE 59 Rzepin–Wroclaw sections, continuation of the E 30 modernisation, from Opole to Kraków, as well as the start of modernisation of the E 75 line from Warsaw via Bialystok to Sokólka, along the pan-European transport corridor I (Rail Baltica). With the use of structural funds from the European Regional Development Fund (ERDF), completion of service improvement between Warsaw and Lódz, is planned. Further proposals for ERDF include improvement of connections between Warsaw, Radom and Kielce, as well as Psary (on the CMK line) and Kraków. Introduction of 200 km/h passenger services on some lines is also planned. Studies into the development of high-speed links

0114305

New turnout installed at Mrozy, on line E 20 between Mińsk Mazowiecki and Siedlce, which is being modernised with the help of ISPA funds (Andrzej Harassek)   1020701

will be carried out; one proposal concerning the connection of Warsaw with Wroclaw and Poznan, via Lódz. The study will have to determine whether the upgrading of existing lines for 200–250 km/h or the construction of an entirely new line would best meet needs.

Electrification

The 11,842 km of the network electrified at 3 kV DC, including nearly 7,900 km of double-track and about 4,000 km of single-track lines (25,289 km track-km), constitutes 60 per cent of all the lines managed by PLK.

Signalling and telecommunication

The majority of station interlockings are still of the electromechanical or relay type. However, computerised equipment is being introduced increasingly, controlling about 4 per cent of turnouts at 37 stations or locations. The three- or four-aspect signalling is in most cases bi-directional; automatic block systems cover 2,661 km of the lines, 115 km of them covered by computerised systems. Increasing numbers of train detection systems based on axle counters are being introduced, replacing track circuit equipment. The eight Centralised Train Control (CTC) centres control 428 turnouts along

368 km of lines. At about 150 locations hot-box detectors are in operation. Increasing numbers of modern automatic protection systems (warning lights and barriers) on level crossings are being installed.

Almost the entire network is covered by simple train to track radio (simplex operation in the 150 MHz waveband). Radio equipment, however, is currently being modernised. In the long term, implementation of the standard European Rail Traffic Management System (ERTMS), including the European Train Control System (ETCS) and GSM-R digital radio system, is planned.

UPDATED

Modernised station at Mrozy, on line E 20 between Mińsk Mazowiecki and Siedlce
(Andrzej Harassek)
1020700

Polish State Railways Joint Stock Company (PKP SA)

Polskie Koleje Państwowe SA
ul Szczęśliwicka 62, PL-00-973 Warsaw
Tel: (+48 22) 474 91 01   Fax: (+48 22) 524 91 02
Web: www.pkp.pl

Key personnel
Supervisory Board
   Chairman: Krzysztof Białowolski
   Vice-Chairman: Jerzy Drygalski
   Board Members: Jacek Barylski,
      Kazimierz Gontarczyk, Stanisław Kogut,
      Jerzy Kędzierski, Marek Tadeusz Krawczyk,
      Wojciech Paprocki, Zygmunt Świrski
Management Board
   President and Chief Executive Officer:
      Andrzej Wach
Members
   Financial Director: Janusz Lach
   Director, Corporate Governance and Privatisation:
      Jacek Bukowski
   Director, Strategy and Organisation:
      Zbigniew Szafrański
   Director, Social Affairs and Promotion:
      Maria Wasiak
Management Board Office
   Tel: (+48 22) 474 91 01   Fax: (+48 22) 474 91 02
Press Spokesman (Public Relations): Anna Wolek
   Tel: (+48 22) 474 92 00   Fax: (+48 22) 474 92 02
   e-mail: rzecznik@pkp.pl
International Co-operation Office
   Tel: (+48 22) 474 91 81   Fax: (+48 22) 474 91 82
   e-mail: kzr@pkp.pl

Political background
In the late 1990s the financial situation of the Polish State Railways (PKP) continued to deteriorate. In 1998 a new structure for the organisation was implemented, which created separate business units covering infrastructure, passenger and freight operations. This was the first stage of PKP restructuring, with an aim to create business-oriented, independent railway companies capable of operating in both domestic and international transport markets. Clear distinctions would have to be made between different fields of activity, separating profitable freight and, potentially, long-distance passenger services, from loss-making regional and suburban services that require support from central governmental or local and regional authorities.

The framework for the further restructuring of PKP was approved by the Polish government in 1999. A parliamentary act of 8 September 2000 covering the commercialisation, restructuring and privatisation of the Polish State Railways state enterprise, provided for three stages of reform:

Commercialisation
The initial step was the transformation of the state-owned PKP enterprise into a joint stock company acting under the name 'Polskie Koleje Państwowe Spółka Akcyjna' (PKP SA or Polish State Railways Joint Stock Company), which entered into the commercial register on 1 January 2001. The State Treasury is the only shareholder of PKP SA, which assumed all legal relationship of the former Polish State Railways and effectively replaced it.

Restructuring
The main objective of the restructuring was to establish PKP SA as a holding company, necessitating transformation of organisation, finance, labour and assets. Accordingly, subsidiaries were established based on the existing business units:
- A joint stock company to manage the rail infrastructure, PKP Polskie Linie Kolejowe Spółka Akcyjna (PKP PLK SA or PKP Polish Railway Lines Joint Stock Company), which commenced activities on 1 October 2001.
- Three train operating companies:
   PKP Intercity Sp z o.o. (PKP Intercity Ltd), operating long-distance, high-quality passenger trains (EC, IC, Ex and EN services);
   PKP Przewozy Regionalne Sp z o. o. (PKP Regional Services Ltd), operating all other passenger services;
   PKP Cargo Spółka Akcyjna (PKP Cargo SA) joint-stock company for freight traffic.
- Three railway companies, operating both their infrastructure, and trains:
   PKP Warszawska Kolej Dojazdowa Sp z oo (PKP WKD or PKP Warsaw Suburban Railway Ltd), operating the 35 km double-track electrified (600 V DC) line between Warsaw and Grodzisk Mazowiecki;
   PKP Szybka Kolej Miejska w Trójmieś´cie Sp z oo (PKP SKM or PKP Rapid Commuter Railway in Tri-City Ltd), serving the so-called Tri-City area of Gdan´sk/Sopot/Gdynia, operating commuter electric trains mainly on dedicated tracks, which it owns, between Gdan´sk and Gdynia, and over PLK tracks further west to Wejherowo; PKP SKM also operates some regional services over PLK lines beyond the Tri-City and Wejherowo area;

   PKP Linia Hutnicza Szerokotorowa Sp z oo (PKP LHS or PKP Metallurgic Broad Gauge Line Ltd), operating freight trains over a broad-gauge (1,520 mm), 395 km single non-electrified line from the Ukrainian border at Hrubieszów to the Silesia industrial region.
- A number of other companies, the most important being:
   PKP Energetyka (PKP Power Supply Ltd), operating power supply systems;
   PKP Telekomunikacja (PKP Telecommunication Ltd), operating the long distance cables and transmission systems, as well as the railway telephone network;
   PKP Informatyka (PKP Informatics Ltd), dealing with information technology systems.
In addition, a number of infrastructure maintenance companies were created from former PKP maintenance units as well as several other auxiliary companies.

A labour restructuring programme, approved by trade unions, has been implemented. The number of employees had fallen to about 140,000 by the end of 2004. This programme is supported by loan agreements with the European Bank for Reconstruction and Development, which is providing a long-term loan of EUR100 million, and with the World Bank, providing a long-term loan of EUR110 million for funding railway restructuring, including social protection measures. All these loans will be repaid by PKP SA.

Privatisation
In the long term, privatisation of PKP SA and its subsidiaries is expected. Excluding PLK SA the restructuring legislation does not impose any restrictions in selling shares in these companies and all potential buyers would be able to buy these on equal terms.

Organisation
PKP SA is the parent company for all the subsidiaries created in 2001. It took over from the PKP state enterprise all its obligations and rights. The main objective of the company is to perform full restructuring and eventually the privatisation of its subsidiaries, excluding PKP PLK SA.

PKP SA does not undertake any transport operations. All the assets necessary for subsidiaries to sustain operations are being transferred, leased or hired to them, generating income for PKP SA.

UPDATED

Pol-Miedz Trans Sp z o.o.

Pol-Copper Trans Ltd
PL-59-301 Lubin
Tel: (+48 76) 847 18 00   Fax: (+48 76) 847 18 09
e-mail: marketing@pmtrans.com.pl
Web: www.pmtrans.com.pl

Key personnel

Chairman of the Board: Marek Markiewicz
Members
  Commercial and Restructuring: Piotr Stryczek
  Logistics: Edward Słodyński

Organisation

Pol-Miedź Trans Sp z o.o. is a member of the
KGHM Polska Miedź SA (Polish Copper) group

of companies involved in mining and processing
copper ore and the production copper products
in the Legnica/Lubin/Głogów area of Lower
Silesia. In the 1970s KGHM built a 25 km single
line linking the mines, ore processing plants and
a sand mine, as well as rail networks within the
KGHM's mines and other plants. In 1997 Pol-Miedź
Trans Sp z o.o. was created as KGHM's transport
division.

Freight operations

The company operates freight trains, mostly of
copper ore and processed ore (concentrates)
between the KGHM plants both over its own
network and on PKP PLK tracks to the copper works
in Głogów. In 2004 the company carried 15 million
tonnes (374 million tonne-km).

Traction and rolling stock

Pol-Miedź Trans has about 50 diesel locomotives,
mostly equivalent to PKP Class SM31, SM42,
SM48. There are also about 800 wagons, mostly
for the transport of copper ore and other bulk
materials such as sand and coal.

NEW ENTRY

Przedsiębiorstwo Transportu Kolejowego i Gospodarki Kamieniem SA (PTKiGK Rybnik)

Railway Transport and Stone Supply Management
Joint Stock Company, Rybnik
ul Kłokocińska 51, PL-44-251 Rybnik
Tel: (+48 32) 739 49 01; 49 71
Fax.: (+48 32) 422 06 92; 02 27
e-mail: ptkigk@ptkigk.com
Web: www.ptkigk.com

Key personnel

Chairman of the Board: Kazimierz Musiolik

Members
  Technology: Grzegorz Przygoda
  Finance: Mieczysław Kuśka
  Owner's Supervision: Tadeusz Wereszczak

Organisation

With a network of 357 km of lines and industrial
branches, this is the largest of the former so-called
'sand railways'. In 2004 some 53 million tonnes (664
million tonne-km) were carried, mainly of coal and
stone. The company is the second largest freight
operator in Poland and in 2003 commenced the
regular transport of coal and other bulk materials

in block trains using its own locomotives over the
PKP PLK lines.
  The company has around 1,120 employees.

Traction and rolling stock

The fleet totals 126 locomotives, comprising two
electrics and 124 diesels of Types TEM2 (SM48) and
SM42 as well as T200, T448 and T419 Czech-built
machines, and some 1,600 wagons.

UPDATED

Przedsiębiorstwo Transportu Kolejowego i Gospodarki Kamieniem Sp zoo (PTKiGK Zabrze)

Railway Transport and Stone Supply Management
Ltd, Zabrze
ul Wolności 337, PL-41-800 Zabrze
Tel: (+48 32) 271 44 41   Fax: (+48 32) 271 50 74
e-mail: holding@ptkigk.pl
Web: www.ptkigk.pl

Key personnel

Chairman of the Board: Zbigniew Pucek

Organisation

The company carried about 34.5 million tonnes
(715 million tonne-km) in 2004, mainly of coal and
stone. In 2003 it commenced regular transport of
coal and other bulk materials in block trains using
its own locomotives over PKP PLK lines.
  The company has around 1,800 employees.

Traction and rolling stock

The company has more than 80 locomotives,
including electrics equivalent to PKP Classes EU07,
ET21 and ET22, 39 main line and 42 shunting
diesels and around 1,300 wagons.

UPDATED

*PTKiGK Classes 4E (EU07) and 3E (ET21) electric locomotives on PKP PLK tracks near Warsaw
(Andrzej Harassek)*   *NEW*/1122846

Rail Polska Sp zoo

ul Willowa 8/10 lok 11, PL-00-790 Warsaw
Tel: (+48 22) 646 54 67; 646 54 68
Fax: (+48 22) 646 54 66
e-mail: railpolska@railpolska.pl
Web: www.railpolska.pl

Key personnel

Chairman of the Board: Edward A Burkhardt

Organisation

Based in Warsaw, Rail Polska Sp zoo was
established as a private train operating company
in Poland in 1999. It is a subsidiary of Rail World Inc,

the company which purchased a 66 per cent
shareholding in Estonian State Railways in 2001.
In 2003 Rail Polska purchased 100 per cent of
the shares in two Polish companies, Kolex
Sp zoo and ZEC-Trans Sp zoo. In August 2003
Kolex and ZEC-Trans obtained licences for rail
freight transport in Poland. ZEC-Trans operated its
first coal train on 13 October 2003, followed by a
second service on 12 November 2003. Rail Polska
was planning to start new operations on other
routes in 2004.

Traction and rolling stock

At the end of 2003 the company owned six TEM2
locomotives (equivalent to PKP Class SM48), five

Class SM42 and three M62 diesels (PKP Class
ST44). Purchase of further engines was expected.
All locomotives are being repainted in a new livery
of Bordeaux red and cream, similar to that adopted
for EWS in the UK and the former Wisconsin
Central in the US.

UPDATED

Sybka Kolej Miejska Sp z o.o.

Rapid Commuter Railway Ltd
ul Wilczy Dół 5, PL-02-798 Warsaw
Tel: (+48 22) 655 43 01   Fax: (+48 22) 655 45 98
e-mail: biuro@skm.warszawa.pl
Web: www.skm.warszawa.pl

Key personnel
Management Board Member: Jerzy Obrębski

Organisation
The company is a joint venture of Warsaw tram authority, Tramwaje Waszawskie Sp z o.o., and Metro Warszawskie Sp z o.o. (Warsaw Metro Ltd), set up to create rapid city-rail services over PKP

PLK lines within the greater Warsaw area. In June 2005 SKM was granted a licence for the transport of passengers and for leasing (or hiring) of traction vehicles.

In December 2005 SKM intends to start its first service between the Warsaw Zachodnia (West) station and Falenica, the last station within greater Warsaw, on the line towards Lublin. In 2006 the company intends to extend the service to Grodzisk Mazowiecki, southwest of Warsaw, to Otwock in the southeast). Further services, due to commence in 2006–07, include: Warsaw Zachodnia–Warsaw Gdańska-Legionowo (north of Warsaw, towards Gdańsk); Warsaw Zachodnia–Piaseczno (south of Warsaw, towards Radom and Kraków over the old line); and possibly Warsaw Wschodnia–Warsaw

Rembertów–Mińsk Mazowiecki (east of Warsaw, towards Terespol).

Traction and rolling stock
The company ordered a first batch of six three-car emus to be built by ZNTK Nowy Sącz. The trains will be completely rebuilt and modernised Class EN57 emus, with new bodyshells with streamlined front ends, new electrical equipment and modernised bogies. The cars' interior will also be entirely new, with more comfortable seats, luggage racks and small tables between seats. An onboard CCTV system will enhance passenger security. All six emus are to be delivered by December 2005.

NEW ENTRY

Portugal

Ministry of Planning, Transport and Territorial Administration

Praca do Comércio, Ala Oriental, P-1100 Lisbon Codex
Tel: (+351 1) 87 95 41   Fax: (+351 1) 86 76 22

Key personnel
Minister: Ferro Rodrigues

Portuguese Railways (CP)

Caminhos de Ferro Portugueses
Calçada do Duque 20, P-1294 Lisbon Codex
Tel: (+351 1) 346 31 81   Fax: (+351 1) 347 65 24

Key personnel
Board of Directors
Chairman: Crisóstoma Teixeira
Members: Eng Vilaça e Moura, Dr Elsa Roncon
    Santos, Dr Silva Rodrigues, Dr Moura Calhão,
    Eng José Espinha, Dr Braancamp Sobral
Directors General
    Infrastructure: Eng Francisco Carapinha
    Engineering and Investment: Eng Nuno Leandro
    Operations: Dr Oliveira Monteiro
    Sintra Line Business Unit: Eng Martins de Brito
Directors
    Planning: Vacant
    Innovation and Development: Eng Tiago Ferreira
    Data Processing: Eng Guimares da Silva
    Legal Affairs: Dr Almeida Coragem
    Finance: Dr Viegas de Barros
    Audit: Eng Vitor Biscaia
    Human Resources: Dr Maria João Tender Arroja
    Investment Management: Dr Tavares Fernandes
Administration Secretariat: Dr Luis Beato
Public Relations: Dr Américo Ramalho
Regional Managers
    Northern Region: A Villaverde
    Southern Region: Eng Alberto Grossinho
    Cascais Line: Eng João Cunha
    Sintra and Circle Lines: Eng Conceição e Silva

Political background
The government announced a radical restructuring of the Portuguese railway sector in 1996, apparently prompted by CP's worsening financial performance. In 1995 CP recorded a deficit of Esc59.961 billion, following deficits of Esc48.37 billion in 1994 and Esc54.275 billion in 1993. As part of the restructuring, the government agreed to assume Esc450 billion of debt owed by the railways.

A new body, REFER (Rede Ferroviária – qv), began taking over responsibility for infrastructure on 1 July 1997, while the remainder of CP was to continue to operate the railway. The bill creating REFER was passed by the parliament in February 1997. Signalling centres were to be the final responsibility assumed by REFER, in January 1999.

Open access rights will encourage new operators to come into the industry, and a regulatory body is to be set up to ensure fair treatment of all companies using the network. Through restructuring, the government hoped to determine which socially necessary services would continue to be subsidised. It was suggested that

regional services might be provided under contract by private companies and more public/private partnerships set up, such as the Mirandela Metro venture between CP and local government on the Tua-Mirandela route (see 'Passenger operations' section below). Operation of commuter services over the 25 de Abril bridge (see 'New lines' section in REFER entry) is to go out to tender.

In 1998 CP was reorganised into five business units: Greater Lisbon Suburban (USGL); Greater Porto Suburban (USGP); Traction & Rolling Stock (UMAT); Interurban & Regional (UVIR); and Freight & Logistics (UTML). The first to come into operation, in February 1998, was UTML, where traffic rose by 27 per cent in 1997. This unit, which operates from 12 to 18 terminals across Portugal, may be privatised within three years.

USGL came into existence in March 1998, managing Sintra, Cascais and Azambuja services, as well as the Lisbon circle (Cintura) line and Barreiro-Setúbal services on the south bank of the Tagus. USGP manage four lines, and became operational in 1998 following completion of certain infrastructure projects.

Outright privatisation of CP is not on the agenda at present.

Organisation
CP has created over 10 subsidiary companies to manage more peripheral activities. Private companies have a stake in some of these subsidiaries, such as those involved in the movement of containers and cars by rail and the sale of advertising space on railway sites. Recently created subsidiaries include SOFLUSA, managing CP's Lisbon-Barreiro ferry service, door-to-door (using passenger trains) parcel carrier TEX and EMEF. The latter undertakes work for CP and other operators at four ex-CP heavy-repair sites with a workforce of 1,930.

CP staff numbers have fallen from 21,037 at the end of 1991 to 13,024 at the end of 1997. This reduction has been achieved in part through early retirement and voluntary redundancy schemes.

Finance
In early 1998 the government partially offset CP's accumulated debt of Esc450 billion by allocating

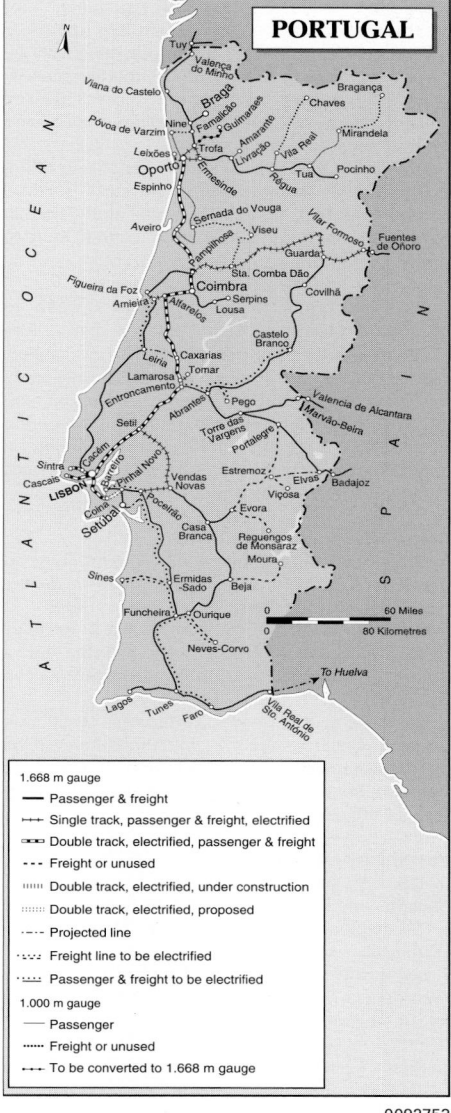

0092753

Diesel locomotives

Class	Wheel arrangement	Power kW	Speed km/h	Weight tonnes	No in service	First built	Builders Mechanical	Engine	Transmission
9001/003	Bo-Bo	425	70	46	3	1959	Alsthom	SACM/MGO	E Alsthom
9004/006	Bo-Bo	440	70	46	3	1959	Alsthom	SACM	E Alsthom
9021/031	Bo-Bo	530	70	46.65	11	1976	Alsthom	SACM	E Alsthom
1001/1006	C	120	41.5	30.4	6	1948	Drewry	Gardner	M Sinclair
1021/1025	B	320	65	36	5	1968	Moyse	Deutz	E Moyse
1051/1068	B	90	38	28.3	13	1955	Moyse	Moyse	E Moyse
1101/1112	Bo-Bo	190	56	41.2	11	1946	GE	Caterpillar	E GE
1151/1186	C	190	58	42	36	1966	Sorefame	Rolls-Royce	H Rolls-Royce
1201/1225	Bo-Bo	480	80	64.7	25	1961	Sorefame	SACM	E B&L
1321/1337	Co-Co	750	120	87	17	1968	Alco	Alco	E GE
1401/1467	Bo-Bo	765	105	64.4	65	1967	Sorefame	EE	E EE
1501/1521	A1A-A1A	1,290	120	111	17	1948	Alco	Alco	E GE
1551/1570	Co-Co	1,290	120	89.7	20	1973	MLW	MLW/Alco	E GE Canada
1801/1810	Co-Co	1,530	140	110.3	10	1968	EE	EE	E EE
1961/013	Co-Co	1,680	120	121	13	1979	MLW	MLW	E GE Canada
1901/13	Co-Co	2,240	100	120	13	1981	Sorefame	SACM	E Alsthom
1931/47	Co-Co	–	120	116	17	1981	Sorefame	SACM	E Alsthom

to the railway shares in Electricidade de Portugal worth Esc151.5 billion, making it the latter company's largest shareholder, although without voting rights. This has effectively boosted CP's capital from Esc128.5 billion to Esc280 billion.

Traffic (million)	1996	1997
Passenger journeys	177.15	178.1
Passenger-km	4,503	4,563
Freight tonnes	7.9	9.3
Freight tonne-km	1,857	2,247

Passenger operations

CP's passenger traffic, in steady decline since the early 1990s, saw a modest increase in 1997. Whilst express and interregional services have grown in popularity, this has not been able to compensate for the decline in regional, suburban and international traffic, largely put down to increased car ownership, an improving road network and a higher standard of living for many sectors of the population.

In 1997 suburban services accounted for 59.9 per cent of total CP passenger-km. Total passenger traffic in terms of passenger-km rose marginally by 1.3 per cent between 1996 and 1997; suburban services achieved an increase of 1.8 per cent in passenger-km, while the figure for long- and medium-distance and regional services was 0.6 per cent.

The first of a fleet of 10 six-car Pendoluso tilting trainsets (see 'Traction and rolling stock' section) began operating between Lisbon and Porto in May 1998, although CP's aim of cutting journey times using this equipment were frustrated by major delays to infrastructure upgrading.

Exclusive rights to operate from Cachão to Mirandela (14 km) on CP's 1,000 mm gauge Tua-Mirandela route have been granted to Mirandela Metro Co. Owned mostly by the municipality of Mirandela, with CP holding the remaining 10 per cent, the company began operating over an initial 4.1 km section from Mirandela in July 1995. Operations with four LRV 2000 railbuses are eventually to be extended to Cachão.

Cross-Tagus link

In December 1997 three groups entered bids to manage cross-Tagus suburban services using the new link described below (see entry for REFER). In mid-1998 the Fertagus consortium, which includes CGEA Transport of France, was selected for a 30-year operating concession. Services commenced in June 1999.

For the first three years, Fertagus is to operate under financial safeguards, but thereafter will be compensated with lower track charges if passenger numbers are lower than expected and with higher fees if traffic exceeds expectations.

Freight operations

In 1998 CP's freight activities were consolidated to become one of the railway's five business units, UTML (Unidade de Transportes de Mercadorias e Logística).

Significant increases in freight traffic were recorded in 1997. Turnover rose 23.5 per cent to Esc14.1 billion, while tonnage grew 17.7 per cent from 7.9 to 9.3 million. Receipts from intermodal traffic increased by 37.1 per cent and tonnage was up 31.5 per cent, while conventional international freight tonnage rose by 58.9 per cent. Tonnage increases for specific commodities included: timber products 71.6 per cent; aggregates 59.7 per cent; iron and steel 46.8 per cent; coal 26.3 per cent; and automotive components 16.7 per cent.

An express service known as CEMI (Comboio Expresso de Mercadorias Internacional) is operated on the Lisbon-Barcelona route in conjunction with RENFE of Spain. Specialising in the movement of palletised consumer goods, CEMI offers door-to-door service with collection and onward distribution by road.

Intermodal operations

Three new regular intermodal services to and from Spain were launched in 1995: an Intercontainer service from Lisbon and Leixões (Oporto) to Santurce (Bilbao); an Intercontainer service from Lisbon and Leixões to Algeciras/Cádiz; and a service for Liscont moving refrigerated containers between

Class 1800 diesel locomotive heading Vila Real de Santo António-Barrreiro passenger service at Tunes (Chris Wilson)   0023361

Class 5600 electric locomotive on Alfa Porto-Lisbon passenger service at Oporto Campanha (Colin Boocock)   0023365

Class 2200 Sintra line emu at Campolida (Colin Boocock)   0023366

Electric locomotives

Class	Wheel arrangement	Power kW	Weight tonnes	No in service	Builders Mechanical	Electrical
2501–2515	Bo-Bo	2,116	72	15	50 c/s Group	50 c/s Group
2551–2570	Bo-Bo	2,116	70.5	20	Sorefame	50 c/s Group
2601–2629	B-B	2,940	78	21	Alsthom	50 c/s Group
5600	Bo-Bo	5,600	88	30	Sorefame	Siemens

Lisbon and Vigo. Lisbon's Beirolas container terminal was replaced by a new facility at Bobadela which opened on 24 July 1995. CP eventually hoped to restructure its freight operations around Bobadela and three other principal terminals at Oporto, Aveiro and Setúbal.

Pecobasa, which owns 10 bimodal Transtrailer bogies, has asked CP to purchase 30 similar bogies to allow the establishment of a Lisbon-Barcelona service.

Traction and rolling stock

In 1998 a separate business unit, UMAT (Unidade de Material e Tração), was established to manage CP's rolling stock fleet. Its remit includes the sale of traction services to external clients.

In 1996, CP's 1,668 mm gauge rolling stock fleet comprised 220 diesel locomotives (including shunters), 82 electric locomotives, 195 emus, 80 dmus, 487 passenger coaches and 3,976 freight wagons. For its 1,000 mm gauge network, CP had at its disposal 10 diesel locomotives, 50 diesel railcars, 15 passenger coaches and 5 freight wagons.

In early 1998 CP was reported to be dissatisfied with levels of traction availability, and was to negotiate new maintenance contracts with EMEF (see 'Organisation' above). Competitive tendering for these services was foreseen if availability did not improve.

Intercity trains

CP's target end-to-end timing of 2 hours 15 minutes for its upgraded Lisbon-Oporto route is based on the use of automatic body-tilting vehicles. In March 1996 an order for 10 six-car Pendoluso trainsets, valued at approximately Esc25 billion, was placed with Fiat (bodyshells and tilting equipment) and Siemens (electrical equipment), with Adtranz Portugal (formerly Sorefame) undertaking final assembly of the fleet; the first five units were to be delivered by mid-1998, in time for Expo '98, with the remainder in service by the end of 1999. However, subsequent manufacturing delays led to the first five trainsets being delivered at a rate of one per month from July 1998. CP negotiated compensation from Fiat Ferroviaria in the form of services and spares valued at Esc1.4 billion.

Meanwhile, upgrading has been undertaken on the existing intercity fleet: 44 passenger

Diesel railcars

Series	Power kW	Speed km/h	Weight tonnes	No in service	First built	Builders Mechanical	Engine	Transmission
9101/103 (NG)	240	70	22	3	1949	Nohab	Scania Vabis	H Lisholm-Smith
9301/310 (NG)	320	70	37	8	1954	Allan	AEC	E Smith
9601/622 (NG)	383	90	64.36	22	1976	Alsthom	SFAC	E Alsthom
0101/115 (BG)	252	100	33.3	12	1948	Nohab	Saab-Scania	H Voith
0301/325 (BG)	360	100	5.5	24	1954	Allan	SSCM	E Smith
0401/419 (BG)	560	110	94.1	19	1965	Sorefame	Rolls-Royce	H Rolls-Royce
0601/0640 (BG)	775	120	110	20	1979	Sorefame	SFAC	H Voith
9701-40 (NG)	720	60	92	10	–	Fiat*		M
9501/502	245	84	30	2	1995	Fiat/Volvo	Volvo	H Voith
9401/406	222	51	30	6	1993	Fiat	Volvo	H Voith

* (acquired from Yugoslav Railways).

Electric railcars

Class	Output kW	Speed km/h	Weight tonnes	No in service	First built	Builders Mechanical	Electrical
2001/2025	1,095	90	117	24	1956	Sorefame	Siemen/AEG/Oerlikon
2051/2074-2082/2090	1,095	90	123.6	33	1956	Sorefame	Siemens/AEG/Oerlikon
2101/2124	1,280	120	132.8	24	1970	Sorefame	Siemens/AEG/Oerlikon
2151/2168	1,280	120	132.8	18	1977	Sorefame	Siemens/AEG/Oerlikon
2201/2215	1,280	120	132.8	15	1984	Sorefame	EFACEC
2301/2342	3,100	120	–	42	1992	Sorefame	Siemens

coaches have been refurbished by CP's EMEF subsidiary at a cost of Esc5 billion. The upgrading features remounting on Y32 bogies, fitting of retention toilets, double-glazing, high-capacity air conditioning and power doors. CP has also installed telephones on the premium 'Alfa' trainsets of the Lisbon-Oporto service and introduced a business car on this route.

Many types in the dmu and emu fleets have also been the subject of refurbishment programmes, including 19 two-car dmus and 34 Cascais line emus. In addition, CP has been studying the acquisition of lightweight diesel railcars to improve the performance and economics of regional and local services.

Cross-Tagus trains

A contract for the supply of 18 double-deck four-car electric multiple-units for operation on the cross-Tagus link (see 'New lines' section in REFER

entry) went to a GEC Alsthom-led consortium that also included CAF. The first trains were delivered in 1998.

A follow-on order for 12 similar sets was placed in December 1997 for Lisbon-Azambuja suburban services.

Suburban trains

Refurbishment of 34 Cascais line trains in Lisbon is being undertaken by EMEF as part of a Esc13.8 billion upgrade programme for the line.

In 1998 Adtranz Portugal and Siemens were awarded a contract to supply a further four four-car emus for the Lisbon-Sintra line for delivery in 1999.

Tenders have also been invited for 22 new emus, 12 of which will be for Cascais services and 10 for suburban services around Oporto. These could be double-deck units, and the eventual contract will include options for a further 22 trains.

Portuguese National Rail Administration (REFER)

Rede Farroviária Nacional

Key personnel

President: Dr Manuel Frasquilho

Gauge: 1,668 mm; 1,000 mm
Route length: 2,576 km; 274 km
Electrification: 1,668 mm gauge, 598 km at 25 kV 50 Hz AC; 25 km at 1.5 kV DC

Political background

REFER began trading as Portugal's new rail infrastructure authority on 1 July 1997. Transfer of track was to be gradual, to be completed by 1 January 1999. Initially, 500 former CP employees were employed by the new company, with 2,000 more due to be taken on in 1998 and a further 3,000 in 1999.

REFER generates income from a combination of track access charges, property rental from operators and other businesses, and government subsidy.

REFER anticipated making a trading loss of Esc2–3 billion for the half-year ending 31 December 1997, with the Finance Ministry due to transfer Esc1.5–3 billion to balance the accounts. As part of its first half-year's trading, REFER invested Esc100 billion, much of which was sourced from the state or from the European Union. Esc120 billion of infrastructure investment has been budgeted for 1998.

The organisation's plan for the period up to 2000 is to spend some Esc600 billion on modernising its network. Half of this figure has already been

allocated, with the cross-Tagus project in Lisbon absorbing Esc130 billion and the modernisation of the Lisbon-Oporto main line accounting for a further Esc200 billion. Overall priority is to be given to this latter line, where contracts for upgrading work on the Braça de Prata-Pampilhosa Da Serra section were awarded in 1997, and for Quitães-Ouar in 1998. Around Esc300 billion is allocated for the suburban networks of Lisbon and Oporto.

REFER has absorbed the two public-sector organisations set up to manage major rail infrastructure schemes in Lisbon – GECAF, which has been responsible for the addition of a rail deck to the cross-Tagus 25 de Abril bridge, and GNFL, the Lisbon Railway Network Office. Also absorbed is GNFP, which was charged with managing rail investment in the greater Oporto area.

Organisation

Functional divisions within REFER's organisational structure cover operations, maintenance, infrastructure engineering, and construction projects, and will report to a board of directors. They are supported by finance and personnel departments. The maintenance division was established in January 1998, while an operations division started functioning in January 1999.

New lines
Cross-Tagus link

Construction of the Esc140 billion 22 km cross-Tagus link for Lisbon suburban traffic between Chelas on the north bank and Fogueteiro on the south bank has been completed, enabling services to commence in June 1998. The project involved the installation of a rail deck to carry 700 tonne trains on the 25 de Abril suspension

bridge, opened in 1967 for road traffic. Between Chelas and Campolide the link makes use of CP's existing Lisbon orbital (Cintura) route, widened to four tracks, but the remainder of the project has involved new construction.

Existing stations at Chelas, Areeiro, Entrecampos, Sete Rios and Campolide are served, and new stations have been built at Alvito, Pragal, Corroios, Foros de Amora and Fogueteiro. It is hoped eventually to extend the link from Fogueteiro via Coina and Penalva to the CP system at Pinhal Novo, thereby providing an alternative to the ferry connection between Lisbon city centre and Barreiro on the south bank of the Tagus, the terminus for passenger trains to and from the south of Portugal. The necessary link to complete this scheme, between Coina and Penalva, will be financed by the state.

International links

In May 1996 representatives of the Portuguese government met with European Commission officials to discuss the possibility of European Union funding for a new international passenger service. Using existing or planned infrastructure and a new connection between the Portuguese and Spanish networks over the River Guadiana, presumably at Vila Real de Santo António, the new service would run between the Spanish cities of La Coruña and Seville via Oporto, Lisbon and Faro.

Various other options have been examined for connecting Lisbon to Europe's developing high-speed network. An upgrade of the Beira Baixa (Entroncamento-Castelo Branco-Guarda) route was proposed in May 1996, whilst CP is known to have considered a direct Lisbon-Irún link. RENFE of Spain has expressed a preference for a route feeding into its emerging Madrid-Barcelona high-speed line.

Improvements to existing lines

Lisbon-Oporto upgrading

The electrified Lisbon-Oporto line (Linha do Norte) generates half CP's income, and its redevelopment became an urgent need with the 1991 completion of a motorway between the two cities. Following a study by the UK consultancy Transmark, in April 1991 the government authorised investment in infrastructure and new rolling stock to allow 220 km/h passenger operation between Lisbon and Oporto, with tilt-body trains covering the 340 km in two hours. A lift of maximum freight wagon axleload to 25 tonnes was also an objective.

On 26 July 1994 CP signed a Esc5.5 billion five-year project and construction management contract for the Linha do Norte upgrading programme with a consortium of ICF Kaiser International, W S Atkins and Fernando Braz Oliveira. A total of Esc190 billion is being spent on infrastructure, which includes renewal of 600 track-km, realignment of 60 track-km for 220 km/h operation and provision of an additional 75 track-km. UIC 60 rail on monobloc sleepers is being installed and the project involves the construction of 160 new bridges and a 1 km viaduct.

Catenary on the Linha do Norte is being renewed and the route resignalled. Level crossings are being eliminated or given automatic protection in conjunction with installation of CTC and the Ericsson ATP system, which will permit a lifting of maximum passenger train speed. Layouts at 42 stations are being remodelled to reduce path conflicts of trunk and local trains, and to avoid speed restriction of non-stopping trains.

REFER has inherited the Linha do Norte upgrading programme and intended to complete it by 2001. However, by 1999 serious delays and unforeseen work were threatening to put back completion until 2010. Costs had also risen substantially. These factors were prompting a renewed assessment of the case for a new high-speed line between the two cities.

Lisbon network redevelopment

Before the establishment of REFER CP allocated Esc3.2 billion for raising capacity on the 27 km Lisbon-Sintra line. The money was to be spent on resignalling the route as far as Cacém with ESTWL90 equipment, eliminating level crossings, renovating stations and remodelling track layouts. It is planned to introduce centralised traffic control, permitting 25 trains an hour in each direction. Four tracks in place of the present two will eventually be in service between Lisbon and Cacém (17.3 km). The route between Campolide and Benfica (3.1 km) has been expanded to four tracks, with trains operating at 4 minute headways at peak hours and serving a new station at Queluz-Massamá.

In addition to minor works to improve interchanges with the Lisbon metro, a modernisation programme costed at Esc60 billion has been proposed for the Lisbon-Cascais route. Level crossing elimination, better bus interchanges and new car parks for rail commuters form the major objectives of the programme, which could be completed by 2002.

Oporto improvements

Recent investment programmes have made provision for a new train maintenance facility at Oporto and the conversion of the 1,000 mm gauge Lousado-Guimarães branch to 1,668 mm gauge. The 1,000 mm gauge Oporto-Póvoa de Varzim and Senhora da Hora-Trofa routes (49.4 km in total) are to form part of a new Esc100 billion light rail system planned for Oporto. A contract to build and initially operate the 68 km network was awarded in early 1997 to Siemens, following competition with GEC Alsthom, a consortium led by Adtranz and a consortium of Spie Batignolles, Bombardier and Ansaldo.

Elsewhere on the CP Oporto network, doubling from two to four tracks between Oporto and Ermesinde is planned. Electrification from Ermesinde eastwards to Marco de Canaveses and double-tracking between Ermesinde and Caíde (completed as far as Valongo in 1995) has been undertaken.

Beira Alta route

Modernisation of REFER's prime international route to Spain, the Beira Alta, was a priority identified by CP. The line runs 201.6 km from Pampilhosa, south of Oporto on the Lisbon main line, to the Spanish border at Vilar Formoso; from here there is a direct Spanish route to the French border via Salamanca and Burgos. The 202 km of single track within Portugal has been electrified, resignalled and has undergone substantial realignment to ease its most severe curves and permit 160 km/h operation. This will entail complete reconstruction of two segments, in addition to level crossing elimination. Completion has cut as much as two hours from Lisbon-Paris passenger train timings, and the Esc38 billion cost of the project has been supported in part by the European Union.

Signalling and telecommunications

In recent years CP has concentrated signalling of its main line network on five electronic control centres, at Oporto, Coimbra, Entroncamento, Lisbon and Setúbal.

Two technologies have been adopted. One is the British SSI (Solid-State Interlocking), which has been supplied by Dimetronic SA of Spain and Westinghouse Brake & Signal. The other is ESTWL90, a German system similar to one adopted by that country's DB AG, which has been furnished by Alcatel's Portuguese subsidiaries, Alcatel SEL and Alcatel Portugal. The local company Efacec assisted in various aspects, including track circuiting, cabling, level crossing automation and installation.

EB Corporation of Norway (now Adtranz) received a CP contract for the delivery of Ericsson-type ATP speed-control equipment for installation on the Lisbon-Oporto line. The contract covers equipment for 300 vehicles. The local partner in this project is Efacec.

Track

Standard rail
1,668 mm gauge: 30–55 kg/m in 8 and 18 m lengths
1,000 mm gauge: 20–36 kg/m in 8 and 12 m lengths
Welded rail
Thermit process is used. Rail used weighs 54, 50, 45, 40 kg/m in 18 and 24 m lengths. The length of continuous welded rail is usually 840 m but occasionally 950 m
Crossties (sleepers)
1,668 mm gauge: 260 × 130 × 2,600 mm, spacing 605 mm
1,000 mm gauge: 230 × 120 × 1,800 mm, spacing 820 to 850 mm
Rail fastening: Screw spikes or bolts. RN flexible fastenings used with welded rail
Filling: Broken stone, gravel or earth
Max curvature
1,668 mm gauge: 5.9° = min radius 300 m
1,000 mm gauge: 29° = min radius 60 m
Longest continuous gradient
1,668 mm gauge: 8.3 km of 1.4% gradient with curves varying from 590 to 1,501 m in radius
1,000 mm gauge: 7.2 km of 2.5% gradient with curves varying from 75 to 500 m in radius
Max gradient
1,668 mm gauge: 1.8% = 1 in 55½
1,000 mm gauge: 2.5% = 1 in 40
Max altitude
1,668 mm gauge: 812.7 m
1,000 mm gauge: 849.7 m
Max axleload
1,668 mm gauge: 19.5 tonnes
1,000 mm gauge: 11 tonnes
Max permitted speed
1,668 mm gauge: 160 km/h
1,000 mm gauge: 80 km/h

Fertagus

Travessia do Tejo SA
Estação do Pragal, Porta 23, P-2800 Almada
Tel: (+351 21) 106 63 00   Fax: (+352 21) 106 63 99
e-mail: info@fertagus.pt
Web: www.fertagus.pt

Key personnel

Managing Director: Cristina Dourado
Commercial Director: Raquel Santos

Gauge: 1,668 mm
Route length: 54 km

Organisation

In July 1999 bus operator Barraqueiro commenced a 30-year concession to run Fertagus suburban rail services linking central Lisbon with Fogueteiro, south of the River Tagus, via the lower deck of the 25 April bridge. Services are operated over infrastructure owned by the Portuguese Rail Infrastructure Authority (REFER).

Traffic (million)	2001	2002	2003
Passenger journeys	14.8	17.5	17.8

Passenger operations

Fertagus serves a 14-station line linking Roma/Areeiro, in central Lisbon, with Setúbal, to which services were extended in October 2004. At Sete

Fertagus double-deck emu at Fogueteiro (Quintus Vosman)

NEW/1115021

Rios and Entrecampos stations, connections are made with CP suburban and Lisbon metro services. At five stations in Almada connections are provided with SulFertagus bus services. Peak-hour services run every 10 minutes between Lisbon and Coina and every 30 minutes between Lisbon and Setúbal. Some 75,000 passengers are carried daily.

Traction and rolling stock

The Fertagus fleet comprises 18 four-car double-deck air conditioned 25 kV AC emus supplied by a consortium led by ALSTOM and including Adtranz (now Bombardier Transportation) and CAF.

UPDATED

Puerto Rico

Ponce & Guayama Railway

Corporación Azucarera de Puerto Rico
Aguirre, Puerto Rico 00608
Tel: (+1 809) 853 38 10

Key personnel
Executive Director: A Martinez

General Superintendent: J Rodriguez
Roadmaster: R Rodriguez
Traffic Manager: J Lopez
Purchasing Manager: R Rivera
Accountant: T Cartagena

Gauge: 1,000 mm
Route length: 96 km

Traction and rolling stock
The railway operates 22 diesel locomotives and
1,280 freight wagons.

Romania

Ministry of Transport

Bd Dinicu Golescu 38, Bucharest 78123
Tel: (+40 1) 617 20 60   Fax: (+40 1) 312 32 05

Key personnel
Minister: Miron Mitrea

Political background
Within the Ministry of Transport, the Railways
Directorate has responsibility for overall regulation
and strategy. The directorate is also responsible
for implementing the changes needed to meet
European Union directives as Romania prepares
for membership of the community. It is also the
body which agrees annual contracts with CFR SA
and CFR-Calatori.

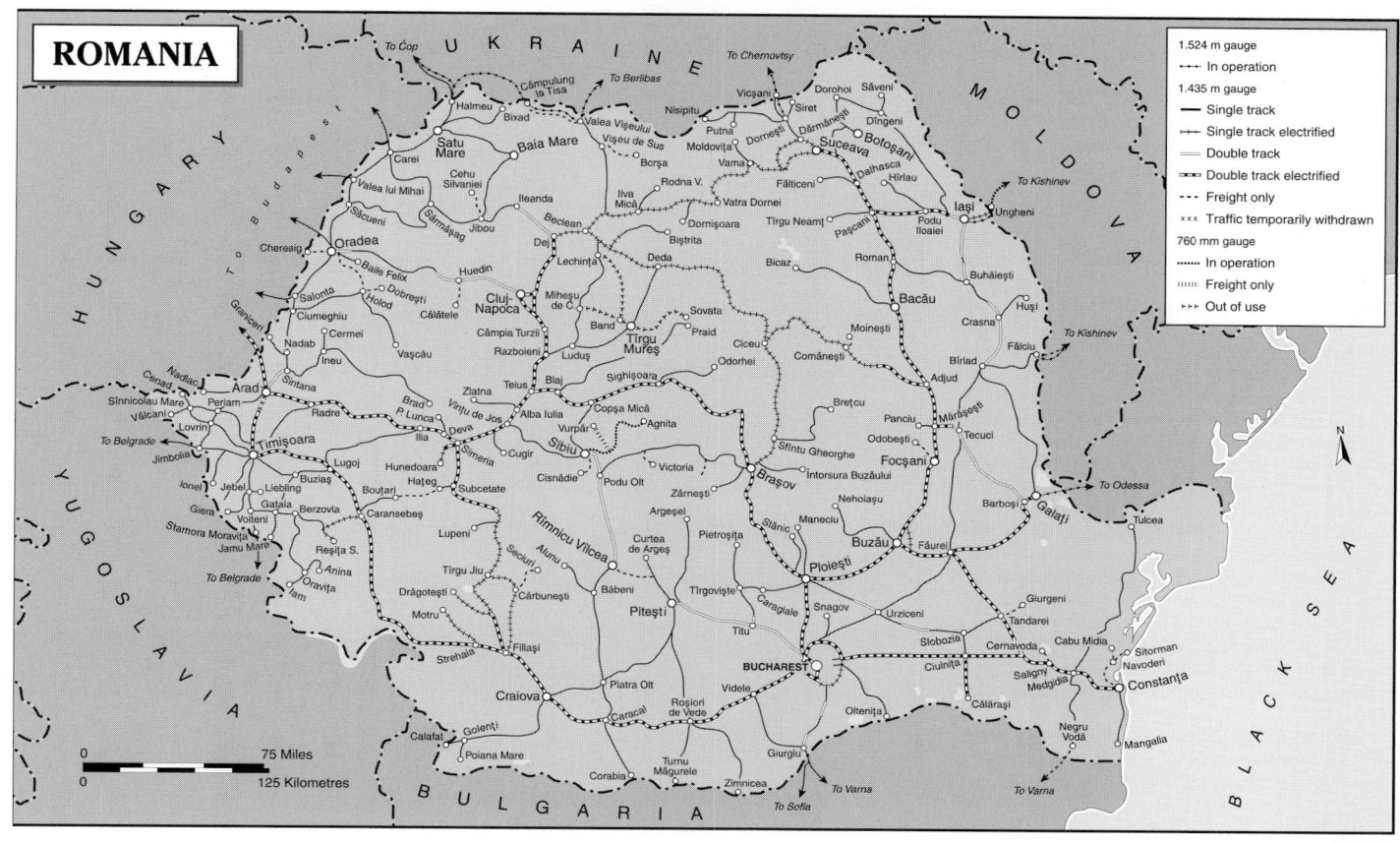

0089970

Romanian Railway Authority (AFR)

Autoritatea Feroviara Romana
Bucharest

Key personnel
General Manager: Ispas Eugen
Chief State Inspector: Dobrescu Lucian

Political background
AFR was established on 1 November 1998 to
oversee the licensing of operators, railway
safety and issues such as conflicts in track access
demands.

CFR Calatori SA

Bd Dinicu Golescu 38, 77111 Bucharest 1
Tel: (+40 1) 222 25 18   Fax: (+40 1) 411 20 54
e-mail: marketing.calatori@cfr.ro
Web: www.cfr.ro

Key personnel
Chairman and Director General: Valentin Bota
Deputy Director General: Vanghele Nacu

Organisation
CFR Calatori is the passenger business created
on the October 1998 restructuring of Romanian
National Railways in accordance with EU Directives.
It is a state-owned joint-stock company. At an
operational level it is managed as eight regions.
There are 62 stations, 12 locomotive depots and
25 coach repair facilities. Of the company's nine
subsidiaries, six are concerned with traction and
rolling stock maintenance and repair and supplies
procurement. The remaining three handle sleeping
and catering car activities and services (CFR
Gevaro SA), ticketing and commercial activities
(Voiaj CFR SA) and parcels, luggage and mail (CFR
Mesagerie SA). In 2004 approximately 19,000 staff
were employed.
   Subsidy for unremunerative services is provided
via passenger service contracts negotiated with
government via the Railways Directorate. A four-
year contract ran from 2001 to 2004.

Traffic (million)	2001	2002	2003
Passenger journeys	113.7	95.6	94.7
Passenger-km	10,966	8,502	8,506

Passenger operations

Around 1,450 services are operated daily. Intercity and express services linking Romania's major centres of population form a key part of CFR Calatori's business, accounting for some 50 per cent of passenger journeys. In addition, on average 36 international trains are run each day. Changes in Romanian industry have contributed to a gradual decline in rail usage.

Traction and rolling stock

In 2003 CFR Calatori operated 365 electric, 501 diesel locomotives and 97 dmus. The electric locomotive fleet was formed of 5,100 kW Class 40/41 machines and their refurbished Class 45 sisters (see below) and Yugoslav-built 3,860 kW Class 43/44 units, some of which have been refurbished by Koncar in Croatia to become Class 46. The diesel fleet comprised 285 Class 60/62 diesel-electrics (see below) and 216 diesel-hydraulic locomotives, mostly Class 80/81. There were also some 3,350 coaches in use, including 151 sleeping/couchette cars and 24 catering vehicles.

In 2002 Siemens Transportation Systems signed a contract to supply 120 Desiro two-car dmus to CFR Calatori. The first 57 units were to be constructed by Siemens, leaving the remaining 62 to be built locally by Astra Vagaone Calatori SA Arad. Designated Class 96 by CFR Calatori, first examples entered service in early 2003.

Siemens has also refurbished 24 Class 40 and 41 5,100 kW electric locomotives in collaboration with Electroputere. The project was financed by an EBRD loan. The modernised locomotives have become Class 45. The modernisation of 57 Class 60 and 62 Sulzer-engined diesel-electric locomotives has also been started. This follows the successful refurbishment in 2000 of two prototypes, installing a General Motors eight-cylinder 710 series engine in place of the original Sulzer unit. Modified locomotives become Class 63, the conversion programme scheduled for completion in August 2006. A third locomotive refurbishment programme covers Class 80 and 81 B-B diesel-hydraulic machines, four of which were modernised by ALSTOM and Faur in 1999, with their original Sulzer engines replaced by Caterpillar or MTU engines to become Classes 82 and 83.

To handle traffic on lightly used regional lines, CFR Calatori has acquired up to 12 ex-DR Class 771 and 772 two-axle railbuses and associated trailer cars from Germany. These have been refurbished in Romania by Marub in Brasov and are designated Class 79.

In addition, CFR Calatori has awarded contracts for the refurbishment of 100 vehicles in its coaching stock fleet, funded by an EBRD loan.

UPDATED

Class 43 electric locomotive with a CFR Calatori local service at Deva (Colin Boocock)   ***NEW***/0585057

Class 96 two-car Siemens Desiro dmu at Bucharest Nord (Colin Boocock)   ***NEW***/0585058

Class 79 diesel railbuses (ex-DR Class 771) acquired from Germany and refurbished for CFR Calatori, seen at Sibiu depot (Colin Boocock)   ***NEW***/0585059

CFR Marfa SA

Societatea Nationala de Transport Feroviar de Marfa
Bd Dinicu Golescu 38, 77111 Bucharest 1
Tel: (+40 1) 638 55 88   Fax: (+40 1) 312 47 00
Web: www.cfrmarfa.cfr.ro

Key personnel

General Manager: Victor Bucureanu
Deputy General Manager: Ion Garoseanu
Deputy General Manager, Finance: Doina Crosman
Traffic Manager: Liviu Raican
Commercial Manager: Sorin Chinde
Operations Manager: Gratian Calin
Technical Manager: George Micu

Organisation

CFR Marfa is the freight business created upon the October 1998 restructuring of Romanian National Railways. Tariffs are deregulated and the company receives no state subsidy. CFR Marfa faces competition from independent operators which have taken advantage of open access rights included in the legislative reforms of 1998.

In 2001 studies into possible privatisation of CFR Marfa were undertaken with the aim of determining the extent and nature of private sector involvement. Subsequently, the government announced its intention to privatise the business based on guidelines to be recommended by the World Bank.

The company is structured as eight regional divisions, with headquarters offices in Bucharest. There are also several subsidiary companies, including:

CFR SSVAC SA, handling the cleaning of tank wagons.

CFR SSVM SA, responsible for cleaning general freight wagons.

CFR Transauto SA, which undertakes road operations related to intermodal and parcels services.

CFR TVM SA, which handles the trans-shipment of freight between standard and broad (1,520 mm) gauge at border points.

In 2004 CFR Marfa employed 21,000 staff.

Traffic (million)	2000	2002
Freight tonnes	71.5	68.1
Freight tonne-km	18,000	17,200

CFR Marfa electric locomotives

Class	Wheel arrangement	Power kW	Speed km/h	Weight tonnes	No in service	First built	Builders Mechanical	Electrical
40/41	Co-Co	5,100	120/160	120-126	354	1965	Electroputere	ASEA, Electroputere
43/44	Bo-Bo	3,400	120/140	78-82	22	1976	Koncar	ASEA, Elin, Koncar, Sécheron

Freight operations

The predominant commodity carried by CFR Marfa is coal, accounting for 40.9 per cent of tonnage in 2003. Petroleum products represented 12.6 per cent of traffic and metals 10.2 per cent. Transit traffic is important, with the Romanian rail network forming part of three European corridors, Corridors IV, VII and IX. In 2000 international traffic accounted for one quarter of tonne-km and is viewed by the company as an area for potential growth.

Rail links exist with Bulgaria, Hungary, Moldova (1,524 mm gauge), Serbia and Ukraine (1,524 mm gauge), while CFR Marfa operates Black Sea train ferry services linking Constanţa and ports in Georgia and Turkey. For these operations CFR Marfa owns two 12,000 tonne train ferries. Both piggyback and container services are operated.

Traction and rolling stock

In 2004, CFR Marfa operated 376 electric, 289 main line diesel locomotives and 261 shunters/trip locomotives. The main line diesel locomotives are 1,545 kW Sulzer-engined Class 60 and 62 diesel-electrics dating from the 1960s, while the lower-powered machines are various diesel-electric and diesel-hydraulic types of up to 930 kW. There were also some 60,000 wagons in the fleet.

UPDATED

CFR SA

Bd Dinicu Golescu 38, 77113 Bucharest 1
Tel: (+40 1) 638 55 88   Fax: (+40 1) 312 47 00

Key personnel
President: Mihai Necolaiciuc

Gauge: 1,435 mm; 1,524 mm; 760 mm
Route length: 10,898 km; 60 km; 427 km
Electrification: 3,888 km at 25 kV AC 50 Hz

Political background
CFR SA is the infrastructure business created on the October 1998 restructuring of Romanian National Railways. As a legally independent, state-owned company, it manages and maintains Romania's national rail network under an indefinite concession. Operations are mainly funded by access charges from CFR Calatori and CFR Marfa (both qv), together with some direct state subsidy and access charges from a small number of independent operators. International loans to fund major investment projects are underwritten by the government. Licensing and regulation of all rail operators is the responsibility of a separate state agency, AFER (qv).

Organisation
The business is managed via a devolved regional structure, supported by headquarters departments overseeing strategic and technical issues.

Improvements to existing lines
In 2001, work was in hand to upgrade a key section of European Corridor IV, between Braşov and Bucharest (160 km). The section from Braşov to Ploieşti also forms part of Corridor IX. The upgrading project includes infrastructure improvements, track renewal and some realignment and upgrading of the catenary. The European Investment Bank is providing a loan covering 75 per cent of the cost of the work; the remaining funding is being provided by the Romanian government.

International loans have also been negotiated to cover much of the cost of modernising the Bucharest–Constanţa line (225 km).

Signalling and telecommunications
Improvements to Corridor IV include the provision by SEL Alcatel of new electronic interlockings at Arad, Braşov, Bucharest and Timişoara. Siemens TS has provided an electronic interlocking at Ploiesti, the first such installation commissioned by the railway.

CFR SA plans to modernise its telecommunications system with an optical fibre network.

Electrification
Most key routes are electrified using the 25 kV AV 50 Hz system. Lines identified by CFR SA for possible future electrification include: Cluj-Napoca–Oradea and beyond to the border with Hungary; Iaşi–Tecuci; and Constanţa–Mangalia.

Russian Federation

Ministry of Railways

Novo-Basmannaya 2, Moscow 107174
Tel: (+7 095) 262 16 28   Fax: (+7 095) 262 90 95

Key personnel
Minister of Railways, Russian Federation:
N E Aksyonenko
First Deputy Ministers: V I Ilyin; M V Ivankov
Deputy Ministers A V Annenkov, E N Vinogradov, S N Gapaev, Y M Gerasimov, S A Grishin, A N Kondratenko, V M Mironov, A S Misharin, V N Morozov, V N Pustovoi, V T Semenov, P A Shevopukov, A Tselko
Heads of Departments
Economics: B M Lapidus
Finance: P G Korotkevich
Freight Operations: (vacant)
Passenger: V N Shataev
Traction: A D Rusak
Freightcar: S S Barbarich
Track and Structures: S A Rabchuk
Safety and Ecology: P S Shanaitsa
Freight and Commercial: Yu M Kosov
Signalling and Train Control: V A Milyukov
Electrification and Power Supply: G B Yakimov
Information Technology: V S Voronin
Personnel and Training: N M Burnosov
Health: O N Sorokin
Technical Policy: V S Nagovitsyn
Railway Restructuring: P K Chichagov
Capital Construction: (vacant)
Clerical: V G Dolzhenko
External Relations: N V Antipov
Wages and Working Conditions: S M Danilov
Statistics: G V Bugrov
Real Estate: A M Vaigel
Legal: A A Melnikov
Internal Organisation: E P Chernyavskii

Russian Railways (OAO RZD)

Rossiiskie Zheleznie Dorogi (RZD)
Novaya Basmannaya 2, Moscow 107174
Tel: (+7 095) 262 99 01
e-mail: info@rzd.ru
Web: www.rzd.ru; www.eng.rzd.ru

Key personnel
President: G Fadeev
First Vice-Presidents: V Yakunin, K Ziabirov
Vice-Presidents: F Andreev, S Babaev, A Belova, N Burnosov, V Gapanovich, S Ivanov, G Kornilov, S Kozyrev, G Kraft, B Lapidus, A Mersianov, S Pegov, I Rotenberg, V Sazonov
Regional Directors
Dalnevostochnaya (Far Eastern) Railway: V A Popov
Muraviyova-Amurskogo ul 20, 680000 Khabarovsk
Tel: (+7 4212) 38 44 00; 38 41 36
Gor'kovskaya (Gorky) Railway: V F Sekhin
Oktyabr'skoy Revolyutsii ul 78, 603011 Nizhnii Novgorod
Tel: (+7 8312) 48 69 00
Web: www.grw.ru
Kaliningradskaya (Kaliningrad): V G Budovskii
Kievskaya ul 1, 263039
Tel: (+7 0112) 58 66 47; 49 26 47; 44 22 75
Krasnoyarskaya (Krasnoyarsk) Railway: V N Suprun
Gorkogo ul 6, 660049 Krasnoyarsk
Tel: (+7 3912) 59 44 00; 59 44 01
Kuibyshevskaya (Kuibyshev) Railway: V G Lemeshko
Komsomolskaya Ploshchad 2–3, 443030 Samara
Tel: (+7 8462) 39 44 00; 39 22 62
Web: www.kbsh.rzd.ru
Moskovskaya (Moscow) Railway: V I Starostenko
Krasnoprudnaya ul 20, 107140 Moscow
Tel: (+7 095) 266 20 50
Web: www.mzd.ru
Oktyabr'skaya (October) Railway: V V Stepov
Ostrovskogo Ploshchad 2, 190011 St Petersburg
Tel: (+7 812) 168 60 40; 168 60 41
Web: www.ozd.rzd.ru
Privolzhskaya (Volga) Railway: M A V Khrapatiy
Moskovskaya ul 8, 410013 Saratov
Tel: (+7 8452) 41 40 13; 41 44 00
Sakhalinskaya (Sakhalin) Railway: M M Zaichenko
Kommunicheskii Prospekt, 693000 Yuzhno-Sakhalinsk
Tel: (+7 4242) 71 44 00; 71 44 48
Severnaya (Northern) Railway: V A Bilokha
Volzhskaya nab 59, 150000 Yaroslavl'
Tel: (+7 0852) 29 44 00; 79 45 84
Web: www.szd.rzd.ru
Severo-Kavkazskaya (North Caucasus) Railway: V B Vorobyov
Teatral'naya Ploshchad 4, 344719 Rostov-on-Don
Tel: (+7 8632) 59 42 33; 59 44 00; 59 50 06

Sverdlovskaya (Sverdlovsk) Railway: S N Shaydulin
Cheluskintsev ul 11, 629013 Ekaterinburg
Tel: (+7 3432) 358 44 00
Web: www.svrw.ru
Vostochno-Sibirskaya (Eastern Siberia) Railway: A G Tishanin
Karla Marska ul 7, 664000 Irkutsk
Tel: (+7 3952) 64 44 00
Yugo-Vostochnaya (South Eastern) Railway: A I Volodko
Revolyutsii Prospect 18, 394621 Voronezh
Tel: (+7 0732) 50 44 60/50 44 50
Fax: (+7 0732) 65 44 50
Yuzhno-Ural'skaya (Southern Urals) Railway: A S Levchenko
Revolyutsii Ploshchad 3, 454000 Chelyabinsk
Tel: (+7 3512) 68 44 00
Zabaikal'skaya (Trans Baikal) Railway: V N Khokhryakov
Leningradskaya ul 34, 672092 Chita
Tel: (+7 3022) 97 43 16
Zapadno-Sibirskaya (Western Siberia) Railway: A V Tselko
Vokzalnaya Magistral ul 14, 630004 Novosibirsk
Tel: (+7 3832) 29 44 00

Gauge: 1,520 mm; Sakhalin: 1,067 mm
Route length: 86,200 km; Sakhalin: 957 km
Electrification: 18,800 km at 3 kV DC; 21,500 km at 25 kV 50 Hz AC

Political background
In the days of the Soviet command economy, SZhD (Soviet Railways) carried over two and a half times the total freight carried by all 16 Class 1 carriers in the US (SZhD's 1990 freight volume was 3,857 million tonnes, 3,717 billion tonne-km). In passenger traffic, SZhD carried about twice as many passenger-km as the railways of France, Germany, Italy and the UK combined (in 1990 SZhD carried 4,273 million passenger journeys, 417 billion passenger-km).

The railways are still by far the most important mode of transport in Russia. RZhD (Russian Railways) accounts for almost half of all passenger-km travelled, and over three-quarters of all non-pipeline tonne-km carried in the country. But with the break up of the Soviet Union in late 1991 and the transition to a market economy, the former

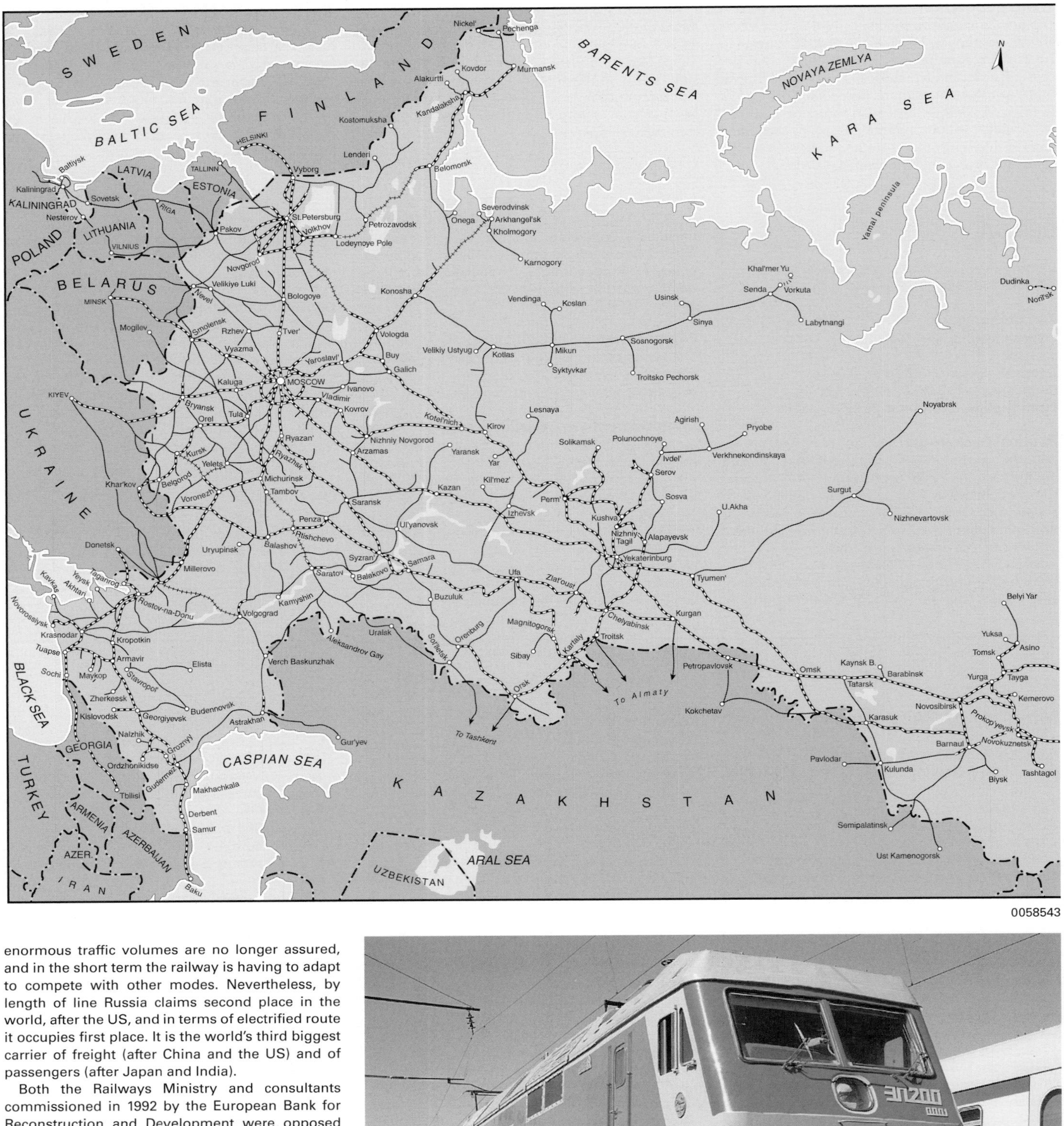

0058543

enormous traffic volumes are no longer assured, and in the short term the railway is having to adapt to compete with other modes. Nevertheless, by length of line Russia claims second place in the world, after the US, and in terms of electrified route it occupies first place. It is the world's third biggest carrier of freight (after China and the US) and of passengers (after Japan and India).

Both the Railways Ministry and consultants commissioned in 1992 by the European Bank for Reconstruction and Development were opposed to railway privatisation, but both accepted that adaptation to the market economy was urgent. The railways were already introducing flexibility to rate fixing, and were establishing premium freight centres, where clients could negotiate the whole transportation process, including pick-up and delivery, storage and documentation, with a single service point. Exploitation of modern information technology has progressively enabled these centres to monitor freight transits for the benefit of clients. Privately owned wagons were encouraged with special rates, although some clients subsequently found those rates too high.

In July 1995 the Federal Railway Law was ratified. This specified that, while non-transport-related facilities might be privatised, the operating railway was not liable to denationalisation. (However, in early 1996 the Railways Minister found it necessary to promise strong opposition to the suggestion that the railways might nevertheless be privatised in response to pressure from Western banks). The Railway Law also forbade strikes of operating workers and allowed the railways to sell off shipments whose transportation had not been paid for. The general

Prototype of Kolomna-built Class EP200 eight-axle AC electric passenger locomotive (Zheleznodorozhnoe Delo/Ivan Khil'ko)

0088204

strengthening of the railways' hand meant that stronger measures were taken against recalcitrant debtors. For a few days in 1995 the railways refused to provide mail cars, and relented only when the Federal Postal Service agreed to pay its enormous debts. However, in subsequent years

some railways were again refusing to handle mail until the postal service paid its debts.

Some ancillary enterprises like equipment supply works have been privatised. In the case of train and station catering services, privatisation brought a decline in an already unappetising performance

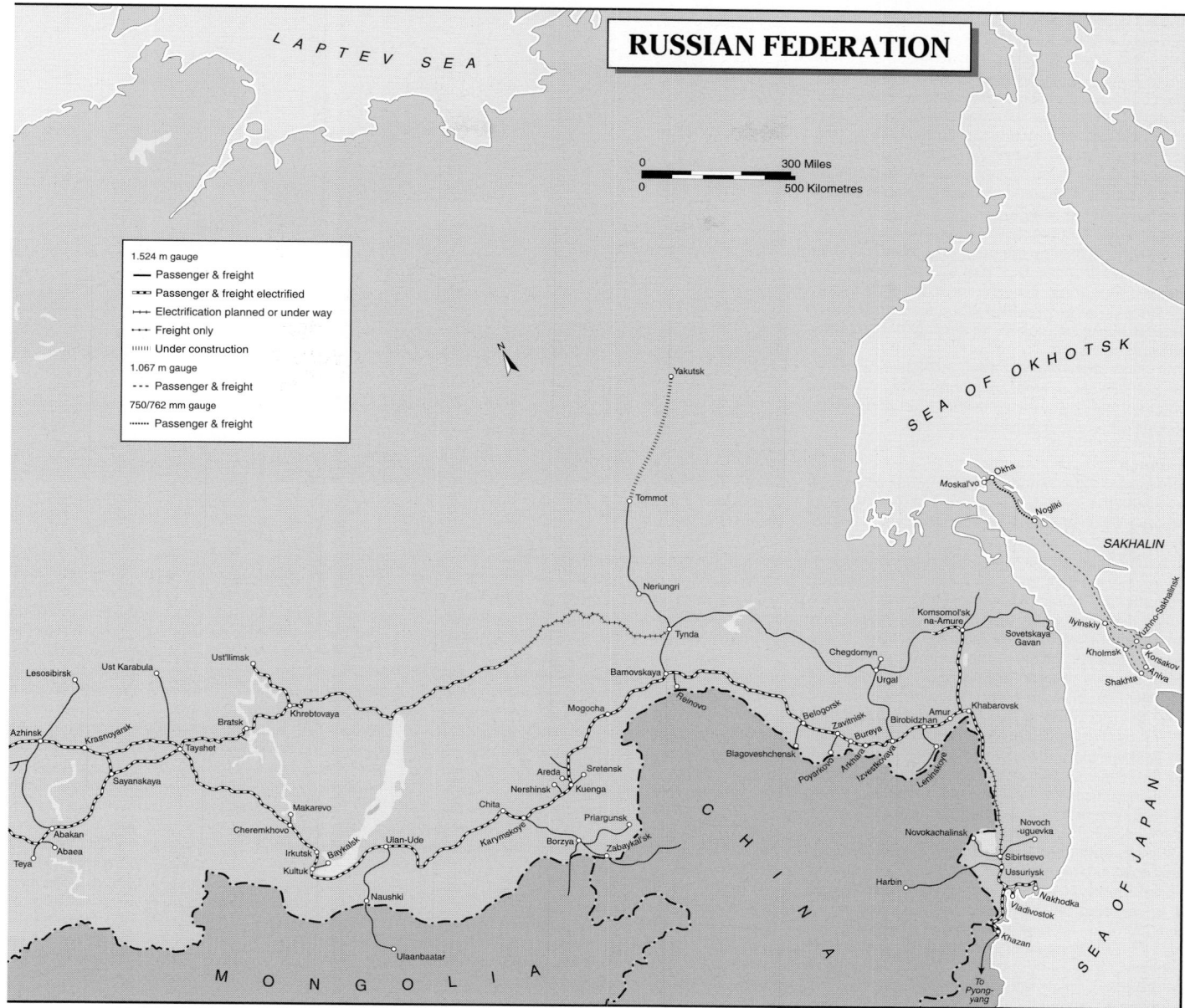

0058544

and the railways in some cases have been allowed to take them back.

After a change of minister in April 1997 (the second such change within a year), the government stated that the immediate task of the railways was to reduce freight tariffs and at the same time pay the specified taxes and contributions to the pension fund. By 2001 pension fund arrears had been almost paid off. Wages were paid promptly, and most overdue taxes settled. The federal taxation service had dropped its practice of despatching squads with sub-machine guns to sequester railway cash at stations and depots, but was still pursuing some railways in the courts. On the Moscow Railway, big cost reductions were obtained by eliminating the night shift wherever possible.

Several detailed proposals for reforming the railways, a so-called 'natural monopoly', were put forward, including one by the Transport Ministry, responsible for non-railway transport, which proposed taking over the railways as well. The Railway Ministry described other competing proposals as simply versions of the derided 'English approach', and it was the Ministry's own proposal that, in early 1998, was favoured by the government. Somewhat clarified, it was finally accepted by the government in 2001, although since 1998 much exploratory work had been done by the Ministry's new Department for Railway Reform. In the first stage (2001–02) the already existing railway-founded passenger operating companies are to settle down, while independent freight operating companies for specialised traffic continue to develop. In the second stage (2003–05) the passenger operating companies will be partially sold off, and subsidiary companies will be formed

to handle general freight. In stage 3 (2006–10) it is expected that genuine competition will appear. Meanwhile, the Ministry of Railways' two roles, as state supervisor of railways and railway operator, will be split; a restructured Russian Federal Railways is expected to be established in 2002. At arm's length from the Ministry, this will be responsible for operation and steadily develop into a holding company responsible for affiliate companies for passenger and freight services, infrastructure, and maintenance services. Initially, the Ministry of Railways valued reform as a means of ending cross-subsidy of passenger by freight traffic, but now appreciates it as a means of raising long-needed capital.

There were 15 independent freight operating companies in early 2001. These were tied to a particular industry: several, notably Linkoil-SPb, BaltTransServis, and Russkii Mir, were oil carriers and the last two seemed likely to acquire their own motive power in 2001. Another big operator was Severstal'trans, established by the Northern Steel company. In addition, there were two early Ministry of Railways companies, GUP-Konteiner for container and Refservis for refrigerator services; the latter was going through difficult times.

Less money from government

In 1994 central government severely restricted its grants for major projects and agreed to maintain finance only for the final tunnel, the Severo–Muiskii, on the Baikul–Amur Magistral (BAM), the Amur bridge reconstruction at Khabarovsk and 130 new passenger vehicles. Overall, by 1995 government investment was about 96 per cent less than it had been three years earlier and this was further

diminished. By 1999 the government was paying only about one-tenth of the Severo–Muiskii tunnel costs. Most capital projects are now financed by RZhD itself, but local authorities sometimes assist, especially with station construction/rebuilding projects and occasionally with the purchase of new rolling stock for commuter services. Some local authorities, providing funds to offset commuter service losses in the expectation that this money would buy new trains, have been disappointed. Meanwhile in 1999 the RZhD introduced new depreciation indices to permit investment to rise from the RUR16 billion of 1998 to RUR21 billion in 1999.

No major line closures have occurred yet, but they have become a possibility: up to 3,000 km of routes were provisionally listed for closure in 1995–96. Some 9,000 km are regarded as warranting closure, and to lessen local opposition it is hoped that some of these will become industrial railways. In 2000 the Krasnoyarsk regional government expressed willingness to take responsibility for the loss-making 259 km Reshoty–Karabuda branch, and it was hoped that other governments would make similar offers.

Organisation

The Russian system had 19 principal railways or regions, including the Kaliningrad and Sakhalin systems. These railways were further subdivided into over 100 divisions. However, a start was made in reducing the number of divisions, with a view to their complete elimination as a management level. By 1999, when the process was temporarily abandoned, about 40 divisions had been eliminated. Merging of the railways was recommended, leading to the formation of five or six big organisations. The BAM Railway was divided between the Far Eastern and East Siberian railways, and the Kemerovo Railway was merged with the West Siberian. However, there was opposition to such amalgamations, largely from local political administrators reluctant to lose their 'own' railway and the tax-base it represents.

In 1999 agreement of local authorities to the elimination of the Krasnoyarsk Railway became unlikely to be forthcoming after the election of the powerful General Lebedev as Governor of Krasnoyarsk. The merging of the Sakhalin Railway into the Far Eastern never took place, although it remains a distant intention. The total number of railways is now 17. These are not financially independent: their incomes are channelled to the Ministry of Railways (which is in effect RzhD) and it decides how much money to allocate back to each railway.

In 1999 there were about 1,600,000 workers on the railway payroll, of whom 1,236,100 were operating workers (compared to 1,521,000 in late 1996). Further reductions are expected.

Finance

The former SZhD ran at a profit. Interest on capital was not charged as an expense, but even so it is thought that the railway made sufficient profit to cover this.

Traditionally, profits from freight operations have been used to cover losses generated by passenger operations. During the early 1990s, with freight traffic declining faster than passenger traffic, this arrangement seemed threatened. But all RzhD regions (except Sakhalin) returned a profit in 1994 with an overall profitability, after subsidies, of 21.9 per cent of income. This fell to 10.2 per cent in 1997 but recovered to 24.5 per cent in 1999. In that year freight operating profit was RUR48,511 million, while there was a RUR17,629 million loss registered by passenger services. The balancing of passenger with freight profits has aroused complaints from Siberia, where passenger traffic is low and freight traffic is high.

Freight tariffs and long-distance passenger fares are now inflation-indexed. Problems and losses occur, however, when the government orders a price-freeze, as happened for the duration of the fourth quarter of 1995. Local authorities are involved in the setting of fares for commuter services, and many of them make a financial contribution. Highway competition, as well as a desire to stimulate traffic, has led to flexible (negotiated) freight tariffs for some commodities, while political pressures have dictated reduced rates for freight moving over 3,000 km in Siberia and the east. Further differentiation of tariffs can be expected. Rates for low-value freight were reduced from one third to one half in 1995, with slight increases for high-value traffic. The railways have made agreements with certain

New Demikhovo-built Class ED4MK emu providing the three-class supplementary-fare Moscow–Ryazan service
(Zheleznodorozhnoe Delo/N Ermolaev) 0085815

regional governments whereby rates on a region's prime produce are reduced (thereby widening the market) in return for prompt payment, price reductions for products and services bought by the railway, and other concessions. However, there have been disappointments. Traffic has not significantly responded to tariff cuts and the Magnitogorsk metallurgical complex, after gaining a substantial rate cut, increased its steel prices to the railways by one third. In 2000 and 2001 there were significant rate increases, but the much-criticised discrimination between international and internal rates was being eliminated.

About 25 per cent of railway income is paid as taxes, compared to 6 per cent in 1991. The railways duly but unenthusiastically make their contribution to the Road Tax.

Passenger operations

Passenger statistics are compromised by the high proportion of commuter passengers who do not purchase a ticket. To solve this problem, turnstile access is being introduced at main stations. At Moscow stations, commuter revenues were said to have increased by twice and even three times after these changes were made. In general, passenger traffic held up better than freight over the 1990s, and in 1999 there was an unexpected recovery in long-distance summer traffic.

Services operating between the republics of the former Soviet Union now undergo lengthy customs inspections at the new state frontiers. For this reason, services from the centre to the Caucasus and the Black Sea are now routed via Voronezh rather than through Ukraine.

Obtaining reservations has become less time-consuming for passengers due to the spread of RzhD's electronic reservation system. The St Petersburg Railway Agency now offers international reservations through the European START-AMADEUS system.

1999 was a good year for safety, with no passenger fatalities in train accidents. Nevertheless passengers perceive rail travel as not entirely safe, largely because of the incidence of on-train crime. Railway police now cover some late-night commuter trains and also selected long-distance services. It is the policy to direct foreign tourists to long-distance trains that are thus protected.

Overall the quality of customer service remains patchy, albeit improving. There have been some fare reductions even though the previous fare level ensured that passenger trains could never make a profit (although long-distance services might become profitable if fare reduction and free travel concessions were less widespread; currently there are over 60 categories of passenger entitled to them). Creation of separate passenger companies has been accepted as a policy and a number of city commuter companies are being developed. On the October Railway, *Transservis* operates

Class ChS7 3 kV DC electric locomotive with a Nizhniy Novgorod–Tyumen express at Yekaterinburg
(Milan Šrámek) 0059400

longer-distance passenger trains and *Transkom* operates commuter services. There is a similar commuter company at Samara (*Samaratransprigorod*) and a variation is being introduced at Moscow, while at Novosibirsk the *Ekspress-prigorod* is a joint-stock company. The West Siberian Railway has established a long-distance passenger subsidiary. Premium service (supplementary fare) trains are becoming more widespread and in early 1999 *Transkom* was demonstrating its first premium-service commuter train. Other railways have since instituted extra-fare local trains, offering the rare (for Russia) division into first, second and third classes. Some do not accept discounted tickets, but are nevertheless well patronised. The prototype ER200 trainset was withdrawn for complete refurbishment and replaced by an identical trainset which performed the Moscow-St Petersburg route twice-weekly.

In December 2000 twice-daily high-speed services were introduced between Moscow and St Petersburg, but this schedule soon relapsed into the previous once-weekly high-speed train. However, in 2001 it was expected that an enhanced service would be restored. Final choice of equipment had not yet been made; a high-speed stainless-steel trainset built by the Tver Works was to undergo trials, the Sokol train was already on trial, and Talgo trains were also a possibility.

Traffic (million)	1997	1998	1999
Passenger journeys	1,595	1,471	1,300
Passenger-km	170,000	152,200	141,062
Freight tonnes	855.8	834.1	947.3
Freight tonne-km	1,096,000	1,006,00	1,204,600

Freight traffic in 2000 grew to 1,373,000 tonne-km. Passenger traffic was reported to have increased by a fifth, but part of this was probably a statistical manifestation of reduced fare evasion.

Main commodities carried by rail (000 tonnes)

	1998	1999
Coal	208,600	227,000
Coke	n/a	n/a
Oil and oil products	n/a	n/a
Manganese and iron ore	72,000	84,700
Ferrous metals	43,700	52,700
Chemical and mineral fertilisers	28,300	33,700
Cement	16,700	19,100
Timber	31,900	42,700
Grain and flour	15,400	15,500
Construction materials	134,000	145,600

Freight operations

The economic changes in Russia, featuring the decline of heavy industry and the emergence of private enterprise road hauliers, have led to a decline in carryings on RzhD. All the main commodity groups saw a decrease in carryings in 1996 compared to 1992.

The 1999 figures may, at last, mark the end of decline. Meanwhile, the railways' share of total common-carrier shipments rose from 75 per cent in 1995 to 86 per cent in 1999, thanks to the decline of water-borne traffic.

The division of the SZhD freight wagon fleet amongst the republics of the former Soviet Union took place amicably, but the new states often suspect their neighbours of sending them wagons in doubtful order that require repair before being returned to the owning system. An automated wagon registration system at frontiers, with charges levied for the use of foreign wagons, was instituted.

Industrial railways

Non-common-carrier lines belonging to particular industries, varying from short factory sidings to large local networks, are said to originate three-quarters of Russian railway freight. Total extent of these lines is probably around 30,000 km. In industrial areas they tend to be amalgamated into industrial railway transport enterprises which have steadily become privatised. One of the largest is the Moscow City Industrial Railway Company which serves 120 sites and owns 50 diesel shunters.

Intermodal operations

As yet, most container services operate 'by train path' rather than 'by timetable'. That is, they run to fixed schedules but only when traffic seems to require them. This irregularity is probably one reason why shippers have shown a certain indifference to container services. But since 1996 there has been a nightly St Petersburg–Moscow container train, and the Moscow–Tallinn (Estonia) service is also regular. Helsinki–Moscow container trains are also showing promise. The Berlin–Moscow container train ran about 500 times in 1995–98. In 1998 the Budapest–Moscow train ran 41 times, but most of its containers were transhipped to and from the highways at the Brest break-of-gauge. On the Trans-Siberian route, which showed great promise in the 1980s, traffic fell with deteriorating service and increasing pilferage. Efforts are being made to improve its attractions and it now provides a secure and swift service (12 days from Pacific port to the Polish frontier), but shippers are slow to return. In 1998, there were 129 container trains operated over this route (47 westward and 82 eastward).

A regular service, carrying motor trucks and their drivers, now links Novorossiysk on the Black Sea with Moscow. Those who criticised it as an inherently uneconomic procedure failed to take into account its main appeal: freedom from the compulsory bribes levied by highway police. In 1998, less than one per cent of traffic was handled in containers, although it was planned to increase this to 16 million tonnes by 2005. The lack of a railway organisation solely interested in container traffic was one brake on development, but in 2000 a new state subsidiary company (*GUP-Konteiner*) was set up, with affiliate companies on each railway.

A new multimodal freight terminal is being built at St Petersburg. Advice on the location, design, specifications, marketing and management of this terminal was provided by Sir Alexander Gibb and Partners of the UK, thanks to European Union funding.

New lines

Moscow–St Petersburg high-speed line

In September 1991 President Yeltsin decreed that a new high-speed route be created between Moscow and St Petersburg (formerly Leningrad).

A special company, RAO VSM, was formed to study, build and operate the high-speed route. Initially, a strong argument for the line was the need to relieve the existing St Petersburg–Moscow railway, but this lost its validity with the drastic fall of freight traffic after 1990.

Class 2TE116 two-section diesel-electric locomotive on a transfer freight at Narva, Estonia (Eddie Barnes)   0088201

Container traffic on the Nakhodka–Brest route (Cniitei Mps)   0088206

The present proposal, which only received final ecological approval in 1999, is for a new line from the outskirts of Moscow to the outskirts of St Petersburg, with existing routes being used for the terminal sections. Initially, trains would run at up to 300 km/h, and later at 350 km/h. The line would be passenger-only and would close for maintenance at night.

RAO VSM is a company independent of Russian Federal Railways even though the October Railway, part of Russian Federal Railways, is a joint founder (the others were the Moscow and St Petersburg governments and the Leningrad regional government). On its formation, the government transferred to the new company controlling shares in state industries about to be privatised. These included rolling stock building plants at Torzhok and Tikhvin. With the successful exploitation of these industries, and its land holdings, RAO VSM might well prosper even if the railway were not built.

Successive railway ministers have been lukewarm in their attitude towards the line (the only minister who was enthusiastic had an unexpectedly short tenure), and there are legitimate doubts about the traffic projections. Meanwhile the Railways Ministry has almost completed the reconstruction of the existing railway with a view to running trains at up to 200 km/h.

Almost all the needed land has been acquired, but work so far done has been limited to clearing the proposed route of wartime land mines and the beginning of the construction of a station/hotel complex as the terminal in St Petersburg. The latter is the responsibility of a British contractor and has been financed by British banks. Meanwhile the Torzhok Works has been building electric commuter trains while Tikhvin, with associate companies, has produced the *Sokol* prototype high-speed train.

In 1999 and 2000, RAO VSM was unable to redeem some of its due bonds. These were guaranteed by the government, which has taken an appropriate proportion of the company's assets in recompense. The prevailing opinion is that the line will not be built until about 2020, and probably not by the existing company. The Railways Ministry is hoping that high-speed lines will also be built in the 2020s between Moscow and the South and Moscow–Brest, in joint private/public ventures. In the meantime, 200 km/h services are anticipated

For details of the latest updates to *Jane's World Railways* online and to discover the additional information available exclusively to online subscribers please visit

jwr.janes.com

between Moscow, Omsk and Novosibirsk over existing lines, the upgrading of which is in course.

Finnish connections

A new 126 km route between Kochkoma and Ledmozero in Karelia, near the Finnish border, was nearing completion in 2001. Constructed by a private company, Gelleflint, the line has been designed to tap the mineral and forestry resources of the region. Gelleflint will interchange with RZhD at Kochkoma and with Finnish State Railways (VR) at the border.

Gelleflint proposes the development of a second line, a 180 km line from Karnogory to Vendinga, which, in combination with the first new line and existing RZhD routes, would establish the so-called No-We (North-West) rail link from northeast Russia to ice-free Finnish ports on the Baltic. A start on construction was proposed for 1997.

In another development, agreement has been reached between VR and RZhD on a 68 km new line from Ruchei Karelski to the border with northern Finland at Alakurtti, with reinstatement of a 6.5 km line on an abandoned trackbed on the Finnish side of the border. This would allow mineral deposits in the Murmansk area to be tapped. Construction was due to begin in 1995.

New link with China

Several Japanese firms have entered joint ventures associated with development of the port of Zarubino, at the southeastern extremity of Russia, below Vladivostok.

On the initiative of the Far Eastern Railway, the 'Golden Link' company was formed to build a line from Kraskino, south of Vladivostok, to link with Chinese Railways. This was finished in 1999 but traffic has failed to develop because it is a private railway, and therefore, as yet, has no legal status.

Another attempt to develop traffic for the Trans Siberian Railway is a link between that railway and South Korea. Russian Railways has offered to help upgrade the North Korean sector of this route.

Dagestan

A new 80 km line avoiding Chechnya has been built in Dagestan. This also reduces the length of the trunk route linking Russia with Azerbaijan.

Yamal peninsula line

Construction of a 400 km line from Vorkuta to gas and oilfields on the Yamal peninsula, which was put on hold in the early 1990s, has restarted; the line was 250 km long by late 1995. This will be the world's northernmost railway.

Amur–Yakutsk Railway

Construction of this 850 km line from Neryungri to Yakutsk has proved difficult, but 380 km have been laid, and completion is expected in 2005.

Improvements to existing lines

Apart from the Moscow–St Petersburg reconstruction, the lines from Moscow to Nizhnii Novgorod and Voronezh are also to be reconstructed. The EBRD is assisting in all three of these projects.

Track renewal and regauging to 1,435 mm gauge on the 47 km main line from Kaliningrad to Mamonova on the Polish frontier was completed in 1992/93 with German financial assistance. Elsewhere on the RZhD system, the condition of bridges requiring renewal or replacement has imposed a large number of speed restrictions.

An important new idea is to develop the Moscow ring railway as an outer circle line of the city metro system. Freight traffic on the ring railway is down by half, while the existing Circle line of the metro is heavily congested. Metro services will be introduced gradually section by section alongside, but separated from, continuing freight traffic. There will be about 30 new stations and interchange points with the existing metro system. If authorised, the first trains could be running by late 2000.

A Russian-Finnish railway accord was signed on 16 April 1996. Projects include upgrading of the St Petersburg–Helsinki line for high-speed operation by 1999 with Pendolino tilting trains. International travel on this line grew by 32 per cent in 1994 and

Class ChS200 8,000 kW DC electric locomotive recently refurbished and modernised by Škoda (Milan Šrámek)   0059401

One of a trial batch of Class EP1 six-axle AC electric passenger locomotives built by Novocherkassk (Zheleznodorozhnoe Delo/Ivan Khil'ko)   0088202

Prototype asynchronous-motored dual-voltage Class EP10 electric locomotive built by a consortium of Adtranz and Novocherkassk (Zheleznodorozhnoe Delo/Ivan Khil'ko)   0088203

is forecast to reach 1 million passengers per year by 2000. The Finnish Railways invested a European Union credit of USD335 million and undertook to find a further USD358 million by 1999. Some USD185–200 million is needed for the Russian section. Through high-speed trains to Turku are planned for 2005.

On the BAM line in Siberia, work on building the Severo–Muiskii tunnel has continued despite lack of cash and was finished in 2001. The BAM has excess capacity, and is in need of greatly increased economic activity in the area. For this reason, two branches to mineral deposits are being constructed. Work has already started on a 65 km line to Udokan which will tap substantial deposits of several minerals. A 320 km branch to the Elginskii coal deposits was begun in 2001 and when finished (some time after 2006) should also bring more traffic to the BAM.

A new 150 km industrial railway is being built in the Urals, linking the mainline railway with the Srednii Timan bauxite mine.

An improvement project has been proposed for railways on the island of Sakhalin in the Pacific, which were built to the 1,067 mm gauge. Freight from the mainland at present is shipped across the Gulf of Tartary on 1,520 mm gauge wagons on eight train ferries, and transhipped on the island into 1,067 mm gauge wagons for onward transit. To allow uninterrupted rail transit, there is a proposal to introduce a bogie changing facility at the port of Kholmsk on the island; clearances would also be enlarged in nine tunnels on the island's main line. For the longer term, regauging of Sakhalin's railways at 1,520 mm and construction of a tunnel to the mainland are under study.

A rather more exotic proposal is a two-tunnel rail link between Japan, Sakhalin and the Siberian mainland.

Traction and rolling stock

In 1993 the rolling stock fleet available for use comprised some 2,653 diesel and 5,043 electric locomotives, 37,940 passenger cars and 548,000 wagons plus over 16,000 multiple-unit coaches.

In 1998, the total (including reserve) freight wagon stock was about 660,000, of which 254,000 were open cars, 95,000 box cars, 74,000 flat cars, and 91,000 tank cars (8-axle tank cars totalled 8,600). The working stock was 385,000 in 1998 and 426,700 in 1999. The working stock of passenger cars was 19,959 in 1999. In 2000, 23 per cent of the operational freight wagons were in private ownership, and this percentage was expected to increase when more favourable tariffs for private-owner wagons were introduced. About 100,000 of the private-owner vehicles are tank wagons.

RZhD's strained finances have continued to restrict deliveries of new rolling stock. Only 29 electric and 35 diesel locomotives were delivered in 1994, and the 1995 programme for new passenger coaches was cut from 980 to 320. Deliveries of electric locomotives were six in 1998 and 20 in 1999, and of diesel locomotives (expressed in sections) 10 and 17. In total 607 passenger cars were delivered in 1999, up from 423 in 1998. In 1997, an order worth EUR48 million was placed with Adtranz for 21 electric locomotives, designated as Class EP10. This Co-Co design has dual power supply and asynchronous motors, and nearly 80 per cent of components will be supplied by Russian factories.

Under an accord between the German and Russian governments reached in 1994, RZhD was due to take delivery of 10 electric locomotives from Adtranz (worth DM40 million) and, at a concessionary price of DM500 million, 500 passenger coaches from former East German builder DWA; 191 new coaches were supplied in 1994.

Also approved by the Railways Ministry was the import of 26 electric locomotives from Škoda, 16 to be bought with hard currency and the remainder to be bartered for supply of electricity to the Czech Republic. In 1998 Škoda was also undertaking refurbishment and upgrading of nine ChS200 3 kV DC electric locomotives. Dating from the late 1970s, the 8,000 kW locomotives was returned to service in 1999 on the Moscow–St Petersburg line, where RhZD plans 200 km/h running. Meanwhile Škoda is co-operating in establishing a repair works

Class TEM18G gas-fuelled diesel shunting locomotive built by Bryansk (Cniitei Mps)   0088207

at Yaroslavl for the several Škoda-built electric locomotive types.

A shortage of passenger locomotives is forcing widespread use of freight traction for passenger services: an average of 800 to 1,200 electric and 605 diesel locomotives were in use in this way in 1995. Loco-hauled stock is being used to cover a shortfall of multiple-units; almost every diesel class appears on such work. A modernisation programme has been agreed for Hungarian-built class D1 dmus using a Zvezda engine and Kaluga electrical equipment.

Recently, numbers of new freight wagons delivered to the railway have fallen well below the number of withdrawals. RZhD acquired 10,400 wagons in 1993 and 7,600 in 1994, falling further to 2,102 in 1998 and 1,557 in 1999. While some new designs are currently being prepared for series production, the cheaper alternative of thoroughly reconstructing older vehicles is also being pursued, with 20,000 such rebuildings scheduled for 1999.

Two Russian sleeping cars have been refurbished in Spain at RENFE's Malaga works as a pilot project for a possible 1,500 carriages.

New locomotive plans

Many of the traction suppliers to the former SZhD were located outside the Russian Federation. Procurement for the RZhD has been reorganised to minimise hard currency expenditure on imports and protect domestic employment.

Škoda in Czechoslovakia was SZhD's principal supplier of passenger electric locomotives in the days of the Comecon trading bloc. The Russian freight electric locomotive production facilities at Novocherkassk have been expanded to produce diversified passenger locomotives as well. The EP1 is a passenger locomotive derived from Novocherkassk's VL65 AC electric. VL65 production has now ended, only 48 units having been built, but 28 of the replacement EP1 were built by the end of 2000. To replace the bigger Czech locomotives the 8-axle 7,200 kW EP200 has been built, to be followed by its DC version, the EP100. The EP201 and EP101 will be similar but designed for 160 km/h instead of 200 km/h. The EP10 is a dual-voltage machine of which 21 are being built by a consortium of Adtranz and Novocherkassk. These are the first Russian locomotives to be equipped with GTO asynchronous technology. The first EP10

Electric locomotives

Class	Wheel arrangement	Output kW continuous/ one hour	Speed km/h	Weight tonnes	First built	Builders Mechanical	Electrical
3 kV DC							
VL8	2 × Bo+Bo	3,760/4,200	80/100	184	1953	N/T	N
VL10	2 × Bo-Bo	4,600/5,360	100	184	1976	N/T	T
VL10u	2 × Bo-Bo	4,600/5,360	100	200	1974	N/T	T
VL11	2 × Bo-Bo	4,600/5,360	100	184	1975	T	T
VL11m	2 × Bo-Bo	NA/6,680	100	184	1987	T	T
VL15	2 × Bo-Bo-Bo	8,400/9,000	100	300	1984	T	T
VL22m	2 × Co+Co	1,860/2,400	80	132	1941	N	N/D
VL23*	Co+Co	2,740/3,150	100	138	1956	N	N
ChS2	Co-Co	3,708/4,200	160	125	1958	Škoda	Škoda
ChS2t	Co-Co	4,080/4,620	160	126	1972	Škoda	Škoda
ChS200	2 × Bo-Bo	8,000/8,400	220	156	1975	Škoda	Škoda
ChS6	2 × Bo-Bo	8,000/8,400	190	164	1979	Škoda	Škoda
ChS7	2 × Bo-Bo	6,160/7,200	180	172	1983	Škoda	Škoda
EP100	Bo-Bo-Bo-Bo	9,600/NA	200	180	1999	Kolomna	Škoda
25 kV AC							
VL60K*	Co-Co	4,050/4,650	100	138	1962	N	N
VL60pk	Co-Co	4,050/4,650	110	138	1965	N	N
VL65	Bo-Bo-Bo	4,680/5,000	120	NA	1992	N	N
VL80K	2 × Bo-Bo	5,920/6,320	110	184	1963	N	N
VL80T	2 × Bo-Bo	5,920/6,320	110	184	1967	N	N
VL80R	2 × Bo-Bo	5,920/6,320	110	192	1967	N	N
VL80S	2 × Bo-Bo	5,920/6,320	110	184/192	1979	N	N
VL85	2 × Bo-Bo-Bo	9,360/10,000	110	288	1983	N	N
VL86F	2 × Bo-Bo-Bo-Bo	NA/10,800	120	288	1985	N	N
ChS4	Co-Co	4,920/5,100	180	123	1965	Škoda	Škoda
ChS4t	Co-Co	4,920/5,100	180	126	1971	Škoda	Škoda
ChS8	2 × Bo-Bo	7,200/NA	180	175	1983	Škoda	Škoda
EP200	Bo-Bo-Bo-Bo	8,000/NA	200	180	1996	Kolomna	N
Dual voltage 3 kV DC/25 kV AC							
VL82m	2 × Bo-Bo	5,760/6,040	110	200	1972	N	N
EP10	Bo-Bo-Bo	7,200/NA	160	132	1998	N	Adtranz

* Also operated in twin- and triple-unit versions.
Abbreviations: D: Dinamo; N: Novocherkassk; T: Tbilisi.

was built in 1998 and is still on trial. This type will be followed by the somewhat similar Russian-built EP2 (DC) and EP3 (AC). It is forecast that over 1,000 units of each of the EP2 and EP3 will be built.

With the 'loss' of diesel loco-building capacity in Ukraine, steps are being taken to raise output from Russia's own plants. At Lyudinovo a new generation of standard diesel-electrics is to be developed in co-operation with GE Transportation Systems of the US, under a technology transfer agreement signed in July 1995. A prototype 3,000 kW TEP400 mixed-traffic diesel locomotive is scheduled from Lyudinovo. Kolomna plans to construct TEP80 and TEP71 passenger diesel locos. Limited imports may continue in the form of 2TE116UP mixed-traffic locomotives from Lugansk. Later additions to the programme include the 3,600 hp TEP100 diesel passenger and the TEM18G gas-fuelled diesel yard locomotive. The prototype of the latter has already been built at Bryansk.

Further co-operation is wanted with foreign firms concerning the latest semiconductor transformers, three-phase traction motors, microprocessor control and frame-hung traction motors. Foreign investment in more joint ventures is also being sought.

A Talgo tilting train was on test on the October Railway in early 1996, including high-speed trials on the Moscow–St Petersburg line.

In 2000 an agreement was signed establishing a joint Russo-Spanish company for the production of Talgo convertible-gauge freight and passenger wagons and coaches. The first gauge-convertible passenger trains were expected to be used for international services to Berlin and Warsaw.

New emu and dmu plans

Short-haul commuter travel is booming at weekends, despite draconian fare increases, because the government has provided suburban land for city inhabitants in a bid to encourage food production and cure urban shortages. More than half of the commuter vehicles are life-expired, or approaching that age, and cannot cope with the still dense suburban traffic.

The former source of emu car supply was the Riga works in Latvia, which Russia now finds expensive. In 1992, 100 emu vehicles were ordered from the Riga works, but the Russians are now developing their own domestic sources of supply.

Emu vehicles have been produced by the Torzhok missile wagon works, while the Tikhvin converted military plant has manufactured motor bogies. Current plans envisage an annual capacity of 750 carriages by 2001.

Novocherkassk produced its first 10-car emu in early 1996 (Class EN1) and the ENZ asynchronous emu was expected to appear in 1999. A lack of funds means that this works' capacity for emu construction remains badly utilised.

The Demikhovo, formerly a builder of narrow-gauge wagons for the mining industry, completed the first of 50 four-car emus in December 1992; Czech builder Škoda has a USD20 million stake in the new concern. Classes ED4 and ED6 were to enter series production in 1996 and 1999 respectively, with Novocherkassk electrical equipment. By 2000 the factory is intended to produce over 1,000 cars per year; 24 10-car trains were produced in 1998 and the Railways Ministry was to acquire about 50 new emu sets in 1999, about half the number urgently required.

The prototype *Sokol* high-speed train has been on trial since 1999. This is a 12-car formation with four dual-current motored vehicles. It is designed to run at up to 250 km/h. It may be purchased for use on the reconstructed St Petersburg–Moscow line.

To replace the Hungarian, Latvian and Czech suppliers of dmus and railcars, the Lyudinovo locomotive plant has begun production of Class DL2 dmu motor cars which, in combination with trailer cars supplied by Tver, results in the DL2B12 12-car dmu. Meanwhile, the Mytischii works has produced Russia's first railbus, the RA1, which has been on trial since early 1998; this is based on that works' metro vehicles.

The DP2 diesel train of completely new design began trials in 2000. It is a joint product of the

Diesel locomotives

Class	Wheel arrangement	Power kW	Speed km/h	Weight tonnes	First Built	Builders Mechanical	Engine	Transmission
TE3	2 × Co-Co	2,944	100	254	1953	V/K/Ko	2 × 2D 100	E ETM
TE7	2 × Co-Co	2,944	140	254	1956	V/K	2 × 2D 100	E ETM
2TE10M*	2 × Co-Co	4,416	100	276	1981	V/K	2 × 10 D100 or M1	E ETM
2TE10L	2 × Co-Co	4,416	100	260	1962	V/K	2 × 10 D100	E ETM
2TE10u	2 × Co-Co	4,412	100	276	1989	V/K	2 × 10 D100M1 or M2	E ETM
2TE10ut	2 × Co-Co	4,412	120	276	1989	V/K	2 × 10 D100M1 or M2	E ETM
2TE10V	2 × Co-Co	4,416	100	276	1974	V/K	2 × 10 D100	E ETM
4TE10S	4 × Co-Co	8,824	100	552	1983	V/K	4 × 10 D100	E ETM
TEM1	Co-Co	736	90	126	1958	B	2D50	E ETM
TEP10	Co-Co	2,208	140	129	1960	V/K	10 D500	E ETM
TEP60	Co-Co	2,208	160	129	1960	Ko	11D45	E ETM
TEP70	Co-Co	2,944	160	135	1973	Ko	2A5D49	E ETM
TEP80	Bo+Bo-Bo+Bo	4,412	160	180	1988	Ko	2-20DG	E ETM
M62**	Co-Co	1,470	100	116.5	1964	V/Ko	14D40	E ETM
2TE116	2× Co-Co	4,500	100	276	1971	V/Ko	2 × 1A5D49	E ETM
2TE136	2 × Bo-Bo+ Bo-Bo	8,832	120	400	1984	V/Ko	120DG(D49)	E ETM
2M62u	2 × Co-Co	2,942	100	252	1987	V/Ko	2 × 14D40	E ETM
TEM2	Co-Co	883	100	120	1960	V/B/P	PDIM or PDI	E ETM
TEM2um	Co-Co	994	100	126	1988	B/P	1PD-4A	E ETM
TEM7	Bo+Bo– Bo+Bo	1,472	100	180	1975	L/Ko	2-2D49	E ETM
ChME2	Bo-Bo	552	80	64	1958	ČKD	65310DR	E ČKD
ChME3†	Co-Co	993	95	123	1964	ČKD	K65310DR	E ČKD
ChME5	Bo+Bo Bo+Bo	1,470	95	168	1985	ČKD	K85310DR	E ČKD

* Also operated in triple-unit version (3TE10M).
** Also operated in twin- and triple-unit versions (2M62, 3M62).
† A variant is classified ChME3T.
Abbreviations: B: Bryansk; K: Kharkov; Ko: Kolomna; L: Lyundinovsk; P: Penza; V: Voroshilovgrad

Electric railcars or multiple-units

Class	Cars per unit	Motor cars per unit	Motored axles/car	Output/motor kW	Speed km/h	First built	Builders Mechanical	Electrical
3 kV DC								
ER1	10	5	4	200	130	1957	Riga/Kalinin	REZ, Dinamo
ER2	10–12	5–6	4	200	130	1962	Riga/Kalinin	REZ
ER22	8	4	4	220	130	1964	Riga	REZ
ER200	10	8	4	240	200	1974	Riga	REZ
ER2R	10–12	5–6	4	240	130	1979	Riga	REZ
ER2T	10–12	5–6	4	235	130	1987	Riga	REZ
ET2	10	6	4	NA	130	1993	Torzhok/Tikhvin	Tikhvin
ED2T	10	5	4	NA	NA	1992	Demikhovo/Tikhvin	Tikhvin
25 kV AC								
ER-9	10	5	4	180	130	1961	Riga/Kalinin	REZ
ER9P	10–12	5–6	4	200	130	1964	Riga	REZ
ER9M	10–12	5–6	4	200	130	1976	Riga	REZ
ER9E	10–12	5–6	4	200	130	1981	Riga	REZ
ER 9T	10–12	5–6	4	200	130	1988	Riga	REZ

Diesel railcars or multiple-units

Class	Cars per unit	Motor cars per unit	Motored axles/car	Power/motor kW	Speed km/h	First built	Mechanical	Builders Engine	Transmission
D1	4	2	2	540	120	1963	Avad/Ganz-Mavag	Ganz-Jendrassik 12VFE17/24	HM
DR1	6	2	2	736	120	1963	Riga	Zvezda M756B	H
DR1P	6	2	2	736	120	1969	Riga	Zvezda M756B	HM
DR1A	6	2	2	736	120	1973	Riga	Zvezda M756B	H
ACh2	1/2	1	2	736	120	1984	Studenka	Zvezda M756B	H

Lyudinovo diesel locomotive works and the Tver passenger coach works.

Other passenger coach developments

Double-deck vehicles are slow to appear, despite earlier agreements with ANF-Industrie (now Bombardier) of France and GEC Alsthom Transporte (now Alstom) of Spain. With the end of German imports, the Tver passenger coach works is now the supplier of long-distance coaches and it was hoped production there would increase to 500 annually in 2000. It is concentrating on high-comfort designs. Meanwhile the Voitovich Works is reconstructing time-expired vehicles and also designing a new range of coaches. In 2000, it was due to deliver the first of 16 new coaches for the *Avrora* Moscow–St Petersburg service. These are to be air conditioned and provided with ecological toilets. The latter are also a feature of vehicles now in the Helsinki–St Petersburg services, which also include Russia's first coaches with wheelchair access.

Signalling and telecommunications

About 63,000 km of the RZhD network are equipped with the automatic block system. Microprocessor-based control centres cover about 25,000 km, and this system is being extended, although much signalling equipment dating from the 1950s and 1960s will need to be replaced at the same time. Several systems of automatic locomotive signalling are in use, but only one is regarded as satisfactory.

Large-scale electronic systems include the *Ekspress* passenger train reservation facility, the *DISPARK* freight car control system and the *DISLOK* network which monitors locomotives and their crews. The laying of fibre optic cables between St Petersburg and the Black Sea, and between Moscow and Vladivostok, is well advanced. The first ground station for satellite communication is being built in the Urals, although earlier experiments were carried out at Krasnoyarsk. The Railways Ministry will have enough capacity to enable its *Transtelekom* subsidiary, already a substantial telephone service provider, to offer a competitive direct-dialling long-distance service to the public.

Electrification

RZhD operates the world's most extensive electrified railway network and about two-thirds of all freight traffic is electrically hauled. Lines around Moscow and in the Urals are electrified on the 3 kV DC system. Most lines electrified since 1956 have been equipped with the 25 kV 50 Hz AC system.

The astonishing pace of electrification between the 1960s and the 1980s has slowed in recent years for economic reasons. In 1990 the 25 kV electrification of the Baikal–Amur Railway from Ust–Kut reached Chara. Two other routes electrified at 25 kV in the Far East were Khrebtovaya–Ust Ilimsk and Khabarovsk–Kruglikovo.

The Trans-Siberian has been electrified as far as Khabarovsk (8,531 km) for many years, and work is proceeding steadily on converting the Khabarovsk–Vladivostok section.

Electrification at 25 kV is proceeding on the North Caucasus Railway. In 1993 wiring was completed of a 50 km section from Krymskaya to Novorossiysy, while 1994 saw the start of work on a further two section, Krasnodar–Tikhoretskaya and Krymskaya–Afimskaya, totalling 180 km. Work on the Tikhoretskaya–Salsk and Likhaya–Morozovskaya sections (290 km) began one year later. After completion it will be possible to employ electric traction to the Volga Railway via Morozovskaya and Kotelnikovo.

In 1997 electrification began on a 158 km section between Sviyazhsk and Tsima, connecting the Gorki and Kuibyshev Railways. The changing geopolitical situation following the break-up of the Soviet Union calls for a further upgrading and electrification of an additional 470 km of north-south lines to avoid passing through Ukraine.

On the South East Railway the Voronezh–Kastornaya route was electrified at 25 kV early in 1998 and work has since switched to the 187 km Elec–Kastornaya–Stariy Oskol section.

In 1996 the 130 km Vologda–Buy line was electrified and work on a westward extension to Cherepovets (124 km) has since been started. Further north, electrification of the Konosha–Milenga–Arkangelsk line (423 km) is in progress and a 126 km section from Nyandoma, north of Konosha, and Plesetskaya was completed in 1998. Both projects employ the 25 kV system.

Electrification at 3 kV DC of the Volkhovstroy–Belomorsk section of the busy Northern Railway's St Petersburg–Petrozavodsk–Belomorsk–Murmansk line is close to completion. In mid-1996 the 121 km Volkhovstroy–Lodeynoye Pole section was completed; an extension northwards to Svir followed in mid-1998. A section from Belomorsk to Idel was energised in August 1997 and a further 165 km south to Medvezhya Gora was completed by the end of 1998. Electrifying the gap between Svir and Medvezhya Gora will eliminate remaining diesel operation over the 1,455 km St Petersburg–Murmansk line, significantly reducing journey times.

Conversion from 3 kV DC to 25 kV AC 50 Hz of the Mineralnye Vody–Kislovodsk (65 km) and Zima–Slyudyanka (453 km) lines was completed in 1996. The latter forms part of the Trans-Siberian route.

Track

A severe backlog in track maintenance work has developed on RZhD in recent years. Ballast is in poor condition in many places. The EBRD report recommended that lighter, more mobile track maintenance equipment and advanced tampers should be used, and advocated a revision of maintenance methods. Western track machines have been purchased and a start made on producing Russian versions of them which initially will be devoted to the Moscow–St Petersburg, Moscow–Smolensk and Moscow–Samara main lines. A trial order was placed for Swedish sleepers with Pandrol clips.

Rails: 75, 65, 50 kg/m
Continuous welded rail: 40,400 km (1999)
Sleepers: Ferro-concrete on 56,800 km (1999)
Sleeper spacing: 1,840/km on plain track, 2,000/km on curves of less than 1,200 m radius
Sleeper dimensions: 2,750 × 250 × 180 (wood); 2,700 × 250 × 193 (concrete)
Max axleload: 23.5 tonnes

UPDATED

Saudi Arabia

Ministry Of Transport

Airport Road, 11178 Riyadh
Tel: (+966 1) 404 30 00; 29 28
Fax: (+966 1) 403 14 01
Web: www.saudinf.com

Key personnel
Minister: Dr Jubarah bin Eid Al-Suraiseri

UPDATED

Saudi Railways Organisation (SRO)

PO Box 36, Dammam 31241
Tel: (+966 3) 871 30 01
Fax: (+966 3) 827 11 30
Web: www.saudirailways.org;
www.saudirailexpansion.com

Key personnel
President: Khalid H Alyahya
Vice-President, Operations: Hamad A Al-Abdulqadir

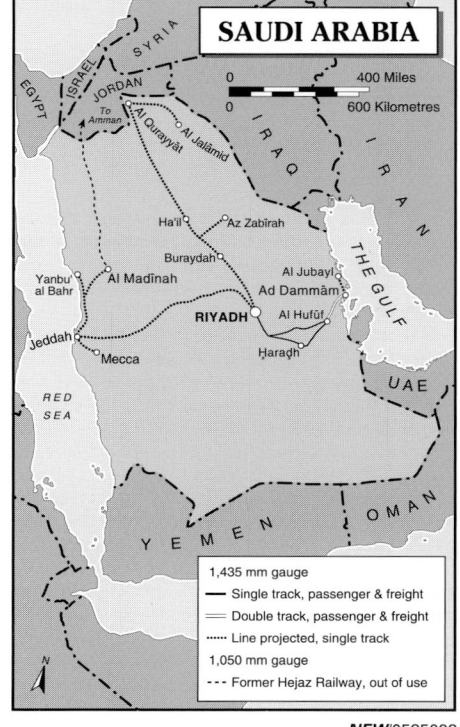

SAUDI ARABIA

0   400 Miles
0   600 Kilometres

1,435 mm gauge
— Single track, passenger & freight
═ Double track, passenger & freight
···· Line projected, single track
1,050 mm gauge
- - - Former Hejaz Railway, out of use

NEW/0585098

Vice-President, Technical Affairs: Ali S Al-Karni
Vice-President, Planning and Budget and Director General, Administration and Finance: Saad Abdulaziz Al-Abbad
General Manager, Human Resources: Mohammed Saleh Al-Qarni
General Manager, Maintenance and Executive Responsible for UIC/International Affairs: Abdullah S Balhaddad
General Manager, Utilities and General Services: Ali Razhan Al-Sheri
Director, Contracting Department: Abdulamir M Al-Sunni

Gauge: 1,435 mm
Route length: 1,392 km (with branch lines & sidings)

Political background
As part of the Saudi Landbridge project (see below), the government expects to grant concessions for freight and passenger services over the entire SRO network, transferring assets and staff to the successful bidders for these contracts.

Organisation
Of the two routes between Dammam and Riyadh, Line 1 (via Haradh) is used by passenger traffic and is engineered for 135 km/h running, while Line 2, the direct route between the two cities, is freight only, with a maximum permitted speed of 60 km/h.

At the end of 2004 SRO employed 1,627 staff.

Traffic	2002	2003	2004
Passenger journeys	913,000	913,000	1,124,000
Passenger-km (million)	294	293	364
Freight tonnes (million)	1.89	1.89	2.55
Freight tonne-km (million)	880	880	1,173

Passenger operations
Passenger services are operated between Dammam and Riyadh.

Freight operations
The principal freight corridor is between the Gulf port of Dammam and Riyadh, Saudi Arabia's largest city and consumer centre. Much of this traffic is containerised: in 2004 SRO handled 283,000 TEUs, container traffic accounting for some 65 per cent of SRO revenue in that year. There are extensive dry port facilities at Riyadh, where there is annual capacity to handle 250,000 TEUs. This facility, and the freight terminals at Al Hufüf and Dammam, have recently undergone modernisation.

New lines
Saudi Landbridge project
In 2005 pre-qualification was due to start of bids for a Build-Operate-Transfer (BOT) concession for the Saudi Landbridge project, an ambitious scheme to provide a rail link for freight and passenger traffic between the Gulf and the Red Sea. This will entail construction of a 950 km line west of the existing Dammam–Riyadh line to Jeddah and a 115 km line northwest from Dammam to the industrial town of Jubail.

In 2004 SNCF International, together with UBS Investment Bank and the National Commercial Bank of Saudi Arabia, was awarded a contract to undertake detailed studies for this scheme and the Western Railway (see below). SRO foresees daily operation of up to 40 double-stack container trains running at 120 km/h and four or five 220 km/h passenger services. Initially the line will be single-track with passing loops but its civil engineering will allow later doubling. It will be diesel operated.

Pre-qualified concessionaires were expected to be invited to bid in mid-2005. A concession award is expected in the third quarter of 2006, with the aim of commencing services on the line in 2010.

Western Railway
Also projected is a 570 km line running north from Jeddah, with a branch to the Red Sea industrial

SRO passenger service headed by GM-built SDL50 locomotive   0058803

town of Yanbu al Bahr, to Al Madinah (Medina) and south to Mecca. As well as extending the reach of the Landbridge scheme, this will handle substantial passenger traffic to the holy cities of Mecca and Medinah. The study contract awarded to SNCF International and its partners for the Saudi Landbridge project also covers the Western Railway.

Northwestern mineral line

Studies by a grouping of Jacobs Engineering and SNC Lavalin were in progress in 2004 into a mineral line of some 1,100 km running northwest from Riyadh via Ha'il to Al Qurayyat, close to the border with Jordan. From here a branch would run east to Al Jalamid, where extensive reserves of phosphates are located. A second branch would be built northeast from near Ha'il to Az Zabirah to serve bauxite deposits. The study into the line followed an earlier evaluation of the scheme by the World Bank.

Improvements to existing lines

Upgrading of the Dammam–Riyadh main line was undertaken in the 1980s. The 140 km of the existing route from Dammam to Al Hufüf were double-tracked and realigned for 120 km/h maximum speed, with continuously welded UIC 60 rail on concrete sleepers. Beyond Al Hufüf a new and direct double-track route of 308 km was built for 150 km/h running. This line is used by passenger trains only. There is a second double-track line from Al Hufüf to Riyadh (416 km), passing through Haradh and Al-Kharj. The old line from Dammam to Al Hufüf and from Al Hufüf to Riyadh via Haradh and Al-Kharj is used for freight only. At present 449 km of the network carries passenger traffic.

New stations were built at Riyadh, Dammam and Al Hufüf. Traffic is controlled by voice orders through the telecommunications system via radio terminals and repeater stations. Some level-crossing automatic barriers are worked by soft-lead batteries recharged by solar power.

Traction and rolling stock

In 2000 the railway was operating 59 diesel locomotives, 58 passenger cars (including nine restaurant cars) and 2,262 freight wagons.

In 2004 SRO ordered eight new diesel locomotives from General Motors. In the same year the railway completed a programme to refurbish much of its existing locomotive fleet.

In 2003 a programme commenced to refurbish the entire SRO passenger fleet. This foreshadowed the arrival of 13 new passenger coaches and four generator/luggage vans for which orders were placed with Samsung Corporation, South Korea, in 2002.

Type of coupler, passenger and freight: AAR Type E
Type of brake: Westinghouse air

Diesel locomotives

Class	Wheel arrangement	Power kW	Speed km/h	Weight tonnes	No in service	First built	Builders Engine	Transmission
1100 (G18W)	Bo-Bo	746	110	67	16	1968	EMD	E EMD
1100 (SW1001)	Bo-Bo	746	110	84	5	1981	EMD	E EMD
2250 (GP38-2)	Bo-Bo	1,678	110	107	1	1973	EMD	E EMD
2000 (GT22CW)	Co-Co	1,492	110	109	3	1976	EMD	E EMD
2000 (SDL38-2)	Co-Co	1,678	110	–	6	1978	EMD	E EMD
3500 (SDL50)	Co-Co	2,611	160	124	6	1981	EMD	E EMD
				125	10*	1985*	EMD	E EMD
				120	7	1998	EMD	E EMD
3600 (CSE 26-21)	Co-Co	2,450	160	126	6	1981	Alco/Francorail	E Jeumont Schneider

* With dynamic braking

Dry port at Riyadh   0087935

Track

All main line track renewals are undertaken with continuously welded UIC 60 kg/m rail on prestressed concrete sleepers with elastic fastenings.

Rail types and weights: 110 lb RE, 115 lb RE, UIC 54, UIC 60

Crossties (sleepers): Prestressed concrete, 215 mm thickness, 2,600 mm long; some wooden

Spacing: 1,667/km

Rail fastenings: Vossloh, ballast cushion of 300 mm

Min curve radius: 565 m

Max gradient: 1.1%

Max permissible axleload: 22 tonnes (25 tonnes in special cases)

UPDATED

Senegal

Ministry of Infrastructure, Equipment and Transport

Ex Camp Lat-Dior Corriche, Dakar
Tel: (+221 823) 83 51; 82 76   Fax: (+221 823) 82 79
Web: http://www.primature.sn

Key personnel
Minister: Mamadou Seck

Société Nationale des Chemins de Fer du Sénégal (SNCS)

PO Box 175, Cité Ballabey, Thiès
Tel: (+221 39) 53 50   Fax: (+221 951) 13 93

Key personnel
Director General: Ibrahima Niang
Directors:
    Infrastructure and Co-ordination: Badara Talla
    Personnel: Ndaraw Fall
    Financial: Serigne Ndiaye

Equipment: Abdoulaye Lo
Operations: Mamadou Gueye
Studies, International Relations,
    Computerisation and Property: Baba Diankha

Gauge: 1,000 mm
Route length: 906 km

Organisation
As one of West Africa's most industrialised countries, Senegal has a railway system comprising two main lines running from Dakar to St Louis and Linguère in the north-east and the border with Mali in the east. The system was originally part of the Federal West African Railway Authority (AOF) before transfer to the Mali Federation in 1960. The disintegration of the Mali Federation caused the division of the former Dakar–Niger system into two networks. The principal line extends 1,286 km from Dakar in Senegal to Koulikoro, the terminus of the railway in the landlocked neighbouring country of Mali.

SNCS was established in November 1989 as successor to the former RCFS; it is an independent semi-public corporation and as such agrees a performance contract each year with the government.

In recent years, the growth of road transport has been a significant challenge for SNCS.

In 1997 the governments of Senegal and Mali retained Canadian consultancy CPCS Transcom to prepare a concession and assist in the selection of a concessionaire to operate international services on the Dakar–Bamako (Mali) line. Expressions of interest were submitted by February 1998 but no further progress was evident until February 2001, when the Senegal and Mali governments revived efforts to select a concessionaire with access to finance to upgrade the infrastructure and commercial expertise to develop traffic.

Passenger operations
There are four international expresses a week from Dakar to Bamako, the capital of Mali, which are

overseen by a joint management board. Typically, these trains carry between 4,000 and 5,000 passengers a month.

A frequent push-pull service operates on the 29 km suburban route from Dakar to Thiaroye and Rufisque. Patronage on this route has boomed, with the service handling some 6 million passengers in a year.

Freight operations

In terms of weight, the principal freight traffic is phosphates, transported by rail to the port of Dakar for export. In 2000, one daily train was serving Taiba, a single weekly train was serving Lam-Lam, and a single daily train was serving Allou Kagne.

In 2000, SNCS was operating two trains of chemical products a day for the chemicals company Sefics between Taiba and Dakar, using wagons owned by Sefics.

In 2000, international traffic on the Dakar–Bamako line totalled 345,704 tonnes, 4.3 per cent down on the 1999 figure of 361,293 tonnes. Freight traffic revenue in 2000 was CFAFr7 billion.

New lines

As part of a larger proposal to convert the Senegalese rail network from metre to standard gauge, the government has conducted studies into possible alignments for the construction of a new line east from Linguère to Faleme and Matan, close to the border with Mauritania, to serve iron ore and phosphate deposits.

Improvements to existing lines

A major aim of the Senegalese government is the conversion of the metre gauge network to standard gauge. This has three objectives: economic development; providing a stimulus to the exploitation of iron ore and phosphates resources in the east of the country; and improving the efficiency of Dakar suburban passenger services. Pre-feasibility studies complete in 2001 with the technical assistance of the government of Pakistan covered: construction of a third line between Dakar and Thiès, including provision for a service to a new international airport; rehabilitiation of the Dakar–Thiès-Saint Louis line, including construction of a new standard gauge line between Thiès and Saint Louis; conversion to standard gauge of the Dakar–Tambacounda–Kidira line; and construction of the new line to the east as detailed in *New lines*.

Traction and rolling stock

The stock of 29 diesel locomotives is mostly French built; French aid is financing Alstom maintenance of the railway's Alsthom AD16B locomotives. In 1996, a protocol was signed between the governments of Senegal and Canada that provided a grant of C$12.5 million to pay for five new locomotives from General Motors Canada; a further C$1 million from Senegalese sources funded acquisition of a pool of spare parts. The locomotive fleet is more than adequate for current SNCS traffic requirements, to such an extent that in 2001 three 1,790 kW machines were leased to RCFM in Mali, in addition to an additional three that were outbased there.

At the end of 2000, the number of passenger cars totalled 102. SNCS has acquired from France 24 stainless steel SNCF 'Mistral' cars of the 1950s and their four generator cars; these are now being used on the international trains to Mali. In 2001, SNCS also had 20 passenger coaches refurbished by Pakistan Railway Carriage Factory. This was in part to provide capacity for traffic generated by the African Cup football competition.

The 'Little Blue Train' suburban service in Dakar was launched in 1987. There are seven five-car trains, with 10 dedicated Type AD16B locomotives as motive power. SNCS has refurbished 41 SNCF

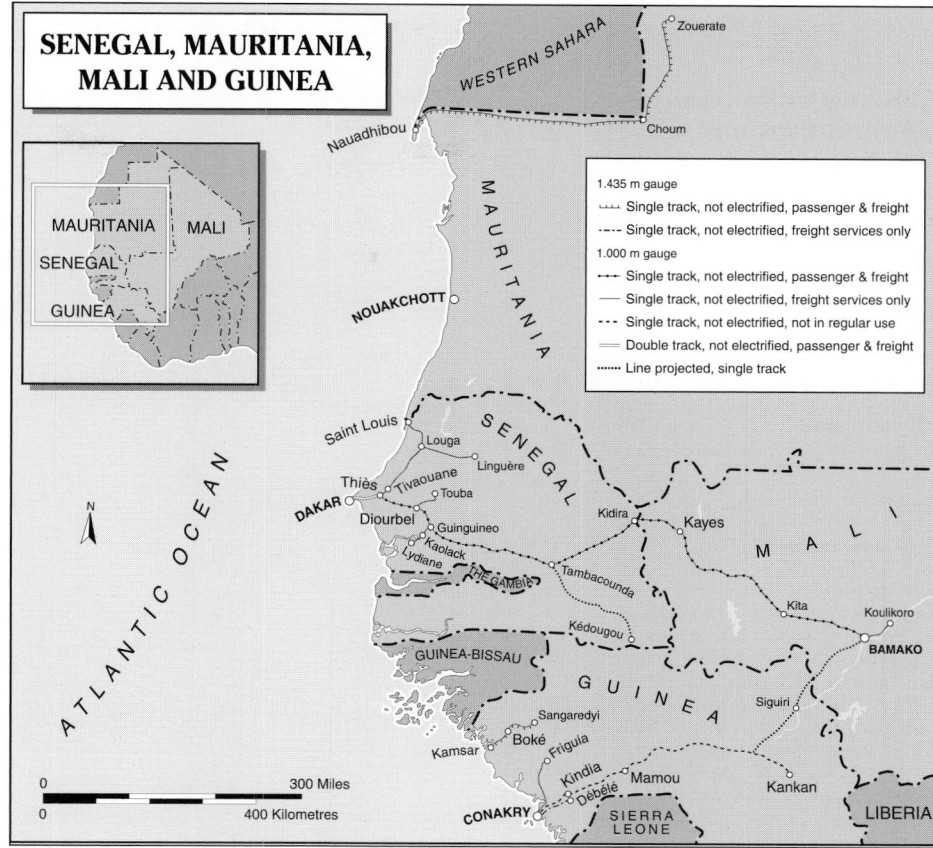

0058514

SNCS Alsthom-built AD16B diesel locomotive at Dakar heading a 'Little Blue Train' push-pull suburban service (Eddie Barnes)   0536198

Type B10t 'Bruhat' passenger cars, and fitted them with ex-freight wagon metre-gauge bogies with their suspension refined. The fleet was further supplemented in 2001 by the supply of 20 coaches from Pakistan as part of the technical co-operation agreement covering studies into standard gauge conversion.

In 2000, the freight wagon fleet totalled 1,015. SNCS and Mali have recently added 140 new wagons to the international pool.

Signalling and telecommunications

Automatic block colourlight signalling is operative on the 70 km of double track between Dakar and

Thiès. Further improvements to signalling and telecommunications systems have been identified as a priority, especially for routes carrying suburban traffic.

Serbia and Montenegro

Ministry of Transport & Communications

Boulevard Avnoj 104, Belgrade

Montenegrin Railways (ZCG)

Železnice Crne Gore a d
Trg Golootočkih žrtava 13, 81000 Podgorica, Montenegro
Tel: (+381 81) 63 34 98   Fax: (+381 81) 63 39 57

Key personnel
Director General: Ranko Medenica

Gauge: 1,435 mm
Route length: 250 km
Electrification: 169 km at 25 kV AC 50 Hz

Background
ZCG was established in January 2003 as a result of the split of the rump of the former Yugoslav Railways (JŽ) into separate companies covering Montenegro and Serbia.

While the electrified trunk route from Belgrade to the Adriatic coast is recognised as a key element of the trans-European transport network, the ZCG section from Vrbnica to Bar is in poor condition as a result of a lack of maintenance and investment. The same applies to much of the network, especially the Podgorica–Niksic line (56 km).

The cross-border section of the Podgorica–Shkodër line linking Montenegro and Albania was reported to be closed in 2003.

Improvements to existing lines
EIB funding of €85 million for an upgrade of the Belgrade–Bar line included €15 million for the ZCG

0554838

section from Vrbica to the coast. This should lead to some improvements by the end of 2004.

Traction and rolling stock
Rolling stock assigned to ZCG following the break up of JŽ comprised 17 electric locomotives, 18 diesel main line and shunting locomotives and six emus. More than half of this fleet was unserviceable in 2003, although tenders were invited in 2003 for the refurbishment of four Class 461 electric locomotives. There were also 68 passenger coaches and 910 wagons.

Serbian Railways (ŽS)

Železnice Srbije
Nemanjina 6, Belgrade 11000
Tel: (+381 11) 361 67 22; 48 11   Fax: (+381 11) 361 67 97

Gauge: 1,435 mm
Route length: approximately 3,800 km
Electrification: 1,195 km at 25 kV 50 Hz AC

Organisation
ŽS was established in January 2003 as a result of the division of the former Yugoslav Railways (JŽ) into two separately administered systems covering Serbia and Montenegro (see separate entry for Montenegrin Railways in the Serbia and Montenegro section of *Railway systems and operators*).

In 2002–03 JŽ and then ŽS continued a process of reconstruction and rehabilitation following heavy damage inflicted on the system by the NATO bombing campaign during the Kosovo crisis in 1999 and the adverse effects of trade sanctions imposed by the UN. A major problem has been a shortage of motive power due to deferred maintenance and a lack of spares. The position has progressively improved as overhauls have been undertaken and some traction has been hired in, including main line diesels from Bulgaria. Sources of funding for network rehabilitation and development have included the European Bank for Reconstruction and Development (EBRD), which in October 2001 agreed a €57 million loan covering locomotive refurbishment and maintenance, track maintenance equipment and funding for a reduction in personnel numbers.

The railways of Kosovo are under separate autonomous administration (see entry in Serbia and Montenegro section of *Railway systems and operators*).

Passenger operations
Passenger business was badly affected by the Kosovo crisis in 1999, but by 2002 many services had been restored despite severe shortages of traction.

Diesel locomotives (including ZCG, Montenegro, fleet)

Class	Wheel arrangement	Power kW	Speed km/h	Weight tonnes	No in service	First built	Mechanical	Engine	Builders Transmission
641-100	Bo-Bo	441	80	62	65	1960	Mávag	Ganz	E Ganz-Mávag
641-300	Bo-Bo	685	80	64	40	1985	Ganz-Mávag	Pielstick	E Ganz-Mávag
642	Bo-Bo	606	80	64	28	1960	DD	MGO	E B&L
643	Bo-Bo	680	80	67.2	14	1967	B&L/DD	MGO	E B&L
645	A1A-A1A	1,820	120	100.2	3	1981	GMF/DD	GM	E GM
661	Co-Co	1,933	112	108/114	99	1961	GM	GM	E GM
662	Co-Co	1,212	120	99	2	1965	DD	MGO/DD	E R Končar/Sever
664	Co-Co	1,617	124	99/103	11	1972	GM	GM	E GM
734*	B-B	478	60	48/54	12	1960	MAK	Maybach GT06	H Voith
666	Co-Co	1,845	122	100	4	1978	GM-USA	GM	E GM

* Ex-German Federal Class 260/261.
B&L: Brissonneau & Lotz; DD: Duro Dakovic; RK: Rade Končar Zagreb·

Electric locomotives (including ZCG, Montenegro, fleet)

Class	Wheel arrangement	Power kW	Speed km/h	Weight tonnes	No in service	First built	Builders Mechanical	Electrical
441	Bo-Bo	3,400/4,080	140	78	96	1970	ASEA/R Končar	ASEA/R Končar
461	Co-Co	5,100/5,400	120	126	97	1972	Electroputere	Electroputere

Diesel railcars or multiple-units

Class	Cars per unit	Motor cars per unit	Motored axles/car	Power/ motor kW	Speed km/h	Cars in service	First built	Mechanical	Engine	Builders Transmission
712	2	1	4	2 × 213	120	15	1980	Duro Dakovic	MAN	H Voith
811	4	2	2	367	118	11	1974	Ganz-Mávag	Ganz-Mávag	M Ganz-Mávag
812	1	1	1	110	90	80	1958	Goša	MAN	M MAN

Electric railcars or multiple-units (including ZCG, Montenegro, fleet)

Class	Cars per unit	Motor cars per unit	Motored axles/car	Output/ motor kW	Speed km/h	Cars in service	First built	Builders Mechanical	Electrical
412	4	2	4	170	120	45	1980	RVR	RVR

Freight operations

Freight traffic suffered heavily as a result of the NATO bombing and UN sanctions. This was exacerbated by the damage inflicted on the country's industrial infrastructure. However, in 2002 some international transit traffic resumed between western Europe and Greece.

New lines

Completion of the 68 km Valjevo-Loznica line, on which work started in 1992, remains an aim of ŽS. The single-track line will incorporate 20 tunnels totalling around 10 km and 69 bridges. The most difficult tunnel, at Trifković (1,719 m), was completed in 1998.

Improvements to existing lines

In the wake of the 1999 Kosovo crisis, rehabilitation of the system was top priority for the former JŽ, with the EBRD and the European Investment Bank (EIB) among institutions to provide funding. Reconstruction and modernisation of the 874 km Yugoslav section of European Corridor X has also been identified as a key development priority. Corridor X (Salzburg-Ljubljana-Zagreb-Belgrade-Niš-Skopje-Thessaloniki, with its Belgrade-Budapest and Niš-Sofia branches) forms the shortest and fastest route between central and western Europe and southeastern Europe and the Middle East. Part of an EIB loan of €85 million finalised in March 2002 was intended to assist with the rehabilitation of sections of Corridor X in Yugoslavia.

Corridor X also features in a joint international project to reduce delays to trains crossing the border into and out of the ŽS system. Based on the UIC's 'Action Border Crossing' (ABC) methodology, steps to reduce stopping times and simplify border formalities were to be implemented at four crossing points: Subotica-Kelebia (MÁV/Hungary); Šid-Tovarnik (HŽ/Croatia); Preševo-Tabanovci (MŽ/Macedonia); and Dimitrovgrad-Dragoman (BDŽ/Bulgaria).

Traction and rolling stock

At the start of 1999 the former JŽ owned 189 electric and 293 diesel locomotives, 45 emu and 106 dmu sets, 722 hauled passenger cars and 17,169 freight wagons. While only one locomotive (a Class 441 electric) is reported lost during the conflict with NATO over Kosovo, just one third of the traction fleet was estimated to be serviceable in 2000. The separation in January 2003 of JŽ into separate Serbian and Montenegrin networks led to 17 electric locomotives, 18 diesel locomotives, six emus, 68 coaches and 910 freight wagons being assigned to ZCG, Montenegro.

Steps taken by the former JŽ to overcome its motive power shortage included sending Class 441 and 461 electric locomotives for overhaul to Electroputere in Romania and hiring in Class 06 diesel-electric locomotives from Bulgaria. An EIB loan agreed in March 2002 was to fund the refurbishment of 60 electric locomotives.

In 1998, JŽ received the first 23 of an order for 100 200 km/h long-distance passenger coaches from local manufacturer Goša. The new vehicles were to be deployed on international services with the introduction of the 1998–99 timetable.

Signalling and telecommunications

About 50 per cent of railway stations are equipped with station relay signalling; automatic block system or interlocking are installed on about 40 per cent of lines, and Automatic Train Control (ATC) on 34 per cent of lines. CTC was commissioned on the Belgrade-Bijelo Polje line (311 km) and between Belgrade and the Montenegrin border and on to Bar (481 km) during 1996, while automatic block is being installed over the Subotica-Belgrade-Niš-Preševo main line, totalling 840 route-km.

Lineside telecommunication cables and dispatching telephone systems with selective dialling and identification are installed on 37 per cent of lines.

Electrification

Electrification plans include the route from Niš to Dimitrovgrad and the Bulgarian border, expected to be completed by 2004.

Track

Rail: UIC 45A is installed over 2,121.7 km; UIC 49 over 1,790 km; and UIC 60 over 146 km
Crossties (sleepers): Wooden, 2,600 × 260 × 160 mm; concrete, 2,400 × 300 × 192 mm
Spacing: 1,670/km
Fastenings: K, SKL-2 and Pandrol
Min curvature radius: 250 m
Max gradient: 2.5%
Max axleload: 22.5 tonnes

Kosovo Railway Enterprise

Hekurudha e Kosoves/Kosovske Zeleznice

Gauge: 1,435 mm
Route length: 330 km

Organisation

The Kosovo Railway Enterprise was set up by UNMIK to administer rail operations following the 1999 war. Initially, all trains were driven by KFOR soldiers, but these duties were due to be transferred to civilian hands by September 2000. Representatives of the railways of Croatia, France and Germany were training local drivers and fitters.

Passenger operations

In 2000, there were two daily return trips between the good station at Pristina and Skopje Volkovo, Macedonia, and on the Kosovo Polje–Mitrovica-Zvecan line. Trains were formed of two or three coaches and were unheated. All were heavily guarded by KFOR troops.

Freight operations

In 2000, freight trains carrying mainly KFOR and humanitarian supplies were operating on the two lines which were seeing passenger services and on the Prizren–Pec, Pristina–Prizren and Kosovo Polje–Leposavic lines.

Improvements to existing lines

In 2000, parts of the system remained unusable due to war damage. Work was under way to repair the tunnel at Sinji, north of Pristina, which for some years has been used as a waste dump and parking area.

Traction and rolling stock

In mid-2000, the Kosovo Railway Enterprise had at its disposal three Yugoslav Railways Class 661 diesel-electrics, two former SNCF Class 63000 diesel-electrics and six former German Rail Class 202 diesel-hydraulics. It was reported that 18 Norwegian State Railways Class Di3 diesel locomotives were to be transferred to Kosovo in September 2000.

The locomotive maintenance facilities at Kosovo Polje were reopened in January 2000.

Slovakia

Ministry of Transport, Post and Telecommunications

Námestie slobodyč. 6, SK-801 05 Bratislava
Tel: (+421 2) 59 49 41 11   Fax: (+421 2) 52 49 47 94
Web: www.telecom.gov.sk

Key personnel

Minister: Pavol Prokopovič
Director, Railways: Dušan Turanovič

NEW ENTRY

Slovakian Republic Railways (ŽSR)

Železnice Slovenskej republiky
Klemensova 8, SK-813 61 Bratislava
Tel: (+421 2) 52 96 15 05; 50 58 70 28
Fax: (+421 2) 52 92 09 08; 52 96 22 96
e-mail: gr@zsr.sk
Web: www.zsr.sk

Key personnel

Chairman of Executive Committee: Oszkár Világi
General Manager: Roman Veselka
General Inspector: Vojtech Štefan Kobetič

General Manager's Office – Under-Secretaries
 Economy: Štefan Hlinka
 Management and Personnel Resources: Ján Žačko
 Technical Development: Ladislav Dimun
 Operations: Ladislav Dimun
General Manager's Office Director: Marcel Haydu
Management Board
 Legal Affairs: Alojz Ceizel
 Control and Investigation: Mária Surovcová
 Strategy: Julius Daubner
 Technical Development and Ecology: Ján Hanúsek
 Investment: Marián Urbánek
 Infrastructure: Daniel Balucha
 Infrastructure Maintenance: Jaroslav Mikla
 Information and Technology: Peter Banič
 Controlling: Anton Kukučka
 Property Management: Ján Prosuch
 Signalling: Ivan Bartoš
 Power and Electrical Engineering:
  Dušan Mitošinka
 Tracks and Buildings: Igor Fedor
 Economy: Rastislav Sirkovský
 Human Resources: Eva Schwarczová
 Operations: Miroslav Barčiak
 Marketing: Milan Nevid'anský
 Social Affairs: Soňa Kubincová
 Protection and Security: Ivan Bernát

Public Relations: Ján Keresteš
Media Contact: Nela Blašková
Regional Administration
 Košice: Ján Juriga
 Trnava: Milan Solárik

Gauge: 1,435 mm; 1,520 mm; 1,000 or 750 mm
Route length: 3,512 km; 100 km; 50 km
Electrification: 807 km at 25 kV 50 Hz AC; 737 km at 3 kV DC; 42 km at 1.5 kV DC; 2 km at 15 kV AC 16²/₃ Hz

Political background

ŽSR is the state-owned company responsible for the infrastructure of the Slovakian railway network. The formal separation of infrastructure and train operations took place on 1 January 2002. More details of the political background to this separation can be found under the entry for Železničná spolčnost as (ZSSK) in the Slovakia section of *Railway systems and operators*. ŽSR will also assume responsibility for managing residual assets and disposing of surplus railway property. Ancillary businesses will be converted in three stages into 17 service companies, which will be offered for sale at a later date. The company's supreme body is the Executive Committee, overseen by the Ministry. At the start of 2004 ŽSR employed 22,106 staff.

In March 2002 ŽSR became a member of the Community of European Railways.

In March 2003 ŽSR joined RailNetEurope (RNE), an association of European rail infrastructure managers.

Privatisation

In a way that mirrors the model adopted in the neighbouring Czech Republic, a process is under that allows private operators to run their own trains on ŽSR tracks. The first company to take advantage of this is a Košice-based steel-producing firm, US Steel Košice, which signed an agreement with ŽSR in December 1999 for the operation of its own block trains over the national network under a licence granted by the Slovak Railway Authority (Štátny dráhový úrad). The monopoly of ZSSK for operating rail services on the Slovakian network came to an end in March 2003, when the system was fully liberalised and open to any operator holding a licence granted by the Ministry of Transport, Post and Telecommunications.

On 1 February 2000 US Steel Košice began operation of three pairs of its own freight trains on a 23 km section of ŽSR's Zvolen–Košice route. Formed of Class 770.5 diesel-electric locomotives and Type Facc and Faccs hopper wagons, the trains convey limestone from a quarry at Turňanad Bodvou to a steelworks at Velká Ida in eastern Slovakia, close to the Hungarian border. During 2004 the results of liberalisation became more evident, with four new domestic operators gaining access to the ŽSR network: LTE Logistik a Transport Slovakia; SŽDS Bratislava; BRKS; and AM-Tuning Topol'cany.

Organisation

In January 2004 ŽSR was split into four regional management centres (OR) which assumed responsibility for traffic control and infrastructure management. A new organisational structure was adopted in January 2005, when the four centres were reduced to just two, based on Trnava and Košice. These two centres control ten subordinate executive units (SŽI).

Finance

Track access charges from both ZSSK and from private operators account for around half of total revenue. In 2003 ZSSK paid Kcs5.67 billion for access to the ŽSR infrastructure for freight traffic and Kcs1.655 billion for passenger services. Charges from private open access operators amounted to Kcs0.024 billion for freight traffic and Kcs0.002 billion for passenger traffic. ŽSR's overall loss in 2003 totalled Kcs956.5 million, one-third of the loss recorded in the previous year.

Finances (Kcs million)	2002	2003
Revenue	18,096	20,685
Expenditure	21,163	21,641
Profit/loss	−3,067	−956

At the start of 2004 ŽSR's assets totalled Kcs74.024 billion, with a total debt burden of Kcs49.868 billion. In January 2004 the Slovak government agreed to take over the railway's historic debt, incurred by the state's inability to reimburse fully the cost of running unprofitable but socially desirable passenger services in the period 1994–2001. Consequently ŽSR wrote off Kcs13.63 billion while the state train operating company, ZSSK, wrote off Kcs7.32 billion.

New lines

In 1994 work began on the construction of a new cross-border alignment from Petržalka, a suburb of the Slovak capital, to Kittsee, in Austria. This partly used the trackbed of the Bratislava–Berg–Vienna line, built in 1914 and closed in April 1945. The new single-track line was built by the 'Rýchla trať' (Fast Line) consortium of construction companies, with the Slovak portion just 1.76 km in length. Construction has also involved electrification at 15 kV AC 16²/₃ Hz as far as Petržalka, the installation of new signalling and safety equipment and the erection of new station buildings at Petržalka and Kittsee.

0576115

The official opening of the line took place on 15 December 1998 and revenue operations began on 7 January 1999, with one pair of fast passenger trains operating between Bratislava–Petržalka and Vienna Südbahnhof. An additional three non-stop return workings each day were introduced from the 1999/2000 timetable change in May 1999, when full electric operation also commenced. These trains are hauled by ÖBB Class 1046 and ZSSK Class 240 electric locomotives. Initially, problems with customs authorities prevented the introduction of new local passenger services between Petržalka and Parndorf; eight pairs of local stopping services using ÖBB Class 4010 and 4020 emus were finally introduced from 1 August 1999. This increased to 14 pairs with the 2003–04 timetable change. Commissioning the new line has reduced journey times betwen the two capitals to 45 minutes compared with the previous best of 1 hour 7 minutes via Devínská Nová Mes and Marchegg.

In 1996 Austria and the Slovak Republic agreed to set up a joint venture to establish a link between Vienna's Schwechat airport and the Ivanka pri Dunaji airport in Slovakia. The so-called Neu Pressburger Bahn, supported by the regional governments of Wien and Niederösterreich, would run through Wolfsthal, Kittsee and Bratislava, partly on existing lines and partly on new construction.

A 14.5 km cross-border connection into Poland was reopened to freight traffic in June 1996.

Up to six pairs of freight trains daily use this line. The Kcs150 million project links Medzilaborce and Lupków. ŽSR introduced two pairs of cross-border passenger trains between Humenné and Sanok, Poland, from June 1999 but these were withdrawn at the 2003–04 timetable change due to poor patronage.

Improvements to existing lines

In the period 1997–2010 ŽSR plans to invest a total of Kcs33.7 billion, of which Kcs21.5 billion will go towards the upgrading of three trunk corridors which are responsible for three-quarters of ZSSK's traffic volume. In mid-1997 ŽSR obtained a loan of Kcs5.8 billion, of which Kcs0.5 billion was intended for modernisation of the 113 km Kúty–Zohor–Bratislava–Galanta section and Kcs2.8 billion for the 46 km Bratislava–Rača–Trnava section.

Modernisation has been undertaken of the 20.2 km line between Čadca and Skalité-Serafínov on the Polish border, which was opened for freight operation in May 1994. The Kcs2.3 billion project involving electrification at 3 kV DC, upgrading 15 bridges and rebuilding the border station at Čadca, under way since 1993, will allow maximum loads for freight trains to be increased from 400 to 800 tonnes and the line speed to be raised from 60 to 100 km/h. Completion was achieved in December 2002 (see 'Electrification').

While some priority ŽSR upgrading programmes are progressing, work on others has been suspended due to a lack of funds, such as double-tracking the Zvolen–Fil'akovo–Rožňava–Košice southern transversal, seen here near Kriváň in September 2003 (Michal Málek)   0580509

Modernisation of the 265 km north-south transversal route is seen as a priority. This route runs from Skalité–Serafínov near the Polish border to the Austrian border at Kittsee (via a new cross-border connection – see 'New lines' section above), via Žilina, Trenčín and Bratislava (European Corridor VI), and forms part of the Warsaw–Vienna international route.

In March 1999 ŽSR embarked on an extensive modernisation programme to upgrade to full European standards its two main international corridors, based on a model adopted by the neighbouring Czech Railways. The complex modernisation will see tracks, overhead catenary and signalling equipment upgraded to allow 160 km/h operation.

The first line to be treated, the 41.7 km Bratislava Rača–Trnava section, forms part of international Corridor V (Trieste–Ljubljana–Budapest–Lvov), running on Slovakian territory via Bratislava, Trnava, Žilina, Košice and Čiena nad Tisou (536.2 km), which is double-track and electrified throughout. Upgrading this first section has been divided into 11 sections at a total cost of Kcs7.9 billion.

The first stage of modernisation began in 1999 on the 8.4 km section between Cífer and Trnava and was completed in September 2000. In August 2003 a second section was completed, Bratislava Rača–Svätý Júr (4.7 km), followed by a third section, Svätý Júr–Pezinok, commissioned in January 2004.

In June 2004 ŽSR commenced works on a second phase of the complex modernisation of the Bratislava Rača–Trnava line, starting with the Šenkvice–Cífer subsection (9.8 km) at a cost of Kcs116.9 million. This stage was to involve the overhaul of 12 bridges and the construction of three new bridges, plus modernisation of six stations on the 12.4 km Bratislava Rača–Trnava section. Work commenced on the last sub-section between Pezinok and Šenkvice at the end of 2004. This involves construction of a completely new elevated alignment near Šenkvice. Total cost of the project is put at €142 million, of which 62 per cent is to come from the state budget and the remainder from the EU's ISPA programme. The whole section between Bratislava Rača and Trnava is due for completion by December 2007. Work is carried out without interruption to traffic, with some trains re-routed to the parallel line via Galanta, Sered' and Leopoldov.

In 2004 preparatory works also began on the next section, Trnava–Nové Mesto nad Váhom (55 km). The first stage of this project involves modernisation of the line between Trnava and Piešt'any (35 km), due for completion at the end of 2007 at a cost of Kcs4.13 billion, of which 85 per cent is coming from the EU's Cohesion Fund.

At the turn of 2004–05 tenders were being called for modernisation of the second stage between Piešt'any and Nové Mesto nad Váhom, estimated at €127.688 million with a similar proportion of EU funding. The following section, Nové Mesto nad Váhom–Púchov (57.5 km), is due for modernisation in 2005–10 at a cost of Kcs7.608 billion.

The double-track 25 kV AC route linking Prague and Budapest which runs from Kúty through Bratislava and Nové Zámky to the Hungarian border at Štúrovo (European Corridor IV) is also to be upgraded for 160 km/h operation. Of the 218 km in Slovak territory, 167 km will have track renewed, with level crossings eliminated and tunnels through Bratislava refurbished. Work on the Kcs19.8 billion project is expected to continue until 2010. In 2001 work began on the first two sections, Devínske Jazero–Zohor (8.3 km), completed in early 2002, and Malacky–Vel'ke Leváre (7.6 km), part of the 70.1 km line from the Czech border at Kúty to the Slovakian capital.

Another project concerns the east-west trunk route from Žilina to Čierna nad Tisou via Košice

(European Corridor V), which is a transit route to the Ukraine. It is already double-track and electrified at 3 kV, but upgrading from 120 to 140 km/h on the 242 km Žilina–Košice section is scheduled for 2006–2010 at a cost of Kcs21.2 billion; the 95 km from Košice to the Ukrainian border at Č ierna would follow, with upgrading for 140 to 160 km/h running at a cost of Kcs12.1 billion.

Work started in the 1970s on the 442 km Bratislava–Levice–Zvolen–Košice southern transversal with the aim of double-tracking and electrifying it throughout, but only slow progress has been made. By the start of 1997, when work stopped for financial reasons, less than half of the route had been doubled, and a similar proportion electrified mainly at 25 kV 50 Hz AC. Track upgrading was complete on the 84 km section west from Košice. ŽSR has divided the line into five sections with the aim of accelerating the programme, but lack of funds means that commencement of work is unlikely before 2007.

In December 2001 ŽSR commenced modernisation of the Slovak-Ukrainian border station at Cierna nad Tisou. Undertaken by ŽS Košice, ŽS Bratislava and ELTRA Košice, the Kcs1.5 billion project is being carried out in three stages and covers: track and overhead catenary renewal at marshalling yards; installation of new telecommunications systems; and environmental protection measures. The first phase covers the 1,520 mm gauge part of the station and is estimated to cost Kcs547 million. The last-mentioned was completed in July 2003.

The 5.4 km 760 mm gauge Trenčianska Teplá-Trenčianske Teplice line in northwest Slovakia underwent heavy maintenance during 2000, involving renewal of the trackbed and catenary masts.

Signalling and telecommunications

Automatic block signalling is in operation on 690 route-km, with another 1,195 route-km covered by semi-automatic block signalling.

Following introduction of automatic block equipment on the 31 km section between Žilina and Považská Bystrica in October 1995, work switched to the 35 route-km from Považská Bystrica through Púchov to the Czech border. This scheme was completed by AŽD Bratislava and formally opened in February 2000 at a cost of Kcs505.9 million.

In December 2004 automatic block equipment was installed on the 68 km single-track line between Kúty and Trnava. The Kcs222.3 million project was undertaken by AŽD Praha.

In December 2002 ŽSR's first Type ESA11 electronic interlocking was installed at the border station of Kúty. The equipment was supplied by AŽD Praha at a cost of Kcs250 million.

On 5 November 2001 Alcatel Slovakia won a Kcs2.6 billion contract to upgrade the ŽSR telecommunications network. Modernisation began in December 2001 and is scheduled for completion by the end of 2004. Work is being carried out in three stages: the first covers Bratislava–Zvolen–Košice–Čierna nad Tisou (completed in May 2003), the second Bratislava–Žilina–Košice (due to be completed in December 2003) and the last Vrútky-Banská Bystrica–Zvolen and Turč ianske Teplice–Prievidza–Nitra–Nové Zámky–Štúrovo/Komárno.

The European Investment Bank is providing Kcs1.5 billion towards the cost of upgrading ŽSR's telecommunications subsidiary, ŽSR Telekomunikácie, which was set up in January 2002.

ŽSR is the first rail operator in Eastern Europe to install GSM-R wireless communication technology. In October 2002 the authority appointed Kapsch CarrierCom as a general supplier of the entire infrastructure, under a contract worth €23 million. A pilot project involves installation of GSM-R technology on part of the international Corridor IV between the Czech border at Kúty and Štúrovo on the Hungarian border via Bratislava and Nové

Zámky and on three short sections from Bratislava to border stations at Marchegg (ÖBB), Kittsee (ÖBB) and Rajka (MÁV).

Electrification

Over 40 per cent of the ŽSR network is electrified, with electric traction in 2002 handling 84.1 per cent of the railway's traffic. The Púchov–Žilina–Košice–Čierna nad Tisou main line with branches, the Plaveč–Prešov–Košice–Čaňa north-south axis and the 1,520 mm gauge Haniska–Trebišov–Kapušány (ŠRT) route are electrified at 3 kV DC, with the 25 kV 50 Hz AC system having been chosen for the remainder of the 1,435 mm gauge electrification. The only junction of the electrification systems is located at Púchov, 45 km southwest of Žilina.

The 1,000 mm gauge High Tatras system is electrified at 1.5 kV DC, while the 750 mm gauge Trenčianska Teplá–Trenčianske Teplice route employs the 600 V DC system.

Electric operation on the sinuous 56 km line between Plaveč and Prešov in the country's far east began on 1 June 1997. Started in May 1995, electrification work at 3 kV DC was carried out by ELTRA Košice at a cost of Kcs594 million. Also on 1 June 1997 electrically powered services commenced over the formerly non-electrified 11 km gap from Čaňa, south of Košice, to the Hungarian border, completely energising this important north-south corridor.

In 1998, work started on the 3 kV DC system of the 20.2 km single-track line between Čadca and the Polish border, which forms part of ŽSR's north-south transit corridor (see 'Improvements to existing lines'). The Kcs2.3 billion project, using a grant of €2.2 million from the EU's Phare programe, has involved electrification, complete upgrading of the trackbed, installation of new signalling equipment and modernisation of the border station at Skalité. Here, a new substation was built. ZSSK plans to use this line, which was only opened for regular freight traffic in September 1995, for extensive north-south freight and intermodal operations between Slovakia and Poland, allowing shorter transit times. It is also using the line for international passenger services. Regular electrically hauled revenue operations over the line began in December 2002. Traffic is controlled by Bombardier Transportation Ebilock interlocking equipment installed at Čadca.

In December 2004 works were formally launched on the 25 kV AC 50 Hz electrification of the 21 km single-track line between Zvolen and Banská Bystrica. Having been postponed several times due to a lack of funds, the Kcs1.098 billion project, of which 75 per cent is coming from EU funds, is being undertaken by a consortium of ŽS Brno, Betamont Zvolen and ELTRA Košice. Due for completion in December 2006, the project also involves resignalling and construction of a new substation at Banská Bystrica.

The next electrification project should see the Zvolen–Fil'akovo line (67 km) wired at 25 kV AC 50 Hz AC by 2010.

Track

Rail: 65 kg/m Type R (1,611 km installed); 60 kg/m Type UIC (27.2 km installed); 49 kg/m Type T and S (2,727 km installed); 44 kg/m Type A (300 km installed)

Max axleload: 22.5 tonnes (1,435 mm gauge); 24.5 tonnes (1,520 mm gauge)

Min curve radius: 350 m (main lines); 200 m (secondary routes)

Sleepers: Wood and concrete spaced 1,640–1,720 per km

Fastenings: Vossloh Sk1 12, 14

Max gradient: 2.7%

UPDATED

Bratislava Regional Railway Authority (BRKS)

Bratislavská Regionálna Kol'ajorá Spoločnost'
Krátka 11, SK-811 03 Bratislava
Fax: (+421 2) 48 28 72 17
e-mail: brks@micronet.sk

Key personnel

Chairman: Dušan Prokop
General Director: Roman Filistein

Political background

The regional authority in the Bratislava area formed this company in April 2003 to take over passenger services on two lines from which ZSSK withdrew trains. It is owned by the Bratislava regional authority (51 per cent), ŽOS Vrútky (20 per cent), Wagon Service Travel (12 per cent), Connex Transport AB (13 per cent) and four local authorities (4 per cent).

Passenger operations

Slovakia's first private rail operator, BRKS began running passenger trains on 7 July 2003 on the Zohor–Záhorská Ves (14 km) line, from which national operator ZSSK withdrew traffic in February the same year. Operations on a section of the Zohor–Plavecký Mikuláš line between Zohor and Rohožnik (24 km) resumed in September 2003. The same day, a direct service using BRKS rolling stock was launched between Zohor and Bratislava (40 km). BRKS subsequently operated nine pairs of weekday-only passenger trains on both routes from Zohor to Záhorská Ves and Rohožnik, with three pairs extended from Zohor to Bratislava. However, the new timetable introduced in August 2004 brought a substantial reduction in passenger services, with four train pairs between Zohor and Záhorská Ves, one between Zohor and Rohožnik and one direct train pair between Záhorská Ves and Bratislava; all weekdays only.

Infrastructure is still owned by ŽSR but since January 2004 BRKS has rented the Zohor–Záhorská Ves and Zohor–Rohožnik lines and intends to purchase them outright at a later date. It also has ambitious plans to build a new line to complete the missing section from Plavecký Mikuláš to Jablonica (15 km), establishing an orbital line around Bratislava.

Freight operations

BRKS has acquired a licence to operate freight trains over the ŽSR network but currently owns no wagons. In August 2004 the company started to run cement trains for Holcim from the Czech/Slovak border at Kúty, on the Breclav–Bratislava line, to Komárom, Hungary, via Bratislava. Frequency was initially three trains per fortnight. In November 2004 BRKS began operating its second freight service, carrying bananas in refrigerated wagons from Kúty or Horní Lideč on the Check/Slovak border to Šaľa.

Traction and rolling stock

In 2005 BRKS passenger services were operated by four of its own elderly Class 830 diesel-electric railcars and one Class 020 trailer, augmented by locomotive-hauled trains consisting of four Type Bp coaches owned by Austrian Federal Railways (ÖBB) during the non-winter period. The Slovakian national operator's high selling price obstructed the acquisition of further Class 830s and 020s. Two Class 2143 diesel locomotives hired from ÖBB since December 2003 were returned in July 2004. The planned purchase of four to six such locomotives as well as that of two redundant Class VT70 dmus

Class 830 diesel railcar, hired from ZSSK, at Záhorská Ves (Quintus Vosman)   0561561

Hired Class 740 and 770 locomotives head BRKS's inaugural freight service at Komárno, Slovakia, in August 2004 (Quintus Vosman)   **NEW**/0585035

from the Graz-Köflacher-Eisenbahn (GKE) in Austria was not realised due to their high selling price.

Freight and some passenger trains are hauled by a fleet of five Class 720.5, 740.4, 742.5 and 770.5 diesel-electric locomotives leased from the Slovakian private companies MAX Cargo and AM-Tuning and from TIMOS in the Czech Republic.

UPDATED

Železničná spoločnost as (ZSSK)

Žabotova 12, SK-811 04 Bratislava
Tel: (+421 2) 55 57 38 34   Fax: (+421 2) 55 57 38 31
e-mail: zsgr@slovakrail.sk
Web: www.slovakrail.sk

Key personnel

Chairman of Supervisory Board: Ján Kotuľa
Chairman of Managing Board and Director General:
  Ondrej Matej
Deputy Chairman of Managing Board and Director,
  Services: Pavol Gábor
Board Member and Director, Commercial:
  Pavol Gallo
Board Member and Director, Operations:
  Jaroslav Bajužik
Management Board
  Manager, Director General's Office: František Stručka
Controlling: Stanislava Slovincová
  Finances: Mária Kováčová
  Marketing: Július Slíž
  Sales Control: Frantošek Vančo
  Facility Management: Vladimír Škvarek
  Strategy: Marián Frko
  Logistics: Anton Fišer
  Control and Inspection: Peter Krcho
  Human Resources: Peter Kliment
  Internal Audit: Ľudmila Šúňová
  Information Technology: Peter Stohl
  Internal and External Relations: Ladislav Martinkovič
  Spokesman and Media Contact: Miloš Čikovský

For key personnel for ZSSK Cargo see separate entry in Slovakia section of *Railway systems and operators*.

Political background

Slovakia has been leading rail restructuring in Eastern Europe since January 2002, when the former state-owned ŽSR was split into a commercial train operating company, ZSSK, and an infrastructure owner, ŽSR. Formally established on 1 January 2002, ZSSK became a joint stock company conducting both the freight and passenger train operations of the former Slovakian Republic Railways (ŽSR). Overseen by a supervisory board, the company's sole shareholder was the Slovak state through the Ministry of Transport, Post and Telecommunications.

Class 411.9 750 mm gauge 600 V DC electric railcar and trailer at Trenčianske Teplice (Quintus Vosman)   0580506

In order to define the relationship between the railway and the state, the Ministry of Transport, Post and Telecommunications drafted an amendment to Act 164/1996, the legislation under which the railway operated. This provided for the establishment of two commercial undertakings to manage rail infrastructure and operate transport services in a profit-oriented market environment. It also defined the extent of state involvement to give both companies a degree of freedom within national transport policies. A state-owned company, Železnice Slovenskej republiky (ŽSR as) (see entry in Slovakia section of *Railway systems and operators*), has retained responsibility for railway infrastructure.

The joint stock commercial and operating enterprise ZSSK was made debt-free and divided into three divisions responsible for: passenger operations (DOP); freight operations (DNP); and rolling stock (DŽKV), overseen by the company's general management based in Bratislava.

Splitting of the freight business
As a result of the continuing restructuring programme, ZSSK was split into two separate companies from 1 January 2005: a joint stock freight company, Železničná spoločnosť Cargo (ZSSK Cargo); and a state-owned passenger operator, Železničná spoločnosť Slovensko (ZSSK). With a 60:40 division of property in favour of ZSSK Cargo, this step formed the initial stage of an ambitious plan to privatise rail freight operation, as approved by the Slovak government in July 2004.

Seven bidders for the freight business from both Slovakia and abroad were shortlisted in September 2004 and the sale should be completed by the end of 2005. The new owner will then have a stake of up to 100 per cent in the new business, which is earning some Kcs1.25 billion annually. The sale is expected to raise Kcs15 billion, which would be used to pay off part of the railway's accumulated debt. The aim of this radical restructuring is to eliminate the cross-subsidising of the loss-making passenger operation from freight revenues and to allow the rail freight operation to become an independent customer-oriented business, in accordance with EU rules.

Apart from the core freight business that should see 49.6 million tonnes carried in 2005, ZSSK Cargo will also concentrate on the lease, maintenance and overhaul of rolling stock. After the division, the company owns some 800 electric and diesel locomotives and a wagon fleet of 18,000. With investment of Kcs534 million for modernisation of the freight wagon fleet and Kcs138 million for overhaul of diesel locomotives, ZSSK Cargo plans a net profit of Kcs 97.6 million in 2005.

The passenger operator, ZSSK, will remain wholly owned by the Slovakian government, which will regulate pricing policy and continue to subsidise the operation of unprofitable passenger services in the public interest. With some 100 electric and diesel locomotives and a fleet of dmus, emus and railcars, its principal role will be providing long-distance and suburban services, with the aim of securing a 22 per cent share of the passenger traffic market by 2010. The process will also mean significant cuts in current staff numbers. At the start of 2005 ZSSK employed 4,744 people and aimed to reduce this to 4,494 by the year-end. ZSSK Cargo's staff stood at 12,256, a figure to be reduced to 12,006 over the same period. Investment should reach Kcs841 million, of which 98 per cent was destined for motive power and rolling stock refurbishment.

Both ZSSK and ZSSK Cargo became members of the Community of European Railways in January 2005.

Privatisation
Ancillary activities such as civil engineering, mechanical engineering, design, catering, and the three major traction and rolling stock workshops at Trnava, Zvolen and Vrútky, were all sold off by the end of 1993 under the privatisation programme. In January 1996 Wagon Slovakia took over from ŽSR as private operator of dining and sleeping cars on international services to 14 European countries.

In a way that mirrors the model adopted in the neighbouring Czech Republic, a process is

Class 812 LUX first class-only two-axle diesel railcar rebuilt from a Class 810 vehicle (Michal Málek)   0580508

Class 754 diesel-electric locomotive heading a ZSSK fast service at Zvolen (Milan Šrámek)   0547054

Class 350 dual-voltage 25 kV AC/3 kV DC electric locomotive at Polárikovo with the 'Csárdás' Budapest–Prague EuroCity service (Quintus Vosman)   0547052

under way of allowing private operators to run their own trains on ŽSR tracks. The monopoly of ZSSK for operating rail services on the Slovakian network came to an end in March 2003, when the system was fully liberalised and open to any operator holding a licence granted by the Ministry of Transport, Post and Telecommunications.

On 22 December 2003 ZSSK decided to lease its subsidiary TLD (Tatranské lanové dráhy), which is responsible for the operation of cable lines in the High Tatras mountain region, to Mountain Holiday & Ski Resorts plc. The UK company has taken on the lease for five years for Kcs150 million, with plans to invest Kcs1 billion on quality and safety improvements. In 2003 TLD recorded a profit of Kcs20 million.

Closure of regional lines

Despite strong opposition from the public and from railway trades unions, including a three-day general strike, ZSSK abandoned passenger traffic on 25 of the country's 36 loss-making regional lines from 2 February 2003. This dramatic step was a consequence of aspects of the government's transport policy: according to the Railway Act, the state must reimburse the cost of running unprofitable passenger services operated in the public interest, but it has not fully done this since the establishment of the independent railway in 1993. In January 2003, the government earmarked support of Kcs4.33 billion from the national budget for ZSSK's public interest services during that year, with Kcs2.3 billion allocated to the infrastructure company, ŽSR. This represented a reduction of Kcs1 billion on the figure expected by ZSSK, a shortfall which when added to the Kcs48 billion debt that burdened the railway left it unable to meet the performance requirements of the timetable. Substitute bus services were introduced on three lines totalling 207 route-km (Trenčín–Chynorany (49 km); Levice–Štúrovo (52 km); and Zvolen–Čata (106 km). This left 22 remaining lines totalling 511 route-km with no passenger rail or substitute bus services.

This step, which ZSSK believes will save up to Kcs468.9 million in 2003, was also expected to bring about a reduction of 443 in the railway's workforce, with further job cuts expected to follow later. It also allowed the withdrawal of 65 obsolete traction units and 45 non-powered passenger vehicles.

Regional authorities were obliged to take over these services or to develop alternative transport services to meet local needs, but they were also short of funds. Despite this, limited passenger services resumed on half of the lines closed in 2003–04, thanks to a major reduction in operating costs by ZSSK and financial support from the regions. The move also reflected an undertaking given by ZSSK management to re-examine the social and environmental impact of the least loss-making closures during the trade union strike of February 2003. Reopening took place in two stages. The first, in June 2003, covered: Žilina–Rajec (operated on behalf of Žilina county since February 2003); Šala–Neded; Trenčín–Chynorany; Úľany nad Žitavou–Zlaté Moravce; Levice–Štúrovo; Šahy–Čata; Hronská Dúbrava–Banská Štiavnica; Plešivec–Muráand Bánovce nad Ondavou–Veľké Kapušany. The second stage, in June 2004, involved: Spišská Nová Ves-Levoča; Zvolen–Šahy; and Vranov nad Topľou-Trebišov.

Finance

In 2004, the final year of the railway's unified freight and passenger activities, ZSSK expected to carry 50.9 million tonnes of freight, maintaining its 22 per cent share of the country's goods transport market. A net profit of Kcs373 million was expected to be recorded, with overall revenues of Kcs16.114 billion. Revenues from passenger traffic were budgeted to reach Kcs2.656 billion, generating a projected loss of Kcs500 million. The government earmarked support of Kcs4.33 billion for unprofitable services run in the public interest, the same sum as that provided in 2003. ZSSK expected total revenues from commercial services provided in 2004 to stand at Kcs24.385 billion against expenditure of Kcs24.373 billion. Of total revenues 66.1 per cent will come from freight traffic, 10.9 per cent from passenger services and 17.8 per cent from the state budget. The largest item of budgeted expenditure is for access to the ŽSR infrastructure network, accounting for 31.3 per cent of ZSSK's costs.

Finances (Kcs million)	2002	2003
Expenditure	27,196	29,516
Total revenues	26,719	24,690
– of which freight revenue	16,111	15,728
– of which passenger revenue	2,186	2,184
Profit/loss	–476	–4,826

Traffic (million)	2001	2002	2003
Passenger journeys	63.47	59.43	51.27
Passenger-km	2,805	2,682	2,316
Freight tonnes	53.59	49.86	50.52
Freight tonne-km	10,929	10,384	10,398

Passenger operations

The political and economic changes that took place in the wake of the 'Velvet Revolution' have resulted in declining passenger numbers. Passenger-km dropped from 6.2 billion in 1989 to 2.3 billion in 2003; road coaches are proving formidable competitors. Reflecting the fall in rail patronage, services in Slovakia were cut by some 10 to 15 per cent in 1993, and further cuts took place in December 1995 and February 2003. ZSSK now accounts for 13 per cent of the country's total passenger journeys.

During the period of the 2004–05 timetable, Slovakia was served by two year-round EuroCity trains on the Budapest–Berlin/Hamburg axis, two Bratislava–Prague EuroCity services and a Warsaw–Budapest via Bratislava EuroCity service.

In December 2003 ZSSK, in collaboration with Austrian Federal Railways (ÖBB), introduced two pairs of through InterCity trains between Košice and Vienna Südbahnhof via the Bratislava–Petržalka/Kittsee border, crossing as an extension of the existing Košice–Bratislava IC connection. In the 2004–05 timetable the Vienna terminus for these trains switched to Westbahnhof to provide better connections with trains for Linz, Salzburg and to Germany and Switzerland. Trains now cover the 549 km journey in 6 hours 32 minutes. Train formation comprises ZSSK's first and second class air conditioned coaches plus second class/service and restaurant cars, all heavily refurbished by ŽOS Vrútky (see below).

To further boost cross-border services between Slovakia and Austria, ZSSK and ÖBB have also introduced frequent EURegio trains between Bratislava and Vienna operating via the Devínska Nová Ves/Marchegg border crossing, augmented by one pair of through ER trains on Vienna–Nitra and Vienna–Nové Zámky routes. By the 2004–05 timetable the number of cross-border passenger trains between the two countries had risen to 55 per day.

In its 2003–04 timetable, ZSSK planned a total of 2,075 passenger services, of which there were 10 EuroCity, eight InterCity, nine expresses, 95 fast trains, 41 semi-fast services, 12 Euregio trains and 1,900 stopping trains.

Freight operations

The political and economic changes in central and eastern Europe have resulted in a major reduction in freight traffic, with the decline of heavy industry, a reduction in north-south transit traffic between the former Eastern Bloc countries and a decline in trade with the former USSR, heavily influencing tonnage figures. Tonne-km more than halved between 1989 and 1995, dropping from 25 to 12 billion. By 1995, the situation had stabilised, with freight revenue rising by 5 per cent in that year to reach Kcs12.8 billion, and rising again in 1998 to Kcs14.172 billion against operating costs of Kcs13.3 billion.

ZSSK still remains Slovakia's principal mode of heavy freight transport, although its market share fell between 1994 and 2002 from 66.3 to 52.9 per cent.

Freight traffic accounted for 63.7 per cent of ZSSK's revenues in 2003. The main commodities transported are iron ore, coal, construction materials, chemicals, metal products and timber. Of this, 36.7 per cent was imported cargo, 25.8 per cent export, 16.6 per cent internal and 20.9 per cent transit business.

In 2003 ZSSK was operating 770 freight trains each day (899 in 2001), of which 108 were regular block trains.

Following an agreement reached between ZSSK and its Austrian counterpart, ÖBB, direct freight trains on the Bratislava–Petržalka–Kittsee, route are cleared by customs and both railways' authorities at Bratislava-vychod marshalling yard. A €5.2 million project has greatly simplified freight flows between both countries since its introduction in April 2002. However it is still necessary to change locomotives at the border station at Bratislava-Petržalka, reflecting the inability of ÖBB's Class 1116 to operate throughout due to technical difficulties.

On 15 September 2003, ŽSR together with Czech Railways (ČD) reintroduced a fast door-to-door parcels service between both countries branded 'Interkurýr'. The service was previously launched in 1996 but was abandoned due to poor results. It is available at 13 ZSSK stations.

Intermodal transport

Intermodal traffic is still at an early stage of development, accounting for just 1.84 per cent of the railway's freight volume at 930,000 tonnes in 2003. Much of the trade with the former USSR was transported in containers; the break-up of the Eastern Bloc resulted in a dramatic fall in intermodal traffic. Of late, ŽSR has been making efforts to rebuild this traffic. In October 1996, a new Kcs100 million terminal was opened at Čierna nad Tisou-Dobrá on the Ukrainian border, where the gauge changes. Two months later a weekly container service was launched between the new terminal and Moscow, Russia.

In November 1996 the Slovak government adopted a plan to develop intermodal traffic, with Combi Slovakia Eurotrans Žilina selected as the national operator. Metrans Danubia, ERS and ČSKD-Intrans are the principal operators in this market.

The 'Tatranexpress' unit container train between Bratislava-Pálenisko and Rotterdam and the 'Tatranexpress II' between Čierna nad Tisou and Rotterdam, both launched in 1997, were discontinued due to poor loadings. Subsequently, in July 1999, a Netherlands-based company,

The first of six Class 840 diesel-electric railcars built by the DMU-GTW consortium, at Vrútky on the occasion of its handover in September 2003 (Quintus Vosman)   0561562

European Rail Shuttles (ERS), in co-operation with local partners ČSKD-Intrans in Prague and SPAP of Bratislava, launched a new container shuttle between Prague Žižkov and Bratislava ÚNS/Pálenisko. Now running three times-weekly between Bratislava-Pálenisko and Mělník-Labe, this provides a direct connection with ERS's daily 'Bohemia Express' service between Mělník-Labe and Rotterdam Maasvlakte/Walhaven. In March 2004 two weekly container shuttles augmented ERS's portfolio in Slovakia, connecting a new terminal at Sládkovičovo with Mělník-Labe, Czech Republic, and Bremerhaven, Germany. Another container shuttle operates between Prague-Uhříněves and Dunajská Streda (Metrans Danubia).

ŽSR now has at its disposal five international intermodal terminals: Bratislava-ÚNS, Čierna nad Tisou; Košice; Ružomberok-Lisková; and Žilina. It plans to open an additional facility at Zvolen. A regional terminal opened at Ružomberok in May 1998. There are also privately owned terminals at Dunajská Streda, Nové Zámky (Metrans Danubia/Ozon) and Sládkovičovo (ERS/Lörincz). Multipurpose combined transport centres are located at Bratislava-Pálenisko and Čierna nad Tisou-Dobrá.

Using funds from the state budget, ŽSR had acquired 152 Type Sgnss container flats by the end of 2000 in a programme to eliminate the 1970s-built Type Pasy vehicles which were barred from international services from the start of 2000. Also meeting RIV international standards are 97 Hungarian-built Type Sgjs container flats and 60 new Type Sdgnss wagons for piggyback operations by Tatra-vagónka. The first *Röllende Landstrasse* piggyback service is planned on the Budapest Haros–Trstená route. The weekday-only trains will consist of 17 Type Saadkms wagons owned by MÁV and a sleeping car for lorry drivers, with an expected journey time of 12 hours. A trial run took place in September 2002.

Traction and rolling stock
At the start of 2002, ZSSK operated 480 electric locomotives (of which 21 were for 1,520 mm gauge), 631 diesel locomotives (of which 23 were for 1,520 mm gauge and five for 1,000 mm), 54 electric multiple-units (of which 30 were for 760 or 1,000 mm gauge), 230 diesel railcars (of which two were for 1,520 mm gauge), 1,500 hauled passenger coaches and 21,200 freight wagons (all 1,435 mm gauge).

In 2003–04 ZSSK planned to invest Kcs3,321 billion on rolling stock modernisation, of which Kcs1.318 billion was to come from the railway's own resources. A €10 million credit is being provided via the Eurofima leasing agency. Plans include the conversion of further Class 810 railcars into new Class 812 to replace ageing Classes 830, 850 and 851, the purchase of six newly-built Class 840 lightweight diesel-electric railcars, modernisation of six Class 735 diesel-electric locomotives into Class 736 and upgrading a further 28 coaches for InterCity trains and 18 for conventional trains (see below). Also due for modernisation are 1,764 freight wagons to meet market demands. ZSSK is also to purchase 22 new covered wagons for the transport of cereals.

ZSSK formally joined the Eurofima financing agency in March 2003, when it bought 0.02 per cent of the shares. Later the shareholding was increased to 0.05 per cent.

Pendolino order cancelled
In April 2002 ZSSK signed an agreement with Czech Railways (ČD) and Alstom Ferroviaria to add to ČD's order for seven Class 680 three-voltage tilting trainsets based on the Pendolino design. ZSSK had ordered four generally similar trainsets for its Bratislava–Žilina–Košice corridor. ZSSK also planned to operate the trains on international services on Prague–Košice, Poprad–Vienna and Berlin–Prague–Budapest routes jointly with ČD's sister units. Both railways had also agreed to proceed jointly to obtain costly approval to operate their fleets over Austrian and German networks. However, in 2003 ZSSK cancelled the order due to a lack of funds and priority is being given to other investments.

Class 425.95 low-floor articulated metre-gauge 1.5 kV DC emu at Štrbské Pleso, in the High Tatras (Edward Barnes) 0547051

Class 405.9 metre-gauge rack railcars at the modernised high-level terminus at Štrba, prior to departure with a service to Štrbské Pleso (Edward Barnes) 0547055

Electric locomotives (1,520 and 1,435 mm gauge)

Class	Wheel arrangement	Output kW Continuous/ one hour	Speed km/h	Weight tonnes	No in service (2003)	First built	Builders Mechanical	Electrical
3 kV DC								
110	Bo-Bo	800/960	80	72	22	1971	Škoda	Škoda
121	Bo-Bo	2,032/2,344	90	88	12	1960	Škoda	Škoda
125.8*	2 × Bo-Bo	4,080/4,680	90	170	21	1976	Škoda	Škoda
131**	2 × Bo-Bo	2 × 2,240/2,500	100	2 × 85	2 × 50	1980	Škoda	Škoda
140	Bo-Bo	2,032/2,344	120	82	12	1953	Škoda	Škoda
163/163.1	Bo-Bo	3,060/3,400	120	85	12/25	1986/92	Škoda	Škoda/ČKD
162	Bo-Bo	3,060/3,400	140	85	8	1992	Škoda	Škoda/ČKD
182	Co-Co	2,790/2,890	90	120	55	1963	Škoda	Škoda
183	Co-Co	2,790/3,000	90	120	40	1971	Škoda	Škoda
25 kV AC								
210	Bo-Bo	880/984	80	72	34	1973	Škoda	Škoda
240	Bo-Bo	3,080/3,200	120	85	98	1968	Škoda	Škoda
263	Bo-Bo	3,060/3,400	120	85	10	1988	Škoda	Škoda/ČKD
Dual-voltage								
350	Bo-Bo	4,000/4,200	160	87.6	18	1974	Škoda	Škoda
363	Bo-Bo	3,060/3,400	120	87	36	1980	Škoda	Škoda/ČKD
362	Bo-Bo	3,060/3,400	160	88	15	1990	Škoda	Škoda/ČKD

* Designed for the Haniska–Matovce 1,520 mm gauge line.
** These locomotives can be operated as single Bo-Bo units when uncoupled.

Electric multiple-units (1,435 mm gauge)

Class	Cars per Unit	Motor cars per unit	Motored axles/car	Output kW Continuous/ One hour	Speed km/h	No in service (2003)	First built	Builders Mechanical	Electrical
3 kV DC									
460	5	2	4	2,000/2,160	110	16	1974	MSV	MEZ
25 kV AC									
560	5	2	4	1,680/1,860	110	8	1970	MSV	MEZ

Electric locomotives
ZSSK has started to fit its fleets of Class 350 and 363 Škoda-built dual-voltage electric locomotives with MIREL microprocessor control to equip them to run at 160 km/h. The first Class 350 upgraded for 160 km/h running on the Bratislava–Prague route was handed over in March 2000, followed by two more later in the same year. The type is heavily employed on the neighbouring ČD.

Since early 1999, a requirement for additional high-speed dual-voltage electric locomotives has led to bogie exchanges between classes. Fifteen Class 363 120 km/h dual-voltage electric locomotives have been fitted with 140 km/h bogies from Class 162 3 kV DC machines and redesignated Class 362. The Class 162 machines which receive 120 km/h bogies have become Class 163.1.

In 2004 ZSSK is to call bids for new three-voltage (3 kV DC/25 kV AC 50 Hz/15 kV AC 16.7 Hz) electric locomotives for international operations. They are to be funded by Eurofima.

Modernisation of the diesel fleet

A programme to modernise ŽSR's diesel locomotive fleet made good progress in 1998, when three new rebuilds emerged. The Class 772 six-axle 111-tonne diesel-electric was reconstructed from a Class 771 by Martinská mechatronická and Vagónka Trenišov. It is powered by a 960 kW Pielstick 8PA4-185M4 engine licence-built by ZTS TEES Martin and was intended for heavy shunting duties. However, series production was not implemented.

The most successful conversion from Class 771 is Class 773, featuring a 1,300 kW Caterpillar engine. Weighing 112 tonnes and intended for heavy shunting and main line freight services, the locomotive has a central cab with lowered hoods and incorporates Theimeg remote-control equipment supplied by Krauss-Maffei. Construction was undertaken by ZŤS-KV Dubnica nad Váhom. Series modernisation began in 1999 and by the end of 2001, 10 locomotives had been completed. ZSSK planned to have 35 of the type operational by 2005.

Another successful conversion is Class 736, rebuilt by ŽOS Zvolen from a Class 735 locomotive. The 64 tonne machine features AC/DC transmission and is powered by a Caterpillar 3512 DITA engine, rated at 990 kW, with a maximum speed of 90 km/h. Two prototypes were produced in 1998 and 2002 to cover freight duties on the Zvolen-Kosiče route and a batch of eight series-built locomotives followed in 2003–04, to be allocated by ZSSK to Zvolen and Nové Zámky depots.

Diesel railcars

Vrútky workshop has undertaken a programme to rebuild Type Baafx railcar trailers into new Class 811 diesel-electric railcars. The Class 811 is used to provide faster and more comfortable services on ŽSR branch lines. Two prototypes powered by a Liaz M1.2C 640F engine emerged late in 1995 and underwent trials on the 56 km route between Kralovany and Trstená. Class 811 is equipped with microprocessor control by NES Nová Dubnica. ŽSR ordered a first series-built batch of ten railcars fitted with a modified Type M1.2c 640s engine of 237 kW, of which six were manufactured by ŽOS Vrútky and four by ŽOS Zvolen. These were completed in 1997–98. In April 1997, a second batch of 17 railcars was ordered from ŽOS Zvolen at a cost of Kcs173.5 million. Delivery was made during 1999. The type has entered service on six regional routes. Prototypes of Type 011 trailers and 912 driving trailers for use with these railcars emerged in late 1996, built in collaboration with ŽOS Zvolen and Správkárna vozů Stúrovo.

The first example of a similar modernisation emerged in September 2001 from ŽOS Zvolen. The new Class 812 lightweight two-axle diesel railcar prototype is a rebuild from a Class 810 railcar. It is powered by a MAN engine rated at 257 kW and has Voith Type DIWA 863.3 hydro-mechanical transmission. The vehicle has a top speed of 88 km/h and seats 53 passengers. It is also equipped with a MIREL microprocessor control system.

The prototype entered revenue operation in June 2002. Following successful trials, series conversions began and by the end of 2003 a total of 25 Class 812 railcars had been completed. Of these, two emerged as a luxury version designated Class 812 LUX. These feature first class only seats, full air conditioning, tinted windows, electro-pneumatic plug doors and vacuum toilets. Seating capacity is reduced to 38, plus space for 32 standees. For 2004–05 a further 30 conversions have been ordered at a cost of Kcs380 million, with ten of these completed by mid-2004. Class 812 is intended to replace Classes 830, 850 and 851 railcars on mountainous lines, including: Poprad Tatry–Tatranská Lomnica/Plaveč; Margecany–Červaná Skala–Brezno; Brezno–Tisovec–Jesenské; Prešov–Bardejov; and Zvolen–Lučenec.

Diesel locomotives (1,520 and 1,435 mm gauge)

Class	Wheel arrangement	Power kW	Speed km/h	Weight tonnes	No in service (2003)	First built	Mechanical	Engine	Builders Transmission	
701	B	147	40	22	9	1957	ČKD	Tatra 930-51	M	ČKD
702	B	147	40	24	2	1968	ZTS Martin	Tatra 930-51	M	ČKD
710	C	302	60 (30)	42	24	1961	ČKD	12V 170DR	H	ČKD
720	Bo-Bo	551	60	61	3	1958	ČKD	6S 310DR	E	ČKD
726	B-B	515	70 (35)	56.6	4	1963	ZTS Martin	K12 170DR	H	ČKD
721	Bo-Bo	551	80	74	121	1963	ČKD	K6S 310DR	E	ČKD
731	Bo-Bo	880	70	72	11	1988	ČKD	K6S 230DR	E	ČKD
735	Bo-Bo	926	90	64	41	1973	ZTS Martin	12PA 4 186	E	ČKD
742	Bo-Bo	883	90	64	82	1977	ČKD	K6S 230DR	E	ČKD
751	Bo-Bo	1,102	100	75	77	1964	ČKD	K6S 310DR	E	ČKD
752	Bo-Bo	1,103	100	74	26	1969	ČKD	K6S 310DR	E	ČKD
753	Bo-Bo	1,325	100	73.2	15	1968	ČKD	K12V 230DR	E	ČKD
750*	Bo-Bo	1,325	100	71.7	47	1991	ČKD	K12V 230DR	E	ČKD
754	Bo-Bo	1,460	100	74.4	25	1975	ČKD	K12V 230DR	E	ČKD
770	Co-Co	993	90	114.6	33	1963	ČKD/Dubnica	K6S 310DR	E	ČKD
771	Co-Co	993	90	115.8	50	1968	SMZ/Dubnica	K6S 310DR	E	ČKD
770.8**	Co-Co	993	90	114.6	11	1983	SMZ/Dubnica	K6S 310DR	E	ČKD
771.8**	Co-Co	993	90	114.6	9	1970	Dubnica	K6S 310DR	E	ČKD
781	Co-Co	1,472	100	116	10	1966	Lugansk	VSZ 14D 40	E	Kharkov
772†	Co-Co	960	90	110	1	1998	Martinská/ Vagónka Trebišov	8PA4-185M4	E	Martinská
773†	Co-Co	1,300	100	112	10	1998	ZŤS-KV Dubnica	Cat 3512 DI-TA/2	E	Siemens
736††	Bo-Bo	990	90	64	2	1998	ŽOS Zvolen	Cat 3512 DI-TA	E	Siemens

Notes:
* A reconstructed Class 753 locomotive with electric 3,000 V heating.
** Designed for the Haniska-Matovce 1,520 mm gauge line.
† Rebuilt from Class 771, equipped with electronic and remote control.
†† Rebuilt from Class 735, equipped with electronic control.

Diesel railcars (1,520 and 1,435 mm gauge)

Class	Cars per unit	Motor cars per unit	Motored axles/car	Power/car KW	No in service (2003)	First built	Mechanical	Engine	Builders Transmission	
811	1	1	2	180/237**	27	1995	VS	Liaz M1.2C640S	E	ČKD†
810	1	1	1	156	100	1975	VS	Liaz ML634	HM	Praga
810.8*	1	1	1	156	2	1982	VS	Liaz ML634	HM	ČKD
812†	1	1	2	257	25	2001	VS	MAN D2866 LUH21	HM	Voith
820	1	1	2	206	7	1963	VS	Tatra T930-4	H	ČKD
830	1	1	2	301	30	1949	VS	ČKD 12V170DR	E	ČKD
840	2††	1	2	420	6	2003	Stadler, ŽOS Vrútky	MTU	E	Bombardier
850	1	1	2	515	25	1962	VS	ČKD 12V170DR	H	ČKD
851	1	1	2	588	14	1967	VS	ČKD 12V170DR	H	ČKD

VS – Vagónka Studénka.
* Designed for the Haniska–Matovce 1,520 mm gauge line.
** First two prototypes equipped with Liaz M1.2C 640F 180 kW engine.
† Rebuilt from Type Baafx trailer at ŽOS Vrútky and ŽOS Zvolen maintenance shops.
†† Plus centre power module.

New emus and dmus

In August 2000, the EMU GTW-Vysoké Tatry consortium, incorporating Stadler Fahrzeuge AG, Adtranz and ŽOS Vrútky, rolled out a prototype of a series of Class 425.95 low-floor articulated emus for ŽSR's 1.5 kV DC metre gauge network in the High Tatras mountains. A fleet of 14 of these aluminium-bodied vehicles was ordered. They are based on Stadler's GTW 2/6 design, replacing obsolete Class 420.95 units built by Tatra Smíchov in 1963–70. Revenue operations began in June 2001, and the whole fleet had entered service by July 2002. Built under licence in Slovakia, each unit consists of two non-powered driving cars and an intermediate powered section. Rated output is 600 kW with a top speed of 80 km/h.

A diesel-electric version of the Class 425.95 emerged from the ŽOS Vrútky plant in March 2003, built by the DMU-GTW consortium comprising Stadler Fahrzeuge AG, Bombardier Transportation Switzerland AG and ŽOS Vrútky. Designated Class 840, the 58.7 tonne vehicles are powered by an MTU 550 kW engine with AC transmission and a top speed of 115 km/h. Each provides seating for 94 passengers (plus 16 tip-up seats) and space for 119 standing passengers. Up to three units can be operated in multiple. Six Class 840 dmus were put into operation at the 2003–04 timetable change, replacing locomotive-hauled formations on the Zvolen–Vrútky–Žilina, Zvolen–Banská Bystrica–Červaná Skala, Žilina–Čadca–Skalité and Zvolen–Košice–Humenné routes.

Passenger coach refurbishment

The private maintenance company, ŽOS Vrútky, has been undertaking coach refurbishment. In conjunction with Hungarian company DVJ Dunakeszi of Budapest, by the end of 1996 the firm

had refurbished 23 160 km/h coaches for use on Bratislava-Košice InterCity trains. Improvements include disc brakes, upgraded interiors with air conditioning, and plug doors. After a three-year break due to a lack of funds, modernisation of a second batch of 26 Hungarian-built coaches was finally ordered in August 1999. Of these, ten are first-class Type Aheer, ten second-class Type Bheer, three first-class/restaurant Type AReer and three second-class/service cars of Type Bdseer. Work commenced at the ŽOS Vrútky workshops in September 1999 and involved close collaboration with DVJ Dunakeszi. Entry into revenue service on international expresses commenced at the end of 2000. Improvements since the refurbishment of the first batch of these vehicles include new anti-skid protection and hydraulic shock-absorbers. In 2001 the former ŽSR and ŽOS Vrútky reached agreement on continuation of a refurbishment programme covering 100 additional coaches, rebuilt from Types A, B, AB, Bh, Bc and BDs and to be completed in four batches by 2006. Eurofima loans are providing 90 per cent of the vehicles' funding. The first batch covered 23 coaches (six Type Apeer first class and 17 Bpeer second class), completed in 2002–03 for both international and domestic IC services. In November 2003 a second stage commenced, incorporating nine first class Type Aeer and 22 second class Type Beer, for completion in 2006. The third and fourth stages will see a total of 13 Type Aeer and 33 class Beer coaches overhauled by 2008.

Freight wagons

As at 30 October 2003 ZSSK owned 18,180 freight wagons, with a further 1,724 vehicles on hire. There are plans to refurbish or purchase new wagons, mainly sliding-roof, sliding-wall, container flats or piggyback types such as Rils, Rilns, Falls, Tadns, Habbins, Hbbillns, Sgnss and Sdgnss to meet the needs of a radically changed market.

In June 1997 the first Type Sdgnss piggyback wagon was manufactured by Tatravagónka Poprad. By the end of 1997, 28 such wagons were delivered with a further 32 following in the first half of 1998 at a total cost of some Kcs300 million. A further batch of 20 Type Sdgnss wagons for container traffic followed in 2000–01.

A prototype of a new Type Rils wagon with fixed ends and sliding roof was purchased in 1998 from ŽOS Trnava. Rebuilt from a Type Res flat wagon, it is intended for carrying bulk and palletised goods. In October 1998 200 new four-axle Type Shimmns steel coil transporters were handed over by Tatravagónka Poprad.

In August 2000, ŽOS Trnava rolled out the first Type Hirrs four-axle twin-unit freight wagon, rebuilt in collaboration with Greenbrier in Germany from two Type Gbkks two-axle wagons. It was also agreed that 204 Type Rils and Rilns wagons with removable tarpaulin roofs and 346 Type Eas high-sided wagons for the transport of substrates would be rebuilt in collaboration with Ahaus-Altstätter Eisenbahn Cargo AG (AAE). The wagons will first be sold to AAE, then leased back and eventually purchased again by ZSSK.

By April 2001, 200 Type Eas 52 and Eas 53 wagons had been upgraded, followed by the modernisation of 250 Type Eas 11 by December of the same year. All three types now have a axleload capacity of 22.5 tonnes and a maximum speed of 100 km/h.

During 2000–01, a batch of 132 Type Habbins, Hbbins and Hbbillns sliding wall wagons rebuilt from older vehicles entered service.

UPDATED

ZSSK Cargo

Železničná spoločnost' Cargo as
Drieňová ulica 24, SK-820 09 Bratislava
e-mail: cargo.o2@zscargo.sk
Web: www.zscargo.sk (under construction)

Key personnel
Chairman of Supervisory Board: Mikuláš Kačaljak
Chairman of Managing Board and Director General:
Pavol Kužma
Deputy Chairman of Managing Board and Director,
Economics: Vladimír Ľupták
Board Member and Director, Commercial:
Peter Klinka
Board Member and Director, Operations:
Ervín Kvašnovský
Board Member and Director, Rolling Stock:
Miroslav Dzurinda
Management Board
Manager, Director General's Office: Jozef Benko
Internal Audit: Peter Pavlík
Internal and External Relations: Silvia Šimonová
Control and Inspection: Miroslav Janáček
Strategy: Ján Simčo
Human Resources: Anton Jaborek
Finance: Ľubomír Húska
Controller: Edita Valentovičová
Marketing: Ján Dančej
Sales: Jozef Melní
Intermodal Traffic: Miroslav Horečný
Traffic: Jiří Jančík
Personnel: Imrich Sloboda
Facility Management: Matilda Tomašovičová
Logistics: Vladimiír Kormaník
Information Technology: Marián Bujňák

Organisation
See entry for Železničná spoločnost' Slovensko as in Slovakia section of *Railway systems and operators*.

NEW ENTRY

Class 773 diesel-electric freight locomotive, rebuilt from Class 771 (Michal Málek) 0580505

Class 240 25 kV AC electric locomotives at the marshalling yard at Devínska Nová Ves (Colin Boocock)
0580507

Two Class 125.8 twin-section 1,520 mm gauge 3 kV DC electric locomotives with an iron ore train at Slaneč (Quintus Vosman) **NEW**/1115020

Class 736 diesel-electric locomotive, rebuilt from Class 735, at Brezno (Quintus Vosman) **NEW**/1115019

Slovenia

Ministry of Transport

Langusova 4, SL-1000 Ljubljana
Tel: (+386 1) 478 80 00   Fax: (+386 1) 478 81 39
e-mail: gp.mzp@gov.si
Web: http://www.goc.si/mpz

Key personnel
Minister: Jakob Presečnik

Slovenian Railways (SŽ)

Holding SlovenskeŽeleznice doo
Kolodvorska ul 11, SL-1506 Ljubljana
Tel: (+386 1) 291 21 00
e-mail: info@slo-zeleznice.si
Web: www.slo-zeleznice.si

Key personnel
Director General: Blaž Miklavčič
Deputy Director General: Andrej Godec
Assistant Director General: Andrej Pagon
Workers' Director: Albert Pavlič
Executive Director, Freight: S Žerjav
Corporate Communications Manager:
  Aleksander Salkič

Gauge: 1,435 mm
Route length: 1,229 km
Electrification: 504 km at 3 kV DC

Political background
Slovenian Railways is part of the former Yugoslav network. Slovenia is a natural junction of transport routes which connect central and western Europe with the Near East or southern Europe. When part of the Yugoslav system, Slovenia's railways transported more than 22 million tonnes of freight and about 21 million passengers in inland and international traffic. However, in the early 1990s, the break-up of Yugoslavia and the ensuing war cut Slovenia's railway links with the former southern republics. This resulted in a drastic fall of Slovenian transport volume. This was reflected also in the reduction of the number of employees, which dropped from 15,206 in 1990 to 11,900 in 1993. By the end of 2002 the number had declined further to 8,794.

In 2003 SŽ became a state-owned holding company with three main divisions responsible for infrastructure and passenger and freight operations.

Traffic (million)	2002	2003
Passenger journeys	14.5	15.1
Passenger-km	749	777
Freight tonnes	16.3	17.2
Freight tonne-km	3,078	3,274

Passenger operations
SŽ operates international passenger trains from the capital Ljubljana to Trieste in northern Italy (165 km) and Graz in Austria (246 km). Slovenia is also served by international trains linking neighbouring countries, such as the 'Simplon Express' (Zagreb–Ljubljana–Trieste–Venice–Geneva) and the Munich–Zagreb overnight service.

Domestically, the most important line is from Ljubljana to Maribor (278 km, two-hourly service), on which Fiat-built Class 310 Pendolino trainsets were introduced during 2000. Branded Inter City Slovenia (ICS), the service replaced the former Želeni vak (green train) operation provided by German-built Class 711 dmus.

A cross-country diesel-worked route runs from Jesenice to Sežana (149 km). For its summer 2000 timetable, SŽ introduced a car-carrying service on this route, eliminating the need for motorists to use the difficult road over the Bohinjsko Pass.

SŽ serves 128 stations.

Freight operations
International freight, both transit and import, accounts for around 90 per cent of SŽ traffic. With TEN Corridors V and X using part of the network, efforts are focused on developing this activity.

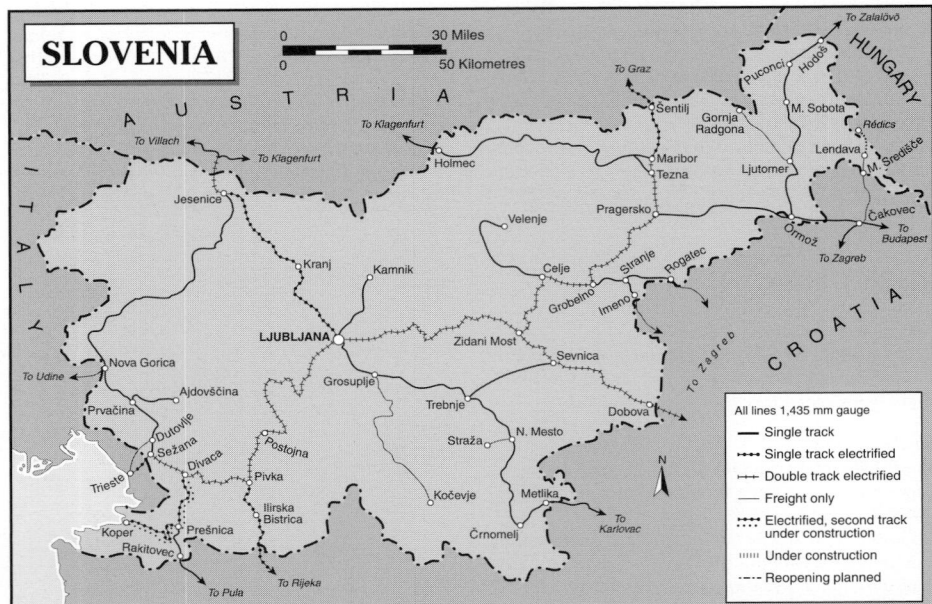

0567915

Class 363 electric locomotive at Zidani Most with a northbound iron ore train from Koper (Philip Wormald)   ***NEW**/0585044*

Diesel locomotives

Class	Wheel arrangement	Power kW	Speed km/h	Weight tonnes	No in service	First built	Builders Mechanical	Engine	Transmission
661	Co-Co	1,440	112	114	3	1961	GM	GM 567	E GM
642	Bo-Bo	452	80	64	17	1960	DD	MGO	E B&L
643	Bo-Bo	507	80	67.2	24	1967	B&L/DD	MGO	E B&L
644	A1A-A1A	1,357	120	90	16	1973	Macosa	GM	E GM
661	Co-Co	1,454	124	112	3	1961, 1973	GM	GM	E GM
664	Co-Co	1,496	124	113	20	1984	DD	GM	E GM
732	C	440	60	44	3	—	Jenbacher Werke	Jenbacher Werke	H Voith

B&L: Brissoneau & Lotz; DD: Djuro Djakovic; GM: General Motors

Electric locomotives

Class	Wheel arrangement	Power kW continuous	Speed km/h	Weight tonnes	No in service	First built	Builders Mechanical	Electrical
342	Bo-Bo	2,280	120	76	22	1968	Ansaldo	Ansaldo
362	Bo-Bo-Bo	3,400	120	112	16	1960	Ansaldo	ASGEN
363	C-C	2,830	125	114	38	1976	Alsthom	Alsthom
541	Bo-Bo	6,000	200	85	20	—*	Siemens	Siemens

* To be delivered 2006–08; three-voltage (3 kV DC/15 kV AC/25 kV AC) locomotives

Class 310 tilting trainset at Zidani Most forming a Ljubljana–Maribor ICS service (Philip Wormald)
NEW/0585045

Freight carryings continue to rebound following the restoration of peace in the Balkan region. The performance in 2003 of 17.2 million tonnes for 3,274 million tonne-km compares with 13.2 million (2,560 million tonne-km) in 1996. SŽ's largest marshalling yard at Zalog outside Ljubljana has been modernised; GEC Alsthom and local supplier Iskra re-equipped the yard to raise sorting capacity from 1,500 to 2,500 wagons daily, using 39 classification tracks. Work was completed in mid-1997.

New international services which commenced from Zalog in 2004 include the 'East-West Rail Shuttle', linking Slovenia with Bologna, Italy, and involving through running with locomotives from each country, and the 'Sava Express', connecting Ljubljana with Serbian Railways at Belgrade. In October 2004 another service, the 'East Express', was established to Istanbul, Turkey.

Intermodal operations

SŽ operates container terminals in Ljubljana, Celje and Maribor and, together with the Port of Koper, the container terminal in Koper.

There are hopes that Koper will become a major entry port for Austria and other parts of central Europe, served by feeder ships from a Mediterranean hub port; this route could save boxes from Asia between five and seven days in transit time, compared to landing at Rotterdam. Koper is only 20 km from the established port

of Trieste in Italy, but operating costs are lower. Accordingly, upgrading infrastructure and increasing capacity of the Divaca–Koper line is a priority for SŽ.

New lines

In May 2001 a 25 km link was constructed from Murska Sobota to the Hungarian border near Hodoš; Hungarian Railways (MÁV) has built a 19 km line on its side of the frontier. The combined link shortens the route for freight running between eastern Europe and the West and the port of Koper in the Adriatic. This substantially increases the importance of SŽ, for this link enables an important part of potential eastern European freight to go by sea (see 'Intermodal operations' above). Some of the funding for the €80 million project was provided by the EU under its PHARE programme.

In February 2001 the Slovenian and Italian governments agreed on the route through their countries of the European Union's Corridor V high-speed line linking Milan–Venice–Trieste–Ljubljana–Budapest–Kiev. Work is expected to start in 2005 with a target completion date of 2015.

Improvements to existing lines

Work in progress in 2004 included an upgrade of the Divaca–Koper route. In the longer term this

includes provision of a second track using a new alignment. EU funding is also being provided to upgrade the line from the Austrian border at Jesenice to Ljubljana and on to Croatia via Dobova, which forms the SŽ portion of Corridor X. The single-track section from Jesenice to Ljubljana is to be doubled.

Work was in progress in 2004 to raise line speeds on the Ljubljana–Maribor section to enable Class 310 tilting trains to operate at 160 km/h.

SŽ also plans to double the Šentilj–Maribor line.

Traction and rolling stock

In 2004 SŽ operated 76 electric and 83 diesel locomotives, three tilting emus, 38 emu and 144 dmu cars. There were also some 160 passenger coaches and around 5,300 freight wagons. Electric locomotives are of French and Italian origin while the bulk of the main line diesel fleet is formed of General Motors and French designs.

In July 2004 SŽ awarded Siemens a contract to supply 20 Class 541 three-voltage electric locomotives based on the Taurus design. Rated at 6,000 kW and mainly intended for freight traffic, the locomotives will be capable of operating cross-border services into Austria, Croatia, Hungary, Germany and Italy. Funding for the locomotives is to be provided by Eurofima. Delivery from Siemens' Munich plant is scheduled for 2006–08. To meet more immediate traffic requirements, in 2004 SŽ signed a contract with Siemens Dispolok covering the hire of four four-voltage Class 189-type electric locomotives from 2005.

During 2000 SŽ received three three-car Class 310 Pendolino tilting trainsets ordered from Fiat Ferroviaria in 1998. They operate intercity services between Ljubljana and Maribor.

Since 2000 SŽ has received from Siemens 10 two-section and 20 three-section 'Desiro' articulated emus for use on its Koper–Postojna and Ljubljana–Zidani Most routes. Designated Classes 312-0 and 312-1 respectively, the vehicles largely replaced Pafawag-built Class 311 and 315 emus. The first five units were built at Siemens plants in Germany, while the remainder were assembled at the Siemens-owned TVT Nova plant in Maribor.

Dmus include 26 two-car Class 713/715 vehicles supplied by German builders in the 1980s and based on DB Class 628.1 and 40 two-car Class 813/814 Fiat-built units dating from the 1970s. The majority of these have undergone extensive refurbishment by TVT in Maribor to become Class 813.100/814.100.

In 2004 SŽ announced plans to procure 100 new four-axle freight wagons for general cargo use. These were to be 85 Type Habbins and 15 Type Habbillns.

Electrification

Principal rail routes in Slovenia (Ljubljana–Maribor, Ljubljana–Jesenice, Ljubljana–Koper/Trieste) are electrified on the 3 kV DC system.

A recent addition to the electrified network has been the line from the Austrian border at Šentilj to Pragersko. Electrification is planned by 2007 of the Pragersko–Ormoz–Murska Sobota line, which forms part of TEN Corridor 5 to Hungary. General modernisation of these sections is also being undertaken, including level crossing elimination and renewal of train control systems.

Track

Rail type and weight (kg/m): UIC 60,54,49,45
Crossties (sleepers)
Wood: 26 × 16 × 260 cm
Concrete: 31-24 × 21-18 × 260 cm
Number per km
   in plain track: 1,660
   in curves: 1,660
Fastenings: K, Pandrol, SKL 2
Min curvature radius: 250 m
Max gradient: 0.26%
Max permissible axleload: 22.5 tonnes

Class 312-1 Desiro emu at Zidani Most after arrival as a local service from Maribor
(Philip Wormald)   *NEW*/0585046   *UPDATED*

South Africa

Department of Public Enterprises

Private Bag X842, Pretoria 0001
Tel: (+27 12) 342 71 11
Fax: (+27 12) 342 72 24

Key personnel
Minister: Jeff Radebe

Department of Transport

Strubenstreet, Forum Building, Pretoria
Tel: (+27 12) 45 72 60
Fax: (+27 12) 461 68 45

Key personnel
Minister: Dullah Omar
Director General of Transport: Khetso Ghordan

Gautrain Rapid Rail Link

Gautrain Project Office
Ten Sixty Six, 12th Floor, 35 Pritchard Street, Corner
Loveday and Harrison Streets, Johannesburg
Tel: (+27 11) 298 49 00
Fax: (+27 11) 298 49 16
Web: corporate.gautrain.co.za

Key personnel
Project Team Chairman: Jack van der Merwe

Political background
The Gautrain Rapid Rail Link is a public-private partnership project developed by the Gauteng provincial government to establish an 80 km fast electrified passenger rail system linking Tshwane (Pretoria) and Johannesburg International Airport (JIA) and Sandton. The scheme aims to improve transport links in South Africa's main economic region, reduce road congestion and stimulate development. Its estimated cost is R7 billion.

Initial planning and pre-feasibility/feasibility studies were undertaken in the late 1990s and early 2000s. A Reference Route was published in 2002, leading to an environmental impact assessment and public consultation.

In July 2005 the Bombela consortium was selected as the preferred bidder from two shortlisted groups that had submitted bids for a Design, Build, Operate and Maintain (DBOM) concession.

The Bombela consortium comprises:
- Bombardier Transportation (consortium leader; core electrical and mechanical systems, including rolling stock)
- Bouygues Travaux Publics (civils)
- RATP Développement (operations)
- Murray & Roberts (rolling stock)
- Loliwe Rail Contractors (civils)
- Loliwe Rail Express (electrical and mechanical systems; maintenance and training)

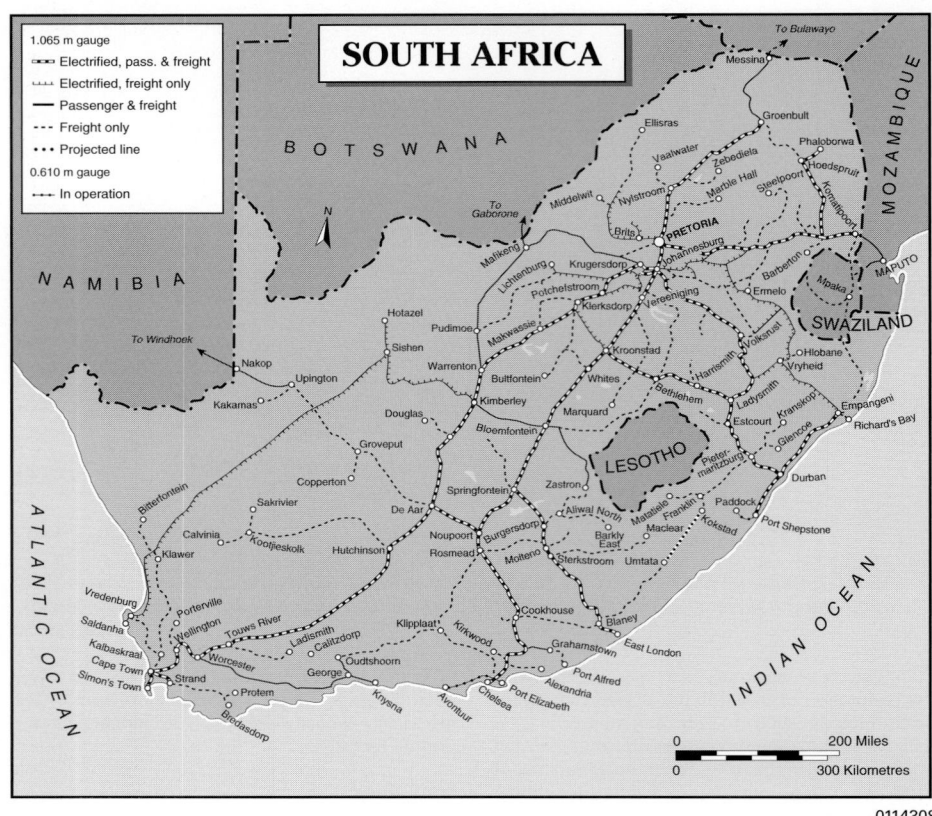

0114308

A final contract award is expected to take place in 2006. Construction is expected to lead to commissioning of the system by 2010, in time for the Soccer World Cup, for which South Africa is the host country.

Organisation
Management of the Gautrain Rapid Rail Link project is led by a Provincial Political Steering Committee, which acts as the main decision-making body. Executive responsibility rests with the Gautrain Project Team, with a Project Review Committee providing specialist advice and guidance.

New lines
The Gautrain Rapid Rail Link provides for an 80 km standard gauge electrified network comprising a north-south line from Hatfield, north of Pretoria, to a southern terminus at Johannesburg Park. From Marlboro, northwest of Sandton, a line will run east to serve JIA. With 10 stations, the network will double as a commuter system and an airport rail link. Interchange with South African Rail Commuter Corporation (SARCC) Metrorail services will be available at Hatfield, Johannesburg Park and Pretoria; improved interchanges are planned at other locations.

A dedicated service is planned between Sandton (Johannesburg) and JIA, with in-town check-in provided at the form er and a journey time of 15 minutes.

The network will employ continuously welded rail on concrete sleepers, with slab track in tunnel sections. Bombardier's Cityflo 250 train control technology will be used.

Traction and rolling stock
The rolling stock that features in the Bombela consortium's preferred bid is Bombardier's Electrostar emu. Manufacture of these will be assigned to the company's Derby, UK plant, with UCW Partnership, a subsidiary of Murray & Roberts, undertaking final assembly in South Africa. It had been estimated that around 20 three- or four-car trainsets would be required for the commencement of services. These would employ either 3 kV DC or 25 kV AC power supply systems and will be capable of operating at up to 160 or 180 km/h.

NEW ENTRY

Metrorail

PO Box 52238 Braamfontein 2017, Johannesburg
Tel: (+27 11) 773 70 91; 70 92
Fax: (+27 11) 773 71 25
e-mail: infoho@metrorail.co.za
Web: http://www.metrorail.co.za

Key personnel
Chief Executive Officer: Honey Mateya

Background
A business unit of Transnet, the state-owned transport organisation, Metrorail operates

commuter rail services under contract to the South African Rail Commuter Corporation (SARCC) (see entry in the South Africa section of *Railway systems and operators*). The contract is a 'pseudo-concession' business agreement intended to pave the way for the eventual concessioning of urban commuter rail operations.

Organisation
Metrorail operates trains owned by SARCC over tracks owned either by SARCC or by Spoornet, the national rail system operator. See entry for SARCC for more details.

The company's Johannesburg headquarters office is complemented by a regional management structure covering five areas of operation: Cape Town, Durban, Eastern Cape, Pretoria and Wits. In 2003 Metrorail employed some 9,200 staff.

Port Shepstone & Alfred County Railway Co Ltd

PO Box 572, 4240 Port Shepstone
Tel: (+27 39) 695 05 20   Fax: (+27 39) 695 05 20

Gauge: 610 mm
Route length: 122 km

Organisation

The Port Shepstone & Alfred County Railway is a narrow gauge line linking Harding and Port Shepstone in Natal that was taken over by private operators when it was abandoned by South African Railways in 1986. In August 2004 the Port Shepstone & Alfred County Railway Co Ltd was placed in final liquidation. Its last train ran on 25 July 2004. There was a possibility that the line's 'Banana Express' tourist service might be reintroduced under a new operator.

UPDATED

South African Rail Commuter Corporation Ltd (SARCC)

Private Bag X2, Sunninghill 2157
Tel: (+27 11) 804 29 00   Fax: (+27 11) 804 38 52
e-mail: info@sarcc.co.za
Web: www.sarcc.co.za

Key personnel

Chief Executive Officer: Eddie Lekota
Group Executives
   Corporate Support and Chief Financial Officer:
      Neli Xaba
   Group Operations Executive, Infrastructure and
      Rolling Stock Development: Brian Jacobs
   Group Operations Executive, Performance
      Compliance, Safety and Security and Service
      Contracts: Enos Ngutshane
   Corporate Relations, Promotions, Information
      Management and Business Intelligence:
      Selomane Maitisa
   Rail Business Strategy, Policy and Network
      Development: Tando Mbikwana
   Chief Executive Officer, Intersite Property
      Management Services: Papi Mphahlele

Subsidiary

Intersite Property Management Services (Pty) Ltd
PO Box 999, Sunninghill 2157
Tel: (+27 11) 804 56 10   Fax: (+27 11) 802 64 75
Web: www.intersite.co.za
Chief Executive: Papi Mphahlele

SARCC electric multiple-units

Class	Cars per unit	Motor cars per unit	Motored axles/car	Output/ motor kW	Speed km/h	No of sets	First built	Builders Mechanical	Builders Electrical
5M*	11–14	3–4	16	220	100	350 (all series)	1957	UCW	GEC
6M	14	8	32	245	110	1 (prototype)	1983	Hitachi	Hitachi
7M	6	6	24	290	110	1 (prototype)	1984	MAN	Siemens/ AEG/BBC
8M	6	6	24	245	110	8	1987	Dorbyl	Hitachi

* Undergoing refurbishment to form Class 10M family.

Gauge: 1,065 mm
Length: operates over 1,150 km, of which 680 km owned, remainder Spoornet
Electrification: 97 per cent of network at 3 kV DC

Political background

SARCC is an agency of the National Department of Transport responsible for the provision of commuter rail services in six metropolitan regions of South Africa. It owns the rolling stock and some of the infrastructure used for these services and manages the operating subsidy. In addition, SARCC specifies service levels and monitors operator performance. Train operations are provided under contract by Metrorail, a business unit of the nationally owned transport undertaking, Transnet.

SARCC's functions also include management of the execution of capital investment projects, upgrading of infrastructure and stations and the co-ordination of public transport with local and provincial government authorities.

The management and development of SARCC's property portfolio, including its stations, is undertaken by a subsidiary, Intersite Property Management Services (Pty) Ltd, which also provides similar services for other bodies.

Traffic (million)	2001–02	2002–03	2003–04
Passenger journeys	466.9	–	482

Passenger operations

SARCC services are operated by Metrorail in six main metropolitan areas around Cape Town, Durban, East London, Johannesburg, Port Elizabeth and Pretoria. Trains serve 470 stations. Services are provided in two classes: Metro and MetroPlus.

New lines

Minor network extensions in Cape Town and East London were under way or planned in 2003.

Traction and rolling stock

In 2004 the rolling stock fleet totalled 4,625 emu cars formed into 375 trainsets. Apart from a few trains built in the 1980s, all stock is based on a design of 1957 vintage. In 2002 deliveries commenced of extensively refurbished examples of the Class 5M family. This followed the conversion of two prototypes, designated Class 10M and 10M1, in 1998 and 1999. Two further prototypes, Class 10M2, were subsequently produced before series refurbishment orders were placed covering 44 motor cars and 132 trailers for delivery by 2003. Contracts were equally shared between two consortia: Transwerk/Bombardier and Union Carriage/Siemens. Those produced by the former are designated Class 10M3 and feature new corrosion-resistant stainless steel bodyshells. They are for Cape Town area services. The Union Carriage/Siemens units, Class 10M4, feature a new Corten steel bodyshell and are for Johannesburg services. Subject to finance SARCC intends to treat the rest of the fleet similarly.

Refurbishment of Class 8M emus also started in 2002.

UPDATED

SARCC (Metrorail) 5M-series emu at Muizenburg on the Cape Town network (Marcel Vleugels)
0567280

Spoornet

South African Railways
Umjanshi House, 30 Wolmarans Street, Johannesburg 2000 (Private Bag X47, Johannesburg 2000)
Tel: (+27 11) 773 87 57   Fax: (+27 11) 773 82 50
e-mail: Johanh@transnet.co.za
Web: http://www.spoornet.co.za

Key personnel

Deputy Managing Director (Transnet):
   M Mkwanazi
Chief Executive: Z Jakavula
General Managers
   Marketing and Sales: (Vacant)
   Information Systems: Dr C R Jardine
   Specialist Businesses: H Evert
Rail and Terminal Services: S Lushaba
Engineering: B J van der Merwe
Finance: H L van der Westhuizen
Human Resources: A Mofokeng
Restructuring and Joint Ventures: R Nair
Planning and Technology: L Petkoon
Executive Business Managers:
   H M Mashele; L Miller

For details of the latest updates to *Jane's World Railways* online and to discover the additional information available exclusively to online subscribers please visit

jwr.janes.com

Executive Managers
COALlink: Peet Cronjé
Orex: H Green
Corporate Affairs: Dr V V Mkhize

Gauge: 1,065 mm; 610 mm
Route length: 19,756 km; 314 km
Electrification: 5,040 km at 3 kV DC; 861 km at 50 kV AC; 2,298 km at 25 kV AC; 9 km dual-current (3 kV DC/25 kV AC)

Political background

Spoornet is the largest division of Transnet, a state-owned company established in 1990 as a result of government policy to commercialise its transport business interests and deregulate the industry in South Africa. As well as its rail business, Transnet has port management, pipeline, road haulage and air carrier divisions. Since the mid-1980s the South African road transport industry has grown considerably to become a major competitor for rail. Initially road was an attractive mode for light manufacturing industries but more recently has made inroads into heavy manufacturing and certain sectors of the mining industry. In response Spoornet has restructured itself to provide integrated freight logistics solutions with the aim of maintaining its leadership in the heavy haul sector and repositioning its other freight and passenger transport businesses.

Organisation

Recent restructuring has seen the centralisation of service planning and control in place of the previous geographical organisation, a move aimed at improving customer service. In 1996 the Two Streams Spoornet programme was introduced, separating the processes of marketing the company's logistics capabilities and its production but retaining customer orientation as the primary goal. Dedicated business units, COALlink and Orex, manage Spoornet's coal and iron ore traffic over the Richards Bay and Sishen-Saldanha lines respectively. The General Freight Services division is grouped into 15 industry-based segments.

Separate units manage Spoornet's Main Line Passenger Services and Blue Train business.

In a bid to help reduce the debt of Spoornet parent Transnet, reported to be R27 billion in 2000, a restructuring plan was announced. This would see the establishment of the company's main operational divisions as separate entities. A further restructuring was expected to follow a re-examination of core and non-core activities.

Since 1991, the Spoornet workforce has dropped from nearly 120,000 to 43,736 in March 1999. The South African Rail Commuter Corporation (Metrorail) business was divested in 1998.

Finances (R million)

Revenue	1997–98	1998–99
Passengers (including commuter)	263.498	284.136
Freight	8,710.006	8,817.751
Other income	573.082	555.369
Total income	9,546.586	9,657.256

Expenditure

Labour	3,739.359	4,140.710
Materials and services	3,878.781	4,406.972
Depreciation	864.903	772.786
Total operating costs	8,483.043	9,320.468
Operating surplus/(shortfall)	1,063,543	336.788

Traffic (million)	1997–98	1998–99
Freight tonnes	186.8	182.7
Freight tonne-km	103,866	102,777
Passenger journeys	5.0	NA
Passenger-km	1,775	1,794

Passenger operations

Spoornet is principally a freight operator, but its Main Line Passenger Services division runs some long-distance passenger trains, including several semi-luxury services. The company also operates the prestigious Blue Train on four routes: Pretoria–Cape Town; Pretoria–Victoria Falls (Zimbabwe); Cape Town–Port Elizabeth (The Garden Route); and Pretoria–Hoedspruit (Valley of the Olifants), Northern Province. Blue Train

Spoornet diesel-electric locomotives

Class	Wheel arrangement	Power kW	Speed km/h	Weight tonnes	No in service	First built	Mechanical	Builders Engine	Transmission
33–000	Co-Co	1,605/1,490	100	91		1965	GE	GE 7 FDL-12	GE 761 A6
33–200	Co-Co	1,640/1,490	100	91	43[1]	1966	EMD	EMD 16-645-E	EMD D 29CC-7
33–400	Co-Co	1,605/1,490	100	91		1968	GE	GE 7 FDL-12	5 GE 761 A6
34–000	Co-Co	2,050/1,940	100	111		1971	GE	GE 7 FDL-12	GE 5 GE 761 A13
34–200	Co-Co	2,145/1,940	100	111		1971	EMD	EMD 16-645-E3	EMD 29B
34–400	Co-Co	2,050/1,940	100	111		1973	GE	GE 7 FDL-12	GE 5 GE 761 A13
34–500	Co-Co	2,050/1,940	100	111	388[2]	1977	GE	GE 7 FDL-12	GE 5 GE 761 A13
34–600	Co-Co	2,245/1,940	100	111		1974	GM SA	EMD 16-645-E3	EMD D 29B
34–800	Co-Co	2,140/1,940	100	111		1978	GM SA	EMD 16-645-E3	EMD 29B
34–900	Co-Co	2,050/1,940	100	111		1979	GE	GE 7 FDL-12	GE 5GE 761 A13
35–000	Co-Co	1,230/1,160	100	82		1972	GE	GE 7 FDL-8	GE 5GE 764-C
35–200	Co-Co	1,195/1,065	100	82	324[3]	1974	EMD/GM SA	EMD 8-645-E3	EMD D 29CCBT
35-400	Co-Co	1,230/1,160		82		1976	GE	GE 7 FDL-8	GE 5GE 764-C1
35-600	Co-Co	1,195/1,065	100	82		1976	GM SA	EMD 8-645-E3	EMD D29 CCBT
36–000	Bo-Bo	875/800	100	72	217[4]	1975	GE	GE 7 FDL-8	GE 5GE 764-C1
36–200	Bo-Bo	875/800	90	72		1980	GM SA	EMD 8-645-E	D29B
37–000	Co-Co	2,340/2,170	100	125	85	1981	GM SA	EMD 16-645E-3B	EMD D31
38–000	Bo-Bo	1,500/780	100	74	49	1993	UCW	CAT 3508	Siemens ABB/6 FRA 5252
91–000	Bo-Bo	52/480	50	44	16	1973	GE	CAT D 379	GE 5GE 778

[1] Total for 33 Class. [2] Total for 34 Class. [3] Total for 35 Class. [4] Total for 36 Class.

Spoornet electric locomotives

Class	Wheel arrangement	Output kW continuous/one hour	Speed km/h	Weight tonnes	No in service	First built	Builders Mechanical	Electrical
3 kV DC								
5E1	Bo-Bo	1,458	97	86	229	1963	UCW	GEC
6E	Bo-Bo	2,252	105	89	42	1970	UCW	GEC
6E1	Bo-Bo	2,252	105	89	702	1969	UCW	GEC
8E	Bo-Bo	704	75	81	82	1983	UCW	50 c/s Group
10E	Co-Co	3,090	90	125	50	1985	UCW	Toshiba
10E1	Co-Co	3,090	90	126	100	1987	UCW	GEC
10E2	Co-Co	3,330	100	126	25	1989	UCW	Toshiba
12E	Bo-Bo	2,252	150	84	5	1963	UCW	GEC
17E	Bo-Bo	2,252	105	90	140	1979	UCW	GEC
18E	Bo-Bo	2,252	105	90	4	1979	UCW	GEC
25 kV AC								
7E	Co-Co	3,000	100	123	67	1978	UCW	50 c/s Group
7EI	Co-Co	3,000	100	125	50	1980	Dorman Long	Hitachi
7E2	Co-Co	3,000	88	126	64	1982	UCW	50 c/s Group
(Series 1 and 2)								
7E3	Co-Co	3,000	100	124	85	1983	Dorbyl	Hitachi
(Series 1 and 2)								
11E	Co-Co	3,900	90	168–172	45	1985	GMSA	GM/ASEA
50 kV AC								
9E	Co-Co	3,840	90	166	31	1978	UCW	GEC
Dual-voltage 3 kV DC/25 kV AC								
14E	Bo-Bo	4,080	140	92	3	1990	SLM	50 c/s Group
14E1	Bo-Bo	4,000	140	97	10	1998	UCW	50 c/s Group

Abbreviations: UCW: Union Carriage & Wagon.

Four General Electric Class 34 locomotives form a heavy freight at Oudtshoorn (Colin Boocock)

0058827

coaching stock has undergone a refurbishment and relaunch programme costing some R70 million. Reports in 1999 suggested the possible sell-off or concessioning of the Blue Train operation.

Freight operations

In 1997–98 Spoornet carried 182.7 million tonnes of freight. This represented a fall of 2.2 per cent on the previous year's figure of 186.8 million tonnes.

In 2000, Spoornet was reported to be seeking strategic equity partners for its COALlink and Orex (iron ore) transport activities.

During FY97-98 a five-year contract worth R400 million was signed by Spoornet and Chrome Resources to convey up to 500,000 tonnes annually of ferrochrome to Richards Bay for export. The Intermodal Wholesale business segment reached agreement with Maersk Lines to convey containers between Gauteng and principal ports.

Signalling and telecommunications

At the beginning of 1999 6,365 route-km was CTC-controlled, 3,530 km mechanically signalled and 609 km under local control.

In 1999 Webb Industries was awarded a contract to install GSM radio equipment on the Richards Bay–Ermelo line.

Electrification

Electrification of the De Aar–Kimberley route is planned.

Traction and rolling stock

At the start of 2000 the railway was operating 2,146 electric and 1,403 diesel locomotives, 1,851 passenger cars (including 72 catering vehicles and 525 miscellaneous including luggage vans, steam-heat coaches), and 123,750 wagons.

In 1999 Class 6E electric locomotives were the subject of a refurbishment programme using Agate traction control electronics supplied by Alstom.

Type of braking: Vacuum and air
Type of coupler in standard use: SAR M-type and S-type (AAR-compatible), AAR E-type and AAR F-type (rotary and non-rotary)

Track

Rail: 60 kg (1,950 km), 57 kg (4,370 km), 48 kg (9,290 km), 40 kg (1,580 km), 30 kg (2,803.3 km) and 22 kg (99.7 km)
Crossties (sleepers)
PY/FY concrete 2,200 × 250 × 200 mm
P2/F4 concrete 2,057 × 203 × 200 mm
Steel 40 and 30 kg 2,060 × 260 × 115 mm
Wood 2,100 × 250 × 125 mm; 2,100 × 250 × 175/185 mm
Fastenings: Pandrol, Fist, E3131 chairs
Spacing: 700 mm: 1,440/km; and 650 mm: 1,538/km
Minimum curvature radius: 90 m
Max gradient: 1.67%
Max axleload: 29 tonnes (locomotives); 26 tonnes (wagons)

Electrically hauled heavy coal train at Three Rivers (Marcel Vleugels)   0099133

Two Class 6E electric locomotives at the head of Spoornet's prestigious 'Blue Train' (Wolfram Veith)   0063787

Spoornet 610 mm gauge Class 91 diesel locomotives on a freight service at Port Elizabeth
(Marcel Vleugels)
0099134

Spain

Ministry of Development

Ministerio de Fomento
Plaza de la Castellana 67, Nuevos Ministerios,
E-28071 Madrid
Tel: (+34 91) 597 70 00   Fax: (+34 91) 597 85 02
Web: www.mfom.es

Key personnel
Minister: Magdalena Álvarez Arza
Infrastructure and Planning Secretary:
    Victor Morlán Gracia
Infrastructure Secretary: Josefina Cruz Villalón

Political background
The Ministerio de Fomento is the government department with responsibility for transport. The appointment of Magdalena Álvarez Arza following the general election in 2004 and changes to the department's remit resulted in a restructuring to focus on core activities of: planning; infrastructure and investment; transport regulation; and economics. The railway directorate reports to the Infrastructure Secretary. The department sponsors the Strategic Infrastructure and Transport Plan (El Plan Estratégico de Infraestructuras y Transporte) covering the period 2005–20. The rail programme supports the Plan's high level objectives to increase economic competitiveness and open up access to public services. Rail investment accounts for around 42 per cent (€103 billion) of projected total investment (€214 billion) in the Plan. It envisages by 2020 increasing the high-speed network from 1,031 to 10,000 km; investing in infrastructure to ensure that 90 per cent of Spain's population lives within 50 km of a station on the high-speed network; developing the existing network to encourage more passenger and freight traffic with an emphasis on adopting the 1,435 mm gauge; further elimination of level crossings; and continuing investment in the conservation and upkeep of railway heritage sites.

UPDATED

Catalan Railways (FGC)

Ferrocarrils de la Generalitat de Catalunya
Av Pau Casals 24, 8è, E-08021 Barcelona
Tel: (+34 93) 366 30 00   Fax: (+34 93) 366 33 50
e-mail: mllevat@fgc.catalunya.net
Web: www.fgc.net

Key personnel
President: Joan Torres i Carol
Directors
    Performance: Albert Tortajada i Flores
    Finance: Lluís Huguet i Viñallonga
    Co-ordination and Expansion:
        Manuel Villalante i Llauradó
    Projects: Pere Calvet i Tordera
    Personnel: Armand Aixut i Freixanet

Gauge: 1,668 mm (Lleida–La Pobla de Segur line); 1,435 mm (Barcelona–Vallès line); 1,000 mm (Llobregat–Anoia line and Ribes–Núria and Montserrat (Abt rack) lines)
Route length: 90 km (Lleida–La Pobla de Segur); 45 km (Barcelona–Vallès); 139 km (Llobregat–Anoia); 13 km (Ribes–Núria) and 5 km (Monistrol–Montserrat)
Electrification: 1,435 mm gauge: 45 km (Barcelona–Vallès) at 1.5 kV DC
1,000 mm gauge: 98 km (Llobregat–Anoia line); 12 km (Ribes–Núria); 5 km (Monistrol–Montserrat); all at 1.5 kV DC

Organisation
FGC operates local railways in the Barcelona area. Its 1,435 mm gauge Barcelona–Vallès line runs from Barcelona's Plaça Catalunya terminus to Sant Cugat where it forks to serve Terrasa and Sabadell. A short branch (2 km) leaves the route at Gràcia to serve Avinguda Tibidabo.

FGC's 1,000 mm gauge Llobregat–Anoia line runs from Barcelona Plaça Espanya, diverging at Martorell to serve Igualada and Manresa (where connection, across town, can be made with RENFE's Barcelona–Lleida service). The station at Montserrat Aeri serves the cable car to Montserrat, and connection can be made at Monistrol de Montserrat with the Montserrat rack railway.

The 1,000 mm gauge Vall de Núria Abt rack line between Ribes de Freser and Núria is isolated from other FGC lines, connecting at Ribes with RENFE's 1,668 mm gauge Barcelona–Puigcerdà–La Tour de Carol route. FGC also operates one cable car system and four funicular railways.

Responsibility for the management and development of the non-electrified 1,668 mm line between Lleida and La Pobla de Segur passed from RENFE to FGC at the beginning of 2005. Passenger services continue to be operated using RENFE's Class 593 units.

At the end of 2003 FGC employed 1,215 staff.

Stadler-built Type GTW 2/6 emu at Monistrol de Montserrat on the recently restored Monistrol–Montserrat metre-gauge rack line (Bryan Philpott)   0567060

Traffic (million)	2002	2003
Passenger journeys	70.5	74.1
Passenger-km	758.2	801.4
Freight tonnes	0.583	0.655
Freight tonne-km	33.6	41.0

Passenger operations
In 2003 the main FGC network carried 74.1 million passengers. Passenger numbers increased on the Barcelona–Vallès line to 54.50 million in 2003 (from 53.58 million in 2002). On the Llobregat–Anoia line passenger numbers increased to 18.47 million in 2003 (from 16.43 million in 2002).

In June 2003 FGC reintroduced services on the 5.2 km rack railway serving the mountain tourist destination of Montserrat. Investment of €56.8 million saw the line rebuilt completely. Services had been withdrawn in 1957. Most trains operate from a new station at Monistrol-Vila, where a 1,000-space car park has been provided, although some connect with FGC services at the nearby Monistrol de Montserrat station on FGC's Llobregat–Anoia line. Five two-car type Beh 2/6 rack versions of Stadler's GTW lightweight low-floor rail vehicle each provide capacity for 200 passengers. During its first 12 months of operation from 12 June 2003 the number of passengers carried was 462,964. The Vall de Núria rack line carried 272,514 passengers in 2003.

Freight operations
FGC is primarily an intensive passenger service operator, but there are significant flows of common and potassium salts along the Llobregat–Anoia line. This traffic originates at mines served by the Súria and Sallent freight-only branches at the northwestern extremity of the line and gains access

Electric multiple-units

Class	Cars per unit	Motor cars per unit	Motored axles/car	Output/ motor kW	Speed km/h	No in service	First built	Builders Mechanical	Electrical
Llobregat–Anoia line									
M211/ T281/T291	3	1	4	276	90	7	1987	Macosa	Alsthom
M211/T291	2	1	4	276	90	3	1987	Macosa	Alsthom
M213/T283	3	2	4	180	90	20	1999	CAF/ Alstom	Adtranz
Barcelona–Vallés line									
M111/T181	3	2	4	276	90	20	1983	MTM/ Alsthom	Alsthom
M112/TM122/ T182	4	3	4	180	90	16	1995	CAF	ABB
Ribes–Núria and Monistrol–Montserrat abt rack lines									
GTW 2/6	2	1	2	300	45	7	2003	Stadler	Bombardier
A	2	2	2	181	37	4	1985	GEC Alsthom/SLM	ABB

Class M213/T283 metre-gauge emu on the Llobregat–Anoia line (Tony Pattison) ***NEW**/1114986*

to the port of Barcelona via another freight-only branch at Sant Boi. A total of 655,092 tonnes were carried in 2003, representing an increase of 12.3 per cent compared to the previous year. This comprised 319,431 tonnes of potassium salts (compared to 237,597 tonnes in 2002) and 335,661 tonnes of common salts (compared to 345,460 tonnes in 2002). Trains load up to 1,200 tonnes gross.

Improvements to existing lines

Extension of the Barcelona–Vallès line at Terrassa, to be constructed in two stages, will create three new stations at UPC-Vallparadis, Nord and Can Roca providing links with the university campus,

RENFE's regional services at a new interchange station, and population growth at Can Roca.

Traction and rolling stock

In 2005 FGC operated 43 emus on the Barcelona-Vallès line. On the Llobregat-Anoia line it operated 30 emus, two dmus, eight diesel locomotives and 180 freight wagons. In addition the Ribes-Núria rack-railway fleet comprised two electric locomotives (265 kW units of 1929 vintage), one diesel (delivered by Stadler/SLM/ABB in 1994) and nine coaches for use with special trains. Regular services were operated by four two-car emus each with 181 kW motors per car delivered in 1985 and

1995 and two Stadler Rail GTW 2/6 two-car units (of the type supplied to the Montserrat rack railway), which joined the fleet in 2003.

In October 2003 FGC ordered 13 three-car emus from an ALSTOM-led consortium that also includes CAF and Bombardier. Completion of the order was scheduled for 2006. These vehicles will be similar to the existing M213/T283 units.

Signalling and telecommunications

The Barcelona–Vallès line has ATP and CTC in operation along its entire length. Installation of ATO is planned for the Barcelona urban area (between Plaça Catalunya, Sarrià and Avinguda Tibidabo), to enable an appreciable increase in train service frequency, especially in the peak hours.

The Llobregat–Anoia line is equipped with the DIMFAP trainstop system between Barcelona and Manresa and between Martorell and Igualada (98 km).

Satellite tracking using the GPS system is now operational on all freight-only lines, and also on the Martorell–Igualada line.

Track

Rail: 1,000 mm gauge: UNE 45 kg/m; UIC 54 kg/m. Abt rack line only: 20 kg/m

1,435 mm gauge: UIC 54 kg/m

Sleepers: Wood 2,400 × 240 × 140 mm*; and 2,000 × 230 × 140 mm

Concrete 2,400 × 300 × 200 mm*; and 1,900 × 260 × 209 mm

Fastenings: Direct or indirect elastic and direct rigid with ribbed plate on wooden sleepers

Min curvature radius: 150 m*; 100 m

Max gradient: 4.4%*; 2.5%

Max axleload: 20 tonne*; 15 tonne

*Barcelona–Vallès line. Other dimensions refer to Llobregat–Anoia line

UPDATED

EuskoTren

Eusko Trenbideak/Ferrocarriles Vascos SA (ET/FV)
Atxuri 6, E-48006 Bilbao
Tel: (+34 94) 401 99 00   Fax: (+34 94) 401 99 01
Web: www.euskotren.es

Key personnel

President: Álvaro Amann Rabanera
Vice-President: Antonio Aiz Salazar
Director General: José Miguel Múgica Peral
Secretary: Jon Arruti Etxebarria
Directors
  Operations: Augustín Menoyo Barcena
  Infrastructure: José Antonio Gorostiza Emparanza
  Organisation and Support Services:
    Silvia Gómez Santos

Gauge: 1,000 mm
Route length: 181 km
Electrification: 176 km at 1.5 kV DC

Political background

Operation of the electrified 1,000 mm gauge railway system in the Basque provinces of Bizkaia (Vizcaya) and Gipuzkoa (Guipúzcoa) was devolved from FEVE to the Basque autonomous community in 1979. ET/FV was created in 1982 and adopted the EuskoTren brand in 1996. EuskoTren remains in public ownership, controlled by the Department for Transport and Public Works of the Basque government, and operates several modes of public transport.

Organisation

A strategic plan, EuskoTren XXI, was launched in 2001. Its principal aims were to increase the number of annual passenger journeys to 39 million and increase annual freight traffic to 2 million tonnes. To implement the plan, a new organisational structure was created in October 2002. An operations business unit responsible for all passenger services, freight, logistics and maintenance, sits alongside separate divisions

for infrastructure, and organisation and support services (including personnel, IT, marketing and purchasing). At the end of 2003 EuskoTren employed 952 people.

Passenger operations

EuskoTren operates the line from Bilbao Atxuri to Donostia (San Sebastián) Amara; between Lasarte Oria and Hendaia (Hendaya) via Amara and Irún (the line known locally as El Topo). An hourly through service operates from 0600 between Atxuri and Amara, with a typical journey time of 2 hours 39 minutes. A limited-stop service (fastest journey time 2 hours 10 minutes) runs twice daily in each direction calling at Bolueta, Durango, Eibar and Zarautz. EuskoTren also provides a passenger service on the Larreineta–Escontrilla funicular railway and between Azpeitia and Lasao at the Basque Railway Museum. Altogether 18.15 million

passengers were carried by rail in 2003, compared with 17.98 million in 2002.

EuskoTren also operates 12 local bus services in Bizkaia and 10 in Gipuzkoa, while the EuskoTran tramway is operated in Bilbao between Atxuri and San Manés, the last-mentioned carrying 1.14 million passengers in 2003. The total number of passengers carried by EuskoTren on its rail, tram, road and funicular services in 2003 was 24.68 million representing an increase of 5.46 per cent compared to 2002.

A separate company, Metro Bilbao (formed in 1993), provides rail services from Bolueta to Plentzia (the line operated formerly by ET/FV) and Urninaga.

Freight operations

EuskoTren transported 154,163 tonnes in 2003 (a decrease of 1.8 per cent from 2002) operating

EuskoTren Class UT-200 emu crossing the border with France at Hendaye (Edward Barnes)
***NEW**/1114975*

258 trains between Ariz-Basauri and Lasarte in collaboration with FEVE. The reduction in 2003 was attributed partly to an industrial dispute affecting output in the iron and steel industry in Asturias.

Improvements to existing lines
EuskoTren XXI
Costed at €973 million, the strategic plan envisages increasing the proportion of double track between Atxuri and Hendaia from 15 to 40 per cent of route-km to increase train frequencies and cut journey times.

Proposals for double-tracking sections between Amorebeita and Durango and Traña and Berriz include relocating the line in a cut-and-cover tunnel in Durango and building new workshops and offices. Initial estimates put the cost of these projects in the region of €137 million. The Iraqi architect Zaha Hadid has been commissioned to design EuskoTren's new station and headquarters at Durango.

A further major doubling project, between Usurbil and Añorga in Gipuzkoa, costed initially at €54 million, involves building seven tunnels and a viaduct, plus two new stations and possibly new maintenance facilities at Lasarte. Doubling of six other sections is planned, along with several new

stations. The plan also includes investment of €172 million in the development of tramway/light rail in seven urban areas, including reinstatement of the tramway between Irún and Hondarribia (Fuenterrabía), the site of Donostia airport, and a 7.6 km tramway for the Basque capital Gasteiz (Vitoria).

Freight improvements centre on reopening of the Urola railway from Zumaia to carry scrap metal to the steelworks at Azpeitia, plus provision of better access and increased siding capacity at the ports of Bermeo and Pasaia and a new freight and intermodal terminal at Kostorbe (Irún), adjacent to the RENFE and SNCF lines. A 'dry port' is to be developed at Amorebieta, to be served by rail from Bermeo.

Traction and rolling stock
In 2005 EuskoTren/EuskoTran operated four electric and two diesel locomotives, 53 electric multiple-units, 19 wagons, two funicular cars, 61 buses and seven trams. In 2004 EuskoTren announced plans to invest €4.1 million in 42 multipurpose 80 km/h freight wagons and a further €36 million in a fleet of up to 12 diesel-electric locomotives for delivery in stages over a 10-year period.

UPDATED

A freight service on the EuskoTren system at Ariz, in this instance behind a FEVE diesel-electric locomotive
(Bryan Philpott)
0576711

Spanish Narrow-Gauge Railways (FEVE)

Ferrocarriles de Vía Estrecha
General Rodrigo 6, Parque de las Naciones, E-28003 Madrid
Tel: (+34 91) 453 38 00   Fax: (+34 91) 453 38 25
e-mail: info@feve.es
Web: www.feve.es
   www.transcantabrico.feve.es

Key personnel
President: Eugenio Dimás Sañudo Aja
Managing Director: Manuel Acero Valbuena
Directors
   General Secretary and Legal Adviser:
      Valentín Pérez García
   Finance and Planning:
      José Carlos Baños Márquez
   Communcations and External Affairs:
      Amador Robles Tascón
   Operations: Juan Carlos Albizuri Higuera
   Infrastructure: Jesús Hallado Arenales

Gauge: 1,000 mm
Route length: 1,268 km
Electrification: 391 km at 1.5 kV DC

Political background
FEVE was set up by the state in 1965 to take control of the country's extensive narrow-gauge railways. Control of several networks has since passed to public operators in the autonomous regions. Investment is provided through the Railway Infrastructure Plan 2000–07. State investment totalling €71.33 million in 2003 was expected to increase to €74.00 million in 2004. Asturias received the highest level of investment (€15.88 million) followed by Castilla y León (€12.02 million) in 2003. The León–Guardo line is independently financed by the autonomous community of Castilla y León.

Organisation
FEVE operates passenger and freight services over an extensive network linking Ferrol, Bilbao and León in the northwest and the 19 km Cartagena–Los Nietos passenger line in Murcia.

FEVE is structured into divisions covering: general secretariat; economics/finance; human resources; infrastructure; operations; traffic management; rolling stock and communications. At the end of 2003 the company employed 1,960 staff in seven autonomous communities. At the

end of 2003 the company employed 19,620 people in seven autonomous communities. The largest concentration of personnel (46 per cent) is based in Asturias.

Traffic (million)	2002	2003
Passenger journeys	12.3	12.3
Passenger-km	230.8	233.6
Freight tonnes	3.2	3.1
Freight tonne-km	479.1	432.7

Passenger operations
Passenger services between Bilbao (Concordia) and León (the line known as La Robla) were reinstated in May 2003, with one through train in each direction taking 7 hours 10 min. Services had been withdrawn over much of the route in 1991. The €41 million project to rebuild the line was financed by the Development Ministry and the autonomous community of Castilla y León and took five years to complete. Investment in 2002 alone amounted to €7.3 million on the last section to undergo track renewal between Arija and Cordovilla. FEVE was reported to be well pleased with patronage, having carried nearly 23,000 passengers in the first eight months of operation.

FEVE's five Metrotrén Asturias lines carried 4.5 million passengers in 2003, the most heavily used being line F6 Oviedo–Infiesto. Metrotrén Asturias promotes integration of FEVE and RENFE passenger networks in the Asturias autonomous community, partly through provision of a unified fare structure.

Operation of the 'Transcantábrico' luxury land-cruise train and other special services represents a significant source of income. Two Transcantábrico trainsets are available for regular scheduled journeys between León and Ferrol. The Transcantábrico is offered for private charter and FEVE also promotes special trains using conventional vehicles, particularly in connection with local and regional festivals.

Freight operations
An increase of 20.5 per cent in freight tonnage to 3.29 million tonnes in 2002 highlighted the strategic importance of FEVE's network to the heavy industry of northwest Spain, though traffic fell by 6.3 per cent in 2003 to 3.09 million tonnes. Coal and its derivatives accounted for 60 per cent of this traffic, with iron and steel products making up 17 per cent. Most freight traffic originated in

Class 1900 electro-diesel locomotive at the head of El Transcantábrico at León (Bryan Philpott)
0567057

Diesel locomotives

Class	Wheel arrangement	Power kW	Speed km/h	Weight tonnes	No in service	First built	Mechanical	Builders Engine	Transmission
1600	Bo-Bo	1,177	90	58	13	1982	MTM	SACM-MGO V16	E Alsthom
1650	Bo-Bo	1,177	90	60	13	1985	MTM, Alsthom	SACM-MGO V16	E Alsthom
1400	B-B	883	60	56	3	1964	Henschel	SACM-MGO V12	H Voith
1500	B-B	772	80	56	15	1965	GECO	Caterpillar	E GECO
1900*	Bo-Bo	1,130	–	60	10	2002	FEVE, Suncove	Caterpillar 3512B	E Siemens

* Electro-diesel locomotives

Diesel railcars or multiple-units

Class	Cars per unit	Motor cars per unit	Motored axles/car	Power/ motor kW	Speed km/h	No in service	First built	Mechanical	Builders Engine	Transmission
2600*	2	2	2	–	80	12	1999	Suncove	–	–
2400	3	2	2	228	80	10	1983	MTM	MAN	E BBC
2400	2	2	2	228	80	17	1985	MTM	MAN	E BBC

* Rebuilt from Class 2300

Electric railcars or multiple-units

Class	Cars per unit	Motor cars per unit	Motored axles/car	Output/ motor kW	Speed km/h	No in service	First built	Builders Mechanical	Electrical
3500	2	1	4	4 × 121	80	2	1981	CAF	AEG/GEE
3500	3	1	4	4 × 121	80	6	1984	CAF	AEG/GEE
3500	3	1	4	4 × 121	80	13	1981	CAF	AEG/GEE
3600*	2	–	–	–	80	12	2000	Suncove	Siemens
3800	2	1	4	4 × 119	80	16	1992	CAF	AEG/GEE

* Rebuilt from Class 2300 dmu

Class 3600 emus at Bilbao Concordia (Edward Barnes)   **NEW**/1115006

Class 1600 diesel-electric locomotives of the first (left) and second series at El Berrón
(Bryan Philpott)   **NEW**/1115007

Class 2600 dmu, rebuilt from a Class 2300 unit, photographed at León (Bryan Philpott)   0567058

Asturias and Cantabria, with the Bilbao area as the principal destination outside the regions.

Improvements to existing lines
Work was completed in 2003 on a €3.2 million project to electrify the short section of line east of Oviedo between Infiesto and Nava. On the same line, the section between El Berrón and Nava is being converted to double track in two stages. A new branch linking Valle Real and Puerto de Raos opened for freight in 2003, while a short cut-off outside Ferrol, completed in April 2004, has bypassed a severe curve and paves the way for track-doubling between Ferrol and Xove. At Gijón a 2.6 km branch is planned to serve the Asturias airport.

Traction and rolling stock
In 2005 FEVE operated 54 diesel locomotives, 49 electric and 39 diesel trainsets, including the new fleet of 10 Class 1900 electro-diesel locomotives. Rolling stock comprised 23 passenger cars and vans and 992 freight wagons. The Transcantábrico fleet comprised a further 27 vehicles and five coaches were dedicated to the Costa Verde tourist train.

FEVE completed the refurbishment of its Class 2400 dmu fleet in 2002 and has introduced its Class 2600 dmus (rebuilt from Class 2300) on commuter services in Castilla and León, Asturias and Murcia. A further five Class 3600 emus (also rebuilt from Class 2300 diesel units) entered service in 2002, bringing the number operational to 12. In addition to 18 intermediate cars being built to lengthen the Class 3500 and 3800 emus, a further six three-car sets of Class 3600 emu are to be built new, and 12 trailer cars ordered to extend the existing trains from two to three cars.

Type of coupler in standard use: Alliance
Type of braking: Air and vacuum

Signalling and telecommunications
Automatic block signalling, automatic train protection and CTC covers all routes.

Electrification
In 2003 electrification was under way of the double-track section between Maliaño and Valle Real, south of Santander.

Track
Rail: 35 kg/m, 45 kg/m and 54 kg/m
Sleepers: Concrete monobloc; and timber 2,000 × 240 × 130 mm
Sleeper spacing: 1,500/km plain track, 1,600/km in curves
Min curvature radius: Main line 100 m
Average curvature radius: 250 m
Max gradient: 3.6% between Cartegena and Los Nietos
Max permissible axleload: 15 tonne
Longest tunnel: 4 km, La Florida, between Gijón and Pola de Laviana

UPDATED

Railway Infrastructure Authority (ADIF)

Administrador de Infraestructuras Ferroviarias (ADIF)
Miguel Angel 21-23, E-28010 Madrid, Spain
Tel: (+34 91) 700 47 00   Fax: (+34 91) 319 61 30
Web: www.infraestructuras-ferroviarias.com

Key personnel

President: Antonio González Marín
General Secretary: María Rosa Sanz Cerezo
Deputy Secretary: Antonio Benavente Jover
Managing Directors
   Operations and Performance:
     Rafael López González
   Legal Services and Affiliates:
     Cristina Borrajo Peláez
   Communications and External Affairs:
     Francisco Javier Sevillano Nacarino
   Planning and Infrastructure Maintenance:
     Manuel Sánchez Doblado
   Projects, Scheduling and Infrastructure
     Construction: Luis María Pérez Fabregat
Finance: Ricardo Bolufer Nieto
Construction: Diego Gómez Sanchez
High-Speed Projects and Construction:
   Vicente Gago Llorente
Security, Human Resources and International
   Relations: Mariano Garrido García

Gauge: 1,668 mm; 1,435 mm; 1,000 mm
Route length: 11,829 km; 953 km; 19 km
Electrification: 1,668 mm gauge: 6,950 km at 3 kV DC
1,435 mm gauge: 953 km at 25 kV 50 Hz AC
1,000 mm gauge: 19 km at 1.5 kV DC

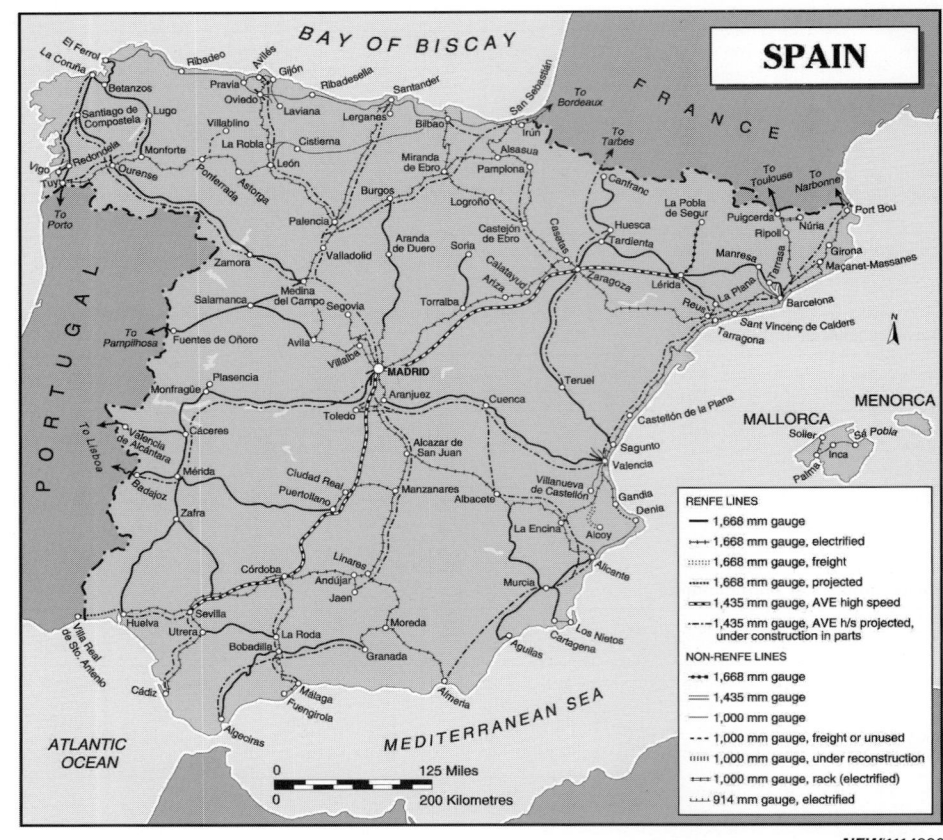

NEW/1114990

Political background

ADIF was created on 1 January 2005 following the implementation of Spain's Railway Sector Law 2004, which provided for the separation of railway operations from infrastructure in line with European Directive 440/91. In practice this means that ADIF is responsible for overseeing the construction and maintenance of the state-owned railway infrastructure, including the development of the high-speed network, management of rail traffic, passenger stations, freight terminals, telecommunication systems and railway heritage.

Heritage

In 2005 ADIF was responsible for the management of 56 disused railway lines and around 1,000 closed station buildings throughout Spain. Many closed lines on both the former national system and the narrow-gauge network of FEVE and the autonomous communities have been transformed into green routes to promote access to the countryside for walkers and cyclists. Disused stations have been put to a variety of uses including youth hostels, restaurants and outdoor centres.

New lines

ADIF has responsibility for overseeing the construction of new lines identified by the Ministerio de Fomento (Ministry of Public Works), their operation and the allocation of track paths and levy of access charges. ADIF's highest profile project is completion of the Madrid–Barcelona–French border 1,435 mm gauge high-speed line.

Madrid–Barcelona

Prioritisation of national transport policy ensured that Spain's first high-speed line between Madrid and Seville was operational in time for the international showcase of Expo '92 in Seville. However, this achievement was at the expense of a drastic budget overspend, which may have influenced the decision to construct the Madrid–Barcelona high-speed line in stages. AVE and Altaria services were introduced over the first section between Puerta de Atocha (currently the Madrid hub of all 1,435 mm gauge routes) and Lleida (via Zaragoza) in 2003. They are expected to be extended to Tarragona in 2006 and Barcelona in 2007.

   The approach to Barcelona is through Tarragona, where a new €40.95 million AVE station, serving the high-speed line and the Barcelona/València/Alicante corridor, and a gauge-changing facility

are under construction. From there the line runs to El Cattlar and Gelida to serve the new AVE station at Barcelona Sagrera, built at a cost of €600 million. A new tunnel, costed at €218 million, links Sagrera with Barcelona Sants. The station at Sants, Spain's busiest in terms of passenger numbers, is itself the subject of a €220 million investment programme providing upgraded facilities including a new platform (six platforms will serve 1,435 mm gauge high-speed lines and eight will serve 1,688 mm gauge tracks), hotel, underground bus station and car park. The high-speed line will continue to the French border and ultimately to Perpignan and Nîmes. Initial Madrid–Lleida

AVE services were provided by S100 trainsets from the Seville high-speed line, restricted to 200 km/h operation. Madrid–Barcelona Altaria services were operated by Class 252 locomotives and Talgo 7 vehicles, changing gauge and locomotive en route. Fastest AVE journey time between Madrid and Lleida was 2 hours 40 minutes. Fastest Altaria journey time between Madrid and Barcelona was 4 hours 35 minutes, compared to around 7 hours previously. A Madrid–Barcelona timing of 2 hours 30 minutes is envisaged on completion of the new line. This is being engineered for a maximum speed of 350 km/h throughout, with a minimum curve radius of 6,000 m, 4.5 m between track centres

AVE trainsets at Seville, terminus of Spain's first high-speed line from Madrid (Edward Barnes)

0567048

and a maximum gradient of 2.5 per cent, subject to eventual commissioning of Level 2 European Train Control Signalling. In the interim, speeds will be raised in stages to 300 km/h with the introduction of Level 1 ETCS.

Other high-speed projects
Construction is planned or under way on other high-speed projects including:

Chamartín–Puerta de Atocha
A new 6.8 km tunnel will link Madrid's principal stations. AVE services will eventually serve Chamartín in addition to Atocha. The tunnel will accommodate twin 1,435 mm tracks.

Madrid–Valencia
Construction of the new line, budgeted around €6 billion, began in 2002. The route diverges east from the Madrid–Seville AVE line about 45 km south of Puerta de Atocha. From Cuenca one line runs direct to Valencia, with another connecting at Albacete with the existing line to La Encina, which will be regauged to 1,435 mm.

Madrid–Toledo
Services over the high-speed line connecting Toledo with Madrid Puerta de Atocha in 25 minutes were scheduled to begin in 2005. The new double-track line diverges west from the existing Madrid–Seville high-speed line 51 km south of Puerta de Atocha and runs for 21 km between Alameda de la Sagra and Toledo. Toledo station has been upgraded at a cost of €700,000. A further 3.9 km spur to the south will provide direct access for trains to and from Ciudad Real, Puertollano and Seville. The project was delivered at a cost of €175 million. In the longer term, direct trains will also link Toledo with new high-speed services from Albacete and Cuenca.

Madrid–Segovia–Valladolid
This project has involved construction of a 28.37 km twin-bore tunnel under the Sierra de Guadarrama along with a 750 m viaduct at Salobral on the Tres Cantos–Colmenar Viejo section at the Madrid end of the project, for which a €43.26 million construction contract was awarded in 2003. A new station for Segovia forms part of the project. ADIF is undertaking research on the test track between Olmeda and Medina del Campo into the development of dual-gauge track with three lines which would enable 1,668 mm gauge vehicles to operate over high-speed lines. Ultimately the test track will be utilised as a high-speed line providing a connection from the Valladolid line to Medina del Campo and on to Zamora and La Coruña.

Córdoba–Málaga
This new line diverges from the Madrid–Seville AVE line at Almodóvar del Río to run for 155 km to Málaga with new stations at Puente Genil (budgeted at €11.65 million) and Antequera-Santa Ana (€8.8 million). High-speed services should operate at 200 km/h between Córdoba and Antequera-Santa Ana from late 2005, with gauge-changing in the vicinity of Antequera-Santa Ana enabling through-running to Granada and Málaga, cutting the Madrid–Málaga journey time to 3 hours 5 minutes. When the Antequera-SantaAna–Málaga

section is commissioned in 2007, the timing will be reduced to 2 hours 15 minutes to match the existing Seville–Madrid schedule. The total budget for the project was €1.68 billion.

León–Asturias
The existing 83 km 1,668 mm gauge line from La Robla (León) to Pola de Lena (Asturias) has 85 tunnels and a 60 km/h speed limit. This project will deliver a new 49.5 km 1,435 mm gauge section requiring 13 tunnels and eight viaducts. Six of the tunnels will be longer than 1 km, the most significant by far a twin-bore 24.6 km tunnel below the Pajares Pass. This is expected to take at least 5 years to construct, with high-speed services not commencing before 2010. Construction is in three sections: Pola de Gordón–Folledo (10.4 km); Folledo–Viadangos (3.9 km); and Viadangos–Telledo (10.3 km). Investment in the project was estimated at €1.06 billion. The line will be engineered for 300 km/h running, including the Pajares tunnel, potentially reducing the journey time between León and Oviedo from 103 to 35 minutes. In March 2005 the Ministerio de Fomento announced that the high-speed line would follow the existing trajectory, suitably upgraded, between Pola de Lena and Gijón, via Oviedo. A Madrid–Oviedo timing of around 3 hours (currently around 5 hours 45 minutes) was envisaged on completion of the new line (and the Madrid–Valladolid high-speed link).

Vitoria–Bilbao–San Sebastián
Extension of the Madrid–Segovia–Vallodolid northeast corridor to Vitoria, Bilbao and San Sebastián will link those cities with Madrid in 1 hour 35 minutes, 2 hours 10 minutes and 2 hours 15 minutes respectively. The Basque 'Y' linking Vitoria, Bilbao and San Sebastián/Irún will provide 159.9 km of new line along with a further 20.2 km between Vitoria and Burgos. The Ministerio de Fomento approved contracts for the first four sections of the Vitoria–Bilbao link (62.6 km) at the end of 2004.

Galicia
Investment of €1.31 billion provides for 123.5 km of new lines; 67.6 km between Santiago de Compostela and Orense and 55.9 km for the Atlantic line.

Zaragoza–Teruel
Construction of the 173.2 km high-speed line between Zaragoza and Teruel is under way between Cuarte de Huerva and Cariñena (34.5 km) and Caminreal and Cella (46.7 km).

Other lines
Madrid suburban tunnel
Work began in 2003 on a €259 million project to build a new tunnel in Madrid for suburban services between Atocha Cercanías and Chamartín. The project is being implemented in three stages and will provide new stations at Puerta del Sol-Gran Vía and Alonso Martínez. This project is separate from the ongoing construction of a 6.8 km tunnel between Puerta de Atocha and Chamartín, without intermediate stations, for use by 1,435 mm gauge high-speed services.

Barajas airport link
Planned investment of €196.7 million will provide a new 8 km suburban rail link, to be constructed in two stages, to Madrid Barajas airport. Journey time from the airport to Chamartín would be 7 minutes and a spur would be provided to permit direct running from the existing Alcalá de Henares line. Public consultation on the proposals ended in April 2005.

Huesca deviation
Investment of €22 million is planned for a 12.3 km line to enable through trains from Zaragoza-Jaca/Canfranc to avoid Huesca. The new line would diverge from the existing line north of Tardienta in a northwesterly direction to rejoin the existing line in the vicinity of Alerre.

Signalling and telecommunications
Following deregulation of telecommunications systems in Spain, ADIF has responsibility for commercial exploitation of excess capacity in the national railway system's 12,000 km fibre optic national trunk network, along with a further 600 km in metropolitan areas.

Track
1,435 mm gauge routes
Rail: UIC 60 kg/m
Sleepers: Type DW monobloc concrete, spaced 1,667/km
Fastenings: Vossloh Type HM
Spacing between tracks: 4.3 m
Max gradient: 1.25%
Min curve radius: In general 0.5°, except stations and 30 km at 0.8°
Max axleload: 22.5 tonnes (up to 200 km/h); 17.2 tonnes (up to 270 km/h)
1,668 mm gauge network
Standard rail
Main lines: 60 kg/m, 54.5 kg/m, 45 kg/m and UIC 54.1 kg/m in 12 and 18 m lengths
Sleepers
Wooden: Mainly creosoted oak, pine, and sometimes beech, 2,600 × 240 × 140 mm for ordinary track. For points, crossings 3, 3.5, 4 and 4.5 m of same width and thickness (centre crossing sleeper being 4,500 × 300 × 14 mm), and for expansion joints 2,600 × 350 × 140 mm. Special sleepers of up to 6.2 m used for diagonals on double track
Reinforced concrete: Type RS or monobloc, thickness 250 mm
Spacing: 1,667/km
Rail fastenings: Screw spikes on wood sleepers and elastic clamps on reinforced concrete sleepers. Elastic fastenings for wood sleepers are also being tested.
Min curve radius: In general 250 m
Max gradient: 4.2% on Ripoll–Puigcerdá line
Longest continuous gradient: 8.27 km of 2%, with 5.85° curves (300 m radius) on 4.84 km
Max altitude: 1,494 m on Ripoll–Puigcerdá line
Max axleload: 22.5 tonnes

UPDATED

Majorca Railways (SFM)

Serveis Ferroviaris de Mallorca
Eusebio Estada, 28 – 1°, E-07004 Palma de Mallorca
Tel: (+34 871) 993 00 00
Fax: (+34 871) 93 00 01
Web: www.tib.caib.es

Key personnel
President: Margarita Isabel Cabrer González
Directors
  Managing: Rafael R Pons Vidal
  Finance: Natalia Lanuza Aguilera
  Infrastructure and Services: Pedro Sintes Ripoll
  Institutional Relations: José Oliver Rebassa
  Integrated Transport: Onofre Plomer Oliver
General Secretary: Carlos Perelló Oliver

Gauge: 1,000 mm
Route length: 77 km

Political background
Responsibility for the 29 km Palma de Mallorca–Inca line was transferred by FEVE to SFM in 1994. In turn, SFM reports to the regional development ministry, which has funded restoration and expansion of the system. SFM services are co-ordinated with those of other modes under Transports de les Illes Balears (tib) branding, which appears on SFM trains.

Passenger operations
Services are operated from Palma to Sa Pobla via Inca (46 km) and from Inca to Manacor (30 km). Following reopening of the Inca–Manacor line in 2003, 88 trains were operated daily on weekdays.

Improvements to existing lines
Projected network developments include: a line linking Sa Pobla–Alcudia–Pollensa; reopening the line to Palma airport and Santanyí; and a new line to Palma University. Construction of an intermodal underground terminal at a new site in Palma has been designed to accommodate the city's expanding public transport needs for the next 50 years. Budgeted at €89.6 million (2004 figure) and financed by the autonomous community and city council, the new station will include five platforms served by 10 tracks; 30 bus stands; and 400 car parking spaces.

Traction and rolling stock
In 1995 SFM acquired four three-car and two two-car dmus from the CAF factory at Beasain Guipúzcoa. They were supplemented by four additional two-car

units procured to coincide with the opening in 2003 of the Inca–Manacor line. Each is powered by two Cummins NTA855R4 engines with Voith hydraulic transmission. Servicing is undertaken at a new maintenance facility established at Son Rullán. A Batignolles/CAF diesel shunter is also retained for works trains.

UPDATED

SFM dmu at Sa Pobla forming a service to Palma (John C Baker)
***NEW**/1114987*

Sóller Railway

Ferrocarril de Sóller SA
Plaza de Espanya 6, E-07100 Sóller, Majorca
Tel: (+34 971) 63 03 01   Fax: (+34 971) 63 12 22
e-mail: info@trendesoller.com
Web: www.trendesoller.com

Key personnel
Managing Director: Miquel Galmés

Gauge: 914 mm
Route length: 28 km
Electrification: 28 km at 1.2 kV DC

Passenger operations
Operating between Palma de Mallorca and Sóller, this privately owned railway carries around 1 million passengers annually, 90 per cent of which are tourists.

Traction and rolling stock
The railway operates one diesel locomotive, four 350 kW electric railcars dating from 1929, 15 trailers and three baggage cars.

UPDATED

Spanish National Railways (RENFE Operadora)

Red Nacional de los Ferrocarriles Españoles
Avenida Pío XII n° 110, E-28036 Madrid
Tel: (+34 91) 300 66 00   Fax: (+34 91) 300 73 36
Web: www.renfe.es

Key personnel
President: José Salguiero Carmona
Managing Directors
  General Secretary and Legal Adviser:
    José Luis Marroquín Mochales
  Transport Services – High-Speed and
    Long-Distance: Abelardo Carillo Jiménez
  Transport Services – Suburban and Regional:
    Javier Pérez Sanz
  Transport Services – Freight and Maintenance:
    Juan Fernández Álvarez
  Planning, Control and Finance: Natalia Garzón
  Human Resources, Security, Quality, Relations
    with ADIF: Ángel Jiménez Gutiérrez
  Communications and External Affairs:
    Manuel Sempere Luján
Corporate Directors
  Security – Train: Antonio Lanchares Asensio
  Security – Personnel: Manuel Rodríguez Simons
  Organisation and Internal Communications:
    Vicente Camarena Miñana
  Reward, Planning – Human Resources:
    José Núñez Blázquez
  Human Resources: José Ángel Méndez González
  Service Quality – Relations with ADIF,
    International: Apolinar Rodríguez Díaz
  Budget Planning: Miguel Minayo de la Cruz
  Management Control and Systems:
    Jesús Solana Madariaga
  Sustainable Development:
    Juan Luis Martín Cuesta
  Purchasing and Resources:
    Pilar Oviedo Cabrillo
  Finance and Administration:
    Miguel Ángel Gutiérrez García
Business Unit Managing Directors
  Long-Distance: Francisco Bonache Córdoba
  High-Speed: María Magdalena Bodelón Alonso
  Regional: Manuel Jesús Simón Peña
  Suburban: Cecilio Gómez-Comino Barrilero
  Freight: Enrique Fernández Diez
  Rolling Stock Maintenance:
    Daniel García Gallego

Political background
Following the general election in March 2004, the incoming PSOE government indicated its intention to implement legislation, enacted by the previous administration, to separate the management of national railway operations from its infrastructure. This resulted in the creation of RENFE Operadora on 1 January 2005 to manage the operational passenger and freight business units and maintain the rolling stock fleet. This was accompanied by the launch of a new RENFE logo. All passenger vehicles would carry the new logo and be lined out in matching purple with an additional stripe in the corporate colour of the particular business unit. Also on 1 January 2005 the newly formed Administrador de Infraestructuras Ferroviarias (ADIF) became responsible for the national system's infrastructure, construction and maintenance.

Organisation
Reporting to RENFE's president, three general directors at board level head the Transport Directorates (high-speed and long-distance, suburban and regional, and freight and maintenance). Five separate divisions (Legal Services; Planning; Control and Finance; Human Resources and Communications; and External Affairs) are headed by directors, also at board level, reporting direct to the president.

Staff levels have continued to fall significantly: at the end of 2003 RENFE employed 30,803 staff compared with 31,422 at the end of 2002 and with 48,884 at the end of 1991.

Finance
State investment in transport covering the period 2005–20 is provided by the Strategic Plan for Infrastructure and Transport (El plan Estratégico de Infraestructuras y Transportes – PEIT) announced by the Ministerio de Fomento in December 2004 and subject to revision every four years. Total investment in the plan was put at EUR241 billion of which EUR103 billion (42.84 per cent) was earmarked for the rail programme. State funding

Newly constructed Class S103 high-speed trainset at Siemens' Krefeld plant (David Haydock)
***NEW**/1122853*

makes up 60 per cent of the total PEIT budget with the remainder to be financed through public and private partnerships.

Traffic (million)	2001	2002	2003
Passenger journeys	466	485	490
Passenger-km	19,191	19,480	19,309
Freight tonnes	25.6	26.3	26.9
Freight tonne-km	11,749	11,660	11,860

Passenger operations

Long distance (Grandes Líneas)

Traffic income (amounting to EUR373.1 million) rose by 0.5 per cent in 2003 compared with the previous year. Total income rose by 0.2 per cent to EUR383.7 million. In 2003 12.8 million passengers were carried; a reduction from 13.2 million in 2002.

The Grandes Líneas product comprises:

Alaris (200 km/h services between Puerta de Atocha and Albacete/València);

Altaria, linking Madrid to a growing number of regional centres including Alicante, Barcelona, Huesca, Logroño, Murcia and Pamplona. Over 475,000 passengers used the Madrid to Murcia Altaria in 2004, its first year of operation, when 3,500 trains ran with 94 per cent punctuality. Altaria services using 1,435 mm gauge and 1,668 mm gauge Class 252 locomotives (52 units are 1,668 mm gauge, 21 are 1,435 mm gauge) and Talgo 7 coaches have operated Puerta de Atocha–Barcelona services over the high-speed line between Puerta de Atocha and Lleida Pirineos since it opened in 2003. Six through trains ran daily in each direction, with a further service each way between Barcelona and Zaragoza Delicias. Line speed was limited to 200 km/h, providing a Madrid–Barcelona journey time of around 5 hours 15 minutes compared to around 7 hours previously. Also in 2003, high-speed services started between Zaragoza and Huesca using Class 594 TRD (Tren Regional Diesel) units, supplemented by one Altaria service in each direction between Puerta de Atocha and Huesca, offering a journey time of 2 hours 40 minutes (compared with around 4 hours 45 minutes previously). Pamplona and Logroño are also served by Altaria services using the high-speed line;

Arco (locomotive-hauled services at up to 200 km/h on the Barcelona/Alicante corridor and some cross-country services);

Diurno (daytime cross-country services using conventional coaches covering distances greater than 500 km with 6–10 hour journey times);

Estrella (overnight services provided by a fleet of 49 sleeping berth coaches and 57 couchette coaches);

Euromed (provided by six 1,668 mm gauge S101 high-speed trainsets on the Barcelona Sants–València/Alicante corridor);

Intercity (160 km/h emu services operating Madrid–País Vasco and Madrid–Andalucía);

Tren Hotel (overnight express services formed of Talgo vehicles operating from Madrid and Barcelona to destinations including Lisbon and Paris).

High-speed services (AVE)

Traffic income (amounting to EUR204.09 million) reduced by 3.1 per cent in 2003 compared with the previous year. Total income decreased by 1.9 per cent to EUR212.79 million. Over the same period, profit decreased by 7.5 per cent from EUR52.04 million to EUR48.16 million. Passengers carried in 2003 totalled 6.0 million, 3.6 per cent fewer than in 2003.

The AVE product comprises:

300 km/h services over the Puerta de Atocha–Seville high-speed line, which opened in 1992; medium-distance 250 km/h services between Puerta de Atocha and Ciudad Real/Puertollano and Seville and Córdoba;

Talgo 200 services between Puerta de Atocha and Málaga;

200 km/h (potentially 350 km/h) services over the Puerta de Atocha–Lleida Pirineos section of the Madrid–Barcelona high-speed line. Services began in 2003 using AVE S100 trainsets from the Seville high-speed line. Four services daily in each direction (in addition to six Altaria services in each direction between Madrid and Barcelona)

Class S102 high-speed trainset forming a Madrid Puerta de Atocha–Lleida Pirineos service at Guadalajara Yebes (Bryan Philpott) **NEW**/1115008

Class 598 diesel-electric tilting dmu (Bryan Philpott) **NEW**/1115010

Class 450 double-deck emu forming a Cercanías service at Pinar de la Rozas (Bryan Philpott)
0576999

Class 333 diesel locomotive and Talgo stock at Valladolid with a Salamanca–Barcelona service
(Bryan Philpott)   0577000

AVE high-speed trainset at Madrid Atocha station (Wolfram Veith)   0567049

Euromed 1,668 mm gauge high-speed trainset forming a Barcelona–Valencia/Alicante service at Tarragona (Marcel Vleugels)   0567051

operated at up to 200 km/h. One train in each direction called at Guadalajara-Yebes, Caltayud and Zaragoza Delicias. Remaining services called only at Zaragoza Delicias. The fastest covered the 492.5 km in 2 hours 40 minutes. In February 2005 the phased introduction of AVE S102 trainsets began with four trains in each direction running to existing schedules.

The AVE 'commitment to punctuality' launched in 1994, 2 years after Madrid–Seville high-speed services began, entitled customers to a full refund of the ticket price should services arrive more than 5 minutes late. Of the 170,000 high-speed services operated between Madrid and Seville during 1994–2005 around 400 trains triggered this compensation. During 2003 the percentage of trains arriving within 3 minutes of schedule was 99.8, the same as in 2002. Of the 18,492 trains operated in 2003, six (0.03 per cent) were delayed for infrastructure-related incidents. The total number of passenger journeys on the Madrid–Seville high-speed line during the period April 1992–April 2005 was put at 58.20 million. The same commitment to punctuality applies to medium-distance services operated over the Madrid–Seville high-speed line. Since October 2003 a 100 per cent refund has been offered to customers on Puerta de Atocha–Lleida Pirineos AVE services for delays exceeding 60 minutes attributed to AVE (50 per cent for delays of 41–60 minutes, reducing to 25 per cent for 21–40 minutes and nil for delays of less than 20 minutes). In 2003 99.7 per cent of these services arrived within 5 minutes of schedule.

Regional (Regionales)

Traffic income amounted to EUR110.51 million in 2003, compared with EUR106.61 million the previous year. Over the same period the number of passenger journeys increased to 26.77 million from 26.32 million. Catalonia generated the most income for the business unit in 2003, followed by Andalusia. These regions produced increases in income of 7.1 and 3.3 per cent respectively in 2003 compared with the previous year. A drop in income was recorded in La Rioja (7.1 per cent), Asturias (1.7 per cent), Aragón (17.0 per cent) and Madrid (0.6 per cent). Asturias continued to produce by far the lowest turnover of all the autonomous communities served by the business unit. The highest percentage increase in revenue in 2003 over the previous year was generated by Murcia at 31.8 per cent, albeit on a modest total traffic income of around EUR0.37 million. The six services operated by the business unit in the autonomous community of Valencia carried 1.96 million passengers in 2004.

Suburban (Cercanías)

The business unit serves 11 metropolitan areas: Asturias (including Gijón and Oviedo), Barcelona, Bilbao, Cádiz, Donastia (San Sebastián, including Irún), Madrid, Málaga, Murcia (including Aguila and Gandía), Santander, Seville and Valencia. Madrid and Barcelona are by far the most heavily used, followed by Bilbao and Valencia.

In 2003 the number of passengers carried increased to 444.84 million compared to 439.6 million in 2002. In 2003 passenger receipts increased to EUR323.3 million compared with EUR310.4 million in 2002. The only regions where passenger receipts were lower than in 2003 were: Bilbao, down 1.3 per cent to EUR11.13 million; Asturias, down 0.1 per cent to EUR6.41 million; and Santander, down 1.1 per cent to EUR0.95 million. Punctuality continued to fall marginally to 98.4 per cent in 2003 from 98.8 per cent in 2002 (arrival within 10 minutes of schedule). The average number of passengers using suburban services in Madrid on weekdays was put at 885,819 at the end of 2004, an increase of 0.6 per cent over the year before.

Freight operations

Traffic income increased by 0.4 per cent to EUR210.08 million in 2003 compared with the previous year. Tonnage increased to 19.10 million in 2003 from 18.82 million in 2002.

Two Alaris trainsets passing Aranjuez forming a Madrid Atocha to Valencia service (Bryan Philpott)
0567050

Class 252 dual-voltage electric locomotive with an Alicante-bound Altaria service formed of Talgo stock at Aranjuez (Bryan Philpott)
0567052

Electric locomotives

Class	Wheel arrangement	Output kW	Speed km/h	Weight tonnes	No in service	First built	Builders Mechanical	Electrical
3 kV DC								
250/250.600	C-C	4,600	140/100	124	40	1982	Krauss-Maffei/ CAF/MTM	BBC
251	B-B-B	4,650	140/100	138	30	1980	Mitsubishi/CAF	Westinghouse
269	B-B	3,100	140/80	88	105	1973	Mitsubishi/CAF	Cenemesa
269	B-B	3,100	160/90	88	157	1973	CAF/Macosa	Westinghouse
3 kV DC/25 kV AC								
252*	Bo-Bo	5,600	220/100	90	73	1991	Krauss-Maffei/ ABB/Henschel/ Meinfesa/CAF	Siemens ABB
1.5 kV DC/3 kV DC								
279	B-B	2,700	130/80	80	15	1967	Mitsubishi/CAF	Mitsubishi/ Cenemesa
289	B-B	3,100	130/80	84	28	1967	Mitsubishi/CAF	Mitsubishi/ Cenemesa

* 52 1,668 mm gauge; 21 1,435 mm gauge.

Diesel locomotives

Class	Wheel arrangement	Power kW	Speed km/h	Weight tonnes	No in service	First built	Builders Mechanical	Engine	Transmission
319–200[1]	Co-Co	1,190	120	110	57	1984, 1994	Macosa	GM 567 C	E WESA
319–300	Co-Co	1,372	140	119	40	1991	Meinfesa	GM 645 E	E GM
319–400	Co-Co	1,372	120	116	10	1992	Meinfesa	GM 645 E	E GM
333[2]	Co-Co	1,875	146	120	93	1974	Macosa	GM 645 E 3	E GM
354	B-B	2,340	180	80	3	1982	Krauss-Maffei	2 × Maybach- Mercedes MD 6652	H Maybach- Mercedes
308	Bo-Bo	520	120	64	39	1966	GE/Babcock & Wilcox	Caterpillar D-398	E GE
309	C	515	50	54	20	1986	MTM	MTU-Bazan 6V396TC13	H Voith
310[3]	Bo-Bo	930	114	78	60	1990	Macosa	GM 645 E3	E GM/Indar
311	Bo-Bo	504	90	80	60	1990	MTM	MTU 8V396TC13	E MTM
321[4]	Co-Co	1,250	120	111	21	1965	Alco	GEC 5G-761	E GEC

[1] Class 354 are Talgo locomotives.
[2] Seven are 1,435 mm gauge.
[3] Includes 26 locomotives modernised and renumbered in the Class 333 300 sub-series and six in the 333 400 sub-series.
[4] Four rebuilt for TMD trials.
[5] Maintenance and infrastructure locomotives.

In País Vasco freight tonnage increased to 5.93 million tonnes in 2004, an increase of 8.3 per cent compared to 5.47 million tonnes for 2003. Container traffic accounted for 2.38 million tonnes, and iron and steel products 1.78 million tonnes in 2004. In Guipúzcoa the freight terminals in the vicinity of Irún handled 2.05 million tonnes. In Vizcaya the Santurtzi freight terminal handled 1.56 million tonnes in 2004 and the steel flow from Asturias to La Naval de Sestao was reinstated after a two year interval.

TMD (Teco de Media Distancia)
RENFE's freight unit has been evaluating the potential market for medium-distance (120–300 km) container traffic. It has been conducting trials with a fixed formation of eight container flats topped and tailed by Class 310 locomotives rebuilt for the purpose at RENFE's Malaga workshops. A daily service has operated on the La Coruña–Orense corridor. Trains load to 720 tonnes with a maximum speed of 110 km/hr. If successful, the concept would be implemented nationally.

International traffic
International traffic increased to 1.88 million tonnes (875 million tonne-km) in 2003 compared to 1.76 million tonnes (798 million tonne-km) in 2002.

Rolling stock operating beyond the French border passes through the wheelset-changing facilities at Irún/Hendaye and Port Bou/Cerbère. RENFE has owned a 20 per cent shareholding in Transfesa, the operator of these facilities, since 1995; SNCF has a shareholding of the same size. Wheelset-changing is preferred where less robust commodities such as new cars, car parts, citrus fruit and finished consumer goods are being carried. Cargoes such as wood, cereals and iron and steel products are usually transshipped, as are containers and swapbodies. Some 4,600 wagons belonging to other European railways are permitted to operate in Spain, and private wagon owners inside Spain (such as Transfesa) have 7,600 vehicles suitable for cross-border operation.

In 1998 RENFE, SNCF, logistics company Decoexsa and TMF founded the Cadaler joint venture to manage flows of iron and steel traffic from the south to the north of Europe. The company is based at Port Bou, where it operates handling equipment and 6,000 m² of warehousing. Logistics, reception and sorting of products takes place here, while additional intermodal distribution centres are located at Valencia and Barcelona. A forecast 25,000 tonnes are expected to be dealt with annually at Port Bou.

Intermodal operations
Traffic income amounted to EUR125.10 million in 2003, an increase of 6.6 per cent from EUR1 17.35 million in 2002. Tonnage increased by 3.9 per cent in 2003 to 7.825 million compared to 7.535 million in 2002.

International services
International traffic increased by 1.6 per cent in 2003 compared to 2.72 million tonnes in 2002. However, a 5.3 per cent increase in imported traffic was offset by a 4.0 per cent reduction in exports compared to 2002. Total tonne-km increased by 4.3 per cent in 2003.

Traction and rolling stock
In 2005 RENFE's fleet comprised: 439 electric locomotives, 275 diesel locomotives (including three Class 354 Talgo locomotives), 184 diesel shunting locomotives, 18 Class S100 AVE trainsets, six Class S101 Euromed trainsets, 10 Alaris trainsets, 584 emus and 142 dmus. A total of 235 passenger coaches (including 106 sleeping cars/couchettes) and 18,563 freight vehicles made up the fleet.

Within the large Class 269 electric locomotive fleet various sub-series have been created according to the type of duty performed. Some examples designated Class 269.7 have been formed into permanently coupled pairs for freight duties. Ten Class 289 machines have been similarly adapted. Class 252 includes 21 units equipped

Electric multiple-units and high-speed trainsets

Class	Cars per unit	Motor cars per unit	Motored axles/car	Output/ motor kW	Speed km/h	No in service	First built	Builders Mechanical	Electrical
1.5 kV DC									
442	2	1	4	524	60	5	1976	MTM	BBC
3 kV DC									
432	3	1	4	290	140	12	1971	CAF/Macosa	Mitsubishi/ Westinghouse
440/ 440R	2/3	1	4	290	140	199	1974	CAF/Macosa	Mitsubishi/ Westinghouse
444	3	1	4		140	9	1980	CAF/Macosa	Westinghouse
446	3	2	4	300	100	167	1989	CAF/MTM/ Macosa	Mits ubishi/ Cenemesa
447	3	2	4	300	120	183	1993	CAF/GEC Alsthom	Siemens/ABB
448	3	1	4	290	160	26	1987	CAF/Macosa	Melco/GEC
450	6	2	4	370	140	24	1994	GEC Alsthom/ CAF	GEC Alsthom
464[3]	2/3/4/5	2	–	–	120	14	2003	ALSTOM/CAF	ALSTOM/ Siemens
451	3	1	4	370	140	12	1994	GEC Alsthom/ CAF	GEC Alsthom
470[1]	3	1	4	290	140	56	1981	CAF/Macosa	Mitsubishi/ Cenemesa
490 (Alaris)	3	2	2	500	220	10	1999	ALSTOM, Fiat	ALSTOM, Fiat
S101 (Euromed)	10	2	4	675	220	6	1996	GEC Alsthom	GEC Alsthom
3 kV DC/25 kV AC									
S100 (AVE)	10	2	4	1,100	300	18	1991	GEC Alsthom	GEC Alsthom
S120[2]	4	2	4	500	250	57	2004	CAF/ALSTOM	ALSTOM
–[2]	9	2	4	1,000	350	22	[4]	Bombardier, Talgo	Bombardier
25 kV AC									
S102	12	2	4	1,000	350	16	2003	Bombardier, Talgo	Bombardier
S103	8	4	4	550	330	26		Siemens, CAF	Siemens
S104	4	2	4	500	250	50	2004	ALSTOM, CAF	ALSTOM

[1] Rebuilt from Class 440.
[2] Variable gauge (1,435/1,668 mm).
[3] Eventual fleet to be 248 units.
[4] Delivery due from 2006.

Diesel multiple-units

Class	Cars per unit	Motor cars per unit	Motored axles/car	Power/ motor kW	Speed km/h	No in service	First built	Mechanical	Builders Engine	Transmission
592	2/3	2	2	213	120	45[1]	1981	Macosa/Ateinsa	MAN	H Voith
593	3	2	2	206	120	22	1981	CAF/BWE	Fiat	M Fiat
594	2	2	2	300	160	23[2]	1997	CAF/Adtranz	MAN	H Voith
596	1	1	2	206	120	23[3]	1997	CAF/BWE	Fiat	M Fiat
598	3	2	2	338	160	21	2004[4]	CAF	–	E –

[1] Includes 27 units modernised for 140 km/h running and renumbered in the 592 200 sub-series.
[2] Includes eight units modified in 2001 and renumbered in the 594 100 sub-series.
[3] Rebuilt from Class 593.
[4] Includes units on order.

with standard-gauge bogies for operation on high-speed lines.

Refurbishment of the Class 446 emu fleet (to be followed by Class 447, 450 and 451) began at RENFE Operadora's Valladolid workshops in 2005 and included upgrading suspension and electrical equipment, along with improvements to air conditioning and internal lighting. The units are expected to remain in service until 2014.

Civia emus
The first of the new generation of Class 464 Civia articulated emus built by ALSTOM, CAF and Siemens for suburban commuter services entered commercial service in 2004 and are being introduced throughout the network. With an aluminium alloy modular design, the units are formed of two, three, four or five coaches accommodating 222, 315, 416 or 517 passengers respectively. The budget for the first build of 14 units was EUR54 million. Ultimately 248 units will operate throughout the suburban network. Particular features to benefit passengers include real-time running information via screens in each coach and an accessibility audit which at the design stage took account of the needs of passengers with disabilities.

High-speed trainset orders
The eight-car S103 'Velaro', a derivative of the German ICE 3 capable of 350 km/h, is under construction by Siemens in collaboration with CAF. Both types are capable of 350 km/h. Sixteen units were initially ordered. In 2004 RENFE placed follow-on orders with Bombardier for 44 power heads to form 22 variable-gauge nine-car versions of the S102 trainset (delivery 2006–08) and with Siemens for 10 more S103 trainsets.

A contract for a further 32 four-car 250 km/h units for the new line was awarded to ALSTOM/ CAF. This was for 20 units of Class S104, the first of which entered service between Seville and Córdoba and between Madrid Puerta de Atocha and Puertollano in December 2004 and January 2005 respectively, along with a further 12 variable-gauge 250 km/h (220 km/h when running on 1,668 mm gauge) units (S120). The first of these began trials with the Grandes Líneas (long-distance) business unit at the end of 2004. In February 2004 a supplementary order was placed with ALSTOM/CAF for 30 and 45 examples respectively of these two types of train for delivery in 2006–08 and 2006–09. The trains built under this later contract will feature a more modern exterior styling than the Pendolino-based units of the initial order.

In the case of each of these high-speed train orders, their manufacturers are also responsible for maintenance for a 14-year period.

Tilting dmus
The first of 21 Class 598 three-car tilting diesel-electric multiple-units built by CAF entered service between La Coruña, Santiago and Vigo in December 2004 and between Seville, Granada and Almería in April 2005.

Class 333 rebuilds
A new variant of Class 333 diesel locomotive was created by rebuilding 32 units during 2002–03. The lifespan of these units (which entered service originally in 1974) would be extended by 20–25 years. The joint project involving the former ALSTOM Albuixech factory at Valencia and RENFE's Villaverde Bajo workshops was delivered at a cost of EUR48 million. Twenty-three of the units designated sub-series 333 300 are in service with Mercancías (freight). The remaining eight units form sub-series 333 400 and have been allocated to Grandes Líneas (long distance passenger) since January 2004 to operate Altaria services between Madrid, Murcia and Cartagena and the Madrid–La Coruña Talgo between Medina Del Campo and La Coruña.

The Talgo fleet
The Talgo Pendular, has a passive body-tilting system designed to permit curve negotiation at a

Class 464 Civia four-car suburban emu at Tudela
(Felipe Arando)
0567053

Class 470 emu forming a Vitorio–Madrid regional express service at Medina del Campo
(Felipe Arando)

20 per cent higher speed than the normal limit without discomfort to passengers. The range includes sleeping-car sets with wheelsets adjustable to gauge-change for international services.

Talgo cars are owned by RENFE but maintained by the builder under contract, which also covers Class 354 diesel-hydraulic locomotives, and specifies high-season coach availability of 97 per cent, dropping to 88 per cent in periods of low demand. In service, each trainset carries a Talgo engineer.

The Talgo-Team consortium's prototype electric locomotive, the first with variable-gauge bogies capable of operating on 1,435 mm and 1,668 mm gauge, commenced trials in 2005. The locomotive has been designed as dual-voltage 25 kV AC/3 kV DC with a maximum speed of 260 km/h. The prototype was expected to be the first of a batch of 44 locomotives to be numbered in the 253.001 series.

0567054   **UPDATED**

Modernised and refurbished Class 333.300 diesel locomotive on an intermodal service at Lozoyuela (Bryan Philpott)
0567055

Class 250 electric locomotive an a bulk freight service at Samper de Calanda (Felipe Arando)
0567056

Class 310 TMD (Teco de Media Distancia) diesel locomotive for short- to medium-distance freight work (Bryan Philpott)
***NEW**/1115009*

Valencia Railways (FGV)

Ferrocarrils de la Generalitat Valenciana
Partida de Xirivelleta s/n, E-46014 Valencia
Tel: (+34 96) 397 65 65   Fax: (+34 96) 397 65 80
Web: www.fgv.es
   www.metrovalencia.com

Key personnel

Managing Director: Marisa Gracia Giménez
Directors
   Management: Antonio Carbonell Pastor
   Operations: Manuel Sansano Muñoz
   Exploitation: Vicente Contreras Bórnez
   Finance: Jesus Cerverón Esteban
   Infrastructure: Crescencio Fernández Holgueras
   Marketing: Jorge Beltrán Oliver
   Organisation and Systems: Antonio Ruiz Cano
   Maintenance: Francisco Garcia Sigüenza
   Personnel: Dionisio García Gómez
   Alicante Representative:
      Damián Uclés Fernández

Gauge: 1,000 mm
Route length: 133 km (Metro Valencia); 93 km (Tram Metropolitan Alicante)
Electrification: 113.5 km at 1.5 kV DC

Political background

FGV, a public company set up by the autonomous community of Valencia, succeeded FEVE as operator of Valencia's 1,000 mm gauge railways. In 2004 the regional administration announced investment of €14.8 billion in its Strategic Plan for Infrastructure (embracing road, rail, infrastructure and telecommunications) covering the period 2004 to 2010. The Strategic Plan envisages Metro Valencia increasing its route length from 133 km to around 200 km and Tram Metropolitan Alicante increasing from 93 km to around 170 km by 2010. At the end of 2003 FGV employed 1,304 people, an increase of 41 over the previous year.

Passenger operations

A total of 53.51 million passengers were carried in 2003, an increase of 6.63 per cent over the previous year.

Valencia

Metro Valencia comprises the integrated Lines 1, 3 and 5 (linking suburban surface lines and intensive underground urban services), the Line T4 tramway and two local bus routes. Metro Valencia carried 51 million passengers in 2003, an increase of 9.6 per cent over the previous year. Line 1 carried 17.2 million passengers in 2003 (an increase of 3.18 per cent over 2002); Line 3 carried 22.4 million (a decrease of 4.53 per cent); Line 4 carried 6 million (a decrease of 4.6 per cent) and Line 5 over which services started on 30 April 2003)

FGV Class 3800 LRV on the Alicante tramway at El Campello (Bryan Philpott)   **NEW**/1114958

Electric railcars or multiple-units

Class	Cars per unit	Motor cars per unit	Motored axles/car	Output/ motor kW	Speed km/h	No in service	Builders Mechanical	Electrical
B&W	3	–	–	–	80	10	Babcock & Wilcox	GEE
3700	2	2	1	220	80	40	CAF	ABB
3900	3	2			80	18	GEC Alsthom	GEC Alsthom
3800 Tram	3	–	–	–	65	24	Siemens	Siemens

Diesel railcars or multiple-units

Class	Cars per unit	Motor cars per unit	Motored axles/car	Power/ motor kW	Speed km/h	No in service	First built	Builders Mechanical	Engine	Transmission
2300	2	1	2	210	80	10	1965	MAN	MAN	M Voith DIWA

carried 5.2 million. Three new stations (at Ayora, Amistat and Aragón) were added to the network when services were introduced over Line 5.

Bus services on two routes carried 0.75 million passengers in 2003.

Alicante

FGV operates Tram Metropolitan Alicante services. Since August 2003 a half-hourly service (0550–2150) with a 20 minute journey time has operated from a new terminal at Puerta del Mar for 14 km over the route of the former Alicante–Denia line as far as El Campello. Departures on the return from El Campello connect with FGV services arriving over the non-electrified line from Denia. A total of 330,369 passenger journeys were made between Alicante and El Campello from 15 August to 31 December 2003. The line between the interchange at El Campello and Denia is earmarked for modernisation.

Improvements to existing lines

Valencia

Construction projects include extension of the new Line 5 under the River Turia along the existing RENFE trajectory towards the airport at Manises and beyond to Ribaoja. A new north-south tramway link to be known as T2 and built partly underground, is planned to link Orriols in the north and Natzaret in the southeast of the city. The line will run underground through the city centre, emerging near Xàtiva, from where it will run on the surface towards the flourishing science and arts centre in the southeast of the city. The first phase of construction from Orriols to Torrefiel was budgeted at €20.5 million and provides six new stations. An extension to Line T4 budgeted at €26 million including six new halts will provide links to La Coma, Terramelar and Valterna. A new station at Bailén, between Jesús and Colón, will provide an interchange between Metro Valencia and RENFE's high-speed and suburban services.

Alicante

Investment of €41.1 million for infrastructure improvements on the Alicante–Denia line was provided in 2002. A €500 million project to develop the Alicante tramway is scheduled for completion in 2007. Completion of the Mercado Central–Finca Adoc section of Line 2 was scheduled for 2005, with construction continuing between Mercado and Plaza de Luceros. Investment of €3.5 million in a new station at La Creueta forms part of the project to extend the tramway from El Campello towards Villajoyasa.

FGV Class 3900 emu at Sant Isidre (Bryan Philpott)   **NEW**/1114957

Traction and rolling stock

Metro Valencia operates 50 electric units on Line 1, comprising: 10 Babcock & Wilcox units (1981); 30 Class 3700 first series units (1987); and 10 Class 3700 second series (1990). A further 18 Class 3900 electric units operate services on Line 3. Metro Valencia's 750 V DC tramway, Line T4, is operated by 24 Siemens-Duewag 3800 series three-car tram units. FGV operates 10 Class 2300 dmus between El Campello and Denia and five tram units between Alicante and El Campello.

Orders were placed in 2003 for nine Class 4100 tram units and 10 Class 4200 units.

In early 2005 FGV announced investment of €60 million in a further 10 units to form Class 4300 to be built by Vossloh at Albuixec, along with investment of €7.21 million in the modernisation and renovation of six of its Class 2300 units for completion in 2006.

FGV also operates four Alsthom ex-FEVE diesel locomotives, principally on permanent way traffic and its Lemon Express tourist train.

Track

Rail: 45 and 54 kg/m
Sleepers: Stedef bibloc concrete 1,800 × 260 × 220 mm with Nabla fastenings; monobloc concrete 1,900 × 240 × 130 mm with HM-Vossloh fastenings; timber 1,900 × 240 × 130 mm
Sleeper spacing: 1,000–1,500/km in plain track; 1,250–1,500/km in curves
Max gradient: 2.4%
Max permissible axleload: 15 tonnes

UPDATED

Sri Lanka

Ministry of Transport

PO Box 588, D R Wijewardana Mawatha, Colombo 10
Tel: (+94 1) 68 73 11; 68 72 12
Fax: (+94 1) 69 45 47

Key personnel
Minister: Felix Perera

Sri Lanka Railway (SLR)

PO Box No 355, Colombo 10
Tel: (+94 1) 43 11 77   Fax: (+94 1) 44 64 90
e-mail: rly@visual.lk

Key personnel
General Manager: Priyal de Silva
Additional General Manager (Administration):
  H M Rupasinghe
Additional General Manager (Operations):
  K A Premasiri
Commercial Superintendent: P H Silva
Chief Engineer (Way and Works): C R Vithanage
Chief Signal and Telecommunications Engineer:
  V S Balasubranamiam
Chief Mechanical Engineer: P P Wijesekera
Chief Engineer (Motive Power): B A P Ariyaratne
Operating Superintendent: GRP Chandratillake
Transportation Superintendent (planning):
  S W Munasinghe
Stores Superintendent: S P Samaranayake
Chief Accountant: A K Theiventhren

Gauge: 1,676 mm
Route length: 1,449 km

Poltical background
SLR is a government department functioning under the Ministry of Transport, Highways and Aviation. The government intends to reconstitute the management structure of SLR, giving it more authority and flexibility in carrying outits maintenance and development programmes. Considerable disruption has been caused to SLR operations by the so-called liberation tigers of

Tamil Eelam, which has been outlawed by the government as a terrorist organisation.

In 1998 a business unit was established to open up the network to private-sector freight operators, who would pay access charges and be responsible for maintenance of infrastructure.

The Railway Management Council was established in 2001 to pursue the objectives of a restructuring programme. Amendments to existing legislation were being considered to facilitate public/private partnerships in the operation of the railway.

Organisation
The SLR network is centred on Colombo, from where nine lines radiate north along the coast to Periyanagavilu, south to Matara and east to the Central Highlands. The 339 km Northern Line branches off the Colombo-Badulla Main Line (290 km) at Polgahawela and crosses north-central and northern provinces to the city of Kankesanturai, although train services beyond Vavuniya have been suspended because of the civil war in the north.

In June 1997 SLR employed 18,070 staff.

The installation of a management information system and a computerised ticketing system feature in SLR's 2001–05 investment programme.

Finances (SLRs million)

Revenues	1999	2000
Passengers	678.5	740.9
Freight	209.6	133.3
Parcels	39.4	32.1
Other	110.9	108.1
Total	1,038.4	1,014.5

Expenditure	1999	2000
Staff/personnel	1,186.6	1,777.0
Materials and stores	78.6	86.5
Fuel/lubricants	742.9	538.9
Pensions/other	182.5	283.4
Total	2,885.6	2,685.8

Traffic (million)	1999	2000
Passenger journeys	83.2	85.9
Passenger-km	3,176	3,430
Freight tonnes	1.160	1.194
Freight tonne-km	94.5	88.2

Passenger operations
Besides commuter services around Colombo and long-distance services on the Main Line and the Coast Line to Matara, intercity trains also operate between other major cities, such as Colombo-Kandy and Colombo-Galle. In all, 170 stations and 155 stops are served.

SLR passenger-km have increased every year since 1991, but in recent years service development has been hampered by inadequate investment. Daily passenger services operated include: 260 Colombo suburban; 35 long-distance; and 21 local.

Freight operations
Freight traffic has suffered from lack of motive power, as the few locomotives which are available are often taken to power passenger trains. In 2001, SLR operated 20 daily freight services, with petroleum products, cement and foodstuff the principal commodities carried.

In early 2001, the Freight Business Unit (FBU) was established to lead a revival in rail freight traffic. An open access policy was also being purchased to facilitate private sector participation with the aim of increasing rail's marketshare from 10 to 50 per cent.

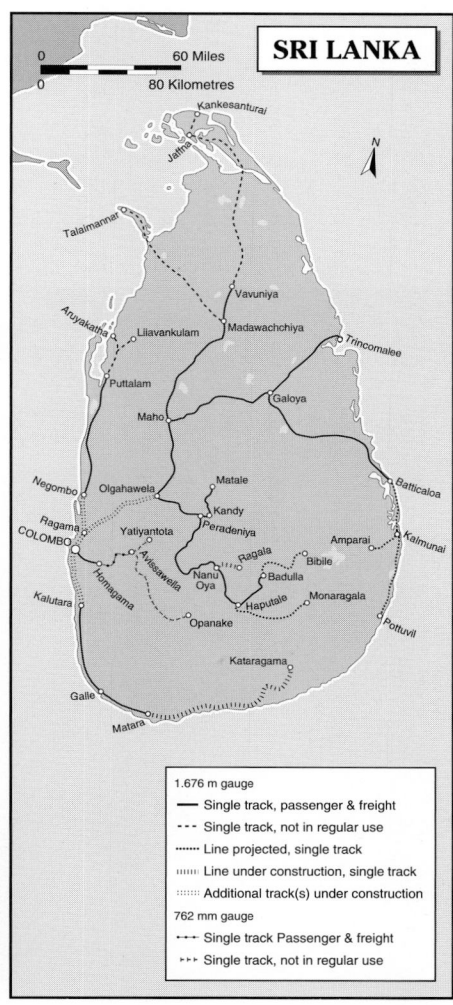

0023136

Sifang-built dmu, one of 15 supplied from 1999   0122770

Intermodal operations

Having performed poorly in recent years, container traffic was to be revived during the period 2001–06. A new facility, the Internal Container Depot, was to be established near Ragama station, 13 km north of Colombo and the line linking the SLR network with Colombo port was to be rehabilitated in 2002. Plans were being pursued to establish container handling yards closer to main export processing zones.

Improvements to existing lines

A continuing track rehabilitation programme underway in 2001 aimed to allow 100 km/h running by 2005. In 2001, negotiations were underway for financial assistance for this programme from Japanese banking institutions. To some extent track maintenance has been mechanised in recent years.

Provision of a third track on the Main Line between Colombo and Ragama has reduced congestion at peak times. Track-doubling between Wadduwa and Aluthgama (17 km) on the Coast Line and between Ragama and Negombo (14 km) on the Puttalam Line was in progress in 2001. Construction of five bridges in connection with further track-doubling was scheduled to commence in 2002, and an additional eight minor bridges were due to be rehabilitated during the period 2002–04.

Construction of a Mass Rapid Transit for the outer Colombo area, upgrading the Colombo–Katunyaka Airport service, extension of the network to link industrial and agricultural zones and the electrification of Colombo suburban services were all the subject of continuing feasibility studies in 2001.

Traction and rolling stock

At the beginning of 2001 SLR operated 129 diesel locomotives, 53 diesel railcars, 1,128 passenger cars and 2,000 freight wagons on 1,676 mm gauge.

During the period 1994–2001, SLR strengthened its fleet by 30 diesel locomotives and 15 dmus for suburban services.

Signalling and telecommunications

Rehabilitation or replacement of the CTC system installed in 1960 has been identified as a major project. In addition, in 2001 SLR planned to install level crossing protection systems at 244 locations.

Phase 2 of the installation by NMA Signalling of colour light signalling on the Coast Line was scheduled for completion by 2003.

Type of coupler in standard use: Auto and screw, ARR standard
Type of braking: Vacuum

Track

Rail: 43 and 39 kg/m (382 km and 866 km)
Crossties (sleepers): Wood (1,248 km) and concrete, spaced 1,550/km; installation of concrete started in 1996
Fastenings: Dog and elastic spikes, Pandrol clips
Min curvature radius: 100.6 m
Max gradient: 2.7%
Max axleload: 16.5 tonnes

Diesel locomotives

Class	Wheel arrangement	Power kW	Speed km/h	Weight tonnes	No in service	First built	Mechanical	Builders Engine	Transmission
M2	A1A-A1A	1,063	80	79	12	1954	GM	GM 12-567C	E GM
M2C	Bo-Bo	1,063	80	79	2	1961	GM	GM 12-567C	E GM
M2D	A1A-A1A	977	80	79	2	1966	GM	GM 12-567E	E GM
M4	Co-Co	1,305	80	97.68	14	1975	MLW	Alco 12-25 103	E Generator GT 581PJ1
M5	Bo-Bo	1,175	80	66	6	1975	Hitachi	MTC-Ikegai 12V652TD11	E Alternator H1-503-Bb
M5A	Bo-Bo	1,175	82	66	1	1993[1]	Hitachi	MTU 396	E Alternator
M5B	Bo-Bo	1,175	82	66	4	1997[1]	Hitachi	Caterpillar 3516	E Alternator
M6	A1A-A1A	1,230	80	85.5	14	1980	Henschel	GM 12-645E	E Generator D 25L
M7	Bo-Bo	746	80	67	15	1981	Brush	GM 08-645E	E BA 1004A/BAE/ 507A
M8	Co-Co	2,400	120	109.8	6	1996	DLW	251-B16-DLW	E Alco
M9	Co-Co	1,340	80	100	10	2000	Alstom	Ruston 12 BK	E Alstom
W1	B-B	857	80	60.55	10	1969	Henschel	Caterpillar 3312	H MTU-Mekydro K 102-1016 PS
W2	B-B	1,173	80	65.3	3	1969	Lokomotivbau	Paxman 16YJXL	H MTU-Mekydro K 182 BU
W3[2]	B-B	857	80	65.3	6	1979	Henschel	Caterpillar	H Voith
Y	0-6-0	410	45	28	20	1969	Hunslet	Rolls-Royce DV 8T	H Rolls-Royce CF 13800
P1		98	32	20.12	2	1950	Hunslet	Ruston Hornsby 6 VPH	H Hunslet axle drive
N1	1-C-1	367	32	41.17	2	1953	Fried Krupp	Deutz 33	H Krupp LIB
N2	B-B	447	32	–	2	1973	Kawasaki	Detroit Diesel 16V71K	H Niigata DBG-138

[1] Date of re-engining.
[2] Class W1 re-engined.

Diesel railcars or multiple-units

Class	Cars per unit	Motor cars per unit	Motored axles/car	Power/ motor kW	Speed km/h	No in service	First built	Mechanical	Builders Engine	Transmission
S3	4	1	4	656	80	6	1959	MAN	MAN L 12V18/21	H Maybach K 104 U
S5	4	2	4	577	80	3	1970	Hitachi	Paxman 8Y JXL	H MTU-Mekydro K 102 UB/55
S6	4	1	4	869	80	8	1975	Hitachi	Paxman 12Y JXL	H MTU-Mekydro K 102 UB
S7	4	1	4	760	80	8	1977	Hitachi	Cummins KTA-2300L	H Hitachi DW 2A
S8	5	1	4	1,005	100	20	1990	Hitachi/ Hyundai	MTU 12V 396 TC13	H
S9	5	1	4	1,015	100	15	1999	Sifang	MTU 12V 396 TC14	E

Hitachi-built dmu at Fort
(Ralph Oakes-Garnett)
0023708

Sudan

Ministry of Transport

PO Box 1300, Khartoum

Key personnel
Minister: Al-Sammani al Cheikh al-Waseilah

UPDATED

Sudan Railways (SRC)

PO Box 1812, Khartoum
Tel: (+249 11) 77 40 09   Fax: (+249 11) 77 06 52

Key personnel
General Manager:
    Eng Omer Mohamed Mohamed Noor
Deputy General Manager, Infrastructure:
    Eng Abdel Rahim M Abdel Rahim
Deputy General Manager, Co-ordination and
    Studies: Gaafar Hub Alla Saeed
Deputy General Manager, Operations:
    Eng Samel Ahmed Samel
Deputy General Manager, Technical:
    Eng Majdi Mohamed Bilal
Deputy General Manager, Finance:
    Abdel Haleem Ahmedi Taha
Regional Managers
    Northern: Mohammed Hamid
    Eastern: Musa Mohammed Musa
    Central: Eng Ali El Tayeb
    Southern: Eng Musa El Goam
    Western: Haydar Ali Mohammed
Managers:
    Traction and Rolling Stock:
        Eng Abdel Majed Basheer
    Tracks and Construction: Eng Bushara Gantoor
    Operation and Traffic: Abdel Wahab Suliman
    Signalling and Telecommunication:
        Adam Ibrahim Shurbaike
    Finance: Mohammed Abdel Rahman Ibrahim
    Personnel: Mahjoob Hassan Hiba
    Planning and Statistics:
        Elsheikh Abdel Mutalab Omer
    Safety: Eng Hassan Bakry Hamad
    Police: Brig Mogadam Habila Abu Zeco
    Internal Auditing: Hayat M Abdel Rahim
    Purchasing and Stores: Ibrahim Bashri Ali
    Legal Administration: Jamal Almobarak Omer
    Purchasing: Ibrahim Bashari Ali
    Public Relations and Executive Affairs:
        Abdel Rahman M Ahmed
    Co-operative Corporation:
        Mirghani Mohammed Ahmed

Gauge: 1,067 mm
Route length: 4,578 km

Political background
The single-track railway used to be the main transport mode in Sudan, but from the late 1970s onwards it has been in serious decline. Some rehabilitation was undertaken with foreign aid during the 1980s famine, but lack of spares and consequent poor availability of motive power severely hampered operations in the early 1990s; the situation eased when 10 new diesel locomotives arrived in January 1995. Operations are also affected by axleload constraints imposed by varying rail weights in use and by the generally poor condition of track. At the beginning of 2005 1,248 km of the network was non-operational, including the 802 km section from Haiya Junction to Sennar via Kassala and Gedaref.

The main line linking the capital Khartoum to Port Sudan (787 km) carries over two-thirds of SRC's traffic and is the only section of the SRC network laid with 90 lb rail.

In 1999 the government announced its intention to privatise freight and passenger operations and some private sector companies were active in 2004, including Al-Bazim (freight only), Um Gamala (freight), Shieku (freight and passenger)

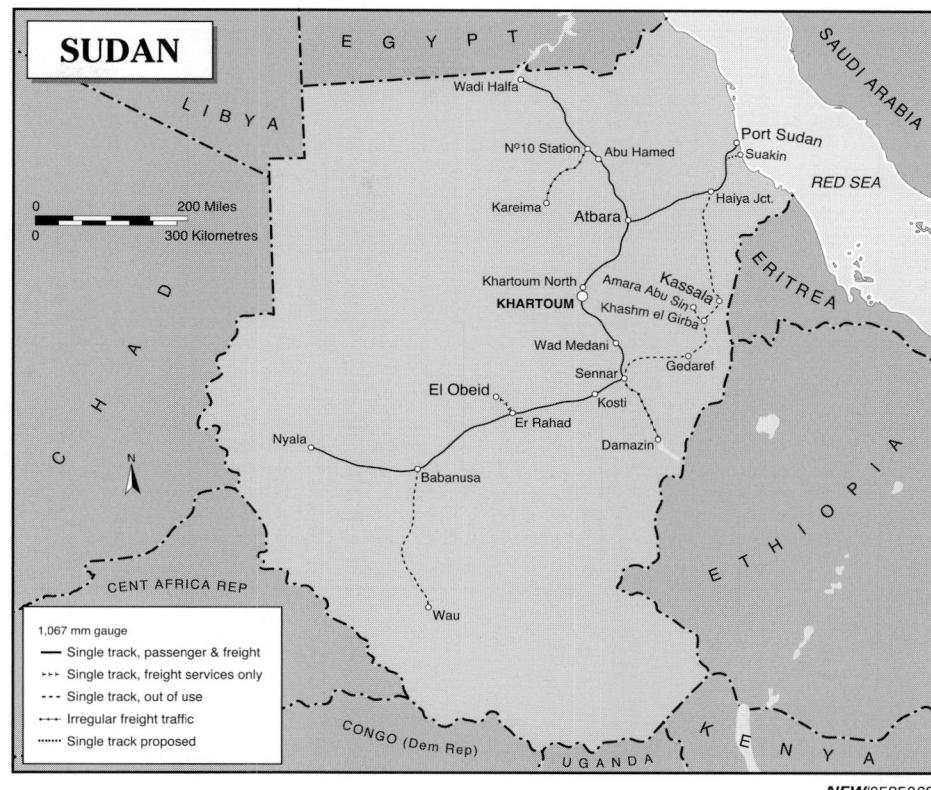

NEW/0585069

and Sudan Free Zones Rail Transport (freight). There was also an SRC-owned company, Sekakion, operating some freight services between Wadi Halfa and Port Sudan and passenger traffic from both Port Sudan and Wadi Halfa to Khartoum and Nyala. All these operate with dedicated locomotives and rolling stock, but these are mostly owned by SRC.

Organisation
At the end of 2003 SRC employed 12,301 staff, down from 13,698 at the end of 2001.

Traffic	2001	2002	2003
Passenger journeys	157,957	145,216	109,216
Passenger-km (million)	78	73	52
Freight tonnes (million)	1.25	1.28	1.27
Freight tonne-km (million)	903	–	889

Passenger operations
There is a weekly train from the capital Khartoum to Wadi Halfa in the north of the country, Port Sudan on the Red Sea and Nyala in Southern Darfur province. Political and operating difficulties have continued to disrupt the network, with many trains cancelled for lack of motive power and substantial portions of the network closed, including the line from Babanusa to Wau. These have all contributed to a decline in passenger traffic.

Freight operations
Port Sudan on the Red Sea and Wadi Halfa on the Nile are both key freight transfer points on the Sudanese network. In recent times, traffic has been boosted by the movement of humanitarian supplies into the Sudanese interior, with the World Food Programme organisation employing SRC and its private sector partners to move aid.

In 1996 crude oil from a new find near El Mujlad in western Sudan began flowing by rail to a refinery at El Obeid; the distance travelled is 466 km and quantities range from 2,000 to 5,000 tonnes a day. Around 1,750 tonnes a day of fuel oil is being transported from the refinery to Khartoum and other towns. More than US$30 million had to be invested in the network for this traffic, which amongst other things covered rehabilitation of 10 locomotives, track repairs and provision of workshop facilities.

New lines
In 2004–05 studies endorsed by the governments of Sudan and Kenya were in progress into a possible independently financed railway linking southern Sudan and Mombasa.

Studies commissioned in 2000 into establishing rail links with the Central African Republic, Chad and Ethiopia had not resulted in any moves to implement these schemes by early 2005.

Improvements to existing lines
In 2003 a Memorandum of Understanding was signed by SRC and a Malaysian turnkey contractor, Lankhorst Pancabumi Contractors Sdn Bhd, covering the double-tracking of the Port Sudan–Khartoum–Sennar line (1,084 km). No moves to implement the scheme had been reported by the end of 2004.

In 2004 work commenced on upgrading sections of the Babanusa–Nyala line, replacing 50 lb rail with 90 lb. Rehabilitation of the Babanusa–Wau line was also planned.

Traction and rolling stock
At the end of 2004 the SRC traction fleet comprised 111 main line diesel locomotives, although only 43 of these were serviceable. There were also 41 shunting locomotives (10 serviceable). By mid-2003 Spoornet, South Africa, had supplied 16 main line diesel locomotives to SRC and was to refurbish 35 of the existing fleet. Also on the stock list were 196 passenger coaches and 4,918 freight wagons, 2,520 of which were out of use.

The Indian consultancy and engineering company RITES secured a contract in 2004 covering the supply to SRC of new diesel locomotives and the refurbishment in Indian Railways workshops of machines from the existing fleet.

In 1996 SRC put into service 200 tank wagons leased from Spoornet of South Africa on the new oil flow from El Mujlad to El Obeid; more than half SRC's 600 tank wagons are employed on this flow.

In 1998 SRC ordered 500 freight wagons and 600 bogies from Wagon Pars of Iran. By March 2004 180 of these had been delivered.

Signalling and telecommunications
Mechanical lower quadrant signalling is used throughout the system with absolute block working between stations. In some cases both points and

Diesel locomotives

Class	Wheel arrangement	Power kW	Weight tonnes	No in service	First built	Mechanical	Builder Engine	Transmission
600	C	335	48	22	1968	Kawasaki	–	H Kawasaki
700	C	375	–	19	1975	Henschel	–	H Voith
1000	Co-Co	1,380	99	8	1960	Vulcan	EE 12CSVT	E EE
1500	A1A-A1A	1,120	78	5	1990	Hitachi	–	E Hitachi
1601	Co-Co	1,230	53	10	1981	Henschel	–	H Voith
1700	Co-Co	1,230	79.2	9	1975	GE	GE 7FDL8	E GE
1710	Co-Co	1,230	79.2	9	1984	GE	GE 7FDL8	E GE
1750	A1A-A1A	1,360	79.2	6	1991	GE	GE 7FDL8	E GE
1800	Co-Co	1,715	na	18	1975	GE	GE 7FDL12	E GE
1850	Co-Co	2,000	na	10	1985	GE	GE 7FDL12	E GE
1860	Co-Co	2,000	na	6	1995	GE	GE 7FDL12	E GE
1900	Co-Co	1,865	76	20	1975	Henschel	EMD 16-645-E	E EMD
1950	Co-Co	1,790	94.1	10	1981	EMD	EMD 16-645-E	E EMD

signals are operated from signalboxes, but in others points are hand-operated individually or from ground frames. The primary communication link is a lineside open pole route, over which SRC serves block instruments and block telephone, selective ringing telephones and a station-to-station telephone and telegraph system. This open route has been vulnerable to theft and destruction.

SRC has an HF radio link for voice communication between major centres throughout the rail network. Now it is considering introduction of radio signalling and solar power.

A central control point at Atbara is responsible for a dispatching system for the Khartoum-Port Sudan main line. Station-to-station VHF radio and digital telecommunications equipment is used in the Atbara area and a teleprinter over a leased line is used for communications between major stations.

Track
Standard rail: Flat bottom, 24.7 kg/m in lengths of 9.14 m and 37.2 kg/m in lengths of 10.97 m; 45 kg/m in lengths of 10.97 and 21 m; joined by fishplates and bolts
Crossties (sleepers): Steel; and wood impregnated under pressure in mixture of creosote and oil (1:1). Concrete used in a few cases as an experiment
Max curvature: Main line 4.5°
Max gradient: 0.66%, except on section in Red Sea Hills between Summit and Port Sudan 1%. Gradient compensation for curves 0.04% per 1° curvature
Max axleload: 75 lb track: 16½ tonnes; 50 lb track: 12½ tonnes
Max speed: 50 lb track: 50 km/h; 75 lb track: 60 km/h
Max altitude: 918.5 m at Summit station on Port Sudan line

UPDATED

Swaziland

Ministry of Works & Communications

PO Box 58, Mbabane
Tel: (+268) 423 21   Fax: (+268) 423 64

Key personnel
Minister: E Magagula

Swaziland Railway (SR)

PO Box 475, Johnstone Street, Mbabane
Tel: (+268) 424 86   Fax: (+268) 450 09

Key personnel
Chairman: L Sithebe
Chief Executive Officer: G J Mahlalela
Directors
    Financial: U Makhubu
    Operating: S Z Ngubane
    Human Resources and Development: M B Mabuza
    Engineering: T Ndlovu

Gauge: 1,067 mm
Route length: 301 km

Political background
In October 2001, the Swaziland government approved the concessioning of SR as a single entity as a public-private partnership with strong local participation.

Organisation
Swaziland's railway was completed in 1964. The main route is from Ka Dake to the Mozambique border at Goba, with lines from Phuzumoya to Lavumisa/Golela and Mpaka to Komatipoort.

A 120 km link from Mpaka to Komatipoort in the eastern Transvaal was opened in 1986; 58 km is in Swaziland and 62 km in South Africa. The line provides a through north-south line to South Africa via the Mpaka-Phuzumoya section of the Ka Dake-Mlawula line and the 95 km Phuzumoya-Golela southern link.

Passenger operations
The twice-weekly 'Trans-Lubombo' service, operated by Spoornet between Durban, Mpaka and Maputo, was discontinued in January 2001. This left tourist trains from South Africa and private charters as the only passenger services using the Swaziland network.

Freight operations
Annual traffic totals around 4 million tonnes, 700 million tonne-km, and includes sugar industry products, minerals and ores from eastern Transvaal, phosphoric acid, fruit, timber and containers.

A major traffic source is Matsapha Dry Port, an inland clearance terminal where export containers are loaded. The success of this facility has led SR to develop plans for a second terminal at Mpaka.

New lines
In 1997 SR unveiled plans to construct a western link with the South African system, from Matsapha via Siphoco to Lothair, terminus of a Spoornet branch from Ermelo. The line, which SR is developing with South African partners, would provide a direct link to Johannesburg and cut the distance from Matsapha by 200 km.

Improvements to existing lines
In 1998 rehabilitation was in progress on SR's west-east corridor, from Matsapha to Siweni and the border with Mozambique.

Traction and rolling stock
Traction is provided by diesel locomotives leased from South Africa's Spoornet; there are three passenger cars and 420 wagons.

Track

	Kadake-Goba	Phuzumoya-Golela
Rail	40 kg/m	48 kg/m (sidings 30 kg/m)
Sleepers	hardwood	concrete
Thickness	127 mm	200 mm
Spacing	814 mm	700 mm
Fastenings	sole plates and coachscrews	Fist BBR
Welded rail	126 km	92 km

Sweden

Ministry for Industry, Employment & Communication

Jakobsgatan 26, SE-103 33 Stockholm
Tel: (+46 8) 405 10 00   Fax: (+46 8) 411 36 16
e-mail: registrator@industry.ministry.se
Web: http://www.sweden.gov.se

Key personnel
Minister: Ulrica Messing

Political background
Separation of infrastructure and operations
In 1988 the state assumed responsibility for upkeep and development of the railway's infrastructure. For the first time in Europe, management of infrastructure and operations was separated, infrastructure becoming the responsibility of Banverket (BV), the National Rail Administration.

Under legislation passed by the Swedish parliament in 1998, track access charges were reduced substantially to improve rail's ability to compete with road transport. The previous charging method has been replaced with a system of variable access fees based on socioeconomic marginal costs.

The 1,053 km Inland Railway from Mora to Gøllivare is a special case. Under a 20-year agreement concluded in 1993, it and three connecting branches were transferred to the control of Inlandsbanan AB (IBAB), a consortium of local authorities along the route.

With effect from 1 January 2001, the business of the former Swedish State Railways (SJ) was split up into six independent companies: SJ AB (passenger services), Green Cargo AB (freight services), Jernhusen AB (real estate), TraffiCare AB (train cleaning and station services), EuroMaint AB (mechanical engineering and maintenance) and Unigrid AB (information technology services). TraffiCare and Unigrid were privatised in 2001.

Phased deregulation
Access to the rail network was liberalised with effect from 1 July 1996. Although SJ AB retains its monopoly on unsubsidised long-distance passenger services, regional transport administrations and the state can now contract

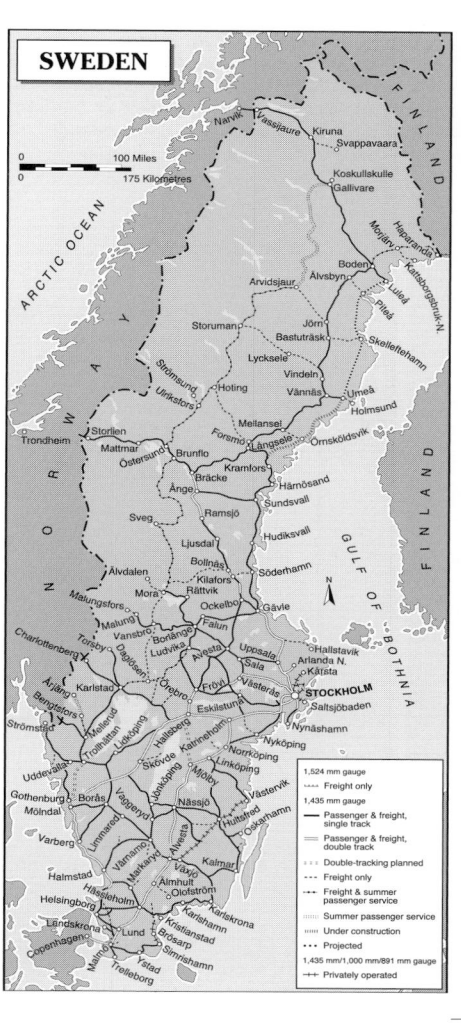

SWEDEN

other operators for subsidised services on any line.

For freight traffic, any licensed operator has the right to run services over any part of the network, though established traffic has priority in track allocation. Responsibility for track allocation and traffic control rests with Banverket.

Further legislation passed in 1998 aims to secure non-discriminatory access to all parts of the rail network by transferring responsibility for industrial sidings, platform structures and passenger information systems to Banverket.

Responsibility for procuring commercially unviable long-distance passenger services was transferred to a new agency, Rikstrafiken, from 1 January 2000.

New legislation that came into force on 1 April 2002 abolished the previous distinction between trunk and regional lines. As a consequence, SJ AB now has operating rights for passenger traffic on all lines belonging to the state railway infrastructure, but no obligation to run services.

SJ AB does not run services that are not commercially viable. These are procured by the state, via Rikstrafiken, and by the 24 regional transport administrations (länstrafikhuvudmän).

0580982

A-Banan Projekt AB

Gamla Brogatan 27, SE-111 20 Stockholm
Tel: (+46 8) 797 98 50   Fax: (+46 8) 797 98 51
e-mail: a-banan@a-banan.lfv.se

Key personnel
Managing Director: Kjell Sundberg

Gauge: 1,435 mm
Route length: 40 km
Electrification: 40 km at 15 kV 16^2/$_3$ Hz AC

Organisation
Established by the Swedish government, A-Banan Projekt AB owns the railway linking central Stockholm with Arlanda Airport. The railway was built by the Arlanda Link Consortium comprising ALSTOM, Mowlem, NCC and Vattenfall. A 40-year concession from 1999 to operate services is held by A-Train AB, which runs trains under the Arlanda Express brand name.

Arlanda Express

A-Train AB
PO Box 130, SE-101 22 Stockholm
Tel: (+46 8) 58 88 90 00   Fax: (+46 8) 58 88 90 01
e-mail: arlandaexpress@atrain.se
Web: http://www.arlandaexpress.com

Key personnel
Chief Executive: Göran Lundgren
Finance Director: Göran Karlsson
Project Director: Torsten Bjuggren
Operations Director: Roman Willenfelt
Marketing Director: Tomas Kreij
Public Relations Manager: Inger Fjordgren
Sales Manager: Maria Bohman

Background
Formerly owned by the Arlanda Link Consortium comprising ALSTOM, Mowlem, Vattenfall and NCC, A-Train AB was sold in 2003 to Macquarie Bank of Australia.

Passenger operations
The direct line between Stockholm Central station and three underground stations at Arlanda airport,

40 km to the north, opened in November 1999 and is owned by A-Banan Projekt AB, a company established by the Swedish government. The link provides travellers with a dedicated airport shuttle (Arlanda Express), as well as allowing regional and long-distance passenger trains to serve the airport terminal.

Under its contract with the government A-Train AB has the right to operate the shuttle service and retain ticket revenues until 2040.

The Arlanda Express service has a maximum speed of 200 km/h. The journey takes 20 minutes and trains depart every 15 minutes, increasing to every 10 minutes at peak times. Check-in facilities

Arlanda Express trainset at Stockholm Central (Quintus Vosman)   0573225

For details of the latest updates to *Jane's World Railways* online and to discover the additional information available exclusively to online subscribers please visit
jwr.janes.com

for air passengers are provided at Stockholm Central station, where luggage can be tagged for collection from designated drop-off points on arrival at the airport.

Platform 1 at Stockholm Central has been converted into a terminal for the Arlanda Express with its own entrances. Taxis and cars can stop outside these entrances and the nearest metro station is 50 metres away.

Three separate rock tunnels have been blasted beneath the airport area. Two of these are single-track tunnels for the Arlanda Express, with two stations serving the airport's four terminals. The third tunnel is double-track and is used by regional and long-distance trains, with a single station beneath the airport's central service area, Sky City. Lifts and escalators provide direct access from the stations to the various terminals.

Traction and rolling stock

The seven airport shuttle trains, built in the UK by ALSTOM, are four-car emus seating 190 passengers. Car floors are entirely at platform level and there is special accommodation for wheelchairs and accompanying passengers.

Banverket

See Swedish National Rail Administration, below.

BK Tåg AB

PO Box 28, SE-571 21 Nässjö
Tel: (+46 380) 55 44 00   Fax: (+46 380) 55 44 44
e-mail: info@bktag.se
Web: http://www.bktag.se

Key personnel

Managing Director: Rolf Torwald
Deputy Managing Director: Tommy Nilsson

Passenger operations

BK Tåg has been operating in this market since 1990 and currently holds contracts with regional transport administrations to run services on the following lines: Hallsberg–Lidköping–Herrljunga in western Sweden, Ystad–Simrishamn in the far south of the country and a group of routes radiating from Nässjö in the southern province of Småland.

In December 1998 a consortium, in which BK Tåg has a 10 per cent stake, won a five-year contract to operate the suburban network in Stockholm from 1 January 2000. The other partner in the consortium is the French company Keolis. A new company, Citypendeln (qv), has been set up to operate this network, which covers 185 route-km and carries over 60 million passengers annually.

In 2001, BK Tåg won a renewed contract to operate a number of routes in the counties of Halland, Jönköping and Småland in southeast Sweden. These include the Jönköping–Värnamo–Alvesta–Växjö line. The five-year contract commenced in August 2002.

Freight operations

In March 1998, BK Tåg began operating container traffic between Karlstad and Gothenburg, in conjunction with the Port of Gothenburg. In addition, the company now transports containers from the Absolut Vodka plant at Åhus to Gothenburg, for onward export to the USA.

Since 1996 the subsidiary company has operated timber traffic over the southern section of the Inlandsbanan between Östersund and Mora.

In March 2000, BK Tåg acquired the business of BSM Järnväg AB, and with it the running of daily block trains carrying liquefied petroleum gas between the Neste Gas terminal at Timrå (near Sundsvall) and the SSAB steelworks at Borlänge.

Class Y1 diesel railcar operated by BK Tåg AB for Jönköpings Länstrafik AB (JLT), photographed at Jönköping (Edward Barnes)   0114221

BK Tåg Class TMX diesel-electric locomotive, acquired from Danish State Railways, with a freight service at Mellerud (Edward Barnes)   0524943

Traction and rolling stock

BK Tåg AB has a fleet of 11 diesel locomotives. Two Class Da electric locomotives are hired from Shortline Väst AB (qv) for the Karlstad–Gothenburg traffic. Haulage for Inlandsbanan timber trains is hired from Inlandsbanan AB (qv). The company owns six Class Y1/YF1 diesel railcars for passenger services to supplement the

rolling stock provided by the regional transport administrations. The company has 65 freight wagons.

Jönköpings Länstrafik has ordered four two-car and six three-car Itino dmus from Bombardier Transportation, for delivery by mid-2003. These will be operated by BK Tåg on the Jönköping–Värnamo and Nässjö–Halmstad lines.

Botniabanan AB

Strandgatan 7, SE-891 33 Örnsköldsvik
Tel: (+46 660) 29 49 00   Fax: (+46 660) 29 49 10
e-mail: info@botniabanan.se
Web: www.botniabanan.se

Key personnel

Chief Executive Officer: Lennart Westberg
Project Manager, Railway Systems:
  Håkan Hellqvist
Communications Manager: Jan Bergman

Political background

Botniabanan AB was founded by the Swedish government (91 per cent) and local authorities (9 per cent) to finance, design and construct a new railway serving the Norrland coast, linking Nyland and Umeå. This followed parliamentary approval in 1997 to build the line, which is intended to provide better communications and improve the economy of a region currently poorly served by rail. The Swedish National Rail Administration (Banverket) is responsible for preliminary evaluation, planning,

design planning and permits. On completion of construction of the line, it will be rented to Banverket, which will undertake operations and maintenance for a 25-year period. The cost of construction is estimated at SKr13.2 billion at January 2005 prices, making the Botniabanan Sweden's largest rail project in modern times. Financing will be provided by loans through the National Debt Office and by possible contributions from the European Union.

New lines

The 190 km Botniabanan (Bothnia Line) will run from a junction with the Sundsvall–Sollefteå Ådalsbanan north of Nyland via Örnsköldsvik, Husum and Nordmaling to Umeå. Designed for possible eventual 250 km/h running, the standard gauge line will be single-track with passing loops and will be electrified at the Swedish standard 15 kV 16²⁄₃ Hz AC system. Initial maximum operating speed will be 200 km/h. Maximum permitted axleload will be 25 tonnes, with a maximum gross train weight of 1,600 tonnes.

The line will feature 15 tunnels totalling 25 km, one of 6 km and one of 5 km, and some 140 bridges, including one of almost 2 km and two of

1 km. The heaviest engineering works will be on the southernmost section between Nyland and Örnsköldsvik. Seven stations are to be constructed at: Örnsköldsvik North; Örnsköldsvik Central; Husum; Nordmaling; Hörnefors; Umeå East; and Umeå Central. These will be closely integrated with other transport modes.

The radio-based ERTMS train control system is to be provided on the Botniabanan, the first planned installation of this technology in Sweden. All road-rail crossings will feature grade separation. GSM-R radio communications equipment will be provided together with pulse code modulation (PCM) transmission technology and a fibre optic telecommunications network.

Work on the Örnsköldsvik–Husum section started in 1999. Construction of the central Örnsköldsvik and Nyland–Örnsköldsvik sections was due to begin in 2003. Work on the Husum–Nordmaling section began in 2004. Construction work between Husum and Umeå is planned to begin during 2005. The entire line is due to be commissioned in 2008.

UPDATED

Citypendeln Sverige AB

Varuvägen 34, SE-125 30 Älvsjö
Tel: (+46 8) 762 27 00   Fax: (+46 8) 762 27 75
e-mail: info@citypendeln.se
Web: http://www.citypendeln.se

Key personnel
Chairman: Tord Hult

Passenger operations
Under a five-year contract with the regional transport administration, Storstockholms Lokaltrafik (SL),

Citypendeln took over the running of trains and stations on the Greater Stockholm suburban network from 6 January 2000. These services were previously provided by Swedish State Railways (SJ).

The majority shareholder in Citypendeln, with a 90 per cent stake, is the French transport group Keolis. The remaining 10 per cent of shares are held by Swedish rail operator BK Tåg AB (qv).

The business has a turnover of €93 million, a workforce of 900, and transported 61 million passengers in its first year of trading. As well as train operations and maintenance, Citypendeln is responsible for managing 50 suburban stations.

Services operate on two lines, totalling 185 route-km: a northwest-southeast line from Bålsta via the city to Västerhaninge and Nynäshamn; and a northeast-southwest line from Märsta via the city to Södertälje and Gnesta. Both lines run in parallel through the city centre.

Traction and rolling stock
Services are operated with a fleet of around 150 Class X1 and X10 emus owned by SL.

Citytunneln

Citytunnelprojektet
Lilla Nygatan 7, Box 4012, SE-203 11 Malmö, Sweden
Tel: (+46 40) 32 14 00   Fax: (+46 40) 32 15 00
e-mail: info@citytunneln.com
Web: www.citytunneln.com

Key personnel
Executive Project Director and Chief Executive Officer: Örjan Larsson
Project Director, Railway: Patrik Magnusson
Project Director, Malmö C: Rolf Dahl
Project Director, Tunnels: Michael Myhré
Manager, Public Relations: Anders Mellberg

Political background
The Citytunnel scheme is a Swedish National Rail Administration (Banverket) project to plan, construct and commission improved rail links to and through Malmö and to the Öresund Fixed Link with Denmark. It is being financed by Banverket, Swedish State Railways (SJ), the City of Malmö, the Skåne (Scania) regional authority and the EU. The project's aims include increasing the competitiveness of public transport by rail in the Skåne region, improving integration in the Öresund region, increasing the competitiveness of the national rail network, reducing environmental problems associated with the existing 'Continental' line to Trelleborg, boosting regional development and increasing the role of Malmö as a regional centre.

Government approval of the scheme was given in 2003 subject to environmental endorsement, given in 2004, and construction started in March 2005. Commissioning is scheduled for 2011. The cost of the project is estimated at SKr8.8 billion at 2001 prices.

New lines
The Citytunnel project provides for construction of 17 km of electrified main line railway, comprising a

Malmö, SWEDEN
Citytunneln Project

To Lund
Malmö Central
Triangeln
To Copenhagen
Öresund Fixed Link
Hyllie
To Ystad
To Trelleborg

New surface line
New line in tunnel
Existing lines
Station

0580228

11 km double-track line from Malmö Central station to the Öresund Fixed Link and 6 km of single line to provide new links to Trelleborg in the south and Ystad in the east. The project also includes construction of a supplementary station at Malmö Central and new stations at Hyllie and Triangeln.

The principal feature of the project is the proposed construction of twin 6 km single-track tunnels under central Malmö. Some 4.5 km of these will be bored, the remainder constructed using the cut-and-cover method. Triangeln station, which will serve the city's commercial and cultural centre, will

be located in the tunnelled section; Hyllie will be at surface level and will serve a new residential and commercial area. Proposals for expanding facilities at Malmö Central include provision of a new station, Malmö C Nedre. This will comprise two underground island platforms served by four tracks immediately north of the existing surface-level facility. Both stations will be served by a new glass concourse.

UPDATED

Green Cargo AB

PO Box 39, SE-17111 Solna
Tel: (+46 8) 762 40 00   Fax: (+46 8) 791 72 21
e-mail: info@greencargo.com
Web: www.greencargo.com

Key personnel
Managing Director: Jan Sundling
Directors
  Finance: Gunnar Andersson
  Operations: Bengt Fors
  Planning: Mats Granath
  Marketing and Sales: Mats Hanson
  Market Support: Olle Wennerstein
  Road: Lars Reinholdsson
  Logistics: Marcin Tubylewicz
  Corporate Communications: Mats Hollander
  Human Resources: vacant
  Information and Communications Technology:
    Kerstin Stenberg

Organisation
Green Cargo AB, formed from the freight division of the former Swedish State Railways, began trading on 1 January 2001. The company is wholly owned by the Swedish state and has a workforce of around 3,350.

The company had a turnover of SKr5.9 billion in 2004 and rail traffic equivalent to 13 billion net tonne-km.

Green Cargo's business is organised into six main divisions: Operations (rail transport), Planning, Road, Logistics, Marketing and Sales, and Market Support.

Freight operations
The company is responsible for 80 per cent of freight transport by rail in Sweden. It provides road transport and intermodal door-to-door solutions that include trains, its own lorries or external partners' lorries. Its third-party logistics operation manages warehousing and cargo processing at

Green Cargo Class Rc4 electric locomotives heading a timber train (Green Cargo)   **NEW**/1114962

Diesel locomotives

Class	Wheel arrangement	Power kW	Speed km/h	Weight tonnes	No in service	First built	Mechanical	Builders Engine	Transmission
T44/ T44R	Bo-Bo	1,235	90	76	115	1968 1983	Nohab Kalmar	GM 12-645E	E GM
V5	C	460	70	48	40	1975	Henschel	Deutz BF12M716	H Voith/ Gmeinder
Z70*	B	333	70	34	50	1990	ABB-Rac	Saab-Scania DSI 14	H Voith L3r 4U2

* Rebuilt from former Class Z65.

Electric locomotives

Class	Wheel arrangement	Output kW	Speed km/h	Weight tonnes	No in service	First built	Builders Mechanical	Electrical
Rc1	Bo-Bo	3,600	135	80	17	1967	Nohab/ASJ/Motala	ASEA
Rc2	Bo-Bo	3,600	135	77	70	1969	Nohab/ASJ/Motala	ASEA
Rc3	Bo-Bo	3,600	160	77	10	1970	Nohab/ASJ/Motala	ASEA
Rc4	Bo-Bo	3,600	135	78	128	1975	Nohab/Kalmar	ASEA
Rm	Bo-Bo	3,600	100	92	6	1977	Nohab	ASEA

Class Rc and Rm, thyristor control.

its logistics centres in Stockholm, Norrköping, Göteborg, and Helsingborg-Köpenhamn. Green Cargo's transport services have been awarded the Good Environmental Choice eco-label from the Swedish Society for Nature Conservation.

The Green Cargo business includes the following subsidiaries and associated companies:
• Nordisk Transport Rail AB (NTR), one of Scandinavia's leading freight forwarding companies.
• TGOJ Trafik AB (see entry in Sweden section of Railway systems and operators), one of Sweden's largest short line freight operators, specialising in trainload traffic for heavy industry.
• Green Cargo A/S, a third-party logistics company in Denmark, 100 per cent owned by Green Cargo AB.
• Hallsbergs Terminal AB, a terminal for intermodal traffic located at Sweden's biggest marshalling yard in Hallsberg. Green Cargo owns 67 per cent of the company.
• CargoNet AS, formerly the freight division of Norwegian State Railways (NSB). Green Cargo took over a 45 per cent stake in the business from 1 January 2002, the remaining 55 per cent remaining with NSB. CargoNet specialises in intermodal transport through a network of terminals in Norway and Sweden, with links to the rest of Europe.

Traction and rolling stock
Green Cargo AB has a fleet of 436 locomotives, 7,700 wagons and 160 road vehicles.

UPDATED

Green Cargo Class T44 diesel-electric locomotive (Green Cargo)
***NEW**/1114963*

Hector Rail AB

Ellagårdsvägen 68, SE-187 45 Täby
Tel: (+46 8) 768 59 08
e-mail: info@hectorrail.com
Web: www.hectorrail.com

Key personnel
Chairman: Ole Kjörrefjord
Managing Director: Mats Nyblom
Production Manager: Sune Hector
Safety and Operations Manager: Leif Gustafsson

Background
Hector Rail AB is an open access freight operator that commenced operations in December 2004. The majority shareholder of the company is Höegh Capital Partners Investments Ltd, a subsidiary of a family-owned Norwegian shipping company whose interests include HAUL, the world's fifth largest car transporting company, with more than 50 ro-ro ships.

Freight operations
Hector Rail's first operation was the haulage of containers for Euroshuttle between Hallsberg and Halden. Services were later to be extended to Hönefoss.

Traction and rolling stock
At the commencement of operations Hector Rail services were handled by two Class Rc3 electric locomotives hired from SJ AB. The company has also acquired six Class EI15 5,400 kW electric locomotives from Norwegian iron ore haulier MTAS. Three of these were to be refurbished by Euromaint and assigned to Euroshuttle traffic, leaving the remaining three as reserve traction for future business opportunities.

NEW ENTRY

IKEA Rail AB

Box 228, SE-260 35 Odäkra
Tel: (+46 42) 25 73 80   Fax: (+46 42) 25 73 80
Web: http://www.ikea.com

Key personnel
Managing Director: Christer Beijbom

Background
This subsidiary of IKEA, the international home furnishings retail chain, was established in April 2001 to develop its own rail services to distribute

the company's products in Europe. Its first train ran on 27 June 2002. IKEA Rail claimed to be the first private company to operate its own freight trains across international borders in Europe. However, in late 2003 the company announced that for commercial reasons its rail operations would cease in January 2004.

Freight operations
At its launch, IKEA Rail was operating five return trains each week between Älmhult, Sweden; and Duisburg, Germany (1,044 km).

Traction and rolling stock
IKEA Rail uses hired-in traction for its services: between Älmhult and Copenhagen, a TGOJ Trafik General Motors-built Class T66 diesel-electric locomotive is used; between Copenhagen and Padborg, Denmark, trains are hauled by a pair of TraXion Class MY diesels; and from Padborg to Duisburg, a pair of RAG Bahn und Hafen Class 145 electric locomotives take over.

Inlandsbanan AB (IBAB)

Köpmangatan 22B, PO Box 561, SE-831 27 Östersund
Tel: (+46 63) 19 44 00
Fax: (+46 63) 19 44 06
e-mail: info@inlandsbanan.se
Web: http://www.inlandsbanan.se

Key personnel
Chairman: Leo Persson
Managing Director: Ulf Johannisson
Marketing Manager: Dan Humble
Finance Manager: Rolf Larsson

Gauge: 1,435 mm
Route length: 1,056 km

Political background
The Inlandsbanan (Inland Railway) extends for 1,053 km (some 10 per cent of total railway route-km in Sweden) from Mora to Gällivare, through some of the most sparsely populated territory in Sweden. After years of declining services and uncertainty over the line's future, parliament agreed in 1992 to lease the

line for 20 years to the 15 district councils along the route.

Inlandsbanan AB (IBAB), a company owned by the 15 councils, assumed responsibility for managing the line in 1993. The addition of three connecting branch lines has brought the total route-km managed by IBAB to 1,096, some 10 per cent of the Swedish rail network.

Following expiry of the initial five-year government funding agreement in 1997, IBAB can now apply for subsidy on an annual basis.

Organisation

IBAB is responsible for passenger services, infrastructure and stations, as well as traffic control and safety. Freight services are run by Green Cargo AB, TGOJ Trafik AB and Inlandsgods AB. Infrastructure maintenance services are procured by competitive tendering. The train operators pay fees to IBAB for access to infrastructure, locomotives and rolling stock.

The board of IBAB has stipulated that the company must operate within the bounds of the finance provided in the transfer agreement and the revenue generated by operations; apart from equity capital, no funding will be required from shareholders. Together with the train operators, IBAB works to develop traffic and to secure the Inlandsbanan as an essential part of the transport system in northern Sweden.

Passenger operations

IBAB has concentrated on developing tourist services, which attracted 22,000 passengers in 2003, an increase of 10 per cent on the previous year. In 1997, IBAB took over the operation and marketing of these services itself, rather than contracting them out as in previous years.

From late June until mid-August, railcars provide daily workings in each direction over both the southern section from Östersund to Mora (322 km) and the northern section from Östersund to Gällivare (746 km). The 'Wilderness Express', a locomotive-hauled train conveying ordinary second-class accommodation, vintage first-class cars and a restaurant car, returned to the route in 1998.

Since 1996, local railway preservation societies have operated a series of steam-hauled specials on different sections of the Inlandsbanan during the summer season.

Freight operations

Timber transport has always been the mainstay of the Inlandsbanan. Freight traffic has grown by 50 per cent in wagonload terms since IBAB took over the line; Green Cargo AB now operates a daily return journey all the way from Östersund to Storuman, whereas previously the policy was

Former SJ Class Y1 diesel railcars forming an IBAB passenger service at Arvidsjaur (Edward Barnes)
0536199

for freight trains to travel the shortest possible distance on the Inlandsbanan.

Inlandsgods AB operates timber trains on the southern section between Östersund and Mora plus peat trains between Sveg and Mora and onward via Banverket lines to Uppsala.

In total 37,000 wagonloads were carried on the line in 2003, corresponding to about 1,100 trainloads.

Improvements to existing lines

Under the 1992 transfer agreement, IBAB is committed to completing upgrading work planned by the former SJ and Banverket covering the 494 route-km to which the former SJ's freight operations were previously confined. In view of the increase in freight traffic, IBAB has upgraded most of the Mora-Arvidsjaur section for an axleload of 22.5 tonnes, the same maximum as Banverket's network, except for the Sver-Brunflo section, where dispensation had already been given for this limit. In the longer term, IBAB is developing a

project jointly with Luleå University to increase the maximum axleload on spiked track to 30 tonnes.

In 2003, IBAB was studying options for the procurement of a traffic management system.

Some of IBAB's network upgrading work has taken the form of employment training projects, operated in conjunction with regional government. It has also been necessary for IBAB, at its own expense, to upgrade several sections where maintenance had been deferred.

Traction and rolling stock

IBAB's diesel motive power comprises three Class Z65 shunters and three Class TMX A1A-A1A locomotives acquired from Danish State Railways in 1994. Passenger services are operated with eight ex-SJ diesel railcars owned by IBAB.

Nine preserved steam locomotives have been acquired for use by railway societies along the line.

Jernhusen AB

Box 520, SE-101 30 Stockholm
Tel: (+46 8) 762 45 00
Fax: (+46 8) 762 33 50
Web: http://www.jernhusen.se

Key personnel

Managing Director: Lasse Jerbéus

Organisation

Formerly SJ Properties, the land and property company created by Swedish State Railways as part of the restructuring resulting from its break-up, Jerhusen AB owns or manages 900 stations, workshops, freight terminals, warehouses and development assets in Sweden. The present company was established in 2001 and remains state owned.

The company acts as station manager for train and bus traffic operators at some 190 locations in Sweden; it manages tenancies at railway properties; it is responsible for optimising the value of property and land no longer needed for railway use; and it has a remit to draw up development plans for stations in major towns and cities.

Malmtrafik i Kiruna AB (MTAB)

SE-981 86 Kiruna
Tel: (+46 980) 710 00
Fax: (+46 980) 109 02
e-mail: info@mtab.com
Web: http://www.mtab.com

Key personnel

Managing Director: Åke Boström

Gauge: 1,435 mm
Route length: 447 km in Sweden (BV), 42 km in Norway (JBV)

Organisation

The Iron Ore Line (Malmbanan) links the mines at Kiruna, Malmberget and Svappavaara with the harbours at Narvik (Norway) and Luleå. Since 1988 operating rights for the Swedish part of the line have rested with the mining company LKAB, which initially contracted Swedish State Railways (SJ) to run the trains on its behalf, together with Norwegian State Railways (NSB).

In 1995 it was decided to establish the ore transport operations as a separate company, Malmtrafik AB (MTAB), originally owned 51 per cent by LKAB, 24.5 per cent by SJ and 24.5 per cent by NSB. MTAB established a Norwegian subsidiary,

Malmtrafikk AS (MTAS), to acquire the former NSB assets and staff.

On 1 July 1996, MTAB and MTAS took over all locomotives, wagons and depot facilities connected with the ore operations. Infrastructure maintenance remains the responsibility of Banverket (Sweden) and Jernbaneverket (Norway).

Since 1 January 2000, MTAB has been a wholly owned subsidiary of LKAB.

The infrastructure authorities are upgrading the line for 30-tonne axleloads, which will allow the operation of 8,160-tonne trains compared with the previous maximum of 5,200 tonnes. Upgrading of the southern section, between Malmberget and Luleå, was completed in 2001.

Work on the northern section, between Kiruna and Narvik, is in progress.

Traction and rolling stock

At January 2002, MTAB and MTAS operated 24 electric line-haul locomotives: 15 Class Dm3 (7,200 kW, ex-SJ), six Class EI15 (5,406 kW, ex-NSB) and three of the new class IORE (see below). In addition, the companies owned 985 ore wagons, together with seven diesel locomotives for shunting and terminal operations and one back-up electric locomotive.

In 1998 the company placed an order with Adtranz (now Bombardier Transportation) for nine electric locomotives to replace classes Dm3 and EI15. Each Class IORE locomotive will comprise two close-coupled six-axle units (Co-Co + Co-Co) with a total power output of 10.8 MW and a maximum tractive effort of 1,200 kN. The first example was delivered in 2000 and underwent extensive testing prior to series delivery in 2002–04.

In 1999 tests commenced with five prototype Series Uno ore wagons from Transwerk of South Africa. A preliminary batch of 68 additional wagons was delivered in December 2000, and MTAB has an option to purchase a total of 209 such wagons.

Upgrading of the line for 30-tonne axleloads will permit a reduction in the fleet to nine locomotives (Class IORE) and around 700 wagons.

Bombardier Transportation Class IORE two-section 10.8 MW electric locomotive
0103974

Rikstrafiken

National Public Transport Agency
PO Box 473, SE-856 06 Sundsvall
Tel: (+46 60) 67 82 50   Fax: (+46 60) 67 82 51
e-mail: registrator@rikstrafiken.se
Web: http://www.rikstrafiken.se

Key personnel

Director General: Kjell Dahlström
Administrative Director: Kerstin Söderland
Chief Procurement Officer: Elisabeth Forslin
Public Relations Officer: Håkan Jacobsson

Political background

Legislation passed by the Swedish parliament in 1998 set up a new agency to promote the development of a nationwide integrated public transport system. Rikstrafiken came into being on 1 July 1999, and on 1 January 2000 it assumed responsibility for procuring long-distance passenger services which SJ AB has declared commercially unviable. It has similar responsibilities in respect of air and ferry services.

Subsidies totalling SKr800 million are allocated to public transport in the state budget, which includes central government support to the regional transport administrations. These funds have now been transferred to Rikstrafiken, but in future will no longer be earmarked for a particular mode. Instead, Rikstrafiken will allocate subsidies to the most appropriate mode on the basis of an overall socioeconomic assessment of public transport requirements.

Commercially unviable long-distance rail services have been subject to competitive tendering since 1993, but it was not until 1999 that private operators succeeded in breaking SJ's long-distance monopoly. Since January 2000, overnight services linking Stockholm and Gothenburg with the far north have been run by Svenska Tågkompaniet AB (qv). Following retendering in 2002, the contract to operate these services for five years from June 2003 was awarded to a new operator, Connex Transport AB. In May 2001 a contract to operate services between Sundsvall and Östersund from June 2002 was awarded to Svenska Tågkompaniet. The remaining state-subsidised services remain in the hands of SJ.

Shortline Väst AB

Per Anders gata 24, SE-521 32 Falköping
Tel: (+46 515) 834 23

Key personnel

Managing Director: Jan-Erik Astonius

Freight operations

Shortline Väst AB was set up in 1994 and has operated feeder freight traffic on the Borås–Herrljunga line (43 km) since January 1995. The company's original service between Herrljunga and Lidköping (55 km) ceased in early 1998.

The company has three electric locomotives, three diesel locomotives and two locotractors. It also provides workshop facilities and has responsibility for shunting duties at Falköping and Borås.

SJ AB

Centralstationen, SE-105 50 Stockholm
Tel: (+46 8) 762 20 00   Fax: (+46 8) 411 12 16
e-mail: sjinfo@sj.se
Web: http://www.sj.se

Key personnel

Chairman: Daniel Johanesson
Managing Director: Jan Forsberg
Directors
  Personnel: Peter Blomqvist
  Production: Jan G Forslund
  Rolling Stock: Jan Kyrk
  Political and International Affairs:
    Magnus Persson
  Marketing and Communications: Gunilla Asker
  Subsidiary Businesses: Christer Jernberg
  Sales: Björn Nilsson
  Long-distance Services: Mikael Wikström
  Götland Region: Gunnar Wulff
  Svealand Region: Mats Wilhelmsson

Organisation

SJ AB, formerly the passenger division of Swedish State Railways, began trading as a limited company on 1 January 2001. The company is wholly owned by the Swedish state and employs around 3,500 people.

The company had a turnover of SKr5.5 billion in 2001. Traffic for the year (excluding contract traffic) totalled 30.6 million passenger journeys and 5.5 billion passenger-km.

Y2K 'Kustpilen' dmu at Lund forming a Malmö–Copenhagen service (David Haydock)   0114259

Electric locomotives

Class	Wheel arrangement	Output kW	Speed km/h	Weight tonnes	No in service	First built	Builders Mechanical	Electrical
Rc3	Bo-Bo	3,600	160	77	32	1970	Nohab/ASJ/Motala	ASEA
Rc6	Bo-Bo	3,600	160	78	100	1982	Hägglunds	ASEA

Diesel railcars or multiple-units

Class	Cars per unit	Motor cars per unit	Motored axles/car	Power/ motor kW	Speed km/h	No in service	First built	Builders Mechanical	Engine	Transmission
Y1	1	1	2	147	130	19	1979	Fiat/ Kalmar	Fiat 8217.12.150	H Fiat SRM
Y1	1	1	2	210	130	17	1979	Fiat/ Kalmar	Volvo THD 102 KB	H Allison
YF1	1	1	2	147	130	4	1981	Fiat	Fiat 8217.12.150	H Fiat SRM
Y2	3	2	2	294	180	14	1990	ABB-Scandia	KHD BF8LV 513CP	M ZFS HP 600
Y2*	3	2	2	310	180	6	1996	Adtranz	Cummins NTAA 855 R7	M ZFS HP 600

* Fitted with catalytic converters and particle filters.

Electric railcars or multiple-units

Class	Cars per unit	Motor cars per unit	Motored axles/car	Output/motor kW	Speed km/h	No in service	First built	Builders Mechanical	Electrical
X1	2	1	4	280	120	94	1967	ASJ/Kalmar	ASEA
X2	7	1	4	815	210	23	1990	ABB	ABB
X2	6	1	4	815	210	3	1994	ABB	ABB
X2-1	5	1	4	815	210	17	1994	ABB	ABB
X9	3	2	2	170	115	5	1960	Carlsson	ASEA
X10	2	1	4	320	140	54	1982	Hägglunds	ASEA
X11	2	1	4	320	140	47	1982	Hägglunds	ASEA
X12	2	1	4	320	160	18	1991	ABB	ABB
X14	2	1	4	320	160	18	1994	ABB	ABB
X31*	3	2	4	480	180	18	1999	Adtranz	Adtranz
X50	2	–	–	–	–	31	2001	Bombardier	Bombardier
X50	3	–	–	–	–	5	2001	Bombardier	Bombardier

* Dual-voltage (15 kV AC/25 kV AC).

SJ AB Class Rc6 electric locomotive waiting at Oslo Central with a return cross-border service to Gothenburg (Wolfram Veith) 0114223

Class X50 emu operated by SJ AB for Tåg i Bergslagen at Borlange (Edward Barnes) 0524944

SJ AB is structured into seven head office departments, a long-distance division, two regional divisions, and divisions for sales, rolling stock, production and subsidiary businesses. Operational and profit responsibility is devolved to the business units.

Traffic (million)	1997	1998	1999
Passenger journeys	106.8	110.9	114.9
Passenger-km	6,814	6,997	7,434

(Data applies to former SJ Passenger Traffic subsidiary)

Passenger operations

SJ AB's product portfolio includes long-distance, regional and local services. Around 20 per cent of business consists of contracted traffic purchased by the public sector through Rikstrafiken (qv) and the 24 regional transport administrations. The remaining 80 per cent of business is made up of commercial traffic on the trunk network.

In recent years SJ has significantly increased its share of the business travel market. On the major business travel routes (Stockholm to Gothenburg, Malmö and Sundsvall) served by X2000 services, SJ has strengthened its position. X2000 services are operated with Adtranz-built Class X2 tilt-body trainsets, designed for 200 km/h operation. X2000 services also operate from Stockholm to Karlstad, Arvika, Falun, Mora, Härnösand, Jönköping, Karlskrona and Helsingborg. The Mälar and Svealand lines serving the populous area west of Stockholm are also operated with Class X2 trainsets. A shortened version of the X2, known as the X2-1, operates on interregional services in the south of Sweden.

New services

The link enabling SJ to provide services to Stockholm's Arlanda airport was opened at the beginning of 2000 and the Öresund Link was opened in July 2000, offering a direct connection to Copenhagen and its airport at Kastrup. The provision of new services to the large airports in Copenhagen and Stockholm are strategically important for SJ, which realises that such effective rail links offer substantial benefits for both train and airline passengers. SJ and the Scandinavian airline SAS are already co-operating with through ticketing, enabling passengers to use one ticket for travel on both modes.

Together with Norwegian State Railways (NSB), SJ is working on a project to develop high-speed services linking Stockholm and Oslo in 4 hours 30 minutes. A new company, Linx AB (qv in *Operators of International rail services in Europe* section) owned jointly by SJ and NSB and based in Gothenburg, has been set up to take over the running of Oslo–Karlstad–Stockholm and Oslo–Gothenburg–Copenhagen services from 2001/2002.

Traction and rolling stock

SJ AB on its formation took over 121 Class Rc3 and Rc6 electric locomotives from Swedish State Railways, as well as most of the dmu and emu fleet.

Of the initial batch of 27 three-car dual-voltage emus supplied by Adtranz to the Swedish and Danish railways for operating services via the Öresund fixed link, 10 units are owned by SJ AB, which has a further eight units on order (Class X31).

In January 2001, six Adtranz Regina two-car emus (Class X50) were introduced on local services from Västerås, operated by SJ AB on behalf of the Västmanland regional transport administration. SJ AB also won the contract to run local services in the Bergslagen region from June 2001, for which 12 Regina and two Itino units are being leased.

In its own right, SJ AB has agreed to lease 11 two-car and five three-car Regina units for operation on Stockholm–Västerås services and a pair of two-car units for Coast-to-Coast services (Gothenburg–Kalmar/Karlskrona).

When current orders are completed, SJ AB will have a total of 36 Regina emus in service on contracted and commercial services.

The first order placed by the new SJ AB, in early 2001, calls for 43 bilevel emus for services in the Mälardalen region such as Stockholm–Västerås. The units, to be supplied by Alstom, will have a top speed of 200 km/h. Of the trains, 27 will be three-car units, seating 319 passengers, while the remaining 16 will be two-car units with 202 seats. Delivery is scheduled to begin in late 2003 and be completed by the end of 2004.

A fleet of 27 hauled coaches built in the 1980s was refurbished in 2001–02 to provide back-up and relief capacity for X2000 services. The refurbished stock, known as BlueX trains, is to operate in six-car sets, offering comparable levels of comfort

to Class X2 trainsets. Although maximum speed was initially limited to 160 km/h, five Class Rc electric locomotives are being modified and reclassified Rc7 to provide haulage at up to 180 km/h.

Svenska Tågkompaniet AB

PO Box 45, SE-971 02 Luleå
Tel: (+46 920) 23 33 33   Fax: (+46 920) 23 33 39
Web: http://www.tagkompaniet.se

Key personnel
Managing Director: Jan Johansson
Operations Director: Sven Malmberg
Marketing Director: Björn Nyström

Passenger operations
Tågkompaniet was set up by a group of senior managers previously employed by Swedish State Railways (SJ) and is jointly owned by its three founder directors and the venture capital company Fylkinvest AB. Operations commenced on 10 January 2000, following the award of a two-year contract to run the state-subsidised overnight services linking Stockholm and Gothenburg with the northern cities of Umeå and Luleå and the Norwegian port of Narvik.

Under a new five-year contract awarded in June 2002, these services are to be taken over by a new operator, Connex Transport AB, from June 2003. Tågkompaniet has lodged an appeal against this decision.

Tågkompaniet's contract also covers passenger services on the Iron Ore line (Luleå–Kiruna–Narvik). In a separate initiative, the company is running a daily return service from Luleå to the Finnish border at Haparanda during the summer months, which has entailed the reopening to passengers of eight stations.

Further expansion came in June 2000, when Tågkompaniet took over the operation of commuter services between Uppsala and Tierp (on the Stockholm–Gävle main line) under a five-year contract with the regional transport administration, Upplands Lokaltrafik.

Tågkompaniet has also won the contract to operate local services for X-Trafik, the regional transport administration in the neighbouring

Tågkompaniet Rc2 electric locomotive with a passenger service at Gällivare (Edward Barnes)

0524945

country of Gävleborg. The five-year agreement, which took effect in June 2001, covers services on the main lines north from Gävle to Ljusdal and Sundsvall.

In May 2001 Tågkompaniet was awarded a contract to run services between Sundsvall and Östersund from June 2002, initially for three years but with an option for a further two.

Traction and rolling stock
Motive power for the services to the far north is provided by 12 Class Rc6 and two Class Rc2 electric locomotives hired from SJ. Initially, a fleet of 93 carriages (including 26 sleeping cars, 30 couchette cars and six cinema cars) was also hired from SJ.

To reduce its dependence on hired stock, the company has since purchased 10 carriages from the Swiss tour operator Reisebüro Mittelthurgau: five ex-DB 'Rheingold' panoramic observation

cars, and five recently refurbished sleeping cars which latterly operated the Nostalgie Istanbul Orient Express land cruise. Tågkompaniet and Mittelthurgau have joined forces to market services using this stock to foreign tourists under the Polar Express name.

Tågkompaniet's motive power and rolling stock is maintained under contract by TGOJ (formerly a subsidiary of SJ) at its Notviken works near Luleå.

The Uppsala–Tierp service is operated with Class X12 emus hired from SJ by Upplands Lokaltrafik. Tågkompaniet has established a small depot at Tierp for cleaning and day-to-day maintenance.

The X-Trafik services are operated with five Regina two-car emus delivered in 2001 by Adtranz Sweden (now Bombardier Transportation). Additional Regina units will be deployed on the Sundsvall–Östersund service from mid-2003.

Swedish National Rail Administration (BV)

Banverket
Jussi Björlings väg 2, SE-781 85 Borlänge
Tel: (+46 243) 44 50 00   Fax: (+46 243) 44 50 09
e-mail: banverket@banverket.se
Web: http://www.banverket.se

Key personnel
Chairman: Mats Hulth
Director General: Bo Bylund
Deputy Director General: Per-Olof Granbom
Directors
  Technical: Rune Lindberg
  Infrastructure Management: Hans Öhman
  Strategy: Lena Ericsson
  Marketing: Lars-Åke Josefsson
  Administration: Mattias Persson
  Finance: Bo Wikström
  Communications: Karin Rosander
  Banverket Production: Claes Sandgren
  Industrial Division: Roger Jonsson
  Banverket Consulting: Björn Östlund
  Information Technology: Rune Lidman
  Banverket Telenät: Lars-Göran Bernland
  Railway Training Centre: Christer Harvigsson
Directors, Independent Administrative Units
  Rail Traffic Administration: Jan Fahlén
  Railway Inspectorate: Over Andersson
Press contact: Jan Erik Kregnes

Gauge: 1,435 mm
Route length: 9,978 km
Electrification: 7,527 km at 15 kV 16⅔ Hz AC

Organisation
Banverket is a state authority responsible for the railway network in Sweden. At the end of 1996 its role was expanded from that of infrastructure provider to one of overall responsibility for the development and co-ordination of the rail system.

Day-to-day scheduling, train control and passenger information functions were fully integrated into Banverket during 2001. An independent unit, the Rail Traffic Administration, decides on operating rights and track allocations, and acts as an arbitrator in the event of disputes. Train traffic control is a division within Banverket.

The state takes full responsibility for the maintenance and upgrading of rail infrastructure, while train operators pay fees to Banverket for using the tracks. Track access charges are based on socio-economic marginal costs and generate around €650 million per annum.

Banverket maintains embankments and tracks, marshalling yards and electrical, signalling and telecommunications installations. It is responsible for the upgrading of existing lines and the construction of new lines.

Banverket's main lines generally allow speeds up to 130 km/h, but a gradual programme is under way to upgrade the most important routes for 160 to 200 km/h. Double or multiple track exists on the following lines:
- Stockholm–Gothenburg
- Stockholm–Malmö
- Stockholm–Gävle and some additional sections onwards to Östersund
- Stockholm–Västerås–Kolbäck

- About 110 km between Gothenburg and Lund
- Frövi–Örebro–Hallsberg

Axleload is normally 22.5 tonnes, but a design standard of 25 tonnes is now applied when upgrading track (for example, by use of UIC-60 rail) and replacing bridges on major lines. The Iron Ore line (Luleå–Kiruna–Narvik) has been upgraded to 30 tonnes axleload.

Finance
Banverket is funded by annual government appropriations. In 2003 Banverket's total volume of operations was SKr12.732 billion (SKr11.2 billion in 2002). Investment in new lines and upgrading, including capital costs, were SKr5.226 billion (SKr4.844 billion), and Banverket received SKr1.956 billion (SKr1.739 billion) for regulatory duties. External contracts generated SKr1.568 billion (SKr1.373 billion).

In February 2004 the government announced that under the Future Plan for Railways it was to invest SKr107.7 billion (€11.7 billion) in capital projects in the railway network during the period 2004–15, as well as SKr38 billion (€4.1 billion) to support operations and maintenance over the same period.

Improvements to existing lines
The rail infrastructure of the Nordic Triangle in Sweden consists of: Stockholm–Järna–Katrineholm–Laxå–Charlottenburg at the Norwegian border; Katrineholm/Järna–Norrköping–Mjölby–Malmö–Öresund fixed link; the Norwegian border at Kornsjö–Göteborg–Ängelholm–Helsingborg–Lund and Ängelholm–Malmö for freight only; and links to ferry ports at Göteborg and Stockholm.

Strängnäs station on Banverket's Stockholm–Eskilstuna line, with a SJ Class X2 tilting trainset (Thomas Fahlander/BV)   1034607

Long-distance transport by rail to and from Sweden consists mainly of freight. Most of this is generated by metal and forest industries located in the central and northern parts of the country and is sent by rail to customers in central Europe and, in combination with shipping, to coastal areas. During the 10 years to 2003 cross-border freight flows overall have been relatively stable. The opening of the fixed link across the Öresund in 2000 changed the Sweden–Denmark transfer point from Helsingborg to Malmö. In the absence of a fixed link across the Fehmar Belt and the presence of weak bridges at Rendsburg in northern Germany and at LilleBelt in Denmark, about half of the heavy southbound freight flow utilises train ferries from Trelleborg to Rostock and Sassnitz.

The pattern of international passenger travel mainly has the capitals of Denmark and Norway as its destinations. During the five years to 2003 preliminary figures indicate that passenger travel by train in Sweden increased by 30 per cent, while freight transport increased by 7 per cent.

Before the Nordic Triangle was selected by the European Union as one of the so-called Essen priority projects, a decision was taken to upgrade about 1,500 km of Swedish main lines from 130 km/h operation to 200 km/h running with tilting trainsets. The programme also included improved accessibility and facilities at most major stations involving platform reconstruction and the installation of lifts and escalators. Lines suffering from insufficient capacity were to be upgraded by doubling or quadrupling. In 1996, when the Swedish rail system was integrated in the Trans-European Network (TEN), about 50 per cent of these improvements had been completed.

Since then, work has continued. The most important project was completion of the fixed link between Sweden and Denmark. Most sections of track between Stockholm and Malmö and Stockholm and Karlstad are now able to sustain operations at 200 km/h, the maximum speed of the X2 tilting trainsets. From Karlstad to the Norwegian border the presence of many curves limits speeds to 160 km/h. About 80 per cent of the Malmö–Göteborg line has at least two tracks and a speed limit of 200 km/h and is prepared for a future increase to 250 km/h.

From 1996 to 2003 €1,400 million has been invested in or assigned to the following Nordic Triangle projects: Stockholm–Charlottenburg – €200 million; Katrineholm-Malmö – €150 million; Göteborg-Malmö – €600 million; and Göteborg-Kornsjö – €90 million. Of this total the European

Union has contributed 4 per cent. In addition, the cost railway share of the Malmö–Copenhagen fixed link has to be added.

These investments have reduced travel times between major cities. For example, the Stockholm–Copenhagen journey, which in 1990 took 8 hours, now takes 5, while the Stockholm–Oslo journey time has been reduced from 6 hours 30 minutes to 5 hours.

A mix of high-speed tilting trains and a substantial flow of slower freight trains south of Mjölby calls for the construction of additional passing loops to minimise delays. Double-tracking of the initial 80 km from Göteborg to Oslo has been started. Work to increase the maximum permissible axle-load from 22.5 to 25 tonnes, the load per metre to 8 tonnes and to implement loading gauge enhancements for wide-bodied containers has also started and will raise freight capacity. Further upgrading of existing tracks to accommodate 250 km/h running is also considered justified in some parts of the Nordic Triangle.

Future Plan for the Railways 2004–15
Some serious bottlenecks remain and Banverket intends to tackle some of these during the period from 2004 to 2015. With only €11.7 billion sanctioned of a requirement estimated at €27 billion, not all of the projects identified by Banverket in its submissions to government for the 2004–15 Future Plan for the Railways investment

plan will be realised over the period. However, several major priority projects will go ahead.

Hallandsås ridge tunnels
Steep gradients over Hallandsås ridge, situated between Halmstad and Ängelholm on the West Coast Line, limit the trailing loads of freight trains, forcing them to be diverted via Hässleholm. This will be resolved by tunnelling to create a modern double-track line. Tunnel construction is due to begin in 2005 for completion by 2012 at a cost of €800 million.

Stockholm and Malmö improvements
The mix of fast and slow trains in the Greater Stockholm and Malmö urban areas is to be solved by the construction of expensive tunnels to separate traffic flows. In Stockholm Banverket plans to construct a tunnel under the centre of the city for commuter services. Construction of the €886 million scheme is due to start in 2006 for completion by 2011.

In Malmö the Citytunneln project (see entry in the Sweden section of *Railway systems and operators*) will bring major improvements to the handling of rail traffic through the city.

Bothnia Line
Construction started in 1999 of the Bothnia Line. This will run from Nyland, north of Sundsvall, to the northern university city of Umeå, a major growth area. A special-purpose company, Botniabanan AB (see entry in the Sweden section of *Railway systems and operators*) has been formed by the government and local authorities to finance, design and construct the 190 km line. Commissioning is scheduled for 2010 at an estimated cost of €1.4 billion.

Northern Bothnia Line
North of Umeå Banverket plans to build a new railway, the Northern Bothnia Line, to complement the existing 270 km Upper Norland line to Luleå. Construction of the first stage of this line, which will follow an alignment closer to the coast, is scheduled to start in 2010. North of Luleå the existing line to the border with Finland at Haparanda is to be modernised, partly realigned and electrified under a €290 million scheme to be implemented over the period 2004–06.

Göteborg airport link
The plan for 2004–15 also includes projected link to Göteborg's airport. This will be the first part of a proposed new Göteborg–Stockholm high-speed link via Borås and Jönköping.

Projects beyond 2015
The TEN network foresees a high-speed Järna–Nyköping–Norrköping–Linköping link, bypassing Mjölby and making it possible completely to separate slow and high-speed trains. It is planned to construct this link in stages. Other lines, especially Göteborg–Malmö, will be upgraded for speeds of over 200 km/h by the installation of pan-European signalling technology.

Banverket's traction fleet includes former Austrian Federal Railways Class 1043 electric locomotives, which are used for hauling works trains (Edward Barnes)   0547799

Signalling and telecommunications

The project to find replacements for 40-year-old relay-based interlockings at small stations, and for traditional line blocking equipment, has reached the design phase. The system is being developed by a Swedish company, Novosignal, and a prototype installation is expected to be made at Gemla station on the Alvesta–Växjö line in 2002.

Since 1997, a new generation of control system for marshalling yards has been successfully in use at Borlänge. The same kind of system, based on an open system concept with PLC and PC hardware and Windows NT software, has now also been introduced at Sävenäs marshalling yard.

A generic specification of functional safety requirements has been drawn up and used for formal verification of some of Banverket's interlocking systems. Further research in this field is in progress.

Telecommunications

The Banverket Telecom system comprises about 20,000 km of metal cable, 13,000 km of fibre optic cable, 85 digital telephone exchanges and 800 radio base stations. There are 26,000 subscriptions, of which 16,000 are used by Banverket or SJ and 10,000 are trackside telephones.

The fibre optic cable is a 12-fibre single-mode cable. Capacity is up to 10Gb/s long-distance and 34 Mb/s short-distance. In addition, wavelength multiplexing (DWDM) is used on major long-distance routes.

Capacity in the fibre optic cable is sold on commercial terms to external customers, of which SJ is one. Other major clients include Tele 2, Comviq, Telenordia and Worldcom, reflecting the deregulated telecommunications market in Sweden.

In July 1998, Banverket placed a contract with Siemens to supply a digital mobile radio communications network. It will employ GSM-R technology, developed to meet the specific requirements of railway operations. Banverket is a member of ERIG (EIRENE Radio Implementation Group), which brings together the railway administrations introducing the EIRENE GSM-R system. The new system, MobiSIR, was inaugurated at the end of May 2001.

Electrification

The Swedish network has 9,400 track-km electrified at 15 kV AC $16\frac{2}{3}$ Hz, with 730 MVA static frequency converters and 640 MVA rotary converters. In 2004 it was intended to install nine more static frequency converters each of 15 MVA. The system is synchronous with the 50 Hz national grid.

The major part of the network employs the booster transformer system but in recent years the auto-transformer system has been installed and by 2004 covered some 600 km. In the northern part of Sweden there is a 1,900 km 132 kV AC $16\frac{2}{3}$ Hz transmission line parallel with the 15 kV system.

A combined auto-transformer and booster transformer system was under consideration in 2004. For some lines with an existing booster transformer system, this can be a simple way to improve the power supply system.

There are several kinds of catenary system in Sweden. The oldest lacks a steady arm or stitch wire and is designed for a maximum speed of 120 km/h. The newest is a modern stitch-wire system and is designed for a maximum speed of 250 km/h. Recently the focus of Banverket has been on punctuality and several projects have been initiated to reduce the number of catenary incidents and dewirements.

Traction and rolling stock

Banverket has a fleet of locomotives for its own operations, including ex-SJ Class T43 diesel-electrics, Vossloh Locomotives Types G1205 and G1206 diesel-hydraulics and three former Austrian Federal Railways Class 1043 electric locomotives similar to the Green Cargo and SJ Rc family.

Track

Standard rail: Type UIC 60 (60 kg/m) or BV 50 (50 kg/m)
On secondary lines: BV 43 (43 kg/m)
Sleepers
Concrete
Type B10: 320 × 222 × 2,500 mm
Type S3: 320 × 250 × 2,500 mm
Wooden
Type 1: 240 × 165 × 2,600 mm
Rail fastenings
Wooden sleepers
On main lines: Hey-Back
On secondary lines: spikes, normally with baseplate
Concrete sleepers: Pandrol E-Type, Hambo, Fist
Sleeper spacing
On main lines: 1,538/km on plain track; 1,538 or 1,667/km in curves of less than 500 m radius
On secondary lines: 1,333/km
Number of sleepers is increased to 2,000/km on the Kiruna-Riksgränsen ore line
Min curve radius: 300 m
Max gradient: 1% on most lines; 2.5% on lines not carrying freight traffic
Max permissible axleload: normally 22.5 tonnes; 25 tonnes on some lines; 30 tonnes on the Kiruna–Riksgränsen ore line

---

Tågåkeriet i Bergslagen AB (Tågab)

Bangårdsgatan 2, SE-681 30 Kristinehamn
Tel: (+46 550) 875 00   Fax: (+46 550) 875 03

Key personnel

Managing Director: Lars Yngström

Freight operations

With over 15,000 wagonloads a year, Tågab is one of Sweden's largest short line operators. Since 1994, the company has run feeder services on two lines in west central Sweden: Hällefors-Filipstad–Kristinehamn (105 km) and Bofors–Degerfors (18 km).

Tågab supplies locomotives and drivers for some Green Cargo trains from Kristinehamn to Hallsberg and Karlstad. Haulage is also provided on occasion for Banverket works trains, and in early 1999 the company signed a five-year contract with Jernbaneverket, the Norwegian infrastructure authority, to provide two locomotives and crews for works trains in Norway.

The company's main shareholders are its principal customers: Ovako Steel and crispbread manufacturer Wasabröd. Kristinehamn council

Former Danish State Railways Tågab Class TMY diesel locomotives at Namsskogen, Norway (Edward Barnes)
0524942

holds a minority stake. Other traffic includes processed timber and liquefied petroleum gas.

Traction and rolling stock

Tågab operates three main line diesel locomotives (ex-Danish State Railways Class MY) and four diesel shunters. In 2002, the fleet was supplemented

by the acquisition of three Class 1043 electric locomotives from Austrian Federal Railways. Their design is similar to that of Green Cargo's Rc2 machines.

The company has its own workshop for heavy maintenance, which also undertakes work for other operators.

TGOJ Trafik AB

Gredbyvägen 3-5, SE-631 21 Eskilstuna
Tel: (+46 16) 17 26 61   Fax: (+46 16) 17 26 01
e-mail: info@tgojtrafik.se
Web: www.tgojtrafik.se

Key personnel

Managing Director: Kjell Färnström
Fiance Manager: Stephan Dagson
Traffic Manager: Ingvar Sandberg
Marketing and Sales Manager: Claes Sörman
Quality Manager: Ingrid Toverland
Administration and Services Manager:
  Ingeborg Jakobsson
Traction Manager: Anders Dahl
Rolling Stock Manager: Leif Johanssen

Background

TGOJ Trafik AB is a subsidiary of Green Cargo, the freight business created by the break-up of the former Swedish State Railways (SJ). SJ took over the company in 1989. Having originally operated mainly over its own lines, TGOJ has since become an open access operator. The former

TGOJ maintenance business now forms part of EuroMaint AB.

Organisation

As well as undertaking its own-account freight operations, TGOJ Trafik hires out traction, rolling stock and train crews via a subsidiary, TGOJ Rental

TGOJ electric locomotives

Class	Wheel arrangement	Output kW	Speed km/h	Weight tonnes	No in service	First built	Builders Mechanical	Electrical
Ma	Co-Co	3,960	100	105	29	1953	Nohab, ASJ, Motala	ASEA
Rc2	Bo-Bo	3,600	135	83	3	1980	Nohab, Falun, Hägglund	ASEA
El 16	Bo-Bo	4,440	140	88	4	1977	Nohab, Hamjern, Strømmen	ASEA

AB, and also has a maintenance business for its own rolling stock.

Freight operations

TGOJ mostly specialises in trainload freight movements for heavy industry, carrying commodities that include steel products and copper ore. Operations are mainly concentrated in central Sweden, although trains are operated to the south and the far north of the country. The company also operates container traffic on behalf of Inter Container Scandinavia AB between several Swedish ports and Eskilstuna, where a 'dry port' was commissioned in 2003.

Traction and rolling stock

At the beginning of 2005 the TGOJ fleet comprised 36 electric locomotives, including some stored, 17 diesel main line and 19 diesel shunting locomotives. The diesel fleet includes two General Motors Class T66 machines delivered in late 2000 and of the same design as the Class 66 locomotives supplied to EWS in the UK. TGOJ also has a fleet of some 800 freight wagons.

Recent acquisitions include three Class 1043 electric locomotives, similar to that of Green Cargo's Rc2 machines, from Austrian Federal Railways, and four ex-Norwegian State Railways Class El 16.

UPDATED

Switzerland

Ministry of Transport, Communications & Power

Bundeshaus Nord, CH-3003 Bern
Tel: (+41 31) 322 57 11   Fax: (+41 31) 322 58 11

Key personnel

Minister: Moritz Leuenberger
Director General, International Railway Office:
  Hans-Rudolf Isliker

Political background

The federal government assumes financial responsibility for Swiss Federal Railways' (SBB) fixed installations, except where directly connected with SBB operations, workshops and traction current supply installations. The state's responsibility is purely financial. In its commercial sector, comprising InterCity passenger, wagonload, container and less-than-wagonload traffic, SBB, as a joint stock company, has managerial freedom and is required to make an annually predetermined contribution to infrastructure costs. The company returned to profitability in 1999.

The 57 so-called private railways (their private shareholders are in fact minimal) are supported by their Cantons. The federal government is statutorily obliged to top up their subsidy in two ways. It stabilises the railways' finances, including capital investment, at levels of contribution that reflect each railway's assessed value to the national or a regional transport system. Thus, support for the Bern–Lötschberg–Simplon (BLS), a trunk system of national importance, or the metre-gauge Rhaetian, socioeconomically vital to the Grisons Canton, is much more generous than the backing for short local lines in less hostile terrain. Secondly, in the case of Alpine territory, railways like the Rhaetian or Bernese Oberland, federal subsidy finances local passenger and freight tariffs based on SBB prices; because of their abnormal upkeep costs such railways charge non-residents much higher rates.

In 1994, the federal council approved revision of the country's Railway Law to put the SBB's Regional services on the same footing as those of the private railways. The Cantons are required to share the costs of all Regional services, whoever provides them, sharing with the federal administration the annual determination of the total amount of subsidy to be provided, and the detail of services to be operated. This may in some instances be in conflict with the 'Bahn 2000' regular-interval service concept (see 'Political background' section in Swiss Federal Railways entry).

Major change is now in progress. New rail legislation has been passed to adapt Swiss railways to the principles of open access for third-party operators, with track charge for infrastructure use, such as are being introduced into surrounding European Union countries. This involves far-reaching changes to the concession principle of ownership and operations which has previously guided Swiss public and private rail competences. The new law came into effect on 1 January 1998.

In 1999, the Bundesrat brought the regulations for the funding of major railway projects into force, backdating them to 1 January 1998. The fund receives the revenues from the flat-rate HGV charge and part of the excise duties on mineral oils (the total amount being SFr200 million in 1998). The flat-rate HGV charge was to be doubled at the start of 2000, and then to be phased out in 2001 in favour of a levy based on actual journeys. There was also to be an increase in VAT. The fund will allow the construction of the new Alpine tunnels, Rail 2000, noise abatement measures, and connections with the European high-speed network.

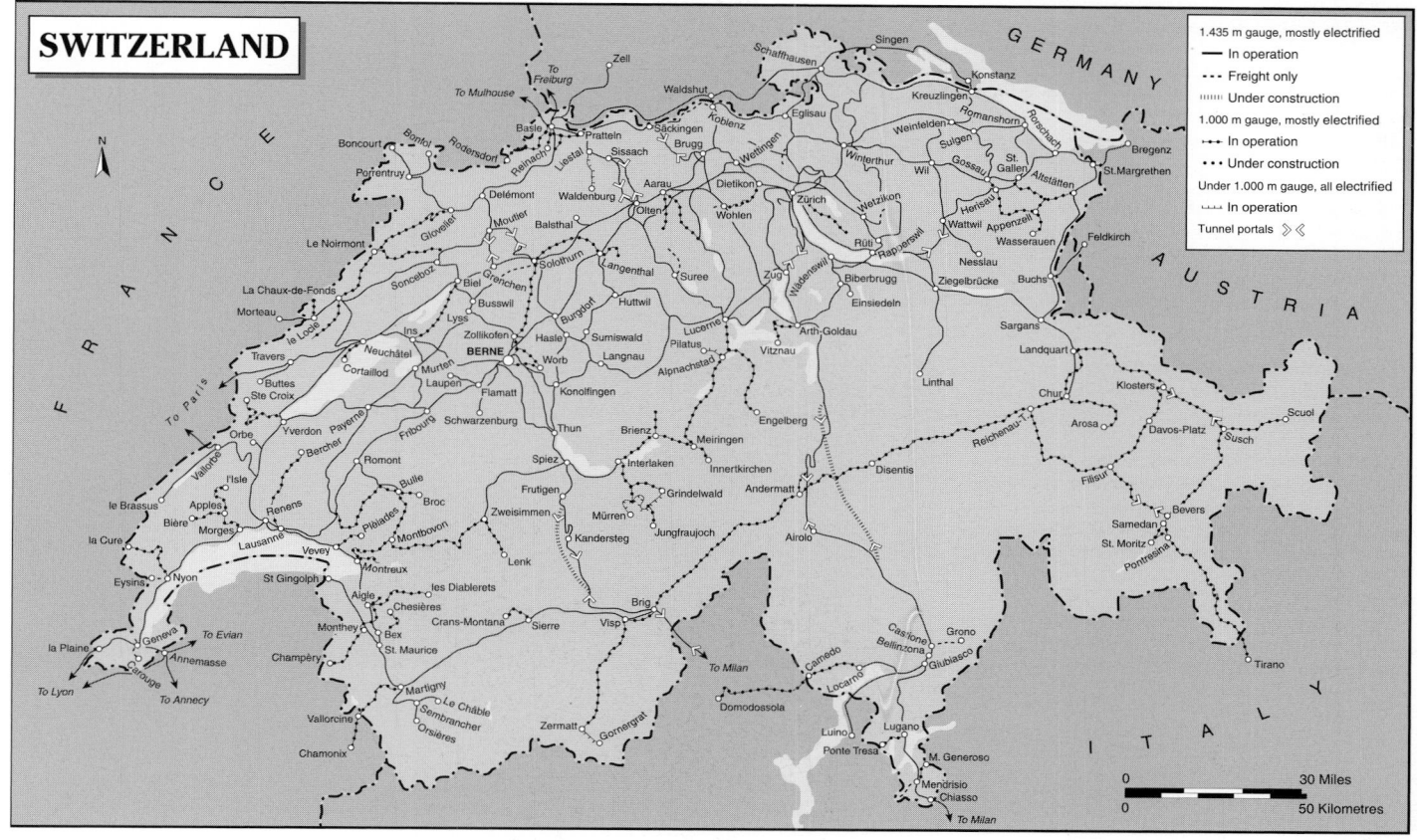

0088734

Appenzell Railway (AB)

Appenzeller Bahnen
Bahnhofplatz 10, CH-9101 Herisau
Tel: (+41 71) 354 50 60   Fax: (+41 71) 354 50 65

Web: http://www.appenzellerbahnen.ch
e-mail: info@appenzellerbahnen.ch

Key personnel

Director: Martin Vogt
  (from 1 July 2004: Dr Hansjüerg Düsel)

Vice-Director, Finance: Urs von Arx
Marketing Manager: Eduard Bühler
Operations Manager: Thomas Hablützel
Infrastructure Manager: Walter Bach
Regional Transport Manager:
  Eduard Bühler

Gauge: 1,000 mm
Route length: 59.76 km (4.15 km of rack rail)
Electrification: 59 km at 1.5 kV DC

Organisation

The system results from the 1988 merger of the Appenzell with the neighbouring St Gallen–Gais–Appenzell Railway. A 32 km line runs from Gossau, on the SBB St Gallen–Zurich main line, to Herisau, Appenzell and Wasserauen, in the Säntis mountain area, and attains a summit of 903 m above sea level at Gonten. A 20 km line extends from St Gallen to Teufen, Gais and Appenzell and is worked by seven rack-equipped electric trainsets. The third, 7 km line connects Gais and Altstätten Stadt and features a ruling gradient of 16 per cent.

Passenger and freight operations

AB's passenger performance in 2002 was 3.2 million passenger journeys, an increase of 1.7 per cent on the previous year. On the freight side, 4,547 tonnes were carried.

Traction and rolling stock

In total the railway owns three locomotives: one steam, one electric and one diesel; 14 three-car electric trainsets; six electric and one diesel railcars plus a driving trailer; and 23 other passenger coaches. AB also operates 38 freight and service wagons and 12 vehicles for transporting standard-gauge wagons.

Signalling and telecommunications

Centralised traffic control (CTC) manages 52 route-km. Automatic block equipment is installed between St Gallen and Appenzell (20 km) and Gossau and Wasseraven (32 km).

Bern–Solothurn Railway (RBS)

Regionalverkehr Bern-Solothurn
PO Box 119, CH-3048 Worblaufen
Tel: (+41 31) 925 55 65   Fax: (+41 31) 925 55 66
e-mail: info@rbs.ch
Web: www.rbs.ch

Key personnel

Director: Hans Amacker
Deputy Director: Hans-Jakob Stricker

Gauge: 1,000 mm; mixed 1,435 and 1,000 mm
Route length: 57 km
Electrification: 47 km at 1.25 kV DC; 10 km at 600 V DC overhead

Organisation

RBS was formed by merger of the Solothurn–Zollikofen–Bern Railway (SZB) and Bern–Worb Railways (VBW). The Bern Zytlogge–Worb Dorf line is electrified at 600 V DC, the voltage of the Bern city tramway network with which both SZB and VBW were connected before the 1960s. At that time SZB's original surface route into Bern was superseded by a new segregated double-track route from Worblaufen which finally tunnelled 1.2 km to a new four-platform terminus beneath the reconstructed Bern main station. The SZB system was then mostly re-electrified at 1.25 kV DC and its route modified to funnel its trains into the new Bern subterranean terminus.

In 2004 RBS employed 300 staff for its railway operations.

RBS also operates feeder bus services over 10 routes, totalling 60 km.

Passenger and freight operations

In 2004 RBS carried 19.5 million rail passengers, a slight increase on the 2003 figure of 19.3 million. Limited amounts of freight are carried including some in standard-gauge wagons carried on RBS transporter wagons.

Improvements to existing lines

Double-tracking is planned of the most heavily used sections of the railway.

Traction and rolling stock

In 2004 RBS operated six electric and two diesel locomotives; 59 emu/light rail cars; nine control trailers; three passenger cars; 12 freight wagons; and 45 transporters for 1,435 mm gauge wagons. Some historic passenger vehicles are retained for special traffic.

During 2001 intermediate low-floor trailers were added to 11 of the fleet of 16 Type Be 4/8 light rail vehicles dating from the 1970s. The lengthened vehicles are designated Type ABe 4/12. RBS has since refurbished these trains.

UPDATED

RBS underground station at Bern, refurbishment of which was completed in 2004 (RBS)

NEW/1143209

Refurbished Class ABe 4/12 emu with part-low-floor centre section (RBS)   *NEW*/1110186

Bière–Apples–Morges Railway (BAM)

Chemin de Fer Bière–Apples–Morges
En Riond Bosson 3, CH-1110 Morges
Tel: (+41 21) 811 43 43   Fax: (+41 21) 801 90 17
e-mail: info@lebam.ch
Web: http://www.swisscraft.ch/bam

Key personnel

Director: P Gaillard

Gauge: 1,000 mm
Route length: 31 km
Electrification: 31 km at 15 kV 16⅔ Hz AC

Organisation

The BAM is located between Lake Geneva and the Jura foothills. From Morges, on the SBB Geneva–Lausanne main line, a 19 km runs north to L'Isle-Mont-La-Ville. A 10 km branch runs west from the intermediate station at Apples to Bière.

Passenger and freight operations

As well as providing regional passenger services, the BAM also handles freight traffic, including the movement of standard gauge vehicles on transporter wagons.

Traction and rolling stock

BAM's rolling stock fleet comprises two Class Ge 4/4 III electric locomotives, one diesel shunter, six railcars, 11 trailers and 48 freight wagons.

BAM Class Be 4/4 electric railcar and trailer near Vufflens-Le-Chateau
(Markéta Šrámková)
0561563

BLS Lötschbergbahn (BLS)

BLS Lötschbergbahn AG
Genfergasse 11, CH-3001 Bern
Tel: (+41 31) 327 27 27   Fax: (+41 31) 327 29 10
e-mail: media@bls.ch
Web: www.bls.ch

Key personnel
Management
  Director: Mathias Tromp
  Infrastructure: Kees van Hoek
  Passenger Traffic: Anna Barbara Remund
  Freight Traffic: Dirk Stahl
  Finance: Hans Flury
  Personnel: Erwin Laetsch
  Secretary General: Thomas Müller
Divisions
  Safety: Eduard Wymann
  Installations: Stefan Huber
  Operations: Walter Flühmann
  Operations, AlpTransit: Markus Barth
  Infrastructure Development: Jean-Piere Kipfer
  Property: Hans Peter Lehmann
  Car-shuttle Traffic: (vacant)
  Regional Rail Passenger Traffic:
    Martin Reichenbach
  Passenger Traffic Development: Beat Luginbühl
  Passenger Traffic Distribution: Hansueli Kunz
  Traction: Thomas Furrer
Workshops: Heinz Schweizer
  Ships: (vacant)
  Marketing: Hans-Peter Ernst
  Corporate Communications: Urs Pfenninger
  Media Relations: Hans Martin Schaer

Route length: 245 km
Electrification: 245 km at 15 kV 16⅔ Hz AC

Organisation

The BLS main line from Frutigen through the Lötschberg Tunnel to Brig was opened in 1913. Certain regional railways with special guarantees from the Canton of Bern were then incorporated in the Lötschberg system, though each retained separate financial and operating identity. The BLS embraces the Spiez–Erlenbach–Zweisimmen line, the Gürbetal–Bern–Schwarzenburg line and the Bern–Neuchâtel line. Much interoperation of rolling stock takes place. In 1997 the company's shareholders approved the merger of all group railway operations into one new company, BLS Lötschbergbahn AG. The group also owns 16 Lake Thun and Lake Brienz ships. The BLS company also owns BLS Alp Transit AG, a subsidiary set up to manage the planning and building of the Lötschberg Base Tunnel, which is to open in 2007.

The BLS system covers the main lines from Thun to Spiez and Interlaken, and from Spiez via the Lötschberg Tunnel to a junction with SBB at Brig. The Lötschberg route is one of Europe's vital international rail links. BLS also owns the Grenchenberg tunnel line between Lengnau and Moutier (MLB), which forms part of the shortest route between Geneva and Basel.

At the end of 2004 BLS employed 1,815 staff.

BLS revenue from passengers in 2003 was SFr131 million, with a further SFr104 million from freight. Total earnings from traffic were SFr235 million and total income was SFr481 million. Total expenditure was SFr463 million with SFr172 million spent on staff and SFr82 million for depreciation.

Under an agreement announced in May 2001 between BLS and SBB, collaboration between the two companies is being enhanced. In December 2004 BLS took over the Bern S-Bahn lines previously worked by SBB, with SBB assuming responsibility for all long-distance traffic of both railways.

Passenger operations

In 2003 the BLS group carried 20.3 million passengers.

Bern Canton supports integration and development of an RER (Regional Express) network (see Swiss Federal Railways entry). For BLS this will principally entail part double-tracking of the Bern–Neuchâtel line, double-tracking throughout of the Bern–Belp line, and provision of additional rolling stock. These requirements are only in part covered by investments planned with credits from federal and cantonal supports for the private railways covering the period up to 2004.

BLS provides the route for the Cisalpino (see entry in *International* section) tilting trains introduced between Basel and Milano in October 1996, bringing substantially reduced journey times.

Car shuttle

BLS runs a push-pull shuttle service for accompanied cars and coaches through the Lötschberg Tunnel between terminals at Kandersteg and Goppenstein. The nine trains are formed of 87 Schlieren and 54 Talbot-built car-carriers, giving an hourly capacity of 600 vehicles.

Freight operations

In July 2001 a new freight company, BLS Cargo AG, was formed from the former freight division of BLS. The German rail freight operator, Railion, holds a 20 per cent stake in the company. Within two years an impressive increase in traffic was achieved: 1,310 net tonne-km in 2003 represented an increase of 126 per cent over the 2001 figure and share of the Swiss rail cargo market of more than 12 per cent. The increase in traffic continued in 2004 thanks to additional transit services via the Lötschberg line and to new transit traffic via the Gotthard route taken up since 2003. Since June 2001 BLS Cargo has also been hauling 'rolling highway' piggyback services between Freiburg/Breisgau and Novara on behalf of RAlpin AG.

Class RABe 'NINA' three-car low-floor emu (BLS)
NEW/1067754

Two BLS Class 465 electric locomotives at Reichenbach with a Novara-bound 'rolling highway' piggyback service (John C Baker) **NEW**/0585038

Electric locomotives

Class	Wheel arrangement	Output kW	Speed km/h	Weight tonnes	No in service	First built	Builders Mechanical	Electrical
Re 485	Bo-Bo	5,600	140	84	20	2002	Bombardier	Bombardier
Re 465	Bo-Bo	6,400	230	84	18	1994	SLM	ABB
Re 420.5	Bo-Bo	4,700	140	80	6	1964	SLM	ABB/Oerlikon/Sécheron
Re 4/4	Bo-Bo	4,980	140	80	35	1964	SLM	ABB
Eea 3/3	C (shunter)	600	75	50	1	1991	SLM	ABB

Electric railcars

Class	Cars per unit	Motor cars per unit	Motored axles/car	Output kW	Speed km/h	No in service	First built	Builders Mechanical	Electrical
RBDe 4/4	3	1	4	1,620	125	22	1982	SIG/SWS/SWP	ABB
RABe 4/8	3	2	4	1,000	140	33*	1998	Bombardier	ALSTOM

* Additional three on order for delivery early 2003.

New lines

The federal council's transalpine tunnel decision of 1989 (see 'New lines' section in the Swiss Federal Railways (SBB) entry) provided for construction of both Gotthard and Lötschberg base tunnels.

The Lötschberg base tunnel will form part of a new 34.5 km route from Frutigen to a junction with the SBB's Rhône Valley main line, with ruling gradient 1 per cent. Like the Gotthard Base Tunnel, the Lötschberg will have clearance for piggybacked 4.2 m high lorries, as also the Simplon Tunnel, following the latter's track-lowering and equipment with a rigid traction current contact system (see 'Improvements to existing lines' section in SBB entry). The federal government will meet the cost, recovering roughly a quarter from road transport fuel taxation.

The twin-bore base tunnel begins at Frutigen. Planned to fork at Km 27, the western arm will emerge, 34.5 km from Frutigen, on the SBB's Rhône Valley main line at Steg, giving a politically important automobile-carrying train service to the Canton Valais, replacing an Alpine motorway connection struck from the programme; it also completes a through route from the Lötschberg line to Lausanne and Geneva. The eastern arm will lead into the Rhône Valley main line at Raron, west of Visp. Tunnel length to Raron Ost will be 34.5 km underground; this will feed traffic for Italy into the approach to the Simplon tunnel. Preliminary work began in 1994 and construction proper began in 1998 after two referenda. Completion of the base tunnel should be achieved in mid-2007.

Traction and rolling stock

In 2004 BLS operated 79 electric locomotives, six electric shunting locomotives, 55 electric railcars, nine electrodiesel shunters, 27 diesel shunters, 44 control trailers, 133 other passenger coaches, 14 baggage cars and some 200 freight wagons (including Lötschberg Tunnel car-carrying shuttle cars). There is also one steam locomotive.

The BLS electric locomotive fleet was supplemented from 2004 by the acquisition from SBB of six Class Re 4/4 II machines. Designated Class Re 420.5 by BLS, they are used mainly on Bern–Neuchâtel regional express services.

In 1996 BLS ordered eight part-low-floor NINA (Niederflur-Nahverkehr) commuter trains from a consortium led by Vevey (since acquired by Bombardier Transportation). Bombardier Talbot supplied bodies and bogies and Holec (subsequently ALSTOM Transport) the electrical equipment. Used for Bern area regional services, the first of the articulated three-section 47.4 m long trains was delivered in late 1998. An order for six similar units for Bern S-Bahn services was placed in 2000, followed by two subsequent contracts for 22 more. In December 2004 33 NINA trains were in operation, leaving the final three units to be delivered in early 2005.

Track

(BLS only)
Rail
SBB IV (UIC 54E), 54 kg/m
Crossties (sleepers): Timber and concrete
Thickness: 150 mm timber, 235 mm concrete
Spacing: 1,666/km

Fastenings
Timber: Ke (bolted spring clips SKL 3)

Min curvature radius
Lötschberg line: 300 m
Other lines: 280 m

Max gradient
Lötschberg line: 2.7%
Other lines: 3.5% (Bern–Scharzenburg)

Max axleload
Lötschberg line: 22.5 tonnes
Other lines: 20 tonnes

UPDATED

Jungfraubahnen

Berner Oberland-Bahnen (BOB); Wengernalp-Bahn (WAB); Jungfraubahn (JB); Schynige Platte Bahn (SPB); Bergbahn Lauterbrunnen Mürren (BLM); Firstbahn (FB); and Harderbahn (HB)
Hardstrasse 14, CH-3800 Interlaken
Tel: (+41 33) 828 71 11   Fax: (+41 33) 828 72 64
e-mail: jb@jungfrau.ch
www.jungfraubahnen.ch

Key personnel

Chief Executive Officer: Walter Steuri
Head of Operations: Christian Balmer
Vice-President, Technical Director: Hans Schlunegger
Vice-President, Marketing and Operations Director: Urs Kessler
Vice-President, Director of Finance: Christoph Seiler
Vice-President, Head of Mountain Offers: Christoph Egger

BOB Type ABt 421.425 articulated low-floor driving trailer built in 2004 by Stadler (Jungfraubahnen) **NEW**/1137532

Jungfraubahnen lines

Railway	BOB	SPB	WAB	JB	BLM
Route length	38.367 km	7.26 km	29.366 km	11.827 km	5.7 km
Gauge	1,000 mm	800 mm	800 mm	1,000 mm	1,000 mm
Max gradient	12%	25%	25%	25%	5%
Minimum curve radius	90 m	60 m	60 m	100 m	40 m
Maximum axleload	20 t	–	11.5 t	16 t	–
Steam locomotives	–	1	–	–	–
Electric locomotives	2	11	8	5	–
Electric railcars	120	–	32	18	4
Passenger cars	41	22	44	18	–
Baggage cars	7	–	–	–	–
Freight wagons	25	7	63	20	4

Diesel locomotives

Class	Wheel arrangement	Power kW	Speed km/h	Weight tonnes	No in service	First built	Mechanical	Engine	Builders Transmission
BOB									
Tm 2/2	B (shunter)	110	30	15	1	1946	Stadler	Saurer	E BBC
HGm 2/2	B (shunter)	296	30	19.5	1	1986	Steck	Deutz	H Steck

Electric locomotives

Class	Wheel arrangement	Output kW	Speed km/h	Weight tonnes	No in service	First built	Builders Mechanical	Electrical
BOB								
1.5 kV DC								
HGe 3/3	Co	295	45	36	2	1914	SLM	BBC
SPB								
1.5 kV DC								
He 2/2	B	220	12	16	11	1910	SLM	BBC/Alioth
WAB								
1.5 kV DC								
He 2/2	B	220	12	16	4	1909	SLM	BBC/Alioth
He 2/2	B	236	12	16.5	2	1926	SLM	BBC/Alioth
He 2/2	B	460	22	16	2	1995	Stadler/SLM	ABB
JB								
1.125 kV AC 3 phase								
Hc 2/2	B	283	18	15	5	1904	SLM	BBC

Electric railcars or multiple-units

Class	Cars per unit	Motor cars per unit	Motored axles/car	Output kW	Speed km/h	Units in service	First built	Builders Mechanical	Electrical
BOB									
1.5 kV DC									
ABDeh 4/4	2	2	4	632	70	2	1949	SLM	BBC
ABeh 4/4 I	4	4	4	1,000	60	4	1965	SLM/SIG	BBC
ABeh 4/4 I	3	3	4	1,000	60	3	1979	SLM/SIG	BBC
ABeh 4/4 II	3	3	4	1,256	70	3	1986	SLM	BBC
WAB									
1.5 kV DC									
BDhe 4/4	11	11	4	440	25	11	1947	SLM	BBC
BDhe 4/4	13	13	4	440	25	13	1963	SLM	BBC
BDhe 4/8	4	4	4	804	28	4	1988	SLM	BBC/Sécheron
BDhe 4/8	4	4	4	880	28	4	2004	Stadler	Bombardier
JB									
1.125 kV AC 3 phase									
BDhe 2/4	6	6	2	440	24	6	1955	SLM	BBC
BDhe 2/4	4	4	2	440	24	4	1966	SLM	BBC
BDhe 4/8	2	2	4	804	27	4	1992	SLM	ABB
Bdhe 4/8	2	2	4	804	27	4	2002	Stadler	Bombardier
BLM									
560 V DC									
BDe 2/4	1	1	2	100	25	1	1913	SIG	Alioth
Be 4/4	3	3	4	208	30	3	1967	SIG	BBC

Organisation

The Jungfraubahnen comprises the Berner Oberland-Bahnen (BOB), the Schynige Platte Bahn (SPB), and the Jungfraubahn Group, with among others the Wengernalp-Bahn (WAB), Jungfraubahn (JB), and Bergbahn Lauterbrunnen-Mürren (BLM) and Firstbahn.

Berner Oberland-Bahnen

BOB operates 38.367 route-km of 1,000 mm gauge from Interlaken Ost to Lauterbrunnen and Grindelwald, electrified at 1.5 kV DC. Sections of route employ Riggenbach rack to cope with maximum gradients of 12 per cent. These four Riggenbach rack sections, with a total length of 4.6 km, were replaced in 2004 with the Von Roll system and with concrete sleepers in place of the previous steel.

Double track has been installed over a 2.5 km section between Wilderswil and Zweilütschinen at a cost of SFr30 million, and curve radius eased from 120 to 200 m over a further 5 km, enabling line speed to be raised from 40 to 70 km/h.

In 2004 BOB introduced a new automatic train protection system (Type ZSI 127) from Siemens, based on the ECTS.

In 2000 the BOB acquired five Type ABt driving trailers from Regionalverkehr Bern-Solothurn (RBS).

BOB's rolling stock fleet includes three ABeh 4/4 II 43-tonne motor coaches and matching BDt control trailers. Each motor coach is powered by four ABB 314 kW series-wound DC motors. The braking system for direct-coupled DC motors is a combined regenerative and resistance brake with automatic changeover according to conditions in the supply system. The added control functions are performed by ABB's MICAS programmable system. Maximum speed is 70 km/h. In 2004 Stadler delivered five articulated low-floor driving trailers, and since the end of 2004 push-pull operation has been usual.

The railway also operates the 7.3 km 800 mm gauge Riggenbach rack Schynige Platte mountain railway (SPB) from Wilderswil, electrified at 1.5 kV DC, and the Bergbahn Lauterbrunnen–Mürren (BLM). The BLM comprises a cable funicular from Lauterbrunnen to Grütschalp and a metre-gauge line on the Lauterbrunnen valley's western wall from Grütschalp to Mürren.

Wengernalp-Bahn

Recording over 2 million passenger journeys annually, WAB operates a 29.366 km 800 mm gauge line running from Grindelwald and Lauterbrunnen to Kleine Scheidegg, 2,060 m above sea level and immediately below the Jungfrau mountain chain. It is electrified at 1.5 kV DC, and employs Riggenbach rack throughout. In 2004 an interlocking system with Frauscher axlecounter track release, specially designed for multi-train operation, was introduced between Grindelwald and Kleine Scheidegg. A control centre is located in Grindelwald-Grund.

The Grindelwald–Kleine Scheidegg section is to be modernised at a cost of SFr47 million. Work will include provision of a double-track section and the modernisation of several stations.

The WAB fleet includes two locomotives for the operation of freight services to the village of Wengen, which is prohibited to road traffic. By the

BOB Type ABeh 4/4 II rack railcar at Zweilütschinen (Jungfraubahnen)
NEW/1107950

WAB Type Bhe 4/8 low-floor railcar with panoramic centre coach (Jungfraubahnen)
NEW/1107952

end of 2004 four Type Bhe 4/8 low-floor panoramic emus were due to be delivered by Stadler. These will provide a much higher passenger capacity than the trains they are to replace.

Jungfraubahn

JB records annually over 1 million passenger journeys. It operates an 11,827 km 1,000 mm gauge line starting from Kleine Scheidegg, which tunnels through the Jungfrau range to attain the highest altitude of any European railway at Jungfraujoch, 3,454 m above sea level, on the ridge between the Jungfrau and Mönch mountains. It employs the Strub rack system and is electrified at 1,125 V 50 Hz three-phase AC.

In 2002 Stadler delivered an additional four Type BDeh 4/8 units similar to those supplied in 1992.

UPDATED

Two JB Type BDhe 4/8 railcars in front of the Eiger (Jungfraubahnen)
NEW/1107951

Jura Railways (CJ)

Chemins de fer du Jura
1 rue Général-Voirol, CH-2710 Tavannes
Tel: (+41 32) 482 64 50   Fax: (+41 32) 482 64 79
e-mail: information@les-cj.ch
Web: www.les-cj.ch

Key personnel

Director: Georges Bregnard
Managers
   Finance and Administration: René Clémençon
   Infrastructure: Yves Cuenin
   Operations: Alain-Pierre Kohler
   Traction and Electrical Installations:
      Theodor Stolz
   Personnel: Christophe Froidevaux
Marketing and Communications: Frank Maillard

Gauge: 1,000 mm; 1,435 mm
Route length: 74 km; 11 km
Electrification: 74 km at 1.5 kV DC; 11 km at 15 kV AC 16.7 Hz

Organisation

Located in northwest Switzerland, the CJ network comprises a 51 km metre gauge line from Glovelier

to Le Chaux-de-Fonds with a 23 km branch from an intermediate station at Noirmont to Tavannes and an 11 km standard gauge line from Porrentruy to Bonfol.

CJ also operates bus services over a 92 km network.

In 2003 the company employed 131 staff for its rail operations.

Passenger and freight operations

In 2003 CJ carried 1.238 million passengers (18.9 million passenger-km) and 108,000 tonnes of freight (2.4 million tonne-km).

CJ Class Stadler-built ABe 2/6 metre gauge emu at Tramelan, with a short train of containerised domestic waste on the right (Milan Šrámek)
0561564

New lines

In December 2001 CJ initiated studies into an extension eastwards of its metre gauge line from Glovelier to Delémont with the aim of improving connections to Basle and eastern Switzerland. Approval of the project was expected in 2005, with completion foreseen in 2006–07.

Lausanne–Echallens–Bercher Railway (LEB)

Compagnie du Chemin de Fer Lausanne–Echallens–Bercher
Place de la Gare 9, Case postale 196, CH-1040 Echallens
Tel: (+41 21) 886 20 15   Fax: (+41 21) 886 20 19
e-mail: admin.leb@leb.ch

Gauge: 1,000 mm
Route length: 24 km
Electrification: 24 km at 1.5 kV DC

Passenger operations

The LEB provides a commuter service between Lausanne and communities to the north of the city, serving some 20 stations. In 2002 2.1 million passengers were carried, 80 per cent of these season ticket holders. Unusually, the railway is not directly linked to the Swiss Federal Railways national network; connections with it in Lausanne are made via light rail and rack metro services. Dining services are also operated and steam specials are run in high season. The railway employs 50 staff.

Freight operations

LEB freight traffic, mostly agricultural produce, is handled by a fleet of road vehicles.

Improvements to existing lines

In May 2000, a 1.1 km tunnelled extension was completed from the LEB's Lausanne terminus of Chauderon to a more conveniently located site at Flon. This has resulted in improved connections with Métro Lausanne–Ouchy (LO), Lausanne–Gare

Traction and rolling stock

At the end of 2003 the metre gauge fleet comprised three electric locomotives, 11 electric railcars, four articulated low-floor emus (Stadler, 2001), eight driving trailers, 10 coaches, one diesel shunter and one electric shunter, 64 freight wagons and 21 service vehicles. The standard gauge fleet consisted of one electric locomotive, two railcars and one driving trailer, two diesel shunters and 22 service vehicles.

UPDATED

LEB emu at Flon station 1020496

(LG) and Métro Ouest (TSOL) urban rail services and will also provide a link with the planned Métro Nord-Est line.

Traction and rolling stock

At the beginning of 2003, LEB operated two single-unit electric railcars and six two-car Class Be 4/8 emus, the latter dating from 1985–1991, as well as two driving trailers and seven trailer cars. One steam locomotive is also retained for special traffic. The maintenance fleet comprises two shunting tractors and seven operational wagons.

Matterhorn Gotthard Bahn AG (MGB)

Nordstrasse 20, CH-3900 Brig
Tel: (+41 27) 927 77 77   Fax: (+41 27) 927 77 79
e-mail: info@mgbahn.ch
Web: www.mgbahn.ch

Key personnel

Chairman: Hans-Rudolf Mooser
Finance, Controlling and Administration:
  Beat Britsch
Marketing and Services: Marcel Mooser
Rolling Stock and Traction: Fernando Lehner
Infrastructure: Willi In-Albon
Strategic Projects: Bernhard Glor
Railway Operations: Peter Rüttimann

Gauge: 1,000 mm
Route length: 144 km, of which 31.9 km Abt rack system
Electrification: 144 km at 11 kV 16²/₃ Hz AC

Organisation

The Matterhorn Gotthard Bahn AG (MGB) parent company was created in January 2003 by the merger of the BVZ Zermatt-Bahn (BVZ) and Furka Oberalp (FO) systems. Equal shares in the management company are owned by BVZ Holding AG and by the Swiss Confederation and cantons. Ownership of the system's track, power supply equipment, control centres, station facilities and workshops is vested in Matterhorn Gotthard Bahn Infrastruktur AG, which is wholly owned by the public sector (the Swiss Confederation and the cantons of Grisons, Uri and Valais). Railway operation and rolling stock ownership rests with Matterhorn Gotthard Bahn Verkehrs AG, 75 per cent of the shares of which are held by BVZ Holding AG and 25 per cent by the Swiss Confederation and cantons.

The combined MGB network comprises a line running northeast from Zermatt via Visp and Brig to Disentis, where a connection is made with the Rhaetian Railway (RhB). From Andermatt a line runs to Göschenen. The former BVZ portion is part adhesion, part Abt rack (four sections Visp–Zermatt, ruling gradient 12.5 per cent). Between Visp and Zermatt the railway climbs 955 m and is single-line throughout, with passing loops. The former FO section extends from Brig to Disentis and Göschenen and also employs the Abt rack system to negotiate maximum gradients of 11 per cent. The summit of the line is at Oberalppass

MGB Class HGe4/4 II electric locomotive with a passenger service at Brig (Brian Webber) 0561565

MGB Class HGe 4/4 II locomotive crossing the Richlerenbrücke, between Andermatt and Realp (MGB)   **NEW**/1110099

(2,033 m), between Andermatt and Disentis. There are 42 stations.

Centrepiece of the route is the 15.44 km Furka Base Tunnel, single track with two crossing loops, completed in 1982. It bypasses the former rack-worked section from Oberwald to Realp which used to close to traffic between October and April each year, lying mostly between 2,000 and 3,000 m above sea level. A private organisation, the Dampfbahn Furka-Bergstrecke AG, has restored the historic route of the 'Glacier Express' between Realp and Gletsch over the Furka Pass and the Rhône glacier, which was closed in 1982, and runs special trains on it.

Gornergrat Railway
MGB also manages the 1,000 m gauge Abt rack Gornergrat–Monte Rosa–Bahnen (GGB) from Zermatt to Gornergrat, 3,089 metres above sea level. This 9.3 km line, electrified at 725 V DC, operates three electric locomotives and 22 electric railcar sets.

Passenger and freight operations
The combined railway expects to carry some 2.5 million passengers and 100,000 tonnes of freight annually. The most notable passenger service is the 'Glacier Express', which runs from Zermatt to Disentis and onward via the RhB to St Moritz.

Shuttle services carrying cars and their occupants operate between Oberwald and Realp via the Furka Base Tunnel. They convey around 180,000 vehicles annually.

To facilitate the bulk transport of domestic refuse from Zermatt, MGB employs the Tuchschmid ACTS container transfer system.

From 2003 the railway commenced moving large volumes of material generated by construction of the Gotthard Base Tunnel, which forms part of the AlpTransit project. This traffic uses the former FO section of line between Disentis and Sedrun, which has been upgraded.

Improvements to existing lines
Development plans include the 'Matterhorn Terminal Täsch' scheme to improve facilities at this location, work on which started in 2004 for completion by 2007, a project to create a full interchange with SBB and improved freight facilities at Visp station and a major realignment of the approach to Brig which in 2012 will see MGB trains diverted into the SBB station.

Traction and rolling stock
The former BVZ fleet comprises nine electric, two diesel-electric and one historic steam locomotive; four diesel tractors; nine electric motor coaches; nine driving trailers; 58 hauled passenger cars; four baggage and mail cars; and 90 freight and service wagons. The electric locomotive fleet is headed by five HGe 4/4 II 1,932 kW electric locomotives similar to those of the former FO fleet and Brünig line of SBB.

The former FO fleet consists of two diesel-electric, one diesel-hydraulic and 16 electric locomotives; five Type BDeh 2/4 and 11 Type Deh 4/4 electric power cars; 79 passenger coaches; four baggage cars; 30 car-carriers; 67 freight wagons; and maintenance vehicles including nine powered units and 11 powered snow-ploughs. Passenger stock includes four panoramic saloons with deep side and roof windows.

In 2000 BVZ ordered two three-car articulated emus from Stadler, with traction equipment supplied by Bombardier Transportation. Designated Class BDSeh 4/8, these entered service between Täsch and Zermatt in 2003.

In 2003 24 new coaches for 'Glacier Express' services were acquired jointly by FO, BVZ and RhB.

In 2002 BVZ took delivery of a Class HGm 2/2 diesel-electric locomotive equipped for the Abt-rack system to enable it also to operate on the GGB line.

UPDATED

Mittelland Regional Railways (RM)

Regionalverkehr Mittelland
Bucherstrasse 1-3, CH-3401 Burgdorf
Tel: (+41 34) 424 50 00
Fax: (+41 34) 424 50 80
e-mail: info@rm-rail.ch
Web: www.regionalverkehr.ch; www.crossrail.ch

Key personnel
Director: Martin Selz
Manager, Central Services: Franz Bieri
Manager, Passenger: Joseph R Zeder
Manager, Logistics and Freight: Otmar Halfmann
Infrastructure Manager: Peter Dübi
Workshops Manager: Ruedi Beutler

RM Stadler GTW 2/8 articulated emu forming a Bern S-Bahn S44 service at Langnau (Philip Wormald)
NEW/1114949

Electric locomotives

Class	Wheel arrangement	Output kW	Speed km/h	Weight tonnes	No in service	First built	Builders Mechanical	Electrical
Re 436	Bo-Bo	4,656	120	80	5	1969	SLM	MFO, BBC
Re 456	Bo-Bo	3,000	125	68	2	1993	SLM	ABB

Electric railcars or multiple-units

Class	Cars per unit	Motor cars per unit	Motored axles/car	Output/ motor kW	Speed km/h	No in service	First built	Builders Mechanical	Electrical
GTW 2/6 526	2	1	2	550	140	6	2003	Stadler	ABB, Traktionssysteme Austria
GTW 2/8 526	3	1	2	550	140	7	2003	Stadler	ABB, Traktionssysteme Austria
RBDe 566	2	1	4	335	125	8	1973	WS	SAAS, ABB, EBT
RBDe 566 II	2	1	4	412	125	13	1984	SWP/SIG	BBC
RBDe 566 II	3	1	4	515	110	3	1966	SIG	BBC

Subsidiary
Crossrail
Am Strackbach, Postfach, CH-3428 Wiler
Tel: (+41 32) 666 36 36
e-mail: info@crossrail.ch
Web: www.crossrail.ch
Contacts: Diana Hennecke; Holger Ulrich

Gauge: 1,435 mm
Route length: 167 km
Electrification: 155.8 km at 15 kV 16²/₃ Hz AC

Organisation
RM was created in 1997 by merger of the Emmentahl-Burgdorf-Thun Railway with railways which had been under its management: the Solothurn-Moutier (SMB); Solothurn-Burgdorf-Hasle-Rüegsau-Konolfingen-Thun and Hasle-Rüegsau-Langnau (EBT); and Ramsei-Sumiswald and Langenthal-Huttwil-Wolhusen (VHB). The RM group of railways connects the north and south of the Canton of Bern, and also reaches out to the hinterland of Lucerne.

In 1999 RM was re-organised into five profit centres covering passenger traffic, freight traffic, infrastructure management, workshops and central services. RM also operates bus services.

In January 2004 RM launched Crossrail, a cross-border freight service (see below) provided in partnership with German operator RAG Bahn- und Hafenbetrieb. Another subsidiary, Cargodrome provides logistics and distribution services from centres in Wiler, Switzerland, and Domodossola, Italy.

Traffic (million)	2001	2002
Passenger journeys	7.6	8.3
Passenger-km	87	94.5
Freight tonnes	–	–
Freight tonne-km	119.2	118.4

Passenger operations
There are eight interchange stations with SBB, and RM connects with and is now part of the Bern S-Bahn system, having operated routes S4 and S44 (Langnau–Burgdorf–Bern) since 1999. Passenger services on minor branches have been replaced by bus (Sumiswald-Wasen in 1994), but the intensification of services on through routes and a growing freight business have required conversion to double track of several critical sections.

Freight operations
A significant development in RM's freight activities occurred in January 2004 when the company launched Crossrail, a subsidiary providing a scheduled cross-border freight linking Dörpen (Duisburg), Germany, with Wilen, Switzerland, and Domodossola in Italy. Crossrail's partner in Germany is RAG Bahn- und Hafenbetrieb, which handles the Switzerland–Germany sector using a Class Re 482 electric locomotive.

Traction and rolling stock
At the beginning of 2004, the RM fleet included seven main line electric locomotives and 37 emus (including vehicles under delivery). The most recent additions to the latter included 13 GTW 2/6 and GTW 2/8 articulated electric railcars from Stadler, deliveries of which commenced in 2003.

In 2004 RM took delivery of a Vossloh G1202 diesel-hydraulic locomotive for trip working for its Crossrail and other freight services.

UPDATED

Montreux–Oberland Bernois Railway (MOB)

Chemin de Fer Montreux–Oberland Bernois
PO Box 1426, Rue du Lac 36, CH-1820 Montreux 1
Tel: (+41 21) 989 81 81   Fax: (+41 21) 963 89 96
e-mail: mob@mob.ch
Web: www.mob.ch
www.goldenpass.ch

Key personnel
President, Managing Board: W von Siebental
Group Director General: Richard Kummrow
Managing Director: M Sandoz
Finance Director: J Brouze
Marketing Manager: H-J Spirgi
Traffic Manager: G Verdan
Way and Works Manager: G Bridevaux
Traction and Workshops Manager: J M Forclaz

Gauge: 1,000 mm; 800 mm
Route length: 85.7 km; 10 km
Electrification: 95.7 km at 860 V DC

Organisation
The MOB Group comprises three electrified railways, funiculars (including the automatic 1.6 km Vevey–Chardonne–Mont Pèlerin Railway), ski lifts, coach operations, travel agencies and hotels and restaurants. The principal railway is the 75.3 km 1,000 mm gauge Montreux–Oberland Bernois (MOB); the Montreux–Territet–Glion–Rochers de Naye (800 mm gauge, 10 km) and Vevey Electric (CEV) (1,000 mm gauge, 10.4 km) railways have an essentially local role.

MOB's main line runs from Montreux via Gstaad to Zweisimmen. Its climb from Lake Geneva to Les Avants and the 2.4 km Col de Jaman Tunnel, at 7.3 per cent ruling grade, is the steepest adhesion line in Switzerland. The summit is 1,269 m above sea level. From Zweisimmen a branch runs to Lenk.

Passenger operations
Starting life as a local cross-country line, MOB began to court the tourist market in 1979 with

Diesel locomotives

Class	Wheel arrangement	Power kW	Speed km/h	Weight tonnes	No in Service	First built	Mechanical	Builders Engine	Transmission
Gm 4/4	Bo-Bo	575	80	44	2	1976	Moyse	Poyaud	E Moyse-Leroy-Sommer
Tm 2/2	B-B	115	33	15	2	1953*	KHD	Deutz	H Voith-Turbo

* Rebuilt 1983–84.

Electric locomotives

Class	Wheel arrangement	Output kW	Speed km/h	Weight tonnes	No in service	First built	Builders Mechanical	Electrical
De 6/6	Bo-Bo-Bo	1,230	55	63	2	1931	SIG	ABB
GDe 4/4	Bo-Bo	1,432	100	50	4	1983	SLM	ABB
Ge 4/4	Bo-Bo	2,400	120	62	4	1995	SLM	ABB

MOB 'Golden Pass Panoramic' service operated as a push-pull formation with a Class Ge 4/4 electric locomotive (MOB)
NEW/0585041

introduction of the 'Panoramic Express'. The 'Super Panoramic Express' followed in 1985, and the 'Crystal Panoramic Express' in 1993. Completely new stock was provided in May 2000, running between Montreux and Lenk as the 'Golden Pass Panoramic', while 1999 saw the introduction of a new trainset on the CEV, operating under the name of 'Train des Etoiles'.

In June 2001 it launched a marketing alliance with BLS Lötschbergbahn and Swiss Federal Railways' metre-gauge Brünig line to promote integrated panoramic rail travel from Montreux to Lucerne via the Bernese Oberland. A development of the MOB's 'Golden Pass Panoramic' service concept, the 'Golden Pass Line' brand name has been adopted and locomotives and rolling stock used by each operator on the group of services carry a gold and white livery. Routes served are: Montreux–Zweisimmen (MOB); Zweisimmen–Interlaken (BLS, standard gauge); and Interlaken–Lucerne (SBB Brünig line). Together these total 250 km.

Traction and rolling stock

MOB operates eight electric and two diesel locomotives, two diesel shunting tractors, 20 motored passenger units, 52 passenger cars and 112 freight wagons, plus numerous service vehicles. Four new Stadler/Adtranz/SLM Class Be 2/6 low-floor electric railcars entered service with the CEV in 1998.

UPDATED

Rhaetian Railway (RhB)

Rhätische Bahn
Bahnhofstrasse 25, CH-7002 Chur
Tel: (+41 81) 288 61 00   Fax: (+41 81) 288 61 01
Web: www.rhb.ch

Key personnel

Director: Erwin Rutishauser
Chief Engineer: Christian Florin
Chief Mechanical Engineer:
    Johannes Georg Bühler
Chief of Finance and Services: Silvio Briccola
Chief of Marketing and Operations:
    Ernst Bachmann

Gauge: 1,000 mm
Route length: 385 km
Electrification: 325 km at 11 kV 16²/₃ Hz AC; 60.7 km at 1 kV DC

Organisation

The Rhaetian Railway is a vital means of communication in the mountainous southeast of Switzerland. Serving the Engadine, the valley of Poschiavo, the Davos area, Arosa and the Grisons Oberland, the railway connects the canton with the SBB network at Chur and Landquart, with the Furka-Oberalp Railway in Disentis/Muster, and with FS of Italy in Tirano. The territory is the most sparsely populated in Switzerland, with a population of some 170,000 averaging 23 per km².

The core of the RhB network is electrified at 11 kV 16²/₃ Hz, but the Bernina Railway from St Moritz to Tirano was electrified at its construction in 1908–10 at 1 kV DC and retains that system. The Bernina is the only Swiss transalpine line that avoids tunnelling, attaining a summit of 2,253 m above sea level at Ospitio Bernina; for 27 km, or 44 per cent of its total distance, it is graded at 7 per cent but is worked entirely by adhesion. The Chur–Arosa Railway of 26.4 route-km, which RhB absorbed in 1943, was converted from a 2.4 kV DC power supply system to the standard 11 kV AC on 1 December 1997; on this line the ruling gradient is 6 per cent.

At the end of 2003 the 13 km Castione–Cama line was closed by RhB.

In 2003 RhB employed 1,497 staff.

Traffic (million)	2002	2003	2004
Passenger journeys	8.752	10.263	9.118
Passenger-km	299.4	321.9	316.1
Freight tonnes	0.785	0.819	0.737
Freight tonne-km	53.9	56.0	52.5

Passenger operations

RhB passenger traffic dropped to 316.1 million passenger-km in 2004 from 321.9 million the previous year.

Freight operations

Main constituents of RhB's considerable freight traffic are construction materials, cement, drinks, chemicals, petroleum products, fuel oil and timber. The wagon fleet includes sliding wall, insulated vans with electric heating and cooling to provide for winter haulage of fresh produce; power is provided by a busline from the train's locomotive.

RhB has developed container and swap-body traffic between Landquart and St Moritz. It has also developed a Bernina line terminal on the Italian frontier at Campocologno, believing that this route has all-weather advantages for traffic from northern Italy.

Class ABe 4/4 electric railcar with a St Moritz–Tirano service (Milan Šrámek)   0552002

During 2000 RhB took delivery of 21 wagons designed to carry ACTS roll-on/roll-off containers. These are used for domestic waste traffic.

New lines

Commercial services began in November 1999 on RhB's 22.3 km cut-off from Klosters to Lavin, which halves journey times between Landquart and the Lower Engadine. Journey times to Scuol-Tarasp from Zurich and Chur have been cut by 135 and 100 minutes to 2 hours 40 minutes and 1½ hours respectively. The cut-off includes the 19.05 km single-track Vereina Tunnel under the Silvretta mountain range. The tunnel has one passing loop and a ruling gradient of 1.5 per cent, and is large enough to accommodate wagons conveying road vehicles up to 4 m in height. Maximum speed through the tunnel is 100 km/h, allowing passage in 17 minutes.

From the Selfranga vehicle-loading terminal at Klosters, a semicircular single-line tunnel on a gradient of 2.5 per cent leads to the Vereina Tunnel portal via a short section of track in the open. The south terminal is at Sagliains, between Susch and Lavin. Each terminal has two loading tracks.

The standard hourly service in each direction is one conventional passenger train and one or two vehicle shuttles; the latter increases to three in peak periods. There are one or two daily freight services.

Traction and rolling stock

In 2003 RhB operated 57 electric, two diesel and two electro-diesel locomotives; 28 electric railcars; 13 diesel and 10 electric shunting tractors; three steam locomotives (used for special trains); 373 passenger coaches (including 10 restaurant cars; 32 baggage and mail cars); and 824 revenue freight wagons, including 73 car-carrying vehicles.

In 2002 and 2003 RhB rebuilt its two Class Gem 4/4 electro-diesel locomotives, equipping each with two Cummins 709 kW engines, new electrical equipment and improved cabs. Their 1 kV DC capability enables them to operate on RhB's Berninabahn, linking St Moritz and Tirano.

In 2000 Stadler supplied 10 Panorama coaches and eight driving trailers for RhB 'Bernina Express' services. In 2003, jointly with the Matterhorn Gotthard Bahn, RhB ordered four new trains, each of six coaches from the same builder for 'Glacier

RhB Class Ge 4/4 III electric locomotive at Thusis with a Chur–St Moritz passenger service (John C Baker)   *NEW*/0585039

Two RhB Class Ge 4/4 II electric locomotives at Domat/Ems with a Lanquart-Pontresina freight
service (John C Baker)   **NEW**/0585040

Express' services. To enter service in 2006, these
comprise five Panorama coaches and one service
vehicle. Also in 2003, RhB ordered 10 additional
Panorama wagons for 'Bernina Express' services
from Stadler, these entering service from 2006.

A refurbishment and modernisation programme
covering 100 coaches, including 60 Type EW II, was
in progress in 2004 at RhB's Landquart workshops
for completion in 2007.

Signalling and telecommunications

The RhB system, mostly single track, is colour light
signalled with automatic block under the oversight
of seven control centres. The majority of passing
loops can be switched for automatic operation
by trains when their stations are unmanned. The
entire system and all power units are equipped
with the Integra 79 ATC system. Ground-to-train
radio has been installed throughout the railway's
network. The RhB has always relied upon Hardy
vacuum brakes for its passenger and freight trains,
enhanced by electric braking on all locomotives.

UPDATED

SBB Cargo

Schweizerische Bundesbahnen Cargo AG
PO Box 4065, Basel
Tel: (+41 51) 229 00 00   Fax: (+41 51) 229 00 01
Web: www.sbbcargo.com

Key personnel

Chairman: Dr Benedikt Weibel
Chief Executive Officer: Daniel Nordmann

Background

Created in 2000, SBB Cargo is the wholly owned
freight subsidiary of the state-owned Swiss
Federal Railways (SBB). Creation of a separate
company to spearhead the railway's freight
business was largely driven by new opportunities
and the competitive environment created by open
access legislation affecting the rail networks of
Switzerland's neighbours. SBB Cargo's response
has been to establish its own subsidiaries in
contiguous countries, as well as partnership
arrangements with foreign-based operators,
to enable it to project its services beyond the
domestic market.

Organisation

Headquartered in Basel, SBB Cargo has three
operational divisions covering commercial
relationships, production and rolling stock
maintenance and support divisions responsible for
business development, communications, finance,
human resources, IT and strategic planning.
Relationships with clients are managed from SBB
Cargo's Customer Service Center at Fribourg. At
the end of 2003 SBB Cargo and its subsidiaries
employed 4,898 staff, down from 5,130 in 2002.

Subsidiaries

- Swiss Rail Cargo Italy srl (SRCI) (100 per cent
  shareholding
- Swiss Rail Cargo Köln (SRCK) (51 per cent)
- ChemOil Logistics AG (51 per cent with
  Transpetrol (49 per cent))
  Entries for SRCI and SRCK appear in the Italy and
Germany sections of *Railway systems and operators*.

SBB Cargo continues to work with its traditional
partners in Austria (Rail Cargo Austria), France
(SNCF Fret), Germany (Railion/DB Cargo) and Italy
(Trenitalia Cargo). It is also acquiring new partners:
since April 2003 FN Cargo (Ferrovie Nord Milano)
in Italy has been operating trains for SBB Cargo on
the Chiasso–Desio route and in October 2003 SBB
Cargo and ERS Railways BV signed an agreement
covering services between Rotterdam and northern
Italy. The company also co-operates in Germany
with Hafen und Güterverkehr Köln (HGK) (see

entry in Germany section of *Railway systems and
operators*), which has a shareholding in SRCK.

Traffic (million)	2001	2002	2003
Freight tonnes	59.0	54.93	54.78
Freight tonne-km	10,534	9,732	9,936

Finance

In 2003 SBB Cargo reduced its annual loss from
SFr96.1 million in 2002 to SFr33.1 million, an
improvement of 65.6 per cent and a significant
step in the company's goal of achieving a positive
result in 2005. At SFr 1.062 billion, traffic revenues
were slightly down on 2002 (SFr1.076 billion) but
operating expenses decreased by 4.4 per cent.

Freight operations

Switzerland's geographic location dictates large
parts of SBB Cargo's activity, with some 50 per cent
of its business by tonne-km (5,017 million in 2003)
accounted for by transit traffic, mostly on the
north-south axis through the Alps to Italy. However,
domestic traffic – especially wagonload – is also a
major source of revenue, with SBB Cargo servicing
650 freight delivery point and 2,450 customers with
private sidings. Overall, the company operates
around 2,300 trains daily.

SBB Cargo provides the following services for
originating traffic:
- Cargo Rail: conventional wagonload
- Cargo Rail Intermodal: domestic wagonload
  intermodal

- Cargo Express: fast wagonload service for
  commodities such as perishables
- Cargo Train Flexi and Cargo Train Fix: short- and
  long-term trainload
- Cargo Domino: domestic door-to-door intermodal
  service, including road collection/delivery
- Intermodal: domestic and international intermodal

Wagonload traffic

In 2003 SBB Cargo implemented a plan aimed
at cutting the losses generated by its wagonload
business while aiming to retain good levels of
service. The first phase of the plan, introduced
in December 2003, brought a major revision of
the timetable for wagonload services. A second
phase initiated in July 2004 saw preliminary
work to adopt remote control shunting more
widely. There has also been a simplification of
day-to-day operations management, with the
number of 'team stations' overseeing wagonload
services reduced from 120 to 43. SBB Cargo is
also upgrading some of its marshalling yards,
with expansion and modernisation of facilities at
Biel/Bienne and Lausanne due for completion by
the end of 2006.

Transalpine traffic

With open access in place in Switzerland,
SBB Cargo faces an increasingly competitive
environment. In particular, Switzerland's other
major transalpine operator, BLS Cargo, together
with its German partner DB Cargo. BLS Cargo's

SBB Cargo Class 620 (Re 6/6) electric locomotive at Mels with a MegaKombi intermodal service
which originated in Austria (John C Baker)   **NEW**/0585054

Diesel locomotives (principal classes)

Class	Wheel arrangement	Power kW	Speed km/h	Weight tonnes	No in service	First built	Builders Mechanical	Engine	Transmission
830 (Em 3/3)	C	450	65	49	29	1959	SLM	SLM	E ABB/Sécheron
Em 831	C	900	60	54	3	1992	RACO	Cummins	H
Bm 840 (Bm 4/4)	Bo-Bo	620	75	72	25	1960	SLM	SLM	E Sécheron
Am 841	Bo-Bo	920	80	80	18	1996	GEC Alsthom	MTU 8V 396	E GEC Alsthom
Am 842.1	B-B	1,100	100	. –	2	2003	Vossloh	–	H Voith
Am 843	B-B	1,500	100	80	37	2002	Vossloh	Caterpillar 3512 B	H Voith
Bm 860 (Bm 6/6)	Co-Co	1,270	75	106	14	1954	SLM	2 × Sulzer 6LDA25	E BBC/SAAS
Am 861	Co-Co	1,440	85	111	6	1976	Thyssen-Henschel	Chantiers de l'Atlantique	E ABB

Electric locomotives (principal classes)

Class	Wheel Arrangement	Power kW	Speed km/h	Weight tonnes	No in service	First built	Builders Mechanical	Electrical
Re 420	Bo-Bo	4,700	140	80	105	1964	SLM	ABB/Oerlikon/Sécheron
Re 421*	Bo-Bo	4,700	140	80	271	1964	SLM	ABB/Oerlikon/Sécgeron
Re 430	Bo-Bo	4,650	125	80	21	1971	SLM	ABB/Oerlikon/Sécheron
Re 474**	Bo-Bo	6,400	140	87	18	2004	Siemens	Siemens
Re 481	Bo-Bo	4,200	140	80	6	2000	Adtranz	Adtranz
482†	Bo-Bo	5,600	140	86	50	2002	Bombardier	Bombardier
Re 484**	Bo-Bo	5,600	140	86	18	2004	Bombardier	Bombardier
Ae 610	Co-Co	4,300	125	120/124	112	1952	SLM	ABB/Oerlikon
Re 620	Bo-Bo-Bo	7,850	140	120	88	1972	SLM	ABB/Sécheron
Ee 930 (Ee 3/3)	C	500	40	45	–	1930	SLM	BBC/MFO/SAAS
Ee 961 (Ee 6/6 II)	Co-Co	730	85	107	10	1980	SLM	ABB

* Some modified to operate on the German network.
** Four-system (15 kV/25 kV AC, 3 kV/1.5 kV DC).
† Two-system (15 kV/25 kV AC).

Class 843 (Type MaK 1700BB) diesel-hydraulic locomotive supplied by Vossloh (Vossloh Locomotives) **NEW**/1036757

Class 484 four-system 5,600 kW electric locomotive (Bombardier Transportation) **NEW**/0585055

Lötschberg route has benefited from upgrading which makes it more suitable for 'Rola' piggyback traffic. The Brenner route from Austria to Italy also provides an alternative north-south freight route through the Alps. In addition, from 2005 the Swiss weight limit for lorries rose from 34 to 40 tonnes, increasing road transport's competitiveness.

SBB Cargo has responded both by strengthening its operational activities and by introducing operating efficiencies. In January 2003 Swiss Rail Cargo Italy (SRCI) was established following the failure of a 2001 project to establish a joint venture with FS Trenitalia Cargo in Italy. SRCI commenced operations in December 2003. The formation of an Italian subsidiary followed the 2002 launch of Swiss Rail Cargo Köln, which has extended SBB Cargo's reach into the important German market and successfully captured business on the Cologne–Basel axis.

To improve the performance of its transalpine services, in December 2003 SBB Cargo applied regular-interval timetabling to most of its trains using the Gotthard and Simplon routes. Under this system, trains depart at their scheduled time with available traffic rather than operating as specific services, improving the utilisation of locomotives and crews. By 2004 some 80 per cent of services were run according to this system.

ChemOil Logistics AG
SBB Cargo has a 51 per cent shareholding in ChemOil Logistics AG, which provides logistics services based on rail for chemical products, liquid gases, mineral oils and bulk and powdered commodities. The Basel-based company's other partner is Transpetrol GmbH, part of the VTG Group.

Basel-Nord gateway project
Land has been purchased by SBB Cargo in Basel that by 2008 will form a 13.7 hectare facility for reforming intermodal freight services. This will operate by using cranes to exchange containers and swapbodies between trains. As well as its primary role of handling international traffic, the yard will also act as a domestic intermodal terminal for the Basel region, replacing the existing Basel-Wolf facility.

Traction and rolling stock
At the end of 2003 SBB Cargo operated 463 main line electric locomotives, 128 shunters/trip locomotives (of which 74 were diesel-powered) and 152 shunting tractors (including 107 diesel, 45 dual-mode). There were 12,171 wagons, supplemented by a further 7,139 privately owned vehicles operated by SBB Cargo.

Electric locomotive acquisitions
In 2004 SBB cargo received the first of 18 Class 474 6,400 kW (on AC supply) dual-voltage electric locomotives from Siemens. Based on the design of the manufacturer's Type ES 64 F4 (Class 189) for Railion, Germany, these are four-system machines capable of operating from 25 kV AC and 3 kV and 1.5 kV DC power supplies as well as the 15 kV AC system used in Switzerland and Germany.

SBB Cargo has also been procuring new electric locomotives from Bombardier. In 2002 the company started taking delivery of 50 5,600 kW Class 482 15 kV/25 kV AC machines similar to the Railion Class 185. Subsequently SBB Cargo ordered 18 examples of a dual-voltage (15 kV AC/3 kV DC) version of this design, equipped to work into Italy. Designated Class 484, the first of these was rolled out in 2004.

Diesel locomotive acquisitions
To modernise its fleet of traction for shunting and trip working, SBB Cargo has been taking delivery of 40 Class 843 1,500 kW diesel-hydraulic locomotives (Type MaK 1700BB) from Vossloh Locomotives, the major part an order for 59 machines that includes 14 for SBB's infrastructure division and five for its passenger business. The type entered service in 2004 and delivery was due to be completed in mid-2005.

Vossloh is also the supplier of two Class 842.1 (Type MaK 1000BB) diesel-hydraulic locomotives hired to operate SBB Cargo services on several lines in Germany and of three Class 840 (Type MaK 2000BB) diesel-hydraulics for its Italian subsidiary, SRCI (see entry in Italy section of *Railway systems and operators*).

Other fleet developments
A batch of 27 Class 420 (ex-Class 4/4 II) electric locomotives has been modified for use on the German network, becoming Class 421. All Class 620

(ex-Class Re 6/6) locomotives are assigned to SBB Cargo, but by the end of 2004 the modern Class 460 electric locomotives had all been allocated to the passenger division.

The six Class 481 electric locomotives are similar to the Railion Class 145 and were originally acquired by the Mittel, passing to SBB in 2002 following the former company's bankruptcy. They operate in Germany.

Wagons
Recent new wagon deliveries or orders have been for sliding-wall vehicles, flats for Cargo Domino swapbodies and Cargo Domino containers.

As well as servicing its own rolling stock, SBB Cargo undertakes maintenance for other operators.

NEW ENTRY

Sihltal–Zurich–Uetliberg Railway (SZU)

Sihltal–Zurich–Uetliberg Bahn
Manessestrasse 152, CH-8045 Zurich
Tel: (+41 1) 206 45 11   Fax: (+41 1) 206 45 10
e-mail: info@szu.ch
Web: www.szu.ch

Key personnel
Chairman: Dr Thomas Wagner
Director: Clemens Schöb

Gauge: 1,435 mm
Route length: 29 km
Electrification: 19 km (Sihltal line) at 15 kV 16⅔ Hz; 10 km (Uetliberg line) at 1.2 kV DC

Organisation
SZU operates an intensive passenger service from Zurich over two routes, one to Sihlbrugg (the Sihltal line) and the other to Uetliberg. These form routes S4 and S10 respectively of the Zurich S-Bahn system. Sister companies operate the Felseneggbahn cable car from

Adliswil to Felsenegg and buses in the Zimmerberg region.

Traction and rolling stock
SZU operates six Class Re 4/4 3,200 kW electric and three diesel shunting/service locomotives, 16 powered emu railcars and 29 trailers and other passenger coaches. The locomotive-powered formations are used on the Sihltal (S4) line.

UPDATED

South Eastern Railway (SOB)

Schweizerische Südostbahn AG
Bahnhofplatz 1a, Postfach, CH-9001 St Gallen
Tel: (+41 71) 228 23 23   Fax: (+41 71) 228 23 33
e-mail: info@sob.ch
Web: www.sob.ch

Key personnel
Director: Dr Guido Schoch
Managers
  Marketing and Services: Ernst Wittmer
  Production (Operations): Heinrich Güttinger
  Finance: Thomas Mangold
  Technical: Frédy Vogler
  Infrastructure: Marcel Latscha

Subsidiary company
Voralpen-Express (jointly with SBB)
c/o SOB, Bahnhofplatz 1a, Postfach, CH-9001 St Gallen
Tel: (+41 71) 228 23 23   Fax: (+41 71) 228 23 33
e-mail: info@voralpen-express.ch
Web: www.voralpen-express.ch

Gauge: 1,435 mm
Route length: 115.1 km
Electrification: 115.1 km at 15 kV 16⅔ Hz AC

Organisation
In December 2001 the SOB and the Bodensee–Toggenburg Railway (BT) merged, more than doubling the route length of this important cross-country system. The original SOB segment runs from Rapperswil to the SBB station at Pfäffikon, and from there to connect with the SBB at Arth-Goldau. Branches serve Wädenswil and Einsiedeln. The former BT line starts at the SBB station at Romanshorn on Bodensee (Lake Constance), joins SBB tracks to St Gallen and then runs on SOB's own line to Wattwil, where it branches to its terminus at Nesslau-Neu St Johann. From Wattwil, however, SOB operates over SBB tracks to Rapperswil. Together, these two portions of the network form a through Romanshorn–St Gallen–Lucerne route.

Passenger operations
The principal service is the hourly 'Voralpen-Express', operated jointly with SBB. This connects Romanshorn, St Gallen, Wattwil, Pfäffikon, Arth-Goldau and Lucerne. The operation and marketing of this branded service is managed by Voralpen-Express, the management organisation established jointly by SOB and SBB.

SOB also runs S-Bahn services – two in the St Gallen area and one as part of the Zurich S-Bahn. Other services are regional, in mostly rural areas, and mainly served by hourly trains.

Traction and rolling stock
In 2005 SOB operated a total of 13 electric locomotives, 22 emus/electric railcars, 17 tractors

for shunting and maintenance duties (including a four-axle diesel unit) and four historic locomotives, including an original steam engine dating from 1910. Noteworthy among the electric locomotives are four Class Re 446, formerly SBB Class Re 4/4 IV, prototypes for 160 km/h passenger services, and eight Class Re 456, the first rectifier engines in Switzerland, built in 1986.

Electric locomotives

Class	Wheel arrangement	Output kW	Speed km/h	Weight tonnes	No in service	First built	Builders Mechanical	Electrical
Re 446*	Bo-Bo	4,960	160	80	4	1982	SLM	BBC
Re 456	Bo-Bo	3,200	130	68	8	1987	SLM	ABB
Re 476**	Bo-Bo	2,920	100	84	1	(1994)	Hennigsdorf	AEG

* Ex-SBB Class Re 4/4 IV.
** Former DR Class 142.

Electric railcars

Class	Cars per unit	Motor cars per unit	Motored axles/car	Output/ motor kW	Speed km/h	Weight tonnes	No in service	First built	Builders Mechanical	Electrical
BDe 576	1	1	4	526	110	72	12	1959-79	SIG	MFO/BBC
RBDe 566	2	1	4	425	125	71	6	1982	FZA	ABB
RBDe 566	2	1	4	425	140	70	4	1995	SWP	ABB
RABe 526*	4	2	2	650	160	118	11	2007	Stadler	ABB

* Vehicles on order.

Class Re 466 electric locomotive heading a Lucerne-bound Voralpen-Express service formed of 'revvivo' coaches near Mogelsberg (SOB)
NEW/1146736

For details of the latest updates to *Jane's World Railways* online and to discover the additional information available exclusively to online subscribers please visit

jwr.janes.com

In 2007 SOB will receive 11 Class RABe 526 low-floor FLIRT emus supplied by Stadler to replace electric railcars dating from the 1960s.

Coaching stock comprises 106 cars, including buffets, driving trailers, special cars for cycle transport and the only complete historic composition surviving from the opening of former BT railway in 1910. All cars are non-smoking, so-called 'fresh-air' vehicles. Locomotive-hauled stock includes the "revivo" stock used for the Voralpen-Express, fully air-conditioned vehicles created by installing new glass-fibre modular interiors into existing bodyshells.

Class RBDe 566 three-car emu forming an Einsiedeln-Rapperswil service (SOB)   *NEW*/1146735

Track

Rail: SBB-profile type I, 46 kg/m
Crossties (sleepers): Steel, concrete, wood, 150 × 260 mm
Spacing: 1,667/km
Fastenings: K and W on wood and steel sleepers. A on steel sleepers, B on concrete sleepers
Min curvature radius: 143 m

Max gradient: 5%
Max axleload: 22.5 tonnes

The former BT section includes the highest railway bridge of Switzerland, the Sitter Viaduct near St Gallen, with a track level 99 m above the river Sitter and built in 1908. With several km at gradients of 5 per cent, the former SOB section is the steepest main line in Switzerland, necessitating a banking locomotive for heavier Voralpen-Express services.

UPDATED

Swiss Federal Railways

Schweizerische Bundesbahnen (SBB AG)
Chemins de Fer Fédéraux Suisses (CFF)
Ferrovie Federali Svizzere (FFS)
Hochschulstrasse 6, CH-3000 Bern 65
Tel: (+41 51) 220 11 11
Fax: (+41 51) 220 42 65
e-mail: railinfo@rail.ch
Web: http://www.rail.ch

Key personnel

Board of Directors
  President: Thierry Lalive d'Epinay
  Chief Executive Officer: Benedikt Weibel
  Finance: Claude-Alain Dulex
  Personnel: Daniel Nordmann
  Passenger Traffic: Paul Blumenthal
  Freight Traffic: Daniel Nordmann
  Infrastructure: Pierre-Alain Urech
  Secretary General: Peter Fülistaler
  Deputy Secretary General: Walter Moser
  Corporate Risk Management:
    Annette Zimmerli
  Head of Communications: Werner Nuber
Passenger Traffic Business
  Director: Paul Blumenthal
  Long-Distance Traffic: Vincent Ducrot
  Finances: Guy Luginbühl
  Business Development: Peter Grossenbacher
  Project Expo 01: Markus Dössegger
  Project ICN: Theo Weiss
  Regional Traffic: Philippe Gauderon
  Customer Services: Peter Lehmann
  Production: Hannes Wittwer; Serge Anet
  Logistics, Personal Security: Buchs Daniel
  Rolling Stock and Maintenance:
    Ferdinando Gianella
  Personnel: Annick Kalantzopoulos
Freight Business (SBB Cargo)
  Director: Daniel Nordmann
  Marketing: Jürg Scheidegger
  Business Development: Hanspeter Vogel
  Information Technology: Burkhard Schulz
  Home Sales: Raphael Waeber
  Foreign Sales: Dirk Broek
  Production: Nicolas Perrin
  Human Resources: Thomas Aebischer
  Customer Services Centre: Samuel Ruggli;
    Dominique Boucrot
Infrastructure Business
  Director: Pierre-Alain Urech
  Rail 2000 and Major Projects: Paul Moser
  Logistics of Works and Purchases: Max Lehmann
  Telecommunications Management:
    Eduard Stiefel
  Development and Technology: Peter Winter
  Network and Path Management, Finances:
    Hans-Jürg Spillmann
  Management of Installation: Erwin Rutishauser
  Management of Maintenance: Reto Burkhardt
  Operational Management: Felix Loeffel
  Property Management: Urs Schlegel
  Personnel: Eric Pétremand

Gauge: 1,435 mm; 1,000 mm
Route length: 2,836 km; 74 km
Electrification: 1,435 mm gauge, 2,836 km at 15 kV 16²/₃ Hz AC; 1,000 mm gauge, 74 km at 15 kV 16²/₃ Hz AC

Political background

Heavy losses during the 1990s prompted the Swiss government to put pressure on SBB to improve productivity and financial results. The railway has had considerable success in this, and returned to profitability in 1999.

Under a reform package approved by the Swiss Council of States in October 1997 and subsequently ratified by the National Council, SBB became a limited company on 1 January 1999, with all shares owned by the state.

At the end of March 1999, the Bundesrat set out the strategic and operational objectives for SBB for the years 1999 to 2000 as required under the new arrangements for the railway. These objectives require the railway to win a greater share of the passenger market and at least to hold its current position *vis-à-vis* road in the goods market. Productivity must improve by 5 per cent a year in both sectors, and infrastructure costs must be reduced by at least 5 per cent.

The workforce will have been cut from 39,000 in 1990 to 28,000 in 2000.

In March 2000, SBB and the rail unions signed a new General Employment Agreement (Gesamtarbeitsvertrag (GAV)) which came into force on 1 January 2001. For the first time in the railway's history, all conditions of work are dealt with in a single agreement between employers and representatives of the employees. Among the main provisions of the new agreement are formalisation of the flexible working agreements established when the 39-hour week was introduced on 1 June 2000, a major expansion of working-together arrangements between management and unions, a pay structure with a performance element, and guaranteed employment.

Bahn 2000

In the 'Bahn 2000', or 'Bahn + Bus 2000' programme, the federal government is financing development of an expanded and closely integrated public passenger transport service nationwide by the next century. For railways, parliament in 1986 approved expenditure of SFr5.4 billion on SBB projects embodied in the plan. At the end of 1991 the federal government budgeted SFr1.3 billion to support the country's 57 private railways' 'Bahn 2000' investments.

SBB's expanded passenger train service plan for 'Bahn 2000' centred on hourly cycling throughout the timetable at critical interchange points. On key routes, however, the 'Bahn 2000' plan doubles InterCity or direct train service from hourly to half-hourly. Realisation of this increase of service frequency began in 1999.

Class 460 electric locomotive with a Basel Chur IC service passing the Rhätische Bahn Station at Zizers (Andrew Marshall)   0088192

The 'Bahn 2000' scheme called for about 130 km of new 200 km/h route, to secure competitive transit times and to cut running time between neighbouring hubs to the 1 hour required. (For details, see under 'New lines' section.)

SBB's major 'Bahn 2000' projects have not proceeded to plan. The initial cost estimate of SFr5.4 billion had, by 1993, swollen to SFr14–16 billion with increased construction industry costs, and through the addition of environmental safeguards enforced by objectors to several schemes.

In 1993, the Bundesrat required SBB to recast the 'Bahn 2000' plan to keep total cost within the 1987 budget, SFr8.1 billion, with subsequent cost inflation. This was exclusive of investment associated with the AlpTransit scheme for new transalpine base tunnels (see 'New lines' section). The revised plan is now expected to cost SFr 6.4 billion. This curtailed programme will achieve most objectives of 'Bahn 2000' by 2005. It will not, however, accommodate the increased traffic expected from completion of the AlpTransit base tunnel plan. With tilt-body technology and double-deck coaches for InterCity services, some costly civil engineering to increase speed and capacity has been saved.

Organisation

In 1997, SBB completed the implementation of a major restructuring intended to bring management decision-making closer to the market and to comply with the provisions of European Union Directive 91/440, even though Switzerland is not an EU member. The new structure provided for a complete separation of infrastructure and operations, and allowed the Swiss network to be opened up to operators other than SBB.

A further step was taken at the start of 1999. The state railway was transformed into a limited liability company wholly owned by the state, headed by a President and with a Chief Executive Officer. Within 'Schweizerische Bundesbahnen SBB' (SBB AG) three divisions have bottom-line responsibility (Passenger Traffic, Freight Traffic and Infrastructure). There is also a Finance Department and a Personnel Department. An Executive Board is made up of the Chief Executive Officer, the heads of passenger, freight and infrastructure divisions and the heads of the finance and personnel departments. Simplification (to achieve a more effective structure) and decentralisation (to bring the railway closer to its customers) continues; an entire layer has been removed by the elimination of the regional directions in Lausanne, Lucerne and Zurich. Regionalverkehr, the regional traffic business, has established a much firmer presence 'on the ground', able to deal directly and easily with the cantons. It produces more than half the total train-km of SBB's passenger traffic operation, amounting to some 2,600 million passenger-km on more than 120 routes in 1998. Over the last five years costs have been reduced by SFr100 million even though the number of train-km produced has gone up by 2.3 per cent. The new regions, with their headquarters, are: Léman (Lausanne), Wallis (Sion), Arc jurassien (Neuenburg), Mittelland (Bern), Zentralschweiz (Lucerne, Tessin (Bellinzona), Nordwestschweiz (Basel), Solothurn-Aargau (Olten), Zürich (Zürich), Nordostschweiz (Winterthur), Säntis-Bodensee (St Gallen), Graubünden-Walensee (Chur). The head of each region is the formal point of contact between SBB and the regional authorities. Grants towards regional traffic are falling – in 2000 at SFr545.7 million they were down SFr82.7 million on the figure for the previous two years – but the service is being expanded, and in the next few years some 50 new stations are to open for regional trains. In order to prepare for the joint-venture freight business with FS (see below), the SBB freight business was converted into a wholly owned limited liability subsidiary, SBB Cargo AG, at the beginning of 2000, as this was a prerequisite for entering into international agreements.

Open access for goods traffic and for the local traffic 'bought' by the cantons also came into force at the beginning of 1999 and SBB has established a one-stop shop to sell the paths. A new timetable gives priority to InterCity and international services, after which paths for sale can be established. SBB's own timetable is already projected forward to 2010.

Finance

SBB had a deficit of SFr496 million in 1995. Energetic measures turned this round to a net

Interlaken Ost-St Gallen service formed of IC2000 bilevel stock and propelled by a Class 460 locomotive at Faulensee (Andrew Marshall)   0109871

profit of SFr120 million in 1999, the first year as a joint-stock company.

Productivity measures being introduced include one-person crews for regional trains and further substitution of buses for trains on poorly used regional services. Stations are being closed or destaffed as their operational facilities are brought under remote control. And a 'lean infrastructure' programme has rigorously embarked upon the removal of all track, points, crossings, and so on, surplus to normal requirements.

SBB's current aim is to improve the results of its freight division annually by SFr290 million up to 2002. To achieve this, productivity must be increased by SFr150 million, which requires a further reduction in personnel.

A federal law of 1987 requires rail and road to satisfy stringent noise emission standards by 2002. Meeting these standards will cost SBB approximately SFr2 billion.

Finances (SFr million)	1998	1999	2000
Income			
Passenger traffic	1,533.7	1,603.6	1,630.3
Goods traffic	979.7	1,012.5	1,081.2
Traffic earnings for services	105.1	113.8	113.8
Traffic earnings/ Infrastructure	14.7	9.6	11.9
Grants for regional services	628.4	557.9	545.7
Grants for piggy-back traffic	110.0	125.0	75.1
Property rentals	233.9	242.7	247.4
Subsidiary earnings	515.4	487.9	530.7
Other income	26.4	7.4	14.7
Internal services	383.0	382.4	456.6
Federal contributions	1,249.0	1,273.0	1,316.0
Reductions in income	(−39.7)	(−64.7)	(−72.5)
Total	5,738.6	5,751.1	5,950.9
Expenditure	*1998*	*1999*	*2000*
Materials	(−324.9)	(−304.0)	(−340.5)
Staff costs	(−3,142.8)	(−2,990.3)	(−2,863.8)
Other costs	(−900.9)	(−959.8)	(−1,231.1)
Depreciation	(−938.3)	(−830.7)	(−879.1)
Inactive investments	(−257.3)	(−304.2)	(−326.4)
Taxation	(−58.9)	(−86.9)	−
Total expenditure	(−5,623.1)	(−5,475.9)	(−5,641.9)
Operational result	115.5	275.2	−
Financial earnings	143.9	108.3	132.8
Financial costs	(−202.8)	(−178.5)	(−343.5)
Sale of real estate	27.6	41.1	34.1
Extraordinary earnings	58.0	83.1	102.4
Extraordinary expenditure	(−163.5)	(−208.9)	(−104.9)
Profit	(−21.3)	120.3	139.9
Traffic (million)	*1998*	*1999*	*2000*
Passenger journeys	266.1	276.0	286.8
Passenger-km	12,484.8	12,615.0	12,835.0
Freight tonnes	52.3	57.3	60.5
Freight tonne-km	9,540.0	9,797.0	10,800.0

Passenger operations

Four out of five of SBB's customers use local services; just one passenger in every five makes

a journey exclusively on long-distance trains. A customer survey showed that in 2000 84.7 per cent of its customers were 'very satisfied' with SBB, a railway on which 94 per cent of trains reach their destination on time or no more than four minutes late.

The cutbacks in 'Bahn 2000' civil engineering investment (see 'Political background' section above) mean that, outside the key Basel–Bern–Zurich triangle, within-the-hour scheduling of InterCity (IC) services between all adjoining pairs of hub stations will not be feasible. Half-hourly service frequency on some routes, such as Basel–Lucerne, may be limited to business peak hours. The objective of connectional interlacing of regional and long-haul trains to offer once-every-daytime-hour service between any pair of SBB stations is affected by cuts in lightly used off-peak trains on many regional services.

The first three 'avec' shops at stations opened in 1999, at Schüpfen, Brügg, and Mettmenstetten. These are a joint venture between SBB, Migros and Kiosk AG, and they sell rail tickets, groceries, and kiosk goods. Open 365 days a year between 06.00 and 20.00 (or longer), they have been very well received by customers. It is planned to develop 50 more over the next few years at middle-sized stations. From the railway's point of view the great advantage is that the most important of the railway's products can be available in these stations without a need for SBB sales staff.

On 1 December 1999 SBB introduced a new national enquiry number, 0900 300 300, available 24 hours a day, to replace the old 157 22 22 number (which closed down at nights). Night service is provided from Zürich, while the other offices in Lausanne, Biel, Bern, Lucerne, Basel, Glaris, St Gallen and Lugano keep to their existing working hours. Four languages are available (English being the fourth), and the cost of the service is SFr1.19 per minute.

SBB is involved with the Swiss Post and the Public Transport Association (VöV) in the so-called 'EasyRide' project. In stages up to 2005 it will become possible to use the entire Swiss public transport system without the need for advance purchase of a ticket. Sensors on each of the 11,000 vehicles of the national public transport system will register each passenger's entry and exit and send the information for calculation of the tariff valid at that time, taking into account any applicable rebates and if the passenger will be invoiced at regular intervals. The system will cost some SFr600 million and it will cut the annual cost of ticketing by some SFr450 million. It will also provide data for timetable compilation and deployment of vehicles. Extension of the system to other modes (for example taxis, ski subscriptions, car-sharing) are provided for at a future stage. The aim is eventually to cover 80 per cent of all mobility. A preliminary study was completed in autumn 1999.

Impuls 97

The June 1997 timetable change saw the introduction of the major train service innovations connected with 'Bahn 2000' and launched under the 'Impuls 97' branding. The first of an initial batch of 58 'InterCity 2000' double-deck coaches entered service on St Gallen–Interlaken services; train service frequency on Zurich–Bern, Zurich–St Gallen

Zurich S-Bahn Class 450 emu at Zürich Altstetten with an Uster service (David Haydock)   0109873

and Bern-Fribourg routes was doubled to half-hourly. Experience has shown that the half-hourly interval timetables are a commercial success.

The 1999 timetable offered an additional 8,000 train-km a day (6,000 in long-distance traffic and 2,000 in local traffic). The routes between Zürich and Lucerne, and Fribourg and Lausanne both received a half-hourly service throughout the day. The main beneficiary of the year's changes was Romandie, as a result of progress on the 'Bahn 2000' construction works: trains can run directly between Lausanne and the Broye again on completion of work on the three tunnels between Palézieux and Lausanne. Delivery of more bilevel coaches means that these can replace older stock on the Basel–Zürich–Chur, Interlaken–Bern–Zürich–St Gallen, and Zürich–Lucerne lines, with the stock displaced being made available for new IC and IR trains on the Lucerne-Basel and Basel-Biel-Geneva lines. The first ICN trains entered service in 2000. The 1999 timetable also saw complete Fly-Baggage-Check-In for almost all airlines on offer at 23 stations, with more due to be added to the list.

SBB entered into partnership with its Austrian and German counterparts ÖBB and DB in the Trans Europe Excellence (TEE) Alliance, which aims by 2005 to offer a common profile, cross-border service. The three railways were initially seeking bids to supply 116 tilting trains for the new through services; 34 of these will be owned by SBB. However, in late 2001 DB withdrew its intention to procure a share of a common fleet of trains, confining its participation in the TEE Alliance to marketing.

Zurich RER

The Zurich RER (Regional Express) service, some 300 route-km, was inaugurated in 1990 (see *Jane's World Railways 1996–97* for details). In 2001, work started on a further expansion phase of the network. This covers track-doubling between Bubikon and Rüti to boost frequencies to Rapperswil, work to enhance capacity on the line to Winterthur, extension of route S3 to Dietikon, and completion of the Zimmerberg tunnel, raising capacity on lines serving Pfäffikon and Zug. The extension of line S3 is due to be completed by 2004; the remaining projects are to be commissioned by 2006.

In a referendum held in September 2001, residents in the Zurich region approved plans for a fourth expansion of the S-Bahn network, at an estimated cost of SFr1.45 billion. To be completed by 2012, the scheme includes construction of a second underground through station beneath Zurich Hbf. A new 4.8 km tunnel from Zurich Hbf to Oerlikon will release capacity on the existing line for long-distance traffic.

Bern RER

May 1998 saw a major expansion of the Bern S-Bahn (first established in 1995). The two existing lines running through the city (S1 Thun-Bern-Fribourg/Laupen and S2 Langnau-Bern-Schwarzenburg) were joined by two new through lines, S3 (Biel-Lyss–Zollikofen–Bern–Ausserholligen–Weissenbühl–

Belp–Seftigen-Thun and S4 (Langnau–Ramsei-Hasle–Rüesgau-Burgdorf–Zollikofen–Bern–Bern Bümpliz Nord). And the regional trains on the line from Bern to Neuchâtel via Kerzers now run under the S5 designation. The 138 stations on the network are served by more than 400 trains a day carrying some 40,000 passengers. As well as SBB, BLS Regionalverkehr Mittelland, the Sensetalbahn, and the RBS network are involved, all operating as equal partners under the S-Bahn brand name (using the symbol of a blue 'S' on a yellow ground) to serve the cantons of Bern, Fribourg, Solothurn, Waadt and Neuenburg. Under the close working agreement between SBB and BLS, the latter will take over SBB's share of Bern S-Bahn operations.

Other S-Bahn operations

Following on from the development of cross-border local services in the Basel area with SNCF, SBB is now to introduce an S-Bahn service across the border into Germany in collaboration with DB. Trains will run between Basel SBB and Offenbach via Basel Bad and Freiburg (Brsg). On the German side this new service will take over the existing Upper Rhine regional traffic. At present the only rail connection between Basel Bad and Basel SBB is provided by long-distance trains.

SBB has been entrusted with the planning of the proposed S-Bahn system for Central Switzerland, on which a decision was expected by the middle of 2001.

Regional operations

In September 2001, SBB announced that it was to form a jointly owned subsidiary with the Mittel–Thurgau Railway (MThB) (qv) to operate regional passenger services over a 550 km network in the

Thurgau, St Gallen and Zurich areas of eastern Switzerland. Named Thurbo AG, the new company was due to commence operations in December 2002. Thurbo was to take over ownership of 10 existing MThB Type GTW 2/6 lightweight emus and order additional similar vehicles (see 'Traction and rolling stock').

Freight operations

Major changes in SBB's freight business were foreshadowed on 30 March 1998 with the signature of an agreement between SBB and Italian State Railways (FS) to establish a joint rail freight business. With headquarters offices in Milan, the new company, Cargo Schweiz Italien GmbH, is to create a 'quality centre' to manage commercial and operational aspects of the two companies' international freight traffic. A second joint venture between SBB and FS undertakes the integration of the two railways' freight operations, creating a separate organisation with dedicated rolling stock, terminals and staff. Josef Egger, previously head of SBB's Central Department for Informatics, moved to the management of the new combined goods operation as Head of Informatics with the task of bringing the different systems of the two railways together. He was heavily involved in the negotiations that led to the establishment of the joint venture.

The two railways hoped to bring the joint operation into life at the start of 2001, but they have since agreed that a step-by-step approach to full union of their freight activities is required because of the much greater complexity of the project than was at first anticipated.

In 2000, SBB freight tonnage increased to a record 60.5 million.

SBB wants to raise the maximum speed of freight trains from 80 to 100 km/h to help obtain greater capacity on the mixed traffic network without new investment by harmonisation of speeds.

A new freight strategy was announced in 2001 under which SBB will aim to improve standards of both service and productivity. This replaces the strategy based on rapid fusion with FS Cargo and on the aim of offering a complete logistics service.

On the broader European front, SBB will seek direct relationships with customers in key geographic markets, will develop its own production in Germany and will seek to further its partnerships with FS (leading to fusion), with Hupac for combined traffic and with HGK as an additional partner in the north. Basel will be developed as a modern, high-performance 'Eurohub' and SBB Cargo will move its headquarters there.

On the domestic front the most important change is the redefinition of SBB Cargo's activities to just those areas where rail can be strong. Four products are envisaged: *CargoRail*, for overnight wagonload traffic; *CargoExpress*, for accelerated transits for single wagons overnight between 55 centres; *CargoClientNet* (working title) for

Classes Re 6/6 and Re 4/4 II locomotives with a train of new light commercial vehicles at Wassen, on the southbound ascent of the Gotthard line (Andrew Marshall)   0109874

Northbound Gotthard line RoLa piggyback service at Göschenen (David Haydock)   0109875

trainload services for major customers between specific centres; and *CargoTrain* (working title), for single-product trainload services. *CargoExpress* was launched in June 2001.

Intermodal operations

SBB is purely a wholesaler of train capacity to intermodal marketing concerns, notably the Swiss company Hupac (for contact details, see UIRR entry in 'Operators of International Rail Services in Europe' section), and the Swiss-based international company Intercontainer (qv), both of which own fleets of intermodal wagons. Hupac deploys a fleet of 345 RoLa ('Rolling Highway') low-floor well wagons for ro-ro movement of complete (tractor and trailer) road trucks. Hupac also operates 25 couchette cars for RoLa trains' trucker crews and 1,750 pocket wagons for carriage of unaccompanied trailers and swapbodies.

Hupac's RoLa trains for accompanied lorries run between Basel and Lugano, between Freiburg-im-Breisgau (just inside Germany) and Lugano or Milan Greco Pirelli, and from Rielasingen to Milan Rogoredo. The RoLa operation, however, cannot be marketed at a cost-covering price and is federally supported out of road-users' petrol/diesel tax.

In April 2001 SBB and Hupac formed a joint venture subsidiary, S-Rail Europe (SRE), which has been registered in Germany as an open access operator. A Class 1116 electric locomotive, acquired by Hupac in 2000, was transferred to SRE and two similar machines were ordered from Siemens. The locomotives are similar to the Austrian Federal's Class 1016/1116 Taurus machines.

A further development in 2001 was the establishment by SBB, BLS and Hupac of the RAlpin joint company to operate RoLa services between Freiburg im Breisgau, Germany and Novara, Italy. Each company holds a 30 per cent shareholding in the venture, with FS Cargo in Italy holding the remaining 10 per cent. Using the Lötschberg route, RAlpin services commenced in June 2001 with an initial four train pairs per day.

Since late 2000 lorry drivers on the Germany–Italy run have been able to put their vehicles on trains and take their rest periods during the rail trip through Switzerland. A 1992 agreement between the EU and Switzerland included the creation of a rolling motorway, set the capacity (105,000 lorry-spaces), and named the terminals (Freiburg im Breisgau, and Novara). A spur was given to the plans by the 1994 Alpine Initiative. In 1998 the federal government sought tenders for operating the Lötschberg–Simplon route, and early in 1999 it was announced that a consortium of SBB, BLS, and Hupac SA had been successful. Revenue of SFr22.4 million in 2001, rising to SFr65 million in 2005, is envisaged. Subsidy will remain constant at about SFr30 million a year. On the infrastructure front, the Lötschberg tunnel is cleared for trains carrying lorries up to 4 m high and 2.4 m wide. In the Simplon area, works have been completed on the Varzo–Preglia section. Works on the Iselle–Domodossola and Domodossola–Novara sections were completed by September 2000. And on the

Varzo–Iselle section one line became available in September 2000, the other in March 2001.

New intermodal systems

The first bimodal operation was launched by the Swiss food distribution company Migros, which annually sends 1.2 million tonnes of freight by rail. Ten Kombitrailers and 20 rail bogies operate from Migros' main Neuendorf centre to St Margrethen.

SBB and the private railways are partners in ACTS SA, a company formed to promote local container traffic. This achieves road-rail transfer without cranage; the Type Rs-x rail wagons, 800 of which are supplied by Tuchschmid AG of Frauenfeld, each have three 20 ft long platforms that can be swung outwards to back up to the rear of a road vehicle chassis equipped with a mechanism to slide a container from one vehicle to the other.

Gotthard route upgrade

The federal government allocated SFr1.46 billion for works, chiefly on the Gotthard route, and on approach routes Basel–Brugg–Arth-Goldau, to expand capacity for transit intermodal traffic pending completion of the new transalpine base tunnels (see 'New lines' section). The aim was to treble the Gotthard route's intermodal piggyback capacity from 160,000 to 470,000 units a year. This embraced unaccompanied semi-trailers, and RoLa movement of highway trucks. There would also be scope for annual throughput of some 330,000 containers and swapbodies.

The works were completed at the end of 1993. As a result, the maximum permissible format of Gotthard piggyback trains rose from 17 to 36 flat wagons with a gross laden weight of 2,000 tonnes. Southbound, single-headed 16-wagon trains run to Dottikon, where they are coupled, so that the locomotive of the rear train becomes the radio-controlled mid-train power. A third locomotive is added to double-head the combined train. Northbound, combined trains are formed during the essential change-of-voltage repowering at the FS-SBB yard in Chiasso. The aim is to power these trains exclusively with the tranche of 75 Class 460 locomotives which SBB expressly ordered for the service. In 1994, the Gotthard route carried 44 piggyback intermodal trains daily.

The clearances of the Gotthard route allow unaccompanied trailers, 3.9 m high at their roof corners, on so-called 'pocket' wagons. If they have deflatable air suspension, 4 m high box trailers can be safely accommodated on the latest type of pocket wagon. But RoLa piggybacked trucks must not be higher than 3.8 m, whereas virtually unrestricted admissibility to RoLa trains of European trucks requires clearance for 4.2 m road vehicle height. That will be built into the new transalpine base tunnels.

New lines
'Bahn 2000' works

The longest stretches of new 200 km/h infrastructure proposed in the original 'Bahn 2000' plan were to be on the Basel-Bern route: one of 34 km between

Muttenz, on Basel's outskirts, and Olten (which entailed tunnels of 4.7 and 12.8 km length, the Adler and Wisenberg respectively); and another of 54 km between the Olten area at Rothrist and Mattstetten, near Bern, with a branch to the Olten-Zofingen–Lucerne line. These give the 22 minute cut in Bern–Basel timings needed to secure under-the-hour timing of direct trains, as between Bern and Lucerne or Zurich. But four-tracking between Basel and Bern would also handle the doubled direct passenger train frequency and anticipated extra international freight traffic via the Bern area following completion of the BLS Lötschberg route double-tracking.

The Rothrist-Mattstetten high-speed route survived the reappraisal of the 'Bahn 2000' project, and work started in 1996. It runs parallel to the N1 motorway and close to the existing railway as far as the region of Herzogenbuchsee, with a branch to an existing secondary line to Solothurn, to be upgraded to a fast route for trunk trains from Biel to Basel or Zurich. At Mattstetten the new line will make an end-on junction with the new Grauholz bypass. The line is expected to open on 19 December 2004.

The 9.5 km Grauholz bypass opened in 1995; it includes a 6.3 km tunnel. Its purpose is to keep Bern–Olten traffic clear of Zollikofen, where the old Olten-Bern and Biel–Bern lines saw some 300 trains a day. With full implementation of 'Bahn 2000' the total would rise to 500, including eight pairs of IC trains in each hour. The bypass starts 4 km out of Bern at Bern-Löchligut. The Grauholz bypass permits half-hourly Bern–Zurich IC service. Further east, the Rupperswil–Aarau segment, including Aarau station and a new tunnel under the town, will become a four-track high-speed section.

Of the planned Muttenz–Olten high-speed segment, the Muttenz-Liestal section (including the Adler Tunnel) opened in December 2000. This section segregates the Hauenstein and Bözberg routes out of Basel. The advantages offered by a continuation through a new Wisenberg Tunnel were held not to be worth the cost, but SBB is now engaged in new studies for an additional tunnel through the Jura chain. A decision is expected in time for the second stage of Rail 2000.

A third stretch of 200 km/h track, 9 km in length, was to be between Zurich Airport and Winterthur. This stretch required boring of the 8.4 km Brüttener Tunnel between Kloten and Winterthur. The entire scheme is now deferred.

Finally, 31 km of 200 km/h line were planned between Vauderens and Villars-sur-Glâne, on the main line from Lausanne to Bern. This, the only 200 km/h project with the sole aim of shortening transit time, was cut back to construction of a new tunnel at Vauderens that is essential to provide adequate clearance for the new double-deck IC trainsets. Construction began in 1998 on the 2.9 km of double track between Vauderens and Prezvers-Siviriez, which includes the 1,975 m tunnel. The section will cost SFr90 million and its completion will allow the existing 1862 tunnel to be abandoned.

Among the SBB's other major 'Bahn 2000' schemes, the short section of the Simplon route still single – the 4.6 km between Salgesch and Leuk in the canton of Wallis – is now being bypassed. The 6.6 km double-track new, 4.2 km in tunnel, will be completed in 2005 and will allow 160 km/h instead of today's 100 km/h. The job is a big one: the railway will be realigned through the Pfynwald, the A9 autobahn will take the alignment of the present T9 cantonal main road and the T9 will take the present railway alignment. The total final cost will be SFr239 million, of which 60 per cent will be borne by the canton and the Federal Roads Department.

In April 2001 the new 10.5 km double-track line from Onnens to St-Aubin via Gorgier was opened and this allows tilting trains to run into Romandie. After this, the next opening expected is of the double-track 10 km new line between Zürich and Thalwil, in 2003, which will relieve the lakeside line of traffic and allow improvements to local services.

New base tunnels

The federal government's AlpTransit plan for new Gotthard and Lötschberg base tunnels, to enable the near-trebling of transalpine capacity for rail-based intermodal traffic, was approved by national referendum in 1992. The federal council adopted the AlpTransit plan, to resist European Union (EU) demands for a transalpine road corridor for 40 tonne trucks, in face of Switzerland's 28 tonne gross loaded weight truck limit, prohibition of night driving, a levy on heavy goods road transit, and noise and emissions controls. Railways claim 68 per cent of the Swiss transit freight market, although substantial diversion of road traffic via France or Austria occurs.

Using rights under the Swiss constitution, a private environmentalist initiative led to a national referendum on transit highway freight in 1994. Against federal government advice, in view of the likely strain on relations with the EU, the country voted to bar Swiss roads to all transit highway freight by 2004. Since neither transalpine base tunnel would be finished by then, rail congestion is inevitable.

Exploratory boring for the Gotthard base tunnel route began in 1995 (the Lötschberg project is dealt with in the BLS Lötschbergbahn entry). It will extend approximately 125 km from Arth-Goldau to the outskirts of Lugano, with the 57 km base tunnel as its centrepiece. The base tunnel, starting at Erstfeld on the northern side, will be the world's longest rail bore; its summit, 550 m above sea level, will be some 600 m below that of the present Gotthard tunnel. The route is described in *JWR 1996–97*.

Clearances will be contoured for piggybacking of road lorries 4.2 m high. The base tunnel itself will be aligned and built so as to simplify its possible elaboration at some future date into the 'Y' form sought by eastern Switzerland. Other schemes exist, subject to financing, to improve access to the Gotthard base tunnel from eastern Switzerland, south Germany and Austria. Use of the Bodensee–Toggenburg line (qv) from St Gallen southwestwards would bring Gotthard traffic from the east and northeast to Pfaffikon, up the SBB's Zurich line to Au and through a new Hirzel Tunnel, to a 'Y' near Zug with a new north-south double-track Zimmerberg Tunnel, to be bored from Thalwil to benefit traffic from Zurich (and the Stuttgart area of Germany) to Arth-Goldau.

The Lötschberg tunnel will be completed in 2006, the Gotthard in 2012. Internal differences in the Swiss Cabinet in 1994 led in 1995 to an independent reappraisal of the cost (about SFr15 billion) and returns, and it was shown that the SBB and BLS could not, as originally planned, pay back the cost from traffic income. Since this was a condition on which the 1992 popular approval was given, a further vote was due to be held on a revised finance package, including increased use of fuel revenues. An agreement reached in January 1998 between Switzerland and the European Union provides for a gradual relaxation of the 28 tonne gross laden weight limit for trucks up to 2005, when a 40 tonne ceiling would be permitted. A transit tax on each lorry and a quota system were considered by the Swiss government sufficient both to provide funding for the two base tunnels and their related works – as well as for other rail infrastructure

investments in Switzerland – and to appease public opinion and the concerns of environmentalists.

In May 1998, a new SBB subsidiary, 'AlpTransit Gotthard AG' was founded in Bern with a capital of SFr5 million to be responsible for the planning and construction of the AlpTransit Gotthard Axis.

In mid-1998, final test results showed that all the strata through which the Gotthard Base Tunnel passes are firm throughout, so that no special problems are to be expected in tunnelling, either by boring machine or by explosive. On 29 November 1998, in two referenda, the Swiss people accepted the so-called FinöV proposals and thus approved the funds to allow a solid financing of major rail projects, allowing work on the two Alpine Base Tunnels to move from the planning phase into the building phase.

Tunnelling began in November 1999 and the Base Tunnel is expected to open by 2012.

Improvements to existing lines

The Rhône Valley main line to the Simplon Tunnel has been upgraded. Complete renewal of track with heavier rail and higher-speed pointwork, realignments at Riddes and Ardon, renewal of overhead equipment, and resignalling accompanied by computer-based control, track-to-train radio and automatic speed control, equip the Martigny-Sion section for 200 km/h. There is provision for solid-state interlockings at Martigny and Sion. A new double-track tunnel will replace the last single-track section, the 5 km in difficult terrain between Salgesch and Leuk. Here, from Leuk to Visp, and (later) through the Simplon Tunnel the speed limit will be 160 km/h.

To achieve clearance for piggybacked trucks of 4.2 m corner height, and to provide for higher passenger train speeds, the Simplon Tunnel's track is being renewed and its bed lowered. At the same time, a rigid traction current conductor replaces traditional catenary. The rigid conductor is closer to the tunnel crown, giving 300 to 400 mm of extra clearance for intermodal piggyback trains. To avoid unacceptable interference with traffic, the works are executed in the least busy periods of the year, covering 3 to 5 km at a time, and the project is expected to take six to eight years in total.

Full seven-car production ICN tilting trainset near Ligerz   0088191

A seven-year reconstruction and upgrading programme for Zürich's main station was announced in April 1998.

In March 1999, work began on the construction of the third line over the 13.5 km between Coppet and Geneva. Completion will be between 2003 and 2005. Provision of this third track will allow the demands of the 'Bahn 2000' timetable to be met by separating out regional services from long-distance traffic.

In the period up to 2005 the approaches to Zürich station are being totally renewed within the scope of the 'Bahn 2000' project. Trains will be sorted at Killwangen-Spreitenbach, Altstetten, Thalwil and Oerlikon, thus allowing crossing movements in the *Hauptbahnhof* throat to be avoided and raising reliability and efficiency. The chosen solution, coupled with the quadrupling of the Wipkingen line, will allow up to 15 parallel movements – necessary if the number of trains is to be raised by 22 per cent in four stages over the next seven years, from 1,350 in 1996 to 1,650. A new four-platform S-Bahn station between Line 3 and the Sihlpost will eliminate problems in the station itself. From 2005 Zürich will have half-hourly services in all directions. At Altstetten Süd SFr176.4 million are to be spent on the improvements, including provision of a single-track S-Bahn underpass.

SBB is currently concerned to establish a more robust power supply and as part of this work it is spending SFr22 million on the 25 km Burgdorf-Wanzwill section of the second Central Switzerland electrical link between its transformer stations in Kerzers and Rupperswil. This is to relieve a 1927 link which is no longer adequate by providing an additional link between the power stations group in Wallis and that in the Gotthard region. The 132 kV connection will also feed the planned substation in Wanzwil which is to feed the 'Bahn 2000' new line between Mattstetten and Rothrist. The link will be built as a 90 per cent joint exercise with other electricity companies. And in mid-September 1998 the Amsteg hydro-electric power station, completely renovated over a five-year period at a cost of SFr460 million, began to supply power. The old Amsteg station produced current for SBB for more than 70 years. For the new one, SBB and the canton of Uri founded in 1992 a Kraftwerk Amsteg KG company with a capital of SFr80 million (90 per cent SBB) to build and operate the new installations. The new equipment generates up to 120 MW.

Traction and rolling stock

By January 2000 SBB's locomotive fleet had been divided between the Passenger and Freight business sectors, with 234 going to the former and 397 to the latter. Depots and personnel are also now similarly assigned to the one business or the other. A decision on main workshops was due to be taken at the end of November 2000.

In 1999, SBB operated 803 standard-gauge electric locomotives; 255 electric power cars; 121 diesel locomotives; 163 electric shunters; and 916 tractors. The rolling stock fleet of 4,392 passenger cars included 53 restaurant, 12 sleeping and 60 couchette cars. Freight wagons totalled 16,111, Swiss Post Office-owned postal cars 561 and baggage cars 333. Over 6,460 privately owned wagons were in use.

Tilting train order

In 1996, SBB ordered 24 nine-car ICN tilting trains from a consortium of Swiss industry in a contract

Rack-equipped Class 101 (HGe 4/4) Brünig line metre-gauge electric locomotive at Lucerne
(LT Peacock)   0088205

worth some SFr500 million. The consortium is led by Bombardier and also involves Alstom. The trains are capable of travelling at 200 km/h and are for use on curvaceous routes; SBB decided to opt for tilting trains to cut down the work required on building new lines for 'Bahn 2000'.

Delivery began in December 1999 and will be completed in 2001. The first route on which the trains ran is the Jura corridor between Lausanne and St Gallen; Geneva to Basel will follow. The tilt trains are marketed under the 'ICN 2000' tag (the bilevel cars run as 'IC 2000').

The tilting trains have a multiple-unit format, with power distributed along the train. Two units can work in multiple at peak times.

In June 2001, SBB ordered an additional 10 seven-car ICN trainsets, with an option on 10 more. Delivery is scheduled in 2004.

SBB also plans to procure an additional 34 tilting trainsets as its contribution to the TEE Alliance project (see 'Passenger operations').

Other developments
In October 2001, SBB Cargo ordered 10 dual-voltage (15 kV AC/25 kV AC) 4,200 kW electric locomotives from Bombardier. These are to be a design similar to that of DB Cargo's Class 185. Construction will take place in Germany, with delivery commencing in February 2002.

SBB Cargo has equipped 10 locomotives of classes Re 460 and Re 4/4 II for working into Germany. This requires the provision of a German-size pantograph, Indusi (and probably LZB), and German train-radio. Up to now only the Re 4/4 II machines 11195 to 11200 have been equipped with pantographs that allow them to work into Lindau. Target station is Ludwigshafen, a centre for chemical traffic.

In 2000 SBB ordered from Stadler 17 GTW 2/8 three-section lightweight emus for services on the Lucerne–Lenzburg line in the Seetal region. Designated Type ABe 520 by SBB, they are to be delivered in 2002 and will be the first three-section examples of the Stadler GTW family. In collaboration with Adtranz, Stadler also supplied 50 Tm 234 service tractors during 2000.

The Thurbo AG subsidiary (see 'Passenger operations') established jointly by SBB and the Mittel-Thurgau Railway (MThB) (qv) plans to acquire an additional 80 GTW-type articulated lightweight emus from Stadler/Bombardier by 2007, supplementing the existing fleet of 10 similar vehicles.

Coach fleet
SBB is investing between SFr260 million and SFr290 million a year up to 2002 in the modernisation of the passenger fleet. Following on from the initial order for 58 air conditioned InterCity 'Bahn 2000' bilevel cars placed in 1993 and the follow-up order for 144 vehicles placed in late 1997, a third series of 48 bilevel InterCity carriages has been ordered at a cost of SFr143 million, to bring this fleet up to 250 units within three years – including, from 2000, vehicles with a bistro section. This order included 19 first class cars, three driving trailers and 26 second class/bistro cars. In place of compartments on the lower level, these last have a bistro-bar with kitchen and office. There are 14 seats and standing room at the bar, and on the upper level there are eight groups of four seats for take-away service. These cars were ordered in place of classic restaurant cars because of the poor economics of the latter and ever-shortening journey times. Also, research has shown that a bistro will appeal to a broader cross-section of the public than a traditional restaurant car.

An order for a fourth series of 70 vehicles, valued at SFr217 million, was placed with an Adtranz/Alstom consortium in February 2001. Deliveries will be made between 2002 and 2004 and will bring the total fleet to 320 cars. No further orders for IC2000 bilevel stock are envisaged.

The standard formation of a push-pull bilevel IC set consists of a Class 460 locomotive, seven intermediate vehicles and a driving trailer, offering a total of 712 seats. Trains can be worked in multiple, allowing division en route to serve two destinations.

The modernisation programme includes SFr248 million to update by 2005 the InterCity fleet of 449 Type IV (EW IV) coaches and the EuroCity fleet. In addition, five baggage cars are to be bought from French National Railways (SNCF), 15 restaurant cars are to be converted for international service, 36 EC coaches are to be rebuilt as driving trailers and baggage compartments are to be built into the first class InterCity cars. The EW IV vehicles are still in good shape after 10 or more years of service and do not yet require major overhaul, so a facelift is to be applied: new furnishings and carpets, a sealed toilet system, electric hand-driers, cycle platforms and a better passenger information system, as well as replacement of the air conditioning and electronic systems. They will be equipped for push-pull operation and 200 km/h running. The EC cars will acquire a closed toilet system. After this work the refurbished vehicles and the new bilevel vehicles will all be fit for use on the new line between Mattstetten and Rothrist.

With push-pull 'InterCity 2000' services in mind, 60 driving trailers were ordered in 1994 and were in widespread service in 1997. Some 300 cars of earlier build, at present limited to 140 km/h, are being modified for 160 km/h operation.

Signalling and telecommunications
Some 99.7 per cent of SBB lines are colourlight signalled with the Swiss standard system. The remaining 0.3 per cent of lines still have mechanical signalling. Some 45 per cent of the system (1,267.2 km) is equipped with automatic block with axle-counters; 26.5 per cent (742.3 km) has automatic block with track circuits; manual block with continuous current serves 28 per cent (788.8 km), and manual block indicators the remaining 0.5 per cent (11.9 km).

SBB has so far not required cab signalling, though this would be useful in the Lakeland areas of the country where mists and fog can make lineside equipment hard to see. However, some of the new lines due to be opened in the next few years will allow speeds at which cab signalling is required. Accordingly a 32 km test section is to be equipped, between Zofingen and Sempach. This is a Gotthard feeder line which suffers from a good many foggy days. Some stations will need alteration, as on safety grounds island platforms will need to be replaced by lineside platforms. 53 vehicles will be equipped with the necessary cab signalling: nine local units, 37 locomotives for goods and passenger traffic, and eight shunting locomotives. The system used will be the world's first ERTMS Level 2 installation in commercial service and is being supplied by Adtranz. The contract covers the radio block centre, the GSM-R radio transmission system, onboard equipment for 66 locomotives and some 200 track beacons (balises). Testing began in May 2000. Signalling

Diesel locomotives

Class	Wheel arrangement	Power kW	Speed km/h	Weight tonnes	No in service	First built	Builders Mechanical	Engine	Transmission
842	Bo-Bo	611	75	66	1	1939	SLM	Sulzer	E ABB
840	Bo-Bo	620	75	72	26	1960	SLM	SLM	E Sécheron
	Bo-Bo	620	75	72	20	1968	SLM	SLM	E Sécheron
	Co-Co	956	75	106	4	1954	SLM	Sulzer	E ABB/ Sécheron
					10	1960			
930	C	326	65	49	5	1959	SLM	SLM	E ABB/ Sécheron
		326	65	49	35	1962	SLM	SLM	E ABB/ Sécheron
863	Co-Co	1,440	85	111	6	1976	Thyssen-Henschel	Chantiers de l'Atlantique	E ABB)
Em 6/6	Co-Co	393	65	104	6	1971	SLM	SLM	E SAAS/ SSB/ABB
831	C	900	60	54	3	1992	RACO	Cummins	H
842	Bo-Bo	1,120	90	80	2	1992	Krupp	MTU	H

Electric locomotives

Class	Wheel arrangement	Speed km/h	Weight tonnes	No in service	First built	Builders Mechanical	Electrical
610	Co-Co	125	120/124	120	1952	SLM	ABB/Oerlikon
410	Bo-Bo	125	57	10	1946	SLM	ABB/Oerlikon
411		125	57	17	1950	SLM	Sécheron
420	Bo-Bo	140	80	273	1964	SLM	ABB/Oerlikon/ Sécheron
430	Bo-Bo	125	80	21	1971	SLM	ABB/Oerlikon/ Sécheron
450	Bo-Bo	130	71	115	1989	SLM	ABB
460	Bo-Bo	100	81	119	1993	SLM	ABB
620	Bo-Bo-Bo	140	120	88	1972	SLM	ABB/Sécheron
930	C	40/50	39/45	118	1928		ABB/Oerlikon/ Sécheron
Ee 6/6	C + C	45	90	2	1952	SLM	ABB/Sécheron
962	Co-Co	85	107	10	1980	SLM	ABB
Eem 6/6	Co-Co	65	104	1	1970	SLM	SAAS
15 kV/25 kV							
Ee 3/3 II	C	45	46	14	1957	SLM	ABB
1.5 kV/3 kV/15 kV/25 kV							
934	C	60	48	10	1962	SLM	Sécheron

New classifications have been reserved for forthcoming locomotives as follows:
    453 Possible three-voltage unit for Geneva and Basel S-Bahn projects
    462 Planned dual-voltage version of Re 4/4 VI

Electric railcars or multiple-units

Class	Cars per unit	Motor cars per unit	Motored axles/car	Output/ motor kW	Speed km/h	No in service	First built	Builders Mechanical	Electrical
15 kV AC									
RAe 2/4	1	1	2	197	125	1	1935	SLM	ABB/MFO/ SAAS/SBB
511	3	3	4	204	125	18	1965	SWP/FFA	SAAS/ABB
512	4	2	4	281	125	4	1976	SWS/SWP/SIG	SAAS
560	2	1	4	412	140	84	1984	FFA/SIG/ SWP/SWA	ABB
540	1	1	4	497	125	80	1959	SIG/SWS	ABB/MFO
536	1	1	4	294	110	2	1952	SLM/SWP	ABB/MFO/SAAS
546	1	1	4	201	75	1	1927	SIG/SWS	SAAS
1.5 kV DC									
Bem 550	2	–	–	600	100	5	1994	SWG-A/S WG-P/SIG	ABB
1.5 kV/3 kV/15 kV/25 kV									
506	6	1	6	577.5	160	3	1961	SIG	MFO

on the whole Olten–Lucerne line is being modernised, involving SFr55.4 million replacement of mechanical installations dating from 1908 to 1932 in Brittnau-Wikon, Reiden, Dagnersellen, Nebikon, Wauwil, Sursee and Nottwill – which are not suitable for the 200 km/h speeds required for the future – by two new ESTW installations, at Dagmersellen and Sursee.

Since April 1998, it has been possible to use train-to-shore radio throughout the 19.803 km length of the Simplon No 1 tunnel. Broadband coverage (160–900 MHz) is available. The total installation project for both tunnels involves the drilling of some 40,000 holes into the tunnel walls. The radio system replaces the telephone system previously in place.

SBB and SNCF have decided to link up their traffic management systems, *Brehat* and *Surf*, thus providing controllers with a picture of operations on both networks and allowing the possibility of more effective intervention in case of problems.

Almost as needy of renewal is the third of SBB's interlockings that are electromechanically controlled, and thus difficult to adapt for remote control from new route-setting panels. SBB is also one of Europe's leading railways in use of Automatic Vehicle Identification, using trackside radio-wave readers and transponder tags, designed by Alcatel-Amtech to UIC standard. The first installations at Erstfeld, on the Gotthard route, and at Italian and French border stations in 1993, were followed by implementation throughout Switzerland, giving much-improved rolling stock control and better customer information.

In common with the Lucerne-Stans-Engelberg Railway (qv), with which it shares tracks between Hergiswil and Lucerne, the Brünig route has been equipped with a ZSL90 system of ATP specially developed for narrow-gauge railways.

A thorough modernisation of the railway telephone service was being undertaken in 2001, affecting all exchanges and all 18,000 handsets.

New equipment was being supplied by Ascom. Total cost of the project is SFr15 million. A project for the modernisation of public information services is running in parallel. This work was expected to cost SFr98 million.

Track
Rail: SBB 1, 46 kg/m (538 km); SBB IV UIC 54E, 54 kg/m (2,382 km); SBB VI UIC 60, 60 kg/m (1,553 km)
Crossties (sleepers): wood (1,765 km); steel (1,096 km); concrete (1,612 km)
Spacing:
In plain track: 1,667/km
Min curvature: 176 m
Max gradient: 4%
Max axleload: 22.5 tonnes

Thurbo AG

Postfach, CH-8280 Kreuzlingen 1
Tel: (+41 512) 23 49 00   Fax: (+49 512) 23 49 90
e-mail: hallo@thurbo.ch
Web: www.thurbo.ch

Subsidiary company
EuroThurbo GmbH
Bahnhofplatz 49, D-78462 Konstanz, Germany
Tel: (+49 7531) 91 51 09

Gauge: 1,435 mm
Route length: 550 km (470 km in Switzerland, 80 km in Germany)

Key personnel
Chairman: Paul Blumenthal
Managing Director: Dr Ernst Boos
Managers
    Finance: Christian Saxer
    Marketing and Sales: Gallus Heuberger
    Transport: Martin Hochreutener
    Production: Alfred Hartmann
    Customer Service: Werner Fritschi
    Communications: Gallus Heuberger
Managing Director, EuroThurbo GmbH:
    Wolf-Dieter Deuschle

Organisation
Thurbo AG was founded in September 2001 jointly by SBB AG and Mittel-Thurgaubahn AG to operate and develop regional services in northeast Switzerland and across the border into southern Germany. Following the bankruptcy of MThB in 2002, SBB assumed a shareholding of 90 per cent, with the remaining 10 per cent held byThurgau canton. Full operations commenced in December 2002.

A subsidiary company, EuroThurbo GmbH, manages business in Germany from offices in Konstanz. From December 2006 EuroThurbo is to take over responsibility for the Konstanz–Singen–Engen 'Seehas' line, which currently forms part of the parent company's network. It is also a joint venture partner with Regentalbahn AG in the Allgäu-Express (ALEX) regional express service in southern Bavaria (see entry in Germany section of *Railway systems and operators*).

Most Thurbo services are operated over SBB and DB tracks but the company owns the former

Class RABe 526 700 (Stadler GTW 2/6 Series 2) electric railcar (Stadler)   **NEW**/1036626

Electric railcars or multiple-units

Class	Cars per unit	Motor cars per unit	Motored axles/car	Output/ motor kW	Speed km/h	No in service	First built	Builders Mechanical	Electrical
15 kV AC									
FLIRT 4/10	4	2	2	250	160	9	2006*	Stadler	ABB Automation
RABe 526 700 (GTW 2/6)	2	1	2	350	140	80	2003**	Stadler	ABB Automation
RABe 526 680 (GTW 2/6)	2	1	2	260	130	10	1998	Stadler	–
ABDe 536	3	1	4	275	100	6	1965	SIG/SWS	BBC, MFO, SAAS
RBDe 566	4	1	4	425	140	4	1994	SIG/SWG	ABB

* On order for Konstanz–Singen–Engen 'Seehas' line
** Delivery to be completed by 2007

MThB main line from Wil via Weinfelden to Kreuzlingen.

At the end of 2004 Thurbo AG employed around 250 staff.

Passenger operations
Since December 2002 Thurbo AG has operated regional services over a mostly electrified 550 km network in an area defined by the Bodensee (Lake Constance) in the east, Koblenz and Winterthur in the west and Uznach in the south. Also covered are cross-border services to Stockach, Singen and Engen, in Germany. In all 15 routes in Switzerland (149 stations and halts) and three in Germany (26) are served.

Traffic (million)	2003
Passenger journeys (Switzerland)	20.6
Passenger-km (Switzerland)	304
Passenger journeys (Germany)	3.4
Passenger-km (Germany)	50

Traction and rolling stock
Much ex-MThB and SBB equipment is being supplemented or replaced progressively by a fleet of 80 Type GTW 2/6 Series 2 lightweight articulated emus supplied by Stadler. All are to be in service by the end of 2007. Their delivery follows that of 10 ex-MThB Stadler GTW 2/6 Series 1 emus, the first electric version of the type to produced. Three Stadler GTW 2/6 diesel railcars procured in 1996–97 are used on the Radolfzell-Stockach branch. Some Thurbo services employ SBB emus/motor cars of Classes RBe 540 and RBDe 560.

In 2004 Thurbo ordered nine FLIRT four-car emus from Stadler for services on the Konstanz–Singen–Engen 'Seehas' line for delivery from 2006.

UPDATED

Transports de Martigny et Régions SA (TMR)

Case Postale 727, CH-1920 Martigny
Tel: (+41 27) 722 20 61   Fax: (+41 27) 722 45 10
e-mail: info@tmrsa.ch
Web: http://www.tmrsa.ch

Key personnel
Director General: A Lugon Moulin

Gauge: 1,000 mm; 1,435 mm
Route length: 19 km; 26 km

Electrification: 19 km at 850 V DC; 26 km at 15 kV AC 16.7 Hz

Organisation
TMR is the result of a decision in 1990 by two railway companies in southeast Switzerland, the metre gauge Martigny-Châtelard (MC) and the standard gauge Martigny-Orsières (MO), to operate under a common management structure while remaining independent companies. A full merger took place in 2000 to create the present company. Subsequently road freight transport, touring coaches and

travel agencies were added to the company's portfolio.

Passenger operations
Services on the former MC line, which sees through trains to Chamonix, France, are operated under the Saint-Bernard Express brand name, while those over the former MO run as the Mont-Blanc Express.

Traction and rolling stock
TMR operates 21 electric railcars on the metre gauge system and 11 on standard gauge lines.

Transports Public Fribourgeois (TPF)

Rue des Pilettes 3, CH-1700 Fribourg
Tel: (+41 26) 351 02 00   Fax: (+44 26) 351 02 90

Key personnel

General Manager: Claude Barraz

Gauge: 1,435 mm; 1,000 mm
Route length: 50 km; 50 km
Electrification: 50 km of 1,435 mm gauge at 15 kV 16²/₃ Hz AC; 50 km of 1,000 mm gauge at 900 V DC

Organisation

Formed in 1942, the former GFM group (Chemins de Fer Fribourgeois Gruyère-Fribourg-Morat) comprised: a 49 km 1,000 mm gauge route from Montbovon MOB via Gruyères to Palézieux (SBB), and from Bulle to Broc; an 18 km 1,435 mm gauge route from Bulle to Romont (SBB); and an isolated 32 km 1,435 mm gauge line from Fribourg via Morat to Anet (BLS-BN).

In July 2000 GFM merged with Fribourg's public transport operator, Transports Publics de Fribourg (TPF), creating the present company.

Passenger and freight operations

In 1995, 2,124,000 passengers were carried (26.5 million passenger-km), and 302,000 tonnes of freight (5.6 million tonne-km). In 1996, traffic declined to 2.0 million passengers (24.9 million passenger-km) and 243,000 tonnes of freight (4.1 million tonne-km).

TPF (former GFM) Type RABDe standard-gauge railcar leaving Fribourg with a freight service (David Haydock)   0109861

Traction and rolling stock

In 1995, GFM's 1,435 mm gauge fleet comprised: 13 electric power cars; two Ae 417 electric locomotives, former DR Class 142 obtained through leasing company Lokoop AG; 13 passenger trailer cars; two electric and five diesel tractors; and four freight wagons.

GFM's 1,000 mm gauge fleet comprised in 1994: two Type GDe 4/4 1,000 kW electric locomotives; 11 electric power cars; 20 other passenger coaches; four electric and two diesel tractors; 45 freight wagons; and 52 transporter bogies for conveyance of 1,435 mm gauge wagons.

Zentralbahn (ZB)

zb Zentralbahn AG
Luzern–Stans–Engelberg Bahn
Postfach 457, CH-6362 Stansstad
Stanserstrasse 2, CH-6362 Stansstad
Tel: (+41 41) 618 85 85   Fax: (+41 41) 618 85 89
e-mail: info@zentralbahn.ch
Web: www.zentralbahn.ch

Key personnel

Managing Director: Josef Langenegger
Managers
    Finance: Kurt Stricker
    Sales and Marketing: Mario Lütolf
    Infrastructure: Martin Röthlisberger
    Production and Workshops: Gerhard Züger
    Personnel: Thomas Studer

Gauge: 1,000 mm
Route length: 90 km
Electrification: 90 km at 15 kV 16²/₃ Hz AC

Organisation

ZB was established in January 2005, the result of a merger of the Luzern-Stans-Engelberg Bahn (LSE) and the SBB Brünigbahn of Swiss Federal Railways. The electrified metre gauge network shares tracks for the 9 km south from Lucerne to Hergiswil. From here, the former LSE line runs southeast via Stans to the Engelberg valley (16 km), while the Brünig line (65 km) runs southwest via Meiringen to Interlaken. Both lines feature Riggenbach rack-operated sections.

Passenger operations

Operations on the former LSE line combine a commuter service for the outskirts of Lucerne with access to the resorts of the Engelberg and Titlis areas. Around 2 million passengers (22 million passenger-km) annually are carried.

Brünig line passenger operations are a combination of regional and tourist trains, the latter including the Golden Pass Panoramic service operated jointly with the BLS Lötschbergbahn and the Montreux-Oberland Bernois Railway.

Improvements to existing lines

On the former LSE line realignment is being undertaken to eliminate the final rack section from

Brünig line Class 101 electric locomotive at Brienz with an Interlaken–Lucerne express service (John C Baker)   **NEW**/0585096

Obermatt into the Engelberg basin, requiring a new 4 km tunnel and easing the gradient from 25 to 10.5 per cent. This will increase line capacity and permit the use of new non-rack rolling stock. Commissioning is planned for 2007.

Traction and rolling stock

At the start of ZB operations in 2005 the former LSE fleet comprised eight BDeh 4/4 electric railcars (736 kW, Schindler/SLM/BBC, 1964) and 41 driving and intermediate trailers; two De 4/4 electric locomotives (894 kW, SLM/BBC, 1942) for Stans–Lucerne–Sachseln commuter services; and five diesel shunters. There were also seven freight wagons, 10 post/baggage cars and 10 service vehicles.

The core of the ex-SBB Brünig fleet is eight Class 101 (Type HGE 4/4 II) rack-equipped electric locomotives (1,932 kW, SLM/ABB, 1989), three

older electric locomotives, survivors of a fleet of 16 Class Deh 4/6 railcars and five electric shunters. There is also a small fleet of diesel shunting locomotives.

In 2004 deliveries commenced from Stadler of ten Class ABe 4/8 SPATZ three-car emus. These were to be used on Lucerne–Giswil, Lucerne–Stans and Lucerne Ost–Meiringen services.

On order from Stadler for ZB in 2005 were three Class ABt three-section driving trailers for Lucerne S-Bahn services. A further order, placed with Stadler by SBB in 2004 before the formation of ZB, was for a rack-equipped diesel-electric locomotive for the Brünig line.

UPDATED

Syria

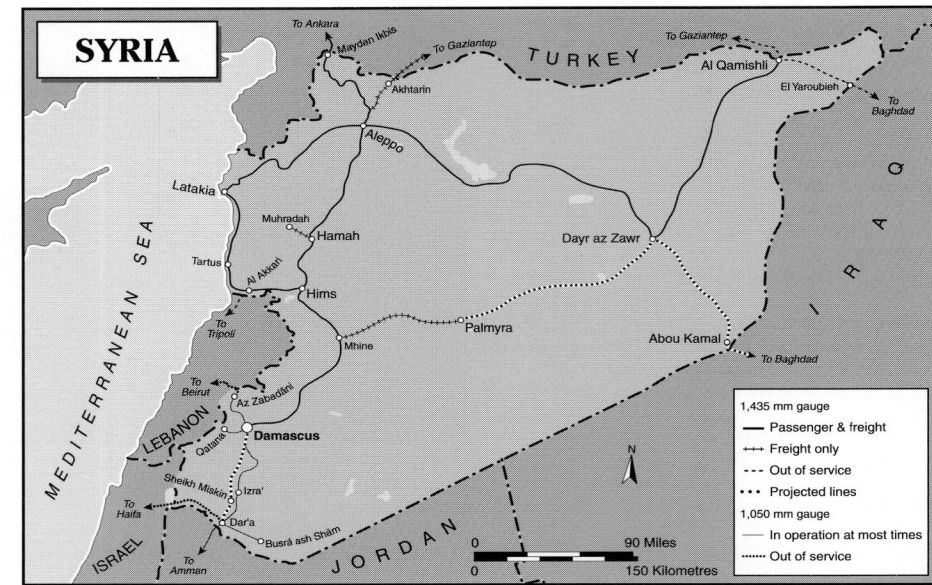

SYRIA

NEW/1114960

Ministry of Transport

PO Box 134, Damascus
Al Jalaa Street, Damascus
Tel: (+963 11) 333 68 01   Fax: (+963 11) 332 33 17

Key personnel
Minister: Mokram Obeid

UPDATED

Chemin de Fer du Hedjaz

PO Box 134, Place Hidjaz, Damascus
Tel: (+963 11) 221 29 50   Fax: (+963 11) 222 71 06

Key personnel
Director General: Ing Salah Ahmad
Head of Director General's Office: Youssef Dakhlallah
Directors
   Operations and Marketing: Ing Hussein Nasser
   Traction and Rolling Stock: Ing Faysal Alnen
   Information Technology, Signalling and
      Communications: Ing Ousama Al-Abiad
   Finance: Mohamed Osman
   Planning: Younes Al-Nasser
   Infrastructure: Ing Wafik Al-Homsy
   Training: Ing Faez Breshe
   Compliance: Rifaat Suleyman

Gauge: 1,050 mm
Route length: 251 km

Political background
In December 2001, the governments of Jordan, Saudi
Arabia and Syria agreed to reconstruct the railway.

Organisation
In addition to its own route length, the CF du
Hedjaz also operates the 67 km narrow-gauge

Damascus–Zerghaya line on behalf of the Syrian
government. Traffic has been at a very low level
for several years. Work on a standard-gauge line
to supersede this line was started but has been
suspended (see CFS entry).

Traffic
In 1999 the railway carried 94,000 passengers and
some 2,000 tonnes of freight.

Improvements to existing lines
Work was in progress in 2004 to realign a 4 km
stretch of the HJR in central Damascus. In a project

intended to alleviate road traffic congestion, one
HJR track and two Syrian Railways tracks will
share a cut-and-cover tunnel. Also planned is
redevelopment of Hedjaz station.

Traction and rolling stock
The railway owns five (of 29) serviceable steam
and three (of nine) diesel locomotives, six railcars,
35 passenger cars and 311 freight wagons.

Chemins de Fer Syriens (CFS)

PO Box 182, Aleppo
Tel: (+963 21) 221 39 00; 221 39 01
Fax: (+963 21) 222 84 80; 222 56 97
e-mail: cfs-syria@net.sy

Key personnel
President and Director General:
   Ing Mohamad Iyad Ghazal
Deputy Directors General: Ing I A M El-Bonn
   Adnan Al-Hussein
Directors
   Rolling Stock and Traction: Ing Ibrahim Mahli
   Movement and Traffic: Ing Ismail Badenkhan
   Fixed Installations: Ing Mahmoud Ismail
   Administrative: Abdul Hakim Badawi
   Financial Affairs: Farreezah Molyess
   Accounts: Ibrahim Na'ana'a
   Planning and Statistics: Ing Ghaleb Katerji
   Marketing: Ing M Kattash
   Signalling and Telecommunications: Ing N Barakat
   Technical: Ing Isa Haji
   Central Region Operations: Ing M Yassin Ghreir
   Eastern Region Operations: Ing Jamal Hasan
   Northern Region Operations: Dr Ing M Tahboub
   Institute of Railways: Ing S Barakat
   Professional Training Centre: Ing Khaled Jneid
   International Relations: H Hamam
   Medical: Dr Ihsan Assy
   Auto-Control: Ahmad Ammaneh
   Studies: Dr Ing Ammar Kaadan
   UIC Officer: M L Srag Al-Din
   Supplies: Abdul Rahman Al-Tayeb
   Legal: B Mokdom
   Assistant Manager, Informatics: Lama Mo'Mar
   Public Relations: Anas Hilali

Gauge: 1,435 mm
Route length: 2,460 km

Political background
Modernisation of the Syrian rail network has been
identified as a priority by the government. In 2003

it planned to invest €9 billion in the system, with
€12 billion annually to be spent in succeeding
years. The longer term development of the
network up to 2020 has been the subject of
studies undertaken with the help of the Japanese
consultancy, Jaika. Proposals include the
construction of new lines for speeds of up to
250 km/h to Iraq, Lebanon, Jordan and Turkey.
   Restructuring of CFS is foreseen, with the state
assuming responsibility for infrastructure and
railway operations placed in the hands of separate
independent business units.

Organisation
All standard-gauge lines in Syria are operated by CFS,
and comprise the lines from the Lebanese border via
Hims and Aleppo to the Turkish border and, in the
northeast, the connecting line between the Turkish
and Iraqi borders. A line runs from the oilfields of Al
Qamishli in the north to the port of Latakia (750 km).
The Hims–Palmyra line was opened to phosphates
traffic (destined for the port of Tartus) in 1980.
   The extension of the railway from Homs
southwards to Damascus (194 km) was opened
in 1983 and the 80 km Tartus–Latakia line in 1992.
   For operational purposes CFS is divided into
three regions: Central, Eastern and Northern.
   At the end of 2003 CFS employed around 12,400
staff.

Traffic (million)	2002	2003	2004
Passenger journeys	1.4	1.9	2.18[1]
Passenger-km	364	527	653[1]
Freight tonnes	5.91	6.39	6.56[1]
Freight tonne-km	1,812	1,882	1,770[1]

[1] 2004 figures show result up to 30 November

Passenger operations
In 2004 traffic continued to recover. This was
achieved despite competition from faster, cheaper
coaches using Syria's developing road network, a
situation exacerbated by passenger rail services
rarely exceeding an average speed of 60 km/h due
to infrastructure constraints.
   Sixteen daily services are operated (20 in
summer), and some overnight trains are run, for
example between Damascus and Latakia. Weekly
international services are run between Damascus
and Tehran via Turkey and Damascus and Istanbul.

Freight operations
Freight traffic also achieved some degree of stability
in 2004 after the 2001 low of 1,491 million tonne-km.
   Freight traffic is strong in bulk freight commodities
such as petroleum products, phosphates, cereals
and cement. Traffic is exchanged with the Turkish
network at Maydan Ikbis. A train ferry link with Greece,
possibly from Latakia to Vólos, has been proposed.

Diesel locomotives

Class	Wheel arrangement	Power kW	Speed km/h	Weight tonnes	No in service	First built	Mechanical	Builders Engine	Transmission
LDE 650	Bo-Bo	485	60	60	9	1968	B&L	SACM	*E* B&L
LDE 1200	Co-Co	895	100	120	11	1972	(Soviet-built)		*E*
LDE 2800	Co-Co	2,090	100	120	80	1974	Voroshilovgrad	5D49	*E* Jaricov
LDE 2800	Co-Co	2,090	100	120	30	1984	Voroshilovgrad	5D49	*E* Jaricov
LDE 1800	Co-Co	1,342	110	90	15	1976	General Electric	7 FDL-8	*E* General Electric
LDE 1800	Co-Co	1,342	135	90	15	1976	General Electric	7 FDL-8	*E* General Electric
LDE 1500	Co-Co	1,120	90	120	25	1985	ČKD	K6S 310 DR	*E* ČKD
AD33C	Co-Co	1,715	120	120	30	1999	Alstom	12RK215	*E* Alstom

New lines

Projects include a new line from Dayr az Zawrto Abou Kemal on the Iraqi frontier (190 km). This is to link up with Iraq's new Baghdad–Husaiba line. Some 40 km of trackbed was under construction in 2002, but no further progress has been reported.

Studies have also been undertaken of a new 203 km line from Palmyra to Dayr az Zawr.

In 1996 preliminary work started on a 1,435 mm gauge line of 101 km from the outskirts of Damascus to Dar'a, near the Jordanian frontier, to supersede the Syrian section of the 1,050 mm gauge Hedjaz Railway. The new line, which was to have a branch from Sheikh Miskin to Suweida, would be engineered for 160 km/h operation. However, construction was suspended pending commencement by Jordan of its part of the scheme.

In 2004 work was in progress to realign access to the capital from the southern section of the Hims–Damascus line. A new Damascus central station is projected.

In 2003 tenders were invited to undertake feasibility studies into a new line netween Damscus and Palmyra.

Improvements to existing lines

Extensive studies conducted since 2000 by the French consultancy, Systra, have focused on the Aleppo–Damascus trunk route. Options studied ranged from raising the maximum speed admissable on the existing line from 120 to 160 km/h to constructing a new high-speed line. While CFS favoured the high-speed option, no decision has been taken.

Over the period 2005–20, CFS plans to refurbish and upgrade the Aleppo–Dayr az Zawr–Al Qamishli main line (745 km), while most other parts of the network were expected to be similarly treated by 2010.

In support of its track improvement plans, in 2003 CFS ordered a Roger 400 diagnostic vehicle from Mer Mec to undertake track geometry and rail profile measurement.

Traction and rolling stock

In 2003 the railway owned 209 diesel locomotives, many of which were reported unserviceable, 478 passenger cars, 33 baggage vans and 5,047 freight

Type AD33C diesel-electric locomotive supplied to CFS by ALSTOM 0088111

wagons. The most recent locomotive deliveries are 30 ALSTOM 1,715 kW Type AD33C 'Prima' Co-Co machines powered by Ruston 12RK215 engines with IGBT control equipment and asynchronous traction motors. Deliveries commenced in 1999.

Poor reliability of the Soviet-built Class LDE 2800 locomotives has led CFS to embark on a repowering programme for 32 of the type using General Electric engines and traction equipment, with new cooling and air filter systems. This programme continued in 2003.

In December 2004 CFS signed a contract with Korean manufacturer Rotem to supply 10 160 km/h intercity diesel trainsets for delivery by the end of 2006. CFS expected to deploy the trains on its main lines from Aleppo to Damascus, Latakia and Al Qamishli.

In 2003 tenders were invited for the supply of 41 freight and 17 passenger diesel locomotives. No resulting orders had been reported by early 2005.

Signalling and telecommunications

On the Latakia-Aleppo-Kamechli line, signalling and telecommunications are of Soviet origin, dating from 1967. Mechanical interlockings control movements at stations, while a relay semi-automatic block system is used between stations. The Aleppo–Homs, Homs–Tartus and Homs–Mahin–Damascus lines are equipped with full relay interlockings at stations and relay semi–automatic block signalling between stations. This was supplied by German Democratic Republic companies in the 1980s.

Track

Rail: 30, 37, 43 and 50 kg/m
Crossties (sleepers): concrete, metal
Spacing:
In plain track: 1,600/km
In curves: 1,840/km
Fastenings: Russe KB, K2 and RN
Min curvature radius: 300 m
Max gradient: 2.5%
Max axleload: 20 tonnes

UPDATED

Taiwan

Ministry of Transportation & Communications

2 Chang-Sha Street, Section 100, Taipei 100
Tel: (+886 2) 23 49 29 00   Fax: (+886 2) 23 49 24 91
Web: http://www.motc.gov.tw

Key personnel
Minister: Lin Ling-san

Taiwan High Speed Rail Corporation (THSRC)

3rd Floor, 100 Hsin Yi Road Section 5, Taipei, Taiwan 110
Tel: (+886 2) 87 89 20 00
Web: http://www.thsrc.com.tw

Key personnel
Chairman: Nita Ing
President: George Liu

Political background

In 1991 revised plans were approved for a US$17 billion, 350 km high-speed line from Taipei to Kao-hsiung. A budget was approved in mid-1995 to cover the initial phases of land acquisition, planning and administration, and it was decided later that a minimum of 40 per cent of the projected cost would be sought from the private sector. Land acquisition started in late 1995 and in September 1997 the Taiwan High Speed Rail Consortium (THSRC) was named preferred bidder to construct and equip the line under a 35-year 'build-operate-transfer' (BOT) contract.

TAIWAN

0 60 Miles
0 80 Kilometres

1,435 mm gauge high-speed line
▭▭▭ Under construction
1.067 m gauge
▛▛▛ Double track electrified
— Single track
┼╍┼ Single track electrified
— Freight only lines
•••• Projected line (approximate alignment)

0576730

THSRC comprises five Taiwanese companies: Continental Engineering Corporation; Evergreen Marine Corporation (Taiwan) Ltd; Fubon Insurance Co Ltd; Pacific Electric Wire & Cable Co Ltd;

and TECO Electric & Machinery Co Ltd. Completion of the project had been planned for July 2003 but this was subsequently revised to October 2005.

New lines

Taiwan's 350 km high-speed railway will be a double-track standard-gauge line extending from the capital city Taipei in the north to the second largest city Kaohsiung in the south. The line is being designed for 350 km/h but initially maximum speeds will be limited to 300 km/h.

The line is to extend for most of the length of the island, linking several of the larger cities located along the densely populated west coast, where about 75 per cent of Taiwan's population lives. The existing transportation facilities in this corridor, consisting of highways, the 1,067 mm gauge existing railway (TRA) and air services, are heavily utilised, and high-speed rail was seen as an obvious way to ensure that growing demand for rapid and convenient travel along the corridor would be met.

Civil engineering works

The project is one of the largest railway construction projects under way in the world. The nature of the topography and difficulties associated with land acquisition have resulted in the railway generally being located on viaduct or in tunnel, with only short sections at grade. The guideway consists of the following:

Viaducts and bridges – 255 km
Mined tunnels (36, including 4 long tunnels) – 39 km
Cut and cover tunnels – 11 km
Embankments and cuttings – 32 km.

A section of viaduct on the high-speed line in Taiwan   0576712

Portal area nearing completion on one of the line's tunnels, more than 30 of which are located in the hilly terrain of the northern section   0576713

On the southern section of the line, where the topography is generally flat, there will be a section of viaduct extending continuously for almost 160 km. This is one of the longest railway viaducts in the world.

For the major civil construction work 11 contracts were signed with international joint ventures between March and May 2000. The twelfth and last civil contract covering a short section at the southern end of the line was let in early 2001. All of these contracts are of the design-and-build type. Construction began in earnest in early 2001 and by the end of 2003 was over 90 per cent complete. Construction work on the three southern contracts covering the 60 km section due to form the track for train and system testing and for staff training was largely completed by this time.

Taiwan is located in an area of high seismic activity so particular attention has been paid to this factor in the design of the civil structures. Other engineering challenges included poor geotechnical conditions at some tunnel sites and the presence of three active earthquake faults that needed to be traversed.

The most northern section of the line, of approximately 20 km through Taipei, is not included in the 12 civil contracts. Here the high-speed line is to be located in an already constructed tunnel that passes through the city centre and presently accommodates the existing 1,067 mm gauge TRA main line. The intention is for the two railways to share this infrastructure, with two tracks each for the high-speed line and TRA.

Stations and depots
The high-speed line will initially have seven stations located at Taipei, Taoyuan, Hsinchu, Taichung, Chiayi, Tainan and Tsoying, on the northern outskirts of Kaohsiung. Provision has been made in the design for five additional stations to be added at appropriate times in the future.

In central Taipei the existing main railway station is to be used. The other six initial stations were designed under four consultancy contracts. Construction contracts were awarded for these in 2002 and by the end of 2003 significant progress had been achieved. Most of these stations are

elevated structures constructed around the guideways built under the civil contracts.

Four depots and maintenance facilities for various activities are being constructed along the route of the line with several more planned in the Taipei area. Contracts for the construction of the four depots and workshops were awarded during early 2003 and construction work on these was well advanced by the end of 2003.

Trackwork
Four trackwork contracts cover the new guideway sections, while a fifth covers the track to be installed in the existing tunnel through central Taipei. All five contracts were awarded during 2002 and construction was progressing well in 2004. While initially both slab and ballasted track were specified for different sections, this was changed and all track is now to be slab apart from that in depots and a short section of main line which will be ballasted. For the most part a Japanese slab track design has been adopted. German 'Rheda' slab track is being used in certain station areas and Sonneville low-vibration track is being installed in the Taipei tunnel to reduce vibration impacts on adjacent buildings. The trackwork contracts are of the design and install type.

The high-speed line is to be double-track standard-gauge, using 60 kg/m rail. The two tracks will be 4.5 m apart.

Core System (Rolling stock and mechanical and electrical equipment)
After intense competition between European and Japanese consortiums, the Core System contracts were awarded in late December 1999 to the Japanese TSC group, which includes Mitsui, Mitsubishi Corporation, Mitsubishi Heavy Industries, Kawasaki and Toshiba. These contracts provide for the supply, installation, commissioning and testing of all rolling stock, traction power supply, signalling and telecommunications equipment.

The line will be electrified at 25 kV 60 Hz AC.

Thirty trainsets are being obtained initially, with options for another 25 sets. The train design is derived from that of the Series 700 Shinkansen train. Each set will consist initially of 12 cars (nine motor cars and three trailers). The first train was rolled out in Japan in January 2004 and was due for delivery to Taiwan later that year. Testing and initial training is to take place over the 60 km section of the main guideway that has been completed at the southern end of the line.

All trains will have ATO and ATP as well as onboard fault diagnosis and monitoring systems.

Train operations
To meet the anticipated passenger demand at acceptable levels of service, several different types of service will be operated with minimum intervals of four minutes between trains.

The fastest services between Taipei and Kaohsiung will include a stop at Taichung and will take 90 minutes for the full journey. The slowest trains, that will stop at all the stations to be built, will take approximately 2 hours 15 minutes for the journey from Taipei to Kaohsiung. A number of services that operate only over the Taipei-Taichung section will also be provided.

Taiwan Railway Administration (TRA)

3 Peiping West Road, Taipei
Tel: (+886 2) 23 81 52 26   Fax: (+886 2) 23 81 13 67
e-mail: railway@railway.gov.tw
Web: www.railway.gov.tw

Key personnel
Director General: T W Hsu
Deputy Directors General: M J Huang; Y N Sheu; C L Shu
Chief Secretary: F N Chen
Chief Engineer: M H Chen

Directors
Transportation: C S Chou
Construction: S C Chen
Rolling Stock: J S Siao
Procurement and Stores: C R Ciou
Planning: C C Liu
Accounting: J L Lin
General Affairs: M J Lai
Personnel: M C Guo
Civil Service Ethics: H Z Hsu
Freight Services: Y H Jang
Catering Services: C B Chen
Employee Training: RT L Chen

Gauge: 1,067 mm
Route length: 1,097 km
Electrification: 685 km at 25 kV 60 Hz AC

Organisation
Structural renovation project
TRA has progressed with reforms in recent years and taken on many major construction and business renovation projects. Projects completed include the double-tracking of the west coast mountain line between Jhunan and Fongyuan and the electrification of the Kaohsiung–Pingtung line. Ongoing projects include track structure renovations, the elevation of the main lines in

urban areas to eliminate level crossings, level crossing protection improvements, and train safety facilities improvements. In the area of business development and reform, TRA has adjusted the pricing structures, upgraded ticketing services, strengthened enquiry services and improved travel and transport facilities. TRA has also diversified to invest in the telecommunication field and in land development to improve its overall financial structure.

Finance

TRA benefits from state funding for its major projects financed by the country's massive accumulation of foreign exchange reserves, which in April 2004 stood at US$227.7 billion. But TRA itself is in deficit, largely because of a burden of excessive staff and pensioners, and of government control of its charges.

Finances (NT$ million)

Revenue	2001	2002	2003
Passenger	16,292	15,673	14,202
Freight	1,183	1,099	1,008
Parcels and mail	42	36	31
Other	3,823	6,293	4,558
Total	21,340	23,101	19,799

Expenditure			
Staff	20,199	18,496	17,917
Materials/services	5,715	5,826	5,250
Depreciation	4,101	4,186	4,390
Financial charges	3,969	4,056	2,193
Total	33,984	32,564	29,750

Traffic (million)	2001	2002	2003
Passenger journeys	186.1	175.3	161.4
Passenger-km	10,037	9,666	8,726
Freight tonnes	12.4	12.1	11.2
Freight tonne-km	985	919	846

Passenger operations

In the passenger sector TRA until recently mainly pursued the development of quality long-distance services with air conditioned rolling stock. These are the most remunerative because, although there is government control of fares, the ceiling imposed on long-distance fares is well above that set for local and commuter travel. However, with the THSRC high-speed line between Taipei and Kaohsiung due to open in late 2005, TRA's focus, at least for the very busy west coast corridor, is expected to change to short- and mid-length journeys of less than 200 km. At present around 70 per cent of journeys are of less than 50 km and around 25 per cent are between 50 and 200 km.

In addition to focusing more on shorter distance travel on the west coast line, TRA is also now concentrating on improving services down the less populated but very scenic east coast line. Electrification and doubling is now complete as far as Hualien and some services taking less than three hours for the Taipei–Hualien journey are now in operation. TRA also introduced faster services in 2004 on the middle section of the east coast line between Hualien and Taitung.

The cross-city tunnel in Taipei, the first section of which opened in 1990, has eliminated traffic congestion caused by level crossings in central Taipei. The tunnel, which has been opened in stages, now extends from west of Panchiao station to Sungshan station in the eastern part of the city. The project has also established a transport centre in the Taipei station area as a foundation for the city's emerging mass transit system.

In 2004 TRA launched a luxury circular train service for tourists, which travels around the island and is called the 'Star of Formosa'.

Freight operations

In the freight sector, bulk commodities account for 60 per cent of the tonnage, with cement, limestone, grain and coal topping the list. Through its Railway Freight Service (RFS), TRA offers a total service, from door-to-door rail-and-truck transits between a dozen main centres to warehousing, responsibility for customs clearance and insurance. RFS has headquarters in Taipei, branch offices in eight other cities and service offices at 69 locations of the rail network.

TRA Class DR E1000 electric power car heading a push-pull formation on the east coast line (Bruce Evans)
0576714

TRA Class DR1000 Nippon Sharyo single-unit diesel railcars on the Pingshi branch (Bruce Evans)
0576715

TRA refurbished Class 300 long-distance emu at Sungshan station, Taipei (Bruce Evan)
0576716

Intermodal operations

TRA has two container terminals of its own at Keelung, one at Chitu and the other at Wutu. In addition, the railway serves the United Container Terminal's installation at Puhsu. TRA is also one of the financial backers of the China Container Terminal Corp and its Wutu Inland Terminal in the Keelung suburbs. The railway has further port container terminals at Tai-chung and Su-ao, and a Taipei area terminal at Cheng-kung.

The Keelung and Kao-hsiung port terminals and the Taipei inland terminal are interconnected by eight dedicated container trains each way daily. TRA has some 600 four-axle flat wagons capable of carrying 40 ft ISO containers.

New lines
Proposed Taipei–Ilan line
With the increased attention being given to the east coast services, feasibility studies have been undertaken into constructing a direct line from Taipei to the east coast city of Ilan. Instead of following the coast line in the same way as the existing line, the new line would pass through the mountain range which forms a formidable physical barrier between the two cities. A new line on such an alignment would for the most part need to be in tunnel and the project would thus have a high capital cost. Initial estimates for the project are of the order of NT$60 billion. The government is studying the possibility of adopting a BOT strategy for implementing the scheme.

Chiang Kai-Shek airport line
In May 1998 the Ministry of Transport & Communications awarded a concession to develop a 35 km rail link from Taipei to Chiang Kai-Shek International airport. The line is envisaged to include 15 stations for commuter traffic. However, in 2004 the project had not progressed as expected and revised implementation plans were being studied.

Improvements to existing lines
East coast line improvements
Following completion of double-tracking of the main line down the western side of the island of Taiwan, the government has funded double-tracking of the route on the eastern side, southward from Su-ao to Hua-lien; and also electrification of this line and its adjoining double-track line to the north. This project was completed in 2003. South of Hua-lien, the east coast line is being upgraded with 50 kg/m rail to Tai-tung and the junction with the new South Coast Link. CTC is to be installed on this section. Target date for completion is 2004.

Taipei cross-city line
The Taipei cross-city tunnel has been opened progressively in stages since 1990 and now extends from Panchiao in the west of the metropolitan area to Sungshan in the east. In 2004 a major project was underway which includes the extension of this tunnel still further eastwards to beyond Nankang. The project also includes the construction of a considerable length of elevated guideway between Hsichih and Wutu. The new construction will eliminate a large number of level crossings, easing traffic congestion in these built-up areas and improving safety for both rail and road users.

Traction and rolling stock
At the end of 2003 the TRA fleet comprised 179 electric and 155 diesel locomotives, 171 dmu vehicles, 66 diesel railcars, 563 emu vehicles, 1,351 passenger coaches and 2,755 freight wagons.

In the mid-1990s delivery was taken of 344 electric multiple-unit cars for Taipei suburban services being built by Daewoo Heavy Industries in South Korea. The cars have Siemens electrical equipment, comprising VVVF propulsion and control systems, and bogies by Adtranz. A further 56 generally similar emu cars were supplied by Rotem in 2002. With electrification of the east coast line to Hualien, these trains are also now seeing service on this route.

In 2004 TRA ordered eight six-car tilting emus from Hitachi for use on its east coast line. Their design is to be similar to that of the Series 885 trainsets operating in Japan.

Other acquisitions include an order for 36 Class DR 1000 single-unit diesel railcars from Nippon Sharyo for suburban and branch line services and 10 Class DR 3100 three-car dmus from the same builder. Two of the DRC units and one dmu were manufactured in Japan, while the remaining vehicles were assembled in Taiwan from knocked-down components.

The push-pull trains made up of UCW-supplied Class E1000 locomotives and matching Hyundai-supplied coaches used for the top quality

TRA Daewoo/Rotem-built Class 600 emus (Bruce Evans)   0576717

Diesel locomotives

Class	Wheel arrangement	Power kW	Speed km/h	Weight tonnes	No in service	First built	Mechanical	Engine	Transmission
								Builders	
R20/R50	A1A-A1A	1,060	100	78	49	1960	EMD	EMD	E EMD
R100	Co-Co	1,237	100	78	39	1970	EMD	EMD	E EMD
R150	Co-Co	1,237	110	89	24	1970	EMD	EMD	E EMD
R180	Co-Co	1,237	110	89	4	1992	EMD	EMD	E EMD
R190	Co-Co	1,237	110	89	46	1992	EMD	EMD	E EMD
S200	A1A-A1A	712	100	65	10	1960	EMD	EMD	E EMD
S300	Bo-Bo	667	75	54	3	1966	EMD	EMD	E EMD
S400	Bo-Bo	825	75	54	4	1970	EMD	EMD	E EMD
R0	Co-Co	1,154	100	85	1	1960	Hitachi	MAN	E Hitachi
DHL100	B-B	900	75	75	16	2003	TRSC	Cummins	H Niigata

Electric locomotives

Class	Wheel arrangement	Power kW	Speed km/h	Weight tonnes	No in service	First built	Mechanical	Electrical
							Builders	
E100	Bo-Bo	2,050	110	72	20	1976	UCW	GEC
E200	Co-Co	2,800	110	96	39	1976	GE	GE
E300	Co-Co	2,800	110	96	39	1976	GE	GE
E400	Co-Co	2,800	130	92	18	1980	GE	GE
E1000	Bo-Bo	–	130	58	64	1996	UCW	GEC

Diesel railcars or multiple-units

Class	Cars per unit	Motor cars per unit	Motored axles/car	Output/ motor kW	Speed km/h	Unit in service	First built	Mechanical	Electrical	Transmission
								Builders		
2050	1	–	–	–	105	7	1971	TRA	Cummins	H
2510	1	–	–	–	110	2	1992	TE	Cummins	H
2700	1	–	–	–	110	21	1966	Tokyu Car	Cummins	H
2800	3	–	–	–	110	15	1982	Tokyu Car	Cummins	H
2900	3	–	–	–	110	5	1987	Hitachi	Cummins	H
3000	3	–	–	–	110	27	1990	Hitachi	Cummins	H
3100	3	–	–	–	110	10	1998	Nippon Sharyo	Cummins	H Nico
1000	1	–	–	–	110	36	1998	Nippon Sharyo	Cummins	H Nico

Electric railcars

Class	Cars per unit	Motor cars per unit	Motored axles/car	Output/ motor kW	Speed km/h	Unit in service	First built	Mechanical	Electrical
								Builders	
100	5	1	4	310	120	11	1979	BREL	GEC
200	3	2	4	116	120	10	1988	Socimi	Brush
300	2	3	4	125	120	8	1986	UCW	GEC
400	2	4	4	125	110	12	1991	UCW	GEC
500	2	3	4	250	110	86	1995	Daewoo	Siemens
600	2	3	3	250	110	14	2002	Daewoo/Rotem	Siemens
[1]	6	3	4	190	130	8	–	Hitachi	Hitachi

[1] Ordered 2004.

long-distance services on the electrified sections of the network have recently been extended from 10 to 12 coaches following the acquisition of additional passenger and snack-bar vehicles.

In addition to new vehicles TRA has been refurbishing older long-distance emus to provide improved service. The first vehicles treated have been the 300 Series dating from 1986. Improvements include new seating, automatic doors and updated information displays. New cabs have been provided on the driving cars.

The most recent diesel locomotives acquired by TRA are 16 Type DHL100 diesel-hydraulic shunting units of 900 kW supplied by Niigata Engineering of Japan in 2002. This series was obtained to replace the 1966-supplied S300 shunters.

For the period from 2005 to 2010 TRA's plans include the purchase of 272 emu and 96 dmu vehicles for commuter services, 84 tilting emu and 54 tilting dmu vehicles for long-distance services, as well as 30 electric and 20 diesel locomotives.

TRA Class DHL100 diesel-hydraulic shunting locomotive supplied by Niigata (Bruce Evans) 0576718

Automatic Train Protection (ATP) systems for commissioning in 2005.

Electrification
The following sections totalling 601 km were energised as at the end of 2003: Keelung–Pingtung (427 km); Patu–Suao (94 km); Hsuhsin–Hualien (80 km).

Type of coupler
Passenger cars: Tight lock automatic AAR-H
Freight cars: AAR E Type automatic
Type of braking: AAR Westinghouse air

Track
Rail: 37 and 50 kg/m; 100 lb/yd
Crossties (sleepers): In tangent track and curves over 600 m radius: 174 × 240 × 2,000 mm
In curves of 300 to 600 m radius: 201 × 240 × 2,000 mm

Spacing
In plain track: 1,760/km (wood), 1,640/km (concrete)
In curves: 1,800/km for radii less than 400 m
Rail fastenings: Pandrol clips and dogspikes
Min curvature radius: 5.82°
Max gradient: 2.5%
Max axleload: 18 tonnes

UPDATED

Signalling
Computer-aided CTC has now been installed around the island covering a total of 903 route-km.

A total of 54 stations have been equipped with electronic interlocking equipment. In 2004 work began to install Bombardier axle-counting and

---

Tajikistan

Ministry of Transport

Nazarshojev 35, 734012 Dushanbe
Fax: (+737 72) 21 44 48

Key personnel
Minister: F Mukhiddinov
Deputy Minister: Abdurahmon Rasulov

Tajik Railways

Tajikskaya Zheleznaya Doroga (TZD)
Nazarshojev 35, 734012 Dushanbe
Fax: (+737 72) 21 44 48

Key personnel
President: M Nuralyev
Chief Engineer: B Shodiyev
Traffic Manager: A Bulugin
Finance Director: M F Narzyev

Gauge: 1,520 mm
Route length: 482 km

Political background
The railway was founded as an independent organisation in October 1994.

Traffic has declined precipitously since independence. The disturbed political situation has resulted in a mass exodus of Russians, and 90 per cent of railway specialist staff have been lost as a result. Further decline will occur if Uzbekistan completes its proposed Angren bypass of the Tajik Railways' northern section. This section was closed by military activities for a week in November 1998.

In 1996, an agreement was made with the Uzbekistan railway covering standardisation, cross-border workings of personnel, and access of Tajik students to Uzbek railway training institutions. Unfortunately this has not led to any noticeable improvement in cross-border operations.

In 1997, the accounts of all 38 enterprises making up the railway were frozen because of overdue tax. They were later unfrozen, and the railway also made inroads into its wages backlog. The average railwayman's wage is five times higher than the national average, which is exceptional for a former Soviet state.

In December 1998, the Asian Development Bank announced a US$21.5 million loan and technical assistance package to assist in the reform of

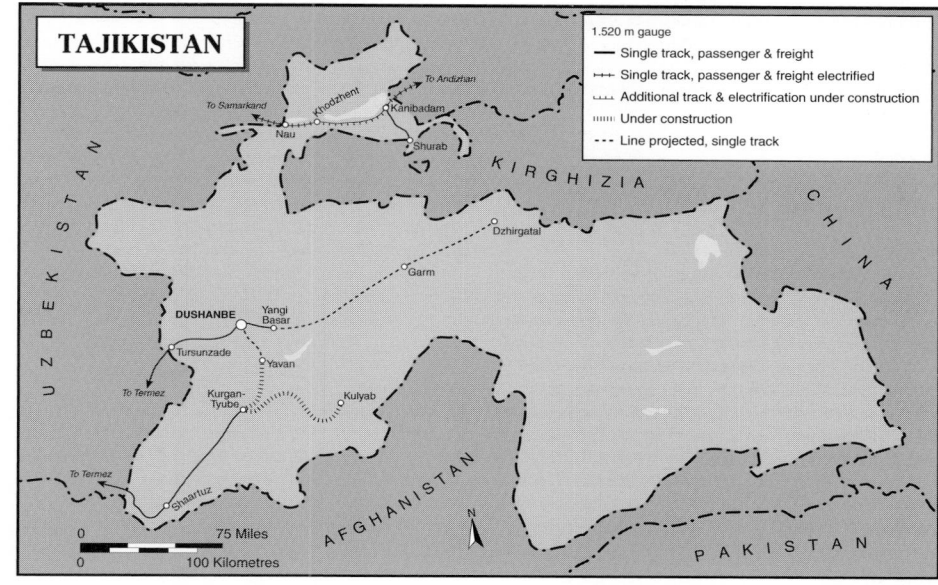

0121645

Tajikistan's transport and energy sectors. Among the programme's aims were the separation of regulatory and operational functions of the railway network and providing state enterprises increased autonomy and incentives to run on commercial lines.

In July 2001, the government announced increased funding for the network to pay for a programme of improvements and to meet debts owed to the Uzbekistan Railway.

Organisation
Formerly part of the USSR's Central Asian Railway, Tajik Railways consists of three lines, each isolated from the others by neighbouring states. In the south there are two branch lines from Termez in Uzbekistan, one to the Tajik capital Dushanbe and Yangi Basar (93 km) and one to Kurgan-Tyube and Vakhsh (220 km). In the north, some 110 km of the east-west Andizhan–Samarkand line runs through Tajikistan. This portion is electrified eastwards as far as Khodzhent (formerly Leninabad), but electrification work on the 59 km section towards Kanibadam has been suspended. A branch from Kanibadam stretches 53 km to Shurab.

The number of operating workers in 1999 was 4,235.

Traffic (million)	1997	1999
Passenger journeys	0.64	0.65
Passenger-km	150.2	61.2
Freight tonnes	0.62	0.90
Freight tonne-km	–	1,300

Passenger operations
In December 1995, Russian Railways refused to handle the Dushanbe–Moscow through service, citing poor quality rolling stock and mass ticketless travel. The through train was reintroduced on a thrice-weekly basis from October 1996, taking 87 hours for the journey.

A weekly Dushanbe–Volgograd train, reputed to carry mainly ticketless travellers in unhygienic conditions, was established during 1997.

New lines
In August 2001, the government authorised planning for a new line from Dushanbe to Dzhirgatal via Garm, with the eventual intention

of extending it to the border with Kyrgyzia. Also planned is a line from Dushanbe to Yavan.

In 2001, work was well advanced on the lines from Kurgan-Tyube to Yaran and to Kulyab.

Improvements to existing lines

Despite a shortage of track components, remedied partly by the lifting of redundant sidings, electrification and double-tracking of the northern line is taking place over busy sections. The first

53 km of this 128 km line were opened in 1998 and the remainder in September 1999. Upgrading continued into 2000, and the bi-weekly passenger train now covers the 128 km in five hours.

Traction and rolling stock

Having initially only 29 locomotives, stationed at the country's only depot at Dushanbe, and finding itself with no track maintenance equipment upon independence, the railway has depended, to some

extent, on equipment loaned by Uzbekistan's railway.

In 1997, the locomotive fleet comprised two double-unit locomotives of Type 2TE10, purchased in 1996; 24 double-unit locomotives of Type 2TE10V/L, of which 16 were serviceable; 10 shunting locomotives of Type TEM2 (all serviceable) and five shunting locomotives of Type ChME3 (of which only one was serviceable). The rolling stock fleet comprised 355 passenger coaches and 2,112 freight wagons.

Tanzania

Ministry of Communications & Transport

PO Box 9144, Dar es Salaam
Tel: (+255 22) 212 22 68   Fax: (+255 22) 211 27 51

Key personnel
Minister: Mark Mwandosya

NEW ENTRY

Tanzanian Railways Corporation (TRC)

PO Box 468, Dar es Salaam
Tel: (+255 51) 211 05 99   Fax: (+255 51) 211 06 00
Web: http://www.trctz.com

Key personnel
Director General: L Mboma
Chief Commercial Manager: Rukia D Shamte
Chief Mechanical Engineer: K A Kisamfu
Chief Civil Engineer: J Mabeyo
Chief of Manpower Development: N N Msoffe
Chief Supplies Manager: A E Munishi
Chief of Finance: S A Riwa
Chief of Corporate Development and Management
   Services: H J Kivina
Chief Signals and Communications Engineer: B Tito

Gauge: 1,000 mm
Length: 2,721 km

Political background

Following the formal break-up of the East African Railways Corporation in 1977, Tanzania set up the independent Tanzanian Railways Corporation to operate the former EAR lines wholly within Tanzania.

By the end of the 1980s, TRC was rundown and in urgent need of investment capital. In a series of World Bank-inspired reforms, the railway's extensive lake-shipping operation was hived off into a separate profit centre, staff numbers have been reduced (down to 8,892 by the end of 2001) and tariffs increased. Some rehabilitation of the system has been possible with funds made available from Canada and the European

Union and the World Bank. In 2002, the Tanzanian government was in the process of privatising TRC by concessioning, with the process expected to be completed by mid-2003.

Traffic (million)	1999	2000	2001
Passenger journeys	0.615	0.631	0.728
Passenger-km	413	–	471
Freight tonnes	1.127	1.165	1.351
Freight tonne-km	1,185	1,210	1,380

Passenger operations

Passenger traffic levels continue to run at less than half the level achieved in the mid-1990s but in 2002 were improving steadily. Bus competition accounts for some of the decline in traffic, and has led to a discontinuation of shuttle services on the Dar es Salaam–Moshi and Dar es Salaam–Tanga lines.

Freight operations

Freight traffic has continued to recover, achieving 1.35 million tonnes (1,380 million tonne-km) in 2001.

Since 1993, TRC has provided dedicated trains for Burundi traffic between Dar es Salaam and its western terminus at Kigoma, where goods are transhipped to road vehicles for the cross-border journey to Burundi. New freight-handling depots at Dar es Salaam and Kigoma provide rapid customs clearance facilities. The Burundi service is seen as the prototype for similar operations serving other neighbouring countries as TRC strives to regain the transit traffic lost at the start of the 1990s. An inland dry port with a capacity of 13,000 TEUs annually has been established at Isaka. This serves Rwandan import and export traffic, as well as cargo to eastern Congo and some to Burundi. A through service to and from Kenya via the Moshi–Taveta–Voi link was restored in 1996 after a 20-year interruption.

New lines

In 1991, Uganda and Tanzania reached agreement in principle for extension into Uganda of TRC's line from the port of Tanga to Arusha. The proposal is to project this line to Musoma on Lake Victoria and install a train ferry to connect Musoma with Uganda's capital, Kampala. A feasibility study,

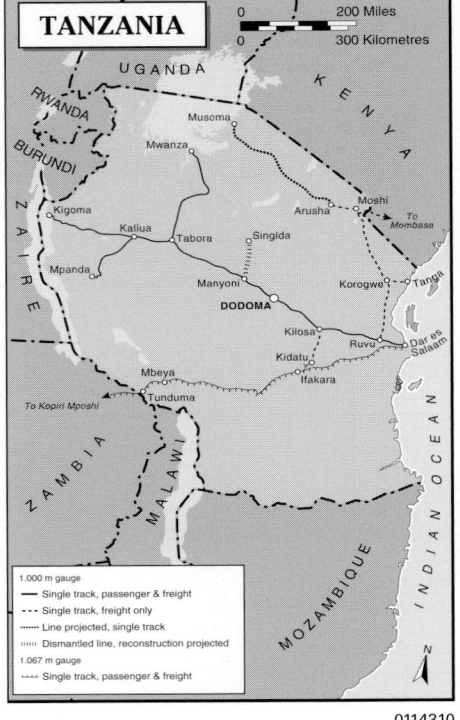

0114310

funded partly by Uganda, was carried out during 1994, and the project has since been adopted by the East African Community.

Traction and rolling stock

In 2001, TRC owned 66 main line diesel locomotives, 148 passenger coaches and 1,970 freight wagons.

Adtranz carried out heavy overhaul on 11 Class 36 diesel shunters, and refurbished 79 passenger coaches at TRC's Dar es Salaam works. A batch of 27 third-class coaches was supplied by India's ICF in 1998–99. During the same period 80 covered bogie wagons were added to the freight fleet.

Type of coupler in standard use: MCA-DA
Type of brake in standard use: Automatic air, Type EST4d

Track
Rail: 55, 60 and 80 lb/yd
Crossties (sleepers): Steel plain track, wood turn-outs
Fastenings: Fish bolts and nuts, fishplates, screw spikes, coach screws, Pandrol
Spacing:
   55 lb/yd: 1,430/km plain, 1,540/km curved track
   60 lb/yd: 1,405/km plain, 1,485/km curved track
   80 lb/yd: 1,402/km plain, 1,482/km curved track
Min curve radius: 8°
Max gradient: 2.5%
Max axleload: 14 tonnes

Diesel locomotives

Class	Wheel arrangement	Power kW	Speed km/h	Weight tonnes	First built	Mechanical	Builders Engine	Transmission
35	C	205	25	36.6	1973	Barclay	Paxman 8 RPHL Mk 7	H Voith L320V
36	C	244	25	36.2	1979	Brush	Ruston-Paxman	E Brush
37	C	295	25	36.2	1985	Henschel	MTU Type 6Y 396	H Voith TC12
64	B-B	559	72	38.3	1979	Henschel	MTU Type EB	H Voith L520-UZ 12V 396 TCII
72	1Bo-Bo1	925	72	68.86	1972	GEC Traction	Ruston-Paxman 8CVST	E GEC
73	Co-Co	1,003	96	72	1975	Varanasi	YDMA4	E Varanasi
87	1Co-Co1	1,370	72	101.4	1966	Eng Elec	Ruston-Paxman 12CVST	E Eng Elec
88	1Co-Co1	1,490	72	110.9	1972	MLW	Alco 251C	E GE Canada

Tanzania-Zambia Railway Authority (TAZARA)

PO Box 2834, Dar es Salaam
Tel: (+255 51) 86 03 40/76 41 91/9
Fax: (+255 51) 86 51 87/86 51 92

Key personnel
Chairman: S Musoma
Co-Chairman: Ronald Makuma
Directors
   Finance: Method A Kashonda
   Corporate Planning: Michael J Ngonyani
   Traffic: J Y Minsula

Technical Services: L M Nsofwa
Regional Manager, Zambia: Morrison S Banda
Regional Manager, Tanzania: Hamis M Teggisa

Gauge: 1,067 mm
Route length: 1,860 km (891 km in Zambia, 969 km in Tanzania)

Diesel locomotives

Class	Wheel arrangement	Power kW	Speed km/h	Weight tonnes	No in service	First built	Mechanical	Builders Engine	Transmission
1A (DFH1)	Bo-Bo	1,000	50	60	14	1971	Chintao	12V 189ZL[1]	SF 2010
1B (DFH-2)	Bo-Bo	2,000	100	80	57	1971	Chintao	12V 180ZL (28 locos)[1]/ MTU 12V 396TC12 (29 locos)	SE 2010 SE2010
DE (U30C)	Co-Co	3,200	100	120	29	1983	Krupp	GE 17 FDL 12HT	E GE
CKD8B	–	2,200	–	–	6	1997	Loric	–	–

[1] Class is being re-engined with MTU units.

Political background

The Tanzania-Zambia Railway (TAZARA) was constructed following an agreement signed in 1967 between the governments of Tanzania, Zambia and the People's Republic of China. Under the agreement, China provided finance and technical services for construction of a railway from Dar es Salaam in Tanzania to Kapiri Mposhi in Zambia, together with equipment, two workshops and other auxiliary facilities. Operation began in 1975.

The loan repayment was to commence in 1983 and to be spread over 30 years, with each country responsible for 50 per cent. Due to their economic problems, however, both countries agreed in 1983 to reschedule the repayment terms; the start of repayment of the main loan was put back 10 years. In 1997, a further loan agreement was signed, with China extending the 30-year contract to provide additional development funding of Y100 million. This was intended to cover staff training and the provision of microwave communications technology on the section of the railway in Zambia.

The Council of Ministers, consisting of three Ministers each from Zambia and Tanzania, is the body established by the two governments to exercise overall control on the railway. All the railway assets are vested in TAZARA, a corporate body whose principal organ is the Board of Directors; this consists of 10 members, five appointed by each government. For operational purposes, the railway is divided into two regions for Tanzania and Zambia, with headquarters at Dar es Salaam and Mpika.

TAZARA's gauge permits through traffic with contiguous railways in Central Africa, in particular Zambia Railways. Its performance, however, has been handicapped by serious problems of traction, rolling stock and track maintenance, of inadequate funds, and the international political strains of the continent. Whereas the railway's designed capacity was 5 million tonnes of freight a year, it has yet to register more than 1.5 million tonnes.

In the 1980s, TAZARA benefited from export/import traffic from Botswana, Malawi, Zaire, Zambia and Zimbabwe seeking to use Dar es Salaam instead of South African ports, but in the 1990s political change in South Africa enabled the landlocked states to resume use of ports in that country. By 1997, TAZARA was in deep trouble, with carryings down to 600,000 tonnes per year. An estimated US$7 million was owed by customers and there were few prospects of being able to collect the money.

In an effort to reverse the decline, the railway adopted a commercialisation policy with an aggressive approach to marketing. Armed police ride with freight trains to deter pilferers. Transit times have been cut, with the fastest freight trains taking four days (as opposed to 10 in the past) to traverse the line. Gantry cranes were being installed at four freight terminals. By 1999, the workforce had been reduced to 5,000 and it was planned to cut this to 3,000 by 2003.

In 1999, it looked increasingly likely that some degree of privatisation would be introduced to TAZARA.

Passenger operations

A long-established passenger service runs the length of the TAZARA line, from Dar es Salaam to Kapiri Mposhi, twice a week. In 1997, a twice-weekly express service, named 'Mukaba' eastbound and 'Kilimanjaro' westbound, was introduced.

There are also three-a-week services running on the Tanzanian and Zambian sections respectively.

Around 1.6 million passenger journeys were recorded in 1998–1999.

Freight operations

Zambian exports, principally copper, have been the mainstay of TAZARA freight tonnage, but much of this is now routed through South Africa. Some traffic diversification has been achieved.

Improvements to existing lines

In a project funded by the Austrian government, long-welding of rail is being installed between Dar es Salaam and Makambako (657 km); this project was more than two-thirds complete in 1999. Completion throughout the route will remove a source of frequent delays due to misaligned or broken rail joints.

Traction and rolling stock

At the start of 1993 TAZARA operated 100 diesel locomotives (many of GE design), 97 passenger coaches (including nine restaurant cars) and 2,230 freight wagons.

In 1997, six CKD8B 2,200 kW diesel locomotives were delivered from Chinese manufacturers, funded by a Chinese government loan of Y200 million. The same loan was to allow the purchase of 80 tank wagons for petroleum products, a breakdown crane, and spares for locomotives and rolling stock.

Signalling and telecommunications

A digital microwave radio system was to be installed over the entire TAZARA system by 2000–01 at a cost of US$500,000.

Type of coupler in standard use: Top action
Type of braking in standard use:
Passenger: Air
Freight cars: Air, vacuum

Track

Rail: 45 kg/m, 12.5 m length
Crossties (sleepers): Prestressed concrete 195 × 208 × 272 mm
Spacing: 1,520/km
Fastening: Electric (spring clip and bolt)
Min curvature: 200 m
Max gradient: 1 in 50
Max altitude: 1,789.43 m, Uyole near Mbeya
Max axleload: 20 tonne

Thailand

Ministry of Transport & Communications

38 Ratchadamneon Nok Avenue, Bangkok 10100
Tel: (+66 2) 281 34 22   Fax: (+66 2) 280 17 14
Web: http://www.motc.go.th

Key personnel

Minister: Suriya Jungrungrerngkij
Permanent Secretary: Srisuk Chandrangsu

State Railway of Thailand (SRT)

Rong Muang Road, Pathumwan, Bangkok 10330
Tel: (+66 2) 220 42 60; 22 04 26 67
Fax: (+66 2) 225 38 01
e-mail: fad_gm@railway.co.th
Web: www.railway.co.th

Key personnel

Governor: Dr Chitsanti Dhanasobhon
Deputy Governors
   Operations 1: Sriyoudh Sirivedhin
   Operations 2: Thavil Samnakorn
   Administration: Bancha Kongnakorn
   Development and Planning:
      Youdtana Tupchareon
   Special Affairs: Suchai Roywirutn
   Advisor to SRT: Teera Rattanavit

Assistant Governors: Prasert Attananda;
   Pijarn Rattanaratree; Sayan Rohitrattana; Viroj
   Treamphongpun
Inspectors   General: Clong   Clongpayaban;
   Chalerm Nontasut; Ithipon Paphavasit
Personnel Manager: Wacharin Teevakul
Traffic Manager: (vacant)
Marketing Manager: Voravuth Mala
Chief Mechanical Engineer: (vacant)
Chief Civil Engineer: (vacant)
Chief Signalling and Telecommunications
   Engineer: Anuwong Sooksiwong
Chief Construction Engineer: Charnchai Anantasate
Directors
   Finance and Accounting: Kallayanee
      Krongboonying
   Information Systems: Yolchai Kemungkorn
   Stores: Kiatpong Muangvuttanant
   Property Management and Development:
      Panthop Malakul Na Ayuthaya
   Internal Auditing: Wayupol Chaisiri
   Special Programme Development:
      Nakorn Chantasorn
Chief, Legal Bureau: Surang Srimeesup
Chief, Medical Bureau: Dr Montree Changtor
Chief, Policy and Planning Bureau:
   Sirima Hiruncharoenvate
Chief, Governor Bureau: Molimas Shatragom

Gauge: 1,000 mm
Route length: 4,071 km

Political background

While restructuring and privatisation of SRT has been government policy since November 1998, no significant moves to implement this had been made by early 2005.

Traffic (million)	2000	2001	2002
Passenger journeys	60.6	56.3	55.7
Passenger-km	10,092	10,321	10,378
Freight tonnes	9.8	9.7	9.9
Freight tonne-km	3,422	3,766	3,898

Passenger operations

In FY2002 SRT carried 55.7 million passengers and achieved 10,378 million passenger-km, decreases compared with the previous year of 1.02 and 0.55 per cent respectively. SRT attributed this result to disruption caused by infrastructure rehabilitation and upgrading and to timetable changes.

Freight operations

In FY2002 SRT transported 9.9 million tonnes of freight for 3,898 million tonne-km, increases of 2.31 and 3.49 per cent respectively. This improved performance was due in part to increased cement traffic in response to the demands of a growing construction industry, to landbridge maritime container traffic from Port Klang and to increased domestic container carryings from Sa Kosi Narai, Surat Thani and Tha Rua Noi terminals.

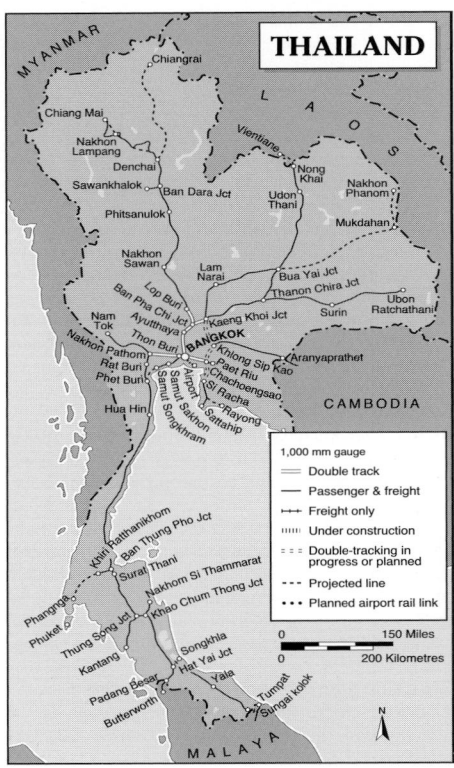

0580983

Class HID diesel locomotive at Ayutthaya with a Ubon Ratchathani–Bangkok passenger service (Marcel Vleugels)   0580259

Japanese-built dmu formation at Donmuang, Bangkok (Bruce Evans)   0580260

SRT has experienced strong growth in both maritime and domestic container traffic   1034657

New lines
New main lines

SRT plans four new lines: Den Chai–Chiang Rai (245 km), in the north of the country; Surat Thani–Phangnga (163 km) in the south; MabTa Phut–Rayong (24 km) in the east; and Bua Yai–Nakhon Phanom (368 km) in the northeast. The design of the first three of these had been completed by 2002, while the last still awaited detailed design. The Royal Decree of Land Acquisition Marking for the Den Chai–Chiang Rai and Surat Thani–Phang Nga lines was implemented in 2001 but no budget had been allocated to build these lines in view of domestic economic conditions. Land acquisition was expected to take two years, with construction of the two lines taking five and four years respectively. However, implementation of these projects, which require substantial investment, has been deferred due to government budget constraints. As a result feasibility studies have to be reviewed and updated before re-submission to government for construction approval. Accordingly, in May 2003 SRT engaged a consultant to update the feasibility study for the Den Chai–Chiang Rai project and to conduct a new study into an extension from Chiang Rai to southern China. These studies were completed in January 2004.

The Thai government has also mooted a 130 km line linking Nakhon Si Thammarat on the east coast with Phangna on the west coast. By acting as a land bridge, this would shorten the sea route between Japan and the Middle East by over 1,000 km and avoid the Strait of Malacca, which is already a target for piracy and is considered vulnerable to terrorist threat.

Airport Link Project

In preparation for the commissioning in 2005 of the Second Bangkok International Airport (SBIA), also known as the 'Suwannabhumi' airport, SRT was assigned to construct a 28 km standard gauge rail link to it from a new station in Bangkok, City Air Terminal. Following feasibility studies and an environmental impact assessment report, a €518 million contract to construct and equip the line was awarded in January 2005 to a consortium led by B Grimm International Ltd and B Grimm Hong Kong Ltd and also including Sino Thai Engineering and Construction plc (Stecon) and Siemens Transportation Systems. Siemens and B Grimm are to be responsible for design, supply, installation and project management of the entire electrical and mechanical systems. The line is expected to be operational in October 2007.

Improvements to existing lines
Double-tracking

With line capacity reaching its limits on various parts of the SRT network, additional tracks are

Sleepers

Type	Untreated hardwood	Creosote-treated softwood	2-block concrete (RS-type)	Monobloc pre-stressed concrete
Dimensions	150 × 200 × 1,900–2,000 mm	150 × 200 × 1,900 mm	1,710 mm long, block 209 × 274 × 600 mm	200 × 260 × 2,000 mm
Spacing	1,430–1,666/km	1,430–1,666/km	1,666/km	1,666/km
Fastenings	Dogspike, Dorken spike or Woodings clip	Dogspike, Dorken spike or Woodings clip	RN clip	Hambo, Fist, Pandrol, Vossloh, Stedef

required to handle an increasing number of trains. As well as from provision in 1999 of a third track between Rangsit and Ban Phachi (61 km), double-tracking on four routes was completed in 2002: Bang Sue–Nakhon Phathom on the Southern line (56 km); Ban Phachi–Lop Buri on the Northern line (43 km); Ban Phachi–Map Kabao on the Northeastern line (44 km). Similar work on the

Hua Mak–Chachoengsao section on the Eastern line (2 × 45 km) was finished in April 2003, apart from signalling and telecommunications work.

In addition, double-tracking of the eastern seaboard line (the Sattahip line), 177 km from Si Racha to the Kaeng Khoi junction of the Nong Khai line via Chachoengsao Junction, is also planned to support the Thai government's

Station improvements in progress at Bang Sue II   1034656

Diesel locomotives

Class	Wheel arrangement	Power kW	Speed km/h	Weight tonnes	No in service	First built*	Mechanical	Builders Engine	Transmission
Davenport	Bo-Bo	375	82	48.12	16	1952	Davenport	Caterpillar D397	E Westinghouse
Davenport	Co-Co	745	92	80	3	1955	Davenport	Caterpillar D397	E Westinghouse
HID	Co-Co	2,135	100	90	22	1993	Hitachi	Cummins KT7A 50-L	E Hitachi
GE	Co-Co	985	103	75	49	1964	GE	Cummins KT 38-L, KT 2300-L	E GE
Alsthom	Co-Co	1,790	95	82.5	49	1975	Alsthom	SEMT 16PA 4V.183	E Alsthom
AHK	Co-Co	1,790	100	82.5	29	1980	Henschel/ Alsthom/ Krupp	SEMT 16PA4 185VG	E Alsthom
ALD	Co-Co	1,790	100	82.5	8	1983	Alsthom	SEMT 16PA4 185VG	E Alsthom
ADD	Co-Co	1,790	100	82.5	19	1985	Alsthom	SEMT 16PA4 185VG	E Alsthom
GEA	Co-Co	1,865	100	86.5	38	1995	GE	Cummins KTA 50-L	E GE
HI	Co-Co	776	70	72	1	1961	Hitachi	MAN W8 V22/ 30 mAuL	E Hitachi
Hunslet	C	180	19.5	30	3	1965	Hunslet	Gardner 8L 3B	H Voith
Krupp	B-B	1,120	90	55	20	1969	Krupp	MTU 12V6 52TB 10, MTU 16V 396TC 14	H Voith
HAS	C	535	58	41.25	10	1986	Henschel	MTU 6V3 96TC12	H Voith
HE	B-B	895	90	52	7	1964	Henschel	MTU 12V 493TB10, 12V 396TC12	H Voith

* As at August 2003.

Diesel railcars or multiple-units

Class	Cars per unit	Motor cars per unit	Motored axles/car	Power/ motor kW	Speed km/h	No in service (cars)**	First built	Mechanical	Builders Engine	Transmission
BPD/ BTD	2	1	2	175 × 2	85	7 + 7	1967	Hitachi	Cummins N855-R2	H Voith
BDP/ BTD	2	1	2	175 × 2	90	23 + 26	1971	Hitachi	Cummins N855-R2	H Voith
BPD/ BTD	3	2	1	165	70	8 + 4	1971	Tokyu	Cummins NHH-220-B-1	H Nico
BPD (THN)	2	2	1	175	100	40	1983	Tokyu Hitachi Nippon	Cummins N855-R2	H Nico
ASR	2	2	1	210	120	20	1991	BREL, UK	Cummins NTA 855 RL	H Voith
BPD	2	2	1	175	100	62	1985	Nippon/ Kinki/ Kawasaki/ Hitachi/Fuji/ Niigata	Cummins N855-R2	H Nico Voith
APN	Intermediate motor car		1	175	100	11	1985	Tokyu	Cummins N855-R1	H Nico
APD	2	2	1	261	120	12	1995	Daewoo	Cummins NTA 855-R1	H Voith
APN	Intermediate motor car		1	261	120	8	1995	Daewoo	Cummins N855-R1	H Voith
APD	2	2	1	261	120	20	1996	Daewoo	Cummins NTA 855-R1	H Voith

BPD = Bogie Power car for Diesel railcar with driving cab
BTD = Bogie Trailer for Diesel railcar with driving cab
ASR = Air-conditioned Sprinter Railcar
APD = Air-conditioned power diesel railcar with driving cab
APN = Air-conditioned power diesel railcar non-driving cab
** As at August 2003.

Eastern Seaboard Area Development Programme. Design of the project has been completed and the construction cost has been estimated at Bt11,889 million. According to traffic demand and studies relating to train operations, the Si Racha–Chachoengsao section (69 km), estimated to cost Bt5,0442 million, should be constructed first in response to forecasts of increasing passenger traffic and freight growth. Submissions relating to this project made to the government in 2002 were still under consideration in 2004. However, SRT statements made in 2003 identified this scheme as the top priority in a list of double-tracking and upgrading projects that would receive some Bt400 billion of government funding, with the aim of completing all by 2007.

Track rehabilitation

Track rehabilitation projects recently completed include: Lop Buri–Chumsaeng (141 km) and Hua Hin–Ban Krut (148 km), completed in October 2000; Chumsaeng–Phitsanulok (108 km), Chai Ya–Thung Song (150 km) and Ban Krut–Chai Ya (244 km), completed in July 2002. Because of safety issues and the need for increased line capacity, track rehabilitation schemes are given very high priority by SRT and are being continuously implemented. Feasibility studies identifying a further 813 km of track to be treated as a priority have been completed and await implementation approval.

For its commuter services, SRT engaged a consultant to produce a master plan and feasibility study into improvement on its network, investigating and confirming the need for investment to improve operations in this sector. Completed by November 2002, it recommended that in a first phase 190 km of track should be doubled and signalling improved at 58 stations. Implemantation approval was awaited in 2004.

Traction and rolling stock

In August 2003 SRT was operating five steam, 234 diesel-electric and 40 diesel-hydraulic locomotives (including shunters), 248 diesel railcars, 1,236 passenger cars (including 263 sleeping, 69 restaurant, and 84 baggage cars) and 7,312 freight wagons. In 2004 it was reported that SRT expected to order 50 diesel locomotives to replace life-expired machines and offer additional traction capacity.

Rolling stock to be supplied by Siemens Transportation Systems for the the Second Bangkok International Airport (SBIA) (see 'Airport Link Project' above) will comprise of five three-car and four four-car emus based on the company's Desiro design.

Type of coupler in standard use:

Locomotives: Type E, AAR-10A
Passenger cars: Type E, AAR-10A automatic, controlled-slack type
Freight wagons: Type E, AAR-10E automatic, controlled-slack type
Diesel railcars: Type E, AAR-10A, automatic tight-lock type

Type of braking in standard use:

Locomotive-hauled stock: vacuum, air and dual-system

Signalling and telecommunications

STR's standard signalling systems are colour light signals, DC track circuit blocks and electric point machines. However, in addition to double-tracking of single-track sections, SRT has also introduced new signalling system, replacing all-relay interlockings with Computer-Based Interlockings (CBIs). The new signalling system must also support immunisation against possible future electrification and provision of Automatic Train Protection (ATP). Sections already

For details of the latest updates to *Jane's World Railways* online and to discover the additional information available exclusively to online subscribers please visit

jwr.janes.com

Newly completed double-tracking between Taling Chan Junction and Nakron Pathom 1034658

provided with or to be equipped with CBI-based systems are:
• Scheme ST1 (Northern line, third track Rangsit–Ban Phachi Junction, 61 km, 11 stations; and Southern line, double-track Bang Sue Junction–Taling Chan Junction, 14.66 km, three stations). Completed in December 2003.
• Scheme ST2 (ST2/N Northern line, double track, Ban Phachi Junction Lop Buri, 43 km, seven stations; ST2/NE Northeastern line, double track, Ban Phachi Junction–Kaeng Khoi–Map Kabao, 44 km, eight stations; ST2/S, Southern line, double track, Taling Chan Junction (excluded)– Nakhon Pathom, 42 km, 10 stations). Contracts signed February 2004.
• Scheme ST3 (Eastern line, double track, Hua Mak–Chachoengsao, 45 km, 10 stations). Contracts signed February 2004.

Track

Rail

Type	Weight (kg/m)
BS 50 R	24.8
BS 60 R & 60 ASCE	29.77
BS 70 R	34.76
BS 70 A & 70 ASCE	34.84
BS 80 A	39.8
BS 100 A	49.6
Others	37, 37.5, 42.5

Min curvature radius: 180 m (turnouts 156 m)
Max gradient: 2.6%
Max altitude: 574.9 m
Max axleload: 15 tonnes

UPDATED

Togo

Ministry of Commerce and Transport

Lomé

Key personnel
Minister: Dedevi Michele Ekue
For map see entry for **Benin**.

Compagnie Togolaise des Mines de Benin (CTMB)

Gauge: 1,000 mm
Track length: approximately 36 km

Organisation
CTMB runs phosphate trains over a distance of 30 km from mines at Hahotoé to the port of Kpémé and on a 6 km branch serving mines at Kpogamé. Loaded trains to Kpémé are run approximately hourly around the clock, transporting around 120,000 tonnes of phosphates per month. Operations are suspended one day per week to enable track maintenance to be carried out. CTMB owns six Alsthom Bo-Bo diesel locomotives, 150 self-discharging phosphate hoppers and a few service vehicles. Traction and rolling stock maintenance is undertaken at a depot near Kpémé.

Main running lines are laid with heavy rail on steel sleepers on crushed stone ballast and are well maintained.

Société Nationale des Chemins de Fer Togolais (SNCT)

PO Box 340, vis-á-vis de la Gare de Lomé, Lomé
Tel: (+228 21) 43 01

Key personnel
President: David de Fanti
General Manager: Yawo Kalepe

Gauge: 1,000 mm
Route length: 532 km

Political background
During 1997–98 the former Réseau des Chemins de Fer du Togo was restructured and its name changed to Société Nationale des chemins de Fer Togolais (SNCT) to facilitate eventual privatisation of both passenger and freight services. Branch lines in the phosphate mining area north of Kpémpé are operated by the Compagnie Togolaise des Mines du Benin (CTMB) (qv).

Organisation
One main route extends from the port of Lomé to Blitta (276 km). There are several branches but many of them are out of service, including the Frontalière Lomé-Togblékové, Côtière Lomé-Aného lines and the short section from Agbonou to Atakamé. A 58 km branch linking Togblékové and Tabligbo, which was opened in 1979 but closed between 1984 and 1990, is used for clinker transport. Of the branches leading to the port of Lomé only the line to Dique Est remains in operation.

Passenger and freight operations
Poor maintenance, due to lack of investment, led to a steady decline in traffic and as a result it was decided to let out railway operations in Togo to a private sector company under a management contract. This resulted in an upturn in traffic with 19 million tonne-km and 9 million passenger-km carried in 1994.

In mid-1998, remaining passenger services between Lomé and Blitta were withdrawn and then reinstated in November of that year, running on Saturdays to Blitta and returning on Sundays. This followed pressure from traders who felt isolated from markets at Lomé. The train is mixed, conveying passengers and commodities such as charcoal and agricultural products.

Freight traffic includes about 0.5 million tonnes of clinker annually on the Tabligbo-Lomé-Digue Est route, conveyed normally in two 20-wagon trains per day each carrying about 1,000 tonnes. Cement trains run regularly on the Lomé-Blitta route.

Traction and rolling stock
SNCT's stock consists of one Soulé railcar, three Canadian-built GM Co-Co diesel electrics, three Alsthom diesels (two B-B and one Bo-Bo), two Henschel B-Bs and two small French-built shunters. One of the two Henschels, both of which had been out of service for some time, has been returned to traffic by robbing its sister of spares. There are 12 coaches, several of which are railcar trailers which are unusable in locomotive hauled formations due to coupler differences. Some vehicles were built locally by Sotometo in 1990. Of some 120 freight wagons, 80 are clinker hoppers.

Signalling and telecommunications
All traffic is radio-controlled, the telephone communications system having been suspended completely.

Track
Rails: Up to 40 kg/m on main lines, lighter on disused branches. Welded rail is in place on some parts of main routes
Sleepers: Mostly steel, on older line sections also wooden, rails are fixed with plates and screws
Ballast: Crushed stone on all lines built after 1960 and on some parts of older lines
Max speed: 35 km/h for railcars, lower for locomotive-hauled trains

Tunisia

Ministry of Transport

Tunis

Key personnel
Minister: Houcine Chouk

Tunisian National Railways (SNCFT)

Société Nationale des Chemins de Fer Tunisiens
PO Box 693, 67 Avenue Farhat Hached, Tunis
Tel: (+216) 71 34 01 66
Fax: (+216) 71 34 85 40

Key personnel
President/Director General:
  Ing Abdel Aziz Chaabane
Deputy Director General: Med Nejib Fitouri
Headquarters Directors
  Management and Finance: Lotfi Ben Fadhl
  Rail Business: Sami Khanfir
Bureau Heads
  Environment and Energy: Faouzia Gardallou
  Development and Strategy: Faika Dali
  Organisation and Information: Hédi Belguith
Directors
  Tunisian Rail Network Unit:
    Abdulsalam Ben Dhiab
  Railway Sector: Sami Khanfir
  International Affairs: Mohamed Chabbi
  Administration and Finance: Mohamed Touj
  Administrative Division: Ahmed Ezzine
  Finance Division: Faouzia El Ghardallou
  Planning and Control: Chazli Al-Kaizani
  Purchasing: Mokhtar Baati
  Main Line Passenger Services: Fayçal Klibi
  Tunis Suburban Services: Abderrahmene Gamha
  Sahel Suburban Services: Hemadi Ben Osman
  Freight Services: Mouldi Zouaoui
  Phosphates Traffic: Djamal Al-Dine Hamza
  Infrastructure Operations and Maintenance:
    Salha Zaidi
  Operations: Jalel Ben Dana
  Property: Khaled Jeddi
  Strategic Development: Abdelhamid Jemmali
  Infrastructure Development: Mokhtar Fennira
  Planning and Operations Control: Chadly
    Guizani
  Communications and External Relations: Jalila
    Houssein
  Information Technology: M Yangui

Gauge: 1,435 mm; 1,000 mm; mixed 1,435/1,000 mm
Route length: 471 km; 1,674 km; 8 km
Electrification: 65 km of 1,000 mm gauge at 25 kV 50 Hz AC

Political background
For the first time, a contract plan was agreed between the government and SNCFT for the period of the Ninth Economic and Social Development Plan (1996–2001). This set specific targets for the railway to achieve by the end of the period. These included: managing the railway under strictly commercial terms, pursuing markets only where rail offers economic and financial advantages compared with other modes, to achieve a balanced financial result after state subsidies are taken into account; improving service quality, comfort, punctuality, information provision and the safe handling of freight traffic, with standards monitored by performance indicators. Under the contract plan the government would provide subsidy for unremunerative services. All rolling stock investments are to be financed by SNCFT.

Under Tunisia's Tenth Economic and Social Development Plan (2002–06), studies were to be undertaken to rationalise the railway's organisation and the size of its management support units. Finance for these and for some infrastructure upgrading were funded by a US$88 million loan from the African Development Bank, signed in 2004.

Organisation
The reorganisation of SNCFT in compliance with its first contract plan with the government led to the adoption of a more commercially oriented structure. This comprises seven business units: two suburban passenger units; main line passenger; phosphates traffic; freight; maintenance; and railway infrastructure. At the end of 2003 SNCFT employed 5,5554 staff.

Operationally SNCFT is divided into three sectors: the Northern standard-gauge sector, covering lines north and northeast of Tunis; the Northern narrow-gauge sector, covering the route south of Tunis to Sfax; and the Southern narrow-gauge sector, responsible for lines south of Sfax. Important equipment differences restrict the cohesion of the two narrow-gauge sectors as a network: rolling stock of the Northern sector (the former Compagnie des Phosphates de Gafsa) employs single buffers with a coupler located underneath, while the Southern sector uses vehicles with two buffers and a coupler between. SNCFT does not employ bogie-changing between narrow and standard gauges.

Connection between Tunisian and Algerian standard-gauge networks is made in the northwest at Ghardimaou while in the future SNCFT will be connected to the system under construction in Libya via a 185 km standard-gauge line linking Gabès and Ras Jedir, effectively completing a trans-Maghreb rail link.

Traffic (million)	2003
Passenger journeys	35.7
Passenger-km	1,242
Freight tonnes	12.2
Freight tonne-km	2,174

Passenger operations
SNCFT operates nine pairs of trains linking Tunis and Sfax (278 km metre gauge), including some running at up to 120 km/h and using refurbished first class and grand comfort coaches only. The Tunis-Ghardimaou standard gauge 'International Line' (211 km) sees five daily return services in each direction.

Diesel-operated push-pull trains serve destinations south of Tunis Ville station, suburban traffic accounting for some 85 per cent of SNCFT passenger journeys 37 per cent of passenger-km. The Sousse–Mahdia metre gauge line is electrified, with about 20 services a day in each direction.

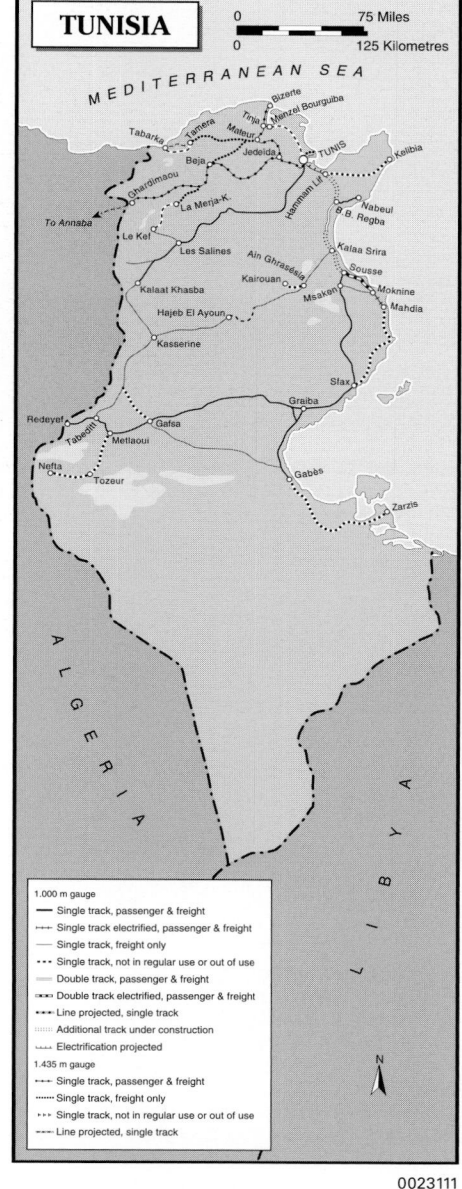

0023111

SNCFT metre-gauge Class 040 GT (General Motors Type GT18) with a southbound express service at Tunis Ville station (Philip Wormald)   ***NEW**/0585047*

SNCFT Class 060 DP (Bombardier Type MX 624) with a Metlaoui–Tunis express service
(Philip Wormald)   **NEW**/0585048

Diesel locomotives

Class	Wheel arrangement	Power kW	Speed km/h	Weight tonnes	No in service	First built	Builders Mechanical	Engine	Transmission
040 DF	Bo-Bo	698	90	59	9	1965	GM-EMD	GM 8-567CR	E GM
060 DH	Co-Co	1,640	100	93	5	1973	GM-EMD	GM 16-645E	E GM
060 DI	Co-Co	1,604	114	90	22	1973	MLW	Alco 251-12	E MLW
040 DJ	Bo-Bo	520				1977	GE	Caterpillar D-379	E GE
040 DK	Bo-Bo	895	116	64	20	1978	Bombardier	Alco 251-6	E MLW
040 DL	B-B	1,323	110	62	10	1981	Ganz-Mávag	Ganz-Pielstick	H Ganz Mávag
040 DM	Bo-Bo	522	114	64	28	1983	GE	Caterpillar D398	E GE
060 DN	Co-Co	1,862	114	89	19	1983	GE	GE FDL12	H GE
040 DO	B-B	1,764	130	64	20	1985	Ganz-Mávag	Ganz-Pielstick	H Ganz Mávag
060 DP	Co-Co	1,764	130	91	20	1984	MLW/ Bombardier	Alco 251-12	H Bombardier
040 GE	Bo-Bo	446	96	49	4	1962	GE	GE	E GE
040 GE	Bo-Bo	515	96	49	4	1965	GE	GE	E GE
060 GR	Co-Co	1,047	100	92	5	1964	GM-EMD	12-567C	E GM
040 GT	Bo-Bo	1,340	–	–	21	1999	GM Canada	GM 8-645E3C	E GM

Freight operations

Phosphates form SNCFT's principal freight commodity, accounting for around 70 per cent of tonnage and nearly three-quarters of tonne-km.

New lines

Several new lines have been proposed, including rebuilding the former Cap Bon line from Fondouk Djedid to Henchir Lebna and extending it to Kelibia. Other schemes include two new north–south lines, one from Borj Mcherga to Kairouan and connections with the prospective Sousse–Kasserine route; and another further west from Gafour to Sidi Bouzid, whence there would be a fork southwest to Gafsa and southeast to Mazouna. A new route around the coast from Monastir to Sfax via Mahdia has also been projected.

Improvements to existing lines

Under its Ninth Plan period, SNCFT undertook double-tracking of the key metre-gauge main line from Tunis southward to Sousse (72 km). Extension

of this line beyond Gabès to Medenine has been projected, with ultimate projection to Tripoli in view. Also undertaken during the plan period was renewal of 26 km of track on the Tunis–Ghardimaou line, together with signalling improvements.

During the Tenth Plan period SNCFT had planned to implement electrification of the 23 km Tunis Ville–Borj Cedria suburban line, although no progress had been reported on this by 2004. The Tunis–Tebourba surburban line has also been identified as a candidate for future electrification.

Reopening of several routes has been proposed, including: the line from Mateur to Tabarka, closed since 1984; the Mateur-Jedeida connection; the Mastouta-Merja link further south; and the line southwest from Sousse to revive a through route from the coast to Kasserine.

Traction and rolling stock

In 2004 SNCFT operated 135 main line and 41 shunting diesel locomotives, six emus and 12 trailers, 268 passenger cars and 4,447 freight

wagons. Many of the dmus supplied to SNCFT in the 1970s by Alsthom and Ganz-Mávag are used as non-powered passenger stock.

All locomotives and passenger cars are capable of bogie change and of operation on either gauge, although in practice this facility is rarely used. Many Tunis area commuter services are operated using diesel locomotives with push-pull coaching stock formations using GT18B diesel-electric locomotives supplied from 1999 by General Motors of Canada.

In 2005 SNCFT signed a contract with French vehicle manufacturer CFD to supply 10 Type AMG two-car metre-gauge dmus.

Sousse–Mahdia electrified services are operated by six three-car Class YZE emus supplied by Ganz-Mavág.

Signalling and telecommunications

A CTC system supplied by Ansaldo Trasporti and located at Tunis Ville station controls the 24 km Tunis–Bordj Cedria section, which handles around 200 trains daily.

Automatic block installation, operative throughout the Tunis–Sousse–Sfax–Gabès single line, with relay interlockings at crossing stations, has also been extended to the sections from Tunis to Djedeeda and Beja, Bir Kassa to Gafour, Sfax to Gabès and Gabès to Gafsa. The section from Sfax to Gafsa and Métlaoui has been experimentally equipped with track-to-train radio. All telephone lines are being progressively placed in ground-level ducts.

Electrification

The Sousse–Mahdia metre-gauge line is electrified at 25 kV AC 50 Hz.

Track

Standard rail
Standard-gauge: Flat bottom, 36–46 kg/m in lengths of 12–18 m
Metre-gauge: Flat bottom, 25–36 kg/m in lengths of 7, 8, 12 m
Welded joints: Thermit welding of rail joints
Crossties (sleepers): Oak impregnated with creosote; metal; concrete RS type
Standard-gauge: 120 × 220 × 2,600 mm
Metre-gauge: 120 × 220 × 2,200 mm
Spacing: 1,500/km
Rail fastenings: Wood sleepers: spikes
Metal sleepers: clips and bolts
Concrete sleepers: special resilient fittings
Filling: Broken stone

Max curvature
Standard-gauge: 7° = min radius 250 m
Metre-gauge: 11.6° = min radius 150 m
Max gradient: 1 in 50
Max altitude: 952 m on Haidra to Kasserine line
Max speed, standard-gauge
Railcars: 100 km/h
Diesel trains: 70 km/h

Max axleload
Standard-gauge: 21 tonnes
Metre-gauge: 18 tonnes

UPDATED

Turkey

Ministry of Transport

Ulâstirma Bakanligi, Ankara
Tel: (+90 312) 550 10 00
Web: http://www.mt.gov.tr

Key personnel
Minister: Binali Yildirim
Under-Secretary: Munir Kutluata
Deputy Under-Secretary, Railways: Mehmet Kutlu

Turkish State Railways (TCDD)

Türkiye Cumhuriyeti Devlet Demiryollari Genel Müdürlüğü Talatpaşa Bulvari, TR-06330 Ankara
Tel: (+90 312) 309 05 15   Fax: (+90 312) 312 32 15
Web: http://www.tcdd.gov.tr

Key personnel
Director General and Chairman of the Board: Süleyman Karaman
Board Members and Deputy Directors General: Fşinasi Kazancioglu, A Kemal Ergüleç

Deputy Directors General: Tayyar Hindistan, Erol Inal
Head of Board of Inspection: Mustafa Çavuioilu
Head of Legal Department: Necdet Alkan
Head of Security Secretariat: Kemal Çiftci
Press and Public Relations Counsellor: Mehmet Ayci
Department Heads
   Permanent Way: Nurullah Tataraçşasigil
   Traction (Acting): Turgut Kumai
   Commercial: Lüfti Özbek
   Financial: Nejat Firat
   Operations: Veysi Kurt

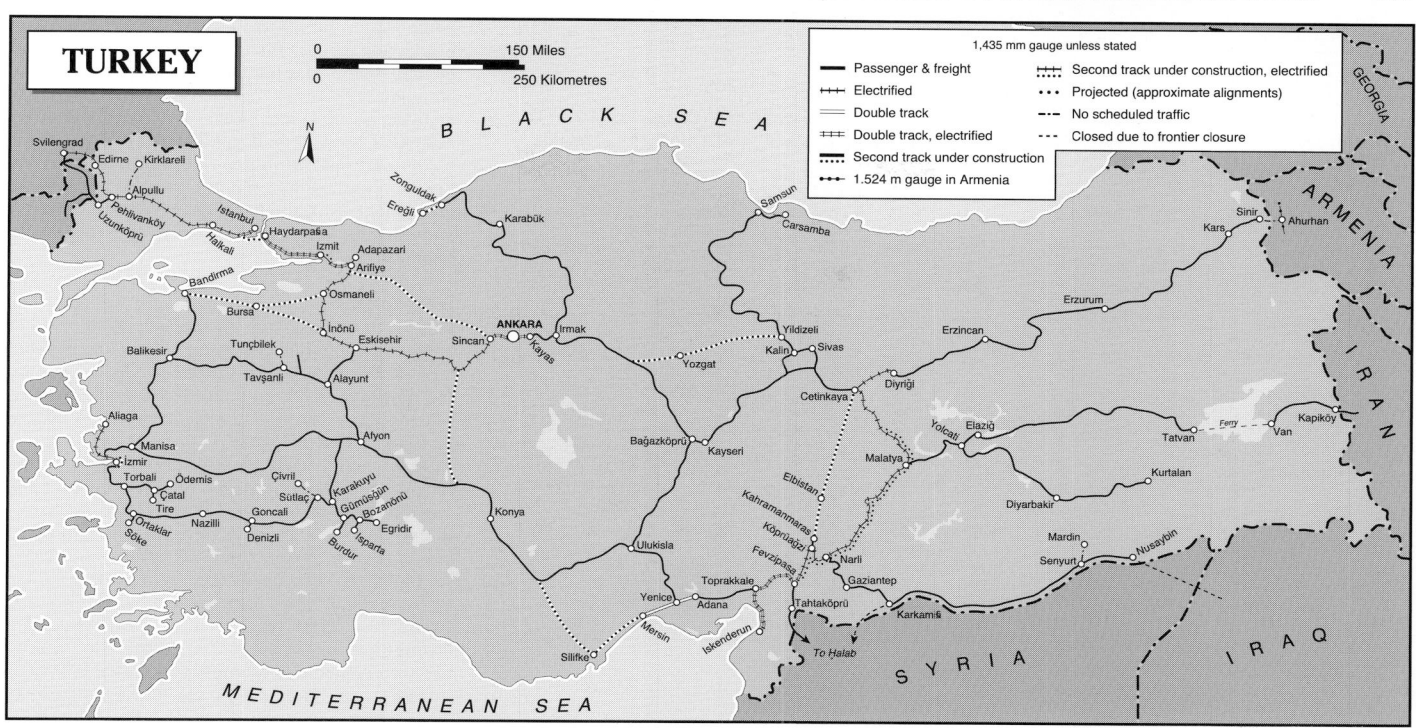

0576731

Health (Acting): Kamil Kahyaoilu
Personnel and Administrative Affairs:
   Hüseyin Öztürk
Purchasing: Ekrem Aksoy
Ports: Emin Tekbai
Construction (Acting): Hayri Varinli
Research, Planning and Co-ordination:
   Ismet Duman
Training and Education (Acting): Murat Senekan
Data Processing: Ersin Tasci
Installations: Mehmet Uras
Real Estate: Suat Altin
Foreign Affairs: Serafettin Deniz
Factories: Nail Adali
Marketing: Nizamettin Aslan

Affiliated companies
Tülomsaş (Locomotive and Motor Corporation of
Turkey)
   Director General: Dr Muammer Kantarci
Tüvasaş (Wagon Industry Corporation of Turkey)

Director General: Ibrahim Ertiryaki
Tüdemsaş (Railway Machinery Industry Corporation
of Turkey)
   Director General: Halil Torun

Gauge: 1,435 mm
Route length: 8,697 km
Electrification: 2,122 km at 25 kV 50 Hz AC

Political background
TCDD restructuring initiatives active in 2004 included securing a grant from the Japanese government to assist in defraying redundancy costs as the railway sought to reduce employment levels. This move corresponded with plans in 2004 to create a legislative structure for the rail market in Turkey matching that of EU countries, to establish a new business unit-based structure for TCDD and to introduce management information systems to provide effective analysis of costs and revenue. Legislation introduced in

June 2003 gave private operators access to the TCDD network.

Several other reform moves have been taken or are planned. From 2003 government initiatives enabled TCDD to shed financial responsibility for its staff hospitals. Work has started on generating revenue from non-operational property assets. Some rationalisation of permanent way workshop facilities has been implemented, while others are to be converted to joint stock companies. In the field of traction and rolling stock maintenance TCDD plans to encourage private sector participation.

Organisation
TCDD is managed by a board of directors comprising a chairman and five members. The chairman is also director general and at the railway's headquarters there are five deputy director generals and 18 departments. In Turkey's provinces TCDD is organised as seven regional directorates. TCDD owns 99.9 per cent of each of three affiliated corporations (see *Key personnel*), which are organised as autonomous general directorates.

TCDD also owns and operates Turkey's seven largest ports, all of which are rail-connected, as well as a train ferry service across Lake Van.

At the end of 2003 TCDD employed 34,526 staff.

Revenue (TL million)	2001	2003
Passenger	63.2	105.1
Freight	293.7	211.7
Ports	–	331.2
Other income	41.2	140
Subsidies	197.7	330.7
Total	595.8	1,118.7

Expenditure (TL million)	2001	2003
Railway personnel	425.2	703.5
Other expenditure	742.0	911.4
Total	1,167.2	1,614.9

Traffic (million)	2001	2002	2003
Passenger journeys	76	73	77
Passenger-km	5,568	5,204	5,878
Freight tonnes	14.6	14.6	15.9
Freight tonne-km	7,561	7,224	8,669

Passenger operations
Passenger business divides roughly half and half (in terms of passenger-km) between suburban and main line. Sirkeci has the most commuter traffic (550 million passenger-km in 2003), followed by Haydarpaşşa (364 million), Ankara (377 million), Basmane (1 million) and Alsancak (3 million).

A feature in the long-haul sector is the development of good quality, first-class-only services. Recent additions to the long-haul rolling

Two Ex-Bosnia-Herzegovina Railways Class E52 Bo-Bo electric locomotives at Istanbul Yenikapi with a freight service (Marcel Vleugels) 0576719

Diesel locomotives

Class	Wheel arrangement	Power kW	Speed km/h	Weight tonnes	No in service	First built	Mechanical	Builders Engine	Transmission
DH 33100	C	450	50	41.2	15	1953	MAK	KTA-1150	H Voith
DH 6500	C	650	60	49.6	6	1960	Krupp	Maybach	H Voith
DH 3600	C	360	50	40.5	3	1968	MAK ELMS	MTU 12V 183	H Voith
DH 3600	C	450	50	40.5	9	1968	MAK ELMS	Cummins	H Voith
DH 7000	C	522	50	51	18	1995	Tülomsaş	Cummins KTA19L	H Voith
DH 9500	B-B	710	40/80	68	26	1999	Tülomsaş	MTU 8V 396	H Voith
DE 21500	Co-Co	1,600	114	111.6	11	1965	GE	GEFDL-12	E GE
DE 24000	Co-Co	1,580	120	112.8	286	1970	Tülomsaş	Pielstick 16PA4-185	E Alsthom
DE 18000	Bo-Bo	1,325	120	80	1	1970	MTE	Pielstick 12PA4-185	E Alsthom
DE 18100	A1A-A1A	1,325	120	87	16	1978	Tülomsaş-MTE	Pielstick 12PA4-185	E Alsthom
DE 22000	Co-Co	1,470	120	117	86	1985	GM	GM 645 E	E GM
DE 11000	Bo-Bo	735	80	68	67	1985	Tülomsaş	MTU 8V 396	E GEC

Electric locomotives

Class	Wheel arrangement	Output kW	Speed km/h	Weight tonnes	No in service	First built	Builders Mechanical	Eletrical
E40000	B-B	2,945	90/130	77	9	1969	50 C/S Group	50 C/S Group
E43000	Bo-Bo-Bo	3,180	90/120	120	45	1987	Tülomsaşş/Toshiba	Toshiba
E52500*	Bo-Bo	3,860	120 or 140	80	20	1970	Koncar	ASEA

*Acquired in 1998 from Bosnia-Herzegovina Railways.

Diesel multiple-units

Class	Cars per unit	Motor cars per unit	Motored axles/car	Power/ motor LP	Speed km/h	No in service	First built	Mechanical	Builders Engine	Transmission
MT 5600	1	1	2	2 × 550	140	11	1992	Tüvasaş	Cummins KTA 19 R	H Voith T 320 RZ
MT 5700	1	1	2	2 × 280	140	30	1993	Fiat	Iveco	H Voith T211R
MT 3000	3	1	2	2 × 150	90	1	1960	Uerdingen	Bussing U10	EM
MT 5500	3	2	4	4 × 145	90	7.5	1961	Fiat	Fiat	H

Electric multiple-units

Class	Cars per unit	Motor cars per unit	Motored axles/car	Output/ motor kw	Speed km/h	No in service	First built	Builders Mechanical	Electrical
E 8000	4	2	2	255	90	21	1955	Alsthom	Alsthom
E 14000	3	1	2	520	119	69	1979	Tüvasaş/50 C/S Group	50 C/S Group

stock include business cars with telephone, telex and data modem equipment, new day cars and sleeping cars with individual shower/WC facilities for each compartment.

For short-haul service TCDD has begun local manufacture of diesel railcars, while new Fiat dmus have been introduced on suburban lines round Izmir pending electrification. It also operates 20 two-car railbus sets bought second-hand from Germany.

Freight operations

The most important commodity is iron ore, at over 4 million tonnes, while coal (over 3.9 million

tonnes) and construction materials (1.9 million tonnes) are also significant. International traffic is around one million tonnes annually.

In a bid to reduce costs TCDD is implementing greater use of block train operations, anticipating that this will reduce shunting costs by around 60 per cent, cut turn-round times by 20 per cent and increase commercial speeds by 300 per cent.

Intermodal operations

Following feasibility studies TCDD has implemented a strategy of creating inland intermodal terminals to relieve the pressure on ports which has resulted from the growth in containerisation and move the

emphasis on handling freight closer to centres of production and demand. By 2004 a facility had already been commissioned at Gaziantep, while planning for terminals at Denizili and Kayseri was at an advanced stage. Similar facilities are planned at Ankara, Kayseri, Konya, Malatya and Mara.

As part of a major ports modernisation programme, partly funded by the European Investment Bank, the facilities at Haydarpaşa, Izmir and Mersin are being provided with container-handling equipment.

New lines

Feasibility studies into a proposed 416 km high-speed passenger link between Ankara and Istanbul have been undertaken. Costed at US$3.5 billion, the new line would be built for 260 km/h running, reducing journey times between the two cities to 2 hours or less.

Construction was due to start in 2004 of the long-planned Bosphorus rail tunnel and its related works, which forms a major electrified suburban rail scheme now known as the Marmaray project. The line will also serve as a key east-west corridor. An initial 13.3 km tunnel section, including 1.6 km of 'immersed tube' construction, is to form the centrepiece of a 76.3 km cross-Bosphorus line linking Halkali and Gebze. The construction cost of the entire scheme is estimated at US$2.7 million.

Other new lines listed by TCDD as priority projects include:

- A 306 km double-track line to bypass Kayseri, on the Ankara–Sivas line, leaving the existing line east of Kirikkale and running via Yozgat and Yildizeli to Sivas.
- Construction of a new line between Polatli, southwest of Ankara, to Afyon as part of a project to upgrade the link between the capital and the port of Izmir (606 km) that also involves upgrading the existing Afyon–Izmir line.
- A 290 km double-track line from Inönü, between Eskesehir and Adapazari, to Bursa and Bandirma.
- A 607 km line south from Ankara via Konya and Silifke to the Mediterranean port of Mersin, mainly for the movement of iron ore.
- A 546 km double-track line in southern Anatolia which would also provide an eventual link with Iraq.
- A 72 km line between Kömürler and Osmaniye to eliminate a bottleneck on the Iskenderun–Sivas iron ore line and provide access to coal mines around Elbistan.
- A 129 km double-track line from Adapazari to serve a steelworks at Ereğli.
- A 285 km line from Cetinkaya via Elbistan to Kahramanmara for iron ore transport and to improve access to the ports of Iskenderun and Mersin.
- A 42 km double-track iron ore line from Zonguldak to Ereğli.

Improvements to existing lines

Priority rehabilitation projects listed by TCDD in 2004 included:

- Modernisation of the 310 km Bandirma–Izmir line, improving connections for both freight and passenger traffic.
- Upgrading the Ankara–Kayseri route, increasing its capacity for coal and iron ore traffic.

Traction and rolling stock

At the end of 2003 the fleet comprised 470 main line and 74 shunting diesel, 74 electric and 50 steam locomotives, 88 emu cars and 49 diesel railcars. In addition, the railway operated 965 passenger coaches, among them 91 sleeping, 116 couchette, and 64 restaurant cars; and 16,070 freight wagons.

In 2004 some Class E52500 locomotives were upgraded by Koncar of Zagreb for 140 km/h operation. At the same time TCDD was reported to have signed a contract with Bosnia-Herzegovina Railways to lease two more with an option on a further eight.

In 2003 TCDD subsidiary Tüvasas commenced production of a three-car prototype of a dmu type intended for service on the domestic network. Subsequent deliveries were to cover three two-car, three three-car and two four-car versions of the

Alsthom Class E8000 (left) and Tüvasas E14000 emus at Sirkeci station, Istanbul (Colin Boocock)

0576720

Ex-Bosnia–Herzegovina Railways Class E52 Bo-Bo electric locomotive at Sirkeci with an Istanbul–Uzunkopru train (Colin Boocock)   0533635

same design as part of a five-year TCDD contract for 168 vehicles.

Type of coupling: Screw; semi-automatic coupler
Type of brake: Knorr and Westinghouse air

Signalling and telecommunications
In 2003 2,449 km of the TCDD network was supervised by CTC.

Electrification
In 2002, contracts were placed with a Turkish consortium to upgrade and electrify 79 km of suburban lines serving Izmir. Routes covered by the scheme are Basmane–Menemen–Aliaga and Alsançak–Cumaovasi.

Track
Rails: 49 kg/m (5,498 km); 46 kg/m (1,554 km); 39 kg/m (1,645 km)
Sleepers: concrete (B55-B58); wooden (2,600 × 160 × 260 mm); steel (2,600 × 240 × 100 mm)
Spacing: 1,612/km
Fastenings: K, HM, N
Min curvature radius: 250 m
Max gradient: 2.7%
Max axleload: 20 tonnes

Class DE 11000 diesel-electric locomotive with an Istanbul–Pythion (Greece) train at Alpullu (Colin Boocock)
0533634

Turkmenistan

Ministry of Transport

Turkmenbashi Shaely 7, 744007 Ashkhabad
Fax: (+993 12) 47 39 58

Key personnel
Minister: Sedar Djepbarov

Turkmenistan State Railway

Türkmenistanyn Döwlet Demir Yoly (TDDY)
Turkmenbashi Shaely 7, 744007 Ashkhabad
Fax: (+993 12) 47 39 58

Key personnel
President: Atdabek Agodzhanov
Chief Engineer: G G Neroubaiski
Directors
  Commercial and Operations: Z T Bakhalov
  Traction: A B Khommadov
  Rolling Stock: D I Nourberdyev
  Passenger Services: S A Dourdyev
  Purchasing: T Y Yazlakov
  Personnel and Infrastructure: G A Veguentchova
  Signalling and Telecommunications:
    A Gueldymouradov
  Finance: S O Perliev
  Technical: G A Sakhatdourdyeva
  Legal: E M Kouziat
  Chief Economist: O Khoudaiberdyev
  External Relations Manager: M K Biachimova

Gauge: 1,520 mm
Route length: 2,440 km all single track; no electrification

Political background
Turkmenistan is investing in its rail network to develop the country's role as a transport hub in Central Asia. Bilateral agreements covering rail freight links have been signed with the governments of Azerbaijan, Georgia and Uzbekistan, and connections with Iran improved in 1996 with the opening of the Sarakhs–Mashhad line. The

Turkmenistan rail system also features in a project supported by the EU to develop a transport corridor between Europe, the Caucasus region and Asia.

Organisation
The network comprises the western portion of the former Soviet Railways Central Asian Railway. Its main route is the 1,141 km

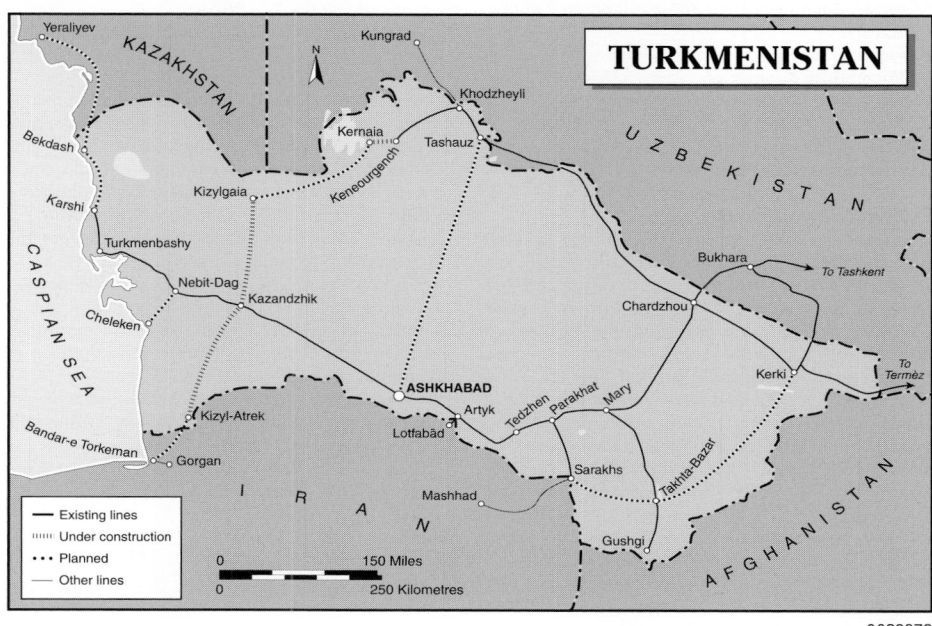

0089973

trans-Caspian line linking the port of Turkmenbashy (formerly Krasnovodsk), the capital Ashkhabad and Chardzhou, near the border with Uzbekistan. A branch from Mary to Kushka (Gushgi) on the Afghan frontier is rarely used because of the unrest in Afghanistan.

In 1997, Japan's Overseas Economic Co-operation Fund granted a US$40 million credit to modernise the Ashkhabad locomotive depot, assimilate new repair technologies, and establish a computerised traffic control system.

Additional funding valued at €35.6 million for modernisation of the network was committed in early 2000 by the Japanese Overseas Economic Co-operation Fund.

Workers engaged in the 'main activity' (railway work) totalled 18,535 in 1998, of which 14,283 were operating workers.

Passenger operations
Passenger traffic in 1995 amounted to 1.9 billion passenger-km. In 1999, 2.3 million passengers were carried (0.7 billion passenger-km). A new fast service was introduced in 1993 from Turkmenbashy to the regional capital of Tashauz, covering the 1,600 km in 36 hours.

Freight operations
Traffic is mainly cotton and other crops, and oil. In 1993, 18.5 million tonnes of freight were carried,

falling to 8.7 million in 1997 and then recovering to 9.8 million in 1999. In that year 7 billion tonne-km were recorded.

New lines
Construction of the 122 km Tedzhen–Sarakhs–Mashhad line, linking Turkmenistan with Iran, has been completed; the line was opened in May 1996.

The branch to Kushka is to be the spring-off point for the proposed trans-Afghan railway, which was the subject of an accord signed between the Pakistan, Afghan and Turkmenistan governments in March 1994. The 800 km route, which has the strong support of the Pakistan government, would link Kushka with Chaman in Pakistan, via Herat, Farah and Kandahar in Afghanistan.

A new link from Chardzhou in a southeasterly direction was opened in early 2000. This line runs along the south side of the Amu Darya river to Kerki where a new bridge has been constructed with a link to the present isolated line at Kerkichi (on the Karshi-Termez (both in Uzbekistan) route). This new river crossing relieves pressure on the life-expired bridge at Chardzhou, for which a replacement is also being planned.

In 1997, a Japanese-led consortium agreed to build a 450 km line from Yeraliyev, in Kazakhstan, to Turkmenbashy (formerly Krasnovodsk). This will provide improved access from western Russia to

the Persian Gulf via Sarakhs and through Iran to the port of Banda' Abbās.

Construction of the 530 km Ashkhabad–Tashauz line began in 2000, with completion expected in 2005.

Traction and rolling stock
The current fleet of locomotives comprises some 233 double-unit locomotives of the 2TE10 derivatives, of which about 120 are available for service. Additionally, there are 98 shunters of types TEM2 and ChME3 of which about 60 are serviceable. Interest has been expressed in a potential re-engining exercise for the 2TE10M locomotives using General Electric (USA) technology.

Hard currency earned from exports of cotton have enabled the railway to purchase spare parts for the rehabilitation at Chardzhou workshops of unserviceable locomotives. The same source has provided funding for new locomotives: in 1995 the railway bought 33 Type 2TE10U two-section diesel-electrics from Lugansk of Ukraine.

Fifteen passenger cars were bought in 1998, but none in 1999. No freight cars were delivered in those years.

Uganda

Ministry of Works, Housing and Communication

PO Box 10, Airport Road, Entebbe
Tel: (+256 42) 32 01 01
Web: www.miniworks.go.ug

Key personnel
Minister: John Nasasira

UPDATED

Uganda Railways Corporation (URC)

PO Box 7150, Nasser Road, Kampala
Tel: (+256 41) 25 80 51   Fax: (+256 41) 34 88 32

Key personnel
Managing Director: D C Murungi
Chief Mechanical Engineer: C Katambira
Chief Civil Engineer: H Akora
Chief Rail Operations Manager: J Okello
Chief Administration Manager: E A Katoroogo
Financial Controller: M B Okella
Assistant Chief Planning and Development Officer: P Oyang
Assistant Chief Signal and Telecommunications Officer: Ruth Kyohairwe
Public Relations Manager: E Zalwango

Gauge: 1,000 mm
Route length: 1,244 km

Political background
Uganda Railways Corporation (URC) was created after the 1977 dissolution of East African Railways (EAR). Since then it has suffered seriously from political dissension between its former partner countries in EAR, from the civil war in its own country and resultant damage, and from a decline in the performance of Ugandan industry and agriculture. These factors have contributed to large sections of the network falling out of use, leaving only the main line from Kampala to the border with Kenya at Malaba (251 km) and the 10 km line from Kampala to Port Bell on Lake Victoria in operation at the beginning of 2004. However, in April 2004 services were restored on the 56 km section from Tororo, on the Kampala–Malaba main line, to Mbale, on the line northeast to Pakwach East that had been closed since 1997 due to the conflict in Uganda. A further section of the line, from Mbale to Soroti, was re-opened in July 2004.

All lines 1,000 mm gauge
— In operation
--- Not in use

NEW/0585060

Long-discussed plans to privatise the network took a major step forward in 2004 when the governments of Uganda and Kenya decided to hand over the operation of their rail networks to private companies as a combined 25-year concession for freight traffic in both countries and a seven-year concession for passenger services in Kenya. The two governments were expected together to hold a 40 per cent stake in the concession. In early 2005 pre-qualification of bidders was in progress. Final bids were due to be submitted in April 2005, with a concessionaire expected to be named in July. The transfer of operations is scheduled for December 2005. Funding to cover restructuring costs associated with concessioning included a US$77 million World Bank loan announced in November 2004.

By 2004 URC staff numbers had fallen to 1,240 compared with more than 5,000 five years earlier.

Traffic	2000	2001	2002
Freight tonnes	772,000	715,000	869,000

Passenger operations
In 1990 traffic amounted to 1.5 million passenger journeys but by 1997 all passenger services had

been suspended. Their revival does not feature in URC privatisation plans.

Freight operations
URC carries around 60 per cent of Uganda's exports and imports, this traffic depending on the railway's links with the networks of Kenya and Tanzania to provide access to the ports of Mombasa and Dar es Salaam. Principal export commodities handled include coffee, cotton and forest products, while inbound goods include cereals, machinery, petroleum products and sugar. In 1999 URC carried 733,000 tonnes of freight compared with 600,000 tonnes in 1998. This upward trend continued in the new millennium. The increase in rail freight volume has been partly attributed to a reduction in maximum permissible loads for road transport, a unique step taken by the Ugandan government and one not attempted by any other African country. Other steps by URC to improve performance include the movement of freight between Nairobi, Kenya, and Kampala in block trains and the provision of a computer-based consignment tracking system.

In 2004 URC in association with Kenya Railways Corporation introduced a direct Kampala–Mombasa

Henschel-built Class 73U diesel-hydraulic locomotive on track maintenance duties near Jinja (James Pattison)

NEW/0585061

container service that eliminated delays at the Uganda-Kenya border, reducing the transit time from 14 to three days. Using 350 dedicated container flats, the service was running three times weekly in each direction.

Train ferry service
More than half of URC's freight tonnage is trans-shipped from/to Lake Victoria wagon ferries. URC owns three such ferries which ply between Port Bell in Uganda, Kisumu in Kenya and Mwanza in Tanzania, with some use of the Jinja terminal for traffic from the east of the country. Kenya Railways Corporation and Tanzania Railways Corporation also each operate one wagon ferry serving Port Bell and Jinja. Each vessel has a capacity of 880 tonnes, enabling 22 wagons with a gross laden weight of 40 tonnes to be carried.

Improvements to existing lines
The condition of the Kampala–Malaba main line was reported as poor in 2004, leading to a reduction in operating speeds and preventing the use of multiple traction for heavier trailing loads. Renewal of track and ballast and replacement of culverts was seen as a priority under the concession arrangements to take effect in 2005.

Reopening in 2004 of the Tororo–Mbale–Soroti (162 km) section of Tororo–Pakwach line (501 km) was seen by URC as a first stage in the eventual restoration of services over the entire route. URC has also studied rehabilitation of the Kampala–Kasese line (333 km), the case for such a project boosted by the prospect of oil reserves in western Uganda.

In 2004 in-motion wagon weighing systems were installed at Kampala and Tororo, a move aimed at reducing derailments and accidents caused by overloaded or unbalanced vehicles. Similar equipment has been installed in Kenya and Tanzania.

Traction and rolling stock
In 2001 URC owned 54 diesel locomotives, 37 of which were serviceable. These are predominantly German-built. There were 1,376 freight wagons (1,252 serviceable). Under a contract signed in 1998, locomotive and rolling stock maintenance and overhaul is carried out by a joint venture of Bombardier Transportation (formerly Adtranz) and Nalukolongo Ltd.

UPDATED

Ukraine

Ministry of Transport

Peremogy Prospekt 14, 01135 Kiev-135
Fax: (+380 44) 216 72 06
Web: www.mtu.gov.ua

Key personnel
Minister: Yevhan Chervonenko

UPDATED

Transcarpathian Railroads

294017 Uzhgorod, Universitetskaja Str 1
Tel: (+380 31) 224 26 62
Fax: (+380 31) 223 78 01

Key personnel
Managing Director: Ivan Ustich

Organisation
Transcarpathian Railroads is a private-sector company set up to develop narrow-gauge railways in southwest Ukraine and neighbouring countries. Its first ambition was to take over the 80 km Beregova–Kusnitsa line run by the Ukrainian Ministry of Transport.

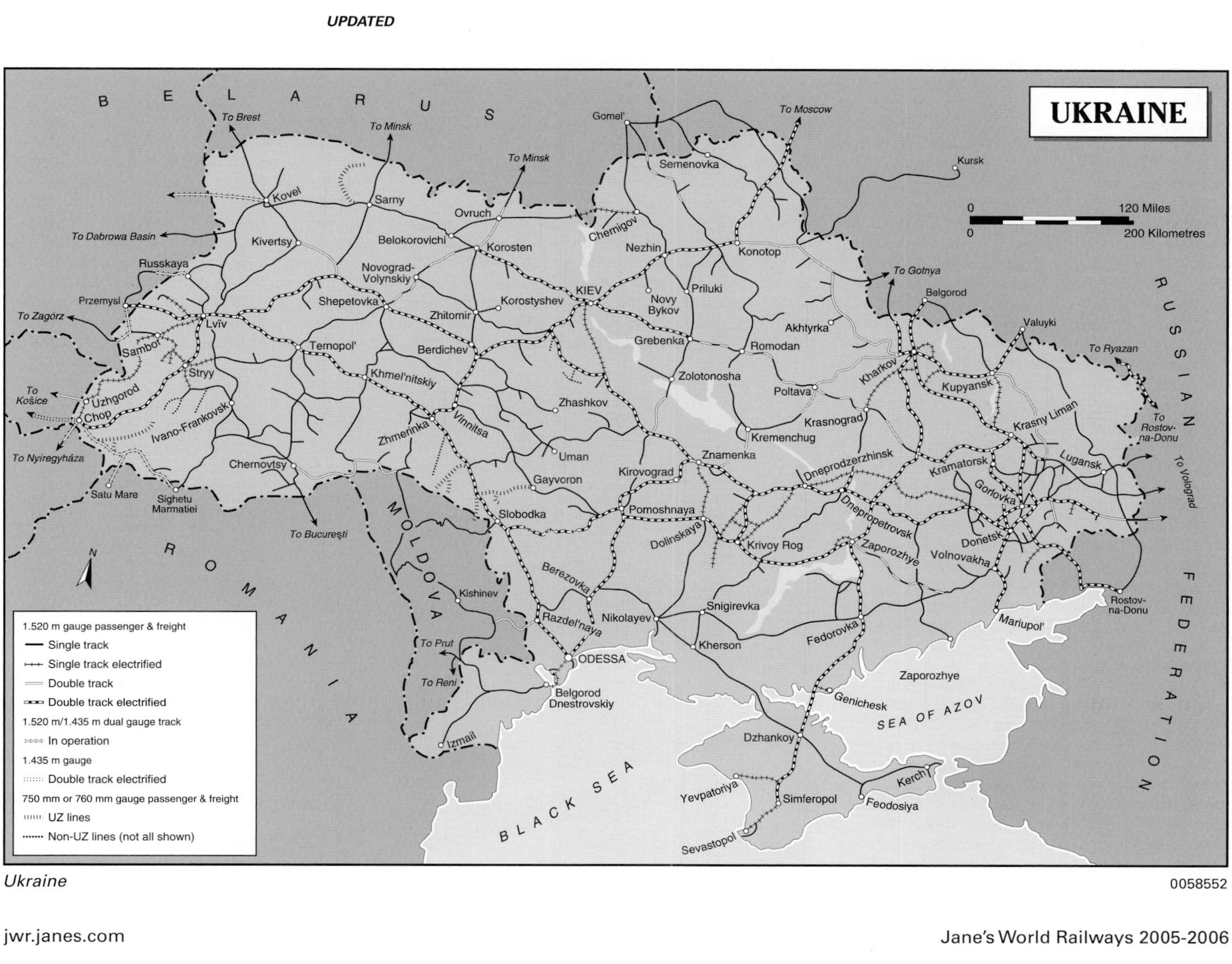

Ukraine

0058552

Ukrainian National Rail Transport Administration (UZ)

Derzhavna Administratsiya Zaliznichnogo Transportu
Ukraïni (Ukrzaliznitsya, UZ)
Vulitsa Lysenko 6, 252601 Kiev-34
Tel: (+380 44) 223 63 05
Fax: (+380 44) 227 65 93; 03 23
Web: http://www.uz.kiev.ua

Key personnel
General Director: G Kirpa

Gauge: 1,520 mm
Route length: 22,473 km
Electrification: 4,320 km at 25 kV 50 Hz AC, 4,930 km at 3 kV DC

Political background

UZ resulted from the 1992 break-up of Soviet Railways (SZD), and consisted of the six railway administrations that served Ukraine: the Donetsk, Lviv (Lvov), Odessa, Dnipro (Dnepr), South West and Southern railways. Its headquarters was established in Kiev and is subordinated to the Railway Department of the Transport Ministry. Of Ukraine's seven neighbours, the Russian Federation, Belarus and Moldova share the same 1,520 mm gauge. Although there are a few short cross-border 1,520 mm gauge lines, most freight is transhipped to standard gauge when it passes to and from the Polish, Slovakian, Hungarian and Romanian railways.

Due partly to the unsettled political situation, and partly to the economic downturn that followed the break-up of the USSR, the UZ administration was faced with a critical situation. It endeavoured to cut costs, but could not always achieve its goals; in 1996 it planned to close 89 low-traffic subdivisions, but local resistance halved that number. An attempt to merge railway divisions aroused wide opposition and may have been one of the factors in the summary dismissal of the General Director in 1997, which came at a time when the apparently excessively reorganised Dnipro Railway was facing possible bankruptcy and there were calls for the establishment of an independent Crimean Railway.

More serious, perhaps, was the growth of crime and corruption and apparently the 1997 upheaval did little to stop this. Early in 2000, following a critical report commissioned by the national security council, there was another hasty change of General Director. The new man, G Kirpa, formerly head of the Lviv Railway, declared his intention to 'disinfect' UZ from railway heads down to cleaners. Five out of the six railway heads, accused of incompetence and financial unreliability and of 'hiding behind the skirts of local administrators,' were dismissed, together with half of the General Director's deputies. It was announced that real money would be required from clients (hitherto only 22 per cent of internal freight charges were paid in ready cash, the rest was payment in kind or credit notes). The host of middlemen and self-proclaimed freight forwarders was to be eliminated, and railway organisations were to be strictly audited and required to use UZ's EkspressBank for all their transactions. In June 2000, apparently as a first fruit of this cleansing, freight traffic increased by six per cent but freight revenue by 47 per cent. Wages began to be paid on time and in full.

UZ had earlier accepted the restructuring principles of EU Directive 91/440 and additionally has been advised by the European Bank for Reconstruction and Development. However, the present intention is to proceed slowly, probably towards transforming UZ into a holding company.

Finance

High inflation (2,500 per cent in 1994 but much less subsequently) has to be taken into account when comparing different years. In 1997, income from freight was 4,026 million karbovanets, 11 per cent more than in 1996. Corresponding figures for passenger income were 652 million and 6.6 per cent more. The share of passenger income had grown considerably from the barely 5 per cent of 1992.

Reluctance or inability of customers to pay their bills has caused problems. In 1997, on the eve of a conscription call-up period, the railway refused to

Class 2M62U diesel-electric locomotive at Zhitomir with a Chop-Kharkov service (Colin Boocock)
0059404

issue military concession tickets until the Defence Ministry agreed to pay its debts.

Since separation from Russia, heavy increases in the costs of energy in the Ukraine – both diesel oil and electricity – have meant that these now form a much higher proportion of the railway's costs than formerly.

The number of workers (excluding those in non-railway activity) was 367,879 in 1999, of which 276,729 were operating workers. A high proportion of staff works less than full time.

Passenger operations

In 1995, passenger traffic stood at 577 million journeys (1994 = 631 million); passenger-km stood at 63.759 billion (1994 = 70.881 billion). By 1999 these figures had fallen respectively to 487 million and 47.6 billion and passenger operations were continuing to return an operating loss.

The railways have no financial support from the state. In the five years 1992–96 only 17 passenger cars were purchased. For long-distance passenger services in 1997, 11,300 vehicles were needed but only 9,000 were serviceable, of which one third were obsolete. Although the Crimean government relieved the railways of land tax, most local governments paid little heed to the government's exhortation to help the railways with passenger subsidies. In 1996, the railways therefore cut some services and put other pressures on local authorities, but with little result. From mid-2000 children under 16 have been carried free. Passenger trains are

increasingly accompanied by transport militia. On the Dnipro Railway, 35 daily passenger trains were so escorted by late 2000.

There is little road competition for passenger traffic; this reflects the high cost of fuel oil and the railway fares structure. In contrast to other former Communist countries, private interurban bus operation has not developed to any great extent, though there is already competition on the Moscow–Kiev–Kishinev (Moldova)–Bucharest–Turkey corridor from Turkish coach operators. Most mail now moves by road.

Freight operations

The major source of freight traffic is the heavily industrialised Donetsk region in the southeast.

Since the break-up of the USSR and SZD, freight traffic has declined by about 50 per cent. In 1995, the railway carried 360 million tonnes (1994 = 408 million); tonne-km stood at 195.762 billion (1994 = 200.423 billion). In 1996, 342.5 million tonnes were carried, of which iron ore amounted to 50 million, and ferrous metals 28.5 million. In 1999, 156.3 billion tonne-km were produced. Tonnes originated totalled 284.2 million, of which coal contributed 82 million.

As with many neighbouring countries, the principal reasons for the decline in traffic are the reduction in industrial output following the loss of traditional markets combined with an upsurge of competition, particularly from road hauliers for international traffic flows.

Class VL80T 25 kV AC electric locomotive at Fastov (Colin Boocock)
0059405

Class ER9M 10-car emu at Fastov, forming a Kiev-bound service (Colin Boocock)   0059406

Thanks to ill-judged tariff increases, much transit traffic was lost in the late 1990s. About a quarter of transit traffic is originated by Intertrans, the forwarder established by the Transport Ministry in 1992.

UZ's Ukrpromzheldortrans (Ukrainian Industrial Railway Transport Combine) operates much, though not all, of the industrial feeder mileage, charging industries directly for work done (but rarely being paid in full). In 2000 it was said to possess 224 diesel locomotives, of which 80 were serviceable. It is being restructured, the first moves being the combination of its 14 industrial railways in Kiev, and a smaller merger at Odessa.

Traction and rolling stock
At the end of 1997, the effective (although not neccesarily operating) rolling stock fleet comprised: 515 passenger electric locomotives, 139 passenger diesel locomotives, 1,349 freight electric locomotives, 1,211 main line diesel freight locomotives, 1,869 diesel shunting locomotives, 1,032 dmu cars and 2,982 emu cars.

The main electric Classes are: VL8, VL10, VL11/11m, VL80 and VL82 families, ChS2, ChS4, ChS7 and ChS8, with emus mainly from the Riga-built ER2 and ER9 groups. The main diesel fleets are: TE3, TE7, TEP60, TEP70, 2TE10 family, M62 family, 2TE116 and 2TE121; dmus are of Classes D1 (Ganz-Mávag), DR1, DR1A and DR1P (Riga).

During 1994, the financial situation had deteriorated to such an extent that freight traffic, wherever feasible, was being routed solely over electrified lines in order to reduce the consumption of diesel oil. More than 20 steam locomotives were returned to service on shunting and trip work, mainly in the Lviv area.

New construction approved by the government for 1998 was 40 dmu cars, eight emu cars and six mixed-traffic DC electric locomotives. It was decided soon after independence that electric locomotives would be built at an engineering works in Dnepropetrovsk, while the existing main line diesel locomotive works at Lugansk would add emu and dmu designs to its product range. The Dnepropetrovsk Electric Locomotive Works (NPO DEVZ) is to build twin-section Bo-Bo freight and passenger designs for both AC and DC operation. The first, the DC freight Type DE1, appeared in 1996 but was seven tonnes overweight. The six mixed-traffic units scheduled for construction in 1998 were to be based on this design and were to be followed in 1999 by an AC version. DC and AC passenger electric prototypes were to appear in 1999 and 2000. Siemens Transportation Systems is supplying three-phase traction equipment based on that used in its Class 152 and 189 locomotives supplied to DB for 21 Class DS3 25 kV AC electric locomotives to be constructed by NPO DEVZ. Construction of a prototype commenced in 2000. The series batch of 20 is to be completed by 2006. The 4,800 kW machines will have a maximum speed of 160 km/h.

To extend the life of ChS4 passenger locomotives for 15 years beyond 2000, the Zaporozhe repair works, with assistance from Škoda, is refurbishing these units. Kharkov Works has produced diesel engines with which to re-engine the Soviet 2TE116 freight design, while Lugansk has produced a passenger version of the latter.

In early 1998 it was announced that Uzbekistan was to participate in a joint venture to manufacture diesel locomotives at Lugansk.

In 1992, 330,000 freight wagons of the Soviet stock were allocated to Ukraine, of which 225,000 were in the operational stock. In January 1996 there were 280,000, of which 100,000 were in the working stock. New acquisitions have been few (only 2,200 wagons were acquired in 1994, for example).

Five coaches were rebuilt with Spanish help in 1996, following which it was decided to obtain assistance from the same source to refurbish coaches older than 20 years to bring them up to European standards. The Dnepropetrovsk repair works commenced refurbishment of 100 coaches in 1999. With some technical assistance from De Dietrich in France, it was expected that the Kryukov wagon works would build new passenger vehicles.

In 1999, 22 passenger coaches were bought, as against 10 in 1998 and nil in 1997. Delivery of freight wagons, however, decreased from 1,416 to 772.

Signalling and telecommunications
The following types of signalling are in use on UZ (with km controlled in brackets): automatic block (13,605 km); interlocking control (3,355); semi-automatic block (7,466); automatic cab signalling (15,113); dispatcher control (8,495).

Electrification
Some 41 per cent of the system is electrified, divided almost equally between 25 kV AC and 3 kV DC. Encouraged by the high price of diesel fuel, the administration adopted a 10-year electrification plan in 1994 which is now in arrears. It envisaged 2,148 km of new electrification. The east-west Russian frontier–Poltava–Kiev line has been the most important objective, as well as the Ternopol–Zhmerinka line. In 1998 the final section of the Moscow–Hungary route was electrified. 187 km were electrified in 1999.

Track
To overcome a shortage of rail, in 1995, each of the six railways in UZ was required to close 200 km of poorly utilised track in order to provide materials for maintaining the main lines.

The Lvov–Kiev main line, part of one of the nominated European transport corridors, is being upgraded to accept 140 km/h running, with the European Bank for Reconstruction and Development providing financial help.

Considerable investment is to be made in increasing line capacity in and around the ports of Odessa and Ilichevsk.

Rail: 65 kg/m (34,620 km); 50 kg/m (11,330 km)
Sleepers: Concrete, wood
Spacing: 1,840/km in plain track, 2,000/km in curves
Min curve radius: 200 m
Max gradient: 1.4%
Max axleload: 22.5 tonnes

United Kingdom

Department for Transport

Eland House, Bressenden Place, London SW1E 5DU
Tel: (+44 20) 79 44 30 00   Fax: (+44 20) 79 44 46 69
Web: www.dft.gov.uk

Key personnel
Secretary of State for Transport: Alistair Darling
Minister for Transport: Tony McNulty
Permanent Under-Secretary: Charlotte Atkins

Director, Rail Strategy and Resource: Mark Lambirth
Director, Rail Implementation and Performance: Vivien Bodnar

Organisation
The privatised railway in the UK is regulated by three autonomous bodies: the Strategic Rail Authority, the Office of Rail Regulation and HM Railway Inspectorate within the Health & Safety Executive.

However, in July 2004 the government announced that it was to dissolve the Strategic Rail Authority, absorbing some of its functions within the Department for Transport and vesting others in Network Rail, the national rail infrastructure company. As part of the same restructuring, responsibility for rail safety was to move from the Health & Safety Executive to the Office of Rail Regulation.

UPDATED

HM Railway Inspectorate Health and Safety Executive

Rose Court, 2 Southwark Bridge, London SE1 9HS
Tel: (+44 20) 77 17 65 02   Fax: (+44 20) 77 17 65 48
Web: www.hse.gov.uk

Key personnel

Director of Rail Safety/Chief Inspector of Railways:
  Dr Allan Sefton
Deputy Chief Inspector/Head of Field Operations
  Division (RI Ops): Linda Williams

Organisation

The Railway Inspectorate's role is to secure the proper control of risks to the health and safety of employees, passengers and others who might be affected by the operation of Britain's railways. It does this by a number of activities including: technical approvals; safety case assessment and acceptance; pro-active programmes of intervention; investigation of railway incidents; giving advice; and administering and enforcing the law. There were 123 inspectors in HM Railway Inspectorate as at January 2005. The Chief Inspector produces an annual report on railway safety in Great Britain which is available on the Internet.

In July 2004 the government announced that is was to transfer the Railway Inspectorate's responsibilities for rail safety from the Health and Safety Executive to the Office of Rail Regulation.

UPDATED

Office of Rail Regulation (ORR)

1 Waterhouse Square, 138-142 Holborn, London EC1N 2TQ
Tel: (+44 20) 72 82 20 00   Fax: (+44 20) 72 82 20 45
e-mail:rail.library@orr.gsi.gov.uk
Web: www.rail-reg.gov.uk

Key personnel

Chairman: Chris Bolt
Chief Executive: Keith Webb
Non-executive Members: Peter Bucks;
  Jeffrey Jowell, QC; Jane May; Jim O'Sullivan;
  Chris Stokes

Political background

The Office of Rail Regulation was established on 5 July 2004 by the Railways and Transport Safety Act 2003. It replaced the Office of the Rail Regulator. As the railway industry's economic regulator, the Office's principal function is to regulate Network Rail's stewardship of the national network. The ORR also licences operators of railway assets, approves agreements for access by operators to track, stations and light maintenance depots, and enforces domestic competition law.

The ORR is led by a Board appointed by the Secretary of State for Transport, under the Chairmanship of Chris Bolt, who has been appointed International Rail Regulator (IRR) by the Secretary of State. The IRR is a statutory office separate from that of the ORR. The IRR licences the operation of certain international rail services in the European Economic Area and access to railway infrastructure in Great Britain for the purpose of the operation of such services.

In July 2004 the government announced its intentions for the regulation of rail health and safety in its White Paper 'The Future of Rail', and the first step in the transfer of that responsibility from the Health and Safety Commission/Executive to the ORR occurred when the Railways Act 2005 received Royal Assent on 7 April 2005. ORR expects to assume its new responsibilities as a combined safety and economic regulator at the end of 2005.

UPDATED

Strategic Rail Authority (SRA)

55 Victoria Street, London SW1H 0EU, UK
Tel: (+44 20) 76 54 60 00   Fax: (+44 20) 76 54 60 10
Web: http://www.sra.gov.uk

Key personnel

Chief Executive Officer: Nick Newton

Political background

The SRA was established by the UK government to provide focus and strategic direction for Britain's railways and to promote their development. It is a non-departmental public body charged with advising government and with implementing government policy for the railway network. Its main roles are: to promote the use of the railway network for the carriage of passengers and freight; to secure the development of the railway network; and to contribute to the development of an integrated system of transport of passengers and freight.

Pending the legislation which enabled its formal establishment, the SRA was set up in mid-1999 as the Shadow Strategic Rail Authority (sSRA), using the powers of the Office of Passenger Rail Franchising (OPRAF) and the British Railways Board (BRB). The functions of both organisations were absorbed by the SRA on 1 February 2001, following Royal Assent of the 2000 Transport Act. In Scotland, the SRA is subject to Directions from the Scottish Minister for Transport. At the same time, the SRA assumed responsibility for the consumer support responsibilities of the Office of the Rail Regulator and the freight and international operations responsibilities of the Department for Transport.

As part of a further restructuring of the UK railway industry, the government announced in July 2004 that it was to dissolve the SRA. Its powers would be transferred to the Department for Transport and Network Rail. No date was given for these changes, which would require parliamentary approval.

Organisation

Passenger services

The SRA assumed responsibility for the management and development of the system of franchises which form the basis of most passenger rail services in Great Britain. Between 1995 and 1997 25 franchises were established based on the lowest bid for subsidy against predictions of little growth. It was subsequently recognised that this arrangement did little to encourage investment, especially as many of the franchises were short: 18 were due to expire by mid-2004.

In response, the sSRA initiated several changes to the franchise replacement process. Most significant among these were proposals, unveiled in June 2000, to amend geographically the structure: of the existing 25 franchises, changes are planned for six, three would be incorporated into other franchises, seven new franchises would be created and nine would remain unchanged. In addition, the sSRA in September 1999 began the process of replacement of existing franchises for which the geographical structure would not significantly change, giving priority to those with less than five years to run. The Strategic Plan included refinements to this process, including combining franchises serving London termini in a move aimed at increasing station capacity and operational flexibility.

The SRA also administers the Rail Passenger Partnership project (RPP) scheme, a government initiative to provide support for innovative proposals aimed at encouraging new rail passengers or at promoting modal shift and integration. Typically, these provide financial support for additional rolling stock and services.

Freight services

Freight operations on the railway network in Great Britain are in the hands of private sector companies, including some third party or own-account operators running services under the provisions of the 1993 Railways Act. Under the 2000 Transport Act, the SRA has a responsibility to promote rail freight. The SRA's freight strategy is designed to support growth through network and access investment and targeted revenue support. It pursues a policy of the retention of existing freight capacity and of the enhancement of both capacity and capability of the network.

Following full vesting of the SRA on 1 February 2001, the organisation took over responsibility for administering rail freight grants in England. This function was previously undertaken by Department of Transport, local government and the Regions. Similar grants in Scotland and Wales are administered by the Scottish Executive and the National Assembly for Wales respectively, in both cases in consultation with the SRA.

In December 1999 the sSRA announced a scheme to award financial incentives totalling up to £6 million for innovative solutions for rail-based logistics. Three awards were announced in June 2000 covering: a two-year trial using a piggyback concept for cement distribution; the development of a freight dmu; and the development of a high-speed flat wagon to convey mini-containers for express parcels traffic on Anglo-Scottish routes.

Infrastructure

In its role of securing the development of the railway network, the SRA plays a key part in reviewing network utilisation and capacity, assessing how this is expected to change and identifying priorities for network enhancement as part of the longer term process of investment planning. The sSRA/SRA has led or participated in several major studies, including: the Review of Rail Capacity in the West Midlands; London North-South Integration and London East-West Study schemes; the South London Metro feasibility study; studies into a new high-speed line linking London and Scotland; and Manchester area study. The SRA also participates in the evaluation of route reinstatement and upgrading schemes proposed by train operators and/or local authorities.

The Strategic Plan published in January 2002 detailed major infrastructure projects that the SRA was backing. These include Crossrail, two cross-city lines in London, for which the SRA has established an eponymous organisation jointly with Transport for London (see 'Crossrail' entry). Also listed in the plan for development are: Great Western Main Line upgrade and a new north-south high-speed line.

Association of Train Operating Companies

3rd Floor 40 Bernard Street, London WC1N 1BY
Tel: (+44 20) 79 04 30 00
Fax: (+44 20) 79 04 30 20
Web: http://www.atoc.org

Key personnel

Chairman: Keith Ludeman
Deputy Chairman: Adrian Shooter
Director General: George Muir

Directors:
Communications: Edward Funnell
Distribution and Marketing Services: David Mapp
Engineering: Rebeka Sellick
Finance: Chris Wade
Policy and Regulation: Alec McTavish
London Support: Paul Smith
Chief Executive, NRES: Chris Scoggins

Political background

The Association of Train Operating Companies (TOCs) was set up to administer schemes that required the joint participation of the 25 passenger TOCs formed out of the old British Rail. ATOC's most important role is to administer the division of monies derived from tickets for journeys which cross TOC boundaries, but it also performs a role as a trade association, acting for the TOCs where there is benefit to them in acting collectively.

Arriva plc

Admiral Way, Doxford International Business Park, Sunderland SR3 3XP
Tel: (+44 191) 520 40 00   Fax: (+44 191) 520 40 01
Web: www.arriva.co.uk

Key personnel

Arriva plc
Chief Executive: Bob Davies
Finance Director: Steve Lonsdale

Arriva Trains Ltd
Finance Director: Peter Telford
Policy Director: Roger Cobbe
Bid Director: Fergus Robertson

Organisation

Arriva is one of the largest transport services organisations in Europe, with operations in Denmark, Germany, Italy, The Netherlands, Portugal, Spain and Sweden, as well as in the UK. Employing some 30,000 staff, the group operates a range of services that includes trains, buses, commuter coaches, demand-response transport and vehicle rental. It operates rail franchises in Denmark, Germany, The Netherlands and the UK.

In the UK Arriva operates the Arriva Trains Wales/Trenau Arriva Cymru franchise under a 15-year contract that commenced in December 2003. This covers interurban, rural and commuter passenger rail services in Wales and the border counties of England. See entry in UK section of *Railway systems and operators*.

Arriva's rail portfolio in mainland Europe has also grown significantly. It owns Noordned (see entry in the Netherlands section of *Railway systems and operators*), which operates integrated rail and bus services in Friesland, and also operates some train services in Groningen. Formerly a joint venture with NS Rail, Noordned became fully owned by Arriva in 2003. In January 2003 Arriva Danmark began to operate two regional rail services in Denmark, serving mid- and north Jutland. Arriva was the first private company to operate rail services tendered by the Danish state. The franchises operate until 2010. See entry for Arriva Tog in the Denmark section of *Railway systems and operators*.

Arriva entered the German market in 2004 with the acquisition of Prignitzer Eisenbahn Gruppe (PEG) (see Germany section of *Railway systems and operators*). PEG runs services in Nordrhein-Westfalen, Brandenburg and Mecklenburg-Vorpommern. In October 2004 Arriva purchased a stake in Regentalbahn AG from the Bavarian state and now has a shareholding of 90 per cent. Regentalbahn runs regional passenger rail services in Bavaria, Thuringia and Saxony. See entry in Germany section of *Railway systems and operators*.

UPDATED

Arriva Trains Northern

Background

The Arriva Trains Northern franchise ended in December 2004, when most of its routes were taken over by the operator of the Northern Rail franchise (see entry in UK section of *Railway systems and operators*). Some of its former longer-distance services form part of the TransPennine franchise.

UPDATED

Arriva Trains Wales

Brunel House, 2 Fitzalan Road, Cardiff CF24 0SU
Tel: (+44 29) 20 43 04 00   Fax: (+44 29) 20 43 02 14
Web: www.arrivatrainswales.co.uk

Key personnel

Managing Director: Peter Strachan
Customer Operations Director: Graeme Bunker
Finance and Commercial Director:
    Richard Roxburgh
Customer Services Director: Ian Bullock
Fleet Director: Ian Papworth
Operations and Safety Director: Tim Bell
Human Resources Director: Dennis Baker

Political background

Consisting mostly of the former Wales & Borders franchise, which in turn was formed of elements of the Wales & West franchise, Valley Lines (Cardiff Railway) routes and Central Trains services to northeast and mid-Wales, the franchise operated as Arriva Trains Wales/Trenau Arriva Cymru (ATW) was created with a geographical profile to match the area administered politically by the Welsh Assembly. Consequently it also includes regional and local services in North Wales previously operated by First North Western. In October 2003 the SRA signed a contract with Arriva plc (see entry in UK section of *Railway systems and operators*) for a 15-year franchise, subject to five-year performance review assessments, which took effect from December 2003. Over the life of the franchise the projected level of public subsidy is £1.68 billion, plus any funding from Rail Passenger Partnership schemes and services funded directly by the Welsh Assembly.

Passenger operations

ATW serves a network of some 4,180 route-km, with a fleet of 116 trains and 1,860 staff. The company operates 235 stations, of which 48 are staffed.

Arriva Trains Wales diesel multiple-units

Class	Cars per unit	Motor cars per unit	Motored axles/car	Power/ motor kW	Speed km/h	Cars in service	First built	Builders Mechanical	Builders Engine	Transmission
142	2	2	1	170	120	46	1985	BREL	Cummins LTA10R	*H* Voith T211r
143	2	2	1	152	120	34	1985	Alexander/ Barclay	Cummins LTA10R	*H* Voith T211r
150	2	2	2	210	120	14	1986	BREL	Cummins NT855R5	*HM* Voith T211r
153	1	1	2	213	120	12	1987*	Leyland	Cummins NT855R5	*HM* Voith T211r
158/0	2	2	2	275/300	145	76	1989	BREL	Cummins NTA855R	*HM* Voith T211r
175	2/3	2/3	2	335	160	–	1999	ALSTOM	Cummins N14	*HM* Voith T21rzze

* Rebuilt 1991.

Class 143 dmu forming a Valley Lines service at Ystrad Mynach (Ken Harris)   0554925

Its main routes are: west Wales to Swansea and Cardiff, through to Shrewsbury and Manchester; Birmingham to Shrewsbury, Aberystywth and Pwllheli; Holyhead to Manchester; Cardiff to Gloucester; the 'Valley Lines' network around Cardiff; and the Swansea-Shrewsbury 'Heart of Wales' line.

On average ATW handles 60,000 passengers each day.

c2c

c2c Rail Ltd
National Express Group, London Lines
Hertford House, 1 Cranwood Street, London
EC1V 9QS
Tel: (+44 20) 74 27 28 09
Web: www.c2c-online.co.uk

Key personnel
Managing Director: Andrew Chivers
Directors
　Finance: Adam Golton
　Operations: Mark Hopwood
　Commercial: David Hamilton

Political background
Under the arrangements put in place by the 1993 Railways Act, the LTS Rail franchise was won by Prism Rail in May 1996. Prism also ran the Cardiff Railway (subsequently absorbed into Wales & Borders Trains and now part of Arriva Trains Wales), Wales & West (now part of Wessex Trains) and West Anglia Great Northern (WAGN) franchises. The franchise was let for 15 years.

LTS Rail was to receive financial support from the Franchising Director, in 1996 prices, of £29.5 million in the first year, declining to £11.2 million in 2010–11.

Corporate developments
In 2000 LTS Rail adopted the branding 'c2c'.

In July 2000 Prism sold its rail operating businesses to National Express Group, which subsequently created its London Lines management structure to enable resources to be shared by c2c and its two other franchises serving the capital, Silverlink and WAGN.

In late 2004 c2c employed 513 staff.

Passenger operations
c2c operates commuter services on the lines east of the capital out of London Fenchurch Street to Tilbury, Southend and Shoeburyness, with the route network extending to 125 km in all and serving 24 stations.

A £50 million investment programme has led to the provision of modern passenger information systems, ticket gates and security systems at most stations.

Traffic (million)	2001–02	2002–03	2003–04
Passenger journeys	27.4	28.8	29.9
Passenger-km	799	826	838

Traction and rolling stock
ATW is an all-diesel-operated business mostly worked with second- and third-generation dmus. For its interurban routes the company employs Class 158 and Class 175 dmus leased from Angel Trains. For local services, it leases Class 142 dmus and Class 153 single cars from Angel and Class 143 and 150/2 dmus from Porterbrook. ATW also leases in Class 37/4 locomotives from EWS and Mk 2 coaches for Rhymney line peak-hour services and some seasonal trains, as well as Class 47s and Mk 2 stock for some North Wales services.

UPDATED

c2c Class 357 emu at Pitsea forming a London Fenchurch Street–Tilbury service (Colin Boocock)
0524947

c2c electric multiple-units

Class	Cars per unit	Motor cars per unit	Motored axles/car	Power/ motor kW	Speed km/h	Cars in service	First built	Builders Mechanical	Electrical
357	4	3	2	250	160	296	1999	Bombardier	Bombardier

Traction and rolling stock
To meet the requirements of the 15-year franchise term, LTS Rail had to acquire stock to replace slam-door units. This process was completed in 2003 when the last of 74 Class 357 four-car Electrostar 25 kV AC emus entered service. Built by Bombardier Transportation (formerly Adtranz), they are leased from Angel Trains (28 Class 357/2) and Porterbrook (46 Class 357/0). From September 2004 five units were subleased to sister company One Great Eastern for London Liverpool Street–Southend Victoria services.

Maintenance of the fleet is undertaken by Bombardier under c2c management control at the operator's East Ham depot, which has benefited from investment in improved facilities.

UPDATED

Central Trains

Central Trains Ltd
102 New Street, Birmingham B2 4JB
Tel: (+44 121) 654 12 00　Fax: (+44 121) 654 12 34
Web: http://www.centraltrains.co.uk

Key personnel
Managing Director: Nick Brown
Business Development Director: Mike Haigh

Political background
Under the arrangements put in place by the 1993 Railways Act, the franchise to run operations on Central Trains was let in March 1997 to National Express Group plc (see entry in UK section of *Railway systems and operators*), a bus and airport operator which also operates the Midland Mainline, Gatwick Express, ScotRail and Silverlink franchises.

The franchise was let for a term of seven years and one month. NEG was to receive from the Franchising Director (now the SRA) and the local

Central Trains Class 170/6 three-car dmu at Nottingham (Ken Harris)
0554702

Central Trains diesel multiple-units

Class	Cars per unit	Motor cars per unit	Motored axles/car	Power/ motor kW	Speed km/h	Cars in service	First built	Mechanical	Builders Engine	Transmission
150/0	3	3	2	210	120	33	1985	BREL	Cummins NT855R5	HM Voith T211r
150/1	2	2	2	210	120	32	1985	BREL	Cummins NT855R5	HM Voith T211r
150/2	2	2	2	210	120	8	1986	BREL	Cummins NT855R5	HM Voith T211r
153	1	1	2	213	120	18	1987*	Leyland Bus	Cummins NT855R5	HM Voith T211r
156	2	2	2	210	120	40	1987	Met-Cam	Cummins NT855R5	HM Voith T211r
158/0	2	2	2	275/300	145	48	1989	BREL	Cummins NTA855R	HM Voith T211r
170/5 & 170/6	2/3	2/3	2	315	160	76	1999	Adtranz	MTU 6R183TD	HM Voith T211r

* Rebuilt 1991.

Central Trains electric multiple-units

Class	Cars per unit	Motor cars per unit	Motored axles/car	Power/ motor kW	Speed km/h	Cars in service	First built	Builders Mechanical	Electrical
323	3	1	4	146	144	78	1993	Hunslet TPL	Holec

authority Centro support, in 1997 prices, of £187.5 million in the first full financial year of the franchise, declining to £132.6 million in 2003–04.

In October 2001, all Central Trains operations in northeast and mid-Wales were handed over to the new Wales & Borders Trains train operating company (see entry in UK section of *Railway systems and operators*).

Under a deal negotiated in March 2002, the Strategic Rail Authority agreed to pay £115 million additional subsidy to Central Trains and its sister company ScotRail over the remaining two years of their franchises to enable them to break even. The parent company, National Express Group, was to contribute £57 million.

In October 2003 the SRA signed a two-year extension with NEG to operate the Central Trains franchise from April 2004 to March 2006. A new franchise was to commence in April 2006 after a competitive tendering process. Subject to further negotiations, a subsidy of £457 million would be paid to NEG over the two-year extension period.

Passenger operations

Central Trains operates services over a 2,182-km route network covering a large part of central England, with many lines extending into South Wales and East Anglia. It also operates an extensive network of local services in the Birmingham/ Wolverhampton conurbation for Centro, the local Passenger Transport Executive.

Traffic (million)	1999–2000	2000–01	2001–02
Passenger journeys	36	36.7	37.0
Passenger-km	1,292.1	1,320.3	1,342
Train-km	30.9	34.3	31.1

Traction and rolling stock

For electric services around Birmingham, Central uses 26 three-car Class 323 emus. For its premier diesel-worked long-distance routes, the business leases Class 170 and 158 dmus from Porterbrook and Angel Trains respectively. Other trains in use include Class 150, 153 and 156 dmus.

The Class 323 emus were the subject of a refurbishment programme in 2003.

Chiltern Railways

Western House, 14 Rickfords Hill, Aylesbury, Buckinghamshire HP20 2RX
Tel: (+44 1296) 33 21 00   Fax: (+44 1296) 33 21 26
Web: www.chilternrailways.co.uk

Key personnel

Chairman: Adrian Shooter
Managing Director: Cath Procter
Directors
  Finance: Tony Allen
  Engineering: Andy Hamilton
  Operations: Richard Maclennan
  Commercial Planning: Mike Bagshaw

Political background

Under the arrangements put in place by the 1993 Railways Act, the Chiltern Railways franchise passed in July 1996 to M40 Trains, a joint venture of the Chiltern's management, the John Laing construction group and venture capitalists 3i. The new owners took action to cut subsidy levels over a seven-year franchise from GBP16.5 million in the first year to GBP3.3 million in the final year, acquire new trains and increase the level of services prevailing when it took over. Subsequently, the 3i shareholdings in M40 Trains were acquired by John Laing, giving it 100 per cent ownership by 2002.

In April 2001 Chiltern Railways announced that it had reached agreement with the Strategic Rail

Chiltern Railways diesel multiple-units

Class	Cars per unit	Motor cars per unit	Motored axles/car	Power/ motor kW	Speed km/h	Cars in service	First built	Mechanical	Builders Engine	Transmission
165/0	2/3	2/3	2	260	120	89	1990	BREL	Perkins 2006-TWH	H Voith T211r
168/0	3	3	2	315	160	20	1998	Adtranz	MTU	H Voith
168/1	3	3	2	315	160	26	2000	Bombardier	MTU	H Voith
168/2	2/3	2/3	2	315	160	15	2003	Bombardier	MTU	H Voith
121	1	1	2	2 × 112	112	1	1965	BR	Leyland	M

Authority to amend its franchise. Chiltern was to receive a one-off subsidy payment in 2000–01 of GBP2.4 million, bringing the total for that year to GBP5.1 million, enabling the company to make investments of GBP11 million in a range of safety and customer service improvements.

In February 2002 the SRA awarded M40 Trains a new 20-year franchise to take effect from July 2003. The extension of the original franchise and its subsequent 20-year extension is linked to a major programme of capacity enhancement schemes. These include reinstating double track between Bicester and Aynho Junction (14 km), completed in 2002, quadruple track at Beaconsfield and between West and South Ruislip, raising the line speed to 160 km/h between Banbury and South Ruislip, signalling improvements and provision of two additional platforms at London Marylebone

station. Improvements in the Birmingham area include the reopening of terminal platforms at Moor Street station, scheduled for 2003. Also planned are transport interchanges at Banbury and High Wycombe, as well as numerous other station improvements.

Passenger operations

Chiltern runs services over 336 km on two routes out of London's Marylebone station: to Aylesbury via Amersham and to Birmingham Snow Hill and Birmingham Snow Hill via High Wycombe. The franchisee's original ambitious revenue growth plans went towards changing the character of the company's operation from one where commuter traffic to London predominates to one where medium-distance traffic, from such places as Leamington Spa to London, is of significance as well. Supporting these growth plans, Railtrack (now Network Rail) restored double track on the 29 km Bicester–Princes Risborough section.

In May 2001 Chiltern extended some services beyond Birmingham Snow Hill to Stourbridge and in 2002 this was extended to serve Kidderminster at peak hours. From December 2004 the service frequency between London and Birmingham was increased to two trains per hour. Chiltern has also taken over services between Leamington Spa and Stratford-upon-Avon.

Traffic (million)	2002–03	2003–04	2004–05
Passenger journeys	12.1	12.8	14.1
Passenger-km	585	635.9	715
Train-km	7.6	7.9	8.3

Traction and rolling stock

Chiltern initially operated a fleet of modern 'Turbo' Class 165 diesel multiple-units owned by the Angel Trains leasing company with 23 two-car and 11

Two Class 168/1 dmus forming a Birmingham Snow Hill–London Marylebone service at Banbury (John C Baker)   0554920

three-car units. Maintenance of the fleet was subcontracted out to the trains' manufacturer, Adtranz (now Bombardier Transportation). Refurbishment of this fleet was undertaken by Bombardier in 2001–04. By 2005 the fleet had grown to 89 Class 165 cars, five two-car units having been transferred from First Great Western Link, and 61 Class 168 cars. In May 2005 Chiltern ordered six additional intermediate powered cars from Bombardier to strengthen existing two-car sets.

In 2002 Chiltern placed in service a refurbished Class 121 single-unit, first-generation railcar to provide a peak-hour feeder service between Aylesbury and Princes Risborough.

UPDATED

Crossrail

Cross London Rail Links Ltd
1 Butler Place, London SW1H 0PT
Tel: (+44 20) 79 41 76 00   Fax: (+44 20) 79 41 77 03
Web: www.crossrail.co.uk

Key personnel
Chairman: Adrian Montague
Chief Executive: Norman Haste
Head of Interface Management: Charles Devereux
Head of Planning: Dr David Anderson
Head of Operations and Development:
   Keith Berryman
Head of Human Resources: Tina Bailey
Head of Public Affairs: Bernard Gambrill
Media Manager: Ian Rathbone

Background
Cross London Rail Links Ltd (CLRL) is an equally owned joint venture, formed in 2001, by the Department for Transport and Transport for London (TfL), the authority responsible for the capital's public transport. It has been allocated £154 million of government funding to undertake feasibility studies into two heavy rail, high-capacity urban passenger rail routes across central London and to acquire parliamentary powers to construct Line 1.

Crossrail Line 1 is a projected east-west line, the core of which would be a new tunnelled alignment linking the City of London and London Docklands commercial districts with Paddington station. In the east the planned line extends to Romford and Shenfield, with a southeastern branch running via the Isle of Dogs to Abbey Wood. In the west, its route terminates at Maidenhead, with an option to serve Heathrow Airport.

Crossrail Line 2 is a proposal for a southwest-northeast route which also requires substantial tunnelling under central London. This would be linked to the southwest London suburban network, running via Victoria, the West End of London and King's Cross/St Pancras to Hackney and London's northeast suburbs.

In July 2003 the Secretary of State for Transport authorised CLRL to consult on Line 1 and report by later that year and in July 2004 to prepare to support a Parliamentary Hybrid Bill. In November 2004 the government announced that it was to put enabling legislation for the project before parliament in 2005.

Crossrail's 'best case scenario' foresees services commencing in the period 2013–16. Line 2 services would commence in 2016 at the earliest. At 2002 prices the estimated cost of the project is £7 billion, with £3 billion for contingencies under the government's 'Green Book' rules.

UPDATED

Direct Rail Services Ltd (DRS)

Kingmoor Depot, Etterby Road, Carlisle CA3 9NZ
Tel: (+44 1228) 40 66 00
e-mail: enquiries@drsl.co.uk
Web: www.directrailservices.com

Key personnel
Managing Director: Neil McNicholas
General Manager, Commercial and Business
   Services: Chris Connelly
Head of Engineering: Ted Cassady

Political background
Under the 1993 Railways Act the government made provision for third party and own-account operators to begin rail freight operations on the Railtrack (now Network Rail) network.

Two traditional customers of the railways initially took advantage of the legislation by setting up their own rail freight operations: National Power, which in 1998 transferred its operations to EWS (see entry in UK section of *Railway systems and operators*), and British Nuclear Fuels Ltd. BNFL established a new subsidiary for the purpose: Direct Rail Services. BNFL subsequently became the Nuclear Decommissioning Authority (NDA), which retains ownership of DRS.

Freight operations
DRS began freight operations in December 1995. Its initial operations centred on BNFL's own requirements, carrying flasks of imported spent nuclear fuel from the port at Barrow about 60 km along the Cumbrian coastline to the BNFL reprocessing plant at Sellafield, and also chemicals from further afield required by the plant. In 1999 DRS took over from EWS the movement of spent nuclear fuel rods travelling between power stations around Great Britain and Sellafield. The growing DRS traction fleet has also found useful employment covering locomotive shortages being experienced by other operators and as emergency traction during

Class 37 and 20 locomotives with a Sellafield-bound nuclear flask train north of Stafford (Ken Harris)   *NEW*/1122854

service disruptions on northwest England and southern Scotland.

In February 2001 DRS commenced a 10-year contract with road haulier and logistics company WH Malcolm to provide rail services between Grangemouth, Scotland, and Daventry International Freight Terminal in the Midlands region of England. The frequency of this service subsequently increased and further flows were added, including Grangemouth–Aberdeen and, in 2005, from Widnes, in northwest England, to Purfleet Thames Terminal in a partnership with Novatrans UK. Further diversification in the company's activities came in 2005 with its provision of traction for vegetation management and test trains for Network Rail and its contractors.

Traction and rolling stock
In 2005 the operational DRS locomotive fleet comprised 15 Class 20/3, four Class 33, 10 Class 37/0 and 37/5, nine Class 37/6, four Class 47/0 and 10 Class 66/4. Delivered in 2003, the Class 66/4 locomotives are mostly used on longer non-nuclear flows such as the Grangemouth–Daventry service.

Trial operations with a batch of four Class 87 electric locomotives made surplus by Virgin West Coast were terminated in 2005, in part due to the increased cost of traction power supplies.

The wagon fleet includes some 60 special-purpose vehicles for the transport of nuclear materials.

UPDATED

English Welsh & Scottish Railway

English Welsh & Scottish Railway Ltd
310 Goswell Road, Islington, London EC1V 7LW
Tel: (+44 20) 77 13 23 00   Fax: (+44 20) 77 13 23 11
Web: http://www.ews-railway.co.uk

Key personnel
Chairman: Carl Ferenbach
Chief Executive: Keth Heller

Chief Operating Officer: Allen Johnson
Directors
   Finance: Patrick Butcher
   Engineering: Stuart Boner
   Planning: Graham Smith
   Human Resources: Rachel Bennett
   Communications: Sue Evans
   Head of Safety: Barry Evans
   Head of Legal: Michelle Davies
   Chief Information Officer: Guy Mason
   Safety Development Manager: Gordon Hunt

Other office
Doncaster CSDC
Lakeside Business Park, Carolina Way, Doncaster DN4 5PN
Tel: (+33 1302) 76 68 01

Political background
English Welsh & Scottish Railway (EWS) is the UK's largest rail freight company. Formed in 1996, the company was initially owned by a consortium led by US regional railroad Wisconsin

Central Transportation Corporation (WCTC), with financial partners Fay Richwhite & Co from New Zealand, Berkshire Partners from the USA and Goldman Sachs from the UK. In October 2001, the purchase of WCTC by Canadian National (CN) was completed. As a result, CN acquired WCTC's 42 per cent shareholding in EWS and nominated three representatives to join the company's board.

EWS operates freight, infrastructure, parcels and passenger charter trains throughout Great Britain, as well as freight services to and from mainland Europe via the Channel Tunnel. Around 8,000 trains are operated each week. It also provides traction and traincrew for scheduled passenger operators and locomotives as standby power. Business has additionally been developed hiring surplus or redundant locomotives for use on construction projects overseas.

While a private company, EWS operates in a regulated industry for which the UK government has broad objectives. The company therefore has dealings with the Office of Rail Regulation and the Strategic Rail Authority, as well as with the Scottish Executive and the Welsh Assembly.

In 2004 the company employed some 6,000 staff.

Organisation

EWS operations are managed from a purpose-built Customer Service Delivery Centre (CSDC) in Doncaster. Staffed by approximately 600 personnel, the CSDC is responsible for customer liaison, train planning, locomotive control, wagon management and performance analysis and improvement.

In July 2004 EWS announced the establishment of two independent subsidiary companies to provide vehicle acceptance and engineering consultancy services both in the UK and in the wider European market. Railway Approvals Ltd is based in Derby and specialises in approvals for new locomotives and rolling stock; Engineering Support Group Ltd provides a range of services including vehicle design, modification and project management.

In September 2001 EWS launched its Rail Industry Services sector. This brought together the previously separate Rail Services and Infrastructure Services units and is responsible for operating trains and providing resources for other railway industry companies and for providing works trains and hauling infrastructure materials for railway infrastructure and engineering companies.

EWS also provides contract maintenance services and facilities for other train operators (see *Manufacturers – Vehicle maintenance equipment and services*).

Passenger operations

Although principally a freight carrier, EWS holds a nationwide operating licence for passenger trains. This allows the company to run more than 1,000 charter trains annually, including steam-hauled trains for tour operators. The company also provides traction, and in some cases traincrew, for

Class 60 diesel locomotive at Cardiff with a train of steel slabs from Margam steelworks to the rolling plant at Llanwern, Newport (Ken Harris) 0583265

scheduled passenger operators, including Arriva Trains Northern, Arriva Trains Wales and ScotRail.

In July 2004 EWS announced that it had taken a 29.9 per cent shareholding in a joint venture company with Danish State Railways (DSB) to bid for the new InterCity East Coast passenger franchise in the UK, due to take effect in 2005.

Freight operations

Bulk freight carried by EWS chiefly comprised coal for electricity generation and industrial use, finished and semi-finished steel products, aggregates for the construction industry and petroleum products.

While there was a significant switch from coal to gas for electricity generation in the UK in the latter part of the last century, the volume of coal moved by rail has held up well. While coal is still carried from domestic deep mines and opencast sources, substantial volumes are also imported via ports which include Avonmouth, Clydeport and Immingham. Industrial users of coal are also supplied by EWS.

Despite a low level of UK road-building activity, EWS's aggregates business continues to develop, with around 16 million tonnes now carried annually.

Changes in UK steel production have had both positive and negative impacts on EWS steel traffic, but with strong world demand for steel business remained stable in 2004. The termination in 2001 of steel production at the Corus steelworks at

Llanwern, Newport, brought to an end the transport of iron ore to the plant from Port Talbot. In contrast, the continuation of rolling at the plant led to the commencement of long-distance flows of steel slabs to Llanwern from Lackenby, in northeast England.

EWS now handles only one surviving flow of iron ore, that between Immingham and Scunthorpe.

Other metals sector traffic includes flows of domestic scrap for steel-making and raw materials and finished products for the aluminium manufacturing industry.

In the petroleum market, rail is still extensively used for a whole range of products, including niche products such as LPG and bitumen. In 2001 the movement by rail of petroleum products from Grangemouth to destinations in Scotland resumed after a gap of eight years.

In addition to moving freight for industrial customers, EWS moves ballast, rail and other construction materials for railway infrastructure engineering work, handling around 80 per cent of Network Rail's business of this type. EWS has also been involved in all the recent major rail infrastructure projects, including the Channel Tunnel Rail Link and upgrade of the West Coast Main Line.

EWS continues actively to develop its Enterprise wagonload business, conveying commodities such as forest products, food and drink and Ministry of Defence stores. Enterprise caters not only for movements within the UK, but also into Europe via the Channel Tunnel.

The movement of cars and car components makes use of both block trains and the Enterprise network, as shippers react to the exacting requirements of this market. Recent successes in this market include contracts to move export versions of Jaguar's X-type model from Halewood (Liverpool) to Southampton and BMW Minis from Oxford to Purfleet, east London.

Royal Mail and parcels traffic

A long-term contract with Royal Mail for the movement of mail by rail was terminated early in 2004 when the two parties failed to reach agreement during renegotiations. However, EWS has successfully established daily high-speed parcels and packages services between Motherwell (Glasgow) and Inverness and the West Midlands and Aberdeen, in both cases offering transit times much better than those available by road. This activity received a boost in January 2004 when express and logistic company DHL awarded the company its first-ever contract entailing the use of rail.

Intermodal operations

EWS is now a major force in the operation of intermodal services in the UK. Channel

Class 66 diesel locomotive with a coal train formed of Type HTA wagons at Oxford with service bound for Didcot power station (Ken Harris) 0583264

Class 66 diesel locomotive piloting a Class 92 dual-voltage electric locomotive at the head of an intermodal service for Bari, Italy, at Paddock Wood (Ken Harris)   0583266

EWS Diesel locomotives

Class	Wheel arrangement	Power kW	Speed km/h	Weight tonnes	No in service	First built	Mechanical	Builders Engine	Transmission
08	C	261	24	49	95	1958	BR	EE 6KT	E EE
09	C	261	43	49	26	1958	BR	EE 6KT	E EE
37/0	Co-Co	1,305	129	105	4	1960	EE	EE 12CSVT	E EE
37/3	Co-Co	1,305	129	107	1	1960	EE	EE 12CSVT	E EE
37/4	Co-Co	1,305	129	107	14	1960	EE	EE 12CSVT	E EE
37/5	Co-Co	1,305	129	107	5	1960	EE	EE 12CSVT	E EE
37/7	Co-Co	1,305	129	119	6	1960	EE	EE 12CSVT	E EE
59/2	Co-Co	2,460	96	121	6	1994	General Motors	GM 645E3C	E GM
60	Co-Co	2,310	96	129	100	1989	Brush	Mirlees MB275T	E Brush
66	Co-Co	2,385	120	127	250	1998	General Motors	GM 710	E GM
67	Bo-Bo	2,385	200	90	30	1999	ALSTOM	GM 710	E GM

EWS Electric locomotives

Class	Wheel arrangement	Power kW continuous	Speed km/h	Weight tonnes	No in service	First built	Builders Mechanical	Electrical
90/0	Bo-Bo	3,360	177	84	25	1988	BREL	GEC
92	Co-Co	5,000	145	126	30	1996	Brush	ABB

Tunnel trains continue to be run under the EWS International banner, but there are increasing intermodal movements from key ports to inland terminals. These take place both as Intermodal Express services and as part of the Enterprise network, and are attracting new business sectors to rail, including white goods and food and drink.

Three of the UK's leading supermarket chains are now regular users of rail.

International traffic

International traffic handled by EWS via the Channel Tunnel includes cars and automotive products, newsprint, china clay and general wagonload freight, serving destinations in Belgium, France, Germany, Italy and Spain. Dedicated intermodal services are operated between the several terminals in the UK and Belgium (Muizen), France (Metz, Paris and Perpignan) and Italy (Bari, Milan, Novara and Turin). After suffering considerable disruption from the efforts of asylum seekers wishing to enter the UK from France, business has recovered: in 2003, the first full year since closure of a controversial asylum seekers' camp at Sangatte, Calais, traffic via the Channel Tunnel increased by 22 per cent over the previous year.

Traction and rolling stock

For its core UK operations, in mid-2004 EWS owned an operational fleet of 416 main line diesel, 97 diesel shunting and 55 electric locomotives. In addition, EWS operates Mendip Rail's fleet of eight privately owned Class 59/0 and 59/1 diesel locomotives used on aggregates trains, plus many privately owned wagons.

The bulk of the diesel locomotive fleet is 250 Class 66 (Model JT42CWR) diesel-electric locomotives built by General Motors between 1998 and 2000. EWS also owns 30 Class 67 high-speed diesel-electric locomotives acquired originally for its mail and parcels businesses but now deployed on a variety of other duties.

The electric locomotive fleet includes 30 Class 92 dual-voltage machines which handle some West Coast Main Line services using a 25 kV AC power supply, and traffic to and through the Channel Tunnel using third rail 750 V DC and 25 kV AC supplies respectively.

EWS has also developed a business supplying for export purposes locomotives surplus to its current requirements. The latest contract, signed in 2004 with French company Fertis, covered the provision for two years of 26 Class 56 and 14 Class 58 2,460 kW locomotives for use on the construction of the LGV Est high-speed line in eastern France. This followed earlier contracts supplying 14 Class 37/7 diesel locomotives to Spain for construction trains in connection with the high-speed line between Barcelona and Madrid and 42 Class 37 locomotives for use during construction of the LGV Méditerranée high-speed line in France. Two redundant Class 58s have also been supplied to Netherlands-based open access freight operator ACTS and other examples of the same type have gone to Spain for use on high-speed line construction projects.

The wagon fleet has been substantially modernised by the acquisition of 2,500 new wagons from Thrall Car. These include 1,145 high-capacity coal wagons, as well as steel carriers, container wagons and open box wagons.

Eurostar (UK) Ltd

Eurostar House, Waterloo Station, London SE1 8SE
Tel: (+44 20) 79 22 61 80   Fax: (+44 20) 79 22 44 24
Web: www.eurostar.com

Key personnel

Chairman: Guillaume Pepy
Chief Executive: Richard Brown
Chief Operating Officer: Jacques Damas
Directors:
  Financial: Ian Nunn
  Commercial: Nick Mercer
  Customer Service: Nicolas Petrovic
  Communications: Paul Charles
  Human Resources: Marc Noaro
  Corporate and Legal Services: Victoria Wilson

Background

Eurostar (UK) Ltd is responsible, in conjunction with the Belgian and French national railways, for the operation of international high-speed passenger services between London, Paris and Brussels via the Channel Tunnel. The company has a shareholding of 32.5 per cent in the Eurostar Group. It was previously known as European Passenger Services Ltd (EPSL).

Eurostar journey times were cut by 20 minutes following the opening in September 2003 of the first phase of the UK's Channel Tunnel Rail Link, seen here where the line crosses the River Medway (Eurostar UK)   0558938

Eurostar electric trainsets

Class	Cars per half-set	Line voltage	Motor cars per unit	Motored axles/car	Output/ motor kW	Speed km/h	No in service	First built	Builders Mechanical	Electrical
373/0	10	25 kV AC 750 V DC 3 kV DC	2	4/2	100	300	22[1]	1994	GEC	GEC/Brush
373/3	8	25 kV AC 750 V DC 3 kV DC	2	4/2	100	300	14[2]	1995	GEC	GEC/Brush

[1] Eurostar (UK) allocation of half-trains for London—Paris/Brussels services. A further 32 half-trains are allocated to France and 8 to Belgium.
[2] Five units for a hire pool for GNER; remainder stored.

In May 1994 EPSL was transferred out of British Rail ownership and became a company owned directly by the government. This was in preparation for the transfer of EPSL to the private sector consortium chosen to build the Channel Tunnel Rail Link, a new £3 billion, 108 km high-speed line from London to the tunnel. Assets worth some £800 million would be transferred, comprising the British share of the fleet of Eurostar trains, Waterloo International station in central London and the North Pole Eurostar maintenance depot in west London.

In June 1994 four private-sector consortia were invited to bid to build the line on a build-and-operate basis: the winning consortium was London & Continental Railways (LCR). LCR was awarded the assets of Eurostar and there was also to have been a cash injection of about £1.4 billion from the government. The government also wrote off £1.3 billion of debt in Eurostar.

LCR took control of EPSL in May 1996 and in October changed the company's name to Eurostar (UK) Ltd. However, in January 1998, LCR's bid to construct the CTRL failed after the government refused additional state funding for the scheme to make good a shortfall in forecast revenues from Eurostar operations. As a result, the operation of Eurostar services passed from LCR, which remains an umbrella company for the construction of the CTRL, to Inter Capital and Regional Rail Ltd, a consortium formed by National Express Group (40 per cent), SNCF (35 per cent), SNCB (15 per cent) and British Airways (10 per cent). The consortium is to operate Eurostar under a management contract valid until 2010. Continuing government subsidy will be required for the immediate future to cover a revenue shortfall against costs.

Passenger operations
Eurostar's principal routes are between London and Paris (up to 16 trains a day in January 2005) and London and Brussels (up to nine trains a day in January 2005). Services on these routes were first introduced in November 1994, using Waterloo International as the London terminal. Some services stop at the intermediate stations at Ashford, Calais Fréthun and Lille. The fastest London–Paris train takes 2 hours 35 minutes, the fastest London–Brussels trip 2 hours 15 minutes.

Since 1994 Eurostar has established itself as the leading carrier from London to Paris and Brussels and has the largest share of the air/rail market on its core routes.

A direct daily train from Waterloo serves Marne-la-Vallée on the LGV Jonction (Paris bypass line), for the nearby Disney complex. Seasonal skiing trains to the French Alps were introduced in 1997, with three Eurostar trains adapted to take 1.5 kV DC supplies from the overhead in the French regions.

In July 2002 Eurostar launched a summer Saturday direct London–Avignon service.

In 2000 plans to introduce services between the Continent and destinations north of London were abandoned as no case could be put forward for operating these on a commercially viable basis. As a consequence, some examples of the Eurostar (UK) Class 373/3 fleet were leased during 2000 to Great North Eastern Railway for domestic traffic.

The opening of Stage 1 of the high-speed Channel Tunnel Rail Link in September 2003 cut journey times from London to the tunnel by 20 minutes. The opening of the second stage in 2007 will further reduce journey times, as well as improving prospects for effective links to Network Rail's East and West Coast Main Lines to serve destinations north of London (see entry for London & Continental Railways) thanks to the use of St Pancras as the new London terminal for Eurostar services. In 2007 the fastest London–Paris journey time will be 2 hours 15 minutes and London–Brussels 1 hour 53 minutes.

Eurostar (UK) ticket sales and seat reservations are handled by the company's ELGAR distribution system, which is accessible to some 4,000 UK travel agents and has links with Galileo (UK), Sabre, Amadeus and Worldspan global distribution systems. The system also has a link to the Association of Train Operating Companies (ATOC) ticket sales system, providing travel industry access to the UK's 25 train operating companies.

Traction and rolling stock
For the London–Paris/Brussels 'Three Capitals' service via the Channel Tunnel, 31 300 km/h Eurostar trainsets were built by GEC Alsthom.

Each Three Capitals train consists of 18 trailers, articulated in two separable rakes and with a 68 tonne power car at each end of the train. The power car has to accommodate three voltages (25 kV on SNCF, 3 kV on SNCB and 750 V in the UK) but within a 17 tonne axleload maximum. Nine sets have been modified to operate from SNCF's 1.5 kV DC system to allow the trains to work through France to Alpine ski resorts.

Each air-conditioned trainset includes first- and standard-class accommodation, two bar-buffet cars and facilities for nursing mothers. Each set is arranged as two identical halves of self-contained power cars and nine trailer halves, easily separable from each other in case of emergency in the Channel Tunnel.

A complete interior refurbishment of the fleet commenced in September 2004.

For the once-intended through services between Paris and the British provinces, a 14-trailer format is used. The failure to implement these services has enabled Eurostar (UK) to lease three sets daily (from a pool of five) to Great North Eastern Railway for use on its London King's Cross–Leeds domestic route. Three of the five units have been repainted in GNER livery.

Eurostar (UK) Ltd's share of the Eurostar trainset fleet is 11 18-trailer and all seven 14-trailer units. Of the remaining 18-trailer sets the SNCF's share is 16 and the SNCB's four. (Details in the table refer to the half trains into which the Eurostar units are separable.) Due to spare capacity in the fleet, some of the sets assigned to SNCF are used on domestic services in France.

Eurostar (UK) retains three Class 37/6 diesel locomotives for emergency standby purposes and general hire. In 2000 the company also put up for sale seven Class 92 dual-voltage electric locomotives originally procured for planned overnight train services.

The traction fleet also includes two Class 73/1 locomotives, adapted with special drawgear for trainset stock movements, and one Class 08 shunter.

UPDATED

FirstGroup plc

Block E, 3rd Floor, Macmillan House, Paddington Station, London W2 1FG
Tel: (+44 20) 72 91 05 05   Fax: (+44 20) 74 36 33 37
Web: http://www.firstgroup.com

Operational headquarters
395 King Street, Aberdeen AB24 5RP
Tel: (+44 1224) 65 01 00   Fax: (+44 1224) 65 01 40

Key personnel
Chief Executive: Moir Lockhead
Chief Operating Officer: Dr Mike Mitchell
Managing Director, Railways: Dean Finch
Managing Director, Buses: David Leeder

Political background
Bus operator FirstBus took advantage of the 1993 Railways Act, winning the Great Eastern Railway franchise and securing a stake in Great Western Holdings, which won the Great Western Trains and North Western Trains franchises. In March 1998, having changed its name to FirstGroup plc, the company acquired the remaining 74.7 per cent shareholding in GWH. Subsequently the company's three franchises were renamed First Great Eastern, First Great Western and First North Western.

In September 2003 the Strategic Rail Authority signed a contract with FGK, a joint venture of FirstGroup plc and Keolis SA, to operate a new Trans-Pennine franchise in northern England. This is formed of portions of the Arriva Trains Northern and First North Western franchises. Operations commenced in early 2004. The franchise runs for eight years with an option to extend it for up to five years.

In November 2003 FirstGroup was named preferred bidder for the Thames Trains franchise. This runs from 1 April 2004 until 31 March 2006, when the franchise territory will be incorporated into the so-called 'Greater Western' franchise which will see existing Great Western, former Thames Trains and Wessex Trains businesses combined into a single franchise. FirstGroup renamed the operation First Great Western Link.

FirstGroup also operates the Tramlink light rail system in Croydon.

In July 2003 it was announced that FirstGroup had acquired GB Railways plc (see entry in UK section of *Railway systems and operators*), subject to shareholder approval. GB Railways held the Anglia Railways franchise, an 80 per cent shareholding in Hull Trains and established GB Railfreight, an open access freight operator. However, the Anglia Railways franchise, restructured also to incorporate the former First Great Eastern and the West Anglia portion of the West Anglia Great Northern franchise, was lost in 2004 to National Express Group.

In July 2004 FirstGroup was awarded a seven-year contract to operate the ScotRail franchise.

First Great Western

First Great Western Trains Ltd
Milford House, 1 Milford Street, Swindon SN1 1HL
Tel: (+44 1793) 49 94 00
Fax: (+44 1793) 49 94 51
Web: www.firstgreatwestern.co.uk

Key personnel
Managing Director: Alison Forster
Commercial Director:
  Glenda Lamont
Finance Director: Ben Caswell
Engineering Director:
  Graham Boot-Handford

Political background
Under the arrangements put in place by the 1993 Railways Act, Great Western Holdings took over the operation of intercity services out of London Paddington terminus in February 1996. GWH was formed as a joint venture between the former British Rail management on the line, the FirstBus

FGW HST forming a London Paddington–Swansea service near Severn Tunnel Junction (Ken Harris)
***NEW**/1114979*

company and the 3i investment company. In March 1998 FirstGroup (renamed from FirstBus) acquired the stakes owned by GWH management and investment companies to become sole shareholder. Subsequently the company was renamed First Great Western (FGW).

Following a commitment by GWH (now FGW) to invest £36 million in new rolling stock, the length of the franchise was extended to 10 years, ending in March 2006. While a subsidy has been paid for the first nine years of the contract, for the final 11-month period a premium payment is to be made.

On expiry of the existing contract in 2006 the routes served by the franchise are to be merged with those of First Great Western Link and Wessex Trains to create a 'Greater Western' franchise.

Passenger operations
FGW operates high-speed services on a 1,368 km route network out of London Paddington to South Wales and the west of England. It also operates a sleeper service out of London Paddington to Devon and Cornwall.

FGW increased the frequency on the London-Bristol route to half-hourly in the summer 1999 timetable and on the London-Cardiff route at the introduction of the summer 2001 timetable.

Semi-fast services using Class 180 dmus were introduced between London Paddington and Exeter via Newbury in 2004.

In 2005 FGW employed some 2,700 staff.

Traffic (million)	2000–01	2001–02	2002–03
Passenger journeys	18.6	19.2	20.2
Passenger-km	2,401.0	2,428	2,556
Train-km	16.8	18.7	16.2

Traction and rolling stock
FGW leases 39 High Speed Train (HST) sets from Angel Trains and five from Porterbrook. It also leases 11 sleeping cars. In 2001 FGW leased from Porterbrook a prototype Class 57/6 locomotive, created by refurbishing a Class 47/8 machine and equipping it with a General Motors 645-series engine. While the prototype was returned to the leasing company, four similar machines also designated Class 57/6 were acquired in 2004 for London Paddington–Penzance sleeper services. In addition, FGW operates 12 Class 08 shunting locomotives.

The entire Angel-owned HST fleet has been refurbished, while the Class 43 power cars have been the subject of a programme of reliability improvements.

First examples of a fleet of 14 Class 180/1 'Adelante' five-car 200 km/h dmus entered regular service in 2002. Subsequently up to 10 of these have been transferred to a sister operator, First Great Western Link, for services between London and Oxford and the line to Worcester.

UPDATED

First Great Western Link

Venture House, 37-43 Blagrave Street, Reading RG1 1PZ
Tel: (+44 118) 908 36 78   Fax: (+44 118) 957 96 48
Web: http://www.firstgreatwestern.co.uk

Key personnel
Managing Director: Alison Forster
Directors
    Commercial Services: Glenda Lamont
    Finance: Ben Caswell
    Operations: Kevin Gale
    Human Resources: John Nolan
    Engineering: Graham Boot-Handford

Political background
Under the arrangements put in place by the 1993 Railways Act, the franchise to run operations on the former Thames Trains network was let in October 1996 to Victory Railways Holdings Ltd, a joint venture between bus operator The Go-Ahead Group plc and the management of Thames Trains. In early 1998, The Go-Ahead Group acquired the 34.8 per cent shareholding in Victory Railways by management and staff to take full control of the company.

The franchise was let for a term of seven years and six months. Victory Railways was to receive from the Franchising Director support, in 1996 prices, of £33.2 million in the first full financial year of the franchise, declining to zero in the last year of the franchise.

In November 2003 the Strategic Rail Authority announced that FirstGroup plc had been selected bidder to take over the franchise after expiry of the existing contract in March 2004. The new franchise would run only until March 2006, when the Thames Trains franchise would be combined with those of Great Western and Wessex Trains to form the so-called 'Greater Western' franchise. On taking over the franchise, FirstGroup renamed it First Great Western Link.

In 2003 the company employed 1,009 staff.

Passenger operations
FGW Link operates suburban services out of London's Paddington station along the Thames valley to Reading and Oxford. The company also runs services between Oxford and Hereford, Oxford and Stratford-upon-Avon, Reading and

FGW Link diesel multiple-units

Class	Cars per unit	Motor cars per unit	Motored axles/car	Power/ motor kW	Speed km/h	Units in service	First built	Mechanical	Builders Engine	Transmission
165/0	2	2	2	260	120	5	1990	BREL	Perkins 2006-TWH	H Voith T211r
165/1	2/3	2/3	2	260	145	36	1992	ABB	Perkins 2006-TWH	H Voith T211r
166	3	3	2	260	145	21	1993	ABB	Perkins 2006-TWH	H Voith T211r

Former Thames Trains Class 166 dmu, now operated by FGW Link, forming an Oxford-bound service near Reading (Ken Harris)
0576721

Basingstoke and Reading and Gatwick Airport. Total route-km operated is 581.

A major recast of services is due to take place in December 2004. This will see Class 180 Adelante dmus, currently operated by First Great Western, introduced on FGW Link services, along with other improvements aimed at significantly increasing capacity.

Traffic (million)	2000–01	2001–02	2002–03
Passenger journeys	36.4	36.5	37.3
Passenger-km	1,012	1,007	1,020
Train-km	13.8	14.0	15.1

Traction and rolling stock
FGW Link uses modern 'Turbo' diesel units built by Adtranz. The fleet consists of Class 165 and 166 dmus leased from Angel Train Contracts. All have been subjected to a refurbishment programme which began in 2000.

First North Western

Background
The First North Western franchise expired in December 2004, much of it becoming part of the Northern Rail franchise, other parts of its network having been transferred to Arriva Trains Wales in 2003 and to the TransPennine franchise in February 2004.

UPDATED

First ScotRail

Caledonian Chambers, 87 Union Street, Glasgow G1 3TA
Tel: (+44 141) 332 98 11   Fax: (+44 141) 335 47 91
e-mail: enquiries@scotrail.co.uk
Web: www.scotrail.co.uk

Key personnel
Managing Director: Mary Dickson
Finance Director: Kenny McPhail
Operations Director: Steve Montgomery
Commercial Director: Gordon Dewar
Engineering Director: Andy Mellors
Human Resources Director: Donald Macpherson
Head of Business Planning: Gerard O'Hanlon
Head of Revenue Protection and Train Planning:
    Jerry Farquharson
Head of Contracts: Mike Price
Deputy Operations Director: Andy Thomas
Head of Marketing: Ellie Newlands
Divisional Operations Manager, Strathclyde
    Services: Jacqui Dey
Sleeper Services Manager: Billy Black
Franchise Partnership Manager: Alan Scott
Customer Services Manager: Pamela Ballantyne
External Relations Manager: John Yellowlees
Communications Manager: Carol Harris

Political background
National Express Group commenced a seven-year franchise to operate ScotRail on 1 April 1997. Responsibility for funding the franchise was transferred to the Scottish Executive from 1 April 2001. In June 2002 the Executive issued Directions and Guidance to the SRA for the re-letting of the

Class 156 dmus in ScotRail (left) and SPT liveries at Glasgow Queen Street   0543288

ScotRail franchise. In October 2003 bids were submitted to the SRA acting as agent for the Scottish Executive. The original franchise expired on 31 March 2004 but a short extension was implemented to mobilise its replacement, which runs for seven years with the option of a three-year extension. In July 2004 FirstGroup was awarded this contract, taking over operations in October of that year.

Passenger operations
First ScotRail operates services on a 3,043 km network. These include: interurban services linking

ALSTOM-built Class 334 'Juniper' emus at Glasgow central forming an SPT-supported service to Ayr (Colin Boocock)   0137115

the five Scottish cities; suburban services around Edinburgh and Glasgow, where the services supported by Strathclyde Passenger Transport comprise the largest suburban network in Great Britain outside London; rural services in Dumfries and Galloway and the West and North Highlands; and sleeper services linking Scotland with London. The company operates 336 stations and serves 354.

A commitment under the original franchise to increase the frequency of services between Edinburgh and Glasgow from half-hourly to 15-minute was implemented six months early, in September 1999. Daytime services were also doubled between Edinburgh and Fife, with an increase in peak-hour capacity funded by Fife Council. An additional 130,000 annual off-peak train-km have been provided on the Glasgow North Electrics network for SPT. New services introduced over and above franchise requirements include: a Tain–Inverness commuter service; a new direct link between Cumbernauld and Falkirk; a new commuter service between Carstairs and Edinburgh; and a new Edinburgh Crossrail route, which opened in June 2002.

A twice-hourly service to Lanark introduced for SPT in January 2003 has some trains routed via the Holytown–Wishaw line, which had not previously seen a passenger service. With support from the Highland Rail Partnership and SPT, the season in 2003 was extended for additional trains which run on summer Saturdays and Sundays on the West Highland lines. A Sunday service between Glasgow and Cumbernauld was introduced in May 2003.

Three new stations opened in 2002 at Beauly, Brunstane and Newcraighall, while Edinburgh Park was commissioned in 2003, taking the total operated by ScotRail to 336. The next new station, at Gartcosh on the SPT network, is to be opened in 2004. Work on reopening the Larkhall–Milngavie route for SPT is due for completion in October 2005. Provision of services on this line is part of the FirstGroup franchise commitment. As an interim

For details of the latest updates to *Jane's World Railways* online and to discover the additional information available exclusively to online subscribers please visit
jwr.janes.com

SPT-liveried Class 170/4 dmu at Glasgow Queen Street station forming a service to Stirling (Colin Boocock) 0570695

First ScotRail diesel multiple-units

Class	Cars per unit	Motor cars per unit	Motored axles/car	Power/ motor kW	Speed km/h	Cars in service	First built	Mechanical	Builders Engine	Transmission
150/2	2	2	2	210	120	24	1986	BREL	Cummins NT855R5	HM Voith T211r
156	2	2	2	210	120	96	1987	Met-Cam	Cummins NT855R5	HM Voith T211r
158/0	2	2	2	275/300	145	80	1989	BREL	Perkins 2006-TWH	HM Voith T211r
170/4	3	3	2	315	160	78*	1999	Adtranz	MTU 6R 183TD13H	H Voith T211rzze

* Plus 87 under delivery or on order in 2004.

First ScotRail electric multiple-units

Class	Cars per unit	Motor cars per unit	Motored axles/car	Power/ motor kW	Speed km/h	Cars in service	First built	Builders Mechanical	Electrical
314	3	2	2	82	120	48	1976	BREL	GEC/Brush
318	3	1	4	268	145	63	1985	BREL	Brush
320	3	1	4	268	120	66	1990	BREL	Brush
334	3	2	2	270	160	120	1999	ALSTOM	ALSTOM

measure SPT plans to introduce an additional Hamilton–Anderston half-hourly service from December 2004. Plans to reopen the Stirling–Alloa–Kincardine line will enable First ScotRail to extend the Glasgow–Stirling service to Alloa. Funding has been awarded by the Scottish Executive to Highland Council for the 'Invernet' package of local services linking Kingussie and Tain with Inverness, commencing in the second quarter of 2005.

The new franchise which started in October 2004 also provides for an extension of Inverness–Aberdeen services to Stonehaven. Other new routes under development are: Newcraighall–Tweedbank (by the Waverley Railway Partnership); Bathgate–Airdrie; and rail links to Glasgow and Edinburgh Airports.

By early 2004 a total of 126 stations on the SPT-sponsored network and 33 in the east and northeast of Scotland were linked by 24-hour online CCTV, help points and long-line public address to customer service centres at Paisley and Dunfermline respectively. A further seven and 18 stations respectively were soon to be added.

Possible additional stations on existing routes are foreseen at Conon Bridge, Gartcosh and Laurencekirk.

Traffic (million)	2000–01	2001–02	2002–03
Passenger journeys	63.2	60.7	57.4
Passenger-km	1,928	1,969	1,915
Train-km	35	37	34.9

Traction and rolling stock

First ScotRail operates only multiple-unit trains, save for on its sleeper services, where Class 90 AC electric locomotives and Class 67 and 37 diesels are hired in from EWS to haul the Mk 3 sleeper vehicles on Anglo-Scottish and internal Scottish legs respectively, and on the Edinburgh–North Berwick route, where Class 90 push-pull formations were introduced in 2004 as a stop-gap measure pending transfer of dmus from elsewhere in the UK.

New trains procured by NEG for ScotRail were of two types: 40 Class 334 'Juniper' three-car emus from ALSTOM, all of which were in service by February 2003; and 26 Class 170/4 'Turbostar' three-car dmus from Adtranz (now Bombardier), 24 of these for the interurban network and two for SPT as compensation for late delivery of Class 334 emus.

The Scottish Executive has funded the procurement of 29 new Turbostar-style three-car trains, the first of which entered service during 2004. Seven of these are for routes supported by SPT. Eight sets are be used to strengthen Edinburgh–Glasgow services, and 12 are to be formed as six-car sets for East of Scotland routes into Edinburgh, replacing the operator's 12 Class 150/2 units, which would be returned off-lease. The final two units were to be used on interurban services. Finance is provided by Porterbrook. Four additional Turbostars are to be transferred to First ScotRail in April 2005 to replace Class 90 operations on the North Berwick line. Additional maintenance facilities for the enlarged fleet are being created by reconstructing and reopening Glasgow Eastfield depot.

A refurbishment programme has been undertaken for the Class 320 emus. ScotRail also secured funding from the SRA's Rail Performance Fund towards a £2.2 million programme for the renewal of certain subsystems and components on its Class 158 fleet. This was completed in 2003. In the same year maintenance of 20 of these dmus used on Highland services was transferred from Edinburgh Haymarket depot to Inverness.

The sleeper fleet, which comprises 53 sleeping cars, 10 lounge cars and 11 club/brake cars, have been the subject of a refurbishment programme which included provision of wheelchair-accessible cabins and the reinstatement of seated accommodation, a facility withdrawn by the former British Rail.

UPDATED

First TransPennine Express

Floor 7, Bridgewater House, 60 Whitworth Street, Manchester M1 6LT
Fax: (+44 161) 228 81 00
Web: www.tpexpress.co.uk

Key personnel

Managing Director: Vernon Barker
Directors
  Commercial: Hugh Clancy
  Operations: Danny Fox
  Fleet: Kenny Scott
  New Trains: Nick Donovan
  Customer Services: Edith Rogers

Background

The TransPennine Express franchise was established by the Strategic Rail Authority to operate passenger services on principal intercity routes in the north of England. It incorporates routes that latterly formed parts of the Arriva Trains Northern and First North Western franchises, serving a 977 route-km 30-station network that includes Manchester, Lancaster, Liverpool and Preston in the west, the Yorkshire communities of Huddersfield,

Class 185 driving car awaiting its final livery at Siemens' Krefeld plant (David Haydock)
NEW/1122852

Leeds and Sheffield, and Middlesbrough, Newcastle and York in the east.

In September 2003 the SRA awarded an eight-year franchise, with an option to extend for up to five further years, to FGK, a consortium of FirstGroup plc and Keolis SA. Operated as First TransPennine Express, the contract commenced in February 2004. FGK is to receive a subsidy of £637 million over the eight-year contract period, commencing with £28.6 million in 2003–04 and rising to £55.8 million in 2011–12.

FM Rail Ltd

RTC Business Park, London Road, Derby DE24 8YB
Tel: (+44 1332) 33 22 22   Fax: (+44 1332) 36 88 88
e-mail: enquiries@fmrail.com
Web: www.fmrail.com

Key personnel
Chairman: Martin Sargent
Deputy Chairman: Rick Edmonson
Managing Director: Bob Gordon
Executive Directors: Roger Bulmer, Andy Lynch
Operations Director: Richard Clark
Non-Executive Director: Marcus Robertson

Background
FM Rail is an independent train operating company established in January 2005 following a merger

Freightliner

Freightliner Group Ltd
3rd Floor, The Podium, 1 Eversholt Street, London NW1 2FL
Tel: (+44 20) 72 00 39 00   Fax: (+44 20) 73 88 64 54
Web: www.freightliner.co.uk

Key personnel
Freightliner Group Ltd
  Chairman: Norman Broadhurst
  Chief Executive: Eddie Fitzsimmons
  Finance Director: Douglas Downie
  Director of Strategy and Company Secretary:
    Robert Goundry
  Head of Safety: Paul Higgins
  Chief Financial Accountant: Paul Naicker
Freightliner Intermodal
  Managing Director: Peter Maybury
  Production Director: John Smith
  Finance Director: Russell Mears
Freightliner Heavy Haul
  Managing Director: Adam Cunliffe
  Director of Coal: Martin Wilkes
  Director of Business Development: Paul Smart
  Director, Contracts: Tony Pritchard

Organisation
Formerly the division of British Rail running deep-sea container services from ports to inland terminals, Freightliner was privatised in 1996 and acquired by a management/staff-led consortium, Management Bid Consortium Ltd, that also included venture capitalists. Since then the company has progessively grown its container services and expanded substantially into the heavy-haul sector of the market, creating Freightliner Heavy Haul Ltd in 1999 to manage the latter business. In 2003 the group employed over 1,400 staff.

Freight operations
Freightliner Intermodal maritime container services link the UK's five major deep-sea ports with seven main inland terminals. Services are also provided at the ports of Belfast and Dublin, Ireland and at six other terminals. The company handles around 25 per cent of maritime containers arriving in or leaving the UK.

Freightliner Heavy Haul is active nationwide in the aggregates, automotive, cement, coal, domestic waste and petroleum products sectors. Over 200 loaded coal trains are operated each week. It is also a major provider of traction for infrastructure trains for Network Rail and has secured contracts from Union Railways (North) to provide haulage of Channel Tunnel Rail Link construction trains, following similar work on the

Passenger operations
Traffic (million)	2004–05
Passenger journeys	14.6
Passenger-km	776.0
Train-km	13.5

Traction and rolling stock
In September 2003 FGK ordered from Siemens Transportation Systems 56 three-car Cummins-powered Class 185 Desiro UK dmus to replace existing equipment used on trans-Pennine routes.

of Fragonset Railways, which provides train operating, traction hire, traction engineering and parts supply services, and Merlin Rail, a provider of train operating, train planning and traincrew. Train operations and capabilities and traction are provided in support of franchise-holding train operators, freight operators and charter train companies. Extensive overhaul and engineering facilities are located at Derby.

Passenger and freight operations
Operations in early 2005 included provision of standby traction and rolling stock for Wessex Trains and the hire of locomotives to GB Railfreight. The company also undertakes rolling stock movements, operation and haulage of inspection trains and train planning, traincrew and traction provision for charter services.

The order was subsequently cut to 51 trains. Delivery is scheduled for 2006. Finance for the new trains is being provided by HSBC Rail. Until their arrival services are operated with 11 Class 175 and 44 Class 158 dmus (23 three-car and 21 two-car). The Class 175 units are sub-leased from Arriva Trains Wales.

New maintenance depots are to be built at Manchester and York, while an existing facility at Cleethorpes is also to be upgraded.

UPDATED

Traction and rolling stock
In early 2005 the operational FM Rail diesel locomotive fleet comprised: one Class 08 shunter; nine Class 31; two Class 33; one Class 45; six Class 47. There was also one Class 73 electro-diesel. The company possesses some 50 additional locomotives which could be returned to operational standards. FM Rail also owns a fleet of around 60 passenger coaches, as well as other vehicles and infrastructure equipment.

NEW ENTRY

Class 66/5 diesel locomotive with a container service at Southampton (Ken Harris)   *NEW*/0585073

Diesel locomotives

Class	Wheel arrangement	Power kW	Speed km/h	Weight tonnes	No in service	First built	Mechanical	Builders Engine	Transmission
08	C	298	24	50	9	1955	BR	EE 6KT	E EE
47	Co-Co	1,550	120	120	7	1963	BR, Brush	Sulzer 12LDA28C	E Brush
57	Co-Co	1,507	120	121	12	1997*	BR, Brush	GM 645-12E3	E Brush
66/5	Co-Co	2,460	120	64	76**	2000	GM	GM 12N-710G3B	E GM
66/6	Co-Co	2,460	105	121	18**	2000	GM	GM 12N-710G3B	E GM
66/9	Co-Co	2,460	120	–	2	2004	GM	GM 12N-710	E GM

* Rebuilt and re-engined by Brush.
** Including examples on order.

Electric locomotives

Class	Wheel arrangement	Power kW continuous	Speed km/h	Weight tonnes	No in service	First built	Builders Mechanical	Electrical
86	Bo-Bo	2,680	120	83	19	1965	BR	AEI
90	Bo-Bo	3,730	120	84.5	11	1987	BREL	GEC

first section of the line. More than 1,000 trains per week are operated.

Traction and rolling stock
In early 2005 Freightliner's diesel locomotive fleet totalled 115 main line units, assigned to dedicated pools for each of its two businesses, and nine

shunters. There were also 30 electric locomotives, all used by Freightliner Intermodal. The bulk of the diesel fleet is formed of General Motors Class 66 locomotives, with examples leased from both HSBC Rail (UK) and Porterbrook. The 12 Class 57 machines are former Class 47s re-engined with General Motors power units and used by Freightliner Intermodal.

The wagon fleet totalled over 1,750 vehicles. Recent acquisitions include up to 400 semipermanently coupled Type FEA container flats from Greenbrier, while some 700 older intermodal vehicles have been refurbished by ALSTOM. For its coal haulage contracts, Freightliner has acquired 350 bogie hopper wagons from Greenbrier. These are leased to the company by Porterbrook Leasing. Greenbrier has also supplied box wagons for aggregates traffic and Marcroft has delivered 44 Type FRA flat wagons for containerised domestic waste.

In January 2002 Freightliner Heavy Haul launched a new type of car-carrying wagon, the Autoflat, for its Autoliner business. The wagon is intended to provide a means of transporting larger cars within the restrictive British loading gauge.

The Freightliner Intermodal road vehicle fleet includes around 180 tractor units and 500 trailers.

UPDATED

Class 66/6 locomotive with an empty petroleum products train from the Midlands to Humber (John C Baker) **NEW**/0585074

Gatwick Express Ltd

Gatwick Express Ltd
52 Grosvenor Gardens, London SW1W 0AU
Tel: (+44 20) 79 73 50 00   Fax: (+44 20) 79 73 50 48
Web: www:gatwickexpress.com

Key personnel
Managing Director: David Stretch
Head of Operations: Dick Whitwell

Political background
The Gatwick Express franchise was won by the National Express Group in 1996 and was the only one let by the Franchising Director that did not require a subsidy from the outset. The franchisee paid a premium (in 1996 prices) of GBP4.6 million in the first year, increasing to GBP22.6 million in 2010–11.

Passenger operations
Gatwick Express operates a shuttle service over the 43 km route between Gatwick Airport (London's second largest – Heathrow is the biggest) and the terminus at Victoria in central London. The service works on a quarter-hourly frequency throughout the day, and hourly throughout the night. The trip takes 30 minutes.

In April 2000 Gatwick Express and Heathrow Express (see entry in UK section of *Railway systems and operators*) formed a marketing alliance, Airport Express, to promote the two companies' services to the travel industry. In December 2000 Stansted Express, part of the West Anglia Great Northern franchise (see entry in UK section of *Railway systems and operators*) joined the alliance.

Traffic (million)	2002–03	2003–04	2004–05
Passenger journeys	4.2	4.5	4.7
Passenger-km	184.0	197.9	227.2
Train-km	2.1	2.55	2.6

Gatwick Express Class 460 trainset approaching Gatwick Airport station (Ken Harris) **NEW**/1122864

Gatwick Express electric multiple-units

Class	Cars per unit	Motor cars per unit	Motored axles/car	Power / motor kW	Speed km/h	Units in service	First built	Builders Mechanical	Electrical
460	8	4	2	270	160	8	2000	Alstom	Alstom

Traction and rolling stock
Gatwick Express operations are undertaken with a fleet of eight eight-car Class 460 750 V DC third rail electric trainsets supplied by ALSTOM to its Juniper design and leased from Porterbrook.

One of the Class 73/2 electro-diesel locomotives formerly employed is retained for depot shunting and standby purposes.

UPDATED

GB Railfreight

15-25 Artillery Lane, London E1 7HA
Tel: (+44 20) 79 04 33 94   Fax: (+44 20) 73 75 25 94
Web: http://www.gbrailfreight.com

Key personnel
Managing Director: John Smith
Executive Director: Max D Steinkopf
Finance Director: David Simons
Commercial Director: Tim Robinson
Business Development Director: Philip McGrath
Fleet Engineer: Mark Jordan

Organisation
GB Railfreight was established as an open access freight-operating subsidiary of the former GB Railways Group plc. The company obtained its train-operating licence in mid-2000, shortly before securing an eight-year contract to haul infrastructure materials and maintenance trains for Railtrack (now Network Rail) to and from

GB Railfreight Class 66/7 locomotive with a Hams Hall (West Midlands)–Felixstowe container train (John C Baker) 0536191

work sites in the east of England. The Railtrack contract commenced on 31 March 2001. In 2002 the company secured contracts for general freight and container flows from Felixstowe to Selby and the West Midlands respectively and in 2003 won contracts covering the movement of containerised gypsum from an East Midlands power station to several processing facilities. Subsequent contracts cover movements of consumer products for a major supermarket chain and further infrastructure services in southern England.

GB Railways

GB Railways Group plc
15-25 Artillery Lane, London E1 7HA
Tel: (+44 20) 74 65 90 13   Fax: (+44 20) 73 75 35 94
e-mail: info@gbrailways.com
Web: http://www.gbrailways.com

Key personnel
Chairman: Lord Sheppard of Didgemere
Deputy Chairman and Chief Executive:
    Jeremy Long
Directors: Michael Schabas; Max D Steinkopf

Background
GB Railways was established to bid for rail franchises let under the 1993 Railways Act,

In 2003 GB Railways was acquired by FirstGroup plc (see entry in UK section of *Railway systems and operators*).

Traction and rolling stock
GB Railfreight operates 17 Class 66/7 locomotives built by General Motors in Canada and financed by HSBC Rail. The company has also acquired six Class 73/2 electro-diesels formerly operated by Gatwick Express. By mid-2004 two of these had been refurbished and returned to traffic for

successfully winning a seven-year three-month contract to run the Anglia Railways (see entry in UK section of *Railway systems and operators*) franchise from January 1997. In September 2000 GB Railways, in a joint venture with Renaissance Railways, launched Hull Trains (see entry in UK section of *Railway systems and operators*) to provide a service between London King's Cross and Hull. GB Railways' shareholding in Hull Trains is 80 per cent. In April 2001 the company's freight subsidiary, GB Railfreight (see entry in UK section of *Railway systems and operators*) commenced operations, making GB Railways the only company to undertake both passenger and freight operations on the Railtrack network.

In December 2000 GB Railways, through its subsidiary GB Railways Eesti AS, signed an

infrastructure work on the electrified third rail DC network in southern England.

In 2004 the wagon fleet consisted on 94 container flats, with a further 93 on order, many of which would be adapted for carrying infrastructure materials.

agreement to purchase the assets of Estonian rail operator Edelaraudtee AS (see entry in Estonia section of *Railway systems and operators*), which operates passenger and freight services in the southwest of the country.

In July 2003 it was announced that FirstGroup plc (see entry in UK section of *Railway systems and operators*) was to acquire GB Railways, subject to shareholder approval.

The Go-Ahead Group plc

3rd Floor, 41-51 Grey Street, Newcastle upon Tyne NE1 6EE
Tel: (+44 191) 232 31 23 (+44 191) 221 03 15
Web: www.go-ahead.com

Key personnel
Group Chief Executive: Chris Moyes
Group Finance Director: Ian Butcher
Chief Executive, Rail: Keith Ludeman

Political background
Go-Ahead Group plc originated from the management buyout in 1987 of the Northern Bus Company. The company remains a major bus operator, especially in London and the south, as well as in northeast England. Go-Ahead also operates two rail franchises created by the break-up of British Rail, with a 65 per cent shareholding in Govia Ltd, the joint venture with the French transport group Keolis, which operates the Thameslink and Southern rail franchises (see

entries in UK section of *Railway systems and operators*).

Go-Ahead also has interests in the aviation ground handling sector through Aviance UK, as well as parking operations nationally under the Meteor name.

UPDATED

Great North Eastern Railway (GNER)

Main Headquarters Building, Station Rise, York YO1 6HT
Tel: (+44 1904) 65 30 22   Fax: (+44 1904) 65 33 92
Web: www.gner.co.uk

Key personnel
Chief Executive: Christopher Garnett
Chief Operating Officer: Jonathan Metcalfe
Directors
    Marketing and Sales: Lysanne McCallion
    Customer Operations: Jim Gilbert
    Production: Richard McClean
    Human Resources: Mike Gooddie
    Finance: Philip Pacey

Political background
Under the arrangements put in place by the 1993 Railways Act, Great Northern Railway (GNER), a subsidiary of the shipping and hotels group Sea Containers, was awarded the franchise for the InterCity East Coast route in March 1996; the trading name of the line was changed later that year to GNER. The franchise was initially to run for seven years, with subsidy declining from GBP 67.3 million in the first year to zero in the seventh year. In January 2002 an agreement was signed with the Strategic Rail Authority to extend the franchise by two years to April 2005.

In March 2005 GNER was awarded a new seven-year contract with an automatic three-year extension subject to agreed performance targets being met. The contract commencement date was 1 May 2005 and foresees GNER paying a premium of GBP1.3 billion over the life of the contract.

An additional feature of this contract will see GNER co-operating with Network Rail to electrify the 24 km Leeds–Hambleton Junction section, increasing capacity and operational flexibility by enabling Leeds to be approached from both

east and west. Estimated cost of this project is GBP70 million. Current electrification only permits access from the west.

Passenger operations
GNER operates intercity services on a 1,473 km network out of King's Cross terminus in London to West Yorkshire, the North East of England and Scotland.

A key feature of the new franchise contract awarded in 2005 is a plan to add 13 extra London–Leeds services by December 2007, creating a day-long half-hourly service between these two cities.

Passenger numbers on this route had increased by 30 per cent in the two years to 2005.

Traffic (million)	2002–03	2003–04	2004–05
Passenger journeys	14.6	15.8	16.7
Passenger-km	3,722	3,939	4,064
Train-km	18.5	18.9	18.8

Traction and rolling stock
GNER operates 31 'Mallard' Class 91-powered electric trains owned by HSBC Rail (UK) Ltd and 10 diesel InterCity 125s (HSTs) owned by Angel Trains. Class 91 electric locomotives have undergone

Refurbished interior of a GNER Mark 4 coach, undertaken by Bombardier (Ken Harris)   **NEW**/1122858

Hired in Class 373 Eurostar trainset forming a GNER London–Leeds service (John C Baker)
NEW/1114953

refurbishment by a consortium of Bombardier and ALSTOM aimed at improving their reliability. In January 2003 Bombardier Transportation was awarded a contract by HSBC Rail (UK) Ltd to provide heavy maintenance services on the Class 91 locomotives for a period of five years, with options to extend this to up to 18 years. Bombardier was also awarded a contract to undertake the refurbishment of the 302 Mark 4 trailer vehicles employed in the Mallard trainsets. This work is scheduled to be completed in October 2005.

Ten HST trailer cars made surplus by Virgin Trains' Voyager/Super Voyager fleet programme were procured from Angel Trains and used to strengthen GNER's HSTs to nine-car sets in 2003. To meet the enhanced service requirements of the new contract awarded in 2005, GNER was to lease three additional HST sets, bringing the fleet of these to 13. The entire fleet was also to undergo interior and external refurbishment, and further work would be undertaken to improve reliability.

From the May 2000 timetable change, GNER introduced Class 373/2 Eurostar sets on domestic services between London Kings Cross and York in a move intended to increase capacity. After infrastructure improvements, these services, branded 'White Rose', were diverted to Leeds. Built for international services from UK regional centres to Brussels and Paris, which were not implemented, the Class 373/2 trains had been stored out of use. The three trains required for daily service are drawn from a pool of four under a 'wet' lease contract with Eurostar (UK) Ltd. Two of the sets in use by GNER carry the company's blue livery, while the other two retain Eurostar colours but without branding.

In 2003 GNER concluded an agreement with EWS (see entry in UK section of *Railway systems and operators*) for the hire of four Class 67 diesel locomotives for standby and rescue purposes. As part of the new franchise contract arrangements agreed in 2005 an additional locomotive was to be hired for this purpose.

GNER also hires four Class 08 locomotives from Wabtec for stock shunting duties.

UPDATED

Heathrow Express

3rd Floor, 30 Eastbourne Terrace, Paddington, London W2 6LE
Tel: (+44 20) 87 50 66 00   Fax: (+44 20) 87 50 66 15
Web: http://www.heathrowexpress.com

Key personnel
Chairman, Heathrow Express and BAA Rail:
  Vernon Murphy
Managing Director: Brian Raven
Head of Operations: Richard Brown

Passenger operations
Heathrow Express provides a dedicated airport express service between London's principal airport, Heathrow, and the rail terminus at Paddington. The company is wholly owned by airport operator BAA. Services are operated without subsidy.

The exit from Paddington is electrified at 25 kV AC. At Heathrow Junction, 19 km out of Paddington, a 7 km line owned by Heathrow Express runs to the airport. The line is double track to a station under the Central Terminal Area, and then single track to a station under Terminal Four. To serve the airport's Terminal 5, on which construction began in 2003, another line is to be built from the CTA, with trains splitting there for Terminals 4 and 5.

The service comprises four trains an hour in each direction travelling at a maximum of 160 km/h, with the Paddington-Heathrow journey taking 15 minutes. Full baggage check-in services are provided at Paddington.

From early 2005 Heathrow Express plans to supplement its non-stop service between the airport and Paddington with a Heathrow Connect half-hourly service with five intermediate stops, for which new rolling stock was ordered in June 2003. This service will be operated as a joint partnership with FirstGroup, the winner of a competition for the former Thames Trains franchise, which covers local services on routes west of Paddington. Until

Heathrow Express Class 332 emu at London Paddington station (Colin Boocock)   0552789

completion of Terminal 5 in 2008, the Heathrow Connect service will call only at the CTA station for Terminals 1 to 3.

Heathrow Express is a member of Airport Express, a sales and marketing alliance that also includes Gatwick Express and Stansted Express, the operators of dedicated rail services to London's other two main airports.

Traction and rolling stock
Rolling stock consists of Class 332 emus (nine four-car and five five-car) supplied by Siemens Transportation Systems, with CAF of Spain undertaking mechanical construction. These vehicles were originally supplied as four-car sets. In 2001 Heathrow Express ordered five additional intermediate trailer cars from CAF to strengthen

members of the existing fleet. These entered service in 2002, enabling five- and nine-car trains to be operated. Siemens also undertakes fleet maintenance on a contract basis at a purpose-built depot at Old Oak Common, west London.

In June 2003 Heathrow Express placed an order with Siemens for four four-car Class 360/2 Desiro UK emus for delivery from late 2004. The contract was later amended to provide for five five-car sets, with the four units originally ordered each to receive an additional trailer car retrospectively and the fifth to be constructed as a five-car set from new. These vehicles are intended for the planned Heathrow Connect stopping service between the airport and Paddington.

Hull Trains

Premier House, Ferensway, Hull HU1 3UF
Tel: (+44 1482) 21 57 45   Fax: (+44 1482) 21 57 45

Key personnel
Managing Director: Mark Leving

Background
Hull Trains is a joint venture established between GB Railways (80 per cent) and Renaissance Railways (20 per cent) to run passenger services

between London and Hull under an initial four-year open access agreement with Network Rail. Services between the two cities commenced in November 2000 and are operated without subsidy.

In July 2003 it was announced that subject to shareholder approval GB Railways was to be acquired by FirstGroup plc (see entry in UK section of *Railway systems and operators*).

Passenger operations
Hull Trains' timetable provides four weekday, three Saturday and one Sunday return services between

London King's Cross and Hull, serving Grantham, Doncaster, Selby and Brough.

Traction and rolling stock
Services are provided using examples of Anglia Railways' fleet of eight Class 170/2 Turbostar dmus.

In September 2002 GB Railways issued letters of intent to procure from Bombardier Transportation four three-car Class 170 dmus in 2004 and four Class 222 200 km/h four-car diesel-electric trainsets by March 2005. To be financed by HSBC Rail, the Class 222 units will be based on the design of vehicles supplied to Virgin CrossCountry and Midland Mainline. The contract for those was finalised in July 2003 and included a 5.3-year agreement covering their maintenance at the Crofton, South Yorkshire, depot being built by Bombardier primarily to maintain the Midland Mainline Class 222 fleet.

Anglia Railways Class 170/2 dmus forming a Hull–London King's Cross service at Claypole, on the East Coast Main Line
(John C Baker)
0547800

Island Line

Island Line Ltd
St John's Road Station, Ryde, Isle of Wight
PO33 2BA
Tel: (+44 1983) 81 25 91   Fax: (+44 1983) 81 78 79
e-mail: comments@island-line.co.uk
Web: http://www.island-line.co.uk

Key personnel
Managing Director: Andrew Haines
General Manager: Steve Wade

Political background
Under the arrangements put in place by the 1993 Railways Act, the franchise to run operations on the 13.6 km railway on the Isle of Wight was let in October 1996 to Stagecoach Holdings plc (see entry in UK section of *Railway systems and operators*), a bus operator. The self-contained Island Line franchise is unique in being the only vertically operated portion, combining train operations and infrastructure, of the former BR network.

The franchise was let for a term of five years. Stagecoach was to receive from the Franchising Director (now the SRA) support, in 1996 prices, of £2.012 million in the first full financial year of the franchise, declining to £1.751 million in the last year.

In 2001 a two-year extension to Stagecoach's franchise was concluded with the Strategic Rail Authority, extending the contract to September 2003. In 2002, a further franchise extension was agreed, extending Stagecoach's operation of the service until February 2007. Conversion of the system to light rail or guided busway operation are among options considered for the line after that, with life-expiry of the 65-year old rolling stock likely to precipitate change.

Island Line electric multiple-units

Class	Cars per unit	Motor cars per unit	Motored axles/car	Power/ motor kW	Speed km/h	Cars in service	First built	Builders Mechanical	Electrical
483*	2	2	2	130	75	12	1938	Met-Cam	GEC

* Built for London Transport, converted for Isle of Wight 1989–90.

Passenger operations
Island Line operates passenger services on a 13.6 km line serving eight stations between Ryde and Shanklin on the eastern coast of the Isle of Wight.

Traffic (million)	2000–01	2001–02	2002–03
Passenger journeys	0.8	0.8	0.8
Passenger-km	5.9	6	6
Train-km	0.30	0.30	0.30

Traction and rolling stock
Island Line is operated with former London Underground electric trains leased from HSBC Rail UK.

London & Continental Railways Ltd

3rd Floor, 183 Eversholt Street, London NW1 1AY
Tel: (+44 20) 73 91 43 00   Fax: (+44 20) 73 91 44 10
Web: www.ctrl.co.uk; www.lcrproperties.com

Key personnel
Executive Chairman: Rob Holden
Finance Director: Mark Bayley
Managing Director, Union Railways (North):
 Alan Dyke
Managing Director, L&C Stations & Property Ltd:
 Stephen Jordan
Managing Director, Eurostar (UK) Ltd:
 Richard Brown

Subsidiary companies
CTRL (UK) Ltd
Union Railways (North) Ltd
Channel Tunnel Rail Link Ltd
London & Continental Stations & Property Ltd
Eurostar (UK) Ltd

Gauge: 1,435 mm
Route length: 113 km (including Section 2, under construction)
Electrification: 113 km at 25 kV AC 50 Hz

Political background
London & Continental Railways (LCR) is the consortium which in 1996 won the build-finance-and-operate Private Finance Initiative contract for the Channel Tunnel Rail Link (CTRL), the high-speed line from the Channel Tunnel to London. The contract includes ownership of the UK element of the tri-national Eurostar service. LCR's shareholders are Ove Arup & Partners, Bechtel Ltd, EDF Energy Ltd, Sir William Halcrow & Partners Ltd,

National Express Group plc, SNCF (French National Railways), Systra-Sofretu-Sofrerail and UBS.

Following a restructuring in 1998 the CTRL project was taken forward as two phased sections with the former Railtrack Group plc. The LCR Group subsequently regained full control of the project when Railtrack sold its CTRL interests in 2002. Section 1 of the CTRL was opened on schedule on 28 September 2003; section 2 is under construction and is planned to open early in 2007. CTRL Section 1 is owned by CTRL (UK) Ltd, which has contracted the operation and maintenance to Network Rail (CTRL) Ltd. The client for Section 2 is LCR subsidiary Union Railways (North) Ltd.

The construction cost of the CTRL is estimated at GBP5.2 billion (inflation-adjusted final cost), of which the government is to contribute GBP3.1 billion. Section 1 is costed at GBP1.9 billion and Section 2 at GBP3.3 billion, including land acquisition costs. The government will receive a share of LCR's cashflow after 2020.

The functions of London & Continental Stations & Property Ltd include the development and regeneration of major brownfield land holdings around St Pancras, Stratford and Ebbsfleet stations.

New lines
In December 1996, parliamentary approval was granted for the building of a new high-speed

A Eurostar trainset during a high-speed test run on the CTRL near Charing, west of Ashford
(Alex Dasi-Sutton)
0567061

Work in progress at St Pancras, London, in July 2004. The eastern half of the station extension, temporarily housing Midland Mainline services, is on the left. Piling work for the western half of the extension is under way (centre), as is refurbishment of the William Barlow arched trainshed in the background **NEW**/1108028

Eurostar trainset crossing the Medway Viaduct on 30 July 2003, making a demonstration run during which a new UK rail speed record of 334.7 km/h was achieved 0558938

route between London and the Channel Tunnel. The new route, then planned to open in 2003, would reduce the journey time between London and Paris via the tunnel by some 35 minutes to around 2 hours 20 minutes and that to Brussels to 2 hours 5 minutes. Subsequent revisions to the construction schedule saw Section 1 completed in 2003 and Section 2 scheduled to be commissioned in 2007.

Construction of Section 1 of the new route (74 km) formally began in October 1998 and was project-managed by Rail Link Engineering (RLE), a joint venture of LCR's four engineering shareholders (Arup, Bechtel, Halcrow and Systra). RLE is also responsible for design, project management and construction of Section 2 (39 km).

Section 1 runs from the Channel Tunnel to Fawkham Junction, near Dartford, connecting with the existing network to Waterloo International terminal. Work on Section 2 of the project began in July 2001. While this is shorter, it is more complex, with extensive tunnelling, two intermediate stations and major works at St Pancras to create a new London terminal.

Route characteristics

Of the new route's 113 km, approximately 25 per cent will be in tunnel. The main structures are the 3.2 km North Downs Tunnel, a 1.2 km viaduct over the Medway valley, 3 km twin tunnels under the Thames, a 1.3 km viaduct at Thurrock and some 17 km of twin tunnels under east London. The line includes junctions with the 750 V DC existing route near Ashford so that international trains from the high-speed line can serve Ashford International station; non-stop trains are able to bypass Ashford. At Southfleet Junction, near Gravesend, a spur forming the west end of Section 1 leaves the CTRL to join the existing network at Fawkham Junction. This allows trains access to the 'classic' network to reach Waterloo terminal on the south side of London pending completion of Section 2, which will take international services into London St Pancras station.

Intermediate stations at Ebbsfleet in north Kent and Stratford in east London are being constructed. These will provide intermediate stops for international and domestic services using the Channel Tunnel Rail Link. On the approach to St Pancras a direct link from the high-speed line to Network Rail's North London Line (giving access to the West Coast Main Line) is being built and there will also be a connection from the terminal to the East Coast Main Line.

At St Pancras, LCR is making full use of the magnificent train shed, designed by the nineteenth century engineer William Barlow of the Midland Railway, and St Pancras Chambers (originally the Midland Grand Hotel), designed by Sir George Gilbert-Scott. Six platforms in the trainshed, extending out onto a new northward extension, will be given over to international services, and platforms are being built on the extension outside the trainshed for Midland Mainline and CTRL domestic services. Two sub-surface platforms will be provided beneath the west side of the station for Thameslink services, to replace the present cramped station at King's Cross Thameslink.

Technical characteristics and operation of the CTRL are mainly based on the principles adopted for the most recently built high-speed lines in France. The line is largely engineered for 300 km/h running over Section 1. On Section 2 the line speed will be 230 km/h on open track and 220 km/h in tunnel sections. Train control is via the TVM430 cab signalling system. The CTRL is electrified using the 25 kV AC 50 Hz system and employs UIC60 rail throughout. Slab track is being installed in the tunnels on Section 2. The line is built to UIC 'C' gauge except for the section through Ashford International station, which complies with 'B+' gauge.

Domestic services

Under the terms of the government's agreement with LCR, a number of paths on the high-speed line are reserved for domestic services from east and north Kent to St Pancras.

In 2003 the SRA initiated consultations regarding domestic services over the new line. These will be operated using 225 km/h dual-voltage emus as part of a redrawn South Eastern network franchise.

UPDATED

Mendip Rail

Mendip Rail Limited, The Pavilion, Torr Works, East Cranmore, Shepton Mallet, Somerset BA4 4SQ
Tel: (+44 1373) 45 67 22   Fax: (+44 1749) 88 01 41
e-mail: alan.taylor@mendip-rail.co.uk

Key personnel
Managing Director: Alan Taylor
Commercial Manager: Alan Freemantle

*Mendip Rail Class 59 locomotive in the latest Foster Yeoman livery near Westbury with aggregates for west London
(Phil Marshall)*
0063646

Freight operations

Mendip Rail provides transportation services in conjunction with English, Welsh and Scottish Railway for two aggregates companies, Foster Yeoman and Hanson Aggregates, to manage the rail movement of stone from the company's quarries in the west of England. Mendip Rail owns nine 2,460 kW General Motors-built Class 59 diesel locomotives, plus a fleet of 413 wagons. In 1998, Mendip Rail conveyed 6.13 million tonnes,

compared with 5.7 million in 1997. In 1999, the company employed 25 staff.

In 1997, one of Mendip's Class 59s was dispatched to Germany, where it works in a joint venture with DB AG (German Railways) on the haulage of stone delivered by ship from Scotland to northwest German ports and then transported by rail for the reconstruction of Berlin.

Recent contracts cover the movement of 2 million tonnes of construction material to

Sevington, Kent for the Channel Tunnel Rail Link, on behalf of Foster Yeoman (October 1999 – October 2001) and 500,000 tonnes of material to Allington, Kent for Hanson Aggregates (March 2000 – March 2001).

In mid-2000, Mendip Rail became a licensed non-passenger train operating company.

Merseyrail Electrics 2002 Ltd

Rail House, Lord Nelson Street, Liverpool, L1 1JF
Tel: (+44 151) 702 25 34   Fax: (+44 151) 702 24 13
Web: www.merseyrail.org

Key personnel
Managing Director: Patrick Verwer
Directors
   Finance: Ian McLaren
   Customer Services: Ingrid van Poelgeest
   Operations: Andrew Heath
   Engineering: Irene Doosje
   Safety: Lesley Cusick
   Human Resources: Alan Haynes

Political background
Under the arrangements put in place by the 1993 Railways Act, the franchise to run operations on the Merseyrail Electrics system was let in January 1997 to MTL Services plc, formerly MTL Trust Holdings Ltd, a group with roots in bus operations in Liverpool. In January 2000, the Arriva bus-operating group made a takeover bid for MTL Services and this subsequently received approval from the Shadow Strategic Rail Authority. Arriva was initially to operate the franchise for a maximum of 12 months under existing terms and conditions, during which period it would be re-let. In 2001 Arriva was granted a two-year extension of the franchise until February 2003.

In 2001 the Strategic Rail Authority and Merseytravel established a methodology to transfer responsibility to Merseytravel for the services covered by this franchise. This resulted in the award in May 2003 of a 25-year contract to operate the network to a joint venture of Serco Group plc and Netherlands-based Ned Rail (Serco/NS joint venture), resulting in the formation of the present company. The contract took effect in July 2003.

In 2005 the company employed around 1,100 staff.

Passenger operations
Merseyrail Electrics runs services on two lines comprising 121 route-km serving 66 stations in the Liverpool suburban area. The Wirral line, extending under the river Mersey, links Liverpool with West Kirby, Ellesmere Port, New Brighton and Chester.

Merseyrail Electrics refurbished Class 508 emu (Merseytravel)   ***NEW**/1141384*

Merseyrail Electrics electric multiple-units

Class	Cars per unit	Motor cars per unit	Motored axles/car	Power/ motor kW	Speed km/h	Cars in service	First built	Builders Mechanical	Builders Electrical
507	3	2	4	82	120	96	1978	BREL	GEC
508	3	2	4	82	120	81	1979	BREL	GEC/Brush

The Northern line links Liverpool with Ormskirk, Kirkby, Southport and Hunts Cross. There are 66 stations. Services are operated at a basic 15-minute frequency.

Traffic (million)	2002–03	2003–04	2004–05
Passenger journeys	24.9	27.8	28.5
Passenger-km	275.2	282.6	291.6
Train-km	5.5	5.9	6.2

Traction and rolling stock
The Merseyrail Electrics system is electrified on the 750 V DC third rail system. The company runs

32 Class 507 and 27 Class 508 emus leased from Angel Train Contracts. A refurbishment programme for the entire fleet, funded by Angel and Merseytravel, was completed by ALSTOM in July 2005.

Tendering for the supply of a new fleet of trains to replace the existing equipment is due to be initiated by 2007.

Maintenance of the fleet is undertaken at Birkenhead North depot.

UPDATED

Midland Mainline

Midland Mainline Ltd
Midland House, Nelson Street, Derby DE1 2SA
Tel: (+44 1332) 26 20 40   Fax: (+44 1332) 26 25 61
e-mail: feedback@midlandmainline.com
Web: http://www.midlandmainline.com

Key personnel
Directors
   Managing: Alan Wilson
   Train Services: Heidi Mottram
   Commercial: Malcolm Brown
   Projects: Peter Garrood
   Finance: Sharon Stone
   Station Services and Human Resources:
      Barry Brown

Political background
Under the arrangements put in place by the 1993 Railways Act, the Midland Mainline franchise was let to the National Express Group (see entry in the United Kingdom section of *Railway*

systems and operators) in April 1996 for a 10-year term.

A subsidy of £16.5 million was to be paid by the Franchising Director in the first year, with decreasing amounts thereafter until Year 5, when the franchise was forecast to turn profitable. In the second half of the franchise there were to be payments to the Franchising Director by the franchisee, with the premium being around £10 million in Year 10.

In August 2000, NEG was awarded a two-year extension of the franchise, until April 2008. As part of the negotiations which resulted in this extension, premium payments due to be made under the original franchise agreement would be retained by Midland Mainline for investment in its network.

Passenger operations
Midland Mainline operates intercity passenger services on a 708 km route network running out of London St Pancras station to Leicester, Derby, Nottingham, Sheffield, Leeds, Burton-on-Trent and Matlock.

Traffic (million)	1999–2000	2000–01	2001–02
Passenger journeys	8.2	8.6	9.0
Passenger-km	1,075.8	1,096.7	1,134.0
Train-km	9.5	11.4	11.4

Improvements to existing lines
In January 2001, Midland Mainline received planning approval to develop a new East Midlands Parkway station close to the M1 motorway and East Midlands Airport. In 2003, land acquisition issues were preventing a start of work on this project.

Traction and rolling stock
Midland Mainline leases 15 HST diesel trainsets from the Porterbrook Leasing Company. These include two additional sets which were cascaded from Virgin CrossCountry in 2002–03. Refurbishment of the fleet was in progress in 2003, including power car re-engining.

In June 1997, the company ordered 13 two-car Class 170/1 Adtranz 'Turbostar' diesel multiple-units with finance from Porterbrook; the deal included maintenance of the new trains by the manufacturer. In October 1997, a further four two-car units of the same type were ordered. In April 2000, 10 intermediate cars were ordered to strengthen some existing Class 170/1 units to three-car sets. Since May 1999, these dmus have been used to provide frequent services to smaller intermediate stations on the route, allowing an increase in frequency to principal stations using HSTs. A hub at Leicester provides an interchange between dmus and HSTs.

In February 2001, Midland Mainline placed an order with Bombardier Transportation for 127 diesel-electric multiple-unit cars based on the design of the Class 220 and 221 Voyager and Super Voyager trains supplied to Virgin CrossCountry. To be designated Class 222 and branded 'Meridian' units, the trains are to be configured as seven nine-car and 16 four-car sets. Finance is being provided by HSBC Rail (UK) Ltd and delivery is due to take place between the second quarter of 2004 and January 2005. The contract with Bombardier Transportation also includes maintenance provision initially for four years, with an option to extend this to 15 years. Maintenance is to be

Midland Mainline HST forming a northbound service near Wellingborough (Ken Harris)   0547057

undertaken at a new depot at Beighton, Sheffield, at a recommissioned facility at Cricklewood, north London, and at Bombardier's Central Rivers depot near Burton-on-Trent. Delivery of

the Class 222 units will enable the Class 170/1 dmus to be transferred to other National Express franchises.

National Express Group plc

75 Davies Street, London W1K 5HT
Tel: (+44 20) 75 29 20 00   Fax: (+44 20) 75 29 21 00
e-mail: info@natex.co.uk
Web: www.nationalexpressgroup.com

Key personnel
Chief Executive: Phil White
Chief Operating Officer: Ray O'Toole
Finance Director: Adam Walker

Trains Division
Midland House, Nelson Street, Derby DE1 2SA
Chief Executive: David Franks

Political background
National Express Group was a coach, bus and airport operation which took advantage of the 1993 Railways Act, winning five out of the 25 passenger franchises on offer, making it the largest single operator. The franchises won were Gatwick Express,

Midland Mainline, Central Trains, North London Railways (now Silverlink) and ScotRail; the group is also a member of the consortium contracted to manage Eurostar international services on behalf of London & Continental Railways, which won the concession to build and operate the Channel Tunnel Rail Link.

In July 2000 National Express announced a takeover bid for Prism Rail, which was operating four franchises: c2c (formerly LTS Rail); Valley Lines (formerly Cardiff Railway); Wales & West; and West Anglia Great Northern (WAGN). The offer valued Prism Rail at £166 million. From October 2001 the Wales & West franchise and certain adjoining franchises were restructured to form two 'managerial units': Wales & Borders Trains and Wessex Trains. As part of this process the Valley Lines franchise was integrated with the new Wales & Borders Trains managerial unit, which National Express ran until this restructured franchise passed to Arriva in 2003 as Arriva Trains Wales.

In December 2003 National Express was named preferred bidder by the SRA for the Greater Anglia franchise, which combines former Anglia and Great Eastern franchises and the West Anglia portion of WAGN (including Stansted Express), all of which serve London's Liverpool Street station. The company took over operation of all these services in April 2004, naming the operation One Railway (see UK section of *Railway systems and operators*).

In February 2004 National Express signed contracts extending its Great Northern portion of WAGN and Wessex Trains franchises until March 2006.

In October 2004 operation of the ScotRail franchise passed to FirstGroup.

The company operates a national call centre for its Trains Division. Located in Sheffield, the centre handles internet ticket sales and oversees development of commercial retailing strategy.

UPDATED

Northern Ireland Railways (NIR)

Northern Ireland Railways Co Limited
Central Station, East Bridge Street, Belfast BT1 3PB
Tel: (+44 1232) 89 94 00   Fax: (+44 1232) 89 94 01
e-mail: feedback@translink.co.uk
Web: http://www.translink.co.uk

Key personnel
Chief Executive: Keith Moffatt
Directors
   Finance: S Armstrong
   Human Resources: Alan Mercer
   Operations: Philip O'Neill
   Mechanical Engineering: Mal McGreevy
Managers
   Railway Services: Seamus Scallon
   Stations: Hilton Parr
   Enterprise Service: Ken McKnight
Marketing Executive: Ciaran Rogan

Gauge: 1,600 mm
Route length: 342 km

Political background
Northern Ireland Railways is one of three operating companies forming part of Translink, the brand name of Northern Ireland Transport Holding Company (NITHCo), a state-owned corporation which manages public transport in the province, including Citybus and Ulsterbus. The chairmanship

of NITHCo is a non-executive government appointed position.

In July 2002 the Northern Ireland Assembly approved a 10-year Regional Transportation Strategy, costed at £3.5 billion. This included

£500 million for the province's railway system, guaranteeing retention of the existing network and providing for some enhancement. Also covered was an investment of £24 million for rolling stock (see *Traction and rolling stock*).

NIR Class 201 diesel-electric locomotive at Belfast Central with an 'Enterprise' service to Dublin, alongside a Class 450 demu
(Colin Boocock)
0573226

Diesel locomotives

Class	Wheel arrangement	Power	Speed km/h	Weight tonnes	No in service	First built	Mechanical	Builders Engine	Transmission
201	Co-Co	2,390	143	112	2	1995	GM	EMD 12-710 G3B	E GM
111	Co-Co	1,678	129	102	3	1980	GM	EMD 12-645E3B	E GM

Diesel railcars or multiple-units

Class	Cars per unit	Motor cars per unit	Motored axles/car	Output/ motor kW	Speed km/h	Units in service	First built	Mechanical	Builders Engine	Transmission
80	3	1	2	175	112	20	1974	BREL	EE 4SRKT	E EE
450	3	1	2	175	112	9	1985	BREL	EE 4SRKT	E EE

Organisation

NIR, Citybus and Ulsterbus are managed by a single Integrated Executive Team, with operations and engineering activities devolved to 23 district offices in the province. Translink employs around 700 staff.

Traffic (million)	2000–01	2001–02	2002–03
Passenger journeys	5.9	6.2	6.3

Passenger operations

NIR operates services within the province from Belfast on routes to Bangor, Larne and Londonderry. It also operates the cross-border intercity 'Enterprise' service jointly with Irish Rail between Belfast and Dublin. Relaunched in October 1997, this service carries around 1 million passengers annually.

Freight operations

A limited freight service operates in Northern Ireland, in the form of cross-border traffic worked to and from Adelaide yard in Belfast by Irish Rail. NIR also operates a same day parcels service between 12 stations within Northern Ireland and to 66 destinations in Ireland in co-operation with Irish Rail.

New lines

A £29 million Belfast cross-harbour rail link incorporating the longest bridge in Ireland was opened in 1994. The route comprises a 2 km single-track line, 1,424 m of it on viaduct, between Yorkgate and Belfast Central stations.

The project represented a significant advance for NIR, as it allowed the railway to consolidate all its passenger activities on Belfast Central station; before construction of the bridge, Larne services terminated at Belfast Yorkgate, on the other side of the river Lagan.

Upgrading to passenger standards of the line from Bleach Green to Antrim to capitalise on the cross-harbour link was completed in June 2001, cutting Belfast–Londonderry journey times by some 30 minutes and permitting operation of through Dublin–Londonderry services without reversal. This led to the abandonment of regular passenger services over the Lisburn–Antrim line in June 2003. However, the line is retained in operational condition for diversionary purposes and could eventually form a component of a Belfast–Lisburn–Antrim–Belfast circle line.

Improvements to existing lines

The 9.7 km Coleraine–Portrush branch was closed for a three-month period during 2000 for complete renewal of track.

Track on the Belfast–Bangor line was renewed in 2001–02, while relaying of the Larne line was in progress in 2003.

An integrated rail and bus station was opened in Bangor in April 2001.

Major refurbishment of Belfast Central station was completed in March 2003.

Traction and rolling stock

NIR has five diesel-electric locomotives, 30 dmus and 36 passenger cars. Two of the locomotives are Class 201 2,390 kW JT42HCW Co-Co General Motors diesel-electric locomotives, NIR's contribution to a pool of four such locomotives jointly maintained with Irish Rail for Enterprise cross-border services. A pool of coaches was also ordered from De Dietrich by the two organisations for this service (see entry for Irish Rail in the Ireland section of *Railway systems and operators*).

In 2001 Porterbrook Leasing concluded an agreement to make available to NIR eight former Gatwick Express vehicles. These were refurbished and regauged in Great Britain prior to transfer to NIR, where they are powered by Class 111 locomotives on Newry–Belfast services.

In 2002 NIR placed an order with CAF in Spain for 23 three-car dmus to replace the Class 80 vehicles. Deliveries are due to begin in 2004, with the whole fleet in service in 2005–06.

Signalling and telecommunications

NIR uses two signalling systems. Centralised traffic control (CTC), with entry-exit colourlight route setting panels, controls 215 km.

Track

Rail: 113 A FB (135 km); 95 lb BH (72 km); 50 kg FB (171 km)
Crossties(sleepers): Prestressed concrete and timber, spaced 1,150/km in plain and 1,300/km in curved track.
Fastenings: Pandrol PR401
Min curvature radius: 180 m
Max gradient: 1.25%
Max axleload: 18 tonnes

Network Rail Ltd

40 Melton Street, London NW1 2EE
Tel: (+44 20) 75 57 80 00   Fax: (+44 20) 75 57 90 00
Web: www.networkrail.co.uk

Key personnel

Chairman: Ian McAllister
Deputy Chairman: Adrian Montague
Chief Executive: John Armitt
Deputy Chief Executive: Iain Coucher
Group Director, Finance: Ron Henderson
Group Director, Government and Corporate Affairs: Victoria Pender
Human Resources Director: Peter Bennett
Projects and Engineering Director: Peter Henderson
Director, Major Projects and Investmen: Simon Kirby
Operations and Customer Services Director: Robin Gisby
Maintenance Director: Richard Fenny
Route Directors
    Anglia: John Wiseman
    Kent: Dave Ward
    London North Eastern: Dyan Crowther
    London North Western: Peter Strachan
    Scotland: Ron McAulay
    Sussex: Andrew Munden
    Wessex: David Pape
    Western: Robbie Burns

Gauge: 1,435 mm
Route length: 16,666 km
Electrification: 3,224 km at 25 kV, 50 Hz AC overhead; 2,047 km at 750 V DC third rail

Political background

Network Rail Ltd in October 2002 became the successor to Railtrack, acquiring the shares of that company after it was placed in administration in October 2001. It is the rail infrastructure authority on the UK mainland, owning the track, stations and signals and controlling the timetabling of trains. Its income mainly comes from access charges for the train paths it sells to passenger and freight train operating companies.

Established by the government in the wake of the Railtrack collapse, Network Rail is a 'not for dividend' company limited by guarantee, meaning that it operates as a commercial business but any surpluses are invested in the safety, reliability and performance of the rail network. The company is owned by 113 member organisations and individuals drawn from a wide range of industry partners and interested parties. These members have similar rights to those of shareholders in a public company except they have no financial stake in the company, nor any rights to dividends or other forms of payment; their role is to monitor the conduct of the business and ensure it maintains high standards of corporate governance.

Network Rail determines its strategy within the regulatory framework of the overall strategy of the Strategic Rail Authority (SRA) (see entry in UK section of *Railway systems and operators*). The Office of Rail Regulation (ORR) (see entry in UK section of *Railway systems and operators*) also sets targets for the UK rail network and these impact on Network Rail's strategy. Following the government White Paper, 'The Future of Rail', during 2005, Network Rail is to take on additional responsibility for overall performance and public reporting, industry planning and small- and medium-size network improvements and tackling service disruption. It will remain accountable to the ORR for stewardship of the rail network.

Organisation

Network Rail's assets include nearly 34,000 km of track, these include 40,000 bridges and tunnels, 9,000 level crossings and 1,000 signal boxes. In 2003 the company implemented the replacement of the six former Railtrack territorial zones into a national, functional structure, allowing the company to focus clearly on the needs of the

Network Rail investments include this high-output ballast cleaning system from Plasser & Theurer (Network Rail)   **NEW**/1115012

passenger train operating companies and freight operators that are the company's customers.

In-house maintenance
A major strategic decision made by Network Rail in October 2003 was to bring rail maintenance in-house and create a single, integrated maintenance operation. When this project was completed in July 2004, some 16,000 maintenance staff, a fleet of over 5,000 road vehicles, a network of training centres and nearly 600 depots had been brought under Network Rail direct control. Benefits delivered by this strategy include: the consistent application of high standards across the network; efficiency savings; the continued improvement in trackside safety standards; and for the first time, complete visibility of asset condition and costs.

Managed stations and property
Network Rail's assets also include around 2,500 stations. Most are leased to train operating companies but 17 major facilities are owned and managed directly by the company. They are: Birmingham New Street; Edinburgh Waverley; Gatwick Airport; Glasgow Central; Leeds City; Liverpool Lime Street, London Bridge (London); London Cannon Street; London Charing Cross; London Euston; London Fenchurch Street; London King's Cross; London Liverpool Street; London Paddington; London Victoria; London Waterloo; and Manchester Piccadilly. Almost 800 million people pass through these 17 stations annually.

In February 2005 day-to-day operational management of Network Rail's managed stations transferred to its Operations and Customer Services function. The majority of station managers now report directly to Area General Managers.

Network Rail's commercial property division, Spacia, is one of the UK's biggest landlords of small to medium-sized enterprises with around 9,000 tenants.

Connections are provided to more than 1,000 freight terminals and 90 maintenance depots are leased to train operators and contractors.

Finance
Improvements in rail performance gathered pace in the first six months of 2004–05 as Network Rail cut delays by 16 per cent and recorded an operating profit. A £20 billion debt issuance programme was launched following the period's end. Turnover increased from £1,501 million to £1,874 million and total operating costs, excluding depreciation and exceptional items, were reduced by £70 million.

In the first half of 2004–05 the company recorded an operating profit of £225 million and a pre-tax loss 0f £34 million, compared to a £233 million loss in 2003–04.

Improvements to existing lines
Network Rail is delivering a number of major projects that are central to its programme to rebuild and enhance the British railway network. These include:

West Coast Route Modernisation
In September 2004 a critical milestone was reached in the modernisation of the West Coast Main Line, which connects London with Birmingham, Glasgow, Liverpool and Manchester, with the introduction of a 200 km/h line speed and faster and more frequent services between London Euston and Birmingham, Stoke-on-Trent, Stockport and Manchester. In 2005 line speed improvements were also close to completion between Crewe, Liverpool and Preston. Work to improve capacity had also begun with four-tracking the Trent Valley route between Tamworth and Armitage. This will result in timetable improvements by December 2008.

Southern Power Supply Upgrade
Network Rail's Southern Power Supply Upgrade (PSU) project is delivering modifications to electrical infrastructure, provision of depot facilities and extended platforms at over 30 stations to support the introduction of more than 2,000 new emu cars across Hampshire, Kent, Surrey and Sussex. The programme is due to be completed by the end of 2005.

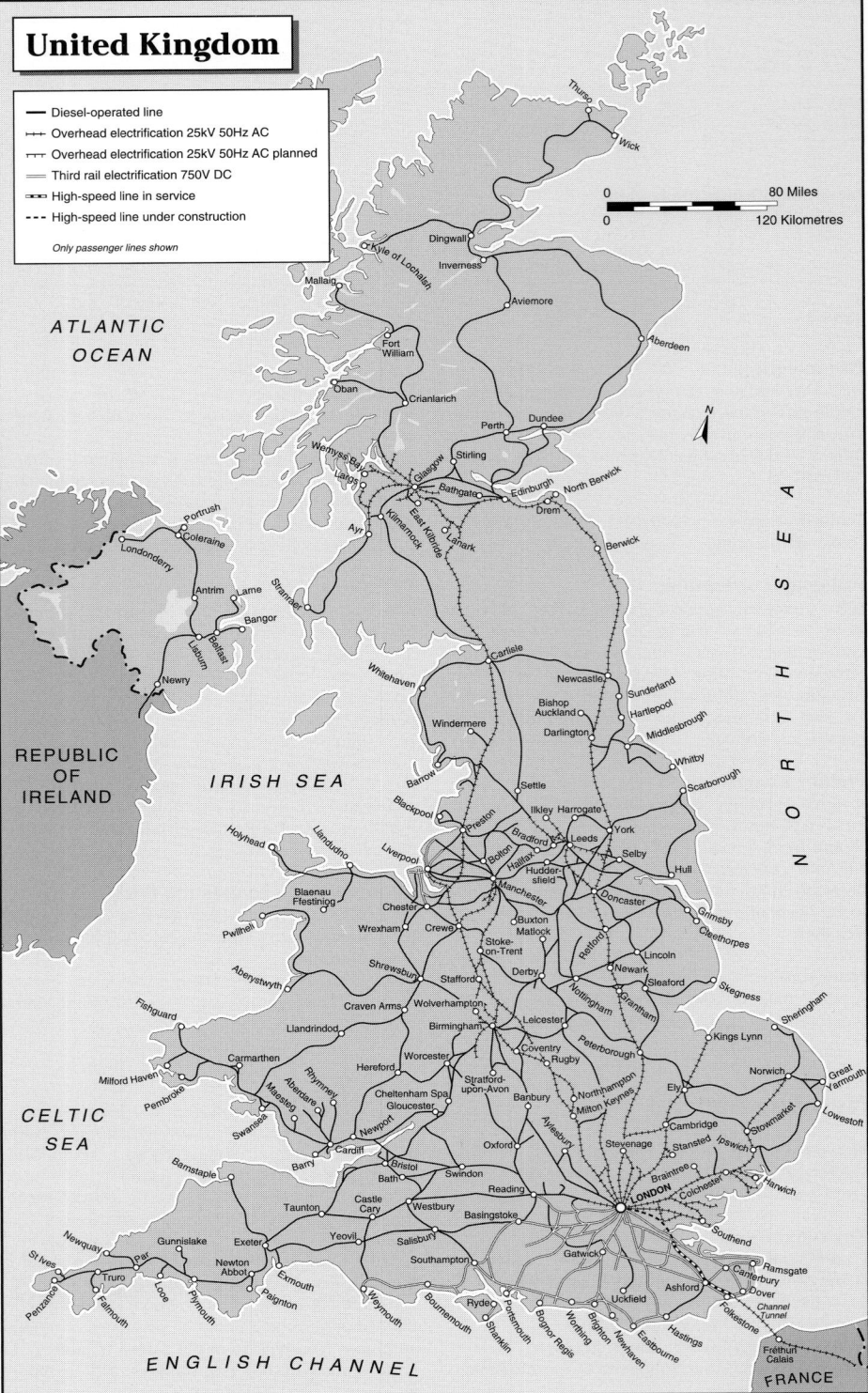

NEW/1114952

Thameslink Programme
The Thameslink Programme is a major upgrade planned for the north-south cross-London Thameslink route, which will improve rail connections in and around the capital by increasing train services, reducing overcrowding and increasing peak-hour capacity. A Transport and Works Act (TWA) order inquiry will reopen in 2005 to resolve outstanding issues. Subject to the outcome of the inquiry, the government is expected to consider the case for investment in the Thameslink Programme as part of its overall spending plans, but implementation is now not likely before 2012.

Global system for mobile communications
A project has been developed to install a new national radio system to enable secure voice and data communications over the whole network. It also supports the proposed adoption of the European Rail Traffic Management System (ERTMS) for train control. Costed at £1.3 billion, the programme is due to be implemented progressively up to 2012.

Maintenance and renewals
Network Rail continues to explore innovative ways to deliver its key role of rebuilding the British railway network. An example is the New Measurement Train (NMT), which can operate at 200 km/h, allowing it to slot in between timetabled high-speed passenger services, replacing manual inspection with mechanised measurement. As well as producing video recordings of the infrastructure and the wheel/rail interface, the train provides 'six foot' gauging data and information on track geometry.

The company has also invested £86 million in state-of-the-art high-output plant, including track relaying and high-output ballast cleaning systems. These significantly increase the company's track renewals capacity and maximise track access slots.

Traction and rolling stock
Network Rail is not a train operator: it mainly procures its haulage and vehicle needs from freight operators, notably EWS, Freightliner and

GB Railfreight. However, the company retains a fleet of traction and rolling stock to meet at least some of its traction needs. These include: a high-speed track recording train, the New Measurement Train (NMT), converted from an InterCity 125 HST trainset; two ex-Gatwick Express Class 73/2 electro-diesel and four Class 31 diesel-electric locomotives for use on infrastructure monitoring and cable laying trains; 32 Windhoff Multi-Purpose Vehicles (MPVs) for functions such as vegetation control and Sandite application; and a variety of older dmu and emu vehicles for Sandite and de-icing work.

The wagon fleet includes a large number of modern Autoballaster hopper wagons and other recently procured ballast carriers.

Signalling and telecommunications
A major milestone was reached in December 2003 with the completion of installation of the Train Protection Warning System (TPWS) throughout the network, within the project budget and to the required timescale. TPWS functions by applying the emergency brakes on a train when the system detects that: a train has passed a signal set at

danger; that a train is travelling at a speed that will prevent it stopping before a signal set at danger; or that a train is approaching a speed restriction or a buffer stop at excessive speed.

Electrification
Of the Network Rail network 31 per cent is electrified, using the 25 kV AC 50 Hz overhead or 750 V DC systems. Most of the latter is concentrated on the former Southern Region network in southeast England.

UPDATED

Northern Rail Ltd

Main Headquarters Building, Station Rise, York YO1 6HT
Web: www.northernrail.org

Key personnel
Chairman: Brian Burdsall
Managing Director: Heidi Mottram
Commercial Director: Chris Kimberley
Engineering Director: Ruud Haket

Political background
The Northern Rail franchise contract covering passenger rail services in northeast England, northwest England, Yorkshire and Humberside was awarded to a Serco-NedRailways joint venture in 2004. Running for 8.75 years commencing in December 2004, the franchise incorporates most of the former Arriva Trains Northern and First North Western franchise areas with the exception of routes transferred to the Arriva Trains Wales and TransPennine Express businesses (see entries in UK section of *Railway systems and operators*). Northern Rail also provides services on routes covered by the Greater Manchester, Tyne & Wear, South Yorkshire and West Yorkshire Passenger Transport Executives, as well as some services in the Merseyside PTE area. Over the full term of the contract Serco-NedRailways will receive £2,434.6 million in public subsidy.

Serco-NedRailways also operates the Merseyrail Electrics franchise (see entry in UK section of *Railway systems and operators*).

Passenger operations
Northern Rail provides interurban, commuter and rural services on a 2,680 km 472-station network covering the whole of northern England and covering major cities that include Leeds, Liverpool, Manchester, Newcastle and Sheffield. It also serves important rural areas, including three National Parks, and coastal towns popular with tourists. Longer-distance interurban services within the area covered by Northern Rail are operated by TransPennine Express, managed since February 2004 by FirstGroup Keolis.

Northern Rail diesel multiple-units

Class	Cars per unit	Motor cars per unit	Motored axles/car	Power/ motor kW	Speed km/h	Cars in service	First built	Mechanical	Builders Engine	Transmission
142	2	2	1	170	120	140	1985	BREL	Cummins LTA10R	H Voith T211r
144	2/3	2/3	1	170	120	56[1]	1986	Alexander	Cummins LTA10R	H Voith T211r
150	2	2	2	213	120	80	1986	BREL	Cummins NT855R5	HM Voith T211r
153	1	1	2	213	120	24	1987[3]	Leyland Bus	Cummins NT855R5	HM Voith T211r
155	2	2	2	213	120	14	1988	Leyland Bus	Cummins NT855R5	HM Voith T211r
156	2	2	2	210	120	92	1987	Met-Cam	Cummins NT855R5	HM Voith T211r
158/0	2/3	2/3	2	275/300	145	–[2]	1989	BREL	Cummins NTA855R/ Perkins 2006-TWH	HM Voith T211r
158/9	2	2	2	275	145	20	1989	BREL	Cummins NT855R1	HM Voith T211r
175	2/3	2/3	2	335	160	–	1999	ALSTOM	Cummins N14RE	HM Voith T211

[1] Includes 10 West Yorkshire PTE cars.
[2] Includes 20 West Yorkshire PTE cars.
[3] Rebuilt 1991.

Northern Rail electric multiple-units

Class	Cars per unit	Motor cars per unit	Motored axles/car	Power/ motor kW	Speed km/h	Cars in service	First built	Builders Mechanical	Electrical
321/9	4	1	4	268	160	12	1991	BREL	Brush
323	3	1	4	146	144	51	1993	Hunslet TPL	Holec
333	4	2	4	175	160	56	2000	CAF	Siemens

Former First North Western Class 175 dmu forming a Manchester Airport–Windermere service at Staveley
(John C Baker)
NEW/0554923

In the livery of former operator First North Western, a Class 323 emu waits at Manchester Piccadilly with a service to Hadfield
(Colin Boocock)
NEW/0585063

At its commencement of operations, Northern Rail employed some 4,700 staff.

Traction and rolling stock

The Northern Rail fleet comprises 232 second- and third-generation dmus inherited from Arriva Trains Northern and First North Western and three small batches of emus: Class 321/9 for the Leeds–Doncaster route; Class 323 for Manchester-based services; and Class 333 for the Leeds–Skipton line.

NEW ENTRY

Siemens-built Class 333 emu forming a Leeds service at Skipton (Colin Boocock)
***NEW**/0536196*

One Railway

One Railway
1 Oliver's Yard, First Floor, 55 City Road, London EC1Y 1HQ
Web: www.onerailway.com

Key personnel

Managing Director: Tim Clarke
Directors
  Finance: Ian Rogers
  Operations and Planning: Mark Phillips
  Customer Services: Andrew MacPherson
  Sales and Marketing: Martin Dean
  Engineering: Steve Rees
  Human Resources: Andrew Meadows
  Projects: Theo Steel
  Head of Corporate Affairs: Jonathan Denby
Business Directors
  Main Line: Richard Clark
  Rural: Clive Morris
  Metro: Perry Ramsey
  West Anglia: Mark Evans
  Stansted Express: Liz Mullen

Head of Safety and Environment: Andy Sanders
Manager, London Terminals: Alan Perry

Political background

As part of a policy adopted to rationalise the number of train operating companies using major London terminals, the Strategic Rail Authority created a so-called Greater Anglia franchise to cover intercity, suburban and regional services radiating east and northeast from the capital. As a result, a seven-year contract to operate this franchise was awarded to London Eastern Railway Ltd, a subsidiary of National Express Group, in January 2004 and took effect on 1 April of the same year under the name One Railway. The contract is subject to an automatic extension of three years if performance and service quality targets are met.

Passenger operations

One Railway services are based on a structure of five businesses, each headed by a director.

Main Line operates longer-distance passenger services between London's Liverpool Street terminus and Ipswich and Norwich and also provides trains to Harwich and on the Clacton line.

Rural embraces local services from Norwich to Great Yarmouth, Lowestoft and Sheringham and to Cambridge, and from Ipswich to Lowestoft and Felixstowe.

Metro covers commuter and regional services between Liverpool Street and Southend, Southminster and Upminster.

West Anglia comprises services run from Liverpool Street to Cambridge, Enfield Town, Hertford East, Stansted and Chingford.

Stansted Express is a dedicated link between Liverpool Street and Stansted Airport.

Most services employ electric traction, although dmus are used on some rural and regional services.

Improvements to be made under the terms of the franchise contract include increasing passenger capacity on Stansted Express services by lengthening trains, the formation of a dedicated business unit to oversee rural services, the introduction of direct services between Liverpool Street and Lowestoft and Peterborough and more services to Stansted Airport. These last-mentioned will include a service from Stratford, location of a future station on the Channel Tunnel Rail Link.

Class 90 electric locomotive propelling a London Liverpool Street–Norwich service formed of ex-Virgin West Coast Mark 3 stock near Manningtree (Ken Harris)
***NEW**/1122863*

Class 170/2 dmu providing a former Anglia Railways Norwich–London Liverpool Street service at Manningtree (Ken Harris)
0570696

One diesel multiple-units

Class	Cars per unit	Motor cars per unit	Motored axles/car	Power/ motor kW	Speed km/h	Cars in service	First built	Mechanical	Builders Engine	Transmission
150/2	2	2	2	210	120	20	1986	BREL	Cummins NT855R5	HM Voith T211r
153	1	1	2	213	120	7	1987	Leyland Bus	Cummins NT855R5	HM Voith T211r
156	2	2	2	210	120	18	1987	BREL	Cummins NT855R5	HM Voith T211r
170/2	2/3	2/3	2	315	160	32	1999	Adtranz	MTU 6R 183TD	HM Voith T211r

One electric multiple-units

Class	Cars per unit	Motor cars per unit	Motored axles/car	Power/ motor kW	Speed km/h	Cars in service	First built	Builders Mechanical	Electrical
315	4	2	2	82	120	172	1980	BREL	GEC
317	4	1	4	247	160	180	1981	BREL	GEC
321/3	4	1	4	268	160	308*	1988	BREL	Brush
322	4	1	4	268	160	20	1990	BREL	Brush
360	4	2	4	194	160	84	2002	Siemens	Siemens

* Includes some units on temporary hire to Silverlink.

Traction and rolling stock

On the London–Norwich route the fleet was in transition in 2005: formations of Mk 2 stock powered in push-pull mode were giving way to 120 Mark 3 coaches displaced from the West Coast Main Line. By mid-2005 the older Class 86/2 electric locomotives that powered these trains had been replaced by 15 Class 90s also displaced from the West Coast Main Line. The Mark 3 vehicles were to undergo refurbishment under a programme initiated in 2005. On its local and regional services One Anglia uses eight three-car and four two-car Class 170/2 'Turbostar' air-conditioned dmus, 10 Class 150/2 two-car dmus, seven Class 153 railcars and nine Class 156 dmus. Three Class 47 diesel-electric locomotives are also hired in as standby traction and for special traffic.

The most recent vehicles in the large emu fleet are 21 four-car Class 360 Siemens 'Desiro UK' emus which entered service in 2003 and are leased from Angel Trains. Nine refurbished Class 317/7 units are dedicated to Stansted Express services. Refurbishment of the Class 315 units was being undertaken by Bombardier in 2005.

UPDATED

Class 360 emu in the colours of former operator, FirstGroup, forming an Ipswich–London Liverpool Street service between Ipswich and Manningtree (Ken Harris)
0583127

Railway Safety

Evergreen House, 160 Euston Road, London NW1 2DX
Tel: (+44 20) 79 04 75 18   Fax: (+44 20) 75 57 90 72
e-mail: enquiries@railwaysafety.org.uk
Web: http://www.railwaysafety.org.uk

Key personnel
Chairman: Sir David Davies
Director: Rod Muttram

Director, Policy and Standards: Aidan Nelson
Director, Safety Management Systems:
 Matt Walter

Background
Railway Safety was established in January 2000 in the wake of the Southall and Ladbrooke Grove fatal accidents as a wholly owned not-for-profit subsidiary of Railtrack plc. The organisation's role is to provide a centre of excellence for all matters relating to UK railway safety and

to help the industry focus on improved safety management. It took over most of the functions of Railtrack's former Safety and Standards directorate.

Sea Containers

Sea Containers Services Ltd
20 Upper Ground, London SE1 9PF
Tel: (+44 20) 78 05 50 00   Fax: (+44 20) 78 05 59 03
Web: http://www.seacontainers.com

Key personnel
President: James Sherwood
Vice-President, Rail: Christopher W M Garnett

Organisation
Sea Containers has its roots in the container leasing industry and has of late diversified into railways. In the UK it operates the luxury Venice–Simplon Orient Express tour train and runs similar services in Asia, Australia and mainland Europe. In 1996 it took over intercity operations on the London–Edinburgh/Aberdeen/ Inverness, and London–Leeds routes, which it groups under GNER Holdings Ltd and is

marketing as the Great North Eastern Railway (see entry in UK section of *Railway systems and operators*).

Silverlink

Silverlink Train Services Ltd
National Express Group, London Lines
Hertford House, 1 Cranwood Street, London EC1V 9QS
Tel: (+44 20) 74 27 28 09
Web: www.silverlink-trains.com

Key personnel
Managing Director, NEG London Lines:
 Andrew Chivers

Directors
 Finance: Adam Golton
 Commercial: Clive Tilley

Political background
Under the arrangements put in place by the 1993 Railways Act, the franchise to run Silverlink (then North London Railways) operations was let in March 1997 to National Express Group plc (NEG).

The franchise was let for a term of seven years and six months. NEG received from the Franchising Director (now SRA) support, in 1997 prices, of

£48.6 million in the first full financial year of the franchise, declining to £16.9 million in 2003–04.

In September 2004 NEG was awarded a contract extending the Silverlink franchise until October 2006. The level of subsidy for this extension was projected at £120.2 million.

Organisation
NEG subsequently created its London Lines management structure to enable resources to be shared by Silverlink and its other franchises serving the capital.

Two Silverlink Class 321 emus forming a London Euston–Northampton service south of Leighton Buzzard (Ken Harris)   0554924

Silverlink electric multiple-units

Class	Cars per unit	Motor cars per unit	Motored axles/car	Power/ motor kW	Speed km/h	Cars in service	First built	Builders Mechanical	Electrical
313[1]	3	2	2	82	120	69	1976	BREL	GEC
321/4	4	1	4	268	160	144	1990	BREL	Brush
350/1	4	2	2	250	160	40	2004[3]	Siemens	Siemens
508[2]	3	1	4	82	120	9	1979	BREL	GEC

[1] Dual-voltage 750 V DC/25 kV AC.
[2] 750 V DC only.
[3] Delivery 2004–05.

Passenger operations

Silverlink Train Services is the name adopted in September 1997 for the former North London Railways. Initially the company operated suburban and regional services out of London's Euston station to Birmingham and on the North London orbital line, the route network extending to 321 km. From September 2004 services on the Birmingham route were cut back to Northampton, with Central Trains assuming responsibility for Northampton–Birmingham services.

Traffic (million)	2001–02	2002–03	2003–04
Passenger journeys	35.6	36.2	38.1
Passenger-km	1000	1,035	1,062.4

Traction and rolling stock

Silverlink's principal route, from Euston to Northampton, is electrified at 25 kV AC overhead; for this line, the company leases 36 Class 321 emus from HSBC Rail UK. Local routes around London are electrified with a mixture of 25 kV AC overhead and 750 V DC third rail, and for these the business leases 23 Class 313 dual-voltage emus from HSBC Rail UK and three Class 508 DC-only emus of generally similar design.

To provide additional capacity, from September 2004 Silverlink employed Class 90 electric locomotives and formations of Mark 3 coaches on some peak-hour Northampton–London Euston services pending delivery of Class 350/1 Desiro UK emus in 2005 by Siemens Transportation Systems. Of an order for 30 of these four-car units, 10 are assigned to Silverlink, the remainder to Central Trains. Financing is by Angel Trains. Siemens' contract for the Class 350/1 units includes full-service maintenance provision for both operators' fleets at a purpose-built depot at Northampton.

On the Bletchley–Bedford and Gospel Oak–Barking diesel-worked branch lines, Silverlink leases seven Class 150/1 dmus.

UPDATED

South Eastern Trains

Friars Bridge Court, 41-45 Blackfriars Road, London SE1 8PG, UK
Tel: (+44 20) 76 20 55 05   Fax: (+44 20) 76 20 55 22
Web: www.setrains.co.uk

Key personnel

Chairman, South Eastern Trains (Holdings) Ltd:
   Doug Sutherland
Managing Director: Michael Holden
Retail Director: Alex Warner
Engineering Director: Phil Verster

Political background

Under the arrangements put in place by the 1993 Railways Act, the franchise to run operations on the former South Eastern Division of British Rail was let to Connex, a wholly owned subsidiary of the French CGEA group, in August 1996. In 1998 CGEA adopted the new name Vivendi, subsequently becoming Veolia Environnement. Connex also ran the neighbouring South Central franchise (see entry in UK section of *Railway systems and operators*) until August 2001.

The franchise was let for a 15-year term. Subsequently the SRA cut back the length of the franchise to run only until 2006. However, at the end of June 2003 the SRA announced that it was to terminate the Connex franchise at the end of that year. This followed rejection by the SRA of a request from Connex for additional funding. Connex subsequently handed in the franchise early, in November 2003, leaving operations to be run by the SRA via a subsidiary, South Eastern Trains, with rail consultancy First Class Partnerships acting as advisors. This arrangement was to apply until the letting of a new Integrated Kent Franchise (IKF) that will include domestic services operated over the Channel Tunnel Rail Link (CTRL). By early 2005 four pre-qualified applicants to run the IKF had been selected, with an award due later this year. The new eight-year contract is scheduled to commence in December 2005, with domestic services over the CTRL due to begin in 2009.

Passenger operations

South Eastern Trains runs services on a 774 km network out of London's Charing Cross and Victoria stations to south London, Kent and East Sussex. It is the biggest of the 25 train operating companies in terms of passenger-km; it moves around 120,000 commuters into London every weekday morning.

Traffic (million)	2001–02	2002–03	2003–04
Passenger journeys	131.4	132.6	132.8
Passenger-km	3,232	3,300	3,296
Train-km	28.2	27.4	28.7

Class 375/6 emu forming a London to Ashford service near Sevenoaks (Ken Harris)   0554703

South Eastern Trains electric multiple-units

Class	Cars per unit	Motor cars per unit	Motored axles/car	Power/ motor kW	Speed km/h	Cars in service	First built	Builders Mechanical	Electrical
421/4	4	1	4	185	145	–[1]	1970	BREL	EE
423	4	1	4	185	145	–[1]	1967	BR	EE
465/0	4	2	4	140	120	388	1992	ABB/Brush	GEC
465/2	4	2	4	140	120	200	1992	GEC	GEC
466	2	1	4	140	120	86	1993	GEC	GEC
375/3	3	2	2	250	160	30	2002	Bombardier	Bombardier
375/6	4	3	2	250	160	120	1999	Bombardier	Bombardier
375/7	4	3	2	250	160	60	2002	Bombardier	Bombardier
375/8	4	3	2	250	160	120	2004	Bombardier	Bombardier
375/9	4	3	2	250	160	108	2004	Bombardier	Bombardier
376	5	4	2	250	160	180[2]	2004	Bombardier	Bombardier
508	3	1	4	82	120	33	1980	BR	GEC

[1] Retirement to be completed in 2005.
[2] Delivery in progress 2005.

Traction and rolling stock
All South Eastern Trains' territory is electrified on the 750 V DC third rail system.

Many of the company's inner-suburban services are run with modern Class 465 and 466 Networker emus (674 vehicles) leased from Angel Trains and HSBC Rail UK. Class 508 emus dating from the 1980s are also employed on some local services.

Progressively Connex had commenced replacing its fleet of over 500 Mk 1 slam-door emus of Classes 411, 421 and 423. These date from the 1950s and 1960s and were used on most longer-distance services. Subsequently orders were placed in several batches for 112 Class 375 (10 three-car and 92 four-car) 'Electrostar' trains from Bombardier Transportation (formerly Adtranz). Revenue-earning services with the type commenced in 2001 and deliveries were completed in 2004. In July 2002 36 examples of a Class 376 five-car derivative of the Electrostar design were ordered for suburban services. They are intended to eliminate remaining slam-door stock by the end of 2005.

Electrostar maintenance has been undertaken by South Eastern Train Maintenance, a joint-venture company originally established by Connex and Bombardier Transportation.

In late 2004 Hitachi was named preferred bidder to supply trains to operate domestic services originating on the classic network and running over the CTRL as defined in the new IKF (see 'Political background' above). These vehicles were to take the form of up to 30 six-car dual-voltage (75 V DC third rail/25 kV AC 50 Hz overhead) emus capable of up to 225 km/h. Their introduction is scheduled for 2009, two years after full commissioning of the CTRL.

UPDATED

Southern

New Southern Railway Ltd
Go-Ahead House, 26-28 Addiscombe Road, Croydon CR9 5GA
Tel: (+44 20) 89 29 86 00   Fax: (+44 20) 89 29 88 64
Web: www.southernrailway.com

Key personnel
Chief Executive, Rail: Keith Ludeman
Directors
    Managing: Charles Horton
    Finance: Bob Mayne
    Commercial: Vince Lucas
    Operations: Chris Burchell
    Human Resources: Nick Mitchell
    Engineering: David Sawyer
    Communications: Samantha Hodder

Political background
Under the arrangements put in place by the 1993 Railways Act, the Network South Central franchise was transferred to London & South Coast Ltd (later renamed Connex), a subsidiary of the French CGEA group, in April 1996. Services were operated under Connex South Central branding, while in 1998 CGEA adopted the new name Vivendi. The franchise was to have run for seven years.

Support payments from the Franchise Director to Connex South Central were to decline from GBP85.3 million in the first year to GBP34.6 million in 2003.

Connex's bid for a replacement franchise to take effect from 2003 was rejected by the Strategic Rail Authority in 2001. Instead, GoVia Ltd, a joint venture of Go-Ahead Group and Keolis SA, which also operates the Thameslink franchise (see entry in UK section of *Railway systems and operators*), took over the existing business on 26 August 2001, compensating Connex for the unfulfilled portion of its contract. In May 2003 GoVia signed a six-year contract to operate the service, the franchise commencing later that month. In May 2004 the name Southern was adopted, replacing the previous South Central.

In 2005 the company employed around 3,500 staff.

Passenger operations
Southern runs services on a 666 km network mostly of 750 V DC electrified lines out of London's Victoria and London Bridge termini to Gatwick Airport, Brighton and other towns on the south coast of England. Services also operate between Brighton and Watford Junction, on the West Coast Main Line, east along the south coast to Ashford and west to Southampton. There is also an extensive suburban commuter operation in south London.

Introduction of new Class 171 dmus has enabled Southern to run hourly through services between Uckfield and London Bridge and was due to do so between Ashford and Brighton from late 2005.

A programme of station improvements, including repainting and new signage, has been implemented.

Southern diesel multiple-units

Class	Cars per unit	Motor cars per unit	Motored axles/car	Power/ motor kW	Speed km/h	Cars in service	First built	Mechanical	Builders Engine	Transmission
171/7	2	2	2	315	160	18	2003	Bombardier	MTU 6R183	H Voith
171/8	4	4	2	315	160	24	2004	Bombardier	MTU 6R183	H Voith

Southern electric multiple-units

Class	Cars per unit	Motor cars per unit	Motored axles/car	Power/ motor kW	Speed km/h	Cars in service	First built	Builders Mechanical	Electrical
377	3	2	2	250	160	84	2002	Bombardier	Bombardier
377	4	3	2	250	160	616*	2002	Bombardier	Bombardier
455	4	1	4	185	120	184	1982	BREL	GEC
456	2	1	2	268	120	48	1990	BREL	Brush

*Total on order; includes 15 dual-voltage (750 V DC/25 kV AC).

Class 170 (since re-designated Class 171/7) dmu on a driver training run at Lewes (Ken Harris)

0583267

Southern Class 377 emu forming a Brighton-bound service at Barnham (Ken Harris)   **NEW**/1122860

Traffic (million)	2002–03	2003–04	2004–05
Passenger journeys	114.9	116.8	127.9
Passenger-km	2,666	2,727	2,914

Traction and rolling stock

Southern runs over 250 four-, three- and two-car emus and 12 dmus (for the non-electrified Uckfield and Hastings–Ashford lines), on lease from the three ex-British Rail rolling stock leasing companies. A small number of surviving slam-door stock built in the 1960s was due to be retired in the third quarter of 2005.

In July 1999 the former franchise holder, Connex Rail, ordered 37 'Electrostar' emus from Adtranz (now Bombardier Transportation) for delivery from 2001. The order was split as 28 three-car (initially Class 375 but subsequently redesignated Class 377) and nine four-car units (Class 377). Subsequently, an additional 30 four-car units were ordered. The new operator, GoVia, confirmed this order in September 2001. In March 2002, GoVia placed a second order with Bombardier for an additional 115 four-car Class 377 emus, bringing the total of the type ordered to 28 three-car and 154 four-car sets (700 cars). Financing for both orders is being provided by Porterbrook Leasing Company.

To replace the ageing Class 205 and 207 demus formerly employed on the non-electrified Uckfield and Hastings-Ashford lines, GoVia received from Bombardier Transportation for 42 Class 171 'Turbostar' dmu cars. Configured as nine two-car and six four-car units and delivered in 2003–05, these are also financed by Porterbrook.

As part of the franchise contract which commenced in May 2003, GoVia announced a rolling refurbishment programme for its fleet of 184 Class 455 emu cars used on south London suburban services. The programme is due to be completed by 2006. Also undergoing refurbishment in 2004 were the 24 Class 456 emus, similarly employed on south London services.

UPDATED

South West Trains Ltd

A subsidiary of Stagecoach Holdings
Friars Bridge Court, 41-45 Blackfriars Road, London SE1 8NZ
Tel: (+44 20) 76 20 52 29   Fax: (+44 20) 76 20 50 15
e-mail: via website
Web: www.southwesttrains.co.uk

Key personnel

Managing Director: Graham Eccles
Operations Director: Stewart Palmer
Commercial Director: Rufus Boyd
Engineering Director: Mac Mackintosh
Finance Director: Andy West
Human Resources Director: Margaret Kay
Customer Service Director: James Burt

Political background

Under the arrangements put in place by the 1993 Railways Act, a franchise to run South West Trains for seven years passed to the bus company Stagecoach (see entry in UK section of *Railway systems and operators*) in February 1996; Stagecoach also operates the Island Line franchise and is a partner in Virgin Trains. Subsidy from the Franchise Director (now the SRA) was set to fall from GBP54.7 million in the first year of operation to GBP40.3 million in the final year, 2003. The franchise was subsequently extended to February 2004 and then by a further three years to 2007. The contract also provides for a further two-year extension subject to the agreement of both parties.

In 2005 the company employed approximately 5,500 staff.

Passenger operations

SWT operates commuter trains in southwest London and long-distance services from London Waterloo to a triangular area of southern England from Brighton in the east to Exeter in the west. Total route-km operated: 991.

Traffic (million)	2002–03	2003–04	2004–05
Passenger journeys	141.1	143.0	161.6
Passenger-km	4,184.3	4,288.4	4,605.8

South West Trains diesel multiple-units

Class	Cars per unit	Motor cars per unit	Motored axles/car	Power/ motor kW	Speed km/h	Units in service	First built	Mechanical	Builders Engine	Transmission
158	2	2	2	275	145	2	1989	BREL	Cummins NTA855R	*HM* Voith T211r
159*	3	3	3	300	145	22	1992	ABB/ Babcock	Cummins NT855R1	*HM* Voith T211r
170	2	2	2	315	160	9	2000	Adtranz	MTU	*HM* Voith

* Built as Class 158 and converted before entering service.

South West Trains electric multiple-units

Class	Cars per unit	Motor cars per unit	Motored axles/car	Power/ motor kW	Speed km/h	Units in service	First built	Builders Mechanical	Electrical
421	3	1	4	185	145	2	1970	BREL	EE
423	4	1	4	185	145	1	1967	BR	EE
442	5	1	4	300	160	24	1988	BREL	EE
444	5	2	4	187	160	45	2002	Siemens	Siemens
450	4	2	4	187	160	110	2002	Siemens	Siemens
455	4	1	4	185	120	91	1983	BREL	GEC
458	4	3	2	270	160	30	1999	ALSTOM	ALSTOM

SWT Class 450 Desiro UK emu at Siemens' fleet maintenance depot at Northam, Southampton (Ken Harris)
0554377

Traction and rolling stock

Most of the area covered by SWT is electrified on the 750 V DC system. SWT operates 302 four- and five-car electric multiple-units owned by all three of the ex-British Rail rolling stock leasing companies, plus 22 three-car Class 159, nine two-car Class 170 and two two-car Class 158 diesel units, which are used on the line from Waterloo to the west of England.

Delivery of 665 Class 444 and 450 Desiro emu cars is complete and slam-door stock has been replaced with the exception of two refurbished units for use on the Lymington branch line and one retained for special duties only. Class 455 emus were undergoing refurbishment in 2005.

SWT also operates two Class 73/1 electro-diesel locomotives for rolling stock movements.

UPDATED

Class 170 dmu forming a Southampton area local service, launched in May 2003
(Ken Harris)
0554376

Stagecoach

Stagecoach Group plc
10 Dunkeld Street, Perth PH1 STW
Tel: (+44 1738) 44 21 11   Fax: (+44 1738) 64 36 48
e-mail: info@stagecoachgroup.com
Web: http://www.stagecoachgroup.com

Key personnel

Acting Non-Executive Chairman: Robert Spiers
Acting Chief Executive: Brian Souter
Executive Director, Rail Operations: Graham Eccles
Business Development Director, Rail: Andy Pitt
Major Projects Director: Allison Ingram
Group Finance Director: Martin Griffiths

Political background

Stagecoach has its roots in the bus industry, having grown fast following bus deregulation in the UK in the 1980s. With that market approaching maturity, it turned its attentions to overseas bus acquisitions and to the opportunities offered by the 1993 Railways Act in the UK. It won the first franchise to be let, South West Trains (qv), and subsequently acquired the Island Line franchise. It also purchased Porterbrook (qv), one of the three rolling stock leasing companies formed with the British Rail rolling stock fleet. Porterbrook was initially sold to a management buyout team and then acquired by Stagecoach in late 1996. In April 2000, Porterbrook was sold to Abbey National for a consideration

valued at £1.4 billion. In 1997, Stagecoach acquired a 26-year operating franchise for the Sheffield-based Supertram light rail system.

In June 1998, Stagecoach announced that it had taken a 49 per cent stake in the Virgin Rail Group for a purchase price of £158 million. The move followed the purchase by Stagecoach of the 59 per cent shareholding by Virgin's venture capital backers and the subsequent transfer of 10 per cent of the equity back to Virgin, giving it a controlling stake.

In April 2001 Stagecoach was named preferred bidder for a new 20-year franchise to operate South West Trains.

Thameslink Rail Ltd

Friars Bridge Court, 41-45 Blackfriars Bridge Road, London SE1 8NZ
Tel: (+44 20) 79 28 51 51
Web: http://www.thameslink.co.uk

Key personnel

Chief Executive: Keith Ludeman
Managing Director: Mark Causebrook
Finance Director: Ken Watson

Political background

Under the arrangements put in place by the 1993 Railways Act, the franchise to run operations on Thameslink was let in March 1997 to Govia, a joint venture combining the British bus operator Go-Ahead Group plc and Via GTI (now Keolis), the French transport group.

The franchise was let for a seven year and one month term. Govia was to receive from the Franchising Director (now the SRA) support, in 1997 prices, of £2.5 million in the first full financial year of the franchise, reversing to a premium payment to the Franchise Director of £28.4 million in 2004. The SRA intends that on expiry of the existing contract, a short-term franchise will be let pending completion of the Thameslink 2000 project (see *Improvements to existing lines*).

In 2002 Thameslink employed some 800 staff.

In August 2001 Govia took over the South Central franchise, with which it shares some section of route, from Connex.

Passenger operations

Thameslink runs cross-London suburban operations through a tunnel linking King's Cross to Blackfriars. Long-distance services run between Bedford and Brighton via Gatwick Airport, and inner-suburban services between Luton and Wimbledon. Total route-km operated over: 203.

Traffic (million)	1999–2000	2000–01	2001–02
Passenger journeys	38	40.2	41.1
Passenger-km	1213.7	1,290.8	1,340
Train-km	11.4	11.4	11.4

Improvements to existing lines

Railtrack's successor, Network Rail, is committed to a major upgrade of the Thameslink lines in a project known as Thameslink 2000. This will see upgrading of the power supply and signalling through the central London tunnels, improvements in track layouts to increase capacity in south London, and a new station adjacent to St Pancras to replace the cramped King's Cross Thameslink station. Successive delays have seen the target completion date slip to 2010, at which point Peterborough, King's Lynn and Letchworth to the north of London, along with Littlehampton,

Eastbourne, Horsham, Dartford and Ashford to the south of London, will be brought into the service net.

Traction and rolling stock

Thameslink operates over 25 kV AC overhead electrified lines to the north of London and 750 V DC third rail in the south. The changeover point is at Farringdon in central London.

The company's rolling stock comprises Class 319 dual-voltage emus leased from Porterbrook; there are 66 four-car units in the fleet. From the end of 1997, these have been refurbished and divided

Thameslink Class 319/4 emu forming a Bedford–Brighton service at Coulsdon (Ken Harris)   0554378

Thameslink Rail electric multiple-units

Class	Cars per unit	Motor cars per unit	Motored axles/car	Power/motor kW	Speed km/h	Cars in service	First Built	Builders Mechanical	Electrical
319/3	4	1	4	247	160	104	1987	BREL	GEC
319/4	4	1	4	247	160	160	1990	BREL	GEC

into two sub-classes: 26 have become Class 319/3 for inner suburban work, while the remaining 40 units are designated Class 319/4 and incorporate two-class accommodation for long-distance work.

During refurbishment the vehicles have been reliveried.

In addition, to meet demand for increased passenger capacity, Thameslink hires two Class 319

emus from South Central and four Class 317 (AC-only) emus from West Anglia Great Northern.

In 1999, Thameslink signed an innovative contract with Porterbrook under which the entire fleet is operated under a contract hire agreement which includes all maintenance. This was claimed to be the first time such an arrangement was used in the UK rail industry.

Virgin Trains

Virgin Rail Group Ltd
North Wing Offices, Euston Station, London NW1 2HS
Tel: (+44 20) 79 04 32 11   Fax: (+44 20) 79 04 32 34
Web: www.virgintrains.co.uk

Key personnel
Chief Executive: Tony Collins
Executive Directors
   Strategic Planning: Chris Tibbits
   Commercial: Graham Leech

Finance: Linda Bell
Managing Director, West Coast: Charles Belcher
Managing Director, CrossCountry: Chris Gibb
Director, Human Resources: Patrick McGrath
Director, Corporate Affairs: Arthur Leathley

Political background
Under the arrangements put in place by the 1993 Railways Act, the franchise to run operations on British Rail's InterCity CrossCountry network was awarded to the Virgin Rail Group in January 1997. The franchise for InterCity West Coast services followed in March 1997. Both were awarded for 15 years.

In October 1998, Stagecoach Holdings acquired a 49 per cent shareholding in the Virgin Rail Group. The remaining 51 per cent is owned by Virgin Management.

Organisation
The two franchises held by the Virgin Rail Group are overseen by an integrated management team and trade collectively as Virgin Trains, although to meet the rules of the Strategic Rail Authority they are legally separate entities.

UPDATED

Virgin CrossCountry

CrossCountry Trains Ltd
Meridian, 4th Floor West, 85 Smallbrook, Queensway, Birmingham B4 4HA
Tel: (+44 121) 654 75 81
Web: www.virgintrains.co.uk

Key personnel
Chief Executive: Tony Collins
Managing Director: Chris Gibb
Director, Finance: Linda Bell
Director, Corporate Affairs: Arthur Leathley

Political background
Under the arrangements put in place by the 1993 Railways Act, the franchise to run operations on British Rail's InterCity CrossCountry division was let in January 1997 to the Virgin Rail Group Ltd, a subsidiary of the Virgin airline company. Virgin also runs the West Coast franchise.

The franchise was originally let for a 15-year term. Virgin was to receive £112.9 million in the

CrossCountry diesel multiple-units

Class	Cars per unit	Motor cars per unit	Power/ motor kW	Speed km/h	Units in service	First built	Mechanical	Builders Engine	Transmission
220	4	4	560	200	34	2000	Bombardier	Cummins	Alstom
221	5	5	560	200	40	2002	Bombardier	Cummins	Alstom
221	4	4	560	200	4	2002	Bombardier	Cummins	Alstom

first full financial year of the franchise (1997 prices) from the Strategic Rail Authority support, reversing to a premium payment from Virgin to the Franchising Director of £10 million in 2011–12. Attempts by the Virgin Rail Group and the SRA to renegotiate the franchise failed to lead to a revised contract and since 2004 the company has operated services under a Letter Agreement. With the SRA's responsibilities transferred to the Department for Transport in 2005, this left the government free to invite new tenders for the franchise.

Passenger operations
Virgin CrossCountry runs long-distance passenger services over a 2,503 km network throughout

Britain, although few of its services reach London. Its route map has the form of a cross intersecting in Birmingham, with northern axes serving Scotland, northeast and northwest England and southern axes serving the west of England, the south coast and South Wales.

Full commissioning of a new fleet of trains in September 2002 initially resulted in a virtual doubling of the number of services operated each day to 215, compared with 110 in 1997, although following discussions with the Strategic Rail Authority, the level of service was cut back in 2003 to levels that posed fewer delivery problems, with trains no longer serving some destinations.

Traffic (million)	2001–02	2002–03	2003–04
Passenger journeys	14.9	17.8	19.2
Passenger-km	2,423	2,577	2,666

Traction and rolling stock
In 2002 Virgin CrossCountry completed commissioning of a new fleet of diesel-electric multiple-units which replaced the locomotive-hauled stock and subsequently the InterCity 125 HST diesel trainsets that it inherited from British Rail. Supplied by Bombardier Transportation, the fleet consists of 40 five-car and four four-car Class 221 Super Voyager tilting demus and 34 four-car Class 220 Voyager non-tilting units. The contract to supply these trains also covers their maintenance until the end of the original 15 year franchise period. To fulfil this requirement, Bombardier constructed a dedicated maintenance facility at Central Rivers, near Burton-on-Trent.

Virgin CrossCountry Class 221 tilting diesel-electric multiple-unit forming a northbound service between Birmingham and Burton-on-Trent (Ken Harris)   **NEW**/1122828

UPDATED

Virgin West Coast

West Coast Trains Ltd
North Wing Offices, Euston Station, London NW1 2HS
Tel: (+44 20 7) 983 80 00
Web: www.virgintrains.co.uk

Key personnel
Chief Executive: Tony Collins
Managing Director: Charles Belcher

Directors
   Finance: Linda Bell
   Corporate Affairs: Arthur Leathley

Political background
Under the arrangements put in place by the 1993 Railways Act, the franchise to run long-distance passenger services on the West Coast main line was let to Virgin Rail Group (see entry in UK section of *Railway systems and operators*), a subsidiary of the airline operator Virgin, in March

1997. Virgin also runs the CrossCountry franchise (see entry in UK section of *Railway systems and operators*).

The franchise was originally let for a 15-year term; it was expected to move into profit in 2002–03, with the SRA support, in 1997 prices, of £76.8 million in the first full year of the franchise, turning to a premium payment of £220.3 million in 2011–12. However, delays and cost over-runs in the project to upgrade the West Coast Main Line, led to a re-negotiation of the contract and from 2003 the business has

Electric trainsets

Class	Cars per unit	Motor cars per unit	Motored axles/car	Power/ motor kW	Speed km/h	Units in service	First built	Builders Mechanical	Electrical
390	9	7	2	425	225*	53	2001	ALSTOM	ALSTOM

* Design speed; operational speed is 200 km/h.

Diesel locomotives

Class	Wheel arrangement	Power kW	Speed km/h	Weight tonnes	No in service	First built	Mechanical	Builders Engine	Transmission
57/3	Co-Co	2,050	153	117	16	2003*	Brush, BR	GM 12-645 F3B	Brush

* Rebuilt from Class 47.

been run under an ongoing contract of agreement pending renegotiation of a new franchise.

Passenger operations

Virgin West Coast runs services on a 1,075 km network out of London's Euston terminus to the West Midlands, the northwest of England, North Wales and central Scotland.

Traffic (million)	2001–02	2002–03	2003–04
Passenger journeys	16.4	15.2	14.9
Passenger-km	3,177	2,897	2,745

Improvements to existing lines

In October 1997 Virgin Rail Group and Railtrack (now Network Rail) jointly announced their agreement of a two-stage scheme to upgrade and re-equip the West Coast main line, with the £5.8 billion cost funded by a combination of track access charges and a revenue-sharing arrangement. A first phase, to have been completed by June 2002, would have allowed Virgin trains to operate between London Euston and Crewe at speeds up to 200 km/h, while a second phase, originally scheduled for commissioning by 2005,

would further increase speeds to 225 km/h. The investment would have allowed Virgin to double the number of paths it uses into and out of Euston station.

The timetable for upgrading the line subsequently slipped and its estimated cost increased substantially. Significant changes in the structure of the UK rail industry in 2001–02 in the wake of Railtrack being placed in administration led to overall responsibility for the West Coast main line upgrade passing initially to the Strategic Rail Authority and subsequently to Network Rail. Together, these changes were contributory factors to the franchise renegotiation referred to above.

In the meantime, under Network Rail management, modernisation of the West Coast main line has continued. A key stage in the upgrading project was reached in September 2004, when 200 km/h services were introduced between London Euston and Manchester. Further work was expected to see the complete London–Glasgow route approved for 200 km/h running in 2005–06. However, a further increase in speed to 225 km/h will necessitate provision of cab signalling, a step which remained deferred in 2005.

Traction and rolling stock

Most of the West Coast main line network is electrified at 25 kV AC 50 Hz. In 2005 Virgin completed commissioning a fleet of 53 nine-car Class 390 Pendolino tilting trainsets supplied by ALSTOM and financed by Angel Trains. These replaced the Class 87 and 90 electric locomotives and Mark 2 and Mark 3 stock previously used on the network. Maintenance of the Pendolino trainsets is undertaken by an ALSTOM subsidiary, West Coast Traincare. For services over the North Wales Coast line beyond Crewe to Holyhead, which is not electrified, Virgin West Coast employs Class 221 diesel-electric tilting trainsets drawn from the fleet of sister operator, Virgin CrossCountry.

To provide haulage of Class 390 trainsets during diversions over non-electrified routes or in exceptional conditions when the overhead power is not available, Virgin leases 16 Class 57/3 diesel-electric locomotives from Porterbrook. Created by Brush Traction by refurbishing and converting Class 47 machines, the Class 57/3s have been repowered with a General Motors 12-645 F3B engine rated at 2,050 kW.

Northbound Virgin West Coast Class 390 Pendolino tilting trainset near Stafford (Ken Harris)

NEW/1122827

UPDATED

WAGN

West Anglia Great Northern Railway Ltd
National Express Group, London Lines
Hertford House, 1 Cranwood Street, London EC1V 9QS
Tel: (+44 20) 79 28 51 51
Web: www.wagn.co.uk

Key personnel

Managing Director: Andrew Chivers
Directors
   Finance: Adam Golton
   Operations: Mark Hopwood
   Commercial: David Hamilton

Political background

Formerly West Anglia Great Northern and still known as WAGN, this operator is the residual portion of a franchise let in December 1996 to Prism Rail plc, a company set up to run rail franchises by a consortium of bus companies. In November 2000 Prism sold its rail operating businesses to National Express Group and Great

Great Northern electric multiple-units

Class	Cars per unit	Motor cars per unit	Motored axles/car	Power/ motor KW	Speed km/h	Cars in service	First built	Builders Mechanical	Electrical
313*	3	2	2	82	120	123	1976	BREL	GEC
317	4	1	4	247	160	52	1981	BREL	GEC
365	4	2	2	140	160	160	1994	ABB	GEC

* Dual-voltage, 25 kV AC, 750 V DC.

Northern now forms part of its London Lines rail operation.

The original franchise was let for a term of seven years and three months. Prism was to receive from the Franchising Director support, in 1996 prices, of GBP52.9 million in the first full financial year of the franchise, reversing to a premium payment of GBP24.8 million in 2004.

On expiry of the franchise at the end of March 2004 the SRA split it. In April 2004 WAGN services operating into and out of London Liverpool Street station became part of the One Railway franchise along with routes formerly operated by Anglia

Railways and First Great Eastern. Remaining services using London King's Cross and Moorgate stations are still known as WAGN and are operated by National Express Group until March 2006 under a two-year extension contract awarded by the Strategic Rail Authority in February 2004. After that date operations were to be integrated with an enlarged Thameslink franchise.

Passenger operations

WAGN operates a 380 km network of suburban (Great Northern Inner) and regional (Great Northern

---

For details of the latest updates to *Jane's World Railways* online and to discover the additional information available exclusively to online subscribers please visit

jwr.janes.com

Outer) services on lines out of London's King's Cross and Moorgate stations, with services reaching as far north as King's Lynn and Peterborough. All services are operated with emus.

Traffic (million)	2004–05
Passenger journeys	37.8
Passenger-km	1,339

Wensleydale Railway plc

35 High Street, Northallerton DL7 8EE
Tel: (+44 1609) 77 93 68
e-mail: admin@wrplc.fsnet.co.uk
Web: http://www.culbard.demon.co.uk

Key personnel
Chairman: Keith Cameron
Chief Executive: Scott Handley
Directors
   Marketing: Ruth Annison
   Commercial: Colin Brown
   Property and Utilities: Andrew Maude
   Engineering: Clive Roberts
   Safety: Mark Flather
   Mechanical Engineering: Steve Deane

Gauge: 1,435 mm
Route length: 35 km

Wessex Trains

2nd Floor, Broadwalk House, Southernhay West, Exeter EX1 1TS
Tel: (+44 1392) 47 31 00
Web: www.wessextrains.co.uk

Key personnel
Managing Director: Alan Wilson
Finance Director: James Griffin
Commercial Director: (vacant)
Director of Operations and Planning: Garry Raven

Political background
Under the arrangements put in place by the 1993 Railways Act, the franchise to run operations on South Wales & West Railway was let in October 1997 to Prism Rail plc, a company set up by a consortium of bus companies. Prism also ran the Cardiff Railway (subsequently Valley Lines), LTS Rail (c2c) and West Anglia Great Northern franchises. The company subsequently changed its name to Wales & West Passenger Trains Ltd.

The franchise was let for a term of seven years and six months. Prism was to receive from the Franchising Director (now the SRA) support, in 1996 prices, of GBP70.9 million for the first full financial year of the franchise, declining to GBP38.1 million in the last year of the franchise.

In July 2000 Prism sold its rail operating business to National Express Group.

In 2001 agreement was reached with the Strategic Rail Authority to extend the franchise to April 2004. In October 2001 former Wales & West services from Wales to some English destinations and within England were combined to form the Wessex Trains managerial unit, pending the eventual awarding of a new franchise contract. Subsequently, the franchise contract with National Express Group was extended to February 2006, by which time routes served by Wessex Trains will be included in a new 'Greater Western' franchise together with lines covered by the existing First Great Western and First Great Western Link (formerly Thames Trains) contracts.

Services in Wales formerly part of the Wales & West franchise and the former Valley Lines franchise are now operated by Arriva Trains (see entry for Arriva Trains Wales in UK section of *Railway systems and operators*).

Passenger operations
Wessex Trains runs services over a 1,394 km network serving 125 stations. It operates interurban Alphaline-branded services from Bristol

Traction and rolling stock
WAGN is exclusively operated by emus operating on the 25 kV AC overhead system, apart from a short 750 V DC third rail section to Moorgate. The company leases 41 Class 313s and examples of Classes 317 and 41 Class 365. In the case of the last-mentioned, the fleet was augmented in 2004 by the transfer from South Eastern Trains of examples of

Background
In May 2003 Wensleydale Railway plc was granted licences by the Rail Regulator covering: the management of infrastructure on a 35 km line in northeast England from Northallerton, on the East Coast Main Line, to Redmire; the operation of passenger and freight trains over it; the operation of stations; and train maintenance. This coincided with the handover of the line to the company on a 99-year lease from Network Rail.

Wensleydale Railway plc was formed in November 2000 by the Wensleydale Railway Association to raise funds by public share issue to reinstate the railway and eventually extend it. Future aims see the line continuing via Hawes to be reconnected in the west to the national network at its former junction with the Settle–Carlisle line at Garsdale.

Two Wessex Trains Class 158 dmus forming a Cardiff–Portsmouth Alphaline service at Southampton (Ken Harris)
0554926

Wessex Trains diesel multiple-units

Class	Cars per unit	Motor cars per unit	Motored axles/car	Power/ motor kW	Speed km/h	Cars in service	First built	Mechanical	Builders Engine	Transmission
143	2	2	1	152	120	16	1985	Alexander/ Barclay	Cummins LTA10R	H Voith T211r
150/2	2	2	2	210	120	50	1986	BREL	Cummins NT855R5	HM Voith T211r
153	1	1	2	213	120	15	1987*	Leyland	Cummins NT855R5	HM Voith T211r
158	2/3	2/3	2	275/300	145	46	1989	BREL	Cummins NTA855R	HM Voith T211r

* Rebuilt 1991.

to Cardiff, Portsmouth, Brighton, Exeter, Plymouth and Penzance and local services to Gloucester, Swindon, Weymouth, Worcester and throughout southwest England.

Traffic (million)	2002–03	2003–04	2004–05
Passenger journeys	9.8	10.2	10.9
Passenger-km	399	409.6	435

Traction and rolling stock
Wessex trains is an all-diesel-operated business worked with second-generation dmus. For its

the type which had themselves been replaced by new 'Electrostar' units.

Units used on Great Northern Outer services have been refurbished to a standard comparable with that of the Class 365s.

UPDATED

Passenger operations
A passenger service running four times a day in each direction between two intermediate stations, Leeming Bar and Leyburn (19 km), was introduced in July 2003, 49 years after passenger trains last ran on the line.

Freight operations
The line sees infrequent freight trains serving Ministry of Defence facilities at Redmire, west of the present limit of passenger working at Leyburn. These are operated by English Welsh & Scottish Railway.

Traction and rolling stock
A Class 107 three-car first-generation dmu was used to launch services in July 2003.

interurban routes, the company employs Class 158 dmus leased from Angel Trains. For local services, it leases Class 153 single cars from Angel and Class 143 and 150/2 dmus from Porterbrook. Peak period relief and seasonal services are provided by Class 31 locomotives and Mark 2 coaches hired from FM Rail.

UPDATED

United States

Department of Transportation

400 7th Street SW, Washington DC 20590
Tel: (+1 202) 366 40 00
Web: www.dot.gov

Key personnel
Secretary: Norman Mineta
Deputy Secretary: (vacant)
Under Secretary, Policy: Jeffrey Shane
General Counsel: Jeffrey Rosen
Assistant Secretaries
  Administration: Vincent Taylor
  Government Affairs: Nicole Nason
  Policy: Emil Frankel
  Budget and Programmes: Linda Combs
Director of Public Affairs: Robert Johnson

Surface Transportation Board

1925 K Street NW, Washington DC 20423-0001
Tel: (+1 202) 565 16 74   Fax: (+1 202) 927 61 07
Web: www.stb.dot.gov

Key personnel
Chairman: Roger P Nober
Vice-Chairman: Wayne O Burkes
General Counsel: Ellen D Hanson
Secretary: Vernon A Williams

Organisation
Created in 1967, the US Department of Transportation (USDOT) is steward of the nation's transport systems and speaks for transport in the Federal government. Its mission is to develop policies and programmes that contribute to provision of safe, efficient and convenient transport at the lowest possible cost. In 2003 there were some 118,000 employees in USDOT's three departmental and 10 operating administrations, including the Federal Railroad Administration (see entry in US section of Railway systems and operators) and the Federal Transit Administration. Also included is the three-member Surface Transportation Board (STB), which replaced the Interstate Commerce Commission (ICC) and assumed its responsibilities in matters such as rates and service issues, restructuring transactions (such as mergers, line sales and abandonments) and related labour matters.

UPDATED

Federal Railroad Administration

1120 Vermont Street NW, Washington DC 20005
Tel: (+1 202) 493 60 14   Fax: (+1 202) 493 64 01
Web: http://www.fra.dot.gov

Regional Administrators of Railroad Safety
Region 1: 55 Broadway, 10th Floor, Room 107, Cambridge, Massachusetts 02142
Tel: (+1 617) 494 23 21

Region 2: Scott Plaza 2, Suite 550, Philadelphia, Pennsylvania 19113
Tel: (+1 610) 521 82 00

Region 3: 1720 Peachtree Road NW, Suite 440 North Tower, Atlanta, Georgia 30309
Tel: (+1 404) 347 27 51

Region 4: 111 N Canal Street, Suite 655, Chicago, Illinois 60606
Tel: (+1 312) 353 62 03

Region 5: 8701 Bedford Euless Road, Suite 425, Hurst, Texas 76053
Tel: (+1 817) 334 36 01

Region 6: 1807 Federal Building, 911 Walnut Street, Kansas City, Missouri 64106-2095
Tel: (+1 816) 426 24 97

Region 7: 650 Capital Mall, Room 7007, Sacramento, California 95814
Tel: (+1 916) 551 12 60

Region 8: Murdock Building, 703 Broadway Street, Suite 650, Vancouver, Washington 98660
Tel: (+1 206) 696 75 36

Key personnel
Administrator (acting): Betty Monro
Chief Counsel: S Mark Lindsey
Directors
  Budget: D J Stadtler

Civil Rights: Carl Ruiz
International Policy: Ted Krohn
Research and Development: Jo Strang

Organisation
The Federal Railroad Administration develops rules and regulations to implement national transportation law affecting railroads. FRA rules apply to some 40 major areas, including driver qualifications; maximum permissible hours of service for railroad personnel; conduct of railroad police; rolling stock safety standards; alcohol and drug testing; and level crossing and signalling system safety.
  FRA also manages research and development projects and contracts to operate FRA's Transportation Technology Center in Pueblo, Colorado with the Association of American Railroads.

Federal Transit Administration (FTA)

Department of Transportation, 400 7th Street SW, Washington DC 20590
Tel: (+1 202) 366 40 40
Fax: (+1 202) 366 98 54
Web: http://www.fta.dot.gov

Key personnel
Administrator: Jennifer L Dorn
Deputy Administrator: Robert Jamison

Organisation
The Federal Transit Administration (FTA) was formerly the Urban Mass Transportation Administration and is the agency responsible for providing federal financial assistance to US cities to improve mass transit. Grants for capital projects, including the acquisition of rolling stock, involve 80 per cent federal and 20 per cent local funding; FTA also provides funds for public transport planning, research and operations.
  FTA has 10 field offices.

Aberdeen & Rockfish Railroad Co

PO Box 917, Aberdeen, North Carolina 28315
Tel: (+1 910) 944 23 41   Fax: (+1 910) 944 97 38
e-mail: aberdeen.rockfish.rr@worldnet.att.net
Web: www.aberdeen-rockfish.com/

Key personnel
President: Edward P Lewis
Vice-President, Traffic: Paul McArdle

Organisation
A&RR has 75 route-km linking Aberdeen with Raeford, Fayetteville and River Terminal, making connections with CSXT in Aberdeen and both CSXT and NS in Fayetteville, North Carolina. The company also owns the Pee Dee River railway in South Carolina, linking McColl, Bennettsville and Marlboro Mills (36 km). Annual average carloadings for both lines are 12,000.

Traction and rolling stock
The company operates nine locomotives (four GP16s; two GP38s; one each of Classes CF7, GP7 and GP18) and 150 freight wagons.

UPDATED

Aberdeen, Carolina & Western Railway Co

102 Depot Street, PO Box 586, Star, North Carolina 27356
Tel: (+1 910) 428 90 30
Fax: (+1 910) 428 99 30
Web: http://www.acwr.com

Key personnel
President: Robert M Menzies II
General Manager: William G Bartosh

Gauge: 1,435 mm
Track length: 225 km

Organisation
The present ACWR was created in 1987 to purchase the NS line from Aberdeen to Star and in 1989 became the contract operator of Norfolk Southern's Charlotte–Gulf line (165 km) under the terms of the NS 'Thoroughbred Short Line Program'. It has 20 employees. Connection is made to CSXT and A&R at Aberdeen, with CSXT and NS at Charlotte, with Winston-Salem Southbound at Norwood and with NS at Gulf. The Charlotte–Gulf line has benefited from a US$2 million improvement programme. Total tonnage carried in 2002 was 1,188,000, generating some 12,000 carloads. Commodities carried include timber and forest products, plastics, construction materials, propane gas and Solite rock products.

Traction and rolling stock
The company owns 12 locomotives of GM-EMD origin.

Alabama & Florida Railway Co

1510 East Three Notch Street, Andalusia, Alabama
36420

A subsidiary of Pioneer Railcorp
1318 S Johansen Road, Peoria, Illinois 61607
Tel: (+1 309) 697 14 00   Fax: (+1 309) 697 53 87
Web: http://www.pioneer-railcorp.com

Key personnel
See entry for Pioneer Railcorp.

Gauge: 1,435 mm
Route length: 123 km

Organisation
Acquired from A&F Inc in 1992, AF runs from
Georgiana, Alabama, where it interchanges with

CSXT, to Geneva. Main commodities carried
are chemicals, forest products and agricultural
products.

Alabama & Gulf Coast Railway

136 North Mount Pleasant Avenue, PO Box 339,
Monroeville, Alabama 36461
Tel: (+1 251) 575 50 08   Fax: (+1 251) 575 19 41
Web: www.railamerica.com

Key personnel
General Manager: Mike Brigham
Business Development: Michael Koile

Gauge: 1,435 mm
Route length: 225 km

Organisation
Acquired by RailAmerica as a result of its takeover
in 2002 of StatesRail, AGR runs from Kimborough,
Alabama, to Pensacola, Florida, and has trackage
rights over NS to Mobile, Alabama. The railway
handles approximately 40,000 carloads annually,
mainly of forest products, chemicals, clay and

food products. Interchange is made with BNSF
at Magnolia, Alabama, via trackage rights from
Kimborough, with CSXT at Hybart via trackage
rights from Atmore and with CSXT at Cantonment,
Florida.

UPDATED

Alabama Railroad Co

RR2, Box 209, Monroeville, Alabama 36460

A subsidiary of Pioneer Railcorp
1318 S Johansen Road, Peoria, Illinois 61607
Tel: (+1 309) 697 14 00   Fax: (+1 309) 697 53 87
Web: http://www.pioneer-railcorp.com

Key personnel
See entry for Pioneer Railcorp.

Gauge: 1,435 mm
Route length: 89 km

Organisation
Acquired from CSX Transportation in 1991,
ALAB runs from Flomaton, Alabama, where
it interchanges with CSXT, to Corduroy. Main
commodities carried are forest products and
cement.

The Alaska Railroad

The Alaska Railroad Corporation
PO Box 107500, Anchorage, Alaska 99510-7500
Tel: (+1 907) 265 23 00   Fax: (+1 907) 265 23 12
Web: http://www.alaskarailroad.com

Key personnel
President and Chief Executive Officer: Pat Gamble
Vice-Presidents
  Chief Operating Officer: Matt Glynn
  Chief Mechanical Officer: Robert Stout
  Corporate Affairs: James B Blasingame
  Real Estate: James Kubitz
  Finance and Administration: Bill O'Leary
  General Counsel: Phyllis C Johnson
Superintendent of Transportation:
  Patrick C Shake
Chief Engineer: Tom E Brooks
Chief Mechanical Engineer: Joshua D Coran

Gauge: 1,435 mm
Length: 846 km

Organisation
The Alaska Railroad (ARR) runs a single-track
main line of 756 km from the ports of Seward on
the Gulf of Alaska, and Whittier on Prince William
Sound, northward through Anchorage and Denali
(formerly McKinley) National Park to Fairbanks,
and eastward to Eielson Air Force Base, with a
branch to Palmer.

The ARR was built in 1923 by the federal
government to open up the state and was
transferred to state ownership in 1985. It is a quasi-
public corporation with a seven-member board of
directors appointed by the Governor of Alaska. Net
earnings in 2001 were US$6.6 million.

Passenger operations
Unique as the only full-service US railroad still
offering both passenger and freight services, ARR
operates passenger service along a large section
of the system. In season (May-September) there
are two scheduled passenger trains with company-
owned passenger and dome cars, connecting with
scenic cruises and other attractions, and dome car
services provided by Princess Tours and Holland
America (Westours). Winter local passenger
service operates on Saturdays/Sundays to link the
state's two major cities. In 2001, the railway carried
472,000 passengers.

Freight operations
Freight connections are made with the US and
Canadian rail systems to the south via Akaska Rail
Marine Service (rail wagons and loose-stowed
freight on barges) weekly from Seattle and CN
Rail's Aquatrain (rail wagons on barges) from
Prince Rupert. In 2001, 7.8 million tons (96,488
carloads) were carried.

Recognising the essential nature of the service
to the state's inhabitants, the Federal Railroad
Administration has in recent years awarded
grants to the railroad for trackwork to improve the
performance of the passenger operations, which
operate at 50 km/h overall.

Alaska Railroad diesel locomotives

Class	Wheel arrange-ment	Transmission	Rated power kW	Max speed km/h	Total Weight Tonnes	No in service (rebuilt)	First built (rebuilt)	Mechanical	Builders Engine	Transmission
1550 (MP15)	B-B	DC Elec	1,120	115	115	4	1978	EMD	EMD 12–645	EMD
2000 (GP38-2/38U)	B-B	AC/DC Elec	1,492	113	116	8	1985	EMD	EMD 16–567	EMD
2500 (GP35/35U)	B-B	DC Elec	1,865	115	118	4	1981	EMD	EMD 16–567	EMD
2800 (GP49)	B-B	AC/DC Elec	2,087	115	113	9	1983	EMD	EMD 12–645	EMD
3000 (GP40-2/40U)	B-B	AC/DC Elec	2,238	115	121	19	1975	EMD	EMD 16–645	EMD
4000 (SD70MAC)	C-C	AC/DC Elec	2,985	–	–	6	1999	EMD	EMD 16–710	EMD

*Alaska Railroad General Motors SD70MAC
locomotive at Anchorage with a passenger
train that includes newly acquired
semi-double-deck coaches
(Wolfram Veith)*
0531908

Traction and rolling stock

In service in 2001 were 52 diesel-electric locomotives, 39 passenger cars, four RDCs and 1,672 wagons. Six locomotives were leased for the summer of 1998 to power infrastructure trains. In 1999–2000, ARR added six new General Motors SD70MAC locomotives to its fleet.

Albuquerque Commuter Rail

Mid-Region Council of Governments
317 Commercial NE, Suite 104, Albuquerque, New Mexico 87102
Tel: (+1 505) 247 17 50   Fax: (+1 505) 247 17 53
Web: www.mrcog-nm.gov

Key personnel

Executive Director, MRCOG: Lawrence Rael
Director of Transportation: Chris Blewett
Projects Planner, Commuter Rail: Tony Sylvester

Organisation

The New Mexico Department of Transportation and the Mid-Region Council of Governments are developing a 74 km commuter route running from Albuquerque southwards to Belen and northwards to Bernalillo over BNSF right-of-way, with start-up service expected to begin by the end of 2005. Following signing of preliminary understandings in September 2004, BNSF granted trackage rights for the operation, and is making track and signal upgrades to accommodate passenger trains. A final agreement will determine access rights to the track and adjacent land for construction of up to seven stations, and will agree a programme of capital improvements. Budgeted cost of the initial operation is USD75 million. A second phase of development will see a northwards extension of 121 km to Santa Fe in operation by 2008.

Passenger operations

In 2005 Herzog Transit Services Inc was selected to operate and maintain the service, which is to be branded 'New Mexico Rail Runner Express'.

Type of coupler in standard use: AAR Automatic Knuckle
Type of braking in standard use: AAR Standard Air

Track

Rail: 115 lb/yd RE-Standard Carbon (57.16 kg/m)
Crossties (sleepers): Treated fir and hardwood 2,045 × 203 × 178 mm

Spacing: 2,019/km
Fastenings: 4- or 6-hole angle bars, steel spikes
Max curvature: 14.5°
Max gradient: 3%
Max axleload: 30.726 tonnes (67,750 lb)

Traction and rolling stock

Operations will commence with a fleet of 10 bilevel cars ordered from Bombardier in October 2004 at a cost of USD22 million, comprising six cab cars and four trailers. The first of these was handed over in June 2005. In January 2005 MRCOG agreed a USD12.7 million contract with Wabtec Corporation's MotivePower subsidiary for supply of five 3,600 hp MPXpress diesel-electric locomotives and spare parts, to be delivered by the end of 2005.

NEW ENTRY

Altamont Commuter Express Authority (ACE)

949 East Channel Street, Stockton, California 95202
Tel: (+1 209) 944 62 24   Fax: (+1 209) 944 62 25
e-mail: info@acerail.com
Web: www.acerail.com

Key personnel

Executive Director: Stacey Mortensen
Director of Operations and Customer Services:
   Gregg Baxter
Accounting Manager: NIla Cordova
Information Technology Manager: Hal Singer
Marketing Manager: Mike Steenburgh
Planning and Capital Projects Manager:
   Brian Schmidt
Operations Manager: Cliff Smith
Facilities and Training Manager: John Vazquez

Political background

The managing agency for the ACE service, which commenced operations in 1998, is the San Joaquin Regional Rail Commission (SJRRC), formed by seven cities and the county of San Joaquin. In turn, the SJRRC, the Almeda Congestion Management Agency and the Santa Clara Valley Transportation Authority subsequently created the Altamont Commuter Express Joint Powers Authority (JPA), which oversees the participation and financial commitment of each member agency. Revenue support comes from a Measure K sales tax in the county of San Joaquin.

Organisation

Altamont Commuter Express (ACE) provides weekday commuter rail services over a 138 km corridor from Stockton, in the richly agricultural San Joaquin Valley, to San Jose, at the heart of the US computer manufacturing industry, helping to relieve three of the most notorious highway bottlenecks in the greater San Francisco area. Services operate over Union Pacific (UP) tracks via

An F40PHM-3C arrives at San Jose with empty coaches for a service to Stockton (Philip Wormald)
NEW/1122825

intermediate station stops at Stockton, Manteca/Lathrop, Tracy, Livermore (2), Pleasanton, Fremont, Santa Clara (2) and San Jose.

Passenger operations

The ACE train currently undertakes three daily runs at peak commuter times. Following a strong start-up in October 1998, ridership grew rapidly to 3,200 daily, but then fell back to no more than 2,000 a day during 2002. This decline, due in part to delays caused by UP freights on the shared route, has forced ACE to drop plans for a fourth daily run as well as longer-term proposals for 12 trips and extension of service to Modesto. The poor ridership also rules out ACE's aim of purchasing its right-of-way from UP. Nevertheless, funding has been made available for some track upgrading work.

Farebox recovery ratio after three years of operation was targeted at between 5 and 60 per cent. The contracted operator is Herzog Transit Services.

Additional bus shuttle services connect ACE with San Francisco's BART metro system at Pleasanton, with Santa Clara Valley Transportation Authority's light rail system at Santa Clara, and with Amtrak's San Jose–Oakland corridor.

Traction and rolling stock

Services are operated with five 3,200 hp (2,385 kW) F40PHM-3Cs from Boise Locomotive and 20 Bombardier gallery cars, including nine cab cars.

UPDATED

Amtrak

National Railroad Passenger Corporation
Washington Union Station, 60 Massachusetts Avenue, New England NE, Washington DC 20002-4285
Tel: (+1 202) 906 38 57   Fax: (+1 202) 906 38 65
Web: www.amtrak.com

Key personnel

President and Chief Executive Officer: David L Gunn

Senior Vice President Operations: William Crosbie
General Counsel & Corporate Secretary:
   Alicia M Serfaty
Chief Financial Officer: David Smith
Inspector General: Fred Weiderhold
Vice Presidents
   Government Affairs and Policy:
      Joseph H McHugh
   Business Diversity: Gerri Mason Hall
   Labor Relations: Joseph M Bress
   Human Resources: Lorraine A Green

Procurement and Materials Management:
   Michael J Rienzi
Strategic Planning and Contract Administration:
   Gilbert Mallery
Customer Services: Emmett Fremaux
Marketing and Sales: Barbara J Richardson
Transportation: Ed Walker
Security: Alfred J Broadbent, Sr
Chief Mechanical Officer: Vince Nesci
Chief Engineer: David Hughes
Chief of Police: Sonya Proctor

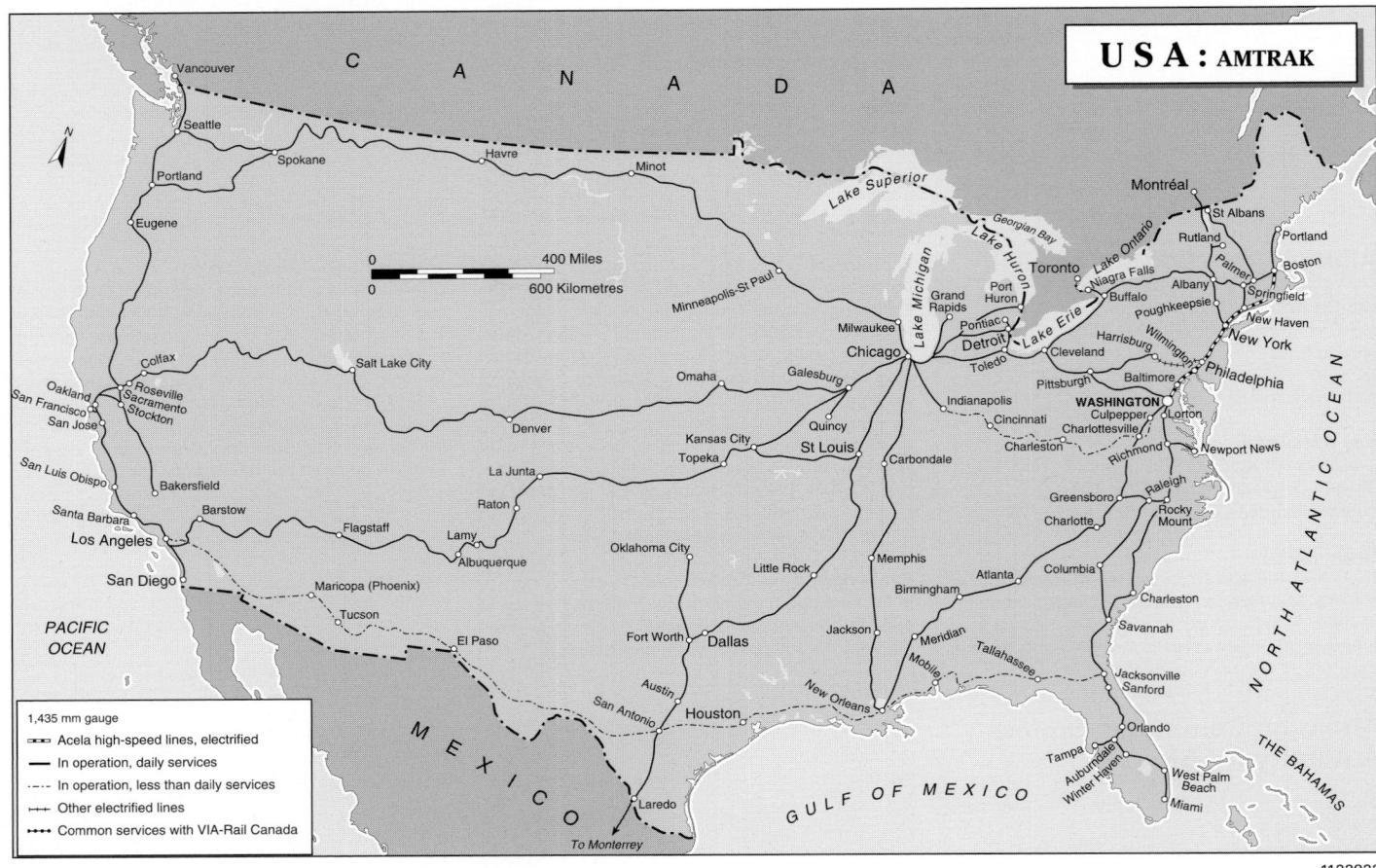

USA : AMTRAK

1122822

Chief Operations Planning Officer: (vacant)
AVP Environmental: Roy Deitchman

Gauge: 1,435 mm
Route length owned: 1,256 km; network operated: 35,200 km
Electrification: 735 km (554.6 km at 12 kV AC 25 Hz; remainder at 25 kV AC 60 Hz)

Political background

Amtrak was created when the Rail Passenger Service Act was enacted in 1970. Services began in May 1971, establishing the first nationwide rail passenger service under one management in the US.

Amtrak's rail passenger service is totally dependent upon the condition of track and related facilities that are owned, designed, maintained and operated by the private freight-hauling railroads. The only exceptions to this are where Amtrak owns its own track: in the Boston-New York-Washington North East Corridor (NEC), on short sections of track elsewhere, and in several major cities where the corporation has acquired passenger terminals.

Amtrak has been supported by federal capital and operating grants, the amount of which is annually budgeted by Congress as part of the overall transportation authorisation. The process generally consists of a joint House and Senate committee reconciling a House figure and a Senate figure with the requests submitted by Amtrak and by the Administration.

For the past several years, Amtrak had been under a Congressional mandate to streamline operations so that operating subsidies could be eliminated by 2002. An unwritten aspect of this intention was that Amtrak should be provided with adequate capital assistance.

In December 1997, President Clinton signed into law the operator's reauthorisation bill, which, contrary to expectations, had been passed by the US Congress during the previous month. The Amtrak Reform and Accountability Act made a further USD2.3 billion in capital improvements available until 2000, as well as providing operating and capital investment funds of up to US$4 billion. The law also relaxed labour agreements to allow Amtrak to contract out certain functions.

Despite these additional funds, Amtrak's condition worsened. The constraint of using the USD2.3 billion on capital operations resulted in many functions that railroads normally consider to be operating costs to be reclassified as capital. The outward impression was that the need for operating funds, and thus achieving self-sufficiency, could be met. In fact, the company was starving for cash. In June 2001, it took out a USD300 million loan secured by Penn Station, New York City, to meet expenses through the end of the 2001 Fiscal Year. For FY2002 Congress appropriated only USD827 million for Amtrak operations despite the continuing cash shortages.

Finance
FY2002-03

President George Warrington left Amtrak in March 2002, and when President David Gunn arrived in May, 2002, it became clear that operational self-sufficiency would not occur. In fact, a cash crisis almost caused the railroad to cease operations in the summer of 2002. An emergency grant as well as a loan from the Department of Transportation provided sufficient cash for operations to continue. For FY2003, Congress appropriated USD1.05 billion. Amtrak had requested USD1.2 billion, an amount that was still heavily influenced by President Warrington's continued belief that self-sufficiency was "just around the corner". The USD1.05 billion came with some several conditions intended to increase the oversight over Amtrak, including continuing to abide by the terms of the June 2002 loan agreement. Unlike previous appropriations, in which the money was directly given to Amtrak, this bill gave the money to the Secretary of Transportation, who was to "approve funding to cover operating losses" of each long-distance route "only after receiving and reviewing a grant request for each specific train route", provided that the financial analysis in each request justifies funding for that route "to the Secretary's satisfaction."

FY2004-05

In preparation for the FY2004 appropriation, Amtrak prepared a five-year Strategic Plan intended to return the railroad infrastructure and equipment to a state of good repair. The total funding needs ranged from USD1.8 billion for FY2004, gradually reducing to USD1.5 billion in FY 2008. The plan submitted to Congress spelled out in detail exactly what equipment would be rebuilt and refurbished, what tracks and interlockings would be replaced in each year, and what signal system components would be rebuilt. Despite the exhaustive detail, and the candour that David Gunn brought to Congress, the Bush Administration only included USD900 million in its FY 2004 request. Congress ultimately appropriated USD1.2 billion for Amtrak, and continued to defer repayment of the 2002 USD100 million loan from the US DoT.

A similar scenario played out in FY2005, with the Bush Administration again proposing only USD900 million for Amtrak, though it stated it would support up to USD1.4 billion if certain reforms were implemented. For FY2005, Amtrak requested USD1.8 billion, an amount justified by the need to keep current trains operating and returning the railroad to a state of good repair. The actual appropriation was USD1.2 billion, with USD709 million for operating support and a requirement that USD492 million be used for capital investment. The appropriation also required that the FY2002 USD100 million DOT Railroad Rehabilitation and Improvement Financing Programme loan be repaid in five equal instalments beginning in FY2005. Due to the lower federal appropriation, several planned programmes had to be scaled back or curtailed. In the aggregate, the FY2005 plan called for infrastructure spending of USD418 million, fleet and facilities spending of USD208 million and other capital expenses of USD80 million. The federal appropriation was supplemented by approximately USD200 million of third-party funding.

FY2006

In FY2006 the Bush Administration proposed no funding for Amtrak, though USD360 million was included to provide continuation of commuter rail services that use the Northeast Corridor (e.g. MBTA, NJT, LIRR, SEPTA, MARC and VRE). The DoT Secretary embarked on a nationwide tour to defend the Administration's proposal arguing that it didn't intend to bankrupt Amtrak, but to merely force the issue of Amtrak reform. Nevertheless, it wasn't until months later that the Administration revealed its reform plan that was basically a repeat of the reforms the Administration called for in the Passenger Rail Investment Reform Act of 2003, that would divide Amtrak into three companies: a private rail infrastructure company that would maintain and operate the infrastructure of the

AEM-7 electric locomotive with a southbound Northeast Corridor service at Newark Liberty International station (Philip Wormald)   *NEW*/1122821

Northeast Corridor under contract to a multi-state 'Northeast Corridor Compact';

a private passenger rail company that would operate trains under contract to states and multi-state contracts;

a 'new' Amtrak that would continue as a government corporation retaining Amtrak's current right to use tracks of the freight railroads and the Amtrak corporate name.

Two months after sending the FY2006 budget to Congress, the Bush Administration issued the 2005 Passenger Rail Investment Reform Act. Though very similar in content to the 2003 Act, it did include conditions for establishing a capital assistance programme with a 50/50 state/federal split that would be made available for states and multi-state compacts to support passenger rail improvements.

In response to the Bush Administration's FY2006 budget proposal, Amtrak requested USD1.82 billion in financial support and tied it to a comprehensive set of strategic initiatives that are aimed at reducing Amtrak's federal subsidy requirement over the next five years. The USD 1.8 billion grant request was divided into six business line areas as follows: Infrastructure Management –SD479 million, NEC Operations – USD28 million, State Corridor Operations – USD166 million, National Long Distance Service – USD537 million, Ancillary Businesses – USD61 million and System Support and Security – USD254 million. For FY2006 Amtrak plans to align financial accounting, planning and management accountability along these five business lines to facilitate future decision-making. The five business lines don't represent a return to the business unit structure, but rather are intended to make the costs and revenues of each segment transparent and separate from other Amtrak costs, such as debt service.

Four key objectives of the grant request and associated reform initiatives were:

• development of passenger rail corridors utilising a 80/20 federal/state matching approach that would have states, rather than Amtrak, lead corridor development initiatives. Amtrak and other interested private operators would be able to bid on the operation of the corridor;

• returning the Northeast Corridor infrastructure to a state of good repair and operational reliability, with phased-in financial responsibility for capital and operating costs assumed on a proportionate basis by all users, including Amtrak, freight and commuter railroads;

• establishment of phased-in financial performance thresholds for Amtrak's existing 15 long-distance trains and any future similar proposed service;

• creation of markets for competition, private commercial participation and industrial reforms in various rail functions.

Whether the required federal legislative reforms will be passed to enable full implementation of Amtrak's strategic reform initiatives remains to be seen. However, Amtrak made the point that it is spending USD1.4 billion annually, an amount insufficient to prevent further deterioration in the condition of its infrastructure and fleet, let alone bring it back to a state of good repair.

Passenger operations

Amtrak's operations cover most of mainland US, but services vary in frequency. The most frequent services are on the Northeast Corridor (NEC) from Washington DC to New York City, where two or more trains operate each hour at speeds up to 225 km/h; and between New York City and Boston at speeds up to 250 km/h. Other corridors with multiple daily trips include New York City to Albany, New York, Washington DC to Richmond, Virginia, routes from Chicago, Illinois, to Detroit, Michigan, Milwaukee, Wisconsin and St. Louis, Missouri; several routes in California (San Diego–Los Angeles–Santa Barbara, San Jose–Oakland–Sacramento, Oakland–Bakersfield); and the Pacific Northwest between Seattle, Washington, and Eugene, Oregon. All national, long distance services are operated at least daily except for routes that have service three times per week.

As part of a 1996 restructuring exercise, Amtrak's activities were reorganised into three strategic business units, namely Northeast Corridor, West Coast and Intercity. These business units were eliminated in October 2002. Operations were divided into seven operating divisions (New England, New York, Mid-Atlantic, Southern Central, Southwest and Pacific). All Division General Superintendents reportto the Operations Vice-President, who, in turn reports to the Senior Vice-President, Operations.

Acela Express high-speed service was introduced in December 2000 with a single Boston–Washington daily round trip.. By the end of 2004 weekday service had increased to 15 Washington–New York round trips, with 10 of those continuing to/from Boston; in addition one New York–Boston round trip was scheduled. The venerable 'Metroliner' service, using refurbished "Concept 2000" Amfleet rolling stock, was scheduled to be eliminated in mid-2005 as availability of Acela Express trainsets improved. Implementation of this plan was delayed for several months when a large number of Acela Express brake discs sidelined the trains until repairs could be made.

Ridership in FY2004 set a new record for the second straight year. Just over 25 million passengers rode Amtrak between October 2003 and September 2004, and increase of over 1 million passengers from FY03. Short-distance and corridor ridership rose 4.3 per cent while ridership on long distance trains rose 3.3 per cent.

Track access

In the Northeast Corridor Amtrak owns its own tracks, but on other routes it has to buy track access from the freight railroads. The original track access contracts between freight railroads and Amtrak expired in April 1996 after a period of 25 years. By September 1996, negotiations had been concluded with several of the host companies and contract extensions agreed with Illinois Central (for 15 years) and Conrail (10 years). An impasse with Burlington Northern Santa Fe was referred to the Surface Transportation Board, then an agreement signed in October 1996 for a 15-year renewal. CSXT and Norfolk Southern have both commented publicly on the 'incompatibility' of freight and passenger services.

Under the track access agreements, Amtrak makes performance incentive payments to host railroads for dispatching its trains over their lines on time. An exponential formula has been adopted, so that at 90 per cent on-time performance, the host railroad has lost 50 per cent of its incentive opportunity.

Northeast Corridor

Amtrak continues to enjoy a dominant presence in the Northeast Corridor. Normal weekday train departures between New York and Boston total 42, a level of service not seen in over 50 years between those two cities. Between New York and Washington there are 97 daily departures. These high levels of service, in conjunction with the popularity of the Acela Express, has resulted in Amtrak's market share between Washington DC and New York City rising to 50 per cent; it now carries more passengers than either of the competing air shuttles. Between Boston and

Amtrak California Type F59PHI locomotive arriving at Fremont-Centerville from San Jose (Philip Wormald)   *NEW*/1122820

Acela Express high-speed tilting trainset on the Northeast Corridor at Newark, Delaware
(David C Warner)   0580261

New York, Amtrak's share of the air/rail market has more than doubled in the past four years to 40 per cent.

Pacific Northwest Rail Corridor
New rolling stock was introduced in January 1999 on the Pacific Northwest Rail Corridor, which extends 746 km between Vancouver, British Columbia and Eugene, Oregon. In 2004 Amtrak Cascades service carried over 600,000 passengers, an increase of 541 per cent over 1993, when the operator began a partnership with the states of Washington and Oregon to develop rail to provide an alternative mode to road. To replace existing leased equipment, two dedicated sets of Talgo passive tilting trainsets were acquired by the state of Washington for Seattle–Portland–Eugene services, and a third set was acquired by Amtrak. The trainsets were ordered in 1996, and delivered in 1998. The 12-car trainsets were designed by Patentes Talgo and assembled locally in Seattle by Pacifica Inc, a public-private partnership of aerospace and marine construction unions. Two additional trainsets were manufactured by Patentes Talgo in a bid to generate a market for its designand they were leased for service on this corridor. Amtrak subsequently purchased one of the trainsets. The fifth trainset, in a unique livery for a planned Las Vegas–Los Angeles route, was purchased in 2003 by the state of Washington and repainted in 2004 to match the other four trainsets, which had also been leased for use on the 'Cascades' service. The end-cars of the 'Cascades' vehicles incorporate glass-fibre streamlined fins intended to harmonise their low-slung appearance with that of the F59PHI locomotives which power the trains. The five trainsets are operated under the 'Amtrak Cascades' branding.

In the later half of 2005 Amtrak and the state of Washington plan to add one additional Seattle–Portland round trip using the Amtrak Cascades equipment. The five trainsets are all named after important northwest mountains. Amtrak's trainsets are named Mt Hood and Mt Olympus. Washington's are named Mt Adams, Mt Baker and Mt Rainier.

The corridor is also served by Amtrak's daily Los Angeles–Seattle 'Coast Starlight'.

State-supported trains (formerly '403(b)' services)
Under the Rail Passenger Service Act of 1970, Amtrak was allowed to add new services to its existing network provided it had the equipment available and the additions would cover their operating costs with no additional federal support. Since this was rarely the case, Section 403(b) of the Act authorised Amtrak to initiate new routes with financial support from a non-Amtrak source. The latter could be a state, a group of states, a regional or local agency, or even an individual with the required financial backing.

For nearly three decades the basic conditions for a 403(b) service were that 70 per cent of long-term losses had to be covered by the requesting state(s). By 2003 Amtrak was operating in a harsher financial environment and the level of support required from states to operate a train rose to 100 per cent of the full, direct operating losses. This was a move that caused anxiety in those states who were keen to introduce such services. Even at this level, there is still a 'hidden' subsidy from Amtrak in the form of unallocated overhead and equipment related costs. In FY2004, Amtrak received USD135 million from 13 states to support certain corridor trains.

Amtrak's new strategic reform initiative (see funding above) calls for a transition beginning in FY2008 for states to contribute 100 per cent of fully-allocated operating losses (excluding interest and depreciation) plus an equipment capital charge. In FY2008 the states would contribute 25 per cent of the costs, rising to 100 per cent in FY2011. Amtrak believes this modification in state support, combined with the Bush Administration's promise of federal matching funds, will encourage the expansion of existing services and the creation of new corridors, such as the comprehensive 'Midwest Regional Rail Initiative' proposed in the late 1990s by a consortium of nine midwest states (Illinois, Indiana, Iowa, Michigan, Minnesota, Missouri, Nebraska, Ohio and Wisconsin).

Keystone Corridor
In 2004 an agreement was reached with the Commonweath of Pennsylvania to return the 167 km Philadelphia–Harrisburg corridor to a state of good repair. This project replaces a similar scheme halted in 2002 when the railroad nearly shut down due to the financial crises. The costs of the USD142 million project will be split 50/50 between Amtrak and the state. Upon completion of the project in 2006, the current two-hour travel time will be cut by over 10 per cent to 105 minutes, and some express trains will make the trip in 90 minutes. The number of daily round trips on the route will increase by nine to 13. Major elements of the project include replacing wooden ties with concrete, replacing all jointed rails with welded ones and improving communications equipment and traction cables. A separate agreement calls for eliminating the remaining three road-rail grade crossing on the route.

Amtral California
Since the mid-1970s California has been a major supporter of intrastate train services. Approval of two major state bond funds in 1990 set the stage for significant improvements in rail. Proposition 108 (Passenger Rail and Clean Air Bond Act) provided USD1 billion in rail bonds, with USD250 million specifically for intercity rail capital projects, and Proposition 116 (Clean Air and Transportation Improvement Act of 1990) provided USD1.9 billion one-time funding for rail projects, including over USD380 million for intercity rail passenger improvement projects. California's state operating support in FY2004 of USD68.3 million represented over half of the total such support Amtrak received. In addition to operating support, the state also provides grants for station and track upgrades and it has purchased 17 diesel locomotives and 83 passengers cars for use on the Pacific Surfliner, Capitol Corridor and San Joaquin routes.

The Pacific Surfliner service between San Luis Obispo and San Diego, a distance of 565 km, began under Amtrak with just two San Diego–Los Angeles round trips. The number of trips has gradually increased over the years and service has been extended north to Santa Barbara, while one daily round trip goes to San Luis Obispo. There are presently 11 weekday and 12 weekend Los Angeles–San Diego round trips, with four weekday and five weekend trips to/from Santa Barbara. In 1999 the service name was changed from San Diegans to Pacific Surfliner, and the route was re-equipped with California-purchased F59PHI locomotives and ALSTOM-built double-deck rolling stock.

The San Joaquin service under Amtrak between Bakersfield and Oakland began in 1974 with one round trip. Second and third round trips were added in 1980 and 1989. The route was re-equipped in the mid-1990s with new diesel locomotives and double-deck 'California Cars'. Six daily round trips

GE-built P42 locomotives at Seattle King Street station after arrival with the 'Empire Builder'
(Brian Webber)   0580262

Type HHP-8 electric locomotive at Queens, New York (David C Warner)   0531914

now operate on the corridor, two to Sacramento and four to Oakland.

Capitol Corridor began at the end of 1991 with three daily round trips between San Jose and Sacramento. Management of the service was transferred to the Capitol Corridor Joint Powers Authority (CCJPA) in 1998. The CCJPA is governed by 16 elected officials from the six member agencies along the 273 km route. Day-to-day management staff is provided by the San Francisco Bay Area Rapid Transit District (BART). Also re-equipped with the California-owned diesels and rolling stock, the levels of service had increased to nine daily round trips by 2001 and there are presently 12 daily round trips (with one extended to Auburn, 35 miles east of Sacramento). Nearly USD80 million has been spent on installing double track and modernizing signalling on the route, owned by the Union Pacific Railroad. The infrastructure improvements and increased service have resulted in annual ridership increasing from 46 million riders in FY1998 (before CCJPA management) to 1.2 million in FY2004.

All three corridors, as well as Amtrak's long distance trains that serve California, are supplemented by an extensive, integrated network of buses, operated by Amtrak using state contract operators. Over half the passengers on the San Joaquin Corridor begin or complete their journey using an Amtrak bus.

North Carolina
In FY2004 the state of North Carolina provided USD2.7million in operating support to Amtrak. The state owns three diesel locomotives and seven passenger cars for use on the Charlotte–Raleigh 'Piedmont'. This train supplements the daily Charlotte–New York 'Carolinian'. These trains use a portion of the state-owned North Carolina Railroad, operated under lease to Norfolk Southern. The Washington–Raleigh–Charlotte route was one of the first nationally-designated high-speed corridors, and for over a decade the state has been working on a 'Sealed Corridor' programme to eliminate grade crossings and instal four-quadrant gates where the crossings cannot be eliminated.

Maine
The Northern New England Passenger Rail Authority (NEEPRA) was formed by the Maine Legislature in 1995 to establish rail service between Boston (North Station) and Portland, Maine. Rail service began in December 2001, with four round trips. Nearly USD50 million was spent to upgrade the tracks, though the owners fought the start of service in a case that went to the Surface Transportation Board. In FY2004, the state of Maine provided USD2.4 million support for the service; NEEPRA oversees the operation. This was the first passenger service in Maine provided by Amtrak, and ended a 36-year gap in through service between Boston and Portland. Though a significant number of the passengers board in the three New Hampshire stations, that state has not provided any funding for the service, leading to a

crisis during FY2005. This highlights the difficulties of running state-supported trains through multiple states, a methodology supported by the Bush Administration (see *Finance* above).

Commuter contracts
In recent years, as its focus has increased on the core business of intercity passenger operations, Amtrak has become more selective in the commuter rail operations it runs for local agencies. In 2003, citing proposed contract conditions that were unfavourable for a company in the severe financial condition in which Amtrak found itself, the company chose not to bid on a renewal of its 16 year-old contract to operate and maintain commuter trains services for the Massachusetts Bay Transportation Authority's 13-route commuter rail system. In July 2005 operation of the Los Angeles region Metrolink commuter trains was to be transferred from Amtrak. Current commuter rail services operated by or for the following agencies are:
Connecticut Department of Transportation (Shoreline East);
Peninsula Corridor Joint Powers Board (Caltrain);
Maryland Department of Transportation (MARC);
North San Diego County Transit Development Board (Coaster);
Virginia Railway Express (VRE);
In addition, in 1999 Amtrak secured a 10-year contract from Central Puget Sound Regional Transit Authority (Sound Transit) to provide maintenance services for the equipment to be used on 'Sounder' commuter services introduced between Seattle and Tacoma late in 1999 and between Seattle and Everett in 2004.

Intermodal operations
During FY2004 Amtrak decided to cease carrying mail and express consignments. After evaluating the business, the company determined it was a marginal addition to the bottom line finances, and that eliminating the negative impacts on passenger train operations resulting from the adding and removing of mail and express equipment from trains, would result in more reliable performance. Except for some limited perishable business from the west coast to the northeast, all mail and

express service ceased by the beginning of FY2005. That service too, is scheduled for elimination once contractual issues are resolved. Amtrak is in the process of disposing equipment associated with these operations (RoadRailer bi-modal vans and associated rail bogies, Material Handling Cars and express boxcars).

As a result of exiting this business, in the third quarter of 2005 service between New York and Chicago and Florida was restructured. The 'Three Rivers' was eliminated, leaving a single Pittsburgh–New York train across Pennsylvania. The southern terminal of the 'Palmetto' was changed from Miami to Savannah, Georgia, and turned into a coach-only train. This change resulted in train service being eliminated to Waldo, Ocala, Wildwood and Dade City, Florida, though motor coach connections to the 'Silver Star' were provided.

Improvements to existing lines
Excluding sections totalling 151 km that are owned by six regional commuter authorities, Amtrak has owned the 735 km Boston-New York-Washington Northeast Corridor route since 1976, including five of its stations: Baltimore, Wilmington, Philadelphia (30th Street), New York (Pennsylvania Station) and Providence. An improvement project (NECIP) on this route was completed in 2001. The last stage to be inaugurated included electrification to Boston (see 'Electrification' section), and upgrading the New York-Washington section to prepare for the introduction of high-speed trainsets.

Fire/Life Safety upgrades to the key tunnels serving New York City's Penn Station are planned, including improved ventilation, emergency egress improvements and increased security.

Traction and rolling stock
At the beginning of 2005 Amtrak's locomotive fleet totalled 323, of which 64 were electrics, 40 Acela Express were power cars (see below), 48 were diesel switchers and the remaining units were line-haul diesel locomotives.

High-performance trainsets
In March 1996 Amtrak placed a contract with the Bombardier/GEC Alsthom (now ALSTOM) consortium for 18 trainsets with active body-tilt, capable of 250 km/h operation, and 15 electric locomotives (see below). In July 1998 a follow-up order for a further two trainsets was announced. Each 'Acela Express' trainset consists of two power cars each developing 4,600 kW, flanking six coaches (one first, four standard and one bistro car) with a seating capacity of 301 per set. Total value of the contract for the rolling stock, three maintenance facilities and maintenance of the trainsets for 10 years was USD1.2 billion (USD611 million for rolling stock only). The operating contract contains significant performance specifications and penalties for failure to meet targets. Following two years of trials at the test track at the Association of American Railroads' Transportation Technology Center in Pueblo, Colorado, and on the Northeast Corridor, revenue service began in late 2000. All trainsets had been delivered by mid-2003. In March 2004 an agreement was reached between Amtrak and the consortium to settle legal disputes associated with the trainset acquisition programme. In addition to a cash settlement, much less than had been requested by the consortium, Amtrak will take over maintenance of the trainsets in October 2006.

Cascades service at White Rock, British Columbia, on the Pacific Northwest Corridor, formed of Talgo passive tilting rolling stock powered by F59PHI locomotives (Quintus Vosman)   0531913

California Corridor service at San Diego, formed of a Type F59PHI locomotive and Surfliner coaching stock (David C Warner) 0531912

Class	Wheel arrangement	Power kW	Speed km/h	Weight short tons	No in service	First built	Builders Mechanical	Engine
Diesel locomotives								
F59PHI	Bo-Bo	2,240	175	–	21	1997	EMD	12–710 G3B
P-32BH	Bo-Bo	2,390	165	129	18	1991	GE	FDL12
P-40	Bo-Bo	2,990	165	127	17	1993	GE	FDL16
P-42	Bo-Bo	3,135	165	127	199	1996	GE	FDL16
P-32AC-DM*	Bo-Bo	2,390	175	127	18	1996	GE	FDL12
12 kV AC electric locomotives								
AEM-7	Bo-Bo	5,200	225	91	22	1980	GM/ASEA	
AEM-7AC**	Bo-Bo	5,200	225	91	29	2000	GM/ASEA	
HHP-8	Bo-Bo	5,970	225	101	15	1999	Bombardier/ ALSTOM	

* Dual-mode: diesel and electric DC.
** Refurbished with AC propulsion by Amtrak and ALSTOM.

Acela Express Trainsets

Class	Cars per unit	Motor cars per unit	Motored axles/car	Power/ motor KW	Speed km/h	Weight short tons	No in service	First built	Builders
HST	8	2	4	–	250		20	2000	Bombardier/ALSTOM

Diesel locomotives

Amtrak received 44 of a new design of diesel locomotive in 1993/4. These units, designated P-40 (and called Genesis I by the builder, General Electric Transportation Systems), are rated at 4,000 hp (2,985 kW)Over half of these units are temporarily stored as locomotive requirements were reduced after Amtrak exited the mail and express business; these numbers are included in the fleet total above. Additional orders for a total of 207 additional similar locomotives rated at 4,250 hp (3,170 kW) were made in the late 1990s. All of these locomotives, designated P42s, were delivered by the end of 2001. In addition, a total of 17 dual-mode Genesis II locomotives (P32AC-DM) were purchased in two batches in the 1990s for service on the Empire Corridor. These 3,200 hp (2,385kW) units feature AC propulsion, and operate off 600 V DC third rail in the Penn Station area.

In 1998 Amtrak ordered 21 F59PHI units from GM, to be assigned to trains in what was then known as the West Coast Business Unit for use on California state-supported services and on Pacific Northwest services. In addition, the state of California owns 15 F59PHIs for use in California service; these units are not included in the Amtrak motive power totals listed above.

All F40 diesel locomotives, varying in age from 16 to 28 years, were retired from active service in 2002. A few units remain in service leased to other operators.

Electric locomotives

Beginning in 2000, Amtrak and ALSTOM commenced a cooperative programme to rebuild the 20-year old AEM-7s using AC propulsion technology. Just over half the fleet was converted before the project was stopped due to budget problems. All AEM-7 overhauls since 2003 have seen these units rebuilt with DC propulsion.

In 1996, as part of the order received by Bombardier and ALSTOM for the high-speed trainsets, Amtrak ordered 15 Type HHP-8 high-powered electric locomotives for use on Northeast Corridor long-distance services. The design is based on that of the Acela Express power cars, and the locomotives are rated at 5,970 kW. All 15 HHP-8s were delivered by 2001.

Refurbishment projects

The first 1991-era series of GE units, the B-32 model, had a mid-life overhaul in the Chicago shops. They are generally used on short-haul routes out of Chicago, and in recent years many have been transferred to yard service.

Cost overruns and contractual problems have plagued a long-running programme to upgrade and repower the existing 1975-vintage RTL Turboliner sets. In FY2004 Amtrak refused to accept any of the three trainsets completed by New York-based Super Steel Schenectady and moved the trainsets to storage in Bear, Delaware.

Beginning in 2000, Amtrak and ALSTOM began a cooperative programme to rebuild the 20-year

old AEM-7s using AC propulsion technology. Just over half the fleet was converted before the project was stopped due to budget problems. All AEM-7 overhauls since 2003 have seen these units rebuilt with DC propulsion.

During FY2005 40 P42, two P32AC-DM and five F59PHI diesel locomotives will be overhauled, and five wrecked P42s will be returned to service.

Locomotive procurement

Amtrak purchased 10 new GP15D (1,120 kW) diesel electric switchers from Motive Power Industries in FY2004 and outsourced the conversion of eight GP40 road diesels into GP38H switchers. These actions will enable the retirement of some switchers that are over 50 years old.

Passenger coaches

Total coaching stock at the start of 2005 was 1,414, including stock of the 20 Acela Express trainsets and Amtrak-owned Talgo equipment. In addition, 83 cars are owned by the state of California, eight cars by the state of North Carolina, and the state of Washington owns 42 Talgo cars. Amtrak also owns slightly more than 200 cars for carrying baggage, express and automobiles.

In FY2004 Amtrak ordered 80 new bi-level auto carriers from Johnstown America. These will be placed in service during 2005 and will replace the auto carriers currently in service on the Sanford–Lorton Auto Train, which are over 30 years old.

The Five Year Strategic Plan calls for the remaining 'Heritage' diners and dormitory cars, both nearly 50 years old, to be retired and replaced with new cars built using the Viewliner vehicle design. Two of the Viewliner prototype cars have been moved to the Wilmington Maintenance Facility and are being used by the Mechanical Department for the prototyping of new interiors.

In FY2005 Amtrak plans to restore 20 wrecked passenger cars to service. Overhauls are planned for over 180 additional pieces of coaching stock. The Superliner remanufacturing programme will continue, with 18 Superliner I sleepers and the conversion of 23 Superliner I smoking coaches to coach/baggage cars. Twenty-eight Horizon coaches and cafes will also be remanufactured.

Signalling and telecommunications
Advanced signalling tests

Amtrak is participating in tests of three different advanced positive train control systems, to be carried out in conjunction with the Federal Railroad Administration. The new systems aim to allow higher operating speeds, provide increased track capacity and improve safety.

On Union Pacific routes in Illinois, the Advanced Train Control System developed in-house by the Association of American Railroads was tested successfully on the Chicago–St Louis route between Dwight and Springfield at speeds up 185 km/h. In Michigan, the Incremental Train Control System developed by Amtrak and Harmon Industries is being tested on the 100 km between New Buffalo and Kalamazoo belonging to Amtrak.

Along the Northeast Corridor, Amtrak has installed a system called Advanced Civil Speed Enforcement System (ACSES). This enforces the maximum authorised speed and all civil speed restrictions, both permanent and temporary. A positive stop is enforced at interlocking home signals. The ACSES system is transponder-based, with wayside transponders providing speed and location information to the train as it moves through the territory. The system is an enhancement to the existing and expanded cab signal system. The ACSES system is presently in service on both main tracks between New Haven and Boston, and on certain stretches between New York and Washington. Long-term plans call for ACSES eventually to be extended to all main tracks between New York and Washington, as funding permits, and it will be required for all trains.

Electrification
Extension to Boston

In total, 482.7 route-km of the Northeast Corridor is electrified at 12 kV 25 Hz AC. From New York to Boston, only the first 120.7 km to New Haven was electrified at 11 kV AC 25 Hz (since converted to

60 Hz) in the early part of the last century and is mostly operable at speeds up to 115 km/h.

With the financial backing of Congress, the US Department of Transportation, and the Coalition of Northeastern State Governors (CONEG) electrification was extended from New Haven to Boston using the 25 kV 60 Hz AC system. Construction began in July 1996 and the project was completed in 2000. Completion of electrification enabled the best New York–Boston journey time to be cut by almost 1 hour.

Track
Rail: 70 kg RE
Crossties (sleepers)
Concrete: Thickness 241 mm, spacing 1,584/km

Wood: Thickness 178 mm, spacing 1,950/km
Fastenings: Concrete ties: Pandrol E2055
 Wood ties: Cut spikes
Min curvature radius: 175 m (10.5°)
Max gradient: 1.9%
Max axleload: 32.88 tonnes

UPDATED

Anacostia & Pacific Company Inc

Suite 335, 53 West Jackson Boulevard, Chicago, Illinois 60604
Tel: (+1 312) 362 18 88   Fax: (+1 312) 362 14 02
e-mail: info@anacostia.com
Web: www.anacostia.com

New York office
535 Fifth Avenue, 33rd Floor, New York, New York 10174
Tel: (+1 212) 687 95 00   Fax: (+1 212) 687 95 01

Key personnel
President: Peter A Gilbertson
Vice-President and Chief Financial Officer:
 Bruce A Lieberman

Organisation
The company is a railroad investment and management firm, with US and foreign interests. It deals in acquisitions, start-ups and restructuring as well as management and management support. Shareholder equity interest varies; it is up to 100 per cent in some cases. The company has formed an affiliate subsidiary, Rail Logistics Services Inc, to

expand into equipment, yard services, and track maintenance. In 2004 principal affiliates consisted of: Anacostia Rail Holdings; Chicago SouthShore & South Bend RR; Louisville & Indiana RR; New York & Atlantic RR; Pacific Harbor Line; and Northern Lines Railway LLC. All these lines are in the USA.

In February 2004 A&P was chosen by the Port of Portland, Oregon, to set up a neutral switching company to provide rail service to port users.

UPDATED

AN Railroad Co

PO Box 250, Port St Joe, Florida 32457
Tel: (+1 850) 229 74 11   Fax: (+1 850) 229 27 55
Web: http://www.rail-management.com

Key personnel
General Manager: R Wayne Parrish

Organisation
The 154.9 route-km former Apalachicola Northern Railroad was acquired by Rail Management Corporation from the St Joe Company in September 2002. ANR links Port St Joe with Apalachicola and Chattahoochee, Florida, where interchange is provided with CSXT. It carries mainly chemicals and forest products, and serves a barge/rail transfer facility.

Traction and rolling stock
It has six locomotives (three SW1500 and three GP15T, all GM-EMD) and 149 freight wagons.

Appalachian & Ohio Railroad

Web: www.watcocompanies.com

Key personnel
President: Rick Webb

Gauge: 1,435 mm
Route length: 254 km

Organisation
CSXT coal-haul lines in its Cowen and Pickens subdivisions in the West Virginia coalfield are leased to Watco, which was to start operations in March 2005. The Cowen subdivision (198 km) runs south from Grafton to Cowen and Bolair, while the

Pickens subdivision is a branch from Hampton to Alexander (56 km). Traffic is mainly coal in unit trains.

NEW ENTRY

Arizona & California Railroad

1301 California Avenue, Parker, Arizona 85344
Tel: (+1 928) 669 66 62   Fax: (+1 928) 669 66 66
Web: http://www.railamerica.com

Key personnel
General Manager: Tanya Cecil

Gauge: 1,435 mm
Route length: 557 km

Organisation
Operated by RailAmerica, the ARZC is a former Santa Fe line linking Cadiz, California, with Matthie, Arizona, via Rice, California, and Parker, Arizona. From Rice, a branch runs south to Ripley.

The railway handles approximately 18,000 carloads annually, commodities carried including cement, asphalt, forest products and steel. Interchange is made with BNSF at Cadiz and at Phoenix, Arizona, via trackage rights between Matthie and Phoenix.

Arizona Eastern Railway Company

PO Box 2200, Claypool, Arizona 85532
Tel: (+1 520) 473 24 47
Fax: (+1 520) 473 24 49
Web: http//www.railamerica.com

Key personnel
General Manager: Forrest Becht

Gauge: 1,435 mm
Route length: 217 km

Organisation
Formerly a Statesrail company and now owned by RailAmerica, the AZER runs southeast through

eastern Arizona from Miami to Bowie. The railway handles approximately 16,000 carloads of copper cathode and related materials annually. Interchange is made with UP at Bowie.

Arkansas–Missouri Railroad Co

306 East Emma Street, Springdale, Arkansas 72764
Tel: (+1 497) 751 86 00
Fax: (+1 497) 751 22 25
e-mail: info@arkansasmissouri-rr.com
Web: http://www.arkansasmissouri-rr.com

Key personnel
President: G Brent McCready
Chief Mechanical Officer: Casey Shepherd

Gauge: 1,435 mm
Route length: 222 km

Organisation
A&M serves northwestern Arkansas, linking Monett, Missouri, with Van Buren and Fort Smith, Arkansas. Connections are made to BNSF at Monett and to UP at Van Buren. Annual carloads totalled 22,900 in 1998; tonnage amounted to 2.3 million. From April to mid-November the railway runs a tourist special from Springdale to Van Buren.

Traction and rolling stock
A&M operates 20 Alco locomotives, comprising one RS1, six T6, one RS232 and 12 C420s. The railroad also operates 850 freight cars and four passenger cars.

Bay Colony Railroad Corporation

420 Washington Street, Braintree, Massachusetts 02184
Tel: (+1 781) 380 35 56   Fax: (+1 781) 380 48 20
Web: http://www.baycolrr.com

Key personnel
Chairman and Chief Executive Officer:
  Gordon H Fay

Senior Vice-President, Marketing:
  Bernard M Reagan
Superintendent, Operations and Maintenance:
  John F Pimentel Sr

Organisation
BCLR operates 204 route-km formed of six unconnected lines in southeastern Massachusetts (four are presently active); these are former Conrail routes which were purchased by the state

of Massachusetts in 1981. It has eight locomotives and 47 freight wagons. The operating centre is Wareham on the Cape Cod line. All connections are to CSXT.

The Bay Line Railroad

PO Box 35098, Panama City, Florida 32412
Tel: (+1 850) 785 46 09
Fax: (+1 850) 747 40 37
Web: http://www.rail-management.com

Key personnel
General Manager: Jerry Hood
Sales and Marketing Manager: Brenda Hatfield

Organisation
The former Atlanta & St Andrews Bay Railroad was purchased by Rail Management Corporation in 1994, when its name was changed. BAYL runs 130 route-km from the Florida port of Panama City northwards to Dothan, Alabama. Connections with CSX and NS are provided at Dothan and with CSX at Cottondale, Florida. Beyond Dothan, BAYL has access rights over some 10 km of CSX trackage to Grimes, whence a 43 km branch runs to Abbeville.

The railroad handles about 28,000 carloads annually, mainly paper and forest products, chemicals and aggregates.

Traction and rolling stock
The rolling stock fleet comprises 13 GM-EMD locomotives (nine GP38, three GP38-2 and a single GP9).

Belt Railway Company of Chicago

6900 South Central Avenue, Chicago, Illinois 60638
Tel: (+1 708) 496 40 00   Fax: (+1 708) 496 40 05
Web: http://www.beltrailway.com

Key personnel
President: Patrick J O'Brien
General Manager, Transportation:
  George A LaValley
Superintendent, Mechanical: Michael S O'Donnell
Assistant Secretary/Treasurer: Pamela S Hagen

Gauge: 1,435 mm
Route length: 45 km

Organisation
The BRC is the innermost of three Chicago-area connecting railroads (see also Indiana Harbor Belt; Elgin, Joliet & Eastern) and is the largest intermediate switching railroad in the USA. Its

13 shares are owned by six of the North American Class I railroads: BNSF, CN, CP, CSXT, NS and UP. It makes connections with other railroads, including all Chicago trunk lines, at 21 locations. BRC's massive Clearing Yard measures 9 km from east to west and carries out classification and blocking for UP (ex-SP), Grand Trunk, CN (ex-Wisconsin Central) and CSX, and services locomotives for several railroads. An average of 8,400 wagons are dispatched daily. Total yard trackage is 480 km. The BRC also serves around 100 online industries.

Since 1990, the BRC has undertaken a rationalisation and modernisation programme including bidirectional signalling, installation of remote signals and switches and the introduction of computerised train dispatching. Since 1990 revenues have reached US$68 million (from US$16.5 million) and the operating ratio has improved to 87.6 per cent. There are some 520 staff. BRC hosts several unit coal train run-throughs, and since mid-1996 it has been cleared throughout for double-stack container trains. Since picking up

some extra traffic diverted from Proviso yard as a result of the UP/CNW merger, the BRC Clearing Yard is virtually at capacity; the company is embarking upon an analysis mapping out a plan for the future, with the participation of the Chicago Operating Rules Association.

BRC should benefit from the proposed expenditure of US$1.5 billion on Chicago area infrastructure improvements under the CREATE (Chicago Area Regional, Environmental and Transportation Efficiency) programme announced in mid-2003. If approved and funded by the federal government, construction could start in 2005 for completion in 2010.

Traction and rolling stock
BRC owns 32 GM-EMD diesel-electric locomotives: six GP10; four MP1500; nine SD40; three SD40M; two SW1200; six SW1500; and two 'slugs'. A major upgrading programme was completed in 2003, with all locomotives equipped for remote control operation which was inaugurated in June 2003.

Bessemer & Lake Erie Railroad Company

PO Box 68, Monroeville, Pennsylvania 15146
Tel: (+1 412) 829 66 00   Fax: (+1 412) 829 66 03
Web: http://www.cn.ca

Organisation
This mineral-haul railroad links the Great Lakes with coal mines and steelworks in western

Pennsylvania, West Virginia and the Ohio River valley. Running some 375 km from the docks at Conneaut on Lake Erie and Wallace Junction, Ohio, to North Bessemer near Pittsburgh, B&LE became part of the Great Lakes Transportation group in 2001 along with the Duluth, Missabe & Iron Range Railway.

In May 2004 the B&LE was acquired by CN, along with other railroad and related interests held by the Great Lakes Transportation group.

Traction and rolling stock
The B&LE has a fleet of 25 locomotives (all EMD six-axle) and 3,070 freight wagons, of which 2,527 in service.

Bethlehem Steel Corporation

Rail Operations Division, Martin Tower, 1170 8th Avenue, Bethlehem, Pennsylvania 18016
Tel: (+1 610) 694 59 37   Fax: (+1 610) 694 33 16
Web: http://www.bethintermodal.com

Key personnel
President: J Michael Zaia
Vice-President, Operations: Patrick R Loughlin
Vice-President, Finance: August N Fix Jr
Vice-President, Business Development:
  Patrick A Sabatino

The BSC operates nine railroads, each of which serves a company mill (the Steelton, Pennsylvania, works, for example, produces rail) and any

receivers/shippers located on line. Each railroad's resident superintendent is responsible for its own business plan and asset management.
  The railroads are:
- Brandywine Valley Railroad, Coatesville, Pennsylvania
  Tel: (+1 610) 383 26 69
- Conemaugh & Black Lick, Johnston, Pennsylvania
  Tel: (+1 814) 533 71 50
- Cumberland Valley Business Park, Chambersburg, Pennsylvania
  Tel: (+1 610) 694 59 74
- Lake Michigan and Indiana, Burns Harbor, Indiana
  Tel: (+1 219) 787 79 88
- Patapsco & Black Rivers, Baltimore, Maryland
  Tel: (+1 410) 388 79 37

- Philadelphia, Bethlehem & New England, Bethlehem, Pennsylvania
  Tel: (+1 610) 694 33 92
- Steelton & Highspire, Steelton, Pennsylvania
  Tel: (+1 717) 986 24 55
- South Buffalo Railway, Lackawanna, New York
  Tel: (+1 716) 821 36 30
- Upper Merion and Plymouth, Conshohocken, Pennsylvania
  Tel: (+1 610) 383 26 69

Total route length of the nine railroads is 528 km. Rolling stock comprises 103 locomotives and 638 freight wagons. Interchange traffic totals around 160,000 carloads annually, while internal traffic amounts to over 170,000 carloads.

Birmingham Southern Railroad Co

PO Box 579, 6200 E J Oliver Boulevard, Fairfield, Alabama 35064
Tel: (+1 205) 783 41 18   Fax: (+1 205) 783 45 07
Web: http://www.tstarinc.com/birmingham

Key personnel
President and Chief Executive Officer:
   Thomas W Sterling

General Manager, Southern Operations:
   J Craig Stepan
General Superintendent: David M Gevaudan

Organisation
The 135 route-km BSRC serves the coal and steel industries in and around Fairfield, Alabama, including the largest integrated steel works in the region. Coal is hauled to Port Birmingham for transfer to barges on the Warrior-Tombigbee Waterway. It is a US Steel/Transtar subsidiary.

Traction and rolling stock
The BSRC operates 37 GM-EMD locomotives (three MP1500, four GP38-2, three SW1000, 12 SW1001, one SW7, three SW9C, 10 SD9R and one slug unit) and 1,118 freight wagons.

Boston & Maine Corporation

A subsidiary of Guilford Rail System
Iron Horse Park, North Billerica, Massachusetts 01862
Tel: (+1 508) 663 11 30   Fax: (+1 508) 663 11 99
Web: http://wwwguilfordrail.com

Key personnel
See entry for 'Guilford Rail System'.

Gauge: 1,435 mm
Length: 2,532 km

Organisation
The principal lines of the Boston & Maine run north and west from Boston through the states of Maine, New Hampshire, Vermont, Massachusetts and in eastern New York state, where it makes connections at Albany and Schenectady with other lines.

In 1983, Guilford Transportation Industries (GTI) took over the B & M. GTI also owns the Maine Central. The railroads preserve their separate identities in some areas (for example equipment), but are being operated as an integral system with common management. As a result of past disputes with the labour force over working practices, management is exercised by another GTI subsidiary, the Springfield Terminal Railway Co.

Since the second quarter of 1995 all shipping documents have carried the Guilford name in order to simplify relations with customers. Guilford is privately held and revenue and performance data are not a matter of public record.

Passenger operations
The Northern New England Passenger Rail Authority (NNEPRA) has endorsed the upgrading of the main line from Plaistow, New Hampshire to Portland, Maine, for passenger operations at speeds of up to 126 km/h. Work began in January 1999 and was expected to take two years.

Maine Central-lettered GP40 at the Boston & Maine yard in East Somerville, Maryland
(F Gerald Rawling)   0021587

Freight operations
About 85 per cent of B & M's freight tonnage is received from connecting lines and two-thirds of it terminates on the system. Forest products from northern New England and Canada predominate.

Intermodal operations
B & M has built a major new intermodal facility called the Devens Inland Port and Distribution Center which opened in late 1993. This facility handles domestic and international containers, piggyback traffic and bulk transfer services; it serves as an extension of the port of Boston. Both the Devens Port and the Ayer automobile facility (see below) are targets of Norfolk Southern in a run-through arrangement which will see NS negotiating reciprocal rights with CP (St Lawrence & Hudson) in order to reach Albany, then a similar arrangement with B&M to serve Massachusetts, since CSX is acquiring Conrail's assets in New England.

Improvements to existing lines
The state of Massachusetts has announced a plan to spend $158 million to raise clearances across the state to accommodate double-stack trains. Phase I has consisted of clearing the section from Devens to Mechanicville, New York, to make connection to CP (St Lawrence & Hudson division).

In 1993, B&M upgraded the Ayer automobile unloading facility and installed 112 lb continuous welded rail on the Worcester main line.

Traction and rolling stock
The railroad operates with a fleet of 45 diesel locomotives of which the largest tranche is 32 GP40/GP40-2 models.

Recent rolling stock acquisitions include 384 boxcars for the transport of paper and paper products and 98 covered hoppers for cement traffic.

Buffalo & Pittsburgh Railroad

Genesee & Wyoming Inc
New York/Pennsylvania Region
1200-C Scottsville Road, Suite 200, Rochester, New York 14624
Tel: (+1 585) 463 33 08   Fax: (+1 585) 463 33 57
Web: www.gwrr.com

Key personnel
Senior Vice-President: David Collins
Vice-President, Marketing and Sales: Ron Klein
Vice President, Transportation: Jason Fuller
Vice-President, Mechanical: Gene Evans
Vice-President, Engineering: David Baer
Director of Marketing and Customer Service:
   Dan Wahle

Organisation
Together with the Rochester & Southern Railroad and the South Buffalo Railway, the Buffalo & Pittsburgh Railroad forms the New York/Pennsylvania Region of Genesee & Wyoming Inc (see entry in USA section of *Railway systems and operators*), and totals some 1,050 route-km. The Buffalo & Pittsburgh runs south from Buffalo, New York, and Erie to a network of lines serving Driftwood in the east, Freeport in the south and, via trackage rights over CSX, to New Castle in the west. The Rochester & Southern Railroad runs south from Rochester, New York, to Dansville and Silver Springs, and is connected to its sister railway via trackage rights over NS metals from Silver Springs to Buffalo. Traffic for the combined system totals some 120,000 carloads annually, with coal,

petroleum products, metals and forest products as the main commodities carried.

Improvements to existing lines
In July 2005 a 26 km section of line between Indiana, Pennsylvania, and Homer City was re-opened after rehabilitation following a long period out of use. Revived to enable supplies of coal to a power station to be moved by rail, the line includes 34 public and private road crossings.

UPDATED

Burlington Northern Santa Fe (BNSF)

2650 Lou Menk Drive, PO Box 96105, Fort Worth, Texas 76161-0057
Tel: (+1 817) 333 20 00   Fax: (+1 817) 352 79 25
Web: http://www.bnsf.com

Key personnel
Chairman, President and Chief Executive Officer:
   Matthew K Rose

Executive Vice-Presidents
   Chief Financial Officer: Tom Hund
   Chief Operations Officer: Carl R Ice
   Chief Marketing Officer: John P Lanigan Jr
   Law, Government Affairs and Secretary:
      Jeffrey R Moreland
Vice-Presidents
   General Corporate Counsel: James H Gallegos
   General Counsel: Paul R Hoferer
   General Tax Counsel: Shelley J Venick
   Senior Regulatory Counsel: Richard E Weicher

Government Relations: A R (Skip) Endres
Treasurer: Linda J Hurt
Controller: Dennis Johnson
Corporate Audit Services: Kenneth J Kempker
Investor Relations: Marsha K Morgan
Corporate Relations: Richard A Russack
Human Resources and Medical:
   Gloria A Zamora
Technology Services and Chief Information
   Officer: Jeffery J Campbell

BNSF coal train at Denver, Colorado, headed by three GM EMD SD70MAC locomotives
(David C Warner)   0554380

Four GE DASH9-44CW power a BNSF intermodal train through Valencia City, New Mexico
(David C Warner)   0554381

Gauge: 1,435 mm
Length: 53,000 route-km

For map, see Union Pacific section.

Organisation

BNSF was created in 1995 from the merger of two major railroads, the Burlington Northern and the Santa Fe. The core of the Santa Fe Railway was a high-speed Chicago–Los Angeles route, with feeder lines in the western and mid-western states. BNSF's territory lies principally to the west of the Mississippi river, serving 28 US states and two Canadian provinces; a major rival is the Union Pacific/Southern Pacific system.

BN itself was the product of a 1970 merger involving the Chicago, Burlington & Quincy; Great Northern; Northern Pacific; and Spokane, Portland & Seattle railways. In 1980, BN bought out the St Louis-San Francisco Railway.

In response to opposition to the UP/SP merger of 1996, UP sold BNSF some 560 km of track and additionally arranged up to 6,100 km in trackage rights to provide competition in areas that would otherwise have had only single-carrier service. The three principal elements of the deal were: allowing BNSF into New Orleans directly; connecting BNSF's Pacific northwest trackage with California's Central Valley and Bay area, via Oregon (Klamath Falls) and northern California to offer a single line haul much shorter than the alternative through Idaho, Wyoming and over the Rockies again; and giving BNSF access to the Denver-Stockton corridor which would otherwise have been a UP monopoly. In addition to existing interchange agreements and trackage rights, in 2002 BNSF entered marketing agreements with Kansas City Southern and Canadian National.

At the beginning of 2003 BNSF employed some 36,000 staff.

Finance

BNSF ended 2002 with US$8,979 million in total freight revenues, down 2 per cent on 2001 on account of weak market conditions for bulk commodities and the generally slack economic situation. Operating income was US$1,656 million against US$1,750 million in 2002, but net income rose 4 per cent to US$760 million.

Revenue ton-km, as 788.8 billion, was 2 per cent down on 2001.

Freight operations

Coal is BNSF's largest single source of freight tonnage, contributing 23 per cent of revenue in 2002 from haulage of 240 million tonnes, down slightly on account of mild winter weather. At US$2,071 million, revenue was 2 per cent down on 2001. Low sulphur coal from the Powder River Basin of Montana and Wyoming accounts for 90 per cent of BNSF's originations. This is hauled to close to 65 electricity generating stations chiefly in the north central, south central and mountain regions, with some hauls to the east and southeast.

A significant new contract was signed with Georgia Power for haulage of some 14 million tonnes a year, starting in January 2004.

BNSF serves a significant area of the major grain-producing regions located in the Midwest and Great Plains and transports large quantities of whole grains to domestic feed lots, major milling centres, and to the Pacific northwest, Gulf and western Great Lakes ports for export. In 2002 agricultural commodities produced 16 per cent of revenues.

BNSF's Industrial Products business area, which produced some 23 per cent of revenue in 2002, encompasses building and construction materials, chemicals and plastics and petroleum, all influenced by the weakness of the economy. In particular BNSF serves the timber-producing regions of the Pacific northwest and the southeast, hauling significant volumes of lumber, plywood and structural panels, wood chips, wood pulp, paper and paper products.

Intermodal operations

The Consumer Products business, which encompasses Intermodal freight, is the number one revenue producer for BNSF, generating US$3.353 billion in 2002 (38 per cent of revenue), much the same as in the previous year. But movements were up in both the international and truckload businesses.

Since 1991 the railroad has had a joint service agreement with road haulier J B Hunt to provide intermodal service between the midwest and Pacific northwest; this partnership continues, along with contracts from other shippers such as UPS, as well as several LTL (Less than TruckLoad) carriers.

In 1996, BNSF, Norfolk Southern and Conrail had formed a domestic intermodal equipment project called NACS (North American Container System), to encourage shippers to use containers in most major markets where there is double-stack service available and taking advantage of the restriction-free interchange of NACS 48 ft containers. By 2002 some 8,000 of these units were available on BNSF.

Intermodal handling capacity received a substantial boost during 2002 in preparation for planned traffic growth on the southern transcontinental main line. BNSF's joint intermodal terminal at the port of Oakland, known as Oakland International Gateway, opened in March 2002, bringing on stream capacity for 250,000 containers annually, while the Commerce automotive facility was converted for intermodal use, providing capacity for a further 180,000 units in the Los Anegeles area. This traffic will also benefit from opening of the 32 km Alameda Corridor, which provides much improved access to the ports of Long Beach and Los Angeles. In Chicago, BNSF commissioned the 251 ha LPC logistics park, with facilities for intermodal, automotive and multi-commodity transfers, along with warehousing and a distribution centre. Completion of phase 1 of the scheme at the end of 2003 will provide capacity for some 450,000 autos annually and will raise Chicago-area container handling capacity to nearly 3 million a year.

In total, including other areas of business, 130 new customer facilities and access links were opened in 2002, among them a further 20 grain shuttle terminals capable of accepting 130-car unit trains.

Improvements to existing lines

BNSF's 2002 capital expenditure was over US$1.358 million, some US$100 million down on 2001; for 2003 a slight increase to US$1.4 billion was forecast.

Maintenance work included laying over 1,100 km of new or reused rail, installation of 2.2 million sleepers and resurfacing of 20,100 km of track.

Traction and rolling stock

At the beginning of 2003 the BNSF locomotive fleet totalled 5,184, including machines on hire. Maintenance and overhaul of some 2,600 locomotives is contracted to GE and GM EMD, with work being carried out at BNSF workshops by BNSF employees. From the beginning of 2003, Alstom Transportation Inc was contracted to maintain the fleet of 434 GM-built SD70MAC units at BNSF's Alliance works in Nebraska.

The freight wagon fleet numbered 88,767 and there were 160 commuter passenger coaches for the routes operated by BNSF for Chicago's Metra (see entry in USA section of *Railway systems and operators*).

Signalling and telecommunications

BN opened its Network Operations Centre (NOC), called the James J Hill Center and designed in co-operation with Union Switch & Signal, in Fort Worth, Texas, in 1995. The centre provides tactical control of four functions:

Business Processes and Functions (that is trains, crews and power) through TSS (Transportation Support Systems);

Information Technology (that is asset management, strategic and tactical planning systemwide);

Human Resources; and

Facility Design.

2002 saw introduction of a new suite of web tools, iPower, which allows customers to plan their freight movements and all associated activities, and to pay bills electronically.

California Northern Railroad

129 Klamath Court, American Canyon, California 94503
Tel: (+1 707) 557 28 68
Fax: (+1 707) 557 29 41
Web: http://www.railamerica.com

Key personnel
General Manager: Bob Jones

Gauge: 1,435 mm
Route length: 410 km

Organisation
Operated by RailAmerica, the CFNR comprises a line from American Canyon, California, north to Tehama, with branches, and an unconnected line from Tracy, southwest of Oakland, to Los Banos. The railway handles approximately 30,000 carloads annually of industrial and agricultural products.

Interchange is made with UP at Davis, Suisin Fairfield, Tehama and Tracy, with the Northwestern Pacific at Schellville and with the Napa Valley line at Rocktram.

Carolina Piedmont Railroad

621 Pond Field Road, Darlington, South Carolina 29540
Tel: (+1 843) 398 98 50   Fax: (+1 843) 398 96 83
Web: www.railamerica.com

Key personnel
General Manager: Lamont Jones
Business Development Manager: Kim Greer

Gauge: 1,435 mm
Route length: 79 km

Organisation
Operated by RailAmerica, the CPDR links East Greenville and Laurens, South Carolina. The railway handles approximately 5,000 carloads annually, carrying commodities which include forest and paper products, food and farm products, chemicals, petroleum and coal. Interchange is

made with NS at Greenville and with CSXT at Laurens.

UPDATED

Carrizo Gorge Railway (CZRY)

8929 Gardena Way, Lakeside, California 92040
Tel: (+1 619) 938 19 43   Fax: (+1 619) 561 43 67
Web: www.carrizogorgerailway.com

Key personnel
President: Gary Sweetwood

Gauge: 1,435 mm
Route length: 241 km

Organisation
This short line connection between San Diego and eastern California was reopened in December 2004 after a 28-year closure as a through route. Formerly operated by contractors as the San Diego & Imperial Valley and later the San Diego & Arizona Eastern (as part of the RailAmerica group), the US section is owned by the San Diego Metropolitan Transit Development Board. The line links

San Diego with Plaster City where there is interchange with UP's main line from Los Angeles to El Paso. The route crosses the Mexican border near Tijuana and runs through Baja California for some 80 km to Lindero, where it re-enters US territory.

CZRY's reopening is important because it provides the only direct route to southeastern US states from the San Diego area and the Tijuana border region. Avoiding the busy Los Angeles railhead will bring considerable time savings for through freight movements, and it is hoped that the route will help divert traffic from the congested ports of Los Angeles and Long Beach to the benefit of San Diego's port. There are also prospects for the railway to handle components and raw materials bound for Baja California's extensive manufacturing industries.

Previously, operation of the line had been problematic because of its international character; contractors were unwilling to invest in rehabilitation of the closed eastern section. Eventually, in 1999 a

group of local entrepreneurs managed to win access rights to the US portion from the San Diego MTDB, at the same time gaining a concession from the Baja California state government for its associated Mexican partner Ferrocarriles Peninsulares del Noreste to operate the Mexican section. This paved the way for investment of some US$5 million to rehabilitate the line sufficiently to allow resumption of through services. Now, a US$10 million grant is likely to become available from California state transportation funds for further upgrading. Initially, CZRY was expected to operate three trains daily, the first of which carried timber bound for furniture manufacturers at Tecate.

Traction and rolling stock
At the start of operations in early 2005, CZRY was operating five diesel-electrics.

NEW ENTRY

Cascade & Columbia River Railroad

901 Omak Avenue, Omak, Washington 98841
Tel: (+1 509) 826 37 52   Fax: (+1 509) 826 38 66
Web: http://www.railamerica.com

Key personnel
General Manager: Buck Workman
Business Development Manager: Tom Hawksworth

Gauge: 1,435 mm
Route length: 209 km

Organisation
Operated by RailAmerica, CSCD is a shortline which runs north through central Washington from Wenatchee, east of Seattle, to Oroville. The railway handles approximately 8,000 carloads annually, mostly of forest and agricultural products. Interchange is made with BNSF at Wenatchee.

Cedar Rapids & Iowa City Railway Co

2330 12th Street SW, Cedar Rapids, Iowa 52404
Tel: (+1 319) 786 36 86   Fax: (+1 319) 398 41 71
Web: http://www.crandic.com

Key personnel
Vice-President and General Manager:
   Paul H Treangen

Organisation
The railway has 85 route-km linking the two places in its name and provides connections with BNSF, CN and Iowa Interstate railroads. Owned by Alliant Energy Resources, it carries some 90,000 carloads annually, mostly of grain and coal. There are 110 employees.

Traction and rolling stock
The railway operates 20 GM-EMD locomotives (two SW9; four SW12; four SW1400; six GP9; and four GP35) and owns or leases 400 wagons.

Central Montana Rail Inc

PO Box 868, Denton, Montana 59430
Tel: (+1 406) 567 22 23   Fax: (+1 406) 567 22 23
e-mail: cmrail@ttc-cmc.net

Key personnel
Chairman: Donald Engellant
Vice-Chairman: Wiley Judeman
Secretary/Treasurer: Larry Barber
General Manager: Carla R Allen

Gauge: 1,435 mm
Route length: 137 km

Organisation
CMRI operates under lease a former BNSF branch which is owned by the state of Montana, linking Geraldine with Denton and Mocassin. Connection is with BNSF at Mocassin, Montana. In 2004 CMRI had seven employees.

Track is 75 ASCE, 90RA, 90RB and 100RE rail. Maximum curvature is 8°, maximum gradient

1.5 per cent and maximum permissible axleload 28.5 tonnes.

Traction and rolling stock
CMRI has six EMD GP9 locomotives and five unpowered Budd diesel railcars from the 1950s used as accommodation on tourist trains. Wagons are supplied by BNSF.

UPDATED

Central Oregon & Pacific Railroad

333 SE Mosher, Roseburg, Oregon 97470
Tel: (+1 541) 957 59 66   Fax: (+1 541) 957 06 86
Web: www.railamerica.com

Key personnel
General Manager: Dan Loveday
Business Development Manager: Tom Hawksworth

Gauge: 1,435 mm
Route length: 723 km

Organisation
Formerly owned by RailTex and operated by RailAmerica since 2000, the CORP comprises two adjoining lines: from Eugene, Oregon, southwest to Coquille; and from Springfield Junction, Oregon, south to Black Butte, in northern California. The railway handles approximately 44,000 carloads

annually, with a diversified traffic base. Interchange is made with UP at Black Butte and with other shortline operators at Eugene, Springfield Junction, White City and Gardiner Junction, Oregon, and at Montague, California.

UPDATED

Central Railroad of Indiana

497 Circle Freeway Drive, Suite 230, Cincinnati, Ohio 45246
Tel: (+1 513) 353 36 14   Fax: (+1 513) 682 46 45
Web: www.railamerica.com

Key personnel
General Manager: Terry Wilson
Marketing and Sales Manager: Mike Klass

Gauge: 1,435 mm
Route length: 137 km

Organisation
Operated by RailAmerica, the CIND links Cincinnati, Ohio, with Shelbyville, Indiana. From Cincinnati, the railway has trackage rights to Springfield, Ohio, and from Shelbyville rights exist to Indianapolis and Frankfort. Approximately 7,000 carloads are handled annually, commodities carried including

cereals, minerals and aggregates, food products, fertilisers and iron and steel.

UPDATED

Central Railroad of Indianapolis

497 Circle Freeway Drive, Suite 230, Cincinnati, Ohio 45246
Tel: (+1 765) 454 79 03   Fax: (+1 765) 454 79 08
Web: www.railamerica.com

Key personnel
General Manager: Al Satunas
Business Development Manager: Mike Klass

Gauge: 1,435 mm
Route length: 137 km

Organisation
Operated by RailAmerica, CERA comprises lines in northern Indiana northwest from Marion to Amboy and west from Marion to Kokomo. Interchange is made with NS at Marion and

Kokomo. Haulage rights exist beyond Kokomo to Frankfort. The railway handles around 3,600 carloads annually.

UPDATED

Chesapeake & Albemarle Railroad

214 North Railroad Street, Ahoshkie, North Carolina 27910
Tel: (+1 252) 332 27 78   Fax: (+1 252) 332 33 25
Web: www.railamerica.com

Key personnel
General Manager: Brad Ovitt
Business Development Manager: Kim Greer

Gauge: 1,435 mm
Route length: 338 km

Organisation
Operated by RailAmerica, CA links Chesapeake, Virginia, with Elizabeth City and Edenton, North Carolina. The railway handles approximately 8,600 carloads annually, consisting mainly of limestone, forest products, grain and scrap metal.

UPDATED

Chicago Rail Link

2728 E 104th Street, Chicago, Illinois 60617-5766
Tel: (+1 773) 721 40 00   Fax: (+1 773) 374 66 05
Web: http://www.omnitrax.com

Key personnel
President: Steve Ward

Organisation
The CRL has 57 route-km. It provides switching and industry service in and around southern Chicago. Its accounts include the Illinois International Port (for export grain), UP's Canal Street and IMX intermodal yards and storage and repair of wagons for several leasing companies. Connections are made with seven Class I railroads and seven

regional or short lines. CRL is an OmniTRAX company and draws from its locomotive fleet, principally GP7s and GP9s.

Chicago SouthShore & South Bend Railroad

505 North Carroll Avenue, Michigan City, Indiana 46360-5082
Tel: (+1 219) 874 90 00
Fax: (+1 219) 879 37 54
Web: http://www.anacostia.com/css

Key personnel
Chairman: Peter A Gilbertson
President: H Terry Hearst
Superintendent: James Thompson
Controller: Lance Werner

Gauge: 1,435 mm
Length: 293 km

Organisation
The railroad provides freight service between Chicago and South Bend, Illinois, hauling mainly steel and coal.

In 1989, the railroad was acquired by Anacostia & Pacific Co Inc, which began operations in January 1990. In early 1991, the new owners completed sale to the Northern Indiana Commuter Transportation District (NICTD) of the 146 km of 1.5 kV DC electrified line used by commuter services between South Bend, Indiana, and the Indiana-Illinois state line. NICTD assumed exclusive responsibility for the passenger service, which runs through to Chicago over Metra tracks, but the CSS&SB retained trackage rights for its freight operations, which total approximately 50,000 wagonloads in a year, with coal accounting for over 50 per cent of freight revenue. The system also serves major

steel and utilities facilities and connects with all transcontinental, regional and local railroads in the Chicago area.

In November 2003 a loading facility was opened in Michigan City at the GAF Materials plant, the lagest producer of roofing materials in the US. This enables CSS&SB to handle the plant's outgoing traffic as well as the inbound raw materials already on rail.

The line is entirely signalled by automatic block. Track is mostly 50 kg RE continuously welded rail on oak sleepers. Maximum curvature is 8°, maximum gradient 2.5 per cent.

Traction and rolling stock
The railroad operates 10 GM-EMD GP38-2 diesel locomotives and 548 freight wagons, some 350 of which are gondolas.

Colorado & Wyoming Railway Co

PO Box 316, Pueblo, Colorado 81002
Tel: (+1 719) 561 63 59
Fax: (+ 1 719) 561 68 37

Key personnel
President: Steve Rowan
Chief Operating Officer: Franklin Lloyd

Organisation
This property is now essentially a switching operation serving a steel mill at Minnequa, near

Pueblo (rail is one product), and connecting to both BNSF and UP. It moves about 15,000 carloads annually. It has 25 km of track, 10 locomotives, including three GP7s, and 397 wagons.

The Columbia Basin Railroad

6 East Arlington, Yakima, Washington 98901
Tel: (+1 509) 453 91 66   Fax: (+1 509) 452 93 49
Web: http://www.cbrr.com

Key personnel

President: Brig Temple
General Manager: Tim Marshall
Senior Locomotive Mechanic: Bob Sluys

Gauge: 1,435 mm
Route length: 135 km

Organisation

Formerly part of the Washington Central Railroad Company (WCRC), which was acquired in 1986 from the former Burlington Northern (BN), the Columbia Basin Railroad (CBRC) was established in 1996 following the merger of BN and Santa Fe to form BNSF. BNSF re-acquired the former WCRC lines and the 135 km that now form CBRC is leased to the Temple family. The locally owned line runs from a junction with BNSF's Spokane–Portland route at Connell to Moses Lake, both in Washington. It carries mainly agricultural products, including grain, potatoes and fertilisers, as well as chemicals and paper products, moving about 31,000 carloads annually.

Traction and rolling stock

In 2002 CBRW operated 12 locomotives: two GP-38s, two SD-9s, four GP-9s and four SW-1200s.

Connecticut Department of Transportation

2800 Berlin Turnpike, PO Box 317546, Newington, Connecticut 06131-7546
Tel: (+1 860) 594 28 00   Fax: (+1 860) 594 34 06

Key personnel

Commissioner: J William Burns
Head of Public Transportation and Rail
   Administrator: Harry P Harris

Passenger operations

Amtrak is contracted to run an 82 km service between New London and New Haven, called 'Shoreline East', serving six intermediate stations.

Service is basically weekday peak-only with two counter (stock positioning) movements; ridership in 2000 was 296,000. Farebox recovery is in the range of 12 per cent. Six 3,000 hp (2,238 kW) diesel locomotives have been delivered from AMF Technotransport to improve service speeds in anticipation of overall corridor speed improvements as the result of electrification (see Amtrak entry in United States section of *Railway systems and operators*).

The Connecticut DoT also oversees and subsidises jointly with the New York Metropolitan Transportation Authority the New Haven line commuter service into New York Grand Central Terminal operated by Metro-North. Connecticut funds approximately 60 per cent of the New Haven line's operating shortfall and 63 per cent of capital costs. Some 20 million passengers board at Connecticut stations annually.

ConnDot has been promoting bus shuttles to its stations to encourage 'reverse' commuting, which are now operating in Greenwich, Stamford, New Haven and Norwalk.

At the end of 2002, the Connecticut Transportation Strategy Board approved a 10-year transport improvement plan which would see both commuter services upgraded. Early purchase of 12 electric locomotives and 40 coaches was recommended to achieve a quick expansion of Metro-North service, while US$96 million would fund extension of the Shoreline East service to New York's Penn station. Other proposals include expansion of New Haven line maintenance depots, a start-up commuter service in the New Haven-Hartford-Springfield corridor, and electrification of the Waterbury and Danbury branches. The plans await approval by the state legislature.

In its investigation of New Haven line capacity problems, the CTSB studied the possibility of designing a bi-level coach that would fit the 4.87 m height of the tunnels leading to Grand Central Terminal. This would be a medium-term palliative whilst more costly infrastructure improvements were designed and implemented.

Traction and rolling stock

In 2002 ConnDot operated eight diesel locomotives and 21 coaches.

A short-term solution to the rolling stock shortage mentioned above was agreed in mid-2004, with the purchase by ConnDot of 38 cars from Virginia Railway Express. These Mafersa-built cars will take over Shoreline East services, allowing that line's cars to be transferred to the New Haven route.

Purchase of up to 340 new emu cars was approved in June 2004, though funding was not expected to be in place before late 2005.

Connecticut Southern Railroad

70 Tolland Street, Building 6, East Hartford, Connecticut 06108
Tel: (+1 860) 291 17 00   Fax: (+1 860) 291 17 03
Web: www.railamerica.com

Key personnel

Acting General Manager: Charles Moore
Business Development Manager: Mark Bromirski

Gauge: 1,435 mm
Route length: 126 km

Organisation

Operated by RailAmerica, CSOR has trackage rights over CSXT from Springfield, Massachusetts to New Haven, Connecticut, providing access to its two otherwise unconnected components. Lines run from Windsor Locks to Suffield and Bradley, and from Hartford to East Windsor Hill and Manchester. The railway handles approximately 16,000 carloads annually, including forest products, metal products, minerals, food and chemicals. Interchange is made with CSXT at West Springfield, Massachusetts, and New Haven, Connecticut.

UPDATED

Ex-LMX Leasing GE C39-8 diesel locomotives with a CSOR service at Berlin, Connecticut (Philip Wormald)   *NEW*/0585056

Copper Basin Railway Inc

PO Drawer 1, Highway 177, Hayden, Arizona 85235
Tel: (+1 520) 356 77 30   Fax: (+1 520) 356 63 04
Web: http://www.rail-management.com

Key personnel

President and Chief Operating Officer:
   L S Jacobson
Operations Superintendent: Bobby Blake
Chief Mechanical Officer: Michael J McGinley

Organisation

CBRY has 112.6 route-km running from Hayden to a connection with UP at Magma, Arizona. It is a member of the Rail Management Corporation group. Its principal traffic is movement of copper ore in unit trains between mines at Race and Asarco's processing plant at Hayden. Annual carloadings are in the order of 110,000.

Traction and rolling stock

It operates 14 GM-EMD locomotives (three GP18, five GP9, four SD39 and two GP39) and 130 wagons.

CSX Transportation Inc (CSXT)

A business unit of CSX Corporation
500 Water Street, Jacksonville, Florida 32202
Tel: (+1 904) 359 32 00   Fax: (+1 904) 359 18 99
Web: www.csxt.com

Key personnel

Chairman, President and Chief Executive Officer,
   CSX Corp: Michael J Ward
Executive Vice-President and Chief Operating
   Officer, CSXT: Tony L Ingram
Executive Vice-President and Chief Financial
   Officer: Oscar Munoz
Executive Vice-President and Chief Commercial
   Officer: Clarence W Gooden
Senior Vice-Presidents
   Coal Service Group: Christopher P Jenkins
   Service Design: Alan P Blumenfeld
   Performance Improvement: Frederick J Favorite
   Law and General Counsel: Ellen Fitzsimmons
   Labour Relations: Robert J Haulter
   President, CSX Intermodal Inc: James R Hertwig

Gauge: 1,435 mm
Route length: approximately 37,000 km

Organisation

CSXT is a rail transport and distribution company operating in 23 US states, the District of Columbia and Ontario and Quebec, Canada. It is one of several companies that make up CSX Corporation. The other surface transport component is: CSX Intermodal Inc, which operates over 500 dedicated trains each week serving a network of 44 intermodal terminals. The holding company, CSX Corporation, is publicly traded on the New York Stock Exchange.

Other transport-related business units are CSX World Terminals, CSX Technology, CSX Real Property, Transflo (materials management and distribution), Total Distribution Services (automotive distribution and storage) and BridgePoint (logistics technology). Non-transport holdings are the Greenbrier resort hotel in White Sulphur Springs, West Virginia, and a majority interest in Yukon Pacific Corporation, which is promoting the Trans-Alaska Gas System.

In 1986 CSXT bid successfully for control of Sea-Land, a major container shipping line. This gave CSXT control of Sea-Land terminals throughout the world. To optimise the scope for synergies of its marine and land intermodal activity, all operations were grouped into a new company, CSX Intermodal.

CSX's domestic container business, CSX Lines, was sold at the beginning of 2003 to a venture known as Horizon Lines formed with the Carlyle Group, allowing CSX to retain an interest while further concentrating on the core rail business. This interest was sold for US$59 million in July 2004 when Horizon was acquired by Castle Harlan, though CSX and one of its affiliates continue as

GE locomotive with a train of auto-racks at Selkirk, New York (David C Warner)   0554638

lessee/sublessor or guarantor on certain vessels and equipment while subleases remain in effect. In February 2005 CSX sold its international terminals business (SL Service Inc) to Dubai Ports International for US$1.142 billion, completing sale of its non-rail operations which began in 2000.

CSXT is the product of the 1980 merger of Chessie System and Seaboard Coast Line, which were co-ordinated into a single system. A multimodal structure was developed to maximise deregulation's opportunities, stress the importance of the train as a link in a total distribution chain, and structure a competitive door-to-door transport system.

In the third quarter of 2004, CSX implemented the first phase of its ONE Plan, including a complete redesign of the operating plan for train movements intended to improve quality of service by reducing handling at yards and terminals, optimising routeing and cutting operating kilometrage. The year also saw completion of a reorganisation that reduced managerial staff by some 900, bringing the total employed in rail operations to around 32,000.

In December 2004 CSX opened its US$8 million Consolidated Training Center in Atlanta, replacing facilities previously provided at six locations. Some 3,500 students are expected to undergo training there each year.

Conrail acquisition

In the Conrail break-up, CSX acquired much of what was once the New York Central plus a shared interest in desirable trackage around Detroit,

New York/New Jersey and in the Pennsylvania/West Virginia coalfields. CSX and NS operate their portions of Conrail under various agreements; CSX's economic interest in Conrail amounts to 42 per cent. With additional trackage acquired from Conrail, the CSX system expanded to some 37,000 route-km. Network benefits included the ability to bypass congestion in Chicago, thus reducing transit times to BNSF and UP destinations by one day, and the use of modern classification facilities at Indianapolis which freed capacity at Cincinnati to speed up shipments to southeastern states.

In late 2004 CSX, NS and Conrail Inc completed a corporate reorganisation of Conrail that gave CSX direct ownership and control of routes and assets that it had previously operated under operating and lease agreements.

Finance

Revenue from the surface transport business increased by US$581 million (8 per cent) in 2004 (a 53-week year), reaching US$8.020 billion, while operating expenses rose 4 per cent to US$7.027 billion. The fourth quarter of 2004 was the 11th consecutive quarter of year-on-year revenue growth, an achievement driven largely by the increased demand for coal. Expenses were significantly affected by the rising cost of diesel fuel during the year, though mitigated somewhat by surcharges; expenditure on fuel was 16 per cent up at US$656 million. The operating ratio (operating expenses as a percentage of income) fell to 87.6 per cent from 91.2 per cent a year earlier. Operating income rose by 53 per cent to US$993 million. Revenue from CSX Intermodal was 5 per cent up, producing operating income of US$152 million, a 38 per cent increase on 2003.

Passenger operations

CSX handles more intercity passenger and commuter operations than any other US freight railroad. As well as Amtrak trains on several routes, it operates or hosts commuter services for Maryland's Mass Transit Administration, Florida Tri-Rail, Massachusetts Bay Transportation Authority, Metro North, NJ Transit, SEPTA and Virginia Railway Express.

Freight operations

Capacity problems and sluggish fleet utilisation combined to slow volume growth at a time when all Class I railroads were coping with record traffic on offer, particularly in the fourth quarter run-up to the holiday season. Several commodities, especially coal and metals, showed potential for additional volume growth had fleet utilisation been better. Also, some customers diverted business to other railroads or to trucks in an attempt to avoid service delays. Average train speed fell by 4 per cent to 32.7 km/h, and on-time arrivals fell by almost 30 per cent to 40.9 per cent. Such were the difficulties with intermodal traffic that in July 2004

A GE AC4400CW locomotive leads a CSXT coal train at Lee Hall, Virginia (David C Warner)   0554637

CSXI cut some 26 trains per week in an effort to consolidate traffic into fewer trains and so make better use of scarce capacity.

In 2004 total carloads rose 3.4 per cent to 5.2 million, continuing the recovery evident since the figure dipped below 5 million in 2002. Two of the railroad's three main revenue sectors recorded volume growth: Merchandise rose 3 per cent and Coal, Coke and Iron Ore 6 per cent, while Automotive declined 4 per cent. In the Merchandise sector (accounting for 60 per cent of revenue and 57 per cent of carloads), revenue rose 8 per cent on account of pricing-up, yield management strategies and fuel surcharges. All markets except Agricultural recorded higher volumes. Metals topped the list due to strong demand at steel mills, growing 9 per cent to produce revenue growth of 17 per cent, while the continued boom in residential construction saw Forest Products up 9 per cent on flat volumes.

Coal, Coke and Iron Ore revenue rose 11 per cent on volume growth of 6 per cent, driven largely by the buoyancy of the Asian export market for coal. In the first quarter, for example, coal traffic was up 10 per cent on the 2003 figure. This sector accounts for about 23 per cent of revenue. Automotive, on the other hand, recorded a further decline in volume, attributed to the continuing decrease in US light vehicle production and short-time working at some plants.

Revenue at CSX Intermodal rose 5 per cent for a 3 per cent increase in volume, with growth recorded in most sectors apart from domestic. There were further strong gains in truck brokerage with implementation of deal space technology to handle pricing, scheduling and capacity reservation.

CSXT continued to attract and participate in industrial development projects, particularly in Florida and Kentucky. The number of schemes rose 20 per cent in 2004 to 145, attracting a total of US$1.5 billion in capital spending.

Improvements to existing lines
Capital investment in infrastructure and rolling stock for 2005 is budgeted at US$1.1 billion.

Traction and rolling stock
At the beginning of 2003 CSXT listed 3,540 diesel locomotives, including leased machines. Among them were GE AC4400 (492 units) and AC6000 (117), GM-EMD SD70MAC (90), SD60 (178), SD40/ SD40–2 (494), GP47 (447), GP40 (469) and GP38–2 (282). At the same date, the wagon fleet totalled around 100,000.

Signalling and telecommunications
Locomotive deliveries during 2004 amounted to 110.

At the beginning of 2003 CSXT awarded a contract to Union Switch & Signal for specification, design and implementation of the proposed Next Generation Dispatch system at its Kenneth C Dufford signalling centre in Jacksonville. NGD will replace the dispatching system installed by US&S in 1988.

Track
Rail type & weight: T-rail, 17.6 kg/m to 69.6 kg/m
Sleepers: Hardwood (101.5 million) 2,600 × 180 × 230 mm; concrete (250,000) 2,600 × 210 (railseat)/ 180 (centre) × 270 mm. Spaced 1,950/km (wood) and 1,640/km (concrete)
Fastenings: Cut spike, wood; Pandrol clips, wood in more than 6° curves and for concrete sleepers
Max curvature: 14°, main line; some branches up to 30°
Max gradients: approx 2.5%, main line; approx 2.9% branch lines
Max permissible axleload: 34.8 tonnes

UPDATED

D & I Railroad Co

PO Box 5829, 300 South Phillips Avenue, Suite 200, Sioux Falls, South Dakota 57117-5829
Tel: (+1 605) 334 50 00   Fax: (+1 605) 334 3656
e-mail: sales@lgeverist.com
Web: http://www.lgeverist.com

Key personnel
President: Jack Parliament
Master Mechanic: Gerald Fox

Organisation
The D&IRR operates 222 route-km of former Milwaukee Road track between Dell Rapids, South Dakota, and Sioux City, Iowa, owned by the state of South Dakota. Its main activities are hauling aggregates, gravel and railroad ballast from its principal shipper and owner of the working assets, the L G Everist Company. A former CNW line from Hawarden to Beresford is now a branch of the D&I and generates as many as 100 grain trains in season. BNSF is the local service provider on two

segments, Sioux Falls to Canton and Elk Point to Sioux City. Interchange is available via BNSF at Sioux Falls, South Dakota, and with BNSF, UP and CN (formerly IC) at Sioux City, Iowa.

Traction and rolling stock
D&IRR operates 12 GM-EMD GP9, two GP20, two GP7, two SD45 and two SD39 locomotives. Rolling stock comprises 250 open-top hoppers and 75 gondolas.

Dakota, Minnesota & Eastern Railroad Corporation

140 N Phillips Avenue, Sioux Falls, South Dakota 57104
Tel: (+1 605) 782 12 00   Fax: (+1 605) 782 12 99
Web: www.dmerail.com

Key personnel
President and Chief Executive Officer:
 Kevin V Schieffer
Senior Vice-President and Chief Financial Officer:
 Kurt V Feaster
Vice-President, Marketing: Lynn A Anderson
Chief Transportation Officer: Richard D Awe
Chief Mechanical Officer: Daniel L Goodwin
Vice-President, Engineering: Steve O Scharnweber

Gauge: 1,435 mm
Route length: 1,825 km

Organisation
DMER, formed in 1986, was the first railroad to be created by a major sale of Chicago & North

Western trackage. It is the second longest US regional railroad, with Winona–Rapid City and Waseca–Mason City main lines as its core.

In May 1996 DMER acquired from UP the 325 km 'Colony line' from Chadron, Nebraska, to Colony, Wyoming, which was also an ex-Chicago & North Western property. The DMER meets this line at Rapid City, South Dakota.

In February 2002 the DMER announced that it had reached agreement to acquire the assets of I&M Rail Link (IMRL), a 2,763 km network to the east linking Chicago with Kansas City and Minneapolis. The acquisition was completed in July 2002, when operation of IMRL passed to be a new wholly owned DMER subsidiary, Iowa, Chicago & Eastern Railroad Corporation (IC&E). Both DMER and IC&E come under the common management of Cedar American Rail Holdings Inc. The two networks interchange at Owatonna, Minnesota.

Freight operations
The DMER handled approximately 65,000 carloads in 2003. Grain is the main commodity, comprising approximately 44 per cent of wagonloads in 2003.

Other important commodities are bentonite, kaolin, cement, metal products, paper, foodstuffs, scrap metal and grain products.

New lines
In June 1997 DMER unveiled a plan for a US$1.4 billion project to extend into the Powder River Basin coalfield in Wyoming. The 965 km line from Winona, Minnesota, to Wasta, South Dakota, would be rebuilt, and a new 450 km link would be built into the coalfields. The plan received the approval of the Surface Transportation Board in January 2002. Completion is expected to take five years.

Traction and rolling stock
DMER operates 66 locomotives, all of GM-EMD origin. The railway also owns or leases around 5,000 wagons, of which 1,225 are of recent manufacture.

UPDATED

Dakota, Missouri Valley & Western Railroad Inc

1131 22nd Street South, Bismarck, North Dakota 58504
Tel: (+1 701) 223 92 82   Fax: (+1 701) 223 00 08

Key personnel
President: Larry C Wood
Executive Vice-President: Jeff Wood

Vice-President: Diane J Wood
Vice-President, Marketing: Dennis Ming
General Manager, Maintenance: Roger C Wood
Transportation Manager: J P Ankenbauer
Manager, Mechanical: Randy Adien

Organisation
DMV&WR, which has 505.7 route-km, was created in 1990 to operate trackage leased from Soo Line (now Canadian Pacific) in North Dakota, principally

from Oakes westward through Bismarck and north to Max. Freight wagons are supplied by CP Rail. Some 18,000 carloads were handled in 2000, mainly agricultural produce.

Traction and rolling stock
DMV&WR operates three GP9 and 10 GM-EMD GP35 locomotives.

For details of the latest updates to *Jane's World Railways* online and to discover the additional information available exclusively to online subscribers please visit
jwr.janes.com

Dakota Southern Railway Company

PO Box 436, Chamberlain, South Dakota 57325
Tel: (+1 605) 734 65 95   Fax: (+1 605) 734 65 95
Web: http://www.sddot.com/fpa/railroad

Key personnel
Chairman: Richard H Huff
President and General Manager: George A Huff IV

Organisation
DSR operates over 305 route-km of 'local option' track owned by the state of South Dakota. The line from Mitchell to Chamberlain and Kadoka generates 1,600 carloads a year, almost entirely grain, and connects at Mitchell to BNSF. DSR has 220 covered hopper wagons on lease.

Traction and rolling stock
DSR has eight locomotives, including one Alco C420, one Alco S3, two GM-EMD SD7, two SD9 and one GE 70-ton unit.

Dallas, Garland & Northeastern Railroad

403 International Parkway, Suite 500, Richardson, Texas 75081
Tel: (+1 972) 808 98 00   Fax: (+1 972) 808 99 00
Web: http://www.railamerica.com

Key personnel
General Manager: Jim Kunz
Business Development Manager: Robin Bergeron

Gauge: 1,435 mm
Route length: 301 km

Organisation
Operated by RailAmerica, DGNO comprises lines running north from Dallas to Sherman and to Trenton using trackage rights between Dallas and Garland (KCS), and Dallas–Irving–Carrolton–Sherman (BNSF). The railway handles approximately 46,000 carloads annually. Interchange is made with KCS and UP at Dallas,

with BNSF at Irving and with a sister company, the Texas Northeastern Railroad, at Sherman and Trenton.

Decatur Junction Railway Co

308 South Chestnut Street, Assumption, Illinois 62510
A subsidiary of Pioneer Railcorp
1318 S Johansen Road, Peoria, Illinois 61607
Tel: (+1 309) 697 14 00
Fax: (+1 309) 697 53 87
Web: http://www.pioneer-railcorp.com

Key personnel
See entry for Pioneer Railcorp

Gauge: 1,435 mm
Route length: 61 km

Organisation
Operated under lease until 31 December 2006 from a consortium of grain dealers, DT comprises two

segments of track in east central Illinois, including 13 km of trackage rights over the Canadian National system through Decatur, Illinois. Main commodities carried are grain, fertilisers and plastics.

Delaware Otsego Corporation

1 Railroad Avenue, Cooperstown, New York 13326
Tel: (+1 607) 547 25 55
Fax: (+1 607) 547 98 34

The Delaware Otsego Corporation is a privately held non-rail holding company whose principal asset is the New York, Susquehanna and Western Railway (qv), a wholly owned subsidiary. The holding company also engages in real estate development to augment the railroad's traffic base. It also owns the defunct Central New York Railroad.

Duluth, Missabe and Iron Range Railway Company

500 Missabe Building, Duluth, Minnesota 55802
Tel: (+1 218) 723 21 15   Fax: (+1 218) 723 21 27
Web: http://www.cn.ca

Organisation
This is the largest iron ore handling railroad in North America, linking taconite plants in Minnesota's Mesabi Iron Range with the ports of Duluth and Two Harbors on Lake Superior, extending to some 340 km. Connections are with BNSF, CN, CP and UP. Ore is transported

either by unit train or lake vessels to steel mills in the Great Lakes region; in 2000 some 520,000 carloads were handled. DM&IR became part of the Great Lakes Transportation group in 2001 along with the Bessemer & Lake Erie Railroad.
   In May 2004 the DM&IR was acquired by CN, along with other railroad-related interests held by the Great Lakes Transportation group.

Traction and rolling stock
It operates 52 locomotives, mainly SD9s and SD38s, and 3,800 freight wagons.

Eastern Alabama Railway

2413 Hill Road, PO Box 658, Sylacauga, Alabama 35150
Tel: (+1 256) 249 11 96   Fax: (+1 256) 249 11 98
Web: www.railamerica.com

Key personnel
General Manager: Mike Brigham
Business Development Manager: Michael Koile

Gauge: 1,435 mm
Route length: 40 km

Organisation
Operated by RailAmerica, EARY is formed of branches from the CSXT Birmingham–Phenix City line in eastern Alabama, principal route Sylcauga to Talladega. Interchange is also made with NS. The railway handles approximately 12,000 carloads annually.

UPDATED

Eastern Idaho Railroad (EIRR)

420 Hansen Street South, Twin Falls, Idaho 83301
Tel: (+1 208) 733 46 86   Fax: (+1 208) 733 17 20
Web: www.watcocompanies.com

Key personnel
General Manager: Jack Lisle
Marketing Manager: Brandon Elliott

Gauge: 1,435 mm
Route length: 483 km

Organisation
Established in 1993, EIRR is one of the largest short line spin-offs of UP trackage. It comprises two unconnected networks in Idaho, the northern lines interchanging with UP at Idaho Falls and the southern lines at Mindoka. Serving a rich

agricultural area, EIRR handles some 35,000 carloads annually of agricultural and industrial produce.

Traction and rolling stock
At the beginning of 2005 EIRR was operating 16 diesel-electrics.

NEW ENTRY

Eastern Shore Railroad

PO Box 312, Cape Charles, Virginia 23310-0312
Tel: (+1 757) 331 10 94   Fax: (+1 757) 331 27 72

Key personnel
President and General Manager: Larry Le Mond
Transportation Superintendent: Ira T Higbee

Organisation
ESHR has 113 route-km. It provides a rail link from the Hampton Roads area to the eastern shore of Virginia by floating wagons across Chesapeake Bay (42 km) on two barges of 18- and 25-wagon capacity. There is interchange with NS at both ends of the route.

Traction and rolling stock
ESHR operates one GM-EMD GP38 and four GM-EMD GP10 locomotives, and 27 freight wagons.

Elgin, Joliet & Eastern Railway Company

1141 Maple Road, Joliet, Illinois 60432
Tel: (+1 815) 740 69 08 Fax: (+1 815) 740 67 29
Web: www.tstarinc.com

Key personnel

President and Chief Executive Officer:
John A Pranaitis

Elkhart & Western Railroad Co

PO Box 1468, Elkhart, Indiana 46515
A subsidiary of Pioneer Railcorp
1318 S Johansen Road, Peoria, Illinois 61607
Tel: (+1 309) 697 14 00 Fax: (+1 309) 697 53 87
Web: http://www.pioneer-railcorp.com

Key personnel

See entry for Pioneer Railcorp

Escanaba and Lake Superior Railroad

One Larkin Plaza, Wells, Michigan 49894
Tel: (+1 906) 786 06 93 Fax: (+1 906) 786 80 12
Web: www.elsrr.com

Key personnel

President: John C Larkin
Vice-President, Marketing: Thomas J Klimek

Farmrail System Inc

PO Box 1750, Clinton, Oklahoma 73601-1750
Tel: (+1 580) 323 12 34 Fax: (+1 580) 323 45 68
Web: http://www.farmrail.com

Key personnel

Chairman, President and Chief Executive Officer:
George C Betke Jr

Florida East Coast Railway Company

Florida East Coast Industries Inc
1 Malaga Street, St Augustine, Florida 32084
Tel: (+1 904) 829 34 21 Fax: (+1 904) 826 23 79
Web: www.fecrwy.com

Key personnel

Chairman, President and Chief Executive Officer:
Robert W Anestis
Vice-Chairman: Robert F MacSwain
Executive Vice-President, Secretary and General
Counsel: Heidi J Eddins
Executive Vice-President and Chief Financial
Officer: Daniel H Popky
Executive Vice-President and President, Florida
East Coast Railway: John D McPherson

Gauge: 1,435 mm
Route length: 859 km

Organisation

FECR is a Class II carrier, one of two wholly owned subsidiaries of Florida East Coast Industries Inc (FECI). The railway's physical plant is 561.6 km of

Fort Smith Railroad Co

A subsidiary of Pioneer Railcorp
1318 S Johanson Road, Peoria, Illinois 61607
Tel: (+1 309) 697 14 00 Fax: (+1 309) 697 53 87
Web: http://www.pioneer-railcorp.com

Key personnel

See entry for Pioneer Railcorp

General Manager, Northern Operations:
James L Neis
Superintendent: David M Gevaudan
Director of Marketing: James H Danzl

Organisation

EJ&E is a US Steel/Transtar property. Its route length is 360 km, and it has 63 locomotives and 4,059 freight wagons. The EJ&E is the outermost of the three belt railroads encircling the Chicago

Gauge: 1,435 mm
Route length: 13 km

Organisation

Acquired from Conrail in 1996 to become the Elkhart & Western Division of the Michigan Southern Railroad (MSO) (see entry in United States section of *Railway systems and operators*), EWR was separated from MSO in 2004. It runs from Elkhart to Mishawaka, Indiana. Interchange

Gauge: 1,435 mm
Route length: 555 km

Organisation

ELSR runs from Green Bay, Wisconsin, and Escanaba, Michigan, to Otonagon, Michigan, on Lake Superior, with branches. It has connections through Green Bay to the CN system. There are several connections in other locations with CN.

Organisation

Farmrail System Inc comprises 558 km of line in western Oklahoma. Farmrail Corporation is lessor/operator for Oklahoma Department of Transportation of the Weatherford–Erick line (132 km) and the Westhorn–Elmer route (143 km), which connect at Clinton. Subsidiary Grainbelt Corporation owns and operates the 283 km Enid–Clinton–Frederick line. Farmrail also has running

main track between Jacksonville and Miami; there is a 145.6 km branch line to Belle Glade on Lake Okeechobee. FECI also owns Flagler Development, a property company which redevelops surplus rail property and other sites.

Finance

Consolidated revenues in 2003 were US$339 million compared with US$301 million in 2002. Income before taxes was US$67 million (US$81 million in 2002).

Freight operations

In addition to intermodal traffic, principal commodities are automobiles, crushed stone, consumer products and foodstuffs. After intermodal (37 per cent), aggregates remain the strongest general commodity, accounting for 33 per cent of revenues. With the construction industry experiencing heavy demand in Florida, aggregates revenue was up 10 per cent in 2003. Income from other commodities slightly increased. Total carloads, at 187.1 million, were up 4 per cent on 2002's result.

Intermodal operations

Historically known for its interline intermodal business, especially through the ports of Fort

Gauge: 1,435 mm
Route length: 35 km

Organisation

Originally a 79 km system for which a 20-year operating lease was acquired by Pioneer Railcorp from Missouri Pacific in 1991, the line was cut back to its present length in 1995, and now links Fort Smith with Fort Chaffee. Connection is made at Fort

---

area. Via 14 interchanges it connects with every Class I operator in the region, serving steel works, petroleum and chemical plants.

UPDATED

is made with Norfolk Southern at Elkhart and EWR has a haulage agreement with CSXT at Fort Wayne, Indiana. Commodities carried include building materials and cement.

Principal traffic is paper, oriented strand board and pulpwood outbound; scrap paper, pulpwood, chemicals and fertilisers inbound; some 30 industries are served. Volume in 2004 was approximately 13,000 carloads.

Traction and rolling stock

ELSR operates 21 locomotives, including several Baldwins. It owns 357 wagons and leases a further 550.

UPDATED

rights between Snyder, Altus and Quanah, Texas. Connections are made to Burlington Northern Santa Fe, Union Pacific and two short lines. Employees: 32; locomotives: 15; annual traffic: 2,500 carloads.

Since 1995, the company has assigned locomotives to a joint-venture subsidiary, Finger Lakes Railway, operating 176 km of lines centred on Geneva in upstate New York.

Lauderdale, Miami and Palm Beach, FEC now originates or terminates 54 per cent of revenues online. Notable in 2003 was continued growth of the 'Hurricane' intermodal service between Miami and Atlanta in co-operation with NS, designed to attract more truck business to the railroad. This was followed in 2003 by 'Nighthawk' services from the New York/New Jersey area, and introduction of express Baltimore and Chicago services. Intermodal revenues, which had been in decline since 1997, began to improve in the second half of 2003 and ended 1.7 per cent up on the year. The late 2002 takeover of intermodal drayage operations was expected to bring efficiency improvements.

Traction and rolling stock

The all-GM-EMD locomotive fleet comprised 74 owned or leased machines in 2004, including some 40 GP40 and GP40-2 units. FECR also operated 2,488 freight wagons and 1,408 highway trailers.

UPDATED

Smith with Kansas City Southern and Union Pacific. Commodities carried include iron and steel, forest products and food and beverages, with around 1 million tons carried annually.

Garden City Western Railway

708 North View Road, Garden City, Kansas 67846
A subsidiary of Pioneer Railcorp
1318 S Johansen Road, Peoria, Illinois 61607
Tel: (+1 309) 697 14 00
Fax: (+1 309) 697 53 87
Web: http://www.pioneer-railcorp.com

Key personnel
See entry for Pioneer Railcorp

Gauge: 1,435 mm
Route length: 72 km

Organisation
Acquired by Pioneer Railcorp in 1999, GCW comprises two branches running from Garden City, Kansas, where interchange is made with Burlington Northern Santa Fe, to Wolf and Shallow Water. Main commodities carried are grain and other agricultural products, fertilisers and foodstuffs.

Gateway Western Railway

114 West 11th Street, Kansas City, Missouri 64105-1804
Tel: (+1 816) 983 13 03   Fax: (+1 618) 624 47 31
Web: http://www.kcsi.com/gwr

Key personnel
President: Don Gill
Vice-President Engineering and Mechanical:
   Paul Fetterman

Organisation
In May 1997, the Surface Transportation Board gave approval for acquisition of the GWWR by the Kansas City Southern Railway Company (qv).

GWWR's route length is 656 km, of which 602 km is solely owned; the remainder is owned jointly or operated under rights. Annual carloadings: 60,000; employees: 263.

The railway was formed in 1990 to take over lines linking Kansas City, St Louis and Springfield, Illinois, and associated branches, formerly owned by the bankrupt Chicago, Missouri & Western. It has a long-term contract with BNSF (ex-Santa Fe) for haulage of that railroad's freight between Kansas City and East St Louis.

In 1995, it took over the Mill Street yard and six locomotives to service customers of the Kansas City Terminal railroad. Also in 1995, the GWWR reached agreement with Conrail to haul the latter's intermodal traffic from St Louis to Kansas City where previously it went by highway. GWWR also owned trackage rights from Springfield to Chicago.

Traction and rolling stock
GWWR operates a fleet of GM-EMD locomotives, comprising 10 GP38 and four GP40 line-haul units and 13 SW1500 switchers.

Genesee & Wyoming Inc

66 Field Point Road, Greenwich, Connecticut 06830
Tel: (+1 203) 629 37 22   Fax: (+1 203) 661 41 06
Web: www.gwrr.com

Key personnel
Chairman and Chief Executive Officer:
   Mortimer B Fuller III
President and Chief Operating Officer:
   Charles N Marshall
Executive Vice-President, Corporate Development:
   Mark W Hastings
Executive Vice-President, Government and
   Industry Affairs: Robert Grossman
Chief Financial Officer: John C Hellmann
Senior Vice-President, General Counsel and
   Secretary: Adam B Frankel
Chief Accounting Officer and Global Controller:
   James M Andres
Senior Vice-Presidents
   Canada: Mario Brault
   Illinois: Spencer D White
   New York & Pennsylvania: David J Collins
   Oregon: Larry Phipps
   GWI Rail Switching Services: James W Benz
   Australia: Mike Mohan
   Mexico: Paul M Victor
   Utah: James N Davis
   Finance and Treasurer: Thomas P Loftus
   Administration and Human Resources:
      Shayne L Magdoff

Organisation
GWI is an operator of shortlines and regional railroads. Its properties in the USA are:
- Allegheny & Eastern RR
- Arkansas Louisiana and Mississippi RR
- Bradford Industrial Rail
- Buffalo & Pittsburgh RR
- Carolina Coastal Railway
- Chattahoochee Industrial RR
- Commonwealth Railway
- Corpus Christi Terminal RR
- Dansville & Mount Norris RR
- First Coast Railroad
- Fordyce & Princeton RR
- Genesee & Wyoming RR
- Golden Isles Terminal RR
- Illinois & Midland RR
- Louisiana & Delta RR
- Pittsburgh & Shawmut RR
- Portland & Western RR
- Rochester & Southern RR
- Salt Lake City Southern RR
- Savannah Port Terminal RR
- South Buffalo Railway
- Talleyrand Terminal RR
- Utah Railway Company
- Willamette & Pacific RR
- York Railway

Through its joint-venture affiliate in Canada, GWI (see Canada section of *Railway systems and operators*), the company operates:
- Québec Gatineau Railway
- Huron Central Railway
- St Laurent & Atlantique Inc (Québec)
- St Lawrence & Atlantic Inc

In Australia, through its joint venture subsidiary with Wesfarmers Ltd, Australian Railroad Group Pty Ltd (ARG) (see in Australia section of *Railway systems and operators*), GWI owns or operates:
- Asia Pacific Transport
- Australia Southern Railroad
- Australia Northern Railroad
- Australia Western Railroad
- Westnet Rail

In Bolivia GWI is a strategic investor in:
- Empresa Ferroviaria Oriental SA (see in Bolivia section of *Railway systems and operators*)

In Mexico GWI operates:
- Ferrocarriles Chiapas–Mayab, SA de CV (FCCM) (see in Mexico section of *Railway systems and operators*)

In 1996 GWI acquired an industrial switching subsidiary, Rail Link Inc, which it has subsequently developed. In 2004 Rail Link was handling rail operations at some 25 locations in 10 states, including port authorities, coal mines and power utilities companies, and was operating seven short line railroads.

In February 2002 GWI acquired Emons Transportation Group, which operated short line railroads in Maine, New Hampshire, Pennsylvania and Vermont, and in Quebec, Canada.

In August 2002 GWI acquired the Utah Railway Company from Mueller Industries Inc.

In 2004 GWI owned or leased approximately 12,800 km of track and operated over an additional 4,800 km under access arrangements. The company employs some 3,000 people and carries about 90 million tonnes annually.

In October 2002 GWI agreed a 15-year lease on BNSF's 122 km line between Salem and Eugene, Oregon, which is contiguous with GWI's Portland & Western Railroad. The agreement was expected to produce an additional 20,000 carloads of traffic annually for GWI.

Traction and rolling stock
At the beginning of 2004 the company operated 345 locomotives in its North American businesses.

UPDATED

Georgia & Florida Railnet

1019 Coastline Avenue, Albany, Georgia 31705
Tel: (+1 229) 435 66 29   Fax: (+1 229) 435 45 71

Key personnel
Vice President and General Manager: WT Hart
Operations Superintendent: D Windham

Organisation
Several formerly independent Georgia shortlines operating in the area between Albany and Valdosta are grouped into GFRR. They are the Atlantic & Gulf Railroad (former Seaboard Coast Line) from Albany to Thomasville and Sylvester; the Georgia Northern from Albany to Moultrie and Sparks; the South Georgia from Adel to the Florida boundary; and the Georgia & Florida from Willacoochee to Valdosta. Principal connections are with CSXT and NS. Total route 210 km; 65 employees; annual carloads 31,000.

Georgia Central Railway

186 Winge Road, Lyons, Georgia 30436
Tel: (+1 912) 526 61 65   Fax: (+1 912) 526 63 99
Web: http://www.rail-management.com

Key personnel
General Manager: Cecil Bowden

Gauge: 1,435 mm
Route length: 275 km

Organisation
A member of the Rail Management Corporation group, GC comprises former CSX trackage from Vidalia, Georgia, to Macon and Savannah, purchased or leased in 1990. Also acquired

was Vidalia–Rhine, later sold to Georgia Department of Transportation and now part of the Heart of Georgia Railroad. There is interchange with CSX at Savannah, with NS at Dublin and Macon, and with HOG at Vidalia. Daily service is provided, traffic including chemicals and forest products.

Traction and rolling stock
There is a fleet of 22 diesel locomotives comprising both GE and EMD units.

Georgia Rail Passenger Authority (GRPA)

276 Memorial Drive, SW, Atlanta, Georgia 30303
Tel: (+1 404) 463 09 65
Web: www.dot.state.ga.us/dot/grpa/index.shtml

Background
GRPA, working in concert with the Georgia Department of Transportation (GDOT), identified seven routes totalling 685 km over which commuter rail service could be implemented. These would run into Atlanta from Athens, Bremen, Canton, Gainesville, Macon, Madison and Senoia.

The first commuter service would run in the Macon Corridor south of Atlanta, linking central Atlanta with Lovejoy (42 km), serving six stations, and forming Phase 1 of what will ultimately be a 166 km route to Macon. Phase 2 would see trains extended by 29 km to Griffin, with two stations,

while Phase 3 would take the service a further 95 km through to Macon, adding four stations. The whole service would operate over existing NS trackage, and the railroad has indicated its willingness to reach an agreement. Cost of developing Phase 1 had been estimated at USD106 million, of which USD43 million was to fund track upgrading for 100 km/h running and provision of additional passing loops. Construction of a basic two-track station in central Atlanta (the so-called Multi-Modal Passenger Terminal) adjacent to MARTA's Five Points metro stop was costed at USD23 million, while the five suburban stations and four park-and-ride lots will cost USD13 million. The four trainsets required for the initial service were to have comprised three or four refurbished bi-level cars to be acquired from Chicago's METRA, hauled by remanufactured 3,000 hp diesel-electric locomotives.

However, changes in the political make-up of the Georgia state senate in 2004–05 resulted in a less

favourable view of the GRPA's proposals and its activities were scaled down. In 2005 negotiations were continuing between the GDOT and NS regarding access for the Lovejoy service, but its implementation was uncertain.

The proposed Macon line forms the first section of the Federally-designated high-speed rail corridor that would see Atlanta linked with Jacksonville, Florida, over existing NS and CSXT rights-of-way. Branches from this alignment would serve Columbus, Albany and Augusta, while an alternative routing might take trains to Savannah. Studies completed in 2003 evaluated options for 'conventional', 'moderate' and 'high-speed' (180 km/h) services in the corridor, costed at USD104 million, USD297 million and USD 393 million respectively. Ridership for the three options was estimated at 89,000, 248,000 and 288,000 trips annually.

NEW ENTRY

Georgia Southwestern Railroad

PO Box 69, 216 Long Drive, Smithville, Georgia 31787
Tel: (+1 229) 846 40 21   Fax: (+1 229) 846 40 25
Web: http://www.gswrr.com

Key personnel
President: Terry Small
Business Development Manager: Anita Horton

Gauge: 1,435 mm
Route length: 302 km

Organisation
Formerly owned by RailTex, GSWR was operated by RailAmerica between February 2000 and March 2002. In March 2002, RailAmerica sold the track to the state of Georgia and the capital stock of GSWR, together with eight locomotives, to a local private operator, which runs the railway on behalf of the state. Parts of the GSWR were then leased or transferred to Georgia Department of Transportation to form the Heart of Georgia Railroad (Vidalia–Americus–Mahrt, Alabama).

The railway comprises lines from Columbus in the northwest of the state to Bainbridge in the

south and via trackage rights to Albany and to White Oak, Alabama, in the west. Approximately 12,000 carloads are handled annually, mainly agricultural and forest products, chemicals, plastics and aggregates. Interchange is made with CSXT at Bainbridge and Cordele, and with NS at Albany, Americus and Arlington.

Improvements to existing lines
Rehabilitation of the Cusseta–Richland–Cuthbert line is in progress in conjunction with Georgia DoT's upgrading of the Heart of Georgia's Preston–Mahrt segment, due for completion in 2004.

Gettysburg & Northern Railroad Co

750 Mummasburg Road, Gettysburg, Pennsylvania 17325
A subsidiary of Pioneer Railcorp
1318 S Johansen Road, Peoria, Illinois 61607
Tel: (+1 309) 697 14 00   Fax: (+1 309) 697 53 87
Web: http://www.pioneer-railcorp.com

Key personnel
See entry for Pioneer Railcorp

Gauge: 1,435 mm
Route length: 40 km

Organisation
Acquired by Pioneer Railcorp from private interests in 2001, the Gettysburg & Northern runs north

from a connection with CSXT at Gettysburg, Pennsylvania, to Mount Holly Springs, where a connection is made with Norfolk Southern. Main commodities carried are paper products, foodstuffs and grain.

Gloster Southern Railroad

PO Box 757, Crossett, Arkansas 71635
Tel: (+1 870) 364 90 00   Fax: (+1 870) 364 45 21

Key personnel
General Manager: Bob Shelley

Gauge: 1,435 mm
Route length: 56 km

Organisation
This short line is owned by the timber and building materials supplier Georgia-Pacific Corporation. It began operations in August 1986 over the

56 km between the Georgia-Pacific plywood mill at Gloster, Mississippi and Slaughter, Louisiana, following abandonment of the branch by its then owner Illinois Central. Today the line provides a connection with CN at Slaughter.

NEW ENTRY

Grand Rapids Eastern

430 East Grove Street, Greenville, Michigan 48838
Tel: (+1 616) 754 00 01   Fax: (+1 616) 754 44 44
Web: www.railamerica.com

Key personnel
General Manager: Terry Wilson
Business Development Manager: Mike Bobic

Gauge: 1,435 mm
Route length: 70 km

Organisation
Operated by RailAmerica, GR links Grand Rapids with Ionia in the east and Marne in the west. The railway handles approximately 3,200 carloads annually, comprising mainly chemicals, paper, farm products and automotive components.

Interchange is made with CSXT and NS at Grand Rapids and with a sister company, the Mid-Michigan Railroad, at Lowell.

UPDATED

Great Miami/US Rail

PO Box 1060, Hamilton, Ohio 45012-1060
Tel: (+1 513) 844 64 10
Fax: (+1 513) 844 69 71
Web: www.great-miami.com

Key personnel
President: Frederick L Stout

Organisation
Great Miami runs two short line railroads, the Great Miami & Western providing switching service

at the Smart Paper Company mill at Hamilton, and the 108 km Great Miami & Scioto running from a CSXT connection at Chillicothe to Red Diamond and Firebrick.

NEW ENTRY

The Great Northwest Railroad, Inc

325 Mill Road, PO Box 1166, Lewiston, Idaho 83501
Tel: (+1 208) 746 50 57

Key personnel
President and Chief Executive Officer: Rick Webb
Manager: Kevin Spradlin

Gauge: 1,435 mm
Route length: 285 km

Organisation
Formerly an operating company for the Union Pacific and BNSF railroads, the former Camas Prairie Railroad was sold in April 1998 to North American RailNet, and in March 2004 was purchased by Watco Companies Inc. It was renamed the Great Northwest Railroad. Annual carloadings are 18,000.

The line leaves the UP at Riparia, Washington, and runs to Lewiston, then crosses the border into Idaho and runs to Kooskia, plus Orofino to Jaypee (out of service). Principal commodities carried are logs and lumber products, while intermodal traffic generated by the river port of Lewiston has also grown.

Traction and rolling stock
The locomotive fleet comprises seven GE B23-7 and one EMD SW1000.

Great Western Railway Company of Colorado

Taylor Avenue Shops, PO Box 537, Loveland, Colorado 80538
Tel: (+1 970) 667 68 83   Fax: (+1 970) 667 17 10
Web: http://www.omnitrax.com

Organisation
GWR serves customers on 178 route-km, with four diesel locomotives. It runs from Loveland to Longmont, Colorado, and also operates the former BNSF line from Greeley to Fort Collins. Interchanges are made with BNSF and UP; the latter brings corn in unit trains to GWR. The railway

is an OmniTRAX company (for further details, see entry for OmniTRAX Inc in the US *Railway systems and operators* section) and undertakes major locomotive repair and rebuilding at Loveland for member companies as well as on contract.

Guilford Rail System

Iron Horse Park, North Billerica, Massachusetts 01862
Tel: (+1 508) 663 11 30
Web: http://www.guilfordrail.com

Key personnel
President: Thomas F Steinger
Executive Vice-President: David A Fink
Vice-Presidents
  Transportation: Sid Culliford
  Engineering: James Patterson

Mechanical: Michael Walsh
Finance: Sonya Clay
Marketing: Richard Willey

Organisation
Guilford is the holding company for several rail assets, namely:
- Boston & Maine Corporation (see entry in the *Railway systems and operators* section)
- Maine Central Railroad (see entry in the *Railway systems and operators* section)

- Portland Terminal Co (see entry in the *Railway systems and operators* section)
- Springfield Terminal Railway.

Gulf, Colorado & San Saba Railway

510 North Bridge, Brady, Texas 76825
Tel: (+1 915) 597 23 59
Fax: (+1 915) 597 35 71

Key personnel
Senior Vice-President: Allan Roach
General Manager: Benny Herrera

Gauge: 1,435 mm
Route length: 108 km

Organisation
This short line started operations in 1993 by acquisition of the abandoned Santa Fe branch from Lometa to San Saba and Brady in Texas.

NEW ENTRY

Heart of Georgia Railroad

908 Elm Avenue, Americus, Georgia 31709
Tel: (+1 229) 924 76 62   Fax: (+1 229) 924 76 65
Web: http://www.hograil.com

Key personnel
President and Chief Executive Officer: Brad Lafevers
Vice-President and Chief Operating Officer:
  Duane Broxterman
Train Operations Manager: Dusty Carnes

Gauge: 1,435 mm
Route length: 170 km

Organisation
HOG started operations in 2000 on lines which had been part of the Georgia Central (Vidalia–Rhine) and Georgia Southwestern (Rochelle–Preston), and which are now leased from Georgia Department of Transportation. Initially operating only between Rochelle and Preston, service was extended eastwards to Vidalia in June 2000

on completion of track rehabilitation. Further work completed in June 2003 saw the Rochelle–Preston section upgraded with 80,000 new sleepers and repairs to 27 grade crossings and 25 bridges. Currently in progress is upgrading of the section between Preston and Mahrt, Alabama, completion of which will see HOG extended to 285 km.

Housatonic Railroad Co Inc

PO Box 1146, Canaan, Connecticut 06018
Tel: (+1 860) 824 08 50   Fax: (+1 860) 824 79 36
Web: http://www.housatonicrailroad.com

Key personnel
President: John R Hanlon Jr
Vice-Presidents:
  Marketing: Rian Nemeroff
  Finance: Robert Finley
  General Counsel: Edward J Rodriguez

Organisation
The Housatonic RR, together with its subsidiary Danbury Terminal RR, operates freight service on 259 km of tracks in the states of Connecticut, New York and Massachusetts; the lines are ex-New Haven, New York Central, Conrail, and Boston & Maine. Connections are made to CSXT at Pittsfield, Massachusetts and Beacon, New York; and to the Boston & Maine at Derby, Connecticut. Paper products, limestone, chemicals and lumber are the principal commodities; total carloadings are 4,500 to 5,000 per year. It

operates a lumber reload centre at Hawleyville, Connecticut.

Traction and rolling stock
The railway operates seven locomotives, including five EMD GP35s and one GP9.

Huron & Eastern Railway

101 Enterprise Drive, Vassar, Michigan 48768
Tel: (+1 989) 823 00 90   Fax: (+1 517) 823 37 94
Web: www.railamerica.com

Key personnel
General Manager: Jack Bixby
Business Development Manager: Mike Bobic

Gauge: 1,435 mm
Route length: 409 km

Organisation
A RailAmerica subsidiary since 1986, HESR comprises a network of shortlines running east of Saginaw, Michigan, to Croswell, Kinde and Harbor Beach. The network has also absorbed the Michigan Central, purchased by RailAmerica in January 2004, running some 150 km from Buena Vista south to Durand. The railway handles approximately 9,500 carloads annually, traffic including agricultural commodities, automotive components, chemicals, sugar beet and sugar products, fertiliser and

aggregates. Interchange is made with CSXT at Saginaw.

In 2004 HESR absorbed the contiguous Saginaw Valley Railway (SGVY), also a RailAmerica subsidiary, which runs east from Saginaw to Marlette and Brown City. This move added 89 km to the HESR system.

UPDATED

Illinois & Midland Railway Co

1500 North Grand Avenue East, Springfield, Illinois 62702
Tel: (+1 217) 788 86 01   Fax: (+1 217) 788 86 60
Web: http://www.gwrr.com

Key personnel
President and General Manager: Spencer White
Executive Vice-President: Raquel Swan
Chief Transportation Officer: Terry Holderread
Chief Mechanical Officer: Hal Bast
Chief Engineer: Allan Johnson

Manager, Marketing and Customer Services:
   Mike Vetter

Gauge: 1,435 mm
Length: 194.7 km

Organisation
Formerly the Chicago & Illinois Midland Railway Company, the I&M acquired its current name following acquisition in 1996 by Genesee & Wyoming. The railway runs south from Peoria, Illinois, via Havana to Springfield and Taylorville. Interchange is made with BNSF, CN, NS and UP, and via the

Gateway Western with CSX Transportation. Coal features prominently in the I&M's traffic, with some 3 million tonnes delivered annually from the Powder River Basin to the Kincaid generating plant. Other commodities carried include agricultural products, building materials and domestic and industrial waste, totalling around 100,000 carloads annually.

Traction and rolling stock
The I&M fleet numbers 17 (four SW1500s; two SD18s; four SD9s; five SD20s; two RS1325s).

Indiana & Ohio Rail System

497 Circle Freeway Drive, Suite 230, Cincinnati, Ohio 45246
Tel: (+1 513) 860 10 00   Fax: (+1 513) 682 46 45
Web: www.railamerica.com

Key personnel
General Manager: Terry Wilson
Marketing and Sales Manager: Greg Dixon

Gauge: 1,435 mm
Route length: 792 km

Organisation
Formerly operated by RailTex and owned by RailAmerica since February 2000, the Indiana & Ohio Rail System comprises two geographically separate railways: the Indiana & Ohio Railway (IORY) is a network of lines centred on Springfield, Ohio, and running north to Diann, Michigan, southeast to Washington Court House and via trackage rights southwest to Cincinnati. From here, using the tracks of a sister company, the Central Railroad of Indiana, and trackage rights, IORY extends its services northwest to Frankfort, Indiana, and beyond. The Indiana & Ohio Central Railroad

(IOCR) runs from Columbus, Ohio, southwest to Logan.
   Together the two railways handle approximately 110,000 carloads annually. Commodities carried include soda ash, limestone, cars, trucks and automotive components, railway equipment, fertiliser, and forest products.

UPDATED

Indiana Harbor Belt Railroad

2721 161st Street, Hammond, Indiana 46323-1099
Tel: (+1 219) 989 47 03   Fax: (+1 219) 989 47 07
Web: www.ihbrr.com

Key personnel
President: Mark D Manion
General Manager: Gary L Gibson
Comptroller: Derek Smith
Director, Business Development and Sales:
   J R Szamatowicz

Organisation
Comprising 252 route-km, Indiana Harbor Belt (IHB) provides industrial switching and transfer services in northwest Indiana and northeast Illinois, especially to several steel plants in Indiana and corn milling plants in Illinois. It provides interchange with 16 other railroads and operates yards at Hammond, Indiana and Riverdale, Illinois. IHB is 51 per cent owned by NS/CSX and 49 per cent by CP. In 2003 traffic totalled 700,000 wagonloads.

Traction and rolling stock
IHB operates 56 locomotives: two GP38-2s; 10 GP40s; two SD38-2s; six SD20s; 11 NW2s; two SW-1200s; and 23 SW-1500 units. The railroad also operates five power boosters, three diesel hump trailers and 1,283 freight wagons, of which 44 per cent are gondolas and 36 per cent coil cars.

UPDATED

Indiana Rail Road Co (INRD)

101 West Ohio Street, Suite 1600, Indianapolis, Indiana 46204
Tel: (+1 317) 262 51 40   Fax: (+1 317) 262 33 14
e-mail: admin@inrd.com
Web: www.inrd.com

Key personnel
President and Chief Executive Officer:
   Thomas G Hoback

Executive Vice-President and Chief Operating
   Officer: John A Rickoff

Organisation
Indiana Rail Road Co (INRD) provides freight service over 249 route-km. In 1985 the company bought 188 route-km serving coalfields in southern Indiana from Illinois Central and has since expanded. Its main line extends southwest from Indianapolis via Bloomington, Indiana, to Newton

and Lis, Illinois. North of Indianapolis, trackage rights exist to Noblesville.

Traction and rolling stock
INRD operates a fleet of 30 diesel locomotives.

UPDATED

Indiana Southern Railroad

PO Box 158, Ashby Yard, Petersburg, Indiana 47567
Tel: (+1 812) 354 80 80
Fax: (+1 812) 354 80 85
Web: www.railamerica.com

Key personnel
General Manager: Charles Fooks

Business Development Managers: Greg Dixon,
   Doug Ernstes

Gauge: 1,435 mm
Route length: 283 km

Organisation
Operated by RailAmerica, ISRR runs southwest from Indianapolis to Evansville, Indiana. The

railway handles approximately 65,000 carloads annually, commodities carried including coal, metal products, petroleum products, grain and grain products, foodstuffs, fertilisers, forest products and alcohol.

UPDATED

Indiana Southwestern Railroad Co

603 Allen Lane, Evansville, Indiana 47710
A subsidiary of Pioneer Railcorp
1318 S Johansen Road, Peoria, Illinois 61607
Tel: (+1 309) 697 14 00
Fax: (+1 309) 697 53 87
Web: http://www.pioneer-railcorp.com

Key personnel
See entry for Pioneer Railcorp

Gauge: 1,435 mm
Route length: 40 km

Organisation
Acquired by Pioneer Railcorp from the Evansville Terminal Railway Company in 2000, ISW runs from

Evansville via Poseyville to Cynthiana, Indiana. Interchange is made with CSXT and Norfolk Southern at Evansville. Main commodities carried are grain, plastics and rail equipment.

Iowa, Chicago & Eastern Railroad Corporation

1910 East Kimberley Road, Davenport, Iowa 52807-2033
Tel: (+1 319) 344 76 00   Fax: (+1 319) 344 77 00
Web: http://www.dmerail.com

Gauge: 1,435 mm
Length: 2,763 km

Organisation
Created in 1997 as I&M Rail Link, this railway serving Illinois, Iowa, Missouri and Minnesota includes a main line linking Minneapolis/St Paul, Chicago and Kansas City. Originally a Milwaukee Road route, it

was merged into the Soo Line in 1986 and later came under the control of Canadian Pacific in 1990.
   In 2002 I&M was acquired by Dakota, Minnesota & Eastern Railroad Corporation (DMER) subsidiary Iowa, Chicago & Eastern Railway. Both railroads come under the common management of Cedar American Rail Holdings Inc. The two networks interchange at Owatonna, Minnesota.

The IC&E connects with all railroads in Chicago, offering a direct intermodal service between Chicago, Davenport, Kansas City and the Twin Cities. Major branch lines serve the rich agricultural areas of northern Iowa and southern Wisconsin. The line also serves a variety of industrial customers, including cereals processors and manufacturers of machinery and agricultural and industrial heavy equipment. Bulk products such as cement and coal are also carried, and a branch to Rockford, Illinois

transports automobile parts to assembly plants and warehouses.

Improvements to existing lines
During 2004 major capital works were in progress on both DMER and IC&E, financed by a loan under the FRA's Railroad Revitalization Improvement Financing programme. Including normal annual expenditure of some US$16 million, the combined capital budget for 2004 was US$65.5 million. Five projects were under way for completion by the end

of 2004: relaying of some 160 track-km with 136 lb/yd rail; installation of 182,000 hardwood sleepers; surfacing of 1,300 km of track; rehabilitation of 146 bridges; and construction of passing loops at six locations.

Traction and rolling stock
The locomotive fleet totals 115, including General Motors SD-40 and SD-45 models, and the railway owns some 2,300 freight wagons.

UPDATED

Iowa Interstate Railroad Ltd (IAIS)

5900 6th Street SW, Cedar Rapids, Iowa 52404
Tel: (+1 319) 298 54 00
e-mail: info@iaisrr.com
Web: www.iaisrr.com

Key personnel
President and Chief Executive Officer:
   Dennis H Miller
Vice-President and Chief Operating Officer:
   Richard Stoeckly
Controller: Lori Frost
Vice-President, Mechanical/Engineering:
   Patrick Sheldon
Vice-President, Sales and Marketing:
   Don Nelson

Gauge: 1,435 mm
Route length: 1,106 km

Organisation
Owned jointly since 1991 by Railroad Development Corporation (see entry in the United States of America section of *Railway systems and operators*) and Heartland Rail Corporation, IAIS was created in 1984 to operate the former Rock Island main line from Omaha, Nebraska, to Bureau, Illinois and on to Chicago using trackage rights over CSXT and Metra lines. A 74 km branch runs south from Bureau to Peoria, providing an outlet for Iowa grain on the Illinois river and to Gulf ports. At the beginning of 2004 RDC purchased Heartland's 80 per cent stake in the railway, assuming 100 per cent ownership.

Traffic consists mainly of cereals, steel, forest products, intermodal, domestic appliances and

chemicals. Some six million tonnes were carried in 2003.

Improvements to existing lines
In 2005 IAIS secured a loan of USD32.7 million from the FRA's Railroad Rehabilitation and Improvement Financing fund. This was destined to finance track improvements between Atlantic, Iowa, and Bureau, Illinois, to accommodate heavier axle loads, as well as other infrastructure improvements.

Traction and rolling stock
At the end of 2002 IAIS owned 35 locomotives, mostly GM-EMD machines, and 1,211 wagons.

UPDATED

Iowa Northern Railway Co

PO Box 640, 122 North Second Street, Greene, Iowa 50636
Tel: (+1 641) 816 58 70   Fax: (+1 641) 816 48 16

Key personnel
General Manager: Mark A Sabin

Organisation
Iowa Northern (224 route-km) connects with CN (Illinois Central, previously CC&P), with the Cedar

Rapids & Iowa City Railroad at Cedar Rapids, with the Iowa, Chicago & Eastern at Plymouth and with Union Pacific at Manly. A 24 km branch runs from Vinton to Dysart. Much of its traffic is interchanged with the latter two railroads. In 2001 the railway carried 17,500 carloads, mainly farm products and fertiliser.

In 1995 the company was bought by the former Iron Road Railways Inc; it is now privately owned.

In October 2003 Iowa Northern took over operation of the 37 km line from Dewar to Oelwein

under contract from Transco Railway Products, which had bought the route from UP. Apart from serving Transco's rolling stock repair shops at Oelwein, the line serves several grain silos.

Traction and rolling stock
Iowa Northern operates 12 locomotives.

Iowa Northwestern Railroad (IANW)

4814 Douglas Street, Omaha, Nebraska 68132
Tel: (+1 402) 558 05 53
Web: www.ianwwr.com

Key personnel
President: John Larkin

Gauge: 1,435 mm
Route length: 60 km

Organisation
This short line runs from an interchange with UP at Mackenzie Junction, near Superior, Iowa, to Allendorf. The railroad was acquired from UP in the 1990s by the Dickinson Osceola Railway Association, which at first leased and later sold the line to IANW. As well as handling freight for

local industries, the railroad offers contract wagon storage facilities, with space for 450 vehicles, and runs a tour train.

Traction and rolling stock
The line is operated with two diesel locomotives.

NEW ENTRY

Kankakee Beaverville & Southern Railroad Co (KBSR)

PO Box 136, Beaverville, Illinois 60912
Tel: (+1 815) 486 72 60
Fax: (+1 815) 486 72 64
e-mail: cfhall@kbsrailroad.com

Key personnel
President: Robert M Garner
Secretary and Treasurer: C F Hall

Organisation
KBSR operates 250 route-km between Kankakee and Danville, and Iroquois Junction to Lafayette, Indiana, and has eight interchanges with other

systems. It hauls about 6,000 carloads annually, mainly grain and agricultural products.

Traction and rolling stock
The rolling stock fleet comprises six diesel locomotives and 284 covered hoppers.

UPDATED

Kansas City Southern Railway Company

PO Box 219335, Kansas City, Missouri 64121-9335
Tel: (+1 816) 983 13 03   Fax: (+1 816) 983 12 97
Web: www.kcsi.com

Key personnel
Chairman, President and Chief Executive Officer:
   Michael R Haverty
Executive Vice-President and Chief Operating
   Officer: Gerald K Davies
Executive Vice-President and Chief Financial
   Officer: Ronald G Russ
Vice-President, Corporate Affairs: Warren K Erdman
Vice-President and Comptroller: Louis G Van Horn
Senior Vice-President, Marketing and Sales:
   Larry O Stevenson

Vice-President and Treasurer: Paul J Weyandt
Associate General Counsel and Corporate
   Secretary: Jay Nadlman
Vice-President and Chief Mechanical Officer:
   James E Fisk
Director, Purchasing: Bob M Bayless

Gauge: 1,435 mm
Length: 5,259 route-km (including Texas-Mexican Railway and GWWR)

Organisation
Kansas City Southern (KCS) is a diversified holding company, the largest holding of which is the Kansas City Southern Railway (KCSR). In 1996, KCS entered into a joint venture with GATX to create Southern Capital Corporation to continue the leasing operations of rolling stock and

maintenance equipment, with KCS as its major client.

KCS serves a 10-state region and connects northern and eastern US railroads to southwestern US states and Mexico. It also offers the shortest rail route between Kansas City and major ports along the Gulf of Mexico. The railway's east-west corridor, the 'Meridian Speedway', provides an effective link between Meridian, Mississippi and Dallas, Texas.

At the end of 2003 KCSR employed 2,670 staff.

NAFTA Rail
KCSR has become a major player in the Mexican market. The roots of this expansion date back to 1995, when the company formed an alliance with Grupo TMM (TMM) to explore joint venture prospects in connection with the North American Free Trade Agreement (NAFTA) cross-Gulf

---

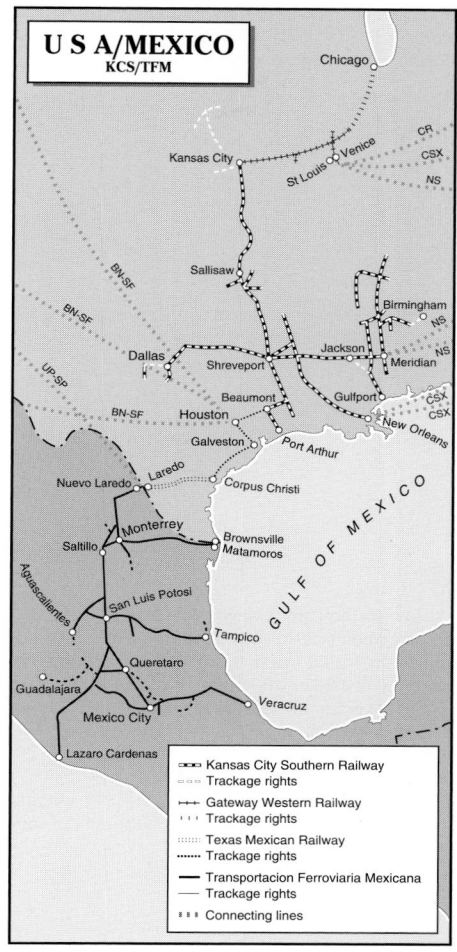

U S A/MEXICO
KCS/TFM

Kansas City Southern Railway
Trackage rights
Gateway Western Railway
Trackage rights
Texas Mexican Railway
Trackage rights
Transportacion Ferroviaria Mexicana
Trackage rights
Connecting lines

0003175

intermodal movements, and privatisation of the Mexican railways. Through the partnership, Mexrail Inc, KCSR acquired a 49 per cent stake in the Texas-Mexican Railway (see entry in US section of *Railway systems and operators*), which operates the primary US-Mexico rail gateway from Corpus Christi to Laredo. Mexrail was sold to TFM in 2002, the proceeds being used to reduce KCS debt. However, in November 2004 KCS received approval from the US Surface Transportation Board to acquire from TFM its entire shareholding in the Texas-Mexican Railway, together with its US portion of the International Rail Bridge at Laredo, Texas.

Of greater significance, though, was the alliance KCS formed with TMM to bid successfully on the 50-year concession to run the Ferrocarril del Noreste, 3,850 km of track which carries some 60 per cent of Mexico's rail freight. The core line is Mexico

City–San Luis Potosi–Monterrey–Nuevo Laredo, with branches to Veracruz, Tampico, Aguascalientes, Metamoros. The operating company is Grupo Transportación Ferroviaria Mexicana, SA de CV (Grupo TFM). One-quarter of the stock initially remained with the Mexican government, KCS/TMM having the option to purchase. This was exercised in 2002, after which KCS owned 46.6 per cent of the shares in Grupo TFM. The success of this expansion was further demonstrated in 2002, when Grupo TFM achieved significantly higher earnings. As a logical conclusion to these developments, the boards of KCS and Grupo TMM announced in April 2003 that the three railways (KCSR, Tex-Mex and TFM) were to be placed under common control of a single holding company, NAFTA Rail, subject to regulatory approval in both the USA and Mexico. As part of the transaction, KCS would change its name to NAFTA Rail. In August 2003 Grupo TMM shareholders voted against the proposals, but negotiations continued with a view to completing the agreement. These negotiations resulted in agreement in December 2004 with TMM to sell its interests in TFM to KCS, subject to the approval of the shareholders of the last-mentioned.

In addition, KCSR created a subsidiary, KCS Transportation Co, which in 1997 acquired the Gateway Western (see entry in US section of *Railway systems and operators*). By linking up the GWWR, the KCS and the Tex-Mex, KCSR has created 'the NAFTA railroad' from Chicago to the Mexico border, while a marketing alliance with CN signed in 1998 extended its reach into Canada. A further alliance signed in 2002 with BNSF gave KCSR access to new grain markets and will provide new opportunities in western states for its chemical and petroleum customers. Other marketing agreements are in place with NS and IC&E.

Further expanding its interests in Latin America, KCS formed a partnership with Mi-Jack Products and a consortium of Panamanian business interests to win a 25-year concession to rehabilitate and operate the 76 km Panama Railroad Company (see entry in Panama section of *Railway systems and operators*) to provide an intermodal alternative to the Panama Canal. The railway was rehabilitated and converted to 1,435 mm gauge in 2001.

Finance
KCSR produced revenues in 2003 of US$581.3 million (2.7 per cent down on 2002). Net income after taxes and interest expenses was US$12.2 million, compared with US$57.27 million in 2002.

Freight operations
Carloadings in 2003 totalled 950,500, up 2 per cent on 2002. Coal continued to lead the commodity sectors at 210,000 carloads, but ended the year with revenue 8.4 per cent down due to reduced inventories at some power stations and poor market conditions. Paper and forest products was the largest sector by revenue (up 8.4 per cent), followed by chemicals

and agricultural and minerals. Cross-border traffic between KCS and TFM grew substantially in both 2002 and 2003, with carloads up 18 per cent and 25 per cent respectively; revenue from cross-border traffic was up 44 per cent in 2003.

KCSR gained access to the port of Mobile, Alabama, at the beginning of 2003, through trackage and haulage rights negotiated with CN. November 2003 saw completion of Phase 2 of expansion of warehousing and transhipment facilities at Jackson, Mississippi, where annual capacity will rise to 3,000 cars by 2006.

The first full year of operation of KCSR's Management Control System computer platform brought significant improvements in operating statistics. Average load-to-load wagon turnround fell from 13.9 days in 2001 to 9.3 days at the end of 2003, while in 2003 alone average train speed rose from 31 km/h to 45 km/h.

Intermodal operations
Intermodal business, chiefly between New Orleans and Dallas, increased modestly for several years but took off in 1994. Growth continued in 1995 and in every year since; revenue in 2003 was up by 5.6 per cent.

KCSR owns and operates six intermodal facilities: Dallas, Texas; Kansas City, Missouri; Sallisaw, Oklahoma; Shreveport and New Orleans, Louisiana; Jackson, Mississippi.

Improvements to existing lines
Around 63 per cent of KCSR's main line track-km is continuously welded rail (cwr). As a result of track improvements, systemwide speeds on the core network have been raised to 85–90 km/h. Expenditure on track and infrastructure rose in 2003 by some 25 per cent to US$50.5 million. Much of the work was concentrated on raising capacity and line speed on the Meridian Speedway route between Dallas, Texas, and Meridian, Mississippi, where four new passing loops were commissioned. Work started in 2004 on a realignment project at Vicksburg, designed to raise line speed by easing curvature, while two track-doubling projects at Jackson, Mississippi, will improve interchange facilities with CN lines.

Traction and rolling stock
Traction and rolling stock at the end of 2003 comprised 487 diesel locomotives (133 owned and 354 leased), of which 423 were main line units, and 13,178 freight cars, of which 2,812 were leased.

Signalling and telecommunications
KCSR has Centralised Traffic Control (CTC) over 1,505 track-km with a centralised computer-aided dispatching system. It has 96 track-km of Automatic Block System (ABS). The entire system is cable-controlled by microwave and/or fibre optic cables.

UPDATED

Kansas & Oklahoma Railroad

1825 West Harry, Wichita, Kansas 67213
Tel: (+1 316) 263 31 13   Fax: (+1 316) 263 55 63
Web: http://www.watcocompanies.com

Key personnel
General Manager: Jim Wineland

Gauge 1,435 mm
Route length: 1,480 km

Organisation
The former Central Kansas Railway was acquired by The Watco Companies from OmniTrax in 2001 and renamed. It comprises a network of ex-Missouri Pacific and ex-Santa Fe trackage

extending westwards from Wichita, Kansas, principally for the transport of outbound grain. About 55,000 carloads are handled annually.

Keokuk Junction Railway Co

17 Water Street, Keokuk, Iowa 56232
A subsidiary of Pioneer Railcorp
1318 S Johansen Road, Peoria, Illinois 61607
Tel: (+1 309) 697 14 00   Fax: (+1 309) 697 53 87
Web: http://www.pioneer-railcorp.com

Key personnel
See entry for Pioneer Railcorp

Gauge: 1,435 mm
Route length: 196 km

Organisation
Acquired by Pioneer Railcorp from KNRECO in 1996, the KJRY runs from Keokuk, Iowa, where interchange is made with Burlington Northern Santa Fe, to LaHarpe, Illinois, where interchange is made with the Toledo, Peoria and Western Railroad. In 2002, KJRY was extended by acquisition of an

additional 19.5 km of line from LaHarpe to Lomax, Illinois, together with 25 km of trackage rights over BNSF's main line from Lomax to Fort Madison, Iowa, where interchange is made with Union Pacific. Main commodity carried is foodstuffs.

Kiamichi Railroad

800 Martin Luther King Boulevard, PO Box 786, Hugo, Oklahoma 74743
Tel: (+1 580) 326 83 57   Fax: (+1 580) 326 66 06
Web: www.railamerica.com

Key personnel
General Manager: Mitch Becker
Business Development: Gary Bradshaw

Organisation

Operated by RailAmerica since 1987, the KRR is an east-west line linking Anthony, Arkansas, and Lakeside, Oklahoma. At Hugo, branches run north to Antlers and south to Paris, Texas. The railway handles approximately 38,000 carloads annually. Traffic includes aggregates, cement, foodstuffs, forest products and paper, grain and grain products, metals, fertiliser and coal. Interchange is made with

BNSF at Lakeside, with KCS at Ashdown, Arkansas, with DQE at Valiant, Oklahoma, and with UP at Durant, Oklahoma, and Hope, Arkansas.

UPDATED

Kyle Railroad Co

PO Box 566, 38 Railroad Avenue, Philipsburg, Kansas 67661
Tel: (+1 785) 543 65 27   Fax: (+1 785) 543 96 46
Web: www.railamerica.com

Key personnel
General Manager: Steve Coomes
Business Development: Vince Como

Gauge: 1,435 mm
Route length: 1,114 km

Organisation

Operated by RailAmerica, KYLE comprises a group of lines linking Limon, Colorado, in the west and Salina and Mahaska, Kansas, in the east. The railway handles approximately 42,000 carloads annually including grain and other agricultural products, coal, asphalt and aggregates. Interchange

with BNSF is made at Courtland, Kansas, with UP at Salina and Colby, Kansas, and at Limon.

UPDATED

Lake Superior & Ishpeming Railroad Co

105 East Washington Street, Marquette, Michigan 49855-4385
Tel: (+1 906) 228 79 79
Fax: (+1 906) 228 79 83
Web: http://www.lsandirr.com

Key personnel
President and General Manager: John F Marshall
Vice-President, Controller: Dewayne D Nygard

Superintendent, Transportation: William J Cooke
Chief Engineer: Robin Trembath

Organisation

Lake Superior & Ishpeming (LS&I) is an 80 km self-contained private railroad which also shares tracks with CN (formerly Wisconsin Central). Built in 1896, LS&I is currently owned by the Cleveland-Cliffs Iron Co, which has an interest in Empire and Tilden, the last two active iron ore mines in the Marquette range.

The railroad handles some 9.5 million tonnes annually. Of this, 8 million tonnes is outbound

ore pellets and the balance is inbound materials, notably bentonite. Coal is also carried. Ore is moved in 11,000 tonne trains and loaded into lake shipping at LS&I dock, Marquette. The dock closes when Lake Superior freezes over, which usually occurs between January and April every year.

Traction and rolling stock
The rolling stock fleet comprises 1,300 wagons and 13 U30C and three U23C locomotives built by GE.

Lancaster & Chester Railway

PO Box 450, Lancaster, South Carolina 29721
Tel: (+1 803) 286 21 00   Fax: (+1 803) 286 41 58
Web: www.landcrailroad.com

Key personnel
President: Steve Gedney

Gauge: 1,435 mm
Route length: 96 km

Organisation

Always independently owned and now controlled by the Springs/Close family, the original L&C railway linking Lancaster and Chester in South Carolina was augmented in 2001 by lease-purchase from NS of its line from Catawba Junction to Chester and Kershaw, bringing the route operated to 96 km. There is interchange with NS at Chester and CSXT at Easy Chester. Traffic is mainly steel, coal, chemicals and building products.

As well as its daily freight operations, the railroad offers freight car leasing, charter of luxury passenger stock, a full-service locomotive workshop and refurbishment of passenger coaches.

Traction and rolling stock
The fleet comprises 11 diesel locomotives, including five GP38-2s.

NEW ENTRY

Laurinburg & Southern Railroad Co

PO Box 1929, Laurinburg, North Carolina 28353
Tel: (+1 910) 276 07 86   Fax: (+1 910) 276 28 53

Key personnel
President: Pete Claussen
General Manager: Rick Pearson

Organisation

The railway owns 45 route-km and also operates the 19 km Red Springs & Northern. It links Raeford with Laurinburg and Johns in North Carolina, carrying some 7,500 carloads annually, mainly grain, fertiliser, soda ash, coal and lime. Connection is direct to CSX at Raeford or via the Aberdeen & Rockfish (for further details, see entry in *Railway systems and operators* section) to NS.

On 31 July 1998, the railway was purchased by Gulf & Ohio Railways System.

Livonia, Avon & Lakeville Railroad (LAL)

PO Box 190B, Lakeville, New York 14480
Tel: (+1 585) 346 20 90   Fax: (+1 585) 346 64 54
www.lalrr.com

Key personnel
Chief Executive Officer: Eugene Blabey
President: William Burt

Gauge: 1,453 mm
Route length: 454 km

Organisation

From its beginning as a 20 km short line in 1964, LAL has developed into a four-line network serving a large area of western New York state and northwestern Pennsylvania. LAL itself operates in Livingston and Monroe counties, linking Rochester with Lakeville and Silver Springs. The railroad's three subsidiaries are: the Ontario Central Railroad, serving Ontario County southeast of Rochester; the Western New York & Pennsylvania from Hornell to Meadville, PA; and B&H Rail Corp, operating in Steuben County, from Corning to Wayland and Hammondsport (the former Bath & Hammondsport Railroad).

Its lines connect with NS, CSX, CP and Rochester & Southern, carrying food products, grain, fertiliser, timber and building materials. The former southern division operating in Steuben County, west of Corning, is now operated as B&H Rail Corp.

Traction and rolling stock
At the beginning of 2005 LAL's fleet comprised 10 diesel-electric locomotives, all Alco units.

NEW ENTRY

Long Island Rail Road

93-02 Sutphin Boulevard, Jamaica Station, Jamaica, New York 11435
Tel: (+1 718) 558 74 00   Fax: (+1 718) 558 82 12
Web: www.mta.info

Key personnel
President: James Dermody
Executive Vice-President: Albert M Cosenza
Vice-Presidents:
Senior Vice-President, Operations:
Raymond P Kenny

Market Development and Public Affairs: Brian P Dolan
Planning, Technology Development and Capital Program Management: John W Coulter Jr
Chief Financial Officer: Nicholas DiMola
General Counsel and Secretary: Mary Mahon

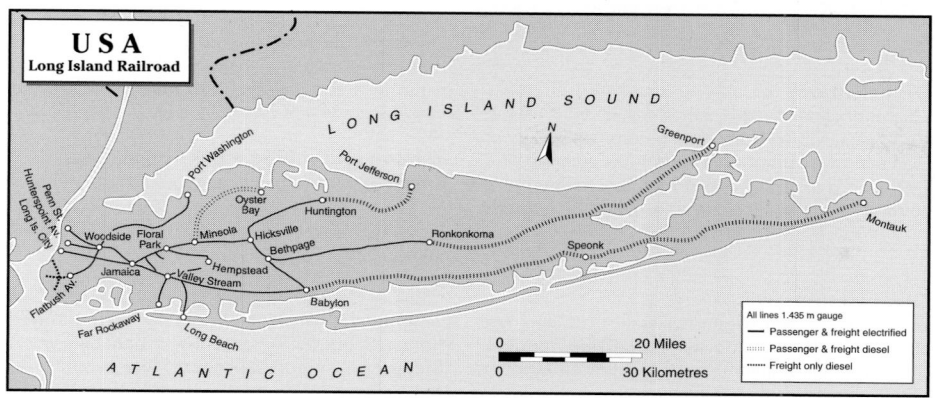

0003167

Design and construction continues for the East Side Access (ESA) Project, a 10-year scheme now costed at US$6.3 billion, which will provide LIRR with its own 8-track four-platform terminal at Grand Central's lower level, scheduled to open in 2012. As well as involving new and reconstructed tunnels between Long Island City and Manhattan, the ESA Project includes improved LIRR passenger access at Grand Central, a new station at Sunnyside Yard, new carriage stabling facilities in Queens and improved traction power supply, signalling and communications systems. Work also includes maintenance facilities at Long Island City, under construction at a cost of US$74.2 million.

Bids were sought in mid-2004 for the 1.8 km tunnel beneath Manhattan which will carry LIRR trains to Grand Central.

Traction and rolling stock
LIRR's most recent acquisitions were 23 General Motors DE30AC diesel-electric locomotives with Siemens AC transmission and 23 DM30AC dual-mode locomotives of similar design, but equipped to operate from LIRR's 750 V DC third-rail supply. Delivered in 2000, the dual-mode locomotives allow through running to New York's Penn station from non-electrified lines. Mitsui supplied 134 air-conditioned double-deck coaches for push-pull operation, in an order valued at US$234 million.

Replacement of the 1968-vintage Budd M-1 emus has continued with some 250 M-7 cars from Bombardier in service by September 2004. MTA now has a total of 978 M-7 cars on order for LIRR and Metro-North (for more details, see Metro-North Railroad Co in the *Railway systems and operators* section, US). Options remain for the supply of an additional 288 cars. The aim is for M-7s to comprise 75 per cent of the LIRR fleet by 2006.

The fleet of around 1,100 passenger vehicles includes 500 Class M-1 multiple-unit cars, operated on inner suburban electrified lines in New York City, Nassau and Suffolk counties, 172 Class M-3 cars acquired in 1984–86 and 250 M-7 cars.

Track
Rail: 100 PS, 112, 115, 119 RE, 130
Sleepers: Wood 7 × 9 in by 8 ft 6 in, spaced 2,983/mile;
concrete 8 × 10 in by 8 ft 6 in, spaced 2,640/mile
Fastenings: Cut spike and Pandrol Fastclips
Max curvature: 6° approx
Max gradient: 2.5%
Max permissible axleload: 41 short tons

UPDATED

Electric railcars or multiple-units

Class	Cars per unit	Motor cars per unit	Motored axles/car	Output/ motor kW	Speed km/h	No in service*	First built	Builders Mechanical
M-1	2	2	4	640	129	500	1968	Budd
M-3	2	2	4	704	129	172	1984	Budd
M-7	2	2	4	–	160	250	2002	Bombardier

* As at September 2004

Executive Director of Capital Program
    Management: Thomas J DeMaria
Chief Transportation Officer: Jane F Dietz
Chief Mechanical Officer: Mark P Sullivan
General Manager, New Fleet: David J Elliot

Gauge: 1,435 mm
Length: 435 km
Electrification: 237 km at 750 V DC third rail

Political background
Long Island Rail Road (LIRR) is a wholly owned subsidiary of the Metropolitan Transportation Authority (MTA), an agency of the State of New York, whose members constitute the railroad's board of directors. In a major reorganisation of the MTA announced in October 2002, LIRR and Metro-North are to be merged as the MTA Railroad.

Finance
LIRR's 2000–04 five-year plan, an element of the New York MTA's budget, calls for a US$2.5 billion capital investment programme which is focusing on rolling stock renewal (US$1 billion), infrastructure maintenance (US$743 million) and stations (US$314 million).

Passenger operations
Passenger journeys in 2003 amounted to 80.9 million, down 3.6 per cent on 2002.

The railroad serves 124 stations in the Long Island counties of Nassau and Suffolk, extending to Greenport and Montauk, each some 150 km from New York, as well as certain communities in eastern Queens. It has nine branches which feed into three western terminals in New York City: Penn Station, Flatbush Avenue (Brooklyn) and Hunterspoint Avenue (open only in peak hours and served only by diesel-powered trains). Penn Station handles 240,000 passengers daily; another 40,000 travel via Brooklyn and Hunterspoint Avenue.

The focal point of the system is the eight-platform Jamaica station, where eight of the nine branches and the three approaches to the New York City terminals converge. During the rush hour, a train movement takes place on average every 30 seconds at Jamaica.

Improvements to existing lines
Following completion of LIRR's US$190 million Penn Station Improvement Project, work started in 2001 on a US$225 million reconstruction of Jamaica station, scheduled to take four years.

Louisiana & Delta Railroad

402 W Washington Street, New Iberia, Louisiana 70560
Tel: (+1 337) 364 96 25   Fax: (+1 337) 369 14 87
Web: http://www.gwrr.com

Key personnel
President: James W Benz
Vice-Presidents
    Operating Services: William A Jasper

Sales and Marketing: C Murray Cook
Short Line Operations: Billy C Eason

Organisation
The railroad is a subsidiary of Genesee & Wyoming's Rail Link short line operation (see entry in United States of America section of *Railway systems and operators*) and operates 14 branches in Louisiana totalling 183 route-km owned and an additional 148 km of trackage rights over BNSF property between Lake Charles and Raceland.

It connects to the BNSF which bought the line into New Orleans as a condition of the UP-SP merger. Annual traffic totals some 12,000 carloads. Principal commodities carried are carbon black, sugar and molasses, pipes, rice and paper products.

Louisville & Indiana Railroad Company

500 Willinger Lane, Jeffersonville, Indiana 47130
Tel: (+1 812) 288 09 40   Fax: (+1 812) 288 49 77
Web: http://www.anacostia.com/lir

Key personnel
Chairman and Chief Executive Officer:
    Peter A Gilbertson
President and Chief Operating Officer:
    John K Secor
General Superintendent: John H Sharp

Gauge: 1,435 mm
Route length: 171 km (owned); 182 km (operated)

Organisation
Louisville & Indiana (LIRC) is a wholly owned subsidiary of Anacostia & Pacific Co Inc (see entry in the United States section of *Railway systems and operators*). Formed in 1994, the LIRC operates 171 km of main line and some secondary track between Indianapolis, Indiana, and Louisville, Kentucky, formerly worked by Conrail. Annual carloadings amount to 16,000 and the railroad employed 40 people in 2002. LIRC connects with CP, CSXT, NS and PAL.

The LIRC main line between Indianapolis and Louisville has been designated as a high-speed rail corridor by the US Department of Transportation.

Improvements to existing lines
Infrastructure works completed in 2003 included installation of some 10,000 new sleepers, surfacing of 69 km of track and redecking three bridges. For 2004, major track rehabilitation was planned at Jeffersonville yard and 3 km of route was to be relaid with 112 lb/yd rail.

Traction and rolling stock
The locomotive fleet consists of 13 locomotives, mostly GP7/GP9 units. Four GP39-2 locomotives were acquired from UP in late 2003 to replace lower horsepower machines.

Madison Railroad

Building 216, 1121 West JPG Woodfill Road, Madison, Indiana 47250
Tel: (+1 812) 273 42 48
Fax: (+1 812) 265 52 51
Web: www.madisonrailroad.com

Key personnel
Chief Executive Officer: Cathy Hale

Organisation
This short line started operations in 1978 and in 1981 was acquired from Penn Central by the City of Madison port authority. It runs from a connection with CSXT at North Vernon 40 km to Madison, south of Indianapolis, carrying mainly polythene,

coal by-products, steel products and scrap. In addition to its main line, MR operates 27 km of track in the industrial park developed on the site of the former Jefferson Proving Ground, offering wagon storage and other facilities.

Traction and rolling stock
Two diesel locomotives are operated.

NEW ENTRY

Maine Central Railroad Company

Subsidiary of Guilford System Industries Rail Division
20 Rigby Road, South Portland, Maine 04106
Tel: (+1 207) 828 64 03   Fax: (+1 207) 828 64 03

Key personnel
See entry for Guilford Rail Systems in the United States of America section of *Railway systems and operators.*

Gauge: 1,435 mm
Length: 1,187 km

Organisation
Maine Central lies within the state of Maine, but another Guilford company, Springfield Terminal Railway, actually operates the system on account of the labour agreements advantageous to Guilford which are in force on the Springfield Terminal Railway.

Guilford has contributed $1.5 million to a $7 million new intermodal facility in Waterville, aimed at capturing truck traffic. Waterville is also the originating/terminating point for the 'DownEast Express' operation started jointly with Conrail.

Maine Central is also being positioned to combine with Boston & Maine and the New Brunswick Southern (see entry in Canada section) to create an import-export business to and from New England via the port of Halifax, Nova Scotia.

Traction and rolling stock
Maine Central operates 73 diesel locomotives, mostly of GM origin.

Manufacturers Railway Co

2850 South Broadway, St Louis, Missouri 63118-1895
Tel: (+1 314) 577 17 49   Fax: (+1 314) 577 31 36
Web: www.anheuser-busch.com/overview/ Railway.htm

Key personnel
President and Chief Executive Officer: Kurt R Andrew
Vice-President and Treasurer: Barbara J Houseworth

Senior Operations Director: Randall J Weitzel
Mechanical Superintendent: Amund L Whittley
Director of Engineering: Kem E Conrad

Organisation
The railway operates over 61 route-km and connects with all major carriers in the St Louis gateway, via the Terminal Railroad of St Louis and Alton & Southern in particular. It is a wholly owned subsidiary of the brewing company Anheuser-Busch. In addition to switching the flagship brewery in St Louis, distributing certain beer brands nationally and hauling grain, for which

the railway maintains a fleet of 282 boxcars and grain hopper cars, the subsidiary secures income from contract wagon and locomotive maintenance, repair and modification.

There is also a subsidiary trucking operation that furnishes cartage (80 purpose-designed trailers) and warehousing services to the parent brewing company and others.

Traction and rolling stock
Operates nine locomotives.

UPDATED

Maryland & Delaware Railroad Co

106 Railroad Avenue, Federalsburg, Maryland 21632
Tel: (+1 410) 754 57 35   Fax: (+1 410) 754 95 28
Web: http://www.mdde.com

Key personnel
President: Eric H Callaway

Vice-President: John C Paredes
General Manager: Nelson J Pearsall III

Organisation
The company operates 191 route-km: three unconnected routes in the Delmarva peninsula, Delaware and Maryland, connecting with the NS line from Wilmington, Delaware to Cape Charles, Virginia. These are the Seaford, Snow Hill and Townsend lines. Interchange with NS

is made at Seaford, Townsend and Frankford, Delaware.

Traction and rolling stock
Eight diesel locomotives (three CF7s, three RS3Ms, one SW900 and one SW8M).

Maryland Midland Railway Inc

PO Box 1000, 40 North Main Street, Union Bridge, Maryland 21791-0568
Tel: (+1 410) 876 03 92
Fax: (+1 410) 775 25 20
e-mail: bdenton@mmidrwy.com
Web: www.mmidrwy.com

Key personnel
Chairman: Harvey L Miller

President and Chief Executive Officer: Paul D Denton
Senior Vice-President, Chief Financial Officer and Chief Operating Officer: David W Bordner
Director, Marketing: Barbara Denton
Manager, Transportation: David B Hart
Manager, Track and Structures: Gary W Smith

Organisation
Maryland Midland operates 108 route-km and connects with CSXT at Highfield and Emory Grove.

In FY2004, the railroad employed 29 staff and handled 14,014 carloads of freight. Cement, stone, aggregates and timber accounted for 95 per cent of traffic.

Traction and rolling stock
Maryland Midland operates 10 diesel locomotives (three GP9s and seven GP38s) and 449 wagons.

UPDATED

Maryland Transit Administration: Maryland Commuter Rail (MARC)

6 St Paul Street, Baltimore, Maryland 21202-1614
Tel: (+1 410) 767 37 78   Fax: (+1 410) 333 04 89
Web: www.mtamaryland.com

Key personnel
MTA Administrator: Robert L Smith
Manager and Chief Operating Officer: (vacant)
Deputy Administrator, Operations: Robert L Mowry
Chief Mechanical Officer: Rex Springston
Chief Transportation Officer: Ira Silverman

Passenger operations
MARC provides Perryville–Baltimore–Washington DC services over Amtrak's Northeast Corridor route (119 km electrified at 12 kV 25 Hz AC) and Baltimore–Washington DC services over 61 km of CSXT. A Frederick/Martinsburg–Washington DC

service is provided over 134 km of CSXT track. Total route is 323 km with 42 stations.

Amtrak operates the Washington–Baltimore–Perryville service with 10 electric locomotives and MARC's newer Japanese-built coaches. Two diesel-powered services are operated for MARC by CSXT. One is over former Baltimore & Ohio tracks between Washington's Union station and Baltimore Camden station. The other is from Washington to Brunswick, Maryland, and Martinsburg, West Virginia. On weekdays, 86 trains are operated; there is no weekend service. Some 6.7 million passengers were carried in 2004, an all-time record and up from the 2003 figure of 6.4 million.

Improvements to existing lines
The MARC network was extended by 21.6 km to Frederick in late 2001, with three daily round trips to Washington. The total cost of the project, including land acquisition, infrastructure improvements,

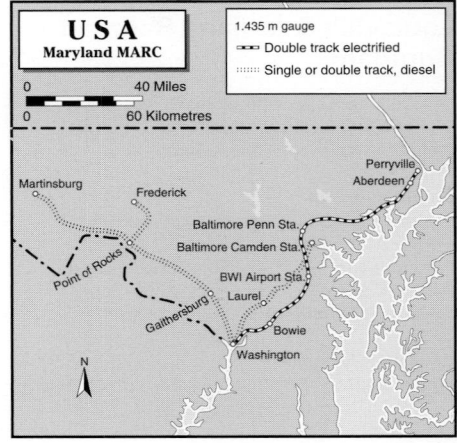

0543472

stations and rolling stock was US$91.3 million. Initial patronage was disappointing, but budget restraints and line capacity prevent early expansion of the service.

Traction and rolling stock
MARC's locomotive fleet consists of four AEM7 electric locomotives, six HHP electric locomotives similar to those supplied by ALSTOM and Bombardier to Amtrak for Northeast Corridor services, 19 GP40WH-2 (3,000 hp/2,238 kW) and six GP39–2 (2,300 hp/1,716 kW) diesel locomotives rebuilt by Morrison Knudsen. There are also 110 coaches, built by Kawasaki and Sumitomo. Twelve gallery cars were acquired from Metra in 2003.

Signalling and telecommunications
CTC covers all of the system. All signalling is owned by Amtrak or CSXT. Amtrak territory between Washington and Perryville is also equipped with cab signalling and ATC.

UPDATED

MARC Type HHP electric locomotive at Baltimore BWI with an evening commuter service from Washington DC (Philip Wormald)
***NEW**/0585043*

Massachusetts Bay Transportation Authority (MBTA)

10 Park Plaza, Room 5720, Boston, Massachusetts 02116
Tel: (+1 617) 222 33 02   Fax: (+1 617) 222 45 39
Web: http://www.mbta.com

Key personnel
General Manager: Michael H Mulhern
Chief Operating Officer: Anne Herzenberg
Treasurer and Controller: Wesley G Wallace
General Manager, Massachusetts Bay Commuter
  Rail Corp: Kevin Lydon

Passenger operations
MBTA provides commuter service over 13 routes extending to 528 km with 116 stations; five serve Boston's North station and eight run to the newly remodelled South station. Much of the network (436 km) comprises ex-Penn Central and Boston & Maine trackage purchased in the 1970s. In 1988, service was extended along Amtrak's Northeast Corridor route to Providence, Rhode Island, and a 37 km extension of the Framingham line to Worcester opened in 1996. Old Colony Railroad services over three routes to Middleboro and Kingston in southeastern Massachusetts started in late 1997, while trains on the restored link from Ipswich to Newburyport commenced in 1998.

MBTA is also the operator of a co-ordinated network of four metro lines, five light rail lines, four trolleybus and 155 bus routes. An integrated fare system is in place allowing use of all modes, including heavy rail, with an inclusive, zone-based monthly ticket. The state of Massachusetts, directed by the governor's office, is looking into prospects for MBTA privatisation, beginning with functional areas such as revenue accounting, payroll and property management.

Operation of the commuter network was contracted to Amtrak in 1986, but the arrangement came to an end in July 2003 when the Massachusetts Bay Commuter Rail Corporation consortium took over. MBCR beat two competitors with a low bid of US$1.07 billion for the five-year operating contract; Amtrak had declined to bid. The new contractor is a consortium comprising Connex North America Inc, Bombardier Transportation and a local consulting company headed by former MBTA general manager James F O'Leary. To smooth the transition, MBTA voted a US$15 million appropriation which was being used mainly for cosmetic improvements to the rolling stock fleet.

MBTA F40PHM-2C locomotive at Plymouth, Massachusetts (David Haydock) 0543490

In 2000 MBTA heavy rail operations were recording some 36 million passenger journeys and 1,150 million passenger-km a year. The heavy rail services recover about 33 per cent of their costs from farebox revenues.

Improvements to existing lines
A five-year investment plan approved in March 2004 aims mainly to upgrade the existing lines and complete reopening of the Greenbush line (see below).

Several network extensions, to both the north and south of Boston, are either under way or the subject of studies, and a batch of measures was approved in May 2004 for improvements to Haverhill, Worcester and Fitchburg line services.

Preliminary work has started on the 28 km Greenbush line, the third Old Colony route, and the project survived proposed budget cuts in early 2003. Extensions to New Bedford and Fall River (69 km), which had been dormant, were revived in mid-2004

when it was agreed to start discussions with CSX on the purchase of 53 km of alignment.

Traction and rolling stock
In 2003 MBTA operated 82 diesel locomotives and around 200 passenger cars. The all-GM locomotive fleet includes 11 F40PHM-2Cs which were undergoing refurbishment by Boise in 2002–04 and 25 F40PH-2C machines subject to similar treatment by Boise in 2000–03. The coaching stock fleet includes vehicles supplied by Kawasaki (bi-level), Bombardier, MBB of Germany and Pullman-Standard. The last-mentioned have been refurbished.

Massachusetts Central Railroad Corporation

2 Wilbraham Street, Palmer, Massachusetts 01069
Tel: (+1 413) 283 59 00
Fax: (+1 413) 283 29 10
Web: www.masscentralrr.com

Key personnel
President: Mike Smith

Organisation
Started operations in 1976 over the 42 km former Conrail branch from Palmer (CSXT and New England Central connections) to South Barre, plus a short branch to Bondsville (5 km). It handles general freight and intermodal traffic at the Palmer inland port opened by the railroad in 1984. MassCentral Transportation Services, a subsidiary set up in 2000, operates drayage service for railroad customers.

Traction and rolling stock
MassCentral operates five diesel locomotives and some 200 freight wagons owned or leased.

NEW ENTRY

Maumee & Western Railroad (MAW)

817 Fifth Street, Defiance, Ohio 43512
Tel: (+1 419) 784 08 89   Fax: (+1 419) 784 05 38

Key personnel
President: Spencer Wendelin

Gauge: 1,435 mm
Route length: 82 km

Organisation
MAW operates portions of the former Wabash (later NS) Woodburn branch in Indiana and Ohio, between Woodburn, Cecil, Defiance and Liberty Center. It acquired the route in 1997 from the bankrupt Indiana Hi-Rail operation with funding provided by the Ohio Rail Development Commission. The entire line was in a poor physical state and the 13 km between Cecil and Defiance is currently out of service due to poor track conditions, preventing operation of a through service and cutting-off several important sources of traffic. About 3,200 carloads are handled annually.

Grants of around US$500,000 from the state of Ohio have helped fund three phases of rehabilitation, under which operations were scheduled to resume between Napoleon and Liberty Center in late 2005. Work included replacement of 3,000 sleepers and installation of some sections of new rail. Phase 3, approved in March 2005, will see the line reopened as a through route, and is expected to generate 960 new carloads.

Traction and rolling stock
Four diesel locomotives are operated.

NEW ENTRY

Metra

Chicago Commuter Rail Service Board
(Northeast Illinois Regional Commuter Railroad Corporation)
547 W Jackson Blvd, Chicago, Illinois 60661-5717
Tel: (+1 312) 322 69 00
Web: www.metrarail.com

Key personnel
Chairman: Jeffrey R Ladd
Executive Director: Philip A Pagano
Deputy Executive Director: G Richard Tidwell
Chief Operations Officer: Vaughn L Stoner
Heads of Department
  Corporate Administration: Michael J Nielsen
  General Development: Jack Groner
  Real Estate and Planning: Patrick McAtee
  Human Resources: Gail Washington
  Treasury and Finance: Frank M Racibozynski
  Materials Management: Paul Kisielius
  Transportation: George Hardwidge
  Mechanical: Richard Soukup
  Engineering: William K Tupper

Gauge: 1,435 mm
Route length: 729 km
Electrification: 99 km at 1.5 kV DC

Passenger operations
Metra is the commuter railroad operating arm of the Chicago Regional Transportation Authority (RTA). Its commuter rail system takes in the six counties of northeast Illinois. Including the operations conducted for Metra by Burlington Northern Santa Fe (BNSF – formerly Burlington Northern) and Union Pacific (UP – formerly Chicago & North Western) on their own infrastructure, the system extends to 729 route-km and covers most of the northeast Illinois region. Trains operate seven days a week over 11 main routes and four branches, serving 228 stations.

BNSF (one route) and UP (three routes) operate commuter services in the Chicago RTA area with Metra-owned equipment and under Metra direction through purchase-of-service contracts. Metra itself owns and runs the former Illinois Central (electric), Milwaukee Road and Rock Island services with its own train crews and equipment; since 1993 it has operated the Norfolk Southern line to Orland Park, now called SouthWest Service, under long-term lease. Metra also operates its North Central service via trackage rights on Canadian National right-of-way. There are also three Heritage Corridor trains each way daily between Chicago and Joliet on the CN route. In addition South Shore Line services operated by Northern Indiana Commuter Transportation District run over Metra tracks from 115th Street to Randolph Terminal.

0583638

Total ridership in 2003 was 76.9 million (projected) for 1,724.5 passenger-km, a fall of 3 and 2.4 per cent respectively on the previous year. Metra has forecast progressive increases to 80.8 million passenger journeys in 2006.

Improvements to existing lines
Metra's 2004 capital programme was set at US$348.1 million. The programme includes: US$43 million for track and infrastructure projects; US$49.1 million for stations and parking; US$123.4 million for rolling stock; US$7 million for support facilities and equipment; US$26.1 million for signalling and telecommunications; US$84.7 million for extensions and service expansion; and US$14.8 million for other investments.

New Start programme
Funding of US$585 million was granted by the Federal Railroad Administration in July 2002 for three projects under the New Start programme of infrastructure works to raise capacity. This covers: an extension of the SouthWest line from Orland Park to Manhattan with three new stations, capacity enhancements and new yard facilities, enabling Metra to increase the number of daily trains from 16 to 30; further double-tracking of the North Central route to Antioch, including five new stations, enabling the service level to be increased from 10 to 22 trains daily; and on the Union Pacific West line a 14.5 km extension from Geneva to Elburn, capacity enhancement work, provision of two new stations and new yard facilities. Completion of the programme is expected in 2005.

STAR Line
Plans for several Metra extensions were consolidated at the beginning of 2003 when the STAR (Suburban Transit Access Route) programme was announced. Recognising the decline in journeys to and from Chicago's central business district, STAR proposes an orbital line serving the northwest suburbs, largely over existing Elgin, Joliet & Eastern and Indiana Harbor Belt trackage. The route would run northwards from Joliet to Hoffman Estates (Prairie Stone) and then east to O'Hare airport, totalling 88 km with 17 stations, 11 of which would provide interchange with existing Metra services. Diesel multiple-units would run a 15-minute all-day service; construction would take 10 to 12 years and cost some US$1.1 billion. A later phase would see the line extended from O'Hare south and east to Midway airport.

SouthEast Service Line
The SouthEast Service Line is a proposal for a new service from La Salle Street station to southern suburbs over CSXT and UP lines to Chicago Heights and Crete. The 56 km line would serve nine new stations.

Other improvements
In early 2003 Metra announced plans to ameliorate pinch-points on UP's Northwest and West lines to allow further service expansion and improve reliability. On the former an extension is proposed of the McHenry branch north to Johnsburg, where a new yard would be provided. At the same time CTC would be introduced. On the UP West line capacity upgrades and signalling improvements are planned, including relocation and renovation of busy A-2 interlocking outside Chicago's Union station to improve operational performance.

There is strong local backing in the state of Wisconsin for extension of Metra service north from Kenosha to Racine and Milwaukee (55 km), but at US$152 million for seven peak-hour return

Metra diesel locomotives

Class	Builder's type	Wheel arrangement	Power kW	Speed km/h	Weight tonnes	No in service	First built	Mechanical	Builders Engine	Transmission
B32A	F40PH	Bo-Bo	2,390	142	118	28	1977	EMD	16-645E3B	E EMD
B32A	F40PH-2	Bo-Bo	2,390	142	118	57	1979	EMD	16-645E	E EMD
B32A	F40PHM-2	Bo-Bo	2,390	142	120	30	1991	EMD	16-645E3B	E EMD
C32A	F40C	Co-Co	2,390	142	165	15	1974	EMD	16-645E	E EMD
–	MP36	Bo-Bo	2,685	142	147	27	2003	Motive Power	16-643F3B	E Wabtec, Motor Coils

Metra electric railcars

Class	Cars per unit	Motor cars per unit	Motored axles/car	Output/ motor kW	Speed km/h	No in service	First built	Builders Mechanical	Electrical
SA2A	1	1	4	120	130	8	1982	Nippon Sharyo	GE
MA3A	1	1	4	120	121	129	1971	St Louis Car	GE
MA3B	1	1	4	120	121	36	1978	Bombardier	GE

trips daily, the scheme may prove too expensive in the current economic climate.

Traction and rolling stock
At the end of 2003 Metra owned 140 diesel locomotives for revenue service, plus 165 electric railcars and 806 push-pull coaches.

Delivery was completed in early 2004 of 26 new Type MP36PH-3S diesel-electric locomotives from Wabtec subsidiary, MotivePower, 15 of which are replacement units. The remainder have been assigned to the North Central Service, which will be expanded from 10 weekday trains to 22 beginning in 2005, when right-of-way upgrades are completed.

Replacement of Metra's oldest cars started with delivery in March 2003 of the first of 300 stainless steel bilevel vehicles from Sumitomo Corporation of America. All due to be delivered by October 2005,

these were ordered in December 2000 to expand the fleet and replace 258 life-expired units built in the 1950s and 1960s that are assigned to the three Union Pacific lines and the BNSF Aurora route. Nippon Sharyo designed and is manufacturing the bodyshells and is performing systems integration. Super Steel Products Corporation is assembling the cars in Milwaukee. In addition, 26 electric cars were ordered from Sumitomo Corp of America in October 2002, starting the process of replacing the St Louis Car 'Highliner' fleet dating from 1971. These cars are being assembled locally by Super Steel, Milwaukee. Two prototype vehicles are due to be delivered in December 2004; delivery of all is scheduled to be completed by February 2006.

Track
Rail: 115, 119, 131, 132, 136 lb/yd
Crossties (sleepers): Wood and concrete
Thickness: 7 × 9 in
Spacing: 1,988/km
Fastenings: Rail anchors
Max curvature: 7° 26'
Max gradient: 1.75%
Max axleload: 65,000 lb

UPDATED

Metrolink

Southern California Regional Rail Authority
700 South Flower Street, Suite 2600, Los Angeles, California 90017-4606
Tel: (+1 213) 452 02 00   Fax: (+1 213) 454 04 25
Web: www.metrolinktrains.com

Key personnel
Chief Executive Officer: David R Solow
Assistant Executive Officer: Steve Wylie
Directors:
Strategic Development and Communications: Stephen H Lantz
Operations: John Kerins
Equipment: William Lydon
Construction and Engineering: Mike McGinley
Finance: Pat Kataura

Gauge: 1,435 mm
Route length: 824 km

Political background
The Southern California Regional Rail Authority (SCRRA) is a joint-powers authority formed in 1991 by five member agencies to develop a commuter rail system for the Los Angeles region. Its members are: Los Angeles County Metropolitan Transportation Authority; Orange County Transportation Authority; Riverside County Transportation Commission; San Bernardino Associated Governments; and Ventura County Transportation Commission. Its activities are governed by a board of 11 directors drawn from these agencies. Between 1990 and 1992 SCRRA purchased right-of-way and operating rights from Southern Pacific, Union Pacific and the Santa Fe Railway Company, whose successors, Union

Pacific and Burlington Northern Santa Fe, retain trackage rights for freight traffic. Services on the first three of the present network of seven lines commenced in 1992.

Operation of the network has been contracted to Amtrak since the beginning, but uncertainty over Amtrak's future led SCRRA to make contingency plans in late 2002. From October of that year SCRRA assumed responsibility for despatching Metrolink trains with Amtrak continuing to operate them pending a tendering process for a new train operating contract. The resulted in a 5-year operating contract being awarded in late 2004 to Connex North America, the US subsidiary of the global Connex group. Worth €77 million, the contract takes effect in July 2005.

Passenger operations
SCRRA services operate under the Metrolink branding over a seven-line, 53-station network. The seven lines are: Antelope Valley (Lancaster–Los Angeles, 123.3 km); Inland Empire-Orange County (San Bernardino–Oceanside, 161.1 km); Orange County (Oceanside–Los Angeles, 140.3 km); Riverside (Riverside–Los Angeles, 94.5 km); San Bernardino (San Bernardino–Los Angeles, 90.4 km); Ventura County (Montalvo–Los Angeles, 114.1 km); and the 91 Line (Riverside–Fullerton–Los Angeles, 99.1 km).

In 2003 137 trains were operated each weekday.

Traffic (million)	2001	2002	2003
Passenger journeys	7.4	7.9	9.1

Traction and rolling stock
The fleet comprises 39 low-emission GM-EMD F59PH or F59PHI locomotives and 139 double-deck coaches bought from Bombardier. Four ex-Amtrak F40PH locomotives were acquired in November 2002 and are scheduled for rebuilding and retrofitting. An additional 12 Bombardier bi-level coaches and two GM-EMD F59PHI locomotives

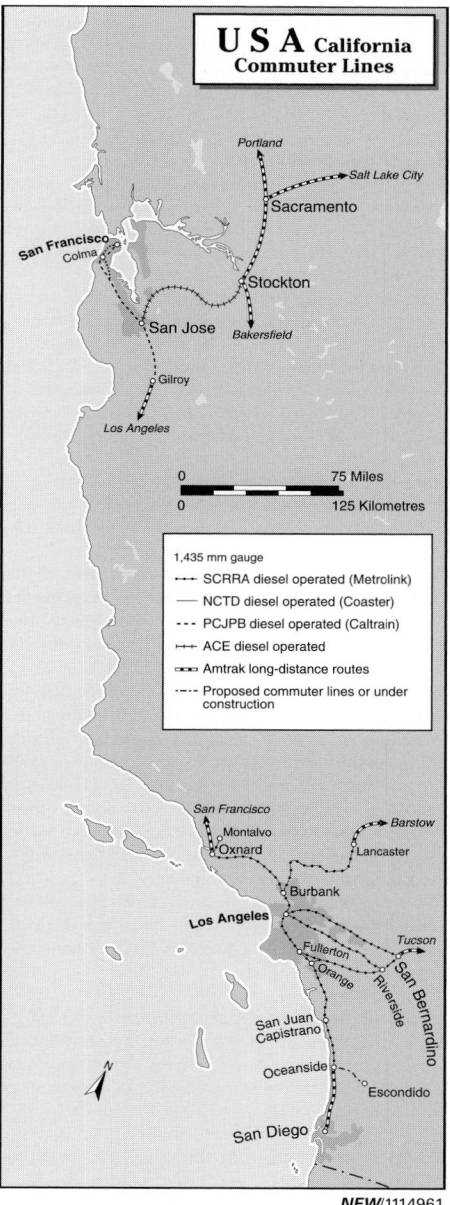

Metrolink F59PHI diesel locomotive and bi-level stock (Metrolink)   **NEW**/1140508

NEW/1114961

were leased from Sound Transit in Seattle, Washington, in late 2004 to help meet growing ridership demand. Invitations to tender for the supply of 35-40 new multilevel coaches were to be released in mid-2005, with deliveries anticipated from late 2007.

In 2003 Bombardier Transportation was awarded a 7-year follow-on contract valued at US$90 million covering maintenance of SCRRA's locomotive and rolling stock fleet at Metrolink's Central Maintenance facility in Los Angeles. The contract includes one 3-year extension option.

Track
Rail: 136, 119, 115
Crossties (sleepers): 83% wood; 17% concrete

Spacing: Wood: 2,006/ km; concrete: 1,630/km
Fastenings: (concrete) Pandrol clip, McKay Safelok
Min curvature radius: 10°
Max gradient: 3%

UPDATED

Metro-North Railroad Co

347 Madison Avenue, 4th Floor, New York, New York 10017
Tel: (+1 212) 340 30 00   Fax: (+1 212) 340 40 37
Web: http://www.mta.info

Key personnel
President and General Manager: Peter Cannito
Vice-Presidents:
  Finance and Administration:
    Genevieve T Firnhaber
  General Counsel: Richard K Bernard
  Human Resources: Gregory Bradley
  Operations: George F Walker
  Planning and Development: Howard Permut
Assistant Vice-President, Operations:
  George F Walker

Gauge: 1,435 mm
Route length: 428 km
Electrification: Hudson and Harlem lines (139 km), 600 V DC third rail; New Haven line (116 km), 11 kV 60 Hz AC overhead

Political background
Metro-North is one of five operating divisions of New York's Metropolitan Transportation Authority (MTA). In a major reorganisation of the MTA announced in October 2002, Metro-North and the Long Island Rail Road are to be merged as the MTA Railroad.

Passenger operations
Metro-North's three main lines that run north out of New York City and east of the Hudson river are the Hudson, Harlem and New Haven lines. These three routes operate out of New York's Grand Central Terminal.

Service on the 158 km, 29-station New Haven line, and its New Canaan, Danbury and Waterbury branches totalling 96 km and 17 stations, is provided by Metro-North under a contract between the Connecticut Department of Transportation (ConnDoT) and MTA/Metro-North. Stamford is the busiest station, boarding almost 5,000 passengers per weekday and receiving 2,100 arrivals. At New Haven connection is made with ConnDoT's Shoreline East service which runs to New London. Capital improvements in Connecticut are funded by ConnDoT, which owns the infrastructure.

Metro-North also provides services west of the Hudson in the New York state counties of Orange and Rockland on the Port Jervis and Passaic Valley lines; trains on these lines run out of the Hoboken terminal on the New Jersey shore of the Hudson. These services are operated by New Jersey Transit (for further details, see New Jersey Transit Rail Operations Inc in the *Railway systems and operators* section, USA) under contract from Metro-North, which assigns part of its rolling stock fleet to NJT. In early 2003, Metro-North took a 49-year lease on the 105 km Port Jervis line from owner Norfolk Southern. This will allow Metro-North to carry out track improvements to raise reliability of the passenger service.

On the three main lines (Hudson, Harlem and New Haven), total passenger journeys in 2002 were 73.2 million. This was the tenth consecutive annual increase.

Improvements to existing lines
An 8 km extension of the Harlem line beyond its current terminus at Dover Plains to Wassaic was opened in 2000, restoring service to a section of route closed in 1971.

Traction and rolling stock
At the end of 2002, resources comprised three electric and 47 diesel locomotives (10 of which

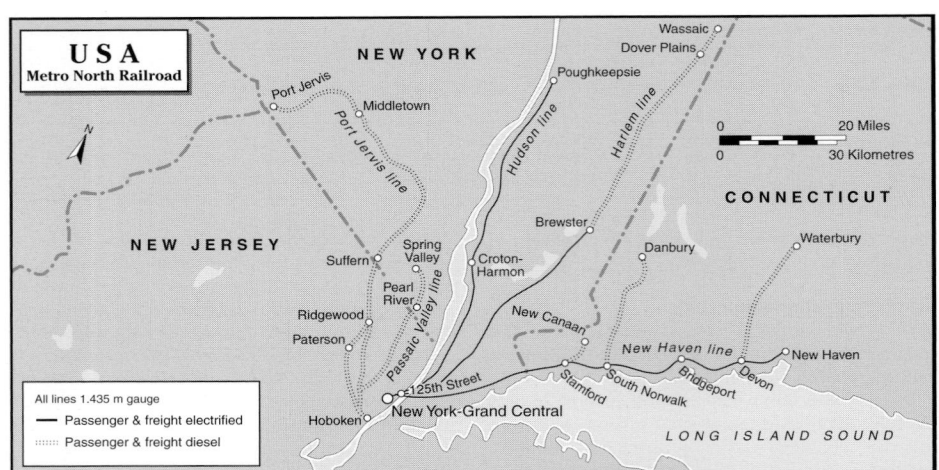

0554470

Metro-North General Electric Type P32AC-DM dual-mode locomotive at Cold Spring, New York, with a New York–Poughkeepsie train (Quintus Vosman)   0536194

Diesel and electrodiesel locomotives (with third-rail pick-up shoes)

Class	Wheel arrangement	Power kW	Speed km/h	Weight short tonnes	No in service	First built	Mechanical	Builders Engine	Transmission
FL9	Bo-A1A	1,490	110	135	8	1957	EMD/MK	GM 567C	E EMD
F10	Bo-Bo	1,490	110	130	3	1957	EMD/MK	GM 567C	E EMD
P32AC-DM	Bo-Bo	2,390	160	Na	20	1995	GE	GE	E GE
GP35[1]	Bo-Bo	1,490	110	131	6	1994 R*	EMD	GM 567	E EMD
GP7u	Bo-Bo	1,120	105	124.5	1	1953	EMD	GM 567 C	E EMD
GP9	Bo-Bo	1,300	105	125.5	1	1955	EMD	GM 567 C	E EMD
GP40	Bo-Bo	2,240	169	128.1	7	1988 92R**	EMD/MK;	GM 16 645	E EMD
FH2[2]							EMD/CR		
BL6	B	559	60	72	2	2000	BMS	Cat 3456	E Kato

* R = year of rebuilding.
** In New Jersey service.

Electric railcars or multiple-units

Class	Cars per unit	Line viltage	Motor cars per unit	Motored axles/car	Output/ motor kW	Speed km/h	No in service	First built	Builders Mechanical	Electrical
M-1	2	600 DC	2	4	110	160	170	1971	Budd	GE
M-2	2	600 DC and 11 kV AC	2	4	120	160	241	1973	GE	GE
M-3	2	600 DC	2	4	120	160	142	1982	Budd	GE
M-4	3	600 DC and 11 kV AC	2	4	120	160	54	1987	Tokyu Car	GE
M-6	3	700 DC and 11 kV AC	3	4	120	160	48	1993	Morrison Knudsen	GE
M-7	2	600 DC	2	2	–	160	–	2002	Bombardier	Bombardier
ACMU	1	600 DC	1	4	75	120	57	1962	Pullman-Standard	GE

M2 emus at Bridgeport, Connecticut (David C Warner)   0543492

are owned by ConnDoT), 722 emus and 183 conventional coaches (of which 40 are owned by ConnDoT). An additional eight locomotives are assigned to west of Hudson services and numbered in with NJ Transit rosters.

The majority of Metro-North's passenger cars serving the third-rail electrified Hudson and Harlem lines and the overhead-electrified New Haven line are ageing emus of Types M-1 and M-3 plus the more recent tranche of 54 M-4 and 48 M-6 three-car emus. Replacement of the M-1 and M-2 fleets has started, with a total of 858 M-7 cars ordered by the MTA for Metro-North and the Long Island Rail Road (for further details, see Long Island Rail Road in the *Railway systems and operators* section, USA). Options exist for a further 408 cars, the purchase from Bombardier of 120 of which was approved by MTA's board of directors in January 2004. The aim is for M-7s to comprise half the Metro-North fleet by 2006.

Non-electrified lines are worked by locomotive-hauled trains, using a variety of diesel traction including Genesis II locomotives equipped with third-rail pick-up shoes for operating into Grand Central Terminal. Seven FL9s have been modernised: they were rebuilt by ABB as FL9AC units with three-phase AC traction motors and a 3,200 hp (2,387 kW) EMD engine.

Signalling and telecommunications
The only remaining segments without cab signalling are the single-track Danbury and Waterbury branches. Metro-North dispatches for Amtrak and freight activity as far as New Haven, Connecticut, and Poughkeepsie, New York.

Track
Rail: 119 RE and 132 RE predominantly, some 140 and 127
Crossties (sleepers): 7 × 9 in × 8 ft 6 in
Spacing: 1,968/km for wood ties; 1,640/km for concrete
Fastenings: Tie plates and cut spike for wood; Pandrol clips for concrete and some wood
Min curvature radius: 109 m
Max gradient: 3%
Max axleload: E72 loading for track design

Michigan Shore Railroad

101 Enterprise Drive, Vassar, Michigan 48768
Tel: (+1 989) 823 00 90
Fax: (+1 989) 823 37 94
Web: www.railamerica.com

Key personnel
General Manager: Bob Dine
Business Development Manager: Mike Bobic

Gauge: 1,435 mm
Route length: 11 km

Organisation
Operated by RailAmerica, the MS is a shortline in the Muskegon, Michigan, area handling approximately 3,200 carloads annually of sand and chemicals. Interchange with CSXT is made at Muskegon Heights.

UPDATED

Michigan Southern Railroad Co

PO Box 239, White Pigeon, Michigan 49099
A subsidiary of Pioneer Railcorp
1318 S Johansen Road, Peoria, Illinois 61607
Tel: (+1 309) 697 14 00
Fax: (+1 309) 697 53 87
Web: http://www.pioneer-railcorp.com

Key personnel
See entry for Pioneer Railcorp

Gauge: 1,435 mm
Route length: 78 km

Organisation
Located in southern Michigan and running between White Pigeon and Coldwater, MSO connects with Norfolk Southern at White Pigeon, Michigan, and via haulage rights with CSXT at Fort Wayne, Indiana. Main commodities carried include scrap metal, paper, fertiliser, aggregates and food products.

Mid-Michigan Railroad

432 East Grove Street, Greenville, Michigan 48838
Tel: (+1 616) 754 00 01
Fax: (+1 616) 754 44 44
Web: www.railamerica.com

Key personnel
General Manager: Jack Bixby
Assistant General Manager: Bob Dine
Business Development Manager: Mike Bobic

Gauge: 1,435 mm
Route length: 176 km

Organisation
Formerly owned by RailTex and since 2000 operated by RailAmerica, MMRR runs from Elmdale to Lowell in Michigan. Approximately 8,200 carloads are handled annually. Interchange is made with CSXT at Paines and Elmdale, with a sister company, Grand Rapids Eastern, at Lowell, and with TSBY at Alma.

UPDATED

Minnesota Commercial Railway

14047 Petronella Drive, Suite 201, Libertyville, Illinois 60048
Tel: (+1 847) 549 04 86
Fax: (+1 847) 549 04 85

Key personnel
Chairman and President: John W Gohmann
Chief Maintenance of Way Officer: Joe Krajcrewski
Chief Mechanical Officer, Cars: John Walsh
Chief Mechanical Officer, Locomotives: Scott Wardrope
Chief Accounting Officer: Galen Miller
Director of Operations: Wayne Hall Jr

Organisation
The company operates over some 190 route-km in the Metro Twin Cities, Minnesota area. In 2001 54,000 revenue units were handled, mainly farm products and waste/scrap material, and the company employed 85 people. The company also operates a warehousing and trucking subsidiary through Commercial Translade of Minnesota. It also owns a heated warehouse in Fridley, Minnesota and specialises in steel transloading, storage and transportation, as well as large bulkier items.

Traction and rolling stock
The fleet consists of 26 diesel locomotives.

Mission Mountain Railroad (MMTR)

730 3rd Street West, Columbia Falls, Montana 59912
Tel: (+1 406) 892 32 93   Fax: (+1 406) 892 32 94
Web: www.watcocompanies.com

Key personnel
General Manager: Norm Brown

Gauge: 1,435 mm
Route length: 60 km

Organisation
MMTR started operations in December 2004 when Watco took over two BNSF branches in Montana under a lease-and-purchase agreement. The north line running from Stryker to Eureka (37 km) is a remnant of the former Great Northern route closed in 1970 on opening of the new Flathead tunnel. The south line is a branch from Columbia Falls to Kalispell (22 km). MMTR expected to handle over 9,000 carloads in its first year of operation, serving 12 customers producing forest products and grain.

NEW ENTRY

Mississippi Central Railroad Co

642 East Van Dorn Avenue, Holly Springs, Mississippi 38635
A subsidiary of Pioneer Railcorp
1318 S Johansen Road, Peoria, Illinois 61607
Tel: (+1 309) 697 14 00   Fax: (+1 309) 697 53 87
Web: http://www.pioneer-railcorp.com

Key personnel
See entry for Pioneer Railcorp

Gauge: 1,435 mm
Route length: 91 km

Organisation
Acquired by Pioneer Railcorp as the Natchez Trace Railroad from Kyle Railways Inc in 1992, MSCI runs from Oxford, Mississippi, to Grand Junction, Tennessee. Interchange is made with Norfolk Southern at Grand Junction and with Burlington Northern Santa Fe at Holly Springs, Mississippi. Main commodities carried include cotton products, steel, forest products, fertiliser and chemicals.

Mississippi Export Railroad Co

PO Box 8743, Moss Point, Mississippi 39562-8743
Tel: (+1 228) 475 33 22   Fax: (+1 228) 475 33 37
Web: http://www.mserailroad.com

Key personnel
President: D Gregory Luce Jr

Organisation
The company owns 67 route-km linking Evanston with Pascagoula, Mississippi, and providing a connection between CN and CSXT lines. MSER's track is of high quality to accommodate unit trains. Facilities include transshipment from river and coastal vessels at MSER's own docks on the Escatawpa river. The company also provides wagon repair and maintenance service, and a mobile diesel repair unit.

Traction and rolling stock
Five GM-EMD locomotives and 262 wagons.

Missouri & Northern Arkansas Railroad

514 North Corner, Carthage, Missouri 64836
Tel: (+1 417) 358 88 00   Fax: (+1 417) 358 09 81
Web: www.railamerica.com

Key personnel
General Manager: Al Satunas, Sr
Business Development Manager: Anita Horton

Gauge: 1,435 mm
Route length: 848 km

Organisation
Formerly owned by RailTex and since 2000 operated by RailAmerica, MNA runs northwest from Diaz, Arkansas, to Pleasant Hill, Missouri, continuing via trackage rights to Kansas City. The railway handles approximately 110,000 carloads annually, commodities carried including coal, foodstuffs and farm products, forest products and paper, fertiliser, minerals, chemicals and steel. Interchange is made with UP at Kansas City, Missouri, and Newport, Arkansas, with BNSF at Aurora, Lamar and Springfield, Missouri and with KCS at Joplin, Missouri.

UPDATED

Mohawk Adirondack & Northern Railroad Corporation

8364 Lewiston Road, Batavia, New York 14020-1245
Tel: (+1 716) 343 53 98   Fax: (+1 716) 343 43 69

Key personnel
President: David J Monte Verde
Vice-President and General Manager:
   Loren Wilmot
Marketing and Chief Financial Officer:
   Charles J Riedmuller

Organisation
Totalling 264 route-km, Mohawk Adirondack & Northern (MA&N) is the largest of five railroads in the states of New York and Pennsylvania that together make up the Genesee Valley Transportation Company (GVTC). The others are Depew, Lancaster & Western; Lowville & Beaver River; and Falls Road Railroad, all in New York; and Delaware-Lackawanna in Pennsylvania. Traffic is mainly in chemicals, fertiliser and grain.
   In 1996 GVTC purchased 73.6 km of track from Conrail with a connection at Lockport, New York. GVTC also purchased an out-of-use yard in Niagara Falls, New York, which has been developed as a business park for customers.

Traction and rolling stock
MA&N operates several Alco models, RS11s, RS18s, displaced from the ex-Central Vermont.

Montana Rail Link Inc

PO Box 16390, Missoula, Montana 59808-6390
Tel: (+1 406) 523 15 00   Fax: (+1 406) 523 14 93
Web: www.montanarail.com

Key personnel
President: Thomas J Walsh
Vice-President, Operations: John L Grewell
Superintendent: Gary Waddell
Chief Engineer: Richard L Keller
Chief Mechanical Officer: Claude van Winkle
Executive Director, Marketing/Customer Service:
   Howard E Nash

Gauge: 1,435 mm
Length: 1,534 route-km (including 110 km BNSF trackage rights)

Organisation
Montana Rail Link (MRL) is a subsidiary of Washington Companies, a diversified natural resources conglomerate. Also owned by Washington Companies is the Southern Railway of British Columbia.
   Formed in 1987 to run Burlington Northern's ex-Northern Pacific main line through Montana as a regional railroad, MRL provides a major corridor between central and southern US states and the Pacific northwest and Canada. Its western end is Spokane, its eastern end is Huntley near Billings; it traverses the Belt Mountains via Bozeman Pass, west of Livingston, at an altitude of 5,561 ft; it crosses the Continental Divide via Mullan Pass, west of Helena, at 5,546 ft. It also provides local service to more than 100 stations. Connection is made with: BNSF at Helena and Laurel, Montana and at Spokane, Washington; with Montana Western at Garrison, Montana; and with Union Pacific at Sandpoint, Idaho.
   MRL is among the larger Class II railroads formed in the last 25 years. Employees numbered 1,000 in 2003. Two-thirds of its route-km are cleared for 96 km/h operation, equipped with CTC and automatic block signalling.
   Construction of new yard facilities at Polson was under way in early 2004, funded by the state of Montana as part of the reconstruction of highway I93.
   Revenue carloads are around 240,000 annually.

Traction and rolling stock
In 2004 MRL rolling stock comprised an all-EMD fleet of 120 diesel-electric locomotives (of which 49 were on lease to BNSF and NS) and 2,100 freight wagons of various types.
   In late 2004 MRL announced that it was to acquire 16 3,200 kW Type SD70ACe diesel-electric locomotives from General Motor EMD to modernise its fleet and reduce the number of units required per train. Delivery was due in mid-2005.

UPDATED

Montreal, Maine & Atlantic Railway (MMA) Ltd

15 Iron Road, Bangor, Maine 04401-9621
Tel: (+1 207) 848 42 00   Fax: (+1 207) 848 42 32
e-mail: info@mmarail.com
Web: www.mmarail.com

Key personnel
Chairman: Edward A Burkhardt
President and Chief Executive Officer:
   Robert C Grindrod
Vice-Presidents
   Engineering and Mechanical: Thomas R Klemm
   Secretary and Treasurer: Fred W Yocum, Jr
   General Manager, Logistics Management
      Systems: Thomas W Flacke
   Marketing: William R Schauer
   Transportation: John C Scott

Gauge: 1,435 mm
Route length: 1,199 km

Organisation
The former Iron Road Railways group was bought out of bankruptcy in early 2003 by Ed Burkhardt's company, Rail World Inc. The new operation brings together the former Bangor & Aroostook, Canadian American, Northern Vermont and Quebec Southern railroads and the Van Buren Bridge Co. MMA connects with eight other railroads (CN, CP, EMRY/NBSR, NECR, ST, SLQ and WACR), and thus provides the most direct route between northern Maine, the port of Saint John (New Brunswick) and Montreal. Its main line also serves the port of Searsport.

An affiliated company, LMS Corp, offers delivery, storage and handling services at Northern Maine Junction and Presque Isle, Bangor, as well as timber trans-loading at Jackman and Van Buren, Maine.

Traction and rolling stock
In 2004 the fleet consisted of 30 locomotives, mostly of GE origin, and 2,835 wagons, some 60 per cent of which are boxcars.

UPDATED

Morristown & Erie/Maine Eastern Railways

PO Box 2206, Morristown, New Jersey 07962
Tel: (+1 973) 267 43 00   Fax: (+1 973) 267 31 38
Web: www.merail.com

Key personnel
President and Chief Executive Officer:
   Wesley R Weis
Vice-President and Chief Operations Officer:
   Gordon R Fuller

Gauge: 1,435 mm
Route length: 400 km

Organisation
M&E operates separate networks in New Jersey and Maine. The original Morristown & Erie Railway was bought out of bankruptcy by a consortium in 1982; today it runs four branch lines in Morris County, New Jersey, totalling 254 km and connected by trackage rights over the Morris & Essex lines of New Jersey Transit. Freight is exchanged with NS at Lake Junction. M&E carries mainly plastics, timber, food products, petroleum derivatives and manufactured goods.

The operation in Maine, known as the Maine Eastern Railroad, started in November 2003 over the former Maine Central routes from Brunswick Junction (Springfield Terminal interchange) northwards to Augusta and eastwards to Rockland. Principal customer is Dragon Cement Inc, which moves cement by rail to Canada by rail or by sea from the port of Rockland to US east coast destinations. Since taking over operations, M&E has developed a strategy for introduction of commuter services and seasonal passenger trains.

A subsidiary, Bayshore Terminal Company, was set up in 1995 to provide a contract switching service at the ConocoPhillips/Infineum Bayway oil refinery, Linden, where some 8,000 wagons are handled annually. Operations were extended in 2002 to the new ConocoPhillips polypropylene plant.

From July 2003 to November 2004, M&E was also contract operator of the Octoraro Railroad (now East Penn Railroad) in Chester County, Pennsylvania, for its owner South East Pennsylvania Transportation Authority.

Traction and rolling stock
M&E operates a total of eight diesel locomotives.

NEW ENTRY

Nashville & Eastern Railroad Corporation

514 Knoxville Avenue, Lebanon, Tennessee 37087
Tel: (+1 615) 444 14 34   Fax: (+1 615) 444 46 82

Key personnel
President: William J Drunsic
General Manager: Craig Wade

Organisation
The NERR is the contracted operator for a four-county rail authority which owns the railroad, extending from Nashville eastwards to Monterey with two branches. The company serves 210 route-km, connecting with CSX at Nashville. Of the 10,000 annual carloads, 80 per cent are inbound and the primary commodities are lumber, plastics and beer.

Traction and rolling stock
Eight ex-CSX U30Bs and one U36B.

Nashville Regional Transportation Authority

501 Union Street, 6th Floor, Nashville, Tennessee 37219
Tel: (+1 615) 862 88 33   Fax: (+1 615) 862 88 40
Web: www.gnrc.org/rta

Key personnel
Executive Director: Eric Beyer

Organisation
Nashville's first commuter rail service, to be known as Music City Star, is expected to start operations between Lebanon and Nashville in late 2005. The 51 km route with six stations will run over trackage owned by Nashville & Eastern Railroad. It forms the East Corridor Alignment of the planned 230 km Middle Tennessee Transit Network which envisages a further four services running into Nashville from Gallatin, Murfreesboro, Franklin and Kingston Springs.

Total cost of the initial phase is a modest US$40 million. Upgrading of track and signalling to accommodate passenger service started in November 2004.

Traction and rolling stock
Two diesel locomotives have been acquired for the start-up service, along with 11 gallery cars from Chicago's Metra.

NEW ENTRY

Nebkota Railway (NRI)

111 North Main, Chadron, Nebraska 69337
Tel: (+1 398) 432 24 87
Fax: (+1 308) 432 32 67

Key personnel
General Manager: Roy Fitzgibbon

Gauge: 1,435 mm
Route length: 162 km

Organisation
This locally owned short line, part of the former Chicago & North Western northern Nebraska main line, started operations in 1994. It runs eastwards from a connection with BNSF at Crawford to Chadron and Merriman. Principal commodities handled are grain, fertiliser and minerals; also hosts tourist trains.

Traction and rolling stock
NRI operates four diesel locomotives.

NEW ENTRY

Nebraska Central Railroad Company

400 Braasch, Suite B, Norfolk, Nebraska 68701
Tel: (+1 402) 371 90 15   Fax: (+1 402) 371 45 88

Key personnel
General Manager: Ken Sidlinger

Organisation
NCRC operates five former UP grain branches, the Stromsberg and Ord lines, totalling 460 route-km. The first branch connects to UP at Grand Island, the second at Central City. NCRC is wholly owned by Rio Grande Pacific Corporation.

Traction and rolling stock
NCRC uses its own GP38s and borrows locomotives from UP when grain volumes reach unit train levels.

New England Central Railroad

2 Federal Street, St Albans, Vermont 05478
Tel: (+1 802) 527 35 00   Fax: (+1 802) 527 34 82
Web: www.railamerica.com

Key personnel

General Manager: Charles Moore
Business Development Manager: Jack Dail

Gauge: 1,435 mm
Route length: 528 km

Organisation

Formerly owned by RailTex and since 2000 operated by RailAmerica, NECR runs south from East Alburg, Vermont, on the border with Canada, through New Hampshire and Massachusetts to New London, Connecticut. The railway handles approximately 34,000 carloads annually, commodities carried including forest and paper products, plastics, copper, coal and flyash. Interchange is made with CN at East Alburg, with CSXT at Palmer, Massachusetts, and with the Providence & Worcester Railroad at New London.

UPDATED

New England Southern Railroad

8 Water Street, Concord, New Hampshire 03301
Tel: (+1 603) 228 85 80   Fax: (+1 603) 228 95 71

Key personnel

President and General Manager: Peter M Dearness

Organisation

The company owns and operates two properties, the NEGS in New Hampshire and the Quincy Bay Railroad in Massachusetts. NEGS averages 2,200 carloads annually, including unit coal trains to the Public Service Co of New Hampshire power station at Bow, and QBRR moves 900. NEGS connects to the Boston & Maine (Guilford) at Manchester, New Hampshire; QBRR connects to NS at Braintree, Massachusetts.

Traction and rolling stock

NEGS owns two GM-EMD locomotives, a GP10 and a GP18.

New Jersey Transit Rail Operations Inc

One Penn Plaza East, Newark, New Jersey 07105-2246
Tel: (+1 201) 491 70 00   Fax: (+1 201) 491 82 18
Web: http://www.njtransit.state.nj.us

Key personnel

Executive Director, NJT: George D Warrington
Vice-President and General Manager, NJT Rail
   Operations Inc: William B Duggan

Gauge: 1,435 mm
Route length, owned: 513 km; Trackage rights: 231 km
Electrification, owned: 108 km at 25 kV 60 Hz AC; 26 km at 12 kV 25 Hz AC; 25 km at 12 kV 60 Hz AC; trackage rights: 93 km at 12 kV 25 Hz AC

Passenger operations

New Jersey Transit (NJT) runs rail, light rail and bus transport throughout the state of New Jersey. Its heavy rail network carried 62.7 million passengers in 2002 over 12 routes running into Newark, Hoboken and New York City. NJT Rail Operations employs train crews, owns rolling stock and much of its network infrastructure (except Amtrak Northeast Corridor route between New York and Trenton) and serves 163 stations.

The extremities of the Port Jervis and Pascack Valley routes extend into New York state and service there is provided under contract to MTA Metro-North (see entry in USA section of *Railway systems and operators*); NJT also runs the Philadelphia–Atlantic City route formerly operated by Amtrak.

Improvements to existing lines

NJT made significant progress with ambitious plans to meet increased demand and facilitate the shift away from commuting by road that is one of the aims of the state of New Jersey's Smart Growth policy. In 2002 nearly a decade of infrastructure work was nearing completion and NJT was able to begin phasing-in a major restructuring of its network and timetables in the run-up to partial opening in late 2003 of the US$450 million Secaucus transfer station.

September 2002 saw completion of the Montclair connection between Montclair and Boontown lines at Bay Street, allowing the start of through running from the Boontown line to New York's Penn station and cutting journey times by 15 to 20 minutes. Revised timetables brought an

New Jersey Transit Type ALP44 electric locomotive with a commuter service at Dover, New Jersey (Eddie Barnes)
0063704

Arrow III electric railcar at Princeton, New Jersey (Quintus Vosman)
0554639

0519656

Diesel locomotives

Class	Wheel arrangement	Power kW	Speed km/h	Weight tonnes	No in service	First built	Mechanical	Builders Engine	Transmission
GP40PH-2	Bo-Bo	2,238	169	133.8	13	1968	EMD	16–645 E3B	E EMD D77
F40PH-2	Bo-Bo	2,238	169	118.6	17	1996R	EMD/CR	16–645 E3B	E EMD D77
GP40FH-2	Bo-Bo	2,238	169	128.1	21	1987–1990[1]	EMD/MK	16–645 E3	E EMD D77
GP40FH-2	Bo-Bo	2,238	169	128.1	19	1994R	EMD/CR	16–645 E3	E EMD D77

[1] Year of rebuild.

Electric locomotives

Class	Wheel arrangement	Output kW	Speed km/h	Weight tonnes	No in service	First built	Builders Mechanical	Electrical
ALP44	Bo-Bo	4,320	200	92.5	32	1990	ABB	ABB
ALP46	Bo-Bo	5,300	160	92.5	29	2002	Bombardier	Bombardier

Electric railcars or multiple-units

Class	Cars per unit	Motored axles/car	Output/motor kW	Speed km/h	No of cars in service	First built	Builders Mechanical	Electrical
Arrow III	1/2				230	1976	GE-AVCO	GE

additional 9,500 seats on the new through services, plus 4,500 elsewhere, providing sufficient capacity to eliminate standing on most services.

Further timetable enhancements have been introduced prior to opening of the Secaucus interchange, on a date which is tied to resumption of PATH metro service from New Jersey to Manhattan following the September 2001 terrorist attack. The loss of PATH service substantially increased passenger numbers on NJT services, leaving insufficient capacity for those who might be attracted to the new Secaucus station. Weekend service was scheduled to start in late 2003, but it is now expected to be early 2004 before restructuring is complete and the new station fully operational.

Ground was broken in 1995 for Secaucus station, a major public transport interchange embracing all of NJT's north New Jersey lines, situated at the point where these lines cross Amtrak's Northeast Corridor route. The multilevel facility, in the Jersey Meadowlands at the heart of a 325,000 m³ office and commercial development, will give connections to Manhattan's Penn station from the three lines in New Jersey that at present terminate in Hoboken. Interchange with services to Newark,

the Jersey Coast, Trenton and Amtrak destinations will also be available via Secaucus. Other works nearing completion include construction of a new storage yard at Morrisville with space for 120 cars, installation of crossovers between Pascack Valley and Bergen County lines north of Secaucus station, expansion of storage sidings at Hoboken to accommodate a further 10 trains and commissioning a new operations control centre at the Meadows maintenance complex.

Capacity at Penn Station
In conjunction with Amtrak and Long Island Rail Road, NJT has implemented projects to increase train frequency into New York City's Penn station, raising capacity through the Hudson River tunnels at peak periods from 19 to 25 trains per hour. Work included the design and construction of a new high-speed signalling system as well as improvements in and around Penn station, where in September 2002 NJT opened its US$105 million 7th Avenue concourse to accommodate increasing passenger numbers. Despite these moves spare capacity in the Hudson River tunnels between Secaucus and Penn station is expected to be exhausted sometime

between 2010 and 2020. Contracts were awarded in mid-2003 for environmental impact studies of a new tunnel which would almost double capacity at a cost of up to US$5 billion. Subject to funding constraints, construction could start in 2010 for completion within six years.

Traction and rolling stock
At the start of 2003 NJT operated 34 electric and 79 diesel locomotives, 230 emu cars and 481 other passenger coaches (including nine diesel locomotives and 34 push-pull cars owned by MTA Metro-North).

Large quantities of new rolling stock have been commissioned or are on order to replace life-expired equipment and expand capacity. Delivery was completed in 2002 of 29 ALP46 multi-current electric locomotives built by Bombardier and based on the design of the German Rail Class 101 machines. A batch of 200 Comet V push-pull cars has come from Alstom Transportation and more new motive power comes on stream in 2004 with delivery of 33 high-horsepower diesel locomotives.

Further growth in patronage towards the end of the decade will be handled by the network's first fleet of bilevel cars, designed to a tightly-configured profile that will allow them to reach Penn station through the Hudson River tunnels. An order for 100 was placed with Bombardier at the end of 2002 at a cost not exceeding US$250 million. With delivery scheduled for 2007–10, the bilevels should be commissioned just as capacity becomes critical once again. NJT has taken options for 176 more, depending on availability of funding.

Track
Rail: New rail standard 132 lb/yd (65.5 kg/m)
Crossties (sleepers): Hardwood
Thickness: Main lines 178 × 229 mm, yards 152 × 203 mm
Spacing: Running lines 1,989/km; yards 1,802/km
Fastenings: 152 mm cut spike, drive-on rail anchors
Min curvature radius: Running lines 194 m, yards 122 m
Max axleload: 20 tonnes at unrestricted speed; locomotives to 32.7 tonnes at restricted speed

New Orleans Public Belt Railroad (NOPB)

4822 Tchoupitoulas Street, New Orleans, Louisiana 70115
Tel: (+1 504) 896 74 00   Fax: (+1 504) 896 74 52
e-mail: info@nopb.com
Web: www.nopb.com

Key personnel
General Manager: Jim Bridger
General Superintendent: Tom Lobello

Gauge: 1,435 mm
Route length: 198 km

Organisation
NOPB is, unusually, a non-profit switching railroad, created to provide impartial service in the New Orleans area. It is owned and operated by the city of New Orleans. The 7 km Huey P Long Bridge, located in Jefferson, Louisiana, is owned and maintained by NOPB and provides access for both road and rail traffic. There is interchange with six

Class 1 railroads, some of which contract switching and local haulage from NOPB.

Traction and rolling stock
Four GM-EMD locomotives (one SW1000 and three SW1500) and four MK-Boise machines (two MK2000 and two MK1500). NOPB also leases 257 boxcars and 65 gondolas for customers' use.

UPDATED

New York & Atlantic Railway

68-01 Otto Road, Glendale, New York 11385
Tel: (+1 718) 497 30 23   Fax: (+1 718) 497 33 64
Web: http://www.anacostia.com/nya

Key personnel
Chairman and Chief Executive Officer:
   Bruce A Liebermann
President: Fred Krebs

Superintendent of Transportation:
   Michael McGinley

Organisation
NYAR has contracted to operate freight services that were previously provided by the Long Island Rail Road; operations began in 1997 over 430 route-km and involve 30 employees. NYAR is a subsidiary of the Anacostia & Pacific Company Inc. Annual carloadings amount to 15,000, including forest

products and paper, aggregates and construction materials and food products. Interchange is made with CP, CSXT, NS and two shortlines.

Traction and rolling stock
NYA operates 11 diesel locomotives and 60 gondola cars.

New York State Department of Transportation (NYSDOT)

Office of Passenger and Freight Transportation
Building 7A, Room 302, Albany, New York 12232
Tel: (+1 518) 457 10 46   Fax: (+1 518) 457 31 83
Web: http://www.dot.state.ny.us

Background
The state of New York contains the entire Empire Corridor New York City–Albany–Buffalo–Niagara Falls and a section of Amtrak's Northeast Corridor. Consequently, the state, through NYSDOT, has acted in partnership with Amtrak to implement programmes to support the

introduction of high-speed rail services on both routes.

Empire Corridor (New York–Albany–Buffalo)
Under TEA-21 legislation, the Empire Corridor was officially designated a high-speed rail route in September 1998. The line is mainly owned by

CSX, but some sections are owned by Amtrak or the Metropolitan Transportation Authority's Metro North Commuter Railroad. The state and Amtrak are sharing equally the cost of the US$200 million Empire Corridor Improvement Program to upgrade the route, with US$140 million of this figure being invested in infrastructure improvements and the remainder in rolling stock upgrades.

Proposed infrastructure investments include: double-tracking the Albany–Schenectady section (29 km); rehabilitating the Livingstone Avenue swing bridge over the Hudson River in Albany; expanding and improving Amtrak's Rensselaer rolling stock maintenance facility in Albany; improvements to track condition and alignment south of Albany and west of Schenectady; and grade crossing improvements.

Proposed rolling stock expenditure covers the refurbishment of seven RTL III Turboliner gas turbine-powered high-speed trainsets originally built under licence to French designs by Rohr Corporation. The first rebuilt trainset entered revenue service in mid-April 2003 and the second was ready for service in May of that year. Refurbishment of the remaining five units was in progress in 2003 at Super Steel Industry's plants near Schenectady. Each trainset comprises two power cars and three trailers, one of which is a café car. While primarily turbine-powered, they are also equipped with third rail collector shoes for operation in New York City tunnels.

Using the refurbished trains on upgraded infrastructure will enable Amtrak to cut 20 minutes from the New York City–Albany (229 km) journey time and 1 hour 30 minutes from existing timings

between New York City and Buffalo (690 km). Empire Corridor ridership is expected to rise from 1.2 million passenger journeys annually to 3 million.

Northeast Corridor
As a contribution to the Northeast Corridor upgrade, NYSDOT has acted in partnership with Amtrak on several projects, including the provision of a connection between the NEC and the Empire Corridor at Pennsylvania Station in New York City and the redevelopment of the James A Farley Post Office Building as the intercity rail component of the restructured terminal.

New York, Susquehanna and Western Railway Corporation

1 Railroad Avenue, Cooperstown, New York 13326
Tel: (+1 607) 547 25 55   Fax: (+1 607) 547 98 34
Web: http://www.nysw.com

Key personnel
President and Chief Executive Officer: W G Rich
Vice-Presidents
　Engineering and Mechanical: Richard J Hensel
　General Manager: Joseph G Senchyshyn
　Treasurer: Frank Quattrocchi
　Marketing and Sales: John Fenton
　General Counsel and Corporate Secretary:
　　Nathan R Fenno

Organisation
New York, Susquehanna and Western (NYS&W) is wholly owned by the Delaware Otsego Corporation

and operates over 800 route-km, of which 320 km is trackage rights. NYS&W covers territory from Syracuse and Utica, New York, through Pennsylvania to northern New Jersey. Between Binghamton and Buffalo, NYS&W has a haulage agreement with CP, using NYS&W supplied power and fuel.

From its junction with CSXT at Syracuse, NYS&W is now the regular route of Sea-Land double-stack container trains from the west coast to the New York/New Jersey area. This was made possible by a renegotiated interchange agreement with the former Conrail and a US$7 million investment in the line between Syracuse and Binghamton. The alternative route via Buffalo and the Norfolk Southern system is still used for CSXI domestic container trains from Chicago. Trains are handled at the CSXI-owned Little Ferry terminal in New Jersey. NYS&W has secured Hanjin's Pacific Rim-Northeast Coast container service, which is handled at a North Bergen terminal, and partners Union Pacific

and Norfolk Southern in a weekly coast-to-coast stack train service from Long Beach, California.

NYS&W serves food-grade transfer facilities for liquid sweeteners at Oakland and Riverdale, New Jersey. The railroad also has an automobile loading facility in North Bergen, operated by CT Services, and brings lumber to the New York/New Jersey metropolitan market via a distribution centre operated by National Distribution in conjunction with Georgia Pacific. These facilities complement an already-strong sugar and plastics bulk transfer business.

Traction and rolling stock
Rolling stock comprises 20 diesel locomotives (three GM-EMD SD70M, four GM-EMD SD45, one GM-EMD F45, four GE B40–8, two Alco C430, one GM-EMD GP38, one GM-EMD GP40, three GM-EMD GP18 and one GM-EMD NW2); 285 freight wagons; and 15 passenger coaches.

Norfolk Southern Corporation

Three Commercial Place, Norfolk, Virginia 23510-2191
Tel: (+1 757) 629 26 00   Fax: (+1 757) 629 23 45
Web: www.nscorp.com

Key personnel
Chairman and Chief Executive Officer:
　David R Goode
President: Charles W Moorman
Vice-Chairmen
　Chief Marketing Officer: L I Prillaman
　Chief Financial Officer: Henry C Wolf
　Chief Operating Officer: Stephen C Tobias
Senior Vice-Presidents
　Public Affairs: John F Corcoran
　Law: Henry D Light, James A Squires
　Operations Planning and Support:
　　John M Samuels
　Energy and Properties: Daniel D Smith
Executive Vice-Presidents
　Finance and Public Affairs: James A Hixon
　Sales and Marketing: Donald W Searle
　Planning and Chief Information Officer:
　　Kathryn B McQuade
　Administration: John P Rathbone
Vice-Presidents
　Human Resources: Thomas H Mullenix
　Business Development: Robert E Martinez
　Customer Service: Deborah H Butler
　Internal Audit: James E Carter, Jr
　Operations Planning and Budget: Terry N Evans
　Real Estate: F Blair Wimbush
　Mechanical: Gerhard A Thelen
　Intermodal Operations: Robert E Huffman
　Public Relations: Robert C Fort
　Taxation: William A Galanko
　Corporate Affairs: H Craig Lewis
　Public Affairs: Bruno Maestri
　Engineering: Gary Woods
　Information Technology: Cindy C Earhart
　Treasurer: William J Romig
　Intermodal Marketing: Michael R McClellan

Labour Relations: Mark R MacMahon
Controller: Marta Stewart
Safety and Environmental:
　Charles J Wehrmeister
Senior General Counsel: Joseph C Dimino
Corporate Secretary: Dezora M Martin

Gauge: 1,435 mm
Length: 34,590 km

Organisation
Norfolk Southern Corporation (NS) is a Virginia-based company that controls a major freight railroad, Norfolk Southern Railway Company. The railway operates 34,600 route-km in 22 eastern states, the District of Columbia, and Ontario. It serves 20 ports and connects with rail partners in the West and in Canada. NS operates the most extensive intermodal network in the East and is the US's largest rail carrier of automotive parts and finished vehicles.

Route structure
The Norfolk & Western and Southern Railway merger of 1982, which created NS, was an end-to-end consolidation. The N&W stretched from Norfolk west to Kansas City and north into the key markets of Chicago, Detroit and Cleveland. The Southern blanketed the southeast, from New Orleans, Mobile and Jacksonville, north to Cincinnati and Washington DC; from the St Louis 'gateway' and Memphis, eastwards to the Atlantic ports of Norfolk, Charleston, Savannah and Brunswick. The railroads connected at 17 points, with major connections at East St Louis, Cincinnati and throughout Virginia and North Carolina.

In 1999 NS and CSX began operating portions of Conrail's 11,000 km system in the northeastern United States through their respective railroad subsidiaries. Under the acquisition agreement, NS obtained the right to operate routes and other assets accounting for 58 per cent of Conrail's 1995 revenues. The ex-Conrail routes and assets operated by NS are leased from the owner, NS subsidiary Pennsylvania Lines LLC (PRR); they increased the

size of the NS service area by some 50 per cent and gave access to the New York metropolitan region and much of the northeast USA, as well as to most major east coast ports north of Norfolk.

In 2004 NS, CSX and Conrail were jointly working to reorganise Conrail and establish direct ownership and control of the last-mentioned.

Finance
Net income for 2003 was US$535 million, up 16 per cent from 2002. Income from railway operations was US$1.06 billion, down 8 per cent over 2002 largely on account of US$107 million in costs incurred under a voluntary separation programme which saw a reduction of 533 in the overall staff complement. At US$6.47 billion, railway operating revenues were 3 per cent up on 2002, while expenses rose 6 per cent to US$5.4 billion.

Operating ratio rose to 83.5 per cent, compared with 81.5 per cent in 2002.

Total capital expenditure of US$798 million was planned for 2003, but actual expenditure amounted to US$720 million, a 4 per cent increase over the slightly depressed 2002 figure. Locomotive purchases rose from the low of 50 in 2002, offsetting reduced spending on signalling and IT equipment.

Spending plans for 2004 included purchase of 100 six-axle locomotives and widespread upgrading of existing traction units (US$178 million), and investment of US$384 million in track and infrastructure improvements. Total capital expenditure for 2004 was budgeted at US$810 million.

Freight operations
Coal
Coal (most of it bituminous), coke and iron ore is NS's largest commodity group as measured by revenue. Reserves of some 7 billion tonnes, much of it low-sulphur, exist in Central Appalachia, where NS originates most of its traffic in this sector. The principal sources are mines in West Virginia, Kentucky, Virginia and Tennessee. Export coal is

A GE Dash 9-40CW locomotive, one of more than 1,000 operated by NS, leads an ex-Conrail machine on a mineral train at La Grange Road, Illinois (Philip Wormald)   **NEW**/0585051

shipped through the NS pier at Lambert's Point in Norfolk, Virginia.

In 2002 coal, coke and iron ore tonnage fell 4 per cent on 2001, while revenues declined by 5 per cent. Some of this loss was clawed back in 2003, with coal tonnage up 1 per cent and revenue up 4 per cent. Utility coal tonnage handled was 130 million tons, against 128 million tons in 2002, while total tonnage rose from 170 to 172 million. The rise in domestic movements was due mainly to a 6 per cent increase in tonnage moving to northeast power stations during the big freeze in early 2003. After a 30 per cent drop in 2001 and a further 18 per cent decline in 2002, export coal tonnage recovered slightly in 2003 with a 9 per cent increase. Exports through the port of Norfolk rose 24 per cent, benefiting from a decline in China's coal exports as production was absorbed by that country's booming domestic demand for energy. Conversely, record steel production in China boosted demand for metallurgical coal and coke imported from the US.

Handling of coal traffic improved as a new coal transportation management system produced better trip and wagon allocation planning.

Agricultural, consumer and government traffic
Agricultural traffic rose by 7 per cent and revenue by 8 per cent, both record figures, thanks to increases in movement of corn, fertilizers, sweeteners and wheat. Following the drought of 2002, there was much restocking of depleted inventories at feed mills

and poultry producers in the northeast, resulting in additional long-haul movements from the Midwest. Government traffic rose by 36 per cent as movement of military vehicles and equipment escalated in the run-up to the Iraq war.

Automotive
After producing its best-ever result in 2002, the Automotive Group saw traffic and revenues decline by 3 per cent in 2003 as US vehicle manufacturers reduced production. Higher revenues were forecast for 2004, with light vehicle production predicted to increase and new business accruing from Honda's Lincoln, Alabama, plant, where a dedicated yard was opened in March 2004 at a cost of US$15 million. A new joint service with UP reduced transit times by two days for vehicle parts moving between the Midwest and Mexican manufacturing plants.

Other business areas
In other business sectors volumes moved only slightly on 2002's figures. Metals and construction traffic fell 1 per cent, for a 1 per cent revenue rise, while chemicals revenue rose 2 per cent on a 1 per cent rise in volume. Improved domestic demand saw paper, clay and forest products traffic rise by 1 per cent, with revenue up 5 per cent.

Intermodal operations
Intermodal traffic grew by 5 per cent to achieve a record volume, stimulated by further service

improvements. Domestic shipments were up 14 per cent and international traffic by 9 per cent, together producing a 5 per cent rise in revenues.

Triple Crown Services Company RoadRailer business volume increased by just 14 per cent, with futher growth constrained by full capacity. More rolling stock was planned for acquisition in 2004.

Traction and rolling stock
At the beginning of 2004 the diesel locomotive fleet totalled some 3,468, while the freight wagon fleet stood at around 100,000 owned and leased. In total, 22 per cent of locomotives and 17 per cent of wagons were leased from NS subsidiary PRR. Traction investments in 2005 were to include 102 2,985 kW diesel locomotives, 52 of these from EMD, the remainder from GE.

Special attention was paid in 2003 to improving fleet utilisation, and both locomotive and wagon utilisation rose by 2 per cent. Fuel consumption was cut by three-quarters of a gallon per wagonload, producing a saving of more than 5 million gallons. Some 5,000 wagons were withdrawn.

Further improvements in fuel efficiency could accrue from a pilot installation of the Locomotive Engineer Assist Display and Event Recorder train handling system developed by New York Air Brake Corp. An 18-month trial of the equipment on 15 GE locomotives is being run in partnership with NYAB, GE and the Federal Railroad Administration research division, which provided a US$615,000 grant.

Signalling and telecommunications
Of the 34,590 route-km operated, some 18,500 km is signalled, of which 13,750 km is equipped with CTC and 4,340 km with automatic block.

Track
In 2003 NS installed 372 km of track and installed 2.8 million new sleepers. Some 8,160 km of track was surfaced and lined.

Rail type and weight: Some 75 per cent of track is laid with rail ranging from 131 to 155 lb/yd (64 to 76.9 kg/m), with 141 lb/yd (69.9 kg/m) the standard. Other lines have 115 lb/yd, 57 kg/m; 100 lb/yd, 49.6 kg/m; and 85 lb/yd, 42.2 kg/m rail
Crossties (sleepers): Wood
Spacing: 1,970/km
Fastenings: 6 in cut spikes
Max curvature: 22°
Max gradient: 4.4%
Max permissible axleload (lb): 71,500±, 36 in diameter wheels; 78,750±, 38 in diameter wheels

UPDATED

North American RailNet Inc

2300 Airport Freeway, Suite 230, Bedford, Texas 76022
Tel: (+1 817) 571 23 56   Fax: (+1 817) 571 23 35

Key personnel
Chairman and Chief Executive Officer:
  Robert F McKenney

President and Chief Operating Officer:
  Roger H Nelson
Executive Vice-President: William E Glavin

Organisation
The company began operations in 1996. Its first property was Nebraska, Kansas & Colorado RailNet Inc, four lines totalling 669 km in its namesake states and operated with five locomotives (two

CF7, two B23–7 and one GP7). Its second property, the 130 km Illinois RailNet, commenced operations in 1997. Other assets include the Mississippi & Tennessee and Alberta RailNet in Canada. The Camas Prairie Railroad was sold to Watco Companies in March 2004.

North Carolina & Virginia Railroad

214 North Railroad Street, Ahoskie, North Carolina 27910
Tel: (+1 252) 332 27 78   Fax: (+1 252) 332 33 25
Web: http://www.railamerica.com

Key personnel
General Manager: Brad Ovitt
Business Development Manager: Kim Greer

Gauge: 1,435 mm
Route length: 87 km

Organisation
Formerly owned by RailTex and since 2000 operated by RailAmerica, NCVA comprises two lines running north from Kelford to Tunis, North Carolina, and to Boykins, Virginia. The railway

handles approximately 2,900 carloads annually, carrying grain, forest products and peanut products. Interchange with CSXT is made at Boykins.

For details of the latest updates to *Jane's World Railways* online and to discover the additional information available exclusively to online subscribers please visit
jwr.janes.com

North Carolina Railroad Company

Suite 100, 2809 Highwoods Boulevard, Raleigh, North Carolina 27604-1000
Tel: (+1 919) 954 76 01
Web: www.ncrr.com

Key personnel
President: Scott Saylor
Public Affairs: Kat Christian

Gauge: 1,435 mm
Route length: 510 km

Organisation
NCRR runs from the port of Morehead City to Raleigh, Durham, Greensboro and Charlotte, North Carolina, over 510 route-km. It carries some 60 daily freights operated by Norfolk Southern, as well as Amtrak's 'Piedmont' and 'Carolinan' passenger trains. All freight operations are undertaken by Norfolk Southern under a long-term agreement.

Following completion in 2002 of track upgrading and installation of CTC between Raleigh and Greensboro, which cut journey time for passenger trains by 20 minutes, approval was given at the start of 2003 for investment of US$24.5 million in similar work between Raleigh and Goldsboro. Continuous-welded rail is being laid, a 10 km length double-tracked, and CTC installed throughout, benefitting current freight and passenger operations. Two commuter operations are at the planning stage. Other infrastructure improvements include replacement of the hurricane-damaged Neuse river bridge at Kinston and a US$5 million rebuild of the Highway 54 bridge at Research Triangle Park, eastern Durham County, including provision of 2 km of double track. Four more sections are being double-tracked during 2005–10, and CTC installed between Garner and Goldsboro.

Traction and rolling stock
NCRR owns no locomotives or rolling stock.

UPDATED

Northern Indiana Commuter Transportation District

33 E US Highway 12, Chesterton, Indiana 46304-3514
Tel: (+1 219) 926 57 44   Fax: (+1 219) 929 44 38
Web: www.nictd.com

Key personnel
Chairman: David L Niezgodski
General Manager: Gerald R Hanas
Chief Operating Officer: Jeffrey S Lowe
Chief Financial Officer: Dario M Brezene
Director, Marketing and Planning: John N Parsons

Gauge: 1,435 mm
Route length: (owned) 109.3 km; (operated) 129.5 km

Passenger operations
Following sale of the Chicago SouthShore & South Bend Railroad's freight operation, Northern Indiana Commuter Transportation District (NICTD) took over full responsibility for its passenger services between Chicago Randolph Street and South Bend. From Chicago to Kensington, operation is over Metra's electrified Illinois Central route. In 1992, a 1 km extension was opened to South Bend Airport.

NICTD owns 109.3 route-km, leases 20.2 route-km and has trackage rights over 22.9 km. Electrification at 1.5 kV DC overhead covers 141.4 km. Ridership in 2004 was 3.5 million; the line continues to attract latent demand such that peak-hour trains average 110 per cent of seat capacity. Income from all sources including fares was US$30 million in 2004.

A major investment study was completed in 2000 and recommended a new service between Chicago, Illinois, and Valparaiso, Indiana, with a second phase to Lowell, Indiana. Power would be supplied by dual-mode locomotives with overhead power from Munster to Chicago and diesel power to a proposed terminal at either Lowell or Valparaiso. Additional feasibility studies were to be undertaken in 2005.

Improvements to existing lines
NICTD's capital investment programme for the near term involves continued improvements to infrastructure, including a US$100 million upgrade of catenary and signalling. Operation of the service continues to be constrained by the 5 km largely single-track section of on-street running in Michigan City, for which a new alignment has long been discussed.

Traction and rolling stock
NICTD operates a fleet of 58 Nippon Sharyo electric railcars and 10 trailers ordered through Sumitomo in three batches delivered in 1982–83, 1992 and 2000. NICTD has built an extension to its Michigan City workshops for mid-life overhauls of the first batch.

The overhaul includes new Toshiba AC propulsion drives. The 2000 order from Sumitomo is also equipped with a similar propulsion system.

UPDATED

Northern Plains Railroad

PO Box 38, Fordville, North Dakota 58321
Tel: (+1 701) 229 33 30   Fax: (+1 701) 229 33 65
Web: www.nprail.com

Key personnel
President: Gregg Haug
Manager, Accounting/Administration:
   Cheryl Harlow

Organisation
Established in 1997, NPR was set up to operate 615 km of leased Canadian Pacific (former Soo Line) track, mostly in North Dakota, commonly referred to as 'the Wheat Lines'. Connections to CP are available at Thief River Falls, Minnesota, or Kenmare, North Dakota. Also operates the Fordville to Harlow section. CP provides the grain cars, sets the rates and markets the services.

Affiliate Northern Plains Rail Services carries out rolling stock maintenance and locomotive painting.

Traction and rolling stock
In 2003 the motive power fleet comprised 10 locomotives, eight GP35s and one each of GP7 and GP9.

UPDATED

North San Diego County Transit District

810 Mission Avenue, Oceanside, California 92054
Tel: (+1 760) 967 28 27   Fax: (+1 760) 722 09 40
e-mail: ekasparik@nctd.org
Web: www.gonctd.com

Key personnel
Executive Director: Karen King
Director of Transportation Services:
   Thomas Lichterman
Manager of Commuter Rail Services:
   Edward Kasparik
Manager of Maintenance-of-Way: Richard Walker
Rail Contract Compliance and System Safety
   Officer: Wayne Penn
Manager of Light Rail Services:
   Walt Stringer

Passenger operations
In February 1995, North San Diego County Transit District (NCTD) began operating its 'Coaster' commuter train service over a 67.2 km San Diego–Oceanside route in California. The service is also operated under the name of San Diego Northern Railway. The line has eight stations, each of which has extensive free parking facilities and connecting bus services. The line is maintained for 145 km/h operation and is equipped with ATS and CTC. Approximately two-thirds of the line remains single-track, although several projects were under way in 2004 to convert sections to double-track.

Amtrak is the current designated operator under a five-year contract. It is also contracted to undertake track maintenance and, under a subcontract with Motive Power Inc to maintain rolling stock.

In 2004 22 Coaster trains were run each weekday and eight on Saturdays. There is no service on Sundays. Coaster services share tracks with 22 daily Amtrak 'Surfliner' intercity services and six BNSF freight trains each day. Ridership averages 5,306 each weekday. In FY 2004 1.429 million passenger journeys were recorded. Trains are push-pull operated, the usual formation comprising

'Coaster' service at Del Mar, California (David C Warner)   0528636

one locomotive, four trailers and a cab control coach.

Full funding approval for the long-proposed US$351.5 million 35 km Oceanside–Escondido 'Sprinter' diesel light rail project was granted in 2003, and 12 Desiro dmus were ordered from Siemens Transportation Systems in 2004. Conversion of this lightly used freight line for passenger service began in 2004, a year later than planned, with services scheduled to start in 2007.

The line, which will have 15 stations, will continue to be used by freight trains during periods when passenger services are not operated.

Traction and rolling stock

Motive power consists of five remanufactured (1995) F40PHM-2C diesel locomotives supplied by Morrison-Knudsen plus two EMD F59PHI locomotives delivered in 2001. Bombardier has supplied 28 bilevel coaches and cab control coaches. All motive power and rolling stock is maintained at a maintenance and stabling facility at Stuart Mesa, 8 km north of Oceanside.

Rolling stock for the Oceanside–Escondido light rail project will take the form of 12 two-car Desiro Classic dmus to be supplied by Siemens Transportation Systems in 2005–06.

UPDATED

North Shore Railroad

356 Priestley Avenue, Northumberland, Pennsylvania 17857
Tel: (+1 570) 473 79 49 Fax: (+1 570) 473 84 32
Web: http://www.nshr.com

Key personnel

Chairman and Chief Executive Officer:
Richard Robey
President: Gary Shields

Vice-President, Operations: Jeff Pontius
Chief Mechanical Officer: Mike Herman

Organisation

NSHR operates eight Pennsylvania short lines and switching carriers: the Juniata Valley (19 km), Lycoming Valley (59 km), Nittany & Bald Eagle (104 km), North Shore (70 km), Shamokin Valley (40 km), Stourbridge (40 km), Union County Industrial (20 km) and Wellsboro & Corning (56 km), five of which are ex-Conrail lines municipally

owned by the SEDA-Cog Joint Rail Authority. There are connections with NS, CP and RBMN.

Traffic is general freight, including aggregates (1.1 million tons in 2003).

Traction and rolling stock

NSHR operates 11 diesel locomotives.

UPDATED

Northstar Commuter Rail

Northstar Corridor Development Authority
Tel: (+1 763) 323 56 92
Web: www.northstartrain.com

Key personnel

Executive Director: Tim Yantos

Political background

Minnesota Department of Transportation and the Northstar Corridor Development Authority, supported by 30 local authorities, are promoting a commuter rail line in the congested Northstar Corridor area of suburban Minneapolis. The corridor runs some 130 km northwest from central Minneapolis to St Cloud along highways 10 and 94.

It is the fastest-growing area of Minnesota, with population up by 20 per cent in the decade to 2000. Highway 10 is heavily congested at peak times and with no road improvements scheduled before 2020, rail is seen as a quick and inexpensive way of coping with further population growth and congestion.

The scheme dates from studies initiated in 1998, which two years later recommended Northstar as the priority rail project in the state's Commuter Rail System Plan. The initial proposal is for a 64.5 km line linking Minneapolis Downtown station with Big Lake over existing BNSF alignment, serving six stations and expected to attract 5,600 riders per day with eight return journeys. Construction cost of Phase I is put at US$265 million, with opening planned for 2008–09.

The 30 counties, cities and townships in the corridor pledged to fund US$44.2 million of the start-up cost and contribute towards operating costs, but the Minnesota state legislature five times refused to approve the necessary US$37.5 million state funding, eventually in February 2005 voting US$10 million to keep the scheme alive. In April 2005 the Governor of Minnesota signed a state construction bond bill that included the initial US$37.5 million for the project, leaving the state to provide a further US$50.8 million in 2006. The federal government is to fund the remainder of the scheme's cost accounting for 50 per cent of the total.

NEW ENTRY

Northwestern Pacific Railroad

4 West 2nd Street, Eureka, California 95501
Tel: (+1 707) 441 16 25
Fax: (+1 707) 441 13 24

Key personnel

Executive Director and Chief Executive Officer:
Dan Hauser
Manager, Passenger Operations: Arthur Lloyd

Organisation

NPRR runs between Eureka and Schellville. It started up in 1996 after the public agency, the North Coast Railroad Authority, purchased the Willets to Schellville section (it had previously acquired the Eureka to Willets section). The previous contract operator, California Northern, was displaced by NPRR but still provides a vital bridge 29 km from Schellville to Suisun City to connect to UP (ex-SP). Carloads for NPRR average 600 per month, mainly timber, grain and aggregates.

Long-term, the public authority is reviewing property south into the area of San Rafael and Tiburon (in Marin County) as a commuter railway opportunity. The right-of-way is in place, as is some out-of-service track. The commuter operator would be the Golden Gate Transit District.

Traction and rolling stock

NPRR operates nine locomotives.

Ohio Central Railroad System

47849 Papermill Road, Coshocton, Ohio 43812
Tel: (+1 740) 622 81 18 Fax: (+1 740) 622 39 41
Web: www.ohiocentral.com

Key personnel

President: William A Strawn
Chief Executive Officer: Jerry J Jacobson
Executive Vice-President: Michael J Connor
Director of Marketing: Marty Pohlod
Chief Engineering Officer: John Dulac
Manager of Motive Power: J A Bozeman
Chief Mechanical Officer: T J Sposato

Superintendent, Freight Cars: J A DeGallo
Chief Accounting Officer: B A Fogle

Gauge: 1,435 mm
Route length: 997 km

Organisation

The Ohio Central Railroad System is the collective name for the Ohio Central, Alliquipa & Ohio River, Columbus & Ohio River, Ohio Southern, Pittsburgh & Ohio Central, Mahoning Valley, Youngstown & Austinstown, Youngstown Belt, Warren & Trumbull and Ohio & Pennsylvania railroads, totalling 732 km. In the fourth quarter

of 2004 a further 265 km of lines were acquired, principally from CSXT.

The company offers repair and rebuild service to other short lines in Ohio and Indiana and to independents such as grain elevator operators.

Traction and rolling stock

At the end of 2004 the fleet consisted of 84 diesel-electric locomotives, 10 steam locomotives, 22 passenger cars, 1,248 freight wagons and 38 service vehicles. The company has bought a number of ex-Class 1 locomotives and given them significant upgrading.

UPDATED

Ohi-Rail Corporation

6200 Salineville Road, Mechanicstown, Ohio 44651
Tel: (+1 330) 738 67 35
e-mail: ohirail@bright.net
Web: www.ohirail.freeshell.org

Key personnel

Chief Executive Officer: Teresa Schiappa
Vice-President and General Manager:
Tom D Barnett

Gauge: 1,435 mm
Route length: 72 km

Organisation

OHIC started service in 1982 on the former Conrail branch from Bayard, Ohio, where today there is a connection with NS, southwards to Minerva and Hopedale, both of which have links to the Wheeling & Lake Erie Railway. Between Bayard and Minerva, track is owned by LWR Inc, while the route thence to Hopedale is owned by and leased from

the state of Ohio. OHIC also operates the switching yard at Minerva. Traffic is mainly hardwood and pine lumber, waste wood, plastics and machinery, and wagons are handled for five clients at Minerva yard.

Traction and rolling stock

The locomotive fleet totals five units, two Alco S-2 and three B23–7Rs (ex-MGA U23B, rebuilt by GE).

UPDATED

OmniTRAX Inc

50 South Steele, Suite 250, Denver, Colorado 80209
Tel: (+1 303) 398 45 00   Fax: (+1 303) 398 450 40
Web: www.omnitrax.com

Key personnel
President: Dwight N Johnson
Chief Operating Officer: Robert Parker
Managing Director: Mike Ogborn
Executive Vice-President, Marketing and Strategic Planning: Alfred M Sauer
Executive Vice-President, Mechanical: Jim Griffiths
Executive Vice-President, Rail: Darcy Brede
Executive Vice-President, Intermodal: Robert M Sleeker
Managing Director: Mike Ogborn
Vice-Presidents
    Financial Planning and Purchasing: Bill Banham
    Safety, Training and Loss Control: Paul E Crawford
    Engineering/Maintenance: Chris Dodge
    Operations, Gulf Region: John E DeWitt
    Southern Operations; Quality Terminal Services

LLC: Jim Caron
Administration and Planning; Quality Terminal Services LLC: Mark Hansen
Regional Vice-President, West Rail: Carl Hollowell
Regional Vice-President, East Rail: Steve Ward
Regional Vice-President, Central Rail: Chuck Littlefield

Organisation
OmniTRAX Inc, an affiliate of the Broe Companies, owns and manages 18 railroads in several US states and three Canadian provinces. It also owns and operates the Port of Churchill, Manitoba. Additionally, OmniTRAX provides intermodal terminal services to Class 1 railways through its subsidiary Quality Terminal Services LLC, customised switching services for major industrial complexes through OmniTRAX Switching Services LLC and locomotive maintenance services to Class 1 railways through OmniTRAX Locomotive Services LLC.

In 2005 OmniTRAX owned and managed the following entities:
    Alabama & Tennessee River Railway (ATR)

Alliance Terminal Railroad (ATR)
Carlton Trail Railway (CTRW) (Canada)
Chicago Rail Link (CRL)
Churchill Marine Tank Farm Company (CMTF)
Fulton County Terminal Railway (FCR)
Georgia & Florida RailNet (GFRR)
Georgia Woodlands RR (GWRC)
Great Western Railway of Colorado (GWR)
Great Western Railway of Iowa (CBGR)
Hudson Bay Port Company (HBPC) (Port of Churchill, Canada)
Hudson Bay Railway (HBRY) (Canada)
Illinois Railway (IR)
Kettle Falls International Railway (KFR)
Manufacturers Junction Railway (MJ)
Nebraska, Kansas & Colorado Railway (NKCR)
Newburgh & South Shore RR (NSR)
Northern Ohio & Western Railway (NOW)
Okanagan Valley Railway (OKAN) (Canada)
Panhandle Northern RR (PNR)

UPDATED

Otter Tail Valley Railroad

200 North Mill Street, Fergus Falls, Minnesota 56357
Tel: (+1 218) 736 60 73   Fax: (+1 218) 736 76 36
Web: http://www.railamerica.com

Key personnel
General Manager: Pam Slifka
Business Development Manager: Robin Bergeron

Gauge: 1,435 mm
Route length: 116 km

Organisation
Operated by RailAmerica, OTVR runs from East Moorhead (Fargo) to southeast Fergus Falls, in western Minnesota. From Fergus Falls, branches extend east to Hoot Lake and west to French. The railway handles approximately 9,000 carloads annually, mostly carrying coal, grain and fertiliser.

Interchange with BNSF is made at Dilworth Yard, Fargo.

Pacific Harbor Line Inc

340 Water Street, Wilmington, California 90744
Tel: (+1 310) 834 45 94   Fax: (+1 310) 513 67 89
Web: www.anacostia.com/phl/phl.htm

Key personnel
Chairman and Chief Executive Officer: Peter A Gilbertson
President: Andrew Fox
Vice-President: Bill Roufs

General Superintendent: M D (Mike) Stolzman
Superintendent, Transportation, Safety and Rules: Russell R Tomren
Chief Engineer: Robert Giannoble
Controller: R Scott Morgan
Manager of Business Development: Don Norton

Organisation
Pacific Harbor Line Inc is an affiliate of Anacostia & Pacific. The ports of Long Beach and Los Angeles selected Anacostia & Pacific to manage and operate

the 128 km of ports-owned trackage, beginning in November 1997. Connections are made to Union Pacific and BNSF. The company employs 120 staff, operates 20 locomotives and handles some 35,000 carloads annually, plus an estimated 750,000 intermodal containers a year.

UPDATED

Paducah & Louisville Railway (PAL)

1500 Kentucky Avenue, Paducah, Kentucky 42003, US
Tel: (+1 270) 444 43 00   Fax: (+1 270) 444 43 88
e-mail: feedback@palrr.com
Web: www.palrr.com

Key personnel
President and Chief Executive Officer: Anthony V Reck
Vice-President and Chief Financial Officer: Tom A Green
Executive Vice-President and General Counsel: J Thomas Garrett

Organisation
The company was formed in 1986 and operates 587 route-km of ex-Illinois central trackage. PAL's main line links Paducah and Louisville; there are three branches, serving Elizabeth, Kevil and Mayfield. Connections are provided with five Class 1 railroads and two shortlines, as well as with four major rail-to-barge complexes on the Ohio and Tennessee rivers. The PAL is a full-service railroad serving some 100 customers. It handles between 150,000 and 225,000 carloads annually. The principal commodity carried is coal, accounting for 65–70 per cent of total traffic volume. Chemicals traffic accounts for 13 per cent of carloads and 44 per cent of revenues. There are some 230 employees.

Traction and rolling stock
In 2004 PAL operated 41 newly refurbished locomotives and a fleet of approximately 1,000 owned or leased wagons.

UPDATED

Palouse River & Coulee City Railroad (PCC)

417 SW Park, Rosalia, Washington 99170-0155
Tel: (+1 509) 523 01 55   Fax: (+1 509) 523 34 04
Web: www.watcocompanies.com

Key personnel
General Manager: Robert Thrall
Marketing Manager: Mark Demers

Gauge: 1,435 mm
Route length: 676 km

Organisation
Watco purchased this group of former UP and BN lines in southeast Washington State in 1992. They comprise three isolated networks, two of which are linked by trackage rights. Routes link Cheney with Coulee City; Marshall with Pullman, Colfax and Hooper Junction; and Wallula with Walla Walla, Dayton and Weston (Oregon). Other branches extend over the border into Idaho. Serving around 70 customers, the lines handle 20 per cent of Washington State's grain shipments, as well as lentils and barley, amounting to some 9,400 carloads annually.

Deferred maintenance had caused concern amongst local industries, and after 18 months of negotiation the Washington State Department of Transportation bought the network in November 2004 for US$8 million. The purchase released grants of up to US$22 million to fund a series of track improvements to be carried out over an eight-year period. At the same time WSDOT concluded a 15-year operating lease with Watco.

Traction and rolling stock
At the end of 2004 the PCC was operating 12 diesel-electrics.

NEW ENTRY

Peninsula Corridor Joint Powers Board (Caltrain)

PO Box 3006, 1250 San Carlos Avenue, San Carlos, California 94070-1306
Tel: (+1 650) 508 62 00   Fax: (+1 650) 508 79 19
Web: www.caltrain.com

Key personnel
Executive Director: Michael J Scanlon
Director, Rail Transportation: Robert L Doty
Manager of Operations: Walt Stringer
Manager of Equipment: Steve Coleman

Organisation
In 1992 the Peninsula Corridor Joint Powers Board, a three-county agency, purchased the 75.6 km Southern Pacific main line from San Francisco to San Jose for US$242.3 million. This is a long-standing commuter route previously operated under the sponsorship of the California Department of Transportation (Caltrans). There were options to purchase additional trackage on three branches (including the Tracy branch and its Dumbarton Bridge over San Francisco Bay) and an additional 45 km of line south to Gilroy. Amtrak was awarded the contract to operate the line; management is provided for the board by the San Mateo County Transit District (SamTrans). Under a three-year contract extension that was approved in 2005, Amtrak will continue to operate the service until June 2009. The agreement, which incorporates incentives for performance improvements includes two one-year options, which would further extend the contract.

Passenger operations
The Peninsula commuter service, marketed as Caltrain, comprises a weekday service of 86 (96 from August 2005) San Francisco–San Jose return trains, including 10 'Baby Bullet' express services employing MP36PH-3C locomotives supplied by MotivePower. Six trains in each direction are extended to serve Gilroy beyond San Jose. Under a 10 year plan, service levels are to rise to over 100 trains a day. Annual journeys totalled 6.6 million in 2003–04, down slightly on the previous year's figure of 6.7 million.

There are 31 stations, including interchanges with Santa Clara County's VTA light rail system at Tamien and Mountain View. Since 1998, the San Francisco terminal has been served by the city's Muni Metro light rail system.

In 2003–04 farebox revenue contributed 30.4 per cent of operating costs, other commercial sources 8.4 per cent; the remainder was covered by subsidy and grants.

Improvements to existing lines
Caltrans' biggest capital project to date, a US$110 million two-year upgrade of parts of the route, started in mid-2002 and was largely complete by mid-2004. The CXT project involves installation

Caltrain Baby Bullet service headed by a MotivePower-built MP36PH-3C locomotive at Burlingame (Philip Wormald)
NEW/1122826

of CTC and bi-directional signalling between Bayshore, Menlo Park and Santa Clara, together with track and signalling improvements and laying of high-speed turnouts at five other locations. As well as increasing line speed, a new four-track section through Lawrence station in Sunnyvale, will allow overtaking by new 'Baby Bullet' limited stop services. Introduced in June 2004, these trains make only four stops, cutting San Francisco–San Jose journey time to 57 minutes.

Two major projects under discussion for years are now going forward. Having several times been rejected on cost grounds, relocation of Caltrain's San Francisco terminal to a more central site is now seen as part of a proposed large-scale rejuvenation of the rundown area around the existing Transbay bus terminal. An extension of almost 2 km would bring Caltrain to a new multi-level interchange which might eventually also accommodate high-speed trains. A draft environmental impact report and statement, published in late 2002, set out route options, plus a new Transbay Joint Powers Authority has been set up to design, build and operate the new terminal. The project moved to the final environmental impact stage in mid-2004.

The second long-term aim is to electrify the route, for which environmental impact statements were prepared in 2003 and public discussions held in mid-2004. This would involve energising some 320 track-km at 25 kV AC 60 Hz, and would initially use new electric locomotives to haul existing stock.

California's financial difficulties placed a question mark over all these schemes, and the proposal for extending Caltrain service from Gilroy to Castroville and Salinas (55 km) by 2006. This is now seen as unlikely to go ahead, as is the proposed extension over existing tracks from Caltrain's Redwood City station over the Dumbarton bridge (see above) to a link with BART at Union City.

A new station at Millbrae provides cross-platform connections with the extension of the BART metro system to San Francisco International Airport, opened in 2003 and replacing the existing shuttle bus. The interchange also connects with SamTrans buses and other shuttle bus services.

Traction and rolling stock
The fleet comprises 20 3,200 hp (2,385 kW) F40PH-2 locomotives from GM-EMD, three F40 locomotives from Boise Locomotive and six MP36PH-3C machines from the latter's successor, MotivePower. The last-mentioned were acquired in 2003 to provide power for Baby Bullet limited stop services. Coaching stock consists of 73 double-deck gallery coaches from Nippon Sharyo, 20 from Sumitomo and 17 from Bombardier. A mid-life overhaul programme of the original coaches has been carried out by ALSTOM Transport Canada. ALSTOM also has a contract covering routine overhaul of the passenger coach fleet. A new maintenance depot is under construction in San Jose, for 2006 opening.

UPDATED

Pennsylvania Southwestern Railroad (PSWR)

1200 Midland Avenue, Midland, Pennsylvania 15059
Tel: (+1 724) 773 27 86   Fax: (+1 724) 773 27 92
Web: www.watcocompanies.com

Key personnel
General Manager: Richard L Van Buskirk

Gauge: 1,435 mm
Route length: 19 km

Organisation
PSWR's main business is serving the J&L Specialty Steel mill at Midland, Pennsylvania, where it connects with NS. Its single diesel-electric switcher handles steel products and scrap.

NEW ENTRY

Peoria & Pekin Union Railway

301 Wesley Road, Creve Coeur, Illinois 61610
Tel: (+1 309) 694 86 00   Fax: (+1 309) 694 86 08

Key personnel
President: P D Feltenstein
Chief Engineer: J H Rada

Organisation
PPUR is a switching line owned among others by CN, NS and UP, operating 194 km of track (128 km owned and 66 km contracted), of which 77 per cent is yard, siding or industrial trackage, in and around Peoria, Illinois.

Traction and rolling stock
PPUR owns seven locomotives and 53 freight cars.

Pinsly Railroad Company

53 Southampton Road, Westfield, Massachusetts
01085
Tel: (+1 413) 568 64 26
Fax: (+1 413) 562 84 60
Web: http://www.pinsly.com

Key personnel
President: Marjorie P Silver

Executive Vice-President: John Levine
General Manager: Bennett Biscan

Pinsly is a holding company for five short line
railroads, totalling 480 km in four states with 125
employees, $14 million in annual revenues, and a
logistics company, Rail Road Distribution Services.
It owns a fleet of 20 locomotives. Rolling stock is
leased. In the Pinsly portfolio are:
  Pioneer Valley Railroad, Massachusetts

Florida Central Railroad
Florida Midland Railroad
Florida Northern Railroad
Arkansas Midland Railroad.

Pioneer Railcorp

1318 S Johansen Road, Peoria, Illinois 61607
Tel: (+1 309) 697 14 00   Fax: (+1 309) 697 53 87
Web: www.pioneer-railcorp.com

Key personnel
Chairman and Chief Executive Officer:
  Guy L Brenkman
President and Chief Financial Officer:
  J Michael Carr
Director of Marketing: Catherine Busch
General Counsel: Daniel A LaKemper
Vice-President of Safety and Compliance:
  Tom S Black
Superintendent of Transportation: Shane D Cullen
Director of Operations Center: Joseph Evans
Chief Agent: Tammy Bridson

Organisation
Pioneer is a holding company formed in 1986 which
owns and/or operates some 856 km of track in 16
locations as follows:
  Alabama Railroad Co
  Alabama & Florida Railway Co
  Decatur Junction Railway Co, Illinois
  Elkhart & Western Railroad, Indiana
  Fort Smith Railroad Co, Arkansas
  Garden City Western Railroad, Kansas
  Gettysburg & Northern Railroad Co, Pennsylvania
  Indiana Southwestern Railway Co
  Kendalville Terminal Railway Co, Indiana
  Keokuk Junction Railroad, Illinois
  Michigan Southern Railroad
  Mississippi Central Railroad Co
  Pioneer Industrial Railway
  Shawnee Terminal Railway Co, Illinois

Vandalia Railroad Co, Illinois
West Michigan Railroad Co, Michigan.
  Pioneer Railcorp also operates the Gettysburg
Scenic Railway tourist line.
  The company owns 95 locomotives in the
1,000 to 1,800 hp (746 to 1,340 kW) range and
over 1,400 wagons, plus materials for reuse or
sale. Pioneer Railroad Equipment Company hires
out and repairs the assets. About one third of the
locomotive fleet is leased to other short lines or
switching operations. Pioneer Railroad Services Inc
provides administrative and management support
functions.

UPDATED

Portland & Western Railroad Inc

Genesee & Wyoming Inc, Oregon Region
650 Hawthorn Avenue SE, Suite 220, Salem,
Oregon 97301
Tel: (+1 503) 365 77 17   Fax: (+1 503) 365 77 87
Web: http://www.gwrr.com

Key personnel
President: Larry Phipps
Vice-Presidents
  Customer Service: Ron D Vincent
  Marketing and Sales: William D Bremer

Mechanical: Jack G Russell
Transportation: Mike Erwin
Engineering: Charles S Kettenring
Manager of Asset Utilization: Tom G Cresswell

Organisation
Together with the Willamette & Pacific Railroad,
the Portland & Western Railroad forms the Oregon
Region of Genesee & Wyoming, and totals some
715 route-km. The network runs northwest from
Portland to Astoria, west to Stimson-Forestex
and south to Salem, Albany and Toledo, with a
connection via trackage rights over UP to Eugene.

Connections are made with BNSF. Traffic amounts
to some 60,000 carloads annually, comprising
mainly paper and forest products, steel, grain,
fertilisers, chemicals and aggregates.

Port Terminal Railroad Association

8934 Manchester Street, Houston, Texas 77012-2149
Tel: (+1 713) 393 65 00
Fax: (+1 713) 393 65 80
Web: http://www.ptra.com

Key personnel
General Manager: Marvin Wells

Organisation
The 51.6 km property, of which 14.5 km are CTC-
equipped, is owned 50 per cent by the UP and
50 per cent by the BNSF. It provides access to the

port of Houston. Fees are assessed on the basis of
use. In a transaction valued at US$90 million, the
railroad replaced all its switchers with 24 units
of type MK1500D from MK Rail during 1996. The
contract includes a 30-year maintenance provision.
  In 2001 the company handled about 500,000
carloads.

Providence & Worcester Railroad Co

PO Box 16551, 75 Hammond Street, Worcester,
Massachusetts 01610
Tel: (+1 508) 755 40 00   Fax: (+1 508) 795 07 48
Web: http://www.pwrr.com

Key personnel
Chairman: Robert H Eder
President: Orville R Harrold
Treasurer and Controller: Robert J Easton
Secretary and General Counsel: Mary A Tanona
Director of Operations: David Fitzgerald
Vice-President, Engineering: P Scott Conti
Chief Mechanical Officer: David Rutkowski

Organisation
Providence & Worcester came into being in 1973 to
operate a 69 km line between its namesake towns
whose stockholders had declined to form part of

the then emerging Conrail. Since then, P&W has
expanded to a 877 km system, partly by buying up
sections of former Boston & Maine and Conrail's
trackage in Connecticut and Rhode Island. Rights
to provide local freight service on the Amtrak
route from Providence to New Haven have been
acquired; P&W also has rights on Metro-North
(MN) from New Haven to New York, over two
MN branches to reach Danbury and over Amtrak to
a link with the New York & Atlantic at Fresh Pond
in Queens, New York. Its principal interchange
points are Worcester and New Haven with CSXT;
Gardner, Massachusetts, with Springfield Terminal;
and New London, Connecticut, with CN, CP and
New England Central. It also owns a deep-water
port facility, the South Quay intermodal terminal at
Wilkesbarre Pier in East Providence, Rhode Island,
and serves two other deep-water ports. In addition,
P&W is involved in the development of the State of
Connecticut's project for restoration and expansion
of rail facilities at the port of New Haven.

Traffic handled in 2003, at around 32,000 carloads,
was virtually unchanged from 2002, with increased
volumes of coal and metal products largely offset
by a decline in shipments of aggregates for the
construction industry.

Traction and rolling stock
P&W owns 31 locomotives, a mix of GE and
GM types, 66 freight cars and five passenger cars.

Puget Sound & Pacific Railroad

PO Box L-2, 501 North 2nd Street, Elma, Washington 98541
Tel: (+1 360) 482 49 94   Fax: (+1 360) 482 39 66
Web: www.railamerica.com

Key personnel
General Manager: Tom Paul
Business Development: Cathrine Martin

RailAmerica, Inc

5300 Broken Sound Boulevard NW, Boca Raton, Florida 33487
Tel: (+1 561) 994 60 15   Fax: (+1 561) 994 46 29
Web: www.railamerica.com

Key personnel
President: Donald D Redfearn
Chief Executive Officer: Charles Swinburn
Executive Vice-President and Chief Operating Officer, North American Rail Group: Joe Conklin
Executive Vice-President and Chief Financial Officer: Michael J Howe
Senior Vice-President and General Counsel: Scott G Williams
Senior Vice-President, Sales and Marketing: Thomas Owen
Senior Vice-President, Corporate Development: Matthew Devine
Senior Vice-President, Eastern Corridor, North American Rail Group: Scott Linn
Senior Vice-President, Western Corridor, North American Rail Group: Tom Schlosser

Corporate background
RailAmerica Inc is the world's largest operator of shortline and regional railways, owning 46 companies operating more than 14,500 route-km in the US and Canada, operating in 26 states and six Canadian provinces. It is a public company, traded on the New York Stock Exchange under the symbol 'RRA'.

The company was formed in 1986 with the purchase of the Huron & Eastern Railway Co (133 km) in Michigan. In November 1992, RailAmerica held its initial public offering, setting the foundation for expansion.

In 1997 RailAmerica acquired its first railroad outside the US with the purchase of the 3,621 km Chilean railway, Ferronor, and later that year expanded into Australia with acquisition of the 6,500 km transcontinental Australian passenger rail service concession.

In 1999 RailAmerica grew rapidly, starting with acquisition of its first railway in Canada with the purchase/lease of the 294 km Esquimault &

Gauge: 1,435 mm
Route length: 241 km

Organisation
Operated by RailAmerica, PSAP comprises a network of lines radiating from Elma, Washington. The railway handles approximately 14,000 carloads annually. Traffic includes forest and paper products, fertiliser and metal products. In addition, freight is carried for the US Navy, which has an interchange

Nanaimo Railway (E&N) on Vancouver Island from Canadian Pacific. It then acquired four major networks: Australia's V/Line Freight railway, now known as Freight Australia; Canada's third largest rail system, RaiLink Ltd; the 966 km Toledo, Peoria & Western Railroad; and RailTex, then North America's largest shortline freight railroad company.

In January 2002 RailAmerica completed the acquisition of StatesRail, another leading US shortline operator, and ParkSierra, an operator of three shortline railroads covering 1,220 km in four western US states.

Weak economic conditions in the US and reduced operating results in Australia caused by the severe drought led the company in December 2002 to implement a number of cost savings and cut about 145 jobs. Then, to reduce debt, it was decided in 2003 to focus more closely on the core North American business, and buyers were sought for the overseas operations. The 55 per cent stake in Ferronor was sold at the beginning of 2004 for US$18.1 million, and agreement was reached in March 2004 for the sale of Freight Australia to Pacific National for US$214 million, subject to regulatory approval. Gary Marino, President and Chief Executive officer since 1994, retired in April 2004. Another US shortline was added to the network in January 2004 with purchase of the Michigan Central Railroad. This has been absorbed into the Huron & Eastern Railway.

On 31 August 2004 the company completed its sale of Freight Australia.

At mid-2004, the North American Rail group employed some 1,750 personnel, and owned or leased 495 diesel locomotives and 7,420 wagons.

Freight operations
RailAmerica undertakes freight operations on the following railways. See also individual entries for each railway.

Canada
Cape Breton & Central Nova Scotia Railway
Central Western Railway
Goderich-Exeter Railway
Lakeland & Waterways Railway
Mackenzie Northern Railway

with PSAP at Shelton. Interchange is also made with BNSF at Centralia and with UP at Blakeslee Junction.

UPDATED

Ottawa Valley Railway
Southern Ontario Railway

US
Alabama & Gulf Coast Railway
Arizona & California Railroad
California Northern Railroad
Carolina Piedmont Railroad
Cascade & Columbia River Railroad
Central Oregon & Pacific Railroad
Central Railroad of Indiana
Central Railroad of Indianapolis
Central Western Railroad
Chesapeake & Albemarle Railroad
Chicago, Fort Wayne & Eastern Railroad
Connecticut & Southern Railroad
Dallas, Garland & Northeastern Railroad
Eastern Alabama Railway
Grand Rapids Eastern
Huron & Eastern Railway
Indiana & Ohio Rail System
Indiana Southern Railroad
Kiamichi Railroad
Kyle Railroad Company
Lahaina, Kaanapali & Pacific Railroad (tourist passenger railway)
Michigan Shore Railroad
Mid-Michigan Railroad
Missouri & Northern Arkansas Railroad
New England Central Railroad
North Carolina & Virginia Railroad
Otter Tail Valley Railroad
Puget Sound & Pacific Railroad
San Diego & Imperial Valley Railroad
San Joaquin Valley Railroad
San Luis & Rio Grande Railroad
South Carolina Central Railroad
Texas Northeastern Railroad
Toledo Peoria & Western Railroad
Ventura County Railroad
Virginia Southern Railroad
West Texas & Lubbock Railroad*

* Assets leased to Permian Basin Railways Inc

UPDATED

Rail Management Corporation

2605 Thomas Drive, Panama City Beach, Florida 32408
Tel: (+1 850) 230 83 31   Fax: (+1 850) 230 88 48
Web: http://www.rail-management.com

Key personnel
Vice-Presidents
  Sales and Marketing: Garland Horton
  Operations: Ed Clark
  Chief Financial Officer: Scott Helms
  Chief Engineer: Doug Davis
  Car Leasing: Barry Waters

Organisation
Created in 1980, RMC owns and operates 14 short line railroads in 11 states, as well as leasing locomotives and rolling stock and provision of trucking services. Its rail operations comprise:
  AN Railway
  Atlantic & Western Railway
  Bay Line Railroad
  Copper Basin Railway
  East Tennessee Railway
  Galveston Railroad
  Georgia Central Railway
  KWT Railway

Little Rock & Western Railway
M&B Railroad
Tomahawk Railway
Valdosta Railway
Western Kentucky Railway
Wilmington Terminal Railroad

Railroad Commission of Texas

Rail Division, PO Box 12967, Austin, Texas 78711-2967
Tel: (+1 512) 463 70 01   Fax: (+1 512) 463 71 53
Web: http://www.rrc.state.tx.us

Key personnel
Director: Jerry L Martin
Special Projects Director: Michael N Jones

Background
A Texas High Speed Rail Authority was created by the Texas legislature's High Speed Rail Act to investigate the possibilities of high-speed rail in Texas, and to grant franchises for the construction, operation, maintenance and financing of high-speed rail routes in Texas, if such routes were determined to be in the public convenience and necessity.

An attempt by the Authority to initiate construction of a TGV route in the early 1990s failed to come to fruition and under Texas law

the Authority's independent existence ended in August 1995, when the enabling legislation was repealed. As a matter of law, any party wishing to build a high-speed rail system in Texas can secure a railroad company charter under Texas railroad statutes and can then utilise the right of eminent domain to acquire right-of-way. The main reason for granting a franchise under the now repealed High Speed Rail Act was to confer exclusive rights to a high-speed developer so that it would be easier to finance a capital-intensive project.

Railroad Development Corporation (RDC)

381 Mansfield Avenue, Suite 500, Pittsburgh, Pennsylvania 15220, US
Tel: (+1 412) 928 07 77 Fax: (+1 412) 928 77 15
e-mail: rdc@rrdc.com
Web: www.rrdc.com

Key personnel
Chairman: Henry Posner III
President: Robert A Pietrandrea
Vice-President, Operations: William J Duggan
Vice-President: Bradley J Knapp
Chief Financial Officer: John F Hensler
Manager, Administration: Kathleen E Przybylski

Organisation
RDC is a railway investment and management company with ownership and management interests in railways in seven countries. In 2005 these were:

América Latina Logística (formerly the Buenos Aires al Pacifico and Ferrocarril Mesopotámico-General Urquiza systems, Argentina) (see entry in Argentina section of *Railway systems and operators*)

Central East African Railways Company (Malawi) (see entry in Malawi section of *Railway systems and operators*)

Nacala Corridor (Mozambique) (see entry in Mozambique section of *Railway systems and operators*)

Eesti Raudtee (Estonia) (see entry in Estonia section of *Railway systems and operators*)

Ferrocarril Central Andino (Peru) (see entry in Peru section of *Railway systems and operators*)

Ferrovías Guatemala (Guatemala) (see entry in Guatemala section of *Railway systems and operators*)

Iowa Interstate Railroad (US) (see entry in US section of *Railway systems and operators*)

With the exception of Iowa Interstate Railroad, RDC has joint venture partners in each country.

In January 2004 RDC acquired Heartland Rail Corporation's 80 per cent holding in the Iowa Interstate Railroad, thereby assuming 100 per cent ownership.

A subsidiary of Iowa Interstate Railroad is Rail Traffic Control Inc, established to provide a broad spectrum of railroad dispatching and transportation consulting services.

In January 2005 a transaction was completed involving RDC and Mozambican and international investors to take over the 914 km Nacala rail and port corridor, which is contiguous with the RDC-led operation in Malawi.

UPDATED

Rarus Railway Co

PO Box 1070, Anaconda, Montana 59711, US
Tel: (+1 406) 563 28 51

Key personnel
President: William T McCarthy

Gauge: 1,435 mm
Route length: 40 km

Organisation
Rarus Railway Co has 40 route-km, the former Butte, Anaconda & Pacific Railway, serving the copper industry and interchanging with UP at Silver Bow, Montana, and with BNSF at Butte, Montana. It handles around 5,000 carloads annually.

Traction and rolling stock
Rolling stock comprises 12 locomotives (one GP7, seven GP9's, two GP38-2's and two GP39-2's) and 67 freight wagons.

UPDATED

Reading & Northern Railroad Co

PO Box 218, Port Clinton, Pennsylvania 19549
Tel: (+1 610) 562 21 00 Fax: (+1 610) 562 19 20
Web: http://www.readingnorthern.com

Key personnel
President: Andrew Muller Jr
Executive Vice-President, Operations:
    Wayne Michel

Organisation
The R&N is a holding company which began operation in 1990. It took over from Conrail a total of 199.5 km, chiefly serving anthracite mines, radiating from Port Clinton, Pennsylvania; connection is via Conrail at Reading. There are two operating subsidiaries: the Blue Mountain Reading & Northern is the freight hauler, the Blue Mountain & Reading is a stand-alone tourist train operation.

In 1996, the Reading & Northern bought Conrail's Lehigh Division, bringing the network to 510 route-km. The railway bought 400 hoppers for coal traffic and in 2002 purchased 25 second-hand locomotives. Traffic hauled in 2002 reached almost 25,000 carloads.

Traction and rolling stock
At the beginning of 2003, the locomotive fleet comprised: four SD38, six SD40-2, one SD45, six SD50, four SW8 and two SW1500, all from EMD; and one B30-a, two C30-7 and one C36-7 from GE. The roster also included two Budd RDC ralicars and some historic locomotives.

Red River Valley & Western Railroad Co

PO Box 608, 116 S 4th Street, Wahpeton, North Dakota 58075
Tel: (+1 701) 642 82 57 Fax: (+1 701) 642 35 34
Web: www.rrvw.net

Key personnel
Chairman: Kent P Shoemaker
President and Chief Executive Officer:
    William F Drusch

Senior Vice-President and General Manager:
    Andrew J Thompson

Organisation
RRVW owns 832 km of former Burlington Northern route in North Dakota, acquired in 1987. Lines run west from Wahpeton to Oakes, then northwards to Jamestown, Carrington, New Rockford and Maddock; from Wahpeton to Breckenridge and Lidgerwood; Wahpeton to Casselton; and Horace–Edgeley. It connects with BNSF and CP. BNSF supplies most of the empty rolling stock to RRVW.

The traffic base is approximately 42,000 carloads annually and includes outbound grain, sugar, corn syrup and feeds; inbound traffic includes steel, coal and fertilisers.

Traction and rolling stock
RRVW operates 14 diesel locomotives: five GP20C; four GP15C; two GP10; two SW1200 and one CF7. There are also two slugs.

UPDATED

Regional Transportation Authority (RTA)

501 Union Street, 6th Floor, Nashville, Tennessee 37219-1705
Tel: (+1 615) 862 88 33 Fax: (+1 615) 880 39 01
e-mail: rta.nrc.org
Web: www.rta-ride.org

Key personnel
Executive Director: Eric C Beyer
Rail Project Manager: Allyson Shumate

Background
RTA is a government-funded agency that has worked closely with the US Federal Transit Administration to develop a commuter rail service based on Nashville and serving mid-Tennessee. Branded 'Music City Star', East Corridor Alignment services are due to commence in late 2005 on a 51 km six-station route east from Riverfront, Nashville, to Lebanon over a single-track line owned by the Nashville and Eastern Railroad Authority. Initially three trains per day in each direction will be operated. At three stations in Nashville, connection will be made with bus services operated by Nashville Metropolitan Transit Authority (MTA).

RTA's longer term plans foresee an eventual network of 232 km, with four further routes from Nashville running southeast to Murfreesboro, south to Franklin, west to Kingston Springs and northeast to Gallatin.

Traction and rolling stock
For the launch of Music City Star services RTA has acquired two EMD F40PH diesel-electric locomotives from Amtrak and 11 bi-level gallery cars from Metra, Chicago.

NEW ENTRY

Saginaw Valley Railway

Organisation

In 2004 the Saginaw Valley Railway, already a RailAmerica subsidary, was absorbed by its neighbouring sister company, the Huron and Eastern Railway (see entry in US section of *Railway systems and operators*).

UPDATED

San Diego & Imperial Valley Railroad

1501 National Avenue, Suite 200, San Diego, California 92113-1029
Tel:(+1 619) 239 73 48   Fax: (+1 619) 239 71 28
Web: http://www.railamerica.com

Key personnel

General Manager: Doug Verity
Business Development Manager: Mike Ortega

Gauge: 1,435 mm
Route length: 262 km

Organisation

Formerly owned by RailTex and since 2000 operated by RailAmerica, SDIY runs south from San Diego to the Mexican border at Tijuana. The defunct portion within Mexico is now leased to the Carrizo Gorge Railway and is being rehabilitated. SDIY's route resumes at Division, California, and runs to Plaster City. A second connected line runs northeast from San Diego to El Cajon. The railway handles approximately 4,400 carloads annually, traffic including liquefied petroleum gas, forest products, grain and foodstuffs.

San Joaquin Valley Railroad

221 North 'F' Street, PO Box 937, Exeter, California 93221
Tel: (+1 559) 592 18 57   Fax: (+1 559) 592 18 59
Web: www.railamerica.com

Key personnel

General Manager: Charles Littlefield
Business Development: Richard McGowan, Bill Tongate

Gauge: 1,435 mm
Route length: 549 km

Organisation

A former SatesRail system now operated by RailAmerica, SJVR comprises a network of lines centred on Bakersfield, Exeter and Fresno, California. The railway handles approximately 32,000 carloads annually, traffic including consumer products, agricultural and food products,

paper, metals and petrochemicals. Interchange is made with BNSF at Fresno and with UP at Fresno and Bakersfield.

UPDATED

San Pedro & Southwestern Railway

796 East Country Club Drive, PO Box 1420, Benson, Arizona 85602
Tel: (+1 520) 586 22 66
Fax: (+1 520) 586 29 99

Key personnel

President: David Parkinson

Gauge: 1,435 mm
Route length: 109 km

Organisation

This former Southern Pacific branch in the extreme south of Arizona, acquired by RailAmerica with its 2002 purchase of StatesRail, was sold in November 2003 to the Arizona Railroad Group. Interchanging with UP's Chicago–Los Angeles main line at Benson, east of Tucson, the railway runs to Fairbank, Curtiss

and Paul Spur near Douglas on the border with Mexico. Once part of a through route to El Paso, Texas, the remaining line now serves the fertiliser and copper industries at Curtiss and Bisbee Junction.

Traction and rolling stock

The railway owns one diesel locomotive; wagons are supplied by UP and other connecting railways.

Seminole Gulf Railway

4110 Centerpoint Drive, Fort Myers, Florida 33916
Tel: (+1 239) 275 60 60   Fax: (+1 239) 275 05 81
Web: http://www.semgulf.com

Key personnel

President and Treasurer: Gordon H Fay
Executive Vice-President: George E Bartholomew
Senior Vice-President, Marketing:
  Bernard M Reagan

Organisation

SGR runs two former CSXT lines in south Florida, extending to 190 route-km. The lines, which are unconnected, run from a link with CSXT at Arcadia to North Naples, and from Oneco to Sarasota and Venice. A dinner train service runs from the colonial-style station at Fort Myers. Traffic is mainly building materials, newsprint, forest products and beer.

Traction and rolling stock

Services are operated with eight GM-EMD GP9 locomotives, 26 freight wagons and seven passenger coaches.

Sound Transit Board

Central Puget Sound Transit Authority
401 S Jackson Street, Seattle, Washington 98014-2826
Tel: (+1 206) 398 51 15   Fax: (+1 206) 398 52 16
Web: http://www.soundtransit.org

Key personnel

Rail Operations Manager: Martin Young

Passenger operations

'Sounder' peak-hour commuter services operated by the Sound Transit Board over BNSF tracks between Tacoma and Seattle (130 km) in the state of Washington started in 1999 and by 2002 was carrying some 3,000 passengers daily. Extensions were planned north from Seattle to Everett and south from Tacoma to Lakewood by 2001, but both have run into financial and environmental difficulties. A 2 km extension to Freighthouse Square in Tacoma opened in September 2003, but was subsequently closed following subsidence of the earthworks. Reopening was planned for late 2004. The implementation of services forms part of the 10-year 'Sound Move' regional transport plan developed by the Central Puget Sound Transit Authority.

Capacity improvements between Tacoma and Seattle, including some double-tracking and

Sound Transit Board General Motors F59PH diesel-electric locomotive and Bombardier-built bilevel coaches at Seattle (David C Warner) 0528629

bidirectional signalling, are due for completion in 2003. This work will allow extra 'Sounder' trains to be run without affecting BNSF's freight operations.

In 2004 funding was allocated by the Puget Sound Regional Council for feasibility studies of a 64 km commuter rail service over BNSF trackage between Snohomish and Renton, known as the Eastside line. A proposal also exists to run a commuter service from Bellingham to Everett (92 km) to connect with Sounder trains.

Traction and rolling stock

An initial fleet of six General Motors F59PH locomotives and 38 bilevel Bombardier coaches was supplied and it is expected eventually to add an additional five locomotives and 20 coaches to the fleet. In 1999, Amtrak was awarded a 10-year programme to provide fleet maintenance services. These are undertaken at the national operator's King Street facility in Seattle.

South Carolina Central Railroad Inc

621 Field Pond Road, Darlington, South Carolina 29540
Tel: (+1 843) 398 98 50 Fax: (+1 843) 398 96 83
Web: www.railamerica.com

Key personnel
General Manager: Lamont Jones
Marketing and Sales Manager: Kim Greer

Gauge: 1,435 mm
Route length: 93 km

Organisation
Formerly owned by RailTex and since 2000 operated by RailAmerica, SCRF comprises two unconnected shortlines in South Carolina: Cheraw–Society Hill and Bishopville–Florence. The railway handles approximately 25,000 carloads

annually. Interchange is made with CSXT at Cheraw and Florence.

UPDATED

South Carolina Division of Public Railways

540 East Bay Street, Charleston, South Carolina 29403
Tel: (+1 843) 727 20 67
Fax: (+1 843) 727 20 05

Key personnel
President: Daniel S Green

Organisation
The state of South Carolina owns and operates three short lines or switching railroads. The East Cooper & Berkeley (28 km) serves BP and Nucor Steel in southern Berkeley County, interchanging with CSXT at Coordesville. Two switching railroads, the Port Utilities Commission of Charleston and the Port Terminal Railroad (20 km), provide links between terminals of the South Carolina State Ports Authority at Charleston, and with the CSXT and NS railroads.

Also owned is the Port Royal Railroad linking Port Royal and Yemassee (42 km), currently closed and awaiting disposal along with the port itself, which is scheduled to close in 2006.

The Public Railways Division also provides technical assistance and consulting service in railway matters to state, local and municipal governments.

NEW ENTRY

South Central Florida Express Inc

Division of United States Sugar Corporation
900 South W C Owen Avenue, Clewiston, Florida 33440
Tel: (+1 863) 983 31 63 Fax: (+1 863) 983 67 73
e-mail: scrr@gate.net
Web: www.ussugar.com/sugar/transportation.html

Key personnel
Vice-President: Sally Conley

Gauge: 1,435 mm
Route length: 254 km

Organisation
As well as its internal network totalling some 190 km, US Sugar owns SCFE, which it purchased in 1994. Formerly the South Central Florida Railroad, the property comprises the former CSXT line from Sebring to Fort Pierce, plus branches to Lake Harbor and Okeelanta. The principal commodities are raw and refined sugar, raw sugar cane and fertilisers. The railroad carries over 70,000 carloads annually.

Traction and rolling stock
SCFE operates 14 diesel locomotives (10 GP11s, two GP16s and two GP18s) and 48 wagons.

UPDATED

South Kansas & Oklahoma Railroad Inc

315 W Third Street, Pittsburg, Kansas 67762
Tel: (+1 620) 231 22 30 Fax: (+1 620) 231 25 68

Key personnel
President: Rock Webb
General Superintendent: Jim Horner

Organisation
SKOL has 800 route-km centred on Cherryville, Kansas. The main line runs southwards to Tulsa, Oklahoma, and its port on the Arkansas river. Other lines radiate east to Columbus and Pittsburg, west to Winfield and Oxford, north to Humboldt and south to Coffeyville. The company, a subsidiary of Watco of Coffeyville, purchased its three lines from Santa Fe in 1990. Its annual traffic is about 42,000

carloads, carrying grain, cement, chemicals, steel and plastics.

Southeastern Pennsylvania Transportation Authority (SEPTA)

1234 Market Street, Philadelphia, Pennsylvania 19107-3780
Tel: (+1 215) 580 40 00
Web: www.septa.org

Key personnel
Chairman: Pasquale Deon
Chief Operating Officer and General Manager: Faye Moore
Chief Officer, Rail Transportation: Richard Hanratty

Gauge: 1,435 mm
Route length: 722 km
Electrification: 722 km at 12 kV 25 Hz AC

Passenger operations
SEPTA operates a multimodal network of public transport services throughout the five-county region of southeastern Pennsylvania. The commuter rail portion consists of 13 electrified lines radiating from Philadelphia. These are operated as eight cross-city routes, serving 153 stations. SEPTA owns about half the route-km; the remainder belongs to Amtrak. The network carried 31 million passengers in 2002.

The present Regional Rail system was formed in 1984 when the commuter lines formerly operated by the Pennsylvania and Reading railroads were linked by a four-track connection beneath central Philadelphia.

SEPTA has extended service over Amtrak's Northeast Corridor to Newark, Delaware, under contract to the Delaware Transportation Authority.

New lines
The MetroRail scheme for a 100 km commuter route in the Schuylkill Valley corridor, put forward by SEPTA and the neighbouring Berks Area Reading Transportation Authority (BARTA), was adopted as 'locally preferred solution' in 2000. Largely using existing right-of-way, a high-quality electrified railway with a 15-min peak-hour service was planned to link Philadelphia with Reading. FTA approval was granted in January 2002 for preliminary engineering studies, but the project languished due to political difficulties. These appeared to be resolved at the end of 2002, but the costly scheme then fell foul of proposed Federal budget cuts and its grant application was rejected in early 2003. SEPTA and BARTA redrafted the proposals in 2004 in an attempt to reduce costs, principally by deferring electrification. Several other short extensions of existing routes have also been proposed from time to time.

Improvements to existing lines
In 2005 work began on the second phase of a major track, communications and signalling improvement programme on SEPTA's principal corridor between Wayne Junction and Glenside.

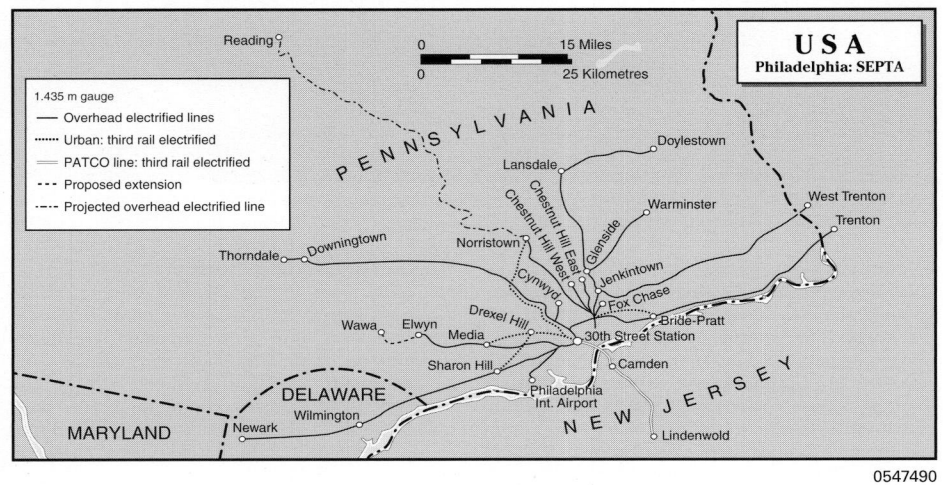

0547490

Electric railcars or multiple-units

Class	Cars per unit	Motor cars per unit	Motored axles/car	Output/ motor kW	Speed km/h	No in service	First built	Builders Mechanical	Electrical
Silverliner 2	1	1	4	110	135	36	1963	Budd	GE
Silverliner 2	1	1	4	110	135	17	1964	Budd	GE
Silverliner 3	1	1	4	110	135	20	1967	St Louis Car Co	GE
Silverliner 4	1	1	4	140	150	47	1974–75	GE	GE
Silverliner 4	2	2	4	140	150	184	1975–76	GE	GE

Traction and rolling stock

The Regional Rail service is operated chiefly by a fleet of 304 Silverliner electric multiple-unit cars.

SEPTA also operates a fleet of 45 Bombardier-built passenger coaches operated as push-pull trainsets with seven AEM7 and one ALP electric locomotives. A refurbishment programme for the Silverliner 4 fleet is due to be completed in 2004.

A prospective order for 104 cars, awarded in early 2004 to the Korean United Transit Systems consortium, was to be tendered again following representations from the unsuccessful bidders.

Type of coupler in standard use: N2A (emus) Tightlock Knuckle (push-pull cars)
Type of braking in standard use: 26C, PS68
Brake air and blended rheostatic (Silverliner 4 cars)

Track
Rail: 140 PS and RE, 132 RE, 131 RE, 130 REHF, 115 RE, 112 RE, 107 NH, 100 PS
Crossties (sleepers): Wood, thickness 175 × 200 mm
Spacing: 530 mm
Fastenings: Pandrol and spike
Min curvature radius: 125 m
Max gradient: 1 in 25

UPDATED

Southern Railroad Company of New Jersey (SRNJ)

PO Box 122, Willingboro, New Jersey 08046
Tel: (+1 609) 871 86 99
Fax: (+1 609) 871 74 32

Key personnel
Vice-President: Tom Collard

Organisation
SRNJ runs four routes in southern New Jersey: from an interchange with CSXT/NS at Winslow Junction to Vineland (26 km) and Atlantic City (51 km), and Atlantic City to Pleasantville (16 km); and from Salem to CSXT/NS at Swedesboro (29 km). It handles some 3,500 carloads annually.

Upgrading and rehabilitation of the Salem line was completed in 2004 following an agreement under which the New Jersey Department of Transportation granted US$600,000 towards the work.

Traction and rolling stock
SRNJ operates six diesel locomotives.

NEW ENTRY

Southwestern Railroad Co

New Mexico Division
PO Box 126, Hurley, New Mexico 88043
Tel: (+1 505) 537 20 04   Fax: (+1 505) 537 26 24

Key personnel
Operations Manager: Ron Lindsey

Texas Division
2 North Main Street, Perryton, Texas 79070
Tel: (+1 806) 435 23 22

Organisation
The New Mexico operation comprises several lines purchased from ATSF in 1990 linking opencast copper mines at Chino and Tyrone with the Phelps-Dodge smelter at Hurley. Phelps-Dodge is the railway's only customer. There is interchange with BNSF at Peruhill. The New Mexico division covers 106 route-km. The Texas division runs the 93 km Perryton–Shattuck (Oklahoma) line in northern Texas.

Traction and rolling stock
The New Mexico division has ten GM-EMD locomotives (four GP7Us, one GP30R, two GP35Rs, one GP40 and two SD45Rs).

Stillwater Central Railroad (SLWC)

123 North Depot Street, Cherryvale, Kansas 67335
Tel: (+1 620) 336 22 91   Fax: (+1 620) 336 27 12
Web: www.watcocompanies.com

Key personnel
General Manager: Jim Horner

Gauge: 1,435 mm
Route length: 420 km

Organisation
Stillwater Central comprises former BNSF routes from Sapulpa to Oklahoma City and Wheatland to Snyder purchased by Watco in 1996 and 2001. The formerly unconnected lines were linked into a single network at the end of 2004 when Watco leased BNSF's line between Wheatland and Oklahoma City, at the same time taking over operation of BNSF's North Yard in Oklahoma City. This allows SLWC to handle BNSF traffic from Oklahoma City to Tulsa. In addition, trackage rights over BNSF between Sapulpa and Tulsa provide a connection with Watco's South Kansas & Oklahoma Railroad (see entry in US section of *Railway systems and operators*). SLWC handles mainly minerals and industrial products, and serves the USA's largest gypsum mine in Barber County. Traffic amounted to some 17,000 carloads in 2003.

In February 2004 the Federal Railroad Administration granted a loan of US$4.6 million to SLWC to refinance existing debt and fund purchase of an additional locomotive.

NEW ENTRY

Tacoma Rail (TMBL)

Tacoma Public Utilities
PO Box 11007, Tacoma, Washington 98411
Tel: (+1 253) 502 88 67
Fax: (+1 253) 922 56 79
Web: www.tacomarail.com

Key personnel
Superintendent: Dennis Dean

Gauge: 1,435 mm
Route length: 328 km

Organisation
This municipally-owned railroad provides service under contract over three divisional operations. The Tidelands Division is a switching service at the port of Tacoma, providing links to Class I railroads. Here the extensive classification and storage yards were expanded in two phases completed in 1998 and 2001.

Tacoma Rail expanded in 1998 with creation of the Mountain Division, linking Tacoma with Chehalis (212 km). This route had been purchased from BN by the City of Tacoma, which contracted operations to Tacoma Rail. The third division, the Capital, started up in November 2004 to provide contract service for BNSF over three of its routes in South Tacoma. Together more than 100,000 train movements are handled annually; there are 90 employees.

Traction and rolling stock
At the beginning of 2005 the fleet comprised 15 locomotives: three of 1,200 hp, four GM MP15AC of 1,500 hp, six of 2,000 hp and two of 3,000 hp.

NEW ENTRY

Tennessee Southern Railroad Co Inc

100 Railroad Street, Mt Pleasant, Tennessee 38474
Tel: (+1 931) 379 58 24   Fax: (+1 931) 379 58 26
Web: http://www.tnsou.com

Key personnel
President: Dennis T Prince
Operations Manager: Tony Brunson

Organisation
TSR operates over 190 route-km. The company bought one of its lines from NS in 1988 and leased the other from CSX in 1989. Routes run from Columbia to Pulaski, Tennessee, and Florence, Alabama, with CSXT interchange at Natco. Traffic is also transferred to barges at the company's docks on the Tennessee river at Florence. It is controlled by Shortlines Inc.

Traction and rolling stock
TSR operates 10 GM-EMD locomotives (two SD18s, three GP30s and five GP9s) and around 50 freight wagons.

Terminal Railroad Association of St Louis

1000 Union Station, St Louis, Missouri 63102
Tel: (+1 314) 539 47 04
Web: www.terminalrailroad.com

Key personnel
President: GT Gates
General Manager: B J Broyles

Organisation
TRRA has 330 track-km, 27 diesel locomotives
(10 GM-EMD SW1200 and 17 GM-EMD SW1500)
and five slugs. It links seven Class 1 and two
switching railroads in the west and southwest of St
Louis and operates an 80-track classification yard at
Madison, Illinois.

UPDATED

Texas & New Mexico Railroad

82 West Broadway Street, Brownfield, Texas 79316
Tel: (+1 806) 637 83 23
Fax: (+1 806) 637 80 74
Web: http://www.iowapacific.com

Key personnel
President: Edwin Ellis

Gauge: 1,435 mm
Route length: 169 km

Organisation
Formerly owned by RailTex and from February
2000 until May 2002 operated by RailAmerica,
the TNMR was then sold to Permian Basin
Railways Inc, a wholly owned subsidiary of
Iowa Pacific Holdings. The line runs north from
Monahans, Texas, to Lovington, New Mexico, and
handles approximately 2,700 carloads annually.
Commodities carried include hazardous waste,
liquefied petroleum gas, chemicals, scrap metal
and minerals. Interchange is made with UP at
Monahans.

Texas Mexican Railway Co

5810 San Bernardo Avenue, Walker Plaza Building,
Suite 350, PO Box 419, Laredo, Texas 78041
Tel: (+1 956) 728 67 00   Fax: (+1 956) 728 67 90
Web: www.tmry.com

Key personnel
President and Chief Executive Officer:
   Mario Mohar
General Manager: Jim Riney
Finance Manager: Rene Robles
Manager, Real Estate and Industrial Development:
   Arturo Dominguez

Gauge: 1,435 mm
Route length: 884 km (including trackage rights)

Organisation
TMR operates from Laredo to Corpus Christi,
Texas. With its line originating in Laredo it connects
with Transportación Ferroviaria Mexicana (TFM)
(see entry in Mexico section of *Railway systems
and operators*). It also controls the US portion
of the International Rail Bridge at Laredo that
provides a connection with the Mexican network.
In 1997 it acquired 589 km of trackage rights from
Corpus Christi to Beaumont, Texas. In 1998 a yard
and intermodal terminal were commissioned at
Laredo.

Corporate development
At the end of November 2004 the US Surface
Transportation Board approved an application
from Kansas City Southern (KCSI) to obtain control
of TMR and of the US portion of the International
Rail Bridge by acquiring these assets from
Mexrail Inc, a subsidiary of Grupo Transportación
Ferroviaria Mexicana (TFM). The transaction was
finalised at the end of 2004. KCSI had owned 49
per cent of TMR since 1995. The move by KCSI to
obtain control of TMR coincided with shareholder
approval of its acquisition of TFM, providing the
US operator with a combined US-Mexican main
line network of some 8,500 km.

Improvements to existing lines
In 2005 TMR secured a loan of USD50 million
from the FRA's Railroad Rehabilitation and
Improvement Financing fund. This was destined
for improvements covering 234 km of track, two
bridges and yards at Corpus Christi and Laredo.

Traction and rolling stock
In 2004 the fleet comprised 38 locomotives and
905 wagons.

UPDATED

Texas Northeastern Railroad

403 International Parkway, Suite 500, Richardson,
Texas 75040
Tel: (+1 972) 808 98 00   Fax: (+1 972) 808 99 00
Web: www.railamerica.com

Key personnel
General Manager: Jim Kunz
Business Development Manager: Robin Bergeron

Gauge: 1,435 mm
Route length: 172 km

Organisation
Formerly owned by RailTex and since 2000
operated by RailAmerica, the TNER comprises
two unconnected lines in northeast Texas: from
Sherman eastwards to Paris; and from Texarkana
westwards to New Boston. The railway handles
approximately 13,800 carloads annually, mainly
carrying consumer products, railroad wagons,
metals and agricultural products. Interchange
is made with UP at Dennison, Texas, with BNSF
at Sherman, Texas, and with a sister company,
the Dallas, Garland & Northeastern Railroad, at
Trenton, Texas.

Timber Rock Railroad (TIBR)

505 West Avenue F, Silsbee, Texas 77656
Tel: (+1 409) 385 66 11
Fax: (+1 409) 386 28 51
Web: www.watcocompanies.com

Key personnel
General Manager: Pat LaCaze

Gauge: 1,435 mm
Route length: 547 km

Organisation
Watco started operating the TIBR in 1998, and
acquired further trackage in 2002 and 2004. Routes
extend from Beaumont, Texas, northwards to
Tenaha and west to Dobbin, with trackage rights
over BNSF thence to Somerville. A branch from
Kirbyville extends over the border into Louisiana
at DeRidder, where there is interchange with KCS.
In addition, there are interchanges with BNSF at
Tenaha, Beaumont, Somerville and Dobbin.
   TIBR bought the Kirbyville to DeRidder line from
BNSF in 1998, incidentally serving Watco's start-up
switching operation located at DeRidder, and took
a 10-year lease on the Kirbyville–Tenaha section
(160 km) in September 2002. A further major
extension to the railroad was completed in July
2004, when TIBR leased the 196 km Silsbee to
Somerville line, again from BNSF. Principal traffic is
forest products and minerals, amounting to some
33,000 carloads a year.

NEW ENTRY

Toledo, Peoria & Western Railroad

1990 East Washington Street, East Peoria, Illinois
61611-2961
Tel: (+1 309) 698 26 00   Fax: (+1 309) 698 26 79
Web: http://www.railamerica.com

Key personnel
General Manager: Alan Satunas

Gauge: 1,435 mm
Route length: 594 km

Organisation
Operated by RailAmerica since 1999, TPW mainly
consists of an east-west line from Logansport,
Indiana, to Peoria, Illinois. The railway handles
approximately 65,600 carloads annually,
commodities carried including automotive
components, chemicals, coal, fertiliser, food
products and steel. Interchange is made with
BNSF, CN, CSXT, UP and other railways at various
locations.

Transkentucky Transportation Railroad Inc

205 Winchester Street, Paris, Kentucky 40361
Tel: (+1 859) 987 15 89

Key personnel

President: C Randall Clark
Manager, Operations: Russell S Rogers

Organisation

The TTI moves trainloads of coal out of the eastern Kentucky and Cumberland Valley coalfields to a rail-to-barge transloading facility at Maysville, Kentucky. It is effectively a short cut for CSXT, which delivers 90–120 wagon trains to TTI at Paris; the latter are moved in two blocks of 60. Capacity is capped at the 6 million tonnes per annum that the transloader can handle.

Traction and rolling stock

At the beginning of 2003 there were 16 diesel locomotives.

Triangle Transit Authority (TTA)

PO Box 13787, Research Triangle Park, North Carolina 27709
Tel: (+1 919) 549 99 99   Fax: (+1 919) 485 74 41
Web: www.ridetta.org

Key personnel

General Manager: John Claflin
Project Manager, Systems: Thomas Janssen

Political background

TTA was established in 1989 to plan, finance, organise and operate a public transport system for the Research Triangle area of North Carolina, serving Durham, Orange and Wake Counties. Its three main programme areas cover regional bus services, a rideshare programme and regional transit planning. Vehicle registration and rental vehicle taxes have been imposed to finance the agency's activities and the first phase of its Regional Transit Plan.

Organisation

Phase 1 of TTA's Regional Rail service covers a 45 km route linking Durham, Research Triangle Park, Morrisville, Cary and Raleigh. The dedicated double-track line is being built on existing alignments owned by North Carolina Railroad and CSX. As originally approved, the line would serve 16 stations but this was cut back to 12 in 2004 in the light of increased construction costs and reduced revenue from taxes. Construction is scheduled to begin in 2006; services are scheduled to begin in late 2008.

Passenger operations

TTA plans foresee a 15 minute service frequency at peak times, reducing to 30 minutes at off-peak times and weekends. Future plans are to operate a 10 minute service frequency at peak times, 20 minute at off-peak and weekends. Services will operate 18 hours per day, seven days a week. Many stations will feature park-and-ride facilities and interchange will be made with other modes.

Traction and rolling stock

In October 2004 TTA selected United Transit Systems (UTS), a consortium of Sojitz Corporation and Rotem Company, to supply 12 two-car dmus, with an option on four additional vehicles. Bodyshells are to be manufactured by Rotem in South Korea, with final assembly taking place at Rotem's facility in Philadelphia to comply with the FTA's Buy America requirements.

NEW ENTRY

Trinity Railway Express

4801 Rock Island Road, Irving, Texas 75061
Tel: (+1 972) 399 89 78   Fax: (+1 972) 313 98 37
Web: www.trinityrailwayexpress.org

Key personnel

Vice-President, Commuter Rail: Kathryn Waters
Chief Operating Officer: Pete Sklannik, Jr

Passenger operations

A joint project of the Fort Worth Transportation Authority (the T) and Dallas Area Rapid Transit (DART), Trinity Railway Express (TRE) is a commuter rail service operating between the central business districts of Dallas and Fort Worth, Texas. The service operates six days a week along the former Rock Island Railroad, now named TRE and forming the 54.7 km TRE/DFW subdivision, with 11 station stops including Dallas and Fort Worth.

The last station that opened, in January 2002, was Fort Worth Intermodal Transportation Center (ITC), to complete the project. The ITC provides an interchange with Amtrak's long-distance services, bus services and, in Dallas, with buses and the city's light rail system. The CentrePort/DFW Airport station provides access to DFW International Airport via a free shuttle bus. TRE,

TRE F59PH-2 locomotives and bilevel coaches   0526355

DART and the T tickets are used on all DART and the T buses, TRE, DART light rail and Airport Shuttles to and from CentrePort/DFW station with a four-hour transfer.

Services are provided under contract by Herzog Transit Services, Inc and right-of-way maintenance is provided under contract by Herzog Contracting Corporation.

In FY2003 TRE carried 2.3 million passengers, compared with 2.14 million the previous year.

Traction and rolling stock

Services are provided by six General Motors F59PH-2 locomotives and 17 ex-GO Transit Bombardier-built bilevel coaches (10 coach cars and seven control cars) and 13 ex-VIA RDC diesel railcars.

UPDATED

Tri-Rail

South Florida Regional Transportation Authority (SFRTA)
800 NW 33rd Street, Suite 100, Pompano Beach, Florida 33064
Tel: (+1 954) 942 72 45   Fax: (+1 954) 788 78 78
Web: www.tri-rail.com; www.sfrta.fl.gov

Key personnel

Chair, SFRTA: Allen Harper
Vice-Chair, SFRTA: Jeff Koons
Executive Director: Joe Giulietti
Deputy Executive Director: Jack Stephens

Organisation

On 1 July 2003 the former Tri-County Commuter Rail Authority, which initiated the Tri-Rail service, became the South Florida Regional Transportation Authority (SFRTA), reflecting plans for expanded rail and bus services throughout the region. However, the commuter rail system continues to operate under the Tri-Rail brand name.

Passenger operations

The Tri-Rail commuter service, funded by the Florida Department of Transport, was introduced over 107 route-km of former CSXT tracks between Miami and West Palm Beach in 1989. An extension north from West Palm Beach to Mangonia Park has made the service 114 km overall, with 18 stations. The state of Florida financed the purchase of the Miami–West Palm Beach line for USD264 million. Amtrak and CSXT continue to share the line; CSXT is responsible for maintenance.

Contract management of Tri-Rail operations is by Herzog Transit Services Inc. Tri-Rail serves all three of the area's international airports which are at Miami, Fort Lauderdale/Hollywood and Palm Beach. Services operate at weekends as well as on weekdays, running at two-hourly and hourly intervals respectively. Transfers to local buses and the Miami metro system have remained free. Connection with the metro is made at Tri-Rail/Metrorail Transfer station.

Implementation of a smartcard fare collection system covering public transport agencies in the South Florida region is scheduled for implementation in late 2006.

Traffic (million)	*2002*	*2003*	*2004*
Passenger journeys	2.6	2.7	2.8

Tri-Rail F40PHL-2 locomotive with bilevel stock (Keith Henderson)   0567689

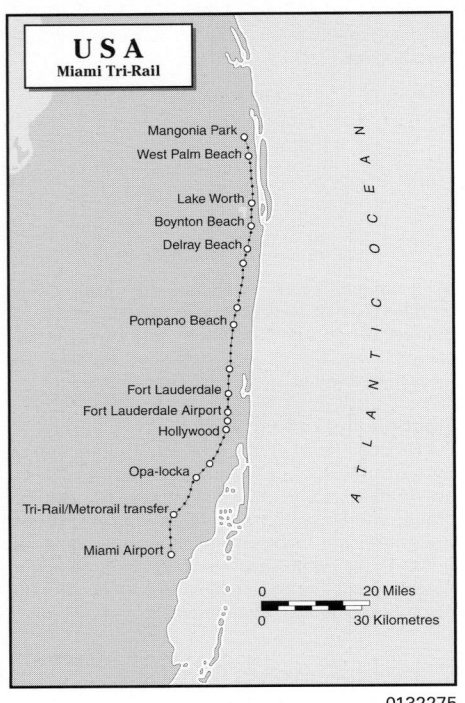

USA
Miami Tri-Rail

Mangonia Park
West Palm Beach
Lake Worth
Boynton Beach
Delray Beach
Pompano Beach
Fort Lauderdale
Fort Lauderdale Airport
Hollywood
Opa-locka
Tri-Rail/Metrorail transfer
Miami Airport

ATLANTIC OCEAN

0   20 Miles
0   30 Kilometres

0132275

Improvements to existing lines

The final stage of Tri-Rail's Double Track Corridor Improvement Program is in progress. Covering 71.3 km, the Segment 5 Project is due to be completed in early 2006. This will complete double-tracking throughout the route and enable Tri-Rail services to run at 20 minute intervals during peak periods. It will also reduce congestion and scheduling conflicts with Amtrak and CSXT traffic.

In 2004 SFRTA initiated public consultation on the Jupiter Corridor Alternatives analysis. This foresees possible extension of the Tri-Rail system some 26 km northwards from its present terminus of Mangonia Park to Jupiter, and possibly beyond into Martin County. The proposals include the construction of a fleet maintenance facility near Mangonia Park. In 2005 attention was focused on the Scripps Transit Feasibility Study, which was to examine the possibility of extending Tri-Rail service from Mangonia Park to a new terminus in the vicinity of Scripps, close to North Central Airport.

A proposal to extend Tri-Rail south from Miami International Airport was abandoned in 2004 in favour of a metro extension.

In 2003 land was acquired for a multimodal passenger interchange at Boca Raton. The first phase of the project, which also includes property developments, was completed in 2005. Other SFRTA property initiatives are planned at Mangonia Park and West Palm Beach stations,

while an administrative/operations facility is to be built at Cypress Park.

In 2005 plans were being developed to create a 'transit-oriented development' to serve the 79th Street Corridor.

Traction and rolling stock

In 2005 Tri-Rail's fleet consisted of 26 double-deck coaches (11 cab cars), five F40PHL-2 locomotives, three F40PHC-2C rebuilt locomotives from Morrison Knudsen, and two ex-Amtrak reconditioned F40s. Six EMD GP49 diesels were acquired from Norfolk Southern in late 2002 and refurbished to provide the extra capacity needed when double-tracking is complete. In addition, in 2004 Tri-Rail ordered a two-car double-deck dmu from Colorado Railcar and a single-deck demonstrator from the same builder following earlier trials with it. The procurement, jointly funded by SFRTA and Florida DoT, followed Tri-Rail trials with the demonstrator and is intended to examine how modern dmus might contribute to low-cost passenger operations. Manufacture of these vehicles is scheduled for 2005.

Signalling and telecommunications

A 900 MHz radio control system has been installed and in 1997 electronically coded rail signalling replaced a system dating from the Second World War.

UPDATED

Tuscola & Saginaw Bay Railway

PO Box 550, 308 West Main Street, Suite 303, Owosso, Michigan 48867-0550
Tel: (+1 989) 725 66 44   Fax: (+1 989) 723 82 26
Web: http://www.tsbrailway.com

Key personnel

Chairman and Chief Executive Officer:
  James E Shepherd

General Manager: James Schell
Vice-President, Operations: Raymond J Robinson

Organisation

TSBR has 690 route-km linking Ann Arbor and Petosky, on Lake Michigan. Traffic is primarily outbound grains, lumber, sand and aggregates. Connections are made to CSX and CN/IC.

Traction and rolling stock

TSBR operates 11 diesel locomotives (two Alco, eight EMD GP35, one GE) and five freight wagons.

Twin Cities & Western Railroad Co (TCW)

2925 12th Street East, Glencoe, Minnesota 55336
Tel: (+1 320) 864 72 00   Fax: (+1 320) 864 72 20
Web: www.tcwr.net

Key personnel

President and Chief Executive Officer:
  William F Drusch
Vice-President and General Manager:
  Mark Wegner

Vice-President, Marketing: Lloyd T Host
General Manager, Operations: Bob Suko
Manager, Transportation Services: Diane M McCall
Director, Mechanical and Maintenance:
  Tim K Jeske

Organisation

TCW has 331 km (226 km owned, 105 km of trackage rights). In 1991 the newly formed company purchased the Soo Line route from Hopkins, on the outskirts of Minneapolis-St Paul, to Appleton, Minnesota, and trackage rights for connection

with all railroads serving the Minneapolis-St Paul conurbation (the Twin Cities). Annual carloadings: 21,000; employees: 65.

Traction and rolling stock

TCW operates nine leased GP20Cs (Caterpillar power plants installed by Generation II locomotives), three CF7 slugs and 350 wagons, mostly hoppers.

UPDATED

Union Pacific Railroad Company

1400 Douglas Street, Omaha, Nebraska 68179-0605
Tel: (+1 402) 544 50 00
Fax: (+1 402) 544 55 72
Web: www.uprr.com

Key personnel

Chairman and Chief Executive Officer:
  Richard K Davidson
President and Chief Operating Officer:
  James R Young
Executive Vice-Presidents
  Finance and Chief Financial Officer:
    Robert M Knight
  Network Design and Integration:
    R Bradley King
  Marketing and Sales: John J Koraleski
  Operations: Dennis J Duffy
Senior Vice-Presidents
  Chief Information Officer: Lynden Tennison
  Law and General Counsel: J Michael Hemmer
  Human Resources: Barbara W Schaefer
  Corporate Relations: Robert W Turner
  Strategic Planning: Charles Eisele
  Director, Public Affairs: John Bromley

Gauge: 1,435 mm
Route length: 53,153 km

Organisation

Union Pacific Railroad Company (UPRR) is the principal business unit of the holding company, Union Pacific Corporation. The company's trucking operations, and Motor Cargo Industries, were sold in November 2003. Union Pacific is a publicly traded company listed on the New York Stock Exchange.

Union Pacific's aggressive acquisitions policy has made it the second mega-system in the US West after the merged Burlington Northern/Santa Fe entity (see entry in USA section of *Railway systems and operators*). It operates 53,000 km in 23 states, of which 43,500 km is owned and the remainder leased or operated under trackage rights. At the end of 2004 UP employed 48,000 people.

In 1997 UP and a consortium of partners were granted a 50-year concession to operate the Ferrocarril Pacifico Norte and Chihuahua–Pacific lines in Mexico, together with a 25 per cent stake in the Mexico City Terminal Company for a price of US$525 million. UP initially held a 13 per cent share in the consortium. An additional 13 per cent ownership was acquired in 1999.

The railroad moved to a new headquarters in Omaha in mid-2004, completed at a cost of US$260 million and bringing under one roof personnel formerly working at ten locations in Omaha and St Louis.

Operating Plan

Union Pacific's Operating Plan has a strong echo of the structure of the former Southern Pacific, acquired in 1996, in that it features discrete routes:
- the I-5, or West Coast, Corridor: California to Seattle;
- the Sunset Route: Los Angeles–Houston–New Orleans;
- the Overland Route: Midwest to Northern California;
- the Golden State Route: Midwest to Southern California;
- the Kansas–Pacific Route: Kansas City to Denver;
- Memphis–Texas–California route;
- Midwest–Arkansas–Texas route.

A redesign of the plan was being undertaken at the beginning of 2005, with implementation of the new Unified Plan scheduled for mid-year. The aim is to further simplify operations, raise average speeds and improve management techniques better to cope with protracted strong demand.

Finance

Union Pacific Corporation recorded a net profit of US$604 million in 2004, down 62 per cent on 2003. Rail operating revenue rose by 5.7 per cent to a record US$12.2 billion, while rail commodity income reached an unprecedented US$11.7 billion, some US$650 million higher than the previous year's figure, itself a record. The operating ratio (operating expenses as a percentage of income) rose to 89.4. Overall carloadings were up 2.4 per cent over 2003, to 9,458,000.

Operating difficulties associated with the year's record demand (see below) combined with substantially increased fuel costs created a challenging financial environment. Operating expenses were up 16 per cent on 2003's figure; higher fuel prices (up 35 per cent), wage and benefit inflation (up 7 per cent) and the ongoing cost of operating difficulties all impacted on expenditure. Operating inefficiencies and the additional costs associated with taking on nearly 5,000 extra personnel amounted to some US$300 million.

Passenger operations

UP runs commuter services between Riverside and Los Angeles under contract to the Southern California Regional Rail Authority (see entry in US section of *Railway systems and operators*).

As a consequence of the takeover of Chicago & Northwestern, UP is contract operator to Metra (see entry in USA section of *Railway systems and operators*) for three Chicago area lines.

Freight operations

UP has a wide traffic mix, with coal from the Powder River Basin and elsewhere, intermodal traffic from Californian ports, and grain from the prairie states being important commodities. Like other Class I railroads, UP experienced unprecedented demand for its services in 2004, driven by vigorous economic growth, strong export markets and higher trucking costs. For the first time in decades, demand for rail service exceeded capacity almost everywhere, and the railroad faced considerable operating difficulties trying to cope. Monthly records for seven-day carloadings were set every month during the year, peaking in November with an all-time weekly record of 194,000 carloads. Strongest growth was experienced in the industrial products, chemicals and intermodal commodity groups.

By business unit and measured in carloadings the year-to-year changes from 2003 to 2004 were: *intermodal* – up 5 per cent; *energy (coal)* – down 1 per cent; *automotive* – up 1 per cent; *chemicals* – up 5 per cent; *agricultural products* – unchanged; *industrial products* – up 2 per cent.

Coal is UP's prime commodity, accounting for 20 per cent of revenue in 2004. Carloadings from the Southern Powder River Basin continued to

GE ES44AC test locomotive at Sacramento, California (Philip Wormald)   **NEW**/1122804

grow, and shipments from mines in Colorado were up 9 per cent due to strong demand for the region's high-BTU coal. Total carloads in 2004 were down 1 per cent on 2003's record figures at 2,172 million.

Movement of general industrial products produced 19 per cent of revenue in 2002, rising by 3 per cent to US$2,035 billion. There was continued strong demand from the building and construction industries for lumber, steel and aggregates, much of it for house-building stimulated by low mortgage interest rates. Total carloadings rose 2 per cent to 1,515,000.

Revenue from chemicals was up 8 per cent, driven by a 5 per cent increase in carloads, while agricultural products revenue saw a 6 per cent rise, though carloadings were flat. In the automotive group, shipments for domestic manufacturers declined on account of lower production and weak demand, but greater import traffic and movement of materials and components helped produce a 1 per cent rise in carloadings and for a 2 per cent increase in revenue. Mexican traffic (included in the above figures) produced US$900 million in revenues in 2004, up 9 per cent, with increased shipments in all commodity groups. Growth in agricultural business revenue was driven by higher wheat exports and beer imports.

Intermodal operations

UP serves all major West Coast and all Gulf Coast ports. Most container traffic runs between West Coast ports and midwestern cities or New York,

but regular double-stack trains also run between Houston and New Orleans. The fastest Chicago–Los Angeles double-stack train schedule is now 50 hours, following the introduction of a premium service. Shippers in the northeast now have better access to Mexican destinations following introduction in 2002 of a new expedited service in partnership with CN and NS which cuts up to three days from previous timings. UP continued working with both US customs officials and Mexican railroads to improve cross-border transit times. The Automated Manifest System is now in operation at all six Mexican gateways, allowing customs officials to pre-clear shipments and so reduce transit times at all six Mexican gateways, is now supplemented by Passport Plus high-security screening.

Intermodal revenues rose 8 per cent in 2004; volume was up by 5 per cent due to an increase in imports from the Far East.

UP's West Coast trains are handled at the Intermodal Container Transfer Facility at Long Beach. SP has a long-term lease until 2036 on the ICTF, which covers 96 ha. An average of 75 trains a week serve the terminal, moving containers mostly for Pacific Rim shippers. The US$100 million facility, a joint venture between UP and the ports of Los Angeles and Long Beach, is 6.5 km from the ports. It has 11 km of track and can load and unload five double-stack trains simultaneously.

ICTF is now complemented by the US$1.8 billion Alameda Corridor project, completed in 2002. Designed jointly by the ports and the city authorities, the 32 km high-capacity rail freight corridor links the former Southern Pacific, Santa Fe and Union Pacific networks with container yards at the dockside, removing trucks hauling containers from city streets, and relieving traffic congestion by eliminating grade crossings on rail lines near the ports.

Improvements to existing lines

Track and infrastructure improvements accounted for US$1.3 billion out of a total capital expenditure of US$1.876 billion in 2004, and similar expenditure was planned for 2005. Some US$295 million is being spent on capacity improvements to the Sunset Route through Arizona, and in the area around North Platte, Nebraska. Some 1,700 km of new and reused rail was installed, along with 4.4 million sleepers.

Traction and rolling stock

At the beginning of 2005, the UP fleet totalled 7,682 freight locomotives, the average age of which was 14.7 years. Among the steps taken to improve operating performance during 2004 was an increase in the number of locomotives on order from 150 to 270, while short-term relief was provided by more than 300 additional leased units. For 2005 delivery, a further 315 locomotives were ordered in September 2004.

UP EMD SD70M and GE C41-8W locomotives head an autorack (car carrier) train at Davis, California (Philip Wormald)   **NEW**/1122803

UP has tested several low-emission technologies, and in April 2005 took delivery of the first diesel-hybrid unit.

In 1999 UP announced a major deal with General Motors to lease 1,000 SD70 4,300 hp (3,285 kW) diesel locomotives. With deliveries completed during 2003, the new locomotives have enabled some 1,500 older machines to be withdrawn, so reducing the number of types operated from 33 to 18. A further SD70M units were ordered from GM-EMD at the beginning of 2002.

UP continues to develop 'distributed power', in which radio-controlled unmanned helper sets are put in mid-train or at the end of long consists, with the result that undesirable 'slack action' is reduced by 50 per cent or better and track capacity can be increased.

In 2002 UP began using remote-controlled locomotives for shunting and in marshalling yards.

Freight wagons

UP's assets at the start of 2005 included over 104,000 revenue-earning freight wagons, either owned or leased, and almost 7,000 service vehicles. During 2005 UP was expecting to add around 4,000 freight cars to the fleet on long or short term leases. At the beginning of 2005 the average age of the wagon fleet was 23 years.

Signalling and telecommunications
Harriman Dispatching Center

Union Switch & Signal has consolidated control of all UP trackage into the Harriman Dispatching Center in Omaha. The most striking feature of the office is its panoramic, video-projected display of the UP system. While this wall display provides summary information, a visual display unit at each dispatcher's console offers a detailed view of operations at any of the 1,800 or so signal control points throughout the system.

The Harriman Center employs Computer-Aided Dispatching System (CADS). Using auto-routeing, dispatchers assign an identity and priority to each train. The computer then takes over and routes trains according to priority, while also automatically determining the meeting and passing of trains on single-track sections.

Transportation Control System

UP's marketing strategy rests on one of the most advanced technical bases in the industry. The heart of UP's service operations, the National Customer Service Center, would be impossible without the Transportation Control System (TCS) and the Automatic Call Directing (ACD) system. TCS schedules and monitors rail operations and performs the accounting function on every item

shipped on the railroad. Web-based systems are now in everyday use, including the UP-designed Bulk Train Planner computer programme to manage unit train operations.

A work order reporting system connects TCS computers directly with UP locomotives, enabling customers to have their data communicated to trains en route. This system was developed in conjunction with ATCS (Advanced Train Control System) and is expected to improve customer service and cut expenses by more than US$20 million a year.

Track
Rail: 133 AREA, 60.3 kg/m
Crossties (sleepers): Wooden, 7 × 9 in × 9 ft
Spacing: 2,019/km
Fastenings: ⅝ × 6¼ in cut track spikes; Portec curve blocks for curvative in excess of 6°
Max curvature: 20°; exceptionally 12°
Max gradient: Main line 2.33%, 4.0% branch lines
Max axleload: 65,750 lb (unrestricted operation), 78,750 lb (restricted)

UPDATED

Union Railroad Co

600 Grant Street, Suite 1887, Pennsylvania 15219-2800
Tel: (+1 412) 433 46 44   Fax: (+1 412) 433 46 45
Web: www.tstarinc.com

Key personnel
President: John C Pranaitis
Vice-Presidents
  Law: Robert N Gentile
  Finance: John A Yokim
General Manager: James L Neis

Chief Engineer and Superintendent, Mechanical:
  R L Janus
Assistant Manager, Signals and Communications:
  Michael D Willby

Organisation
URR operates over a 105 km main line serving steelworks and associated industries in the Monongahela Valley, plus some 320 km of yard track and private sidings. It interchanges with five other railroads: B&LE, CSXT, NS, W&LE and B&P. URR also operates Duquesne Wharf on the Monongahela River just south of Pittsburgh, where

coal and other bulk commodities are transferred from river barges for local distribution by rail. With docking for up to 98 vessels, Duquesne Wharf has capacity for 8.5 million tonnes annually.

It is a Transtar (see entry in United States of America section of *Railway systems and operators*) property and became the first railroad in the US to be certified under ISO 9002 quality standards.

Traction and rolling stock
URR has 33 diesel locomotives and 603 wagons.

UPDATED

Upper Merion and Plymouth Railroad Co

PO Box 404, Conshohocken, Pennsylvania 19428
Tel: (+1 610) 828 75 36; (+1 610) 383 22 37
Fax: (+1 610) 828 67 90; (+1 610) 383 24 40

Key personnel
President: Gary R Shields
Controller: John W Jankowski

UMPR, a Bethlehem Steel railroad, has 17.7 route-km; it operates three GM-EMD locomotives (one SW9, SW1 and NW2) and has 3,246 freight

wagons. It handles between 6,000 and 9,000 loads a year, connecting to NS at Consohocken.

Another Bethlehem subsidiary, the Brandywine Valley Railroad, handles 100,000 loads and empties annually, plus provides 30,000 revenue switches to online customers, connecting with both NS and CSXT.

US Steel Company/Blackstone Partners/Transtar Inc

Web: http://www.tstarinc.com

The one-time US Steel Company rail properties are now jointly owned by US Steel and Blackstone

Partners, a venture capital firm. Each road is operated by an on-site team with separately registered rolling stock and managed by Transtar Inc, which is located on the Union Railroad. The properties are:
  Birmingham Southern Railroad
  Elgin, Joliet & Eastern Railroad

Fairfield Southern Railroad
Lake Terminal, Ohio, Railroad
McKeesport Connecting Railroad
Union Railroad

Utah Railway Company (URC)

Genesee & Wyoming Inc, Utah Region
4692 North 300 West, Suite 220, Provo, Utah 84604
Tel: (+1 801) 221 74 60   Fax: (+1 801) 221 74 62
Web: www.gwrr.com

Key personnel
President and General Manager: James N Davis
Vice-President, Finance and Administration:
  Scott Tucker
Superintendent: Tim Ercanbrack
Chief Engineer: William Callor
Maintenance of Way/Mechanical Supervisor:
  Scott Cox
Director, Sales and Marketing: Barry Olsen

Gauge: 1,435 mm
Route length: (owned) 72 km; (trackage rights) 605 km

Organisation
Acquired in August 2002 from Mueller Industries Inc by Genesee & Wyoming Inc, the Utah Railway Company (URC) operates a 677 km network that includes trackage agreements with UP between Provo, Utah, and Grand Junction, Colorado, and with BNSF between Provo and Ogden, Utah. The company also serves industrial customers in the Salt Lake City area through trackage rights with the Utah Transit Authority. The principal commodity carried is coal, especially low-sulphur material to power stations in Utah and Nevada, with 60,000

carloads handled annually. Switching services are also provided by URC for BNSF in the Salt Lake City, Provo and Ogden areas. URC also owns the Salt Lake City Southern Railroad (40 km), serving around 30 industries between Salt Lake City and Draper.

Traction and rolling stock
UTAH operates a fleet of 23 diesel locomotives, mostly General Motors SD40s.

UPDATED

Utah Transit Authority

PO Box 30810, Salt Lake City, Utah 84130-0810
Tel: (+1 801) 262 56 26   Fax: (+1 801) 287 45 40
Web: www.rideuta.com

Key personnel

General Manager: John Inglish
Director of Rail Services: Paul O'Brien
Commuter Rail Project Manager: Steve Meyer

Organisation

In September 2002 UTA purchased 282 km of rights-of-way from Union Pacific, stretching southwards from Brigham City through Ogden and Salt Lake City, to Provo and Payson, with the intention of preserving the alignment for future use as commuter or light rail routes. UTA's first commuter route will follow part of this alignment between Pleasant View, north of Ogden, and Salt Lake City (71 km), with nine stations. In mid-2004 the project entered the pre-construction phase, having been allocated USD9 million in preliminary funding, and in mid-2005, preparatory work such as relocation of utilities and procurement of materials, was initiated. Service is scheduled to start in 2008 with diesel locomotives hauling bilevel coaches. The cost of the project is estimated at USD582 million.

Traction and rolling stock

In 2004 UTA acquired 30 gallery cars at minimal cost from Metra, Chicago. These were to be refurbished and made ADA-compliant. Further rolling stock acquisitions were planned.

UPDATED

Vandalia Railroad Co

609 West Main Street, Vandalia, Illinois 62471
A subsidiary of Pioneer Railcorp
1318 S Johansen Road, Peoria, Illinois 61607
Tel: (+1 309) 697 14 00   Fax: (+1 309) 697 53 87

Key personnel

See entry for Pioneer Railcorp.

Gauge: 1,435 mm
Route length: 4 km

Organisation

Acquired by Pioneer Railcorp in 1994, VRRC serves industries at Vandalia, Illinois, carrying steel, plastics, wheat and fertilisers. It connects with CSXT.

Ventura County Railroad

333 Ponoma, Port Hueneme, California 93041
Tel: (+1 805) 488 65 46
Fax: (+1 805) 488 65 17
Web: http://www.railamerica.com

Key personnel

General Manager: Doug Verity
Business Development Manager: Mike Ortega

Gauge: 1,435 mm
Route length: 21 km

Organisation

Operated by RailAmerica, VCRR is a network of freight-carrying shortlines in and around Port Hueneme, California. The railway handles approximately 3,100 carloads annually. Interchange is made with UP at Oxnard.

Vermont Rail System

One Railway Lane, Burlington, Vermont 05401
Tel: (+1 802) 658 25 50   Fax: (+1 802) 658 25 53
Web: www.vermontrailway.com

Key personnel

President: David W Wulfson
Executive Vice-President and Treasurer:
   Lisa W Wulfson
Vice-Presidents: Jerome M Hebda

Marketing: Ed Fitzgerald
Intermodal: Eric Moffett

Organisation

The Vermont Rail System comprises five short line railroads totalling 370 km:
   Vermont Railway Inc
   Green Mountain Railroad Corporation
   Clarendon & Pittsford Railroad Company
   Washington County Railroad Company
   New York & Ogdensburg Railway.

Connections are made with other carriers at: Whitehall, New York (Canadian Pacific); Burlington and Bellows Falls, Vermont (New England Central); Bellows Falls (Springfield Terminal Railroad); Palmer, Massachusetts (CSX Transportation, via NECR); and Newport, Vermont (Montreal, Maine & Atlantic Railroad). Tourist trains are also operated.

The railroad operates 14 locomotives and carries around 23,000 carloads annually.

UPDATED

Virginia Railway Express (VRE)

1500 King Street, Suite 202, Alexandria, Virginia 22314
Tel: (+1 703) 684 10 01
Fax: (+1 703) 684 13 13
e-mail: gotrains@vre.org
Web: www.vre.org

Key personnel

Chief Executive Officer: Dale Zehner
Deputy Chief Executive Officer: Jennifer Straub
Director, Finance: Donna Boxer
Director, Construction and Facilities:
   Sirel Mouchantaf
Superintendent, Operations, Safety and Security:
   David A Snyder
Manager, Operations Support: Dennis E Larson
Manager of Customer Services: Wendy Lemieux
Manager, Public Affairs: Mark Roeber
Manager of Marketing: Ann King

Organisation

Virginia Railway Express (VRE), a partnership of the Northern Virginia Transportation Commission (NVTC) and the Potomac and Rappahannock Transportation Commissions, began commuter operations in 1992 over two lines into Washington Union station, from Fredericksburg (87 km, owned by CSXT) and Manassas (53 km, owned by NS), serving 18 stations in northern and central Virginia and terminating at Union Station in Washington, DC. Operating aspects of VRE are overseen by the VRE Operations Board, consisting of seven commissioners, three each from NVTC and PRTC and the Director of the Virginia Department of Rail and Public Transportation.

Passenger operations

In 2004 VRE was operating 32 trains per day. Passenger journeys totalled 3.5 million in FY2004, up from 3.3 million in FY2003 and continuing a trend that saw traffic grow by 16 per cent annually for the preceding four years – an average that is about four times the national figure.

Farebox recovery stands at about 66 per cent. This relatively high ratio enables VRE to provide services such as free parking; a guaranteed ride home outside train times; a 'Security Blanket' daycare programme to cover extra costs when trains are late; and free ride certificates for late-running services.

Improvements to existing lines

The VRE Operations Board has awarded a USD28 million contract to Abernathy Construction Corp to

Virginia Railway Express FP59 locomotive with Bombardier bilevel stock   1036912

build a new double-track bridge across Quantico Creek, eliminating a single-track bottleneck on the Fredericksburg line. The project is part of a USD66million capital improvement programme on the CSX-owned tracks used by VRE that will permit operation of more trains.

In 2005 VRE was also implementing parking improvements to add capacity to the system and extending platforms to accommodate longer trains.

Traction and rolling stock
VRE motive power consists of 21 diesel locomotives (10 GP39-2C, five GP40P-2C, four F40 and two FP59) and 119 coaches (13 Kawasaki bilevel, 50 Gallery bilevel, 38 Mafersa single-level, 18 leased Bombardier bilevel vehicles). In May 2005 VRE awarded a contract to Sumitomo Corporation of America in conjunction with its partner, Nippon Sharyo Ltd, to supply 11 bilevel cars, with an option on 50 more. Delivery is scheduled for 2006–07.

In addition to the expansion work detailed above, VRE will need more coaches and locomotives to expand service on its existing lines and for a proposed expansion to Gainesville-Haymarket or Spotsylvania and Fauquier counties. The agency has applied for federal and state funding to purchase 61 double-deck cars and 20 locomotives, as well as funding for increased car parking provision and to lengthen station platforms to handle longer trains – the current maximum is six cars. Current growth patterns indicate VRE could be carrying up to 20,000 weekday riders by 2010.

UPDATED

Virginia Southern Railroad

PO Box 12, Keysville, Virginia 23947
Tel: (+1 804) 736 88 62   Fax: (+1 804) 736 99 68
Web: http://www.railamerica.com

Key personnel
General Manager: Brad Ovitt
Business Development Manager: Kim Greer

Gauge: 1,435 mm
Route length: 121 km

Organisation
Formerly owned by RailTex and since 2000 operated by RailAmerica, VSRR runs south from Burkeville, Virginia, to Oxford, North Carolina. The railway handles approximately 6,800 carloads annually, commodities carried including coal, food products and forest products. Interchange is made with NS at Burkeville and at Oxford.

Washington State Department of Transportation

Transportation Building, PO Box 47300, Olympia, Washington 98504-7300
Tel: (+1 360) 705 79 01   Fax: (+1 360) 705 68 21
Web: www.wsdot.wa.gov

Key personnel
Secretary of Transportation: Douglas B MacDonald
Rail Programme Manager: Ken Uznanski

Passenger operations
Amtrak 'Cascades' services on the Pacific Northwest Rail Corridor are operated by Amtrak with support from the states of Washington and Oregon. Services cover 746 km between Vancouver, British Columbia and Eugene, Oregon. Patronage has grown each year for a decade, reaching 603,000 in 2004, up from 590,000 in 2003.

In 1999 new equipment was introduced on Amtrak Cascades services, replacing leased rolling stock. Two dedicated Talgo passive, tilting trainsets were acquired by the state of Washington and two by Amtrak. The 12-car sets were designed by Patentes Talgo and assembled in Seattle by Pacifica Inc, a public-private partnership of aerospace and marine construction unions. An additional trainset was manufactured by Talgo in a bid to generate a market for its design; for the four years up to 2003 this set has been leased for use on Amtrak Cascades services, and was purchased outright by WSDoT in November 2003. The end-cars of the Cascades vehicles incorporate fibre-glass streamlined fins intended to harmonise their appearance with that of the Amtrak F59PHI locomotives that power the trains.

A new daily service between Seattle and Bellingham, Washington, began in 1999, complementing the existing Seattle–Vancouver service. It is intended to extend the Bellingham service to Vancouver when resources to complete significant track work in Canada are secured. The necessary USD14 million has to be provided by

Amtrak Cascades service near Bellingham Wash (Steven J Brown)   *NEW*/1146776

either the British Columbia provincial government, the Canadian federal government or local transportation agencies.

In 2000 a second daily Seattle–Portland round trip was extended south to Eugene, Oregon, a segment sponsored by the state of Oregon.

A station stop at Tukwila, Washington, was added in 2001, a new station at Everett was completed in 2002, and reconstruction of the historic King Street terminal in Seattle started in 2003. In 2004 another stop was added at Oregon City, Oregon, and a new station also opened at Mt Vernon, Washington.

In 2003 the Washington State Legislature appropriated funds for a significant number of capital projects to be completed during following years, including corridor improvement schemes such as installation of high-speed crossovers, easing curvature and station improvements. Completion of these projects will provide opportunities for reduced Amtrak Cascades journey times and increased service frequencies.

UPDATED

West Michigan Railroad Co

15 Industrial Drive, Paw Paw, Michigan 47079

A subsidiary of Pioneer Railcorp
1318 S Johansen Road, Peoria, Illinois 61607
Tel: (+1 309) 697 14 00
Fax: (+1 309) 697 53 87
Web: http://www.pioneer-railcorp.com

Key personnel
See entry for Pioneer Railcorp.

Gauge: 1,435 mm
Route length: 24 km

Organisation
Acquired by Pioneer Railcorp as the Kalamazoo, Lakeshore & Chicago Railroad in 1995, WMI runs from Harford to Paw Paw, Michigan. Interchange is made with CSXT at Paw Paw. Main commodities carried are food products and beverages.

West Texas & Lubbock Railroad

821 West Broadway Street, Brownfield, Texas
79315
Tel: (+1 806) 637 83 23   Fax: (+1 806) 637 80 74
Web: http://www.iowapacific.com

Key personnel
President: Edwin Ellis
General Manager: Curtis Goodin

Wheeling & Lake Erie Railway

100 E First Street, Brewster, Ohio 44613
Tel: (+1 330) 767 34 01
Fax: (+1 330) 767 70 25
Web: http://www.wlerwy.com

Key personnel
Chief Executive Officer: Larry R Parsons
President and Chief Operations Officer: Steve Wait
Vice-Presidents
    Chief Financial Officer: Michael A Mokodean
    Law: William A Callison
    Transportation: James I Northcraft
    Marketing, Sales, Stone, Timber, Paper:
        Larry D Wood
    Marketing, Sales, Coals, Chemicals, Petroleum:
        Richard Elston
    Engineering: Kasey O'Connor

Wichita, Tillman & Jackson Railway Co, Inc

4420 West Vickery Boulevard, Suite 110, Fort Worth,
Texas 76107
Tel: (+1 817) 737 72 88   Fax: (+1 817) 732 26 10

Key personnel
Chairman, President and Chief Executive
    Officer of Rio Grande Pacific Corporation:
        Richard D Bertel
Supervisor: Edmond Yelle

Winchester & Western Railroad Co

Virginia Division
PO Box 264, 126 East Piccadilly Street, Winchester,
Virginia 22601
Tel: (+1 540) 662 26 00   Fax: (+1 540) 667 36 92
e-mail: wwrail@visuallink.com (Virginia Division
    only)

Key personnel
President: W P Light
Agent: Sarah D Vessella

Wisconsin & Southern Railroad Co

PO Box 9229, 5300 N 33rd Street, Milwaukee,
Wisconsin 53209-0229
Tel: (+1 414) 438 88 20   Fax: (+1 414) 438 88 26
Web: http://www.wsorrailroad.com

Key personnel
President and Chief Executive Officer:
    William E Gardner
Secretary and Treasurer: Lucy Stone-Gardner
Vice-President, Financial: T J Karp
Vice-President, Sales and Marketing: J V Lombard
Superintendent of Transportation: H M McConville
Superindendent, Maintenance of Way:
    B M Meighan

Gauge: 1,435 mm
Route length: 167 km

Organisation
Operated by RailAmerica from 1995 until 2002,
WTLR runs from Lubbock, Texas, southwest to
Seagraves and west to Whiteface. In May 2002,
Permian Basin Railways Inc, a wholly owned
subsidiary of Iowa Pacific Holdings, signed a
long-term operating lease for the line. The railway

Organisation
A wholly owned subsidiary of the Wheeling
Corporation, the WLER was formed in 1990 from
the sale by Norfolk Southern of lines that once
formed the original Wheeling & Lake Erie and the
Akron Canton & Youngstown and Pittsburgh &
West Virginia Railroads. It operates over
1,100 km. Later, on the break-up of Conrail, WLER
gained track access rights to Toledo, Ohio, and over
CSXT from Lima to Carey, together adding some
180 km to the network. The railway carries about
130,000 carloads annually, with gross revenue of
US$50 million.
    WLER also operates the 30 km Akron Barberton
Cluster Railway, which provides switching services
to industries in Akron, Ohio.
    During 1994, the railway restructured financially,
and received a US$2.4 million investment from the
state of Ohio; in 1996 the Stark County Development

Organisation
The WT&J operates over 163.5 route-km. The
railroad consists of two segments, one from the UP
at Wichita Falls, Texas (with connections to BNSF)
running northwest to Altus, Oklahoma, and the
other connecting Waurika, Oklahoma, on the Union
Pacific with Walters, Oklahoma. Track in Texas is
leased from the UP; in Oklahoma it is leased from
the state. Annual carloadings are 7,700; wheat and
sand are the main commodities.
    WT&J is part of a holding company, Rio Grande
Pacific Corporation, that also has the Nebraska

New Jersey Division
PO Box 1024, Burlington Road, Bridgeton,
New Jersey 08302
Tel: (+1 856) 451 64 00   Fax: (+1 856) 451 70 16

Key personnel
Trainmaster: M T Luczkiewicz
General Agent: F A Winkler

Organisation
W&WR has 164 route-km of unsignalled trackage.
The New Jersey Division links Vineland with
Dorchester and Mauricetown. The Virginia Division

Organisation
The company has 1,030 route-km, in three line
clusters, including 418 km acquired with the
purchase of the Wisconsin & Calumet Railroad
from Chicago West Pullman Transportation
Corporation in 1992. Wisconsin & Southern
(WSOR) connects with CN (Wisconsin Central),
Union Pacific (formerly CNW) and Canadian Pacific
at several locations. In addition, the railroad enjoys
trackage rights over Metra between Fox Lake and
Cragin and thence over Belt Railway of Chicago to
reach BRC's Clearing Yard to connect with eastern
and southern railroads.
    In November 2003 WSR completed purchase
from Canadian Pacific of the 52 km Waterloo Spur
linking Madison and Watertown, Wisconson, which
it had leased since 1998.

handles approximately 4,000 carloads annually,
commodities carried including cotton and cotton
products, sodium sulphate, chemicals, fertiliser,
scrap metal and steel. Interchange is made with
BNSF at Lubbock and with UP at Amarillo, Texas.

Board used a US$10 million state loan to build
Neomodal, a North East Ohio intermodal facility at
Masillon, which is targeted at a large market within
a 180 km radius focusing on consumer perishables
inbound and manufactured goods outbound. WLER
interchanges with CN, CSX and NS. In 1995, after an
interval of 10 years, the railway reactivated the former
Norfolk & Western ore dock at Huron, Ohio, for the
purpose of moving taconite to a Wheeling-Pittsburgh
Steel Company mill at Steubenville, Ohio.

Traction and rolling stock
WLER has 44 locomotives (seven SD45, 14 SD40
and 23 GP35) and some 1,200 freight wagons.
The company's Brewster workshops offer repair,
refurbishment and fabrication services for other
railroads, with covered space for 50 wagons.

Central and the Idaho, Northern and Pacific, each of
which is locally managed.

Traction and rolling stock
WT&J has six GM-EMD GP7 locomotives.

links Gore and Winchester with Hagerstown in
Maryland. Tonnage moved in 2001 was 1.0 million,
compared with 1.12 million in the previous year.

Traction and rolling stock
The New Jersey Division operates with seven
GM-EMD GP9 locomotives and one road slug. The
Virginia Division operates with six GP9s and one
road slug. Winchester & Western has 624 freight
wagons.

In 2001 52,000 carloads were recorded. The
railway carries a varied range of commodities,
including grain and grain products, coal, timber,
paper and paper products, fertiliser, chemicals and
aggregates.

Traction and rolling stock
WSOR has an all-GM fleet of 28 locomotives of
several models, and 786 freight wagons. The
company has closed its Northern Railcar workshop
in Cudahy, Wisconsin, that specialised in rebuilding
and maintaining privately owned passenger and
freight vehicles and in custom-painting. This
work is now carried out at its Horicon, Wisconsin,
shops.

Wyoming Colorado Railroad Inc

PO Box 1876, Ogden, Utah 84402
Tel: (+1 801) 621 53 11   Fax: (+1 801) 393 77 33

Key personnel
President: David L Durbano

Yadkin Valley Railroad Co

PO Box 1218, Rural Hall, North Carolina 27045
Tel: (+1 336) 969 6055   Fax: (+1 336) 969 91 68
Web: http://www.gorailways.net

Key personnel
President: H Peter Claussen
General Manager: Todd Burchette

Vice-President, Operations: Gregory L Kissel
General Manager: L Brown

Organisation
WCRI operates three lines serving the timber
industry in Wyoming and Oregon, extending to
75 km and carrying about 1,000 carloads annually.

Organisation
This property was first sold by Norfolk Southern in
1989 as an element of NS's 'Thoroughbred' short line
leasing initiative, then acquired in 1994 by Gulf &
Ohio Railways. The YVRR totals 160 route-km
based on Rural Hall, South Carolina, comprising
routes to Mount Airy and North Wilkesboro.
Annual carloadings are 15,000; corn, coal and

Connections are made with UP at Ontario, Oregon,
and Laramie and Wolcott Junction, Wyoming. It has
seven GM-EMD locomotives.

soyabean meal are the three primary commodities.
Interchange is with NS.

Traction and rolling stock
Five GM-EMD GP9 and four GM-EMD GP10
rebuilds.

Uruguay

Ministry of Transport & Public Works

Rincón 561, 8to piso, Montevideo
Tel: (+ 598 2) 96 05 09; 95 70 13
Fax: (+598 2) 96 28 83

Key personnel
Minister: L Cáceres Behrens

State Railways Administration (AFE)

Administración de Ferrocarriles del Estado
PO Box 419, Calle La Paz 1095, Montevideo
Tel: (+598 2) 94 08 05
Fax: (+598 2) 94 08 47

Key personnel
President: V Vaillant
Vice-President: Dr F Caride B
Directors: O Lopez Balestra, J C Hernandez,
   E Silveira
Secretary General: V Varela
General Manager: M Anastasia
Deputy General Manager: H Chapuis
Directors
   Legal Affairs: Dr R Jimenez
   Audit: E Garcia
   Special Projects: A Santos
Managers
   Finance and Accounting: G Leva
   Human Resources: J Ceriani
   Operations: H Riccardi
   Rolling Stock: J Lopez Baggi
   Infrastructure: G Tettamanti
   Communications: A Lujambio

Gauge: 1,435 mm
Route length: 2,073 km

Political background

In October 1991, legislation was passed which
effectively reduced the role of AFE to that of
track authority with responsibility for maintaining
track and co-ordinating operations. It was to be
financed via taxation and by charging fees to
operators. The rump of AFE has been permitted
to compete with private operators in providing
services, although the future of the country's
railways will depend on private companies owning
and running their own motive power and rolling
stock.
    Of the current network, some 461 km are closed
and a further 460 km only in partial operation.
Track quality has been seriously prejudiced by the
financial cuts of 1988 when substantial numbers
of permanent way staff were shed. Total AFE staff
numbers fell from 10,000 in 1987 to 2,100 at the
start of 1994.

Finance

By 1997 annual revenue stood at around US$3
million. Recent policy has emphasised tighter cost
control, resulting in a reduction of US$1 million in
overall operating costs. This will enable subsidy to
be cut by US$45 million.
    AFE's projected investment strategy for 1994–97
had US$30 million earmarked for track repairs
and US$11.8 million to be spent on rolling stock.
The remainder was reserved for the purchase
of intermodal equipment. During 1996, out of
a total investment budget of US$6.53 million,
US$4.16 million was to be spent on major
track improvements, US$1.55 million on freight
wagon refurbishment and US$0.35 million on
communications.

Passenger operations

In 1988 passenger services were withdrawn as an
economy measure. In March 1990 a new political
administration set out to restore passenger
services by encouraging AFE to undertake joint
ventures with the private sector. Between 1990
and 1992 several 'charter' trains were run and in
January 1993 regular passenger services were
restored on the 118 km Tacuarembó-Rivera line in
the north of the country.
    Suburban services resumed between
Montevideo and 25 de Agosto on 25 August
1993. By November of that year, the four trains
running in each direction were carrying over
1,200 passengers a day on the 64 km route. In
1997 476,900 passengers were carried. These two
restored passenger services were managed by the
private company Luxtol SA, which was responsible
for ticket sales, onboard service and the cleaning
of trains and stations. AFE received a fixed amount
per train-km which covered the hire of train crews
and rolling stock.
    Expansion and extension of the existing 25 de
Agosto services to San José de Mayo and Florida
was proposed. However, the return of the anti-rail
political party in 1996 had a dramatic effect. The
Ferrotransporte company managed by Luxtol had its
August 1993 concession withdrawn. AFE itself took
over the running of all trains in January 1997 and
suspended indefinitely the deep rural Tacuarembó-
Rivera train due to its lack of profitability. Although
upgrading work on the 118 km route had resulted
in a 2 hours 40 minutes end-to-end timing in
1993, by 1996 this had lengthened to 3 hours
45 minutes, despite faster railcars having replaced
diesel locomotives and coaches. Intense political
lobbying resulted in the Tacuarembó-Rivera train
being reinstated on 3 March 1997. Furthermore, the
timetabled four trains a day on the Montevideo-25
de Agosto commuter service were increased to five
following the reinstatement of the Peñarol-Sayago
feeder service.
    A further suspension of the Tacuarembó-Rivera
service occurred on 31 December 1998 but trains
resumed running in April 1999.

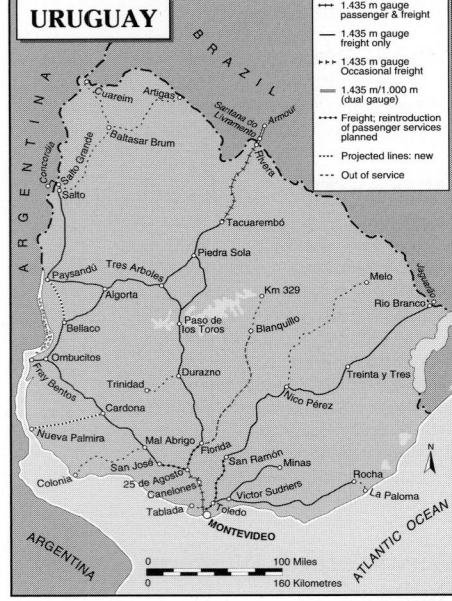

0023116

Freight operations

Around 1.5 million tonnes of freight were carried in
1997, compared with an average of 1 million tonnes
in the preceding 10 years.
    In early 1998, the Compañía Uruguya de Cemento
Portland acquired 28 Czech-built second-hand
hopper wagons to convey clinker over the 118 km
Verdum–Montevideo route using AFE traction.
    AFE and the Ministry of Transport and Public
Works have invited expressions of interest for
a concession to operate timber trains between
Montevideo and both Rivera and Blanquillo. The
railway expects its timber traffic to increase and
is investing in track improvements accordingly:
its 10 new GE locomotives were purchased with
this timber traffic in mind, and were expected to
enable AFE to increase train lengths by 50 per cent
and modify the pattern of its freight operations.
New sidings to serve industrial customers, silos,
containers and handling equipment for containers
and pallets are also the subject of investment.

Intermodal operations

AFE's investment strategy for 1994–97 made
provision for the purchase of two container cranes
(US$0.56 million) and 150 bulk containers (US$0.9
million). By August 1994, 44 containers had been
purchased for US$350,000.

New lines

Recent planning objectives have included
development of the Littoral line and extension of
the Central line northwards to a point known as

Km 441 from its present terminus at Km 329. In conjunction with the latter project, US$3 million was spent on a new rail bridge across the Río Negro, but the scheme was later abandoned and the bridge converted for use by road traffic.

AFE has been reported as studying the possibility of constructing a new branch from the Montevideo–Fray Bentos line at Cardona to the recently expanded port of Nueva Palmira. The branch would be 70 km long.

Improvements to existing lines
In 1997 AFE embarked on a two-and-a-half year track renewal programme. The use of 50 kg/m rail and new sleepers has cut derailments by 40 per cent. However, line speeds of 20–60 km/h are still prevalent. Funding for the project is achieved through productivity improvements and cost savings. By mid-1997, repairs were under way on 1,339 km of the 1,926 km operational network. A further 422 km were due for upgrading. The Uruguay army undertook repairs on the 250 km San José–Mercedes route, which had been closed since 1994, and work was reported to have switched to the 98 km Reboledo–Nico Pérez line. The World Bank has provided US$20 million to renovate

587 km to meet the needs of forestry products traffic; this has been matched by US$20 million from the government.

Electrification of the Montevideo–Progreso (26.4 km) section of AFE's route to 25 de Agosto was proposed as part of an LRT scheme under development in 1996. The upgraded infrastructure would be used by dual-mode LRVs similar to those adopted by the German cities of Karlsruhe and Saarbrücken.

In 1998 AFE called for tenders for a light rail line, to be known as 'Tren de la Costa', linking Montevideo and El Pinar. Ten consortia had already expressed an interest in developing the scheme.

Traction and rolling stock
At the start of 1993 the locomotive fleet comprised 24 Alco/GE and 18 Alsthom 825 hp (615 kW) diesel-electrics, seven diesel-hydraulics for shunting duties and four steam locomotives. In November of that year, AFE took delivery of 10 1,800 hp (1,340 kW) Type C18-7i diesel-electric locomotives built by GE Canada. The investment programme for 1994–97 made provision for the rebuilding of three shunting locomotives.

Passenger services are operated with refurbished equipment, namely six 200 hp (150 kW) Brill

60 railcars of 1936 vintage and 10 Fiat-Materfer coaches. The latter are hauled by Alsthom or newer GE locomotives. AFE has begun refurbishing four Ganz-Mavag diesel trainsets; the first was returned to service in September 1994. Its fleet of passenger equipment also comprises five ex-DB Uerdingen railbuses and 97 other passenger coaches (including 21 wooden-bodied vehicles and 10 Allan trailers used for charter trains), of which half have latterly been unserviceable. In early 1998, AFE was reported to be considering leasing five Class 593 dmus from RENFE in Spain in a package which would include maintenance.

At the start of 1993, AFE operated 2,413 freight wagons. In 1998 delivery was expected from Czech builders of 150 wagons for timber or container traffic and 150 low-sided wagons.

Track
Rail: 20 kg/m (529 km); 30–40 kg/m (1,361 km); 40–50 kg/m (969 km); 50 kg/m or heavier (132 km)
Sleepers: Steel and timber
Min curve radius: Generally 500 m; less than 500 m over 97 km

Líneas Férreas Uruguayas (LFU)

Guaviyú 2941, 11800 Montevideo
Tel: (+598 2) 203 67 42   Fax: (+598 2) 203 67 42
Swiss project office:
Fahrplancenter, Tellstrasse 45, CH-8400 Winterthur, Switzerland
Tel: (+41 52) 213 12 20   Fax: (+41 52) 213 12 20

Key personnel
Uruguay: Marcelo Benoit; Esteban Martínez; Alvaro Saavedra
Switzerland: Peter Lais; Samuel Rachdi

Political background
In 1988 the State Railways Administration (AFE) withdrew all rail passenger services throughout Uruguay. In 1993 services were reinstated on two short sections of line, Montevideo–25 de Agosto and Tacuarembó–Rivera, leaving all remaining towns and cities without access to passenger rail transport. Many of these towns are not served

by the main national roads network and some are inaccessible during adverse weather. Líneas Férreas Uruguayas (LFU), a Uruguayan-Swiss consortium, plans to reintroduce passenger services over as many lines as possible. In 1999 LFU had received network access approval from AFE for modest charges and was to operate without government subsidy. The consortium was seeking low-cost second-hand lightweight rolling stock. In Uruguay the consortium was finalising the most appropriate form that the operating company should take.

Passenger operations
In a first phase, the proposed fleet of railbuses would operate services on three types of route: lines with no parallel road; lines between major communities; and suburban lines around Montevideo and in the Salto/Paysandú area. Three daily long-distance services are planned: one from Rivera, in the north of the country, to Salto in the west and connecting at Tres Arboles with a service to Montevideo; one

serving the densely populated Montevideo-Paso de los Toros corridor, complementing the service from Tres Arboles, which takes the same route; and a third possible weekly service between Montevideo and Treinta y Tres. Proposed short-haul services include Salto–Paysandú, Rio Branco–Treinta y Tres, Treinta y Tres–José Pedro Varela, Florida–Sarandí del Yí and Montevideo–La Floresta. The consortium has also expressed an interest in taking over the Tacuarembó–Rivera service should AFE wish to withdraw from it.

LFU was aiming at a commencement of services by the end of 1999.

A second planned phase of service development by LFU foresees the possible procurement of 20 to 25 X4500 'Caravelle' railcars from France for deployment on long-distance services and to equip a reintroduction of services from Montevideo to Fray Bentos, Colonia, Melo, Rocha and Rivera and between Sarandí del Yí and Islas de la Paloma. In this phase LFU would also take over responsibility for track maintenance over lines no longer used by AFE.

Uzbekistan

Ministry of Transport

ul T Shewchenko 7, 700061 Tashkent
Fax: (+7 3712) 59 52 51

Key personnel
Head of Rail Transport: Normat Ermetov

Uzbekistan Railway (UTY)

Uzbekistan Temir Yullari
T Shevchenko 7, 700061 Tashkent
Tel: (+7 3712) 32 44 00
Fax: (+7 3712) 59 52 51

Key personnel
President: Ravshan Zakhidov

Gauge: 1,520 mm
Route length: 3,950 km
Electrification: 620 km at 25 kV AC

Organisation
This network comprises the greater part of the former Soviet Railways Central Asian Railway, centred on the artery from Chardzhou in Turkmenistan to Bukhara, Samarkand, Dzhizak and Tashkent. It became an independent entity from January 1992. The northwest and far-eastern sections could only be accessed through the neighbouring states of Turkmenistan and Tajikistan respectively which, given the political tensions in the whole region, complicated the railway's operations considerably.

Due mainly to clients' continued failure to pay their debts, the railways were in severe financial difficulties in 1999. The labour force remained excessive, and electric locomotives were reported, on average, to spend only seven per cent of their day actually hauling trains.

However, in March 2001 a presidential order converted UTY into an open share company, with the aim of introducing competition and encouraging investment.

The railway is divided into five regional administrations: Tashkent, Ferganar, Bukhara, Aral and Karshi.

The labour force (excluding workers in non-rail activity) was 45,223 in 1999, compared to 53,842 in 1997. Operating workers in 1999 totalled 29,406.

Passenger operations
The passenger business stood at 20.4 million journeys in 1993, for 5.9 billion passenger-km. By 1999, passenger journeys had declined to 13.3 million and passenger-km to 1.9 billion. A number of passenger trains were withdrawn in 1999. To cut costs further, remaining passenger services in the capital were concentrated on the Central station.

The extra costs associated with catering for this growth led to an increased financial loss for the railway, as a result of which fares were raised substantially in early 1995. Local authorities contributed towards the cost of running new high-quality trains between major cities. Some trains were leased to their staffs, and were said to be very presentable.

For details of the latest updates to *Jane's World Railways* online and to discover the additional information available exclusively to online subscribers please visit
jwr.janes.com

No new coaches had been purchased for the Central Asian Railway in the two years before the break-up of the USSR, and subsequently the lack of repair facilities has exacerbated the shortage of passenger stock. Steps were taken in 1998 to alleviate this situation (see 'Traction and rolling stock').

Freight operations

Freight traffic has slumped following the break-up of the Soviet Union, with freight tonnage down from 90.8 million in 1991 to 57 million in 1993. An expected recovery in 1994 failed to materialise, with tonnage falling to 40 million. However, in 1995 volume increased to 46 million tonnes following improved international transits. But by 1997 it was down to 41 million. In 1999, 41 million tonnes were again carried, but tonne-km were 13.9 billion, down from 16.5 billion in 1997. The railway carries nine-tenths of the total common-carrier freight of the republic. The dominant traffics are cotton and construction materials.

New lines

From early 2001 freight services to the northwest began to use the new 341 km cut-off route Uchkuduk-Misken-Nukus, thereby avoiding transit across Turkmenistan. The 115 km Misken-Sultanuizdag line was completed in 1999, linking the Aral region with central Uzbekistan. There is also a project to link the Karshi-Kitab and Termez-Denau lines in the south, again bypassing Turkmenistan.

Work on this 124 km Guzar-Kumkurgan line began in 1996.

In the east, a bypass is proposed from Angren to the Kokand-Namangan line, hitherto accessed from Khavast through Tajikistan. However, the very difficult terrain makes any early start on this project unlikely.

In 1997, construction began of the 50 km Shavat-Dzhurmurtau line.

A works for track products has been opened, producing some 400 categories of material formerly imported.

Improvements to existing lines

The modernisation, including electrification, of the main trunk route between Bukhara, Samarkand and Tashkent is continuing. In 1998 a loan worth US$100 million was made available by the Asian Development Bank to upgrade 320 km of this route from Chengel'dy (Kazakhstan) to Tashkent and Samarkand. Then, in late 2000, the Bank approved a loan to help modernise the 341 km Samarkand-Bukhara-Khodzha Davlet main line, with additional investment from UTY.

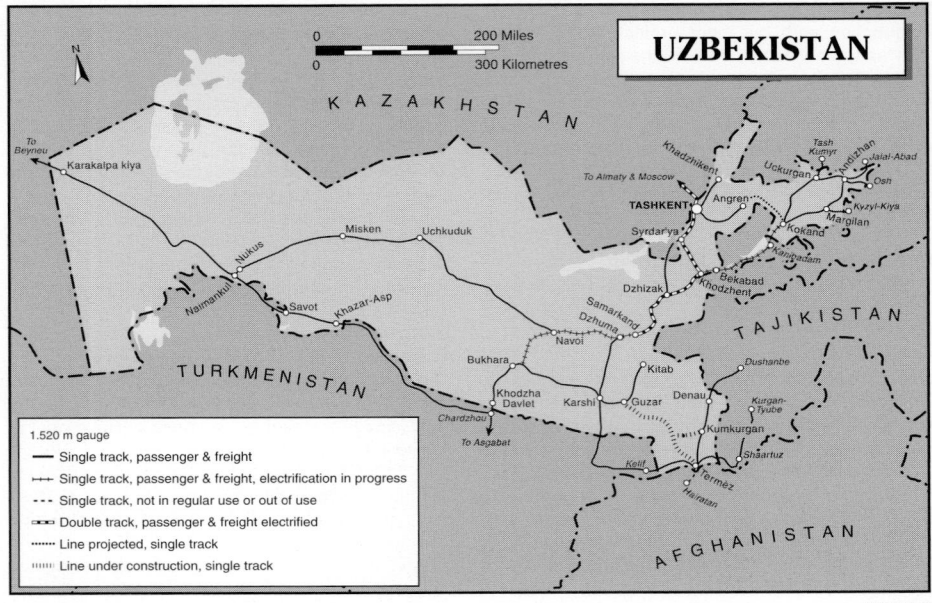

0525468

Traction and rolling stock

The current traction is as follows:
Diesel (main line) locomotives 2TE10/3TE10 variants: 1,082 units (480 ½ double and triple locomotives);
Diesel (shunting) locomotives TEM2/chME3: 313 locomotives (all single units);
Electric (main line) locomotives 3VL80/VL80, 2VL60/VL60: 173 units (84 double and triple locomotives); electric multiple-units ER2, ER9: 90 units (45 sets).

The ER2 units (24 two-car units) are stored out of use.

In early 1998, the government of Uzbekistan announced that it was to participate in a joint venture with Ukraine to manufacture diesel locomotives at the latter country's Lugansk plant and in 2001 negotiations were proceeding for the acquisition of 30 Skoda-built electric locomotives.

Five new coaches were acquired in 1998 and a Russian-built Demikhovo emu has been on trial. In the same year an additional 25 vehicles were ordered from Bombardier Transportation's DWA subsidiary in German, using Japanese finance. A Japanese-financed passenger car depot at Tashkent should also ease the situation and this facility may also build new passenger stock.

No new passenger cars were received in 1999, but there were 100 new freight cars (50 in 1998, nil in 1997). Passenger coaches total 1,450 (many are

no longer serviceable). Freight wagons total 32,500 (many are no longer serviceable).

In 2001, the European Bank for Reconstruction and Development granted a loan to assist with the refurbishment and rehabilitation of diesel locomotives. Also planned was the construction of a new locomotive repair shop in Tashkent, with some funding provided by the Overseas Economic Co-operation Fund of Japan.

Signalling and communications

Provision of fibre optic communications technology for the 630 km section from Bukhara to Keles, on the border with Kazakhstan, forms a key feature of a five-year modernisation programme commenced in 2000.

Electrification

Apart from the 354 km between Tashkent and Samarkand, a few short industrial lines have been electrified. Some routes in the Fergana basin were electrified in 1993, including Khavast-Bekabad. In 1999 the Samarkand-Marakand and Misken-Sultannuisdag lines were energised, which brought electrified mileage to 620 km. Another 1,340 km are to be electrified over the longer term.

This includes the 275 km section between Samarkand and Bukhara, electrification of which was in progress in 2002.

Venezuela

Ministry of Transport & Communications

Caracas

Key personnel
Minister: C Zaa

Venezuelan State Railways (Ferrocar)

Instituto Autónomo Ferrocarriles del Estado
PO Box 146, Avenida Lecuna, Parque Central, Torre Este, Piso 45, Caracas 1010
Tel: (+58 2) 509 35 00/1
Fax: (+58 2) 574 70 21

Key personnel
President: Ing Juan Carlos Hiedra Cobo
Vice-President: O Ramirez Osio
Vice-President, Legal Affairs: C Salazar
Vice-President, Marketing: Antonio Zapata
Vice-President, Internal Accounting: L Morales

Managers
    Construction: Olegario Braga
    Operations: C Alberto Bueñano
    Administration and Finance: I Antunez
    Planning and Budgets: G Vanorio
    Personnel: Mercedes Polo Mimo
    Property: R Gosselain

Gauge: 1,435 mm
Route length: 434 km

Political background

Recent legislation pertaining to Venezuela's ambitious railway construction programme has made provision for private funding to be used alongside public money.

Organisation

The Ferrocar system consists of the 176 km Puerto Cabello–Barquisimeto route with branches from Yaritagua to Acarigua and Morón to Tucucas.

Freight traffic, amounting to 422,541 tonnes (54.5 million tonne-km) in 1997, consists of grain, fertiliser, sand and sugar cane waste (bagasse), often moving via private sidings serving industry.

Ferrocar has in the past operated passenger services on the 173 km Puerto Cabello–Barquisimeto line, but at last report no passenger services were operating.

At the end of 1997, Ferrocar employed nearly 500 people.

Finance

Ferrocar's provisional investment budget for 1996 amounted to Br14.963 billion, including Br1.341 billion for major track improvements, Br567 million for workshops and repair facilities and Br13.055 billion for new lines, of which Br12.493 billion was to pay for civil works on the Cua–Puerto Cabello project.

New lines
National Railways Plan

Venezuela's revised National Railways Plan aims to create a 3,447 km system of lines of local, regional and national importance by 2020. The programme involves the eventual creation of four interconnecting systems that are known as the Central Western, Central Region, Eastern Plains and Western railways.

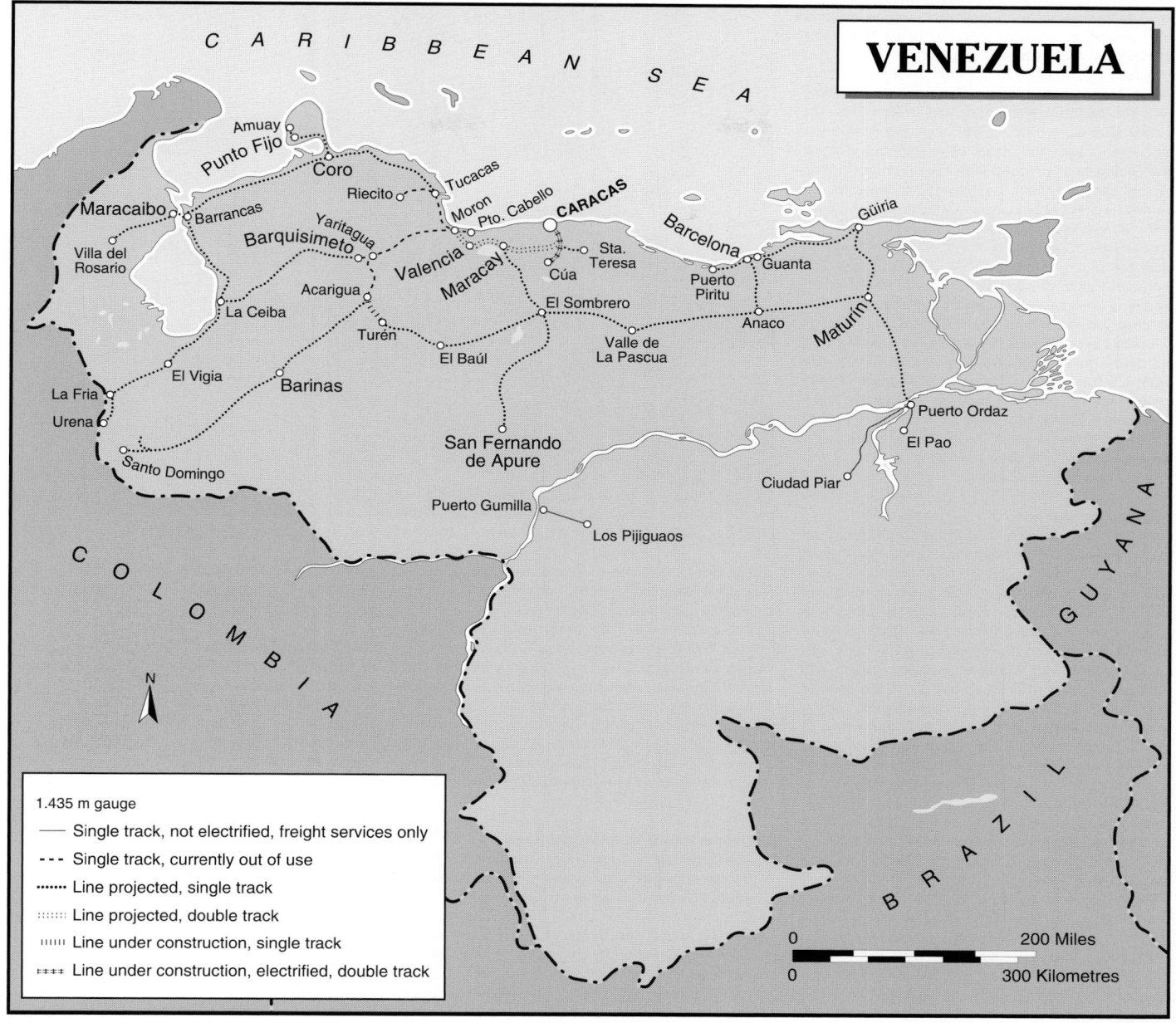

VENEZUELA

1.435 m gauge

— Single track, not electrified, freight services only

--- Single track, currently out of use

······ Line projected, single track

∷∷∷∷ Line projected, double track

ıııııı Line under construction, single track

⊬⊬⊬⊬ Line under construction, electrified, double track

0561638

Central Western Railway

Ferrocar's existing routes form the core of this system. Extension of the branch leaving the main line at Morón beyond its present terminus at Tucacas to Riecito, to a total length of 109 km, had been completed in early 1998 but not inaugurated. Construction, costed at a total of Br2.3 billion, has been funded by Ferrocar and the national oil company PDVSA, as the line will carry phosphates from deposits at Riecito to a petrochemical works at Morón. This traffic is eventually expected to amount to some 1.2 million tonnes a year, and Ferrocar sees potential for tourist services in the Tucacas and Chichiriviche areas.

A proposed 285 km extension would take the Riecito branch to Coro, Puerto Cardón and Amuay for oil industry traffic. This has been costed at Br15.5 billion for infrastructure and Br8.5 million for rolling stock. A total of Br10 million was allocated in 1996 to studies for a 185 km Yaracal–Coro route.

In 1991, work began on a 45 km extension from Acarigua to Turén, for which financing had been approved. The majority of the civil engineering work had been completed by early 1994, but further funding was required before tracklaying and terminal construction could begin. Costed at Br0.7 billion, the Acarigua–Turén extension is expected to carry some 500,000 to 800,000 tonnes of freight (mostly agricultural produce) a year with potential for traffic to grow by 20 per cent a year.

From Turén, a 115 km extension is planned to El Baúl. From El Baúl, a further eastward extension would connect the Central Western system with the proposed Eastern Plains network at El Sombrero.

Central Region Railway

The major Ferrocar enterprise now under way is the long-planned new east-west Central Region system, also known as the Central Trunk. This will extend from Caracas to Puerto Cabello via Cúa, Maracay and Valencia, and will link the capital and the central region of the country with the Central Western trunk at Puerto Cabello. It has been designated a project of national importance. At Puerto Cabello the Central Trunk would also connect with the Central Western, enabling the latter to rail coal and minerals to the Caracas conurbation.

After a call for tenders in 1989, an Italian-Japanese-Venezuelan consortium, Contuy Medio, led by Cogefar-Impresit SpA of Milan, won in 1991 the competition to build the first 43 km from Caracas to Cúa, serving the expanding new town of Tuy Medio. A contract valued at US$800 million was eventually signed with the Venezuelan government on 29 May 1996, whereby the consortium agreed to meet 50 per cent of the construction cost. Ferrocar's 1996 investment budget made provision for a Br12.493 billion advance payment for civil works on the Caracas–Cúa route, expected to open in 2001 and initially to carry some 64,000 passengers per day, rising to 110,000 per day by 2007.

This first stage traverses difficult terrain, with a difference of 623 m in level between its extremities, and will involve a ruling gradient of 2.3 per cent and 10.6 km of tunnelling, including one bore of 6.8 km. The line will be electrified at 25 kV 60 Hz AC, and laid with a single track later to be doubled.

Maximum speeds of 120 km/h will be exploited by 13 Japanese-built four-car emus consisting of two motor cars enclosing two trailers, with a capacity for 448 passengers. CTC will be located in Caracas and there will be two intermediate stations. The Caracas terminus will be at Mercado for interchange with Line 3 of the city's metro system, currently under construction.

The second stage of the Central Region system is the 176 km Cúa–Puerto Cabello section. Private finance was expected to fund the project, costed at Br52.7 billion for civil works and Br12.6 billion for rolling stock; Ferrocar allocated Br213 million to further studies in 1996. The Cúa–Puerto Cabello route is to be engineered for 180 km/h passenger operation, but is also expected to carry heavy freight traffic once Venezuela's national network emerges. Ferrocar expects to eventually handle 96,000 passengers and 139,000 tonnes of freight between Caracas and Puerto Cabello every day. A construction contract, possibly as part of an operating concession, was due to be awarded in early 1998. Groups from Canada, France and Mexico were reported to have expressed an interest.

Eastern Plains Railway

A proposed Eastern Plains system will be formed of south-north and east-west axes together totalling 938 km. The 394 km south-north route would start at Puerto Ordaz on the River Orinoco, connecting with the existing 141 km route to Ciudad Piar operated by Ferrominera for the transport of iron ore. From Puerto Ordaz, the new line heads north through Maturín for a deep sea port at Puerto Guiria on the Gulf of Cariaco, in the state of Sucre. Costed at Br32.5 billion for civil works and

Br4.3 billion for rolling stock, the south-north axis is expected by Ferrocar to carry 12 to 21 million tonnes of iron ore a year.

In 1995, Br45 million was spent on surveying work for the Puerto Ordaz–Maturín–Guacarapo component (342 km) of the project, allocated Br75 million for further studies in 1996. In 1995, Ferrocar concluded agreements with the states of Anzoategui, Bolívar and Monagas to proceed with the south-north axis of the Eastern Plains system (with branches to Barcelona and Guanta on the Caribbean) to be built either as a mixed private/state venture or by offering concessions. Expressions of interest were sought in 1996 by mining company CVG Bauxilum (see Los Pijiguaos Railway) to construct a route from Ciudad Guayana (east of Puerto Ordaz) to Puerto Guiria or another Caribbean port.

The 544 km east-west line would run from Maturín via Zaraza and El Sombrero to a junction with the Central Trunk at Cagua. Traffic would consist mainly of agricultural produce moving westwards from the eastern plains, estimated to be in the region of 8 million tonnes a year. The Maturín–Cagua route has been costed at Br45.3 billion for civil works and Br6.5 billion for rolling stock.

Western Railway

This system's principal traffic would be coal and phosphates from deposits in the southwest of Venezuela, moving to tidewater for export. International links have also been considered and, in May 1996, Ferrovias of Colombia announced proposals for a new 126 km route from Cúcuta across the border to La Concha and La Ceiba on Lake Maracaibo in Venezuela. The proposed Western Railway's initial 116 km section from La Fría to La Concha would carry some 3 million tonnes a year, and has been costed at Br10.6 billion for civil works and Br0.9 billion for rolling stock. Construction of this route has been deemed a priority and was to be publicly funded to the tune of 50 to 80 per cent. In 1995, Br31.3 million was spent on initial studies for the Urena–La Fría–La Concha–La Ceiba route, including geotechnical and hydrological work, and further La Frí–La Ceiba (243 km) studies received Br198 million in 1996.

The Western Railway would also include three other routes. The first is El Vigía–La Ceiba–Barquisimeto (417 km), connecting the La Fría–La Concha–La Ceiba route with the Central Western system and costing Br33.2 billion for civil works and Br6.9 billion for rolling stock. This line

Ferrocar GM-EMD GP15-1 diesel-electric locomotive at Barquisimeto (Eddie Barnes)   0021603

would carry 7.9 milion tonnes a year according to Ferrocar's forecasts. Much of this traffic would be coal and agricultural products.

A second line, 472 km long, would head north from La Fría to a new port on the Gulf of Venezuela, known as Puerto Nuevo, via Maracaibo and the coal mining area of Guasare. Civil works are costed at Br41.9 billion. The third route runs for 446 km from Santo Domingo via Barinas to the southern branch of the Central Western at Turén. This route would carry 10 million tonnes of traffic a year, principally coal and phosphates. Civil works have been costed at Br38 billion.

Improvements to existing lines

In anticipation of heavier traffic expected to be generated by the extension of its branches (roughly twice the present amount) and the construction of other connecting trunk routes, upgrading work was undertaken on the track of the Puerto Cabello–Barquisimeto main line of the Central Western Railway in the early 1990s. However, in late 1997 Ferrocar admitted that track

had badly deteriorated, and that US$67 million had been assigned to fund an 18 month upgrading programme.

Traction and rolling stock

At the start of 1998, Ferrocar operated 17 diesel locomotives, namely six GM-EMD 1,300 kW GP9s, four GM-EMD 1,100 kW GP15-1s and two 110 kW shunting locomotives. Other rolling stock comprised 14 passenger coaches and 264 freight wagons.

Type of coupler in standard use: US Type E
Type of braking in standard use: Air

Track

Rail: ASCE 49 kg/m and UIC 60 kg/m
Sleepers: Dywidag concrete, 2,500 × 227 × 300 mm
Spacing: 1,670/km, 1,336/km in curves
Min curve radius: 800 m
Max gradient: 1.4%
Max permitted axleload: 31.75 tonnes
Max speed: 70 km/h

Ferrominera

CVG Ferrominera Orinoco CA
PO Box 399, Puerto Orduz, Bolívar State
Tel: (+58 86) 30 31 11   Fax: (+58 86) 30 36 56

Key personnel

General Superintendent, Piar Division: E Carabello
Superintendents
   Pao Division: H Brazon
   General Shops: M Aro G
   Track and Structures: D Massiah R
General Supervisor: J Diaz

Gauge: 1,435 mm
Route length: 196 km

Freight operations

Ferrominera, a state-owned company, operates two railways in eastern Venezuela principally for the transport of iron ore to the River Orinoco. The line linking Ciudad Piar in the Cerro Bolívar range with Puerto Ordaz on the river is 141 km long, and that connecting El Pao with Palua on the Orinoco 55 km long. Annual carryings amount to some 30 million tonnes of freight.

Traction and rolling stock

Ferrominera operated a fleet of 36 diesel locomotives and 1,587 freight wagons in 1990. In 1996, Freios Knorr of Brazil was contracted to supply the company with 100 braking systems to be fitted to new wagons ordered from Transimpex of Bulgaria.

Los Pijiguaos Railway

CVG Bauxilum
Carretera Nacional Caicara-Puerto Ayacucho, Los Pijiguaos, Bolívar State
Tel: (+58 2) 572 16 20
Fax: (+58 2) 993 76 85

Key personnel

President: Pedro Mantellini
Executive Vice-President: José L Garcia G
General Manager: Gustavo Quintero
Mineral Handling Manager: Oscar Portes
Traffic and Railway Maintenance Superintendent:
   Juan Carlos Fermin

Gauge: 1,435 mm
Route length: 52 km

Freight operations

CVG Bauxilum operates a 52 km railway linking bauxite deposits at Los Pijiguaos with the Orinoco river port of Gumilla. Opened in 1989, the line carried 5 million tonnes in 1995.

Improvements to existing lines

A new mineral handling system has been installed which will allow the railway to carry 5.3 million tonnes of bauxite a year. The system incorporates an automatic loading station and an automatic car dumper, both with a capacity of 3,600 tonnes an hour, and allows six trains carrying 2,500 tonnes each at 60 km/h to be operated in a period of 8 hours.

Traction and rolling stock

At the end of 1995, the railway operated five diesel-electric locomotives, including GM-EMD SD38-2

and SW900 types, and 119 freight wagons, mostly 90 tonne gondolas.

Type of coupler in standard use: F150
Type of braking in standard use: 26L, 24RL, 6BL

Track

Rail: AREA 132 RE, 54.75 kg/m
Sleepers: Dywidag concrete, 2,500 × 227 × 300 mm
Spacing: 1,667/km
Fastenings: RN 300
Min curve radius: 36°
Max gradient: 1%
Max permissible axleload: 30 tonnes

Vietnam

Ministry of Communications and Transport

80 Tran Hung Dao, Hanoi
Tel: (+84 4) 25 20 79

Key personnel
Minister: Buy Danh Luu
Railways Director: Nguyen Hieu Liem

Vietnam Railways (DSVN)

Duong Sat Viet Nam
118 Duong le Duan, Hanoi
Tel: (+84 4) 942 44 01   Fax: (+84 4) 942 49 98
e-mail: vr.hn.irstd@fpt.vn

Key personnel
Director General: Dr Nguyen Huu Bang
Deputy Directors General: Vuong Dinh Khanh,
    Phan Van Gian, Dang Duc Thinh, Vu Xuan Hong,
    Tran Phuc Tien

Gauge: 1,435 mm; 1,000 mm; mixed gauge
Route length: 178 km; 2,169 km; 253 km

Political background
After a long period of stagnation, DSVN underwent reorganisation in 1994–95, following the government's agreement to separate infrastructure and operating costs from the beginning of 1995. The railway, freed from responsibility for upkeep and renewal of its infrastructure, nevertheless continues to manage day-to-day execution of these tasks and retains control over rolling stock maintenance and procurement. It is required to cover its operating costs as well as remitting 10 per cent of revenues to the government as a track access charge. In return, the government shoulders all costs associated with track, signalling, telecommunications and structures.

In January 2002, the government published a development and investment strategy for the railway network covering the period up to 2020. This foresees the reorganisation of DSVN to become a state-owned corporation.

Organisation
The most important route is the main line linking Hanoi with Ho Chi Minh (Saigon) (1,726 km); other routes radiate from Hanoi to the port of Haiphong (96 km) and Quan Trieu (mixed-gauge, 54 km), while two lines run to the border with China – northwest to Lao Cai and northeast to Dong Dang (mixed-gauge). A 1,435 mm gauge line runs from Quan Trieu to Kep and Halong.

Hanoi has a western orbital route, built in the 1980s, which diverges from the main line to the north at Dong Anh, crosses the Red River on the Thang Long bridge, and joins the southern main line to Ho Chi Minh (Saigon) at Van Dien.

Operationally, the railway is divided into three divisions based on Hanoi, Da Nang and Ho Chi Minh.

Traffic (million)	1999	2000	2001
Passenger journeys	9.3	9.8	10.6
Passenger-km	2,727	3,200	3,426
Freight tonnes	5.0	6.1	6.3
Freight tonne-km	1,398	1,901	2,000

Passenger operations
By 1994, passenger traffic had fallen below 8 million journeys annually, with some 80 per cent of the passenger-km logged by the Hanoi–Ho Chi Minh (Saigon) route. Traffic on this artery is expected to grow to more than 14 million journeys a year, a renaissance which started in 1994 with the commissioning of a new premium service, the Reunification Express, which cut the transit time by half to 36 hours. By 2001, the number of passenger journeys had grown strongly to 10.6 million.

Further growth is expected on the two routes to the Chinese border. At Dong Dang, a metre-gauge DSVN train connects twice-weekly with a 1,435 mm Chinese Railways service to Pinxiang. On the 530 km all-metre-gauge route from Hanoi via Lao Cai to Kunming in China, an inaugural direct service ran in April 1997.

In 2002, there were 250 stations on the DSVN network.

Freight operations
Recent years have seen considerable growth in freight traffic. Tonne-km in 1998 were 1.3 billion; this figure rose to 2 billion in 2001.

There is considerable potential for increased traffic with China, both over the 1,435 mm gauge route via Dong Dang and the metre-gauge line that provides a through connection from the port of Haiphong to Kunming. On the Dong Dang route (on which trans-shipment is required at the gauge change at the border) international services resumed in 1996, after having been broken off during a border dispute in 1979.

Intermodal operations
A joint venture between VSDN and New Zealand interests known as Rail Express opened up the route between the port of Haiphong and Hanoi with a US$5 million investment package. Just over US$3 million was contributed to this by the venture's New Zealand partner, Minzr Containers Ltd, a company established by engineering conglomerate McConnell International which is drawing in rail expertise from Tranz Rail.

VSDN has improved the track between the two cities and an on-dock rail terminal has been built in Haiphong. Rail Express obtained its operating licence in August 1996 and started running three container trains a day (total capacity 60 TEU) between Haiphong and the capital during 1997. More powerful locomotives and new wagons are needed if the venture is to expand.

New lines
DSVN's development programme for the period 2001–2010 foresees implementation of the following schemes:
- construction of a new 43 km line linking Yen Vien, Pha Lai and Co Thanh
- construction of a new 132 km line linking Saigon and Loc Ninh
- construction of a 19 km elevated railway between Yen Vien and Van Dien, on the east-west corridor
- construction of a 9 km elevated railway to serve a mass transit system for Ho Chi Minh

The development programme for the period 2010–2020 includes four new line projects: Saigon–Vung Tau: Saigon–My Tho–Can Tho; Dong Ha–Lao Bao; and Vung Ang–Tan Ap. Also included is preparatory work for a proposed standard gauge high-speed line from Hanoi to Saigon.

Improvements to existing lines
Plans for upgrading the existing network have been set out by DSVN in development programmes covering the periods 2001–2010 and 2010–2020. During the first of these periods, the focus of attention was to be on:
- strengthening the infrastructure on the central portion of the network
- strengthening bridges and repairing tunnels
- measures to protect against landslips on the Western line
- sleeper replacement on the Northern line
- the mechanisation of track maintenance
- modernisation or elimination of level crossings
- modernisation and upgrading of the telecommunications network
- establishing effective links with ports
Specific project in the programme include:

Hanoi–Ho Chi Minh
- upgrading of 10 major bridges
- realignment of the Trang Bom–Binh Trieu section
- boring a new 10 km tunnel through the Hai Van pass

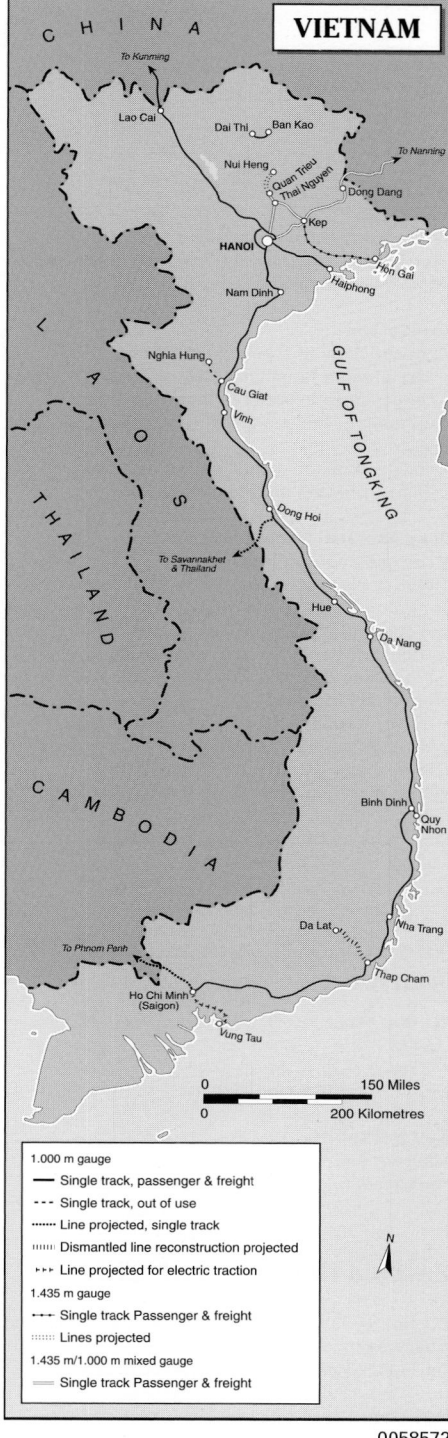

VIETNAM

0058572

- modernisation of signalling and telecommunications for the entire 1,725 km line
- rehabilitation of terminals at Da Nang and Nha Trang and the construction of 16 new stations to replace facilities at existing sites
- general line upgrading, improving the superstructure, widening the trackbed and undertaking flood prevention measures
- installation of level crossing protection systems

Hanoi–Lao Ca
- track doubling of the Bac Hong–Yen Bai section and general upgrading

Hanoi–Lang Son
- rehabilitation and upgrading of the entire 167 km line, including track and signalling, to provide a four-hour journey time

Hanoi–Haiphong (96 km)
- improvements and upgrading to increase line speed to 100–110 km/h

Diesel locomotives

Class	Wheel arrangement	Power kW	Speed km/h	Weight tonnes	No in service	First built	Mechanical	Builders Engine	Transmission
D4H	Bo-Bo	290	50	24	190[1]	1968	Kambarka	1D12-400	H
D4H	Bo-Bo	325	50	24	1	1997	Dong Co	368 MTU 12V 183	H
D5H[2]	Bo-Bo	347	50	40	13	1968	Walkers	CAT D-355	H
D9E	Bo-Bo	670	100	52	28	1963	GE	D 379	E
D11H[3]	Bo-Bo	–	–	–	(58)	1979	23 August	–	H
D12E	Bo-Bo	736	80	56	40	1985	ČKD	K6S 230DR	E
D13E	Co-Co	895	100	72	12	1984	DLW	251-D	E
D18E	Co-Co	1,472	100	84	16	1982	CMI	LTR 240-C	E

[1] 11 locomotives operational on 1,435 mm gauge.
[2] Purchased second-hand from Queensland Railways 1993–96.
[3] Whole fleet of 58 out of service.

Kep–Cai Lan

• construction of a passenger station at Ha Long and a freight terminal at Cai Lan, together with a connecting line between these two locations
• upgrading the standard gauge Kep–Chi Linh section (37 km)
• construction of a dual-gauge link between Co Thanh, Chi Linh and Ha Long (84.9 km)

Dong Anh–Thai Nguyen (55 km)

• conversion to standard gauge

Also included in the 2001–2010 development programme were improvements to passenger stations and freight facilities along the east-west corridor, including at Hanoi Central station.

Projects featuring in the 2010–2020 development programme include:

• track doubling on the Hanoi–Vinh and Saigon–Nha Trang sections of the Hanoi–Ho Chi Minh line
• electrification of the 96 km Hanoi–Hai Phong line

• realignment of the 156 km Yen Bai–Lao Cai section of the Hanoi–Lao Cai line
• works to establish urban railway networks in Hanoi and Ho Chi Minh

Signalling and telecommunications

Operations are controlled by tokenless semi-automatic block signalling, and token block working. In 2002, semi-automatic block signalling was being installed on the Hanoi–Ho Chi Minh route and is planned for other lines. The same route was the subject of a contract placed in 2000 with an Alstom-Alcatel CIT consortium which included an upgrade of telecommunications facilities.

Traction and rolling stock

At the end of 2002, the standard gauge fleet comprised 14 diesel-hydraulic and three diesel-electric locomotives, six steam locomotives, eight coaches and 332 freight wagons. The metre gauge fleet comprised 240 diesel-hydraulic and

113 diesel-electric locomotives, 35 steam locomotives, 897 coaches and 4,027 freight wagons. At that time, 38 diesel locomotives were reported to be unserviceable.

Recent deliveries include 15 1,000 kW, 251 D-engined diesel-electric Co-Co locomotives, similar to Indian Railways' Type YDM4, from India's Projects & Equipment Corp (PEC); and from Belgian industry came 16 1,300 kW, 84 tonne diesel-electric Co-Cos with Cockerill engines. ČKD Praha has delivered 20 736 kW, 56-tonne Type DEV-736 diesel-electric locomotives for metre-gauge. Other diesel traction includes GE and Russian-built Type TY 300 kW units, and 13 locos bought second-hand from Queensland Rail in 1994.

Procurements planned in the period from 2001 to 2010 foresee the acquisition of 60 to 100 metre gauge locomotives of 900 to 1,340 kW and eight to 10 standard gauge locomotives. Also planned is the establishment of facilities to construct over the same period 500 to 800 passenger coaches and 200 to 300 freight wagons, including container flats and vehicles for perishables. The railway also plans to establish a locomotive overhaul facility and to re-organise and re-equip its locomotive and rolling stock maintenance depots, with priority to be given to those in Hanoi and Ho Chi Minh.

Track

Rail: 43 kg/m rail is installed over 2,274 track-km; 38 kg/m rail is installed over 260 track-km
Sleepers: concrete (1,157 km); steel (1,103 km); wood (482 km)
Min curve radius: 90 m
Max gradient: 1.7%
Max permissible axleload: standard gauge – 16 tonnes; metre gauge – 14 tonnes

---

Zambia

Ministry of Communications & Transport

PO Box 50065, Block 33, Fairley Road, Ridgeway, Lusaka
Tel: (+260 1) 25 14 44; 25 17 40
Fax: (+260 1) 25 17 95; 25 32 60

Key personnel

Minister: Abel Chambeshi
Permanent Secretary: Bob Samakai
UPDATED

Zambia Railways Ltd (ZRL)

PO Box 80935, Corner of Buntungwa Street and Ghana Avenue, Kabwe
Tel: (+260 5) 22 22 01; 9   Fax: (+260 5) 22 44 11

Key personnel

Chairman: P S Chamunda
Managing Director: Robert Crawford
Directors
   Technical Services: Igwa U Sichula
   Traffic and Marketing: Chris C Musonda
   Finance: Goodson M Moonga
   Personnel: Angela Malawo
General Managers
   Workshops: Bedford Lungu
   Central: H C K Nyimbili
   South: Baxton Siwila
   North: Kingston Mkandawire
Chief Civil Engineer: Yubya Mwanawina
Chief Mechanical and Electrical Engineer:
   Webster Mutambo
Chief Signal and Telecommunications Engineer:
   P C Lumumbe
Managers
   Traffic: David Mwaliteta
   Finance: A L Fernando
   Marketing: Frank Kangwa
   Purchasing and Stores: Francis Zulu
   Passenger Services: Luke Mwanza
   Freight Services: Hilary Mphuka

Gauge: 1,067 mm
Route length: 1,266 km

Political background

Formerly part of Rhodesia Railways (RR), Zambia Railways was segregated as an autonomous system in 1976. It comprises the old RR system north of the Victoria Falls Bridge, to which was added in 1970 the 164 km Zambesi Sawmills Railway from Livingstone to Mulobezi. RR remained a legal entity until 1996, when its assets were eventually divided between ZRL and National Railways of Zimbabwe.

Since its independence, ZRL has been handicapped by the political crises in the region and the problems of some neighbouring railways,

which have clouded definition of the land-locked country's rail routes to the sea ports with uncertainty. Rail outlets are of critical importance to Zambia's copper industry, which generates 90 per cent of its exports. The TAZARA system's operating difficulties (see Tanzania entry) have restricted the potential of its route to Dar-es-Salaam, originally envisaged as Zambia's primary export rail route, and the Benguela Railway to Lobito in Angola has been affected by the unrest to the west. Assignment of copper traffic to the TAZARA or Victoria Falls routes is decided by the government.

Since 1984, ZRL has been a limited company, with freedom to set its own tariffs, though it is still subject to a fuel oil tax, the revenues from which support road infrastructure.

1114989

ZRL is one of the poorest performers in the country's transport sector, and this has led to the government of Zambia seeking foreign aid in order to implement a comprehensive restructuring plan. Funding was agreed with the Swedish International Development Agency (SIDA) in 1997, the terms of which required the board of directors of ZRL to recruit and select a new management company to take over the railway for a period of up to 2.5 years. A consortium consisting of the Swedish company Hifab and German rail consultants DE-Consult was selected for this role and staff of the company duly took over the leading management positions in March 1998.

Before the new company commenced its work, staffing levels had been reduced from 8,500 in 1993 to 5,190 in 1997. Major problems experienced related almost exclusively to the very poor cashflow situation and constant lack of funds for investment, particularly in staff severance payments, infrastructure rehabilitation and traction and rolling stock repair and refurbishment.

In March 2000, the Zambian government approved in principle plans to privatise ZRL. In August 2001, the Zambia Privatisation Agency invited bids for a long-term concession to operate the ZRL system and manage its assets. This followed agreement with the World Bank to provide a loan to fund selective rehabilitation of track, locomotives and freight wagons. The closing date for bids was 7 December 2001.

Passenger operations

In 1997, ZRL was running a daily service over the 851 km between Livingstone and Kitwe (augmented by a thrice-weekly service on the Livingstone–Lusaka section of the route), two daily return trips between Livingstone and Victoria Falls (13 km) and a twice-weekly service between Livingstone and Mulobezi (163 km). For service details on the TAZARA line see Tanzania entry.

Passengers benefited from introduction of refurbished coaching stock in 1991, but this did little to attract extra patronage. Traffic figures for 1997 were 1.26 million passenger journeys, 267 million passenger-km – little more than half what they had been a decade before.

General Electric Co-Co diesel-electric locomotive at Livingstone with a ZRL passenger service (Oubeck.com)   0114661

Freight operations

Around two-thirds of all freight carried is copper, but this has been vulnerable to road competition. In addition to the constraints on performance due to fallible equipment, traffic growth has been restricted by several factors: slow turn-round of wagons at customers' sidings and on neighbouring railways; inflation and the depreciation of the Kwacha; and unsettled politics and strife in neighbouring countries. Traffic figures for the year to 31 March 1998 were 1.4 million tonnes – less than half what they had been only five years previously. However, a modest recovery was achieved in 1999, when the railway was expected to carry 1.8 million tonnes.

Traction and rolling stock

In 1997, ZRL owned 66 diesel locomotives (44 operational, 35 in use), mostly comprising General Electric U20C 1,330 kW units, plus 12 General Electric U15C 1,016 kW shunters. ZRL's most recently acquired traction is a batch of 15 Type GT36CU-MP locomotives from GM Canada, delivered from 1993.

The wagon fleet in 1997 totalled 6,345, of which 3,345 were in use. There were 77 passenger cars, including six sleepers, 19 standard class, 36 economy class, three buffet/dining and six snack cars. Other recent deliveries were 100 bogie covered wagons from BN and 25 container flats from BREC, Belgium, while Braithwaite of India was supplying 20 sulphuric acid tank wagons in 1996–97.

Type of coupler in standard use: Alliance automatic, bottom-operated, Contour 10A
Type of brake in standard use: Vacuum

Signalling and telecommunications

The 851 km in total from Livingstone to Ndola and the Copperbelt branch section between Ndola and Kitwe are controlled by CTC (centralised traffic control) with multiple-aspect colourlight signals. Sections outside CTC territory are worked on the token block system or, in the Copperbelt, on the train staff system. The CTC system was installed during 1961–64 and utilised open-wire carrier circuits along the line or rail. Operating from a centre at Kabwe, the CTC has now been renewed with all-electronic apparatus by Siemens AG at a cost of US$15.8 million. A total of 61 relay interlockings are remote-controlled from Kabwe, while a further six stations have locally controlled relay interlockings.

The overhead line carrier system has been replaced by radio transmission because of theft and vandalism of wires, generators and batteries, with resultant interruption of the new CTC. The African Development Bank funded the US$9.14 million conversion which was finished in 1990.

Other recent projects include installation of a Mitsui digital multiplex microwave radio system between Ndola and Livingstone and upgrading of a computer mainframe. A start was made with purchase of hand-held radios for communication between yard staff and signalboxes. Installation of mains electricity was completed at a further 15 stations, replacing diesel generators.

Track

Rail: 45.13 kg/m
Crossties (sleepers): Wood, 127 mm thickness; concrete, 200 mm thickness; spaced 1,340/km in plain track, 1,400/km in curves, for concrete, 1,400/km in both cases for wood
Fastenings: Coach screw, clip and spring washer (triple coil) for wood, Pandrol for concrete sleepers
Min curve radius: 8.7°
Max gradient: 1.75%
Max axleload: 15.25 tonnes

ZRL General Motors GT36CU-MP diesel-electric locomotive at Kapiri Mposhi with a freight service (Oubeck.com)   0114660

Njanji Railways

PO Box, Lusaka

Key personnel
Managing Director: C Mayatwa

Gauge: 1,067 mm
Route length: 16 km

Organisation

This operation, started-up by the state copper mining company and now run independently, established a cross-city commuter service from Chilenje to George in Lusaka in 1990, largely over existing tracks. The two diesel railbuses originally employed proved inadequate for the demand, and now the service is provided by two locomotives and 10 coaches hired from ZR.

The success of the limited service offered has stimulated plans for extension over ZR's main line to serve Ngwerere in the north and Chilanga to the south.

It is possible that ZR will take over the management of the line, which would then be run as a self-contained business unit.

Zimbabwe

Ministry of Transport and Communications

PO Box Cy 595, Kaguvi Building, Causeway, Harare
Tel: (+263 4) 70 06 93   Fax: (+263 4) 70 82 25

Key personnel
Minister: Witness Mangwende

National Railways of Zimbabwe (NRZ)

PO Box 596, Bulawayo
Tel: (+263 9) 36 35 21   Fax: (+263 9) 36 35 43

Key personnel
General Manager: Munesu Munodawafa
Directors
    Finance: Patrick Bondai
    Operations and Marketing: Mazikhethela Banana
    Technical Services: Lewis Mukwada
Chief Internal Auditor: David Sithole
Finance Manager: Lochard D Mkandla
Chief Traffic Manager: Dennis Nyoni
Manager, Strategic Planning: Bernard Dzawanda
Chief Engineer, Traction and Rolling Stock:
    Norman Marange
Human Resources Manager: (vacant)
Chief Engineer, Infrastructure: Herbert Dzinotyiwei
Manager, Supplies and Stores: Kainos P Magunda
Corporate Affairs Manager: Misheck Matanhire
Manager, Safety and Environmental Services:
    Claris Pfute
Security Services Manager: Frank Msutu
Information Technology Manager:
    Simbarashe Mutikani
Public Relations Manager: Fanuel Masikani
Manager, Buisness Development:
    Chenjerai Nziramasanga
Manager, Eastern Area:
    Christopher Madzimbamuto
Manager, Southern Area: Petros Gwizi
A/Business Manager, Passenger Services:
    Ephraim Choto

Gauge: 1,067 mm
Route length: 3,077 mm (includes 318 km BBR line)
Electrification: 313 km at 25 kV 50 Hz AC

Political background
NRZ was created in 1967 out of Rhodesia Railways (RR), the northern portion of which became Zambia Railways. The old RR remained a legal entity until 1996, when division of its assets between the two countries was finally agreed.

NRZ continues to be adversely affected by fuel shortages. A lack of foreign exchange has also had an impact on the procurement of spares for traction and rolling stock and for infrastructure fittings such as electrical equipment. Matters are compounded by theft and vandalism. In 2004 a programme to rehabilitate the network and its equipment was initiated, with a planned completion date of August 2005.

Traffic	2002
Passenger journeys	359,000
Passenger-km (million)	–
Freight tonnes (million)	8.1
Freight tonne-km (million)	–

Passenger operations
There is a daily train between the two largest cities, the capital Harare and Bulawayo, and tourist services between Bulawayo and Victoria Falls. The daily service between Harare and Mutare on the Beira Corridor route was relaunched with new rolling stock in 1999. There are also services between Somabhula and Chiredzi in the southeast of the country.

In July 2001 NRZ launched a daily commuter service between Marimba and Harare. This is operated jointly with Zimbabwe United Passenger Services Company (Zupco), which provides feeder bus services. This was followed later in the same

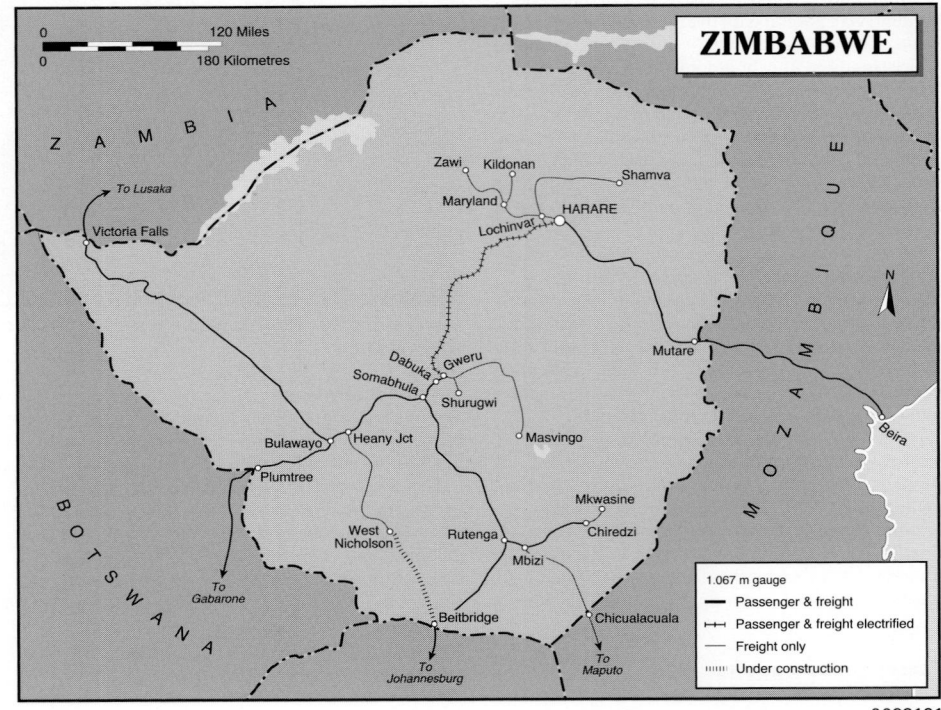

0023121

year on the Ruwa and Tafara routes in Harare and the Luveve/Mabvuku route in Bulawayo. It was expected that additional routes would be added.

Freight operations
Economic conditions and the deteriorating fabric of the network and its equipment have led to a continuing downward trend in freight movement, with only 8.1 million tonnes lifted, compared with more then 18 million tonnes in 1998.

Intermodal operations
In recent yars, almost 90 per cent of the country's tobacco crop has been exported in containers, along with tea, coffee, graphite, hides and skins, nickel and tin. Containerised imports for the Harare area's extensive industrial development include machinery, lubricants, bricks, iron and steel. NRZ's two container terminals are Dabcon, at Dabuka near Gweru, and Locon, at Lochinvar near Harare.

New lines
Freight services over the 350 km Beitbridge–Bulawayo line commenced in July 1999. Developed as a private-sector 'Build, Operate and Transfer' (BOT) scheme, the BBR provides a direct link from South Africa to Bulawayo and on to Zambia and to Mutare, close to the border with Mozambique. As a result of the opening of the line, journey times between Bulawayo and Durban have been cut from 25 to four days.

A 30-year concession to operate and maintain the line was acquired by the Bulawaya–Beitbridge Railway Company, in which a 15 per cent stake is held by NRZ, 40 per cent by the New Limpopo Bridge Company, and the remainder by Gensec, Nedcor, Old Mutual and Sanlam. Spoornet of South Africa is contracted to run train services.

Traction and rolling stock
At the beginning of 2000 NRZ owned 30 electric, 42 steam and 152 diesel locomotives; 325 passenger cars (including 17 buffet/dining, and 117 sleeping cars) and 11,387 freight wagons.

At the end of 2004 NRZ stated that it had refurbished 22 diesel locomotives; work on at least six of these funded by Freight Forwarders of Zimbabwe in a move intended to increase capacity. There were also reports that a loan from China was to fund the supply of 10 locomotives, eight commuter trains and three sets of intercity rolling stock.

Type of coupler in standard use: Automatic centre buffer coupler, Alliance No 1 and 2 heads
Passenger cars: 5 × 5 in, 5½ × 5 in shanks
Freight wagons: 8 × 6 in, 7 × 5 in shanks
Type of braking in standard use: Vacuum on passenger and freight stock except for liner train wagons, which are on direct release airbrakes

Signalling and telecommunications
A new CTC (centralised traffic control) supervisory and remote-control system supplied by Siemens

Diesel locomotives

Class	Wheel arrangement	Power kW	Speed km/h	Weight tonnes	No in service	First Built	Mechanical	Builders Engine	Transmission
DE 6	Co-Co	1,559	116	90.84	9	1966	GE-USA	GE 7-FDL-12 (re-engined 1998)	E GE-USA
DE 9	Bo-Bo	732	103	61.3	64	1975	GE (U11B)	D 3512 D11 JWAG Caterpillar (re-engined 1992–94)	E GE
DE 10	Co-Co	1,678	107	94.35	61	1982	GM USA/Canada	12-645E3 B (GM)	E GM
DE 11	Co-Co	2,508	102	113.998	13	1992		16-645E3B	E

Electric locomotives

Class	Wheel arrangement	Output kW	Speed km/h	Weight tonnes	No in service	First built	Builders Mechanical	Electrical
EL1	Co-Co	2,400	100	114	30	1983	ZECO/SGP	50 c/s Group

Zus was commissioned on the Victoria Falls–Bulawayo section in 1993. However, its usefulness was reduced by continuing dependence on a vulnerable copper-wire pole route for communications. CTC now extends to around 1,250 route-km.

Radio control is employed between Harare and Mutare (the 300 km Zimbabwean section of the Beira Corridor export route to the Indian Ocean), replacing the signalling system dating from 1960.

Electrification

The 313 km route between Harare and Gweru is electrified on the 25 kV 50 Hz AC system.

NRZ has more Class EL1 electric locomotives than current traffic levels demand, so locomotive availability is maintained by cannibalising mothballed machines.

Track

Rail: UICE Class B standard 54 kg/m (313 km of main line only); remainder of main line BS45 kg/m; branch lines BS45, 40 and 30 kg/m. A new standard rail, BS90A, has been adopted for all future use

Crossties (sleepers)

Type:	Thickness (under seat)
Concrete	226 mm
Hardwood	115 mm
Steel	10 mm; 13 mm

Spacing: 1,429/km
Min curvature radius: Main line 600 m, branch lines 400 m
Max gradient: 1 in 80 (main line)
Max axleload: 18.6 tonnes

UPDATED

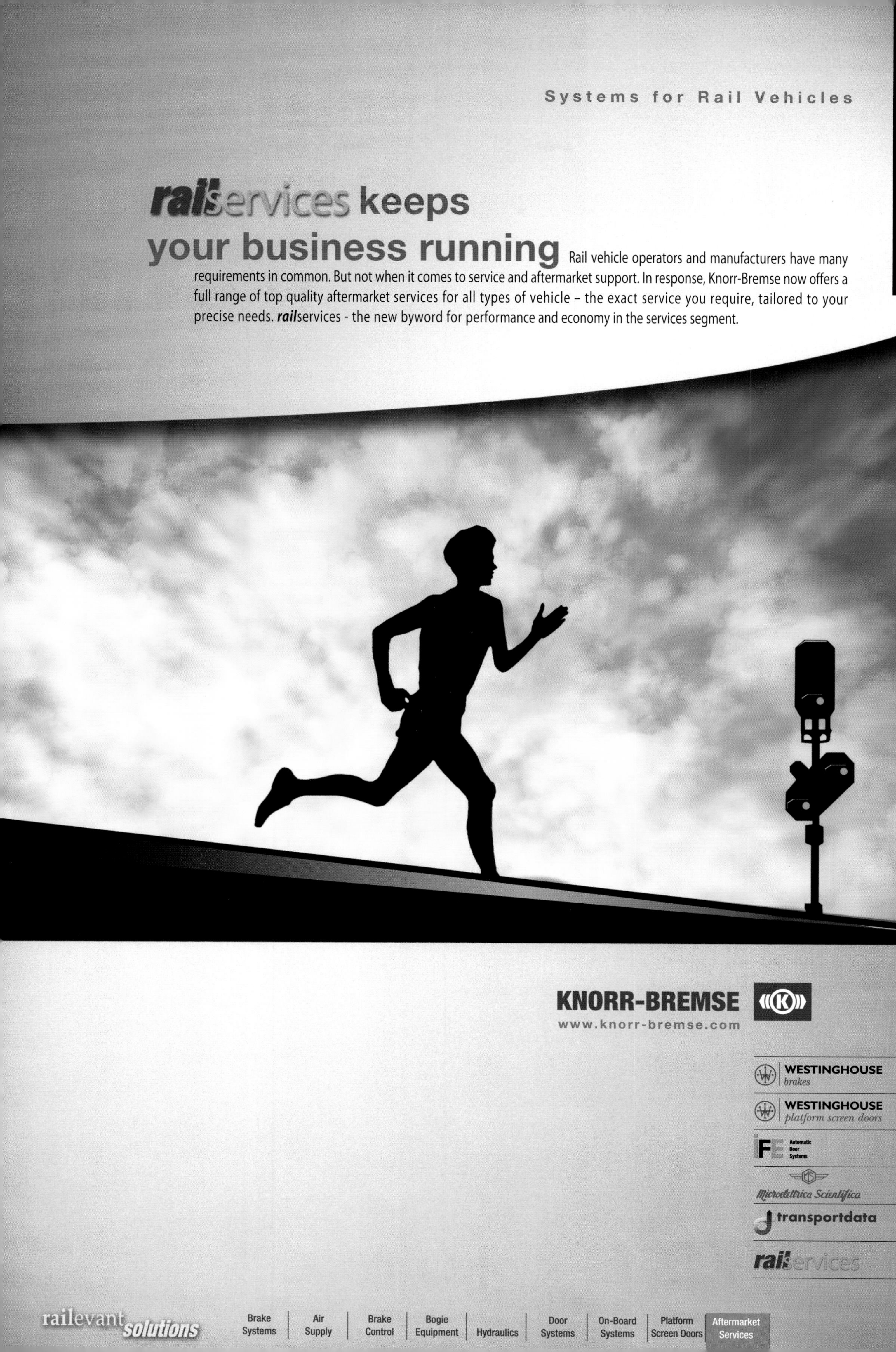
Systems for Rail Vehicles
railservices keeps your business running
Rail vehicle operators and manufacturers have many requirements in common. But not when it comes to service and aftermarket support. In response, Knorr-Bremse now offers a full range of top quality aftermarket services for all types of vehicle – the exact service you require, tailored to your precise needs. railservices - the new byword for performance and economy in the services segment.
KNORR-BREMSE
www.knorr-bremse.com
WESTINGHOUSE brakes
WESTINGHOUSE platform screen doors
IFE Automatic Door Systems
Microelettrica Scientifica
transportdata
railservices
railevant solutions
Brake Systems
Air Supply
Brake Control
Bogie Equipment
Hydraulics
Door Systems
On-Board Systems
Platform Screen Doors
Aftermarket Services

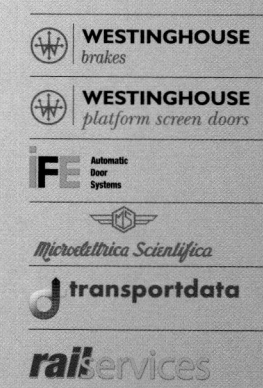

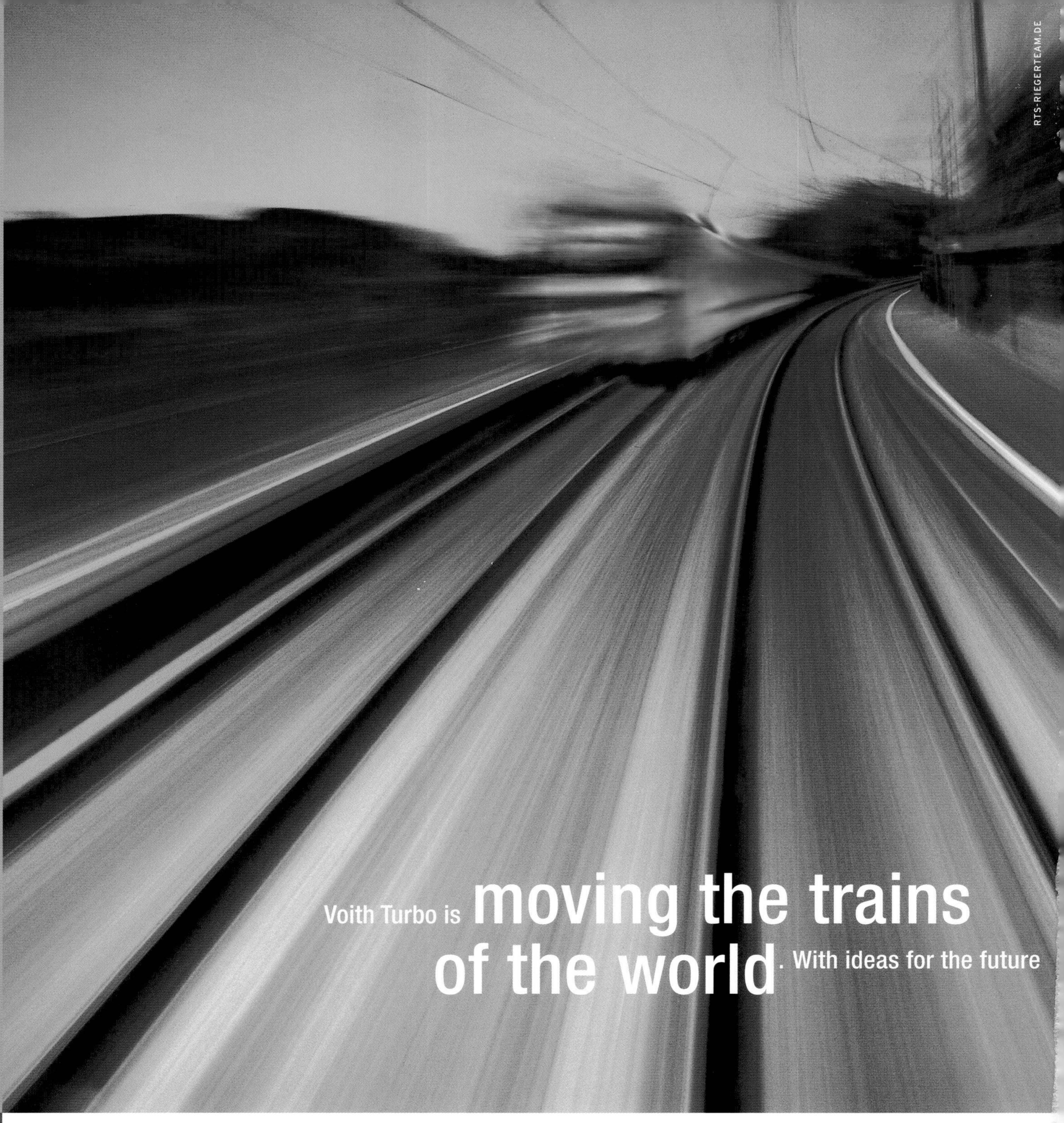
RTS-RIEGERTEAM.DE
Voith Turbo is moving the trains of the world. With ideas for the future

Industry | Marine | Rail | Road

Voith Turbo is a specialist in driveline and control solutions, cooling systems and Scharfenberg® couplers. Almost every freight or passenger train in the world has Voith Turbo equipment installed – for higher speed, higher comfort and lower noise levels. Our competence in the design of systems is based on decades of experience.

This enables us to find individual solutions to every possible task, however complex and challenging it may be. We accept responsibility for our solutions, anywhere in the world, so that our customers can be confident even with their most ground-breaking projects.

www.voithturbo.com

Voith Turbo

VOITH

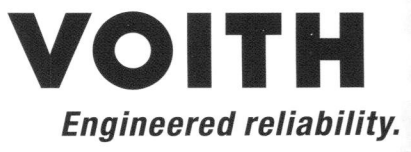

Engineered reliability.

MANUFACTURERS AND SERVICES

Locomotives and passenger vehicles
Diesel engines, transmission and fuelling systems
Electric traction equipment
Passenger coach equipment
Freight vehicles and equipment
Brakes and drawgear
Bogies and suspension, wheels and axles, bearings
Simulation and training systems
Signalling and communications systems
Passenger information systems
Revenue collection systems and station equipment
Electrification contractors and equipment suppliers
Cables and cable equipment
Permanent way components, equipment and services
Freight yard and terminal equipment
Vehicle maintenance equipment and services
Turnkey systems contractors
Information technology systems
Rolling stock leasing companies
Consultancy services

LOCOMOTIVES AND PASSENGER VEHICLES

Alphabetical listing

Alan Keef Limited
Alcan Mass Transportation Systems
Alna Yusoki-Yohin Co Ltd
Alstom Transport
Ansaldobreda SpA
Astra Vagoane Călători SA Arad
Bharat Earth Movers Ltd (BEML)
BHEL – Bharat Heavy Electricals Ltd
BMZ
Bombardier Transportation
Brookville Equipment Corporation
Brush Traction
Bumar-Fablok Ltd
CAF – Construcciones y Auxiliar de Ferrocarriles SA
CFD Bagneres
Changzhou Diesel & Mining Locomotive Plant
China Northern Locomotive and Rolling Stock Industry (Group) Corporation (CNR)
China South Locomotive and Rolling Stock Industry (Group) Corporation (CSR)
Chittaranjan Locomotive Works
ČKD Vagonka as
ČMKS holding as
CNR Beijing 'February 7th' Locomotive Works
CNR – Changchun Railway Vehicles Co Ltd
CNR Dalian Locomotive and Rolling Stock Works
CNR Datong Electric Locomotive Co Ltd
Cockerill Maintenance & Ingenerie SA (CMI)
Colorado Railcar Manufacturing LLC
CostaRail Srl
CSR Qishuyan Locomotive and Rolling Stock Works (QS)
DEVZ – Dnepropetrovsk State Electric Locomotive Factory
Diesel Locomotive Works (DLW)
EDI Rail
E G Steele & Co Ltd
Electro-Motive Division (EMD)
Electroputere
Fahrzeugtechnik Dessau AG
FIREMA Trasporti SpA
Fuji Car Manufacturing Co
Fuji Heavy Industries

Ganz Motor Kft
GE Transportation Systems
Gevisa SA
GLG Gmeinder Lokomotiven- und Maschinenfabrik GmbH
Hanjin
Hitachi Ltd
Hunslet-Barclay Ltd
INKA
The Integral Coach Factory (ICF)
Interlok Bahnconsulting Schmidtendorf
Interlok Ltd
Jessop
Kaniewski ZUR
Kawasaki Heavy Industries Ltd
Alan Keef Limited
Keller Elettromeccanica SpA
Kinki Sharyo Co Ltd
Kolomna
Končar – Električne Lokomotive dd
Luganskteplovozstroj
Lyudinovo Locomotive Works JSC
MaLoWa Bahnwerkstatt Ltd
Man Ferrostaal AG
Metrowagonmash Joint Stock Company
Mitsubishi Electric Corporation
Moës
Motive Power Inc
Motive Power Pty Ltd
Muromteplovoz
National Railway Equipment Company
NEVZ
Newag GmbH & Co KG
Niigata Transys Co Ltd
Nippon Sharyo Ltd
Pakistan Railways Carriage
Parry People Movers Ltd
PEC Ltd
PESA Bydoszcz SA
Power Machines Group
Price
Qualter Hall & Co Ltd
RailPower Technologies Corporation

Rail Services International SA (RSI)
RCF – Rail Coach Factory
Relco Locomotives Inc
Republic Locomotive
Rocafort Ingenieros
Rotem Company
RSD (A division of DCD-Dorbyl (Pty) Ltd)
RVR
Saalasti Oy
SAN Engineering & Locomotive Co Ltd
Santa Matilde
Schöma
SCIF
SEMAF
Siemens AG
Škoda Transportation sro
Stadler Rail Group
Tafesa SA
Talgo
Talgo Oy
The Integral Coach Factory (ICF)
Tokyu Car Corporation
Toshiba Corporation
Transwerk
Tülomsas
Tüvasas
Unilokomotive Ltd
Union Carriage & Wagon Co (Pty) Ltd
United Goninan
Ventra Locomotives Ltd
Villares
Vossloh Locomotives GmbH
Wabtec Rail Ltd
Wagon Pars
Winpro AG
Zeco
Zephir SpA
Zhuzhou Electric Locomotive Works
ŽOS Vrútky as
Zwiehoff GmbH

Company listing by country

AUSTRALIA
EDI Rail
Motive Power Pty Ltd
United Goninan

BELGIUM
Cockerill Maintenance & Ingenerie SA (CMI)
Moës
Rail Services International SA (RSI)

BRAZIL
Gevisa SA
Santa Matilde
Villares

CANADA
Bombardier Transportation
RailPower Technologies Corporation

CHINA
Changzhou Diesel & Mining Locomotive Plant
China Northern Locomotive and Rolling Stock Industry (Group) Corporation (CNR)
China South Locomotive and Rolling Stock Industry (Group) Corporation (CSR)
CNR Beijing 'February 7th' Locomotive Works
CNR – Changchun Railway Vehicles Co Ltd
CNR Dalian Locomotive and Rolling Stock Works
CNR Datong Locomotive Works

CSR Qishuyan Locomotive and Rolling Stock Works (QS)
Zhuzhou Electric Locomotive Works

CROATIA
Končar – Električne Lokomotive dd

CZECH REPUBLIC
ČKD Vagonka as
ČMKS holding as
Škoda Transportation sro

EGYPT
SEMAF

FINLAND
Saalasti Oy
Talgo Oy

FRANCE
Alstom Transport
CFD Bagneres

GERMANY
Fahrzeugtechnik Dessau AG
GLG Gmeinder Lokomotiven- und Maschinenfabrik GmbH
Interlok Bahnconsulting Schmidtendorf
MaLoWa Bahnwerkstatt Ltd

MAN Ferrostaal AG
Newag GmbH & Co KG
Schöma
Siemens AG
Transrapid International GmbH & Co KG
Vossloh Locomotives GmbH
Zwiehoff GmbH

HUNGARY
Ganz Motor Kft

INDIA
Bharat Earth Movers Ltd (BEML)
BHEL – Bharat Heavy Electricals Ltd
Chittaranjan Locomotive Works
Diesel Locomotive Works (DLW)
The Integral Coach Factory (ICF)
Jessop
PEC Ltd
RCF – Rail Coach Factory
SAN Engineering & Locomotive Co Ltd
Ventra Locomotives Ltd

INDONESIA
INKA
Wagon Pars

IRELAND
Unilokomotive Ltd

Company listing by country–*continued*

ITALY
Ansaldobreda SpA
CostaRail Srl
FIREMA Trasporti SpA
Keller Elettromeccanica SpA
Zephir SpA

JAPAN
Alna Yusoki-Yohin Co Ltd
Fuji Car Manufacturing Co
Fuji Heavy Industries
Hitachi Ltd
Kawasaki Heavy Industries Ltd
Kinki Sharyo Co Ltd
Mitsubishi Electric Corporation
Niigata Transys Co Ltd
Nippon Sharyo Ltd
Tokyu Car Corporation
Toshiba Corporation

KOREA, SOUTH
Hanjin
Rotem Company

LATVIA
RVR

MOROCCO
SCIF

NEW ZEALAND
Price

PAKISTAN
Pakistan Railways Carriage

POLAND
Bumar-Fablok Ltd
Interlok Ltd
Kaniewski ZUR
PESA Bydoszcz SA

ROMANIA
Astra Vagoane Călători SA Arad
Electroputere

RUSSIAN FEDERATION
BMZ
Kolomna
Lyudinovo Locomotive Works JSC
Metrowagonmash Joint Stock Company
Muromteplovoz
NEVZ
Power Machines Group

SLOVAKIA
ŽOS Vrútky as

SOUTH AFRICA
RSD (A division of DCD-Dorbyl (Pty) Ltd)
Transwerk
Union Carriage & Wagon Co (Pty) Ltd

SPAIN
CAF – Construcciones y Auxiliar de Ferrocarriles SA
Rocafort Ingenieros
Tafesa SA
Talgo

SWITZERLAND
Alcan Mass Transportation Systems

Stadler Rail Group
Winpro AG

TURKEY
Tülomsas
Tüvasas

UKRAINE
DEVZ – Dnepropetrovsk State Electric Locomotive Factory
Luganskteplovozstroj

UNITED KINGDOM
Brush Traction
E G Steele & Co Ltd
Hunslet-Barclay Ltd
Alan Keef Limited
Parry People Movers Ltd
Qualter Hall & Co Ltd
Wabtec Rail Ltd

UNITED STATES
Brookville Equipment Corporation
Colorado Railcar Manufacturing LLC
Electro-Motive Diesel (EMD)
GE Transportation Systems
MotivePower Inc
National Railway Equipment Company
Relco Locomotives Inc
Republic Locomotive

ZIMBABWE
Zeco

Alcan Mass Transportation Systems

(Alcan Alesa Engineering Ltd)
PO Box 1250, Max Högger-Strasse 6, CH-8048 Zurich, Switzerland
Tel: (+41 43) 497 44 22   Fax: (+41 43) 497 45 85
Web: www.alcan-masstransportation.com

Key personnel
Managing Director: Frédéric Zufferey
Commercial Director: Giorgio Destefani

Background
Alcan Mass Transportation Systems is a member of the Engineered Products Business Group of Alcan Inc, which operates plants in 37 countries. The Business Unit Mass Transportation Systems is the marketing and sales organisation representing the Alcan group of companies in the transportation market worldwide.

Products
Supply of aluminium semis (large extrusions up to 800 mm width, sheets), structural sub-assemblies (such as finished floor elements, side walls and crash modules) as well as 2D composite components produced by the different Alcan plants and partner companies.

Contracts
Alcan Mass Transportation Systems has entered into co-operation agreements with more than 30 rolling stock manufacturers.

Recent contracts include low-floor regional train sets for Austrian, Danish, French, Greek, Italian, Slovakian, Spanish, Swiss and US railways; suburban train sets for Barcelona, Helsinki and

Stadler's regional diesel train type GTW for Swiss Federal Railways with bodyshells built to Alcan's hybrid system
0546801

New generation of suburban trains for RENFE (Civia) built by CAF in Zaragoza and Beasain (Spain) using large aluminium extrusions and crash modules from Alcan
NEW/1047984

Suburban trains of the X'trapolis family built by ALSTOM La Rochelle (France) and Ballarat (Australia) using light-weight interior components from Alcan
NEW/1047982

TGV Duplex – ALSTOM's high-speed intercity double-deck train using large aluminium extrusions from Alcan
NEW/1135342

Melbourne; double-deck coaches for Czech, Finnish, French, Italian and Swiss railways; tilting trains for Czech, German, Italian, Portuguese, Slovenian, Spanish, Swiss and UK railways; high-speed trains for French, German, Italian, Spanish and US railways; Trasrapid Maglev train for the airline line Shanghai.

UPDATED

Talgo's high-speed train AVE Class S102 for Renfe's new Madrid—Barcelona line. The bodyshells are built using large aluminium extrusions and crash modules from Alcan   ***NEW***/1135341

Alna Yusoki-Yohin Co Ltd

4-5 Higashi Naniwa-cho 1-chome, Amagasaki 660-8572, Japan
Tel: (+81 6) 401 72 83   Fax: (+81 6) 401 61 68
Web: http://www.alna.co.jp

Key personnel
President: N Yamazawa
Senior Director, Engineering: T Kaihara
Managing Director, Sales and Production: K Torao

Products
Aluminium, mild steel and stainless-steel electric railcars, passenger coaches and light rail vehicles.

ALSTOM Transport

Worldwide Headquarters of Transport Sector:
48 rue Albert Dhalenne, F-93482 Saint-Ouen Cedex France
Tel: (+33 1) 41 66 90 00   Fax: (+33 1) 41 66 96 66
Web: www.transport.alstom.com

Key personnel
President, Transport Sector: Philippe Mellier
Chief Operating Officer: Gérard Blanc
Chief Financial Officer: Roland Kientz
Regional Senior Vice-Presidents:
   North Europe: Terence Watson
   South Europe: Charles Carlier
   Asia Pacific: Marc Chatelard
   NAFTA: Roelof van Ark
   Iberian-Americas: Antonio Oporto
Senior Vice-President Human Resources:
   Bruno Guillemet
Senior Vice-President Legal: Fred Einbinder
Senior Vice-President Technical: François Lacôte
Senior Vice-President Communications: Patrick Bessy
Senior Vice-President Public Affairs:
   Maurice Benassayag
Senior Vice-President Business Excellence:
   Jean-Michel Geffriaud
Senior Vice-President Rolling Stock Engineering:
   Marc van Damme
Senior Vice-President Rolling Stock Manufacturing:
   Frank Lecoq
Senior Vice-President Train Life Services:
   Antonio Moreno
Senior Vice-President Infrastructure: Alain Goga
Senior Vice-President Transport Information
   Solutions: Michel Marien
Senior Vice-President Components: Roberto Sestini
Senior Vice-President Systems: Yves Bourbin

Background
ALSTOM, floated on the Paris, London and New York stock exchanges in June 1998, has a presence in over 70 countries and employs 75,000 people. For each of its main markets, ALSTOM offers a complete range of systems, components and services covering design and manufacture as well as commissioning and long-term maintenance. The company also has experience in large-scale systems integration and the management of turnkey projects.

In March 1999, ALSTOM and ABB announced their decision to merge their power generation activities in a 50-50 joint company to be known as ABB ALSTOM Power.

The company is organised into five sectors: power generation, power service, power conversion, rail transport and marine systems.

Following the terms agreed by the EC's acceptance of ALSTOM's financing package in 2004, in June 2005 ALSTOM and United Group Ltd signed a binding agreement for the sale of ALSTOM's transport activities in Australia and New Zealand. ALSTOM's transport operations in Australia and New Zealand include engineering and maintenance support, road and rail infrastructure projects and the provision of professional services and systems to the transport industry throughout Australia and New Zealand.

The Transport Sector of ALSTOM employs nearly 25,000 people. It has production facilities in more than 20 countries. ALSTOM reports that one in every four metro trains in the world, in over 40 major cities, is an ALSTOM product: nearly 550 of ALSTOM-built TGV trainsets are in operation in seven countries.

ALSTOM has acquired or formed the following companies:

In July 2002, ALSTOM announced the signing of a joint venture agreement with German Rail (DB AG) to undertake locomotive renovation and leasing at a DB site at Stendal, Germany. Subject to approval by national and European regulatory authorities, operations by the venture, in which ALSTOM will hold a 51 per cent stake, was expected to start in September 2002. In the UK, ALSTOM acquired Railcare in 2001; it specialises in heavy overhaul and refurbishment of locomotives, dmus and emus and main line coaches. With facilities at Coventry, Springburn (Glasgow) and Wolverton, Railcare has become part of ALSTOM's Transport Services business as ALSTOM Railcare Ltd.

In November 2000, ALSTOM and General Motors Electro-Motive Division (EMD) announced the formation of a joint venture, ALSTOM EMD Services, to provide locomotive maintenance services worldwide.

In Italy, ALSTOM acquired a 51 per cent shareholding in Fiat Ferroviaria in October 2000. As well as the high-speed tilting trainsets for which it has become best known, Fiat Ferroviaria's product capabilities include dmus and emus, metro cars, LRVs, diesel and electric locomotives, passenger coaches, bogies and components.

In Canada, ALSTOM acquired Telecite Inc in 1999. Founded in 1986, Telecite specialises in developing, manufacturing and marketing real-time information and communication systems for public transport and other applications.

In France, ALSTOM increased its shareholding in De Dietrich Ferroviaire from 17.5 per cent acquired in 1995 to 68.75 per cent in 1998 and 98.75 per cent in January 1999.

In the UK, ALSTOM acquired Wessex Traincare in 1998; it specialises in heavy overhaul and refurbishment of main line coaches, and electrical multiple-units.

The SASIB railway signalling business was acquired in 1998. It is based in Italy and has a presence in Europe along with GRS in the US.

Mafersa's stainless steel technology was bought by ALSTOM in 1997. Mafersa, Brazil, was founded in 1944 and its products include cars for suburban, metro and long-distance services, LRVs, car bodyshells, wheels and axles. It also undertakes the refurbishing and replacement of the main components of passenger cars.

Konstal, Poland, was bought in 1997. Its product range includes tramcars, LRVs and metros.

Also in 1997, an agreement was signed with the Fidelity and Deposit Company of Maryland to take possession of the Hornell facilities, New York, which were previously owned by the American Passenger Rail Car Company, LCC (Amerail).

The Canadian company AMF (now ALSTOM Canada Inc) was acquired in 1996 from Canadian National railway undertaking. It specialises in the maintenance and refurbishment of rolling stock, locomotives and coaches.

Linke-Hofmann-Busch (LHB) in Germany, became a subsidiary company in 1995. Its products include metro cars, low-floor LRVs and conventional

Eurostar high-speed trainset   ***NEW***/0585243

tramway rolling stock. It also delivers emus, dmus and ICE cars.

ALSTOM Transport was set up in 1995 following the acquisition of the Romanian company, Faur. It builds rail vehicles in Romania and refurbishes rolling stock.

CMW, based in Brazil, was bought in 1995. It produces a range of signalling equipment for urban, suburban and national railway network management and security.

In October 2003 a restructuring of the company saw the establishment of four customer-focused regional businesses (Northern Europe, Southern Europe, the Americas and Asia-Pacific). These are supported by a single operations group which manages ALSTOM Transport's design, engineering and manufacturing facilities.

Products

ALSTOM designs, manufactures, tests and commissions rolling stock for commuter, metro and light rail applications; as well as single- or double-deck, electric or diesel-electric trainsets for suburban, regional high-speed and very-high-speed use. ALSTOM also produces locomotives and freight wagons and markets traction and electromechanical subsystems to third-party wagon and carriage builders.

The company also designs, configures, integrates, tests and commissions rail transport network signalling systems for urban and main line rail infrastructure authorities. These include traffic management, traffic control, maintenance diagnostics and planning solutions, passenger information facilities and automatic driving systems.

These products and systems are also marketed by ALSTOM within a range of rail transport system solution packages, which may include: project management, project financing, civil works, lifetime maintenance, and initial or long-term transport service operation. All this offering is backed up by a worldwide service organisation offering customers a range of service packages from basic warranty and parts to 'lifetime support'.

Locomotives

ALSTOM has developed a new family of locomotives, named Prima. Prima locomotives can be equipped with each of the relevant train control systems to enable them to run on all electrified lines in Europe and are available in both diesel and multivoltage electric versions. The design is claimed to cut traction costs by up to 30 per cent, through its ONIX technology. It is offered as the world's first locomotive to use an IGBT traction system. A 4,200 kW 140 km/h multivoltage demonstrator has been produced.

Passenger vehicles

ALSTOM has developed six comprehensive families of passenger rolling stock to address the rail transport market, from LRVs and metros through commuter, regional and intercity trains, to high-speed and very-high-speed trains. Optionic Design© is the name given to ALSTOM's product design process, which integrates from the very beginning a wide range of configurations, based on international customers' evolving needs around the world.

The six families are: Citadis™ (light rail vehicles for city operation), Metropolis™ (metro cars for city operation), X'trapolis™ (commuter trainsets), Coradia™ (regional and intercity trainsets),

Passenger version of the Prima Type DE43C diesel-electric locomotive supplied to Islamic Iranian Republic Railways   0554536

Pendolino™ (high-speed trainsets) and TGV (very-high-speed trainsets).

All these families use service-proven equipment, bogies and subassemblies such as ONIX™ Drive and Agate train control, produced by the many ALSTOM units around the world. These factors contribute to lower LCC and higher reliability and availability.

Citadis features include:
- modularity, with a choice of car dimensions, high or low floor
- improved vision and comfort, with 35 per cent more glass surfaces compared to a traditional tramway
- high safety levels for drivers and passengers, with reinforced cab structure and bodyshell structure easy maintenance thanks to centralised spare parts supply, with plug-in, pull-out layout of equipment interfaces
- quick repairability with easy-to-exchange windows and panelling.

Metropolis is the modular mass transit solution and addresses the different needs of customers worldwide, from the most traditional metro to the most sophisticated driverless version. Choices of train dimensions, train configurations, multiple-unit operation, train-monitoring architecture and manned or unmanned operation mean that customers are able to select design parameters to create the right metro for their network. ALSTOM also offers tailor-made metro solutions for networks with specific design requirements, such as those of Paris, London and New York.

X'trapolis is a high-capacity commuter train. Like the Citadis and Metropolis families, it offers flexibility in train dimensions and configurations. Other features include: electrical multiple-units and passenger coaches, single- and double-decks, easy maintenance and upgrading to deal with different passenger flow requirements.

X'trapolis is designed to meet all track gauge, vehicle gauge and voltage requirements.

Coradia addresses the increasing trends of urban and outer-city development. Coradia regional and

intercity trains are modular flexible products, designed to provide high standards of comfort, safety and performance. The Coradia family holds the record for the fastest dmu and double-deck emu (200 and 220 km/h respectively).

The Coradia Duplex double-deck train concept with distributed power has a new bodyshell based on a system of laser-cut interlocking pieces. In this way, the structure is flexible and can adapt to any specific gauge dimensions or body shapes for technical and aesthetic purposes. It can operate at up to 220 km/h in dmu, emu or locomotive-operated push-pull version.

The Pendolino family operates at maximum speeds in the 200–270 km/h range. Most of this family are tilting trains but a non-tilting version has also been developed to suit high-speed shuttle-type operations.

The TGV family of very-high-speed trains includes single-deck, double-deck and the new AGV (Automotrice à Grande Vitesse). They are all based on the articulated trainset concept, which provides high security for passengers and drivers alike. The TGV Duplex features an aluminium bodyshell, high capacity and speeds of up to 320 km/h. The AGV can travel up to 350 km/h and has the advantage of distributed power throughout the trainset. A tilting TGV is also possible.

The Pendolino and TGV product families can all be built as tilting trains, using ALSTOM's field-proven, third-generation Tiltronix technology. For the TGV, Tiltronix would be applied to optimise performance if the train were required to leave dedicated high-speed routes and operate on traditional lines. ALSTOM's industrial unit in Savigliano, Italy, is the company's centre for Tiltronix technology.

TGV is an SNCF trademark.

Contact points
Brazil
ALSTOM Transporte Ltda
Al Campinas 463-4 Floor,
01404-902 São Paulo
Tel: (+55 11) 30 69 07 22
Contact: Juliana Souza

France
ALSTOM Transport – La Rochelle Plant
avenue du Commandant Lysiak, 17440 Aytre
BP 359, F-17001 La Rochelle Cedex
Tel: (+33 5) 46 51 30 00
Contact: Eric Lenoir

ALSTOM Transport – Valenciennes Plant
rue Jacquart – BP 45, F-59494 Petite Forêt
Tel: (+33 3) 27 14 18 00
Contact: Eric Lenoir

ALSTOM DDF
6 route de Strasbourg, F-67110 Reichshoffen
Tel: (+33 3) 88 80 25 00
Contact: Eric Lenoir

Pendolino Alaris non-tilting high-speed emu for RENFE, Spain   0554537

Coradia LINT two-car regional dmu operated by Syntus, Netherlands　　0554539

Germany
ALSTOM LHB GmbH
Linke-Hofmann-Busch-Strasse 1, D-38239 Salzgitter
Tel: (+49 53) 41 21 05
Contact: Sabine Gross

Italy
ALSTOM Ferroviaria SpA
Via O Moreno 23, I-12038 Savigliano (CN), Italy
Tel: (+39 0172) 71 83 08　Fax: (+39 0172) 71 83 06
Contact: Manuela Bozzolan

ALSTOM Transport SpA
Via Nazario Suro 38, Sesto San Giovanni, Milan, Italy
Tel: (+39 02) 248 82 62 02　Fax: (+39 02) 240 25 52
Contact: Manuela Bozzolan

Poland
ALSTOM Konstal SA
Ul Katowicka 104/41 500 Chorzow
Tel: (+48 32) 349 10 00
Contact: Michal Prochownik

Spain
ALSTOM Transporte – Barcelona Plant
Crta – B140 de Sta Perpetua a Mollet del Valles km 7.5, E-08120 Sta Perpetua de Mogoda, Barcelona
Tel: (+34 93) 575 50 00
Contact: Andres Lopes-Morancho

UK
ALSTOM Transport UK
Newbold Road, Rugby, CV21 2NH, UK
Tel: (+44 1788) 54 56 25　Fax: (+44 1788) 54 64 40
Contact: Malcolm Cowling

US
ALSTOM Transportation Inc
353, Lexington Avenue Suite 800, New York, New York 10016
Tel: (+1 212) 557 72 62
Contact: Tod Harvey

ALSTOM Transport (ex-Amerail)
1 Transit Drive, Hornell, New York 14843
Tel: (+1 607) 324 45 95
Contact: Tod Harvey

Systems solutions, systems integration and project financing
Project management, systems integration and supply packages. These packages are custom-designed from a set of modules which start with the rolling stock and signalling system and extend to embrace the remaining electrical and mechanical systems, the civil works, project finance, lifetime maintenance and even initial or long-term transport service operation.

ALSTOM Transport Systems
48 rue Albert Dhalenne, F-93482 Saint-Ouen Cedex
Tel: (+33 1) 41 66 90 00　Fax: (+33 1) 41 66 96 66
Senior Vice-President International Systems
　Operations: Yves Bourbin

ALSTOM Transport Systems Infrastructure Business Unit
33 rue des Bateliers, F-93404 Saint Ouen Cedex, France
Tel: (+33 1) 40 10 62 62　Fax: (+33 1) 40 10 60 60
Senior Vice-President, International Systems
　Infrastructure: Alain Goga

Traction equipment and subsystems
ALSTOM supplies fully integrated propulsion system packages, traction equipment and support services for modern railway and urban transportation vehicles. It offers a range of products, services and expertise in transport electronics, and in electrical and mechanical engineering and establishes partnerships with its customers to support them throughout the life cycle of their equipment.

Contact points
Traction equipment, power modules and switchgear:
ALSTOM Transport SA
50 rue du Dr Guimier, PO Box 4, F-65600 Semeac, Tarbes
Tel: (+33 5) 62 53 41 21

Onboard electronic equipment:
ALSTOM Transport SA
11-13 avenue de Bel Air, F-69627 Villeurbanne Cedex
Tel: (+33 4) 72 81 52 00

Motors:
ALSTOM Transport
7 av de Lattre de Tassigny, BP 49, F-25290 Ornans Cedex
Tel: (+33 3) 81 62 44 00

Bogies and dampers:
ALSTOM Transport
1 rue Baptiste Marcet, PO Box 42, F-71202 Le Creusot Cedex
Tel: (+33 3) 85 77 60 00

Transport services:
ALSTOM Train Services
PO Box 248, Leigh Road, Washwood Heath, Birmingham B8 2YF
Tel: (+44 121) 695 36 00 Fax (+44 121) 327 56 31

Maintenance, equipment and services section
Maintenance, renovation and spare parts:
ALSTOM Transport Train Life Services
48 rue Albert Dhalenne, F-93482 Saint-Ouen, Cedex, France
Tel: (+33 1) 41 66 90 00　Fax: (+33 1) 41 66 96 66
Senior Vice-President Train Life Services:
　Antonio Oporto

Signalling systems:
ALSTOM Transport Information Solutions
48 rue Albert Dhalenne, F-93482 Saint-Ouen, Cedex, France
Tel: (+33 1) 41 66 90 00　Fax: (+33 1) 41 66 96 66
Senior Vice-President Transport Information
　Solutions: Michel Marien

See main entry in Signalling, communications and traffic control equipment

Contracts
Argentina: ALSTOM received an order from Métrovias SA for the supply of 16 Metropolis trainsets for Line A of the Buenos Aires metro system. The order calls for the manufacture of 80 stainless steel metro cars, to be equipped with ONIX traction systems and built in ALSTOM's Brazil unit.

Australia: ALSTOM signed contracts in 2000 with Metrolink for the supply and maintenance of new trams for the Yarra Trams network of Melbourne. Under the terms of the contracts, ALSTOM was to provide 36 Citadis 300 low-floor tramsets. ALSTOM is also responsible for maintaining the new fleet over 15 years. Deliveries commenced in 2001.

In 1999 ALSTOM Transport won an order to supply and maintain 58 three-car X'trapolis suburban trains for the Connex Trains Melbourne network. Deliveries commenced in 2002.

Belgium: SNCB (Société Nationale des Chemins de Fer Belges) placed an order in 1999 for 35 sets of six double-deck passenger coaches with a consortium composed of ALSTOM and Bombardier Transportation. The order is shared equally between the two consortium members. The six-car sets are configured in pairs to form 220 km/h locomotive-hauled formations to boost intercity services to and from Brussels.

Brazil: Companhia Paulista de Trens Metropolitanos (CPTM), the public sector organisation responsible for suburban railway transport in São Paulo State, selected the Sistrem Consortium, led by ALSTOM, to build its new 9 km line from Capão Redondo to Largo Treze. The Sistrem Consortium was formed to undertake the project on a turnkey basis. Within it, ALSTOM is responsible for the overall technical management of the project, systems integration, provision from its factories in Brazil and France of eight six-car Metropolis trainsets, the signalling system, operations control centre and various items of electrical and mechanical equipment for the stations. Other members of the consortium are Bombardier Transportation, CAF and Siemens TSO.

The new line is the first phase of a project which it is planned to extend as far as Chacara Klabin station (Vila Mariana region) in the future. ALSTOM supplied 48 Metropolis metro cars to CPTM, the public transport operator in São Paulo. The metros were manufactured in Brazil. These 48 Metropolis cars are part of an ALSTOM turnkey project and are destined for the Line 5 extension of the São Paulo network which was inaugurated in October 2002.

Canada: ALSTOM was awarded an order by VIA Rail, Canada for the supply of 139 former Nightstock coaches. VIA Rail uses these coaches for overnight services, operating on its flagship

X'trapolis emu for Connex Trains, Melbourne, Australia　　0576722

Class 390 Pendolino high-speed tilting trainset for Virgin West Coast, UK   0554542

Class BB 27000 Prima dual-voltage electric freight locomotive supplied to SNCF Fret, France

0554543

selected for the supply of 16 trainsets of eight cars each.

Also in October 2004 the Ministry of Railways of the People's Republic of China chose ALSTOM for the supply of regional trains and locomotives. The contract specifies that 60 regional trains are to be supplied and built in China by ALSTOM working in partnership with the Changchun Railway Company. ALSTOM will supply three complete trainsets, six trainsets in kits, and equipment for the remaining 51 trainsets from its sites in Savigliano, Italy and La Rochelle in France. The trainsets are scheduled to enter service in 2007.

In June 2004 ALSTOM signed a co-operation agreement with Datong Electric Locomotive Co, part of China Northern Locomotive & Industry (Group) Corporation (CNR), covering the manufacture of electric locomotives based on the Prima range. The agreement allows the two companies to compete for future freight and passenger locomotive orders in the Chinese market.

In October 2003 ALSTOM was awarded a contract by the Shanghai Mass Transit Yangpu Line Development Co to supply 28 six-car Metropolis trainsets for the 23.4 km 24-station Yangpu metro line. The design of the vehicles, to be delivered between 2005 and 2007, will be similar to that of those being supplied by ALSTOM for the city's Xinmin line. Some 70 per cent of the manufacturing content will be sourced in China.

In 2002 ALSTOM, in partnership with Nanjing Puzhen Rolling Stock Works, was selected by Nanjing Metro Company Ltd to supply 20 six-car Metropolis metro trains for Phase the city's North-South Line scheme. The first train is to be built at ALSTOM's plant in Valenciennes, France, with the remainder of the contract fulfilled by the Nanjing Puzhen Rolling Stock Works. Deliveries are due to take place between May 2002 and June 2006.

ALSTOM has been awarded a contract by the Shanghai Mass Transit Pearl Line Development Co Ltd for the supply of 28 six-car Metropolis trainsets, worth some Euro203 million, for Line 3 of the city's metro. The new line links the southeast to the north of the city. The trains' lightweight aluminium cars will be equipped with ALSTOM's ONIX 1500 drive propulsion system based on IGBT technology. A significant portion of work relating to the contract will be carried out in China by several factories, including the Shanghai ALSTOM Transport Electrical Equipment Company Ltd and Nanjing Puzhen Rolling Stock Works.

A contract for the new Shanghai tramway, the Xinmin line, covers the supply of 152 Metropolis cars, with an option for a further 148. This contract was between JIUSHI, a company representing the Municipality of Shanghai, and SATCO, a joint venture between ALSTOM (with a 40 per cent stake) and Shanghai Electric Corporation (SEC) (60 per cent).

Czech Republic: ALSTOM is a member of a consortium supplying to ⏚ seven three-voltage seven-car tilting trainsets (Series 680) based on the

routes between Halifax, Montreal and Toronto. The vehicles entered service in 2002.

Chile: In April 2005 ALSTOM was awarded three contracts by Santiago Metro as part of its extension programme for line 2 north Recoleta Avenue. The first contract covers the supply of 42 metro cars and a two-year supervision of maintenance of the capital's metro. Santiago Metro exercised the option included in this contract for an additional 43 metro cars and a two-year supervision of maintenance of these cars.

In 2002 ALSTOM won a turnkey contract to equip the new 32 km Line 4 of the Santiago metro system. This includes the supply of 60 three-car Metropolis stainless steel-bodied metro trains. Commissioning of the system is scheduled for the end of 2005.

Also in 2002 Metro Regional de Valparaiso (Merval) selected ALSTOM to supply 28 two-car emus to provide main line suburban services between Valparaiso, Viña del Mar and Limache. The trains are part of ALSTOM's X'trapolis range of suburban trains. Deliveries are scheduled for 2005. The contract also covers maintenance of the trains.

ALSTOM has signed a contract with Metro SA, operator of the Santiago metro, for the supply of rolling stock intended to strengthen the existing fleet. ALSTOM is to supply 92 cars (10 eight-car trains, a seven-car trainset, and five trailer cars to be added to existing trainsets). The order includes the maintenance of the new trainsets for two years. The new trains will use the latest technology, including Agate Media, ALSTOM's passenger information system. The vehicles are designed with walkthrough gangways, which allow passengers to circulate freely throughout the train. The new vehicles will allow Metro SA to cope with rising traffic and to meet the needs linked to the extension of Lines 5 and 2 of its network. First

deliveries were scheduled for early 2002 at a rate of one train per month. This contract is a follow-on to the order placed by Metro SA in 1997, which was for six seven-car trainsets.

China: In October 2004 the Shanghai Shetong Holdings Group chose the ALSTOM consortium for the supply of Metropolis trains for the extension of Shanghai's metro line 1. The consortium, comprising SATCO (a joint venture of ALSTOM Transport and Shanghai Electric Corporation) and CSR Nanjing Puzhen Rolling Stock Works, was

Class 680 Pendolino high-speed tilting trainset for Czech Railways   0554545

M12N double-deck emu for SNCF 'Transilien' Paris area suburban services   0554548

design of Italian Railways Class ETR 460 Pendolino trainsets. Delivery was made in 2003.

Denmark: In February 2005 ALSTOM was awarded a contract by Lokalbanen, a Danish public transport operator, for the supply of 27 two-car Coradia LINT 41 regional diesel railcars for services in the greater Copenhagen region. The first delivery will take place in autumn 2006, the final delivery is scheduled for autumn 2007.

In 2002, Arriva Tog AS ordered 30 two-car Coradia LINT regional dmus for its Jutland franchise. The order includes an option for five additional vehicles. Delivery is scheduled for 2004.

Egypt: In September 2000 an order was received from Egyptian National Railways to supply 30 MTU-powered 900 kW Type GA900 AS diesel shunting locomotives. Delivery commenced in 2004.

Finland: Options were exercised in 2002 to supply to VR-Group an additional eight of a possible 15 six-car Pendolino electric tilting trainsets to supplement the 10 Class Sm3 trains already in service or in production. Delivery of the first trainset is scheduled for 2004.

At the same time, options were taken up by VR-Group to order 20 of an additional 40 Class Sm4 two-car aluminium-bodied suburban emus similar to 10 examples delivered by the former Fiat Ferroviaria in 1999. Named 'City Train' by the operator, deliveries began in 2004.

France: In June 2005 SNCF notified ALSTOM of an order for six double-deck trainsets for use in it's TER regional express fleet. This is a follow-on order that corresponds to an option in a contract signed in November 2000 and follows on from an order for 66 cars announced on 21 June 2005. Delivery of the first order is expected to begin at the end of 2006.

In March 2005 ALSTOM was awarded a contract by SNCF for the upgrade of 48 existing Corail regional cars. In addition, there are options for the transformation of up to 30 further regional cars.

In January 2005, as part of the opening of the LEA line, ALSTOM was awarded a contract for the supply of 10 tramsets, identical to the 47 low-floor tramsets already delivered by ALSTOM since December 2000 for Lyon's tram lines 1 and 2.

In September 2004 Communauté de l'Agglomération de Nice Côte d'Azur (CANCA) ordered 20 Citadis trams for its new tramway in Nice. The firm order also includes an option for a further eight and an option to subsequently increase capacity of the 20 trams. In July 2004 ALSTOM was named as supplier of trams for a new light rail system in Le Mans. Le Mans Métropole was to order 23 30 m Citadis vehicles for the 14 km system, with delivery commencing in late 2006. The line is due to be commissioned in 2007.

In February 2004 SNCF ordered 60 dual-voltage 4,200 kW Prima electric locomotives for passenger services in the Ile de France. Delivery is scheduled for 2006. The locomotives are similar to 210 freight locomotives ordered by SNCF from ALSTOM. In January 2004 SNCF confirmed an order for seven dual-voltage TGV duplex trainsets and 15 sets of eight trailers. The order is part of a master agreement signed in October 2000 covering the supply of 82 TGV duplex trainsets by the partnership of ALSTOM and Bombardier. This order will bring the

fleet of SNCF TGV duplex trainsets to 104. In January 2003 SNCF ordered 30 power cars for the TGV Est line. These will be coupled with trailers from existing TGV Réseau trainsets. This will release power cars from the TGV Réseau trainsets to be integrated with the 15 trailer sections of the order detailed above. In December 2003 the Paris transport authority, RATP, ordered 70 40 M Citadis trams Maréchaux Sud line in the south of Paris. Delivery is scheduled to be completed by 2006.

In July 2003 Strasbourg public transport operator CTS ordered 35 45 m Citadis trams to cover extensions to the city's network. Delivery is scheduled to begin in 2005. The contract includes an option on 12 additional vehicles.

In 2003, Grenoble's transport authority, SMTC, ordered 35 40 metre bi-directional Citadis trams for the third line of its light rail network. Delivery is scheduled for 2005.

Also in 2003 SITRAM, the transport authority for the metropolitan area of Mulhouse, ordered 20 Citadis trams, with an option for seven more, for its new TramTrain network. As well as serving the city, these vehicles will operate over mainline tracks in the Thur Valley. Delivery was scheduled for 2005.

The Communauté Urbaine de Bordeaux (CUB) has placed firm orders for 70 trams placed in batches of 38 and 32 vehicles, covering 56 40 m and 14 30 m vehicles. Delivery of the first tram took place in 2002. When commercial service begins in Bordeaux in 2003, the Citadis became the first tramway to operate catenary-free in certain areas on its route. From the inception of this project, one of the customer's primary concerns was that its historic old city and city centre not be marred by unsightly overhead wires. The trams will rely on a brand new and innovative technology for powering the trams by ground, called Innorail. Innorail is a system for powering trams at ground level, freeing networks from unsightly overhead catenary. Innorail uses a segmented third rail to transmit electricity to the tram. The high-performance system is completely safe: power control units, buried under the track at 22 m intervals, receive radio signals from an antenna on the tram. The control unit activates the segment only when the tram passes over and completely covers it.

In December 2002, French National Railways (SNCF) notified ALSTOM of its decision to order 86 Duplex double-deck emus for use in its TER regional express fleet. The follow-on order corresponds to the second option in a contract signed in November 2000 and is part of an investment programme to be carried out over several years and financed by regional authorities, for 629 new cars. This order, which calls for 25 three-car trainsets and 11 intermediate cars, brings the number of cars ordered in the framework so far to 327.

An order placed in March 2002 for 40 vehicles brought to 299 the number of Coradia ATER regional railcars ordered by SNCF. The vehicles are built in Reichshoffen, France.

In 2002 SNCF placed a contract with ALSTOM to supply 10 Type Z2N four-car double-deck emus for services in the Ile-de-France region. Delivery was scheduled for 2003.

In 2002 SNCF placed with ALSTOM an order to supply 29 Coradia Z TER three-car regional emus, bringing to 51 the number of vehicles of this type ordered by the operator.

The Paris Transport Authority (RATP) is to order 161 metro trains (805 metro cars) from the consortium comprising ALSTOM, Bombardier Transportation and Technicatome. The trains are destined to replace some 40 per cent of the RATP's current metro fleet, beginning in 2005. Designated MF 2000, the new trains will be composed of five cars, three of which will be powered by ONIX traction systems. Similar equipment has also been supplied by ALSTOM to the Washington, DC metro. Construction is to be undertaken at Valenciennes. A pre-series train is to be delivered for testing on Line 2 of the Paris Métro in December 2003. Full production is scheduled to begin by mid-2004 and the first series train for Line 2 will be delivered by December 2005. Deliveries for Line 5 will begin in mid-2008, and for Line 9 in mid-2011. Final deliveries are scheduled for December 2015.

In July 1998 SNCF ordered 120 Prima locomotives with options on 180 more. Each has a power rating of 4,200 kW and a maximum speed of 140 km/h. The contract was subsequently amended to cover the supply of 180 Class BB 27000 dual-voltage (25 kV AC/1.5 kV DC) locomotives, 29 Class BB 37000 three-current machines additionally equipped to operate from the German and Swiss 15 kV AC systems and one three-current Class BB 37500 locomotive additionally equipped to operate from the Belgian 3 kV DC power supply.

By 2000 ALSTOM had completed deliveries of an order to supply 80 TER2N two-car double-deck emus for the Nord-Pas-de-Calais, Provence-Côte d'Azur and Rhône-Alpes regions. SNCF subsequently notified ALSTOM of its decision to order 205 new-generation TER2N double-deck regional emu cars. This first firm order comes within the framework of an investment programme for 629 new cars, to be carried out over several years and financed by regional authorities. The new distributed power vehicle, part of the Coradia Duplex product family, will be able to operate at up to 160 km/h. The order calls for 72 trainsets, composed of two to five cars, according to service requirements. The number of seated places will be 220 in two-car and 576 in five-car versions. First deliveries began in December 2003.

Class Z 21500 three-car regional emu supplied to SNCF   0554549

TGV Duplex high-speed trainset for SNCF France 0583279

The Paris Transport Authority (RATP) notified ALSTOM in 2000 of its decision to order 13 Citadis trams, destined for existing or projected tramway lines in the Ile-de-France region. This firm order is part of a programme involving a total of 60 vehicles needed to strengthen services on tramway lines T1 (Saint Denis-Bobigny) and T2 (La Défense-Issy Val de Seine). Delivery of the new vehicles began in March 2002.

Germany: In 2005 ALSTOM was awarded lead consortiums tram contracts by the cities of Braunschweig, Darmstadt and Gera. HEAG mobilio GmbH, the transport operatior in Darmstadt, has chosen an ALSTOM-led consortium for the delivery of 18 low-floor trams. Braunschweiger Verkehrs-AG has chosen an ALSTOM/Bombardier consortium for the delivery of 12 low-floor trams and Geraer Verkehrsbetrieb GmbH has awarded ALSTOM a contract for the delivery of six low-floor trams with an option for an additional six.

In November 2003 a consortium of ALSTOM and Bombardier Transportation was awarded a contract by S-Bahn Hamburg GmbH to supply nine three-car dual-voltage 474.3 emus and to retrofit 33 similar but DC-only existing units for dual-voltage operation. Delivery is scheduled for 2006–07.

In September 2003 Deutsche Bahn AG ordered 27 two-car Coradia LINT dmus for services on DB Regio Lower Saxony's network in the Harz-Weser area. The trains are due to enter service in 2005.

In May 2003 Deutsche Bahn AG ordered 60 additional Class 423 four-car emus from ALSTOM. In total, DB AG has ordered 456 of these trains, more than 300 of which have already been delivered.

In 2003 DB Regionalbahn Wesfalen ordered 28 Coradia LINT two-regional dmus for services on the Sauerland network and around Siegen. The same contract also covered the upgrading of 19 existing Coradia LINT Class 640 single-car units to match the equipment and comfort levels of the new trains. The contract is scheduled for completion at the end of 2004.

In 2002 Hessische Landesbahn and the Westerwaldbahn ordered 28 new Coradia LINT dmus. The order includes ten single-unit and 18 two-car vehicles that allow multiple operation and flexible use. The trainsets will be delivered between April and November 2004. The vehicles will then start regular passenger service on the four lines of the Westerwaldnetz around Limburg.

In 2002 ALSTOM was awarded a contract by Hamburger Hochbahn AG for the manufacture of a sixth series of Type DT4 underground vehicles. The new four-unit trainsets will be delivered between March 2004 and April 2005.

In 2002 Landesnahverkehrsgesellschaft Niedersachsen (LNVG) placed an order for 16 Coradia LINT two-car regional dmus for NordWestBahn services in northwest Germany. Delivery was scheduled for 2003. This contract includes an option for 30 additional vehicles and follows an earlier order for 23 trains of this type. A 15-year service agreement covers fleet maintenance.

In 2001 Regionalbahn Kassel GmbH in co-operation with Kasseler Verkehrsgesellscahft placed a contract with ALSTOM to supply 28 Regio Citadis dual-voltage and diesel-hybrid tram-trains for suburban services around Kassel. Intended to operate over both heavy rail main lines and the city's tram system, the order comprises two vehicle types: 18 dual-voltage (600/750 V DC/15 kV AC 16 2/3 Hz); and 10 diesel-hybrid (600/750 V DC/diesel-electric). Delivery of the all-electric vehicles is scheduled to start in 2003, with the diesel-hybrid variant following in 2005.

Iran: ALSTOM is supplying RAI with 100 3,200 kW Prima DE 43C AC diesel-electric locomotives for passenger and freight traffic. Delivered from March 2002, the locomotives are equipped with MAN B&W Ruston diesel engines and ONIX asynchronous drive systems. The first 20 locomotives are being manufactured at ALSTOM's factory in Belfort, eastern France. The remaining 80 units are being assembled in Iran by Wagon Pars, ALSTOM's local partner for the contract.

Ireland: ALSTOM has won orders to supply 26 Citadis tram sets, with an option for an additional 14, for Dublin's LUAS LRT line. The low-floor vehicles, which use the ONIX 800 propulsion system with asynchronous traction motors, are approximately 30 m long, 2.4 m wide, and can carry up to 170 passengers. Delivery of the order began in October 2001 and was completed in May 2002. A further order was placed in August 2000 for 14 40 m partial low-floor vehicles for Dublin's Line B.

Israel: In 1998 Israel Railways (IR) commissioned three new Bo-Bo diesel-electric shunters as part of its traction fleet modernisation programme. The 920 kW centre-cab machines were supplied by ALSTOM Spain, and are similar to locomotives delivered to Spanish National Railways and Swiss Federal Railways. IR has also taken delivery of 10 Bo-Bo and eight Co-Co Prima 2,460 kW main line diesel-electric locomotives from the same builder against a June 1996 contract and in July 2000 placed an order for an additional 10 Bo-Bo locomotives. They are powered by General Motors engines.

Italy: In April 2004 Gruppo Torinese Trasporti (GTT) ordered 10 three-car Coradia emus for regional services in the Turin area. Delivery was scheduled for 2005–06.

In March 2004 Trenitalia placed an order with ALSTOM for 12 seven-car Pendolino dual-voltage high-speed tilting trainsets for delivery by 2007.

Six Minuetto trainsets were delivered to Trenitalia in September 2004. Trenitalia and the Italian regions have ordered a total of 139 Minuetto three-car regional trainsets (72 emus and 67 dmus). These orders fall within the framework of a multi-year investment programme that foresees the acquisition of 200 trains from ALSTOM's Coradia family.

In 2002 Ferrovia Trento-Malè SpA ordered 12 two-car Coradia regional emus for services in the Trento province of northern Italy. Delivery was scheduled for 2004.

In January 2001, ALSTOM was awarded a contract by FS (Trenitalia) for 37 three-car articulated Coradia trains (20 emus and 16 dmus) for regional services. The trains will be assigned to Trenitalia (14 emus, 10 dmus) and the provinces of Arezzo (four emus), Salerno (three emus) and Trento (six dmus). Deliveries were arranged to begin in 2004. Options cover the supply of an additional 90 emus and 74 dmus.

ALSTOM won an order, in consortium with Costaferroviaria, to supply 10 emus, including an option for six, for the Rome—Viterbe line.

Luxembourg: Luxembourg National Railways (CFL) authorised SNCF in 2001 to notify ALSTOM on its behalf that it will place an order for new-generation double-deck electrical multiple units. This order comes within the framework of a sales agreement schedule for 629 Coradia Duplex vehicles, signed by SNCF in 2000. The first firm order of this agreement, financed by the Regions of France, was for 205 vehicles. The new order for 36 vehicles (12 three-car trainsets), destined for CFL, was confirmed to ALSTOM in 2001. ANF Industries (Bombardier Transport) will participate in the manufacture of these trains.

Netherlands: In April 2004 HTM Personenvervoer NV awarded ALSTOM an order for 50 RegioCitadis dual-voltage three-car tram-trains for services on RandstadRail, a light rail link connecting the Hague, Zoetermeer and Rotterdam. Delivery is scheduled for 2006.

KTX high-speed trainset for Korail, Korea 0583280

For details of the latest updates to *Jane's World Railways* online and to discover the additional information available exclusively to online subscribers please visit

jwr.janes.com

Coradia double-deck emu for SJ, Sweden 0583281

ALSTOM secured a contract to supply 23 two-car Coradia Lint lightweight dmus to Syntus, a consortium formed by Netherlands Railways (NS Reizigers), the Dutch bus company VSN, and French-based public transport operator Cariane Multimodal International. The trains are used in the eastern Netherlands province of Gelderland, where a five-year concession to operate services was won by Syntus.

Rotterdam city transport authority RET selected ALSTOM to supply 60 new trams for its three Tram-Plus lines, which serve suburban areas not covered by the city's metro. ALSTOM will supply its full low-floor, unidirectional Citadis tram, featuring four double doors on one side of each vehicle. The new fleet began service in 2003.

Poland: Warsaw Metro chose ALSTOM for the delivery of the new rolling stock for the extended line which is part of the extension programme carried out by the Polish capital. Metro Warszawskie, the operator of the Polish metro, has now put into service the last six-car trainset of a total contract of 18 new Metropolis trainsets from ALSTOM.

ALSTOM won an order to upgrade the Bytom to Katowice tramway for the Tramway Communication Company of Katowice in Silesia. As well as provision of new vehicles, the turnkey order includes the refurbishment of the rail infrastructure and stations on the existing 20 km Line 6/41. The trams, which will be supplied by the company's Polish subsidiary ALSTOM Konstal, will be fitted with ONIX traction drives.

Singapore: ALSTOM won an order to supply the Singapore Land Transport Authority (LTA) with 25 six-car Metropolis trainsets for its new North East Line. This was the largest electromechanical contract awarded within the current LTA development programme. It covers the design, implementation, production, and on-site testing of the trainsets. The driverless signalling and control system will also be supplied by ALSTOM.

In 2002 ALSTOM was confirmed as supplier of the third phase of the city's 34 km Circle Line system, which when completed will employ 26 three-car Metropolis trainsets. Deliveries of these are due to commence in 2004.

In 2000 LTA ordered a further 42 driverless Metropolis cars for the Marina Line (MCL).

South Korea: ALSTOM won a contract for the development of a new 400 km high-speed rail link between Seoul and Pusan. This project involves a substantial amount of local input and ALSTOM is managing the technology transfer and co-production needed to meet these objectives. In 1994, 46 TGV trainsets were ordered to serve this line, with production split between ALSTOM and its Korean partners. Commercial service began in April 2004.

Spain: In November 2004 Ferrocarrils de la Generalitat Valenciana (FGV) ordered 10 four-car trains from ALSTOM. Delivery is scheduled to begin in 2006. In March 2004 Metropolitano de Tenerife awarded ALSTOM a contract to supply 20 30 m Citadis trams for the island's first tram line. Delivery is scheduled for 2006. The contract includes options for maintenance for the purchase of up to 13 additional trams and for extending the 30 m vehicles to 40 m.

In February 2004 Spanish National Railways (RENFE) placed a contract with ALSTOM and CAF for the supply and maintenance of 30 high-speed shuttle trains and 45 variable-gauge high-speed trains. The Class S 104 shuttle trains are of similar design to the Pendolino Alaris trains previously ordered by RENFE. Delivery is scheduled for 2006–08 and 2006–09 respectively.

In October 2003 Ferrocarrils de la Generalitat de Catalunya (FGC) placed an order with a consortium of ALSTOM, CAF and Bombardier Transportation for 13 three-car metre-gauge emus for Barcelona area suburban services. Their design is similar to that of 20 units supplied in the late 1990s.

Also in October 2003 Spanish National Railways (RENFE) ordered 40 Civia suburban 3 kV DC emus from ALSTOM. The aluminium alloy-bodied articulated units, which feature low-floor entrances, are to be delivered as 20 four-car and 20 three-car sets. The contract includes undertaking the vehicles' maintenance.

In April 2003 ALSTOM won a contract to build nine light rail trains from FGV, the transportation

authority of the Spanish region of Valencia. The light trains will operate in both tram mode, with a top speed of 70 km/h, and train mode, with a top speed of 100 kph. The bi-directional trainsets will be 37 m long, with passenger capacity of 99 seated and 204 standing. The vehicles are for the Tren-Tram service from Alicante to Altea, (about 50 km). Trip time between Alicante and Altea will be less than an hour. Delivery was scheduled to begin in May 2005 and to be complete by October 2005.

In 2002 Autoritat del Transport Metropolita SpA (ATM) awarded ALSTOM a contract to supply 50 five-car Metropolis driverless trainsets for the 41.3 km Line 9 of the Barcelona metro system. ALSTOM's share of the order is 86 per cent; Ansaldo is to supply bogies and auxiliary converters. Delivery of the vehicles is due to take place from 2004 to 2007.

Sweden: In 2002 Storstockholms Lokaltrafik (SL) placed an order with ALSTOM to supply 55 six-car Coradia Lirex regional emus for operation in the greater Stockholm area. The contract includes an option for 50 additional trains. Deliveries were due to begin in 2005.

In 2000, ALSTOM was awarded a contract by SJ AB to supply 113 Coradia Duplex double-deck emu cars to be formed into two- and three-car sets. Delivered between 2003 and 2004, the trains were destined for inter-regional services in the Lake Mälaren region.

Switzerland: In March 2004 Cisalpino, a joint venture between Swiss Federal Railways and the Italian national operator Trenitalia, ordered 14 seven-car Pendolino three voltage tilting trainsets for delivery between 2007 and 2008.

As part of a turnkey contract awarded in 2002 by the Administration of Switzerland's Vaud canton, ALSTOM is to supply 15 two-car driverless rubber-tyred metro trainsets for a new 6 km linking Ouchy with the district of Epalinges, Lausanne. The Urbalis train control system will be employed. Commissioning of the system is due to take place in 2007.

As part of the Intercity Neigezüge SBB consortium, in 2001 ALSTOM received a contract to supply Tiltronix tilting technology and equipment for 10 additional ICN seven-car tilting trainsets for Swiss Federal Railways (SBB). This latest order will bering to 44 the number of ICN trainsets orderedby SBB.

Thailand: In September 2000 as part of the Nippon-Euro Subway Consortium, ALSTOM was selected by Bangkok Metro Company Ltd to become a partner in a 25-year concession to design, build, operate and maintain the 20 km Blue Line of the Bangkok metro. ALSTOM is to supply 21 three-car Metropolis trainsets, with an option on an additional four sets. The company is also responsible for overall management of E&M supply, systems integration, signalling, communications systems and provision of depot/workshop equipment.

Tunisia: In June 2004 the Société du Métro de Tunis (SMLT) made a provisional contract award to ALSTOM covering the supply of 30 Citadis low-floor trams. The 30 m bi-directional vehicles will both strengthen the existing fleet on SMLT's 32 km network and serve two planned 7 km extensions to be commissioned in 2006.

UK: ALSTOM delivered the first completed tilting Pendolino train for the West Coast route in 2001. Virgin Trains ordered 53 Pendolinos. Completion of deliveries was due to made in 2004. They are built and tested for 225 km/h, but will run at a lower speed until the deferred infrastructure enhancement envisaged in Phase II of the West Coast upgrade is implemented.

US: In 2002 a consortium of ALSTOM and Kawasaki received an order from the Metropolitan Transportation Authority of New York City Transit (MTA-NYC) to supply 660 heavy rail subway cars. To be delivered in 2006–07, the new cars will replace existing Type R32, R38, R40 and R42 vehicles.

Also in 2002, ALSTOM was awarded a contract by the Washington Metropolitan Transportation Authority (WMATA) to supply 62 heavy rail subway cars for its Blue Line extension to Largo Town Centre. The contract includes an option to supply

Coradia Duplex emu for SNCF, France 0583285

Coradia 'Minuetto' for Trenitalia, Italy   **NEW**/0583286

120 cars for WMATA's Orange Line extension and for general traffic growth.

In 2000, ALSTOM supplied Amtrak with eight five-car double-deck trainsets for services in the Southern and Central Coast regions of California.

In 2001 New Jersey Transit Corporation (NJT) voted to award ALSTOM a contract for the design and manufacture of a fleet of 33 new diesel-electric passenger locomotives, spare parts and an option for an additional five locomotives. NJT also has

the option to award an additional 42 locomotives in the future.

Uzbekistan: 15 Prima 1,520 mm 25 kV gauge 4,200 kW electric locomotives have been delivered.

Venezuela: ALSTOM, as part of the FRAMECA consortium, was awarded a turnkey order from the Caracas metro authority, CAMC, for the 5.5 km Line 4 of the city's metro system. In addition to supplying 44 metro cars and the signalling system, ALSTOM will have carried out electrification of the line and provided a complete fire protection system.

In 2002 as part of the same FRAMECA consortium, ALSTOM won a turnkey contract from CAMC to equip an extension to Line 3 of the Caracas metro. This included the supply of seven six-car metro trainsets.

UPDATED

Ansaldobreda SpA

Via Argine 425, I-80147 Naples, Italy
Tel: (+39 081) 243 11 11   Fax: (+39 081) 243 26 98; 99
Web: www.ansaldobreda.it

Key personnel

President: Fausto Cutuli
Chief Executive Officer: Roberto Assereto
Industrial Managing Director: Francesco Schirripa
Executive Vice-President, Commercial:
  Claudio Mannucci
Executive Vice-President, Engineering:
  Carlo Pellegrini

Background

Ansaldobreda is the Finmeccanica Transportation Sector company responsible for designing and manufacturing railway and mass transit vehicles. The company is a merger of the Ansaldo Trasporti unit, which produces electronic drives and vehicle-borne equipment, with Breda Costruzione Ferroviarie, a leading rail vehicle manufacturer.

Products

Ansaldobreda supplies railway and mass transit vehicles applying three different construction technologies: aluminium, stainless steel and carbon steel. The range of its products varies from high-speed trains, electric and diesel locomotives, electric and diesel locomotives and diesel multiple-units, all types of passenger coaches single- and double-deck, freight wagons, bogies, metro cars, low-floor light rail vehicles.

Contracts

In 2004 Ansaldobreda won a contract for 12 high-speed trains that will connect the Netherlands and Belgium.

Forty Sirio LRVs for Gothenburg. Each LRV is 29 m long and is single-ended, seating 83 with standing room for 96. Delivery is expected by 2005; 35 trams for the Greek city of Athens to be in operation during the Olympic Games in 2004. The company is building more than 130 similar trams for the cities of Milan, Naples, Sassari and Bergamo. The rolling stock fleet of 19 driverless vehicles with IGBT inverter control was supplied for a 15 km automated LRT system in Copenhagen with 15 stops, linking Norreport in the centre with Orestaden and Lergravsparken.

Ansaldobreda Class IC/4 dmu for Danish State Railways (DSB)   0580477

Mock-up of front section of Ansaldobreda three-car articulated emu for the Circumvesuviana Railway, Italy (Ken Harris)   **NEW**/1115018

Ansaldobreda Class BM72 emu for Norwegian State Railways (NSB)   0530158

Ansaldobreda ETR 500 high-speed trainset for FS   0530159

Ansaldobreda E402 electric locomotive for FS   0530160

Ansaldobreda TAF emu for North Milan Railways (FNM)   0530161

In 2003 Moroccan Railways (ONCFM) ordered 18 four-car double-deck emus, with an option on six more, based on the design of the TAF emus supplied to FS Trenitalia and North Milan Railway. Delivery is scheduled for 2006.

In 2000, Danish State Railways placed an order with Ansaldobreda for 83 Class MG (IC/4) four-car high-speed dmus. Delivery was due to commence in 2003. An option exists for 67 additional units. A new contract for 23 Class IC2 dmus was awarded in 2002.

Oslo Sporveier has taken delivery of 17 articulated LRVs with IGBT inverter drive. An option exists for 13 more. They have low floors and seat 96 with 122 standing. Each car is mounted on four bogies, each having two asynchronous traction motors; top speed is 80 km/h and floor height 350 mm. The cars are of the same type as those supplied for Line 1 of the Midland Metro.

Z1-type coaches for Italian Railways (FS); amenity coaches for Eurotunnel; panoramic coaches for BVZ, FO and MOB, Switzerland; trailer vehicles for ETR 500 high-speed trainsets for FS; and double-deck emu trailer cars for FS and North Milan Railway.

Traction equipment for 73 E652 chopper-controlled locomotives for Italian Railways (FS) has been supplied, also 60 ETR 500 high-speed trainsets for FS (supplied by the Trevi Consortium, of which Ansaldo Trasporti is a member); and 120 E402 electric locomotives with inverter drives and asynchronous traction motors for FS, with 80 equipped for operation at 3 kV DC and 25 kV 50 Hz AC and 24 for operation at 3 kV DC, 25 kV AC, 15 kV AC.

In 1997, Norwegian State Railways (NSB) Norway awarded Ansaldo Trasporti a contract for the delivery of 36 emus for local traffic.

UPDATED

Astra Vagoane Călători SA Arad

1-3 Petru Rareş, RO-2900 Arad, Romania
Tel: (+40 257) 23 62 10   Fax: (+40 257) 25 81 68
e-mail: astra.calatori@astrac.rdsar.ro
Web: www.astracalatori.ro

Key personnel
Administrator: Alexandru Truta
General Manager: Dan Micalacian
Production Director: Gheorghe Vărşăndan
Commercial Director: Raj Epuran
International Relations Director: Romulus Nosner

Background
The company was formed in 1998 by splitting it from freight wagon and passenger coaches manufacturer Astra Vagoane Arad SA.

Products
Certified ISO 9001 for the design, manufacture and refurbishment of passenger coaches and rail urban passenger vehicles. Passenger coaches for international and domestic traffic, metro cars, dmus, emus, light rail vehicles. Refurbishment is also undertaken and the company has renovated coaches for Romanian Railways, alone and in co-operation with Alstom.

Astra is licensed to build 200 km/h Corail coaches.

UPDATED

Z1 UIC 200 km/h passenger coach AVA 200 – Corail licence   0142356

CNR Beijing 'February 7th' Locomotive Works

1 Yanggingzhuang, Changxindian, Beijing 100072, China
Tel: (+86 10) 83 30 60 01   Fax: (+86 10) 83 30 37 36

Key personnel
Contact: Wang Dongming

Background
Member of China North Locomotive and Rolling Stock Industry (Group) Corporation.

Products
Diesel locomotives; diesel engines; hydraulic transmissions; fuel injectors; fuel injection pumps; cardan shafts; and bogies.

The BJ series of diesel-hydraulic locomotives is produced in 1,990 kW B-B and 3,980 kW B-B versions. The B-B has been in series production since 1975. Both models employ the Type 12V240ZJ-1 12-cylinder 1,000 rpm engine, which is also manufactured in the works. The B-B weighs 92 tonnes, has a starting tractive effort of 23.1 t, a continuous tractive effort of 16.27 t at 24.3 km/h and a maximum speed of 120 km/h.

The works also manufactures the Model DF7 Co-Co diesel-electric locomotive, for heavy shunting. It is fitted with a four-stroke 12-cylinder Vee engine of Type 12V240ZJ-1, exhaust turbocharged with intercooling, which has a rating of 1,470 kW at 1,000 rpm. The transmission is AC/DC alternator, employing silicon rectifiers.

The Model DF7B Co-Co diesel-electric locomotive produced for freight traffic has a rating of 1,840 kW at 1,000 rpm and is equipped with rheostatic braking.

To meet the needs of industrial and mining industries the works produces the Model GK1E diesel locomotive with a hydrostatic transmission system. The locomotive is powered by a six-cylinder Vee diesel engine, which has a 240 mm bore and a stroke of 260 mm, producing a maximum rating of 1,000 kW at 1,100 rpm.

Bharat Earth Movers Ltd (BEML)

BEML Soudha, 23/1 4th Main, SR Nagar, Bangalore 560 027, India
Tel: (+91 80) 222 44 58
Fax: (+91 80) 229 19 80
e-mail: techrnd@vsnl.com
Web: http://www.bemlindia.com

Key personnel
Chairman and Managing Director:
  Shri V RS Natarajan
Director, R&D: V S Venkatanathan
Executives Director: M P Sriram
Deputy General Manager (R&D): P Bayya Reddy

25 kV AC emu for Indian Railways   0527276

Products
Electric multiple-units, diesel railbuses, railcoaches.

Contracts
Recent contracts for Indian Railways include the supply of a lightweight, two-axle, 1,676 mm gauge diesel railbus and electric multiple-units with 25 kV AC traction equipment.

Lightweight passenger coaches of integral welded steel construction of all types including sleeper coaches, day travel coaches, treasury vans, postal vans, brake and luggage vans and motor-cum-parcel vans. The division has supplied over 12,000 coaches of different types for Indian Railways. Coaches have also been exported to Bangladesh and Sri Lanka.

BEML has been selected by Delhi Metro to manufacture 45 rakes (180 coaches) through technology transfer from a Japanese/Korean consortium.

Two-axle 1,676 mm-gauge diesel railbus for Indian Railways   0527275

Bharat Heavy Electricals Limited (BHEL)

Transportation Business Department
Integrated Office Complex
Lodhi Road, New Delhi 110003, India
Tel: (+91 11) 24 36 93 77; 51 79 33 00
Fax: (+91 11) 24 36 94 23
e-mail: opb@bheldindustry.com
Web: www.bhel.com

Key personnel
General Manager: O P Bhutani

Products
Complete rolling stock including a variety of traction machines, traction controls, traction transformers and locomotive components for diesel locomotives, AC locomotives, AC/DC locomotives, demus, AC emus, DC emus and AC/DC emus (three-phased drives). Complete AC and

AC/DC locomotives, battery powered road locomotives and OHE test cars. Diesel shunting locomotives range from 261-1,939 kW (350-2,600 hp) and special purpose vehicles including diesel electric tower cars, rail-cum-road vehicles, dynamic track stabilisers, ballast cleaning machines.

UPDATED

BMZ

JSC Bryansk Engineering Works
ulica Ulyanova 26, 241 015 Bryansk, Russian Federation
Tel: (+7 832) 55 86 73; 00 30   Fax: (+7 95) 203 33 95

Key personnel
Senior Marketing Manager: Natali Skrobova
Marketing Manager: Dmitri Melnichuk

Products
Diesel shunting locomotives, generator vans.

Bombardier Transportation

Management Office
Saatwinkler Damm 43, D-13627 Berlin, Germany
Tel: (+49 30) 383 20   Fax: (+49 30) 38 32 20 00
Web: www.transportation.bombardier.com

Key personnel
President: André Navarri
Chief Operating Officer: Wolfgang Tœlsner
Vice-President, Communications: Linda Coates

Vice-President, Strategy, Markets and Product Planning: Trung Ngo
Vice-President and Chief Procurement Officer: Pierre Attendu

Bombardier Transportation
North America
1101 Parent Street, Saint-Bruno, Québec J3V 6E6, Canada
Tel: (+1 450) 441 20 20   Fax: (+1 450) 441 15 15
President, North America: William Spurr

Bombardier Transportation Mainline & Metros
Am Rathenaupark, D-16761 Hennigsdorf, Germany
Tel: (+49 33) 330 28 90   Fax: (+49 33) 33 02 89 20 88
President, Mainline & Metros: Olof Persson

Bombardier Transportation Light Rail Vehicles
Donaufelder Strasse 73-79, A-1211 Vienna, Austria
Tel: (+43 1) 251 10   Fax: (+43 1) 25 11 08
President, Light Rail Vehicles: Walter Grawenhoff

Bombardier Transportation Locomotives & Freight
Brown-Boveri Strasse 5, PO Box 8384, CH-8050 Zürich, Switzerland
Tel: (+41 1) 318 33 33   Fax: (+41 1) 318 27 27
President, Locomotives & Freight:
    Edmund Schlummer

Bombardier Transportation Total Transit Systems
1501 Lebanon Church Road, Pittsburgh, Pennsylvania 15236 USA
Tel: (+1 412) 655 57 00   Fax: (+1 412) 650 64 86
President, Total Transit Systems: Ray Betler

Bombardier Transportation Propulsion and Controls
Brown-Boveri Strasse 5, CH-8050 Zürich, Switzerland
Tel: (+41 1) 318 33 33   Fax: (+41 1) 318 15 43
President, Propulsion and Controls:
    Åke Wennberg

Bombardier Transportation Rail Control Solutions
10 Church Street, Reading RH1 2SQ, UK
Tel: (+44 118) 953 80 00   Fax: (+44 118) 953 84 83
President: Josef Doppelbauer

Bombardier Transportation Services
Litchurch Lane, Derby DE24 8AD, UK
Tel: (+44 1332) 34 46 66   Fax: (+44 1332) 26 64 72
President, Services: Rik Dobbelaere

Bombardier Talent dmu for Deutsche Bahn, Germany   0558259

ICE-tilting train for Deutsche Bahn, Germany   0116578

Bombardier NINA emu for BLS, Switzerland   0116583

Bombardier Transportation, Bogies
Siegstrasse 27, D-57250 Netphen, Germany
Tel: (+49 271) 70 20   Fax: (+49 271) 70 22 22
President: Robert Wassmer

Bombardier Transportation, London Underground
Projects
Litchurch Lane, Derby DE24 8AD, UK
Tel: (+44 1332) 34 46 66   Fax: (+44 1332) 251 635
President: T C Chew

Australia
John Ince
Tel: (+61 3) 97 94 22 33
Fax: (+61 3) 97 06 92 19

Austria
Bernhard Rieder
Tel: (+43 1) 25 11 01 78
Fax: (+43 1) 25 11 05 31

Belgium
André Detollenaere
Tel: (+32 50) 40 11 11   Fax: (+32 50) 40 18 40

Brazil
Serge van Themsche
Tel: (+55 11) 37 48 97 00
Fax: (+55 11) 37 48 99 31

Canada
Hélène Gagnon
Tel: (+1 450) 441 81 56
Fax: (+1 450) 441 30 90

China
Jian Wei Zhang
Tel: (+86 10) 85 29 68 00
Fax: (+86 10) 85 29 91 09

Denmark
Kirsten Petersen
Tel: (+45) 86 42 53 00
Fax: (+45) 86 41 45 64

Finland
Christoffer Enckell
Tel: (+35 81) 02 22 62 10
Fax: (+35 81) 022 20 67

France
Jean-Pierre Hulot
Tel: (+33 1) 53 45 84
Fax: (+33 1) 53 45 84 83

Germany
Ulrich Bieger
Tel: (+49 30) 38 32 11 78
Fax: (+49 30) 38 32 20 20

Hungary
Janos Ujhelyi
Tel: (+36 27) 54 21 47
Fax: (+36 27) 39 01 43

India
Mahesh Kumar Ahuja
Tel: (+91 11) 618 67 94
Fax: (+91 11) 618 66 51

Israel
Yossy Daskal
Tel: (+972 3) 612 11 06
Fax: (+972 3) 612 11 07

Italy
Luis Ramos
Tel: (+35 12) 14 96 95 54
Fax: (+35 12) 14 96 93 94

Korea
Darryl Sailor
Tel: (+65) 65 49 71 50
Fax: (+65) 65 49 72 92

Latin America (except Brazil and Mexico)
Francisco Garcia
Tel: (+613 3) 84 31 00
Fax: (+613 3) 84 52 40

Malaysia
Darryl Sailor
Tel: (+65) 65 49 71 50
Fax: (+65) 65 49 72 92

Mexico
Alejandro Gutierrez Marcos
Tel: (+52 55) 50 93 77 00
Fax: (+52 55) 50 93 77 51

Norway
Tom Korstad
Tel: (+47) 63 80 96 58
Fax: (+47) 63 80 97 76

Poland
Janusz Kucmin
Tel: (+48 71) 356 25 96
Fax: (+48 71) 355 57 31

Portugal
Luis Ramos
Tel: (+35 12) 14 96 95 54
Fax: (+35 12) 14 96 93 94

Russia
Christian-David Mueller
Tel: (+7 09) 57 75 18 30/35
Fax: (+7 09) 57 75 18 32

Bombardier emu for Öresund Link services between Denmark and Sweden   0558262

Voyager demu for Virgin Trains, UK   0116284

Bombardier BiLevel commuter car for Sound Transit Board, Seattle   0116575

Spain
Luis Ramos
Tel: (+35 12) 14 96 95 54
Fax: (+35 12) 14 96 93 94

Sweden
Christian H Schmidt
Tel: (+46 21) 31 78 17
Fax: (+46 21) 13 51 32

Switzerland
Alfred Ruckstuhl
Tel: (+41 1) 318 27 33
Fax: (+41 1) 318 30 80

Taiwan
Michael Chung
Tel: (+886 2) 87 88 16 29 Ext. 32
Fax: (+886 2) 87 88 16 09

Turkey
A Birol Altan
Tel: (+90 312) 26 64 70
Fax: (+90 312) 2266 64 52

UK
Neil Harvey
Tel: (+44 1332) 26 64 70
Fax: (+44 1332) 26 64 72

US
Maryanne Roberts
Tel: (+1 215) 639 14 44
Fax: (+1 215) 639 37 24

Background
Serving a diverse customer base around the world, Bombardier Transportation is the global leader in the rail equipment manufacturing and servicing industry. Its wide range of products includes passenger rail vehicles and total transit systems. It also manufactures locomotives, bogies, propulsion and controls and provides rail control solutions.

Bombardier Transportation's revenues for the fiscal year ended Jan. 31, 2004 amount to $6.9 billion US (6.1 billion Euros). It is a unit of Bombardier Inc., a global corporation based in Canada, world-leading manufacturer of innovative transportation solutions, from regional aircraft and business jets to rail transportation equipment. Bombardier Inc.'s revenues for the fiscal year ended Jan. 31, 2004 totalled $15.5 billion US and its shares are traded on the Toronto Stock Exchange (BBD).

Bombardier has withdrawn from the freight car business and in December 2004 sold its Saxon facility in Niesky, Germany and also sold its interests in the Mexico-based freight car manufacturing joint venture with the Greenbrier Companies.

Products
Bombardier Transportation offers a full range of rail vehicles for urban and mainline operation as well as modernization of rolling stock and operations and maintenance services. Products include

Bombardier AGC high-capacity regional express train for SNCF, France   0583231

M7 emu for Long Island Rail Road, New York, US   0528510

metro cars, light rail vehicles/trams, single and double-deck electric multiple units (EMUs), diesel multiple units (DMUs) and coaches; tilting trains and high-speed trains. Bombardier Transportation also supplies complete transportation systems, from high-capacity urban transit systems to automated people movers. Moreover, Bombardier Transportation offers electric and diesel locomotives; propulsion and controls; rail control solutions; bogies and freight cars.

Bombardier, FLEXITY, MOVIA, FICAS, BiLevel, Electrostar, Turbostar, Regina, Talent and TRAXX, are trademarks of Bombardier Inc or its subsidiaries.

SkyTrain is a trademark of BC Transit Corp.

Contracts
Australia: In October 2004 Bombardier was awarded a contract to supply the city of Adelaide with nine light rail vehicles of the Bombardier FLEXITY Classic family for the Glenelg tramway. The delivery is scheduled to start at the end of 2005.

In a joint venture with Australian train manufacturer EDI Rail, Bombardier was awarded a contract in November 2004 by the Queensland Government to supply 24 three-car emus to QR. Sixteen of these are for Gold Coast and other interurban services, with the remaining eight destined for increased suburban operations.

In 2002 Bombardier Transportation was awarded a contract from the Victoria State Government and V/Line to provide passenger rolling stock for regional services in Victoria, Australia. This

Bombardier Turbostar dmus for UK operators 0092252

Bombardier Electrostar emu for South Central, UK 0558264

was a follow-on order to a contract awarded to Bombardier Transportation in November 2001 and includes the manufacture of nine two-car diesel units (dmus) and a 15-year maintenance agreement. This represents a total of 38 two-car dmus and an agreement for maintenance of these trains for a period of 15 years. The Vlocity 160 stainless steel trains have been designed in Australia. Bombardier Transportation's facility in Derby, UK will provide the bogies; its Västerås, Sweden site will supply the train control system. The trains will be manufactured at Bombardier's Dandenong facility in Victoria.

Also in 2002 Bombardier Transportation and Australian train manufacturer EDI-Rail announced that their joint venture received an order from the Western Australian government to build 31 three-car electric commuter trains for the southern leg of the Perth Urban Rail Development (PERD) project. The contract includes 15 years of maintenance and the construction of a maintenance and stabling facility. Bombardier Transportation's scope is to design the railcars and supply propulsion and electrical systems. EDIRail will be supplying bodyshells and bogies.

Austria: In December 2004 Bombardier received an order from Wiener Linien, Vienna's local operator, for the construction of 38 low-floor light rail vehicles (LRVs). The contract also includes an option for a further 42 LRVs. Delivery of the LRVs for Line U6 is scheduled for the period between the end of 2006 and mid-2008. Bombardier received an order for four additional three-car light rail vehicles for the Viennese operator Wiener Lokalbahnen AG. This order is an execution of an option from a previous contract for six vehicles which have been in operation since 2000.

In November 2003 Austrian Federal Railways ordered 60 four-car Talent emus for regional and commuter services. This follows a 2001 order for 40 four-car and 11 three-car Talent emus from Bombardier Transportation and ELIN EBG Traction.

Belgium: In September 2003 the Brussels transport authority, STIB, awarded Bombardier a contract to supply 46 bi-directional FLEXITY Outlook low-floor trams for delivery between 2005 and 2007. In 2001 De Lijn ordered 10 low-floor intermediate cars to equip existing vehicles, two further batches of 11 cars each were ordered in 2001 and 2002.

In 1999, SNCB (Belgium National Railways) ordered 210 double-deck coaches from Bombardier Transportation in consortium with ALSTOM. In October 2004 the same consortium was awarded a contract for 70 additional coaches of the same type, together with an order to modify 19 Class 27 electric locomotives to work with the new vehicles.

Delivery of this second order is scheduled to take place between October 2006 and February 2008.

Canada: In March 2005 Bombardier received a firm order from the Great Toronto Transit Authority (GO Transit) for 10 additional Bombardier BiLevel commuter rail vehicles. The order will add to GO Transit's existing fleet of 385 Bombardier-built commuter railcars. In March 2004 the Greater Toronto Transit Authority (GO Transit) awarded Bombardier Transportation a contract to supply 10 new-generation BiLevelcommuter coaches for delivery from late 2004. Bombardier Transportation had already delivered 361 BiLevel suburban vehicles to this operator.

In September 2003 Agence Métropolitaine de Transport (AMT) placed an order with Bombardier for 22 aluminium-bodied BiLevel coaches for its Montreal/Dorion to Rigaud line. Deliveries were completed in June 2005.

In 2002 Bombardier Transportation participated in the official ceremony to launch the fully automated 20-km Millennium Line, an extension of the Vancouver Sky-Train system. At 49 km, Sky-Train is the longest fully automated driverless transit system in the world.

China: In 2005 Bombardier received an order for 20 eight-car high-speed trains for China. In February 2005 Bombardier Transportation and its joint venture partners, Power Corporation of Canada and China South Locomotive and Rolling Stock Industry (Group) Corporation, through their joint venture Bombardier Sifang Power (Qingdao) Transportation Ltd (BSP), along with consortium partner Sifang Locomotive and Rolling Stock Co Ltd, received an order from the Ministry of Railways of China (MOR) for the production and delivery of 361 cars to be used for new train line services to Lhasa in Tibet. Delivery is scheduled to take place between December 2005 and May 2006.

In October 2004 the Bombardier Sifang power (Qingdao) Transportation Ltd (BSP) joint venture (Bombardier Transportation, Power Corporation of Canada, China South Locomotive and Rolling Stock Industry (Group) Corporation) was awarded a contract by the Chinese Ministry of Railways for 20 eight-car high-speed trainsets. Bogies for the 200 km/h vehicles will be manufactured at Bombardier's Siegen plant in Germany and some propulsion equipment will be produced at Västerås, Sweden. Final assembly will be undertaken in China, with delivery scheduled for 2006–07.

In February 2003 Changchun Bombardier Railway Vehicles Ltd (CBRC), a joint venture between Bombardier Transportation and Changchun Car Company, announced a follow-on order from Shenzhen Metro Corporation for three six-car Moviatrainsets for the first phase of Line 1 of the Shenzhen metro. This followed an order placed with CBRC in November 2001 for 114 similar metro cars (19 six-car trains) for the same customer. The first train of the original order will be manufactured in Hennigsdorf, Germany with the remaining 18 trains to be built at the CBRC plant in Changchun, Jilin Province in northeast China. Bombardier's facility in Västeras, Sweden will provide the propulsion. The metro trainsets are planned to be delivered between early 2004 and the second half of 2005. Revenue service is scheduled to commence in December 2004.

In December 2002 CBRC received an order from Shanghai Metro Operation Company for 60 metro cars configured as 10 six-car sets for use on Shanghai Metro Line 1. Delivery is due between June 2004 and May 2005.

In August 2000 Guangzhou Metro Corporation placed an order with CBRC for 26 six-car Movia trainsets to serve Line 2 of the city's metro network. The first two units were manufactured in Germany, leaving the remaining 24 to be assembled at the Changchun plant in China. The first trainset from this order was formally handed over in December 2003. In October 2004 an order was placed with CBRC for 48 additional cars (eight six-car trainsets) for delivery in 2006 to equip Lines 1 and 2.

Bombardier 'Xinshisu' tilting train, China 0524868

In 1999 Bombardier Transportation received an order to build 300 intercity coaches at one of its joint venture manufacturing facilities in China, Bombardier Sifang Power (Qingdao). The first 22 coaches from this order were delivered to the Shanghai Railways Administration Bureau in January 2003 and were part of a batch of 70 destined for this enterprise.

Croatia: In 2003 Bombardier Transportation received an order from Croatian State Railways (HZ) for eight two-car 'Regio-Swinger' dmus with tilting technology. Delivery commenced in 2004.

Denmark: In 2001, an order was received from UK-based Porterbrook Leasing Company Ltd for 42 double-deck passenger coaches to be leased to Danish State Railways (DSB) for services between Copenhagen and Kalundborg. The order comprised 33 intermediate trailers and nine driving cars. In 2002, a further order was received from Porterbrook for 25 additional double-deck coaches. These vehicles are leased by DSB for use on Copenhagen Lolland/Falster and Copenhagen Seeland routes.

Other contracts fulfilled in Denmark include 48 IR4 Flexliner emus for DSB and 13 Flexliner RL2D dmus for several independent Danish railways.

France: SNCF ordered 100 AGC trains in 2004 and 48 AGCs in 2005. In December 2004 Bombardier Transportation received an order from the Communauté Urbaine Provence Métropole (CUMPM) for the delivery of 26 bi-directional Bombardier FLEXITY Outlook trams. The new 100 per cent low-floor trams will enter service on the new tram network of Marseille that opens at the beginning of 2007. Deliveries are scheduled between September 2006 and April 2007.

By February 2003 Bombardier Transportation had received firm orders for 279 high-capacity Type AGC (Autorail Grande Capacité) regional trains against a 2001 contract placed by SNCF covering the supply of up to 500 of these vehicles. The AGC design provides for diesel, electric or dual-powered versions of the train. Deliveries commenced in 2004.

In January 2003 the Paris Transport Authority (RATP) placed a contract with a consortium of Bombardier Transportation and ALSTOM Transport for 14 five-car MI2N double-deck emus for the city's suburban network. The order brought the number of vehicles of this type ordered by RATP to 215. Bombardier's share in the contract covers the production of 42 intermediate cars. The vehicles entered service in 2004.

Also in January 2003 a consortium of Bombardier Transportation and ALSTOM received an order from SNCF to supply 25 three-car TER2N NG emus for services in the Rhône-Alpes and Lorraine regions and 11 intermediate cars to be added to two-car trainsets already ordered for the Rhône-Alpes region. Bombardier is to be responsible for the manufacture of 25 vehicles and 86 motor bogies. Delivery is scheduled to begin in March 2006. This

Class ET 474 suburban emu for S-Bahn Hamburg by a consortium that includes Bombardier

0567265

brought the number of trains of this type ordered to 327; options exist for 302 additional vehicles.

In October 2002 Bombardier Transportation received an order from ALSTOM Transport to supply six trailer cars and 12 unpowered bogies for Type ZTER emus for SNCF.

In 2001 a consortium of Bombardier Transportation and ALSTOM received an order from SNCF for 18 TGV Duplex double-deck high-speed trainsets. These represent part-confirmation of an option on 60 vehicles included in the contract placed in 2000 (see below). Bombardier's share of the contract, which was fulfilled between March 2004 and March 2005, covers the manufacture of two first class and one second class car and six carrying bogies for each 10-car trainset.

In 2001 Bombardier received an order from RATP, the Paris transport authority, for 161 five-car Type MF-2000 metro trains in consortium with ALSTOM Transport and Technicatome.

In June 2004 Bombardier Transportation was awarded an initial contract by the Communauté Urbaine Marseille Provence Métropole (MPM) for the engineering work and the conception of the industrial design as well as the production of a 1:1 mock-up of a new bi-directional tram of the Bombardier FLEXITY Outlook series. The next step will entail the order of 25 to 30 series vehicles.

Germany: In September 2004 Dresden Urban Transport Operator DVB AG placed an order with Bombardier to supply 20, 30 m three-section

FLEXITY Classic trams for delivery in 2006–07. In July 2004 Connex placed a follow-up contract with Bombardier for nine three-car Talent dmus, bringing to 40 the number of these vehicles ordered by this operator. Intended for use on the Marschbahn line between Itzehoe and Husum, the trains will be delivered in late 2005.

In May 2004 Berliner Verkehrsbetriebe (BVG) awarded Bombardier a contract to supply 20 additional Class HK four-car metro trainsets. This followed the supply of four pre-production trainsets supplied to BVG in 2001. Delivery is scheduled for 2006–07, with manufacturing taking place at Bombardier's Hennigsdorf facility.

In November 2003 Bombardier received a contract from the Rhein-Main transport group (RMV) for 22 Itino two-car diesel-hydraulic trains for delivery in 2005. The vehicles will be used on the Odenwaldbahn.

In October 2003 Bombardier won a contract to supply 12 FLEXITY Classic 45 m low-floor trams for service in Leipzig. The contract includes an option for 12 additional vehicles.

ATI ordered three 3-car Talen dmus in 2004, six additional emus Class 423 were ordered by DB AG in 2004. Bombardier received an order from Nordwestbahn for the delivery of three 3-car Talen dmus in 2003. PEG Westmünsterland/ATI ordered six 3-car Talen dmus in 2003.

In July 2003 Deutsche Bahn AG (DB) awarded Bombardier Transportation a contact to supply 298 double-deck coaches for service in Bavaria and Schleswig-Holstein. The order is for 78 driving trailers and 220 conventional trailers. Delivery is scheduled for 2004–06. Included in the contract is an option on 300 additional vehicles.

In May 2003 a consortium of Bombardier Transportation and ALSTOM LHB received a contract from DB for 60 Class 423 emus for S-Bahn suburban services. Bombardier's participation as consortium leader is to supply propulsion and electrical equipment and to undertake interior fitting, final assembly, testing and commissioning. Deliveries started in 2004. The order brought to 456 the number of trains of this type ordered by DB.

In February 2003 a consortium of Bombardier Transportation and Siemens Transportation Systems was awarded a contract by the Munich transport authority, Stadtwerke München, to supply eight six-car trainsets for the city's metro system. This followed an earlier contract placed in 1997 for 10 trainsets of the same type. Bombardier is to manufacture the carbodies and assemble the trains at its Hennigsdorf site, while the company's Siegen plant will supply bogies. Siemens is the general contractor and is responsible for all electrical equipment.

In January 2003 an order was received from the Local Transport Authority of Lower Saxony (LNVG)

Bombardier C20 metro trainset employing FICAS construction technology for SL, Stockholm

0567395

CP 2000 suburban emu for CP, Portugal   0567391

Acela high-speed train for Amtrak, US   0583228

dmus, converting an option on an earlier order for 75 units. DB also ordered an additional 117 double-deck suburban vehicles. So far, Deutsche Bahn has ordered over 1,000 vehicles of this type.

In 2001 a contract was awarded by the transport authority of Halle (Hallesche Verkehrs AG, Germany) for the supply of 30 low-floor FLEXITY Classic trams. Deliveries commenced in spring 2004.

Also in 2001 Bombardier Transportation was awarded a contract by the Bonn transport authority (SWB) for 15 high-floor trams, in partnership with Kiepe Elekrik.

Bombardier is to deliver 69 FLEXITY Swift low-floor light rail vehicles of a new generation to the Cologne transport authority (KVB). They will be in operation from 2005 onwards.

Deliveries commenced in 2000 of a DB Cargo order for 400 Class 185 dual-voltage electric locomotives, derived from the Class 145 4,000 kW single-voltage design which was rolled out in 1999. Eighty Class 145 locomotives have been delivered to DB and two have been procured by chemicals company BASF for open access operations. In 2001, deliveries commenced to DB of 31 Class 146 locomotives, a derivative of Class 145 for regional passenger services.

In 2000 DB took up an option for an additional 46 Class 612 two-car 'Regio-Swinger' tilting dmus, bringing to 250 the number of the type and its Class 611 predecessor ordered by this operator.

Also in 2000 DB received the last examples of an order for 110 Class 423 four-car emus for suburban services and was taking delivery of an order for 160 Class 424 low-floor four-car S-Bahn emus being supplied by a consortium of Bombardier and Siemens.

In October 2003 Bombardier Transportation received an order for 90 single-deck coaches from Connex. The coaches will be in operation on the Nord-Ostsee-Bahn.

Greece: National operator OSE has received 25 DE2000 diesel-electric locomotives which are capable of later conversion to electric traction. Bombardier is also a member of a consortium which received a contract placed by the Athens-Piraeus Electric Railways Company for 40 three-car metro trainsets.

Israel: In 2002 Bombardier Transportation received an order from the Ports and Railways Authority-Israel Railways (Tel Aviv) to deliver a further six four-car double-deck trains, each consisting of one generator car and three trailer cars. In 2004 Bombardier Transportation received a follow-on order of 54 additional double-deck coaches.

Since 2001, 32 three-car dmus based on the Danish IC3 design have been ordered by IR.

Italy: In April 2005 Bombardier Transportation received an order from Trenitalia for the supply of 100 Type E464 electric locomotives. This follows an initial order for 50 units placed in 1996 followed by three additional options for 90, 100 and 48 units received in 1999, 2001 and 2003 respectively. Deliveries of this new set of 100 locomotives are scheduled to take place between September 2005 and November 2007. Production of the new locomotives will overlap with the delivery schedule of the previous contract, scheduled to

for eight Class 146.1 15 kV AC electric locomotives and 40 double-deck passenger coaches. This was a follow-on contract to a 2002 order for 10 similar locomotives and 66 coaches. To be delivered as two batches between July and October 2003 and July and October 2005, the trains will provide Metronom regional express services on Hamburg–Bremen and Hamburg–Uelzen routes. The locomotives are to be manufactured at the company's Kassel plant, while the coaches will be produced at Görlitz. 78 additional double-deck coaches were ordered in 2005.

In December 2002 Bombardier Transportation received an order from Bremer Strassenbahn AG for 20 FLEXITY Classic 35.4 m low-floor trams for delivery from September 2005 to April 2007. Traction equipment is to be supplied by Kiepe Elektrik.

In December 2002 a consortium of Bombardier Transportation and ALSTOM LHB was awarded a contract by Hamburger Hochbahn AG (Hochbahn) to manufacture 15 Type DT4 four-car trains for the city's metro system. The order is the sixth from Hochbahn for vehicles of this type, for a total of 126 trainsets. Bombardier will be responsible for traction and electrical equipment. Delivery is scheduled between March 2004 and April 2005.

Also in December 2002 Karsdorfer Eisenbahn (KEG) ordered eight Blue Tiger diesel-electric locomotives which were delivered between July 2003 and April 2004. A Blue Tiger was also ordered by Mindener Kreisbahn and delivered in May 2003.

In 2002 Bombardier Transportation received an order to manufacture 60 FLEXITY Classic low-floor trams for the city of Frankfurt/Main, Germany. Also in 2002, a consortium comprising Bombardier Transportation, Siemens and ALSTOM received an order from DB for the delivery of 28 seven-car ICE T high-speed trains.

In October 2002 Bombardier Transportation received an order from the DB for the production of 65 double-deck passenger coaches. The 54 trailer cars and 11 driving cars, were manufactured at Bombardier Transportation's site in Görlitz, Germany. They are operated in the area of Berlin-Brandenburg and in Mecklenburg-West Pomerania.

In 2001 a consortium of Bombardier Transportation Systems and Siemens Transportation Systems received an order from DB to supply 40 Class 425.2 four-car emus for delivery between April and November 2003. The trains are used for Rhein-Neckar commuter services. DB ordered 20 additional emus Class 425.2 in 2002 and 2003.

In 2001 Bombardier Transportation received an order from the Dresden transport authority, DVB, for 20 FLEXITY Classic 45 m low-floor trams, in 2003 additional 12 vehicles were ordered.

In 2001 a DB subsidiary, DB Regionalbahn Rheinland GmbH, ordered 26 two-car Talent

Bombardier Regina emu, Sweden   0583229

end by September 2006. To date, Bombardier Transportation has delivered 240 (out of 388) Type E453 locomotives to Trenitalia. Bombardier also received an order from Trenitalia for the upgrade of 60 ETR 500 electric power-heads.

In December 2003 Trenitalia awarded Bombardier a contract for 48 Class E464 TRAXX single-cab electric locomotives for regional passenger services. This brought to 288 the number of locomotives of this type ordered by Trenitalia. Delivery of this latest contract is scheduled for 2005–06.

In June 2003 Bombardier Transportation was awarded an order by the regional operator Ferrovia Emilia Romagna (FER) for the supply of three electric locomotives, type E464, including spare parts. The contract, which also includes an option for three further units. The three vehicles are scheduled to be delivered by mid-2005. To date, Bombardier Transportation has already delivered some 150 units in Italy.

In 2002 Italian Railways (Trenitalia) placed a contract with Bombardier Transportation for 42 Class EU11 6 MW 3 kV DC Bo-Bo electric locomotives.

Luxembourg: In February 2003 Bombardier Transportation received an order from Luxembourg Railways (CFL) for 85 double-deck passenger coaches (67 trailers and 18 driving trailers). To be manufactured at the company's Görlitz plant in Germany, the vehicles are due to be delivered between November 2004 and September 2005.

Mexico: In 2003 in consortium with CAF, Bombardier Transportation finalised a contract with Mexico City's transit authority, Sistema de Transporte Colectivo-Metro (STC), for the supply of 405 subway cars for STC's Line 2. Delivery commenced in 2004 and is due to continue through to mid-2006.

Netherlands: In June 2004 Bombardier Transportation was awarded a contract from RET, the public transport operator of Rotterdam, to design and build 21 Bombardier FLEXITY Swift vehicles. The contract includes an option for up to 21 additional vehicles. The new vehicles are scheduled to be delivered between January and November 2008.

Poland: In 2001 Bombardier Transportation, in consortium with Kiepe Elektrik, was awarded a contract from Krakow's Transport Authority (MPK), Poland for the supply of 12 FLEXITY Classic low-floor trams. The transaction carries an option for 18 additional trams. Delivery of the vehicles took place in 2003. A total of 14 low-floor trams of the same design are in revenue service in Krakow since December 1999.

Portugal: The new Porto Metro system commenced operation in December 2002. Bombardier, as a partner in the Normetro Consortium, responsible for the delivery of full turnkey, was awarded the contract in 1998 and also supplies 72 FLEXITY Outlook 100 per cent low-floor trams.

In 2000 the company, in consortium with Siemens, received an order to supply 22 four-car emus to Portuguese Railways, for Oporto commuter service and 12 five-car emus to Cascais line.

Romania: In 2002, deliveries of 108 Movia began. The vehicles have been ordered for Line 2 of the Bucharest metro by Metrorex RA. In 2005 Bombardier received an order to deliver an additional 120 Movia cars.

South Africa: As part of a DBOM concession for the Gautrain Rapid Rail Link, for which the Bombardier-led Bombela consortium was named preferred bidder in 2005, the company is to supply a fleet of Electrostar emus. Manufacture of these will be undertaken at Bombardier's Derby, UK, facility, with final assembly taking place in South Africa by UCW Partnership.

Spain: In April 2004 a consortium of Bombardier Transportation and Patentes Talgo received from RENFE an order for 44 variable gauge power cars for 22 nine-car train sets to be used on Spain's high-speed lines. Delivery is scheduled for 2006–8.

In 2001, a contract was awarded by Spanish National Railways (RENFE) for the construction of 16 high-speed trains for services on the standard-gauge line linking Madrid, Barcelona and the French border. Train configuration will comprise 12 trailer cars and power cars. The order followed

Bombardier Talent emu for ÖBB, Austria)   0583230

AVE S 102 very high-speed power car for RENFE, Spain   **NEW**/0585278

the development by a consortium with Talgo of a prototype trainset, which completed trials at speeds up to 350 km/h.

Sweden: In March 2005 Bombardier Transportation received an additional order from Region Skåne for the production and delivery of eight three-car emus. Region Skåne previously ordered a total of 32 Oeresund trains. In 2004 Bombardier received an order from Västtrafik for the delivery of two 2-car ITINO dmus. Three 2-car ITINO dmus were ordered by Tarnsitio for the operator Vämlandstrafik also in 2004.

In March 2003 Bombardier Transportation received an order for three 3-car Regina emus from the Swedish local transport authority Västtrafik. In September 2002 10 two-car units of this type were ordered by the leasing company, Transitio. Together these contracts brought the number of vehicles ordered of this type to 68 trains (148 cars).

In 2001 the Stockholm Public Transportation Authority (SL) placed a contract with Bombardier Transportation to supply an additional 70 three-car Vagn 2000 metro units between September 2002 and early 2004. This brought to 270 the number of Vagn 2000 units ordered by SL. Production is undertaken at Bombardier sites at Kalmar and Västerås, Sweden, with bogies manufactured at its Derby, UK, plant.

In 2001 additional orders were received from Skånetrafiken for seven three-car Contessaintercity trainsets for Öresund Link services between Sweden and Denmark, bringing to 56 the number of trains of this type ordered by operators in the two countries.

In 2000 the first of nine two-section 10,800 kW 360-tonne three-phase electric locomotives for the LKAB iron ore mining company was handed over. Designed for use on the 540 km Luleå Kiruna Narvik line, the locomotives are based on the Octeon platform developed by the former Adtranz.

Switzerland: In May 2003 SBB Cargo ordered 18 Class 482 (Type 185) dual-system (15 kV AC/3 kV DC) electric locomotives for cross-border freight operations. Delivery was due from October 2004 to March 2005. This was a follow-on order to a contract placed in September 2002 for 40 similar locomotives for delivery from December 2002.

In 2003 and 2004 BLS ordered 21 intermediate cars for BLS Jumbo.

In January 2003 Swiss Federal Railways awarded a consortium of Bombardier Transportation and ALSTOM Transport a contract to supply 51 IC 2000 double-deck intercity coaches (30 driving trailers and 21 second class coaches), which were delivered between December 2003 and June 2004. The same contract covered the refurbishment of 16 IC 2000 restaurant cars. Production of the new vehicles took place at Bombardier's Pratteln plant in Switzerland.

In 2002 Bombardier Transportation received an order from the Geneva transport authority for the supply of 21 low-floor FLEXITY Outlook trams. The new tram for Geneva is based on a proven technology which features 100 per cent low-floor interior and conventional wheel set bogies.

Also in 2002, Bombardier Transportation has been awarded a contract by BLS Cargo AG in Switzerland to deliver 10 dual-voltage freight locomotives. The order is valued at approximately €30 million. Delivery of the locomotives, which BLS Cargo AG will put in service for the cross-border traffic between Germany and Switzerland, took place in 2002-03. In Switzerland, 10 freight locomotives of the same type are already in service with SBB Cargo (the Swiss Federal Railways). Various private railway operators also operate freight locomotives of the same successful type; these include Deutsche Bahn, with more than 70.

In 2001, a consortium of Bombardier Transportation and ALSTOM Transport was

Meridian demu for Midland Mainline, UK   **NEW**/0585276

awarded a contract by BLS Lötschbergbahn to supply 18 three-car NINA 15 kV AC emus for service in the regional network on the Bern area. Deliveries were made in 2003-04. This followed an earlier order for eight similar units which had been in service since 1999.

In 2001, an order was placed by Swiss Federal Railways with a consortium led by Bombardier Transportation and including ALSTOM Transport for 10 seven-car Class ICN tilting trainsets, with an option for an additional 10 trains. This followed earlier orders for 24 trainsets of this type, the first of which commenced commercial service in 2000. In 2001, the option on 10 additional trains was taken up for delivery in 2004.

In 2000 orders were received from three Swiss regional operators, BLS Lötschbergbahn, Transports de Martigny et Région and Transports Régionaux Neuchâtelois for 11 NINA low-floor emu trainsets. Earlier the company had supplied eight three-car NINA trainsets to BLS Lötschbergbahn for Bern area services.

In October 2003 the Zurich transport authority (VBZ) placed an order for further 68 Cobra trams, following the thorough testing of six preproduction trial vehicles which were delivered in 2001. In total 74 vehicles have been ordered.

Taiwan: In June 2003 Bombardier Transportation finalised a contract with Kung Sing Engineering Corporation (KSECO) to supply electrical and mechanical works for the 15 km Neihu Line rapid transit system in Taipei. The contract includes the supply of 202 vehicles, the first 60 of which are due to be delivered by mid-2007, when the first 10 km of the line is due to be commissioned.

Turkey: In 2002 a consortium of Bombardier Transportation and the engineering group Yapi Merkezi of Turkey was awarded a contract by the Eskisehir Greater City Municipality to supply a 14.2 km light rail transit system, including the delivery of 18 FLEXITY Outlook 100 per cent low-floor trams.

In 2001 the Istanbul Transportation Company ordered 55 FLEXITY Swift low-floor light rail vehicles. In 2000, the Izmir Light Rail Transit system started operation. Bombardier was also involved in the construction of a light rail system for the city of Adana which included the delivery of 36 light rail vehicles.

UK: In May 2005 Bombardier Transportation was awarded a contract from Docklands Light Railway, London, for the design and build of 24 automatically guided light rail cars, together with an option for a further nine vehicles. The new vehicles are scheduled to be delivered between May 2007 and September 2008. In March 2005 Bombardier Transportation received an order from Angel Trains for the delivery of 36 TRAXX locomotives, consisting of 26 Bombardier TRAXX F140 MS multi-system locomotives and 10 F140 DC locomotives under a framework agreement that

offer Angel Trains the possibility of ordering up to 100 locomotives subject to the same general conditions. Delivery of the DC locomotives is scheduled between October 2006 and March 2007, with the multi-system locomotives following between January 2007 and September 2007. The locomotives will be manufactured at Bombardier's facilites in Kassel, Germany.

In March 2004 Line One of the Nottingham Express Transit (NET) light rail system was commissioned. Bombardier Transportation, as a member of the Arrow Light Rail Ltd concession company, has supplied 15 low-floor light rail vehicles as part of the turnkey rail system for the City of Nottingham.

In April 2003 Bombardier Transportation received contracts from the Metronet consortium, of which the company is a member, covering the supply of 1,738 metro cars for London Underground's Victoria Line and its sub-surface lines. In the early stages of the turnkey project, Bombardier is to build two pre-production trains for the Victoria Line by 2006 to be followed by 47 eight-car trains for the same line. Bombardier is also to refurbish existing District Line rolling stock.

In September 2002 GB Railways Group plc issued letters of intent to Bombardier Transportation to purchase four Class 222 four-car high-speed (200 km/h) dmus for its subsidiary, Hull Trains. The order was confirmed in July 2003. Financed by HSBC Rail, they are due to be delivered in mid-2005.

In 2002 Bombardier Transportation won two contracts from Connex South Eastern (now South Eastern Trains): an order for 36 five-car Class 376

suburban Electrostaremus for services in south London; and an order for 228 additional Class 375 Electrostar emu cars complementing the previous Connex order for 210 cars (55 trains). Both orders include maintenance agreements.

Electrostar emus have been ordered by or supplied to three UK rail operators: c2c Rail Ltd (74 four-car 25 kV AC units); Connex South Eastern (now South Eastern Trains) (45 four-car and 10 three-car, all 750 V DC); and Southern (formerly South Central) (154 four-car, 28 three-car all 750 V DC).

In 2005 Bombardier received an order for six intermediate cars for the Turbostars of Chiltern Railways.

Bombardier Transportation also received an order for the provision of 45 Turbostar diesel-multiple-unit cars for Porterbrook Leasing Company Ltd in the UK, part of Abbey National Treasury Services plc.

In 2002 HSBC Rail UK Ltd, on behalf of the UK train operator Midland Mainline, placed an order with Bombardier Transportation for 127 high-speed diesel-electric cars. Technically similar to the Voyager units supplied for the Virgin Rail Group, the Meridian units will operate express services between London and Derby, Nottingham and Sheffield. Delivery commenced in 2004.

Also in 2002 UK operator Govia confirmed an additional order for 460 Class 375/377 Electrostar 750 V DC emu cars for services on its Southern (formerly South Central) network. Financed by Porterbrook Leasing Company Ltd, the contract brought to 700 the number of cars of this type ordered for this operator. Assembly takes place at Bombardier's Derby plant in the UK.

In May 2001 Virgin Rail Group took delivery of the first Voyager diesel-electric multiple unit. Bombardier Transportation is responsible for the design, supply and maintenance of 78 trainsets, comprising 352 diesel-electric cars, of which 216 are equipped with Bombardier's tilting technology.

US: In December 2004, Metro Transit of Minneapolis/St Paul opened the completed Hiawatha light rail line, served by Bombardier FLEXITY LRVs. In June 2004 services began on the first 13 km phase of the line. Bombardier is manufacturing a total of 27 LRVs for Metro Transit. Carbody shells are being manufactured at Bombardier Transportation facilities in Mexico with final assembly and testing taking place in Plattsburgh, New York. The first vehicle was delivered in March 2003.

In October 2004 the Mid-Region Council of Governments and the New Mexico Department of Transportation placed an order with Bombardier Transportation for the supply of 10 Bombardier BiLevel commuter rail vehicles for a new service between Belen and Bernalillo, New Mexico. The coaches are being built at Bombardier's Thunder Bay, Ontario, plant for delivery in 2005.

In April 2004 the Metropolitan Transportation Authority/Metro-North Railroad, New York, awarded Bombardier Transportation a contract to supply 120 Type M-7 emu commuter cars, exercising an option that brought the number of cars ordered by this operator and the Metropolitan Transportation

Bombardier Class 185 TRAXX for locomotive for Deutsche Bahn, Germany   **NEW**/0585277

Authority/Long Island Rail Road to 978. Bodyshells are being manufactured in La Pocatière, Québec, and assembly is taking place in Plattsburgh.

In December 2002 NJ Transit awarded Bombardier Transportation a contract to supply 100 stainless steel-bodied multi-level commuter coaches, with options for up to 176 additional vehicles. Six 'Pilot Cars' are to be delivered in July 2005 for a six-month test period. Delivery of the remaining coaches will begin after the testing is completed. Carbodies are to be manufactured at the company's La Pocatière plant in Quebec, with final assembly taking place at the Plattsburgh, New York facility.

In 2002 Bombardier Transportation received orders for its Bombardier of BiLevel commuter cars from four North American transportation authorities: the North San Diego County Transit District; the Central Puget Sound Regional Transit Authority; the San Joaquin Regional Rail Commission; and the Fort Worth Transportation Authority. Together, these contracts were for 12 cab cars and 18 trailer coaches that were manufactured at the company's plants at Thunder Bay, Ontario, and Barre, Vermont, and delivered during 2003.

In 2001, an order was received from the Metropolitan Transportation Authority/New York City Transit for 350 Type R142 metro cars. The contract represented the conversion of an existing option for 200 cars included in a 1997 order for 680 cars, and a new order for 150 cars required by NYCT to meet increased levels of ridership. Deliveries of all 1,030 cars were completed in 2003.

The Port Authority of New York and New Jersey awarded Bombardier Transportation a contract for the design, construction, operation and maintenance of the New York JFK International Airport Automated Light Rail System. The contract called for the turnkey design and construction of the driverless light rail system and the supply of 32 vehicles.

Venezuela: In 2002 Bombardier Transportation, through participation in the Frameca consortium, was awarded a contract for the manufacture of 56 motor bogies and 28 carrying bogies for the extension of the Caracas metro. Deliveries were started in 2003. This is a follow-on order to previous orders awarded since 1977.

Developments
With regard to product development, the emphasis at Bombardier Transportation is on innovative product families featuring common platforms, proven technologies, and increased standardisation and modularisation. One of the most recent examples is the Bombardier TRAXX MS multi-system freight locomotive, the first of which was presented to Swiss Federal Railways (SBB) in May 2004. In addition to the significant benefits customers derive from commonality, this newest member of the TRAXX locomotive family also is able to operate across bordering countries that have different rail infrastructures. For instance, the 18 TRAXX MS electric locomotives for SBB Cargo can operate on both the alternating-current network in Switzerland and the direct-current network in neighbouring Italy.

In early 2004, Bombardier delivered the first AGC (Autorail Grande Capacité) high-capacity regional trainset to the French National Railways (SNCF). The versatile AGC family of multiple units can run on diesel fuel, electricity or a combination of the two, and will be used to renew regional express (TER) train fleets operating across France. SNCF, on behalf of the French regions, has placed firm orders for 379 AGC trains, the most recent for 100 train in January 2005. Bombardier delivered the first AGC trainset to the customer just 26 months after Notice to Proceed, a significant milestone reflecting the focus on responsiveness and customer satisfaction.

The Bombardier MOVIA family of metro vehicles also provides a practical expression of the platform concept. The outstanding economics of the flat-pack MOVIA cars, designed for easy assembly in the local regions where they will operate, make them an ideal solution for emerging markets in eastern Europe and elsewhere. In January 2005 Bombardier Transportation received an order from Metrorex in Romania to supply 20 six-car MOVIA trainsets for the Bucharest Metro.

Bombardier FLEXITY light rail vehicles represent another innovative product family. FLEXITY vehicles made their North American debut in 2005 when the first light rail transit line serving Minneapolis/St Paul, US, initiated revenue service. Noteworthy orders during the year included 26 bi-directional FLEXITY Outlook low-floor trams for Marseille, France, and another 30 for Valencia, Spain.

Bombardier double-decker coaches are also highly popular with European cities and regions that need to maximize passenger capacity on limited trackways. Germany is a good example, where in 2005, Bombardier delivered the 1,500th double-decker coach to Deutsche Bahn AG.

32 high-speed power heads were delivered on schedule to Spanish National Railways (RENFE) for the AVE S 102 high-speed trains. RENFE also ordered 44 additional Bombardier power heads designed to reach maximum speeds of 250 km/h. Both Spanish contracts involved Bombardier Transportation and consortium partner Patentes Talgo.

In China, the Bombardier Sifang Power (Qingdao) Transportation Ltd (BSP) joint venture received an order from the Ministry of Railways of China for the production and delivery of 20 eight-car high-speed trainsets.

UPDATED

Brookville Equipment Corporation

PO Box 130, 175 Evans Street, Brookville, Pennsylvania 15825, USA
Tel: (+1 814) 849 20 00   Fax: (+1 814) 849 20 10
e-mail: bec@brookvilleequipment.com
Web: http://www.brookvilleequipment.com

Key personnel
President: Dalph S McNeil
Vice-President, Production: Larry J Conrad

Director of Marketing: Tim Raffeinner
Sales: Chris Rhoades

Products
Haulage equipment for mines, tunnelling and industrial applications.

The company produces 4- to 20-ton battery-powered locomotives, 4- to 45-ton diesel-powered locomotives and 45- to 150-ton diesel-electric locomotives for marshalling yard operations. In 2002 Brookville introduced bogie assemblies for PCC-style streetcars and trams.

Support equipment includes diesel- or battery-powered haulage tractors, maintenance vehicles and personnel carriers for rail or road use.

Brookville also rebuilds/remanufactures existing locomotives to original OEM specifications and undertakes the restoration and modernisation of PCC-style streetcars.

Brush Traction

PO Box 17, Loughborough LE11 1HS, UK
Tel: (+44 1509) 61 70 00   Fax: (+44 1509) 61 70 01
e-mail: sales@brushtraction.com
Web: http://www.brushtraction.com

Key personnel
General Director: J M G Bidewell
Sales Manager: P L Needham
Engineering Manager: A Haworth
Production Manager: I G Hall
Purchasing Manager: J Marshall

Background
Brush Electrical Engineering was founded in 1889. Today the company is a major supplier of electric propulsion equipment and locomotives and is able to undertake refurbishment of complete vehicles or components. The company is a member of the FKI Group.

Products
Diesel electric and electric locomotives for mainline and shunting applications, also battery-electric service locomotives for metro systems; complete AC and DC propulsion packages, including traction motors and control equipment; locomotive servicing, refurbishment and upgrade.

Contracts
Channel Tunnel Shuttle locomotives for Eurotunnel, Class 92 electric locomotives for Railfreight distribution and dual-mode battery-electric locomotives for Hong Kong MTRC. Brush Traction is also undertaking a number of major locomotive mid-life overhaul and repowering projects, examples including contracts with Porterbrook Leasing for the repower and re-engineering of Class 47 locomotives, creating Class 57 type. The Class 57 locomotives have 500 kW electric train supply facility and are used for rescue and diversionary working of passenger trains.

The mid-life overhaul and upgrading to 7.0 MW of 39 Eurotunnel Shuttle locomotives is also being undertaken at the Brush Traction works, over a seven-year period.

Bumar-Fablok Ltd

Fabryka Maszyn Budowlanych i Lokomotyw 'Bumar-Fablok' SA
ul Fabryczna 16, PL-32-500 Chrzanów, Poland
Tel: (+48 32) 624 66 66   Fax: (+48 32) 623 29 25
e-mail: info@fablok.com.pl
Web: www.fablok.com.pl

Key personnel
President and Managing Director:
  Piotr Majcherczyk

Marketing and Development Director:
  Andrej Smolana
Economy and Finance Director: Maria Pitala

Products
Modernisation, repair and general overhaul of diesel locomotives type 6Da, modernised diesel locomotives type 6Da/R; production driving axle-sets for electrical and diesel locomotives, motor bogies for power cars, mechanical gears, cylindrical gears, braking system components for rail vehicles, and brake adjusters.

UPDATED

CAF – Construcciones y Auxiliar de Ferrocarriles SA

Padilla 17 – 6°, E-28006 Madrid, Spain
Tel: (+34 91) 435 25 00   Fax: (+34 91) 436 03 96
e-mail: export.caf@caf.es
Web: www.caf.es

Key personnel

President and Chief Executive Officer:
  J M Baztarrica
Managing Directors: A Arizcorreta, A Lergarda
Contact: J Esnaola

Works

Beasain
J M Iturrioz 26, E-20200 Beasain, Spain
Tel: (+34 943) 88 01 00   Fax: (+34 943) 88 14 20

Irún
Calle Anaca 13, E-20301 Irún, Spain
Tel: (+34 943) 61 33 42   Fax: (+34 943) 61 81 55

Zaragoza
Av de Cataluña 299, E-50014 Zaragoza, Spain
Tel: (+34 976) 76 51 00   Fax: (+34 976) 57 26 48

Offices

CAF Argentina SA
Pte Luis Sáenz Peña 31° – 7° piso, 1110 Capital
Federal Buenos Aires, Argentina
Tel: (+51 11) 43 83 20 06   Fax: (+54 11) 43 81 48 37
e-mail: cafadministracion@cafarg.com.ar

CAF Brasil Industria e Comercio
Rua Pedroso Alvarenga 58 conj 52, CEP 04531-000
São Paulo, Brazil
Tel: (+55 11) 31 67 17 20   Fax: (+55 11) 30 79 87 62
e-mail: cafsaopaulo@cafbrasil.com.br

CAF Mexico SA de CV
Prolongación Uxmal 988, Col Sta Cruz Atoyac,
03310 Mexico DF, Mexico
Tel: (+52 55) 56 88 75 43   Fax: (+52 55) 56 88 11 56
e-mail: cafmex@prodigy.net.mx

CAF USA Inc
1401 K Street NW, Suite 803, Washington DC
20005-3418, US
Tel: (+1 202) 898 48 48   Fax: (+1 202) 216 89 29
e-mail: cafusa@cafusa.com

Products

Electric, diesel-electric and diesel-hydraulic
locomotives; multiple-units and hauled coaches;
metro cars; light rail vehicles/trams.

Contracts

Belgium: In 2004 Brussels Metro operation
STIB ordered 15 six-car trainsets for delivery in
July 2006.

Driving car from CAF-built Class 2900 dmu for Irish Rail, on display at the UITP Mobility & City Transport Exhibition 2003 in Madrid (Ken Harris)   0554141

Brazil: In 2002 CAF was awarded a turnkey contract
to construct and operate for 20 years a metro
system in Salvador de Bahía. The project includes
the supply of rolling stock.

A consortium led by CAF, including ALSTOM and
the former Adtranz, supplied 30 four-car emus to
CPTM, São Paulo.

China: In collaboration with the former Adtranz,
CAF supplied 184 aluminium-bodied emu cars to
MTRC Hong Kong for its Airport Express (11 eight-
car trains) and Lantau line commuter services
(12 eight-car trains).

Ireland: In 2003 an order was placed for nine
additional trains of the same type.

In November 2002 Irish Rail awarded CAF a
contract to supply 67 locomotive-hauled coaches,
with an option on 24 more. Intended mainly for
Dublin-Cork services, the vehicles are to be a mix of
first, standard and buffet coaches as well as driving
trailers equipped with generators for train power
supply. The new vehicles are due to enter service
in October 2005.

In 2000 Irish Rail placed a contract with CAF to
supply 15 Class 2900 four-car dmus.

Italy: In December 2002, CAF was awarded a
EUR230 million contract to supply 33 trains to
the Rome Metro. This was extended by six new
trains with final delivery of the units in 2007. The
order includes maintenance work on rolling stock
supplied. CAF is creating a subsidiary in Rome to
perform the task.

In 2001 CAF signed a contract increasing the
number of air conditioned six-car trains ordered by
Comune di Roma for the capital's metro network
from 33 to 45.

Mexico: In October 2002 CAF was awarded a share
of a contract to supply new trains for the Mexico
City metro. The company is to produce 17 nine-car
trains of a 45-train order and will also supply some
electrical components and running gear for the
entire contract.

Portugal: In collaboration with ALSTOM, CAF
undertook construction of 36 cars of an order for 18
four-car double-deck emus for Lisbon cross-Tagus
services.

Spain: In 2004 MINTRA (Madrid, Infraestructuras
del Transporte) awarded CAF a EUR609.8 million
design contract. The order will comprise 90 trains
made up of 432 narrow-gauge cars. A further 14
motor cars will be broad-gauge types that will
be coupled to units currently providing services
on line 8 (airport). These trains form part of an
entire line of CAF's own products for metropolitan
railways.

In February 2004 the Spanish National Railways
(RENFE) board of directors awarded the CAF/
ALSTOM consortium 30 250 km/h shuttle-service
trains for EUR353 million and 45 250 km/h shuttle-
service trains equipped with the variable-gauge
system for EUR584 million. Delivery of both types
is scheduled to begin in 2006.

Serveis Ferroviaris de Mallorca placed a contract
with CAF in 2003 for 10 two-car dmus for the
island's metre gauge network.

CAF is a member of a consortium that includes
ALSTOM and Bombardier which in 2003 secured a
contract to deliver 13 three-car emus to Ferrocarrils
de la Generalitat de Catalunya (FGC). Their design
will be similar to that of 20 Class 213 vehicles
supplied to the same operator against a 1995
contract and will serve the Llobregat–Anoia metre
gauge line.

In December 2002 CAF secured a contract to
supply 33 trains for Line 5 of the Barcelona metro.
Deliveries commenced in 2004.

In 2001 RENFE awarded a consortium of CAF
and ALSTOM a contract to supply 12 AVE high-
speed trainsets for the Madrid-Barcelona line. The
trains are to be equipped with CAF's Brava gauge
changing system to enable them to operate on
both the standard-gauge high-speed lines and the
broad-gauge (1,668 mm) of the classic Spanish
network.

Civia emu for RENFE   ***NEW**/1135654*

In 2001 RENFE awarded CAF a contract to supply 21 three-car tilting dmus for regional services. Delivery was due to commence in 2003. These follow earlier deliveries of 16 Flexliner-type dmus supplied to RENFE for regional services.

In 2001 CAF, in consortium with ALSTOM, was awarded a contract by RENFE to supply 20 Alaris non-tilting four-car 270 km/h regional emus for use on the Madrid-Barcelona high-speed line.

Recent emu contracts for RENFE include batches of 29 and 46, Class 447 air conditioned emus.

Recent locomotive contracts for RENFE include the manufacture of 15 Class 252 high-speed electric locomotives and the reconstruction of four Class 269.600 electric locomotives for 200 km/h operation.

The most recent deliveries for the Madrid metro system have involved Type S-8000 three-car trainsets with IGBT technology. Earlier deliveries to Metro de Madrid included 44 S-600 and 72 Series 2000 trainsets.

The latest type to be supplied to the Barcelona metro system is the four-car Type S-550, which employs IGBT technology. Earlier deliveries were of the Series 2100 type.

UK: In 2002 CAF received an order from the Northern Ireland government for 23 three-car dmus for service on the Northern Ireland Railways network from 2004.

In 2000 CAF commenced delivery of 16 Class 333 three-car 25 kV AC emus for Northern Spirit (now

Class S120 dual-voltage auto gauge-changing emu for RENFE   *NEW*/1135653

Arriva Trains Northern). Subsequently, orders were placed for 16 additional trailer cars to strengthen all sets to four-car units.

Acting as a general contractor, CAF, together with Siemens Transportation Systems, built 14 25 kV AC Class 332 four-car emus for Heathrow Express. In 2001, orders were placed for an additional five trailers to strengthen five of these trains.

US: Recent contracts include: the supply of 192 metro cars for WMATA, Washington; supply of 40 LRVs to Sacramento Regional Transit District; and the supply of 28 new trams and the refurbishment of 55 existing vehicles for PATCO, Pittsburgh.

UPDATED

CNR Changchun Railway Vehicles Co Ltd

5 Qingyin Road, Changchun 130062, Jilin, China
Tel: (+86 431) 790 23 01   Fax: (+86 431) 293 87 40

Key personnel
Director of Factory: Ma Shukun

Background
Changchun Railway Vehicles Co Ltd is the largest Chinese manufacturer of passenger coaches and urban transit vehicles.

Products
Development, design, production, sales, installation, refurbishment, maintenance and aftersales service for metro cars; electric multiple-units for main line applications; sleeping cars, dining cars, mail vans, generator vans and passenger coaches.

Metro cars include DK20 and the DK9 with chopper voltage control. All cars are motored (4 × 86 kW) and designed to operate in multiples of two, four or six.

Contracts
The latest metro design is the DK20 wide-body set, consisting of two driving motor cars and four non-driving trailer cars, for 750 V DC third-rail operation; each car is 19 m long, 3.51 m high and 2.8 m wide.

CFD Bagneres

9–11 rue Benoît Malon, F-92150 Suresnes, France
Tel: (+33 1) 45 06 44 00   Fax: (+33 1) 47 28 48 84
e-mail: cfd@cfd.fr
Web: www.cfd.fr

Key personnel
President: P Esnault
Sales Manager: A Gonzalez

Principal subsidiaries
Conception Ferroviaire & Développement

Products
Diesel locomotives from 150 to 2,000 kW, dmus.

UPDATED

Changzhou Diesel & Mining Locomotive Plant

100 Xinshi Road, Changzhou, 213002 Jiangsu, China
Tel: (+86 519) 667 27 45   Fax: (+86 519) 660 04 43
Web: http://www.js.cei.gov.cn

Key personnel
Managing Director: Xie Yintang
Chief Engineer: Xu Di'an

Marketing Director: Wang Chenxian
Production Director: Wang Jianxun
Export Sales Director: Wu Leping

Background
Changzhou Diesel and Mining Locomotive Plant was created in 1964 and is located between Shanghai and Nanjing. The plant has obtained ISO 9001 quality certification.

Products
Diesel locomotives from 60 to 750 kW. Electric locomotives for mining applications from 3 to 55 tonnes.

UPDATED

China Northern Locomotive and Rolling Stock Industry (Group) Corporation (CNR)

11 Yangfangdian Road, Haidian District, Beijing 100038, China
Tel: (+86 10) 51 86 23 70
Fax: (+86 10) 51 86 23 74
e-mail: loriciec@cnrgc.com.cn
Web: www.cnrgc.com

Key personnel
President: Cui Dianguo
Vice-President: Zhao Guangxing
President of LORIC Import & Export Corporation Ltd:
  Cao Guobing

Vice-President of LORIC Import & Export
  Corporation Ltd: Chen Dayong
Senior Engineer of LORIC Import & Export
  Corporation Ltd: Yang Xiangjing

Subsidiaries
The corporation comprises 18 production plants, two research institutes and one trading company.

Production plants
CNR Beijing February 7th Locomotives Works
1 Yanggingzhuang, Changxindian, Fengtai District, Beijing 100072, China
Tel: (+86 10) 83 30 60 01
Fax: (+86 10) 83 30 37 36
Legal Representative: Wang Dongming

CNR Beijing Nankou Locomotive & Rolling Stock Machinery Works
Daobei, Nankou Town, Changping District, Beijing 102202, China
Tel: (+86 10) 51 01 33 61   Fax: (+86 10) 69 77 18 09
Legal Representative: Song Zhigui

CNR Changchun Locomotive Works
2155 Kaixuan Road, Changchun 130052, Jilin, China
Tel: (+86 431) 615 77 76   Fax: (+86 431) 613 44 44
Legal Representative: Zhang Xiuchen

CNR Changchun Railway Vehicle Co Ltd
5 Qingyin Road, Changchun 130062, Jilin, China
Tel: (+86 431) 790 23 01   Fax: (+86 431) 293 87 40
Legal Representative: Ma Shukun

CNR 'China Star' high-speed trainset   1048665

CNR Lanzhou Locomotive Works
49 Wuwei Road, Qilihe District, Lanzhou 730050,
Gansu, China
Tel: (+86 931) 298 54 61   Fax: (+86 931) 286 72 79
Legal Representative: Chen Beiqun

CNR Mudanjiang Locomotive and Rolling Stock
Works
55 Locomotive Road, Yangming District,
Mudanjiang 157013, Heilongjiang, China
Tel: (+86 453) 896 82 30   Fax: (+86 453) 633 10 08
Legal Representative: Xia Binhai

CNR Qiqihar Railway Rolling Stock (Group) Co
Ltd (QRRS)
10 Zhinghua East Road, Qiqihar 161002,
Heilongjiang, China
Tel: (+86 452) 293 83 34   Fax: (+86 452) 251 44 64
e-mail: zsjqrrsintl@qrrs.com.cn
Web: www.qrrs.com.cn
Legal Representative: Wei Yan

CNR Shenyang Locomotive & Rolling Stock Co Ltd
75 Kunshan West Road, Huanggu District,
Shenyang 110035, China
Tel: (+86 24) 86 41 32 57   Fax: (+86 24) 86 40 87 30
Legal Representative: Miao Huangsheng

CNR Type CKD9 diesel-electric locomotive for Pakistan Railways   0583232

CNR luxury coaches   0552546

CNR Dalian Locomotive & Rolling Stock Work
51 Zhongchang Street, Shahekou District, Dalian
116022, Liaoning, China
Tel: (+86 411) 419 82 54   Fax: (+86 411) 465 42 45
Legal Representative: Tan Chengxu

CNR Datong Electric Locomotive Co Ltd
1 Qianjin Street, Datong 037038, Shanxi, China
Tel: (+86 352) 509 08 78
Fax: (+86 352) 509 09 84
Legal Representative: Shi Xiaoding

CNR Harbin Railway Rolling Stock Co Ltd
82 Tongjiang Street, Daoli District, Harbin 150018,
Heilongjiang, China
Tel: (+86 451) 84 61 34 56   Fax: (+86 451) 84 65 66 57
Legal Representative: Wu Yaobin

CNR Jinan Locomotive & Rolling Stock Works
73 Huaichun Street, Huaiyin District, Jinan 250022,
Shangdon, China
Tel: (+86 531) 718 60 24   Fax: (+86 531) 795 75 48
Legal Representative: Zhang Chongzhi

CNR electric locomotives

Type	Wheel arrangement	Line voltage	Output (kW) continuous	Speed (km/h)	Weight tonnes	First built	Builders
SS3B	Co-Co	25 kV/50 Hz	4,350	100	138	1993	Datong, Taiyuan
SS4G	2(Bo-Bo)	25 kV/50 Hz	6,400	100	184	1994	Datong, Dalian
SS6B	Co-Co	25 kV/50 Hz	4,800	100	138	2000	Datong
SS7C	Bo-Co-Bo	25 kV/50 Hz	4,800	132	125	1998	Datong
SS7D	Bo-Co-Bo	25 kV/50 Hz	4,800	170	126	1999	Datong
SS7E	Co-Co	25 kV/50 Hz	4,800	170	126	2001	Datong
Sky Shuttle	Bo-Bo	25 kV/50 Hz	4,800	200	82	2002	Datong
SSJ3	Co-Co	25 kV/50 Hz	7,200	120	150	2004	Dalian

Interior of sleeping car by CNR   0552547

CNR electric railcars and multiple-units

Class	Cars per unit	Line voltage	Motor cars per unit	Wheel arrangement	Output (kW) per motor	Speed (km/h)	Weight (tonnes) per car	No in service	First built	Builders
DK20	6	750 V DC	6	Bo-Bo	86	80	35	42	1994	Changchun
DK28/31	6	750 V DC	6	Bo-Bo	180	80	Mc, M37.5, T30.5	114	1998	Changchun
DKZ2	7	750 V DC	7	Bo-Bo	132	80	M35.8, Mc37.4, Tc31.5	144	1998	Chanchun
DKZ3	7	750 V DC	5	Bo-Bo	132	80	M35.8, Mc 37.4,	77	2000	Chanchun
KDZ2/Z5	6	25 kV/50 Hz	3	Bo-Bo	180	120	M54, T52	12	1999	Changchun
DDJ1	7	25 kV/50 Hz	1	Bo-Bo	1,000	210	M84, Tc52, T48/52	7	1999	Chanchun
DJJ1	7	25 kV/50 Hz	1	Bo-Bo	1,225	220	M78, Ms52, Mc50, T48	7	2000	Changchun
DL6W	1	750 V DC	1	Bo-2-Bo	455	60	3.544	25	2000	DalianInstitute
FG	4	1.5 kV DC	2	Bo-Bo	170	100	M36, T32	40	2002	Dalian
DK32/34	4	750 V DC	2	Bo-Bo	180	80	M37, T29	136	2002	Changchun
China Star	11	25 kV/50Hz	2	Bo-Bo	1,225	270	M78, T44/48	11	2002	Datong Changchun
Tianjin	4	1,500 V DC	2	Bo-Bo	200	100	Mcp39.4, T33.4	12	2003	Changchun
Wuhan	4	750 V DC	2	Bo-Bo	180	100	M35.5, Tc31.5	8	2003	Changchun
Guangzhou	6	1,500 V DC	4	Bo-Bo	220	100	A35, B/C38	8	2003	Changchun
Shanghai	6	1,000–1,500 V	4	Bo-Bo	220	80	M38.3	310	2004	Changchun
Changbaishan	9	25 kV/50 Hz	6	Bo-Bo	265	210	M52.5, T53	9	2003	Changchun
QkZ2	4	1,500 V	4		105	80	McL28.6, M:27.6	76	2004	Changchun

CNR diesel railcars and multiple-units

Class	Cars per unit	Motor cars per unit	Motored axles per motor car	Transmission	Rated power (kW) per motor	Speed (km/h)	Weight (t) per car	First built	Builders
SYZ25	4	6	4	AC/DC	91	120	Mc72, T68	1998	Tangshan
NZJ2	12	2	Co-Co	AC/DC	600	180	M135, T53.8	2000	Dalian
Harbin	7	2	B-2	Hydraulic	–	140	M84, T43	2000	Changchun
Putian	1	1	A1A-A1A	AC/DC	600	160	M111	2002	Dalian
160 km/h	5	4	1A-2, 2-2, 2-A1	Hydraulic	–	160	M64.3, T53	2003	Tangshan

CNR diesel locomotives

Type	Wheel arrangement	Transmission	Power (kW)	Speed (km/h)	Weight tonnes	First built	Diesel engine	Builders
DF4C	Co-Co	AC/DC	2,650	100	138	1985	16V240ZJC	Dalian
DF4D	Co-Co	AC/DC	2,940	145/100	138	1996	16V240ZJD	Dalian
DF4DJ	Co-Co	AC/DC/AC	2,940	145	138	2000	16V240ZJD	Dalian
DF5B	Co-Co	AC/DC	1,840	100	135	1996	12V240ZJF	Dalian
DF5D	Co-Co	AC/DC	1,470	100	135	1999	8V240ZJ	Dalian
DF6	Co-Co	AC/DC	2,940	118	138	1989	16V240ZJD	Dalian
DF7C	Co-Co	AD/DC	1,470	100	135	1991	12V240ZJ6D	Feb 7th
DF7C2	Co-Co	AC/DC	1,840	100	138	2000	12V240ZJ6F	Feb 7th
DF7D	Co-Co	AC/DC	1,840	100	138	1999	12V240ZJ6C	Feb 7th
DF7E	Co-Co	AC/DC	1,840	100	138	1999	12V240ZJ6B	Feb 7th
DF7F	Co-Co	AC/DC	2,650	100	2 × 138	2000	EQ16V240ZJ	Feb 7th
DF7J	Co-Co	AC/DC/AC	1,730	100	135	2003	12V240ZJ6E	Feb 7th
DF8B	Co-Co	AC/DC	3,820	100	138	2000	12V280ZJ	Dalian
DF10F	2 Co-Co	AC/DC	2 × 2,200	140	2 × 123	1996	12V240ZJD	Dalian
DF10F1	Co-Co	AC/DC	1,840	140	120	2002	12V240ZJD-2	Dalian
DF11	Co-Co	AC/DC	3,820	170	138	2000	12V280ZJ	Dalian
GK1E31	B-B	Hydraulic	1,000	80/40	92	1999	EQ6240ZJ	Feb 7th
GKD1	Bo-Bo	AC/DC	990	80	84	1990	6240ZJ	Dalian
CKD4A	Co-Co	AC/DC	2,940	100	111	2002	16V240ZJD	Dalian
CKD7	Bo-Bo-Bo	AC/DC	1,250	90	76	1993	CAT3516	Dalian
CKD8A	Co-Co	AC/DC	1,985	100	96	1996	12V240ZJD	Dalian
CKD8B	Co-Co	AC/DC	2,200	100	120	1997	12V240ZJD	Dalian
CKD8C	Co-Co	AC/DC	1,840	180	120	2001	12V240ZJD-2	Dalian
CKD9	Co-Co	AC/DC	2,650/1,840	100	140	2003	16V240ZJD1	Dalian
CK1E	C-C	Hydraulic	970	50	75	2003	CAT3508B	Feb 7th

CNR Type SS7E electric locomotive 0595032

CNR Type CK1E diesel-hydraulic locomotive for Vietnam **NEW**/1135830

CNR metro trainset
0595030

CNR Shenyang Railway Brake Factory
120 Nanliuzhong Road, Tiexi District, Shenyang 110023, Liaoning, China
Tel: (+86 24) 62 07 12 55   Fax: (+86 24) 25 86 21 14
Legal Representative: Jia Shiyuan

CNR Taiyuan Locomotive & Rolling Stock Works
10 Jiefang North Road, Taiyuan 030009, Shanxi, China
Tel: (+86 351) 264 94 50   Fax: (+86 351) 304 95 63
Legal Representative: Zhang Yili

CNR Tangshan Locomotive & Rolling Stock Works
3 Changqian Road, Fengrun District, Tangshan 063035, Hebei, China
Tel: (+86 315) 324 24 08   Fax: (+86 315) 324 16 12
Legal Representative: Wang Run

CNR Tianjin Locomotive & Rolling Stock Machinery Works
1 Nankou East Road, Hebei District, Tianjin 300232, China
Tel: (+86 22) 26 27 02 68   Fax: (+86 22) 26 27 12 34
Legal Representative: Zong Baoquan

CNR Yongji Electric Machine Factory
18 Dianji Street, Yongji 044502, Shanxi, China
Tel: (+86 359) 807 51 62   Fax: (+86 359) 807 52 90
Legal Representative: Dong Yu

CNR Xi'an Rolling Stock Works
Jianzhang Road, Sanqiao Town, Xi'an 710086, Shannxi, China
Tel: (+86 29) 236 91 26   Fax: (+86 29) 451 24 50
Legal Representative: Sun Kai

Research institutes
CNR Dalian Locomotive Research Institute
49 Zhongchang Street, Shahekou District, Dalian 116021, Liaoning, China
Tel: (+86 411) 460 10 10   Fax: (+86 411) 460 16 17
Legal Representative: Zhang Yan

CNR Sifang Rolling Stock Research Institute
231 Ruichang Road, Sifang District, Qingdao 266031, Shandong, China
Tel: (+86 532) 499 27 10   Fax: (+86 532) 499 29 61
Legal Representative: Ren Yujun

Trading company
LORIC Import & Export Corporation Ltd
11 Yangfangdian Road, Haidian District, Beijing 100038, China
Tel: (+86 10) 51 86 23 69   Fax: (+86 10) 51 86 23 74
Legal Representative: Zhao Guangxing

Background

CNR was established in October 2000 as a result of the division into two regional groups of the former China National Railway Locomotive and Rolling Stock Industry Corporation (LORIC). It is one of the world's leading suppliers of a complete range of systems, components and services for locomotive and rolling stock covering design, manufacture, refurbishment, maintenance and leasing. CNR has a presence in over 30 countries and regions and employs 102,908 people.

The annual output is 280 electric locomotives, 350 diesel locomotives, 1,650 passenger coaches, 30 multiple unit train-sets, 700 urban railway vehicles, 20,000 for freight wagons and 150 railway cranes. Its locomotive, coach and wagon manufacturing plants are among the largest in the world.

Products

CNR undertakes the design, manufacture, testing, commissioning and maintenance of locomotives and rolling stock, including: electric locomotives, diesel-electric and diesel-hydraulic locomotives from 280 to 7,200 kW for main line and shunting

CNR Type CKD 8C diesel-electric locomotive for Malaysia   0595031

CNR Type DF7C modular diesel-electric shunter   **NEW**/1135827

CNR Type DF4DJ AC drive diesel-electric locomotive   0595033

duties; dmus and emus for urban, suburban and regional transport; tilting and high-speed trains; trams and light rail vehicles; metro cars; and passenger coaches.

Electric locomotives

The SS range of electric locomotives features microprocessor control and anti-slip braking. The SS4G and SS3B eight-axle double units are suitable for heavy haul freight service. The SS7 series is used for high-speed passenger service. The SS7E is the modular mass transit solution which addresses the need for higher speeds. The 'Sky Shuttle' passenger locomotive is equipped with AC three-phase equipment. Type DJ3 is a high-power AC propulsion electric locomotive to be used to boost main line freight services in China.

Diesel locomotives

DF family

The DF range of diesel-electric locomotives, with a four-stroke turbocharged diesel engine and AC/DC transmission, are the main types of main line locomotive for both passenger and freight traffic. They run on three-axle bogies with roller bearings. The traction motors are fully suspended with hollow shaft quill drive.

Types DF4D, DF8B are designed for heavy-duty freight trains, Types DF4D, DF10F and DF11 for high-speed passenger service and Types DF5 and DF7 for shunting work. Type DF4DJ is equipped with AC traction equipment and IGBT converters.

GK family

The GK range of diesel locomotives are intended for shunting and industrial applications. Many have hydraulic transmission and have a B-B configuration to suit tight curves.

CK family

The CK series diesel locomotives are designed for markets other than China and feature a lower axle load; examples are in operation on the narrow-gauge railways of Malaysia, Myanmar, Nigeria, Tanzania and Vietnam.

Passenger coaches

All types of passenger coaches including single- or double-deck seating coaches, sleeping cars and restaurant cars, as well as mail vans, luggage vans and generator vans, with air-conditioning or normal ventilation. They were delivered to Iran, Malaysia, Myanmar, North Korea and Pakistan. The 25T family is a new product range used in Chinese Railways' fifth speed acceleration. Capable of operating at up to 160 km/h, it is the latest in a series of designs first built in 1992.

Metro trainsets, multiple-units and trams/LRVs

CNR products address increasing trends of urban and suburban development. Its regional and intercity trains are flexible products, designed to provide high standards of comfort, safety and performance. The 'China Star' holds the Chinese speed record for an emu of 321.5 km/h.

Travelling cranes

CNR produces an extensive range of travelling cranes, with lifting capacities from eight to 160 tonnes and capable of hauled speeds of up to 120 km/h. As well as supplying the domestic Chinese market, CNR has also delivered cranes to Myanmar, Pakistan and Tanzania.

CNR railway crane

Model, (lift capacity (t)), factory name
N151 (15t), China, Qiqihar
N603 (60t), Burma, Qiqihar
N1002 (100t), China, Lanzhou
N1004 (100t), China, Lanzhou
N1005 (100t), Tanzania, Lanzhou
N1601 (160t), China, Qiqihar
NS1252 (125t), China, Qiqihar
NS1601 (160t), China, Qiqihar
BT8.3 (8t), Pakistan, Lanzhou

CNR Model NS1252 diesel railway crane 0595034

CNR Type QKZ2 monorail car for Chongqing ***NEW**/1135821*

Contracts

China: From 2000 to 2004, CNR supplied 757 electric locomotives, 1,640 diesel locomotives, 7,409 passenger coaches and 74,440 freight wagons. Most of them were used by the Ministry of Railways (MOR).

In 2004 CNR delivered 380 Type 25T coaches to the MOR for the fifth acceleration of passenger services.

In December 2004 CNR received a contract with Beijing Metro to supply 192 metros for Line 5. They will be put into service in 2006 before the 2008 Olympic Games. CNR delivered 174 VVVF metro cars to Beijing Metro for Fuxingmen-Bawangfen Line before 2001.

In June 2004 CNR delivered the first metro trainsets to extension Line 1 of the Shanghai Subway Company. The Shanghai Subway purchased 10 six-car, broad aluminium-bodied trainsets from CNR.

Also in 2004 CNR delivered two catenary operation cars to Taiwan.

In 2002 CNR delivered 136 LRVs to the Beijing LRT system and received a contract from Tianjin Binhai LRT to supply 114 stainless steel-bodied rapid transit vehicles. Delivery began in October 2003. Another order was received in March 2004 to supply 25 four-car aluminium-bodied metro trainsets for the Line 1.

In October 2003 CNR began to deliver rapid transit vehicles for the Wuhan LRT system. All 12 Type B aluminium-bodied four-car vehicles were completed before October 2004.

In August 2000 in co-operation with Bombardier, CNR signed a contract with Shenzhen Metro to supply 22 trainsets (132 metros) for the first phase. In March 2004 CNR delivered the first trainset to the Shenzhen Metro system. Delivery is scheduled to be completed in February 2006.

In November 2002, as a main contractor, CNR won a contract for 21 monorail trainsets (84 rolling stock) with technical co-operation from Hitachi, Japan for the first phase of the Jiaochangkou–Xinshanchun line of the Chongqing system LRT. The first monorail trainset was put into operation in December 2004.

Congo: In 2005 CNR received a contract to supply one diesel-hydraulic locomotive.

Iran: By 2000 CNR had completed deliveries of an order to supply 217 metro cars for Tehran Urban & Suburban Railways. CNR was subsequently awarded further contracts for metro cars: 105 in 2002, 77 in 2004 and 119 in November 2004; 30 double-deck coaches in November 2004.

Iraq: In April 2002 50 model DF10FI 1,840 kW diesel-electric locomotives were delivered to Iraq under a tender contract approved by the United Nations.

Malaysia: In November 2001 two Type CKD8C 1,840 kW diesel-electric locomotives were delivered to Malayan Railway (KTM). CNR was awarded a further order by KTM for 20 three-phase propulsion diesel-electric locomotives in November 2003. They are to be metre gauge Co-Co machines with a 20 tonne axleload.

North Korea: Two model GKD4A 2,940 kW Co-Co diesel-electric locomotives were delivered to North Korea in July 2002. The diesel engine type is 16V240ZJD.

Pakistan: A first batch of eight 2,450 kW diesel-electric locomotives was delivered in June 2003 to Pakistan Railways (PR), followed by a second batch formed of seven 1,840 kW diesel-electric locomotives in July of the same year. Delivery of these 15 locomotives was to be followed by the local construction of 54 further machines supplied by CNR in CKD or SKD form under a technology transfer contract awarded in November 2002. This covered the supply of 44 2,450 kW and 25 1,840 kW 1,676 mm gauge locomotives. They are equipped with VVVF/CVCF transmission.

A second batch of 26 coaches was handed over to PR in August 2003, bringing to 40 the number of vehicles delivered. They were supplied under a contract which was signed in November 2001 for a total of 175 coaches, including a further 135 supplied as CKD kits and technology transfer. These comprise 52 semi-cushioned berth air-conditioned sleeping cars, 10 air-conditioned cushioned-berth sleeping cars, 87 semi-cushioned seating coaches, 16 generator cars and 10 brake vans. To meet Pakistani conditions, the interiors and seating are designed to be comfortable and bright. Track gauge is 1,676 mm and the carbody width 3,250 mm.

Vietnam: Vietnamese coal mining and phosphor mining companies selected CNR to supply eight metre gauge diesel-hydraulic locomotives. Six were delivered from 2003 to 2004; the remaining two were due to enter service in 2005.

UPDATED

China South Locomotive and Rolling Stock Industry (Group) Corporation (CSR)

11 Yangfangdian Road, Haidian District, Beijing, China
Tel: (+86 10) 63 98 47 70 Fax: (+86 10) 63 98 47 66
e-mail: csrft@csrgc.com.cn
Web: http://www.csrgc.com.cn

Key personnel
General Manager: Zhao Xiaogang

Background
CSR was established in 2000 as a result of the division into two regional groups of the former China National Railway Locomotive and Rolling Stock Industry Corporation (LORIC). The corporation comprises 25 state-owned enterprises employing around 112,000 staff. Some facilities are listed as part of CSR and of its northern China counterpart, China North Locomotive and Rolling Stock Industry (Group) Corporation (CNR).

Subsidiaries
CSR comprises 16 manufacturing plants and eight holding companies. There are also three mutual shareholding subsidiaries.

Works
CSR Beijing Feb 7th Rolling Stock Works
Zhangguozhuang Fengtai District, Beijing 100072, China
Web: http://www.eqc.com.cn

CSR Beijing Locomotive & Rolling Stock Machinery Works
West Changping Railway Station, Beijing 102249, China
Web: http://www.cpgc.com

CSR Chengdu Locomotive & Rolling Stock Works
Erxian Bridge, North Road, Chengdu 610057, Sichuan, China
Web: http://www.cdjcc.com

CSR Guiyang Rolling Stock Works
Baiyun District, Guiyang 550017, Guizhou, China
Web: http://www.southhuiton.com

CSR Luoyang Locomotive Works
Qiming East Road, Luoyang 471002, Henan, China
Web: http://www.lylw.co.cn

CSR Meishan Rolling Stock Works
Dongpo, Meishan 620032, Sichuan, China
Web: http://www.msrsco.com

CSR Nanjing Puzhen Rolling Stock Works
Puzhen, Nanjing 210032, Jiangsu, China
Web: http://www.njpzclc.com

CSR Qishuyan Locomotive & Rolling Stock Works
Qishuyan, Changzhou 213011, Jiangsu, China
Web: http://www.qscn.com

CSR Shijiazhuang Rolling Stock Works
Front Street of Rolling Stock Works, Shijiazhuang 050000, Hebei, China
Web: http://www.sjzclc.com

CSR Tongling Rolling Stock Works
Phoenix Mountain Avenue, Tongling 244142, Anhui, China
Web: http://www.tlclc.com

CSR Wuchang Rolling Stock Works
Peace Avenue, Wuchang, Wuhan 430062, Hubei, China
Web: http://www.chinacool168.com

CSR Wuhan Jiang'an Rolling Stock Works
Jiang'an District, Wuhan 430012, Hubei, China
Web: http://www.csrgc-ja.com

CSR Xiangfan Diesel Locomotive Works
Xiangyang District, Xiangfan 441105, Hubei, China
Web: http://www.csr-xfjc.com

CSR Zhuzhou Electric Locomotive Works
Tianxin, Zhuzhou 412001, Hunan, China
Web: http://www.gofront.com
(see Zhuzhou Electric Locomotive Works entry in *Locomotives and powered/non-powered passenger vehicles* section)

CSR Zhuzhou Rolling Stock Works
Hetang District, Zhuzhou 412003, Hunan, China
Web: http://www.csr-zrsw.com

CSR Ziyang Locomotive Works
Ziyang 641301, Sichuan, China
Web: http://www.zyloco.com

Holding companies
Beijing Railway Industry Trade Co
11 Yangfandian Road, Haidan District, Beijing 100038, China

CSR Marketing & Leasing Corp
11 Yangfandian Road, Haidan District, Beijing 100038, China

CSR Qishuyan Locomotive & Rolling Stock Technology Research Institute
Qishuyan, Changzhou 441047, Jiangsu, China
Web: http://www.jcqys.com

CSR Sifang Locomotive & Rolling Stock Co Ltd
Laoshan District, Qingdao 266101, Shandong, China
Web: http://www.cqsf.com

CSR Xiangfan Traction Motor Co Ltd
Changhong North Road, Xiangfan 441047, Hubei, China
Web: http://www.xqdj.com

CSR Zhuzhou Electric Locomotive Research Institute
Shifeng District, Zhuzhou 412001, Hunan, China
Web: http://www.zelri.com

Sifang Locomotive & Rolling Stock Co Ltd
Sifang District, Qingdao 266031, Shandong, China
Web: http://www.csrsf.com

South Huitong Co Ltd
Baiyun District, Guiyang 550017, Guizhou, China
Web: http://www.southhuiton.com

Mutual shareholding subsidiaries
CNR Dalian Locomotive Research Institute
CNR Sifang Rolling Stock Research Institute
LORIC Import & Export Corp Ltd

Products
Diesel and electric locomotives, high-speed trainsets, dmus and emus, metro trainsets and passenger coaches, including:

High-speed trainsets
China Star 25 kV AC 50 Hz power cars: 4,800 kW, 78 tonnes, maximum speed 270 km/h.

Diesel locomotives
DF8B Co-Co diesel-electric freight locomotive: 3,100 kW, 138 or 150 tones, maximum speed 100 km/h. A variant of this design for high-altitude operation has been developed for service in Tibet.

Electric locomotives
DJ2 25 kV AC 50 Hz Bo-Bo electric passenger locomotive: 4,800 kW, 84 tonnes, maximum speed 200 km/h.
SS8 25 kV AC 50 Hz Bo-Bo electric passenger locomotive: 3,600 kW, 88 tonnes, maximum speed 240 km/h.

Emus
25 kV AC 50 Hz 160 km/h 3,200 emu with AC propulsion.

Contracts
Export contracts include the supply of high-powered 25 kV AC 50 Hz electric locomotives to Uzbekistan.

Chittaranjan Locomotive Works

Chittaranjan, 713331, Burdwan District, West Bengal, India
Tel: (+91 341) 252 55 22   Fax: (+91 341) 252 61 53
e-mail: dgp_secyclw@sancharnet.in
Web: http://www.clwindia.com

Key personnel
Chief Executive Officer: Surinder Jain
Chief Electrical Engineer: A K Vohra
Chief Mechanical Engineer: P K Chatterjee
Controller of Stores: C Ilangovan
Financial Adviser/Chief Accounts Officer:
   Sri Harish Chandra
Chief Personnel Officer: Sunil Bajpai
Head of Public Relations: C N Jha

Background
CLW was the first production unit to be set up within Indian Railways. Established in 1950, CLW has turned out 2,351 steam locomotives of five different classes and seven classes of diesel locomotives.

A special feature of the plant is its in-house manufacturing facility for traction motors, control equipment and heavy steel castings.

Products
Electric locomotives; traction motors, castings. The present product mix consists of the following types of locomotives:

WAG7: This is a broad gauge machine rated at 3,730 kW and can haul 4,500 t of freight up a 1 in 200 incline. It is for 25 kV operation, with tap changer control and DC traction motors, and has a starting tractive effort of 42 t and a continuous tractive effort of 24 t. It has 630 kW Type HS15250A traction motors and is designed for a speed of 100 km/h. It has a dual brake system with rheostatic braking and can be operated in multiple units.

WAP4: This is a broad gauge machine rated at 3,730 kW and can haul a maximum of 26 coaches at a maximum speed of 140 km/h. It operates on 25 kV, with tap changer control and DC traction motors, and has a starting tractive effort of 30.8 t and a

CLW WAP-5 three-phase locomotive for passenger service   0063353

CLW WAG-7 25 kV AC high-adhesion Co-Co locomotive for Indian Railways freight service   0063351

continuous tractive effort of 19 t. It has HS1525OA 630 kW traction motors and flexi-coil Co-Co steel bogies. It is suitable for hauling air-braked stock and is designed for multiple-unit operation.

WAG9/WAG9H: CLW manufactures this three-phase 4,500 kW broad gauge locomotive with GTO drive and microprocessor control through a technology transfer agreement from Bombardier Transportation (formerly Adtranz). CLW assembled 15 with SKD/CKD components from Bombardier Transportation (formerly Adtranz). The first locomotive was handed over in 1998. It operates on 25 kV and has a starting tractive effort of 47 tonnes. It is designed for a maximum speed of 100 km/h. The locomotives are intended for heavy haulage work, due to their higher adhesion, and have regenerative braking and a microprocessor-based fault diagnostic system. Maintenance is reduced through the use of electronic controls and three-phase AC traction motors.

WAP5: CLW is a manufacturer of these three-phase 4,500 kW broad gauge electric passenger locomotives with technology acquired from Bombardier Transportation (formerly Adtranz Switzerland). The locomotive is designed to haul 26-coach trains at a maximum speed of 160 km/h.

WAP7: This is a 4,500 kW three-phase broad gauge machine with a top speed of 140 km/h.

ČKD Vagonka as

1 máje 3176/102, CZ-70931, Czech Republic
Tel: (+420 59) 747 71 11   Fax: (+420 59) 747 71 90
e-mail: info@vagonka.cz
Web: www.vagonka.cz

Key personnel
Chairman of the Board of Directors and General
    Manager: Miroslav Kupec
Sales Director: Ivo Gurnak

Background
ČKD Vagonka was founded in January 2000 and at present has its headquarters and workshops in Ostrava. The company is a direct successor of the traditional railway passenger vehicles production plant which was established at Studénka in 1900, and has key capabilities in development, design, technology and production of railway passenger vehicles.

It is wholly owned by Škoda Holding as.

Products
The main operation is the development, design and manufacture of double-deck electric multiple-units using aluminium bodyshells; diesel-electric and diesel-hydraulic railcars and multiple-units; light regional vehicles; railcars; trailers and coaches.

A supplementary operation consists of repairs and modernisation of vehicles, delivery of spare parts and manufacturing co-operation with foreign partners.

Contracts
Recent contracts include the supply of double-deck Class 471 emus for Czech Railways, designed for rapid suburban traffic using 3 kV DC nominal voltage. Four-axle railcars for Finnish Railways (VR Ltd) intended for traffic on secondary non-electrified tracks. Electric multiple-units for suburban transport on electrified tracks with 3 kV DC nominal voltage and 25 kV AC 50 Hz, optionally for regional transport. Repairs and modernisation of passenger cars. Aluminium alloy constructions of sub-assembly parts for passenger vehicles.

UPDATED

ČKD Vagonka double-deck Class 471 emu for Czech Railways 1066406

ČKD Vagonka Dm 12 diesel railcar for VR-Group, Finland 1066405

ČMKS holding as

ČMKS Holding as, Drahelick 2083, CZ-28803 Nymburk, Czech Republic
Tel: (+420 325) 53 13 35   Fax: (+420 325) 53 13 35
e-mail: cmks@cmks.cz
Web: www. cmks.cz

Key personnel
Chairman of the Board, General Director:
    Ing Josef Barta
Chief Designer, Rail Vehicles: Ing Pavel Matys

Subsidiary companies
ČMKS Jihlavská lokomotivní společnost sro (Jihlava, CZ)
ČMKS Lokomotivy as (Česká Třebová, CZ)
ČMKS Letohradske strojirny, sro (Letohrad, CZ)
ČMKS – Polska Sp zoo (Gliwice, Poland)
ČMKS – Zeleznicni vlecky, sro (Nymburk, CZ)

Ex-Czech Railways Class 753 diesel-electric locomotive refurbished by CMKS Holding for FNM Cargo, Italy
***NEW**/1141381*

For details of the latest updates to *Jane's World Railways* online and to discover the additional information available exclusively to online subscribers please visit
jwr.janes.com

Background

ČMKS holding as is a private company involved in transport engineering, in particular diesel-electric freight and shunting locomotives. The company's main focus is on the design, development, manufacture, overhaul, refurbishment and upgrading of diesel-electric traction vehicles, activities in which ČMKS has many years' experience.

Products

New diesel-electric locomotive manufacturing: shunting locomotive Class 709.4, intended for light and medium-heavy shunting service. The locomotive is powered by Caterpillar C 15 diesel engine with power output of 392 kW, diesel power transmission (AC/DC), Bo' wheel arrangement, axle load 18 t and max speed 80 km/h. The locomotive was presented at the Innotrans 2004 trade fair in Berlin and is due to be certificated for service in Germany and other EU countries. The first 10 Series 709.4 will be delivered to Serbia in 2005–2006 (ZTP Belgrade/Serbian Railways).

A modified 22 t axle load Series 709.4 shunting locomotive will also be manufactured.

Refurbishment, re-engining and modernisation of locomotives: refurbishment, re-engining, part and full modernisation of two, three, four and six axle diesel-electric and electric locomotives and special rail vehicles. Components overhaul and refurbishment. A modular system is employed for diesel-electric modernising with the option of a new Caterpillar diesel engine available in outputs from 300 to 1,500 kW. This programme includes modernised locomotives for Czech Railways, Czech private operators and foreign private operators and sidings.

Contracts

In 2004 ČMKS won a tender for the overhaul of diesel-electric and electric locomotives for Czech Railways up to 2008.

In 2004 ČMKS also delivered a total of 38 modernised main line locomotives Class 752.5 and 753.7 for private operators in Italy.

UPDATED

Class 709.4 new-build two-axle diesel-electric shunter by CMKS Holding   *NEW*/1141379

Class 724 modernised diesel-electric locomotive
NEW/1141380

Cockerill Maintenance & Ingenerie SA (CMI)

Avenue Greiner 1, B-4100 Seraing, Belgium
Tel: (+32 4) 330 24 33   Fax: (+32 4) 330 25 45
e-mail: locos.diesel@cmi.be
Web: www.cmi.be

Key personnel
President: B Serin
Sales Director: J F Levaux

Contact: D Vanham (Manager, Locomotives & Diesel Engines)
Locomotives Sales: M Metz

Products
Shunting locomotives 185 to 1,000 kW, on two, three or four axles, for all gauges and types of track. Among its latest products is a shunting/branch line locomotive featuring hydrostatic transmission.

Maintenance, rehabilitation and retro-fitting of shunting locomotives; radio remote-control.

UPDATED

Colorado Railcar Manufacturing LLC

1011 14th Street, Fort Lupton, Colorado 80621, USA
Tel: (+1 303) 857 10 66   Fax: (+1 303) 857 42 09
e-mail: sales@coloradorailcar.com
Web: www.coloradorail.com

Key personnel
President: Tom Rader
Vice-President, Sales: Tom Janaky

Products
Passenger rail vehicles. The company has developed and built a prototype of a commuter dmu that is fully compliant with Federal Railroad Administration and APTA structural safety requirements. Each steel-bodied power car

Colorado Railcar single-deck prototype dmu demonstrator, subsequently ordered by SFRTA
NEW/0585219

is equipped with two Detroit Diesel Series 60 440 kW engines with Voith T212 hydrodynamic transmission and Voith KE553 final drive. A 130 kW Deutz water-cooled diesel engine provides power for auxiliaries such as heating and air conditioning. Vehicle configuration proposals provide for up to three intermediate trailer cars, for streamlined and non-streamlined power cars and for double-deck power cars and trailers.

Colorado Railcar also manufactures passenger coaches and has specialised in low-volume contracts for operators of tourist services, supplying single-level and bi-level dome cars to customers that include Rocky Mountaineer Railtours, Canada, the Alaska Railroad and Princess Cruises and Tours.

Recent developments have included a new double-deck dmu which raises the seating capacity from 90 to 188 passengers. The bilevel passenger car has 200 seats in a low-floor, ADA accessible configuration.

Contracts

Florida Regional Authority (SFRTA formerly Tri-Rail), representing the FRA and Florida DOT in a joint procurement agreement, approved the contract for the dmu demonstration project and authorised the acquisition of three commuter railcars from Colorado Railcar Manufacturing. The purchase includes the existing single-level dmu which entered service in February 2004, the newly designed double-deck dmu and the new double-deck companion commuter coach. Production completion is scheduled for the end of March 2005.

In late 2004 Colorado Railcar was awarded a contract by the Grand Canadian Railtour Co, Canada, for a dmu railcar and two single-level dome cars for Rocky Mountaineer tourist passenger services. The vehicles are due to enter service in 2006.

UPDATED

CostaRail Srl

Viale IV Novembre, I-23845 Costa Masnaga (LC), Italy
Tel: (+39 031) 86 94 11   Fax: (+39 031) 85 53 30
e-mail: hkastner@costarail.it

Key personnel

Commercial Manager: Heinz Kastner

Background

Formerly Costaferroviaria, the company went into temporary receivership and was taken over by Rail Services International Group, for which RSI Italia SpA is the operative leader. In October 2004, CostaRail Srl was established, re-launching all existing Costaferroviaria activities, including orders, trademarks, patents and projects.

Products

Electric multiple-units, passenger coaches, freight wagons and bogies.

Contracts

CostaRail has restarted the construction of three battery powered locomotives for a tourist train in Aosta Valley, Italy, following the interuption during the takeover of Costaferroviaria by RSI.

As part of a consortium, CostaRail has been awarded a contract by Italian State Railway, Trenitalia, for the construction of 39 double-deck passenger coaches.

Repair of a fire damaged electric locomotive for Skoda of Ferrovie Nord Milano.

UPDATED

CSR Qishuyan Locomotive and Rolling Stock Works (QS)

Qishuyan, Changzhou City, Jiangsu Province 213011, Changzhou City, China
Tel: (+86 519) 877 17 11   Fax: (+86 519) 877 03 58
Web: http://www.qscn.com

Key personnel

President: Gu Mingkang
Export Directors: Jiang Ping; Wang Weiping
Export Manager: Li Guohua

Background

CSR Qishuyan Locomotive and Rolling Stock Works (QS), a member of China Southern Locomotive and Rolling Stock Industy Corporation (Group), is the main production enterprise for high-powered diesel locomotives for trunk lines of the Chinese Railways network and also produces freight wagons. The company was established in 1905, and now covers an area of 1,760,000 m² with over 10,000 employees (of whom 3,000 are technicians).

QS leads diesel engine development, design, research and domestic manufacturing and is the centre for researching high-power diesel locomotive technology in China.

QS researched and manufactured the first 160 km/h Dong Feng 11 quasi-high-speed diesel passenger locomotive in 1992. This was tested at up to 186 km/h, a record for Chinese Railways at that time. This type of diesel locomotive has made a major contribution to raising operational speeds on Chinese Railways.

In 1997, QS researched and manufactured the 3,680 kW Dong Feng 8B heavy duty freight diesel locomotive. It is now the main type of diesel locomotive for heavy duty freight on Chinese Railways.

In 1999, co-operating with Nanjing Puzhen Rolling Stock Works, QS researched and manufactured the 180 km/h NZJ1 quasi-high-speed diesel trainset, which comprises two power cars and nine trailers. QS Works took charge of the research and manufacture of the two power cars. This was tested at up to 199.4 km/h, a new record for diesel traction on Chinese Railways. In the same year, QS also developed and successfully manufactured the Dong Feng 8B diesel locomotive using bogies with rolling bearing traction motor suspension.

In 2000, QS manufactured the 280 diesel engine first Series products, the Type GKD2 1,470 kW diesel-electric shunting locomotive.

In 2001, QS manufactured Dong Feng 8B diesel locomotives using radial bogies, the Dong Feng IIZ special diesel locomotive with a 16V280ZJB engine, nominally rated at 4,705 kW.

In 2002, the trial manufacture was completed of an AC transmission diesel locomotive using the 16V280ZJB diesel engine, developed in co-operation with the Austria AVL Company. This is rated at 4,410 kW. The GK2C diesel-hydraulic shunting locomotive has also been developed, as well as the plateau locomotive specially designed for Qinghai–Tibet lines, which has been put into use for service testing.

Products

Diesel-electric locomotives, locomotive parts and components, overhaul and repair of locomotives and railways wagons.

Dong Feng 11
Track gauge: 1,435 mm
Wheel arrangement: Co-Co
Max speed: 170 km/h
Axle load: 23 tonnes
Max starting tractive effort: 245 kN
Continuous tractive effort: 160 kN
Min curvature radius: 145 m
Max height (rail surface to top end): 4,736 mm
Max width: 3,304 mm
Length over couplers: 21,250 mm

Dong Feng 8B
Track gauge: 1,435 mm
Wheel arrangement: Co-Co
Max speed: 100 km/h
Axle load: 23/25 tonnes
Max starting tractive effort: 520 kN
Continuous tractive effort: 340 kN
Min curvature radius: 145 m
Max height (rail surface to top): 4,736 mm
Max width: 3,304 mm
Length over couplers: 22,000 mm

Contracts

In 2002, five DSVN type diesel locomotives were exported to Vietnam. As well as producing locomotives for the domestic market, QS also exports its products.

CNR Dalian Locomotive and Rolling Stock Works

51 Zhongchang Street, Shahekou District, 116022 Liaoning, China
Tel: (+86 411) 419 82 54
Fax: (+86 411) 465 42 45
Web: http://www.dloco.com

Key personnel

Contact: Tan Chengxu

Background

Member of CNR (see CNR entry in *Locomotives and powered/non-powered passenger vehicles* section). Dalian has formed a joint venture company with Toshiba, Dalian Toshiba Locomotive Electrical Equipment Co Ltd, to develop AC transmission systems for locomotives.

Products

Diesel locomotives of various power ratings for passenger, freight and shunting applications.

The DF4D passenger diesel locomotive is newly introduced and has a 16V240ZJD diesel engine, maximum power output of 2,940 kW. The locomotive is for main line passenger work and has a maximum speed of 145 km/h, and can haul 20 coaches.

The DF4D freight diesel locomotive has been developed on the basis of the DF4D passenger locomotive. It has an improved gear ratio (63:14) and a maximum starting tractive effort (480.5 kN) for heavy haul work in China.

The DF5 diesel electric shunting locomotive has a maximum power rating of 1,210 kW and an 8240ZJ diesel engine with three-phase AC/DC transmission and an axle loading of 22.5 tonnes. It is for shunting and industrial work.

The DF5B locomotive, based on the DF5 locomotive, is powered by the 12V240ZJF diesel engine with a maximum power rating of 1,840 kW. The locomotive is narrow, making it suitable for mining operations.

The DF6 locomotive is a new generation of high-power and high-performance units, built in co-operation with other firms. Its power unit, the 16V240ZJD has been improved in association with Ricardo Consulting Engineers, UK, while the electric transmission has been developed in co-operation with General Electric Co, USA. It features advanced technology including micro-processor control, diagnostic display system, dynamic brakes with three-stage expansion and anti-slip/slide devices. It has a maximum power rating of 2,940 kW.

The DFI0F passenger diesel locomotive is a new model for mainline passenger operation with a maximum speed of 160 km/h. It is in Co-Co axle configuration and has an axle loading of 20 t. The engine is predominantly used on main line work.

The CKD7 locomotive has been specially designed and built for main line work on Myanmar Railways. It has a CAT 3516 diesel engine as its power unit with a maximum power rating of

1,250 kW, and features two cabs and Bo-Bo-Bo arrangement. It has a dual vacuum-air brake system and rheostatic brake system.

The CKD8A locomotive, designed for Nigerian Railways, is equipped with the advanced 12V240ZJG diesel engine and has a maximum power rating of 1,800 kW. The turbocharger and fuel injection system can be specified from manufacturers other than Dalian. Passenger and freight work on narrow gauge (1,067 mm).

The CKD6 locomotive is also for Nigerian Railways. It is powered by the 6240ZJ diesel engine and has a maximum power rating of 1,200 kW. The

wheel arrangement A 1 A-A 1 A, and the axle load is 12.5 t, with a tare weight of 75 t.

The CKD8B locomotive, with a 2,200 kW power rating, which is powered by the 12V240ZJD diesel engine, has been designed specially for the Tanzania-Zambia Railway.

The GK diesel-hydraulic shunting locomotive is for both shunting and light haulage. This locomotive, powered by the 6240ZJ diesel engine with a power rating of 990 kW, has a hydraulic transmission and hydraulic reversing instead of the original electric transmission mode which is cited as being lower in cost.

CNR Datong Electric Locomotive Co Ltd

1 Qianjin Street, Datong 037038, Shanxi, China
Tel: (+86 352) 509 08 78   Fax: (+86 352) 509 09 84

Key personnel
Contact: Shi Xiaoding

Background
Datong ABC Castings Company Ltd (a Sino-American joint venture, Datong Works, owns 60 per

cent of the shares). Datong Electric Locomotive Co Ltd is a member of CNR and CSR (see CNR and CSR entries in *Locomotives and powered/non-powered passenger vehicles* section).

Products
SS3 passenger and freight electric locomotives; SS4/SS7 electric locomotives; DF4 diesel-electric freight locomotive; DT20 diesel-electric shunting locomotive; DF4, DF8 combined piston of steel crown and aluminium skirt; ND5 combined piston of steel crown and aluminium skirt for 7FDL diesel

engines; forged aluminium piston and piston pins for diesel engines; bogies and parts for locomotives and rolling stock.

Contracts include supply of 20 DF4 diesel-electric locomotives, 44 SS3 electric locomotives and five SS7 electric locomotives for the Ministry of Railways.

DEVZ – Dnepropetrovsk State Electric Locomotive Factory

13 Orbitalnaya Street, Dnepropetrovsk 49868, Ukraine
Tel: (+380 56) 258 51 33; 42 35; (56) 773 26 78
Fax: (+380 56) 773 29 17; 773 26 77

Key personnel
General Director: Chumask Valery Victorovych

Background
The firm was established in 1937 as Ukraine's major locomotive and repair works. In 1958 it switched to the manufacture of industrial electric locomotives.

Products
Main line electric locomotives, mining electric locomotives, industrial narrow-gauge electric locomotives, repair, emergency renovation, electric traction components, starters for diesel locomotives, wheelsets.

The DS freight locomotive is for operation on 3 kV overhead lines and 1,520 mm gauge. It has a maximum power rating of 6,250 kW at a speed of 100 km/h.

Contracts
Locomotives have been supplied for operation on rail lines and open-cast mines in Russia, Kazakhstan, Mongolia and other countries.

Developments
In 1993–1995, under the programme of Ukraine's own enterprise encouraging diesel locomotive production, DEVZ produced a prototype model. Designs have been produced for advanced four- and eight-axle locomotives.

DEVZ DS Class eight-axle freight locomotive   0092266

DEVZ prototype advanced four-axle locomotive
0092267

Diesel Locomotive Works (DLW)

Varanasi 221 004, (UP), India
Tel: (+91 542) 227 05 51; 55
Fax: (+91 542) 227 06 03; 227 01 38
e-mail: gmdlw@satyam.net.in
Web: www.diessellocoworks.com

Key personnel
General Manager: Ravindra Sharma
Chief Mechanical Engineer:
  Pratap Srivastava
Chief Project Manager/TOT: S M Verma
Controller of Stores: R Ramesh

Chief Design Engineer: Anoop Kumar
Financial Adviser and Chief Accounts Officer:
  M S Bhatia
Chief Marketing Manager: R P Mishra
Chief Mechanical Engineer/Production:
  Hemant Katiyar
Chief Public Relations Officer: Dinesh Shukla

Background
A production unit of Indian Railways, DLW was set up in collaboration with Alco, USA, in 1961 and commenced manufacturing in 1963. It has so far supplied 4,660 locomotives, mainly to Indian Railways but also including 321 locomotives

for non-railway customers such as power plants, port trusts and industrial users. DLW has recently had its quality standards upgraded to ISO 9001–2000, having been awarded ISO 9002 certification in 1997. It has also been awarded ISO 14001 and integrated ISO 9001: 14001 certification.

Products
Diesel-electric locomotives for 1,676 mm (broad gauge) and metre gauge; diesel engines of 1,000 to 3,100 kW and components for diesel locomotives. The company also manufactures high-capacity generator sets.

The following Alco/GM engine locomotive types have been produced or are in production:

Alco design:
WDG3A: 1,676 mm gauge 2,300 kW (3,100 hp) Co-Co for main line freight service
WDM3C: 1,676 mm gauge 2,445 kW (3,300 hp) Co-Co for main line freight service
WDM3D: 1,676 mm gauge, 2,445 kW (3,300 hp) Co-Co for main line mixed service
WDM2: 1,676 mm gauge 1,940 kW (2,600 hp) Co-Co for main line mixed service
WDM3A: 1,676 mm gauge 2,310 kW (3,100 hp) Co-Co for main line mixed service
WDM6: 1,676 mm gauge 1,000 kW (1,350 hp) Bo-Bo for shunting and main line service
WDM7: 1,676 mm gauge 1,565 kW (2,150 hp) Co-Co for main service
WDP1: 1,676 mm gauge 1,715 kW (2,300 hp) Bo-Bo for main line passenger service
WDS6: 1,676 mm gauge 1,000 kW (1,350 hp) Co-Co for shunting service
YDM4: 1,000 mm gauge 1,000 kW (1,350 hp) Co-Co for main line mixed service

General Motors design:
WDG4: 1,676 mm gauge 3,100 kW (4,000 hp) Co-Co for main line freight service.
WDP4: 1,676 mm gauge 3,100 kW (4,000 hp) Co-Co for main line passenger service.

By 2004 export orders totalled 78 locomotives, including machines to Bangladesh, Malaysia, Myanmar, Sri Lanka, Tanzania and Vietnam. In addition, 40 Type YDM4 locomotives manufactured by DLW were operating on a lease basis in Malaysia and Tanzania.

In 1995, Indian Railways signed a technology transfer contract with General Motors (EMD), USA, to manufacture 3,100 kW (4,000 hp) state-of-the-art AC-AC microprocessor-controlled Type GT46MAC Class WDG4 freight locomotives, as well as 12- 16- and 20-cylinder diesel engines, produced at DLW. The first series of 31 ordered from General Motors (EMD) are working on South Central Railways. Indigenous manufacture of these locomotives has also been taken up at DLW and over 55 locomotives have so far been manufactured.

The agreement with General Motors (EMD) also covers a Class WDP4 160 km/h passenger version of the GT46MAC design, which have entered production at DLW.

With the acquisition of AC technology from GM/EMD, DLW is offering Alco and GM technology worldwide.

DLW also manufactures spares for diesel locomotives including 6-, 12- and 16-cylinder Alco diesel power packs, cylinder blocks and engine components. An automated cylinder liner chrome plating facility was commissioned in 1996.

UPDATED

DLW Class WDP4 1,676 mm gauge, 3,100 kW main line locomotive for passenger service
***NEW**/1139352*

DLW Class WDM3D 1,676 mm gauge, 2,460 kW main line locomotive for mixed traffic service
1139350

DLW Class WDP1 1,676 mm gauge 1,715 kW main line locomotive for passenger service
***NEW**/1139353*

DLW Class WDG4 1,676 mm gauge 3,100 kW main line locomotive for freight service
***NEW**/1139351*

EDI Rail

2B Factory Street, Granville, New South Wales 2142, Australia
Tel: (+61 2) 96 37 82 88   Fax: (+61 2) 96 37 67 83
e-mail: sales@edirail.com.au
Web: www.edirail.com.au

Key personnel
Chief Executive Officer: Guy Wannop
Executive General Manager, Freight: Danny Broad
Executive General Manager, Passenger:
  David Williamson

Background
A division of Downer EDI Ltd, EDI Rail is the result of a merger of Clyde Engineering and the rail activities of Walkers Ltd.

Works
EDI Rail has manufacturing, maintenance or design facilities in: Bathurst, Granville, Kooragang Island and Cardiff, New South Wales; Darwin, Northern Territory; Forrestfield and Nowergup, Western Australia; Rockhampton, Gladstone, Maryborough and Brisbane, Queensland; Newport and Geelong, Victoria; Port Augusta, Dry Creek, Whyalla, Port Lincoln, South Australia.

Products
Diesel locomotives, dmus, emus, including tilting trains, LRVs.

Contracts
In 2003 EDI Rail was awarded a contract by Pacific National to supply 13 2,420 kW PN Class 1,067 mm gauge diesel-electric locomotives for east coast freight services into Queensland. Their design is similar to that of QR's 4,000 Class.

In March 2003 EDI Rail signed a contract to supply to QR, Queensland, 11 4000 Class 2,424 kW 1,067 mm gauge diesel-electric locomotives for delivery in 2004–05. This brought to 49 the number of this type ordered by QR.

In January 2003 EDI Rail announced orders from Asia Pacific Transport (APT) for locomotives and freight wagons for use on the new Alice Springs to Darwin standard gauge rail link. The four FQ Class diesel-electric locomotives were due to be delivered by the end of 2003. The contract also covers provision of maintenance services (see entry for EDI Rail in the *Vehicle maintenance equipment and services* section).

In December 2002 the State Rail Authority of New South Wales (SRA) signed a Stage 2 contract to supply 15 additional four-car Millennium double-deck emus for the Sydney suburban network. Delivery is scheduled for 2004–05. The order follows an earlier (Stage 1) contract for 81 Millennium cars to be delivered by the end of 2003. A Stage 3 option covers 60 additional cars.

In 2001 Western Australian Government Railways ordered 31 three-car emus from a consortium of EDI Rail and Bombardier for Transperth suburban services. Deliveries are due to commence in early 2004.

Earlier passenger vehicle contracts include: two diesel-powered and two electric-powered 1,067 mm gauge tilt trains for QR; Airtrain, IMU and SMU emus for QR; STAR LRVs for Kuala Lumpur.

UPDATED

Millennium double-deck emu by EDI Rail for SRA's CityRail suburban services in Sydney   0594642

Diesel Tilt Train supplied by EDI Rail to QR (Brian Webber)   0554900

4000 Class diesel-electric locomotive supplied by EDI Rail to QR
(Brian Webber)
0099111

E G Steele & Co Ltd

25 Dalziel Street, Hamilton ML3 9AU, UK
Tel: (+44 1698) 28 37 65   Fax: (+44 1698) 89 15 50
e-mail: egsteelecoltd@btinternet.com

Key personnel
Managing Director: David Steele
Logistics Manager/Director, Export Sales: Ian Hood
Quality Manager: Cameron Gibson

Products
Locopulsor shunting machine, a single-wheel vehicle capable of moving freight wagons weighing 160 to 200 tonnes on straight level track. It can also move wagons in curves, split a line of wagons and handle a wagon on a turntable.

UK agents for Trackmobile road/rail shunters.

UPDATED

Electro-Motive Diesel (EMD)

9301 West 55th Street, La Grange, Illinois
60525, US
Tel: (+1 708) 387 60 00; 61 07
Web: www.emdiesels.com

Key personnel
President and Chief Executive Officer:
John Hamilton

Canadian Manufacturing Facility
PO Box 5160, London, Ontario N6A 4N5, Canada
Tel: (+1 519) 452 50 00

Licensees
EMD has locomotive manufacturing licence
agreements with the following companies:
Clyde Engineering Company Pty Ltd, Granville,
New South Wales, Australia
Hyundai Rolling Stock Company, Seoul, South
Korea
ALSTOM Transporte SA, Valencia, Spain
Turkish State Railways, Ankara, Turkey

Background
In April 2005 General Motors Corporation sold its
Electro-Motive Division to a group of investors
led by Greenbrier Equity Group LLC and Berkshire
Partners LLC. The transaction covered EMD's North
American and international locomotive supply
businesses, its power, marine and industrial
products, all of its locomotives maintenance
contracts worldwide and its manufacturing facilities
at La Grange, Illinois, and London, Ontario.

Products
Diesel-electric freight and passenger locomotives
from 746 to 4,476 kW (1,000 to 6,000 hp);
remanufacturing and rebuilding services;
locomotive leasing and contract maintenance.

SD90MAC: 4,476 kW (6,000 hp) AC-driven freight
locomotive. It is offered as a 1:2 replacement for
the SD40-2 locomotive. The H-engine has a 16 per
cent improvement in fuel efficiency over that of
the SD40-2 16-645E3B engined locomotive and
it has a starting tractive and braking capability of
170,000 lbs continuous, 200,000 lbs starting
tractive effort and 115,000 lbs braking. The SD90MAC
has integrated cab electronics, EM2000 advanced
computer, electronic air brakes, automatic parking
brakes, WhisperCab and a patented radial bogie.

GT46MAC: 2984 kW (4,000 hp) heavy haul/
high-speed AC traction locomotive, in production
since 1993, fitted with 16-710G3B engine. Starting
tractive effort 549 kN and braking 270 kN. A 12 per
cent fuel efficiency improvement is cited over the
645E series. It has the EM2000 advanced computer
system and HTSC bogie. Maintenance intervals
are 90 days.

JT42CWR: 2,238 kW (3,000 hp) DC traction freight
locomotive with 12N-710G3B-EC turbocharged
engine, geared for a top speed of 124 km/h. A
12 per cent fuel efficiency improvement is cited
over the 645E series. It has the EM2000 advanced

SD90MAC locomotive for Union Pacific   0003462

SD80MAC locomotive for Conrail   0003463

computer system and HTSC bogie. Maintenance
intervals are 90 days.

GP15D/GP20D: 1,119/1,679 kW(1,500/2,250 hp)
shunting locomotives with EMD170 electronic
fuel injection twin-turbocharged engines built by
Caterpillar. Fitted with EM2000 control technology.

F59PHI: 2,238 kW (3,000 hp) DC traction,
streamlined passenger locomotive with
12-710G3B-EC turbocharged diesel engine and top
speed of 176 km/h (110 mph), WhisperCab which
isolates crew from noise and vibration, GP-SS
bogies. Fitted with EM2000 control technology.

DE30AC/DM30AC: 2,238 kW (3,000 hp) AC traction
passenger locomotive with 12-710 turbocharged
engine with electronic fuel injection geared to top
speed of 160 kph (100 mph), bolsterless bogie,
integrated cab electronics, EM2000 advanced
computer and aerodynamic body. Dual-mode
version can operate off third rail.

*Functionally Integrated Railroad Electronics
(FIRE®):* an advanced locomotive management
system.

UPDATED

Electroputere

SC Electroputere SA Holding
Calea Bucureşti 144, RO-1100 Craiova, Romania
Tel: (+40 251) 43 77 00   Fax: (+40 251) 43 77 30
e-mail: electroputere@electroputere.ro
Web: http://www.electroputere.ro

Key personnel
General Manager: Mondea Cornel
e-mail: cmondea@electroputere.ro

Background
Electroputere was founded in September 1949,
manufacturing equipment for the energy industry
and railway transport. In 1990 Electroputere was
separated into seven trading companies and in
1994 became SC Electroputere SA Holding. Now
the company is made up of four manufacturing
divisions, a general services division and the
following factories: tools, devices checkers, foundry
and forge and repairing and modernisation for
tools and engineering installations.

Products
Diesel-electric Co-Co locomotives from 1,570 to
3,000 kW; electric Co-Co locomotives of 5,100 kW;
electric shunting locomotives of 1,250 kW.

Fahrzeugtechnik Dessau AG

Am Waggonbau 11, D-06844 Dessau, Germany
Tel: (+49 340) 253 70
Fax: (+49 340) 253 71 05
e-mail: zentrale@fahrzeugtechnik-dessau.de
Web: http://www.fahrzeugtechnik-dessau.com

Key personnel
Chairman of the Supervisory Board:
Jürgen Jantzen
Board of Directors: Dr Ing Joachim Pfannmüller
Eberhard Mann
Walter Schulz
Sales: Christain Dziubiel
Public Relations: Karin Pursche

Subsidiary companies
Fahrzeug-Outfit Gmb
HFTD Metall- und Komponentenbau GmbH Dessau
Gedack Metallschlauchservice GmbH

Products
Car bodies and subassemblies for rail vehicles;
prototypes and small series of complete vehicles;

driver's cab modules; rail vehicle doors; rail vehicle drive systems; repair, welding, supplier auditing, construction supervision and testing, production rig development and construction and technical calculation services; and anti-corrosion, blasting and painting and external finishing services for rail vehicles (via the Fahrzeug-Outfit subsidiary).

Contracts
Rail vehicles: Dortmund 2 vehicles (H-Bahngesellschaft); Type Rimms 660 wagons (DB Cargo AG); sliding wall wagons (Bombardier/Greenbrier); Metropolitan intercity trainsets (DB AG); Type Faln 128 wagons (Bombardier); prototype and pre-series Class 670 double-deck railbus (Bombardier/DB AG). Underframe modules for: Combino LRVs (Siemens); cars for Karlsruhe trams (Siemens); subassemblies for GT6 trams for Jena (Bombardier); roof sections for Class 101, 145, 152 and 185 electric locomotives (Bombardier and Siemens); LIREX dmu (ALSTOM); aluminium assemblies for ICE 3, VT 605 and VT 605 trainsets; trams for Magdeburg (ALSTOM); various cab modules, including Guangzhou metro (Bombardier); drive systems for Talent dmus for NSB, Norway (Bombardier), DB Class VT 612 dmus (Cummins/Adtranz), DB Class VT 642 dmus (MTU/Siemens).

FIREMA Trasporti SpA

Headquarters
Via Provinciale Appia, Località Ponteselice, I-81100 Caserta, Italy
Tel: (+39 0823) 09 71 11   Fax: (+39 0823) 46 68 12
Web: http//www.firema.it

Key personnel
Chairman: A de Benerdictis
Managing Director: L Rigno
Commercial Manager: S d'Arminio
Marketing Manager: M Fantini

Commercial and technical offices
Via Triboniano n 220, I-20156 Milan, Italy
Tel: (+39 02) 23 02 02 23   Fax: (+39 02) 23 02 03 00

Products
Electric locomotives, high-speed trainsets, electric multiple-units, railcars, advanced guided transit systems, metro rolling stock and light rail vehicles.

Contracts
Contracts include mechanical parts for 20 single-voltage (3 kV DC) and 23 dual-voltage (3 kV DC and 25 kV 50 Hz AC) E402 electric locomotives for Italian Railways (FS); electrical equipment for 26 E652 electric locomotives for FS; and mechanical parts for 12 E652 locomotives for FS.

Other contracts include the supply of 20 power car bodyshells, 56 first-class and 38 second-class light-alloy trailers, 40 trailer bogies, 12 sets of AC traction motors and 60 auxiliary static converters for 30 ETR 500 high-speed trainsets for Italian Railways (FS); 20 power cars, 77 trailers, 30 motor bogies, 154 trailer bogies and electronic equipment for 30 dual-voltage (3 kV DC and 25 kV 50 Hz AC) ETR 500 trainsets for FS; six Type E82 two-car 3 kV DC emus for SEPSA, Naples (Cumana and Circumflegrea Railways); five Type E84A 3 kV DC emus for COTRAL (Rome–Viterbo); 104 power cars and 64 trailers for 72 four-car double-deck emus for FS and North Milan Railway (FNME); and four two-car emus for Circumetnea Railway. Firema has supplied 16 articulated low-floor LRVs for the Midland Metro in Birmingham and 30 double-articulated low-floor LRVs for Oslo Sporveier. Electrical equipment for both types supplied by Ansaldobreda.

Recent orders include 25 diesel-hydraulic Bo-Bo locomotives Class D146 for Italian State Railways; 1 diesel-hydraulic locomotive Type D146 for Autorità portuale di Marina di Carrara, deliveries from the end of 2001; 100 bogie hopper wagons Type Talns for DB Cargo, deliveries from April 2000.

TAF double-deck emu in service with FS   0063359

Firema D146 diesel-hydraulic locomotive   0089812

Fuji Car Manufacturing Co

Fuji Car Manufacturing Co Ltd
3 Shoho Building, 2-2-3 Nishishinsaibashi, Chuo-ku, Osaka 542, Japan
Tel: (+81 6) 213 27 11
Fax: (+81 6) 213 40 71

Key personnel
Export Sales: M Miyama

Products
People mover systems (FAST) and passenger coaches.

Fuji Heavy Industries

Fuji Heavy Industries Ltd
Transportation & Ecology Systems Division
Subaru Building 7-2, Nishishinjuku 1-chome, Shinjyuku-ku, Tokyo 160, Japan
Tel: (+81 3) 33 47 24 92   Fax: (+81 3) 33 47 24 75

Main works
Utsunomiya Manufacturing Division
1-11, Yonan 1-chome, Utsunomiya, Tochigi 320, Japan

Key personnel
General Manager, Overseas Group: Hideo Ueno

Products
Diesel trainsets, tilting dmus, passenger coaches and railcars.

Ganz Motor Kft

Postafiók 263, H-1475 Budapest, Hungary
Tel: (+36 1) 210 11 50   Fax: (+36 1) 313 48 09
e-mail: ganzmoto@mail.datanet.hu

Key personnel

Managing Director: Péter Pétervári
Sales Director: Martinkó András
Technical Director: Orsovai József

Background

Ganz Motor Kft, established in 1990 as a successor of the former Ganz Mávag, is a manufacturer of diesel and gas engines. In 2004 its partner company Ganz Vagon Kft, following reorganisation, merged into Ganz Motor Kft bringing with it the ability to manufacture complete rolling stock vehicles. It also owns the company Ganz Holding Rt.

Products

Electric and diesel railcars, trams and LRVs, metro cars, passenger coaches, diesel-electric and diesel-mechanical locomotives, trailer vehicles for electric multiple-units.

Diesel engines: Ganz-Pielstick PA4V 185VG (6, 8, 12, 16 and 18 cylinders); Ganz-Jendrassik VFE 17/24.

UPDATED

GE Transportation Systems

2901 East Lake Road, Erie, Pennsylvania 16531 USA
Tel: (+1 814) 875 22 40   Fax: (+1 814) 875 59 11
Web: http://www.getransportation.com

Key personnel

President and Chief Executive Officer:
   John Krenicki, Jr
General Managers
   Locomotive Commercial Operation:
      Michael Abrams
   Global Services Operation: David Tucker
   Global Supply Chain Management:
      Bill Fitzgerald
   Engineering Operation: Steve Gray

Affiliates

A Goninan & Co Ltd, Australia (licensee)
Dorbyl Transport Products, South Africa (licensee)
GE Transportation Systems Global Signaling, USA (joint venture)
GE Transportation Systems South America, Brasil (joint venture)
PT GE Locomotif, Indonesia (joint venture)

Background

GE Transportation Systems is a global technology leader and supplier to the railroad, transit, marine, stationary and mining industries. GE provides freight and passenger locomotives, propulsion and auxiliary power systems for transit vehicles, motorised drive systems for mining trucks and drill platforms, marine and stationary engines, product replacement parts and major components for GE and other platforms, information-based service applications like remote monitoring and diagnostics as well as information technology solutions to its customers.

GE Transportation Systems Global Signaling is a provider of on-board and wayside signalling and communications systems, train inspection systems, crossing warning systems and services for the global railroad and transit industry. (See GE Transportation Systems Global Signaling entry in *Signalling and communications systems* section).

Products and services

Locomotive overview
GE locomotives include the high-performance GE AC6000 CW™ and GE AC4400 CW™ locomotives, which use AC technology for higher horsepower,

GE AC4400 CW™ Locomotive for Ferrocarril del Sureste, SA de CV   0092472

lower operating costs and fewer maintenance needs. The GE line also includes reliable DC propulsion DASH 7™, DASH 8™ and DASH 9™ locomotives. With performance-enhancing upgrades, customers can further improve tractive effort, reduce maintenance costs, lower fuel consumption and emissions, and achieve other efficiencies in these models. The GENESIS® Series locomotives are lightweight aerodynamic units designed for passenger service. In 2003 GE launched its Evolution Series diesel locomotive range aimed at providing increased fuel efficiency and more demanding emission standards.

Global Services
GE Transportation Systems has made a significant investment in product enhancements, digital technologies and web-based delivery systems to put solutions directly into customers' hands. With business resources and six Sigma tools, GE Transportation Systems demonstrates customer-centricity by working on key issues for customers, from improving primary business processes like product reliability, to improving profitability and value for customers, while sharing best practices.

Information-based communications management
Wireless communications technology operating on the LOCOCOMM™ platform is designed to deliver improved asset utilisation and productivity. Applications such as PinPoint™ asset tracking

and Expert on Alert™ diagnostics provide real-time location and condition information for improved locomotive management decisions. The LOCOCOMM™ platform is a rugged micro-processor that integrates with locomotive control and train-line systems to provide critical information through the most efficient and reliable communication mode available using OptiPath™ gives operational and maintenance managers the capability to make real-time locomotive management decisions and shift repairs on-train. Through digital service delivery, web-based applications manage maintenance workflow with greater speed, efficiency and accuracy.

GE Transportation Systems offers a comprehensive range of genuine and enhanced compatible parts and major components for GE and other platforms in the railroad, marine and stationary industries. Using the company website, customers can check part availability and place orders online.

Locomotive enhancements
(http://www.gelocoenhancements.com)
In-service locomotives can benefit from the latest in GE Transportation Systems' productivity and performance improving technologies. Products such as auto engine start/stop, battery job, battery saver, controlled stop, fuel saving packages and handbrake warning systems can improve fuel efficiency, reduce operating costs and standardise older fleets. Multiple emissions solutions are available for locomotives as well as marine and stationary engines for GE and other platforms.

Competitive products
(http://www.gecompetitiveproducts.com)
GE Transportation Systems offers a full range of new and rebuilt locomotive, marine, stationary and industrial service replacement parts and major components for GE and other platforms, including EMD. Continued expansion of high technology products and services includes broadened capability for products like engines, traction motors, alternators and generators, turbochargers, power assemblies, electronics and accessories, such as small rotating devices, cylinder assemblies, moulded cables and other electrical and mechanical parts.

Repair services
(http://www.getsrepairservices.com)
GE Transportation Systems' Repair Services offers a one-stop solution for locomotive wrecks, repairs, remanufactures or overhauls for GE and other platforms.

GE GENESIS® Series passenger locomotive for Amtrak's Empire Corridor   0092473

Locomotive modernisations
(http://www.gebrightstar.com)
Selective introduction of new technology can deliver great productivity from in-service locomotives and can extend existing asset life by as much as 15 to 20 years. The microprocessor-based BrightStar™ Control System from GE Transportation Systems can provide self-test and on-board diagnostic capabilities to improve adhesion, reliability and availability of older locomotives.

Solutions to improve product performance
Digitisation, GE's initiative for more efficient customer collaboration including E-Business, is re-energising the business and providing new opportunities to deliver faster, more accurate and more valuable services to its customers. Online orders and quotes, product catalogue and value calculators for product upgrades are provided on the company's websites.

Locomotive Products
GE AC6000 CW™ (6,000 hp, 4,474 kW), GE AC4400 CW™ (4,400 hp, 3,281 kW) and the Blue Tiger™ family (1,119–2,984 kW) diesel locomotives feature AC traction motors regulated by separate, computer-controlled inverters. The AC traction control system utilises the latest Insulated Gate Bipolar Transistor (IGBT) technology for improved reliability and packaging.

DASH 9™ and DASH 8™ diesel-electric locomotive propulsion systems use DC traction motors in full-time parallel configuration supplied by rectified, three-phase traction alternators (computer-regulated adhesion process). All models are microprocessor-controlled. Operator interface is made through computer displays that allow access to diagnostics and integrated ancillary systems information.

DASH 7™ diesel models include the Constant Horsepower Excitation Control (CHEC) and the speed-based adhesion control (Sentry) system with automatic wheel diameter calibration.

GENESIS® Series passenger locomotives are available either as straight diesel models with DC propulsion or as dual-mode with AC propulsion, able to operate from a third-rail electric supply or by using diesel power in non-electrified zones.

Evolution Series diesel locomotives utilise GE's latest GEVO engine range to provide increased efficiency while meeting new US standards on emissions due to come into effect in 2005.

Electric locomotives utilise DC traction motors in full-time parallel configuration supplied by phased thyristor control and fast, analogue-regulated adhesion control.

GE service experts based in Erie, Pennsylvania, USA, use advanced communications and diagnostics tools to monitor locomotive fleets around the globe   0092474

Locomotive Engines & Components
GE offers two diesel engine families: the GE 7HDL™ 16-cylinder, 4,474 kW and the GE 7FDL™ 16-cylinder, 3,281 kW. The GE 7FDL 16-, 12-, and 8-cylinder can also be set at various power ratings. In 2003 the company launched its GEVO engine range, which in its 12-cylinder version produces the same 3,281 kW output as its 16-cylinder predecessor.

All AC, DASH 9, DASH 8 and DASH 7 models feature a three-phase traction alternator feeding a full-wave bridge traction rectifier. The GMG-type alternator is used on 4,474 traction kW (AC) and 3,281 traction kW (AC, DASH 9 and DASH 8) models.

Most AC, DASH 9, DASH 8, and DASH 7 models include a single stator traction alternator output with sufficient capacity to accommodate all volt-amp requirements through the entire speed range.

All AC, DASH 9, DASH 8, and DASH 7 models feature traction motors connected in a full-time parallel configuration in the propulsion mode to provide consistent propulsion behaviour, to enhance speed-tractive effort characteristics, and to eliminate motor transition contactor maintenance. The DC traction motor family consists of GE 752, GE 793, GE 794, GE 761 and GE 764 models. The

AC traction motor family consists of GEB 13 and GEB 15 models.

Locomotive Technologies
GE's AC locomotives feature direct, air-cooled, easily replaceable phase module inverter systems (single inverter per axle). The GE AC traction motor includes a low-slip capability. IGBT technology delivers more reliability with the added benefits of less weight and volume. The units are air-cooled, as with the former GTO technology, yet have significantly reduced component count and complexity.

Electronic fuel injection (EFI) promotes fuel savings, reduces emissions, and lowers maintenance costs. Split cooling improves fuel efficiency and reduces engine temperatures.

HiAd™ (high adhesion) bogies feature a low-weight transfer design with 10-year overhaul intervals. The GE steerable bogie is a three-axle, three-motor, high-adhesion class bogie designed to reduce rail forces in curves. The steerable bogies allow locomotives to negotiate previously prohibited curved track.

GE Transportation Systems Global Signaling designs advanced electronic train and railway products and systems such as LOCOTROL® distributed power, EPx™ Direct Braking and Universal Control Valve™, TrainTalk™ intra-train communications, Precision Dispatch™, Navigator Dispatch™, Precision Train Control™, PINPOINT™ locomotive tracking and asset management system, and train defect detection systems. These and other GE technologies are integrated into GE microprocessor-based locomotives through an open-architecture control system.

Locomotive Leasing
Locomotive Management Services (LMS) manages a variety of locomotive lease fleets including DASH 8 locomotives. The mission of LMS is to provide highly reliable and productive locomotives for lease to its customers on terms that are tailored to support the rail lines. Lease periods range from short-term operating leases to long-term commitments. Full-time (for a specified time period) leases over a multiyear commitment can be ordered.

TranShopNet.com™, developed by GE Transportation Systems, is an online marketplace that links buyers and sellers of equipment, parts and supplies
0092475

Gevisa SA

Transit Area, Av Mofarrej, 592 CEP 05311-000, São Paulo SP, Brazil
Tel: (+55 11) 838 25 60; 858 25 03
Fax: (+55 11) 838 25 70; 25 00

Key personnel
Commercial Director: Ronald H Moriyama
Transit Division Manager: Arnaldo Adoglio Júnior
Marketing Manager: Mário Calvani
Director, South American Operations, GE
    Transportation Systems: Marcelo Mosci
Sales Manager: Carlos E Teixeira

Other offices
Praca Papa Joao XXIII, 28 Cidade Industrial, I-32210-100 Contagem MG, Brazil
Tel: (+55 31) 369 33 33   Fax: (+55 31) 369 33 34
e-mail: faleconosco.getrans@trans.ge.com
Web: http://www.getransportation.com.br

Products
Electric traction motors, auxiliary power and control equipment for metro cars, emus and light rail vehicles; motor generator sets for emus; rehabilitation of control equipment and traction motors.

Rolling stock: diesel-powered locomotives; refurbishment and repair; remanufacture/reconstruction; maintenance; painting/livery and spare parts. Traction/control: (diesel) complete traction package, engines, components, traction motors, generators, mechanical equipment, gears/shafts/couplings and turbochargers.

GLG Gmeinder Lokomotivenfabrik GmbH

Anton Gmeinder Strasse 5, D-74821 Mosbach, Baden, Germany
Tel: (+49 6261) 674 70   Fax: (+49 6261) 674 72 19
e-mail: info@glfg.de
Web: www.gmeinder-lokomotivenfabrik.de

Key personnel
Managing Directors: Norbert Wuddel
Head of Rail Traction Division: Helmut Eifler

Background
In 2003 private rail operator Transport & Logistik AG and other investors acquired a 100 per cent shareholding in Gmeinder locomotives.

Products
The current manufacturing programme comprises: two- and three-axle diesel shunting locomotives; and four-axle diesel shunting and main line locomotives. These can be supplied with hydrodynamic (Voith), diesel-electric or hydrostatic transmission. Electronic control

systems produced by prominent suppliers can also be incorporated.
GLG also offers locomotives for hire (Classes V60, V100 and 232).

UPDATED

Hanjin

Hanjin Heavy Industries Co Ltd
118, 2 Ka, Namdaemunro, Chung-ku, Seoul, South Korea
Tel: (+82 2) 728 54 41/54 20
Fax: (+82 2) 755 09 28/756 54 55
Web: http://www.hhic.co.kr

Key personnel
President: Young Soo Song
Executive Vice-President: Soo Bu Lee
General Manager: Byung Chun Choi
General Manager: Moo Yeong Jeong

Products
Electric and diesel multiple-units.
    Recent contracts include 24 emu cars for the Bundang line in 1997; 402 emu cars for Seoul, Lines 7 and 8, in 1997; 336 emu cars for Pusan Line 2; 216 emu cars for Taegu Subway Agency No 1 Line which opened in 1997; 24 cars for the Kwacheon line; 20 emu cars for Seoul Metropolitan Subway Corporation; 36 cars for KNR Korea; 146 cars for the Seoul-Pusan high-speed line project (1994–2001).
    A manufacturing facility is now in operation at Sang-Ju in Kyoung Sang North Province for construction of high-speed trains.

Hanjin emu for Pusan Urban Transit Authority 0063360

Hitachi Ltd

Transportation Systems Division
6 Kanda Surugadai 4-chome, Chiyoda-ku, Tokyo 101-8010, Japan
Tel: (+81 3) 32 58 11 11   Fax: (+81 3) 32 58 52 30
Web: www.hitachi-rail.com

Key personnel
Chief Operating Officer of Transportation Systems:
    Gaku Suzuki
General Manager, Transportation Systems Sales
    Division: Chiaki Ueda
General Manager, Transport Management and
    Control Systems Division: Kazuo Kera
General Manager Rolling Stock System Division:
    Toshihide Uchimura

Main works
Transportation Equipment
Kasado Works: 794 Higashitoyoi, Kudamatsu City, Yamaguchi Pref 744-8601
Mito Works: 1070 Ichige, Hitachinaka-shi, Ibaraki Pref 312-8506
Hitachi Works: 1-1, 3-chome, Saiwai-cho, Hitachi City, Ibaraki Pref 317-0073

Series 683 for JR West Railway Company 0134966

Kokubu Works: 1-1, 1-chome, Kokubu-cho, Hitachi City, Ibaraki Pref 316-8501

Subsidiary
Hitachi Europe Ltd
European Headquarters
Whitebrook Park, Lower Cookham Road, Maidenhead SL6 8YA, UK
Tel: (+44 1628) 58 50 00   Fax: (+44 1628) 58 53 73
e-mail: enquiries@hitachi-eu.com
Web: www.hitachi-eu.com

Background
Hitachi is one of the world's leading global electrical engineering companies generating 1 per cent of Japan's GDP. Hitachi manufactures and markets a wide range of products, including computers, semiconductors, consumer products, rolling stock and power and industrial equipment. Hitachi has 913 subsidiaries, including 276 overseas corporations. Hitachi's share of the Japanese rolling stock market is approximately 40 per cent and it has delivered vehicles into all market sectors, including Shinkansen, 'Limited Express', commuter emu and metro trains.

Hitachi Europe is a wholly owned subsidiary of Hitachi Ltd with operations in 15 countries across Europe, Asia and Africa.

Products
Hitachi is a main supplier for Shinkansen, Monorail System, Linear Metro System, Maglev System, tilting train, signalling and sub-station system, including: emus and dmus for city, urban and regional networks, monorail cars, linear induction motor powered metro trains for small-bore tunnels; lightweight alluminium and stainless steel and steel body shells; AC propulsion system using IGBT inverter; auxiliary power supply; air conditioning; ATC, ATO and automatic train diagnostics systems. Hitachi has developed an aluminium car body train with friction stir welding and module construction.

Contracts
Monorail systems: Sentosa Express for Sentosa Development Corporation, Singapore; Okinawa Monorail, Japan; Series 2000 for Tokyo Monorail, Japan; Series 1000 for Tokyo Tama Inter City Monorail, Japan; Monorail cars for Chongqing, China IGBT propulsion systems; IGBT inverters for Beijing Urban Railway Construction Project.

High-speed trains: rolling stock, control system for Taiwan High Speed Rail Corporation.

Limited Express: Tsukuba Express for Metropolitan Intercity Railway Company, Japan; Series 257 Limited Express for JR East, Japan; Series 683 Limited Express for JR West; Series 885 Limited Express for JR Kyushu, Japan.

Commuter emu: KL3-97 stainless steel emu for Jabotabek suburban network in Jakarta, Indonesia; transit vehicle for Metropolitan Atlanta Rapid Transit Authority (MARTA) in USA; Series 20000 for Seibu Railway Company, Japan; Series 815 for JR Kyushu, Japan.

Metro emu: Series 12000 linear motor propulsion emu for Transportation Bureau of Monorail System; Series 2000 for Tokyo Monorail, Tokyo Metropolitan Government; Series 1000 for Tokyo Tama InterCity Monorail, Japan.

Series 257 for JR East   0134968

Hitachi's Series 815 emu for Kyushu Railway Company, Japan   0098048

Current customers include: Beijing, Taiwan Railway Administration, Sentosa Development Railway Corporation, Chongqing Rail Transit General Corporation, MPS, Hokkaido Railway Company, East Japan Railway Company, Central Japan Railway Company, West Japan Railway Company, Kyushu Railway Company, Teito Rapid Transit Authority, Transportation Bureau of Tokyo Metropolitan Government, Seibu Railway, Tokyo Corporation, Tobu Railway, Kinki Nippon Railway, Japan, UK, supply of 30 high-speed 'A' trains to operate domestic services on the Channel Tunnel Rail Link (CTRL).

UPDATED

Hunslet-Barclay Ltd

Caledonia Works, West Langlands Street, Kilmarnock KA1 2QD, UK
Tel: (+44 1563) 52 35 73   Fax: (+44 1563) 54 10 76
e-mail: mail@hbltd.co.uk
Web: http://www.hunsletbarclay.co.uk

Key personnel
Board of Directors:
    Chairman (Non Executive): H Kuebel
    Managing Director (Executive): John Flowers
    Non-Executive: S Bauer
Executive Management:
    Production: A Cuthbertson
    Finance: B Connell

Senior Management:
    Train Sales: R Edmond
    Wheelset Sales: M Douglas
    Locomotive Sales and Site Services: T Clare
    Train Engineering: S Scott

Background
Hunslet-Barclay is a leading company in railway engineering services, part of Waagner-Biro Holding AG, Austria. Hunslet-Barclay is based in Kilmarnock. The company has five core business areas: trains, locomotives, wheels, spares and repairs, and site services. In addition, the company has a Hugh Smith machine tools division.

Products
All Hunslet products are designed and built by Hunslet-Barclay Ltd in Kilmarnock.

The range includes design and manufacture of railbuses; conversion and repair of powered and non-powered passenger vehicles; diesel-hydraulic, diesel-electric and trolley/battery and electric surface and underground/metro locomotives; fully flame-proofed locomotives. Locomotives supplied range from 3 to 120 t and 50 to 1,000 kW to suit any rail gauge. Included are radio remote-controlled and robot locomotives incorporating programmable logic controller systems and wheel traction control.

Specialist products include metro locomotives and rack-and-pinion locomotives.

INKA

Head office and factory
T (Persero) Industri Kereta Api (PT INKA)
Jalan Yos Sudarso No 71, Madiun 63122, Indonesia
Tel: (+62 351) 45 22 71   Fax: (+62 351) 45 22 75
e-mail: sekretariat@inka.web.id
Web: www.inka.web.id

Representative office
Arthaloka building, 3rd Floor, Jalan Jend, Sudirman
Kav 2, Jakarta
Tel: (+62 21) 251 44 24   Fax: (+62 21) 251 44 23
e-mail: inkajkt@cbn.net.id

Key personnel
President: Ir Roos Diatmoko
General Manager, Railway Rolling Stock Division:
  Suryanto
Finance and Administration Director:
  Drs Udin Supriatman
General Manager of Technology:
  Ir M Harsan Badawi
Engineering Manager: Ir Gunesti Wahyu
Design Manager: Ir Indarto Wibisono
Marketing Manager: Ir M Dedi Tarmidi
Business Development Manager:
  Ir Muchlis Budiman
Procurement Manager: Soedjito Taathadi

Background
PT INKA was originally established in 1981 as a state owned company, transforming from Indonesian State Railway's steam locomotive maintenance shop.

Products
Assembly and renovation of freight wagons (300 units a year), passenger coaches (120 units a year), diesel and electric railcars (40 units a year), bogies (200 units a year).

Locomotives (in collaboration with GE Transportation, 15 units a year). Various special vehicles, including track motor and inspection cars and amusement park trains.

Contracts
In 2005 INKA will manufacture two AC/DC diesel-electric locomotives in co-operation with GE Transportation. This new locomotive type will replace older DC/DC diesel-electric locomotives,

Electric Jakarta area commuter emu by INKA, with Toshiba VVF inverter control (INKA)   **NEW**/0585220

with an expected requirement for up to 50 examples.

INKA has supplied two emus for commuter service in Jakarta and the surrounding area. They feature a lightweight stainless steel bodyshell and AC traction control with IGBT VVVF inverter with vector control from Toshiba and KBGM electro-pneumatic brake system from Knorr Bremse. After two years in service, these trains are achieving 95 per cent availability. In 2005–07, in co-operation with a European supplier and financed by KfW, INKA will fulfil a follow-on order for 10 sets of similar vehicles.

During 2003, INKA delivered 55 refeer container wagons and two power-generating coaches to KTMB Malaysia. The container wagon is specially designed to be used for the Thailand–Malaysia rail network (landbridge), mainly for carrying fresh meat and vegetables.

In March 2005 INKA was to sign a contract agreement with Bangladesh Railways for the supply of 50 broad-gauge passenger coaches.

INKA has also delivered two track motor vehicles to Leighton Asia for a double-track project in Indonesia.

New developments
INKA anticipates that Indonesian Railways will place an order to develop a new design of diesel-electric multiple-units using the latest VVVF inverter technology and underfloor engine for the mountains and busy Jakarta–Bandung corridor.

UPDATED

The Integral Coach Factory (ICF)

Ministry of Railways, Chennai 600 038, Tamil Nadu, India
Tel: (+91 44) 26 26 30 91   Fax: (+91 44) 26 26 31 11
e-mail: coachnet@vsnl.com
Web: www.icf.gov.in

Key personnel
General Manager: Shivendra Kumar
Financial Advisor and Chief Accounts Officer:
  S Ram Mohan
Secretary: P Suresh
Public Relations: J Baktavatchalam

Background
ICF was established in 1955 in collaboration with the Swiss Car & Elevator Manufacturing Co. The agreement ended in 1961, and the company is now controlled by the Ministry of Railways. With a staff of 13,000, ICF is capable of producing 1,000 coaches a year and since inception has produced 34,982 coaches of nearly 200 types and has exported coaches, bogies and spares to 11 countries.

ICF is ISO-9001 accredited and ISO 14001 for 'Environmental Management System'.

Products
ICF produces many different types of electric multiple units, metro cars and diesel railcars. The company offers in-house design and development facilities and can manufacture and supply any type of coach.

During the production year (2004–2005) ICF has produced three new types of coaches which include: luxury coaches for the Deccan Odyssey Tourist train; AC/DC dual voltage emu rake and the aerodynamically profiled dmu. Of these, the Deccan Odyssey coaches include features such as variable room temperature in individual cabins,

ICF D-2000 four-car diesel-electric multiple-unit
0126211

ICF chair-cum-guard car for Intercity 'Jan Shatabdi' train being introduced by IR (KK Gupta)   0558277

Internet and mobile facilities onboard, gymnasium and conference car. The AC/DC dual voltage emu rake for the Mumbai suburban transit system will ensure the smooth transition from 1,500 V DC traction to 25 kV AC traction. The demu has better aesthetics, adhesion and fuel efficiency.

Contracts
Within (2004–2005) exported 11 stainless steel shells to Malaysian Railways.

UPDATED

AC/DC emu for Bombay suburban section
(K K Gupta)
0126504

Interlok Bahnconsulting Schmidtendorf

Bruchsaler Strasse 3, D-10715, Berlin, Germany
Tel: (+49 177) 302 42 53
Fax: (+49 177) 993 02 42 53
e-mail: office@interlok.info
Web: http://www.interlok.info

Key personnel
General Sales Manager: Hermann Schmidtendorf
General Manager Rail Department:
   Uwe Liedecke

Products
Remanufactured, modernised standard gauge passenger/sleeping/dining coaches.

Trade agent for Interlok Ltd and Kaniewski ZUR – Poland, Mitteldeutsche Gesellschaft für Metallhandel und Anlagenverwertung, Germany.

Contracts
24 overhauled, completely modernised passenger and bistro coaches for long distance InterRegio services in northern Germany.

Interlok Ltd

Warsztatowa 8, PL-64-920 Pila, Poland
Tel: (+48 67) 213 20 68
e-mail: pilapoland@interlok.info
Web: http://www.pila.interlok.info/indexe.htm

Key personnel
General Manager: Henryk Palczewski
General Manager Technology: Marek Furtacz

Background
Interlok Ltd was founded in January 1989 as a service/export joint venture between Polish State Locomotive Repair Workshop ZNTK Pila and German private trader HF Schmidtendorf. In 1999 it became fully privatised by additional shareholders from France, Germany, Greece, Poland and Switzerland. It maintains co-operation from former state partner ZNTK, now ZNLS (Diesel Locomotive Repair Workshop).

Products and services
Steam locomotives, steam boilers and other components of all gauges, new and full overhaul, modernisation, historical reconstruction and replicas; steam boiler parts are massively forged,

then welded or riveted; stays are welded or threaded; fabrication of new steam cylinders welded construction; standard gauge diesel locomotives: SP32 1,015 kW BoBo modernised with 12 V MTU engine; narrow gauge diesel locomotives; standard gauge railbuses, new construction and remanufactured.

Contracts
Recent contracts include 15 Interlok Liliput 03-002 steam locomotives for Leipzig leisure park railway. Three 2 ft 6 in 365 kW Lyd2 diesel locomotives for St Kitts Scenic Railway in the Caribbean.

NEW ENTRY

Interlok's Liliput steam locomotive for Leipzig Leisure Park, Germany
0579329

Jessop

Jessop & Company Ltd, Calcutta
63 Netaji Subhas Road, Calcutta 700001, India
Tel: (+91 33) 243 20 41/243 34 20
Fax: (+91 33) 243 16 10

Works
Dum Dum Works, 21 & 22 Jessore Road, Calcutta 700 028, India

Tel: (+91 33) 551 99 22/59 92
Fax: (+91 33) 551 28 68

Key personnel
Managing Director: A K Sur
Director (Engineering and Commercial):
   P C Bhattacharya
Director (Production): P K Mukherjee
Secretary: R D G Raghavan
Senior Manager (Exports): Amit Ghosh

Products
Rolling stock including electric multiple-unit coaches, LRVs, passenger coaches and freight wagons.

Contracts include 175 electric multiple-unit coaches for Indian Railways. A total of 51 such coaches was delivered to Indian Railways in 1993–94.

Kaniewski ZUR

ul Roszarnicza 17, PL-63-230 Witaszyce, Poland
Fax: (+48 62) 740 12 46
e-mail: kaniewski@interlok.info
Web: http://kaniewski.interlok.info

Key personnel
General Manager: Marek Kaniewski

Products
Diesel locomotives: remanufactured standard gauge; new construction 73 kW WLP50 diesel

locomotive for industry and leisure parks with Polish Leyland licensed engine for 20 to 40 in narrow gauge; second-hand narrow gauge freight cars; new construction of narrow gauge mining/tunnel building cars, narrow gauge passenger cars, hand pump cars; equipment for railway workshops: lifting jacks, turntables, railway cranes.

Contracts
Recent contracts include seven 1,470 kW standard gauge ST44 diesel locomotives for Prignitzer Eisenbahn PEG Ltd, Germany.

Kaniewski ZUR 73 kW WLP50 industrial diesel locomotive
0579331

Kawasaki Heavy Industries Ltd

Rolling Stock Company
World Trade Center Building, 4-1 Hamamatsu-cho 2-chome, Minato-ku, Tokyo 105-0116, Japan
Tel: (+81 3) 34 35 25 72   Fax: (+81 3) 34 35 21 57
Web: www.hki.co.jp

Works
Hyogo Works
1-18 Wadayama-dori 2-chome, Hyogo-ku, Kobe 652-0884, Japan

Kawasaki Rail Car, Inc
Yonkers Plant
29 Wells Avenue, Building #4, Yonkers, New York 10701, USA

Kawasaki Motors Manufacturing Corp, USA
Lincoln Plant
6600 Northwest 27th Street, Lincoln, Nebraska 68524, USA

Key personnel
President, Rolling StockCompany: Masashi Segawa
General Manager, Marketing and Sales Division: Akira Hattori
Deputy General Manager, Marketing and Sales Division: Ryoshi Hirano
Senior Manager, Overseas Marketing Department: Hiroshi Murao

Kawasaki's R143 subway car for MTA New York City Transit   1032908

Background
Established in 1906, including other business segments such as aerospace, shipbuilding and consumer products, Kawasaki continues to contribute to the development and modernisation of railway transportation and rolling stocks. The newest Lincoln plant, established in 2002, is equipped with state-of-the-art manufacturing facilities and is the only rolling stock manufacturing plant in North America covering the full process from fabrication of carbody to final assembly and bogie manufacturing.

Kawasaki is expanding its overseas rolling stock operations to keep pace with the growing demands worldwide.

Products
High-speed trainsets; electric and diesel railcars; MRT vehicles; monorail vehicles; automated guideway transit systems; passenger coaches; electric, diesel-electric, diesel-hydraulic locomotives; freight cars; complete bogies sets; platform screen doors.

Contracts
Current or recent contracts include: 126 SMRT emus for Singapore's Land Transport Authority (LTA) (delivered 1999–2000); as a member of the IKK Consortium with Itochu Corp and Kinki Sharyo, 250 rapid transit emus to Kowloon-Canton Railway Corporation (2001–03); as a member of Taiwan Shinkansen Corporation, 360 high-speed trainset cars to Taiwan High Speed Rail Corporation (THSRC) (2000 and ongoing); 321 rapid trainset emus to Taipei's Department of Rapid Transit Systems (DORTS) in Taiwan (2003 and ongoing); 46 towing locomotives to Panama Canal Authority (ACP, formerly Panama Canal Commission) (1999–2003); 520 R142 subway cars (1999–2002) and 212 R143 subway cars (2001–03) to MTA New York City Transit (NYCT); 107 bilevel passenger coaches to Virginia Railway Express (VRE) (1999).

Most recent contracts include ongoing projects to supply 54 towing locomotives to ACP; 80 R142A subway cars and 260 R160B subway cars to NYCT; 28 bilevel passenger coaches to MBTA; in association with Sifang, 480 cars of 200 km/h emus to China and 340 PA-5 cars to Port Authority Trans-Hudson (PATH) of New York and Jersey.

UPDATED

Alan Keef Limited

Lea Line, Ross-on-Wye HR9 7LQ, UK
Tel: (+44 1989) 75 07 57   Fax: (+44 1989) 75 07 80
e-mail: sales@alankeef.co.uk
Web: http://www.alankeef.co.uk

Key personnel
Chairman: Alan Keef
Managing Director: Patrick Keef

Products
Design, manufacture and refurbishment of light railway equipment for the industrial and leisure markets. Products include narrow gauge diesel locomotives for underground and surface use, steam locomotives, associated rolling stock and trackwork. Sole suppliers of spare parts for Simplex, Ruston, Lister and Planet locomotives.

Contracts
Overhaul of three steam locomotives for Irish tourist railways. Manufacture of 'Puffing Billy' locomotive, steam outline diesel locomotive and seven carriages for various UK tourist railways.

Keller Elettromeccanica SpA

Zona Industriale I-09039, Villacidro, Cagliari, Italy
Tel: (+39 070) 933 62 02   Fax: (+39 070) 933 62 44
e-mail: info@keller.it
Web: http://www.keller.it

Key personnel

Chairman: Piero Mancini
Board members: Sergio Zanarini; Paolo Piombini;
   Giovanni Cappietti

Other office

Sales department
via Volturno, 10/12, I-50019 Sesto Fiorentino,
Firenze, Italy
Tel: (+39 055) 34 08 80   Fax: (+39 055) 34 08 81
e-mail: commerciale@keller.it

Background

Established in March 2000, Keller Elettromeccanica
is a private limited company owned by the Busi

Impianti SpA group, Bologna, and the Ciet SpA
group, Arezzo.

Products

Design and manufacture of passenger coaches
and special purpose vehicles, including: UIC
Z1 passenger coaches; measure coaches for
diagnostic train substructures and superstructures;
railway maintenance vehicles; luggage coaches
manufacture for UIC Z1 diagnostic train.

Kinki Sharyo Co Ltd

The Kinki Sharyo Co Ltd
2-6-41 Inada-Uemachi, Higashi-Osaka City, Osaka
577-8511, Japan
Tel: (+81 6) 67 46 52 40
Fax: (+81 6) 67 45 51 35
e-mail: sharyo@kinkisharyo.co.jp
Web: www.kinkisharyo.co.jp

Key personnel

President: Koichi Sakurai
Managing Director, Chief Operating Officer:
   Nobumitsu Kurokawa
Director, Sales: Akio Yamamoto

Subsidiary company

Kinki Sharyo (USA) Inc/Kinki Sharyo International,
LLC
45 Shawmut Road, Canton, Massachusetts 02021,
USA
Tel: (+1 617) 949 24 40
Fax: (+1 781) 828 80 25

Background

Kinki Sharyo Co Ltd is a subsidiary of the Kinki
Nippon Railway.

Products

Electric multiple-units for main line, commuter,
rapid transit, metro and light rail vehicles, double-
deck passenger coaches.

Contracts

Recent orders include: Series 321 commuter emus
for JR West, low floor LRVs for Phoenix, USA and
low floor LRVs for Seattle, USA; commuter emus
for KCRC, Hong Kong (Ma On Shan Line); Series
E257 Limited Express emus for JR East; Series 700
Shinkansen for JR West; commuter emus for Kinki
Nippon Railway; LRVs for Dallas, and low-floor LRVs
for New Jersey, Santa Clara Valley Transportation
Authority, Phoenix and Seattle, USA.

UPDATED

Kinki Sharyo commuter emu for KCRC, Kong Kong, supplied in 12-, seven- and four-car versions
0134173

*Kinki Sharyo low-floor LRV for Santa Clara
Valley Transportation Authority*
0548178

Kolomna

JSC Kolomensky Zavod,
Partizan Street, 42 Kolomna, Moscow Region
140408, Russia
Tel: (+7 0966) 15 52 22; 15 11 17
Fax: (+7 0966) 15 17 96
e-mail: kolomzavod@kolomna.ru
Web: www.kolomnadiesel.com

Marketing and Sales Department

Tel: (+7 0955) 156 94 13
Fax: (+7 0955) 721 37 55

Key personnel

General Director: V N Vlasov
Technical Director: V A Shelemet'yev
Marketing and Sales Director: A A Samoukin

Background

Originally founded in 1863 as JSC Kolomensky
Zavod, the plant was the first to start building
bridges for railway and municipal transport and
was the first of a few to start serial production of
steam locomotives.

Today the company, which employs
approximately 9,000 people, is a multifunctional

modern enterprise that produces diesel
locomotives, electric locomotives, electric power
stations, diesel engines and diesel-generator sets.

Products

Main line passenger diesel-electric locomotives:
TEP70; TEP70BC with electric power supply to the
cars of the train, rated at 2,984 kW (4,000 hp) and
rated speed 160 km/h; 2TE70 double section main
line freight locomotives rated at 3,043 kW (4,080 hp)
in one section and designed speed 110 km/h; EP200
passenger AC electric locomotive rated at 8,000 kW
and designed speed 200 km/h; modern four-stroke

medium-speed supercharged diesel engines; Type D42, dimensions CHN 30/38, produced in in-line version (4-, 6-, 8- cylinders) power range 450–2,500 kW; Type D49, dimensions CHN 26/26 produced in 4- and 6- cylinders in-line and in 8-, 12-, 16- and 20- cylinders in Vee-configuration within the power range 370–5,300 kW.

Diesel engines and diesel-generator sets are used for: building new locomotives; re-powering rolling stock; spare parts for the engines Type D49, D42, 40DM, 14D40 and 11D45 for diesel and electric locomotives. Production is certified and meets the requirements of the European ecological standards regarding harmful matter (NO, CO, HC) in exhaust gas and emissions.

Services also include technical assistance during commissioning of the plant's production and during warranty period and maintenance of the service life; spare parts supply; overhaul of the engines in a specialised workshop, equipped with all the necessary equipment; routine repairs on site.

UPDATED   *Kolomna TEP70 BC diesel electric passenger locomotive*   *NEW*/1143492

Končar – Električne Lokomotive dd

Končar-Electric Locomotives Inc
Velimira Škorpika 7, HR-10090 Zagreb, Croatia
Tel: (+385 1) 349 69 59; 349 69 50
Fax: (+385 1) 349 69 60; 349 69 70
e-mail: uprava@koncar-ellok.hr
Web: www.koncar.hr/ellok

Key personnel
President: Jusuf Crnalić
Members of Board: Vesna Boinović-Grubić
  Željko Šakič
Marketing and Sales Manager:
  Zvonimir Cvijin

Products
Electric locomotives for AC and DC traction; electric multiple-units; trams; shunting locomotives; light rail vehicles; parts, components and systems for electric traction.

UPDATED

Class E52 Bo-Bo 25 kV AC 50 Hz electric locomotive refurbished for Turkish State Railways   0122615

Class 46 Bo-Bo 25 kV AC 50 Hz electric locomotive refurbished for CFR Calatori, Romania
0122617

Luganskteplovozstroj

Lugansk Diesel Locomotive Works
ulica Frunze 107, 348 002 Lugansk, Ukraine

Products
Diesel locomotives.

Lyudinovo Locomotive Works JSC

1 K Liebknecht Street, Lyudinovo, Kaluga region 249400, Russia
Tel: (+7 08444) 201 20; 252 59
Fax: (+7 08444) 252 59

Key personnel
General Director: Peter F Baum
Executive Director: Sergey M Fomin
Chief Engineer: Anatoly I Gerasimov
Import and Export, Sales: Nicolai N Denisov

Products
Diesel-hydraulic and diesel-electric locomotives 597 to 2,984 kW (800 to 4,000 hp) for industrial, shunting and main line applications. Spare parts for locomotives. The product range includes: TGM4B 600 kW diesel-hydraulic shunter; TGM6D 833 kW diesel-hydraulic shunter; TGM 1000 1,000 kW diesel-hydraulic shunting and short-haul locomotive; TEM7A 1,470 kW Bo+Bo-Bo+Bo diesel-electric shunting and short-haul locomotive; TERA1 3,063 kW Bo+Bo-Bo+Bo diesel-electric main line locomotive; DL2 diesel-powered passenger trainset.

Lyudinovo LW also makes rail lubricating machines, rotary snow ploughs, track maintenance cars, dmus, narrow-gauge trains and tank wagons.

UPDATED

MaLoWa Bahnwerkstatt Ltd

Hauptstrasse 10, D-06308 Benndorf, Germany
Tel: (+49 34) 772 255 88 Fax: (+49 34) 772 255 89
e-mail: malowa@interlok.info
Web: http://www.interlok.info/indexe.htm

Key personnel

General Sales Manager: Gerhard Kellner
General Manager Finances: Klemens Peukert
General Manager Technology: Günter Vorwerg

Products

Diesel locomotives, steam locomotives, freight cars, passenger coaches, shunting, mining and battery locomotives, all gauges: general overhaul, repairs, modernisation in own workshop facilities, sale. Rent via partner company Adam&MaLoWa Lokvermietung Ltd. Trade agent for Interlok Ltd, Pila/Poland.

Contracts

Recent contracts include: Malowa steam locomotive for Delmenhorst/Harpstedt Railway (Germany), in co-operation with associated company Interlok Ltd, Poland.

Malowa steam locomotive for Delmenhorst/Harpstedt Railway, Germany　　　0579332

MAN Ferrostaal AG

Hohenzollernstrasse 24, D-45128 Essen, Germany
Tel: (+49 201) 818 01 Fax: (+49 201) 818 28 22
e-mail: info@manferrostaal.com
Web: www.manferrostaal.com

Key personnel

Head of the Department of Infrastructure and
　Transport Systems: Helmut Julius

Background

Re-named from Ferrostaal AG, MAN Ferrostaal AG is a member of the MAN AG group.

Products

Main line, shunting and mining locomotives with diesel-hydraulic, diesel-electric and electric traction. Electric or diesel-electric railcars, railbuses, light rail vehicles and other multiple-units for urban public transport; motor and trailer bogies; rolling stock components including wheels, axles, wheelsets, bearings, suspension parts and couplers.

Passenger coaches including special designs; inspection and service trolleys for track and overhead maintenance and track installation; motor and trailer bogies for freight wagons, passenger coaches, locomotives, railcars; rolling stock components such as wheels, axles, assembled wheelsets, bearings, suspensions, couplers and electrical equipment.

UPDATED

Metrowagonmash Joint Stock Company

4 Kolontsov str, Mytishchi, 141009 Moscow, Russian Federation
Tel: (+7 095) 581 12 56 Fax: (+7 095) 581 12 56
e-mail: mail@metrowagonmash.ru

Web: http://www.mwm.ru
　http://www.metrowagonmash.ru
　http://www.metrowagonmash.com

Key personnel

General Manager: J A Gulko
Technical Director: J P Soldatov
Deputy General Manager: A A Andreyev

Subsidiaries

Spola Spare Parts Production Co (Ukraine)
Vyshnevolotski Machine Building Plant

Products

Metro cars, railbuses, dmu, bogies, spare parts for metro cars.

Surface metro trainset by Metrowagonmash　　　1034508

Metro car saloon by Metrowagonmash　　　1034509

Mitsubishi Electric Corporation

Mitsubishi Denki Building, 2-3 Marunouchi 2-chome,
Chiyoda-ku, Tokyo 100, Japan
Tel: (+81 3) 32 18 34 30
Fax: (+81 3) 32 18 28 95
Web: http://www.melco.co.jp/society/traffic/

Key personnel

President: Tamotsu Nomakuchi
Executive General Manager (Public Utility Systems Group): Takaaki Kijima

Subsidiaries

Australia
Mitsubishi Electric Australia Pty Ltd

348 Victoria Road, Rydalmere, New South Wales 2116, Australia
Tel: (+61 2) 96 84 77 77

Mexico
Melco de Mexico SA de CV
Mariano Escobedo No 69, Tlanepantla 54030, Edo de Mexico
Tel: (+52 5) 390 73 44

Electric locomotives equipped by Mitsubishi Electric since 1990

Class	Railway	Wheel arrangement	Line voltage	Rated output (kW) continuous	Max speed (km/h)	Weight (t)	No in service	Year first built	Builders Mechanical parts	Electrical equipment
EF 500	JR Freight	Bo-Bo-Bo	20 kV AC/ 1.5 kV DC	6,000	120	100.8	1	1990	Kawasaki Heavy Industry	Mitsubishi Electric
EF 210	JR Freight	Bo-Bo-Bo	1.5 kV DC	3,390	110	100.8	30	1995	Kawasaki Heavy Industry	Mitsubishi Electric
EF 510	JR Freight	Bo-Bo-Bo	20 kV AC/ 1.5 kV DC	3,390	110	100.8	1	2002	Kawasaki Heavy Industry	Mitsubishi Electric

Singapore

Mitsubishi Electric Asia Pte Ltd
307 Alexandra Road #05-01/02, Mitsubishi Electric Bldg, Singapore 159943
Tel: (+65) 64 73 23 08

UK

Mitsubishi Electric Europe BV
15th floor, Leon House, 223 High Street, Croydon, Surrey CR0 9XT, UK
Tel: (+44 20) 86 86 95 51

USA

Mitsubishi Electric Power Products Inc
1211 Avenue of the Americas, 43rd floor, New York, New York 10036, USA
Tel: (+1 212) 704 66 73

Products

Complete electric locomotives, diesel-electric locomotives, traction motors, VVVF inverters, main transformers, rectifiers, drive gears, flexible couplings, brake systems, static inverters.

Contracts

Recent contracts include: electrical equipment for 192 emus, New York MTA, Long Island Rail Road (USA), May 1999; electrical equipment for 250 emus, KCRC West/East rail, Hong Kong, August 1999; electrical equipment for 104 emus, MTRC C651 Tsuen-Kwan-O extension, Hong Kong; electrical equipment for 328 emus, Seoul Metro Line 6, Korea; electrical equipment for 240 emus, Delhi Metro Railway Corporation, (India), May 2001, electrical equipment for 126 emus, Atliko Metro (Greece), Feb 2002.

Moës

Moteurs Moës SA
62 Rue de Huy, B-4300 Waremme, Belgium
Tel: (+32 19) 32 23 52   Fax: (+32 19) 32 34 48

Key personnel

Chairman: C Froidbise
Managing Director: R Thirion
Sales Director: J Antoine

Products

Narrow-gauge diesel-hydraulic and diesel-mechanical locomotives (3 to 30 tonnes, 10 to 190 kW) for mine railways.

MotivePower Inc

Address

4600 Apple Street, Boise, Idaho 83716, US
Tel: (+1 208) 947 48 00   Fax: (+1 208) 947 48 20
Web: www.wabtec.com

Key personnel

General Manager: Mark S Warner

Background

MotivePower Inc is a subsidiary company of Wabtec Corporation, the latter created in 1999 as a result of the 1999 merger of Westinghouse Air Brake Co and MotivePower Industries Inc.

Products

Passenger diesel locomotives, switching locomotives, liquefied natural gas locomotives, remanufactured locomotives.

Contracts

Contracts for new-build locomotives include the following:
   USA: In January 2005 the Mid-Regional Council of Governments in Albuquerque, New Mexico, awarded the company a contract to supply five MPXpress® diesel locomotives for a start-up commuter rail service. Delivery is scheduled for late 2005.
   In 2004 MotivePower Inc was awarded a contract by Amtrak for 10 Type MP15B switching locomotives for delivery the same year.
   In 2003 MotivePower Inc delivered six Type MP36PH-3C MPXpress® diesel locomotives to Peninsula Corridor Joint Powers Board for Caltrain's new 'Baby Bullet' express services between San Francisco and San Jose.
   In 2002 Chicago commuter operator Metra received the first of 26 Type MP36PH-3S 2,685 kW diesel electric locomotives. The streamlined machines feature a 500 kW head-end power supply.

NEW ENTRY

Motive Power Pty Ltd

PO Box 666, Toongabbie, New South Wales, Australia
Tel: (+61 2) 98 96 71 81   Fax: (+61 2) 98 96 71 16

e-mail: info@motivepower.com.au
Web: http://www.motivepower.com.au

Products

Specialised rail vehicles from mainline adhesion locomotives and rolling stock, to narrow gauge sugar cane machines. Some of these units operate underground in heavy duty mining applications.

Muromteplovoz

Murom Diesel Locomotive Works
ulica Filatova 10, 602 200 Murom, Vladimir region, Russian Federation
Tel: (+7 095) 291 31 68   Fax: (+7 09234) 443 03
e-mail: mteplo@cl.murom.ru
Web: http://www.cl.murom.ru

Key personnel

Director General: V Kharitinov
Director, Export Sales and Marketing: E Tretyakov

Products

Diesel shunting locomotives.

Muromteplovoz Type TGM23D diesel shunting locomotive
0103977

For details of the latest updates to *Jane's World Railways* online and to discover the additional information available exclusively to online subscribers please visit
jwr.janes.com

National Railway Equipment Company

14400 South Robey Street, PO Box 2270, Dixmoor, Ilinois 60426, USA
Tel: (+1 708) 388 60 02   Fax: (+1 708) 388 24 87

International Division
908 Shawnee Street, Mt Vernon, Illinois 62864, USA
Tel: (+1 618) 241 92 70   Fax: (+1 618) 241 92 75

Key personnel
President: Lawrence J Beal
Vice-Presidents: Wilfred A Burrows,
   Patrick C Frangella
Vice President International Sales: John Tooke

Products
National Railway Equipment Company has been engaged in the remanufacture of diesel-electric locomotives for nearly 20 years. Locomotives ranging from 447 to 2,686 kW in four- or six-axle configurations can be supplied.

In addition to complete locomotives, NREC supplies components including diesel engines, electrical rotating equipment and air compressors. Each is rebuilt to Original Equipment Manufacturer (OEM) specifications before dispatch.

NREC's inventory includes over 200 diesel locomotives which can be rebuilt to customer specifications.

Contracts include the design and remanufacture of 14 EMD SD39-2M locomotives for Ferrocarril Del Pacifico SA, Santiago, Chile; the rebuilding of U10B locomotives for Ferrovias, Colombia; the rebuilding and supply of EMD locomotives for Boke Trading, Guinea.

Recent USA contracts include 33 EMD SD40-2 locomotives for the Union Pacific Railroad and work for the Canadian National Illinois Central Railroad, Burlington Northern Santa Fe Railroad, the Kansas City Southern Railway company and the Gateway Western Railroad.

NEVZ

Locomotive Building Plant of Novocherkassk (NEVZ)
7 Mashinostroiteley Str, 346 413 Novocherkassk, Rostov region, Russian Federation
Tel: (+7 86352) 344 46   Fax: (+7 86352) 338 38

Key personnel
General Director: Noskov Alexander Leonidovich
Director of Production: Podust Sergey Fedorovich
Technical Director:
   Sukhokon Vladimir Timofeyevich
Chief of Foreign Economic Department:
   Budkov Alexander Markovich

Products
Main line passenger and cargo electric locomotives, AC-powered electric trains.

UPDATED

Newag GmbH & Co KG

Ripshorster Strasse 321, D-46117 Oberhausen, Germany
Tel: (+49 208) 86 50 30   Fax: (+49 208) 865 03 20
e-mail: info@newag.de
Web: www.newag.de

Key personnel
Managing Director: C Kohl
Technical Director: M Hanke
Sales Director: R Franz

Products
Diesel-hydraulic locomotives from 150 to 900 kW for gauges 750 to 1,676 mm. The remanufacture of diesel-electric and diesel-hydraulic locomotives of 20 to 2,240 kW.

UPDATED

Niigata Transys Co Ltd

F 9-7, Yaesu 2-chome, Chuo-ku, Tokyo
Tel: (+81 3) 62 14 28 77   Fax: (+81 3) 62 14 28 71
Web: www.niigata-transys.com

Background
Niigata Transys Co Ltd was formed in 2003 by Ishikawajima-Harima Industries Co Ltd (IHI) following the insolvency of Niigata Engineering Co Ltd.

Key personnel
President: M Sekine

Products
Manufacture and sale of rolling stock. Design, manufacture and construction of guided railway systems.

UPDATED

Nippon Sharyo Ltd

Head office
1-1 Sanbonmatsu-cho, Atsuta-ku, Nagoya 456-8691, Japan
Tel: (+81 3) 36 68 33 30   Fax: (+81 3) 36 69 02 38
Web: www.n-sharyo.co.jp

Overseas contact
Marunouchi Central Building, 1-9-1 Marunouchi, Chiyoda-ku, Tokyo 100-0005, Japan
Tel: (+81 3) 66 88 67 94   Fax: (+81 3) 66 88 68 10

Key personnel
President, Director: Kazuhisa Matsuda
Managing Director and General Manager, Rolling
   Stock Division: Katsuyuki Ikushima
Business Contact General Manager, International
   Sales Department: Masataka Nakajima

Subsidiary company
Nippon Sharyo USA Inc
600 Third Avenue, 28th Floor, New York, New York 10016, USA
Tel: (+1 212) 949 22 28   Fax: (+1 212) 949 22 29

Products
Electric and diesel multiple-unit cars, LRVs, automated guideway transit cars, monorail cars, bogies for urban and suburban trainsets, high-speed trains.

Contracts
Between 2004 and 2005, 78 emu cars for Yokohama City Municipal Transit Bureau were produced and delivered. Various emu cars for Japanese domestic railways are in production.

Nippon Sharyo with Sumitomo Corporation is completing deliveries of 300 stainless steel push-pull gallery cars for Metra, Chicago, US.

Nippon Sharyo stainless steel gallery emu for Metra, Chicago, US
NEW/1143110

The deliveries commenced in 2002 and following this 26 new gallery emus for Metra are being produced.

UPDATED

Nippon Sharyo 300 Series stainless steel emu for Nagoya Railway, Japan
0526573

Pakistan Railways Carriage

Head Office, Pakistan Railways, Headquarters Office, Lahore
Tel: (+92 42) 920 17 69; 630 80 97
Fax: (+92 42) 920 17 69; 630 80 97

Main works address
Sector I-11, Islamabad, Pakistan
Tel: (+92 51) 44 13 45; 44 27 76; 44 41 62
Fax: (+92 51) 44 27 76

Key personnel
Chairman, Pakistan Railways:
    Masood A Dhar
Managing Director, Mechanical Works:
    Abdul Wahid Khan
Deputy Chief Mechanical Engineer:
    Asad Ahsan
Works Manager: Jalal-ud-Din

Products
Passenger coaches.
    The factory has so far produced over 1,698 passenger coaches, including 223 exported to Bangladesh. It is currently carrying out mid-term refurbishment of passenger coaches for Pakistan Railways.

UPDATED

Parry People Movers Ltd

Overend Road, Cradley Heath, West Midlands B64 7DD, UK
Tel: (+44 1384) 56 95 53   Fax: (+44 1384) 63 77 53
e-mail: info@parrypeoplemovers.com
Web: www.parrypeoplemovers.com

Key personnel
Chairman: John Parry
Projects Manager: Caspar Lucas

Background
Parry People Movers Ltd was formed in 1992 to develop a light rail transport system conceived by the engineering company JPM Parry& Associates Ltd.

Products
Vehicle capacities are from 30 to 170 passengers and vehicles can be coupled to increase capacity. Low-cost light trams and railcars for all normal track gauges. Smaller vehicles are intended for use on lightly-used urban tramways and other lines in tram-type operations. The larger vehicles in the range are suitable for use either as intermediate size trams or as railcars for branch lines of heavy rail networks.
    Power supplied by flywheel energy storage system, recharging either from an efficient on-board engine (LPG, diesel or other fuel) or from intermittent concealed conductor rail at stations requiring no continuous electrical distribution. The use of energy storage reduces the instantaneous power requirement of the engine. A hydrogen fuel cell option is being developed. Emission-free operation is possible for all or part of a route. Energy-efficient operation is assured by the vehicles' low weight and consistency of power demand. No overhead wires are required. The vehicles' safe electrics and freedom from complex infrastructure make them acceptable for indoor use.
    Vehicle passenger capacity from 30 to 170 people; vehicles can be coupled to increase capacity. All formats are available with modern or heritage exterior styling. Vehicles and their interior styling is to customer's specification and meets all necessary regulations. Complete powered chassis supplied for fitting to separately procured bodywork. The standard range consists of: PPM 30 – narrow-body urban tram with low-floor access for up to 30 passengers. PPM 35 – standard width railbus with low-floor access for up to 35 passengers. PPM 50 – short-range light railcar for up to 50 passengers, suitable for light tramways with low-floor access, or marginal branch line and spurs to existing rail networks with level access at standard platform height. PPM 80 – bogie tramcar/railcar with low- or high-floor for up to 80 passengers, suitable for urban tramways and railway branch lines. PPM 100 – paired single-ended railcars with low- or high-floor for up to 100 passengers, suitable for light rail or metro-style operation on tramways and railway branch lines. PPM 170 – paired single-ended bogie tramcars/railcars with high- or low-floor access for up to 170 passengers, suitable for railway branch lines and urban tramways.

UPDATED

PPM 50 Railcar at Stourbridge Junction station (Phil Evans)
1033406

PPM 35 Railbus by Parry People Movers (Tony Pattison)
0043468

PEC Ltd

A Government of India Enterprise
Hansalaya, 15 Barakhamba Road, New Delhi
110 001, India
Tel: (+91 11) 331 63 72; 36 19; 47 23
Fax: (+91 11) 331 52 79; 47 97; 36 64
Telex: 031-65199 PEC-IN

Cable: PECOIND
e-mail: burman@peclimited.com

Key personnel
Chairman: A K Srivastava
Director: Manjit Yinayak
General Manager: S S Roy Burman

Products
Diesel-electric, diesel-hydraulic, electric, industrial
and mining locomotives; spares for locomotives
passenger coaches.

Diesel-electric locomotives and spares have
been exported to Tanzania and to Vietnam.

PESA Bydoszcz SA

Zygmunta Augusta 11 Street, PL-85-082 Bydgoszcz
Tel: (+48 52) 518 02 48
Fax: (+48 52) 518 52 39
e-mail: pesa@pesa.pl
Web: www.pesa.pl

Key personnel
President and General Director: Tomasz Zaboklicki
Directors:
    Production and Technical: Zenon Duszynski
    Marketing and Development: Zygfryd Zurawski
    Financial: Robert Swiechowicz
Proxy Deputy Director and Head of Production:
    Andrzej Karwasz
Head of Marketing, Passenger Coaches: Jerzy Berg
Head of Development: Andrzej Ciupa

Background
PESA Bydoszcz SA formerly traded as ZNTK
Bydoszcz SA. PESA's holding company is Pojazdy
Szynowe PESA Bydgoszcz Spólka Akcyjna Holding.

Products
Lightweight diesel railcars; air conditioned open
and compartment coaches; sleeping cars. The
company also undertakes refurbishment and
modernisation of locomotives, passenger coaches
and trams.

The Type 214M 'Partner' diesel-hydraulic railcar
was introduced in 2001 to provide a low-cost

Class 620M Partner diesel railcar by PESA for Ukrainian Railways (Ken Harris)   ***NEW**/0585018*

vehicle for regional lines. Built on a former coach
underframe with new bogies, the part-low-floor
(30 per cent) single-car unit is powered by a
500 kW power pack. Maximum speed is 120 km/h,
axleload 14 t.

Open and compartment passenger coaches are
built under licence to Ukrainian designs.

Contracts
In 2004 the Ukrainian National Railway Transport
Administration (UZ) ordered 70 diesel railcars for
delivery by the end of 2008. The contract is for 64
vehicles to be used for passenger services and six
to be used as inspection cars.

NEW ENTRY

Power Machines Group

(Energomachexport + LMZ + Electrosila + ZTL + KTZ)
25A Protopopovsky per, 129090 Moscow, Russian
Federation
Tel: (+7 095) 725 27 63   Fax: (+7 095) 688 79 90
e-mail: mail@power-m.ru
Web: www.power-m.ru

Key personnel
General Director: Evgeny Yakovlev
Sales Director (Machinery building):
    Alexander Zhigalov
Advertising Manager: Natalia Kuznetsova

Background
Power Machines unites the leading Russian
power equipment enterprises and supplies

railway and transport equipment produced by
Kalugaputmash, Luganskteplovoz, Muromteplovoz,
Railway Repair-Mechanical Plant of Sverdlovsk,
Kolomensky Zavod, Metrowagonmash, Vyksa Steel
Works.

Products
Shunting and main line locomotives.

UPDATED

Price

A & G Price
Division of CPD Engineering Ltd
Beach Road, Thames, New Zealand
Tel: (+64 7) 868 60 60
Fax: (+64 9) 309 28 19
e-mail: contracts@agprice.co.nz

Key personnel
Chief Executive Officer: Grant Burnett
Company Financial Officer: Ian Findlay
Commercial Manager: Neil Howe
Production Manager: Peter Yates
Sales Manager: Bill Lovell
Industrial Product Sales Manager: Peter Feran
Estimating Engineer: Barry Ingle

Products
Passenger coach manufacture, modification and
repair.

Qualter Hall & Co Ltd

PO Box 8, Johnson Street, Barnsley S75 2BY, UK
Tel: (+44 1226) 20 57 61
Fax: (+44 1226) 28 62 69
Web: http://www.qualterhall.co.uk

Key personnel
Managing Director: George Orton
Manufacturing Director: Keith Richardson

Background
Qualter Hall is a part of Waagner-Biro group of
companies, Austria.

Products
Fabricated structures for rolling stock including
locomotive superstructures; passenger and
freight vehicles; narrow-gauge underground
transportation systems including cabs, bolsters
and underframes. The company has shotblast

and painting facilities. Complete design, supply,
installation and commissioning of rope hauled
transportation systems.

Contracts
The company has built 59 shuttle locomotive
superstructures for the Channel Tunnel, weighing
approximately 33 tons each; various steel components
for main line multiple-units.

VERIFIED

RailPower Technologies Corporation

777 Dunsmuir Street, Suite 1118, PO Box 10443,
Vancouver, British Columbia V7R 1K4, Canada
Tel: (+1 604) 904 00 85
Web: www.railpower.com

Key personnel
President and Chief Executive Officer: Jim Maier
Chief Financial Officer and Controller:
    Alain Voisin
Executive Vice-President, Coporate Development:
    Simon Clarke
Manager, Customer Engineering: Ray Cousineau

Manager, Manufacturing and Technology
    Development: Mark Klag

Background
In 2005 RailPower signed assembly and supply
agreements with Super Steel Products Corp,
of Milwaukee, Wisconsin, and Super Steel

Schenectady, Inc, of Glenville, New York, in a move intended to respond more effectively to the orders from the US east coast market.

Products

Hybrid switching locomotives which for traction employ custom-designed led acid batteries that are kept charged by a small generator driven by a 90 kW diesel-fuelled Isuzu engine.

A prototype of the 1,492 kW (2,000 hp) Green Goat locomotive has been built on the premises of Southern Railway of British Columbia. It is based on a conventional GP9 frame and bogies and features a low-profile hood that houses the batteries, generator and diesel engine. The locomotive weighs 140 short tons and has a maximum speed of 32 km/h. RailPower claims that this configuration offers a reduction of 30 to 45 per cent in fuel consumption compared with an equivalent conventional diesel-electric locomotive, and reduces the emission of harmful pollutants by some 90 per cent.

After a trial service period with Union Pacific in Chicago, the Green Goat prototype was transferred in July 2003 to Pacific Harbor Line, which serves the ports of Los Angeles and Long Beach, for further evaluation.

In 2003 a prototype of a second design, the Green Kid, was under construction. This is a lower powered machine (760 kW/1,000 hp) weighing 124 short tons and with a top speed of 24 km/h. Unlike the Green Goat, it is a driverless, remote-controlled locomotive. Battery charging is powered by a 50 kW Isuzu engine.

GG20B Green Goat hybrid shunting locomotive by RailPower Technologies Corporation

NEW/0585013

Developments

RailPower has patented a concept for a Compressed Integrated Natural Gas Locomotive (CINGL) that employs gas turbine technology. The company claims that it would offer freight operators savings of 25 to 33 per cent in life-cycle costs and reduce nitrous oxide (NOx) emissions by more than 99 per cent.

RailPower is also developing technologies for fuel cell propulsion, using water to produce hydrogen as a fuel source.

UPDATED

Rail Services International SA (RSI)

Head office
Direction Internationale 38, rue de la Convention, F-94270 Le Kremlin, Bicêtre, France
Tel: (+331) 53 14 17 30   Fax: (+33 1) 53 14 17 49
e-mail: info@railsi.com
Web: www.railsi.com

Key personnel

Managing Director: Philippe Aloyol

Subsidiaries

RSI Austria
Werkstätte und Wäscherei, Domaniggasse, 2, A-1100 Vienna, Austria
Tel: (+43 1) 617 77 71 12   Fax: (+43 1) 617 77 71 28
e-mail: info@railsi.at
Managing Director: Reinhard Rössler

RSI Belgium
Vaartblekersstraat, 29, B-8400 Oostende, Belgium
Tel: (+32 59) 56 18 80   Fax: (+32 59) 70 23 75
e-mail: info@railsi.be
Managing Director: Jan Baert

RSI Italia SpA
Via Sesto San Giovanni, 9, I-20126 Milan, Italy
Tel: (+39 02) 66 14 02 01   Fax: (+39 02) 66 10 09 61
e-mail: info@railsi.it
Managing Director: Renato Mantegazza
Unit Manager Milan: Guido Sarzilla
Unit Manager Rome: Pasquale Grieco

CostaRail Srl
Viale 4 Novembre, I-22041 Costamasnaga, Como, Italy
Tel: (+39 031) 86 94 11   Fax: (+39 031) 85 53 30
e-mail: hkastner@costarail.it
Commercial Manager: Heinz Kastner

RSI Netherlands
Onderhoudspost Watergraafsmeer, Kruislaan, 254, NL-1098 Amsterdam SM
Tel: (+31 20) 557 66 30   Fax: (+31 20) 557 88 18
e-mail: info@railsi.nl
Managing Director: Jan Baert
Unit Manager: Bart Janssen

Background

Formerly the railway maintenance business of Compagnie des Wagons Lits, RSI is an independent company handling the maintenance, overhaul, repair and refurbishment of rail vehicles in several European countries. In 2004 the company acquired the former Costaferroviaria SpA in Italy, renaming the company CostaRail Srl.

Products

Development, design, engineering and technical assistance for fitting out passenger car interiors. RSI's Study and Design Department works on new projects such as interiors and technical specifications for passenger cars. RSI carries out maintenance, overhaul and major refurbishment of all kinds of rolling stock at its four workshops (Milan, Oostende, Rome and Vienna) or at units in 20 European locations.

Customer support for railway operators, day-to-day operations, wheelset maintenance, repair, warranty, mobile operations at customer's plant and supply of spare parts.

UPDATED

RCF – Rail Coach Factory

Kapurthala 144602, Punjab, India
Tel: (+91 181) 245 83 56
Fax: (+91 181) 245 70 91
e-mail: cplercf@rediffmail.com
Web: www.rcf.nic.in

Key personnel

General Manager: Yashpal Gupta
Chief Mechanical Engineer: O P Agarwal
Chief Electrical Engineer: Kulbhushan
Controller of Stores: Harish Gulati
Financial Adviser and Chief Accounts Officer:
   Gurdev Singh
Chief Planning Engineer: Kul Bhushan
Chief Design Engineer: S K Aggarwal

Interior of a stainless steel-bodied coach for Indian Railways built by RCF under a technology transfer agreement with ALSTOM

1033662

Background

Established in 1986, RCF is a coach manufacturing unit of Indian Railways. It has manufactured around 15,400 passenger coaches including self-propelled passenger vehicles. RCF is equipped with CAD centre and CNC machines to undertake design and manufacture of bogies, shell (both with stainless steel and corten steel), FRP interiors as per customer requirements. Its annual production of approximately 1,200 coaches for Indian Railways includes both broad gauge and metre gauge coaches and also approximately 20 per cent air-conditioned coaches. It also serves non-railway customers, such as defence organisations, the Department of Posts and other research organisations. RCF is an ISO-9001 and ISO-14001 certified unit.

Products

Diesel-electric multiple-units (metre gauge); electric multiple-units (broad gauge); self-propelled accident relief trains (broad gauge); air-conditioned/non-air-conditioned chair cars/ sleeper coaches (1,000 mm gauge and 1,676 mm gauge). Also designed and manufactured special purpose vehicles including: refrigerated vans for transportation of perishables, double-deck coaches; coach container flats (with air suspension springs), high-capacity parcels vans; track recording cars; overhead equipment testing cars; oscillograph cars and inspection cars.

Developments

Since 1998, high-speed air-conditioned chair and generator cars designed and manufactured by RCF have been in Indian Railways Inter City Express service. Tested at up to 180 km/h, the vehicles feature disc brakes, flexicoil secondary suspension, UIC drawgear and panoramic windows.

Six-car metre gauge diesel-electric multiple-unit by RCF for Indian Railways   1033661

Recently under technology transfer agreement, RCF has acquired technology from ALSTOM LHB (body shell) and ALSTOM FIAT, Switzerland (bogies) to manufacture stainless steel bodied coaches, which have been successfully tested up to 180 km/h. AC chair cars (78 passenger seating capacity), sleeper cars and power cars to this new design have already been manufactured by RCF and these coaches have been in service on the Inter City Rajdhani Express train between Mumbai and New Delhi. Other variants like non-AC sleepers cars, AC/AC coaches in the self-generating version have also been successfully designed and manufactured

by RCF. The design includes superior features like FRP panelling, controlled discharge toilet system, improved sound and thermal insulation, and CBC with anti-climbing features.

RCF has developed permanent magnet 12.5 kW alternators for successful manufacture of self generating metre gauge AC coaches and high-speed ALSTOM LHB design self-generating coaches. These alternators are very compact, light and more energy-efficient.

UPDATED

Relco Locomotives Inc

113 Industrial Avenue, Minooka, Illinois 60447-0058, US
Tel: (+1 815) 467 30 30
Fax: (+1 815) 467 30 39
Web: www.relcolomotives.com

Key personnel

Vice-President of Marketing: Eric C Bachman

Products

Diesel-electric shunting and main line locomotives for sale or lease. The company can also perform full service maintenance and heavy repairs throughout

the US and leases most of its units with full maintenance contracts. Relco offers remanufacture of motive power to factory-new specifications. Also performs aftermarket repowering with new, high-technology, low-emission engines and alternator/ generator sets. Other products and services include new and remanufactured parts and components, supply/install/service locomotive radio remote-control systems and export capabilities. Relco is fully certified under the American Association of Railroads M-1003 quality assurance programme.

Contracts

Contracts include the supply of locomotives and/or contract services to over 150 industrial companies, railways and government agencies.

UPDATED

Diesel-electric locomotive remanufactured by Relco
0518978

Republic Locomotive

Republic Transportation Systems Incorporated
Suite 101, 131 Falls Street, Greenville, South Carolina 29601-2825, USA
Tel: (+1 864) 271 40 00   Fax: (+1 864) 233 21 03

Works

1861 West Washington Road, Greenville, South Carolina 29601, USA

Key personnel

President and Chief Executive Officer:
  Hugh B Hamilton Jr
Manufacturing Manager: Mike Dixon
Director of Engineering: Tim Armstrong

Principal affiliates

Republic Locomotive Works Inc
Republic Group Inc
Republic Raileasing Inc

Products

New or remanufactured locomotives for railway, industrial and passenger applications.

Republic has successfully tested the RD20 model diesel-electric locomotive, developed by Republic and Detroit Diesel. Available is a switching locomotive with AC traction motors.

Rocafort Ingenieros

Rocafort Ingenieros SL
Pl El Regás, C/Ciencia 25, E-08850 Gavá (Barcelona), Spain
Tel: (+34 93) 633 39 10

Fax: (+34 93) 662 94 50
e-mail: rocafort@rocafort.net

Key personnel

President, General Manager: Gerardo Rocafort
Technical Manager: Daniel Dedieu

Products

Systems for general interiors and refurbishment of railway vehicles, technology transfer, toilet modules and closed WC systems, touristic trains.

UPDATED

Rotem Company

Headquarters
Landmark Tower, 837-36, Yeoksam-dong, Gangnam-gu, Seoul 135-937, South Korea
Tel: (+82 2) 21 12 82 94
Fax: (+82 2) 21 12 98 73
Web: www.rotem.co.kr

Key personnel
Vice Chairman and Chief Executive Officer:
 Soon-Won Chung
Senior Executive Vice-President: Yeo-Sung Lee
Executive Vice-President: Jae-Hong Kim

Works
Changwon plant
#85 Daewon-dong, Changwon, Gyeongsangnam-do
641-808, Korea
Tel: (+82 55) 273 13 41   Fax: (+82 55) 273 17 41
Fax: (+82 55) 273 17 41

Uiwang plant/central research and development centre
#462-18 Sam-dong, Uiwang, Gyeonggi-do 437-718, Korea
Tel: (+82 31) 460 14 01   Fax: (+82 31) 460 17 94
Fax: (+82 31) 460 17 81

Technical research institute
#80-10 Mabook-ri, Guseong-eup, Yongin, Gyeonggi-do 449-910, Korea
Tel: (+82 31) 288 11 14
Fax: (+82 31) 284 021 29

Background
Established in 1964 when Daewoo Heavy Industry started manufacturing rolling stock, followed by Hyundai and Hanjin Heavy Industry a few years later. In 1999, the three companies were consolidated into KOROS. Hyundai Motors Group acquired the share of Daewoo and former company name KOROS became Rotem. Rotem is now an affiliate of Hyundai Motors Group and has its headquarters in Seoul and two factories in Uiwang and Changwon. Uiwang has a capability of manufacturing 500 emus per year and has the capability to manufacture electric equipment such as traction motors, SIV inverters etc. The research and design centre is also located in Uiwang. The Changwon factory has a capability to manufacture 700 emus per year. Rotem has an annual capacity to manufacture approximately 1,200 emus. Certifications such as the ISO 9001 certificate for Quality, 14001 for environment and 18001 for occupational health and safety management have been acquired at all three sites.

Products
Rotem manufactures emus, demus, light rail vehicles, high-speed trains, Maglev, passenger coach, locomotives, electrical equipment and subsystems for the railways industry. It has experience in developing rail systems such as E & M turnkey systems, engineering services such as feasibility studies in rail transportation and also arranges financing for certain projects.

Recently, Rotem has applied RAMS (Reliability, Availability, Maintainability, Safety) to vehicles supplied to Hong Kong and China and received incentives from MTRC, Hong-Kong Subway Company.

Rotem successfully developed the new high-speed train system which can reach speeds of 350 km/h and over. The carbody is made from aluminum to increase speed as compared to the KTX, Korea's express train which is made from steel. Rotem's new high-speed train will become commercial on the Honam line in 2007.

Geared for middle and long-distance trips, Rotem has been developing and supplying dmus. The push-pull type Korean National Railroad dmu, delivered in 1988 to 2000, with a 2,954 kW 3,960 hp) engine operates at the speed of 150 km/h for commercial service. Rotem can design and manufacture other types of dmu according to various clients needs, providing flexibility to the basic formation and power distribution and/or changin the location of power units (under frame or upper frame).

Rotem concept emu for urban applications (Ken Harris) ***NEW**/0585024*

Rotem two-car maglev vehicle (Ken Harris) ***NEW**/0585025*

Rotem diesel electric locomotive 0092237

Rotem's urban transport, Maglev system is a magnetically levitated vehicle powered by linear induction motors running on an elevated guideway. Made from lightweight aluminum and mounted on three magnetically levitated air-sprung bogies whcih glide above the elevated guideway. The bogies are designed to wrap the rail, making a safe train operation.

Contracts

Ireland: Rotem was awarded a contract for the supply of 120 demus to Ireland in 2005. The trains will be operated on 1,947 km of wide gauge track owned by IE and will be manufactured according to European safety standards.

Iran: Rotem will provide 120 dmus and transfer technology to the Iran Khodro Rail Transport Industries Co (IRICO). 24 units will be delivered to IRICO by 2006 and the remainder will be produced in Iran, with the help of Rotem's technology, in 2008.

Korea: Rotem received an order from Korean National Railroad (KNR) for the supply of 120 commuter emus for the KNR main lines. Initially, 60 emus will be delivered in December 2005 and the remainder are scheduled for delivery in June 2006.

In 2004, Rotem delivered 18 KNR electric locomotives and received an additional order for 17 electric locomotives at the end of 2005.

In January 2005, a consortium led by Rotem and several engineering and civil construction companies, won a tender for the construction of Seoul Metro Line 9. Through the Seoul Metro 9 project, Rotem, in addition to being a vehicle supplier, is aiming to be a global rail system turn-key supplier.

Seoul Metropolitan Subway Corporation (SMSC) ordered 69 metro trains for the replacement

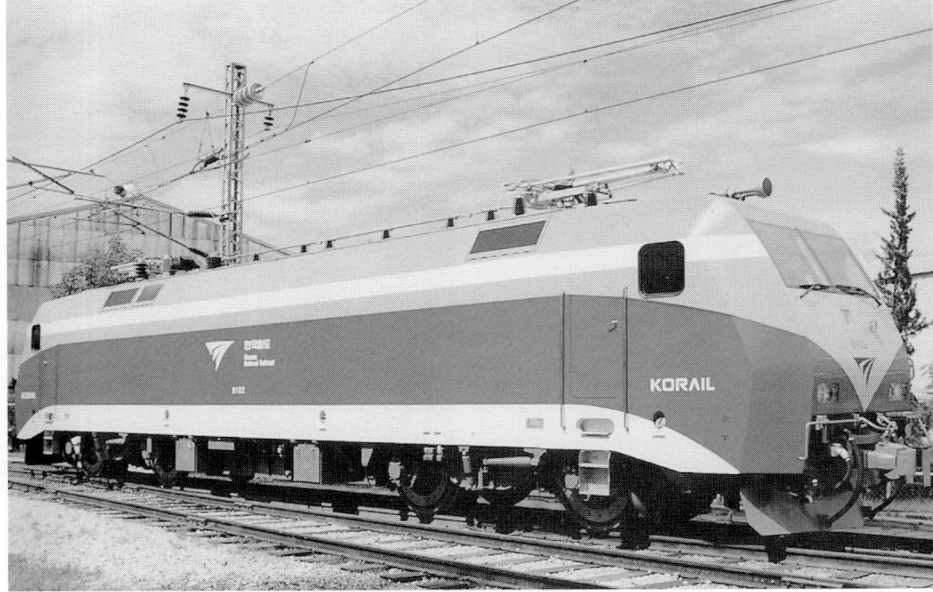

Rotem KNR Type 8100 (DEL 01) locomotive   0092238

line 2 trains. In addition Rotem has received refurbishment orders from domestic and overseas customers totalling approximately 1,200 units and these outdated train sets will be upgraded and well equipped to allow safe and comfortable passenger service.

US: The Triangle Transit Authority (TTA) awarded a contract of 32 dmus to United Transit System, of which Rotem is a member. Rotem (US) will provide a minimum of 24 Federal Railroad Administration

safety complient railcars, configured as 12 married pairs, and an option for eight additional vehicles configured as four additional married pairs. The 12-station, 28 mile Regional Rail Transit System from Durham to the Research Triangle Park, Cary, and Raleigh, North Carolina, is scheduled to begin operation in 2008.

UPDATED

RSD (A division of DCD-Dorbyl (Pty) Ltd)

PO Box 229, Boksburg 1460, South Africa
Tel: (+27 11) 914 14 00
Fax: (+27 11) 914 38 85
Web: http://www.dcd-dorbyl.com

Key personnel
Executive Director: Carl Rehder
Business Manager: Norman Taylor
Divisional Financial Manager: Mark Charters

Products
Funkey mining and industrial locomotives, GE diesel electric mainline and industrial locomotives, freight wagons, couplers and drawgear, renewal parts and servicing.

RVR

Riga Carriage Building Works
201 Brivibas Gatve, Riga LV-1039, Latvia
Tel: (+371 2) 37 25 21
Fax: (+371 7) 55 52 19; 82 83 96

Key personnel
President: Janis Anderson
Vice-President, Financial Director: Velerij Novarro
Vice-President, Technical Director: Robert Reingardt
Commercial Director: Sergey Chigorin
Sales and Marketing Manager: Vadim Maximov

Products
1,435 and 1,520 mm gauge diesel and electric (AC and DC) multiple-units for local services.

Saalasti Oy

Juvan teollisuukatu 28, FIN-02920 Espoo, Finland
Tel: (+358 9) 251 15 50   Fax: (+358 9) 25 11 55 10
e-mail: info@salaasti.fi
Web: www.saalasti.fi

Key personnel
Managing Director: Timo Saalasti
Director of Transport Machinery Department:
  Teijo Saalasti

Products
Saalasti specialises in building vehicles and systems to customer requirements. Typical products include: diesel shunting locomotives and shunting robots; automatic shunting yard systems; rail maintenance vehicles; ganger's trolleys; shunting couplings; snow ploughs.

The OTSO shunting locomotives range from 30 tonne, two-axle through to 100 tonne, four-axle locomotive. Locomotives are equipped, for example, with Caterpillar engine and Voith Turbo fully hydraulic power transmission. Airbrake components including anti-slip and anti-skid devices are supplied from Knorr-Bremse. Locomotives can be equipped with radio-control, automatic Vapiti couplers and snow ploughs. The OTSO shunting robots range from 30 tonne, two-axle through to 100 tonne four-axle

Saalasti Oy Otso-robot   1109857

robots. Shunting robots are equipped with a Deutz water-cooled engine and hydrostatic power transmission. Shunting robots are radio controlled but can be equipped with automatic driving systems if required by the customer. Other possible accessories include a train brake

controlling system with hydraulic compressor, snowplough and snowbrush. Saalasti Ganger's Trolley is a multipurpose vehicle which can be equipped for many kind of rail maintenance work. An underfloor engine enables effective use of space and a variable design. The trolley can also

Examples of Saalasti diesel shunting locomotives and Ganger's Trolley

	OTSO-robot 50	OTSO-robot 100	OTSO 4	OTSO 8 & 10	Ganger's Trolley
Weight (t)	50	100	40–44	80–90	25–40
Power (kW)	126	220	350	700–1,400	180–315
Drawbar pull (kN)	50	180	130	210–270	60–120
Speed (km/h)	4 or 8	4–8	25	Up to 80	Up to 100
Axle arrangement	B	B+B	B	B'B'	B
Rail gauge (mm)	1,435/1,524	1,435/1,524	1,435/1,524	1,435/1,524	1,435/1,524

be equipped with a crane, personal lifting device, snowplough and various hydraulic accessories.

Contracts

Recent contracts include deliveries of three OTSO-robot 100 automatic shunting locomotives to Metsä-Botnia and UPM-Kymmene mills in Finland for wood unloading. A track maintenance vehicle to Railcare AB in Sweden.

UPDATED

SAN Engineering & Locomotive Co Ltd

PO Box 4802, Whitefield Road, Whitefield, Bangalore 48, Karnataka, India
Tel: (+91 80) 845 22 71-76
Fax: (+91 80) 845 22 60; 845 31 95
e-mail: inderm@san-engineering.com
　srinidhi@san-engineering.com
Web: http://www.san-engineering.com

Key personnel
Managing Director: Milind S Thakker
Chief Executive: Inder Mahadevan

Principal subsidiary
Engineering Products Division
Plots 1 & 10, Hebbal Industrial Area, Belwadi Post, Mysore 571 106, India

Products
Diesel-hydraulic and diesel-electric locomotives, self propelled rail vehicles such as emus, dhmu, UTV and OHE cars. Transmissions for locomotives and industrial applications, heavy duty cardan shafts for locomotives and industrial applications, powered and non-powered final drives, gears and gearboxes.

Locomotives manufactured are used by core sector industries such as cement plants, thermal plants, oil refineries, petrochemical complexes, steel plants, fertilizer plants and construction sites. Locomotives are tailored to customers needs.

The company also manufactures flameproof locomotives for underground coal mines.

Santa Matilde

Rua Frei Caneca 784, São Paulo, CEP 01307-000, Brazil
Tel: (+55 11) 41 58 34 85　　Fax: (+55 11) 41 58 34 87
e-mail: mtfrail@uol.com.br

Works
Rua Isaltino Silveira 768, CEP 25804-020, Tres Rios, Rio de Janeiro

Key personnel
Managing Director: Edson Silva Ferreira
Finance Director: Fernando Jose Pimentel Duarte
Operations Director: Manoel Mendes
Marketing & Export Director:
　Eduardo Hubert K Monteiro

Products
Emus; passenger coaches.

Contracts
Contracts include refurbishment of Series 160 and Series 401/431 stainless steel emus for CPTM, and refurbishment of two Series 500 and five Series 800 stainless steel emus for Flumitrens.

Schöma

Christoph Schöttler Maschinenfabrik GmbH
PO Box 1509, D-49345 Diepholz, Germany
Tel: (+49 5441) 99 70
Fax: (+49 5441) 997 44
Web: www.schoma.de

Key personnel
Chairman: Ing C Schöttler
Sales Manager: Ing R P Bogs

Products
Standard- and narrow-gauge diesel-hydraulic locomotives for shunting duties and mining, works and tunnel construction systems in operation worldwide. The company has extended its product range to locomotives of 80 t weight and 600 kW output.

Sales have included 14 CFL-500 VR works locomotives for London Underground Ltd's Jubilee Line Extension; 12 CEL-500R battery-electric locomotives for the North-East Line MRT

in Singapore; various two-, three- and four-axled (Bo-Bo type) works locomotives of 22 t axle-load.

UPDATED

Schöma battery-electric service locomotive supplied for the North-East Line in Singapore　　*NEW*/1143739

Schöma shunting locomotive of 80 tonnes supplied to a cement plant in South Korea　　*NEW*/1143738

SCIF

Société Cherifienne du Matériel Industriel et Ferroviaire
PO Box 2604, Allée des Cactus, Aïn-es-Sebaa, Casablanca, Morocco

Tel: (+212 2) 35 39 11
Fax: (+212 2) 35 09 60

Key personnel
Deputy Director General: M Lahkim

Products
Passenger cars.

SEMAF

Société Générale Egyptienne de Matériel des Chemins de Fer
Ein Helwan, Cairo, Egypt
Tel: (+20 2) 78 23 58; 78 21 77; 78 27 16; 556 27 16
Fax: (+20 2) 78 84 13

Key personnel

Chairman: Eng T El-Maghraby
Technical Manager: Eng A Rahik
Commercial/Financial Manager: A Farid

Works Manager, Coach and Metro: Dr Eng L Melek
Works Manager, Wagon and Bogie:
    Eng El-Sherbini

Products

Power cars, passenger cars, railcar/trailers, trams and metro cars.

Contracts

Include trams for Cairo, Helwan, Heliopolis and Alexandria. Production capability totals more than 100 tramcars per year.

SEMAF has assembled 72 cars for Line 2 of the Cairo metro, under a contract awarded jointly with Kinki Sharyo. An initial batch of 18 cars was supplied complete from Japan. It has also assembled a further 12 tramcars for Alexandria from parts supplied by Kinki Sharyo.

Siemens AG

Transportation Systems

PO Box 3240, D-91050 Erlangen, Germany
Tel: (+49 9131) 72 02 38   Fax: (+49 9131) 72 45 98
e-mail: transportation.systems@siemens.com
Web: www.siemens.com/transportation

Corporate headquarters

Siemens AG
Transportation Systems
PO Box 3240, D-91050 Erlangen, Germany
Tel: (+49 9131) 7-0

Key personnel

President, Transportation Systems Group:
    Hans M Schabert
Group Vice-Presidents: Alfred Frank; Joern F Sens;
    Friedrich Smaxwil

Works

Austria
Siemens Transportation Systems GmbH & Co KG
Leberstrasse 34, A-1110 Vienna
Tel: (+43 51) 70 70   Fax: (+43 51) 70 75 15 95

Siemens Transportation Systems GmbH & Co KG
Eggenberger Strasse 31, A-8021 Graz
Tel: (+43 51) 70 70   Fax: (+43 51) 70 75 35 08

Czech Republic
Siemens sro
Transportation Systems
Evropska 33a, CZ-160 00 Prague 6
Tel: (+420 233) 03 33 03   Fax: (+420 233) 03 11 12

Germany
Siemens AG
Transportation Systems
Duisburger Strasse 145, D-47829 Krefeld
Tel: (+49 2151) 45 00   Fax: (+49 2151) 450 12 14

Siemens AG
Transportation Systems
Krauss-Maffei-Strasse 2, D-80997 Munich
Tel: (+49 89) 889 90
Fax: (+49 211) 88 99 33 36

US
Siemens Transportation Inc
7464 French Road, Sacramento, California 95828
Tel: (+1 916) 681 30 00   Fax: (+1 916) 681 30 06

Test centre
Siemens AG
Transportation Systems
Friedrich-List-Allee 1, D-41844 Wegberg-Wildenrath, Germany
Tel: (+49 2432) 97 00   Fax: (+49 2432) 97 02 00

Organisation

Siemens Transportation Systems (TS) is organised as seven business divisions based on the products or services supplied:
Locomotives: electric locomotives; diesel-electric locomotives; special purpose locomotives; operating leases (Dispolok); and refurbishment
Trains: high-speed and intercity trains; commuter and regional trains; passenger coaches
Mass Transit: metro vehicles; suburban trains; trams; light rail vehicles; components
Rail Automation (see entry in *Signalling and communications systems* section): signalling

Eurorunner diesel electric locomotive for Kowloon-Canton Railway Corporation   **NEW**/0585298

and control systems; interlockings; automatic train control systems; signalling systems and components; telecommunications systems; and communications systems for mass transit and main line systems
Electrification (see entry in *Electrification contractors and equipment suppliers* section): products and systems for main line and mass transit contact lines; and products and systems for main line and mass transit traction power supplies
Turnkey Systems (see entry in *Turnkey systems contractors* section): turnkey systems for mass transit and main line systems
Integrated Services (see entry in *Vehicle maintenance equipment and services* section): maintenance; spare parts; training; documentation; diagnostic services; and consultancy

In September 2004 Siemens TS employed 17,900 staff.

Subsidiary companies

Siemens Transportation Systems SAS (France) (100 per cent*)
Siemens Transportation Systems Inc (USA) (100 per cent)

Mass Transit; Locomotives, Trains
Siemens Transportation Systems GmbH & Co KG, Austria (100 per cent*)
Siemens Dispolok GmbH (Germany) (100 per cent)
Technisches Gemeinschaftsbüro GmbH (TGB), Germany (100 per cent)
Siemens kolejová vozidla sro (SKV), Czech Republic (67.05 per cent)

Class 189 electric locomotive for German Rail   0525400

Diesel and electric locomotive production at the Siemens TS facility at Munich, with ÖBB Class 2016 'Eurorunner' diesel-electrics and DB Class 189, OSE Class 120 and SBB Class 474 'Eurosprinter' electrics undergoing final fitting out (Ken Harris)   0585010

Siemens Traction Equipment Ltd (STEZ), China (51 per cent*)

Rail Automation
messMa GmbH, Germany (100 per cent)
Siemens Signalling Ltd, China (70 per cent)
ESTEL Rail Automation SPA, Algeria (51 per cent)

Electrification/Turnkey Systems
ERL Maintenance Support Sdn Bhd (E-MAS), Malaysia (51 per cent)
Transrapid International GmbH & Co KG (TRI), Germany (50 per cent)

Integrated Services
Siemens Rail Services Pty Ltd (SRS), Australia (100 per cent)
Dienstleistungsgesellschaft für Kommunikation des Stadt- und Regionalverkehrs mbH (DKS), Germany (51 per cent)
NERTUS Mantienimiento Feroviario SA, Spain (51 per cent)

* shareholding held by Siemens AG parent company in each country

Background

In October 2001, Siemens acquired the passenger vehicle manufacturing interests of Czech rolling stock manufacturer ČKD Dopravni Systémi, forming these into a wholly owned subsidiary, Siemens kolejová vozidla sro. Its main products are vehicles for urban and main line transport applications, including metro cars.

In January 2001, Siemens 'Dispolok' leasing and rental business was established as a wholly owned subsidiary, Siemens Dispolok GmbH (see entry in *Rolling stock leasing companies* section).

Products

Design, development and manufacture of main line electric locomotives; diesel-electric locomotives; high-speed trainsets; diesel and electric tilting and non-tilting trainsets; emus; dmus with electric, hydraulic or mechanical transmission; passenger coaches; lightweight diesel or electric railcars; suburban and metro cars; light rail vehicles and trams, fully automated steel-wheeled metro systems (AGT); fully automated rubber-tyred metro systems (Val); Ultra Low Floor trams (ULF); locomotive and rail vehicle refurbishment. The current product portfolio includes:

Eurosprinter
Designed as a universal high-performance three-phase electric locomotive primarily for European networks, the Eurosprinter can be supplied with a continuous rating of 4,200–6,400 kW and top speeds of up to 230 km/h. Siemens has received orders for the type both from national railway companies and from open access operators.

Eurorunner
The Eurorunner is a multi-purpose main line diesel-electric locomotive aimed primarily at European markets and is designed to be supplied in power outputs ranging from 1,500 to 4,200 kW. The first order for the type was for a 2,000 kW 140 km/h version from Austrian Federal Railways (see below).

Velaro
This latest development in the family of high-speed trains initially designed for the German Rail network as ICE features distributed power rather than the power car and trailers concept of earlier ICE units. As well as supplying examples of the type to DB AG, Siemens has won high-speed train orders from NS, Netherlands, and RENFE, Spain, where it is designated Velaro E by Siemens.

ICE T
Tilting version of the ICE high-speed train, supplied to DB AG.

Venturio
The Venturio family has been developed for intercity and inter-regional applications where operations in the 160 to 250 km/h speed range are required. A modular platform enables trains to be supplied in tilting or non-tilting, electric or diesel-electric versions in formations from three to nine cars.

Desiro
This modular family of articulated dmu and emu vehicles is intended for suburban, commuter and regional traffic. Diesel versions can be supplied in diesel-electric or -mechanical versions. Customers include: DB AG (dmu Class VT 642 and emu Classes 424–426); SŽ, Slovenia (see below); Express Rail Link, Malaysia (see below); ÖBB, Austria (see below); MÁV, Hungary (see below); and CFR, Romania (see below). A non-articulated version has been developed for the UK market and has attracted orders from several operators, including a diesel version for TransPennine Express.

Avanto/S70
The Avanto/S70 is a hybrid design of low-floor light rail vehicle intended to operate both on city centre tram routes and over outlying main line suburban networks. The vehicle concept is configured to allow multi-voltage operation and provision is made to incorporate a diesel engine to permit operations on non-electrified routes. The first order for the type was received from SNCF, France, in July 2002.

AGT
AGT is a fully automated steel-wheeled metro system. The system has been ordered by VAG Nuremberg for its Rubin metro project.

VAL
The VAL rubber-tyred fully automated metro system is already proven in several cities worldwide. Recent contracts are for systems in Rennes, Roissy-Charles de Gaulle and Toulouse, France, and Turin, Italy.

ULF
The ULF ultra-low floor modular tramcar features the lowest entrance height in the world. Low-floor throughout, the entrance height is 197 mm. The Vienna transit operator Wiener Linien placed the first order for 150 ULFs in 1997 and a second one for another 150 in 2004.

CargoMover
The CargoMover is an innovative fully automatic diesel-powered freight-carrying vehicle developed for regional lines and distances of up to 150 km. Radar sensors are used to detect relative speeds and distances from objects, providing data to the vehicle's control system. Movements are regulated from conventional control centres using electronic interlocking systems and GSM-R communications technology. A prototype was demonstrated in 2002.

Transrapid
With Thyssen Krupp and Transrapid International, Siemens is a joint partner in the development and supply of the Transrapid magnetic levitation high-speed transit system. Siemens supplies traction technology, power supply, operations control and communications systems for Transrapid projects.

A key element of Siemens' product strategy is the company's Wildenrath Test Center in northwest Germany. Commissioned in January 1997, this

Driving car for Type M1 metro trainset for Prague at the Siemens SKV plant (Ken Harris)   0585011

Type SD70 MAC diesel-electric locomotive for Burlington Northern Santa Fe, built by General Motors with Siemens AC traction equipment   0525401

facility includes five test tracks, the longest 6.083 km and four of them dual-gauge (standard/metre), together with a capability to energise the overhead power supply systems at various voltages. These features enable the company to undertake exhaustive test running and mileage accumulation of all types of vehicle.

Contracts

Recent contracts include:

Australia: In 2002 Siemens delivered the first of 59 Combino trams to M>Tram, Melbourne. The order comprises 38 three-section and 21 five-section vehicles. Siemens also received a contract to supply 72 three-car vehicles for Bayside Trains, Melbourne (now Connex Trains Melbourne).

Austria: In June 2004 a consortium of Siemens and Elin EBG was awarded a contract by the Vienna urban operator Wiener Linien to supply 150 additional Ultra-Low-Floor (ULF) trams, with an option for 150 more. Siemens share of the contract is about 70 per cent by value. Delivery of the vehicles, which will join 150 similar trams in service in Vienna, will take place from Siemens' plant in the city from 2006 to 2012.

In May 2003 Siemens was awarded a contract by Austrian Federal Railways (ÖBB) for 20 two-car Desiro dmus based on the DB Class LVT 642, with an option on 120 additional trainsets.

In September 2002 Siemens was awarded a contract by the municipal authorities of Vienna to supply 25 six-car metro trains based on the company's MoMo platform. This first order was placed within the framework of an agreement signed in 1998 which includes an option for 60 additional trains and followed the successful testing of a prototype train introduced in 2001. Delivery of the trains, destined to work on Lines U1 and U2 of the Vienna system, will take place in 2005–08.

Deliveries continued in 2003 of orders for 400 Class 1016 and 1116 (dual voltage) 6.4 MW Eurosprinter electric locomotives for ÖBB.

In January 2002 Siemens handed over the first of 70 Class 2016 'Hercules' 1,600 kW diesel-electric locomotives to ÖBB. Placed in November 1998, the contract included an option on 80 similar machines, which are members of the Siemens Eurorunner family of locomotives.

Belgium: In December 2001 Siemens announced that it had won a contract from the Flemish transport operator De Lijn to supply 47 Hermelijn-type part-low-floor trams for services in Antwerp (30 single-ended vehicles) and Ghent (17 double-ended). The order followed earlier deliveries to De Lijn of 45 similar vehicles.

Bulgaria: In January 2005 Siemens and Bulgarian State Railways (BDZ) signed a contract for the delivery of 25 diesel-powered regional multiple units and their subsequent maintenance

for the period of seven years. The contract also includes an option for the delivery of 25 electrically powered trains.

Canada: In December 2004 Siemens received an order from the transit company of the city of Calgary in Alberta, for 33 Type SD 160 light rail transit (LRT) vehicles. It will expand the existing fleet to a total of 149 vehicles. Delivery of the first RT units is due to start in early 2006.

China: In December 2004 Siemens in partnership with CSR Zhuzhou Electric Locomotive Works secured a contract to supply 180 two-section 9,600 kW electric locomotives to the Chinese Ministry of Railways. A development of the Class DJ1 (see below), the 184 tonne machines will be delivered between mid-2006 and the end of 2007.

Initial deployment will be on coal trains on the Datong–Qinhuangdao line.

In May 2003 Siemens, in cooperation with Zhuzhou Electric Locomotive Works (ZELW), won a contract to supply 40 three-car trainsets to Guangzhou Metro Co. Intended for the city's new metro Line 3, the trains will be delivered from September 2005 to mid-2007.

In March 2002 Siemens, in co-operation with ZELW, signed a contract for 28 six-car metro trainsets. Shipment started in early 2004 and will continue until mid-2006.

In January 2002 Siemens announced an order for five Eurorunner diesel-electric locomotives for the Kowloon Canton Railway Corporation, Hong Kong. These were delivered in 2003.

In 2001, through its joint-venture company Siemens Traction Equipment Ltd, Zhuzhou (STEZ), Siemens completed the first of 20 Class DJ1 three-phase 6,400 kW eight-axle 25 kV AC electric freight locomotives for Chinese Railways. The first three machines were built at the company's Graz, Austria, plant; the remaining 17 were to be constructed in Zhuzhou.

Czech Republic: In September 2003 Czech Railways placed orders with Siemens for 26 200 km/h passenger coaches for delivery from 2005. Their design at Siemens Prague facility will involve extensive participation by Czech suppliers.

Follow-up orders from Prague Metro for 26 five-car Type M1 metro trainsets were in production at Siemens' Prague plant in 2004. This followed the earlier supply of 22 similar five-car sets, for which Siemens provided traction equipment.

France: In February 2004 a consortium of Siemens and ALSTOM won a contract from French National Railways (SNCF) to supply 400 84 tonne 1,600 kW diesel-electric freight locomotives for delivery from 2006 to 2015. Siemens is to supply propulsion and control systems and manufacture 130 bodyshells at its Munich plant. ALSTOM is to manufacture mechanical equipment and assemble the locomotives at its Belfort facility in France.

In July 2002 Siemens received an order from SNCF for 15 dual-voltage Avanto light rail 'tram-train' vehicles for use on an 8 km line between

ICE T high-speed electric tilting trainset for German Rail   0525402

Class 450 Desiro UK emu for South West Trains   0525403

Aulnay-sous-Bois and Bondy, in the eastern suburbs of Paris. Delivery was due in 2004. The contract includes an option on 20 similar vehicles.

In May 2001 Siemens subsidiary Matra Transport International (now Siemens Transportation Systems SAS) was awarded a turnkey contract by SMTC, Toulouse, to supply and equip the city's second fully automated metro line, the 16 km Line B. The contract included the supply of 35 VAL 208 driverless metro vehicles. The line was scheduled to open in 2007.

Germany: In February 2003 the Munich municipal authority, Stadtwerke München, ordered eight additional Type C metro trains for delivery in 2005–06. The order represents confirmation of an option on a 1997 contract for 10 trains, the first of which entered service in November 2002. Carbodies and running gear are supplied by Bombardier Transportation.

In September 2002 Siemens handed over to DB Cargo the first of 100 Class 189 four-system electric freight locomotives.

In June 2002 Siemens, as a member of a consortium led by Bombardier Transportation, secured a follow-on order for in total 23 Type GT 8-100 D/2S-M dual-voltage (750 V DC/15 kV AC) light rail vehicles for VBK, Karlsruhe. Siemens is responsible for the mechanical portions of the contract, as well as the high-voltage equipment and dual-voltage technology. The new vehicles will bring the number of dual-voltage LRVs used on Karlsruhe's 'tram-train' network to 122.

By May 2002 orders from Germany for the Eurosprinter 6.4 MW electric locomotive had been received from German Rail (DB AG) for 25 machines designated Class 182 and Siemens Dispolok for 38 locomotives for hire or lease.

In May 2002 as leader of the ICT2 consortium, which also includes ALSTOM Transport and Bombardier Transportation, Siemens signed a

Desiro emu for Slovenian Railways   0525404

contract to supply to DB AG 28 seven-car Class 411 tilting ICE high-speed trainsets. Siemens was to be responsible for the supply of electrical equipment and the production of 91 vehicles, with delivery scheduled for June 2004–February 2006.

In December 2001, Siemens announced that in consortium with Bombardier Transportation it had won a contract to supply to DB AG 40 Class 425

four-car emus for service on the Verkehrsverbund Rhein-Neckar regional network from the end of 2003. Orders for 20 additional four-car emus out of an option were placed in September 2002.

In November 2001 Siemens announced an order from Verkehrs AG (VAG), Nuremburg, to equip the new U3 automated metro line in the city and to retrofit the existing VAG U2 metro lines for automatic operation. The contract includes the supply of 30 Type DT 3 driverless trainsets, 16 for service on Line U3 and 14 for Line U2. Line U3 was to be commissioned in 2006 and Line U2 converted to automatic operation by the end of 2007.

In August 2001 SWU, the municipal operator in Ulm, ordered eight additional five-section single-ended metre-gauge Combino light rail vehicles which were delivered in 2003. The order included an option on two more vehicles. At the same time, Rheinbahn, Düsseldorf, exercised its option to procure 15 Type NF 8 five-section Combinos, adding these to 36 seven section NF 10 vehicles previously ordered from Siemens.

Other German operators to order Combino low-floor trams include: Stadtwerke Augsburg (29 seven-section single-ended metre-gauge), EVAG, Erfurt (36 five-section uni-directional, 12 three-section uni-directional); FVAG, Freiburg (18 seven section bi-directional metre-gauge); Stadtwerke Nordhausen (four three-section uni-directional metre-gauge, three three-section bi-directional metre-gauge with additional diesel-electric propulsion equipment – dual system); ViP, Potsdam (16 five-section uni-directional).

Greece: In 2003 a Siemens-led consortium that also included local supplier Hellenic Shipyards completed the first of 20 five-car 25 kV AC articulated emus for the Hellenic Railways Organisation (OSE). As a member of a consortium with Adtranz (now Bombardier Transportation), in August 2000 Siemens delivered the first of 40 three-car metro trains for Athens-Piraeus Electric Railways (ISAP). Siemens was responsible for the vehicles' traction, train control and safety systems.

Hungary: In April 2003 the Budapest public transport authority, BKV, placed an order with a consortium led by Siemens for 40 nine-section 53 m trams for Lines 4 and 6 of the city's network.

In May 2002 Siemens was awarded a contract by Hungarian State Railways (MÁV) to supply 13 two-car Desiro dmus based on the DB Class LVT 642. Delivery took place in September 2003.

In September 2001 Siemens announced orders for 15 Type ES 64 U2 Eurosprinter dual-voltage (25 kV AC/15 kV AC) electric locomotives from operators in Hungary. Ten have been supplied to MÁV and five to the private operator, Györ-Sopron-Ebenfurt-Vasút (GySEV). Delivery was made

Emu for Keihin Express Railway   0525405

Aerial view of Siemens' Wildenrath Test Centre   0525406

Siemens Desiro UK Class 360 emu supplied for operation by FirstGroup, now operated by One Railway   ***NEW***/0585300

in 2002, the two batches becoming Class 1047 and 1047.5 respectively.

India: In May 2005 Siemens received a notification of award of a contract for the Mumbai Railway Vikas Corporation (MRVc) for propulsion system and electrical equipment for emu rolling stock to be built at the Integral Coach Factory, Chennai. The emus will be in operation on both of Mumbai's suburban arterial networks, the Central and Western Railway.

Iran: In September 2000 Siemens signed a contract with Iranian Islamic Republican Railways for the supply of 20 four-car diesel-hydraulic trainsets. The contract also includes provision for technology transfer, with 15 units being manufactured locally in Iran.

Italy: In November 2001, as part of the VAL 208 Torino GEIE consortium, Siemens received an order for 46 automated four-car VAL 208 trainsets for the planned driverless metro in Turin. Assembly was to be carried out at Siemens' plants in the Czech Republic, with delivery scheduled for 2004–05.

Malaysia: In April 2002 the Express Rail Link system connecting Kuala Lumpur with its new international airport was commissioned. The line was built by a Siemens-led consortium and is served by 12 four-car Desiro articulated 25 kV AC emus. Eight of these are used for KLIA Express airport services, while the remaining four are assigned to high-density suburban services on the new line.

Netherlands: In March 2001 Siemens TS and Siemens Netherlands received a follow-on order from GVBA Amsterdam for 60 five-section uni-directional Combino trams. They were to join 95 similar vehicles ordered in March 2000; four of these were to be supplied as bi-directional vehicles. Delivery of the combined orders ran from 2001 to 2004.

Norway: In July 2003 Oslo's urban transport operator, AS Oslo Sporveier, placed an order with Siemens for 33 three-car metro trainsets. Delivery is to run from the end of 2005 to 2008.

Portugal: As part of a turnkey contract, in August 2002 Siemens won an order to supply 24 five-section Combino LRVs to Metro Transportes do Sul to equip a new light rail system linking Almada and Seixal, south of Lisbon.

In February 2000 a Siemens-led consortium that included Adtranz (now Bombardier Transportation) secured a contract to supply 22 four-car and 12 five-car emus for commuter services in the Oporto and Lisbon suburban areas respectively.

Siemens was to supply electrical and electronic equipment. Deliveries were due to be completed by 2004. The contract includes an option on 20 additional units.

Romania: In April 2002 Siemens was awarded a contract by CFR Calatori to supply 120 two-car Desiro dmus based on the DB Class LVT642. Of these, 57 were to be supplied complete from Germany, with the remaining 63 to be assembled in Romania. First examples entered service in 2003.

Russia: In April 2005 Siemens signed a contract for the development of high-speed trains for Russia. The delivery contract is scheduled for signing at the end of 2005. RZD intends to place orders for sixty of these trains following an agreement reached in December 2004.

Slovenia: In July 2004 Siemens was awarded a contract by Slovenian Railways to supply 20 Class 541 three-system EuroSprinter electric locomotives. Assembly is to take place at Siemens' Munich plant in 2006–08.

Spain: In March 2001 Siemens announced that it had won a contract to supply Spanish National Railways (RENFE) with 16 trainsets, based on the

ICE 3 operated by DB, for service on the Madrid–Barcelona high-speed line. Designated the Velaro E model by Siemens and AVE S 103 by RENFE, delivery commenced in 2004. In March 2004 RENFE placed a further order for 10 additional trains of this type.

Switzerland: In March 2003 Siemens received an order from Swiss Federal Railways (SBB) for 35 four-car double-deck 15 kV AC emus for Zurich suburban services. Delivery is scheduled for 2005–08.

Combino trams have been supplied to BVB, Basel (28 seven-section single-ended metre-gauge), and SVB, Berne (15 five-section single-ended metre-gauge).

Taiwan: In August 2001 Siemens announced that it had secured a contract from Kaohsiung Rapid Transit Corporation to equip two metro lines in the city, including the supply of 42 three-car trainsets. The system was due to be commissioned in 2007.

Thailand: As part of a turnkey contact awarded to Siemens in January 2002 to construct, equip and maintain the 20 km Bangkok Blue Line for Bangkok Metro Co Ltd, the company is supplying 19 three-car trainsets. The first vehicle was rolled out in Vienna in July 2003.

UK: In September 2003 Siemens received an order from the First Group/Keolis consortium, FGK, to supply 56 (subsequently reduced to 51) Desiro three-car dmus for the new TransPennine franchise in northern England. The trains are due to enter service in 2006.

In June 2003 Heathrow Express ordered four four-car Class 360/2 Desiro 25 kV AC emus for an expansion of services between London Paddington station and Heathrow Airport. This was subsequently revised to cover the supply of five five-car trains. Delivery of the first of these, initially to be supplied as four-car sets, commenced in 2004.

In 2002 deliveries commenced of an original order for 785 Desiro UK emu cars for South West Trains, with finance provided by Angel Trains. The trains were to be configured as 45 five-car sets for express services (Class 444), and 100 four-car and 32 five-car sets for suburban services (both Class 450). Deliveries were due to be completed by 2004. Subsequent changes to the contract led to the Class 450 order being amended to 110 four-car trains, with 24 further four-car trains built as Class 350/1 AC-only units for services on the UK's West Coast Main Line currently operated by Central Trains and Silverlink. The SWT contract included an option on 321 additional cars to be delivered by 2009.

In July 2000 Angel Trains, a UK-based rolling stock leasing company, ordered 21 Desiro UK Class 360 four-car 25 kV AC emus for use on First Great Eastern services in eastern England. All were in service in 2004.

USA: In June 2004 Siemens received an order from the North County Transit District (NCTD) for 12 Desiro Classic two-car dmus to serve the 35 km Oceanside-Escondido regional passenger service

Siemens Desiro DMU for DB AG   ***NEW***/0585299

in Southern California. This was Siemens' first multiple-unit order from the USA. Production of the vehicles is to take place in Germany for delivery in 2005–06.

In May 2002 Siemens received orders for LRV from operators in three US cities: three additional Avanto/S70 vehicles were ordered to complement an earlier order for 15 to equip a 12 km transit system being built by the company in Houston under a turnkey contract (see below); 11 Avanto/S70 LRVs were ordered for service in San Diego and 22 SD 460 units were delivered in 2004 to the St Louis transit network.

In November 2001 Massachusetts Bay Transportation Authority (MBTA) placed an order with Siemens for the supply of 94 metro vehicles for its Blue Line, deliveries commenced from the beginning of 2004. A locally based associate company, Transportation and Transit Associates Inc, was to take over final assembly of the vehicles under Siemens supervision. The contract also covers the refurbishment of existing vehicles.

In March 2001 Siemens announced that it had secured a turnkey contract to build and equip the first light rail line in Houston, Texas, for the Metropolitan Transit Authority of Harris County. Completed in December 2003, the 12 km line is served by 15 Type S 70 low-floor LRVs.

Venezuela: In July 2000 Siemens secured a turnkey contract to construct and equip the first 6.9 km phase of Line 1 of a light rail system for Maracaibo, including the supply of 12 LRVs.

Vietnam: In co-operation with Vossloh, in October 2004 Siemens was awarded a contract by Vietnam Railways to supply 16 Type AR15 1,500 kW diesel-electric locomotives. The 81 tonne three-phase metre-gauge Co-Co locomotives will be manufactured at Siemens' Munich plant for delivery in 2006.

UPDATED   Siemens suburban emu for Connex Melbourne   *NEW*/0585294

Škoda Transportation sro

Tylova 1/57, CZ-31600 Plzeň, Czech Republic
Tel: (+420 378) 13 50 00
Fax: (+420 378) 13 64 55
e-mail: transportation@skoda.cz
Web: www.skoda.cz/transportation

Key personnel
Chief Executive Officer: Tomas Krsek
Chief Operating Officer: Michal Korecky
Finance Director and Executive: Pavel Novotný
Manufacturing Director: Ales Jedlicka
Technical Director: Ladislav Krivanec
Purchasing Director: Pavel Vokoun

Products
Škoda Transportation sro has built over 5,500 electric locomotives, many of them for the railways of the former Soviet Union, Bulgaria and Poland as well as those of the Czech Republic and Slovakia. At present, Škoda Transportation sro concentrates on both building and refurbishment of electric locomotives, emus, metro trainsets, low-floor tramcars.

Škoda's refurbished Russian Type 81-71 for Prague metro has IGBT traction control (Milan Šrámek)   0116303

Contracts
In March 2004 a contract between Czech Railways and Škoda DT was signed for the purchase of 20 new 3-system electric locomotives of Type 109E. These locomotives are primarily intended for EC/IC passenger and fast freight trains for the Czech Republic, Slovakia, Germany, Austria, Poland and Hungary as well as the corridors of the European Union which are electrified at 3 kV dc, 25 kV/50 Hz and 15 kV/16.7 Hz.

In January 2004 Škoda signed a contract for the supply of 20 30 m five-sectional low-floor unidirectional tramcars Type 14 T for the Czech

Škoda's Type 10 T1 LRV for Sound Transit, Tacoma, Washington, US
(Milan Šrámek)
0567151

capital Prague. This included an option for a further 40 tramcars, with supplies scheduled to commence by 2005.

In 2004 Škoda won a contract to supply six 29.5 m five-section bi-directional low-floor trams Type 06T to Cagliari, Italy. Delivery is scheduled for 2006. In 2005, Škoda was awarded with a contract to supply eight 30 m five-sectional low-floor tramcars type 16 T for Wroclaw, Poland. Contracts for emus include the supply of electrical equipment and bogies for double-deck emus for Czech Railways (ČD). Delivery of these emus was scheduled from 1998 through to present. These are produced in association with ČKD Vagónka as (a member of Skoda Holding group) for which Škoda is supplying the electrical components and bogies.

At present, Prague metro cars are being reconstructed and refurbished. The work includes installation of modernised drive and communications systems. All Škoda refurbished metro trains are also equipped with new cabs and passenger-friendly interiors.

In late 2001 Škoda rolled out a prototype of a new five-car Type 81-55 metro trainset for Kiev, Ukraine. This combines mechanical parts from the Vagonmash factory in Russia with the latest traction and pneumatic equipment by Škoda. The asynchronous drive uses IGBT traction inverters. Škoda also signed a contract for the supply of five metro trainsets for Metro Kazan with an option for a further five. These trainsets will be built together with Vagonmash, which will supply mechanical parts, while Škoda will supply electrical and pneumatic equipment.

Developments

Škoda Transportation sro recently finished tests of the new five-section low-floor tramcar Type 05 T, its constructional features could be used as the basic platform for the complete family of new trams. Škoda has also finished tests of the new asynchronous metro car type 6 Mt with a stainless steel carbody and proven asynchronous traction drive.

UPDATED

Škoda's three-system electric locomotive for Czech Railways   1036645

Škoda's asynchronous-motored metro trainset
with stainless steel carbody
NEW/1036646

Stadler Rail Group

Stadler Rail AG
Bahnhofplatz, CH-9565 Bussnang, Switzerland
Tel: (+41 71) 626 21 20   Fax: (+41 71) 626 21 28
e-mail: stadler.rail@stadlerrail.ch

Stadler Bussnang AG
Industriestrasse 4, CH-9565 Bussnang, Switzerland
Tel: (+41 71) 626 20 20   Fax: (+41 71) 626 20 21
e-mail: stadler.bussnang@stadlerrail.ch

Stadler Altenrhein AG
Park Altenrhein für Industrie und Gewerbe, CH-9423 Altenrhein, Switzerland
Tel: (+41 71) 858 41 41   Fax: (+41 71) 858 41 42
e-mail: stadler.altenrhein@stadlerrail.ch

Stadler's low-floor streetcar (one-direction)
Helsinki (HKL), Finland
NEW/1146518

Stadler Pankow GmbH
Lessingstrasse 102, D-13158 Berlin, Germany
Tel: (+49 30) 91 91 1616   Fax: (+49 30) 91 91 21 50
e-mail: stadler.pankow@stadlerrail.de

Stadler Weiden GmbH
Zur Centralwerkstätte 11, D-92637 Weiden, Germany
Tel: (+49 961) 634 66-0   Fax: (+49 961) 63 46 69 95
e-mail: stadler.weiden@stadlerrail.de

Media Relations Stadler Rail AG
Silvia Bär
Tel: (+41 71) 626 20 34
e-mail: silvia.baer@stadlerrail.ch
Web: www.stadlerrail.com

Key personnel

President and Chief Executive Officer, Stadler Rail Group: Peter Spuhler
Chief Operating Officer, Stadler Rail Group: Michael Daum
Director Marketing and Sales, Stadler Rail Group: Peter A Jenelten
Managing Director, Stadler Bussnang: Peter Spuhler
Managing Director, Stadler Altenrhein: Hans Kubat
Managing Director, Stadler Pankow: Michael Daum
Managing Director, Stadler Weiden: Thomas Clasen

Background

The Stadler Rail Group includes, in addition to Stadler Bussnang AG, Stadler Altenrhein Ag in Switzerland, Stadler Pankow GmbH and Stadler Weiden GmbH in Germany. Over 1,200 co-workers are employed within the group. Stadler Rail Group's most well-known vehicle family, in addition to the GTW articulated railcar (380 trains sold), is the FLIRT (163 trains sold) together with the Regio Shuttle RS1 (354 trains sold).

In July 2004 Stadler handed over its repair, overhaul and refitting activities in Switzerland to Winpro.

The increase in production of articulated multiple units and rack railways, combined with the FLIRT multiple-unit for regional railways, has prompted Stadler to increase capacity at the Bussnang plant.

Products

Dmus, emus, trams and light rail vehicles, narrow-gauge rail cars, rack-rail cars and passenger coaches.

UPDATED

Stadler's low-floor mutliple unit FLIRT, Basel　　　*NEW*/1146517

Stadler's Type Regio Shuttle RS1 low-floor railcar, Erfurter Industrie Bahn (EIB), Germany
NEW/1146516

Stadler's articulated GTW dmu, Vinschgerbahn (Val Venosta), Italy
NEW/1146515

Stadler Type GTW 2/6 articulated electric railcar built for Thurbo AG (Switzerland)
1036626

Tafesa SA

Carretera de Andalucía Km 9, E-28021 Madrid, Spain
Tel: (+34 91) 798 05 50   Fax: (+34 91) 798 09 61
e-mail: tafesa@tafesa.com
Web: www.tafesa.com

Key personnel

Group Chairman: Paloma Fernández-Souza Faro
General Manager: J Eliecer Esteban Gutierrez
Technical Director: Luis Peromarta Calvo
Marketing and Commercial Manager:
  Jesús Montes Chinchón

Products

Manufacture and maintenance of passenger coaches, maintenance vehicles and powered passenger vehicles.

UPDATED

Talgo

Patentes Talgo SA
C/Gabriel Garía Marquéz 4, E-28230 Las Rozas, Madrid, Spain
Tel: (+34 91) 631 38 00
Fax: (+34 91) 631 38 93
e-mail: marketing@talgo.com
Web: www.talgo.com

Key personnel

Chairman: J L Oriol
Sales and Communications Director:
  Jesús Aranda Bayona
Financial Manager: J L Rhodes
Technical Manager: J L López Gómez
Production Manager: S Vallejo

Subsidiary companies

Talgo America, Inc
505 Fifth Avenue South – Suite 180, Seattle, Washington 98104, US
Tel: (+1 206) 748 61 40
Fax: (+1 206) 748 61 47
e-mail: info@talgoamerica.com
Web: www.talgoamerica.com

Talgo LRC, LLC
704 E Gallatin, Livinston, Montana 59047, US
Tel: (+1 406) 222 42 00
Fax: (+1 406) 222 76 79
e-mail: sales@talgolrc.com
Web: www.talgolrc.com

(Talgo) TTA, LLC
7940 Route 415, Bath, New York 14810, USA
Tel: (+1 607) 776 47 91
Fax: (+1 607) 776 74 96
e-mail: tta@ttallc.com
Web: www.ttallc.com

Talgo (Deutschland) GmbH
Revaler Strasse 99, D-10245 Berlin, Germany
Tel: (+49 30) 238 80 00
Fax: (+49 30) 23 88 00 11
e-mail: info@talgo.de
Web: www.talgo.de

Talgo Oy (see Talgo Oy entry in *Locomotives and powered/non-powered passenger vehicles* section)

Elektroniikkatie 2, FIN-90570 Oulu, Finland
Tel: (+358 8) 870 69 00
Fax: (+358 8) 870 69 70
e-mail: sales@talgo.fi
Web: www.talgo.fi

Background

Patentes Talgo acquired Transtech in 1999 to form Talgo-Transtech, now Talgo Oy. Transtech was the railway division of Finnish steel manufacturer Rautaruukki.

Products

Lightweight low-centre-of-gravity passenger coaches employing the Talgo system of suspension and wheel guidance, designed to permit higher curving speed without passenger discomfort or undue track wear (powered and non-powered). Also passenger coaches, electric locomotives, diesel-hydraulic and diesel-electric locomotives and LRVs.

Each Talgo low-floor vehicle (except trainset end vehicles) is carried on a single pair of half-axles with independent wheels. The design includes coaches with automatic variable-gauge axles and bogies for through running between Spain and France and coaches with pendular (passive tilt) suspension.

Automatic variable-gauge axles and bogies for freight wagons; modular pit lathes for wheels and disc brakes; automatic equipment for dynamic measurement of wheel characteristics.

Contracts include Talgo trains for RENFE; Washington State Department of Transportation, US; Amtrak, US; and German Railways (DB AG).

In March 2001 a consortium of Talgo and Adtranz was awarded a contract by RENFE, Spain, to supply 16 Talgo 350 high-speed electric trainsets (see 'Developments' section) for the line linking Madrid, Barcelona and the French border. The standard-gauge 25 kV AC 50 Hz trains are to be formed of two power cars and 12 trailers. Talgo is to undertake supply of the trailers and participate in developing the train's general concept.

Developments

Talgo has produced a prototype of its Talgo XXI diesel-powered tilting trainset. Designed for operation at speeds of up to 220 km/h, the train comprises two power cars and three trailers. Production versions will be formed of two power cars and up to 12 trailers. These will be produced in two versions – club class, with seating for 29 passengers, and coach, accommodating 40. A bistro car also features in the configuration options.

Power cars are produced by Krauss-Maffei, and incorporate a 1,500 kW MTU diesel engine and Voith hydraulic transmission. The train concept includes optional provision of the Talgo RD automatic gauge-changing system. This allows through operations using the standard-gauge domestic high-speed line and cross-border running into France. A 'BT' type designation for the power car is a reference to a unique wheel arrangement, which provides one two-axle traction bogie at the leading end and a single Talgo Pendular tilting axle at the rear, this shared with the first trailer car.

Newly developed is an improved sealed bodyshell design to reduce noise levels and pressure peaks when trains pass, and a repositioning of the air conditioning units from within the bodyshell to the underfloor area, reducing the centre of gravity and providing extra seating space. Axleloading is 17.5 tonnes. Other features include reclining and rotating seats, allowing their reversal so as to face direction of travel, a GPS-driven system for passenger announcements and power points for laptop computers.

The Talgo 350 high-speed electric trainset has been developed jointly by Talgo and Bombardier Transportation. This comprises two 25 kV AC 50 Hz power cars and up to twelve 13 metre trailers of Talgo's Series 7 'new generation' design, employing the company's established articulated, single-axle concept. Each 20 metre power car has a continuous rating of 4,000 kW, with all axles powered by asynchronous traction motors. The power cars are supplied by Adtranz, which is responsible for the train's traction system, control equipment and onboard electronics. Maximum design speed is 350 km/h, with a 17 t axle-load.

UPDATED

Talgo Oy

Elektroniikkatie 2, FIN-90570 Oulu, Finland
Tel: (+358 8) 870 69 00
Fax: (+358 8) 870 69 70
e-mail: sales@talgo.fi
Web: http://www.talgo.fi

Key personnel

Managing Director: Tapani Tapaninaho
Sales Director: Matti Haapakangas
Sales Manager: Matti Asikainen

Background

Formerly Transtech, the railway division of Finnish steel manufacturer Rautaruukki, the company was acquired by Patentes Talgo in 1999.

Products

Electric trainsets, emus, passenger coaches, and light rail vehicles, locomotives, locomotive bodyshells.

Impression of double-deck sleeping car by Talgo Oy for VR Group Ltd, Finland
1036799

Contracts

In 2003 Talgo Oy received an order from VR Group Ltd for 20 double-deck sleeping cars for delivery in 2005–06.

Developments

In 2002 Talgo launched its Talgo 22 concept, a double-deck emu design based on the parent company's proven technology of individually mounted and steered wheels with articulated body sections. The vehicle's design is based on the use of traction equipment by ELIN EBG Traction GmbH. In May 2004 the company announced that it was to build a test train to prove the Talgo 22 concept.

UPDATED

Talgo double-deck intercity coach for VR Finland
0089427

Computer graphic of Talgo 22 concept double-deck emu
1036800

Tokyu Car Corporation

Head Office
Overseas Project, Planning Department, Railway Division:
3-1 Ohkawa, Kanazawa-ku, Yokohama-city 236-0043, Japan
Tel: (+81 45) 785 30 09
Fax: (+81 45) 785 65 50

Sales Department
4-1-1 Taishido, Setagaya-ku, Tokyo 154-0004, Japan
Tel: (+81 3) 54 31 10 91
Fax: (+81 3) 54 31 10 59
Web: www.tokyu-car.co.jp

Key personnel
Chairman: Takahisa Tozawa
President: Takeo Momose
Executive Vice-President: Yasuo Ajima

Series E3-1000 mini-Shinkansen in service on the Yamagata line for East Japan Railways　0079819

Recent Tokyu electric multiple-unit

Class (Railway's own designation)	Cars per unit	Line voltage	Motor cars per unit	Motored axles per motor car	Rated output per motor (kW)	Max speed (km/h)	Weight (t) per car	Total seating capacity	Length per car (mm)	No in service	Rate of acceleration (km/h/s)	Year first built	Builders Mechanical parts	Electrical parts
Yokoha Minato-Mirai emu Y 500	8	DC 1,500 V	4	4	190	120	26.0 (cab car) 32.0–33.0 (motorcar) 24.5	48 (total 141) (cab car) 51 or 54 (total 152) (middle car)	20,200 (cab car) 20,000 (middle car)	48	3.3	2003	Tokyu Car	Hitachi
Shikoku Railway 5000 emu	1	DC 1,500 V	1		Trailer	130	41	First class: 36, ordinary class: 36	20,860	6		2003	Tokyu Car	
E23 emu (suburban use)	10 or 15	DC 1,500 V	4 or 6		Trailer (first class)	120	34.5 to 35.4	First class: 90	20,500	Before debut		2004	Tokyu Car	
E231 emu (suburban use)	10 or 15	DC 1,500 V	4 or 6	4	95	120	26.8 to 27.9 (cab car), 28.8 to 29.9 (motor car) 23.0 to 23.5 (trailer car)	36 to 43 (total of 138 to 143) (cab car), 54 to 60 (total 162) (middle car)	20,000		2.5	2000	Tokyu Car, Kawasaki Heavy Industries, East Japan Railways	Hitachi, Mitsubishi Electric, Toshiba, Toyo Denki
8520 (Irish Rail) emu	4	DC 1,500 V	2	4	140	110	34.2, 34.5 (cab car) 39.7 (motorcar)	40	20,130 (cab car) 20,000 (motorcar)	40	3.3	2004	Tokyu Car	Toshiba

Y 500 Series emu for Yokohama Minato-Mirai Railway   1036650

Double-deck, first class E231 Series emu for East Japan Railways   1036649

Works

Yokohama Plant (Railway rolling stock and heavy duty trailer)
3-1 Ohkawa, Kanazawa-ward, Yokohama-city 236-0043, Japan

Wakayama Plant (Railway track turnout, container and logistics equipment)
770-8 Oaza-Kitaseida, Uchita-town, Naka-county, Wakayama-prefecture 649-6402, Japan

Products

Electric and diesel railcars, passenger coach and light rail vehicles. Recent products include:

Yokohama Minato-Mirai Y 500 series for subway and commuter use
Yokohama Minato-Mirai line is a new subway approximately 4.1 km from Yokohama to Motomachi-Chuukagai (China-town) via Minato-Mirai 21 area of Yokohama harbor new town. The new line's train has run to Shibuya terminal on Tokyu's Touyoko line since 1 February 2004. Tokyu will finally manufacture 150 vehicles of it's emu 5000 Series and Y 500 Series. The Y500 Series is based on Tokyu Corporation's emu 5000 commuter Series resulting in lower costs due to the common design of car body construction, interior and electrical equipment.

East Japan Railway's E 231 Series for commuter and suburban use
In 2000, Tokyu Car and East Japan Railway developed JR East's E231 series as a standard suburban and commuter vehicle and has since manufactured 450 vehicles. In 2004 Tokyu Car manufactured a double-deck first class vehicle for the E231 which will be used in the metropolitan area.

Double-deck first and ordinary Class 5000 Series emu for 'Marine Line'
Tokyu Car has built up to six double-deck vehicles for the rapid service train 'Marine Liner' which was released in October 2003. These vehicles are based on JR East's double-deck first class E217 Series. The train conventionally runs for 58 minutes per 71.7 km between Okayama and Takamatsu with maximum velocity of 130 km/h.

Iarnród Éireann (Irish Rail) Class 8520 Series emu
Iarnród Éireann (IÉ) Class 8520 Series emu is based on the Class 8510 emu which was delivered to IÉ in 2001. An air conditioning system has been added to the vehicle. The new Series (Class 8520) is built with maximum priority on vehicle safety as well as accessibility for handicapped passengers. The car body has been constructed to comply with British fire safety standard BS6853 cat. Ib. and has improved crashworthiness. Closed Circuit Television (CCTV) has been installed and there is arrangement for wheelchair space.

Contracts

East Japan Railways and other private railway firms have ordered various vehicles for super express and commuter trains. Typical orders include:
JR East's E257-500 Series emu for Limited Express, 20 vehicles; JR East's E231 Series emu for suburban use double-deck trailer for first class, 100 vehicles; JWR East's E231 Series emu for

Class 8520 emu for Irish Rail   1036648

Series E257 trainset for JR East   0122823

Double-deck, first and ordinary class 5000 Series emu for Marine Liner, Shikoku Railway   1036647

suburban use, 140; JR East's E531 Series emu for suburban use, 50 vehicles; Keihin Electric Express Railway's 1000 Series emu for commuter use, 12 vehicles; Keio Electric Railway's 9000 Series

emu for commuter use, 8 vehicles; Nankai Electric Railway's 2300 Series emu for commuter use, 6 vehicles; Odakyu Electric Railway's 3000 Series emu for commuter use, 44 vehicles; Bureau of

Transportation, Tokyo Metropolitan Government 10–300 Series emu for subway use, 108 vehicles.

UPDATED

Toshiba Corporation

Railway Projects Department
Toshiba Building, 1-1 Shibaura 1-chome, Minato-ku, Tokyo 105-8001, Japan
Tel: (+81 3) 34 57 49 24
Fax: (+81 3) 54 44 92 63
Web: www.toshiba.co.jp

European office
Toshiba International (Europe) Ltd
71-79 Station Road, West Drayton, Middlesex UB7 7LT, UK
Tel: (+44 1895) 42 74 00   Fax: (+44 1895) 44 94 93

Key personnel
President and Chief Executive Officer:
  Atsutoshi Nishida

Vice-President, Transportation Systems Division:
  Takio Ooyama
Senior Manager, Railway Projects Dept: Koji Toda

Products
Electric, diesel-electric and diesel-hydraulic locomotives; electric traction equipment; auxiliary power supply systems; air conditioning systems.

UPDATED

Transwerk

Transwerk, Business Development Section
Transwerk Park, Lynette Street, Kilner Park, Pretoria 0186, South Africa
Tel: (+27 12) 842 50 30   Fax: (+27 12) 842 59 78

Products
Specialist in refurbishment and other services for developing countries including CKD kits for locomotive refurbishment.

Contracts include the continuing refurbishment of 3,700 locomotives, 3,000 coaches and 4,600 commuter cars and 150,000 freight wagons in South Africa.

Tülomsas

The Locomotive and Motor Corporation of Turkey
Ahmet Kanatli Caddesi, TR-26490 Eskisehir, Turkey
Tel: (+90 222) 225 99 56
Fax: (+90 222) 225 72 72; 57 57
e-mail: tulomsas@tulomsas.com.tr
Web: http://www.tulomsas.com.tr

Key personnel
Managing Director: D Zeki Daloglu

Assistant General Managers: Galip Pala,
  Cengiz Özan, Fatih Turan, Haluk Akova
Head of Marketing: Erol Çetin

Products
Diesel-hydraulic and diesel-electric shunting and main line locomotives (545 to 1,790 kW); electric locomotives (3,200 kW); special purpose freight wagons including tank wagons, refrigerated hopper wagons and wagons for ore transport, tracked tanks and containers; diesel engines with

outputs of 1,865, 2,200 and 2,400 hp; traction motors, motor generator sets, control equipment for electric locomotives, maintenance and repair work; bogies for locomotives and freight wagons; self-propelled rail vehicles for infrastructure maintenance including car equipped with hydraulic crane, catenary maintenance and inspection car.

Tüvasaş

Türkiye Vagon Sanayii AS
Turkish Wagon Industry Corporation
Milli Egemenlik Caddesi 123, Adapazari, Turkey
Tel: (+90 264) 275 16 60
Fax: (+90 264) 275 16 79
e-mail: info@tuvasas.com.tr
Web: http://www.tuvasas.com.tr

Background
An affiliate company of Turkish State Railways (TCDD), Tüvasas was originally a passenger coach repair facility. It commenced vehicle manufacturing in 1975.

Products
Passenger coaches, dmus, emus and LRVs. Coaching stock production centres on the TVS

2000 family of vehicles, which has been produced in open, compartment, restaurant car, sleeping car and couchette versions for TCDD. The company has also produced the Sakarya single-unit railcar, which is powered by a Cummins 410 kW engine with Voith transmission, participated in the manufacture of LRVs for Bursa (in co-operation with Siemens TS) and has developed concepts for a 160 km/h three-car dmu.

Unilokomotive Ltd

Oranmore, Co Galway, Ireland
Tel: (+353 91) 79 08 90   Fax (+353 91) 79 08 46
e-mail: talbracht@unilok.ie
Web: http://www.unilok.ie

Key personnel
Chairman: Owen McConn
Sales Director: Tom Albracht

Products
The Unilok range of road/rail shunting locomotives and special-purpose shunting and inspection vehicles. Uniloks are in service worldwide and are available in most rail gauges and coupler types. A wide range of diesel engines may be fitted according to customer preference. Engines may be air- or water-cooled from 60 to 110 kW. Uniloks are designed and built to the customer's specifications and can be supplied to haul loads up

to 1,500 tonnes. Maximum speed on road and rail is 30 km/h. A wide range of optional equipment can be fitted, including snow-plough, road sweeper, hydraulic crane, front-end loader and radio remote control.

Union Carriage & Wagon Co (Pty) Ltd

PO Box 335, Nigel 1490, South Africa
Tel: (+27 11) 814 44 11   Fax: (+27 11) 814 51 56
e-mail: union@ucw.co.za
Web: http://www.ucw.co.za

Works
Marievale Road, Vorsterkroon, Nigel 1490

Key personnel
Chief Executive: L Taljaard
Financial Manager: G Blewitt
Contracts Manager: D Ward
Engineering Manager: P Watts
Manufacturing and Media: E Wills

Union Carriage push-pull Bo-Bo 25 kV electric locomotive for the Taiwan Railway Administration
0063373

Products

Electric, diesel-electric, diesel-hydraulic and diesel-mechanical locomotives from 50 to 180 t; railcars; electric and diesel-electric multiple-units; main line passenger coaching stock; specialised freight vehicles.

Contracts

64 push-pull locomotives for Taiwan Railway Administration; 66 emu cars for KTM Malaysia's Kuala Lumpur suburban network; 56 passenger coaches for National Railways of Zimbabwe.

Union Carriage three-car emu for Malaysia
0063372

Union Carriage main line economy-class coach for NRZ
0063694

United Goninan

PO Box 33000, Hamilton, New South Wales 2303, Australia
Broadmeadow Road, Broadmeadow, New South Wales 2292, Australia
Tel: (+61 2) 49 23 50 00   Fax: (+61 2) 49 23 50 01
Web: www.unitedgoninan.com.au

Key personnel

Chief Operating Officer: John McLuckie

Background

A member of the United Group, the company changed its name from A Goninan & Co Ltd in February 2001.

Products

Diesel-electric locomotives. United Goninan is a licensee of GE Transportation Systems, USA. Diesel locomotive refurbishment.

Powered single- and double-deck vehicles, light rail vehicles.

Passenger coaches and passenger coach refurbishment.

Contracts

Contracts include: two two-car and one three-car 200 km/h Prospector dmus for the Western Australian Government Railways Commission; 12 GE Dash 9 diesel-electric locomotives to Pilbara Rail Co, in conjunction with GE Transportation Systems; 14 dmus and 41 double-deck emus for RailCorp, 9 AC diesel-electric locomotives for QR.

UPDATED

Prospector dmu designed by United Goninan for Western Australian Government Railways   1037002

Ventra Locomotives Ltd

(A Division of the Dukes Retreat Ltd)
Works and Administration Office
Plot No 23-B, IDA, Patancheru, 502319, Medak District Andhra Pradesh, India
Tel: (+91 8455) 24 06 70; 06 71; 05 63; 06 22
Fax: (+91 8455) 24 11 28
e-mail: hyd2_ventra@sancharnet.in
Web: www.seekandsource.com/ventra

Key personnel

Deputy General Manager: S Henry Samuel

Other offices

Bangalore, Chennai, Calcutta, Nagpur, Panta, Lucknow, Mumbai, New Delhi.

Products

Diesel-hydraulic shunting locomotives; battery locomotives for mining, tunnelling and construction applications; various self-propelled rail vehicles such as rail cars, track maintenance vehicles, transfer cars and other rail equipment; mine cars. Diesel-hydraulic locomotives cover special types such as flame-proof and also remote-controlled locomotives. The powered vehicles feature in the range of 4 to 1,120 kW (5 to 1,500 hp) of any rail gauge and up to 140 t weight which can be adopted to the needs of any industry.

UPDATED

Villares

Equipamentos Villares SA
Rua Alexander Levi 202, 01520 São Paulo, Brazil
Tel: (+55 11) 279 33 09   Fax: (+55 11) 270 05 11

Key personnel

Managing Director: M Silveira
Manufacturing Director: J Cassio Daltrini

Products

Multiple-unit trainsets; diesel locomotives under EMD licence; electric locomotives under ALSTOM licence; diesel shunting locomotives.

Vossloh Locomotives GmbH

PO Box 9293, D-24152 Kiel
Falckensteiner Strasse 2, D-24159 Kiel, Germany
Tel: (+49 431) 39 99 21 95
Fax: (+49 431) 39 99 22 74
e-mail: vertrieb.kiel@vl.vossloh.com
Web: www.vossloh-locomotives.com

Works

Service-Zentrum Moers, Baerler Strasse 100,
D-47441 Moers, Germany
Tel: (+49 2841) 14 04 10
Fax: (+49 2841) 14 04 50

Key personnel

Board of Directors: Andreas Hopmann
Chief Executive Officer: Dr Georg Hauschild

Subsidiary companies

Vossloh Locomotives has a 100 per cent
shareholding in Locomotion Service GmbH to
maintain locomotives owned by Angel Train Cargo
Ltd and others.

Background

The Kiel works started production in 1920 of the
benzene-driven railcar and in 1925 the first diesel
locomotives. In 1947 the Deutsche Bundesbahn
DB placed its first order for Diesel locomotives
with 31 units. In 1971 the Kiel works celebrated
the delivery of 1,000 units for DB. In 1998 Siemens
Schienenfahrzeugtechnik GmbH, formerly Krupp
MaK Maschinenbau, was acquired by Vossloh AG
and renamed Vossloh Schienenfahrzeugtechnik
GmbH. MaK Maschinenbau was acquired by Vossloh
AG and renamed Vossloh Schienenfahrzeugtechnik
GmbH (VSFT). Since then it has become a member
of the Vossloh Group.

In April 2003, the entire Vossloh Group adopted
a uniform corporate design and VSFT became
Vossloh Locomotives GmbH. Annual production is
80 to 130 locomotives.

Products

Diesel-hydraulic and diesel-electric locomotives
ranging from 390 to 2,700 kW for all duties, gauges,
speeds for both main line and industrial service.
Rail vehicle components including bogies and
cooling systems. Maintenance, rehabilitation and
repowering of locomotives (in Moers).

Vossloh Locomotives has delivered its vehicles
in recent years to state and private railways as well
as to leasing companies like Angel Trains Cargo
and Mitsui, within Europe: G 1206 (1,570 kW) to
BV (Sweden), MK 600 (390 kW) to DSB (Denmark),
HLD 77 (1,150 kW) to SNCB (Belgium), Rh 2070
(738 kW) to ÖBB (Austria), Am 843 (1,500 kW) to
SBB (Switzerland), G 1206 (1,500 kW) to SNCF
(France). G 400 B (390 kW), G 1206 and G 2000
BB (2,240 kW) locomotives are already on duty
on private railways in Austria, Germany, Italy,
Luxembourg and Switzerland. Full certification is
in preparation for Austria, Belgium, Italy and the
Netherlands.

In 2002 Vossloh Locomotives completed its range
of diesel-hydraulic locomotives with the G 1000 B-B
(1,100 kW) and the G 1700 B-B (1,700 kW) models.

Developments

The new, main-line MaK 2000-4 BB, with a diesel
engine output upgraded to 2,700 kW. This makes
this one of the world's most powerful, four-axle,
single-engine, diesel-powered locomotives.

It can be delivered with a soot filter like all
locomotives of the Class AM843 (79 units) for
Switzerland.

UPDATED

*Vossloh Locomotive Type G 2000-4 BB
2,700 kW diesel-hydraulic locomotive*
(Ken Harris)
***NEW**/0585012*

*Vossloh Locomotives G2000 BB 2nd and 3rd Series for Austria, Belgium, France, Germany, Italy,
Netherlands and Poland, MaK 2000BB*
1036758

Vossloh Locomotives G 1000BB for France, Germany and Italy
1036756

Vossloh Locomotives Class AM843 (Type G 1700 BB) for SBB Cargo, BLS and SERSA Switzerland
1036757

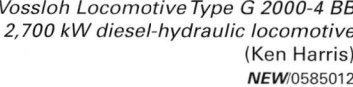

Wabtec Rail Ltd

PO Box 400, Doncaster Works, Hexthorpe Road, Doncaster DN1 1SL, UK
Tel: (+44 1302) 34 07 00   Fax: (+44 1302) 32 13 49
Web: www.wabtec.com

Key personnel
Managing Director: John Meehan
Engineering Director: Mike Roe
Finance Director: Robert Johnson
Operations Director: Chris Weatherall
Commercial Manager: Paul Robinson

Background
Wabtec Rail undertakes the overhaul of passenger rolling stock, main line locomotives, bogies, wheelsets, air brake equipment, hydraulic dampers and buffers design in addition to the manufacture of shunting locomotives and freight wagons.

As part of the Wabtec Corporation, Wabtec Rail also supplies to the UK rolling stock owners and maintainers, composite brake blocks and pads, Wabtec Railway Electronics train data recorders and electronic equipment, Cardwell TMX braking systems and Wabtec air brake equipment.

Products
Wabtec Rail undertakes the heavy maintenance, refurbishment and conversion of passenger rolling stock, diesel and electric multiple units. This includes the overhaul of the rolling stock bogies, couplers, gangways, braking equipment and control systems. Passenger rolling stock work undertaken includes interior refurbishments, conversions and rolling stock modernisation. Wabtec Rail also undertakes the repainting and re-livery of rolling stock. Modern painting booths are able to accommodate 23 m vehicles and apply all types of surface coatings, including water based systems.

Wabtec also performs heavy overhaul, maintenance and refurbishment of main line locomotives and shunters, also designing and manufacturing industrial shunting locomotives. The company owns the design rights to the Sentinel, Valiant, Vanguard and Steelman ranges of locomotives and is able to offer the rebuilding and modernisation of these locomotives.

Contracts
Recent contracts include the conversion of the track recording coach and the lab 5 coach, including complete train assembly and testing for Network Rail's Network Measurement Train (NMT).

Current contracts include C4 and C6 heavy maintenance contracts for Angel Trains and Portbrook Leasing, the installation of Wabtec Railway Electronics data recorders for Angel Trains and vehicle repainting for Arriva Trains and Bombardier Transportation. Main line locomotive works have included the refurbishment and modernisation of Class 20 locomotives for Direct Rail Services (DRS), power unit bearing renewal on Class 60 locomotives for English Welsh and Scottish Railway (EWS), Class 66 locomotive cab sound proofing modification for EWS and Freightliner and the installation of TPWS systems on Class 92 locomotives for EWS.

UPDATED

Wagon Pars

Wagon Pars Company
10 Azarshar Street, South Kheradmand Avenue, Tehran, 15846, Iran
Tel: (+98 21) 884 83 30/83 39
Fax: (+98 21) 884 83 38

Works
Km 4, Tehran Rd Arak
Tel: (+98 861) 330 46 50   Fax: (+98 861) 339 99

Key personnel
Marketing Executive: Reza Esfahlani

Products
Diesel locomotives, including the ME10 diesel-electric locomotive for shunting and other work for speeds up to 100 km/h. Eighty locomotives are being built under a technology transfer deal with the former GEC Alsthom (now ALSTOM).

Also builder of coaches and wagons (see *Freight vehicles and equipment*).

Winpro AG

Postfach, CH-8401 Winterthur, Switzerland
Tel: (+41 52) 262 11 77
Fax: (+41 52) 262 03 03
e-mail: info@winpro-ag.ch
Web: http://www.winpro-ag.ch

Background
Winpro is the successor to the long-established former SLM company.

Products
Special rail vehicles including maintenance equipment. Recent output includes the Type Tm 234 diesel-powered two-axle maintenance vehicle for Swiss Federal Railways. The company also manufactures bogies and undertakes vehicle overhaul, repair and retrofit activities.

Zeco

ZECO (1996)
PO Box 1874, Bulawayo, Zimbabwe
Tel: (+263 9) 789 31   Fax: (+263 9) 722 59; 789 31
e-mail: ZECO@acacia.samara.co.zw

Works
38 Josiah Chinamano Road, Belmont, Bulawayo, Zimbabwe

Key personnel
Managing Director: M A Da Silva
Technical Director: K A Meth
Project Engineer: R McGann
Division Managers
  Rolling Stock: P Birnie
  Structural: M Sciscio
  Erection: S C Braddock

Products
Construction of diesel-electric, diesel-hydraulic and electric locomotives and rolling stock; overhaul and rebuilding of steam, diesel-electric, diesel-hydraulic and electric locomotives and rolling stock.

Zephir SpA

85 Via Salvador Allende, I-41100 Modena, Italy
Tel: (+39 059) 25 25 54   Fax: (+39 059) 25 37 59
e-mail: zephir@zephirspa.com
Web: http://www.zephirspa.com

Key personnel
Managing and Sales Director: Vittorio Cereghini
Purchasing Manager: Fabrizio Della Rovere
Production Manager: Daniele Bergamini

Products
Rail/road shunting tractors. The Zephir range includes the IL Locotrattore for wagon handling which is able to travel on roads as well as on rails by means of four pneumatic tyres combined with four 400 mm railway-type flanged steel wheels. The latter are fitted to two hydraulic axles that are lowered by the operator once the machine has been driven on and aligned to the rails.

The tractor has a strong steel chassis and body structure, and is equipped with buffers at either end to allow the machine to operate in both directions.

The Lok series comprises 12 models with drawbar pulls from 2,500 to 23,000 kg and towing capacities up to 4,600 t. Optional equipment offered includes automatic coupling/uncoupling controlled from the driving position; airbraking system for wagons; remote control for rail operations; anti-explosion systems; and one or two cabs.

VERIFIED

Zhuzhou Electric Locomotive Works

Tianxin, Zhuzhou, Hunan, China 412001
Tel: (+86 733) 844 48 48; 13 51
Fax: (+86 733) 843 10 03; 23 99
e-mail: csrft@csrgc.com.cn

Key personnel
Director: Zhao Xiaogang
Executive President: Li Zhixuan

Corporate background
Zhuzhou Electric Locomotive Works (ZELW) was founded in 1936. Zhuzhou is a member of CSR (see CSR entry in *Locomotives and powered/non-powered passenger vehicles* section).

Zhuzhou owns a 49 per cent shareholding in a joint venture company established with Siemens TS, Siemens Traction Equipment Ltd, to develop electric traction technology.

Products
Four-, six- and eight-axle electric locomotives for passenger and freight operation, with an 80 per cent share of the Chinese electric locomotive market.

In recent years, ZELW has become a major supplier to the Chinese high-speed rail and heavy haulage field. It has supplied 158 Type SS8 locomotives which has increased the average passenger transportation speed on Chinese Railways by 50 per cent.

The locomotive reached a record speed of 240 km/h during a test run. The latest generation of DDJ emus is designed for high-speed operation with a top speed of 200 km/h.

ŽOS Vrútky as

Dielenska Kružná, SK-038 61 Vrútky, Slovakia
Tel: (+421 842) 420 51 01
Fax: (+421 842) 428 15 95
e-mail: zos-vrutky@zos-vrutky.sk
Web: http://www.zos-vrutky.sk

Key personnel
Director for Business and Production:
  Ladislav Bonda

Background
Formerly the Vrútky workshops of the Czechoslovak and subsequently Slovakian Republic Railways (ŽSR), ŽOS Vrútky became a joint stock company in 1994. ŽSR retains a shareholding of 34 per cent; the remainder is held by private shareholders.

Products
Emus and dmus for 1,000 and 1,435 mm gauge. In a consortium with Stadler Fahrzeuge of Switzerland and Daimler Chrysler Rail Systems (now Bombardier Transportation), the company produced 14 1.5 kV DC Class 425.95 metre-gauge emus for ŽSR's High Tatras network. The vehicles are based on Stadler's GTW 2/6 design. The company is currently modernising Class 350 electric locomotives which are able to operate on international corridors in both 3 kV DC and 25 kV 50 Hz AC, at speeds of up to 160 km/h.

Zwiehoff GmbH

Tegernseestrasse 15, D-83022 Rosenheim, Germany
Tel: (+49 8031) 21 96 01   Fax: (+49 8031) 21 96 03
e-mail: info@zwiehoff.com
Web: www.zwiehoff.com

Key personnel
Managing Director: Gerd Zwiehoff
Export Manager: Elisabeth Schoemer

Products
Road/rail vehicles with a shunting capacity of up to 3,000 tonnes or equipped as multipurpose vehicles; self-propelled Mini Shunter with a shunting capacity of up to 150 tonnes; self-propelled Maxi Shunter with a shunting capacity of up to 200 tonnes; forklift truck-propelled wagon shunter with a shunting capacity of up to 300 tonnes.

UPDATED

For details of the latest updates to *Jane's World Railways* online and to discover the additional information available exclusively to online subscribers please visit
jwr.janes.com

DIESEL ENGINES, TRANSMISSION AND FUELLING SYSTEMS

Alphabetical listing

Alstom Power
Caterpillar Inc
CFD Industrie
China Northern Locomotive and Rolling Stock Industry (Group) Corporation
China South Locomotive and Rolling Stock Industry (Group) Corporation (CSR)
Cockerill Mechanical Industries SA (CMI)
Cummins Engine Company Ltd, Rail Marketing
Dalian Locomotive & Rolling Stock Works
David Brown Engineering Ltd
Detroit Diesel
Electric-Motive Diesel (EMD)
Eminox Ltd
Freudenberg Schwab GmbH
Ganz-David Brown Getriebetechnik GmbH
General Electric Transportation Systems
Giro Engineering Ltd
Gmeinder Getriebe und Maschinenfabrik GmbH

Haynes Corporation
Hitachi Ltd
Honeywell Serck
Hygate Transmissions Ltd
Isotta Fraschini Motori SpA
Iveco SpA
Kim Hotstart
Kolomna
Komatsu Diesel Co Ltd
MAN Nutzfahrzeuge Aktiengesellschaft
MAN B&W Diesel Ltd
MTU Friedrichshafen GmbH
NICO
Niigata Power Systems Co Ltd
OMT SpA
Paulstra
PSI (Peaker Services Inc)
Regulateurs Europa
SAN Engineering & Locomotive Co Ltd

SEMT Pielstick
SILSAN
Socofer
Spicer Gelenkwellenbau GmbH & Co KG (GWB)
Tülomsas
Turbomeca
Twiflex Ltd
Twin Disc Incorporated
UCM Reşiţa – Machine Building Company
Unipar Inc
Voith Safeset AB
Voith Turbo GmbH & Co KG
Volvo Penta
ZF Friedrichshafen AG
ZF Padova

Company listing by country

BELGIUM
Cockerill Mechanical Industries SA (CMI)

CHINA
China Northern Locomotive and Rolling Stock Industry (Group) Corporation
China South Locomotive and Rolling STock Industry (Group) Corporation (CSR)
Dalian Locomotive & Rolling Stock Works

FRANCE
CFD Industrie
Paulstra
SEMT Pielstick
Socofer
Turbomeca

GERMANY
Freudenberg Schwab GmbH
Gmeinder Getriebe und Maschinenfabrik GmbH
MAN Nutzfahrzeuge Aktiengesellschaft
MTU Friedrichshafen GmbH
Spicer Gelenkwellenbau GmbH & Co KG (GWB)
Voith Turbo GmbH & Co KG
Volvo Penta
ZF Friedrichshafen AG

HUNGARY
Ganz-David Brown Getriebetechnik GmbH

INDIA
SAN Engineering & Locomotive Co Ltd

ITALY
Isotta Fraschini Motori SpA
Iveco SpA
OMT SpA
ZF Padova

JAPAN
Hitachi Ltd
Komatsu Diesel Co Ltd
NICO
Niigata Power Systems Co Ltd

ROMANIA
UCM Reşiţa – Machine Building Company

RUSSIAN FEDERATION
Kolomna

SWEDEN
Voith Safeset AB

TURKEY
SILSAN
Tülomsas

UNITED KINGDOM
Alstom Power
Cummins Engine Company Ltd, Rail Marketing
David Brown Engineering Ltd
Eminox Ltd
Giro Engineering Ltd
Honeywell Serck
Hygate Transmissions Ltd
MAN B&W Diesel Ltd
Regulateurs Europa
Twiflex Ltd

UNITED STATES
Caterpillar Inc
Detroit Diesel
Electric-Motive Diesel (EMD)
General Electric Transportation Systems
Haynes Corporation
Kim Hotstart
PSI (Peaker Services Inc)
Twin Disc Incorporated
Unipar Inc

Alstom Power

Napier Turbochargers
PO Box 1, Lincoln LN2 5DJ, UK
Tel: (+44 1522) 58 47 97
Fax: (+44 1522) 58 49 02
e-mail: napier.sales@power.alstom.com

Key personnel
Director: David Green

Products
Napier turbochargers for new-build and retrofit rail applications. Current products include the NA256 model developed as a retrofit for the SA08 4DP turbochargers fitted to the Paxman Valenta engine. The company also offers parts supply, service support, overhauls, retrofits and training services.

Caterpillar Inc

100 NE Adams Street, Peoria, Illinois 61629, USA
Tel: (+1 309) 578 37 81   Fax: (+1 309) 578 73 29
e-mail: cat_power@cat.com
Web: http://www.cat-engines.com

Works
Engine Products Division
PO Box 610, Mossville, Illinois 61552-0610, USA

Key personnel
Vice-President, Engine Products Division:
  James J Parker
Manager, Worldwide Locomotive Business:
  Charles L Wills

Principal subsidiary companies
Perkins Engines Company Ltd
MaK Motoren GmbH
F G Wilson (Engineering) Ltd
Solar Turbines Inc

Products
Diesel engines 3.7 to 15,000 kW (5 to 20,100 hp), for traction, electric power generation and maintenance-of-way equipment.

Caterpillar is one of the world's leading manufacturers of diesel and spark-ignited engines for a wide range of applications. With Perkins Engines and CM (Caterpillar MaK) engines, Caterpillar manufactures a range of engines from 3.7 to 15,000 kW at plants in USA, Europe, India and Japan. Currently, parts are distributed through 26 Caterpillar distribution centres in 10 countries and a worldwide network of 207 independent dealers in 1,800 locations.

Caterpillar has been a supplier of diesel engine power to the rail industry since the mid-1930s. The company's 3400, 3500 and 3600 Families of engines are well-suited in size, power and performance for locomotive traction engines. The 3400 Family powered generator sets provide Head End Power (HEP) in passenger locomotives and the full range of Caterpillar's small and medium-sized engines can be found in maintenance-of-way equipment. Caterpillar also offers several engine models for use in underground, mining, tunnelling and other environmentally sensitive applications. Caterpillar's powertrain core competency is exhibited by a broad offering of mechanical transmissions for the rail industry. The company can custom fit engines, transmissions, torque converters, controls and all required attachments to suit customer applications.

Advanced engineering features are offered by Caterpillar to meet emission standards worldwide. These include electronic controls, electronic unit injectors and other technologies that dramatically reduce engine emissions. Caterpillar reports that exhaust emission levels are well within the emission regulations set by the United States Environmental Protection Agency for non-road and locomotive applications. Caterpillar diesel locomotive engines meet UIC 623 or ERRI 2003 emission limits.

The 3600 Family of engines is available in speeds of up to 1,000 rpm for main line locomotives. Model 3606 and 3608 engines are being used to repower Class 232-type locomotives operated by several railways including German Railways (DB). These repowering projects utilise existing traction generators.

The 3500 Family of engines is used to power diesel-electric and diesel-hydraulic shunting and main line locomotives. The largest users are in Zimbabwe, Austria, Germany, Canada and the USA. In 2000, the range of power ratings for the fully electronic 3500B was further expanded. These engines continue to offer reduced fuel consumption, reduced emissions and system diagnostics. Operators of locomotives with 3500B engines include German Railways (DB), FEVE (Spain), Romanian National Railways (SNCFR), and Chinese Railways (MOR). Additional units are on order for several European customers.

High-speed versions of the 3500 engine, packaged with a two-bearing traction alternator and dual auxiliary alternator on an integral welded base, provide an alternative to completely replacing old and inefficient equipment. The unique four-point, rubber isolated mounting system can be installed in many existing carbodies. Caterpillar has started shipping packages using the 3500B unit to EMD (General Motors) for a new range of shunting locomotives. Vossloh Locomotives GmbH (VSFT) completed the sale of the first of 20 MaK G2000 locomotives with Caterpillar 3516B engines rated at 2,240 kW (3,000 hp).

Low-speed 3500 Series engines offer a practical way to repower old locomotives that were previously powered by Caterpillar 6.25 in bore engines. These engines continue to be installed in locomotives in South America and South Africa.

The 3400 Family engines are used to power diesel-electric and diesel-hydraulic shunting locomotives. They are available with mechanical or electronic fuel systems and ratings up to 783 kW at 2,100 rpm. Currently, these engines are being sold into traction, electric power generation and maintenance-of-way applications. Recent contracts include 60 3412E engines rated at 746 kW (1,000 hp) for new locomotives for Austrian Federal Railways; 30 additional locomotives are on order. Cat 3400 generator set engines provide 50 and 60 Hz Head-End Power (HEP) in passenger locomotives and power wagons. The 3412 engine is a standard offering from EMD (GM) and MotivePower in their passenger locomotives for North America.

General specifications
Cylinders: Removable wet-type cylinder liners of hardened cast iron. Alloy cast-iron cylinder heads with water directors and removable injectors. The 3500 and 3600 Family utilises individual cylinder heads for serviceability.
Fuel injection: Caterpillar-designed unit injectors and pump-and-line systems on all direct injection models. Standard arrangements utilise replaceable full-flow fuel filters.
Aspiration: All models utilise turbocharging and aftercooling (intercoolers).
Fuel: No 2 diesel fuel oil (ASTM specification D396).
Lubrication: Full pressure systems including gear-type pumps, replaceable full-flow filters and water-cooled oil coolers.
Cooling systems: Centrifugal-type circulating pumps with thermostatically controlled bypass. All systems are capable of pressurisation and designed to use antifreeze.
Starting: Air or electric.

Model		3612	3608	3606	3516B HD	3516B	3512B	3508B	3412E	3456	3406E	3176C	3306
Bore & stroke	mm	280 × 300	280 × 300	280 × 300	170 × 215	170 × 190	170 × 190	170 × 190	137 × 152	140 × 171	137 × 165	125 × 140	121 × 152
	(in)	(11 × 11.8)	(11 × 11.8)	(11 × 11.8)	(6.7 × 8.5)	(6.6 × 7.5)	(6.7 × 7.5)	(6.7 × 7.5)	(5.4 × 6)	(5.5 × 6.75)	(5.4 × 6)	(4.9 × 5.5)	(4.75 × 6)
No of cylinders		V12	18	16	V16	V16	V12	V8	V12	16	16	16	16
Turbocharged		Yes	Yes	Yes	Yes	Yes	Yes	Yes	Yes	Yes	Yes	Yes	Yes
Aftercooled		Yes	Yes	Yes	Yes	Yes	Yes	Yes	Yes	Yes	Yes	Yes	Yes
Locomotive	bhp	5,300	3,300	2,650	3,000	2,600	1,950	1,300	1,050	660	575	425	270
Rating	kW	3,950	2,460	1,980	2,240	1,940	1,455	970	783	492	429	317	201
	rpm	1,000	1,000	1,000	1,800	1,800	1,800	1,800	2,100	2,100	2,100	2,100	2,200
UIC rated	kW	4,330	2,710	2,300	2,460	2,130	1,600	1,065	860	540	470	350	220
Power	rpm	1,000	1,000	1,000	1,800	1,800	1,800	1,800	2,100	2,100	2,100	2,100	2,200
Weight (dry)	kg	25,100	20,400	16,700	7,500	7,500	6,080	4,310	2,435	1,353	1,300	932	970
	(lb)	(55,220)	(44,880)	(36,740)	(16,500)	(16,500)	(13,400)	(9,500)	(5,365)	(2,979)	(2,867)	(2,050)	(2,140)
Length	mm	4,120	4,540	3,720	3,366	3,366	2,675	2,135	1,723	1,532	1,532	1,287	1,270
	(in)	(162)	(179)	(147)	(133)	(133)	(105)	(84)	(68)	(60)	(60)	(51)	(50)
Width	mm	1,720	1,780	1,780	1,443	1,443	1,443	1,443	1,143	968	968	832	812
	(in)	(68)	(70)	(70)	(57)	(57)	(57)	(57)	(45)	(38)	(38)	(33)	(32)
Height	mm	2,741	2,741	2,631	1,980	1,980	1,839	1,839	1,319	1,375	1,375	1,287	1,160
	(in)	(108)	(108)	(104)	(78)	(78)	(72)	(72)	(52)	(54)	(54)	(51)	(46)

CFD Industrie

9-11 rue Benoît Malon, F-92150 Suresnes, France
Tel: (+33 1) 45 06 44 00   Fax: (+33 1) 47 28 48 84
e-mail: cfd@cfd.fr
Web: http://www.cfd.fr

Key personnel
President: François de Coincy
General Manager: P Esnault
Sales Manager: M Hallet

Products
Mechanical transmissions for mainline and shunting locomotives; power shift gearboxes (up to 900 hp input power); eight-speed hydromechanical 'Asynchro' transmissions; and locomotive final drives.

China Northern Locomotive and Rolling Stock Industry (Group) Corporation

11 Yangfangdian Road, Haidian District, Beijing 100038, China
Tel: (+86 10) 51 86 23 70   Fax: (+86 10) 51 86 23 74
e-mail: loriciec@cnrgc.com.cn
Web:www.cnrgc.com

Works addresses and key personnel
Chairman and Managing Director: Wang Tai-Wen
President: Cui Dianguo
Vice-President: Zhao Guangxing
President of LORIC Import & Export Corporation Ltd: Cao Guo-Bing
Vice-President of LORIC Import & Export Corporation Ltd: Chen Dayong
Senior Engineer of LORIC Import & Export Corporation Ltd: Yang Xiang-Jing

CNR hydrodynamic transmission   0583233

Products
CNR builds diesel engines, hydrodynamic transmissions, gearboxes, cardan shafts, turbochargers, intercoolers, injection pumps, fuel injectors, governors, water pumps and oil pumps. The annual output of diesel engines is 500 with a rated power of over 1,000 kW and motor 6,000 with a rated power of over 32 kW.

The engine range includes Series 240 and 280 from 810 kW to 4,440 kW. These units have direct injection, four-stroke with turbocharging and intercooling.

Cylinder block: All cast iron for Series 280, cast welded steel for Series 240. Removable wet-type cylinder liners of hardened cast iron.

Cylinder heads: Unit alloy cast iron cylinder with double bottom decks; water directors and removable injectors. Valve seats are replaceable corrosion-resistant insets.

Pistons: Thin-walled nodular iron, steel-crowned combined aluminium-skirted or steel-crowned combined iron-skirted.

Fuel system: Unit multi-hole type injector and plunger pumps on all models of the direct injection type. With mechanical or electronic fuel system, camshaft-actuated fuel injectors provide accurate metering and timing.

Aspiration: All models utilise turbochargers and intercoolers.

Lubrication: Large-capacity gear pump provides pressure lubrication to all bearings and oil supply for piston cooling, replaceable full-flow filters and water-cooled oil coolers.

Cooling system: Gear-driven centrifugal water pump. Large-volume water passages provide even flow of coolant around cylinder liners, valves and injectors.

Contracts
Locomotives with 12V240ZJ and 16V240ZJ diesel engines are running in Iran, Iraq, Malaysia, Nigeria, North Korea, Pakistan and Tanzania.

UPDATED

CNR turbocharger   0595035

CNR 12V280ZJ diesel engine   0552543

CNR governor   0583234

CNR diesel engines

Type	No of Cylinders	Bore mm	Stroke mm	Swept volume litres	UIC rated power kW	Full-rated speed rpm	BMEP bar	Dimensions (mm) length	width	height	Dry weight kg	First built	Builder
4240ZJ	4	240	275	49.76	810	1,000	19.53	3,102	1,710	2,647	10,995	2000	Dalian
6240ZJD	6	240	275	74.64	1,200	1,000	19.29	3,902	1,723	2,571	12,515	1998	Dalian
EQ6240ZJ	6	240	275	74.64	1,100	1,000	17.70	4,226	1,760	2,500	12,000	1999	Feb 7th
8240ZJ	8	240	275	99.52	1,320	1,000	16.00	4,827	1,735	2,624	15,900	1984	Dalian
8V240ZJ	8	240	275	99.52	1,620	1,000	19.53	3,454	1,775	2,697	13,875	1999	Dalian
12V240ZJ6D	V12	240	275	149.29	1,620	1,000	13.40	4,378	1,780	2,726	16,700	1998	Feb 7th
12V240ZJ6E	V12	240	275	149.29	2,400	1,000	19.30	4,300	1,780	2,760	17,100	2000	Feb 7th
12V240ZJ6L	V12	240	275	149.29	2,400	1,000	19.30	4,300	1,780	2,760		2003	Feb 7th
12V240ZJD	V12	240	275	149.29	2,400	1,000	19.29	4,298	1,790	2,642	16,300	1992	Dalian
12V240ZJG	V12	240	275	149.29	2,206	1,000	17.73	4,298	1,790	2,642	16,300	1996	Dalian
16V240ZJC	V16	240	275	199.05	2,942	1,000	17.75	5,339	1,790	3,085	23,230	1984	Dalian
16V240ZJD	V16	240	275	199.05	3,240	1,000	19.53	5,339	1,790	3,085	22,600	1986	Dalian
16V240JE	V16	240	275	199.05	3,680	1,000	22.20	5,339	1,790	3,085	23,600	1993	Dalian
EQ12V280ZJ	V12	280	285	210.60	2,900	1,000	16.50	5,700	1,800	2,800	21,000	2003	Feb 7th
12V280ZJ	V12	280	320	236.45	4,440	1,000	22.98	5,757	1,817	2,859	25,000	2000	Dalian

China South Locomotive and Rolling Stock Industry (Group) Corporation (CSR)

11 Yangfangdian Road, Haidian District, Beijing, China
Tel: (+86 10) 63 98 47 70   Fax: (+86 10) 63 98 47 66
e-mail: csrft@csrgc.com.cn
Web: http://www.csrgc.com.cn

Key personnel

General Manager: Zhao Xiaogang

Background

See China Southern Locomotive and Rolling Stock Industry (Group) Corporation (CSR) entry in *Locomotives and powered/non-powered passenger vehicles* section.

Works

See China Southern Locomotive and Rolling Stock Industry (Group) Corporation (CSR) entry in *Locomotives and powered/non-powered passenger vehicles* section.

Products

The diesel engine range includes:
Type 12V180ZJ engine: four-stroke, 12-cylinder vee, turbocharged and intercooled; rated power 993 kW at 1,500 rpm.
Type 16V240ZJC engine: four-stroke 16-cylinder vee, turbocharged and intercooled; rated power 2,940 kW at 1,000 rpm.

NEW ENTRY

Cockerill Mechanical Industries SA (CMI)

Avenue Greiner 1, B-4100 Seraing, Belgium
Tel: (+32 4) 330 24 33   Fax: (+32 4) 330 25 45
e-mail: locos.diesel@cmigroupe.com
Web: www.cmigroupe.com

Products

240CO Series of diesel enginesdeveloping 1,170 to 3,120 kW (1,590 to 4,240 hp) at 1,000 rpm, it is in use worldwide for rail traction, marine propulsion and power generation duties.

Aftermarket activities: spare parts and technical assistance.

UPDATED

Cummins Engine Company Ltd, Rail Marketing

Royal Oak Way South, Daventry NN11 5NU, UK
Tel: (+44 1327) 88 60 00
Fax: (+44 1327) 88 61 15
e-mail: cabo.customerassistance@cummins.com
Web: www.cummins.com

Key personnel

Manager Locomotive Engine Business:
   Steve P Dodman

Head Office

PO Box 3005, Columbus, Indiana 47202-3005, USA
Tel: (+1 812) 377 33 55
Fax: (+1 812) 377 30 82

Subsidiary companies

Fleetguard (filtration systems)
Holset (turbochargers)
Nelson Industries (exhaust systems)
Diesel ReCon (remanufactured engines and components)

Products

Diesel engines and locomotives, railcars and auxiliary power. The engine range includes the QSK19-R and N14-R with capacities of 19 and 14 litres respectively. Both are horizontal for underfloor installation in railcars and dmus.

Developments

Electronic controllers have been introduced to meet emissions legislation and to improve performance.

The latest technology QSK45L and QSK60L are now in service.

UPDATED

Locomotive engines

Engine	Maximum rating hp	rpm	Displacement litres (m³)	Bore and stroke mm (in)	No. of cylinders
M11-L	250	2,100	10.8 (661)	125 × 147 (4.92 × 5.79)	6
M11-L1	330	2,100	10.8 (661)	125 × 147 (4.92 × 5.79)	6
M11-L2	350	2,100	10.8 (661)	125 × 147 (4.92 × 5.79)	6
N14-L	360	2,100	14 (855)	140 × 152 (5.5 × 6)	6
N14-L1	425	2,100	14 (855)	140 × 152 (5.5 × 6)	6
N14-L2	475	2,100	14 (855)	140 × 152 (5.5 × 6)	6
KTA19-L	600	2,100	19 (1,150)	159 × 159 (6.25 × 6.25)	6
QSK19-L1	650	2,100	19 (1,150)	159 × 159 (6.25 × 6.25)	6
QSK19-L2	700	2,100	19 (1,150)	159 × 159 (6.25 × 6.25)	6
QSK19-L3	750	2,100	19 (1,150)	159 × 159 (6.25 × 6.25)	6
VTA28-L3	725	2,100	28 (1,710)	140 × 152 (5.5 × 6)	12
VTA28-L2	800	2,100	28 (1,710)	140 × 152 (5.5 × 6)	12
N14E-R	400	2,100	14 (855)	140 × 152 (5.5 × 6)	6 (horiz)
N14E-R3	450	2,100	14 (855)	140 × 152 (5.5 × 6)	6 (horiz)
N14E-R3	518	2,100	14 (855)	140 × 152 (5.5 × 6)	6 (horiz)
QSK19-R	757	2,100	19 (1,150)	159 × 159 (6.25 × 6.25)	6 (horiz)
QSK19-R1	650	2,100	19 (1,150)	159 × 159 (6.25 × 6.25)	6 (horiz)

Railcar engines

Engine	Maximum rating hp	rpm	Displacement litres (m³)	Bore and stroke mm (in)	No. of cylinders
N14E-R	400	2,100	14 (855)	140 × 152 (5.5 × 6)	6 (horiz)
N14E-R3	450	2,100	14 (855)	140 × 152 (5.5 × 6)	6 (horiz)
N14E-R3	518	2,100	14 (855)	140 × 152 (5.5 × 6)	6 (horiz)
QSK19-R	757	2,100	19 (1,150)	159 × 159 (6.25 × 6.25)	6 (horiz)
QSK19-R1	650	2,100	19 (1,150)	159 × 159 (6.25 × 6.25)	6 (horiz)

Underfloor auxiliary engines

Engine	Maximum rating hp	rpm	Displacement litres (m³)	Bore and stroke mm (in)	No. of cylinders
4B3.9-GR	66	1,800 (1,500 also avail)	3.9 (239)	102 × 120 (4.02 × 4.72)	4
4B3.9-GR3	78	1,800 (1,500 also avail)	3.9 (239)	102 × 120 (4.02 × 4.72)	4
4BT3.9-GR1	86	1,800 (1,500 also avail)	3.9 (239)	102 × 120 (4.02 × 4.72)	4
6B5.9-GR	80	1,500 (1,800 also avail)	5.9 (359)	102 × 120 (4.02 × 4.72)	6
6BT5.9-GR1	135	1,800 (1,500 also avail)	5.9 (359)	102 × 120 (4.02 × 4.72)	6
6BT5.9-GR2	166	1,800 (1,500 also avail)	5.9 (359)	102 × 120 (4.02 × 4.72)	6

Cummins QSK78-L engine rated at 2,240–2,600 kW
0569845

Dalian Locomotive & Rolling Stock Works

51 Zhongchang Street, Shahekou District, Dalian, China
Tel: (+86 411) 460 20 43   Fax: (+86 411) 460 10 93
Web: http://www.dloco.com

Key personnel
For a full list of personnel, see Dalian entry in *Locomotives and powered/non-powered passenger vehicles* section.

Background
See Dalian entry in *Locomotives and powered/non-powered passenger vehicles* section.

Products
Diesel engines and diesel generating sets. The 240ZJ series of diesel engines features a cast welded engine block, an alloy modular cast-iron crankshaft, parallel connecting rods, wet cylinder liners, steel-crowned and aluminium-ringed pistons (or steel crowned, iron-ringed pistons), individual injection pumps, closed-type injection equipment, gas turbochargers and a constant speed and constant power hydraulic governor.

This range of engines can be equipped with various kinds of governors, fuel injectors, turbochargers and generators according to the number of cylinders and cylinder arrangement to provide a range of power outputs to meet various applications.

The 16V240ZJC model is a development of the 16V240ZJB which has been in production for many years. The power rating of the new engine has been increased by improvements in the design and construction of the crankshaft, connecting rods and cylinder liners.

The 16V240ZJD diesel engine has been developed by co-operation in technology with Ricardo Consulting Engineers, UK. Many of the main components, such as the crankshaft and cylinder head, have been improved. New parts including ABB VTC 254-13 turbochargers, Bryce FCVAB fuel injection pumps, NTDLB injectors and NOVA Swiss high pressure pipes, Glacier main journal and connecting rod bearing shells and a Woodward PGEV governor have been introduced.

The 16V240ZJE diesel engine, being improved, is a new generation based on 240ZJD in co-operation with Ricardo Consulting Engineers. It has maximum power rating of 3,300 kW and is designed for heavy haulage work.

Developments
The new high power 280ZJ series diesel engines are under the course of development in co-operation with another company.

David Brown Engineering Ltd

Park Road, Huddersfield HD4 5DD, UK
Tel: (+44 1484) 46 55 00   Fax: (+44 1484) 46 55 86
e-mail: sales@davidbrown.com

Key personnel
Managing Director: C Reed
Sales and Marketing Director and Project Manager: N Crossley

Subsidiaries
Ganz-David Brown Transmissions Kft, Hungary
Hygate Transmissions, UK

Background
Part of Textron Power Transmission.

Products
Complete design, consultancy and manufacture of all types of gears and gearbox assemblies for railway applications.

Spiral and hypoid bevel, spur and helical gears for spares to complete systems for light rail, mainline locomotives and multiple-units. The company also overhauls, designs new, or redesigns and upgrades existing gearboxes to meet customer requirements.

Contracts
Main line gearbox and coupling assemblies for China; LRV Eurotram gearboxes for Strasbourg phase 2 and for Milan; gearbox assemblies for Seoul Metro and spare gear sets for UK, Sweden, Pakistan, Holland and Korea.

Developments
High-speed coupling is being developed.

Detroit Diesel

Detroit Diesel Corporation
13400 Outer Drive West, Detroit, Michigan 48239-4001, USA
Tel: (+1 313) 592 50 00   Fax: (+1 313) 592 75 80
Web: http://www.detroitdiesel.com

Key personnel
Chief Executive Officer and Chairman: Roger S Penske
President and Chief Operating Officer: Ludvik F Koci
Vice-President, International: Paul A Moreton
Vice-President, Construction and Industrial Sales: Jeffrey Sylvester

Products
Diesel and alternative fuel engines from 5 to 10,000 hp. Generator sets.

Electric-Motive Diesel (EMD)

Electric-Motive Diesel
9301 West 55th Street, LaGrange, Illinois 60525, US
Tel: (+1 708) 387 60 00
Web: www.emdiesels.com

Key personnel
President and Chief Executive Officer: John Hamilton

Background
In April 2005 General Motors Corporation sold its Electro-Motive Division to a group of investors led by Greenbrier Equity Group LLC and Berkshire Partners LLC. The transaction covered EMD's North American and international locomotive supply businesses, its power, marine and industrial products, all of its locomotives maintenance contracts worldwide and its manufacturing facilities at La Grange, Illinois, and London, Ontario.

Products
Electro-Motive Division first developed the Model 567 diesel engine in 1938 when it began locomotive manufacture at LaGrange, Illinois, USA.

To provide increased hp and greater efficiency, the Model 645 engine was introduced in mid-1965. The major change in the Model 645 over the 567 was the increase in cylinder liner bore from 216 mm (8½ in) to 230 mm (9¹⁄₆ in), the stroke remaining at 254 mm (10 in).

The turbocharged 645E3B engine introduced in 1979 and the turbocharged 645F3 engine introduced in 1981 were a result of the search for increased product reliability, performance and fuel economy. With increased hp and fuel economy these engines could haul more tonnage at the same speed or the same tonnage at a higher speed than their predecessors.

The 710G series of engines is an evolutionary development of GM/EMD's turbocharged, uniflow-scavenged, two-stroke cycle engine. The 16-cylinder 710G3B is rated at 4,250 hp at 900 rpm for locomotive applications and has a displacement of 710 in³ per cylinder. This series is the result of a succession of improvements to the engine. From 1989 to 1991, for example, the fuel efficiency of the 710G was increased by 1.5 per cent. Greater displacement and an advanced turbocharger give the 710G the capacity for significant increases in hp.

Full load fuel consumption of the model 710G3B engine is down 11.5 per cent from the 1980 model and 4.9 per cent from the 1983 model in the 645 range. Among the ways fuel efficiency has been increased is improved turbocharger aftercoolers and fuel injectors, and low-restriction liner intake ports.

As compared to the 645 series engines, the 710 features a longer stroke and added displacement which led to these structural improvements in the engine: Model G crankcase; larger diameter plunger injectors; larger diameter crankshaft; new camshaft; longer cylinder liner; and longer piston and rod assembly. The 710G design also increased the overall dimensions: the new engine is 1⁵⁄₈ in higher and 4⁵⁄₈ in longer.

A recent innovation is the Functionally Integrated Railroad Electronics (FIRE®) is an advanced locomotive management system.

Model		16-645FB	16-710G
Bore	in	9.06	9.06
Stroke	in	10	11
Displacement	in³	645	710
Cylinder spacing	in	16⁵⁄₈	16⁵⁄₈
Bank angle		45	45
Compression ratio		16.0:1	16.0:1
Engine speed	rpm	950	900
bhp		3,800	4,250

UPDATED

Eminox Ltd

North Warren Road, Gainsborough DN21 2TU, UK
Tel: (+44 1427) 81 00 88   Fax: (+44 1427) 81 00 61
e-mail: john.perry@eminox.com
Web: http://www.eminox.com

Key personnel
Sales Manager, Rail: John Perry
Head of Sales and Marketing: Paul Priestley

Products
Design and manufacture of stainless steel exhaust systems for use on diesel multiple-units and railcars, including exhaust after-treatment; catalyst and particulate filters.

Freudenberg Schwab GmbH

Postplatz 3, D-16761 Hennigsdorf, Germany
Tel: (+49 3302) 206 20   Fax: (+49 3302) 20 62 77
e-mail: info@freudenberg-schwab.de
Web: http://www.freudenberg-schwab.de

Key personnel

Chief Executive Officers: Dr Detlef Cordts;
    Peter Kofmel; Jörg Sost

Ganz-David Brown Getriebetechnik GmbH

Orczy Strasse 46-48, H-1089 Budapest, Hungary
Tel: (+36 1) 210 11 50   Fax: (+36 1) 334 03 64
e-mail: gdb@mail.datanet.hu
Web: http://www.gdb.hu

Key personnel

General Manager: József Fáy
Marketing and Technical Manager: Istvàn Lörincz
Chief Engineer: László Boronkai
Production Manager: Sándor Szluk
Engineering Manager: Laszló Meszaros

General Electric Transportation Systems

2910 East Lake Road, Erie, Pennsylvania 16531,
USA
Tel: (+1 814) 875 59 11   Fax: (+1 814) 875 64 87

Key personnel

For a full list of personnel, see General Electric
Transportation Systems entry in *Locomotives
and powered/non-powered passenger vehicles*
section.

Products

Diesel engines and complete electric
transmissions.

Series HDL (16-cylinder, Vee) engine
Type: four-stroke cycle, turbocharged and aftercooled (intercooled).
Cylinders: Bore 250 mm (9.84 in), Stroke 320 mm
(12.60 in).
Fuel Injection: Individual injection and fuel pumps
with EFI control.
Turbocharger: Two, single stage.
Lubrication and cooling: Forced full flow filtered
oil to all bearings and pistons, gear-type engine-
driven pump.

Series FDL (8-, 12- and 16-cylinder Vee) engines
Type: four-stroke cycle, turbocharged and aftercooled (intercooled).

Quality Manager: Leo S E Lang
Marketing Director: Bernd Werner

Products

Vibration control: rubber/metal elastic elements,
standard as well as specially designed. Range
includes elastic mountings for traction equipment
and components for primary and secondary

Products

Hydrodynamic and hydromechanical power
transmissions for diesel railcars and locomotives.
Axledrive units with helical and spiral bevel gears
for railcars, locomotives and light rail vehicles.
Wholly suspended axledrive units for monomotor
bogies for metro trainsets. Helical gear pairs for
high-speed locomotives and railcars, spiral bevel
gear pairs for locomotives, railcars and light rail
vehicles; drives for low-floor LRVs.

suspension, resilient couplings linking motor,
gearbox and axle.

Contracts

Recent contracts include: 160 complete powered
wheelsets for railbuses, 20 hydrodynamical
power transmissions for dmus and 800 sets of
traction gears for 'Trainitalia'; 113 hydrodynamic
gearboxes have been delivered to CIS railways for
modernisation of dmus.

Current engine specifications

Model	7FDL8	7FDL12	7FDL16	7HDL16
No of cylinders	8	12	16	16
Output (UIC) standard	2,000	3,300	4,100	6,250
Stroke cycle	4	4	4	4
Cylinder arrangement	45° Vee	45° Vee	45° Vee	45° Vee
Bore	228.6 mm *(9 in)*	228.6 mm *(9 in)*	228.6 mm *(9 in)*	250 mm *(9.84 in)*
Stroke	266.7 mm *(10½ in)*	266.7 mm *(10½ in)*	266.7 mm *(10½ in)*	320 mm *(12.60 in)*
Compression ratio	12:7:1	12:7:1	12:7:1	15.2:1
Idle speed	385 rpm	385 rpm	385 rpm	350 rpm
Full-rated speed	1,050 rpm	1,050 rpm	1,050 rpm	1,050 rpm
Firing order	1R-1L-2R-2L-4R-4L-3R-3L	1R-1L-5R-5L-3R-3L-6R 6L-2R-2L-4R-4L	1R-1L-3R-3L-7R-7L-4R-4L 8R-8L-6R-6L-2R-2L-5R-5L	1R-1L-3R-3L-7R-7L-5R-5L 8R-8L-6R-6L-2R-2L-4R-4L
Turbocharger	One	One	One	Two
Engine dimensions				
Height (excluding stack)	2,191 mm *(86¼ in)*	2,289 mm ½*(90⅛ in)**	2,289 mm *(90⅛ in)*	2,611 mm *(102.80 in)*
Length (overall)	3,264 mm *(1281/2 in)*	4,051 mm *(159½ in)*	4,902 mm *(193 in)*	4,984 mm *(196.24)*
Width (overall)	1,734 mm *(68¼ in)*	1,740 mm *(68⅜in)*	1,740 mm *(68⅜in)*	1,700 mm *(66.93 in)*
Weight (dry)	12,200 kg *(27,000 lb)*	15,900 kg *(35,000 lb)*	19,700 kg *(43,500 lb)*	23,800 kg *(52,500 lb)*

Cylinders: Bore 229 mm (9 in), Stroke 267 mm
(10-1/2 in) Swept volume, 668 cu in per cylinder.
Individual unitised cast cylinder with renewable
liner and head.
Fuel injection: Individual injectors and fuel pumps.
Electronic fuel injection (EFI) available.
Turbocharger: One, single stage, exhaust-driven
(no gear drive to crankshaft).
Lubrication: Forced, full-flow, filtered oil to all bearings
and pistons, gear-type engine-driven pump.

Cooling system: Forced circulation water cooling
of cylinders, turbocharger and intercoolers. The
water passages are external of the crankcase and
mainframe.

Giro Engineering Ltd

Talisman, Duncan Road, Park Gate, Southampton
SO31 7GA, UK
Tel: (+44 1489) 88 52 88   Fax: (+44 1489) 88 51 99
e-mail: giro@giroeng.com
Web: www.giroeng.com

Key personnel

Managing Director: J P Williams
Commercial Director: C R Galley

Commercial Manager: J P Tarleton
    jpt@giroeng.com

Products

Design and manufacture of sheathed and
unsheathed fuel injection pipes for diesel engines
from 3.73 to 18,650 kW (five to 25,000 hp) to SOLAS
regulations. Giro holds Type approvals from LR,
ABS, BV and MoD(N) and is an approved supplier
to the British MOD (Navy). Giro is LRQA audited to
ISO 9001:2000.

Contracts

UK customer, EWSR. Significant business has also
been conducted in China and Australia.

UPDATED

Gmeinder Getriebe und Maschinenfabrik GmbH

Anton Gmeinder Strasse, 5, D-74821 Mosbach,
Baden, Germany
Tel: (+49 6261) 674 70   Fax: (+49 6261) 674 72 19
e-mail: info@glfg.de
Web: www.gmeinder-lokomotivenfabrik.de

Key personnel

Managing Directors: Norbert Wuddel; Werner Kilp
Head of Rail Traction Division: Helmut Eifler

Background

In 2004 private rail operator Transport & Logistik
AG acquired a 51 per cent shareholding in
Gmeinder.

Products

Axle drives for lightweight rail vehicles, dmus and
emus, tilting trainsets; axle drives for locomotives
with cardan shaft drive; special vehicles such as
cranes. Vehicles equipped include the RegioShuttle
lightweight dmu, DB Class VT 628 dmu and VT 611/
612 tilting dmus.

UPDATED

Haynes Corporation

3581 Mercantile Avenue, Dept WS, Naples, Florida
34101, US
Tel: (+1 239) 436 15 78   Fax: (+1 239) 643 53 11
e-mail: mailtosales@haynesnesco.com
Web: www.haynesnesco.com

Key personnel

Vice-President Sales and Marketing: Greg Schultz
Sales Engineer: Kirk Leutz
Director of Manufacturing: Dan Neely
Sales and Marketing Co-ordinator:
  Brandie J Dixon

Background

Haynes Corporation was founded in 1960 in
Michigan. In 1972 the company acquired Adeco
Products, Inc and in 1976 the Busch-Sulzer Division
of Nordberg. Later Haynes finalised acquisition
of the exclusive manufacturing and distribution
rights of the American Bosch diesel fuel injection
APF product line from United Technologies. In 1988
Haynes acquired the Bendix diesel fuel injection
product line which included the Naples, Florida
manufacturing facility.

Haynes continued its expansion to include a
west coast operation after acquiring Hatch & Kirk
Fuel Injection Systems, formerly Kordy Colyer. The
Norwalk, California facility, which specialises in
Detroil Diesel, Cummins, and EMD fuel injection
components now operates as Haynes Fuel Injection
Corporation.

Products

Fuel injection systems and components.

Developments include latest in CNC machines,
CAD/CAM and MRP system integrating marketing,
manufacturing, purchasing, inventory control,
engineering and accounting.

NEW ENTRY

Hitachi Ltd

Overseas Marketing Department
Transportation Systems Sales Division
6 Kanda Surugadai 4-chome, Chiyoda-ku, Tokyo
101-8010, Japan
Tel: (+81 3) 32 58 11 11   Fax: (+81 3) 32 58 52 30
Web: www.hitachi-rail.com

Key personnel

Chief Operating Officer of Transportation Systems:
  Gaku Suzuki
General Manager, Transportation Systems Sales
  Division: Chiaki Ueda
General Manager, Transport Management and
  Control Systems Division: Kazuo Kera

General Manager Rolling Stock System Division:
  Toshihide Uchimura

Products

Complete control equipment, electric transmissions
and hydraulic transmissions.

UPDATED

Honeywell Serck

Warwick Road, Birmingham B11 2QY, UK
Tel: (+44 121) 766 66 66   Fax: (+44 121) 766 60 14
e-mail: john.g.brookes@honeywell.com

Key personnel

Sales Director: John Brookes
Technical Sales Manager: Paul Egerton

Background

Honeywell Serck is part of Honeywell Thermal
Systems and a subsidiary of the Honeywell
Corporation.

Products

Integrated cooling modules for cooling oil, water
and air systems for diesel locomotives.

Contracts

Recent contracts include: redesign of the engine
cooling group for South West Trains' Class 159
dmus (UK); charge air cooling equipment for
Countrylink (Australia) XPT power cars re-engined
with MAN B&W Paxman 12VP185 engines;
radiator elements for Sri Lanka Railway Class M7
locomotives and transmission oil coolers for the
same operator's Hunslet shunting locomotives.

Hygate Transmissions Ltd

Part of the David Brown Group (a Textron company)
Lower Bristol Road, Bath BA2 3EB, UK
Tel: (+44 1225) 33 40 00   Fax: (+44 1225) 31 85 82
e-mail: sales.hygate@davidbrown.com
Web: http://www.davidbrown.com/products/rail

Key personnel

Managing Director: C Reed
Sales and Marketing Director: N Crossley
Technical Sales Manager: N Antrobus

Products

Design and manufacture of gearbox assemblies
and loose gears for main line, urban railway,
mass transit systems and LRVs. Single and double
reduction parallel shaft gearboxes and spiral bevel
gearboxes. Double engagement gear couplings
for rail applications. Redesign and upgrading of
existing transmissions. Consultancy, maintenance,
overhaul and repair services.

Contracts

Recent contracts include: locomotive gear sets
for Belgium and Syria; gearboxes for Øresund
Link and regional emus in Sweden and emus for
Gatwick Express, ScotRail and SouthWest Trains
in the UK; metro car gearboxes for Seoul Lines
6 and 7 and Pusan Line 2 in Korea; and gearboxes
for trains in Milan and Rome.

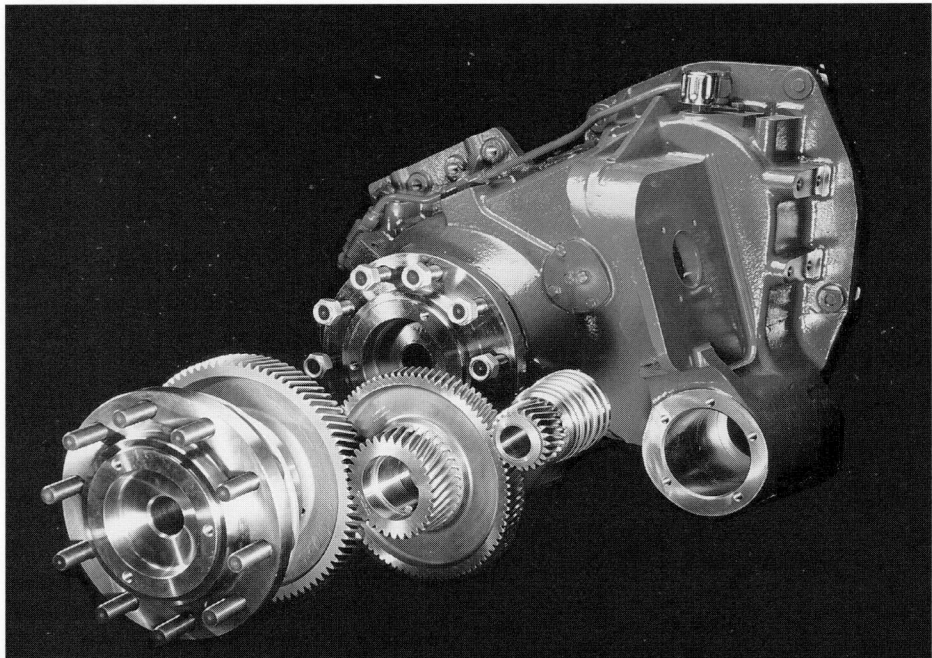

*Hygate double-reduction gearbox for Eurotram LRV stub axle mounted and carrying rail wheel and
disc brake assembly as fitted for Strasbourg Phase 2*   0063623

Isotta Fraschini Motori SpA

Via le F de Blasio, I-70123 Bari, Italy
Tel: (+39 080) 534 52 53   Fax: (+39 080) 531 10 95
e-mail: isottafraschini@isottafraschini.it
Web: www.isottafraschini.it

Key personnel

Chief Executive Officer: Edgardo Bertoli
Managers:
  Research and Development: Lucio D'Amelio
  Commercial: Umberto Brandini
  Operation: Lucio D'Amelio

  Purchasing: Anatonio Barulli
  After Sales Service and Spares: Ezio Ferluga

Background

Isotta Fraschini is a member of the Fincantieri group.

Products

Diesel engines for rail traction, as well as for
marine and industrial applications.

IF 1300 family
Developing up to 50 kW per cylinder, the IF 1300
is a four-stroke engine produced in 6-, 8- and
12-cylinder vee versions, with turbocharged,
turbocharged and intercooled options offering
power ratings from 140 to 600 kW at 1,500–
2,300 rpm.

IF 1700 family
Developing up to 147 kW per cylinder, the IF 1700
is a four-stroke engine produced in 8-, 12- and
16-cylinder vee versions, with naturally aspirated,
turbocharged and turbocharged and intercooled
options offering power ratings from 275 to
2,200 kW at 1,500–1,800 rpm.

NEW ENTRY

Iveco SpA

Engine Business Unit
Lungo Stura Lazio 49, I-10156 Turin, Italy
Tel: (+39 011) 007 51 81   Fax: (+39 011) 007 62 75
Web: http://www.iveco.com

Subsidiaries

Iveco LA
Brazil
Engine Division CO
Rua Almeda da Serra 222, Vale do Sereno, 34000-000
Nova Lima (MG)
Tel: (+55 31) 32 86 37 32   Fax: (+55 31) 32 86 37 35

China
Engine Division
PR Head Office, 10/F Jinling Hotel, World Trade
Center, 2 Hanzhong Road, 210005 Nanjing
Tel: (+86 25) 471 09 81   Fax: (+86 25) 470 11 05

France
Iveco France SA
Division Moteurs Iveco Aifo
50 rue Ampère, BP 103, F-69685 Chassieu Cedex
Tel: (+33 4) 72 47 22 22   Fax: (+33 4) 78 90 59 90

Germany
Iveco Magirus AG
Dieselmotoren Vertrieb

Heiner Flieschmann Strasse 9, D-74172 Neckarsulm
Tel: (+49 7132) 97 69 18   Fax: (+49 7132) 97 69 40

India
Iveco NV
Engine Division
Liaison Office – India, Hotel Taj Palace, Suite 131,
First Floor, Sardar Patel Marg, 110021 New Delhi
Tel: (+91 98) 104 038 82   Fax: (+91 11) 410 29 80

Italy
Iveco Aifo SpA
Viale dell'Industria 15/17, I-20010 Pregnana
Milanese, Milan
Tel: (+39 02) 95 51 01   Fax: (+39 02) 93 59 00 29

Sweden
Engine Division
Lergöksgatan 12, SE-421 50 Västra Frölunda
Tel: (+46 31) 49 24 50   Fax: (+46 31) 49 24 57

United Kingdom
Iveco UK Ltd
Engine Division
Road One, Industrial Estate, Winsford CW7 3QP
Tel: (+44 1606) 54 10 27   Fax: (+44 1606) 54 11 24

United States of America
Iveco Motors of NA
245 E North Avenue, Carol Stream, Illinois 60188-2021

Tel: (+1 630) 260 50 60
Fax: (+1 630) 260 50 63

Products

Diesel engines and power packs for rail vehicles
ranging from 200 to 840 kW.

The Vector 8 Euro 3 power pack features a 560 kW
supercharged and cooled common rail injection
engine with mechanical transmission designed
for underfloor installation in dmus and railcars. It
is equipped with a 65 kVA 400 V 50 Hz auxiliary
generator.

The Cursor range of six-cylinder in-line engines
features C78 ENT (221 kW at 2,200 rpm), C10
ENT (295 kW at 2,100 rpm) and C13 ENT (368 kW
at 1,900 rpm) versions. The C78 ENT model is
available in horizontal or vertical versions, the C10
ENT and C13 ENT as vertical models only. The EUI
injection system is employed.

The Vector range of 90 'vee' engines features
V06 ENT (420 kW at 2,000 rpm), V08 ENT (560 kW
at 2,000 rpm) and V12 ENT (840 kW at 2,000 rpm)
models. The common rail injection system is
employed.

Kim Hotstart

Kim Hotstart Mfg Co
East 5723 Alki, Spokane, Washington 99212, USA
Tel: (+1 509) 534 61 71   Fax: (+1 509) 534 42 16
e-mail: sales@kimhotstart.com
Web: www.kimhotstart.com

Key personnel

Vice-President and General Manager:
   Rick Robinson

Director, Sales and Marketing: Terry Judge
Manager, Industrial Sales, Locomotive Division:
   Jason Barnes
International Sales Manager: Trond Liaboe

Products

Winter layover heating systems for diesel and
petrol engines. Available in 240 and 480 V AC three
and single phase; special voltages on request.
Heating systems designed to heat coolant only or
combination of coolant and oil. Models range from
11 to 36 kW plus the patented Diesel Driven Heating
System (DDHS) which includes a three-cylinder

Lister-Petter water-cooled diesel engine with a
water-circulation pump, plus a belt-driven 72 V DC
alternator. The system operates independently of
other locomotive systems.

Heavy-duty 50 A trickle chargers for locomotive
applications. Available from 12 to 96 V DC.

Contracts

Contracts include supply of systems to BNSF (2003,
2005), MotivePower (2003).

UPDATED

Kolomna

Kolomensky Joint Stock Company
Partizan Street, 140408 Kolomna, Moscow Region,
Russia
Tel: (+7 0966) 13 84 44; 13 81 11
Fax: (+7 0966) 15 47 44
e-mail: kolomzavod@kolomna.ru
Web: www.kolomnadiesel.com

Key personnel

General Director: V N Vlasov
Technical Director: V A Shelemet'yev
Marketing Director: E N Gunchenko

Products

D49 series of diesel engines for locomotives. The
D49 series comprises four-stroke, turbocharged
engines with 8, 12, 16 and 20 cylinders in
Vee-versions.

D49 diesel engines and generating sets are in
service in Europe and countries in southeast Asia,
Africa, and South America as well as northern

| Engine type | Output | | Dimensions (mm) | | | Weight (kg) | Type of locomotive |
	kW	hp	Length	Width	Height		Diesel engine
8 VD49	500–1,760	680–2,390	3,370–3,720	1,665	2,290–2,810	10,070	TGM 6V, TGM 8
12 VD49	1,100–2,650	1,500–3,600	3,920–4,030	1,665	2,745–3,030	13,600–14,400	2TE 116, TEM 7A
16 VD49	1,470–3,700	2,000–5,030	4,900	1,655	3,070	17,460–18,500	2TE 116, 2TE 121, TEP 70
20 VD49	3,700–5,000	5,030–6,800	6,270	1,665	3,190	20,900	TE 136, TEP 80

Output, specific fuel consumption and weight values are for ISO standard reference conditions.

regions of Russia. Recent contracts include the
supply of diesel engines to Germany and Iran.

D49 diesel four-stroke turbocharged for locomotives
Bore: 260 mm
Stroke: 260 mm
Speed: min 750–1,000 rpm
Mean effective pressure: MPa 1.1–1.96
Mean piston speed: 6.8–8.6 m/s
Specific fuel consumption: g/kW/h 190 +5%
8, 12, 16 and 20 cylinder versions in Vee-
configuration. A gas-oil version of the 16 VD49 is
also produced, with ratings in the 2,470–2,600 kW
(3,360–3,530 hp) range.

UPDATED

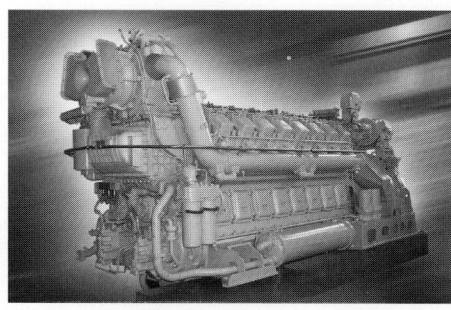

Kolomna Type D49 CHN26/26 16 cylinder diesel
engine in Vee-configuration   1143493

Komatsu Diesel Co Ltd

2-3-6, Akasaka, Minato-ku, Tokyo 107-8414, Japan
Tel: (+81 3) 55 61 29 61   Fax: (+81 3) 55 61 29 67
e-mail: yasuo_okamoto@kdl.komatsu.co.jp
Web: http://www.komatsu.com/kdl

Works

400 Yokokura Shinden, Oyama-shi, Togichi 323-8558

Key personnel

Director, Sales Development & Application
   Department: Y Okamoto

Products

Diesel engines, torque converters, transmissions,
hydraulic pumps, hydraulic motors and hydraulic
control valves.

The engine range features light weight and
compactness, making it suitable for use in tilting
rolling stock, good cold-start capability, easy
maintenance, low fuel consumption (through
measures including improved intake port
configuration and high pressure injection system)
and multifunction electronic governor.

The K-ATOMiCS (Komatsu Advanced
Transmission with Optimum Modulation Control

Komatsu SA12V170 engine   0099146

System) automatic transmission system eliminates shift shock and prevents torque cut-off. A hydraulic retarder and a wheel spin control device can be installed. Maintenance costs are reduced by changing the durable rubber couplings by the engine flywheel.

Komatsu started manufacturing diesel engines for railcar applications about 10 years ago and has gained a major share of the Japanese market. Recently the company has developed a new high-performance engine for locomotive use, the SDA12V170. Three units were produced in late 1999 for a new locomotive design developed for JR Freight, the DF200, or 'Eco-Power Red Bear', as it is popularly known.

The SDA12V170 develops 1,325 kW at 1,800 rpm. It is equipped with a dual-circuit aftercooler system which enhances combustion efficiency and power output while offering reduced fuel consumption and increased endurance through the use of Komatsu original cast pistons. Other features of the engine in the DF200 application include a starter system that controls current for both vehicle and engine control and engine condition-monitoring.

The SA6D140HE Komatsu is a low emission engine which meets the Euro2 standard. It is equipped with a common rail injection system, variable injection timing, re-entrant piston, high swirl port and high efficiency turbocharger.

UPDATED

Komatsu KTF3335A torque converter   0023832

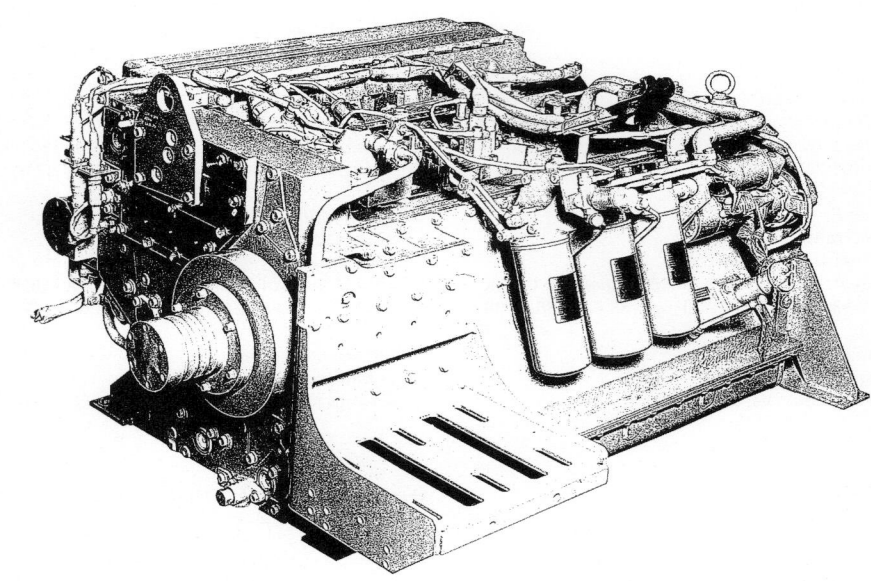

Komatsu SA6D140-HE engine
1108046

Komatsu railcar engines

	Type	Cyl. No	Aspiration	Bore mm	Stroke mm	Displacement cc	Maximum rating kW (hp)	rpm	Continuous rating kW (hp)	per/min	Length mm	Width mm	Height mm	Weight kg
S6D125-H	Horizontal	6	T	125	150	11,040	221 (296)	2,200	199 (266)	2,200	1,376	1,179	770	990
SA6D125-HE[2]	Horizontal	6	T/A	125	150	11,040	294 (395)	2,200	265 (395)	2,200	1,376	1,179	770	1,150
SA6D125-HD	Horizontal	6	T/A[1]	125	150	11,040	342 (459)	2,100	309 (414)	2,100	1,680	1,215	792	1,250
SA6D140-HE[2]	Horizontal	6	T/A	140	165	15,240	431 (577)	2,000	386 (518)	2,000	1,624	1,257	757	1,700
SA6D140-HD	Horizontal	6	T/A[1]	140	165	15,240	475 (636)	2,000	427 (572)	2,000	1,484	1,329	757	1,670

[1] Dual Circuit Aftercooler.

[2] Engine models suffixed by 'E' meet the Euro2 standard.

Torque converter specifications

Model	Max Input Power kW (hp)	rpm	Element	Torque Converter Stage	Phase	Weight kg	Overall Dimension Length mm	Width mm	Height mm	Type	Retarder (Option) Absorbed Power kW (hp)	per/min
KTF3335A	257 (345)	2,200	3	1	2	800	1,008	777	710	Hydraulic	147 (203)	2,000
KTF3345A	331 (444)	2,000	3	1	2	980	1,120	883	757	Hydraulic	147 (203)	2,000

Komatsu locomotive engines

	Type	Cyl. No	Aspiration	Bore mm	Stroke mm	Displacement cc	Maximum rating kW (hp)	rpm	Continuous rating kW (hp)	per/min	Length mm	Width mm	Height mm	Weight kg
S6D125	Vertical	6	T	125	150	11,040	221 (296)	2,200	188 (252)	2,200	1,350	699	1,134	880
SA6D125	Vertical	6	T/A	125	150	11,040	276 (370)	2,200	236 (316)	2,200	1,350	699	1,135	900
SA6D140	Vertical	6	T/A	140	165	15,240	405 (542)	2,100	346 (464)	2,100	1,515	780	1,210	1,230
SA6D170	Vertical	6	T/A	170	170	23,150	552 (740)	2,000	471 (632)	2,000	1,938	998	1,685	2,440
SA12V170	Vee	12	T/A	170	170	46,340	1,104 (1,480)	2,100	942 (1,262)	2,100	2,220	1,203	1,752	4,800
SDA12V170	Vee	12	T/A[1]	170	170	46,340	1,545 (2,071)	1,800	1,324 (1,775)	1,800	2,220	1,203	1,752	4,900
SA16V170	Vee	16	T/A	170	170	61,740	1,472 (1,972)	2,100	1,251 (1,677)	2,100	2,783	1,410	1,698	6,336

[1] Dual Circuit Aftercooler.

MAN Nutzfahrzeuge Aktiengesellschaft

Business Unit Engines

Vogelweiherstrasse 33, D-90441, Nueremberg, Germany
Tel: (+49 911) 420 20 02
Fax: (+49 911) 420 19 32
e-mail: industrialengines@mn.man.de
Web: www.man-nutzfahrzeuge.de

Key personnel

Director, Business Unit: Arnd Loettgen
Director, Engines and Components: Kurt Heuser
General Manager, Industrial Engines: Wolfgang Kuntze

Products

MAN manufactures various engine types for installation in railcars and locomotives based on MAN's truck engine range. These are four-stroke, direct-injection, water-cooled diesel engines especially developed for the demands of railway operation.

For railway applications, MAN Nutzfahrzeuge Aktiengesellschaft manufactures two engine series meeting the current exhaust emission levels UIC II and Euromot 2. Based on the D2876 LUE6XX, MAN offers three ratings 301, 338 and 375 kW, each rated at 2,000 rpm. These ratings are accredited to UIC II. Based on the D2842 LE6xx, MAN offers following ratings 662, 588 and 500 kW each rated at 2,100 rpm as well as for repowering 480 kW at 1,550 rpm. This V12-cylinder is built in an underfloor version with a maximum height of 860 mm. The ratings 662 and 588 kW meet the UIC II emission level as well as EURO 2.

MAN Nujtzfahrzeuge AG also manufactures gas engines for LPG in a horizontal configuration. Besides the naturally aspirated 151 and 177 kW six-cylinder engines in the D 28 family, a turbocharged and intercooled version rated 231 kW is also available. First installation of a gas engine in a light railcar was done on the island of Usedom, Germany.

For auxiliary drives and generating sets MAN produces six-cylinder in-line engines rated from 47 to 313 kW and V-8, V-10 and V-12 cylinder engines from 325 to 587 at 1,500 to 1,800 rpm.

UPDATED

MAN B&W Diesel Ltd

Bramhall Moor Lane, Hazel Grove, Stockport
SK7 5AQ, UK
Tel: (+44 161) 483 10 00   Fax: (+44 161) 487 14 65
e-mail: sales@manbwltd.com
Web: http://www.manbw.com

Key personnel

Managing Director: Rolf Studte
Senior Vice-president Sales: Charles Foulkes
Finance Director: Tony Blagbrough
Sales Director, Traction: Chris Gallagher
Publicity Officer: Wendy Morris

Principal subsidiaries

MAN B&W Diesel Ltd, Regulateurs Europe
MAN B&W Diesel Australia Pty Ltd
MAN B&W Diesel Canada Ltd
MAN B&W Diesel (India) Ltd

Products

A range of modern diesel engines for various
rail traction duties. MAN B&W Diesel engines
power freight and passenger locomotives, both
conventional and high-speed, which are suitable

for new locomotives and repowering applications.
The engine ranges are designed to meet applicable
rail authority type tests, including known future
emissions requirements.

RK215

The RK215T is a medium-speed (900–1,000 rpm)
four-stroke turbocharged and intercooled high
power-to-weight engine with a bore of 215 mm
and a stroke of 275 mm. For rail applications, it is
produced in six-cylinder in-line and eight-, 12- and
16-cylinder vee configurations covering a power
range from 1,065 to 3,160 kW.

RK270

The RK270T is a medium-speed (900–1,000 rpm)
four-stroke turbocharged and intercooled engine
with a bore of 270 mm and a stroke of 305 mm. For rail
applications, it is produced in six- and eight-cylinder
in-line and 12-, 16- and 20-cylinder vee configurations
covering a power range from 1,875 to 6,875 kW.

RK280

The RK280 is a medium-speed (1,000) four-stroke
turbocharged and change-cooled engine with
a bore of 280 mm and a stroke of 330 mm. It is

produced in 12-, 16- and 20-cylinder configuration
giving a continuous rated power of up to
9,000 kW.

VP185

The VP185 is a high-speed (1,800 rpm) four-stroke
turbocharged, intercooled and aftercooled engine
with a bore of 185 mm and a stroke of 196 mm.
For rail applications, it is produced in 12- and
16-cylinder vee configurations developing 1,860
and 2,800 kW respectively.

Contracts

More than 140 RK215 engines have been supplied
to equip ALSTOM-built Prima diesel-electric
locomotives for Iran, Syria and Sri Lanka. Two
12RK215T engines have also been supplied
for shunting locomotives for use at the port of
Qinhuangdao, China. 12VP185 engines have been
supplied for high-speed train repowering projects
in Australia (Countrylink) and the UK (Midland
Mainline).

UPDATED

MTU Friedrichshafen GmbH

D-88040 Friedrichshafen, Germany
Tel: (+49 7541) 90 26 91   Fax: (+49 7541) 90 39 49
e-mail: rail@mtu-online.com
Web: www.mtu-online.com

Key personnel

Senior Manager Rail Applications: Matthias Vogel

Background

MTU Friedrichshafen in Germany and Detroit
Diesel in the US, two DaimlerChrysler companies,
have consolidated their off-highway activities.
With their joint range of products, supplemented
by Mercedes-Benz engines, a worldwide leading
supplier of engines and systems has been set up

DMUs like this Siemens Desiro unit for DB AG are equipped with MTU PowerPack complete drive systems   **NEW**/0585228

Railway traction

Power Creator

*High and medium-speed
4-stroke Diesel engines*

*Unitary power ratings from
1 500 to 6 000 kW*

support
service
performance
industrial partnership
research

S.E.M.T. PIELSTICK

Always going further

S.E.M.T. Pielstick . LE RONSARD PARIS NORD 2 • 22 AVENUE DES NATIONS • BP 84013 VILLEPINTE • 95931 ROISSY CH DE GAULLE CEDEX (FRANCE)
Tel. + 33 (0)1 48 17 63 00 • Fax + 33 (0)1 48 17 63 49 • e-mail : sales_traction@pielstick.com • www.pielstick.com

covering marine, rail, mining, construction and industrial, defence and agriculture applications as well as power generation. The joint product programme includes compact diesel engines with a power output ranging from 30 to 9,100 kW, gas engines with ratings between 410 and 1,320 kW and gas turbines of 3 to 30 MW. It is complemented by electronic governing, monitoring and control systems.

With the rail industry, which is served by the brand MTU, the company has a long and close partnership. Since 1950, MTU and its predecessor companies have delivered more than 14,000 units for rail vehicles. The engines are installed in a large variety of applications and services with some of them operating under the most severe environmental conditions.

Products

Engines used for shunting/switching locomotives with frequent load changes, for multi-purpose and main-line services. For diesel railcars and high-speed trains, the company has gained a capability to design and supply complete drive systems (MTU-PowerPacks®).

Contracts

Recent contracts include: first orders for Powerpack complete drive systems from Russian train builder, Metrowaggon Masch; Series 4000 engines for Siemens Eurorunner locomotives for ÖBB, Dispolok GmbH and KCRC, Hong Kong; 400 8V 4000 engines for a DB programme to re-power Class 290 locomotives; 52 12V 4000 engines for a SNCF programme to re-power Class BB66000 locomotives; and first orders, from Vossloh Locomotives, for the 20-cylinder version of the Series 4000 engine.

Developments

MTU has developed a 20-cylinder version of its Series 4,000 with a power output of 3,000 kW, the new engine is the most powerful high-speed diesel unit on the locomotive market.

Several units of the new Euro 3 Powerpack were supplied for field testing in railcars. The new model is powered by the new 6H 1,800 engine and complies with the Euro 3 standards.

UPDATED

Engines for railcars

Engine	Speed rpm	Output (UIC-rated power) kW	Output (UIC) mhp	Length mm	Width mm	Height mm	Mass	Exhaust emission
Engines for railcars								
6H 1800 R80	1,900	315	480	1,480	1,415	715	1,000	EPA Tier 2, UIC II, EURO II
PowerPacks for rails and their content								
Engines for push-pull trains and locomotives								
8V 4000 R41R	1,500	850	1,140	2,015	1,650	1,985	4,660	UIC II
8V 4000 R41	1,800	1,000	1,360	2,015	1,650	1,985	4,660	UIC II
8V 4000 R41L	1,860	1,100	1,495	2,015	1,650	1,985	4,660	UIC II
12V 4000 R41R	1,380	1,850	1,500	2,505	1,635	2,080	6,190	UIC II
12V 4000 R41	1,800	1,500	2,040	2,505	1,635	2,080	6,190	UIC II
12V 4000 R41L	1,860	1,650	2,245	2,505	1,635	2,080	6,190	UIC II
16V 4000 R41R	1,500	1,700	2,280	2,875	1,635	2,020	7,410	UIC II
16V 4000 R41	1,800	2,000	2,720	2,875	1,635	2,020	7,410	UIC II
16V 4000 R41L	1,860	2,200	2,990	2,875	1,635	2,020	7,410	UIC II
20V 4000 R42	1,800	2,700	3,670	3,630	1,470	2,055	9,450	UIC II
20V 4000 R42L	1,800	3,000	1,800	3,630	1,470	2,055	9,450	UIC II

DB Class 290 diesel-hydraulic locomotive re-powered with 8V 4,000 engine   0580242

NICO

Niigata Converter Company Ltd
405-3, Yoshinocho 1-chome, Omiya, Saitama 330-8646, Japan
Tel: (+81 48) 652 80 69   Fax: (+81 48) 652 87 19
e-mail: h-gocho@niigata-converta.co.jp
Web: http://www.niigata-converta.co.jp

Key personnel
President: S Takeuchi
Managing Director: T Hayashi
General Manager, International Operation: A Ochiai

Main works
Kamo plant: Gejyo Bo 405, Kamo-City, Niigata 959-1391, Japan

Omiya plant: 405-3 Yoshinocho 1-chome, Omiya, Saitama 330-8646, Japan
Muikamachi plant: Kawakubo 1095-1, Muikamachi, Minami-Uonuma, Niigata 949-6603, Japan

Principal subsidiaries
Niigata Converter Co Ltd, Europe Office, Beursplein 37, NL-3011 AA Rotterdam, Netherlands
Tel: (+31 10) 405 30 89   Fax: (+31 10) 405 50 67

Products
Single-stage torque converters, three-stage torque converters, power shift transmissions, hydraulic couplings for engines rated from 37 to 895 kW (50 to 1,200 hp).
Model TACN-22-1600 two-speed forward, two-speed reverse power shift transmission with Type 8 single-stage torque converter for 261 kW (350 hp) diesel railcars.
Model TACN-22-2000 two-speed forward, two-speed reverse power shift transmission with three-stage torque converter for 331 kW (444 hp) express diesel railcars.
Model TACN-33-3000 three-speed forward, three-speed reverse power shift transmission with Type 8 single-stage torque converter for 485 kW (650 hp) express diesel railcars.

Niigata Transys Co Ltd

F 9-7, Yaesu 2-chome, Chuo-ku, Tokyo, Japan
Tel: (+81 3) 62 14 28 77
Fax: (+81 3) 62 14 28 71
Web: www.niigata-transys.com

Key personnel
Administration Department: Hiroshi Goto

Subsidiary
NICO Precision Co Inc
Oaza Kawakubo 1095-1, Muikamachi, Minami-Uonuma-gun, Niigata, Japan

Background
The company was founded in 2003 as part of the restructuring of the former Niigata Engineering Co Ltd. Ishikawajima-Harima Heavy Industries Co Ltd (IHI) owns a shareholding of 70 per cent.

Products
Diesel engines for rail traction use up to 2,000 hp.

12V16FX
Type: 12-cylinder Vee, water-cooled, four-stroke turbocharged and charge air-cooled.

Cylinders: Bore 165 mm. Stroke 185 mm. Swept volume 3.96 litres per cylinder.

DMF18HZ
Type: 6-cylinder horizontal in-line, water-cooled, four-stroke turbocharged and charge air-cooled.

Cylinders: Bore 150 mm. Stroke 165 mm. Swept volume 2.92 litres per cylinder.

General specifications for both models
Cylinders: Monobloc cast-iron cylinder block and crankcase, removable cast-iron liners with integral water jacket. Cast-iron cylinder heads secured by studs.
Fuel injection: Bosch-type injectors and Bosch pump.
Lubrication: Forced feed.
Starting: Electric starter.

UPDATED

OMT SpA

Via Ferrero 67/A, I-10090 Casine Vica, Rivoli (TO), Italy
Tel: (+39 011) 957 53 54   Fax: (+39 011) 957 54 74
Web: http://www.omt-torino.com

Products
Diesel fuel injection equipment.

Paulstra

61 rue Marius Aufan, F-92305 Levallois-Perret, France
Tel: (+33 1) 40 89 53 31
Fax: (+33 1) 47 58 75 16
e-mail: auto.levallois@hutchinson.fr
Web: www.paulstra-vibrachoc.com

Key personnel
Commercial Director: Laurent Poirier

Products
Tetraflex coupling of 4,000 Nm torque capacity, part of a family of power transmission couplings developed by Paulstra. Used between electric motors and gearboxes, they are characterised by a reduced axial thickness and a radial misalignment capacity of several millimetres. The torque range available is between 2,000 and 8,000 Nm with a maximum speed of 3,000 to 3,500 rpm.

UPDATED

PSI (Peaker Services Inc)

8080 Kensington Court, Brighton, Michigan 48116-8591, USA
Tel: (+1 810) 437 41 74   Fax: (+1 810) 437 82 80
Web: http://www.peaker.com

Key personnel
President: Richard R Steele
Marketing and Sales: Vance Shoger, Frank Boatwright
Sales Engineers: Kim Stone, Terry Warrick

Products
Diesel engine rebuilding and maintenance; exchange service for diesel engine components and personnel training.
PSI is an independently owned company specialising in medium-speed, large diesel engines manufactured by the Electro-Motive Division of General Motors Corporation. Services include engine conversions and upgrading, unit exchange components, engine overhaul (including repair of case and pan sections, power assemblies, pumps and governors), field repairs, service contracts, application studies, crankcase line boring and repairs, locomotive inspections and evaluations and personnel training in locomotive maintenance and repair.

Regulateurs Europa

Port Lane, Colchester CO1 2NX, UK
Tel: (+44 1206) 79 95 56   Fax: (+44 1206) 79 26 85
e-mail: uksales@regulateurs-europa.com

Netherlands facility:
1e Energieweg 8, NL-9301 LK Roden, Netherlands
Tel: (+31 5050) 198 88   Fax: (+31 5050) 136 18
e-mail: sales@regulateurs-europa.com

Background
Regulateurs Europa is a subsidiary of MAN B&W Diesels Ltd.

Products
Diesel engine management and governor control systems, including load control and engine de-rates; remote monitoring of locomotives via cellular telephone or satellite communications; refurbishment and installation of locomotive control systems. Products include both mechanical-hydraulic and microprocessor-based governing systems.

SAN Engineering & Locomotive Co Ltd

PO Box 4802, Whitefield Road, Whitefield, Bangalore 48, Karnataka, India
Tel: (+91 80) 845 22 71-76
Fax: (+91 80) 845 22 60; 845 31 95
e-mail: inderm@san-engineering.com
Web: http://www.san-engineering.com

Key personnel
See entry in *locomotives and powered/non-powered passenger vehicles* section.

Principal subsidiary
Engineering Products Division
Plots 1 & 10, Hebbal Industrial Area, Belwadi Post, Mysore 571 106, India
Tel: (+91 821) 40 24 30; 40 23 21
Fax: (+92 821) 40 23 21
e-mail: mysore@san-engineering.com

Products
Transmissions for locomotives and industrial applications, heavy duty cardan shafts for locomotives and industrial applications, powered and non-powered final drives, gears and gearboxes.

SEMT Pielstick

Le Ronsard Paris Nord 2, 22 avenue des Nations, BP 84013 Villepinte, F-95931 Roissy CDG Cedex, France
Tel: (+33 1) 48 17 63 00
Fax: (+33 1) 48 17 63 49
e-mail: sales_traction@pielstick.com
Web: www.pielstick.com

Key personnel
Chairman of the Board of Management: Pierre Bousseau
Sales Director: Jean-Luc Cavellat

Background
Founded in 1946, the company is owned 66.6 per cent by MAN B&W and 33.4 per cent MTU Friedrichshafen.

Products
High-and medium-speed diesel engines for main line passenger and freight locomotives, including the PA4 (8 to 16 cylinders, 1,325 kW to 2,650 kW) and PA6 (8 to 16 cylinders, 2,800 kW to 6000 kW) series.

PA4-200 Series
Type: 8, 12 and 16 Vee (90°), 4-cycle, water-cooled, with variable geometry precombustion chamber (VG).
Cylinders: Bore 200 mm. Stroke 210 mm. Swept volume 6.6 litres per cylinder. Wet liners, individual cast-iron cylinder heads.
Pistons: Cast-iron pistons with steel crown cooled by pressure lubricating oil fed through connecting rod and piston pin into an annular chamber level with top compression ring.
Fuel injection: Monobloc injection pump inside the Vee, controlled by electronic governor.

Superchargers: Exhaust gas turbochargers, one for 8- and 12-cylinder engine, two 16 cylinders, between cylinder banks. Air coolers arranged on timing gear side.
Cooling: Water pumps fitted on timing gear end of frame.
Starting: Either electrically or by compressed air.

PA5-255 Series
Type: 6 and 8 cylinders in-line, 12 and 16 cylinders Vee, supercharged, water-cooled.
Cylinders: Bore 255 mm. Stroke 270 mm. Swept volume 13.79 litres per cylinder. Wet liners directly mounted in the crankcase, without cooling jackets. Individual cast-iron cylinder heads. Single combustion chamber. Direct injection.
Fuel injection: Direct injection by means of injectors of the multihole type. Individual injecting

	PA4-200 VGA		PA5-255				PA6B-280		
No of cylinders	12(V)	16(V)	6(L)	8(L)	12(V)	16(V)	8(L)	12(V)	16 (V)
Turbocharged	Yes	Yes	Yes	Yes	Yes	Yes	–	–	–
Supercharged	–	–	–	–	–	–	Yes	Yes	Yes
Charge air-cooled	Yes	Yes	Yes	Yes	Yes	Yes	Yes	Yes	Yes
Power rating (UIC) (hp)	2,400	3,600	1,955	2,610	3,915	5,220	3,760	5,640	7,500
Engine speed (rpm)	1,500	1,500	1,000	1,000	1,000	1,000	1,000	1,000	1,000
Piston speed (m/s)	10.5	10.5	9	9	9	9	11	11	11
Bmep (bar)	20.5	20.5	19	19	19	19	–	–	–
Weight (dry) (kg)	8,000	10,000	10,500	13,500	17,000	22,200	18,000	27,000	34,000
Length (mm)	2,630	3,630	3,590	4,400	4,060	5,140	4,820	4,480	5,270
Width (mm)	1,600	1,680	1,300	1,310	1,980	2,070	1,450	1,700	1,700
Height (mm)	1,975	1,875	2,255	2,440	2,620	2,870	2,755	2,890	2,890

pump housed in the crankcase, directly controlled by the camshafts. Injection controlled by hydraulic speed governor.

Turbochargers: Two per engine, driven by a turbine on the exhaust gas, and housed in the centreline of the engine above each end of the crankcase. Air cooler at supercharger outlets, housed above the middle of the crankcase, and fed by the low temperature water circuit.

The PA5 engine has successfully run 100 hours UIC tests (12 PA5 V).

PA6B-280 Series
Type: 6 and 8 cylinders in-line, 12 and 16 cylinders Vee, supercharged, water-cooled.

Cylinders: Bore 280 mm. Stroke 330 mm. Swept volume 20.3 litres per cylinder. Wet liners directly mounted in the crankcase, with cooling jackets. Individual cast-iron cylinder heads. Single combustion chamber. Direct injection.

Fuel injection: Direct injection by means of injectors of the multihole type. Individual injecting pump housed in the crankcase, directly controlled by the camshafts. Injection controlled by hydraulic speed governor.

Turbochargers: Driven by a turbine on the exhaust gas, and housed above the end of the crankcase. Air cooler at supercharger outlets, housed above the middle of the crankcase and crossed by a special waterline.

Cooling: Two water pumps of the centrifugal type, driven by the timing train, one for jacket and cylinder head line, the other for air cooler and lube oil line.

360 hour UIC test
The 12-cylinder 12PA6V-280 engine has officially run its 360 hour UIC locomotive test in accordance with ORE regulations.

UPDATED

SILSAN

Silindir ve MotorElemanlari san ve Tic, AS
Mersin Yolu Uzeri 10, PK 127 Carsi, TR-01210 Adana, Turkey
Tel: (+90 322) 441 00 12   Fax: (+90 322) 441 00 86

Key personnel
Managing Director: Mehmet Bacaksizlar
Sales and Marketing: Caghan Bacaksizlar

Products
The company manufactures reciprocating diesel and petrol engine components for rail and road vehicles.

Socofer

7 boulevard Louis XI, Zone Industrielle du Menneton, BP 0507, F-37205 Tours Cedex 3, France
Tel: (+33 2) 47 39 28 24
Fax: (+33 2) 47 37 63 04
e-mail: info@socofer.com
Web: www.socofer.com

Key personnel
President: Pierre Hallé
Director, Rail Division: Bertrand Hallé
Sales and Marketing Manager: Fabrice Thierry

Products
Engine retrofitting kits supplied pre-assembled complete with power unit and all associated

equipment. Kits are designed to fit with the existing vehicle structure.

Contracts
Provision of retrofit engine packages for 85 SNCF Classes X2100 and X2299 diesel railcars; 50 engine retrofitting kits based on an MTU 12V4000 R41 power unit for SNCF Class BB66000 locomotives; two engine retrofitting kits based on a Wartsila UD30 V12 R5D power unit for BB63500 diesel locomotives operated by the Bouches de Rhône departmental unit.

NEW ENTRY

Handover of the first SNCF Class BB66000 locomotives to receive an engine retrofit kit from Socofer
0573983

Spicer Gelenkwellenbau GmbH & Co KG (GWB)

Werk I/ Westendhof 5-9, D-45143 Essen, Germany
Werk II/ 2.Schnieringstrasse 49, 45329 Essen, Germany
Tel: (+49 201) 812 40   Fax: (+49 201) 812 44 59
e-mail: industrial@dana.com
Web: http://www.gwb-essen.de

Key personnel
General Manager: Lars Christoph Schäfer

Background
GWB is a subsidiary of the US-based DANA Corporation.

Products
Driveshaft systems and components for locomotives and rail vehicles. The range extends from standard driveshafts for torques starting at 1.5 kNm to ultra-heavy duty units rated for 15,000 kNm and beyond.

Tülomsas

The Locomotive and Motor Corporation of Turkey
Ahmet Kanatli Cad, TR-26490 Eskisehir, Turkey
Tel: (+90 222) 225 99 56   Fax: (+90 224) 225 72 72
e-mail: tulomsas@tulomsas.com.tr
Web: http://www.tulomsas.com.tr

Key personnel
For a full list of personnel, see Tülomsas entry in *Locomotives and powered/non-powered passenger vehicles* section.

Products
Diesel engines with outputs of 1,865, 2,200 and 2,400 hp.

Turbomeca

F-64511 Bordes, France
Tel: (+33 5) 59 12 50 00
Fax: (+33 5) 59 53 15 12

Key personnel
President: Jean-Bernard Cocheteux
Executive Vice-President: Jean-Louis Chenard
Marketing Manager: Bernard Watier
Turbomeca Engine Corporation

Subsidiary
Turbomeca Engine Corporation
2709 Forum Drive, Grand Prairie, Texas 75052, USA
Tel: (+1 214) 641 66 45

Products
There are two turbines currently being marketed for railway use. The first is the Makila TM-1600. This engine is used for both traction motor drive as well as direct propulsion. The TM-1600 has a speed of 6,500 rpm and a power output of 1,050 kW (1,600 hp).

It is a two-shaft engine and weighs 410 kg with a length of 1,850 mm.

The Astazou is a single-shaft 1,500 rpm engine with an output of 450 hp and 330 kW. It measures 1,500 mm and weighs 300 kg. It is used for running hotel power.

A third engine developed with Volvo and Ulstein is available. Called Eurodyn, it has a power of 2.5/3 mW and a efficiency of 34 per cent. Its lightness makes it suitable for high-powered locomotives.

Twiflex Ltd

104 The Green, Twickenham TW2 5AQ, UK
Tel: (+44 20) 88 94 11 61
Fax: (+44 20) 88 94 60 56

Key personnel
Sales and Marketing Director:
    Jonathan P Cooksley
Technical Director: Peter Wood

Background
Twiflex Limited is owned by Hay Hall Group.

Products
Advanced braking technology, industrial disc brakes, Layrub and Laylink flexible shafts and couplings; industrial and marine disc brakes and Flexi-clutch couplings.

Both Layrub and Laylink couplings incorporate compressed cylindrical rubber blocks. The Laylink coupling carries these blocks in links, while the Layrub coupling carries them in a carrying plate. The use of these couplings and flexible shafts allows large amounts of angular and axial misalignment to be accommodated; it also absorbs shock, controls vibrations and simplifies close coupling in confined spaces. The units need no servicing or lubrication and can cater for very high operating speeds and transmission of high power without loss.

Twin Disc Incorporated

1328 Racine Street, Racine, Wisconsin 53403-1758, USA
Tel: (+1 262) 638 40 00   Fax: (+1 262) 638 44 82
Web: http://www.twindisc.com

Key personnel
Chairman, Chief Executive Officer: M E Batten
President, Chief Operating Officer: M H Joyce
Vice-President, Marine Marketing and Distribution:
    H C Fabry
Commercial Manager, Production Systems and
    Operations: J H Batten

Principal subsidiaries
Twin Disc International SA, Chaussée de Namur 54, B-1400 Nivelles, Belgium
Tel: (+32 67) 88 72 11   Fax: (+32 67) 88 73 33

Twin Disc (Pacific) Pty Ltd, PO Box 442, Virginia, Queensland 4014, Australia
Tel: (+61 7) 32 65 12 00   Fax: (+61 7) 38 65 13 71

Twin Disc (South Africa) Pty Ltd, PO Box 40542, Cleveland 2022, South Africa
Tel: (+27 11) 626 27 14   Fax: (+27 11) 626 27 17

Twin Disc (Far East) Ltd, PO Box 155, Jurong Town Post Office, Singapore 9161
Tel: (+65) 62 61 89 09   Fax: (+65) 62 64 20 80

Products
Universal joints; hydraulic torque converters, power shift transmissions and controls suitable for locomotives and railcars.

UCM Reşiţa SA – Machine Building Company

1 Golului Street, RO-320053 Reşiţa, Romania
Tel: (+40 255) 21 71 11   Fax: (+40 255) 22 30 82
e-mail: contact@ucmr.ro
Web: www.ucmr.ro

Key personnel
President and General Director: Adrian Chebutiu
Operational Director: Sorin David
Commercial Director: Dan Petrescu

Background
Resita Works was established in 1771, originally as a factory of metal manufacturing and machining and has evolved over the years into its present manufacturing status.

2,240 kW diesel engine produced by UCM Resita under ALCO licence.
NEW/1112234

For details of the latest updates to *Jane's World Railways* online and to discover the additional information available exclusively to online subscribers please visit
jwr.janes.com

R 251 series engines

No of cylinders	6	8	12	16
rpm range	400–1,100	400–1,000	400–1,100	400–1,100
Max, full load (HP)	700–1,520	1,000–1,720	1,500–3,040	1,900–3,960
Compression ratio	(12.5) 11.5 to 1	(12.5) 11.5 to 1	(12.5) 11.5 to 1	(12.5) 11.5 to 1
Bore & stroke (mm)	228 × 267	228 × 267	228 × 267	228 × 267
Displacement (dm³)	65.7	87.6	131.4	175.2
Turbocharged	Yes	Yes	Yes	Yes
Aftercooled	Yes	Yes	Yes	Yes
Fuel system type	Direct	Direct	Direct	Direct
Heat dissipation rate				
Oil (kcal/Hp.min)	1.2	1.2	1.2	1.2
Water (kcal/Hp.min)	5.5	5.5	5.5	5.5
Starting system	air motor	–	–	air motor
Cum/start, average	7	7	7	7
Pressure range (bar)	6–10	6–10	6–10	6–10
Weight – dry (tons)	11.2	12.0	14.7	19.3
–wet (tons)	12.0	12.7	15.5	21.0

Based on Alco licence. MDO operation also available.

LDS/LDSR series engines

No of cylinders	6	12	12
rpm range	350 – 750	350 – 750	350 – 750
Max, full load (HP)	1,250	2,100	2,500
Compression ratio	11.25 to 1	11.25 to 1	11.25 to 1
Bore & stroke (mm)	280 × 360	280 × 360	280 × 360
Displacement (dm³)	133	266	266
Turbocharged	Yes	Yes	Yes
Aftercooled	Yes	No	Yes
Fuel system type	Direct	Direct	Direct
Heat dissipation rate			
Oil (kcal/Hp.min)	1.3	1.8	1.3
Water (kcal/Hp.min)	5.6	4.4	5.6
Starting system	1	2	2
Cum/start, average	7	7	7
Pressure range (bar)	6–10	6–10	6–10
Weight – dry (tons)	10.5	28.0	29.3
–wet (tons)	11.3	29.5	30.8

[1] Based on Sulzer licence. External system with traction generator or dyna-starter. Air motor on special request.
[2] DC generator included. Engine start by DC generator. Air motor on special request.

Products

High-speed diesel engines for locomotives, under Sulzer licence with power range between 634 kW (850 hp) and 1,865 kW (2,500 hp) and Alco licence with power range between 522 kW (700 hp) and 2,984 kW (4,000 hp); spare parts for R251, and LDS/LDSR engines; repairs and refurbishment for R251, Alco 251 and LDS/LDSR engines.

UPDATED

Unipar Inc

7210 Polson Lane, Hazelwood, Missouri 63042, USA
Tel: (+1 314) 521 81 00   Fax: (+1 314) 521 80 52

Key personnel

Executive Vice-President: Dennis McClure
Sales Manager: Mark Cleveland

Products

New and remanufactured replacement power assemblies for GM/EMD locomotives. Component parts for Alco, GE and GM locomotives.

VERIFIED

Voith Turbo GmbH & Co KG

PO Box 1930, D-89509 Heidenheim, Germany
Tel: (+49 7321) 37 28 62   Fax: (+49 7321) 37 71 10
e-mail: info.voithturbo@voith.com
Web: http://www.voithturbo.com

Key personnel

President, Chief Executive Officer and
  Vice-President: Peter Edelmann
General Manager, Turbo Transmissions:
  Manfred Lerch
General Manager, Axle Drives: Arno Hoepner
General Manager, Service Centre: Karl Sing
General Manager Scharfenberg® Couplers:
  Martin Wawra

UK subsidiary

Voith Turbo Ltd
6 Beddington Farm Road, Croydon CR0 4XB, UK
Tel: (+44 20) 86 67 03 33   Fax: (+44 20) 86 67 04 03
e-mail: turbo.uk@voith.com
Manager, Rail Products: A L Morris

Background

Voith has been in business since 1869 and subsidiary company Voith Turbo has been established in the UK since 1962.

Products

Hydrodynamic transmissions and retarders, torque converters and automatic hydromechanical transmissions. Final drive reduction gearboxes for mechanical or electrical drives in locomotives, metro cars and light rail vehicles. DIWA hydromechanical transmissions for light rail vehicles. Limited-slip differential device for metro cars and light rail vehicles. Voith Turbo provides drive solutions for rail vehicles. Customer-specific drive systems, extending from the technical design, configuration and construction to application and testing created using technology innovative systems solutions and services.

Voith automatic hydrodynamic transmissions are designed specifically for installation in rail vehicles. The basic components are drainable hydraulic torque converters and fluid couplings, which in combination provide tractive effort over a wide speed range in an efficient and cost-effective manner. The filling characteristics ensure wear-free, smooth shifting without interruption of tractive effort. Turbo-reversing transmissions used in shunting locomotives extend the principle to allow direction shifting in a similar manner. All non-reversing turbo transmissions can be fitted with a wear-free retarder.

Voith final drives are available in bevel and spur gear configurations, from axle-hung drives to bogie or body-suspended quill-shaft drives, with single or double reduction.

Scharfenberg® coupler has developed into one of the most important and successful coupling systems since its invention and introduction in 1903. Scharfenberg is an internationally recognised brand name and over 300,000 couplers have been supplied worldwide.

Development

The LS 640 re V2 split turbo transmission with an output of 2,700 kW is in development.

Voith Turbo has developed the L620 re V2 turbo transmission with an input power of 2,300 kW for main line diesel locomotives   0105887

Voith Turbo Safeset AB

Rönningevägen 8, SE-824 34 Hudiksvall, Sweden
Tel: (+46 650) 54 01 50
Fax: (+46 650) 54 01 65
e-mail: info.safeset@voith.com
Web: www.safeset.com

Background

Since 1992, the company has been wholly owned by J M Voith.

Products

Torque limiting couplings, including the Safeset® coupling for drivelines in high-speed trains, which in event of cardan shaft failure protects the driveline from over-torque.

The system has been supplied to Fiat Ferroviaria for ETR-series Pendolino tilting trainsets, to ALSTOM for Virgin Rail Group's West Coast Main Line Pendolinos in the UK and to AEA Technology for retrofitting on Class 91 electric locomotives operated by GNER in the UK.

UPDATED

Volvo Penta

Volvo Penta Central Europe GmbH
Redderkoppel 5, D-24159 Kiel, Germany
Tel: (+49 431) 399 42 07   Fax: (+49 431) 399 41 55
e-mail: vp.90040hm@memo.volvo.se
Web: http://www.volvo.com

Key personnel
Business Manager, Rail: Helmut Möller

Products
RailPac unit with Volvo DH10A horizontal diesel engine and Volvo Powertronic VT1605PT five-speed automatic transmission.

Volvo Penta engines are the power units in the 'Cargosprinter' Windhoff freight dmu, now in operation with DB Germany. Each power car has two Rail Pac engines, the front one driving the rear axle of the front powered bogie and the engine at the rear driving the front axle of the rear bogie. Each Rail Pac comprises a Volvo DH10-A360 engine rated at 265 kW, at 2,050 rpm, and meeting Euro-2 regulations. The drive is through a Volvo Powertronic Pt 1650 five-speed automatic gearbox which includes converter and retarder as well as electronic torque limiter.

The Volvo Penta DH10A-360 horizontal diesel engine is used to power the DWA LVT/S 64-seat railcar.

MAV Hungary is took delivery of Rail Pac engines for re-powering of Railcargo units in 1998/99.

Volvo Penta RailPac unit with DH10A diesel engine   0063655

ZF Friedrichshafen AG

ZF Friedrichshafen AG
D-88038 Friedrichshafen, Germany
Tel: (+49 7541) 770   Fax: (+49 7541) 77 90 80 00
e-mail: postoffice@zf.com
Web: http://www.zf.com

Key personnel
Supervisory Board:
  Chairman: Wolf Hartmut Prellwitz
  Vice-Chairman, Employee-elected
    representative: Johann Kirchgässner
Board of Management:
  Chief Executive Officer: Siegfried Goll
    Paul Ballmeier, Uwe Berner, Michael Paul,
    Wolfgang Vogel

Business Units
Car Driveline Technology, ZF Getriebe GmbH, Saarbrücken, Germany; car chassis technology, ZF Lemförder Fahrwerktechnik AG & Co KG, Lemförder, Germany; commercial vehicles and special driveline technology, ZF Friedrichshafen AG, Friedrichshafen, Germany; off-road driveline technology and axle systems, ZF Passau GmbH, Passau, Germany; powertrain and chassis components, ZF Sachs AG, Schweinfurt, Germany; steering technology, ZF Lenksysteme GmbH, a joint venture of ZF Friedrichshafen AG und Robert Boch GmbH, Schwäbisch Gmünd, Germany.

Background
Shareholders include Zeppelin Foundation, Friedrichshafen, Germany 93.8 per cent; Dr Jürgen Ulderup Foundation, Lemförder, Germany 6.2 per cent.

Products
Rail driveline technology, including transmissions for locomotives, driveline units for underground trains, city rail vehicles, and metro trains as well as input kits for regional trains (electric and diesel railcars).

In collaboration with ZF Sachs AG, ZF Bahntechnik GmbH can supply diesel railcar engine manufacturers with complete systems, comprising clutches, transmissions and electronic control units. This is especially important when using transmissions in rail vehicles such as diesel railcars, where each car has its own driveline: in this case, a highly developed electronic control unit has to perfectly synchronise several transmissions in line with gear change operations to the millisecond.

The company is developing and producing new transmission systems in co-operation with suppliers of wheels, engines, axles and brakes. This ensures that all cost and delivery advantages can be offered by one supplier. For example, ZF Bahntechnik has supplied the new IC4 intercity train for Danish State Railways with the AS Rail Transmission, as well as the clutch, prop shaft, axle reversing transmission and torque support.

ZF also produces powertrain and suspension components for rail vehicles. The product range of dampers for rail vehicle applications includes primary and secondary dampers (vertical and horizontal), yaw dampers, dampers for engines, superstructures and articulated roof structures.

ZF Sachs has further developed Pneumatic Damping Control (PDC) dampers for use in freight wagons. Pneumatically controlled proportional damper valves progressively adapt the damping to the respective load condition via the brake pressure.

This system is especially suitable as a horizontal damper system for major load differences. In their function as vertical dampers, PDC dampers also improve comfort in pneumatically damped rail vehicles. This allows for higher speeds at a lower driveload.

Complete rubber-metal modules and individual components for vibration damping in rail vehicles in close co-operation with the leading locomotives, tram and wagon manufacturers. ZF Boge Elastmetall plays an active part in future-oriented developments like monorail technology (Transrapid), tilting technology (Neitec), high-speed trains (ICE), and low-floor tram technology (Combino).

Complete heavy damping systems are developed in close co-operation with ZF Sachs. The bearings of these systems are suitable for any application. ZF Boge Elastmetall offers elaborate fully machined rubber-metal parts, components and ready-to-install modules from one source, due to the high vertical range of manufacture.

Special transmissions and high-quality gears for rail vehicles. The product range includes cylindrical and cone-shaped geometry. Supplies for all kinds of trains are provided, including high-speed trains, including the Italian ETR 500 and the French TGV, underground trains and trams.

ZF Padova is equipped for spare parts production to customer specifications and drawings. New components can be produced by reverse engineering based on a sample. A technical department, set up especially for these tasks, offers customer support and has the expertise to develop new transmissions according to requirements and technical conditions.

ZF Padova SpA

Via Penghe N 48, I-35030 Caselle di Selvazzano, Padua, Italy
Tel: (+39 049) 829 93 11
Fax: (+39 049) 829 95 50; 95 60; 95 70
e-mail: special-products@zf.com
Web: http://www.zf.com

Key personnel
President: H G Harter
Managing Director: Roland Heil
Product Manager: Adriano Giuriati

Background
ZF Padova is owned by ZF Friedrichshafen AG based in Germany.

Products
Special transmissions and high-quality gears for rail vehicles. The product range includes cylindrical and cone-shaped geometry. Supplies for all kinds of trains are provided, including high-speed trains, such as the Italian ETR 500 and the French TGV, underground trains and trams.

ZF Padova is equipped for spare parts production to customer specifications and drawings. New components can be produced by reverse engineering based on a sample. A technical department, set up especially for these tasks, offers customer support and has the expertise to develop new transmissions according to requirements and technical conditions.

Contracts
ZF Padova is a certified supplier for Trenitallia, SNCF, SNCB and other leading rail vehicle companies and manufacturers.

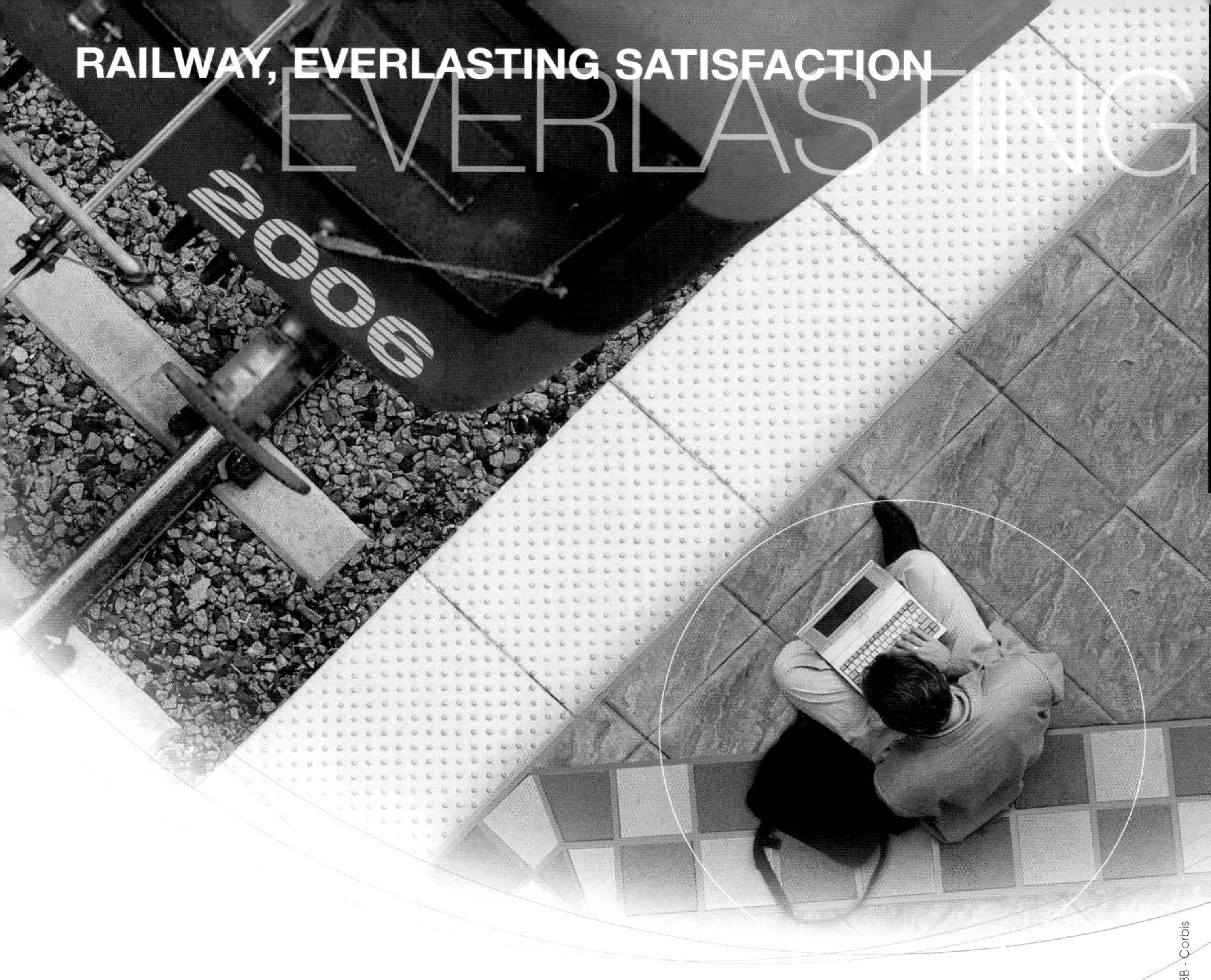

RAILWAY, EVERLASTING SATISFACTION

EVERLASTING

2006

Electric Traction Equipment

© Copyright 2004 ABB - SYGMA n° 05480 B. Photos : SBB - Corbis

ABB, A RELIABLE ENERGY

Partner of railway manufacturers and operators, ABB transforms, regulates and distributes power wherever railway goes.

Developing a comprehensive offer and high performance solutions, ABB contributes actively to the progress of comfort and security. Offering its human competencies and technological expertise to the railway industry, ABB takes part in its development over the five continents.

With equipment reliability, passenger safety and environmental protection as its top priorities, ABB remains loyal to its 100 year old tradition of contributing to world class rail systems.

ELECTRIC TRACTION EQUIPMENT

Alphabetical listing

3M Electronics Markets Materials Division, Europe
ABB
A K Fans
ALSTOM Transport
Ansaldobreda SpA
APS Electronic AG
Behr Industry GmbH & Co, KG
Bharat Heavy Electricals Ltd (BHEL)
Bombardier Transportation
Brecknell Willis & Co Ltd
Brush Traction
China Northern Locomotive & Rolling Stock
    Industry (Group) Corporation(CNR)
China South Locomotive and Rolling Stock Industry
    (Group) Corporation (CSR)
Clyde Engineering
Contransys Pvt Ltd
Crompton Greaves Ltd
Daewoo Heavy Industries Ltd
Dynex Semiconductor
EFACEC Sistemas de Electrónica SA

EG & G Rotron
ELIN EBG Traction GmbH
EVPÜ
Faiveley Transport SA
Ferraz Shawmut
FIREMA Trasporti SpA
Freudenberg Schwab GmbH
Fuji Electric Systems Co Ltd
Ganz Transelektro Traction Electric Ltd
Gardner Denver Ltd
GE Transportation Systems
Hall Industries Inc
Hitachi Ltd
Hyundai Heavy Industries Co Ltd
InfoSystems GmbH
KMT-teknikka Oy
Končar – Elecktrične Lokomotive dd
Lechmotoren GmbH
Lekov AS
Mitsubishi Electric Corporation
Morio Denki Co Ltd

National
Parizzi
Permali Gloucester Ltd
Production SpA
RDS Technology Ltd
Riga Electric Machine Building Works
Rotem Company
SAFT
Schunk Bahntechnik GmbH
Schunk Kohlenstofftechnik GmbH
Sécheron SA
Siemens AG
Škoda Trakči motory sro
SPII
Stemmann-Technik GmbH
Toyo Denki Seizo KK
Traktionssysteme Austria GmbH
Transportation Products Sales Co Inc
Tülomsas
Vossloh Kiepe GmbH

Company listing by country

AUSTRALIA
Clyde Engineering

AUSTRIA
ELIN EBG Traction GmbH
Schunk Bahntechnik GmbH
Traktionssysteme Austria GmbH

CHINA
China Northern Locomotive & Rolling Stock
    Industry (Group) Corporation (CNR)
China South Locomotive and Rolling Stock Industry
    (Group) Corporation (CSR)

CROATIA
Končar – Elecktrične Lokomotive dd

CZECH REPUBLIC
Lekov AS
Škoda Trakči motory sro

FINLAND
KMT-teknikka Oy

FRANCE
ALSTOM Transport
Faiveley Transport SA
Ferraz
SAFT

GERMANY
3M Electronics Markets Materials Division, Europe
Behr Industry GmbH & Co, KG
Freudenberg Schwab GmbH

InfoSystems GmbH
Lechmotoren GmbH
Schunk Kohlenstofftechnik GmbH
Siemens AG
Stemmann-Technik GmbH
Vossloh Kiepe GmbH

HUNGARY
Ganz Transelektro Traction Electric Ltd

INDIA
Bharat Heavy Electricals Ltd (BHEL)
Contransys Pvt Ltd
Crompton Greaves Ltd

ITALY
Ansaldobreda SpA
FIREMA Trasporti SpA
Parizzi
Production SpA
SPII

JAPAN
Fuji Electric Co Ltd
Hitachi Ltd
Mitsubishi Electric Corporation
Morio Denki Co Ltd
Toyo Denki Seizo KK

KOREA, SOUTH
Daewoo Heavy Industries Ltd
Hyundai Heavy Industries Co Ltd
Rotem Company

LATVIA
Riga Electric Machine Building Works

PORTUGAL
EFACEC Sistemas de Electrónica SA

SLOVAKIA
EVPÜ

SWITZERLAND
ABB
APS Electronic AG
Bombardier Transportation
Sécheron SA

TURKEY
Tülomsas

UNITED KINGDOM
A K Fans
Brecknell Willis & Co Ltd
Brush Traction
Dynex Semiconductor
Gardner Denver Ltd
Permali Gloucester Ltd
RDS Technology Ltd

UNITED STATES
EG & G Rotron
GE Transportation Systems
Hall Industries Inc
National
Transportation Products

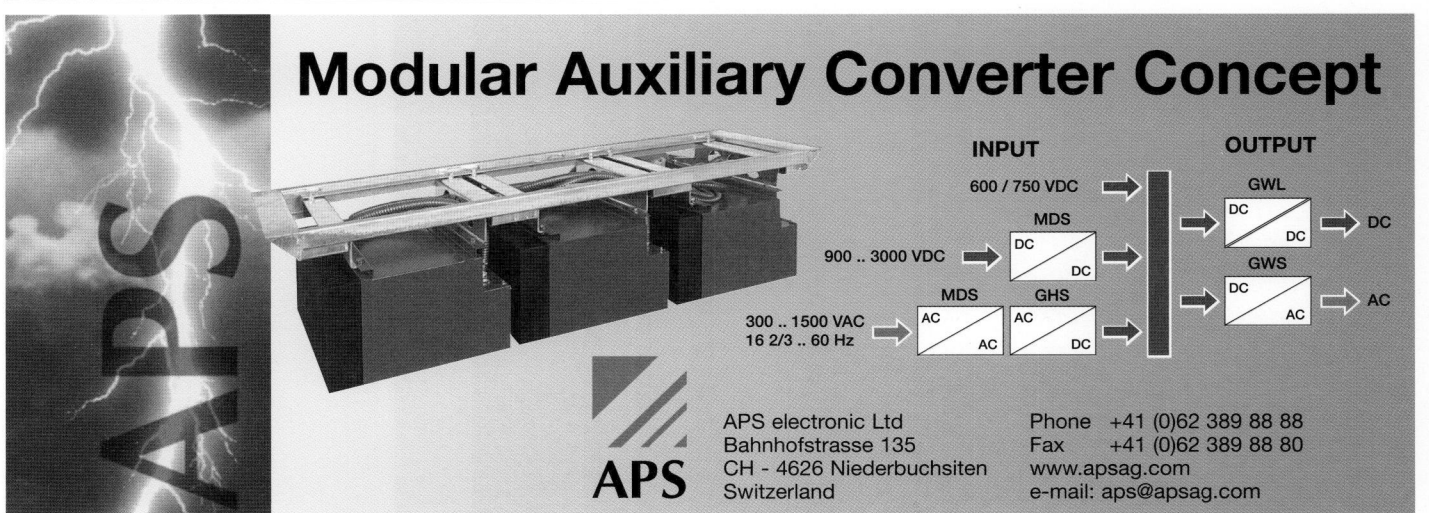

Modular Auxiliary Converter Concept

INPUT
600 / 750 VDC
900 .. 3000 VDC
300 .. 1500 VAC
16 2/3 .. 60 Hz

MDS
MDS   GHS

OUTPUT
GWL   DC
GWS   AC

APS electronic Ltd
Bahnhofstrasse 135
CH - 4626 Niederbuchsiten
Switzerland

Phone   +41 (0)62 389 88 88
Fax   +41 (0)62 389 88 80
www.apsag.com
e-mail: aps@apsag.com

3M Electronics Markets Materials Division, Europe

c/o 3M Germany
Carl Schurz Strasse 1, D-41453 Neuss, Germany
Tel: (+49 2131) 14 24 17   Fax: (+49 2131) 14 27 49
e-mail: pbreloer@mmm.com
Web: www.3m.com

Key personnel
European Business Manager: P Breloer

Head office
3M Center
St Paul, Minnesota 55144, USA
Tel: (+1 651) 736 38 95   Fax: (+1 651) 736 60 41
Web: www.mmm.com/

Vice President, EMMO: Jesse Singh
Global Business Director, Specialty Fluids:
  Craig Schwartz

Products
Fluorinert dielectric liquids for converter system direct contact cooling; coolants for converters, systems (FC-72, FC-3284, FC-87).

NEW ENTRY

ABB

Rail Customer Segment
Rue des Sablières 4-6, Case postale 2095, CH-1211 Geneva 2, Switzerland
Tel: (+41 58) 586 22 11   Fax: (+41 58) 586 20 21
e-mail: info.abb.secheron@ch.abb.com
Web: www.abb.com/railway

Products
ABB offers a complete range of components and subsystems for both rolling stock and fixed installations.

Traction transformers for single- and multi-system operation (15kV 16.7 Hz and/or 25 kV 50 or 60 Hz, other voltages and frequences for example, 11 kV 25 Hz, and also for DC operation as 750 V, 1.5 kV and 3kV) in a power range up to 10 MVA and with underframe, vertical or roof-mounted installation.

Traction power and auxiliary converters, battery chargers; traction motors and generators, auxiliary motors; power semiconductors, from 300 to 12 kV and 200 V to 8.5 kV; communication equipment; surge arrestors; turbo-chargers; low voltage components, relays, connectors; sub-station switchgear for both indoor and outdoor installations and single- or two-phase operation; high power converters and rectifiers for frequency conversion; power quality equipment, static VAR compensation; sub-station transformers; auxiliary and booster transformers; protection and control; high-voltage equipment; power supply systems; compact substations.

A K Fans

A K Fans Ltd
32-34 Park Royal Road, London NW10 7LN, UK
Tel: (+44 20) 89 61 68 88   Fax: (+44 20) 89 65 06 01
e-mail: sales@akfansltd.btinternet.com

Key personnel
Managing Director: D C Moore
Financial Controller: R Brazell
Operations Manager: J M Brinkworth
Chief Engineer: A McArthur
General Sales Manager: G A Anderson
Rail Sales Manager: P Heapy

Products
Fan for rail traction applications; types include axial, centrifugal, mixed flow and crossflow, with AC or brushless DC motors. Applications include cooling of traction motors converters, auxiliary inverters, brake resistors, compressors and transformers. Cooling systems comprising matched fan and heat exchanger assemblies.

Contracts
Contracts include cooling underbody equipment on trains for Arlanda, Incheon, Manila, Adana and Gardermoen.

UPDATED

ALSTOM Transport

48 rue Albert Dhalenne, F-93482 Saint-Ouen Cedex, France
Tel: (+33 1) 41 66 90 00   Fax: (+33 1) 41 66 96 66
Web: www.transport.alstom.com

Contact points
Traction drives, power modules, switchgear
ALSTOM Transport
50 rue du Dr Guimier, BP 4, F-65600 Semeac, Tarbes, France
Tel: (+33 5) 62 53 41 21   Fax: (+33 5) 62 53 40 01
Contact: Eric Lenoir Traction drives, auxiliary converters

ALSTOM Transport
11-13 avenue du Bel Air, F-69627 Villeurbanne Cedex, France
Tel: (+33 4) 72 81 52 00   Fax: (+33 4) 72 81 52 87
Contact: Eric Lenoir

Motors
On-board electronic systems
ALSTOM Transport
7 av de Lattre de Tassigny, BP 49, F-25290 Ornans Cedex
Tel: (+33 3) 81 62 44 00   Fax: (+33 3) 81 62 44 01
Contact: Eric Lenoir

Motors
ALSTOM Transport
50/52 rue Cambier Dupret 6001, Charleroi, Belgium
Tel: (+32 71) 44 54 11   Fax: (+32 71) 44 57 82
Contact: Jacques Toppet

ALSTOM Transport Netherlands BV
PO Box 3021, NL-2980 DA, Ridderkerk, Netherlands
Tel: (+31 180) 45 28 57   Fax: (+31 180) 45 28 60
Contact: Mathilde De Winter

ALSTOM Transportation Inc
1 Transit Drive, Hornell, New York 14843
Tel: (+1 607) 324 45 95   Fax: (+1 607) 324 23 68
Contact: Jennifer Luo
Manhattan office

ALSTOM Transportation Inc
353 Lexington Avenue, Suite 1100, New York, New York 10016, US
Tel: (+1 201) 692 53 20

Traction drives, auxiliary converters, switchgear, motors, transformers
ALSTOM Transport Ltd
F – 4, East of Kailash, New Delhi 110 065, India
Tel: (+91 11) 628 77 16; 77 47; 77 48
Fax: (+91 11) 628 77 15
Contact: Sudha Gupta

Traction drives, auxiliary converters, switchgear
Shanghai ALSTOM Transport Electrical Equipment Co Ltd
Room 915, Electric Power Building, 430 Xu Jia Hui Road, 200025 Shanghai, China
Tel: (+86 21) 64 72 51 57   Fax: (+86 21) 64 72 96 62
Contact: Lu Yin

Products
The company's strategy for traction equipment centres on its IGBT propulsion system known as ONIX (Onduleur à Integration exceptionnelle). The know-how in advanced IGBT power modules, Agate control electronic systems and ONIX asynchronous traction motors provides a compact, lightweight, integrated traction system, which is in reliable service around the world. ONIX is suitable for all line voltages (including the latest 3,000 V). In the ONIX range is the 350, for electric and trolleybuses (up to a line voltage of 400 V), 800 for metros and LRVs (up to a line voltage of 900 V), 1,500 for metros, emus, locomotives and high-speed trains (up to a line voltage of 2,000 V) and 3,000 V for emus, locomotives and high-speed trains (up to a line voltage of 3,000 V).

ONIX's modularity enables quick and easy maintenance and its flexibility allows a choice of cooling system to suit the local operating environment.

More than 5,500 units have been sold worldwide, representing 60 per cent of the world IGBT market.

ALSTOM's electric and electronic sub-systems include: electronic systems featuring the AGATE integrated family of traction control (AGATE Control), auxiliary control (AGATE Aux), Train Control Management Systems (AGATE Link), passenger information systems (AGATE Media, see Passenger Information Systems section), as well as depot monitoring systems.

A modular range of asynchronous, synchronous and DC motors, which are lightweight, compact and can be readily adapted to specific needs.

Auxiliary converters based on the modular CARA auxiliary power conversion and distribution packages.

A range of AC and DC switchgear, circuit breakers and cab equipment.

Contracts
Details of ALSTOM-built rolling stock contracts for which traction equipment has been ordered or supplied appear in the *Locomotives and powered/ non-powered passenger vehicles* section.
Australia: ALSTOM won an order from Clyde Engineering to supply the traction equipment (80 ONIX 1500 IGBT inverter drives plus the new generation of AGATE train management and passenger information systems), auxiliary power supplies and train operating system, as well as a comprehensive maintenance package, for a fleet of 20 four-car Millennium double-deck train sets. The trains are operated by CityRail on Sydney's suburban rail network.

In November 2002 ALSTOM won a contract for the supply of equipment (ONIX traction systems, AGATE train management systems and CARA auxiliary converters) for a further 15 four-car Millennium trainsets.
Brazil: ALSTOM has won a turnkey order from Metrofor for a metro system in Fortaleza, north eastern Brazil. The order covers 10 four-car trainsets, which incorporate the new ONIX 3000 drive system supplied by ALSTOM's plant in Tarbes, France.

For the Rio de Janeiro and Salvador metros, ALSTOM will respectively supply 20 TU ONIX 3 kV and 20 TU ONIX 152 HP modules.
Chile: In July 2002 ALSTOM was selected to supply new rolling stock for Line 4 of the Santiago metro. This contract covers 120 motor cars and 60 trailer cars (60 trainsets). Two TCU ONIX 172 MP modules will equip each motor car. The new ONIX standard case with one supervisor will be used to manage the traction equipment.
China: In 1999 Shanghai Mass Transit Pearl Line Development awarded ALSTOM a contract covering the supply of new rolling stock for Shanghai Metro Line 3. 112 ONIX 1500 traction drives will equip the 28 six-car trainsets. In addition 76 ONIX 15000 modules will equip 38 four-car trainsets for the Shanghai Xin Min Line under a contract awarded by Shanghai Modern Rail Transit Co Ltd in 2000.
France: In July 2001 ALSTOM, as part of a consortium, won a contract to supply the first of a new type MF 2000 metro car fleet to the Paris public transport authority, RATP. RATP expects to order a further 160 five-car sets over a period of 15 years. For each train, ALSTOM will provide a traction system composed of three elements: one ONIX™ 172 MP inverter modular element driving four traction motors and associated speed probes; one ONIX 172 standard modular traction case including a safety panel specifically designed to comply with RATP's requirements and to be integrated to the supervisor module and one line filter inductor separately mounted to the train. In all ALSTOM will supply RATP with 480 traction elements.

A contract with RATP, the Paris public transport authority, for the refurbishment of MP89, MP73,

ONIX 800 propulsion set for Virgin Trains Pendolino tilting trainset   0114492

MF67 and MS61 metro vehicles includes the supply of 2,500 ATM electromagnetic contactors.

Germany: In September 2000 ALSTOM received an order from Siemens to supply Type 22CB vacuum circuit breakers with silicon insulators and earthing switches to equip 100 Class 189 locomotives for DB AG.

India: Mitsubishi Electric (MELCO) awarded a contract to ALSTOM for the design, supply and testing of traction equipment for new rolling stock destined for the Delhi metro, now under construction. ALSTOM will manufacture, assemble and test MELCO-designed traction equipment: traction motors, AC control cases, inverter cases, converter/inverter cases, and auxiliary converter cases. ALSTOM's facility in Coimbatore, in southern India, will carry out case assembling, wiring and testing. Other ALSTOM units in India will be responsible for the remainder of the equipment. Delivery of the equipment was due to begin in early 2002 to run through to November 2004. Testing of completed trains will be carried out in India. This order comes within the framework of a contract placed with the Mitsubishi/MELCO/KOROS consortium in May 2001 for 240 new metro cars destined for Delhi's first two metro lines.

Mexico: ALSTOM received an order from Bombardier Concarril and CAF for 13 six-car trainsets for Line A of the Mexico City metro. This was ALSTOM's first complete order for its new-generation ONIX traction system with Agate electronic power control.

Netherlands: Recent contracts include: 10 trolleybuses for Arnhem; 15 trolleybuses for Solingen; 13 hybrid vehicles for Eindhoven; 36 articulated metro and rapid transit vehicles for Rotterdam; refurbishment of 71 articulated metro and rapid transit vehicles for Rotterdam and 2 electric motor car units for Netherlands Railways; 316 auxiliary power supply units for the Netherlands Railways.

In 2001 ALSTOM signed a contract for the supply and commissioning of electrical equipment for Netherlands Railways. The contract was awarded by Bombardier Transportation, which signed a major refurbishment contract with Dutch Railways (NS Reizigers) in July covering a major modernisation of 60 three-car suburban units. Each renovated train with two 1,500 V DC chopper installations and two static converters. The chopper installation replaces the resistor-controlled drive. Two prototypes, delivered in 1996, have already proved successful.

Russia: Moscow Metro has placed an order with ALSTOM for 40 ONIX propulsion systems. The cars are a new design of light metro developed together by Moscow Metro, Metrowagonmash (the leading Russian metro car builder) and ALSTOM. They will operate initially on the new line under construction for the CITY Business centre in Moscow, and later on further lines planned in Moscow's districts. The ONIX IGBT propulsion which will power these new

cars has already been tested on Moscow's heavy Yauza-type metro cars with three prototype drives.

South Africa: ALSTOM was selected by Spoornet to supply and install its Agate control electronics in 99 locomotives as part of its locomotive refurbishment programme.

Spain: In 2003 ATM, the Metropolitan Transportation Authority, awarded ALSTOM a contract to supply 50 five-car trainsets for service on Barcelona Metro Line 9. These trains will be equipped with 200 ONIX 152 HP traction drives.

Switzerland: In consortium with Bombardier Transportation, ALSTOM in December 2001 received a contract to provide the complete electrical installation for 18 three-car Nina emus for BLS Lötschbergbahn. Equipment to be supplied includes: high-voltage current collection and distribution equipment; auxiliary power supply converters and equipment; and ONIX propulsion equipment. The contract follows earlier orders for 14 similar emus.

UK: In 2003 Bombardier awarded ALSTOM a contract to supply traction and equipment for Midland Mainline Class 222 Meridian demus. The supply includes: 127 traction systems (ONIX 750 V, force air cooled); 127 alternators; 254 traction motors; 254 speed sensors and 381 brake resistors.

ALSTOM Transport has supplied Bombardier Transportation with ONIX traction drive systems (352 ONIX IGBT inverter drives including ALSTOM's standard AGATE microprocessor system) and auxiliary equipment for the fleet of 352 Class 220/221 demu cars operated by Virgin CrossCountry. Under the terms of the contract ALSTOM is responsible for providing all spares for the trains until the end of the franchise in 2012.

US: In November 2002 ALSTOM won an order for R160 subway cars for New York City: 660 vehicles configured as either four-car or five-car units. NYCT has two options to purchase further R160 cars, the first for 660 vehicles, and the second for 380 or 420 vehicles. ALSTOM and Kawasaki will build the vehicles. The ALSTOM propulsion system is similar to that provided for the R142 fleet currently entering service. Also won earlier in 2002 was a contract to equip 62 new vehicle trains for Washington Metropolitan Area Transit Authority 6000 Series, with an option for 120 further vehicles. The propulsion system is the same as ALSTOM is supplying for the 2000/3000 series refurbishment. The complete refurbishment of 364 WMATA 2000/3000 Series vehicles is currently underway. The contract, won in 2000, includes replacing the existing DC propulsion systems with advanced ONIX 172MP traction drives. ALSTOM successfully completed the refurbishment of 490 subway cars for Chicago Transit Authority five months ahead of schedule. New IGBT auxiliary inverters were built for the fleet. ALSTOM is supplying ONIX 812 traction systems and CARA auxiliary converters for 70 LRV sets to train builder Kinki Sharyo for

a follow-on order by Santa Clara Valley Transit Authority (SCVTA). Some of the 30 LRV sets from an initial order are in revenue operation. Very similar traction and auxiliary packages are also being manufactured for Kinki Sharyo and NJT (New Jersey Transit Authority). This follow-on order for 26 sets will supplement the 45 LRVs now in revenue service. ALSTOM is also supplying CAF with traction and auxiliary systems for 80 vehicles ordered by Sacramento Regional Transit District.

Venezuela: In 2003 CAMC Caracas Metro awarded a contract for seven new six-car trainsets for the Line 3. They will be equipped by 28 ONIX 172 MP type traction drives.

UPDATED

Ansaldobreda SpA

425 Via Argine, I-80147 Naples, Italy
Tel: (+39 081) 243 11 11   Fax: (+39 081) 243 26 98
Web: www.ansaldobreda.it

Key personnel
President: Fausto Cutuli
Chief Executive Officer: Roberto Assereto
Industrial Managing Director: Francesco Schirripa
Executive Vice-President, Commercial:
    Claudio Mannucci
Executive Vice-President, Engineering:
    Carlo Pellegrini

Products
Electric propulsion equipment for mainline, urban and suburban railway vehicles with AC and DC traction motors; electronic converters and controls; auxiliary equipment; planning, design and management methodologies for public transport; sales, assembly, start-up servicing.

UPDATED

APS Electronic AG

Bahnhofstrasse 135, CH-4626 Niederbuchsiten, Switzerland
Tel: (+41 62) 389 88 88   Fax: (+41 62) 389 88 80
e-mail: aps@apsag.com
Web: http://www.apsag.com

Key personnel
Managing Directors: Urs Christen, Heinrich Naegelin

Products
Auxiliary converters; battery chargers.
    Over the past 10 years many contracts have been carried out for railway, metro, tramway and trolleybus operators in Europe.

Behr Industry GmbH & Co, KG

Heilbronner Strasse 380, D-70469 Stuttgart, Germany
Tel: (+49 711) 896 20 11   Fax: (+49 711) 896 30 75
e-mail: behrindustry@behrgroup.com
Web: www.behrindustry.behrgroup.com

Key personnel
Sales Director: Ruediger Wanner
    e-mail: ruediger.wanner@behrgroup.com
Sales Manager: Christoph Adolff
    e-mail: christoph.adolff@behrgroup.com

Products
Cooling for electrically driven vehicles: converter cooling (oil/water), transformer cooling (oil), cooling of driving E-motors (water) for locomotives and multiple units. Installation: underfloor, on the roof or inside the vehicle. Components or complete systems. Solutions for new vehicles, refurbishment, overhaul of existing systems. Electronic cooling: watercooled cooling plates for semiconductors in converters, air/air heat exchangers for electronic racks and cabinets.

UPDATED

Bharat Heavy Electricals Ltd (BHEL)

Bhopal, 462 022 India
Tel: (+91 11) 461 65 44   Fax: (+91 11) 462 94 23

Key personnel

For a full list of personnel, see Bharat Heavy Electricals Ltd (BHEL) entry in *Locomotives and powered/non-powered passenger vehicles* section.
BHEL
Bangalore, 560 026, India
Tel: (+91 80) 852 01 92   Fax: (+91 80) 806 61 01 37

Key personnel

General Manager: V K Bhatnagar

Products

Electrical propulsion equipment for electric (AC and DC), diesel-electric and battery rail vehicles, including choppers, AC inverters and traction motors; equipment for traction substations. BHEL is the major supplier of electrical equipment to the locomotive and coach building works of Indian Railways and has supplied equipment to private rolling stock builders.

Power converter and controls for three-phase locomotives, AC-powered emus and AD/DC dual-voltage emus.

Bombardier MITRAC DR 3600 Drive for high power application   0558267

Bombardier Transportation

Propulsion and Controls
Brown-Boveri Strasse 5, CH-8050 Zurich, Switzerland
Tel: (+41 1) 318 33 33   Fax: (+41 1) 318 15 43
Web: www.mitrac.bombardier.com

Key personnel

President: André Navarri
Chief Operating Officer: Wolfgang Tœlsner
Vice-President, Communications: Linda Coates
Vice-President, Strategy, Markets and Product Planning: Trung Ngo
Vice-President and Chief Procurement Officer: Pierre Attendu

Products

Bombardier Transportation is the leading supplier for complete propulsion, train control and management system and related services, both to internal railcar manufacturing divisions, as well as to other railcar manufacturers. Being a single-source supplier for the entire scope of propulsion and control, Bombardier Transportation ensures full integration of the functionality and turnkey solutions. *MITRAC* propulsion systems include converters, traction motors, gears and auxiliary power systems supplies for all types of railway applications. *MITRAC* train control and management systems (TCMS) are used for automation, communication and diagnosis including train network, data communication onboard the train and from train to ground as well as train functionality (including processor units and driver's display units).

MITRAC and FLEXITY are trademarks of Bombardier Inc or its subsidiaries

Developments

Bombardier's developments help guarantee higher levels of performance, increased safety and reliability, improved life cycle cost and reduced energy consumption. Bombardier's *MITRAC* Energy Saver, for example, is an onboard solution which allows energy savings of up to 30 per cent and catenary free operation in section up to 1,000 m. Bombardier's train adapted, Internet Protocol-based MITRAC Train Control and Management Systems is based on an open and standardised information protocol. Data transference will be possible on a much larger scale, data access will be easier and faster and information exchange with other infrastructures will be quicker. This new technology will enable applications for better fleet and service management as well as better service for passengers.

Projects

Belgium: Brussels T3000 1, 46 trams; Angel Trains Cargo, 26 multi-system freight locomotives, Angel Trains Cargo 10 3 kV DC freight locomotives.
China: Ministry of Railways of China, 40 high speed trains; Shenzen Metro Corporation, 18 MoviaAmetro cars; Shanghai Metro Operation Company; Guangzhou Metro Line 2, 26 MOVIA trains; Qishuyan Locomotive and Rolling Stock, 76 diesel-electric locmotives.
France: SNCF, additional 100 AGC vehicles (total of 1,174 AGC vehicles); Eurotunnel, 37 Type ESL V locomotives.
Germany: DB AG, 298 Class ET 423 emus; DB AG, 250 two-system Class BR185 electric locomotives; DB Regio, 24 BR146 locomotives; LNVG, additional 78 double-deck coaches and 9 locomotives (total 184 double-deck coaches and 27 TRAXX P160 AC electric locomotives; Frankfurt, 60 FLEXITY Classic trams; Halle, 30 FLEXITY Classic trams; Rhein Neckar, 36 Variobahn vehicles; Converter units for Stuttgart, 27 trams Class sdt 8,10 114; Dresden, 20 FLEXITY Classic trams.
India: Supply of propulsion components for locomotives.
Iran: Tehran/North Ext Line 1, 15 Metro trains.
Israel: Israel Railways, 54 double deck coaches.
Italy: Metro Rome, 198 metro cars for Trenitalia Cargo, 42 Class Eu11 locomotives and 100 Class E464 locomotives.
Luxembourg: CFL, 85 multivoltage system cars.
Russian: Novocherkassk Locomotive Works, 9 Class EP10 dual-voltage electric passenger locomotives.
Spain: Metro Madrid, additional 20 Class s/2000 metro cars, Metro Madrid, additional 47 Class 8000 trains, Renfe, AVE S/102 high-speed power heads, Renfe, 44 HSP 250 high-speed multi system power head with variable gauge, FGC, 13 emu cars/Series 213 extension.
Sweden: Regina emus for various operators in Sweden. Class OUT emus for SJ and DSB, Stockholm Public Transport Authority, Class C20/C21 metro trains.

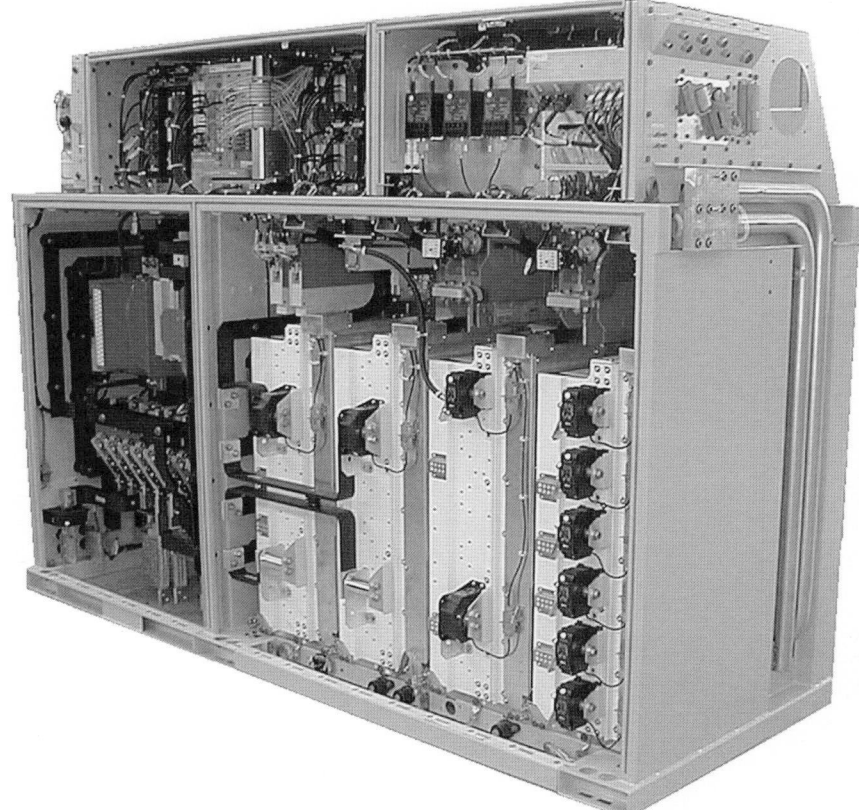

Bombardier MITRAC TC 3300 Traction Converter for high power application   **NEW**/0585270

Switzerland: SBB Cargo, 18 multi-system freight locomotives RE484 in operation, Zurich, 69 Cobra trams.

Turkey: Eskisehir, 18 FLEXITY Outlook trams.

UK: London Underground sub-surface-lines and the Victoria Line, 1,738 metro cars, c2c, South Eastern trains and GoVia Southern (formerly South Central), 1614 Electrostar trains since 1997.

US: Atlanta, 118 metro cars, New York City Authority, 80 metro carsets (Kawasaki R142 and R143 series); Dallas-Fort Worth Airport, 64 Innovia Class 20 people movers; San Francisco Bay Area Rapid Transit, 439 carsets.

UPDATED

Brecknell Willis & Co Ltd

Member of the Fandstan Electric Group
PO Box 10, Tapstone Road, Chard TA20 2DE, UK
Tel: (+44 1460) 649 41   Fax: (+44 1460) 661 22
e-mail: mail@brecknell-willis.co.uk
Web: www.brecknell-willis.co.uk

Key personnel
Managing Director: Tony White
Sales Manager, Trainborne Equipment:
  Andrew Hales
Third Rail, Commercial Operations Manager:
  David Bailey
Overhead Systems, Chief Engineer: David Hartland

Overseas office
Brecknell Willis Taiwan

Products
Current collection and power distribution equipment for the transport sector. This includes the design, manufacture, supply and installation of complete systems.

The product groups are pantographs and third rail current collectors; conductor rail systems; light rail overhead systems and automatic gas tensioning equipment for overhead systems.

Contracts
Conductor rail systems for the Jubilee Line Extension Project and the Northern Line Upgrade (London Underground Ltd), Copenhagen, Ankara and Taipei Metro systems. Overhead contact systems for the Dublin Luas, Midland Metro LRT system, Vancouver, Manchester Metro and refurbishment of Blackpool Tramway. Aluminium composite rail for DB, Berlin and Merseyrail UK.

Current collector systems for all new UK emu vehicles, West Coast Main Line Pendolino and current collection equipment for Shanghai Maglev; current collectors for the new vehicles for Hong Kong, Delhi Metro, KL-Monorail.

Supplied Eurostar and Channel Tunnel Shuttle trainborne current collectors as well as all standard high-speed pantographs for 25 kV electrified line operation in the UK. Light rail pantographs were supplied for the Strasbourg, Manchester, Birmingham, Sheffield and Tyne & Wear trams and shoegear for Glasgow, Amsterdam and Taipei Metros.

UPDATED

Brecknell Willis emu pantograph for KCRC, Hong Kong 0077582

Dockland Light Railways Shoegear, Brecknell Willis
NEW/1135121

Brecknell Willis Copenhagen conductor rail *NEW*/1135120

Brecknell Willis, Dublin Luas – overhead lines *NEW*/1135122

Brush Traction

PO Box 17, Loughborough LE11 1HS, UK
Tel: (+44 1509) 61 70 00   Fax: (+44 1509) 61 70 01
e-mail: sales@brushtraction.com
Web: www.brushtraction.com

Key personnel

General Manager: J M G Bidewell
Sales Manager: P L Needham
Engineering Manager: A Haworth
Production Manager: I G Hall
Purchasing Manager: J Marshall

Background

Brush Electrical Engineering was founded in 1889 and today the company is a major supplier and refurbisher of electric propulsion equipment. The company is a member of the FKI Group.

In June 2005 Brush Traction entered into a licence agreement with RailPower Technologies Corp, Vancouver, Canada, to be the exclusive Green Goat® Series locomotive manufacturer for the UK and Ireland rail markets. Brush Traction will also have the exclusive rights for service and maintenance of the hybrid locomotives in these countries.

Products

Electrical propulsion equipment including control equipment, traction motors, transformers and electrical auxiliaries for electric multiple-units, metro cars and light rail vehicles utilising both DC and AC traction motors.

Contracts

Recent contracts include the supply of chopper propulsion equipments to Bombardier, for operation on the Docklands Light Railway, London.

UPDATED

China Northern Locomotive and Rolling Stock Industry (Group) Corporation (CNR)

11 Yangfangdian Road, Haidian District, Beijing 100038, China
Tel: (+86 10) 51 86 23 70   Fax: (+86 10) 51 86 23 74
e-mail: loriciec@cnrgc.com.cn
Web: www.cnrgc.com

CNR Model JF214 synchronous main and auxiliary alternator   0552544

Model ZD106S asynchronous traction motor by CNR   0552545

CNR high power rectifier diodes and thysistors
NEW/1135822

Key personnel

Chairman and Managing Director: Wang Tai-Wen
President: Cui Dianguo
Vice-President: Zhao Guangxing
President of LORIC Import & Export Corporation Ltd: Cao Guo-Bing
Vice-President of LORIC Import & Export Corporation Ltd: Chen Dayong
Senior Engineer of LORIC Import & Export Corporation Ltd: Yang Xiang-Jing

Products

Development, design, engineering, production, sales, installation, refurbishment, maintenance and after sales service for all types of electric propulsion equipment (traction motors, drive gears, flexible couplings, chopper controllers, main transformers, rectifiers, main alternators); auxiliary electrical equipment (converters, inverters, motor-alternators); train control equipment.

Traction motors are produced with outputs from 32 to 1,200 kW and alternators from 400 to 5,000 kVA.

Contracts

CNR traction and electrical equipment has been sold to customers in Canada, Egypt, Iran, Malaysia, Switzerland and the US.

UPDATED

China South Locomotive and Rolling Stock Industry (Group) Corporation (CSR)

11 Yangfangdian Road, Haidian District, Beijing, China
Tel: (+86 10) 63 98 47 70   Fax: (+86 10) 63 98 47 66
e-mail: csrft@csrgc.com.cn
Web: http://www.csrgc.com.cn

Key personnel

General Manager: Zhao Xiaogang

Background

CSR was established in 2000 as a result of the division into two regional groups of the former China National Railway Locomotive and Rolling Stock Industry Corporation (LORIC). The corporation comprises 24 state-owned enterprises employing around 116,000 staff. Some facilities are listed as part of CSR and of its northern China counterpart, China North Locomotive and Rolling Stock Industry (Group) Corporation (CNR).

Works

See China South Locomotive and Rolling Stock Industry (Group) Corporation (CSR) entry in *Locomotives and powered/non-powered passenger vehicles* section.

Products

Electric traction systems and products, including: Type GTA24A3 synchronous main alternator, rated current 5,300 A, maximum voltage 1,080 V, maximum speed 1,000 rpm; Type JF205A synchronous main alternator, rated power 1,130 kVA, rated current 3,000/1,007 A, rated voltage 216/644 V, rated speed 1,800 rpm; Type TQFR-3000E synchronous main alternator, rated power 2,911 kVA, rated current 3,955/2,183 A, rated voltage 425/770 V, rated speed 1,000 rpm.

Clyde Engineering

Factory Street, Granville, PO Box 73, New South Wales 2142, Australia
Tel: (+61 2) 96 37 82 88   Fax: (+61 2) 98 97 21 74
A member of the Evans Deakin Industries Group

Key personnel

For a full list of personnel, see Clyde Engineering entry in *Locomotives and powered/non-powered passenger vehicles* section.

Products

Traction motors.

Contransys Pvt Ltd

16 Hare Street, Calcutta 700 001, India
Tel: (+91 33) 22 48 23 91   Fax: (+91 33) 22 48 93 82
e-mail: contransys@gems.vsnl.net.in

Key personnel

Managing Director: Y K Daga
Chief Executive: O P Dokania

Products

Pantographs and trolleypoles for LRVs and suburban transit systems and modified versions for other requirements; heavy-duty pantographs for railways for various catenary voltages up to 25 kV, custom-designed and produced for any catenary arrangement; air-brake compressor with integral motor suitable for light rail systems. Door systems, semi-automatic and automatic for railways, metro systems and mass systems.

Contransys Type IR-03H pantograph   0089075

Crompton Greaves Ltd

6th floor, CG House, Dr Annie Besant Road, Prabhadevi, Mumbai 400 025, India
Tel: (+91 22) 423 77 77
Fax: (+91 22) 423 77 88

Rail Transportation Systems Division
Vandhna 11, Tolstoy Marg, New Delhi 110 001, India
Tel: (+91 11) 331 21 47; 373 04 45
Fax: (+91 11) 331 43 60; 335 21 34
e-mail: rtsdm7@mantraonline.com; harsh@mail.cgl.co.in

Key personnel

Managing Director: S M Trehan
General Manager, Rail Transportation Systems Division: A K Raina
All India Marketing Manager, Rail Transportation Systems Division: Harsh Dhingra

Products

AC and DC traction motors; traction alternators; AC and DC auxiliary motors; brushless alternators; locomotive transformers; power converters and auxiliary converters; control electronics for locomotives; surge arrestors; rotary converters; static inverters. Complete electrics for demus and Diesel-Electric Tower Cars (DETC).

Developments

Three-phase electric equipment for locomotives is under development.

Contracts

A four-wheel powered tower wagon has been supplied to the Indian Railway Board.

Daewoo Heavy Industries Ltd

PO Box 7955, Daewoo Centre Building 23nd Fl, 541, 5-Ga, Namdaemun-Ro, Jung-Gu, Seoul, South Korea
Tel: (+82 2) 726 31 79; 31 82
Fax: (+82 2) 726 31 86; 756 26 79

Key personnel

For a full list of personnel, see Daewoo Heavy Industries Ltd entry in *Locomotives and powered/non-powered passenger vehicles* section.

Products

Traction motors (AC, DC), static inverters (IGBT, GTO, PTR), propulsion equipment (VVVF, chopper, rheostatic control) and TCMS (Train Control & Monitoring System).

Recent contracts include technical licences wtih Magatherm Electronics Ltd, India, for the supply of four types of static inverters for Indian Railways.

Dynex Semiconductor

Doddington Road, Lincoln LN6 3LF
Tel: (+44 1522) 50 05 00
Fax: (+44 1522) 50 00 20
e-mail: power_solutions@dynexsemi.com
Web: www.dynexsemi.com

Key personnel

President and Chief Executive Officer:
    Dr Paul Taylor
Chief Financial Officer: Bob Lockwood
Vice-President, World-Wide Sales and Marketing:
    Dr Nick Mallinson

Products

Power semiconductor devices: thyristors, diodes, transistors, IGBTs, gate turn-off thyristors, power modules, and air, oil, water and phase change cooling assemblies. These products may be used for onboard or track side applications.

Contracts

Contracts include Eurostar, TGV Nord, Sybic, Metro Interconnexion (RER), France; Hong Kong MTRC;

London Underground Ltd's Jubilee Line, Class 325 emus for Royal Mail, Networker Class 465 in the UK; Seoul Metro, South Korea; and locomotives for Taiwan.

UPDATED

EFACEC Sistemas de Electrónica SA

Apartado 3078 1018, P-4471-907 Moreira da Maua, Portugal
Tel: (+351 22) 940 20 00   Fax: (+351 2) 948 54 28
e-mail: se@efacec.pt
Web: http://www.efacec.pt

Products

DC and AC traction motors, transformers.

EG & G Rotron

55 Hasbrouck Lane, Woodstock, New York 12498, USA
Tel: (+1 845) 679 24 01   Fax: (+1 845) 679 18 78

Subsidiary company

EG & G Rotron UK
Coronation Road, High Wycombe HP12 3TP, UK
Tel: (+44 1494) 45 16 61   Fax: (+44 1494) 45 24 25

Key personnel

President and General Manager: Peter Stewart
Business Element Manager, Transportation:
    Norm Smith
Marketing Manager, Transportation:
    Keith Hallenbeck

Products

Maintenance-free brushless DC motors, blowers, pumps, fans and blowers for bus, LRV and locomotive applications. Also HVAC products, including boost pumps, ventilators and fuel pumps.

ELIN EBG Traction GmbH

Cumberlandstrasse 32-34, A-1141 Vienna, Austria
Tel: (+43 1) 899 90 22 87   Fax: (+43 1) 899 90 38 62
e-mail: contact@elinebgtraction.at
Web: http://www.elinebgtraction.at

Key personnel

Chair: Peter Rauter
Head of Sales and Marketing: Christian Serjannis

Background

ELIN EBG Traction GmbH is a member of the VaTech Group.

Products

Development, design, manufacture, marketing, sales and installation of electric traction for main line locomotives, railcars, light rail vehicles, trams, metro trainsets, mining and industrial locomotives.

ELIN EBG Traction produces traction transformers, force-ventilated and water-cooled three-phase asynchronous traction motors, water-cooled and air-cooled IGBT transistor traction inverters, microprocessor-based traction and vehicle control systems, IGBT transistor traction inverters for mining locomotives and auxiliary power supply inverters for locomotives, railcars and passenger coaches.

Contracts

Vehicles for which ELW has supplied equipment include: Saarbrücken dual-system LRVs, LRVs for Portland and Tacoma, Lodz Cityrunner low-floor trams, Rome trams and the Škoda-Inekon Astra low-floor tram.

EVPÜ

Elektrotechnický Vyskumný a Projectový Ústav Trencianska 19 SK-018 51 Nov Dubnica, Slovakia

Tel: (+421 42) 443 21 61; 440 92 24; 440 92 22
Fax: (+421 42) 443 42 52
e-mail: trade@evpu.sk
Web: http://www.evpu.sk

Key personnel

Contacts: Dipl Ing Frantisek Jankovic;
    Ing Marcel Pcola

Products

Electric traction control equipment for locomotives, including three-phase systems; electric traction control equipment for trams, trolleybuses and metro cars; transformers; electrical components for rail vehicles and trolleybuses; battery charging equipment.

Faiveley Transport SA

143 boulevard Anatole France, Carrefour Pleyel, F-93285 Saint-Denis Cedex, France
Tel: (+33 1) 48 13 65 00   Fax: (+33 1) 48 13 66 47
e-mail: info@faiveley.com
Web: www.faiveley.com

Key personnel

Chairman and Chief Executive Officer:
    Robert Joyeux
Financial Director: Sven Schopp
General Manager: Pierre Sainfort
Communications Manager: Edmond Ballerin

Background

The Faiveley group completed the acquisition of SAB WABCO in November 2004.

Products

Pantographs, high-voltage switching, auxiliary converters, master controllers.

Recent developments include the CX family of pneumatically cushioned pantographs, comprising the AX and CX designs. The AX can operate where other pantographs are extended at speeds up to 220 km/h; it has been fitted to SNCF Class BB22200 locomotives. The AX is fitted with a regulator that ensures a constant pressure is maintained against the contact wire, at all speeds, using aerodynamic devices.

The CX operates at speeds up to 320 km/h. The CX is controlled by a microprocessor which varies the contact force (via a servo-valve in the pantograph's pneumatic system) in line with the train's speed and direction of travel, the position of the pantograph and the type of catenary overhead. The CX also features an automatic drop device, a low-friction spring box suspension and a collection head with independent wear strips.

UPDATED

Ferraz Shawmut

1 rue Jean Novel, F-69626 Villeurbanne Cedex, France
Tel: (+33 4) 72 22 66 11   Fax: (+33 4) 72 22 67 13

Works

28 rue Saint Philippe, F-69003 Lyon, France
70 avenue de la Gare, PO Box 18, F-38290 La Verpilliere, France
rue Vaucanson, F-69720 St Bonnet de Mure, France

Key personnel

Marketing Manager: M Renart
Export Sales Manager: H Behr
Commercial Director: J Brenet

Principal subsidiaries

Fouilleret, Ferraz Corporation (USA)
Nihon Ferraz (Japan)

Products

Earth return current units; brush-holders for electric traction motors; current-collecting device on live rail; fuses with very high breaking capacity for protection of power semiconductors; shoe fuses; automatic earthing device with large short-circuit capability; resistors, disconnectors and switches.

FIREMA Trasporti SpA

Headquarters
Via Provinciale Appia, Località Ponteselice, I-81100
Caserta, Italy
Tel: (+39 0823) 09 71 11   Fax: (+39 0823) 46 68 12
Web: www.firema.it

Key personnel
Chairman: Gianfranco Fiore
Chief Executive Officer: Roberto Fiore
Operations Director: Maurizio Russo
Commercial Manager: Sergio d'Arminio M
Marketing Manager: Agostino Astori

Commercial and technical offices
Via Triboniano n 220, I-20156 Milan, Italy
Tel: (+39 02) 23 02 02 23   Fax: (+39 02) 23 02 03 00

Products
Electromechanical and electronic (chopper and
inverter) traction equipment for mainline and
suburban, metro and light rail applications. Traction
motors for AC and DC equipment, main generators
for diesel-electric locomotives.

Contracts
Contracts include electrical equipment for E652
and E402 locomotives for Italian Railways (FS);
electrical equipment, including traction motors,
for double-deck emus for FS/North Milan Railway;
auxiliary static converters for ETR 500 high-speed
trainsets for FS; and traction equipment for emus
for Circumetnea Railway and remote control for
Type Z1 coaches.

UPDATED

Freudenberg Schwab GmbH

Postplatz 3, D-16761 Hennigsdorf, Germany
Tel: (+49 3302) 206 20   Fax: (+49 3302) 20 62 77
e-mail: info@freudenberg-schwab.de
Web: http://www.freudenberg-schwab.de

Key personnel
Chief Executive Officers: Dr Detlef Cordts;
   Peter Kofmel; Jörg Sost
Quality Manager: Leo S E Lang
Marketing Director: Bernd Werner

Products
Precision seals: radial shaft seals, complete sealing
kits for electric motors, O-rings, seals and elastic
materials.
   (See also *Bogies and suspension, wheels and
axles, bearings, Passenger coach equipment,
Freight vehicles and equipment, Brakes and
drawgear* and *Diesel engines, transmission and
fuelling systems* sections).

Fuji Electric Systems Co, Ltd

Gate City Ohsaki, East Tower 11-2, Osaki 1-chome,
Shinagawa-ku, Tokyo 141-0032, Japan
Tel: (+81 3) 54 35 70 46   Fax: (+81 3) 54 35 74 23
e-mail: info@fesys.co.jp
Web: http://www.fesys.co.jp

Key personnel
Managing Director, Electrical Systems Company
   Group: H Itou
General Manager, Transportation Systems Sales
   Department: K Kimura

Background
The Transportation Systems Sales Department of
what was previously Fuji Electric Co Ltd, is part of
the newly divided Fuji Electric Systems Co, Ltd.

Products
Traction motors with VVVF control, static
auxiliary power supply (SIV) systems, power
supply equipment; computer-based supervisory
remote-control equipment; water-cooled silicon
rectifiers. GIS (Gas Insulated Switchgear) and mini

Singapore MRT train using Fuji Electric inverter control systems   0098044

high-speed circuit breakers; moulded transformers;
control systems incorporating electric power
management, station office apparatus control,
data management and disaster prevention
management.

Ganz Transelektro Traction Electric Ltd

PO Box 250, H-1243 Budapest, Hungary
Kir ly u. 163, H-1061 Budapest, Hungary
Tel: (+36 1) 432 88 50   Fax: (+36 1) 262 36 38
Web: http://www.traction-ganztrans.hu

Background
Formerly a state enterprise, the company was
privatised in 1991 as Ganz Ansaldo Electric Ltd, with
the Italian company initially taking a shareholding
of 51 per cent but increasing this to 99.99 per cent
in 1998. In 1999 Ansaldo divested its interest,
leading to the establishment of the present private
sector company in 2000.

Products
Electric traction equipment for heavy rail
vehicles, trams and trolleybuses, including
control equipment, IGBT-based DC choppers for
trolleybuses, IGBT-based auxiliary converters and
IGBT-based inverters for mass transit and heavy
rail vehicles, pantographs, braking resistors.

Contracts
Recently the company has supplied electric
traction equipment for tram refurbishment
projects in Budapest, Miskolc and Riga (Latvia),
as well as for new vehicles for Debrecen, Hungary.
Equipment has also been supplied for Ganz-Solaris
trolleybuses operating in Riga, Rome and Tallinn
and for Astra-Ikarus vehicles for Bucharest.

Gardner Denver Ltd

Chequers Bridge, Gloucester GL1 4LL, UK
Tel: (+44 1452) 33 83 38   Fax: (+44 1452) 33 83 07
e-mail: indsales@belliss.co.uk

Key personnel
Managing Director: C Barker
Product Manager: A Davis

Background
Gardner Denver Ltd was previously called
Hamworthy Compressor Systems Ltd.

Products
Pantograph air compressor: a compact air-cooled
compressor direct-coupled to DC motor (0.7 kW).
Constructed in heavy-duty cast iron, this

compressor offers maximum scope for choice
of mounting position on a locomotive. Charging
rate: 126 litres/min over pressure range 0 to 75 psi
(0 to 5 bar). Maximum operating pressure: 100 psi
(7 bar). Dimensions: 46 × 26 × 27 cm; weight with
motor, 43 kg.

UPDATED

GE Transportation Systems

2901 East Lake Road, Erie, Pennsylvania 16531,
USA
Tel: (+1 814) 875 34 57   Fax: (+1 814) 875 59 11
Web: http://www.getransportation.com

Key personnel
For a full list of personnel, see GE Transportation
Systems entry in *Locomotives and powered/non-
powered passenger vehicles* section.

Products
GE offers two diesel-engine families: the GE
7HDL™ 16-cylinder, 4,474 kW and the GE 7FDL™
16-cylinder, 3,281 kW. The GE 7FDL 16-, 12-, and
8-cylinder can also be set at various power ratings.
All AC, DASH 9, DASH 8 and DASH 7 models
feature a three-phase traction alternator feeding a
full-wave bridge traction rectifier. The GMG-type
alternator is used on 4,474 traction kW (AC) and
3,281 traction KW (AC, DASH 9 and DASH 8)
models.
   Most AC, DASH 9, DASH 8 and DASH 7 models
include a single stator traction alternator output
with sufficient capacity to accommodate all
volt-amp requirements through the entire speed
range, according to GE Transportation. All AC,
DASH 9, DASH 8 and DASH 7 models feature
traction motors connected in a full-time parallel
configuration in the propulsion mode to provide
consistent propulsion behaviour, to enhance
speed-tractive effort characteristics, and to
eliminate motor transition contactor maintenance.
The DC traction motor family consists of GE 752,
GE 793, GE 794, GE 761 and GE 764 models. The
AC traction motor family consists of GEB 13 and
GEB 15 models.

Locomotive technologies
GE's AC locomotives feature direct, air-cooled,
phase module inverter systems (single inverter
per axle). The GE AC traction motor includes a
low-slip capability. IGBT technology delivers more
reliability with the added benefits of less weight
and volume. The units are air-cooled, as with the
former GTO technology, yet have significantly
reduced component count and complexity.
   Electronic Fuel Injection (EFI) promotes
fuel savings, reduces emissions, and lowers
maintenance costs. Split cooling improves fuel
efficiency and reduces engine temperatures.

HiAd™ (high adhesion) bogies feature a low-weight transfer design with 10-year overhaul intervals. The GE steerable bogie allows locomotives to negotiate previously prohibited curved track.

GE Transportation Global Signaling designs advanced electronic train and railway products and systems such as LOCOTROL® distributed power, EPX™ Direct Braking and Universal Control Valve™, TrainTalk™ intra-train communications, Precision Dispatch™, Navigator Dispatch™, Precision Train Control™, PINPOINT™ locomotive tracking and asset management system, and train defect detection systems. These and other GE Technologies are integrated into GE microprocessor-based locomotives through an open-architecture control system.

Hall Industries Inc

514 Mecklem Lane, Ellwood City, Pennsylvania 16117, USA
Tel: (+1 724) 752 20 00   Fax: (+1 724) 758 15 58
e-mail: service@hallind.com
Web: http://www.hallind.com

Key personnel
Contacts: Dick Harrison, Scott Kennedy

Products
Pantographs for light rail vehicles, emus and locomotives for 750–3,000 V DC and 25 kV AC power supply systems, master controllers, traction control equipment.

VERIFIED

Hitachi Ltd

Overseas Marketing Department
Transportation Systems Sales Division
6 Kanda Surugadai 4-chome, Chiyoda-ku,
Tokyo 101-8010, Japan
Tel: (+81 3) 32 58 11 11   Fax: (+81 3) 32 58 52 30
Web: www.hitachi-rail.com

Key personnel
Chief Operating Officer of Transportation Systems:
   Gaku Suzuki
General Manager, Transportation Systems Sales
   Division: Chiaki Ueda
General Manager, Transport Management and
   Control Systems Division: Kazuo Kera
General Manager, Rolling Stock System Division:
   Toshihide Uchimura

Products
Propulsion systems, auxiliary power supply, ATP/ATO equipment, bogie and air conditioning equipment.

In view of the increasing use of asynchronous motor propulsion systems, Hitachi has been developing VVVF inverters for trainsets of 1,500 V DC railways.

The small size of Hitachi VVVF inverters has been achieved by use of 4,500 V high-voltage GTO thyristors. Direct digital control provides accuracy in constant speed control, slip-skid correction control and start control on up-gradients. The signalling system is protected from noise and electromagnetic interference by extensive shielding and optimised wiring layout.

Contracts
Hitachi was awarded a contract to upgrade drive units for Type 8M commuter emus from Southern African Rail Commuter Corporation (SARCC).

In December 2004 Hitachi was awarded a contract from the Beijing Subway Authority, as part of a consortium consisting of Yongji Electric Machine Factory and Sumitomo Corporation, which includes the supply of 96 sets of IGBT VVVF inverter propulsion systems.

UPDATED

Hyundai Heavy Industries Co Ltd

Hyundai Building Main 15F, 140-2, Gye-dong, Jongno-gu, 110793 Seoul, South Korea
Tel: (+82 2) 746 75 31   Fax: (+82 2) 746 76 48
e-mail: railway@hhi.co.kr
Web: www.hyundai-elec.co.kr

Key personnel
Contact: Myoung-yong Shim
Tel: (+82 2) 746 75 31
e-mail: erail999@hhi.co.kr

Products
Traction and control: diesel traction motors and generators. Electric: traction motors, power converters and auxiliary converters. Emu traction motors, VVVF inverters, SIV and TCMS. Control: train information management systems, AC traction motors. Static inverters, train control management systems and traction transformers. Electrification: substations, GIS, transformers, switch gears, rectifiers. Power supply and distribution, SCADA.

UPDATED

InfoSystems GmbH

Uellendahler Strasse 437, D-42109 Wuppertal, Germany
Tel: (+49 202) 709 50   Fax: (+49 202) 709 51 37
e-mail: info@wtal.infosystem.de
Web: http://www.infosystem.de

Key personnel
Managing Director: Dr Rolf-Dieter Krächter

Background
InfoSystems GmbH was founded in 1997 with the companies Brose, Krueger and Wandel & Goltermann.

Products
A range of matrix displays such as InfoProfil, LED, InfoProfil LCD and InfoProfil DOT, stationary advertisement and information systems, audio systems, electronic rear view mirrors and video surveying cameras.

KMT-teknikka Oy

PO Box 116, FIN-38701 Kankaanpää, Finland
Tel: (+358 2) 573 12 39   Fax: (+358 2) 573 22 80
e-mail: esa.pyoria@kmt.fi
Web: http://www.kmt.fi

Key personnel
Managing Director: Esa Pyöriä
Key Account Manager: Aki Tuononen

Products
Assembly of electric traction equipment, switchgear, pre-assembled electrical centres and control consoles; fabrication and manufacture of cabinets and enclosures in steel, stainless steel and aluminium for rail applications.

Contracts
Pre-assembly of main converter units for VR Class Sr2 electric locomotives built by Bombardier Transportation; assembly of control consoles for Class Sr2 locomotives; electrical distribution boxes, electrical control centres, heating elements and catering service lifts for VR Type ICS double-deck coaches built by Talgo; electrical control centres and neutral section sensor boxes for VR Pendolino trainsets built by Alstom.

Končar – Elecktrične Lokomotive dd

Končar-Electric Locomotives Inc
Velimira Škorpika 7, HR-10090 Zagreb, Croatia
Tel: (+385 1) 349 69 59   Fax: (+385 1) 349 69 60

e-mail: upravakoncar-ellok.hr
Web: www.koncar.hr/koncar/ellok/

Key personnel
President: Jusuf Cmalić
Members of Board: Vesna Boinović Grubić;
   Željko Šakić
Marketing and Sales Manager: Zvonimir Cvijin

Products
Light rail vehicles; parts, components and systems for electric traction.

UPDATED

Lechmotoren GmbH

Suedliche Roemerstrasse 12-16, D-86972 Altenstadt, Germany
Tel: (+49 8861) 71 00   Fax: (+49 8861) 71 01 80
e-mail: lechmotoren@compuserve.com
Web: http://www.lechmotoren.de

Products
Converters, generators up to 1.5 kV, internal combustion generator sets, AC motors, asynchronous/synchronous motors, electronic components for urban transport applications.

Lekov AS

Jirotova 375, CZ-336 01 Blovice, Czech Republic
Tel: (+420 379) 20 71 11; 20 71 62
Fax: (+420 379) 20 72 01; 20 72 02
e-mail: lekov@lekov.cz
Web: www.lekov.cz

Key personnel
Chief Executive Officer: Michal Ovsjannikov
Accounting Manager: Alena Zoubková
Sales Manager: Jan Ovsjannikov
Technical Manager: Tomáš Lorenc
Production Manager: Stanislav Zoubek
Quality Manager: Radek Hrubý

Background
Faively Transport took a 75 per cent shareholding in Lekov in December 2002.

Products
Components for trolleybuses, trams, metros and railways. These include high-speed circuit breakers, electromagnetic and electropneumatic contactors, master controllers, protection relays, reversers, disconnecting switches, earthing switches, pantographs, resistors and magnetic valves, trolleybus current collectors.

Contracts
Contracts include delivery of electrical equipment to Škoda for the refurbishment of Russian Railways Type CS200 and 82 E9 CS locomotives. Supply of: electrical units to Czech Railways; switches to GE Rail (USA); electrical devices to the cities of Ostrava, Pilsen, Liberec, Brno, Bratislava; electrical devices for Czech Pendolino CDT 680 manufactured by ALSTOM Ferroviaria, Italy; electrical devices for Fret and Prima locomotives for ALSTOM Transport, France; supply of sets of reversers for locomotives of type BR 189 manufactured by Siemens AG, Germany; trolleybus current collectors for Boston, Ostrava, Vilnius, and Winterthur; electrical equipment for AGC, Bombardier France; pantographs for Prague trams; switches for Mitsubishi, Japan.

UPDATED

Mitsubishi Electric Corporation

Mitsubishi Denki Bldg, 2-3 Marunouchi 2-chome, Chiyoda-ku, Tokyo 100, Japan
Tel: (+81 3) 32 18 34 30   Fax: (+81 3) 32 18 28 95
Web: http://www.melco.co.jp/society/traffic/

Key personnel

President: Tamotsu Nomakuchi
Executive General Manager (Public Utility Systems Group): Takaaki Kijima

Products

Electric propulsion equipment (traction motors, VVVF inverters, chopper controllers, main transformers, rectifiers, drive gears, flexible couplings), brake systems, auxiliary electrical equipment (static inverters). Air conditioning systems, train control equipment (ATC, ATO, ATS). Integral control systems, communication systems, substation systems (DC and AC) and equipment, station depot and inspection equipment, magnetic and super conduction magnets.

Contracts

Recent contracts include: electrical equipment for 192 emus, New York MTA, Long Island Rail Road (USA), May 1999; electrical equipment for 250 emus, KCRC West/East Rail, Hong Kong, August 1999; electrical equipment for 104 emus, MTRC C651 Tsuen-Kwan-O extension, Hong Kong; electrical equipment for 328 emus, Seoul Metro Line 6, Korea; electrical equipment for 240 emus, Delhi Metro Railway Corporation, (India), May 2001; electric equipment for 126 emus, Attiko Metro (Greece) February 2002.

Morio Denki Co Ltd

34-1 Tateishi 4-chome, Katsushika-ku, Tokyo 124-0012, Japan
Tel: (+81 3) 36 91 31 81   Fax: (+81 3) 36 92 13 33
Web: www.morio.co.jp

Main works

2 Natooka, Ryugasaki City, Ibaragi Pref 301-0845, Japan

Key personnel

President: S Yamagata
Senior Managing Director: K Miura

Subsidiaries

Shanghai Morio Denki Co, Ltd (China)

Products

Control equipment, including master controllers, switch boxes, distribution boards, junction boxes and conductor switches. Digital speedometers.

UPDATED

National

National Electrical Carbon Corporation
PO Box 1056, Greenville, South Carolina 29602, USA
Tel: (+1 864) 458 77 77   Fax: (+1 864) 281 01 80

Key personnel

General Manager: M Cox
Sales Director: D Klas
Director, International Sales: J T Tidswell
Director, International Marketing: K Osman

Principal subsidiaries

Fulmer Company Inc Export, Pennsylvania, USA
National Electrical Carbon BV, Hoorn, Netherlands
National Electrical Carbon Canada, Mississauga, Ontario, Canada
National Electrical Carbon Limited, Sheffield, UK

Products

Carbon brushes for all traction, commutator, motor generator and auxiliary equipment, carbon brush holders and wheel flange lubricators.

Parizzi

Parizzi (a Fiat Ferroviaria company)
Elettromeccanica Parizzi SpA
Via C Romani 10, I-20091 Bresso, Milan, Italy

Tel: (+39 02) 66 52 31
Fax: (+39 02) 614 04 08

Products

Energy converters: mono- and poly-current static converters fed by the primary power line to supply power for onboard auxiliary systems in three-phase 50 Hz AC (input voltages: 3,000 V DC, 1,500 V DC, 1,500 V AC, 50 Hz, 1,000 V AC 16⅔ Hz. Power: 40 to 85 kVA); mono- and poly-current static battery chargers fed by the primary power line (input voltages: as for auxiliary systems above, plus 600, 750 and 1,500 V DC. Power: 1, 5 to 14 kW); inverters and converters; UPS.

Traction equipment: variable frequency and voltage inverters for the supply and control of asynchronous traction motors for diesel-electric locomotives and electric trainsets; AC traction motors.

Components: electrical and electropneumatic components for railcars and locomotives; automatic starters; controllers and reverse switches; electrical connectors for multiple traction; various transducers.

Electromechanical devices: battery chargers operated from the axle of the bogie of railway vehicles.

Electronic systems: microprocessor-based control and diagnostic systems; centralised tachometric/tachographic and event-recording systems; electronic anti-slide devices to adapt the braking force to the instantaneous adhesion condition of railway vehicles; electronic anti-slip devices for locomotives; bivalent anti-slide/slip devices.

Permali Gloucester Ltd

Bristol Road, Gloucester GL1 5TT, UK
Tel: (+44 1452) 52 82 82   Fax: (+44 1452) 50 74 09
e-mail: sales@permali.co.uk
Web: www.permali.co.uk

Key personnel

Managing Director: A J T King
Technical Director: David Tudor
Sales and Marketing Manager: Fraser Rankin

Products

Halogen-free laminates for shoe beams, shoe arms and arc barriers for third rail current collection systems.

Contracts

Contracts include the supply of shoe beams, shoe arms and arc barriers to London Underground Ltd and shoe beam and shoe arm assemblies for EWS Class 92 locomotives.

UPDATED

Production SpA

Zona Asi Sud, I-81025 Marcianise (CE), Italy
Tel: (+39 0823) 82 16 41   Fax: (+39 0823) 82 13 25
e-mail: info@productionspa.com
Web: http://www.productionspa.com

Products

Pantographs for rail vehicles, trams and trolleybuses.

RDS Technology Ltd

Cirencester Road, Minchinhampton, Stroud GL6 9BH, UK
Tel: (+44 1453) 73 33 00   Fax: (+44 1453) 73 33 11
e-mail: info@rdstec.com
Web: http://www.rdstec.com

Key personnel

Managing Director: R Danby
Engineering Director: P Nelson
OEM Business Director: J Athawes

Products

True Ground Speed Sensor (TGSS) for slip and traction control.

Riga Electric Machine Building Works

31 Ganibu dambis, LV-1005 Riga, Latvia
Tel: (+371 7) 38 13 50
Fax: (+371 7) 733 41 33

Products

Electric control sets for 3 kV DC and 25 kV AC trainsets; electrical control sets for passenger vehicle lighting and power supply; 100 to 280 kW DC traction motors for 3 kV DC emus; |repair and refurbishment of DC motors; low voltage DC motors up to 20 kW for forklifts.

Rotem Company

Headquarters

Landmark Tower, 837-36, Yeoksam-dong, Gangnam-gu, Seoul 135-937, South Korea
Tel: (+82 2) 21 12 82 94   Fax: (+82 2) 21 12 98 73
Web: www.rotem.co.kr

Key personnel

Vice Chairman and Chief Executive Officer: Soon-Won Chung
Senior Executive Vice-President: Yeo-Sung Lee
Executive Vice-President: Jae-Hong Kim

Background

Established in 1964 when Daewoo Heavy Industry started manufacturing rolling stock, followed by Hyundai and Hanjin Heavy Industry a few years later. In 1999, the three companies were consolidated into KOROS. Hyundai Motors Group acquired the share of Daewoo in and former company name KOROS became Rotem. Rotem is now an affiliate of Hyundai Motors Group and has its headquarters in Seoul and two factories in Uiwang and Changwon. Uiwang has a capability of manufacturing 500 emus per year and has the capability to manufacture electric equipment such as traction motors, SIV inverters etc. The research and design centre is also located in Uiwang. The Changwon factory has a capability to manufacture 700 emus per year. Rotem has an annual capacity to manufacture approximately 1,200 emus. Certifications such as the ISO 9001 certificate for Quality, 14001 for environment and 18001 for occupational health and safety management have been acquired at all three sites.

Products

Rotem supplies fully integrated propulsion system packages, traction equipment, auxiliary power supply systems, support services for modern railway and urban transport vehicles. It manufactures electrical equipment for emus, electric locomotives and light rail vehicles, including PWM converter control, VVVF inverter control, chopper control equipment using power semiconductor device such as IGBT, IPM and GTO technology, AC and DC traction motors, main transformer and auxiliary power supply, train control and monitoring systems and onboard signalling systems. Recently Rotem has developed integrated propulsion system packages, traction equipment and train control systems for the Korean high-speed train.

Rotem's electrical equipment meets a broad range of the applications for diversified rolling stocks. The electrical equipment shop provides quality and reliability in various products such as traction motors, train control and monitoring systems, signalling systems, main inverter and static inverters for use on electric cars and locomotives.

Contracts

GTO VVVF inverter AC drive systems have been supplied for Seoul metro line 7 and 8 and also more than 200 sets of chopper control system for Seoul metro line 3 and 4. Recently the VVVF inverter system using IGBT or IPM devices have been supplied to local authorities such as Daegu, Busan, Seoul metro and the Incheon International Airport Railroad (IIAR).

Rotem has been manufacturing various traction motors for electric locomotives and cars as well as high-speed trains. Rotem supplied more than 6,000 sets of AC/DC traction motors for Korean National Railroad (KNR) and local authorities such as Gwangju, Daegu, Busan, Daejeon, Seoul metro, IIAR and Taiwan Railway Administration (TRA).

Rotem supplied more than 1,600 sets of static inverters using IGBT, GTO and power transistors for Seoul metro lines, KNR and overseas railway authorities in Taiwan, Philippines and Greece.

Recent orders include 180 kVA IGBT static inverters for Attiko Metro II in Greece and 190 kVA IGBT static inverters for IIAR.

Rotem supplied more than 750 sets of train monitoring and control system (TCMS) or train monitoring and control system (TCMS) or train monitoring system (TMS) to several domestic and overseas authorities. Currently Rotem is designing more than 450 of these sets for Seoul metro line 9, Brazil Rio, Brazil Salvador, Iran and Ireland.

UPDATED

SAFT

Advanced and Industrial Battery Group
12 rue Sadi Carnot, F-93170 Bagnolet, France
Tel: (+33 1) 49 93 19 18   Fax: (+33 1) 49 93 19 64
Web: http://www.saftbatteries.com

Key personnel
Chairman: John Searle
Managing Director, IBG: Bertrand Olivesi
Marketing and Sales Director, IBG:
   Fred-Erik Hapiak
Marketing Manager: Michael Lippert
Communications Director: Jill Ledger
   Tel: (+33 1) 49 93 17 77
   e-mail: jill.ledger@saft.alcatel.fr

Products
Saft Ni-Cd batteries, pocket plates or sintered/plastic bonded electrode types are for emergency supply and security purposes. All items are lightweight and compact and are available in stainless steel, flame-retardant or standard plastic containers.

SAFT offers an integrated battery assembly for onboard power for railway networks by combining advanced Ni-Cd batteries with custom-made containers, reducing weight and volume.

A new range of Ni-Cd batteries, sintered/Pbe, improve weight and volume by 40 per cent. Matrics (MRX batteries) are built for trams, emus, electric locomotives and light rail.

Contracts
The company has supplied rail and bus companies in 56 countries and contracts include supply to the Pendolino tilting trainsets and passenger coaches for SNCB, RENFE, Taipei and Pakistan. Recent contracts include: supply of MRX batteries for Turin trams, SNCF's new AGC regional express trains, AVE S103 high-speed train in Spain built by Siemens.

Schunk Bahntechnik GmbH

Aupoint 23, A-5101 Bergheim bei Salzburg, Austria
Tel: (+43 662) 45 92 00   Fax: (+43 662) 459 20 01
e-mail: office@schunk-group.at
Web: http://www.schunk-group.com

Subsidiary companies
Schunk Kohlenstoff technik GmbH
Schunk Metall und Kunststoff GmbH

Key personnel
Director: F Rabacher

Products
Electric traction equipment, pantographs for locomotives, high-speed trains, LRVs and trams, earthing contacts and ground switches.

Schunk Kohlenstofftechnik GmbH

Rodheimer Strasse 59, D-35452 Heuchelheim, Germany
Tel: (+49 641) 60 80   Fax: (+49 641) 608 17 48
e-mail: infobox@schunk-group.com
Web: www.schunk-group.com

Key personnel
Public Relations Manager: Andrea Gossel

Products
Carbon sliding strips for main line and urban rail vehicle pantographs, third and fourth rail current collector shoes, trolleybus and trolley pole systems, carbon brushes, brush holders, pantographs (high-speed and urban rail vehicles), earthing contacts, contactor sets, cam gear switchers, third rail pantographs, foil connectors.

UPDATED

Sécheron SA

Routes des Moulières 5, CH-1217 Meyrin, Geneva, Switzerland
Tel: (+41 22) 739 41 11   Fax: (+41 22) 739 48 11
e-mail: info@secheron.com
Web: www.secheron.com

Key personnel
Chief Executive Officer: Claude Durand
PBU Components Director: Jo Murer
PBU Substations Director: Dominique Jamet
PBU Electronics Director: Peter Stauffer
Marketing Manager: Gilbert Lile
Corporate Sales Manager: Jimmy Cuche

Subsidiary companies
Sécheron Tchéquie sro, Prague, Czech Republic

Products
Switching and protection devices (DC High-Speed Circuit Breakers HSCB); AC and DC power and

Sécheron AC vacuum circuit breakers   *NEW*/1136189

Sécheron DC High-Speed Circuit-Breaker (HSCB)   1136193

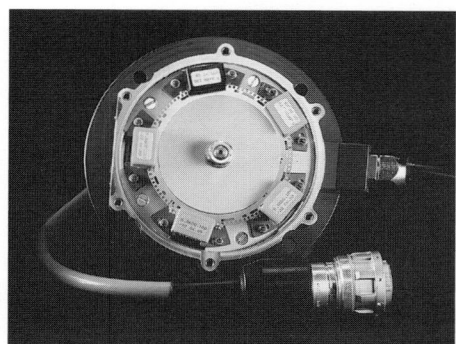

Sécheron pulse generator   *NEW*/1136190

auxiliary contactors, AC vacuum circuit-breakers, disconnectors, disconnection and changeover switches, earthing switches, master controllers, wheel flange lubricators, electric couplers, electronic speed measuring and data recording systems, speed sensors, multifunctional speed display.

Recent projects include: various components for LIRR M7; various components for AGC for SNCF; various components for locomotives for Chinese Railways.

UPDATED

Siemens AG

Transportation Systems
Train Division
Duisburger Strasse 145, D-47820 Krefeld, Germany
Tel: (+49 2151) 450   Fax: (+49 2151) 450
e-mail: trains@siemens.com
Web: www.siemens.com/transportation

Corporate headquarters
Siemens AG
Transportation Systems
PO Box 3240, D-91050 Erlingen, Germany
Tel: (+49 9131) 7-0

Key personnel
President, Transportation Systems Group:
Hans M Schabert
Group Vice-Presidents: Alfred Frank;
Joern F Sens; Friedrich Smaxwil
Heads of Trains Division:
Dr Dietrich G Moeller, Jochen Wiessner

Products
Major products include: traction power supply systems and equipment and auxiliary power

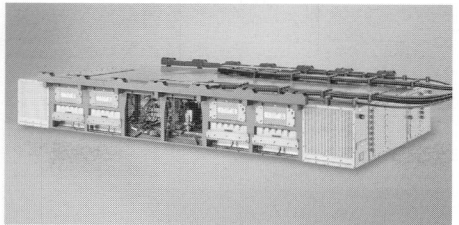

Siemens traction package for dmus for India
NEW/0585224

Siemens AC traction motor   *NEW*/0585225

supply systems for diesel and electric locomotives and passenger rail vehicles of all types; microprocessor controllers; equipment cabinets; test and diagnostic equipment.

Contracts
Recent or current contracts include:
India: In May 2005 Siemens received a notification of award of a contract from the Mumbai Railway Vikas Corporation (MRVc) for propulsion system and electrical equipment for emu rolling stock to be built at the Integral Coach Factory, Chennai. The emus will be in operation on both of Mumbai's suburban arterial networks, the Central and Western Railway.

In January 2004 Siemens was awarded a contract by the Indian Railway Board to supply complete electrical equipment for 70 three-car dual-voltage (1.5 kV DC/25 kV AC 50 Hz) emus to be used on Central Railway and Western Railway services in the greater Mumbai area. The scope of supply covers pantographs, transformers, traction converters and auxiliary power converters, traction motors, train control equipment and passenger information systems, all to be produced locally. Mechanical assembly of the vehicles will be undertaken by Integral Coach Factory (ICF) in Chennai.
Spain: In June 2002 Siemens received an order from Spanish National Railways (RENFE) for

electrical equipment for 40 Civia regional emus, comprising 31 five-car and nine three-car trains. The contract includes the supply of the first 3 kV DC IGBT converter to be supplied to the rail market. Deliveries of the trains, to be built in Spain by CAF, are scheduled to take place between 2004 and 2009. Siemens also supplied electrical equipment for the first 14 Civia emus, the first of which entered service in 2003.
Ukraine: In February 2000 Siemens announced a contract to supply three-phase AC traction and control equipment for a prototype Class DS 34.8 MW electric locomotive to be built by GP NPK Electrovozostroeniya for the Ukrainian national railway company (Ukrzaliznizija). Following successful testing of this prototype, equipment was ordered in April 2004 for 100 additional machines to be supplied between 2004 and 2009.

NEW ENTRY

Škoda Electric sro

Tylova 1/57, Plzeč, CZ-316 00, Plzeč, Czech Republic
Tel: (+420 378) 11 80 27   Fax: (+420 378) 11 74 18
e-mail: electric@skoda.cz
Web: www.skoda.cz/electric

Background
Previously named Škoda Trakčí motory, the company is now an integral part of Škoda Electric sro which is one of the major companies of the industrial group, Škoda.

Products
Asynchronous three-phase water-cooled AC traction motors up to 1,600 kW; DC traction motors up to 1,000 kW. Components of AC motors including coils and stators. General repairs and overhauls, service and maintenance, spare parts and motors or components of motors for special applications.

UPDATED

SPII

SPII SpA
Via Volpi 37, I-21047, Saronno VA, Italy
Tel: (+39 02) 962 29 21   Fax: (+39 02) 960 96 11

Key personnel
Principal: Dr Ing Roberto Foiadelli

Products
Electromechanical components for rail, tram and metro vehicles, including driver's control panels, master controllers, switch panels, circuit breakers, rotary switches, relays and isolators. Newly introduced are driver consoles, master controllers and switch panels for locomotives.

Other equipment by SP11 includes pneumatic selection switches, rotary switches for connecting batteries, controllers and multipole switches.

Contracts include complete driving control units for E412, E402, ETR500, ETR460, ETR470, ETR480, E652 and E633 locomotives.

Stemmann-Technik GmbH

PO Box 1460, D-48459 Schüttorf, Germany
Tel: (+49 5923) 810   Fax: (+49 5923) 811 00
e-mail: info@stemmann.de
Web: http://www.stemmann.de

Background
Stemmann-Technik is a member of the Fandstand Electric Group.

Products
Standard pantographs for light rail vehicles, underground tramways and suburban transit systems per German DIN specification 43,187, and modified versions to suit customer requirements;

Prototype Class DS 3 electric locomotive for Ukrzaliznizija, Ukraine, with AC traction and control equipment supplied by Siemens   0580308

heavy-duty pantographs for trunk railroads for all catenary voltages (up to 25 kV) in conventional (diamond) and contemporary single-arm style; pantographs for industrial locomotives, custom-designed to suit any catenary arrangement. Frost earthing contacts for railway vehicles. Third rail shoegear for metro vehicles. Stinger system overhead conduction rail current collection for depot installations.

Third rail current collection equipment; Stinger roof-mounted conductor rail systems for installation in depots and workshops.

Contracts

DSA 380 pantographs for Talgo 350 high-speed trainsets for the line linking Madrid, Barcelona and the French border; DSA pantographs for the French AGV prototype high-speed trainset; Fb 215 third rail shoegear and AB 433A ground contacts for Series M1 Prague metro trainsets; Stinger roof-mounted catenary for Athens metro depot.

VERIFIED

Toyo Denki Seizo KK

Toyo Electric Manufacturing Co Ltd
No.1 Nurihiko Building, 9-2 Kyobashi 2-chome, Chuo-ku, Tokyo 104-0031, Japan
Tel: (+81 3) 35 35 06 41   Fax: (+81 3) 35 35 06 50

Key personnel

President: Kunio Kai
Vice-President: Motonobu Matsubara
Managing Director: Keisuke Tanaka
Director: Kenzo Terashima

Works

Yokohama: 3-8 Fukuura 3-chome, Kanazawa-ku, Yokohama 236, Japan

Products

Electrical equipment for electric multiple-units, electric and diesel-electric locomotives, light rail vehicles, rubber-tyred vehicles and maglev systems. Propulsion equipment, including induction motors, DC traction motors, drive gear units, VVVF inverter control systems, master controllers, high-speed circuit breakers, unit switches and microprocessor-controlled electronic devices. Auxiliary power supply equipment, including static inverters and converters and brushless motor-alternators. Current collection devices, door actuators, train information control systems and speedometers.

UPDATED

Traktionssysteme Austria GmbH

Brown Boveri Strasse 1, A-2351 Wiener Neudorf, Austria
Tel: (+43 2236) 81 18-0   Fax: (+43 2236) 811 82 37
e-mail: office@traktionssysteme.at
Web: http://www.Traktionssysteme.at

Key personnel

Executive Directors:
   Dr Günter Eichhübl Mag Robert Tencl

Background

Formery ABB Antriebssysteme GmbH, in 2002 the company acquired a majority shareholding in Elin EBG Moteren GmbH.

Products

Liquid-cooled and air-cooled AC traction motors in a power range from 45 to 1,600 kW. Synchronised and asynchronous traction generators with a power of approximately 400 kVA up to approximately 4,000 kVA.

Transportation Products Sales Co Inc

61 Cepi Drive, Suite B, Chesterfied, Missouri 63005, US
Tel: (+1 314) 532 11 44   Fax: (+1 314) 532 14 82
e-mail: tpscerms@tpsc-arms.com
Web: www.tpscarms@tpsc-arms.com

Key personnel

President: Walter J Winzen
Vice-President Operations, Director of Engineering:
   Sid Bakker
Vice-President of Sales: Clay Gillette

Products

Batteries by GNB Technologies for starting diesel-electric locomotives. DC power systems for rail application electronics.

NEW ENTRY

Tülomsas

The Locomotive & Motor Corporation of Turkey
Ahmet Kanatli Cad, TR-26490 Eskisehir, Turkey
Tel: (+90 222) 225 99 56
Fax: (+90 225) 72 72; 57 57
e-mail: tulomsas@tulomsas.com.tr
Web: http://www.tulomsas.com.tr

Key personnel

Managing Director: D Zeki Daloglu
Assistant General Managers: Galip Pala,
   Cengiz Özan, Fatih Turan, Haluk Akova
Head of Marketing: Erol Çetin

Products

Traction motors, motor generator sets, control equipment for electric locomotives, bogies.

Vossloh Kiepe GmbH

Bublitzer Strasse 28, D-40599 Düsseldorf, Germany
PO Box 130540, D-40555 Düsseldorf, Germany
Tel: (+49 211) 749 70   Fax: (+49 211) 749 73 00
e-mail: info@kiepe-elektrik.com
Web: www.vossloh-kiepe.com

Subsidiaries

Vossloh Kiepe Corp, Ottawa, Canada

Key personnel

General Managers: Thomas Weber, Jürgen Textor
Key Account Manager: Wolfram Huober
Marketing Director and Public Relations: M Mixa

Background

In Europe Vossloh Kiepe GmbH is one of the leading manufacturers of electrical equipment for trolleybuses, light rail vehicles, tramways, subways and their modernisations.

Kiepe was established in 1906 as a family business. It has been a subsidiary of the following companies: ACEC (1972), ALSTOM (1998), AEG Rail Systems (1993) and Schaltbau Group (1996). In 2003 it became a member of the Vossloh Group.

Products

Complete electrical equipment for trolleybuses, light rail vehicles, tramways and subways rated DC 600/750/1,500 V.

Range

Includes latest technology in three-phase IGBT AC (direct pulse inverter) and DC chopper power electronics controlled by microprocessor technology, with regenerative braking, built-in diagnosis interface, and roll-back inhibitors. Some recent modernization orders even involved historic contactor banks or rotating pedal controllers.

Contracts

Electrical equipment for 59 new high-floor trams for Cologne, 48 low-floor bi-articulated trolleybuses with Kiepe double-axle-drive for Geneva, 188 low-floor and 40 articulated low-floor trolleybuses for Vancouver, the modernisation of 18 PCC tramcars with Kiepe IGBT direct pulse inverter traction equipment for Philadelphia, USA, and of 46 articulated low-floor units for Geneva; also three-phase inverter drives, databus and on-board diagnostic systems. The same equipment was delivered for 15 high-floor LRVs for Bonn. 69 sets of equipment for a new flow-floor LRV will be delivered for Cologne between 2005 and 2007.

Recent contracts also include sets of equipment for 40 trams for Budapest, a further 20 low-floor trams for Bremen, 15 for Düsseldorf, 28 for Schwerin, 12 for Krakow in Poland, 18 for Graz in Austria and prototypes for Taiwan and Istanbul.

28 dual-voltage LRVs (tram-trains) are being supplied for Saarbrücken; these can operate on 750 V DC in the city and on 15 kV interurban lines.

Kiepe supplies components and subassemblies for most metro and LRV systems in Germany, and undertakes installation and wiring of electrical equipment in manufacturers' works.

Recently and currently supplying three-phase equipment for electric trolleybuses in: Arnhem, Netherlands; Bologna, Parma, Milan and Modena, Italy; Bern, Biel, Montreux, Lausanne, Lucerne and Fribourg, Switzerland; Lyon, France; Innsbruck, Salzburg and Linz, Austria; Quito, Ecuador; Athens, Greece; Budapest, Hungary; Minsk, Belarus; Riga, Latvia; Merida, Venezuela; Solingen, Eberswalde and Esslingen, Germany; and Bergen in Norway.

UPDATED

PASSENGER COACH EQUIPMENT

Alphabetical listing

3M Europe SA
Aim Aviation (Henshalls)
Air International Transit Pty Ltd
Air Mixing
Air Mixing SAS
Airscrew Ltd
A K Fans Ltd
Albatros Corporation
Albright International
Alna Yusoki-Yohin Co Ltd Koki Ltd
ALSTOM Transport SA
Altro floors
American Seating Company
ApATeCh Electro
Apricot Technology AG
Astra Vagoane Călători Arad SA/Arad
Atlas International Ltd
Autoroche Industrie
AVE Rail Products
Avery Dennison
Behr Industry
Baier + Köppel GmbH + Co
R G Bayham Ltd
Bekaert Progressive Composites
BG Teknik Orhus
Bode
Bonar Floors Ltd
Borcad
Brot Technologies SA
captron Electronic gmbh
Carrier Sütrak Transportkälte GmbH
Carrier Transicold
China Northern Locomotive & Rolling Stock
    Industry Group Corporation (CNR)
Cleff
Clerprem SpA
CMC
Compin
Concargo Ltd
ContiTech Luftfedersysteme GmbH
Craig & Derricott Ltd
Cressall Resistors Transit Division
Cromweld Steels Ltd
Dansk Dekor-Laminat A/S
Deans Powered Doors Ltd
Deutsch Relays Inc
Dewhurst plc
DKS
EAO AG
Ebac Industrial Products Ltd
EC-Engineering Oy
Ederena Concept
EFACEC Sistemas de Electrónica, SA
Egetæpper A/S
EKE-Electronics Ltd
Ellamp Interiors SpA
Ellcon National Inc
ELNO
ESW-EXTEL Systems Wedel
EVAC International Ltd
LPA Excil Electronics Ltd
Fahrzeugausrüstung Berlin GmbH (FAGA)
FAINSA – Fabricación Asientos Vehículos
    Industriales SA
Faiveley Transport SA
FASI Seating Systems
Fel SA
Ferranti Technologies Ltd
Ferraz Shawmut
Ferro
Ferro International A/S
Fersystem
Fiberline Composites A/S
Fine Products SA

FISA Srl
Fischer Industries Pty Ltd
Flachglas Wernberg GmbH
Floormaster-Interfloor
Freudenberg Schwab GmbH
Fuji Electric Co Ltd
FY – Composites Oy
Gabriel A/S
GAI-Tronics
Georg Eknes Industrier A/S
Georges Halais SA
GEZE GmbH
Giumma SpA
GKN Aerospace Transparency Systems Kings
    Norton Ltd
Grammer AG
Grupo Antolin
Gummi Metall Technik GmbH (GMT)
Hekatron Vertriebs GmbH
Hexcel Composites
Hepworth Rail International
Hodgson and Hodgson Group Ltd
John Holdsworth & Co Ltd
Huber + Suhner AG
Honeywell Serck
Howden Industrial (Howden Buffalo)
Hübner GmbH
Icon Northern Rubber Co Ltd
IMI Norgren GmbH
International Metals Reclamation Co dbac
    (INMETCO)
International Nameplate Supplies Ltd
iQR
ITT Veam, LLC
Joyce-Loebl Ltd
Jupiter Plast
Kidde
Franz Kiel GmbH & Co KG
Kay-Metzeler Ltd
Kleeneze Sealtech Ltd
Knorr-Bremse GmbH
Konvekta AG
KV Ltd
Lazzerini
Lenord, Bauer & Co GmbH
Liebherr-Verkehrstechnik Frankfurt
LPA Channel Electric Ltd
Lumikko Oy
Mafelec
Marl International Ltd
Mayser GmbH & Co KG
Merak Sistemas Integrados de Climatización
Microelettrica Scientifica SpA
Microphor
Mobile Climate Control Corporation
Monogram Systems
Morio Denki Co Ltd
Mors Smitt
M S Relais
MTB Equipment Ltd
Nabtesco Corporation
Narita Manufacturing Ltd
The Network Connection
Neu Systems Ferroviaires
Orvec International Ltd
Paltechnica
Parker Pneumatic
Pascal International AB
Pars Komponenty sro
People Seating Ltd
Percy Lane Products
Permali Gloucester Limited
Phoenix Traffic Technology GmbH
Pickersgill-Kaye Ltd

Pintsch Bamag Antriebs-und Verkehrstechnik
    GmbH
PIXY AG
Polarteknik PMC Oy AB
Portaramp
Powernetics Ltd
Power-One AG
Powertron Converters Ltd
Rail Interiors SpA
RailTronic AG
Reidler Decal
Rex
RICA
Robert Wagner
Rohm and Haas (UK) Ltd
Ruspa Officine SpA
Safenet
SAFT
Saint-Gobain Sully
Saint-Gobain Sully NA, Inc
SBF Spezialleuchten Wurzen GmbH
Schaltbau GmbH
Schaltbau Holding AG
Schlegel Swiss Standard AG
Secheron
Seira Elettronica Industriale srl
Selectron Systems AG
Semco Vacuumteknik A/S
SEPSA
Seratec Verkehrstechnik
SERINOX
Siemens Rolling Stock Electronics
Slingsby Aviation Ltd
SMA Technologie AG
Sofanar
Soprano
Specialty Bulb Co Inc
SPS Isoclima SpA
Stone India Ltd
Stone International
Stratiforme Industries
Supersine Duramark Limited
Svend A Nielsen A/S
TBA Textiles Ltd
Technical Resin Bonders
Techni-Industrie SA
Técnicas Modulares e Industriales SA (Temoinsa)
Teknoware Oy
Telephonics
th-contact AG
Thermo King Corporation
Tiflex Ltd
Time 24 Ltd
TODCO Inc
Toshiba Corporation
Toyo Denki Seizo KK
Transit Control Systems
Transmatic Inc
Trelleborg Woodville Rail
Trevira GmbH
TrioPlast
UniControls AS
Vapor Corporation
Vapor Rail Inc
Vapor Stone UK Limited
Vogelsitze GmbH
Wabtec Corporation
Walter Mäder Aqualack GmbH
Weserland Sitzsysteme GmbH
Westcode Inc
Widney Transport Components (Pty) Ltd
Winstanley
XP plc
Yutaka Manufacturing Co Ltd

Company listing by country

AUSTRALIA
Air International Transit Pty Ltd
Fischer Industries Pty Ltd
iQR

BELGIUM
3M Europe SA
Bekaert Progressive Composites
FASI Seating Systems

CANADA
International Nameplate Supplies Ltd
Mobile Climate Control Corporation
Stone International
Vapor Rail Inc

CHINA
China Northern Locomotive & Rolling Stock
    Industry Group Corporation (CNR)

CZECH REPUBLIC
Borcad cz sro
Pars Komponenty sro
Thermo King Corporation
UniControls AS

DENMARK
BG Teknik Orhus
Dansk Dekor-Laminat A/S
Egetæpper A/S
Ferro International A/S
Fiberline Composites A/S
Gabriel A/S
Jupiter Plast
Semco Vacuumteknik A/S
Svend A Nielsen A/S

FINLAND
EC-Engineering Oy
EKE-Electronics Ltd
EVAC International
FY – Composites Oy
Lumikko Oy
Polarteknik PMC Oy AB
Teknoware Oy

FRANCE
ALSTOM Transport SA
Autoroche Industrie
Brot Technologies SA
Compin
Ederena Concept
ELNO
Faiveley Transport SA
Fels SA
Ferraz Shawmut
Fersystem
Georges Halais SA
Grupo Antolin
Mafelec
MS Relais
Neu Systems Ferroviaires
SAFT
Saint-Gobain Sully
SERINOX
Sofanar
Soprano
Stratiforme Industries
Techni Industrie SA
TrioPlast

GERMANY
Baier + Köppel GmbH + Co
Behr Industry
Bode
captron Electronic gmbh
Carrier Sütrak Transportkälte GmbH
Cleff
ContiTech Luftfedersysteme GmbH
DKS
EVAC International Ltd
EWS-EXTEL Systems Wedel
Fahrzeugausrüstung Berlin GmbH (FAGA)

Flachglas Wernberg GmbH
Freudenberg Schwab GmbH
GEZE GmbH
Grammer AG
Gummi Metall Technik GmbH (GMT)
Hekatron Vertriebs GmbH
Hübner GmbH
IMI Norgren GmbH
Konvekta AG
Lenord, Bauer & Co GmbH
Liebherr-Verkehrstechnik Frankfurt
Mayser GmbH & Co KG
Phoenix Traffic Technology GmbH
Pintsch Bamag Antriebs-und Verkehrstechnik GmbH
SBF Spezialleuchten Wurzen GmbH
Schaltbau GmbH
Schaltbau Holding AG
Siemens Rolling Stock Electronics
SMA Technologie AG
Trevira GmbH
Vogelsitze GmbH
Walter Mäder Aqualack GmbH
Weserland Sitzsysteme GmbH

INDIA
Stone India Ltd

ISRAEL
Paltechnica

ITALY
Air Mixing
Air Mixing SAS
Clerprem SpA
Ellamp Interiors SpA
FISA Srl
Giumma SpA
Lazzerini
Microelettrica Scientifica SpA
Rail Interiors SpA
RICA
Ruspa Officine SpA
Seira Elettronica Industriale srl
SPS Isoclima SpA

JAPAN
Alna Yusoki-Yohin Co Ltd Koki Ltd
Fuji Electric Co Ltd
Morio Denki Co Ltd
Nabtesco Corporation
Narita Manufacturing Ltd
Toshiba Corporation
Toyo Denki Seizo KK
Yutaka Manufacturing Co Ltd

NETHERLANDS
Avery Dennison
Mors Smitt

NORWAY
Georg Eknes Industrier A/S

PORTUGAL
EFACEC Sistemas de Electrónica, SA

ROMANIA
Astra Vagoane Călători Arad SA/Arad

RUSSIAN FEDERATION
ApATeCh Electro

SOUTH AFRICA
Widney Transport Components (Pty) Ltd

SPAIN
Albatros Corporation
CMC Interiors
FAINSA – Fabricación Asientos Vehículos Industriales
    SA
Fine Products SA
Merak Sistemas Integrados de Climatización
SEPSA
Técnicas Modulares e Industriales SA (Temoinsa)

SWEDEN
Pascal International AB
Safenet

SWITZERLAND
Apricot Technology
EAO AG
Huber + Suhner AG
PIXY AG
Power-One AG
RailTronic AG
Rex
Schlegel Swiss Standard AG
Secheron SA
Selectron Systems AG
Seratec Verkehrstechnik
th-contact AG

UNITED KINGDOM
A K Fans Ltd
Aim Aviation (Henshalls)
Airscrew Ltd
Albright International
Altro Floors
AVE Rail Products
R & G Bayham Ltd
Bonar Floors Ltd
Cole
Concargo Ltd
Craig & Derricott Ltd
Cressall Resistors Transit Division
Cromweld Steels Ltd
Deans Powered Doors Ltd
Dewhurst plc
Eagle Ottawa Callow & Maddox
Ebac Ltd
Excil Electronics Ltd
Ferranti Technologies Ltd
Ferro
Floormaster-Interfloor
GAI-Tronics Ltd
Hepworth Rail International
Hexcel Composites
Hodgson and Hodgson Group Ltd
John Holdsworth & Co Ltd
Honeywell Serck
Howden Industrial (Howden Buffalo)
Icon Northern Rubber Co Ltd
Joyce-Loebl Ltd
Kay-Metzeler Ltd
Kidde
Kleeneze Sealtech Ltd
KV Ltd
LPA Channel Electric Ltd
Marl International Ltd
MTB Equipment Ltd
The Network Connection
Orvec International Ltd
Parker Pneumatic
People Seating Ltd
Percy Lane Products
Permali Gloucester Limited
Pickersgill-Kaye Ltd
Portaramp
Powernetics Ltd
Powertron Converters Ltd
Rohm and Haas (UK) Ltd
Slingsby Aviation Ltd
Supersine Duramark Limited
TBA Textiles Ltd
Technical Resin Bonders
Thorn Transport Lighting
Tiflex Ltd
Time 24 Ltd
Trelleborg Woodville Rail
Vapor Stone UK Limited
Winstanley
XP plc

UNITED STATES
American Seating Company
C&H Chemical
Carrier Transicold

Company listing by country — *continued*

Deutsch Relays Inc
Ellcon National Inc
International Metals Reclamation Co dbac
    (INMETCO)
ITT Veam, LLC
Microphor

Monogram Systems
Reidler Decal
Saint-Gobain Sully NA, Inc
Specialty Bulb Co Inc
Telephonics
TODCO Inc

Transit Control Systems
Transmatic Inc
Vapor Corporation
Wabtec Corporation
Westcode Inc

For details of the latest updates to *Jane's World Railways* online and to discover the additional
information available exclusively to online subscribers please visit
jwr.janes.com

3M Europe SA

Hermeslaan 7, B-1831 Diegem, Belgium
Tel: (+32 2) 722 45 00   Fax: (+32 2) 722 45 11

Key personnel
Marketing Communication Europe: Nicole N Roy
Commercial Graphics Marketing: Severine Soufflet

Products
Bonding systems and adhesive to replace traditional rivets, bolts, and screws in train interiors; anti-graffiti and anti-scratching films; graphics for rail vehicle liveries, interiors and advertising.

Aim Aviation (Henshalls)

Aim Aviation (Henshalls) Ltd
Abbot Close, Oyster Lane, Byfleet KT14 7JT, UK
Tel: (+44 1932) 35 10 11   Fax: (+44 1932) 35 27 92

Key personnel
Chairman and Chief Executive: J C Smith
Managing Director: K Robinson
Financial Director: M J Davis
Marketing Director: M Eyre

Products
Catering equipment for rolling stock, products for carriage interiors.

Recent contracts include the supply of wall-mounted grilles for the Channel Tunnel trains, and the supply of toilet cubicles for the Channel Tunnel nightstock.

Air International Transit Pty Ltd

PO Box 6605, Blacktown Business Centre, Blacktown, New South Wales 2148, Australia
Tel: (+61 2) 98 30 71 00   Fax: (+61 2) 96 72 10 18
e-mail: kallen@airinter.com.au
Web: http://www.airinter.com.au

Key personnel
Executive General Manager: Stephen G Betts
General Manager, Business Development:
   Keith Allen

Principal subsidiaries
Air International Transit US Inc
Air International Transit UK Ltd

Products
Air conditioning and ventilation systems for rolling stock including saloon and locomotive cabs and buses. Unit configurations include integrated roof mounts, split systems, wall/side mounts. Auxiliary power supplies. System controls using programmable logic control or relay logic with diagnostic facilities.

Contracts
Contracts include supply of equipment for LRVs in Hong Kong, USA, Philippines and Turkey, emus in Australia, Hong Kong, India and UK, double-deck trains in Australia, USA and locomotive cabs for Indonesia, Australia and Eurotunnel.

Air Mixing

Via Di Papina 8, I-20043, Arcore MI, Italy
Tel: (+39 0) 61 70 65   Fax: (+39 0) 61 73 87
e-mail: air.mixing.service@interbusiness.it

Products
Ventilation systems for passenger vehicles. The Air Mixing system is composed of flexible tubing, manufactured in a special airtight fabric, in which a series of calibrated perforation is made according to experimented mathematical models. This creates, along the duct, both a uniform repartition of the air flow and a repartition that benefits certain distinct zones.

Air Mixing SAS

Via Papina 8, I-20043 Arcore (MI), Italy
Tel: (+39 039) 61 70 65   Fax: (+39 039) 61 73 87
e-mail: info@airmixing.it

Products
Air distribution ducting for rail vehicle air-conditioning systems manufactured in an airtight fabric. The ducting is supported by aluminium tracking within a fibreglass cable run. The product range includes curves, T-joints and reductions. The ducting meets current European fire-protection norms and has an estimated life of over 25 years.

NEW ENTRY

Airscrew Ltd

111 Windmill Road, Sunbury-on-Thames TW16 7EF, UK
Tel: (+44 1932) 76 58 22
Fax: (+44 1932) 76 10 98
e-mail: mail@airscrew.co.uk
Web: http://www.airscrew.co.uk

Key personnel
Sales Manager, Rail Products: Peter Heapy
Marketing Director: Bryan Hiscock

Associate company
Aircontrol Technologies Limited

Products
Cooling fans and systems for locomotives, dmus and emus. Applications include transformer cooling units, converter cooling modules, traction motor blowers and brake compressor cooling fans.

Heating and ventilation units for rail vehicle cabs and saloons.

Contracts
Contracts include supply of traction motor blowers for UK Class 92 locomotives, Eurostar common block cooling, networker brake resistor cooling and equipment cask cooling for West Coast Main Line, UK.

A K Fans Ltd

32-34 Park Royal Road, London NW10 7LN, UK
Tel: (+44 20) 89 61 68 88   Fax: (+44 20) 89 65 06 01
e-mail: sales@akfansltd.btinternet.com

Key personnel
For a full list of personnel, see A K Fans Ltd entry in *Electric traction equipment* section.

Products
Fans for rail traction applications; axial, centrifugal, mixed flow and crossflow with AC or brushless DC motors. Fan systems and allied equipment including cab heaters and cab ventilation units.

Contracts include fans for saloon heaters on London Underground Jubilee and Northern Line and cab air re-circulation systems for Victoria Line.

Albatros Corporation

c/o Ruiz de Alarcón, 13, E-28014 Madrid, Spain
Tel: (+34 91) 532 41 81   Fax: (+34 91) 522 76 97
e-mail: albatros@albatros-sl.es
Web: www.albatros-sl.es

Works
Merak
Pol Ind 'La Estación', C/Gavilanes 16, E-28320 Pinto, Madrid, Spain
Tel: (+34 91) 495 90 00   Fax: (+34 91) 691 09 97

Albatros also has factories located in Barcelona (Spain), Alcázar de San Juan (Spain), Albany (USA) and Shanghai (China).

Key personnel
Chairman: Nicolás Fúster
Human Resources Director: Patricia Fúster
National Market Director: Félix Ramos
Finance and Control Director: Luis Gil
Marketing and Business Director: Enrique Galavis
Information Technology Director: Andrés Morales
Legal Advisor: José Maria Navia - Osorio
Safety and Security Area: Pedro de la Antonia
Industrial Operations: Arturo Delgado
Converters Area: Antonio Sosa
Merak Managing Director: Julio Rey
AUS Managing Director: Valero Torrelles
ARTS Managing Director: Carlos Rico

Background
Albatros is formed by the following companies:

Merak (formerly known as Stone Ibérica SA) which specialises in the design and manufacture of air-conditioning, ventilation and heating systems for trains, tramways and metros worldwide.

SEPSA, dedicated to the design and subsequent manufacture of electronic equipment for railway vehicles, specialising in two products: static converters and information and control systems.

CMC Interiors, which designs and manufactures modular components for railway vehicles, both externally and internally.

Albatros Alcázar SA which manufactures and maintains all the corporate products. Albatros also has factories located in Albany (US) and Shanghai (China).

Products
Cab and saloon air-conditioning and heating and ventilation systems and equipment for new-build and vehicle refurbishment projects; electronic and microelectronic control equipment; static converters; onboard passenger information and data communications systems; vehicle interior systems and components; inverter ballasts for train lighting applications; static speed regulators for asynchronous motors; pressure wave protection systems; battery charger protection relays. Onboard security cameras, video information systems for passenger entertainment and communication.

Contracts
Vehicles for which air-conditioning equipment has been supplied include: Heathrow Express (UK); New York JFK International Airport automated light metro (USA); Boston Green Line light rail (USA); Chicago Transit Authority metro refurbishment (USA); Airport Express (Hong Kong, China).

UPDATED

Albright International

125 Red Lion Road, Tolworth, Surbiton KT6 7QS, UK
Tel: (+44 20) 83 90 53 57   Fax: (+44 20) 83 90 19 27
Web: http://www.albright.co.uk

Also at
Evingar Trading Estate, Whitchurch RG28 7BB
Tel: (+44 125) 689 30 60

Key personnel
Joint Managing Directors: N Bedggood, A Catt
Sales, Home and Export: P Pickworth
Quality Assurance & Technical Manager:
   Peter Gigance

Products
DC solenoid switches, contactors and battery disconnecting switches. Albright contactors are used to switch the following equipment: cab and corridor lighting, windscreen demister, auxiliary compressor, pantograph heater, saloon half lighting, toilet trace heating, battery isolation.

The contactors are manufactured in five basic ranges, extending from 80 A to 600 A continuous rating and are available in normally open, normally closed and changeover configurations.

Magnetic blowouts can be fitted to most of the range to allow safe operation at voltages in excess of 48 V.

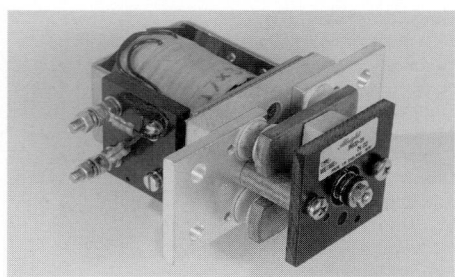

Contactor by Albright International   0023842

Operating coils can be wound to suit voltages from 6 to 240 for continuous or intermittent operation.

Emergency battery disconnect switches are made in single-pole and double-pole variants with current ratings of 100 A to 250 A. These switches are capable of rupturing full-load battery currents in an emergency.

Alna Yusoki-Yohin Co Ltd
Koki Ltd

A subsidiary of the Hankyu Corp
4-5 Higashi Naniwa-cho 1-chome, Amagasaki-City, 660 Hyogo Pref, Japan
Tel: (+81 6) 401 72 83   Fax: (+81 6) 401 61 68
e-mail: soumu@alna.co.jp
Web: http://www.alna.co.jp

Key personnel
President: N Yamazawa
Senior Director, Engineering: T Kaihara
Managing Director, Sales and Production: K Torao

Products
Aluminium, mild steel and steel electric railcars and passenger coaches; light rail vehicles.

ALSTOM Transport SA

48 rue Albert Dhalenne, F-93482 Saint-Ouen Cedex, France
Tel: (+33 1) 41 66 90 00   Fax: (+33 1) 41 66 96 66
Web: www.transport.alstom.com

Key personnel
President, Transport Sector: Philippe Mellier
Chief Operating Officer: Gérard Blanc
Chief Financial Officer: Roland Kientz
Regional Senior Vice-Presidents:
  Asia Pacific: Marc Chatelard
NAFTA: Roelof van Ark
Iberian-Americas: Northern Europe:
  Terence Watson

Contact points
ALSTOM Transport
Rue Cambier Dupret 50-52, B-6001 Charleroi, Belgium
Tel: (+32 71) 44 54 11   Fax: (+32 71) 44 57 82

Products
DC, AC and multivoltage static converters designed for the supply of rolling stock auxiliaries (AC and DC), such as air conditioning, heating and ventilation units, battery chargers and power equipment cooling.

ALSTOM Transport SA
11-13, avenue de Bel Air, F-69627 Villeurbanne Cedex, France
Tel: (+33 4) 72 81 52 00 Fax (+33 4) 72 81 52 87
Products
Onboard railway electronic systems for train management, propulsion and auxiliary control and passenger information.

ALSTOM Transport Inc
1 Transit Drive, New York 14843, US
Tel: (+1 607) 324 45 95   Fax: (+1 607) 324 45 68

UPDATED

Altro Floors

Works Road, Letchworth SG6 1NW, Hertfordshire, UK
Tel: (+44 1462) 48 04 80   Fax: (+44 1462) 48 00 10
e-mail: info@altro.co.uk
Web: http://www.altro.co.uk

Other offices
Australia
Australian Safety Flooring Pty Ltd, Transit Division
Tel: (+61 3) 97 64 56 66

Canada
Altro –Transit Division Canada
Tel: (+1 800) 565 46 58

Germany
Altro GmbH, Transit Abteilung
Tel: (+49 40) 51 94 90

Sweden
Altro Nordic AB, Transit Abteilung
Tel: (+46 40) 31 22 00

UK
Altro Limited, Transit Division
Tel: (+44 1462) 48 04 80

USA
Altro –Transit Division USA
Tel: (+1 800) 382 03 33

Products
Transflor® flooring for buses and coaches. The floor material is made of PVC with silicon carbide, quartz, glass scrim and aluminium oxide chips on a non-woven backing. It can have welded joints for a continuous surface to prevent water ingress. The range includes: thickness of between 2 and 2.5 mm; up to 2 m wide; roll lengths up to 20 m; weights from 2.5 kg/m².

Contracts
Flooring has been supplied to Duewag, Neoplan, Van Hool, MAN, Aabenraa, Kässbohrer, LHB, Scania, Säffle and DAB Silkeborg.

American Seating Company

Transportation Products Group
401 American Seating Center, Grand Rapids, Michigan 49504, USA
Tel: (+1 616) 732 64 06   Fax: (+1 616) 732 64 91

Key personnel
President, Transportation Products:
  Dave McLaughlin
Manager, OEM Sales & Service: Bruce Wright

Products
Passenger seating for bus, metro, rapid transit and trams; aftermarket service; parts on all American Seating models; seating and securement systems for special service vehicles.

Developments
Lightweight modular seating has been introduced with interchangeable parts.

NEW ENTRY

ApATeCh Electro

Zhukovsky Street 2, Building 131, Dubna, 141980 Moscow Region, Russian Federation
Tel: (+7 09621) 257 92   Fax: (+7 09621) 234 92
e-mail: electro@dubna.ru

Products
Glass-fibre plastic seating shells, and plastic interior panels and window frame casings for new and refurbished passenger rail vehicles.

Apricot Technology AG

Postfach 218, CH-6331 Hünenberg, Switzerland
Tel: (+41 41) 784 41 11   Fax: (+41 41) 784 41 30
e-mail: office@apricot.ch

Key personnel
Chief Executive Officer: Domenico Fontana

Subsidiary
Apricot Technology GmbH
Müngstener Strasse 10, D-42285 Wuppertal, Germany
Tel: (+49 202) 76 96 20   Fax: (+49 202) 769 62 22
e-mail: office@apricot.de
Web: http://www.apricot.de

Products
Onboard video surveillance systems, infra-red passenger counting systems, vehicle interior lighting systems, onboard audio and visual passenger information and entertainment systems.

Atlas International Ltd

Merrington Lane, Spennymoor DL16 7UR, UK
Tel: (+44 191) 301 31 15   Fax: (+44 191) 301 31 10

Key personnel
Logistics Manager: T Burton

Associate company
Thorn Transport Lighting

Products
Lighting components for rail vehicles; inverter ballasts for fluorescent lighting 24, 36, 52, 70/72 and 110 V DC, AC; ballasts for various voltages; lampholders and other lighting accessories including luminaires and subassemblies of standard or special types; AC and DC converters for low-voltage tungsten halogen lamps for 12 V reading lights and locomotive headlights; inverter ballasts for fluorescent lamps to the European standard, and luminaires using compact fluorescent lamps.

An extended range of Atlas brand electronic lighting inverter ballasts has been introduced to operate at 15 to 40 W 26 to 38 mm tubes. Advances in production technology have enabled them to offer these compact electronic inverters in voltages covering 24 to 110 V DC. They meet RIA, European (EN), French (NF), International (UIC) and associated standards for rolling stock.

Astra Vagoane Călători Arad SA/Arad

1-3 Petru Rareş, R-2900 Arad, Romania
Tel: (+40 257) 23 62 10   Fax: (+40 257) 25 81 68
e-mail: astra.calatori@astrac.rdsar.ro
Web: http://www.astracalatori.ro

Key personnel
See entry in *Locomotive and powered/non-powered passenger vehicles* section.

Products
Doors, windows, seats, luggage racks, toilet cubicles, sidewalls.

Autoroche Industrie

3 rue de la Cotonnière, F-14000 Caen, France
Tel: (+33 2) 31 74 73 13   Fax: (+33 2) 31 74 75 98
e-mail: info@autoroche.com

Subsidiary
ABL Lights
50 Golf Club Boulevard, Mosinee, Wisconsin 54455, USA
Tel: (+1 715) 693 15 30   Fax: (+1 715) 693 15 34
Web: http://www.abllights.com

Products

Exterior lighting systems and indicators for rail vehicles.

AVE Rail Products

Derby Carriage Works, Litchurch Lane, Derby DE24 8AP, UK
Tel: (+44 1332) 25 75 00   Fax: (+44 1332) 37 19 50
e-mail: admin@averail.co.uk
Web: www.averail.co.uk

Key personnel

Managing Director: M P Thompson
General Manager: S J Ollier
Sales Manager: T Clews

Background

In 2005 Transport Design International UK (TDI), joined AVE Rail Products to form a new interior design division, TDI-AVE Design and Engineering Services. AVE supplies complete rail vehicle interiors, components and electrical systems and has supplied nearly 2,000 vehicles including West Coast Main Line Pendolinos, Alstom's Juniper vehicles, Coradia Class 180 fleet and Bombardier's Electrostars. TDI's portfolio, developed over the last 20 years, covers flagship projects in both light and heavy rail around the world.

The company is now able to offer full international design and low cost supply using its TDI-AVE Design and Engineering Service Division and its newly formed Chinese joint venture, Changzhou Evergreen AVE Ltd.

Products

Complete interior packages for rail vehicles. The company's expertise includes turnkey interior packages, cab and cabin modules, doors, exterior panels and structures, electrical services, powder coating, driver vigilance systems, train data recorders and onboard CCTV systems and the use of advanced manufacturing materials.

UPDATED

Avery Dennison

Graphics Division
Rijndijk 86, PO Box 118, NL-2394 ZG Hazerswoude, The Netherlands
Tel: (+31 71) 342 15 00   Fax: (+31 71) 342 15 38
Web: http://www.averygraphics.com

Products

Anti-graffiti film for rail vehicles; glass protection film; adhesive colour vinyl film for livery applications and vandal and graffiti resistant.

Baier + Köppel GmbH + Co

Präzisionsapparatefabrik, Beethovenstrasse 14, D-91257 Pegnitz/Bayern, Germany
Tel: (+49 9241) 72 90   Fax: (+49 9241) 729 50
e-mail: beka@beka-lube.de;
  beka@beka-max.de
Web: http://www.beka-lube.de

FluiLub lubrication nozzle positioned to deliver lubricant to the wheel flange of a light rail vehicle 0126646

Products

FluiLub vehicle-mounted wheel flange lubrication systems.

R & G Bayham Ltd

Rutherford Road, Daneshill West Industrial Estate, Basingstoke RG24 8PG, UK
Tel: (+44 1256) 46 49 11   Fax: (+44 1256) 46 43 66
e-mail: sales@bayham.demon.uk
Web: http://www.bayham.demon.co.uk

Key personnel

Chairman: A Madigan
Managing Director: C Balment
Export Sales: B Sudworth

Subsidiary

The Ranger Instrument Company Ltd

Products

R & G Bayham provides a comprehensive range of tank contents measurement solutions for rail service. Instruments are available from simple mechanically operated direct reading contents gauges to remote reading systems with integral alarm facilities. Models are manufactured for specific applications including fuel oil contents monitoring and coolant and water tank level measurement. Combinations gauges can be supplied with local dial indication plus outputs for management systems including level warning and engine shutdown signals.

Behr Industry GmbH & Co, KG

Heilbronner Strasse 380, D-70469 Stuttgart, Germany
Tel: (+49 711) 896 20 11   Fax: (+49 711) 896 30 75
e-mail: behrindustry@behrgroup.com
Web: www.behrindustry.behrgroup.com

Key personnel

Sales Director: Ruediger Wanner
  e-mail: ruediger.wanner@behrgroup.com
Sales Manager: Wilfried Sonnenfeld
  e-mail: wilfried.sonnenfeld@behrgroup.com

Products

Complete air conditioning systems for passenger coaches, trams and commuter and subway trains; complete air conditioning systems for locomotive cabs and driving units. Capabilities include design of air ducting and air circulation in the vehicle, power supply for climate control and associated power electronics.

UPDATED

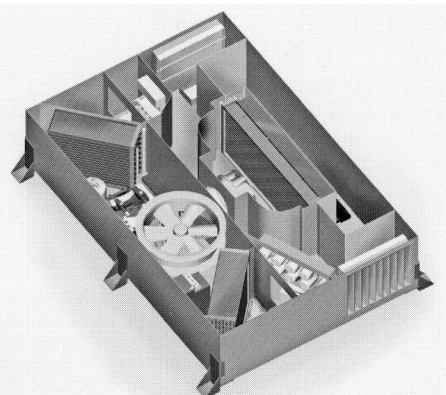

HVAC module of a railway car air conditioning system 0114662

Bekaert Progressive Composites

Industriepark De Bruwaan 2, B-9700 Oudenaarde, Belgium
Tel: (+32 55) 33 30 11   Fax: (+32 55) 33 30 50
e-mail: info@composites.com
Web: www.bekaertcomposites.com

Subsidiary company

Bekaert Composites Spain
Aritzbidea 67, E-48100 Munguia (Vizcaya), Spain
Tel: (+34 94) 674 03 16   Fax: (+34 94) 615 61 14

Products

Components and modules in fibre-reinforced composites for rail vehicle applications. Both pultrusion and filament winding production processes are used.

UPDATED

BG Teknik Orhus

Grenåvej 148, DK-8240 Risskov, Denmark
Tel: (+45) 87 41 80 10   Fax: (+45) 86 17 44 44
e-mail: bg@bg-dk.dk
Web: http://www.bg-dk.dk

Key personnel

Contact: Benny Godsk Jensen

Products

Vehicle front and tail light systems; LED interior lighting systems; windscreen wiper and washer systems; power supply systems for lighting and wiper systems.

Bode

Gebrüder Bode & Co GmbH
Ochshäuserstrasse 14, D-34123 Kassel, Germany
Tel: (+49 561) 500 90   Fax: (+49 561) 559 56
e-mail: bode@bode-kassel.com

Key personnel

Managing Directors: Klaus-Peter Schwarz
  Rainer Wicke
Sales Manager: Jürgen Holz

Background

Bode is a member of the Schaltbau Group.

Products

Electric and pneumatic door systems for LRVs, metro and high-speed vehicles, outswing plug doors and inswing plug doors; pressure-sealed doors for high-speed trains; ramp systems; step systems; door controls.

Contracts include supply of door systems to Hong Kong MTRC airport link; Amtrak Northeast Corridor; ICE and ICT, Germany; Transrapid, Germany; ICN, Switzerland; and Kuala Lumpur metro.

Bonar Floors Ltd

High Holborn Road, Ripley DE5 3NT, UK
Tel: (+44 1773) 74 41 21   Fax: (+44 1773) 74 41 42
Web: http://www.bonarfloors.com

Products

Floorcoverings for rail vehicles, including Flotex Transport, which is fully waterproof and has been designed to trap dust mites and other allergens.

Borcad cz sro

Fryčovice 673, CZ-739 45 Fryčovice okres Frýdek-Mistek, Czech Republic
Tel: (+420 558) 64 06 31   Fax: (+420 558) 66 80 87
e-mail: borcad@borcad.cz
Web: http://www.borcad.cz

Key personnel

Chairman: Ing Ivan Boruta
Managing Director: Milan Burgár
Technical Director: Ing Petr Mil
Finance Manager: Ing Milan Homza
Commercial Manager: Ing Miroslav Havelka

Products

Seats for high-speed trainsets, long-distance and regional trains and restaurant cars; beds and couchettes for sleeping cars.

Contracts

These include the supply of seating for the Czech Railways Class 680 Pendolino trainset and for Slovakian Railways Class 840 regional diesel railcars.

Brot Technologies SA

4 rue de la Fauvette, F-95100 Argenteuil, France
Tel: (+33 1) 34 10 79 76   Fax: (+33 1) 34 10 79 79
e-mail: rail@brot.fr
Web: http://www.brot.fr

Products

Windscreen wiper systems for rail vehicles. A recent addition to the product range is the Brot 4, an electro-pneumatic model which features electronic control.
VERIFIED

captron Electronic gmbh

Bodenseestrasse 129, D-81243 Munich, Germany
Tel: (+49 89) 889 69 50   Fax: (+49 89) 88 96 95 55
e-mail: info@captron.de
Web: http://www.captron.de

Products

CHT series wall- and glass-mounted pushbutton sensors for buses and rail vehicles; HWT series wall- and pole-mounted pushbutton stop request sensors; associated connector systems.

Carrier Sütrak Transportkälte GmbH

Heinkelstrasse 5, D-71272 Renningen, Germany
Tel: (+49 7159) 92 32 02   Fax: (+49 7159) 92 31 08
e-mail: oliver.wels@carrier.utc.com
Web: http://www.carrier.com

Key personnel

Marketing and Distribution: Oliver Wels

UK office

Sütrak UK Ltd
24-25 Saddleback Road, Westgate Industrial Estate, Northampton NN5 5HL
Tel: (+44 1604) 58 14 68   Fax: (+44 1604) 75 81 32
Sales Director: Jeremy Smith

USA office

Sütrak Corporation
6899 East 49th Avenue, Commerce City, Colorado 80022
Tel: (+1 303) 287 27 00   Fax: (+1 303) 286 10 05

Products

Air conditioning equipment for rail vehicles, people movers and buses; roof-mounted, integrated modular and integrated compact systems, all available with R134A refrigerant.

The AC136 is quieter and more efficient with less servicing. The roof-mounted air conditioning AC136 has been specially created for city and intercity buses. It is available as three variants; with cooling and heating, air treatment only (without condenser) and as a ventilation unit only, with heating but without cooling. The cooling output ranges from 24 to 32 kW, depending on the specification and the heating output ranges to 45.6 kW.

Carrier Transicold

Carrier Refrigeration Operations (Division of Carrier Corporation)
PO Box 4805, Carrier Parkway, Building TR20, Syracuse, New York 13221, USA
Tel: (+1 315) 432 64 34   Fax: (+1 315) 432 72 18

Key personnel

President: Nick Pinchuk
Vice-President, Transport Air Conditioning:
  Lex van der Weerd

Regional Sales Offices

North America:
Carrier Transicold-Transport A/C
715 Willow Springs Lane, York, Pennsylvania 17402, USA
Tel: (+1 717) 767 65 31   Fax: (+1 717) 764 04 01

Latin America:
Carrier Transicold-Mexico
Tezoquipa No 142 Col. La Joya, Deleg. Tlalpan 14090 Mexico DF
Tel: (+52 5) 573 55 55; 655 05 07
Fax: (+52 5) 655 21 02

Europe/Middle East/Africa:
Carrier-Sütrak, Heinkelstrasse 5, D-71272 Renningen, Germany
Tel: (+49 7159) 92 31 00   Fax: (+49 7159) 92 31 08

Asia Pacific:
Carrier Transicold-APO
12 Gul Road, Singapore 629343
Tel: (+65) 68 62 00 98 Fax (+65) 68 62 32 86

Business units

Transport Air Conditioning Group
Transport Refrigeration and Air Conditioning

Products

Air conditioning and heating systems for rail and bus applications, roof-mounted, rear-mounted or in-vehicle; air conditioners for small vehicles; components including compressors, evaporators and heater coils, open drive and semi-hermetic compressors; and replacement components.

China Northern Locomotive & Rolling Stock Industry Group Corporation (CNR)

11 Yangfangdian Road, Haidian District, Beijing 100038, China
Tel: (+86 10) 51 86 23 70   Fax: (+86 10) 51 86 23 74
e-mail: loriciec@cnrgc.com.cn
Web:www.cnrgc.com

Key personnel

Chairman and Managing Director: Wang Tai-Wen
President: Cui Dianguo
Vice-President: Zhao Guangxing
President of LORIC Import & Export Corporation Ltd:
  Cao Guo-Bing
Vice-President of LORIC Import & Export
  Corporation Ltd: Chen Dayong
Senior Engineer of LORIC Import & Export
  Corporation Ltd: Yang Xiang-Jing

Products

Design, manufacture and supply of a range of toilet cubicles, air spring, doors, chairs, sidewalls and complete interior packages for rail vehicles.

NEW ENTRY

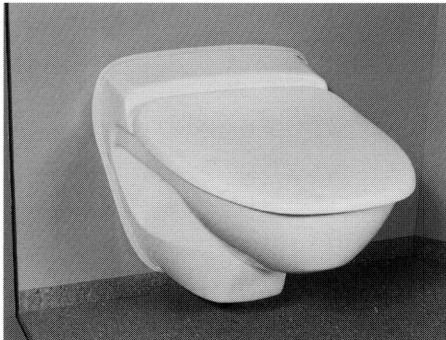

CNR EVAC2000E compact vacuum toilet
NEW/1135832

Cleff

Carl Wilhelm Cleff GmbH & Co KG
Postfach 260180, D-42243 Wuppertal, Germany
Tel: (+49 202) 64 79 90   Fax: (+49 202) 647 99 88
e-mail: marketing@cleff-wpt.de

Products

Ventilation and lighting systems for buses and coaches; windows for rail vehicles; interior doors; interior fittings.

Clerprem SpA

Via Bianche 10, I-36010 Carre (VI), Italy
Tel: (+39 0445) 86 97 00   Fax: (+39 0445) 86 97 77
e-mail: info@clerprem.com

Products

Seating systems for urban, suburban, regional and high-speed trains.

CMC

Pol. Ind Can Roca, C/. Mar del Japon, n° 3, E-08130 Sta Perpetua de Mogoda, Barcelona, Spain
Tel: (+34 93) 544 66 66   Fax: (+34 93) 544 82 19
e-mail: cmc@cmc.albatros-sl.es

Key personnel

General Manager: Antonio Lillo
Industrial Manager: Miguel Aliseda

Background

CMC is a member of the Albatros Group.

Products

Concept, design, fabrication and maintenance service of complete interiors for new and refurbished passenger rolling stock.
VERIFIED

Compin

1, rue du Guesclin, ZI de Netreville, PO Box 1804, F-27018 Evreux, France
Tel: (+33 2) 32 33 92 21   Fax: (+33 2) 32 33 92 29
e-mail: commercial@compin.com
Web: www.compin.com

Key personnel

President: Patrick Balloffet

Products

Seating systems and accessories for urban, suburban, regional and high-speed trains.
UPDATED

Concargo Ltd

Old Mixon Crescent, Weston-super-Mare BS24 9AH, UK
Tel: (+44 1934) 62 82 21   Fax: (+44 1934) 41 76 23
e-mail: sales@concargo.co.uk
Web: http://www.concargo.co.uk

Key personnel

General Manager: Andrew Morris

Products

Glass reinforced plastics components including side panels, window surrounds, skirts, ceiling panels, third-rail covers, seats, consoles, cab fronts. Materials include phenolic, epoxy and Class 1 polyester resin systems using the hand lay and resin transfer moulding process.

Contracts

Moulded components for London Underground Jubilee, Piccadilly and Northern lines; Turbostar and Electrostar interior and exterior mouldings.
VERIFIED

ContiTech Luftfedersysteme GmbH

Vahrenwalder Strasse 9, D-30165 Hanover, Germany
Tel: (+49 511) 938 13 04
Fax: (+49 511) 938 938 13 05
e-mail: railway_suspension_part@ls.contitech.de
Web: http://www.contitech.de

US office
ContiTech North America Inc
136 Summit Avenue, Montvale, New Jersey 07645
Tel: (+1 201) 930 06 00 (ext. 101)
Fax: (+1 201) 930 00 50
e-mail: rkremmeicke@contitech-usa.com
Product Manager: Rainer Kremmeicke

Key personnel
Customer Management, Rolling Stock:
   Manfred Hunze

Background
ContiTech Profile GmbH is a member of the ContiTech Group, part of Continental AG.

Products
Door and window sealing profiles.

Craig & Derricott Ltd

Hall Lane, Walsall Wood, Walsall WS9 9DP, UK
Tel: (+44 1543) 37 55 41   Fax: (+44 1543) 45 26 10
Fax: (+44 1543) 36 16 19 (Direct Sales)
e-mail: info@craigandderricott.com
Web: www.craigandderricott.com

Key personnel
Managing Director: Gordon Barraclough
Engineering and Production Director:
   Andy Dolman
Sales Director: Paul Cranshaw

Products
Design, manufacture and supply of a range of switchgear and custom-designed control panels adapted to meet the control and safety requirements of the rail industry, control desks, uncouplers, control panels, equipment cases, switches, communication panels, safety-related equipment and wiring harnesses. Specifically including: shunting control panels; customised rotary switches and isolators; rotary switches featuring high-security key locks; and mushroom-headed push-buttons used as both emergency stops and passenger alarm switches. Also supplied is a range of limit and reed switches suitable for mounting in or around all rail equipment.

Low-profile push-button and LED indicator units, master drivers key switches, passenger alarm handles, drum switches and underframe starting switches.

Craig & Derricott's passenger talk-back unit   **NEW**/1136294

Contracts
Supply of uncouplers and all the communication panels within the Virgin Pendolino trains, supply of switch panels into Virgin CrossCountry. Supply of uncouplers and earthing switches into Bombardier Electrostar Vehicles, supply of passenger communications handles for Hong Kong MRTC, supply of switches for drivers cabs into Bombardier, Siemens and London Underground, supply of passenger emergency handles into London Underground.

UPDATED

Cressall Resistors Transit Division

Transit Division, Peacock Way, Melton Constable NR24 2BZ, UK
Tel: (+44 1263) 86 05 81
Fax: (+44 1263) 86 14 17
e-mail: transit@cressall.com
Web: http://www.cressall.com

Key personnel
General Manager: Eric Williams
Finance Director: Dennis Cummins
Sales and Marketing Manager: Roger Mason
Principal Engineer: Russell Everett
Manufacturing Manager: Mark Dent
Purchasing Manager: John Crick

Background
Parent company is Cressall Resistors, Evington Valley Road, Leicester, LE5 5LZ.

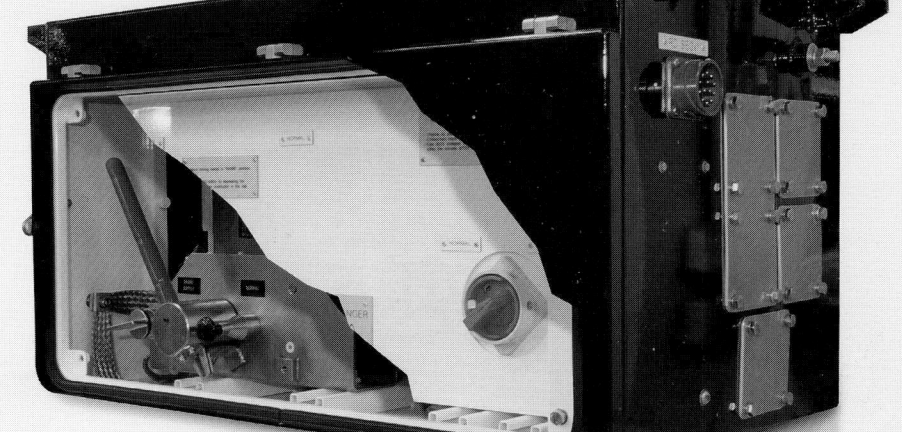

Craig & Derricott's shoegear earthing switches   0526803

Products

Power resistors for transport applications, including naturally cooled and forced cooled roof or under-car-mounted resistors, designed to meet the load requirements of each vehicle. Infra-red temperature monitoring and protection system offering remote indication and status.

Cromweld Steels Ltd

The Old Vicarage, Tittensor, Stoke-on-Trent, Staffordshire ST12 9HY, UK
Tel: (+44 1782) 37 41 39   Fax: (+44 1782) 37 33 88

Key personnel
For a full list of personnel, see Cromweld Steels Ltd entry in *Freight vehicles and equipment* section.

Subsidiary
Cromweld Steels
PO Box 1500, Cornelius, North Carolina 28031, USA
Tel: (+1 704) 896 81 14   Fax: (+1 704) 896 81 15

Products
3CR12 ferritic stainless steel. This material is claimed to offer a low stable cost stainless-steel option for passenger coach and light rail vehicles.

Dansk Dekor-Laminat A/S

Grønlandsveg 197, DK-7100 Vejle, Denmark
Tel: (+45 7642) 82 82   Fax: (+45 7582) 71 21
e-mail: dd@dandekor.dk
Web: http://www.dandekor.dk

Products
Etronit-M high-pressure and compact laminates, DanDekor real veneer laminates and Alunit lightweight construction material for rail vehicle interiors. Applications include wall and ceiling panels, tables, partitions and door panels.

Deans Powered Doors Ltd

PO Box 8, Borwick Drive, Beverley HU17 0HQ, UK
Tel: (+44 1482) 86 81 11   Fax: (+44 1482) 88 18 90
Web: http://www.deans-doors.co.uk

Key personnel
Managing Director: Derek Skidmore
Sales and Marketing Director: Malcolm Phillips
Technical Director: P Spencer

Background
Deans is part of Manganese Bronze Holdings Plc, a leading group in the automotive and transport industry. The group is organised into two divisional sectors – the Vehicle Division and the Components Division. Group products range from taxi-cabs and automotive castings, to metal powders and sinter products.

Products
Powered doors, powered access ramps, handrail and handrail fittings, microprocessor controllers and diagnostics for city buses, coaches and LRVs.

Range includes inward-opening glider doors: one conventionally gasket glazed, the other featuring flush-bonded glazing. Both doors are driven by reliable, cost-effective pneumatic actuators. The flush-bonded door is fitted with Deans' active draught seal, which lifts out of the way when the door is opened. The new seal is effective on angled floors up to the DDA (Disability Discrimination Act, UK) limit of 5°.

Deans' range also includes PowerLeaf and PowerSlide access ramps, together with a manually operated alternative. There are six PowerSlide models to choose from, one for each of the most popular chassis designs currently in use in the UK.

Developments
In 2001 an outward-opening minibus door, with flush-bonded glazing, was announced. The new door, an outward-opening plug door, features profiled flush-bonded glazing. It is designed for use on the latest generation of low-floor minibuses, welfare vehicles and small coaches. Its electric rotary actuators operate from the bottom, rather than above the door.

Deutsch Relays Inc

55 Engineers Road, Hauppauge, New York 11788, USA
Tel: (+1 631) 342 17 00   Fax: (+1 631) 342 94 55
e-mail: info@deutschrelays.com
Web: http://www.deutschrelays.com

Key personnel
President: Tom Sadusky
Account Manager, Railway Market:
   Eugene Agresta
Engineering Manager: Keith Gaedje
Purchasing Manager: Warren Stricoff

Overseas offices
Compagnie Deutsch Orleans
22 rue des Chaises, F-45140 St. Jean de la Ruelle, Cedex, France
Tel: (+33 2) 38 70 45 00   Fax: (+33 2) 38 70 45 99
Web: http://www.compagnie-deutsche.com
Division Manager: Gerome Avelange

Compagnie Deutsch GmbH
Fraunhoferstrasse 11B, D-82152 Martinsried, Germany
Tel: (+49 89) 899 15 70   Fax: (+49 89) 857 46 84
e-mail: info@compagnie-deutsch.de
Division/Marketing Manager: Sebastian Bendak

Deutsch Italia SRL
Sede e uffici, I-20158 Milan, P le Lugano, 9, Italy
Tel: (+39 02) 39 32 30 08   Fax: (+39 02) 646 41 95
Sales Engineer: Walter Merlini

Products
Hermetically sealed relays, sockets and solid-state timers/time delay relays which are designed for reliable operation under severe environmental conditions. Relays, sockets and timers are available in a variety of terminations and mounting styles for the railway industry such as PC board, panel, track and wiring harness installations. Relays are designed for long life switching of low to medium current levels in 1, 2, 3, 4 and 6 pole form C and form Z configurations. All relays can be included with internal voltage suppression directly on the coil of the relay. Deutsch's design also features a single pivoting armature switching whereby all movable pole weld resistant contacts switch together and 'non-overlapping', 'back-check' or 'force-guided' principles can be capitalised on. Environment characteristics for the critical safety relays include a wide temperature range: –40°C to +125°C, vibration (any axis): 20 g 2,000 Hz and shock (any axis): 200 g 6MS. Applications include logic interface in automatic and manual train control systems, sensor and actuator interface, lighting, braking and other control and sensing systems.

New developments
Deutsch Relays has released a new micro contactor called the MCT 110. The MCT 110 series is a 1 cu in contactor that switches 10 amps resistive at

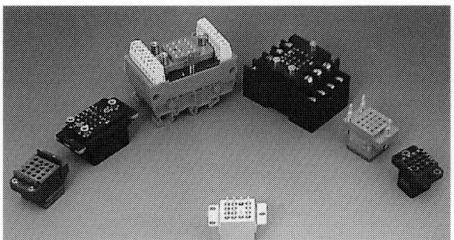

Deutsch Relays railway sockets   0120204

Deutsch Relays railway sockets   1048812

110 V DC for 100,000 cycles. Applications include emergency braking systems.

Deutsch Relays also offers DEST, a hermetically sealed voltage sensor. The DEST is able to detect voltage spikes according to specified voltage thresholds. Applications include connecting the DEST up to a battery to avoid power loss.

Contracts
Contracts include relays, sockets and timers, designed for: MTA LIRR M-7 emu project – Bombardier Transportation; Minneapolis LRV project – Bombardier Transportation; Port Authority New York/New Jersey JFK Airport LRV 'AirTrain' – Bombardier Transportation; Amtrak Northeast Corridor (Acela) project – Bombardier/ALSTOM; Dallas Area Rapid Transit (DART) project – Kinki Sharyo; VTA LRV project – Kinki Sharyo; NJT Hudson Bergen LRV - Kinki Sharyo; Washington DC Metro (WMATA) project – ALSTOM.

Dewhurst plc

Inverness Road, Hounslow TW3 3LT, UK
Tel: (+44 20) 86 07 73 00
Fax: (+44 20) 85 72 59 86
e-mail: railsales@dewhurst.co.uk
Web: http://www.dewhurst.co.uk

Key personnel
Rail Product Manager: John Harris

Principal subsidiary companies
Dupar Controls, Canada
The Fixture Company, USA
LiftStore, UK
Australian Lift Company, Australia

Products
Push-button controls and indicators to meet the requirements of retrofit and new rolling stock vehicles. Standard ranges of vandal-resistant push-buttons, keypads and push-button control panels for internal and external passenger doors, vestibule doors, emergency call panels, drivers' cab controls, guard stations and crew access.

Platform TR and RA signal boxes and signalling control panels.

Contracts
Contracts include LED bodyside status indicators on the new generation of Adtranz passenger vehicles.

Other contracts include passenger door push-buttons for Heathrow Express and Stockholm LRVs. Driver's cab controls for MTRC Lantau Airport extension, Hong Kong. Upgrade to LED illumination of passenger controls on the Mk IV intercity fleet and status indicators on

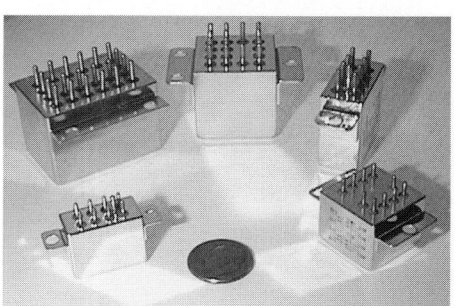

Deutsch Relays hermetically sealed relays in standard metal enclosure   0087707

London Underground's Hammersmith, Bakerloo, Metropoliton, Central and Circle Lines. Connex header panels with illumination.

VERIFIED

DKS

Dienstleistungsgesellschaft für Kommunikationsanlagen des Stadt- und Regionalverkehrs mbH
Oskar Jäger Strasse 155, D-50825 Cologne, Germany
Tel: (+49 221) 954 44 20   Fax: (+49 221) 95 44 42 23
e-mail: info@dks-koeln.de
Web: http://www.dks-koeln.de

Key personnel
Technical Managing Director:
   Dipl Ing Christian Doering
Commercial Managing Director: Ralf Kochs

Products
SIUS onboard monitoring and recording systems for rail vehicles, trams and LRVs and buses. The ruggedised systems can be equipped for video or digital recording and can be installed in both newbuild and retrofit applications. A driver's monitor can also be provided.

EAO AG

Tannwaldstrasse 88, CH-4601, Olten, Switzerland
Tel: (+41 622) 80 92 95
e-mail: info@eao.com
Web: www.eao.com

Key personnel
Contact: Roland Schmid

Other offices
Austria
Tel: (+43 1) 259 96 23   Fax: (+43 1) 259 96 23
e-mail: sales.ede@eao.com

Belgium
Tel: (+32 3) 777 82 36   Fax: (+32 3) 777 84 19
e-mail sales.ebl@eaocom

China
Tel: (+852 27) 86 91 41
e-mail: sales@ehk@eao.com.hk

France
Tel: (+33 1) 64 43 37 37   Fax: (+ 33 1) 64 43 37 48
e-mail: sales.ese@eao.com

Germany
Tel: (+49 201) 858 70   Fax: (+49 201) 858 72 57
e-mail: sales.ede@eao.com

Japan
Tel: (+81 3) 54 01 09 53   Fax: (+81 3) 54 01 09 68
e-mail: sales.esj@eao.com

Netherlands
Tel: (+31 78) 653 17 00   Fax: (+31 78) 653 17 99
e-mail: sales.enl@eao.com

Sweden
Tel: (+46 8) 683 86 60   Fax: (+46 8) 724 29 12
e-mail: sales.esw@eao.com

Switzerland
Tel: (+41 62) 388 95 00   Fax: (+41 62) 388 95 55
e-mail: sales.ech@eao.com

UK
Tel: (+44 1444) 23 60 00   Fax: (+44 1444) 23 66 41
e-mail: sales.euk@eao.com

USA
Tel: (+1 203) 877 45 77   Fax: (+1 203) 877 36 94
e-mail: sales.eus@eao.com

Background
EAO is a global manufacturer of high-quality human machine interface products and solutions,

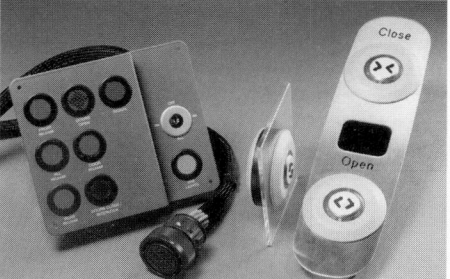

EAO's panel combination   0087708

from switches, key pads and keyboards to complete custom built control panels.

Founded in 1947, EAO's range of target industries include: transportation; machinery; telecommunications; process control; lifting and moving and automative.

EAO has nine specialised sales and customer service centres around the world and a network of trained specialist agents and representatives in more than 50 countries. EAO is certified and managed according to international standards ISO9001:2000, ISO 14001 and VDA6.1.

Products
Driver's cab and door control switch components and customised HMI panels. Components include: push-buttons, rotary switches, emergency stop switches and crew and access switches. Passenger operated buttons include halo illuminated push buttons for door control and toilets.

Contracts
EAO supplies products to many train manufacturing companies on a global scale, including ALSTOM Transport, CAF, Firema, Kawasaki, Kinki Sharyo, Rotem, Siemens Transportation, Skoda, Stadler, Talgo and Vossloh. Typical examples of end user projects using EAO materials include: AGC & TER2NNG (SNCF); NINA (BLS); FLIRT (SBB); West Rail emu (KCRC); Desiro (ÖBB); Desiro (AngelTrains); X40 (SJ); IC4 (DSB); AM96 (SNCB); CL 471 (CD); BR 152 (DB); Electrostar (UK); Class 455 (UK); Class 465 (UK).

UPDATED

Ebac Industrial Products Ltd

St Helen Trading Estate, Bishop Auckland, County Durham DL14 9AD, UK
Tel: (+44 1388) 66 44 00   Fax: (+44 1388) 66 25 90
e-mail: comm_info@ebac.com
Web: www.ebacuk.com

Key personnel
Managing Director: Stephen Lilly

Principal subsidiaries
Ebac Industrial Products Inc
704 Middle Ground Blvd, Suite A-2, Newport News, Virginia 23606-2528, USA
Tel: (+1 757) 873 68 00   Fax: (+1 757) 873 36 32
(Enquiries to Ebac Ltd, UK)

Ebac Industrial Products Ltd
Miraustra 64-66, D-13509 Berlin, Germany
Tel: (+49 30) 435 57 23   Fax: (+49 30) 43 55 72 40

Products
Design, manufacture, supply and overhaul of air conditioning, heating and ventilation systems for all types of rail vehicles. Heat pump technology; microprocessor-based controls and diagnostics; chilled water systems; low refrigerant content, R134a systems; roof-mounted, underframe-mounted and split systems.

UPDATED

EC-Engineering Oy

Santaniitynkatu 12A, FIN-04250 Kerava, Finland
Tel: (+358 9) 294 03 27   Fax: (+358 9) 294 03 28

e-mail: ec-engineering@ec-engineering.fi
Web: http://www.ec-engineering.fi

Key personnel
Managing Director: Tapio Ollanketo

German office
EC-Engineering Oy
Büro Deutschland
Postfach 2029, D-32779 Lage, Germany
Tel: (+49 5232) 784 33   Fax: (+49 5232) 781 33
e-mail: ec-vs@t-online.de
Contact: Frhr Horst v Schleinitz

Background
EC-Engineering Oy is a member of the Finnish Transportation Expertise Network (Finten) association.

Products
Custom-designed sandwich panels and vacuum-formed GRP components for passenger rail vehicles, including complex shapes and products for heavy wear conditions. Applications include: window and wall elements; interior stairs for double-deck coaches; elements for vestibules.

Contracts
Roof fairings for Pendolino trainsets (Talgo); exterior roof and wall panels for RegioShuttle dmu (Bombardier); floor panels for Class ET 423 emu (Alstom); and wall and window panels, interior dividing walls and roof panels for ICS double-deck coaches (Talgo).

Ederena Concept

Avenue du Parc, F-40230 St Vincent de Tyrosse, France
Tel: (+33 5) 58 77 46 46   Fax: (+33 5) 58 77 46 45
e-mail: sales@ederena.com
Web: http://www.ederena.com

Products
Ederena designs and manufactures high-performance composite/mechanical subassemblies by sandwich bonding materials.

EFACEC Sistemas de Electrónica, SA

Av Eng Frederico Ulrich, PO Box 31, P-4470 Maia, Portugal
Tel: (+351 2) 941 36 66   Fax: (+351 2) 948 54 28

Products
Static converters.

Contracts
Recent contracts include supply of static converters for Corail-type coaches for CP.

Egetæpper A/S

Industrivej Nord 25, DK-7400 Herning, Denmark
Tel: (+45) 97 11 88 11   Fax: (+45) 97 11 95 80
e-mail: ege@ege.dk
Web: http://www.egecarpet.com

Products
ege range of flooring textiles for rail vehicles.

EKE-Electronics Ltd

Piispanportti 7, FIN-02240 Espoo, Finland
Tel: (+358 9) 61 30 33 08   Fax: (+358 9) 61 30 33 00
e-mail: electronics@eke.fi
Web: www.eke.com

Key personnel
Managing Director: Anssi Laakkonen
Development: Jyrki Keurulainen
Sales: Joni Juuth

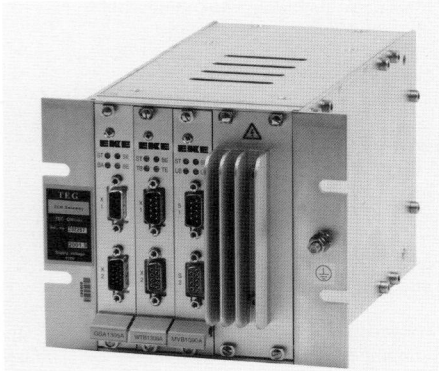

EKE-Trainnet® Gateway   0134728

Customer Projects: Sami Hienonen
Service and Technical Support: Mika Linden
Marketing Manager: Katja Urpelainen

Products

EKE-Trainnet® TCN standard-based integrated train management systems. Products include a comprehensive set of modules, components and tools for complete train monitoring, diagnostics and control systems. System solutions include EKE-TMS train management systems and TCN gateways, EKE-TDR event recorders with protected memory and EKE-MMI user interface with display and keypad.

Contracts

Recent contracts include: TMS, TCN gateway and remote control and user interface for Bombardier Transportation, TCN gateway and remote control and user interface for Bombardier Transportation (FAGA, Israel double-deck project) 2002–2003; TDR Event Recorder for Korean Railroad Research Institute/Rotem (Korean G7 High-Speed Train), 2002; TCN gateway for Zhuzhou Electric Loco. Res. Institute (China) (ZELRI, High-Speed Train project), 2002; TMS, TCN gateway and remote control, TMS, TDR and driver's interface for Walkers Pty Ltd (Cairns Tilt Train, Australia), 1999–2002; WTB Trainbus for Bombardier Transportation (Virgin CrossCountry dmus, UK), 1999–2001.

UPDATED

Ellamp Interiors SpA

Via Rossini, 7, I-21020 Bodio Lomnago Varese, Italy
Tel: (+39 0332) 94 37 11   Fax: (+39 0332) 94 37 65
e-mail: info@elleampgroup.com
Web: http://www.ellampgroup.com

Products

Design and manufacture of interior modules, components and fabrications for main line and urban rail vehicles, buses and coaches in aluminium alloy, composite materials, high-pressure laminate postforming and reinforced fibreglass. Products include air conditioning channels, handrails, interior panels, lighting systems and luggage racks.

Contracts

Projects undertaken recently include products for Pendolino-series tilting trainsets (ALSTOM Transport, Fiat Ferroviaria) and Talent dmus for Norway (Bombardier Transportation).

Ellcon National Inc

50 Beechtree Boulevard, PO Box 9377, Greenville, South Carolina 29604-9377, USA
Tel: (+1 864) 277 50 00   Fax: (+1 864) 277 52 07
Web: http://www.ellcon.com

Key personnel

Chairman: E P Kondra
President: Douglas E Kondra
Executive Vice-President: R A Nitsch
Treasurer: L F D'Alessio

Licensee

Gregg Company Ltd
15 Dyatt Place, PO Box 430, Hackensack, New Jersey 07602-0430, USA

Sales representatives

Canada
Pandrol Canada Ltd
8310 Cote de Liesse Road, Suite 100, Montreal, Quebec H4T 1G7

Mexico
Piedras Negras 35, Club de Golf 'La Hacienda', Atizapan de Zaragoza, Edo De, Mexico

Products

Ball-style straight cocks, ball-style angle cocks, retaining valve, stainless-steel ball valves, electronic air brake system, 'Norson' pneumatic discharge gates, handbrakes, bogie- mounted brakes, empty load brakes, slack adjusters, branch pipe tees.

Body-mounted brake cylinders, dirt collector cut out chock.

ELNO

BP 46-17 rue Jean Pierre Timbaud, F-95100 Argenteuil, France
Tel: (+33 1) 39 98 44 44   Fax: (+33 1) 39 98 44 46
e-mail: sales@elno
Web: http://www.elno.fr/

Key personnel

Chairman: Philippe Bertin
General Manager: Jacques Fedon
Export Manager: Gabriel Grosjean

Subsidiary company

Deutsche Elno, Germany

Products

Audio and video systems for passenger vehicles. Audio components including driver/guard handsets and microphones, loudspeakers, power amplifiers, passenger alarm units and remote/central communications and alarm management units. Active noise reduction system.

Contracts

Contracts include supply of equipment to Eurostar, TGV Atlantique (ALSTOM and Bombardier), SNCF/RATP, SNCFB and rail operators in UK.

ESW-EXTEL Systems Wedel

Industriestrasse 33, D-22876 Wedel, Germany
Tel: (+49 4103) 60 36 71   Fax: (+49 4103) 60 45 03
e-mail: sales@esw-wedel.de
Web: http://www.esw-wedel.de

Key personnel

Marketing Manager: Rolf Forstmann

Products

Electric body tilting equipment.

EVAC International Ltd

Purotie 1, FIN-00380 Helsinki, Finland
Tel: (+358 9) 50 67 61   Fax: (+358 9) 50 67 63 33

Key personnel

Managing Director: Matti Tanska

EVAC companies

EVAC Oy
(address above)
Managing Director: Olli Björkqvist

EVAC AB
SE-295 39 Bromölla, Sweden
Tel: (+46 456) 485 00   Fax: (+46 456) 279 72
Managing Director: Nils Andersson

EVAC GmbH
Hafenstrasse 32A, D-22880 Wedel, Germany
Tel: (+49 41) 03 91 68 0   Fax: (+49 41) 03 91 68 90
Managing Director: Nils Andersson

Envirovac Inc
1260 Turret Drive, Rockford, Illinois 61115-1486, USA
Tel: (+1 815) 654 83 00   Fax: (+1 815) 654 83 06
Managing Director: Robert Schafer

EVAC Vacuum Systems (Shanghai) Co Ltd (EVSS)
Unit E, 13F, Jiu Shi Fu Xing Mansion, 918 Huai Hai Road (M), Shanghai, 200020, China
Tel: (+86 21) 64 15 95 80   Fax: (+86 21) 64 15 75 50
Managing Director: Nils Andersson

Products

Vacuum toilet and sewage handling systems.

LPA Excil Electronics Ltd

Ripley Drive, Normanton WF6 1QT, UK
Tel: (+44 1924) 22 41 00
Fax: (+44 1924) 22 41 11
e-mail: sales@excil.co.uk
Web: http://www.excil.co.uk

Key personnel

Managing Director: Phil Burns
Sales Manager: David Burley
Technical Manager: John Hesketh

Background

LPA Excil Electronics is a member of the LPA Group plc.

Products

Passenger vehicle interior lighting systems; electronic lighting inverters; control and monitoring equipment.

Contracts

Interior lighting for London Underground Ltd Piccadilly line metro cars and luminaires for Northern and Jubilee line stock.

Actuator by ESW-EXTEL Systems   0092271

Fahrzeugausrüstung Berlin GmbH (FAGA)

A subsidiary of Deutsche Waggonbau AG
Wolfener Strasse 23, D-12681 Berlin, Germany
Tel: (+49 30) 93 64 20   Fax: (+49 30) 93 64 23 02

Key personnel
Managing Director: Reinhard Schwarzenau

Products
Power supply systems, DC and AC generators, rectifiers, inverters, high-voltage equipment boxes in single and multitension design, switch cabinets, transistorised fluorescent ballasts, train/vehicle control systems, controlling and regulating and diagnostic devices.

FAINSA – Fabricación Asientos Vehículos Industriales SA

Calle Horta s/n, E-08107 Martorelles (Barcelona), Spain
Tel: (+34 93) 579 69 70   Fax: (+34 93) 570 18 38
e-mail: fainsa@fainsa.com

Key personnel
President: Juan Singla
Managing Director: Rafaél Roldán
Commercial Manager: Francesc Puig
Export Manager: Marc Vidal

Subsidiary companies
MTP Equipment (UK Sales)
FAINSA Corporation, USA

Products
Passenger seating for railway vehicles, LRVs and metro cars, including berths for sleeping cars.
Recent developments include: design of a rotatable seat for high-speed train applications. Establishing new anti-vandalism systems; new production facilities in Plattsburg, New York; signature of several technology transfer agreements.

Contracts
Recent contracts for RENFE in Spain include: D-200 vehicle refurbishment programme; seats for TRD dmus; Class 440 refurbishment; seats for Class 447 emus; and second-class seats for TALGO vehicles. FAINSA has also equipped refurbishment projects for FEVE, Spain.
Other contracts: dmu refurbishment for Chile; refurbishment of regional trains for TGOJ, Sweden; first- and second-class seats for ŽSR, Slovakia; seats for Sacramento and Washington metro cars; seats for passenger coaches for New Jersey; dmu seats for SNCB, Belgium; seats for Bombardier's Itino projects; second-class seats for ENR, Egypt; seats for regional emus for the UK; tram seats for vehicles for Kassel, Milan, Prague and Turin; sleeping car berths for Greece.

Faiveley Transport SA

Head Office, International Division
Carrefour Pleyel, 143 boulevard Anatole France, F-93285 Saint-Denis Cedex, France
Tel: (+33 1) 48 13 65 00   Fax: (+33 1) 48 13 66 47
e-mail: info@faiveley.com
Web: www.faiveley.com

Main works
Electromechanical Division
Les Yvaudières, 75 avenue Yves Farge, F-37705 Saint-Pierre-des-Corps, France
Tel: (+33 2) 47 32 55 55
Fax: (+33 2) 47 44 80 24

Production Centre, Electromechanics and Air Conditioning Division
ZI, 1 rue des Grands Mortiers, F-37705 Saint-Pierre-des-Corps, France
Tel: (+33 2) 47 32 55 55
Fax: (+33 2) 47 63 19 31

Electronics Division
rue Amélia Earhart, ZI du Bois de Plante, PO Box 43, F-37700 La-Ville-aux-Dames, France
Tel: (+33 2) 47 32 55 55   Fax: (+33 2) 47 32 56 61

Background
The Faiveley group completed the acquisition of SAB WABCO in November 2004.

Key personnel
Chairman and Chief Executive Officer: Robert Joyeux
Financial Director: Sven Schopp
General Manager: Pierre Sainfort
Communications Manager: Edmond Ballerin

Subsidiaries
Faiveley Española SA & Transequipos
Autovia Reus Km 5, Apartado 525, E-43080 Tarragona, Spain
Tel: (+34 977) 03 00 02   Fax: (+34 977) 03 00 10
e-mail: info-esp@faiveley.net

Faiveley Italia SpA
Via Caduti del Lavoro 16/18, Località Ferlina, I-37012 Verona, Italy
Tel: (+39 045) 671 75 42   Fax: (+39 045) 671 75 35
e-mail: fyitalia@tin.it

Equipfer Faiveley Ltda
241 Rua Major Paladino, Via Leopoldina, São Paulo SP, 05307-000 Brazil
Tel: (+55 11) 36 47 04 00   Fax: (+55 11) 36 47 04 10
e-mail: info-brasil@faiveley.com

Faiveley UK Ltd
Unit 10, Ninian Industrial Park, Ninian Way, Tamworth B77 5DE, UK
Tel: (+44 1827) 26 28 30   Fax: (+44 1827) 26 28 31
e-mail: info-uk@faiveley.com

HFG – HVAC Faiveley GmbH & Co KG
Industriestrasse 60, D-04435 Schkeuditz, Germany
Tel: (+49 34204) 853 00   Fax: (+49 34204) 853 02
e-mail: info@faiveley.de

Shanghai Faiveley Railway Technology Co Ltd
1481 Gong He Xin Road, Shanghai 200072, China
Tel: (+86 21) 56 62 58 04   Fax: (+86 21) 56 62 55 00
e-mail: info-china@faiveley.com

Faiveley Far East Ltd
Unit B, 16F, Guangdong Investment Tower, 148 Connaught Road Central, Sheungwan, Hong Kong, China
Tel: (+ 852) 28 61 17 88   Fax: (+852) 28 61 17 44
e-mail: info-asia@faiveley.com

Faiveley Rail Inc
213 Welsh Pool Road, Pickering Creek Industrial Park, Exton, Pennsylvania 19341, US
Tel: (+1 610) 524 91 10   Fax: (+49 610) 524 91 90
e-mail: info-usa@faiveley.com

LEKOV as
Jirotova 375 – CZ-336 01 Blovice, Czech Republic
Tel: (+420 371) 52 22 40   Fax: (+420 371) 52 21 81
e-mail: lekov@lekov.cz

Products
Air conditioning, intercirculation gangways, internal doors, access doors, odometry/tachometry systems and event recorders, video surveillance systems, services (renovate, maintain, install, advise).

UPDATED

FASI Seating Systems

Clerprem Benelux, Rue des Ateliers no 10, B-7850 Petit Enghien, Belgium
Tel: (+32 2) 395 58 30
Fax: (+32 2) 395 64 96

Products
Seating systems.

Fels SA

2 rue J M Jacquard, F-67400 Illkirch Graffenstaden, France
Tel: (+33 3) 88 67 10 60   Fax: (+33 3) 88 67 33 10
e-mail: fels@fels.fr
Web: www.fels.fr

Products
Electrical contacts; special purpose-made connectors.

UPDATED

Ferranti Technologies Ltd

Cairo House, Waterhead, Oldham OL4 3JA, UK
Tel: (+44 161) 624 02 81   Fax: (+44 161) 624 52 44
e-mail: sales@ferranti-technologies.co.uk
Web: http://www.ferranti-technologies.co.uk

Key personnel
Managing Director: T C Scuoler
Finance Director: F Brinksman
Commercial Director: K R Mills
Operations Director: G J Lowe
Business Development Director: S R Warren

Products
Distance/Velocity Measurement Device (DVMD) utilising non-contact Doppler radar sensing and featuring integral processing electronics. Train-mounted applications include traction control, slip/slide protection, odometry and speed measurement; track-located applications for sensing speed and length of passing rolling stock.
Design, development and production of control electronics to customer specifications; manufacture and repair of third-party electronic and electromechanical assemblies.
Range of power conversion equipment, including transformer rectifier units, inverters, power supplies and other equipment for low-power auxiliary functions.
UV laser cable marking and manufacture of cable looms; environmental testing (UKAS-approved).

Contracts
Contracts include a DVMD device selected by ALSTOM Transport Service UK for integration into the traction control package of an overseas refurbishment programme.
Production for Alstom Traction Ltd, following previous design and development contracts, of a family of key electronic subsystems for use on the Juniper generic train.

Ferranti Technologies distance/velocity measurement device 0023846

Ferraz Shawmut

1 rue Jean Novel, F-69626 Villeurbanne Cedex, France
Tel: (+33 4) 72 22 66 11   Fax: (+33 4) 72 22 67 13

Key personnel
For full personnel listing, see Ferraz Shawmut entry in *Electric traction equipment* section.

Products
Earth return current units and associated resistors to prevent current flowing through bearing of axleboxes and associated resistors; fuses with

very high breaking capacity for DC/AC converter protection and for heating circuits protection.

Ferro

Ferro (Great Britain) Ltd, Powder Coatings Division
Westgate, Aldridge WS9 8YH, UK
Tel: (+44 1922) 74 13 00   Fax: (+44 1922) 74 13 27

Key personnel
General Manager: A J Pitchford
UK Sales Manager: M F Haines
Export Director: M Davies
Transportation Market Manager: A Phillips

Products
Powder coatings for the rail industry for passenger vehicle interiors, station fittings and signage, cladding, trunking and switchgear assemblies.
Range includes: Bonalux AG 2000 anti-graffiti, 491 series polyesters, 4620 series epoxies, all fire resistant to BS 476 Pys 6 and 7 (Class 1) and smoke emission to BS 6853.
Ferro coatings have been supplied to London Underground Ltd's Central, Piccadilly, Northern and Jubilee lines and Hong Kong MTRC, Virgin Trains, First Great Western, Midland Mainline and specified by major rolling stock manufacturers.

Ferro International A/S

Tirsbækvej 15, DK-7120 Vejle Ø, Denmark
Tel: (+45 75) 89 56 11   Fax: (+45 75) 89 59 37
e-mail: info@ferro-int.dk
Web: www.ferro-int.dk

Key personnel
Managing Director, Daily Manager and Export
  Manager: Torben Lerris

Products
Complete powered and automated doors for trains and buses, sliding interior doors for passenger vehicles, including electric and pneumatic doors operated by push-buttons or automatically by sensors; curved electric sliding doors for toilet modules and compartments.

Contracts
Recent contracts include: interior doors and gangway doors for VR, Talgo-Transtech, Bombardier and EVAC AB, Sweden; the supply of interior sliding doors to Bombardier (M6); and curved automatic doors for EVAC.
UPDATED

Fersystem

Parc d'activitiés de Conneuil, 1 avenue Léonard de Vinci, PO Box 30, F-37270 Montlouis-sur-Loire, France
Tel: (+33 2) 47 45 19 45   Fax: (+33 2) 47 45 11 34
e-mail: fersystem@wanadoo.fr
Web: www.fersystem.com

Key personnel
President: Jean Chapouthier

Products
Doors for rail vehicles and fire barriers.

Contracts
Contracts include supply of door systems and fire barriers for Eurostar in partnership with Westinghouse Brakes & Signals.
UPDATED

Fiberline Composites A/S

Nr Bjertvej 88, DK-6000 Kolding, Denmark
Tel: (+45) 70 13 77 13   Fax: (+45) 70 13 77 14

e-mail: fiberline@fiberline.com
Web: www.fiberline.com

Key personnel
Sales Manager: Stig Krogh Pedersen
Technical Solution Manager: Peter Kjaer
Market Co-ordinator: Lisbeth Malmskov

Products
Lightweight corrosion-resistant GRP profiles for rail vehicles.

Contracts
Contracts include the supply of exterior panels for Talent dmus built by Bombardier Transportation for German Rail (DB AG) and panels for the Swedish Regina and Öresund trains produced by Bombardier Transportation AB.
UPDATED

Fine Products SA

Polígono El Sequero, parcela 21, E-26509 Agoncillo, La Rioja, Spain
Tel: (+34 941) 43 70 32   Fax: (+34 941) 43 71 85
e-mail: commercial.fine@fer.es
Web: http://www.fineproducts.es

Key personnel
Transport and Energy Commercial Manager:
  Guillermo Del Rio Navarro

Products
Emergency evacuation ramps for passenger rail vehicles; access ramps for mobility-impaired passengers; control desks; equipment cabinets; coupler fairings; luggage stacks; vehicle side skirts.

FISA Srl

Zona industriale Rivoli
I-33010 Osoppo, Udine, Italy
Tel: (+39 0432) 98 60 71   Fax: (+39 0432) 98 60 86
e-mail: fisa@fisaitaly.com
  info@fisaitaly.com (sales)

Key personnel
Contact: Irene Dolzani

Products
Passenger and driver seating for buses and rail vehicles including double-deck.
UPDATED

Fischer Industries Pty Ltd

13 Whiting Street, Atarmon, New South Wales 2064, Australia
Tel: (+61 2) 94 36 06 11   Fax: (+61 2) 94 38 24 35
e-mail: info@fischerind.com.au
Web: www.fischerind.com.au

Key personnel
Director: Peter Fischer
General Manager: Warren Hocking

Products
Events recorders/data loggers, vigilance systems, Train Management Systems (TMS), analogue and digital meters, door controllers, inverters/converters.
UPDATED

Flachglas Wernberg GmbH

Nuernberger Strasse 140, D-92533 Wernberg-Koeblitz, Germany
Tel: (+49 9604) 480   Fax: (+49 9604) 483 97
Web: http://www.flachglas.de

Key personnel
Sales Director: Martin Rädel

Products
Module supplier of glass for special applications: low-E insulating glass; solar control glass; noise-reducing insulating glass; toughened glass; laminated glass; bullet-resistant glass; screen-printed glass; all with and without glazing system, flat and curved.

Developments
New solutions for emergency windows include heated glass and laminated and toughened complete windows.

Floormaster-Interfloor

Edinburgh Road, Heathhall, Dumfries DG1 1QA, UK
Tel: (+44 1387) 26 95 51   Fax: (+44 1387) 25 01 87
Web: http://www.interfloor.com

Products
Fire-safe, low-smoke, low-toxicity floor coverings for rail vehicle applications, including Floormaster 9200 M2F1 for surface stock and Floormaster Plus 9300 for use in metro vehicles.

Freudenberg Schwab GmbH

Postplatz 3, D-16761 Hennigsdorf, Germany
Tel: (+49 3302) 206 20
Fax: (+49 3302) 20 62 77
e-mail: info@freudenberg-schwab.de
Web: http://www.freudenberg-schwab.de

Key personnel
Chief Executive Officers: Dr Detlef Cordts;
  Peter Kofmel; Jörg Sost
Quality Manager: Leo S E Lang
Marketing Director: Bernd Werner

Background
Freudenberg Schwab GmbH is jointly owned by Freudenberg & Co (51 per cent) and Schwab Holding AG (49 per cent). The company collaborates closely with Schwab Schwingungstechnik AG in Switzerland.

Product
Noise reduction and vibration control systems and components especially multi-layer springs, cone springs, axle type bushes, ultra bushes, spherical bearings, buffers, spring seats, elastic coupling elements, hydrobushings, active vibration absorbers.
The company has worked with: ALSTOM, Bombardier Transportation, ContiTech, Deutsche Bahn AG, A Friedr Flender AG, Phoenix AG, Scharfenbergkupplung, SAB WABCO BSI Verkehrstechnik Products, Siemens, Voith Turbo, Vossloh Schienenfahrzeugtechnik, WATTEEUW Power Transmission Co and ZF Bahntechnik.

Fuji Electric Systems Co, Ltd

Gate City Ohsaki, East Tower 11-2, Osaki 1-chome, Shinagawa-ku, Tokyo 141-0032, Japan
Tel: (+81 3) 54 35 70 46   Fax: (+81 3) 54 35 74 23
e-mail: info@fesys.co.jp
Web: http://www.fesys.co.jp

Key personnel
Managing Director, Electrical Systems Company
  Group: H Itou
General Manager, Transportation Systems Sales
  Department: K Kimura

Background
The Transportation Systems Sales Department of what was previously Fuji Electric Co Ltd, ispart of the newly divided Fuji Electric Systems Co, Ltd.

Products

Traction motors with VVVF control; static auxiliary power supply systems; linear motor-drive door systems for rail vehicles; converter-inverter for Shinkansen.

FY – Composites Oy

Nosturikatu 7, FIN-37150 Nokia, Finland
Tel: (+358 3) 342 99 00   Fax: (+358 3) 342 99 14
e-mail: jaakko.barsk@fy-composites.com
Web: http://www.fy-composites.com

Key personnel

Managing Director: Jaakko Barsk

Products

Composite parts for coaches. On-train automatic public address system which uses a Global Positioning System receiver to make announcements at predetermined locations. The system uses a CD-ROM for extra capacity in the recorded multilingual announcements. Finnish Railways has been operating the system since 1995 and has over 225 modules in service.

Contracts

FY-Composites Oy produced a series of 14 different composite parts for the Intercity double-decker coaches for the Finnish Railways.

Gabriel A/S

Hjulmagervej 55, DK-9000 Aalborg, Denmark
Tel: (+45) 96 30 31 00   Fax: (+45) 98 13 25 44
e-mail: mail@gabriel.dk
Web: http://www.gabriel.dk

Key personnel

Contact: Bent Illum

Products

Textiles for rail vehicle interiors with high performance in flammability- and abrasion-resistance.

GAI-Tronics

A division of Hubbell Limited
Brunel Drive, Stretton Business Park, Burton-on-Trent DE13 0BZ, UK
Tel: (+44 1283) 50 05 00   Fax: (+44 1283) 50 04 00
e-mail: sales@gai-tronics.co.uk
Web: www.gai-tronics.co.uk

Key personnel

Business Unit Manager: Graham Lines
Financial Director: Toby Balmer
Engineering Manager, Applications:
   Richard Rumsby
Manufacturing Director: Mark Bradford
Commercial Manager: Roger Goodall
Marketing Consultant: Nicole Ireland
Special Projects Manager: Steve Smith

Other offices

GAI-Tronics Corporation, USA
Tel: (+1 610) 777 13 74   Fax: (+1 610) 775 65 40
Web: www.gai-tronics.com

GAI-Tronics Srl, Italy
Tel: (+39 02) 48 60 14 60   Fax: (+39 02) 458 56 25

GAI-Tronics Corporation, Malaysia
Tel: (+60 3) 89 45 40 35   Fax: (+60 3) 89 45 46 75

Background

Established in 1964, Gai-Tronics is a major provider of specialised telecommunications for both UK and worldwide railways, manufacturing weather- and vandal-resistant communication equipment.

Products

Onboard communications and information systems including passenger announcement system; crew communications; driver/control centre radio communications; emergency driver/passenger communications; disabled persons communications; on-train entertainment; audible warning and prerecorded digitised special messages.

The products and systems are developed for use in hazardous or hostile areas and are vandal and weather resistant.

UPDATED

Georg Eknes Industrier A/S

N-5913 Elkangervåg, Norway
Tel: (+47) 56 35 75 00   Fax: (+47) 56 35 75 01
e-mail: transit@eknes.no
Web: http://www.eknes.no

Products

Hispek range of seating systems for rail vehicles.

Georges Halais SA

5 rue Ambroise Croizat, BP 203, F-95190 Goussainville, France
Tel: (+33 1) 30 18 00 10   Fax: (+33 1) 30 18 00 11

Key personnel

Chief Executive Officer: Yves Andreux

Products

Components for passenger car interiors including: stainless steel fittings, ceilings, interior doors and walls, table lamps, water heaters, lighting fixtures, handles and handrails, hinges, luggage racks, blinds, locks, slots for seat reservation cards, litter bins, ashtrays, picture frames, tables, folding tables and skirts.

Contracts

Contracts include TGV (SNCF), MP89 (RATP), MI2N (SNCF/RATP) and TER2N (SNCF/RATP), Virgin.

GEZE GmbH

PO Box 1363, D-71226 Leonberg, Germany
Tel: (+49 7152) 20 30   Fax: (+49 7152) 20 33 10
e-mail: vertrieb.services.de@geze.com
Web: http://www.geze.com

UK subsidiary

GEZE UK Ltd
Bleinheim Way, Fradley Park, Lichfield, WS13 8SX
Tel: (+44 1543) 44 30 00   Fax: (+44 1543) 44 30 01
e-mail: geze.uk@geze.com
Web: http://www.geze-uk.com

Products

Window and door systems (single, double and telescopic); closing mechanisms, electromechanical or electropneumatic drives for single- and double-leaf doors; actuators; boarding and alighting equipment for buses trams and trains. RWA and safety technology glass systems.

Giumma SpA

Via Pian Masino 1A, I-16011 Arenzano, Genova, Italy
Tel: (+39 010) 91 32 51   Fax: (+39 010) 911 10 65
e-mail: giumma@giumma.com
Web: http://www.giumma.com

Products

Prefabricated interior packages in self-extinguishing resin with reinforced glass fibre for rail vehicles, including toilet modules, wall panels and gangways.

Contracts

Recent contracts include the supply of toilet modules to ALSTOM (Fiat Ferroviaria) for Pendolino trainsets for Finnish Railways, Italian State Railways, Portuguese Railways, Slovenian Railways and Virgin Trains in the UK.

GKN Aerospace Transparency Systems Kings Norton Ltd

Eckersall Road, Kings Norton, Birmingham B38 8SR, UK
Tel: (+44 121) 606 41 00   Fax: (+44 121) 606 41 91
e-mail: pilkaero@aol.com

Key personnel

General Manager: R A Harper

Products

Design and manufacture of heated/unheated, curved/flat, framed/unframed impact-resistant transparencies for railway and transit industries.

Grammer AG

Köferinger Strasse 9-13, D-92245 Kümmersbruck, Germany
Tel: (+49 9621) 88 00
Fax: (+49 9621) 88 01 30
e-mail: info@grammer.com
Web: www.grammer.com
Chairman and Chief Executive Officer: Heinz-J Otto
Director, Seating Systems: Peter Nagel

Associate company

Lazzerini & Co Srl
Via Toscana, I-60030 Monsano (AN), Italy
Tel: (+39 0731) 602 61   Fax: (+39 0731) 604 49
e-mail: cml.lazzerini@fastnet.it

Works

Amber, Kummersbruck

Products

Suspension driver seats, passenger seats and other passenger coach equipment. Three-point seat belts, tables and interior fittings and the passenger seat range includes the Comfort reclining seat with arm rest and leather trim.

UPDATED

Grupo Antolin

Boulevard Blaise-Pascal, F-42230 Roche-la-Molière, France
Tel: (+33 4) 77 50 53 00   Fax: (+33 4) 77 50 53 01
e-mail: marilyn.poncon@grupoantolin.com
Web: http://www.grupoantolin.com

Head office

Ctra Madrid-Irün, Km 244,8, E-09007 Burgos, Spain
Tel: (+34 947) 47 77 00   Fax: (+34 947) 48 48 08

Key personnel

Managing Director: René Picout
Sales Director: André Bocchi

Products

Passenger and driver seats for railway equipment and other public transportation. The latest railway seat to be introduced is the Linea, which is designed to allow travel in the conventional sitting position (with a 17° reclined backrest) or in a semi-horizontal position (with a reclining backrest from 17 to 25°). The seat consists of an integrated head restraint, central/left and right armrests and a reversible and adjustable foot made of cast aluminium.

Gummi Metall Technik GmbH (GMT)

5 Liechtersmatten, D-77815 Bühl, Germany
Tel: (+49 7223) 80 40
Fax: (+49 7223) 210 75
e-mail: info@gmt-gmbh.de
Web: www.gmt-gmbh.de

Key personnel

Manager: S Engstler

Sales offices

Austria
GMT Gummi Metall Technik Ges mbH
Gewerbestrasse A-65082 Grödig
Tel: (+43 6246) 74 00 60   Fax: (+43 6246) 740 06 22
Contacts: Robert Weber; Joseph Fellinger

Belgium
GMT Belgium SA/NV
165 Chaussée de Louvain, B-5310 Eghezée
Tel: (+32 81) 81 14 40   Fax: (+32 81) 24 40
e-mail: info@gmt-belgium.be
Web: www.gmt-belgium.be
Contact: Jean De Corte

France
GMT France Sarl
ZI Ste Agathe, Rue Paul Langevin, BP 10049,
F-57192 Florange
Tel: (+33 3) 82 59 33 90   Fax: (+33 3) 82 59 33 99
e-mail: gmt.france@wanadoo.fr
Contact: Didier Pouchèle

Ireland
GMT Ireland Ltd
Clifden, Co Galway
Tel: (+353 95) 213 82   Fax: (+353 95) 217 04
e-mail: gmtirl@iol.ie
Contact: Ivor Duane

Italy
Via Magenta, 77/14 A, I-20017 Rho (MI)
Tel: (+39 02) 93 26 10 20   Fax: (+39 02) 93 26 10 90
e-mail: info@pantecnica.it
Contacts: Davide Fatigat; Flavia Fatigati

Malaysia
GMT Gummi Metall Technik (M) SDN BHD
Industrial Estate, PO Box 82, 33000 Kuala Kangsar/
Perak
Tel: (+60 5) 776 17 42   Fax: (+60 5) 776 57 00
e-mail: rmtsb@jaring.my
Contact: Ahmad Zamzuri Bin Abd

Netherlands
GMT Benelux BV
Rudolf Dieselweg 14, PO 3298, NL-5928 RA Venlo
Tel: (+31 77) 387 25 56   Fax: (+31 77) 382 44 91
e-mail: info@gmt-benelux.nl
Contact: Mijnheer Jan Lagewaard

Switzerland
GMT Gumeta AG
Kautschuk-Werk, Buchrainstrasse 2, CH-6030
Ebikon
Tel: (+41 440) 17 17   Fax: (+41 440) 50 60
Contact: Mrs Engstler

UK
GMT Rubber Metal Technic Ltd
The Sidings, Station Road, Guiseley, Leeds LS20 8BX
Tel: (+44 1943) 87 06 70   Fax: (+44 1943) 87 06 31
e-mail: andrew@gmt.gb.com
Contact: Andrew Melville

USA
GMT International Corp
PO Box 117 Villa Industrial, Villa Rica, Georgia
30180, USA
Tel: (+1 770) 459 57 57   Fax: (+1 770) 459 09 57
e-mail: gmt@gmt-international.com
Contact: Heiko Beutner

Products
Primary and secondary suspensions (cones, bolsters, axle springs, chevrons, side bearers, roller springs), bushes, reaction and traction rods, lateral buffers, centre pivot, all bogies applications. Floor suspensions, resilient wheels (all existing systems).

Contracts
ICE and ICE II high-speed trains, Germany; DWA (Bombardier) railbus project. Clients include ALSTOM, Bombardier (Adtranz), SNCF, SNCB, DB, Siemens.

UPDATED

Hekatron Vertriebs GmbH

Brühlmatten 9, D-79295 Sulzburg, Germany
Tel: (+49 7634) 50 02 64   Fax: (+49 7634) 50 03 23
e-mail: export@hekatron.de
Web: http://www.hekatron.com

Background
Hekatron Vetriebs GmbH is a member of the Swiss Securitas Group.

Products
Smoke switches for fire protection systems in passenger and freight rolling stock and locomotives.

Hepworth Rail International

Hepworth House, Brook Street, Redditch B98 8NF, UK
Tel: (+44 1527) 601 46   Fax: (+44 1527) 668 36
e-mail: bhepworth@b-hepworth.com
Web: http://www.b-hepworth.com

Key personnel
Chief Executive: J P Eddy

Subsidiary companies
Dudleys Screenwipers (address as above)
Hepworth Marine International

Products
Windscreen wash/wipe systems for road and rail vehicles and marine applications.

VERIFIED

Resilient components for rail vehicles

DIN ISO 9001
Certified Quality

 **® GUMMI · METALL · TECHNIK · GMBH**
P.O.Box 1253 · 77802 Bühl · Germany · E-Mail: info@gmt-gmbh.de
Tel. +49/72 23/804-0 · Fax +49/72 23/2 10 75 · www.gmt-gmbh.de

USA · England · Belgium · France · Malaysia · Switzerland · Austria · Ireland · India

Hexcel

Duxford, Cambridge CB6 2JR, UK
Tel: (+44 1223) 83 83 70   Fax: (+44 1223) 83 85 64
e-mail: communications@hexcel-eu.com
Web: http://www.hexcel.com

Key personnel
Head of Sales and Marketing, Europe:
    Thierry Merlot
Communications Manager: Rachel Atkinson

Products
Hexlite® sandwich panels for interiors and structures; honeycomb cores (metallic and non-metallic); honeycomb energy absorbers; epoxy and phenolic prepregs; reinforcement fabrics, Redux® film adhesives.

Contracts
Structural flooring for Bombardier IC2000 carriages, sandwich panels for Alstom Pendolino interiors, energy absorbers for TGV trainsets and prepregs for SNCF.

Hodgson and Hodgson Group Ltd

Winnington Hall Mews, Northwich CW8 4DU, UK
Tel: (+44 1606) 765 93   Fax: (+44 1606) 743 15
e-mail: hodgsons@easynet.co.uk
Web: www.acoustic.co.uk

Key personnel
Chairman: G Balshaw-Jones
Managing Director: J Roberts
Technical Director: N Grundy
Sales Director: Margaret Narburgh
Commercial Director: P Rollinson
Export Sales Manager: E Fitzpatrick

Products
Thermal and acoustic component services for bus and railway traction units and rolling stock. Thermal and acoustic products for associated buildings. Design, manufacture and supply of finished products or components direct to site or the production line.

Contracts
Recent projects include: Waterloo Eurostar Terminal (buildings), St Petersburg Rail Terminal (buildings), Barratt Housing Project (trackside development), Eurotram (complete vehicle), Europa Transrapid (complete vehicle), MTRC Hong Kong (complete vehicle), Arlanda, Stockholm (complete vehicle), Gatwick Express (complete vehicle), Juniper, Turbostar and Electrostar (complete vehicle), West Coast Main Line (complete vehicle), First Bus, Mellor Vancraft, Optare and Marshalls (exhaust jacketing, moulded engine compartment and interior panelling).

UPDATED

John Holdsworth & Co Ltd

Shaw Lodge Mills, Halifax HX3 9ET, UK
Tel: (+44 1422) 43 30 00   Fax: (+44 1422) 43 33 00
e-mail: info@holdsworth.co.uk
Web: http://www.holdsworth.co.uk

Key personnel
Sales Director: Michael Holdsworth
Sales: Richard Field, Peter Hobson, Mike Formby

Subsidiary companies
Holdsworth North America Inc
Holdsworth Australasia Pty Ltd
Happich-Holdsworth GmbH

Products
Manufacturers of transport seating fabrics for the rail, bus, coach, airport and ferry markets worldwide, incorporating a bespoke design and styling service. The 7,000 series of fabrics has been developed to meet BS6853:1999 railway specification.

Contracts
Supply of seat fabrics for Virgin CrossCountry and West Coast Main Line vehicles; the Desiro trains for South West Trains, all London Underground Lines plus other projects within the UK and worldwide. The company has reached a new market in South Korea by supplying fabrics for Korean National Railways and Incheon Metro.

VERIFIED

Honeywell Serck

Warwick Road, Birmingham B11 2QY, UK
Tel: (+44 121) 766 66 66   Fax: (+44 121) 766 60 14
e-mail: sind@serckht.co.uk

Key personnel
For a full list of personnel, see Honeywell Serck entry in *Diesel engines, transmission and fuelling systems* section.

Corporate background
Honeywell Serck is part of Honeywell Thermal Systems and a subsidiary of the Honeywell Corporation.

Products
Integrated cooling modules air conditioning systems for dmus, emus and passenger coaches.
    Recent contracts include: redesign of the air conditioning system in South West Trains (UK) Class 159 dmus.

Howden Industrial UK (Howden Buffalo)

Old Govan Road, Renfrew, Glasgow PA4 8XJ, UK
Tel: (+44 141) 885 75 00   Fax: (+44 141) 885 75 55
Web: www.howden.com

Key personnel
General Manager: Ian Thompson

Other offices
Howden Industrial
9 rue d'Estienne d'Orves, F-92503 Rueil-Malmaison Cedex, France
Tel: (+33 1) 47 10 11 06   Fax: (+33 1) 47 10 11 19
Sales Manager: Alexandre Graziani

Howden Industrial (Turbewerke Meissen Howden)
Naundorfer Strasse 4, D-01640 Coswig, Germany
Tel: (+49 35) 239 40   Fax: (+49 35) 239 42 65
Sales Manager: Peter Godermeier

Products
Cooling fans for locomotive engines, transformers, inverters and dynamic braking systems.

UPDATED

Huber + Suhner AG

Polymer Components
CH-9100 Herisau, Switzerland
Tel: (+41 71) 353 41 11   Fax: (+41 71) 353 44 44
e-mail: info@hubersuhner.com
Web: http://www.hubersuhner.com

Key personnel
Contact: Rudolf Wenger

Subsidiaries
Huber + Suhner (Australia) Pty Ltd, Australia
Huber + Suhner (Canada) Ltd, Canada
Huber + Suhner (Hong Kong) Ltd, China
Huber + Suhner France, France
Huner + Suhner GmbH, Germany
Huber + Suhner (Singapore) Pte Ltd, Singapore
Huber + Suhner (UK) Ltd, UK
Huber + Suhner Inc, USA

Products
Low-fire, low-smoke profile materials for train windows and doors.

Hübner GmbH

Agathofstrasse 15, D-34123 Kassel, Germany
Postfach 101920, D-34019 Kassel, Germany
Tel: (+49 561) 99 80   Fax: (+49 561) 998 15 15
e-mail: info@hubner-germany.com
Web: http://www.hubner-germany.com

Key personnel
Executive Chairman: Reinhard Hübner
General Manager: Günter Schwind
Director Train Passageway Design: Ongo Britzue

Products
Gangway systems for metros and commuter trains, regional trains, passenger trains and super high-speed rail vehicles. Folding and corrugated bellows for articulated buses, railway vehicles, boarding bridges and special requirements; vehicle articulation systems; rail vehicle gangways; moulded rubber parts; rubber profiles.
    Hübner has supplied gangways for Class 158, 165, 168, Turbostar, Electrostar DB class 643 emus, Juniper and Pendolino emus, ICE cars, Berlin metro, Strasbourg metro and metros in Cologne, Amsterdam and Rotterdam.

Icon Northern Rubber Co Ltd

Retford DN22 6HH, UK
Tel: (+44 1777) 71 43 00
Fax: (+44 1777) 70 97 39
e-mail: info@iconpolymer.com
Web: www.northernrubber.co.uk

Key personnel
Group Chief Executive: R Gogerty
Group Sales and Marketing Director: R Karssiens

Products
Gangway diaphragms, inter-car gap protection systems and other fabric-reinforced rubber products. Fire-resistant kick straps on ticket gates, inter-car protection mouldings. Carriage interior covings.

Contracts
Contracts include inter-car gap protection mouldings for ALSTOM rolling stock for London Underground Northern and Jubilee Lines; fire-resistant kick strips on WCL ticket gates for London Underground.

UPDATED

IMI Norgren GmbH

Actuator Division
Bruckstrasse 93, D-46519 Alpen, Germany
Tel: (+49 2802) 490   Fax: (+49 2802) 68 04
e-mail: info@norgren-herion.de
Web: http://www.norgren.com

Background
IMI Norgren GmbH is part of the IMI Group.

Products
Pneumatic actuators for rail vehicle applications, including: internal and external door operation; door step control; seat adjustment; toilet operation; water preparation; pantograph retraction; brakes; de-coupling equipment; retractable mirrors; and body tilting systems.
    Clients include Alstom Transport, Bode, Bombardier Transportation, DB AG, IFE, PFA and Stadler-Pankow.

International Metals Reclamation Co dbac (INMETCO)

One INMETCO Drive, Ellwood City, Pennsylvania 16117, US
Tel: (+1 724) 758 55 15
e-mail: ahardies@inco.com
Web: www.inmetco.com

Key personnel
President: Stephen Heddle
Sales Manager: Al Hardies

Products
Recycling of batteries used in locomotives and rolling stock, radios, phones and computers.

UPDATED

International Nameplate Supplies Ltd

1420 Crumlin Road, London, Ontario N5V 1S1, Canada
Tel: (+1 800) 565 35 09   Fax: (+1 519) 455 44 09
e-mail: sales@inps.on.ca
Web: http://www.inps.on.ca

Key personnel
Director: David Gibson

Products
Dead-front graphics (panels which have a dull finish, and hidden lights; they are commonly found on cars for the indicating lights). The company has the technology to form small runs of plastic without incurring the enormous tooling required for injection moulding. It also engraves serial plates, wire and hydraulic markers, warning and informational signs; hot stamping of customer cards, wire markers, parking and expiration decals. Cutting of stencils and legends, corporate logos, painting patterns, vehicle identification, and numbering vehicles. Silk screen printing of warning and information nameplates, rate decals and corporate logos.

Stainless steel etching, aluminium anodising, screen printing and four-colour process and reverse screened polycarbonate.

Contracts
Has primarily supplied General Motors Coach and Motor Coach Industries. Currently INPS is the sole supplier of locomotive graphics services to General Motors EMD division, CSX Transportation and Amtrak. INPS also engineers, manufactures and supplies National Steel Car, Bombardier Transportation, ALSTOM, New Flyer, Nova Bus, Orion Bus, Thomas Built Bus and other companies. Advanced Thermo Dynamics Ltd (ATC) is affiliated to IRR and INPS through joint ownership. ATC is involved in both transit, transportation and military projects and products that involve heating and cooling (HVAC) and generator set manufacture and design.

iQR

Railcentre 1, 305 Edward Street, Brisbane, Queensland, Australia 4000
GPO Box 1429, Brisbane, Queensland 4001
Tel: (+61 7) 32 32 33 90   Fax: (+61 7) 32 35 33 46
e-mail: sales@iqr.com.au
Web: www.iqr.com.au

Key personnel
General Manager: Michael Walsh
Marketing and Business Development Manager:
   Peter Harris
Sales Manager: Youfa Chen

Background
iQR was previously Queensland Rail Consultancy Services.

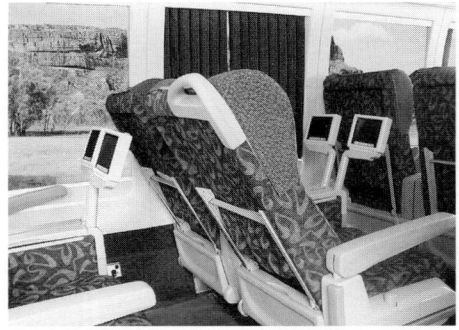

iQR's Train Information Entertainment Systems (TIES)   *NEW*/1140307

Products
iQR's Train Information and Entertainment System (TIES) is an audio-visual information and entertainment system. TIES is designed specifically for rail applications where the system is distributed throughout multiple passenger cars and can be tailored for client's specific requirements. The passenger video screens can be mounted overhead, on seat-backs or in seat armrests. The system supports Media File Server, DVD, satellite TV, driver's-view camera and computer generated information.

Contracts
QR's Cairns and Rockhampton tilt trains have train information and entertainment systems installed and operating.

NEW ENTRY

ITT Veam, LLC

Division of Litton Systems Inc
100 New Wood Road, Watertown, Connecticut 06795, USA
Tel: (+1 860) 274 96 81   Fax: (+1 860) 274 49 63

Key personnel
General Manager: Mike Hansen

Products
Electrical, optical and pneumatic connectors for trainline, brake systems, air conditioning, speed sensing, communications, lighting, automatic coupling and traction motor applications.

Joyce-Loebl Ltd

390 Princesway, Team Valley Trading Estate, Gateshead NE11 0TU, UK
Tel: (+44 191) 420 30 00   Fax: (+44 191) 420 30 30
e-mail: andy.kevins@joyce-loebl.com
Web: www.joyce-loebl.net

Key personnel
Director, Transport Systems: Mike Wade

Products
Train-borne digital CCTV systems.

UPDATED

Jupiter Plast A/S

Bakkedraget 1 DK-4793 Bogø, Denmark
Tel: (+45) 55 89 33 33   Fax: (+45) 55 89 33 66
e-mail: jupiter@jupiter.as.dk
Web: http://www.designinsite.dk

Products
Passenger vehicle components in composite materials including toilet modules, door leaves and seat frames.

Contracts
Jupiter has been involved with supplying the component parts on the following train projects: toilet cabins for ICE (Germany), exterior cabin doors

and sofas for IC3 (Denmark), complete partitions, window frames and complete flooring for OUT (Sweden/Denmark), complete partitions and flooring for Regina (Sweden), front hoods, exterior roof and walls and toilet cabins for Regio Shuttle (Germany), toilet cabins for LVT 642 (Germany), toilet cabins for Stadler (Switzerland), toilet cabins for EMG 312 (Austria), toilet cabins for M6 (Belgium), toilet cabins for ITINO (Germany).

Kidde Graviner Ltd

Mathisen Way, Colnbrook, Slough SL3 0HB, UK
Tel: (+44 1753) 68 32 45   Fax: (+44 1753) 68 51 26
Web: http://www.kiddegraviner.co.uk

Key personnel
Vehicle Sales Manager: Ronnie Drugan

Products
Fire protection systems for locomotives, dmus and rolling stock, including halon replacement gaseous fire suppression systems using FM-200, FE-13 & CO_2; AFFF (Aqueous Film-Forming Foam), Powder and Water.

Systems have been supplied for Class 169, 170, 175/1, 180 and 200 dmus, high-speed locomotives (Eurostar, Amtrak Flyer and ETR500); freight locomotives (Class 66 and 92); new product developments include passenger coach fire detection and control systems, along with Halon replacement with Argonite for Class 91s.

New systems developed for passenger vehicles also include emergency braking control and discharging, utilising the latest sanding technology.

Franz Kiel GmbH & Co KG

Nürnberger Strasse 62, D-86720 Nördlingen, Germany
Tel: (+49 9081) 210 30   Fax: (+49 9081) 210 31 51
e-mail: info@kiel-sitze.de
Web: www.kiel-sitze.de

Key personnel
Managing Director: Gerhard Hellweg

Products
Seating for buses, coaches, main line and light rail vehicles, including mother-and-child seat.

UPDATED

Kleeneze Sealtech Ltd

Ansteys Road, Hanham, Bristol BS15 3SS, UK
Tel: (+44 117) 947 51 49   Fax: (+44 117) 960 01 41
e-mail: enq@ksltd.com
Web: http://www.ksltd.com

Key personnel
Commercial Director: David Love
Export Manager: David Eggleden
Marketing Manager: Mrs Fitton

Subsidiaries
Record Industrial Brushes (sister company)
Kullen GmbH (parent company)

Products
Carriage door brushstrip seals. Rodent brush, internal panel seals, pillar seals, cab door seals. Draft proofing and pest control sealing for buildings.

Kay-Metzeler Ltd

Wellington Road, Bollington, Macclesfield SK10 5JJ, UK
e-mail: info@kay-metzeler.co.uk
Web: www.kay-metzeler.co.uk

Background
Kay-Metzeler Ltd is a subsidiary of British Vita plc.

Products

Transprotect™ foam for use in rail vehicle seating. The material conforms to leading European fire protection standards, with resistance to ignition, low heat release, low smoke and toxic gas emission, together with physical properties that offer wide-ranging design potential.

NEW ENTRY

Knorr-Bremse GmbH

Division IFE Automatic Door Systems
Patertal 20, A-3340 Waidhofen/Ybbs, Austria
Tel: (+43 7442) 51 50   Fax: (+43 7442) 515 13
e-mail: doors_vk@ife-doors.com
Web: www.ife-doors.com

Key personnel

General Manager: Dipl Ing Thomas Feser
Sales Director, Export and Marketing:
 Ing Wolfgang Steiner

Background

Previously called IFE Industrie-Einrichtungen Fertigungs AG, in 1997 Knorr-Bremse Systems for Rail Vehicles GmbH took a 49 per cent shareholding. The company fully merged into Knorr-Bremse in August 2002 and changed its name to Knorr-Bremse GmbH, Division IFE Automatic Door Systems.

Products

Plug, pocket and sliding door systems for rail vehicles; external sliding doors, pocket doors, inside swing doors; movable steps; door control equipment.

UPDATED

Konvekta AG

Am Nordbahnhof 5, D-34613 Schwalmstadt, Germany
Tel: (+49 6691) 760   Fax: (+49 6691) 761 71
e-mail: info@konvekta.com
Web: http://www.konvekta.com

Products

Air-conditioning systems for buses and rail vehicles.

KV Ltd

Presley Way, Crownhill, Milton Keynes MK8 0HB, UK
Tel: (+44 1908) 56 15 15   Fax: (+44 1908) 56 12 27
e-mail: sales@kvautomation.co.uk
Web: http://www.devicelink.com

Key personnel

Managing Director: A R Cersell
Export Sales Director: A C Hough
European Sales Director: I R Davies
Technical Director: H M Stoneman
Financial Director: B Kentish

Subsidiary

Instruments and Movements Ltd

Products

Pneumatic control components and modular systems for control of passenger, driver cab and internal coach doors, pneumatic components and systems for use on rail vehicles.

Lazzerini

Lazzerini & Co Srl
Via Toscana 1, I-60030 Monsano (AN), Italy
Tel: (+39 0731) 602 61   Fax: (+39 0731) 604 49
e-mail: cml.lazzerini

Background

Lazzerini is a subsidiary of Grammer AG (qv).

Products

Seating for passenger vehicles.

Lenord, Bauer & Co GmbH

Dohlenstrasse 32, D-46145 Oberhausen, Germany
Tel: (+49 208) 996 30   Fax: (+49 208) 67 62 92
e-mail: info@lenord.de
Web: http://www.lenord.de

Agents in UK

Motor Technology Ltd
Motec House, Chadkirk Industrial Estate, Romiley, Stockport, Cheshire SK6 3LE, UK
Tel: (+44 161) 427 36 41   Fax: (+44 161) 427 13 06
e-mail: sales@motec.co.uk

Products

Speed sensors and encoders for rail vehicles, applicable to motor speed monitoring systems; wheel slip protection systems; traction control systems; and door movement control systems.

VERIFIED

GEL 247 MiniCoder in a typical application scanning a toothed wheel to detect traction motor speed 0126826

Liebherr-Verkehrstechnik GmbH

Liebherr Strasse 1, A-2100 Korneuburg, Austria
Tel: (+43 2262) 60 20   Fax: (+43 2262) 25 01
e-mail: info@lvf.liebherr.com
Web: www.liebherr.com

Products

Air conditioning, heating and ventilating equipment for locomotives and passenger, multiple-unit and rapid transit stock; split systems; unitary equipment; underfloor-mounted, ceiling integrated or rooftop installation. Microprocessor-based electronic controls for HVAC equipment with integrated diagnostic and control features for all pneumatic and electrical components of a coach.

Contracts

Together with Angel Trains Ltd, Liebherr implemented the air cycle air conditioner as a built-in, underframe-mounted unit, based on virtually the same thermodynamic principles and main components as the version used in ICE (InterCityExpress) trains.

Other contracts have included the development of a roof-mounted unit for ALSTOM LHB's technology demonstrator train and the development and supply of pressure-sealed compact air conditioning units for high-speed trains such as the German ICT and the Virgin high-speed train for the West Coast Main Line, UK; the heating and ventilation system for Austrian Rail double-deck cars and the air-conditioning system for the TER 2N operated by SNCF, France. The double-deck car air conditioning system family has recently been expanded with the acquisition of a contract for Zurich's S-Bahn commuter trains.

Many Liebherr air conditioners have been installed in railcars for metropolitan and suburban systems, including the Siemens Combino family, the Bombardier Incentro, GT8N and R 3.3, and the ALSTOM Regio Citadis. Subway trains in Berlin and Munich also utilise Liebherr heating, ventilation and air conditioning equipment.

UPDATED

LPA Channel Electric Ltd

Bath Road, Thatcham RG18 3ST, UK
Tel: (+1635) 86 48 66   Fax: (+1635) 86 91 78

Channel Electric Beaufort blower for upgrading air conditioning units 0063682

Key personnel

Managing Director: George Renshaw
Marketing Manager: Alex Burt
Commercial Director: Chris Antysz

Background

Previously called Channel Electric Equipment, LPA Channel Electric is an ISO 9002 registered company and is a member of LPA Group plc.

It supplies electrical and electro-mechanical products.

Products

Beaufort blower for air conditioning units. This fan system incorporates a single maintenance-free brushless DC motor. The Beaufort unit delivers airflow rates up to 36,000 litres/min and has a design life of eight years. Other products include: relays, connectors, circuit breakers, switches, contactors, fans, motors, blowers, harnesses, assemblies, enclosures and stud terminals.

UPDATED

Lumikko Oy

PO Box 304, FIN-60101 Seinäjoki, Finland
Tel: (+358 6) 420 40 00   Fax: (+358 6) 414 19 21
e-mail: lumikko@lumikko.com
Web: http://www.lumikko.com

Key personnel

Production Manager: Kari Saikkonen

Products

Lumikko HVAC devices for locomotives, passenger coaches and other rail vehicles; systems are capable of operation in temperatures ranging from −40 to +40°C.

Contracts

Contracts include: the supply of over 500 cab air conditioning units for locomotives, dmus and emus operated by Finnish Railways (VR-Group Ltd); air conditioners for 12 VR-Group restaurant cars; refrigeration and freezing equipment for restaurant cars manufactured by Talgo Oy.

Mafelec

F-38480 Chimilin, France
Tel: (+33 4) 76 32 07 33   Fax: (+33 4) 76 32 54 11
Web: http://www.mafelec.fr

Products

Automatic railway control components including man-machine interface, visualisation and display, connection, safety, engineering and electronics.

Marl International Ltd

Marl Business Park, Ulverston LA12 9BN, UK
Tel: (+44 1229) 58 24 30   Fax: (+44 1229) 58 51 55
e-mail: sales@marl.co.uk
Web: http://www.marl.co.uk;
 www.marlrailproducts.com

Products

Door pressels; illuminated LED gauge rings for applications such as rail vehicle speedometer and brake gauges.

UPDATED

Mayser GmbH & Co KG

Örlinger Strasse 1-3, D-89072 Ulm, Germany
Tel: (+49 731) 206 10   Fax: (+49 731) 612 22
Web: www.mayser.de

Key personnel

Contact: Deirdre Duffy-B

Products

Safety products for passenger rail vehicles and buses, including powered door gap detection systems, Mayser Miniature Safety Edges for vehicle doors to detect trapped objects and Mayser Safety Mats, used in buses and rail vehicles for passenger detection and passenger counting.

UPDATED

One example of safe obstacle detection for public transport users from Mayser NEW/1143619

Merak Sistemas Integrados de Climatización

Ruiz de Alarcón no 13-3°, E-28014 Madrid, Spain
Tel: (+34 91) 532 41 81   Fax: (+34 91) 522 76 97
e-mail: albatros@albatros-es.es

Key personnel

President: Nicolás Fúster
Managing Director: Julio Rey
Commercial Director: Manuel V Rey
Engineering Director: Pablo Bronchalo

Background

Merak formerly traded as Stone Ibérica SA and is a member of the Albatros Group.

Products

Design, manufacture and maintenance of air conditioning equipment, heating and ventilation equipment, electronic and microprocessor controls, inverter ballasts for fluorescent tubes, and static speed regulator for asynchronous motors.

Contracts

Recent contracts include: NJT (New Jersey Transit) 2003, Metro Madrid 3000 and 9000 2004, Metro Barcelona L1, L2 and L3 2004, Metro Brussels 2004,

Shanghai Line 3 (2002), Guangzhou Line 2 (2002), Singapore Nel (2002), Marina Line (2003), Talgo 350 Spain (2003), CTA 2200–2600 Chicago (2001), WMATA 2K/3K/5K/6K Washington (2002), Metro Madrid (2002), Heathrow Express (2000), ICE-350 Spain (2003), M-7 New York (2002), supply of air conditioning equipment for Vancouver Train (1999) Canada, Talgo Trains S-7 (1999) Spain, JFK Train (1999) USA, Flumitrens (1999) Brazil, DMU Belgica (1999), Northern Spirit (1999), USA, Oporto Metro (2000), Shanghai Metro (2000), Oporto Suburban Train, M-7 Train, USA.

UPDATED

Microelettrica Scientifica SpA

Via Alberelle 56/58, I-20089 Rozzano (MI), Italy
Tel: (+39 02) 57 57 31   Fax: (+39 02) 57 51 09 40
e-mail: info@microelettrica.com
Web: www.microelettrica.com

Key personnel

Managing Director: Dr Marco Boldrini

Products

Standard and custom-designed contactors, digital electronic protection relays and power resistors for braking and other rail applications.

UPDATED

Microphor

452 East Hill Road, Willits, California 95490, USA
Tel: (+1 707) 459 55 63   Fax: (+1 707) 459 66 17
e-mail: info@microphor.com
Web: www.microphor.com

Key personnel

Vice-President, Operations: Mark Natalizia
Customer Service: Brian Banzhaf
Marketing and Sales Manager: Walter Hess

Background

Microphor is part of the Wabtec Corporation.

Products

Macerator and air-assisted flush toilets/waste retention and treatment systems (32 ounce, one quart and two quart per flush) for passenger vehicles and locomotives; thermoelectric refrigerators and ice boxes, low-temperature protection systems, air compressors, freeze dump valves, ditch lights and accessories for locomotives; custom-manufactured components in plastic and sheet metal for locomotives, passenger vehicles and freight wagons.

New developments include synthetic filter media for onboard sewage treatment.

Contracts

Contracts include the supply of toilets and waste retention/treatment systems to Metro-North, Amtrak, MARC, New Jersey Transit, Amerail, GE Transportation Systems and Electro-Motive Division, General Motors Corporation.

Products are supplied to customers in Australia, Canada, China, Mexico, South Africa, Vietnam and UK.

UPDATED

Mobile Climate Control Corporation

80 Kincort Street, Toronto, Ontario M6M 5G1, Canada
Tel: (+1 416) 242 64 06   Fax: (+1 416) 242 64 06
Web: http://www.MCCII.com

Subsidiaries

Industrial Division
2050 Drew Road, Mississauga, Ontario L5S 1N3, Canada
Tel: (+1 905) 405 00 04   Fax: (+1 905) 405 99 94

USA East
2200 Dywer Avenue, Utica, New York 13501, USA
Tel: (+1 315) 738 15 00   Fax: (+1 315) 738 19 19

USA West
426 Winnebago Avenue, Fairmont, Minnesota 56031, USA
Tel: (+ 1 507) 238 27 83   Fax: (+1 507) 238 41 51

MCC Europe
AB Baldersgatan 24, Box 96, S-761 21 Norrtälje, Sweden
Tel: (+46 176) 20 78 00   Fax: (+46 176) 20 78 10

Soprano
Parc Technologique de l'Isle d'Abeau, 40 rue condorcet, Vaulx Milieu F-38090, France

Products

Air conditioning in passenger coaches. The company has research and development facilities and laboratories in Ontario, Canada.

Monogram Systems

A Zodiac company
800 West Artesia Boulevard, Compton, California 90224-9057, USA
Tel: (+1 310) 638 84 45   Fax: (+1 310) 638 84 58
e-mail: fkaviani@monogram.zodiac.com

Key personnel

President: Mike Rozenblatt
Vice-President, Sales and Marketing: James R Durso
Sales Manager: F Kaviani

Products

Waste collection systems for passenger vehicles. Water flush, vacuum waste collection systems capable of 72-hour full retention. Self-contained, single-position modular vacuum toilets. Waste compactors for café cars and buffet cars.

Contracts

Contracts include vacuum waste collection systems for 280 vehicles of Amtrak's Superliner I fleet, and the supply of vacuum waste collection systems to Bombardier of Canada for 38 Amtrak Superliner II coach cars. Other orders include 80 sets of full-retention vacuum waste systems for Amtrak's Horizon coach fleet and 300 sets of toilets for its Amfleet I refurbishment programme. Monogram is supplying systems for 35 RTLIII Turboliners for Super Steel, New York and 160 cars for the Metro North M2 rehabilitation.

Morio Denki Co Ltd

34-1 Tateishi 4-chome, Katsushika-ku, Tokyo 124-0012, Japan
Tel: (+81 3) 36 91 31 81   Fax: (+81 3) 36 92 13 33
Web: www.morio.co.jp

Key personnel

President: S Yamagata
Senior Managing Director: K Miura

Products

Fluorescent ceiling lights, headlights and tail-lights, destination display systems, heating systems, door-operating switches.

UPDATED

Mors Smitt

PO Box 7023, NL-3502 KA, Utrecht, Netherlands
Vrieslantlaan 6, NL-3526 AA, Utrecht, Netherlands
Tel: (+31 30) 288 13 11
Fax: (+31 30) 289 88 16
e-mail: sales@nieaf-smitt.nl
Web: www.nieaf-smitt.nl/railway

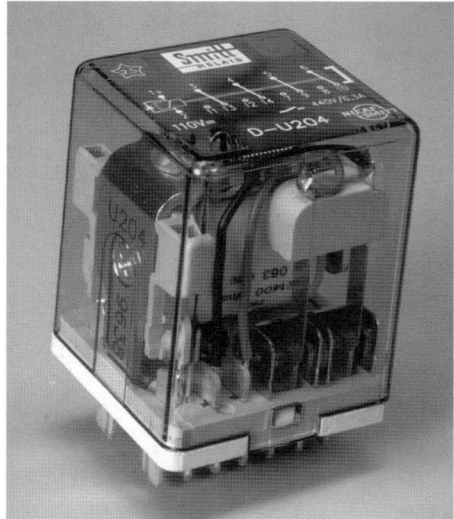

Mors Smitt D-U204 relay 0122493

Key personnel
Managing Director: Arne J Wijnmaalen
Area Sales Managers: Arjan B Mann,
 Cho-Sa Hu, Igor Cubleac
Product Market Manager Railway:
 Erwin K G Veldman

Products
Circuit protection components: hydraulic-magnetic circuit breakers, ground fault breakers and remotely operated circuit breakers; electrical control solutions: form, fit and function electrical control solutions for new built and retrofit/refurbishment applications; electronic time delay relays: delay on drop-out (instantaneous contacts), delay on pull-in (instantaneous contacts), delay on pull-in and/or drop-out (selectable by dip switch), delay on pull-in or drop-out pulse controlled, one shot on pull-in or drop-out, threshold with delay on drop-out and flashing; electronic timers: delay on drop-out, delay on pull-in, one shot on pull-in or drop-out, threshold with delay on drop-out and flashing; instantaneous relays: dry circuit, latching, miniature, PCB and safety-critical; measuring and monitoring relays: voltage monitoring, current monitoring and analogue value monitoring; panel indicators: analogue indicators for speed, voltage, power and current; protection relays: high voltage, frequency and current monitoring; track safety equipment: high voltage electrical monitoring and measuring equipment; accessories: relay sockets, retaining clips.

UPDATED

MS Relais

Tour Rosny 2, Avenue du Général de Gaulle, Rosny sous Bois, F-93118, France
Tel: (+33 1) 48 12 14 40   Fax: (+33 1) 48 55 90 01
e-mail: sales@morssmittrelais.com
Web: www.morssmittrelais.com

Key personnel
Sales Manager France: P Vigier
Sales Manager North and South America: C Acard
Area Sales Manager: G Morgado
Product Market Manager: E Veldman

Products
Railway relays, protection traction relays, panel indicators, relay panels, rehabilitation relay panels, circuit breakers.

Contracts
International contracts include Bombardier, ALSTOM, Siemens and Kawasaki. Projects include: COMET 5, LIRR M7, Mexico subway, TGV, WMATA 6K. End users: AMTRAK, KCRC, LIRR, NJT, NYCTA, RATP, RENFE, Shanghai Metro, SNCF, STM, TTC, VIA Rail, WAMATA.

NEW ENTRY

MTB Equipment Ltd

7-9 Barton Road, Water Eaton Industrial Estate, Bletchley, Milton Keynes MK2 3HX, UK
Tel: (+44 1908) 37 95 21   Fax: (+44 1908) 27 06 04
e-mail: info@mtb-equipment.com
Web: http://www.mtb-equipment.com

Key personnel
Managing Director: John Mainwaring
Sales Director: Tony Berrington

Products
Manufacturers of rail vehicle seats and suppliers of accessories for driving cabs including driver's night blinds and sun visors. Bus and coach seating, roller blinds, vehicle interior equipment.

Nabtesco Corporation

Railroad Products Company, 9-18 Kaigan 1-chome, Minato-ku, Tokyo 105-0022, Japan
Tel: (+81 3) 54 70 24 01   Fax: (+81 3) 54 70 24 24
e-mail: takashi_koyama@nabtesco.com
Web: www.nabtesco.com

Key personnel
Company President: Koshiro Yoshida
Company Vice-President: Masanori Kawanishi
General Manager, Sales and Marketing:
 Yukiyasu Fujimoto
General Manager, Overseas Marketing:
 Takashi Koyama

Background
Nabtesco Corporation was previously called NABCO Ltd.

Products
Plug and sliding door systems; platform doors (manufactured and supplied by Nabtesco Corporation, NABCO Company); pressure-sealing systems for passenger coaches; windscreen wiper motors.

Contracts
Recent contracts include the supply of automatic sliding door systems to JR East for E231 and E531 commuter trainsets, to Changi Airport Automated People Mover System in Singapore, Miami Airport Automated People Mover system in USA. Nabtesco supplied over 28,700 sets of automatic electric sliding door systems (including the products licensed by Faiveley) for various Japanese operators including JR East commuter trainsets.

UPDATED

Narita Manufacturing Ltd

20-12 Hanaomote-cho, Atsuta-ku, Nagoya 456-0033, Japan
Tel: (+81 52) 881 61 91   Fax: (+81 52) 881 67 48
e-mail: sinarita@narita.co.jp
Web: http://www.narita.co.jp

Key personnel
President: Masatoshi Narita
Executive Director (Export Sales and Marketing):
 Shuichi Narita
Executive Director, General Affairs and Quality
 Assurance: Haruo Narita

Products
Vestibule diaphragms, gangways, rubber bellows, inter-car barriers, door leaves, driving consoles, fuel and water tanks, interior panels, air ducts for railway vehicles and platform door leaves.

Contracts
Recent contracts include the supply of gangway systems for Taiwan High Speed Train and Series 800 Shinkansen (JR-Kyushu), for Guangzhou Metro Line 2, for Melbourne Bayside Trains (now Connex Trains Melbourne), for Chongqing monorail, door leaves for Taiwan High Speed Train

Narita sound proof gangway for EMU of Melbourne Bayside on testing device 0533656

and Series 800 Shinkansen (JR-Kyushu), passenger cars for Metra USA, and for Series 01 Metro train of TRTA (Tokyo Rapid Transit Authority), inside panels of passenger cars for Taiwan High Speed Train, inter-car barriers for Nagoya Municipal Transportation Bureau.

VERIFIED

The Network Connection

The Mill, Lodge Lane, Derby DE1 3HB, UK
Tel: (+44 1332) 20 21 72
e-mail: uk@tncx.com
Web: http://www.projectrainbow.com

Key personnel
Managing Director: Stephen J Ollier
Commercial Director: Philip Campbell
Sales and Marketing Director: Julian Burrell
Engineering Director: Dominic Newton

Parent company
Global Technologies Ltd
1911 Chestnut Street, Suite 120 Philadelphia, Pennsylvania 19103, USA
Tel: (+1 215) 972 81 91   Fax: (+1 215) 972 81 83

Products
Services include hardware provision and integration into the rail vehicle, service support for the agreed life of product, automated content loading via integrated telecommunications systems, content provision, revenue generation management.

UPDATED

Neu Systèmes Ferroviaires

PO Box 2026, 70 rue du Collège, F-59700 Marcq en Baroeul, France
Tel: (+33 3) 20 45 65 46   Fax: (+33 3) 20 45 64 98
e-mail: mail@neu-nsf.com
Web: http://www.neu-nsf.com

Key personnel
Chief Executive Officer: Guy Leblon
Commercial Manager: Franck Vinchon
Export Assistant: Caroline Dufour

Subsidiaries
Atelier Neu Systems Ferroviaires
ZI Neuville en Ferrain, Voie Nouvelle, rue de Reckem, F-599960 Neuville en Ferrain

Conestra
Trupbacher Strasse 26a, D-57072 Siegen, Germany
Tel: (+49 271) 372 05 03; 372 05 05
Contact: Karl Heinz Brull

Neu SF Polska
ul Krucza 28, PL-00522 Warsaw, Poland
Tel: (+48 22) 622 84 71; 622 84 72

Background
Neu Systèmes Ferroviaires was founded in January 1991 and is a member of the NEU Group.

Products
Air-conditioning, heating and ventilation equipment, control systems for HVAC, exhaust fans, cooling fans, high-pressure fans for adjustment of pressure inside coach.
    Contracts include heating and ventilation equipment for Rotterdam metro (168 units), Z2N driving cabs for SNCF (800 units), NINA cars for BLS Switzerland (Bombardier Transportation, 24 units).
    Recent contracts: Rotterdam metro for Bombardier; Z2N driving cabs for SNCF; NINA/BLS; Metro VAL 208, Lille for Matra; TVR Nancy for Alstom de Dietrich; Warsaw metro for AISTOM Konstal; metro VAL 208 Rennes for Siemens and Citadis for ALSTOM.

UPDATED

Orvec International Ltd

Malmo Road, Hull HU7 0YF, UK
Tel: (+44 1482) 87 91 46   Fax: (+44 1482) 62 53 25
e-mail: service@orvec.com
Web: http://www.orvec.com

Key personnel
Managing Director: Graham Stonehouse
Sales Director: Richard Heath

Products
Bespoke and customised antimacassars, headrest covers, disposable headrest covers, tray mats (including non-slip), blankets, amenity packs, pillows and pillow covers.

VERIFIED

Paltechnica

Kibbutz Nitzanim D N, Evtach IL-79290, Israel
Tel: (+972 8) 672 10 80; 10 81   Fax: (+972 8) 672 97 71
Web: www.paltechnica.co.il

Key personnel
President and Chief Executive Officer: Avner Gazit
Business Development Manager: Richard Sarfati

Background
Paltechnica is affiliated to the KGM group.

Products
Seating systems for trains and buses.

Contracts
Paltechnica seating systems have been supplied for Alstom coaches and Bombardier Transportation IC3 dmus supplied to Israel Railways.

UPDATED

Parker Pneumatic

Parker Hannifin plc
Pneumatic Division, Walkmill Lane, Bridgtown, Cannock, Staffordshire WS11 3LR, UK
Tel: (+44 1543) 45 60 00
Fax: (+44 1543) 45 60 01

Key personnel
Application Manager: Brian Umney

Products
Parker manufactures pneumatic systems with a range of applications in rolling stock, including seat adjustment cylinders (Italian ETR 500 high-speed train), pneumatic circuit blocks for onboard toilet flush (French TGV), door opening control panel

(ETR 500), valves and cylinders for heating and ventilating.

Pars Komponenty sro

Butovická ul, (areál TVS), CZ-742 13 Studénka, Czech Republic
Tel: (+420 556) 45 50 00   Fax: (+420 556) 45 50 10
e-mail: info@parskomponenty.cz
Web: www.parskomponenty.cz

Key personnel
Director: Vladimír Vyhlídal
Technical Director: Schreier Jirí
Financial Director: Martina Mazancová
Production Director: René Krístek

Background
The company was originally founded in 1999 following the merger of Komponenty sro and Pars Holding sro Sumperk.
    Pars Komponenty develops and manufactures parts for the mass transportation industry.

Products
Subsystems and components for passenger rail systems and buses, including: exterior and interior door systems; windows; luggage racks; interior walls; lifting platforms for passengers with limited mobility.

NEW ENTRY

Pascal International AB

Box 33, SE-280 10 Sösdala, Sweden
Tel: (+46 451) 660 80   Fax: (+46 451) 603 70
e-mail: contact@pascal-system.com
Web: http://www.pascal-system.com

Products
DUX spring technology for passenger rail vehicle seating. The system comprises a self-contained spring unit consisting of multiple rows of springs mounted into channels of non-woven fabric.
    Pascal FR meets fire resistance standards to BS 5852 – Crib 7, UIC 564-2, ISO 6941 and M1.

VERIFIED

People Seating Ltd

Unit 9, Washington Street Industrial Estate, Netherton, Dudley DY2 9RE, UK
Tel: (+44 1384) 25 71 24   Fax: (+44 1384) 24 21 06
Web: www.peopleseating.co.uk

Key personnel
Managing Director: David J Poston
Sales and Marketing Co-ordinator: Kaye Blunt

Products
Passenger and driver seating, interior trim components. Speciality seat refurbishment upgrade.

UPDATED

Percy Lane Products

Lichfield Road, Tamworth B79 7TL, UK
Tel: (+44 1827) 638 21   Fax: (+44 1827) 31 01 59
e-mail: main@percy-lane.co.uk
Web: http://www.percy-lane.co.uk

Key personnel
Executive Chairman: G H Fowler
Managing Director: P S Wright
Sales Director: J W Whetton
Business Development Director: N Greenhalgh
Purchasing Manager: L Clarkson
Commercial Manager: D J Knight

Products
Windows, sashes, impact-resistant windscreens, luggage racks, detrainment devices, Beclawat range of products; aluminium fabrications, including gangway frames, doors.

Contracts
Contracts include PBKATGV2N side windows (for Alstom) bodyside glazing for MP89 stock for Paris metro (for Alstom); detrainment device for London Underground Northern Line (for Alstom); bodyside glazing for Connex vehicles (for Bombardier Transportation); emergency exit doors and detrainment ramp for Singapore SMRT (for Alstom); bodyside windows for Virgin CrossCountry (for Bombardier) and for Docklands Light Railway vehicles. Detrainment steps for London Underground Piccadilly Line cars.

Developments
The company has introduced a new purpose-designed vehicle access ramp for the disabled and a range of flush bodyside windows.

VERIFIED

Permali Gloucester Limited

Bristol Road, Gloucester GL1 5TT, UK
Tel: (+44 1452) 52 82 82   Fax: (+44 1452) 50 74 09
e-mail: sales@permali.co.uk
Web: http://www.permali.co.uk

Key personnel
Managing Director: A J T King
Technical Director: David Tudor
Sales and Marketing Manager: Fraser Rankin

Products
Complicated moulded composites to both laminate and sandwich form. Available in a range of resin systems for optimised fire and structural performance.

VERIFIED

Phoenix Traffic Technology GmbH

Hannoversche Strasse 88, D-21079 Hamburg, Germany
Tel: (+49 40) 76 67 27 78   Fax: (+49 40) 766 77 27 78
e-mail: michael.groth@phxtt.de
Web: www.phoenix-ag.com

Key personnel
Chief Engineer: Peter Eckworth
Engineer: Jörg Frohn
Export Sales: Peter Wist
Sales Manager: Michael Groth

Products
Doorseals, windowseals and anti-vibration parts.

UPDATED

Pickersgill-Kaye Ltd

Pepper Road, Hunslet, Leeds LS10 2PP, UK
Tel: (+44 113) 277 55 31   Fax: (+44 113) 276 02 21
e-mail: sales@pkaye.co.uk
Web: http://www.pkaye.co.uk

Key personnel
Managing Director: Peter Murphy
Sales Manager: Harry Griffiths

Products
Lock assemblies for internal and external applications; passenger emergency alarm handles and emergency talkback units; K-Tex emergency window breaker device; LED indicators for door status, driver's desk.

VERIFIED

Pintsch Bamag Antriebs-und Verkehrstechnik GmbH

PO Box 100420, D-46524 Dinslaken, Germany
Tel: (+49 2064) 60 20   Fax: (+49 2064) 60 22 66
e-mail: info@pintschbamag.de
Web: www.pintschbamag.de

Key personnel

Managing Director: Dr Rolf-Dieter Krächter
Strategic Concerns: Ulrich Nagorski
Head of Business Unit: Thomas Milewski
Export Manager: Peter Bunzeck

Background

Pintsch Bamag is a member of the Schaltbau Group.

Products

Railway vehicle equipment for high-speed, regional and local traffic. Door systems, door step systems, door control systems, door attachments, diagnosis systems, MVB bus system interfaces. All kinds of power supply systems (single- and multi-voltage), battery chargers. Current collection systems, lighting components (top and rear lights), heating devices.

UPDATED

PIXY AG

Bruggerstrasse 37, CH-4500 Baden, Switzerland
Tel: (+41 56) 221 72 10   Fax: (+41 56) 21 72 59
e-mail: sales@pixy.ch
Web: http://www.pixy.ch

Key personnel

Managing Director: Mark Meier
Chief Executive Officer: Franz Steuri
Sales and Marketing: Markus Koller
Chief, Development: Wolf Liebert
Chief, Production: Lukas Gautschi

Background

The company became independent of the Sécheron Group following a management buyout in 2002 and is ISO 9001 certified.

Products

Development, manufacture and sale of PC-based visualisation equipment with and without software, for use in the toughest environments. Based on components developed in-house, its displays are used on mobile and stationary platforms in the railway field. Applications include ETCS (the European Train Control System), diagnostics, controls and data loggers.

UPDATED

Polarteknik PMC Oy Ab

Klaavolantie 1, FIN-3270 Huittinen, Finland
Tel: (+358 2) 560 15 00   Fax: (+358 2) 56 85 01
e-mail: jouni.saarnia@polarteknik.com
Web: www.polarteknik.com

Head Office

Mestarintie 7, PO Box 24, FIN-01731 Vantaa, Finland
Tel: (+358 9) 87 80 80   Fax: (+358 9) 87 80 81 80
e-mail: info@polarteknik.com

Key personnel

Divisional Manager: Jouni Saarnia

Background

Previously Berendsen PMC Oy Ab, Pimatic, the company is owned by Segulah AB.

Products

Pimatic automatic interior door systems; pressure-sealed gangway doors; interior doors; fire barrier doors; electro-pneumatic and electric-powered door gear for rail vehicles; pneumatic bus actuators.

Contracts

Contracts include the supply of interior door systems for Bombardier Transportation projects, including Electrostar emus for the UK, and for double-deck intercity coaches supply to VR, Finland, by Talgo, ALSTOM Sweden and Talgo Spain high-speed trains.

UPDATED

Portaramp

Roman House, Roman Way, Rison Way Industrial Estate, Thetford IP24 1HT, UK
Tel: (+44 1842) 82 14 00   Fax: (+44 1842) 82 14 01
e-mail: ramps@globalnet.co.uk
Web: http://www.portaramp.co.uk

Key personnel

Production Manager: Keith Jones

Products

Portable wheelchair ramp, to provide access to rail vehicles for wheelchair users. Available in four standard lengths and fully DIPTAC (Disabled Persons Transport Advisory Committee) compliant; manufactured from lightweight strong aluminium alloy; folds in half for compact easy storage.

Developments

Pneumatically powered integrated bus ramp developed in conjunction with Norgren pneumatics. The ramp has been designed to be DIPTAC compliant.

Powernetics Ltd

Jason Works, Clarence Street, Loughborough LE11 1DX, UK
Tel: (+44 1509) 21 41 53   Fax: (+44 1509) 26 24 60
e-mail: jag@powernetics.co.uk
Web: www.powernetics.co.uk

Key personnel

Managing Director: Satish Chada
Financial Controller: Bob Lawson
Engineering Director: Nilesh Chouhan
Operations Manager: Konrad Chada
Business Development Manager: Aran Chada
Sales Manager: Jim Goddard
Purchasing Manager: Russel Roughton
QA Manager: Gordon Anderson
Production Manager: Andy Guigno

Products

Independent private limited company with 30 years' experience in design, manufacture, test, installation, commissioning and maintenance of auxiliary, standby/emergency AC/DC power supply systems for the rail industry in trackside, tunnel, switchroom, REB and trainborne environments.

RailPower UPS systems for signalling, radio, telecoms and level crossing applications. Static inverters for trainborne HVAC and domestic catering applications. Heavy duty battery chargers to BR1875Std for lineside signalling. DC PSUs for NRN driver/train radio, secondary door locking and track circuit actuation applications.

Powernetics is a BS EN ISO9001 accredited company, with Network Rail product acceptance for RT/E/PS00007 (PA0765) and is link-up supply registered (ID No 15400) for 17 separate categories.

System design and integration of UPS and associated switchgear/power monitoring within REBs and switchrooms for SSI, FSP and PSP applications.

Contracts

Derby PSB/Life Extension Works – 5 × RailPower UPS 540 V 30–60 kVA for Railtrack Midlands Zone 2000/500 k Horsham area resignalling (HARSTL2K) – 6 × RailPower UPS/REBs 400 V 4–40 kVA for Railtrack Southern Zone 2001/500 k.

East Anglia signalling upgrade – 20 × RailPower 240/400 V HV UPS for Railtrack East Anglia Zone 2002/500 k NRS supply RailPower BR1875Std

battery chargers to IMCs throughout Network Rail 1994–2003 onwards 1,000–3,000 k.

Chiltern Lines – Project Evergreen – 2 × RailPower UPS/REBs 650 V 20 kVA for Railtrack Midlands Zone 2002–2003/250 k.

Cherwell Valley re-signalling project, Leamington Spa retro-fit 30 kVA UPS and PEB for Network Rail Midlands Region 2002–2003/100 k.

Quintrell Downs re-signalling project, 5 kVA RailPower UPS for Network Rail Great Western Regional 2003–2004/15 k.

Three Bridges and Eastleigh 4 × 5 kVA RailPower UPS for Network Rail Southern Region – 62 k – 2004 (UPS for TD's).

Sheerness PSB – 1 × 3 kVA RailPower UPS for Sherbourne Resignalling Project on Network Rail Southern Region – 15 k – 2004 (UPS for TDR).

UPDATED

Power-One AG

Ackerstrasse 56, CH-8610 Uster, Switzerland
Tel: (+41 1) 944 82 16   Fax: (+41 1) 944 80 11
e-mail: info@power-one.com
Web: http://www.power-one.com

Key personnel

President: Hans Grüter
Product Manager: Claude Abächerli

Background

Power-One products were formerly marketed under the Melcher name.

Products

DC-DC converters, inverters, battery chargers – 80 product families with output power in a range of 1 to 1,000 W. All Power-One power products are ISO 9001 certified. Miniature-size switching regulator.

UPDATED

Powertron Converters Ltd

Glebe Farm Technical Campus, Knapwell, Cambridge CB3 8GG, UK
Tel: (+44 1223) 72 20 00   Fax: (+44 1223) 72 20 50
e-mail: sales@powertron.co.uk
Web: www.powertron.co.uk

Key personnel

Chairman: Miles Rackowe
Managing Director: Mike Carter
Technical Director: Andy Dickeson

Background

Powertron Converters Ltd is an internationally trading company recently formed as a continuation of the activities of Powertron Ltd, which was founded in 1971.

Powertron is a founder member of the Power Supply Manufacturers Association, and a member of the Railway Industry Association.

Products

High-reliability switch mode power supplies and DC-DC converters in the power range 5 W to 1 kW. A principal area of activity is DC-DC converters for use on railway rolling stock. Applications include lighting, communications, brake monitoring equipment, fire protection equipment and train management systems.

UPDATED

Rail Interiors SpA

Via Carrara, I-04013 Tre Ponti (Latina), Italy
Tel: (+39 0773) 44 41 (+39 0773) 44 47 44
e-mail: info@railinteriors.com
Web: www.railinteriors.com

Background

Established in 1972, Rail Interiors is a division of Aviointeriors SpA.

Products

Driver's seats, passenger seating, decorated laminates and modules for vehicle interiors, toilet modules, luggage stacks, racks and shelves, walls, end panels, interior doors, ceiling panels, heating and ventilation channels, vestibule modules, galley and restaurant car modules and lighting systems. The company's interior panels and modules make extensive use of GRP, phenolic prepregs-honeycomb core-cured, contoured cold-bonded honeycomb panels and aluminium extrusions.

Contracts

Rail Interiors' customers include Alstom Transport, Ansaldobreda, Bombardier Transportation, Costaferroviaria (now CostaRail Srl), FS Trenitalia, MerMec and VR Group, Finland.

UPDATED

RailTronic AG

Postfach, Fabrikstrasse 10, CH-8370 Sirnach, Switzerland
Tel: (+41 71) 969 37 73   Fax: (+41 71) 969 37 74
e-mail: info@railtronic.com
Web: www.railtronic.com

Key personnel

Chief Executive Officer: Ulrich Plathner
General Manager Business Development and
    Marketing: Pieter de Ruijter

Products

Electronic ballasts, timetable illuminators, high-voltage charging and power supply systems and complete interior lighting and ceiling systems for urban and main line rail vehicles.

UPDATED

Reidler Decal

The Reidler Decal Corporation
1 Reidler Road, Industrial Park, St Clair, Pennsylvania 17970, USA
Tel: (+1 717) 429 18 12   Fax: (+1 717) 429 15 28

Key personnel

President: Richard Reidler
Vice-President, Operations: Barry Frey

Products

Rail-Cal™ decals for rail vehicles, featuring heavy adhesive for pitted surfaces and a double-baking process for chemical and solvent protection; prismatic delineators to make rail vehicles more visible in poor light conditions and full graphics wrap programs.

Rex Articoli Tecnici SA

Via Catenazzi 1, CH-6850 Mendrisio, Switzerland
Tel: (+41 91) 640 50 50   Fax: (+41 91) 640 50 55
e-mail: sales@rex.ch
Web: http://www.rex.ch

Key personnel

General Manager: M Favini

Products

Design and manufacture of rubber and elastic thermoplastic products for the rail industry.

RICA

Via Podgora 26, I-31029 Vittorio Veneto (TV), Italy
Tel: (+39 0438) 91 01
Fax: (+39 0438) 91 22 36; 91 22 72; 91 03 26
e-mail: rica@zoppas-industries.com
Web: www.rica.it

Background

RICA is a member of the Zoppas Industries Group.

Products

Heating systems for rail vehicles, including: finned element duct heaters; air conditioning duct heaters; insulator mounted heaters; horizontal upflow heater units; wall and floor heating elements; cab heater units; vertical convection heaters; toilet heater units and boilers; waste tank heaters; windscreen defrosting systems; flexible heater systems; coach entrance floor heating systems; locomotive anti-slip system heating systems; and pipe and exhaust tank heating systems.

UPDATED

Robert Wagner

PO Box 1604, D-42477 Radevormwald, Germany
Tel: (+49 2195) 70 04; 70 05   Fax: (+49 2195) 10 19
e-mail: mail@robertwagner.de
Web: http://www.robertwagner.de

Key personnel

Managing Director: Reinhold Wagner

Products

Hinges for doors, window locks and mechanisms, water valves and associated equipment, interior fittings for rail vehicles.

Rohm and Haas (UK) Ltd

Powder Coatings Division
Westgate, Aldridge, West Midlands WS9 8YH, UK
Tel: (+44 1922) 74 13 45   Fax: (+44 1922) 74 13 27

Key personnel

Site Director: Tony Pitchford
European Product Manager: Richard Norris
Rail Industry Manager: Alan Phillips

Products

Powder coatings for the rail industry for use in passenger vehicle interiors, on station fittings and signage, and on electrical trunking and switchgear assemblies. The product range comprises: Bonalux AG 2000 anti-graffiti powder coatings; 491 Series polyester powder coatings; and 4620 Series epoxy powder coatings. All are fire-resistant to BS 476 Parts 6 and 7, smoke and toxic fume emission complying with BS 2583. Impress, a new range of innovative powder coatings for 'in-mould' application to plastic composite, for example: rolling stock seating.

Contracts

Powder coatings have been supplied to London Underground Ltd (Central, Jubilee, Piccadilly and Northern Lines), MTRC Hong Kong, First Great Western, Midland Mainline, Virgin Trains and to leading rolling stock manufacturers.

Ruspa Officine SpA

Via Cristoforo Colombo 2, I-10070 Robassomero, Italy
Tel: (+39 011) 923 41 11   Fax: (+39 011) 923 41 06
e-mail: info@ruspa.com
Web: www.ruspa.com

Key personnel

President: R Ruspa
General Manager: L Ruspa
Chief of Sales Department: C Ruspa

Products

Interior styling equipment, including magazine nets, ashtrays, handles, glass holders, food trays, lighting, and stainless steel wheel covers; seat systems for bus and rail applications; foot rests; arm rests.

UPDATED

Safenet AB

Beragården, SE-24010 Dalby, Sweden
Tel: (+46 46) 20 33 43   Fax: (+46 46) 20 33 47
e-mail: info@safenet.se
Web: http://www.safenet.se

Products

Manufacture of pneumatic and electric door opening systems and moveable steps for trains, buses, ships and buildings.

SAFT

Advanced and Industrial Battery Group
12 rue Sadi Carnot, F-93170 Bagnolet, France
Tel: (+33 1) 49 93 17 69   Fax: (+33 1) 49 93 19 50
Web: http://www.saftbatteries.com

Key personnel

Chairman: John Searle
Managing Director, IBG: Bertrand Olivesi
Marketing and Sales Director, IBG: Fred-Erik Hapiak
Marketing Manager: Michael Lippert
Communications Director: Jill Ledger
    Tel: (+33 1)49 93 17 77
    e-mail: Jill.ledger@saft.alcatel.fr

Products

Ni-Cd batteries, pocket plates or sintered/plastic bonded electrode types are for emergency supply and security purposes. All items are lightweight and compact and are available in stainless steel, flame-retardant or standard plastic containers.

SAFT offers an integrated battery assembly for onboard power for railway networks by combining advanced Ni-Cd batteries with custom-made containers, reducing weight and volume.

A new range of Ni-Cd batteries, sintered/Pbe, improve weight and volume by 40 per cent. Matrics (MRX batteries) are built for trams, emus, electric locomotives and light rail.

Contracts

The company has supplied rail and bus companies in 56 countries and contracts include supply to the Pendolino tilting trainsets and passenger coaches for SNCB, RENFE, Taipei and Pakistan. Recent contracts include: supply of MRX batteries for Turin trams, SNCF's new AGC regional express trains, AVES 103 high-speed train in Spain built by Siemens.

UPDATED

Saint-Gobain Sully

16 route d'Isdes, F-45600 Sully-sur-Loire, France
Tel: (+33 2) 38 37 30 00   Fax: (+33 2) 38 37 30 40

Key personnel

Marketing Manager: Guy Pajot

Products

Locomotive windscreens. Traction fitted with Sully windscreens includes the Thalys TGV, the ETR 500 high-speed train, Series 700 Skinkansen (Japan) and Acela trainsets for Amtrack (USA).

Saint-Gobain Sully NA, Inc

2175 Kumry Road, PO Box 70, Trumbauersville, Pennsylvania 18970-0070, USA
Tel: (+1 215) 536 03 33   Fax: (+1 215) 536 68 72
e-mail: douglasjones@msn.com

Key personnel

General Manager: Scott Switzer
Sales Manager, Railroad: Douglas Jones

Background

Saint-Gobain Sully NA is part of the Saint-Gobain Corporation.

Products

Windscreens and other glass for rolling stock including frames and safety glazing with high-impact resistance and good optical quality, wire grid heating elements for glazing, solar protection.

Contracts

Contracts include supply of glazing to Bombardier Transportation for Northeast Corridor; for New York CTA R142 and R142A cars; Los Angeles metro; St Louis metro.

UPDATED

SBF Spezialleuchten Wurzen GmbH

Badergraben 16, D-04808 Wurzen, Germany
Tel: (+49 3425) 905 15   Fax: (+49 3425) 90 51 62
e-mail: info@sbf-spezialleuchten.de
Web: http://www.sbf-spezialleuchten.de

Key personnel

Managing Director: Hans D Sehn
Sales Manager: Fritz Strobelt

Products

Light fittings for passenger car interiors and exteriors.

Schaltbau GmbH

Klausenburger Strasse 6, D-81677 Munich, Germany
Tel: (+49 89) 93 00 50   Fax: (+49 89) 93 00 53 50
e-mail: contact@schaltbau.de
Web: www.schaltbau.de

Works

Industrie Strasse 12, D-84149 Velden/Vils, Germany
Dietmar-von-Ayst Strasse 10, D-94501 Aldersbach, Germany

Key personnel

Managing Directors: Hans Kudlacek, Dirk Konrad

Subsidiary companies

Schaltbau Electric America Inc
705 Interchange Boulevard, Newark, Delaware 19711, USA
Tel: (+1 302) 266 05 00   Fax: (+1 302) 266 77 47
e-mail: info@schaltbau.com

Xi'an Schaltbau Electric Corporation Ltd
6 Golden Flower South Road, Xi'an 710048, China
Tel: (+86) 292 22 47 22
Fax: (+86) 292 22 47 24

TA Technologies
10 rue Désiré Granet, F-95104 Argenteuil, France
Tel: (+33 1) 39 98 49 49   Fax: (+33 1) 39 81 92 64
e-mail: cna@easynet.fr

Products

High-voltage equipment for passenger coaches; electrical equipment for diesel-hydraulic railcars; electrical components for locomotives, LRVs and passenger coaches.

UPDATED

Schaltbau Holding AG

PO Box 801540, D-81615 Munich, Germany
Tel: (+49 89) 93 00 50   Fax: (+49 89) 93 00 53 50
e-mail: schaltbau@schaltbau.de
Web: www.schaltbau.de

Key personnel

Board Members: Dr Jürgen Cammann,
   Waltraud Hertreiter

Background

Schaltbau Holding AG is a holding company for several companies in the railway equipment market.

Subsidiary companies

Schaltbau GmbH
Gebrüder Bode GmbH & Co GKh
Pintsch Bamag Antriebs- und Verkehrstechnik GmbH

UPDATED

Schlegel Swiss Standard AG

IG Park, CH-9423 Altenrhein, Switzerland
Tel: (+41 71) 858 45 45   Fax: (+41 71) 858 45 90
e-mail: info@schlegel.ch
Web: www.schlegel.ch

Key personnel

Managing Director: Daniel Niederer
Business Development and Marketing:
   Roman Hächler

Background

Schlegel Swiss Standard took over the passenger coach interiors business of Schindler Technik AG.

Products

Seats; interior fittings including luggage racks and wall-mounted tables; interior doors and door systems; self-contained submodules for all kinds of passenger coaches, walls, ceilings and other interior components and complete subsystems.

Schlegel Swiss Standard offers complete interiors.

Contracts

Recent contracts include the supply of lighweight seats for Swiss Federal Railways (SBB), featuring wooden frames and other biodegradable/recyclable components.

UPDATED

First-class seat by Schlegel Swiss Standard
0063688

Sécheron SA

Route des Moulières 5, CH-1217 Meyrin, Geneva, Switzerland
Tel: (+41 22) 739 41 11   Fax: (+41 22) 739 48 11
e-mail: info@secheron.com
Web: www.secheron.com

Key personnel

Chief Executive Officer: Claude Durand
PBU Components Director: Jo Murer
PBU Substations Director: Dominique Jamet
PBU Electronics Director: Peter Stauffer
Marketing Manager: Gilbert Lile
Corporate Sales Manager: Jimmy Cuche

Subsidiary companies

Sécheron Tchéquie, spol. sro, Prague, Czech Republic

Products

DC High-Speed Circuit Breakers (HSBC); AC and DC power and auxiliary contactors, AC vacuum circuit breakers, disconnection and changeover switches, single-arm pantographs, master controllers, electronic speed measuring and data recording systems Hasler TELOC®, Hasler Optical Pulse Generators, multifunctional display systems, wheel flange lubricators, automatic centre-couplings.

Complete system engineering, installation and setting up of modular control electronics and associated traction/auxiliary components.

Recent projects include: various components for M7 LIRR; various components for AGC for SNCF; various components for locomotives for Chinese Railways.

NEW ENTRY

Selectron Systems AG

Bernstrasse 70 CH-3250 Lyss, Switzerland
Tel: (+41 32) 387 61 61   Fax: (+41 32) 387 61 00
e-mail: info@selectron.ch
Web: www.selectron.ch

Background

Selectron Systems is a member of the Schneider Electric Group.

Products

Complete control systems for rail vehicles, including TCMS (Train Communication Management Systems), including MAS-Traffic open control and communication system, including train bus and vehicle bus. This integrates traction and propulsion monitoring, braking, vehicle speed control, power supply, lighting, heating, ventilation, air conditioning, toilets, doors and other functions.

Selectron Systems also supplies control components and undertakes hardware manufacturing and engineering for vehicle subsystems and for stationary applications.

Selectron Systems control technology is in service with operators in Austria, Poland, Germany, Netherlands, Switzerland and others.

Developments

Selectron has developed the MAS 73x controller family, designed to achieve compact dimensions and a high number of I/O connections. The modules have a robust steel housing and comply with EN 50155 standards.

UPDATED

Semco Vakuumteknik A/S

Semco Vakuumteknik A/S
PO Box 157, Svendborgvej 226, DK-5260 Odense S, Denmark
Tel: (+45) 63 15 33 00   Fax: (+45) 63 15 33 01
e-mail: man@semco-vt.dk
Web: www.semcovakuumteknik.dk

Key personnel

Managing Director: Bent Clausen
Sales Manager: Morten Arndal Nielsen

Products

Vacuum toilet systems; complete toilet compartments; toilet system components, effluent tanks, water tanks, sensors and spares.

Contracts

Contracts include 210 toilet systems for Bombardier Transportation for M6 coaches for SNCB, 239 vacuum toilets for Bombardier Transportation for TER '2N'NG, 294 vacuum toilets for ALSTOM for A-TER and Z-TER, 249 toilet modules for Ansaldobreda for IC4, 505 toilet modules for Siemens AG, VT642 – Desiro and vacuum toilets for Australia, Korea, Russian Federation, US and many other countries.

UPDATED

SEPSA

Sistemas Electrónicos de Potencia SA
Albatros 7 and 9, (Pol Ind) La Estación, E-28320 Pinto, Madrid, Spain
Tel: (+34 91) 495 70 00   Fax: (+34 91) 495 70 60
Web: http://www.sepsa.es

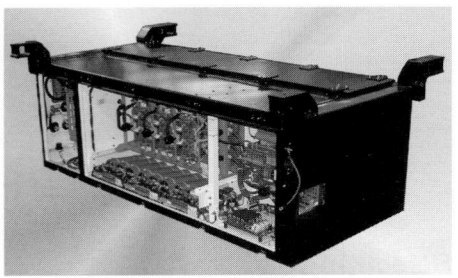

SEPSA 65.5 kVA static converter for Washington Metro 0125399

Key personnel
President: Nicolas Fuster
General Manager: Felix Ramos
Commercial Director: Antonio Sosa
Technical Director: Carlos de la Viesca

Background
SEPSA is a member of the Albatros Group (qv).

Products
Static converters with large power range up to 400 kVA. DC and AC inputs and multivoltage outputs to feed auxiliary rail equipment such as air conditioning, heating, compressors and lighting. Power electronic equipment for railway applications includes converters, inverters, choppers, rectifiers and battery chargers. Microprocessor control, high-switching frequency and IGBT technology.

Passenger information systems, public address, station announcers and displays (IRIS). Monitoring and Controlling System (PLC) to drive both auxiliary equipment and traction. CESIS crash event recording equipment.

Contracts
Recent contracts include equipment for R-142 in New York, M-7 for LIRR, WMATA 5000 and WMATA 2000/3000 RATP, vehicles for Washington, Paris, London Underground, Madrid Metro S7000/8000 and Talgo Series 7 vehicles.

SERINOX

route de Sainte Marguerite, PO Box 70, F-63307 Thiers Cedex, France
Tel: (+33 4) 73 80 22 01   Fax: (+33 4) 73 80 84 73
Web: http://www.serinox.fr

Key personnel
General Manager: Philippe Furodet

Products
Steel components for toilet systems and toilet compartments, including retention tanks, toilet bowls, litter bins, door handles and wash basins.

Siemens Rolling Stock Electronics

Siemens AG
TS LM E
Krauss-Maffei-Strasse 2, PO Box 500340, D-80973 Munich, Germany
Tel: (+49 89) 88 99 33 00   Fax: (+49 89) 88 99 28 50
e-mail: info@siemens-k-systems.com
Web: www.k-modular.com

Key personnel
Managing Board: Dr Volker Kefer; Bruno Flad
Rolling Stock Electronics: Dr Albin Oberhofer;
   Ralf Ruck;

Background
A division of the Transportation Systems group of Siemens AG.

Products
Modular, rail-proven microprocessor-based locomotive control, command and monitoring systems (KM-DIREKT, K-DIREKT and K-MODULAR) with integrated online diagnostics (K-MEMO)

suited to safe (SIL 3) radio remote control; integrated and stand-alone wheelslip and skid control and protection systems (K-MICRO and K-MICRO compact) for all kinds of rail vehicle; rugged contact- and wear-less active speed sensors (KMG-2H) and LCD displays (K-MONITOR); speed control/regulation and automation systems (KM-PROFA); sanding gear (KM-1sp and ED) and crash elements.

UPDATED

Seira Elettronica Industriale srl

Via Ca' del Bosco 1, I-37046 Minerbe (Verona), Italy
Tel: (+39 0442) 64 09 55   Fax: (+39 0442) 64 02 49
e-mail: info@seiraelettronica.it
Web: http://www.seiraelettronica.it

Products
Static converters for auxiliary power supplies for locomotives and rail vehicles.

Seratec Verkehrstechnik

Hühnerstrasse 66, CH-3123 Belp, Switzerland
Tel/Fax: (+41 31) 819 45 66
e-mail: info@seratec.ch
Web: http://www.seratec.ch

German office
Brückenstrasse 16, D-86153 Augsburg, Germany
Tel: (+49 821) 301 10   Fax: (+49 821) 333 80
e-mail: info@seratec.de

Products
Data recording systems for rail vehicles and buses. Products include journey data recording equipment, tachometers/speed recorders, speedometers, driver vigilance systems and journey data analysis systems.

Slingsby Aviation Ltd

Kirbymoorside, York YO62 6EZ, UK
Tel: (+44 1751) 43 24 74   Fax: (+44 1751) 43 11 73
e-mail: sales@slingsby.co.uk
Web: www.slingsby.co.uk

Key personnel
Sales and Marketing Manager: Rob Collinson

Background
Slingsby Aviation is a member of the Cobham plc group.

Products
Emergency detrainment systems in composite and metallic materials.

UPDATED

SMA Technologie AG

Hannoversche Strasse 1-5, D-34266 Niestetal, Germany
Tel: (+49 561) 952 20   Fax: (+49 561) 952 21 00
e-mail: railtech@sma.de
Web: www.sma.de

Key personnel
Managing Director, Railway and Solar Technology:
   Günther Cramer
Sales Managers: Birgit Wilde; Dirk Wimmer

Products
Single- and multi-voltage auxiliary power supplies as well as stand-alone battery chargers and inverters, customised power electronics and microcomputer solutions for rolling stock.

Contracts
Recent contracts have included the auxiliary power supplies for the tram-train RegioCitadis Kassel and

the commuter train Coradia Lirex Stockholm. The trains are manufactured by ALSTOM LHB. SMA also supplies the auxiliary power supplies for the new metro trains for Kaohsiung manufactured by Siemens. Major refurbishment contracts include new inverters for air conditioning for the Bvmz 185 of DB AG as well as for the BordBistro coaches of DB AG. Multiple-voltage auxiliary power supplies for double-deck sleeping coaches of City Night Line manufactured by Bombardier.

Due to existing production facilities in the People's Republic of China, SMA has won numerous contracts to supply auxiliary power supplies to carbuilders in China. SMA manufactures, for example, the stand-alone battery chargers for Guangzhou Line 2 and Shenzhen Line 1. The trains are manufactured by Bombardier in China.

UPDATED

Sofanor

94 rue Valériani, F-59920 Quiévrechain, France
Tel: (+33 3) 27 22 76 00   Fax: (+33 3) 27 22 76 22
e-mail: contact@sofanor.fr
Web: www.sofanor.fr

Background
Now independent, Sofanor was formerly a subsidiary of Bombardier Transportation, having been acquired in the 1960s by vehicle builder ANF Industrie.

Products
Design and manufacture of interior systems and fitting for new vehicles and refurbishment projects. These include: ceiling modules; air-conditioning ducting; luggage racks; lighting fittings; gangway doors; fire-break partitions; seats; vandal-resistant upholstery; and electrical cubicles.

Contracts
Sofanor customers include ALSTOM, Bombardier, Siemens, SNCB Belgium, SNCF France and RATP Paris. Projects include equipment or fittings for: AGC dmu/emu, TER 2N emu, MI2N emu, Z2N dmu, TGV Atlantique/Duplex, Lille VAL 208 (France); Meridian and Voyager demus (UK); AM96 emu/l11 coach (Belgium); Eurostar and Thalys (international).

NEW ENTRY

Soprano

Parc Technologique de l'Isle d'Abeau
40 rue Condorcet, F-38090 Vaulx Milieu, France
Tel: (+33 4) 74 82 27 27   Fax: (+33 4) 74 82 27 04
Web: http://www.soprano.fr
e-mail: soprano@soprano.fr

Key personnel
Chief Executive Officer: Marianne Mannerheim

Background
Soprano is a subsidiary of Mobile Climate Control Industries Inc (Toronto, Canada)

Products
Development, engineering, design, manufacturing, supply, refurbishment and support services for rail vehicles (passenger coaches and driver's cabins).

Range includes climate control HVAC (heating, ventilation and air conditioning) systems, onboard programmable logic controllers, train communication and control and fault diagnosis onboard power supply, static converters, certified to ISO 9001.

Specialty Bulb Co Inc

The Specialty Bulb Co Inc
80 Orville Drive, Bohemia, New York 11716-0231, USA
Tel: (+1 631) 589 33 93   Fax: (+1 631) 563 30 89
e-mail: info@bulbspecialists.com
Web: http://www.bulbspecialists.com

Key personnel

President: Judith Beja

Vice-President, Technical: Caden Zollo

Vice-President, Sales and Marketing: Edie Muldoon

Products

Major supplier to rail and mass transit industry of USA and European lamps for car, signal, headlights and other applications.

SPS Isoclima SpA

Via E Mattei, I-35046 Saletto (PD), Italy

Tel: (+39 0429) 895 44   Fax: (+39 0429) 89 92 94

e-mail: info@sps.it

Web: http://www.isoclima.net

Key personnel

General Manager: Alberto Bertolini

Works

Via Enico Mattei I, I-35046 Aletto (PD), Italy

Tel: (+39 0429) 895 44   Fax: (+39 0429) 89 92 94

Products

Windscreens: flat, curved, heated, framed with antispall, high-impact-resistant for high-speed trains, locomotives, emus, dmus, LRVs, trams and buses; glass for light covers, interior glazing and partitions.

Stone India Ltd

16 Taratalla Road, Calcutta 700 088, India

Tel: (+91 33) 24 01 46 61/668   Fax: (+91 33) 24 01 48 86/24 01 34 51

e-mail: mktg.kolkata@stoneindia.co.in

e-mail: gautamghosh@stoneindia.co.in

Web: www.stoneindia.co.in

Key personnel

Managing Director: A Mondal

Chief Marketing Officer: Gautam Ghosh

National Sales Manager: R K Ganeshan

Products

Train lighting alternators, air conditioning, pantographs for electric locomotives and emus.

UPDATED

Stone International

A member of Vapor Group

10655 Henri Bourassa West, Montreal, Quebec H4S 1A1, Canada

Stone UK Limited (Stone International)

Unit 9, Crossways Business Park, Stone Marshes, Dartford, DA2 6QG, UK

Tel: (+44 1322) 28 93 23   Fax: (+44 1322) 28 92 82

Managing Director: Anthony J Walsh

Stone Safety Service

240 South Main Street, South Hackensack, New Jersey 07606, USA

Tel: (+1 201) 489 02 00   Fax: (+1 201) 489 93 62

Vice-President: Vincent Mirandi

Products

Air conditioning, heating, pressure ventilation and temperature control equipment, static inverters, battery chargers, alternators, DC motors.

Stratiforme Industries

26 route Nationale, F-59235 Bersee, France

Tel: (+33 3) 20 84 90 10   Fax: (+33 3) 20 59 28 00

Web: http://www.stratiforme.com

Technoforme

26 route Nationale, F-59235 Bersee, France

Tel: (+33 20) 849 00   Fax: (+33 20) 84 90 22

Compreforme

BP 8, F-21402 Chatillon sur Seine, France

Tel: (+33 80) 91 09 85   Fax: (+33 80) 91 40 61

Products

Polyester thermohardening plastics to manufacture phenolic resin moulded pieces with in-mould gelcoat coating. Stratiforme has two production units, at Bersee and Chatillon sur Seine, with a testing laboratory at Chatillon.

Moulded pieces for driver cabins, front ends, driver consoles, toilet cubicles, stairways, side panels, seats and interior liner panels.

Front ends for the French, Spanish, Italian, Korean and Eurostar TGV high-speed trains; urban and suburban rolling stock.

Underground passenger cars for Paris, Caracas, New York, Taipei, Jacksonville, Chicago, Atlanta, San Francisco; tramways and streetcars for Lille, Brussels, Grenoble, Cologne and Saarbrücken; and interior fittings for Intercity trains.

Stratiforme moulded structural units are used in Talent dmus   0085461

Supersine Duramark Limited

Freemantle Road, Lowestoft, Suffolk NR33 0EA, UK

Tel: (+44 1502) 50 12 34   Fax: (+44 1502) 58 35 44

e-mail: sales@ssdm.co.uk

Web: http://www.ssdm.co.uk

Products

System Deco self-adhesive laminates for use on trains as an alternative to paint, with anti-graffiti options. Glass protection film applied to windows to protect them from vandalism. SSDM On-line, developed by SSDM's electronic media company, Clear Interactive Ltd. The Internet-based system allows for artwork approvals and instant ordering around-the-clock.

Svend A Nielsen A/S

Gillelejevej 30, Esbønderup, DK-3230 Graested, Denmark

Tel: (+45) 48 39 88 88   Fax: (+45) 48 39 88 98

e-mail: info@san.as.com

Web: http://www.san-as.com

Products

Driver's cab and passenger compartment heating systems for rail vehicles and buses.

NEW ENTRY

TBA Textiles Ltd

PO Box 40, Rochdale OL12 7EQ, UK

Tel: (+44 1706) 64 74 22   Fax: (+44 1706) 75 02 56

e-mail: info@tbatextiles.co.uk

Web: http://www.tbatextiles.com

Key personnel

Managing Director: Dr A V Ruddy

Products

Fireblocking and anti-vandal fabrics for passenger transport seating. Fire resisting/insulating liners for use within the bodywork. Moulded fire-resistant seat pans.

Technical Resin Bonders

12 Clifton Road, Huntingdon, Cambridgeshire PE29 7EN, UK

Tel: (+44 1480) 44 74 00   Fax: (+44 1480) 41 49 92

e-mail: sales@trbonders.co.uk

Web: www.trbonders.co.uk

Key personnel

Sales Manager: Robert Hodgson

Background

Formerly part of Ciba Geigy, Technical Resin Bonders is part of the Bondsword Group.

Products

Lightweight structural and decorative honeycomb panels for vehicle new build and refurbishment programmes, including: aluminium honeycomb floors; external doors; internal doors; partitions; draught screens; table tops; ceiling panels; lower body sides; tables for passengers with disabilities; toilet modules; catering modules; overhead lockers; train skirts; energy absorbers; driver protection panels and cladding panels.

UPDATED

Techni-Industrie SA

ZI de la Chambrouillère, F-53960 Bonchamp-les-Laval, France

Tel: (+33 2) 43 59 23 80   Fax: (+33 2) 43 59 23 89

e-mail: techni@techni-industrie.fr

Web: http://www.techni-industrie.fr

Key personnel

Chairman: Gérard Lelasseux

Commercial and Technical Manager: Dan Diaconu

Products

Components for powered passenger vehicles including front ends and chassis for LRVs and bodyshell parts for TGV high-speed trainsets. Locomotive cabs and chassis.

Técnicas Modulares e Industriales SA (Temoinsa)

Polígono Industrial Congost, Avenida San Juliá 100, E-08400 Granollers, Barcelona, Spain

Tel: (+34 93) 860 92 00   Fax: (+34 93) 860 92 13

e-mail: tmi@temoinsa.com

Web: www.temoinsa.com

Key personnel

Chairman: Alvaro Colomer Castellano

Subsidiaries Manager: Miguel De Sagarra Romeo

Chief Executive Officer: Mercé Sala Schnorkowski

Strategic Vice-President: Fernando Delgado Leiria

Products

Design, engineering, manufacture, supply and technical assistance for the fitting of complete

---

For details of the latest updates to *Jane's World Railways* online and to discover the additional information available exclusively to online subscribers please visit

jwr.janes.com

passenger coach interiors with fully developed modular systems, including air conditioning, heating and ventilation, vacuum toilet systems; high-technology composites, control systems, passenger information systems and electric cabinets.

Turnkey projects for complete interiors of new vehicles and modernisation/upgrading works.

Contracts

Contracts include: toilet modules and systems destined for Long Island for Bombardier (Canada), air conditioning saloon coaches for Bombardier/Kioleides (Greece), Railcars for Bombardier (Germany), different types of complete coaches for OSE (Greece), Sirio Trams for Ansaldobreda (Italy), interiors and air conditioning for emus destined for Denmark for Ansaldobreda (Italy), saloon coaches and double-decker coaches for Trenitalia (Italy), complete interiors for s/9000 Metro Madrid coaches for Ansaldobreda (Italy), 1st and 2nd class saloon coaches, buffet and suburban coaches for ENR and SEMAF (Egypt), complete refurbishment of various passenger coaches for SNCF (France), toilet modules for Civia coaches for ALSTOM and CAF (Spain), toilet modules and systems for S/213 coaches for ALSTOM and CAF (Spain), toilet modules and systems for saloon coaches for ONCF (Morocco).

Other contracts have included fitting out of s/440 units for RENFE (Spain), s/444 units destined for Chile by RENFE (Spain), D-160 coaches destined for Chile by RENFE (Spain), double-decker coaches for ALSTOM (Spain) and for CP Tagus Crossing (Portugal), toilet modules destined for Finland by ALSTOM (France), interior components for saloon coaches for Talgo (Spain), metro coaches destined for Copenhagen, Atlanta and Madrid for Ansaldobreda (Italy), bistro coaches for Amtrak (US), Connex Line toilet modules for Bombardier (UK), toilet modules for Virgin CrossCountry Bombardier (France), coaches for WCML for ALSTOM (UK), air conditioning for Liebherr (Austria), saloon, sleeper coaches and toilet modules for BSP (China).

UPDATED

Teknoware Oy

Ilmarisentie 8, FI-15200 Lahti, Finland
Tel: (+358 3) 88 30 20   Fax: (+358 3) 883 02 40
e-mail: sales@teknoware.fi
Web: www.teknoware.fi

Key personnel

Export Manager: (Austria, Czech Republic, Croatia, Germany, Hungary, Romania, Slovakia, Slovenia, Switzerland, Yugoslavia): Jeroen Hinnen
Export Director (Rest of World): Paul Steurs

Products

Interior lighting electronics and lighting systems; inverters/ballasts and fluorescent lighting systems for LRVs and passenger coaches. Optiline, optimal optic fibre solutions for night lighting and special effect lighting for public transport vehicle interiors.

Recent developments include an increased range of halogen converters and dimmable lighting inverters/ballasts, as well as halogen and LED spots and readings lights.

UPDATED

Telephonics

Telephonics Corporation
815 Broad Hollow Road, Farmingdale, New York 11735, USA
Tel: (+1 631) 549 60 62   Fax: (+1 631) 549 60 18

Key personnel

Vice-President, Business Development: Philip Greco
Manager, Business Development: Norbert Trokki

Products

Communication, information and surveillance systems for mass transit: integrated car communication and passenger information systems; wayside communication and central control systems; closed circuit television systems for wayside, stations and vehicles; train line multiplexers; network controller and vehicle monitoring.

Contracts

Vehicle communication, train multiplexer, network controller for New York City Transit subway cars.

Vehicle communication, radios and door observation CCTV for South Eastern Pennsylvania Transit Authority.

Vehicle communication system for Massachusetts Bay Transit Authority light rail.

Vehicle communication, health monitoring and vehicle CCTV at Hudson Bergen light rail transit system.

Integrated wayside communication system for Newark APM.

Vehicle communication and passenger entertainment for Caltrans.

Developments

Train line multiplexer for transport of vehicle controls and digital audio using EI standards; car network controller is used to convert LON to propulsion/brake commands; public address and information systems based on internet technology.

---

th-contact AG

Schönmattstrasse 6, CH-4153 Reinach, Switzerland
Tel: (+41 61) 716 75 75   Fax: (+41 61) 711 77 67
e-mail: info@th-contact.ch
Web: http://www.th-contact.ch

France
550 av du Général De Gaulle, BP 203, F-88106 Saint-Dié, Cédex
Tel: (+33) 329 56 23 82   Fax: (+33) 329 56 10 42
e-mail: the-contact@wanadoo.fr
Web: http://www.th-contact.fr

UK
Unit 9, Lion Industrial Park, Northgate Way, Aldridge, Walsall WS9 9RL
Tel: (+44 1922) 45 22 12   Fax: (+44 1922) 45 54 17
e-mail: sales@th-contact.co.uk
Web: http://www.th-contact.co.uk

USA
th-contact inc
2121 South Oneida, Suite 515, Denver, Colorado 80224
Tel: (+1 303) 757 62 00   Fax: (+1 303) 757 68 00
e-mail: info@th-contact.com
Web: http://www.th-contact.com

Background

In December 2001, a co-operation agreement was announced between th-contact AG and German-based Escha Bauelemente GmbH.

Products

Switching devices and indicator lamps for public transport.

Thermo King Corporation

314 West 90th Street, Minneapolis, Minnesota 55420, USA
Tel: (+1 952) 887 22 00   Fax: (+1 952) 887 25 29
Web: http://www.thermoking.com

Key personnel

Vice-President, North America: P Smith

Subsidiaries

Thermo King Asia Pacific
Hong Kong
Tel: (+852) 25 07 91 00   Fax: (+852) 28 27 51 59

Thermo King do Brasil
Campinas, Brazil
Tel: (+55 19) 745 64 00   Fax: (+55 19) 245 11 06

Thermo King Europe
Brussels, Belgium
Tel: (+32 2) 714 57 11   Fax: (+32 2) 714 57 12

Thermo King Europe
Kolin, Czech Republic
Tel: (+420 321) 75 71 11   Fax: (+420 321) 75 71 70

Background

Thermo King Corporation is part of the Ingersoll-Rand Company.

Products

Heating, ventilation and air conditioning units (HVAC) for buses and light rail vehicles. Range includes roof-mounted and integral systems, front-mounted systems, heating convection systems and small bus air conditioning systems, backed up by global service network.

Developments

Electrical HVAC unit designed for any type of bus, including electric, hybrid or diesel/gas engine driven.

Tiflex Ltd

Tiflex House, Liskeard, PL14 4NB, UK
Tel: (+44 1579) 32 08 08
Fax: (+44 1579) 32 08 02
e-mail: treadmaster@tiflex.co.uk
Web: www.tiflex.co.uk

Key personnel

Managing Director: N A Spearman
Sales and Marketing Director: A Tuffield
Product Manager: Barry Curtis

Products

Treadmaster smoke and fire-resistant floor coverings, slip-resistant floor coverings, for rail and road vehicles and for buildings; specialist stair nosings.

Contracts include the supply of floorings for the Heathrow Express trainsets, Midland Metro LRVs and MoD naval vessels.

UPDATED

Time 24 Ltd

Unit 69, Victoria Road, Burgess Hill RH15 9TR, UK
Tel: (+44 1444) 25 76 55   Fax: (+44 1444) 25 90 00
e-mail: sales@time24.co.uk

Key personnel

Managing Director: David Shore
Finance Director: Mark Willifer
Sales and Marketing Director: Chris Young
Technical Director: David Shergold
Purchasing Manager: Martin Arter

Products

Traction and brake controllers; cable assemblies, harnesses and looms; cab desks; project management. Body end cubicles, drivers' cupboards, lighting systems and vestibules.

TODCO Inc

7167 Route 353, Cattaraugus, New York 14719-9537, USA
Tel: (+1 716) 373 62 04
Fax: (+1 716) 373 64 72
Web: http://www.todco.com

Products

Doors, floors, walls, windscreens, ceilings, seating and other interior modules for rail vehicles and buses. Panels are produced without seals, glazing or furniture and in primed-only condition. Recent products include the Partner door, which is capable of being operated in sliding-plug configuration.

Toshiba Corporation

Railway Projects Department
Toshiba Building, 1-Shibaura 1-chome, Minato-ku,
Tokyo 105-8001, Japan
Tel: (+81 3) 34 57 49 24   Fax: (+81 3) 54 44 92 63

Key personnel

President and Chief Executive Officer:
  Tadashi Okamura
Vice-President: Takio Ooyama
Senior Manager, Railway Projects Department:
  Koji Toda

Products

Heating, ventilation and air conditioning
equipment; AC and DC electrification equipment.

UPDATED

Toyo Denki Seizo KK

Toyo Electric Manufacturing Co Ltd, No 1 Nurihiko
Bldg, 9-2 Kyobashi, 2-chome, Chuo-ku, Tokyo
104-0031, Japan
Tel: (+81 3) 35 35 06 41
Fax: (+81 3) 35 35 06 50

Key personnel

President: Kunio Kai
Vice-President: Motonobu Matsubara
Managing Director: Keisuke Tanaka
Director: Kenzo Terashima

Products

Door operating systems and equipment, including
door actuators, opening/closing switches and
control systems.

UPDATED

Transit Control Systems

11451 North Ocean Circle, Anaheim, California
92806, USA
Tel: (+1 714) 234 30 30

Key personnel

President: Peter J Anello
Vice-President, Marketing and Contracts:
  Jimmie C Collins
Chief Engineer: J Kiel

Products

Drivers' consoles, manual controllers, fault
monitoring, controls, communications, heating
and air conditioning subsystem equipment.
TCS also supplies locomotive heating and air
conditioning systems.

Contracts

Contracts include: Heating, Ventilation and Air
Conditioning (HVAC) equipment for MBTA Red Line
cars; drivers' consoles and control equipment for Los
Angeles; HVAC and communication equipment for
Septa Norristown cars; drivers' consoles for NYCTA;
knife switch assembly for St Louis LRVs; drivers'
consoles control and communication equipment for
BART's C2 cars; trainsets and the HVAC overhaul of
WMATA's 298 Rohr cars; and manual controllers for
Atlanta's MARTA CQ312 cars.

Transmatic Inc

6145 Delfield Industrial Drive, Waterford, Michigan
48329, USA
Tel: (+1 248) 623 25 00
Fax: (+1 248) 623 28 39
Web: http://www.transmaticgroup.com

Key personnel

President: O K Dealey Jr
Vice-President, Sales and Marketing: M T Hoffman
Vice-President, Environmental Systems:
  D Scott McConnell

UK subsidiary

Transmatic Europe Ltd
Unit 2, City Park Industrial Estate, Gelderd Road,
Leeds LS12 6DR, UK
Tel: (+44 113) 279 99 89
Fax: (+44 113) 279 41 27
e-mail: sales@transmatic.co.uk
Managing Director: Terry Calnon

Products

Interior lighting and advertising coving for buses
and urban transit vehicles, multipurpose lighting/
air conditioning duct modules, surface-mounted
fluorescent lighting, destination sign lighting,
interior cleaning systems for buses and rail
vehicles.

Trevira GmbH

Philipp-Reis-Strasse 2, D-65795 Hattersheim,
Germany
Fax: (+49 69) 305 163 42
e-mail: treviracs@fra.trevira.com
Web: www.treviracs.com

Key personnel

Head Marketing and Sales Home Textiles:
  Dr Ulrich Girrbach,

Products

Flame-retardant fibres and filament yarns for
interior textiles such as upholstery.

UPDATED

TrioPlast

ZI A – 262, Avenue Georges Washington, BP 556,
F-62411, Bethune, Cedex, France
Tel: (+33 321) 63 23 73   Fax: (+33 321) 56 63 69
Web: http://www.trioplast.fr-trioplast@trioplast.fr

Key personnel

Managing Director: Jacques Duhoo

Products

Developing and manufacturing of tables, floor and
doors, produced in composite materials for the
railway industry and more specifically for high-
speed train passenger cars.

Contracts

With most of the French high-speed trains (TGV)
and ALSTOM and Bombardier.

UniControls AS

Křenická ulice-2257, CZ-100 00 Prague 10, Czech
Republic
Tel: (+420 2) 72 01 14 11   Fax: (+420 2) 74 81 44 75
e-mail: unic@unicontrols.cz
Web: www.unicontrols.cz

Key personnel

Marketing Manager: Marian Belosovic

Products

Communications and control systems for trains
and rail vehicles, including: train communication
network, wire-train-bus equipment; multivehicle
communications equipment; vehicle communication
devices; and driver's cab equipment and displays.

Contracts include train communications and
control systems for refurbished Russian-built
Prague metro stock; train communications
system for St Petersburg metro stock; train and
vehicle communications system and driver's cab
equipment for Czech Railways Class 471 emus;
automatic train control system for Czech Railways
Class 680 tilting trainsets; driver's cab equipment
for refurbished Class 772 locomotive for Slovakian
Railways (ZSSK).

UPDATED

Vapor Rail Inc

10655 Henri-Bourassa West, Saint-Laurent, Quebec
H4S 1A1, Canada
Tel: (+1 514) 335 42 00
Fax: (+1 514) 335 42 31
Web: http://www.vaporrail.com
http://wabtec.com

Key personnel

Vice-President, Marketing and Business
  Development: Michel Germain

Background

A member of the WABCO Transit Group, a Wabtec
company.

Products

Design and manufacture of complete door systems
for passenger rail vehicles, related accessories and
peripherals, including accessibility aids.

Vapor Stone UK Limited

2nd Avenue, Centrum 100, Burton-on-Trent
DE14 2WF, UK
Tel: (+44 1283) 74 33 00   Fax: (+44 1283) 74 33 33
Web: http://www.wabtec.com

Key personnel

Managing Director: Anthony J Walsh
Commercial: Rob Turner

Background

A Wabtec company.

Products

Design and manufacture of door systems and
HVAC equipment for passenger transit vehicles
plus upgrades and maintenance.

Vogelsitze GmbH

Kleinsteinbacher Strasse 44, D-76228 Karlsruhe,
Germany
Tel: (+49 721) 470 20   Fax: (+49 721) 470 21 70
e-mail: info@vogelsitze.de
Web: www.vogelsitze.de

Key personnel

Chief Operating Officer: Dr Robert Kocsis

Products

Seating.

UPDATED

Wabtec Corporation

1001 Air Brake Avenue, Wilmerding, Pennsylvania
15148, USA
Tel: (+1 412) 825 10 00   Fax: (+1 412) 825 10 19
Web: www.wabtec.com

Key personnel

Chairman, President and Chief Executive Officer:
  William E Kassling
Vice-Chairman of the Board: Emilio A Fernandez
  Board of Directors: Robert J Brooks,
  Kim G Davis, Lee B Foster II,
  Michael W D Howell, James P Miscoll,
  James V Napier
Senior Vice-President, Chief Financial Officer:
  Alvaro Garica-Tunon
Vice-President, Investor Relations: Tim Wesley

Products

Electronic data recorders, end-of-train units, speed
indicators, alertness devices, onboard computers,
fault monitors/annunciators, fuel level indicators;
electromechanical and electropneumatic door
actuators for passenger vehicles; air conditioning
equipment for passenger vehicles.

UPDATED

Walter Mäder Aqualack GmbH

Gewerbepark 40, D-59069 Hamm, Germany
Tel: (+49 2385) 935 60 Fax: (+49 2385) 93 56 49
e-mail: info@maeder-aqualack.de

Products
Water-borne and solvent-borne paints and coatings for rail vehicles of all types.

Weserland Sitzsysteme GmbH

Immenweg 19-21, D-31582 Nienburg, Germany
Tel: (+49 5021) 960 99 15 Fax: (+49 5021) 960 99 18
e-mail: info@wesersitz.de
Web: http://www.wesersitz.de

Key personnel
Managing Director: Walter Anklam

Products
Design, development and manufacture of seating systems for main line and urban rail vehicles. The company also undertakes the refurbishment and cleaning of rail vehicle seating and interiors.

Westcode Inc

1372 Enterprise Drive, West Chester, Pennsylvania 19380, USA
Tel: (+1 610) 738 12 00 Fax: (+1 610) 696 74 20
Web: http://www.westcodeus.com

Key personnel
Chairman and Chief Executive Officer:
 Edward J Widdowson

Subsidiary company
Westcode (UK) Ltd
PO Box 1582, Chippenham SN15 3ZR, UK
Tel: (+44 1249) 78 34 56 Fax: (+44 1249) 78 36 66
Managing Director: David A Thompson

Products and services
Manufacture and refurbishment of rail vehicle heating, ventilation and air conditioning (HVAC) systems; design, manufacture and overhaul of train door systems; design and manufacture of pneumatic systems, including brake control and air supply equipment and test rigs.

Contracts
HVAC equipment for urban operators in Puerto Rico, Taipei and Toronto; refurbishment of HVAC equipment for 439 cars for BART, San Francisco; HVAC equipment for ScotRail, UK and IE, Ireland; over 40,000 door systems for operators in North America; test equipment for NYCTA, New York and TRTC, Taipei.

Widney Transport Components (Pty) Ltd

Widney Transport Components (Pty) Ltd
PO Box 17291, Rondhort, 1457, Gauteng, South Africa
Tel: (+27 11) 864 48 04 Fax: (+27 11) 908 18 56
Web: www.shatterprufe.co.za

Key personnel
Managing Director: G Scott
Export Marketing Director: K Luyt
General Manager: F G Murray
Sales and Marketing Manager: A Booi

Products
Windows for passenger vehicles, including sliding, hopper and double-glazed types; doors for metro coaches, general carriage fittings, door windows for metro coaches and drop windows (MKIV).

Contracts
Contracts have included supply of bus sliding glass.

UPDATED

Winstanley

Winstanley & Co (Kings Norton) Ltd
Racehorse Road, Pinvin Industrial Estate, Pershore WR10 2DG, UK
Tel: (+44 1386) 55 22 78 Fax: (+44 1386) 55 64 36
e-mail: trevor.clews@winstanleyco.co.uk

Key personnel
Chairman: John Foley
Managing Director: Jerry King
Commercial Director: Trevor Clews

Subsidiary companies
Portable Balers Ltd
Adcon Modular MRF Systems

Products
Design, manufacture and finishing of aluminium, steel and stainless steel fabrications, primarily for the railway industry. Factoring of products made in composites or wood and the supply of complete railway vehicle interiors and ancillary equipment to the major rolling stock passenger coach equipment.

Trelleborg Woodville Rail

Hearthcote Road, Swadlincote, DE11 9DX, UK
Tel: (+44 1283) 22 11 22
Fax: (+44 1283) 21 97 68
e-mail: joanne.cresswell@trelleborg.com
Web: www.trelleborg.com

Key personnel
Managing Director: David Semple
General Manager: Mike Hawkins
Business Unit Manager: John Blackham
Technical Manager: Nigel Bailey

Background
Trelleborg Woodville Rail, formerly Woodville Polymer Engineering, is part of Trelleborg AB, within the Trelleborg Sealing Solutions division.

Products
Design, development and manufacture of gangways and flexible treadplates; rail accessories fabrication; and industrial engineering services.

Contracts
Recent contracts include: Virgin CrossCountry, UK; MTRC TKO Line, Hong Kong; Vancouver Sky Train, Canada; Juniper emus, UK; Øresund Link emus, Sweden/Denmark; Delhi Metro, India; Manila Metro, Philippines; Prospector Avon Link, Western Australia; Circle Line, Singapore; Shanghai Metro Pearl Line, China; Gwang Ju Metro, South Korea; Attiko Metro, Greece; Hunter Valley, New South Wales, Australia; Daegu Line 2, South Korea; Seoul Metropolitan, South Korea.

UPDATED

XP plc

Horseshoe Park, Pangbourne RG8 7JW, UK
Tel: (+44 118) 984 55 15 Fax: (+44 118) 984 34 23
e-mail: sales@xpplc.com
Web: www.xpplc.com

Key personnel
Managing Director: Steve Robinson
Technical Director: Gary Bocock
Divisional Director (Railways): Richard Bartlett

Products
Electronic power supplies and DC/DC converters that meet UK and European standards RIA12 and 13 for rail applications.

UPDATED

XP modular DC/DC converters that meet UK and European standards 0003469

Yutaka Manufacturing Co Ltd

1-18-17 Kitakojiya Ota-ku, Tokyo 144-0032, Japan
Tel: (+81 3) 37 41 41 31 Fax: (+81 3) 57 05 70 65
e-mail: hideo.kamei@yutaka-ss.co.jp
Web: www.yutaka-ss.co.jp/

Key personnel
President and Chief Executive Officer:
 Yasuyuki Maki

Products
Jumper cable connectors; high- and low-voltage connectors; multicontact connectors for power input and output; automatic train coupling/uncoupling systems; various kinds of control equipment for rolling stock.

NEW ENTRY

FREIGHT VEHICLES AND EQUIPMENT

Alphabetical listing

ABRF Industries
Alna Koki Company Ltd
ALSTOM LHB GmbH
ALSTOM Transport
Arbel
Azovmash
Babcock Rail
Bharat Bhari Udyog Nigam Ltd (BBUNL)
Bharat Wagon & Engineering
BMZ
Bradken Rail
Braithwaite
Burn Standard
CAF – Construcciones y Auxiliar de Ferrocarriles SA
Cattaneo
China Northern Locomotive and Rolling Stock Industry (Group) Corporation
China South Locomotive and Rolling Stock Industry (Group) Corporation (CSR)
Cimmco International
CNR Qigihar Railway Rolling Stock (Group) Co Ltd (QRRS)
CostaRail Srl
Cromweld Steels
Daewoo Heavy Industries Ltd
Dalian Locomotive and Rolling Stock Works
W H Davis Ltd
E G Steele & Co Ltd
EKA Limited
Fabryka 'Wagon' SA
Fabryka Wagonów Gniewczyna
FIREMA Trasporti SpA
FM Industries
FreightCar America Inc
Freudenberg Schwab GmbH
Fuji Car Manufacturing Co Ltd

Ganz Vagon Kft
Goša
Graaff
The Greenbrier Companies
The Gregg Company
Gunderson
Gunderson-Concarril
Hanjin Heavy Industries Co Ltd
Hindusthan Engineering & Industries Ltd (HEI)
Holland Company LP
Hyundai
INKA
Interlok Bahnconsulting Schmidtendorf
iQR
IRECO LLC
Jessop
K & M
Kawasaki Heavy Industries Ltd
Keller Elettromeccanica SpA
K Industrier AB
KT Steel
Kryukovsky
Lyudinovo Locomotive Works JSC
MAN Ferrostaal AG
Marcroft Engineering Ltd
MÁV
Modalohr
More Wear Industries
National Steel Car Ltd
Nippon Sharyo Ltd
Oleo International Ltd
Pakistan Railways Carriage Factory
Palfinger Bermüller GmbH
PEC Ltd
PESA Bydoszcz SA
Power Machines Group

Price
Procor Ltd
Progress Rail Services Corporation (Locomotive and Railcar Services)
Qishuyan
Rolanfer Matériel Ferroviaire SA
Rotem Company
RSD (DCD-Dorbyl (Pty) Ltd)
SCIF
Sambre et Meuse
Siemens AG
Ray Smith Group plc
Société Générale Egyptienne de Matériel des Chemins de Fer (SEMAF)
STAG AG
Tabor Szynowy Opole SA
Tafesa SA
Talgo Oy
Tatravagónka AS
Texmaco Ltd
Transwerk
TrentonWorks Ltd
Trinity Difco
Tuchschmid AG
Tülomsas
Union Tank
United Goninan
Unity Railway Supply Co Inc
Wabash National
Wabtec Rail Ltd
Waggonbau Brüninghaus GmbH
Wagon Pars
WagonySwidnica SA
Windhoff Bahn-und Anlagentechnik GmbH
Zimbabwe Engineering Ltd
Zhuzhou

Company listing by country

AUSTRALIA
Bradken Rail
iQR
United Goninan

CANADA
National Steel Car Ltd
Procor Ltd
TrentonWorks Ltd

CHINA
China Northern Locomotive and Rolling Stock Industry (Group) Corporation
China South Locomotive and Rolling Stock Industry (Group) Corporation (CSR)
CNR Qigihar Railway Rolling Stock (Group) Co Ltd (QRRS)
Dalian Locomotive and Rolling Stock Works
Qishuyan
Zhuzhou

EGYPT
Société Générale Egyptienne de Matériel des Chemins de Fer (SEMAF)

FINLAND
Talgo Oy

FRANCE
ABRF Industries
Arbel
Modalohr
Rolanfer Matériel Ferroviaire SA
Sambre et Meuse

GERMANY
ALSTOM LHB GmbH
Freudenberg Schwab GmbH

Graaff
Interlok Bahnconsulting Schmidtendorf
MAN Ferrostaal AG
Palfinger Bermüller GmbH
Siemens AG
Waggonbau Brüninghaus GmbH
Windhoff Bahn-und Anlagentechnik GmbH

HUNGARY
Ganz Vagon Kft
MÁV

INDIA
Bharat Bhari Udyog Nigam Ltd (BBUNL)
Bharat Wagon & Engineering
Braithwaite
Burn Standard
Cimmco International
Hindusthan Engineering & Industries Ltd (HEI)
Jessop
KT Steel
PEC Ltd
Texmaco Ltd

INDONESIA
INKA

IRAN
Wagon Pars

ITALY
CostaRail Srl
FIREMA Trasporti SpA
K & M
Keller Elettromeccanica SpA

JAPAN
Alna Koki Company Ltd

Fuji Car Manufacturing Co Ltd
Kawasaki Heavy Industries Ltd
Nippon Sharyo Ltd

KOREA, SOUTH
Daewoo Heavy Industries Ltd
Hanjin Heavy Industries Co Ltd
Hyundai
Rotem Company

MEXICO
Gunderson-Concarril

MOROCCO
SCIF

NEW ZEALAND
Price

PAKISTAN
Pakistan Railways Carriage Factory

POLAND
ALSTOM Transport
Fabryka 'Wagon' SA
Fabryka Wagonów Gniewczyna
PESA Bydoszcz SA
Tabor Szynowy Opole SA
WagonySwidnica SA

RUSSIAN FEDERATION
BMZ
Lyudinovo Locomotive Works JSC
Power Machines Group

SERBIA & MONTENEGRO
Goša

Company listing by country—*continued*

SLOVAKIA
Tatravagónka AS

SOUTH AFRICA
RSD (DCD-Dorbyl (Pty) Ltd)
Transwerk

SPAIN
CAF – Construcciones y Auxiliar de Ferrocarriles SA
Tafesa SA

SWEDEN
K Industrier AB

SWITZERLAND
Cattaneo
STAG AG
Tuchschmid AG

TURKEY
Tülomsas

UKRAINE
Azovmash
Kryukovsky

UNITED KINGDOM
Babcock Rail
Cromweld Steels
W H Davis Ltd
E G Steele & Co Ltd
EKA Limited
Marcroft Engineering Ltd
Oleo International Ltd
Ray Smith Group plc
Wabtec Rail Ltd

UNITED STATES
FM Industries
FreightCar America Inc
The Greenbrier Companies
The Gregg Company
Gunderson

Holland Company LP
IRECO LLC
Progress Rail Services Corporation (Locomotive and Railcar Services)
Trinity Difco
Union Tank
Unity Railway Supply Co Inc
Wabash National

ZIMBABWE
More Wear Industries
Zimbabwe Engineering Ltd

ABRF Industries

Ateliers Bretons de Réalisations Ferroviaires Industries
PO Box 19, ZI rue Lafayette, F-44141 Châteaubriant Cedex, France
Tel: (+33 2) 40 81 19 20   Fax: (+33 2) 40 28 02 02

Key personnel
General Manager: Jean-Luc Remondeau
Commercial and Technical Director:
  Gérard Gueguin
Commercial Engineer: Jean-Pierre Cadiou

Products
Freight wagons, including 'Easiloader' curtain hood wagon for general merchandise, tank, container and hopper wagons.

Alna Koki Company Ltd

4-5 Higashi Naniwa-cho 1-chome, Amagasaki 660-8572, Japan
Tel: (+81 6) 401 72 83   Fax: (+81 6) 401 61 68

Key personnel
For full personnel listing, see Alna Koki Company Ltd entry in *Locomotives and powered/non-powered passenger vehicles* section.

Products
General-purpose freight wagons; low-floor wagons; tank wagons; dump wagons.

ALSTOM LHB GmbH

Linke-Hofmann-Busch-Strasse 1, D-38239 Salzgitter, Germany
Tel: (+49 5341) 90 00   Fax: (+49 5341) 900 49 14
Web: www.transport.alstom.com

Key personnel
Freight Product Line Director: Günter Köhler

Products
Development, design, engineering and manufacture of freight wagons of all types (to UIC specifications), refurbishment and repair (authorised repair shop), self-discharging hopper wagons (four-/six-axle), covered wagons (tarpaulin or sliding hood), large-volume sliding-wall wagons, wagons with special shock protection (against shunting impacts), covered and closed double-deck car-carriers for high-value.

UPDATED

Closed double-deck car carrier unit during assembly by ALSTOM   0554540

Special-purpose tank wagon by ALSTOM   0554546

Faals 151 bulk iron ore carrier, Germany
NEW/0585241

ALSTOM's CeSa wagon
NEW/0585242

ALSTOM Transport

ALSTOM Kontal SA (ex Konstal SA)
Ul Katowicka 104, PL-415-00 Chorzow, Poland
Tel: (+48 32) 241 10 51   Fax: (+48 32) 241 33 97
Web: www.transport.alstom.com

Key personnel
Contact: Teresa Klusek
Freight Product Line Director: Günter Köhler

Products
Freight wagons, including self-discharging hopper wagons; four-axle, six-axle and eight-axle pocket wagons; tank wagons; and specialised wagons for the transport of extra-heavy and extra-long loads, including hollow-platform wagons for transport of extra-long loads.
For further product information see ALSTOM LHB entry in Freight vehicles section.

Contracts
Contracts included delivery of three, 100 freight wagons to First Union Rail Corporation, the leasing division of First Union National Bank.

UPDATED

Arbel Fauvet Rail (AFR)

140 rue du Paradis, F-59500 Douai, France
Tel: (+33 3) 27 93 49 49   Fax: (+33 3) 27 96 07 43
Web: http://www.a-f-r.fr

Key personnel
President and Managing Director: Bruno Flocco
Production Director: Didier Chevalet
Commercial Director: Jean Jomeau

Subsidiaries
Lormafer – maintenance, repair and refurbishment of goods wagons.

Products
Design, manufacture and refurbishment of a wide range of freight wagons and tank containers. Wagon range includes container, open high-sided, covered, tank, hopper, car-carrying (double-deck), shuttle (for Eurotunnel), coil-carrying and chemical tank containers.

Azovmash

1 Mashinostroitelei Pl, 87535 Mariupol, Ukraine
Tel: (+380 629) 56 08 53
Fax: (+380 629) 38 30 31; 38 54 37
e-mail: info@azovmash.com.ua;
   ves@azovmash.com.ua

Key personnel
Chairman: Alexander Savchuk

Background
Originally one of the leading heavy engineering undertakings in the former Soviet Union, Azovmash became a joint stock company in 2000. As well as its rail vehicle building activity, the company produces heavy equipment for the steelmaking industry, mining equipment, cranes and heavy lifting equipment, aircraft refuelling systems, armoured fighting vehicles and consumer products.

Products
Tank wagons of various types, including vehicles for the carriage of petroleum products, hazardous products, foodstuffs, solidifying products and liquefied gases; wagons for bulk products, such as cement carriers and hoppers; open wagons; tank containers.

Babcock BES

Rosyth Business Park, Rosyth, Fife KY11 2YD, UK
Tel: (+44 1383) 41 21 31   Fax: (+44 1383) 41 77 74
Web: http://www.babcockbes.co.uk

Key personnel
Managing Director: Douglas Lindsay
General Manager: Andy Somerville

Background
Babcock BES is a subsidiary of Babcock International Group plc.

Products
Manufacture of freight wagons. In September 2002, Babcock unveiled its Mega 3 piggyback wagon, which commenced trial running on the UK network in 2003 as part of an initiative sponsored by the Strategic Rail Authority to encourage road traffic to rail.

Bharat Bhari Udyog Nigam Ltd (BBUNL)

26 Raja Santosh Road, Alipore, Calcutta 700 027, India
Tel: (+91 33) 24 79 55 35   Fax: (+91 33) 24 79 70 46
e-mail: bbuni@vsni.net.in
Web: http://www.bbuni.com

Key personnel
Chairman and Managing Director: Satish C Gupta

Organisation
Bharat Bhari Udyog Nigam Ltd (BBUNL) is a public sector holding company under the administrative control of the Union Ministry of Heavy Industries and Public Enterprises, Government of India. Its subsidiary units are: Burn Standard Co Ltd (BSCL); Braithwaite & Co Ltd (BWT); Bharat Wagon & Engineering Co Ltd (BWEL) and BBJ Construction Co Ltd (BBJ). BBUNL also acts as the marketing, design and development organisation for these subsidiary units.

Bharat Wagon & Engineering

Bharat Wagon & Engineering Company Ltd (A Government of India undertaking)
C Block, 5th Floor, Maurya Lok, Dak Bungalow Road, Patna 800 001, India
Tel: (+91 612) 22 66 99   Fax: (+91 612) 22 21 47

Key personnel
General Manager: S P Singh

Background
A Government of India undertaking and subsidiary of Bharat Bhari Udyog Nigam Ltd (BBUNL).

Products
Freight wagons.

BMZ

JSC Bryansk Engineering Works
Ulica Ulyanova 26, 241 015 Bryansk, Russian Federation
Tel: (+7 832) 55 86 73; 00 30   Fax: (+7 95) 203 33 95

Key personnel
For full personnel listing, see BMZ entry in *Locomotives and powered/non-powered passenger vehicles* section.

Products
Freight wagons, including grain hoppers and refrigerated wagons. Specialised wagons for the iron and steel industry. Refrigerated containers.

Bradken Rail

2 Maud Street, Mayfield West, New South Wales 2304, Australia
PO Box 105, Waratah, New South Wales 2298, Australia
Tel: (+61 2) 49 41 26 77   Fax: +(61 2) 49 41 26 61
e-mail: rail@bradken.com.au
Web: www.bradken.com.au

Background
Previously ANI Railway Transportation Group. Bradken's rail business structure consists of three strategic business units: freight rolling stock, express parts and maintenance.

Products
The Express Parts business unit supplies spare and renewed parts for freight wagons, passenger vehicles and locomotives, as well as drawgear, track and signal equipment, bogie and brake components. The company stocks thousands of off-the-shelf line items.

UPDATED

Braithwaite

Braithwaite & Company Ltd
5 Hide Road, Calcutta 700 043, India
Tel: (+91 33) 439 79 62; 74 13; 74 15; 67 27; 66 13
Fax: (+91 33) 439 56 07; 76 32
e-mail: braith.company@gems.vsnl.net.in

Branch office
74 Janpath, New Delhi 110 001
Tel: (+91 11) 372 31 44

Works
Clive Works, 5 Hide Road, Calcutta 700 043
Angus Works, PO Angus, Dist Hooghly, West Bengal
Victoria Works, P-61, CGR Road, Calcutta 700 043

Marketing Design and Development Office
Bharat Bhari Udyog Nigam Ltd (BBUNL) (Parent company)
26 Raja Santosh Road, Alipore, Calcutta 700 027, India

Azovmash four-axle tank wagon for petroleum products   0552542

Braithwaite coal hopper wagon for Vietnam Railways 0002416

Braithwaite 24-wheel special-purpose wagon with 182 tonnes payload 0002417

Tel: (+91 33) 24 79 55 35 Fax: (+91 33) 24 79 70 46
e-mail: bbuni@vsni.net.in
Web: http://www.bbuni.com

Key personnel
Managing Director: P K Mukherjee
General Manager (Marketing and Projects):
A K Battacharyya
Manager (Marketing): M K Chakraborty
Deputy Manager Marketing: S K Basu

Background
A Government of India undertaking and subsidiary of Bharat Bhari Udyog Nigam Ltd (BBUNL).

Products
Railway wagons of different gauge systems (all types- general or special purpose), fabricated steel structures for railway bridges, cast steel bogies and centre buffer couplers. All products are ISO 9000 certified.

Burn Standard

Burn Standard Co Ltd
10-C Hungerford Street, Calcutta 700 017, India
Tel: (+91 33) 247 10 67; 17 62; 17 72
Fax: (+91 33) 247 17 88

Works
20-22 Nityadhan Mukherjee Road, Howrah 711 101
Burnpur Works, Burnpur 713 325
Tel: (+91 33) 660 26 01; 5 Fax: (+91 341) 20 85 30

Marketing Design and Development Office
Bharat Bhari Udyog Nigam Ltd (BBUNL) (Parent company)
26 Raja Santosh Road, Alipore, Calcutta 700 027, India
Tel: (+91 33) 24 79 55 35 Fax: (+91 33) 24 79 70 46
e-mail: bbuni@vsni.net.in
Web: http://www.bbuni.com

Key personnel
Chairman: Shri A K Mohapatra
Managing Director: R P Singh
Director, Engineering: I C Sinha

Background
A Government of India undertaking and subsidiary of Bharat Bhari Udyog Nigam Ltd (BBUNL).

Principal subsidiaries
Bharat Brakes & Valves Ltd
22 Gobra Road, Calcutta 700 014
Reyrolle Burn Ltd
99 Dr Abani Dutta Road, Howrah 711 101

Products
Freight wagons.

CAF – Construcciones y Auxiliar de Ferrocarriles SA

Padilla 17 – 6°, E-28006 Madrid, Spain
Tel: (+34 91) 435 25 00 Fax: (+34 91) 436 03 96
e-mail: export.caf@caf.es
Web: www.caf.es

Key personnel
President and Chief Executive Officer:
J M Baztarrica
Managing Directors: A Arizcorreta, A Lergarda
Contact: J Esnaola

Work
Beasain
J M Iturrioz 26, E-20200 Beasain, Spain
Tel: (+34 943) 88 01 00 Fax: (+34 943) 88 14 20

Irún
Calle Anaca 13, E-20301 Irún, Spain
Tel: (+34 943) 61 33 42 Fax: (+34 943) 61 81 55

Zaragoza
Av de Cataluña 299, E-50014 Zaragoza, Spain
Tel: (+34 976) 76 51 00 Fax: (+34 976) 57 26 48

Offices
CAF Argentina SA
Pte Luis Sáenz Peña 31° – 7° piso, 1110 Capital Federal Buenos Aires, Argentina
Tel: (+51 11) 43 83 20 06 Fax: (+54 11) 43 81 48 37
e-mail: cafadministracion@cafarg.com.ar

CAF Brasil Industria e Comercio
Rua Pedroso Alvarenga 58 conj 52, CEP 04531-000 São Paulo, Brazil
Tel: (+55 11) 31 67 17 20 Fax: (+55 11) 30 79 87 62
e-mail: cafsaopaulo@cafbrasil.com.br

CAF Mexico SA de CV
Prolongación Uxmal 988, Col Sta Cruz Atoyac, 03310 Mexico DF, Mexico
Tel: (+52 55) 56 88 75 43
Fax: (+52 55) 56 88 11 56
e-mail: cafmex@prodigy.net.mx

CAF USA Inc
1401 K Street NW, Suite 803, Washington DC 20005-3418, USA
Tel: (+1 202) 898 48 48 Fax: (+1 202) 216 89 29
e-mail: cafusa@cafusa.com

Products
Freight wagons. Designs according to UIC, AAR or individual specifications.
Contracts include 200 pocket wagons and 223 container wagons for Intercontainer and 53 hopper wagons for Israel.

UPDATED

Ferriere Cattaneo SA

Via Ferriere 12, CH-6512 Giubiasco, Switzerland
Tel: (+41 91) 850 91 91
Fax: (+41 91) 850 91 92
e-mail: fcsa@ferrierecattaneo.ch
Web: www.ferrierecattaneo.ch

Key personnel
Managing Director: Aleardo Cattaneo
Technical Director: Hans Tandetzki
Engineering Director: Erik Fregni
Freight Wagon Team Manager: Eugenio Moro

Products
Freight wagons.
Product range includes freight wagons for intermodal transport such as: the double pocket wagon T3000 for the transport of semitrailers (also Mega type included) and containers. Eight axle ultra low floor wagon (rolling highway concept) for the transport of lorries; the double Mega-wagon for the transport of semitrailers, swapbodies for ss-traffic up to 18 tonne/axle; container wagons suitable for the lateral transhipment type Cargo Domino. Other wagons can be manufactured according to customer specification.

UPDATED

China Northern Locomotive and Rolling Stock Industry (Group) Corporation (CNR)

11 Yangfangdian Road, Haidian District, Beijing 100038, China
Tel: (+86 10) 51 86 23 70 Fax: (+86 10) 51 86 23 74
e-mail: loriciec@cnrgc.com.cn
Web: www.cnrgc.com

Key personnel
Chairman and Managing Director: Wang Tai-Wen
President: Cui Dianguo
Vice-President: Zhao Guangxing
President of LORIC Import & Export Corporation Ltd: Cao Guo-Bing
Vice-President of LORIC Import & Export Corporation Ltd: Chen Dayong
Senior Engineer of LORIC Import & Export Corporation Ltd: Yang Xiang-Jing

Products
Development, design, engineering, production, sales, installation, refurbishment, maintenance and after sales service for freight wagons including covered wagons; open-top wagons for coal, ore, steel and timber; hopper wagons for grain, ore, fertiliser; flat wagons; container flat wagons; depressed centre flat wagons; double-deck flat

CNR Type C80 aluminium-bodied open-top coal hopper 1048666

CNR Freight wagons

Type	Model	Use	Tare weight (t)	Loading capacity (t)	Effective capacity (m³)	Year first built	Builders
Covered wagons	P64A		25.3	58.0	135.0	1996	Qiqihar
	P64GK		24.0	60.0	135.0	2002	Qiqihar
	P65		22.8	58.0	135.0	1998	Qiqihar
	Pxy		26.5	73.0	158.0	2003	Qiqihar
Open-top wagons	C64		22.5	61.0	73.3	1991	Qiqihar
	C64K		23.0	61.0	73.3	2002	Qiqihar
	C76C		22.9	76.0	82.0	1998	Qiqihar
	C80[1]	Aluminium alloy body	20.0	80.0	87.0	2003	Qiqihar
Hopper wagons	K13NK	Ballast hopper	23.6	95.0	36.0	2002	Taiyuan
	L18	Grain	25.0	67.5	85.0	1996	Qiqihar
	C32	Coal	24.5	75.3	106.0	2002	Qiqihar
	C35	Grain	24.7		86.5	2000	Qiqihar
	HFE	Grain			105.0	2004	Qiqihar
Flat wagons	C3		19.5	80.0		1999	Qiqihar
	C5	Five-pack	61.0	190.0		2004	Qiqihar
	C10	Double-deck	22.0	78.0		2001	Qiqihar
	N17A		19.7	60.0		1996	Qiqihar
Tank wagons	G11J	Caustic soda	20.6	63.0	66.4	1992	Xi'an
	G17B	Viscous	21.0	62.0	69.7	1998	Xi'an
	G70K	Light oil	40.9	40.3	96.0	2002	Xi'an
	GY95SK	Liquefied petroleum gas	38.9	40.3	96.0	2002	Xi'an
	GY95AK	Liquefied petroleum gas				2002	Xi'an
Heavy-duty loads wagons	D12	Depressed centre	119.0	120.0		1994	Habin
	D18A	Depressed centre	226.0	180.0		1991	Habin
	D30A	Schnabel wagon	226.0	300.0		1997	Qiqihar
	D32	Depressed centre		320.0		2003	Qiqihar
	D38	Schnabel wagon		380.0		1995	Qiqihar
Containers	J24A	Liquid tank	4.17	26.3	24.86	1997	Xi'an
	1D	Oil, chemicals	2.365	12.0	7.5	1997	Xi'an

[1] Aluminium alloy body.

wagons; tank wagons for all types of liquids and chemicals; tipper wagons; ballast wagons; cabooses and hoists; as well as liquid tank containers.

The annual output is over 20,000 vehicles, most of them 1,435 mm gauge. Thousands of freight wagons in various track gauges have been delivered customers in countries and regions which include Albania, Australia, Botswana, Brazil, Cuba, Hong Kong, Myanmar, Nigeria, Sri Lanka, Taiwan, Tanzania, Vietnam and Zambia,

Contracts

Australia: In March 1999 CNR received an order from ATN-Access, Australia, to supply three Type C3 25-tonne axleload container flat wagons. The bogies are of a diagonal linked side-frame design with cushioned bearing saddles, providing for exceptional ride quality and minimal flange wear. They accept the standard AAR wheel set of 250 kN axle force, as used in Australia. The cushioned bearing saddles allow for steering and, as a consequence, the wagons run very freely and quietly.

In 2000 Australian customers subsequently awarded CNR a contract for 50 Type C3 container flat wagons with 85-tonne loading capacity and 44 C-35 grain hopper wagons, in 2001 for one Type C37 ballast wagon, three Type C10A double-stack container well wagons and 50 Type C32 coal hopper wagons, and in 2002 for 105 Type C35 grain hopper wagons.

In 2004 CNR was awarded a three new contracts to supply 29 Type C5 five-pack articulated wagons and 75 Type C3 TEU skeletal container wagons.

Brazil: In October 2003 CNR won a contract with CVRD, Brazil, to supply 200 grain hopper wagons. They were delivered in January 2004.

China: CNR delivered over 2,000 Type C80 tub coal open-top aluminium-bodied wagons for MOR in 2004.

CNR delivered 12 ballaster wagons to Taiwan in January 2004.

UPDATED

CNR Type C-10 double-stack container flat wagon for Australia 0583235

Model C76C tub coal open-top wagon for CNR 0552548

Model C37 ballast hopper wagon for Australia by CNR 0595039

CNR Type C5 five-section articulated flat wagon
NEW/1135831

China South Locomotive and Rolling Stock Industry (Group) Corporation (CSR)

11 Yangfangdian Road, Haidian District, Beijing, China
Tel: (+86 10) 63 98 47 70   Fax: (+86 10) 63 98 47 66
e-mail: csrft@csrgc.com.cn
Web: http://www.csrgc.com.cn

Key personnel
General Manager: Zhao Xiaogang

Background
CSR was established in 2000 as a result of the division into two regional groups of the former China National Railway Locomotive and Rolling Stock Industry Corporation (LORIC). The corporation comprises 24 state-owned enterprises employing around 116,000 staff. Some facilities are listed as part of CSR and of its northern China counterpart, China North Locomotive and Rolling Stock Industry (Group) Corporation (CNR).

Works
See China South Locomotive and Rolling Stock Industry (Group) Corporation (CSR) entry in *Locomotives and powered/non-powered passenger vehicles* section.

Products
Freight wagons of various types, including: general-purpose open wagons with a 23-tonne axle-load; 100-tonne gross aluminium-bodied hopper wagons; Type P64 covered wagons with 1-tonne axle-load; Type D2 depressed centre flat wagon; and Type C62A 82-tonne gross all-steel open wagon.

Contracts
A recent contract covers the supply of 1,300 freight wagons to Pakistan Railways. Other export orders were for iron ore carriers for Brazil and flat wagons for Tazara, Tanzania/Zambia.

Cimmco International

A division of Cimmco Birla Ltd
D-180 Okhla Phase 1, New Delhi 110 020, India
Tel: (+91 11) 331 43 83; 84; 85
Fax: (+91 11) 332 07 77; 372 35 20

Main works
Wagon Division, Bharatpur 321 001, Rajasthan, India

Key personnel
Chairman: S Birla
President: R Upadhaya
General Manager: M P Gupta
Marketing Manager: G Sodhi

Products
Design and manufacture of freight wagons. The company manufactures wagons to meet special material handling applications and has supplied over 30,000 wagons in India and abroad. The range includes: covered wagons, bottom and side discharge wagons; tank wagons with heating arrangements for transport of all types of liquid.

CNR Qiqihar Railway Rolling Stock (Group) Co Ltd (QRRS)

10 ZhongHuaDongLu Road, Qiqihar 161002, China
Tel: (+86 452) 293 84 72; 293 84 99
Fax: (+86 452) 251 67 23
e-mail: grrsintl@qrrs.com.cn
Web: www.qrrs.com.cn

Key personnel
Chairman of Board of Directors and General Manager: Wei Yan
Chief Engineer: Yu Lianyou
International: Liu Dezeng; Ms Zhang Xianbin

Background
QRRS is a member of CNR.

Products
Freight wagons with speeds of up to 120 km/h, including P64GK, and P65 covered wagons, Type C3 and C10A and articulated container wagons supplied to Australia; Type D38 380 tonne special heavy duty wagons; Type L18 grain hopper wagons supplied to Chinese Railways; Type C35-100 tonne hopper wagons supplied to Australia; Type HFE 100 tonne grain hopper wagons supplied to Brazil; Type C32 120 tonne coal hopper wagons supplied to Australia.

Equipped with rotary couplers and high-capacity draft gear and other fittings.

UPDATED

CostaRail Srl

Viale 4 Novembre, I-22041 Costamasnaga, Como, Italy
Tel: (+39 031) 86 94 11
Fax: (+39 031) 85 53 30
e-mail: hkastner@costarail.it

Key personnel
Commercial Manager: Heinz Kastner

Background
Formerly Costaferroviaria, the company went into temporary receivership and was taken over by Rail Services International Group, for which RSI Italia SpA is the operative leader. In October 2004, CostaRail Srl was established, re-launching all existing Costaferroviaria activities, including orders, trademarks, patents and projects.

Products
Freight wagons.

Orders have included 520 BA655/656 23.5 tonne/ axle bogies and 400 BD682/683 22.5 tonne/axle bogies and 200 Shimmns wagons for DB Cargo, 475 car transport wagons for STVA and 52 hoppers and 108 flat wagons for Spanish infrastructure operator GIF.

UPDATED

Cromweld Steels

Cromweld Steels Ltd
The Old Vicarage, Tittensor, Stoke-on-Trent ST12 9HY, UK
Tel: (+44 1782) 37 41 39
Fax: (+44 1782) 37 33 88
e-mail: enquiries@cromweld.com
Web: http://www.cromweld.com

Key personnel
Contacts: Jacqueline Redman, Neil Cooper

Subsidiary
Cromweld Steels
PO Box 1500, Cornelius, North Carolina 28031, USA
Tel: (+1 704) 896 81 14
Fax: (+1 704) 896 81 15

Products
Supplier of 3CR12, a low-cost stainless steel for wagon construction. Typical applications include structural frames and subframes, flooring, walls and outer skinning/cladding, as well as applications for the construction of rail infrastructure.

Daewoo Heavy Industries Ltd

PO Box 7955, Daewoo Centre Building 22nd Floor, 541, 5-Ga, Namdaemun-Ro, Jung-gu, Seoul, South Korea
Tel: (+82 2) 726 31 79; 31 82
Fax: (+82 2) 726 31 86; 756 26 79

Key personnel
For full personnel listing, see Daewoo Heavy Industries Ltd entry in *Locomotives and powered/ non-powered passenger vehicles* section.

Products
Freight wagons.

Daewoo has supplied more than 10,000 freight wagons to operators in over 30 countries.

Recent contracts include the supply of 117 container flat wagons and 24 ballast hopper wagons for Korean National Railroad.

Dalian Locomotive and Rolling Stock Works

51 Zhongchang Street, Shahekou District, Dalian, 116022, China
Tel: (+86 411) 460 20 43   Fax: (+86 411) 465 42 45
Web: http://www.dloco.com

Key personnel
For full personnel listing, see Dalian Locomotive and Rolling Stock Works entry in *Locomotives and powered/non-powered passenger vehicles* section.

Background
See Dalian Locomotive and Rolling Stock Works entry in *Locomotives and powered/non-powered passenger vehicles* section.

Products
Freight wagons, including Type C64 and C62B gondola wagons.

Recent contracts include the supply of 900 C64 and 100 C62B gondola wagons for Sino Railways Construction Development Centre, People's Republic of China.

W H Davis Ltd

PO Box 3, Langwith Junction, Mansfield NG20 9SA, UK
Tel: (+44 1623) 74 26 21   Fax: (+44 1623) 74 44 74
e-mail: maco.burg@whdavis.co.uk
Web: http://www.whdavis.co.uk

Key personnel
Chairman: D Sharpe
Sales Director, Wagons: M S Burge
Company Secretary: M A Jackson
Sales Director, Containers: D G Bradley

Products
Freight wagons, containers and swapbodies.

Contracts
Contracts include the supply of rolling stock to Kenya Railways, Bardon London Ltd, Cleveland Potash, Marcon Topmix, Indonesia and Network Rail.

VERIFIED

E G Steele & Co Ltd

25 Dalziel Street, Hamilton ML3 9AU, UK
Tel: (+44 1698) 28 37 65   Fax: (+44 1698) 89 15 50
e-mail: egsteelecoltd@btinternet.com

Key personnel
Managing Director: David Steele
Logistics Manager/Director, Export Sales: Ian Hood
Quality Manager: Cameron Gibson

Products
Suppliers of spare parts for freight wagons operating in the UK, especially wagons of a continental UIC design which are able to operate in most European countries. Stock of 300 different items of UIC parts include: drawgear, coupling systems, springs, manganese wear plates and bogie parts.

UPDATED

EKA Limited

Valkyrie House, 38 Packhorse Road, Gerrards Cross, SL9 8EB, UK
Tel: (+44 1753) 88 98 18　Fax: (+44 1753) 88 00 04
e-mail: eka@ekalimited.com
Web: ekalimited.com

Key personnel
Managing Director: W O Forster
Sales Liaison Manager: J E Fadelle

Products
The EKA 'Stevedore' side-loading semi-trailer is ideally suited to handling ISO containers at lightly used road/rail transfer yards where the cost of expensive fixed gantries or heavy forklift trucks cannot be justified. There is also no need for separate vehicles for transporting containers by road. With the 'Stevedore', one man can collect a container at a transfer yard and deliver to a destination where it can be grounded for emptying or stacked for storage. Individually controlled hydraulic stabilisers ensure total stability and ease of levelling. There are versions for 20 and 40 ft long ISO containers weighing up to 28 tonnes.

EKA Simple Rail Transfer Equipment (SRTE), comprises a demountable sideloader which unloads 20 ft ISO containers/flatracks from rail wagons alongside. EKA has supplied 24 SRTE units to the Ministry of Defence, UK, and one SRTE to the Ministry of Defence, Ireland. Several Stevedores have also been supplied in the UK.

UPDATED

Fabryka 'Wagon' SA

ul Wroclawska 93, PL-63-400 Ostrøw Wielkopolski, Poland
Tel: (+48 62) 595 39 13　Fax: (+48 62) 591 27 29
e-mail: mr@fabryka-wagon.pl; zntk@osw.pl
Web: http://www.fabryka-wagon.pl

Background
Established in 1920, Fabryka 'Wagon' SA is the former Ostrøw rolling stock repair workshop of Polish State Railways.

In mid-2004 the company was reported to be in receivership, with three potential buyers bidding for part or all of the business.

Products
Freight wagons, including coal hoppers, bottom discharge wagons, container flats, pocket wagons for piggyback traffic and sliding wall covered wagons. Freight wagon components are also produced.

NEW ENTRY

Fabryka Wagonów Gniewczyna

Gniewczyna 591, PL-37 203 Gniewczyna, Poland
Tel: (+48 16) 648 83 64　Fax: (+48 16) 648 85 87
e-mail: admin@gniewczyna.pl
Web: http://www.gniewczyna.pl

Products
Open wagons, self-unloading mineral wagons; telescopic hood coil and steel carriers; telescopic hood general cargo wagons; pocket wagons for semi-trailers; container flats; ACTS-type container wagons.

The company also undertakes repairs to wagons of all types.

FIREMA Trasporti SpA

Headquarters
Via Provinciale Appia, Località Ponteselice, I-81100 Caserta, Italy
Tel: (+39 0823) 09 71 11　Fax: (+39 0823) 46 68 12
Web: http//www.firema.it

Key personnel
Chairman: A de Benerdictis
Managing Director: L Rigno

Commercial Manager: S d'Arminio
Marketing Manager: M Fantini

Commercial and technical offices
Via Triboniano n 220, I-20156 Milan, Italy
Tel: (+39 02) 23 02 02 23　Fax: (+39 02) 23 02 03 00

Products
Freight wagons. FIREMA has completed a prototype 'Twist Wagon', designed to simplify the loading and unloading of 'rolling highway' trains. The wagon deck pivots around its centre, and integral end ramps allow lorries to be driven on or off without the need for a special platform.

Recent orders include 110 Type VFaccs bogie hopper wagons, 20 Type DDm double-deck car-carrying wagons for Italian Railways (FS) and 100 bogie hopper wagons for DB Cargo.

FM Industries

FM Industries
(a Division of Progress Rail Services)
8600 Will Rogers Blvd, Fort Worth, Texas 76140, USA
Tel: (+1 817) 293 42 20　Fax: (+1 817) 551 58 10

Key personnel
President: James Kingerski

Parent company
Progress Rail Services (qv)

Products
Hydraulic end-of-wagon cushioning devices.

FreightCar America Inc

2 North Riverside Plaza, Suite 1250, Chicago, Illinois 60606, USA
Tel: (+1 814) 533 50 00　Fax: (+1 814) 533 50 10
e-mail: webmaster@freightcar.net
Web: www.freightcaramerica.com

Key personnel
Chairman, President and Chief Executive Officer: John E Caroll Jr
Vice-President, Marketing: Edward J Whalen
Vice-President, Engineering: James D Hart

Background
Previously called Johnstown America Corporation, Freight Car America Inc is a freight wagon builder, operating manufacturing facilities in Johnstown, Pennsylvania, Danville, Illinois and in Roanoke, Virginia. Its subsidiary, Chicago-based JAIX Leasing Company provides lease financing and rail wagon management for new and rebuilt wagons.

Products
Manufacture and repair of freight wagons, including open top aluminium gondola and hopper wagons used for the haulage of coal, spine wagons for intermodal traffic and steel and aluminium, open and covered hopper wagons used for the haulage of grain, fertiliser, iron ore, minerals and other bulk products.

FreightCar America, Inc has also introduced an aluminium vehicle carrier wagon used to haul automobiles and small trucks as well as a specialised spine wagon used to haul steel slabs and a transverse trough spine wagon used to haul steel coils.

UPDATED

Freudenberg Schwab GmbH

Postplatz 3, D-16761 Hennigsdorf, Germany
Tel: (+49 3302) 206 20　Fax: (+49 3302) 20 62 77
e-mail: info@freudenberg-schwab.de
Web: http://www.freudenberg-schwab.de

Key personnel
Chief Executive Officers: Dr Detlef Cordts; Peter Kofmel; Jörg Sost
Quality Manager: Leo S E Lang
Marketing Director: Bernd Werner

Products
Vibration control: rubber/metal elastic elements, standard as well as specially designed. Include resilient bushes, spherical bearings, buffers, conical mountings, V-mountings, machine mountings, rectangular mountings and hydraulic bushes.

Innovative products for vibration control include hydraulic mountings and bushes for speed-dependent stiffness, such as axle guide bearings for rail vehicles on tightly curved track, and systems for reduction of structure-borne noise such as equalisation of vibrations from wheel flats.

Fuji Car Manufacturing Co Ltd

3 Shoho Building, 2-2-3 Nishishinsaibashi, Chuo-ku, Osaka 542, Japan
Tel: (+81 6) 213 27 11　Fax: (+81 6) 213 40 71

Key personnel
For full personnel listing, see Fuji Car Manufacturing Co Ltd entry in *Locomotives and powered/non-powered passenger vehicles* section.

Products
Freight wagons, ladle wagons and bogies.

Ganz Motor Kft

Postafiók 263, H-1475 Budapest, Hungary
Tel: (+36 1) 210 11 50　Fax: (+36 1) 313 48 09
e-mail: ganzmoto@mail.datanet.hu

Key personnel
Managing Director: Péter Pétervári
Sales Director: Martinkó András
Technical Director: Orsovai József

Products
Specialised freight wagons. Recent deliveries include 22 sliding wall wagons for Delacher & Co, Austria; 20 similar wagons for Györ-Sopron-Ebenfurt Railway (GySEV), Hungary; 30 Type Saadkms ROLA intermodal wagons for Hungarian State Railways (MÁV), manufactured under

FreightCar America open top hopper wagon for hauling coal　　0129561

licence from Waggonfabrik Talbot of Germany; 82 coil transport wagons manufactured for Adtranz-Siegen, Germany with 200 more on order and 60 Type Shimms telescopic roof steel coil transport wagons for MÁV Hungary. Three twin-car covered car transport wagons have been delivered to Austria Car Rail Logistics of Austria, running under the GySEV branding.

Goša

Goša Holding Corporation
Equipment and Vehicle Industries
Kolarčeva 8, 11000 Belgrade, Federal Republic of Yugoslavia
Tel: (+381 11) 63 36 50; 63 30 31
Fax: (+381 11) 63 45 31

Products

Tank wagons; refrigerated wagons and special purpose wagons.

Graaff

Graaff Transportsysteme GmbH
Heinrich-Nagel Strasse 1, D-31008 Elze, Germany
Tel: (+49 5068) 182 15
Fax: (+49 5068) 182 81
e-mail: aduchatsch@graaff-transportsysteme.de

Key personnel

Financial Director: Wolfgang Hassepass
Production Director: Günter Homes
Sales Director: Eberhard Miehlke

Products

Stainless steel hopper wagons 96 m³ and 90 m³ for the transport of cereals, sugar according to hygienic regulations. Tanks wagons for chemical products; container wagons; coil wagons; insulated and refrigerated wagons with shock-absorbing systems for load protection and electric heating and air ventilation, sliding doors; tarpaulin-covered wagons; 31 m car carriers; insulated and refrigerated containers, low-floor container carrying wagons; swab body designs for short-sea; specialised containers for telecommunication; sandwich panels. Insulated customised sandwich panels and kits for truck bodies.

Developments

New stainless steel bogie hopper wagons covering HACCP regulations for single block train transport designs for the transport of grain and in a specially modified version for sugar. Clients include SBB and DB.

Two-axle refrigerated sliding door wagon for the transport of chilled food and temperature-sensitive goods. Upgraded 45 ft refrigerated swap-body designed especially for short-sea crossings. New German-registered tank wagon designed for operation between Spanish broad gauge and standard gauge.

The Greenbrier Companies

One Centerpointe Drive, Suite 200, Lake Oswego, Oregon 97035, US
Tel: (+1 503) 684 70 00   Fax: (+1 503) 684 75 53
Web: www.gbrx.com

Key personnel

Chairman of the Board: Alan James
President and Chief Executive Officer:
   William A Furman
Senior Vice-President, Marketing and Sales:
   Robin D Bisson
Senior Vice-President, Chief Financial Officer:
   Larry G Brady

Subsidiaries

Gunderson-Concarril SA de CV
Domicilio Conocido, Zona Industrial, Cd Sahagun, Edo de Hgo, CP 43990 Mexico
Gunderson Inc
4350 Northwest Front Avenue, Portland, Oregon 97210, US
Tel: (+1 503) 972 57 00   Fax: (+1 503) 972 59 86

Worldwide offices
Greenbrier Europe Office
Welterstrasser 60 D-57072, Siegen, Germany
Tel: (+49 271) 70 22 53   Fax: (+49 271) 741 28 94

Greenbrier London Office
Bonnie Gillespie Greenbrier Europe
25 Hasker Street, London SW3 2LE, UK
Tel: (+44 20) 75 84 18 98

Background

Greenbrier manufactures freight wagons at Gunderson Inc (Portland, Oregon), TrentonWorks Ltd (Nova Scotia, Canada), WagonySwidnica (Poland), Greenbrier-Concarril (Mexico). In January 2000, Greenbrier acquired the German-based freight wagon manufacturing interests of Adtranz from DaimlerChrysler.

In 2004 Greenbrier acquired Bombardier Transportation's interest in Greenbrier-Concarill LLC and its manufacturing subsidiary Gunderson Concarril SA de CV. The transaction has resulted in Greenbrier owning 100 per cent of these companies. Gunderson-Concarril will continue to lease a portion of Bombardier's Sahagun facility and production of freight cars will continue at the same location.

UPDATED

Graaff insulated sliding door wagon   1066400

The Gregg Company

The Gregg Company Ltd
15 Dyatt Place, Hackensack, New Jersey 07601, USA
Tel: (+1 201) 489 24 40   Fax: (+1 201) 592 02 82

Key personnel
Chairman: RT Gregg

Products

Covered and open wagons, mineral wagons, refrigerator wagons, tank wagons, wagons for transport of road vehicles, container wagons, bulk powder wagons, flat wagons, ballast and aluminium wagons, conventional and self-steering bogies for freight and passenger vehicles.

VERIFIED

Graaff grain hopper wagon
1066399

Gunderson Inc

A subsidiary of The Greenbrier Companies
4350 Northwest Front Avenue, Portland, Oregon
97210, US
Tel: (+1 503) 972 57 00   Fax: (+1 503) 972 59 86

Key personnel
Chairman: C Bruce Ward
President: L Clark Wood
Chief Engineer: Gary S Kaleta

Products
Freight wagons. Conventional freight wagons
include box cars, centre-partition lumber cars, flat
cars and gondolas. Principal intermodal products
are the Maxi-Stack III, the Husky-Stack family.

The Maxi-Stack III uses 125-short ton bogies
and can handle 20, 24, 40, 45 or 48 ft containers in
all wells, as well as any size container from 20 to
53 ft on the top level. The Husky-Stack is a stand-
alone version of the double-stack wagon, designed
for very heavy container loads up to 166,000 lb
(75,300 kg).

Conventional car types include box car wagons
for paper and wood industries, refrigerated
wagons for frozen foods, gondolas, covered
hopper wagons, woodchip wagons, flat wagons
and other speciality cars. Gunderson also repairs
and refurbishes freight wagons, wheels and axles.

Other wagon products include the heavy capacity
Maxi-Stack® III double-stack wagon, Maxi-Stack®
AP (all-purpose) double-stack wagon and the Auto-
Max® high-capacity automobile carrier wagon.

Gunderson builds ocean-going barges at its
marine facility at the same location in Portland,
Oregon.

NEW ENTRY

Gunderson-Concarril

Gunderson-Concarril SA de CV
A subsidiary of The Greenbrier Companies Inc
Domicilio Conocido, Zona Industrial, Cd Sahagun,
Edo de Hgo, CP 43990 Mexico

Mexico sales office
Greenbrier de Mexico, S de RI de CV
Boulevard Manuel Avila Camacho, 340 Piso 11
Col Lomas de Chapultepec, Mexico, DF 11000
Tel: (+52 5) 202 49 25   Fax: (+52 5) 202 53 87
Web: www.gbrx.com

Key personnel
General Manager: Victor Silva

Background
Greenbrier-Concarill LLC was established in 1998 as
a joint venture between The Greenbrier Companies
and Bombardier Transportation. Gunderson-Concarill

Hanjin bulk cement wagon　　0002425

SA de CV is the manufacturing subsidiary through
which rail freight cars are produced for the North
American market. In 2004 Bombardier sold its
50 per cent shareholding in the company to the
Greenbrier Companies giving Greenbrier 100 per
cent ownership of both Greenbrier-Concarril and its
subsidiary Gunderson Concarril.

Products
Freight wagons. Conventional freight wagons
include box, flat and gondolas. Gunderson-
Concarril's mill gondola car is designed, engineered
and built to meet the demands of hauling heavy
loads such as steel pipe and steel scrap. It features
extra strong ends, covers, side posts, top chords,
and steel decking. This car meets AAR specifications
for construction and interchange requirements.

UPDATED

Hanjin Heavy Industries Co Ltd

118 Namdaemunro-2-Ga, Chung-ku, Seoul, South
Korea
Tel: (+82 2) 728 54 41; 54 20
Fax: (+82 2) 755 09 28; 756 54 55

Key personnel
For full personnel listing, see Hanjin Heavy Industries
Co Ltd entry in *Locomotives and powered/non-
powered passenger vehicles* section.

Products
Freight wagons, including covered wagons,
flat wagons, hopper wagons, gondola cars, ore
wagons, box cars and tank wagons. Rolling stock
components.

Contracts include 22 cement hopper wagons
for Thailand in 1997. Deliveries overall have passed
the 5,000 mark.

Hindusthan Engineering & Industries Ltd (HEI)

Mody Building, 27 Sir RN Mukherjee Road, Calcutta
700 001, India
Tel: (+91 33) 22 48 01 66
Fax: (+91 33) 22 48 19 22; 22 43 56 07
e-mail: hindus@cal2.vsnl.net.in

Works
Santragachi Plant (SP)
PO Box Jagacha 711 311, District Howrah, West
Bengal, India

Key personnel
Executive Director: M L Lohia

Products
Freight wagons, including open wagons, covered
wagons, tank wagons, intermodal wagons, special
purpose wagons including container flat wagons
and bottom discharge wagons for the transport of
cement, coal and fertiliser.

UPDATED

Holland Company LP

1000 Holland Drive, Crete, Illinois 60417-2120,
USA
Tel: (+1 708) 672 23 00
Fax: (+1 708) 672 01 19
e-mail: postmaster@hollandco.com
Web: http://www.hollandco.com

Key personnel
President: Phil Moeller
Vice-President, Rail Mechanical Group:
　Len O'Kray
General Manager, Track Testing Services:
　Robert Madderom
International Sales Manager: Billy Hedrick
General Manager, Mobile Welding Operations:
　Mark Rovnyak
Controller: Frank Francis
General Manager, Equipment Division:
　Robert Norby

Gunderson-Concarril mill gondola wagon　　0063721

Products

Anti-wear components for wagons; product protection systems for steel coils and paper rolls, load securement systems for commercial and military vehicles; container securement locks for securing intermodal containers to rail wagons and for double-stacking.

Hyundai

Hyundai Precision & Ind Co Ltd
Hyundai Building, 140-2 Gye-Dong, Chongro-ku, Seoul, South Korea
Tel: (+82 2) 719 06 49   Fax: (+82 2) 719 07 41

Key personnel

For full personnel listing, see Hyundai entry in *Locomotives and powered/non-powered passenger vehicles* section.

Products

General freight wagons; special freight wagons.

INKA

Head office and factory

PT (Persero) Industri Kereta Api (PT INKA)
Jalan Yos Sudarso No 71, Madiun 63122, Indonesia
Tel: (+62 351) 45 22 71
Fax: (+62 351) 45 22 75
e-mail: Sekretariat@inka.web.id
Web: www.inka.web.id

Representative Office

Arthaloka building, 3rd Floor, Jalan Jend, Sudirman Kav 2, Jakarta
Tel: (+62 21) 251 44 24
Fax: (+62 21) 251 44 23
e-mail: inkajkt@cbn.net.id

Key personnel

President: Ir Roos Diatmoko
General Manager, Railway Rolling Stock Division:
   Suryanto
Finance and Administration Director:
   Drs Udin Supriatman
General Manager of Technology:
   Ir M Harsan Badawi
Engineering Manager: Ir Gunesti Wahyu
Design Manager: Ir Indarto Wibisono
Marketing Manager: Ir M Dedi Tarmidi
Business Development Manager:
   Ir Muchlis Budiman
Procurement Manager: Soedjito Taathadi

Products

Freight wagons. Up to 300 units are produced annually.
   Indonesian Railways has taken delivery of 158 coal hopper wagons; 70 ballast hopper wagons have been ordered by State Railways of Thailand.

UPDATED

Interlok Bahnconsulting Schmidtendorf

Bruchsaler Strasse 3, D-10715, Berlin, Germany
Tel: (+49 177) 302 42 53
Fax: (+49 177) 993 02 42 53
e-mail: office@interlok.info
Web: http://www.interlok.info

Key personnel

General Sales Manager: Hermann Schmidtendorf
General Manager Rail Department: Uwe Liedecke

Products

Remanufactured, modernised standard gauge freight cars (UIC, RIV) mainly from former Eastern Europe, sale, rent, such as ballast cars, dump cars Fakks 418Vg platform cars; remanufactured modernised standard gauge passenger/sleeping/dining coaches; rails, switches, sleepers, second-hand

Interlok Bahnconsulting Type 418Vg side-tipping wagon for Prignitzer Eisenbahn Cargo (PEC) Ltd, Germany
0579330

for standard gauge; second-hand and new for narrow gauge; second-hand railway maintenance machines. Trade agent for Interlok Ltd and Kaniewski ZUR, Poland, Mitteldeutsche Gesellschaft für Metallhandel und Anlagenverwertung, Germany.

Contracts

20 418Vg dump cars for Prignitzer Eisenbahn Cargo (PEC) Ltd, Germany.

iQR

Railcentre 1, 305 Edward Street, Brisbane, Queensland, Australia 4000
GPO Box 1429, Brisbane, Queensland 4001
Tel: (+61 7) 32 32 33 90   Fax: (+61 7) 32 35 33 46
e-mail: sales@iqr.com.au
Web: www.iqr.com.au

Key personnel

General Manager: Michael Walsh
Marketing and Business Development Manager:
   Peter Harris
Sales Manager: Youfa Chen

Background

iQR was previously Queensland Rail Consultancy Services.

Products

iQR designs, manufactures and supplies a range of specialised containers to increase the efficiency of the haulage of coal, minerals and containerised and bulk freight. iQR offers a range of different size segregation units which enables the carriage of multiple types of dangerous goods within a single road or rail transportation unit. iQR's light powder containers feature a proprietary unloading system which is compatible with standard discharge equipment and enables powders such as cement and fly ash to be carried in non-pressurised containers, offering increased carrying capacity. iQR can also custom design fit-for-purpose containers to clients' requirements.

NEW ENTRY

IRECO LLC

805 Golf Lane, Bensenville, Illinois 60106, USA
Tel: (+1 630) 741 01 55   Fax: (+1 630) 595 06 46

Key personnel

President and Chief Executive Officer:
   Harold R O'Connor
Vice-President: Robert S Grandy
Executive Vice-President: Robert Holden

Products

Components for freight wagons including: AEI tag bracket, adjustable air brake hose support and bottom rod safety support, autorack door opener, brake head adapter, bulkhead safety

cable, centrebeam cable assembly, container guide, cotter lock and guard (self-locking), draft key retainers, edge protector (lumber) hatch cover lock (combination hinge-lock and lock with hinge and retainer hook), hitch lockout device, lading anchors (various). Pipe anchors (various), bulkhead webbing and securement systems.

Jessop

Jessop & Company Ltd, Calcutta
63 Netaji Subhas Road, Calcutta 700 001, India
Tel: (+91 33) 243 20 41; 243 34 20
Fax: (+91 33) 243 16 10

Key personnel

For full personnel listing, see Jessop entry in *Locomotives and powered/non-powered passenger vehicles* section.

Products

Freight wagons, including tank wagons and covered wagons. Diesel breakdown cranes.

K & M

K & M Industrie Metalmeccaniche srl
Via Ugo La Malfa 6, I-90146 Palermo, Italy
Tel: (+39 091) 16 80 05 11
Fax: (+39 091) 16 80 05 16
e-mail: km@telegest.it

Corporate background

The company previously traded as Keller SpA.

Products

Freight wagons including high-capacity wagons; wagons with sliding doors; intermodal wagons; infrastructure maintenance vehicles.

Kawasaki Heavy Industries Ltd

World Trade Center Building, 4-1 Hamamatsu-cho 2-chome, Minato-ku, Tokyo 105, Japan
Tel: (+81 3) 34 35 25 88
Fax: (+81 3) 34 35 21 57; 34 36 30 37

Key personnel

President, Rolling Stock Company:
   Masashi Segawa
General Manager, Marketing and Sales Division:
   Akira Hattori
Deputy General Manager, Marketing and Sales
   Division: Ryoshi Hirano
Senior Manager, Overseas Marketing Department:
   Hiroshi Murao

Products

Freight wagons.

UPDATED

Keller Elettromeccanica SpA

Zona Industriale I-09039, Villacidro, Cagliari, Italy
Tel: (+39 070) 933 62 02   Fax: (+39 070) 933 62 44
e-mail: info@keller.it
Web: http://www.keller.it

Key personnel
For a full list of personnel, see Keller Elettromeccanica SpA entry in *Locomotives and powered/non-powered passenger vehicles* section.

Products
Freight wagons including: Type Kgps platform wagons; Type Vfaccs hopper cars; Type Laadkks flat wagons; Type Tadgs hopper wagons; Type Habillss wagons with sliding light alloy walls.

K Industrier AB

Stora Varvsgatan 14, SE-211 19 Malmö, Sweden
Tel: (+46 40) 34 80 00   Fax: (+46 40) 34 87 75
Web: www.kindustrier.se

Key personnel
President: Björn Widell
Executive Vice-Presidents
   Head of Engineering: Peter Linde
   Production: Karl-Gustav Andersson
   Head of Finance: Nils-Arne Nilsson

Background
K Industrier AB formerly traded as Kockums Industrier AB.

Products
Freight wagons, including covered, steel-carrying, copper-carrying wagons and container flats. A recent product is the Type Hiqqrrs-vw011 two-section box wagon with automatic electrically powered vertically sliding sides and a 25-tonne axleload.

K Industrier has most recently introduced the newly developed Type Hiqqrrs-vw041 covered freight wagon adapted for the European market. Both wagon sides can be opened to full height.

UPDATED

K T Steel

K T Steel Industries Pvt, Ltd
9 Altamount Road, Bombay 400 026, India
Tel: (+91 22) 386 35 03; 386 04 34
Fax: (+91 22) 363 46 53

Key personnel
Chairman: T K Gupta
Executive Director: V R Gupta

Products
Rolling stock. In addition to supplying wagons for the home market, K T Steel Industries has built wagons for Iranian State Railways (400 ore wagons); South Korea (120 covered wagons); Sudan (120 covered wagons); and Sri Lanka. The company specialises in the manufacture of special purpose wagons including 21,000- and 40,000-litre milk tank wagons; and pressurised tank wagons.

Kryukovsky

Kryukovsky Railway Car Building Works
139 I Pridhodko Street, Kremenchung 315307, Ukraine
Tel: (+38 5366) 695 05; 697 95; 611 01
Fax: (+38 532) 50 14 21; (+38 5366) 611 01;
   (+38 44) 295 72 03
e-mail: root@kvsz.poltava.ua

Products
Vehicles for 1,435 and 1,520 mm gauge with a cargo capacity of up to 75 tonnes, including: open wagons; hopper wagons; tank wagons; flat wagons for intermodal traffic.

Type Hiqqrrs-vw041 wagon by K Industrier (K Industrier)   ***NEW**/0585026*

Bogie hopper wagon with 72 tonne capacity by Kryukovsky   0087656

Open wagon with 70 tonne capacity by Kryukovsky   0087655

Lyudinovo Locomotive Works JSC

1 K Liebknecht Street, Lyudinovo, Kaluga region 249400, Russian Federation
Tel: (+7 084 44) 201 20; 252 59
Fax: (+7 084 44) 201 20; 252 59

Key personnel
For full personnel listing, see Lyudinovo Locomotive Works JSC entry in *Locomotives and powered/non-powered passenger vehicles* section.

Products
Tank wagons.

MAN Ferrostaal AG

Hohenzollernstrasse 24, D-45128 Essen, Germany
Tel: (+49 201) 818 01   Fax: (+49 201) 818 28 22
e-mail: info@manferrostaal.com
Web: www.manferrostaal.com

Key personnel
Head of the Department of Infrastructure and Transport Systems: Helmut Julius

Background
Re-named from Ferrostaal AG, MAN Ferrostaal AG is a member of the MAN AG group.

Products
Freight wagons, including special designs; inspection and service trolleys for track and overhead maintenance and track installation; bogies for freight wagons, rail-mounted cranes and special trolleys; rolling stock components such as wheels, axles, assembled wheelsets, bearings, suspensions and couplers.

UPDATED

Marcroft Engineering Ltd

Whieldon Road, Stoke-on-Trent, ST4 4HP, UK
Tel: (+44 1782) 84 40 75   Fax: (+44 1782) 84 35 79
e-mail: sales@marcroft.co.uk
Web: http://www.marcroft.co.uk

Key personnel
Managing Director: Robert McNeil
Commercial Director: Richard Simons
Senior Commercial Assistant: Debbie Timmis

Background
Previously part of the European Rail Division of VTG Rail UK Limited, Marcroft Engineering was subject to a management buyout in February 2004.

Products
Marcroft Engineering provides a broad range of general repair, refurbishment and modification services on all types of freight wagons.

MÁV

MÁV Debreceni Jármüjavtó kft
Faraktár út 67, H-4034 Debrecen, Hungary
Tel: (+36 52) 34 68 00   Fax: (+36 52) 31 47 25

Key personnel
Managing Director: Imre Kerékgyártó

MÁV container wagon   0002427

Associate company
Dunakeszi Wagon Building & Repair Works
See Adtranz, *Locomotives and powered passenger vehicles* section

Products
Freight wagons including 10-axle low-floor large goods vehicle (LGV) carrier and tank wagons.

Modalohr

Lohr Industrie SA
29 rue du 14 juillet, BP1 – Hangenbieten, F-67838 Tanneries Cedex, France
Tel: (+33 3) 88 38 98 00
Fax: (+33 3) 88 38 06 36
Web: http://www.modalohr.com; http://www.lohr.fr

Works
ZI de Duppigheim, F-67120 Duppigheim, France.

Key personnel
President: Philippe Mangeard

Products
Modalohr is a low-floor wagon concept developed by Lohr Industrie for the piggyback movement by rail of standard road freight vehicles within existing loading gauges in France and neighbouring countries (UIC GB1 gauge). The design of the wagon features a bed which pivots to facilitate loading and unloading at a diagonal angle. Wagons are formed as articulated pairs running on standard freight vehicle bogies. The length of articulated road vehicles necessitates the separation of the road tractor unit from its semi-trailer after loading. The Modalohr wagons are dependent on dedicated terminal facilities. Most notably these take the form of equipment mounted between the rails of the terminal siding to activate the pivoting of the rail wagon beds for loading and unloading. The terminal requires a paved or asphalt loading area of a height that aligns with the Modalohr wagon bed.

A pilot service using Modalohr vehicles was due to start during 2003. Operated by CME, a partnership between SNCF Participations and Lohr Industrie, this would link terminals at Aiton, near Bourg St Maurice, in the French Alps, and Obassano, near Turin (175 km).

More Wear Industries

PO Box 2199, 5510 Harare, Zimbabwe
Tel: (+263 4) 62 17 31
Fax: (+263 4) 62 17 37
e-mail: morewear@web.co.zw

Key personnel
Chief Executive: M N Komo
Marketing and Sales Executive: A Kapota

Products
Freight wagons, high-sided wagons, tankers, flat-deck container wagons, box wagons, Gloucester-rubber suspension bogies, ride control bogies and railway spares.

National Steel Car Ltd

PO Box 2450, Hamilton, Ontario L8N 3J4, Canada
Tel: (+1 905) 544 33 11   Fax: (+1 905) 547 40 69
e-mail: hnicholson@steelcar.com
Web: http://www.steelcar.com

Key personnel
Chairman and Chief Executive Officer: G Aziz
Chief Operating Officer: D Elliott
Executive Vice-President, Marketing and Sales: H Nicholson

Products
Freight wagons; industrial and mining wagons.

Contracts
Supply of covered hopper wagons for the transport of grain, plastic pellets and potash; centre-beam flat wagons for packaged timber; coil wagons; 53 ft double-stack wells; 60 ft high-cube boxcars; 360 ton heavy duty flat wagons.

UPDATED

Nippon Sharyo Ltd

Head office
1-1 Sanbonmatsu-cho, Atsuta-ku, Nagoya 456-8691, Japan
Tel: (+81 3) 36 68 33 30   Fax: (+81 3) 36 69 02 38
Web: www.n-sharyo.co.jp

Overseas contact
Maruncuchi Central Building, 1-9-1 Maruncuchi, Chiyoda-ku, Tokyo 100-0005, Japan
Tel: (+81 3) 66 88 67 94   Fax: (+81 3) 66 88 68 10

Key personnel
President: Kazuhisa Matsuda
Managing Director and General Manager, Rolling Stock Division: Katsuyuki Ikushima
Business Contact General Manager, International Sales Department: Masataka Nakajima

Products
Freight wagons of all types including tank wagons, international and domestic containers and intermodal wagons, heavy-duty carriers and pressure vessels.

UPDATED

Oleo International Ltd

Grovelands, Longford Road, Exhall, Coventry CV7 9NE, UK
Tel: (+44 2476) 64 55 55   Fax: (+44 2476) 36 42 87

Key personnel
Managing Director: S B Gelderd
Commercial Director: C C Brown

Products
Supplier of hydraulic side buffers to major European railways. Oleo specialises in gas hydraulic buffer capsules with strokes of 105, 110 and 150 mm. All products comply with the relevant UIC standards. New products include crash buffers capable of 400 Kj and 800 Kj energy absorption per buffer. The 400 Kj buffer complies with OTIF and RID standards. Oleo also produces hydraulic draft gear. The DA4463 complies with AAR standards. All products are suitable for freight wagons, passenger coaches and locomotives.

UPDATED

Pakistan Railways Carriage Factory

Sector I-11, Khayaban-e-Sir Syed, Islamabad, Pakistan
Tel: (+92 51) 86 02 53; 86 03 49
Fax: (+92 51) 86 15 50

Key personnel
For full personnel listing, see Pakistan Railways Carriage Factory entry in *Locomotives and powered/non-powered passenger vehicles* section.

Products
Freight wagons for any gauge.
Contracts include 34 wagons each with a capacity of 67.1 tonnes for Bangladesh.

Palfinger Bermüller GmbH

Georg-Wimmer-Ring 25, D-85604 Zorneding-Pöring, Germany
Tel: (+49 8106) 309 90   Fax: (+49 8106) 30 99 29
e-mail: info@mobiler.de
Web: http://www.mobiler.info

Key personnel
Chief Executives Officer: Wolfgang Bermüller

Products
The Mobiler lorry-mounted transhipment system for transferring containers and swapbodies to and from rail vehicles. Transfer is achieved by two hydraulically driven shaft traversers controlled by an electronic Programmable Logic Controller (PLC).

Contracts
The Mobiler system has been adopted by Rail Cargo Austria, Railion (DB Cargo AG) and SBB Cargo, Switzerland. It is also a feature of Siemens' CargoMover automated freight-carrying rail vehicle.

PEC Ltd

A Government of India enterprise
Hansalaya, 15 Barakhamba Road, New Delhi 110 001, India
Tel: (+91 11) 331 63 72; 36 19; 55 08; 33 51; 47 23
Fax: (+91 11) 331 52 79; 36 64
e-mail: burman@peclimited.com

Key personnel
For full personnel listing, see PEC Ltd entry in *Locomotives and powered/non-powered passenger vehicles* section.

Products
Freight wagons for various gauges; spares. PEC has exported over 9,300 wagons to various countries such as Uganda, Tanzania, Zambia, Hungary, Sri Lanka, Myanmar, Bangladesh, South Korea, Malaysia, Nigeria, Poland, Sudan, Iran, Yugoslavia and Vietnam.

PESA Bydoszcz SA

Zygmunta Augusta 11 Street, PL-85-082 Bydoszcz
Tel: (+48 52) 518 02 48   Fax: (+48 52) 518 52 39
e-mail: pesa@pesa.pl
Web: www.pesa.pl

Key personnel
President and General Director: Tomasz Zaboklicki

Background
PESA Bydoszcz SA formerly traded as ZNTK Bydoszcz SA.

Products
Freight wagons, including: dry bulk wagons; side-discharge wagons; side-tipping wagons; silo wagons; tank wagons; and timber carriers.

NEW ENTRY

Power Machines Group

(Energomachexport + LMZ + Electrosila + ZTL + KTZ)
25A Protopopovsky per, 129090 Moscow, Russia
Tel: (+7 095) 725 27 63   Fax: (+7 095) 688 79 90
e-mail: mail@power-m.ru
Web: www.power-m.ru

Key personnel
General Director: Evgeny Yakovlev
Sales Director (Machinery building): Alexander Zhigalov
Advertising Manager: Natalia Kuznetsova

Background
Power Machines unites the leading Russian power equipment enterprises and supplies railway and transport equipment produced by Kalugaputmash, Luganskteplovoz, Muromteplovoz, Railway Repair-Mechanical Plant of Sverdlovsk, Kolomensky Zavod, Metrowagonmash, Vyksa Steel Works.

Products
Freight wagons. The range available includes four- and six-axle freight vehicles for 1,435, 1,520 and 1,676 mm gauges; box wagons, flat wagons for containers; open-top wagons; tank wagons; dumpers; and hopper wagons.

UPDATED

Price

A & G Price
Division of CPD Engineering Ltd
Beach Road, Thames, New Zealand
Tel: (+64 7) 868 60 60   Fax: (+64 9) 309 28 19
e-mail: contracts@agprice.co.nz

Key personnel
Chief Executive Officer: Grant Burnett
Company Financial Officer: Ian Findlay
Commercial Manager: Neil Howe
Production Manager: Peter Yates
Sales Manager: Bill Lovell
Industrial Product Sales Manager: Peter Feran
Estimating Engineer: Barry Ingle

Products
Freight wagon manufacture, modification and repair, supply of components, points and crossings.

Procor Ltd

2001 Speers Road, Oakville, Ontario L6J 5E1, Canada
Tel: (+1 905) 827 41 11   Fax: (+1 905) 827 09 13
Web: http://www.procor.com

Key personnel
President: Ronald W Way
Vice-President, Leasing: Roger Tipple
Vice-President, Rail Devices: M C Parker
Director of Engineering: J S McKechnie
Fleet Manager: N Dachuk
Manager, Business Development: Doug Reece
Manager, Communications: Y Amor

Background
Procor is a member of the Marmon Group.

Products
Tank and special-purpose freight wagons.
Procor designs and manufactures tank wagons and freight wagons for a great variety of products, for lease to shippers in Canada. The company operates and maintains the largest (over 22,000 vehicles) fleet of privately owned freight and tank wagons in Canada.

Progress Rail Services Corporation (Locomotive and Railcar Services)

1600 Progress Drive, PO Box 1037, Albertville, Alabama 35950, USA
Tel: (+1 256) 593 12 60
Fax: (+1 256) 593 12 49
Web: http://www.progressrail.com

Key personnel
Senior Vice-President: Jackie Nesmith

Subsidiary companies/divisions
Facilities are located in 25 states, Canada and Mexico.
Cushioning Division (manufacturer of railcar cushioning devices)
Progress Vanguard Corporation (axles)
Railcar Repair Division (wagon repairs)
Wheel Shop Division (wheel reconditioning)
Parts Division (bolsters, sideframes, air cylinders)
Locomotive and Transit Products Division (axles, traction motors, wheelsets)

Products
Reconditioned and new wagon parts including wheels, axles, bolsters and bogie sideframes; hydraulic cushioning units, wagon and locomotive repair facilities. Also wagon and locomotive leasing and repair.

Qishuyan

Qishuyan Locomotive and Rolling Stock Works
Qishuyan, Changzhou City, Jiangsu Province, People's Republic of China
Tel: (+86 519) 877 17 11   Fax: (+86 519) 877 03 58

Key personnel
For full personnel listing, see Qishuyan entry in *Locomotives and powered/non-powered passenger vehicles* section.

Products
Freight wagons and components.

Rolanfer Matériel Ferroviaire SA

6 rue Thomas Edison, BP 60022, F-57971 Yutz, Cedex, France
Tel: (+33 3) 82 59 56 56
Fax: (+33 3) 82 82 05 77
e-mail: commercial@rolanfer-mf.com

Key personnel
Chairman: Y Henry
Export Sales and Marketing Manager: Laurence Courvalet

Products
Design and construction of freight wagons.

Type Feelrrs four-section side-discharge wagon by PESA for PKP Cargo (Ken Harris)   **NEW**/0585027

Construction of six-axle flat wagons for steel industry customers for the transport of heavy and hot products. Modification of existing wagons on request.

UPDATED

Rotem Company

Headquarters
Landmark Tower, 837-36, Yeoksam-dong, Gangnam-gu, Seoul 135-937, South Korea
Tel: (+82 2) 21 12 82 94
Fax: (+82 2) 21 12 98 73
Web: www.rotem.co.kr

Key personnel
Vice Chairman and Chief Executive Officer:
   Soon-Won Chung
Senior Executive Vice-President: Yeo-Sung Lee
Executive Vice-President: Jae-Hong Kim

Background
Established in 1964 when Daewoo Heavy Industry started manufacturing rolling stock, followed by Hyundai and Hanjin Heavy Industry a few years later. In 1999, the three companies were consolidated into KOROS. Hyundai Motors Group acquired the share of Daewoo in 2001 and the former company KOROS became Rotem. Rotem is now an affiliate of Hyundai Motors Group and has its headquarters in Seoul and two factories in Uiwang and Changwon. Uiwang has a capability of manufacturing 500 emus per year and has the capability to manufacture electric equipment such as traction motors, SIV inverters etc. The research and design centre is also located in Uiwang. The Changwon factory has capacity to manufacture 700 emus per year. Rotem has an annual capacity to manufacture approximately 1,200 emus. Certifications such as the ISO 9001 certificate for Quality, 14001 for environment and 18001 for occupational health and safety management have been acquired at all three sites.

Products
Rotem manufactures gondola cars, box cars, ore wagons, petroleum tank wagons, ballast hoppers, refrigerated wagons, high speed gas containers for transport of ammonia and polypropylene, and other specially designed vehicles, including double-deck car-carriers.

UPDATED

RSD (DCD-Dorbyl (Pty) Ltd)

PO Box 229, Boksburg 1460, South Africa
Tel: (+27 11) 914 14 00   Fax: (+27 11) 914 38 85
Web: http://www.dcd-dorbyl.com

Key personnel
Executive Director: Carl Rehder
Business Manager: Norman Taylor
Divisional Financial Manager: Mark Charters

Products
Freight wagons, tank wagons including dry bulk and LPG tank wagons.

Contracts
Contracts include 100 covered wagons for Ghana Railway Corporation, two LPG tank wagons for Hydro-Congo, Brazzaville, Congo-Brazzaville, and 150 ore wagons for SNIM in Mauritania.

Sambre et Meuse

Railway Division
2 rue des Usines, F-59750 Feignies, France
Tel: (+33 3) 27 69 69 69   Fax: (+33 3) 27 69 69 89

Key personnel
Managing Director: Daniel Pain
Sales and Marketing Manager: Jean Jomeau

Products
RailTrailer bimodal system. The RailTrailer has been designed to be as close to a standard trailer as possible with a profile which can fit in the A loading gauge. The bogie/trailer interface allows for independent roll and pitch movement of two consecutive trailers which reduces the stress on each trailer. Various body versions are available: the prototype is a curtain-sided trailer and other versions under design include a flatbed for containers, plywood or rigid steel walls; refrigerated and a tank configuration.

UPDATED

SCIF

Société Cherifienne du Matériel Industriel et Ferroviaire
PO Box 2604, Allée des Cactus, Aïn-es-Sebaa, Casablanca, Morocco
Tel: (+212 27) 35 39 11   Fax: (+212 27) 35 09 60

Key personnel
For full personnel listing, see SCIF entry in *Locomotives and powered/non-powered passenger vehicles* section.

Products
Freight wagons.

Siemens AG

Transportation Systems
PO Box 3240, D-9105 Erlangen, Germany
Tel: (+49 9131) 7-0
e-mail: cargomover@siemens.com
Web: www.siemens-cargomover.com

Key personnel
President, Transportation Systems Group:
   Hans M Schabert
Group Vice Presidents: Alfred Frank, Joern F Sens, Friedrich Smaxwil

Background
The CargoMover system has been developed in collaboration with the Institut für Schienenfahrzeugtechnik (Institute of Rail Vehicle Technology) of the RWTH technical college in Aachen, the University of Braunschweig and various divisions of Siemens.

Products
Developed to bring to rail the economies of road freight transport, CargoMover is an automated, driverless diesel-powered vehicle for conveying consignments of up to 60 tonnes, especially those in containers and swapbodies. Siemens claims that the vehicle is especially suitable for high-volume routes of between 50 and 150 km, where conventional rail freight is not competitive.

For movement control and operational safety, the cabless vehicle takes advantage of ETCS Level 2 technology, with S21 balises and GSM-R communications. In the vehicle's test phase this system has proved effective. Detection of the track and obstacles on it is achieved by three sensor systems: an array of five radar sensors to provide data on relative speeds and distances from objects; a video camera and pivoted laser scanner for track detection; and a second laser scanner that sweeps the area in front of the vehicle at a height of 50 cm, enabling the vehicle to react to obstacles on or next to the track. Information from the three systems is compiled in a data fusion computer which supplies the signals for vehicle control. At a speed of 30 km/h the vehicle can stop in a distance of around 7 m if an obstacle is detected. In normal operation its top speed is 80 km/h.

For the transfer from road vehicles of containers and swapbodies, CargoMover employs the Mobiler system manufactured by Palfinger Bermüller GmbH.

UPDATED

Ray Smith Group plc

Fengate, Peterborough, PE1 5XG, UK
Tel: (+44 1733) 639 36
Fax: (+44 1733) 34 70 90

Key personnel
Group Sales and Joint Managing Director:
   D Browning

Products
Demountable systems for all weights of chassis. Ground loader bodies for police, fire and MOD requirements. Cantilever tail-lifts from 500–3,000 kg capacity.

Société Générale Egyptienne de Matériel des Chemins de Fer (SEMAF)

PO Box, Cairo, Egypt
Tel: (+20 2) 556 21 77; 555 00 37
Fax: (+20 2) 556 40 96; 555 00 37

Siemens automated CargoMover vehicle

0569807

Key personnel

For full personnel listing, see Société Générale Egyptienne de Matériel des Chemins de Fer (SEMAF) entry in *Locomotives and powered/non-powered passenger vehicles* section.

Products

Freight wagons.

STAG AG

Industriestrasse, CH-7304 Maienfeld, Switzerland
Tel: (+41 81) 303 58 00   Fax: (+41 81) 303 58 99
e-mail: office@stag.net
Web: http://www.stag.net

Key personnel

Managing Director: Christian Gloor
Sales and Marketing: Josef Doller

Principal subsidiaries

STAG GmbH
D-66740 Saarlouis, Germany

STAG Sarl
F-67100 Strasbourg, France

Products

Bulk transport wagons with pneumatic discharge and unloading equipment and components.

Contracts

Contracts include supply of 180 wagons for China, 75 wagons for SBB, Switzerland and 140 wagons for TPI Bangkok, Thailand.

VERIFIED

Tabor Szynowy Opole SA

Zaklad Zastal Wagony – Joint Stock Company
PL-65-119 Zielona Góra, Poland
Tel: (+48 68) 328 45 52   Fax: (+48 68) 328 46 50
e-mail: wagony@zastal.pl
Web: www.taborszynowy.com.pl

Key personnel

President, Tabor Szynowy Opole SA:
  Andrzej Świerczek
Plant Director, Zaklad Zastal Wagony:
  Franciszek Robert Siwulski
Commercial Director: Janina Król
Technical Director: Piotr Koltuniewicz

Products

Freight wagons manufactured according to UIC standards, including: four-axle coal wagons (Eanos), self-discharge wagons for aggregates, ore, cereals, bulk materials (Falns), covered wagons (Hbbins, Tanoos), specialised wagons (Shmmns), flat wagons. Freight wagons for operation on 1,520 mm (CIS) gauge; other wagons according to customer technical documentation; wagon main assemblies (body, frame and so on) and spare parts; wagon refurbishment, steel structures and technological equipment.

Contracts

Standard Eanos coal wagons for PKP, Poland (1998); self-discharging Falns wagons for PKP, Poland (1998/9); self-discharging Tanoos wagons for Bombardier Transportation DWA, Germany (1998/99); 500 Shmmns hot-rolled steel coil carriers for SNCB, Belgium (2001/02); 240 Hbbins-tt covered wagons with sliding protective cover and movable internal partition walls for DB Cargo AG (2002); 10 Hbbins-tt covered wagons for Bombardier Transportation (Bahntechnologie) Germany GmbH & Co (2003); over 200 self-discharging Tanoos wagons for Bombardier Transportation (Bahntechnologie) Germany GmbH & Co (2003).

UPDATED

Tafesa SA

Carretera de Andalucía Km 9, E-28021 Madrid, Spain
Tel: (+34 91) 798 05 50   Fax: (+34 91) 798 09 61

e-mail: tafesa@tafesa.com
Web: www.tafesa.com

Key personnel

Group Chairman: Paloma Fernández-Souza Faro
General Manager: Eliecer Esteban Gutierrez
Technical Director: Luis Peromarta Calvo
Marketing and Commercial Manager:
  Jesús Montes Chinchón

Products

All types of freight rolling stock; special containers; bogies; covered wagons; car transporter wagons; dual-purpose container wagons; piggyback wagons; air discharge cement wagons; ventilated wagons with sliding doors; power line masts; vans; bimodal systems.

UPDATED

Talgo Oy

Elektroniikkatie 2, FIN-90570 Oulu, Finland
Tel: (+358 8) 870 69 00   Fax: (+358 8) 870 69 70
e-mail: sales@talgo.fi
Web: http://www.talgo.fi

Key personnel

Managing Director: Tapani Tapaninaho
Sales Director: Matti Haapakangas
Sales Manager: Matti Asikainen

Products

Freight wagons, including: car-carrying wagons; sliding wall and sliding hood wagons; intermodal wagons; hopper wagons; special wagons including molten sulphur wagons; special wagons for steelworks, timber wagons and wine-carrying wagons.

Contracts

In 2003 Talgo Oy received an order from VR Ltd for 15 closed car carriers (160 km/h) for delivery 2004–05.

Tatravagónka AS

Štefánikova 887/53, SK-058 01 Poprad, Slovakia
Tel: (+421 52) 711 27 00   Fax: (+421 52) 711 21 56
e-mail: frantisek.hudak@tatravagonka.sk
Web: www.tatravagonka.sk

Key personnel

General Director: Július Vachmansky
Sales Director: František Hudák
Marketing: Miroslav Štephán

Products

Freight wagons, including open, covered, flat, tank, container, intermodal, car transporters, coil steel, bulk powder, hopper and specialised types. Bogies for freight wagons, types Y 25, Y 33, Y 37, three-axle bogies with 1,000 mm gauge. Passenger coach body shells.

Type Laes 559 articulated car transporter by Tatravagónka AS 0552468

New products include: Sgmrss 80, light container wagon; Shimmnss, coil steel wagon; Laaps timber wagon, and Zacens 40 m² insulated tank wagon for liquid sulphur.

Contracts

Customers include Slovakian Railways (ZSSK), Czech Railways (ČD), DB Cargo, CFL, AAE, ALSTOM Transport, VTG Hamburg, METRANS CZ Prague.

UPDATED

Texmaco Ltd

6th Floor, Birla Building, 9/1 RN Mukherjee Road, Calcutta 700 001, India
Tel: (+91 33) 22 42 43 83; 22 48 01 35
Fax: (+91 33) 22 42 58 33
e-mail: texprs@ca13.vsnl.net.in

Main works

Belgahria, Calcutta 700 056, India
Tel: (+91 33) 25 41 16 31; 12 04
Fax: (+91 33) 25 41 24 52; 24 26

Key personnel

President: R Maheshwari
Senior Vice-President (Engineering and Marketing)
  Rolling Stock: A K Sinha
Senior Vice-President (Steel Foundry): D H Kela

Products

Freight wagons, bogies and couplers. Also railway points and crossings.

UPDATED

Transwerk

PO Box 15912, LynnEast 0039, South Africa, Transwerk Park, Lynette Street, Kilner Park, Pretoria
Tel: (+27 12) 391 13 04   Fax: (+27 12) 391 13 71
e-mail: johnmat@iafrica.com
Web: http://www.transwerk.co.za

Works

Boemfontain, Durban (Bayhead), Johannesburg (Germiston), Port Elizabeth (Uitenhange), Pretoria (Koedespoort).

Key personnel

Corporate Marketing Manager: John Mathew

Background

Transwerk is a subsidiary of Transnet, the state-owned South African transport undertaking.

Products

Freight wagons of various types, including conventional and bottom-discharge coal and iron ore carriers, cement wagons and tank wagons. Annual production capacity is for around 500 vehicles.

NEW ENTRY

TrentonWorks Ltd

A subsidiary of the Greenbrier Companies Inc
PO Box 130, Trenton, Nova Scotia B0K 1X0, Canada
Tel: (+1 902) 752 15 41   Fax: (+1 902) 755 32 44

TrentonWorks 27 m flat wagon 0063728

Key personnel
President: Richard McKay
Engineering Manager: Glenn MacDonald

Products
Freight wagons, including box cars for paper and wood, centre partition lumber cars, flat wagons and gondolas.

Conventional wagon types include refrigerated wagons, covered hopper wagons, woodchip wagons, flat wagons and other speciality wagons.

Trenton builds a 27 m intermodal flat wagon in 70 and 100 tonne configurations for auto-rack service according to TTX Company's specification.

Trinity Rail Group

One Tower Lanes, Suite 2900, Oakbrook Terrace, Illinois 60181, USA
Tel: (+1 630) 571 59 01   Fax: (+1 630) 571 57 22
Web: http://www.trinityrail.com

Key personnel
Chief Executive Officer: Michael E Flannery

Subsidiaries
Trinity Rail Operations
One Tower Lanes, Suite 2900, Oakbrook Terrace, Illinois 60181, USA
Tel: (+1 630) 571 59 02   Fax: (+1 630) 571 57 93
President: Martin Graham
e-mail: martin.graham@trinityrail.com

Trinity Rail Components and Repair Inc
2525 Stemmons Freeway, Dallas, Texas 75207, USA
Tel: (+1 214) 631 44 20
President: Patrick Wallace
e-mail: pat.wallace@trinityrail.com

Trinity Rail GmbH
Brunngasse 4, CH-8400 Winterthur, Switzerland
Tel: (+41 52) 269 32 42   Fax: (+41 52) 269 32 43
e-mail: winterthur@trinityraileurope.com
President: H Christian Schmalbruch

Background
A subsidiary of Trinity Industries Inc, Trinity Rail Group is a result of the merger in 2001 of Trinity Industries' railcar operations and Thrall Car Manufacturing Co, combining both the North American and European businesses of those two companies.

The group comprises three main units: Trinity Rail Operations, which manufactures wagons at plants in North America; Trinity Rail Components and Repair, which consists of the group's axle manufacturing businesses, repair activities and service part supply units; and Trinity Rail GmbH, covering the group's European wagon-related activities. Two other units, Trinity Industries Leasing Co and Trinity Rail Market Services, together handle vehicle leasing and financing services.

Works
North America
USA: Georgia, Illinois, Montana, Ohio, Oklahoma, Pennsylvania and Texas
Mexico: two plants

Europe
Astra Vagaone, Arad, Romania
TVS Vagonka Studénka, Studénka, Czech Republic

Other facilities
Europe
ICPVA, Arad, Romania (engineering and testing)
Rail Project, Poprad, Slovakia (engineering)
Thrall Europa, York, UK (service parts)
Wagonmarket, Poprad, Slovakia (marketing and sales)

Products
North America
Covered hopper wagons for products such as cereals, cement and plastic pellets, box cars, open hoppers and rotary gondolas for coal traffic, intermodal and non-intermodal flat wagons, steel gondolas and coil wagons, special vehicles for automotive products. Trinity also claims to be North America's leading manufacturer of tank wagons.

Europe
Intermodal wagons, flat wagons, tank wagons, steel products wagons, covered and open hoppers, box and open wagons. Y-25 Lsd1, Y-25 SLsd1 and Y-33 bogies, buffers, drawgear, brake shoe holders, bogie and wagon components.

Tuchschmid AG

Kehlhofstrasse 54, CH-8501 Frauenfeld, Switzerland
Tel: (+41 52) 728 81 11   Fax: (+41 52) 728 81 00
e-mail: info@tuchschmid.ch
Web: http://www.tuchschmid.ch

Key personnel
Manager: Richard Nägeli
Manager, Intermodal Transport Systems:
   Derrick Stormink
Manager, High Rack Storage: Urs Kern

Products
ACTS system of container transfer, developed in conjunction with Swiss Federal Railways and road hauliers. The system dispenses with independent transfer machines and enables the driver of a road vehicle to achieve a road-rail transfer or vice-versa on his own. The system employs special flat wagons equipped with rotating guideways and road chassis equipped with a tilting frame and chain mechanism to slide the containers on and off the wagons.

Rotating guideways in various configurations to suit customers' needs including 16.5 tonnes for old-type flat wagons and 21/30 tonnes for new-type flat wagons.

Tülomsas

The Locomotive and Motor Corporation of Turkey
Ahmet Kanatli Cad, TR-26490 Eskisehir, Turkey
Tel: (+90 222) 224 99 56   Fax: (+90 222) 225 72 72
e-mail: tulomsas@tulomsas.com.tr
Web: http://www.tulomsas.com.tr

Key personnel
For full personnel listing, see Tülomsas entry in *Locomotives and powered/non-powered passenger vehicles* section.

Products
Special-purpose freight wagons including tank wagons, refrigerated hopper wagons and wagons for the transport of ore, tracked tanks and containers.

Union Tank

Union Tank Car Co
A member of The Marmon Group of companies
175 West Jackson Boulevard, Chicago, Illinois 60604, USA
Tel: (+1 312) 431 31 11
Fax: (+1 312) 431 50 03

Key personnel
President: Frank D Lester
Senior Vice-President, Marketing and Sales:
   William L Snelgrove
Senior Vice-President and Controller:
   Mark J Garrette
Operations: Louis A Kulekowskis
Vice-President, Fleet Management:
   Wiliam R Constantino

Products
Steel, stainless steel and aluminium tank wagons for liquids and compressed gases; covered hopper wagons for plastic pellets and resins.

United Goninan

PO Box 33000, Hamilton, New South Wales 2303, Australia
Broadmeadow Road, Broadmeadow, New South Wales 2292, Australia
Web: www.unitedgoninan.com.au

Key personnel
Chief Operating Officer: John McLuckie

Products
Freight wagons, including bulk hoppers, tank wagons, flat wagons and container transport wagons.

Contracts
Recent contracts include the supply of iron ore wagons for BHP Billiton, coal wagons for QR.

UPDATED

Unity Railway Supply Co Inc

805 Golf Lane, Bensenville, Illinois 60106, USA
Tel: (+1 630) 595 45 60
Fax: (+1 630) 595 06 46

Key personnel
Chairman: Harold R O'Connor
President: Robert S Grandy
Executive Vice-President, Financial: Robert Holden

Products
Components for freight wagons including: roller bearing adapters; airbrake hose supports; batten bars; brake levers, bottom rods, jaws and eyes; brake pins; brake release rods; brake steps; steel castings, centre pins; draft key retainers; draft lugs; end platforms; camcar fasteners; pipe bracket/dust strainer filters; standard and split wedge friction castings; grab irons; handbrakes, handwheels and bell cranks; journal lubricators; ladders; lading anchors; lumber corner protectors; train

23,000 gallon capacity tank wagon by Union Tank Car for general-purpose service   0002430

line pipe anchors; gravity-type outlet gates; plug door guides; running boards; reusable shipping containers; sill steps; striker castings; trainline trolley shackles; blue flag and barricade warning lights; and wheel chocks; bottom rod safety supports; brake shoe keys; pedestal liners; trailer and container locks; webbing-systems and winches.

UPDATED

Wabash National

Wabash National Corporation, RoadRailer Division
PO Box 6129, Lafayette, Indiana 47903, USA
Tel: (+1 771) 449 57 45   Fax: (+1 771) 449 54 74
Web: http://www.wabashnational.com

Key personnel
President and Chief Executive Officer, Wabash National: William P Greubel
President, RoadRailer and Senior Vice-President, Marketing, Wabash National: Lawrence J Gross
Chief Operating Officer: Richard J Giromini
Vice-President and General Manger: Bryan K Langford
Vice-President, Engineering: Rodney Ehrlich

Products
Design, production, licensing, marketing and sales of the RoadRailer® Mark V intermodal system for both North America and international markets through the RoadRailer division.

The RoadRailer Mark V system consists of a detachable two or three-axle rail bogie supporting specially designed truck trailers. All Mark V rail bogies can be used to carry any RoadRailer trailer, no matter what the length, height or type. All RoadRailer trailers use their air suspension to lower and raise the road wheels, eliminating the need for terminal cranes. RoadRailer trailers are joined together with special couplers to form trains (up to 125 units in length or 4,800 short tons in trailing weight), or at the end of conventional trains. RoadRailer trains are coupled to conventional wagons by means of the CouplerMate® bogie.

For the North American market, DuraPlate® dry vans are available in 48, 53 and 57 ft lengths, in both standard and high-cube versions. The high-cube van uses smaller tyres to provide an interior height of up to 118 in at the nose and 121 in at the rear of the trailer. The AutoRailer® van is a special type of high-cube trailer that has a raisable deck for carrying up to six full-size cars. A new variant of the AutoRailer has a split-deck design, making it possible to carry trucks and cars or light commercial vehicles. Other dry vans are the 48 and 28 ft sheet and post trailers. The 48 ft van is for mail service and has large side doors. The 28 ft van is called a PupRailer trailer, has a single axle, pintle hook and roll-up door.

Sleeper-carrying wagon designed and manufactured for Jarvis Rail by Wabtec Rail   1037654

The ReeferRailer® trailer is a refrigerated RoadRailer and is available in 48 and 53 ft lengths. For the international market, RoadRailer trailers are available in a wide range of configurations and all trailers built for Europe are compatible with each other, permitting co-ordinated service.

Current RoadRailer operators include Triple Crown, Swift Transportation, Schneider National, Amtrak and Burlington Northern Santa Fe. Nearly 6,000 units are in service in the USA. RoadRailer trailers are also in operation in Brazil, Thailand, China, India, Germany, France, Italy and Australia, and prototypes have been tested in the UK, Thailand, China and India.

Wabtec Rail Ltd

PO Box 400, Doncaster Works, Hexthorpe Road, Doncaster DN1 1SL, UK
Tel: (+44 1302) 34 07 00   Fax: (+44 1302) 32 13 49
Web: www.wabtec.com

Key personnel
Managing Director: John Meehan
Engineering Director: Mike Roe
Finance Director: Robert Johnson
Operations Director: Chris Weatherall
Commercial Manager: Paul Robinson

Background
Wabtec Rail undertakes the overhaul of passenger rolling stock, main line locomotives, bogies, wheelsets, air brake equipment, hydraulic dampers and buffers design in addition to the manufacture of shunting locomotives and freight wagons.

As part of the Wabtec Corporation, Wabtec Rail also supplies to the UK rolling stock owners and maintainers, composite brake blocks and pads, Wabtec Railway Electronics train data recorders and electronic equipment, Cardwell TMX braking systems and Wabtec air brake equipment.

Products
Wabtec Rail undertakes the design and manufacture of freight wagons, also the conversion, overhaul and refurbishment of wagons.

Contracts
Wabtec Rail has supplied 240 High Output Ballast (HOBS) wagons to Network Rail. The HOBS wagons give railway infrastructure companies an efficient means of depositing ballast to either side of the track or between the rails. In addition, radio remote control means that one man can control the ballast delivery for a whole train.

Wabtec Rail has designed and built sleeper carrying wagons for operation with the Jarvis Rail-owned Track Renewal Train. The sleeper-carrying wagons are designed to accommodate the train's sleeper delivery gantries whilst maximising the number of sleepers that can be carried within the minimum length of train. Additionally Wabtec Rail has converted large numbers of wagons for English Welsh & Scottish Railway (EWS). This has included converting HAA coal hopper wagons to MHA open box-bodied wagons, TTA tank wagons to MTA open box-bodied wagons and HEA coal hopper wagons to MEA open-bodied wagons.

UPDATED

Wabash National Triple Crown Services RoadRailer trailer with DuraPlate® RoadRailer trailers on Mark V intermediate bogies   0063729

Waggonbau Brüninghaus GmbH

Brüninghausstrasse 1, D-58239 Schwerte, Germany
Tel: (+49 2304) 68 90   Fax: (+49 2304) 68 91 51
e-mail: info@waggonbaubrueninghaus.de
Web: http://www.waggonbaubrueninghaus.de

Key personnel
General Manager: Bernhard van Engelen
Sales and Marketing Manager: Peter Shwarzwald
Procurement Manager: Hubert Baumgart

Bogie tank wagon for chlorine transport produced by Waggonbau Brüninghaus for VTG-Lehnkering 0546910

Background

Waggonbau Brüninghaus has been a wholly owned subsidiary company of Karsdorfer Eisenbahngesellschaft (KEG) since September 2001. For further details of KEG, see KEG entry in the *Railway systems and operators* section, Germany.

Products

Manufacturer of railway vehicles and especially tank wagons. Design of new vehicles as well as reconstruction of small series up to single vehicles. New designs include four-axle pressure tank wagon manufactured using special steel which yields low weight and operations down to –40°C and up to 33 bar, bogie tank wagons for chlorine transport for VTG-Lehnkering; four-axle tipping wagons; and special vehicles for the transport of hazardous materials.

Wagon Pars

Wagon Pars Company
10 Azarshar Street, South Kheradmand Avenue, Tehran, 15846, Iran
Tel: (+98 21) 884 83 30; 83 39
Fax: (+98 21) 884 83 38

Works address
Km 4, Tehran Rd, Arak
Tel: (+98 861) 330 46 50   Fax: (+98 861) 339 99

Key personnel

Marketing Executive: Reza Esfahlani

Products

Freight wagons including cement and powder wagon, LPG tank wagon, hopper wagon, covered wagon and four-axle wagon.

WagonySwidnica SA

A subsidiary of The Greenbrier Companies Inc
ul Strzclinska 35, Swidnica, PL-581-00 Poland

Warsaw sales office
ul Widok 8, 8th floor, PL-000-23 Warsaw, Poland
Tel: (+48 22) 690 6810   Fax: (+46 22) 690 6811

Greenbrier Europe office
Welterstrasser 60 D-57072, Siegen, Germany
Tel: (+49 271) 271 70 22 53   Fax: (+49 271) 741 28 94

Greenbrier London office
25 Hasker Street, London SW3 2LE, UK
Tel: (+44 20) 75 84 18 98

WagonySwidnica 105 m³ tank wagon 0063730

Key personnel

President: Tom Peczerski
Chief Engineer: Dionizy Studzinski

Products

Freight wagons. Conventional freight wagons include tank wagons, gondolas and flat wagons.

WagonySwidnica has gained approval from Deutsche Bahn AG, Germany.

The 105 m³ tank wagon for transport of liquefied hydrocarbons has a maximum load capacity of 56 tons. Tank wagons for transport of products such as mineral oils, LPG and ammonia are also built at the plant.

New products include the East-West Tank Wagon which incorporates both automated and link and pin couplers, in addition to a dual-valve system to cope with differences in load/discharge practice in Russia and West Europe. Also, a lighter weight/increased-volume tank wagon is being designed, based on the US stub sill tank wagon.

Windhoff Bahn-und Anlagentechnik GmbH

PO Box 1963, D-48409 Rheine, Germany
Tel: (+49 5971) 580
Fax: (+49 5971) 582 09
e-mail: info@windhoff.de
Web: http://www.windhoff.de

Key personnel

Board Members: Herbert Liessem,
   Georg Vennemann
Finance Director: Helmut Gielians
Sales Directors: Dr Martin Hindersmann,
   Uwe Dolkemeyer, Gerd Heitmeier
Technical Director: Juergen Auschner
Purchasing Manager: Stefan Berkemeyer

Products

Cargo transport: freight diesel multiple-unit for low volume freight movements.

Zimbabwe Engineering Ltd

PO Box 1874, Bulawayo, Zimbabwe
Tel: (+263 9) 789 31   Fax: (+263 9) 722 59
e-mail: zeco@acacia.samara.co.zw

Key personnel

For full personnel listing, see Zimbabwe Engineering Ltd entry in *Locomotives and powered/non-powered passenger vehicles* section.

Products

Construction, overhaul and rebuilding of freight wagons. Supply of complete bogies, bogie components and all spares for freight wagons.

Zhuzhou

Rolling Stock Works
Xinhua East Road, Zhuzhou, Hunan, China
Tel: (+86 733) 840 32 84
Fax: (+86 733) 840 32 84; 31 34

Key personnel

Director: Liu Yuzhou

Products

Freight wagons including gondola cars, hopper wagons, flat wagons, heavy-duty wagons, tankers, wagon components and bogies.

Contracts include salt wagons for Botswana Railways, steel gondolas for Tanzania and Zambia and ballast wagons for Hong Kong.

For details of the latest updates to *Jane's World Railways* online and to discover the additional information available exclusively to online subscribers please visit
jwr.janes.com

BRAKES AND DRAWGEAR

Alphabetical listing

Abex Rail
Acieries de Ploërmel
ADES Technologies
Amsted Rail International
Anchor Brake Shoe LLC
ASF-Keystone Incorporated
ASM Spencer Moulton SAS
Atlas Copco Construction + Mining Ltd
Becorit GmbH
Bremskerl Reibbelagwerke Emerling & Co KG
Buffalo Brake Beam Co
S A Buhlmann NV
Carbone Lorraine
Cardwell Westinghouse WABCO
China Northern Locomotive and Rolling Stock
    Industry (Group) Corporation
Cimmco International
Cobra Brake Shoes
Cobreq
Comet Industries Inc
Cometna
Conbrako (Pty) Ltd
William Cook Rail
DAKO-CZ as
Dellner Couplers AB
Delta Rail
domnick hunter Limited
Ellcon National Inc
Escorts Ltd
European Friction Industries Ltd
Federal-Mogul (FERODO) Ltd

FMI – FM Industries Inc
Forges de Fresnes
Frenoplast Bulhak i Cieślawski SP J
Frenos Calefaccion y Señales
Freudenberg Schwab GmbH
Fritex
Futuris Brakes International
GE Harris
Graham-White Manufacturing Co
Greysham and Co
Greysham International
Hanning & Kahl GmbH & Co KG
Honeywell Bremsbelag (Jurid)
IB Italian Brakes SpA
ICER Brakes SA
Jarret
Kamax SA
Karl Georg
Keystone
Keystone Bahntechnik GmbH
Knorr-Bremse AG
Knorr-Bremse Rail Systems (UK) Limited
Kovis doo
KYB – Kayaba Industry Co Ltd
LAF – Les Appareils Ferroviaires
Lesjöfors AB
Meridian Rail LLC
Metcalfe Railway Products Ltd
Metpro
Miner
Mitsubishi Electric Corporation

Multi-Service Supply
MZT Hepos AD
Nabtesco Corporation
Newag GmbH & Co KG
Oleo International Ltd
Paulstra
Peddinghaus Group
Poli Costruzione Materiali Trazione SpA
Progressive Engineering (AUL) Ltd
Le Réservoir
SAB WABCO
SEE
SMC Pneumatics
Stabeg
Standard Car Truck Co
Stone India Ltd
Sumitomo Metal Industries Ltd
Svendborg Brake A/S
Textar
ThyssenKrupp GfT Gleistechnik GmbH
TMD Friction
Triax-YSD
Ueda
Unity Railway Supply Co Inc
Voith Turbo Rail Systems Ltd
Voith Turbo Scharfenberg GmbH & Co KG
Wabtec Corporation
Westinghouse Saxby Farmer Ltd
Yutaka Manufacturing Co Ltd

Company listing by Country

AUSTRALIA
Futuris Brakes International
SMC Pneumatics

AUSTRIA
Stabeg

BELGIUM
S A Buhlmann NV

BRAZIL
Cobreq

CHINA
China Northern Locomotive and Rolling Stock
    Industry (Group) Corporation

CZECH REPUBLIC
DAKO-CZ as

DENMARK
Svendborg Brake A/S

FRANCE
Acieries de Ploërmel
ADES Technologies
ASM Spencer Moulton SAS
Carbone Lorraine
Delta Rail
Forges de Fresnes
Jarret
LAF – Les Appareils Ferroviaires
Paulstra
Le Réservoir
SEE

GERMANY
Abex Rail
Becorit GmbH
Bremskerl Reibbelagwerke Emerling & Co KG
Freudenberg Schwab GmbH

Hanning & Kahl GmbH & Co KG
Honeywell Bremsbelag (Jurid)
Karl Georg
Keystone Bahntechnik GmbH
Knorr-Bremse AG
Newag GmbH & Co KG
Peddinghaus Group
Textar
ThyssenKrupp GfT Gleistechnik GmbH
Voith Turbo Scharfenberg GmbH & Co KG

INDIA
Cimmco International
Escorts Ltd
Greysham and Co
Greysham International
Stone India Ltd
Westinghouse Saxby Farmer Ltd

ITALY
IB Italian Brakes SpA
Poli Costruzione Materiali Trazione SpA

JAPAN
KYB – Kayaba Industry Co Ltd
Mitsubishi Electric Corporation
Nabtesco Corporation
Sumitomo Metal Industries Ltd
Ueda
Yutaka Manufacturing Co Ltd

MACEDONIA
MZT Hepos AD

POLAND
Frenoplast Butmak I Cieślawski SP J
Kamax SA

PORTUGAL
Cometna

RUSSIAN FEDERATION
Fritex

SLOVENIA
Kovis doo

SOUTH AFRICA
Conbrako (Pty) Ltd
Metpro

SPAIN
Frenos Calefaccion y Señales
ICER Brakes SA

SWEDEN
Dellner Couplers AB
Lesjöfors AB
SAB WABCO

UNITED KINGDOM
Atlas Copco Construction + Mining Ltd
William Cook Rail
domnick hunter Limited
European Friction Industries Ltd
Federal-Mogul (FERODO) Ltd
Knorr-Bremse Rail Systems (UK) Limited
Metcalfe Railway Products Ltd
Oleo International Ltd
Progressive Engineering (AUL) Ltd
TMD Friction
Voith Turbo Rail Systems Ltd

UNITED STATES
Amsted Rail International
Anchor Brake Shoe LLC
ASF-Keystone Incorporated
Buffalo Brake Beam Co
Cardwell Westinghouse WABCO
Cobra Brake Shoes
Comet Industries Inc

Company listing by Country—*continued*

Ellcon National Inc
FMI – FM Industries Inc
GE Harris
Graham-White Manufacturing Co

Keystone
Meridian Rail LLC
Miner
Multi-Service Supply

Standard Car Truck Co
Triax-YSD
Unity Railway Supply Co Inc
Wabtec Corporation

Abex Rail

Lütticher Strasse 565, D-52074 Aachen, Germany
Tel: (+49 241) 712 83   Fax: (+49 241) 712 52
e-mail: info@abexrail.com

Key personnel
Managing Director: KW Kever

Main works
Rütgers Rail Italia, Avellino, Italy

Products
Composition and sintered brake shoes and disc
brake pads.
VERIFIED

Acieries de Ploërmel

PO Box 103, F-56804 Ploërmel, France
Tel: (+33 2) 97 73 24 70   Fax: (+33 2) 97 74 03 90
e-mail: ap.ctrochu@wanadoo.fr

Key personnel
Managing Director: Jean-Luc Lancelot
Sales Manager: Alain Noblet

Background
Acieries de Ploërmel is a member of the Amsted
Group.

Products
Brake block holders, buffers.

ADES Technologies

13 rue Edouard Martel, ZI de la Chauvetière,
F-42100 St Etienne, France
Tel: (+33 4) 77 59 59 23   Fax: (+33 4) 77 80 95 64
e-mail: business@ades-technologies.com

Key personnel
Managing Director: Jean-Louis Modrin
Commercial Director: Jean-Michel Pagnerre

Products
Design, manufacture and maintenance of hydraulic
and pneumatic systems, including: Auxim cock and
isolating valves for compressed air applications;
Raflex fittings for metal tubing to SNCF STM 820 A
Norm and UIC 803 35 OR standards; Auxim
internal safety devices for tank wagons. The company's
Someplan department provides its Systemier concept
for the design and assembly of modules and panels
from preformed tubes or manifold blocks.

Contracts
ADES Technologies is supplying ALSTOM,
Bombardier, Siemens and SNCF.
UPDATED

Amsted Rail International

ASF-Keystone Incorporated
1700 Walnut Street, Granite City, Illinois 62040, USA
Tel: (+1 618) 452 21 11   Fax: (+1 618) 452 71 55

Key personnel
Vice-President: Steve Becker

Products
Cast three-piece bogies, couplers, twin packs, draft
gears and cushions, discharge gates, sideframes
and bolsters, low profile centre plates, draft sills
and coil springs.

Anchor Brake Shoe LLC

1920 Downes Drive, West Chicago, Illinois 60185,
USA
Tel: (+1 630) 293 11 10   Fax: (+1 630) 293 71 88

Key personnel
Chairman: Richard A Mathes
President: Jack M Payne
Plant Manager: James Quattrone
Field Service Manager: Joseph H Samolowicz
Quality Control Supervisor: Michael Tatera

Products
Composition brake shoes for locomotives and
freight cars.

ASF-Keystone Incorporated

ASF-Keystone/Amsted Industries International
200 West Monroe Street, Suite 2301, Chicago,
Illinois 60606, USA
Tel: (+1 312) 372 53 84   Fax: (+1 312) 372 82 30

Key personnel
Vice-President: Steve Becker

Products
Cast three-piece bogies, couplers, twin packs, draft
gears and cushions, discharge gates, sideframes
and bolsters, low profile centre plates, draft sills,
coil springs.

Atlas Copco Compressors Ltd

Swallowdale Lane, Hemel Hempstead HP2 7HA, UK
Tel: (+44 1442) 26 12 01   Fax: (+44 1442) 23 47 91
e-mail: gba.info@uk.atlascopco.com
Web: www.atlascopco.co.uk

Key personnel
Managing Director: Geert Follens

Products
Rotary screw compressors for electric and diesel
locomotives, railcars, LRVs and tramcars.
UPDATED

ASM Spencer Moulton SAS

21 rue de la Gare, F-45330 Malesherbes, France
Tel: (+33 2) 38 32 72 29   Fax: (+33 2) 38 34 73 42
e-mail: gjoly@asmtrac.com
Web: www.asmtrac.com

Key personnel
Sales Manager: Guy Joly

Products
Drawgear, buffers, rubber-to-metal bonded
components and suspension systems.
UPDATED

Becorit GmbH

Rumplestrasse 6-10, D-45659 Recklinghausen,
Germany
Tel: (+49 2361) 66 66   Fax: (+49 2361) 66 67 40
e-mail: info@becorit.de
Web: http://www.becorit.de

Products
Manufacturers of disc brakes and composition brake
shoes for all rail vehicle applications. A wide range
of products includes organic disc brake pads and
sintered brake pads for use with high-speed trainsets
(ICE 1, 2 and 3, TGV) and other passenger vehicles
(IC vehicles, emus), and composition brake shoes for
use with freight wagons (K, L and LL shoes).

Bremskerl Reibbelagwerke Emerling & Co KG

Brakenhof 7, D-31629 Estorf-Leeseringen, Germany
Tel: (+49 5025) 97 80   Fax: (+49 5025) 97 81 10
e-mail: bk@bremskerl.de
Web: http://www.bremskerl.de

Key personnel
Managing Director: Herr Gramatke
Sales Director: Herr Wolf
Technical Director: Herr Hering

UK office
Bremskerl (UK) Ltd, Unit 2, Stable Yard, Windsor
Bridge Road, Bath BA2 3AY, UK
Tel: (+44 1225) 44 28 95   Fax: (+44 1225) 44 28 96
e-mail: online@bremskerl.uk.com
General Manager: Chris Prior

North American office
PO Box 965, Arlington Heights, Chicago, Illinois
60006-0965, USA

Products
Asbestos-free organic and sintered metal disc
brake pads and wheel tread brake blocks.

Contracts
Customers include: DB, SNCF, FS, ÖBB, SJ, VR,
RATP, SBB and MTA Los Angeles.

Buffalo Brake Beam Co

400 Ingham Avenue, Lackawanna, New York
14218-2536, USA
Tel: (+1 716) 823 42 00   Fax: (+1 716) 822 38 23
e-mail: bbb@brakebeam.com
Web: http://www.brakebeam.com

Key personnel
Chairman and Chief Executive Officer:
   Richard G Adams
President and Chief Operating Officer:
   Garold L Stone Jr
Vice-President, Sales: Christopher F Adams
Director of Engineering: Louis E Bobsein
Director of Sales: Christopher F Adams

Products
Wagon brake beams, unit side-frame wear plates
(steel or plastic), brake rod connectors, brake shoe
keys and coupler carrier wear plates.

S A Buhlmann NV

Leuvensesteenweg 31, B-1932 St Stevens-Woluwe,
Belgium
Tel: (+32 2) 711 20 30   Fax: (+32 2) 720 20 64
e-mail: buhlman@skynet.be

Key personnel
Chairman: Olivier Buhlmann
Manager, Rolling Stock Department:
   Frederic Collier

Products
Railway, including light rail, for: hydraulic brakes,
electric and electromagnetic; automatic couplers.

Carbone Lorraine

Brake division
41 rue Jean Jaurès, BP 148, F-92231 Gennevilliers,
France
Tel: (+33 1) 41 85 43 65   Fax: (+33 1) 41 85 43 63
e-mail: herve.mace@carbonelorraine.com
Web: www.carbonelorraine.com

Key personnel
Managing Director: Luc Themelin
Sales and Marketing Managers: Hervé Macé,
Plant Manager: Loïc Lelièvre

Principal subsidiaries
Carbone of America Corp
400 Myrtle Avenue, Boonton, New Jersey 07005,
USA
Tel: (+1 973) 334 07 00

Carbono Lorena SA
Av Francisco Monteiro 1701, 09400-000, Ribeirao
Pires-SP, Brazil
Tel: (+55 112) 46 62 33

Carbone Lorraine
Apollo Building, Room 703, 1440 Yan'An Road,
Shanghai 200040, China
Tel: (+86 21) 62 49 68 36

Carbone Lorraine Korea Co, Ltd
Eden Bldg, 4 Fl-1579-1, Seocho Dong, Seocho-Ku,
Seoul, South Korea
Tel: (+82 2) 598 00 71

Le Carbone K K
6 F, Shinkuju Royal Bldg, 7-21-1, Nishi-Shinkuju,
Shinkuju-Ku, Tokyo 160-0023, Japan
Tel: (+81 353) 32 53 61

Le Carbone Lorraine Australia Pty Ltd
PO Box 196, 75 Sparks Avenue, Fairfield, Victoria
3078, Australia
Tel: (+61 3) 94 89 24 55

Le Carbone (SA) Pty Ltd
CNr Commando and Wright Street, Industria, West
Johannesburg, South Africa
Tel: (+27 11) 474 00 00

Products
Low-noise, sintered metal substitutes for cast-iron
brake shoes; low-friction coefficients sintered metal
brake shoes C10, medium friction coefficients
sintered metal brake shoes C17, for hot and wet
and icy effective braking; high-friction coefficients
sintered metal brake shoes C30, for wagons and
locomotives speeding over 100 km/h; sintered
metal disc brake pads for very high-speed trains;
dynamometer braking simulations; customised
friction material for specific applications.

Contracts
Recent contracts include Korean High Speed Train,
Eurostar, as well as French National Railways TGV
for sintered metal brake pad supply.
    Also the new regional train from Bombardier,
(160 km/h) high-capacity Type AGC.

UPDATED

Cardwell Westinghouse WABCO

8400 South Stewart Avenue, Chicago, Illinois
60620-1794, USA
Tel: (+1 708) 655 52 00   Fax: (+1 708) 655 52 02

Key personnel
Vice-President and General Manager:
   Mark Van Cleave

Background
Cardwell Westinghouse has become a subsidiary
of WABCO (qv)

Products
Friction, rubber-friction and hydraulic friction
draftgear; handbrakes; automatic slack adjusters.
Vacuum brake equipment and components, Alliance
couplers, AAR standard-type Alliance couplers,
MCA couplers, automatic centre buffer couplers,
enhanced screw couplers, screw couplings.

China Northern Locomotive and Rolling Stock Industry (Group) Corporation (CNR)

11 Yangfangdian Road, Haidian District, Beijing
100038, China
Tel: (+86 10) 51 86 23 70   Fax: (+86 10) 51 86 23 74
e-mail: loriciec@cnrgc.com.cn
Web: www.loriciec@cnrgc.com.cn

Key personnel
Chairman and Managing Director: Wang Tai-Wen
President: Cui Dianguo

CNR Model 120 distribution valve   0595037

CNR Model MT-3 draft gear   0552549

Vice-President: Zhao Guangxing
President of LORIC Import & Export Corporation
   Ltd: Cao Guo-Bing
Vice-President of LORIC Import & Export
   Corporation Ltd: Chen Dayong
Senior Engineer of LORIC Import & Export
   Corporation Ltd: Yang Xiang-Jing

Products
Development, design, engineering, production,
sales, installation, refurbishment, maintenance
and after-sales services for all types of brakes and
drawgear.

UPDATED

Cimmco International

A division of Cimmco Birla Ltd
Prakash Deep, 7 Tolstoy Marg, New Delhi 110 001,
India
Tel: (+91 11) 331 43 83; 84; 85
Fax: (+91 11) 332 07 77; 372 35 20

Key personnel
For full personnel listing, see Cimmco International
entry in *Freight vehicles and equipment* section.

Products
Vacuum brake equipment and components;
automatic centre-buffer couplers, including AAR

types E and F, Alliance II and high-tensile couplers;
enhanced centre-buffer couplers for freight
wagons; ABC couplers for locomotives; and MCA
and PH type couplers for coaches.

Cobra Brake Shoes

Railroad Friction Products Corporation
PO Box 1349, Laurinburg, North Carolina 28353,
USA
Tel: (+1 910) 844 97 10   Fax: (+1 910) 844 97 33

Key personnel
Vice-President and General Manager: F J Grejda
Director, Sales and Marketing: L R Charity
Product Manager: Michael F Griffin

Products
Composition brake shoes and disc pads.

Cobreq

Cia Brasileira de Equipamentos
Praia do Flamengo 200, 9° Andar, Flamengo 22210,
Rio de Janeiro, Brazil
Tel: (+55 21) 285 22 33
Fax: (+55 21) 285 70 60

Main works
Rua Tupi 293, Caixa Postal 54, Vila Maria, Indaiatuba,
13300-001 São Paulo, Brazil
Tel: (+55 192) 75 31 33   Fax: (+55 192) 75 71 29

Key personnel
Sales Director: R Darigo

Products
Non-metallic composition brake shoes and brake
pads for railroad vehicles.

Comet Industries Inc

4800 Deramus Avenue, Kansas City, Missouri
64120, USA
Tel: (+1 816) 245 54 00   Fax: (+1 816) 245 54 35

Key personnel
President: Steve Woodson
Chief Financial Officer: Mike Klinock
Rail Products Operations Manager: Randall Haan
Director of Business Development: John Killian

Products
New and used wagon parts; reconditioned and
second-hand bolsters/sideframes; distributor of
Kohler power equipment; voice date installation
and maintenance.

Cometna

Copanhia Metalurgica Nacionel
Rua Marechel Gomes Dacosta, Famoes P-2675
Odivelas, Lisbon, Portugal
Tel: (+351 1) 933 31 39   Fax: (+351 1) 933 31 43

Key personnel
President: Jose Bissaia Barreto

Products
Knuckle-type couplers for locomotives and rolling
stock.

Conbrako (Pty) Ltd

PO Box 4018, Luipaardsvlei 1743, Transvaal, South
Africa
Tel: (+27 11) 762 24 21   Fax: (+27 11) 762 65 35
e-mail: conbrako@mweb.co.za

Key personnel
Managing Director: R G Child

Products
Air and vacuum brakes; drawgear; snubbers, handbrakes.

William Cook Rail

Cross Green, Leeds LS9 0SG, UK
Tel: (+44 113) 249 63 63   Fax: (+44 113) 249 13 76
e-mail: admin@cook-catton.co.uk
Web: http://www.william-cook.co.uk

Key personnel
Managing Director: Kevin Grayley
Field Sales Manager: David Walshaw
Also at
Parkway Avenue, Sheffield S9 4WA
Tel: (+44 114) 273 01 21   Fax: (+44 114) 275 25 08
Sales Director: Trevor Stephenson

Products
Steel brake discs. Couplers and coupler assemblies, including Drophead Buckeye, Alliance and Tightlock types.

DAKO-CZ as

Budovatelů 323, CZ-53843 Třemošnice, Czech Republic
Tel: (+420 469) 61 71 11   Fax: (+420 469) 61 71 15
e-mail: dakocz@dako-cz.cz
Web: http://www.dako-cz.cz

Key personnel
Chairman: Ing Václav Lebeda
Commercial Director: Milan Polák

Background
DAKO-CZ, previously known as ČKD-DAKO, changed its name in August 2001.

Products
Manufacturer of brake devices and components for rail vehicles; development, manufacture and sales of hydraulic and mechanical devices and components for both civil and military equipment.
   Principal products include DAKO complete compressed-air brake systems for both freight wagons, passenger cars and locomotives conforming with UIC standards; compressed-air brake devices for high-speed traffic; brake systems for trams.
   DAKO-CZ is IS0 9001 certified.

Dellner Couplers AB

Vikavägen 144, SE-791 95 Falun, Sweden
Tel: (+46 23) 76 54 00
Fax: (+46 23) 76 54 10
e-mail: info@dellner.se
Web: www.dellner.com

Key personnel
Managing Director: C Nicolin
Technical Director: D Ernst
Deputy Managing Director: L Wicklander
Manager, Aftersales and Nordic countries:
  T Westbom
International Sales Manager: R Danielsson
Marketing Director: H Gustafsson

Dellner Couplers France
10 Passage Ronsin, F-77300 Fontainebleau, France
Tel: (+33 1) 64 69 55 21   Fax: (+33 1) 64 69 55 28
e-mail: dellnercouplers1@hotmail.com

Dellner Kupplungen GmbH
Stahlstrasse 4a, D-42281 Wuppertal, Germany
Tel: (+49 202) 50 40 26   Fax: (+49 202) 50 60 21
e-mail: dellner@dellner.de

Dellner Couplers Inc
8334-H Arrowridge Boulevard, Charlotte, North Carolina 28273, USA
Tel: (+1 704) 527 21 21   Fax: (+1 704) 527 21 25
e-mail: ttarantino@dellnercouplers.com
Web: www.dellner.com

Dellner Couplers Sp zoo
Osada Kolejowa 12, PL-81-220 Gdynia, Poland
Tel: (+48 586) 28 54 81   Fax: (+48 586) 28 54 80
e-mail: dellner@dellner.com.pl

Products
Supply of automatic couplers, semi-permanent couplers, side buffers, semi-trailer joints, adapters, hatches, snow gaiters and other front-end parts for rail vehicles.

Contracts
Couplers being supplied to Washington metro, Santa Clara VTA and NJJ, USA, Munich metro and Porto metro.

UPDATED

Delta Rail

1 Rue Roussel, F-92250 La Garenne Colombes, France
Tel: (+33 1) 42 42 11 44
Fax: (+33 1) 42 42 11 16

Key personnel
President: Yves Daunas
Marketing: Maire Collins

Products
Rolling stock buffers, shock-absorbers, suspension air springs, platform buffers.

---

domnick hunter Limited

Dukesway, Team Valley Trading Estate, Gateshead NE11 0PZ, UK
Tel (+44 191) 402 90 00   Fax (+44 191) 482 62 96
e-mail: info@domnickhunter.com
Web: www.domnickhunter.com/railways

Key personnel
Business Manager: Graham S Leach

Subsidiary companies
Subsidiaries in the following countries: Australia, Benelux, Brazil, Canada, China, Czech Republic, Denmark, France, Germany, India, Indonesia, Italy, Japan, Malaysia, Norway, Poland, Singapore, Spain, Sweden, Thailand and USA.

Products
Compressed air systems for railway filtration applications; compressed air railway dryers and brake systems. Applications also include: pneumatic door systems; air suspension systems; pantograph operation; track cleaning; and paint spraying.

UPDATED

Ellcon National Inc

50 Beechtree Boulevard, PO Box 9377, Greenville, South Carolina 29604-9377, USA
Tel: (+1 803) 277 50 00
Fax: (+1 803) 277 52 07

Key personnel
For full personnel listing, see Ellcon National Inc entry in *Passenger coach equipment* section.

Products
Handbrakes, slack adjusters and bogie-mounted brakes.

Escorts Ltd

Railway Equipment Division
Plot No 115, Sector 24, Faridabad 121 005, India
Tel: (+91 129) 544 22 74   Fax: (+91 129) 544 22 75
Web: http://www.escortsgroup.com

Key personnel
Vice-President and Business Head: Devraj Singh
Head, Railway Equipment Division:
   Krishna Havaldar
   e-mail: ksh@hotmail.com
Head, Marketing: Sunil Khanna
   e-mail: sunil.khanna@escortsred.com
Head, Exports: Sunil Jain
   e-mail: suniljain59@redifmail.com

Products
Complete airbrake systems for locomotives, freight and passenger vehicles, overhead equipment inspection cars and diesel cranes. Electropneumatic brake systems for metro cars and electric multiple-units. Vacuum control valves. Composite brake blocks. Automatic and semi-permanent centre buffer-couplers for emus and metro cars. Rail fastening systems and vulcanised rubber components.
   The products incorporate technologies from world renowned companies like Schaku and Knorr-Bremse of Germany, General & Railways Supplies and Vulcanite of Australia and ICER of Spain.
   Escorts is the prime supplier to Indian Railways, delivering over 50,000 air brake sets, 2,500 sets of electropneumatic brake equipment and 15,000 centre buffer-couplers. Products are also exported to over 15 countries.
   New products under development or validation include a bogie-mounted brake system for freight wagons.
   Escorts (Railway Equipment Division) is certified to ISO 9002.

European Friction Industries Ltd

Enterprise House, 6/7 Bonville Road, Brislington, Bristol BS4 5PF, UK
Tel: (+44 117) 971 48 37   Fax: (+44 117) 971 65 78
e-mail: rail@efiltd.co.uk

Key personnel
Rail Business Manager: Kevin Alexander
   e-mail: Kevin@efiltd.co.uk

Products
Composite friction materials for braking for all types of rail vehicles, including tram, heavy freight, high-speed and mining applications; OE and replacement pads and blocks.

UPDATED

Federal-Mogul (FERODO) Ltd

Chapel-en-le-Frith, High Peak SK23 0JP, UK
Tel: (+44 1298) 81 15 98   Fax: (+44 1298) 81 15 80
e-mail: fpgrailenquiries@eu.fmo.com

Key personnel
Director of Operations: T M Saxby
Commercial Manager: H Lavender

Products
High-performance, cost-effective friction brake materials for all types of rail vehicles, including high-speed, passenger and light rail vehicles. Composite disc brake pad materials including Sinter Metal Pads. Low-friction 'L' and 'LL' blocks and high-friction 'K' blocks are supplied, covering a wide spectrum of braking requirements.

FMI – FM Industries Inc

8600 Will Rogers Boulevard, PO Box 40555 Fort Worth, Texas 76140, USA
Tel: (+1 817) 293 42 20   Fax: (+1 817) 551 58 01

Key personnel
President: James Kingerski
Senior Vice-President, Operations: M S Dew
Vice-President, Sales and Marketing: T W Howe
Vice-President Engineering: R N Hodges

Products
FreightMaster/Freight-Saver/Hydra-Buff brand of end-of-car and centre-of-car hydraulic cushioning devices.

FMI offers new and reconditioned gas return cushioning devices designed to eliminate high in-train forces. FMI's EOC devices are shipped precharged with nitrogen gas and gagged short of full extension for ease of installation. A sight glass to verify the hydraulic fluid level is provided in all new and reconditioned devices.

Forges de Fresnes

80 rue Pasteur, PO Box 11, F-59970 Fresnes-sur-Escaut, France
Tel: (+33 3) 27 25 92 22　Fax: (+33 3) 27 26 17 27

Key personnel
Manager: J M Deramaux

Products
Forged brake beam assemblies and other forgings for bogies, passenger coaches and freight wagons.

Frenoplast Bulhak i Cieślawski SP J

Korpele 75 – Strefa, PL-12-100 Szczytno, Poland
Tel: (+48 89) 624 97 54
Fax: (+48 89) 624 97 55
e-mail: info@frenoplast.pl
Web: www.frenoplast.pl

Products
Disc pads for rail vehicles; centre plate linings and shoes; composition brake inserts; clutch linings for anti-slip devices.

Frenoplast manufactures friction materials for railway braking systems which have been supplied to many European countries. The company offers composite brake pads conforming to UIC 541-3, as well as K, L and LL composite blocks. Further products include centre pivot and side bearer liners for freight bogies.

UPDATED

Frenos Calefaccion y Señales

Sociedad Española de Frenos, Calefacción y Señales SA
Calle Nicolás Fúster 2, E-28320 Pinto, Madrid, Spain
Tel: (+34 91) 691 00 54　Fax: (+34 91) 691 01 00
e-mail: sefrenos@frenos.dobytec.es

Key personnel
Board Member: Nicolás Fúster Junquera
General Director: Miguel Angel Martín Jiménez
Commercial Director: Agustín Lagartos Ruano
Manufacturing Director: Eugenio Blázquez

Products
Airbrake products, vacuum brakes and high-voltage heating; compressed air production and treatment; all types of brake control units; electronic anti-skid, anti-spin and brake control systems (analogue and microprocessor); bogie equipment such as cylinders, discs, block brake units; hydraulic brakes, magnetic track brakes and vacuum toilets.

Contracts include the supply of brake equipment for: 16 emus for Metro Madrid, 33 emus and 154 wagons for RENFE, 80 demus for Belgium and 13 emus for Metro Bilbao.

Freudenberg Schwab GmbH

Postplatz 3, D-16761 Hennigsdorf, Germany
Tel: (+49 3302) 206 20　Fax: (+49 3302) 20 62 77
e-mail: info@freudenberg-schwab.de
Web: http://www.freudenberg-schwab.de

Key personnel
Chief Executive Officers: Dr Detlef Cordts;
　Peter Kofmel; Jörg Sost
Quality Manager: Leo S E Lang
Marketing Director: Bernd Werner

Background
Freudenberg Schwab GmbH is jointly owned by Freudenberg & Co (51 per cent) and Schwab Holding AG (49 per cent). The company collaborates closely with Schwab Schwingungstechnik AG in Switzerland.

Products
Vibration control: rubber/metal elastic elements, standard as well as specially designed. Includes resilient mountings for brake calipers and compressors.

Fritex

79 Sovetskaya ul, 150003 Yaroslavl, Russia
Tel: (+7 0852) 25 42 84　Fax: (+7 0852) 25 47 10
e-mail: sbt@fritex.yaroslavl.ru
Web: http://www.fritex.yaroslavl.ru

Products
Brake shoes in asbestos and composite materials, including shoes capable of operation in temperatures ranging from +50 to –60°C. Products are certified to ISO 9001.

Futuris Brakes International

6 Wenban Street, Wetherill Park, New South Wales 2164, Australia
Tel: (+61 2) 87 84 84 00　Fax: (+61 2) 87 84 84 90

Key personnel
Managing Director: Mark Kuenzi
Marketing and Sales/Business Development
　Manager: Chris Katakouzinos

Background
Futuris is the trading name of FIP Pty Ltd and is a member of the Wabtec Corporation.

Products
High, medium and low-friction non-asbestos composite brake shoes for railway braking applications. Direct cast iron replacement using composite technology.

UPDATED

GE Harris

GE Harris Railway Electronics LLC
407 John Rhodes Boulevard, Melbourne, Florida 32934, USA
Tel: (+1 407) 242 51 74　Fax: (+1 407) 242 50 19
e-mail: rjohns@ge-harris.com
Web: http://www.geharris.com

Key personnel
See *Signalling and Telecommunications Systems*.

Products
The LOCOTROL® EB integrates the controls of distributed power and locomotive electronic air brake into one package, providing five-fold reliability gain and automated diagnostics. The modular designed LOCOTROL EB electronic air brake has fewer line-replaceable units and offers higher level diagnostics, enabling fast ready track service. The LOCOTROL EB system provides the performance advantages of distributed power and locomotive electronic braking.

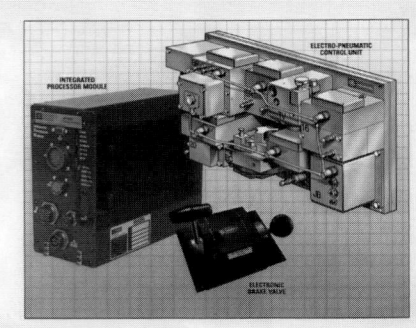

LOCOTROL EB system　0041484

Graham-White Manufacturing Co

Sales Division, PO Box 1099, 1242 Colorado Street, Salem, Virginia 24153-1099, USA
Tel: (+1 540) 387 56 20　Fax: (+1 540) 387 56 39
e-mail: sbruce@grahamwhite.com
Web: www.grahamwhite.com

Key personnel
President and Chief Executive Officer:
　James S Frantz
Vice-President, Marketing: W Stewart Bruce Jr
International Customer Service Representative:
　Beth Beckner
Manager Field Service: Clayton Abbott

Products
Pneumatic and electropneumatic devices for locomotives and powered passenger vehicles such as regenerative air dryers, locomotive sanding systems, air filters, automatic drain valves, air check valves (one- and two-way), solenoid valves, horn valves, coalescing air filters, bell ringers, analogue air gauges, air test fittings for air gauges and pressure switches, electric timers, mirror and windshield wing combinations, locomotive cab awnings and ventilators for locomotive cabs. SafeSet™ locomotive parking brake, electric motor driven; E-Bell® electronic bell; new-generation 994 air dryer system, which keeps compressed air clean, dry and oil-free, with six-year intervals between dryer overhauls. It makes use of Graham-White's patented self-adjusting purge valve and plug-in mounting bracket.

The company offers full remanufacturing services for all of its products as well as 26-L, 30 CDW, D-24 equipment and freight car brake valves from two locations, Salem, Virginia and Carson City, Nevada.

The company is now ISO 9001:2000 accredited.

Developments
GW is to release a new automated single car test device, 750 Series. Also to be released, an alternative to the SafeSet electric parking brake, GW's 'lever style' with the introduction of the 'wheel style' electric locomotive parking brake.

UPDATED

Greysham and Co

7249 (1/1) Roop Nagar, Delhi 110 007, India
Tel: (+91 11) 23 97 37 46; 23 97 38 54; 23 97 39 89;
　23 97 53 68; 23 94 24 63
Fax: (+91 11) 23 93 10 21; 23 94 08 92
e-mail: greysham@nda.vsnl.net.in
Web: http://www.greyshamco.com

Key personnel
Managing Director: Mohan Singh
Executive Director: Subodh Singh
General Manager (Operations):
　Pradeep Kumar Gupta
Export Manager: ST Ghosh

Products
The company is ISO 9002 accredited.

Braking systems – complete air brake systems for freight and passenger vehicles; OHE (overhead inspection) cars, dmu (motor and trailer) cars, C3W distributor valves to SAB WABCO design; EST distributor valve; automatic load sensing device; empty/load change-over valves; bogie-mounted brake systems and brake cylinders; reservoirs and rubber hoses.

Vacuum brake equipment, including 'E' and 'F' type cylinders; Prestall cylinders; QSA valves; couplings; release valves; couplings and similar equipment.

LED signal lights for railway traffic, non-powered water coolers for passenger coaches; semi-automatic locks for flat wagons.

A load-sensing device has been introduced and a shock-absorber range is being developed.

Greysham International

1/1-A, Man Singh Place, Roop Nagar, Delhi 110 007, India
Tel: (+91 11) 23 84 93 81   Fax: (+91 11) 23 84 30 68
e-mail: sales@greysham.com
Web: www.greysham.com

Key personnel
Managing Director and Chairman: Govind Singh
Joint Managing Director: Sanjeev Singh
Export Manager: V B Arya

Associate company
Greysham (International) Pvt Ltd

Products
Air brake systems for freight and passenger vehicles including: distributor valves, automatic empty/load device, load sensing device, brake cylinders, centrifugal dirt collector, end cocks, isolating cocks, guards emergency valve, hose couplings, passenger emergency signal device, passenger emergency relay valve, reservoirs, slack adjusters, non-asbestos brake blocks type 'L' and 'K', pipe and pipe fittings, brake cylinder with in-built slack adjuster.

Vacuum brake equipment and spares 'E' and 'F' type vacuum brake cylinders, prestall cylinders, DA valve, release valve, hose couplings, reservoirs, QSA valve, dummy couplings, brake gear pins and bushes.

ISO 9001-2000 certified, accredited by NQAQSR, New Zealand.

UPDATED

Hanning & Kahl GmbH & Co KG

Rudolf Diesel Strasse 6, D-33813 Oerlinghausen, Germany
PO Box 1342, D-33806 Oerlinghausen, Germany
Tel: (+49 5202) 70 76 00   Fax: (+49 5202) 70 76 29
e-mail: info@huk.hanning.com
Web: www.hanning-kahl.de

Key personnel
General Manager: Wolfgang Helas
Brake Division Manager: Dietrich Radtke
LRT Division Manager: Christian Schmidt
Service Division Manager: Peter Spilker
Sales Manager, Brakes: Jürgen Stammeier
Sales Manager, LRT: Hans-Joachim Pässler
Sales Manager, Services/LRT: Joachim Zehn
Sales Manager, Services/Brakes: Martin Epp

Products
Brake Division: Electrohydraulic brake systems; spring-applied actuators, active callipers, hydraulic-power units, hydraulic emergency release units, electronic brake control systems with slide protection, track brakes, and filter and flushing units.

Service Division: Services and testing and measuring equipment for track brakes and hydraulic brake systems.

Contracts
Contracts include the supply of equipment to operators in Potsdam, Helsinki, Chemnitz, Bielefeld, Kassel, St Etienne, Bucharest and FVG Delijn.

UPDATED

Honeywell Bremsbelag (Jurid)

Honeywell Bremsbelag GmbH (Jurid Products)
PO Box 1201, D-21504 Glinde, Germany
Tel: (+49 40) 727 10   Fax: (+49 40) 72 71 24 08

Main works
Glinder Weg 1, D-21509 Glinde/Hamburg, Germany

Key personnel
Managing Director: G Kasper
Sales Director: R Drummond
Sales Manager, Railway Products: Peter Franz

Products
Composition brake blocks; disc brake pads; friction plates; sintered brake blocks and disc brake pads for heavy-duty rail brakes; data acquisition equipment and instrumentation for rail brake system evaluation.

UPDATED

IB Italian Brakes SpA

Via Ponti di Napoli, I-80036 Palma Campania (NA), Italy
Tel: (+39 081) 47 11 98   Fax: (+39 081) 827 70 31
e-mail: info@italianbrakes.com
Web: http://www.italianbrakes.com

Products
Synthetic brake pads and shoes for main line and urban rail vehicles, passenger coaches and freight wagons.

NEW ENTRY

ICER Brakes SA

Poligono Industrial Agustinos, C/G, s/n, E-31013 Pamplona, Spain
Tel: (+34 948) 32 16 40   Fax: (+34 948) 28 63 67
e-mail: railtech@icerbrakes.com
Web: http://www.icerbrakes.com

Key personnel
Chairman: Victor Ruiz Rubio
Managing Director: Juan Miguel Sucunza
Railway Product Manager: Blanca Iturralde

Products
High (K-blocks), medium (L-blocks), low (LL-blocks) and special friction asbestos-free composition brake blocks, disc brake pads and friction plates, including UIC-approved, for coaches, high-speed coaches, locomotives, freight cars, emus, LRVs and metro cars.

ICER Brakes is currently working on the ERS project (Euro Rolling Silent) for replacing cast iron blocks in all the wagons within Europe. The main objective is to reduce the rolling noise produced by the high roughness of the wheel, created by the cast iron blocks while braking and the braking noise. ICER Brakes has already successfully replaced cast iron blocks in Portugal, where they are no longer used.

Jarret

198 avenue des Grésillons, F-92602 Asnières Cedex, France
Tel: (+33 1) 46 88 16 20   Fax: (+33 1) 47 90 03 57
e-mail: contact@jarret.fr

Key personnel
Chairman and General Manager: Bruno Domange
Managing Director: Antoine Domange

Products
Buffers; shock-absorbers for automatic couplers; buffer stops.

Kamax SA

ul Zielona 2, PL-37-220 Kanczuga, Poland
Tel: (+48 16) 648 76 82   Fax: (+48 16) 648 76 82
e-mail: kamax@kamax.com.pl
Web: http://www.kamax.com.pl

Products
Elastomeric shock absorbers for rail vehicle buffers; complete buffers; absorbers for automatic couplers.

Karl Georg

Karl Georg Bantechnik GmbH & Co KG
Rheinstrasse 15, D-57638 Neitersen, Germany
Tel: (+49 26) 81 80 80   Fax: (+49 26) 818 08 21
e-mail: K.Georg.Bahntechnik@T-Online.de

Key personnel
Managing Directors:
    Michael Schnaufer George W Hoffmann
Sales Manager: Achim Geyer

Background
Karl Georg is a member of the Amsted Group.

Products
Karl Georg produces buffers using a hot-extruding process. Products include: side buffers in all categories according to UIC 526-1; longstroke buffers according to UIC 526-3; passenger coach buffers according to UIC 528; drawgear; buffers for industrial equipment; parts for automatic couplers; hot-extruded components from 20 to 350 kg.

Keystone

Keystone Industries Inc
3420 Simpson Ferry Road, PO Box 456, Camp Hill, Pennsylvania 17001-0456, USA
Tel: (+1 717) 761 36 90   Fax: (+1 717) 763 99 17

Key personnel
President: G W Hoffman
Vice-President, International Marketing:
    WT Malinowski
Vice-President and Chief Technical Officer:
    M P Scott

Products
Hydraulic end-of-car and centre-of-car cushioning unit with strokes from 120 to 500 mm for freight wagon, locomotive and passenger coach applications. Friction steel and elastomeric draft gear for freight wagons, locomotives and passenger coaches. KeyGard elastomeric pads for railroad and industrial applications. Keystone hopper outlet sliding gates.

Contracts include the supply of Keystone products to UP, SP, BNSF, CSX, NS, Conrail, CN and CP and railways in Asia, Australia, Africa and South America, in addition to General Motors, GE Transportation Systems, Thrall, TTX, Johnstown America, TrentonWorks, Transcisco Industries, Goninan and Dorbyl.

Keystone Bahntechnik GmbH

Rheinstrasse 15, D-57638 Neitersen, Germany
Tel: (+49 2681) 80 80   Fax: (+49 2681) 808 21
e-mail: keystone@krec.de
Web: www.krec.de

Key personnel
Managing Directors: Lothar Gintze,
    Michael Aherne, Achim Geyer
Head of Technical Department: Matthias Wagner

Background
Keystone Bahntechnik is a result of the merger between Karl Georg Bahntechnik/Neitersen, Eisenbahntechnik Halberstadt/Halberstadt and 'KG' Ringfeder Bahntechnik/Krefeld. It is a member of the Amsted Industries Group, USA, and is the main office of Keystone Europe.

Products

Buffers using a hot-extruding process. Includes buffers in all UIC 526-1 categories, longstroke buffers according to UIC 526-3, passenger coach buffers according to UIC 528, drawgear, draft gears, connection bars including damper systems, crash buffers and crash-car-protection-systems, friction springs, hydraulic energy absorption dampers, buffers for industrial equipment, parts for automatic couplers, and hot-extruded components from 20 to 350 kg.

UPDATED

Knorr-Bremse Rail Systems (UK) Limited

Westinghouse Way, Hampton Park East, Melksham SN12 6TL, UK
Tel: (+44 1225) 89 87 00   Fax: (+44 1225) 89 87 01
e-mail: wbl.sales@westbrake.com
Web: www.knorr-bremse.co.uk

Key personnel

Managing Director: Paul R Johnson
Engineering Director: Jason Abbott
Marketing and Sales Director: Peter C Johnson
Projects Director: Danny Lee
Manufacturing Director: Martyn T Perkins
Finance Director: Stephen Thomas

Background

Knorr-Bremse Rail Systems (UK) Limited is a member of Knorr-Bremse group. In March 2005 the company moved from Chippenham to a purpose-built factory in Melksham. At the same time the company changed its name from Westinghouse Brakes to Knorr-Bremse Rail Systems (UK) Ltd to reflect its enhanced sales and service role within the global Knorr-Bremse Group.

Products

Brake systems comprising air and vacuum brake equipment, electro-pneumatic brake systems with digital or analogue control for metro and commuter passenger trains. Equipment includes rotary and reciprocating air compressors, air treatment equipment, drivers brake and brake/traction controllers, wheelslide protection equipment, sanding systems, warning equipment, air suspension control equipment, brake actuation equipment, tread and disc brake equipment and magnetic track brakes.

A comprehensive product support service is available providing equipment repair, overhaul and long-term maintenance and spare parts supply for the UK and Irish markets, which is for the complete Knorr-Bremse product range. In addition to braking systems and associated equipment, other products supported within the 'on-board systems' division include IFE train doors, Merak HVAC, Frensistemi toilet systems and passenger information systems, and Microelettrica power electrics components.

Contracts

In September 2004 Westinghouse Brakes was awarded over GBP100 million of braking systems and maintenance business by Bombardier Transportation for its new London Underground trains. The initial contract is for the supply of the entire brake system for 376 new deep tube fleet metro cars being built by Bombardier Transportation for Metronet Rail use on the Victoria Line. The braking system incorporates the latest technology from Knorr-Bremse Sfs, including EP2002 Distributed Brake Control and the VV120T Oil-Free Compressor. The second stage of the contract includes the supply of braking systems to a further 1,362 Bombardier cars for Metronet's sub-surface lines, for example Circle, District, Hammersmith and City, and Metropolitan Lines. Westinghouse Brakes will also provide maintenance and service support for the brake control, bogie brake and air supply equipment.

Other recent contracts include: brake systems for Desiro, Electrostar, Juniper and Coradia multiple-unit vehicle platforms, Arriva Trains Northern trains, 'Voyager' CrossCountry and Midland Mainline

Knorr-Bremse Type WZK pneumatic brake caliper units   0567153

Knorr-Bremse oil-free compressor   0567154

'Meridian' demus, Kuala Lumpur Airport Express trains; Docklands Light Railway, Shanghai Pearl Line and Shanghai Xin Min metro trains; Class 66 locomotives and wheelslide equipment for the First Great Western high-speed trains.

UPDATED

Knorr-Bremse AG

Moosacher Strasse 80, D-80809 Munich, Germany
Tel: (+49 89) 354 70   Fax: (+49 89) 35 47 27 67
Web: www.knorr-bremse.com

Key personnel

Chairman: Dieter Wilhelm
Executive Board: H H Thiele, Peter Riedlinger; Jens Theuerkorn

Knorr-Bremse Systeme für Schienenfahrzeuge GmbH
(address as Knorr-Bremse AG)
Managing Directors: Dr Frank Gropengiesser, Wolfgang Schlosser; Dr Albrecht Köhler; Rolf Graf

Works

Berlin and Munich

Subsidiary companies

transportdata AG, Munich, Germany, Knorr-Bremse Electronic GmbH, Munich
Knorr-Bremse GesmbH, Mödling, Austria
Oerlikon-Knorr Eisenbahntechnik AG, Niederhasli, Switzerland
Knorr-Bremse GmbH, Mödling, Austria
Dr techn J Zelisko GmbH, Mödling, Austria
Knorr-Bremse GmbH Division IFE Automatic Door Systems, Waidhofen an de Ybbs, Austria
Westinghouse Brakes (UK) Ltd, Chippenham, UK
Freinrail Systèmes Ferroviaires SA, Reims Cedex, France
Frensistemi Srl, Florence, Italy
Sociedad Espanola de Frenos, Calefaccion y Señales SA, Pinto/Madrid, Pinto (Madrid), Spain
Knorr-Bremse Nordic Rail Services AB, Lund, Sweden
Knorr-Bremse Systemy dla Kolejowych Srodków Lokomocji PL Sp Zoo, Katowice, Poland
New York Air Brake Corporation, Watertown, USA
Knorr Brake Corporation, Westminster, USA
Knorr Brake Ltd, Kingston, Canada

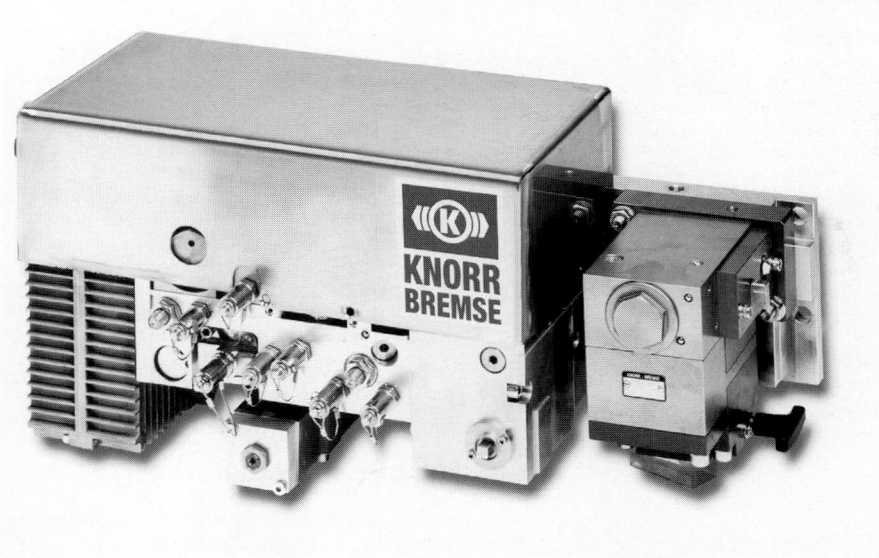

Knorr-Bremse modular configurable EP Compact (R) brake control system for suburban and intercity trains

NEW/0585082

Knorr-Bremse Sistemas para Veículos Ferroiários Ltda, São Paulo, Brazil
Knorr Bremse SA Pty Ltd, Kempton Park, South Africa
Knorr-Bremse India Private Ltd, Faridabad, India
Knorr-Bremse Rail Systems Japan, Tokyo, Japan
Knorr-Bremse Far East Ltd Rail Systems Division, Hong Kong, China
Knorr-Bremse Brake Equipment (Shanghai) Co Ltd, Shanghai, China
Knorr-Bremse Australia Pty Ltd, Granville, Australia

Products
Knorr-Bremse Systems for Rail Vehicles projects, develops, manufactures and supplies brake systems and on-board systems for all kinds of rail vehicles worldwide. This comprises brake systems and on-board systems for locomotives and freight cars with AAR-technology and locomotives, freight and passenger cars, dmus and emus with UIC-technology. The product range also includes brake and on-board systems for tramways, metros, people movers and special vehicles. The product range of brake systems comprises air supply systems (piston- and screw compressors, oil-free compressors, air dryers, oil filters, condensate collectors, complete air-supply systems), brake control systems (carrier systems, control units, brake controller, valves, ESRA platform, sensor systems for wheel slide control, pressure management and load control), bogie equipment (brake discs and pads, brake callipers, block brake units, track brakes, eddy-current brakes) and hydraulic systems (hydraulic units, brake actuators, suspension systems). The product range of on-board systems comprises automatic door systems, platform screen doors, dampers, toilet systems, windscreen wipers, sanding equipment, passenger information systems, traffic control and transport telematics systems. A comprehensive after-sales package is offered. The broad spectrum of different services from urban traffic enterprises up to national operating companies covers the entire life cycle of the products of Knorr-Bremse Systems for Rail Vehicles.

Contracts
Recent contracts include: brake equipment for Virgin Trains West Coast Main Line (UK); KCRC and MTRC, Hong Kong (China); Vienna metro (Austria); Lisbon metro (Portugal); Long Island Rail Road and Amtrak Acela (USA).

UPDATED

Kovis doo

Velika Dolina 37, SL-8261 Jesenice na Dolenjskem, Slovenia
Tel: (+386 7) 457 31 00
Fax: (+386 7) 495 73 32
e-mail: kovis@siol.net; info@kovis.si
Web: www.kovis.si

Key personnel
Sales and Marketing: Aleksander Gajski

Products
Brake discs for European manufacturers and operators of rail vehicles. Produced under UIC regulations, Kovis brake discs can be used for all types of rolling stock, for passenger and freight vehicles as well as for diesel-powered trains, metro and tram cars. The products range comprises wheel and axle-mounted brake discs, available in grey cast or graphite cast iron. The friction rings can be divided or non-divided. Kovis also produces bearing boxes for various rail vehicle manufacturers and operators. The company has also developed a bearing box for the Y25t, with an axle load of 22.5 t and 25 t. The bearing boxes can be used in passenger traffic as well as for freight. The main product range of brake discs and bearing boxes, brake lining support (brake pad holders) are produced in dimensions from 350 up to 400 cm², with or without cross positioned bolt holder pin, according to customer needs.

UPDATED

KYB – Kayaba Industry Co Ltd

World Trade Center Building, 2-4-1, Hamamatsu Cho, Minato Ku, Tokyo 105-6111, Japan
Tel: (+81 3) 34 35 35 751   Fax: (+81 3) 34 36 29 07
Overseas business department:
Tel: (+81 3) 34 35 35 81   Fax: (+81 3) 34 36 29 08
Web: http://www.kyb.co.jp

Key personnel
President: Tadahiko Ozawa

European representation
Gadelius Europe AB
Box 859, SE-301 18 Halmstad, Sweden
Tel: (+46 35) 18 20 71   Fax: (+46 35) 18 20 71
e-mail: europe@gadelius.com
Web: http://www.gadelius.com

Products
Caliper brakes; tread cleaning devices.

LAF – Les Appareils Ferroviaires

Les Appareils Ferroviaires
55 Rue du Bois Chaland, CE 2928 Lisses, F-91029 Evry Cedex, France
Tel: (+33 1) 69 11 93 26   Fax: (+33 1) 69 11 93 27
e-mail: laf@cimgroupe.com

Key personnel
General Manager: Alain Lovambac
Sales Department: C Danel
Technical Department: S Franco

Products
Couplers and draftgear, including automatic and knuckle couplers, special transition couplers, draw hooks and screw couplings, ring (friction) springs for draft and buffing gear.

Lesjöfors AB

SE-524 92 Herrljunga, Sweden
Tel: (+46 513) 220 00   Fax: (+46 513) 230 24
e-mail: info@lesjoforsab.com
Web: http://www.lesjofors.com

Products
Springs for rail vehicle braking systems. Companies supplied include SAB Wabco.

Meridian Rail LLC

1200 Corporate Drive, Birmingham, Alabama 35242, USA
Tel: (+1 205) 991 03 84   Fax: (+1 205) 972 13 86
Web: www.meridianrail.com

Key personnel
President: Rick Turner
Vice-President, Operations: Bill Holcomb
Vice-President, Sales and Marketing: Frank Cristelli

Background
Meridian Rail's predecessor was ABC-NACO, which had filed for Chapter 11 in November 2001, and its assets sold, in January 2002. ABC-NACO re-emerged as Meridian Rail in January 2002.

Products
Wheelset inventory management and distribution of new and used wheelsets to railroads and maintenance centres in the US and Mexico.

Meridian Rail Reconditioning Services (MRS) is a full service reconditioning business certified under AAR M-1003, with technical certification under M-212 for couplers and yokes and M-300 for brake beams. In addition, MRS carries a number of freight car component parts for resale through its facility in Chicago and through wheel shops located throughout the US.

UPDATED

Metcalfe Railway Products Ltd

Tolletts Farm, Leek Old Road, Sutton, Macclesfield, Cheshire SK11 0HZ, UK
Tel: (+44 1260) 25 23 29   Fax: (+44 1260) 25 34 13
e-mail: info@metcalferailprodltd.co.uk
Web: http://www.metcalferailprodltd.co.uk

Key personnel
Chairman and Managing Director:
  Richard H Metcalfe
Office Manager: Nigel Boyle

Products
UIC-approved Oerlikon-type air brake equipment for locomotives, freight wagons and passenger vehicles, together with associated equipment. SAB-type slack adjusters, tread brake equipment and weigh valves. Brake cylinders, hose couplings, anti-frost devices. UIC- and ISO-approved. Manuals and on-site training courses.

Contracts

Metcalfe Railway Products Ltd is continuing its exclusive agreement with MZT Hepos AD for the supply of railway braking equipment.

UPDATED

Metpro

PO Box 911-810, Rosslyn 0200, South Africa
Tel: (+27 12) 541 61 10   Fax: (+27 12) 541 35 28
Web: http://www.de.dorbyl.com

Key personnel

Works Manager: Arrie Meyer
Financial Manager: Shaun Bosch

Background

Metpro is a division of DCD-Dorbyl (Pty) Ltd.

Products

Composition brake blocks.

Miner

Miner Enterprises Inc
International Sales
PO Box 471, 1200 East State Street, Geneva, Illinois 60134, USA
Tel: (+1 630) 232 30 00   Fax: (+1 630) 232 30 55

Key personnel

President: K C Jurasek
Vice-President: M L McGuigan
Marketing Managers: C Vanbutsele

Foreign Licensees

Australia: Bradken Rail, Maud Street, Waratah, Newcastle, New South Wales 2298

Foreign Sales Agents

In Argentina; Austria; Brazil; Chile; Colombia; Egypt; Greece; India; Israel; Italy; Mexico; Netherlands; South Africa; Switzerland; Turkey; and Venezuela.

Products

Draftgear; constant-contact side bearings; TecsPak buffer and traction springs; buffers; discharge systems for bulk commodities; brake beams.

TecsPak is a heavy-duty elastomeric spring package based on Hytrel.

The Miner TF-880 and SL-76 draftgear, developed for use in 24⅝ in standard pockets, are approved in accordance with AAR Specification M-901E.

The Crown SE draftgear is an all-steel draftgear engineered for heavy unit train applications.

Miner Constant Contact Side Bearings are engineered to reduce bogie hunting on freight wagons.

The AutoMEC air cylinder-activated door mechanism is designed primarily for coal hopper wagon unloading and is adaptable for 8 to 20 door openings.

The Autolok II discharge gate for hopper wagons features the BackLOK positive locking mechanism, ledgeless clear opening, sealed door and low operating torque. It meeets all AAR S-233 specification requirements.

Mitsubishi Electric Corporation

Mitsubishi Denki Bldg, 2-3 Marunouchi 2-chome, Chiyoda-ku, Tokyo 100, Japan
Tel: (+81 3) 32 18 34 30   Fax: (+81 3) 32 18 90 48
Web: http://www.melco.co.jp/society/traffic/

Key personnel

President: Tamotsu Nomakuchi
Executive General Manager (Public Utility Systems Group): Takaaki Kijima

Products

Airbrake equipment for locomotives, electric multiple-units, light rail vehicles, rubber-tyred vehicles and monorails.

Multi-Service Supply Division

The Buncher Co
Ferry Street and Avenue 'C', Leetsdale, Pennsylvania 15056, USA
Tel: (+1 412) 741 15 00   Fax: (+1 412) 741 33 20
e-mail: multiservice@buncher.com
Web: http://www.buncher.com

Key personnel

President: Thomas Balestrieri
General Manager: Robert Lewis
Sales and Marketing Director, Marketing: Paul Bittner
International Sales Representative: Mark Jackovic

Products

Airbrake systems and components for locomotives, freight wagons and passenger coaches, repair and reconditioning.

VERIFIED

MZT Hepos AD

Pero Nakov bb, 1000 Skopje, Former Yugoslav Republic of Macedonia
Tel: (+389 02) 254 97 80; 254 97 91
Fax: (+389 02) 254 98 51; 254 98 48
e-mail: mzthepos@on.net.mk
Web: www.hepos.com.mk

Key personnel

General Manager: Vlado Atanasovski
Deputy General Manager: Stojče Smileski
Sales Director: Dušan Popović

Background

MZT HEPOS was established in 1953 with the purchase of the licenses from Oerlikon (for various driver's brake valves, distributer type Est, electro-pneumatic valves). SAB (for DRV slack adjusters, weighing valves, changeover boxes), and STABEG (for brake cylinders). Although originally designed to produce components and devices for the former Yugoslav Railways and wagon and locomotive builders, the company has grown to be one of the biggest brake equipment producers in southern and eastern Europe. All products are manufactured to applicable International Railway Standards (UIC) under strict ISO 9001 quality conditions. MZT HEPOS has been part of the POLi Group (Italy) since September 2003.

Products

Complete pneumatic and brake systems for locomotives, passenger and freight vehicles; brake

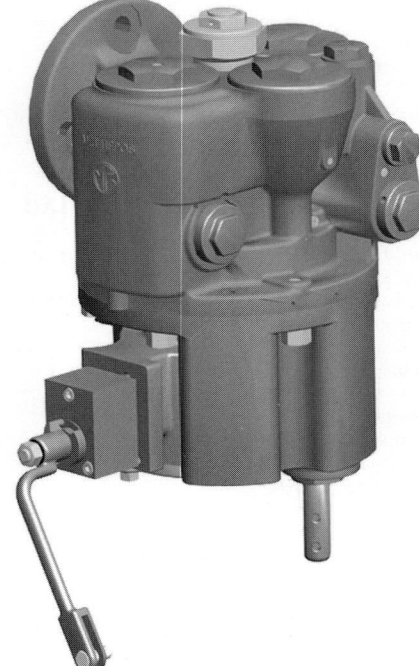

MZT-HEPOS distributor valve   **NEW**/1113740

equipment (UIC approved) including pneumatic, electropneumatic, hydropneumatic systems and electronic components; driver's brake valves, distributors, disc brakes and tread brake actuators, brake cylinders, slack adjusters, load brake valves, auxiliary pneumatic equipment, end cocks, hoses and coupling heads, air dryers, brake panels, windscreen washers, wheelslide protection devices, diagnostic and test stands.

New products include: distributor valve MH3f HBG310, the main control unit for pneumatic systems of railway vehicles, developed and manufactured by MZT HEPOS and approved by the UIC in 2003. All vehicles that run with the MH3f HBG300 unit also carry the brake system, MH, which is an acronym of MZT HEPOS.

UPDATED

Nabtesco Corporation

Railroad Products Company, 9-18 Kaigan 1-chome, Minato-ku, Tokyo 105-0022, Japan
Tel: (+81 3) 54 70 24 01   Fax: (+81 3) 54 70 24 24
e-mail: takashi_koyama@nabtesco.com
Web: www.nabtesco.com

Key personnel

Company President: Koshiro Yoshida
Company Vice-President: Masanori Kawanishi
General Manager, Sales and Marketing: Yukiyasu Fujimoto
General Manager, Overseas Marketing: Takashi Koyama

Background

Nabtesco Corporation was previously called NABCO Ltd.

Products

Airbrake systems; air compressors, composition brake shoes; automatic slack adjusters; air suspension levelling valves and automatic brake testing systems.

Plug and sliding door systems; platform doors (manufactured and supplied by Nabtesco Corporation, NABCO Company); pressure-sealing systems; windscreen wiper motors.

NABCO brake systems are operating on over 49,700 emu cars of various Japanese operators and on the Shinkansen high-speed trainsets of the JR Group companies.

Contracts

Recent contracts include the supply of braking systems to the People's Republic of China for 224 metro cars for Beijing Urban Transit Railway Co Ltd, 96 metro cars for Beijing subway Jington Development Co Ltd, 48 cars for Wuhan Rail Transit Co Ltd and 84 monorail cars for Chongqing Rail Transit General Corporation, 116 cars for Tianjin Binhai Mass Transit Development Co Ltd and to Taiwan for 321 metro cars for Taipei Rapid Transit Corporation.

UPDATED

Newag GmbH & Co KG

PO Box 120355, D-46117 Oberhausen, Germany
Tel: (+49 208) 86 50 30   Fax: (+49 208) 865 03 20
e-mail: info@newag.de

Key personnel

For full personnel listing, see Newag GmbH & Co KG entry in *Locomotives and powered/non-powered passenger vehicles* section.

Products

Composition brake shoes, shoe carriers, disc brake pads. Newag has developed a range of non-asbestos composition brake shoes and can now supply appropriate friction materials with friction coefficients between 0.15 and 0.35 μm for any desired brake shoe. The same applies to Newag disc brake pads, which can be supplied with friction coefficients between 0.25 and 0.35 μm.

Oleo International Ltd

Grovelands, Longford Road, Exhall, Coventry CV7 9NE, UK
Tel: (+44 2476) 64 55 55   Fax: (+44 2476) 36 42 87

Key personnel
For full personnel listing, see Oleo International Ltd entry in *Freight vehicles and equipment* section.

Products
Supplier of hydraulic side buffers to major European railways. Oleo has introduced specialised long-stroke side buffers to its range. By increasing the stroke from the traditional 105 to 150 mm, the energy capacity is increased by up to 50 per cent. The Type 5SC-150 has been developed to meet the requirements of provisional UIC 526-3. Long-stroke hydraulic side buffers give additional protection to delicate goods.

Also available are the standard Type 4EC-80–105 and 5SC-105 (high static resistance) side buffers. The company also offers the Type 3RCA in combination with a friction spring, supplied by Ringfeder.

The Oleo Hydraulic Draftgear Type DA 4463 (Hycon) has an energy absorption capacity of 300,000 ft lb.

Products include the Oleo Hydraulic Drawspring, an adaptation of the Oleo side buffer capsule to protect drawhooks from snatch damage, particularly on long trains.

Energy-absorbing Anticlimber Side Buffers for passenger stock are on the Heathrow Express trainsets and Class 465 emus in the UK.

UPDATED

Paulstra

Railway Department
61 rue Marius Aufan, F-92305 Levallois-Perret, France
Tel: (+33 1) 40 89 53 31   Fax: (+33 1) 47 58 75 16
e-mail: auto.levallois@hutchinson.fr
Web: www.paulstra-vibrachoc.com

Key personnel
Commercial Director: Laurent Poirier

Subsidiary companies
Stop-Choc GmbH
Stop-Choc UK
Vibrachoc España
Paulstra Industries Inc, USA

Products
Buffers; elastic couplings; primary springs, air spring secondary suspension, axle bushes, engine mounts, sandwich mounts and pads, chevrons, bogie pivot bushes, rubber/metal secondary suspension, gangway seals, floor tiles and oil seals.

UPDATED

Peddinghaus Group

Carl Dan Peddinghaus GmbH & Co KG
Mittelstrasse 64, D-58256 Ennepetal, Germany
Tel: (+49 2333) 79 60   Fax: (+49 2333) 79 63 88
e-mail: cdp-en@t-online.de
Web: http://www.peddinghaus-group.de

Products
A supplier of high quality steel and aluminium forgings and components including screw couplings and hooks.

Poli Costruzione Materiali Trazione SpA

Via Fontanella 11, I-26010 Camisano, Cremona, Italy
Tel: (+39 0373) 77 72 33   Fax: (+39 0373) 77 72 29
e-mail: info@polibrakes.com
Web: www.polibrakes.com

Key personnel
Chief Executive Officer: Francesco Poli
Technical Director: Thomas Alm
Technical Export Sales Manager: Dott Giuseppe Poli
Purchasing Manager: Pier Antonio Biondi

Principal subsidiary
MZT HEPOS AD
Pero Nakov bb, 1000 Skopje, Macedonia

Products
Complete brake systems and their components for any kind of railway vehicle, high-speed train, passenger car, freight car, locomotive, track maintenance vehicle; tread brake units; axle-mounted and wheel-mounted brake discs by sectors; electro-magnetic track brakes; resilient wheels; wheelsets. All products comply with UNI EN ISO 9001:2000 quality system, with the new EC regulations and one UIC-approved.

UPDATED

Progressive Engineering (AUL) Ltd

Unit 5, Progressive Business Parts, Groby Road, Audenshaw, Manchester M34 5HT, UK
Tel: (+44 161) 371 04 40   Fax: (+44 161) 371 04 44
e-mail: johnw@progressive-eng.demon.co.uk
Web: http://www.progressive-eng.demon.co.uk

Key personnel
Managing Director: J Williams
Chairman: C Williams
Works Manager: P Moss

Products
Design and manufacture of valves, mountings and auxiliary fittings, air brake dump valves, anti tow-away interlocks. Subcontract machining including T-bar section.

Le Réservoir

Le Réservoir SA
rue Eugène Sue, PO Box 1139, F-03103 Montluçon Cedex, France
Tel: (+33 4) 70 03 47 47   Fax: (+33 4) 70 03 77 03

Key personnel
Director: Jean Claude Mardele
Commercial Director: Pierre Poncie
Export Manager: Clotilde Aufaure

Products
Air vessels for brake and suspension systems for trains and buses.

Contracts
Le Réservoir's air vessels are mounted on the French TGV high-speed train, TGV double-deck TER new generation (regional train) and for export on the Eurostar TGV, KTGV in Korea and the metro trains in Mexico.

UPDATED

SAB WABCO

Roskildevägen 1, PO Box 193, SE-201 21 Malmö, Sweden
Tel: (+46 40) 35 04 60   Fax: (+46 40) 30 38 03
e-mail: info@sabwabco.com
Web: www.sabwabco.com

Background
Formerly owned by Cardo Rail AB, the acquisition of SAB WABCO by the Faiveley Transport group was completed in November 2004.

Key personnel
Chief Executive Officer: Robert Joyeux
Chief Operating Officer: Pierre Sainfort
Product Line Manager 'Brake and Couplers': Mats Svensson
Product Line Manager 'Electronics': Jean-Pierre Guy
Product Line Manager 'Doors': Harald Zinggrebe
Product Line Manager 'Wheels': Michael Walter
Product Line Manager 'After-Market and Services': Mario Padovani
European Profit Centers: Ulrich Giesen
International Profits Centers (outside Europe) and Business Development: Marc Chocat
Chief Financial Officer: Etienne Haumont

Subsidiaries
SAB WABCO AB, Sweden
SAB WABCO Nordic AB, Sweden
SAB WABCO International AB, Sweden
SAB WABCO SA, France
SAB WABCO NV, Belgium
SAB IBERICA SA, SAB WABCO SA, Spain
SAB WABCO Products Ltd, UK
SAB WABCO Ltd, UK
SAB WABCO SpA, Italy
SAB WABCO do Brazil SA, Brazil
SAB WABCO CS sro, Czech Republic
SAB WABCO D&M Engineering Ltd, Australia
SAB WABCO BSI Verkehrstechnik Products GmbH, Germany
SAB WABCO BSI Verkehrstechnik GmbH, Germany
Gutehoffnungshütte Radsatz GmbH, Germany
SAB WABCO KP GmbH, Germany
SAB WABCO Korea Ltd, South Korea
SAB WABCO India Ltd, India
SAB WABCO Polska sp zoo, Poland

Products
Braking systems for locomotives, passenger and freight vehicles, including LRVs and guided vehicles with rubber tyres; UIC approved automatic airbrakes to the requirements of most railway administrations; electropneumatic, electrohydraulic, electromechanical and all-electric brake systems; vacuum and combined brake systems.

Air generation: air supply equipment, reciprocating and screw compressors, air treatment devices.

Tread brakes/disc brakes: automatic slack adjusters, variable load devices; friction pair devices, tread brakes, brake discs, calliper assemblies, disc brake actuators with spring, hydraulic or mechanically operated parking; friction materials; electromagnetic track brakes.

Wheel products: solid wheels, wheelsets, resilient and low-noise wheels, running gear.

Other: UIC-approved microprocessor-controlled wheelslide protection devices, anti-slip and speed controls; automatic test equipment for brake controls; automatic computer-controlled systems for marshalling yards; transit car coupling systems.

Recent contracts for high-speed trains include supply of complete brake systems for Pendolino trains, TGV South Korea, TGV France, brake equipment for X2000 Sweden and wheels for ICE Germany.

Brake systems and brake equipment for dmus and emus have been supplied to: Australia, Brazil, France, Germany, Hungary, Italy, Netherlands, Spain, Sweden, Taiwan and UK. Also wheels for ET474 cars, Germany.

Complete brake systems and brake equipment for freight wagons have been supplied to many railways around the world.

Complete brake systems for container wagons among others have been supplied to AAE, RoadRailer and NS.

Electrohydraulic brake systems have been supplied for LRVs and trams for Cologne, Rome, and Grenoble, electrohydraulic brake systems and wheels to Strasbourg, and wheels to Val de Seine. Electromechanical systems have been supplied to Hungary and Czech Republic, and wheels have been supplied for Darmstadt, Nuremberg, Vienna and Manila.

Electropneumatic brake systems have been delivered to the Paris metro, Rome metro, Santiago metro, Teheran metro and Tehran metro and electropneumatic brake systems and wheels to the Stockholm metro.

For locomotives, complete brake systems have gone to France, Germany and Italy.

UPDATED

SEE

Société Européenne d'Engrenages
5 rue Henri Cavallier, PO Box 716, F-89107
Saint-Denis-les-Sens, France
Tel: (+33 3) 86 95 62 00   Fax: (+33 3) 86 95 62 41

Key personnel
President: J P Fontaine
Commercial Director: P Davion

Products
Disc brakes, hangers and hydraulic calipers.
   SEE designed and manufactured the high-performance disc brake mounted on SNCF's TGV-Atlantique high-speed train (300 km/h speed in service).
   The company has introduced a wheel flanged disc brake provided with flexible steel wheel fixation.

SMC Pneumatics

Transport Division, 18 Hudson Avenue, Castle Hill, NSW 2154, Australia
Tel: (+61 2) 935 48 22 22   Fax: (+61 2) 96 34 77 64

Key personnel
For full personnel listing, see SMC Pneumatics entry in *Electric traction equipment* section.

Products
Air dryers and filters, drain valves, pressure and flow control valves, EP valves.

Stabeg

Stabeg Apparatebau GmbH
Reinlgasse 5-9, A-1140 Vienna, Austria
Tel: (+43 1) 92 26 28   Fax: (+43 1) 92 61 66

Products
Airbrake equipment for locomotives, coaches, freight wagons; air springs; drawgear; buffer and drawgear; side buffers.

Standard Car Truck Co

865 Busse Highway, Park Ridge, Illinois 60068, USA
Tel: (+1 847) 692 60 50   Fax: (+1 847) 692 62 99

Key personnel
For full personnel listing, see Standard Car Truck Co entry in *Bogies and suspension*, wheels and axles, bearings section.

Products
High and low-friction Anchor composition brake shoes for locomotives and freight wagons.

Stone India Ltd

16 Taratalla Road, Calcutta 700 088, India
Tel: (+91 33) 24 01 46 61/668
Fax: (+91 33) 24 01 48 86/24 01 34 51
e-mail: mktg.kolkata@stoneindia.co.in
e-mail: gautamghosh@stoneindia.co.in
Web: www.stoneindia.co.in

Key personnel
Managing Director: A Mondal
Chief Marketing Officer: Gautam Ghosh
National Sales Manager: R K Ganeshan

Products
Airbrake systems for locomotives, passenger coaches and freight wagons. Slack adjusters for coaches and wagons, tread brake units for locomotives, brake cylinders, brake cylinders with integral slack adjuster, air dryers for diesel and electric locomotives and for electric and diesel multiple units.

UPDATED

Sumitomo Metal Industries Ltd

8-11 Harumi 1-Chome, Tokyo 104-6111, Japan
Tel: (+81 3) 44 16 62 71
Fax: (+81 3) 44 16 67 90
e-mail: skr@sumitomometals.co.jp
Web: www.sumitomometals.co.jp/e/index.html

Works address
Osaka Steel Works, 1-109 Shimaya 5-chome, Konohana-ku, Osaka 554-0024, Japan

Key personnel
Senior Managing Executive Officer: Yasutaka Toya
Senior Managing Executive Officer: Kaoru Goto
Managing Executive Officer: Mitsunori Okada
General Manager: Shiuji Morinobu
Senior Manager: Yukinari Akimoto

Products
Couplers; draftgear and steel castings for the railway industry; bogies (see *Bogies and suspension, wheels and axles, bearings* section).

UPDATED

Svendborg Brake A/S

Jernbanevej 9, DK-5882 Vejstrup, Denmark
Tel: (+45) 63 25 52 55
Fax: (+45) 62 28 10 58
e-mail: sb@svendborg-brakes.dk
Web: http://www.svendborg-brakes.dk

Products
Hydraulic and pneumatic disc brakes, electronic products for brakes, hydraulic and pneumatic control devices for brakes.

Textar

Textar Kupplungs-und Industriebeläge GmbH
Industriestrasse 7, D-57577 Hamm/Sieg, Germany
Tel: (+49 2682) 70 82 34   Fax: (+49 2682) 70 81 50

Products
Non-asbestos friction materials, mainly based on organic binders.

ThyssenKrupp GfT Gleistechnik GmbH

Altendorfer Strasse 120, D-45143 Essen, Germany
Tel: (+49 201) 188 37 65
Fax: (+49 201) 188 37 57
Web: www.thyssenkruppgleistechnik.de

Key personnel
Contact: H Weiб

Background
ThyssenKrupp GfT Gleistechnik GmbH is a subsidiary of ThyssenKrupp Services AG, Düsseldorf, Germany.

Products
Brake discs.

NEW ENTRY

TMD Friction

PO Box 18, Cleckheaton, BD19 3UJ, UK
Tel: (+44 1274) 85 40 00
Fax: (+44 1274) 85 40 01

Key personnel
Managing Director: Bob Cornish

Products
Asbestos-free low- and high-friction composition brake blocks for passenger and freight applications. Asbestos-free disc pads for a variety of applications.

Contracts
Recent contracts include the supply of disc brake pads for Virgin West Coast tilting trains and high-friction brake blocks for the Singapore metro.

Triax-YSD

401 North 8th Street, Benton Harbour, Michigan 49022, USA
Tel: (+1 269) 925 00 11   Fax: (+1 269) 925 00 00
e-mail: sales@triax-ysd.com
Web: http://www.triax-ysd.com

Key personnel
President: William Mundinger
General Manager: Glenn L Wright

Products
Solid-truss brake beams and bogie-mounted braking systems including custom designs.

Ueda

Ueda Brake Co Ltd
10-19 Tomobuchi-cho 2-chome, Miyakojima-ku, Osaka, Japan
Tel: (+81 6) 921 29 71   Fax: (+81 6) 921 29 75

Key personnel
President: Takafumi Ueda
Marketing Director: Tadayoshi Akiyama
Senior Manager, Engineering: Akio Yoshioka

Products
Composite brake shoes to prevent reduction of the adhesion coefficient between wheel and rail during operation in wet weather.

Unity Railway Supply Co Inc

805 Golf Lane, Bensenville, Illinois 60106, USA
Tel: (+1 630) 595 45 60   Fax: (+1 630) 595 06 46

Key personnel
Chairman: Harold R O'Connor
President: Robert S Grandy
Executive Vice-President, Financial: Robert Holden

Products
Air brake hose supports; batten bars; roller bearing adapters; handbrakes; hand wheels and bell cranks; brake levers and parts; cast steel draft lugs and centre plates; centre pins; end platforms (grating type); fasteners, filters, pipe brackets/dust strainers; friction castings (standard and split wedge); gravity discharge outlet gates; handholds (grab irons); journal lubricators; ladders; lumber corner protectors; plug door guides; running boards (grating type); side posts; sill steps; warning lights (blue flag and barricade); wheel chocks.

UPDATED

Voith Turbo Rail Systems Ltd

Unit 7, Silverdale Industrial Centre, Silverdale Road, Hayes UB3 3BP, UK
Tel: (+44 20) 88 61 21 31   Fax: (+44 20) 88 69 17 26
e-mail: turbo.uk@voith.com

Key personnel
Managing Director: W L 'Chip' Boyd
General Manager: Paul Mallows
Sales Contact: Roger Everest

Background
In October 2002, Radenton Scharfenberg and Voith Turbo merged to become Voith Turbo Rail Systems Ltd.

Products

'Wedgelock' autocouplers and bar couplers; anticlimbers; Scharfenberg autocouplers and bar couplers; Tightlock autocouplers; Albert couplers; shock-absorbing front hoods.

Contracts

Recent contracts include the supply of autocouplers for trainsets in use on London Underground Ltd's Northern line. Recent contracts include orders from Great Western Holdings and ScotRail.

Voith Turbo Scharfenberg GmbH & Co KG

PO Box 311157, D-38231 Salzgitter, Germany
Tel: (+49 5341) 21 02   Fax: (+49 5341) 21 42 02
Web: www.voithturbo.de

Key personnel

General Manager: M Wawra
Sales Director: H Costard
Technical Director: S Kobert

Principal licensees and subsidiary companies

Australia: Voith Turbo Scharfenberg Pty Ltd
11 Sleigh Place, Wetherill Park, 2164, New South Wales, Australia
Spain: CAF
UK: Voith Turbo Rail Systems Ltd (qv)
France: Voith Turbo Scharfenberg GmbH & Co KG
Succursale Couplematic, 21 rue de Clichy, F-93584 Saint Ouen Cedex, France

Background

The company name changed from Scharfenbergkupplung GmbH & Co KG to Voith Turbo Scharfenberg GmbH & Co KG in September 2002.

Products

Automatic multifunction couplers, semi-permanent couplers and drawgear for light rail, rapid transit, trams, metros, powered and non-powered passenger vehicles, freight wagons, locomotives, automated guideway transit, mountain railways; special couplers for shunting vehicles, ladle cars, cranes; adaptor couplers, electric couplers; impact protection modules, complete front noses, articulations.

Contracts

Recent contracts include automatic and semi-permanent couplers for high-speed trains including: Talgo 350 and Velaro E, Spain; TGV, France; ICE 3, Germany; Chinese metros such as Guangzhou, Shenzen, Nanjing and Shanghai, also Delhi Metro, India, Singapore Circle Line and North-East Line, Vienna subway, Austria; Oresund OTU trains, Sweden; Rubin Nürnberg, Germany; V'locity and PURD Perth, Australia; the AVE S 102 (Talgo 350, Spain) plus delivery of the complete front nose including driver's cabin.

UPDATED

Wabtec Corporation

1001 Air Brake Avenue, Wilmerding, Pennsylvania 15148, US
Tel: (+1 412) 825 10 00   Fax: (+1 412) 825 10 19
Web: www.wabtec.com

Key personnel

Chairman, President and Chief Executive Officer:
   William E Kassling
Vice-Chairman of the Board: Emilio A Fernandez
Board of Directors: Robert J Brooks, Kim G Davis,
   Lee B Foster II, Michael W D Howell,
   James P Miscoll, James V Napier
Senior Vice-President, Chief Financial Officer:
   Alvaro Garcia-Tunon
Vice-President, Investor Relations: Tim Wesley

Principal subsidiaries

Cardwell Westinghouse Company
FIP Pty Ltd
Jinwu Control Systems
Microphor
Milufab Inc
Motive Power
Pioneer Friction Ltd
Stone Air
Vapor Bus International
Vapor Europe
Vapor Rail
Vapor Stone UK Limited

Products

Complete brake system components for locomotives, powered and non-powered passenger vehicles and freight wagons; pneumatic, electropneumatic, electrohydraulic, electromechanical and electronic (pneumatic and hydraulic) brake equipment; mechanical and electric reciprocating and rotary air compressors; bogie-mounted brake cylinders (actuators), tread and disc brake units, brake discs, automatic slack adjusters, track brakes; composition (non-metallic, asbestos-free) brake shoes (blocks) and disc brake pads; fully automatic mechanical, pneumatic and electrical couplers for LRVs, metro trainsets and electric multiple-units; friction-type, all-metal and rubber/metal draft gears, draft gears for freight service; geared handbrakes; electronic data recorders, end-of-train units, speed indicators, alertness devices, onboard computers, fault monitors/annunciators, fuel level indicators; electromechanical and electropneumatic door actuators for passenger vehicles; air conditioning equipment for passenger vehicles.

UPDATED

Westinghouse Saxby Farmer Ltd

17 Convent Road, Entally, Calcutta 700014, India
Tel: (+91 33) 22 44 71 61
Fax: (+91 33) 22 44 71 65
e-mail: wsfedp@cal2.vsnl.net.in

Key personnel

Managing Director: A N Dutta
Director, Finance and Marketing: D K De Sarker
Executive Director: C K Chaki

Products

Electropneumatic brake system including regenerative brake blending equipment for emu coaches; air and vacuum brake system for locomotives, passenger coaches, diesel rail cranes, special purpose railway vehicles and tramcars; semi-permanent and automatic centre buffer couplers for emu coaches and centre buffer couplers for passenger coaches.

Contracts

Recent contracts include the supply of 104 sets of electropneumatic brake equipment and 166 sets of semi-permanent centre buffer couplers to Indian Railways.

UPDATED

Yutaka Manufacturing Co Ltd

1-18-17 Kitakojiya Ota-ku, Tokyo 144-0032, Japan
Tel: (+81 3) 37 41 41 31
Fax: (+81 3) 57 05 70 65
e-mail: hideo.kamei@yutaka-ss.co.jp
Web: www.yutaka-ss.co.jp/

Key personnel

President and Chief Executive Officer:
   Yasuyuki Maki

Products

Automatic train coupling/uncoupling systems.

UPDATED

For details of the latest updates to *Jane's World Railways* online and to discover the additional information available exclusively to online subscribers please visit
jwr.janes.com

Bogies and Suspension, Wheels and Axels, Bearings

What you see is what you get.

The World's Highest Standards In Railroad Wheel Sets

LOCOMOTIVE • TRANSIT • LIGHT RAIL • FREIGHT • INDUSTRIAL • HISTORICAL

In a bind?

Relax.

Stop fretting and call
Glenn: (814) 684-8484,
or catch him at glenn@orxrail.com
ORX • One Park Ave • Tipton, PA 16684 • USA

ORX®

The World's Highest Standards In Railroad Wheel Sets

LOCOMOTIVE • TRANSIT • LIGHT RAIL • FREIGHT • INDUSTRIAL • HISTORICAL

BOGIES AND SUSPENSION, WHEELS AND AXLES, BEARINGS

Alphabetical listing

Acieries de Ploërmel
ALSTOM Transport
American Koyo Corporation
AMPEP plc
Amsted Rail International
Amurrio Ferrocarrill y Equipos SA
Ansaldobreda SpA
Arbel Fauvet Rail (AFR)
ASM Spencer Moulton SAS
ASF
Astra Vagoane Călători Arad SA/Arad
Ateliers de Braine-le-Comte et Thiriau Réunis SA
AWS
Bharat Heavy Electricals Ltd (BHEL)
Bochumer Verein Verkehrstechnik GmbH
Bombardier Transportation
Bonatrans AS
Bradken Rail
Brenco, Incorporated/ QBS
British Springs
Buckeye Steel Castings Co
Bumar-Fablok Ltd
CAF – Construcciones y Auxiliar de Ferrocarriles SA
CFD Industrie
China Northern Locomotive and Rolling Stock
    Industry (Group) Corporation
Cimmco International
ČKD Kutná Hora as
CNR Qiqihar Railway Rolling Stock (Group) Co Ltd
    (QRRS)
Cockerill Forges & Ringmill SA
Comet Industries Inc
Cometna
ContiTech Luftfedersysteme GmbH
CostaRail Srl
Curtis-Wright Antriebstechnik GmbH
Delimon GmbH
Dellner Dampers AB
Delta Rail
Devol
EDI Rail
ELH Eisenbhanlaufwerke Halle GmbH & Co KG
Escorts Ltd
Extel Systems Wedel (ESW)
FAG Kugelfischer AG & Co KG
Federal-Mogul Corporation
Federnwerke J P Grueber GmbH & Co KG

Ferraz Shawmut
FIREMA Trasporti SpA
Freudenberg Schwab GmbH
Fuchs Lubricants (UK) plc
Ganz Motor Kft
Gummi Metall Technik GmbH
Greysham (International) Pvt Ltd
Griffin
Gutenhoffnungshütte Radsatz GmbH
Hindusthan Engineering & Industries Ltd (HEI)
IBG Monforts GmbH & Co
The Integral Coach Factory (ICF)
IMS UK
Issels
INKA
Kinex-ZVL A/S
KLW – Wheelco SA
K Industrier AB
Koni BV
Koyo Seiko Co Ltd
Kryukovsky
KYB – Kayaba Industry Co Ltd
Lesjöfors AB
Langen & Sondermann
Lord Corporation
Lubricant Consult GmbH
Lucchini Group
Magnus LLC
MAN Ferrostaal AG
Meridian Rail LLC
MIBA Gleitlager AG
MSA Mediterr Shock-Absorbers SpA
Multi-Service Supply Division
MWL Brasil Rodas & Eixos Ltda
Nippon Sharyo Ltd
Nizhnedneprovsky
NMB – Minebea UK Ltd
NSK Ltd
NSK UK Ltd
NTN Toyo
ORX
Peddinghaus Group
Penn
Phoenix Traffic Technology GmbH
Poli Costruzione Materiali Trazione SpA
Power Machines Group
Probotec Limited

Puzhen Rolling Stock Works of Ministry of
    Railways
CNR Qiqihar Railway Rolling Stock (Group) Co Ltd
    (QRRS)
Radsatz Ilsenburg
Railko Ltd
REBS Zentralschmiertechnik GmbH
Ringrollers
Rotem Company
RSD (DCD-Dorbyl (Pty) Ltd)
RTM SA
R W Mac Co
SAB WABCO
Sambre et Meuse
Silvertown UK Ltd
SKF Ab
Standard Car Truck Co
Standard Steel, LLC
Steel Authority of India Ltd
Stojírna Oslavany spol sro
Sumitomo Metal Industries Ltd
Superior Graphite Co
Swasap Pty Ltd
Tafesa SA
Talgo
Tatravagónka AS
Techlam
Technical Service and Marketing Inc (TSM)
Techni-Industrie SA
Tenmat Ltd
ThyssenKrupp GfT Gleistechnik GmbH
Timken Rail Service
Toyo Tire & Rubber Co Ltd
Trelleborg Industrial AVS
Tülomsas
United Goninan
Uralvagonzavod
Valdunes
Vibratech Inc
Vogel
Vossloh Locomotives GmbH
Wabtec Rail Ltd
Wheel & Axle
William Cook Rail
Woodhead Shock Absorbers
ZF Sachs AG
Zimbabwe Engineering Ltd (Zeco)

Company listing by country

AUSTRALIA
Bradken Rail
EDI Rail
United Goninan

AUSTRIA
MIBA Gleitlager AG

BELGIUM
Ateliers de Braine-le-Comte et Thiriau Réunis SA
Cockerill Forges & Ringmill SA
RTM SA

BRAZIL
MWL Brasil Rodas & Eixos Ltda

CANADA
Bombardier Transportation

CHINA
China Northern Locomotive and Rolling Stock
    Industry (Group) Corporation
Puzhen Rolling Stock Works of Ministry of
    Railways
CNR Qiqihar Railway Rolling Stock (Group) Co Ltd
    (QRRS)

CZECH REPUBLIC
Bonatrans AS
ČKD Kutná Hora as
Stojírna Oslavany spol sro

FRANCE
Acieries de Ploërmel
ALSTOM Transport
Arbel Fauvet Rail (AFR)
ASM Spencer Moulton SAS
CFD Industrie
Delta Rail
Sambre et Meuse
Techlam
Techni-Industrie SA
Valdunes

GERMANY
AWS
Bochumer Verein Verkehrstechnik GmbH
ContiTech Luftfedersysteme GmbH
Delimon GmbH
ELH Eisenhanlaufwerke Halle GmbH & Co KG
Extel Systems Wedel (ESW)
FAG Kugelfischer AG & Co KG
Federnwerke J P Grueber GmbH & Co KG

Freudenberg Schwab GmbH
Gummi Metall Technik GmbH
Gutenhoffnungshütte Radsatz GmbH
IBG Monforts GmbH & Co
Langen & Sondermann
Lubricant Consult GmbH
MAN Ferrostaal AG
Peddinghaus Group
Phoenix Traffic Technology GmbH
Radsatz Ilsenburg
REBS Zentralschmiertechnik GmbH
Thyssen Krupp
Vogel
Vossloh Locomotives GmbH
ZF Sachs AG

HUNGARY
Ganz motor Kft

INDIA
Bharat Heavy Electricals Ltd (BHEL)
Cimmco International
Escorts Ltd
Greysham (International) Pvt Ltd
Hindusthan Engineering & Industries Ltd (HEI)
INKA

Company listing by Country—*continued*

The Integral Coach Factory (ICF)
Steel Authority of India Ltd
Wheel & Axle

ITALY
CostaRail Srl
Ansaldobreda SpA
FIREMA Trasporti SpA
Lucchini Group
MSA Mediterr Shock-Absorbers SpA
Poli Costruzione Materiali Trazione SpA

JAPAN
Koyo Seiko Co Ltd
KYB – Kayaba Industry Co Ltd
Nippon Sharyo Ltd
NSK Ltd
NTN Toyo
Sumitomo Metal Industries Ltd
Toyo Tire Rubber Co Ltd

KOREA, SOUTH
Rotem Company

NETHERLANDS
Koni BV

POLAND
Bumar-Fablok Ltd Ltd

PORTUGAL
Cometna

ROMANIA
Astra Vagoane Călători Arad SA/Arad
Mecanoexportimport SA

RUSSIAN FEDERATION
Power Machines Group
Uralvagonzavod

SLOVAKIA
Kinex-ZVL A/S
Tatravagónka AS

SOUTH AFRICA
Ringrollers
RSD (DCD-Dorbyl (Pty) Ltd)
Swasap Pty Ltd

SPAIN
Amurrio Ferrocarrill y Equipos SA
CAF – Construcciones y Auxiliar de Ferrocarriles SA
Tafesa SA
Talgo

SWEDEN
Dellner Dampers AB
K Industrier AB
Lesjöfors AB
SAB WABCO
SKF Ab

SWITZERLAND
Curtis-Wright Antriebstechnik GmbH
KLW – Wheelco SA

TURKEY
Tülomsas

UKRAINE
Kryukovsky
Nizhnedneprovsky

UNITED KINGDOM
AMPEP plc
British Springs
Devol
Fuchs Lubricants (UK) plc
IMS UK
NMB – Minebea UK Ltd

NSK UK Ltd
Probotec Limited
Railko Ltd
Silvertown UK Ltd
Tenmat Ltd
Timken Rail Service
Trelleborg
Wabtec Rail Ltd
William Cook Rail
Woodhead Shock Absorbers

UNITED STATES
American Koyo Corporation
Amsted Rail International
ASF
Brenco, Incorporated/ QBS
Buckeye Steel Castings Co
Comet Industries Inc
Federal-Mogul Corporation
Griffin
Lord Corporation
Magnus LLC
Meridian Rail LLC
Multi-Service Supply Division
ORX
Penn
R W Mac Co
Standard Car Truck Co
Standard Steel, LLC
Superior Graphite Co
Technical Service and Marketing Inc (TSM)
Vibratech Inc

ZIMBABWE
Issels
Zimbabwe Engineering Ltd (Zeco)

Acieries de Ploërmel

PO Box 103, F-56804 Ploërmel, France
Tel: (+33 2) 97 73 24 70
Fax: (+33 2) 97 74 03 90
e-mail: ap.ctrochu@wanadoo.fr

Key personnel
For full personnel listing, see Acieries de Ploërmel entry in *Brakes and drawgear* section.

Background
Acieries de Ploërmel is a member of the Amsted Group.

Products
Bogie components for TGV trainsets, pendular trains, LRVs and metro trains.

Recent contracts include parts for bogies for ALSTOM's UK Juniper emu contract; for Bombardier for the Cologne metro and for Croydon Tramlink, UK; for ALSTOM for the Citadis contract in France and for new ALSTOM projects of Metropolis and MF2000.

ALSTOM Transport

1 rue Baptiste Marcet, PO Box 42, F-71202 Le Creusot Cedex, France
Tel: (+33 3) 85 73 60 00
Fax: (+33 3) 85 73 67 99
e-mail: laurent.bellot@alstom.transport.com
Web: www.transport.alstom.com

Key personnel
Managing Director: Guillaume Mehlman
Contact: Eric Lenoir

Joint venture
ALSTOM Qingdao Railway Equipment Co
231 Ruichang Road, Qingdao 266031, China
Tel: (+86 532) 499 16 05
Fax: (+86 532) 499 21 02
Contact: Lu Yin
Tel: (+86 10) 64 10 62 88   Fax: (+86 10) 64 10 62 65

Products
ALSTOM designs and produces bogies and dampers, for supply through ALSTOM Transport directly to operators or through alternative car builders. These components are accompanied by worldwide service support.

Applications extend from urban tram and metros to locomotives and very high-speed trains.

DISPEN dampers are specifically engineered to railway standards and serve bogie requirements in both primary and secondary suspensions (vertical, horizontal and anti-yaw damping) as well as inter-car applications.

Annual production capacity of bogies and dampers is 1,500 and 40,000 respectively.

Key orders and deliveries since 2000 include Shanghai, Warsaw, NYCT New York, and Singapore. Other recent contracts include: TGVs and freight locomotives for SNCF, emus for Melbourne and trams for Lyon, Bordeaux, Dublin and Rotterdam.

DISPEN Damper business was particularly active with a sharp increase in European and Chinese orders, and strong demand in service support activities.

UPDATED

American Koyo Corporation

Division of KCU Corporation of USA
29570 Clemens Road, Westlake, Ohio 44145, USA
Tel: (+1 216) 835 10 00
Fax: (+1 216 835) 93 47

Key personnel
General Manager: Yoshio Yabuno
Vice-President, Sales: Ray Normandin
OEM Sales Manager: Roger Lewis
After-Sales Manager: Don Kishton
OE Industrial Marketing Manager: Dale Neumann

Products
ABU-type journal roller bearings.

AMPEP plc

Strode Road, Clevedon BS21 6QQ, UK
Tel: (+44 1275) 87 60 21
Fax: (+44 1275) 87 84 80
e-mail: mail@ampep.co.uk
Web: www.ampep.co.uk

Key personnel
Industrial Business Manager: Andrew Vicarage

Background
The company was formed in 1963 and in 1988 became part of the SKF Group.

Products
Self-lubricating PTFE/glass fibre-lined plain bearings. AMPEP bearings have been used in rail bogies for over 20 years. Applications include tilt mechanisms, torsion bars, damper attachment points, swing links, brake mechanisms, steering linkages, valve linkages, auto- coupler and tailpin linkages.

UPDATED

Amsted Rail International

200 West Monroe Street, Suite 2301, Chicago, Illinois 60606, USA
Tel: (+1 312) 372 53 84
Fax: (+1 312) 372 82 30
Web: http://www.amsted.com

Key personnel
President: Glen F Lazar

Products
Griffen Wheel Co pressure-poured cast steel low-stress wheels.

Amurrio Ferrocarrill y Equipos SA

Maskuribai 10, E-014709 Amurrio, Alava, Spain
Tel: (+34 945) 89 16 00
Fax: (+34 945) 89 24 80
e-mail: aferreq@sea.es
Web: http://www.amufer.es

Key personnel
General Manager: J M de Lapatza
Joint Manager: L M de Lapatza
Director of Engineering: V Ruiz
Technical Manager: R Sanabria
Sales Manager: J M Gutierrez
Quality Manager: J A Garcia

Products
Bearings and axleboxes for locomotives and wagons; adaptors; wheelsets, bogies, centre pivots; buffing gear, traction motor yokes and frames; frames and bolsters for bogies and other rolling stock parts.

Ansaldobreda SpA

425 Via Argine, I-80147 Naples, Italy
Tel: (+39 081) 243 11 11   Fax: (+39 081) 243 26 98
Web:www.ansaldobreda.it

Key personnel
President: Fausto Cutuli
Chief Executive Officer: Roberto Assereto
Industrial Managing Director: Francesco Schirripa
Executive Vice-President, Commercial:
    Claudio Mannucci
Executive Vice-President, Engineering:
    Carlo Pellegrini

Products
Motor and trailer bogies for locomotives, high-speed trainsets, metro trainsets, light rail vehicles, passenger coaches and freight wagons.

Contracts
Contracts include motor and trailer bogies for light rail vehicles for San Francisco and Boston; motor and trailer bogies for metro trainsets for Atlanta, Copenhagen, San Francisco and Los Angeles; motor and trailer bogies for double-deck emus for Italian Railways (FS); motor and trailer bogies for ETR 500 high-speed trainsets for FS; bogies for freight wagons; and motor bogies for E 402B electric locomotives for FS.

UPDATED

Arbel Fauvet Rail (AFR)

140 rue du Paradis, F-59500 Douai, France
Tel: (+33 3) 27 93 49 49   Fax: (+33 3) 27 96 07 43
Web: http://www.a-f-r.fr

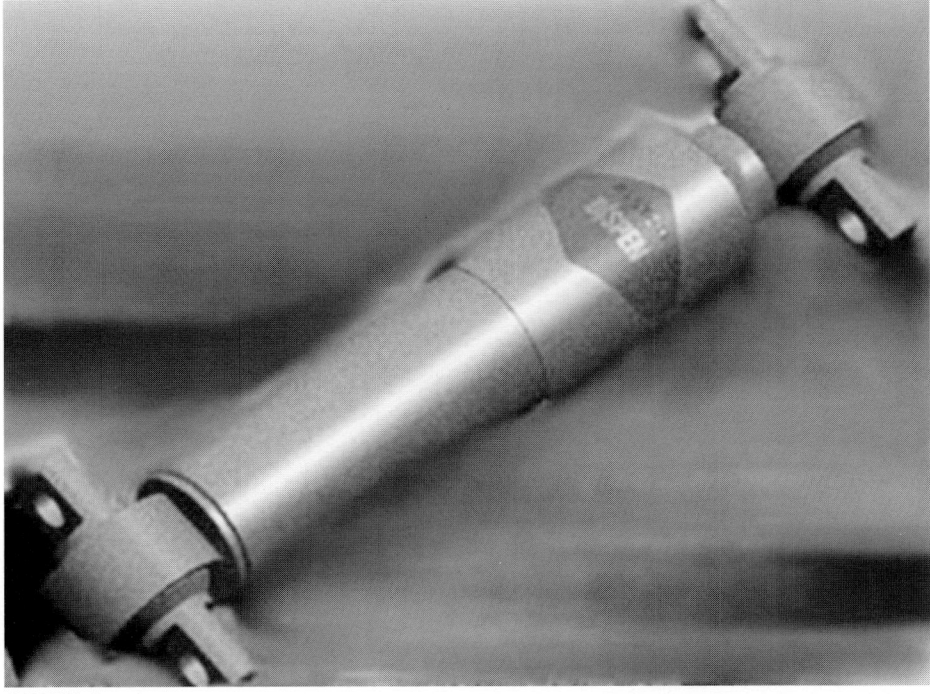

DISPEN damper   0525424

Key personnel

President and Managing Director: Bruno Flocco
Production Director: Didier Chevalet
Commercial Director: Jean Jomeau

Products

Bogies for freight wagons including AFR 25 (25/25.5 tonnes), Y25 Lsd 1 (22.5 tonnes), Y33 A2 (20 tonnes), and Eurobogie (22.5 tonnes for Eurotunnel) types.

ASF

American Steel Foundries/Amsted Industries International (a division of Amsted Industries Inc)
200 West Monroe Street, Suite 2301, Chicago, Illinois 60606, USA
Tel: (+1 312) 372 53 84
Fax: (+1 312) 372 82 30

Key personnel

For full personnel listing, see ASF entry in *Brakes and drawgear* section.

Products

ASF designs, tests, manufactures and markets cast steel components and hot-wound coil products for both the rail and road industry. Freight car components include side frames, bolsters, bogies, draft sill end castings, couplers, rotary couplers, yokes, articulated connectors, slack-free drawbars, AAR and low-profile centre plates, snubbing packages (Ride Control, Super Service Ride Control, Ridemaster, Super Service Ridemaster), AR-1 self-steering trucks, strikers, draft lugs, AAR coils and control coils.

ASM Spencer Moulton SAS

21 rue de la Gare, F-45330 Malesherbes, France
Tel: (+33 2) 38 32 72 29
Fax: (+33 2) 38 34 73 42
e-mail: gjoly@asmtrac.com
Web: www.asmtrac.com

Key personnel

Sales Manager: Guy Joly

Products

Suspension systems, including safety springs, chevrons and primary/secondary suspension bondings; drawgear packs, buffing springs.

UPDATED

Astra Vagoane Călători Arad SA/Arad

1-3 Petru Rareş, R-2900 Arad, Romania
Tel: (+40 257) 23 62 10
Fax: (+40 257) 25 81 68
e-mail: astra.calatori@astrac.rdsar.ro
Web: http://www.astracalatori.ro

Key personnel

For full personnel listing, see Astra Vagoane Călători Arad SA/Arad entry in *Locomotive and powered/non-powered passenger vehicles* section.

Products

Bogies for metro cars, passenger coaches, freight wagons and special purpose vehicles. Production includes 200 km/h Y32 bogies manufactured under Alstom/De Dietrich licence. Ring-type springs for traction and dumpers.

Ateliers de Braine-le-Comte et Thiriau Réunis SA

Rue des Frères Dulait 14, B-7090 Braine-le-Comte, Belgium
Tel: (+32 67) 56 02 11
Fax: (+32 67) 56 12 17

Key personnel

Commercial Manager: A Lejeune
Commercial Manager, Railway Division: R Brohée

Products

Welded bogies, three-axle bogies.

AWS

Achslagerwerk Stassfurt GmbH
An der Liethe 5, D-39418 Stassfurt, Germany
Tel: (+49 3925) 96 03
Fax: (+49 3925) 96 04 05
e-mail: info@aws-tec.de
Web: http://www.aws-tec.de

Key personnel

Managing Director: Heinz-Jürgen Luig
Chief Buyer and Sales Manager: Georg Lohmann

Works address

PO Box 1153, D-39401 Stassfurt, Germany

Products

Axlebox housings for rail vehicles including locomotives, railcars, passenger cars and freight wagons to UIC and GOST specification and for the standard Type Y25 freight bogie.

Bharat Heavy Electricals Limited (BHEL)

Transportation Business Department
Integrated Office Complex
Lodhi Road, New Delhi 110003, India
Tel: (+91 11) 24 36 93 77; 51 79 33 00
Fax: (+91 11) 24 36 94 23
e-mail: opb@bheldindustry.com
Web: www.bhel.com

Key personnel

General Manager: O P Bhutani

Products

Fabricated steel bogies for freight and passenger locomotives.

Contracts

Contracts include 150 fabricated bogies for Indian Railways.

UPDATED

Bochumer Verein Verkehrstechnik GmbH

PO Box 101466, D-44714 Bochum, Germany
Tel: (+49 234) 689 10
Fax: (+49 234) 689 15 80
e-mail: info@bochumer-verein.de
Web: www.bochumer-verein.de

Key personnel

Head of Sales and Marketing: Andreas Dal Canton
Head of Design and Calculation:
  Dipl Ing Franz Murawa
Head of Quality Management:
  Dipl Ing Michael Ditzler

Products

Rolling stock components, wheels/wheelsets and axles. Production programme also includes rubber cushioned wheels for urban traffic, low-floor axles with independent wheels for low-floor cars, stress balanced wheels with noise absorbers for high-speed trains and other passenger cars.

UPDATED

Bombardier Transportation

Bogies
Siegstr. 27, D-57250 Netphen, Germany
Tel: (+49 271) 70 20
Fax: (+49 271) 70 23 96
Web: www.transportation.bombardier.com

Key personnel

President, Bogies: Robert Wassmer

Products

Bombardier Transportation has the most comprehensive bogie portfolio in the world, offering bogie solutions for all market segments and types of rolling stock: trams, metros, commuter and regional trains, long-distance, high-speed and locomotives. An average of around 5,000 bogies are manufactured per year, making Bombardier a world leader in this sector.

Bombardier Transportation's design centre for bogies are in Switzerland and in the UK and manufacturing sites are in Germany and in France. Frame production is mainly centralised in its own plant in Poland and in Hungary.

In the near future, Bombardier Transportation intends to introduce its new 'mechatronic bogie'. Mechatronic means the reduction of the mechanics of a complex system to the absolute minimum by the introduction of active, multifunctional elements. These are flexibly driven by integrated, model-based controllers and form an integral part of an information network. Condition monitoring is automatically provided by these systems.

The bogies portfolio is divided into three bogie families: small for trams and light rail vehicles, dekum for metros, commuter and regional trains, long distance and high-speed trains and large for locomotives.

Bombardier BM 1000 Bogies for low-floor trams (S Family)   *NEW*/0585275

Examples of the Bombardier bogie portfolio for different applications are: HVP tilting bogie for high-speed diesel trains; BlueTiger bogie for six-axle locomotives; Mechatronic bogie; MD 523 for trailer coaches; Series 3 for narrow-gauge electric trains; B 5000 for Intercity trains and BM 1000 for low-floor trams.

UPDATED

Bombardier HVP tilting bogie for high-speed diesel trains (M Family)   **NEW**/0585271

Bonatrans AS

Bezručova 300, CZ-735 94 Bohumín, Czech Republic
Tel: (+420 59) 708 23 04
Fax: (+420 59) 708 28 05
e-mail: info@bonatrans.cz
Web: www.bonatrans.cz

Key personnel
Managing Director: Pavel Lazar
Commercial Director: Vilém Balcárek
Purchasing Director: Jaroslav Sedlák
Technical Director: Radim Zima
Production Director: Jan Kusněř
Financial Director: Ladislav Sitko

Bombardier MD 523 bogie for trailer coaches (M Family)   0558271

Background
The company previously traded as ŽDB AS, changing its name to Bonatrans AS in April 1999.

Products
Manufacture and delivery of Wheelsets, solid wheels, resilient wheels, axles, tyres and wheel centres for passenger coaches, high-speed trains, light rail and city vehicles, metro cars, locomotives, and freight wagons.

Contracts
Bonatrans is one of the biggest suppliers of wheelsets for freight transport vehicles in Europe with deliveries including Tatravagónka Poprad, Trinity Rail and Greenbriar.

UPDATED

Bombardier Series 3 bogie for narrow gauge (M Family)   0558272

Bradken

2 Maud Street, Mayfield West, New South Wales 2304, Australia
PO Box 105, Waratah, New South Wales 2298, Australia
Tel: (+61 2) 49 41 26 77
Fax: (+61 2) 49 41 26 61
e-mail: rail@bradken.com.au
Web: www.bradken.com.au

Background
Bradken's rail business structure consists of three strategic business units: freight rolling stock, maintenance and express parts.

Products
Bradken offers a range of bogie designs to suit standard, broad, narrow and one-metre gauge railways, with axle load capacities ranging from 9¼ to 40 tonnes and a variety of wagons to suit customer specifications including coal wagons, bulk grain wagons, fuel tank wagons, ballast wagons and container wagons.

UPDATED

Bombardier B 5000, powered bogie for Intercity trains (M Family)   **NEW**/0585272

Brenco, Incorporated/QBS

Brenco, Incorporated/Amsted Rail International
PO Box 389, Petersburg, Virginia 23804, USA
Tel: (+1 804) 86 31 11   Fax: (+1 804) 861 69 89
Web: http://www.amsted.com

Key personnel
Vice-President of International Sales and Sourcing:
Arun Dhir

Products
Tapered roller bearings.

NEW ENTRY

Bombardier Mechatronic bogie   **NEW**/0585274

British Springs

Stanmore Industrial Estate, Bridgnorth
WV15 5HR, UK
Tel: (+44 1746) 76 18 55   Fax: (+44 1746) 76 73 98
e-mail: sales@british-springs.co.uk
Web: http://www.british-springs.co.uk

Key personnel
Director: P Stone

Background
British Springs is a member of the United Industries
plc group.

Products
Rail vehicle suspension springs, including multi-
leaf, parabolic and coil springs. The company also
undertakes customised spring design and spring
refurbishment.

NEW ENTRY

Buckeye Steel Castings Co

2211 Parsons Avenue, Columbus, Ohio 43207, USA
Tel: (+1 614) 444 21 21   Fax: (+1 614) 445 20 84

Key personnel
President and Chief Executive Officer:
   Joe W Harden
Vice-President, Sales: Jeffrey E Laird
Director, Engineering: J R Downes
Manager, Product Engineering: R Polley

Products
Cast-steel four-wheel bogie side frames, bogie
bolsters, wagon couplers, draft yokes, centre plates,
sill centre braces, draft sill ends, six-wheel bogies,
span bolsters, and other castings for railroad
wagons. Running gear for railroad passenger
cars and mass transit rail vehicles. Buckeye Steel
Castings is a major supplier to railroads, railcar
builders and railcar repair shops.

Bumar-Fablok Ltd

Fabryka Maszyn Budowlanych i Lokomotyw
'Bumar-Fablok SA
ul Fabryczna 16, PL-32-500 Chrzanów, Poland
Tel: (+48 32) 624 66 66   Fax: (+48 32) 623 29 25
e-mail: info@fablok.com.pl
Web: www.fablok.com.pl

Key personnel
President and Managing Director:
   Piotr Majcherczyk
Marketing and Development Director:
   Andrej Smolana
Economy and Finance Director: Maria Pitala
Production and Sales Director: Leszek Tylutki

Products
Axles, axle gears and pinions for EMD and GE
locomotives, wheel centres, bogies for freight cars
type Y25 Lsd 1/Y25 Ls (s)1, driving bogies for diesel
locomotives type 1LNa, PGears, braking system
components, tyres for heavy and light rail vehicles,
cylindrical toothed wheels, tyred and monoblock
wheelsets with wheels from 360 to 1,130 mm
diameters. ISO 9002 and AAR certified.

UPDATED

CAF – Construcciones y Auxiliar de Ferrocarriles SA

Padilla 17 – 6°, E-28006 Madrid, Spain
Tel: (+34 91) 435 25 00   Fax: (+34 91) 436 03 96
e-mail: export.caf@caf.es
Web: www.caf.es

Key personnel
President and Chief Executive Officer:
   J M Baztarrica
Managing Directors: A Arizcorreta, A Lergarda
Contact: J Esnaola

Beasain
J M Iturrioz 26, E-20200 Beasain, Spain
Tel: (+34 943) 88 01 00
Fax: (+34 943) 88 14 20

Irún
Calle Anaca 13, E-20301 Irún, Spain
Tel: (+34 943) 61 33 42
Fax: (+34 943) 61 81 55

Zaragoza
Av de Cataluña 299, E-50014 Zaragoza, Spain
Tel: (+34 976) 76 51 00
Fax: (+34 976) 57 26 48

Offices
CAF Argentina SA
Pte Luis Sáenz Peña 31° – 7° piso, 1110 Capital
Federal Buenos Aires, Argentina
Tel: (+51 11) 43 83 20 06
Fax: (+54 11) 43 81 48 37
e-mail: cafadministracion@cafarg.com.ar

CAF Brasil Industria e Comercio
Rua Pedroso Alvarenga 58 conj 52, CEP 04531-000
São Paulo, Brazil
Tel: (+55 11) 31 67 17 20
Fax: (+55 11) 30 79 87 62
e-mail: cafsaopaulo@cafbrasil.com.br

CAF Mexico SA de CV
Prolongación Uxmal 988, Col Sta Cruz Atoyac,
03310 Mexico DF, Mexico
Tel: (+52 55) 56 88 75 43
Fax: (+52 55) 56 88 11 56
e-mail: cafmex@prodigy.net.mx

CAF USA Inc
1401 K Street NW, Suite 803, Washington DC 20005-
3418, USA
Tel: (+1 202) 898 48 48
Fax: (+1 202) 216 89 29
e-mail: cafusa@cafusa.com

Padilla 17, E-28006 Madrid, Spain
Tel: (+34 91) 575 64 03
Fax: (+34 91) 576 81 08
e-mail: export.caf@nexo.es
Web: www.caf.es

Products
Bogies, axles and wheel assemblies. CAF also
produces the Integral Intelligent Tilting System
(Sistema Inteligente e Basculatión Integral, or SIBI)
and the Brava automatic gauge-changing bogie
system.

UPDATED

CFD Industrie

9-11 rue Benoît Malon, F-92150 Suresnes, France
Tel: (+33 1) 45 06 44 00
Fax: (+33 1) 47 28 48 84

Key personnel
Full full personnel listing, see CFD Industrie
entry in *Locomotives and powered/non-powered
passenger vehicles* section.

Products
Locomotive bogies for axleloads up to 22 tonnes
and speeds up to 120 km/h. Passenger-coach
bogies for speeds up to 140 km/h.

China Northern Locomotive and Rolling Stock Industry (Group) Corporation

11 Yangfangdian Road, Haidian District, Beijing
100038, China
Tel: (+86 10) 51 86 23 70
Fax: (+86 10) 51 86 23 74
e-mail: loriciec@cnrgc.com.cn
Web: www.cnrgc.com

Key personnel
Chairman and Managing Director: Wang Tai-Wen
President: Cui Dianguo
Vice-President: Zhao Guangxing
President of LORIC Import & Export Corporation
   Ltd: Cao Guo-Bing
Vice-President of LORIC Import & Export
   Corporation Ltd: Chen Dayong
Senior Engineer of LORIC Import & Export
   Corporation Ltd: Yang Xiang-Jing

Products
Development, design, engineering, production,
sales, installation, refurbishment, maintenance and
after-sales services for all types of bogies.
   Many bogies have been delivered to Iraq, South
Korea, Thailand and Vietnam.

UPDATED

UPDATED   CNR Type ZK6 wagon bogie   *NEW*/1135834

CNR coach bogie

0595038

Cimmco International

A division of Cimmco Birla Ltd
Prakash Deep (6th Floor), 7 Tolstoy Marg, New
Delhi 110,001, India
Tel: (+91 11) 331 43 83; 84; 85
Fax: (+91 11) 332 07 77; 372 35 20

Key personnel
Full full personnel listing, see Cimmco International
entry in *Freight vehicles and equipment* section.

Products
Cast-steel bogies for passenger and freight
wagons.

ČKD Kutná Hora as

CZ-284 49 Kutná Hora, Czech Republic
Tel: (+420 327) 50 61 11   Fax: (+420 327) 50 65 87
e-mail: sales@ckdkh.cz
Web: http://www.ckdkh.cz

Background
Founded in 1967, the company was transformed
from a state-owned enterprise to a joint stock
company in 1995 as part of the Czech Republic's
privatisation reforms.

Products
Y25, Y33 and TF25 bogies; bogies to special designs
for main line and light rail vehicles; cast and
welded bogie frames and components.

UPDATED

Cockerill Forges & Ringmill SA

PO Box 65, B-4100 Seraing 1, Belgium
Tel: (+32 4) 330 35 35   Fax: (+32 4) 337 79 02
e-mail: cfr@cfr.be

Works
Main Cockerill Site, Seraing, Belgium

Key personnel
Chief Executive Officer: Marc Theunissen
Marketing and Sales: Raymond Rauw

Products
Steel tyres for all types of railway, light rail,
tramway and metro rolling stock.
UPDATED

Comet Industries Inc

4800 Deramus Avenue, Kansas City, Missouri
64120, USA
Tel: (+1 816) 245 94 00   Fax: (+1 816) 245 94 61

Key personnel
President: Steve Woodson
Chief Financial Officer: Mike Klinock
Rail Products Operations Manager: Randall Haan
Director of Business Development: John Killian

Products
Rail-associated bearings and components.

Cometna

Companhia Metalurgica Nacionel
Rua Marechal Gomes Dacosta, Famoes P-2675
Odivelas, Lisbon, Portugal
Tel: (+351 1) 933 31 39   Fax: (+351 1) 933 31 43

Key personnel
President: Jose Bissaia Barreto

Products
Cast-steel bogies for locomotives and freight
wagons.

ContiTech Luftfedersysteme GmbH

Vahrenwalder Strasse 9, D-30165 Hanover,
Germany
Tel: (+49 511) 938 13 04
Fax: (+49 511) 938 938 13 05
e-mail: railway_suspension_part@ls.contitech.de
Web: http://www.contitech.de

US office
ContiTech North America Inc
136 Summit Avenue, Montvale, New Jersey 07645
Tel: (+1 201) 930 06 00 (ext. 101)
Fax: (+1 201) 930 00 50
e-mail: rkremmeicke@contitech-usa.com
Product Manager: Rainer Kremmeicke

Key personnel
Customer Management, Rolling Stock:
   Manfred Hunze

Background
ContiTech Profile GmbH is a member of the
ContiTech Group, part of Continental AG.

Products
Suspension products for rail vehicles including
primary springs; rubber metal conical springs
with integrated hydraulic damping system; rubber
metal layer springs; rolling rubber springs for
axle suspension; secondary airbags; air spring
suspension systems.
   Vehicle programmes employing ContiTech
suspension components include: Alstom's LIREX
prototype; Amtrak Acela trainsets; DB AG VT605
and ICE 3 trainsets; TGV-Duplex trainsets; and
SNCF X-TER emus.
   The ContiTech product range also includes
Clouth rolling rubber and combination springs.

*ContiTech spring systems fitted to a second-
generation ICE high-speed trainset bogie*
0103971

CostaRail Srl

Viale IV Novembre, I-23845 Costa Masnaga (LC), Italy
Tel: (+39 031) 86 94 11   Fax: (+39 031) 85 53 30
e-mail: hkastner@costarail.it

Key personnel
Commercial Manager: Heinz Kastner

Background
Formerly Costaferroviaria, the company went
into temporary receivership and was taken over
by Rail Services International Group, for which
RSI Italia SpA is the operative leader. In October
2004, CostaRail Srl was established, re-launching
all existing Costaferroviaria activities, including
orders, trademarks, patents and projects.

Products
Electric multiple-units, passenger coaches, freight
wagons and bogies.

Contracts
Development and construction of 36 bogies for the
Y36P2, for completion and fitting during scheduled
maintenance.
   Construction of 20 bogie frames for the F7195C,
awarded by Italian State Railway, Trenitalia.

NEW ENTRY

Curtiss-Wright Antriebstechnik GmbH

Badstrasse 5, CH-8212 Neuhausen am Rheinfall,
Switzerland
Tel: (+41 52) 674 65 22   Fax: (+41 52) 674 66 09
e-mail: info@cwat.ch
Web: http://www.cwat.ch

Key personnel
Personnel Manager: Harry Zai

Products
Electromechanical active tilt system for rail
vehicles.

Contracts
ICN tilting high-speed trainsets for Swiss Federal
Railways.

Delimon GmbH

Postfach 102052, D-40011 Düsseldorf, Germany
Tel: (+49 211) 777 40   Fax: (+49 211) 777 42 10
e-mail: info@delimon.de

Background
Delimon GmbH is a member of Vesper Corporation.

Products
Railjet wheel flange lubrication systems for high-
speed main line, metro and light line rail vehicles.

Dellner Dampers AB

Box 51, SE-642 22 Flen, Sweden
Tel: (+46 157) 243 63   Fax: (+46 157) 243 69
e-mail: dampers@dellner.se
Web: http://www.dellner-dampers.se

Background
Formerly Precima Development AB, Dellner
Dampers AB became a member of the Dellner
Group in 2003.

Products
Dampers for railway rolling stock, including:
primary and secondary suspension; roll dampers;
yaw dampers; secondary couple dampers; and
displacement-sensitive dampers.
   Customers supplied include: Alstom, Bombardier
Transportation, Dellner Couplers, Siemens,
and TGOJ.

Delta Rail

1 Rue Roussel, F-92250 La Garenne Colombes,
France
Tel: (+33 1) 42 42 11 44   Fax: (+33 1) 42 42 11 16

Key personnel
For full personnel listing, see Delta Rail entry in
Brakes and drawgear section.

Products
Shock-absorbers, suspension air springs.

Devol

Devol Engineering Limited
Clarence Street, Greenock PA15 1LR, UK
Tel: (+44 1475) 72 53 20   Fax: (+44 1475) 78 78 73
e-mail: sales@devol.com
Web: www.devol.com

Key personnel
Rail Products Manager: Martin Wainwright
Internal Sales: Mitchell Farquhar
Operations Director: G Stark

Subsidiary company
Devol Moulding Services Ltd

Products

Design and manufacture of thermoplastic components for the rail industry. Devlon is an exclusive company range, a tough and resilient thermoplastic produced in-house by either monomer casting or extrusion. Devlon was first introduced into the UK rail industry in 1973, the initial application being tread brake rigging bushes fitted to passenger vehicles. Devlon is interchangeable with steel bushes, thus avoiding the need to replace any current pins and housings and increasing the maintenance period. More common applications include disc and tread brake bushes, suspension bearings, gangway fascia wear plates, bogie centre pivot liners and wear pads.

UPDATED

EDI Rail

28 Factory Street, Granville, New South Wales 2142, Australia
Tel: (+61 2) 96 37 82 88   Fax: (+61 2) 96 37 67 83
e-mail: sales@edirail.com.au
Web: www.edirail.com.au

Key personnel

Chief Executive Officer: Guy Wannop
Executive Genera Manager, Freight: Danny Broad
Executive General Manager, Passenger:
   David Williamson

Background

A division of Downer EDI Ltd, EDI Rail is the result of a merger of Clyde Engineering and the rail activities of Walkers Ltd.

Products

Bogies for locomotives, freight wagons and passenger vehicles.

UPDATED

ELH Eisenbhanlaufwerke Halle GmbH & Co KG

Hans-Dietrich-Genscher, Strasse 34, D-06188 Queis, Germany
Tel: (+49 346) 02 55 1-0   Fax: (+49 346) 025 51 30
e-mail: information@elh.de
Web: www.elh.de

Key personnel

Manager: Michael Schnaufer
Sales: Rainer Schmalenberg
Design: Detlef Scholdan

Background

ELH Eisenbhanlaufwerke Halle GmbH & Co KG was created in 1998, emerging from the repair and maintenance works of the Deutsche Bahn AG in Halle/Saale. In 1999 the company moved into a new works site at Halle-Queis, with a total area of 20,000 m².

Products

Bogies for freight and passenger vehicles as well as for special rail vehicles for rail track construction and maintenance. ELH offers axle loads from 18 up to 25 tonnes.

NEW ENTRY

Escorts Ltd

Railway Equipment Division
Plot 115, Sector 24, Faridabad 121 005, India
Tel: (+91 129) 544 22 75; 528 02 86
Fax: (+91 129) 23 21 48; 528 30 69; 526 42 86
e-mail: cgtech@asp-escorts@gndel.global.net.in

Key personnel

For a full list of personnel, see Escorts Ltd entry in *Brakes and drawgear* section.

Products

Shock-absorbers for locomotives, emus, passenger coaches and diesel cranes. Escorts is the prime supplier to Indian Railways, having supplied over 400,000 shock-absorbers.

New products under development include Uniflow-type shock-absorbers with adjustable damping.

Extel Systems Wedel (ESW)

Gesellschaft für Ausrüstung mbH
Industriestrasse 33, D-22876, Wedel, Germany
Tel: (+49 4103) 60 36 71   Fax: (+49 4103) 60 45 03
e-mail: info@esw-wedel.de
Web: http://www.esw-wedel.de

Products

neicontrol-E® eletromechanical tilting technology for regional trains. NEICODRIVE-E® tilting technology for high velocity trains and actuators for pantograph tilting.

FAG Kugelfischer AG & Co KG

Business Unit Spindle Bearings
PO Box 1260, D-97419 Schweinfurt, Germany
Tel: (+49 9721) 91 39 78
Fax: (+49 9721) 91 37 88
e-mail: rail_transport@fag.de
Web: www.fag.de

Key personnel

Division Manager: Dr Raimund Abele

Works

Georg-Schäfer Strasse 30, D-97421 Schweinfurt, Germany

Products and services

FAG products for rail vehicles include bearings for various types of axleboxes (also light alloy housings and axleboxes equipped with speed-, temperature- and vibration sensors); cartridge units; insulated traction motor bearings; transmission bearings; test rigs for wheelsets with mounted bearings; specific Arcanol greases and mounting/dismounting devices.

Applications include locomotives, freight cars and passenger coaches, commuter train vehicles, subways and tramways.

UPDATED

Federal-Mogul Corporation

Powertrain Systems (formerly known as Glacier Clevite)
5037 North State Route 60, McConnelsville, Ohio 43756, USA
Tel: (+1 740) 962 42 42   Fax: (+1 740) 962 82 02

Key personnel

Operations Manager: Ted McConnell
General Manager, Global Marketing:
   Joseph J Vauter
Manager Railroad Replacement: David A Comer

Products

Main, con-rod and flanged main journal bearings, thrust washers, cam bearings and various other shapes for accessory applications. Heavy wall innovations include tri-metal steel-backed cast-copper lead bearings, precision-plated overlays, nickel dam and the delta wall which allows the loading capacity to be raised without increasing the size of the bearing. An OE supplier to the world's locomotive engine PASSENGER COACH EQUIPMENT, Federal-Mogul also produces bearings for other large engine, compressor and aircraft applications.

Federnwerke J P Grueber GmbH & Co KG

PO Box 600131, D-58137 Hagen, Germany
Tel: (+49 2331) 965 60   Fax: (+49 2331) 96 56 56
e-mail: grueberfedern@grueber.de
Web: http://www.grueber.de

Products

Alloy leaf and parabolic springs for rail vehicles; alloy thermoformed springs; buffer and conical springs.

Ferraz Shawmut

1 rue Jean Novel, F-69626 Villeurbanne Cedex, France
Tel: (+33 4) 72 22 66 11   Fax: (+33 4) 72 22 67 13

Key personnel

For full personnel listing, see Ferraz Shawmut entry in *Electric traction equipment* section.

Products

Earth current units to prevent current flowing through bearings of axleboxes and associated resistors; current collecting device on live rail and its associated shoe fuse box.

FIREMA Trasporti SpA

Headquarters

Via Provinciale Appia, Località Ponteselice, I-81100 Caserta, Italy
Tel: (+39 0823) 09 71 11   Fax: (+39 0823) 46 68 12
Web: www.firema.it

Key personnel

Chairman: Gianfranco Fiore
Chief Executive Officer: Roberto Fiore
Operations Director: Maurizio Russo
Commercial Manager: Sergio d'Arminio M
Marketing Manager: Agostino Astori

Commercial and technical offices

Via Triboniano n 220, I-20156 Milan, Italy
Tel: (+39 02) 23 02 02 23   Fax: (+39 02) 23 02 03 00

Products

Motor and trailer bogies for high-speed trainsets and locomotives. Contracts include the supply of 70 motor bogies for E402 locomotives; 194 trailer bogies and 30 motor bogies for ETR 500 high-speed trains, all for Italian Railways.

UPDATED

Freudenberg Schwab GmbH

Postplatz 3, D-16761 Hennigsdorf, Germany
Tel: (+49 3302) 206 20   Fax: (+49 3302) 20 62 77
e-mail: info@freudenberg-schwab.de
Web: http://www.freudenberg-schwab.de

Key personnel

Chief Executive Officers: Dr Detlef Cordts;
   Peter Kofmel; Jörg Sost
Quality Manager: Leo S E Lang
Marketing Director: Bernd Werner

Background

Freudenberg Schwab GmbH is jointly owned by Freudenberg & Co (51 per cent) and Schwab Holding AG (49 per cent). The company collaborates closely with Schwab Schwingungstechnik AG in Switzerland.

Products

Vibration control: rubber/metal elastic elements, standard as well as specially designed. Range includes elastic mountings for traction equipment and components for primary and secondary suspension, resilient couplings linking motor, gearbox, axle and other mountings, resilient bushes, spherical bearings, buffers, conical mountings, V-mountings, machine mountings, rectangular mountings and hydraulic bushes.

Innovative products for vibration control include hydraulic mountings and bushes for speed-dependent stiffness, such as axle guide bearings for rail vehicles on tightly curved track, and active absorbers for reduction of structure-borne noise such as equalisation of vibrations from wheel flats.

(See also Freudenberg Schwab Gmbh entries in *Brakes and drawgear* and *Passenger coach equipment* sections.)

UPDATED

Fuchs Lubricants (UK) plc

New Century Street, Hanley, Stoke-on-Trent ST1 5HU, UK
Tel: (+44 8701) 20 04 00   Fax: (+44 1782) 20 20 72
e-mail: contact-uk@fuchs-oil.com
Web: www.fuchslubricants.com

Key personnel
Managing Director: R Halhead
General Sales Manager: David Atkin
Railway Accounts Manager: Peter Baker

Products
Manufacture and supply of a full range of lubricants for the railway industry including: diesel engine oils; transmission fluids; hydraulic oils, special greases for wheel-flange and switch-plate applications; biodegradable products for most applications, including engine oils, gear oils, hydraulic oils and grease. The company is also a supplier of industrial cleaning products, degreasing fluids, hygiene products and anti-freeze for engine cooling systems.

UPDATED

Ganz motor Kft

Postafiók 263, H-1475 Budapest, Hungary
Tel: (+36 1) 210 11 50   Fax: (+36 1) 313 48 09
e-mail: ganzmoto@mail.datanet.hu

Key personnel
Managing Director: Péter Pétervári
Sales Director: Martinkó András
Technical Director: Orsovai József

Products
Fabricated bolsterless bogies with mechanical or air-sprung secondary suspension for speeds of up to 200 km/h and for 1,000 to 1,676 mm gauges.

Motor bogies for railcars and multiple-unit vehicles with bevel-gear axleboxes or electric traction motor drives.

Greysham (International) Pvt Ltd

1/1-A, Man Singh Place, Roop Nagar, Delhi 110 007, India
Tel: (+91 11) 23 84 93 81
Fax: (+91 11) 23 84 10 68; 23 84 30 68
e-mail: sales@greysham.com
Web: www.greysham.com

Key personnel
Managing Director and Chairman: Govind Singh
Joint Managing Director: Sanjeev Singh
Export Manager: V B Arya

Products
Axlebox assemblies.
ISO 9001-2000 accredited by NQAQSR, New Zealand.

UPDATED

Griffin

Griffin Wheel Co/Amsted Rail International
200 West Monroe Street, Suite 2300, Chicago, Illinois 60606, USA
Tel: (+1 312) 853 56 96   Fax: (+1 312) 346 33 76
Web: http://www.amsted.com

Key personnel
Manager International Wheel Sales and Operations:
Leslie G Wood

Products
Cast steel low-stress wheels.

Gummi Metall Technik GmbH

5 Liechtersmatten, D-77815 Bühl, Germany
Tel: (+49 23) 80 40
Fax: (+49 23) 210 75
e-mail: info@gmt-gmbh.de
Web: www.gmt-gmbh.de

Key personnel
Manager: S Engstler

Sales offices
See Gummi Metall Technik GmbH entry in *Passenger coach equipment* section.

Products
Rubber and rubber-metal elements, such as axle springs, rolling rubber springs, bushes, cone mountings, ball joints and bonded springs. Resilient wheel systems.

UPDATED

Gutenhoffnungshütte Radsatz GmbH

Postfach 110226, D-46122 Oberhausen, Germany
Tel: (+49 208) 692 43 50
Fax: (+49 208) 692 43 55
e-mail: gr.radsatz@sabwabco.com

Background
Gutenhoffnungshütte Radsatz GmbH, formerly part of the Cardo BSI Rail Group, became part of the Faiveley Transport group in November 2004 with the latter's acquisition of SAB WABCO.

Products
Driven and non-driven wheelsets for locomotives, dmus, emus, metro cars, light rail vehicles, passenger coaches and freight wagons; running gear components and systems for low-floor light rail vehicles; independent wheel axles; independent wheel units; EEF non-driven wheel pairs; solid wheels; tyred wheels; rubber-sprung wheels; wheelset axles; track wheels; sound absorbers. Wheelsets are also manufactured for special vehicles for the rail and steel industries.

UPDATED

Hindusthan Engineering & Industries Ltd (HEI)

Mody Building, 27 Sir RN Mukherjee Road, Calcutta 700 001, India
Tel: (+91 33) 22 48 01 66
Fax: (+91 33) 22 48 19 22; 22 43 56 07
e-mail: hindus@cal2.vsnl.net.in

Works
Bamunari Plant (BP), National Highway No 2, Bumunari 712 205, District Hooghly, West Bengal, India

Key personnel
Executive Director: M L Lohia

Products
Cast steel two and three-axle bogies, locomotive bogies fitted with suspension arrangements, axleboxes, high-tensile centre coupler both transition and non-transition type and high capacity draft gear. The company is approved by the Association of American Railroad under AAR M-1003 and AAR M-210.

HEI exports under license agreement from Standard Car Truck Company to the US, Australia, South Korea and other parts of the world.

UPDATED

IBG Monforts GmbH & Co

An der Waldesruh 23, D-41238 Mönchengladbach, Germany
Tel: (+49 2166) 86 82 40   Fax: (+49 2166) 86 82 44
e-mail: info@ibg-monforts.de
Web: www.ibg-monforts.de

Key personnel
Managing Director: Klaus Sasserath

Products
Slide bearing systems for applications such as bogie air springs.

Contracts
Contracts include the supply of bearing systems for ICE-2 high-speed trainsets, VT611 tilting dmus and double-deck coaches for DB AG as well as for ICT trainsets.

UPDATED

The Integral Coach Factory (ICF)

Ministry of Railways, Chennai 600 038, India
Tel: (+91 44) 26 26 31 11   Fax: (+91 44) 26 28 47 56
e-mail: coachnet@vsnl.com
Web: http://www.icf.gov.in

Key personnel
For a full list of personnel, see The Integral Coach Factory (ICF) entry in *Locomotives and powered/non-powered passenger vehicles* section.

Products
All-coil all-steel high-speed bogies with bogie-mounted brakes and conventional brakes; bogies with pneumatic suspension system.

IMS UK

Arley Road, Saltley, Birmingham B8 1QX, UK
Tel: (+44 121) 326 10 00
Fax: (+44 121) 326 10 10
e-mail: ims.millsales@ims-group.com
Web: www.ims-uk.com

Key personnel
Financial Director: J Christie
Division Directors: C Beal; A Hance; J Symes;
  J Murray
Export Manager: B Christie
Purchasing Manager: M Hole

Other offices
Canklow Meadows Industrial Estate
West Bawtry Road, Rotherham, South Yorkshire S60 2XN, UK
Tel: (+44 1709) 78 80 00   Fax: (+44 1709) 78 80 30
e-mail: imsuk.north@ims-group.com
Arley Road, Saltley, Birmingham B8 1QX, UK
Tel: (+44 121) 326 10 00   Fax: (+44 121) 326 10 10
e-mail: ims.millsales@ims-group.com

Fyne Avenue, Righead Industrial Estate, Bellshill, Lanarkshire ML4 3JY, UK
Tel: (+44 1698) 74 69 22   Fax: (+44 1698) 74 69 97
e-mail: imsuk.millsales@ims-group.com
Carrwood Road, Chesterfield Trading Esate, Sheep Bridge, Chesterfield, Derbyshire S41 9QB, UK
Tel: (+44 1246) 26 45 00   Fax: (+44 1246) 26 45 50
e-mail: imsuk.energy@ims-group.com

Products
Stockholding and distribution of speciality steels. Mill representation in the UK for Valdunes railway wheels, axles, brake discs, assembled wheelsets

and forged safety-critical parts; RTM (Roues et Trains Montés) a subsidiary of Valdunes, offering wheelset overhaul in addition to railway wheels, axles and assembled parts; CFR (Cockerill Forges & Ringmill) tyres for rail vehicles and trams; LAF (Les Appareils Ferroviaires) automatic couplers for freight and passenger vehicles.

UPDATED

INKA

Head office and factory
T (Persero) Industri Kereta Api (PT INKA)
Jalan Yos Sudarso No 71, Madiun 63122, Indonesia
Tel: (+62 351) 45 22 71   Fax: (+62 351) 45 22 75
e-mail: sekretariat@inka.web.id
Web: www.inka.web.id

Representative office
Arthaloka building, 3rd Floor, Jalan Jend, Sudirman Kav 2, Jakarta
Tel: (+62 21) 251 44 24   Fax: (+62 21) 251 44 23
e-mail: inkajkt@cbn.net.id

Key personnel
President: Ir Roos Diatmoko
General Manager, Railway Rolling Stock Division: Suryanto
Finance and Administration Director: Drs Udin Supriatman
General Manager of Technology: Ir M Harsan Badawi
Engineering Manager: Ir Gunesti Wahyu
Design Manager: Ir Indarto Wibisono
Marketing Manager: Ir M Dedi Tarmidi
Business Development Manager: Ir Muchlis Budiman
Procurement Manager: Soedjito Taathadi

Background
PT INKA was originally established in 1981 as a state owned company, transforming from Indonesian State Railway's steam locomotive maintenance shop.

Products
Assembly and renovation of freight wagons (300 units a year), passenger coaches (120 units a year), diesel and electric railcars (40 units a year), bogies (200 units a year).
   Locomotives (in collaboration with GE Transportation, 15 units a year). Various special vehicles, including track motor and inspection cars and amusement park trains.

NEW ENTRY

Issels

F Issels & Son Ltd
PO Box 2199, Bulawayo, Zimbabwe

Products
Bogies including the A3 Ride Control bogie pack for use on railways in Africa, north of the Limpopo.

UPDATED

K Industrier AB

Stora Varvsgatan 14, SE-211 19 Malmö, Sweden
Tel: (+46 40) 34 80 00   Fax: (+46 40) 34 87 75
Web: www.kindustrier.se

Key personnel
President: Björn Widell
Executive Vice-Presidents
   Engineering: Peter Linde
   Production: Karl-Gustav Andersson
Head of Finance: Nils-Arne Nilsson

Background
K Industrier AB formerly traded as Kockums Industrier AB.

Products
Bogies for freight vehicles, including the Y25 TTV, with a 25-tonne axleload.

UPDATED

KINEX-KLF a/s

Kukučinová 2346, SK-024 01 Kysucké Nové Mesto, Slovakia
Tel: (+421 41) 420 10 40   Fax: (+421 41) 420 18 70
e-mail: frantisek.vilk@kinex-klf.sk
Web: http://www.kinex-klf.sk

Key personnel
Managing Director: Igor Kováč
Commercial Director: Tibor Ďurajka
Railway Segment Manager: Frantisek Vilk

Background
KINEX-KLF a/s was previously called KLF-ZVL a/s.

Products
Single-row ball bearings, single-row cylindrical roller bearings and special bearings for axle box arrangement of freight wagons, passenger coaches, electric and diesel locomotives; for transmission shafts; for traction motors and generators, compressor motors; and axle boxes for 22.5 and 25 tonnes axle load.

UPDATED

KLW – Wheelco SA

Via Calloni 1, PO Box 45, CH-6907 Lugano, Switzerland
Tel: (+41 91) 986 58 00   Fax: (+41 91) 986 58 01
e-mail: info@klw-wheelco.ch
Web: www.klw-wheelco.ch

Key personnel
General Manager: Michael Is'kov

Products
Wheels, tyres, wheelsets, bogies and bogie compartments; rolling stock components.

UPDATED

Koni BV

Langeweg 1, PO Box 1014, NL-3260 AA Oud-Beijerland, Netherlands
Tel: (+31 186) 63 55 00   Fax: (+31 186) 63 56 05
e-mail: p.bakkeren@koni.nl
Web: www.koni.com

Key personnel
Chairman and Managing Director: R Vrenken
Marketing and Sales Director: H Kromhout

Other offices
Ste Koni – France
BP no 9, F-06271 Villeneuve-Loubet Cedex, France

Koni Germany
Rheinstrasse 96, D-56285 Ransbach-Baumbach, Germany

Koni BV
Postbox 1014, NL-3260 AA, Oud-Beijerland, Netherlands

Koni North America
1961-A International Way, Hebron, Kentucky 41048, USA
Tel: (+1 859) 586 41 00   Fax: (+1 859) 334 33 40
Web: www.koni-na.com

Products
In-house research, development, engineering and manufacturing of dampers for buses, trailers, trains, metros, passenger coaches and freight wagons, locomotives, high-speed trains and pantographs; electrically controlled dampers.

UPDATED

Koyo Seiko Co Ltd

5-8 Minamisemba 3-chome, Chuo-ku, Osaka 542, Japan
Tel: (+81 6) 245 60 87   Fax: (+81 6) 244 08 14

Key personnel
President: Hiroshi Inoue
Vice-President: Takatoyo Uematsu
Senior Executive Director: Fumio Morishita
Executive Director: Hajime Hori

Products
Axlebox cartridge-type cylindrical and tapered sealed journal roller bearings; open-type cylindrical, tapered and spherical roller bearings and ball-bearings.
   Traction motor bearings, including insulated cylindrical roller bearings and ball bearings. Standard bearings are also available.
   Gear bearings, including tapered, cylindrical and spherical roller bearings and ball-bearings.
   Also ball spline-type universal drive bearings.
   Koyo cartridge-type tapered roller bearings with seals are used on the Shinkansen Series 500 Nozomi axle journals.

Kryukovsky

Kryukovsky Railway Car Building Works
139 I Pridhodko Street, Kremenchung 315307, Ukraine
Tel: (+38 5366) 695 05; 697 95; 611 01
Fax: (+38 532) 50 14 21; (+38 5366) 611 01;
   (+38 44) 295 72 03
e-mail: root@kvsz.poltava.ua

Two-axle freight wagon bogie by Kryukovsky

0087657

Products

Two-axle bogies for freight wagons; wheelsets, including products to UIC 812-3 and ISO 1005/3 standards for passenger coaches and freight cars; wheels and axles; equipment for strengthening wheels.

KYB – Kayaba Industry Co Ltd

World Trade Center Building, 2-4-1, Hamamatsu Cho, Minato Ku, Tokyo 105-6111, Japan
Tel: (+81 3) 34 35 35 751   Fax: (+81 3) 34 36 29 07
Overseas business department:
Tel: (+81 3) 34 35 35 81   Fax: (+81 3) 34 36 29 08
Web: http://www.kyb.co.jp

Key personnel

President: Tadahiko Ozawa

European representation
Gadelius Europe AB
Box 859, SE-301 18 Halmstad, Sweden
Tel: (+46 35) 18 20 71   Fax: (+46 35) 18 20 71
e-mail: europe@gadelius.com
Web: http://www.gadelius.com

Products

Caliper brakes; tread cleaning.

Lesjöfors AB

SE-524 92 Herrljunga, Sweden
Tel: (+46 513) 220 00   Fax: (+46 513) 230 24
e-mail: info@lesjoforsab.com
Web: http://www.lesjofors.com

Products

Suspension springs and stabiliser equipment for rail vehicles. Companies supplied include Adtranz and Bombardier Transportation.

Langen & Sondermann

Langen & Sondermann Federnwerk
Bergkampstrasse 57, D-44534 Lünen, Germany
Tel: (+49 2306) 75 05 70   Fax: (+49 2306) 576 72
e-mail: post@langen-sondermann.de
Web: http://www.langen-sondermann.de

Key personnel

Managing Director: Jürgen Kohl
Sales Director: Heinz Hermann
Technical Manager: Dipl Ing Dieter Schmidt

Products

Conventional (trapezoidal) leaf springs, parabolic springs (also design and development), coil springs (up to metal diameter of 80 mm); TKS-springs® (coil springs with progressive compression characteristics); design and development of steel springs.

VERIFIED

Lord Corporation

Mechanical Products Division, 2000 West Grandview Boulevard, PO Box 10040, Erie, Pennsylvania 16514-0040, USA
Tel: (+1 814) 868 54 24   Fax: (+1 814) 868 31 09

Key personnel

Regional General Manager, International Business: Eric Ravinowic
Manager, International Business Development: John Fisher

Principal subsidiary

Lord SA
Immeuble Strategy Center, 1-, Rue des Gaudines, F-78100, Saint Germain-en-Laye, France
Tel: (+33 3) 90 20 34 70   Fax: (+33 3) 90 20 39 39

Products

LC-Pads for roller bearing adaptors, designed to reduce lateral forces, accommodate motion without wear, reduce rail wear and eliminate adaptor crown wear on self-steering bogies.

V Springs (chevron springs) for the primary suspension system. Applications include primary suspension for rapid transit bogies, locomotive bogies, mining cars and maintenance equipment.

Bolster mounts to accommodate lateral movement of locomotive bolsters.

Dyna-Deck for loading-shock protection. Designed to absorb longitudinal shocks and movements up to 12 in.

Lubricant Consult GmbH

Gutenbergstrasse 13 D-63477 Maintal, Germany
Tel: (+49 6109) 765 00
Fax: (+49 6109) 76 50 51
e-mail: webmaster@lubcon.com
Web: http://www.lubcon.com

Subsidiaries

Austria
Büro Linz, Pollheimer Strasse, 16 A-4020, Linz
Tel: (+43 73) 267 16 30   Fax: (+43 73) 267 16 30
e-mail: info-austria@lubcon.com

Czech Republic/Slovakia
LUBCON sro
Diouhá 783 CZ-76321, Slavicin
Tel: (+420 636) 34 36 18   Fax: (+420 636) 34 20 09
e-mail: info-czechrepublic@lubcon.com

France
LUBCON SARL
Alpespace – La Pyramide F-73800, Montmélian
Tel: (+33 4) 79 84 38 60   Fax: (+33 4) 79 84 38 61
e-mail: france@lubcon.com

Italy
LUBCON Italia SRL
Via Trieste, 16 1-21040 Sumirago (VA)
Tel: (+39 0331) 90 87 62   Fax: (+39 0331) 90 87 54
e-mail: info-italy@lubcon.com

Poland
LUBCON Polska Sp zoo
Al. Lotników Polskich 1, PL-21-045 Swidnik
Tel: (+48 81) 468 82 60   Fax: (+48 81) 468 82 61
e-mail: lubcon@lubcon.com.pl

Slovenia
LUBCON doo
Pod gradom 2 SLO-2380, Slovenj Gradec
Tel: (+386 2) 883 19 86   Fax: (+386 2) 883 19 87
e-mail: info-slovenia@lubcon.com

Switzerland
LUBCON Lubricant Consult AG
Im Leisibühl 35 CH-8044, Zurich
Tel: (+41 1) 882 30 37
Fax: (+41 1) 882 30 38
e-mail: lubcon-ch@bluewin.ch

USA
TURMO LUBRICATION, Inc
760 36 Street SE, Grand Rapids, Michigan 49548
Tel: (+1 616) 247 01 29   Fax: (+1 616) 247 08 01
e-mail: turmolube@aol.com

Products

Lubcon biologically degradable special lubricants for the wheel/rail system to counteract wear and noise.

Lucchini Group

Headquarters
Lucchini SpA
Via Oberdan 1/a, I-25128 Brescia, Italy
Tel: (+39 030) 399 21
Fax: (+39 030) 399 25 17
e-mail: commerciale@lucchini.com
Web: www.lucchini.com

Key personnel

President: Giuseppe Lucchini
Managing Director: Giovanni Gillerio
Corporate Commercial Director: Giovanni Bajetti

Background

Lucchini Group focuses on design and manufacture of a wide range of steel products for the railway sector. Rail and components for permanent way are produced at Lucchini Piombino. Manganese cast steel monobloc crossing for railway switches are produced at Bari Fonderie Meridionali (BFM). Wheels, axles, wheelsets are produced at Lucchini Sidermeccanica Lovere and its subsidiary companies: Lucchini UK, Lucchini Sweden and Lucchini Poland.

Subsidiary companies

Lucchini Sidermeccanica SpA
Via G. Paglia 45, I-24065 Lovere (BG), Italy
Tel: (+39 035) 96 35 66   Fax: (+39 035) 96 35 52
e-mail: rollingstock@lucchini.it
Chairman and Managing Director: Erder Mingoli
General Manager: Augusto Mensi
Sales Manager: Roberto Forcella

Lucchini UK – Wheel Systems Division
Ashburton Road West, Trafford Park, Manchester M17 1GU, UK
Tel: (+44 161) 872 04 92   Fax: (+44 161) 872 28 95
e-mail: salesuk@lucchini.co.uk
Chairman: Erder Mingoli
General Manager: Ian Dolman
Sales Manager: Chris Fawdry

Lucchini Sweden
Bruks Gatan, Box 210, SE-73523 Surahammar, Sweden
Tel: (+46 22) 03 47 00   Fax: (+46 22) 03 47 80
e-mail: info@lucchini.se
Chairman: Erder Mingoli
Managing Director and Sales Manager: Lennart Nordhall

Lucchini Poland
Ul Kasprowicza, 132, PL-01949 Warsaw, Poland
Tel: (+48 22) 835 89 20   Fax: (+48 22) 835 08 33
e-mail: info@lucchini.pl
Chairman: Erder Mingoli
General Manager: Krzysztof Laskowski
Sales Manager: Albert Siekierka

Products

Lucchini Sidermeccanica SpA, Lovere Works: design and manufacture of wheels, tyres, axles and assembly of wheelsets complete with axleboxes, brake discs and drive units.

Lucchini Sidermeccanica SpA, LMF Works: overhaul and full refurbishment of wheelsets including axleboxes, bearings, disc brakes and drive units. Production of axleboxes for railway and mass transit systems.

Lucchini UK – Wheel Systems Division, Manchester Works: production of all types of railway wheels and assembly of complete wheelsets for trucks, carriages and locomotives.

Lucchini Sweden, Surahammar Works: wheels, tyres, axles and wheelsets with axleboxes and drive units.

Lucchini Poland, Warsaw Works: wheels, forged axles and wheelsets complete with axleboxes and drive units.

UPDATED

Magnus LLC

PO Box 1029, Fremont, Nebraska 68026-1029, USA
Tel: (+1 402) 721 95 40   Fax: (+1 402) 721 23 77
e-mail: dburns@magnus-farley.com
Web: http://www.magnus-farley.com

Key personnel

President: J E Macklin
Sales Manager: David Burns

Products

High-leaded bronze bearings for traction motor application; tin bronze bearings and special analysis; solid journal bearings; proprietary centrifugal castings, horizontal and vertical; babbitt bearing linings, centrifugal, static and plated; Statistical Process Control (SPC); robotic material handling and integrated machining; electric induction furnaces and spectrographic analysis.

MAN Ferrostaal AG

Hohenzollernstrasse 24, D-45128 Essen, Germany
Tel: (+49 201) 818 01   Fax: (+49 201) 818 28 22
e-mail: info@manferrostaal.com
Web: www.manferrostaal.com

Key personnel
Head of the Department of Infrastructure and Transport Systems: Helmut Julius

Background
Re-named from Ferrostaal AG, MAN Ferrostaal AG is a member of the MAN AG group.

Products
Motor and trailer bogies for all types of rolling stock; wheels, axles, assembled wheelsets, bearings and suspensions.

UPDATED

Meridian Rail LLC

1200 Corporate Drive, Birmingham, Alabama 35242, USA
Tel: (+1 205) 991 03 84   Fax: (+1 205) 972 13 86
Web: www.meridianrail.com

Key personnel
President: Rick Turner
Vice-President, Operations: Bill Holcomb
Vice-President, Sales and Marketing: Frank Cristelli

Background
Meridian Rail's predecessor was ABC-NACO, which had filed for Chapter 11 in November 2001, and its assets sold, in January 2002. ABC-NACO re-emerged as Meridian Rail in January 2002.

Products
Wheelset inventory management and distribution of new and used wheelsets to railroads and maintenance centres in the US and Mexico.
   Meridian Rail Reconditioning Services (MRS) is a full service reconditioning business certified under AAR M-1003, with technical certification under M-212 for couplers and yokes and M-300 for brake beams. In addition, MRS carries a number of freight car component parts for resale through its facility in Chicago and through wheel shops located throughout the US.

UPDATED

MIBA Gleitlager GmbH

Dr Mitterbauer Strasse 3, A-4663 Laakirchen, Austria
Tel: (+43 7613) 254 10   Fax: (+43 7613) 42 57

Key personnel
President, Sales: Mag Dr Wolfgang Litzlbauer
Vice-President, Sales: Johann Jandrasits

Products
Engine bearings and bushes for diesel locomotives.

MSA Mediterr Shock-Absorbers SpA

Via G A Valentini 127, I-93100 Caltanissetta, Italy
Tel: (+39 0934) 57 53 60   Fax: (+39 0934) 57 53 73
Works
Stada Statale 457 No 37/39, I-14033 Castell' Alfero, AT, Italy
Tel: (+39 0141) 20 47 57   Fax: (+39 0141) 20 46 70

Products
Shock-absorbers for rail vehicles, including specialised applications.

Multi-Service Supply Division

The Buncher Co
Ferry Street and Avenue 'C', Leetsdale, Pennsylvania 15056, USA
Tel: (+1 412) 741 15 00   Fax: (+1 412) 741 33 20
e-mail: multiservice@buncher.com
Web: http://www.buncher.com

Key personnel
For full personnel listing, see Multi-Service Supply Division entry in *Brakes and drawgear* section.

Products
Roller bearings; couplers; yokes, draft gears; wheel and axle assemblies; complete bogie assemblies with or without brake rigging.

VERIFIED

MWL Brasil Rodas & Eixos Ltda

Rodovia Vito Ardito, s/n Km I – CP 189-CEP, 12280-000 Cacapava, São Paulo, Brazil
Tel: (+55 12) 36 54 74 00
Fax: (+55 12) 36 54 74 10
e-mail: mwlbrasil@mwlbrasil.com.br
Web: www.mwlbrasil.com.br

Key personnel
President: Aparecido Nobuo Terazima
Industrial Director: Domingos Miniccuci
Financial Director: João Aquino Carvalho Junior

Background
Recently known as Mafersa SA. MWL Brasil Rodas & Exios purchased the technology and licence of Mafersa in November 1999 and manufactures its products under the MWL name.

Products
Forged rail vehicle wheels, axles, gear blanks, wheelsets and crane wheels, sheave wheels.

UPDATED

Nippon Sharyo Ltd

Head office
1-1 Sanbonmatsu-cho, Atsuta-ku, Nagoya 456-8691, Japan
Tel: (+81 3) 36 68 33 30   Fax: (+81 3) 36 69 02 38
Web: www.n-sharyo.co.jp

Overseas contact
Maruncuchi Central Building, 1-9-1 Maruncuchi, Chiyoda-ku, Tokyo 100-0005, Japan
Tel: (+81 3) 66 88 67 94   Fax: (+81 3) 66 88 68 10

Key personnel
President: Kazuhisa Matsuda
Managing Director and General Manager, Rolling Stock Division: Katsuyuki Ikushima
Business Contact General Manager, International Sales Department: Masataka Nakajima

Products
Powered and trailer bogies for all types of rolling stock, including tilting and self-steering bogies.

UPDATED

Nizhnedneprovsky

Nizhnedneprovsky Tube Rolling Plant
21 Solotov St, Dnepropetrovsk, 320060, Ukraine
Tel: (+380 0562) 20 73 01; 20 73 90
Fax: (+380 0562) 27 16 43; 23 05 45

Key personnel
Vice Director of Foreign Economic Relations: Mikhail Staroseletsky
Head of Communications: Svetlana Chernikova

Background
Nizhnedneprovsky Tube Rolling Plant has been manufacturing wheels and tyres for over 60 years.

Products
Manufacturer and supply of more than 50 sizes of wheels and tyres for emus, LRVs, metrocars, trainsets, locomotives and freight wagons. Steel for the wheels and tyres is produced in the company's open-hearth furnaces and is refined for resistance to fatigue and brittleness.

Contracts
Include supply of wheels and tyres to 36 countries including fromer CIS countries, Bangladesh, China, France, Germany, India, Italy and Tanzania.

NMB – Minebea UK Ltd

Doddington Road, Lincoln LN6 3RA, UK
Tel: (+44 1522) 50 09 33   Fax: (+44 1522) 69 64 85

Key personnel
General Manager and Director: Mark N Stansfield
Sales Manager: A E Morton

Background
NMB – Minebea UK Ltd was previously called Rose Bearings Ltd.

Products
Rod end and spherical bearings to aerospace, railway, automotive and power generation industries. All types from military standard to customer bespoke. Specialist in coupling bearings for articulated PSV, including seal and gasket technology.

UPDATED

NSK Ltd

Nissei Building, 1-6-3 Ohsaki, Shinagawa-ku, Tokyo 141, Japan
Tel: (+81 3) 37 79 71 20   Fax: (+81 3) 37 79 74 33
Web: www.nsk.com

Key personnel
President: T Sekiya
Managing Director: K Moriya

Products
Axleboxes, journal bearings, cylindrical and spherical roller bearings, ball-bearings and pillow units.

UPDATED

NSK UK Ltd

Northern Road, Newark, Nottinghamshire NG24 2JF, UK
Tel: (+44 1636) 60 51 23   Fax: (+44 1636) 60 27 75
e-mail: info-uk@nsk.com
Web: www.eu.nsk.com

Background
Previously called NSK-RHP UK, the company has changed its name to NSK UK Ltd.

Products
Traction motor bearings, axlebox bearings, ball and roller bearings.

Developments
Axle bearing technology for high-speed bullet trains.

UPDATED

NTN Toyo

NTN Toyo Bearing Co Ltd
3-17 chome Kyomachibori, Nishi-ku, Osaka, Japan
Tel: (+81 6) 443 50 01

US office
American NTN Bearing Manufacturing
Corporation
1600 East Bishop Court, Mount Prospect, Illinois
60056, USA
Tel: (+1 708) 298 75 00   Fax: (+1 708) 699 97 44

Products
Journal roller bearings of all types.

ORX

One Park Avenue, Tipton, Pennsylvania 16684,
USA
Tel: (+1 814) 684 84 84   Fax: (+1 814) 684 84 00
e-mail: orx@orxrail.com
Web: http://www.orxrail.com

Key personnel
President: Glenn Brandimarte
Sales Manager: Rocky Pacifico

Products
Axles, wheelsets and bogies.
ORX is the OEM (Original Equipment
Manufacturer) supplier for GM DD locomotives,
Los Angeles Red Line cars, New York City R142
cars and Amtrak's Acela Express trains.

Peddinghaus Group

Carl Dan Peddinghaus GmbH & Co KG
Mittelstrasse 64, D-58256 Ennepetal, Germany
Tel: (+49 2333) 79 60
Fax: (+49 2333) 79 63 88
e-mail: cdp-en@t-online.de
Web: http://www.peddinghaus-group.de

Products
A supplier of high quality steel and aluminium
forgings and components including railway axles
and coupling shafts.

Penn

Penn Machine Company (a member of the Marmon
Group)
106 Station Street, Johnstown, Pennsylvania
15905, USA
Tel: (+1 814) 288 15 47
Fax: (+1 814) 288 22 60
e-mail: pmcsales@pennmach.com
Web: www.pennmach.com

Key personnel
Vice-President and General Manager:
  H Karl Wiegand
Vice-President, Manufacturing and Engineering:
  Thomas Redvay
Vice-President, Transportation Sales: Richard E Trail
Sales Engineer, Transportation Sales: Peter Basile

Principal subsidiary
Penn Locomotive Gear Company
310 Innovation Drive, Blairsville, Philadelphia
15717, USA
Tel: (+1 724) 459 03 02   Fax: (+1 724) 459 48 69

Sales office
210 Pine Street, Carnegie, Pennsylvania 15106, USA
Tel: (+1 412) 279 44 60
Fax: (+1 412) 279 44 65

Licensing agreements
Penn Machine Company has agreements with
ZF Hurth and Bochumer Verein of Germany to
manufacture passenger vehicle gearboxes and
resilient wheels in the USA.

Products
Resilient wheels; axles; pinions; gears; journal
boxes; complete gearboxes; gearbox components;
pinions, gears and shafts for diesel-electric
locomotives, primarily General Motors, GE
Transportation Systems and Alco types.

Contracts
Contracts include the supply of resilient wheels
and axles to Bombardier for LRVs for Minneapolis;
LRVs for Breda in Boston; and to Kinki Sharyo for
LRVs in Phoenix and Seattle; gearbox overhaul
contracts with WMATA, ALSTOM and Baltimore.

UPDATED

Phoenix Traffic Technology GmbH

Hannoversche Strasse 88, D-21079 Hamburg,
Germany
Tel: (+49 40) 76 67 27 78
Fax: (+49 40) 76 67 27 78
e-mail: michael.groth@phxtt.de
Web: www.phoenix-ag.com

Key personnel
Chief Engineer: Peter Eckworth
Engineer: Jörg Frohn
Export Sales: Peter Wist
Sales Manager: Michael Groth

Subsidiary companies
Phoenix Benelux BVBA
Tel: (+32 3) 206 74 20
Sales Manager: René Van Schaik

Phoenix Distribution Sarl
Tel: (+33 4) 72 46 09 74
Sales Manager: Michel Marrot

Phoenix Rubber Asia Pacific Pte Ltd
Tel: (+65) 68 96 23 19
Manager: David Sim

Phoenix Elastomerprodukte Vertriebsges mbH
Tel: (+43 1) 259 56 24 14
Sales Manager: Heinz Tichy

Phoenix (GB) Ltd
Tel: (+44 121) 522 08 83
Sales Manager: Kieron Moloney

Phoenix Rubber Iberica SL
Tel: (+34 93) 477 27 20
Sales Office: Núria Mullor

Phoenix North America Inc
Tel: (+1 732) 346 53 53
Sales Manager: Peter Tiedemann

Products
Rubber and airspring systems for primary and
secondary bogie suspensions. Auxiliary rubber
elements for railway applications.

UPDATED

Poli Costruzione Materiali Trazione SpA

Via Marconi 3, I-26014 Romanengro (CR), Italy
Tel: (+39 0373) 27 01 26   Fax: (+39 0373) 72 90 97
e-mail: info@polibrakes.com
Web: http://www.polibrakes.com

Key personnel
For full personnel listing, see Poli Costruzione
Materiali Trazione SpA entry in *Brakes and
drawgear* section.

Products
Gear reductors, resilient wheels, tyres, manufacture
and overhaul of complete wheelsets for traction
units and trailers.

Probotec Limited

Cambrian House, 6 Charnwood Court, Parc
Nantgarw, Cardiff CF15 7QZ, UK
Tel: (+44 1443) 84 82 50
Fax: (+44 1443) 84 82 51
e-mail: uk@probotec.com
Web: www.probotec.com

Key personnel
Chairman: S C Harris
Chief Executive Officer: Claude Elsen
Sales Director, UK and Nordic countries:
  Nick Hughes
Finance Director: E W Pulman

Other offices
Sven Hauser, 26, Boulevard Royal L-2649
Luxembourg
Tel: (+35 2) 22 99 99 22 28
Fax: (+35 2) 22 99 99 24 00
e-mail: europe@probotec.com
Web: www.probotec.com
Director Worldwide Business Development:
  Irmhild Saabel

Czech Republic office
Probotec sro
Masarykova 645, CZ-284 12 Kutna Hora, Czech
Republic
Tel: (+42 032) 173 49 60
Fax: (+42 032) 173 49 69
e-mail: czech@probotec.com
General Manager: Jiri Sechovec

Probotec Nordic Countries AB
Fabriksvägen 9, S-18632 Vallentuna, Sweden
Tel: (+46 8) 51 18 77 57
Fax: (+46 8) 51 18 77 67
Director Nordic Region: Rolf Torwald

Background
In October 2002, Probotec Ltd, formerly Powell
Duffryn Rail, acquired the European business of
Meridian Rail, which included the acquisition of
the Axle Motion and Unitruck European product
lines. The company changed its name to Probotec
in May 2004.
  After financial difficulties prompted Probotec
Limited to call in administrators, the company's
assets were sold in April 2005 to the UK's leading
rail freight operator, English Welsh & Scottish
Railway (EWS). EWS was to ensure continuity of
supply of Probotec products to all its customers.

Products
Bogies for freight wagons including the TF (Track
Friendly) range of bogies and suspensions. Wheels,
axles and wheelsets for freight and passenger
stock.
  Range includes: *TF25SA* single axle suspension,
UNItruck single axle suspension, the *Axle Motion
AM* and *TF25* track friendly bogies.

UPDATED

Power Machines Group

(Energomachexport + LMZ + Electrosila + ZTL + KTZ)
25A Protopopovsky per, 129090 Moscow, Russian
Federation
Tel: (+7 095) 725 27 63   Fax: (+7 095) 688 79 90
e-mail: mail@power-m.ru
Web: www.power-m.ru

Key personnel
General Director: Evgeny Yakovlev
Sales Director (Machinery building):
  Alexander Zhigalov
Advertising Manager: Natalia Kuznetsova

Background
Power Machines unites the leading Russian power
equipment enterprises and supplies railway and
transport equipment produced by Kalugaputmash,
Luganskteplovoz, Muromteplovoz, Railway Repair-
Mechanical Plant of Sverdlovsk, Kolomensky
Zavod, Metrowagonmash, Vyksa Steel Works.

Products

Bogies for passenger and freight vehicles; primary and secondary suspension units; rolled- steel wheels, tyres; axles and wheelsets; and Type SA-3 automatic coupling devices.

UPDATED

Puzhen Rolling Stock Works of Ministry of Railways

No 5 Longhu Lane, Puzhen, Nanjing, China
Tel: (+86 75) 85 24 24; 85 47 03
Fax: (+86 75) 85 24 24; 85 26 55

Key personnel

Factory Director: Song Zu Yu
Chief Engineer: Wang Wui Sheng

Products

Roller bearings.

Roller bearings featuring an oblique block frame and solid cage have been introduced. Sections of this bearing can be interchanged with existing bearings.

CNR Qiqihar Railway Rolling Stock (Group) Co Ltd (QRRS)

10 ZhongHuaDongLu Road, Qiqihar 161002, China
Tel: (+86 452) 293 84 72; 293 84 99
Fax: (+86 452) 251 67 23
e-mail: qrrsintl@qrrs.com.cn
Web: www.qrrs.com.cn

Key personnel

Chairman of Board of Directors and General
  Manager: Wei Yan
Chief Engineer: Yu Lianyou
International: Liu Dezeng; Ms Zhang Xianbin

Background

QRRS is a member of CNR.

Products

High-speed freight bogie for 160 km/h operations; Type K1 bogie (120 km/h) with four-rod linkage of QRRS's patent; Type K2 bogie (120 km/h) with linkage below bolster; Barber S-2-HD bolster and side frame, and Barber S-2-M gauge bogie with authorisation from Standard Car Truck Co, US.

UPDATED

Radsatz Ilsenburg

Radsatzfabrik Ilsenburg GmbH
Schmiedestr 16/17, D-38871 Ilsenburg, Germany
Tel: (+49 394) 529 30
Fax: (+49 394) 529 32 05
e-mail: info@rafil-gmbh.de
Web: http://www.rafil-gmbh.de

Key personnel

General Managers: Dipl Ing Jorg Villmann,
  Dipl Kaufm Müller
Sales Manager: H Böhme

Products

Wheels, wheelsets, brake pulleys and axles.

A special development is a lightweight monobloc wheel, proved in German Railways (DB AG) service. It has an axleload from 22.5 to 25 tonnes and is capable of a maximum speed of 350 km/h.

Also available is a wheelset that can change gauge from 1,435 to 1,524 mm and 1,668 mm.

Contracts include the supply of wheelsets to Greenbrier, Bombardier, DB AG, ÖBB, Preussag Stahl, ZF Hurth, Tatra Vagónka Poprad, ALSTOM, Siemens SGP, SBB, SJ, and NeDtrain.

Railko Ltd

Boundary Road, Loudwater, High Wycombe HP10 9QU, UK
Tel: (+44 1628) 52 49 01   Fax: (+44 1628) 81 07 61
e-mail: info@railko.co.uk
Web: http://www.railko.co.uk

Key personnel

Managing Director: S Little
Sales Director: Dr G D Wells
Sales Manager, Railways: B Kirby

Products

Composite thermoset bearings and thermoplastic bearings. Applications include centre pivot liners, side bearer liners, torsion bar bearings, friction dampers, brake gear bushes, door slides, sliding gangway liners and axlebox guides.

VERIFIED

REBS Zentralschmiertechnik GmbH

Duisburger Strasse 115, D-40885 Ratingen, Germany
Tel: (+49 2102) 930 60   Fax: (+49 2102) 93 06 40
e-mail: info@rebs.de
Web: http://www.rebs.de

Products

Wheel flange lubrication systems for trains, metro vehicles, trams and LRVs.

Ringrollers

PO Box 504, Springs 1560, South Africa
Tel: (+27 11) 362 66 70   Fax: (+27 11) 815 28 05
e-mail: ringsales@rail.dcd-dorbyl.com
Web: http://www.dcd-dorbyl.com/ringrollers

Key personnel

General Manager: Stephen Nel
Financial Manager: Ashley Padayachee
Technical Manager: Martin de Lange
Manufacturing Manager: Nic Richter
Sales Manager – Africa: Fred Burr-Dixon
Export Marketing Manager: Roberto Gaspari
Human Resources Officer: Benjamin Motloba

Background

A division of DCD-Dorbyl Ltd, agents for Lucchini in Italy.

Products

Forged railway tyres, flanges and rectangular rings manufactured to a variety of specifications in carbon, alloy and stainless steel. It has expanded its mainstream forging activities to include the provision of heat treatment and machining facilities.

Rotem Company

Headquarters
Landmark Tower, 837-36, Yeoksam-dong, Gangnam-gu, Seoul, 135-937, South Korea
Tel: (+82 2) 21 12 82 94   Fax: (+82 2) 21 12 98 73
Web: www.rotem.co.kr

Key personnel

Vice Chairman and Chief Executive Officer:
  Soon-Won Chung
Senior Executive Vice-President: Yeo-Sung Lee
Executive Vice-President: Jae-Hong Kim

Background

Established in 1964 when Daewoo Heavy Industry started manufacturing rolling stock, followed by Hyundai and Hanjin Heavy Industry a few years later. In 1999, the three companies were consolidated into KOROS. Hyundai Motors Group acquired the share of Daewoo in 2001 and the former company KOROS became Rotem. Rotem

Rotem bolsterless bogie with air suspension for the Seoul metro   0099151

is now an affiliate of Hyundai Motors Group and has its headquarters in Seoul and two factories in Uiwang and Changwon. Uiwang has capacity for manufacturing 500 emus per year and has the capability to manufacture electric equipment such as traction motors, SIV inverters etc. The research and design centre is also located in Uiwang. The Changwon factory has capacity to manufacture 700 emus per year. Rotem has an annual capacity to manufacture approximately 1,200 emus. Certifications such as the ISO 9001 certificate for Quality, 14001 for environment and 18001 for occupational health and safety management have been acquired at all three sites.

Organisation

Rotem manufactures a complete range of bogies suitable for high-speed trains, diesel, electric locomotives, emus, dmus, passenger coaches and all types of wagons. To verify bogie performance, CAD, CAE and other simulation techniques are applied from the initial stage of 3-D digital mock-up design, finite element analysis, vehicle dynamic simulation, static and fatigue load testing of bogie frames, roller rig running tests of the complete bogie, riding quality tests and dynamic stress tests.

UPDATED

RSD (DCD-Dorbyl (Pty) Ltd)

PO Box 229, Boksburg 1460, South Africa
Tel: (+27 1) 914 14 00
Fax: (+27 1) 914 38 85
Web: http://www.de.dorbyl.com/rsd

Key personnel

For full personnel listing, see RSD (DCD-Dorbyl (Pty) Ltd) entry in *Locomotives and powered/non-powered passenger vehicles* section.

Products

Locomotive, coach and freight wagon bogies, including the Scheffel High Stability bogie.

RTM SA

Roues et Trains Montés SA
Quai Greiner, porte 2, B-4100 Seraing, Belgium
Tel: (+32 4) 338 83 00
Fax: (+32 4) 338 82 29

Key personnel

Managing Director: P Marchettini

Background

A subsidiary of Usinor-Sacilor.

Products

Wheelsets, loose wheels (monobloc and tyred), loose axles; rewheeling and overhaul of wheelsets.

R W Mac Co

PO Box 56, Crete, Illinois 60417, USA
Tel: (+1 708) 672 63 76; 81

Key personnel

President: R W MacDonnell
Vice-President, Sales and Exports: J K MacDonnell

Products

Car-safe freight wagon bolster supports; locomotive bogie bolster supports; EMD and GE Journal box wear plates (steel); custom-made locomotive and freight wagon parts.

SAB WABCO

Roskildevägen 1, PO Box 193, SE-201 21 Malmö, Sweden
Tel: (+46 40) 35 04 60   Fax: (+46 40) 30 38 03
e-mail: info@sabwabco.com
Web: www.sabwabco.com

Background

Formerly owned by Cardo Rail AB, the acquisition of SAB WABCO by the Faiveley Transport group was completed in November 2004.

Key personnel

Chief Executive Officer: Robert Joyeux
Chief Operating Officer: Pierre Sainfort
Product Line Manager 'Brake and Couplers':
  Mats Svensson
Product Line Manager 'Electronics':
  Jean-Pierre Guy
Product Line Manager 'Doors': Harald Zinggrebe
Product Line Manager 'Wheels': Michael Walter
Product Line Manager 'After-Market and Services':
  Mario Padovani
European Profit Centers: Ulrich Giesen
International Profits Centres (outside Europe) and
  Business Development: Marc Chocat
Chief Financial Officer: Etienne Haumont

Products

Air suspension control equipment; resilient wheels for main line vehicles, noise reducing wheels for LRVs, laser measurement tools for wearing condition of wheels.

UPDATED

Sambre et Meuse

Railway Division
2 rue des Usines, F-59750 Feignies, France
Tel: (+33 3) 27 69 69 69   Fax: (+33 3) 27 69 69 89

Key personnel

Managing Director: Daniel Pain
Sales and Marketing Manager: Jean Jomeau

Products

Design, testing and manufacture of three-piece bogies for gauges from 1,000 to 1,600 mm; rigid cast-steel bogies Types Y25, Y33, Y39 and VNH for 1,435 mm and other European gauges, for speeds up to 160 km/h and 25 tonne axleloads. Parts for welded bogies including spring supports, pivots and suspension caps.

UPDATED

Silvertown UK Ltd

Horninglow Road, Burton upon Trent, Staffordshire DE13 0SN, UK
Tel: (+44 1283) 51 05 10   Fax: (+44 1283) 51 00 52
e-mail: sales.enq@silvertown.co.uk
Web: www.silvertown.co.uk

Background

Silvertown UK Ltd, previously a member of the Unipoly group, was acquired by The Icon Polymer Group in January 2005. An agreement was also concluded to dispose of its manufacturing site in Burton upon Trent, with further news to be announced on the future location of the business.

Products

Silentbloc elastomeric components and assemblies including: resilient rubber/metal bushes; ball joints; anti-vibration and shock mountings; flexible couplings; laminated spring and link assemblies; suspension systems and anti-roll bar systems.

Silentbloc also provides a comprehensive re-manufacturing service for products requiring refurbishment during normal service life.

Andre elastomeric bearings, moulded from natural rubber and incorporating steel plates to reduce compression strains.

Contracts

Supply of Silentbloc suspension components to the Euroshuttle Consortium – Locomotives (ESCL) for bogies for locomotives to run HGV services in the Channel Tunnel.

Silvertown won a first contract of a £multi-billion infrastructure project in Libya, the country's first railway, designed to run all along the Mediterranean coast from the Tunisian border to Egypt. The contract is for Andre elastomeric structural bearing to isolate 48 bridges on the first 200 km section linking Tripoli with Misurata.

UPDATED

SKF AB

Hornsgatan 1, SE-415 50 Gothenburg, Sweden
Tel: (+46 31) 337 14 32
Fax: (+46 31) 337 17 77
e-mail: egon.ekdahl@skf.com
Web: www.railways.skf.com

Key personnel

President and Chief Executive Officer:
  Tom Johnstone
Finance: Tore Bertilsson
Business Development: Henrik Lange
HR & Sustainability: Eva Hansdotter
Technology Development: Henning Wittemeyer
Communication: Lars G Malmer
Director, Railway Business Unit: Egon Ekdahl

Background

Founded in 1907 the business is organised into five divisions: industrial, automotive, electrical, service, aero and steel. Each division serves a global market, focussing on its specific customer segments.

SKF employs 38,600 employees across 76 production sites in 22 countries.
ISO 14001 certified.

Products

Axleboxes for all types of rolling stock; axlebridges for low-floor vehicles, sealed and greased, double row taper bearing units, speed and temperature sensors; bogie monitoring systems; traction motor and transmission bearings; motor support housings and roller bearings; self-lubricating plain bearings; electrically insulated bearings; maintenance, mounting/dismounting equipment; bearing refurbishment.

Contracts

Locomotives and multiple unit applications:
Australia: Explorer diesel railcars.
Austria: 1012, 1014, 114 and 1822 electric locomotives.
Finland: Sr2 electric locomotives.
France: TML, MI 2N, M10N, Z2 and X-TER trains.
Germany: 101, 145 and 152 electric locomotives.
  LVT 642 and 611 tilting trains, suburban electrical and diesel units.
Italy: E 402, E 412 and E 464 electric locomotives. TAF trains.
Netherlands: DM 90 and IRM.
Norway: El 18 electric locomotives.
Spain: S 252 electric locomotives.
Sweden: X- 15 and Oeresund trains.
Switzerland: Lok 2000.
UK: Class 58, 60 and 67 locomotives.
  Class 165, 315, 317, 318, 319, 320, 321, 322, 323, 357, 442, 456, 465 and 466.
USA: Diesel electric locomotives manufactured by EMD.

UPDATED

Standard Car Truck Co

865 Busse Highway, Park Ridge, Illinois 60068, USA
Tel: (+1 847) 692 60 50   Fax: (+1 847) 692 62 99
Web: http://www.sctco.com

Key personnel

Chairman and Chief Executive Officer:
  Richard A Mathes
Chief Operating Officer/President: Dan Schroeder
Vice-President, International: David J Watson
Vice-President, Sales & Service: Mark Pace
Secretary/Treasurer: Donald J Popernik

Subsidiary companies

Anchor Brake Shoe Co
Barber Spring
Triangle Engineered Products Co
Sancast Inc
Durox Company

Products

Cast steel bogies (AAR three-piece type) incorporating load-sensitive Barber Stabilized suspension and control elements.

The Barber Stabilized bogie utilises vertical damping forces directly related to the vehicle's load. A wide range of capacities and track gauges can be accommodated.

The same damping features are offered in self-steering radial bogies. The advantages of self-steering can also be enjoyed with existing bogies through the use of the Barber Frame Bracing retrofit.

Standard Steel, LLC

500 N Walnut Street, Burnham, Pennsylvania 17009, USA
Tel: (+1 717) 248 49 11   Fax: (+1 717) 248 80 50

Key personnel

President and Chief Executive Officer:
  Michael J Farrell
Senior Vice-President Sales and Operations:
  John M Hilton
Senior Vice-President Finance and Administration:
  Dana L Patterson

Principal subsidiaries

Standard Steel, Burnham and Latrobe, Pennsylvania, USA

Background

Standard Steel, LLC was previously named Freedom Forge Corporation.

Products

Various types of wheels, axles and mounted assemblies (freight, diesel and transit). Box lids.

Steel Authority of India Ltd

Durgapur Steel Plant, Durgapur, West Bengal, India
Tel: (+91 343) 830 00   Fax: (+91 343) 823 17

Key personnel

For full personnel listing, see Steel Authority of India Ltd entry in *Permanent way equipment* section.

Products

Forged wheels, axles and wheelsets.
Recent contracts include supply of rails, wheels, axles and wheelsets to Indian Railways and rails to Bangladesh Railways.

Stojírna Oslavany spol sro

Padochovsk 31, CZ-664 12 Oslavany, Czech Republic
Tel: (+420 546) 49 22 11; 49 22 61
Fax: (+420 546) 49 29 99; 42 34 87
e-mail: mihalisko@st.os.cz

Key personnel

Managing Director: Ing Ivo Šrámek
Manufacturing Director: Milan Vaněček
Technical and Sales Director: Ing Jozef Mihalisko
Finance Director: Ing Zdeněk Vévoda

Products

Shock absorbers for rail vehicle bogie damping systems. Dampers are also manufactured for rail switches.

Sumitomo Metal Industries Ltd

8-11 Harumi I-Chome, Chu-ku, Tokyo 104-6111, Japan
Tel: (+81 3) 44 16 62 71   Fax: (+81 3) 44 16 67 90
e-mail: skr@sumitomometals.co.jp
Web: www.sumitomometals.co.jp/e/index.html

Key personnel

Senior Managing Executive Officer: Yasutaka Toya
Senior Managing Executive Officer: Kaoru Goto
Managing Executive Officer: Mitsunori Okada
General Manager: Shiuji Morinobu
Senior Manager: Yukinovi Akimoto

Works

Osaka Steelworks, 5-1-109 Shimaya, Konohana-Ku, Osaka 554-0024

Products

Wheels, tyres, axles, wheelsets, bogies, air springs, gear units, brake discs, automatic couplers, draftgear.

Sumitomo has developed a bolsterless bogie with 40 per cent fewer components and weighing some 15 per cent less than conventional designs. Better running performance through curves is claimed. Also developed linear induction motors for the Osaka and Tokyo mini-metro systems.

Contracts

Numerous types of powered and trailing bogies have been supplied to Japanese private railways and metro systems. Sumitomo's market share of bogies for Japanese private railways and metros is put at over 70 per cent.

UPDATED

Superior Graphite Co

10 S Riverside Plaza, Suite 1600, Chicago, Ilinois 60606, USA
Tel: (+1 312) 559 29 99   Fax: (+1 312) 559 90 64
Web: www.superiorgraphite.com

Key personnel

Chairman: Peter R Carney
President and Chief Executive Officer:
  Edward O Carney
Senior Vice-President Operations, Sales and
  Marketing: Wesley C Krueger
Senior Vice-President Finance: Ronald G Pawelko

Overseas offices

Sweden
PO Box 1300, Sundsvall
Tel: (+46 60) 13 41 88
Fax: (+46 60) 13 41 28

Canada
PO Box 20015, John Galt Postal Station, Cambridge, Ontario N1R 8C8
Tel: (+1 519) 650 16 08
Fax: (+1 519) 650 18 03

Products

Graphite-based rail lubricants.

Lubricants have been supplied to Indian Railways and rail systems in the US including Burlington Northern Railroad.

UPDATED

Swasap (Pty) Ltd

PO Box 366, Germiston 1400, Gauteng, South Africa
Tel: (+27 11) 873 66 66
Fax: (+27 11) 873 76 88
e-mail: swasap@bongroup.com
Web: http://www.bongroup.com/swasap

Main works

Rinkhals Street, Industries East, Germiston, Gauteng, South Africa

Key personnel

Managing Director: Derek Anderson
Sales Director: Welcome Ngeju
Contracts Controller: Gerrie Nel

Products

Axles and wheelsets.

Contracts

Contracts include supply of 2,676 axles for several US metro rail systems as well as exports to the UK, Australia, Canada and other global customers.

VERIFIED

Tafesa SA

Carretera de Andalucía Km 9, E-28021 Madrid, Spain
Tel: (+34 91) 798 05 50
Fax: (+34 91) 798 09 61
e-mail: tafesa@tafesa.com
Web: www.tafesa.com

Key personnel

Group Chairman: Paloma Fernández-Souza Faro
General Manager: J Eliecer Esteban Gutierrez
Technical Director: Luis Peromarta Calvo
Marketing and Commercial Manager:
  Jesús Montes Chinchón

Products

Design, testing and manufacture of bogies.

UPDATED

Talgo

Patentes Talgo SA
C/Gabriel García Márquez 4, E-28230 Las Rozas, Madrid, Spain
Tel: (+34 91) 631 38 00; 27 00
Fax: (+34 91) 631 38 99
e-mail: marketing@talgo.com
Web: http://www.talgo.com

Key personnel

For full personnel listing, see Talgo entry in *Locomotives and powered/non-powered passenger vehicles* section.

Subsidiary companies

For a full list of subsidiary companies see Talgo entry in *Locomotives and powered/non-powered passenger vehicles* section.

Products

Talgo RD automatic wheelset gauge-changing system, enabling freight wagons and Talgo passenger vehicles to operate between railway networks with track gauges ranging from 1,000 to 1,668 mm. The system comprises two elements: wheelsets in which the position of the wheels on the axles can be unlocked to enable them to be moved to conform to the gauge of the destination railway; fixed gauge-changing installations which include side-bearers to take the weight of each vehicle as wheelset adjustment takes place.

UPDATED

Tatravagónka AS

Štefánikova 887/53, SK-058 01 Poprad, Slovakia
Tel: (+421 52) 72 32 75
Fax: (+421 52) 72 17 32
e-mail: market@vapop.sk
Web: www.tatravagonka.sk

Key personnel

General Director: Július Vachmansky
Sales Director: Frantisek Hudák
Marketing: Miroslav Stephán

Products

Bogies for freight wagons, types Y 25, Y 33, Y 37, three-axle bogies with 1,000 mm gauge.

Contracts

Customers include: ČSD, PKP, MÁV, DR, VTG, DB, AAE,ČD, ŽSR, DWA, LHB,KRUPP, Hoesch, Zastal Zelona Gora, ZNTK, Gniewczyna, WTR, Transwaggon, Thrall Car, FW Ostrow, Wielkopolski, Slovak Republic, Czech Republic, SNCF, Duro Dakovič, Nik Kioleides.

UPDATED

Techlam

1 rue de l'Industrie, BP 6, F-68701 Cernay Cedex, France
Tel: (+33 3) 89 75 30 82   Fax: (+33 3) 89 39 89 62
e-mail: info@techlam.org

Products

Flexible mechanical systems and elastomer/rubber components, including: anti-roll bar components for TGV high-speed trainset motor bogies; Orgal steering system for VAL 208 automated rubber-tyred metro vehicles; and elements for inter-car articulation for low-floor light rail vehicles.

Techni-Industrie SA

ZI de la Chambrouillère, F-53960 Bonchamp-lès-Laval, France
Tel: (+33 2) 43 59 23 80
Fax: (+33 2) 43 59 23 89
e-mail: techni@techni-industrie.fr
Web: http://www.techni-industrie.fr

Key personnel

Chairman: Gérard Lelasseux
Commercial and Technical Manager: Dan Diaconu

Products

Bogies for metro trainsets and freight wagons.

Tenmat Ltd

Ashburton Road West, Trafford Park, Manchester M17 1RU, UK
Tel: (+44 161) 872 21 81
Fax: (+44 161) 872 75 96
e-mail: info@tenmat.com
Web: http://www.tenmat.com

Key personnel

Marketing Officer: Colin Stansfield
Technical Services Manager: Dave Hill

Products

Ferobestos and Feroform wearing and bearing materials for centre pivot liners, side bearer pads, brake pivot bushes and other applications; non-asbestos arc and heat-resistant insulation boards; millboards and intumescents for thermal insulation and fire protection.

ThyssenKrupp GfT Gleistechnik GmbH

Altendorfer Strasse 120, D-45143 Essen, Germany
Tel: (+49 201) 188 37 65   Fax: (+49 201) 188 37 57
Web: www.thyssenkruppgleistechnik.de

Key personnel
Contact: H Weiß

Background
ThyssenKrupp GfT Gleistechnik GmbH is a subsidiary of ThyssenKrupp Services AG, Düsseldorf, Germany.

Products
Wagon components; wheel sets, wheels, axles, wheel tyres.

NEW ENTRY

Timken Rail Service

16 Quorn Way, Grafton Industrial Estate, Northampton NN1 2PN, UK
Tel: (+44 1604) 62 70 15
Fax: (+44 1604) 63 64 54

Head office
The Timken Company
1875 Dueber Avenue, SW, Canton, Ohio 44706, USA
Tel: (+1 330) 471 30 00

UK head office
British Timken Ltd
Main Road, Duston, Northampton NN5 6UL, UK
Tel: (+44 1604) 75 23 11
Canton, Ohio 44706, USA

Key personnel
President, Rail: Vinnie Dasari (Canton)
General Manager, Duston Bearing Plant:
  William Kelleher (Duston)
Director, European Rail and Passenger Systems:
  Keith P Kruger (Quorn Way)
Manager, Operations and Engineering:
  Christian Moly (Quorn Way)

Manufacturing plants
Australia, Brazil, Canada, China, France, Italy, Poland, South Africa, UK and USA.

Products and services
Tapered roller bearings; AP and SP tapered roller bearing cartridge units; and complete axleboxes and motor suspension units.
  Design, manufacture and supply of tapered roller bearings and ancillary equipment covering transmissions, axleboxes, traction motor suspension units and other equipment, such as cooling fans and screw compressors.
  Reconditioning and remanufacturing of tapered roller bearings for rail applications.

Trelleborg Industrial AVS

1 Hoods Close, Leicester LE4 2BN, UK
Tel: (+44 116) 267 03 00
Fax: (+44 116) 267 03 01
e-mail: industrialavs.uk@trelleborg.com
Web: http://www.trelleborg.com

Key personnel
Managing Director: Leif Olsson
Sales Manager: Tony Carter
Public Relations: David Somerlad

Background
Trelleborg Industrial AVS is the amalgamation of Metalastik Limited and Novibra, Sweden.

Products
Design and manufacture of rubber-to-metal bonded components for anti-vibration and suspension systems for bus, truck, rail and other industrial and engineering applications.

Contracts
Supply of rubber chevron springs to refurbish 109 of the Mafersa-built rolling stock fleet in use on the 47.1 km São Paulo metro, Brazil. The springs are rubber/metal laminated units fitted between the wheel axles and bogie frame, designed to absorb high levels of shock and vibration.

Technical Service and Marketing Inc (TSM)

10765 Ambassador Drive, Kansas City, Missouri 64153, USA
Tel: (+1 816) 891 65 44   Fax: (+1 816) 891 93 29

Key personnel
For full personnel listing, see Technical Service and Marketing Inc (TSM) entry in *Diesel engines, transmission and fuelling systems* section.

Products
Onboard flange lubricator systems with electronic control.
  Contracts include the supply of flange lubricator systems to Burlington Northern Santa Fe, USA, maintained and serviced by TSM Services.

Toyo Tire & Rubber Co Ltd

Mejironakano Building, 2-17-22 Takada, Toshima-ku, Tokyo 718544, Japan
Web: http://www.toyo-rubber.co.jp

European distributor
Gadelius Europe AB
Halmstad, Sweden
Tel: (+46 35) 18 20 73   Fax: (+46 35) 18 20 71
e-mail: lars.winslott@gadelius.com
Web: http://www.gadelius.com

Products
Air springs and anti-vibration components for rail vehicles.

NEW ENTRY

Tülomsas

The Locomotive & Motor Corporation of Turkey
Ahmet Kanatli Cad, TR-26490 Eskisehir, Turkey
Tel: (+90 222) 224 99 56   Fax: (+90 222) 225 72 72
e-mail: tulomsas@tulomsas.com.tr
Web: http://www.tulomsas.com.tr

Key personnel
For full personnel listing, see Tülomsas entry in *Locomotives and powered/non-powered passenger vehicles* section.

Products
Bogies for locomotives and freight wagons.

United Goninan

PO Box 33000, Hamilton, New South Wales 2303, Australia
Broadmeadow Road, Broadmeadow, New South Wales 2292, Australia
Tel: (+61 2) 49 23 50 00   Fax: (+61 2) 49 23 50 01
Web: www.unitedgoninan.com.au

Key personnel
Chief Operating Officer: John McLuckie

Products
Bogies for locomotives, passenger vehicles and freight wagons.

Contracts
Design and manufacture of fabricated bogies for dmus and double-deck emus for Railcorp, New South Wales, Australia.

UPDATED

Uralvagonzavod

622006 Nizhny Tagil, Russian Federation
Tel: (+7 83435) 23 17 74; 23 01 97
Fax: (+7 83435) 23 34 92; 23 03 57

Products
Bogies.

Valdunes

Immeuble International, Bâtiment A2, 2 rue Stephenson, F-78180 Montigny-le-Bretonneux, France
Tel: (+33 1) 39 30 84 84   Fax: (+33 1) 39 30 84 75
Web: http://www.valdunes.com

Key personnel
Chief Executive Officer: Jean-Pierre Auger
Commercial Manager: Christian Pignerol

Products
Forged wheels, including low-stress wheels, sound-dampening wheels and wheels for light rail vehicles and heavy-haul applications; axles; powered and trailer wheelsets; brake discs; gear blanks.

Contracts
Contracts include the supply of wheels and wheelsets to operators, traction and rolling stock builders and private owners in over 60 countries.

Vibratech Inc

11980 Walden Avenue, Alden, New York 14004-9790, USA
Tel: (+1 716) 937 79 03   Fax: (+1 716) 937 46 92
e-mail: dcovelli@vibratech.com
Web: http://www.vibratech.com

Key personnel
Sales Manager: Daniel Covelli

Products
Vibration and motion damping systems.
  Customers include Adtranz, Gevisa, ALSTOM, RSD Dorbyl, General Electric, Kawasaki, Siemens, Sumitomo and Bombardier.

Vibratech telescopic shock-absorbers to control yaw motion in locomotive, freight and passenger rail applications 0063736

Vogel

Willy Vogel Aktiengesellschaft
Motzener Strasse 35/37, D-12277, Germany
Tel: (+49 30) 72 00 20   Fax: (+49 30) 72 00 21 11
e-mail: info@vogel-berlin-de
Web: http://www.vogelag.com

Key personnel
Director, International Sales: Vincent F Warnecke

Background
Willy Vogel AG has been acquired by SKF, from previous owners Hannover Finanz Group, a private equity firm.

Products
Vogel wheel flange lubrication systems for rail vehicles.

UPDATED

Vossloh Locomotives GmbH

PO Box 9293, D-24152 Kiel
Falckensteiner Strasse 2, D-24159 Kiel, Germany
Tel: (+49 431) 39 99 21 05
Fax: (+49 431) 39 99 22 74
e-mail: vertrieb.kiel@vl.vossloh.com
Web: www.vossloh-locomotives.com

Works

Service-Zentrum Moers, Baerler Strasse 100, D-47441 Moers, Germany
Tel: (+49 2841) 14 04 10   Fax: (+49 2841) 14 04 50

Key personnel

Board of Directors: Andreas Hopmann
Chief Executive Officer: Dr Georg Hauschild

Products

Design and manufacturing of two- and three-axle bogies for hydraulic and electric drive according to European and US standards.

UPDATED

Wabtec Rail Ltd

PO Box 400, Doncaster Works, Hexthorpe Road, Doncaster DN1 1SL, UK
Tel: (+44 1302) 34 07 00   Fax: (+44 1302) 32 13 49
Web: www.wabtec.com

Key personnel

Managing Director: John Meehan
Engineering Director: Mike Roe
Finance Director: Robert Johnson
Operations Director: Chris Weatherall
Commercial Manager: Paul Robinson

Background

Wabtec Rail undertakes the overhaul of passenger rolling stock, main line locomotives, bogies, wheelsets, air brake equipment, hydraulic dampers and buffers design in addition to the manufacture of shunting locomotives and freight wagons.

As part of the Wabtec Corporation, Wabtec Rail also supplies to the UK rolling stock owners and maintainers, composite brake blocks and pads, Wabtec Railway Electronics train data recorders and electronic equipment, Cardwell TMX braking systems and Wabtec air brake equipment.

Products

Wabtec Rail is a major overhauler of wheelsets, with the ability to re-profile, re-wheel and re-axle a wide range of locomotive, passenger stock and freight wagon wheelsets. Wabtec Rail's modern wheelset overhaul facility is equipped with full ultrasonic and magnetic particle axle testing equipment and is accredited to overhaul all wheelset types.

In addition, Wabtec Rail has a dedicated bogie overhaul facility undertaking the overhaul of diesel multiple-unit, electric multiple-unit, passenger coach, locomotive and freight wagon bogies.

UPDATED

Wheel & Axle

Wheel & Axle Plant (Indian Railways)
Yelahanka, Bangalore 560 064, India
Tel: (+91 80) 846 03 49; 20 45
Fax: (+91 80) 846 03 67

Key personnel

General Manager: Gopal Krishna Malhotra
Chief Mechanical Engineer: Rajneesh Dubey

Products

Cast-steel wheels, axles and wheelsets.
Contracts include the supply of 948 wheels and 200 axles to Progress Rail Services, USA.

William Cook Rail

Cross Green, Leeds LS9 0SG, UK
Tel: (+44 113) 249 63 63
Fax: (+44 113) 249 13 76
e-mail: admin@cook-catton.co.uk
Web: http://www.william-cook.co.uk

Key personnel

For full personnel listing, see William Cook Rail entry in *Brakes and drawgear* section.

Products

Bogie castings, including frame castings, centre castings, brackets, axleboxes, traction motor casings and suspension parts.

Woodhead Shock Absorbers

Church Street, Ossett WF5 9DL, UK
Tel: (+44 1924) 27 35 21
Fax: (+44 1924) 27 61 57; 26 46 56

Key personnel

Chairman: S Beyazit
Director: A C Kart

Products

Dampers (primary and secondary) for rolling stock, bogies and door actuating systems. Refurbishment facility for rolling stock and bogie dampers.

UPDATED

ZF Sachs AG

Center of Excellence
Rail Vehicle Damping
Bogestrasse 50, D-53783 Eitorf, Germany
Fax: (+49 2243) 122 80
e-mail: rail.technology@zf.de
Web: www.zfsachs.com

Key personnel

General Manager: Thomas Memmesheimer

Works

Schweinfurt, Germany

ZF Sachs dampers for rail vehicles   0122673

Background

ZF Sachs AG is owned by ZF Friedrichshafen AG based in Germany.

Products

Powertrain and suspension components for rail vehicles. The product range of dampers for rail vehicle applications includes primary and secondary dampers (vertical and horizontal), yaw dampers, dampers for engines, superstructures and articulated roof structures.

ZF Sachs has further developed PDC (Pneumatic Damping Control) dampers for use in freight wagons. Pneumatically controlled proportional damper valves progressively adapt the damping to the respective load condition via the brake pressure. This system is especially suitable as a horizontal damper system for major load differences. In their function as vertical dampers, PDC dampers also improve comfort in pneumatically damped rail vehicles. This allows for higher speeds at a lower driveload.

UPDATED

Zimbabwe Engineering Ltd (Zeco)

PO Box 1874, Bulawayo, Zimbabwe
Tel: (+263 9) 789 31   Fax: (+263 9) 722 59

Key personnel

For full personnel listing, see Zimbabwe Engineering Ltd (Zeco) entry in *Locomotives and powered/non-powered passenger vehicles* section.

Products

Bogies, bogie components and spares for freight wagons.

SIMULATION AND TRAINING SYSTEMS

Alphabetical listing

Alion Science and Technology
Corys TESS
Dornier GmbH
DST
EBIM SA
Indra

Inovex Digital Training
Kolmex sa
Krauss-Maffei Wegmann GmbH & Co KG
Oktal
Orthstar Inc
PC-Rail Software

STERIA Group
Systra
Tata
Transurb Technirail SA
Vossloh Information Technologies York Ltd

Company listing by country

BELGIUM
Transurb Technirail SA

FRANCE
Corys TESS
EBIM SA
Oktal
STERIA Group
Systra

GERMANY
Dornier GmbH
DST
Krauss-Maffei Wegmann GmbH & Co KG

INDIA
Tata

POLAND
Kolmex sa

SPAIN
Indra

UNITED KINGDOM
Inovex Digital Trainineg
PC-Rail Software
Vossloh Information Technologies York Ltd

UNITED STATES
Alion Science and Technology
Orthstar Inc

For details of the latest updates to *Jane's World Railways* online and to discover the additional information available exclusively to online subscribers please visit
jwr.janes.com

Alion Science and Technology

185 Admiral Cochrane Drive, Annapolis, Maryland 21401, USA
Tel: (+1 410) 573 71 40   Fax: (+1 410) 573 73 10
e-mail: mjones@alionscience.com
Web: www.alionscience.com

Key personnel

President and Chief Executive Officer:
   Bahman Atefi
Senior Vice-President: Lynn Cumberbatch
Program Manager Transport Technology:
   Mel Jones

Background

Alion Science and Technology Corporation (Alion) is a 100 per cent employee-owned, independent, for-profit, research and development company.

Headquartered in McLean, Virginia, US, Alion has more than 1,900 employees at major offices and laboratories in 35 cities worldwide, providing technology services to government and commercial customers. With experts in modelling and simulationg, wireless communication, defense operational support, information technology and transport technology, Alion in particular, has 20 years experience in providing driver simulators, hardware, and associated software to an international customer base.

Products

Alion Science and Technology's Transport Technology Group is a worldwide supplier of railway industry-accepted training products for operations and maintenance. Incorporating the latest advances in rail technology, and producing the industry's most detailed simulator software models, Alion's simulation modelling has become the accepted standard of quality for the rail industry worldwide. Each system is customised with the features and functions to meet specific user needs and full customer service is provided for all products.

UPDATED

Alion Science and Technology's simulator displaying the approach to St Pancras Station in London, UK   0547061

Alion Science and Technology's CoachSim™ train driver simulator   1048525

Corys TESS

74 rue des Martyrs, F-38027 Grenoble, Cedex 1, France
Tel: (+33 4) 76 28 82 00   Fax: (+33 4) 76 28 82 11
e-mail: coryscom@corys.fr
Web: http://www.corys.com

Key personnel

Managing Director: Frank Chevalley
Transport Business Unit Manager:
   Damien Convert

UK office

Rail Training International (RTI)
35 Old Queens Street, London SW1 9JD
Tel: (+44 20) 73 40 09 00   Fax: (+44 20) 72 33 34 11

US office

Corys TESS Inc
9507 South Kedzie Avenue, Evergreen Park, Illinois 60805, USA
Tel: (+1 708) 422 22 20   Fax: (+1 708) 422 12 25

Background

Corys TESS (Training & Engineering Support Systems) has over 25-years experience in simulation and training. Since 1997, Tractebel Engineering have had a majority holding of Corys TESS.

Products

Complete range of turnkey railway solutions from computer-based training, and training consultancy to full motion replica train-driving simulators.

Contracts

Recent train simulator contracts include SNCFT Tunisia, VAG Nurnberg Germany, Indian Railways, NSB Norway, Transperth Australia and Northern Ireland Railways.

Corys TESS has train driving simulator references on all continents with production facilities in Grenoble, France and Chicago, USA. Through its subsidiary Rail Training International, Corys TESS can supply training solutions comprising the integration of the training technology within the training programme.

UPDATED

Corys TESS in-house computer generated image for Connex, UK   0594130

Dornier GmbH

D-88039 Friedrichshafen, Germany
Tel: (+49 7545) 891 47   Fax: (+49 7545) 887 72
e-mail: simulation.training@dornier.eads.net
Web: http://www.simulation-training.com
   http://www.eads.net

Key personnel

Marketing Manager (Driving Simulation):
   Jürgen Rau

Background

Dornier is a corporate unit of EADS.

Corys TESS dynamic full scope simulator with panoramic view for RET, Netherlands   0594129

Dornier ICE simulator for the Deutsche Bahn AG   1032286

Dornier ICE simulator cabin with projected visual system   1032287

Products

Dornier provides driving simulators for all kinds of wheeled vehicles (trucks, buses and cars) as well as for railway vehicles, including locomotives, trams and light rail vehicles and metro trains.

Simulators for both wheeled and railway vehicles including driver's cabin, visual system, image generation, vehicles dynamics, motion system, sound system, instructor station and observer station for trainees.

DST

Deutsche System-Technik GmbH
PO Box 450262, D-28296 Bremen, Germany
Tel: (+49 421) 428 70
Fax: (+49 421) 40 46 60

Key personnel

Chairmen and Chief Executive Officers:
 Hans-Jörg Zobel, Bruno Jacobi
Director, International Sales and Marketing:
 Dr Christian Wetzel

Products

Driving and signalling simulators.

EBIM SA

ZI St Joseph, F-04012 Manosque Cedex, France
Tel: (+33 4) 92 72 18 66   Fax: (+33 4) 92 87 31 86

Key personnel

Chairman: M Giusti
Managing Director: J J Valade
Assistant Managing Director: B Giusti
Business Manager: A Nassif

Products

Static and dynamic driving simulators; computer-assisted training systems for railway personnel. French National Railways (SNCF) has 42 EBIM multimedia computer-assisted training stations in service at 36 training centres. The training stations are used to familiarise personnel with different types of rolling stock (TGV high-speed trainsets, Corail passenger coaches and suburban trainsets) and train them in operational, safety and fault finding/rectification procedures.

Recent contracts include the supply of a static Eurostar driving simulator and a static Class 92 locomotive driving simulator to EPS, UK; two static driving simulators to Eurotunnel; and three dynamic (Eurostar, Class 92 and TGV-Réseau) driving simulators to SNCF, France.

Indra

Mar Egeo 4, Pol Industrial no 1, E-28830 San Fernando de Henares, Madrid
Tel: (+34 91) 626 90 61; 86 00
Fax: (+34 91) 626 90 76
Web: http://www.indra.es

Products

Computer-based simulator systems for train drivers, station masters, traffic controllers and maintenance staff. Driver simulators can be supplied in fixed or moving positions, with a 45 to 180° field of view and specific or generic driving cabs and visual databases. Traffic control simulators can be supplied to train staff in the handling of individual station traffic, operating in local or remote mode, and as a CTC post, and can be configured to feature signals, track circuits, levers, points and level crossings. Integrated simulator systems combining driving and traffic control can also be supplied.

In collaboration with Madrid bus operator EMT, Indra has also developed an urban bus driver simulator system. Features include a driving position with interchangeable elements for eight different vehicle types, ticket vending machine and operational help systems, a motion system, surround field of more than 180° with exterior and interior rear-view mirrors, virtual city features, varying traffic, passenger and pedestrian models, environmental sound generation and an instructor station.

Inovex Digital Training

Tasman House, Lamport Drive, Heartlands Business Park, Daventry NN11 5YH, UK
Tel: (+44 1327) 31 06 54
Fax: (+44 1327) 31 00 67
e-mail: phil.creaney@inovex.net
Web: http://www.inovex.net

Key personnel

Chief Executive Officer: Dr Urs Guggenbuehl
Sales and Marketing Director: Phil Creaney
Design and Development Manager: David Horne

Other office

Inovex Digital Training
Steinackerstrasse 34, Basserdorf CH-8302, Kloten, Switzerland
Tel: (+41 43) 255 57 00   Fax: (+41 43) 255 57 01
e-mail: gabriela.bloesch@inovex.net
Web: http://www.inovex.net

Products and services

Inovex designs and develops training solutions including computer-based training (CBT) and e-learning programmes employing software-based simulation techniques. The company has wide experience working with the Swiss rail industry including driver training on the new advanced train control systems ERTMS, where it developed a computer-based course simulating the man/machine interface employing a series of typical user scenarios.

Inovex is part of the UK rail industry's Link-Up database qualifying under three categories: train driver training, road vehicle services driver training and technical training.

NEW ENTRY

Kolmex sa

Grzybowska 80/82, PL-00-844 Warsaw, Poland
Tel: (+48 22) 661 50 00
Fax: (+48 22) 620 93 81
e-mail: kolmexsa@kolmex.com.pl
Web: http://www.kolmex.com.pl

Key personnel

Chairman: Andrzej Nalecz

Products

Railway simulator for driver training.

Krauss-Maffei Wegmann GmbH & Co KG

Krauss-Maffei-Strasse 11, D-80997 Munich, Germany
Tel: (+49 89) 81 48 20   Fax: (+49 89) 81 40 49 23
e-mail: info@kmweg.de
Web: http://www.kmweg.de

Key personnel

Director: Dieter Weber
Sales and Projects: Robert Bodner
Commercial Affairs: Friedemann Neuhäuser
Executive Sales Manager: Hermann J Bidermann

Products

Planning, development, manufacturing, integration and marketing of training and simulations for civilian and non-civilian applications. With its leading role in the field of driving simulation, the products in that field cover the complete spectrum for the installation of integrated training concepts including: Computer Based Training (CBT), part task trainers and full mission simulators.

The simulation systems include the complete train model and KMW has modelled locomotives and trains of many of the major locomotive manufacturers worldwide including Siemens, Bombardier and ALSTOM. The simulation of any track logics, signalling systems or train control systems such as ATO/ATP, the German standard LZB. The simulation of train diagnostic system displays. A sound simulation system which includes the simulation of a train radio system. The simulation of any operational and/or technical malfunctions that can online be released by the instructor (on- or offline) and a highly sophisticated instructor's station software (MMI) which enables the instructor to operate the simulator.

Oktal

Immeuble Aurelien 2, 2 rue Boudeville, F-31100 Toulouse, France
Tel: (+33 5) 62 11 50 10   Fax: (+33 5) 62 11 50 29
e-mail: oktal@oktal.fr
Web: http://www.oktal.fr

Paris office
55 rue Thiers, F-92100 Boulogne Billancourt, Paris
Tel: (+33 1) 46 94 97 80

Associated company

Technirail, Belgium

Products

Train driving simulators, including visual and motion systems.

Orthstar Inc

Airport Corporate Park, PO Box 459, Big Flats, New York 14814, USA
Tel: (+1 607) 562 21 00
Fax: (+1 607) 562 21 10
e-mail: info@orthstar.com
Web: http://www.orthstar.com

Key personnel

Chairman, Chief Executive Officer and Director of Engineering: James E Orsillo
President and Chief Financial Officer: Joseph E Strykowski
Executive Vice-President and Chief Operating Officer: Carolyn B Spencer

Background

Orthstar Inc, a systems and software engineering organisation, bought Hughes Training rail simulation business in 1997.

Products

Locomotive and train simulators, metro car simulators, GPS tracking systems and parts task trainers. Development of real-time on-train and wayside products for the rail industry.

Contracts

Orthstar clients include ALSTOM, the NJT Comet 5 data car network traffic parametric simulation/study and the NJT Comet 5 data car network integration test plan; advanced automated train system for BART as a strategic partner with GE Global Signal (Harmon); train simulator visual server upgrades for Metro North Railroad, Canadian National, Norfork Southern; computerised metro door control systems (software) for MARTA, NYCT (R142A and R143), NJT (Hudsen Bergen) as a strategic partner with Vapor Corporation; software/system support for EPIC and EPIC II locomotive air brake system as a strategic partner with WABTEC (WABCO); automation of Netherlands Spoorwegen (Dutch railway) Amsterdam Station; V & V System/software testing of aspects of the GE Transportation AC 6000 locomotive and BART computerised door system; marshalling yards for NS-Conrail; hump yards for Union Pacific (Fresco Rail); metro train control system for MARTA; locomotive preventive maintenance for Burlington Northern Santa Fa Railroad; transit car simulators for SEPTA and Metro North; high-speed passenger train simulator for Swedish Railroad; locomotive and train simulators for Canadian National NS-Conrail, Grand Trunk Western, UP-SP; computerised train control system and train traffic simulator for Netherland Spoorwegen.

TCSim system of KMW, at research facilities of German Railways in Munich-Freimann   0143311

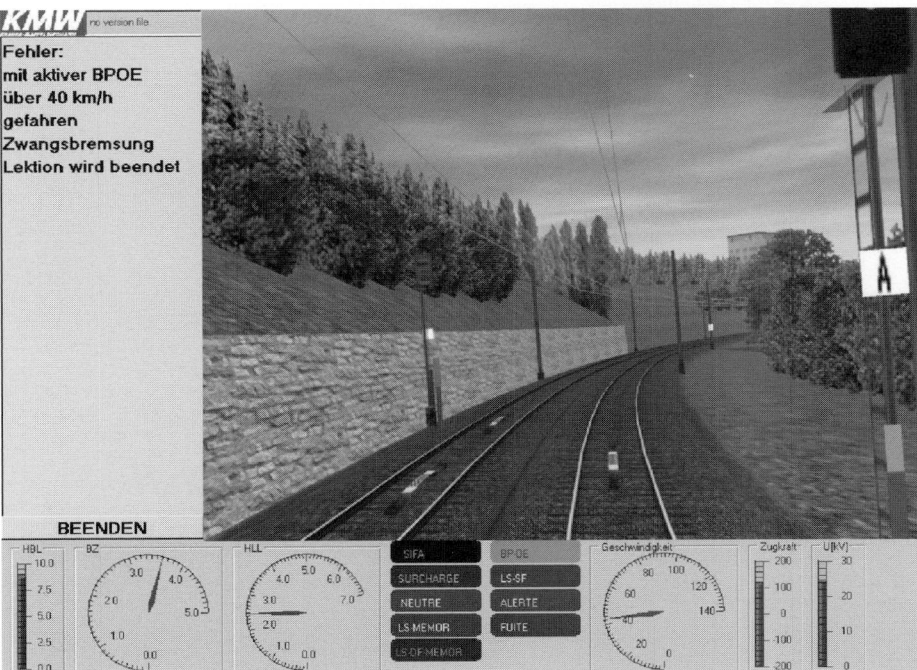

KMW training device with partial simulation as delivered to Luxembourg Railways (together with a full mission simulator and a CBT system)   0143312

PC-Rail Software

PO Box 27, Cromer NR27 9RL, UK
Tel: (+44 1263) 51 40 47
e-mail: info@pcrail.co.uk
Web: www.pcrail.co.uk

Key personnel

Principle Consultant: John D Dennis

Products

PC-Rail range of railway signalling simulations for use with Windows 95 or later (including 98, ME, XP, NT and 2000).

UPDATED

STERIA Group

46 rue Camille Desmoulins, F-92782 Issy les Moulineaux, France
Tel: (+33 1) 34 88 63 99
Fax: (+33 1) 34 88 60 15
e-mail: jacques.lafay@steria.fr
Web: www.steria.fr

Key personnel

See STERIA Group entry in *Signalling and communications systems* section.

Products

Traffic control, model networks, scheduling, equipment and power consumption control, passenger information, ticketing, simulation, high-reliability systems (Atelier B).

Contracts

Recent contracts executed and obtained – metro controller training simulator for RATP Paris; Météor Line 14, Line 13.

UPDATED

Systra

5 avenue du Coq, F-75009 Paris, France
Tel: (+33 1) 40 16 61 00
Fax: (+33 1) 40 16 61 04
e-mail: systra@systra.com
Web: http://www.systra.com

Key personnel

Chairman: Michel Cornil
President: Philippe Citroën
Vice-President, Finance and Legal Affairs:
  Jean-Claude Roynier
Vice-President, Human Resources:
  Jean-Marie Champigny
Vice-President, Europe: Jean-Pierre Orsi
Vice-President, Asia Pacific:
  Jean-Christophe Hugonnard
Vice-President, America, Africa and Middle-East:
  Alain Estève
Vice-President, France: Gérard Chaldoreille

Subsidiaries

MVA Ltd, UK
MVA, Hong Kong
Systra Consulting, USA
Canarail, Canada
Sotec Ingénierie, France

Other subsidiaries

Citilabs, UK, USA
Mexistra, Mexico
Sofrecade, Chile
Eurometudes, Romania
Semto, Reunion Island
Systra, Philippines
Systra, Shanghai, China

Background

Systra SA is an international engineering company, a subsidiary of SNCF and RATP (36 per cent each).

Capabilities

Systra offers consulting and engineering services for all rail and urban transport infrastructure and systems (high-speed trains, conventional rail, mass transit, AGT, LRT, BRT). Capabilities include: transport planning and organisation, from master plan development to system feasibility analysis, institutional organisation, design.

UPDATED

Tata

Tata Electronic Development Services
A division of Tata Electric Companies
Bombay House, 24 Homi Mody Street, Mumbai 400 001, India
Tel: (+91 22) 56 65 82 82
Fax: (+91 22) 56 65 81 60
e-mail: tec@tata.com
Web: http://www.tata.com

Key personnel

General Manager, Operations: S Gnyana Sundar
Senior Manager, Projects & Marketing:
  Col J Banerjee
Senior Manager, Simulators and Manufacturing:
  V Kishore Kumar

Background

Tata Power was formed as a result of an amalgamation in April 2000 of three companies, Tata Hydro-Electric Power Supply Co, Andhra Valley Power Supply Co and Tata Power Co.

Products

Replica locomotive driving simulators. Tata simulators feature an exact replica of the locomotive cab, a six-axis motion base, visual display replayed from video discs and sound replay using digitised real sounds stored by computer.

A driver console for the WAG-9 locomotive supplied by Adtranz Sweden is available; also the subsystems of WAG-9 locomotives.

Transurb Technirail SA

60 rue Ravenstein, Bte 18, B-1000 Brussels, Belgium
Tel: (+32 2) 548 53 11
Fax: (+32 2) 513 94 19
e-mail: info@transurb.com
Web: http://www.transurb.com

Key personnel

Chief Executive: Patrick Steyaert
Commercial Manager: Etienne Deblon
Export Manager: Stefaan Van de Kelder
Training Manager: P Vandenbroek

Background

Transurb Technirail SA is part of the Société Nationale des Chemins de Fer Belges, Belgian State Railways (SNCB). The company was founded in 1995 as Technirail and merged with Transurb Consult in 1999 to form Transurb Technirail SA.

Products and services

Production of simulators for drivers of locomotives, metro and bus systems, maintenance systems and electric motion platforms.

Products include: full simulators for rail drivers and virtual stations which are PC-based with software generated instruments presented on-screen.

Vossloh Information Technologies York Ltd

Jervaulx House, 6 St Mary's Court, Blossom Street, York YO24 1AH, UK
Tel: (+44 1904) 63 90 91
Fax: (+44 1904) 63 90 92
e-mail: info@vity.vossloh.com
Web: www.vit.vossloh.com

Key personnel

Managing Director: Roland F Albert
Head of International Business Development:
  Alastair Dick
UK Business Development Manager: David Clarke

Background

Vossloh Information Technologies York Ltd, formerly Comreco Rail Ltd, is a subsidiary of Vossloh Information Technologies GmbH, Kiel, Germany, which is part of the Vossloh Group.

Projects

Provision of consultancy, planning and management of rail operations, exploitation of infrastructure capacity and development of operations planning processes for a wide range of clients including Network Rail, First ScotRail, South West Trains, EWS, NSB, SJ, Strategic Rail Authority, London Underground, Israel Railways, Grand Central Trains and Arriva. Vossloh Information Technologies York Ltd is ISO 9001/2000 accredited.

Contracts

Recent contracts include: supply of TrainPlan, RailPlan, ResourcePlan and Resource Manager to Norwegian State Railways; supply of UK-wide integrated train planning systems and planning database with TrainPlan to Network Rail; supply of TrainPlan and AccessPlan to Banverket; supply of TrainPlan and ResourcePlan to Israel Railways and First ScotRail (UK); supply of TrainPlan and RailPlan to Indonesian Railway and Romanian State Railways; supply of Train Plan to Green Cargo, SJ, JBV, Amtrak and New Jersey Transit; supply of RailPlan and PowerPlan to Kowloon–Canton Railway Corporation (Hong Kong).

UPDATED

SIGNALLING AND COMMUNICATIONS SYSTEMS

Alphabetical listing

Alcatel Transport Automation Solutions
Alpha Zaicon Technology Inc
ALSTOM Transport
AMEC SPIE Rail (UK)
Ametek
Andrew
Ansaldo Segnalamento Ferroviario SpA
Ansaldo Signal Finland Oy
Ansaldo Signal NV
Ansaldo Signal Sweden AB (ANSAB)
Ansaldo Signal UK Ltd and Ireland
Antenna Specialists
APD Communications
AREX
ARINC Inc
Atron Electronic GmbH
AŽD Praha sro
Bombardier Transportation
Bosch – Robert Bosch GmbH
BP Solar Ltd
S A Buhlmann NV
Cadex Electronics Inc
Cattron-Theimeg
cdsrail
China Railway Signal & Communication
    Corporation (CRSC)
Chloride Power Protection
Collis Engineering Ltd
Computer Products Professional bv
Crompton Greaves Ltd
CSEE Transport
David Clark Company
Deuta-Werke GmbH
Dialight
Dimetronic SA
EFACEC Sistemas de Electrónica SA
Erico Inc
Fels SA
Fiber Options

Frequentis GesmbH
GAI-Tronics Ltd
Ganz Transelektro Traction Electric Ltd
GE Rail
GrantRail Ltd
Hanning & Kahl GmbH & Co KG
Harrington Generators International
Henry Williams Group
Hitachi Ltd
Isolux
Kapsch Group
Kruch Gesmbh & Co KG
Kyosan Electric Mfg Co Ltd
Leach International Europe SA
Lincoln Industries
LPA Channel Electric Ltd
Luxram Lighting Ltd
Marconi Communications Limited
Marl International Ltd
Maxon Europe
Mer Mec SpA
Mirror Technology
Morio Denki Co Ltd
Motorola Canada Ltd
Motorola (Schweiz) AG
Nippon Signal Co Ltd
Nortel Networks Germany GmbH & Co
NRS – National Railway Supplies
Peek Traffic BV
Pintsch Bamag Antreibs – und Verkehrstechik
    GmbH
Powernetics Ltd
Progressive Engineering (AUL) Ltd
Quest
Railroad Signal International
Rails Company
Safetran Systems Corporation
Sagem SA, Defence and Security
SchlumbergerSema Group

Sécheron SA
Selectra srl
Sistemas Electrónicos de Potencia SA (SEPSA)
Siemens AG
Siemens Switzerland
Signal & Systemn Technik (SST)
Signal House Ltd
Specialty Bulb Co Inc
Springboard Wireless Networks Inc
Stein GmbH
STERIA Group
STS Signals
TagMaster AB
Teknis Electronics Pty Ltd
Telephonics Corporation
Thales Communications Ltd
Thales Telecom Services
Thoreb AG
Tiefenbach GmbH
Toshiba Corporation
TransCore
Transmitton Ltd
Transportation Products Sales Co Inc
Trivector System AB
UniControls AS
Universal Power Systems
Union Switch & Signal Asia Pacific Region
Union Switch & Signal Inc (US&S)
Vialis NMA
Vossloh Information Technologies Malmö AB
Wabtec Railway Electronics
Western-Cullen-Hayes Inc
Westinghouse Rail Systems Ltd
Westinghouse Saxby Farmer Ltd
Zelisko Elektrotechnik und Elektronik GesmbH
Zenitel

Company listing by country

AUSTRALIA
Teknis Electronics Pty Ltd
Union Switch & Signal Asia Pacific Region

AUSTRIA
Frequentis GesmbH
Kapsch Group
Kruch Gesmbh & Co KG
Zelisko Elektrotechnik und Elektronik GesmbH

BELGIUM
S A Buhlmann NV
Zenitel

CANADA
Alpha Zaicon Technology Inc
Cadex Electronics Inc
Motorola Canada Ltd
Springboard Wireless Networks Inc

CHINA
China Railway Signal & Communication
    Corporation (CRSC)

CZECH REPUBLIC
AŽD Praha sro
UniControls AS

FINLAND
Ansaldo Signal Finland Oy

FRANCE
Alcatel Transport Automation Solutions
ALSTOM Transport
CSEE Transport
Fels SA

Leach International Europe SA
Sagem SA, Defence and Security
STERIA Group

GERMANY
Atron Electronic GmbH
Bosch – Robert Bosch GmbH
Deuta-Werke GmbH
Hanning & Kahl GmbH & Co KG
Nortel Networks Germany GmbH & Co
Pintsch Bamag Antreibs – und Verkehrstechik
    GmbH
Siemens AG
Signal & Systemn Technik (SST)
Stein GmbH
Tiefenbach GmbH

HUNGARY
Ganz Transelektro Traction Electric Ltd

INDIA
Crompton Greaves Ltd
Westinghouse Saxby Farmer Ltd

ITALY
Ansaldo Segnalamento Ferroviario SpA
Mer Mec SpA
Selectra srl

JAPAN
Hitachi Ltd
Kyosan Electric Mfg Co Ltd
Morio Denki Co Ltd
Nippon Signal Co Ltd
Toshiba Corporation

NETHERLANDS
Ansaldo Signal NV
Computer Products Professional bv
Peek Traffic BV
Vialis NMA

POLAND
AREX

PORTUGAL
EFACEC Sistemas de Electrónica SA

SPAIN
Dimetronic SA
Isolux
Sistemas Electrónicos de Potencia SA (SEPSA)

SWEDEN
Ansaldo Signal Sweden AB (ANSAB)
TagMaster AB
Thoreb AB
Trivector System AB
Vossloh Information Technologies Malmö AB

SWITZERLAND
Andrew
Luxram Lighting Ltd
Motorola (Schweiz) AG
Sécheron SA
Siemens Switzerland

UNITED KINGDOM
AMEC SPIE Rail (UK)
Ansaldo Signal UK Ltd and Ireland
APD Communications

Company listing by country—*continued*

Bombardier Transportation
BP Solar Ltd
cdsrail
Chloride Power Protection
Collis Engineering Ltd
Dialight
GAI-Tronics Ltd
GrantRail Ltd
Harrington Generators International
Henry Williams Group
LPA Channel Electric Ltd
Marconi Communications Limited
Marl International Ltd
Maxon Europe
Mirror Technology
NRS – National Railway Supplies

Powernetics Ltd
Progressive Engineering (AUL) Ltd
SchlumbergerSema Group
Signal House Ltd
STS Signals
Thales Communications Ltd
Thales Telecom Services
Transmitton Ltd
Universal Power Systems
Westinghouse Rail Systems Ltd

UNITED STATES
Ametek
Antenna Specialists
ARINC Inc
Cattron-Theimeg

David Clark Company
Erico Inc
Fiber Options
GE Rail International
Lincoln Industries
Quest
Railroad Signal
Rails Company
Safetran Systems Corporation
Specialty Bulb Co Inc
Telephonics Corporation
TransCore
Transportation Products Sales Co Inc
Union Switch & Signal Inc (US&S)
Wabtec Railway Electronics
Western-Cullen-Hayes Inc

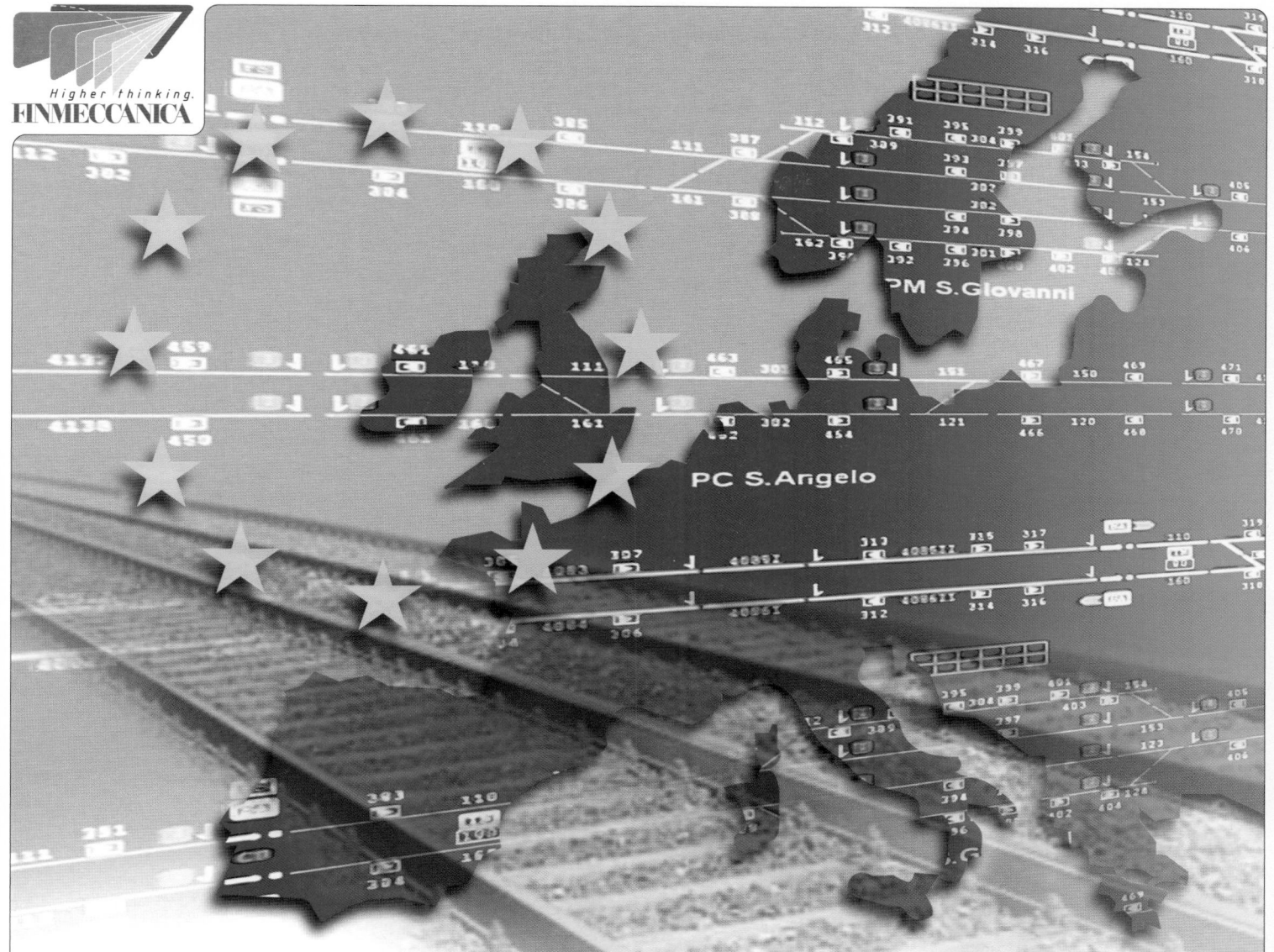

[ERTMS the shape of things to come in Europe]

ERTMS, the European Railway Traffic Management System, is changing the shape of Europe's rail networks

ERTMS is being applied to the different national signalling systems to provide seamless interoperability and control

across the European rail network. Ansaldo Signal is currently installing the signalling and train control systems for

75 per cent of the contracts signed to date on European ERTMS/ETCS level 2 high speed lines. We're a leader in

the development of signalling, Automatic Train Control (ATC), and train supervision systems. Whether it's a wayside

signalling project or a complete, integrated system that employs the latest advancements such as ERTMS,

we have the experience and resources to provide innovative solutions. Choose Ansaldo Signal as your partner.

As the shape of signalling continues to change, you can count on us to lead the way. **www.ansaldo-signal.com**

ANSALDO SIGNAL THE NETHERLANDS • ANSALDO SEGNALAMENTO FERROVIARIO ITALY • ANSALDO SIGNAL UK • ANSALDO SIGNAL IRELAND • ANSALDO SIGNAL SWEDEN • ANSALDO SIGNAL FINLAND • CSEE TRANSPORT FRANCE • UNION SWITCH & SIGNAL INC USA • UNION SWITCH & SIGNAL PTY AUSTRALIA • UNION SWITCH & SIGNAL PVT INDIA • UNION SWITCH & SIGNAL SDN MALAYSIA

GLOBAL RESOURCES
LOCAL RESPONSE

ANSALDO SIGNAL

Alcatel Transport Automation Solutions

10 rue Latécoère, Vélizy F-78141, France
Tel: (+33 1) 30 77 17 30
Fax: (+33 1) 30 77 12 68
Web: www.alcatel.com/tas

Key personnel
President: Jean-Pierre Forestier
Vice-President, Business Development Main Lines:
   Robert Mattenberger
Vice-hyphen;President, Business Development
   Urban Rail: John D Mills
Director, Communication: Bénédicte Massin

Subsidiary companies
Alcatel TAS Austria
Scheydgasse 41, A-1210 Vienna
Tel: (+43 127) 722 57 79
Fax: (+43 127) 722 36 24
Managing Director: Alfred Veider
Communication Manager: Christian Studnicka

Alcatel TAS Canada
1235 Ormont Drive, Weston, Ontario, M9L 2W6
Tel: (+1 416) 742 39 00   Fax: (+1 416) 742 11 36
Managing Director: Walter Friesen
Communication Manager: Roger Fradgley

Alcatel TAS China
32F Times Square, 500 Zhangyang Road, Pu Dong
200122 Shanghai
Tel: (+ 86 21) 68 60 45 45 11 90
Fax: (+ 86 21) 58 36 83 00
Managing Director: Guy Sellier

Alcatel TAS France
1 rue Ampère, Massy, PO Box 56, F-91302, France
Tel: (+33 1) 69 76 90 02   Fax: (+33 1) 69 76 90 01
Managing Director: Gérard Avice

Alcatel TAS Germany
Holderaeckerstrasse 10, D-70499 Stuttgart
Tel: (+49 711) 82 14 65 00
Fax: (+49 711) 82 14 67 19
Managing Director: Hans Leibbrand
Communication Manager: Hans-Jürgen Krehle

Alcatel TAS Portugal
Estrada Malveira da Serra 955, P-2750-782 Cascais
Tel: (+351 21) 485 91 52   Fax: (+351 21) 485 91 12
Managing Director: Joao Araujo
Communication Manager: Joao Salgueiro

Alcatel TAS Spain
5 Ramirez de Prado, E-28045 Madrid
Tel: (+34 91) 330 95 70   Fax: (+34 91) 330 95 76
Managing Director: Anastasio Gallego
Communication Manager: Ramon Mayorga

Alcatel TAS Switzerland
Friesenbergstrasse 75, CH-8055 Zürich
Tel: (+41 1) 465 35 00   Fax: (+41 1) 465 23 19
Managing Director: Franz Stampfli

Alcatel TAS UK
2N140 Christchurch Way, Greenwich, London
SE10 0AG
Tel: (+44 20) 84 65 12 11   Fax: (+44 20) 84 65 14 51
Managing Director: John D Mills

Alcatel TAS USA
5700 Corporate Drive, Suite 300, Pittsburgh,
Pennsylvania 15237
Tel: (+1 412) 366 88 14
Fax: (+1 412) 366 88 17
Managing Director: John Brohm

Background
Alcatel Transport Automation Solutions (TAS)
employs 2,750 people and has main offices in
Austria, Canada, China, France, Germany, Portugal,
Spain, Switzerland, UK and USA.

Products
Alcatel offers a host of proven main line railway
and urban transit solutions to move people
safely through the network in a comfortable,
reliable and cost-effective way. Drawing on wide

international experience in all aspects of transport
automation, Alcatel is one of the leading suppliers
of signalling and train control, train routing, control
centres and integrated rail communications
technologies.

For main line operators, Alcatel offers AlTrac
(in compliance with ERTMS/ETCS specifications)
for train control, NetTrac network management,
LockTrack electronic interlocking system, and
FieldTrac field equipment (axle counters, point
machines, signals). For urban line operators,
Alcatel offers proven SelTrac® communications-
based train control, NetTrack MT network
management, and LockTrac electronic interlocking
system solutions.

Contracts
China: Alcatel has been selected to provide its
SelTrac® communications-based train control
(CBTC) system for Guangzhou Metro Line 3 with
moving-block technology for system-wide safety,
reliability and availability. The 36.3 km long
Guangzhou Metro Line 3, linking three suburbs
and crossing the Zhujiang River, is scheduled to
begin operations in 2006. Guangzhou, the capital
of Guangdong Province and the largest city in
southern China, is implementing an aggressive
plan to build new metro lines over the next
decade.
Finland: Alcatel has been selected to equip the
new section of the Finnish double-tracked
line, Kerava–Lahti, with its innovative control,
supervision and operation systems for interlocking
zones. This contract should be completed by
March 2006.
France: Alcatel has been awarded a contract
to provide a complete Radio Communications-
Based Train Control (CBTC) solution to the Régie
Autonome des Transports Parisiens (RATP) for
metro-line 13 crossing Paris from Porte de Châtillon
to Saint-Denis and Asnières. The contract is for four
years.
Hungary: Alcatel has been awarded two contracts
by Hungarian Railways to equip the Budapest–
Cegléd–Szolnok main line with modern electronic
interlocking systems and to upgrade the line
between Budapest and the Austrian border with
the standardised European Train Control System
(ETCS).
Romania: October 2004, Compania Nationala
Caile Ferate Romane (CN CFR), awarded Alcatel a
contract to equip railways in the southern region of
Romania with its electronic interlocking solutions
to replace the old relay-based interlocking systems.
Alcatel will modernise 11 stations, mainly located
in the Pan-European Transport Corridor IV, crossing
the country in the south and linking the cities of
Constanta and Lugoj.
Spain: Alcatel has been awarded three contracts
by the Transport Ministry of Spain/GIF to deploy the
European standard ETCS (European Train Control
System) Level 1 and 2 for the three new sections
of the Spanish high–speed lines, Lerida–Barcelona,
Segovia–Valladolid and Madrid–Toledo. Alcatel
will also equip the Lerida-Barcelona and Segovia–
Valladolid sections with a fixed communications
network.
Switzerland: Alcatel has been awarded a contract
by BLS AlpTransit AG for the engineering,
installation and delivery of railway security and
signalling systems for the 34.6 km Lötschberg base
tunnel, as well as for the Frutigen and Visp railway
stations in Switzerland. The Lötschberg tunnel
is the core of the Lötschberg base line between
Frutigen and St German which forms an important
part of the Swiss rail corridor through the Alps from
north to south.
Turkey: Alcatel has been awarded a contract
to supply signalling and telecommunication
systems to the Turkish State Railways (TCDD)
for the first renovation phase of the Ankara–
Istanbul Railway Line Rehabilitation Project.
Alcatel's solutions, including an ETCS system,
will ensure traffic control and safety along the
250 km double line section and bi-directional
operations between the Inönü and Essenkent
stations.
UK: Alcatel has been selected by Tube Lines
to re-signal the London Underground Jubilee
and Northern Lines with SelTrac®, its proven

communications-based train control technology
incorporating NetTrac MT central control. This
contract is for a term of seven years.
USA: Alcatel has been selected to provide open-
standards wireless Communications-Based Train
Control to USA's first fully automated, line-haul,
urban monorail rapid transit system in Las Vegas. It
was due to be inaugurated in July 2004.

UPDATED

Alpha Zaicon Technology Inc

40 Hearthstone Crescent, Toronto, Ontario
M2R IG3, Canada
Tel: (+1 416) 633 47 17   Fax: (+1 416) 633 47 10
e-mail: alpha@istar.ca
Web: http://www.alpha-zaicon.com

Products
PIES (Platform Intrusion Emergency Stop) system
based on sensor panels linked to MID-5000
controllers which integrate directly with SCADA
and ATC control systems.

VERIFIED

Alstec Ltd

Cambridge Road, Whetstone, Leicester LE8 6LH, UK
Tel: (+44 116) 275 07 50   Fax: (+44 116) 275 07 87
e-mail: cust.enquiries@alstec.com
Web: http://www.alstec.com

Key personnel
Senior Sales Engineer: Peter Webster

Products
Telecontrol technology for rail applications,
including the Telecode TDM time division
multiplexer for: transmission of CCTV level
crossing controls and indications; transmission
of controls and indications between signalboxes
and remote interlockings; multi-pair signalling
cable replacement; transmission of emergency
alarms; power protection; and block and bell cable
replacement systems. Alstec also undertakes EMC
testing.

ALSTOM Transport

Information Solutions
48 rue Albert Dhalenne, F-93482 Saint-Ouen Cedex,
France
Tel: (+33 1) 41 66 90 00   Fax: (+33 1) 41 66 96 66
Web: www.transport.alstom.com

Key personnel
Senior Vice-President, International Product Line
   Management: Michel Marien

Australia
ALSTOM Transport
Transportation Division, 373 Horsley Road,
Millperra, New South Wales 2214
Tel: (+61 2) 97 74 74 44   Fax: (+61 2) 97 74 74 44

Belgium
ALSTOM Transport
PO Box 4211, B-6000 Charleroi, Belgium
Tel: (+32 71) 44 54 11   Fax: (+32 71) 44 57 06

Brazil
ALSTOM Transport
Av. Octaviano, Alves de Lima 1480, Casa Verde,
02501 000, São Paulo, Brazil
Tel: (+55 11) 38 55 62 00   Fax: (+55 11) 38 55 63 95
Managing Director: Francisco Amigo

Canada
ALSTOM Transport
3330 De Miniac Street, St Laurent, Quebec, Canada
H4S IY4
Tel: (+1 514) 333 08 88   Fax: (+1 514) 333 04 96

Denmark
ALSTOM Transport
Priorparken 530, DK-2605 Brøndby, Denmark
Tel: (+45) 43 43 84 00   Fax: (+45) 43 43 84 01

Egypt
ALSTOM Transport
Street 201, No 13 Maadi, Cairo, Egypt
Tel: (+20 2) 352 80 73   Fax: (+20 2) 354 34 38

France
ALSTOM Transport
11-13 avenue du Bel Air, F-69627 Villeurbanne, France
Tel: (+33 4) 72 81 52 00   Fax: (+33 4) 72 81 52 87

Greece
ALSTOM Transport
6 Parnassou Street, GR-151 24 Marouissi, Greece
Tel: (+30 1) 614 31 85   Fax: (+30 1) 614 31 87

Hong Kong
ALSTOM Transport
9/F New Kowloon Plaza, 38 Tai Kok Road, Hong Kong
Tel: (+852) 23 90 19 01   Fax: (+852) 23 81 86 80

Hungary
ALSTOM Transport
Lehel út 3/b 1 Floor, H-1062 Budapest, Hungary
Tel: (+36 1) 359 98 77   Fax: (+36 1) 359 98 70

Italy
ALSTOM Signaling SpA
Via di Corticella 87/89, I-40128 Bologna
Tel: (+39 051) 416 31 11   Fax: (+39 051) 416 35 94

ALSTOM Signaling Telecommunication
Via dell Elettronica 17, I-37139 Verona, Italy
Tel: (+39 045) 851 05 22   Fax: (+39 045) 851 05 30

Netherlands
ALSTOM Transport
Algemene Sein Industrie, 100 Moeder Teresalaan, PO Box 3333, NL-3502 GH Utrecht, Netherlands
Tel: (+31 30) 292 96 11   Fax: (+31 30) 294 76 21

South Africa
ALSTOM Transport
Railway Signaling South Africa, PO Box 4583, Germiston South 1411, South Africa
Tel: (+27 11) 902 77 20   Fax: (+27 11) 902 90 20

Spain
ALSTOM Transport
Apolonio Morales 13A, E-28036 Madrid, Spain
Tel: (+34 91) 343 17 70   Fax: (+34 91) 350 99 95

UK
ALSTOM Transport
Borehamwood Industrial Park, Rowley Lane, Borehamwood WD6 5PZ, UK
Tel: (+44 20) 89 53 99 22   Fax: (+44 20) 82 07 59 05

US
ALSTOM Transport
150 Sawgrass Drive, Rochester, New York 14620, US
Tel: (+1 716) 783 20 00   Fax: (+1 716) 783 20 88

Background

Following the terms agreed by the EC's acceptance of ALSTOM's financing package in 2004, in June 2005 ALSTOM and United Group Ltd signed a binding agreement for the sale of ALSTOM's transport activities in Australia and New Zealand. ALSTOM's transport operations in Australia and New Zealand include engineering and maintenance support, road and rail infrastructure projects and the provision of professional services and systems to the transport industry throughout Australia and New Zealand.

Products

Covers main line and mass transit. For main line, ALSTOM offers ATLAS for high-speed lines, intercity and regional lines as well as for low-density lines, freight-only lines and single-track operations. This new-generation signalling system applies the latest technologies for train detection, interlocking, route management and speed control. ATLAS implements the new European Rail Traffic Management System (ERTMS) which sets interoperability standards for train control throughout the continent.

ALSTOM claims a leading position in the market for ERTMS projects and products, with more orders in more countries than its competitors. The modular ATLAS system is available at three ERTMS levels, ATLAS 100, 200 and 300, while ATLAS 400 offers a low-cost solution for low- to medium-density lines.

In 2005 ALSTOM is active in several projects, some involving interoperability, some to provide test tracks:

- An on-board ERTMS Level 1 (ATLAS 100) signalling equipment contract for CFL, Luxembourg
- An ERTMS Level 1 contract for Zaragoza-Huesca line, Spain, as well as the equipment of various types of trains in ERTMS Levels 1 & 2 (ATLAS 100 & 200)
- The ERTMS Level 1 contract for the airport link, Athens, Greece
- Onboard ERMTS Level 2 signalling equipment for the Czech Pendolinos provided by ALSTOM
- The Betuwe Route project in the Netherlands, involving the development of a complete signalling system based on ERTMS Level 2 (ATLAS 200)
- The West Coast Main Line in the UK, ALSTOM has supplied TASS (Tilt Authorisation and Speed Supervision System) for the whole infrastructure and the entire fleet of 53 ALSTOM Pendolino trains and trains from other suppliers
- In Italy, ALSTOM is involved in Rome-Naples ERTMS Level 2 as well as further projects for infrastructure, such as Bologna-Firenze. ALSTOM is involved in equipping various types of trains, such as Pendolinos and New Pendolinos, Cisalpino and ETR500 high-speed trains all in ERTMS Level 2 (ATLAS 200)
- In Switzerland, ALSTOM is equipping the Mattstetten-Rothrist line (Bern-Olten) as well as over 500 vehicles with ERTMS Levels 1 & 2 (ATLAS 100 & 200)
- In France, ALSTOM is designing Euro-KVB which is going to keep KVB functionalities and provide a smooth transition to ERTMS Levels 1 & 2 (ATLAS 100 & 200)
- In Belgium, ALSTOM is supplying ATLAS 200 (ERTMS Level 2) for the new high-speed Lines 3 & 4 of the SNCB network

ALSTOM's ATLAS-based Advanced Civil Speed Enforcement System (ACSES) has been successfully tested on the Transportation Test Center test track in Pueblo, US and is looked at favourably for installation on Amtrak's Northeast Corridor from Washington BC to Boston. ACSES is designed to control and enforce train movements at various speeds, enforce civil speed restrictions (including curves, bridges and poor track conditions), provide positive train stopping at interlockings and protect work crews with temporary speed restrictions. ACSES generates and enforces profiles for various train types as required by Positive Train Control (PTC) and Communications-based Train Control (CBTC). The system integrates track transponder transmission systems, mobile communication radios and onboard computers, speed sensors and aspect display units.

For mass transit operations, ALSTOM offers the URBALIS solution for heavy metro and suburban networks as well as for light rail systems.

These systems are modular and adaptable and can be designed to follow changes in railway design.

With the contract for provision of a fully integrated signalling system to the NEL metro, ALSTOM is offering to this first fully automatic heavy metro state-of-the-art technology for automatic train control, electronic interlocking, and automatic train supervision.

ALSTOM has developed five product categories:

- Smartway: conventional signalling products covering basic needs including train detection, points equipment, level crossing equipment, signals, relays and accessories.
- Smartlock: mainly electronic interlock for railway stations.
- Advantik: ATC systems for main line operation.
- Mastria: ATC for mass transit operation including ATP, ATO and all attached facilities. The ATC system can be fully integrated with interlocking and train supervision functions.
- Iconis: Automatic Train Superversion (ATS) also called Integrated Electronic Control Center System, integrating all activities required for train control, train supervision or monitoring, management functions, SCADA requirements, maintenance or alarm centralisation and supervision.
- AGATE e-Media: passenger information systems for trains and stations.

Other contracts

Belgium: In November 2003 a consortium of ALSTOM and Siemens Transportation Systems won a contract from Belgian National Railways (SNCB) to supply new electronic control centres and signalling equipment. ALSTOM's scope of supply covers Smartlock electronic interlocking equipment.

Brazil: In July 2005 Companhia do Metropolitano de São Paulo awarded an ALSTOM-led consortium a contract for the supply, installation and testing of signalling and control equipment for the extension of São Paulo's metro line 2 (green line).

In November 2002, Porto Alegre metro operator Trensurb commissioned a new control centre supplied by ALSTOM. The centre covers 33.8 km and 17 stations on the system. Under the same contract, ALSTOM also established two local control posts and provided signalling systems for them and refurbished three other posts.

Chile: In April 2005 ALSTOM was awarded a contract by Santiago Metro for signalling equipment and automatic train control as part of its extension programme for line 2 north Recoleta Avenue. URBALIS Automatic Train Control equipment features in a turnkey project secured by ALSTOM in July 2002 to equip the 32 km Line 4 of Santiago's metro network.

China: The Kowloon-Canton Railway Corporation (KCRC) of Hong Kong awarded ALSTOM a contract for supply of a signalling system for its East Rail line. The order has two aspects: a signalling system for the new 9 km two-station East Rail line extension; and a capacity upgrade of the entire East Rail line to allow up to 27 trains to operate on the line per hour. The East Rail line is already in operation with an ALSTOM signalling system. The upgrade will be made possible by integrating ALSTOM's AXONIS Automatic Train Control and Signalling system. This will increase line capacity and enhance line safety without requiring any major modification to rolling stock or infrastructure.

France: In January 2005 ALSTOM received a contract, as part of the opening of the LEA line, to supply (within a consortium led by Cegelec Centre Est) automation and signalling equipment for the 26 crossings on this line.

Greece: As part of a turnkey contract placed in 2002 by Ergose, a subsidiary of the Hellenic Railways Organisation, ALSTOM is to supply the signalling system for a new 30 km line linking the centre of Athens with a new airport at Spata. This will comply with the ERTMS European standard and will incorporate ALSTOM's Atlas system.

India: ALSTOM Transport Ltd (India) has signed a contract with Delhi Metro Rail Corporation for the design and supply of an integrated automatic train control, signalling and telecommunication system for the operator's metro project currently under construction in Delhi.

Indonesia: As a member of a consortium led by Itochu, ALSTOM in September 2002 was awarded a contract by the Indonesian Ministry of Communication and Transport's Directorate General of Land Communications for a turnkey signalling and communications system for the upgrading and doubling of the Cikampek Cirebon line in Jakarta. Under the contract, which is due to be completed by December 2004, ALSTOM's Atlas signalling and communications system is to be installed at eight stations.

Italy: Signalling equipment formed the major part of a contract awarded in November 2003 to an ALSTOM-led consortium by RFI, the country's rail infrastructure authority. The equipment is to cover the line between Milan and Bologna and includes Smartlock electronic interlocking technology, renovation of five relay-based interlockings, provision of an automatic train protection system and the installation of a telecommunications system at two stations.

Korea, South: In 2002 as a member of the IKFC consortium, ALSTOM was awarded a turnkey contract by the Incheon Internationa Airport Railroad Co Ltd that includes the supply of signalling equipment. The line is due to be commissioned at the end of 2005.

Netherlands: In October 2003, in consortium with its partner Holland Railconsult, ALSTOM won a contract from ProRail to supply its ATLAS 200 train control and signalling system for the Betuwe dedicated freight line linking Rotterdam with the German border. The system is ALSTOM's implementation of ERTMS Level 2. The line is due to be commissioned in 2007.

Spain: In February 2005 Mintra (Madrid Infraestructuras del Transporte), chose an ALSTOM–led consortium for the supply and maintenance of signalling equipment for Madrid's four new light rail lines.

Switzerland: In 2002 ALSTOM secured a contract from Swiss Federal Railways to supply components from its ATLAS system to equip the Mattstetten–Rothrist line according to European Train Control System (ETCS) Levels 1 and 2 standards. Final delivery and commissioning were scheduled for November 2004.

Taiwan: In October 2003, as part of a wider infrastructure contract awarded by the Department of Rapid Transit Systems, ALSTOM was asked to supply signalling equipment for extensions to the Orange and Blue lines of Taipei's metro network.

Vietnam: In September 2002 an ALSTOM led consortium that also includes Alcatel CIT was awarded a contract by Vietnam Railways to modernise signalling and telecommunications facilities on the Hanoi Ho Chi Minh City line. ALSTOM was to modernise signalling at three stations on the Hanoi Vinh section using its Smartlock interlocking and Smartway trackside equipment for train protection. Training for Vietnam Railways signalling engineers is also covered by the contract.

UPDATED

AMEC SPIE Rail (UK)

Stephenson House, 2 Cherry Orchard Road, Croydon CR9 6JA, UK
Tel: (+44 20) 86 67 36 66
Fax: (+44 20) 86 67 27 03
e-mail: atkinsonb.amecrail@ems.rail.co.uk
Web: http://www.amec.com/rail

Key personnel
Managing Director: John Moss
Business Development Director: Barry Atkinson
Commercial Director: Bill Daly
Financial Director: Richard Hunt
Operations Director: Kevin Beauchamp
Personnel Director: Alan Barnes
Technical Director: Bruce Littlewood

Other offices
Head office
Pole Edison
Parc Saint Christophe, F-95861, Cergy Pontoise, Cedex, France
Tel: (+33 1) 34 22 50 00   Fax: (+33 1) 34 22 51 26

Europe: AMEC SPIE Rail (Benelux) Abay Rail (Belgium); AMEC SPIE Rail SET GmbH (Germany); AMEC SPIE Rail, Cariboni Impianti SRL (Italy)
UK: AMEC SPIE Rail Training; AMEC SPIE Rail (Southern); AMEC SPIE Rail (Western); AMEC SPIE Rail (North East); AMEC SPIE Rail (North West); AMEC SPIE Rail (Scotland)

Background
Following the acquisition of 41.6 per cent of SPIE SA (France) by AMEC plc, a UK-based railway infrastructure company was formed called AMEC SPIE Rail Systems Ltd. In 2003 AMEC exercised its option to purchase the remaining share of SPIE SA, which has a 50 per cent interest in ETF, the French track renewals company.

Services
AMEC SPIE Rail provides operators with state-of-the-art control centres, high capacity data transmission

systems, passenger information systems and specialised telecommunications networks.

AMEC SPIE Rail is currently implementing the new GSM-R Telecommunication system in several European countries.

NEW ENTRY

Ametek

Ametek Panalarm Division
7401 North Hamlin Avenue, Skokie, Illinois 60076, USA
Tel: (+1 312) 675 25 00   Fax: (+1 312) 675 30 11
Web: http://www.panalarm.com

Key personnel
General Manager: Roger Piegza
Sales Manager: Frank Kay
Export Sales Manager: Ervin Whitfield
Transportation Marketing Manager:
   Rick Schoneman

Products
Panalarm mosaic tile graphic displays, available in illuminated and non-illuminated form, and capable of printing with any symbols to customer requirement: incorporating push-button or switch controls, indicator lights, digital readout displays as required; annunciators; event recorders.

Contracts include supply of displays to Union Switch and Signal (1998), Metro North Railroad (1998), Harmon Industries (1998) and Metro Dade Transit (1999).

Andrew

Andrew Kommunikationssysteme AG
Bächliwis 2b, CH-8184 Bachenbülach, Zurich, Switzerland
Tel: (+41 1) 863 73 52   Fax: (+41 1) 863 73 56

Key personnel
Managing Director: R J Fiedler
Systems Engineering: H H Junge, M Kalt
Operations: T Clachers

Products
Design, manufacture and installation of radio communications systems for tunnels and buildings; Heliax coaxial cables, Radiax leaky feeder cables, Panel antennas, unidirectional and bidirectional amplifiers, cell extenders for GSM and PCN.

Ansaldo Segnalamento Ferroviario SpA

Headquarters
Via Paulo Mantovani, 3–5
1-1651, Genoa, Italy
Tel: (+039 010) 655 2550   Fax: (+39 010) 655 21 03
e-mail: contact@asf.ansaldo.it

Key personnel
Chief Executive Officer: S De Luca
Vice-President, Sales and Marketing: S Pidatella
Communications Manager (Contact): D De Palma

Background
Ansaldo Segnalamento Ferroviario is part of Ansaldo Signal group, controlled by Finmeccanica.

Products and Services
Cab equipment, train management system, control equipment.

Infrastructure
Signalling and train control, complete train control systems, subsystems, interlockings, automated driving, remote control, panels/display, track circuits/axle counters, line-side signal, power suppliers, level crossing protection, train health monitoring, installation/management.

Data information systems: remote control system, Centralised Traffic Control (CTC), Supervisory

Control And Data Acquisition (SCADA), station management system, automatic train supervision, data systems, diagnostic warning systems, train control scheduling, train management/vehicle location, marshalling yard control, automated shunting, management information systems, simulator/test equipment, tools and software.

Operation support systems
Decision and co-ordination supports, railway traffic simulators, interactive systems for the drawing of railway timetables, interactive systems for train positioning.

Traffic control
CTC for low traffic and main lines, main station management systems, integrated electronic management systems for electrification, yard automation systems, SCADA, communication management systems, Office Control Centre (OCC) for mass transit.

Train control system for (ERTMS/ETCS) European Railways Traffic Management System.

Wayside signalling equipment
Computer-based and all-relay interlocking systems, automatic block systems, data transmission equipment, signals, track circuits, switch machines, power supply equipment, relays.

Onboard automation and equipment
Automatic Train Control (ATC), Automatic Train Protection (ATP), Automatic Train Operation (ATO), Automatic Train Supervision (ATS), train to wayside data transmission, train diagnostic systems, end-of-train equipment.

Service and maintenance
Training of personnel, after sales service.

Contracts
Belgium: Pilot line for the ERTMS/ETC level 1 on the high-speed line N96.

Brazil: Supply of the signalling system and computer based interlockings for São Gabriel – via Norte section and São Gabriel depot of Belo Horizonte metro.

France: ERTMS/ETCS wayside subsupplier of Eurobalise and LEU on LGV Est between Vaires-sur-Marne and Baudrecourt.

Hungary: ERTMS/ETCS wayside subsupplier of Eurobalise and BTM on Vienna–Budapest Corridor Five.

Italy: Design, supply, installation and commissioning for the Italian high–speed lines (Turin–Milan, Milan–Bologna, Bologna–Florence, Rome–Naples) of the centralised traffic control system; central control of the electric traction installations and auxiliary services; ERTMS/ETCS level 2; Computer-Based Interlockings (CBI); remote terminal units for the local management and the data collection relevant to station systems and auxiliary services (such as meteo and seismic monitoring, premises monitoring, hot box detection system, switch point heating).

75 CBIs are in progress, including: Turin Lingotto, Voghera stations, Naples and Genoa junctions; Metano-Malles, Tirrenica and Rome–Pantano lines.

Automatic Train Protection, SCMT performed with ERTMS components. Design, supply, installation and commissioning of six new centralised traffic control systems SCC, for the main line and junctions including Genoa, Naples, Venice area junctions, and the Adriatic, Tirrenica and Bologna–Brennero lines.

UK: Improvement of capacity and increase of performance of the railway facilities between Slade Lane Junction and Cheadle Hulme with associated modifications in the fringe signalling controlled areas at Macclesfield, Wilmslow, Preston, Northenden and Manchester Piccadilly, in the area south of Manchester. Extension of the first contract that was limited to the Cheadle Hulme area. The extension provides for a complete resignalling of the line and includes the supply of a new ACC system controlling the stations from Wilmslow (included) to Sandbach (included) as well as the associated modifications in the fringe signalling controlled areas at Crewe, Stoke and Greenbank.

UPDATED

Ansaldo Signal Finland Oy

An Ansaldo Signal company, controlled by Finmeccanica
PO Box 35, Teknobulevardi 3-5, FIN-01531 Vantaa, Finland
Tel: (+358 9) 25 17 83 15   Fax: (+358 9) 25 17 83 16
e-mail: marko.ankerman@ansaldosignal.fi

Key personnel
Managing Director: Leif Daniels
Project Manager: Marko Ankerman

Background
Ansaldo Signal Finland Oy is a subsidiary of Ansaldo Signal Sweden AB.

Products
L10 000 Automatic Train Protection (ATP). L12 000 modular ATP and Automatic Train Operation (ATO) system for railway, metro and automated transportation systems. L15 000 onboard train control system for special application, such as Specific Transmission Module (STM) for ERTMS. Represents Ansaldo Signal group in Finland for ERTMS products, Traffic Control Centres (TCC), monitoring devices (including hot box detectors and hot wheel detectors) and general railway/metro signalling products such as point machines, track circuits and signals.

Contracts
Finland: Microlok II® interlocking for Jyväskylä-Pieksämäki line. RHK (Finnish Rail Administration). ATP balises and encoders for Finnish National Railway network, RHK. L 15 000 Specific Transmission Module (STM) for ERTMS operation on Finnish national network, RHK.

NEW ENTRY

Ansaldo Signal NV

The World Trade Center, Schiphol Boulevard 267, NL-1118 BH, Schiphol, Netherlands
Tel: (+31 20) 405 98 41   Fax: (+31 20) 405 98 41
e-mail: info@ansaldo-signal.com
Web: www.ansaldo-signal.com

Key personnel
Chief Executive Officer: Roberto Gagliardi
Vice-President, Business Development:
  Emmanuel Viollet
Communications Manager (Contact):
  Paco Cabeza-López

Background
Ansaldo Signal NV, a Finmeccanica company, is a global supplier of signalling, control and automation systems for mass transit, conventional and high-speed railways. Ansaldo Signal is the leader of a worldwide, ten-company group, employing over 3,000 people. It is able to provide local engineering solutions and apply advanced, safe and reliable technologies.

Subsidiary companies
Ansaldo Segnalamento Ferroviario (Italy)
CSEE Transport (France)
Union Switch & Signal Inc (US)
Union Switch & Signal Pty (Australia)
Union Switch & Signal Pvt (India)
Union Switch & Signal Sdn (Malaysia)
Ansaldo Signal UK (UK)
Ansaldo Signal Ireland (Ireland)
Ansaldo Signal Sweden (Sweden)
Ansaldo Signal Finland (Finland)

Products and services
Operation support systems: Decision and co-ordination support, railway traffic simulators, interactive systems for the drawing of railway timetables, interactive systems for train positioning.
Traffic control: Centralised Traffic Control (CTC) for low traffic and main lines, main station management systems, integrated electronic management systems for electrification, yards automation systems. Supervisory Control And Acquisition (SCADA), communication management system, Office Control Centre (OCC) for mass transit.
Train control system for ERTMS/ETCS (European Railways Traffic Management System).
Wayside signalling equipment: Computer-based and all-relay interlocking systems, automatic block systems, data transmission equipment, signals; track circuits, switch machines, power supply equipment, relays.
Onboard automation and equipment: Automatic Train Control (ATC), Automatic Train Protection (ATP), Automatic Train Operation (ATO), Automatic Train Supervision (ATS), train to wayside data transmission, train diagnostic systems, end-of-train equipment.
Service and maintenance: Training of personnel, after sales services.

UPDATED

Ansaldo Signal Sweden AB (ANSAB)

PO Box 8142, Gunnebogatan 22, SE-163 08 Spånga, Sweden
Tel: (+46 8) 621 95 00   Fax: (+46 8) 621 14 24
e-mail: info@ansaldosignal.se
Web: www.ansaldosignal.se

Key personnel
Managing Director: Leif Daniels
Marketing Manager: Lars Wennerholm
Technical Manager: Bertil Sjöbergh
Product Manager: Lucas Orve
Sales Manager: Ulf Hammarbäck

ATP interrogator and ATP transponders 0561207

ATP-equipped train and ATP transponders 1027185

ATP for Monorail, Kuala Lumpur, Malaysia *NEW*/1137137

Traffic Control Centre for Northern Sweden, BV Sweden   **NEW**/1137135

Background
Ansaldo Signal Sweden AB is a subsidiary of Ansaldo Signal NV.

Products
L10 000 Automatic Train Protection (ATP). L12 000 modular Automatic Train Protection (ATP) and Automatic Train Operation (ATO) system for railway, metro and automated transportation systems. L15 000 onboard train control system for special applications, such as Specific Transmission Module (STM) for ERTMS. Represents Ansaldo Signal group on Scandinavian and Baltic market for ERTMS products, Traffic Control Centres (TCC), monitoring devices (including hot box detectors and hot wheel detectors) and general railway/metro signalling products such as point machines, track circuits and signals.

Contracts
Finland: Microlok II® interlocking for Jyväskylä–Pieksämäki line. RHK (Finnish Rail Administration). ATP balises and encoders for Finnish National Railway network, RHK. L 15 000 Specific Transmission Module (STM) for ERTMS operation on Finnish national network, RHK.
Malaysia: L 12 000 ATP for Kuala Lumpur Monorail.
Norway: L 15 000 Specific Transmission Module (STM) for ERTMS operation on Norwegian national network, JBV (Norwegian National Railway Administration).
Sweden: Microlok II® interlocking and L12 000 ATP for Roslagsbanan line, SL (Stockholm Local Transport). L 10 000 ATP for X40 regional trains, SJ (Swedish State Railways). L 10 000 ATP for X60 commuter trains, (SL). Traffic Control Centre for northern Sweden, BV (Swedish National Railway Administration). L 15 000 Specific Transmission Module (STM) for ERTMS operation on Swedish national network, BV.
Sweden/Denmark: L 10 000 combined Swedish/Danish ATO for the Öresund link.

UPDATED

Business Development Director: Derek Collins
Director of Engineering: Ian McCullough
Commercial Director: Maurizio Rossi
Director of Safety Assurance: Andy Harrison

Ireland Office
Ansaldo Signal Ireland Ltd
Monavalley, Tralee, Co Kerry, Ireland
Tel: (+353) 667 12 44 11
Fax: (+353) 667 12 44 13
e-mail: jlaide@ansaldo-signal.ie
General Manager, Ireland: Jimmy Laide
Technical Sales Manager: Francis Foley

Products
Design, engineering and integration of signalling, control and automation for rail transport infrastructure. ASUK has full access to the entire product catalogue of the Ansaldo Signal Group including advanced computer based interlockings, train detection and positioning systems, automatic block system, automatic train control, operation and traffic control and wayside signalling equipment. Also, products that have been developed for ERTMS/ETCS interoperability standards.

Contracts
Ireland: Supply to Iarnród Éireann of Mini CTC for the Tralee Line, cabinets for the Network Rail West Coast project and upgrades to various level crossings.
Fabrication and fit out of a range of equipment cabinets along with the provision of level crossing control systems, CTC, signalling equipment.
Study, in association with ASIR, for the replacement of the Irish Rail Data Transmission System (DTS).
UK: Application of the ASF interlocking and associated products in the south Manchester and adjacent areas.

NEW ENTRY

Ansaldo Signal UK Ltd and Ireland
An Ansaldo Signal Company, controlled by Finmeccanica

Headquarters
Euston Tower, 286 Euston Road, London NW1 3AT, UK
Tel: (+44 20) 78 74 62 20
Fax: (+44 20) 78 74 62 62
e-mail: asukmarketing@ansaldo-signal.co.uk

Key personnel
Chief Executive Officer, Ansaldo Signal UK Ltd:
  Derek Hume

Antenna Specialists
31225 Bainbridge Road, A, Cleveland, Ohio 44139-2281, USA
Tel: (+1 440) 349 84 00
Fax: (+1 440) 349 84 07

Key personnel
President: F Kim Goryance
Marketing Director: Carol Broniman
International Director: Jo Virant

Products
Two-way communications antenna systems; cellular equipment.

UPDATED

APD Communications
16 Shenley Pavilions, Chalkdell Drive, Milton Keynes MK5 6LB, UK
Tel: (+44 1482) 80 83 00
Fax: (+44 1908) 50 74 91
e-mail: marketing@apdcomms.com
Web: www.apdcomms.com

Background
APD is a specialist provider of mobile information, resource location and control room solutions to the public sector organisations and businesses with field-based operatives. ADP is part of Lynx Group.

Products
APD's mobile information platform connects its clients mobile workforces to any enterprise application, including e-mail, intranet, customer database, job despatch and contractor management, using any network and bearer (GSM, GPRS, TETRA, private mobile radio, Mobitex) and delivers information to any in-vehicle or hand-held mobile data terminal or Personal Digital Assistant (PDA). APD also provides a vehicle managment solution, INCA™, which allows the user to track and locate assets, as well as offer lone-worker safety capabilities, and CORTEX, an integrated communications control system for the emergency services.
Clients for APD include emergency services, local authority and other organisations including Metropolitan Police, London Underground, Anglian Water and Swedish Police.

UPDATED

AREX
PL 80-454 Gdańsk, ul. Nad Stawem 5, Poland
Tel: (+48 58) 344 35 40
Fax: (+48 58) 344 35 39
e-mail: marketing@arex.pl
Web: http://www.arex.pl

Key personnel
Founder and General Director:
  D Phil Andrzej Darski
Technical Manager: Krzysztof Lutowicz

Products
DIMaC-EK control and communications system. The system is provided especially for installation in unstaffed stations and is supervised from one dispatcher's centre. The DIMaC-EK system incorporates systems to control electric switch point heating; light; electric power switch control; hydrotechnical tunnel services and lighting; electrical heating cubicle; electric power measurement; pump service; intruder alarm.
These systems are independent and can operate in automatic mode with full diagnostics of the equipment, with possibility of remote change-over and monitoring. Each is provided with measuring, actuating and control equipment. Microprocessor plant controllers operating in accordance with the preset algorithms are provided as control equipment. The controllers are fitted in switchboards and control cabinets.

ARINC Inc
2551 Riva Road, Annapolis, Maryland 21401-7465, USA
Tel: (+1 410) 266 40 00
e-mail: corpcomm@arinc.com
Web: www.arinc.com

Key personnel
Vice-President Marketing and Sales:
  Stephen E Means

Background
As well as serving the surface transport market, ARINC Inc provides communications, systems engineering and integration solutions for airports, aviation, defence and government.

Products

Communications-based train control systems for main line and urban networks; Positive Train Control (PTC) systems; train control centre integration.

Contracts

In July 2003 ARINC commissioned a train control system featuring ATC for a 22 km extension to the Gold Line of the Los Angeles metro system. As well as regulating train movements, the system also controls traction power supplies. Similar systems were also being provided for the systems Blue Line (35 km) and Green Line (32 km) to create a centralised system to control and monitor the entire Los Angeles metro network.

ARINC has also acted as systems engineering team leader for the North American Joint Positive Train Control (NAJPTC) programme and is providing technical support to the NAJPTC programme to implement PTC on a high-speed corridor in Illinois.

NEW ENTRY

Atron Electronic GmbH

Am Ziegelstadel 12 + 14, D-85570 Markt Schwaben, Germany
Tel: (+49 8121) 934 20   Fax: (+49 8121) 93 42 77
e-mail: vertrieb@atron.de
Web: www.atron.de

Key personnel
Marketing: Christine Baumgart

Products
Automatic Vehicle Location and Control (AVLC) systems based on spontaneous data transmission that communicates deviations from the planned status to central control, optimising bandwidth use of systems such as GSM and Tetra/Tetrapol.

UPDATED

AŽD Praha sro

Žirovnická 2, Prague 10, CZ-10617, Czech Republic
Tel: (+420) 267 28 74 37   Fax: (+420) 272 65 61 59
e-mail: mno.rsp@azd.cz
Web: www.azd.cz

Key personnel
General Director: Zdeněk Chrdle
Executive Director: František Jeckel

Services
Research, development, design, manufacture, installation, sales and servicing of signalling, shunting, information, telecommunication and automation technologies for railway transport.

Products
Electronic signalling and interlocking systems, level crossing safeguard systems, ATP and ATO systems, wayside equipment (point machines, locks, track circuits, axle counters, light signals), installation of telecommunications and information networks, cabling (metallic or FO cables).

UPDATED

Bombardier Transportation

Rail Control Solutions
St Giles House, 10 Church Street, Reading, Berkshire, RG1 2SD, UK
Tel: (+44 118) 953 80 00   Fax: (+44 118) 953 80 09
Web: www.transportation.bombardier.com

Key personnel
President, Rail Control Solutions:
 Josef Doppelbauer

Metro do Porto control room   *NEW*/0585268

Products

Development, design, engineering, production, sales, installation, maintenance, after-sales service of, and customer support for, rail control and signalling systems for main line, ERTMS and mass transit.

Bombardier INTERFLO offers main line solutions individually tailored to customers' needs, encompassing integrated operations control systems and computer based interlocking systems, as well as automatic train protection and wayside equipment. INTERFLO also provides ERTMS/ETCS (The European Rail Traffic Management System/European Train Control System) solutions for ERTMS Level 1 and ERTMS Level 2.

Bombardier CITYFLO provides complete mass transit solutions for all types of mass transit such as trams, light rail vehicles and metros, to suit various operating modes and customers' needs. These include cab signalling, semi-automatic train operation and unattended train operation.

Bombardier's product portfolio encompasses a wide range of products available as stand-alone products, or as part of Bombardier INTERFLO and CITYFLO solutions, products include:

EBI Screen control rooms, traffic management systems for efficient and economic management of main line and mass transit networks

EBI Lock computer-based interlockings, which supervise and control wayside objects such as signals, point machines and level crossings

EBI Com radio block centre, compiles information from interlockings and trains in its control area and sends movement authorities and other information to individual trains, taking into account a safe distance to the train ahead

EBI Link ATC (automatic train control) wayside equipment, providing either fixed or variable data from track to train utilising balises

EBI Track train detection, a range of equipment consisting of train location systems, jointless track circuits, coded jointless track circuits and axle counters, providing train detection and track-to-train communication

EBI Switch point machines, a range of point machines that includes sleeper-integrated point machines, conventional end-of-sleeper machines and machines mounted in a recess between the rails

EBI Light signals, a range of optical signals including colour light multi-aspect signals, fibre optic searchlight signals, fibre optic alpha numeric signals and tunnel signals

EBI Gate level crossings, providing barriers and signals

EBI Cab ATC onboard equipment, which supports the driver and continuously supervises the speed of the train. The EBI Cab onboard system warns the driver if the maximum permitted speed is exceeded and activates the brakes in the event

of danger. The system guarantees the highest possible level of safety

EBI Star SatNav onboard units, providing enhanced transport timeliness and improved utilisation of locomotives by reducing standstill and empty runs, offering punctuality, cost-effectiveness and environmental protection

EBI Tool design and maintenance, which generates site data automatically allowing engineering to be achieved quickly and efficiently.

In 1995 European railways unanimously selected Bombardier Transportation's magnetic transponder (Balise) technology for track-to-train communication proposed for the Euro-Balise.

Contracts

Benelux: In April 2005 Bombardier Transportation and Arcadis Infra BV signed a memorandum of understanding that will assist in the progress of the delivery of signalling projects in the Netherlands, Belgium and Luxemburg. This agreement builds upon the successful co-operation of the companies on the ongoing Bev21 ERTMS Pilot and Amsterdam–Utrecht resignalling projects.

Denmark: In 2002, Bombardier Transportation won a contract from Banestyrelsen to supply three relay interlockings, numerous EBI Lock computer-based interlockings and a train control system for Copenhagen's Circle Line.

Finland: In 2002, Bombardier Transportation developed a stationary-based passenger information system concept for travel centres. Today, 10 main line stations are equipped with systems controlling LCD-based concourse displays and arrival/departure displays.

The real-time automatic passenger information system for the coast line was integrated with the traffic control and SCADA system for southern Finland and today offers real-time information on various displays and public address announcements at 37 main line stations.

In 1997, Bombardier Transportation delivered the first section of the nationwide ATC system to the Kirkkonummi–Turku coast line. By the end of 2003, 3,158 km of the Finnish rail network and the rolling stock had been equipped with ATC.

In 1993 an integrated traffic control and traction power SCADA system for southern Finland was commissioned. The system is now controlling 37 main line stations and 16 traction power feeder and switching stations on the coast line from Helsinki–Turku and main line from Helsinki – Riihimaki.

Germany: In March 2005 Bombardier Transportation received an order to supply EBI Lock 500 computer-based interlocking (CBI) systems to Deutsche Bahn (DB Netz AG). This new contract follows the previous order awarded to Bombardier in 2003 to supply EBI Lock 950 CBI systems to the Deutsch Bahn's pilot project on the Mannheim–Rheinau line, in South West Germany.

Greece: Hellenic Railway (OSE) has ordered centralised traffic control to increase capacity between Athens and Larissa. In addition, Bombardier Transportation is delivering upgraded EBI Gate level crossing systems to the railways around Athens.

Italy: In March 2005 Bombardier Transportation was awarded two contracts from Italferr, the engineering subsidiary of the Italian State Railways, to supply its EBI Lock 950 computer-based interlocking, an automatic block and telecommunications system for the Italian railway network. Deliveries are scheduled to end in 2007. Italferr awarded Bombardier Transportation two contracts for the implementation of an INTERFLO 200 solution comprising railway traffic management and safety systems, EBI Lock computer-based interlockings, and automatic block on the Pozzuolo–Treviglio line. Under the contracts, Bombardier Transportation will supply, install, test and commission four EBI Lock computer-based interlockings. The new interlockings will control a total of 333 signals, 161 point machines and 169 track circuits.

Bombardier Transportation received a further contract from Italferr for the implementation of EBI Lock computer-based interlockings, on the Milan junction. This technology will be applied to Pioltello, Melzo Scalo and the double junction of Pozzuolo; the automatic block on the Pioltello – D.B. Pozzuolo section; telecommunication equipment on the rail stations and sections on the Milano–Treviglio line.

In 2002, Rete Ferroviaria Italiana awarded Bombardier Transportation a contract to provide Eurobalises and LEUs for approximately 1,500 km of FS rail network, covering the area Turin, Milan, Rome and Ancona. This is the first phase of a project that covers around 45 per cent of all the lines to be equipped.

Latvia: Bombardier Transportation's first project in Latvia for Latvian Railways, includes the design, supply, installation, test and commissioning of EBI Lock computer-based interlockings for Latvia's major railway junction, Riga Passenger Station, and the supply of 140 EBI Switch point machines.

Lithuania: Bombardier Transportation was awarded a contract by Lithuanian Railways for the design, installation and commissioning of a complete EBI Lock computer-based interlocking system with connection to an EBI Screen control room and installation of EBI Link ATC wayside equipment and telecommunication along the 120 km line from Vilnius–Kleipeda.

Netherlands: As part of the Bev21 programme, Pro Rail awarded Bombardier Transportation a contract to develop the Bev21 System and gain Dutch national safety acceptance of both wayside and onboard systems that support the ERTMS/ETCS Standards at levels 1, 2 and STM and dual signalling, for both ERTMS equipped and conventionally equipped trains.

Following on from the progress of the project, Pro Rail awarded Bombardier Transportation a further contract to design, supply, install, test and commission an ERTMS Level 2 dual signalling safety system for the Amsterdam – Utrecht track doubling project, thereby upgrading the line.

In March 2005 Bombardier Transportation successfully completed cross exchange tests between its INTERFLO 450 ERTMS Level 2 and both ALSTOM and Siemens ERTMS equipment on the Bev 21 ERTMS lines in the Netherlands. In April 2005 Bombardier Transportation and NedTrain Consulting, an independent engineering and consulting office established by NedTrain, entered into an agreement for the development and distribution of ATB-AG Specific Transmission Modules (STMs) and ATB-EG Phase 5 stand alone systems. Used in conjunction with ETCS onboard equipment, these products make it possible for rolling stock to run not only on the new ERTMS equipped cross border lines, but also on the existing Dutch domestic railway network.

Bombardier Transportation was also awarded a contract by Railinfrabeheer for the delivery of 1,200 EBI Switch point machines. The deliveries are scheduled to take place from May 2002 to May 2007.

Philippines: Bombardier Transportation designed, manufactured, tested and delivered a complete CITYFLO 250 solution for Line 3 of the LRT in Manila. This fully integrated system includes EBI Lock computer-based interlockings, EBI Screen control room and EBI Cab ATC onboard equipment, guaranteeing a low cost solid and reliable solution. Bombardier Transportation was also awarded a 5 year maintenance contact for Line 3, which includes training.

Poland: Bombardier Transportation supplied Polish Railways (PKP) with 33 EBI Lock computer-based interlockings. These are installed on the E-20 and E-30 main lines. In addition, Bombardier Transportation has delivered 220 EBI Gate level crossings as well as EBI Light signals, EBI Switch point machines and track circuits to the whole PKP network.

The E-20 and E-30 main lines are part of the transportation corridors linking Eastern and Western Europe. The E-20 main line was the first modernised line in 1994 and 90% of the signalling equipment was supplied by Bombardier Transportation.

Portugal: Bombardier Transportation was awarded the contract for the delivery of the full turnkey project for Metro do Porto, including the project management, in-house design, installation and commissioning of a CITYFLO 250 solution.

The Metro do Porto CITYFLO 250 solution supplied consists of a traffic management system including automatic vehicle regulation (AVR), 17 EBI Lock computer-based interlockings, EBI Cab ATC onboard equipment, EBI Link ATC wayside equipment, EBI Light signals, EBI Switch point machines, EBI Gate level crossings and EBI Track train detection. In total, Metro do Porto covers 70 km of track, 50 km of which is converted rail tracks with an additional 20 km of newly built tracks, stopping at 66 stations with 72 vehicles supplied by Bombardier Transportation.

Romania: The Bucharest Metro Authorities, Metrorex, awarded Bombardier Transportation a contract for the in-house design, installation, commissioning and supply of the state-of-the-art CITYFLO 350 mass transit signalling solution for Line 2. The CITYFLO 350 solution includes EBI Lock computer-based interlockings, EBI Screen control room, EBI Switch point machines, EBI Cab ATC onboard equipment with jointless audio frequency coded track circuits and EBI Cruise ATO. Building on the success of the Line 2 contract Bombardier Transportation was awarded a further signalling contract from Metrorex to supply an additional CITYFLO 350 mass transit solution for Bucharest Metro Lines 1 and 3. Within this order Bombardier will be responsible for the design, supply, testing and commissioning of the CITYFLO 350 solution, which encompasses an EBI Screen control room, EBI Lock 950 computer-based interlockings and wayside equipment that includes EBI Track train detection track circuits, balises and EBI Switch point machines.

Russian Federation: Bombardier Transportation has supplied EBI Lock computer-based interlockings for the St Petersburg–Moscow line. A further 26 stations have been equipped with EBI Lock, with orders received to supply an additional 15 stations with EBI Lock computer-based interlockings.

South Africa: As part of a DBOM concession for the Gautrain Rapid Rail Link, for which the Bombardier-led Bombela consortium was named preferred bidder in 2005, the company is to supply CITYFLO 250 train control technology and communications systems.

South Korea: Bombardier Transportation was awarded a contract by a consortium led by Taejung Electric Construction Co to supply INTERFLO 250 ERTMS Level 1 to the Korean National Railroad (KNR). The system consists of Bombardier EBI Cab ATC onboard equipment, EBI Link ATC wayside equipment consisting of lineside electronic units (LEUs) and balises for the lines from Seoul to Busan and from Daejon to Mokpo, totalling 760 km. The final delivery is scheduled to take place in December 2006.

The Busan Urban Transport Authority (BUTA), through LGIS, awarded Bombardier Transportation a contract for the in-house design, installation and commissioning of a CITYFLO 350 solution on Busan Line 2. The CITYFLO 350 solution supplied includes EBI Lock computer-based interlockings, EBI Switch point machines, EBI Cruise ATO and EBI Cab ATC onboard equipment with EBI Track train detection jointless audio frequency coded track circuits.

Spain: Metro Madrid awarded a contract to Bombardier Transportation to supply its CITYFLO 450 solution for the renovation of Lines 1 and 6. Metro Madrid needed to upgrade the capacity of both lines without extensive changes to the infrastructure and it was felt that only a communication based train control (CBTC) system could achieve this. Furthermore the system should have the ability to evolve into a fully driverless system. The CITYFLO 450 solution from Bombardier Transportation fulfilled all these demands, being a state of the art communication based train control (CBTC) system featuring moving block (MB) operation with the driver (semi-automatic train operation, or STO) using radio communication between train and wayside. The system can either operate as just a moving block system for new installations, as an overlay system for installations already in service or as a moving block system with a fixed block track circuit based fall-back system, as in the case of Metro Madrid. Furthermore the CITYFLO 450 can be upgraded to a CITYFLO 650 fully driverless system. The orders for Metro Madrid Lines 1 and 6 include the supply of onboard equipment for 117 trains, EBI Lock computer based interlockings, EBI Screen control rooms, EBI Track jointless track circuits and EBI Switch 700 hydraulic point machines as well as LED (Light Emitting Diode) type signals Madrid Metro line 1 is

EBI Switch point machine   **NEW**/0585267

EBI Lock computer-based interlocking   **NEW**/0585266

ERTMS/ETCS Level 2 system was accepted by Swiss Federal Railways, SBB, in November 2003.

Bombardier Transportation developed the trackside equipment and the train-borne equipment. 63 vehicles of five different types were equipped to run on the Olten–Lucerne line, a 35 km double track with nine stations and a line speed of 140 km/h, capable of running up to 57,000 trains per year and transporting more than 3,000,000 passengers and 240,000 tonnes of goods.

The INTERFLO 450 solution is proven to be reliable and safe, meeting the standards set by SBB. The test results show outstanding performance levels of 99.6 per cent availability, reaching standards above conventional systems.

Thailand: Bombardier Transportation received a contract from the State Railway of Thailand to supply and install an INTERFLO 200 signalling system, consisting of EBI Lock computer-based interlocking systems, EBI Link ATC wayside equipment and associated communications systems on all four new double-track routes into and out of Bangkok, covering 174 km of double- and triple- track and 33 stations.

Taiwan: Bombardier Transportation was awarded a contract by the Taiwan Railway Administration, for a complete INTERFLO 250 ERTMS Level 1 ATP system which included ATP computers for 832 locomotives, 14,000 balises, and 2,000 encoders. The order also included 13 simulators for driver training and systems for train data presentation including driver panels.

UK: In April 2005 Bombardier Transportation was awarded a contract from Strathclyde Passenger Transport to replace the Glasgow Subway's existing centralised control centre (CTC) originally delivered by Bombardier in 1997. The system, part of Bombardier's CITYFLO mass transit solutions, is to be delivered in July 2006.

Bombardier Transportation is a member of the Metronet consortium, which is currently undertaking a major programme to modernise the London Underground network's signals. Additionally, Bombardier Transportation is currently delivering an upgraded control room solution for the East London line of the Underground. The first line of the network was commissioned in 2004.

In 2003 Bombardier Transportation delivered the final elements of the re-signalling contract at Leeds station. Based on Solid State Interlocking (SSI), signalling for the remodelled track layout is now managed from an Integrated Electronic Control Centre (IECC) at York, which is the largest in the UK.

As part of the Arrow Light Rail Ltd Consortium, Bombardier Transportation provides the light rail signalling and control solution, CITYFLO 150, for the Nottingham Express Transit (NET), Line 1 light rail system. Bombardier Transportation has supplied a range of passenger facilities that include: passenger information displays on trams and at tram stops, closed-circuit television cameras that transmit images to the systems control centre, passenger emergency help points and public address systems.

Bombardier Transportation has also equipped the control centre with a radio system, private automatic branch exchange (PABX) and an auxiliary/traction SCADA.

Greater Manchester's Metrolink is at the cutting edge of public transport, with more and more passengers using the system each year. Bombardier Transportation was awarded the contract to supply a CITYFLO 150 solution for a fully integrated signalling and control system that includes the supply of an EBI Screen control room, passenger information displays, automation system, audio public address, passenger help points, and closed circuit television security surveillance. In addition, Bombardier Transportation fitted six trams with automatic tram stop, public address, passenger emergency call, tram radios and vehicle recognition system.

US: The South Eastern Pennsylvania Transport Authorities (SEPTA) awarded Bombardier Transportation a contract for the in-house design, installation support and commissioning of a CITYFLO 450 solution for the loop around the city. The CITYFLO 450 solution supplied includes the EBI Com radio block centre, EBI Cab ATC onboard equipment and the EBI Screen control room. The CITYFLO 450 solution was chosen as the cab-signalling system can be adapted to manually operated vehicles running in mixed traffic and in a transit tunnel.

16.7 km long, has 27 stations and for this line 68 trains have to be equipped with ATP and ATO. Madrid Metro line 6 is 23.5 km long and has 27 stations. For this line 49 trains have to be equipped with ATP and ATO. As a continuation of the ERTMS Level 1 line from Albacete – Villar de Chinchilla, Bombardier was awarded a further contract by the Spanish Rail Infrastructure Administrator (ADIF) to install ERTMS Level 1 wayside equipment on the extension of this ERTMS Level 1 line, from Villar de Chincilla – La Encina, linking with the EBI Cab 900 area from the Mediterranean Corridor. The ERTMS Level 1 wayside equipment supplied includes 333 EBI Link 2000 Eurobalises and 53 EBI Link 2000 lineside electronic units (LEUs), all of which interface with existing interlocking systems. With the 50 km extension of the double track line it will be possible to drive under ATP for more than 500 km from Albacete to Barcelona via Valencia.

Barcelona Metro awarded Bombardier Transportation a contract for the in-house design, installation and commissioning of a CITYFLO 350 solution for Lines 1 and 3. The CITYFLO 350 solution supplied includes EBI Lock computer-based interlockings, EBI Cab ATC onboard equipment with EBI Track train detection jointless audio frequency coded track circuits and EBI Cruise ATO.

The Basque Government awarded Bombardier Transportation a full turnkey contract for the in-house design, installation and commissioning of a CITYFLO 350 solution for Bilbao Metro Lines 1 and 2. The CITYFLO 350 solution supplied for Lines 1 and 2 includes EBI Lock computer-based

interlockings, EBI Cab ATC onboard equipment with approximately 350 jointless audio frequency coded track circuits and EBI Cruise ATO.

Bombardier Transportation was awarded the Madrid automatic people mover (APM) project in 2001, representing the first fully automated transit system in Spain. The CITYFLO 550 solution consists of an EBI Screen control room and EBI Cab ATC onboard equipment. 19 fully automated CX-100 vehicles will operate in a pinched loop configuration with each vehicle fitted with CCTV capability for monitoring safety conditions onboard.

Sweden: In April 2005 Bombardier Transportation won a signalling contract from Botniabanan AB to produce the ERTMS system specifications as the first phase in introducing the new system on the Bothnia Line. It is intended that Bombardier will subsequently supply its state-of-the-art INTERFLO 450 main line signalling solutions, providing ERTMS/ETCS Level 2 functionality for the entire Bothnia Line, from Nyland to Gimonäs. Work commenced in January 2005 and is scheduled to be completed in late 2009. Bombardier Transportation has won orders to supply the Swedish Railway Administration, Banverket, with 160 EBI Lock computer-based interlockings and seven EBI Screen control rooms. Currently there are contracts to upgrade the EBI Screen control rooms.

Switzerland: Bombardier Transportation is the first supplier in the industry to have successfully proven in full commercial operation its INTERFLO 450 ERTMS Level 2 technology on its pilot line in Switzerland. The world's first commercial

In 2003, the Bay Area Rapid Transit (BART) system in San Francisco, California commissioned its link to San Francisco Airport. Bombardier Transportation supplied the proven fixed block signalling system, CITYFLO 450. In 2003, Bombardier Transportation commissioned the completely refurbished automatic people mover system at the Seattle Tacoma Airport in Seattle, Washington. The project involved a complete replacement of the original 1970s vehicle fleet and signalling system, accomplished whilst keeping the system operating during peak periods. Bombardier Transportation's CITYFLO 650 communication-based moving block technology was selected largely due to its suitability for overlay applications. In 2003, Bombardier Transportation supplied the versatile CITYFLO 650 solution for the SFO AirTrain automatic people mover system. The CITYFLO 650 solution supplied includes the EBI Com radio block centre, EBI Cab onboard continuous ATP and ATO system and the EBI Screencontrol room and supervisory system.

UPDATED

Bosch – Robert Bosch GmbH

Zitadellenweg 34, D-13578 Berlin, Germany
Tel: (+49 711) 81 10   Fax: (+49 711) 811 66 30

UK subsidiary
Bosch Telecom Ltd
PO Box 98, Broadwater Park, North Orbital Road, Denham UB9 5HJ
Tel: (+44 1895) 83 44 66   Fax: (+44 1895) 83 85 48

Products
Mobile radio communications, including radio transceivers, trunked radios and base stations; satellite systems; security techniques; traffic control techniques.

Bosch has been involved in pioneer work on the digital PMR standard and the DISCO SR 440 has a digital/analogue dual function system with encryption.

Range of trunked radios in the VHF and UHF bands; Dikos 210 digital communications system; Flexplex XMP1 combined PCM transmission and cross-connect system; intercom systems for platform and trackside operations.

BP Solar Ltd

PO Box 191, Chertsey Road, Sunbury-on-Thames TW16 7XA, UK
Tel: (+44 1932) 77 95 43   Fax: (+44 1932) 76 26 86
Web: http://www.bpsolar.com

Key personnel
Managing Director: Michael Pitcher
Sales Manager, Railway Systems: Peter Carter

Principal subsidiaries
BP Solar Arabia, Saudi Arabia
BP Solar Australia, Australia
BP Solar España, Spain
BP Solar Inc, USA
Tata BP Solar, India

Products
Solar electric photovoltaic power systems to supply signalling, telecommunications, track circuits and lighting. PV solar cell materials and PV cell materials convert light into electricity.

Contracts include KTM Malaysia, Perumka (Indonesia), Ghana Railways, Indian Railways, Mozambique Railways, Kenya Railways and Saudi Rail and an IMWp multicrystalline roof atria system has been installed into stations for Deutsche Bahn.

S A Buhlmann NV

Leuvensesteenweg 31, B-1932 St Stevens-Woluwe, Brussels, Belgium
Tel: (+32 2) 711 20 30   Fax: (+32 2) 720 20 64
e-mail: buhlman@skynet.be

Key personnel
For full personnel listing, see A Buhlmann NV entry in *Brakes and drawgear* section.

Products
Point setting and point controllers; track circuit equipment and accessories; level crossing gates, barriers and warning signals.

Cadex Electronics Inc

22000 Fraserwood Way, Richmond, British Columbia V6W 1J6, Canada
Tel: (+1 604) 231 77 77   Fax: (+1 604) 231 77 55
e-mail: info@cadex.com
Web: http://www.cadex.com

Key personnel
President and Chief Executive Officer:
  Isidor Buchmann
Accounts Manager, Export: Richard Janzen
Marketing Specialist: Andrew Green

Products
Cadex C7000 series battery analysers service and restore batteries as used in portable communications equipment, computers and data acquisition devices; battery-specific adapters for all common batteries; multiple batteries serviced with the FlexArm™; user-selectable programmes for all battery needs such as the QuickTest™ programme which supports all common battery chemistries. Cadex analysers print service reports and battery labels and interfacing with a PC is possible with BatteryShop™. Cadex Universal Conditioning Chargers (UCCs) supplied in one-, two- and

Cadex Universal Conditioning Chargers six-bay model   0593063

six-bay models. The chargers accommodate Lithium Ion/Polymer, Nickel Cadmium and Nickel Metal Hydride batteries.

Cattron-Theimeg

58 West Shenango Street, Sharpsville, Pennsylvania 16150-1198, USA
Tel: (+1 412) 962 35 71
Fax: (+1 412) 962 43 10
e-mail: mail@cattron.com
Web: http://www.cattron-theimeg.com

Key personnel
President and Chief Executive Officer:
  James C Robertson
Chief Technology Officer: Carl Verholek

Principal subsidiaries
Cattron-Theimeg (UK) Ltd
Riverdene Industrial Estate, Molesey Road, Hersham, Walton-on-Thames KT12 4RY, UK
Tel: (+44 1932) 24 75 11
Fax: (+44 1932) 22 09 37
General Manager: Ian Martin

Cattron-Theimeg Canada, Ltd
150 Armstrong Avenue, Units 5-6, Georgetown, Ontario, Canada
Tel: (+1 905) 873 94 40   Fax: (+1 905) 873 94 49
General Manager: Bill Goldie

Cattron-Theimeg Africa (Pty) Ltd
25 O'Rielly Merry Road, Rynfield, Benoni, South Africa
PO Box 15444, Farrarmere, Benoni, Gauteng 1518, South Africa
Tel: (+27 11) 849 57 17   Fax: (+27 11) 425 59 38

Cattron-Theimeg Europe GmbH
Krefelder Strasse 423-425, D-41066 Mönchengladbach, Germany

Cadex Universal Conditioning Chargers one-bay model   0593064

Cattron-Theimeg DR Series portable remote-control system used to control ballast wagon doors
0114664

Products

Radio and infra-red portable control systems and related products allowing operators to control equipment from a safe distance. Cattron-Theimeg® portable remote controls can be used to control overhead cranes, monorails, hoists, shunting locomotives, underground mining locomotives, railcar movers, ballast wagon doors and many other types of equipment.

cdsrail

1570 Parkway, Solent Business Park, Fareham PO15 7AG, UK
Tel: (+44 1489) 57 17 71   Fax: (+44 1489) 57 15 55
e-mail: sales@cdsrail.com
Web: http://www.cdsrail.com

Products

Asset monitoring systems for railways infrastructure. The company's Trackwatch system provides event and condition monitoring for: relay and electronic/solid state signal events; signal interlockings; level crossings; point machines; clamp locks; point heaters; signalling power supplies and UPS systems; flood monitoring; track circuit condition monitoring; hot axlebox detectors; axle counters; and equipment status.

The Pointwatch system has been developed to provide early warning of impending points failure. In a typical installation it uses non-invasive, rugged sensors to monitor motor current, hydraulic air pressure, operating force and changeover time.

Analysis of data and alarms can be monitored from a single workstation using Trackwatch Analysis software or at client workstations attached to a Trackwatch Sentinel master supervisory system database server.

Infrastructure asset monitoring systems have been supplied by cdsrail to railways in India, Netherlands, Poland, Singapore, South Africa and UK.

China Railway Signal & Communication Corporation (CRSC)

111 Zao Jia Cun, Fengtai, Beijing, China
Tel: (+86 1) 381 62 89   Fax: (+86 1) 381 62 89

Key personnel

President: Yu Xiaomang
Chief Engineer: Wang Jinchen
Deputy Chief Engineer: Lu Jiasheng
General Manager, International Co-operation
    Department: Lu Delian

Products

All-relay interlockings, automatic and tokenless block, ATC equipment, CTC, cab signalling equipment, track circuits, relays, point machines, communications and telephone systems.

Chloride Power Protection

Barton Park, Eastleigh, SO50 6RZ, UK
Tel: (+44 23) 80 64 98 38   Fax: (+44 23) 80 61 08 52
e-mail: abi.linnartz@chloridepower.com
Web: http://www.chloridepower.com

Other offices

Australia, Brazil, France, Germany, Italy, Portugal, Singapore, Spain, Thailand, Turkey and the USA.

Products

Supply of electrical power supply protection systems.

Collis Engineering Ltd

Salcombe Road, Meadow Lane Industrial Estate, Alfreton, Derbyshire DE55 7RG, UK
Tel: (+44 1773) 83 32 55   Fax: (+44 1773) 52 06 93
e-mail: sales@collis.co.uk
Web: www.collis.co.uk

Key personnel

Managing Director: Peter Roberts
Contracts General Manager: Pete Savage
Technical Manager: Masoud Jafari
Sales and Marketing Manager: David Eades

Products

Design, supply and installation of railway signalling structures, posts, cantilevers and gantries. Mechanical signalling equipment designed, supplied and installed, as well as embankment platforms. Railway civil engineering, including full geotechnical and topographical surveying, foundation design and installation.

Contracts

Recent contracts include structures supplied for West Coast in Scotland (Carillion), structures and civils works for Jarvis (Mold & Crewe/Preston) and mechanical signals, installed for Network Rail. Also mechanical signalling contract for National Railway Supplies (NRS).

UPDATED

Computer Products Professional bv

Biezenvijer 4, NL-3297 GK Puttershoek, Netherlands
Tel: (+31 78) 676 29 99   Fax: (+31 78) 676 52 02
Web: http://www.compprof-RTP.com

Key personnel

Director: Ing W G Bouwmeester

Products

Remote control units for sensing and control of railway switches and signals for train tracking. Control units for mimic panels. Alpha-numeric indicators.

Crompton Greaves Ltd

1 Dr V B Gandhi Marg, Mumbai 400 023, India
Tel: (+91 22) 202 80 25

Rail Transportation Systems Division
5G, Vandhna, 11 Tolstoy Marg, 110 001 New Delhi, India
Tel: (+91 11) 331 70 75; 373 04 45
Fax: (+91 11) 331 70 75; 332 43 60
e-mail: dhingra_cgl@mantraonline.com

Key personnel

Managing Director: S M Trehan
General Manager, Rail Transportation Systems
    Division: A K Raina
All India Marketing Manager, Rail Transportation
    Systems Division: Harsh Dhingra

Products

Signalling relays, point machines, axle counters, route relay interlocking systems, solid state signalling and data loggers.

Developments

Integrated power supply, AFTC and digital axle counters.

CSEE Transport

An Ansaldo Signal company, controlled by Finmeccanica
4 avenue du Canada, F-91944 Les Ulis, France
Tel: (+33 1) 69 29 65 65   Fax: (+33 1) 69 29 07 07
Web: www.csee-transport.com

Key personnel

Chief Executive Officer: Georges Dubot
Chief Financial Officer and Member of the Board:
    Dominique Athanassiadis
Vice-President, Operations:
    Jean Jacques Jarjanette
Vice-President, Sales: Gérard Bonnas

Vice-President, Business Development:
    Gilles Pascault
Vice-President, Development: Pierre Advani
Vice-President, Manufacturing: Francis Cornet
Vice-President, Human Resources:
    Catherine Porret

Background

CSEE Transport is a wholly owned subsidiary of Ansaldo Signal NV.

Works

BP 13, ZAC des Portes de Riom, F-63201 Riom Cedex, France

Principal subsidiaries

CSEE Trasp. Hong Kong Ltd, Hong Kong
Sinelbras Sinalizacao e Electr. Brasilera Ltd, Brazil
Beijing CS Signal Controlling System Ltd, Beijing
Acelec SA, France
Equipos de Control y Segnalizacion SA (Ecosen), Venezuela
Frameca, France

Branches

CSEE Transport in Spain
CSEE Transport in Portugal
CSEE Transport in the UK

Products

Products for high-speed, mass transit and conventional lines including:

TVM systems
Automatic Train Control (ATC), Automatic Train Protection (ATP) and Automatic Train Operation (ATO); driving-assistance systems grouping a complete range and several generations. All these systems, TVM 300, TVM 430, TVM SEI (interlocking) up to the TVM ERTMS system, conform to the railway standards of European inter-operability.

Supervision
Operation Control Center (OCC): these interconnected computerised systems ensure the centralised management of traffic. They allow the network operators to optimise the use of rolling stock and energy.

Access Control Gates, PAC: they provide a more reliable supervision of user movement and better access, in both transport systems and public buildings.

Hot-box and hot-wheeled detectors, track circuits, point machines and electronic treadles.

Contracts

LGV Est ERTMS level 2, high-speed line
Mediterranean TGV: ATC and signalling equipment (high-speed line)
Korean TGV: ATC and signalling equipment (high-speed line)
Channel Tunnel–London: ATC, supervision, signalling equipment (high-speed line)
Madrid–Barcelona: ATC and signalling equipment (high–speed line/ERTMS)
Vienna–Budapest: Installation and testing of the new system ERTMS/ETCS level 1
Zhengzhou–Beijing–Shenyang: ATC and signalling equipment (high-speed technology)
Paris Metro: Maintenance of the way-side equipment and aid to driving + Piragam system
Lisbon Metro: Signalling, ATC and supervision
Newcastle Metro: OCC system
Hong Kong Metro/Lantau Airport Express: Control Centre and supervision
Hong Kong Metro/Station Management System: Supervision system

UPDATED

David Clark Company

360 Franklin Street, Worcester, Massachusetts 01604, USA
Tel: (+1 508) 751 58 00   Fax: (+1 508) 753 58 27
e-mail: sales@davidclark.com
Web: www.davidclark.com

Key personnel
Vice-President: Richard M Urella
International Sales Manager: James E Comer
Sales Manager: Mark E Gardell

Products
Locomotive crew intercom systems featuring noise-reducing headsets to provide hands-free intercom, music and radio transmitting and reception with alarms integrated into the intercom system. Noise attenuating headsets and radio interference adapters for mobile and portable radios.

Contracts
Recent contracts include supply of equipment to Burlington Northern Santa Fe, Quebec North Shore and Labrador, Canadian Pacific, Canadian National and Amtrak.

UPDATED

Deuta-Werke GmbH

Paffrather Strasse 140, D- 51465 Bergisch Gladbach, Germany
Tel: (+49 2202) 95 80   Fax: (+49 2202) 95 81 45
e-mail: support@deuta.de
Web: http://www.deuta.com

Key personnel
Manager Marketing and Sales: Wolfgang Fabek
Export Manager: Joachim Baumann

Principal subsidiaries
Shanghai Deuta Electronic & Electrical Equipment Co Ltd
Deuta America Corporation, St Louis, Missouri 63146, USA

Products
Sensors: pickups, AC generators, electronic/electric pulse generators, opto-electronic generators, microwave sensors.

Indicators: electric indicators, electric meters, eddy current tachometers, panel-mounted clocks, electric and mechanical counters, modular driver's cab indicators, multi-function displays, digital indicators.

Incident recorders: electric incident recorders, short-distance incident recorders, digital storage cassettes, evaluation software for data recorders.

System components: central distance and speed measuring units, electronic control units, multi-function modules.

IT solutions: speed, distance and position systems for train navigation and protection, tracing and tracking information systems for passenger and freight, train information systems, network databank systems for fleet management and diagnostics.

Embedded systems: Deuta embedded PC boards and components are suitable for use over a wide temperature range (–40° to +85°C) and in environments exposed to vibration and shock. All-in-one Geode EPC – compact EPC board with low-power processor, onboard graphics and environment controller, field bus functionality and an integrated power supply. Especially suitable for use in mobile applications for process visualisation, terminals (MMI), data acquisition units. DSP module, highly integrated DSP module with a programmable, eight-channel, 14/16 bit ADC, including data pre-processing and CAN bus controller. For universal usage in industrial measuring and control systems. Combination of the DSP module with an EPC enables inexpensive implementation of an all-in-one solution with integrated MMI interfaces and industrial field bus functionality.

Driver desks: Deuta-Werke develops driver's desks as complete systems with latest technologies such as integrated CAN bus networks.

Dialight

Exning Road, Newmarket CB8 0AX, UK
Tel: (+44 1638) 66 23 17   Fax: (+44 1638) 56 04 55

Key personnel
For full personnel listing, see Dialight entry in *Passenger coach equipment* section.

Products
Level crossing signal lights.

Dimetronic SA

Avda de Castella 2, Parque Empresarial, E-28830 San Fernando de Henares, Madrid, Spain
Tel: (+34 91) 675 42 12   Fax: (+34 91) 756 21 15
e-mail: marketing.dimetronic@invensys.com

Key personnel
Managing Director: Carlos Manzano
International Markets Director: Luis Garcia

Principal subsidiary
Dimetronic Portugal

Background
Dimetronic is part of Invensys Rail Group.

Products
Signalling systems; electronic interlockings, CTC, ATP, ATO, ATC, ERTMS, train describer, automatic route setting, cab signalling and traffic regulation equipment.

Contracts
Most of the projects are of a turnkey nature. Contracts include Westrace electronic Interlockings and CTC systems for RENFE in Spain and Lisbon metros; ATP/ATO, CTC and Westrace for Madrid metro, Barcelona, Manila, Valencia and Bucharest metro; SSI and CTC for REFER (Portugal); and Westrace electronic interlocking for FGC Spain.

EFACEC Sistemas de Electrónica SA

Signalling Systems Division, Av Eng Ulrich, PO Box 31, P-4470 Maia, Portugal
Tel: (+351 2) 941 36 66   Fax: (+351 2) 948 54 28

Products
Signalling and telecommunications systems and components, including electronic interlocking, automatic train speed control, track circuits, level crossing equipment including barrier machines, and ATP panels, encoders, recording and evaluation units. Installation, repair and maintenance of signalling equipment.

Erico Inc

34600 Solon Road, Solon, Ohio 44139, USA
Tel: (+1 440) 248 01 00   Fax: (+1 440) 309 89 11
e-mail: info@erico.com
Web: http://www.erico.com

Key personnel
Division Manager: Dan Johnston
Marketing Manager: Mitchell Bednarek
Export Sales Manager: Phil Graham

Principal Subsidiary Companies
In Australia, Brazil, Chile, France, Germany, Hong Kong, Mexico, Netherlands, South Africa and UK.

Products
An engineered system of bonding, grounding and electronic protection products: power and signal bonds, track circuit connectors; Cadweld® exothermic bonding process; ground enhancement material; electronic protection panels, electronic protection devices.

Fels SA

2 rue J M Jacquard, F-67400 Illkirch Graffenstaden, France
Tel: (+33 3) 88 67 10 60   Fax: (+33 3) 88 67 33 10
e-mail: fels@fels.fr
Web: www.fels.fr

Products
Electrical contacts; special connectors.

UPDATED

Fiber Options

80 Orville Drive, Suite 102, Bohemia, New York 11716, USA
Tel: (+1 631) 567 83 20
Fax: (+1 631) 567 83 22
e-mail: info@fiberoptions.com
Web: http://www.fiberoptions.com

Key personnel
President: John Collins
Founder: Bob Delia
Director of Marketing: Jerry Jacobson

Products
Fibre optic communication systems.

Deuta-Werke's MFT 1 terminal technology   1026868

Frequentis GesmbH

Wolfganggasse 58-60, A-1120 Vienna, Austria
Tel: (+43 1) 81 15 00
Fax (+43 1) 811 50 13 19
e-mail: marketing-group@frequentis.com
Web: http://www.frequentis.com

Key personnel

Managing Director: Sylvia Bardach

Background

The core business of Frequentis is air traffic control but recently the company has started to build new business in the fields of command and control systems, maritime systems and TETRA (Terrestrial Trunked Radio).

Products

Telecommunication systems for rail and public transport systems. Products include the Dicora-S GSM-R dispatcher terminal.

Contracts

In 2002 Frequentis secured its first contract in the rail market. This involved the supply of a telecommunications network to Railtrack plc (now Network Rail), UK, for its Rail Traffic Control Centre covering much of the West Coast Main Line. As well as the latest digital broadband interfaces, the contract covered provision of interfaces to existing systems, including analogue train radio, railway telephones and emergency telephone systems.

GAI-Tronics

A division of Hubbell Limited
Brunel Drive, Stretton Business Park, Burton-on-Trent DE13 0BZ, UK
Tel: (+44 1283) 50 05 00
Fax: (+44 1283) 50 04 00
e-mail: sales@gai-tronics.co.uk
Web: www.gai-tronics.co.uk

Key personnel

Business Unit Manager: Graham Lines
Financial Director: Toby Balmer
Engineering Manager: Richard Rumsby
Manufacturing Director: Mark Bradford
Commercial Manager: Roger Goodall
Marketing Consultant: Nicole Ireland
Special Projects Manager: Steve Smith

Other offices

GAI-Tronics Corporation, USA
Tel: (+1 610) 777 13 74   Fax: (+1 610) 775 65 40
Web: www.gai-tronics.com

GAI-Tronics Srl, Italy
Tel: (+39 02) 48 60 14 60   Fax: (+39 02) 458 56 25
Web: www.gai-tronics.co.uk

GAI-Tronics Corporation, Malaysia
Tel: (+60 3) 89 45 40 35   Fax: (+60 3) 89 45 46 75
Web: www.gai-tronics.co.uk

Background

Established in 1964, Gai-Tronics is a major provider of specialised telecommunications for both UK and worldwide railways, manufacturing weather- and vandal-resistant communication equipment.

Products

Metal-bodied weather-resilient telephones; SMART self-monitoring and reporting telephone systems. Network Rail certified (PA05/570) telephones for trackside and level crossings. Illuminated crossing telephones.

UPDATED

Ganz Transelektro Traction Electric Ltd

PO Box 250, H-1243 Budapest, Hungary
Király u. 163, H-1061 Budapest, Hungary
Tel: (+36 1) 432 88 50
Fax: (+36 1) 262 36 38
Web: http://www.traction-ganztrans.hu

Corporate background

Formerly a state enterprise, the company was privatised in 1991 as Ganz Ansaldo Electric Ltd, with the Italian company initially taking a shareholding of 51 per cent but increasing this to 99.99 per cent in 1998. In 1999 Ansaldo divested its interest, leading to the establishment of the present private sector company in 2000.

Products

Centralised traffic control systems, signalling systems for stations, block interlocking equipment, track circuit equipment and axle-counters, level crossing systems, automatic train control equipment, control systems, panels and mimic boards, point machines.

GE Rail

Global Signalling
2712 S. Dillingham Road, PO Box 600, Grain Valley, Missouri 64029, US
Tel: (+1 816) 650 31 12   Fax: (+1 816) 650 63 29
Web:www.getransportation.com

Key personnel

General Manager: Kevin Caponecchi
Marketing Leader: Peter Thomas
Sales Leader: Andrew Zaborny
Transit Sales Manager: Patrick McKenna
European Sales Manager: Olof Kjelberg

Subsidiaries

Global Signalling – UK
The Maltings, Hoe Lane, Ware SG12 9LR, UK
Tel: (+44 1920) 46 22 82   Fax: (+44 1920) 46 07 02

Train Control Systems
321 SE AA Highway, Blue Springs, Missouri 64014, US
Tel: (+1 816) 229 60 55   Fax: (+1 816) 229 61 22

Advanced Communication Systems Division
40 Pond Park Road, Hingham, Massachusetts 02043 4371, US
Tel: (+1 781) 740 02 00   Fax: (+1 781) 740 02 02

DJR Installation & Maintenance
3800 Ten Oaks Road, Suite B, PO Box 305, Glenelg, Maryland 21737 0305, US
Tel: (+1 410) 442 17 06   Fax: (+1 410) 442 29 71

Global Signalling – Germany
Bruchstrasse 79A, Bad Durkheim D-67098, Germany
Tel: (+49 63) 22 94 78 25   Fax: (+49 63) 22 94 78 25

Globall Signalling – Italy
21 Via Pietro Fanfani, I-50127 Firenze, Italy
Tel: (+39 055) 423 41   Fax: (+39 055) 438 68

Products

Previously called GE Transportation, GE Rail, a division of General Electric Company, comprises aircraft engines, rail, marine and off-highway business units. The company is one of the world's leading manufacturers of products and services for freight and passenger locomotives, railway signalling and communication systems. As a leader of innovative signalling and communication solutions for the rail freight and transit industry, GE

Transportation designs and manufactures a product and service line including wayside and vehicle-carried components, engineering and railway management services, training, rail-highway level crossings, transit/commuter signalling and train control, communications, control centres, traction motors, traction power systems and information technology solutions.

UPDATED

GrantRail

1 Carolina Court, Lakeside, Doncaster DN4 5RA, UK
Tel: (+44 1302) 79 11 00   Fax: (+44 1302) 79 12 00
Web: http://www.grantrail.co.uk

Key personnel

Chief Executive: Gren Edwards

Background

GrantRail is a joint venture between the Corus Rail Group and Volker Stein Rail and Traffic.

Capabilities

Capabilities cover: support for permanent way renewals, enhancements and maintenance; modifications and renewals of signalling construction; and installation of new signalling.

Hanning & Kahl GmbH & Co KG

Rudolf Diesel Strasse 6, D-33813 Oerlinghausen, Germany
PO Box 1342, D-33806 Oerlinghausen, Germany
Tel: (+49 5202) 70 76 00   Fax: (+49 5202) 70 76 29
e-mail: info@huk.hanning.com
Web: www.hanning-kahl.de

Key personnel

General Manager: Wolfgang Helas
Brake Division Manager: Dietrich Radtke
LRT Division Manager: Christian Schmidt
Service Division Manager: Peter Spilker
Sales Manager, Brakes: Jürgen Stammeier
Sales Manager, LRT: Hans-Joachim Pässler
Sales Manager, Services/LRT: Joachim Zehn
Sales Manager, Services/Brakes: Martin Epp

Products

LRT Division
Point controls, signalling systems, level crossing safety devices, single-line track safety devices, track circuits, mass detectors, vehicle reporting system, radio control, electronic data recorder and accessories including insulated guard rail tie bars, rail termination boxes and contact systems.

Point setting mechanisms
LRT Division: Point controls, signalling systems, level crossing safety devices, single-line track safety devices, track circuits, mass detectors, vehicle reporting systems, radio control, electronic data recorders and accessories including guard rail tie bars, rail boxes and contact systems. Point setting mechanisms for all gauges and types of rail, manual point mechanisms, electric point mechanisms with magnetic, motor, electrohydraulic or central control.

All point mechanisms can also be set manually and are available with a tongue detector and a mechanical double-interlocking device for tongues in open and closed positions.

Train to Wayside Communication Systems (TWC), LED signalling devices, passenger information systems, vehicle management systems, depot management systems, point management systems (WEDIS), route diagnosis systems (FADIS), point heaters, point heater controllers.

For details of the latest updates to *Jane's World Railways* online and to discover the additional information available exclusively to online subscribers please visit

jwr.janes.com

Service Division: Services and testing and measuring equipment for point setting mechanisms and controllers, signalling installations, TWC systems.

Contracts
Recent contracts include projects in Calgary, Salt Lake City, Dallas, San Diego, San Jose, Manchester, Birmingham, Croydon, Sheffield; Rome, Turin, Milan; cities in Germany, Switzerland, Austria, Belgium, Netherlands, Norway, Sweden, Finland, Melbourne and Hong Kong.

UPDATED

Harrington Generators International

Ravenstor Road, Wirksworth, Derbyshire, DE4 4FY, UK
Tel: (+44 1629) 82 42 84   Fax: (+44 1629) 82 46 13
e-mail: sales@harringtongen.co.uk
Web: http://www.harringtongen.com

Key personnel
Managing Director: Peter Harrington
Sales Manager: Barry Kimber

Products
Static and mobile generators for standby power applications for signalling systems.

Henry Williams Group

Dodsworth Street, Darlington DL1 2NJ, UK
Tel: (+44 1325) 46 27 22   Fax: (+44 1325) 24 52 20
e-mail: sales@hwilliams.co.uk
Web: http://www.hwilliams.co.uk

Key personnel
Managing Director: Alan W D Puddick
Production Director: George R Fenley
Sales Engineers: Bryan Blareau; Jon Heslop
Sales Co-ordinator: Stan Thompson

Products
Henry Williams has the capability for the design, manufacture and installation of level crossing systems and small signalling schemes.

Contracts
Projects include REBs for Rugby resignalling, location cases for RETB project and crossings for Tilbury Docks and West Burton power station.

Hitachi Ltd

Overseas Marketing Department
Transportation Systems Sales Division
6 Kanda Surugadai 4-chome, Chiyoda-ku, Tokyo 101-8010, Japan
Tel: (+81 3) 32 58 11 11
Fax: (+81 3) 32 58 52 30
Web: www.hitachi-rail.com

Key personnel
Chief Operating Officer of Transportation Systems: Gaku Suzuki
General Manager, Transportation Systems Sales Division: Chiaki Ueda
General Manager, Transport Management and Control Systems Division: Kazuo Kera
General Manager Rolling Stock System Division: Toshihide Uchimura

Products
Total railway traffic management system including data acquisition and passenger information service, Automatic Train Control (ATC), Automatic Train Operation (ATO), Autonomous Train Integration (ATI) train communication networks.

UPDATED

Isolux

Isolux Wat SA
Alcocer 41, E-28021 Madrid, Spain
Tel: (+34 91) 796 30 00 (+34 91) 798 37 70
e-mail: sistemas@isolux.es
Web: http://www.isolux.es

Key personnel
Director, Control & Systems Division: Miguel Angel Tapia
Commercial Director Control & Systems Division: José Ignacio Martinez

Products
Hot wheel detector; station signalling and operation system; centralised control of installations in stations; video surveillance and control of stations.
Contracts include hot wheel detectors and signalling systems for RENFE, and CTC for RENFE. Video surveillance has been supplied for FGC Barcelona metro and CTC for Madrid metro.

Kapsch Group

Wagenseilgasse 1, A-1121 Vienna, Austria
Tel: (+43 1) 81 11 10   Fax: (+43 1) 81 11 18 88
Web: http://www.kapsch.headoffice@kapsch.net

Subsidiaries
Austria
Austria Telecommunication GmbH
Triester Strasse 70, A-1102 Vienna
Tel: (+43 1) 60 50 10   Fax: (+43 1) 605 01 32 01
Web: http://www.austria.telecommunication.at@kapsch.net

Bulgaria
Kapsch AG Bulgaria
ul Sitnjakovo N: 5A, BG-1124 Sofia
Tel: (+359 2) 46 82 00   Fax: (+359 2) 46 82 00
e-mail: vvkapsch@mbox.infotel.bg

Chile
Combitech Traffic Systems AB
Av Las Condes 11400 Suite 32, 668 0002 Vitacura, Santiago
Tel: (+56 2) 217 53 25   Fax: (+56 2) 217 53 10
e-mail: ctschile@terra.cl

Croatia
Kapsch Telecom doo
Kneza Mislava 1, HR-10000 Zagreb
Tel: (+385 1) 466 48 46   Fax: (+385 1) 461 79 32
Web: http://www.kapsch.telecom.hr@kapsch.net

Czech Republic
Kapsch Telecom spol sro
Opletalova 1015/55 CZ-110 00 Praha 1
Tel: (+420 2) 21 46 63 11   Fax: (+420 2) 22 24 42 88
Web: http://www.kapsch.telecom.cz@kaptsch.net

Germany
Kapsch Telecom GmbH
Stuttgarterstrasse 61, D-71554 Weissach im Tal
Tel: (+49 7191) 355 90   Fax: (+49 7191) 35 59 60 80
Web: http://www.kapsch.telecom.de@kapsch.net

Hungary
Kapsch Telecom Kft
Bocskai út 77-79, H-1113 Budapest
Tel: (+36 1) 209 21 10   Fax: (+36 1) 209 21 11
Web: http://www.kapsch.telecom.hu@kapsch.net

Poland
Kapsch Telecom Sp zoo
ul Sniadeckich 1, PL-02-785 Warszawa
Tel: (+48 22) 544 60 00   Fax: (+48 22) 544 60 05
Web: http://www.kapsch.telecom.pl@kapsch.net

Koltel Sp zoo
ul Ludwikowo 1, PL-85-502 Bydgoszcz
Tel: (+48 52) 322 86 93   Fax: (+48 52) 322 78 84
Web: http://www.koltel.pl@kapsch.net

Russia
OOO Kapsch NIIShA Tel
ul Nizhegorodskaja 27, RF-109029 Moskwa

Tel: (+7 095) 262 82 74
Fax: (+7 095) 262 04 43
Web: http://www.kapsch.niishatel.ru@kapsch.net

Slovak Republic
Kapsch Telecom spol sro
Bratislava Business Center, Plynárenská 1, SK-821 09 Bratislava
Tel: (+421 7) 53 41 83 00   Fax: (+421 7) 53 41 83 01
Web: http://www.kapsch.telecom.sk@kapsch.net

Slovenia
Kapsch Telecom doo
Kotnikova 12, SI-1000 Ljubljana
Tel: (+386 1) 300 84 70   Fax: (+386 1) 300 84 79
Web: http://www.kapsch.telecom.si@kapsch.net

Ukraine
Kapsch Telecom TOV
Bocskai ut 77-79, H-1113 Budapest
Tel: (+36 1) 209 21 10   Fax: (+36 1) 209 21 11
Web: http://www.kapsch.telecom.hu@kapsch.net

Products
Closed-circuit radio communication systems for main and ancillary lines, KS2000 closed-circuit telecommunications system and GSM-Railway (GSM-R) for high-speed trains.
GSM-R is a uniform radio communication system developed by Kapsch.
The GSM core network is the dominant mobile communications platform and is the basis for over 20,000 roaming agreements in the world. The multirate codec AMR has recently been developed for GSM and it will also be the standard codec for UMTS.

Kruch Gesmbh & Co KG

Pfarrgasse 87, A-1230 Vienna, Austria
Tel: (+43 1) 616 31 65   Fax: (+43 1) 616 31 68
e-mail: office@kruch.at
Web: http://www.kruch.at

Products
Signal masts; signal lights; mast-mounted switch cabinets.

Kyosan Electric Mfg Co Ltd

4-2 Marunouchi 3-chome, Chiyoda-ku, Tokyo 100-0005, Japan
Tel: (+81 3) 32 14 81 36   Fax: (+81 3) 32 11 24 50
Web: www.kyosan.co.jp

Head office
29-1, Heian-cho, 2-chome, Tsurumi-ku, Yokohama, Japan
Tel: (+81 45) 501 12 61   Fax: (+81 45) 15 61

Key personnel
President: T Nishikawa
Director and Executive Officer, Export Sales: Kazuo Hinata
General Manager, Overseas Department: F Takenouchi

Subsidiary
Taiwan Kyosan Co Ltd, Taichung, Taiwan

Products
Centralised Traffic Control (CTC) system, Automatic Train Protection (ATP) system, Automatic Train Operation (ATO) system, Automatic Train Stop (ATS) system, automatic block signal equipment, solid-state interlocking equipment, relay interlocking equipment, level crossing signal and gate information display system.

Contracts
Contracts include supply of systems for the automated people mover for Hong Kong airport at Lantau, Singapore LRT Sengkang and Punggol line signalling systems and solid state interlocking system for Seoul Rolling Stock Depot.

UPDATED

Leach International Europe SA

2 rue Goethe, F-57430 Sarralbe, France
Tel: (+33 3) 87 97 98 97
Fax: (+33 3) 87 97 84 04
e-mail: jmsigaud@leachint.fr
Web: http://www.leachintl.com

Key personnel
Commercial Director: J M Sigaud
Export Sales Manager: P Saunders

Products
Relays, time delay relays and timers, contactors, switchlights, panels and keyboards, switching and control equipment, electromechanical and electronic power distribution assemblies.

Lincoln Industries

8021 National Tumpike, Louisville, Kentucky 40214, USA
Tel: (+1 502) 368 65 65
Fax: (+1 502) 367 14 84

Key personnel
Vice-President, Sales: Eddie Ramer

Background
Lincoln Industries is a division of Progress Rail Services Corporation.

Products and services
Level crossing safety devices, signalling hardware, relay cases and housings, cantilevers, signal masts and bases, switch layout and accessories, inspection and severe storm damage repair.

LPA Channel Electric Ltd

Bath Road, Thatcham RG18 3ST, UK
Tel: (+44 1635) 86 48 66
Fax: (+44 1635) 86 91 78

Key personnel
Managing Director: George Renshaw
Marketing Manager: Alex Burt
Commercial Director: Chris Antysz

Background
Previously called Channel Electric Equipment, LPA Channel Electric is an ISO 9002 registered company and is a member of LPA Group plc.

It supplies electrical and electro-mechanical products.

Products
Terminal junction module assemblies and other components for telecommunications use; relays, connectors, circuit breakers, switches, contactors, fans, motors, blowers, harnesses, assemblies, enclosures and stud terminals.

Luxram Lighting Ltd

PO Box, Calendariaweg 2A, CH-6405 Immensee, Switzerland
Tel: (+41 41) 854 44 44
Fax: (+41 41) 854 44 50
e-mail: info@luxram.com
Web: http://www.luxram.com

Key personnel
General Manager: F J Naegeli
Production Manager: H Ullrich

Products
Lamps for railway applications, including signal lamps.

Recent contracts include the supply of lamps to SBB, Luxembourg Railway Company, SNCB, MÁV, Croatian State Railway, KTM, Malaysia, Melio Nedellin and Swedish Rail.

Marconi Communications Limited

Strategic Networks
New Century Park, PO Box 146, Coventry CV3 1LQ, UK
Tel: (+44 24) 76 56 55 00
Fax: (+44 24) 76 56 58 88
e-mail: marketing.transportation@marconi.com
Web: http://www.comms.marconi.com

Key personnel
Managing Director, Strategic Networks:
   Neil Sutcliffe
Sales Director, Strategic Networks:
   Graham Outterside
Contact: S Meade

Products
Design, supply, maintenance and management of complex transportation communication networks. Has specialist knowledge in system design, integration, implementation and project management, incorporating the following communication systems: PDH and SDH digital transmission systems, optical fibre, copper distribution and signalling cable systems, telephone systems, PABX, fixed and track-to-train CCTV, analogue and digital public address, passenger information displays, VHF/UHF mobile radio, passenger alarm and help point systems and a comprehensive Station Management System (SiMS) platform.

SiMS includes control of all communications equipment on a station, in addition to tunnel telephone systems, ticketing machines, SCADA and a comprehensive decision support package.

An example of integration capability is the development of an integrated communications system, controlling telecommunications, public address, CCN and passenger information displays via a single screen.

Marconi Communications CNS (Customer Network Services) targets communication networks and control systems in the mass transport sector.

Contracts
Contracts include London Underground Ltd Jubilee Line Extension (prime contractor for Contract 204); London, Tilbury and Southend resignalling scheme (Railtrack); Docklands Light Railway; Beijing Metro Line 1 modernisation; and London Underground Ltd Northern Line rolling stock communications infrastructure and maintenance.

Strategic Networks is providing communications for Hong Kong's Lantau Airport rail link and for Midland Metro Line 1.

Marl International Ltd

Marl Business Park, Ulverston LA12 9BN, UK
Tel: (+44 1229) 58 24 30   Fax: (+44 1229) 58 51 55
e-mail: sales@marl.co.uk
Web: http://www.marl.co.uk;
   www.marlrailproducts.com

Trackside signal using LED technology, developed by Marl International in conjunction with Signal House Ltd   0546911

Products
Lumarled LED signal lamps; 238 Series bipolar LED lamps for signal applications; Solo filament bulb replacements; LED bulkhead lamps for use as buffer stop lights.

VERIFIED

Maxon Europe

Maxon House, Maxted Close, Hemel Hempstead HP2 7EG, UK
Tel: (+44 1442) 26 77 77
Fax: (+44 1442) 21 55 15
e-mail: sales@maxon.co.uk
Web: http://www.maxon.co.uk

Key personnel
Sales and Marketing Manager: John French

Subsidiary companies
MRSA (Maxon France, Paris)
MISA (Maxon Iberia, Madrid)

Products
VHF and UHF hand-held radios, mobile radios, all with a wide area of paging.

Contracts include the supply to Polish State Railways of SL70 hand-held portable radios.

NEW ENTRY

Mer Mec SpA

Via Oberdan 70, I-70043 Monopoli (BA), Italy
Tel: (+39 080) 887 65 70
Fax: (+39 080) 887 40 28
e-mail: mermec@mermec.it
Web: www.mermec.it

Key personnel
Managing Director: Vito Pertosa
Marketing and Sales, Executive Director:
   Luca Ebreo
Marketing Manager: Pietro Stama
International Sales Manager: Carlo Evangelisti
Domestic Sales Manager: Mario Girolami
After Sales and Commissioning Manager:
   Giuseppe Aurisicchio
Quality Assurance Manager: Mauro Simone
Research and Development Programme Manager:
   Patrizia Sforza

Products
Design, development and manufacturing of signalling systems compliant with ERTMS/ETCS standards. Retrofit of onboard signalling equipment.

Design and development of measuring and diagnostic systems for signalling and telecommunication networks monitoring.

UPDATED

Mirror Technology

Unit 4, Redwood House, Orchard Trading Estate, Toddington GL54 5EB, UK
Tel: (+44 1242) 62 15 34
Fax: (+44 1242) 62 15 29
e-mail: mirtec@aol.com
Web: http://www.mirrortechnology.co.uk

Key personnel
Sales Director: Malcolm Robertson
Production Director: R M J Chambers

Products
Platform mirrors for driver-only operation, surveillance and as an aid to safety by providing rearward visibility from the driver's cab; vandal-resistant polycarbonate pedestrian subway mirrors.

Contracts

UK train operating companies supplied include Chiltern Railways, South Central, Docklands Light Rail; Lisbon Metro, Manchester, Medellin Metro, Metrolink, Midland Metro, Sheffield Supertram, Thames Trains and West Anglia Great Northern and most main line operators.

Morio Denki Co Ltd

34-1 Tateishi 4-chome, Katsushika-ku, Tokyo 124-0012, Japan
Tel: (+81 3) 36 91 31 81   Fax: (+81 3) 36 92 13 33
Web: www.morio.co.jp

Key personnel
President: S Yamagata
Senior Managing Director: K Miura

Products
Automatic Train Control and Automatic Train Stop units, station override prevention systems, event recorders, monitoring systems and alarm systems.

UPDATED

Motorola Canada Ltd

Communications Division
3125 Steeles Avenue East, North York, Ontario M2H 2H6, Canada
Tel: (+1 416) 499 14 41

Products
Two-way portable radios; the PHD 5000 Trackside Communications Network, a self-contained and integrated voice and data system, interconnecting management, control offices, field staff and vehicles on the move.

Motorola (Schweiz) AG

Fabrikstrasse 8, CH-4614 Haegendorf, Switzerland
Web: http://www.motorola.com

Background
Motorola took over the railway communications business of Ascom Radiocom AG in 1999.

Products
Railway communications systems, including: track-to-train radio systems; tunnel radio systems; portable, mobile and fixed radio communications systems. Recent products include the MSTR track-to-train mobile station, designed for cross-border rail operations.

Hand-held portable UHF and Band Three trunked radios; base station equipment. Band Three Radio Electronic Token Block (RETB); Field Radio Unit (FRU); Band Three mobile National Rail Network (NRN); Band Three trunked Starnet system.

Nippon Signal Co Ltd

1-1 Higashi-Ikebukuro, 3-chome, Toshima-ku 170-6047 Tokyo, Japan
Tel: (+81 3) 59 54 45 47   Fax: (+81 3) 59 54 45 51
e-mail: info@signal.co.jp
Web: www.signal.co.jp

Key personnel
President: K Nishimura

Products
Integrated Traffic Control by Computer (ITC); Centralised Traffic Control System (CTC); Automatic Program Route Control System (PRC); Automatic Train Control System (ATC); Automatic Train Operation (ATO); Automatic Train Stop System (ATS); train detection equipment; transponder; maintenance management system;

relay interlocking equipment; train information processing system; electronic interlocking equipment; electronic block system; railway crossing monitor system; Train Navigation System (TNS); other signal equipment; universal traffic management system; centralised area traffic control system; local controllers; traffic signal-integrated street lights; ultrasonic vehicle detector; vehicle-type classifying detector; image processing vehicle detector; road traffic information system; traffic information board; traffic flow data counter.

UPDATED

Nortel Networks Germany GmbH & Co

Hahnstrasse 37-39, D-60528 Frankfurt am Main, Germany
Tel: (+49 69) 669 70   Fax: (+49 69) 97 11 11
e-mail: euroinfo@nortelnetworks.com
Web: http://www.nortelnetworks.com

Key personnel
Assistant Business Development GSM-R:
   Leonie Geray

Subsidiary companies
Offices in Canada, Europe, Asia-Pacific, Caribbean, Latin Amercia, Middle East, Africa and USA.

Products
Missioncritical telephony and IP-optimised networks supplied to customers in 150 countries; ASCI (Advanced Speech Call Items) for high-speed trains. GSM-R base transceiver station (BTS) for trackside coverage under adverse evironmental conditions; specially-adapted radio algorithms to allow error-free voice and data communications at train speeds up to 500 km/h. Nortel is a member of the MORANE consortium (Mobile Radio for Railway Networks in Europe).

NRS – National Railway Supplies

Gresty Road, Crewe CW2 6EH, UK
Tel: (+44 1270) 53 30 00   Fax: (+44 1270) 53 39 56
e-mail: commercial@natrail.com
Web: www.natrail.com

Other offices
Leeman Road, York YO26 4ZD, UK
Tel: (+44 1904) 52 22 93   Fax: (+44 1904) 52 26 96

Key personnel
Managing Director: Graham Jackson
Commercial Director: David A Kierton
Marketing Manager: Dave Tilmouth

Background
NRS is a wholly owned subsidiary of the Unipart Group of Companies.

Products
NRS provides a logistics and supply chain management service to the UK rail industry including the management of customer warehouses, the distribution of 23,000 products and the repair and recondition of a wide range of products. NRS also designs, develops and manufactures a range of new products for infrastructure and rolling stock.

Contracts
Contracts include Network Rail, BBRISL, Carillion Rail, AMEC Spie Rail, Jarvis Rail, Westinghouse Rail Systems and London Underground. Manufacture, production, servicing, repair and distribution of signalling, track maintenance and telecoms products, infrastructure and permanent way. Also design and assembly of signalling, TPWS and signalling location cases.

UPDATED

Peek Traffic BV

Basicweg 16, PO Box 2542, NL-3800 GB Amersfoort, Netherlands
Tel: (+31 35) 689 17 77
Fax: (+31 35) 689 18 50
e-mail: info@peektraffic.nl
Web: http://www.peektraffic.nl

Key personnel
Business Manager, Railway Products:
   J R Opperman
Export Manager: A Koopmans

Products
Train identification and information systems for metros and light rail systems.

Pintsch Bamag Antriebs- und Verkehrstechnik GmbH

PO Box 100420, D-46524 Dinslaken, Germany
Tel: (+49 2064) 60 20   Fax: (+49 2064) 60 22 66
e-mail: info@pintschbamag.de
Web: www.pintschbamag.de

Key personnel
Managing Director: Dr Rolf-Dieter Krächter
Strategic Concerns: Ulrich Nagorski
Export Manager Signalling: Ulrich Rink
Head of Business Unit: Hans-Jürgen Dröttboom

Background
Pintsch Bamag is a member of the Schaltbau Group.

Products
Level crossing protection systems and components, computer controlled level crossing protection (all types of supervision: main signal indication, locally and remote monitoring), train detection devices, trackside signals (bulbs and LED technology), roadside signals (flashing or static light), barriers (different types), audible alarm devices, radar obstacle detection (danger area warning system), 4-quad protection; fibre-optic speed indicators.

UPDATED

Powernetics Ltd

Jason Works, Clarence Street, Loughborough, LE11 1DX, UK
Tel: (+44 1509) 21 41 53   Fax: (+44 1509) 26 24 60
e-mail: jag@powernetics.co.uk
Web: www.powernetics.co.uk

Key personnel
Managing Director: Satish Chada
Financial Controller: Bob Lawson
Engineering Director: Nilesh Chouhan
Operations Manager: Konrad Chada
Business Development Manager: Aran Chada
Sales Manager: Jim Goddard
Purchasing Manager: Russel Roughton
QA Manager: Gordon Anderson
Production Manager: Andy Guigno

Products
Independent private limited company with 30 years' experience in design, manufacture, test, installation, commissioning and maintenance of auxiliary, standby/emergency AC/DC power supply systems for the rail industry in trackside, tunnel, switchroom, REB and trainborne environments.

System design and integration of UPS and associated switchgear/power monitoring within REBs and switchrooms for SSI, FSP and PSP applications.

RailPower UPS systems for signalling, radio, telecoms and level crossing applications.

UPDATED

Progressive Engineering (AUL) Ltd

Unit 5, Progressive Business Parts, Groby Road, Audenshaw, Manchester M34 5HT, UK
Tel: (+44 161) 371 04 40
Fax: (+44 161) 371 04 44
e-mail: johnw@progressive-eng.com
Web: http://www.progressive-eng.com

Key personnel
Managing Director: J Williams
Chairman: C Williams
Works Manager: P Moss

Products
Hot axlebox detection system.

Quest

Quest Corporation
12900 York Road, North Royalton, Ohio 44133, USA
Tel: (+1 440) 230 94 00
Fax: (+1 440) 582 77 65
e-mail: questcor@aol.com

Key personnel
President: Kurtis S Wetzel
Sales and Marketing Manager: Daniel J Donovan

Products
Hot bearing simulators; hot wheel simulators; hand-held line pressure testers; locomotive warning strobes; locomotive headlight/crossing light controls.

Railroad Signal International

15110 East Pine Street, Tulsa, Oklahoma 74116, USA
Tel: (+1 918) 234 15 22
Fax: (+1 918) 234 15 29
e-mail: railroadsignal01@aol.com
Web: http://www.railroadsignalinc.com

Key personnel
President: Eddie Burns

Products
Engineering and signal design, installation and 24-hour maintenance of signal systems, level crossing signals, turnkey capability, manufacture and distribution of signal parts and components; repair of printed circuit cards; remanufacture of switch/point machines, relays, gate mechanisms and other signal equipment, installation and maintenance of hot box detectors, bridge detectors, drag detectors and AEI reader systems, supply of train flashing rear end services.

UPDATED

Rails Company

101 Newark Way, Maplewood, New Jersey 07040-3393, USA
Tel: (+1 973) 763 43 20
Fax: (+1 973) 763 25 85
e-mail: rails@railsco.com
Web: http://www.railsco.com

Key personnel
President: G N Burwell
Vice-President: J Maldonado
Secretary/Treasurer: M Kinda
Sales: J Vertun, L Ford, D Burwell, P Hest

Products
Batteries for signalling and communications applications.

UPDATED

Safetran Systems Corporation

An Invensys company
2400 Nelson Miller Parkway, Louisville, Kentucky 40223, USA
Tel: (+1 800) 626 27 10
Fax: (+1 502) 244 74 44
Web: http://www.safetran.com

Key personnel
Vice-President, Marketing, Sales and Service: John J Paljug
General Manager, Marketing and Technology: John T Sharkey
Sales and Service Engineer: Bob Shaw

Products
Signalling, communications, warehouse services and maintenance-of-way equipment for railway and mass transit applications.

Products include: crossing warning products; power supplies; wayside signal products; freight yard products; communications services; maintenance-of-way transit systems housings and foundations.

Safetran's new wayside communications controller/field protocol device (WCC/FPD) maintenance utilities for Windows 98/NT are designed to be used by customers requiring maintenance and diagnostic access to their WCC/FPD equipment from local, remote, or dial-in Windows 98/NT workstations.

Sagem SA, Defence and Security

Le Ponant de Paris, 27 rue Leblanc, F-75512 Paris, Cedex, France
Tel: (+33 1) 40 70 63 63
Fax: (+33 1) 40 70 66 40
Web: http://www.sagem.com

Key personnel
President: Grégoire Olivier

Background
Part of Group Sagem.

Projects
Driver assistance systems: platform and train doorway monitoring, onboard repetition of wayside signalling, train detectors/treadles.

SchlumbergerSema Group

Rail Control Systems
143/149 Farringdon Road, London EC1R 3AD, UK
Tel: (+44 20) 78 30 44 44
Fax: (+44 20) 72 78 05 74

Affiliated and associated companies
Belgium
SchlumbergerSema
Rue de Stalle 96 Stallestraat, B-1180
Tel: (+32 2) 333 55 11
Fax: (+32 2) 333 55 22
e-mail: info-request@be.sema.com

SchlumbergerSema Global Services
Raketstraat 98 rue de la Fusee, B-1130
Tel: (+32 2) 724 92 90
Fax: (+32 2) 724 92 92
e-mail: request@sgs.be.sema.com

Brazil
SchlumbergerSema
Rua Alexandre Dumas, 1711, Bloco 12-1, andar Sao Paulo, CEP 04717004
Tel: (+55 11) 34 44 76 00

China
SchlumbergerSema
6th floor, Lido Office Tower, Lido Place, Jichang Road, Beijing 10004

France
SchlumbergerSema
50 avenue Jean Jaurés, BP 62012 F-92542 Montrouge
Tel: (+33 1) 46 00 66 67

Schlumberger Limited Paris
42, rue Saint Dominique, F-75007 Paris
Tel: (+33 1) 40 62 10 00

Netherlands
SchlumbergerSema Informatica
Van Houten Industriepark 11, PO Box 143, NL-1380, AC Weesp
Tel: (+31 294) 23 95 00
Fax: (+31 294) 23 95 01

Mexico
SchlumbergerSema
Ejercito Nacional 425, Colonia Granada, Delegacion, Miguel Hidalgo, CP 11520, Mexico DF
Tel: (+52 55) 52 63 30 00

Germany
Dreieich
Otto-Hahn Strasse 36, D-63303 Dreieich
Tel: (+49 69) 23 83 30 00

Italy
SchlumbergerSema Rome, Via Riccardo Morandi 36, I-00050
Tel: (+39 6) 83 07 42 01

Spain
SchlumbergerSema
Avinguda Diagonal 210-218, Barcelona E-08018
Tel: (+34 93) 486 18 18

SchlumbergerSema
Albarracin 25, E-28037 Madrid
Tel: (+34 914) 40 88 00

UK
Schlumberger Cambridge Research
High Cross, Madingly Road, Cambridge CB3 0EL
Tel: (+44 1223) 32 52 00

SchlumbergerSema
4 Triton Square, Regent's Place, London NW1 3HG
Tel: (+44 207) 830 44 47

Schlumberger House
Buckingham Gate, Gatwick Airport, West Sussex RH6 0NZ
Tel: (+44 1293) 55 66 55

USA
Schlumberger Austin Technology Centre
8311 Ranch Road 620 N, Austin, Texas 78726-4010

SchlumbergerSema
6399 South Fiddler's Green Circle, Suite 600, Greenwood Village, Colorado 80111
Tel: (+1 303) 741 84 00

SchlumbergeriCenter
1325 South Dairy Ashford, Houston, Texas 77077-2307
Tel: (+1 281) 285 13 00

Schlumberger Solutions Center
SchlumbergerSema
5599 San Felipe, Houston, Texas 77056-2724
Tel: (+1 713) 513 20 00

Schlumberger Data Management Center
(in the CMS Energy Building)
5444 Westheimer, Houston, Texas 77056
Tel: (+1 713) 513 20 00

Schlumberger Limited New York
153 E 53rd Street, 57th Floor, New York NY 10022
Tel: (+1 212) 350 94 00

Schlumberger Doll Research
36 Old Quarry Road, Ridgefield, Connecticut 06877-4108
Tel: (+1 203) 431 50 00

Schlumberger Reservoic Completic Center
14910 Airline Road, Rosharon, Texas 77583-1590
Tel: (+1 281) 285 52 00

Schlumberger Sugar Land Campus
The Forum
210 Schlumberger Drive, Sugar Land, Texas 77478
Tel: (+1 281) 285 85 00

Background
In 2001, Schlumberger Limited acquired Sema plc and combined it with part of its former transaction business and other acquisitions. The company is now called SchlumbergerSema and is one of two business segments of Schlumberger Limited.

Products
Design and supply of computer systems; electronic signalling control equipment; automatic route-setting systems; passenger information systems; rail control systems consultancy.

The main product supplied is the Integrated Electronic Control Centre (IECC). SchlumbergerSema Group developed the IECC and it is now the standard control system used for major resignalling schemes in the UK.

Sécheron SA

Routes des Moulières 5, CH-1217 Meyrin, Geneva, Switzerland
Tel: (+41 22) 739 41 11
Fax: (+41 22) 739 48 11
e-mail: info@secheron.com
Web: www.secheron.com

Key personnel
Chief Executive Officer: Claude Durand
PBU Components Director: Jo Murer
PBU Substations Director: Dominique Jamet
PBU Electronics Director: Peter Stauffer
Marketing Manager: Gilbert Lile
Corporate Sales Manager: Jimmy Cuche

Subsidiary company
Sécheron Tchéquie, spol. sro, Prague, Czech Republic

Products
Microprocessor-controlled Hasler TELOC® On-Train Monitoring and Recording (OTMR) systems, speed and distance measuring systems, Hasler optical pulse generators for axle or gearbox mounting, modular cab display systems for ATC and ATP applications.

Recent orders include the supply of Hasler TELOC® systems; ETCS SBB with ALSTOM; Attiko Greece with Melco; ETCS for Talgo 350 with Bombardier; ETCS for Korean National Railway.

UPDATED

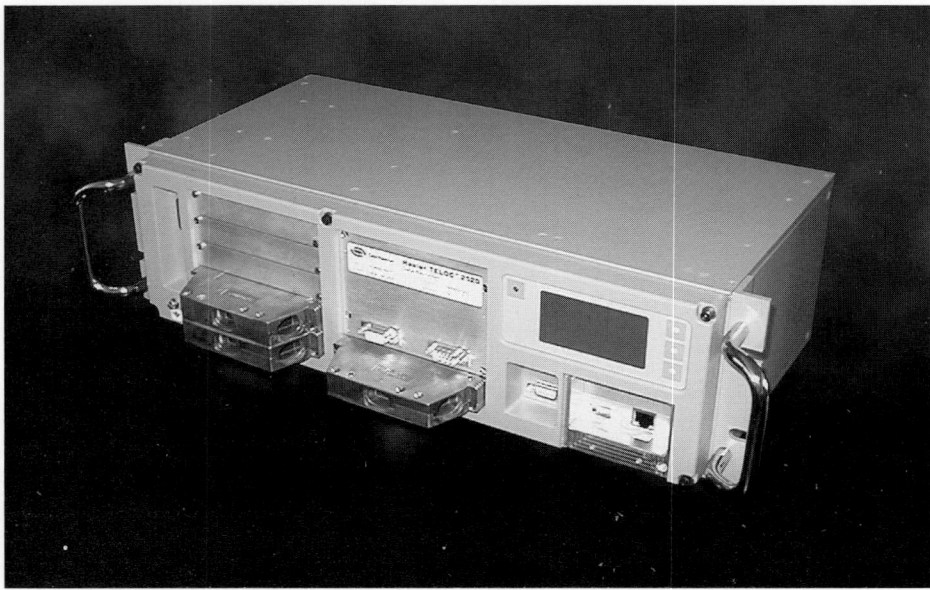

Sécheron Hasler TELOC® 2500 system   **NEW**/1136192

Selectra srl

Via delle Nazioni 5, I-44100 Ferrara, Italy
Tel: (+39 0532) 74 80 00
Fax: (+39 0532) 74 44 55
e-mail: info@selectra.org
Web: http://www.selectra.org

Products
OptoCross opto-electronic system for monitoring and controlling level crossing safety.

Sistemas Electrónicos de Potencia SA (SEPSA)

Albatros 7 and 9, (Pol Ind) La Estación, E-28320 Pinto, Madrid, Spain
Tel: (+34 91) 691 52 61
Fax: (+34 91) 691 39 77
Web: http://www.sepsa.es

Key personnel
For full personnel listing, see Sistemas Electrónicos de Potencia SA (SEPSA) entry in *Passenger coach equipment* section.

Background
SEPSA is a member of the Albatros Group (qv).

Products
Automatic Train Protection (ATP) systems (CESARES).

Siemens Switzerland

Siemens Switzerland Ltd, Transportation Systems
Industriestrasse   42,   CH-8304   Wallisellen, Switzerland
Tel: (+41 585) 58 55 85   Fax: (+41 585) 58 05 01
e-mail: ts@siemens.ch
Web: www.siemens.ch/ts

Key personnel
Managing Director: Willy Gehrer
Manager, Export Department, South East Asia: A Hefti

Products
Electronic, hybrid and relay interlockings; intermittent and continuous ATP and ATC systems, including vehicle and trackside equipment; CTC and electronic remote-control systems; block and track vacancy proving equipment (axle counters, track circuits, last vehicle detection); point locking equipment including point machines and point locking system; signals and indicators; safety relays.

Swiss Railways (SBB) has awarded Siemens a contract to build and operate a GSM-R (GSM-Railway) pilot system to provide mobile radio communication on trains. The first test journeys are to be carried out along the 36 km long pilot stretch between Zofingen (Canton Aargau) and Sempach (Canton Lucerne), where the GSM-R technology is being tested in accordance with the European EIRENE requirements. The aim of this pilot project is to test the suitability of the GSM-R system for nationwide implementation in Switzerland. SBB is the second European railway company to implement GSM-R into railway traffic with Siemens as technology supplier. In 1998 Siemens started to install a nationwide turnkey GSM-R system including switching systems and base stations in Sweden for the railway infrastructure company Banverket.

GSM-R was designed to meet the railway-specific communication requirements and is also planned as a communications platform for information services, in the first instance for train personnel and later to provide information for passengers. For safety reasons GSM-R works in a different frequency spectrum from traditional GSM.

UPDATED

Siemens AG

Transportation Systems
Rail Automation Division
PO Box 3327, D-38023 Braunschweig, Germany
Tel: (+49 531) 22 60   Fax: (+49 531) 226 64
e-mail: rail-automation@siemens.com
Web: www.siemens.com/transportation

Corporate headquarters
Siemens AG
Transportation Systems
PO Box 3240, D-91050 Erlangen, Germany
Tel: (+49 9131) 7-0

Key personnel
President, Transportation Systems Group: Hans M Schabert
Group Vice-Presidents: Alfred Frank; Joern F Sens; Friedrich Smaxwil
Heads of Rail Automation Division: Andreas Busemann; Dr Roland Alter

Products and services
Major product categories include:
Signalling: relay-based and electronic interlockings; radio-based train control systems; intermittent and continuous Automatic Train Control (ATC) and Automatic Train Protection (ATP) systems; hump yard systems; level crossing systems; and signalling components.

Operations control systems: centralised train control systems; train describer systems; control centres; Man-Machine Interface (HMI) systems; and marshalling yard equipment.

Rail communications: ACCS control centre software; transmission technologies (OTN, ATM, SDH, SONET, IP); CCTV surveillance; public address and passenger information; clock systems; help point and information systems; PABX; SCADA; emergency and security systems; GSM-R infrastructure; cab radio; analogue and digital radio systems; systems integration; project management; RAMS.

Major service categories include:
Simulation of railway operation; optimised operational planning and control; training; project management; life-cycle costing; systems integration; testing before commissioning using the Braunschweig Test Center; installation, commissioning and maintenance; and financing.

Contracts
Recent or current contracts include:
Belgium: In a consortium with ALSTOM Transport Siemens was awarded a contract in November 2003 by Belgian National Railways (SNCB) to provide new electronic control centres and signalling equipment. Siemens is to supply its Rail Automation Technology control room equipment and is responsible for project management of the contract.

Canada: In June 2002 Siemens received an order from Canadian National for an operations control system to handle and monitor all operations on the company's 12,000 km network. Equipped with 46 operator consoles and located in Montreal, the system was due to be commissioned by the end of 2005.

China: In November 2002 a consortium of Siemens and the Nanjing Research Institute of Electronic Technology was awarded a contract to provide signalling and control equipment for Nanjing's first metro line. Scheduled to be inaugurated in mid-2005, the 39 km 16-station line is the first element of a planned six-line, 139 track-km metro and light rail network. Signalling equipment for the first line will feature Sicaselectronic interlocking equipment and the LZB 700 M automatic train control system.

In August 2002 Siemens was awarded a contract to provide the signalling and train control systems for the first two metro lines to be built in Shenzen. Covering 20 km and 18 stations, the two lines are due to be commissioned in 2005. The contract includes the supply of Sicas electronic interlocking equipment and the LZB 700 M automatic train control system.

In August 2001 Siemens secured a contract from KCRC, Hong Kong, to supply and install signalling and control equipment and passenger information systems for 4 km of new line and the existing 30 km network of the Tuen Mun light rail system.

Also in August 2001 Siemens won an order to supply a Vicos operations control system, Sicas electronic interlocking and ZUB 200 intermittent ATC system for the new 17 km Xinmin Line LRT system in southern Shanghai. Trial operations began in 2003.

In October 2000 Siemens was awarded a contract by Guangzhou Metro Corporation and Chinese Railways to plan, supply and commission a complete signalling and operations control system for Line 2 of the Guangzhou metro. The 23 km 20-station line was commissioned in 2004.

France: In February 2004 Siemens was awarded a contract by the Paris public transport authority, RATP, to supply train control and route protection systems for Paris Métro Lines 3, 5, 9, 10 and 12. This covers technology for transmitting data between trackside and trains for all five lines, onboard equipment for trains on Lines 3, 10 and 12 (112 Type MF67 metro cars) and trackside equipment for Lines 5 and 9 (34.2 km). Phased completion is to extend from 2007 to 2009.

In December 2003 an extension to the fully automated Paris Métro Line 14 between Bibliothèque François Mitterand and Gare Saint Lazare was commissioned, featuring train control, signalling and communications systems supplied by Siemens. A southern extension of the line to the future Olympiades station is planned for commissioning in 2007.

Germany: In June 2004 the 15 km rail link to Cologne-Bonn airport was inaugurated. As well as supplying significant aspects of the signalling and safety system, including an electronic interlocking, Siemens was responsible for provision of a tunnel radio system, the first such application for DB, a GSM-R-based train radio system and a cable-bound railway telephone system. Contracts also covered a passenger information system and station and tunnel fire alarm systems.

In February 2002 Siemens was awarded a contract by German Rail (DB AG) to supply and install a new interlocking system at Braunschweig main station. To be controlled from an operations centre in Hanover, the Type EI S interlocking will be based on Siemens' Simis C microcomputer technology. Work involves renovation of 294 signals and 78 switches, as well as provision of derailers and track vacancy detection sections.

In January 2002 DB AG placed a contract with Siemens to upgrade the main interlocking at Frankfurt. The project also features the use of a Type EI S interlocking and Simis computers, and includes renovation of 315 signals, 331 switches, 41 electronically decentralised switches and 306 track vacancy detection sections.

In March 2004 Siemens completed a contract placed in 2001 to modernise the signalling system at Magdeburg, providing Simis C interlockings. The project included provision of three sub-centres

SICAS computer cabinet for the Cologne–Mülheim interlocking 0525407

Desiro railcar equipped as a demonstrator for Siemens' Trainguard ETCS family **NEW**/0585295

Control centre by Siemens for Hanover's mass transit system **NEW**/0585222

Control centre by Siemens for the Houston light rail system   NEW/0585223

and eight decentralised control computers, and the modernisation of 441 signals and 312 switches and derailers.

Greece: In October 2003, as a member of a consortium that also includes Greek construction company Aktor ATE and the German track construction company H F Wiebe GmbH & Co KG, Siemens was awarded a contract by ERGA OSE to supply electronic interlockings, an operations control system, switch mechanisms and axle counters for a new 110 km double-track line linking Athens with Kiata, near Corinth. The contract also covered the supply and installation of trackside components for the ETCS (Level 1) train control system.

Hungary: In January 2005 Siemens was awarded a contract by Budapest's transit authorities (BKV Rt) for the supply of a new control, singalling and safety system to the M2 metro line.

India: In September 2004 Siemens received an order from Delhi Metro Rail Corporation to supply signalling and communications systems for its 23.5 km Line 3. Equipment to be supplied includes Vicos OC 100 and 501 operations control systems, Sicas ECC electronic interlockings, LZB 700 continuous train control system and S 700 K point machines. Revenue services on the line are planned to start in early 2006.

Malaysia: In February 2001 Siemens received an order from Mitsui, acting as general contractor, to supply signalling and train control equipment for the upgraded line between Rawang and Ipoh. Siemens was to be responsible for system engineering, supplying and commissioning the CTC system, signalling and point operating equipment.

Netherlands: In 2002 a general agreement was signed with Railinfrabeheer covering the provision of its Simis W electronic interlocking technology to upgrade Elektronische Beveiliging Siemens (EBS) interlockings on the Netherlands rail network. A first phase of the project covers several interlockings on the Vleugel route, south of Utrecht, where work is due to be completed by 2011.

Portugal: As part of a turnkey contract signed in August 2002 to establish a 13 km light rail system between Almada and Seixal, south of Lisbon, Siemens is to supply signalling and operations control equipment.

Romania: In January 2005 Siemens received an order from Romanian National Railway, CFR for the supply of type Simis W electronic interlockings. These interlockings are intended for seven stations to the northwest of Bucharest.

Spain: In March 2003 Siemens, in a consortium with Dimetronic SA, won a contract to equip Barcelona's automatic driverless Metro Line 9 (41.4 km) with a train protection and operations control system. The entire line is due to be commissioned by 2008.

UK: The Dorset Coast resignalling scheme was completed by Siemens in December 2003. Covering 26 route-km, the system employs SIMIS W interlocking, the Vicos OC111 train control system and AzSM(E) axle-counters.

USA: In July 2000 Siemens won an order from MTA, New York, to upgrade and expand the communications system used by New York City Transit. A goal of the project is to permit an unrestricted exchange of information among the 188 stations and three traffic control centres of the NYCT system. Additional functions for voice and data communications and a CCTV system were due to be added at later stages of the project. The contract also included maintenance of the system for five years from its commissioning in April 2004.

Other orders for signalling technology form part of turnkey contracts, detailed in the Turnkey systems contractors section.

UPDATED

Signal & Systemn Technik (SST)

Halsschlag 12, D-56427 Siershahn, Germany
Tel: (+49 2623) 608 60
Fax: (+49 2623) 60 86 60
e-mail: mail@sst.ag
Web: http://www. Sst.ag

Products
Hot box and hot wheel detectors, flat wheel detectors, run-time monitoring of pantographs, car clearance detection, wind and airflow monitoring, weather monitoring stations, rock fall, vehicle fall and landslide detection.

Services
Network integration and remote maintenance.

Contracts
Deutsche Bahn; AVE, Spain; Rheinbraun; Austrian Federal Railways; Spanish National Railways RENFE; Swiss Federal Railways SBB, CFF, FFS; French National Railways SNFC; Nederlandse Spoorwegen; State Rail Authority of New South Wales; Ferrovie; RailTrack.

Signal House Ltd

Signal House, Cherrycourt Way, Stanbridge Road, Leighton Buzzard LU7 4UH, UK
Tel: (+44 1525) 37 74 77   Fax: (+44 1525) 85 09 99
e-mail: sales@signalhouse.demon.co.uk
Web: www.collis.co.uk

Key personnel
Managing Director: Peter Roberts
Operations Director: P Hobbs
Technical Director: J Wareing
Senior Sales Engineer: D Farrington
Sales and Marketing Manager: D Eades

Associate company
Collis Engineering Ltd

Products
Conventional and LED colour light signals; fibre optic signalling indicators; trackside enclosures; Planlite 5000 series switch heaters; oil and electric lamps for signalling and other application; terminal blocks and fuse holders.

UPDATED

Specialty Bulb Co Inc

The Specialty Bulb Co Inc
80 Orville Grove, Bohemia, New York 11716-0231, USA
Tel: (+1 631) 589 33 93   Fax: (+1 631) 563 30 89
e-mail: info@bulbspecialists.com
Web: http://www.bulbspecialists.com

Key personnel
President: Judith Beja
Vice-President, Technical: Caden Zollo
Vice-President, Sales and Marketing:
  Edie Muldoon

Products
Lamps for signal applications.

Springboard Wireless Networks Inc

5100 Orbitor Drive, Mississauga, Ontario L4W 4Z4, Canada
Tel: (+1 905) 238 52 55   Fax: (+1 905) 212 20 04
e-mail: info@springboardwireless.com
Web: http://www.springboardwireless.com

Key personnel
President, Chief Executive Officer: Herman Chang
Vice President Marketing and Sales: Dave Fisher

Products
Communications-based Train Control (CBTC). Springboard can provide train control suppliers and rail operators with a one-stop shopping point for data communications requirements – from R & D to system engineering to installation. Expertise in designing custom radio frequency (RF) modems, routers and protocol converters.

Wireless solutions for rail applications include code line replacement, highway grade crossings and positive train separation. The RailPath system is a wireless network that has been specifically developed for CBTC applications. RailPath has been designed to allow for compatibility with the world's leading train control systems. Springboard's RailPath was field tested in CBTC demonstrations on New York City Transit's Culver Line (January to July 1999). For these NYCT demonstrations, Springboard was partnered with Alcatel Transport Automation, ALSTOM Signaling and Matra Transport International/Union Switch & Signal.

The company also provides integrated radio-dispatch solutions for major rail clients such as CN, CP, and TTC. InterTalk, the company's standards based communication control system, integrates wide area mobile radio and telephone networks.

Stein GmbH

Stahlgruberring 36, D-81829 Munich, Germany
Tel: (+49 89) 427 19 00
Fax: (+44 89) 427 190 19
e-mail: office@steingmbh.com
Web: http://www.steingmbh.com

Products

Bi-control remote radio control system for locomotives. bi-control employs a two-computer system, which monitors itself automatically during operation and processes all commands in fail-safe mode. Up to 23 radio remote controls may work in the same area on one frequency. The system is tested and approved by TÜV Rail and the Eisenbahn-Bundesamt in Germany.

VERIFIED

STERIA Group

46 rue Camille Desmoulins, F-92782 Issy les Moulineaux, France
Tel: (+33 1) 34 38 60 00   Fax: (+33 1) 34 88 60 15
Web: www.steria.fr

Key personnel

Railway Department Manager: Jean Charles Tarlier
Transportation Sales Manager: Jacques Lafay
Transportation Communication Representative:
   François Lemoine Ninet

Products

Traffic regulation and monitoring systems, road pricing and supervisory systems, railway supervisory and monitoring systems; passenger information systems, data processing, management control, training simulators and programmes, computer-aided diagnosis, traffic simulation; ticketing systems.

Contracts

Supervisory Control System/OCC for RATP: metro lines 1,2,3,5,6,7,8,9,10,11,12,13; Paris Rapid Mass tTransit (RER A & B); Lyons Metro – Line's A,B,C; CFF/SBB Swiss Federal Railways. Railway equipment supervision for Lei T Fü Deutsche Bahn (Germany).

UPDATED

STS Signals

Doulton Road, Cradley Heath B64 5QB, UK
Tel: (+44 1384) 85 85 21
Fax: (+44 1384) 56 77 10
e-mail: signals@sts-international.co.uk
Web: www.sts-switchgear.com

Key personnel

Chairman: D Miller
Business Manager: T Warrington
Commercial Manager: W Thompson

Products

Signalling equipment ground-based, including miniature signalling relays, shelf relays, colour light signals, key token instruments, circuit controllers and electric lever locks.
   For onboard use, the equipment range covers TPWS and AWS systems, driver safety pedals and wiring assemblies.

UPDATED

TagMaster AB

Electrum 410, SE-164 40 Kista, Sweden
Tel: (+46 8) 632 19 50   Fax: (+46 8) 750 53 62
e-mail: info@tagmaster.se
Web: http://www.tagmaster.se

Key personnel

President: Magnus Rehn
Vice-Presidents
   Marketing and Sales: Christian Nordberg
   Information Solutions/Research and Development:
      Mikael Willgert
   Production: Peter Beijar
   Finance and Administration: Marika Falk

Products

Long-range Radio Frequency Identification (RFID) systems for rail and public transport applications, providing position and identification data. System components comprise fixed ID programmable identification tags and the TagMaster Reader, which reads tag information and offers various interfaces for onward data transmission. Systems can be configured with either tags or readers on moving vehicles according to data requirements. Applications include maintenance and operational or quality control, high-precision event triggering and traffic and passenger information.

Contracts

Users include Netherlands Railways, measuring wheel quality and axleloads, London Underground, enabling an onboard radio system automatically to change talk-group when moving between different areas of coverage, and the Pearl Line light rail system, using TagMaster RFID technology for train identification for traffic and passenger information purposes.

Teknis Electronics Pty Ltd

258 Halifax Street, Adelaide, South Australia 5000, Australia
Tel: (+61 8) 82 23 54 11
Fax: (+61 8) 82 23 54 99
e-mail: info@teknis.net
Web: http://www.teknis.net

Key personnel

Managing Director: K Bladon

Products

Electronic control systems, SCADA systems. DDU (Driver Display Unit), as in-cab train driver's display for signalling and train orders. WCM (Wheel Condition Monitor) for analysing wheel condition and detecting wheel faults causing damage to track structure. In-motion train weighing at line speed. WMS (Wayside Monitoring System) integrating the WCM and other industry standard sensors and instrumentation into a single data/communications system.

Telephonics Corporation

815 Broadhollow Road, Farmingdale, New York 11735, USA
Tel: (+1 516) 549 60 62
Fax: (+1 516) 549 60 18

Key personnel

Vice-President, Business Development:
   Philip Greco
Manager, Business Development: Norbert Trokki

Products

Communication, information and surveillance systems for mass transit: integrated car communication and passenger information systems; wayside communication and central control systems; closed circuit television systems for wayside, stations and vehicles; train line multiplexers; network controller and vehicle monitoring.

Contracts

Vehicle communication, train multiplexer, network controller for New York City Transit subway cars.
   Vehicle communication, radios and door observation CCTV for South Eastern Pennsylvania Transit Authority.
   Vehicle communication system for Massachusetts Bay Transit Authority light rail.
   Vehicle communication, health monitoring and vehicle CCTV at Hudson Bergen light rail transit system.
   Integrated wayside communication system for Newark APM.
   Vehicle communication and passenger entertainment for Caltrans.

Developments

Train line multiplexer for transport of vehicle controls and digital audio using EI standards; car network controller is used to convert LON to propulsion/brake commands; public address and information systems based on internet technology.

Thales Communications Ltd

Newton Road, Crawley RH10 9TS, UK
Tel: (+44 1293) 51 88 55   Fax: (+44 1293) 41 64 00
e-mail: kenm@rmel.com

Key personnel

Transport Sales Manager: Ken McFarland

Background

Thales Communications Ltd was formerly Redifon MEL.

Products

Platform door interface to ensure platform door and train door synchronisation; TPWS, preventing trains going through danger signals; track-to-train, data communications, solutions in a range of technologies; TramCom, a two-way data communication system between trams and trackside equipment.

Thales Telecom Services

Phoenix House, Station Hill, Reading RG1 1NB, UK
Tel: (+44 118) 908 60 00
Fax: (+44 118) 908 63 85
Web: http://www.thales-ts.com

Key personnel

Chief Executive Officer: Peter Batley
Finance Director: Stuart Palmer
Sales, Marketing and Strategy Director:
   Peter Saunders

Background

Thales Telecom Services was formed in April 2002 by the merger of its project and maintenance organisations, Thales Translink and Thales Fieldforce. The company is part of the Thales Group formed by the merger of Thomson-CSF and Racal. Formerly trading as Racal Telecom, the company's origins lie in its acquisition of the former BRT private telecommunications network during the privatisation of British Rail in the 1990s.

Products

The business of Thales Telecom Services principally consists of the provision of specialist operational and business telecommunications services to the UK railway industry. Its core network comprised 17,000 km of trunk cable, of which 4,000 route-km are made up of fibre-optic cable. During 1997 the former Racal upgraded the network by creating a resilient SDH backbone network, which includes a further 1,200 route-km of fibre-optic cable. Facilities provided include national fixed-line telecommunications systems in support of trackside telephony, signalling control systems, CCTV and mainframe computer connections. The company also provides facilities management for Network Rail's National Radio Network, which provides driver to shore communications.
   Thales Telecom Services also installs and maintains telecommunications systems. The comprehensive service includes national cable and fibre networks, PABXs and call routing equipment, radio and VSAT systems, customer information systems, customer premises equipment, local and wide area networks, transmission systems and structured cabling. The company offers a nationwide service with support on call 24 hours a day, seven days a week. The company also has service partner status with Ericsson.

Contracts

Thales Telecom Services is the leading partner in CityLink Telecommunications Ltd, a special purpose company formed to operate, maintain and renew London Underground's operational radio and transmission systems. The 20-year contract has been let on a Private Finance Initiative (PFI) basis. Its contribution to the project includes the design, installation, testing and commissioning of a new network-wide SDH transmission system with CCTV, and to manage the implementation of a new digital Tetra radio system. Also for London Underground, Thales Telecom Services developed in-cab CCTV, a system designed to enhance the safety of passengers on station platforms by enabling drivers to view the whole platform clearly before arrival, while the train berths and on departure, all from within the cab.

Thoreb AB

Gruvgatan 37, SE-421 30 Västra Frölunda, Sweden
Tel: (+46 31) 734 39 00   Fax: (+46 31) 734 39 10
e-mail: info@thoreb.se
Web: www.thoreb.se;
      www.true-realtime.com

Key personnel
Managing Director: Michael Sigvardsson
Marketing Directors: Johan Kallvik; Esbjorn Lif
Chief Financial Officer: Thord Brynielsson

Products
Automatic vehicle location, passenger counting, data communication, passenger information displays, communication systems for public transport vehicles and multiplex electrical systems for heavy vehicles. Real time passenger information systems.

NEW ENTRY

Tiefenbach GmbH

PO Box 91 13 60, D-45538 Sprockhoevel, Germany
Tel: (+49 2324) 70 54   Fax: (+49 2324) 70 51 14
e-mail: info@tiefenbach.de
Web: www.tiefenbach.de

Key personnel
Managing Director: Peter Jochums
Sales Manager: Juergen Burghoff
Technical Manager: Walter Pyschny

Background
Tiefenbach is a member of the Hauhinco Group.

Subsidiary company
Tiefenbach GmbH
Transportation Technology
1325 Evans City Road, Evans City, PA 16033, USA
Tel: (+1 724) 789 70 50
Fax: (+1 724) 789 70 56
Web: www.tiefenbachgmbh.com

Products
Point control equipment: complete control equipment for automation and signalling of depots and marshalling yards; axle counter systems; track vacancy detection and level crossing switching with axle counter systems based on Tiefenbach's double wheel sensor.

A recently introduced product is the train positioning and information system, ALOIS, developed jointly with the Universität der Bundeswehr, Munich (AGIS/ikv). Intended to maximise the efficiency and traction and rolling stock utilisation of both marshalling yards and regional rail systems, ALOIS employs GPS technology to provide locomotive or train position data. This is transmitted to a central control unit to meet operational train control, data management and operations archiving requirements. Users of the system include BASF at its Ludwigshafen facility and German regional operator EVB.

Contracts
Complete systems design and supply of depot signalling works at Cottbus (DB AG), Cologne

(DB AG), Antwerp (SNCB), marshalling yard control and signalling works at Kassel (DB AG), Hamm (DB AG), Gdansk, Poland; complete systems design and supply of axle counter systems for SOB-Südostbahn, Switzerland (main line); AKN Eisenbahn AG suburban feeder service at Hamburg, Germany. Level crossing control system at Ford Motor Company, Detroit. Axle counter system, Duluth USA; Constant warning time systems at Detroit for Conrail (main line).

UPDATED

Toshiba Corporation

Railway Projects Department
Toshiba Building, 1-1 Shibaura 1-chome, Minato-ku, Tokyo 105-8001, Japan
Tel: (+81 3) 34 57 49 24
Fax: (+81 3) 54 44 92 63
Web: www.toshiba.co.jp

Key personnel
President and Chief Executive Officer:
   Atsutoshi Nishida
Vice-President, Transportation Systems Division:
   Takio Ooyama
Senior Manager, Railway Projects Dept: Koji Toda

Products
Automatic train control equipment; automatic train stop equipment; electrical indicating and train describing equipment; centralised traffic control and remote-control systems; marshalling yard equipment, including retarders.

UPDATED

TransCore

19111 Dallas Parkway, Suite 300, Dallas, Texas 75287-3106, USA
Tel: (+1 972) 397 81 97   Fax: (+1 972) 733 64 86
Web: www.transcore.com

Key personnel
President and Chief Executive Officer:
   John Worthington
Chief Operating Officer: John Simler
Executive Vice-President, Marketing: Dick Blackwell
Executive Vice-President: John Foote
Executive Vice-President and Chief Technical
   Officer: Kelly Gravelle
Chief Scientist and Executive Vice-President:
   Dr Jerry Landt

Products
Amtech® brand radio frequency technology for automatic train control. Automatic equipment identification; automatic equipment monitoring systems; automatic train positioning systems; automatic train location; automatic train separation. Activation of audio and visual annunciation.

UPDATED

Transmitton Ltd

Ashby Park, Ashby-de-la-Zouch LE65 1JD, UK
Tel: (+44 1530) 25 80 00   Fax: (+44 1530) 25 80 08
e-mail: sales@transmitton.co.uk
Web: www.transmitton.co.uk

Key personnel
Managing Director: I Wright
Sales Director: E Turnock
Financial Director: J Blackwell

Products
Transmitton provides a range of integrated control and asset management solutions to the rail industry. These include; integrated station management systems, network management systems, real-time passenger information systems, remote condition monitoring, traction power SCADA, asset management systems.

The fastflex Remote Terminal Unit (RTU) is modular in design and suits a wide range of applications for stand alone control, point-to-point and remote communications. Transmitton's cromos software offers real-time control and asset management technologies. Using the core technologies, Transmitton has successfully implemented a wide range of solutions within the rail market including: Integrated Station Management (ISM), Network Management Systems (NMS), Real-time Passenger Information Systems (RTPIS), Remote Condition Monitoring (RCM) and Traction Power SCADA (Supervisory Control and Data Acquisition).

ADT Rail Systems is a Transmitton company which specialises in the specification, design and manufacture of these systems and provides performance data through its on track integrated suite of monitoring products.

UPDATED

Transportation Products Sales Co Inc

618 Cepi Drive, Suite B, Chesterfields, Missouri 63005, US
Tel: (+1 314) 532 11 44
Fax: (+1 314) 532 14 82
e-mail: tpscarms@tpsc-arms.com
Web: www.tpscarms@tpsc-arms.com

Key personnel
President: Walter J Winzen
Vice-President Operations, Director of Engineering:
   Sid Bakker
Vice-President of Sales: Clay Gillette

Products
Batteries by GNB Technologies for back-up duties in signalling and telecommunications systems. DC power systems for rail electronics.

UPDATED

Trivector System AB

Åldermansgatan 13, SE-227 64 Lund, Sweden
Tel: (+46 42) 38 65 00
Fax: (+46 42) 38 65 25
e-mail: info@trivector.se

Key personnel
Managing Director: Klas Odelid
Marketing Manager: Ola Fogelberg
Project Manager: Anders Månsson

Products
RASC (RAdio Signal Control) system for assigning priority to buses at crossings regulated by signals. This system consists of the software contained in Triveco 8 together with the component system IRU-SS, an intelligent receiver unit contained in the control apparatus. It has been installed at 40 intersections in Malmö, Sweden and is to be installed in the city of Karlstad in the future.

UniControls AS

Křenická ulice-2257, CZ-100 00 Prague 10, Czech Republic
Tel: (+420 2) 72 01 14 11
Fax: (+420 2) 74 81 44 75
e-mail: unic@unicontrols.cz
Web: www.unicontrols.cz

Key personnel
Marketing Manager: Marian Belosovic

Products
Communications and control systems for trains and rail vehicles, including: Train Communication Network-Wire Train Bus (TCN-WTB) equipment; multi-vehicle communications equipment; vehicle communications devices; and drivers cab equipment and displays.

Contracts

Train communications and control systems for refurbished Russian-built Prague metro stock; train communications system for St Petersburg metro stock; train and vehicle communications system and drivers cab equipment for Czech Railways Class 471 emus; automatic train control system for Czech Railways Class 680 tilting trainsets; drivers cab equipment for refurbished Class 772 locomotive for Slovakian Railways (ZSSK).

UPDATED

Universal Power Systems

Weldon Road, Loughborough LE11 5RN, UK
Tel: (+44 1509) 26 11 00
Fax: (+44 1509) 26 11 48
e-mail: sales@upsltd.com
Web: http://www.upsltd.co.uk

Key personnel

General Manager, Rail Division: Chris Smith

Products

Uninterruptible power supply systems.

Union Switch & Signal Asia Pacific Region

An Ansaldo Signal company, controlled by Finmeccanica
Asia Pacific Regional Headquarters
39 Harvey St North, Eagle Farm, Brisbane, Queensland 4009, Australia
Tel: (+61 7) 38 68 93 33
Fax: (+61 7) 32 68 22 19
e-mail: info@ansaldo-signal.com.au

Key personnel

Chief Executive Officer, Asia Pacific Region:
    Lyle Jackson
General Manager, Australia: Graham Russell
General Manager, Malaysia: Peter Costello
General Manager, India: S Lahiri
Director of Sales and Marketing, Asia Pacific
    Region: Martin Sossay Raj
Director of Finance, Asia Pacific Region:
    Miranda Mathews
Director of Safety and Quality, Asia Pacific Region:
    Les Brearley

Regional offices

Union Switch & Signal Pty Ltd
Brisbane: Tel: (+61 7) 38 68 93 33
    Fax: (+61 7) 32 68 22 19
Sydney: Tel: (+61 2) 98 79 37 77
    Fax: (+61 2) 98 79 61 55
Newcastle: Tel: (+61 2) 49 62 95 60
    Fax: (+61 2) 49 62 95 42
Perth: Tel: (+61 8) 92 56 00 00
    Fax: (+61 8) 92 56 11 99
Karratha: Tel: (+61 8) 91 44 13 12
    Fax: (+61 8) 91 44 11 27
    e-mail: info@ansaldo-signal.com.au

Union Switch & Signal Malaysia Sdn Bhd
Kuala Lumpur: Tel: (+60 3) 40 45 80 55
    Fax: (+60 3) 40 45 89 90
    e-mail: info@ansaldo-signal.com.au

Union Switch & Signal India Pvt Ltd
Bangalore: Tel: (+91 80) 552 57 37
    Fax: (+91 80) 552 57 32
Calcutta: Tel: (+91 33) 22 83 60 31
    Fax: (+91 33) 22 83 60 31
    e-mail: info@ansaldo-signal.com.au

Products

Design, engineering and manufacture of signalling, communications, control and automation systems for railway and mass transit industries. US&S has a product portfolio that includes the most advanced capabilities in high-speed lines, ERTMS/ETCS, computer based interlocking, automatic block system, automatic train control,

operation and traffic control and wayside signalling equipment.

Contracts

Australia: Processor-based interlocking project, advanced train communications, Perth suburban Automatic Train Protection (ATP) extension, Westrail; West Angelas rail project, West Angelas voice radio communications, North Ltd; system expansions for heavy rail system in the Pilbara (Western Australia), Pilbara Rail. Signalling and ATP systems, South West Metro, Perth; safeworking systems, Botswana Railways.
India: Bangalore Cantonment Station, South Western Railways; ERTMS Southern sector, Indian Railways; 29 stations Processor Based Interlocking (PBI), South Eastern Central Railways; 14 stations PBI, Southern Railways, 37 stations PBI; 16 stations PBI, Bangladesh Railways.
Malaysia: Signalling, ATP and train control system and communications and SCADA system for Kuala Lumpur Monorail, Malaysian Monorail Technology; Rawang Ipoh double track project, KTM Berhad; signalling and ATP system for rail link Port Klang to Westport on Pulau Indah, KTM Berhad, signalling and ATP systems for Kuala Lumpur Central station, KTM Berhad.

NEW ENTRY

Union Switch & Signal Inc (US&S)

An Ansaldo Signal NV company
Systems & Research Center
1000 Technology Drive, Pittsburgh, Pennsylvania 15219-3120 USA
Tel: (+1 412) 688 24 00   Fax: (+1 412) 688 23 99
e-mail: info@switch.com
Web: http://www.switch.com

Key personnel

President and Chief Executive Officer: Ken Burk
Vice-President, Marketing and Product
    Development: Jeremy Hill
Director, International Marketing: Dan Antonucci
Director, Railroad Sales & Customer Service:
    George Rudge
Director, Transit Systems Sales: John Fisher

Works

645 Russell Street, Batesburg, South Carolina 29006, USA
Tel: (+1 803) 532 44 32   Fax: (+1 803) 532 29 40

Other offices

Union Switch & Signal International Co
Rousseau No 14-3er Piso, Col Nueva Anzures, Mexico DF 11590, Mexico
Tel: (+52 5) 254 18 17   Fax: (+52 5) 254 42 91
Managing Director: Lorenzo Simonini

US&S International Project Co
Hungguk Life Building, 5th Floor, 6-7 Su Nae Dong, Pundang, Korea
Tel: (+82 31) 719 07 61; 2
Fax: (+82 31) 719 07 63
Acting Managing Director: YY Wang

Products

Design and manufacture of signalling and train control systems for passenger, freight, high-speed and mass transit railways including: vehicle-mounted signalling, positioning and monitoring systems; marshalling yard components and management systems; automation and integration technologies for dedicated and multimodal traffic management.

Vialis NMA

Vialis NMA Railsystems bv
PO Box 318, NL-3700 AH Zeist, Netherlands
Tel: (+31 30) 698 38 00   Fax: (+31 30) 698 38 09
e-mail: info@vialis.nl
Web: http://www.vialis.nl

Vialis NMA Railway Signalling bv
PO Box 70, NL-1723 ZH Noord-Scharwoude, Netherlands
Tel: (+31 226) 33 67 00   Fax: (+31 226) 33 67 12
e-mail: sales@nmarail.nl
Web: http://www.vialis.nl

Background

NMA Railsystems bv and its sister company Vialis NMA Railway Signalling bv are subsidiaries of Vialis, part of the Koninklijke Volker Wessels Stevin nv (KVWS) group.

Products

Vialis NMA Railsystems bv develops, assembles and overhauls power systems and signalling components for level crossing, signalling and switch systems and also develops and implements entire systems. Vialis NMA Railway Signalling manufactures electrical components for signalling systems, including signals, switch components and level crossing protection systems.

Recent product developments include: the application of LED technology to railway signals; the introduction of a maintenance-free, vandal-resistant level crossing barrier mechanism; a monitoring system to detect faults in level crossing half-barriers; an in-sleeper low-maintenance point switch system; and an integrated safety system incorporating axle-counters, automatic vehicle identification and other detection systems to enable trains of different types, including LRVs, to utilise the same sections of track.

Vossloh Information Technologies Malmö AB

Emilstorpsgatan 23-25, PO Box 16108, SE-200 25 Malmö, Sweden
Tel: (+46 40) 671 65 00   Fax: (+46 40) 671 65 99
e-mail: info.malmoe@vitm.vossloh.com
Web: www.vit.vossloh.com

Key personnel

Managing Director: Torsten Bergh
Board of Directors: Reinhold Hundt,
    Désirée Gruchot, Heinz-Holger Neustock,
    Machiel Baaijens
Head of Marketing: Dr Marion Behr

Background

Vossloh Information Technologies Malmö AB is a subsidiary of Vossloh Information Technologies GmbH, Kiel, Germany, which is part of the Vossloh Group.

Products

Electronic interlocking system for small and medium-sized stations (Alister): Alister is a vital interlocking system based on industrial components. It has been developed in full compliance with CENELEC standards and fulfils the requirements of safety level SIL 4. As standard industrial components are used, life cycle costs can be kept low and compatible replacement parts are available.

Electronic interlocking system for marshalling yards (AlisterCargo): AlisterCargo is an electronic interlocking system based on industrial components and designed especially for marshalling yards, sidings and industrial railways. The flexibility of its standard industrial components, which are already in used in many different industrial applications, ensures easy modification and extension of the system's functions. It can be connected to existing signals and operations monitoring and control systems through standard interfaces.

UPDATED

Wabtec Railway Electronics

21200 Dorsey Mill Road, Germantown, Maryland 20876, USA
Tel: (+1 301) 515 20 00   Fax: (+1 301) 515 21 00
Web: http://www.wabtec.com

M6W Locomotive event recorder with wireless and memory card download port 0134169

Key personnel

President: Robert Haag
Vice-President Marketing and Sales:
　MarkT Kramer

Background

Wabtec Railway Electronics formerly Pulse Electronics, is a division of Wabtec Corporation. WRE was founded in 1977 and its first product was the locomotive event recorder, Train Trax, which is currently installed worldwide and is a standard on many US railroads.

Products

Train Trax solid-state locomotive even recording system, a flexible, virtually maintenance-free, fully FRA-compliant recording system designed to improve operational safety and performance by monitoring and recording key channels from the locomotive; TrainLink II End of Train Telemetry System, which continuously monitors key conditions on the last car including brake pipe pressure, end of train motion, battery condition and marker light status and communicate the information to the lead locomotive while adding end of train emergency braking capability; locomotive speed indicators with both MPH and KPH models; Iso-Amp Speed Isolation Amplifier an alternative to the Axle Generator or passive sensors mounted on the locomotive drive gear; Train Sentry III Engineman Alertness Device, a solid-state system designed to enhance the safe operation of railway vehicles by monitoring the alertness of the engineman; FuelLink fuel measurement system, which incorporates solid-state electronics and advanced pneumatics for determining locomotive fuel levels accurate to +/– one per cent; Q-Tron's QEG 1000 Electronic autostart system, which reduces engine fuel consumption resulting in significant fuel savings by safely shutting down and restarting the locomotive engine during appropriate idle periods; LocoTemp II cooling system controller, which reliably monitors and controls locomotive coolings systems temperature, providing 'New Locomotive' technology to a non-microprocessor fleet; Q-Tron QTRAC 1000 locomotive traction control system, a standalone unit that monitors and controls locomotive wheel slip by reducing excitation to the main AC (or DC) generator when wheel slip is detected; Wayside Device Monitor (WDM-450) vigilant highway crossing monitoring system, which provides continuous real-time monitoring and event recording of the status of grade crossing installations; electronically controlled pneumatic braking system.

Western-Cullen-Hayes Inc

2700 West 36th Place, Chicago, Illinois 60632-1682, USA
Tel: (+1 773) 254 96 00　Fax: (+1 773) 254 11 10
e-mail: wch@wch.com
Web: http://www.wch.com

Key personnel

President: Ronald L McDaniel
Vice-President: Barbara Gulick
Sales Manager: Carl J Pambianco
Customer Service Manager: Bill Crain
Systems Application: RodneyYourist

Products

Railway safety signals and accessories, gate arms, railway crossing signals, industrial crossing warning systems, flashing light signals: incandescent and LED, bells: AC-DC and electronic, switch lamps and targets, bumping posts: fixed, sliding and hydraulic; wheel stops, chocks, switch point guards, programmable yard switch machine, track drills, rail benders, rail tongs, journal and hydraulic jacks, derails: sliding, hinged and portable and accessories, derail operators: ELDO, Delectric and solar powered, blue flags, wagon re-railers, locomotive revolving lights and warning bells, and other custom designed equipment for railroad, transit and industrial applications.

VERIFIED

Westinghouse Rail Systems Ltd

PO Box 79, Pew Hill, Chippenham SN15 1JD, UK
Tel: (+44 1249) 44 14 41　Fax: (+44 1249) 65 23 22
e-mail: wrsl.marketing@invensys.com
Web: www.westinghouserail.co.uk

Key personnel

Managing Director: J A Cotton
Sales & Marketing Director: J E Clark
Head of Projects: Alistair McPhee
Head ofTechnology: Charles Riley
Operations Director: Steve Barry
Finance Director: Andy Bryon
Head of Human Resources: Richard Drury

Associated companies

Invensys Rail Systems, UK
Westinghouse Signals, Australia
Foxboro Transportation, Australia
Dimetronic, Spain
Safetran Systems Corporation, USA
Burco Services, USA

Background

Westinghouse Rail Systems Ltd is a subsidiary of Invensys plc.

Products

Westinghouse Rail Systems produces a comprehensive range of railway signalling and train control systems. These include computer-based interlocking, electronic interlocking, relay interlocking, automatic train control, control and management centres, passenger information systems, train describers, traction power tele-control and SCADA systems.

　Products include LED signals, colourlight signals, position-light junction route indicators, theatre

Westinghouse Rail Systems WESTLOCK computer-based interlocking system 0538399

and stencil route indicators and position-light shunt signals, electric point machines, jointed and jointless track circuits, apparatus cases, remote control, indication and data acquisition equipment, safety plug-in relays and solid-state interlocking modules.

TBS100 is the new transmission-board signalling system for urban mass transit railways. This may be configured to provide Automatic Train Operation (ATO), fixed and moving-block Automatic Train Protection (ATP) and Automatic Train Supervision (ATS) systems. The system can be overlaid on existing signalling allowing minimum disruption to operations during installation. TBS provides greater capacity, better regulation, improves operation and enhances infrastructure performance.

WESTRACE is a second-generation safety processor which has been developed by Westinghouse Rail Systems and its three associate signalling companies to satisfy a range of safety applications to worldwide standards. Applications range from simple wayside interlockings to complete Centralised Traffic Control (CTC). The equipment can provide trainborne safety processing for ATP systems, or can be configured as a solid-state highway crossing controller.

WESTRONIC is a microprocessor-based data handling and transmission system designed specifically for railway applications. It is a modular system that can be configured to provide any specific combination of signalling, train control or supervisory facilities, train description, traction power telecontrol, station plant supervision and monitoring, panel processing and so on.

WESTCAD is a comprehensive PC-based scaleable signalling control and display system which can be configured to suit any customer requirement.

Westinghouse Rail Systems SCADA technology provides management centre technology with capability for the full integration of numerous operational systems including traction power telecontrol, supervision of power supply sub-station, environmental control, public address with CCTV, fare collection and tunnel ventilation, centralised clock systems, optical fibre transmission and fixed and mobile radio.

FUTUR is the Invensys Rail group's versatile ERTMS system tested and proven on Spanish main line installation to Levels 1 and 2. This is a highly robust system, configured to the latest European SRS, and capable of being developed to suit low-cost and UGTMS applications.

Customer support includes routine maintenance, fault attendance, diagnosis and repair services provided to suit customer needs. Maintenance and support services include maintenance monitoring, diagnostic and logging systems, control and allocation of spares and maintenance logistic support, together with repair and refurbishment of electronic, electric and electromechanical equipment, including service-exchange facilities.

Contracts

Westinghouse Rail Systems has been awarded two London Underground signalling contracts by Bombardier Transportation. The contracts cover the provision of signalling, train control and train supervision systems as part of a Bombardier contract with the Metronet consortium. The Bombardier contract with Metronet encompasses rolling stock supply, refurbishment and maintenance, signalling upgrade projects for Infraco BCV and Infraco SSL under the London Underground Public Private Partnership (PPP) scheme.

UPDATED

Westinghouse Saxby Farmer Ltd

17 Convent Road, Entally, Calcutta 700014, India
Tel: (+91 33) 22 44 71 61
Fax: (+91 33) 22 44 71 65
e-mail: wsfedp@cal2.vsnl.net.in

Key personnel

Managing Director: A N Dutta
Director, Finance & Marketing: D K De Sarker
Executive Director: C K Chaki

Products

Signalling equipment including route relay and point interlocking systems and signalling relays of all types.

Contracts

Recent contracts include the supply of route relay and interlocking system for seven stations and 36,900 nos of different types of signalling relays to Indian railways.

UPDATED

Zelisko Elektrotechnik und Elektronik GesmbH

Steinfeldergasse 12, 43-45, A-2340 Mödling, Austria
Tel: (+43 2236) 40 60
Fax: (+43 2236) 40 62 99
e-mail: info@zelisko.com
Web: http://www.zelisko.com

Key personnel

Managing Director: Dr Wolfgang Widl
Sales Director: Gerhard Brychta

Corporate background

Zelisko is a wholly owned subsidiary of the Knorr-Bremse Group.

Products

Safety equipment for level crossings including half and full barriers, with acoustic signals (the barrier drives are controlled by battery-powered 24 V DC shunt motors powered by nickel-iron batteries with constant charge from the mains electricity); railway signals in fibre optic and LED displays; alphanumeric LED displays for passenger information; telecommunications equipment.

Zenitel

Research Park Zelik
Pontbeek 63, B-1731 Zelik, Belgium
Tel: (+32 2) 370 53 11
Fax: (+32 2) 370 51 19
e-mail: info.belgium@zenitel.biz
Web: http://www.saitrh.com

Key personnel

Managing Director and Chief Executive Officer: Patrick De Groote

Background

Previously known as Sait-Devlonics and part of the Sait-Stento Group.

Products

Turnkey implementation of HF, VHF, UHF, SHF fixed and mobile telecommunication networks; communication and control systems for public transport, including use of radio in underground and other difficult environments.

PASSENGER INFORMATION SYSTEMS

Alphabetical listing

Albatros Corporation
ALSTOM Transport
Annax Anzeigesysteme GmbH
Apricot Technology AG
Conrac GmbH
Data Display Co Ltd
Densitron Ferrograph Limited
ELNO SN
Firema Trasporti SpA
Focon Electronic Systems A/S
GAI-Tronics Ltd
Globe Transportation Graphics
Gorba AG
Halo Graphics
Hanover Displays Limited
ICL
Ikusi – Ángel Iglesias SA
Info Systems GmbH
Infotec Limited
Inova Corporation

Insta Visual Solutions Oy
Interalia Inc
Jasmin Simtec
Joyce-Loebl Ltd
KeTech
KJ GmbH Installationen
KRONE-REW GmbH
LTS
Lumino Licht Elektronik GmbH
Meister Electronic GmbH
Metra Blansko AS
Mitron Oy
Morio Denki Co Ltd
Moser-Baer SA
Net Display Systems bv
Omega Electronics SA
PAGE
Pallas Informatik A/S
Postfield Systems
Ruf Telematik AG

SA ViewCom
Sistemas Electrónicas de Potencia SA (SEPSA)
Siemens Transit Telematic Systems AG
Siemens Transportation Systems
SLE
Socel Visioner
Solari di Udine SpA
STERIA Group
Sysco SpA
Techspan Systems Ltd
Telephonics Corporation
Televic NV
Trivector System AB
UniControls AS
Vaughan Harmon Systems Ltd
Voice Perfect Ltd
Vossloh Information Technologies Karlsfeld GmbH
Vultron International Ltd

Company listing by country

BELGIUM
Televic NV

CANADA
Interalia Inc

CZECH REPUBLIC
Metra Blansko AS
UniControls AS

DENMARK
Focon Electronic Systems A/S
Pallas Informatik A/S
SA ViewCom

FINLAND
Insta Visual Solutions Oy
Mitron Oy

FRANCE
ALSTOM Transport
ELNO SN
SLE
Socel Visioner
STERIA Group

GERMANY
Annax Anzeigesysteme GmbH
Conrac GmbH
Info Systems GmbH

KJ GmbH Installationen
KRONE-REW GmbH
Lumino Licht Elektronik GmbH
Meister Electronic GmbH
Siemens Transportation Systems
Vossloh Information Technologies Karlsfeld GmbH

IRELAND
Data Display Co Ltd

ITALY
Firema Trasporti SpA
Solari di Udine SpA
Sysco SpA

JAPAN
Morio Denki Co Ltd

NETHERLANDS
Net Display Systems bv

SPAIN
Albatros Corporation
Ikusi – Ángel Iglesias SA
PAGE
Sistemas Electrónicas de Potencia SA (SEPSA)

SWEDEN
LTS
Trivector System AB

SWITZERLAND
Apricot Technology AG
Gorba AG
Moser-Baer SA
Omega Electronics SA
Ruf Telematik AG
Siemens Transit Telematic Systems AG

UNITED KINGDOM
Densitron Ferrograph Limited
GAI-Tronics Ltd
Halo Graphics
Hanover Displays Limited
ICL
Infotec Limited
Jasmin Simtec
Joyce-Loebl Ltd
KeTech
Postfield Systems
Techspan Systems Ltd
Vaughan Harmon Systems Ltd
Voice Perfect Ltd
Vultron International Ltd

UNITED STATES
Globe Transportation Graphics
Inova Corporation
Telephonics Corporation

Albatros Corporation

c/o Ruiz de Alarcón, 13, E-28014 Madrid, Spain
Tel: (+34 91) 532 41 81   Fax: (+34 91) 522 76 97
e-mail: albatros@albatros-sl.es
Web: www.albatros-sl.es

Works
Merak
Pol Ind 'La Estación', C/Gavilanes 16, E-28320 Pinto,
Madrid, Spain
Tel: (+34 91) 495 90 00   Fax: (+34 91) 691 09 97
Albatros also has factories located in Barcelona
(Spain), Alcázar de San Juan (Spain), Albany (USA)
and Shanghai (China).

Key personnel
Chairman: Nicolás Fúster
Human Resources Director: Patricia Fúster
National Market Director: Félix Ramos
Finance and Control Director: Luis Gil
Marketing and Business Director: Enrique Galavis
Information Technology Director: Andrés Morales
Legal Advisor: José Maria Navia – Osorio
Safety and Security Area: Pedro de la Antonia
Industrial Operations: Arturo Delgado
Converters Area: Antonio Sosa
Merak Managing Director: Julio Rey
AUS Managing Director: Valero Torrelles
ARTS Managing Director: Carlos Rico

Background
Albatros is formed by the following companies:
    Merak (formerly known as Stone Ibérica SA)
which specialises in the design and manufacture of
air-conditioning, ventilation and heating systems
for trains, tramways and metros worldwide.
    SEPSA, dedicated to the design and subsequent
manufacture of electronic equipment for railway
vehicles, specialising in two products: static
converters and information and control systems.
    CMC Interiors, which designs and manufactures
modular components for railway vehicles, both
externally and internally.
    Albatros Alcázar SA which manufactures and
maintains all the corporate products. Albatros also
has factories located in Albany (US) and Shanghai
(China).

Products
Telematics for onboard security and comfort
including security cameras and video information
systems for passenger entertainment and
communication.

UPDATED

ALSTOM Transport

Information Solutions
48 rue Albert Dhalenne, F-93482 Saint-Ouen Cedex,
France
Tel: (+33 1) 41 66 90 00   Fax: (+33 1) 41 66 96 66
Web: www.transport.alstom.com

Key personnel
Managing Director: Angelo Guercioni

Canada offices
503 Levy Street, St Laurent, Quebec, Canada H4R 2N9
Tel: (+1 514) 333 08 88   Fax: (+1 514) 333 04 96
Managing Director: Angelo Guercioni

3330 De Miniac Street, St Laurent, Quebec, Canada
H4S 1Y4
Tel: (+1 514) 333 08 88   Fax: (+1 514) 333 04 96
Managing Director: Angelo Guercioni

Products
Real-time onboard passenger information systems for
rolling stock: AGATE e-Media passenger information
systems for mass transit and railway stations.
    ALSTOM acquired Telecite Inc in 1999. The
Montreal-based Telecite becomes ALSTOM's
worldwide centre of excellence in Advanced
Traveller Information Systems (ATIS).
    The company specialises in conceiving,
deploying, maintaining and marketing real-time
information and communication systems for

public transportation and other applications. The
continued pursuit of innovation, coupled with
system engineering and integration, has allowed
ALSTOM (formerly Telecite Inc) to refine and expand
its offerings to include LCD passenger information
displays, public address, CCTV surveillance, seat
reservation system and at-seat audio and video
on demand with full interactive Intranet/Internet
HTML content as part of an integrated multi-modal
system.
    The acquisition of Telecite further strengthens
ALSTOM's presence in North America. Recent
acquisitions have expanded its rolling stock,
traction equipment, signalling systems, field
service and turn-key project management
capabilities in the US and Canada, complementing
its already strong presence in Mexico.
    ALSTOM's co-operation with Telecite over
the years leading to acquisition has brought
orders from Europe, Latin America and Asia. For
example, ALSTOM chose Telecite as its partner to
supply real-time advanced traveller information
systems to the Mass Transit Railway Corporation
(MTRC) in Hong Kong. It was installed on
104 cars (eight trains, of 13 cars each) and each
car has four displays. Other important references
of Telecite are the two French contracts for
the supply of passenger information systems
to the new TER 2N emu (SNCF) and the future
MF2000 metro trainset (RATP).

UPDATED

Annax Anzeigesysteme GmbH

Wettersteinstrasse   18,   D-82024   Taufkirchen,
Germany
Tel: (+49 89) 614 43 60   Fax: (+49 89) 614 436 63
e-mail: vertrieb@annax.de
Web: www.annax.de

Sales office
Kurmainzer   Ring   51,   D-63834   Sulzbach/Main,
Germany
Tel: (+49 89) 614 436 30   Fax: (+49 89) 614 436 81

Products
Information display systems for external and
internal onboard vehicle applications and for
stationary use at bus and tram stops and in
stations. Annax offers displays emplying LCD, LED
and BiLED technologies.
    The stationary displays can be configured to
operate with the Annax-DECT wireless transmission
system which receives vehicles location and
running information and supplies this as real time
information at bus and tram stops.

NEW ENTRY

Apricot Technology AG

Postfach 218, CH-6331 Hünenberg, Switzerland
Tel: (+41 41) 784 41 11   Fax: (+41 41) 784 41 30
e-mail: office@apricot.ch

Key personnel
Chief Executive Officer: Domenico Fontana

Subsidiary
Apricot Technology GmbH
Müngstener   Strasse   10,   D-42285   Wuppertal,
Germany
Tel: (+49 202) 76 96 20   Fax: (+49 202) 769 62 22
e-mail: office@apricot.de
Web: http://www.apricot.de

Products
Onboard   and   fixed   passenger   information
systems using dot, flip, roller, LCD, LED and VFD
displays; onboard and fixed 'infotainment' systems
combining passenger information, entertainment
and advertising; digital audio announcement
systems;   control   equipment   for   passenger
information systems.

Conrac GmbH

Lindenstrasse 8, D-97990 Weikersheim, Germany
Tel: (+49 7934) 10 10   Fax: (+49 7934) 10 11 02
e-mail: marketing@conrac.de
Web: www.conrac.de

Key personnel
General Managers: Walter Hammel,
    Ulrich Brueggermann
Director Sales: Ingo Richter
Sales Manager Information Display Systems:
    Toni Lang
Sales Manager Transport Solutions: Klaus Schipper
Manager Conrac Asia: Kee Sek Huat
Manager Sales Office Italy: Bernhard Gemassmer
Marketing Communications: Petra Ollhoff
Sales Manager Conrac MENA: Binu Sharma

Subsidiaries
Conrac France SARL
47 route de Baillon, F-95270 Chaumontel, France
Tel: (+33 1) 30 35 06 34   Fax: (+33 1) 34 71 29 26
e-mail: info@conracfrance.fr

Conrac Asia Display Products Ltd
82 Genting Lane # 05-04, 349450 Singapore
Tel: (+65) 67 42 79 88   Fax: (+65) 67 47 39 33
e-mail: sales@conrac-asia.com
Web: www.conrac-asia.com

Conrac MEXA FZE
DAFZA Dubai Airport Free Zone, 3E G06 East Wing 3,
Dubai, United Arab Emirates
Tel: (+971 4) 299 40 09   Fax: (+971 4) 299 55 87
e-mail: info@conrac.ae
Web: www.conrac.ae

Conrac Sales Office Italy
Via R R Jaribalidi 30, I-00145 Rome, Italy
Tel: (+39 06) 45 43 92 02   Fax: (+39 06) 45 43 91 79
e-mail: info@conrac.it
Web: www.conrac.it

Products
Intelligent information display systems for railway
applications   (hardware   and   software).   Special
customised solutions such as IP protection, front
access   facilities,   climate   control;   large-screen
colour plasma and TFT displays; large-screen CRT
monitors, large-screen multimedia displays; split-
flap, LED and LCD boards. Special customised
solutions such as IP54 protection for outdoor

*Conrac IP54 protected large screen TFT/LCD
displays, Zug, Switzerland*   1066397

Conrac double-sided LCD/TFT indicator (4 x 30 inch)   *NEW*/1137109

applications, front access facilities and climate control. Different models for connection to standard PC systems or with key display functions from a central server in the network. The large plasma screens and TFT and LCD displays feature robust aluminium housings and laminated safety glass with high efficiency anti-glare treatment. Designed for reliable operation in landscape and portrait mode applications, the PD-Series features a high quality direct digital interface and gamma correction circuitry. Conrac is an ISO 9001 certified company.

Developments
New developments include 32 and 40 inch in high-resolution colour LCD/T FT displays in 16.9 format. Latest products are single-and double-sided TFT indicators in various configurations.
UPDATED

Data Display Co Ltd

Deerpark Industrial Estate, Ennistymon, Co Clare, Ireland
Tel: (+353 65) 707 26 00   Fax: (+353 65) 707 13 11
e-mail: sales@data-display.com
Web: www.data-display.com

Key personnel
Managing Director: Kevin Neville
Marketing Manager: Paul Neville

Subsidiary companies
Data Display UK Ltd
Tel: (+44 23) 92 24 75 00
e-mail: sales@datadisplayuk.com

Data Display USA
Tel: (+1 631) 218 21 30
e-mail: salesinfo@ddusa.com

Data Display Netherlands
Tel: (+31 78) 684 05 04
e-mail: info@data-display.nl

Data Display Dublin
Tel: (+353 65) 707 26 00
e-mail: maryhow@data-display.com

Data Display France
Tel: (+33 1) 43 03 75 00
e-mail: infos@datadisplayfrance.com

Data Display Portugal
Tel: (+351 21) 910 67 60
e-mail: datadisplayportugal@mail.telepac.pt

Poletech, Sweden
Tel: (+46 171) 41 45 90
e-mail: info@poltech.se

Data Display Australia
Tel: (+61 3) 95 87 85 77
e-mail: mclark@datadisplay.com.au

Products
Data Display designs and manufactures electronic passenger information displays for the railway market. It provides a comprehensive range of platform, concourse and on-train displays. Technologies include LED, LCD, electromechanical, TFT and monitors.

Contracts
Supply of LED passenger information displays to London Underground, Denver Light Rail (USA), Mersey Rail, London Line, Heathrow Express, Denver Bus (USA), Wales & Borders Trains (now Arriva Trains Wales) and VTA (USA). Further railway projects include Thales (Hong Kong), DRIMS (Netherlands), Paris Metro (France), Lisbon Metro, Dublin Area Rapid Transit as well as London Underground Central and Jubilee Lines (UK).
UPDATED

Densitron Ferrograph Ltd

New York Way, New York Industrial Park, Newcastle-upon-Tyne NE27 0QF, UK

Tel: (+44 191) 280 88 00   Fax: (+44 191) 280 88 10
e-mail: ferrograph@ferrograph.com
Web: http://www.ferrograph.com

Key personnel
Business Development Manager: Steve Higginson

Products
Passenger information displays, LED, LCD, CRT monitor and plasma technologies.

Most recent developments include a bus stop display using GPRS technology for wireless operation, a digital rear projection display available in 72, 84 and 100 in diagonal screen size units and a 40 in TFT display designed for hostile environments such as railway and bus stations, displaying a full 40 in, 16:9 format display with flat screen technology.

ELNO SN

17 rue Jean Pierre Timbaud, F-95100 Argenteuil, France
Tel: (+33 1) 39 98 44 44   Fax: (+33 1) 39 98 44 46
e-mail: sales@elno.fr

Key personnel
For full personnel listing, see ELNO SN entry in *Passenger coach equipment* section.

Products
Audio and video systems for passenger vehicles.

FIREMA Trasporti SpA

Headquarters
Via Provinciale Appia, Località Ponteselice, I-81100 Caserta, Italy
Tel: (+39 0823) 09 71 11   Fax: (+39 0823) 46 68 12
Web: www.firema.it

Key personnel
Chairman: Gianfranco Fiore
Chief Executive Officer: Roberto Fiore
Operations Director: Maurizio Russo
Commercial Manager: S d'Arminio
Marketing Manager: Agostino Astori

Commercial and technical offices
Via Triboniano n 220, I-20156 Milan, Italy
Tel: (+39 02) 23 02 02 23
Fax: (+39 02) 23 02 03 00

Products
Onboard computers; automatic voice announcement of next stop, public address for buses and trains.
UPDATED

Focon Electronic Systems A/S

Damvangz, PO Box 269, DK-6400 Sønderborg, Denmark
Tel: (+45) 73 42 25 00   Fax: (+45) 73 42 25 01
e-mail: focon@focon.dk
Web: www.focon.com

Key personnel
Managing Director: Niels-Henrik Hedegaard
Senior Managers
    Sales: Jens Moldrup
    Research and Development:
        Lars Bo Kjong-Rasmussen
    Projects and Quality: Per Viggo Rasmussen
    Finance, Administration and IT:
        Henrik Kock Clausen
    Customer Support and Logistics: Lars Jansson
Manager, Purchasing: Birgit Elvardt Bader
Manager, Marketing: Bent Sloth Lave
Service Manager: Jan Østergaard Pedersen

Background
Focon is part of the Mark IV IDS Group.

Products

High performance passenger information systems including: audio-visual and communication systems for onboard applications, passenger entertainment systems, public address, alarm, crew communications and talk-back systems, automatic seat reservation systems. Video surveillance and full-colour graphic displays.

Contracts

Intercity trains: AVE S103 (Spain), Icx (Germany), IC4 (Denmark), CP2000 (Portugal), IR4/IC3 (Denmark, Sweden, Israel), Regina (Sweden), OUT (Denmark, Sweden), Pendoluso (Portugal), ICE (Germany), X2000 (Sweden). Metro: London Underground Victoria Line Upgrade (UK), D-stock (UK), London Underground Jubilee Line (UK), Bucharest Metro (Romania), C20/C21 (Sweden), S-Train Copenhagen (Denmark). Regional Trains: IC2 (Denmark), X60 (Sweden), SGM III (Denmark), Itino (Sweden, Germany). Light Rail: Satra (Brazil), Metro do Porto (Portugal), A32 (Sweden, Netherlands, Turkey), Strasbourg (France).

UPDATED

GAI-Tronics

A division of Hubbell Limited
Brunel Drive, Stretton Business Park, Burton-on-Trent DE13 0BZ, UK
Tel: (+44 1283) 50 05 00   Fax: (+44 1283) 50 04 00
e-mail: sales@gai-tronics.co.uk
Web: www.gai-tronics.co.uk

Key personnel

Business Unit Manager: Graham Lines
Financial Director: Toby Balmer
Engineering Manager: Richard Rumsby
Manufacturing Director: Mark Bradford
Commercial Manager: Roger Goodall
Marketing Consultant: Nicole Ireland
Special Projects Manager: Steve Smith

Other offices

GAI-Tronics Corporation, USA
Tel: (+1 610) 777 13 74   Fax: (+1 610) 775 65 40
Web: www.gai-tronics.com

GAI-Tronics Srl, Italy
Tel: (+39 02) 48 60 14 60   Fax: (+39 02) 458 56 25

GAI-Tronics Corporation, Malaysia
Tel: (+60 3) 89 45 40 35   Fax: (+60 3) 89 45 46 75

Background

GAI-Tronics was established in 1964 to provide communications equipment to the mining industry. It has now expanded into the provision of specialised communications equipment for road and rail transportation industries.

Products

Help point telephone: hands-free use, weather- and vandal-resistant. Allows easy access to information or emergency assistance. Option for remote monitoring, programming and maintenance.

UPDATED

Globe Transportation Graphics

7127 Rutherford Road, Baltimore, Maryland 21244, USA
Tel: (+1 410) 685 67 50   Fax: (+1 410) 752 88 28
e-mail: info@globegrafix.com
Web: www.globegrafix.com

Key personnel

Regional Sales Manager, West: Joe Pollard
Regional Sales Manager, North: Shawn Beight
Regional Sales Manager, South: Ron Owens

Products

Globe Transportation Graphics interior and exterior durable TransGrafix® markings, decals and signage

can be customised, are graffiti- and chemical-proof, weather- and vandal-resistant and available with raised letters and braille. Globe sub-surface prints on the most durable and impervious of surface materials and utilises adhesives of unprecedented strength and bond. TransGrafix® also features High Performance Photo-Luminescence (HPPL) to glow safely in a power failure.

Contracts

Globe has sold TransGrafix® to the passenger rail markets in Hong Kong, Australia, Turkey, France, Russian Federation, Switzerland, Singapore and Canada.

NEW ENTRY

Gorba AG

Sandackerstrasse, CH-9245 Oberbüren, Switzerland
Tel: (+41 71) 955 74 74   Fax: (+44 71) 951 96 74
e-mail: info@gorba.com
Web: http://www.gorba.com

Key personnel

President: Daniel Fäh
Administration: Cornelia Eberli

Subsidiary company

Gorba GmbH
Osterbooksweg 69, D-22869 Schenefeld b. Hamburg, Germany
Tel: (+49 40) 41 59 70   Fax: (+49 40) 41 45 97 33
e-mail: gerkens@gorba.com

Products

Audio and visual passenger information systems including dot matrix, LED, LCD, roller-blind, Gorbastop in-vehicle signs, GPS-synchronised clocks, Gorbaplan roller-blind/LCD variable route displays and Gorbaterm variable information indicator systems for stationary applications.

Halo Graphics

Osborne House, Station Road, Burgess Hill RH15 9EN, UK
Tel: (+44 1444) 24 77 17   Fax: (+44 1444) 87 02 20
e-mail: sales@halogroup.co.uk
Web: http://www.halogroup.co.uk

Key personnel

Managing Director: Peter Low
Sales Director: John Veasey
Commercial Director: Bob King

Products

Livery for rail rolling stock (passenger and freight), internal and external signs and notices; station nameplates and signing.

Contracts

Contracts in the UK have included manufacture and application of livery to 73 four-car emus for Great Eastern Railway; manufacture and application of livery to 30 four-car emus for Connex South Eastern; manufacture and supply of livery to Connex South Central; manufacture and supply of Thameslink livery for Wessex Traincare; supply of decals and notices for South West Trains; supply of decals for power cars and passenger cars for Great Western; ScotRail decals and notices for Railcare Ltd; supply of Midland Mainline decals to Adtranz; supply of Virgin Decals to Bombardier Transportation; supply of livery for English Welsh & Scottish Railway; supply of livery and notices for Silverlink.

Hanover Displays Limited

Unit 24, Cliffe Industrial Estate, Lewes BN8 6JL, UK
Tel: (+44 1273) 47 75 28   Fax: (+44 1273) 40 77 66
e-mail: sales@hanoverdisplays.com
Web: http://www.hanoverdisplays.com

Key personnel

Managing Director: D G Williams
International Sales Manager: A Williams

Subsidiary

Hanover Sarl
Bureau 124, Tertia 3000, rue Henri Matisse, F-59300 Aulnoye-les-Valenciennes, France
Tel: (+33 3) 27 25 36 99   Fax: (+33 3) 27 25 30 90

Products

Electronic information display systems using electromagnetic flip-dot and LED technologies for use in vehicles and fixed installations including bus stops and stations. The systems provide large memory capacity with simple reprogramming facilities for route network updating. An autonomous driver's controller is provided, but the signs can also be slaved to onboard electronic equipment such as ticket machines.

A flexible version of the company's flip-dot sign specifically for new rail vehicles with steeply inclined curved screens has been introduced. The sign bends to the shape of the screen and fixes against it without the need for a second window.

ICL

Eskdale Road, Winnersh, Wokingham RG41 5TT, UK
Tel: (+44 1189) 63 44 96   Fax: (+44 1189) 63 44 07

Key personnel

Sales Manager, Rail: Gary Mills

Products

Passenger information systems using Fujitsu plasma screen technology.

Ikusi – Ángel Iglesias SA

PO Box 1320, E-20080 San Sebastian, Spain
Tel: (+34 943) 44 88 00
Fax: (+34 943) 44 88 20
e-mail: sb@ikusi.com
Web: http://www.ikusi.com

Key personnel

Deputy Director, Public Information Systems:
   Imanol Saenz
Marketing Manager: Nora Iglesias

Products

Displays, monitors and automated public address, audio and video entertainment systems; video monitoring by CCTV of passenger coach interior and exterior, with integrated alarm systems; mobile phone coverage on trains and on stations.

Info Systems GmbH

Uellendahler Strasse 437, D-42109 Wuppertal, Germany
Tel: (+49 202) 709 50   Fax: (+49 202) 709 51 02
e-mail: info@wtal.infosystem.de
Web: http://www.infosystems.de

Key personnel

Managing Directors: F Khavand, K Bämer
Sales Manager: H Spahn
Export Sales Manager: M Scharowsky

Background

Info Systems is a member of the Schaltbau Group.

Products

Passenger information systems for buses and trains; components for suburban road and rail traffic systems.

The FZG 400 remote-control unit generates, stores and transmits data for the information of driver, passengers and ground-based central control. It can be operated either as an autonomous control and information system or as part of a computer-controlled central control system. Separate input and output channels are available for data exchange and the vehicle bus for flow of data inside the vehicle.

The company's LED full-matrix indicator for vehicle interior display of stopping points allows 98 items of information with a total of 2,040 letters. The unit is supplied with the names of stopping points via an IBIS remote-control unit.

Units have been supplied to DB AG, Belgium (Brussels metro and De Lijn, Antwerp).

Infotec Limited

The Maltings, Tamworth Road, Ashby de la Zouch, LE65 2PS, UK
Tel: (+44 1530) 56 65 02 Fax (+44 1530) 56 01 11
e-mail: sales@infotec.co.uk
Web: http://www.infotec.co.uk

Background

Infotec Limited has established a technology partnership with Transmitton Limited, providing for the latter company to supply integrated control and real-time technology to provide complete passenger information systems.

Products

Manufacturers of electronic display systems, including: dot matrix LCD displays, LED and plasma displays, plasma replacement systems for LED displays, dot matrix LCD clocks, display control systems, and elements for audio-visual bus stop information display systems.

Contracts

Recent contracts to provide dot matrix LCD display systems for Network Rail's Manchester Piccadilly station, UK, and LED and plasma displays for London Underground Limited, c2c Rail, Silverlink Trains and WAGN, UK, and LED and plasma displays for Midland Mainline, UK.

Inova Corporation

110 Avon Street, Charlottesville, Virginia 22902, USA
Tel: (+1 434) 817 80 00 Fax: (+1 434) 817 80 02
e-mail: info@inovacorp.com
Web: www.inovatransit.com

UK office

City Tower, Level 4, 40 Basinghall Street, London EC2V 5DE, UK

Products

Products include: LightLink™, a robust, multi-modal passenger information system which displays real-time schedule and status information, including next train countdowns, arrivals and departures, special operating hours, route changes, delays and service interruption alerts. Lightlink can deliver information through a wide range of display options including Variable Message Signs (VMS) (manufactured by Inova), video monitors and walls, plasma displays, public address systems, wireless devices, the Internet and interactive voice response systems.

UPDATED

Insta Visual Solutions Oy

Sarankulmankatu 20, FIN-33900 Tampere, Finland
Tel: (+358 20) 771 71 11 Fax: (+358 20) 711 79 20
e-mail: ivs@insta.fi
Web: www.ivs.fi

Key personnel

Director, Business Operations: Kari Pulkkinen

Subsidiary companies

Switzerland
Insta Visual Solutions (Switzerland) GmbH

Schulhausstrasse 21, CH-6318 Walchwil
Tel: (+41 41) 759 02 72 Fax: (+41 41) 759 02 71
e-mail: ivs@ivs.fi

Background

Insta Visual Solutions Oy is a member of the Instrumentointi Group.

Products

Video and entertainment systems, passenger announcement systems, crew communication systems, internal and external displays.

Contracts

Include: bodyside and interior information displays, entertainment system and automatic audio information system for ICS double-deck intercity coaches supplied by Talgo to VR-Group, Finland; exterior bodyside information displays for Virgin UK West Coast Pendolino trainsets supplied by Alstom; entertainment systems for Pendolino trainsets supplied to VR-Group by Fiat/Transtech and by Alstom.

UPDATED

Interalia Inc

4110 79th Street NW, Calgary, Alberta T3B 5C2, Canada
Tel: (+1 403) 288 27 06 (ext. 105)
Fax: (+1 403) 288 59 35
e-mail: bcormack@interalia.ca
Web: www.interalia.com

Key personnel

President and Chief Executive Officer: John Trester
Chief Financial Officer: Garth Hunter
Chief Operations Officer and Manager Special Products including Transportation Division: Robert Cormack

Subsidiaries

Interalia Communications Inc, USA
Interalia Communications Ltd, Europe, Middle East and Africa

Products

Passenger information systems including onboard and platform announcers:

Commander – platform announcer and sign controller offering full live broadcast and local recording capability

Trouncer – onboard and platform announcer/sign controller

Transit voice – platform digital voice announcement system for smaller platforms.

Vehicle monitoring: *Transit Owl* – for monitoring train equipment and security in remote waysides.

Passenger help point systems: *IBBISS* – Provides fire, emergency and information reporting and communication to transportation system passengers.

Contracts

Commander – 27 installations at the London Underground. Trouncer installations through Geofocus at the Chicago Metra, Capital Corridor Oakland, Salt Lake City, and Ft Lauderdale. Trouncer installation with Trak Com Wireless at Vancouver Skytrain and Detroit People Mover. Interalia installations at Calgary Transit, Edmonton Transit, Warrington, Poole and Manchester in the UK, and Cologne Arena in Germany.

UPDATED

Jasmin Simtec

Sellers Wood Drive, Bulwell, Nottingham NG6 8UX, UK
Tel: (+44 115) 916 51 65 Fax: (+44 115) 916 14 91
e-mail: paul.corcoran@jasmin.plc.uk
Web: http://www.jasmin.plc.uk

Key personnel

Marketing Director: Paul Corcoran

Products

Passenger information systems for on and off-train use, including front-of-train and internal information displays, closed-circuit television, audio systems, full train management systems and on-station passenger information systems.

Joyce-Loebl Ltd

390 Princesway, Team Valley Trading Estate, Gateshead NE11 0TU, UK
Tel: (+44 191) 420 30 00 Fax: (+44 191) 420 30 30
e-mail: mike.wade@joyce-loebl.com

Key personnel

Director, Transport Systems: Mike Wade

Products

Integrated electronic onboard passenger information systems including: seat reservation displays; front of train displays; side of train displays; audio systems; side of train displays; and digital CCTV security systems.

KeTech

KeTech House, 35-37 Tavistock Place, Bedford MK40 2RZ, UK
Tel: (+44 1234) 33 08 00 Fax: (+44 1234) 33 08 01
e-mail: jek@ketech.com
Web: http://www.ketech.com

Key personnel

Director: John Kearney
General Manager (London Operations Centre): Paul Clause

Products

Integrated communications and control systems for passenger information systems, including: KeCIS, providing control of single-station or network-wide information displays; and KeSMS, providing total communications systems control of customer information; integrated station management; public address; closed circuit television; passenger help points; station radios; SCADA; clocks; fire alarm and security interfaces; telephones; door entry; access control, networking; digital voice announcement; and breakdown broadcast messaging.

KJ GmbH Installationen

Fechnerstrasse 29, D-01139 Dresden, Germany
Tel: (+49 351) 89 45 60 Fax: (+49 351) 894 561 00
e-mail: info@kj-gmbh.de
Web: www.kj-gmbh.de

Key personnel

Managing Directors: Maik Juppe; Uwe Köhn

Products

Stationary passenger information display systems using LCD technology and related control systems.

Contracts

In 2004 KJ installed a major new passenger information system at the DB station in Halle, Germany.

NEW ENTRY

KRONE-REW GmbH

Knorrstrasse 119, D-80807 Munich, Germany
Tel: (+49 89) 351 80 39 Fax: (+49 89) 354 19 45

Key personnel

Board of Directors: Hans-Dieter Grunwald, Hans-Karl Mucha, Leo Meyboom

Principal subsidiaries
KRONE (Australia) Tech Pty Ltd
POB 335, Wyong, New South Wales 2259, Australia
KRONE (UK) Tech Ltd
Runnings Road, Kingsditch Trading Estate, Cheltenham GL51 9NQ, UK
KRONE GmbH
Kroneplatz, A-2521 Truman/NÖ, Austria

Products
Passenger information and advertising signage systems, split flaps, video, LCD, automatic control and specialised interfaces. Passenger information systems are installed at railway stations in Austria, Denmark, Finland, Germany, Italy, Netherlands, Norway, Spain, Sweden and the UK. The company developed a liquid crystal transflective display for use on board the DD- IRM and SM 90 trainsets for NS.

LTS

Linné Trafiksystem AB
FO Petersons gata 28, SE-412 31 Västra Frölanda, Sweden
Tel: (+46 31) 89 69 60   Fax: (+46 31) 49 44 65
Web: http://www.linnedata.se

Key personnel
Managing Director: Bengt Rodung
Sales Director: Stefan Bertling

Products
Turnkey supplier of intelligent passenger information systems.
   The Linaria/Dynamite system has been installed in 30 cities. Passenger information systems have been supplied to SL Stockholm and Gothenberg local operators.

Lumino Licht Elektronik GmbH

Europark Fichtenhain A8, D-47807 Krefeld, Germany
Tel: (+49 2151) 819 60   Fax: (+49 2151) 819 63 59
e-mail: info@lumino.de
Web: http://www.lumino.de

Products
Dynamic LED-based passenger information systems for station platforms and tram stops. Displays can be configured to show text, graphics and logos and are produced to display images in green, red and yellow and at variable levels of brightness. Formats include single- or double-sided and single- or multiline displays. Dynamic audio announcements and clock systems can also be integrated with the systems, which are supplied with all software and hardware.

Meister Electronic GmbH

Kölner Strasse 39-45, D-51149 Cologne, Germany
Tel: (+49 2203) 17 01 20   Fax: (+49 2203) 17 01 30
e-mail: customerrelations@meisterelectronics.com
Web: http://www.meisterelectronics.com

Key personnel
Chief Executive Officer: Fritz E Meister
Chief Financial Officer: René Stoffels
Chief Operating Officer: Frank Krüger

Products
Onboard multimedia systems, electronic destination signs, digital announcement and intercom systems; video surveillance systems, refurbishment of onboard communication systems.

Metra Blansko AS

Porici 24, CZ-678 49 Blansko, Czech Republic
Tel: (+420 506) 49 41 15   Fax: (+420 506) 49 41 45
e-mail: metra@metra.cz
Web: http://www.metra.cz

Products
On-train interior and exterior dot matrix and LED display systems.

Contracts
Include provision of information systems for Czech Railways Class 471 double-deck emu vehicles and for new metro cars for Prague.

Mitron Oy

PO Box 113, Tiilenyojankatu 5, FIN-30101 Forssa, Finland
Tel: (+358 3) 424 04 00   Fax: (+358 3) 435 53 21
e-mail: feedback@mitron.fi
Web: http://www.mitron.fi

Key personnel
Director: Heino Ruottinen
Project Manager: Kimmo Ylander

Products
Optical information systems for buses and trains including LCD, mosaic, dot-matrix and LCD-mosaic displays; platform and station displays.
   Customers include VR Finland, Helsinki City Tram department.

Morio Denki Co Ltd

34-1 Tateishi 4-chome, Katsushika-ku, Tokyo 124-0012, Japan
Tel: (+81 3) 36 91 31 81   Fax: (+81 3) 36 92 13 33

Key personnel
President: S Yamagata
Managing Director and General Manager, Marketing Division: K Miura

Products
Information systems featuring LED, liquid crystal and plasma display. Train destination displays.

Contracts
Recent contracts include the supply of passenger information systems for Taiwan High Speed rail cars and 700 Shinkansen high-speed trainsets.

UPDATED

Moser-Baer SA

Export Division
Ch du Champ-des-Filles 14, CH-1228 Plan-les-Ouates, Switzerland
Tel: (+41 22) 884 96 11   Fax: (+41 22) 884 96 90
e-mail: export@mobatime.com
Web: www.mobatime.com

Key personnel
Managing Director: Urs Moser
Export Division Manager: J C Zgraggen
Export Sales Manager: T Fric

Products
Industrial timing equipment: modular master clock systems, time distribution to computers/networks (NTP server), mains frequency supervision, radio and satellite (GPS) time receivers, digital and analogue secondary clock systems, self-setting Mobaline digital and analogue secondary clock systems, façade clocks, tower and decorative clocks.

UPDATED

Net Display Systems bv

Luchthavenweg 31, NL-5657 EA Eindhoven, Netherlands
Tel: (+31 40) 266 11 77   Fax: (+31 40) 266 11 78
e-mail: info@nds-nl.com
Web: http://www.nds-nl.com

Key personnel
Managing Director: Louis van Geldrop
Sales Director: Jan Kosters
Product and Marketing Manager: Arthur Damen

Affiliated company
Internet Display Services

Products
Public Area Display System (PADS). The system supports the following display devices: TVs, VGA monitors or plasma's (4:3 and 16:9), LEDs, LCDs, split-flap board and video walls. Remote Monitor Control (RMC) additional software package for PADS enabling the user to inspect what is actually displayed, obtain information about the operation of the system and install new software from a desktop PC.

Omega Electronics SA

PO Box 6, rue des Prés 149, CH-2500 Bienne, Switzerland
Tel: (+41 32) 343 37 77   Fax: (+41 32) 343 38 00
e-mail: PIS@omega-electronics.ch
Web: http://www.omega-electronics.ch

Key personnel
Managing Director: H Kayal
Sales: M Schumacher

Products
Passenger information systems. Flap and LCD display with white characters on a blue background; also three-colour LED display for general use, CRT, Plasma and TFT monitors.

Contracts
Recent contracts include supply of systems for Leipzig main station, Germany, Zurich airport main station, Fribourg bus station, several Swiss and German railway stations and many others.

VERIFIED

PAGE

Avda de La Industria, 24, E-28760, Tres Cantos, Madrid, Spain
Tel: (+91 807) 39 99   Fax: (+91 807) 18 04
e-mail: page@pagetelecom.com
Web: http://www.pagetelecom.com

Products
Integrated passenger information systems; information displays; automatic centralised public address systems; automatic announcement systems; onboard communications systems, CCTV.

Contracts
PAGE has developed and installed an automatic passenger information system for RENFE's Madrid suburban network.

Pallas Informatik A/S

Allerød Stationsvej 2D, DK-3450 Allerød, Denmark
Tel: (+45) 48 10 24 10   Fax: (+45) 48 10 24 01
e-mail: pallas@pallas.dk
Web: http://www.pallas.dk

Key personnel
Managing Directors: Svend Vitting Anderson, Karsten Funder

Products
Tele and Radio base Information systems for Trains (TRIT), comprising train running information displays on platforms, a central TRIT server, a TRIT computer on all trains and a communications system providing links with all other units in the system. The system is based on data supplied by the infrastructure operator, the train operator's timetable, the operational management disposition

of equipment, current train running information and train layout, including seat availability.

Contracts

Since 1998 Pallas Informatik has worked with Danish State Railways to develop and install a TRIT system for the national rail network.

Postfield Systems

11 Old Barn Lane, Kenley CR8 5AU, UK
Tel: (+44 20) 86 45 97 60   Fax: (+44 20) 86 60 18 04
e-mail: sales@postfield.co.uk
Web: www.postfield.co.uk

Key personnel

Directors: J F Coward; L E Hardy
Sales Contact: Wendy Coward

Products

Manufacturer of hi-tech information software and visual display equipment, which specialises in both customer information and data display. The company supplies the rail industry and has considerable experience in the design, development, fabrication, assembly, project management, commissioning and maintenance support for the following: Customer Information Systems (CIS), CCTV and Public Address (PA).

Contracts

Survey, design, supply and testing of Doo mirrors and CCTV for the re-opened Larkhall branch line.

Supply and installation of CCTV surveillance system; installation of customer information system at six stations on the Silverlink metro network. Supply and manufacture of Bailey Concepts housings for Irish Rail and ScotRail.

UPDATED

Ruf Telematik AG

Rütistrasse 12, CH-8952 Schlieren, Switzerland
Tel: (+41 1) 733 84 00   Fax: (+41 1) 733 83 00
Web: http://www.ruf.ch

Key personnel

Marketing: Rolf Hess

Products

VisiWeb onboard dynamic passenger information systems; station departure/arrival displays; multimedia passenger interior displays and acoustic announcements.

Contracts

Ruf Telematik passenger information systems are installed on Swiss Federal Railways' FLIRT emus and ICN tilting trainsets, on trains operated by the Matterhorn Gotthard Bahn and on Basle Euro-Airport bus shuttle, providing dynamic details of flight departures.

SA ViewCom A/S

Kløvervej 40, DK-7190 Billund, Denmark
Tel: (+45) 72 19 34 00   Fax: (+45) 72 19 34 01
e-mail: mail@saviewcom.com
Web: http://www.saviewcom.com

Sales office

Gefionsweg 7 DK-3400 Hillerød, Denmark
Tel: (+45) 72 19 35 00   Fax: (+45) 72 19 35 01

Key personnel

President: Jette Kruhøffer
Vice-President, Sales and Marketing:
  Povl Moustgaard Knudsen
Financial Manager: Linda Petersen
Quality Manager: Kristian Sigaard
Research and Development Manager:
  Kaare Thorsteinsson
Manager of Logistics: Lie Rask

Background

A wholly owned subsidiary of Kirkbi A/S, an associated company in the Lego Group, SA ViewCom was founded in 1999 as a merger of Modulex ViewCom and ScanAcoustic. In 2002 the company announced a strategy that would see it concentrating on serving the rail market.

Products

Onboard train solutions including: visual passenger information systems; public address; onboard intercom, passenger alarm units; digital audio announcements; passenger entertainment; GPS or tacho positioning; seat reservation; GSM communications; integration of door signals; CCTV.

Contracts

SA ViewCom has supplied its LON-based passenger information system PILON to: Express Rail Link (Malaysia) Desiro emus; NSB Class 73 Signatur and 73B Agenda emus (Norway); SL Class X10 emus (Sweden); and Angel Trains Class 350, One (First Great Eastern) 360, and South West Trains Class 444 and 450 Desiro emus (all UK). CCTV system for DSB including platform surveillance, a kiosk alarm system, PIR sensors and a remote control centre. Siemens TS subsystems for more than 210 four- and five-car trains.

UPDATED

Siemens Transit Telematic Systems AG

Industrieplatz 3, CH-8212 Neuhausen, Switzerland
Tel: (+41 585) 55 11 11   Fax: (+41 585) 55 11 12
e-mail: info.tts@siemens.com
Web: http://www.siemens-tts.ch

Key personnel

Chief Executive Officer, Senior Management:
  Hans-Peter Schär

Background

Previously called Häni-Prolectron AG, the company changed its name to Siemens Transit Telematic Systems AG in 2001 and is a subsidiary of Siemens Schweiz AG.

Products

ALVC systems, integrated planning software, onboard computers, radio and wayside equipment, dynamic passenger information systems, traffic signal pre-emption systems, mobile fare management systems; electronic components for local and long-distance transport.

System consulting, system integration, project management, integration and support, commissioning, maintenance and training/ instruction.

Siemens Transportation Systems

Transportation Systems

Rail Automation Division
PO Box 3327, D-38023 Braunschweig, Germany
Tel: (+49 531) 22 60
Fax: (+49 531) 226 42 64
e-mail: rail-automation@siemens.com
Web: http://www.siemens.com/transportation

Corporate headquarters

Siemens AG
Transportation Systems
PO Box 3240, D-91050 Erlangen, Germany
Tel: (+49 9131) 7-0

Key personnel

President, Transportation Systems Group:
  Hans M Schabert
Group Vice-Presidents: Alfred Frank, Joern F Sens,
  Friedrich Smaxwil
Heads of Rail Automation Division:
  Andreas Busemann, Dr Roland Alter

Products

Passenger information systems for mass transit and main line railways, including the Vicos® 1 dynamic system.

UPDATED

Sistemas Electrónicas de Potencia SA (SEPSA)

Albatros 7 and 9, (Pol Ind) La Estación, E-28320 Pinto, Madrid, Spain
Tel: (+34 91) 691 52 61   Fax: (+34 91) 691 39 77
Web: http://www.sepsa.es

Key personnel

For full personnel listing, see Sistemas Electrónicas de Potencia SA (SEPSA) entry in *Passenger coach equipment* section.

Background

SEPSA is a member of the Albatros Group (qv).

Products

Onboard passenger information systems comprising public address systems, automatic station announcement systems and internal (LED or dot matrix) and external (LED or electromagnetic) information display units.

SLE

avenue Emmanuel Pontrémoli, Nice La Plaine 1, Building F4, F-06200 Nice, France
Tel: (+33 4) 92 29 60 30
Fax: (+33 4) 92 29 60 31
e-mail: sle@sle-fr.com
Web: www.sle-fr.com

Key personnel

Company Manager and Sales and Marketing
  Director: Wilfrid Rouger

Background

SLE is a subsidiary of the Mark IV Industries group.

Products and services

Real-time information and automatic vehicle location systems (real-time location and tracking of vehicles through GPS or microwave beacons); statistics tools for travel time of vehicles; real-time information systems; information and vocal announcements systems; onboard video surveillance systems; bus stop information signs; onboard display signs; external destination signs (LED or LCD); driver keyboards.

Contracts

Hampshire County Council (UK): real-time information and automatic vehicle location systems (location and forecasting of estimated times of arrival at bus stops, WEB, WAP, SMS services); RATP (Paris, France) voice announcement systems for 3,500 buses; London (London Bus Systems) real-time information and automatic vehicle location systems (regulation, location and forecasting of estimated times of arrival at bus stops); Neuchâtel (Switzerland) real-time information and automatic vehicle location systems (regulation, GPS location); Lyon (SYTRAL, France) real-time information and automatic vehicle location systems with microwave beacon location for tramways; Avignon and Vitrolles (France) real-time information and automatic vehicle location systems with GPS location; Lansing and Cincinnati (US) onboard voice announcement system.

NEW ENTRY

Socel Visioner

702 rue du Pont Rouge, BP 3, F-59236 Frelinghien, France
Tel: (+33 3) 20 48 81 78   Fax: (+33 3) 20 48 85 10
e-mail: visionor@visionor.fr
Web: www.visionor.fr

Products

Fixed and onboard passenger information systems and displays. Displays employ CRT, LCD, LED, Plasma, TFT and VFD technologies.

NEW ENTRY

TFT bus stop information display by Socel Visioner   **NEW**/0585287

Solari di Udine SpA

Via Gino Pieri 29, I-33100 Udine, Italy
Tel: (+39 0432) 49 71   Fax: (+39 0432) 48 01 60
e-mail: info.solari@solari.it
Web: http://www.solari.it

Key personnel

Chairman: Massimo Paniccia
Vice-President: Arduino Paniccia
Export Products Sales Director:
   Dino Domeneghetti
Systems Sales Director: Alberto Vazzoler
Products Sales Director: Alberto Zuliani
Technical Director: Giorgio Segatto

Products

Passenger and staff information display systems using monitors and display boards (flaps: LED; TA; LCD; plasma; 16/9 - 4/3 monitors); master and slave clocks; automatic announcement systems; time and attendance recording systems; advertising display systems; automatic information systems; access control systems. Provision of systems on a turnkey basis; staff training and organisation of maintenance.

Contracts

Recent contracts include projects at 20 railway stations in Morocco, SNCF (France), NJT Secaucus (USA), Torino (Italy), Milan Airport FIDS (Italy).

STERIA Group

46 rue Camille Desmoulins, F-92782 Issy les Moulineaux, France
Tel: (+33 1) 34 38 60 00   Fax: (+33 1) 34 88 60 15
Web: www.steria.fr

Key personnel

See STERIA Group entry in *Signalling and communications systems* section.

Products

Include traffic control, model networks, scheduling, equipment and power consumption control, passenger information, ticketing, simulation, high-reliability systems (Atelier B).

Contracts

Contracts executed and obtained include PCC – central control stations on the Parisian RER rapid transit system for RATP Paris; PCS – public transportation passenger and staff information system for RATP Paris; 3615 RATP - public teletext information server; RIS – passenger information system for DB AG. Traveller Information System for RAPT-Metronic and PCS; SNCF 3615; CFF/SBB Swiss Railways.

UPDATED

Sysco SpA

Via di Vannina 78, I-00156 Rome, Italy
Tel: (+39 06) 41 20 60 21
Fax: (+39 06) 410 05 10
e-mail: info@syscosrl.it
Web: http://www.syscosrl.it

Key personnel

President and Chief Executive Officer:
   Vincenzo Manzini
Marketing Manager: Bruno Angius
Technical Manager: Romano Mariani

Works

Via di Vannina 78, I-00156 Rome, Italy

Products

An ISO 9001 certified company, Sysco SpA supplies turnkey passenger information systems and peripheral equipment for railway stations, airports, bus and ferry terminals. It designs and supplies LED peripheral panels, with standard serial or ethernet digital interface, for any specific application. It makes split-flap information boards with 100, 60 and 35 mm character heights and widths with 1, 2, 4, 6, 8 modules; 40 and 60 flaps available. Standard mechanical assemblies and up to IP65 protection grade enclosures for demanding applications are offered.

The Sysco application software includes stand-alone and distributed multistation packages in Windows 2000/NT® system environment, operating in fault tolerant server configuration.

Other products include LCD/TFT and plasma video display terminals with its own powerful graphic generators and text-to-speech digital audio broadcast sub-systems.

Contracts

Recent contracts include the supply to Ferrovie Nord Milano of the LED general boards and platform indicators for the new station of Milano Domodossola and the supply to the Italian rail infrastructure authority (RFI) of 20 turnkey passenger information systems which integrate the technology standard recently defined by RFI for the LED and the LCD/TFT peripheral equipment.

Techspan Systems Ltd

Jarvis House, Griffin Lane, Aylesbury, HP19 8BP, UK
Tel: (+44 1296) 67 30 00   Fax: (+44 1296) 67 30 02
e-mail: techspan@jarvis-uk.com
Web: www.techspan.co.uk

Key personnel

General Manager: Jim Smith
Finance Director: Louise Martin
Business Development Manager: Darren Smith
Technical Director: Jeremy Hinton
Manufacturing Director: John Stottor

Background

Techspan is part of the Jarvis Group of companies.

Products

Design, supply, installation and commission of Variable Message Signs (VMS). Also a product range including LED and electro-mechanical equipment used for highways applications, with plasma, CRT monitor displays and LED being used for Customer Information Systems (CIS). Intelligent controllers such as VDG's VMS drivers, signalling data loggers and telematic decoders. Techspan supplies complete control systems and bespoke software, or can integrate its products with existing IT networks. Techspan can manage and maintain the equipment 24 hours a day all year round providing a full support service.

UPDATED

Telephonics Corporation

815 Broad Hollow Road, Farmingdale, New York 11735, USA
Tel: (+1 631) 549 60 62   Fax: (+1 631) 549 60 18

Key personnel

For a full list of personnel, see Telephonics Corporation entry in *Passenger coach equipment* section.

Products

Integrated digital communications systems for rail vehicles, including public address, passenger/crew intercom, automated announcements, radio communications. Other products include passenger entertainment, CCTV and control centre equipment.

Contracts

Train line multiplexer, network controller and vehicle communication systems for MTA New York R142 and R143 212 metro cars, vehicle communication, radio network and CCTV door observation for SEPTA Philadelphia, vehicle communication for MBTA Boston, vehicle communication, health monitoring and vehicle CCTV for Hudson-Bergen LRVs, integrated wayside communication for Newark APM and vehicle communication/passenger entertainment for Caltrans, USA.

Sysco's split flap modules for information boards   0092477

Televic NV

Leo Bekaertlaan 1, B-8870 Izegem, Belgium
Tel: (+32 51) 30 30 45
Fax: (+32 51) 31 06 70
e-mail: sales@televic.com
Web: http://www.televic.com

Subsidiaries
Televic SA
ZI de la Pilaterie, Actiparc, Bâtiment A, 19 rue
Ladrie, F-59650 Villeneuve d'Ascq, France
Tel: (+33 3) 28 33 88 10   Fax: (+33 3) 28 33 88 11

Televic NV
Piet Mertenshof 14, NL-4882 BB Zundert,
Netherlands
Tel: (+31 76) 597 93 41   Fax: (+31 76) 597 94 89

Televic UK Ltd
Woolpit Business Park, Unit 19A, Woolpit, Bury St
Edmunds IP30 9UP, UK
Tel: (+44 1359) 24 42 62   Fax: (+44 1359) 24 25 79

Products
Onboard passenger information systems
incorporating public address and intercom facilities;
passenger at-seat entertainment and 'infotainment'
systems; control systems for onboard systems.
Capabilities include research and development,
project management, production and total quality
management.

Trivector System AB

Åldermansgatan 13, SE-227 64 Lund, Sweden
Tel: (+46 42) 38 65 00   Fax: (+46 42) 38 65 25
e-mail: info@trivector.se

Key personnel
Managing Director: Klas Odelid
Marketing Manager: Ola Fogelberg
Project Manager: Anders Månsson

Products
TriTrans, a module-based information system
for public transport which operates in real time
and serves all parties involved such as planners,
dispatchers, drivers and passengers. TriTrans is
based on a network of onboard vehicle computers
and a traffic control centre where information
is stored and processed to provide the different
user groups with the most reliable information.
IVIS (In-Vehicle Information System) contains the
complete set of information system components
which include vehicle computer Triveco 8, speech
modules, message displays and passenger
counter and is currently in use in the Skania
region of southern Sweden; Karlstad, Sweden;
and Oslo, Norway. IntraInfo, a system for real-time
information at bus or train stops which provides
travellers with information in real time on train
and bus departure and arrival times. The system
contains the computer unit IRU-IS, together with
whatever message displays and monitors the
customer requires. It is in use in Karlstad and soon
to be in use in Dalarna.

UniControls AS

Křenická ulice-2257, CZ-100 00 Prague 10, Czech
Republic
Tel: (+420 2) 72 01 14 11   Fax: (+420 2) 74 81 44 75
e-mail: unic@unicontrols.cz
Web: www.unicontrols.cz

Key personnel
Marketing Manager: Marian Belosovic

Products
LED-based passenger information systems.

UPDATED

Vaughan Harmon Systems Ltd

The Maltings, Hoe Lane, Ware SG12 9LR, UK
Tel: (+44 1920) 44 33 00   Fax: (+44 1920) 46 07 02
e-mail: sales@vaughanharmon.com
Web: http://www.vaughanharmon.com

Head office
Harmon Industries Inc
Blue Springs, Missouri, USA

Key personnel
For full personnel listing, see Vaughan
Harmon Systems Ltd entry in *Signalling and
communications systems* section.

Products
Passenger information systems; train reporting
systems; staff information systems; timetable
creation and control systems.

Voice Perfect Ltd

103 Friern Barnet Road, London N11 3EU,UK
Tel: (+44 20) 82 11 32 11   Fax: (+44 20) 83 68 08 88
e-mail: tech@voiceperfect.co.uk
Web: www.voiceperfect.co.uk

Key personnel
Commercial Director: Nick Hallett
Technical Director: Rufus Potter

Products
Digital voice announcers: PC-local customer
address system (PC-LCAS); PC-Discrete Customer
Address System (PC-DCAS); PC-Scheduled
Customer Announcing System (PC-SCAS).

Voice transfer systems: Talk-2 and variants;
Talk-2 voice intercom and door release; GateLine
Approach Point system (GLAP).

Voice Perfect offers supply, installation, user
and management training, commissioning and
administrative support, together with preventative
and corrective maintenance of its products and is
ISO 9002 accredited.

Over 100 PC-DCAS are now installed on London
Underground stations either as stand-alone or
integrated with help point systems. Both systems
feature a playlist of the user's compilation for
'fixed' or unchanging messages, users' own
recordings and announcements compiled from
fragments of variables.

3-Factors software and repertoire structure is
the engine for production of both seamless and
believable variable announcements selected from
a cascading series of icons by the user. Other
languages can be accommodated as a pre-recorded
human voice is used. DCAS is for Section XII
underground sites, while LCAS is a version with
the focus on more local pre-recorded material for
non-Section XII stations. PC-SCAS uses a scheduler
as its compilation engine. 3-Factors still provides
seamless variable messages, augmented by the
capability to broadcast differing input from remote
signalling sources up to 12 separate destination
zones. This can be linked to a CIS server to replicate
and augment visual messages and be integrated into
single-screen solutions. Talk-2 in up to five different
variants provides a customised solution to new or
retrofit secure ticketing window situations. Around
500 units are installed in London Underground
ticket offices. The same principle is extended to
station control rooms where the supervisor is able
to conduct a two-way conversation with a customer
outside, protected by a locked door until the release
button is activated. GLAP boxes are equipped with
Talk-2 circuitry and peripherals modified to suit a
small, enclosed environment.

Contracts
Upgrading PC-DCAS to PC-DCAS and Talk-2 on
Jubilee Line Extension, PC-DCAS, LCAS and Talk-2
on all Metronet SSL stations, GLAB at selected sites.
PC-SCAS and CIS on 17Tube Lines SMS installations
as year one and two of the seven and a half year PPP
refurbishment project.

UPDATED

Vossloh Information Technologies Karlsfeld GmbH

PO Box 1320, D-85751 Karlsfeld,
Dieselstrasse 8, D-85757 Karlsfeld, Germany
Tel: (+49 8131) 907 53 00
Fax: (+49 8131) 907 51 10
e-mail: info@vitk.vossloh.com
Web: www.vit.vossloh.com

Key personnel
Managing Directors: Reinhold Hundt,
    Werner Malcherek
Head of Sales: Albert Bastius
Head of Marketing: Dr Marion Behr

Background
Vossloh Information Technologies Karlsfeld GmbH
is a subsidiary of Vossloh Information Technologies
GmbH, Kiel, Germany, which is part of the Vossloh
Group.

Products
Fully automatic networked stationary and mobile
customer information systems and products for
railways, urban public transport and airports.
Monitoring and control of customer information
allows the operators to control groups of stations
or an entire urban network. A user-friendly fault
detection concept ensures an optimal degree
of reliability and functionality for service and
maintenance. Real-time arrival data allows precise
passenger information. This information can be
shown on displays available in split-flap, LCD, LED,
BiLED, TFT, Plasma and CRT display technology. It
will also be used on innovative display technologies
such as electronic ink.

UPDATED

Vultron International Ltd

City Park Industrial Estate, Gelderd Road, Leeds
LS12 6DR, UK
Tel: (+44 113) 263 03 23
Fax: (+44 113) 279 41 27
e-mail: sales@vultron.co.uk
Web: www.vultron.co.uk

Key personnel
Managing Director: John Moorhouse
Project Manager: Paul Kiley
Purchasing Manager: Rod Giles
Director of Operations: Robert D Lemmen

Products
Vultron International Ltd is a manufacturer and
supplier of electronic information display systems,
including talking information systems that translate
passenger information text into speech using a
speech conversion software, enabling operators to
address the needs of visually impaired passengers.
Vultron's manufacturing facility is capable of
producing information displays in electromagnetic,
LCD and LED technologies. The in-house design
and manufacturing skills and range of technologies
enable the company to offer customer-tailored
products.

Contracts
Vultron's LCD clocks have been installed at
Glasgow's Partick and Queen Street stations.
Recently display boards, using the new LCDs, have
been supplied to West Yorkshire PTE for eight of its
largest bus stations.

Developments
Recent developments in products for the bus
market have seen the introduction of a range of
LCD products for use at bus stops and bus shelters.
These displays, which utilise a graphic panel LCD,
can be used to display scheduled and/or real-time
bus information.

UPDATED

Thoreb AB

Gruvgatan 37, SE-421 30 Västra Frölunda, Sweden
Tel: (+46 31) 734 39 00
Fax: (+46 31) 734 39 10
e-mail: info@thoreb.se
Web: www.thoreb.se;
   www.true-realtime.com

Key personnel
Managing Director: Michael Sigvardsson
Marketing Directors: Johan Kallvik; Esbjorn Lif
Chief Financial Officer: Thord Brynielsson

Products
Automatic vehicle location, passenger counting,
data communication, passenger information
displays, communication systems for public
transport vehicles and multiplex electrical
systems for heavy vehicles. Real time passenger
information systems.

NEW ENTRY

For details of the latest updates to *Jane's World Railways* online and to discover the additional
information available exclusively to online subscribers please visit
jwr.janes.com

REVENUE COLLECTION SYSTEMS AND STATION EQUIPMENT

Alphabetical listing

4P Mobile Data Processing
Abberfield Technology
ADT Security Systems
Almex Information Systems
Ascom AG
ASK
Atron Electronic GmbH
Automaten Technik Baumann GmbH
Automatic Systems
BelBim AS
Bemrose Booth Ltd
Burall InfoSmart Ltd
BZA
Casas M, SL
Comelta SA
Cubic Transportation Systems
Dassault Automatismes et Télécommunications
DocuSystems Inc
Elgeba Gerätebau GmbH
ERG Transit Systems
Faiveley Transport SA

FIREMA Trasporti SpA
Floormaster-Interfloor
Fujitsu Systems
The Gates Rubber Co
GFI-Genfare
Guhl & Scheibler
Gunnebo Entrance Control SpA
Hering-Bau
Höft & Wessel
Howe Green Ltd
IBM
ICA Traffic GmbH
IER
Indra
Intec Ltd
Kaba Gilgen AG
Klüssendorf Produkte und Vertriebs GmbH
L-3 Communications Corporation
Maexbic SA
Magnadata International Ltd
Magnetic Autocontrol

MEI
Motorola
Nabtesco Corporation
Narita Manufacturing Ltd
Newbury Data
Nippon Signal Co Ltd
Omron Corporation
Philips Semiconductors
Regazzi SA
Sadamel Ticketing Systems
Scheidt & Bachmann GmbH
SchlumbergerSema
Shere Limited
Siemens Transportation Systems
STERIA Group
Takamisawa Cybernetics Co Ltd
Thales
Toshiba Corporation
Toyo Denki Seizo KK
Wanzl Metallwarenfabrik GmbH
Zelisko Elektrotechnik un Elektronik GesmbH

Company listing by country

AUSTRALIA
Abberfield Technology
ERG Transit Systems

AUSTRIA
Philips Semiconductors
Zelisko Elektrotechnik un Elektronik GesmbH

BELGIUM
Automatic Systems

CANADA
ADT Security Systems

FRANCE
ASK
Dassault Automatismes et Télécommunications
Faiveley Transport SA
IER
SchlumbergerSema
STERIA Group
Thales

GERMANY
Atron Electronic GmbH
Automaten Technik Baumann GmbH
Elgeba Gerätebau GmbH
Hering-Bau
Höft & Wessel
ICA Traffic GmbH
Klüssendorf Produkte und Vertriebs GmbH

Scheidt & Bachmann GmbH
Siemens Transportation Systems
Wanzl Metallwarenfabrik GmbH

ITALY
4P Mobile Data Processing
FIREMA Trasporti SpA
Gunnebo Entrance Control SpA

JAPAN
Nabtesco Corporation
Narita Manufacturing Ltd
Nippon Signal Co Ltd
Omron Corporation
Takamisawa Cybernetics Co Ltd
Toshiba Corporation
Toyo Denki Seizo KK

KOREA, SOUTH
Intec Ltd

SPAIN
Casas M, SL
Comelta SA
Indra
Maexbic SA

SWITZERLAND
Ascom AG
Guhl & Scheibler
Kaba Gilgen AG

MEI
Regazzi SA
Sadamel Ticketing Systems

TURKEY
BelBim AS

UNITED KINGDOM
Almex Information Systems
Bemrose Booth Ltd
Burall InfoSmart Ltd
Floormaster-Interfloor
Fujitsu Systems
Howe Green Ltd
IBM
ICL
Magnadata International Ltd
Newbury Data
Shere Limited

UNITED STATES
BZA
Cubic Transportation Systems
DocuSystems Inc
GFI-Genfare
L-3 Communications Corporation
Magnetic Autocontrol
Motorola

4P Mobile Data Processing

Viale della Regione Veneto 26, I-35127 Padova, Italy
Tel: (+39 049) 806 98 11   Fax: (+39 049) 806 98 43
e-mail: webinfo@4p-online.com
Web: http://www.4p-online.com

Key personnel
Managing Director: Silvano Mansutti

USA office
7400 Oxford Avenue, Philadelphia, Pennsylvania 19111-3095

Product
Hand-held computer with card reader for outdoor operations.

Abberfield Technology Pty Ltd

32 Cross Street, Brookvale, New South Wales 2100, Australia
Tel: (+61 2) 99 33 28 44   Fax: (+61 2) 99 38 34 62
e-mail: contact@abberfield.com.au
Web: http://www.abberfield.com.au

Key personnel
Managing Director: John M Colyer
Sales: Ann Stokes

UK Technical Support Office
Abberfield (Europe) Ltd
4 Andover Street, Sheffield S3 9EG
Tel: (+44 114) 272 71 08   Fax: (+44 114) 272 71 08

Products
Ticket vending machines and ticket validators; design and manufacture of ticketing systems.

ADT Security Systems

2815 Matheson Boulevard East, Mississauga, Ontario L4W 5J8, Canada

Products
Access control, closed circuit system television, fire alarm systems and intruder alarm systems.

Almex Information Systems

Metric Group
Metric House, Love Lane, Cirencester GL7 1YG, UK
Tel: (+44 1285) 65 14 41   Fax: (+44 1285) 65 06 33
e-mail: info@almex.demon.co.uk

Key personnel
Managing Director: Marcus Burton
Sales Team: Ashley Bailey
   Alistair Aitken
   Sophie Fitzpatrick
   Claudia Johnston

Principal subsidiaries
Allwood Brighton Office Centre
2 Brighton Road, 3rd Floor, Clifton, New Jersey 07012, USA
Tel: (+1 201) 777 59 69
General Manager: James Meany

Almex GmbH
Kuehnstrasse 71, D-22045 Hamburg, Germany
Tel: (+49 40) 66 99 22 20
Managing Director: Klaus Schiering

Background
Almex Information Systems is a member of the Hanover-based Höft & Wessel group.

Products
Ticketing and revenue collection systems for railway and light rail applications; ticket issuing machines, portable ticket issuing machines, magnetic ticket validators, automatic vending machines; contact and contactless smartcards.

Ascom AG

Belpstrasse 37, CH-3000 Bern 14, Switzerland
Tel: (+41 31) 999
Web: http://www.ascom.com

Key personnel
Head of Business Unit: Stefan Kalt
Marketing: Daniel Burkhalter

Associate companies
Ascom Autelca AG
Wobstrasse 201, CH-3073 Gümligen, Bern, Switzerland
Tel: (+41 31) 999 61 11   Fax: (+41 31) 999 68 11
Head of Unit: Christopher Franzen
Sales: Leo Muff
Marketing: Robert Engel

Ascom Monétel SA
rue Claude Chappe, F-07503 Guilherand-Granges, France
Tel: (+33 4) 75 81 41 41   Fax: (+33 4) 75 81 42 00
Head of Unit: Eric Jean
Sales and Marketing: Robert Coste

Products
Automatic revenue collection systems including stationary and onboard ticket vending machines, ticket office machines, driver consoles, access gates and validators for rail, metro, tramway and bus operators. Ascom's fare collection systems are designed for use with magnetic tickets (TFCO and TFC1 formats) or with contactless smartcards (memory type A and microprocessor type B).

Collection and processing of all relevant data, revenue collection in multi-operator transport networks and clearing of cash and cashless payments are covered by customised data management systems.

Contracts
Recent contracts include the supply of equipment to DB AG, SBB, ÖBB, NS, MTRC Hong Kong, rail operators in UK, New Jersey Transit, RATP Paris, Mexico City, Meddelin (Colombia), Kuala Lumpur, Brasilia, Toulouse, Goiania (Brazil), Adelaide, Nice, Montpellier, Porto and Malaysia.

ASK

15 rue du Louvre, F-75001 Paris, France
Tel: (+33 1) 42 33 64 15   Fax: (+33 1) 42 33 64 14
e-mail: info@ask.fr
Web: http://www.ask.fr

Factory
15 Traverse des Brucs, F-06560 Sophia Antipolis, France
Tel: (+33 4) 97 21 40 00   Fax: (+33 4) 92 38 93 21

Offices
Suite 908B, Lippo Center, Tower 1 – 89 Queensway, Hong Kong, China
Tel: (+852) 29 18 90 33   Fax: (+852) 25 21 26 26

155 Post Road East, Suite 3, Westport, Connecticut 06880, USA
Tel: (+1 203) 226 96 70   Fax: (+1 203) 698 30 28

Products
Contactless smartcards and related equipment for public transport revenue collection. Products include contact and contactless smartcards, contactless memory cards, contactless paper tickets, contact and contactless smartcard readers and handheld readers. Support services include project and service management and the development and production of customised smartcards.

Atron Electronic GmbH

Am Ziegelstadel 12 + 14, D-85570 Markt Schwaben, Germany
Tel: (+49 8121) 934 20   Fax: (+49 8121) 93 42 77

e-mail: vertrieb@atron.de
Web: www.atron.de

Key personnel
Marketing: Christine Baumgart

Products
Smartcard terminals; AFC systems; stationary and mobile ticket machines.

UPDATED

Automatic Systems

Avenue Mercator 5, B-1300 Wavre, Brussels, Belgium
Tel: (+32 10) 23 02 11   Fax: (+32 10) 23 02 02
e-mail: asmail@automatic-systems.com
Web: http://www.automatic-systems.com

Key personnel
Chairman of the Board: Joël Favé-Lesage
Commercial Director: John De Winter
Financial Director: Frank Harel
Projects and Public Transport Director:
   Benoît Lordet
Marketing Director: Jean-Pierre Leleu
Engineering and Costing Director:
   Etienne Bertrand
Project Development Director: Maxime Boulvain
Manufacturing Director: Christophe Tribouillard
Quality Manager: Eric Vanderelst
Security Manager: Jean-Pierre Deketelaere
Communication Manager: Dominique Gilbart

Background
Automatic Systems is a member of the IER, a subsidiary of the French group Bolloré.

Subsidiary companies
Benelux
Automatic System sa
5, avenue Mercator, B-1300 Wavre, Brussels
Tel: (+32 10) 23 02 11   Fax: (+32 10) 23 02 02
e-mail: asmail@automatic-systems.com

Canada – Montreal
Automatic Systems America Inc
4005, Boulevard Matte Local D, Brossard, Québec J4Y 2P4
Tel: (+1 450) 659 07 37   Fax: (+1 450) 659 09 66
e-mail: sales@automatic-systems.ca

China
Automatic Systems China
Kuen Yang Plaza # 1101, 798 Zhao Jia Bang Road, Shanghai, China 200030
Tel: (+86 21) 64 66 82 05   Fax: (+86 21) 64 73 68 06
e-mail: assales@ieschina.com

France
Automatic Systems sa
18 rue de l'Estérel, Silic BP 518, F-94623 Rungis, Cedex, Paris
Tel: (+33 1) 56 70 07 07   Fax: (+33 1) 56 70 07 08
e-mail: asmail@automatic-systems.fr

Germany
Automatic Systems sa
Kinzigstrasse 56, D-75210 Keltern
Tel: (+49 7236) 93 26 43   Fax: (+49 7236) 93 27 92
e-mail: info@auto-sys.de

Spain
Automatic Systems Española sa
C/Vallés, 52-54 E-08820, El Prat de Llobregat, Barcelona
Tel: (+34 93) 478 77 55
Fax: (+34 93) 478 67 02
e-mail: asemadrid@automatic-systems.net

UK
Automatic Systems Equipment UK Ltd
Two Brabazon Court, Borman, Tamworth, Staffordshire B79 7TA
Tel: (+44 1827) 31 35 39
Fax: (+44 1827) 31 35 40
e-mail: sales@automaticsystems.co.uk

Automatic Systems' high-speed gates, POR365, installed by Bay Area Rapid Transit (BART), San Francisco, USA   0114854

USA

Automatic Control Systems Inc
8 Haven Avenue, Suite 205, Port Washington, New York 11050
Tel: (+1 516) 944 94 98   Fax: (+1 516) 767 34 46
e-mail: marketing@automaticsystems.net

Products
Design and manufacture of access control toll and security equipment for vehicles and pedestrians; automatic rising barriers, tripod turnstiles, high security gates, safety rotating drums, unguarded gates, automatic exit doors, high-speed gates, special tripod turnstiles for buses and uni- or bidirectional counting system to count pedestrians in public areas. The modular public transport range of equipment includes the TGH800 series high security door, TGF820 series high-speed door, TGD830 series bi-directional door, TGT850 series tripod turnstile, TGE860 series exit-only door and TGL87 series lock door.

Contracts
Recent contracts include: an agreement with Scheidt & Bachmann USA Inc for the supply of over 700 pedestrian access devices for the Massachusetts Bay Transport Authority (MBTA). In February 2003, in close partnership with Scheidt & Bachmann, it won the contract for the replacement of all the MBTA automatic fare collection (AFC) systems. This project includes fare vending machines, fares boxes, fare media validating equipment, automatic gates, station information centres, outlet sales, handheld card readers, high-speed magnetic encoding machines, smart card initialization machines, photo ID equipment and central processing systems.

Automatic Systems supplied equipment for fee collection in underground networks in Singapore, Brasilia, Buenos Aires, Warsaw, Barcelona, Stockholm, in railway networks in Portsmouth, Southampton, Epsom, Chelmsford, Romford, UK and in the light rail system of Tunis. The installation of 200 PNG382 in partnership with Sainco Traffico for the Catalan region of Catalogne, Spain; South West Trains, London; Connex, London. The delivery of 70 turnstiles for Irish Rail, Ireland and the delivery of more than 1,200 barriers from 8 to 13.5 m for SNCB Belgian National Railways, Belgium.

Automaten Technik Baumann GmbH

An der Bahn 11, D-92706 Luhe-Wildenau, Germany
Tel: (+49 9607) 922 20   Fax: (+49 9607) 922 22 35
e-mail: info@automatentechnik-baumann.de
Web: http://www.automatentechnik-baumann.de

Key personnel
Sales: Jörg Zeitler

Products
Ticket-vending machines for the issue of pre-printed tickets using coins or debit cards; Klüssendorf- type ticket cancelling machines; chipcard reloading and vending machines; parking 'pay and display' machines.

BelBim AS

Bulgurlu Mahir Libadiye Cad No 27, TR-81190 Üsküdar, Istanbul, Turkey
Tel: (+90 216) 521 70 90   Fax: (+90 216) 521 73 74
e-mail: info@belbim.com.tr
Web: http://www.belbim.com.tr

Key personnel
General Manager: Hasan Haki
Project Co-ordinator: Hamza Ozturk

Background
BelBim is affiliated to Istanbul Municipality and was established in 1987.

Products
Advice on design, development, production of electronic and computer automation projects, traffic signalling equipment and other utility issues for both public and private institutions.

A smartcard electronic ticket named AKBIL was developed for Istanbul. The city has 12 million inhabitants and five million passengers a day, and has a steamship, motorboat, ferry, suburban railway, light rail transport, trams, buses and private minibuses. Until AKBIL was introduced, passengers used over 45 different tickets, tokens, passes and cards. AKBIL stands for Electronic Ticket and Fare Collection System for the Intermodal Public Transport System.

The system is based on touch-based technology and the electronic ticket is recharged at automatic refilling machines. Each ticket has a coloured plastic handle about 5 cm long with transponder. It has a minimum operational life of six years and is water resistant.

The fare collection system now comprises:
- 3,600 validators, bus type
- 85 validators, tram type
- 752 turnstiles (seabus, steamships, subways, suburban railway system)
- 99 selling/refilling machines for AKBIL
- 68 automatic refilling machines for AKBIL
- 342 vending machines, equipped with AKBIL validators
- various station data collection computers
- transmission equipment
- database management software.

Bemrose Booth Ltd

Stockholm Road, Sutton Fields, Hull HU7 0XY, UK
Tel: (+44 1482) 82 63 43   Fax: (+44 1482) 37 13 86
Web: www.bemrosebooth.com

Key personnel
Director: Tim Cammack
Business Manager: Lisa Precious

Background
Bemrose Booth Ltd was formed by a management buyout in July 2000 from Bemrose Corporation plc. The company was acquired in December 2003 by Appleton, a US manufacturer of value-added paper products based in Appleton, Wisconsin.

Products
Printing of specialised secure tickets including related controlled stationery for transit auto parking applications.

Contracts
Recent contracts include a five-year deal to single source supply the UK rail network's requirements for tickets and controlled stationery.

UPDATED

Burall InfoSmart Ltd

Cromwell Road, Wisbech PE14 0SN, UK
Tel: (+44 1945) 46 81 00   Fax: (+44 1945) 46 70 95
Web: www.burall-infosmart.com

Key personnel
Managing Director: R J Duddin
Sales and Marketing Director: M G Moorey
Business Development Director: L S Faulkner

Products
Low-coercivity and high-coercivity magnetic stripe cards, tickets and tokens, in plain and thermal paper, card, laminate and plastic; pre-encoding is a speciality. Contactless chip cards including high-frequency proximity cards to the Philips Mifare® and ISO 14443A/B standards. Full bureau service, software and systems for colourcard personalisation, numbering and programming.

Developments
Card personalisation products and services certified compatible with the emerging ITSO standard in the UK. Small-memory, disposable contactless tickets.

NEW ENTRY

BZA

Suite 230, 8466 North Lockwood Ridge, Sarasota, Florida 34243, USA
Tel: (+1 941) 351 67 97   Fax: (+1 941) 351 95 12

Products
Stored-value smartcards, mobile ticket-vending unit, smartcard loading terminals, smartcard reader.

Casas M, SL

PO Box 1.333, Poligono Santa Rita, E-08755 Castellbisbal, Barcelona, Spain
Tel: (+34 93) 772 46 00
Fax: (+34 93) 772 21 30
e-mail: casas@casas.net
Web: http://www.casas.net

Key personnel
Export Manager: Umberto Ferrari Cassina
Marketing: Rosa Maria Casas

Products
Compas range of modular seating in aluminium, with or without backs, up to six seats in length. Impact and scratch-resistant material is used. Suma range in steel has seating for up to six. Units can be joined together.

Casas seats are fire, impact and scratch resistant.

Contracts
Provision of seating at Barajas Airport in 1997 and 1998, Antaiya International Airport in Turkey, Rio de Janeiro International Airport and Belem International Airport in Brazil in 1998 and recently in Cadorna railway station, Milan, Italy.

Comelta SA

Avda Parc Tecnológic, 4, E-08290 Cerdanyola del Vallès, Barcelona, Spain
Tel: (+34 91) 582 19 91   Fax: (+34 91) 582 19 92
e-mail: infocom@comelta.es
Web: http://www.comelta.es

Key personnel
Technical Director: Mark Rocky
Sales Manager: Maria Delmar Vías

Background
Comelta is owned by Avanzit.

Products
Ticketing machines: self-service terminals, TIC, Serviticket, Smallticket.

Cubic Transportation Systems

Cubic Corporation
5650 Kearny Mesa Road, San Diego, California 92111, USA
Tel: (+1 619) 268 31 00   Fax: (+1 619) 571 99 87
Web: www.cubic.com

Key personnel
Chairman: Walter C Zable
President and Chief Executive Officer:
  Richard Johnson
Chief Operating Officer: Dave Lapczynski

Principal subsidiaries
Cubic Transportation Systems Ltd, European Headquarters, UK
Cubic Nordic, a branch of Cubic Transportation Systems Ltd
Cubic Transportation Systems Australia Pty Limited, Australia
Cubic Transportation Systems GmbH, Germany
Cubic Transportation Systems (Beijing) Co, Ltd

Background
The company has major operating units in the UK; Australia; China; Denmark; and Tennessee, New York and Washington DC, USA. Cubic Transportation Systems is one of two major business segments of Cubic Corporation. The company has recently expanded its Card Services Group to help transport agencies leverage their smartcard-based fare collection systems by linking them with other transportation-related and third-party products and services, including parking, tolls, taxis, event ticketing, concessions, retail and security access control.

Products
Design, manufacture, installation and maintenance of electronic ticketing and automatic fare collection systems, including magnetic and contactless smartcard systems for advanced fare collection applications. Cubic has developed Nextfare™ Central System, a configurable suite of software module designed using industry standards, open platforms and commercial off-the-shelf applications.

Nextfare supplies the core smartcard transaction processing; financial operations including revenue clearing, settlement and reporting; a debit/credit account gateway for payments and customer service database support.

Range
Provision of turnkey systems for automatic fare collection for public transport including bus, bus rapid transit, light rail, commuter rail, heavy rail, ferry and parking. Cubic solutions and services include system design, central computer systems, equipment design and manufacturing, device-level software, integration, test, intallation, warranty, maintenance, computer hosting services, call centre services, card management and distribution services, financial clearing and settlement, multi-application support and outsourcing services.

Contracts
Active projects include: Atlanta, Baltimore, Bangkok, Brisbane, Edmonton, Houston, New Jersey, New York, London, Scandinavia, Singapore, South Florida, Vancouver, Virginia and Washington DC.

UPDATED

Dassault Automatismes et Télécommunications

9 rue Elsa Triolet, ZI Les Gâtines, PO Box 13, F-78373 Plaisir Cedex, France
Tel: (+33 1) 30 81 27 53   Fax: (+33 1) 30 81 25 21
Web: http://www.dassault-at.fr

Key personnel
Chairman and Chief Executive Officer:
  Jean-Paul Vautrey
Executive Vice-President: Daniel Pelletier
Director, Transport Department:
  Jean-Pierre Fournier

Principal subsidiaries
Dassault Electronique Inc, New York, USA
DAT Telecommunications Equipment (Beijing) Co Ltd
ICS, Madrid, Spain
Business Relations Office, Hong Kong

Products
Automatic ticketing and access control systems for transport operators in urban, air, sea and rail applications. Range includes ATB printers, boarding gates and card readers, self-service ticketing and self check-in devices.

The BPR 640 boarding pass printer is designed for processing magnetic boarding passes for travel agencies or check-in desks. This compact machine allows ticket validation, reading, encoding and printing with the possibility of using three different bins for paper load. The main characteristics of this printer are its size (220 × 220 × 400 mm), plug and play architecture and easy maintenance.

The BGD 320 boarding gate device is capable of decoding/encoding magnetic tickets for all the boarding passes issued from ATB printers.

Dassault AT also offers AFC systems for rail operators including ticket vending and office machines, control gates and validators and central management systems which collect traffic data.

Dassault AT started supplying ticketing systems for RATP/Paris Metro over 30 years ago. Since then, the company has extended its market throughout the world to cities including Madrid, London, Cairo, Los Angeles and Calcutta.

Dassault AT has developed a range of contactless ticketing products which have been implemented in several locations in Europe.

DocuSystems Inc

8700 Waukegan Rd, Morton Grove, Illinois 60053, USA
Tel: (+1 847) 583 76 07   Fax: (+1 847) 583 12 47
e-mail: vheaton@docusysinc.com
Web: http://www.docusystems.net

Key personnel
Vice-President, Sales and Marketing:
  Vincent J Heaton

International office
Tring Business Park, Upper Icknield Way, Tring HP23 4JX, UK
Tel: (+44 1442) 82 40 11   Fax: (+44 1442) 82 85 31
Director of Sales: A Heseltine

Products
Magnetically striped and bar coded tickets and cards, and other security printed products, on paper and plastic.

Elgeba Gerätebau GmbH

Eudenbacher Strasse 10-12, D-53604 Bad Honnef, Germany
Tel: (+49 2224) 828 50   Fax: (+49 2224) 802 94
e-mail: info@elgeba.de
Web: http://www.elgeba.de

Key personnel
Managing Director: Bodo Faber
Sales Manager: Peter Stegmayer

Products
Ticket cancelling, vending and printing machines, information displays, master control units, EDP revenue systems, special purpose machines, cashless payment systems (magnetic or smartcards).

ERG Transit Systems Ltd

247 Balcatta Road, Balcatta Western Australia 6021, Australia
Tel: (+61 8) 92 73 11 00   Fax: (+61 8) 93 44 36 86
e-mail: info@au.ergtransit.com
Web: http://www.erggroup.com

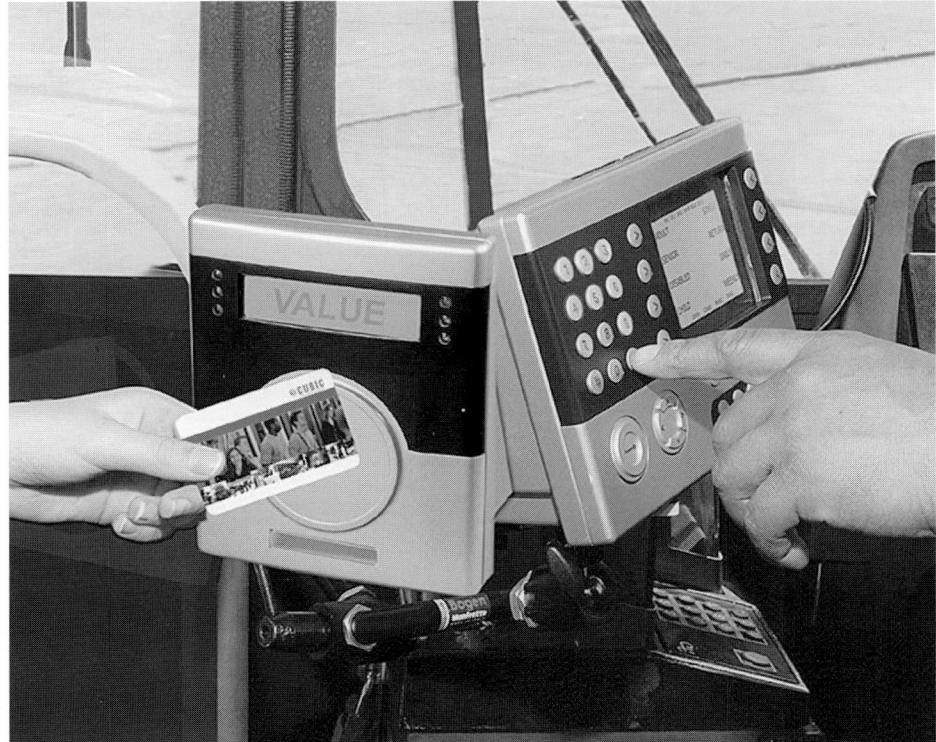

Cubic's mobile ticketing machine   1065753

Key personnel

Chief Executive Officer: Dr Allan Sullivan
Chief Operating Officer: Terry O'Leary
General Manager Asia Pacific: Rob Noble

Principal subsidiaries

ERG Transit Systems (Eur) NV
Kleine Kloosterstraat 23, B-1932 Zaventum, Belgium
Tel: (+32 2) 722 89 11   Fax: (+32 2) 720 87 94
e-mail: info@be.ergtransit.com
Chief Executive Officer: Franky Carbonez

ERG Transit Systems (USA) Inc
Suite 900, 1800 Sutter Street, Concord, California 94520, USA
Tel: (+1 925) 686 82 00   Fax: (+1 925) 686 82 20
e-mail: info@us.ergtransit.com
General Manager: Mike Nash

Background

ERG Transit Systems specialises in the design, development, supply and operation of integrated fare management and software systems for public transport. The company pioneered the use of magnetic tickets and contactless smartcards in public transport.

Products

Fare collection products and systems for all modes of transport using all types of ticket technology including paper, magnetic stripe and contactless and dual interface smartcards.

Range includes: advanced on-board devices such as ticket issuing machines, validators, contactless smartcard readers, portable inspection and ticket issuing devices; on-station equipment such as platform validators, add value machines, gate control units; data communications equipment and software; back-office software for small to medium size and for complex multi-modal transit systems; and central clearing house software.

Developments

ERG has developed MASS, a central computer processing system that is capable of managing a smartcard database, financial reconciliation and management data for both transit and other card applications. ERG has developed the TP5000 ticket processor which is designed for issuing paper tickets and processing contactless smartcards on buses and trams. This new ticket issuing machine is based on ERG's successful TP4000 and includes many of its proven components such as the high-speed thermal printer, quick-disconnect mounting cradle and durable plastic mouldings. The TP5000 incorporates a large graphical display, a power microprocessor, a Global Positioning System and an in-built wireless local area network.

Contracts

In the first quarter of 2003 ERG won three major ticketing contracts to implement smartcard based integrated ticketing systems for the cites of Sydney, Australia; Seattle, US and Stockholm, Sweden.

In April 2002, Singapore's ez-link smartcard ticketing system went into full service. The system was developed by ERG, and is one of the largest integrated smartcard based transit systems in the world, with more than 22,000 readers in place across the five transit operators covering bus, rail and light rail.

The Singapore project included the installation of ERG's central clearing house and data processing network which is currently processing over four million transactions per day. ERG is implementing the largest smartcard contract in the US to design, build, operate and maintain the TransLink® fare payment system in the San Francisco Bay area. In June 2002, ERG completed the commissioning of the entire San Francisco Muni Metro system. Phase 1 of the project implementation encompassing the six largest operators in the area and involving selected buses, light, medium and heavy rail and ferries, was completed in the first quarter of 2003. ERG established a central clearing house to process transaction and settle payments on a daily basis between the operators and the participants in the TransLink® scheme.

In March 2001, ERG and SchlumbergerSema were awarded a contract to lead the implementation of a new fare collection system for the city of Bordeaux. The first phase of the project comprising the city's 700 buses went live in May 2003 and the system will be fully installed prior to the inauguration of Bordeaux' new tramway in September 2003. ERG and SchlumbergeSema also won a contract to supply the city of Grenoble with a new ticketing system comprising both magnetic tickets and smartcards. Phase 1 of the project went live in January 2002 and the full system rollout was completed in April 2003.

ERG is also implementing smartcard based automated fare collection systems in cities such as Manchester, UK; Gothenburg, Sweden; Rome, Italy; Oslo, Norway and Toronto, Canada.

Faiveley Transport SA

Carrefour Pleyel, 143 boulevard Anatole France, F-93285 Saint-Denis Cedex, France
Tel: (+33 1) 48 13 65 00   Fax: (+33 1) 48 13 66 47
e-mail: info@faiveley.com
Web: www.faiveley.com

Background

The Faiveley group completed the acquisition of SAB WABCO in November 2004.

Key personnel

Chairman and Chief Executive Officer:
  Robert Joyeux
Financial Director: Sven Schopp
General Manager: Pierre Sainfort
Communications Manager: Edmond Ballerin

Products

Platform screen doors.

UPDATED

FIREMA Trasporti SpA

Headquarters

Via Provinciale Appia, Località Ponteselice, I-81100 Caserta, Italy
Tel: (+39 0823) 09 71 11   Fax: (+39 0823) 46 68 12
Web: http//www.firema.it

Key personnel

Chairman: A de Benerdictis
Managing Director: L Rigno
Commercial Manager: S d'Arminio
Marketing Manager: M Fantini

Commercial and technical offices

Via Triboniano n 220, I-20156 Milan, Italy
Tel: (+39 02) 23 02 02 23   Fax: (+39 02) 23 02 03 00

Products

Automatic fare collection systems employing magnetic or paper tickets and/or contactless smartcards; integrated ticketing and station automation.

Contracts

Contracts include the supply of automatic fare collection systems for eight stations on Rome Metro Line A; 10 stations on the Rome Metro Line B and entry/exit gates for four stations on Milan's Passante Ferroviario.

Floormaster-Interfloor

Edinburgh Road, Heathhall, Dumfries DG1 1QA, UK
Tel: (+44 1387) 26 95 51   Fax: (+44 1387) 24 09 00
Web: www.interfloor.com

Products

Fire-safe, low-smoke, low-toxicity floor coverings for rail vehicle applications, including Floormaster 9200 M2F1 for surface stock and Floormaster Plus 9300 for use in metro vehicles.

UPDATED

Fujitsu Systems

26 Finsbury Square, London EC2A 1SL, UK
Tel: (+44 870) 242 79 98   Fax: (+44 870) 242 44 45
e-mail: askfujitsu@services.fujitsu.com
Web: http://www.fujitsu.com

Key personnel

Account Director, Rail: Richard Dickson
Business Systems, Rail: Jon Wellings
Business Development, Rail: Carol Jones

Products

Rail Journey Information Service (RJIS), an integrated solution for rail journey information providing timetable, fares, routeing guide and other supplementary information. RJIS has been accredited by the Association of Train Operating Companies (ATOC).

Station Terminals for Advanced Rail retailing (STAR), a new ticketing issuing system, accredited by the Rail Settlement Plan (RSP). STAR is fully integrated with RJIS and offers a comprehensive and integrated rail journey enquiry and ticket issuing system. It also streamlines the capture of warrant and voucher information for submission to the Travel Trade and Warrants Services (TTWS). STAR is also designed to be fully compatible with the new chip and PIN technologies which are designed to make credit and debit card transactions more secure.

VERIFIED

GFI Genfare

751 Pratt Boulevard, Elk Grove Village, Illinois 60007, USA
Tel: (+1 847) 593 88 55   Fax: (+1 847) 593 18 24
e-mail: kim.green@gfi.gensig.com
Web: http://www.gfigenfare.com

Key personnel

President: James A Pacelli
Vice-President, Sales and Marketing:
  Kim Richard Green

Products

Design, manufacture, sales, installation and maintenance of Automatic Fare Collection (AFC) systems. GFI's fare collection products include: electronic validating and registering fareboxes; electronic fare gates for underground railways; magnetic card processing systems; electronic smartcard processing systems; passenger processing systems; revenue vaults; ticket and token vending equipment; data collection and reporting systems. GFI's line of audio equipment and systems provide automated stop announcement systems.

GFI Genfare's Odyssey validating farebox is now available, a modern transit farebox, providing coin and note validation and smaller than other systems.

Contracts

GFI's fare collection systems are installed in over 200 cities across North America and are in use at more than 85 per cent of the largest North American transit agencies. Over 40,000 electronic registering fareboxes are in service at major transit bus operations; also automatic faregates and vending equipment are installed in many of the largest rail (subway) systems in the US.

GFI has supplied fare collection equipment and systems to major US transit agencies in cities including Atlanta, Boston, Chicago, Cincinnati, Cleveland, Columbus, Dallas, Denver, Detroit, Kansas City, Los Angeles, Louisville, Memphis, Miami, Milwaukee, Minneapolis, New Jersey, New Orleans, New York, Philadelphia, Pittsburgh, Portland and Washington DC. GFI has supplied systems throughout Canada.

Guhl & Scheibler

Pfeffingerring 201, CH-4147 Aesch, Switzerland
Tel: (+41 61) 756 20 20   Fax: (+41 61) 756 21 00
e-mail: mail@guhl-scheibler.ch
Web: http://www.stralfordsgroup.com.ch

Key personnel

Innovation Manager: Merckell Patrick
e-mail: mailto:patrik.merckell@guhl-scheibler.ch

Products

Various machine-produced tickets and ticket machines.

Gunnebo Entrance Control SpA

Via A Volta 15, I-38015 Lavis (Trento), Italy
Tel: (+39 0461) 24 89 00   Fax: (+39 0461) 24 89 71
e-mail: metro@gunneboentrance.com
Web: www.gunneboentrance.com

Head Office

Gunnebo Entrance Control AB
SE-590 93 Gunnebo, Sweden
President: Lars Proos
Market Communication Manager:
  Johan Holmgvist

Key personnel

Managing Director: Michele Maistri
Gennebo Metro General Manager: Leo M Detassis
Domestic Sales Director: Mauro Bonetto

Background

Gunnebo Entrance Control SpA, part of the Gunnebo Group is a fast-growing international security group with over 100 companies in 32 countries and sales to a further 100 markets through agents and distributors. Gunnebo Entrance Control is one of the leading specialists in entrance control solutions.

Products

The Italdis product line offers a wide selection of standardised, functional entrance control solutions. This includes basic tripod turnstiles to more advanced speed gates, full height turnstiles and high performance anti-return gates and security manlocks.

Contracts

Recent contracts include: Transadelaide Light Rail, Australia; Montreal Metro, Montreal, Canada; Shanghai Metro, China; Shenzhen Metro, China; Transmilénio Mass Transit System, Bogotà, Colombia; Toulouse Metro, Toulouse, France; KCRC West Rail, Hong Kong; MTRC, Hong Kong; Tehran Metro, Iran; SBME Metro Milan, Italy; Genova Metro, Italy; Napoli Railways, Naples, Italy; TMB Barcelona Metro, Spain; SL Stockholm Metro, Sweden; Bursa Ray, Bursa, Turkey.

UPDATED

Hering-Bau

Hering GmbH & Co KG
Neuländer 1, D-57299 Burbach, Germany
Tel: (+49 2736) 272 61   Fax: (+49 2736) 272 36
e-mail: gruppe@hering-bau.de
Web: http://www.hering-bau.de

Key personnel

Director: Dipl Ing Annette Hering

Products

Seating, waiting shelters, permanent and temporary prefabricated station platforms, modular toilets for stations.

Höft & Wessel AG

Rotenburger Strasse 20, D-30659 Hannover, Germany
Tel: (+49 511) 610 20   Fax: (+49 511) 610 24 11
e-mail: info@hoeft-wessel.de
Web: http://www.hoeft-wessel.de

Key personnel

Managing Directors: Michael Höft, Rolf Wessel, Peter Claussen
Marketing Communications Manager:
  Nicole Funck

Products

Development and production of ticketing systems, electronic payment, devices for mobile data acquisition, internet terminals, parking systems and telematics.

Range

The HW9096 is a portable terminal for use by staff in trains and buses. It accepts cashless payment means.

The H4290/95 is a fixed vending machine with a touch-screen or keys, payment with or without cash.

The HW4220 is a cashless smartcard vending machine, able to act also as an information terminal.

The HW4560 proximity card reader is a smartcard read/write system.

The H4225 mobile mini-ticket pillar is a mobile vending machine for use in vehicles and accepts electronic cards.

The HW4240 information pillar is a fixed information terminal and is operated with electronic cards and operated by means of a touch-sensitive screen.

The HW4510 ferry unit is a portable ticket sales unit for outdoor use and can take cashless payment means.

The HW4581 inspector unit is a portable device for checking electronic tickets.

HW Agent is management software for personnel and terminals, sales records and submission of cash data from electronic cards.

Howe Green Ltd

12 Merchant Drive, Mead Lane Industrial Estate, Hertford SG13 7BH, UK
Tel: (+44 1992) 55 43 88   Fax: (+44 1992) 58 46 12
e-mail: info@howegreen.co.uk
Web: http://www.howegreen.co.uk

Key personnel

UK Sales Director: Richard Centa

Products

Access and duct covers used in concourses, platforms and service areas where sealing and good surface appearance are important. These conform to EN124 where applicable. Also Visedge hatch covers for incorporation in the floors of rolling stock and a patented bonding system for floor covering.

These products are also made under licence by Arden Architectural Specialties Inc, USA.

NEW ENTRY

IBM

IBM Global Travel & Transportation Industry Solutions Unit
1 New Square, Bedfont Lakes, Feltham TW14 8HB, UK
Tel (+44 20) 88 18 40 00   Fax: (+44 20) 88 18 54 37

Key personnel

EMEA Travel and Transportation Industry Leader:
  Maria-José Gomez Martin
  Tel: (+34 91) 397 71 36
Global Segment Executive, Rail, Freight and Logistics: Gregory L Smith
  Tel: (+1 404) 921 55 91

Products

Self-service kiosks for rail ticketing; Internet booking systems for railways, safety training with IBM e-learning solution, wireless solutions for railways.

ICA Traffic GmbH

Walter Welp Strasse 25, D-44149 Dortmund, Germany
Tel: (+49 231) 917 04 40   Fax: (+49 231) 17 13 83
e-mail: info@ica-traffic.de
Web: http://www.ica-traffic.de

Key personnel

Sales Director: Jörg Metzger

Products

Automatic ticket vending systems, including the Dualis family of systems developed to handle smartcards and credit/debit cards and incorporating touchscreen technology.

VERIFIED

IER

3 rue Salomon de Rothschild, PO Box 320, F-92156 Suresnes Cedex, France
Tel: (+33 1) 41 38 60 00   Fax: (+33 1) 41 38 62 00
e-mail: wier-contact@ier.fr
Web: www.ier.fr

Key personnel

President and CEO: Edmond Marchegay
Senior Vice-President, Industrial Affairs:
  Jacques Bouillon
Senior Vice-President, Worldwide Operations:
  Christophe Lamoine
Senior Vice-President, Marketing: Jean-Louis Natta
Senior Vice-President, Worldwide Sales:
  Jean-Pierre Sany
Communication Director: Emmanuelle Mussard

Subsidiaries

Automatic Systems SA
Avenue Mercator 5, B-1300 Wavre, Belgium
Tel: (+32 10) 23 02 11   Fax: (+32 10) 23 02 02

Automatic Systems America Inc
4005 Boulevard Matte, Brossard, Quebec J4Y 2P4, Canada
Tel: (+1 450) 650 07 37   Fax: (+1 450) 659 09 66

Automatic Systems SA
Kuen Yang Plaza #1101, 798 Zhao Jia Bang Road, Shanghai 200030, China
Tel: (+86 21) 64 73 67 92   Fax: (+86 21) 64 73 68 06

Automatic Systems SA
Kinzigstrasse 56, D-75210 Keltern, Germany
Tel: (+49 7236) 93 26 43   Fax: (+49 7236) 93 27 92

Automatic Systems Espanola SA
Calle Vallés 52-54, E-08820 El Prat de Llobregat, Spain
Tel: (+34 93) 478 77 55   Fax: (+34 93) 478 67 02

Calle Antracita n° 7, 4ª planta, E-28045 Madrid, Spain
Tel: (+34 91) 659 07 66   Fax: (+34 91) 654 23 07

Automatic Systems Equipment
Two Brabazon Court, Borman, Tamworth B79 7TA, UK
Tel: (+44 1827) 31 35 39   Fax: (+44 1827) 31 35 40

Automatic Control Systems Inc
8 Haven Avenue, Suite 205, Port Washington, New York 11050, USA
Tel: (+1 516) 944 94 98   Fax: (+1 516) 767 34 46

Background

IER is a subsidiary of the Bolloré Group.

Products

Design, manufacture and marketing of ticket terminals for the major air, rail and sea networks. Printers, readers, electronic ticketing terminals and self-service ticket machines.

Automatic Systems (AS), subsidiary of IER, designs, manufactures and markets a full range of access control equipment.

UPDATED

Indra

Avda de Bruselas Alcobendas, 35 E-28108 Madrid, Spain
Tel: (+34 91) 480 73 00   Fax: (+34 91) 480 73 33
Web: www.indra.es

Key personnel
Commercial Manager: Alberto Calvo
Export Manager: Eneko de Irala

Products
Access control equipment for ISO and Edmonson machines; automatic barriers, gates and turnstiles; automatic ticket machines; ticket office machines; magnetic and contactless smartcard technology; credit card prepayment systems; computerised management and system control; urban, intercity and multimodal software applications; passenger control and flow regulation.

Contracts
Latest contracts include: Pearl Line (Shanghai Metro, China); LRT (Tiansin Light Railway, Thailand); RATP (Paris Metro, France); Attiko (Athens Metro, Greece); STIB (Brussels Metro, Belgium); TRAM (Tranvía of Barcelona, Spain); Merval (Metro de Valparaiso, Chile).

UPDATED

Intec Ltd

192 Pang I-Dong, Song Pa-Ku, Seoul, 138-050 South Korea
Tel: (+82 2) 34 34 40 00   Fax: (+82 2) 34 34 41 70
e-mail: jyulbest@unitel.co.kr

Products
Ticket machines; access control cards; smartcard and contactless systems and machines; AFC systems.
Contracts include provision of an AFC system for metro and KNR lines in Seoul.
Intec has developed a non-touch prepaid card. It is not a smartcard but contains monetary value and has been developed in conjunction with LG Credit Card Services.

Kaba Gilgen AG

Freiburgstrasse 34, CH-3150 Schwarzenburg, Switzerland
Tel: (+41 31) 734 41 11   Fax: (+41 31) 734 44 75
e-mail: info@kgs.kaba.com
Web: www.kaba-gilgen.ch

Key personnel
Sales Director: Konrad Zweifel
Director ADP, Automatic Doors for Public Transport: Hans Krähenbühl

Products
Platform screen door systems, platform gate door systems, access control systems, automatic pedestrian doors.

UPDATED

Klüssendorf Produkte und Vertriebs GmbH

Zitadellenweg 20 D-F, D-13599 Berlin, Germany
Tel: (+49 30) 35 48 10   Fax: (+49 30) 35 48 12 59
e-mail: info:kpv.de
Web: http://www.kpv.de

Key personnel
Managing Director: F Vandepoele
Manager, Sales Promotion: W Burghausen

Products
Ticket machines and validators/cancellers, including microprocessor-based units; counter ticket printers.

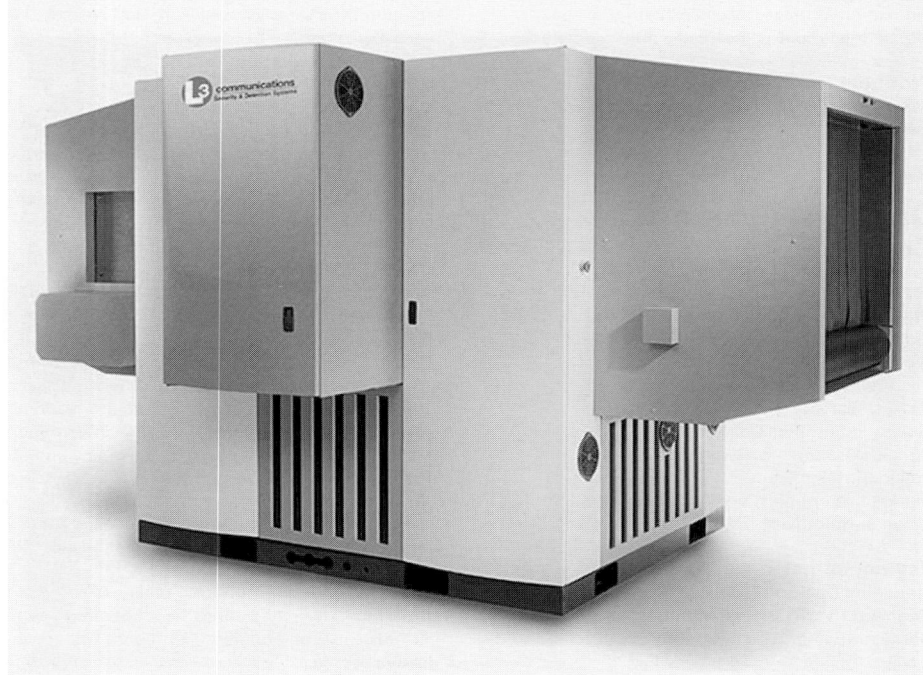

L-3 Communications MVT baggage screening system   0580688

L-3 Communications Corporation

600 Third Avenue, New York, New York 10016, USA
Tel: (+1 212) 697 11 11   Fax: (+1 212) 682 95 53
Web: http://www.l-3com.com

Key personnel
General Manager and Senior Vice-President: Allen Barber
Director of Marketing: Douglas Stevenson

Products
Rail baggage screening systems for explosives detection. L-3's MVT (Multi-View Tomography) system is based on applications widely used at airports and is capable of scanning up to 1,800 bags per hour. The MVT features a large tunnel to accommodate maximum-size baggage. Operator workstations incorporate a touchscreen interface.

Development
In May 2004 the MVT was commissioned at the station at New Carrollton, Maryland, by the US Transportation Security Administration for its Transit and Rail Inspection Pilot (TRIP) programme, the country's first rail passenger baggage screening programme.

NEW ENTRY

Maexbic SA

Ctra C 17, km 26.3, Centro Commercial Sant Jordi, E-08480 L'Ametlla de Vallès, Spain
Tel: (+34 938) 43 24 00   Fax: (+34 938) 43 21 02
e-mail: maexbic@maexbic.es
Web: http://www.maexbic.es

Products
Fixed automatic ticket vending machines capable of selling and recharging contactless smartcards, onboard electronic ticket vending machines with built-in smartcard readers, onboard smartcard readers/cancellers, fixed magnetic and smartcard validators, multi-journey card cancellers, management software for revenue and journey analysis.

Magnadata International Ltd

Norfolk Street, Boston PE21 6AF, UK
Tel: (+44 1205) 31 00 31   Fax: (+44 1205) 31 26 12

e-mail: sales@magnadata.co.uk
Web: www.magnadata.co.uk

Key personnel
Managing Director: R Colclough
Sales Director: A Laidlaw
Business Development and Marketing Director: Paul Johnson

Overseas sales offices
Magnadata USA Inc
100 Route 70, Suite 9, Lakewood, New Jersey 08701, USA
Tel: (+1 732) 901 93 99   Fax: (+1 732) 901 71 71
e-mail: jbonannomagusa@optonline.net

Magnadata Pty Ltd
Sydney, Australia
Tel: (+61 2) 96 23 74 00   Fax: (+61 2) 98 33 10 86
e-mail: gregg@magnadata.com.au

Products
Magnetic striped tickets and low cost smart cards for automatic fare collection systems. These can be supplied in a number of formats (cut single, fan-folded, reel to dimensional requirement) on a variety of materials (paper, plastic, paper/plastic sandwich), including thermally coated materials. Both low- and high-coercivity magnetic striped tickets can be supplied along with chips conforming to both ISO 14443A and B.
Numerous security features can be incorporated into the ticket design, including anti-photocopying inks, UV inks and security backgrounds. Magnadata also manufactures ATB tickets, and can supply plastic thermal material for medium- and long-term magnetic transport tickets and also plastic smartcards.

Contracts
Magnadata has supplied some of the world's first projects with ticketing solutions. Supply of the first countrywide smart project in the Netherlands with low-cost smart tickets.
Supply of dual cards with both magnetic and smart technologies in low-cost format to Porto for Euro 2004.
In conjunction wtih Z-Card, Magnadata produced the first ticket to include an underground map to support the 2012 London Olympic bid.
Magnadata gained ITSO (International Transport Smartcard Organisation) Approval for the Ik and 4K Mifare product.

UPDATED

Magnetic Autocontrol

Magnetic Automation Corporation
3160 Murrell Road, Rockledge, Florida 32955, USA
Tel: (+1 321) 635 85 85   Fax: (+1 321) 635 94 49
e-mail: info@ac.magnetic-usa.com
Web: http://www.magnetic-autocontrol.com

Products

Magnetic Autocontrol access control systems. Barriers can be connected to control systems which include light barriers and readers for code cards, coins, tickets and fingerprints. Barriers are produced in a variety of versions and all models can operate in 'open' mode.

MEI

Mars Electronics International Inc
Geneva Branch
PO Box 2650, CH-1211 Geneva 2, Switzerland
Tel: (+41 22) 884 05 05   Fax: (+41 22) 884 05 04
e-mail: meicustomerservice@effem.com
Web: http://www.meiglobal.com

Key personnel

Industry Manager: Reinhard Banasch
Marketing Manager: Serge Guillod
Business Development Manager: Maurice Reber
Sales Manager: Ruedi Lüthi

Associate companies

USA
Mars Electronics International, 1301 Wilson Drive, West Chester, Pennsylvania 19380
Tel: (+1 610) 430 25 00   Fax: (+1 610) 430 26 94

UK
Mars Electronics International
Eskdale Road, Winnersh Triangle, Wokingham RG41 5AQ
Tel: (+44 118) 969 77 00   Fax: (+44 118) 99 44 64 12

Products

Development and manufacture of banknote validators under the Sodeco® Cash Management Systems brand. These systems are designed for incorporation into ticket vending machines.

The new range of compact high-speed Sodeco® BNA5 validators is now complete and will accept from four to 60 banknotes inserted in all four directions in which it is possible to present them. Three main products are offered.

The Sodeco® BNA 52/54 is a validator with an escrow facility for up to 15 notes, a stacker and a security cash-box with a capacity of 1,000 banknotes or an optional 2,000 note high-capacity cash-box on the latest model. The BNA product has become the standard specification bill acceptor for most mass transit applications worldwide.

The Sodeco® BNA 51/54 is a validator with a stacker and is compatible with the above cash-boxes. The Sodeco® BNA 50 is a validator only, with an optional stainless steel drawer-box with a 100 to 400 note capacity.

All MEI products are designed to offer the highest performance in terms of security against counterfeit notes and fraudulent manipulation. MEI's specification for these machines includes a high acceptance rate over time, combined with the low jam rate, easy servicing and maintenance. The robust construction is suited for outside use and the units are fully compatible with Windows NT/95® environment for those users who wish to network their TVMs. Money collection is both simple and secure with the principle of cash-box exchange avoiding the need for contact with the cash until the box is emptied. Money is also protected by electronic security measures in addition to the more conventional locks and keys. All Sodeco® validators are CE marked and UL approved.

MEI also manufactures the CashFlow® range of change-givers and coin mechanisms for the acceptance of coins in unattended situations and the provision of coin-based change.

MEI offers a worldwide supply and service operation with 'just-in-time' deliveries and technical back-up, plus a research and development

programme that ensures new notes can be accepted as soon as they are issued. MEI states that its manufacturing sites are ISO 9001 certified.

Recent installations of Sodeco® validators include The Long Island Railroad, New York; Metro North, New York; MTA Baltimore; WMATA, Washington DC; Blue Line, Los Angeles; BC Transit, Vancouver; former British Rail companies, UK; SRA City Rail Sydney; Berliner Verkehrsbetrieb; SBB Switzerland; KCRC/MTRC Hong Kong and KTM Malaysian Railways; STIB; NYCTA; San Diego Trolley among many others.

Motorola

Motorola Worldwide Smartcard Solutions Division (WSSD)
1301 East Algonquin Road, 5th Floor, Schaumburg, Illinois 60196, USA
Tel: (+1 847) 576 69 31
Web: http://www.motorola.com

Key personnel

Vice-President and General Manager:
   Francois Dutray

Products

Motorola Inc provides embedded electronic and integrated communications solutions. Motorola's Worldwide Smartcard Solutions Division (WSSD) provides complete multi-application smartcard system solutions including smartcards, application development, systems integration and operations management. The company's platforms allow organisations quickly to deploy and build value-added smart card applications in areas such as transit, access control, campus, government and healthcare.

Nabtesco Corporation

Railroad Products Company, 9-18 Kaigan 1-chome, Minato-ku, Tokyo 105-0022, Japan
Tel: (+81 3) 54 70 24 01   Fax: (+81 3) 54 70 24 24
e-mail: takashi_koyama@nabtesco.com
Web: www.nabtesco.com

Key personnel

Company President: Koshiro Yoshida
Company Vice-President: Masanori Kawanishi
General Manager, Sales and Marketing:
   Yukiyasu Fujimoto
General Manager, Overseas Marketing:
   Takashi Koyama

Background

Nabtesco Corporation was previously called NABCO Ltd.

Products

Platform doors (manufactured and supplied by Nabtesco Corporation, NABCO Company).

NEW ENTRY

Narita Manufacturing Ltd

20-12 Hanaomote-cho, Atsuta-ku, Nagoya 456-0033, Japan
Tel: (+81 52) 881 61 91
Fax: (+81 52) 881 67 48 (General Affairs)
e-mail: sinarita@narita.co.jp
Web: http://www.narita.co.jp

Key personnel

President: Masatoshi Narita
Executive Director (Export Sales and Marketing):
   Shuichi Narita
Executive Director, General Affairs and Quality
   Assurance: Haruo Narita

Products

Platform door leaves.

VERIFIED

Newbury Data Recording Ltd

Premier Park, Road One, Winsford, Cheshire CW7 3PT, UK
Tel: (+44 1606) 59 34 24   Fax: (+44 1606) 55 83 83
Web: http://www.newburydata.co.uk

Key personnel

Managing Director: Alan J Phillips
Product General Manager: Ashley Bailey
Marketing Executive: Philippa Molyneux

Products

Flexstore is a hand-held AB ticket reader, designed to read and display information held on magnetic stripe tickets and to download that data into a central computer for passenger and ticket analysis purposes. Currently supplied to European Passenger Services for that company's Eurostar trains.

Flexfare is a modular booking office ticketing system, comprising a terminal, receipt printer and ISO-sized card ticket issuer. Fully configurable to customer's requirements.

Nippon Signal Co Ltd

1-1, Higashi-Ikebukuro 3-chome, Toshima-ku 170 6047, Tokyo, Japan
Tel: (+81 3) 59 54 46 78   Fax: (+81 3) 59 54 45 78
e-mail: info@signal.co.jp
Web: www.signal.co.jp

Key personnel

President: K Nishimura

Products

Automatic gate; automatic ticket vending machine; automatic fare adjustment machine; data processing equipment; station controller; coupon vending machine; automatic coupon vending machine; automatic pre-paid card vending machine; centralised card encoder; ticket issuing machine for staff; centralised monitoring equipment.

Nippon Signal Co Ltd also makes systems for contactless IC rewritable card read/writers and the RFID item management system.

UPDATED

Omron Corporation

Omron Tokyo Building, 3-4-10 Toranomon, Minato-Ku, Tokyo 105-0001, Japan
Tel: (+81 3) 34 36 72 64
Fax: (+81 3) 34 36 70 54

Key personnel

Manager, Advanced Modules Business Company, Global AFC Systems Sales and Marketing:
   K Yokochi

Products

Complete automatic fare collection systems for magnetic striped ticket and contactless smart card and token, including ticket issuing machines, barrier equipment, validators/cancellers, and fare adjustment machines.

A contactless smartcard has been developed for opening automatic gates. Lithium batteries are not used – the card has an induction system for its power supply.

Philips Semiconductors

Mikronweg 1, A-8101 Gratkorn, Austria
Tel: (+43 31) 24 29 93 45
e-mail: info.bli@philips.com
Web: www.semiconductors.philips.com

Key personnel

Marketing Manager Transport: Jason Hitipeuw
Communications Manager: Alexander Tarzi

Background

With an estimated market share of 80 per cent, billions of card transactions and a total of over 400 million ICs shipped, Philips Semiconductors MIFARE® interface technology claims to be the defacto industry standard for contactless and dual interface proximity smart card schemes.

An open standard, fully-compliant with ISO 14443 A, MIFARE®-based contactless and dual interface smartcard and reader products are available from a variety of suppliers, with an independent certification authority guaranteeing compatibility. This ensures that supplies will continue to meet the rapidly growing demands and, with its commitment to contactless smartcard IC technology, Philips Semiconductors delivers a complete portfolio of MIFARE® ICs.

Products

Philips Semiconductors offers a unique, total capability in IC-based identification with a portfolio covering all smartcard and RFID applications.

With MIFARE®, Philips Semiconductors fully supports an open platform strategy ranging from low- to high-end IC products, thus being the first to provide all required components to set up a 100 per cent contactless ticking system. In line with the international standard for contactless smartcards ISO 14443 A, the MIFARE® Interface Platform covers ICs for smartcards and read/write terminals: Contactless intelligent memory card ICs, such as MIFARE® Classic for contactless multi-application smartcards. Sophisticated high-security contactless microcontroller ICs for dual interface smartcards allow to securely combine contactless applications with contact ones, such as banking, mobile communications and secure network access on a single smartcard. For the reader infrastructure, Philips offers a unique family of MIFARE® cost-effective, single-chip reader ICs for high-volume applications, which are easy to design-in.

MIFARE® has been successfully installed in public transport schemes all over the world. In China 26 cities run transport systems based on MIFARE®, along with other systems established worldwide including: London Underground, UK; Warsaw, Poland; Izmir, Turkey; Mumbai, India; Trondheim, Norway; Santiago, Chile; Sao Paulo, Brazil; Seoul, Pusan, Korea; and Moscow, Russia.

UPDATED

Regazzi SA

Via alle Gerre
Zona Industriale, CH-6596 Gordola, Switzerland
Tel: (+41 91) 735 66 00   Fax: (+41 91) 735 66 99
e-mail: info@regazzi.ch
Web: http://www.regazzi.ch

Key personnel

Sales Manager: Barry Gibson

Products

Stainless steel seating and waste bins for stations.

Contracts

Regazzi has recently supplied seating and waste bins for 620 Swiss railway stations.

Sadamel Ticketing Systems

73 rue du Collège, CH-2300 La Chaux-de-Fonds, Switzerland
Tel: (+41 32) 968 07 70   Fax: (+41 32) 968 08 85
e-mail: info@sadamel.ch
Web: http://www.sadamel.ch

Key personnel

Chairman: Roger Cattin
Managing Director: Louis-George Lecerf
General Director: Roger Cattin
Development Director: Jerôme Froidevaux
Production Director: Daniel Courtet
Sales Manager: Daniel Eberhard

Products

Automatic ticket vending machines suitable for paper, magnetic and contactless tickets with payment by coins, banknotes and bank cards. Automatic fare collection management system including monitoring of networked ticket vending and validating equipment. Passenger-operated automatic vending machines. Onboard or counter-based automatic ticket vending machines, ticket cancelling units. Coins recycling unit.

Contracts

Contracts include Swiss Federal Railways (SBB): 120 ticket vending machines (counter-based with touchscreen facility); Verkehrsbetriebe Luzern (VBL): 230 stationary ticket vending machines and onboard vending machines for 55 buses and six sales points.

Transports Publics Neuchâtelois (TN) to fit retrofit stationary ticket vending machines with the Swiss Cash Card (electronic purse) Portugese Railways (USGL and USGP) for a complete automatic fare collection system with 220 ticket vending machines and 320 magnetic card validators. All the machines and validators are networked to a management system (Sadagest). The latest contract received is from Satu/Oeras for a contactless system with access gates.

Scheidt & Bachmann GmbH

PO Box 201143, D-41211 Mönchengladbach, Germany
Tel: (+49 2166) 26 65 50   Fax: (+49 2166) 26 66 99
e-mail: admin@scheidt-bachmann.de
Web: www.scheidt-bachmann.de

Key personnel

Managing Director: Ratthias Augustguicek
Head of Fare Collection Systems Dept:
   Christoph Poos
Marketing Manager, Europe: Manfred Feiter
Head of Business Development: Frithjof Struye

Background

The company was founded in 1872 by Friedrich Scheidt and Carl Bachmann and remains privately owned.

Products

Automatic fare collection and ticket vending machines for stationary and onboard applications; point of sales; central computer systems for accounting; data provision and technical administration of associated fare collection equipment. Complete smartcard systems.

Recent products include the FAA-2000/C (cashless) and the FAA-2000/M (cards and coins) ticket vending machines incorporating the main features of the FAA 2000 family of machines. The FAA-2000/C and the FAA 2000/M can be used in onboard or stationary applications, will validate electronic tickets and can serve as an information point. The FAA-2000/M can also encode and distribute smart cards. The FAA 2000 machines can also be used as multimedia terminals, providing advertising and marketing potential, or as an information platform displaying information on unique local events or common local data.

UPDATED

SchlumbergerSema

50 avenue Jean Jaurès, BP 62012, F-92542 Montrouge Cedex, France
Tel: (+33 1) 47 46 79 50   Fax: (+33 1) 47 46 68 66
e-mail: marand@montrouge.ts.slb.com
Web: http://www.slb.com

Subsidiary

SchlumbergerSema Test & Transactions
Ferndown Industrial Estate, Wimborne BH21 7PP
Tel: (+44 1202) 85 09 25   Fax: (+44 1202) 85 09 03

Asia office

SchlumbergerSema Singapore
Tel: (+65) 746 63 44
e-mail: schew@singapore.asia.slb.com
Contact: Sally Chew

Background

In 2001, Schlumberger Limited acquired Sema plc and combined it with part of its former transaction business and other acquisitions. The company is now called SchlumbergerSema and is one of two business segments of Schlumberger Limited.

Products

The range comprises smartcard systems and the Addams, DAC, Discobb and TVM ticket dispensers.

Addams: A multidestination ticket dispenser with an emphasis on security, since it automatically prints a financial control ticket and will not allow any ticket to be issued with the door open. Modular, it provides a choice of destination, 30 possible fare structures and choice of class.

DAC: An automatic ticket book dispenser with a dual application. It receives money which is deposited by drivers (coins, notes and cheques) and also supplies them with tickets 24 hours a day. Security is an important feature as a receipt is issued for each transaction which is performed.

Discobb: A dispenser and validator for single tickets. All parameters can be modified (ticket type, display, fares).

TVM: A ticket vending machine which is fully automated for ticketing and fare collection.

Installation of a complete ticketing solution for Tramtrack Croydon Ltd, the operating company of Croydon Tramlink has been carried out. The automatic fare collection system is based on a remote Java-server and is designed to reduce the average transaction time to less than 15 seconds. This helps to maximise revenues by reducing queuing time and increasing passenger throughput, particularly at peak times.

The Croydon Tramlink project is a light rail public transport system in the south of London linking Croydon with Wimbledon, Beckenham and New Addington. Tickets sold by the ticket vending machines will be magnetically encoded upon issue. This will make them compatible with the ticketing systems used by London Underground and local rail services.

The automatic fare collection system consists of 78 SchlumbergerSema High-Flow traffic ticket vending machines, but also includes a fibre-optic local area network (LAN) with a Java-based server for remote monitoring and control of the whole ticket vending machine network including comprehensive analysis of all passenger traffic and revenue. The LAN feeds a hub supporting communication and application servers, a workstation and printers. This system enables alarms, warnings and transaction details to be sent to the central computer and commands, operational settings and fare variations to the ticket vending machines.

SchlumbergerSema has installed some 90,000 terminals in over 30 countries.

Shere Limited

4 Bridge Park, Merrow Industrial Estate, Guildford, GU4 7BF, UK
Tel: (+44 1483) 55 74 00   Fax: (+44 1483) 55 74 01
e-mail: briscoen@shere.com
Web: http://www.shere.com

Products

Kiosk systems integrating ticket printers, card readers, touchscreens, pin pads, scanners, UPS, sensors, alarms and desk top ticketing systems.

Siemens Transportation Systems

Dept VT27 Automatic Fare Collection
PO Box 910220, D-12414 Berlin, Germany
Tel: (+49 30) 386 512 46; 510 32; 513 21
Fax: (+49 30) 61 74 10 32

e-mail: internet@ts.siemens.de
Web: http://www.siemens.com/ts

Key personnel
General Manager: Volker Rind
Sales (Germany): M Netka
Sales (other countries): J Janssen

Products
Automatic fare collection systems; complete turnkey systems for integrated ticketing and fare collection, based on magnetic cards and contact and contactless smartcards; system components, including smartcard readers, ticket vending machines, ticket gates and ticket office machines.

Recent contracts include BTSC Bangkok, Thailand 1997; VRS Cologne/Bonn and BVG Berlin.

CHIPTICKET contactless and dual-interface smartcard system for public transport; secure contactless card terminal.

STERIA Group

46 rue Canille Desnoulins, F-92782, Issy Les Noulineaux, Cedex 9, France
Tel: (+33 1) 34 88 60 00   Fax: (+33 1) 34 88 60 15
Web: www.steria.fr

Key personnel
See STERIA Group entry in *Signalling and communications systems* section.

Products
Ticketing including clipcards and contactless smartcards, suitable for multimodal transport, simulation, high-reliability systems (Atelier B).

Contracts
Ticketing contracts with Easyride SBB (Switzerland); SNCF and RATP.

UPDATED

Takamisawa Cybernetics Co Ltd

Nakano Heiwa Building, 48-5, 2-Chome Chuo, Nakano-ku, Tokyo 164-0011, Japan
Tel: (+81 3) 53 71 33 61   Fax: (+81 3) 53 71 33 59
e-mail: export_dept@tacy.co.jp
Web: www.tacy.co.jp

Key personnel
Manager of Trading Department: J Tada

Works
Nagano Factory No 1
525 Kitagawa, Saku-City, Nagano, 384-0304, Japan
Takamisawa Service Co Ltd
Takamisawa Mex Co Ltd

Products
Design and manufacture of automatic fare collection systems and equipment, including automatic ticket vending machines, automatic gates, fare adjustment machines and ticket printers; AFC-related currency and card handling unit.

Customers include all Japan railway companies and operators in some Asian countries.

UPDATED

Thales

Transport & Energy activities
Centre du Bois des Bordes, PO Box 57, F-91229 Brétigny-sur-Orge Cedex, France
Tel: (+33 1) 69 88 52 00   Fax: (+33 1) 60 84 82 30
e-mail: info@thales-transportservices.com
Web: www.thalesgroup.com

Key personnel
Chairman and Chief Executive Officer:
  Jean-Louis Olié
Vice-Chairman: Tim Cavanagh

Sales and Marketing Director: Ian Woodroofe
Communication Manager: Anne-Lydie Bladier
Transport Activities:
  Managing Director: Bruno Cohades
Energy Activities:
  Managing Director: Denis Chedeville

Background
Thales is an international electronics and systems group, serving defence, aerospace, services and security markets. The group employs 61,500 people worldwide and generated revenues of €10.6 billion in 2003.

Products and Services
Thales is one of the world's leaders in secure integrated fare collection systems. As a leader in Asia, Thales integrates systems for public transport and energy activities such as integrated fare collection, train supervision and control, road toll collection, traffic and fleet management, public parking systems, telematics and electronic fund transfer solutions. It also offers supervision of electricity, gas and oil transport systems, air navigation messages switching systems as well as a wide range of services that cover the whole life cycle of systems and products.

Thales enables transport operators worldwide to improve passenger traffic flow and deliver better client information and management systems. Thales has developed high-performance management systems that are based upon open architecture. These are flexible and scalable in order to optimise customer investment. The systems integrate contactless technology, which allows transport operators to reduce maintenance costs and facilitate cash transactions. They can manage complex season ticket schemes and ensure high passenger throughput. Thales systems also enable travellers to access several transport modes and services with the same card, and the transaction can be carried out with the card still contained in the travellers wallet, saving time. Systems are designed for multi-application cards and guarantee the security and reliability of the system and equipment.

Contracts
Thales has been selected as an integrator of secure city card systems for public transport by the city authorities of Bangkok, Cairo, Calcutta, Kuala Lumpur, Madrid, Manila, Mexico City, New York, Pusan, Seoul, Strasbourg and Oslo.

Thales is a market leader in integrated ticketing systems in Asia and is a global frontrunner in electronic ticketing system integration and in supervision and control systems.

Today Thales systems are in service in some of the world's biggest cities, including Hong Kong, New Delhi, Paris, London, Singapore, Taipei and Guangzhou. In the Netherlands, Thales, within the East West consortium, provides its smartcard technology and experience to implement a nationwide electronic ticketing system.

NEW ENTRY

Toshiba Corporation

Railway Projects Department
1-1 Shibaura, 1-chome, Minato-ku, Tokyo 105-8001, Japan
Tel: (+81 3) 34 57 49 24   Fax: (+81 3) 54 44 92 63
Web: www.toshiba.co.jp

Key personnel
President and Chief Executive Officer:
  Atsutoshi Nishida
Vice-President, Transportation Systems Division:
  Takio Ooyama
Senior Manager, Railway Projects Dept: Koji Toda

Products
Automatic fare and toll collection systems based on customised units from a basic range.

UPDATED

Toyo Denki Seizo KK

Toyo Electric Manufacturing Co Ltd
No 1 Nurihiko Bldg, 9-2 Kyobashi, 2-chome, Chuo-ku, Tokyo 104-0031, Japan
Tel: (+81 3) 35 35 06 41   Fax: (+81 3) 35 35 06 50

Key personnel
President: Kunio Kai
Vice-President: Motonobu Matsubara
Managing Director: Keisuke Tanaka
Director: Kenzo Terashima

Products
Ticket issuing systems for suburban and rapid transit railway systems; automatic ticket issuing machines and fare adjusting equipment.

Toyo Denki's ticket issuing system for suburban and rapid transit systems can issue magnetic tickets for automatic gates. It can also calculate fares for complex urban networks where different routeings are possible and there is much interline traffic between different operators.

UPDATED

Wanzl Metallwarenfabrik GmbH

Passenger Handling Services
Hasberger Strasse 13, D-87757 Kirchheim/Schwaben, Germany
Tel: (+49 8266) 29 92 87   Fax: (+49 8266) 29 92 90
e-mail: airport@wanzl.de
Web: http://www.wanzl-airport.com

Key personnel
Business Unit Manager: Ralf Aubele

Products
Luggage trolleys. The Travel 300 and Euro-Spinter models feature a coin deposit system and integrated braking to prevent runaways.

Contracts
The Travel 300 model has been authorised for use by German Rail (DB AG).

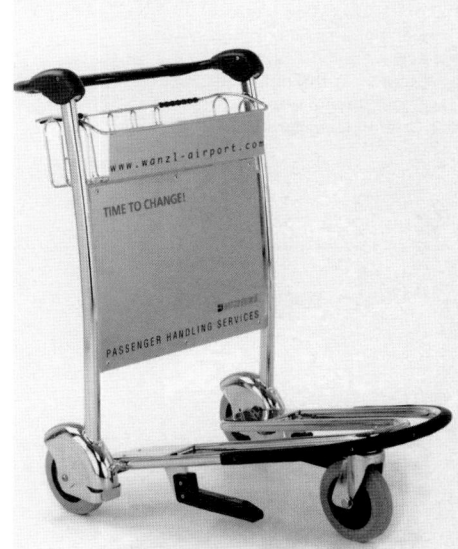

Wanzl Travel 300 luggage trolley, approved for use by DB AG (Wanzl Metallwarenfabrik GmbH)
0569529

Westinghouse Platform Screen Doors

Westinghouse Brakes (UK) Ltd
Foundry Lane, Chippenham SN15 1JB, UK
Tel: (+44 1249) 44 20 01   Fax: (+44 1249) 44 24 41
e-mail: psd@westbrake.com
Web: www.platformscreendoors.com

Westinghouse platform screen doors, Farrer Park Station, Singapore (Ken Seet, Photographer and Poh Siew Wah, Artist)   1035173

Westinghouse platform screen doors, Guangzhou Metro   0552976

Westinghouse platform screen doors, Changi Airport   0552975

Key personnel

Managing Director: Paul R Johnson
Engineering Director: Jason Abbott
Marketing and Sales Director: Peter C Johnson
Projects Director: Danny Lee
Business Development Manager: Colin Fullalove

Background

Westinghouse Brakes (UK) Ltd is part of the Knorr-Bremse Group.

Products

Platform screen door systems, platform edge doors and platform safety gates capability.

Westinghouse Platform Screen Doors designs and installs railway-based automatic platform screen doors. Systems are tailored to suit individual customer specifications for new and retrofit applications, combining safety, reliability and architecturally pleasing designs. Asset management capabilities include comprehensive after-sales services, training and maintenance programmes.

Contracts

Projects worldwide include:
China: Guangzhou Metro Lines 2 and 4, Shenzhen Metro Lines 1 and 4, Hong Kong International Airport.
Denmark: Shenzhen Metro.
France: Toulouse Line B, Roissy Line 1.
Malaysia: Kuala Lumpur LRT2.
Singapore: North South Line, East West Line, North East Line, Changi Airport Line, Marina/Circle Line.
UK: London Underground Jubilee Line.

UPDATED

Zelisko Elektrotechnik un Elektronik GesmbH

Steinfeldergasse 12, A-2340 Mödling, Austria
Tel: (+43 2236) 40 60
Fax: (+43 2236) 40 62 99
e-mail: info@zelisko.com
Web: http://www.zelisko.com

Key personnel

For a full list of personnel, see Zelisko Elektrotechnik un Elektronik GesmbH entry in *Signalling and communications systems* section.

Products

Ticket printers; automatic ticket vending machines; fare collection software; technology for the integration of magnetic cards or smartcards.

ELECTRIFICATION CONTRACTORS AND EQUIPMENT SUPPLIERS

Alphabetical listing

Allied Insulators (Group) Ltd
ALSTOM Transport
Aluminium Inductors Ltd
AMEC SPIE Rail (UK)
Ampcontrol Pty Ltd
ApATeCh Electro
Balfour Beatty Rail GmbH
Barclay Mowlem Construction Ltd
Bharat Earth Movers Ltd (BEML)
Benkler AG
Brecknell, Willis & Co Ltd
Brush Transformers Ltd
Carbone of America (LCL) Ltd
Cembre SpA
CostaRail Srl
Crompton Greaves Ltd
Cuadrelec SA
Ebo Systems
EFACEC Sistemas de Electrónica SA
Elpro BahnstromAnlagen GmbH
Ensto Sekko Oy
Ferraz SA

Firema Trasporti SpA
FKI Switchgear
Flury
Fuji Electric Co Ltd
Furrer + Frey AG
Galland Sarl
Geismar
Greysham (International) Pvt Ltd
HEI
Hitachi Ltd
Hyundai
Insul-8 Corporation
A Kaufmann AG
Kershaw Manufacturing
Kruch GesmbH & Co KG
Kummler & Matter AG
Lerc SA
Merlin Gerin Brasil SA
Mer Mec SpA
Muromteplovoz
nkt cables GmbH
Paul Keller Engineering Ltd

Pfisterer Srl
Plasser & Theurer
Powernetics Ltd
RPG Transmissiom Ltd
SAE (India) Ltd
SAFT
SDCEM
Sécheron SA
Sefag AG
Selectra srl
Siemens AG
Sirti SpA
South Wales Transformers Ltd
SPIE Enertrans
Supertek Enterprise Inc
Toshiba Corporation
Total Power Solutions
Transmitton Ltd
TrendRail Ltd
Ultra Electronics Ltd PMES
Wild & Grunder AG
Windhoff Bahn-und Anlagentechnik GmbH

Company listing by country

AUSTRALIA
Ampcontrol Pty Ltd
Barclay Mowlem Construction Ltd
Kruch GesmbH & Co KG

AUSTRIA
Plasser & Theurer

BRAZIL
Merlin Gerin Brasil SA

CANADA
Carbone of America (LCL) Ltd

FINLAND
Ensto Sekko Oy

FRANCE
Ebo Systems
Ferraz SA
Galland Sarl
Geismar
Lerc SA
SAFT
SDCEM
SPIE Enertrans

GERMANY
Balfour Beatty Rail GmbH
Elpro BahnstromAnlagen GmbH
nkt cables GmbH
Siemens AG
Windhoff Bahn-und Anlagentechnik GmbH

INDIA
Bharat Earth Movers Ltd (BEML)
Crompton Greaves Ltd
Greysham (International) Pvt Ltd
HEI
RPG Transmissiom Ltd
SAE (India) Ltd

ITALY
Cembre SpA
CostaRail Srl
Firema Trasporti SpA
Mer Mec SpA
Pfisterer Srl
Selectra srl
Sirti SpA

JAPAN
Fuji Electric Co Ltd
Hitachi Ltd
Toshiba Corporation

KOREA, SOUTH
Hyundai

PORTUGAL
EFACEC Sistemas de Electrónica SA

RUSSIAN FEDERATION
ApATeCh Electro
Muromteplovoz

SPAIN
Cuadrelec SA

SWITZERLAND
Benkler AG
Flury
Furrer + Frey AG
A Kaufmann AG
Kummler & Matter AG
Paul Keller Engineering Ltd
Sécheron SA
Sefag AG
Wild & Grunder AG

UNITED KINGDOM
Allied Insulators (Group) Ltd
ALSTOM Transport
Aluminium Inductors Ltd
AMEC SPIE Rail (UK)
Brecknell, Willis & Co Ltd
Brush Transformers Ltd
FKI Switchgear
Powernetics Ltd
South Wales Transformers Ltd
Total Power Solutions
Transmitton Ltd
TrendRail Ltd
Ultra Electronics Ltd PMES

UNITED STATES
Insul-8 Corporation
Kershaw Manufacturing
Supertek Enterprise Inc

Allied Insulators (Group) Ltd

PO Box 17, Milton, Stoke-on-Trent ST2 7EE, UK
Tel: (+44 1782) 53 43 21
Fax: (+44 1782) 54 58 04
e-mail: sales@alliedgroup.co.uk
Web: http://www.alliedgroup.com

Key personnel

Managing Director: D R Perrin
Commercial Director: R G Shenton
Sales & Marketing Director: M Pettigrew

Principal subsidiary companies

Allied Insulators Ltd
Doulton Insulators Ltd
Hopyard Foundries Ltd

Products

Insulator assemblies for feeder transmission, tracked overhead transmission, third rail systems, pantograph support and switching apparatus.

Current contracts include the supply of overhead catenary insulators to Railtrack, London Underground (third rail). Recent contracts include supply of equipment to Hong Kong, Taiwan and South Africa.

ALSTOM Transport

System Infrastructure Business Unit
33 rue des Bateliers, F-93404 Saint Ouen, Cedex, France
Tel: (+33 1) 40 10 62 62
Fax: (+33 1) 40 10 60 60

Key personnel

President, Transport Sector: Philippe Mellier
Chief Operating Officer: Gérard Blanc
Chief Financial Officer: Roland Kientz
Senior Vice-Presidents:
    International Product Line Development:
        Alain Goga
    Asia Pacific: Marc Chatelard
    Southern Europe: Charles Carlier
    Americas: Francis Jelensperger
    Northern Europe: Terence Watson

Electrification SpA
10 Via Lago dei Tartari, I-00012 Guidonia, Rome, Italy
Tel: (+39 06) 0774 37 74 85
Fax: (+39 06) 0774 35 34 30

ALSTOM Transport Service (PanChex)
48 rue Albert Dhalenne, F-93482 Saint-Ouen Cedex, France
Tel: (+33 1) 41 66 86 09
Fax: (+33 1) 41 66 92 70

Services

The Infrastructure Business unit of ALSTOM Transport Systems offers solutions at the system or subsystem level for power generation and distribution including: AC and DC traction substations; overhead facilities; contact lines or catenaries; third-rail or at-level integrated supply system; SCADA; auxiliary power supply; track laying; maintenance workshops; communications; signalling (tramways); electromechanical equipment in station; and electronic guidance systems for buses.

Its scope includes design, procurement, installation, commissioning, technical assistance, maintenance and training.

The electrification activities of ALSTOM cover power supply production, distribution and control of the traction current.

ALSTOM provides AC power distribution networks from high-voltage to medium-voltage conversion to feed traction substations and low-voltage station utilities.

Conversion is to AC single-phase current, primarily for main line railways:

- 25 kV 50 Hz for conventional railways
- 15 kV 16⅔ Hz for Rail Link
- 2 x 25 kV 50 Hz for high-speed trains

Conversion to DC current in six-pulse or 12-pulse rectification:

- 3,000 V for conventional railways
- 1,500 V for suburban trains and urban mass transit systems with overhead catenary
- 750 V for mass rail transit metro systems through third rail
- 750 V for light rail transit systems and trolleybuses through overhead contact wire.

More than 1,100 rectifier units have been supplied by ALSTOM, supplying more than 3.3 million MW.

ALSTOM provides in-line traction current distribution:

- Third rail (or fourth rail) for mass transit systems, mainly 750 V.
- Overhead catenary for 25 kV AC high-speed trains, single-phase AC for conventional railways and for DC systems in 3,000, 1,500 and 750 V light rail transit systems.

More than 17,000 km of overhead wire has been supplied by ALSTOM. Overhead contact lines have been supplied for urban tramways, trolleybuses or guided buses.

ALSTOM provides innovative power electronics solutions, and control systems for local and remote traction substations. The company has developed a new concept for urban light transit systems (tramways and rubber-tyred vehicles with electric propulsion) to minimise the impact on the environment in the heart of a city and to provide traction power supply avoiding overhead contact wires.

ALISS (At Level Integrated Supply System) is a hidden, reliable, maintenance-free flexible system using static safety-redundant IGBT commutation, to provide vehicles with traction power, only on adequate sectors with dynamic identification under the vehicle assemblies. A diagnostic and monitoring management signalling system ensures communication with the centralised control centre. After successful reliability and performance laboratory tests, on-site demonstration with real loaded vehicles took place from mid-2000 on ALSTOM's Aytré plant's test tracks. PanChex overhead line protection system is a trackside asset protection system that monitors the interaction of each passing train with the overhead line system. PanChex is a licensed technology from AEA Technology Rail Ltd.

Contracts

25 kV: TGV France (SNCF) and TGV Korea (KHRC)
25 kV: Algeria (SNTF), Portugal (CP), Costa Rica
16⅔ Hz: Arlanda Express airport rail link
3,000 V: Brazilian railways (Fepasa, CBTU), Morocco (ONCFM), Italy (FS)
1,500 V: Cairo metro line 1 (NAT), La Paz Pantitlan (Covitur, Mexico), Dublin (CIE) for Howth Bray line, Serpong and Tangerang for Indonesia (Jabotabek), Hong Kong Lantau airport rail link
750 V third-rail metros: Caracas lines 1 to 4 (CAMC), Athens lines 2 and 3, Mexico (STC and Covitur), Santiago de Chile (Metro SA), Cairo line 2 (NAT), Istanbul, Ankara, Lyon, Marseille
750 V overhead: tramways and trolleybuses (contact wire): Azteca Xochimilco (Mexico), Manchester Metrolink
Tramways in France: Rouen, Grenoble, Montpellier, Lyon, Nantes, Lille, Bobigny
Trolleybuses: Belo Horizonte, Lyon, Nancy, Grenoble, Marseille, Mexico
600–750 V: London Underground, Manchester, Hong Kong tramways, Mexico trolleybuses, New York, Chicago.

Developments

Modular Traction Substations
To avoid the installation of unsightly metallic shelters in towns, ALSTOM has developed a modular traction substation, the use of which means that storage of materials on site and long construction periods are no longer necessary. Increasing worldwide demand for light metros or tramway systems, the growing trend towards standardisation of substations and the search for the most competitive overall price are all factors which prompted ALSTOM to develop the modular substation.

A complete substation is created, using production distribution and traction current control equipment. This consists of two concrete shelters easy to transport by road, each one comprising premises equipped with access doors for operational personnel as well as for possible handling of equipment. Once the concrete slab is constructed, installation is fast. All cable links are prepared for a quick assembly on site and only an easy connection to the electrical network is required. Tests carried out in factory conditions using reduced voltage, allow immediate commissioning.

UPDATED

Aluminium Inductors Ltd

29 Lower Coombe Street, Croydon CR0 1AA, UK
Tel: (+44 181) 680 21 00
Fax: (+44 181) 681 15 77

Key personnel

Managing Director: Barry Martindale

Products

Transformers, inverter-based supply systems.

A power conversion system designed by R-R Industrial Controls, Gateshead, UK, has a new design of transformer to cut weight and boost efficiency.

In association with R-R Industrial Controls advanced high-density inverter-based supply systems for trains are being developed to cope with both alterations in rail traction DC supply environment and the provision of reliable power for emergency battery chargers, three-phase fan motors for air conditioning and motors for air compressor braking systems.

The systems make use of IGBT switching circuits.

AMEC SPIE Rail (UK)

Stephenson House, 2 Cherry Orchard Road, Croydon CR9 6JA, UK
Tel: (+44 20) 86 67 36 66
Fax: (+44 20) 86 67 27 03
e-mail: atkinsonb.amecrail@ems.rail.co.uk
Web: http://www.amec.com/rail

Key personnel

Managing Director: John Moss
Business Development Director: Barry Atkinson
Commercial Director: Bill Daly
Financial Director: Richard Hunt
Operations Director: Kevin Beauchamp
Personnel Director: Alan Barnes
Technical Director: Bruce Littlewood

Other offices

Head office
Pole Edison
Parc Saint Christophe, F-95861, Cergy Pontoise, Cedex, France
Tel: (+33 1) 34 22 50 00   Fax: (+33 1) 34 22 51 26

Europe: AMEC SPIE Rail (Benelux) Abay Rail (Belgium); AMEC SPIE Rail SET GmbH (Germany); AMEC SPIE Rail, Cariboni Impianti SRL (Italy)
UK: AMEC SPIE Rail Training; AMEC SPIE Rail (Southern); AMEC SPIE Rail (Western); AMEC SPIE Rail (North East); AMEC SPIE Rail (North West); AMEC SPIE Rail (Scotland)

Background

Following the acquisition of 41.6 per cent of SPIE SA (France) by AMEC plc, a UK-based railway infrastructure company was formed called AMEC SPIE Rail Systems Ltd. In 2003 AMEC exercised its option to purchase the remaining share of SPIE SA, which has a 50 per cent interest in ETF, the French track renewals company.

Services

Design and construction of catenary systems for high-speed main line and urban systems including provision of traction power supply systems.

Contracts

Include provision of catenary systems for TGV high-speed lines in France. Renovation of overhead catenary systems in southern Italy and design and construction of Phase 1 of the UK's Channel Tunnel Rail Link. The electrificiation of the Betuwe route in the Netherlands. The catenary upgrade for the Brussels–Gent high-speed line. The upgrade of the new catenary works in east Germany. The catenary system for the Essen–Wuppertal region express line. Catenary works for the Nürnberg–Ingolstadt section of the ICE (high-speed) Berlin to Munich line.

NEW ENTRY

Ampcontrol Pty Ltd

16 Old Punt Road, Tomago, New South Wales 2322, Australia
Tel: (+61 2) 49 61 90 00
Fax: (+61 2) 49 61 90 09
e-mail: corporate@ampcontrol.com.au
Web: www.ampcontrol.com.au

Key personnel

National Sales Manager: Gary Hillier
Business Development Manager: Peter Hogg

Products

Traction control switchgear; traction power control systems; traction substation switchrooms (AC and DC); specialised transformers and switchgear; rail communications.

Contracts

220 intelligent field telephones for the NSW Rail Infrastructure Corporation (RIC). Solar powered trackside wireless phones as well as electronic magneto and CB types supplied to the NSW Rail Infrastructure Corporation (RIC).

UPDATED

ApATeCh Electro

Zhukovsky Street 2, Building 131, Dubna, 141980 Moscow Region, Russian Federation
Tel: (+7 09621) 257 92
Fax: (+7 09621) 234 92
e-mail: electro@dubna.ru

Products

Plain and ribbed rod insulators in composite materials for overhead contact line systems for main line and urban railways.

Balfour Beatty Rail GmbH

Power Systems
Garmischer Strasse 35, D-81377 Munich, Germany
Tel: (+49 89) 743 19 02   Fax (+49 89) 74 31 92 50
e-mail: info.powersystems@bbrail.com
Web: www.balfourbeatty.com

Key personnel

Managing Director: Manfred Leger

Subsidiary companies

Balfour Beatty GmbH
Brown Boveri Strasse 1, A-2351 Wiener Neudorf, Austria
Tel: (+43 2236) 90 40 06 00
Fax: (+43 2236) 90 40 06 20

Balfour Beatty Rail Ltda
Av Maria Coelho Aguiar 215, Bloco B – 3° Andar, São Paulo, SP 05804-900, Brazil
Tel: (+55 11) 37 48 63 89
Fax: (+55 11) 37 48 99 31

Balfour Beatty Rail srl
Centro Direzionale Milano Oltre 2, Palazzo Cedri, Viale Europe 24, I-20090 Milan, Italy
Tel: (+39 02) 26 92 70 89
Fax: (+39 02) 26 92 70 60

Balfour Beatty Rail Sdn Bhd
Lot 1E & 1G, 1st Floor, 2 Tasik Ampang, Jalan Hulu Kelang, 68000 Ampang, Selangor Darul Ehsan, Malaysia
Tel: (+60 3) 42 52 73 66   Fax: (+60 3) 42 52 40 88

Balfour Beatty Rail (Iberica) SA
C/San Josà Artesano 12-14, Poligono Industrial, E-28108 Alcobendas (Madrid), Spain
Tel: (+34 91) 657 91 19   Fax: (+34 91) 657 91 25

Balfour Beatty Rail AB
Kopparbergsvàgen 6, Box 413, SE-721 08 Vàsteras, Sweden
Tel: (+46 21) 15 44 00
Fax: (+46 21) 41 24 90

Background

In August 2005 Balfour Beatty announced its intention to acquire SBB, the specialist German signalling contractor. SBB will become part of Balfour Beatty Rail Power Systems.

Products and services

Railway electrification products and services, covering complete system design, supply, installation, construction management and commissioning for main line, mass transit and light rail systems.

Catenary system products and services include: main line catenary systems for speeds of up to 400 km/h; mass transit overhead contact line systems; mass transit conductor rail systems; catenary components and equipment; project management; commissioning; and customer support.

Power supply system products and services for AC systems include: $16^2/_3$ Hz and 50 Hz substations; static converters; boosters and auto-transformer stations; station control technology/protection systems; outdoor cabling systems and indoor installation. For DC systems these include: substations; diode rectifiers; voltage-controlled rectifiers; and switchgear.

For both AC and DC systems Balfour Beatty Rail provides: network control and telecontrol systems; project management; commissioning; and customer support.

Contracts

Current or recent contracts include:
Brazil: Electrification equipment for Line 5 of the São Paulo metro for CPTM.
Bulgaria: Electrification of the 131 km Dupnitza Kulata line for Bulgarian State Railways.
China: Electrification of the Harbin Dalian line in consortium with Siemens Transportation Systems. The project covers some 950 km of double-track line.
Germany: The electrification system for the 70 km Nuremberg Ingolstadt high-speed line for Deutsche Bahn AG.
Greece: Electrification of the Piraeus Athens Thessaloniki main line for Ergose SA.
Italy: The electrification of the Rome Naples high-speed line for FS SpA and TAV SpA.

Malaysia: The electrification system for the STAR LRT system in Kuala Lumpur.
Portugal: The electrification system for the Metro do Porto light rail project.
Sweden: Electrification of the Øresund Fixed Link between Sweden and Denmark.
Turkey: The electrification system for the Adana LRT project.

UPDATED

Barclay Mowlem Construction Ltd

Building 3, Level 2, 20 Bridge Street, Pymble, New South Wales 2073, Australia
Tel: (+61 2) 98 55 16 00
Fax: (+61 2) 98 55 16 20
Web: www.barclaymowlem.com

Key personnel

Managing Director: David Hudson
Director and General Manager, Rail: Bill Killinger
Commercial Manager, Rail: Bob Cooke
Manager, Rail Australia: Graeme Spragg
Manager, Rail Hong Kong: Bill Hardy

Background

Barclay Mowlem is a subsidiary of John Mowlem Co plc.

Subsidiary company

Austrak Pty Ltd

Services

Barclay Mowlem delivers premium services in railway infrastructure design, trackwork, high-speed rail, mass transit systems and depot construction, electrification, signalling, maintenance, renewals and plant and equipment supply throughout Australia, New Zealand and South East Asia.

The rail group is currently operational in Australia, Hong Kong, Malaysia, Singapore, Taiwan and Thailand.

UPDATED

Bharat Earth Movers Ltd (BEML)

BEML Soudha, 23/1 4th Main, SR Nagar, Bangalore 560 027, India
Tel: (+91 80) 222 44 58   Fax: (+91 80) 229 19 80
e-mail: techrnd@vsnl.com
Web: http://www.bemlindia.com

Key personnel

For full personnel listing, see Bharat Earth Movers Ltd (BEML) entry in *Locomotives and powered/ non-powered passenger vehicles* section.

Four-wheeled overhead equipment inspection car for Indian Railways   0527277

Eight-wheeled overhead equipment inspection car for Indian Railways 0527278

Products

Overhead equipment inspection cars, in eight-wheeled and four-wheeled versions for periodic inspection and maintenance of overhead equipment on electrified rail routes. These self-propelled cars are self-contained with workshop, storage facilities, staff cabins and elevating platform and are equipped for repairs to overhead equipment and erecting catenary and contact wires.

Contracts

Recent contracts include the supply of 30 eight-wheeled cars and 10 four-wheeled cars (1,676 mm gauge) for Indian Railways.

Benkler AG

Nordstrasse 1, CH-5612 Willmergen, Switzerland
Tel: (+41 56) 618 72 00
Fax: (+41 56) 618 72 99
e-mail: info@benkler.ch
Web: http://www.benkler.ch

Corporate background

Benkler AG is a member of the Sersa Group (qv).

Services

Overhead line maintenance and construction equipment and services; cable construction.

Brecknell Willis & Co Ltd

Member of the Fandstan Electric Group
PO Box 10, Tapstone Road, Chard TA20 2DE, UK
Tel: (+44 1460) 649 41   Fax: (+44 1460) 661 22
e-mail: mail@brecknell-willis.co.uk
Web: www.brecknell-willis.co.uk

Key personnel

Managing Director: Tony White
Sales Manager, Trainborne Equipment:
   Andrew Hales
Third Rail, Commercial Operations Manager:
   David Bailey
Overhead Systems, Chief Engineer: David Hartland

Products

Current collection and power distribution equipment for the transport sector. This includes design, manufacture, supply, installation, commissioning of complete systems and after sales service.

The product groups are pantographs and third rail current collectors; conductor rail systems; aluminium/stainless composite conductor rail and accessories, light rail overhead systems, automatic gas tensioning equipment, spring boxes and ground return units for railways, metro, light rail systems and tramways.

Contracts

Conductor rail systems for the Jubilee Line Extension Project and the Northern Line Upgrade (London Underground Ltd), Copenhagen, Ankara and Taipei Metro systems. Overhead contact systems for the Dublin Luas, Midland Metro LRT system, Vancouver, Manchester Metro and refurbishment of Blackpool Tramway. Aluminium composite rail for DB, Berlin and Merseyrail UK.

Current collector systems for all new UK emu vehicles, West Coast Main Line Pendolino and current collection equipment for Shanghai Maglev; current collectors for the new vehicles for Hong Kong, Delhi Metro, KL-Monorail.

Supplied Eurostar and Channel Tunnel Shuttle trainborne current collectors as well as all standard high-speed pantographs for 25 kV electrified line operation in the UK. Light rail pantographs were supplied for the Strasbourg, Manchester, Birmingham, Sheffield and Tyne & Wear trams and shoegear for Glasgow, Amsterdam and Taipei Metros.

UPDATED

Brush Transformers Ltd

PO Box 20, Loughborough LE11 1HN, UK
Tel: (+44 1509) 61 14 11   Fax: (+44 1509) 61 05 50
e-mail: sales@btl.fki-eng.com
Web: www.fki-eng.com

Key personnel

Managing Director: D I Woolhouse
Operations Director: M W Gunn
Sales and Marketing Manager: P J Coe

Background

Brush Transformers Limited is part of the FKI Group of international manufacturing companies, with over 75 operating units servicing many diverse market sectors.

Products

Distribution and power transformers in the range of 2 to 60 MVA up to 145 kV, dry-type 100 kVA to 4 MVA 15 kV, traction repair and flameproof transformers and flameproof switchgear. External, bolt-on, high-speed resistor type up to 500 A, 66 kV fully insulated.

After-service support provides customers with repair and refurbishment of traction transformers, industrial power transformers, tap-changers and flameproof equipment. Applications include power utilities, major contractors, the oil and gas industry, as well as petrochemical, rail, steel and coal industries.

UPDATED

Carbone of America (LCL) Ltd

225 Harwood Boulevard, Dorion, Quebec J7V 1Y3, Canada
Tel: (+1 450) 455 57 28   Fax: (+1 450) 455 50 52

Key personnel

General Manager and Customer Liaison:
   Marc Charlebois

Products

Current collection systems, including mechanism, collector shoes and current limiting shoe fuses and ground return units for metro systems. Also supply of wayside switchgear, 1,000 A to 5,000 A, load break and transfer switches.

Cembre SpA

Via Serenissima 9, I-25135 Brescia, Italy
Tel: (+39 030) 369 21   Fax: (+39 030) 336 57 66
e-mail: info@cembre.com
Web: www.cembre.com

Principal subsidiary

Cembre Ltd
Fairview Industrial Estate, Kingsbury Road, Curdworth, Sutton Coldfield B76 9EE, UK
Tel: (+44 1675) 47 04 40   Fax: (+44 1675) 47 02 20
e-mail: sales@cembre.co.uk

Subsidiary companies

Cembre AS
Fossner Senter, N-3160 Stokke, Norway
Tel: (+47) 33 36 17 65   Fax: (+47) 33 36 17 66
e-mail: cembre@cembre.no

Cembre España SL
Calle Llanos de Jerez 2, Coslada, E-28820 Madrid, Spain
Tel: (+34 91) 485 25 80   Fax: (+34 91) 485 25 81
e-mail: info@cembre.es

Cembre GmbH
Taunusstrasse 23, D-80807 Munich, Germany
Tel: (+49 89) 358 06 76   Fax: (+49 89) 35 80 67 77
e-mail: info@cembre.de

Cembre Inc.
Raritan Center Business Park
70 Campus Plaza II, Edison, New Jersey 08837, USA
Tel: (+1 732) 225 74 15   Fax: (+1 732) 225 74 14
e-mail: salesus@cembre.com

Cembre Sarl
22 avenue Ferdinand de Lesseps, F-91420 Morangis, France
Tel: (+33 1) 60 49 11 90   Fax: (+33 1) 60 49 29 10
e-mail: info@cembre.fr

Products

Drilling machines for rail web and wooden sleepers; rail bush contact kits; related accessories.

UPDATED

For details of the latest updates to *Jane's World Railways* online and to discover the additional information available exclusively to online subscribers please visit

jwr.janes.com

CostaRail Srl

Viale 4 Novembre, I-23845 Costamasnaga, Italy
Tel: (+39 031) 86 94 11   Fax: (+39 031) 85 53 30
e-mail: hkastner@costarail.it

Key personnel
Commercial Manager:
  Heinz Kastner

Background
Formerly Costaferroviaria, the company went into temporary receivership and was taken over by Rail Services International Group, for which RSI Italia SpA is the operative leader. In October 2004, CostaRail Srl was established, re-launching all existing Costaferroviaria activities, including orders, trademarks, patents and projects.

Products
ASTRIDE road/rail vehicle which is suitable for maintenance of electric overhead lines and is equipped with a platform fitted to a crane providing a wide range of access. The vehicle has a railway system which takes the power from the engine of the road vehicle (modified Iveco 150). The front axle is driven and the rear axle is trailing. The wheels are 500 mm in diameter and are fitted with leaf springs and shock-absorbers in order to obtain good contact with the rail even at high speeds. Traction is hydromechanical and the braking system utilises disc brakes. Also, the vehicle is provided with service, emergency and parking brakes.

Maximum speed is more than 50 km/h, while at low speeds it is possible to use a remote control from the inspection basket.

ASTRIDE, fitted by Permaquip, is in service in the UK.

UPDATED

Crompton Greaves Ltd

I Dr V B Gandhi Marg, Mumbai 400 023, India
Tel: (+91 22) 202 80 25

Rail Transportation Systems Division
5-E Vandhna 11, Tolstoy Marg, New Delhi 110 001, India
Tel: (+91 11) 331 70 75; 373 04 45
Fax: (+91 11) 331 70 75; 332 43 60
e-mail: dhingra_cgl@mantraonline.com
Web: http://www.cromptongreaves.com

Key personnel
Managing Director: S M Trehan
General Manager, Rail: A K Raina
All India Marketing Manager, Rail Transportration
  Systems Division: Harsh Dhingra

Products
Traction transformers, SF6 gas interrupters/circuit breakers, lightning arrestors, turnkey electrification contracts.

Cuadrelec SA

C/Primavera 1-2-3, Polígono Industrial Las Monjas, E-28850 Torrejón de Ardoz, Madrid, Spain
Tel: (+34 91) 656 3 77
Fax: (+34 91) 656 37 41
e-mail: cuadrelec@sistelcom.com
Web: http://www.cuadrelec.com

Products
Traction rectifiers for 3.3 kV, 1.5 kV and 750 V DC power supply systems.

Ebo Systems

Boulevard d'Europe, PO Box 10 F-67211, Obernai, Cedex, France
Tel: (+33 3) 88 49 50 51   Fax: (+33 3) 88 49 50 14
e-mail: info@ebo-systems.com
Web: http://www.ebo-systems.com

Key personnel
Managing Director: Ulrich Pelz

Products
FRP/GRP cable management systems; cable trays, ground ducts, cable ladders; fixing and supporting material.

VERIFIED

EFACEC Sistemas de Electrónica SA

Av Eng Frederico Ulrich, PO Box 31, P-4470 Maia, Portugal
Tel: (+351 2) 941 36 66   Fax: (+351 2) 948 54 28

Products
Electrification equipment, including traction substations (1,500 V DC and 25 kV 50 Hz AC) and associated telecontrol systems, catenary systems.

Elpro BahnstromAnlagen GmbH

Marzahner Strasse 34, D-13053 Berlin, Germany
Tel: (+49 30) 98 61 22 53   Fax: (+49 30) 98 61 22 51
e-mail: manfred.pietzker@elpro.de
Web: http://www.elpro.de

Products
Equipment for AC and DC electrification systems, including substations, switchgear, overhead lines, control and distribution systems.

Ensto Sekko Oy

Head Office
PO Box 51, FIN-06101 Porvoo, Finland
Tel: (+358 204) 76 21   Fax: (+358 204) 76 27 70
e-mail: utility.networks@ensto.com
Web: www.ensto.com

UK office
Ensto UK Ltd
Regus House, George Curl Way, Southampton, Hampshire SO18 2RZ
Tel: (+44 2380) 30 20 15   Fax: (+44 2380) 30 20 15
e-mail: ian.strachan@ensto.com
Sales Manager: Ian Strachan

Key personnel
Managing Director: Marjut Haverinen
Product Manager: Veijo Vilenius

Products
Aluminium cantilevers and components; composite, ceramic and glass insulators; crimp and screw-type fittings; hot-dipped galvanised steel parts.

Ensto has a solution for lightweight corrosion resistant OLE cantilever assemblies which uses a patented reinforced tubular section with solid ends, in combination with heat tempered aluminium clamps and composite insulators.

Contracts
Recent contracts include VR Track, Eltel Networks, Finland; Banverket, SL-Infrateknik, Sweden; Jernbaneverket, Norway; DSB, Denmark; West Coast Main Line, Balfour Beatty, UK and SNCB, Belgium.

UPDATED

Ferraz Shawmut

1 rue Jean Novel, F-69626 Villeurbanne Cedex, France
Tel: (+33 4) 72 22 66 11
Fax: (+33 4) 72 22 67 13

Key personnel
For full personnel listing, see Ferraz Shawmut entry in *Electric traction equipment* section.

Products
AC protistor fuses for the internal protection of the AC/DC and/or AC/DC/AC substation converters, large power filters and auxiliary circuits; disconnectors to isolate substation converters; automatic fast-acting earthing device with large short-circuit capability, which can be either bi- or unidirectional.

FIREMA Trasporti SpA

Headquarters and main facilities
Via Provinciale Appia, Località Ponteselice, I-81100 Caserta, Italy
Tel: (+39 0823) 09 71 11   Fax: (+39 0823) 46 68 12
Web: http//www.firema.it

Key personnel
Chairman: Gianfranco Fiore
Chief Executive Officer: Roberto Fiore
Operations Director: Maurizio Russo
Marketing and Commercial Manager:
  Agostino Astori

Commercial and technical offices
Via Triboniano n 220, I-20156 Milan, Italy
Tel: (+39 02) 23 02 02 23   Fax: (+39 02) 23 02 03 00

Products
Substation equipment for AC and DC electrification; converting group and auxiliary transformers; solid-state rectifiers; high-/medium-voltage switchgear with high-speed circuit breakers, auxiliary and protection relays; minor parts and maintenance.

Work continues on development of diagnostics and automatic maintenance systems.

Flury

Arthur Flury AG
Fabrikstrasse 4, CH-4543 Deitingen, Switzerland
Tel: (+41 32) 613 33 66   Fax: (+41 32) 613 33 68
e-mail: aflury@bluewin.ch
Web: http://www.aflury.ch

Key personnel
President: Adrian Flury
Managing Director: Jürg Zwahlen

Products
Components for overhead electrification systems, including section insulators and phase breaks; messenger wire and contact wire insulators; earthing equipment; terminals, suspension clamps, connecting clamps and feeder clamps.

Fuji Electric Co Ltd

Gate City Ohsaki, East Tower, 11-2, Osaki 1-chome, Shimagawa-ku, Tokyo 141-0032, Japan
Tel: (+81 3) 54 35 70 46   Fax: (+81 3) 54 35 74 23
e-mail: info@ffesys.co.jp
Web: www.fesys.co.jp

Key personnel
Managing Director, Electrical Systems Company
  Group: H Itou
General Manager, Transportation Systems Sales
  Department: K Kimura

Background
The Transportation Systems Sales Department of what was previously Fuji Electric Co Ltd, is part of the newly divided Fuji Electric Systems Co, Ltd.

Products
Power supply equipment: computer-based remote supervisory control equipment; water cooling silicon rectifiers, SF6 gas circuit breakers and mini high-speed circuit breakers; moulded transformers; total control systems including electric power management, station office apparatus control, data management and disaster prevention management.

UPDATED

Furrer + Frey AG

PO Box 182, Thunstrasse 35, CH-3000 Berne 6, Switzerland
Tel: (+41 31) 357 61 11   Fax: (+41 31) 357 61 00
e-mail: adm@furrerfrey.ch
Web: www.furrerfrey.ch

Key personnel

Chief Executive Officer, Export and Marketing:
B Furrer
Executive Officer, Construction Department for
Railway Electrification: F Friedli
Executive Officer, Electrification and Design
Products: R Marti
Executive Officer, Export Department: R D Brodbek
Consulting Officer: U Wili

Works

PO Box, Eisenbahnstrasse 62–64, CH-3645 Gwatt

Subsidiary

UP AG, Berne

Products

Design, manufacture and installation of overhead contact lines for railways, up to 25 kV AC. Aerial surveys for electrification projects.

Specialist equipment includes overhead contact lines for track railways, tram and light rail systems; overhead conductor rails; movable conductor rail for depots and maintenance facilities. The provision of software for electrification projects; consultancy.

Contracts

1999–2005 fixed and movable overhead conductor rail installations in depots and maintenance facilities in Switzerland, Germany, Denmark, Sweden, Norway, Italy and Hong Kong; 1998–2004 electrification of Zimmerbergtunnel on section Zurich–Thalwil for Swiss Federal Railways within project Rail 2000, with design speed 200 km/h; moveable conductor rail installations on five swing and bascule bridges on the northeast corridor section between Boston and New Haven for Amtrak; 2001–2005 new catenary installation in the reconstructed railway station Chur of Rhaetian Railways (narrow gauge); 2003–2004 supply of patented new catenary tunnel support in very narrow standard gauge Grenchenbergtunnel; 2004 conductor rail for Line 2 of Guangzhou Metro; 2003–2004 overhead conductor rail installation for speeds 200 km/h in Sittenbergtunnel (2 km), Austria; 2001–2005 supply of 400 supports for narrow tunnels to Norwegian Rail Administration; 2003–2004 planning, material supply and installation works in the tunnel between Salgesch and Leuk on double track section of Swiss Federal Railway (SBB), 5 km.

UPDATED

Galland Sarl

20 rue de l'Insurrection Parisienne, F-94600 Choisy-le-Roi, France
Tel: (+33 1) 46 80 25 72   Fax: (+33 1) 46 80 83 42
e-mail: info@j-galland.com
Web: http://www.j-galland.com

Key personnel

Managing Director: Denis Galland
Chairman: Dominique Bec
Technical Director: Philippe d'Huy
Export Manager: Jacques Milhem

Products

Catenary equipment including catenary from 750 V to 25 kV; normal, reinforced and flexible gantries; lightweight section insulators from tramway to high-speed train (tested at 270 km/h); spring and pulley tensioning devices; isolating and selector switches, with and without earth (new European standard); tramway equipment – catenary, anchorage, section insulators, Kevlar terminations, delta suspension, tensioning devices, isolator and selector switches.

Geismar

113 bis avenue Charles-de-Gaulle, F-92200 Neuilly sur Seine, France
Tel: (+33 1) 41 43 40 40   Fax: (+33 1) 46 40 71 70
e-mail: geismar@geismar.com
Web: www.geismar.com

Works

5 rue d'Altkirch, F-68006 Colmar Cedex, France
Tel: (+33 3) 89 80 22 11
Fax: (+33 3) 89 79 78 45
e-mail: colmar@geismar.com

Key personnel

Publicity: Patrick Lambert

Products

Geismar provides equipment to erect, maintain and inspect AC and DC catenary lines; a large variety of dedicated tools; a range of overhead line components; overhead line unrolling/renewal equipment: skids, wagons and trains; access units and vehicles: elevators, road-railers, maintenance and inspection motorcars; electronic measuring/recording: hand-held devices and onboard systems.

Geismar also provides tracklaying and maintenance equipment.

UPDATED

Greysham (International) Pvt Ltd

1/1-A, Man Singh Place, Roop Nagar, Delhi 110 007, India
Tel: (+91 11) 23 84 93 81
Fax: (+91 11) 23 84 30 68
e-mail: sales@greysham.com
Web: www.greysham.com

Key personnel

Managing Director and Chairman: Govind Singh
Joint Managing Director: Sanjeev Singh
Export Manager: V B Arya

Products

Overhead fittings for railway electrification projects.
ISO 9001-2000 certified, accredited by NQAQSR, New Zealand.

UPDATED

FKI Switchgear

Switchgear Works, Castleton, Rochdale OL11 2SS, UK
Tel: (+44 1706) 63 20 51
Fax: (+44 1706) 67 42 36
e-mail: sales@st.fki-et.com
Web: www.fki-et.com

Key personnel

Managing Director: Roy Milward
Engineering Director: Steve Lane
Sales and Marketing Director: Steve Dymond
Marketing Manager: Kevin Lynch

Background

FKI Switchgear brings together Whipp & Bourne and Hawker Siddley Switchgear Limited, two successful UK switchgear companies. Whipp & Bourne is a long-established major manufacturer of heavy-duty electrical switchgear, the original company being formed in 1903. The company has large experience in the production of electrical switchgear, specifically designed to meet the system requirements of AC and DC electrification for controlling the power supplies to both heavy and light rail transit systems together with metro systems.

www.furrerfrey.ch

Furrer+Frey AG
Overhead contact line engineering
Design, manufacturing, installation
Thunstrasse 35, P.O. Box 182
CH-3000 Berne 6, Switzerland
Telephone   + 41 31 357 61 11
Fax   + 41 31 357 61 00

Furrer+Frey®
Overhead contact lines

Products and services

Design, engineering, manufacturing, installation and after-sales service of switchgear for electrical distribution in heavy/light rail transit systems, metro systems and other industries. Products include: high-speed DC circuit breakers for ratings up to 12,000 A, 3,000 V DC and 255 kA short circuit; medium voltage vacuum switchgear for ratings up to 3,150 A, 12 kV and 50 kA short circuit; microprocessor controlled auto reclosers with ratings up to 630 A, 38 kV and 12 kA short circuit; associated substation switchgear for AC and DC electrification projects is also available.

Developments

'Lightning' incorporating the high-speed NDC circuit breaker is the latest edition to FKI's range of DC switchgear products. This product provides a new DC switchgear solution by using the patented magnetic actuator technology. Other products include the patented, low energy, single coil magnetic actuator which allows for reductions in the number of total parts, delivering system wide benefits such as reduced maintenance and an extended working life of all switchgear. Whipp and Bourne has various switchgear products in service with traction authorities worldwide.

UPDATED

HEI

Hindusthan Engineering & Industries Ltd
Mody Building, 27 Sir RN Mukherjee Road, Calcutta 700 001, India
Tel: (+91 33) 248 01 66
Fax: (+91 33) 248 19 22; 220 26 07
e-mail: hindus@cal2.vsnl.net.in

Works

Insulators & Electrical Company (IEC)
1/8 New Industrial Area, Mandideep 462,046, District Raisen, Madhya Pradesh, India

Key personnel

President IEC: B D Tulsian

Products

Manufacture and supply of 25 kV solid core insulators.

Hitachi Ltd

Overseas Marketing Department
Transportation Systems Sales Division
6 Kanda Surugadai 4-chome, Chiyoda-ku, Tokyo 101-8010, Japan
Tel: (+81 3) 32 58 11 11
Fax: (+81 3) 32 58 52 30
Web: www.hitachi-rail.com

Key personnel

Chief Operating Officer of Transportation Systems:
  Gaku Suzuki
General Manager, Transportation Systems Sales
  Division: Chiaki Ueda
General Manager, Transport Management and
  Control Systems Division: Kazuo Kera
General Manager Rolling Stock System Division:
  Toshihide Uchimura

Products

Traction substation equipment for AC and DC electrification projects; diode and thyristor rectifiers; power regenerative inverters; transformers; AC and DC switchgear; control and protection devices; computerised systems – substation supervisory remote-control systems, automatic car diagnosis system, station management system, security system.

UPDATED

Hyundai

Electro-Electric Systems
Hyundai Building, 140-2 Kye-dong, Chongro-ku, Seoul 110-793, South Korea
Tel: (+82 2) 746 75 30   Fax: (+82 2) 746 74 79
e-mail: rolling@hhi.co.kr
Web: http://www.hhi.co.kr

Products

Transformers; rectifiers; switchgear; SCADA systems.

Insul-8 Corporation

10102 F Street, Omaha, Nebraska 68127, USA
Tel: (+1 402) 339 93 00   Fax: (+1 402) 339 96 27
e-mail: rprell@insul-8.com
Web: www.insul-8.com

Key personnel

President: Lon Miller
Project Manager: Richard Prell

Other offices in Canada, Australia and Manchester, UK

Background

Insul-8 is part of the Delachaux Group, Gennevilliers, France.

Products

Conductor rail systems up to 6,000 A. Specialises in conductor rail with stainless steel on aluminium extrusion and can be used for overhead cranes and specialised rail for transit systems.
  Supply of overhead stinger systems for maintenance facilities.

Contracts

Las Vegas Monorail, Dallas/Forth Worth Airport, Lisbon-Oerias Airport, Toronto Airport.

Developments

Contact rails with choice of amperage range to meet system demands.

UPDATED

A Kaufmann AG

Pilatusstrasse 2, CH-6300 Zug, Switzerland
Tel: (+41 41) 711 67 00   Fax: (+41 41) 859 16 01
e-mail: info@kago.ch
Web: www.kago.com

Products

KAGO specialist engineering products for railways, including a complete range of non-screwed rail contact clamps, discs and strips for electrical rail connections (return current, signalling circuits, earthing), cable fastenings for rails and sleepers, special welding electrodes for copper welding; complete range of screwing, welding and grounding fittings; self-tapping sleeper screws for concrete, steel or wooden sleepers; sleeper spring clips; high-voltage insulations.
  Recent development: KAGO axle counter box.

Services

Track returns and earthing wires; earthing poles; heavy-duty and special mountings and drillings; suspensions for radiating cables.

UPDATED

Kershaw Manufacturing

PO Box 244100, Montgomery, Alabama 36124, USA
Tel: (+1 334) 387 91 00
Fax: (+1 334) 215 75 51
Web: http://www.kershawusa.com

Key personnel

Vice-President and COO: G Reg Valley
Vice-President of Sales and International Sales
  Manager: Phil Brown

Products and services

Vehicles to cut safely trees or branches growing in the path of overhead lines, power-lines and catenary. Rubber-tyred or rail-mounted, one-man operation, with reach up to 21.2 m.

Kruch GesmbH & Co KG

Pfarrgasse 87, A-1230 Vienna, Austria
Tel: (+43 1) 616 31 65
Fax: (+43 1) 616 31 68
e-mail: office@kruch.at
Web: http://www.kruch.at

Products

Suspension clamps and hanger clamps for bearer cables and overhead contact wires.

Kummler & Matter AG

Hohlstrasse 176, CH-8026 Zürich, Switzerland
Tel: (+41 44) 247 47 66
Fax: (+41 44) 247 47 47
e-mail: kuma@kuma.ch
Web: www.kuma.ch

KAGO rail contact clamps (inset) and cable connection in progress

0006271

Key personnel

President and Chief Executive Officer:
  Daniel Steiner
Finance Director: Michel Villoz
Product Development Director: Reto Hügli
Project Engineering Installation Director:
  René Kopp
Purchasing Manager: H P Villringer
Area Sales Managers: Rodolfo Middlemann,
Reto Hugli, Willy-Urs Brassel, André Eichhorn

Products

Overhead contact line equipment for light rail,
branch lines, suburban and main line railways, and
trolleybuses. Engineering and feasibility studies.
Installation and supervision of installation.

UPDATED

Lerc SA

Chemin des Hamaïdes, PO Box 119, F-59732 Saint
Amand les Eaux Cedex, France
Tel: (+33 3) 27 22 85 50   Fax: (+33 3) 27 22 85 05
e-mail: commercial@lerc.fr
Web: http://www.lerc.fr

Key personnel

Managing Director: J Mourey
Technical Director: Y Foissac
Sales Manager: F Romet
Export Marketing Manager: Vincent Lernoud

Subsidiary company

Janssen Engineering

Products

A range of insulators with silicon shed shells
for railway, tram and metro lines; bushing and
insulating systems; composite insulators for
energy (transport and distribution). Production and
refurbishment of composite bushings. Synthetic
insulators for catenaries, insulated steady arms
for urban transport, insulated consoles for urban
transport, composite parts processed by pultrusion,
moulding and winding.

VERIFIED

Merlin Gerin Brasil SA

Av Brigadeiro Faria Lima 2003, 14th andar, 01451-001
São Paulo SP, Brazil
Tel: (+55 11) 816 45 00
Fax: (+55 11) 813 09 43

Main works

Av da Saudade s/n, 13171-320, Sumaré, SP, Brazil

Key personnel

For full personnel listing, see Merlin Gerin Brasil
SA entry in *Electric traction equipment* section.

Products

Traction chopper control system; high and low-
voltage switchboards; self-inductive coils, resistors
and static converters, traction rectifiers; control
boards; auxiliary switchboards and low-voltage
rectifiers; track circuits; power and signalling mimic
panels; relay racks; data transmission; boards;
low-, medium- and high-tension equipment.

Mer Mec SpA

Via Oberdan 70, I-70043 Monopoli (BA), Italy
Tel: (+39 080) 887 65 70
Fax: (+39 080) 887 40 28
e-mail: mermec@mermec.it
Web: www.mermec.it

Key personnel

Managing Director: Vito Pertosa
Marketing and Sales, Executive Director:
  Luca Ebreo

Marketing Manager: Pietro Stama
International Sales Manager: Carlo Evangelisti
Domestic Sales Manager: Mario Girolami
After Sales and Commissioning Manager:
  Guiseppe Aurisicchio
Quality Assurance Manager: Mauro Simone
Research and Development Programme Manager:
  Patrizia Sforza

Services

Design, development and manufacturing of
measuring/diagnostic systems and vehicles
for railway infrastructure monitoring. Mer
Mec patented overhead line (OHL) equipment
represents state-of-the-art technology in no-contact
systems for contact wire's wear and geometry
measurements. These can be installed on any
railway cars, can operate in all weather conditions
at speeds of up to 350 km/h and are provided with
software application for real-time/off-line data
processing and analysis. Mer Mec products include
OHL geometry, contact wire wear, pantograph
interaction, arcing measurement and video
inspection with automatic flaw detection.

Design and development of highly modular
software solutions (RAMSYS® – Railway Asset
Management System) supporting enhanced
data analysis for optimised maintenance and
renewal planning of railway infrastuctures.
Design, engineering and manufacturing of
vehicles/equipment dedicated to overhead line
maintenance.

Contracts

Mer Mec has won contracts to supply diagnostic
systems/vehicles for OHL monitoring in
Hong Kong, Italy, Norway, South Korea, Spain,
Sweden and Switzerland.

UPDATED

Muromteplovoz

Murom Diesel Locomotive Works
ulica Filatova 10, 602 200 Murom, Vladimir region,
Russian Federation
Tel: (+7 095) 291 31 68   Fax: (+7 09234) 443 03
e-mail: mteplo@cl.murom.ru
Web: http://www.cl.murom.ru

Key personnel

Director General: V Kharitinov
Director, Export Sales and Marketing: E Tretyakov

Products

Equipment for the erection and maintenance of AC
and DC overhead catenary systems.

nkt cables GmbH

Schanzenstrasse 6-20, D-51063 Cologne, Germany
Tel: (+49 221) 676 38 66   Fax: (+49 221) 676 24 22
e-mail: infoservice@nktcables.com
Web: www.nktcables.com

Products

Copper-magnesium and silver-alloy contact wires.

UPDATED

Paul Keller Engineering Ltd

Hochbordstrasse 9, CH-8600 Dübendorf, Switzerland
Tel: (+41 1) 821 40 27   Fax: (+41 1) 821 45 40
e-mail: pkag@pkag.ch
Web: http://www.pkag.ch

Subsidiaries

Berninastrasse 4, CH-5430 Wettingen, Switzerland
Tel: (+41 56) 438 08 88
Fax: (+41 56) 438 08 89

Bahnhofstrasse 37, CH-7302 Landquart,
Switzerland
Tel: (+41 81) 330 66 60   Fax: (+41 81) 33 06 66 10

Huggenberg, CH-8354 Hofstetten, Switzerland
Tel: (+41 52) 364 16 76   Fax: (+41 52) 364 16 56

Via Collina 54, CH-6612 Ascona, Switzerland
Tel: (+41 91) 791 63 01   Fax: (+41 91) 791 58 85

Alte Steinhauserstrasse 33, CH-6330 Cham,
Switzerland
Tel: (+41 41) 740 23 33
Fax: (+41 41) 74 06 33 30

Products

Power distribution networks; traction power supply
for railways; site surveying; corrosion protection;
structures and catenaries; control systems;
telecommunication systems.

*Muromteplovoz Type ARV-1 self-propelled overhead line maintenance vehicle with lifting and
rotating cradle*   0103973

FUM catenary erector　　　　　　　　　　　　　　　　　　　0598372

Pfisterer Srl

Articoli Elettrici Speciali
Via Sirtori 45d, I-20017 Passirana di Rho (MI), Italy
Tel: (+39 02) 931 58 11
Fax: (+39 02) 93 15 81 37
e-mail: zorzan@pfisterer.it
Web: www.pfisterer.it

Key personnel
Managing Director: Ing Luciano Femminis
Export Manager: Ing Lorenzo Mosna

Other offices
Pfisterer
Kontaktsysteme GmbH & Co KG
Rosenstrasse 44, D-73650 Winterbach, Germany
Tel: (+49 7181) 70 05　Fax: (+49 711) 301 21 97
Web: www.pfisterer.de

Pfisterer GmbH
A-1091 Wien, Augasse 17, Austria
Tel: (+43 1) 31 76 53 10　Fax: (+43 1) 317 65 31 12

Sefag Export
CH-6102 Malters/Luzern, Werkstrasse 7, Switzerland
Tel: (+41 41) 499 72 72　Fax: (+41 41) 497 22 69

Upresa SA
E-08025 Barcelona, Calle Industria 90-92, Spain
Tel: (+34 93) 436 47 01　Fax: (+34 93) 436 77 01
Web: www.upresa.es

Upresa SA
Arroya Fontarròn 39, E-28030, Madrid, Spain
Tel: (+34 91) 430 51 51　Fax: (+34 91) 437 39 10
Web: www.upresa.es

Solikap Acessòrios Eléctricos SA
Rua da Rainha 340, Apartado 2075, 4407 S Felix da
Marinha, Codex, Portugal
Tel: (+351 22) 762 47 14 64　Fax: (+351 22) 762 47 39
e-mail: solikap.ruiribeiro@netc.pt

Products
Silicon rubber composite insulators for railway,
tram and metro lines; clamps and earthing devices;
equipotential reversible cable hanger for use on
Italian Railways (FS) high-speed lines electrified
at 25 kV 50 Hz AC; voltage detectors, tensioning
system Tensorex®, a mechanical pull regulation
system which can handle continuous tension on
the contact wire specially designed for narrow
spaces like tunnels.

PU+ clamp; Ambrosius, automatic controlling
system for contact wire and catenary tensioning,
designed for the automatic remote control of
the tensioning performance; Phantom Tensorex,
automatic tensioning device for tramway and
railway contact lines, not visible, with low impact
on the environment.

UPDATED

Plasser & Theurer

Johannesgasse 3, A-1010 Vienna, Austria
Tel: (+43 1) 51 57 20　Fax: (+43 1) 513 18 01
Web: showroom.creative.co.at

Main works
Pummererstrasse 5, A-4021 Linz/Donau, Austria

Products
Since 1981 Plasser & Theurer has developed and
built machines for the installation and maintenance
of overhead wires and the associated equipment.
The particular advantage highlighted by Plasser &
Theurer catenary renewal machines is that contact
wire and carrying cable are installed with the
final tension as well as in the correct stagger.
The unavoidable post-tensioning necessary
when using traditional methods of installation
can be dispensed with entirely, which enables a
considerable increase in renewal output, reports
the company.

The benefits of the system include: immediate
use of the reopened line at full speed after
renewal; no closure of adjacent tracks; low staff
requirements; high-quality work; and renewal of a
complete section in one track possession.

With appropriate equipment of the machines,
return current circuits on the outer side of the
masts or feeders at the top of the masts can be
removed and installed.

The FUM 100.051 is equipped with reeling-off
devices to install one carrying cable and two
contact wires in one operation.

From the complete catenary renewal system to
a variety of sizes of motor tower car for inspection
and maintenance, individual solutions are adapted
to meet the requirements of the specific railway
administration.

UPDATED

Powernetics Ltd

Jason Works, Clarence Street, Loughborough
LE11 1DX, UK
Tel: (+44 1509) 21 41 53
Fax: (+44 1509) 26 24 60
e-mail: jag@powernetics.co.uk
Web: www.powernetics.co.uk

Key personnel
Managing Director: Satish Chada
Financial Controller: Bob Lawson
Engineering Director: Nilesh Chouhan
Operations Manager: Konrad Chada
Business Development Manager: Aran Chada
Sales Manager: Jim Goddard
Purchasing Manager: Russel Roughton
QA Manager: Gordon Anderson
Production Manager: Andy Guigno

Products
Independent private limited company with
30 years' experience in design, manufacture, test,
installation, commissioning and maintenance of
auxiliary, standby/emergency AC/DC power supply
systems for the rail industry in trackside, tunnel,
switchroom, REB and trainborne environments.

System design and integration of UPS and
associated switchgear/power monitoring within
REBs and switchrooms for SSI, FSP and PSP
applications.

UPDATED

RPG Transmissiom Ltd

29&30 Community Commercial Centre, Basant Lok,
Vasant Vihar, New Delhi 110057, India
Tel: (+91 11) 614 26 55; 614 58 01
Fax: (+91 11) 614 63 40
e-mail: chari@rpgtl.rpgms.ems.vsnl.net.in

Main works address
PO Box 96, Jabalpur-482001, India

Key personnel
Executive Director: S C Khanna
General Manager (Projects & Marketing):
D Luthra
Chief Manager (Projects & Marketing): A K Das

Principal subsidiary
KEC International Ltd, Mumbai, India

Products
Design, supply and erection of 3 kV DC and
25 kV 50 HZ AC overhead equipment, sub-stations,
booster transformer stations; telecommunication
cabling for railway electrification.

Contracts include electrification of 1,050 track
km at 25 KV 50 Hz AC for Indian Railways; design,
supply and erection of 25 kV 50 HZ AC overhead
equipment, booster, transformer stations,
switching stations in the Nidadavolu-Annavaram
section of SC Railway, India, completed in May
1997. Three contracts obtained for a total of 1,050
track km in the Kharagpur–Bhubaneswar section of
the of SE Railway, India, in 1998.

SAE (India) Ltd

29-30 Community Commercial Centre, Basant Lok,
Vasant Vihar, New Delhi 110 057, India
Tel: (+91 11) 688 26 55; 58 01
Fax: (+91 11) 611 11 90; 688 59 58

Main works
PO Box 96, Jabalpur 482001, India

Key personnel
President: Dr R K Dwivedi
Vice-President, Finance: Y L Madan
General Manager: P Varma
General Manager, Marketing: D Luthra
Chief Manager, Projects: A K Das

Products
Design, supply and erection of 3 kV DC and
25 kV 50 Hz AC overhead equipment; substations;
booster transformer stations; telecommunications
cabling for railway electrification.

Contracts include electrification of 300 track-km
at 25 kV 50 Hz AC for Indian Railways.

SAFT

Advanced and Industrial Battery Group
12 rue Sadi Carnot, F-93170 Bagnolet, France
Tel: (+33 1) 49 93 19 18　Fax: (+33 1) 49 93 19 64
Web: http://www.saftbatteries.com

Key personnel
For full personnel listing, see SAFT entry in *Electric
traction equipment* section.

Products

Saft Nife Ni-Cd batteries, with pocket or low maintenance sintered plastic bonded electrodes for supplying energy to all fixed electrification equipment.

Recent contracts include the supply of batteries to Singapore metro, Pakistan Railways, CP Rail.

SDCEM

Société Dauphinoise de Constructions Electro Mécaniques
10 allée de La Grange, F-38450 Vif, France
Tel: (+33 4) 76 72 76 72
Fax: (+33 4) 76 72 46 26

Key personnel

Managers: Claude Yvetot; Alain Plirai;
    François Mees; Gérard Dubois

Products

Catenary and switch disconnectors for 1.5/3 kV DC, 25 kV and 15 kV 16²/₃ Hz AC with manual and electrical operating mechanisms; 25 to 330 kV disconnectors for substations.

Sécheron SA

Route des Moulières 5, CH-1217 Meyrin, Geneva, Switzerland
Tel: (+41 22) 739 41 11
Fax: (+41 22) 739 48 11
e-mail: info@secheron.com
Web: www.secheron.com

Key personnel

Chief Executive Officer: Claude Durand
PBU Components Director: Jo Murer
PBU Substations Director: Dominique Jamet
PBU Electronics Director: Peter Stauffer
Marketing Manager: Gilbert Lile
Corporate Sales Manager: Jimmy Cuche

Subsidiaries

Sécheron Tchéquie, spol sro, Praha, Czech Republic

Products

DC traction power substations and ancillary equipment including system engineering and network computer simulation; solid-state rectifiers and inverters; harmonic filters; DC switchgear; DC high-speed circuit breakers, isolating and changeover switches; electronic protection relays; microprocessor-based remote-control and protection systems.

Contracts

Recent contracts include the design, supply and installation of DC switchgear and transformer/rectifier groups for Singapore Circle Line; DC switchgear for Tram SA in Athens; DC switchgear for Wuhan, Tianjin, Nanjing and Beijing Ba-Tong Lines in China; DC switchgear for Daegu Line 2 in Korea; rectifiers and DC switchgear for Bordeaux, Lyon and Orléans tramways in France.

UPDATED

Sefag AG

Werkstrasse 7, CH-6102 Malters, Switzerland
Tel: (+41 41) 499 72 72   Fax: (+41 41) 497 22 69
e-mail: connect@sefag.ch

Key personnel

Managing Director: H Wicki
Insulators Manager: W Fluri
Components (Domestic Market) Manager:
    W Bachmann
Production Manager: W Wipfli
Marketing and Sales Manager: M Peter
Business Services: H Wicki
Manager, Transmission Lines: W Huiber
Components (International Manager): J Fries
R & D and Marketing Manager Insulators:
    F Schmuck

Products

Silcosil composite insulators with silicon sheds as suspension, dead-end and post insulators, and as special insulators for tunnels and high-speed routes.

Contracts

Contracts include the supply of insulators to Swiss Federal Railways, Austrian Federal Railways, Bern-Lötschberg-Simplon Railway, and other railways in Switzerland and abroad.

UPDATED

Selectra srl

Via delle Nazioni 5, I-44100 Ferrara, Italy
Tel: (+39 0532) 74 80 00   Fax: (+39 0532) 74 44 55
e-mail: info@selectra.org
Web: http://www.selectra.org

Products

OptoCross opto-electronic systems for the high-speed contactless measurement of catenary

parameters, including height, stagger and thickness; PantoStat system for computerised pantograph control and catenary interaction measurement. The systems are designed to be mounted on any type of rail vehicle and can function at speeds of up to 360 km/h.

Siemens AG

Transportation Systems

Electrification Division
PO Box 3240, D-91050 Erlangen, Germany
Tel: (+49 9131) 70
Fax: (+49 9131) 72 83 64
e-mail: railway.electrification@siemens.com
Web: www.siemens.com/transportation

Corporate headquarters

Siemens AG
Transportation Systems
PO Box 3240, D-91050 Erlangen, Germany
Tel: (+49 9131) 7-0

Key personnel

President, Transportation Systems Group:
    Hans M Schabert
Group Vice Presidents: Alfred Frank; Joern F Sens;
    Friedrich Smaxwil
Heads of Electrification Division: Peter Schraut;
    Michael Duttenhofer;

Products and services

Supply and installation of traction power supply systems for mass transit and main line railways; supply and installation of contact lines for mass transit and main line railways.

Contracts

Recent contracts include:
China: In August 2001, a Siemens-led consortium with Barclay Mowlem received a contract covering the electrification of KCRC's East Rail Extension Project in Hong Kong. The contract included the supply of electrification equipment for the new line and project management.
Germany: In February 2001, Siemens was awarded a contract by DB Netz AG to install Germany's first gas-insulated switchgear at Griebnitzsee, near Potsdam. The substation equipment is insulated with inert sulphur hexafluoride.

In January 2001, Siemens-led consortium that includes Elpro BahnstromAnlagen GmbH secured a contract to install complete electrical equipment for 30 rectifier substations on the Berlin S-Bahn system. The contract was completed by the end of 2003.
Iran: In Mashhad, northeastern Iran, TS EL has delivered power supply substations and catenary equipment for a 20 km double-track light rail line. The network will be commissioned in 2006–07.
Thailand: In January 2005 Siemens, together with consortium partners B Grimm and Sino Thai Engineering and Construction plc (STECON) signed a contract with State Railway of Thailand for the construction of the 28 km Suvarnabhumi Airport Rail Link and the City Air Terminal. Both Siemens and B Grimm are taking responsibility for the design, supply, installation and project management of the whole electrical and mechanical system.
UK: In April 2003 Siemens, in a joint venture with AMEC SPIE Rail Systems Ltd, signed a seven-year framework agreement to upgrade the traction power supply system in the Wessex area of southern England, covering 750 V DC lines from London to Bournemouth and Weymouth. Siemens' role in the project covers the design, manufacture and supply of new DC substations.

Other orders for electrification technology form part of turnkey contracts, detailed in the Siemens Transportation Systems entry in the *Turnkey systems contractors* section.

Sécheron DC switchgear

NEW/1136191

UPDATED

Trackside installations for 750 V DC traction power upgrade for Network Rail, UK   **NEW**/0585227

Harbin-Dalian 25 kV AC electrification in China, equipped and installed by Siemens   **NEW**/0585226

Products

Design, construction and maintenance of railway infrastructure including: trackwork, catenary systems, power supplies and electromechanical systems for railways including high-speed rail, metros, tramways and other transport systems.

Recent contracts include Channel Tunnel, RATP (Paris metro), TGV Atlantique, TGV Nord, Cairo metro and Heathrow Express.

Supertek Enterprise Inc

2231 Colby Avenue, Los Angeles, California 90064, USA
Tel: (+1 310) 444 11 55   Fax: (+1 310) 444 11 64
e-mail: SPKDAVID@aol.com

Key personnel
President: David Chang

Korean address
Supertek Inc
1070-12 Hwagok Hong, Kangsen Gu, Seoul, South Korea
Tel: (+82 2) 696 19 90   Fax: (+82 2) 698 82 64
e-mail: DollyDoh@hitel.net
Director: J S Doh

Products
Complete overhead systems, substation systems, signal systems, communication systems, AFC systems.

Toshiba Corporation

Railway Projects Department
Toshiba Building, 1-Shibaura 1-chome, Minato-ku, Tokyo 105-8001, Japan
Tel: (+81 3) 34 57 49 24   Fax: (+81 3) 54 44 92 63
Web: www.toshiba.co.jp

Key personnel
President and Chief Executive Officer:
    Atsutoshi Nishida
Vice-President, Transportation Systems Division:
    Takio Ooyama
Senior Manager, Railway Projects Dept: Koji Toda

Products
SCADA (Supervisory Control and Data aquisition) for power supply systems. AC and DC electrified systems, power transmission and distribution systems. SCADA for facilities.

UPDATED

Sirti SpA

Via Pirelli 20, 20100 Milan, Italy
Tel: (+39 6) 22 54 54 22   Fax: (+39 6) 22 55 54 30

Key personnel
Marketing & Sales, Transport Sector: Giuseppe Celli

Products
Telecommunication, electrification and signalling systems for railways and metros.

South Wales Transformers Ltd

Newport Road, Blackwood, Gwent NP12 2XP, UK
Tel: (+44 1495) 23 21 00   Fax: (+44 1495) 23 21 32
e-mail: sales@swt.fki-et.com
Web: http://www.fki-et.com

Key personnel
Managing Director: A L Williams
Sales and Marketing Manager: P J Coe
Finance Director: J M Smith
Tendering Manager: P Hurndall

Background
South Wales Transformers, formerly South Wales Switchgear Limited (Transformers), began transformer manufacturing in its Treforest factory in 1945 as an extension to its well-established switchgear and cable manufacturing activities, and by 1964 it was manufacturing transformers up to 15 MVA. Most recently the company has become part of the FKI Group of international manufacturing companies, having previously belonged to Hawker Siddeley Group and then the BTR Group. FKI has over 75 operating units servicing many diverse market sectors.

Products
Design, manufacture and supply of: liquid cooled transformers up to 5 MVA (33 kV); transformer substations up to 2.5 MVA (22 kV); earthing/auxiliary transformers and reactors; fabricated transformer accessories and booster transformers.

SPIE Enertrans

Parc Saint-Cristophe, F-95861 Cergy-Pontoise Cedex, France
Tel: (+33 1) 34 22 56 08   Fax: (+33 1) 34 22 62 76

Key personnel
Operations Director: M Fortuné
Business Development Manager: R Zampieri

Total Power Solutions

77 Micklegate, York YO1 6LJ, UK
Tel: (+44 1904) 54 15 55
Fax: (+44 1904) 54 15 56

Corporate background
Total Power Solutions is a joint venture between GrantRail and Total Power Solutions Ltd.

Products and services
Design, supply and installation of 25 kV AC overhead line equipment.

Capabilities include: project planning and management; consultancy services; supply and installation construction works; new works commissioning; isolation planning and implementation services; design and data verification.

Transmitton Ltd

Ashby Park, Ashby-de-la-Zouch LE65 1JD, UK
Tel: (+44 1530) 25 80 00   Fax: (+44 1530) 25 80 05
e-mail: sales@transmitton.co.uk
Web: www.transmitton.co.uk

Key personnel

Managing Director: I Wright
Sales Director: E Turnock
Financial Director: J Blackwell

Products

Transmitton provides a range of integrated control and asset management solutions to the rail industry. These include; integrated station management systems, network management systems, real-time passenger information systems, remote condition monitoring, traction power SCADA, asset management systems.

The fastflex Remote Terminal Unit (RTU) is modular in design and suits a wide range of applications for stand alone control, point-to-point and remote communications. Transmitton's cromos software offers real-time control and asset management technologies. Using the core technologies, Transmitton has successfully implemented a wide range of solutions within the rail market including: Integrated Station Management (ISM), Network Management Systems (NMS), Real-time Passenger Information Systems (RTPIS), Remote Condition Monitoring (RCM) and Traction Power SCADA (Supervisory Control and Data Acquisition).

ADT Rail Systems is a Transmitton company which specialises in the specification, design and manufacture of these systems and provides performance data through its on track integrated suite of monitoring products.

UPDATED

TrendRail Ltd

Unit 11, Brindley Road, St Helens WA9 4HY, UK
Tel: (+44 1744) 85 11 00   Fax: (+44 1744) 85 11 22
e-mail: info@trendrail.com; j.taylor@trendrail.com
Web: http://www.trendrail.com

Key personnel

Managing Director: Jack Taylor
Company Secretary: Margaret Taylor
General Manager: Marion Smith
Production Manager: Gary Birchall

Products

Railway engineering and electrification products/ catenary wire fittings; small parts steelwork including arcing horns, angles, fabrications, plates, links and balance weights and fasteners. ISO 9002 accredited, Investors in People and Link-up approved.

Contracts

Contracts include West Coast Main Line refurbishment, 2000–2002; spares for maintenance to all regions, Channel Tunnel Rail Link, light rail projects (Hong Kong).

VERIFIED

Ultra Electronics Ltd

PMES
Armitage Road, Rugeley, WS15 1DR, UK
Tel: (+44 1889) 50 33 00   Fax: (+44 1889) 57 29 13
e-mail: enquiries@ultra-pmes.com
Web: http://www.ultra-pmes.com

Key personnel

Managing Director: A M Freer
Marketing Director: J S Greenhalgh
Engineering Director: J P Wardale
Sales Manager, Transit Systems: T W Boston
Public Relations: L M John

Principal subsidiary company

PMES (Asia) Pte Ltd, Singapore

Background

Previously Foster Rectifiers, and THORN EMI Electronics, Ultra Electronics Ltd PMES has over 40 years' experience of design and manufacture of transformer rectifier systems for traction applications.

Products and Services

Supplier of high-quality DC power systems. Equipment has been supplied to the Network Rail (UK), London Underground and overseas transit authorities.

Equipment can be supplied as 'stand alone' units, although contracts are often for turnkey projects including complete DC substations design and supply. Scopre of supply may include system design, integration, co-ordination, project management, equipment supply, installation and commissioning. Scope of equipment supply includes modular buildings, transformers, rectifiers, AC, DC and LV switchboard, SCADA, auxiliary equipment, power cables and civil works.

Current major projects

AMEC for Docklands Light Railway – London City Airport extension: turnkey project incuding provision of new modular substations.

Network Rail (UK) – Power supply upgrade project: contract for the provision of four new modular substations (within tight timescales) as part of the renewal of the South Zone DC power supply system.

Ansaldo Trasporti – Dublin Light Rail (Luas) project: design and supply of 15 900 V, 750 DC indoor air natural rectifiers complete with Open Circuit Arm Detectors.

Balfour Beatty Rail for Network Rail –Tyne & Wear Metro Sunderland Extension: complete project including provision of new modular substations.

ABB for London Underground Ltd, Asset Renewal: design, supply and commissioning of 2,000 kW 630/ 750 DC air natural rectifiers with LNAN transformers complete with Open Circuit Arm Detectors.

Wild & Grunder AG

Kelamttstrasse 10, CH-6403 Küssnacht, Switzerland
Tel: (+41 41) 850 60 20   Fax: (+41 41) 850 60 26

Key personnel

Engineer: Roland Ferrari

Products

Computer system for mapping information on lineside structures, such as location of overhead support poles and the date on which they were installed.

Windhoff Bahn-und Anlagentechnik GmbH

PO Box 1963, D-48409 Rheine, Germany
Tel: (+49 5971) 580   Fax: (+49 5971) 582 09
e-mail: info@windhoff.de
Web: www.windhoff.de

Key personnel

Board Members: Herbert Liessem,
Georg Vennemann
Finance Director: Helmut Gielians
Sales Directors: Dr Martin Hindersmann,
Uwe Dolkemeyer,
Technical Director: Juergen Auschner
Purchasing Manager: Stefan Berkemeyer

Products

Installation and maintenance of catenary systems: placing of foundations, mast mounting, installation of catenary wires, catenary maintenance, catenary measuring, complete catenary reconstruction.

UPDATED

CABLES AND CABLE EQUIPMENT

Alphabetical listing

Adaptaflex
AEI Cables
Agro AG
ALSTOM Transport Electrification SpA
Andrew
ApATeCh Electro
BCM Contracts Ltd
BICC
Brand-Rex Ltd
Carrier Khéops Bac
Drallim Telecom

DSG-Canusa UK Ltd
Ebo Systems
Fels SA
Ferranti Technologies Ltd
Harting Ltd
Hellermann
HPP GmbH
Huber + Suhner AG
Hypertac
Icore International Ltd
R Legrand SA

LPA Industries
Marconi Communications
Multi-Contact (UK) Ltd
Nexans
Pirelli Cables Ltd
PMA UK Ltd
Racal Telecom
Simclar International Corporation
Tyco Electronics UK Ltd
Weidmüller Interface GmbH & Co
Yutaka Manufacturing Co Ltd

Company listing by country

FRANCE
Carrier Khéops Bac
Ebo Systems
Fels SA
Hypertac
R Legrand SA
Nexans

GERMANY
HPP GmbH

ITALY
ALSTOM Transport Electrification SpA

JAPAN
Weidmüller Interface GmbH & Co
Yutaka Manufacturing Co Ltd

RUSSIAN FEDERATION
ApATeCh Electro

SWITZERLAND
Agro AG
Andrew
Huber + Suhner AG

UNITED KINGDOM
Adaptaflex
AEI Cables
BCM Contracts Ltd
BICC
Brand-Rex Ltd
Drallim Telecom
DSG-Canusa UK Ltd
Ferranti Technologies Ltd

Harting Ltd
Hellermann
Icore International Ltd
LPA Industries
Marconi Communications
Multi-Contact (UK) Ltd
Pirelli Cables Ltd
PMA UK Ltd
Racal Telecom
Tyco Electronics UK Ltd

UNITED STATES
Simclar International Corporation

Adaptaflex

Station Road, Coleshill, Birmingham B46 1HT, UK
Tel: (+44 1675) 46 82 00   Fax: (+44 1675) 46 49 30
Export: Tel: (+44 1675) 46 82 46
Fax: (+44 1675) 46 49 44
e-mail: sales@adaptaflex.co.uk
Web: www.adaptaflex.com

Key personnel

Marketing Manager: Martyn Turner
UK Sales Manager: Dave White
Export Sales Managers: Steve Bradley

Products

Adaptasteel: metallic-based conduit system for cable management applications.
Adaptalok/Adaptaseal: non metallic-based conduit systems.
   All products available with approvals from UK railway companies.
   Contracts include supply of equipment to London Underground, SNCF, Bombardier and ALSTOM.

UPDATED

AEI Cables

Durham Road, Birtley, Chester-le-Street DH3 2RA, UK
Tel: (+44 191) 410 31 11   Fax: (+44 191) 410 83 12
e-mail: sales@aeicables.co.uk
Web: www.aeicables.co.uk

Key personnel

UK Sales Director: Jim Duffy
Export Sales Director: David Fraser

Background

AEI Cables is a subsidiary of TT electronics plc.

Products

Rolling stock products include both elastomeric power cables and thin wall control and instrumentation cables, manufactured to both UK national and international specifications.
   AEI Cables is a major supplier of infrastructure cables in the UK, where a full range of signalling cables, power cables, jumper cables and earthing conductors are manufactured, as well as both mineral and soft-skinned fire performance cables. A complete range of low-smoke, zero-halogen cables is also supplied for use in trackside stations and buildings, both above and below ground.

Contracts

AEI holds approvals from London Underground, Network Rail (both UK) and a number of major equipment manufacturers and has cables in service across the world, including the Far East, Africa, the Americas and Europe.

UPDATED

Agro AG

Korbackerweg 7, CH-5502 Hunzenschwil, Switzerland
Tel: (+41 62) 889 47 47   Fax: (+41 62) 889 47 50

Products

Cable seals for shielded cables which cut out electromagnetic interference.

ALSTOM Transport Electrification SpA

10 Via Lago dei Tartari, I-00012 Guidonia, Rome, Italy
Tel: (+39 0774) 37 74 85   Fax: (+39 0774) 35 34 30

See also ALSTOM entry in *Locomotives and powered/non-powered passenger vehicles* section.

Key personnel

Managing Director: Massimino Colombo

Products

Cable, fibre optic and radio communications systems employing base band, FDM and PCM technology.

Andrew

Andrew Kommunikationssysteme AG
Bächliwis 2b, CH-8184 Bachenbülach, Zurich, Switzerland
Tel: (+41 1) 863 73 52   Fax: (+41 1) 863 73 56

Key personnel

See *Signalling and telecommunications* section.

Products

Heliax coaxial cables, Radiax leaky feeder cables.

ApATeCh Electro

Zhukovsky Street 2, Building 131, Dubna, 141980 Moscow Region, Russian Federation
Tel: (+7 09621) 257 92   Fax: (+7 09621) 234 92
e-mail: electro@dubna.ru

Products

Ground cable duct systems in composite materials.

BCM GRC Limited

Unit 22, Civic Industrial Park, Whitchurch SY13 1TT, UK
Tel: (+44 1948) 66 53 21   Fax: (+44 1948) 66 63 81
e-mail: info@bcmgrc.com
Web: www.bcmgrc.com

Key personnel

Directors: T F Jordan; M P O'Conner
Sales Manager, Rail: Mike Camp

Background

BCM GRC Limited was previously called BMC Contracts Limited.

Products

A range of lightweight and easily installed cable channel and its accessories – which includes both elevated and ground systems. Complying with UK Railtrack Group standards the glass fibre reinforced concrete (GRC) cable channel system includes not only both new and replacement channel – available in sizes which correspond to the internal dimensions of existing concrete channel – but also a full range of accessories such as channel bends, joint boxes, T-pieces, UTX chambers and junction boxes, point covers and joint bay covers.
   Together these allow continuous cable runs to be installed either at ground level, over bridges, or through tunnels, with simple connection to each other or to existing cable troughs.
   The elevated cable channel, for example, has been specially designed to be installed above ground level by using tailor-made channel supports, brackets and security clamps. The system also includes easily installed H-piece locking devices which lock the channel lids to the channel to ensure that the system remains tamper-proof.
   Apart from being easily installed and maintained, the BCM cable channel system also offers advantages by using GRC as a manufacturing material. These include a high resistance to fire and the emission of toxic fumes, lightweight yet high strength, an ability to be worked with standard hand tooling without damage or breakage, plus other rot-proof, rust-proof, and shatter-proof characteristics.
   The material's lightweight characteristics – it weighs approximately one quarter the weight of concrete – plus the channel's ease of installation, mean that a six man team should be able to lay

at least 600 m of channel in one day – estimated at least twice that possible using a concrete alternative.

UPDATED

BICC

BICC Cables Ltd
Chester Business Park, Chester CH4 9PZ, UK
Tel: (+44 1244) 68 84 00   Fax: (+44 1244) 68 84 01

Main works

Leigh Industrial Cables, Leigh Works, Leigh WN7 4HB, UK
Tel: (+44 1942) 67 24 68   Fax: (+44 1942) 67 99 83

Key personnel

Director and General Manager: A C Unsworth
Market Sector Manager, Mining and Railways:
   A Greenwood

Products

Signalling and control cables.
   Contracts include the supply of signalling cables to London Underground Ltd for the Central Line resignalling scheme.
   BICC Components has developed a new connector which eliminates the need for time-consuming trackside crimping of cables. It is produced by the company's Flexo Products unit and has been specially developed for the Flexo Paulve signalling detectors on the Paris express metro RER system and national railways, SNCF. BICC Components is part of BICC Cables.

Brand-Rex Ltd

Viewfield Industrial Estate, Glenrothes, Fife KT6 2RS, UK
Tel: (+44 1592) 77 21 24   Fax: (+44 1592) 77 53 14
e-mail: jdonnel@brand-rex.com
Web: http://www.brand-rex.com

Other offices

Argentina, Australia, Brazil, Colombia, France, Germany, Hong Kong, Italy, Singapore, Spain, United Arab Emirates and USA.

Key personnel

Managing Director: Ian Mack

Products

Brand-Rex is a specialist copper and fibre wiring systems manufacturer supplying power, control, data and communication cables, using irradiation cross-linking technology, to the rail mass transit industries. The company makes structured wiring systems for data networks and cabling solutions for military and other industries. All Brand-Rex optical fibre cables contain fibres manufactured by Corning Inc.

Carrier Khéops Bac

Group Compagnie Deutsch
Z1 Sud, boulevard Pierre Lefaucheux, F-72027 Le Mans, Cedex 2, France
Tel: (+33 2) 43 61 45 45
Fax: (+33 2) 43 61 45 01
e-mail: ckb@compagnie-deutsch.com
Web: http://www.compagnie-deutsch.com

Key personnel

Managing Director: Denis Plantey
Finance Director: Jean-Claude Thieury
Marketing Manager: Patrick Roseleur
Public Relations: Agnes Favalelli

Products

Electrical connectors for signalling, power and control supply for metro, suburban rail, high-speed trains and LRT systems; fibre optic systems, coaxial connection systems.

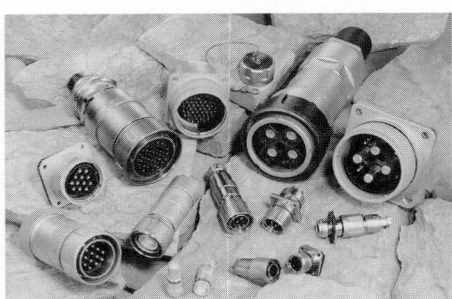

Carrier Khéops CMC series of connectors
0087658

Contracts

Equipment has been supplied for metro systems in Santiago, Mexico, Caracas and Singapore, and for tramways in Grenoble and St Etienne and M6 cars for Korean TGV, AGC, MF 2000, NJT locomotives, WMATA Metro cars.

Developments

CMR series, new concept for modular power connection.

Drallim Telecom

A division of Drallim Industries Ltd
Brett Drive, Bexhill-on-Sea TN40 2JP, UK
Tel: (+44 1424) 21 66 11   Fax: (+44 1424) 21 66 36
e-mail: email@drallim.com
Web: http://www.drallim.com

Key personnel

Chief Executive: R B McBrien
Marketing Director: M F Dawson

Products

Pressurisation equipment for telephone cables and waveguides; microprocessor telephone network alarm monitoring systems for pressurised cables and security alarm systems.

Contracts include the supply of pressurisation equipment complete with automatic monitoring to the Derby/Birmingham area in the UK and the supply and installation of cable pressurisation units at 11 sites between London Euston and Bletchley for Railtrack, UK, with staff training and continuing spares support.

DSG-Canusa UK Ltd

Bergstrand House, Parkwood Close, Broadley Industrial Park, Roborough, Plymouth PL6 7EZ, UK
Tel: (+44 1752) 20 98 80   Fax: (+44 1752) 20 98 50
Web: http://www.dsgcanusa.com

Key personnel

Managing Director: S Hill
UK Sales Manager: Nigel Westermann

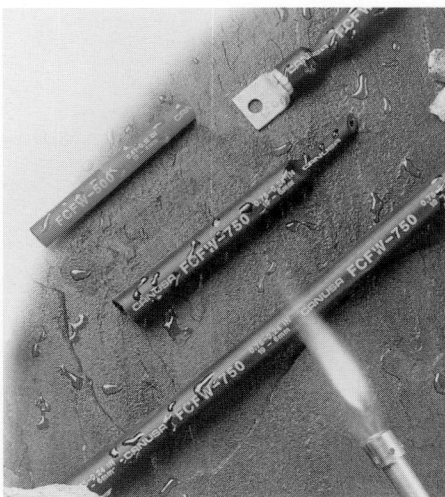

Canusa Systems flame-retardant heavy wall heatshrink tubing
0021682

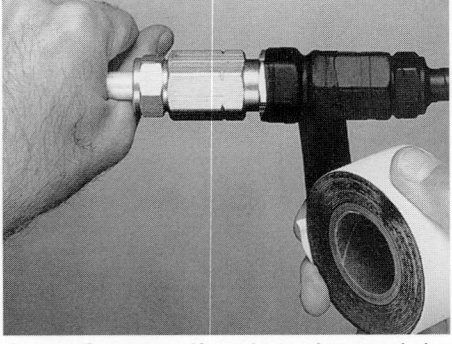

Canusa Systems self-amalgamating tape being applied to an aluminium connector
0021681

Products

Heat-shrinkable products for cable protection, jointing and insulation. Major supplier in the supply of zero halogen heat-shrinkable tubings.

DSG-Canusa products are extensively specified throughout the railway industry.

Ebo Systems

Boulevard d'Europe, PO Box 10, F-67211 Obernai Cedex, France
Tel: (+33 3) 88 49 50 51   Fax: (+33 3) 88 49 50 14
e-mail: info@ebo-systems.com
Web: http://www.ebo-systems.com

Key personnel

Manager: Ulrich Pelz

Contact points

France, Portugal, Spain, Belgium, Luxembourg
Ebo Systems
32/34 Avenue Salvador Allende, F-93800 Epinay/Seine, France
Tel: (+33 1) 49 71 55 80   Fax: (+33 1) 49 71 55 85
Contact: Bernard Morvan
e-mail: info.fr@ebo-systems.com
Web: http://www.ebo-systems.com

Germany
Ebo Systems GmbH
Zum Gurrterstal, D-66440 Blieskastel
Tel: (+49 6842) 92 19 10   Fax: (+49 6842) 92 19 19
Contact: Emst Messenzehl
e-mail: ebosystemsD@t-online.de
Web: http://www.ebo-systems.com

Italy
Ebo Systems Srl
Via Guido Rossa 5/7, I-Cumo (BG)
Tel: (+39 035) 46 20 88   Fax: (+33 035) 46 20 89
Contact: Antonio Rondelli
e-mail: info.it@ebo-systems.com
Web: http://www.ebo-systems.com

Switzerland
Ebo Systems AG
Tambourstrasse 8, CH-8833 Samstagern
Tel: (+41 1) 78 78 78 7   Fax (+41 1) 78 87 79 9
Contact: René Fontolliet
e-mail: info.ch@ebo-systems.com
Web: http://www.ebo-systems.com

Head office, Middle East and other countries
See main address
Key Account Manager: Pascal Muller

Products

FRP/GRP cable management systems: cable trays, ground ducts, cable ladders; supports for trays, ducts and ladders.

FRP/GRP profiles for light structures, gratings, hand rails, stairs and ladders with safety cages.

Contracts include UK-France Channel Tunnel (1999); Vienna metro (1989–97); Berlin metro (1990–95); Swedish Railways (1991–95); Norwegian Railways; Czech Railways (1996); and French Railways.

VERIFIED

Fels SA

2 rue J M Jacquard, F-67400 Illkirch Graffenstaden, France
Tel: (+33 3) 88 67 10 60   Fax: (+33 3) 88 67 33 10
e-mail: fels@fels.fr
Web: www.fels.fr

Products

Electrical contacts; special connectors.

NEW ENTRY

Ferranti Technologies Ltd

Cairo House, Waterhead, Oldham OL4 3JA, UK
Tel: (+44 161) 624 02 81   Fax: (+44 161) 624 52 44
e-mail: sales@ferranti-technologies.co.uk
Web: http://www.ferranti-technologies.co.uk

Key personnel

See *Passenger coach equipment* section.

Products

UV laser cable marking and manufacturer of cable looms.

Harting Ltd

Caswell Road, Brackmills Industrial Estate, Northampton NN4 7PW, UK
Tel: (+44 1604) 76 66 86   Fax: (+44 1604) 70 67 77

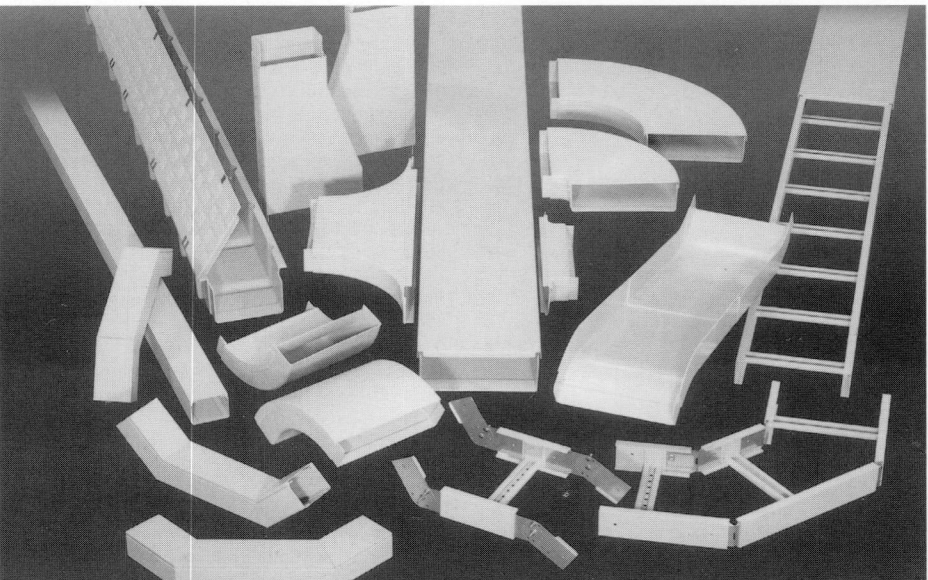

Ebo Systems FRP/GRP cable tray system
0063797

Key personnel
Managing Director: R Wood
General Manager: D T Franklin

Products
Heavy-duty electrical connectors for rail and other transport applications. The range includes hoods, housings, inserts and contacts, all selected to offer environmental resistance and EMI/FRI protection. Fibre optic devices can also be supplied for critical control applications where immunity from electromagnetic interference is required.

Hellermann

Hellermann Insuloid
Leestone Road, Wythenshawe, Manchester M22 4RH, UK
Tel: (+44 161) 998 85 51   Fax: (+44 161) 945 37 08

Key personnel
European Product Manager: Tony Orme

Products
Flame-retardant ties in nylon or stainless steel.
Contracts include the supply of low-toxicity flame-retardant ties to London Underground and SNCF.

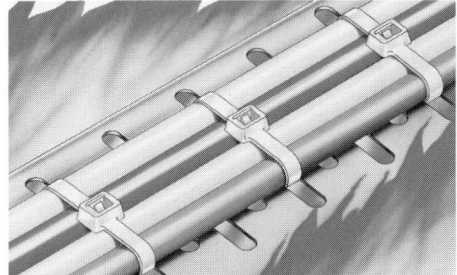

Hellermann Insuloid flame-retardant and high-temperature cable ties   0021683

HPP GmbH

Schweriner Strasse 5, D-23970 Wismar, Germany
Tel: (+49 3841) 22 87 22
Fax: (+49 3841) 22 87 23
e-mail: blickehpp@aol.com
Web: http://www.hpp-gmbh.com

Products
Plastic cable protection systems, including cable ducts, GRP support systems and cable protection pipes.

Huber + Suhner AG

Wire & Cable Division
CH-8330 Pfäffikon ZH, Switzerland
Tel: (+41 1) 952 22 11
Fax: (+41 1) 952 26 50
e-mail: info@hubersuhner.com
Web: http://www.hubersuhner.com

Fibre Optics Division
CH-9100 Herisau, Switzerland
Tel: (+41 71) 353 41 11
Fax: (+41 71) 353 46 47

Key personnel
Contact, Wire & Cable Division: Susann Kuchen
Contact, Fire Optics Division: Stephan Grimm

Subsidiaries
Huber + Suhner (Australia) Pty Ltd, Australia
Huber + Suhner (Canada) Ltd, Canada
Huber + Suhner (Hong Kong) Ltd, China
Huber + Suhner France, France
Huber + Suhner GmbH, Germany
Huber + Suhner (Singapore) Pte Ltd, Singapore
Huber + Suhner (UK) Ltd, UK
Huber + Suhner Inc, USA

Products
Design, development, manufacture and supply of complete cable systems for all rail vehicle and bus applications; supply of cable system components. Cables using copper, fibreoptic, coaxial and hybrid materials are supplied. The fibreoptic product range includes cables, connectors and assembly and installation.

Hypertac

31 rue Isidore Maille, F-76410 Saint Aubin lès Elbeuf, France
Tel: (+33 2) 32 96 91 76   Fax: (+33 2) 32 96 91 70
e-mail: info@hypertac.fr
Web: http://www.hypertac.com

Background
Hypertac is part of Smiths Interconnect, a division of Smiths Group. Hypertac designs and manufactures connectors for electrical and electronic applications.

Products
Design, development and manufacturing of electrical connection systems for the rail traction industry.
All Hypertac Interconnect solutions have machine-turned round pins in the male half and the special Hypertac hyperboloid socket contact in the female half. This socket has a number of spring wires, set at a small angle to the centre line giving the special hyperboloid contact. As the pin enters the socket, the spring wires are deflected within their elastic limit. There is a very smooth wiping action and as result, minimal wear.
Products include rectangular connectors Ihs/Ihf series including modular connectors, rack and panel connectors and amovable contacts, as approved by SNCF and RATP.
Circular couplers include three-pole connectors, power couplers, car-to-car connection.
The Hypertac group has provided interconnect products for more than 40 years for mass and rapid transit projects throughout the world, including TGV, Eurostar, Mass Transit Railway Hong Kong, London Underground and Seoul subway, using its engineering expertise to design connectors for applications such as signalling equipment, power supplies, AC/DC converters, lighting, communications such as ERTMS systems (RF/HF link) and auxiliary equipment.

Icore International Ltd

Leigh Road, Slough SL1 4BB, UK
Tel: (+44 1753) 57 41 34   Fax: (+44 1753) 87 36 24
e-mail: information@icore.co.uk
Web: http://www.icoregroup.com

Key personnel
Product Manager, Fluid Systems:
  Christine Wickings
European Sales Manager: Gary Evans
Industrial Sales Manager: Aidan Butler

Products
Icore designs, manufactures and supplies wired interconnect systems, including lightweight and heavy duty zero halogen conduit systems, EMC protection for wired harnesses, specialised high power connectors and custom designed solutions. Typical applications include: intercarriage signal and power jumpers and carriage to bogie links; automatic coupling systems, sensors and braking systems.

R Legrand SA

BP 3, F-69270 Couzon au Mont d'Or, France
Tel: (+33 4) 78 22 26 96   Fax: (+33 4) 78 22 62 02
e-mail: nbonin@btp-legrand.com
Web: http://www.btp-legrand.com

Key personnel
Export Executive: Nicolas Bonin

Products
Prefabricated modular cable pits made of either reinforced concrete or recycled/recyclable polystyrene. Pits can be supplied in 20 cm concrete sections or 25 cm polystyrene sections.

LPA Industries

PO Box 15, Tudor Works, Debden Road, Saffron Walden CB11 4AN, UK
Tel: (+44 1799) 51 28 00   Fax: (+44 1799) 51 28 28
e-mail: sales@lpa-ind.demin.co.uk
Web: www.lpa-industries.co.uk

Key personnel
Chief Executive Officer: Peter Pollock
Sales and Marketing Director: Kevin Comfoot
Production Director: Graham Clark
Exports Manager: Derek Orley

Background
LPA Industries is a member of LPA Group plc and is ISO 9001 and 9002 certified.

Products
Cable glands, cleats, trays, plugs, sockets and other electrical terminations and connectors.

UPDATED

Marconi Communications

Marconi Communications Limited
New Century Park, PO Box 53, Coventry CV3 1HJ, UK
Tel: (+44 1203) 56 55 00   Fax: (+44 1203) 56 58 88
e-mail: pete.hallard@marconicoms.com
Web: http://www.marconicomms.com

Key personnel
See *Signalling and telecommunications* section.

Products
Optical fibre, copper distribution and signalling cable systems.

Multi-Contact (UK) Ltd

3 Presley Way, Crownhill, Milton Keynes MK8 0ES, UK
Tel: (+44 1908) 26 55 44   Fax: (+44 1908) 26 20 80
e-mail: uk@multi-contact.com
Web: www.multi-contact.com

Key personnel
Managing Director: Selwyn Corns
Commercial Manager: Kim Symms-Fahey
Technical Manager: Charlie Galiszewski

Main works
Multi-Contact AG Basel
Stockbrunnenrain 8, CH-4123 Allschwil 1, Switzerland
Tel: (+41 61) 306 55 55   Fax: (+41 61) 306 55 56

Products
Special-purpose high-current connectors. Typical rail applications include: power supply connections for preheating coaches; high-current connectors in modular static converters; automatic train couplers with control leads; plug-in power connector valves; and plug-in magnetic coils for linear maglev systems.
Flexible single-core cables; PVC, silicone, TPE and Teflon-insulated 0.10 to 95 mm².

UPDATED

Nexans

16 rue de Monceau, F-75008 Paris, France
Tel: (+33 1) 56 69 84 00   Fax: (+33 1) 56 69 84 84
Web: www.nexans.com

Key personnel
Chairman and Chief Executive Officer:
Gérard Hauser
Chief Financial Officer: Frédéric Vincent
Senior Corporate Vice-President, Human Resources:
François Saint-Dizier
Senior Corporate Vice-President, Communications:
Véronique Guillot-Pelpel
Executive Vice-President, Rest of the World Area:
Bruno Thomas
Executive Vice-President, Strategic Operations:
Pascal Portevin
Executive Vice-President, Europe Area: Yvon Raak
Executive Vice-President, North America and Asia
Area: Michel Lemaire

Products
For rolling stock: data and energy cables and
components. These include: standard and thinwall
cables with or without shielding; databus cables;
UIC cables; optical fibre cables; special power
cables; sheathed single-core and multi-core
cables; high-voltage cables and systems; complete
harnesses with connections and sealing ends;
jumper cables for carrying control and power
functions between cars; signalling and data
transmission cables. Virtually all cables are zero-
halogen.
For infrastructure: LANs for communications
and control functions; overhead catenary cables
and low-, medium- and high-voltage cables for
track feed, traction and equipment; signalling
cables; copper, fibre and leaky/radiating
telecommunications cables; halogen-free fire-
performance cables. Nexans also provides
customised engineering, turnkey installation and
maintenance services.

UPDATED

Pirelli Cables Ltd

PO Box 6, Leigh Road, Eastleigh SO50 9YE, UK
Tel: (+44 870) 513 31 43   Fax: (+44 23) 80 29 54 65
Web: http://www.pirelli.co.uk

Key personnel
Managing Director: Martyn Coffey
Business Director, Utilities: Michael Simms
Director, General Market: Grayson Spurlock
Manufacturing Director, Energy Cables:
Paul Davies
Chief Engineer: Doug Gracias

Products
Power, telecommunications cables and cable
accessories. Pirelli Cables makes a wide range
of power cables, from track feeder cables to
high-voltage cables for track power supplies. The
company has introduced low-smoke zero halogen
cables for power, signalling and communication
purposes. These cables, designated LSOH (Low-
Smoke Zero Halogen), are specifically designed
for metro systems and were developed in close
collaboration with London Underground Ltd. LUL
has been supplied with optical fibre communication
cables sheathed with LSOH material for in-tunnel
use, and with X-Flam 15 for use above ground.
Telecommunication cables for railway use
include range of multipair designs to Railtrack
specifications, specially screened designs for use
with 25 kV AC overhead systems and optical fibre
cables.
Cable glands, terminations and cable accessories
are available from Pirelli's components unit.

UPDATED

PMA UK Ltd

Unit 4, Imperial Court, Magellan House, Walworth
Industrial Estate, Andover SP10 5NT, UK
Tel: (+44 1264) 33 35 27   Fax: (+44 1264) 33 36 43
e-mail: sales@pma-uk.com
Web: http://www.pma-uk.com

Key personnel
Country Manager: Joy Levett
Financial Controller: Leanne Smith
Regional Sales Manager (North): Mark Fletcher
Regional Manager South: Donna Clash
Key Account Manager: Annabella Webb

Head office
PMA-AG
Aathalstrasse 90, CH-8610 Uster, Switzerland
Tel: (+41 1) 905 61 11   Fax: (+41 1) 905 61 22
e-mail: info@pma.ch
Web: http://www.pma.ch
Chairman: Ernst Schwarz
Worldwide Sales Manager: Rene Grivaz
Financial Director: Dieter Grabher
Marketing Assistant: Irene Freitag

Subsidiary companies
Australia, Austria, Belgium, Denmark, Finland,
France, Germany, UK, Hong Kong, Hungary,
Israel, Italy, Japan, Netherlands, Norway, Poland,
Portugal, Singapore, Spain, Sweden, Taiwan,
Turkey, Canada and USA.

Products
Nylon-based halogen-free cable protection; PMAfix
cable connectors. VAM is a specially modified
nylon 6 conduit with self-extinguishing and
flammability characteristics particularly for internal
carriage zones.

Developments
A new product range – Light Line which can
be combined with the complete PMA standard
product range.

Contracts
Contracts include supply of VAM conduit to London
Underground Northern line, cable protectors to
Siemens, ALSTOM and Bombardier Transportation.

VERIFIED

*PMAFLEX VAM conduit and PMAfix male
connectors with IP68 metal thread for increased
mechanical requirements*   0021685

Racal Telecom

Phoenix House, Station Hill, Reading, RG1 1NB,
UK
Tel: (+44 118) 908 60 00   Fax: (+44 118) 908 64 54

Key personnel
See *Signalling and telecommunications* section.

Products
Racal Telecom's business principally consists
of the provision of operational and business
telecomunications services to the railway industry
in the UK. In March 1996, Racal Telecom's core
network comprised 11,000 route-km of trunk cable,
of which some 4,000 route-km were made up of
fibre optic cable. During 1997, Racal upgraded
Racal Telecom's network by creating a resilient
SDH backbone network which included a further
1,200 route-km of fibre optic cable.

Simclar Corporation

9114 58th Place, Suite 500, Kenosha, Wisconsin
53144, USA
Tel: (+1 262) 653 99 99   Fax: (+1 262) 653 95 09
e-mail: simclar@execpc.com
Web: http://www.simclar.com

Head office
Simclar International Ltd
Pitreavie Business Park, Dunfermline KT11
8UN, UK
Tel: (+44 1383) 73 51 61   Fax: (+44 1383) 73 99 86
e-mail: sales@simclar.com

Products
Cable harness assemblies for traction equipment,
cabs, high-voltage and low-voltage applications,
data communications systems and earthing
equipment.

Tyco Electronics UK Ltd

Faraday Road, Dorcan, Swindon SN3 5HH, UK
Tel: (+44 1793) 52 81 71   Fax: (+44 1793) 57 25 16
Web: www.tycoelectronics.com

Main offices
Tyco Electronics Corporation, 300 Constitution
Drive, Menlo Park, California 94025, US
Tel: (+1 650) 316 33 33   Fax: (+1 650) 316 21 13

Key personnel
Worldwide Marketing Manager for Rail:
Vito Provenzano
Marketing Communications Manager:
Reinhard Fritzsche

Background
Previously traded as Raychem Ltd.

Products
Products based on plastics, metals and chemicals
including: low fire hazard wire and cable, heat-
shrinkable zero-halogen tubing and moulded parts,
wire marking systems, electrical interconnection
devices, freeze protection of contact rails, points
and crossings, brake and diesel fuel lines;
temperature and condensation control; HV and
LV cable accessories; electrical harness sealing
products, electrical connector adapters.
Computerised marking with TMS System 90
equipment reduces time and effort spent on
electrical installations for locomotives.

UPDATED

Weidmüller Interface GmbH & Co

AnderTalle 89, D-33102 Paderborn, Germany
Tel: (+49 5252) 96 00   Fax: (+49 5252) 96 01 16
e-mail: weidmueller@weidmueller.de
Web: http://weidmueller.de

Key personnel
Owner: W Schubel
Chief Executive Officer: Thomas Hagan
Managing Director, ConNect: Ch. Bönsch
Head of Transportation: Mikhael Hoenig
International Project Manager: Stephen Ward

Subsidiaries
Australia, Austria, Bahrain, Belgium, Brazil,
Bulgaria, Canada, China, Croatia, Czech Republic,
Denmark, Finland, France, Germany, Greece, Hong
Kong, Hungary, India, Israel, Italy, Japan, South

Korea, Former Yugoslav Republic of Macedonia, Malaysia, Mexico, Netherlands, Norway, Pakistan, Poland, Portugal, Romania, Singapore, Slovakia, Slovenia, South Africa, Spain, Sweden, Switzerland, Taiwan, Thailand, Turkey, UK, Ukraine and USA.

Products

Connectors and connection equipment including terminal blocks and terminal strips, HD connectors and terminal connectors for printed circuits.

Installation: wire, cables, terminals and instruments, enclosures, manual and automatic tools, ferrules and mounting trails.

Electronic interface modules, relay interface and PLC interfaces, power supply.

I/O components: decentralised I/O, field bus, SCADA and control software, industrial PC.

Customers include ALSTOM, Bombardier and Siemens.

Yutaka Manufacturing Co Ltd

1-18-17 Kitakojiya Ota-ku, Tokyo 144-0032, Japan
Tel: (+81 3) 37 41 41 31   Fax: (+81 3) 57 05 70 65
e-mail: hideo.kamei@yutaka-ss.co.jp
Web: www.yutaka-ss.co.jp/

Key personnel

President and Chief Executive Officer:
Yasuyuki Maki

Products

Jumper cable connectors; high- and low-voltage connectors; multicontact connectors for power input and output.

UPDATED

"Big grinding trains are economical on long sections of open track, but what I also need is a more versatile machine for complicated trackwork.

Who can I turn to for advice?"

"Speno"

Speno has developed a fleet of innovative machines that can be adapted to a network's rail grinding needs.
The concept is modular and allows machine combinations to suit specific grinding applications, job sizes and track possessions.

Talk to Speno!

SPENO INTERNATIONAL SA
26, Parc Château-Banquet
POB 16
1211 Geneva 21, Switzerland
Tel: +41 22 906 46 00
Fax: +41 22 906 46 01
e-mail: info@speno.ch

www.gottwald.com

Competence on rails

Whatever you need — track-maintenance cranes, breakdown cranes ... — we have the technology and experience to ensure you get the tailored solution you need.

Gottwald railway cranes — throughout the world

Gottwald Port Technology GmbH • Postfach: 18 03 43 • 40570 Düsseldorf, Germany
Phone: +49 211 7102-0 • Fax: +49 211 7102-651 • info@gottwald.com • www.gottwald.com

GOTTWALD
port technology

PERMANENT WAY COMPONENTS, EQUIPMENT AND SERVICES

Alphabetical listing

A & K Railroad Materials Inc
Abetong Teknik
Abloy Security Ltd
AB Strängbeton
Abtus Ltd
Alfred McAlpine
ALH Rail Coatings Ltd
ALSTOM Transport
AMECA Engineering srl
Amec Spie Rail Systems Ltd
American Railroad Curvelining Corporation (ARC)
Amey Rail Ltd
Amurrio Ferrocarril y Equipos SA
ApaTeCh
Arbil Ltd
Arcelor
Aspen Aerials Inc
Aspen Equipment Company
Atlantic Track
Atlas Copco Construction + Mining Ltd
BAC Corrosion Control Limited
Balfour Beatty Rail Ltd
R Bance & Co Ltd
Banverket Industridivisionen
Banestyrelsen Service Division
Barclay Mowlem Construction Ltd
Beilhack Systemtechnik und Vertriebs GmbH
Bharat Earth Movers Ltd (BEML)
Benkler AG
Benntec Systemtechnik GmbH
Bomag GmbH
Bumar Ltd
Burn Standard
BWG – Gesellschaft mbH & Co KG
Camlock Lifting Clamps Ltd
Carillion Rail
CBI Engineering A/S
CDM nv/sa
Cemafer
Cembre Ltd
Century Group Inc
Chemetron
China Northern Locomotive and Rolling Stock Industry (Group) Corporation
China South Locomotive and Rolling Stock Industry (Group) Corporation (CSR)
Cimmco International
CNR Qiqihar Railway Rolling Stock (Group) Co Ltd (QRRS)
Colbond Geosynthetics
Colebrand International Limited
Collis Engineering Ltd
Cometi
Comsa SA
Conbrako
Contec GmbH
Contitech Transportbandsysteme GmbH
Cooper & Turner
Corus Cogifer
Corus Rail Products
Corus Railway Infrastructure Services
Cowans Sheldon
Créabéton
CXT Rail Products
DBT GB Ltd
Delimon GmbH
Delkor Pty Ltd
Derby Rubber Products
Desec Ltd
DISAB Vacuum Technology AB
Drouard
Edgar Allen Ltd
Edilon BV
Edmund Nuttall Ltd
Eichholz
Electrolux Construction Products UK
Elektro-Thermit GmbH and Co KG
Enzesfeld-Caro Metallwerke AG
ESAB AB, Welding Equipment
ESCO Equipment Service Co
Eurailscout Inspection and Analysis BV

Exel Oy
'EXIM' Verkehrs, – Hafen– und Umwelttechnik GmbH & Co Vertriebs KG
FAB-RA-CAST
Fassetta mécanique
Faur SA
Ferotrack Engineering Ltd
Findlay, Irvine Ltd
First Engineering Ltd
Framafer
Geismar
Gemco George Moss Ltd
GERB Vibration Control Systems
Getzner Werkstoffe GmbH
Gleisbaumechanik Brandenburg
Gottwald Port Technology GmbH
GrantRail Ltd
Greenwood Engineering
Grinaker Duraset
Guided Ultrasonics (Rail) Ltd
Hall Rail
Harrington Generators International
Hanning & Kahl GmbH & Co KG
Harsco Track Technologies Ltd
Harsco Track Technologies Ltd
Harsco Track Technologies Pty Ltd
Heat Trace Limited
Hindusthan Engineering & Industries Ltd
Henry Williams Limited
H F Wiebe GmbH & Co KG
Hi-Force Hydraulics Ltd
Hioma-aine Oy
Holdfast Level Crossings Ltd
Holland Company LP
Huck International Inc
IAD Rail Systems
Interlok Bahnconsulting Schmidtendorf
Interep SA
IPA
JacksonEve Infrastructure Services
JAFCO
Jarret
Jarvis Rail
JC Bamford Excavators Ltd
JEZ Sistemas Ferroviarios
Kalyn Siebert
A Kaufmann AG
Kelsan Technologies Corporation
Kershaw
Kier Rail
KIHN SA
Kirow Leipzig AG
Kloos Oving bv
Knox Kershaw Inc
K Industrier AB
Konepaja Mankinen Ky
Laser Rail Ltd
Leica Geosystems AG
Lindapter International
Linsinger Maschinenbau GesmbH
Loram Maintenance of Way, Inc
Lord Corporation
Lucchini Group
LUKAS
Lyudinovo Locomotive Works JSC
MAN Ferrostaal AG
Matisa Matériel Industriel SA
Max Bögl
Mer Mec SpA
metronom Gesellschaft für Industrievermessung mbH
Metrum Information Storage Limited
Mitsukawa Metal Works Co Ltd
Modern Track Machinery Canada Ltd
moklansa GmbH
Moog GmbH
Mossboda Trä AB
Mowlem Railways
MTH Praha AS
Muromteplovoz
NARSTCO

NEU International
Newag GmbH & Co KG
Newt International
Nordco
NRS – National Railway Supplies
ØDS-Caltronic A/S
Oleo International Ltd
ORTEC Gesellschaft für Schienentechnische Systeme mbH
Orton/McCullough
Osmose Railroad Division
Outreau Technologies
Palfinger Europe GmbH
Pandrol Rail Fastenings
Peddinghaus Group
Pennsylvania Steel Technologies
Permali Gloucester Limited
Pettibone
Pfleiderer track systems
Phoenix AG
Pintsch Aben BV
Plasser & Theurer
Pohl Corporation
Polysafe Level Crossing Systems Ltd
Portec
Pouget
Power Machines Group
Presto Products Company
Progress Rail Services (Engineering and Track Services)
CNR Qiqihar Rolling Stock (Group) Co Ltd
Quest Corporation
Racine Railroad Products Inc
Railquip Inc
Rails Company
Railtech International
Railtech Schlatter Systems
RailWorks Corporation
Ranalah
A Rawie GmbH & Co KG
Rex
Rexquote Ltd
RICA
RMC Rail Products
RMS Locotec
Robel
Rosenqvist Rail Tech AB
Rotabroach Ltd
Rotamag Track Equipment
Salient Systems Inc
Sateba
Schlatter AG, HA
Schreck-Mieves GmbH
Schweizer Electronic AG
Schwihag Gesellschaft für Eisenbahnoberbau mbH
SECO-RAIL
Selectra srl
Semperit
Sersa Group Management AG
SGB Group
S G Technologies GmbH
Siemens Switzerland
Sika Ltd
SILF Srl
H J Skelton (Canada) Ltd
H J Skelton & Co Ltd
Slab Track Systems International
SMIS
Socofer
Sonatest Ltd
Speno International SA
Sperling Railway Services Inc
Sperry
Spie
Spitzke GmbH
SPX Fluid Power
SRS Rail Vehicles AB
Steel Authority of India
STRAIL Verkehrssysteme
Strukton Railinfra BV
System Bahnbau International (SBI)

Alphabetical listing—*continued*

Tarmac Precast Concrete Ltd
Techni-Métal Systèmes
Teknikum Oy
Tensol Rail
Thermit Australia Pty
Thermit Welding (GB) Ltd
Thermon Manufacturing Co
ThyssenKrupp GfT Gleistechnik GmbH
Tiefenbach GmbH
Tie & Track Systems Inc
Tiflex Ltd
Tipco Inc
TKL Rail
Tokimec Rail Techno Inc

Travipos
Trinity Difco Inc
TSO SA
TSTG Schienen Technik GmbH
Tülomsas
Türk + Hillinger GmbH
Turkington Precast Concrete
Ultra Dynamics
Unit Rail Anchor Company Inc
VAE Aktiengesellschaft
VAE Nortrak
VAE UK Ltd
Vaia Car
Voestalpine Railpro BV

Vortok International
Vossloh Cogifer
Vossloh Fastening Systems GmbH
Vossloh Infrastructure Services
Wabtec Rail Ltd
Wacker (GB) Ltd
Leonhard Weiss GmbH and Co KG
Western-Cullen-Hayes Inc
Willy Vogel AG
Windhoff Bahn-und Anlagentechnik GmbH
Zöllner GmbH
Zweiweg Schneider GmbH & Co KG

Company listing by country

AUSTRALIA
Barclay Mowlem Construction Ltd
Delkor Pty Ltd
Derby Rubber Products
Gemco George Moss Ltd
Harsco Track Technologies Pty Ltd
Thermit Australia Pty
TKL Rail

AUSTRIA
Enzesfeld-Caro Metallwerke AG
Getzner Werkstoffe GmbH
Linsinger Maschinenbau GesmbH
Palfinger Europe GmbH
Plasser & Theurer
Semperit
VAE Aktiengesellschaft

BELGIUM
CDM nv/sa

CANADA
Kelsan Technologies Corporation
Modern Track Machinery Canada Ltd
H J Skelton (Canada) Ltd
Tipco Inc
VAE Nortrak

CHINA
China Northern Locomotive and Rolling Stock Industry (Group) Corporation
China South Locomotive and Rolling Stock Industry (Group) Corporation (CSR)
CNR Qiqihar Rolling Stock (Group) Co Ltd

CZECH REPUBLIC
MTH Praha AS

DENMARK
Banestyrelsen Service Division
CBI Engineering A/S
Greenwood Engineering
ØDS-Caltronic A/S

FINLAND
Desec Ltd
Exel Oy
Hioma-aine Oy
Konepaja Mankinen Ky
Teknikum Oy

FRANCE
ALSTOM Transport
Cowans Sheldon
Drouard
Fassetta mécanique
Framafer
Geismar
Interep SA
Jarret
NEU International
Outreau Technologies
Pouget
Railtech International
Railtech Schlatter Systems
Sateba
SECO-RAIL
Socofer

Spie
Techni-Métal Systèmes
TSO SA
Vossloh Cogifer
Vossloh Infrastructure Services

GERMANY
Beilhack Systemtechnik und Vertiebs GmbH
Benntec Systemtechnik GmbH
Bomag GmbH
BWG – Gesellschaft mbH & Co KG
Cemafer
Contec GmbH
Contitech Transportbandsysteme GmbH
Delimon GmbH
Eichholz
Elektro-Thermit GmbH and Co KG
'EXIM' Verkehrs–, Hafen– und Umwelttechnik GmbH & Co Vertriebs KG
GERB Vibration Control Systems
Gleisbaumechanik Brandenburg
Gottwald Port Technology GmbH
H F Wiebe GmbH & Co KG
Hanning & Kahl GmbH & Co KG
Interlok Bahnconsulting Schmidtendorf
Kirow Leipzig AG
LUKAS
MAN Ferrostaal AG
Max Bögl
metronom Gesellschaft für Industrievermessung mbH
moklansa GmbH
Moog GmbH
Newag GmbH & Co KG
ORTEC Gesellschaft für Schienentechnische Systeme mbH
Peddinghaus Group
Pfleiderer track systems
Phoenix AG
A Rawie GmbH & Co KG
Robel
Schreck-Mieves GmbH
S G Technologies GmbH
Slab Track Systems International
Spitzke GmbH
STRAIL Verkehrssysteme
System Bahnbau International (SBI)
ThyssenKrupp GfT Gleistechnik GmbH
Tiefenbach GmbH
TSTG Schienen Technik GmbH
Türk + Hillinger GmbH
Vossloh Fastening Systems GmbH
Leonhard Weiss GmbH and Co KG
Willy Vogel AG
Windhoff Bahn-und Anlagentechnik GmbH
Zöllner GmbH
Zweiweg Schneider GmbH & Co KG

INDIA
Bharat Earth Movers Ltd (BEML)
Burn Standard
Cimmco International
Hindusthan Engineering & Industries Ltd
Steel Authority of India

ITALY
AMECA Engineering srl
Cometi

IPA
Lucchini Group
Mer Mec SpA
RICA
Selectra srl
SILF Srl
Vaia Car

JAPAN
Mitsukawa Metal Works Co Ltd
Tokimec Rail Techno Inc

LUXEMBOURG
KIHN SA

NETHERLANDS
Colbond Geosynthetics
Edilon BV
Eurailscout Inspection and Analysis BV
Kloos Oving bv
Pintsch Aben BV
Strukton Railinfra BV
Voestalpine Railpro BV

POLAND
Bumar Ltd

ROMANIA
Faur SA

RUSSIAN FEDERATION
ApaTeCh
Lyudinovo Locomotive Works JSC
Muromteplovoz
Power Machines Group

SOUTH AFRICA
Conbrako
Grinaker Duraset

SPAIN
Amurrio Ferrocarril y Equipos SA
Arcelor
Comsa SA
JEZ Sistemas Ferroviarios
Travipos

SWEDEN
AB Strängbeton
Abetong Teknik
Banverket Industridivisionen
DISAB Vacuum Technology AB
ESAB AB, Welding Equipment
Inexa Profil AB
K Industrier AB
Mossboda Trä AB
Rosenqvist Rail Tech AB
SRS Rail Vehicles AB

SWITZERLAND
Benkler AG
Créabéton
A Kaufmann AG
Leica Geosystems AG
Matisa Matériel Industriel SA
Rex
Schlatter AG, HA
Schweizer Electronic AG

Company listing by country—*continued*

Schwihag Gesellschaft für Eisenbahnoberbau mbH
Sersa Group Management AG
Siemens Switzerland
Speno International SA
Tensol Rail

TURKEY
Tülomsas

UNITED KINGDOM
Abloy Security Ltd
Abtus Ltd
Alfred McAlpine
ALH Rail Coatings Ltd
Amec Spie Rail Systems Ltd
Amey Rail Ltd
Arbil Ltd
Atlas Copco Construction + Mining Ltd
BAC Corrosion Control Limited
Balfour Beatty Rail Ltd
R Bance & Co Ltd
Camlock Lifting Clamps Ltd
Carillion Rail
Cembre Ltd
Colebrand International Limited
Collis Engineering Ltd
Cooper & Turner
Corus Cogifer
Corus Rail Products
Corus Railway Infrastructure Services
DBT GB Ltd
Edgar Allen Ltd
Edmund Nuttall Ltd
Electrolux Construction Products UK
Ferotrack Engineering Ltd
Findlay, Irvine Ltd
First Engineering Ltd
GrantRail Ltd
Guided Ultrasonics (Rail) Ltd
Hall Rail
Harrington Generators International
Harsco Track Technologies Ltd
Heat Trace Limited
Henry Williams Limited

Hi-Force Hydraulics Ltd
Holdfast Level Crossings Ltd
IAD Rail Systems
Infrasoft Ltd
JacksonEve Infrastructure Services
JAFCO
Jarvis Rail
JC Bamford Excavators Ltd
Kier Rail
Laser Rail Ltd
Lindapter International
Metrum Information Storage Limited
Mowlem Railways
Newt International
NRS – National Railway Supplies
Oleo International Ltd
Pandrol Rail Fastenings
Permali Gloucester Limited
Polysafe Level Crossing Systems Ltd
Ranalah
Rexquote Ltd
RMC Rail Products
RMS Locotec
Rotabroach Ltd
Rotamag Track Equipment
SGB Group
Sika Ltd
H J Skelton & Co Ltd
SMIS
Sonatest Ltd
SPX Fluid Power
Tarmac Precast Concrete Ltd
Thermit Welding (GB) Ltd
Tiflex Ltd
Turkington Precast Concrete
Ultra Dynamics
VAE UK Ltd
Vortok International
Wabtec Rail Ltd
Wacker (GB) Ltd

UNITED STATES
A & K Railroad Materials Inc
ABC-NACO Inc

American Railroad Curvelining Corporation (ARC)
Aspen Aerials Inc
Aspen Equipment Company
Atlantic Track
Century Group Inc
Chemetron
CXT Rail Products
ESCO Equipment Service Co
FAB-RA-CAST
Harsco Track Technologies Ltd
Holland Company LP
Huck International Inc
Kalyn Siebert
Kershaw
Knox Kershaw Inc
Loram Maintenance of Way, Inc
Lord Corporation
NARSTCO
Nordco
Orton/McCullough
Osmose Railroad Division
Pennsylvania Steel Technologies
Pettibone
Pohl Corporation
Portec
Presto Products Company
Progress Rail Services (Engineering and Track Services)
Quest Corporation
Racine Railroad Products Inc
Railquip Inc
Rails Company
RailWorks Corporation
Salient Systems Inc
Sperling Railway Services Inc
Sperry
Thermon Manufacturing Co
Tie & Track Systems Inc
Trinity Difco Inc
Unit Rail Anchor Company Inc
Western-Cullen-Hayes Inc

For details of the latest updates to *Jane's World Railways* online and to discover the additional information available exclusively to online subscribers please visit
jwr.janes.com

Abloy Security Ltd

2-3 Hatters Lane, Croxley Business Park, Watford WD18 8QY, UK
Tel: (+44 1923) 25 50 66 Fax: (+44 1923) 65 50 01
e-mail: smockett@abloysecurity.co.uk
Web: www.abloysecurity.co.uk

Key personnel
Marketing Communications Manager: Julie Day

Background
Abloy Security Ltd is a member of the Assa Abloy Group.

Products
A range of high security cylinder systems, mechanical and electric locks, padlocks, cam and furniture locks, panic exit hardware and standalone access control products to the UK and Ireland.

NEW ENTRY

A & K Railroad Materials Inc

1505 South Redwood Road, PO Box 30076, Salt Lake City, Utah 84130, USA
Tel: (+1 801) 974 54 84 Fax: (+1 801) 972 20 41

International office
5206 FM 1960 West Suite 103, Houston, Texas 77069, USA
Tel: (+1 281) 893 39 08 Fax: (+1 281) 893 83 71
Web: http://www.akrailroad.com

Key personnel
Chairman: K W Schumacher
President: M H Kulmer
Vice-President, International: Julian Polit
Manager, International Sales: Alfredo Sansores

Products
Rail (new and used), rail accessories, track tools, trackwork materials, welding, continuous welded rail, complete switches, frogs, anchors, bolts, spikes, lockwashers, gauge rods, ties, sleepers, hand track tools and other track materials.

Abetong Teknik

PO Box 24, SE-351 03 Växjö, Sweden
Tel: (+46 470) 965 00 Fax: (+46 470) 160 81
e-mail: abetong@.com
Web: http://www.abetong.com

Main works
Växjö

Key personnel
Managing Director: Ulf Malmqvist

Subsidiaries
Swetrak AS, Estonia
UAB Swetrak, Lithuania

Products
Prestressed concrete sleepers for main lines and turnouts; prefabricated grade crossings; production equipment and technical services for the manufacture and design of concrete sleepers.

The company has supplied its know-how to 34 different factories worldwide; together these plants have now produced over 40 million sleepers.

The company has designed concrete sleepers for heavy-duty turnouts to suit traffic specifications of 37 tonnes maximum and 30 tonnes nominal axleloads at up to 80 km/h, and annual gross train tonnages of more than 50 million. Abetong has also developed special sleepers for high-speed lines.

UPDATED

AB Strängbeton

Box 5074, SE-131 05 Nacka, Sweden
Tel: (+46 8) 615 82 00 Fax: (+46 8) 641 66 70

Products and services
Prestressed concrete sleepers. Services provided include: total track construction planning; feasibility studies for sleeper production; supply of complete plant for concrete sleeper manufacture, with technology transfer; project planning, management and control; and technical training.

Abtus Ltd

Falconer Road, Haverhill CB9 7XU, UK
Tel: (+44 1440) 70 29 38 Fax: (+44 1440) 70 29 61
e-mail: info@abtus.co.uk
Web: http://www.abtus.co.uk

Key personnel
Managing Director: Russell Owen
Technical Manager: Ashley May

Principal subsidiary
Tergor Electronics Ltd

Products
Sighting, void detection, track and overhead line measurement (digital and analogue), bond drilling and track slewing equipment; self-powered track maintenance vehicles; design, consultancy, full repair and recalibration.

Contracts
Recent contracts include the supply of standard and purpose-made equipment to Railtrack (now Network Rail), London Underground Ltd, Docklands Light Railway, British Rail Infrastructure Services, all major European railway systems and Hong Kong.

VERIFIED

Alfred McAlpine

Exchange House, Kelburn Court, Leacroft Road, Birchwood, Warrington WA3 6SY, UK
Tel: (+44 1925) 85 80 00 Fax: (+44 1925) 85 80 99
Web: http://www.alredmcalpine.com

Key personnel
General Manager: Jeff Boden

Organisation
Alfred McAlpine provides a wide variety of services within the rail sector, from complex major projects to discrete schemes for individual clients. The services include engineering works on the railway, ranging from structures maintenance/renewals and embankment stabilisation through to effective management and delivery of complex multi-discipline schemes.

Through their Assurance Case for works on Railtrack infrastructure, Alfred McAlpine undertakes works both for Railtrack directly and for rail industry clients wishing to carry out construction, renewal or maintenance works on the railway.

ALH Rail Coatings Ltd

Carolina Court, Lakeside, Doncaster DN4 5RA, UK
Tel: (+44 1302) 79 11 00 Fax: (+44 1302) 79 12 00

Works
Hebden Road, Scunthorpe DN15 8DT, UK
Tel: (+44 1724) 84 87 65 Fax: (+44 1724) 84 87 65

Key personnel
Managing Director: Bernard Stell
Technical Director: Malcolm Davies
Sales Director: Robert Venell

Background
ALH Rail Coatings is a joint venture between GrantRail and Hyperlast.

Products
Using its expertise as a formulator and manufacturer of polyurethane and epoxy resin systems, ALH supplies a patented precoated rail system, offering high electrical insulation properties combined with reduced noise and vibration characteristics.

The system was chosen by Laing Civil Engineering and GrantRail Ltd for street running sections of the Midland Metro project and the Phase 2 extension of the Manchester Metrolink light rail system.

ALSTOM Transport

Infrastructure Operations
48 rue Albert Dhalenne, F-93482 Saint-Ouen Cedex, France
Tel: (+33 1) 41 66 90 00 Fax: (+33 1) 41 66 96 66
Web: www.transport.alstom.com

Key personnel
Senior Vice-President Infrastructure: Alain Goga

Services
ALSTOM Transport Infrastructure product line covers the design, procurement and installation of electromechanical fixed infrastructure and its maintenance.

Products
Infrastructure Operations offers comprehensive solutions at the system or subsystem level for power generation and distribution including AC and DC traction substations; overhead facilities; contact lines or catenaries; third rail or at-level integrated supply system; auxiliary power supply; track laying; maintenance workshops; communications; signalling (installations); electrical and mechanical equipment in stations; and electronic guidance systems for buses.

ALSTOM has considerable experience relating to urban track for mass transit and light rail/tramway systems, both for ballasted and concrete-bed applications, and steel-wheeled or rubber-tyred vehicles.

Through its expertise in concrete, ALSTOM has developed a new concept in urban track laying. APPITRACK™ is a new automatic process for the construction of track on a concrete bed which avoids the use of sleepers. The method involves slip-form concreting derived from motorway and airport runway construction.

The track is automatically constructed directly on the project alignment using automatic plate and pin inserts. The process reduces erection time, costs and nuisance impact on the urban environment and allows higher noise and vibration requirements to be met. The APPITRACK™ process uses computerised monitoring aids for accuracy of execution and allows:
- speed of execution and limitation of disturbance (noise, vibrations);
- cost savings combined with a high performance level.

Contracts
Ireland: In February 2004 an ALSTOM-led consortium was awarded a five-year contract, with an extension option for a further five years, to maintain the infrastructure of Dublin's Luas tram network from later that year.
France: In November 2004 SYTRAL, the transport union of the Lyon metropolitan area, chose ALSTOM for the supply and installation of tram infrastructure for LEA, the eastern line of the Lyon agglomeration. The contract covers the supply and installation of a 2,150 m double concrete track, a 13,300 m double ballast track and a 1,700 m single track in the depot.
Spain: ALSTOM Transport Infrastructure designed and installed the substations, electrical installations and control systems – 11 substations 2 x 25 kV/400 kV/2 x 60 MVA and 54 autotransformer post 2 x 15 MVA – for the Madrid–Barcelona high-speed line. A confidence in ALSTOM which has been renewed for the Madrid–Alicante and Cordoba-Malaga lines.

Thailand: For the Bangkok Blue Line, ALSTOM Infrastructure laid 54 km of metro lines within 18 months. The contract also covered the design and installation of the entire tunnel ventilation system, the environmental control system, fire protection/detection and drainage for three stations.

Additional contracts
ALSTOM has taken responsibility for the full electromechanical fixed infrastructure in metro projects including Caracas line 1, 2, 3 and 4, Cairo line 1 and 2, Athens 2 and 3, Santiago line 5 and Istanbul.

ALSTOM has also supplied electromechanical fixed equipment for the Paris and Toulouse VAL systems and the Saint-Etienne, Rouen, Orleans, Lyon, Bordeaux, Grenoble and Barcelona tramway systems.

UPDATED

AMECA Engineering srl

Via G Di Vittorio 4, I-42025 Cavriago di Reggio Emilia (RE), Italy
Tel: (+39 0522) 94 11 17; 94 16 65
Fax: (+39 0522) 94 19 02
e-mail: info@amecaengineering.com
Web: http://www.amevaengineering.com

Products
Railway maintenance equipment, including: portal cranes for turnout maintenance; ballast cleaners; lifting machines; motorised and non-motorised trolleys; jib cranes; road-rail vehicles; hoists for overhead line maintenance; rail clip installers and removers; rail renovation systems; exhaust gas blowers for tunnels.

AMEC SPIE Rail (UK)

Stephenson House, 2 Cherry Orchard Road, Croydon CR9 6JA, UK
Tel: (+44 20) 86 67 36 66   Fax: (+44 20) 86 67 27 03
e-mail: geoff.waite@amec.com
Web: www.amec.com/rail

Key personnel
Managing Director: John Moss
Business Development Director: Barry Atkinson
Commercial Director: Bill Daly
Financial Director: Mark Riley
Operations Director: Kevin Beauchamp
Head of Personnel: Christine Hardy
Technical Director: Bruce Littlewood
Business Support Director: Paul Harding
Communications and Media Manager:
   Sandie Rudd

Other offices
Head office
Pole Edison
Parc Saint Christophe, F-95861, Cergy Pontoise, Cedex, France
Tel: (+33 1) 34 22 50 00
Fax: (+33 1) 34 22 51 26
Europe: AMEC SPIE Rail (Benelux) Abay Rail (Belgium); AMEC SPIE Rail SET GmbH (Germany); AMEC SPIE Rail, Cariboni Impianti SRL (Italy)
UK: AMEC SPIE Rail Training; AMEC SPIE Rail (Southern); AMEC SPIE Rail (Western); AMEC SPIE Rail (North East); AMEC SPIE Rail (North West); AMEC SPIE Rail (Scotland)

Background
Following the acquisition of 41.6 per cent of SPIE SA (France) by AMEC plc, a UK-based railway infrastructure company was formed called AMEC SPIE Rail Systems Ltd. In 2003 AMEC exercised its option to purchase the remaining share of SPIE SA, which has a 50 per cent interest in ETF, the French track renewals company.

Services
Design, supply, installation, testing and commissioning of all types of railway infrastructure, including catenary systems, power supply and electromechanical packages.

Contracts
AMEC SPIE Rail has designed, built and commissioned some of the world's most prestigious railway projects. These include: CTRL; major sections of the French TGV network such as the Atlantique and Nord lines; Cairo Metro; Denmark's Storebaelt fixed crossing and LRT/MRT systems in Europe and other world markets.

UPDATED

American Railroad Curvelining Corporation (ARC)

137 Hollywood Avenue, Little Neck-Douglaston, New York 11363-1110, USA
Tel: (+1 718) 224 11 35   Fax: (+1 718) 279 22 01
e-mail: renef@erols.com

Key personnel
President: René A Fiechter

Subsidiaries
Bondarc Division

Products
ARC Corporation creates equipment for maintaining the railroad track. The *Roll-ordinator* measures the track, the *Curveliner* converts the data into numbers for the *Throwmeters*, allowing the gang to realign the curve. The Bondarc division pioneered epoxy for the railroads, from bonding rails to their rail joints, Bondarc developed a technique giving a traction resistance of over 1 million lb, allowing glued insulated joints to be installed. The New York subways used large quantities of Bondarc to repair station steps and floors.

Amey Rail Ltd

1 Redcliff Street, Bristol BS1 6QZ, UK
Tel: (+44 117) 934 88 36   Fax: (+44 117) 934 85 56

Key personnel
Managing Director: Richard Entwhistle
Managing Director, Infrastructure: Stephen Peat
Operations Director: Richard Adams
Marketing Manager: Peter Burton

Background
Amey Railways is a subsidiary of Amey plc, previously British Rail Western Infrastructure Maintenance Unit.

Services
Amey Railways undertakes rail infrastructure maintenance and construction and renewal work, including permanent way, signalling and civil engineering. Activities also include mechanical and electrical engineering, telecommunications work and planning and project management. The company also provides rail safety training.

Contracts
Recent Railtrack contracts include infrastructure maintenance in the Bristol, Chilterns, Exeter, Newport, Reading and Thames areas, plain line track renewals in the Great Western Zone, rerailing in the Cardiff Valleys and pointwork renewals at Bristol. Signalling projects include a three-year Great Western Zone renewals programme, Guildford area resignalling and a scheme in the Taff Valley. Civil engineering projects include seven structure renewals, Great Western Zone tunnel repairs and station regeneration schemes.

Amurrio Ferrocarril y Equipos SA

Maskuribai 10, E-01470 Amurrio, Alava, Spain
Tel: (+34 945) 89 16 00   Fax: (+34 945) 89 24 80
e-mail: afferq@sea.es

Key personnel
For full personnel listing, see Amurrio Ferrocarril y Equipos SA entry in *Bogies and suspension, wheels and axles, bearings* section.

Products
Points, crossings, movable-frog crossings for high-speed crossovers, turnouts, manganese steel frogs, expansion joints, insulated rail joints, turntables, buffers, height gauges, rerailers.

ApaTeCh

PO Box 388, Zhukovsky City, 140180, Moscow Region, Russian Federation
Tel: (+7 095) 556 42 47   Fax: (+7 095) 911 00 19
e-mail: ushakov@tsagi.rssi.ru

Key personnel
Director: Andrey Ushakov

Products
Insulating fishplates using metal/composites to provide electrically insulated joints between rails of all types. The fishplates are supplied with a guaranteed service life of three years on lines handling train speeds of up to 200 km/h. They provide electrical resistance of over 100 ω and are designed for installation in environments with a temperature range of −60° to +80°C. Fishplates are also produced for use with UIC-60 rail for speeds of up to 300 km/h.

Arbil Ltd

Lifting Gear Centre, Foundry Lane, Fishponds Trading Estate, Bristol BS5 7XH, UK
A member of the Raymond Bills group of companies
Tel: (+44 117) 965 31 43   Fax: (+44 117) 965 86 07

Key personnel
Manager: Dave Vale

Products
Rail thimbles. Four standard models are available with fixed or swivel eye, manual or hydraulically operated.

The MK2 has a working load limit of 3 tonnes on lifting points of vertical roller pins. Features include improved design of tapered rolling pins for ease of replacement, hydraulic cylinder fitted with safety valve in case of hose failure as part of standard specification and heat-treated wearing parts.

Arbil MK2 rail thimble   0006273

Arcelor

Apartado 520, (Edificio Energías 2a P), E-33200 Gijón, Asturias, Spain
Tel: (+34 98) 518 71 67   Fax: (+34 98) 518 75 43
e-mail: fernando.sainz-varona@arcelor.com
Web: http://www.aceralia.es

Key personnel
Marketing Manager: D Fernando Sáinz Varona

Background

Arcelor is the result of a merger of three major European steel producers. Aceralia, Arbed and Usinor, claimed to have created the biggest steel company in the world.

Products

Arcelor produces a wide range of sizes and steel grades in rails, especially for high-speed and heavy-haul lines including UIC 60, UIC 54, RN 45, TR 45, TR 57, TR 68, Vignole 46, 100A, 90A, 80A.

VERIFIED

Aspen Aerials Inc

4303 West 1st Street, PO Box 16958, Duluth, Minnesota 55816-0958, USA
Tel: (+1 218) 624 11 11   Fax: (+1 218) 624 17 14

Key personnel

Vice-President: John W Stubenvoll

Products

Aspen A-30 bridge inspection unit which can reach 9m under bridges. An aerial platform, it operates either side of the truck, within a 2.4 m width when rotated, and does not require outriggers. It features an articulating fourth boom and interchangeable platforms. These can be either a three-person inspection bucket with 272 kg capacity, or a 9 m long, 454 kg capacity maintenance platform.

VERIFIED

Aspen Equipment Company

Railroad Division
9150 Pillsbury Avenue South, Minneapolis, Minnesota 55420, US
Tel: (+1 952) 888 25 25
Fax: (+1 952) 656 71 59
e-mail: jgallo@aspeneq.com
Web: www.aspenequipment.com

Key personnel

President: Steven Sill
Vice-President and General Manager: Tom Cherne
Vice-President, Production: Jerry Neises
Treasurer: Mike Lindberg
National Sales Manager: John Gallo

Other offices

Aspen Equipment Company
613 South East Magazine Road, Ankeny, Iowa 50021, US

Aspen Equipment Company
10922 Sapp Brother Drive, Omaha, Nebraska 68138, US

Aspen Aerials, Inc
4303 West 1st Street, Duluth, Minnesota 55807, US
Web: www.aspenaerials.com
Vice-President: John Stubenvoll
Manufacturer of underbridge inspection units

Products

Aspen Equipment Company is a truck equipment installer and manufacturer supplying hi-rail trucks in a variety of configurations. Representing over 40 lines of equipment from various manufacturers including cranes, hi-rail gear, service bodies, aerials, welders, air compressors, winches and hydraulic tool systems. The company also manufactures custom-built service bodies and offers on-staff engineering, design and specification writing services. The company also has hi-rail trucks available for long and short-term rentals. Its subsidiary, Aspen Aerials, Inc, specialises in the manufacture of underbridge inspection units for rail and non-rail applications.

UPDATED

Atlantic Track

Atlantic Track and Turnout Co
270 Broad Street, PO Box 1589, Bloomfield, New Jersey 07003, USA
Tel: (+1 973) 748 58 85   Fax: (+1 973) 748 45 20
e-mail: info@atlantictrack.com
Web: http://www.atlantictrack.com

Main works

St Clair Industrial Park, RD No 3, PO Box 360, Pottsville, Pennsylvania 17901, USA

Key personnel

President and Chief Executive Officer:
  Peter Hughes
Export and Domestic Sales Director:
  Charlie Killeen

Products

All new ASCE, AREA, ARA-A, ARA-B rail sections produced; full line of relay rail and special trackwork; track accessories including switch materials, maintenance tools and insulating material. Third-rail transit products including glass fibre insulators, coverboard and brackets.

Atlas Copco Construction + Mining Ltd

PO Box 79, Swallowdale Lane, Hemel Hempstead HP2 7HA, UK
Tel: (+44 1442) 22 21 00   Fax: (+44 1442) 23 44 67
e-mail: john.fitzpatrick@atlascopco.com
Web: http://www.atlascopco.co.uk

Key personnel

Executive: John Fitzpatrick

Products

Self-contained power tamper/drill and pneumatic equipment, hydraulic breakers.

VERIFIED

BAC Corrosion Control Limited

Stafford Park 11, Telford TF3 3AY, UK
Tel: (+44 1952) 29 03 21   Fax: (+44 1952) 29 03 25
e-mail: bac@bacgroup.com
Web: http://www.brightbond.com

Products

Bright-Bond pin brazing systems for track bonding for signalling purposes, the connection of heater strips and earth connections to masts carrying overhead power supply lines.

There are two main types of Bright-Bond unit: the BB3, with the capacity to produce approximately 150 connections per charge; and the portable BB2 unit, for service and maintenance operations, giving approximately 50 connections per charge. A wide range of ancillary equipment is also supplied, including several types of brazing gun, specially designed bonds and cables, batteries and accessories.

VERIFIED

Bright-Bond pin brazing system for track bonding 0109479

Balfour Beatty Rail Ltd

7 Mayday Road, Thornton Heath, CR7 7XA, UK
Tel: (+44 20) 86 84 69 22
e-mail: info@bbrail.com
Web: www.bbrail.com

Key personnel

Group Managing Director: Jim Cohen
Chief Operating Officer: Andy Rose
Marketing and Planning Manager: Sean Pang

Background

Balfour Beatty Rail is an international leader in rail engineering projects. Employing 7,000 staff, the company services the full range of rail infrastructure activities from high-speed rail to mass rapid transit, covering inter-city, rural and commuter routes, for both public and private railways. Balfour Beatty Rail also works in partnerships and alliances.

In August 2005 the company acquired the Pennine Group, the UK ground engineering specialists.

Capabilities

Balfour Beatty Rail can deliver large, complex, multi-disciplinary projects using its professional project management skills throughout the whole cycle. The company undertakes work from the initial planning phase, the identification of project requirements and interfaces through to construction, testing and commissioning and handover of the works.

Balfour Beatty Rail has the capability to supply all types of track systems; high-speed main line, heavy haul, urban mass transit and commuter systems, and light rail transits, working on new systems as well as re-modelling and renewals of existing lines. In addition to contracting skills, the company is an international designer and manufacturer of switches and crossing, special trackwork and associated products.

The company has experience in the design and construction of ballasted track, track on concrete including advanced techniques for noise and vibration problems and has developed its own patented embedded slab track system.

Balfour Beatty Rail provides a total power and electrification capability for all types of new railway projects as well as upgrading, renovating and converting existing systems. The company has a large, competent and diverse capability for power and electrification projects ranging from overhead lines, AC and DC power supply to conductor rail systems.

The company also has in-house capability to undertake turnkey signalling and rail telecommunications solutions including design, planning, application engineering, installation, commissioning and testing. It can support established technology, ranging over a wide variety of signalling systems from high-speed lines, low-speed freight lines, metro systems and level crossing and has an in-house design capability based through the UK which now also includes Solid State Interlocking (SSI) suite based at Derby.

Balfour Beatty Rail can offer asset management including management systems, software and tools to facilitate data acquisition, perform inspection and monitoring and support decision making and planning. It can also support remedial activities through its own resources and wide range of equipment and plants. Balfour Beatty Rail specialises in the care of switches, rails, track systems, signalling systems, current collection systems and power supply systems.

Principal subsidiaries

Balfour Beatty Rail Infrastructure Ltd
7th Floor, Russell Square House, 10-12 Russell Square, London WC1B 5EH, UK
Tel: (+44 20) 70 79 47 00   Fax: (+44 20) 70 79 47 01
Web: www.bbrail.com

Key personnel
Managing Director: Eric Prescott
Marketing Manager: Liz Murray-Leslie

Services
Balfour Beatty Rail Infrastructure Services Ltd undertakes all types of infrastructure inspection, maintenance and renewal, covering track, signalling and overhead and third rail power supply systems on main line and urban rail networks.

Balfour Beatty Rail Plant Ltd
PO Box 5065, Raynesway, Derby DE21 7ZQ, UK
Tel: (+44 1332) 66 14 91
Fax: (+44 1332) 28 82 22
Web: www.bbrail.com

Key personnel
Managing Director: Keith Fidler
Engineering and Safety Director: Raymond Reed
Finance Director: Kate Busman
Fleet Services Director: Jeff Bussey
Plant Services Director: David Watson
Business Development Manager: Steve MacIver
Head of Engineering, Health, Safety, Quality and Environment: Dave Elias
Quality and Environment Systems Manager: Tim Russell
Safety Advisor: Peter Turner
Business Development: Lynne Sherlock

Services
Balfour Beatty Rail Plant Ltd has three operating units: the Rail Plant Unit provides mechanised on-track equipment services, including tampers for plain line and for switches and crossings, trams, rail cranes, tracklayers, ballast regulators, ballast cleaners and gophers. The Raynesway Plant Unit provides road/rail equipment, general plant maintenance and hire services for portable tools, accommodation, safety equipment, communications and CCTV systems plus design and manufacture of specialised road/rail equipment. The Fleet Services Unit undertakes transport fleet maintenance and fleet management and maintains Balfour Beatty's 7,000-vehicle fleet.

Balfour Beatty Rail Power Systems GmbH
For further information refer to Balfour Beatty Rail GmbH entry in *Electrification contractors and equipment suppliers* section

Balfour Beatty Rail Projects Ltd
For further information refer to Balfour Beatty Rail Projects Ltd entry in *Turnkey systems contractors*

Balfour Beatty Rail Technologies Ltd
For further information refer to Balfour Beatty Rail Technologies Ltd entry in *Consultancy* section

Balfour Beatty Rail Track Systems Ltd
Osmaston Street, Sandiacre, Nottingham NG10 5AN, UK
Tel: (+44 115) 921 82 18   Fax: (+44 115) 921 82 38
e-mail: phil.bean@bbrail.co.uk
Web: www.balfourbeatty.com

Key personnel
General Manager: Keith Churm
Business Development Manager: Philip Bean

US subsidiaries
Balfour Beatty Rail Systems Inc
1024 Route 519, Suite 300, Eighty Four, Pennsylvania 15330, US
Tel: (+1 724) 225 61 55   Fax: (+1 724) 228 81 13

Balfour Beatty Rail Maintenance, Inc
Marta Track Constructors Inc
Metroplex Corporation

UPDATED

R Bance & Co Ltd

Cockcrow Hill House, St Mary's Road, Surbiton KT6 5HE, UK
Tel: (+44 20) 83 98 71 41
Fax: (+44 20) 83 98 47 65
e-mail: admin@bance.com
Web: www.bance.com

Bance Alumicart inspection/haulage vehicle   0536547

Bance 2 Diesel with two loaded trailers   0125198

Key personnel
Managing Director: R Bance
Sales Director: G Smales
Technical Director: B Steel

Products
Continuous rail flaw detection vehicles; rail moving vehicles for confined environments; emergency response vehicles; motorised trolleys; Alumicart inspection/haulage vehicles; trolleys and skates; tapered rail joint shims for maintaining jointed track; impact wrenches, sockets and augers; rail disc cutters 12 and 14 in; rail drills; tampers; hydraulic tools; track measurement gauges; platform measurement gauges; rechargeable lamps (worksite and emergency).

Services
Ultrasonic rail flaw detection; continuous rail depth recording; continuous platform gauge recording.

UPDATED

Banverket Industridivisionen

Box 67, SE-57121 Nässjö, Sweden
Tel: (+46 380) 724 00
Fax: (+46 380) 725 00
Web: http://www.banverket.se

Key personnel
General Manager: Göthe Persson

Services
Banverket Industrial Division is an offshoot of the Swedish National Rail Administration set up to tender for contract work. It has had some success in obtaining contracts outside its parent company.

Recent projects include: the West Coast line, Sweden, the Danish part of the Öresund Link, the coast-to-coast part of the Öresund Link, the Gardermoen project, Norway and the Arlanda Link, Sweden.

Banestyrelsen Service Division

Vanløse Allé 89, DK-2720 Vanløse, Denmark
Tel: (+45) 82 34 45 02
e-mail: service@servicebane.dk

Services
Maintenance and construction operations within: track works; overhead catenary systems; power systems; interlocking systems; telecommunication.

Barclay Mowlem Construction Ltd

Building 3, Level 2, 20 Bridge Street, Pymble, New South Wales 2073, Australia
Tel: (+61 2) 98 55 16 00   Fax: (+61 2) 98 55 16 20
Web: www.barclaymowlem.com

Key personnel
Managing Director: David Hudson
Director and General Manager, Rail: Bill Killinger
Commercial Manager, Rail: Bob Cooke
Manager, Rail Australia: Graeme Spragg
Manager, Rail Hong Kong: Bill Hardy

Background
Barclay Mowlem is a subsidiary of John Mowlem Co plc.

Subsidiary company
Austrak Pty Ltd

Services
Barclay Mowlem delivers premium services in railway infrastructure design, trackwork, high-speed rail, mass transit systems and depot construction, electrification, signalling, maintenance, renewals and plant and equipment supply throughout Australia, New Zealand and South East Asia.

The rail group is currently operational in Australia, Hong Kong, Malaysia, Singapore, Taiwan and Thailand.

UPDATED

Beilhack Systemtechnik und Vertriebs GmbH

PO Box 100155, D-83001 Rosenheim, Germany
Tel: (+49 8035) 98 40   Fax: (+49 8035) 98 42 30
e-mail: info@beilhack.de
Web: http://www.beilhack.de

Key personnel

Managing Partners
Susanne Potocnik (Finance)
Joachim Schwope (Sales, Distribution)
Bernd Tobie (Technology)

Products

Production and supply of self-propelled, propelled and vehicle-mounted snow-clearing equipment and vehicles, including blowers, ploughs, rail sweeping equipment, flangers and ice cutters; multipurpose infrastructure maintenance vehicles, including self-propelled units with tipper, crane and mower equipment; feasibility studies and consultancy and telematic systems (hard- and software).

Bharat Earth Movers Ltd (BEML)

BEML Soudha, 23/1 4th Main, SR Nagar, Bangalore 560 027, India
Tel: (+91 80) 222 44 58   Fax: (+91 80) 229 19 80
e-mail: techrnd@vsnl.com
Web: http://www.bemlindia.com

BEML spoil disposal unit for Indian Railways   0527280

BEML tracklaying equipment for Indian Railways   0527279

Key personnel

For a full key personnel listing, see Bharat Earth Movers Ltd (BEML) entry in *Locomotives and powered/non-powered passenger vehicles* section.

Products

Tracklaying equipment, including tracklayers and spoil disposal units. The BEML self-propelled diesel-hydraulic tracklayer is designed to move on an auxiliary track. It has hydraulically operated grippers for handling concrete sleepers and rails from wagons and panel assemblies of concrete sleepers and rails. Maximum lifting capacity is nine tonnes.

BEML's spoil disposal unit is designed for use with ballast cleaning machines for the reception, storage and unloading of spoil generated during ballast screening. Equipped with a hydraulically operated horizontal conveyor inside its hopper and a slew conveyor at one end, the spoil disposal unit can unload itself into similar units coupled together, wagons on an adjacent track or on to the slope of the formation.

Contracts include the supply of 29 tracklayers and five spoil disposal units to Indian Railways.

Benkler AG

Nordstrasse 1, CH-5612 Willmergen, Switzerland
Tel: (+41 56) 618 72 00
Fax: (+41 56) 618 72 99
e-mail: info@benkler.ch
Web: http://www.benkler.ch

Corporate background

Benkler AG is a member of the Sersa Group (qv).

Services

Rail grinding; track inspection; overhead line maintenance and construction.

Benntec Systemtechnik GmbH

Walter-Geerdes-Strasse 10-12, D-28307 Bremen, Germany
Tel: (+49 421) 43 84 90
Fax: (+49 421) 438 49 90
e-mail: info@benntec.de
Web: http://www.benntec.de

Key personnel

Sales Manager: R Beck

Products

RailCheck, automatic recognition and marking of fault takes place online during the inspection run. A track allocation system serves to allocate exactly the position of the fault to the data of the track.

HeadCheck, developed specially for the detailed automatic inspection of running edges of the rails. The system inspects the edges for fine surface cracks.

CrackCheck, a digital recording system that serves to inspect the automatic condition of ballastless tracks and concrete sleepers. The finest crack structures will be recognised online at an inspection speed between 50 and 100 km/h.

VegetationCheck, determines automatically the degree of vegetation on the track grid as well as on the track and in the surrounding areas. The data is recorded depending on the configuration at an inspection speed up to 100 km/h.

RailScan, a mobile digital colour image processing system for the visual inspection of the railroad track, the track environment and the overhead, the general condition of the track.

Bomag GmbH

Hellerwald, D-56154 Boppard, Germany
Tel: (+49 6742) 10 00
Fax: (+49 6742) 30 90
e-mail: info@bomag.de
Web: http://www.bomag.de
Part of United Dominion company

Key personnel

President: Lothar Wahl
Senior Vice-Presidents
Marketing and Sales: Dr Kay Mayland
Dr Kay Mayland
Finance: Martin Ochotta
Vice-President, Sales: Joachim Untiedt

Principal subsidiaries

In Austria, Canada, France, Japan, Jordan, Singapore, UK and USA.

Products

Tampers, vibrating plates (single direction and reversible), single and double drum rollers (hand guided), trench compactors, tandem rollers, combi rollers, single drum rollers (including asphalt versions), pneumatic tyred rollers, towed vibratory rollers, sanitary landfill compactors, asphalt surface recycler, soil stabiliser, hardware and software for compaction control of permanent way including data processing of all measured data, high-frequency internal vibrator.

Bumar Ltd

Al Jana Pawa II nr 11, PL-00-828 Warsaw, Poland
Tel: (+48 22) 620 46 65   Fax: (+48 22) 654 70 16
e-mail: b2@phzbumar.com.pl
Web: http://www.phzbumar.com.pl

Key personnel

For a full list of key personnel see Bumar Ltd
entry in *Bogies and suspension, wheels and axles,
bearings* section.

Products

Spring rail fastening for heavy and light rail
applications; dog spikes.

Burn Standard

A subsidiary of: Bharat Bhari Udyog Nigam Ltd
10-C Hungerford Street, Calcutta 700 017 India
Tel: (+91 33) 247 10 67; 17 62; 17 72
Fax: (+91 33) 247 17 88

Key personnel

See *Freight vehicles and equipment* section.

Products

Points and crossings, sleepers, fishplates, bridge
girders.

BWG – Gesellschaft mbH & Co KG

PO Box 305, Wetzlarer Strasse 101, D-35510
Butzbach, Germany
Tel: (+49 6033) 89 21 20   Fax: (+49 6033) 892 61 20
e-mail: d.schluck@bwg-wbg.com
Web: http://www.bwg.cc

Key personnel

Chief Executive Officer: Thomas Kabtbrenner
Chief Financial Officer: Udo Haubmaun
Chief Technology Officer: Hubertus Hohne
Export Manager: D Schluck

Products

Development and manufacture of turnouts and
permanent way systems made of flat bottom
and grooved rails, services for permanent way
construction.

Contracts

Contracts include Amtrak/NJT; DORTS, Taipei;
Konkan Railway, India; DB AG, NSB Norway and
VR Finland.

Camlock Lifting Clamps Ltd

Knutsford Way, Sealand Industrial Estate, Chester
CH1 4NZ, UK
Tel: (+44 1244) 37 53 75   Fax: (+44 1244) 37 74 03
e-mail: sales@camlok.co.uk
Web: http://www.camlok.co.uk
   http://www.liftingclamps.com

Products

Rail handling equipment including: MR multi-rail
clamps; MRS multi-rail clamps grabs; CR and SCR
rail clamps; RA rail anchors and RP rail pulling
clamps.

Developments

Camlok has recently designed a rail clamp that
can perform similar operations to the standard
CR rail clamp but with a hydraulic open and close
mechanism. Groups of CRH rail clamps can be
combined to lift sections of rail up to 180 m in length.

Carillion Rail

Gloucester House, 65 Smallbrook Queensway,
Birmingham B5 4HP, UK
Tel: (+44 121) 345 15 13   Fax: (+44 121) 345 15 18

e-mail: lesdeakin@gtrm.co.uk
Web: http://www.carillionrail.com

Key personnel

Managing Director: Paul Kirk
Commercial Director: Mike Hawe
Director of Projects: Bob Collard and Mark Cutler

Background

Carillion Rail is a trading name of GT Railway
Maintenance Ltd and is wholly owned by
Carillion plc. Carillion Rail was formally launched
in April 2002 as a new rail business combining
the resources of GTRM, Centrac and Carillion
Infrastructure, and has three business groups:
Projects; Infrastructure Maintenance; and Rail
Services. GTRM Ltd was previously owned by a
joint venture between Alstom and Carillion plc
(formerly Tarmac Construction Ltd). Formerly
known as Central Infrastructure Maintenance Co,
GTRM Ltd was one of the British Rail Infrastructure
Service units, sold off to Alstom and Carillion in
1996 as part of the British Rail privatisation.

Carillion Rail has taken over Swedish Rail
Systems Entreprenad (SRSE), the Swedish-based
rail infrastructure maintenance specialist which
also has interests in Norway, Denmark and Finland.
SRSE currently has major contracts in Sweden with
Malmo City Tunnel and Inlandsbanen. It has new
assets worth about SKr80 million.

Services

The activities of Carillion Rail embrace the disciplines
of permanent way, civil engineering, signalling,
telecommunications, power supplies, OLE and plant
and consultancy. All projects incorporate design,
installation, testing and commissioning.

The Infrastructure Maintenance Group is
responsible for maintenance and repair work on
many major routes throughout the UK railway
network. This includes the West Coast Main Line
from London Euston to Scotland and much of
the central band of England and Wales. The group
offers expertise in permanent way maintenance and
inspection; signalling maintenance and inspection;
electrical distribution equipment maintenance and
inspection; off-track maintenance and inspection,
including vegetation management; rapid
response and repair works; telecommunications
maintenance and inspection; overhead line and
third rail equipment maintenance and inspection,
closed branch lines maintenance and inspection
and station and platform regeneration and
refurbishment.

The Rail Services Group, headed by the Rail
Services Directors, contains the following units:
Rail Plant; Eurailscout GB and Rail Testing. Rail Plant
provides on-track machines, having a full range
of machine maintenance plant, track renewals
plant, electrification plant, wiring trains and an
autoballaster. Eurailscout GB offers innovative
solutions to track inspection and analysis, asset
management. A joint venture between Carillion
plc and Europool BV, Eurailscout offers services
throughout the UK, including track recording and
track analysis using software programmes such
as IRIS. Rail Testing provides a comprehensive
range of welding, testing and calibration services
to deliver to both internal and external customers.
The Calibration and Test Centre in Crewe provides
an extensive calibration and repair service for
instrumentation and tools used within the rail
industry which require calibration in accordance
with the current standards.

CBI Engineering A/S

Færgeparken 21, DK-3600 Frederikssund, Denmark
Tel: (+45) 47 31 33 88   Fax: (+45) 47 31 22 33
e-mail: cbi@cbi.dk
Web: http://www.track-collector.com

Products

Track Collector system manufactured in fibreglass-
reinforced plastic for installation in areas of track
that require protection from pollution by oil and
other contaminants. Options include skid-safe
surfaces and grating.

CDM nv/sa

Reutenbeek 9-11, B-3090 Overijse, Belgium
Tel: (+32 2) 686 15 60
Fax: (+32 2) 687 35 52
e-mail: general@cdm.be
Web: http://www.cdm.be

Products

Elastic materials and systems for track noise and
vibration isolation for high-speed, main line, metro
and light rail infrastructure. Products include rail
support pads and strips, under-baseplate pads and
under-sleeper mats, ballast mats and embedded
rail systems.

Cemafer

Cemafer Gleisbaumaschinen und Geräte GmbH
Ihringer Landstrasse 3, PO Box 1327, D-79206
Breisach, Germany
Tel: (+49 7667) 905 90   Fax: (+49 7667) 90 59 59
e-mail: cemafer@cemafer.com
Web: http://www.cemafer.com

Key personnel

General Manager: A Wagner

Products

Power wrenches, coach-screwing machines, rail
drills, rail saws, sleeper drills, sleeper adzing and
drilling machines, rail grinding equipment, rail
benders, light tampers, inspection trolleys, trailers,
portal cranes, hand tools, electric generators
(portable), gauges, jacks, rail cutting machines,
rail stripping machines, sleeper boring machines,
sleeper placing machines, spanners, spike drivers
and extractors, and tracklaying equipment.

Cembre Ltd

Fairview Industrial Estate, Kingsbury Road,
Curdworth, Sutton Coldfield B76 9EE, UK
Tel: (+44 1675) 47 04 40   Fax: (+44 1675) 47 02 20
Web: www.cembre.com

Associated companies

Cembre SpA
Via Serenissima 9, I-25135 Brescia, Italy
Tel: (+39 030) 369 21   Fax: (+39 030) 336 57 66
e-mail: info@cembre.com
Web: www.cembre.com

Cembre A/S
Fossner Senter, N-3160 Stokke, Norway
Tel: (+47) 33 36 17 65   Fax: (+47) 33 36 17 66
e-mail: cembre@cembre.no

*Rail mounted support trolley for Cembre
sleeper drills*   0567156

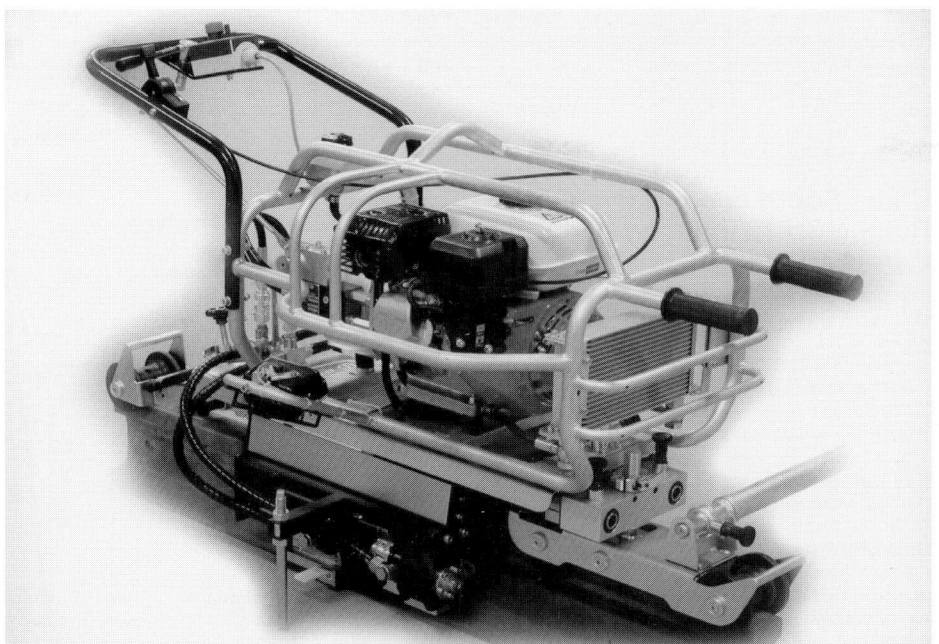

Machine for insertion/extraction of Pandrol rail clips from Cembre   0567155

Century's custom concrete grade crossing turnout   1047988

Cembre España SL
Called Llanos de Jerez, 2 Pol Ind de Coslada,
E-28820 Madrid, Spain
Tel: (+34 91) 485 25 80   Fax: (+34 91) 485 25 81
e-mail: info@cembre.com

Cembre GmbH
Taunusstrasse 23, D-80807 Munich, Germany
Tel: (+49 89) 358 06 76   Fax: (+49 89) 35 80 67 77
e-mail: info@cembre.com

Cembre Inc
Raritan Center Business Park,
70 Campus Plaza II, Edison, New Jersey 08837, USA
Tel: (+1 732) 225 74 15   Fax: (+1 732) 225 74 14
e-mail: salesus@cembre.com

Cembre Sarl
22 avenue Ferdinand de Lesseps, F-91420
Morangis, France
Tel: (+33 1) 60 49 11 90   Fax: (+33 1) 60 49 29 10
e-mail: info@cembre.fr

Products
Drilling machines for rail web and wooden sleepers; rail bush contact kits; related accessories; rail-mounted.

A recent development is the automatic Pandrol clip machine which is capable of the simultaneous insertion or extraction of clips on both sides of the rail, or on one side only.

UPDATED

Century Group Inc

PO Box 228, Sulphur, Louisiana 70664-0228, USA
Tel: (+1 800) 527 52 32 ext 118
Fax: (+1 800) 887 21 53
e-mail: sales@centurygrp.com
Web: http://www.centurygrp.com

Key personnel
President/CEO, Railroad Products Division:
   Rusty Vincent
Vice-President, Sales and Marketing, Railroad
   Products Division: Jerry McCoombs

Products
Full-depth concrete grade crossings in North America. Manufactured of high strength reinforced concrete, the Century crossings are durable, safe, economical and simple to install. The versatile concrete grade crossings are manufactured to fit any size rail and are compatible with all major types of rail fastening systems. The crossings are manufactured for curves, turnouts, diamond crossings, devil strips and many other applications

and with all crossings comes an innovative elastomeric flangeway filler for the safety of pedestrian and vehicular traffic and also to protect the track structure from contaminants.

The HDPE Enviropan system is a state-of-the-art railroad spill collection system to assist railroads, military facilities, light rail transit and industry to protect the environment. It is a high impact puncture- and tear-resistant closed drain system which minimises exposure at pan and cross drain connections. The Enviropan modular lightweight construction allows for fast installation, eliminating railroad track downtime.

Contracts
Recent contracts include: Railroad crossings and spill collection systems for Metropolitan Atlanta Rapid Transit (MARTA) 2003; Burlington/Northern Santa Fe railroad 2002; Union Pacific Railroad 2003; Kansas City Southern Railroad 2003; Canadian National Railroad 2003; Conoco Phillips 2003; Amtrak 2003; and Dallas Area Rapid Transit (DART) 2003.

Chemetron

Chemetron Railway Products Inc
8021 National Turnpike, Louisville Kentucky 40214, USA
Tel: (+1 502) 368 65 62   Fax: (+1 502) 367 14 84

Main works
5600 Stillwell Street, Kansas City, Missouri 64120, USA

Key personnel
Assistant Vice-President, Welding and Equipment
   Sales: Larry J Taylor

Century's HDPE Enviropans   1029025

Century's concrete pedestrian crossing at Light Rail Transit station   1047987

Background

Chemetron Railway Products Inc is a division of Progress Rail Services Corporation.

Products

Electric flash-butt rail welding plants including rail welders, rail end polishers, base grinders, rail saws, automatic rail straighteners and rail pushers. Rail trains, rail wagons and miscellaneous rail handling equipment including turnkey design of rail welding plants. Contract welding service of rail into continuous welded rail in any of standard, alloy or head-hardened rails. Ergonomically safe manually operated switch stands.

Chemetron rail welding machines operate using AC or DC power. Systems are solid-state and can be controlled with various levels of automation. Production capabilities are in excess of 25 welds/h for all rail sizes. Transportable/mobile flash-butt rail welding plants are also available, including truck-mounted road/rail in-track units.

Chemetron operates plants in Canada and the USA and provides equipment and technical advice to Asia, Australia, Mexico and South America.

China Northern Locomotive and Rolling Stock Industry (Group) Corporation (CNR)

11 Yangfangdian Road, Haidian District, Beijing 100038, China
Tel: (+86 10) 51 86 23 70   Fax: (+86 10) 51 86 23 74
e-mail: loriciec@cnrgc.com.cn
Web: www.cnrgc.com

Key personnel

Chairman and Managing Director: Wang Tai-Wen
President: Cui Dianguo
Vice-President: Zhao Guangxing
President of LORIC Import & Export Corporation Ltd: Cao Guo-Bing
Vice-President of LORIC Import & Export Corporation Ltd: Chen Dayong
Senior Engineer of LORIC Import & Export Corporation Ltd: Yang Xiang-Jing

Products

Development, design, engineering, production, sales, installation, refurbishment, maintenance

CNR Model JXK18-9/196 battery-powered AC drive tunnel construction locomotive   1048669

CNR Model TY3 catenary inspection vehicle
1048670

and after-sales services for all types of catenary maintenance vehicles, tunnelling locomotives and long rail transportation vehicles.

NEW ENTRY

China South Locomotive and Rolling Stock Industry (Group) Corporation (CSR)

11 Yangfangdian Road, Haidian District, Beijing, China
Tel: (+86 10) 63 98 47 70
Fax: (+86 10) 63 98 47 66
e-mail: csrft@csrgc.com.cn
Web: http://www.csrgc.com.cn

Key personnel

General Manager: Zhao Xiaogang

Background

CSR was established in 2000 as a result of the division into two regional groups of the former China National Railway Locomotive and Rolling Stock Industry Corporation (LORIC). The corporation comprises 24 state-owned enterprises employing around 116,000 staff. Some facilities are listed as part of CSR and of its northern China counterpart, China North Locomotive and Rolling Stock Industry (Group) Corporation (CNR).

Works

See CSR entry in *Locomotives and powered/non-powered passenger vehicles* section.

Products

Railway cranes and test vehicles. The range of cranes includes:

Type N151 15T diesel-hydraulic self-propelled 57-tonne crane with a lifting capacity of 15 tonnes with outriggers and 10 tonnes without.

Type N1601 160T diesel-hydraulic self-propelled 92-tonne crane with a lifting capacity of 160 tonnes.

Type NZS0631 63T self-propelled 88-tonne self-propelled telescopic crane with a lifting capacity of 63 tonnes.

Cimmco International

Prakash Deep, 7 Tolstoy Marg, New Delhi 110001, India
Tel: (+91 11) 331 43 83; 331 43 84; 331 43 85
Fax: (+91 11) 332 077; 372 35 20

Key personnel

Chairman: S Birla
President: R Upadhaya
General Manager: M P Gupta
Marketing Manager: G Sodhi

Products

Permanent way materials including: cast-iron, pressed steel and concrete sleepers; elastic rail fastening system; rigid fasteners; points and crossings; rail anchors; fishplates, nuts and bolts; track tools and various types of spikes.

CNR Qiqihar Railway Rolling Stock (Group) Co Ltd (QRRS)

10 ZhongHuaDongLu Road, Qiqihar 161002, China
Tel: (+86 452) 293 84 72; 293 84 99
Fax: (+86 452) 251 67 23
e-mail: qrrsintl@qrrs.com.cn
Web: www.qrrs.com.cn

Key personnel

Chairman of Board of Directors and General Manager: Wei Yan
Chief Engineer: Yu Lianyou
International: Liu Dezeng; Ms Zhang Xianbin

Background

QRRS is a member of CNR.

Products

Diesel-hydraulic railway cranes (15, 63, 100, 125 and 160 tonnes) equipped with fixed or telescope beam.

UPDATED

Colbond Geosynthetics

PO Box 9600, Werstervoortsedijk, NL-6800 TC Arnhem, Netherlands
Tel: (+31 26) 366 46 00   Fax: (+31 26) 366 58 12
e-mail: geosynthetics@colbond.com
Web: http://www.geosynthetics.colbond.com

Key personnel

General Manager: Axel Poscher
Sales Manager: Blair Rawes
Marketing Manager: Wim Voskamp
Research and Development Manager: Willem Gevers

Products

The company's products include: track components, Enkadrain® for structural drainage; Enkamat® for erosion control; Armater® for erosion control; Enkagrid® for soil improvement, and Colbonddrain® for soil impovement.

Colebrand International Limited

162-168 Regent Street, London W1B 5TD, UK
Tel: (+44 20) 74 39 10 00
Fax: (+44 20) 77 34 33 58
e-mail: enquiries@colebrand.com
Web: www.colebrand.com

Key personnel

Managing Director: K N Tusch

Main Works

CXL Factory, Goodshawfold Road, Rossendale BB4 8QF, UK

Products

Design and manufacture of the Lock-Up Device (LUD), a specialised component that has wide ranging application to both new and existing bridges and which is particularly applicable to railway bridges. LUDS can transform a normally free moving joint into a temporarily fixed joint. It is activated under the application of a transient load, remaining rigid whilst the load is applied before reverting to its free moving passive state. The LUD provides a means of sharing loads within a structure by re-distributing horizontal forces within the structure away from the directly loaded sub-structure and into adjacent structural elements. This enhances the overall load carrying capacity and makes the LUD ideally suited for application on railway bridges. The LUD has been used successfully for many years on high-speed main line and mass transit railway bridges in the UK and overseas.

UPDATED

Collis Engineering Ltd

Salcombe Road, Meadow Lane Industrial Estate, Alfreton, Derbyshire DE55 7RG, UK
Tel: (+44 1773) 83 32 55   Fax: (+44 1773) 52 06 93
e-mail: sales@collis.co.uk
Web: www.collis.co.uk

Key personnel

Managing Director: Peter Roberts
Contracts General Manager: Pete Savage
Technical Manager: Masoud Jafari
Sales and Marketing Manager: David Eades

Products

Supply of points fittings and fixing plates to the railway industry. Bespoke permanent way equipment is manufactured when standard 'off-the-shelf- solutions' are not applicable. Structural steelwork is designed and manufactured in-house and installed on request.

UPDATED

Cometi

Costruzioni Meccaniche Tiberine
Zona Industriale Fiumicello 19, I-52037 Sansepolcro (AR), Italy
Tel: (+39 0575) 74 42 11   Fax: (+39 0575) 74 42 24

Products

On-track machines for the maintenance of permanent way and for the installation and maintenance of overhead power supply systems for main line and light rail systems.

Comsa SA

Edificio Numancia 1, Calle Viriato 47, E-08014 Barcelona, Spain
Tel: (+34 93) 430 15 152   Fax: (+34 93) 405 13 30
Web: http://www.comsa.com

Subsidiaries

Intraesa
(address as parent company)
Tel: (+34 93) 430 49 44   Fax: (+34 93) 439 17 69

Travipos SA
Calle Irlanda del Norte s/n, Poligono Industrial Constanti, Sector Norte, E-43120 Constanti, Tarragona, Spain
Tel: (+34 977) 29 65 53   Fax: (+34 977) 29 65 53
Joint venture with Pfleiderer Verkehrstechnik GmbH & Co KG

Services

Construction and maintenance of high-speed, conventional and mass transit railway infrastructure including metro and light rail; construction of railway installations such as traction and rolling stock maintenance depots and marshalling yards.

Contracts include infrastructure maintenance for the Madrid-Seville high-speed line, construction and maintenance work on the Alicante-Barcelona Euromed line and construction of the Valencia light rail system. Overseas projects include rolling stock and maintenance facilities for Kuala Lumpur's STAR light rail system.

Conbrako

PO Box 4018, Luipaardsvlei 1743, Transvaal, South Africa
Tel: (+27 11) 762 24 21   Fax: (+27 11) 762 65 35

Products

Track jacks.

Contec GmbH

In den Eichen, D-56244 Ötzingen-Sainerholz, Germany
Tel: (+49 2666) 952 00
Fax: (+49 2666) 83 74
e-mail: info@contec-group.com
Web: http://www.contec-group.com

Products

Switch machines; track wiring systems; control systems; lineside cabinets; electrically and petrol-driven rail drilling equipment; rail bonding systems; fastenings.

Services

Civil engineering; cable laying; signal gantry assembly; track and signalling planning; depot design; signalling cable installation.

Contitech Transportbandsysteme GmbH

Clouth Cologne Plant
Niehler Strasse 102-116, D-50733 Cologne, Germany
Tel: (+49 221) 777 36 24; 36 97
Fax: (+49 221) 777 37 00

Key personnel

Managing Director: Norbert Martin

Subsidiary

Tradegal Lda
Av Visconde Valmor 69-5°, Andar, P-1050 Lisbon, Portugal
Tel: (+351 1) 795 90 82; 90 89
Fax: (+351 1) 795 90 97

Products

Sub-ballast mats for vibration control; protective mats (Clouth-ASM®) for waterproof coatings of bridges and structures; rolling rubber springs (Clouth Rollfeder®) for primary and secondary suspension for rail vehicles; resilient track fastenings; elastomeric bridge bearings; mass-spring systems; Clouth Oil-Ex® elastomeric mat to absorb liquid hydrocarbons, such as oil, lubricants of low viscosity, motor fuels and organic solvents.

Cooper & Turner

Sheffield Road, Sheffield S9 1RS, UK
Tel: (+44 114) 256 00 57
Fax: (+44 114) 244 55 29
e-mail: sales@cooperandturner.com
Web: www.cooperandturner.com

Key personnel

Directors: Paul Cook, Alan White

Products

Fish-bolts, track-bolts, screwspikes, crossing-bolts, Renlok locknut, insulated fishplate kits, HSFG bolts and nuts.

UPDATED

Corus Cogifer

Hebden Road, Scunthorpe DN15 8XX, UK
Tel: (+44 1724) 86 21 31
Fax: (+44 1724) 29 52 43
e-mail: info@coruscogifer.com
Web: http://www.coruscogifer.com

Key personnel

Managing Director: Ian Lindsay
Business Development Manager: David Walters

Capabilities

Corus Cogifer is one of the leaders in the supply of railway switches and crossings and related track components in the UK. The company is a 50/50 joint venture between Corus and Vossloh Cogifer. The business designs, manufactures and assembles switches, crossings and railway track layout systems for the railway maintenance and renewals market.

VERIFIED

Corus Rail Products

UK

Moss Bay, Derwent House, Workington CA14 5AE, UK
Tel: (+44 1900) 643 21
Fax: (+44 1900) 84 24 00

France

Commercial
2 avenue du President Kennedy, F-78100 Saint-Germain-en-Laye, France
Tel: (+33 1) 39 04 63 00   Fax: (+33 1) 39 04 63 44

Mill
164 rue du Marechal Foch, F-57705 Hayange Cedex, France
Tel: (+33 3) 82 57 45 04   Fax: (+33 3) 82 57 45 41
e-mail: rail@corusgroup.com
Web: www.muchmorethanrail.com

Key personnel

Managing Director: Jon Bolton
Director: Gerard Glas
Commercial Director, International: Hubert Dabas
Commercial Director, UK: Geoff Suitor
Technical Director: Daniel Boulanger

Background

Corus manufactures rail products at facilities in the UK and France. Supported by comprehensive quality and testing procedures and advanced logistical systems, the business supplies rail networks worldwide. Commercial and technical teams are supported by local representatives in over 85 countries with extensive experience in international trade. Its customers benefit from expertise in high-speed lines, heavy haul, urban transport, logistics and establishing finance packages.

Products and services

The product portfolio includes: heavy rail (flat bottom, bullhead, conductor) in all grades, also head hardened, grooved, stainless steel clad, corrosion resistant, asymmetric, light bridge and crane rail, long welded rail strings up to 220 m, steel sleepers, fishplates and baseplates.

Corus is also able to offer a wide range of railway infrastructure services.

UPDATED

Corus Railway Infrastructure Services

Headquarters
PO Box 298, York YO1 6YH, UK
Tel: (+44 1904) 45 46 00   Fax: (+44 1904) 45 46 01
e-mail: rail@Corusgroup.com
Web: www.muchmorethanrail.com

Key personnel

Managing Director: Jon Bolton
Director, Railway Infrastructure Services:
   David Marsden
Business Development Director: Jeremy Blake
Technical Director: Craig Scott
Projects Director: Bill Clark

Capabilities

Corus provides customers with railway infrastructure services, including: railway design, technical and operations consultancy, modular systems and track maintenance and renewal. It employs over 1,000 people with offices in York, London, Birmingham, Manchester, Rotherham and Doncaster, the business is focused on delivering solutions to enable better performance of the railway.

Corus also manufactures and supplies rail products from facilities in the UK and France.

UPDATED

Cowans Sheldon

The Clarke Chapman Group Ltd, PO Box 9, Salt Meadows Road, Gateshead HE8 1SW, UK
Tel: (+44 191) 477 22 71   Fax: (+44 191) 477 10 09
Web: www.cowanssheldon.co.uk

Key personnel

General Manager: Les Richardson
Business Sales Manager: Martin Howell

Products

Diesel-electric and diesel-hydraulic railway breakdown, general purpose and tracklaying cranes from 12 tonnes to 250 tonnes lifting capacity. Rail delivery and recovery systems and modular maintenance systems.

Refurbishment, life extension and upgrade of existing cranes as well as track maintenance machines.

UPDATED

Créabéton

Créabéton Matériaux SA
Industrie Nord 2, CH-3225 Müntschemier, Switzerland
Tel: (+41 32) 312 98 50
Fax: (+41 32) 312 98 88
e-mail: muentsche@creabeton.ch
Web: http://www.tribeton.ch

Corporate background
Créabéton Matériaux SA is a member of the Vigier group.

Products
Tribeton concrete sleepers for standard and narrow gauge railways; concrete products for railway civil engineering applications.

CXT Rail Products

2420 N Pioneer Lane, Spokane, Washington 99216, USA
Tel: (+1 509) 924 63 00
Fax: (+1 509) 927 02 99
e-mail: info@cxtinc.com
Web: www.lbfoster.com

Key personnel
President: Alec Bloem
Vice-President, Marketing, Engineering:
    Derek Firth
Contact: Desiree Mendoza

Works
2420 N Pioneer Lane, Spokane, Washington 99216
15708 E Marietta, Spokane, Washington 99216
710 E US Highway 30, Grand Island, Nebraska 68801

Grand Island Tie Plant
710 E US Highway #20, PO Box 1808, Grand Island, Nebraska 68801
Tel: (+1 308) 382 54 00
Fax: (+1 308) 382 32 50

Background
CXT Rail Products is a division of L B Foster Company.

Following the award of the Union Pacific Railroad contract, CXT will expand and modernise its Grand Island, Nebraska plant and build a new facility in Tucson, Arizona to accommodate the contract's requirements.

Products
Prestressed concrete sleepers for track and turnouts; prefabricated buildings and precast concrete grade crossing panels.

CXT has developed geometric design capabilities for turnout layouts; for track, with a facility for gauge widening; for tangent sleeper development, and for standard track sleepers.

Contracts
January 2005, CXT was awarded a long-term contract (through 2012) for the supply of prestressed concrete railroad sleepers to the Union Pacific Railroad (UPRR).

Concrete sleepers supplied to the Calgary LRT, MTA Baltimore, Vancouver, Utah, Los Angeles, New Jersey, Denver, Portland and Southern California Regional Rail Authority, UP, BNSF and many other heavy-haul railways.

Standard and curved concrete level crossing panels supplied to the UPRR, Burlington Northern Santa Fe Railway and other mainline railroads through North America as well as light rail transit systems in California, Oregon, Utah and Washington State.

UPDATED

DBT GB Ltd

Hallam Fields Road, Ilkeston DE7 4BS, UK
Tel: (+44 115) 951 25 00
Fax: (+44 115) 932 96 83
e-mail: info@dbtgb.com
Web: www.dbt.de

Products
Developed jointly with Jarvis Fastline Ltd, DBT GB manufactures the RBE-1 rapid ballast excavator, a road-rail crawler-mounted excavator developed primarily to remove spent ballast from the trackbed during renewals and transfer it to rail wagons on an adjacent track via conveyors mounted on the vehicle. Powered by a 150 kW diesel engine, the excavator employs rail wheels to travel to the site of work and is controlled by one operator. The RBE-1 also has bulk material handling applications at freight terminals.

UPDATED

Delimon GmbH

Postfach 102052, D-40011 Düsseldorf, Germany
Tel: (+49 211) 777 40
Fax: (+49 211) 777 42 10
e-mail: info@delimon.de

Background
Delimon GmbH is a member of Vesper Corporation.

Products
Stationary track lubrication systems for all track types, including grooved rail and Vignoles.

Delkor Pty Ltd

75 Hutchinson Street, St Peters, New South Wales 2044, Australia
Tel: (+61 2) 95 50 51 11
Fax: (+61 2) 95 50 56 25

e-mail: delkor@delkor.com.au
Web: http://www.delkor.com.au

Key personnel
Managing Director: Peter Herbert
Technical Manager: Peter Schonstein
Sales Manager: George Stamboulis

Products
Elastic rail fasteners, including noise-reducing fasteners for bridges and tunnels; ballast mats for ballasted track; floating slab systems.

Contracts
The supply of noise-reducing elastic rail fasteners for the Sydney Harbour Bridge, Australia; and rail fasteners for the Tsing Ma Bridge, Hong Kong.

Derby Rubber Products

84 Derby Street, Silverwater 2128, Australia
Tel: (+61 2) 96 48 49 11
Fax: (+61 2) 96 48 47 83
Web: http://www.derby rubber.com.au

Other office
Derby Rubber Products (Europe) BV
c/ Tavro Business Centre, President Kennedylaan 19,
NL-2517 JK, The Hague, Netherlands
Tel: (+31 70) 360 09 79
Fax: (+31 70) 345 51 99

Key personnel
Managing Director: Stephen Sheppard
Technical Manager: Clinton Miller
General Manager: David Martin

Products
Rubber broom elements for ballast regulating machines. Also rolling stock extrusions for doors, windows, suspension, brake and air conditioning systems.

Desec Ltd

FIN-39700 Parkano, Finland
Tel: (+358 3) 448 34 42; (+358 5) 260 95 50
Fax: (+358 3) 448 34 43; (+358 5) 260 94 48
e-mail: desec@vip.fi
Web: http://www.desec.com

Key personnel
Managing Director: Seppo Koivisto
Director International Sales: Einari Venäläinen

Desec Tracklayer (TL) laying turnouts

0553300

Turnout transport wagons   0559121

Products

Turnout replacement machines; trolleys for turnout and track laying; lifting devices for track maintenance; supply of turnout transport wagons; railway cranes.

Electrolux Construction Products UK

Oldends Lane Industrial Estate, Stonedale Road, Stonehouse GL10 3SY, UK
Tel: (+44 1453) 82 03 05; 82 03 06
Fax: (+44 1453) 79 18 31
Web: http://www.partner.industrial.com; http://www.dimas.com

Key personnel
Product Manager: Peter Waldron
Marketing Co-ordinator: Maggie Wakefield

Background
Electrolux Construction Products UK is a member of the Electrolux Group.

Products
Handheld power tools for rail maintenance and installation. The product range includes the Partner K1250 rail models with 119cc/2-stroke petrol-powered units, available in 14 or 16 in versions with a new model which has an integral cut-off control that automatically immobilises the power unit when the rail clamp unit is removed.

DISAB Vacuum Technology AB

Box 170, SE-241 23 Eslöv, Sweden
Tel: (+46 46) 413 55 43 00
Fax: (+46 46) 413 55 43 01
e-mail: info@disab.se
Web: http://www.disab.se

Products
Railvac air-vacuum excavator system for cleaning cable culverts and drainage ditches and other rail infrastructure applications.

Drouard

Parc Saint-Christophe, F-95865 Cergy-Pontoise Cedex, France
Tel: (+33 1) 34 22 50 00
Fax: (+33 1) 34 22 62 29

Key personnel
Director: D Mallet
General Manager: J Lemercier

Products
Tracklaying and associated works (including TGV); manufacture of concrete sleepers; maintenance and renovation of track and ballast for rail, metros and tramways.

Edgar Allen Ltd

PO Box 42, Shepcote Lane, Sheffield S9 1QW, UK
Tel: (+44 114) 244 66 21
Fax: (+44 114) 242 68 26
Web: http://www.edgar-allen.co.uk

Other office
Whitburn Road, Bathgate, West Lothian EH48 2RB, UK
Tel: (+44 1506) 65 23 41
Fax: (+44 1506) 63 13 31

Key personnel
Directors:
  Managing: R A Laird
  Financial: C Murphy
  Operations: D Eyre
Managers:
  UK Sales: D Day
  Export Sales: T Grindle
  Quality Assurance: I Grant

Background
Edgar Allen is part of the Mowlem Group.

Products
Design and manufacture of trackwork, switches and crossings for railways, mass transit, tramways, docks and harbours and steel works.
The company is a major supplier to Railtrack of switches and crossings in manganese and other steels. It also supplies trackwork to North America to AREA specifications and to other countries to UIC specifications.

Edilon BV

Nijverheidsweg 23, PO Box 1000, NL-2031 CN Haarlem, Netherlands
Tel: (+31 23) 531 95 19   Fax: (+31 23) 531 07 51
e-mail: mail@edilon.nl
Web: http://www.edilon.com

Key personnel
Managing Director: A J Houck
Sales Director: W P Schram

Principal subsidiaries
Edilon International BV, St Denis, France
Edilon Corkelast SA, Spain
Edilon GmbH, Munich, Germany

Products
Specialised adhesives and elastomers for permanent way applications: Edilon Corkelast, a resilient pourable elastomer for embedded rail and embedded single block systems; Edilon Dex range, specialised epoxy-based products for timber sleeper preservation and glued insulated rail joints; and Edilon acoustic web blocks for noise reduction from track. In co-operation with the Silent Bridge Group, Edilon has introduced the integrated steel Silent Bridge concept with optimal noise and vibration concepts.

Edmund Nuttall Ltd

St James House, Knoll Road, Camberley GU15 3XW, UK
Tel: (+44 1276) 634 84   Fax: (+44 1276) 660 60
e-mail: headoffice@edmund-nuttall.co.uk
Web: http://www.edmund-nuttall.co.uk

Key personnel
Managing Director: Peter B Brooks

Services
Civil engineering contractors serving both private and public sectors for large- and small-scale projects in many sectors.

Contracts
Include: five-year alliance for infrastructure renewals with Railtrack's North West Zone; involvement in a slope stabilisation contract with Railtrack throughout its Scottish Zone; signalling installation for Railtrack and London Underground, £110 million for extension of Docklands Light Railway, London; various aspects of work on Luton Parkway Station, UK; construction of a large culvert to support a four-track mainline railway (Paddington to South Wales and southwest England).

Eichholz

Eichholz GmbH & Co KG
Bahnhofstrasse 17, D-97922 Lauda-Königshofen, Germany
Tel: (+49 93) 43 50 60
e-mail: info@eichholz.de
Web: http://www.eichholz.de

Services
Track building, platforms, large machines, civil engineering, rail grinding and underground construction. Engaged in major new building and extension projects for Deutsche Bahn and also the high-speed railway lines of Hanover–Berlin, Cologne–Frankfurt and Nuremberg–Munich.

Elektro-Thermit GmbH and Co KG

Chemiestrasse 24, D-06132 Halle, Germany
Tel: (+49 345) 779 56 00   Fax: (+49 345) 779 57 70
e-mail: info@elektro-thermit.de
Web: http://www.elektro-thermit.de

Key personnel
Managing Directors: Dr Henri Cohrt, Dr Frank Kuster

Products
Thermit rail welding equipment and consumables, heating devices and accessories, hydraulic equipment, measuring devices, insulated and non-insulated fishplate joints; Vortok system.

VERIFIED

Enzesfeld-Caro Metallwerke AG

Fabrikstraße 2, A-2551 Enzesfeld, Austria
Tel: (+43 2256) 811 45   Fax: (+43 2256) 813 40
e-mail: austroroll@caro.at
Web: http://www.austroroll.at

Products

Austroroll – switch point rollers.

UPDATED

ESAB AB, Welding Equipment

SE-695 81 Laxa, Sweden
Tel: (+46 584) 810 00   Fax: (+46 584) 41 17 21
e-mail: engineering.laxa@esab.se

Key personnel

Sales Manager: Johnny Sundin
Product Manager: Sylve Antonsson

Products

Fixed and mobile flash-butt welding machines for rail; equipment for welding automation in rolling stock production; equipment for building-up welding of wheels; equipment for hardfacing and repair of rail profiles.

UPDATED

ESCO Equipment Service Co

117 Garlisch Drive, Elk Grove Village, Illinois 60007, USA
Tel: (+1 847) 758 98 60   Fax: (+1 847) 758 98 61
e-mail: escoequip@aol.com

Key personnel

President: Thomas Y Gehr
Executive Vice-President: Tom Dickey

Products

Abrasive cut-off and grinding stone; rail welding; rail saws and drills, rail ultrasonic testing; hydraulic track tools; magnetic pick-up devices; rail and tie tongs; rail fastening machines and maintenance-of-way machines; solar-powered switch machines, hy-rail cranes.

UPDATED

Eurailscout Inspection and Analysis BV

Postbox 349, NL-3800 AH Amersfoot, Stationsplein 325, Netherlands
NL-3818 LE Amersfoot, Netherlands
Tel: (+31 33) 469 70 00   Fax: (+31 33) 469 70 50
e-mail: info@eurailscout.com
Web: http://www.eurailscout.com

Key personnel

Directors: Dr Peter Hanspach, Anton Weel

Other offices

Büro Berlin, Markgrafendamm 24, Haus 16, D-10245 Berlin, Germany
Tel: (+49 30) 29 38 08 50   Fax: (+49 30) 29 38 08 51

Eurailscout GB Ltd
Inspection & Analyses
Derwent House, rtc Business Park, London Road, Derby DE24 8UP, UK
Tel: (+44 1332) 26 21 98

Services

A joint venture formed by GSG Knape Gleissanierung (Germany) and Strukton Railinfra (Netherlands), Eurailscout undertakes inspection, measurement and analysis of track geometry, rail profile, rail surface, and overhead line using a UST 96 inspection vehicle for ultrasonic inspection of tracks with great precision at high speed. It is specifically used to detect rail defects and has been awarded all appropriate approvals. Measurements are carried out at speeds up to 100 kph. An additional feature is the eddy current measuring system with eight special eddy current probes.

Eurailscout also works with hand-held equipment to conduct important measurements on rail geometry and switches and for ultrasonic rail inspection. The SGMT5 is a mobile instrument for the measurement, recording and presentation of the rail geometry and switch parameters. The MT95 ultrasonic hand-held equipment is primarily used to inspect critical spots such as welds, switch points and rail joints. The head checker is a mobile eddy current measurement system which is provided with a GPS-System.

Exel Oy

PO Box 29, FIN-52701 Mäntyharju, Finland
Tel: (+358 15) 346 11
Fax: (+358 15) 346 12 16

International sales and technical support

Kolmark Ltd
Irisviksv 34H2, FIN-02230 Esbo, Finland
Tel: (+358 9) 88 15 81 40
Fax: (+358 9) 88 15 81 45
e-mail: jan.kolster@kolmark.fi

Key personnel

Managing Director: Jan Kolster

Products

Insulated rail joints for jointed and welded track; composite profiles and tubing for power line support.

'EXIM' Verkehrs-, Hafen- und Umwelttechnik GmbH & Co Vertriebs KG

PO Box 1406, D-35004 Marburg, Germany
Tel: (+49 6421) 810 01; 003   Fax: (+49 6421) 853 53
e-mail: eximgmbhuc@aol.com

Key personnel

General Manager: A J Frangoulis

Products

Sleepers, turnouts, rolling stock materials, permanent way equipment, switch point machines, fastenings, glued joints, welding equipment, workshop machines.

NEW ENTRY

FAB-RA-CAST

FAB-RA-CAST™ is a division of Horizon Manufacturing Inc
23820 Lee Baker Drive, Southfield, Michigan 48075, USA
Tel: (+1 248) 354 71 85   Fax: (+1 248) 354 71 85

Key personnel

President: John Cook

Products

Concrete railway crossings systems; rubber pour-in flangeway filler material. Contracts include the Tri-Met light rail system in Portland, Oregon.

Fassetta mécanique

36 boulevard de la Gare, F-13713 La Penne S/Huveaune, Cedex, France
Tel: (+33 4) 91 87 70 30
Fax: (+33 4) 91 87 70 39
e-mail: fasmec@fassetta.com
Web: www.fassetta.com

Key personnel

General and Export Manager: Frederic Fassetta
Technical Sales Manager: Bernard Rousset

Products

Track construction and maintenance equipment including switch relaying equipment. Special machines designed and produced on request. Tamping machines, re-sleepering machines, track geometry, recording and inspection trolleys.

UPDATED

Faur SA

Basarabia Boulevard 256, Bucharest 3, R-73249, Romania
Tel: (+40 1) 255 15 13   Fax: (+40 1) 255 00 71

Key personnel

Development and General Manager: Victor Vieru
Economic Manager: Dumitru Ghinea
Image, Human Resources, Strategy Manager: Livia Niculescu
Sales and Purchasing Manager: Vasile Diaconescu
Production Manager: Mihai Baldea

Products

Track maintenance machines and spare parts for track maintenance machines.

Recent contracts include the supply of track maintenance machines to the Russian Federation.

Together with Plasser & Theurer, Austria, production of eight 09- 32CSM track maintenance machines for SNCFR Romania.

Ferotrack Engineering Ltd

377 Kilburn High Road, London NW6 2QN, UK
Tel: (+44 20) 76 24 01 03   Fax: (+44 20) 76 24 89 79

Works

Willow Vale, Davenham Road, Oakley MK43 7SZ, UK

Key personnel

Managing Director, Export Sales: I O Schwarz
Director, Production and Design: K T Birchall

Products

Electric pad point and crossing heaters; cartridge heaters; clamp lock heaters; control cabinets; plus a major range of 996 Type Ni/Cd rechargeable batteries with integral electronic charger.

Findlay, Irvine Ltd

Bog Road, Penicuik, Midlothian EH26 9BU, UK
Tel: (+44 1968) 67 12 00   Fax: (+44 1968) 67 12 37
e-mail: sales@findlayirvine.com
Web: www.findlayirvine.com

Key personnel

Chairman: Colin Stewart
Managing Director: Colin Irvine
Sales Manager: Tom Findlay

Products

Electronic control systems for railway switch heating and monitoring.

The Icelert 407M controller is a switch heating controller utilising the latest technology designed specifically for rail switch heating. It uses a variety of sensors to determine the possibility of the formation of ice. These include two rail-mounted temperature sensors (one on an unheated section of rail, one on a heated section), a precipitation (rain or snow) sensor and an optional blown-snow detector.

The control unit has a digital display readout which indicates all the adjustable parameters such as 'set point' temperatures, 'delay' on switch-off of heating, 'hysteresis' levels and fault conditions. There are also LEDs to indicate when 'set points' have been reached and when precipitation and faults have been detected. Self-testing, manual override facilities and monitoring outputs have also been incorporated.

The use of two cold rail set points and two hot rail set points (one if precipitation is detected and the other if precipitation is not detected) gives a total of four set points. The Icelert 407M automatically uses the 'wet' set points if precipitation or snow is detected and the 'dry' set points otherwise. This enables considerable cost savings to be made in dry conditions by not switching the heat on until the temperature drops to a lower level than if the precipitation is detected.

Findlay, Irvine also produces a monitoring datalogger system for use in conjunction with the Icelert 407M accepting analogue, digital and 20 mA loop inputs from a wide range of proprietary sensors.

Contracts

Contracts include the supply of Icelert controllers to Polish Railways; and remote switch heating monitoring systems and weather monitoring control units for Network Rail (Railtrack), UK.

UPDATED

First Engineering Ltd

8th Floor, Buchanan House, 58 Port Dundas Road, Glasgow G4 0HG, UK
Tel: (+44 141) 335 34 01   Fax: (+44 141) 335 36 08
e-mail: enquiries@firstengineering.co.uk
Web: www.firstengineering.co.uk

Key personnel

Chief Executive: Janette Anderson
Commercial Director: John Cowie

Background

Formerly known as Scotland Infrastructure Maintenance Co, First Engineering was one of the British Rail Infrastructure Services Units sold off as part of the British Rail privatisation. The company was sold in 1996 to a management buyout team known as TrackAction.

Subsequently the company was acquired by Babcock International Group plc.

Services

Railway infrastructure construction, maintenance, renewals and consultancy, in addition to property management and maintenance service from its facilities divisions.

UPDATED

Framafer

Société Française de Construction de Matériel Ferroviaire
77 rue de la Gare, F-57803 Bening-lès-Saint-Avold, France
Tel: (+33 3) 87 29 22 00   Fax: (+33 3) 87 81 50 63

Key personnel

Sales Manager: Philippe Crovisier

Background

Framafer is a subsidiary of Plasser & Theurer.

Products

Automatic track levelling, tamping and lining machines, ballast cleaners and ballast regulators, track relaying systems, sleeper changing machines, brush-cutters, ditch-cleaners, track measuring cars, flash-butt welding machines, multipurpose track maintenance machines.

UPDATED

Geismar

113 bis avenue Charles-de-Gaulle, F-92200 Neuilly sur Seine, France
Tel: (+33 1) 41 43 40 40   Fax: (+33 1) 46 40 71 70

e-mail: geismar@geismar.com
Web: www.geismar.com

Works

5 rue d'Altkirch, F-68006 Colmar Cedex, France
Tel: (+33 3) 89 80 22 11
Fax: (+33 3) 89 79 78 45
e-mail: colmar@geismar.com

Key personnel

Publicity: Patrick Lambert

Products and services

Tracklaying and maintenance equipment and services. Permanent way tools. Hand-held machinery, including rail drills, saws, disc cutters, grinders, benders, weld shears, tensors, pre-heaters, strikers, descalers, lifters, loaders; sleeper drills, benders, adzers, plug drivers; fastening machines, fishbolters, coachscrewers, impact wrenches, elastic clip and spike inserters and extractors.

Heavy equipment, including gantries for laying and replacing track panels and switches, threaders, slewers, sleeper-changers, tampers and regulators.

Transport and maintenance vehicles, including inspection and flying gang trolleys, heavy-duty track cars, shunters, trailers, and railway excavators.

Turnkey plants and workshop machinery for rail welding, reprofiling or machining, timber sleeper machining and impregnation, steel sleeper reclamation.

Measuring instrumentation, including manual gauges and devices, hand-pushed or self-propelled or onboard electronic systems.

Geismar also supplies equipment to install, maintain and inspect AC and DC overhead catenary.

UPDATED

Gemco George Moss Ltd

PO Box 136, Mount Hawthorn 6016, 461-465 Scarborough Beach Road, Osborne Park, Western Australia 6017, Australia
Tel: (+61 8) 94 46 88 44   Fax: (+61 8) 94 46 34 04

Key personnel

For a full list of personnel, see Gemco George Moss Ltd entry in *Locomotives and powered/non-powered passenger vehicles* section.

Products

Track maintenance machinery, such as resleepering machines, rail-handling cranes, sleeper-handling machines, ballast scarifiers (linear), spike pullers, track jacks; tracklayers; rail flash-butt welding equipment.

The company also offers rail flaw detection, track recording equipment and track management systems.

GERB Vibration Control Systems

GERB Schwingungsisolierungen GmbH & Co KG
Silviastrasse 21, D-45131 Essen, Germany
Tel: (+49 201) 266 04 20   Fax: (+49 201) 266 04 50
e-mail: rail@gerb.com
Web: http://www.gerb.com

Head office and works

Roedernallee 174-176, D-13407 Berlin (Reinickendorf), Germany
Tel: (+49 30) 419 10
Fax: (+49 30) 419 11 99

Products

Floating slab track systems and other products for the control of track and structural noise and vibration for high-speed, conventional, metro and light rail systems.

VERIFIED

Getzner Werkstoffe GmbH

Head Office
Herrenau 5, PO Box, A-6706 Bürs, Austria
Tel: (+43 5552) 20 10   Fax: (+43 5552) 20 18 99
e-mail: sylomer@getzner.at
Web: www.getzner.at/werkstoffe

Key personnel

Managing Director: R Pfefferkorn
Sales Director: P Burtscher
Advertising Manager: Roland Loacker

Associated company

Getzner Werkstoffe GmbH
Nördliche Münchner Strasse
27a, D-82031 Grünwald, Germany
Tel: (+49 89) 693 50 00   Fax: (+49 89) 69 35 00 11

Am Borsigturm 11, D-13507 Berlin, Germany
Tel: (+49 30) 40 50 34 00   Fax: (+49 30) 40 50 34 35

Getzner Werkstoffe GmbH
Middle East Regional Office
Abdul-Hameed Sharaf Strasse, 114, Rimawi Center, Shmeisani, PO Box 961 303, Amman 11196, Jordan
Tel: (+96 26) 560 73 41   Fax: (+96 26) 569 73 52
e-mail: geme@go.com.jo

Nihon Getzner K K
Shinjuku Park Tower, 30th Floor, 3-7-1 Nishi-Shinjuku, Shinjuku-ku, Tokyo 163-1030, Japan
Tel: (+81 3) 53 26 30 30   Fax: (+81 3) 53 26 30 01
e-mail: sylomer@getzner.at

Products

System solutions with elastic materials: Sylomer, Sylodyn and Sylodamp (all polyurethane materials), used for noise and vibration reduction in track and structure construction, sleeper pads, ballast mats for subways, light rail and main line

Sylomer® mass spring system on a main line bridge in Germany   0101700

track; elastic bearings for track slabs; resilient baseplate pads and resilient rail pads.

Getzner Werkstoffe can also supply tailor-made orders to customer's requirements.

UPDATED

Gleisbaumechanik Brandenburg

Am Sudtor, D-14774 Brandenburg-Kirchmoser, Germany
Tel: (+49 3381) 804 44 32
Fax: (+49 3381) 804 43 81
e-mail: gleisbaumechanik@t-online.de
Web: http://www.gleisbaumechanik.de

Products
Self-propelled rail, signalling and catenary inspection and maintenance vehicles.

Recent products include: the Train Control Testcar (TCT) for testing European Train Control Systems installations; and the Series 711.1 catenary inspection and maintenance vehicle for high-speed lines.

Series 711.1 self-propelled catenary inspection and maintenance vehicle in display at InnoTrans 2002 (Ken Harris) 0536546

Gottwald Port Technology GmbH

PO Box 18 03 43, D-40570 Düsseldorf, Germany
Tel: (+49 211) 710 20   Fax: (+49 211) 710 26 51
e-mail: info@gottwald.com
Web: www.gottwald.com

Key personnel
Chief Executive Officer: Dirk Kiessling
Chief Technical Officer: Dr Mathias Dobner

Background
Formerly Mannesmann Dematic AG Gottwald, the company changed its name in 2002 to Gottwald Port Technology GmbH, and is owned by the investment company Kohlberg Kravis Roberts & Co (KKR) and Siemens.

Products
Manufacture and supply of railway cranes for single- and multi-track lines including: breakdown cranes with telescopic and fixed booms; track-laying and bridge-building cranes; universal cranes.

The main features of Gottwald's three varieties of railway crane are: high lifting capacities with moderate crane tare weight; optimum transport and job operation dimensions; wide working ranges taking into account maximum wheel and

Gottwald GS 130.06TT crane, Germany *NEW*/0585259

axle loads; ergonomic operating conditions and operating safety; low maintenance requirements; Gottwald bogies provide high running qualities in train formation and under own power; high towing speed for fast arrival on site; crane travelling self-propelled speed up to 100 km/h with the model GS 150.14 TR; fully hydrostatic drive for precise and powerful travel and working motions.

UPDATED

GrantRail

1 Carolina Court, Lakeside, Doncaster DN4 5RA, UK
Tel: (+44 1302) 79 11 00   Fax: (+44 1302) 79 12 00
e-mail: sales@grantrail.co.uk
Web: http://www.grantrail.co.uk

Key personnel
Chief Executive: Gren Edwards

Background
GrantRail is a joint venture between the Corus Rail Group and Volker Stein Rail and Traffic.

Associate companies
ALH Rail Coatings Ltd (50 per cent)
Total Power Solutions (50 per cent)

Capabilities
Construction, renewal and maintenance of LRT, metro, underground and heavy rail systems. Complete systems are offered for LRT projects, covering planning, supply of rail and installation.

Recent investments in equipment include a Kirow 810 C (UK) crane, three Matisa B45 tamping machines and a Matisa R24 ballast regulator.

Contracts
Major customers are Network Rail (Railtrack), London Underground, Manchester Metrolink and Midland Metro. As well as fulfilling numerous industrial and freight facilities contracts, GrantRail has also developed its own signalling and electrification divisions.

Greenwood Engineering

H J Holst Vej 3-5C, DK-2605 Brøndby, Denmark
Tel: (+45) 36 36 02 00   Fax: (+45) 36 36 00 01
e-mail: trine@greenwood.dk
Web: www.greenwood.dk

Products
Development and production of measuring equipment including the MiniProf Wheel for measuring wheel profiles, the MiniProf Switch for measuring railhead profiles in switches and crossings and the MiniProf Brake for measuring brake discs.

UPDATED

MiniProf Rail, measuring railhead profile 0134496

GrantRail ballast regulator 0561661

Grinaker Duraset

PO Box 751752, Gardenview 2047, South Africa
Tel: (+27 11) 454 28 15   Fax: (+27 11) 454 17 13
e-mail: chris@gcl.co.za
Web: http://www.gcl.co.za/branches

Works
PO Box 365, Brakpan 1540, South Africa
Tel: (+27 11) 813 23 40   Fax: (+27 11) 813 42 22

Key personnel
Executive Director: J C Havinga
Managing Director: C A Visser
Product Manager: K Burger

Products
Prestressed concrete railway products including sleepers, level crossing slabs, turnouts and electrification masts.

Guided Ultrasonics (Rail) Ltd

17 Dover Beck Close, Nottingham NG15 9ER, UK
Tel: (+44 1623) 49 10 93   Fax: (+44 1623) 49 10 93
e-mail: info@gu-rail.com
Web: www.guided-ultrasonics.com/rail

Key personnel
Managing Director: M Russell

Products
G-Scan rail flaw detection system employing low-frequency ultrasonic guided wave technology. The system is suited to testing applications that include: aluminothermic welds; flash-butt welds; wheelburns and squats; areas of base corrosion including level crossings and tunnels; and gauge corner cracking and rolling contact fatigue.

NEW ENTRY

Hall Rail

Lyons Industrial Estate, Hetton-le-Hole DH5 0RF, UK
(+44 191) 526 21 14   Fax: (+44 191) 517 01 12
e-mail: enquiries@hallrail.co.uk
Web: www.hallrail.com

Key personnel
Managing Director: Reg Hall
Operations Director: Russell Tapping
Business Development Manager: Mike Andrews
Procurement and Safety Manager: Graeme Kitto
Production Manager: Trevor Canning
Production Support and Quality Approval
Manager: Ashley Hodgson

Products
Track products, including: switches and crossings for main line, urban, industrial and narrow gauge railways; buffer stops; components for switches and crossings; AWS magnets.

Contracts
Hall Rail has supplied rail operators and infrastructure maintenance companies in Australia, Ireland, Nigeria, Sudan and Thailand, as well as in the UK.

NEW ENTRY

Hanning & Kahl GmbH & Co KG

Rudolf Diesel Strasse 6, D-33818 Oerlinghausen, Germany
PO Box 1342, D-33806 Oerlinghausen, Germany
Tel: (+49 5202) 70 76 00   Fax: (+49 5202) 70 76 29
e-mail: info@huk.hanning.com
Web: www.hanning-kahl.de

Key Personnel
General Manager: Wolfgang Helas
Brake Division Manager: Dietrich Radtke
LRT Division Manager: Christian Schmidt

Hanning & Kahl's point setting mechanism
0088387

Service Division Manager: Peter Spilker
Sales Manager, Brakes: Jürgen Stammeier
Sales Manager, LRT: Hans-Joachim Pässler
Sales Manager, Services/LRT: Joachim Zehn
Sales Manager, Services/Brakes: Martin Epp

Products
LRT Division: points mechanisms for all gauges and types of rail with magnetic, motor or electrohydraulic drive; manual setting mechanisms. Point setting mechanisms are also available with a tongue detector and mechanical double-interlocking for tongues in open and closed positions.

Point controllers, depot controllers, signalling systems for point controllers, level crossing safety devices, single-line track safety devices, mass detectors, vehicle reporting systems, radio control, electronic data recorders and accessories including guard rail tie bars, rail boxes and contact systems.

Service Division: Services and testing and measuring equipment for point setting mechanisms and controllers, signalling installations, TWC systems.

Contracts
Contracts include the supply of point-setting mechanisms to Calgary (Canada); Salt Lake City, Dallas, San Diego, San Jose (USA); Manchester, Birmingham, Croydon and Sheffield (UK); Rome, Turin, Milan (Italy); Melbourne (Australia); Hong Kong; and cities in Germany, Switzerland, Austria, Belgium, the Netherlands, Norway, Sweden and Finland.

UPDATED

Harrington Generators International

Ravenstor Road, Wirksworth, Derbyshire, DE4 4FY, UK
Tel: (+44 1629) 82 42 84   Fax:(+44 1629) 82 46 13
e-mail: sales@harringtongen.co.uk
Web: http://www.harringtongen.com

Key personnel
Managing Director: Peter Harrington
Sales Manager: Barry Kimber

Products
Power generators for welding applications and audible track warning systems.

Harsco Track Technologies Ltd

2401 Edmund Road, Box 20, Cayce-West Columbia, South Carolina 29171-0020, USA
Tel: (+1 803) 822 91 60   Fax: (+1 803) 822 81 07
e-mail: rnewman@harscotrack.com
Web: www.harscotrack.com

Harsco rail grinder
1067230

Harsco 20-stone grinder
1067229

Harsco track renewal system 1067233

Harsco new track construction machine **NEW**/1030859

Harsco's RGH20C rail grinder **NEW**/1030861

Harsco stone blower tamper **NEW**/1030863

Key personnel
President: G Robert Newman
Vice-President, Sales and Marketing: Donald Benza
Director, Sales North and South America:
Jonathan Reilly
Senior Director, Contracting: Stephen Byers

Subsidiaries
Australia
Harsco Track Technologies, Harsco Corporation
4 Strathwyn Street, PO Box 5287, Brendale,
Queensland 4500, Australia
Tel: (+61 7) 32 05 65 00   Fax (+61 7) 32 05 73 69
Managing Director: Kim Harley
Sales Manager: G A Twilley
Service Manager: P R Hibberson

UK
Harsco Track Technologies
Unit 1, Chewton Street, Eastwood, Nottingham
NG16 3HB
Tel: (+44 1773) 53 94 80   Fax: (+44 1773) 53 94 81
e-mail: httuk@harsco.com
Director Quality and Safety Case: L A Withers
Stoneblower Operations Director: A Wardle
Grinding Operations Manager: S Gear

USA
Harsco Track Technologies, Harsco Corporation
415 North Main Street, PO Box 415, Fairmont,
Minnesota 56031-0415
Tel: (+1 507) 235 33 61   Fax: (+1 507) 235 73 70
200 South Jackson Road, Ludington, Michigan
49431
Tel: (+1 231) 843 34 31   Fax: (+1 231) 843 48 30

Background
Harsco acquired Pandrol Jackson in 2000.

Products/Services
A single source supplier for railway track
maintenance equipment. Hy-Rail® Road-Rail guide
wheel attachments, production/switch and spot
tampers, specialised utility track vehicles, switch
and crossing undercutters, production and spot
tie exchangers, spike drivers/pullers, tie cranes,
tie plugging machines, track construction and
renewal equipment and rail grinding equipment.
Harsco Track Technologies also provides contract
services specialising in track renewal, tracklaying
and rail grinding. The Tie Masters Service Program
performs tie renewal operations on a per tie contract
basis. Remanufacturing Services provides in-plant
remanufacturing and upgrading of customer
machines.

Contracts
Indian Railways, track renewal trains; rail grinder,
Mauritania; 2nd switch and crossing rail grinder,
Germany; narrow gauge tamping machine, Peru;
12 crane cars for New York City Transit Subway
System, USA; three-year agreement to operate and
maintain the HTT rail grinders previously sold to
Network Rail in the UK.

UPDATED

Harsco Track Technologies Ltd

Harsco Track Technologies
Unit 1, Chewton Street, Eastwood, Nottingham
NG16 3HB
Tel: (+44 1773) 53 94 80   Fax: (+44 1773) 53 94 81
e-mail: httuk@harsco.com

Key personnel
Director Quality and Safety Case: L A Withers
Stoneblower Operations Director: A Wardle
Grinding Operations Manager: S Gear

Background
Now part of Fairmont Tamper, Harsco Track
Technologies previously traded as Permaquip.

Products
Design and manufacture of tools, plant and vehicles
for infrastructure maintenance and construction.
Products, including those of Fairmont Tamper, are
available for sale, lease or hire.

Harsco Track Technologies high-capacity trolley supplied to New Jersey Transit   0006290

Land Rover converted to road/rail by Harsco Track Technologies   0006291

Sleeper handling: hand tools and plant for moving, machining and changing wooden and concrete sleepers, including a portable sleeper squarer/spacer.

Rail handling: Ironman and powered rail pullers for the mechanised handling of rail and Continuously Welded Rail (CWR) within or outside track possessions; rail threaders, rail joint straighteners and specialised rail carrying vehicles.

Stressing CWR: hydraulic tensors for all-weather control of stressing CWR, including obstructionless types with lightweight power packs; weld trimmers for track welding and accessories; self-contained mobile welding workshops for CWR maintenance.

Track maintenance and construction: tampers, ballast regulators, maintenance grinders, complete track relaying trains; Permaclipper for production rate installation and removal of Pandrol fastenings, including the Mk V Permaclipper adaptable for use with Pandrol Fastclips; slewing machines, lightweight Ironman for manual movement of points and crossings; hydraulic spike extractors.

Materials and personnel transport: a wide range of vehicles ranging from manual trolleys to road/rail trucks; on-track vehicles including tug units and general purpose Tramms; specialist vehicles for metro systems and customised personnel carriers.

Structures and overhead line maintenance: access platforms mounted on road/rail or on-track vehicles, featuring cantilever extension, creep control for driving from the elevated platform, electrical/hydraulic Power Take Off (PTO) and auxiliary lighting.

Road/rail trucks: a wide range of road vehicles can be adapted for road/rail use with the Fairmont hi-rail system of bolt-on rail guidance wheels.

Fittings for road/rail trucks include cranes, access platforms, tipper bodies, rail carrying frames, drum carriers, crew cabs and trailers. General purpose road/rail trucks available for hire in the UK, including a 17 tonne GVW unit with crane and rail carrying facility.

Contract support: spares and service support; operator and contract support for specialist contract work such as stressing, joint straightening, weed control, ballast cleaning and sleeper squaring.

UPDATED

Harsco Track Technologies Pty Ltd

PO Box 5287, 4 Strathwyn Street, Brendale, Queensland 4500, Australia
Tel: (+61 7) 32 05 65 00   Fax: (+61 7) 32 05 73 69
e-mail: sales@harscotrack.com.au
Web: www.harscotrack.com

Key personnel
Managing Director: D K Harley
Sales Manager: G A Twilley
Contract and Service Manager: P R Hibberson

Products
The complete range of Fairmont, Tamper and Jackson track maintenance equipment including tamping machines, stoneblowers, ballast regulators, ballast undercutters, sleeper renewal equipment, track renewal and construction trains, rail grinders, hy-rail equipment, track and overhead geometry vehicles, utility vehicles and a comprehensive range of support equipment.

UPDATED

Heat Trace Limited

Tracer House, Cromwell Road, Bredbury, Stockport SK6 2RF, UK
Tel: (+44 161) 430 83 33   Fax: (+44 161) 430 86 54
e-mail: webenquiry@heat-trace.com
Web: http://www.heat-trace.com

Key personnel
Chairman: N Malone
Managing Director: A Pearson
Director of Technology: J O'Connor

Products
Rail and switch point heating systems.

Hindusthan Engineering & Industries Ltd (HEI)

Mody Building, 27 Sir RN Mukherjee Road, Calcutta 700 001, India
Tel: (+91 33) 22 48 01 66
Fax: (+91 33) 22 48 19 22; 22 43 56 07
e-mail: hindus@cal2.vsnl.net.in

Works
Tiljala Plant (TP)
38 Tiljala Road, Calcutta 700 039, West Bengal, India

Key personnel
Executive Director: M L Lohia

Products
Turnouts (all types and angles), crossings (swing nose, welded and fabricated), diamond (single and double slips), switches (symmetrical and asymmetrical), derailing switches, expansion joints, stretcher bars, clamp locks and spring setting devices for switches, steel sleepers, track fittings, guard rails, cut spikes, track bolts, frog and switch bolts, contract design and engineering services.

UPDATED

Henry Williams Limited

Dodsworth Street, Darlington DL1 2NJ, UK
Tel: (+44 1325) 46 27 22   Fax: (+44 1325) 24 52 20
e-mail: info@hwilliams.co.uk
Web: http://www.hwilliams.co.uk

Key personnel
Managing Director, Forgings and Fabrications:
 Barrie Hope
Sales Engineers: Lee Taylor, Bryan Blareau,
 Jon Heslop

Products
Manufacture of a range of track and signalling products. Forged and fabricated railway equipment including forged, and rolled fishplates; rail anchors and clips; switch clamps; lineside apparatus cases; signalling cranks and rods; switch levers; track tools and gauges; maintenance escalator trolleys; and specialist fabrications and machined parts.

Design, manufacture and installation of level crossing systems; small signalling schemes and domino signalling panels.

UPDATED

For details of the latest updates to *Jane's World Railways* online and to discover the additional information available exclusively to online subscribers please visit
jwr.janes.com

H F Wiebe GmbH & Co KG

Im Finigen 8, D-28832 Achim, Germany
Tel: (+49 4202) 98 70
Fax: (+49 4202) 98 71 00
e-mail: info-wiebe-achim@wiebe.de
Web: http://www.wiebe.de

Key personnel
Chairman of the Board: Hermann Wiebe, Sen
Managing Directors: Thorsten Bode,
    Norbert Modersitzki Werner Zitz

Services
All aspects of new construction, conversion or maintenance whether for rolling stock, platform facilities or stations; GBM track construction machines; SDS safety with latest equipment in FA and AWS (track possession and automatic warning systems) engineering; BLP construction supervision and logistics.
    Wiebe also undertakes building engineering, tunnel construction and building construction.

Hi-Force Hydraulics Ltd

Bentley Way, Daventry NN11 5QH, UK
Tel: (+44 1327) 301 00 00
Fax: (+44 1327) 70 65 55
e-mail: sales@hi-force.com
Web: http://www.hi-force.com/hi-force

Key personnel
Directors: Chris Jones, John Taylor
Sales Manager: Ronnie Birch

Products
Petrol-engined hydraulic pumps for rail stressing equipment, hydraulic rail joint pusher for insulated joint maintenance, hydraulic nut splitters, wagon rerailing equipment, low-height telescopic jacks, hydraulic tools, presses, pumps and cylinders.

VERIFIED

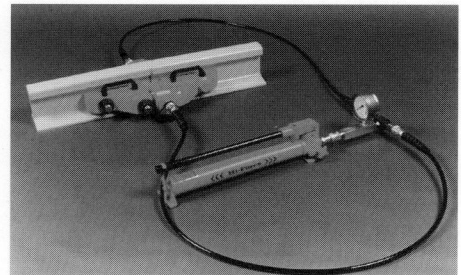

Hi-Force Hydraulics HRJP65 rail joint pusher
0064325

Hioma-aine Oy

PO Box 133, FIN-06151 Porvoo, Finland
Tel: (+358 19) 265 40 00
Fax: (+358 19) 265 40 82
e-mail: abrasives@hioma.fi
Web: www.hioma.fi

Key personnel
Managing Director, Exports & Marketing:
    Stefan Nymark

Main works office
Jakarintie 451, FIN-07320 Jakari, Finland

Products
Grinding wheels in resin bond for rail profiling and grinding of rail joint weldings.

Contracts
Recent contracts include: VR-Rata Oy, Finland, 2002; Banverket, Sweden, 2002; Verktoy og Maskin AS, Norway, 2002.

UPDATED

Holdfast Level Crossings Ltd

Brockenhurst, Chedworth, Cheltenham, GL54 4AA, UK
Tel: (+44 1242) 57 88 01   Fax: (+44 1285) 72 07 48
e-mail: peter@railcrossings.co.uk
Web: http://www.railcrossings.co.uk

Key personnel
Managing Director: Peter Coates Smith

Products
Full depth rubber level crossing systems to suit any rail, sleeper and situation. Features include: very fast and low cost installation; high antiskid values; stable and durable panels.

Holland Company LP

1000 Holland Drive, Crete, Illinois 60417-2120, USA
Tel: (+1 708) 672 23 00   Fax: (+1 708) 672 01 19
e-mail: postmaster@hollandco.com
Web: www.hollandco.com

Key personnel
President: Phil Moeller
Vice-President Rail Mechanical Group: Len O'Kray
General Manager Track Testing Services:
    Robert Madderom
International Sales Manager: Billy Hedrick
General Manager Mobile Welding Operations:
    Mark Rovnyak
Controller: Frank Francis
General Manager Equipment Division:
    Robert Norby
General Manager, MOW Sales: Kevin Flaherty

Products
Sales and contracting of electric flash-butt welding personnel and equipment; rail and road mobile welders, portable on-site welding plants; track measurement services, including geometry, rail wear and track strength measurement. Products also include the Intelliweld® fully digitised flash-butt welding control system. Holland can supply all support equipment or provide a turnkey operation.
    Supply of rail welding equipment, rail and Continuous Welded Rail (CWR) handling and processing equipment, track testing services equipment and railroad maintenance-of-way equipment. Products include Holland's mobile welder in-track flash-butt welding machine, containerised welding units, railpullers, welding plant equipment, rail grinders and polishers, CWR unloader units, railtrains, pick-up units and TrackSTAR®.

UPDATED

Huck International Inc

Huck Fasteners-HQ
PO Box 27207, Tucson, Arizona 85726, USA

UK address
Huck International Ltd
Unit C, Stafford Park 7, Telford TF3 3BQ
Tel: (+44 1952) 29 00 11   Fax: (+44 1952) 29 04 59
Vice-President: John Coles

Key personnel
Executive Vice-President, International Operations:
    Robert S Levine

Products
Fastening systems, including lockbolts for use in rail joints and rolling stock.

IAD Rail Systems

63 + 64 Gazelle Road, Weston-super-Mare, BS24 9ES, UK
Tel: (+44 1934) 42 70 00
Fax: (+44 1934) 42 70 20
e-mail: head-office@iadrailsystems.com
Web: www.iadrailsystems.com

IAD Rail Systems' HPSA switch actuation
NEW/1110185

Key personnel
Managing Director: Trevor Brown
Director of Marketing: Mick Ledger

Background
IAD Rail Systems is the division of Claverham Group Ltd supplying the needs of the rail industry. The company is a wholly owned subsidiary of Hamilton Standard Corporation, itself, part of the United Technologies Corporation.

Products
Switch and crossing actuation systems. The company developed its High Performance Switch Actuator (HPSA), which incorporates reliable actuation, locking and detection technologies in close co-operation with the UK rail industry. The expertise developed was extended to in-sleeper torsional back-drives to create the company's PowerLink backdrive, which when combined with HPSA provides a high reliability total system. Suitable for line speeds of up to 250 km/h, the system is claimed to be easy to install and to require no scheduled maintenance. Through tamping is also possible.
    The first production installation of HPSA took place in the UK at a location in south London on Railtrack's Southern Zone late in 2000. Since then over 200 systems have been installed.

UPDATED

Interlok Bahnconsulting Schmidtendorf

Bruchsaler Strasse 3, D-10715, Berlin, Germany
Tel: (+49 177) 302 42 53   Fax: (+49 177) 993 02 42 53
e-mail: office@interlok.info
Web: http://www.interlok.info

Key personnel
General Sales Manager: Hermann Schmidtendorf
General Manager Rail Department: Uwe Liedecke

Products
Remanufactured, modernised standard gauge freight cars (UIC, RIV) mainly from former Eastern Europe, sale, rent, such as ballast cars, dump cars Fakks 418Vg, platform cars; remanufactured modernised standard gauge passenger/sleeping/dining coaches; rails, switches, sleepers, second-hand for standard gauge; second-hand and new for narrow gauge; second-hand railway maintenance machines. Trade agent for Interlok Ltd and Kaniewski ZUR, Poland, Mitteldeutsche Gesellschaft für Metallhandel und Anlagenverwertung, Germany.

Contracts
Second-hand rail for Volkswagon (VW) Transport Ltd, Germany.

Interep SA

rue de l'Industrie, F-43110 Aurec/Loire, France
Tel: (+33 4) 77 35 20 21   Fax: (+33 4) 77 35 26 17
e-mail: sales@interep.fr
Web: www.interep.fr

Key personnel
Managing Director: Dr Francis Joachim
Sales and Marketing Manager:
    Philippe Charbonnier
Research and Development Manager:
    Jean-Philippe Montagnon
Financial Manager: Patrice Cusin
Technical Manager: Laure Walter

Products

Microcellular rubber foams to reduce vibrations, for use as a ballast mat or under the sleeper for conventional ballasted tracks (marketed under the name Caoutchouc Mousse); also pad under block or baseplate for non-ballasted tracks.

Contracts

Microcellular mats under floating slabs have been supplied for tram lines in Turin, Italy. Microcellular rubber pads for non-ballasted tracks have been supplied to Bilbao, Spain, RATP Paris (Eole and Météor) and Athens, Greece.

Developments

A microcellular rubber mat with advanced damping properties Type 43–45 is suitable for metro systems. An in-house dynamic test machine for controls and simulation has been developed.

UPDATED

IPA

Industria Prefabbricati E Affini
Via Provinciale per Trescore, I-24050 Calcinate (BG), Italy
Tel: (+39 035) 442 30 77   Fax: (+39 035) 442 32 05

Works

Via Don P Bonetti 45, I-24060 Gorlago (BG), Italy
Tel: (+39 035) 95 10 66   Fax: (+39 035) 95 24 14

Key personnel

Executive: Enzo De Biasio

Principal subsidiary

Ipabras
Via Oswaldo Pinto Martins 260, 06900 Embû-Guacu, São Paulo, Brazil

Products

Prestressed concrete sleepers for track and switches; prestressed concrete slabs for track, switches, bridges, level crossings and tramways; noise barriers.

JacksonEve Infrastructure Services

West Offices, City Business Centre, Station Rise, York YO1 6HT, UK
Tel: (+44 1904) 61 33 61   Fax: (+44 1904) 61 29 36
Web: www.jacksoneve.co.uk

Key personnel

Managing Director: Graham Reid
Business Development Manager:
   Chris Buckingham

Corporate background

A member of The Peterhouse Group, JacksonEve Infrastructure Services was formed in 2001 by combining the activities of Jackson Rail (civil engineering and permanent way contracting), Eve NCI (cable installation), Eve Rail (signalling and power networks) and Eve Utility Services (utilities infrastructure).

Services

Civil engineering, permanent way maintenance and renewal, signalling, traction power supplies, telecommunications and cable installation.

NEW ENTRY

JAFCO

JAFCO Tools Ltd
St Paul's Road, Wood Green, Wednesbury WS10 9QX, UK
Tel: (+44 121) 556 77 00   Fax: (+44 121) 556 77 88

Key personnel

Managing Director: Jane Antill

Products

Track maintenance hand tools, both insulated with fibreglass handles and non-insulated. Supplier to the major UK track operators/contractors. Registered to ISO 9002.

Jarret

198 avenue des Grésillons, F-92602 Asnières Cedex, France
Tel: (+33 1) 46 88 16 20   Fax: (+33 1) 47 90 03 57
e-mail: contact@jarret.fr

Key personnel

Chairman and General Manager: Bruno Domange
Managing Director: Antoine Domange

Products

Shock-absorbers for bridge protection; protection against earthquakes for railway bridges, end-of-track buffers.

Jarvis Rail

Jarvis House, Toft Green, York Y01 6JZ, UK
Tel: (+44 1904) 71 27 12   Fax: (+44 1904) 71 37 10
e-mail: marketing@jarvis-uk.com

Key personnel

Managing Director: Tony Cunningham
Chief Operating Officer: Kevin Hyde
Commercial Director: Martin Brazier
Business Development Director: Doug Gillespie

Services

Responsible for track renewals and maintenance, Jarvis is one of the largest railway engineering organisations in the UK with a workforce of over 5,000. It owns approximately 60 per cent of the UK's on-track plant. Regional offices across the UK are supported by a network of over 70 depots at key points on the infrastructure.

Contracts

Contracts include Railtrack maintenance work contracts for Central, East Coast Main Line and Liverpool, Merseyside and North Wales areas involving some 3,530 km of railway infrastructure and the track upgrading of the West Coast Main Line, valued at £550 million.

JC Bamford Excavators Ltd

JCB
Rocester ST14 5JP, UK
Tel: (+44 1889) 59 03 12   Fax: (+44 1889) 59 34 55
e-mail: gordon.henderson@jcb.com
Web: http://www.jcb.com

Key personnel

Director, UK Sales: Gordon Henderson

Products

JCB JS175W Roadrail road-rail vehicle. The 22 tonne unit features a heavy duty JS200W chassis to support independently driven wheels and four independent stabilisers. Lifting capacity with stabilisers deployed is 4,400 kg at 4.5 m reach and over 3,200 kg without stabilisers. Maximum rail speed is 23 km/h. Attachments which can be fitted to the Roadrail's light duty, triple-articulated boom include: rail 'indig' bucket for excavation between rails; hedge flails; ballast brushes; rail thimbles; hydraulic rail beams; and rail clip insertion and removal machines.

JEZ Sistemas Ferroviarios

Arantzar s/n, E-01400 Llodio, Alava, Spain
Tel: (+34 94) 672 00 92
Fax: (+34 94) 672 00 92
e-mail: infor@jez.es
Web: http://www.jez.es

Key personnel

President: Txaber de Errazti

Background

JEZ Sistemas Ferroviarios is a member of the VAE group of companies.

Products

Specialises in points and crossings including high-speed turnouts.

Kalyn Siebert

Kalyn Siebert Inc
PO Box 758, Gatesville, Texas 76528, USA
Tel: (+1 817) 865 72 35
Fax: (+1 817) 865 72 34
e-mail: wesc@kalyntx.com
Web: http://www.kalynsiebert.com

Key personnel

Sales Manager: Wes Chandler

Products

Road trailers for the transport of rail vehicles, with hydraulically operated folding goose-necks for loading and unloading.

KAGO easily mounted electrical rail contact with vibration-resistant cable fixing
0093216

A Kaufmann AG

Pilatusstrasse 2, CH-6300 Zug, Switzerland
Tel: (+41 41) 711 67 00   Fax: (+41 41) 859 16 01
e-mail: info@kago.com
Web: www.kago.com

Products

KAGO specialist engineering products for railways, including a complete range of non-screwed rail contact clamps, discs and strips for electrical rail connections (return current, signalling circuits, earthing), cable fastenings for rails and sleepers, special welding electrodes for copper welding; complete range of screwing, welding and grounding fittings; self-tapping sleeper screws for concrete, steel or wooden sleepers; sleeper spring clips; high-voltage insulations.

Recent development: KAGO axle counter box.

Services

Track returns and earthing wires; earthing poles; heavy-duty and special mountings and drillings; suspensions for radiating cables.

UPDATED

Kelsan Technologies Corporation

1140 West 15th Street, North Vancouver, British Columbia V7P 1M9, Canada
Tel: (+1 604) 984 61 00   Fax: (+1 604) 984 34 19
e-mail: info@kelsan.com
Web: www.kelsan.com

Key personnel

Group Vice-President: Richard Jarosinski

Background

Kelsan Technologies Corporation was acquired by US based Portec Rail Products, Inc in 2005.

Subsidiaries

Kelsan Technologies Corp, Australia and Asia
3/1 Trenoweth Street, Brunswick, Victoria 3056, Australia
Tel: (+61 3) 93 83 38 50   Fax: (+61 3) 93 76 66 55
e-mail: cdale@kelsan.com

Dispotel, France
12 rue d'Oran, F-75018 Paris, France
Tel: (+33 1) 42 51 14 14   Fax: (+33 1) 42 57 10 97
e-mail: info@dispotel.fr
Web: www.dispotel.fr

Kelsan Technologies Corp, Europe (Stone Office)
Unit 1, 14G-Whitebridge Estate, Stone, Staffordshire ST15 8LQ, UK
Tel: (+44 1785) 28 66 11   Fax: (+44 20) 85 80 98 44

Kelsan USA
PMB # 158, 4756 University Village Plaza NE, Seattle, Washington 98105-502, USA
Tel: (+1 604) 984 61 00   Fax: (+1 604) 984 61 02

Products

Keltrack® water-based friction modifier to provide positive friction between wheel and rail, reducing noise, corrugation growth, lateral forces, rail and wheel wear. The product is designed for application using the Portec Rail Products Inc Protector IV trackside top-of-rail applicator. An interlocking solid stick lubricant is used to extend flange life.

Kelsan solid stick LCF lubricant, an interlocking lubricant used to extend flange life.

UPDATED

Kershaw

Kershaw Manufacturing Co
PO Box 244100, Montgomery, Alabama 36124, USA
Tel: (+1 334) 215 10 00   Fax: (+1 334) 215 75 51
Web: http://www.kershawusa.com

Key personnel

Vice-President and Chief Operating Officer:
  Greg Valley
Vice-President of Sales and International Sales
  Manager: Phil Brown

Background

Kershaw Manufacturing Co is a division of Progress Rail Services Corporation.

Products and services

Established in 1944. Design and manufacture of machines for 914; 1,000; 1,067; 1,435; 1,520; 1,600 and 1,676 mm track gauges: ballast regulators (120 to 275 hp); ballast undercutters/cleaners (200 to 750 m³/h); machines for sleeper or rail replacement; snow/ice and sand removers; one-man operated railcar-loading-ramps (20 t/axle); road/rail locomotive cranes (60 to 150 t); on/off-track tree, high branches and brush cutting/clipping vehicles; customised motor cars.

Kier Rail

Tempsford Hall, Station Road, Sandy, Bedfordshire SG19 2BD, UK
Tel: (+44 1767) 64 01 11   Fax: (+44 1767) 64 17 62
e-mail: matt.crabtree@kier.co.uk
Web: http://www.kier.co.uk

Key personnel

Chairman: Paul Sheffield
General Manager: Matthew Crabtree

Works

Euston Station, Barnby Street, London NW1 2RS, UK

Background

Kier Group. Kier Rail is a division of Kier Construction.

Products

Management and construction of railway-related projects including trackwork, signals, earthworks, mechanical and electrical works, land remediation, catenary works, building and civil engineering and planned maintenance.

Contracts

Contracts include: Railtrack station regeneration programme, Southern Zone, in an extended management contract, 1996–2000; West Coast Traincare including upgrade of maintenance depots on West Coast Main Line for tilting trains, 2000. CTRL contracts 103, 250 and 303, Greenburn Open Cast Coal Rail Link and Poole Station Redevelopment.

UPDATED

KIHN SA

17 rue de l'Usine, L-3701 Rumelange, Luxembourg
Tel: (+352 56) 477 11   Fax: (+352 56) 58 54
e-mail: kihn@pt.lu

Key personnel

Manager: Jean Pierre Allegrucci
Export Sales Engineer: Jose Chartier

Products

Engineering and supply of turnouts, points and crossings, crossovers and junctions, trackwork combinations, monobloc and welded frogs, diamond crossings, expansion joints, glued insulated joints and special layouts for urban transport systems, main line railways and industrial network.

K Industrier AB

Stora Varvsgatan 14, SE-211 19 Malmö, Sweden
Tel: (+46 40) 34 80 00   Fax: (+46 40) 34 87 75
Web: www.kindustrier.se

Key personnel

President: Björn Widell
Executive Vice-Presidents
  Engineering: Peter Linde
Production: Karl-Gustav Andersson
Head of Finance: Nils-Arne Nilsson

Background

K Industrier AB formerly traded as Kockums Industrier AB.

Products

Include the new MPSV-2 multipurpose vehicle. It is a flexible vehicle, equipped with a quick-lock system for interchangeable cargo platforms. It can be adapted for different purposes such as transportation of materials and staff as well as overhead line work and clearing activities on or around the track area. A turntable allows 180° rotation, easily controlled by the driver. The MPSV-2 can pull 400 tonnes of wagons with up to 18 axles. It is powered by a 350 kW diesel engine and has a hydraulic power unit for other equipment.

UPDATED

Kirow Leipzig AG

Spinnereistrasse 13, D-04179 Leipzig, Germany
Tel: (+49 341) 495 30   Fax: (+49 341) 477 32 74
e-mail: info@kirow.de
Web: www.kirow.de

Key personnel

Managing Director: L Koehne
Technical Director: KT Giele
Marketing Director: J Kühn

Kelsan's Keltrack® Trackside Transit top-of-rail liquid friction modifier, aimed at reducing squeal noise and increasing rail life
NEW/0585288

Kirow track and bridge KRC 810 construction crane for Amtrak 0533655

Laser Rail's LaserPeT, designed to measure metro networks 1067755

Products

Cranes for bridge and track construction, breakdowns and accidents. The latest model is the KRC 1200 railway crane.

Contracts

Recent contracts include supply of KRC 1200 cranes to Weiss Track Construction, Germany; Swietelsky Track Construction, Austria; Euroswitch, Switzerland; Voker Stevin Track Construction, Netherlands; State Railway of Maroc, Morocco; French National Railways (SNCF), France; Hering-Bau Track Construction, Germany; Balfour Beatty, UK; Hungarian National Railways (MÁV), Hungary; KRC 810 cranes to Grant Rail Construction, UK; Balfour Beatty, UK; Spitzke Track Construction, Germany; DBG (DB Duisburg), Germany; Wiebe Skandinavia, Germany.

UPDATED

Kloos Oving bv

West-Kinderdijk 24, NL-2953 XW Alblasserdam, Netherlands
PO Box 3, NL-2960 AA Kinderdijk, Netherlands
Tel: (+31 78) 691 40 00   Fax: (+31 78) 691 45 42
e-mail: info@kloos-oving.s3c.nl
Web: www.kloos-oving.nl

Key personnel

General Manager: J van Houwelingen

Products

Design, development, construction and delivery of standard and custom-built track materials for main line, metro, light rail systems and heavy-duty systems including: turnouts, crossings, points, expansion joints and special constructions.

Contracts

Recent contracts include the supply of track materials to operators in Egypt, Germany, Indonesia, Iran and the Netherlands.

UPDATED

Knox Kershaw Inc

11211 Trackwork Street, Montgomery, Alabama 36117, USA
Tel: (+1 334) 387 56 69   Fax: (+1 334) 387 45 54
e-mail: sales@knoxkershaw.com
Web: http://www.knoxkershaw.com

Key personnel

President and Chief Executive Officer:
  Knox Kershaw
Marketing: Jaky Felix

Products

Ballast regulators, tie cranes, yard cleaners, rail utility vehicles, switch undercutters, railway cranes, repair kits, improved performance kits, equipment for sleeper and rail replacement, sand and snow removal, vegetation control and various other maintenance operations.

Konepaja Mankinen Ky

Tehtaankatu 9, FIN-11710 Riihimäki, Finland
Tel: (+358 19) 76 42 60   Fax: (+358 19) 72 30 99
e-mail: info@mankinen.fi
Web: www.mankinen.fi

Key personnel

Contact: Mika Mankinen

Products and services

Repair, overhaul and spare parts manufacture of track maintenance machines. The company trademark is ENERCO.

UPDATED

Laser Rail Ltd

Fitology House, Smedley Street East, Matlock DE4 3GH, UK
Tel: (+44 1629) 76 07 50   Fax: (+44 1629) 76 07 51
e-mail: info@laser-rail.co.uk
Web: www.laser-rail.co.uk

Key personnel

Managing Director: David M Johnson
Executive Director: Alison B Stansfield
Engineering Director: Steve Ingleton
Non-Executive Director: Hugh Fenwick
Metro Division Engineering Director: Bob Silver

Background

Laser Rail, established in 1989, is a technical consultancy supplying complete turnkey projects for rail infrastructure and metro networks. It combines experienced engineers with the latest high-technology laser-gauging systems and software to measure and analyse clients' infrastructure and provide new and innovative vehicle designs.

Products

Measuring systems: Laser Gauging Vehicle (LGV), a road-rail vehicle for measuring structure profiles accurately and at speed. Laser Profiling Trolley (LaserPeT), a portable version of the LGV used at walking speed and designed to measure metro infrastructure. LaserSweep™, a portable measuring device for measuring structure profiles in areas where larger systems are uneconomical to use. Product approval received from Network Rail and London Underground.

Software: ClearRoute™, Stress Route™, DesignRoute™ software for vehicles and infrastructure management.

Laser Rail's LaserSweep, enables surveying in isolated locations 1067756

Laser Rail's Laser Gauging Vehicles measuring the East Lancashire Railway   1067757

Lindapter's rail clip   0092275

Services: Track geometry measurement and design, route assessment for existing and new rolling stock, interoperability, feasibility studies for existing and new routes, design of new rolling stock. Risk assessment.

Laser Rail has a formal research and development group to develop core technologies associated with its activities. Projects being undertaken include infrastructure measurement, monitoring and analysis, vehicle/track interaction technology and intelligent video systems. Laser Rail also provides software training and certification to various levels of competency and this can be supplied as part of the overall support package. Laser Rail operates in the UK, Europe and Australasia.

UPDATED

Leica Geosystems AG

Europa Strasse 21, CH-8152 Glattbrugg, Switzerland
Tel: (+41 1) 809 33 11   Fax: (+41 1) 810 79 37
e-mail: info.swiss@leica-geosystems.com
Web: http://www.leica-goesystems.ch

Products
Railway surveying systems in collaboration with Amberg Measuring Technique Ltd. The GRP3000 model is based on the TGS3000 track gauging trolley and measures superelevation, gauge and positioning. The integrated Profiler 100 provides automatic clearance profile measurements and offers real-time comparison with theoretical clearance envelopes in the field, while the Leica TPS Total Station system provides seamless communications. The system also provides three-dimensional documentation of track geometry and environment.

Weighing less than 60 kg, the GRP5000 system is designed for the kinematic survey of clearance profiles and the measurement of track geometry. Applications include the documentation of trackside structures.

Lindapter International

Lindsay House, Brackenbeck Road, Bradford BD7 2NF, UK
Tel: (+44 1274) 52 14 44   Fax: (+44 1274) 52 11 30
e-mail: enquiries@lindapter.com
Web: www.lindapter.com

Key personnel
Technical Support Manager: Michael Knight

Other offices
Lindapter SA
Paris Nord II, 14 rue de la Perdix, F-95700 Roissy, France
Tel: (+33 1) 48 17 87 90   Fax: (+33 1) 48 17 87 99
Contact: Jaques Babault

Lindapter GmbH
Ernestinestr. 67, D-45141 Essen, Germany
Tel: (+49 201) 21 47 78   Fax: (+49 201) 29 06 14
Contact: Sabine Reimann

Background
Lindapter International was formed in 1934 by Henry Lindsay to market traditional hook bolt adapters and has been part of the Victualic Group, Glynwed International and in 1999 was taken over by Tyco International Ltd.

Products
Holdfast adjustable rail clips; the Soft clip holds rails in precise alignment while the Hard clip prevents vertical rail movement. A Spring clip also caters for rail wave while holding the rail down. A Type BR clip suits flat bottom or bridge rails up to an 8° slope. The Temporary Support System supports and insulates running rails while essential repair work is being carried out.

Contracts
Manchester Piccadilly station reroofing. Greenwich station transport interchange (connections in roof structure), London Underground rerailing project. Other contracts include Rome metro, Channel Tunnel and the East Coast Main Line, UK.

UPDATED

Linsinger Maschinenbau GesmbH

Dr Linsinger Strasse 24, A-4662 Steyrermühl, Austria
Tel: (+43 7613) 88 40   Fax: (+43 7613) 88 40 38
e-mail: maschinenbau@linsinger.com
Web: www.linsinger.com

Key personnel
Shareholder: F Weingärtner
Managing Director: H Knoll
Technical Manager, Development Department: W Neubauer
Marketing: H Knoll; Hans-Peter Bartmann

Leica GRP5000 railway infrastructure survey system   0572379

Linsinger SF03-FFS rail profiling train for Alpha Rail Team *NEW*/0585290

German office

Linsinger Rail Milling
Rathaus Strasse 33, D-97922 Lauda-Königshofen
Tel: (+49 9343) 50 94 47   Fax: (+49 9343) 50 94 48
e-mail: rail-milling@linsinger.com

Background

Linsinger is a subsidiary of the Weingärtner Group.

Products

Mobile and stationary rail processing machines (both milling and grinding systems) and rail measuring systems.

Recent products include the SF03-FFS mobile rail reprofiling train that undertakes both milling and grinding in a single pass. In operation, the vehicle works at speeds of 600 to 1,500 metres per hour. It can travel between work sites at up to 100 km/h.

Contracts

Early customers for the SF03-FFS vehicle included Alpha Rail Team GmbH & Co KG of Nuremburg, a subsidiary of the Swiss Sersa group, and DB subsidiary Deutsche Gleis-und Tiefbau GmbH.

UPDATED

Loram Maintenance of Way, Inc

PO Box 188, 3900 Arrowhead Drive, Hamel, Minnesota 55340, USA
Tel: (+1 763) 478 60 14   Fax: (+1 763) 478 22 21
e-mail: sales@loram.com
Web: http://www.loram.com

Key personnel

Vice-President, Marketing: P J Homan
Manager, International: T L Smith

Products

Loram manufactures equipment and provides a full-service approach to railroad maintenance services for heavy haul, passenger/freight and transit railroads. Customers can opt either to purchase or lease, or contract one of Loram's leased equipment crews. Products include a full range of self-propelled rail grinders, self-propelled ditch cleaners; self-propelled shoulder ballast cleaners, L Series of self-propelled transit grinders and rail mounted vacuum excavators.

VERIFIED

Lord Corporation

Mechanical Products Division, 2000 West Grandview Blvd, PO Box 10040, Erie, Pennsylvania 16514-0040, USA
Tel: (+1 814) 868 54 24   Fax: (+1 814) 868 31 09

Key personnel

For a full list of personnel, see Lord Corporation entry in *Bogies and suspension, wheels and axles, bearings* section.

Products

Elastomeric direct fixation fasteners. Manufactured in a variety of designs for new and existing transit systems, they are installed at grade, below grade and on elevated structures, and are fully tested and qualified by user transit authorities. The mid-range direct fixation fastener vertical spring rate ranges from 100,000 to 300,000 lb/sq in, the low-range (soft) direct fixation fastener from 60,000 to 90,000 lb/sq in, and the mid-range special trackwork fastener's vertical spring rate is comparable to that of the mid-range direct fixation fastener. Rail clamping systems are available for aerial and rigid installations.

Lucchini Group

Lucchini SpA
Via Oberdan 1/a, I-25128 Brescia, Italy
Tel: (+39 030) 399 25 66
Fax: (+39 030) 39 17 82
e-mail: rails@lucchini.com
Web: www.lucchini.com

Key personnel

President: Giuseppe Lucchini
Managing Director: Giovanni Gillerio
Corporate Commercial Director: Giovanni Bajetti

Subsidiary company

Lucchini Piombino SpA
Lucchini SpA
Via Oberdan 1/a, I-25128 Brescia, Italy
Tel: (+39 030) 399 25 66
Fax: (+39 030) 39 17 82
e-mail: rails@lucchini.com
Web: www.lucchini.com
President and Managing Director:
   Giovanni Schinelli
Sales Manager: Priamo Priami

Works

Piombino Works (Lucchini SpA)
V.le della Resistenza 2, I-57025 Piombino Livorno, Italy

Lucchini Sidermeccanica SpA
Via G Paglia 45, I-24065 Lovere (BG), Italy
Tel: (+39 035) 96 35 66
Fax: (+39 035) 96 35 52
e-mail: rollingstock@lucchini.com
President and Managing Director: Erder Mingoli
General Manager: Augusto Mensi
Sales Manager: Roberto Forcella

Bari Fonderie Meridionali – BFM (Lucchini Sidermeccanica SpA)
Via Tommaso Columbo 7, I-70123 Bari, Italy
Tel: (+39 080) 582 71 11   Fax: (+39 080) 534 41 35
e-mail: salesbfm:lucchini.com
President and Managing Director: Erder Mingoli
General Manager: Giacomo Pondrelli
Sales Manager: Igor Mariani

Products

Lucchini Piombino: Rails from 27 to 70 kg/m in various steel grades. Maximum rail lengths: 108 m for unwelded rails, 144 m for welded.

Bari Fonderie Meridionali (BFM) Bari Works: Manganese cast steel monobloc crossings for switches; mechanical moulded cast steel components.

UPDATED

LUKAS

LUKAS Hydraulic GmbH & Co Kg
Weinstrasse 39, D-91058, Erlangen, Germany
Tel: (+49 9131) 69 80   Fax: (+49 9131) 69 83 94

Key personnel

Sales Manager: Herr U Kirchner

Products

Development and manufacture of hydraulic rerailing equipment for rolling stock.

LUKAS rerailing equipment allows for precise lifting and rerailing to within 1 mm. LUKAS also makes equipment for uprighting overturned rolling stock. It also makes a pulling device for pulling apart rolling stock, either from each other or from tunnel/bridge walls.

LUKAS rescue tools and pneumatic lifting bags are in worldwide use.

Lyudinovo Locomotive Works JSC

1 K Liebknecht Street, Lyudinovo, Kaluga region 249400, Russia
Tel: (+7 084) 442 01 20; 252 59
Fax: (+7 084) 442 01 20; 252 59

Key personnel

For a full list of personnel, see Lyudinovo Locomotive Works JSC entry in *Locomotives and powered/ non-powered passenger vehicles* section.

Products

Rail lubricating machines, rotary snow ploughs track maintenance cars.

MAN Ferrostaal AG

Hohenzollernstrasse 24, D-45128 Essen, Germany
Tel: (+49 201) 818 01   Fax: (+49 201) 818 28 22
e-mail: info@manferrostaal.com
Web: www.ferrostaal.com

Key personnel

Head of the Department of Infrastructure and
   Transport Systems: Helmut Julius

Background

Re-named from Ferrostaal AG, MAN Ferrostaal AG is a member of the MAN AG group.

Products

Permanent way materials: rail of all grades, light and heavy vignole rail, crane rail, grooved rail; wooden, concrete and steel sleepers; rail fastening systems and individual components for ballasted and slab tracks, resilient and rigid clips, sole plates, sleeper screws, spring washers, anti-creep rail anchors, sleeper anchoring devices; switch and crossing systems of various types, expansion joints, insulated joints; turntables; sliding buffer stops; rail welding materials; plastic components, screw dowels, plastic dowels for reconstruction of wooden and concrete sleepers, rail pads; elastomers for reducing groundborne noise, ballast mats, bearings for floating and slab tracks, resilient pads.

UPDATED

Matisa Matériel Industriel SA

Case Postale, CH-1023 Crissier 1, Switzerland
Tel: (+41 21) 631 21 11   Fax: (+41 21) 631 21 68
e-mail: matisa@matisa.ch
Web: www.matisa.ch

Key personnel
Managing Director: Rainer von Schack
Technical Director: Jörg Ganz
Marketing Director: Jörg Marbach

Principal subsidiaries
France
Matisa SA
Offices and workshop: 9 rue de l'Industrie, ZI Les Sablons, F-89100 Sens
Tel: (+33 3) 86 95 83 35   Fax: (+33 3) 86 95 36 94
e-mail: matisa.@matisa.fr
President: Michel Chevrery
Sales of Matisa group products in France, Benelux and French-speaking Africa. After-sales service, spare parts and overhaul of Matisa machines in France.

Germany
Matisa Maschinen GmbH
Kronenstrasse 2, D-78166 Donaueschingen
Tel: (+49 771) 15 80 63   Fax: (+49 771) 15 80 64
e-mail: matisa.donaueschingen@t-online.de
General Manager: F Wernick
Sales of Matisa group products in Germany, after-sales service and spare parts of Matisa machines operating in Germany.

Italy
Matisa SpA
Via Ardeatina km 21, I-00040 Pomezia/Santa Palomba (Rome)
Tel: (+39 06) 91 82 91   Fax: (+39 06) 91 98 45 74
e-mail: matisa@matisa.it
General Manager: Eng J Berga
Sales of Matisa group products and after-sales service, spare parts and overhaul of Matisa machines in Italy.

Japan
Matisa Japan Co Ltd
Shiba Building, 5-16-7 Shiba, J-Minato-Ku, Tokyo 108
Tel: (+81 3) 34 54 75 61   Fax: (+81 3) 34 54 75 63
e-mail: matisatokyo@gol.com
General Manager: H Otake
Sales of Matisa group products and after-sales service, spare parts and overhaul of Matisa machines operating in Japan.

Spain
Matisa Matériel Industriel SA
Sucursal Española
Avda de Brasil 17, Piso 11F, E-28020, Madrid
Tel: (+34 91) 556 12 80   Fax: (+34 91) 556 68 79
e-mail: matisa@accessnet.es
Works: Estacion de Grinon, E-28970 Grinon, Madrid
Commercial Director: E Puertas
Technical Director: S Gonzalez
Sales of Matisa group products in Spain; after-sales service, spare parts and overhaul of Matisa machines operating in Spain and Portugal.

UK
Matisa (UK) Ltd
PO Box 202, Scunthorpe DN15 6XR
Tel: (+44 1724) 85 93 96   Fax: (+44 1724) 85 44 90
e-mail: matisa@matisa.co.uk
General Manager: F Messerli
Sales of Matisa group products and after-sales service, spare parts and overhaul of Matisa machines operating in the UK.

Products
Manufacture and sale of track construction and maintenance machinery, including tamper-leveller-liners (continuous, conventional, combined for points and crossings and plain track); regulators with and without hoppers; tracklaying and track renewal trains; ballast cleaners; track and catenary measuring, recording and analysis vehicles; track and catenary service vehicles.

UPDATED

Max Bögl

Industriegebiet Schlierferheide/Bögl
PO Box 11 20, D-92301 Neumarkt, Germany
Tel: (+49 91 81) 90 90   Fax: (+49 91 81) 90 50 61

Products
Slab track system.

Mer Mec SpA

Via Oberdan 70, I-70043 Monopoli (BA), Italy
Tel: (+39 080) 887 65 70   Fax: (+39 080) 887 40 28
e-mail: mermec@mermec.it
Web: www.mermec.it

Key personnel
Managing Director: Vito Pertosa
Marketing and Sales, Executive Director:
  Luca Ebreo
Marketing Manager: Pietro Stama
International Sales Manager: Carlo Evangelisti
Domestic Sales Manager: Mario Girolami
After Sales and Commissioning Manager:
  Guiseppe Aurisicchio
Quality Assurance Manager: Mauro Simone
Research and Development Programme Manager:
  Patrizia Sforza

Products
Design, development and manufacturing of measuring/diagnostic systems and vehicles for railway infrastructure monitoring. Mer Mec track measuring systems adopt no-contact optoelectronic technologies. They can be installed on any railway car (locomotive, passenger coach, freight wagon, dedicated vehicle). Systems are provided with software for real-time or off-line (unattended systems) data processing and analysis. They operate in all weather conditions at speeds of up to 350 km/h. Mer Mec product portfolio of diagnostic/measuring systems/vehicles for permanent way monitoring includes: track geometry, rail profile, rail corrugation and video inspection with automatic recognition of defects.

The study of interactive forces between wheel and rail, such as oscillatory motion affecting the vehicle or asymmetric wear pattern affecting the wheel tread profile, has been introduced to evaluate risk of derailment, level of comfort offered on board and to ensure rolling stock achieves full service life potential. Mer Mec offers a set of different systems specifically developed for railway needs: wheel-rail interaction, lateral and vertical forces applied by the wheel on the rail, vehicle accelerations, wheel-rail contact, angles of contact and equivalent conicity, wheel-profile. Design and development of highly modular software solution (RAMSYS® Railway Asset Management System) supporting enhanced data analysis for optimised maintenance and renewal planning of railway infrastructure.

Design, engineering and manufacturing of vehicles/equipment dedicated to permanent way maintenance.

Contracts
Mer Mec has won contracts to supply diagnostic systems/vehicles for permanent way monitoring in Brazil, France, Italy, South Korea, Spain, Switzerland, Syria.

UPDATED

metronom Gesellschaft für Industrievermessung mbH

Hauptstrasse 17-19, D-55120 Mainz, Germany
Tel: (+49 6131) 96 25 70
Fax: (+49 6131) 962 57 18
e-mail: info@metronom.de
Web: http://www.metronom.de

US office
17672 Laurel Park Drive North, Suite 400, Livonia, Missouri 48152, USA
Tel: (+1 734) 384 59 91
Fax: (+1 734) 384 59 93
e-mail: infor@metronomna.com
Web: http://www.metronomna.com

Products and services
Railway measuring technology, including: systems for measuring gauge, camber, contact wire height and tunnel profiles; systems for detecting and surveying loading gauge restrictions; systems for inspecting contact wires; video documentation and inspection systems; video measuring systems; stereoscopic measuring systems; testing and calibration software; mobile 3D measuring services. Also undertaken are: studies; training; technical consultancy; individual measuring and evaluation software; contract development of railway measuring systems; development and realisation of complete measuring trains.

Projects include development, integration and support for the measuring technology installed in a complete loading gauge restriction measuring vehicle (LIMEZ II).

Metrum Information Storage Limited

Oxford Street, Long Eaton NG10 1JR, UK
Tel: (+44 115) 972 09 49   Fax: (+44 115) 946 19 53
e-mail: enquiries@metrum.co.uk
Web: http://www.metrum.co.uk

Products
Measurement and monitoring equipment for the railway industry, covering all aspects of railway track parameter recording and analysis. With equipment ranging from lightweight portable single-operator trolleys to self-propelled measuring frames, Metrum provides a series of instruments that measure ride comfort and monitor vibration. The company also designs and produces a variety

Archimede high-speed diagnostic train supplied by Mer Mec to RFI, Italy   0142176

The MacKarT multipurpose maintenance vehicle   0114663

of inspection vehicles and trailers for transporting maintenance crews and equipment to site.

A recently developed product is the MacKarT, an easily transported multipurpose maintenance vehicle, which can be lifted on and off the track by four personnel in less than 10 minutes. The cart comprises a main chassis weighing approximately 110 kg, a battery pack comprising 12 70 Ah sealed lead acid traction batteries each weighing 25 kg, and a range of tops which can be specifically designed for a wide range of applications. Two standard tops are available: a nine-man carriage weighing 110 kg and a flat top with a single driving position weighing approximately 60 kg.

Mitsukawa Metal Works Co Ltd

21 Harima-cho-nijima, Kako district, Hyogo Pref, Japan
Tel: (+81 794) 35 22 88   Fax: (+81 3) 32 84 03 61

Key personnel
Sales Director: K Kukuda

Products
Rail fastenings, steel sleepers and forged crossings.

Modern Track Machinery Canada Ltd

5926 Shawson Drive, Mississauga, Ontario L4W 3W5, Canada
Tel: (+1 905) 546 12 11   Fax: (+1 905) 564 12 17
e-mail: sales@mtmgeismar.com

Key personnel
President: Daniel Geismar
General Manager: Michael J Byrne

Background
MTM Canada is a subsidiary of Geismar Corporation of Paris, France.

Principal subsidiary
Modern Track Machinery Inc (US Affiliate)
1415 Davis Road, Elgin, Illinois 60123-1375, USA
Tel: (+1 847) 697 75 10   Fax: (+1 847) 697 01 36
General Manager: Al Reynolds

Products
Railway maintenance equipment (hydraulics, gas, high frequency including saws, drills, grinders, impact wrenches, spike drivers, pullers, tie inserters) tampers, motor trolleys, material and rail handling equipment, road/rail cranes/excavators.

Contracts
Sales to all North American railways, transits and contractors. Export to central and south America, Africa and Asia.

UPDATED

moklansa GmbH

Webershohl 53, D-44319 Dortmund, Germany
Tel: (+49 231) 27 15 91   Fax: (+49 231) 27 15 93
e-mail: info@moklansa.de
Web: http://www.moklansa.de

Sales representation
VTEC
Westfalendamm 172, D-44141 Dortmund, Germany
Tel: (+49 231) 56 55 99 50
Fax: (+49 231) 56 55 99 55
e-mail: info@vtec-gmbh.com
Web: http://www.vtec-gmbh.de

Products
Electronically controlled rail lubrication systems for heavy and light rail systems including grooved and Vignoles-type rail. Power source options include current transformer, battery or solar.

Moog GmbH

Brückenuntersichtsgeräte + Hocharbeitsbühnen
D-88693 Deggenhausertalm Im Gewerbegebeit 8, Germany
Tel: (+49 75) 55 93 30   Fax: (+49 75) 559 33 66
French Office:
Pont Acces, 2 rue de Condant, F-67480 Forstfield (Strasbourg)

Products
Bridge inspection equipment for road and railway bridges. The design was originally for electric trains and allows lowering and folding up of the platform without shutting off the electric current in overhead wires. The tower system can be extend above the bridge railing with a telescopic arm.

MOOG 1200 T and 1500 T bridge lifts can be mounted on a vehicle or self-propelled railcar, has two points of rotation for all round mobility, horizontal reach of 12 to 15 m, the bucket arm has a rotation of 2 × 180° with a working area for several people. Also available are dual track version, railway vehicle unit, pier inspection unit.

Mossboda Trä AB

SE-718 92 Frövi, Sweden
Tel: (+46 581) 720 30
Fax: (+46 581) 720 36

Products
Pressure-treated and unimpregnated pine sleepers. Customers include Banverket and railway construction companies.

Mowlem Railways

Head Office
White Lion Court, Swan Street, Isleworth, Middlesex TW7 6RN, UK
Tel: (+44 20) 85 68 91 11   Fax: (+44 20) 88 47 48 02
Web: www.mowlem.com

Key Personnel
Managing Director: Colin Graidage

Background
Founded in 1998, Mowlem Railways is the specialist rail division of John Mowlem & Company plc.

Services
Design, construction and maintenance of light urban transit.

Contracts
Within the UK, the construction and maintenance of the Manchester Metro Link. The design and build of the Docklands Light Railway: Lewisham Extension, which is maintained by Mowlem through Private Finance Initiative (PFI) concession. A three-year infrastructure maintenance contract on the Croydon Tramlink system. The company was previously engaged in trackworks on a sub-contract basis during the original construction of the Tramlink.

UPDATED

MTH Praha AS

Kandertova 1a/1131, CZ-180 00, Praha 8, Libeň, Czech Republic
Tel: (+420 2) 84 09 32 02   Fax: (+420 2) 84 09 32 81
e-mail: mth@mth.cz
Web: www.mth.cz

Key personnel
President: Martin Pinl
General Director: Petr Wagenknecht
Director, Sales: Pavel Türk

Background
MTH Praha AS is a member of the Cimex Group.

Indian joint venture company
MTH India PVT Ltd
409 Chiranjeev Tower, 43, Nehru Place, New Delhi, 110 019, India
Tel: (+91 11) 51 60 83 35   Fax: (+91 11) 51 61 18 10
e-mail: arun@technipimpex.com

Key personnel
Managing Director: Suresh Kumar Mandal
Marketing Director: Arun Kumar Mandal

Products
Manufacture of permanent way maintenance machines and equipment including ballast cleaners, ballast wagons and ballast distribution systems; formation and rehabilitation machines; track stabilisers; catenary maintenance vehicles; ballast compactors; brush cutters; track rotary snow ploughs; track recording cars; motor cars and power units; hand tools.

Contracts
Contracts include the supply of maintenance equipment to Austria, Canada, Croatia, Czech Republic, India, Russia, Slovakia and US.

Recent contracts have included 30 APV vehicles for overhead catenary maintenance for Italian Railways.

UPDATED

Muromteplovoz

Murom Diesel Locomotive Works
ulica Filatova 10 602 200 Murom, Vladimir region, Russian Federation
Tel: (+7 095) 291 31 68   Fax: (+7 09234) 443 03
e-mail: mteplo@cl.murom.ru
Web: http://www.cl.murom.ru

Muromteplovoz Type AGS-1 self-propelled multipurpose infrastructure maintenance vehicle

0103968

Muromteplovoz Type RN-04 rail-end clearance maker   0103969

Key personnel

For a full list of personnel, see Muromteplovoz entry in *Locomotives and powered/non-powered passenger vehicles* section.

Products

Self-propelled track maintenance vehicles; ballast cleaners; hydraulic jacks; rail straighteners; clearance makers.

NARSTCO

Plant 300 Ward Road, Midlothian, Texas 76065, USA
Tel: (+1 972) 775 55 60
Fax: (+1 972) 775 55 14

Key personnel

Vice-President, Marketing: John Fox

Products

Design, manufacture and supply of steel track and turnout sleepers. Steel sleepers are made from hot-rolled bar sections: H-section are for heavy-haul work, and M-section for lighter traffic. Steel turnout sleeper sets are available as full, partial and blank, all complete with fastenings. Insulated and non-insulated packages are available to any custom length. Also supply of steel sleeper fastening systems.

NEW ENTRY

NEU International

PO Box 4039, 70 rue du Collège, F-59704 Marcq en Baroeul, France
Tel: (+33 3) 20 45 64 35
Fax: (+33 3) 20 35 65 99
e-mail: railway@neu-international.com
Web: www.neu-international.com

Other offices

PO Box 488, Paoli, Pennsylvania 19301-0488, USA
Tel: (+1 610) 725 04 01
Fax: (+1 610) 725 04 02
e-mail: mail@neu-inc.com

PO Box 406, Thomson Road Post Office, Singapore 915714
Tel: (+65) 64 30 66 86
Fax: (+65) 65 52 71 31
e-mail: mail@neu-pte.com.sg

Products

The rail/road Vaktrak is designed to meet all types of tracking cleaning requirements for train, light rail, in-station, in-tunnel or outside (with grass or pavement) on ballast or on concrete. The Vaktrak module consists of free modules which can be adapted to the requested cleaning configuration and can be installed on existing rolling stock. NEU has also developed the NEU CET system, an automatic and hygenic way to empty chemical and/or retention toilet tanks. This system can be fixed installation or a mobile unit depending on the depot configuration and the train availability. NEU International product range also includes train washing plants (drive through or gantry systems). Completely automatic and flexible for train washing, they are designed to suit all machine profiles at the requested speed.

UPDATED

Newag GmbH & Co KG

Ripshorster Strasse 321, D-46117 Oberhausen, Germany
Tel: (+49 208) 86 50 30   Fax: (+49 208) 865 03 20
e-mail: info@newag.de
Web: www.newag.de

Key personnel

Managing Director: C Kohl
Technical Director: M Hanke
Sales Director: R Franz

Products

Newag SVM slab track laying system; SVM 98 long welded rail unloading and laying system; SE 2002 fully automated rail fastening machine for Vossloh rail fasteners; FE 2003 integrated rail/road truck for overhead catenary matinenance, inspection and measuring; Jumbo railroader for track and catenary maintenance; railroader truck U 300 L 3 for catenary maintenance; railcars for track and catenary maintenance; Metrostar underground and LRT cleaning and maintenance vehicles; SWG ballast undercutting and resleepering attachment for rail/road excavators and cranes; ASG ballast tamping attachment for rail/road excavators and cranes.

Newag reconditioned tamping machines; ballast regulators; dynamic track stabilisers; ballast cleaners; flash butt welders; gantry cranes originally manufactured by Plasser & Theurer, MATISA, Donelli and others, fully refurbished and modernised at Newag's specialised on track machine overhaul workshop for track construction and maintenance with aftersales service worldwide.

MK tool monobloc ultra wear resistant tamping lines integrally hardfaced with tungsten carbide hard metal tiles for Plasser and Matisa tampers. MK tool wear parts based on tungsten carbide technology for renewal trains, ballast cleaners, sub formation and formation stabilising and reinforcement machines, ballast regulators.

Developments

The slab track laying machine Newag SVM has been developed and successfully employed for the Frankfurt–Cologne high-speed line. It is a further development of the SVM 98 long welded rail unloading and laying machine. The fully automatic Newag SE 2002 rail fastening assembly machine, based on a Newag Jumbo heavy rail/road truck, has been commissioned on the Frankfurt–Cologne high-speed line too. This machine, fitted with a satellite type mounting and assembly unit for continuous non-stop action, allows the complete and automatic installation of Vossloh elastic rail

Fully automated rail fastening machine for Vossloh rail fasteners based on a Newag Jumbo Railroader truck   *NEW*/0585229

Newag Metrostar underground and LRT cleaning and maintenance vehicles   ***NEW**/0585233*

Refurbished Plasser production tamper 08/32 U for track work on high-speed line in Spain
***NEW**/0585236*

fastening systems skl 12, 14 and 16 at a speed of approximately 700 metres per hour with only one operator acting as a surveyor. The Newag SE 2002 also features a torque measuring and recording system with real-time display, print out and floppy disc recording.

A new product line, Newag Metrostar, has been introduced as a cost-efficient, maintenance and cleaning vehicle specifically developed for metros and LRTs. Bangkoki MRTA received a tunnel cleaning and inspection vehicle along with a rerailing and inspection vehicle for its new line in 2003/2004.

UPDATED

Newt International

1 Orion Court, Rodney Road, Portsmouth PO4 8SZ, UK
Tel: (+44 23) 92 73 00 60   Fax: (+44 23) 92 73 00 70
e-mail: sales@lizard.co.uk
Web: http://www.lizard.co.uk

Products
Rail flaw detection system employing the Lizard® field gradient camera.

Developed as an alternative to ultrasonic testing, the capabilities of the field gradient camera are claimed to include detection of poor surface fusion in welds; high-level accuracy and repeatability; no limitations imposed by any surface geometry or crack size; crack depth and length/indication of high-speed inspection; suitable for use on non-conductive coatings; surface cleaning not required; arrays are able to 'see' through coatings in excess of 8 mm; arrays available in both manual and electronically scanned versions; the 2-D planar array with asymmetric excitation can be configured to inspect almost any 3-D shape; no couplant required; direct contact with the rail is not essential; use of multimode software; database of records; improved asset management planning, for example condition monitoring; detection of defects on new rail; monitoring of growth of existing defects; detection and monitoring of profile changes; detection of changes in the condition of the surface.

Nordco

245 W Forest Hill Avenue, Oak Creek, Wisconsin 53154, USA
Tel: (+1 414) 766 21 80   Fax: (+1 414) 766 23 79
e-mail: sales@nordco.com

Key personnel
President: Bruce Boczkiewicz
Vice-President: Steve Wiedenfeld

Products
Self-propelled adzers, rail drills, hydraulic spiking machines by hydraulic spike pullers, tie drills, gauging machines, rail lifters, track inspectors, screw spiking machines, anchor removers, anchor applicators. Tie remover/inserter, tie plugger, ballast regulator, snow clearing machines, brush cutters.

Model NETP-I tie plugging machine is a high-production two-person machine able to fill spike holes while operating in rail or regauging gangs. It has a 400 gallon capacity of non-foam polyurethane compound, known as SpikeFast.

Trailblazer Model BC60 Brushcutter has a 30 ft reach from the track centre on each side, hydrostatic drive, four-speed manual transmission and a speed of 35 mph.

Tie Remover Inserter Plate Positioner (TRIPP) can remove and insert ties from either side of the main frame. The machine will mechanically position tie plates in the gauge section of the rail on the first, second or third tie behind the tie being removed and the gripper is adjustable to eliminate damage to new ties being installed.

G2 ballast regulator has a rear-mounted engine, uses a hydraulically positioned ballast plough and can travel at 35 mph using a four-wheel hydrostatic shaft drive system.

VERIFIED

NRS – National Railway Supplies

Gresty Road, Crewe CW2 6EH, UK
Tel: (+44 1270) 53 30 00   Fax: (+44 1270) 53 39 56
e-mail: commercial@natrail.com
Web: www.natrail.com

Nordco G2 ballast regulator   0093218

Other offices
Leeman Road, York YO26 4ZD, UK
Tel: (+44 1904) 52 22 93   Fax: (+44 1904) 52 26 96

Key personnel
Managing Director: Graham Jackson
Commercial Director: David A Kierton
Marketing Manager: Dave Tilmouth

Background
NRS is a wholly owned subsidiary of the Unipart Group of companies.

Products
NRS provides a logistics and supply chain management service to the UK rail industry including the management of customer warehouses, the distribution of 23,000 products and the repair and recondition of a wide range of products. NRS also designs, develops and manufactures a range of new products for infrastructure and rolling stock.

Contracts
Contracts include Network Rail, BBRISL, Carillion Rail, AMEC Spie Rail, Jarvis Rail, Westinghouse Rail Systems and London Underground. Manufacture, production, servicing, repair and distribution of signalling, track maintenance and telecoms products, infrastructure and permanent way. Also design and assembly of signalling, TPWS and signalling location cases.

UPDATED

ØDS-Caltronic A/S

Kroghsgade 1, DK-2100 Copenhagen Ø, Denmark
Tel: (+45) 35 26 60 11   Fax: (+45) 35 26 50 18
e-mail: ods@oedan.dk
Web: http://www.odegaard.dk

Key personnel
Managing Director: John Ødegaard
Director: Ulrik Danneskiold-Samsøe
Business Manager: Ove Ramkow-Pedersen

Background
ØDS-Caltronic A/S is a company in the Ødegaard & Danneskiold-Samsøe group.

Product
Wheel monitoring systems for online detection of train wheel defects such as flats. The system is modular and can be combined with proprietary AVI wagon identification equipment.

Oleo International Ltd

PO Box 216, Grovelands Estate, Longford Road, Exhall, Coventry CV7 9NE, UK
Tel: (+44 2476) 64 55 55
Fax: (+44 2476) 36 42 87

Key personnel
Managing Director: S B Gelderd
Commercial Director: C C Brown

Products
Oleo produces a wide range of long stroke hydraulic buffers suitable for mounting on fixed or sliding end-stops. Applications include freight yards, steelworks and passenger terminals.
On sliding friction end-stops the need for continual resetting of the friction elements is eliminated, and the hydraulic buffers absorb all of the impact energy at low speeds. Typically a 400 tonne train may be arrested by a pair of 800 mm stroke buffers at 6 km/h without causing the end-stop to slide. Initial and final jerk forces are also eliminated. These buffer units are available for all types of rail operation from LRVs to heavy freight.

UPDATED

ORTEC Gesellschaft für Schienentechnische Systeme mbH

Eigelstein 10-12, D-50668 Cologne, Germany
Tel: (+49 221) 120 69 60   Fax: (+49 221) 12 06 96 66
e-mail: info@ortec-gmbh.de
Web: http://www.ortec-gmbh.de

Key personnel
Managing Director: Hermann Ortwein

Products
Vibration-insulating rail fasteners, Whisper Rail continuous elastic rail embedment material, noise insulation material, ISOLast insulating embedment material, Loadmaster for reduction of ground and structure-borne vibrations.

Contracts
Contracts include reconstruction of 8 km of tramway in Budapest with ISOLast; 26 km of Whisper Rail for the Wuppertal monorail; 1 km of Whisper Rail for Brunswick.

Orton/McCullough

Orton/McCullough Crane Company
1244 East Market Street, PO Box 830, Huntington, Indiana 46750, USA
Tel: (+1 260) 356 79 00   Fax: (+1 260) 356 79 02

Key personnel
Chairman: J F McCullough

Products
Cranes and heavy lifting gear.

VERIFIED

Osmose Railroad Division

PO Box 8276, Madison, Wisconsin 53708, USA
Tel: (+1 608) 221 22 92   Fax: (+1 608) 221 06 18
Web: www.osmose.com

Key personnel
President: Harry Holekamp
Products Manager: David Ostby

Products
TIE-GARD® preservative gel to prevent decay of sleeper plate areas.

Services
The engineered repair of concrete, steel and timber railroad bridges.

UPDATED

Outreau Technologies

rue Pierre Curie, BP 119, F-62230 Outreau, France
Tel: (+33 3) 21 99 53 00   Fax: (+33 3) 21 99 53 03
www.outreautech.com

Key personnel
Managing Director: Daniel Pain
Sales and Marketing Manager: Jean Jomeau

Background
Outreau Technologies is a subsidiary of Manoir Industries.

Products
Cast monobloc, manganese steel crossings to AREA and UIC specifications; welded crossings; cast cradles for movable point crossings; cast bodies and tongues for tram systems; track components for metro systems.

Contracts
Recent contracts include equipment supplied to MTRC's Lantau line in Hong Kong, to

Balfour Beatty for the MRT Changi airport extension in Singapore, Corus Cogifer for the Milan–Rome high-speed line; SNCF, RATP, SNCB; Network Rail; Deutsche Bahn AG; Ferrovie dello Stato; Israel Railways; Kloos Oving; SNIM Mauritania; Nederlandse Spoorwegen; NSB; ONCF Morocco; CP Portugal; SBB/CFF/FFS; Amtrak; Banverket; Burlington Northern; and Banestyrelsen.

UPDATED

Palfinger Europe GmbH

Palfinger Europe GmbH
Moosmühlstrasse 1, A-5203 Köstendorf/Salzburg, Austria
Tel: (+43 6216) 76 60 54 05   Fax: (+43 6216) 77 63
e-mail: h.gollegger@palfinger.com
Web: www.palfinger.com

Key personnel
Sales and Marketing Manager, Railway Cranes: Heinrich Gollegegger

Products
Cranes (1 to 75 tonne/m lifting moment), aerial access platforms for catenary inspection construction and maintenance.

UPDATED

Palfinger railway aerial platform　　0021665

Pandrol Rail Fastenings

63 Station Road, Addlestone, Weybridge KT15 2AR, UK
Tel: (+44 1932) 83 45 00   Fax: (+44 1932) 85 08 58
e-mail: info@pandrol.com
Web: http://www.pandrol.com

Key personnel
Managing Director: G M Lodge
Chief Operating Officer: J Beal-Preston

Principal subsidiaries
Pandrol Asia Pacific Office, Perth, Australia
Pandrol Australia Pty Ltd, Blacktown, Australia
Pandrol Avaux SA, Anderlues, Belgium
Pandrol Fixacoes Ltda, Brazil
Pandrol Canada Ltd, Edmonton, Canada
Promorail SA, Paris, France
Pandrol Rail Fastenings, Germany
PT Pandrol Indonesia, Jakarta, Indonesia
Pandrol Italia SpA, Teramo, Italy
Pandrol Daewon Ltd, Seoul, Republic of Korea
Pandrol South Africa

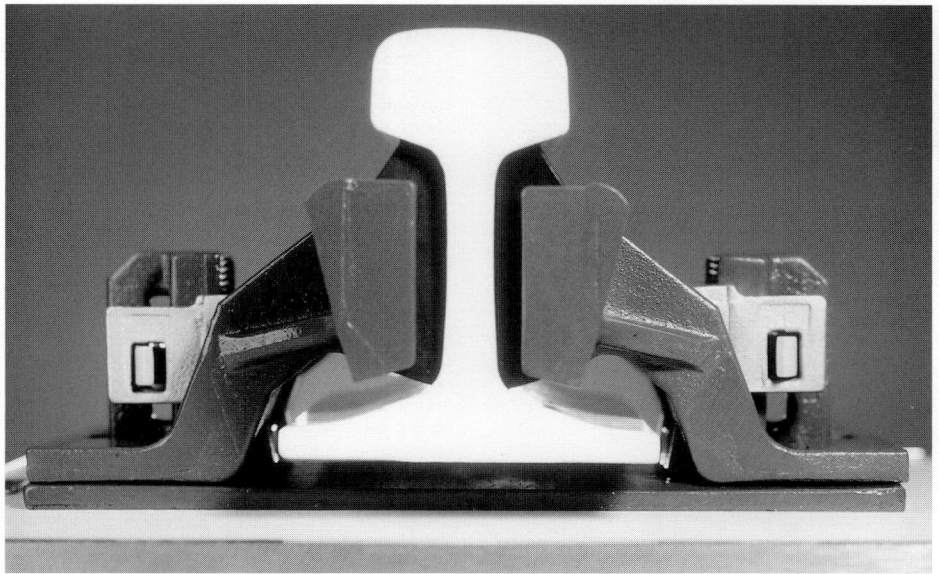

Pandrol VANGUARD system   0526814

Pandrol VIPA-SP assembly   0089430

Pandrol UK Ltd, Worksop, UK
Vortok International Ltd, Addlestone, Surrey, UK
Pandrol USA LP, Bridgeport, New Jersey, USA

Products

Design and manufacture of rail fastening systems and associated installation equipment; resilient rail pads; resiliently supported direct fixation (DF) systems; Vortok Coils for restoring worn screwspike holes (see entry for Vortok International).

Pandrol continues its research into the dynamic behaviour of track to increase understanding of the relationship between forces in track and component performance, and of the generation of noise and vibration. Pandrol has developed a new range of fastening designs for specific applications, including the Pandrol Fastclip® system, which is designed for low-cost installation and maintenance, and the new track support systems designed for application in areas which are sensitive to noise and vibration, including Pandrol VIPA-SP, the latest product in the Pandrol VIPA range incorporating FASTCLIP® and VANGUARD.

Peddinghaus Group

Carl Dan Peddinghaus GmbH & Co KG
Mittelstrasse 64, D-58256 Ennepetal, Germany
Tel: (+49 2333) 79 60
Fax: (+49 2333) 79 63 88
e-mail: cdp-en@t-online.de
Web: http://www.peddinghaus-group.de

Products

A supplier of high quality steel and aluminium forgings and components including maintenance-free roller slide plates; lubrication-free clamp lock devices; lubrication-free point locks; valve rods; checkrail plates and fishplates.

Pennsylvania Steel Technologies

215 South Front Street, Steelton, Pennsylvania 17113, USA
Tel: (+1 717) 986 20 00   Fax: (+1 717) 986 27 00

Key personnel

Chairman and Chief Executive Officer:
   Robert S Miller
General Manager Operations: David P Wirick
General Manager Commercial: Kirkland H Gibson, Jr

Products

Rail, including 80 ft long rail in three grades (in-line head-hardened, medium hardness and standard), t-rail, crane rail and contact rail sections. Also adjustable boltless braces.

Permali Gloucester Limited

Bristol Road, Gloucester GL1 5TT, UK
Tel: (+44 1452) 52 82 82   Fax: (+44 1452) 50 74 09
Web: http://www.permali.co.uk
e-mail: sales@permali.co.uk

Key personnel

Managing Director: A J T King
Technical Director: David Tudor
Sales and Marketing Manager: Fraser Rankin

Products

Permaglass MER components include shoebeams and arc boxes/barriers for a range of electrically powered units.

Permaglass COMP E2 is another material developed for use as an insulating rail joint.

VERIFIED

Pettibone

Pettibone Corporation
Railroad Products Group
5401 W Grand Avenue, Chicago, Illinois 60639, USA
Tel: (+1 312) 745 94 96   Fax: (+1 312) 237 37 63

Main works

Pettibone Ohio, 6917 Bessemer Avenue, Cleveland, Ohio 44127, USA

Key personnel

President: Larry Klumpp
General Manager: T E Hitesman

Products

Switches, switch-points, frogs, crossings, switch stands, guardrails, rail fasteners, compromise joints, switch plates, mobile maintenance-of-way and material handling equipment.

Pfleiderer track systems

Ingolstaedter Strasse 51, D-92318 Neumarkt, Germany
Tel: (+49 9181) 281 36   Fax: (+49 9181) 286 46
e-mail: tracksystems@pfleiderer.com
Web: www.pfleiderer-track.com

Key personnel

Executive Vice-President: Hans Bachmann

Background

Since 1954 Pfleiderer has produced concrete main-track and turnout sleepers for railways and urban traffic in Germany and around the world. As one of the market leaders in Germany, Pfleiderer offers engineering, production, supply, logistics and quality management. Pfleiderer has production plants in Germany, Hungary, the Netherlands, Romania and Taiwan and annually produces more than two million main-track sleepers, and over 500,000 linear metres of turnout sleepers.

Pfleiderer's RHEDA 2000® ballastless track system   0580519

Pfleiderer's GETRAC A1 system features a sleeper 2.60 m long with normal contact width **NEW**/1146564

ATD-G – the Green Track 0580521

Pfleiderer track systems responsible for design and engineering of the line sections on the new Taiwan High Speed Rail (THSR) equipped with RHEDA 2000® **NEW**/1146563

Pfleiderer track systems, in partnership with Royal BAM NV, is directly responsible for project management, engineering and quality assurance for the HSL-Zuid high-speed line **NEW**/1146562

In August 2005 Pfleiderer AG signed an agreement covering the sale of Pfleiderer track systems to Vossloh AG. The transaction was subject to authorisation by the supervisory board of both companies and to approval by the competition authorities in relevant markets.

Products

Pfleiderer offers an extensive portfolio of innovative track systems for installation on ballast, asphalt and concrete for a great number and variety of applications.

The RHEDA 2000® ballastless track system, currently the most advanced development in the RHEDA family. The two most essential modifications over the previous models consist of the following: employment of an especially adjusted bi-block lattice-truss sleeper; merging of the infill concrete and the reinforced trough slab to a homogeneous track-support layer. RHEDA 2000® is well suited for implementation on embankments, over bridges, in tunnels as well as for turnout zones due to its end-to-end systems engineering. In areas sensitive to vibrations it can also be implemented in conjunction with a mass-spring system.

The GETRAC® ballastless track model features direct support of the track panel on an asphalt supporting layer. The GETRAC®A3 is the latest and highest-performance product in the GETRAC® line. The wide sleepers that are part of the system reduce the extent of loads transmitted into the sub-grade.

Contracts

Since 1991, more than 700 km of ballastless track have been installed in Germany. Additional major projects are currently in progress, including the line between Nuremberg and Ingolstadt, Germany and the major North-South link for Berlin.

The construction of the line from Nuremberg to Ingolstadt, part of the new high-speed link between Nuremberg and Munich has taken place completely as ballastless track. The RHEDA 2000® system will be installed in the south and central project section of the line, with a total length of approximately 75 km. In addition, one of the current rail-infrastructure projects in Europe is the construction of the new Dutch high-speed line, HSL-Zuid between Amsterdam and the Belgian border. The RHEDA 2000® will be implemented over the entire line. Construction of the track itself will take place by a joint-venture company founded for this purpose, with the Dutch Royal BAM Group and Pfleiderer track systems as partners. The company RHEDA 2000 vof is responsible for project planning, engineering, construction and quality management.

Pfleiderer has succeeded in entering the Asian market with the new construction of the high-speed line between Taipei and Kaohsiung, in Taiwan. In addition to the use of Japanese ballastless technology on the open section of the track, the RHEDA 2000® system was installed on the demanding sections near the railway stations and at the turnout zones. The majority of the 125,000 main-track sleepers came from local production facilities that were built and operated under the direction of Pfleiderer track systems. Production is now complete, and turnover of the line for operations was scheduled for 2005.

UPDATED

Phoenix AG

Hannoversche Strasse 88, D-21079 Hamburg, Germany
Tel: (+49 40) 76 67 20 05; 27 79
Fax: (+49 40) 76 67 29 43
e-mail: info@phoenix-ag.com
Web: www.phoenix-ag.com

Key personnel
Chief Engineer: Bernd Pahl
Sales Manager: Thomas Barschke

PROGRESS AND PERFORMANCE

The name Plasser & Theurer is synonymous with highly developed and innovative machines for the laying and maintenance of track on railways all over the world. Besides its outstanding technological achievements, Plasser & Theurer has always endeavoured to find solutions together with the customer and to be a reliable, long-term partner. Many decades of experience, up-to-date know-how and the resulting excellent quality are reflected in more than 13,000 Plasser & Theurer track maintenance machines supplied to 103 countries around the world.

Plasser & Theurer

Plasser & Theurer I Export von Bahnbaumaschinen Gesellschaft m.b.H. I A-1010 Wien I Johannesgasse 3 I Tel. (+43) 1 515 72 - 0 I Telefax (+43) 1 513 18 01

The RHEDA CITY ballastless sytem is based on the same functional principles as are all systems in the RHEDA family. The top layer can be executed in multilayer courses of asphalt, concrete orpaving blocks   **NEW**/1146567

Products

Elastomer trackbed matting: CentriCon and Megiflex rail fasteners; rubber groove-sealing sections for safety of rails in workshops and other pedestrian areas; rubber boots with pads; continuous rail seating; and noise-absorbing material.

UPDATED

Pintsch Aben BV

PO Box 63, NL-3600 AB Maarssen, Netherlands
Tel: (+31 346) 55 27 72
Fax: (+31 346) 55 43 93
e-mail: info@pintschaben.com
Web: www.pintschaben.com

Key personnel

Managing Director: Gert J Bosscher
Account Manager: Hans Hendriksen

Background

Pintsch Aben BV is a subsidiary of Pintsch Bamag, a member of the Schaltbau Group.

Products

Point heating systems (electric and propane gas), point heating components, solid-state snow detectors; local and regional control and communication technology (data transmission, supervision centres).

UPDATED

Plasser & Theurer

Johannesgasse 3, A-1010 Vienna, Austria
Tel: (+43 1) 515 72-0
Fax: (+43 1) 513 18 01
e-mail: werbung@plassertheurer.co.at
Web: showroom.creative.co.at

Main works

Pummererstrasse 5, A-4021 Linz/Donau, Austria

Principal subsidiaries

Australia, Brazil, Canada, Denmark, France, Germany, Hong Kong, India, Italy, Japan, Mexico, Poland, South Africa, Spain, UK and US.

Products

Automatic track levelling, lining and tamping machines; universal tamping machines for tracks, switches and crossings; dynamic track stabilisers; ballast consolidating machines; ballast regulators; ballast cleaning machines; ballast vacuum excavators; formation rehabilitation machines; track laying and relaying machines; gantry cranes; rail grinding and planing machines; mobile rail welding machines, track recording cars; railway motor vehicles; catenary maintenance and inspection cars; catenary renewal trains; railway cranes; special machines and lightweight equipment for track maintenance.

The Plasser &Theurer 08 and 09 Series of tamping machines offers a range of equipment to meet the most varied conditions and requirements.

The 09 Series machines travel forward in a continuous motion while the lifting, lining and tamping units, positioned on a separate underframe, work in cyclic action.

The Dynamic Tamping Express 09-3X is a continuous action three-sleeper tamping machine of the 09-3X Series with incorporated dynamic track stabilisation. The UNIMAT series are machines for plain track, switches and crossings. The latest development is the UNIMAT 08-475. Beside tiltable tamping lines and a three-point lifting device the UNIMAT 08-475 offers the ability to swing out the

outer tamping units up to 3.2 m from the centre of the machine. This means that the departing line at switches can be treated simultaneously (four-rail tamping). The pivoting suspension of the tamping units permits easy adaptation to slanting sleepers.

The UNIMAT 09-4S machines combine the advantages of a continuous action plain line tamping machine with the features of the latest generation of switch tamping machines. Based on the working principle of the 09 Series, the UNIMAT 09-4S can be equipped with a stabilising trailer similar to the Dynamic Tamping Express 09-3X.

Plasser & Theurer states that the range of ballast distributing and profiling machines in the PBR, SSP and USP series and the BDS Ballast Distribution and Storage System cover every type of operation. The BDS system allows MFS material conveyor and hopper units to be coupled between the ballast placement/ploughing section and the sweeping/ballast transfer section to increase storage capacity.

The DGS 62 N (DynamicTrack Stabiliser) produces stabilisation of the track following tamping work to increase the resistance against lateral displacement of the track without altering the track geometry. This eliminates speed restrictions. The RM series offers various sizes of undercutter cleaners for plain track and points. The RM 95-700 represents the new compact class of ballast cleaning machine with two screening units for a high output and the possibility to add new ballast. The VM 170 Jumbo vacuum scraper/excavator removes material such as ballast where obstacles may be present. The EM series offers various sizes of track measuring and recording cars for measuring at speeds up to 250 km/h. For measuring at speeds over 120 km/h, Plasser &Theurer has developed and built vehicles which are equipped with non-contact measuring systems. A laser-based non-contact catenary measuring system can also be provided. The EM-SAT 120 track survey car enables fully mechanised surveying of the actual track geometry using a laser reference chord. The EM-SAT 120 provides high precision measuring and cost-efficiency. The new system, EM-SAT with GPS, will allow further cost savings. Reference points on the masts will be replaced by absolute GPS measurement.

The APT 500 introduced mobile flash-butt welding to the track. Plasser & Theurer welding machines have been built in a variety of designs (mobile, self-propelled, on-track welding machines, on-/off-track welding machines, container-mounted units). Closure welds below neutral rail temperature can be performed using the APT 600 S equipped with a high-capacity rail tensioning device.

The SBM 250 rail rectification (planing) machines re-profile side and head-worn rails in situ. The GWM 250 is a rail grinding machine with two grinding units per rail and six grinding stones per unit; the GWM 550 is a machine with five grinding units.

Plasser &Theurer DynamicTamping Express 09-3X   0598374

Plasser & Theurer EM-SAT 120 with GPS   0598373

Both machines can work on plain track, switches and crossings. The formation rehabilitation machines PM 200 R, AHM 800 R and RPM Series have various applications of ballast recycling. The SUM-SMD and SVM and HUZ Series (high-speed track laying and relaying machines of modular design) are the present day models, which can be adapted for any conditions. The SUZ 500 UVR is a machine for track relaying, new laying and removal with a sleeper laying unit that works in two-sleeper cycle to increase the working speed.

The latest development on this area is the RU 800 S, a combined renewal and ballast cleaning machine.

The WM system is designed for relaying switches and crossings, while the WTW system can transport the pre-assembled switches and switch sections from place of manufacture to the worksite. The MSW, MTW and FUM and CEM series were built for the maintenance, inspection and renewal of catenary at high output. The BCR bridge inspection and repair vehicle enables working operation, particularly at sections of bridges with awkward access.

To meet the demands of customers, a wide range of special machines (for example, sand removal machines) for track maintenance and track work, is available, including all kinds of motor vehicles and single or twin-jib heavy railway cranes as well as lightweight track maintenance equipment.

UPDATED

Pohl Corporation

PO Box 13613, Reading, Pennsylvania 19612, USA
Tel; (+1 610) 926 54 00
Fax: (+1 610) 926 18 97

Key personnel
President: Walter Pohl

Products
Rail from 12 lb ASCE to 175 Crane Rail and related accessories; special trackwork and switch components; spikes; steel sleepers (ties) for narrow-gauge track for the mining industry; New Century® switch stands and replacement parts.

Polysafe Level Crossing Systems Ltd

King Street Industrial Estate, Langtoft, Peterborough PE6 9NF, UK
Tel: (+44 1778) 56 05 55
Fax: (+44 1778) 56 07 73
e-mail: sales@polysafe.co.uk
Web: http://www.polysafe.co.uk

Products
Steel-framed polymer concrete level crossing systems for main line, urban and industrial railway applications.

Portec

Portec Rail Products Inc
PO Box 38250, 900 Freeport Road, Pittsburgh, Pennsylvania 15238-8250, USA
Tel: (+1 412) 782 60 00   Fax: (+1 412) 781 10 37
Web: www.PortecRail.com

Key personnel
Group Vice-President, Railway Maintenance Products: Richard Jarosinski
Group Vice-President, Canadian and UK operations: Konstantinos Papazoglou
Vice-President and General Manager, Railway Maintenance Products: Bruce Wise
Vice-President of Finance and Chief Financial Officer: Michael Bornak

Subsidiaries
Railway Maintenance Products Division
900 Old Freeport Road, Box 38250 Pittsburgh, Philadelphia 15238-8250 US
Tel: (+1 412) 782 60 00   Fax: (+1 412) 782 10 37
e-mail: RMPsales@PortecRail.com
Vice-President and General Manager: Bruce Wise

Salient Systems, Inc
4330 Tuller Road, Dublin, Ohio 43017, US
Tel: (+1 614) 792 58 00   Fax: (+1 614) 792 58 88
e-mail: information@salientsystems.com

Kelsan Technologies Corporation
Kelsan Technologies Corporation
1140 West 15th Street, North Vancouver, British Columbia, Canada V7P 1M9
Tel: (+1 604) 984 61 00   Fax: (+1 604) 984 34 19
e-mail: info@kelsan.com
Web: www.kelsan.com

Canada
Portec Rail Products Ltd
2044 32nd Avenue, Lachine, Quebec H8T 3H7
Tel: (+1 514) 636 55 90   Fax: (+1 514) 636 57 47
e-mail: CANADAsales@PortecRail.com
Vice-President and General Manager: Gerry Clark

UK
Portec Rail Products (UK) Ltd
Vauxhall Industrial Estate, Ruabon, Wrexham LL16 6UY
Tel: (+44 1978) 82 08 20   Fax: (+44 1978) 82 14 39
e-mail: portec@portec.co.uk
Web: www.portecrail.com

Products and services
Standard and insulated rail joints, rail and flange lubrication systems; rail friction measuring systems; lubrication consulting and contracting services;

geocomposite track materials for environmental protection from rail curve greases; locomotive refuelling point mats; track components including rail anchors, gauge plate insulators, switch rod insulators, rail separation temporary repair devices, switchglides, switch point protectors and rail rollover protection devices.

In partnership with Environmental Lubricants Manufacturing, Inc (ELM), Portec Rail also supplies Soy Trak™, a soybean-based high-performance rail curve lubricant, claimed to be safe for the environment.

Contracts
Contracts have included: supply of switchglide to Netherlands Rail; supply of lubricators to Irish Rail and Bulgarian Railways.

UPDATED

Pouget

PO Box 69, 6 allée du Val du Moulin, F-93240 Stains, France
Tel: (+33 1) 48 26 62 12   Fax: (+33 1) 48 22 37 15
e-mail: pouget.rail@wanadoo.fr

Key personnel
General Manager: Robert Pouget

Products
Low-noise rail (standard and rail-grooved rail), rail-corrugation reprofiling unit, hydraulic coach-screwing and fishbolting machines, sleeper drills, disc rail-cutting machines, rail saws, rail drills, portable vibrating tampers, rail grinders, rail loaders, light ballast cleaners, sleeper adzing machines, scaffolding on track, light trolleys, manual and motorised transport platforms, crawlers-conveyors, rail-welding equipment, measurements and control tools, lighting equipment, hand tools, jacks.

UPDATED

Power Machines Group

(Energomachexport + LMZ + Electrosila + ZTL + KTZ)
25A Protopopovsky per, 129090 Moscow, Russian Federation
Tel: (+7 095) 725 27 63   Fax: (+7 095) 688 79 90
e-mail: mail@power-m.ru
Web: www.power-m.ru

Key personnel
General Director: Evgeny Yakovlev
Sales Director (Machinery building): Alexander Zhigalov
Advertising Manager: Natalia Kuznetsova

Background
Power Machines unites the leading Russian power equipment enterprises and supplies railway and transport equipment produced by Kalugaputmash, Lugansteplovoz, Muromteplovoz, Railway Repair-Mechanical Plant of Sverdlovsk, Kolomensky Zavod, Metrowagonmash, Vyksa Steel Works.

Products
Ballast cleaning machines; tamper-leveller-liner machines; tracklaying cranes and gantries; snow-plough equipment; rail welding equipment; portable powered machines for tamping, rail cutting, drilling and grinding; maintenance railcars.

UPDATED

Presto Products Company

PO Box 2399, Appleton, Wisconsin 54912-2399, USA
Tel: (+1 920) 738 11 18   Fax: (+1 920) 739 12 22
e-mail: info@prestogeo.com
Web: http://www.prestogeo.com

Products

Geoweb® perforated cellular confinement system. Applications include: stabilisation of track sub-base; slope and channel protection; earth retention; and surface and base stabilization in terminal yards and port facilities.

Progress Rail Services (Engineering and Track Services)

1600 Progress Drive, PO Box 1037, Albertville, Alabama 35950, USA
Tel: (+1 256) 593 12 60   Fax: (+1 256) 593 12 49
Web: http://www.progressrail.com

Key personnel

Senior Vice-President: David Roeder

Subsidiary companies/divisions

Facilities located in 25 states, Canada and Mexico. The Engineering and Track Services Unit of Progress Rail Services Corporation includes:
Rail and Trackwork Division
(rail crossings, turnouts, track components)
Rail Welding Division
(rail welding)
DAPCO Rail Services
(ultrasonic rail flaw detection)
Signals Division
(level crossing safety devices and signal equipment)
Maintenance of Way Division
(All types of maintenance of way equipment).

Products and services

New and relay rail; custom switches and crossings; rail welding; new and used maintenance-of-way equipment; grade crossing safety devices and signal equipment and mobile rail pick-up crews.

Quest Corporation

950 Keynote Circle, Brooklyn Heights, Cleveland, Ohio 44131, USA
Tel: (+1 216) 398 94 00   Fax: (+1 216) 398 77 65

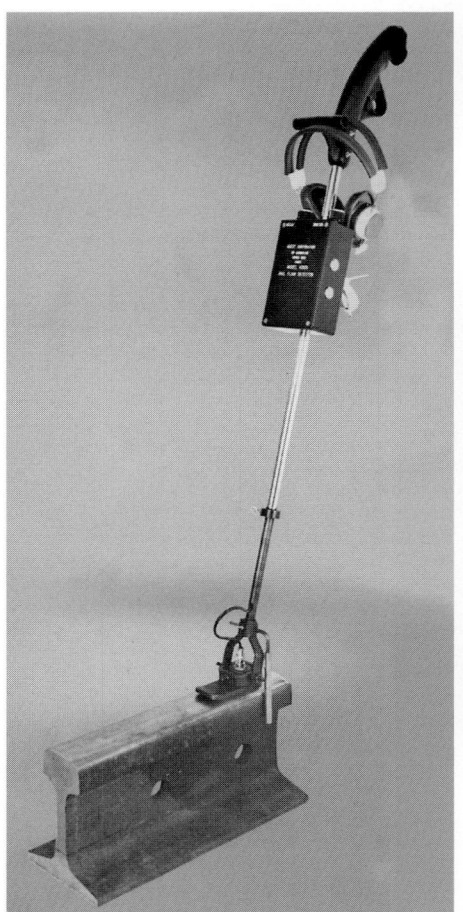

Quest track flaw detector   0021668

Key personnel

For full list of personnel, see Quest Corporation entry in *Signalling and communications systems* section.

Products

Hand-held ultra-sonic rail flaw detectors; electronic weighing systems and controls.

Racine Railroad Products Inc

PO Box 4029, 1524 Frederick Street, Racine, Wisconsin 53404, USA
Tel: (+1 414) 637 96 81   Fax: (+1 414) 637 90 69

Key personnel

President: S J Birkholz
Chief Engineer: D Brenny
Operations Manager: Jim Hilliard
Sales and Marketing Manager: J Syltie
Service Manager: R Rhodes
Controller: G Harmann
Customer Service: L Powell, P Degen

Products

Automatic anchor applicators; dual anchor spreaders and adjusters; dual clip applicators; Pandrol Fastclip applicators/removers; anchor removers; clip setters/applicators; abrasive saws; electric and petrol-driven profile grinders; reciprocating hacksaws; rail drills.

Railquip Inc

3731 Northcrest Road, Suite 6, Atlanta, Georgia 30340, US
Tel: (+1 770) 458 41 57; (+1 800) 325 02 96
Fax: (+1 770) 458 53 65
e-mail: sales@railquip.com
Web: www.railquip.com

Key personnel

President: Helmut Schroeder
Sales Director: Paul Wojcik
Sales Manager: Bill Schroeder
Treasurer and Office Manager: Debbie Fox

Products

Maintenance and supply of: hydraulic track jacks; Azobe hardwood ties; road/rail-equipped Unimog multipurpose vehicle; CNC controlled rail bending machine; Gottwald tracklaying cranes with horizontally extendable boom up to 150 tons (capacity) and 250 tons terrain cranes; Hi-rail equipment trailers; track vacuum refuse collector; gasoline-powered vacuum cleaner back-pack-mounted or with two-wheeled refuse container; cable channels; laser track measuring device; rail aligner; gasoline engine impact wrench; tamper machine and rail grinding machines; automatic rail welding machine.

Contracts

Recent contracts include: Metro Maracaibo, Maracaibo, Venezuela; Tamper track jacks, rail

Railquip's hi-rail equipped trailer with rerailing containers, bogie assembly stand and maxi railcar mover   *NEW*/1143022

floor detector, hand tampers, rail carts, hi-rail equipped trailers, diesel powered unimog; rerailing equipment for Amtrak, New York; rerailing equipment for New Jersey Transit; Maxi railcar movers for Metromara.

UPDATED

Rails Company

101 Newark Way, Maplewood, New Jersey 07040-3393, USA
Tel: (+1 973) 763 43 20   Fax: (+1 973) 763 25 85
e-mail: rails@railsco.com
Web: http://www.railsco.com

Key personnel
President: G N Burwell
Vice-President: J Maldonado
Secretary/Treasurer: M Kinda
Sales: J Vertun, L Ford, D Burwell, P Hest

Products
Rail anchors, switch point locks, switch heaters (propane and natural gas), lubricators, compressors, controls, snow detectors, wheel stops and car retarders.

UPDATED

Railtech International

119 avenue Louis Roche – BP 152, F-92231 Gennevilliers Cedex, France
Tel: (+33 1) 46 88 17 00   Fax: (+33 1) 46 88 17 01
e-mail: management@railtech.fr
Web: http://www.railtech.fr

Key personnel
Chairman and Chief Executive: Jean-Pierre Colliaut
Engineering Director: Didier Bourdon
Sales and Marketing Manager: Gilles du Fou

Products
Aluminothermic rail welding and track maintenance equipment; third rail electrification.

Railtech Schlatter Systems

119 avenue Louis Roche – BP 152, F-92231 Gennevilliers Cedex, France
Tel: (+33 1) 46 88 17 30   Fax: (+33 1) 46 88 17 40
e-mail: management@railtech.fr
Web: www.railtech.fr

Key personnel
Managing Director: P Cathou
Executive Director: Jean-Pierre Mornac

Products
Stationary and mobile flash-butt welding machines and ancillary equipment. Rail welding plants.

UPDATED

RailWorks Corporation

6225 Smith Avenue, Suite 200 Baltimore, Maryland 21209-3613, USA
Tel: (+1 410) 580 60 00

Key personnel
Chairman and Chief Executive Officer: Ab Rees
President and Chief Operating Officer: Jim Kimsey

Subsidiaries
Railworks Products and Services Group
Tel: (+1 412) 325 02 02
e-mail: wdonley@railworks.com
President: William R Donley
Vice-President of Business Development:
  George Caric

Railworks Track Systems Group
Tel: (+1 801) 366 93 39
President: Robert D Wolff

Railworks Transit Systems Group
(+1 914) 323 30 12
President: C William Moore

Pacific Northern Rail Contractors Inc
(a RailWorks Company)
Tel: (+1 604) 850 91 66
President: Henry Braun

Background
In September 2001, RailWorks Corporation and its operating subsidiaries in the US voluntarily filed for protection under chapter 11 of the US Bankruptcy Code.

Services
Integrated rail system services and products, active in new construction, rehabilitation, track repair and maintenance, signalling, communications, electrical and other track-related systems plus rail products manufacturing and supply.

Projects
New York City Transit Authority (NYCTA) White Plains – Phase II signal rehabilitation project, White Plains, New York.
New Jersey Transit (NJT) Morrisville Train Storage Yard, Falls Township Philadelphia: construct 12 new storage yard tracks and to install related crossovers, turnouts, electrical switch heaters and yard lighting. Modification of existing tracks, rehabilitation of two existing yard tracks, yard drainage, accesses roads and paving.
Minnesota Constructors Team, Hiawatha Light Rail Line, Minneapolis: extending from downtown Minneapolis, through Minneapolis, through the Minneapolis–St Paul International Airports to the Mall of America in Bloomington. The completed project will be approximately 11.4 miles long.
TransLink (Vancouver), Millennium Line Expansion of the Vancouver Skytrain, Vancouver, Canada: Linear induction motor and power rail installation, including the installation of 40,000 lineal metres of LIM (Linear Induction Motor) reaction rail and 80,000 linear metres of a side-mounted power rail including thermal expansion joints, power feeds, isolation joints, and coverboard. Also, core drilling and grouting of 120,00 support studs for the LIM rail and core drilling 60,000 holes to anchor the power rail insulator brackets.
New York City Transit Authority (NYCTA) Carnarsie Line, New York: complete installation of a new Communications Based Train Control System (CBTC) for 13 route miles of the Canarsie Line subway in association with Matra Transport International and Union Switch & Signal.
NYCTA, New York: Rehabilitate the train control signal system for 5.1 miles of track on the Flushing Line subway. Also to furnish and install eight escalators at five locations in the boroughs of Brooklyn and Manhattan.
Miami International Airport, Miami, Florida: Electrical and guideway construction services including automatic train control systems and 8,500 ft of concrete guideway in association with Sumitomo.
Panama Canal Railway, a venture between Kansas City Southern Industries and Lanco International: Reconstruction and expansion of track and facilities involving a 44-mile stretch of track, running parallel to the Canal from Colon to Balboa.
Vancouver Wharves, British Columbia, Canada: Rail yard construction services.
Long Island Rail Road, Long Island: Rehabilitation and improvements of the diesel yard.
IPSCO Steel, Mobile: Steel mill expansion project involving 30,000 crossties and additional switch ties.
US Lime and Minteral-Arkansas Lime, Batesville: The design/build project involves the removal and construction of 15,000 ft of track including three turnouts, grade crossing and bridgework.
Tuscola and Saginaw Bay Railroad, Michigan: Refurbishment project involving 20,000 crossties for the 12.5-mile branch line.

Ranalah

Ranalah Moulds Ltd
New Road, Newhaven BN9 0EH, UK
Tel: (+44 1273) 51 46 76   Fax: (+44 1273) 51 65 29

Key personnel
Directors: J H Layfield, P F Phillips
Technical Director: P F Phillips
Technical/Design Manager: S Orwin

Products
Design of prestressed concrete sleepers, moulds and equipment; design and manufacture of equipment to suit any type of sleeper production; factory layout design, plant supply and commissioning; full operational training, initial management assistance. Turnkey service for supply of sleeper production and manufacturing plant.

Contracts
Contracts include supply of sleeper plants to Hong Kong, Philippines, Thailand, Georgia, Turkey, Sudan, Malaysia, Bulgaria, Europe and the UK.

A Rawie GmbH & Co KG

Dornierstrasse 11, D-49090 Osnabrück, Germany
Tel: (+49 541) 91 20 70   Fax: (+49 541) 912 07 10
e-mail: info@rawie.de
Web: http://www.rawie.de

Key personnel
Managing Director, Marketing: J Fründ
Export Manager: N L C Pratt
Project Engineer: T Riesopp
Senior Design Engineer: H Klose

Products
Fixed and friction buffer stops; fixed and/or friction buffer stops with hydraulic or elastomeric cylinders; friction, fixed and folding wheel stops; specialist track endings to customer requirements, including folding buffer stops and buffer stops with integral loading ramps.

Contracts
Include the supply of buffer stops to Lisbon Metro, Portugal; MTRC, Hong Kong; Boston, USA; Bielefeld, Germany; and Guangzhou, China. Heathrow Express, Kuala Lumpur, Star and Putra Lines, North East Line Singapore NEL and Changi Extension, High-speed line Taipei/Kaohsiung, Metro Kaohsiung, Red Line.

UPDATED

Rex

Rex Articoli Tecnici
Via Catenazzi 1, CH-6850 Mendrisio, Switzerland
Tel: (+41 91) 640 50 50   Fax: (+41 91) 640 50 55
e-mail: sales@rex.ch
Web: http://www.rex.ch

Key personnel
General Manager: M Favini

Products
Swisscross modular rubber level crossing system suitable for various track gauges; rail pads; microcellular pads; rubber boots for concrete sleepers; heatable rubber elements for station platforms.

Rexquote Ltd

Broadgauge Business Park, Bishops Lydeard, Taunton TA4 3BU, UK
Tel: (+44 1823) 43 33 98   Fax: (+44 1823) 43 37 25
e-mail: sales@rexquote.co.uk
Web: http://www.rexquote.co.uk

Products
Road-rail conversions and products including: road-rail hydraulic excavators (roadrailers);

road-rail hydraulic crane/excavators (superailers/ megarailers); road-rail dumpers and road-rail access platforms; road-rail tracked vehicles; trailers; rail attachments and one-off engineering projects.

VERIFIED

RICA

Via Podgora 26, I-31029 Vittorio Veneto (TV), Italy
Tel: (+39 0438) 91 01
Fax: (+39 0438) 91 22 36; 91 22 72; 91 03 26
e-mail: rica@zoppas.com
Web: www.rica.it

Background
RICA is a member of the Zoppas Industries Group.

Products
Switch-point and rail de-icing systems.

UPDATED

RMC Rail Products

Aston Church Road, Saltley, Birmingham B8 1QF, UK
Tel: (+44 121) 327 08 44
Fax: (+44 121) 327 75 45
e-mail: sleepers@rmc.co.uk

Key personnel
General Manager: Nick Gainsford
    e-mail: nick.gainsford@rmc.co.uk
Sales Manager: Andy Carey
    e-mail: andy.carey@rmc.co.uk
Works Manager: Stuart Neil
    e-mail: stuart.neil@rmc.co.uk
Quality Manager: Paul Crowther
    e-mail: paul.crowther@rmc.co.uk

Other offices
St Helen Auckland, Bishop Auckland, DL14 9AJ, UK
Tel: (+44 1388) 60 39 31   Fax: (+44 1388) 45 00 56

Products
Pretensioned concrete monobloc sleepers; reinforced concrete sleepers; concrete cable troughs, crossing bearers, platform units; plus general precast products for station regeneration and building work.

Contracts
Half of the annual requirement of concrete sleepers used by Network Rail is supplied by RMC, which is also a large supplier of concrete crossing bearers and cable troughing.

UPDATED

RMS Locotec

Rail Management Services
Vanguard Works, Bretton Street, Dewsbury WF12 9BJ, UK
Tel: (+44 1924) 46 50 50
Fax: (+44 1924) 46 54 22
e-mail: sales@rmslocotec.com
Web: www.rmslocotec.com

Key personnel
Managing Director: Lawrence Crossan
Commercial Manager: Derek Webb
Production Development Engineer: Peter Briddon

Services
Maintenance of track, structures and fixed equipment; short- and long-term locomotive hire, maintenance and overhaul; safety management and training; engineering consultancy; rail operations and facilities management.

UPDATED

Robel

Robel Bahnbaumaschinen GmbH
Industriestrasse 31, D-83395 Freilassing, Germany
Tel: (+49 8654) 60 90   Fax: (+49 8654) 60 91 00
e-mail: info@robel.info
Web: www.robel.info

Key personnel
Managing Director: Erwin Stocker

Products
Heavy machinery: track vehicles, rail handling vehicles, track laying machines for panels, sleepers and switches, rail loading and transportation equipment, catenary maintenance systems, special wagons and trailers.

Small machinery: clipping machines, rail drills, rail cutters, rail grinders (profile, web and switch), rail lifting devices, rail clamps for temporary joints, rail benders, plain line and switch blade, power wrenches, carrying tongs for sleepers and rail, track jacks, mechanical and hydraulic, track lifting and slewing machines, material transport trolleys (1 and 12 tonnes) lightweight tamping units, site illumination systems, track measuring devices, full range of special railway hand tools.

Contracts
A contract with Pandrol Rail Fastenings based in the UK, for the supply of two grinding machines type 13.48 and a clipping machine type 34.01.

In 2002, a contract was signed with German Rail (DB) for 27 track vehicles. Several machines have already been delivered and are in use and the last of the machines will be delivered in 2006.

UPDATED

Rosenqvist Rail Tech AB

Box 334, SE-824 27 Hudiksvall, Sweden
Tel: (+46 650) 165 05   Fax: (+46 650) 165 01
e-mail: anders.rosenqvist@rosenqvist-group.se
Web: www.rosenqvistsrt.se

Key personnel
Managing Director: Anders Rosenqvist

Products
Rail fastening machines and attachments; sleeper laying and replacement attachments; technical consulting and engineering services; and training.

UPDATED

Rotabroach Ltd

Imperial Works, Sheffield Road, Tinsley, Sheffield S9 2YL, UK
Tel: (+44 114) 221 25 10   Fax: (+44 114) 221 25 63
e-mail: info@rotabroach.co.uk
Web: http://www.rotabroach.co.uk

Key personnel
Director and General Manager: R Stych
Managing Director: M Duffin
Sales and Marketing: N E Price
Sales Director: P W Dorsett
Technical Director: B Wood

Products
Petrol-driven rail drill for remote locations, designed with an integral two-stroke engine and petrol tank. It weighs 18 kg and can be operated and carried by one man.

Micro bonder weighing 12.3 kg for bonding hoe and signal drilling applications. It is ready for use quickly by slipping the clamping unit beneath the rail and locking it with a single lever to produce two bonding holes with either a drill up to 12 mm diameter or a Rotabroach cutter up to 22 mm diameter. The machine indexes from one centre to the other on a horizontal slideway.

Hydraulic rail drill can be used for cutting holes up to 38 mm diameter through rails up to 25 mm

thick. It can be used on high tensile steel rails where high torque and low running speeds are required for cutting and because it is quiet when operational can be used at night in urban areas.

High-frequency rail drill with a 135 V or 265 V 200 Hz electric motor; can be used for fish plate holes, stretcher rails and bonding hole applications.

Electric rail drill used by railways around the world. It weighs 21 kg and comes in two variations, the RD 171M 110 V and RD 173M 230 V.

Contracts
Recent contracts include the supply of equipment to SBB, ČD, Japan, Singapore, Hong Kong and Australian National Railways.

Rotamag Track Equipment

41 Catley Road, Darnall, Sheffield S9 5JF, UK
Tel: (+44 114) 291 10 20   Fax: (+44 114) 261 31 86
e-mail: bobsenior@bryar.co.uk
Web: www.rotamag.co.uk

Key personnel
Managing Directors: Vic Archer
Operations Director (Contact): Bob Senior

Background
Rotamag Rail is part of the Bryar Group.

Products
Rail drilling machines; rail broaching cutters; rail jacks; rail bonding equipment; permanent way hand tools; third rail shrouds; rail trolleys; weld shears; and track-handling equipment.

Contracts
Contracts include the supply of rail bonding equipment, drilling machines and pulling tools for London Underground Ltd, Glasgow Underground, SNCB and SJ.

UPDATED

Salient Systems Inc

4330 Tuller Road, Dublin, Ohio 43017-5008, USA
Tel: (+1 614) 792 58 00   Fax: (+1 614) 792 58 88
Web: http://www.salientsystems.com

Key personnel
President and Chief Executive Officer:
    Harold Harrison
Executive Vice-President and Chief Financial
    Officer: S A Harrison
General Manager and Sales Director: Dana Earl

Products
Microprocessor-based wayside inspection and detection instrumentation for monitoring track/ train dynamics at selected wayside sites. Salient's Mk II Wheel Impact Load Detector (WILD) will report and identify flat or out-of-round wheels causing excessive impact to track and structures.

Sateba

Tour Ariane
5 place de la Pyramide, F-92088 Paris La Defénse, Cedex, France
Tel: (+33 1) 46 53 29 00   Fax: (+33 1) 46 53 29 10
e-mail: sateba.paris@sateba.com
Web: www.sateba.com

Key personnel
General Manager: D Vallet

Works
Chalon sur Sâone, Tours La Rick, Charmes, France Morocco, Portugal and UK.

Background
Sateba Système Vagneux is a public limited company which in 2004 had sales of €42 million. Main shareholder is Bonna Sabla.

The company manufactures under licence in Egypt, Greece, Italy, Korea, Mexico, Morocco, UK and Venezuela.

Products

Design and manufacture of Vagneux system of concrete sleepers, prestressed concrete sleepers for turnouts and SAT S312 solution for slab track; design and commissioning of sleeper manufacturing plants. Also technical studies, assistance and staff training.

Contracts

Include tramways at Lille, Nantes, Strasbourg, Grenoble, Rouen, St Etienne, Geneva, Lyon, and Montpellier, France.

Supply to Belgium, Brazil, Portugal, Switzerland and the US.

UPDATED

Schlatter AG, HA

Brandstrasse 24, CH-8952 Schlieren, Switzerland
Tel: (+41 1) 732 71 11; 72 00
Fax: (+41 1) 732 72 02
e-mail: email@schlatter.ch
Web: www.schlatter.ch

Key personnel

Senior Engineer: Juerg Wahrenberger
Product Manager: Robert Schniegler

Subsidiary companies

Railtech Schlatter Systems SAS (joint venture)

Products

Stationary and mobile flash-butt welding equipment.

UPDATED

Schreck-Mieves GmbH

Krückenweg 113, D-44225 Dortmund, Germany
Tel: (+49 231) 710 82 59
Fax: (+49 231) 71 15 88
e-mail: julia.schreck@schreck-mieves.de
Web: http://www.schreck-mieves.de

Key personnel

Product Marketing: Julia Schreck

Products and services

Planning, construction and maintenance of complete track systems; supply of turnouts and track materials; EKOS switch tongue roller systems for main line and light rail applications; EWOS lubrication-free switch locking systems; EMA mobile track inspection and electronic measurement equipment and services; switch inspection services; wheel-rail interface investigation and analysis; consultancy services on track and switch maintenance, and management of track systems.

Schweizer Electronic AG

Industriestrasse 3, CH-6260 Reiden, Switzerland
Tel: (+41 62) 749 07 07
Fax: (+41 62) 749 07 00
e-mail: info@schweizer-electronic.ch
Web: http://www.schweizer-electronic.ch

Products

Safety equipment and systems for permanent way engineering staff and track workers.

Recent products include the Minimel 95 system in which train movements are detected by hand-held or rail-mounted switches, or from the signalling centre, and relayed by radio a control centre. This then activates devices in the work area, which emit audible and visual warnings.

Schwihag Gesellschaft für Eisenbahnoberbau mbH

Lebernstrasse 3, CH-8274 Tägerwilen, Switzerland
Tel: (+41 71) 666 88 00   Fax: (+41 71) 666 88 01
e-mail: info@schwihag.com
Web: www.schwihag.com

Key personnel

Managing Director:
Dipl Betriebswirt Karl-Heinz Schwiede
Sales Director: Dr Ing Roland Buda
Technical Director: Frank Meyer

Background

Schwihag Gesellschaft für Eisenbahnoberbau mbH was founded in 1971 and has ISO 9001 certification.

Research and development, design, production and marketing of progressive components for railway switches and crossing assemblies.

Products

IBSR and IBRR rail anchoring systems for turnouts, including clips, baseplates and slide plates.

Roller slide plates for lubrication-free operation, available for full and shallow depth switch assemblies. Boltless check rail support for continuous check rails in curved track and opposite crossings. Modular sleepers with an integrated switch actuating system. Hollow steel sleepers for cable crossings.

Equipment has been supplied to the MTRC Lantau Airport Railway, KCRC Hong Kong, SMRT Singapore, TRC Taipei, Canadian National, TTC Toronto and Network Rail, UK as well as to many other major European national railway companies including DB, RENFE, RFI Italy, SNCB and SNCF.

UPDATED

SECO-RAIL

Espace Lumière, Bâtiment 1, 6 rue Emile Pathè, F-78403 Chatou Cedex, France
Tel: (+33 1) 30 09 83 00   Fax: (+33 1) 30 09 83 01
e-mail: info@seco-rail.com

Background

SECO-Rail is part of the Colas group.

Services

Track laying; track renewal; track maintenance.

VERIFIED

Selectra srl

Via delle Nazioni 5, I-44100 Ferrara, Italy
Tel: (+39 0532) 74 80 00
Fax: (+39 0532) 74 44 55
e-mail: info@selectra.org
Web: http://www.selectra.org

Products

OptoCross opto-electronic system for the high-speed contactless measurement of geometric and dynamic track parameters.

Semperit

Semperit Technische Produkte GmbH & Co KG
Triester Bundesstrasse 26, A-2632 Wimpassing, Austria
Tel: (+43 2630) 31 05 46   Fax: (+43 2630) 31 05 38
Web: www.semperit.at

Key personnel

Divisional Managers: Dr Günther Mai, Peter Horn

Products

Rubber and plastic track items, including rail pads, elastic sleeper supports and plastic fastenings.

Elastomer profiles, including track trough sealing sections.

UPDATED

Sersa Group Management AG

Churerstrasse 104, CH-8808 Pfäffikon, Switzerland
Tel: (+41 55) 420 30 31   Fax: (+41 55) 420 30 32
e-mail: info@sersa-group.com
Web: http://www.sersa.ch

Key personnel

President: Konrad Schnyder

Subsidiary companies

Sersa SA
Brauerstrasse 126, CH-8004 Zurich, Switzerland
Tel: (+41 43) 322 23 23   Fax: (+41 43) 322 23 99

Sersa GmbH
Kaulsdorfer Strasse 209, D-12555 Berlin, Germany
Tel: (+49 30) 565 46 60   Fax: (+49 30) 56 54 66 15

Sersa (UK) Ltd
Sersa House, Auster Road, Clifton Moor Industrial Estate, Clifton Moor, York YO30 4XA, UK
Tel: (+44 1904) 47 99 68   Fax: (+44 1904) 47 99 70

Benkler AG
Eurogleis
Euroswitch
Gleis- und Bautechnologie AG
Klenk Verwaltungs GmbH
Metrag AG
Parachini SA
SIT GmbH

Services

Track construction; construction of track substructures; planning industrial track systems; track maintenance; operation, hire and maintenance of track construction and maintenance machinery and works locomotives; rail grinding; rail welding; track cleaning; resinification of sleeper screw holes; overhead line construction and maintenance; cable construction.

SGB Group

Harsco House, Regent Park, 299 Kingston Road, Leatherhead, Surrey KT22 7SG, UK
Tel: (+44 1372) 38 13 00
e-mail: info@sgb.co.uk
Web: http://www.sg.co.uk

Background

SGB is a member of Harsco Corporation Access Services Group.

Link trolley for the BOSS Zone 1 access system
0114665

Road-rail mobile elevating work platform from SGB on the Channel Tunnel Rail Link 0576759

Products

Access equipment for infrastructure construction and maintenance. Recent products include the road/rail boom and rail trolley, developed specifically to provide access for hazardous work in the rail sector. It has been developed in partnership with SGB's sister company, Harsco Track Technologies. The SGB rail boom was recently used by ElecTrack Installations, to install overhead power lines on Section 1 of the Channel Tunnel Rail Link project. The SGB Rail Trolley is a rail chassis system for the SGB BoSS GRP scaffold tower, produced for use in Zone 1 areas. The tower is non-sparking, non-conductive and resistant to a range of potentially damaging substances such as oils, solvents, lacquers and petrol.

S G Technologies GmbH

D-66687 Wadern-Büschfeld, Germany
Tel: (+49 6874) 691 09   Fax: (+49 6874) 691 59
e-mail: sgtechnologies.com
Web: http://www.sgtechnologies.com

Key personnel

General Managers: Achim Pietsch,
 Dr Wolfgang Guth
Manager, Moulded Parts Department:
 Werner Berens
Sales Manager, Rail Products: Werner Koch

Subsidiaries

SaarGummi Iberica, Madrid, Spain
SaarGummi Czech, Cerveny Kostelec, Czech Republic
SGGS, India
SaarGummi Americas, Ann Arbor, USA
SaarGummi Tennessee, Pulaski, USA
SaarGummi Quebec, Canada
SaarGummi do Brasil, Bahia, Brazil

Products

Products for ballasted and ballastless track, including: solid rubber boots and baseplate pads; rail and microcellular pads of foamed synthetic rubber with closed cellular structure.

Contracts

Recent contracts include the supply of components to Deutsche Bahn AG, Hanover-Berlin, Hanover–Würzburg and Cologne–Frankfurt new high–speed Intercity III line, Leighton RSA (West Rail, Hong Kong), Channel Tunnel Rail Link ballastless high-speed lines and Nuremberg-Ingolstadt, Hamburg–Berlin new high–speed lines.

UPDATED

Siemens Switzerland

Siemens Switzerland Ltd, Transportation Systems
Industriestrasse 42, CH-8304 Wallisellen, Switzerland
Tel: (+41 585) 58 55 85   Fax: (+41 585) 58 05 01
e-mail: ts@siemens.ch
Web: http://www.siemens.ch/ts

Key personnel

Managing Director: Willy Gehrer
Manager, Export Department, South East Asia:
 A Hefti

Products

Block and track vacancy proving equipment (axle counters, track circuits, last vehicle detection); point locking equipment including point machines and point locking system; signals and indicators; safety relays.

UPDATED

Sika Ltd

Watchmead, Welwyn Garden City AL7 1BQ, UK
Tel: (+44 1707) 39 44 44   Fax: (+44 1707) 32 91 29
e-mail: sales@uk.sika.com
Web: www.sika.co.uk

Key personnel

Managing Director: D Bratt
Projects Manager: P Gair

Products

Sika Rail resilient rail-fixing systems are a combination of tough elastic, polyurethane reaction-curing binders and compressible fillers that absorb vibration and reduce noise. They are available in a variety of grades to ensure suitability with differing load-bearing requirements.

Sika also provides specialist construction products including high-specification concrete admixtures, jointing systems, mortars and grouts, adhesives and bonding agents, waterproofing, corrosion inhibitors, concrete repair and protective coatings, single-ply roofing membranes and high performance industrial and commercial flooring systems.

Contracts

Contracts within the UK include: Base-plate bedding material for direct fixation system for London Underground Northern Line; Heathrow Express (track fixation at Paddington station platforms and at Semi-Outside Double Slip (SODS) crossover, Heathrow Airport); Leeds Portal Link, Liverpool; Newcastle Metro; Docklands Light Railway; The Channel Tunnel Rail Link, Clerkenwell. Structural strengthening: Church Street Cast Iron Bridge, Telford; North Harrow Bridge; Worston Bridge; The Approach Bridge at Humber Sea Terminal; Ealing Road Bridge; LUL Edgeware Road Station, London; Birmingham Airport cable train, Hook-a-Gate Bridge and Gypsy Lane Flyover, A38 Bristol; Rail fixing and grouting: Tate & Lyle Crane Rails, London; Rugby Maintenance Depot; Newcastle Metro, Docklands Light Railway and Liverpool Portal Link Tunnel. Roofing: Stockport railway station, Burnley bus station and Salisbury railway station. Industrial flooring: Arriva Bus Depot, Birkenhead; Turner Powertrains; The Channel Tunnel Rail Link, Contracts 230 and 240; Northern Traincare Facility, Ealing Common Wash Depot and Bombardier Transportation in Kent. Concrete repair and protection: Kirklees bus station car park.

UPDATED

SILF Srl

Via Romagnosi 60, I-29100 Piacenza, Italy
Tel: (+39 0523) 33 85 85   Fax: (+39 0523) 38 51 24
e-mail: silf@writeme.com
Web: http://www.silfsrl.com

Works

SS 10, Via Emilia Pavese 2, I-29010 Sarmato Piacenza, Italy
Tel: (+39 0523) 88 62 00   Fax: (+39 0523) 88 78 20

Key personnel

Managing Director: C Luisa

Products

Light and heavy rail sections; crane rail and grooved rail for tramways; guardrail; wooden, concrete and steel sleepers; plates; clips; nuts and bolts; spikes; anchor bolts; screwspikes; fishplates and baseplates; buffer stops; frogs; points and crossings; turnouts.

Contracts

Contracts include supply of equipment to Fiat, Condo-Metro, Rome, Nova SpA, Milan and MetroRoma SpA.

H J Skelton (Canada) Ltd

165 Oxford Street E, London, Ontario N6A ITA, Canada
Tel: (+1 519) 67 99 18 00
Fax: (+1 519) 679 01 93; 434 47 87

Key personnel

General Manager: Peter Fraser
Sales Director: Geoffrey Richey

Associated company

TKL Rail, Australia

Products

Supplier of a wide variety of track components, special trackwork, sliding rail expansion joints, switch machines, sliding rail buffer stops, Icosit polyurethane/cork grout for undersealing grooved rail and injected pads for direct fixation. Specialises in LRT in-street applications, also railway rail (both T and grooved), crane rails and turnouts. Rawie sliding friction buffer stops to suit a range of applications. Range includes high-speed buffer stops bumping posts to stop trains at up to 56 km/h and also for low-floor LRVs. Contec switch machines and track wiring systems. Range of rails, points, crossing and special trackwork.

Contracts

Five-year supply contract for TTC Toronto for all-manganese frogs, crossing and points for tram tracks; special trackwork for Calgary LRT; Icosit polyurethane grout for Tri-Met Portland, Memphis and Salt Lake City; screw spikes and washers for CN Rail and CP.

H J Skelton & Co Ltd

9 The Broadway, Thatcham RG19 3JA, UK
Tel: (+44 1635) 86 52 56
Fax: (+44 1635) 86 57 10
e-mail: info@hjskelton.com
Web: http://www.hjskelton.co.uk

Key personnel

Director: J W G Smith

Products

Agents for Rawie sliding friction buffer stops, fixed stops and wheel stops throughout the UK and Ireland. Railway rails and crane rail to British, European and North American (BS, UIC and ASTM) standards. Special trackwork in flat bottom and grooved rails.

Contracts

Recent contracts include the supply of Rawie friction element buffer stops for the Channel Tunnel Rail Link, St Pancras; Iarnród Éireann, Ireland; and Nottingham Express Transit.

VERIFIED

Slab Track Systems International

Web: http://www.sts-international.com
e-mail: infor@sts-international.com

Germany:
Max Bögl Bauunternehmung GmbH & Co KG
Industriegebiet Schlieferheide/Bögl, PO Box 1120,
D-92301 Neumarkt, Germany
Tel: (+49 91) 81 90 90   Fax: (+49 91) 81 90 50 61

Netherlands:
Grimbergen Engineering & Projects BV
Bedrijfsweg 23-25, PO Box 145, NL-2400 AC Alphen
aan den Rijn, Netherlands
Tel: (+31 172) 43 27 21   Fax: (+31 172) 44 42 21

Products
The Bögl solid railbed system, comprising
prefabricated panels, connected longitudinally and
installed on a hydraulically bonded bed.

Socofer's Sandite module for Network Rail   ***NEW**/1112525*

SMIS

Surrey Materials Inspection Systems
Alan Turing Road, Surrey Research Park, Guildford,
Surrey, GU2 5YF, UK
Tel: (+44 1483) 50 66 11   Fax: (+44 1483) 56 31 14
e-mail: materials@smis.co.uk
Web: http://www.smis.co.uk

Key personnel
General Manager: Dr N MacCuaig

Products
Automated and manual equipment for railway non-
destructive testing and infrastructure monitoring
applications; vehicle-based inspection systems
for infrastructure monitoring including ultrasonic track inspection, overhead line inspection,
track geometry, structure gauging, as well as
a range of depot maintenance equipment. The
BR2000 is claimed to be the lowest noise flaw
detector available for axle inspection and the only
such instrument to be certified by BR. The R56
digital flaw detector for track, wheels and other
applications was developed in collaboration with
BR for manual inspection of track and is certified
by Railtrack.
Recent contracts include the supply to Indian
Railways of two sets of instrumentation to monitor
the condition of overhead catenary and the
installation of a major online ultra-sonic inspection
system for the Steel Authority of India Ltd.

Socofer

7 boulevard Louis XI, Zone Industrielle du
Menneton, BP 0507, F-37205 Tours Cedex 3, France
Tel: (+33 2) 47 39 28 24   Fax: (+33 2) 47 37 63 04
e-mail: info@socofer.com
Web: www.socofer.com

Key personnel
President: Pierre Hallé
Director Railway Division: Bertrand Hallé
Sales and Marketing Manager: Fabrice Thierry

Products
Railway infrastructure vehicles, including road-
rail inspection vehicles and self-propelled track
and overhead line inspection and maintenance
vehicles. Track maintenance: railhead waterjetting,
Sandite application and track weed spraying.
Infrastructure maintenance: handling cranes,
personnel lifting baskets and platforms.

Contracts
Recent contracts include 30 LORRIC road-rail
overhead line inspection and maintenance
vehicles for SNCF, France; 138 track maintenance
and inspection vehicles for Syrian Railways; self-
propelled overhead line maintenance vehicles for
INEO SCLE and Vossloh Infrastructure services;
64 track maintenance modules for Network Rail;

LORRIC road-rail overhead line inspection for SNCF   ***NEW**/1112526*

1 weedspray module for Network Rail CTL line;
3 waterjet modules for SNCF.

NEW ENTRY

Sonatest Ltd

Dickens Road, Old Wolverton, Milton Keynes MK12
5QQ, UK
Tel: (+44 1908) 31 63 45   Fax: (+44 1908) 32 13 23
e-mail: sales@sonatest-plc.com
Web: www.sonatest-plc.com

Key personnel
Managing Director: M S Reilly
UK Sales Director: Kelvin Cook
Operations Manager: Carol Stevenson

Principal subsidiary
Sonatest Inc
4734 Research Drive, San Antonio, Texas, US
Tel: (+1 210) 697 03 35   Fax: (+1 210) 697 07 67
e-mail: sonatest@sbcglobal.net
Vice-President: R Sidney

Products
Ultrasonic digital flaw detectors and thickness
gauges. This includes the special application
Railscan 125 ultra-sonic portable flaw detector
for rail testing; Powerscan 400 for axle testing;
ultrasonic transducers for testing rail, wheels and
axles, including special probes used in systems
for Thermit weld testing; rail tester trolleys and
equipment for the Bance rail cart. Sonatest also
supplies equipment for x-ray, phased array and
eddy current testing. Also available are systems
and robotic methods of NDT. Current research and
development projects are focused on dry scanning
systems and concrete testing.

Contracts
Equipment has been supplied to Applied
Inspection, Balfour Beatty Rail Maintenance, SNCF,
GTRM, London Underground Ltd, MTR Corporation
of Hong Kong and Serco Ltd. Current research and
development projects are focused on dry scanning
systems and concrete testing.

UPDATED

Speno International SA

PO Box 16, 26 Parc Château-Banquet, CH-1211
Geneva 21, Switzerland
Tel: (+41 22) 906 46 00   Fax: (+41 22) 906 46 01
e-mail: info@speno.ch
Web: www.speno.vhost.virtua.ch

Key personnel

Managing Director: J J Méroz
Marketing Managers: D Mor; R Koller
Finance Manager: G Chignol
Technical Manager: L Palmieri
Purchasing Manager: G Tochon
Production Manager: J Neumayer
Maintenance Manager: G Stranczl
Manager, External Affairs: W Schöch
Quality Assurance Manager: G Ferioli

Principal subsidiaries

Speno Rail Maintenance Australia Pty Ltd
168 Campbell Street, Belmont, Western Australia
6104, Australia
Tel: (+61 8) 94 79 14 99   Fax: (+61 8) 94 79 13 49

Nippon Speno KK
Resona Gotanda Building 5F, 1-23-9 Nishigotanda,
Shinagawa-Ku, Tokyo 141-0031, Japan
Tel: (+81 3) 34 95 71 61   Fax: (+81 3) 34 95 71 62

Products

Design, development and manufacture of machines
for in-track rail maintenance, including rectification
by grinding, rail measurement and detection of
internal rail flaws.

There are 150 Speno machines in service around
the world, engaged on high-speed and conventional
lines, and also in the demanding conditions of heavy-
haul and metropolitan traffic. Speno machines are
operated by railway authorities, intermediaries or by
Speno International itself on a service basis.

Speno rail grinders are grouped in series, depending
on production requirements. There are typically series
of machines with eight, 16, 24 or 48 units. Machines
can be coupled to produce intermediate numbers
of units. Specialised versions exist for operation
on plain track and in the complicated conditions of
switches and crossings. Optional equipment permits
underground working. The machines have multiple
applications. They can treat rail deformation in both
the longitudinal and the transverse planes as well
as damaged surface condition. Speno rail grinders
are suitable for preventative and corrective rail
maintenance strategies.

Speno measurement equipment provides
continuous recording of longitudinal and transverse
rail profiles. There are two main versions,
distinguished by operating speed. The high-
speed equipment, fitted on independent vehicles,
supplies data for network-wide asset monitoring
and maintenance planning. The second version is
incorporated into rail grinders. Its role is to facilitate
machine operation by computer assistance, and to
check and record work progress for management.

Speno rail flaw detection units include rail
vehicles that run non-stop while reporting
potential flaws for subsequent confirmation, and
rail/road vehicles that can stop for immediate

confirmation where local conditions permit.
The non-stop operating speeds of Speno rail
flaw detectors are exceptionally high. Detection
is assured by continuous automatic analysis of
flaw information combined with graphic display.
Location of potential flaws is facilitated by paint
marking of the track.

Speno International runs an extensive research
and development programme both in the design
and the use of its products.

UPDATED

Sperling Railway Services Inc

4313 Southway Street SW, Canton, Ohio 44706,
USA
Tel: (+1 330) 479 20 04   Fax: (+1 330) 479 20 06
e-mail: info@sperlingrailway.com
Web: www.sperlingrailway.com

Key personnel

President: Fred Sperling
Vice-President, Sales and Marketing:
    Warren Stryffeler

Products

Maintenance of way equipment, related items and
railway track signs.

UPDATED

Sperry

Sperry Rail Service Inc
46 Shelter Rock Road, Danbury, Connecticut 06810,
USA
Tel: (+1 203) 791 45 00   Fax: (+1 203) 797 84 17
e-mail: info@sperryrail.com; aveitch@sperryrail.com
Web: www.sperryrail.com

Key personnel

President: Gary S Klein
Managing Director, International Business:
    Alastair Veitch
Director of International Operations:
    Robert Crocker
Director of Sales: Glenn Rooney
Director of Engineering: R Mark Havira

Principal subsidiaries

Sperry Rail International (UK) Ltd
Trent House, RTC Business Park, London Road
DE24 8UP, UK
Tel: (+44 1332) 26 25 65

Background

Founded in 1928, the company's single purpose was to
improve railway safety by focusing on rail flaw detection.
The company has built its reputation by investing in
research and development and continuously improving
ultrasonic and induction testing techniques.

Products and services

Sperry Rail Service is a leader in ultrasonic and
induction rail flaw detection. The company provides

a comprehensive range of rail flaw detection
equipment that includes a variety of railbound and
hi-rail vehicles and portable hand testing equipment,
all based on Sperry's Roller Search Unit (RSU)
technology, and incorporating advanced pattern-
recognition methods. The company's growing fleet
currently comprises more than 155 test vehicles that
are active throughout the world. Sperry can provide
tailored solutions to customers' operations needs,
including advanced data management processes.
Sperry's current development activities include
increased test speeds, improved rail life management
and enhanced defect detection.

UPDATED

Spie

Société Spie (Service Voies Ferrées)
Parc Saint-Cristophe, Edison 3, F-95861 Cergy-
Pontoise Cedex, France
Tel: (+33 1) 34 22 50 02; 3   Fax: (+33 1) 34 22 57 24
Web: http://www.spie.fr

Key personnel

Operations Director: D Mallet
Export Manager: J Lemercier

Products

Track welding and maintenance equipment.

Spitzke GmbH

Warmensteinacher Strasse 60, D-12349 Berlin,
Germany
Tel: (+49 30) 762 90 90
Fax: (+49 30) 76 29 09 90; 76 29 09 91
e-mail: berlin@spitzke.de
Web: http://www.spitzke.de

Electrical engineering/contact lines division
Markgrafendamm 24/Haus 16, D-10245 Berlin,
Germany
Tel: (+49 30) 293 46 50   Fax: (+49 30) 29 34 65 99

Key personnel

Managing Director: Waldemar Münich

Subsidiary

Spitzke Spoorbouw BV
NL-3633 CZ Vreeland, Netherlands
Tel: (+31 294) 23 02 45   Fax: (+31 294) 23 02 45

Services

Planning, development, construction and
maintenance of track facilities for main line, metro
and light rail networks, including conductor rail
installation and maintenance, cable installation and
laying, electrical engineering and the installation
and maintenance of overhead power supply lines,
and civil engineering work such as the construction
of new platforms. A logistics department oversees
the planning and implementation of materials
supply to track construction sites.

Spitzke equipment includes: four tamping
machines; three rapid ballast levelling machines;
one 45 tonne crane; two ballast bed cleaning
machines; two V100 and two LDK 1250 locomotives;
and various excavators, bulldozers and wheel
loaders.

SPX Fluid Power

Spilsby Road, Romford, Essex RM3 8SB, UK
Tel: (+44 1708) 33 65 00   Fax: (+44 1708) 33 65 43
e-mail: briancannon@fluidpower.spx.com

Corporate background

SPX Fluid Power is the trading name of Smith
Industries Hydraulics Company, part of Smiths
Industries plc.

Products

Condition-monitored in-sleeper Rail Clamp Lock
Point system; level crossing barriers; hydraulic

Speno RR16-M rail grinder   0546912

power packs for train stops, brakes, passenger welfare devices and permanent way maintenance equipment; and jointing machines for signal cables.

Countries to which equipment has been supplied include Australia, the Czech Republic, Hong Kong, India, Ireland, Malaysia, Singapore and the UK.

SRS Rail Vehicles AB

PO Box 11178, SE-16 111 BRomma, Sweden
Tel: (+46 8) 799 67 00   Fax: (+46 48) 28 43 33
e-mail: srs.bromma@srs.scancem.com
Web: www.srsrv.com

Key personnel
Managing Director: Alf Göransson

Subsidiary
SRS Rail Ltd
Bembridge, Hook Heath Avenue, Woking, Surrey GU22 0HN, UK
Tel: (+44 1483) 77 38 25   Fax: (+44 1483) 77 38 30
Web: www.srsrailuk.co.uk

Products
Road-rail vehicles in standard, modular or customised designs for: track maintenance; welding; recovery; bridge inspection; overhead line inspection and maintenance. Switch and track panel replacement systems. Clicomatic rail lubricators.

UPDATED

Steel Authority of India

Steel Authority of India Ltd
International Trade Division
13th Floor, Hindustan Times House, 18-20 Kasturba Gandhi Marg, New Delhi 110 001, India
Tel: (+91 11) 332 73 16; 335 57 33
Fax: (+91 11) 332 10 18; 371 27 74

Main works
Bhilai, Madya Pradesh
Durgapur Steel Plant, West Bengal

Key personnel
Chairman: Arvind Pande
Commercial Director: A K Singh
Executive Director, International Trade Division: B N Jha

Products
Rails of various types.

STRAIL Verkehrssysteme

Gummiwerk Kraiburg Elastik GmbH
Göllstrasse 8, D-84529 Tittmoning, Germany
Tel: (+49 8683) 70 10   Fax: (+49 8683) 70 11 26
e-mail: info@strail.de
Web: http://www.strail.com

Products
Modular level crossing surfaces for main line railways, metros, light rail and industrial sidings. Products include: STRAIL, for heavy duty rail and road applications; innoSTRAIL, for applications such as farm crossings and track access platforms; pedeSTRAIL, for personnel crossings; and STRAILprofile, a rail seal for bitumen level crossings. STRAIL/Kraiburg also manufactures elastic rail profiles to enable the trackwork of urban railways to be harmonised with its environment.

Strukton Railinfra BV

Westkanaaldijk 2, NL-3542 DA Utrecht
PO Box 1025, NL-3600 BA Maarssen, Netherlands
Tel: (+31 302) 240 72 00   Fax: (+31 302) 248 66 01
Web: www.struktonrailinfra.com

Key personnel
Managing Directors: Ir A Schoots, Ir G J Vos
Public Relations: Drs I W J van Dam-Aaldijk
Marketing: J J Gozens (Strukton Railinfra Development & Technology)

Background
Strukton Railinfra is part of Strukton Groep NV.

Services
Innovation, engineering, maintenance management, information systems, monitoring systems, product development, project management, maintenance breakdown repair, renewal, new construction, power supply, underground infrastructure, cabling, telecommunications, rolling stock, traction, production and maintenance of high output equipment, production welding, competence development and marketing monitoring.

UPDATED

System Bahnbau International (SBI)

Bessemerstrasse 42b, D-12103 Berlin, Germany
Tel: (+49 30) 75 48 73 78   Fax: (+49 30) 75 48 73 79
e-mail: wolf@sbi-ffc.de

Key personnel
Technical Managing Director: Wolf-Dietrich Rommel
Commercial Managing Director: Roland Österlein
Head of Technical Department: Günther Wolf

Associate company
System-Bahnbau International
Brunnenstrasse 36, D-74504 Crallsheim, Germany
Tel: (+49 79) 332 43   Fax: (+49 79) 332 60

Background
SBI is a joint venture between Leonard Weiss GmbH & Co and ILBAU Deutschland GmbH.

Products and services
Construction of Feste Fahrbahn Crailsheim system ballastless track in the variations: 3.2 m wide for high-speed traffic; 2.4 m wide for conventional rail applications; and a Low Cost Track (LCT) option for speeds of up to 120 km/h. Contracts include the Hannover-Berlin high-speed line (two sections of 17 and 580 km); 4 km of urban track in Munich; and 3 km of urban track in Stuttgart.

Tarmac Precast Concrete Ltd

Tallington Factory, Tallington, Stamford PR9 4RL, UK
Tel: (+44 1778) 38 10 00   Fax: (+44 1778) 34 80 41
e-mail: tarmacprecastenquiries@tarmac.co.uk
Web: www.tarmacprecast.co.uk

Key personnel
Director/General Manager: Nigel Claxton
Commercial Manager: Paul Whitham
Director: Howard Taylor
UK Sales Manager, Rail: Philip Slinn
Customer Services Manager, Rail: David Evans

Principal subsidiaries
RCC
Charcon Tunnels

Products
Approved precast products for the rail industry. The product range includes sleepers, bearers and other bespoke products.

UPDATED

Techni-Métal Systèmes

Parc d'activités de Fiancey, F-26250 Livron, France
Tel: (+33 4) 75 85 85 30   Fax: (+33 4) 75 85 85 35
e-mail: techni.metal@wanadoo.fr
Web: http://www.techni-metal.fr

Key personnel
Managing Director: Bernard Mouton
General Manager: Christian Berger

Background
Techni-Métal Systèmes is a subsidiary of Techni-Métal Entreprise SA.

Products
Equipment for railway and tunnel construction, including rubber-tyred or rail-mounted self-propelled gantries for applications such as rail handling, road-rail multi-purpose vehicles, self-propelled cranes, handling equipment for track panels, sleepers and tunnel segments, equipment for slab track construction and hoists and access platforms.

Contracts
Equipment has been supplied to Cogifer and Sols Bétonnage for the Lyons metro, Ferreira for the Lisbon metro, Cogifer for the Turin metro, SNFC, France and to the Italian-Thai joint venture for railway construction in Thailand.

Teknikum Oy

PO Box 13, Nokiankatu 1, FIN-38210 Vammala, Finland
Tel: (+358 3) 519 11   Fax: (+358 3) 511 34 54
Web: http://www.teknikum.com

Key personnel
Managing Director: Kalevi Vuorisalo
Marketing Director: Juha Myllärinen
Secretary: Sinikka Sisto

Products
Level crossing components including Teknicross rubber level crossing systems, rubber EVA and cork rubber rail pods, insulators, rail screw sleeves and rail rubber for flange grooves.

Tensol Rail SA

CH-6776 Piotta, Switzerland
Tel: (+41 91) 873 66 11   Fax: (+41 91) 873 66 10
e-mail: tensolrail@tensolrail.com
Web: http://www.tensolrail.com

Sales office
Tensol Rail SA, Glutschbachstrasse 49, CH-3661 Uetendorf
Tel: (+41 33) 346 50 28   Fax: (+41 33) 346 50 29

Key personnel
Manager, Railway Maintenance: J Marfurt

Background
Tensol Rail is a member of the TrackNet Holding group.

Products
Rail fastenings, points, crossings; racks and rack turnouts for cog-wheel railways; turnouts; turntables and transfer tables.

Rail welding and grinding and permanent way equipment.

Thermit Australia Pty

170 Somersby Falls Road, Somersby, New South Wales Australia 2250
Tel: (+61 2) 43 40 49 88   Fax: (+61 2) 43 40 40 04
e-mail: info@thermit@com.au

Key personnel
Managing Director: Ron Moller
Financial Manager: Megan Barr
Marketing Manager: Bryan Pieper
Services Manager: David Harnot
Technical Manager: Paul Radmann

Subsidiary company
Rail Track Maintenance, New Zealand

Products

Thermit brand welding materials; hydraulic rail shears; rail weld grinders; glued and mechanical insulated rail joints; training of Thermit welders; removable level crossing systems; Kryorit ballast stabilisation; waterproof bridge membranes.

Contracts

Recent contracts include welding materials and level crossing systems for the Alice Springs-Darwin railway, and welding materials for the Taiwan High Speed Rail project.

UPDATED

Thermit Welding (GB) Ltd

87 Ferry Lane, Rainham RM13 9YH, UK
Tel: (+44 1708) 52 26 26   Fax: (+44 1708) 55 38 06
e-mail: tcc@thermitwelding.co.uk
Web: http://www.thermitwelding.co.uk

Key personnel

Director, Operations: T C Clifton
Director, Technical Sales: R S Johnson

Principal subsidiary

Thermitrex (Pty) Ltd, Boksburg, South Africa

Products

Aluminothermic rail welding products and rail welding, training and inspection; insulated rail joints; ballast stabilisation.

Contracts

Recent contracts include Hong Kong Lantau Airport Railway; Heathrow Express; Jubilee Line Extension; Kuala Lumpur LRT and Midland Metro and Croydon Tramlink.

Thermon Manufacturing Co

Commercial Products Division
100 Thermon Drive, PO Box 609, San Marcos, Texas 78667-0609, USA
Tel: (+1 512) 396 58 01   Fax: (+1 512) 754 24 31

Key personnel

President and Chief Executive Officer:
   Mark Burdick
Support Services Manager: Sandra Michaewicz

Worldwide offices

Thermon UK Ltd
Seventh Avenue, Gateshead Tyne and Wear NE11 0JW, UK
Tel: (+44191) 499 49 00

Thermon France
18 rue du Marais, Montreuil F-93100, France
Tel: (+33 1) 48 70 42 90

Thermon Deutschland
Raiffeisenstrasse 45, Olpe D-57462, Germany
Tel: (+49 2761) 938 30

Thermon Far East
3rd Floor, Recruit Yokohama Building, Yokohama 221-0056, Japan
Tel: (+81 45) 461 03 73

Vorkauf Sociedad Anónima, Commandante Franco, 3 Madrid E-28016, Spain
Tel: (+34 91) 359 17 12

Products

Heat tracing systems for rail infrastructure applications, including: third rail heating systems; point heating systems; including rail web and switch rod heaters; heating systems for people mover and rapid transit guideways; light rail vehicle pantograph heaters; heating systems for station platform surfaces; heating systems for depots and fuelling and servicing installations.

ThyssenKrupp GfT Gleistechnik GmbH

Altendorfer Strasse 120, D-45143 Essen, Germany
Tel: (+49 201) 188 37 65   Fax: (+49 201) 188 37 57
Web: www.thyssenkruppgleistechnik.de

Key personnel

Contact: H Weiβ

Background

ThyssenKrupp GfT Gleistechnik GmbH is a subsidiary of ThyssenKrupp Services AG, Düsseldorf, Germany.

Products

Development, manufacture and sale of products and services for railway superstructures. Project planning and consultation, high quality rails, turnouts, track material, rolling stock, integrated logistic support.

Contracts

Annual supply to the Deutsche Bahn AG of approximately 100,000 tons of track, 80,000 steel Y ties and up to 150 switches.

ICE high-speed line Cologne–Frankfurt, JIT supply of track, unloading, laying, welding including commissioning documentation.

High-speed Channel Tunnel Rail Link, UK, construction of provisional work and side track out of 324 m pieces of track for the continuous supply of the London Channel Tunnel construction.

Taiwan high-speed line supply of track.

NEW ENTRY

Thyssen Krupp Y-steel sleeper system   0105890

Tie & Track Systems Inc

12300 South New Avenue, Lemont, Illinois 60439, USA
Tel: (+1 630) 257 60 04   Fax: (+1 630) 257 59 44
e-mail: information@ttsties.com
Web: http://www.ttsties.com

Products

Steel sleepers for track and pointwork applications. These are of a self-compacting design intended to direct ballast migration to the maximum load points beneath each sleeper. Custom designs can be developed for specific applications.

Tiefenbach GmbH

PO Box 91 13 60, D-45538 Sprockhoevel, Germany
Tel: (+49 2324) 70 54   Fax: (+49 2324) 70 51 14
e-mail: info@tiefenbach.de
Web: http://www.tiefenbach.de

Key personnel

Managing Director: Peter Jochums
Sales: Juergen Burghoff
Export Sales Manager (Contracts): Achim Weirather

Background

Tiefenbach is a member of the Hauhinco Group.

Products

Point control equipment: complete control equipment for automation and signalling of depots and marshalling yards; axle counter systems; track vacancy detection and level crossing switching with axle counter systems based on Tiefenbach's double wheel sensor.

Contracts

Complete systems design and supply of depot signalling works at Cottbus (DB AG), Cologne (DB AG), Antwerp (SNCB), marshalling yard control and signalling works at Kassel (DB AG), Hamm (DB AG), Gdansk, Poland; complete systems design and supply of axle counter systems for SOB-Sudostbahn, Switzerland (main line); AKN Eisenbahn AG suburban feeder service at Hamburg, Germany, level crossing systems for Conrail, USA and Ford Motor Company, Detroit.

Tiflex Ltd

Tiflex House, Liskeard, PL14 4NB, UK
Tel: (+44 1579) 32 08 08   Fax: (+44 1579) 32 08 02
e-mail: trackelast@tiflex.co.uk
Web: www.tiflex.co.uk

Key personnel

Managing Director: N Spearman
Sales and Marketing Director: A Tuffield
Product Specialist: T R Smith

Products

Rail pads, baseplate pads, undersleeper pads, ballast mats, floating slab track bearings, anti-vibration track support materials.

Contracts include the supply of trackform bearings for the Tsing Ma Bridge, Hong Kong, floating slab track bearings for the Jubilee Line Extension, London, undersleeper pads for the Sabadell tunnel in Spain and anti-vibration materials for the refurbishment of the Bucharest metro.

UPDATED

Tipco Inc

1 Coventry Road, Bramalea, Ontario L6T 4B1, Canada
Tel: (+1 905) 791 98 11   Fax: (+1 905) 791 49 17
Web: http://www.tipcopunch.com

Key personnel

President: John Ferrone
Manager, Track Bit Department: B Crichton

Products

Flat beaded track bits.

VERIFIED

TKL Rail

Thompsons, Kelly & Lewis Pty Ltd
26 Faigh Street, Mulgrave, Victoria 3170, Australia
Tel: (+61 3) 95 62 07 44   Fax: (+61 3) 95 62 28 16

Works

5 Parker Street, Castlemaine, Victoria 3450, Australia

For details of the latest updates to *Jane's World Railways* online and to discover the additional information available exclusively to online subscribers please visit

jwr.janes.com

Key personnel
Sales Manager, Railway Products: E A Smith
Manager, Railway Products: W G Kinscher
Manager, Sales and Marketing: A Grage
General Manager: W J Coulter

Principal subsidiary
Davies & Baird

Products
Trackwork, turnouts, points and crossings, switches; steel turnout sleepers, insulated/non-insulated; and a range of accessories.

Contracts include the supply of tramway trackwork for Hong Kong; crossings for MTRC, Hong Kong; turnouts for the Jabotabek track layout improvement project, Indonesia; turnouts for State Railway Authority of New South Wales, Australia; and points and crossings for Westrail and Victoria Public Transport Corporation, Australia; turnouts for the Sydney LRT system, turnouts for BHP Iron Ore Nelson Point yard and Jimblebar mine, Australia; five-year contract for tramway trackwork for Toronto Canada and turnouts for the State Railway of Thailand Track Rehabilitation Project, Phases 1 and 2; turnouts for upgrading the Forrestfield Yard, Perth, Australia; turnouts for upgrading the Mount Isa line, Queensland, Australia.

Tokimec Rail Techno Inc

2-16-46, Minami Kamata, Ota-ku, Tokyo 144-8551, Japan
Tel: (+81 3) 37 32 70 61   Fax: (+81 3) 37 32 70 50
Web:www.tokimec.co.jp

Key personnel
President: Takashi Kamiya

Products
Ultra-sonic rail inspection cars, ultrasonic rail flaw detectors, portable ultra-sonic rail flaw imager; switch profile gauges, portable rail section measuring devices, expansion gap gauge, 'DataDepot' system (non-contact, high-speed communications system).

NEW ENTRY

Travipos

Calle Irlanda s/n, Sector Norte, E-43120 Constanti (Tarragona), Spain
Tel: (+34 977) 29 65 53   Fax: (+34 977) 29 65 35

Background
Travipos is a joint venture of Comsa SA and Pfleiderer Verkehrstechnik GmbH & Co KG (see entries in *Permanent way components, equipment and services section*).

Products
Pre-stressed concrete sleepers for high-speed and main line railways and for sidings.

Trinity Difco Inc

PO Box 238, Findlay, Ohio 45839, USA
Tel: (+1 419) 422 05 25   Fax: (+1 419) 422 12 75

Key personnel
For a full list of personnel, see Trinity Difco Inc entry in *Freight vehicles and equipment* section.

Products
Rolling stock for permanent way construction and maintenance duties including air-operated side-dump wagons; air-operated drop-end side-dump wagons; and Ballaster and Auto-Ballaster systems.

The Auto-Ballaster system provides automated ballasting using remote valves located at one end of the wagon to control the gate operation. Radio remote controls are available as an option. Ballast flow, which is adjustable, can be directed to the centre or either side of the track. Power comes from the locomotive air compressor; a separate line provides air to each car in the train. The gates, which open and close with enough power to shear limestone ballast, are made of heavy-steel plates with a shape and motion that tends to push the ballast back into the wagon when closing. The gates can also be operated manually.

TSO SA

Chemin du Corps de Garde, Zone Industrielle, PO Box 8, F-77501 Chelles Cedex, France
Tel: (+33 1) 64 72 72 00   Fax: (+33 1) 64 26 30 23
e-mail: info@tso.fr

Key personnel
President of the Executive Board: Emmanuèle Perron
Export Manager: Claude Petit

Products
Track construction and maintenance for main line and metro systems; slab track construction; aluminothermic and flash-butt welding; conductor rail system, overhead catenary system.

UPDATED

TSTG Schienen Technik GmbH

Kaiser-Wilhelm-Strasse 100, D-47166 Duisburg, Germany
Tel: (+49 203) 522 46 93   Fax: (+49 203) 522 46 94
e-mail: info@tstg.de
Web: www.tstg.de

Key personnel
Chairman: Peter Sokolowsky
Board of Management: Karl Ebner
   Hans Pfeiler

Products
Manufacturer of rail worldwide including: flat-bottom rail, grooved rails, crane rails, rails for switches and crossings, msc rails, steel-sleepers; up to 120 m without welding.

UPDATED

Tülomsas

The Locomotive and Motor Corporation of Turkey
Ahmet Kanatli Cad, TR-26490 Eskisehir, Turkey
Tel: (+90 222) 224 99 56
Fax: (+90 222) 225 57 57; 72 72
e-mail: tulomsas@tulomsas.com.tr
Web: http://www.tulomsas.com.tr

Key personnel
Managing Director: D Zeki Daloglu
Assistant General Managers: Galip Pala, Cengiz Özan, Fatih Turan, Haluk Akova
Head of Marketing: Erol Çetin

Products
Self-propelled rail vehicles for infrastructure maintenance including car equipped with hydraulic crane, catenary maintenance and inspection car.

Türk + Hillinger GmbH

PO Box 242, D-78503 Tuttlingen, Germany
Tel: (+49 7461) 701 40
Fax: (+49 7461) 70 14 10

Key personnel
President: Erich Hillinger
Managing Director: Eberhard Härter

Principal subsidiaries
Türk + Hillinger GmbH
Dorotheenstrasse 22, D-9102 Limbach/Oberfrohna, Germany
Tel: (+49 3722) 718 90   Fax: (+49 3722) 71 89 16

Türk + Hillinger Hungaria Kft
Arany J u 2, H-3350 Kal, Hungary
Tel: (+36 36) 48 70 53   Fax: (+36 36) 48 70 53

Products
Electric point-heating systems: these consist of flat, tubular heaters with a chrome-nickel-steel connection housing or a complete watertight connecting cable. All clamps and springs for installation at different rail profiles are available.

Contracts include the supply of point heaters for DB AG, SBB and SNCF.

Turkington Precast Concrete

James Park, Mahon Road, Portadown BT62 3EH, County Armagh, UK
Tel: (+44 28) 38 33 28 07   Fax: (+44 28) 38 36 17 79
e-mail: info@turkington-precast.com
Web: http://www.turkington-precast.com

Key personnel
Managing Director: Jim McKeag
Executive Director: Trevor Turkington
Sales and Marketing Manager-Precast: David Hamilton

Products
Prestressed concrete railway sleepers, railway cable troughing.

Contracts
Supplier to Northern Ireland Railways and Iarnród Éireann in Ireland, supplier to Network Rail, UK.

Ultra Dynamics

Upperfield Road, Kingsditch Trading Estate, Cheltenham GL51 9NY, UK
Tel: (+44 1242) 70 79 00   Fax: (+44 1242) 70 79 01
e-mail: sales@ultradynamics.demon.co.uk

Ultra retractable retarders in marshalling yard at Hallsberg, Sweden   1034519

Key personnel

For a full list of personnel, see Ultra Dynamics entry in *Freight vehicles and equipment*.

Products

Spring frog switch dampers for controlling the return of the wing rail on spring frogs. The high-performance and heavy-duty design has applications for both original switch manufacturers and main line retrofit purposes.

The dampers are supplied with fasteners for easy replacement and are double-sealed with low-friction seals with high-speed high-pressure rating.

Unit Rail Anchor Company Inc

2604 Industrial Street, Atchison, Kansas 66002, USA
Tel: (+1 913) 367 72 00   Fax: (+1 913) 367 05 59
Web: http://www.unitrail.com

Key personnel

President: Paul T Ciolino
International Sales: Carol Hale

Products

Unit rail anchors, spring and drive-on; reclamation and remanufacture of rail anchors; E-Z Wrench spring anchor applicator tool.

VAE Aktiengesellschaft

Head office
Rotenturmstrasse 5-9, A-1010 Vienna, Austria
Tel: (+43 1) 53 11 80   Fax: (+43 1) 53 11 82 22
e-mail: marketing@vae.co.at
Web: http://www.vae-ag.com

Key personnel

President Marketing, Sales (International):
  Mag Dr Marc Kaddoura
President Technical, Engineering:
  Dipl Ing Johannes Rainer Oswald
President Finance, Controlling: Werner Saringer

Principal subsidiaries

BWG, Gesellschaft mbH & Co KG, Germany
HBW Light Rail BV, Netherlands
JEZ Sistemas Ferroviarios SL, Spain
Transwerk Perway (Pty) Ltd, South Africa
UAB VAE Legetecha, Lithuania
VAE Africa (Pty) Ltd, South Africa
VAE Apcarom SA, Romania
VAE Eisenbahnsystem & GmbH, Austria
VAE Italia Srl, Italy
VAE Nortrak Cheyenne Inc, USA
VAE Nortrak Inc, USA
VAE Nortrak Ltd, Canada
VAE Railway Systems Pty Ltd, Australia
VAE Riga SIA, Latvia
VAE Sofia OOD, Bulgaria
VAE UK Ltd, UK
VAMAV Vasúti Berendezések Kft, Hungary
Welchenwerk Wörth GmbH, Austria

Products

Layouts, turnouts and turnout components, crossings with moveable points, switch devices, hydraulic switch operating system, monitoring systems, safety systems, locking systems, setting support systems, bearings/attachments, intelligent turnout systems, layout, turnouts and components made of grooved rails, setting devices for grooved rail turnouts, cast manganese crossings, built-up crossings, hot box and brake detection systems, rail movement joints, segmented sleepers, synthetic sleepers, steel sleepers, hollow steel sleepers, insulating joints, diamond crossings, buffer stops, check rail, coach-screws, transition rail.

VAE Nortrak Ltd

16160 River Road, Richmond, British Columbia V6V 1L6, Canada
Tel: (+1 604) 273 30 30   Fax: (+1 604) 273 89 27
Web: www.nortrak.com

Key personnel

President: Al Tuningley
Chief Financial Officer: Eduard Peinhopf

Other offices

VAE Nortrak North America Inc
3930 Valley East Industrial Drive, Birmingham, Alabama 35217, US
Tel: (+1 205) 854 28 84   Fax: (+1 205) 854 28 85

VAE Nortrak North America Inc
1740 Pacific Avenue, Cheyenne, Wyoming 82007-1004, US
Tel: (+1 307) 778 87 00   Fax: (+1 307) 778 87 77

VAE Nortrak North America Inc
2705 South State Street, Chicago Heights, Illinois 60411-4894, US
Tel: (+1 708) 757 65 68   Fax: (+1 708) 757 68 14

VAE Nortrak North America Inc
2300 South Freeway, Pueblo, Colorado 81004, US
Tel: (+1 316) 284 00 88   Fax: (+1 316) 284 00 95

VAE Nortrak North America Inc
405 West Street, Newton, Kansas 67114, US
Tel: (+1 316) 284 00 88   Fax: (+1 316) 284 00 95

Background

VAE Nortrak is a member of the VAE group of companies.

Products

Supply and manufacture of track materials including RBM, SSGM, welded heel manganese frogs, welded spring manganese frogs, jointless and boltless manganese frogs, movable point frogs and vario frogs; AREA and asymmetrical switches, guardrails, transition rails and related components; new relay rail, crane rail, tie plates and joint bars; screwspikes, cut spikes, track-bolts, frog and switch-bolts; track tools. Contract design and engineering services.

UPDATED

VAE UK Ltd

Head office
Sir Harry Lauder Road, Portobello, Edinburgh EH15 2QA, UK
Tel: (+44 131) 550 22 97   Fax: (+44 131) 550 26 60
e-mail: jim.gemmell@vae.co.uk
Web: www.vae-ag.com

Key personnel

Managing Director: Jim Gemmell
Manufacturing Manager: Jimmy O'Neill

Other office

Sales and Technical office
VAE UK Ltd
Enterprise House, Carlton Road, Worksop S81 7QF, UK
Tel: (+44 1909) 50 51 54   Fax: (+44 1909) 53 23 04
Technical/Sales Manager: Kevin Ball

Background

VAE UK is a member of the VAE group of companies.

Products

Manufacturer of switch, crossing and railway systems, servicing renewals and maintenance markets.

UPDATED

Vaia Car

Via Isorella 24, I-25012 Calvisano (BS), Italy
Tel: (+39 030) 968 62 61   Fax: (+39 030) 968 67 00
e-mail: vaiacar@vaiacar.it
Web: http://www.vaiacar.it

Products

Permanent way maintenance equipment, including: road-rail mobile flash butt welders, cranes, excavators and radio-controlled tractive units; cranes; rail, switch, track panel and sleeper handling equipment; tampers; sleeper replacement equipment.

Voestalpine Railpro BV

Nieuwe Crailoseweg 8, PO Box 888, NL-1200 AW Hilversum, Netherlands
Tel: (+31 35) 688 96 00   Fax: (+31 35) 688 96 66

Background

In October 2002, Railpro BV's majority shareholding was taken over by the Austrian company Voestalpine Bahnsysteme GmbH and the name changed to Voestalpine Railpro BV. In 2001, Voestalpine Railpro (previously Railpro BV) linked up with Kummler+Matter (suppliers of catenary systems) and Larnifil (supplier of contact and suspension wires).

Products

Suppliers of all materials required in railway infrastructure work, acting as a stockist for contractors. The company can arrange transport to the worksite by road, water or rail; it operates a fleet of 2,200 rail wagons.

Developments

As a result of the conversion of catenary voltage from 1,500 to 25 kV there is a demand for new types of catenary systems. Voestalpine Railpro entered an agreement with VD Leegte Metaal for the production of welded overhead support arms and enables the company to supply fully engineered and complete catenary systems, both for train, tram and metro systems, and for light rail.

Recently Voestalpine Rail BV concluded an agreement with Strukton for the supply of catenary components for the Amsterdam–Utrecht project.

Vortok International

Units 6-7 Haxter Close, Belliver Industrial Estate, Roborough, Plymouth PL6 7DD, UK
Tel: (+44 1752) 70 06 01
Fax: (+44 1752) 70 23 53
e-mail: sales@vortok.co.uk
Web: http://www.vortok.demon.co.uk

Key personnel

Contact: David Townsend

Principal subsidiary

Multiclip Company Ltd

Products

Permanent way maintenance equipment and components, including: the Vortok Coil, for the rehabilitation of loose screws in wood sleepers; temporary sign board supports for securely and safely placing signs near the rail without ballast penetration; clip-on insulators for the prevention of track circuit signal failure by items passing under both rails; insulated block joint trimmers, portable and self-powered grinders for deburring rail ends at block joints; VERSE, a non-destructive method of measuring stress-free rail temperature; adjustable block spacers, enabling worn check block to be moved without removal from the track; and rigid safety barriers, fitted to the foot of the adjacent open line to enable green zone working at higher train speeds.

Vortok is a supplier to most European railway companies.

Vossloh Cogifer

Headquarters
54 avenue Victor Hugo, BP 56606, F-92566 Rueil Malmaison, Cedex, France
Tel: (+33 1) 55 47 73 00
Fax: (+33 1) 55 47 73 92
e-mail: contact@vcsa.vossloh.com
Web: vcsa.vossloh.com

Key personnel

President: James Sanders
Managing Director: Claude Schwartz
Deputy Managing Director: Guy Delorme

Business Development Director:
  Marc-Antoine de Dietrich
Deputy Managing Director, Signalling Department:
  Jean Louis Binder
Sales Managers: Freddy Sudol, GerardThorez

Works

Reichshoffen – Points and Crossings Department
BP 1, F-67110 Reichshoffen, France
Tel: (+33 3) 88 80 86 80   Fax: (+33 3) 88 09 67 33
e-mail: contact@vcsa.vossloh.com

Reichsoffen – Signalling Department
4 rue d'Oberbronn, Reichshoffen BP 02, F-67891
Niederbronn Cedex, France
Tel: (+33 3) 88 80 85 00   Fax: (+33 3) 88 80 85 18
e-mail: system@vcsa.vossloh.com

Fére en Tardenois – Points and Crossings
Department
Zone Industrielle, F-02130 Fère en Tardenois,
France
Tel: (+33 3) 23 82 58 88   Fax: (+33 3) 23 82 71 99
e-mail: contact@vcsa.vossloh.com

Subsidiaries

Eav-Durieux
138 rue d'Anderlues, B-7141 Carnières, Belgium
Tel: (+32 64) 43 14 00   Fax: (+32 64) 45 86 32
e-mail: durieux-eay@bmedi.be

Vossloh Cogifer Finland
Telakkatie 18, FIN-25570, Teijo, Finland
Tel: (+35 8) 27 36 60 10   Fax: (+35 8) 27 36 60 20
e-mail: cogifer.teijo@vcfi.vossloh.com

Jacquemard – Avr
389 rue des Frères Lumière, ZI de Molina la
Chazotte, F-42650 Saint Jean de Bonnefonds,
France
Tel: (+33 4) 77 47 68 68   Fax: (+33 4) 77 47 68 69
e-mail: michel.cuminetti@vcsa.vossloh.com

Vossloh Laeis GmbH
Ruwerer Strasse 21, D-54292 Trier, Germany
Tel: (+49 651) 55 80   Fax: (+49 651) 558 15
e-mail: vla.vossloh.com

Vossloh Cogifer Italia
Uffici di Milano, Via Pregnanza 32, I-20010
Cornaredo (MI), Italy
Tel: (+39 02) 93 56 58 77   Fax: (+39 02) 93 56 20 16
e-mail: salesoffice@vsscogifer.it

Kihn SA
17 rue de l'Usine, BP 20, L-3701 Rumelange,
Luxembourg
Tel: (+352) 564 77 11   Fax: (+352) 56 58 54
e-mail: contact@kihn.com

Vossloh Cogifer Norway
Storgata 21, PO Box 292, N-2001 Lilleström,
Norway
Tel: (+47) 64 84 35 90   Fax: (+47) 64 84 35 99
e-mail: cogifer@vcno.vossloh.com

Cogifer Polska
UL Ludwikowo 2, PL-85-502 Bydgoszcz, Poland
Tel: (+48 52) 322 52 24   Fax: (+48 52) 322 46 76
e-mail: kzn@vcpo.vossloh.com

Futrifer SA
Edificio Coopali, Rua Jose Afonso, n° 4C, ler Andar,
Espaço H, P-1600-130 Lisbon, Portugal
Tel: (+351 217) 20 05 80; 81; 82; 83; 84
Fax: (+351 217) 20 05 89
e-mail: lisboa@futrifer.pt

Amurrio Ferrocarriles y Equipos
Maskuribai 10, E-01470 Amurrio (Alava), Spain
Tel: (+34 945) 89 16 00   Fax: (+34 945) 89 24 80
e-mail: aferreq@sea.es

Vossloh Nordic Switch System
Södr Grev Rosengatan 1, Box 1502, SE-701
15 Örebro, Sweden
Tel: (+46 19) 17 25 00   Fax: (+46 19) 17 24 50
e-mail: contact@vcn.vossloh.com

ATO
31st floor, Italthai Tower, 2034/132 New Petchburi
Road, Bangkapi, Huaykwang, Bangkok 10320,
Thailand
Tel: (+66 2) 716 14 19   Fax: (+66 2) 716 14 20
e-mail: frederic.milliet@cogifer.co.th

Corus Cogifer
Hebden Road, Scunthorpe DN15 8XX, UK
Tel: (+44 1724) 86 21 31   Fax: (+44 1724) 29 52 43
e-mail: info@coruscogifer.com

J S Industries
104, Amarchand Sharma Complex, Sadar Patel
Road, Secunderabad 500 003 A P, India
Tel: (+91 40) 27 80 52 26   Fax: (+91 40) 27 80 52 27

Vossloh MIN Skretnice
Saraferska 2, 18000 Nis, Serbia
Tel: (+381 18) 58 17 35   Fax: (+381 18) 58 17 38
e-mail: company@min-bss.co.yu

Swedish Rail System
Förskeppsgatan 8, Box 1512-27100 Ystad, Sweden
Tel: (+46 411) 694 00   Fax: (+46 411) 183 08

Products

Design, manufacture and installation of switches
and crossings for high-speed railways, metro
systems, light rail systems and main line and
suburban rail networks. Products include: moveable
manganese frogs; countered switches and
crossings; manganese frogs with welded legs; and
special forgings for switch rails; switch mechanisms;
clamp lock systems; traffic detectors; and the Paulvé,
mechanically driven points detector. Support
services include: diagnostic reports on points in
service; inspection; use and maintenance training;
technology transfer; and financial services.

UPDATED

Vossloh Fastening Systems GmbH

Vosslohstrasse 4, D-58791 Werdohl, Germany
Tel: (+49 2392) 520
Fax: (+49 2392) 523 75
e-mail: info@vfs.vossloh.com
Web: www.vossloh.com

Key personnel

Managing Directors: James N Sanders;
  Dr Georg Hauschild
Sales Manager: Dipl Ing Dirk Vorderbrück
Technical Sales: Winfried Bösterling
Overseas Business Office, Regional Sales Managers,
  Düsseldorf: Joachim Spors; Dirk Pfeiffer

Works

Vossloh Werdohl GmbH
Vosslohstrasse 4, D-58791, Werdohl

Overseas business office, Düsseldorf
Tel: (+49 2102) 490 90   Fax: (+49 2102) 490 94

Products

Rail fastening systems for both ballast-bed and slab
substructures compatible with all climatic zones
and load profiles ranging from heavy-load to high-
speed lines. Logistic services for tracks and special
trackworks, noise protection for railways, plastic
cable duct systems, surface coating (according to
the Dacromet process).

Developments

Vossloh has recently developed the Sk1 24, which
is the first clamping plate available for the ribbed-
plate permanent way that permits the use of 'soft
intermediate bearings' in this track segment.

UPDATED

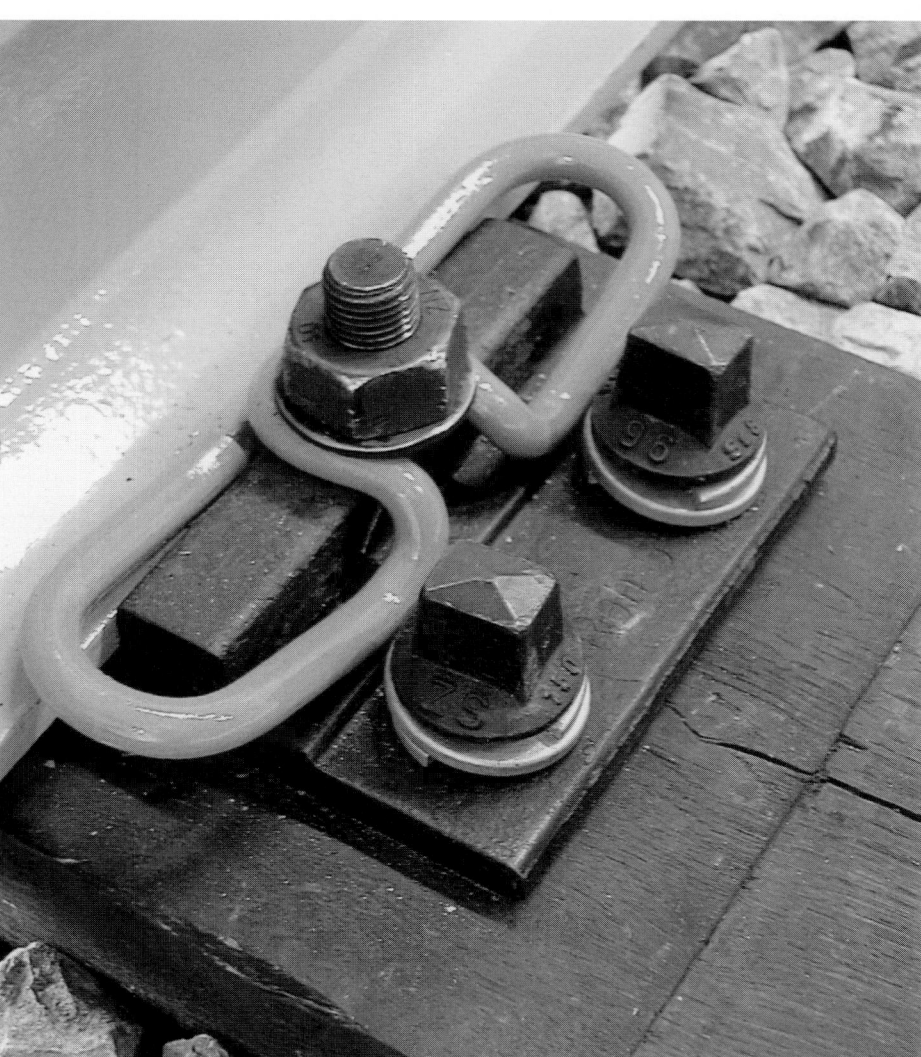

Vossloh's Sk1 24, first tension clamp for ribbed-plate tracks, enabling the use of 'soft buffer bearings' in this section of the tracks (Vossloh)   *NEW*/0585289

Vossloh Infrastructure Services

267 Chaussée Jules César – ZI, BP 62 F-95250,
Beachamp, France
Tel: (+33 1) 30 40 59 00
Fax: (+33 1) 30 40 59 20

Key personnel
Chairman and Chief Executive Officer: Henri Dehe
Managing Director, Major Projects:
    Alain Montgaudon
Managing Director, Administration and Finance:
    Didier Gaudin
General Secretary: Lionel Peraud

Other offices
Industrial siding division
Agence de Nantes
ZI des Acacias, F-44260 Savenay
Tel: (+33 2) 28 01 68 28
Fax: (+33 2) 28 01 68 29

Agence de Rouen
57 rue Rouget de Lisle, BP 44, F-76140 Le Petit
Quevilly Cedex
Tel: (+33 2) 35 72 29 55   Fax: (+33 2) 35 73 64 77

Agence de Dunkerque
Centre de Travaux du Littoral
Rue de Meuninck, ZI de Petite Synthe, F-59640
Dunkerque Cedex
Tel: (+33 3) 28 24 12 15   Fax: (+33 3) 28 60 61 32

Agence de Lille
20 rue Félix Faure, F-59872 Saint-André Cedex
Tel: (+33 3) 28 38 83 50   Fax: (+33 3) 20 40 97 35

Agence de Florange
4 rue d'Alsace, F-57190 Florange
Tel: (+33 3) 82 59 84 50
Fax: (+33 3) 82 59 92 09

Agence de Strasbourg
18 rue de Brest, BP 18, F-67026 Strasbourg Cedex
Tel: (+33 3) 88 39 41 78   Fax: (+33 3) 88 39 77 54

Agence de Bordeaux
140 avenue du Maréchal Leclerc, F-33130 Bègles
Tel: (+33 5) 56 85 96 32   Fax: (+33 5) 56 49 35 10

Agence de Sète
63 quai J-J. Bosc, F-34200 Sète
Tel: (+33 4) 67 74 96 94   Fax: (+33 4) 67 46 00 93

Agence de Marseille
40 boulevard de l'Europe, F-13127 Vitrolles
Tel: (+33 4) 42 89 25 87
Fax: (+33 4) 42 89 54 78

Agence de Lyon
51 rue de Collières, BP 268, F-69802 Saint-Priest
Cedex
Tel: (+33 4) 72 23 64 87
Fax: (+33 4) 37 25 59 43

Agence de Toulouse
Zone Industrielle d'En Jacca
9 Chemin de Garrabot, BP 142, F-31774 Colomiers
Cedex
Tel: (+33 5) 34 55 27 10
Fax: (+33 5) 61 78 16 37

Major projects division
Agence de Bordeaux
Tramway de Bordeaux
40 rue Promis, F-33100 Bordeaux
Tel: (+33 5) 57 54 12 60
Fax: (+33 5) 57 54 12 59

Agence de Clermont
Tramway de Clermont-Ferrand
24 boulevard Etienne Clémentel, F-63100 Clermont-
Ferrand
Tel: (+33 4) 73 74 91 10
Fax: (+33 4) 73 25 40 04

Sonora n° 85, Col. Roma, 06700 Mexico DF
Tel: (+52 55) 55 53 24 44   Fax: (+52 55) 55 53 38 43

Catenary division
Division Caténaires
14 bis, Vieux Chemin de Paris, BP 41, F-94192
Villeneuve St. Georges Cedex
Tel: (+33 1) 56 87 21 10
Fax: (+33 1) 55 47 73 94

Subsidiaries
SOLUXTRAFER
Agence de Carnières, 138 rue d'Anderlues, B-7141
Carnières, Belgium
Tel: (+32) 64 43 24 60
Fax: (+32) 64 43 24 70

Agence d'Athus
7 rue du Terminal, B-6791 Athus, Belgium
Tel: (+32) 63 38 29 90
Fax: (+32) 63 38 29 99

ETF Head Office
2 rue Saint Pétersbourg, F-75008 Paris, France
Tel: (+33 1) 53 04 95 10
Fax: (+33 1) 53 04 95 20
e-mail: etf@etf.com

North Area Management
267 chaussée Jules César ZI, BP 70, F-95250
Beauchamp, France
Tel: (+33 1) 30 40 59 59
Fax: (+33 1) 30 40 59 58

South Area Management
51 rue de Collières, BP 268, F-69802 Saint-Priest
Cedex, France
Tel: (+33 4) 72 23 64 80
Fax: (+33 4) 78 20 62 44

Dehe Bahnbau GmbH
Ruwerer Strasse 21, D-54292 Trier, Germany
Tel: (+49 651) 55 80
Fax: (+49 651) 558 15

SOLUXTRAFER – Head Office
2-4 Route de Longwy, Rodange, Luxembourg
Tel: (+352) 504 60 21
Fax: (+352) 504 60 22 00

Cogimex
Sonora n° 85, Col Roma, 06700 Mexico DF
Tel: (+52 55) 55 53 24 44
Fax: (+52 55) 55 53 38 43

DDL
2-4 Route de Longwy, Rodange, Luxembourg
Tel: (+352) 504 87 51   Fax: (+352) 504 87 52 00

High Output Ballast (HOBS) wagon supplied by Wabtec Rail to Network Rail   1037653

Cogifer TF Thailand Branch
31st Floor, Italthai Tower, 2034/132 New Petchburi
Road,
Bangkapi, Huaykwang, Bangkok 10320, Thailand
Tel: (+66 2) 716 14 19
Fax: (+66 2) 716 14 20

Services
Track laying for railways, high-speed railways,
metros, automatic metros (VAL), tramways (on rails
and rubber tyre), industrial sidings, track renewal.
    Catenary design, studies, manufacturing,
installation and commissioning.

Wabtec Rail Ltd

PO Box 400, Doncaster Works, Hexthorpe Road,
Doncaster DN1 1SL, UK
Tel: (+44 1302) 34 07 00
Fax: (+44 1302) 32 13 49
Web: www.wabtec.com

Key personnel
Managing Director: John Meehan
Engineering Director: Mike Roe
Finance Director: Robert Johnson
Operations Director: Chris Weatherall
Commercial Manager: Paul Robinson

Background
Wabtec Rail undertakes the overhaul of passenger
rolling stock, main line locomotives, bogies,
wheelsets, air brake equipment, hydraulic
dampers and buffers design in addition to the
manufacture of shunting locomotives and freight
wagons.
    As part of the Wabtec Corporation, Wabtec Rail
also supplies to the UK rolling stock owners and
maintainers, composite brake blocks and pads,
Wabtec Railway Electronics train data recorders
and electronic equipment, Cardwell TMX braking
systems and Wabtec air brake equipment.

Products
Wabtec Rail has supplied 240 High Output Ballast
(HOBS) wagons to Network Rail. The HOBS wagons
give railway infrastructure companies an efficient
means of depositing ballast to either side of the
track or between the rails. In addition, radio remote
control means that one man can control the ballast
delivery for a whole train.

UPDATED

Wacker (GB) Ltd

Lea Road, Waltham Cross, Hertfordshire EN9 1AW, UK
Tel: (+44 1992) 70 72 00   Fax: (+44 1992) 70 72 01
e-mail: jane.carter@en.wackergroup.com
Web: www.wackergroup.com

Key personnel
Managing Director: Andrew Howells
National Sales Manager: Clive Downham

Products
Petrol-powered tie tamper with remote ignition cut-out system; triple-plate ballast compaction set, comprising three vibrating plates, allowing single passes covering the full width of the track bed up to 2.6 m.
    Contracts include supply of Wacker machines to railway operators in Europe.

UPDATED

Leonhard Weiss GmbH and Co KG

Leonhard-Weiss-Strasse 22, D-73037 Goeppingen, Germany
Tel: (+49 7161) 60 12 41   Fax: (+49 7161) 60 14 07
e-mail: gleisbau@leonhard-weiss.de
Web: www.leonhard-weiss.de

Key personnel
Chief Executive Officer: Dr Martin Werner
Director, International Projects: Markus Hofmann

Products and services
Among other products and services within Leonhard Weiss GmbH and Co the Track Construction department offers a track reconstruction and maintenance programme.
    It offers track renewal trains for wood, steel and concrete sleepers, track and bridge construction cranes, computerised track survey systems, tampers and similar equipment on a turnkey rental or lease-to-own basis.
    The product and services range also includes the laying of concrete and permanent slab track for commuter, urban and high-speed applications. All the above can include on-site training and knowledge transfer.

UPDATED

Leonhard Weiss 'on track' tamping machine for maintenance of plain line, switches and crossings   0583221

Western-Cullen-Hayes Inc

2700 West 36th Place, Chicago, Illinois 60632-1682, USA
Tel: (+1 773) 254 96 00   Fax: (+1 773) 254 11 10
e-mail: wch@wch.com
Web: www.wch.com

Key personnel
President: Robert L McDaniel
Vice-President: Barbara Gulick
Sales Manager: Carl J Pambianco
Customer Service Manager: Bill Crain
Systems Application: Rodney Yourist

Principal subsidiary
Hayes Plant Western-Cullen-Hayes Inc
120 North 3rd Street, Box 756, Richmond, Indiana 47374, USA

Products
Railway safety track appliances; bumping posts, fixed, sliding and hydraulic; wheel stops and chocks; switch point guards; yard switch machines; WCHT-72 programmable and Solar Tech switch machines; track drills; rail benders; rail tongs; journal and hydraulic jacks; sliding, hinged and portable derails; derail operators; Eldo, DeLectric and solar-powered derail operators; blue flags; and other custom-designed equipment for railway and industrial applications.

UPDATED

Willy Vogel AG

Motzener Strasse 35/37, D-12277 Berlin, Germany
Tel: (+49 30) 72 00 20   Fax: (+49 30) 72 00 21
e-mail: info@vogel-berlin.de
Web: www.vogelag.com

Key personnel
Director International Sales: Andreas Breuer
Press, Marketing: Gotz Mehr
    Tel: (+49 30) 72 00 21 09

Background
Willy Vogel AG has been acquired by SKF, from previous owners Hannover Finanz Group, a private equity firm.

Main subsidiaries
Vogel Japan Ltd, Osaka, Japan
Vogel Lubrication, Inc, Newport News, Virginia, USA
Vogel Nederland BV, Enschede, the Netherlands
Willy Vogel Belgium BVBA, Mechelen, Belgium
Willy Vogel Ibérica, SA, Spain
Berger Vogel srl, Milan, Italy
Vogel France SAS, Saumur, France
Willy Vogel Hungary Kft, Biatorbègy, Hungary

Products
Wheel flange lubrication systems for rail vehicles.

UPDATED

Windhoff Bahn- und Anlagentechnik GmbH

PO Box 1963, D-48409 Rheine, Germany
Tel: (+49 5971) 580
Fax: (+49 5971) 582 09
e-mail: info@windhoff.de
Web: www.windhoff.de

Key personnel
Board Members:
    Herbert Liessem, Georg Vennemann
Finance Director: Helmut Gielians
Sales Directors: Dr Martin Hindersmann, Uwe Dolkemeyer
Technical Director: Juergen Auschner
Purchasing Manager: Stefan Berkemeyer

Products
Construction and maintenance of rail tracks: Multi-Purpose Vehicles (MPV), vehicles with crane or excavator, attachments (ballast broom, tamping unit), rail grinding and finished vehicles. Sales and service of SRS Products in Germany: catenary vehicles, bridge/tunnel inspection vehicles, track welding vehicles, recovery vehicles, flexible vehicles, special vehicles.

UPDATED

Zöllner GmbH

Signal System Technologies
Zur Fähre 1, D-24143 Kiel, Germany
Tel: (+49 431) 702 71 11
Fax: (+49 431) 702 72 02
e-mail: signal@zoellner.de
Web: http://www.zoellner.de

Key personnel
General Manager, Operations: Ulrich Matthieson

Products
ATWS Autoprowa radio-based warning systems for protecting personnel on track worksites. The modular system typically comprises an SSE2 main control unit, radio remote controls, Type WGL flashing lamp, Type WGH horn, F 300 treadle rail contacts for train detection, and cables and cable remote controls. The system can be configured for bidirectional operation on two tracks. The WGH horn system measures ambient noise levels and adjusts the warning volume level appropriately. An independent rechargeable battery provides power for the system.
    Zöllner also produces the Autoprowa Light ZAL warning system for spot worksites. In addition, the company provides inspection and training services related to its systems.

Zweiweg Schneider GmbH & Co KG

Postfach 60, D-42791 Leichlingen, Germany
Tel: (+49 2174) 790 95
Fax: (+49 2174) 79 09 70
e-mail: info@zweiweg.de
Web: http://www.zweiweg.de

Key personnel
For full personnel list, see Zweiweg Schneider GmbH & Co KG entry in *Freight yard and terminal equipment* section.

Products
Track-guidance rollers which convert a road vehicle into a rail vehicle. Besides its use in creating a shunting unit (the resultant tractive power equals approximately that of a 20 tonne locomotive), the device also permits use of the Daimler-Benz Unimog truck on rails as a working unit with various supplementary equipment.
    A Zweiweg Unimog model ZW 82S provided with a steam-jet can be employed for points cleaning.

Kirow track-laying crane used by Leonhard Weiss GmbH & Co   0524953

For example, another unit equipped with a loading crane (and at the same time as a shunting unit) can haul up to 25 loaded wagons. For winter operation a rotary snow-plough or a drum-type snow-plough can be fitted, permitting effective snow removal on rails as well as on the road.

Two special units are available for the construction and maintenance of catenary: a Zweiweg Unimog with hydraulic lifting platform and the Zweiweg road-railer with working platform, which was initially constructed for the Netherlands Railways. The Zweiweg Unimog

vehicles are also available for broad-gauge lines.

Recent contracts include: 22 vehicles to Japan; 14 vehicles for Federal German Railways and four ballast ploughs to Amey Fleet Services, UK.

FREIGHT YARD AND TERMINAL EQUIPMENT

Alphabetical listing

ABC Rail Products Corporation
A Rawie GmbH & Co
Aldon Company Inc
Babcock & Wilcox Española SA
Blatchford Transport Equipment
Bosch – Robert Bosch GmbH
Caillard Cowans Sheldon
Cattron-Theimeg
Central Power Products
Control Chief Corp
E G Steele & Co Ltd
Gottwald Port Technology GmbH
Hegenscheidt-MFD GmbH & Co KG
Kalmar
Kalmar Industries BV

KE Kranbau Eberswalde
Kershaw
Kocks Krane International GmbH
Hans Künz GesmbH
Kyosan Electric Mfg Co Ltd
Liebherr Container Cranes Ltd
Mitsubishi Heavy Industries Ltd
Noell
Oleo International Ltd
Pfister
Railweight
Safetran Systems Corporation
Schenk Process GmbH
Siemens Switzerland
Siemens Transportation Systems

Sika Ltd
Swing Thru International Ltd
Telemotive
Thyssen
Trackmobile, Inc
Tuchschmid Enterprises AG
UCA Railroad Equipment
Ultra Dynamics
Unilokomotive Ltd
Vollert GmbH & Co KG Anlagenbau
Wabtec Rail Ltd
Weighwell
ZAGRO Bahn- und Baumaschinen GmbH
Zweiweg Schneider GmbH & Co KG
Zwiehoff GmbH

Company listing by country

AUSTRIA
Hans Künz GesmbH

BELGIUM
UCA Railroad Equipment

FRANCE
Caillard Cowans Sheldon

GERMANY
A Rawie GmbH & Co
Bosch – Robert Bosch GmbH
Gottwald Port Technology GmbH
Hegenscheidt-MFD GmbH & Co KG
KE Kranbau Eberswalde
Kocks Krane International GmbH
Noell
Pfister
Schenk Process GmbH
Siemens Transportation Systems
Thyssen
Vollert GmbH & Co KG Anlagenbau
ZAGRO Bahn- und Baumaschinen GmbH
Zweiweg Schneider GmbH & Co KG
Zwiehoff GmbH

IRELAND
Liebherr Container Cranes Ltd
Unilokomotive Ltd

JAPAN
Kyosan Electric Mfg Co Ltd
Mitsubishi Heavy Industries Ltd

NETHERLANDS
Kalmar Industries BV

NEW ZEALAND
Swing Thru International Ltd

SPAIN
Babcock & Wilcox Española SA

SWEDEN
Kalmar

SWITZERLAND
Siemens Switzerland
Tuchschmid Enterprises AG

UNITED KINGDOM
Blatchford Transport Equipment
E G Steele & Co Ltd
Oleo International Ltd
Railweight
Sika Ltd
Ultra Dynamics
Wabtec Rail Ltd
Weighwell

UNITED STATES
ABC Rail Products Corporation
Aldon Company Inc
Cattron-Theimeg
Central Power Products
Control Chief Corp
Kershaw
Safetran Systems Corporation
Telemotive
Trackmobile, Inc

ABC Rail Products Corporation

Track Products Division, 200 S Michigan Avenue, Chicago, Illinois 60604, USA
Tel: (+1 312) 322 03 60   Fax: (+1 312) 322 03 77

Key personnel
For a full personnel listing, see ABC Rail Products Corporation entry in *Brakes and drawgear* section.

Products
Yard control systems.

A Rawie GmbH & Co

Dornierstrasse 11, D-49090 Osnabrück, Germany
Tel: (+49 541) 91 20 70   Fax: (+49 541) 912 07 36
e-mail: info@rawie.com

Key personnel
For a full list of personnel, see A Rawie GmbH & Co entry in *Permanent way components, equipment and services* section.

Products
Buffer stops with integral loading ramps; friction, fixed and folding (manual or motorised) wheel stops; fixed and friction buffer stops; specialist track endings to customer requirements.

Aldon Company Inc

3410 Sunset Avenue, Waukegan, Illinois 60087, USA
Tel: (+1 847) 623 88 00; 01
Fax: (+1 847) 623 61 39
e-mail: e-rail@aldonco.com

Key personnel
President: J R Ornig
General Manager: J A Shelton

Products
Wheel blocks, wheel chocks, warning signs, portable friction rail skids, wagon stops, bumping posts, electric and pneumatic wagon shakers, winch-type wagon pullers, electric wagon haulers, power wagon movers, wagon door wrenches and pullers, automatic bulk wagon gate openers and retarders.

Babcock & Wilcox Española SA

PO Box 294, Alameda Recalde 27, E-48009 Bilbao, Spain
Tel: (+34 94) 424 17 61   Fax: (+34 94) 423 70 92

Key personnel
For full list of personnel, see Babcock & Wilcox Española SA entry in *Locomotives and powered/non-powered passenger vehicles* section.

Products
Mechanical handling equipment; container cranes; portal cranes; dockside cranes; shipyard portal cranes; giant shipyard portal cranes; polar cranes for nuclear power facilities; overhead travelling cranes of all kinds; ingot mould stripper and foundry ladle cranes; ship discharging machinery; wagon tipplers.

Blatchford Transport Equipment

A division of Herbert Pool Ltd
95 Fleet Road, Fleet GU51 3PJ, UK
Tel: (+44 1252) 62 04 44   Fax: (+44 1252) 62 22 92

Key personnel
Managing Director: Nigel Pool

Products
Mechanical handling systems for intermodal transport, including truck/trailer-mounted side loaders; trailer and rail-mounted cranes; self-propelled rail-mounted cranes; heavy-duty cranes for terminal use; low-temperature application container handlers.

The Blatchford side loader is available for fixed mounting to vehicle chassis or sliding mounting to trailer chassis for handling 20 to 45 ft long ISO containers.

The Blatchford T-lift trailer and rail-mounted container and swapbody cranes use a patented lifting and cross-transfer system with capacities up to 36 tonnes.

Bosch – Robert Bosch GmbH

Zitadellenweg 34, D-13578 Berlin, Germany
Tel: (+49 711) 81 10   Fax: (+49 711) 811 66 30

Products
Radio systems for shunting operations.

Caillard Cowans Sheldon

PO Box 1368, Place Caillard, F-76065 Le Havre Cedex, France
Tel: (+33 2) 35 25 81 31   Fax: (+33 2) 35 25 11 41
e-mail: caillard@caillard.fr
Web: http://www.caillard.fr

Key personnel
Managing Director: K Bayram
Head of Railway Cranes Department:
   F Elbaroudi

Products
Diesel-electric and diesel-hydraulic railway breakdown cranes with lifting capacities from 60 tonnes to 250 tonnes. Recent contracts include an oder for four diesel-hydraulic breakdown cranes for Indian Railways.

Cattron-Theimeg

58 West Shenango Street, Sharpsville, Pennsylvania 16150-1198, USA
Tel: (+1 724) 962 35 71
Fax: (+1 724) 962 43 10
e-mail: mail@cattron.com
Web: http://www.cattron-theimeg.com

Key personnel
See entry in *Signalling and communications systems* section.

UK subsidiary
Cattron-Theimeg (UK) Ltd
Riverdene Industrial Estate, Molesey Road, Hersham, Walton on Thames KT12 4RY
Tel: (+44 1932) 24 75 11
Fax: (+44 1932) 22 09 37
General Manager: Ian Martin

Products
Radio and infra-red cordless control systems for railway equipment including shunting locomotives, overhead cranes, wagon movers and ballast wagon doors.

Central Power Products

Central Manufacturing
4116 Dr Greaves Road, PO Box 777, Grandview, Missouri 64030, USA
Tel: (+1 816) 767 03 00
Fax: (+1 816) 763 07 05
e-mail: central@birch.net
Web: http://www.shuttlewagon.net

Key personnel
President and Chief Executive Officer: John L Ying
Vice-President, Sales and Engineering:
   Jack Highfill
Vice-President, Operations: Ed Harbour
Customer Support: Joe Eaves

Central Power Products Shuttlewagon road/rail tractor in service with Eurostar, Belgium   0134161

600 Series Shuttlewagon for heavy pulling applications   1026739

New Be Series Shuttlewagon   1026737

Products

Shuttlewagon®: mobile road/rail wagon mover of 24,000 to 55,000 lb drawbar pull (eight models are available); maintenance of way equipment; RYD-A-RAIL® road-to-rail conversion units.

Contracts

Contracts include supply of mobile wagon movers to Norfolk Southern Railroad, BART San Francisco, Long Island Railroad, Metro North Railroad, CSX, Union Pacific, Amtrak, Chicago Transit Authority and Eurostar, Belgium.

VERIFIED

Control Chief Corp

PO Box 141, Bradford, Pennsylvania 16701, USA
Tel: (+1 814) 362 68 11   Fax: (+1 814) 368 41 33
Web: http://www.controlchief.com

Key personnel

Chief Executive Officer/President: Douglas S Bell

Products

Control Chief manufacturers a broad range of industrial wireless remote control systems. Each remote control system is custom-engineered for rail applications.

E G Steele & Co Ltd

25 Dalziel Street, Hamilton ML3 9AU, UK
Tel: (+44 1698) 28 37 65   Fax: (+44 1698) 89 15 50
e-mail: egsteelecoltd@btinternet.com

Key personnel

Managing Director: David Steele
Logistics Manager/Director, Export Sales: Ian Hood
Quality Manager: Cameron Gibson

Products

Locopulsor shunting machine, a single-wheel vehicle capable of moving wagons weighing 160 to 200 tonnes on straight level track. It can also move wagons in curves, split a line of wagons and handle a wagon on a turntable.

The company is UK agent for Trackmobile road/rail shunting equipment. It offers a range of shunters capable of moving loads from 10 to 2,000 tonnes.

UPDATED

Gottwald Port Technology GmbH

PO Box 18 03 43, D-40570 Düsseldorf, Germany
Tel: (+49 211) 710 20
Fax: (+49 211) 710 26 50
e-mail: info@gottwald.com
Web: www.gottwald.com

Key personnel

Chief Executive Officer and Chief Financial Officer: Dirk Kiessling
Chief Technical Officer: Dr Mathias Dobner

Background

Formerly Mannesmann Dematic AG Gottwald and then Siemens Dematic AG, the company changed its name in 2002 to Gottwald Port Technology GmbH, and is owned by the investment company Kohlberg Kravis Roberts & Co (KKR).

Products

Manufacture and supply of railway cranes for single- and multi-track lines including: breakdown cranes with telescopic and fixed booms; track-laying and bridge-building cranes; universal cranes.

The main features of Gottwald's three varieties of railway crane are: high lifting capacities with moderate crane tare weight; optimum transport and job operation dimensions; wide working ranges taking into account maximum wheel and axle loads; ergonomic operating conditions and operating safety; low maintenance requirements; Gottwald bogies provide high running qualities in train formation and under own power; high towing speed for fast arrival on site; crane travelling self-propelled speed up to 100 km/h with the model GS 150.14 TR; fully hydrostatic drive for precise and powerful travel and working motions.

UPDATED

Hegenscheidt-MFD GmbH & Co KG

Hegenscheidt Platz, D-41812 Erkelenz, Germany
Tel: (+49 2431) 860
Fax: (+49 2431) 864 66
e-mail: hegenscheidt.mfd@nshgroup.com
Web: http://www.hegenscheidt-mfd.de

Gottwald GS 140.09 H breakdown, India   *NEW*/0585258

Key personnel
Managing Director: Dr Winfried Büdenbender
Sales Director: Markus von Reden

Products
Portable hydraulic rerailing systems in aluminium alloy; hydraulic equipment for rapid track clearing after serious accidents; road/rail vehicle with rerailing equipment.

Kalmar

SE-341 81 Ljungby, Sweden
Tel: (+46 372) 260 00   Fax: (+46 372) 263 90

Key personnel
Managing Director: Jonas Suaufesson
Vice-President, Corporate Communications:
 S E Petterson
Subsidiary conmpanies in Austria, France, Germany, Hong Kong, Netherlands, Norway, Spain, Singapore and USA

UK subsidiary
Kalmar UK Ltd
Siskin Drive, Coventry CV3 4FJ, UK
Tel: (+44 1203) 83 45 00   Fax: (+44 1203) 83 45 23

Key personnel
Managing Director: J Arkell
Financial Controller: E Pook
Parts and Service Director: Keith Snow

Products
IC engine and electric counterbalance lift-trucks up to 90 tonnes capacity; IC engine and electric side loaders from 2 to 15 tonnes capacity; IC engine reachstackers for container and intermodal handling.

Kalmar Industries BV

PO Box 5303, Doklaan 22, NL-3008 AH Rotterdam, Netherlands
Tel: (+31 10) 294 66 66   Fax: (+31 10) 294 67 77
Web: http://www.kalmarind.com

Key personnel
Managing Director: J K Lukumaa
Sales Manager: K Derks

Products
Container quay cranes; rail-mounted container stacking cranes; automated stacking cranes; rail-mounted harbour cranes; straddle carriers.

KE Kranbau Eberswalde

Heegermühler Strasse 64, D-16225 Eberswalde, Germany
Tel: (+49 3334) 620   Fax: (+49 3334) 62 23 08
e-mail: info@kranbau-eberswalde.de
Web: http://www.kranbau-eberswalde.de

Products
Multi-purpose cranes, slewing cranes and container gantry cranes, operated worldwide.

UPDATED

Kershaw

Kershaw Manufacturing Co Inc
PO Box 244100, Montgomery, Alabama 36124-4100, USA
Tel: (+1 334) 387 91 00   Fax: (+1 334) 215 75 51
Web: http://www.kershawusa.com

Key personnel
Vice-President and COO: Greg Valley
Vice-President of Sales and International Sales
 Manager: Phil Brown

Background
Kerhsaw Manufacturing is a division of Progress Rail Services Corporation.

Products and services
Rail/road yard cleaners; road/rail cranes (60 to 150 t) for handling containers, wagons and locomotives; one-man operated loading ramps for rubber-tyred or rail-mounted vehicles.

Kocks Krane International GmbH

Weserstrasse 64, D-28757 Bremen, Germany
Tel: (+49 421) 660 10   Fax: (+49 421) 660 13 67
e-mail: info@kockskrane.de
Web: www.kockskrane.de

Products
Container gantry cranes; rail-mounted container yard cranes.

UPDATED

Hans Künz GesmbH

Gerbestrasse 15, A-6971 Hard, Austria
Tel: (+43 5574) 688 30   Fax: (+43 5574) 68 83 19
e-mail: sales@kuenz.com
Web: http://www.kuenz.com

Products
Container cranes and spreaders for intermodal container terminals.

Development
With three other companies, Künz was selected by Rail Cargo Austria to participate in the development of the Interoperable Horizontal Transshipment (Inhotra) programme to construct a container storage and retrieval system that separates unloading from storage. The programme aims to reduce the costs and train turn-round times at intermodal terminals.

Kyosan Electric Mfg Co Ltd

4-2 Marunouchi 3-chome, Chiyoda-ku, Tokyo 100-0005, Japan
Tel: (+81 3) 32 14 81 36   Fax: (+81 3) 32 11 24 50
Web: www.kyosan.co.jp

Head office
29-1, Heian-cho, 2-chome, Tsurumi-ku, Yokohama, Japan
Tel: (+81 45) 501 12 61   Fax: (+81 45) 15 61

Key personnel
President: T Nishikawa
Director and Executive Officer, Export Sales:
 Kazuo Hinata
General Manager, Overseas Department:
 F Takenouchi

Subsidiary
Taiwan Kyosan Co Ltd, Taichung, Taiwan

Products
Automatic wagon haulage systems.

UPDATED

Liebherr Container Cranes Ltd

Killarney, Co Kerry, Ireland
Tel: (+353 64) 702 00   Fax: (+353 64) 316 02; 327 35
e-mail: sales@lcc.liebherr.com

UK sales office
Liebherr Great Britain Ltd, Travellers Lane, Welham Green, Hatfield AL9 7HW, UK
Tel: (+44 1707) 26 81 61   Fax: (+44 1707) 26 16 95

Key personnel
Directors: Pat O'Leary, Reinger Geiler, John Coffey, Andreas Boehm
Sales and Marketing Manager: Gerry Bunyan

Products
Liebherr manufactures rail-mounted container handling cranes and RTGs for ship-to-shore terminals, railway and trucking terminals and storage yards. Sizes, speeds and safe working loads to meet all international tenders and customers' specific requirements.

UPDATED

Mitsubishi Heavy Industries Ltd

Steel Structure Systems International Operations Dept
16-5, Konan 2-Chome, Minato-Ku, Tokyo 108-8215, Japan
Tel: (+81 3) 67 16 40 71   Fax: (+81 3) 67 16 58 07

Key personnel
Manager: Susumu Maeda

Products
Straddle carrier and gantry cranes.

Noell

Noell Stahl-und Machinenbau GmbH
Division NHV, Alfred Nobel Strasse 20, D-97080 Würzburg, Germany
Tel: (+49 931) 903 12 69   Fax: (+49 931) 903 10 16
e-mail: johannc.wu@noelle.de

Key personnel
Sales and Marketing Director: Bernd Vossnacke
Sales Director: V Schuessler

Principal subsidiaries
Noell Inc
2411 Dulles Corner Park, Suite 410, Herndorn, Virginia 22071, USA
Manager: Manfred Kohler
e-mail: mkohler@noellcrane.com

Preussag Noell
China Merchants Mechanical Engineering Co Ltd
Floor 16B, Heng Tong Hua Yuan Hubin Bei Lu, Xiamen 361012, People's Republic of China
Tel: (+86 592) 511 28 59   Fax: (+86 592) 511 30 29
e-mail: hessey@pncm.com

Peiner France
124 rue Nationale, Stiring Wendel, F-57600 Forbach, France
Manager: Roger Poliwoda

Products
Rail-mounted and rubber-tyred container gantry cranes and straddle carriers.
 Contracts include the supply of 10 rail-mounted gantry cranes for APL Los Angeles, USA; and additional rubber-tyred gantry crane for Rail Combi, Sweden; and straddle carriers for Portnet, South Africa, Hessenatie Antwerpen, Belgium and VIT Norfolk, USA.

Oleo International Ltd

PO Box 216, Grovelands Estate, Longford Road, Exhall, Coventry CV7 9NE, UK
Tel: (+44 2476) 64 55 55   Fax: (+44 2476) 36 42 87

Key personnel
For a full list of personnel, see Oleo International Ltd entry in *Freight vehicles and equipment* section.

Products
50 Series, 70 Series and 700 Series long-stroke hydraulic buffers (250–2,400 mm stroke) available for mounting on either rigid or sliding stop structures, providing effective emergency impact protection for railway rolling stock.

Pfister

Pfister Waagen GmbH
Stätzlinger Strasse 70, D-86165 Augsburg, Germany
Tel: (+49 821) 794 90   Fax: (+49 821) 794 92 45
e-mail: marketing@waagen.pfister.de
Web: http://www.pfister.de

Products
Sirius dynamic in-motion weighrail vehicle weighing system; Pluto modular rail weighbridges.

Railweight

A trading name of Avery Berkel Ltd
Hurstfield Industrial Estate, Hurst Street, Reddish, Stockport SK5 7BB, UK
Tel: (+44 161) 431 51 55   Fax: (+44 161) 433 13 56
e-mail: sales@railweight.co.uk
Web: http://www.railweight.co.uk

Key personnel
General Manager: Campbell Deas
Sales: M Burgess; G Villalon
Customer Services: R Tyson

Products
Weighing systems for all types of rolling stock, including locomotives, freight wagons and passenger-carrying vehicles. For general freight weighing applications, Weighline provides in-motion weight data of axles, bogies, full draft or unit train, for many different train combinations. For workshop maintenance weighing applications, Weighline provides static or in-motion weight data for locomotives and rolling stock. It is used for quality control procedures and adjusting wheel/bogies suspensions to ensure balanced loading. Weighline High Speed, the latest version, is capable of in-motion weighing at speeds up to 70 km/h. It can provide valuable data including side-to-side and end-to-end imbalance loading, at main line speeds.

Integrated management systems are available which automatically capture vehicle number and weight data and detect out of balance loads, wheel flats and other characteristics.

Contracts
Recent contracts include the supply of equipment to: RSA, Australia; Shen Hua Mines, China; POSO Steel, South Korea; and RJB Mining, UK.

Safetran Systems Corporation

2400 Nelson Miller Parkway, Louisville, Kentucky 40223, USA
Tel: (+1 800) 626 27 10   Fax: (+1 502) 244 74 44
Web: http://www.safetran.com

Key personnel
For full list of personnel, see Safetran Systems Corporation entry in *Signalling and communications* section.

Products
Marshalling yard communication systems; dispatcher communication systems.

Schenk Process GmbH

Landwehrstrasse 55, D-64293 Darmstadt, Germany
Tel: (+49 61) 561 32 29 87   Fax: (+49 61) 51 32 27 54
e-mail: pr.process@schenk.net
Web: http://www.schenk-process.de

Products
Multirail rail vehicle weighing system based on transducers installed in a special concrete sleeper. Multirail can be used both as a 'LegalWeight' dynamic weighing system in legal-for-trade freight traffic and as a 'WheelLoad' testing station in maintenance depots. Weight data is process by a PC-based system. The system is claimed to achieve an accuracy of within 0.5 per cent. It can also precisely identify wheel flats. Installation into existing track can be achieved in a few hours.

When combined with Schenk's 'WheelScan' wheel diagnostic system, Multirail can provide load distribution status of trains running at speeds of up to 240 km/h.

Siemens Switzerland

Siemens Switzerland Ltd, Transportation Systems
Industriestrasse 42, CH-8304 Wallisellen, Switzerland
Tel: (+41 0) 585 58 55 85   Fax: (+41 0) 585 58 05 01
e-mail: ts@siemens.ch
Web: http://www.siemens.ch/ts

Key personnel
See entry in *Signalling and communications* section.

Products
Domino mosaic panels for marshalling yard control systems; alphanumerical keyboard control.

UPDATED

Siemens Transportation Systems

Rail Automation Business Sector
PO Box 3327, D-38023 Braunschweig, Germany
Tel: (+49 531) 226 28 88   Fax: (+49 531) 226 48 88
e-mail: rail-automation@siemens.com
Web: http://www.siemens.com/ts

Key personnel
Business Sector Executive Management:
    Helmut Heike
Business Sector Technical Manager: Helmut Heike
Business Sector Commercial Manager:
    Bertram Boronowski

Products
Planning, control and monitoring of humping operations in marshalling yards; Automatic Vehicle Identification systems; electrically operated points; freight management and dispatching systems.

Marshalling yard operation and freight traffic solutions
Systems for marshalling yard operation (flat type and gravity type); freight management and dispatching systems; identification systems; management and consulting services for freight traffic.

Marshalling yard systems comprise a multicomputer system in hot standby operation that can be extended with locally controlled systems.

The marshalling yard system is designed for hump yards (for example with variable hump speed and computer-controlled retarder) and shunting areas. It reduces the time and costs for shunting operations and improves shunting quality. These systems offer high reliability and easy handling in combination with accurate slowing down even on curved tracks.

Freight management and dispatching systems include client server systems with standard relational databases to administer train, wagon and freight data. It reduces the need for shunting manpower. Identification and location systems aid in net-wide optimisation for handling freight wagons and trains.

Marshalling yard systems have been installed in Vienna and Villach (ÖBB), Bologna and Milan (FS), Munich (DB), Ludwigshafen (industry – BASF), Hamburg Harbour Railway, Antwerp (SNCB), Limmattal (SBB) and Kijfhoek (NS).

Customers for freight management and dispatching systems include Vienna (ÖBB), Limmattal and Chiasso (SBB), Hamburg Maschen (DB) and Bologna (FS).

Sika Ltd

Watchmead, Welwyn Garden City AL7 1BQ, UK
Tel: (+44 1707) 39 44 44   Fax: (+44 1707) 32 91 29
e-mail: sales@uk.sika.com
Web: www.sika.co.uk

Key personnel
For a full list of personnel, see Sika Ltd entry in *Permanent way components, equipment and services* section.

Products
Sika Rail elastic rail fixing systems are a combination of elastic, durable, reaction curing binders and compressible fillers that absorb vibration and redistribute eccentric loading ensuring that no compressive stresses are developed leading to edge failure.

Sika provides specialist construction products including high-specification concrete admixtures, jointing systems, mortars and grouts, adhesives and bonding agents, waterproofing, corrosion inhibitors, concrete repair and protective coatings, single ply roofing membranes and industrial or commercial flooring systems.

UPDATED

Swing Thru International Ltd

PO Box 1148, Dunedin, New Zealand
Tel: (+64 3) 471 84 60   Fax: (+64 3) 471 80 99
e-mail: sales@swingthru.co.nz
Web: www.swingthru.com

Swing Thru STKC-35T unit unladen

1020440

35,000 kg (77,000 lb) lift capacity Swing Thru units 1020439

Key personnel

General Manager: Geoff Kid

Products

Swing Thru is an innovative self-loading truck- or trailer-mounted container handling system. It has the ability to load and unload containers from either side of the vehicle and tranship containers across the host vehicle in a single movement. The contain handling systems are available in configuration from 10 ft to 48 ft in length and weighing up to 35,000 kg (77,000 lb).

The Swing Thru can be positioned between a truck and rail wagon and used as a mobile crane. The versatility of the Swing Thru unit allows a full container to be placed on one side and an empty container to be picked up from the opposite side without having to reposition the vehicle it is mounted on. The Swing Thru can also transport containers from the railhead and deliver them direct to the customer's premises.

UPDATED

Telemotive

Telemotive Industrial Controls
175 Wall Street, Glendale Heights, Chicago, Illinois
60139-1985, USA
Tel: (+1 630) 582 11 11   Fax: (+1 630) 582 11 94

Thyssen TKS multipiston retarder system, left, activated, right, free flow mode 0021674

e-mail: info@telemotive.com
Web: http://www.telemotive.com

Key personnel

Marketing Manager: Ken Bird

Products

Radio remote-control systems for locomotives, wagon movers, cranes and other freight handling systems.

Thyssen

Thyssen Umformtechnik & Guss GmbH
PO Box 28 11 44, D-47241 Duisburg, Germany
Tel: (+49 203) 73 22 80
Fax: (+49 203) 73 22 96

Foreign Sales Agency
Siemens Aktiengesellschaft
Transportation Systems Group
Goods Transport Division (VT 1C)
Ackerstrasse 22, D-38126 Braunschweig, Germany
Tel: (+49 531) 226 25 50   Fax: (+49 531) 226 41 04

Products

Marshalling yard retarders and wagon-moving equipment including the Thyssen TKG multipiston retarder system. The company has supplied retarders

for more than 60 German Railways' (DB AG) marshalling yards, and to German industry and railway companies abroad. The Hamburg Maschen shunting yard alone incorporates as many as 136 Thyssen retarders.

Contracts include the supply of seven primary two-rail beam retarders and 40 secondary single-rail beam retarders for Antwerp Noord marshalling yard, Belgium; 10 primary two-rail beam retarders and 43 secondary single-rail beam retarders for Kijfhoek marshalling yard, Netherlands.

Trackmobile, Inc

1602 Executive Drive, La Grange, Georgia
30240-5751, USA
Tel: (+1 706) 884 66 51
Fax: (+1 706) 884 03 90
e-mail: trackmobile@trackmobile.com
Web: www.trackmobile.com

Key personnel

President: Jack W Kennedy
Executive Vice-President, Worldwide Marketing:
   James R Codlin

Products

Manufacture of bimodal (road/rail) mobile rail vehicle movers. The Trackmobile® range comprises seven different models with tractive effort from 7,076 kg to 25,000 kg. Features include automatic weight transfer couplers. Various gauge and coupler configurations are available.

Trackmobile, Inc has recently introduced as standard equipment on Magnum Series models, high output rotary screw air compressors for service and train brakes.

UPDATED

Tuchschmid Enterprises AG

Kehlhofstrasse 54, CH-8501 Frauenfeld, Switzerland
Tel: (+41 52) 728 81 11
Fax: (+41 52) 728 81 00
e-mail: info@tuchschmid.ch

Key personnel

For a full list of personnel, see Tuchschmid Enterprises AG entry in *Freight vehicles and equipment* section.

Products

COMPACTTERMINAL is a low-cost intermodal terminal. The system is modular and suitable for all sizes from installations for small throughputs to high-performance freight distribution centres. Hercules power-operated transfer equipment for the ACTS road-rail container handling system.

UCA Railroad Equipment

(UCA bvba)
30 Vaartkai, B-2170 Antwerp (Merksem), Belgium
Tel: (+32 3) 646 78 76; 641 66 90
Fax: (+32 3) 646 78 72; 641 66 99
e-mail: info@uca.be
Web: http://www.uca.be

Key personnel

Managing Director: Yves Radermecker

Products

Road-rail shunting vehicles for use in yards and terminals. The range comprises two groups of models: the eight to 18 tonne E10/E12/E14 and RB10/RB12/RB14/RB16/RB18 range is a rigid vehicle designed to handle trailing loads of 600 to 1,600 tonnes on track gauges from 1,435 to

UCA RB10-RB18 series road-rail shunter (Ken Harris) 0528638

1,610 mm; the 20 to 28 tonne RB20 to RB28-4 range is an articulated unit designed for loads of 2,000 to 3,000 tonnes, on 1,435 gauge track. The two ranges are powered by Perkins 79 or 88 kW intercooled and Cummins 128 kW turbocharged aftercooled engines respectively.

Ultra Dynamics

Upperfield Road, Kingsditch Trading Estate, Cheltenham GL51 9NY, UK
Tel: (+44 1242) 70 79 00   Fax: (+44 1242) 70 79 01
e-mail: sales@ultradynamics.demon.co.uk

Key personnel
Managing Director: M Lane
Marketing Manager: S Middleton
Chief Project Engineer: Andrew Carnan

Principal subsidiary
Ultra Dynamics Inc
1110A Claycroft Road, Columbus, Ohio 43230-6625, USA
Tel: (+1 614) 759 90 00
Fax: (+1 614) 759 90 46
e-mail: sales@ultradynamics.com
Web: http://www.ultradynamics.com

Products
Ultra retarders are speed-sensitive units bolted to the inside of rails at strategic intervals along the track and these can be installed on the hump, in the switching area and in the classification tracks to provide the required wagon speed control.

Wagons are retarded accurately from the switching area to a safe buffing speed in the sidings. In the case of automatic couplers, wagon speeds are controlled between the specified bandwidth to ensure coupling takes place. The retarder units are preset during manufacture to the required speed control conditions of a particular marshalling yard. Noise levels are low and no exterior power source is required.

Retarders can be configured as High Capacity or Trackmaster models. Retarders are available with a retraction facility for use in yards where the operations require considerable resorting of trains.

The company also manufactures emergency stopping systems and safety systems and is BS EN ISO9001 approved.

Contracts
Contracts include installations in Copenhagen and Padborg marshalling yards (DSB); Vienna and Villach in Austria (ÖBB); Nuremberg and Zwickau, Germany (DB); Hallesberg, Sweden (SJ); Limmattal, Switzerland (SBB); Keiyo Railways, Japan, Murata marshalling yard and several yards of the Union Pacific, CP and BNSF rail companies in North America.

Unilokomotive Ltd

Oranmore, Galway, Ireland
Tel: (+353 91) 79 08 90   Fax: (+353 91) 79 08 46
e-mail: talbracht@unilok.ie
Web: http://www.unilok.ie

Ultra retarders in Zwickau marshalling yard, Germany 1034524

Key personnel
See entry in *Locomotives and powered/non-powered passenger vehicles* section.

Products
The Unilok range of road/rail wagon movers and shunting locomotives. Models are classified according to maximum drawbar pull up to 12,500 kg. A range of diesel engines may be fitted with outputs from 70 to 150 hp (52 to 112 kW) providing maximum speeds up to 30 km/h on road and rail. All machines are available with hydrostatic drive transmissions and a range of optional equipment including radio remote control, snow-plough, and hydraulic crane.

Uniloks are in service worldwide and are available in all gauges and coupler types.

Vollert GmbH + Co KG Anlagenbau

D-74185 Weinsberg, Germany
Tel: (+49 7134) 522 29   Fax: (+49 7134) 522 22
e-mail: info@vollert.de
Web: www.vollert.de

Products
Shunting equipment of various types; wagon transfer cars, radio-controlled diesel, battery or electric robot shunters of varying sizes and power including models capable of moving trains of 100 to 10,000 tonnes in temperatures between −50°C and +50°C; remote-controlled functions include disengagement of couplings and an infinitely variable traction speed for accurate wagon positioning at discharge points.

UPDATED

Wabtec Rail Ltd

PO Box 400, Doncaster Works, Hexthorpe Road, Doncaster DN1 1SL, UK
Tel: (+44 1302) 34 07 00   Fax: (+44 1302) 32 13 49
Web: www.wabtec.com

Key personnel
Managing Director: John Meehan
Engineering Director: Mike Roe
Finance Director: Robert Johnson
Operations Director: Chris Weatherall
Commercial Manager: Paul Robinson

Background
Wabtec Rail undertakes the overhaul of passenger rolling stock, main line locomotives, bogies, wheelsets, air brake equipment, hydraulic dampers and buffers design in addition to the manufacture of shunting locomotives and freight wagons.

As part of the Wabtec Corporation, Wabtec Rail also supplies to the UK rolling stock owners and maintainers, composite brake blocks and pads, Wabtec Railway Electronics train data recorders and electronic equipment, Cardwell TMX braking systems and Wabtec air brake equipment.

Products
Wabtec Rail can provide shunting locomotives, either on a short- or long-term contract hire basis. The locomotives are supplied on a fully maintained basis, with back-up by engineers to attend promptly to any breakdowns.

Locomotives available include 0-4-0 and 0-6-0 'Sentinel' type industrial shunting locomotives up to Class 08 locomotives. Locomotives can be provided fitted with a remote control system if required by the customer.

UPDATED

Weighwell

Unit 5, Thornes Trading Estate, Wakefield WF1 5QN, UK
Tel: (+44 1924) 29 98 90
Fax: (+44 1924) 20 02 97

e-mail: sales@weighwell.co.uk
Web: www.Weighwell.com

Key personnel
Managing Director: P Horsfall
General Manager: D Morfitt
Sales Consultant: Ron Wood
Business Development Manager: A Johnson

Products
In-motion and static rail vehicle weighing systems, Portable Train weighing equipment, trade approved and non-trade approved. The Network Rail approved Weighwell Portable Train weigher (PTW1) is for in-motion weighing. The Network Rail approved Weighwell Portable Train Weigher (PTW2) is for static weighing of individual wheel weights of rail vehicles. Both the PTW1 and PTW2 can be carried by two persons to any site and installed in minutes with no interuption to train movements.

UPDATED

ZAGRO Bahn- und Baumaschinen GmbH

Mühlstrasse 11-15, D-74906 Bad Rappenau, Germany
Tel: (+49 7266) 916 80   Fax: (+49 7266) 91 68 25
e-mail: info@zagro.de
Web: http://www.zagro.de

Key personnel
Managing Director and Sales Manager:
  Wolfgang Zappel

Products
Road vehicles equipped with ZAGRO track guiding equipment for railway operation. A variety of mounted implements ensures multiple use and efficiency. Designed for quick derailing and rerailing. All drive functions can be radio-controlled.

The range of road vehicles equipped with ZAGRO railway guide wheels includes the Mercedes-Benz Unimog for servicing, maintenance and transport duties on standard, broad and narrow-gauge systems; Mercedes-Benz Sprinter-type for use as a personnel carrier and for servicing and maintenance duties; Renault Traffic 4 × 4 for duties including inspection and maintenance from the track or alongside. The ZAGRO forklift shunter is platform-driven, with a haulage capacity of 300 tonnes.

The ZAGRO Mini-Shunter and Maxi-Shunter have a maximum tractive effort of 200 tonnes. Both can be powered by petrol, gas or diesel engines.

The required thrust force is transmitted from its wheel flanges to the wagon wheel. This and the continuously controlled hydraulic drive ensure safe braking of wagons. Wagons can be shunted in both directions without removing the machine. Wagons can be shunted in both directions without removing the machine.

All vehicles can be supplied for use on narrow, standard or broad gauge railways.

VERIFIED

Zweiweg Schneider GmbH & Co KG

Postfach 60, D-42791 Leichlingen, Germany
Tel: (+49 2174) 790 95   Fax: (+49 2174) 79 09 70
e-mail: zweiweg@t-online.de
Web: http://www.zweiweg.de

Key personnel
Chief Executive Officer: Josef Wagner
Managing Director: Walter Wagner
Technical Director: Jörge Lange
Chief Engineers: Rashed Jarrar; Michael Meyer
Sales Managers: Werner Gassen; Ingo Vogt

Products
Daimler-Chrysler Unimog rail/road vehicles with track guidance device. Applications include shunting, maintenance and overhead construction tasks, as well as track cleaning, cleaning grooved rails, tunnel cleaning, rescue, tower work,

snow-ploughs, brush cutters, ballast ploughs and cranes. Zweiweg Schneider also supplies a range of trucks with hydrostatic drives (bogie and single-axle) for multiple rail applications.

Clients supplied include BASF, Bayer, Siemens, DB, NS, Schlatter, ATM Milan, GTRM, Amey Fleet Services, STIB, JR Central, JR East and the Tokyo metro.

Zwiehoff GmbH

Tegernseestrasse 15, D-83022 Rosenheim, Germany
Tel: (+49 8031) 21 96 01   Fax: (+49 8031) 21 96 03
e-mail: info@zwiehoff.com
Web: www.zwiehoff.com

Key personnel
Managing Director: Gerd Zwiehoff
Export Manager: Elisabeth Schoemer

Products
Road/rail vehicles with a shunting capacity of up to 3,000 tonnes or equipped as multipurpose vehicles; self-propelled Mini Shunter with a shunting capacity of up to 150 tonnes; self-propelled Maxi Shunter with a shunting capacity of up to 200 tonnes; forklift truck-propelled wagon shunter with a shunting capacity of up to 300 tonnes.

UPDATED

Zwiehoff Maxi Shunter 0064334

VEHICLE MAINTENANCE EQUIPMENT AND SERVICES

Alphabetical listing

Alfred Kärcher GmbH & Co KG
ALSTOM Transport
Alzmetall
ASC Industries
Astra Vagoane Călători Arad SA/Arad
Atlas
BBM
Bingham Rail
Blaschke Umwelttechnik
Bombardier Transportation
Bradken Rail
BvL Oberflächentechnik GmbH
CAE Cleaning Technologies plc
CAF – Construcciones y Auxiliar de Ferrocarriles SA
CAM Industries Inc
Ceccato SpA
CESPA
Cimmco International
Clyde Materials Handling
Corade
Cragg Railcharger
Crous Chemicals GmbH
Delaware Car Company
Dawson-Aquamatic
Deutsche Bahn AG
Devonport Royal Dockyard Ltd
EDI Rail
E G Steele & Co Ltd
Emanuel srl
Ematech
English Welsh & Scottish Railway Ltd
Eurogamma srl
EuroMaint AB
Eurostar (UK) Ltd
EuroTrac GmbH Verkehrstechnik
Fabryka 'Wagon' SA
Fabryka Wagonów Gniewczyna SA
Ferifos
Walter Finkbeiner GmbH
Fraunhofer Institute for Non-Destructive Testing

GE
Gevisa S/A
Gilardoni SpA
Gmeinder Lokomotiven- und Maschinenfabrik GmbH
Greenwood Engineering A/S
Gregomatic AG
Hegenscheidt-MFD GmbH & Co KG
Hi-Force Hydraulics
Holland Company LP
Hovair Systems Ltd
Hunslet-Barclay Ltd
HYWEMA Lifting Systems
IBEG Maschinen-und Gerätebau GmbH
INKA
Instron
Interlok Ltd
International Technical Services
iQR
Kambre TTM AB
Alfred Kärcher GmbH & Co KG
Keller Elettromeccanica SpA
Končar-Electric Locomotives Inc
Lloyds Somers
MAN Ferrostaal AG
Marcroft Engineering Ltd
Maschinen- und Stahlbau Dresden
McConnell Research Enterprises Pty Ltd
Mechan Ltd
Metronet REW Ltd
MotivePower Inc
AB Ph Nederman & Co
NedTrain
Nencki AG
Neuero Technology GmbH
NEU International
Orval
Pars nova as
PESA Bydoszcz SA
Pfaff-silberblau Verkehrstechnik GmbH & Co KG
Portec Rail Products Inc

Proceco Inc
Rail Passenger Services Inc (RPS)
Railquip Inc
Rail Services International SA (RSI)
Railway Projects Limited (RPL)
Rescar Companies
RMS Locotec
Rolanfer Matériel Ferroviaire SA
Ross & White Company
Rosvagonmash
Rotary Lift
Rotem Company
SAFOP SpA
SEFAC
Siemens Transportation Systems
Simmons
Smith Bros & Webb Ltd
Sogema Engineering
SSI Corporation
Stertil-Koni
SweMaint
Talgo
Talgo Oy
Thrall
Toshiba Corporation
TraffiCare AB
Trenitalia SpA
Ultrasonic Sciences Ltd
United Goninan
Üstra
Vanjax Sales Pvt Ltd
VIA Rail Canada
Von Roll BETEC
Vossloh Locomotives GmbH
Wabtec Rail Ltd
Wesurail Limited
Whiting Corporation
Windhoff Bahn-und Anlagentechnik GmbH
ŽOS Trnava as
ŽOS Vrútky as
ŽOS Zvolen as

Company listing by country

AUSTRALIA
Bradken Rail
EDI Rail
iQR
McConnell Research Enterprises Pty Ltd
United Goninan

BRAZIL
Gevisa S/A

CANADA
Bombardier Transportation
International Technical Services
SSI Corporation
VIA Rail Canada

CROATIA
Končar-Electric Locomotives Inc

CZECH REPUBLIC
Pars nova as

DENMARK
Greenwood Engineering A/S

FINLAND
Talgo Oy

FRANCE
ALSTOM Transport
Ferifos
NEU International
Orval
Rail Services International SA (RSI)
Rolanfer Matériel Ferroviaire SA
SEFAC
Sogema Engineering

GERMANY
Alzmetall
Blaschke Umwelttechnik
BvL Oberflächentechnik GmbH
Deutsche Bahn AG
EuroTrac GmbH Verkehrstechnik
Walter Finkbeiner GmbH
Fraunhofer Institute for Non-Destructive Testing
Gmeinder Lokomotiven- und Maschinenfabrik GmbH
Hegenscheidt-MFD GmbH & Co KG
HYWEMA Lifting Systems

IBEG Maschinen-und Gerätebau GmbH
Alfred Kärcher GmbH & Co KG
MAN Ferrostaal AG
Maschinen- und Stahlbau Dresden
Neuero Technology GmbH
Pfaff-silberblau Verkehrstechnik GmbH & Co KG
Siemens Transportation Systems
Üstra
Vossloh Locomotives GmbH
Windhoff Bahn-und Anlagentechnik GmbH

INDIA
Cimmco International
Vanjax Sales Pvt Ltd

INDONESIA
INKA

ITALY
BBM
Ceccato SpA
CESPA
Corade
Emanuel srl

For details of the latest updates to *Jane's World Railways* online and to discover the additional information available exclusively to online subscribers please visit
jwr.janes.com

Company listing by country—*continued*

Eurogamma srl
Gilardoni SpA
Keller Elettromeccanica SpA
SAFOP SpA
Trenitalia SpA

JAPAN
Toshiba Corporation

KOREA, SOUTH
Rotem Company

NETHERLANDS
Ematech
NedTrain
Stertil-Koni

POLAND
Fabryka 'Wagon' SA
Fabryka Wagonów Gniewczyna SA
Interlok Ltd
PESA Bydoszcz SA

ROMANIA
Astra Vagoane Călători Arad SA/Arad

RUSSIAN FEDERATION
Rosvagonmash

SLOVAKIA
ŽOS Trnava as
ŽOS Vrútky as
ŽOS Zvolen as

SPAIN
CAF – Construcciones y Auxiliar de Ferrocarriles SA
Talgo

SWEDEN
EuroMaint AB
Kambre TTM AB
AB Ph Nederman & Co
SweMaint
TraffiCare AB

SWITZERLAND
Crous Chemicals GmbH
Gregomatic AG
Nencki AG
Von Roll BETEC

UNITED KINGDOM
Atlas
Bingham Rail
CAE Cleaning Technologies plc
Clyde Materials Handling
Dawson-Aquamatic
Devonport Royal Dockyard Ltd
E G Steele & Co Ltd
English Welsh & Scottish Railway Ltd
Eurostar (UK) Ltd
Hi-Force Hydraulics
Hovair Systems Ltd
Hunslet-Barclay Ltd
Lloyds Somers
Marcroft Engineering Ltd
Mechan Ltd
Metronet REW Ltd
Railway Projects Limited (RPL)
RMS Locotec
Smith Bros & Webb Ltd
Thrall
Ultrasonic Sciences Ltd
Wabtec Rail Ltd
Wesurail Limited

UNITED STATES
ASC Industries
CAM Industries Inc
Cragg Railcharger
Delaware Car Company
GE
Holland Company LP
Instron
MotivePower Inc
Portec Rail Products Inc
Proceco Inc
Rail Passenger Services Inc (RPS)
Railquip Inc
Rescar Companies
Ross & White Company
Rotary Lift
Simmons
Whiting Corporation
Proceco Inc
Rail Passenger Services Inc (RPS)
Railquip Inc
Rescar Companies
Ross & White Company
Rotary Lift
Simmons
Whiting Corporation

ALSTOM Transport

Service Business
48 rue Albert Dhalenne, F-93482 Saint-Ouen, France
Tel: (+33 1) 41 66 90 00   Fax: (+33 1) 41 66 96 66
Web: www.transport.alstom.com

Key personnel

President, Transport Sector: Philippe Mellier
Chief Operating Officer: Gérard Blanc
Chief Financial Officer: Roland Kientz
Senior Vice-President Train Life Services:
   Dominique Pouliquen
Regional Senior Vice-Presidents:
   Asia Pacific: Marc Chatelard
   Southern Europe: Charles Carlier
   Northern Europe: Terence Watson
   NAFTA: Roelof van Ark
   Iberian-Americas: Antonio Oporto

Contact addresses

Brazil
ALSTOM Transport do Brazil
Lapa Unit, Avenida Raimundo Pereira de Magalhães,230, CP 05092-901, São Paulo
Tel: (+55 11) 863 21 31   Fax: (+55 11) 260 02 24

Canada
ALSTOM Canada Inc
11012 MacLeod Trail, SE, Suite 850, Calgary, Alberta T2J 6A5
Tel: (+1 403) 278 84 98   Fax: (+1 403) 278 94 83

France
ALSTOM Transport
11-13 avenue du Bel Air, F-69627 Villeurbanne
Tel: (+33 4) 72 81 52 00   Fax: (+33 4) 72 81 52 87

ALSTOM Transport
Porte Magenta, 1 rue Baptiste Marcet, BP 42, F-71202 Le Creusot
Tel: (+33 3) 85 73 60 00   Fax: (+ 33 3) 85 73 67 99

ALSTOM Transport
Parc d'activités Lavoisier, rue Jacquard – BP 45, F-59494 Valenciennes
Tel: (+33 3) 27 14 18 00   Fax: (+33 3) 27 14 18 83

ALSTOM Transport
50, rue du Dr Guinier – BP 4, Semeac, F-65600 (Tarbes)
Tel: (+33 5) 62 53 41 21   Fax: (+33 5) 62 53 40 01

Germany
ALSTOM Transport
Linke-Hofmann-Busche-Strasse 1, D-38239 Salzgitter
Tel: (+49) 53 41 90 00   Fax: (+49) 534 19 00 69 43

ALSTOM Lokomotiven Service GmbH
Tangermuender Strasse 23A, D-39576 Stendal, Germany
Tel: (+49 3931) 254 00   Fax: (+49 3931) 256 00

Hong Kong
ALSTOM Transport Hong Kong Ltd
Room 910/912, New Kowloon Plaza, 38 Tai Kok Tsui Road, Kowloon, Hong Kong
Tel: (+852 26) 94 28 17   Fax: (+852 26) 88 57 94

Italy
ALSTOM Ferroviaria SpA
Via Ottavio Moreno 23, Savigliano, I-12038 Cuneo
Tel: (+39) 01 72 71 81 11   Fax: (+39) 01 72 71 83 06

Mexico
ALSTOM Transporte
Norte 45 No 919, Col Industrial Vallejo, 02300 Mexico DF
Tel: (+52 5) 719 08 70   Fax: (+52 5) 719 08 96

ALSTOM Transporte
Av Mario Colín s/n, Col. Valle Ceylán, Tlalnepantla, 54150, Estado de Mexico
Tel: (+52 5) 390 10 03   Fax: (+52 5) 390 50 32

ALSTOM Transporte SA
Prol. Miguel Alemán s/n, Col Ferrocarrilera, 91120, Xalapa, Veracruz
Tel: (+52 28) 40 01 29   Fax: (+52 28) 40 08 65

ALSTOM Transporte
Av Manuel Barragán #4850, Col Miguel Hidalgo, 64290, San Nicolás, Nuevo León
Tel: (+52 8) 351 55 56   Fax: (+52 8) 351 95 69

Romania
ALSTOM Transport Bucharest
256 Blvd Basarabia, Sector 3, 73249 Bucharest
Tel: (+401 255) 47 00   Fax: (+401 255) 66 11

Spain
ALSTOM Transporte SA
Paseo de la Castellana 257, 5 Plta, E-28046 Madrid
Tel: (+34 91) 334 57 00   Fax: (+34 91) 334 57 21

UK
ALSTOM Transport Service UK
PO Box 248, Leigh Road, Washwood Health, Birmingham B8 2YF
Tel: (+44 121) 695 36 00   Fax: (+44 121) 695 39 40

ALSTOM Transport Service UK
Campbell Road, Eastleigh SO50 5ZB
Tel: (+44 23) 80 62 40 01   Fax: (+44 23) 80 62 40 05

ALSTOM Transport Service UK
Channel Way, Preston PR1 8XL
Tel: (+44 1772) 55 35 71   Fax: (+44 1772) 53 33 66

ALSTOM Railcare Ltd
3 Ibstock Road, Coventry CV6 6NL

US
ALSTOM Signaling Inc
150 Sawgrass Drive, Rochester, New York 14620
Tel: (+1 716) 783 20 00   Fax: (+1 716) 274 87 77

ALSTOM Transportation Inc
1 Transit Drive, Hornell, New York 14843
Tel: (+1 607) 324 45 95   Fax: (+1 607) 324 45 68

ALSTOM Transport
650 Warrenville Road, Suite 200, Lisle, Illinois 60532
Tel: (+1 630) 719 34 10   Fax: (+1 630) 719 34 60

Venezuela
ALSTOM Transport Service
Av Francisco de Miranda, Editico Banco del Orinoco, PH La Floresta, Caracas 1060
Tel: (+58 2) 285 81 81   Fax: (+58 2) 284 98 98

Background

Train Life Services business of ALSTOM Transport provides a comprehensive range of service options for passenger and freight rolling stock, track and infrastructure. This applies to ALSTOM and non-ALSTOM manufactured equipment. ALSTOM's previous transport activities in Australia and New Zealand were to be disposed of in line with the terms agreed by the EC's acceptance of ALSTOM's financing package in 2004.

Train Life Services

Total TrainLife Management©
Service Business of ALSTOM Transport has created distinct service-product ranges within its Total TrainLife Management concept. These can be provided individually or combined into comprehensive support packages. Principal service-products include:

Maintenance
This provides operators and owners with a maintenance service regardless of the original manufacturer. ALSTOM can take its people and expertise to the operators' premises and assume control and ownership of the facilities and manpower, or it can complement the operator's existing resources through a maintenance management function. ALSTOM also offers the option of maintaining rolling stock at existing ALSTOM depots, where the servicing of 'passing traffic' can be accommodated.

In addition, it provides technical support services where it identifies performance improvements to enhance a network's availability.

Renovation and modernisation
As well as routine, heavy overhaul and repair activities, ALSTOM's renovation service offers interior retrofits and technology upgrades which together deliver 'new vehicle' features, for both ALSTOM-built and other manufacturers' rolling stock.

Parts supply
ALSTOM offers a worldwide supply chain management service. This is designed to provide a flexible response to different customer needs. ALSTOM has the ability to source replacement parts for all makes of rolling stock, from the latest equipment to vehicles that have been in service for more than 30 years. ALSTOM can also provide turnkey warehousing, procurement and inventory management.

Contracts

Maintenance contracts linked to new train deliveries are listed in the *Locomotives and powered/non-powered passenger vehicles* section. Other recent contracts secured or in progress include:
France: In February 2004 a consortium led by ALSTOM including Cannes La Bocca Industries (CLBI) was awarded a contract by SNCF to refurbish 71 Class Z2 emus by 2008. Refurbishment work will be undertaken in CLBI and SNCF workshops with ALSTOM acting as project manager and supplying major equipment.

In August 2003 ALSTOM signed a three-year contract with French National Railways (SNCF) covering the maintenance of cooling units for power modules of the operator's fleet of TGV high-speed trainsets. An option would extend the contract for two years and bring to 4,000 the number of cooling units treated.

Maintenance facility for Citadis trams in Barcelona   *NEW*/0585262

ALSTOM undertakes the maintenance of AVE high-speed trainsets operated by RENFE, Spain

0525427

The Klaw rail wheel horizontal lift　　0109480

Germany: In August 2002 Railion awarded ALSTOM Lokomotiven Service an order for the renovation of five diesel-hydraulic locomotives. In addition to renovating diesel locomotives for other companies, ALSTOM Lokomotiven Service has its own renovated pool of approximately 300 middle-power diesel locomotives in Stendal for sell or lease.

Ireland: In May 2002, ALSTOM was awarded a 15-year contract by Ireland's Railway Procurement Agency (RPA) to maintain the fleet of 40 Citadis trams supplied by the company for Dublin's light rail system.

Romania: In November 2003 ALSTOM was awarded a 15-year contract by Metrorex to maintain the 400-car Bucharest metro fleet for a period of 15 years from January 2004. Under the contract ALSTOM was to take over three Metrorex maintenance depots, a workshop and associated staff.

Spain: A contract placed by RENFE in February 2004 with a consortium of ALSTOM and CAF for the supply of 30 high-speed shuttle trains and 45 variable-gauge high-speed trains includes the provision of comprehensive maintenance services for a period of 14 years from the introduction of the trains in 2006. In October 2002, ALSTOM was awarded a contract by Spanish National Railways to undertake the maintenance for 14 years of 24 trains in its high-speed fleet. The contract covers 18 AVE high-speed trainsets used on the Madrid–Seville line, six Euromed trainsets used on the Barcelona–Valencia–Alicante line and 21 Class 252 electric locomotives which power 200 km/h push-pull services on the Madrid–Seville and Madrid–Barcelona routes.

In August 2000, ALSTOM received a contract from Spanish National Railways to refurbish 30 Class 333 diesel-electric locomotives. Barcelona LRT – 25-year maintenance contract for rolling stock and infrastructure.

Sweden: In March 2004 SJ awarded ALSTOM a four-year contract covering the maintenance of 43 two- and three-car double-deck Coradia emus from July 2004. The contract includes a three-year extension option. The work will be carried out in workshops at Västerås.

UK: In 2002, ALSTOM secured a contract to refurbish 94 metro vehicles operated by Docklands Light Railway Ltd, London. To be carried out at the company's Wolverton plant, the work was completed by late 2004.

In July 2001, ALSTOM was awarded a contract by Anglia Railways to renovate 120 Mark 2 locomotive-hauled coaches owned by HSBC Rail (UK) Ltd. Undertaken at ALSTOM's Eastleigh facility between 2001 and 2004, the work includes full bogie overhaul.

ALSTOM holds a long-term maintenance contract with Virgin Trains covering the operator's West Coast Main Line rolling stock. In February 1999, Virgin transferred all its maintenance activities to a dedicated company, West Coast Traincare, since absorbed within ALSTOM Transport Service Ltd. This is responsible for: managing maintenance depots; employing all maintenance staff; and maintaining and cleaning trains. Initial arrangements covered the existing train fleet; as

Virgin Trains new fleet of 53 nine-car Pendolino tilting trainsets is progressively delivered, ALSTOM is assuming responsibility for their commissioning and maintenance.

US: In December 2002 Alstom was awarded a contract by the Metropolitan Atlanta Rapid Transit Authority (MARTA) to overhaul 120 metro cars of its fleet of 248. A subsequent contract, let in February 2003, covers the overhaul of 118 additional cars. The work was to be undertaken at ALSTOM's Hornell, New York plant.

In November 2002, ALSTOM won a contract from Burlington Northern Santa Fe to undertake the maintenance of 434 diesel-electric freight locomotives for a 12-year period. In its execution of the contract, ALSTOM was to apply condition-based maintenance techniques based mainly on performance and reliability analyses of each piece of equipment and subsystem.

The Board of the Washington Metropolitan Area Transit Authority (WMATA) has awarded ALSTOM a contract valued at EUR370 million to overhaul 364 cars of its 764-car rapid transit fleet. ALSTOM is completely to refurbish the vehicles, which were built between 1983 and 1988, replacing the existing propulsion systems with its ONIX traction drives, and install an advanced cab signalling system. Work is to be carried out at ALSTOM's Hornell and Rochester, New York, facilities. Deliveries were scheduled from August 2002 to June 2005.

UPDATED

Alzmetall

Alzmetall Werkzeugmaschinenfabrik und Giesserei Friedrich GmbH & Co
Harald Friedrich Strasse 2-8, D-83352 Altenmarkt/ Alz, Germany
Tel: (+49 8621) 880　　Fax: (+49 8621) 882 13
Web: www.alzmetall.de

Key personnel
Managing Director: R Ilg
Export Director: H Christis

Products
Drilling machines; machining centres; special purpose machines.

UPDATED

ASC Industries

1406 West 175th Street, East Hazelcrest, Illinois 60429-1820, USA
Tel: (+1 708) 647 49 00　　Fax: (+1 708) 799 60 71
Web: http://www.ascindustries.com

Key personnel
Vice-President Sales and Marketing: Tony Fastuca

Products
The Klaw rail wheel handling device. This features three adjustable arms and is ideal for lifting 36, 40, and 42 inch wheels. Its multiple pick points allow for horizontal to vertical and vertical to horizontal movement. The three-prong system keeps the load steady while lifting and allows it to be manipulated with better control.

Astra Vagoane Călători Arad SA/Arad

1-3 Petru Rareş, RO-2900 Arad, Romania
Tel: (+40 257) 23 62 10
Fax: (+40 257) 25 81 68
e-mail: astra.calatori@astrac.rdsar.ro
Web: http://www.astracalatori.ro

Key personnel
For a full list of personnel see Astra Vagoane Călători Arad SA/Arad entry in *Locomotives and powered/ non-powered passenger vehicles* section.

Background
The company was formed in 1998 by splitting it from freight wagon and passenger coaches manufacturer Astra Vagoane Arad SA (qv).

Products
Certified ISO 9001 for the design, manufacture and refurbishment of passenger coaches and rail urban passenger vehicles. Passenger coaches for international and domestic traffic, metro cars, dmus, emus, light rail vehicles. Refurbishment is also undertaken and the company has renovated coaches for Romanian Railways, alone and in co-operation with Alstom.

Astra is licensed to build 200 km/h Corail coaches.

Atlas

Atlas Engineering Company
12 Croydon Road, Caterham CR3 6QB, UK
Tel: (+44 1883) 34 76 35
Fax: (+44 1883) 34 56 62
e-mail: atlas.hp@btinternet.com
Web: http://www.atlasengineering.co.uk

Key personnel
Sales Director: P J Hines
Finance Director: M Prockter

Products
Mobile railway lifting jack (up to 35 tonnes capacity); wheel profile trueing machines; crank axle turning machines; jacks; screwing machines; underfloor wheel trueing machines; double wheel lathes; hydraulic wheel presses.

Recent contracts include the supply of equipment to: JFK Airport, USA; Railtrack, UK; and Adana metro, Turkey.

BBM

Officine Mecchaniche BBM SpA
Via Mottinello 141, I-36028 Rossano Veneto (VI),
Italy
Tel: (+39 0424) 54 44 00   Fax: (+39 0424) 54 03 72
e-mail: info@bbm.it
Web: http://www.bbm.it

Products

Vehicle lifting and handling equipment; wheelset
presses; testing equipment. Products comply with
ISO 9000 requirements.

Bingham Rail

10 Heather Court, Shaw Wood Way, Doncaster DN2
5YL, UK
Tel: (+44 1302) 36 68 82   Fax: (+44 1302) 36 68 83
e-mail: info@trainwash.co.uk
Web: http://www.trainwash.co.uk

Products

Train washing equipment designed, manufactured
and installed in either 'Brush' or 'Flail' type to
clean the vehicle bodysides, eaves, roofs and
front end/rear of end of the cabs. Under chassis
washing designed, manufactured and installed to
suit clients' requirements, either semi-automatic or
fully automatic.

Blaschke Umwelttechnik

Industriestrasse 13, D-86405 Meitingen, Germany
Tel: (+49 8271) 816 90   Fax: (+49 8271) 81 69 40
e-mail: info@hblaschke.com
Web: http://www.hblaschke.com
    http://www.hb-umwelttechnik.de

Products

Modular rail vehicle exhaust extraction systems
for maintenance depots and workshops. Various
swivel-arm designs are employed, enabling an
operator to position extraction ports precisely
using a remote control device.

Contracts

Among clients to which Blaschke has supplied
exhaust extraction systems are: Deutsche Bahn AG
and various local operators in Germany, Belgian
National Railways and Rhätische Bahn, Switzerland.

Bombardier Transportation

Services
Litchurch Lane, Derby DE24 8AD, UK
Tel: (+44 1332) 34 46 66   Fax: (+44 1332) 26 64 72

North America
1101 Parent Street, Saint-Bruno, Québec J3V 6E6,
Canada
Tel: (+1 450) 441 20 20   Fax: (+1 450) 441 15 15
Web: www.transportation.bombardier.com

SG2 metro cars refurbished for RET Rotterdam (Ken Harris)   0116288

Key personnel

President, Services: Rik Dobbelaere
Vice-President Operations and Maintenance, Total
    Transit Systems: Michael Shaman
Vice-President, Services, North America: Mike Hardt

Services

Bombardier Transportation's complete range of
support services is designed to meet the rapidly
changing needs of the rail industry worldwide. The
company's portfolio includes full train and fleet
maintenance, materials and logistics solutions,
both vehicle overhaul and modernisation and
component re-engineering, as well as operations
and maintenance of complete transit systems. While
working in partnership with operators, Bombardier
Transportation develops long-term support
packages to meet individual client requirements.
This can include customer-supported management
information systems, supply and training of staff
and consulting services. This comprehensive
portfolio and experience offers customers a
range of benefits, including increased availability
for a given fleet through balanced maintenance,
increased reliability through monitoring and
upgrades, lower life cycle costs, full support parts
management and above all safe trains.

As examples of providing hands-on support
from the beginning of the vehicle life Bombardier
has a number of parts supply contracts around the
world. These include consignment stock agreements
whereby stocks of Bombardier-owned parts are
stored at the clients' premises awaiting drawdown:
examples include BVG, the Berlin public transport
authority, Öresund Trains (OTU) and the Gardemoen
Airport Shuttle in Norway. In the UK Bombardier is
supplying kits of parts to several train operators:
these kits have been developed with the operator
to provide the specific self-contained package of
parts that are required for specific maintenance
operations. For Gardemoen in Norway Bombardier
provides a vendor-managed inventory of parts. These
different arrangements are optimised to provide
maximum availability of parts at minimal cost to the
operator. The company is developing Internet parts
ordering which will further simplify and lower the
costs of operators' ordering processes.

Bombardier operates and maintains automated
people mover systems at a number of airports
including Atlanta, Denver, Frankfurt, Houston,
Miami, Newark, Rome and San Francisco, and
provides maintenance and technical assistance
at others such as Dallas/Fort Worth International
Airport. On behalf of transportation authorities,
Bombardier is operating and maintaining the fully
automated Las Vegas Monorail, AirTrain JKF in
New York City, and the future Yong-In LRT upon its
completion in 2009.

*Bombardier Transportation maintains GO
Transit rolling stock at the Willowbrook, Toronto
facility*   0099502

Bombardier refurbished passenger coaches for Hungary   0524877

Bombardier maintenance facility in Central Rivers, UK 0558273

Other maintenance examples include locomotives in Italy and in Germany, where the LNVG operation is supported by Bombardier Transportation.

Bombardier has a range of facilities around the world, where it is modernising vehicles and re-engineering components.

Contracts

Australia: The contract from the Victoria State Government and V/Line to provide passenger rolling stock for regional services in Victoria, Australia, includes a 15-year maintenance agreement for a total of 38 V/locity two-car dmus.

Provision of maintenance for 15 years, including the construction of a maintenance and stabling facility, was a feature of a contract placed in 2002 with a joint venture of Bombardier Transportation and EDI-Rail to supply new emus for Perth area suburban services. From mid-2004 31 three-car emus, with an option for 10 more, are to be delivered to the Western Australian Government for the southern leg of the Perth Urban Rail Development project.

Canada: In Toronto, Bombardier is performing locomotive and coach maintenance services for GO Transit under a three-year option to a six-year contract awarded in December 1996. Services include supplying management, labour and materials; managing train movements in the yard; and maintaining 385 Bombardier-built BiLevel* commuter coaches and 45 General Motors F59PH locomotives.

Bombardier also provides maintenance services for the Ottawa-Carleton Regional Transit Commission (OC Transpo) in Ottawa, Ontario, under a recently awarded three-year extension to an existing contract. The scope of work on this contract includes providing management, labour, material and technical expertise to maintain a fleet of state-of-the-art, Bombardier-built diesel multiple units (dmus).

China: In September 2002 Bombardier Transportation was awarded a contract by Shanghai Metro Operation Corporation to undertake a 10 year overhaul on 96 metro cars between 2002 and 2005.

Germany: Locomotive re-engineering and modernisation contracts in Germany include: equipping 80 Class 298 diesel locomotives with radio remote control and MICAS-L control technology for DB Cargo; modernising 16 MaK diesel locomotives for HGK Cologne; modernising four Class V100.4 and two Class 60 diesel locomotives for various German operators. Additionally refurbishment of two Class 232 2,460 kW diesel locomotives for Schauffele; and the supply of four Class 293 diesel locomotives for GSG, a rail infrastructure company.

Bombardier also has a 15 year contract with LNVG providing fleet maintenance for 18 locomotives, 88 double-deck coaches and 18 driving van trailers.

Hungary: Bombardier is the majority owner of Bombardier MAV Kft, the rail coach manufacturing plant located in Dunakeszi. Over the last decade more than 600 coaches have been modernised at this site.

Italy: In June 2005 Bombardier received an order from Trenitalia (Italian Railways) for the upgrade of 60 ETR 500 electric power-heads.

Norway: In September 2002 Bombardier Transportation received an order from Flytoget AS and Norwegian State Railways (NSB) to supply spare parts for five years for the fleet of Class BM71 high-speed emus that operate between Oslo and Gardermoen Airport.

South Africa: Bombardier won a contract for the modernisation and overhaul of 45 Class 11E electric locomotives for Spoornet: this includes upgrading the control technology with the MITRAC system. The locomotives are used to haul heavy coal trains between the Mpumalanga coalfields and Richards Bay.

Spain: In February 2004 a consortium of Bombardier Transportation and Patentes Talgo was awarded a 14-year contract to maintain RENFE's 16 Class S 102 AVE high-speed trainsets, deliveries of which commenced in 2003.

Sweden: Bombardier has maintenance contracts for approximately 192 emu cars being operated in Sweden.

In January 2005, Bombardier received an order from Affarsverket Statens Jarnvagar (Swedish State Railways) for the heavy maintenance of 16 locomotives, nine emus and components including bogies, axles and traction motors.

UK: In May 2005 Bombardier announced the award of a contract received from Porterbrook for the refurbishment of 110 passenger coaches for National Express Group's Great Anglia franchisee One. In November 2004, Bombardier received a contract from HSBC for the refurbishment of 244 emus, also for use by One. This is the second contract for Class 315 refurbishment that Bombardier has received from HSBC. Four Class 315 units had previously been modified by Bombardier under a contract awarded in 2002. In June 2004 Bombardier announced the award of a contract by Porterbrook Leasing Ltd to undertake the refurbishment of 91 Class 455 four-car emus operated by South West Trains. The work is being carried out at the company's Ashford, Kent, facility between 2004 and 2007.

In January 2003 Bombardier Transportation received a contract from HSBC Rail (UK) Ltd covering the heavy maintenance of 31 Class 91 high-speed electric locomotives operated by Great North Eastern Railway. The contract covered a five-year period with an option to extend it for additional five-year periods up to a maximum of 18 years.

In September 2002 HSBC Rail (UK) Ltd awarded Bombardier Transportation a contract to refurbish 302 Mark 4 passenger coaches used by Great North Eastern Railway. Completion of the programme is scheduled for 2005.

Bombardier Transportation has a 14-year contract with Virgin Trains to maintain 352 Class 220/221 demu cars. The fleet is maintained at a purpose-built facility at Central Rivers, near Burton upon Trent, UK and at other sites.

The 99-year Croydon Tramlink concession in the UK has been awarded to Tramtrack Croydon – a consortium with Bombardier as a key member. Bombardier was awarded the contract to manufacture the 24 trams and is now maintaining them.

Bombardier has built 296 emu cars for c2c, part of the National Express Group. Bombardier is maintaining these vehicles at the East Ham Depot.

In 2002, Bombardier Transportation secured an initial four-year maintenance contract related to an order for 127 Class 222 diesel-electric multiple-unit cars for Midland Mainline. A new depot is to be built in South Yorkshire, and use will also be made of two other re-commissioned or existing facilities. The contract includes an option to extend it to 15 years.

A 20-year materials supply and technical support agreement was signed in March 2002 by Bombardier Transportation and Govia as part of a contract to supply 700 Class 377 Electrostar emu cars for services on the Southern (formerly South Central) franchise in southern England.

Bombardier Transportation maintains the Croydon trams in the UK 0558274

Bombardier's Refurbishment Centre at Derby, UK, with a Mark 3 coach and Class 315 emu undergoing modernisation, both for One Railway (Ken Harris)   ***NEW**/0585279*

As part of the Metronet consortium Bombardier will be refurbishing District Line vehicles. Bombardier has a rolling contract to supply 1,738 new cars for four of Metronet's six lines by the end of 2015. During 2007, the company will take over all maintenance of the Victoria Line and sub surface line fleets to the end of the 30 year PPP contract, subject to periodic review.

US: On 1 July 2003, as part of the Massachusetts Bay Commuter Railroad Company consortium, Bombardier Transportation began a five year services contract to carry out maintenance for the Massachusetts Bay Transportation Authority (MBTA)'s commuter rail system. The contract includes the supply of management, labour and materials and materials management to maintain the MBTA's fleet of 80 locomotives, 377 single-level and bi-level coaches, and 30 work train vehicles.

In 1998, Bombardier was awarded a three-year contract, followed by a two-year option, to provide maintenance services for the Metrolink commuter rail fleet of the Southern California Regional Rail Authority (SCRRA). In 2003, the SCRRA awarded Bombardier a new seven-year contract, with a three-year option, commencing 1 July 2003. The scope of work is to supply management, labour and materials; manage train movements in the yard; and maintain 146 Bombardier-built BiLevels*, 37 General Motors locomotives, and work train vehicles.

As a member of the Southern New Jersey Rail Group, Bombardier was awarded a contract to design, build, operate and maintain a unique 34-mile turnkey system that operates between Trenton and Camden, New Jersey. This project is the first application of a diesel-electric light rail transit system operating on an existing freight corridor in the US. Passenger and freight service is time-separated, with freight operating only at night. Bombardier has supplied 20 articulated diesel-electric light rail vehicles, signalling, communications, traffic control, central control, start-up and testing, and is operating and maintaining the 34-mile commuter system for 10 years. Bombardier is also in charge of train dispatching, including freight trains. Revenue service began on 14 March 2004.

As leader of the Bombardier ALSTOM Consortium, Bombardier built 20 high-speed trainsets for Amtrak, as well as its turnkey maintenance facilities in Boston, New York and Washington, DC where all service and inspection activities for the fleet takes place. Bombardier is providing management services of all maintenance activities until October 2006.

* Trademarks of Bombardier.

UPDATED

Bradken Rail

2 Maud Street, Mayfield West, New South Wales 2304, Australia
PO Box 105, Waratah, New South Wales 2298, Australia
Tel: (+61 2) 49 41 26 77
Fax: +(61 2) 49 41 26 61
e-mail: rail@bradken.com.au
Web: www.bradken.com.au

Background
Bradken's rail business structure consists of three strategic business units: freight rolling stock, express parts and maintenance.

Services
MainteNet is the service, maintenance and refurbishment business unit of Bradken, enabling the group to provide a total freight rail package to the industry by adding maintenance, service and refurbishment to the capital build and spare parts products ranges of its freight, rolling stock and express parts business units.

MainteNet services include: rolling stock refurbishment, bogie refurbishment, drawgear refurbishment, running maintenance, in-field service, wagon conversions, structural repairs and collision damage repairs. All services are tailored to meet OEM, AAR, ROA and other rail systems standards.

UPDATED

BvL Oberflächentechnik GmbH

Grenzstrasse 16, D-48488 Emsbüren, Germany
Tel: (+49 5903) 951 60
Fax: (+49 5903) 951 90
e-mail: bvl-oberflaechentechnik@t-online.de
Web: http://www.bvl-oberflaechentechnik.de

Key personnel
Managing Directors: Wilhelm van Lengerich; Bernhard Sievering

Products
Cleaning systems for bogies and wheelsets.

CAE Cleaning Technologies plc

Mount Street, Bradford BD3 9SN, UK
Tel: (+44 1274) 72 93 41
Fax: (+44 1274) 37 07 99
e-mail: sales@cae-ct.co.uk
Web: http://www.caeclean.com

Key personnel
Managing Director: David G Holmes
Sales Director: Alan R Holmes
Sales and Marketing Manager: Sakeb Zahoor

Corporate background
CAE Cleaning Technologies plc is part of the CAE Cleaning Technologies Group.

Products
Bogie washers and wheelset washers (high-pressure aqueous); bearing washers; traction motor washers; axle gear washers; ultrasonic cleaning systems.

Contracts include supply of equipment to DSB Denmark, Adtranz, the Hellenic Railways Organisation and London Underground.

CAF – Construcciones y Auxiliar de Ferrocarriles SA

Padilla 17 – 6°, E-28006 Madrid, Spain
Tel: (+34 91) 435 25 00
Fax: (+34 91) 436 03 96
e-mail: export.caf@caf.es
Web: www.caf.es

Key personnel
President and Chief Executive Officer: J M Baztarrica
Managing Directors: A Arizcorreta, A Lergarda
Contact: J Esnaola

Works
Beasain
J M Iturrioz 26 E-20200 Beasain, Spain
Tel: (+34 943) 88 01 00   Fax: (+34 943) 88 14 20

Irún
Calle Anaca 13, E-20301 Irún, Spain
Tel: (+34 943) 61 33 42   Fax: (+34 943) 61 81 55

Zaragoza
Av de Cataluña 299, E-50014 Zaragoza, Spain
Tel: (+34 976) 76 51 00   Fax: (+34 976) 57 26 48

Offices
CAF Argentina SA
Pte Luis Sáenz Peña 31° – 7° piso, 1110 Capital Federal Buenos Aires, Argentina
Tel: (+51 11) 43 83 20 06   Fax: (+54 11) 43 81 48 37
e-mail: cafadministracion@cafarg.com.ar

CAF Brasil Industria e Comercio
Rua Pedroso Alvarenga 58 conj 52, CEP 04531-000 São Paulo, Brazil
Tel: (+55 11) 31 67 17 20   Fax: (+55 11) 30 79 87 62
e-mail: cafsaopaulo@cafbrasil.com.br

CAF Mexico SA de CV
Prolongación Uxmal 988, Col Sta Cruz Atoyac, 03310 Mexico DF, Mexico
Tel: (+52 55) 56 88 75 43   Fax: (+52 55) 56 88 11 56
e-mail: cafmex@prodigy.net.mx

CAF USA Inc
1401 K Street NW, Suite 803, Washington DC 20005-3418, USA
Tel: (+1 202) 898 48 48   Fax: (+1 202) 216 89 29
e-mail: cafusa@cafusa.com

Services
Maintenance of emus, dmus, metros, locomotives and other rail vehicles: scheduled and unscheduled maintenance, repairs, modifications and upgrades, supply of materials, operations support, staff training, technical assessments, computerised support.

Refurbishment and modernisation of rail vehicles.

Contracts
Contracts include maintenance of vehicles for the Cercanías and Trenes Regionales divisions of RENFE, Spain, for metro systems in Barcelona, Madrid and Mexico City, emus used by MTRC, Hong Kong, for its Airport Express and Tung Chung Line services.

Refurbishment contracts include work on 3 kV DC emus for RENFE, Spain, dmus for FEVE and emus for ET/FV, both Spain, metro cars for STC, Mexico, and Metrovías, Argentina, suburban emus for Rio de Janeiro and tank wagons for Saltra.

UPDATED

CAM Industries Inc

215 Philadelphia Street, Hanover, Pennsylvania 17331, USA
Tel: (+1 717) 637 59 88
Fax: (+1 717) 637 93 29
e-mail: sales@camindustries.com
Web: www.camindustries.com

Key personnel
President: Charles A McGough III

Products
Complete equipment for electric traction motor repair workshops including undercutters, handling equipment, test equipment and universal armature machines.

Contracts
BART, San Francisco, US.

UPDATED

Ceccato SpA

Via Selva Malolo, 5/7, I-36041, Alte Di Montecchio Maggiore (VI), Italy
Tel: (+39 0444) 70 84 11
Fax: (+39 0444) 69 55 44
e-mail: info@ceccato-carwash.it
Web: www.ceccato.it

Key personnel
Managing Director: Piero Rizzon
Technical Director: Paolo Brovedani
Sales Director: Alessandro Maturo

Products
Complete plants for the external washing of surface trains, trams, underground trains and associated rolling stock. Units can be configured with a high number of modular groups; each one designed to perform a specific washing function on different shapes and surfaces of the vehicles.

UPDATED

CESPA

Costruzioni Elettromeccaniche Spavone
Via Luigi Volpicella 145, I-80147 Napoli, Italy
Tel: (+39 081) 752 48 63
Fax: (+39 081) 559 05 61
e-mail: cespaiet@tin.it
Web: http://www.cespaitaly.com

Key personnel
Technical Director: Massimo Spavone
Managing Director and General Manager:
  Luigi Spavone

Products
Repair and workshop machinery; lifting equipment; workshop bogies for supporting vehicles; mobile lifting jacks for rail and bus applications; testing platforms for bogies; elevated platforms for access to vehicles; rain testing plants; electric motor maintenance equipment; underfloor lifts for bogies and wheelset mounting equipment.
   Recent contracts have included the supply of equipment to Ansaldo Trasporti, Adtranz, Breda, Fiat Ferroviaria, Firema and Italian Railways.

Cimmco International

Prakash Deep, 7 Tolstoy Marg, New Delhi 110001, India
Tel: (+91 11) 331 43 83; 384; 385
Fax: (+91 11) 332 07 77; 372 35 20

Key personnel
For a full list of personnel, see Cimmco International entry in *Freight vehicles and equipment* section.

Products
Machinery and equipment for manufacture and maintenance of rolling stock.

Clyde Materials Handling

Shaw Lane Industrial Estate, Doncaster DN2 4SE, UK
Tel: (+44 1302) 32 13 13
Fax: (+44 1302) 55 44 00

Key personnel
Sales Manager: Jeff A Buston

Products
Automated sand filling systems for locomotives and heavy and light rail vehicles.

Corade

Via P Toselli 81 I-50144, Florence Fl, Italy
Tel: (+39 055) 322 71
Fax: (+39 055) 32 27 27
e-mail: info@corade.it
Web: http://www.corade.it

Products
Painting booths and under chassis air blowing systems for railways and urban vehicles. Corade also provides a wide range of design and consulting services for developing and supplying equipment for rail vehicle depots and workshops.

Cragg Railcharger

4972 Highway 169 North, New Hope, Minnesota 55428, USA
Tel: (+1 612) 537 37 02   Fax: (+1 612) 537 37 78

Key personnel
For a full list of personnel, see Cragg Railcharger entry in *Signalling and communications systems* section.

Product
Cragg ST-2L, a digital display device to verify hot bearings and wheel defects, using a non-contact thermometer.

Crous Chemicals GmbH

Im Feld 4, Härkingen, CH-4624, Switzerland
Tel: (+41 62) 398 39 39   Fax: (+41 62) 398 39 38
e-mail: info@crous-chemicals.com
Web: http://www.crous-chemicals.com
   http://www.anti-graffiti.net

Products
Graffiti removal products and systems for rail vehicle exteriors and interiors, as well as for fixed installation; graffiti protection products and systems; graffiti removal services.

Delaware Car Company

2nd and Lombard Streets, Wilmington, Delaware 19899, USA
Tel: (+1 302) 655 66 65   Fax: (+1 302) 655 71 26

Key personnel
President: Harry E Hill
Chief Engineer: J Winter
Mechanical Superintendent: L J Reed
Vice-President, General Manager: T J Crowley
Marketing and Special Projects Engineer:
  S F Rogowski

Products
Refurbishment, repair and assembly of passenger rolling stock, including metro, suburban and commuter cars. Specialities include stainless steel parts fabrication and bogie repair and overhaul.

Contracts
Maryland Commuter Railraod (MARC) – structural repair to four MARC III bi-level cars and structural repair to one Marc II coach; Chicago Transit Authority (CTA), structural repair to one subway car; New Jersey Transit – installation of new and upgraded cab signal equipment on the entire fleet of MUs, cab cars and locomotives, cab signal equipment supplied by US&S; North Carolina DOT – structural repairs and refurbishment of nine commuter coaches, combine car and lounge cars; SE Pennsylvania Transit Authority (SEPTA), structural repair to one Silverliner IV commuter car.

UPDATED

Dawson-Aquamatic

Gomersal Works, Gomersal, Cleckheaton BD19 4LQ, UK
Tel: (+44 1274) 87 34 22   Fax: (+44 1274) 87 49 30

Key personnel
Managing Director: B J Turner
Divisional Manager: P Barnett

Products
Design, manufacture and installation of drive-through washing and brushing systems for railcars, ranging from the simplest detergent/water wash-up to fully automatic installations for daily detergent washing and periodic removal of oxides and staining by acidic solutions; supporting control systems, water storage, effluent treatment and water recycling systems; railway workshop cleaning plant including bogie washing installations.

Deutsche Bahn AG

Beriech Spezialwerke
Fahrzeugbau Halberstadt, Agustenstrasse, D-38820 Halberstadt, Germany
Tel: (+49 39) 415 20   Fax: (+49 39) 412 41 42
e-mail: Fahrzeugbau-HBS@t-online.de

Services
Modernisation and refurbishment of rail vehicles.

Devonport Royal Dockyard Ltd

Plymouth PL1 4SG, UK
Tel: (+44 1752) 55 29 08   Fax: (+44 1752) 55 41 00
e-mail: john.small@devonport.co.uk
Web: http://www.devonport.co.uk

Key personnel
Group Manager, Rail Support: Geoff Buck
Sales and Marketing Manager, Rail Support:
  John Small

Services
Repair of locomotive power units, including turbochargers, pumps and injectors; repair of power and trailer bogies; passenger coach refurbishment and modification; refurbishment of rail infrastructure vehicles; and repair of electrical equipment, including traction motors, alternators and generators.

Contracts include: repair of Class 43 (HST) power cars for UK operators, as well as overhaul of traction motors, generators and alternators; repair of Class 80 power units for Northern Ireland Railways; overhaul of Trac Gopher infrastructure vehicles; conversion work on catering cars; and fitting secondary door locks to HST coaches.

EDI Rail

28 Factory Street, Granville, New South Wales 2142, Australia
Tel: (+61 2) 96 37 82 88   Fax: (+61 2) 96 37 67 83
e-mail: sales@edirail.com.au
Web: www.edirail.com.au

Key personnel

Chief Executive Officer: Guy Wannop
Executive General Manager, Freight: Danny Broad
Executive General Manager, Passenger:
   David Williamson

Background

A division of Downer EDI Ltd, EDI Rail is the result of a merger of Clyde Engineering and the rail activities of Walkers Ltd.

Works

EDI Rail has manufacturing, maintenance or design facilities in: Bathurst, Granville, Kooragang Island and Cardiff, New South Wales; Darwin, Northern Territory; Forrestfield and Nowergup, Western Australia; Rockhampton, Gladstone, Maryborough and Brisbane, Queensland; Newport and Geelong, Victoria; Port Augusta, Dry Creek, Whyalla, Port Lincoln, South Australia.

Services

Complete maintenance support for rail vehicles of all types; overhaul of locomotives, passenger trains, freight wagons, diesel engines and traction equipment; vehicle modernisation and refurbishment; bogie and wheelset maintenance.

Contracts

Recent contracts include:
   Maintenance of FQ Class diesel locomotives and freight rolling stock for Asia Pacific Transport for a ten-year period from the end of 2003.
   Maintenance for 15 years of SRA's Millennium emus used on CityRail suburban services in Sydney, with an option to extend the term of the contract to 35 years.
   Refurbishment of some 300 Comeng-built emus for M>Train, Melbourne.

NEW ENTRY

E G Steele & Co Ltd

25 Dalziel Street, Hamilton ML3 9AU, UK
Tel: (+44 1698) 28 37 65
Fax: (+44 1698) 89 15 50
e-mail: egsteelecoltd@btinternet.com

Key personnel

Managing Director: David Steele
Logistics Manager/Director, Export Sales: Ian Hood
Quality Manager: Cameron Gibson

Services

Repair of private wagons, hirers, suppliers of spare parts and shunting equipment. E G Steele also has two breakdown units which can travel at short notice to any breakdown situation which may occur in Scotland or North East England. The company maintains a fully equipped van fitted with generator sets, welders' sets, air brake test rigs together with all the other necessary tooling required.

UPDATED

Emanuel srl

Via Marconi 3, I-40011 Anzola Emilia (BO), Italy
Tel: (+39 051) 73 26 52   Fax: (+39 051) 73 40 01
e-mail: info@emanuel1899.com
Web: http://www.emanuel1899.com

Products

Hydraulic and electro-mechanical lifting systems and equipment for heavy-rail and light-rail vehicles.

Ematech

PO Box 8093, NL-3503 RB Utrecht, Netherlands
Tel: (+31 30) 246 91 60   Fax: (+31 30) 246 91 76
e-mail: info@ematech.nl
Web: http://www.ematech.nl

Corporate background

Ematech has been an autonomous subsidiary of NS (Netherlands Railways) since 1996.

Key personnel

General Manager: Ing Joost Kruiswijk
Engineering Manager: Ing Edwin de Groot
Sales and Marketing Manager: Paul Matlung

Products

Overhaul, maintenance and upgrade of traction motors, converters, generators and ancillary equipment. Calibration of safety equipment.
   Life-cycle management programmes: ISO 9002 qualified.
   Recent contracts for clients in the Netherlands include work for NS Reizigers, Railion, RET Rotterdam, Connexxion and GVB Amsterdam. Non-Dutch customers include: Italian Railways; ATM Milan; Docklands Light Railway, London, and HSBC in the UK; MIVB, Brussels; and Essener Verkehrsbetriebe GmbH and Stadtwerke Bielefeld GmbH in Germany. Safety equipment calibration has been undertaken for Railpro in the Netherlands.

English Welsh & Scottish Railway Ltd

310 Goswell Road, Islington, London EC1V 7LW, UK
Tel: (+44 20) 77 13 23 00   Fax: (+44 20) 77 13 23 11
Web: http://www.ews-railway.co.uk

Key personnel

Chief Executive: Philip Mengel
Commercial Engineering Manager: Keith Miller

Services

The UK's leading rail freight operator, EWS also offers traction and rolling stock maintenance facilities and expertise for other operators. Under a contract placed by train builder Bombardier Transportation, EWS undertakes routine maintenance on examples of the Class 220 Voyager and Class 221 Super Voyager high-speed diesel-electric multiple-unit fleet operated by Virgin CrossCountry. By early 2002, EWS had established maintenance facilities for these trains within its existing depots in Bristol (Barton Hill), Gatwick (Three Bridges), London (Old Oak Common), Newcastle (Tyne Yard) and Southampton (Eastleigh).

Eurogamma srl

Via di Castelpulci snc I-50018 Scandicci (FI), Italy
Tel: (+39 055) 72 20 01   Fax: (+39 055) 722 00 20
e-mail: info@eurogamma.com

Products

Mobile lifting jacks for rail vehicles. Jacks can be grouped to lift a single vehicle or an entire train, with a modular design control unit which can be reprogrammed to synchronise the number of jacks in use. Products are ISO 9001-certificated.

EuroMaint AB

PO BOx 1555, SE-171 29 Stockholm, Sweden
Tel: (+46 8) 762 51 00   Fax: (+46 8) 762 32 05
Web: www.euromaint.se

Key personnel

Managing Director, Chief Executive Officer:
   Jonas Samuelson
Sales and Marketing Director: Jonas Samuelson

Background

Formerly SJ Engineering, the business sector responsible for traction and rolling stock maintenance for Swedish State Railways. EuroMaint AB was established in 2001 as part of the restructuring resulting from the break-up of the state railway company. Three of the companies which formed EuroMaint 2001 (RPL RailPartsLogistics, TGOJ and TrainMaint) were merged into EuroMaint in January 2003 and the operation organised into three departments: marketing and sales, production, and material supply. The group of companies forming EuroMaint includes TrainTech Engineering, based in Solna, which specifies maintenance requirements, devises maintenance programmes, provides documentation and acts as an independent consultant for vehicle procurement. EuroMaint also holds a 33 per cent shareholding in SwedeRail AB.

Services

Traction and rolling stock maintenance, overhaul, repair, refurbishment, component purchasing, spare parts supply and vehicle procurement. EuroMaint's principal clients are SJ AB and GreenCargo AB, but the company has also secured contracts from other train operators in Sweden as well as from other industrial partners.

NEW ENTRY

Eurostar (UK) Ltd

North Pole International, Mitre Way, London W10 6AT, UK
Tel: (+44 20) 89 64 70 89
Fax: (+44 20) 89 64 70 07

Key personnel

Head of Engineering Production: David Bailey

Services

Workshop facilities including wheel lathe, bogie drop, vehicle wash, simultaneous lift and LDA (Lavatory Discharge Apron). Services include technical engineering, technical support, commissioning engineering, safety case, document scrutiny, training and languages.

Contracts

Contracts include Eurostar servicing and maintenance.

VERIFIED

EuroTrac GmbH Verkhrstechnik

Diedrichstrasse 5, D-24143 Kiel, Germany
Tel: (+49 431) 70 69 66
Fax: (+49 431) 706 96 89
Web: http://www.eurotrac-vt.de

Key personnel

Managing Directors: Dr Wolfgang Rösch;
   James N Sanders

Background

EuroTrac GmbH is part of the Vossloh AG group.

Products and services

Services include: workshop planning and logistics; organisation of vehicle maintenance and repair; technical maintenance support.
   Products include: FlexBase® modular workshop, FlexTrac® and Flexbridge® modular maintenance

FlexTrac® modular workshop track system (Innotrans 2002) (Ken Harris)   0526802

facility track and vehicle lifting systems, all of which can be dismantled, adapted or relocated according to changing user needs; FlexTime®, an IT solution to the management of vehicle maintenance; and FlexKit®, an IT-based product intended to optimise component-based maintenance.

Fabryka 'Wagon' SA

ul Wroclawska 93, PL-63-400 Ostrøw Wielkopolski, Poland
Tel: (+48 62) 595 39 13   Fax: (+48 62) 591 27 29
e-mail: mr@fabryka-wagon.pl; zntk@osw.pl
Web: http://www.fabryka-wagon.pl

Background
Established in 1920, Fabryka 'Wagon' SA is the former Ostrøw rolling stock repair workshop of Polish State Railways.

Products
Repair, refurbishment and modernisation of passenger coaches, including restaurant and sleeping cars; repair, refurbishment and modernisation of freight wagons and their subassemblies and components.

Fabryka Wagonów Gniewczyna SA

PL-37-203 Gniewczyna, Poland
Tel: (+48 16) 648 83 64   Fax: (+48 16) 648 85 87
e-mail: admin@gniewczyna.pl;
    market@gniewczyna.pl
Web: http://www.gniewczyna.pl

Products
Repairs to freight wagons of all types.

Ferifos

ZI du Ventillon, F-13270 Fos-sur-Mer, France
Tel: (+33 4) 42 11 30 00
e-mail: jacques_harauchamps@ferifos.fr
Web: http://www.ferifos.fr

Key personnel
Director General: Jacques Harauchamps

Background
Ferifos is a member of Groupe Ermewa SA (see Groupe Ermewa SA entry in *Rolling stock leasing companies* section).

Services
Overhaul, maintenance and refurbishment of freight wagons and their components.

Walter Finkbeiner GmbH

Alte Poststrasse 9/11, D-72250 Freudenstadt, Germany
Tel: (+49 7441) 40 31
Fax: (+49 7441) 877 78
e-mail: info@finkbeiner-lifts.com
Web: http://www.finkbeiner-lifts.com

Key personnel
Managing Director: Gerhard Finkbeiner

Products
Mobile lifting equipment including mobile column lifts for rail vehicles such as streetcars and wagons: *EHB 707K*, capacity per column 7,500 kg, *EHB 710K* capacity per column 10,000 kg; each column with integrated electro-hydraulic drive, double load securing, chassis support bracket, easy to manoeuvre by one person, fully synchronised for operation of maximum 24 columns.

VERIFIED

Fraunhofer Institute for Non-Destructive Testing

Fraunhofer Institute for Non-Destructive Testing
Fraunhofer Institut Zerstörungsfreie Prüfverfahren
Universität Gebäude 37, D-66123 Saarbrücken, Germany
Tel: (+49 681) 930 20
Fax: (+49 681) 93 02 59 01
e-mail: info@izfp.fhg.de
Web: http://www.izfp.fhg.de

Key personnel
Managing Director: Prof Dr Michael Kröning
Director, Applications Centre:
    Dipl-Ing Bernd Rockstroh

Products and services
Automated ultrasonic wheelset testing such as test and measurement technology, including specification, design, development and manufacturing and continuing through all service cycles, including operations and maintenance.

Policymakers and the Deutsche Bahn respond to the high density of German business transportation by implementing specific measures to enhance profitability, efficiency, availability and safety of railroad operations, including non-destructive testing of railroad wheels, complete wheelsets and rails.

GE

GE Transportation Systems
2901 East Lake Road, Erie, Pennsylvania 16531, USA
Tel: (+1 814) 875 22 40   Fax: (+1 814) 875 59 11
Web: http://www.getransportation.com

Key personnel
See main entry in *Locomotives and powered/non-powered passenger vehicles* section.

Services
GE Global locomotive maintenance service maintains several thousand locomotives in comprehensive service programmes. GE Transportation Systems provides customised

Walter Finkbeiner EHB707K-12 lifting equipment   0538336

maintenance solutions ranging from field technical support to full turnkey maintenance facility operations. These services feature high performance through reliability-centred maintenance programmes, continuous training, supply chain integration and dedicated engineering support.

Gevisa SA

Transit Area, Av Mofarrej, 592 CEP 05311-000, São Paulo SP, Brazil
Tel: (+55 11) 838 25 60; 858 25 03
Fax: (+55 11) 838 25 70; 25 00

Key personnel
Commercial Director: Ronald H Moriyama
Transit Division Manager: Arnaldo Adoglio Júnior
Marketing Manager: Mário Calvani
Director, South American Operations, GE
  Transportation Systems: Marcelo Mosci
Sales Manager: Carlos E Teixeira

Services
Rolling stock overhaul and workshop services, including: refurbishment; repair; remanufacturing and reconstruction; maintenance; painting; supply of spares.

Gilardoni SpA

Via Arturo Gilardoni 1, I-23826 Mandello del Lario (Lecco), Italy
Tel: (+39 0341) 70 52 82
Fax: (+39 0341) 70 52 83
e-mail: gx@gilardoni.it
Web: http://www.gilardoni.it

Products
Non-destructive testing systems for rail applications, including ultrasonic testing of rail vehicle wheels.

GLG Gmeinder Lokmotivenfabrik GmbH

Anton Gmeinder Strasse 5, D-74821 Mosbach, Baden, Germany
Tel: (+49 6261) 674 70
Fax: (+49 6261) 674 72 19
e-mail: info@glfg.de
Web: www.gmeinder-lokomotivenfabrik.de

Key personnel
Managing Directors: Norbert Wuddel; Werner Kilp
Head of Rail Traction Division: Helmut Eifler

Background
In 2004 private rail operator Transport & Logistik AG acquired a 51 per cent shareholding in Gmeinder.

Services
Modernisation and refurbishment, re-engining, general overhaul, maintenance and repair of diesel locomotives.

UPDATED

Greenwood Engineering A/S

H J Holstvej 3-5C, DK-2605 Brøndby, Denmark
Tel: (+45) 36 36 02 00
Fax: (+45) 36 36 00 01
e-mail: trine@greenwood.dk
Web: www.greenwood.dk

Products
Development and production of measuring equipment including the MiniProf Wheel for measuring wheel profiles, the MiniProf Switch

MiniProf Wheel measuring wheel profile 0134497

for measuring railhead profiles in switches and crossings and the MiniProf Brake for measuring brake discs.

UPDATED

Gregomatic AG

Zwydenweg 14, CH-6052 Hergiswil NW, Switzerland
Tel: (+41 41) 630 32 78   Fax: (+41 41) 630 32 79
e-mail: info@gregomatic.com
Web: http://www.gregomatic.com

Products
Vacuum washing system for vehicle interiors.

Hegenscheidt-MFD GmbH & Co KG

Hegenscheidt Platz D-41812 Erkelenz, Germany
Tel: (+49 2431) 860   Fax: (+49 2431) 864 66
e-mail: hegenscheidt.mfd@nshgroup.com
Web: http://www.hegenscheidt-mfd.de

Key personnel
Managing Director: Dr Winfried Büdenbender
Sales Director: Markus von Reden

Subsidiaries
Hegenscheidt-MFD Corporation
6255 Center Drive, Sterling Heights, Michigan 48312, USA
Tel: (+1 586) 274 49 00
Fax: (+1 586) 274 49 16
e-mail: info_mfd@hegenscheidtmfd.com

Hegenscheidt Australia Pty Ltd
Suite 11, Hightpoint Business Centre, 3374 Pacific Highway, PO Box 775, Springwood Queensland 4127, Australia
Tel: (+61 7) 33 87 77 11
Fax: (+61 7) 33 87 77 71
e-mail: hegenscheidt@ozemail.com.au

Background
In 1995, the activities of Wilhelm Hegenscheidt GmbH and Hoesch Maschinefabrik Deutschland GmbH were merged as Hegenscheidt-MFD GmbH, a subsidiary of Vossloh AG.
In 2001, a joint venture was established between Niles-Simmons and Hegenscheidt-MFD.

Products
Underfloor wheel lathes for machining wheelsets *in situ* ; above-floor wheel lathes for universal machining of wheelsets; hydraulic rerailing systems for lifting and repositioning vehicles; special equipment for machining engine crankshafts; wheelset diagnostic systems.

UPDATED

Hi-Force Hydraulics

Bentley Way, Daventry NN11 5QH, UK
Tel: (+44 1327) 30 10 00   Fax: (+44 1327) 70 65 55
e-mail: sales@hi-force.com
Web: http://www.hi-force.com/hi-force

Key personnel
Directors: Chris Jones, John Taylor
Sales Manager: Ronnie Birch

Products
High-pressure hydraulic tools, including wagon door straighteners; jacks and presses; torque tools and accessories; pumps; crimping tools and nutsplitters.

Contracts
Recent contracts include the supply of hydraulic jacks to London Underground Ltd.

VERIFIED

Holland Company LP

1000 Holland Drive, Crete, Illinois 60417-2120, USA
Tel: (+1 708) 672 23 00   Fax: (+1 708) 672 01 19
e-mail: postmaster@hollandco.com
Web: www.hollandco.com

Key personnel
President: Phil Moeller
Vice-President Rail Mechanical Group: Len O'Kray
General Manager Track Testing Services:
  Robert Madderom
International Sales Manager: Billy Hedrick
General Manager Mobile Welding Operations:
  Mark Rovnyak
Controller: Frank Francis
General Manager Equipment Division:
  Robert Norby
General Manager MOW Sales: Kevin Flaherty

Services
Wagon cleaning, light wagon repair, train loading, bulk material transfer. Locomotive fuelling and service are available through its M-BAR-D RailTech, Inc division.

UPDATED

Hovair Systems Ltd

A division of British Turntable Company Ltd
Emblem Street, Bolton, UK
Tel: (+44 1204) 52 56 26
Fax: (+44 1204) 38 24 07
e-mail: info@turntable.co.uk
Web: www.hovair.co.uk

Key personnel
Managing Director: John Entwistle
Technical Director: David Houghton

Background
In April 2004, British Turntable Company Ltd bought the assets of Hovair Systems and transferred its design and manufacturing capability to the Head Office in Bolton, Lancashire. The Hovair company and brand name continues to operate as a division of British Turntable Company Ltd.

Products
Air film load-handling equipment. The system, which uses air bearing technology, has a multidirectional capability providing low resistance to motion and no damage to floors. It allows carriages to be moved within the workshop area unconstrained by rails. Complete carriages, with or without bogies, can be manoeuvred throughout the whole workshop, by only two men, allowing for removal of finished stock from a production or repair line.
Carriages or trainsets can be rotated removing the need for turning facilities and, in manufacturing

areas, can be used for movement between build, paint and fitting out. Complete carriages can be stored close together and, when required, taken to the single incoming rail for transfer to the rail network.

UPDATED

Hunslet-Barclay Ltd

Caledonia Works, West Langlands Street, Kilmarnock KA1 2QD, UK
Tel: (+44 1563) 52 35 73   Fax: (+44 1563) 54 10 76
e-mail: mail@hbltd.co.uk
Web: http://www.hunsletbarclay.co.uk

Key personnel
Board members
   Chairman: H Kuebel
   Managing Director: John Flowers
   Non-Executive Director: S Bauer
Executive Managers
   Production: A Cuthbertson
   Finance: B Connell
Senior Managers
   Train Sales: R Edmond
   Wheelset Sales: M Douglas
   Locomotive Sales and Site Services: T Clare
   Train Engineering: S Scott

Background
Hunslet-Barclay is a part of the Waagner-Biro group of companies, Austria.

Services
Overhaul and refurbishment of shunting, tunnelling and mining locomotives, passenger rolling stock and infrastructure maintenance vehicles; overhaul and refurbishment of wheelsets for freight wagons and locomotive-hauled passenger coaches; overhaul of rolling stock bearings.

HYWEMA Lifting Systems

Wuppertaler Strasse 134-148, D-42653 Solingen, Germany
Tel: (+49 212) 257 70   Fax: (+49 212) 257 71 00
Web: www.hywema.de

Key personnel
General Manager: D Paul
Sales: R Heidtmann
Purchase: G Greupner

Products
Mobile wheel lifts, specially made for all commercial vehicles of 24 tonnes up to 60 tonnes capacity.

Mobile lifting system developed by HYWEMA for use during the manufacture of high-speed trainsets   0129422

The model RG has been built for repairs, servicing, cleaning or inspection of buses, commercial and special vehicles.

UPDATED

IBEG Maschinen-und Gerätebau GmbH

Stettiner Strasse 9, D-45770 Marl, Germany
Tel: (+49 2365) 510 30
Fax: (+49 2365) 126 47
e-mail: info@ibeg.com
Web: http://www.ibeg.com

Products
Bogie and wheelset maintenance equipment: axle parallelism measurement systems; rail vehicle load measurement systems; laser wheel profile measurement systems; wheel circumference measurement systems; bogie noise measurement systems.

INKA

Head office and factory
T (Persero) Industri Kereta Api (PT INKA)
Jalan Yos Sudarso No 71, Madiun 63122, Indonesia
Tel: (+62 351) 45 22 71
Fax: (+62 351) 45 22 75
e-mail: sekretariat@inka.web.id
Web: www.inka.web.id

Representative office
Arthaloka building, 3rd Floor, Jalan Jend, Sudirman Kav 2, Jakarta
Tel: (+62 21) 251 44 24   Fax: (+62 21) 251 44 23
e-mail: inkajkt@cbn.net.id

Key personnel
President: Ir Roos Diatmoko
General Manager, Railway Rolling Stock Division:
   Suryanto
Finance and Administration Director:
   Drs Udin Supriatman
General Manager of Technology:
   Ir M Harsan Badawi
Engineering Manager: Ir Gunesti Wahyu
Design Manager: Ir Indarto Wibisono
Marketing Manager: Ir M Dedi Tarmidi
Business Development Manager:
   Ir Muchlis Budiman
Procurement Manager: Soedjito Taathadi

Background
PT INKA was originally established in 1981 as a state owned company, transforming from Indonesian State Railway's steam locomotive maintenance shop.

Products
Assembly and renovation of freight wagons (300 units a year), passenger coaches (120 units a year), diesel and electric railcars (40 units a year), bogies (200 units a year).
   Locomotives (in collaboration with GE Transportation, 15 units a year). Various special vehicles, including track motor and inspection cars and amusement park trains.

NEW ENTRY

Instron

Corporate Headquarters
100 Royall Street, Canton, Massachusetts 02021-1089, USA
Tel: (+1 781) 575 50 00

Key personnel
General Managers: Malcolm Buchanan,
   Norman Smith
Marketing Manager: Bernd Schlichtenbrede
Market Services Manager:
   Brigitte Iffländer-Wiegmann
Technical Manager: Klaus J Vedder

European Headquarters
Coronation Road, High Wycombe, Buckinghamshire HP12 3SY, UK
Tel: (+44 1494) 46 46 46   Fax: (+44 1494) 45 61 23

Subsidiaries
Instron Structural Testing Systems
28700 Cabot Drive, Suite 100 Novi, Michigan 48377, USA
Tel: (+1 248) 553 46 30   Fax: (+1 248) 553 68 69

HYWEMA mobile lift-jack for rail vehicle (BVG Berlin)   0536548

Instron Canada
975 Fraser Drive, Unit 1, Burlington, Ontario L7L
4X8, Canada
Tel: (+1 800) 461 91 23   Fax: (+1 905) 639 86 83

Instron de Mexico
Bulevar V Carranza 4120, Local 20, Col Villa
Olimpica, CP 25230 Saltillo, Coah, Mexico
Tel: (+52 84) 44 39 14 19   Fax: (+52 84) 44 39 14 20

Instron Deutschland
Landwehrstrasse 65, D-64293, Germany
Tel: (+49 6151) 39 17-0   Fax: (+49 6151) 391 75 00

Instrumentacion y Servicos Tecsis Ltda
Avenida Holanda 1248, Casilla 50/9 Correo 9,
Providencia, Santiago, Chile
Tel: (+56 2) 205 13 13   Fax: (+56 2) 225 07 59

Products

Testor hardness testing machines, spring testing
machines, universal testing machines for tensile,
compression, shear and bending tests; pendulum
impact testing machines; special purpose testing
machines to customer specification, automatic
testing machines and installations.

Interlok Ltd

Warsztatowa 8, PL-64-920 Pila, Poland
Tel: (+48 67) 213 20 68
e-mail: pilapoland@interlok.info
Web: http://www.pila.interlok.info/indexe.htm

Key personnel

General Manager: Henryk Palczewski
General Manager Technology: Marek Furtacz

Products and services

Steam locomotives, steam boilers and other
components of all gauges, new and full overhaul,
modernisation, historical reconstruction and
replicas; steam boiler parts are massively forged,
then welded or riveted; stays are welded or
threaded; fabrication of new steam cylinders
welded construction; diesel standard gauge
locomotives: SP32 1,015 kW BoBo modernised
with 12 V MTU engine; narrow gauge diesel
locomotives; standard gauge railbuses, new
construction and remanufactured.

NEW ENTRY

International Technical Services

400 Queens Avenue, London, Ontario N6B 1X9,
Canada
Tel: (+1 519) 439 23 62   Fax: (+1 519) 675 18 68
e-mail: info@ITSrail.com
Web: http://www.ITSrail.com

Key personnel

For a full list of personnel, see International Technical
Services entry in *Consultancy services* section.

Capabilities

EMD locomotive specialists; EMD engine rebuild;
locomotives and freight vehicle maintenance; air
brake systems specialists; locomotive rebuilding
and repowering; welding services (engine repair).

iQR

Railcentre 1, 305 Edward Street, Brisbane,
Queensland, Australia 4000
GPO Box 1429, Brisbane, Queensland 4001
Tel: (+61 7) 32 32 33 90   Fax: (+61 7) 32 35 33 46
e-mail: sales@iqr.com.au
Web: www.iqr.com.au

Key personnel

General Manager: Michael Walsh
Marketing and Business Development Manager:
  Peter Harris
Sales Manager: Youfa Chen

Maintenance on QR's tilt train   ***NEW***/1140311

Background

iQR was previously Queensland Rail Consultancy
Services.

Services

iQR's workforce utilise the latest technology
to provide an extensive range of high quality
maintenance services for rolling stock, including:
locomotives, wagons, passenger carriages,
electric, suburban and interurban multiple units.
All workshops are Quality Certified to ISO 9002.

iQR can also help organisations develop whole-
of-life maintenance programmes that suit specific
rollingstock or total system operations and can assist
in the development of maintenance facilities, including
concept, design and performance specifications.

Locomotives: iQR can complete the overhaul,
component change out, maintenance and repair
as well as remanufacture back to OEM standards
for diesel electric locomotives. This includes:
de-wheeling and steam cleaning, repair or
replacement of wear surfaces, bushes, springs and
brake systems, complete re-assembly.

Wagons: modification and maintenance of
all types of wagons, including coal, ore, and
refrigeration containers. Services include: overhaul,
complete component change out, such as doors,
door mechanisms, shedder shields, couplers and
draft gear. Repair of accident damaged wagons,
overhaul of bogies, testing of airbrakes, state of the
art equipment to ensure component reliability to
OEM standards, air conditioning and refrigeration.

Contracts

Sistem Transit Aliran Ringan Sdn Bhd (STAR),
Malaysia: inspection, testing and acceptance of 90
sets of light rail vehicles manufactured at Walkers
Ltd, Maryborough for parts one & two of the Kuala
Lumpur STAR project.

Westrail (New Metro Rail), Perth Western
Australia: inspection of five new emu two-car sets
of rail passenger vehicles during manufacture at
Walters Ltd, Maryborough. Inspection of 31 new
emu three-car sets of rail passenger vehicles
during manufacture at EDI Rail (formerly Walkers
Ltd, Maryborough).

NEW ENTRY

Kambre TTM AB

PO Box 7221, SE-187 13 Täby, Sweden
Tel: (+46 8) 54 44 04 25   Fax: (+46 8) 54 44 04 29
e-mail: kambre@kambre.se
Web: http://www.kambre.s

Key personnel

Chairman: Otto Suensson
Managing Director: Karl-Axel Kambre

Principal subsidiaries

Halltoap Gruppen AB

Products

Train washing machines, featuring brush and
brushless washing equipment; specialised
washing machines for cleaning the upper
and lower surfaces of raked train noses; train
interior cleaning systems. Bus maintenance
systems, including washing and interior cleaning
systems.

Alfred Kärcher GmbH & Co KG

Alfred-Kärcher-Strasse 28-40, D-71364 Winnenden,
Germany
Tel: (+49 7195) 140   Fax: (+49 7195) 14 22 12
e-mail: info@kaercher.com
Web: http://www.kaercher.com

Key personnel

Managing Directors: Hartmut Jenner,
  George Metz, Dr Bernhard Graf

Principal subsidiary

Karcher (UK) Ltd
Karcher House, Beaumont Road, Banbury, Oxon
OX16 1TB, UK
Tel: (+44 1295) 75 20 82   Fax: (+44 1295) 75 21 03
e-mail: enquiries@karcher.co.uk

Products

Kärcher manufactures high pressure cleaners
such as the CHH 8000 OptiWash. The product
range includes vacuum cleaners, steam cleaners,

Fully automated washing plant for DB AG ICE high-speed trainsets installed by Kärcher at Frankfurt/ Höchst 0103637

Transit jacking bogie system supplied by Somers Railway Engineering to Alstom 0105893

cleaning robots, sweepers, scrubber-driers, cleaning agents, brush-type vehicle washers and water treatment plants.

Services
Project planning, architecture, implementation and turnkey handover. Kärcher offers assistance with financing issues.

Keller Elettromeccanica SpA

Zona Industriale I-09039, Villacidro, Cagliari, Italy
Tel: (+39 070) 933 62 02   Fax: (+39 070) 933 62 44
e-mail: info@keller.it
Web: http://www.keller.it

Key personnel
For a full list of personnel, see Keller Elettromeccanica SpA entry in *Locomotives and powered/non-powered passenger vehicles* section.

Services
Maintenance, overhaul, conversion and refurbishment of all types of passenger coaches and freight vehicles. Projects include: conversion of self-service coaches into Ristobar coaches; overhaul of UIC-X coaches into IR trailer coaches; overhaul of Gran Comfort coaches; maintenance of UIC-X coaches; refurbishment of Ale 801–940 Le 108 trains; overhaul of various types of wagons; cyclic overhaul of couchette coaches.

Končar-Electric Locomotives Inc

Velimira Škorpika 7, HR-10090 Zagreb, Croatia
Tel: (+385 1) 349 69 59   Fax: (+385 1) 349 69 60
e-mail: upravakoncar-ellok.hr
Web: www.koncar.hr/koncar/ellok/

Key personnel
President: Jusuf Cmalić
Members of Board: Vesna Boinović-Grubić; ŽeljkoŠakič
Marketing and Sales Manager: Zvonimir Cvijin

Services
Repair, reconstruction, refurbishment and modernisation of locomotives, dmus, emus and trains.

UPDATED

Lloyds Somers

Atlas house, 4/6 Bedwell Lane, Sutton Coldfield, West Midlands B74 4AB, UK
Tel: (+44 870) 197 55 00
Fax: (+44 870) 197 55 99
e-mail: sales@lloydsbritish.com
Web: www.somers-handling.com

Key personnel
General Manager: Stephen Reece
Sales Manager: Alec Foley

Products
Mobile and fixed rail vehicle lifting jacks; traversers; scissor lift tables; bogie rotators; bogie drop systems; bogie turntables; bogie turning systems; rail drop systems; all types of workshop equipment.

UPDATED

MAN Ferrostaal AG

Hohenzollernstrasse 24, D-45128 Essen, Germany
Tel: (+49 201) 818 01
Fax: (+49 201) 818 28 22
e-mail: info@manferrostaal.com
Web: www.manferrostaal.com

Key personnel
Head of Industrial Equipment and Systems Division: Helmut Julius

Background
Re-named from Ferrostaal AG, MAN Ferrostaal AG is a member of the MAN AG group.

Services
Refurbishment of heavy and light rail vehicles.

UPDATED

Marcroft Engineering Ltd

Whieldon Road, Stoke-on-Trent, ST4 4HP, UK
Tel: (+44 1782) 84 40 75
Fax: (+44 1782) 84 35 79
e-mail: sales@marcroft.co.uk
Web: http://www.marcroft.co.uk

Key personnel
For a full list of personnel, see Marcroft Engineering Ltd entry in *Freight vehicles and equipment* section.

Background
For full background information, see Marcroft Engineering Ltd entry in *Freight vehicles and equipment* section.

Services
Overhaul and refurbishment of freight rolling stock and infrastructure vehicles; vehicle damage repair and modifications; wheelset overhaul; provision of spares for air brake systems; refurbishment, repair and cleaning of tank containers.
Preparation and painting of road and rail vehicles of all types.

Class 1061 Bo-Bo-Bo 3 kV DC electric locomotive refurbished for Croatian Railways 0122620

Maschinen- und Stahlbau Dresden

Postfach 27 01 45, D-01171 Dresden, Germany
Tel: (+49 351) 423 40
Fax: (+49 351) 423 41 03
e-mail: info@msd-dresden.de
Web: http://www.msd-dresden.de

Key personnel
Managing Director: Dr Ing Martin Herrenknecht
Managing and Sales Director:
 Dipl-Ing Jürgen Bialek
Head of Production: Raimund Schäfer

Corporate background
MSD is a subsidiary of Herrenknecht AG.

Products
Traversers; locomotive and wagon turntables; vehicle lifting systems; wheelset handling systems.
 Customers include: Bombardier Transportation; Deutsche Bahn AG; FGV, Valencia; Volkswagen, Hanover.

150 tonne MSD traverser system installed at the Opladen workshops of Deutsche Bahn AG 0131018

McConnell Research Enterprises Pty Ltd

68 Bond Street West, Mordialloc, Victoria 3195, Australia
Tel: (+61 3) 95 87 22 66
Fax: (+61 3) 95 80 78 48
e-mail: maxsteam@ozemail.com.au
Web: http://www.maxi-stream.com.au

Key personnel
Managing Director: Peter McConnell
Director and Sales Manager: John McConnell

Products
Maxi-Steam fixed and truck-mounted vehicle interior cleaning systems. Clients supplied include Melbourne's Public Transport Commission, Westrail and KTM in Malaysia.

Mechan Ltd

Thorncliffe Park, Chapeltown, Sheffield S35 2PH, UK
Tel: (+44 114) 257 05 63
Fax: (+44 114) 245 11 24
e-mail: admin@mechan.co.uk
Web: www.mechan.co.uk

Key personnel
Managing Director: A Scripps
Engineering Director: G L Cofield

Products
Mechan Ltd manufactures a complete range of rail vehicle lifting jacks with the following capacities: 6- and 10-tonne jacks for trams, light rail and underground trains; 15- and 20-tonne jacks for rail coaches and multiple-units; 25- and 35-tonne jacks for locomotives and on-track plant; 45 tonne plus for special applications.
 Bogies and wheelset handling and storage; design and manufacture of lifting and handling equipment for all kinds of modules; general lifting with swing jib cranes, mobile gantries, hoist units, standard scissor tables and chain slings. Mechan manufactures SMART depot personnel protection systems.

Contracts
During 2003 Mechan completed and commissioned a contract for Siemens at its new Northam Depot in Southampton, UK. This comprised of a bogie drop/change unit with the latest distributed control system and remote diagnostics, a depot personnel protection system linked to Sofis train identification and in turn all linked to Mechan Depot Manager™, an information retrieval system, which displays the depot status on a plasma screen in the supervisor's office. The company also won the contract for two Wheelset drop/change units and four depot personnel protection systems through AMEC for South Central Railways.

UPDATED

Metronet REW Ltd

130 Bollo Lane, Acton, London W3 8BZ, UK
Tel: (+44 20) 79 18 66 66
Fax: (+44 20) 79 18 65 99
e-mail: sales@r-e-w.co.uk
Web: http://www.metronetuk.com

Key personnel
Sales Manager: Chris Darrall
REW Manager: John Copeman

Background
A subsidiary of Metronet Rail SSL Ltd. REW has 75 years' experience in rolling stock and signal equipment maintenance.

Services
Repair and overhaul of railway compressors and control systems, traction motors including on- and off-site capability for all railway equipment, electronics, clocks and passenger seating, signalling equipment, wheelsets and gearboxes.

Contracts
Maintenance contracts with London Underground and also train companies including: Eurostar, Arriva, Jarvis, South West Trains, Serco Metrolink (Manchester), Strathclyde Passenger Transport.

MotivePower Inc

Address
4600 Apple Street, Boise, Idaho 83716, US
Tel: (+1 208) 947 48 00
Fax: (+1 208) 947 48 20
Web: www.wabtec.com

Key personnel
General Manager: Mark S Warner

Background
MotivePower Inc is a subsidiary company of Wabtec Corporation, the latter created in 1999 as a result of the 1999 merger of Westinghouse Air Brake Co and MotivePower Industries Inc.

Services
Locomotive overhauls; fleet maintenance; frame and body fabrication and collision repairs; locomotive remanufacturing; electrical upgrades; locomotive testing.

Contracts
Recent locomotive overhaul contracts include work on 47 F40PH-2C locomotives for MBTA, Boston, and on nine F59PHI locomotives for Caltrans.

NEW ENTRY

AB Ph Nederman & Co

Sydhamnsgatan 2, SE-252 28 Helsingborg, Sweden
Tel: (+46 42) 18 87 00   Fax: (+46 42) 14 79 71
e-mail: support@nederman.se
Web: http://www.nederman.com

Products
Norfi automated exhaust extractor system for locomotive and diesel-powered rolling stock maintenance and servicing facilities. The system comprises an electrically driven, movable extraction arm and drive carriage mounted on a high-level suction rail. The extraction is controlled by a hand-held radio transmitter; E-Pak vacuum system for extracting workshop fumes.

NedTrain

PO Box 2167, KTT-9, NL-3500 GD Utrecht, Netherlands
Tel: (+31 30) 300 46 01   Fax: (+31 30) 300 46 48
e-mail: info@nedtrain.nl
Web: www.nedtrain.nl

Key personnel
Chief Executive Officer: J P B Huberts
Director Business and Development:
 A J M Spaninks
Financial Director: G Taute

Subsidiary companies
NedTrain Consulting
PO Box 2016-KTT 6, NL-3500 GA Utrecht, Netherlands
Tel: (+31 30) 300 47 00   Fax: (+31 30) 300 48 00

Ematech
PO Box 8093, NL-3503 RB Utrecht, Netherlands
Tel: (+31 30) 246 91 60   Fax: (+31 30) 246 91 76

Background
NedTrain supports operators and owners of rolling stock during the entire life cycle of their fleets, optimising performance.

Services
Service and maintenance, upgrading and overhauling, refurbishment, consulting, components and damage repair of railway rolling stock, both passenger and freight. Refurbishment, overhaul, modification and damage repair of rolling stock. Refurbishment and overhaul of rolling stock systems and components, including bogies, wheelsets, traction motors, brake components and automatic couplers.

Contracts
Contracts include the maintenance and service of NS Reizigers, NS International, Syntus and NoordNed passenger rolling stock and maintenance of Railion locomotives. Other contracts include major refurbishment and upgrading projects of NS Reizigers rolling stock, modification of GM (class 66) locomotives and maintenance and overhaul of freight wagons for DB Cargo, VTG, AAE, GE Capital, GATX.

UPDATED

Nencki AG

Anlagen-Fahrzeugbau, Gaswerkstrasse 27, CH-4901
Langenthal, Switzerland
Tel: (+41 62) 919 93 93
Fax: (+41 62) 919 93 90
e-mail: info@nencki.ch
Web: http://www.nencki.ch

Key personnel
Managing Director: Dr Sepp Káppeli
Sales Manager: A Gerber

Products
Bogie and axle exchanging facilities, lifting tables, scissor platforms and lifting jacks; hydraulic bogie test stands, working platforms for paint shops and washing plants. Side and front access platforms. Rail handling equipment such as bending and straightening presses. Self-propelled rail cars with working platforms and cranes for catenary wire installation.

UPDATED

Twin sand pump for Ferromex, Manzanillo, Mexico to fill high capacity sand boxes 1033688

CET Installation for SNCF at Châtillon, France for TGV trains 1033690

NEU International's Vaktrak cleaning train for the Munich metro 1033689

Neuero Technology GmbH

Neuerostrasse 1, D-49324 Melle, Germany
Tel: (+49 5422) 60 70
Fax: (+49 5422) 60 72 10
e-mail: info@neuero-tec.de
Web: www.neuero-tec.de

Key personnel
Managing Director: Dipl Ing Bernhard Uhlen
Sales Director: Dipl Ing Heinrich Wöstefeld

Products
Lifting jacks, underfloor lifting plants, bogie lifting stands, lifting trucks and tables, shunting vehicles, dismantling devices for wheelsets and bogies, mobile handling equipment, turntables, maintenance platforms, auxiliary bogies, lifting and turning devices. Measuring and testing equipment including devices for measuring rolling stock, axles and wheelsets; test benches for metal, rubber and cylindrical springs, shock-absorber test benches, bogie assembly and test benches; washing machines for trains and bogies.

Contracts
Recent contracts include: underfloor lifting plants for Metro Shanghai, China; Metro Nanjing, China; Delhi Metro, India; London Underground, UK; Martha, Atlanta, USA; Bogestra, Bochum, Germany; Docklands London, UK; Birse Metro, London, UK. Lifting jacks for DB AG Germany; DSB, Denmark; SJ AB Stockholm, Sweden. Elevated tracks with split rails for DB Hamburg, Germany; DB Frankfurt, Germany. Bogie storage system for DB Krefeld, Germany; lifting jacks and turntables for Metro de Maracaibo, Venezuela; different maintenance equipment for Charlotte Area Transit System, US; roof working platforms for DB Frankfurt, Germany; DB Kassel, Germany.

UPDATED

NEU International

PO Box 4039, 70 rue du Collège, F-59707
Marcq en Baroeul, France
Tel: (+33 3) 20 45 65 09   Fax: (+33 3) 20 45 65 99
e-mail: railway@neu-international.com
Web: www.neu-international.com

Other offices
NEU Inc
PO Box 488, Paoli, Pennsylvania 19301-0488, USA
Tel: (+1 610) 725 04 01
Fax: (+1 610) 725 04 02
e-mail: mail@neu-inc.com

NEU Solids Handling Pte Ltd
PO Box 406, Thomson Road Post Office, Singapore 915714

Tel: (+65) 64 30 66 86   Fax: (+65) 65 52 71 31
e-mail: mail@neu-pte.com.sg

Products
NEU International develops specific solutions for safety, cleanliness and reliability of all types of railway networks and depots (train, tramway, metro).

Among these NEU International offers the automatic sand feeding system: Neu Distri Sand (NDS) for efficient and reliable locomotive sandbox filling. This installation is suitable for all types of railway depots and all sizes of sandboxes (train, light rail or freight locomotive). The rail/road Vaktrak is designed to meet all types of tracking cleaning requirements for train, light rail, in-station, in-tunnel or outside (with grass or pavement) on ballast or on concrete. The Vaktrak module consists of free modules which can be adapted to the requested cleaning configuration and can be installed on existing rolling stock. NEU has also developed the NEU CET system, an automatic and hygenic way to empty chemical and/or retention toilet tanks. This system can be fixed installation or a mobile unit depending on the depot configuration and the train availability.

NEU International's train washing plant 1033687

The NEU International product range also includes train washing plants (drive through or gantry systems). Completely automatic and flexible for train washing, they are designed to suit all machine profiles at the requested speed.

Contracts
References include: NYCT New York – Amtrak, Ferrocarril Mexicano, Amey McAlpine, New Jersey Transit, Üstra Hanover, MUNI San Francisco, RENFE Barcelona, FS Italy, SMSC Seoul, RATP Paris, Deutsche Bahn, SNCF, Lisbon Metro, KHRC Seoul, SWM Munich.

UPDATED

Orval

Ateliers d'Orval
Route de l'Ombrée, Orval, F-18200 Saint Armand-Montrand, France
Tel: (+33 2) 48 96 07 39
Fax: (+33 2) 48 96 50 97
e-mail: dg.atelorval@wanadoo.fr
Web: http://www.ermewa.com

Key personnel
Director General: Pierre-Yves Gessé

Background
Orval is a member of Groupe Ermewa SA (see Groupe Ermewa SA entry in *Rolling stock leasing companies* section).

Pars nova as

Žerotíinova 1833/56, CZ-787 01 Šumperk, Czech Republic
Tel: (+420 583) 36 51 11
Fax: (+420 583) 36 54 10
e-mail: pars@parsnova.cz
Web: www.parsnova.cz

Key personnel
General Manager: Ing Tomáš Ignačák

Background
Formerly the State Railways workshops of the former Czechoslovak State Railways, Pars nova commenced commercial activities in 1993 as Pars DMN sro Sumperk, which at this time, remained state-owned. The current name was adopted in August 2000, when Pars nova became a joint-stock company.

Services

Modernisation, refurbishment, overhaul and crash repair of: diesel locomotives; electric locomotives; dmus; railcars; trams and light rail vehicles, passenger coaches and inspection and maintenance vehicles. Overhaul and repair of: diesel engines; electric traction equipment; wheelsets; batteries.

UPDATED

PESA Bydoszcz SA

Zygmunta Augusta 11 Street, PL-85-082 Bydgoszcz
Tel: (+48 52) 518 02 48
Fax: (+48 52) 518 52 39
e-mail: pesa@pesa.pl
Web: www.pesa.pl

Key personnel

President and General Director: Tomasz Zaboklicki
Directors
  Production and Technical: Zenon Duszynski
  Marketing and Development: Zygfryd Zurawski
  Financial: Robert Swiechowicz
Proxy Deputy Director and Head of Production:
  Andrzej Karwasz
Head of Marketing, Passenger Coaches:
  Jerzy Berg
Head of Development: Andrzej Ciupa

Background

PESA Bydoszcz SA formerly traded as ZNTK Bydoszcz SA.

Services

Modernisation, refurbishment and repair of diesel locomotives, passenger coaches, trams and freight wagons.

NEW ENTRY

Pfaff-silberblau Verkehrstechnik GmbH & Co KG

Äussere Industriestrasse 18, D-86316 Friedberg, Derching, Germany
Tel: (+49 821) 780 16 60
Fax: (+49 821) 780 16 69
e-mail: verkehrstechnik@pfaff-silberblau.de
Web: www.pfaff-silberblau.de

Key personnel

Managing Director: Peter Zeller
Chief Engineer: Rudolf Eichner

Background

Pfaff-silberblau has over 135 years experience in the industry and is recognised for its range of lifting equipment.

  Pfaff-silberblau Verkehrstechnik is the specialist within the group for executing turnkey projects for its clients within the railway industry.

Products

Lifting systems and jacks for rail vehicles. Lifting systems are available for rail vehicles (trams, metros, railways) built to the customer's specification and to the loading capacity required. The systems are fully lowerable to below ground level. The lifting of different types of vehicles and various lengths can be achieved by means of one lifting arrangement.

  Standard duty lifting jacks are produced with capacities from 5,000 to 10,000 kg. Heavy duty lifting jacks are produced with capacities from 10,000 to 50,000 kg. Direct drive to the spindle is by a worm gear reduction box with automatic lubrication system. These lifting jacks can be delivered stationary, mobile on concrete floor, or on a manually or electrically driven auxiliary rail system, with fixed or mobile load supports.

  Also supplied: lifting-turning devices, underfloor bogie- and wheelset lowering systems (drop tables), accommodation bogies, rescue bogies, bogie disassembly wagons, bogie handling systems, turntables, hydraulic lifting tables, working platforms and spinning posts.

Contracts

Lifting jacks and underfloor lifting installations have been installed in Belgium, Canada, China, Denmark, France, Germany, Greece, Hong Kong, India, Ireland, Italy, Japan, Korea, Malaysia, Netherlands, Portugal, South Africa, Spain, Turkey and the US.

UPDATED

Portec Rail Products Inc

Railway Maintenance Products Division, Box 38250, 900 Old Freeport Road, Pittsburgh, Pennsylvania 15238-8250, USA
Tel: (+1 412) 782 60 00
Fax: (+1 412) 782 10 37
e-mail: RMPsales@PortecRail.com
Web: www.PortecRail.com

Key personnel

Group Vice-President: Richard Jarosinski

Products

Portec Rail Products spot wagon and locomotive repair systems allow for repair of wagons and locomotives in an efficient central area, with wagon-handling achieved by 'rabbits' under push-button control. The systems include integrated jacking, job cranes, hosereels and other accessories. The systems operate on the basic principle of moving the wagons to the men and materials, rather than have men carry materials to wagons needing repairs. Advantages claimed include increased labour efficiency, reduction in wagon hire costs, and savings of 50 to 90 per cent in switch engine hours.

  Portec Rail Products locomotive drop table is a single-axle hydraulic unit for changing locomotive wheel/axle/traction motor assemblies. The work table is mounted on a transporter.

  Portec Rail also offers a line of environmental products suited to the needs of rail operator repair shops.

UPDATED

Proceco Inc

14790 St Augustine Rd, Jacksonville, Florida 32258, USA
Tel: (+1 904) 886 02 00
Fax: (+1 904) 886 02 32

Principal subsidiary

Proceco Industrial Machinery Ltd
7300 Tellier Street, Montréal, Québec H1N 3T7, Canada
Tel: (+1 514) 254 84 94
Fax: (+1 514) 254 81 84

Products

Machinery washers for items such as bogies, engine blocks and cylinder heads.

  Locomotive traction motor and main generator (alternator) spray washing and vacuum drying system, vacuum-pressure impregnating devices, traction motor stripping and assembly device, traction motor remanufacturing transfer line.

Railquip Inc

3731 Northcrest Road, Suite 6, Atlanta, Georgia 30340, US
Tel: (+1 770) 458 41 57; (+1 800) 325 02 96
Fax: (+1 770) 458 53 65
e-mail: sales@railquip.com
Web: www.railquip.com

Key personnel

President: Helmut Schroeder
Sales Director: Paul Wojcik
Sales Manager: Bill Schroeder
Treasurer and Office Manager: Debbie Fox

Products

Maintenance equipment for railcar and locomotive workshops; portable hydraulic rerailing equipment and containers; underfloor and above-ground car hoists; body stands; turntables; spinning posts; bogie assembly and test stands; portable hydraulic rerailing equipment; emergency truck for locked axles; mobile train wash; power lift bags; compressed air stations; wheel presses and wheel lathes; mechanical wheel handlers; railcar movers; laser track measuring devices; track aligners; mobile and stationary waste removal systems; hydraulic track jacks; road/rail-equipped trailers; track vacuum refuse collectors; plastic cable channels; stationary roof and floor access platforms; CNC controlled rail bending machines; mobile roof access platforms; bogie hoists; transfer tables.

Contracts

Recent contracts include: maintenance equipment for Metro Maracaibo, Maracaibo, Venezuela; maintenance facility car hoists, turntables and body stands for Charlotte Area Transit Systems, Charlotte, North Carolina; pump-out wagons for Massachusetts Bay Transport Authority; portable hydraulic rerailing equipment for Amtrak, Boston, Massachusetts; two sets of mobile car hoists for Amtrak, Boston Massachusetts; Maxi railcar

Railquip's Maxi railcar mover   *NEW*/1143023

mover for Sara Lee, Paris, Texas; forklift truck propelled railcar mover for Trinity Industries, Dallas, Texas.

UPDATED

Railway Projects Limited (RPL)

Lisbon House, 5-7 St Mary's Gate, Derby DE1 3JA, UK
Tel: (+44 1332) 34 92 55   Fax: (+44 1332) 29 46 88
e-mail: sales@railwayprojects.co.uk
Web: www.railwayprojects.co.uk

Key personnel
Managing Director: Ian Duffy
Financial Director: Sue Llanos
Engineering Director: Jim Thomson
Production Director: Jez Ward
Sales and Marketing Director: Kelvin Roberts

Services
Railway Projects Limited is an engineering services company offering a turnkey approach for rolling stock maintenance, refurbishment, modification, overhaul and installation projects. RPL specialises in on-site working and is ISO 9002 certified and Link Up approved.

UPDATED

Rescar Companies

Rescar Incorporated
1101 31st Street, Suite 250, Downers Grove, Illinois 60515, USA
Tel: (+1 630) 963 11 14   Fax: (+1 630) 963 63 42
e-mail: marketing@rescar.com

Key personnel
Chief Executive Officer: Joe Schieszler
President: Gus Schieszler Senior Vice-President and Chief Financial Officer: Joe Schieszler
Senior Vice-Presidents: Steven L Brown; Richard P Hoffman; Marvin Hughes

Principal subsidiaries
Rescar Inc
Rescar Industries Inc
Transtek Solutions Ink

Products
Repair and maintenance of freight wagons. Rescar operates repair workshops at Longview, Orange and Channelview, Texas; Du Bois, Pennsylvania; Cedar Rapids, Iowa; Gordon, Georgia; Elk Mills, Maryland; Chicago, Illinois; Washington, Indiana; and Hudson, Colorado. Other services include mobile operations, servicing at the customer's plant, wagon shunting services and rail car cleaning facilities; locomotive servicing and fleet management.

RMS Locotec

Rail Management Services
Vanguard Works, Bretton Street, Dewsbury WF12 9BJ, UK
Tel: (+44 1924) 46 50 50   Fax: (+44 1924) 46 54 22
e-mail: sales@rmslocotec.com
Web: www.rmslocotec.com

Key personnel
Managing Director: Lawrence Crossan
Commercial Manager: Derek Webb
Production Development Engineer: Peter Briddon

Services
Maintenance and overhaul of diesel-electric and diesel-hydraulic shunting locomotives.

NEW ENTRY

Rolanfer Matériel Ferroviaire SA

6 rue Thomas Edison, BP 60022, F-57971 Yutz, Cedex, France
Tel: (+33 3) 82 59 56 56
Fax: (+33 3) 82 82 05 77
e-mail: commercial@rolanfer-mf.com

Key personnel
Chairman: Y Henry
General Manager: Jean-Claude Schmitz
Export Sales and Marketing Manager: Laurence Courvalet

Services
Overhaul and repair of second-hand freight wagons.

VERIFIED

Ross & White Company

1090 Alexander Court, PO Box 970, Cary, Illinois 60013-0970, USA
Tel: (+1 847) 516 39 00   Fax: (+1 847) 516 39 89
e-mail: sales@rossandwhite.com
Web: www.rossandwhite.com

Ross & White Company pressure washing equipment 0125199

Ross & White Company pressure washing equipment *NEW*/1112487

Key personnel
President: Jeffery A Ross
Vice-President: Roy A Schuetz

Products
Design, manufacture and installation of train washing and companion water reclamation systems; sand handling equipment; Buck Cyclone cleaners for passenger coach interiors; brush scrubbing and pressure washing equipment for passenger coach exteriors.

Contracts
Include the installation of a coach washing facility for Metra, USA; development and installation of a car progression train washing system for Chicago Transit Authority, USA; development and installation of a gantry-type moving washing system for Chicago Transit Authority; design, supply and installation of a three-lane train washing system for MBTA, USA; and design and installation of six-train washing systems for WMATA, USA.

UPDATED

Rosvagonmash

1a Sokoljnicheslij Val, Box 30, 107113 Moscow, Russian Federation
Tel: (+7 095) 269 06 36; 269 20 88
Fax: (+7 095) 269 58 82

Services
Repair and refurbishment of passenger rail vehicles. Recent contracts include the refurbishment of Russian Railways Class ER1 and ER2 emus.

Rotary Lift

2700 Lanier Drive, Madison, Indiana 47250, USA
Tel: (+1 812) 273 16 22   Fax: (+1 812) 273 65 02
e-mail: userlink@rotarylift.com
Web: www.rotarylift.com

Key personnel
President: Michael Jobe
Vice-President, Sales and Marketing: Matt Webster
Managing Director (Europe): Eric Howlett
Tel: (+49 6251) 58 26 40   Fax: (+49 6251) 58 26 40

Background
Rotary Lift's parent company is Dover Corporation.

Products
Vehicle lifting equipment and productivity service tools for maintenance and repairs; lifting capacities 2,727 to 58,968 kg (6,000 to 130,000 lb).

UPDATED

Rotem Company

Landmarktower,   837-36,   Yeoksam-dong, Gangnam-gu, Seoul, 135-937, South Korea
Tel: (+82 2) 21 12 82 94   Fax: (+82 2) 21 12 98 73
Web: www.rotem.co.kr

Key personnel
Vice Chairman and Chief Executive Officer:
   Soon-Won Chung
Senior Executive Vice-President: Yeo-Sung Lee
Executive Vice-President: Jae-Hong Kim

Background
Established in 1964 when Daewoo Heavy Industry started manufacturing rolling stock, followed by Hyundai and Hanjin Heavy Industry a few years later. In 1999, the three companies were consolidated into KOROS. Hyundai Motors Group acquired the share of Daewoo in 2001 and the former company KOROS became Rotem. Rotem is now an affiliate of Hyundai Motors Group and has its headquarters in Seoul and two factories in Uiwang and Changwon. Uiwang has

capacity for manufacturing 500 emus per year and has the capability to manufacture electric equipment such as traction motors, SIV inverters etc. The research and design centre is also located in Uiwang. The Changwon factory has capacity to manufacture 700 emus per year. Rotem has an annual capacity to manufacture approximately 1,200 emus. Certifications such as the ISO 9001 certificate for Quality, 14001 for environment and 18001 for occupational health and safety management have been acquired at all three sites.

Services
Logistics systems include the design and construction of turnkey maintenance and overhaul workshops with E&M facilities and computerised maintenance information systems. Rotem provides full fleet maintenance services for new and existing rolling stock technology and experience in maintenance service. Rotem has a high-level rolling stock availability and reliability for its customer to focus only on key task operations.

Contracts
Rotem has completed the manufacture and delivery of KTX (Korean Train Express) and is also providing maintenance services for KTX to KORAIL in Pusan Depot.

UPDATED

Rail Passenger Services Inc (RPS)

PO Box 26381, Tucson, Arizona 85726, USA
Tel: (+1 520) 747 03 46   Fax: (+1 520) 747 03 78

Key personnel
President: Peter M Robbins
Vice-President and Chief Financial Officer:
   William B Pickeral

Principal subsidiary
Arizona Rail Car Inc
PO Box 26381, Tucson, Arizona 85726, USA
Tel: (+1 520) 748 17 86   Fax: (+1 520) 747 03 78

Products
Heavy passenger coach and freight wagon rebuilding and modernisation. The company specialises in one-off customised passenger coach refurbishments.
   In addition, the company also operates the 'Sierra Madre Express' tourist train in Mexico. See entry for Fenomex in the Mexico section of Railway Systems.

Rail Services International SA (RSI)

Head office
Direction Internationale 38, rue de la Convention, F-94270 Le Kremlin, Bicêtre, France
Tel: (+331) 53 14 17 30   Fax: (+33 1) 53 14 17 49
e-mail: info@railsi.com
Web: www.railsi.com

Key personnel
Managing Director: Philippe Aloyol

Subsidiaries
RSI Austria
Domaniggasse, 2, A-1100 Vienna, Austria
Tel: (+43 1) 617 77 71 12   Fax: (+43 1) 617 77 71 28
e-mail: info@railsi.at
Managing Director: Reinhard Rössler

RSI Belgium
Vaartblekersstraat, 29, B-8400 Ostend, Belgium
Tel: (+32 59) 56 18 80   Fax: (+32 59) 70 23 75
e-mail: info@railsi.be
Managing Director: Jan Baert

RSI Italia SpA
Via Sesto San Giovanni 9, I-20126 Milan, Italy
Tel: (+39 02) 66 14 02 01   Fax: (+39 02) 66 10 09 61
e-mail: info@railsi.it

Managing Director: Renato Mantegazza
Unit Manager Milan: Guido Sarzilla
Unit Manager Rome: Pasquale Grieco

CostaRail Srl
Viale 4 Novembre, I-22041 Costamasnaga, Como, Italy
Tel: (+39 031) 86 94 11   Fax: (+39 031) 85 53 30
e-mail: hkastner@costarail.it
Commercial Manager: Heinz Kastner

RSI Netherlands
Onderhoudpost Watergraafsmeer, Kruislaan 254, NL-1098 Amsterdam SM
Tel: (+31 20) 557 66 30   Fax: (+31 20) 557 88 18
e-mail: info@railsi.nl
Managing Director: Jan Baert
Unit Manager: Bart Janssen

Background
Formerly the railway maintenance business of Compagnie des Wagons Lits, RSI is an independent company handling the maintenance, overhaul, repair and refurbishment of rail vehicles in several European countries. In 2004 the company acquired the former Costaferroviaria SpA in Italy, renaming the company CostaRail Srl.

Services
Development, design, engineering and technical assistance for fitting out passenger car interiors. RSI's Study and Design Department works on new projects such as interiors and technical specifications for passenger cars. RSI carries out maintenance, overhaul and major refurbishment of all kinds of rolling stock at its four workshops (Milan, Ostend, Rome and Vienna) or units in 20 European locations.
   Customer support for railway operators, day-to-day operations, wheelset maintenance, repair, warranty, mobile operations at customer's plant and supply of spare parts.

Contracts
RSI has signed maintenance contracts with CFF, FS, NS, ÖBB, SNCB, SNCF, ICF and Eurotunnel.

UPDATED

SAFOP SpA

Corso L Zanussi 55, I-33080 Porcia (PN), Italy
Tel: (+39 0434) 59 77 11   Fax: (+39 044) 92 25 83
e-mail: safop@safop.com
Web: www.safop.com

Subsidiaries
SAFOP USA Inc
3790 Commerce Court, Suite 100, Wheatfield, New York 14120, US
Tel: (+1 716) 213 00   Fax: (+1 716) 693 98 69

SAFOP UK
12 Aysgarth Road, Newsome, Huddersfield, West Yorkshire HD4 6QY, UK
Tel: (+44 1484) 30 72 32
e-mail: safopuk@ntlworld.com

Products
Wheelset reprofiling systems, including: underfloor lathes; single and tandem lathes; portal lathes; and universal lathes.

NEW ENTRY

SEFAC

Société d'Estampage et de Forge Ardennes Champagne
1 rue Andre Compain, F-08800 Montherme, France
Tel: (+33 3) 24 53 01 82
Fax: (+33 3) 24 53 29 18
e-mail: vjolliot@sefac.fr
Web: www.sefac.fr

Key personnel
President: Emmanuel de Rohan Chabof
Engineering Manager: E Letellier
Rail Sales Manager: Vincent Jolliot

Principal subsidiaries

SEFAC SA
Camino de Rejas, Nave 10, E-28820 Coslada, Spain
Tel: (+34 91) 672 36 12   Fax: (+34 91) 672 33 96
e-mail: commercial@sefac.es
Manager: P Maigre

Products

Electromechanical lifting systems with mobile or fixed columns, capacity per column from 5 to 40 tonnes. SEFAC lifting systems allow the lifting of power cars, LRVs, trainsets and wagons for maintenance work to be carried out.

Contracts

Contracts include the supply of lifting systems to railway operators including SNCF, SNTF, FGC, Ferrovías, TRA, SNCB, OCTRA and CP; metro operators in Hong Kong, Baltimore, Lisbon, Paris, Rotterdam and Cairo; and to rolling stock manufacturers including ALSTOM, Bombardier-ANF and Fiat.

UPDATED

Siemens AG

Transportation Systems

Integrated Services Division
Mozartstrasse 33b, D-91052 Erlangen, Germany
Tel: (+49 9131) 72 38 12   Fax: (+49 9131) 72 11 63
e-mail: service@ts.siemens.de
Web: www.siemens.com/transportation

Corporate Headquarters
Siemens AG
Transportation Systems
PO Box 3240, D-91050 Erlangen, Germany
Tel: (+49 9131) 7-0

Key personnel

President, Transportation Systems Group:
  Hans M Schabert
Group Vice-Presidents: Alfred Frank, Joern F Sens,
  Friedrich Smaxwil
Heads of Integrated Services Division:
  Arne Kleversaat, Thorsten Sponholz

Services

Integrated Services was established in December 1999, offering service and maintenance provision for: rolling stock; infrastructure; signalling and communications systems; and power supply systems.

Rolling stock service provision includes: routine vehicle maintenance; spare parts supply; and bogie overhaul. Infrastructure service provision includes: maintenance; spare parts supply; and diagnostic services. The Charter Rail service is based on using the customer's own staff, which are sub-contracted at a fixed price, with Siemens assuming responsibility for maintenance planning, spares provision and services with guaranteed costs and availability.

Integrated Services also offers consulting, training, documentation and financing services.

Developments

In July 2002 Siemens announced that it had formed a joint venture subsidiary with Spanish National Railways (RENFE) to maintain and repair a fleet of 30 Class 440 and 20 Class 447 emus operating in the greater Barcelona area. Siemens holds a 51 per cent shareholding in the company, Nertus Mantenimiento Feroviario SA. Maintenance activities were to be undertaken at a RENFE-owned facility in Barcelona.

In January 2002 Siemens acquired a 50 per cent shareholding in Leipziger Infrastruktur Betriebe (LIB), which was formerly the infrastructure maintenance subsidiary of Leipzig's transit authority, Leipziger Verkehrsbetriebe (LVB). This followed the earlier acquisition by Siemens of a 50 per cent shareholding in the authority's rolling stock maintenance subsidiary, Leipziger Fahrzeug-Service Betriebe (LFB). Both ventures, now acting as IFTEC GmbH & Co KG, are regarded as pioneering steps in public-private partnerships in Germany.

Siemens has a shareholding of 51 per cent in ERL Maintenance Support Sdn Bhd (EMAS), which provides repair and maintenance services for the rolling stock and infrastructure of the Express Rail Link airport line in Kuala Lumpur Malaysia. The contract provides for transfer of technology to Siemens' partner, ERL, within three years of the start of operations, which took place in April 2002.

In September 2000 Siemens launched its Internet-based Rail Mall spare parts ordering system, enabling operators of light rail vehicles supplied by the company to obtain parts in most countries in Europe within 24 hours.

Contracts

Current contracts include:

Argentina: Rolling stock maintenance for Metrovías, Buenos Aires, providing maintenance and refurbishment of 64 vehicles.

Australia: Siemens received contracts from National Express Group Australia covering full service provision for 59 Combino five-and three-section Melbourne Swanston Combino LRVs from 2003 to 2017 and 62 three-car Melbourne Bayside metro trains from 2003 to 2020. The franchise has since been taken over by Connex Trains Melbourne.

Brazil: In December 2002 Siemens concluded an agreement with Companhia de Trens Metropolitanos (CPTM), São Paulo, to provide maintenance services for 10 Class 3000 emus for five years.

Czech Republic: In September 2004 Siemens signed a Memorandum of Understanding with the Prague transit authorities covering the maintenance of 48 Type M1 metro trains for a period of 14-years. The fleet includes six trains to be delivered by Siemens in 2005-6.

Denmark: From January 2003 Siemens took over complete maintenance of 13 Class EG 3100 25 kV/15 kV AC electric freight locomotives operated by Railion Denmark. The contract, which initially runs for five years with an option for a one-year extension, includes a 'Mobile Service' facility which provides for maintenance and service of the locomotives when they operate outside Denmark.

Egypt: Siemens is responsible for the maintenance of Egyptian National Railways' 350 km Cairo–Baharya freight line, covering track, signalling and communications systems.

Germany: Under a contract covering the period 2002-07 Siemens is responsible for maintenance of the complete infrastructure (signalling, telecommunications, electrification system) and vehicles of the Düsseldorf Sky Train AGT system, supplied as a turnkey project.

Siemens has a separate contract covering the maintenance of signalling equipment (1996–2006) and electrification system (1998-2003) of an industrial railway system operated by Vattenfall Europe.

Siemens was responsible to DB AG for maintenance of the signalling and operations control system of the upgraded Berlin–Magdeburg high-speed line for the period 1996–2003.

Hungary: In conjunction with a contract to supply 40 Combino low-floor trams to the Budapest urban transport authority, BKV, Siemens concluded a two-year agreement to provide maintenance services following the vehicles' delivery in 2005–06.

India: In February 2003 Siemens took over the maintenance of traction converters and traction control units (Siemens' scope of supply) for 31 GT46 six-axle diesel-electric locomotives operated by Indian Railways.

Poland: In January 2003 Siemens announced a Charter Rail 12-year agreement to provide maintenance services for 14 Combino trams that it was to supply to MPK, Poznan, from November 2003. The contract includes a five-year extension option.

Spain: Under a contract signed in July 2001 Siemens is to be responsible for the maintenance for 14 years of 16 new Velaro E high-speed trainsets which the company is to supply for RENFE's 625 km Madrid–Barcelona line.

Siemens has a contract covering maintenance of telecommunications equipment on the Madrid-Seville line for the period 1992-2007.

Under a five-year agreement with Spanish National Railways, Siemens is responsible for substations serving the 471 km Madrid Seville high-speed line, and for maintenance of the power supply and telecommunications systems.

Thailand: Having supplied and equipped the 23.1 km BTS mass transit system in Bangkok, Siemens is responsible for maintenance, covering 105 metro vehicles, track, signalling, operations control systems, power supply, safety systems, stations and all technical equipment for the period 2000–05.

As part of a turnkey contract Siemens is to maintain for 10 years the Bangkok Blue Line of Bangkok Metro Co Ltd, which commenced operations in 2004.

UK: In August 2004 Siemens received a contract from National Express Group (NEG) covering maintenance and spare parts supply for 20-years for 30 Class 350/1 Desiro emus employed on commuter services at the southern end of the West Coast Main Line. The contract also covers the creation of a new depot at Northampton. In May 2004 Siemens was awarded a contract by NEG to maintain 21 Desiro UK Class 360 emus for seven years with a possible extension for three more years. The trains are operated by NEG's One Eastern Railway franchise. Maintenance will be undertaken at Ilford, east London.

South West Trains Class 450 emu at Siemens' fleet maintenance depot, Southampton, UK
(Ken Harris)
0554722

Siemens has secured rolling stock maintenance contracts for several other urban and main line operators in the UK. These include: Arriva Trains Northern, servicing Class 333 emus supplied by Siemens, using a depot in Leeds; Heathrow Express, for which Siemens constructed and operates a dedicated depot in west London to maintain the operator's fleet of Siemens-supplied Class 332 emus; South Yorkshire Supertram, maintaining 25 trams at a purpose-built depot in Sheffield. In addition, Siemens has a 20-year contract starting in 2003 covering the maintenance of a fleet of 665 Desiro Class 444 and 450 emu cars being supplied to South West Trains. This is being undertaken at a depot at Southampton commissioned in 2003.

In October 2004 Heathrow Express renewed its contract with Siemens to cover maintenance of its fleet until 2023.

Other orders for infrastructure and rolling stock maintenance services form part of turnkey contracts, detailed in the *Turnkey systems contractors* section.

UPDATED

Simmons

Simmons Machine Tool Corporation
1700 North Broadway, Albany, New York 12204, USA
Tel: (+1 518) 462 54 31   Fax: (+1 518) 462 03 71
e-mail: smt@smtgroup.com
Web: http://www.smtgroup.com

Key personnel
Chairman and Chief Executive: Hans J Naumann
President: John O Naumann
Finance Director: David A Simonian

Subsidiary
Simmons-Stanray   Wheel   Truing   Machine
Corporation
(address as above)

Affiliated company
Niles-Simmons Industrieanlagen GmbH
Zwickauer Strasse 355, D-09117 Chemnitz, Germany

Products
Design and manufacture of equipment and machines for manual and automated railway and transit wheel workshops, including: underfloor wheel profiling machines; special machines for wheel, axle and wheelset maintenance; CNC grinding centres; CNC vertical turning and boring centres. Simmons also provides engineering design and layout services for complete wheel shops.

Smith Bros & Webb Ltd

Britannia Works, Arden Forest Industrial Estate, Alcester, Warwickshire B49 6EX, UK
Tel: (+44 1789) 40 00 96   Fax: (+44 1789) 40 02 31
e-mail: sbw@vehicle-washing-systems.co.uk
Web: http://www.vehicle-washing-systems.co.uk

Key personnel
Managing Director: John P Bennett
Sales and Marketing Manager: Rob Everton
Service and Installation Manager: Bob Smith
Technical Manager: Tony Appleton

Products
Britannia automatic washing systems for main line and metro trains. Options available include: detergent or acid application; front, rear, roof, valence and skirt washing; blow-drying; pure water final rinse; train speed indication; water recycling and effluent treatment. After-sales service offered.

Smith Bros & Webb train washing system for First Great Western at Bristol, UK   0087659

Recent contracts include train washing installations for Croydon Tramlink, West Midlands Metro and First Great Western (Bristol, UK) and two systems for MTRC, Hong Kong.

Sogema Engineering

Zone Industrielle Roubaix Est, Rue de la Papinerie, BP 62, F-59452 Lys Lez Lannoy, Cedex, France
Tel: (+33 3) 20 66 10 70   Fax: (+33 3) 20 66 10 71
e-mail: contact@sogema-engineering.com
Web: www.sogema-engineering.com

Products
Bogie and wheelset maintenance equipment, including: bogie presses; wheel presses; test benches; bogie turning equipment; bearing removal tools; spring calibration presses; and wheelset measuring benches.

NEW ENTRY

SSI Corporation

1650 Bonhill Road, Mississauga, Ontario L5T 1C8, Canada
Tel: (+1 905) 795 92 74   Fax: (+1 905) 795 13 50
e-mail: info@ssiwash.com
Web: http://www.ssiwash.com

Key personnel
President: Seymour Techner

Products
Modular train and bus washing systems; vehicle interior vacuum cleaning systems; dryers.

NEW ENTRY

Stertil-Koni

PO Box 23, NL-9288 ZG Kootstertille, Pays-Netherlands
Tel: (+31 512) 33 44 44   Fax: (+31 512) 33 44 30
e-mail: info@stertil.nl
Web: http://www.stertil.nl

Subsidiary
Stertil BV
1900 Benhill Avenue, Baltimore, Maryland 21226, USA
Tel: (+1 410) 355 71 00
Fax: (+1 410) 355 71 70
e-mail: lifts@stertil-koni.com
Web: http://www.stertil-koni.com

Products
Hydraulic lifting systems for rail vehicles. Systems have been supplied to urban rail systems in Brussels, Calgary, Graz, Innsbruck, Jacksonville (on behalf of Bombardier Transportation) and Linz.

SweMaint

Utbyvägen 151, SE-415 07 Gothenburg, Sweden
Tel: (+46 31) 10 36 00
Fax: (+46 31) 10 36 69
e-mail: info@swemaint.se
Web: www.swemaint.se

Key personnel
Managing Director: Hakan Fredriksson

Services
Freight   wagon   maintenance,   repair   and
refurbishment.

UPDATED

For details of the latest updates to *Jane's World Railways* online and to discover the additional information available exclusively to online subscribers please visit
jwr.janes.com

Talgo

Patentes Talgo SA
C/Gabriel García Marquéz 4, E-28230 Las Rozas,
Madrid, Spain
Tel: (+34 91) 631 38 00
Fax: (+34 91) 631 38 99
e-mail: marketing@talgo.com
Web: http://www.talgo.com

Key personnel

For a full list of personnel, see Talgo entry
in *Locomotives and powered/non-powered
passenger vehicles section.*

Subsidiaries

For a full list of subsidiaries see Talgo entry
in *Locomotives and powered/non-powered
passenger vehicles* section.

Products

Design and manufacture of machinery and
equipment for rolling stock wheelset examination
and maintenance, including pit lathes with CNC
control or a hydraulic copying device for loads of
18, 26 or 30 tonnes per wheelset, with accessories
for machining brake discs *in situ*; hunting cars for
positioning rolling stock, bogies or wheelsets up to
300 tonnes in weight at speeds up to 5 km/h; and
equipment for dynamic measurement of wheel
parameters, including diameter, flange height and
thickness, to determine turning programmes for
CNC-controlled lathes; equipment for wheel defect
detection; manual gauges for measuring wheel
parameters.

Contracts

Recent contracts include the supply of underfloor
wheel lathes to: TBA, Argentina; DB Reise &
Touristik, Germany; Lisbon metro, Portugal.

Talgo Oy

Elektroniikkatie 2, FIN-90570 Oulu, Finland
Tel: (+358 8) 870 69 00   Fax: (+358 8) 870 69 70
e-mail: sales@talgo.fi
Web: http://www.talgo.fi

Key personnel

Managing Director: Tapani Tapaninaho
Sales Director: Matti Haapakangas
Sales Manager: Matti Asikainen

Services

Rail vehicle refurbishment.

Contracts

In 2003 Talgo Oy and Helsinki City Transport
closed an agreement on the refurbishment of the
bodyshells of 20 articulated trams. The agreement
includes an option for an additional order for
the refurbishment of 10–20 bodyshells during
2005–07.
   Talgo Oy also received a contract from Helsinki
City Transport (HKL) for the refurbishment of 39
metro trains. This was to be caried out at Talgo's
Otanmäki works during 2003–08.

Thrall

Thrall Europa Service Parts
Holgate Park, 156 Holgate Road, York YO24 4FJ, UK
Tel: (+44 1904) 75 60 00
Fax: (+44 1904) 75 60 06
e-mail: ServiceParts@ThrallEurope.co.uk

Key personnel

For a full list of personnel, see Thrall entry in *Freight
vehicles and equipment* section.

Services

Supply of spare parts and components for the
manufacture, repair, conversion and overhaul
of freight wagons, with a capability to distribute
products throughout Europe.

Toshiba Corporation

Railway Projects Department
Toshiba Building, 1-1, Shibaura 1-chome, Minato-
ku, Tokyo 105-8001, Japan
Tel: (+81 3) 34 57 49 24   Fax: (+81 3) 54 44 92 63
Web: www.toshiba.co.jp

Key personnel

President and Chief Executive Officer:
   Atsutoshi Nishida
Vice-President, Transportation Systems Division:
   Takio Ooyama
Senior Manager, Railway Projects Dept: Koji Toda

Products

Multipurpose test equipment for traction motors;
automatic test equipment for silicon rectifiers;
battery trucks for transporting carbodies; battery
tractors interchangeable with both steel and
rubber-tyred wheelsets for moving carbodies in
shops or yards.

UPDATED

TraffiCare AB

Box 340, SE-101 26 Stockholm, Sweden
Tel: (+46 8) 762 35 00
Web: http://www.trafficare.se

Key personnel

Managing Director: Tommie Wikström

Background

Formerly a subsidiary of Swedish State Railways,
Trafficare was privatised in 2001.

Services

TraffiCare's 'Ready Train' service covers vehicle
cleaning and preparation, train formation, logistics
and minor repairs. Clients include SJ AB, A-Train,
Tågkompaniet and Linx.

Trenitalia SpA

Italian State Railways
Rolling Stock Technology Department (UTMR)
Viala S Lavagnini 58, I-50129 Florence, Italy
Tel: (+39 055) 48 90 02
Fax: (+39 055) 46 12 66

Services

With more than 5,000 employees, UTMR is
responsible for the maintenance of Italian
State Railways (FS) traction and rolling stock.
It also undertakes overhaul, refurbishment and
conversion projects using the 13 facilities that form
the company's Major Repairs Workshops (OGR)
division. These resources and capabilities are
offered to other operators.

Ultrasonic Sciences Ltd

Unit 4 Springlakes Industrial Estate, Deadbrook
Lane, Aldershot, Hants, GU12 4UH, UK
Tel: (+44 1252) 35 05 50
Fax: (+44 1252) 35 04 45
e-mail: info@ultrasonic-sciences.co.uk
Web: www.ultrasonic-sciences.co.uk

Key personnel

Chairman: J B Kennelly
Sales Director: C S Gartside

Products

Automated and semi-automated ultrasonic
testing systems for manufacturing plant and
in-service inspection. Applications include testing
of machined solid axles and forged or cast wheels,
in situ inspection of hollow axles from the bore,
and inspection of wheeltreads in maintenance or
services workshops.

Contracts

Supply of systems for Lucchini (1999), SNCF
(2000), Bombardier Transportation (2001 and 2004),
ALSTOM (2002), Korean Railways (2004) and Indian
Railways (2001, 2002, 2004).

UPDATED

United Goninan

PO Box 33000, Hamilton, New South Wales 2303,
Australia
Broadmeadow Road, Broadmeadow, New South
Wales 2292, Australia
Tel: (+61 2) 49 23 50 00
Fax: (+61 2) 49 23 50 01
Web: www.unitedgoninan.com.au

Key personnel

Chief Operating Officer: John McLuckie

Services

Rail vehicle maintenance; locomotive and
passenger vehicle overhaul, upgrading and
refurbishment.

Contracts

Include maintenance of City Rail fleet of 1,450
double-deck emus for Railcorp NSW; maintenance
of 120 emus for MTR Hong Kong at the Tseung
Kwan O depot; maintenance of 95 dmus for
TransAdelaide at the Adelaide Maintenance Centre;
maintenance of 120 diesel-electric locomotives for
Pacific National at Spotswood Depot Melbourne;
maintenance of 23 diesel-electric locomotives
and 360 freight vehicles for Bluescope Steel at
Port Kembla; maintenance of Indian Pacific, Ghan
and Overlander long-distance passenger trains for
Great Southern Railways at Keswick Maintenance
Centre, Adelaide.

UPDATED

Üstra

Hannoversche Verkhersbetriebe AG
Werkstätten Schienenfahrzeuge, Friedrich Lehner
Weg 1, Postfach 2540, D-30025 Hanover, Germany
Tel: (+49 511) 16 68 25 65   Fax: (+49 511) 16 68 21 98

Background

Üstra is the operator of Hanover's public transport
system, including its 115 km light rail network. Its
workshops offer services commercially to other
urban and main line rail operators and to vehicle
and subsystems manufacturers.

Services

Rail vehicle inspection, maintenance, overhaul,
refurbishment, subsystems assembly and
overhaul.

Contracts

Transport operators featuring in Üstra's list of
customers include: BVG, Berlin; Stadtwerke Bielefeld;
VGF, Frankfurt am Main. Manufacturer clients
include: Alstom LHB; Bombardier Transportation;
Siemens; and ZF Hurth Bahntechnik.

Vanjax Sales Pvt Ltd

343 Sidco Industrial Estate, Ambattur, Chennai
600 098, India
Tel: (+91 44) 26 25 46 67; 48 75; 53 00
Fax: (+91 44) 26 52 90 78
e-mail: vanjaxsale@sancharnet.in
Web: www.vanjax.com

Key personnel

Chairman: Fakhruddin Vanak
Managing Director: Daniel F Vanak
Joint Managing Director: Juzar F Vanak
Vice-President, Commercial: Nafisa F Vanak

Vanjax hydraulic head and toes lifting jack with 10 t capacity   **NEW**/1142738

Products

Hydraulic portable floor cranes, workshop presses, trolley jacks, hydraulic jacks and cylinders, hydraulic pullers, pipe benders, bolt tensioners and hydraulic pumps/power packs.

The access platform has an articulated boom with telescopic extensions and a 180° rotating platform. The units capacity is 230 kg, with a maximum raised working height of 22 m. The platforms can be truck mounted. The towable access platforms can be custom made to order with 360° slewing by means of hydraulic cylinder rack and pinion or hydraulic motor (continuous slewing), with a capacity of 160 kg, 227 kg and 250 kg and a maximum height from six to 21 m.

The telescopic platform has a capacity of 500 kg with a maximum raised height of 13 m and can be towable or mounted on special vehicles.

The most recently patented design is the hydraulic head and toe lifting jack (VXTTJ-10/230), with a capacity of 10 tonnes.

UPDATED

VIA Rail Canada

Equipment Maintenance Business and Maintenance Services
201 Ash Avenue, Montreal, Quebec H3K 3K2, Canada
Tel: (+1 514) 934 75 45   Fax: (+1 514) 934 75 20
e-mail: alan_mackenzie@viarail.ca
Web: http://www.viarail.ca

Key personnel

Manager, Planning and Logistics: Alan MacKenzie

Services

VIA Rail Canada is the country's national long-distance passenger train operator. It offers the following services to other operators: inspections, maintenance and repair of rolling stock; daily and weekly train servicing; inspection and overhaul of bogies; repair and requalification of dampers and other components; wheel profiling; bearing replacement; automatic train washing; fluid analysis; vibration analysis; testing and calibration of test instruments; assembly of wiring harnesses and equipment storage.

In addition, the following professional services are offered through a skilled management and technical team: operations planning; repair feasibility studies; project management; technical studies and performance specifications; industrial engineering and productivity studies; vehicle performance testing.

VIA has maintenance centres at Montreal, Winnipeg and Vancouver and train servicing locations at Toronto and Halifax.

Von Roll BETEC

Edenstrasse 20, Post Box, CH-8045 Zurich, Switzerland
Tel: (+41 1) 204 31 11   Fax: (+41 1) 204 31 12
Web: http://www.vonroll.ch

Key personnel

Managing Director: Dr Peter Schildknecht
Marketing Manager: Marc Bickel

Background

Von Roll BETEC took over the production and aftersales support of underfloor wheelset profiling machines from Kellenberger in 1998.

Products

Underfloor wheel profiling equipment including wet grinding machines wheelset measuring systems.

Vossloh Locomotives GmbH

PO Box 9293, D-24152 Kiel
Falckensteiner Strasse 2, D-24159 Kiel, Germany
Tel: (+49 431) 39 99 21 95
Fax: (+49 431) 39 99 22 74
e-mail: vertrieb.kiel@vl.vossloh.com
Web: www.vossloh-locomotives.com

Works

Service-Zentrum Moers, Baerler Strasse 100, D-47441 Moers, Germany
Tel: (+49 2841) 14 04 10
Fax: (+49 2841) 14 04 50

Key personnel

Board of Directors: Andreas Hopmann
Chief Executive Officer: Dr Georg Hauschild

Products

Fleet management and maintenance. Full service or maintenance programmes adapted to the client's specific needs for Vossloh Locomotives GmbH as well as for diesel freight and shunting locomotives from other provenances.

UPDATED

Vossloh Locomotives' Service Centre at Moers, Germany, showing MaK 2000BB and MaK 1206 types from the manufacturer's range (David Haydock)   **NEW**/1122859

Wabtec Rail Ltd

PO Box 400, Doncaster Works, Hexthorpe Road, Doncaster DN1 1SL, UK
Tel: (+44 1302) 34 07 00   Fax: (+44 1302) 32 13 49
Web: www.wabtec.com

Key personnel

Managing Director: John Meehan
Engineering Director: Mike Roe
Finance Director: Robert Johnson
Operations Director: Chris Weatherall
Commercial Manager: Paul Robinson

Background

Wabtec Rail is one of the UK's leading railway rolling stock engineering companies undertaking the overhaul of passenger rolling stock, main line locomotives, bogies, wheelsets, air brake equipment, hydraulic dampers and buffers design in addition to the manufacture of shunting locomotives and freight wagons.

As part of the Wabtec Corporation, Wabtec Rail also supplies to the UK rolling stock owners and maintainers, composite brake blocks and pads, Wabtec Railway Electronics train data recorders and electronic equipment, Cardwell TMX braking systems and Wabtec air brake equipment.

Products

Wabtec Rail undertakes the maintenance, overhaul, repair and refurbishment of all types of railway rolling stock, encompassing locomotives, passenger rolling stock, diesel multiple units, electric multiple units and freight wagons.

In 2005 Wabtec Rail secured a contract from UK leasing company Angel Trains Ltd covering the interior refurbishment of 288 Class 317 25 kV AC emu cars. The trains are operated by West Anglia Great Northern on services out of London's Kings Cross station.

Contracts

Current contracts include passenger rolling stock overhaul and heavy maintenance for Angel Trains and Porterbrook Leasing, freight wagon overhaul and conversion for EWS and GE Rail, bogie overhauls for Railpart, Maintrain, Siemens and Freightliner, wheelset overhauls for EWS and Railpart, air brake equipment overhauls for EWS, Railpart, Bombardier Transportation and ALSTOM.

Wabtec Rail's Fleetcare division maintains fleets of freight wagons operating throughout the UK. This work is undertaken either at the customer's premises or by mobile teams and is supported by its main Doncaster Works where engineers are able to produce maintenance specifications and undertake safety performance monitoring. Wabtec Rail's extensive fleet maintenance database is able to provide customers with vehicles maintenance histories, fleet availabilities, failure trend analysis and component overhaul and repair records.

UPDATED

Wesurail Limited

21-22 Auster Road, Clifton Moor, York YO30 4XA, UK
Tel: (+44 1904) 69 25 44   Fax: (+44 1904) 69 25 66
e-mail: wesurail@aol.com

Key personnel

Managing Director: Alan Sherlock
Design Engineer Train Washer: Ken Dews
Design Engineer CET's: Dave Perry

Products

Automatic controlled emission toilet systems; mobile CET bowsers; automatic train wash systems.

Contracts

West Coast Traincare, five train wash systems, five CET systems, fuelling system; Israel Railways, automatic CET system, Delhi Metro, automatic train wash system; SW Trains, four CET systems; Siemens, one CET, one trainwash, one water recycling; Irish Rail, one trainwash.

NEW ENTRY

Whiting Corporation

26000 Whiting Way, Monee, Illinois 60449-8060, USA
Tel: (+1 800) 255 85 94   Fax: (+1 708) 587 20 01
e-mail: info@whitingcorp.com
Web: www.whitingcorp.com

Key personnel

President: J L Kahn
Vice-President, Sales and Marketing: Alan J Burke
Director of Marketing: Stuart J Lipsteuer
Manager, Transportation Sales: Dave Cunningham
Manager Heavy Rail Equipment Sales:
   Jim Thompson

Products

Custom designed and pre-engineered overhead travelling cranes for all types of critical applications. The maintenance cranes are long-life and low maintenance, built in capacities up to 800 t for a single hook and potentially cab, floor or remote controlled. Conventional/shallow pit car hoists, body hoists/supports, bogie repair hoists, transfer tables, bogie/vehicle turntables, portable electric jacks, wagon/train progression systems, drop tables, overhead and gantry cranes, traction motor dollies, truck locomotive and wagon/car turntables, rip jacks, sanding cranes.

UPDATED

Windhoff Bahn-und Anlagentechnik GmbH

PO Box 1963, D-48409 Rheine, Germany
Tel: (+49 5971) 580   Fax: (+49 5971) 582 09
e-mail: info@windhoff.de
Web: www.windhoff.de

Key personnel

Board Members: Herbert Liessem, Georg Vennemann
Finance Director: Helmut Gielians
Sales Directors: Dr Martin Hindersmann, Uwe Dolkemeyer,
Technical Director: Juergen Auschner
Purchasing Manager: Stefan Berkemeyer

Products

Rail systems: elevated tracks; train/rail car lifting system. Plants and equipment for handling, maintenance and repair of rail cars and dedicated assemblies: exchange of wheelsets and bogies, bogie measuring stand, mobile works platforms, roof access platforms, traverses, turntables, wheel scales. Equipment for handling and transport of components: bogies transport and storage, wheelset transport and storage.

UPDATED

ŽOS Trnava as

Koniarekova 19, SK-917 21, Slovakia
Tel: (+421 33) 556 71 11   Fax: (+421 33) 556 72 04
e-mail: marketing@zos.sk
Web: http://www.zos.sk

Services

Overhaul, maintenance, repair and conversion of freight wagons and passenger coaches; wheelset maintenance, including wheel reprofiling; overhaul of braking systems.

ŽOS Vrútky as

Dielenska Kružná, SK-038 61 Vrútky, Slovakia
Tel: (+421 842) 420 51 01
Fax: (+421 842) 428 15 95
e-mail: zos-vrutky@zos-vrutky.sk
Web: www.zos-vrutky.sk

Background

Formerly the Vrútky workshops of the Czechoslovak and subsequently Slovakian Republic Railways (ŽSR), ŽOS Vrútky became a joint stock company in 1994. ŽSR retains a shareholding of 34 per cent; the remainder is held by private shareholders.

Services

ŽOS Vrútky carries out servicing and general overhaul of electric locomotives, emus, passenger coaches and special cars. The company also carries out repairs to components including bogies, wheelsets, traction and auxiliary motors, traction transformers and other electric and electronic rotating and atypical machinery, transformers of various types and designs.

Contracts

In 2004 the company was undertaking the modernisation of Class 350 dual-voltage electric locomotives.

UPDATED

ŽOS Zvolen as

Môtovská cesta 259/11, SK-960 03 Zvolen, Slovakia
Tel: (+421 45) 530 21 11   Fax: (+421 45) 532 05 26
e-mail: zoszv@zoszv.sk
Web: www.zoszv.sk

Background

Formerly the Zvolen workshops of the Czechoslovak and subsequently Slovakian Republic Railways (ŽSR), ŽOS Zvolen was privatised in 1995.

Services

Modernisation, refurbishment, overhaul and repair of diesel locomotives, railcars and passenger coaches.

Contracts

Recent contracts include: modernisation of 36 Class 812 diesel railcars; modernisation of 11 Class 736 diesel-electric locomotives for Slovakian Railways (ZSSK); conversion and re-engining of four Class 740 locomotives with a Caterpillar Series 3412 627 kW power unit to become Class 724 for US Steel Košice; similar conversion of two Class 770 locomotives with a Caterpillar Series 3512 1,455 kW to become Class 774 for the same customer.

Conversion and re-engining of one Class 740 locomotive with a Caterpillar Series 3508 DITA 970 kW power unit to become Class 744 for Slovnaft Bratislava.

UPDATED

TURNKEY SYSTEMS CONTRACTORS

Alphabetical listing

Alcatel Canada Inc
ALSTOM Transport
Ansaldo Trasporti Sistemi Ferroviari
Balfour Beatty Rail Projects Ltd
Bombardier Transportation
Chemetron Railway Products Inc

Dimetronic SA
Interinfra
Marconi Communications
Parsons Brinckerhoff Inc
Ranalah
Rotem Company

SAT
Scitel Telematics Ltd A
Siemens Transportation Systems
Técnicas Modulares e Industriales (Temoinsa)
Transrapid International GmbH & Co KG
Transys Projects Ltd

Company listing by country

CANADA
Alcatel Canada Inc
Bombardier Transportation

FRANCE
ALSTOM Transport
Interinfra
SAT

GERMANY
Siemens Transportation Systems
Transrapid International GmbH & Co KG

HUNGARY
Scitel Telematics Ltd A

ITALY
Ansaldo Trasporti Sistemi Ferroviari

KOREA, SOUTH
Rotem Company

SPAIN
Dimetronic SA
Técnicas Modulares e Industriales (Temoinsa)

UNITED KINGDOM
Balfour Beatty Rail Projects Ltd
Marconi Communications
Ranalah
Transys Projects Ltd

UNITED STATES
Chemetron Railway Products Inc
Parsons Brinckerhoff Inc

Alcatel Canada Inc

Transport Automation
1235 Ormont Drive, Weston, Ontario M9L 2W6, Canada
Tel: (+1 416) 742 39 00   Fax: (+1 416) 742 11 36

Key personnel

General Manager: Walter Friesen
Director, Business Development: Kevin D Fitzgerald

Projects

Turnkey signalling systems for Stockholm metro, Quebec Cartier Railway, Toronto Transit Commission, Walt Disney World, Hong Kong MTRC and KCRC West Rail and East Rail Extension, San Francisco MUNI, and London Docklands Light Railway.

ALSTOM Transport

Systems Business
48 rue Albert Dhalenne, F-93482 Saint-Ouen Cedex, Paris, France
Tel: (+33 1) 41 66 90 00   Fax: (+33 1) 41 66 96 66
Web: www.transport.alstom.com

Key personnel

President, Transport Sector: Philippe Mellier
Chief Operating Officer: Gérard Blanc
Chief Financial Officer: Roland Kientz
Senior Vice-Presidents:
　Systems Business, International Product Line Management: Laurent Troger
　Regional Asia Pacific: Marc Chatelard
　Southern Europe: Charles Carlier
　Americas: Francis Jelensperger
　Northern Europe: Terence Watson

Background

Systems Business in its current form was created in 1998 following the acquisition of Cegelec, the electrical contracting arm of the Alcatel ALSTOM group.

Services

Systems Business of ALSTOM Transport offers global public transportation solutions including turnkey management, execution of infrastructure packages, interfaces with civil works and rolling stock.

The business addresses urban transit systems, suburban lines and main lines, and can be involved in projects at the individual subsystem level, or at the full turnkey level.

Systems Business acts as main contractor for metro and light rail transit systems, airport rail links and fully automatic metros and houses all the competencies necessary for the development of a project concept, its detailed design, and its delivery.

The Infrastructure Business unit of ALSTOM Transport Systems offers solutions at the system or subsystem level for including power generation and distribution. It includes AC and DC traction substations, overhead lines, surface contact third rail, at-level integrated supply system, SCADA, auxiliary power supply, track laying, maintenance workshops, communications, signalling (tramways), electrical and mechanical equipment in-station.

Its scope embraces design, development, installation, financing, managing, commissioning, technical assistance, maintenance and training.

Contracts

Brazil: In October 2003, as a member of the Via Amarela consortium, ALSTOM was awarded a contract by Companhia Metropolitano de São Paulo to equip Line 4 (the Amarela Line) of the city's metro system. The scope of the contract covers electrical and mechanical infrastructure, including power supply and electrical distribution systems, telecommunications equipment and auxiliary systems such as fire detection, pumping and lighting. The line is due to be commissioned in 2007.

Chile: In July 2002, ALSTOM was selected to supply rolling stock, Urbalis™ automatic train control equipment and maintenance services for the 32 km north-south Line 4 of Santiago's metro system.

Egypt: Egypt's National Authority for Tunnels (NAT) has awarded ALSTOM, as leader of the Interinfra consortium, a turnkey contract for the extension of Line 2 (Phase 2C). The 2.5 km surface line will run south of Giza Suburban station to El Mounib, adding two new stations. The turnkey contract is composed of civil works, electromechanical equipment and power supply and low-voltage equipment. ALSTOM's contribution includes power supply, signalling and Automatic Train Operation (ATO), and a centralised control system, as well as overall project management of the turnkey contract.

France: In June 2000 the Urban Community of Greater Bordeaux (CUB) chose a consortium led by ALSTOM to carry out the contruction of three tram lines that would serve the most densely populated areas of the urban and near-suburban areas of Bordeaux, supplying the 70 Citadis, the largest tram fleet in France. ALSTOM is directing the laying of track and the surfacing of the tram platforms. A contract option includes complete maintenance of all of the equipment supplied by the consortium. The system entered commercial service in 2004.

Greece: In 2002 as a member of a consortium which also includes Greek companies, ALSTOM was awarded a turnkey contract by Ergose, a subsidiary of the Hellenic Railways Organisations, to supply and construct electrical and mechanical elements of a new line linking the centre of Athens with a new airport at Spata. The line will comprise 10 km of existing infrastructure, which will be upgraded, and 20 km of new line. ALSTOM is responsible for turnkey project management and design and engineering, and will supply the substations, signalling, catenary and track. The signalling system will comply with the ERTMS European standard and will incorporate ALSTOM's Atlas system. The new line was due to be commissioned in 2004.

The metro system was built as a turnkey project by the Olympic Metro Consortium, led by Siemens (Germany) and Interinfra (France), of which ALSTOM is the main shareholder, comprising 21 German, French and Greek companies.

Korea, South: In 2002 as a member of the IKFC consortium that also included Korean rolling stock builder Rotem, ALSTOM was awarded a turnkey contract by the Incheon International Airport Railroad Co Ltd to undertake the project management of, and supply equipment for, a new 60 km line linking the centre of Seoul with Incheon International Airport. ALSTOM and its Korean-based company Eukorail, which was created initially for the country's high-speed rail project, are responsible for project management, system engineering and integration and the supply of signalling equipment. The line is due to be commissioned at the end of 2005.

Poland: ALSTOM has also won an order to upgrade the Bytom to Katowice tramway for the Tramway Communication Company of Katowice in Silesia. As well as provision of new vehicles, the turnkey order includes the refurbishment of the rail infrastructure and stations on the existing 20 km Line 6/41. The trams, which will be supplied by the company's Polish subsidiary ALSTOM Konstal, will be fitted with ONIX traction drives.

Singapore: Singapore's Land Transport Authority (LTA) has awarded the ALSTOM/STE consortium an order worth EUR170 million, for the second phase of construction of its automatic Circle metro line. ALSTOM's share of this contract is valued at EUR123 million. ALSTOM will supply its AXONIS™ automatic metro system for this line.

This new order is an extension of the Circle Line Phase 1 project, awarded to ALSTOM on a turnkey basis in December 2000. The first stage covers 5.6 km and has six stations. This second section will extend the Marina line by 5 km and add five new stations. Subsequent contracts placed with ALSTOM covered construction of the remainder of the line, which will eventually cover 34 km and circle the island of Singapore, making it the world's longest automatic metro line.

For this second phase, ALSTOM will supply seven three-car Metropolis™ trainsets, the signalling system and related infrastructure and will be responsible for the overall management of the project. The delivery of the first trainsets was scheduled for 2004, with commercial service commencing in 2006.

In addition to the Circle Line, ALSTOM also supplied and is supplying the Northeast Line with its Axonis automatic metro system. The Northeast Line is also a turnkey contract for which ALSTOM supplied the rolling stock, signalling and information solution, with expertise in structuring, financing, designing, constructing, testing and commissioning integrated rail systems. The Northeast Line is the largest fully automatic metro in the world and entered commercial service in 2003.

Spain: As part of the TramMet consortium, ALSTOM was awarded contracts to build and equip Lines 1 and 2 of the Barcelona light rail system. As well as supplying 19 and 18 Citadis tramsets respectively for the new lines, ALSTOM is responsible for system engineering, traction power supply substations, telecommunications, ticketing, signalling, workshop equipment and project management of the electrical and mechanical package. ALSTOM holds a shareholding of 25 per cent in TramMet, which holds 15-year operating and maintenance concessions on the system. The first phase was commissioned in March 2004.

Taiwan: In consortium with China Technical Consultant IT, in October 2003 ALSTOM was awarded a contract by the Department of Rapid Transit Systems (DORTS) covering infrastructure for extensions to Taipei's Orange and Blue metro lines. The scope of the contract covers project

Singapore's Northeast Line automated metro system   ***NEW**/0585260*

management, system integration, signalling, half of the track work and maintenance depot equipment. **Venezuela:** ALSTOM, as part of the FRAMECA consortium, has been awarded a turnkey order from the Caracas metro authority, CAMC, for the 5.5 km Line 4 of the city's metro system, which was scheduled to enter service in 2002. In addition to supplying 44 metro cars and the signalling system, ALSTOM was to carry out electrification of the line and provide a complete fire protection system.

UPDATED

Ansaldo Trasporti Sistemi Ferroviari

Via Argine 425, I-80147 Naples, Italy
Fax: (+39 081) 243 25 70

Genoa office
Via dei Pescatori 35, I-16129 Genoa, Italy
Fax: (+39 010) 655 20 28

Key personnel
Chief Executive: Sante Roberti

Background
Ansaldo Trasporti Sistemi Ferroviari continues the business activities of the former Ansaldo Trasporti SpA System Business Unit following a break-up of the company in April 2001. These cover the design, construction, testing, commissioning, system engineering, project management, maintenance and project financing of subsystems or complete main line and urban electrified mass transit systems as the main contractor or partner. The company employs around 300 staff.

Projects
Include: Copenhagen driverless metro, Denmark; Dublin light rail lines A and C, Ireland; Midland Metro, UK; Manchester Metrolink, UK; Lima metro, Peru; Genoa LRT and trolleybus systems, Italy; Sassari LRT, Italy; Naples metro line 6; Rome metro; Circumvesuviana and other regional lines, Italy; Rome Naples high-speed railway, Italy; and Trieste innovative stream system, Italy.

Balfour Beatty Rail Projects Ltd

Head office
B203 Midland House, Nelson Street, Derby DE1 2SA
Tel: (+44 1332) 26 26 66   Fax: (+44 1332) 26 22 95
e-mail: info.bbrp@bbrail.com

Key personnel
Managing Director: Rob Boulger
Business Development Director: Peter Kehoe
Business Development Manager: Dave Mackay
Commercial Director: Keith Hampson
Engineering, Safety and Assurance Director:
  Nick Dunne
Finance Director: Rory Mitchell
General Manager: Paul Copeland
General Manager Signalling: Simon Stockwell
Human Resources Director: Ian Home
Operations Director: Paul Holland
Special Tracks Systems: Vernon Turnbull

Background
In August 2005 Balfour Beatty acquired Pennine Group, the UK ground engineering specialists.

Products
Design, supply, installation, testing and commissioning of railway infrastructure, including track, overhead line systems, traction power supplies and telecommunications for high-speed, mixed traffic, heavy haul, light rail and mass transit systems. The company provides multidisciplinary project implementation including project management design and construction.

Contracts
Current contracts include the design, supply and installation of multi-disciplinary rail infrastructure

works for Heathrow Terminal 5 and Watford–Bletchley, UK; the design, supply and installation of overhead line and power supply on the West Coast Main Line, UK; supply, installation and commissioning of the entire section of new track including the power rail for the first steel-wheeled tram on the Santiago metro system, Chile.

Previous contracts have included the design, supply and installation of trackwork, signalling, telecommunications, overhead line, civil works and power supply of the Euston remodelling, London UK; design, supply and installation of the overhead line equipment and the supply and installation of trackwork for the Hong Kong Mass Transit, Lantau and Airport Railway Project; design and construction of the Changi Airport Line, Singapore.

UPDATED

Bombardier Transportation

Total Transit Systems
1501   Lebanon   Church   Road,   Pittsburgh, Pennsylvania, US
Tel: (+1 412) 655 57 00   Fax: (+1 412) 655 58 60
Web: www.transportation.bombardier.com

Key personnel
President, Total Transit Systems: Raymond T Betler

Products and Services
Working in partnership with major international, regional and local civil engineering and construction companies, as well as local suppliers, Bombardier Transportation develops, designs, integrates, installs and delivers the industry's broadest range of technologies – from large-scale urban transit systems to automated people movers.
- Automated people mover systems
- Automated monorail systems
- Automated Advanced Rapid Transit (ART)
- Light rail transit systems
- Metro/rapid transit systems
- Operations and maintenance services
  Bombardier, CITYFLO, Innovia and CX-100 are trademarks of Bombardier Inc or its subsidiaries.
  SkyTrain is a trademark of BC Transit Corp.

Contracts
Canada: The Millennium Line, an extension to Vancouver's fully automated SkyTrain system, opened in August 2002. Bombardier Transportation completed the design, supply and installation of the electrical and mechanical systems for the 20.5 km line. Bombardier's scope of work included automatic train control and communication systems, power supply and distribution system, trackwork, power rail, platform and guideway intrusion detection systems, system engineering and integration, and testing and commissioning for 20.5 km of dual-track guideway. Under a previous contract, Bombardier also supplied 60 ART MK II SkyTrain vehicles, employing Bombardier's LIM technology. Bombardier had previously supplied the original Vancouver Expo Line, as well as 150 ART MK I vehicles.

China: In May 2005 Bombardier was awarded a contract for a 2-km automated people mover system for the Beijing Capital International Airport. Bombardier will be responsible for project management, systems engineering and integration, testing and commissioning, in addition to the design and supply of the 11 CX-100 vehicles, the Bombardier CITYFLO 550 automated train control technology, communications systems, platform screen doors, switches, and equipping the maintenance facility. As part of the Airport's major expansion project, the new APM system will operate on dual-guideway between two stations and will serve the new national terminal, designated T3. Designed to carry 4,100 passengers per hour per direction, the CX-100 vehicles will be capable of operating as single-car trains or in consists of up to four-car trains. The system is scheduled to open in late 2007.

Korea, South: The Young-In LRT consortium, of which Bombardier is the lead member, was awarded a 35 year Build-Transfer-Operate (BTO) concession contract for a fully automated 18.5 km ART system by the city of Young-In, Republic of Korea. Intended to serve 15 stations on mostly elevated double-track guideway, the new line will link the Seoul subway, via Young-In City, and will terminate in Everland, one of the world's most popular theme park. Bombardier is responsible for the design and supply of 30 driverless ART LIM powered vehicles, the Bombardier CITYFLO 650 automatic train control technology, communications systems, project management, systems engineering and integration, testing and commissioning, as well as up to 30 years of operation and maintenance services. Bombardier will lead the integration of the scope of work with Daelim Industrial Co, who will be the co-lead for the design-build portion of the contract. The Yong-In system is scheduled to enter revenue service in 2009 and is projected to carry 13 million passengers annually.

Portugal: In 1998 Bombardier Transportation, as a member of the Normetro Consortium, was responsible for the delivery of a full turnkey, 71.6 km light rail system under contract to the city of Porto, Portugal. Bombardier is responsible for the design, supply, installation and testing and commissioning of 72 light rail vehicles, signalling system, and depot/workshop equipment. When completed, the system will feature 66 above ground and 11 underground stations. Phase 1 opened in December 2002, Phase II in 2004 with the entire network scheduled to open in 2006.

South Africa: In 2005 the Bombardier-led Bombela consortium was named preferred bidder for a concession to design, build, operate and maintain the Gautrain Rapid Rail Link. Planned for completion in 2010, the 80 km electrified standard-gauge system will link Johannesburg, Tshwane (Pretoria) and Johannesburg International Airport, serving 10 stations. Subject to final contract, expected in 2006, Bombardier will be responsible for the core electrical and mechanical systems, including a fleet of Electrostar emus, CITYFLO 250 train control technology, power supply and distribution systems, communication systems, automatic fare collection and trackwork and maintenance equipment. Also provided by the company will

Bombardier advanced rapid transit system for New York's JFK International Airport, US   0583062

Fully automated advanced rapid transit system for Vancouver SkyTrain, Canada 0583070

An automated people mover system for San Francisco International Airport was completed by Bombardier in 2003 under a turnkey contract that includes operation and maintenance for three years 0583065

Bombardier light rail system for Eskişehir, Turkey **NEW**/0585265

be project management, systems engineering and integration, testing and commissioning.

Spain: Bombardier Transportation is supplying the first application of automated people mover technology in Spain at the Barajas International Airport in Madrid. The system will connect a new midfield terminal with a new satellite terminal. The project includes the supply of 19 vehicles, 2.7 km of underground guideway running surface, and the CITYFLO automatic train control and power distribution systems. The system is scheduled to begin operation in early 2006.

Taiwan: Bombardier Transportation is supplying a 15-km rapid transit system in the city of Taipei. As prime subcontractor to Kung Sing Engineering Corporation (KSECO), a Taiwanese construction company, Bombardier is designing and supplying all of the system-wide electrical and mechanical elements for the new Neihu Line, an extension of the existing Muzha Line. The contract includes 202 rubber-tired vehicles and the ATC retrofit of 102 existing vehicles, as well as upgrading the Muzha Line Control Centre. Following the deployment of Bombardier *CITYFLO* 650 automatic train control technology, both the new fleet and the original fleet will be capable of operating interchangeably on both lines. Passenger service is scheduled to begin in 2008.

Turkey: In 2002, a consortium consisting of Bombardier Transportation and engineering group Yapi Merkezi of Turkey, was awarded a contract to supply a 15 km LRT system for the City of Eskisehir. Bombardier was responsible for the design, supply and installation of the turnkey system, including power supply, communications, traffic light and switch control systems, as well as operations and maintenance support for the first year of operations. Eighteen 100 per cent low-floor Bombardier FLEXITY Outlook trams were also provided. The network is designed to handle around 170,000 passengers per day and opened in 2004.

UK: As a member of the Arrow Light Rail Ltd Concession Company, Bombardier Transportation was responsible for the landmark 30.5 year contract to design, build, operate and maintain the Nottingham Express Transit (NET) Line 1 light rail system. Bombardier's scope for the 14 km turnkey system included 15 low-floor light rail vehicles, project management, system engineering and integration, power supply and distribution system, signalling and system control, communications and security systems, ticketing equipment and depot maintenance equipment. The system opened in March 2004.

US: At Dallas/Fort Worth International Airport, Bombardier supplied the first application of its new Bombardier Innovia technology, with CITYFLO 650 automatic train control technology. The system connects the airport's existing terminals with a new terminal and a new parking garage. The 16 km dual-track guideway system includes 64 vehicles and a five-year system-wide maintenance contract. The system opened in May 2005.

In 2001, Bombardier was awarded the contract to expand its people mover system at George Bush Intercontinental Airport in Houston. The new extension connects the present system, originally supplied and operated by Bombardier, to the new international terminal complex. Ten new Bombardier CX-100 vehicles and 11 new switches were supplied for the 0.9 km extension as well as a new off-line maintenance facility with carwash, and a new power distribution system. The system uses CITYFLO 550 automatic train control technology and opened ahead of schedule in January 2005.

Bombardier Transportation, as the lead member of the Las Vegas Monorail Team, supplied its Monorail System in the heart of the resort corridor. Designed according to urban transit safety standards, the system links eight major resort properties and the Las Vegas Convention Center. The 36 cars operate in nine four-car trains and represent the latest innovation in monorail technology. Bombardier was responsible for providing all the electrical and mechanical elements of the system including design and supply of 36 monorail cars, overall project management, automatic train control, communications systems, power supply and distribution systems, automatic fare collection systems, guideway and guidance switching systems, system engineering and integration, platform doors for seven stations, testing and commissioning, training and manuals, and up to 15 years of operations and maintenance services. The system entered revenue service in 2004.

In 1999, Bombardier Transportation, as a member of the Southern New Jersey Rail Group, was awarded a contract to design, build, operate and maintain a unique 55 km (34 mile) turnkey system to operate between Trenton and Camden. Commissioned in 2004, this project – called the River LINE – was the first application of a diesel light rail transit system on an existing freight corridor in the US. Bombardier supplied all of the electrical and mechanical system elements including 20 diesel multiple-units and is providing 10 years of operations and maintenance.

Also in 1999, Bombardier was chosen to supply a replacement people mover system to Seattle-Tacoma International Airport. The original system, supplied to Seattle-Tacoma International Airport in 1973, consists of two transit loops connecting the north and south satellites with the main terminal. A separate shuttle system runs the length of the main terminal, linking the satellite loops, serving main terminal ticketing and providing access to the parking garages. For the new system, Bombardier supplied and installed 21 Bombardier CX-100 vehicles, an overlay of the Bombardier CITYFLO 650 train control technology, central control equipment, a power distribution system and station doors. The North Loop opened for passenger service in May 2003; the South Loop and Shuttle lines opened ahead of schedule in November 2003.

As part of the AirRail Transit Consortium, Bombardier Transportation supplied a fully automated rapid transit system for the John F Kennedy International Airport, New York, US. Under contract to the Port Authority of New York and New Jersey, using a design-build-operate and maintain approach, the consortium was responsible for the turnkey design and construction of the driverless light rail system, including 32 ART MK II LIM-powered vehicles, as well as operations and maintenance for a period of up to 15 years. The system opened in December 2003.

Bombardier supplied an APM system to the San Francisco International Airport that links a new international terminal with existing domestic terminals, parking garages and the airport's rapid transit station. The system includes 38 CX-100 vehicles, nine stations and 10km of elevated guideway, as well as operation and maintenance services. The system opened for passenger service in March 2003.

UPDATED

Chemetron Railway Products Inc

177 West Hintz Road, Wheeling, Illinois 60090, USA
Tel: (+1 847) 520 54 54   Fax: (+1 847) 520 63 73
e-mail: cttsales@aol.com

Key personnel
See main entry in *Permanent way components, equipment and services* section.

Projects
Turnkey design of rail welding plants. Contract welding service of rail into continuous welded rail in any of standard, alloy or head-hardened rails.

Dimetronic SA

Avda de Castilla 2, Parque Empresarial, E-28830 San Fernando de Henares, Spain
Tel: (+34 91) 675 42 12   Fax: (+34 91) 756 21 15
e-mail: marketing@dimetronic.es

Projects
Most of the projects contracted are of a turnkey nature.

Interinfra

Parc Dhalenne, 2 rue Albert Dhalenne, F-93400 St Ouen Cedex, France
Tel: (+33 1) 41 66 84 15   Fax: (+33 1) 41 66 84 62

Key personnel
Chairman and Chief Executive Officer:
    Charles Carlier
Deputy Managing Director: Henry Bussery
General Secretary: Bertrand Dupuy

Projects
Turnkey contracts for the supply of railway and rapid transit systems outside France.

Marconi Communications

Marconi Communications Limited
New Century Park, PO Box 53, Coventry CV3 1HJ, UK
Tel: (+44 24) 76 56 55 00   Fax: (+44 24) 76 56 58 88
e-mail: pete.hallard@marconicoms.com
Web: http://www.marconicomms.com

Key personnel
See main entry in *Signalling and communications systems* section.

Projects
Marconi Communications Strategic Networks provides communication solutions for the rail, metro and light rail market worldwide. Marconi Communications designs, supplies, maintains and manages complex transportation communication networks.

The business has specialist knowledge in system design, integration, implementation and project management of specialist communication projects.

An example of the company's integration capability is the development of a communications system controlling telecommunications public address, CCTV and passenger information displays via a single screen.

Marconi Communications (Customer Network Services) is a whole life service provision business targetting communication networks and control systems in the transportation sector.

Strategic networks is providing communications for Hong Kong's Lantau Airport rail link, London Underground Ltd Jubilee Line Extension and Northern Line rolling stock communications infrastructure and maintenance, and for Midland Metro Line 1.

Parsons Brinckerhoff Inc

Headquarters
One Penn Plaza, New York, New York 10119, USA
Tel: (+1 212) 465 50 00   Fax: (+1 212) 465 50 96
e-mail: pbinfo@pbworld.com
Web: http://www.pbworld.com

Key personnel
President: Thomas J O'Neill
Chairman: Morris S Levy
Controller: Richard A Schrader

Projects
Parsons Brinckerhoff is providing EMCOR Rail with mechanical and electrical services design, tunnel ventilation design, safety assurance and RAMS for contract 588 on the Channel Tunnel Rail Link Project (CTRL). This is a four-year project as part of an integrated EMCOR/Rail/Rail Link Engineering project team.

As a subcontractor for the Tren Urbano system, San Juan, Puerto Rico, Parsons Brinckerhoff is providing programme management and design services for the systems and test track turnkey contract; the company is also involved in the West Coast modernisation project for Network Rail, and Thameslink 2000.

Ranalah

Ranalah Moulds Ltd
New Road, Newhaven BN9 0EH, UK
Tel: (+44 1273) 51 46 76   Fax: (+44 1273) 51 65 29

Key personnel
See main entry in *Permanent way components, equipment and services* section.

Projects
Turnkey service for supply of sleeper manufacturing plant.

Contracts
Contracts include the supply of sleeper plants to Hong Kong, Philippines, Thailand, Georgia, Turkey, Sudan, Malaysia; also to Bulgaria, Europe and the UK.

Rotem Company

Headquarters
Landmarktower, 837-36, Yeoksam-dong, Gangnam-gu, Seoul, 135-937, South Korea
Tel: (+82 2) 21 12 82 94
Fax: (+82 2) 21 12 98 73
Web: www.rotem.co.kr

Key personnel
Vice Chairman and Chief Executive Officer:
    Soon-Won Chung
Senior Executive Vice-President: Yeo-Sung Lee
Executive Vice-President: Jae-Hong Kim

Background
Established in 1964 when Daewoo Heavy Industry started manufacturing rolling stock, followed by Hyundai and Hanjin Heavy Industry a few years later. In 1999, the three companies were consolidated into KOROS. Hyundai Motors Group acquired the share of Daewoo in 2001 and the former company KOROS became Rotem. Rotem is now an affiliate of Hyundai Motors Group and has its headquarters in Seoul and two factories in Uiwang and Changwon. Uiwang has capacity for manufacturing 500 emus per year and has the capability to manufacture electric equipment such as traction motors, SIV inverters etc. The research and design centre is also located in Uiwang. The Changwon factory has capacity to manufacture 700 emus per year. Rotem has an annual capacity to manufacture approximately 1,200 emus. Certifications such as the ISO 9001 certificate for Quality, 14001 for environment and 18001 for occupational health and safety management have been acquired at all three sites.

Projects
Currently Rotem is engaged in metro rail system turnkey projects and light rail system turnkey projects in Korea and overseas countries. Rotem's experience in the total rail system technology is based on 30-years of rolling stock production and supply and rail systems interface experience between vehicles and wayside systems. Rotem supplies electrical and mechanical (E&M) systems and holds full responsibility in project management and systems engineering/integration. Rotem also provides operations and maintenance (O&M) services ranging from light rail to metro rail systems with efficient management and high maintenance technology. Rotem also maintains good partnerships with major international and domestic civil construction companies and E&M subsystem suppliers.

Contracts
Rotem is participating in the Kimhae LRT project in Korea. The project is expected to start commercial service in January 2010. Rotem will provide detailed design, supply and installation of the E&M system including the supply of 50 cars (LRT), signalling systems, electric systems, telecommunication systems, AFC systems, platform screen doors, elevator and escalator maintenance facilities and systems engineering.

UPDATED

SAT

Part of the SAGEM Group
Network and Telecommunications Division
11 rue Watt, F-75626 Paris Cedex 13, France
Tel: (+33 1) 55 75 75 75
Fax: (+33 1) 55 75 30 94

Key personnel
See also main entry in *Signalling and communications systems* section.

Projects
Engineering and network design and turnkey networks.

Scitel Telematics Ltd A

Árbóc u 6, 3rd Floor, H-1133 Budapest, Hungary
Tel: (+361 359) 98 77   Fax: (+361 359) 98 70
e-mail: marketing@transport.alstom.hu

Key personnel
See main entry in ALSTOM Signaling *Signalling and communications systems* section.

Projects
Turnkey projects comprising design, supply, installation, commissioning and maintenance of signalling, telecommunications and vehicle identification systems, train traffic monitoring and control systems from stations to regions with schedule planning, staff assignment and passenger information.

Siemens AG

Transportation Systems
Siemens AG
Transportation Systems
PO Box 910220, D-12414 Berlin, Germany
Tel: (+49 30) 386 50
Fax: (+49 30) 386 514 31
e-mail: turnkey.transportation@siemens.com
Web: www.siemens.com/transportation

Corporate headquarters
Siemens AG, Transportation Systems, PO Box 3240, 91050 Erlangen, Germany
Tel: (+49 9131) 7-0

Key personnel
President, Transportation Systems Group:
  Hans M Schabert
Group Vice Presidents: Alfred Frank; Joern F Sens;
  Friedrich Smaxwil
Heads of Turnkey Systems Division: Erich Kaeser;
  Dr Ewald Feidner

Services
Turnkey projects including: new construction or refurbishment of main line systems; intercity and high-speed systems; commuter and express rail links; mass transit systems; light rail systems; automated guided transit systems; and maglev systems. Capabilities include:
- development of technical concepts including operational concepts, line plans, specifications, RAMS and EMC studies;
- provision or procurement of: civil engineering; infrastructure; traction power supply and distribution; control systems; signalling and safety systems; telecommunications systems; rolling stock; automated fare collection systems; trackwork; building services including fire alarm and fire-fighting systems; lifts and escalators; signage; UPS; testing, commissioning and training services; and operation and maintenance;
- project management functions include project planning; project control; project supervision; and system integration and co-ordination.
Siemens also arranges financing of turnkey projects.

Projects
Recent or current projects include:

China: As an equal partner in the Transrapid International consortium, Siemens is participating in the fulfilment of a contract to build and equip a 30 km magnetic levitation line linking Shanghai's Pudong international airport with the city's Lujiazui financial district. The system was commissioned in December 2003, with maglev trains operating at speeds of up to 430 km/h to complete the journey in seven minutes.

France: In May 2001 Matra Transport International (now Siemens Transportation Systems SAS) was awarded a contract by SMTC, Toulouse, to supply and equip its second automated metro line, the 16 km 20-station Line B. The contract includes the supply of 35 Type VAL 208 vehicles. The company is working with the Toulouse metro operator, SMAT, as system integrator on the project, which is due to be completed in 2007.

Germany: For the 204 km Cologne Rhein/Main high-speed line, which was commissioned in 2002, Siemens acted as consortium leader for the equipment technology group, which covered: project management; electronic interlocking; overhead contact line; telecommunications and remote monitoring systems; and the implementation of a tunnel rescue system.

In November 2001 Siemens was awarded a contract by Verkehrs AG (VAG), Nuremberg, to equip the authority's Line U3, then under construction, for automated, driverless operation employing Siemens' AGT system. In addition, Siemens was to retrofit Line U2 for similar operation. The contract included the supply of 30 Type DT 3 two-car trainsets, 16 for Line U3 and 14 for Line U2. The automatic train protection and operations control system was to be designed and manufactured at Siemens' Braunschweig facility.

Italy: In November 2001, as a member of the project company Val 208 Torino GEIE, Siemens received an order from SATTI, Turin (now Gruppo Trasporti Torinese), to supply 46 Val 208 automated metro trainsets. This followed the award of a contract in July 2000 to Siemens subsidiary Matra Transport International (now Siemens Transportation Systems SAS) to supply and equip a 9.6 km 15-station Val automatic rapid transit system for the city.

Malaysia: In April 2002 the Express Rail Link system connecting Kuala Lumpur with its new airport, was commissioned. As leader of the SYZ Consortium, Siemens was responsible to the line's concessionaire, ERL SB, for track, signalling and train control systems, traction power supply and overhead line equipment, telecommunications systems, E & M construction, SCADA equipment, depot and workshop facilities and rolling stock. The last-mentioned took the form of 12 articulated four-car Desiro ET emus. Siemens is also the majority partner in EMAS, which is responsible for maintenance of the entire system.

Netherlands: In 2001, as a partner in the 'Infaspeed' concession, Siemens won a contract to supply the power supply and distribution system, the ETCS Level 2 signalling system, communications systems and ancillary equipment, as well as maintenance over a 25-year period, for the new high-speed line between Amsterdam and the Belgian border (HSL-Zuid). Siemens is also leading systems integration for the high-speed link.

Portugal: In August 2002 Siemens secured a turnkey contract to build and equip an initial 13 km phase of a light rail system for Metro Transportes do Sul (MTS) linking the communities of Almada and Seixal, south of Lisbon. MTS has a concession to establish the system and operate it for 27 years. Siemens' contract covers: the supply of 24 five-section Combino low-floor LRVs; complete signalling and operations control equipment; a communications system; traction power supply equipment and overhead contact lines; and the equipment for a vehicle maintenance and repair facility. A 20 km extension to the system is projected.

Shanghai maglev system installed by the Transrapid International consortium, of which Siemens is a member   ***NEW**/0585297*

Siemens TS led a turnkey consortium to build and equip the Express Rail Link airport railway in Kuala Lumpur and is a majority partner in the company that maintains it   ***NEW**/0585283*

Taiwan: In August 2001 Siemens won a contract from the Kaohsiung Rapid Transit Corporation (KRTC) to supply signalling equipment, traction power supply equipment and 42 three-car trainsets for the Red Line (28 km, including 9 km elevated, 23 stations) and Orange Line (14 km, 14 stations) of the city's metro system. Siemens was also to be responsible for project management and systems integration of the electromechanical portion of the project. Services are due to commence in 2007.

Thailand: In January 2005 Siemens, together with consortium partners B Grimm and Sino Thai Engineering and Construction plc (STECON) signed a contract with State Railway of Thailand for the construction of the 28 km Suvarnabhumi Airport Rail Link and the City Air Terminal. Both Siemens and B Grimm are taking responsibility for the design, supply, installation and project management of the whole electrical and mechanical system including trackwork, rolling stock, signalling system, power supply, communication systems, automatic fare collection, tunnel equipment, depot and workshop equipment, check-in facilities as well as baggage handling system.

In January 2002 Siemens announced that it had signed an agreement with Bangkok Metro Corporation Ltd for the supply and maintenance of the Thai capital's first metro system, a 20 km line with 18 stations. The project entails supply of the line's complete infrastructure, including signalling, power supply, communications and depot equipment, as well as the manufacture of 19 three-car trainsets to serve the line. The contract also covers project management and maintenance of the line over a 10-year period. The line was due to be commissioned in 2004.

USA: In March 2001 Siemens was awarded a contract by the Metropolitan Transit Authority of Harris County to construct Houston's first light rail system. Siemens was responsible for project management, planning, delivery, installation and commissioning of the 11 km line, which was commissioned in January 2004. This included the manufacture and supply of 18 S 70 LRVs, and of the line's signalling, control and traction power supply systems.

Venezuela: In July 2000 Siemens announced that it had won a turnkey contract to build the initial phase of the first light rail line in Maracaibo. Siemens was to supply infrastructure equipment, including signalling and communications systems, traction power supply equipment, ticketing systems and a maintenance depot for the 6.9 km line, which will connect the city with its airport and will eventually total 14 km. In addition, the company was to supply 12 LRVs to serve the line.

UPDATED

Técnicas Modulares e Industriales (Temoinsa)

Polígono Industrial Congost, Avenida San Juliá 100, E-08400 Granollers, Barcelona, Spain
Tel: (+34 93) 860 92 00   Fax: (+34 93) 860 92 13
e-mail: tmi@temoinsa.com
Web: http://www.temoinsa.com

Key personnel
Chairman: Alvaro Colomer
Subsidiaries Manager: Miguel de Sagarra
Chief Executive Officer: Mercé Sala Schnorkowski
Front Office Manager: Antonio Fábregas

Projects
Turnkey projects for complete interiors of new vehicles and refurbishment.

Transrapid International GmbH & Co KG

Pascalstrasse 10f, D-10587 Berlin, Germany
Tel: (+49 30) 39 84 30   Fax: (+49 30) 39 84 35 99
e-mail: transrapid.international@tri.de
Web: http://www.transrapid.de

Two Transrapid vehicles in front of depot on the Shanghai system   1036751

Other offices
Transrapid International-USA, Inc
400 Seventh Street NW, Fourth Floor, Washington DC 20004, USA
Tel: (+1 202) 969 11 00   Fax: (+1 202) 969 11 03
e-mail: info@transrapid-usa.com
Web: http://www.transrapid-usa.com

Transrapid International PR China Representative Office
Chaoyangmenwai Avenue 16, 22/F, China Life Tower, Chaoyang District, 100020 Beijing, China
Tel: (+86 10) 85 25 29 99   Fax: (+86 10) 85 25 21 70
Web: http://www.transrapid.com.cn

Key personnel
Managing Directors: Hans-Juergen Petersen (speaker), Hans Georg Raschbichler, Horst Engels

Background
In May 1998, Transrapid International (TRI) was founded as a joint company located in Berlin. Siemens and ThyssenKrupp each have an equal share in the company.

The activities of TRI include marketing the Transrapid System worldwide, the system engineering for projects, and the preparation of offers in cooperation with the parent companies, Siemens and ThyssenKrupp. At the realisation phase, TRI takes over the project management and during later operation, the maintenance management.

The TRI markets the Transrapid System worldwide, specifically in Mainland China, Germany, Netherlands and the USA.

Products
The Transrapid System is a track-bound mass transportation system, suitable for long-distance service, regional networks, and airport-to-city connection at speeds of more than 500 km/h.

The Transrapid system comprises aircraft-style trains running on a dedicated track which can be installed at grade or elevated.

Transrapid vehicles are levitated and guided without contact using magnetic force and a synchronous longstator linear motor for propelling and braking. Transrapid vehicles consist of two to 10 sections and are capable of carrying over 1,100 passengers or 175 tons of cargo. The Transrapid system can cope with up to 10 per cent gradients and a cant of 12° (16° in special cases) in curves.

Transrapid vehicles can accelerate to 300 km/h in less than 100 seconds (4.2 km) and a two-minute station stop only adds four minutes to the total travel time.

The 31.5 km test track and facility has been in operation in northeast Germany since 1984.

The Transrapid maglev system in Shanghai successfully completed its maiden trip on 31 December 2002. The world's first commercial Transrapid line started operating in Shanghai in December 2003. The first maglev system within Germany is coming closer to realisation: the Transrapid route between Munich central station and the Franz-Josef Strauss airport is now in the planning stage and services are scheduled to commence in 2009/10.

UPDATED

Transys Projects Ltd

2 Priestley Wharf, Holt Street, Aston Science Park, Birmingham B7 4BN, UK
Tel: (+44 121) 359 77 77   Fax: (+44 121) 359 18 11
e-mail: info@transysprojects.ltd.uk
Web: www.transysprojects.ltd.uk

Key personnel
Managing Director: Jeremy J Ashley
Sales and Marketing Director: Kevin Lane
Engineering Director: Karl J Barras
Financial Director: Dorothy J Lidster

Background
Transys Project Ltd is an independent company.

Capabilities
Multi-disciplined engineering consultancy and turnkey solutions provider, covering all aspects of passenger rail vehicles and their related support services. This covers mass transit vehicles, light rail vehicles and tramcars as well as main line diesel and electric multiple units and passenger coaches. Certified to BS EN ISO 9001 with 'link-up' accreditation in 26 relevant areas.

Specific capabilities include: complete turnkey service for traction and rolling stock; refurbishment and enhancements; modifications and reliability improvements; roll out of modifications and installations such as TPWS, OTMR, GSMR, ERTMS, New WSP and train sander equipment.

Projects
Turnkey installation of on-train monitor record for six UK fleets. Turnkey installation of over 700 sets of trainborne sanders to combat low adhesion in leaf-fall seasons. Turnkey installation of new public address and power outlets for computers. Installation of train protection warning systems. Complete control system rewire. Turnkey installation of new wheel-slide protection equipment. Numerous projects for reliability and passenger comfort enhancement.

UPDATED

INFORMATION TECHNOLOGY SYSTEMS

Alphabetical listing

Andersen Consulting
Bentley Systems Inc
Cadex Electronics Inc
Cityway SA
Com-Net Software Specialists Inc
Datastream Systems (UK) Ltd
Dilax (International) AG
EDB Unigrid AB
European Rail Software Applications (ERSA)
Fleet Software
Fuel Conservation Technologies
Goal Systems
ICL
Infodev Inc
InfoMill Ltd
InfoVision Systems Ltd

Infrasoft Corporation
Infrasoft Ltd
Innovata
Interautomation
Institut für Regional-und Fernverkehrsplanung (IRFP)
Laser Rail Ltd
Maxwell Soft Park Ltd
MultiModal Applied Systems
Onerail
PAFEC Ltd
Pallas Informatik A/S
PC-Soft GmbH
PSI Transportation GmbH
Rail Management Consultants
Railtech Solutions Ltd

RMI
Sabre Inc
Science Systems Group
Siemens Transit Telematic Systems AG
Socratec GmbH
Spear Technologies Inc
STERIA Group
Systra Consulting Inc
Timera Inc
Trivector System AB
Union Pacific Technologies
VIPS AB
Vossloh Information Technologies GmbH
Vossloh Information Technologies York Ltd
ZT

Company listing by country

AUSTRALIA
Onerail

CANADA
Cadex Electronics Inc
Fuel Conservation Technologies
Infodev Inc

DENMARK
Pallas Informatik A/S

FRANCE
Cityway SA
European Rail Software Applications (ERSA)
STERIA Group

GERMANY
Institut für Regional-und Fernverkehrsplanung (IRFP)
Interautomation
PC-Soft GmbH
PSI Transportation GmbH
Rail Management Consultants

Socratec GmbH
Vossloh Information Technologies GmbH

INDIA
Maxwell Soft Park Ltd

SPAIN
Goal Systems

SWEDEN
EDB Unigrid AB
Trivector System AB
VIPS AB

SWITZERLAND
Dilax (International) AG
Siemens Transit Telematic Systems AG

UNITED KINGDOM
Datastream Systems (UK) Ltd
Fleet Software
ICL
InfoMill Ltd

InfoVision Systems Ltd
Infrasoft Ltd
Laser Rail Ltd
PAFEC Ltd
Railtech Solutions Ltd
Science Systems Group
Vossloh Information Technologies York Ltd

UNITED STATES
Andersen Consulting
Bentley Systems Inc
Com-Net Software Specialists Inc
Infrasoft Corporation
Innovata
MultiModal Applied Systems
RMI
Sabre Inc
Spear Technologies Inc
Systra Consulting Inc
Timera Inc
Union Pacific Technologies
ZT

Andersen Consulting

100 S Wacker Drive, Suite 1070, Chicago, Illinois 60606, USA
Tel: (+1 312) 507 29 00   Fax: (+1 312) 507 79 65
Web: http://www.ac.com

Key personnel
See main entry in *Consultancy services* section.

Capabilities
Andersen Consulting Transportation and Travel Services Group services include IT strategy, customer service, network design, information systems, operations, service delivery, service design and financial/accounting systems.

Projects
Andersen Consulting Transportation and Travel Services include many of the world's major railway operators.

Bentley Systems Inc

685 Stockton Drive, Exton, Pennsylvania 19341-0678, USA
Tel: (+1 610) 458 50 00   Fax: (+1 610) 458 10 60
Web: http://www.bentley.com

Key personnel
Chief Executive Officer: Greg Bentley

International headquarters:
Bentley Systems Europe
Wegalaan 2, NL-2132 JC Hoofddorp, Netherlands
Tel: (+31 23) 556 05 60   Fax: (+31 23) 556 05 65

Products
Software solutions for architecture, engineering and construction. Rail-specific software includes InRail, part of Bentley's InRoads suite, which provides track design and layout functions applicable to high-speed, conventional heavy rail or light rail systems. Functionality includes: regression points; curvature diagrams; horizontal regression analysis; vertical regression analysis; slew diagrams; a horizontal connection editor; layout of turnouts; cant; interactive geometry; alignment by elements; coordinate geometry; and feature-based digital terrain model generation.

Bentley's portfolio also includes: EED (Elementary Electrical Diagrams) Software, from which was developed EED Signal Relay, which is used to support the maintenance of existing signalling systems and in the design of new signalling schemes; and MicroStation V8, which allows users to create 3-D models of permanent assets.

Projects in which Bentley software solutions have been used include the Madrid–Seville high-speed line in Spain, the Alameda Corridor project linking Los Angeles with Long Beach, California, and London Underground's Jubilee Line Extension.

Cadex Electronics Inc

22000 Fraserwood Way, Richmond, British Columbia V6W 1J6, Canada
Tel: (+1 604) 231 77 77   Fax: (+1 604) 231 77 55
e-mail: info@cadex.com
Web: http://www.cadex.com

Key personnel
President and Chief Executive Officer:
   Isidor Buchmann
Accounts Manager, Export: Richard Janzen
Marketing Specialist: Andrew Green

Products
Cadex C7000 series battery analysers service and restore batteries as used in portable communications equipment, computers and data acquisition devices; battery-specific adapters for all common batteries; multiple batteries serviced with the FlexArm™; user-selectable programmes for all battery needs such as the QuickTest™ programme which supports all common battery chemistries. Cadex analysers print service reports and battery labels and interfacing with a PC is possible with BatteryShop™. Cadex Universal Conditioning Chargers (UCCs) supplied in one-, two-and six-bay models. The chargers accommodate Lithium Ion/Polymer, Nickel Cadmium and Nickel Metal Hydride batteries.

Windows® application software is available to obtain data from up to 120 C7000 series analysers. Battery test results, inventory status and performance graphs are stored in a database from which print reports can be generated.

Cityway SA

Parc du Golf, Bâtiment 7, F-13856 Aix en Provence Cedex 03, France
Tel: (+33 4) 42 37 18 40   Fax: (+33 4) 42 39 45 15
e-mail: info@cityway.fr
Web: http://www.cityway.fr

Key personnel
Managing Director: Laurent Briant
Sales Manager: Olivier Marrone
Office Manager: Chloé Spano

Background
A majority shareholding in Cityway is held by Connex, the transport subsidiary of Vivendi Environnement.

Products
Transport information management tools, including: Go@T, a multimodal trip planner search engine; Transinfo, a website design tool for public transport operators; Know Now, providing real-time traffic disruption management; Contakt, a call centre management tool; Rapido, a management tool for timetable and stop-points mapping; and Guid'Edit, for editing bus guides and timetables.

Com-Net Software Specialists Inc

3728 Benner Rd, Miamisburg, Ohio 45342, USA
Tel: (+1 937) 859 63 23   Fax: (+1 937) 859 75 11
Web: http://www.comnet-fids.com

Key personnel
Systems Integration Account Executive:
   Linda Palmer

Products
Provision of turnkey integrated information display systems, including real-time information display.

Com-Net's WTI (Windows on Transit Information System) is an integrated information display system based on experience from more than 250 installations.

Other products include video monitors, LED and LCD display systems and outdoor LCD display systems.

Datastream Systems (UK) Ltd

1210 Parkview, Arlington Business Park, Theale, RG7 4TY, UK
Tel: (+44 1189) 65 77 42   Fax: (+44 1189) 65 77 41
e-mail: info@dstm.co.uk
Web: http://www.datastream.net

Head office
50 Datastream Plaza, Greenville, South Carolina 29605, USA
Tel: (+1 864) 422 50 01   Fax: (+1 864) 422 50 00
Web: http://www.datastream.net

Products
Asset management systems for rail applications, covering both infrastructure and rolling stock. Users of the company's MP5i system include: London Underground Ltd; Glasgow Underground; Angel Trains; FirstGroup; Banestyrelsen, Denmark; EMEF, Portugal; Netherlands Railways; BASF, Germany; and RATP, Paris.

Contracts
A contract with Romanian Railways covered the provision of an MP5i system to manage 250 infrastructure sites on the country's network as part of the World Bank-funded IRIS project.

Dilax (International) AG

Fidlerstrasse 2, CH-8272 Ermatingen, Switzerland
Tel: (+41 71) 663 75 75   Fax: (+41 71) 663 75 76
e-mail: info@dilax.ch
Web: http://www.dilax.ch

Key personnel
Contact: Iris Bährle

Subsidiary company
Dilax Intelcom GmbH
Schillerstrasse 3, D-10625 Berlin, Germany
Tel: (+49 30) 77 30 92 40   Fax: (+49 30) 77 30 92 50

Products
Design manufacture and supply of automatic systems for passenger counting and operations analysis. Systems are supplied for both fixed and vehicle-mounted applications. Operators supplied with Dilax and automatic passenger counting systems include: Hamburger Hochbalm; S-Bahn Hamburg and AKN-Eisenbahn, Germany; SBB and TPG, Switzerland; and Muni, San Francisco, USA.

EDB Unigrid AB

Box 40, SE-171 11 Solna, Sweden
Tel: (+46 8) 762 50 00   Fax: (+46 8) 762 40 40
Web: http://www.edbunigrid.com

Key personnel
Managing Director: Björn Nilsson

Background
Formerly a division of Swedish State Railways and established as a limited company. The Norwegian IT group EDB Business Parter acquired EDB Unigrid AB in November 2001, since when it has been the platform for further expansion in the Swedish market.

Services
EDB Unigrid AB offers services with IS outsourcing. The portfolio of services includes the operation of applications, client terminals, networks, central servers and mainframe, related to business-critical systems and applications.

European Rail Software Applications (ERSA)

84 route de Strasbourg, BP 273, F-67504 Haguenau Cedex, France
Tel: (+33 3) 88 07 15 50   Fax: (+33 3) 88 07 15 51
e-mail: info@ersa-france.com
Web: http://www.ersa-france.com

Key personnel
Technical Director: Patrick Deutsch

Products
European Rail Software Applications is a 60 per cent subsidiary of ERS (qv in *International Railway Associations and Agencies* section), specialising in the development and adaptation of simulation software for railway applications. ERSA also maintains and supports other software packages.

Fleet Software

3 Newton Business Centre, Thorncliffe Park Estate, Chapeltown, Sheffield S30 4PH, UK
Tel: (+44 114) 257 16 00   Fax: (+44 114) 257 16 09

Key personnel
Managing Director: John Rands

Products

TACT, software for rolling stock maintenance in the UK. It records individual parts performance and management of reliability issues. TACT software has been approved for use across the networks of National Express Group and Prism. TACT was first introduced to Supertram, Gatwick Express and in 1996 to Heathrow Express. It has now been extended to rolling stock manufacturers.

TACT technology was originally developed around the privatised bus industry when engineering costs became an issue.

Fuel Conservation Technologies

3 Jacks Round, Stouffville, Ontario L4A 1L6, Canada
Tel: (+1 416) 985 68 63
e-mail: fctechnologies@sympatico.ca
Web: http://www.fueloptimizer.ca

Key personnel

President: Phil Reid

Products

Locomotion Management Systems is a suite of programmes incorporating the following modules: Fuel Optimizer, which combines locomotive engine performance data with map characteristics for the route to be travelled to optimise fuel consumption when selecting traction; Train Simulator, which determines fuel consumption for trains by using predefined Train Route Records; Trip Records, which provides for the storage of trip data and fuel used; Train Manager, which is used to edit Train Route records; and Train Scheduler, used for the scheduling and dispatching of trains and for determining fuel consumption and tax payable on it. Mapping data of all railroads in North America has been compiled to the Fuel Optimizer and Train Simulator modules to function.

Goal Systems

Julio Camba 1, 3 Ofic 2, E-28028 Madrid, Spain
Tel: (+34 91) 725 30 00   Fax: (+34 91) 725 56 08
e-mail: goal@goalsystems.com
Web: http://www.goalsystems.com

Key personnel

Research and Development Team Leader:
   Pasquale Iannelli

Products

IT solutions for bus and rail transport resource planning and optimisation. Software products include:

   GoalBus, with basic modules covering generation of bus timetables and services and optimum assignment of drivers to shifts;

   GoalDriver, with basic modules covering drivers' timetable planning and replanning of drivers' timetable periods, including responses to incidents;

   GoalRail, with basic modules covering timetable planning and rolling stock utilisation, driver and supervisor planning, allocation of rolling stock, drivers and supervisors, replanning of rolling stock, drivers and supervisors.

   Supplementary modules are available for each product to expand the scope of the system.

ICL

West Avenue, Kidsgrove, Stoke-on-Trent ST7 1TL, UK
Tel: (+44 1782) 78 14 44   Fax: (+44 1782) 78 14 55

Key personnel

Business Manager, Rail: Richard Betts

Works

Eskdale Road, Winnersh, Wokingham RG41 5TT

Products

Railway journey information systems, revenue collection and allocation systems, ATB2 ticket printers, materials and purchasing management systems, hand-held devices for penalty fares issue and other similar uses, possession management systems, outsourced and managed IT services, IT systems design and build.

   Contracts include London Transport PASS ticketing system, London Underground Possession Management, London Underground penalty fares, materials and purchasing, strategic consulting and IT operations, Gardermobanen IT ticketing, train movements, personnel, finance and information systems, Eurostar departure control system.

   Train companies in the UK are installing a high-technology journey information system which will provide improved customer service, giving more accurate and up-to-date information on train times, routes and fares. The train companies have signed a contract with ICL to develop a new Rail Journey Information Service (RJIS) which will be at the heart of the integrated transport network of the future.

   The system can also supply information about station and other facilities necessary to integrated transport, such as the availability of bus links, cycle facilities, disabled access, taxis and car parking drop-off points and detailed information about other public transport services such as tram and ferry timetables.

Infodev Inc

PO Box 1222 HV, Quebec QC, G1R 5A7, Canada
Tel: (+1 418) 681 35 39   Fax: (+1 418) 681 12 09
e-mail: info@infodev.ca

Key personnel

Vice-President and Director, Research and
   Development: Pierre Deslauriers
Chief Executive Officer: Alain Miville de Chene
Transit Sales: Sandra Howlett

Other offices

Infodev USA
7373 Newcrest Circle, Las Vegas, Nevada 89147-4935, USA
Tel: (+1 702) 889 67 01   Fax: (+1 702) 889 03 80
e-mail: info@infodev.ca
Web: http://www.infodev.us

Transit Sales
Tel: (+1 418) 681 35 39   Fax: (+1 418) 681 12 09
e-mail: info@infodev.ca

Products

Production of Automatic Passenger Counting systems (APC) and Automatic Vehicle Location systems (AVL) using directional optical sensors and also a GPS satellite positioning system applied to vehicles in order to identify and track their positions.

InfoMill Ltd

Lynton Mill House, Lynton Street, Derby DE22 3RW, UK
Tel: (+44 1332) 29 35 19   Fax: (+44 1332) 29 68 45
Web: www.infomill.com

Key personnel

Managing Director: Jonathan Ralphs
Sales and Marketing Director: Holger Levey

Other offices

North America
35 Corporate Drive, 4th Floor, Burlington, Massachusetts 01803
Tel: (+1 781) 685 49 02   Fax: (+1 781) 685 46 01

Europe
Pentagon House, Sir Frank Whittle Road, Derby DE21 4XA
Tel: (+44 1332) 25 31 70   Fax: (+44 1332) 29 53 60

Asia Pacific
Level 20, Tower 2, Darling Park, 201 Sussex Street, Sydney, NSW 2000
Tel: (+61 2) 90 06 16 58   Fax: (+61 2) 90 06 10 10

Capabilities

InfoMill supports the service management community with its unique parts identification technology running on PCs, handheld computers and smart phones. InfoMill's flagship product, PartsArena provides the field engineer with their technical support information, including illustrated parts catalogues with a user interface. This helps the engineer to quickly and accurately select the required part. PartArena passes this selection to the enterprise's chosen service management system for ordering. PartsArena works with most leading ERP, CRM and service management solutions and allows updates to be distributed across the service arena.

   InfoMill has built up a close relationship with a number of key partners in the service management, mobile communications and technology sectors.

NEW ENTRY

InfoVision Systems Ltd

Slack Lane, Derby DE22 3FL, UK
Tel: (+44 1332) 34 71 23
Fax: (+44 1332) 34 51 10
e-mail: info@infovision.co.uk
Web: http://www.infovision.co.uk

Key personnel

Chairman and Chief Executive: Peter Crawford
Systems Director: Stuart Reece
Sales and Marekting Director: Tony Preece
Financial Director: John Stride
Operations Director: Dean Taylor
Business Development (Rail): Steve Nicks

Capabilities

DRUID™ software enabling fast and intelligent manipulation of complex data which can be delivered on CD-ROM and hard disk or network-based media.

   On-line systems for publishing data via internet, intranet and worldwide web.

   Electronic documentation comprising technical authorship, technical sources and driveline in common formats.

   Multiple and bespoke software development services.

DRUID Equinox™: advanced asset management system suitable for all industry sectors including transportation and utilities. Provides a seamless interface to DRUID™ and other applications.

In-Tend™: manages and monitors the progress of the tendering process from the initial set up of a project, through to monitoring and renewal of the awarded contract.

Projects

Supply of DRUID™ and related documentation services to Midland metro (1999), Siemens/Northern Spirit Class 333 (1999), Bombardier/Virgin Trains Class 220/221 (2000) and Siemens/Heathrow Express (2000). Supply of In-Tend™ to large public sector organisation (2002).

Infrasoft Corporation

99 Rosewood Drive, Suite 150, Danvers, Massachusetts 01923, USA
Tel: (+1 978) 777 99 88
Fax: (+1 978) 777 52 59
Web: http://www.infrasoft-civil.com

Works

Infrasoft Ltd, North Heath Lane, Horsham RH12 5QE, UK

Key personnel
Chief Executive Officer: Rick Fiery
Vice-President, Engineering: Stan Fenton
Managing Director, Infrasoft Ltd/Vice-President of
   International Channels & Marketing: Jim Paton

Products
MXRAIL, an engineer-friendly tool for rapid design
of rail projects from new design to infrastructure
maintenance and renewal.

Infrasoft Corporation

900 Cummings Center, Suite 312 T, Beverly,
Massachusetts 01915, USA
Tel: (+1 978) 927 20 33   Fax: (+1 978) 927 10 04
e-mail: usinfo@infrasoft-civil.com
Web: http://www.infrasoft-civil.com

Key personnel
Chief Executive Officer (Infrasoft Corporation):
   Rick Fiery
Chief Operating Officer (Infrasoft Corporation):
   Bryan Taylor
Director and Senior Vice-President Marketing
   Infrasoft Corporation and Managing Director
   (Infrasoft Ltd): Jim Paton
Vice-President Engineering (Infrasoft Corporation):
   Stan Fenton

Subsidiaries
Australia
Infrasoft PTY Ltd
303 Burwood Highway, East Burwood, Victoria
3151, Australia
Tel: (+61 3) 98 03 55 22   Fax: (+61 3) 98 87 90 10
e-mail: info@infrasoft.com.au

The Netherlands
Infrasoft bv
Bijster 11, NL-4817 HZ Breda, Netherlands
Tel: (+31 76) 531 85 85   Fax: (+31 76) 531 85 95
e-mail: beneluxinfo@infrasoft-civil.com

India
Infrasoft Limited
55 Community Center (1st Floor), East of Kailash,
New Delhi 110 065, India
Tel: (+91 11) 643 87 81   Fax: (+91 11) 643 87 85
e-mail: indiainfo@infrasoft-civil.com

UK
Infrasoft Ltd
North Heath Lane, Horsham RH12 5QE, UK
Tel: (+44 1403) 25 95 11   Fax: (+44 1403) 21 77 46
e-mail: info@infrasoft.civil.com
Web: http://www.infrasoft-civil.com

Products
MXRAIL is a Microsoft Windows-compliant design
and analysis system used for high-speed to light
rail, switch and crossings, alignment matching,
cant design, drainage, and automated drawing
production.

Innovata

2800 Vista Ridge Drive, Suwanee, Georgia
30024, USA
Tel: (+1 770) 614 49 00
Fax: (+1 770) 714 41 31
Web: http://www.innovata-LLC.com

Key personnel
General Manager, Europe: Richard Thorne
   Tel: (+44 1582) 63 50 18

Capabilities
Innovate collects, processes and distributes travel
data and services, initially as Dittler Brothers,
a speciality print company, and since 1998 as a
separate, independent company under the Innovata
name. Innovata supplies accurate and reliable
travel data in comprehensive databases that
encompass airline schedules, hotel information,
car rental locations and passenger rail schedules.

Its customer base includes airlines, airports, hotel
chains, travel agents, Internet travel portals and
websites, US government agencies plus a variety
of travel-related consultancy companies and
aviation-related manufacturers.

VERIFIED

Interautomation

Interautomation Deutschland GmbH
Ollenhauerstrasse 98, D-13403 Berlin, Germany
Tel: (+49 30) 412 20 87
e-mail: wolfgang.doerks@interautomation.de

Key personnel
Sales Manager, Vehicle Technology:
   Wolfgang Dörks

Products
Automatic passenger counting, door controlling,
time management and access control systems
for urban transit systems, regional and suburban
trains, buses and depots. Products were
originally developed and marketed by Pronova
Elektronik, which was acquired by Interautomation
Deutschland on 1 January 2000.

Institut für Regional- und Fernverkehrsplanung (IRFP)

Schützengasse 16, D-01067 Dresden, Germany
Tel: (+49 351) 470 68 19   Fax: (+49 351) 476 81 90
e-mail: info@irfp.de
Web: http://www.irfp.de

Products
FBS (Fahrplan-Bearbeitiungs-System) PC-based
train scheduling and timetable planning software
for main line and regional railway networks.

Laser Rail Ltd

Fitology House, Smedley Street East, Matlock DE4
3GH, UK
Tel: (+44 1629) 76 07 50   Fax: (+44 1629) 76 07 51
e-mail: info@laser-rail.co.uk
Web: www.laser-rail.co.uk

Key personnel
Managing Director: David M Johnson
Executive Director: Alison B Stansfield
Engineering Director: Steve Ingleton
Non-Executive Director: Hugh Fenwick
Metro Division Engineering Director: Bob Silver

Background
Laser Rail, established in 1989, is a technical
consultancy supplying complete turnkey projects
for rail infrastructure and metro networks.
It combines experienced engineers with the
latest high-technology laser-gauging systems
and software to measure and analyse clients'
infrastructure and provide new and innovative
vehicle designs.

Products
Measuring systems: Laser Gauging Vehicle (LGV),
a road-rail vehicle for measuring structure profiles
accurately and at speed. Laser Profiling Trolley
(LaserPeT), a portable version of the LGV used
at walking speed and designed to measure
metro infrastructure. LaserSweep™, a portable
measuring device for measuring structure profiles
in areas where larger systems are uneconomical to
use. Product approval received from Network Rail
and London Underground.

Software: ClearRoute™, Stress Route™,
DesignRoute™ software for vehicles and
infrastructure management.

Services: Track geometry measurement and
design, route assessment for existing and new
rolling stock, interoperability, feasibility studies

for existing and new routes, design of new rolling
stock. Risk assessment.
   Laser Rail has a formal research and development
group to develop core technologies associated with
its activities. Projects being undertaken include
infrastructure measurement, monitoring and
analysis, vehicle/track interaction technology and
intelligent video systems. Laser Rail also provides
software training and certification to various levels
of competency and this can be supplied as part of
the overall support package. Laser Rail operates in
the UK, Europe and Australasia.

UPDATED

Maxwell Soft Park Ltd

1/3/3 Millennium Business Park, MIDC Mahape,
Navi Mumbai 400 701, India
Tel: (+91 20) 566 18 50   Fax: (+91 20) 566 18 51
e-mail: pune@maxwell-india.com
Web: http://www.maxwell-India.com

Key personnel
Managing Director: Mohan Borole

Products
TOMIS (Train Operations Management and
Information System) covering: movement tracking;
freight wagon accounting; passenger coach
accounting; locomotive accounting; crew control;
fixed assets information; commercial functions;
and analytical processing. TOMIS is a consultancy-
based system which includes detailed system and
solutions based on: preset methods of operations;
level of automation; service rules; management
perception; evaluation of solutions; improvement
sought; existing problems and bottlenecks; and
financial commitment.

MultiModal Applied Systems

125 Village Boulevard, Suite 270, Princeton, New
Jersey 08540, USA
Tel: (+1 609) 419 98 00   Fax: (+1 609) 419 96 00
Web: http://www.multimodalinc.com

Products
MultiRail is an integrated PC software system
designed for the development and maintenance
of railway operating plans including network
design, timetables, equipment planning, crew
management, traffic analysis, wagon blocking,
and trip planning. MultiRail is available in versions
that are optimised for either freight or passenger
railways. The product has many graphical tools
including a graphical network and track builder,
editable string lines, and graphical crew and
equipment management. The freight version
supports the analysis of train sizes, yard workloads,
traffic routings, wagon schedules and provides
many types of system level statistics.
   MultiRail is used by every Class I railroad in
North America and is installed with railways
in Europe, Asia, South America, Africa and
Australia.

VERIFIED

Onerail

Level 1, 263 Liverpool Street, Darlinghurst, Sydney
2010, NSW, Australia
Tel: (+61 02) 93 39 12 22
Fax: (+61 02) 93 26 01 99
e-mail: rail@onerail.com
Web: http://www.onerail.com

Key personnel
Chief Executive Officer: Grant Holmes
Chief Operating Officer: Ross Holland
Marketing Director: Matthew Stewart
Director of Professional Services and Support:
   Darryl Garbutt

Corporate background

Onerail has its head office in Sydney Australia and has its technology and development office in Toronto Canada.

Products

Founded in 1996, Onerail offers rail operators four key products: the Orion Reservation System; Orion Select; Odyssey Distribution Systems and Your Odyssey.

Onerail's Orion reservation system is engineered for high-volume transaction and mission critical reservations and ticketing environments. The system's contemporary architecture fully supports a wide range of functionality, including: reservations, inventory control, pricing and ticketing, train consists, scheduling, seat and berth allocations and feeds to major financial systems. Orion has been developed in a Graphical User Interface (GUI) environment.

In addition to the features of Orion, Orion Select integrates rail, tour and accounting functionality into one system.

The Onerail Odyssey Distribution technology enables travel agents and consumers to make timetable, schedule, seat availability, fare enquiries and reservations on-line. Connection to Odyssey is available via: all travel agent distribution channels; the consumer direct via www.onerail.com; or the rail operator as their online booking engine.

Each client's Odyssey system is a tailored version of the Odyssey Distribution Solution, enabling the rail operator to brand its online booking tool.

In 2002, users of Onerail technology include: Via Rail (Canada), Amtrak (USA), railways of New South Wales (Australia), Queensland Rail (Australia), Venice Simplon Orient Express (Europe), Great Southern Pacific Express (Australia), Eastern & Oriental Express (Singapore), Road to Mandalay (Myanmar), The British Pullman (UK) and the Northern Bell (UK).

PAFEC Ltd

Strelley Hall, Nottingham NG8 6PE, UK
Tel: (+44 115) 935 70 55   Fax: (+44 115) 935 70 64
e-mail: peter.roberts@pafec.com
Web: http://www.pafec.com

Key personnel

Sales Manager: Peter Roberts

Products

Electronic document management (EDM) for the rail industry including EDM system capable of managing document handling such as archiving, visual browsing, document search techniques, audit trails and document distribution.

Pallas Informatik A/S

Allerød Stationsvej 2D, DK-3450 Allerød, Denmark
Tel: (+45) 48 10 24 10   Fax: (+45) 48 10 24 01
e-mail: pallas@pallas.dk
Web: http://www.pallas.dk

Key personnel

Managing Directors:
Svend Vitting Anderson, Karsten Funder

Products

Fleet Management, an Internet-based software programme developed to provide detailed information to management, station and workshop personnel. The system provides a real-time overview of trains in service, a parameterised view of train characteristics and a parameterised selection of trains requiring attention. A change of detail level enables the user to retrieve data pertaining to a specific train and the system can be integrated with existing vehicle workshop systems.

PC-Soft GmbH

Adolf Hennecke Strasse 37, D-01968 Senftenberg, Germany
Tel: (+49 3573) 707 50   Fax: (+49 3573) 70 75 19

e-mail: info@pcsoft.de
Web: www.pcsoft.de

Key personnel

Consultant: Christoph Baum

Products

Provision of software engineering and consultancy services in the field of railway systems.

The product software suite vips® consists of different modules, providing solutions for computer-based management of transport processes, vehicles and infrastructure. As an open system, vips® provides interfaces that can be integrated with other software systems such as ERP-systems like SAP/R3. The vips® software suite includes vips®/d – which supports the logistics process by centralised planning, controlling and monitoring the flow of information and goods. VIPS® Carsis supports planning, controlling and monitoring of all activities of maintenance, diagnostics as well as warranty management. vips®/r supports monitoring of straining and condition of all tracks, overhead contact lines as well as the safety features and equipment. vips®/r includes the following features: detailed reports of inspection data, generating trends and prospects, controlling of critical values, exposure analysis for every single rail or switch.

These software products are available as client/server systems and run on various operating systems, including Unix and Windows platforms.

UPDATED

PSI Transportation GmbH

Hohenzollerndamm 150, D-14199 Berlin, Germany
Tel: (+49 30) 897 148 33   Fax: (+49 30) 897 148 90
Web: http://www.psitrans.de

Background

PSI Transportation GmbH is a subsidiary of PSI AG.

Products

The PSITraffic software suite includes operations management, operations planning and dynamic passenger information systems. It also includes freight transport management, transport network planning and cargo logistics chain systems.

Contracts

PSI Transportation has supplied software systems to DVB, Dresden, ZOB GmbH, operator of Hamburg's central bus station, the Hamburg S-Bahn, Rostocker Strassenbahn AG, Rostock, the Süd-Thüringen-Bahn and GVB, Amsterdam.

Rail Management Consultants

Lister Strasse 15, D-30163 Hanover, Germany
Tel: (+49 511) 33 69 95 00
Fax: (+49 511) 33 69 95 99
e-mail: info@rmcon.de
Web: http://www.rmcon.de

Products

RailSys 2.0 timetable and infrastructure management software. Developed to operate on PCs running Windows NT, 2000 or XP, the software suite comprises four modules: Infrastructure Manager; Timetable Manager; Simulation Manager; and Evaluation Manager. RailSys was developed in co-operation with the Institute of Transport, Railway Construction and Operation of the University of Hanover.

Railtech Solutions Ltd

The Connect Centre, Kingston Crescent, Portsmouth PO2 8QL, UK
Tel: (+44 23) 92 65 63 00   Fax: (+44 23) 92 65 63 01
e-mail: enquiries@railtech-solutions.co.uk
Web: www.railtech-solutions.co.uk

Key personnel

Technical Director: Ruby Choudhury

Products

RailTech Solutions provides an extensive range of systems that have been developed to provide a complete service management solution supporting the needs of rapid emergency response to faults, incidents or defects that occur on or to any aspect of infrastructure, asset or service. F2000 is a Windows-based web-enabled fault management system for railway infrastructure, which embraces fault logging, tracking and reporting and incorporates an ability to import from or export to other databases. The system was developed to replace earlier infrastructure fault reporting and management systems employed in the UK rail market.

Modules and system enhancements under development include: the F2000 System Console, which provides control centres with the ability to monitor different F2000 parameters in real time, integrating with supported telephone systems; integration with vehicle tracking; integration with digital voice recorders; the Work Information Management Module (e-WIMS), designed to manage planned or scheduled maintenance work against assets; and the MF2000 mobile fault management system, which enables faults to be managed by mobile engineers using handheld devices.

UPDATED

RMI

Railcar Management, LLC
dba RMI, providers of RailConnect
1819 Peachtree Road, Suite 300, Atlanta, Georgia 30309, US
Tel: (+1 404) 355 67 34   Fax: (+1 404) 352 88 14
e-mail: marketing@railcarmgt.com
Web: www.railcarmgt.com

Key personnel

Chairman: J Peter Kleifgen
President: Charles 'Mac' Purdy
Vice-President, Business Development: Karl Knauff
Vice-President, Marketing: Paul Pascutti

Products

RMI is an independent provider of accurate, reliable, comprehensive and secure rail information services to the transport industry. RMI's services are accessed via the Internet through RailConnect®, its web-based portal. RailConnect® is a suite of proprietary, integrated applications that enables railroads to manage their business on a day-to-day basis and provide a higher level of customer service and better communications with other railroads and transport companies. Services include transportation management, revenue management, equipment, shipper freight and fleet management services and related executive information systems.

Customers include nearly 300 railroads that use RailConnect® to run their day-to-day operations. More than 75 per cent of short line and regional railroads in the US, Canada and Mexico utilise one or more of RMI's services. Currently, RMI processes over five million carloads annually through its RailConnect® suite of services.

UPDATED

Sabre Inc

4255 Amon Carter Boulevard, Fort Worth, Texas 76155, USA
Tel: (+1 817) 963 64 00
Web: http://www.sabre.com

Subsidiary companies

Sabre Inc, France
11-13 Avenue de Friedland, 75008 Paris, France
Tel: (+33 1) 53 53 55 00   Fax: (+33 1) 53 53 55 00

Sabre UK Marketing Ltd
Trinity Square, 23-59 Staines Road, Hounslow, TW3 3HE, UK
Tel: (+44 20) 88 14 42 00   Fax: (+44 20) 85 77 49 39

Capabilities

Information technology solutions including customised software development and software products, transaction processing, consulting and information technology outsourcing.

Customers include Amtrak, USA; Belgium Railways (SNCB); Eurostar UK; German Federal Railways (DB AG); London Underground; Société Nationale des Chemins de Fer Français (SNCF); Swiss Rail (SBB); Taipei Rapid Transit.

The SABRE Group has signed a seven-year agreement with London Underground Ltd to modernise and maintain its train and crew scheduling system. The system will streamline the scheduling process for London Underground's 480 trains, 267 stations and 1,250 daily train staff.

Science Systems Group

23 Clothier Road, Brislington, Bristol BS4 5SS, UK
Tel: (+44 117) 971 72 51   Fax: (+44 117) 972 18 46

Key personnel

Managing Director: Dr M D Love
Operations Director: B T Evans
Business Development Director: Peter J M Turner
Business Development Executive: Richard C Jones

Capabilities

Software development and project management for train control systems, control centres, simulation, passenger information and automatic fare collection.

Projects

Work has been carried out on London Underground Central Line and the Jubilee Line Extension; Seoul subway system; trainer's interface software (LUL); Docklands Light Railway, London.

Siemens Transit Telematic Systems AG

Industriestrieplatz 3, CH-8212 Neuhausen, Switzerland
Tel: (+41 585) 55 11 11   Fax: (+41 585) 55 11 12.
e-mail: info.tts@siemens.com
Web: http://www.siemens-tts.ch

Key personnel

Chief Executive Officer, Senior Management:
   Schär Hans-Peter

Background

Previously called Häni-Prolectron AG, the company changed its name to Siemens Transit Telematic Systems AG in 2001 and is a subsidiary of Siemens Schweiz AG.

Products

ALVC systems, integrated planning software, onboard computers, radio and wayside equipment, dynamic passenger information systems, traffic signal pre-emption systems, mobile fare management systems; electronic components for local and long-distance transport.

System consulting, system integration, project management, integration and support, commissioning, maintenance and training/instruction.

Socratec Telematic GmbH

Im Gewerbepark D 29, D-93059 Regensburg, Germany
Tel: (+49 941) 46 02 10   Fax: (+49 941) 460 21 11
e-mail: info@socratec.de
Web: www.socratec.de

Background

Founded in 1996, Socratec is a leading specialist in a variety of applications for traffic telematics on the basis of GPS and the Internet. Its competence is based on more than one decade of experience in the sector of satellite navigation.

In 2004 the company became Socratec Telematic GmbH.

Products

FleetView, onboard computer and software for tracking and monitoring all types of vehicles including buses and trams. The FleetView System also supports the exchange of messages between control centre and vehicle.

FreightView, autonomous on-board computer for ideal tracking and tracing of trailers, freight wagons and in intermodal transportation. The autonomous, battery-powered systems allows the tracking of the transported goods independently of the vehicle.

Fleet View XS, a very small box for tracking and tracing vehicles or objects via the Internet.

UPDATED

Spear Technologies Inc

436 14th Street, Suite 200, Oakland, California 94612, US
Tel: (+1 510) 267 33 33   Fax: (+1 510) 267 33 44
e-mail: marketing@speartechnologies.com
Web: www.speartechnologies.com

Key personnel

Chief Executive Officer: Mike Thomas
Vice-President, Marketing: Ken Voss

Products

Spear Technologies provides enterprise asset management software solutions for rail operators. The Spear 3i™ assists rail operators to improve service, safety and economic performance. The Spear 3i™ is designed specifically for commuter and freight rail fleets, facilities and maintenance of way.

Developments

The company has introduced a new software product designed for rail infrastructure asset management. Infrastructure Manager™ is a fully integrated application of the Spear 3i™ suite of Enterprise Asset Management (EAM) software applications for transport maintenance and materials management.

UPDATED

STERIA Group

46 rue Camille Desmoulins F-92782, Issy les Moulineaux, France
Tel: (+33 1) 34 38 60 00   Fax: (+33 1) 34 88 60 15
Web: www.steria.fr

Key personnel

See STERIA Group entry in *Signalling and communications* section.

Products

Include traffic control, model networks, scheduling, equipment and power consumption control, high reliability systems (Atelier B).

Contracts

Recent contracts executed and obtained include PCC – central control stations on the Parisian RER rapid transit system for RATP Paris; PCS – public transportation passenger and staff information system for RATP Paris; 3615 RATP – public teletext information server; RIS – passenger information system for DB AG; safety critical software B Method for RATP-Météor and SNCF; maintenance management system for LTA Singapore, North East Line, Circle Line, Taiwan High Speed Railway Corp and SNCF-SYSDAM; resource management for

RATP – @llegr@; car park management system for Toulouse (park and ride).

UPDATED

Systra Consulting Inc

2 Whipple Place, Suite 302, Lebanon, New Hampshire 03766-1356, USA
Tel: (+1 603) 448 02 00   Fax: (+1 603) 448 17 50
e-mail: info@railsim.com
Web: www.systraconsulting.com

Key personnel

President: Albrecht P Engel
Senior Vice Presidents: Dennis Fordham,
   David Thurston, Ruby Siegel
Vice-President, Rail Operations Analysis/Simulation:
   F William Lipfert, Jr

Products

RAILSIM® V7 Simulation Software Suite, simulation/engineering package – SYSTRA's RAILSIM v7 is a family of rail network modelling and analysis software applications that run on Windows 2000 and XP. RAILSIM packages are customised according to the licensee's needs. For example, the simulation/engineering package contains: RAILSIM network simulator, RAILSIM editor, RAILSIM track profile generator, RAILSIM Train Performance Calculator (TPC), RAILSIM rolling stock libraries, RAILSIM headway calculation, safe braking distance calculation and signal design add-ons. RAILSIM output includes graphical plots and text reports compatible with any Windows plotter or printer. Reports can be stored in AutoCAD® DXF file or comma-delimited text format, ready for incorporation into reports, spreadsheets and engineering drawings. Other typical RAILSIM license packages include: Base Licence, LFA Package, Simulation/Engineering package.

UPDATED

Timera Inc

5775 Flatiron Parkway, Suite 110, Boulder, Colorado 80301, USA
Tel: (+1 303) 444 37 58   Fax: (+1 303) 444 97 49
e-mail: lw_tatro@pstechno.com

Key personnel

Vice-President, Marketing: L Wayne Tatro Jr

Corporate background

Timera is part of the Fenix group of companies, a wholly owned subsidiary of Union Pacific Corporation, which serves as a holding company for four UP technology firms.

Products

Formerly PS Technology, Timera supplies System for Crew Assignment Tracking (SCAT) workforce management software. Customers include most Class 1 freight railroads in North America and several commuter lines. The software product suite addresses workforce management needs by minimising time required for scheduling and assigning staff to shifts. Products are designed automatically to fill vacancies with qualified staff, schedule and monitor holiday leave, maintain assignments, rosters, records, details of employee status and history and calculate gross pay.

Trivector System AB

Åldermansgatan 13, SE-227 64 Lund, Sweden
Tel: (+46 42) 38 65 00   Fax: (+46 42) 38 65 25
e-mail: info@trivector.se

Key personnel

Managing Director: Klas Odelid
Marketing Manager: Ola Fogelberg
Project Manager: Anders Månsson

Products

VEMOS (Vehicle Monitoring System) is a comprehensive information system for a public transportation system. It works in real time and contains a traffic control system, a central system and a depot system. The complete system is in operation in the city of Karlstad, Sweden and is to be in operation in the Dalarna region in the future. RAPP (Route Analysis Programme Package) is a system for analysing route-time data.

Union Pacific Technologies

7930 Clayton Road, St Louis, Missouri 63117, USA
Tel: (+1 314) 768 68 00
e-mail: DABOCK@notes.up.com
Web: http://www.uptweb.com

Products and services

A subsidiary of Union Pacific Corporation, Union Pacific Technologies (UPT) specialises in developing computer systems for the transportation industry. It has had a major role in extending communications/data/management systems into former SP territory and in interfacing with BNSF over the 6,400 km of trackage and haulage access. For 1997–1998 the installation of TCS on the SP was a priority. Since 1993 it has installed a yard management programme at the 17 largest yards on the National Railways of Mexico (FNM) system and installed a computer upgrade at the FNM central office in Mexico City.

With the onset of privatisation in Mexico, UPT foresees a large market for systems to connect the several new private properties that are evolving. UPT has taken on a Mexican partner to offer its products in Spanish language versions throughout Latin America. To explore the international market, UPT has joined with an IBM subsidiary, Integrated Systems Solutions Corporation.

Wisconsin Central has also adopted TCS as its primary management tool, and UPT is also implementing a version of the system in the UK for the English, Welsh and Scottish Railway.

VIPS AB

FO Petersons Gata 28, SE-421 31 Västra Frölunda, Sweden
Tel: (+46 31) 89 69 40   Fax: (+46 31) 47 86 01

Key personnel

Managing Director: Bo Sahlström

Products

Supplier of PC-based strategic planning systems mainly for public transport but also for private transport.

More than 70 VIPS systems are installed worldwide with bus, tram, metro and heavy rail operators.

Vossloh Information Technologies GmbH

Edisonstrasse 3, D-24145, Kiel, Germany
Tel: (+49 431) 248 14 88   Fax: (+49 431) 248 15 01
e-mail: info@vit.vossloh.com
Web: www.vit.vossloh.com

Key personnel

Managing Directors: James Sanders;
   Reinhold Hundt
Head of Sales and Project Management:
   Hermann Becker
Head of Development: Heinz-Holger Neustock
Head of Commercial Affairs: Désirée Gruchot
Head of Marketing: Dr Marion Behr

Background

Vossloh Information Technologies GmbH develops information and communications technology for rail transport companies. The company, with its head office in Kiel, Germany is part of the Vossloh Group.

Vossloh's display based on electronic ink technology for stationary and mobile customer information systems (Vossloh)   ***NEW***/0585221

Products

Operations control systems for national, regional and metropolitan railways

OCCs for railways and public transport enable rail transport companies to centralise monitoring and control of both rail traffic and technical installations in stations. Concentration and automation of operations management tasks can produce considerable increases in cost efficiency. Vossloh IT's control systems help transport companies to ensure that rail traffic operations are safe and punctual.

Transport information systems

The Operations Information System of Vossloh IT is an intelligent, networked information system, which provides up-to-date information on trains, timing and routes to infrastructure owners, train operating and transport companies and passengers. The system co-ordinates connections and ensures that resources are made available at the right time.

Operations and interlocking simulator for planning and training purposes

Vossloh's operations and interlocking simulator BEST enables detailed simulations of the interlocking during the early planning stages. The user interface is identical to the display that operators find during their daily work. All interlocking functions, specific applications, monitor pictures and the entire timetable with all defined train types are displayed. With the help of the simulation, the user can assess the actual operational process in the station and on the line (irrespective of the original interlocking) already upon completion of the planning process and will furthermore increase project safety. Once the planning phase has been completed, the user can continue making efficient use of BEST as a training simulator. In this function the simulator enables detailed simulations of the interlocking.

Contracts

September 2004, German Rail's (DB AG) one thousandth IC/EC passenger railcar was fitted with Vossloh's mobile passenger information system as part of German Rail's vehicle upgrade project. By the end of 2004 1,198 IC passenger railcars and 141 locomotives were fitted with the Vossloh passenger information systems and the new Vossloh UIC data bus, allowing for the first time, trainwide data transfer and passenger information on freely coupled passenger railcars.

Developments

Planned for commercialisation in 2005, Vossloh has presented the next generation of passenger information displays using the latest in electronic ink display technology in co-operation with E Ink Corporation, USA. These displays can be easily integrated into existing and new passenger information systems.

UPDATED

Vossloh Information Technologies York Ltd

Jervaulx House, 6 St Mary's Court, Blossom Street, York Y24 1AH, UK
Tel: (+44 1904) 63 90 91
Fax: (+44 1904) 63 90 92
e-mail: info@vity.vossloh.com
Web: www.vit.vossloh.com

Key personnel

Managing Director: Roland F Albert
Head of International Business Development:
   Alastair Dick
UK Business Development Manager:
   David Clarke

Background

Vossloh Information Technologies York Ltd, formerly Comreco Rail Ltd, is a subsidiary of Vossloh Information Technologies GmbH, Kiel, Germany, which is part of the Vossloh Group.

Services

Provision of consultancy, planning and management of rail operations, exploitation of infrastructure capacity and development of operations planning processes for a wide range of clients including Network Rail, First ScotRail, South West Trains, EWS, NSB, SJ, Strategic Rail Authority, London Underground, Israel Railways, Grand Central Trains, and Arriva. Vossloh Information Technologies York Ltd is ISO 9001/2000 accredited.

UPDATED

ZT

Zeta-Tech Associates Inc
900 Kings Highway North, PO Box 8407, Cherry Hill, New Jersey 08002, USA
Tel: (+1 609) 779 77 95
Fax: (+1 609) 779 74 36
e-mail: zetatech@zetatech.com
Web: http://www.zetatech.com

Key personnel

President: Dr Allan M Zarembski
Vice-President of Costing and Economic Analysis:
   Randolph R Resor
Director of Marketing: Jim Blaze
Director of Training and Field Engineering:
   Donald Holfeld
Director of Engineering Analysis: Joseph W Palese
Senior Engineer: Pradeep K Patel
Manager, Engineering Systems: John Webster
Project Engineers: Sunil Kondapalli, Leonid Katz
Senior Programmers: Nick Forte, Maciej Gorny
Office Administrator: Kim Corrigan
Administrative Assistant: Katy White

Capabilities

Zeta-Tech Associates is a technical consulting and applied technology company directed at the railway and transportation industries. Its expertise covers:

Track and track systems covering fasteners and fastener systems; sleepers; track strength; track buckling; track maintenance; and track geometry.

Vehicle/track interaction; freight wagon systems; inspection and measurement systems; fatigue design and analysis of structures; applied economics; technical marketing; computer simulation and modelling; transportation cost analysis.

Operations analysis including train simulation modelling freight and passenger equipment; and benefit analysis of improved operations, equipment, advanced train control systems.

Costing including development of detailed operating costs, cost allocation and life cycle costing.

Technical training for all areas of the railway/transit industry, including needs assessment, training material development, training delivery and training evaluation.

Maintenance management, comprising the development and application of computer software for use in forecasting component failure and planning of maintenance requirements, including database development and track component degradation/failure modelling and prioritisation of maintenance activities.

Custom software development including integrated graphic facilities management, geographical information systems, component failure modelling, clearance simulation and other rail-specific software.

For details of the latest updates to *Jane's World Railways* online and to discover the additional information available exclusively to online subscribers please visit

jwr.janes.com

ROLLING STOCK LEASING COMPANIES

AAE

Ahaus Alstätter Eisenbahn Holding AG
Poststrasse 6, PO Box 856, CH-6301 Zug, Switzerland
Tel: (+41 41) 727 20 50   Fax: (+41 41) 727 20 75
e-mail: ole.nygaard@aae.ch
Web: http://www.aae.ch/

Key personnel
Managing Director: DR Eckhart Lehmann
Deputy Managing Director, Operations:
   Markus Vaerst
Deputy Managing Director, Finance:
   Mark Stevenson
Technical Director: Dr Johannes Nicolin
Sales and Marketing: Ole Nygaard

Vehicles
AAE has over 10,000 freight wagons rented to state railway operators and private companies, including DB AG, SBB, NS, SNCB, SJ, DSB, NSB, CFL, ÖBB, ČD, MÁV, ŽSR, Intercontainer/Interfrigo, Hupac, Novotrans, Cemat and Nordwaggon. AAE is a member of UIC, RIV and BCC and its fleet includes covered wagons, flat wagons, pocket wagons and container wagons.

Algeco SA

16 avenue de l'Opera, F-75040 Paris Cedex 01, France
Tel: (+33 1) 42 86 23 00
Fax: (+33 1) 42 97 41 59

Key personnel
Assistant Director, Exploitation: Michel Bernard

Vehicles
Tank and special purpose wagons; ISO tank containers for hazardous products.

Angel Trains Cargo

Frankrijklei 121, B-2000 Antwerp, Belgium
Tel: (+32 3) 470 27 00   Fax: (+32 3) 470 27 09
e-mail: cargo@angeltrains.com
Web: http://www.angeltrains.com

Key personnel
General Manager: André Bloemen

Background
Angel Trains Cargo, previously called Locomotion Capital, was established in July 2000 to provide operating leases for freight locomotives in Europe. The company is part of the Angel Trains Group, a subsidiary of the Royal Bank of Scotland Group plc, with a 10 per cent shareholding by Vossloh AG.

Vehicles
By late 2002, Angel Trains Cargo owned or had ordered more than 70 locomotives. These included 50 Vossloh Schienenfahrzeugtechnik G 1206 or G 2000 diesel-hydraulic machines and around 20 Class 145 electric locomotives from Bombardier Transportation. Customers included: DL Cargo (Belgium); SNCF Fret (France); and RAG Bahn und Hafen, Rail4Chem and Connex Cargo (Germany).

Angel Trains Ltd

22nd Floor, Portland House, Stag Place, London SW1E 5BH, UK
Tel: (+44 20) 75 92 05 00
Fax: (+44 20) 75 92 05 20
e-mail: feedback@angeltrains.com;
   international@angeltrains.com
Web: www.angeltrains.com

Key personnel
Chairman: Leith Roberton
Managing Director: Haydn Abbott
Corporate Development Director: John Vale
Engineering Director: Tim Dugher
Finance Director: George Lynn
Director International: Tim Jackson
Legal Director: Louise Oddy
Director UK: Peter Rigby
Head of Communications: Jane Adley

Angel Trains Ltd

Train operating company (TOC)	Class	Quantity cars/ locomotives	TOC total
Arriva Trains Wales			
	142	30	
	153	5	
	158	74	
	175	70	
			179
c2c			
	357	112	112
Central Trains			
	150	74	
	158	36	
			110
Chiltern Railways			
	165	89	89
English Welsh & Scottish Railway			
	66	250	
	67	30	
			280
First Great Western			
	180	70	
	HST power car	85	
	HST trailer	311	
			466
First Great Western Link			
	165	88	
	166	63	
			151
First ScotRail			
	156	96	
	314	48	
			144
GNER			
	HST power car	23	
	HST trailer	97	
			120
Great Northern			
	317	48	48
Merseyside PTS			
	507	96	
	508	81	
			177
Midland Mainline			
	HST power car	5	
	HST trailer	16	
			21
Northern Rail			
	142	158	
	150	58	
	153	12	
	156	56	
	333	64	
			348
One (London Eastern Railway)			
	317	240	
	360	84	
			340
Silverlink Trains			
	150	14	
	350[1]	120	
	508	9	
			143
South Eastern Trains			
	423	200	
	465	200	
	466	86	
	508	36	
			522

Angel Trains Ltd

Train operating company (TOC)	Class	Quantity cars/ locomotives	TOC total
South West Trains			
	158	4	
	442	120	
	444	225*	
	450	440*	
			789
Southern			
	421	83	83
Virgin West Coast			
	390	477	477
Virgin CrossCountry			
	220	136	
	221	216	
			352
Wessex Trains			
	153	13	
	158	28	
			41

* Includes units on order/under delivery in 2005.

Background
Angel Trains was established in April 1994 within British Rail and in 1995 it was transferred to the UK government in preparation for its sale to the private sector. Angel Trains is now 100 per cent owned by The Royal Bank of Scotland Group plc.

Angel Trains is one of Britain's three train leasing companies. It provides much of Britain's rail industry with rolling stock and is now expanding its business both in the UK and in other countries. It is active in the procurement, financing, ownership, maintenance and asset management of new and used equipment. Its key role is in the introduction of new capital and ideas to railway operations.

In 2000, Angel Trains formed Locomotion Capital in a joint venture with Vossloh (10 per cent). In 2003 the company was renamed Angel Trains Cargo. The company leases freight locomotives. In 2002, Angel Trains announced the establishment of Angel Trains Europa GmbH to handle passenger rolling stock procurement and leasing in the wider European market.

Traction and rolling stock
Angel Trains has been investing £2.4 billion in new trains for UK operators, including: £365 million for 280 new freight locomotives for English Welsh & Scottish Railway, the UK's largest-ever order for freight traction; £593 million for 53 225 km/h tilting trains for the West Coast Main Line, built by Alstom and to be operated by Virgin Trains; £55 million for new Class 333 emus for Arriva Trains Northern (now Northern Rail); £78 million for new dmus for First North Western (now Arriva Trains Wales); £74 million for dmus for First Great Western; £75 million for Desiro UK emus to First Great Eastern (now One) and £640 million for Desiro UK emus for South West Trains.

Non-UK traction and rolling stock
Investments in passenger rolling stock for use in mainland Europe originally made from the UK and subsequently managed by the company's Angel Trains Europa subsidiary included by late 2002: 30 Siemens Desiro VT 642 two-car dmus for Danish State Railways and Connex Regiobahn subsidiaries LausitzBahn, NordWestBahn and the Ostmecklenburgische Eisenbahn; 29 Alstom Coradia LINT 41 two-car dmus for Arriva Tog, Denmark; and 11 three-car Bombardier Talent VT 643.3 dmus for a possible contract from an operator in Germany.

UPDATED

Armita

Armita Nederland BV
Apollolaan 109, NL-1077 AN Amsterdam, Netherlands
Tel: (+31 20) 673 61 17   Fax: (+31 20) 673 58 57

Key personnel
Manager: H M Endstra

Vehicles
Tank cars: 420.

ARR Rail Rent

Transportmittel Vermietungs GesmbH
Kunigundbergerstrasse 40, A-2380 Perchtoldsdorf,
Austria
Tel: (+43 1) 865 66 85
Fax: (+43 1) 865 66 85 91

Key personnel
Managing Directors: Gernot Schwayer,
 Dr Helmut Breit, Johannes Hansbart

Vehicles
Short and long-term leasing of freight wagons. The
company owns a fleet of over 1,000 wagons.

VTG Rail UK Ltd

4a Berkeley Business Park, Wainwright Road,
Worcester, WR4 9FA, UK
Tel: (+44 1905) 75 09 00
Fax: (+44 1905) 75 09 19

Key personnel
Managing Director: J K Jagger
Tank Fleet Manager: P M Lugg
Bulk Fleet Manager: I R Shaw
Fleet Engineering Manager: N Day

Background
VTG Rail UK Ltd, formerly trading as CAIB UK Ltd,
is now part of the VTG group of companies.

Vehicles
The UK fleet numbers approximately 2,500 wagons
with newly manufactured vehicles being added as
part of a continuous renewal process. Vehicle types
currently available include two- and four-axle tank
wagons for Class A and B petroleum products,
liquefied gases including chemical gases, liquid
chemicals (stainless and line mild steel), and
chemical slurries; two- and 4-axle hopper wagons
for carrying aggregates, chemicals, grain and
other granular and powdered materials; two- and
four-axle open and closed box wagons for use
in aggregates, rail infrastructure projects, scrap
steel and waste transportation and four-axle steel
coil carrying wagons; wagons for intermodal
operations, including the Advanced Container
Handling System (ACTS) operated by the lorry
driver.
   In addition, a number of wagons are registered
for international transport enabling operations
to continental destinations. The VTG Rail UK
fleet is maintained by Marcroft Engineering Ltd
and benefits from the support of group related
rail workshops across Europe. VTG Rail UK can
supply a full range of associated services including
loading and unloading trains, on-site shunting
operations, track maintenance and can make
train operating arrangements as required by the
customers.

UPDATED

CFCL Australia Pty Limited

PO Box 6406, North Sydney, New South Wales
2060, Australia
Tel: (+61 1) 70 85 99 10 20
Fax: (+61 1) 70 85 99 40 70
e-mail: mcgeemj@aol.com
Web: http://www.crdx.com/cfcl.australia.htm

Key personnel
Manager: Mike McGee

Organisation
CFCLA Australia is owned by Chicago Freight Car
Leasing Co of USA.

Services
The company is involved in the leasing and
provision of locomotives and rolling stock to the
industry.

Vehicles
The company purchased 13 former Australia
National EL Class 2,240 kW locomotives and leases
them to others. They were overhauled by Goninans,
Western Australia, to Dash 8 technical standards
and had gearing altered to improve their suitability
for freight work. Five were leased to Austrac (qv)
for Sydney–Melbourne freight hauls in 1999 while
others are regularly seen on Melbourne–Adelaide
freights.
   CFCL Australia has also purchased 50 new
container flat wagons from China and has rebuilt
another 22 locally.

CGTX Inc

15th Floor, 1600 Boulevard René Lévesque Ouest,
Montréal, Québec H3H 1P9, Canada
Tel: (+1 514) 931 73 43   Fax: (+1 514) 931 55 34

Key personnel
President and Chief Executive Officer: J C Leger
Vice-President and Treasurer: Jacques Poulin
Vice-President, Marketing and Sales: R A Podsiadlo
Vice-President, Engineering/Fleet Maintenance:
 G Sinclair
Director, Fleet Maintenance: G Cooper

Products
Lessors of railway rolling stock in Canada: tank
wagons and freight wagons. The company has
maintenance workshops at Montréal, Red Deer and
Moose Jaw.

Vehicles
8,300.

Chicago Freight Car Leasing Co

1 O'Hare Centre, Suite 7000, 6250 N River Road,
Rosemont, Illinois 60018, USA
Tel: (+1 847) 318 80 00   Fax: (+1 847) 318 80 45
e-mail: tom@crdx.com

Key personnel
President: F R Sasser
Senior Vice-President, Marketing and Sales:
 T F Kuklinski

Products
New and rebuilt freight wagons of all types; leasing
services.

Vehicles
7,000.

VERIFIED

Convoy

Convoy-Contigas BV
Apollolaan 109, NL-1077 AN Amsterdam, Netherlands
Tel: (+31 20) 673 61 17
Fax: (+31 20) 673 58 57

Key personnel
Director: W Endstra

Vehicles
180 tank wagons.

Dyrekcja Eksploatacji Cystern Sp zoo (DEC)

ul Twarda 30, PL-00-831 Warsaw, Poland
Tel: (+48 22) 622 05 05; 697 91 94
Fax: (+48 22) 697 91 95
e-mail: marketing@decyst.com.pl
Web: http://www.decyst.com.pl

Key personnel
Managing Director: Stefan Garus
Commercial Director: Tadeusz Kościelak

Background
DEC is wholly-owned by GATX, based in the USA.

Products
DEC is the owner of 11,500 tank wagons for the
transport of light and heavy petroleum products,
liquefied gases, chemicals, molasses, liquid
foodstuffs and other products. Facilities are
situated all over Poland, mainly on the premises
of refineries, fuel and reloading depots, at
pipeline terminals and at railway border crossing
points.
   DEC divisions have been authorised by
Polish Railways and the Railway Technical
Supervision to carry out repairs. The rolling
stock repair plant in Ostróda is the production
plant within DEC. Its main responsibility is to
provide general overhauls and inspection of
wagons. This plant is also adapted to carry out
modernisation of existing stock. It handles the
assembly of new types of wagons and freight
wagon bogies.

Tank wagon owned by DEC   0063720

E G Steele & Co Ltd

25 Dalziel Street, Hamilton ML3 9AU, UK
Tel: (+44 1698) 28 37 65   Fax: (+44 1698) 89 15 50
e-mail: egsteelecoltd@btinternet.com

Key personnel

Managing Director: David Steele
Logistics Manager/Director, Export Sales: Ian Hood
Quality Manager: Cameron Gibson

Vehicles

Currently a fleet of approximately 50 wagons which
consists of: 50 tonne pressure discharge powder
wagons, 45 tonne petroleum wagons, and 45 tonne
stainless steel sulphuric acid wagons.

UPDATED

Ermewa

Groupe Ermewa SA
7 rue du Mont-Blanc, CH-1211 Geneva 1,
Switzerland
Tel: (+41 22) 906 04 21   Fax: (+41 22) 906 04 96
e-mail: info@ermewa.com
Web: http://www.ermewa.com

Paris office

Ermewa SA
Le Stratège, 172 rue de la République, F-92817
Puteaux Cedex, France
Tel: (+33 1) 49 07 25 31
Fax: (+33 1) 49 07 25 35

Key personnel

Chief Executive Officer: Josef Küttel
Director, Wagons: Pierre Messulam
Chief Financial Officer: Alain Stocker

Paris office

Director General and Chief Executive Officer:
   Michel Van Wymeersch
Deputy Director General: Xavier Ducluzeau
Technology Director: Jean Idiart
Finance Director: François de Castelnau
Secretary General: Monique Roquain

Ermewa-Sati

(address as *Paris office* above)
Tel: (+33 1) 49 07 26 00   Fax: (+33 1) 47 73 88 10
e-mail: wagons@ermewa-sati.com
Chief Executive Officer: Pascal Varin

Ermewa SA

(address and tel/fax as for Groupe Ermewa SA
above)

Ermewa GmbH

Friedrich-Ebert-Damm 143, D-22047 Hamburg,
Germany
Tel: (+49 40) 694 00 03   Fax: (+49 40) 694 00 02
e-mail: h.vossen@ermewa.de
Managing Director: Heiner Vossen

Locatransports

56 rue de Rome, F-75008 Paris, France
Tel: (+33 1) 42 93 24 75   Fax: (+33 1) 45 22 64 36

EVS

(address as *Paris office* above)
Tel: (+33 1) 49 07 23 41   Fax: (+33 1) 490 72 38 40
President and Director General: Bruno Dambrine

See also entries for Ferifos and Orval in the *Vehicle
maintenance equipment and services* section.

Background

The Ermewa Group specialises in the transport and
logistics of bulk products and has 40 subsidiaries
in 20 countries. As well as being active in the
freight vehicle leasing market, the group undertakes
freight forwarding, maritime transport, container
rental and wagon repair and refurbishment.
Since April 2003 the company's shares have been
owned by Investors in Private Equity (IPE) (50.4
per cent) and French National Railways (SNCF)
(49.6 per cent).

Vehicles

Ermewa owns around 18,000 specialised freight
wagons, most of them for the transport of bulk
products. The group also owns some 20,000 tank
containers and 10,000 mini-tanks and Intermediate
Bulk Containers (IBCs).

First Union Rail

6250 River Road, Suite 5000, Rosemont, Illinois
60018, USA
Tel: (+1 847) 318 75 75   Fax: (+1 847) 318 75 88
Web: http://www.firstunionrail.com

Key personnel

President: Jack Thomas
Vice-President, Marketing: Rich Seymour
Vice-President, Operations: Rick Grossman
Vice-President, Finance: Lori Heissler

Products

Comprehensive fleet management services for
railcar owners: approximately 70,000 railcars
in fleet. Provides fleet management services,
innovative lease packages, railcar portfolio
acquisition, sale and lease-back programmes on all
types of railroad equipment. First Union Rail also
has longer-term financing available through single
investor and leverage leases.

VERIFIED

GE Capital Rail Services

161 North Clark, 7th Floor, Chicago, Illinois
60601, USA
Tel: (+1 870) 443 78 37

Key personnel

President and Chief Executive Officer:
   R W Speetzen
Executive Vice-President, Sales: K Schneider

Background

GE Capital Services, USA, bought Cargowaggon
in 1997. It has become part of GE Capital's Rail
Services business, based in Chicago, Illinois.

Products and services

GE Capital Rail Services offers a range of rail
vehicles and leasing solutions worldwide. The fleet
spans a variety of railcar equipment, including
covered hopper wagons, tank wagons, boxcars and
pressure differential wagons and for industries such
as agriculture, forest products, utilities, petroleum/
chemicals, auto, steel and consumer goods.
   Leasing solutions include per-diem, fixed,
operating and finance leases, sales/leasebacks
and structured financial products. Also repair,
maintenance and administrative services through a
network of service centres. The European business,
GE Rail Services, provides of rail transport services.
Through its offices in England, France, Germany,
Italy and Sweden, it provides rail vehicle equipment
and related services to railways and other transport
providers throughout Western and Central Europe.

General American
Transportation Corp (GATX)

500 W Monroe Street, Chicago, Illinois 60661, USA
Tel: (+1 312) 621 62 00   Fax: (+1 312) 621 66 36
Web: http://www.gatx.com

Key personnel

President: D Ward Fuller
Senior Vice-President: D Stephen Menzies
Vice-President and Chief Financial Officer:
   D J Schaffer

Subsidiaries

Dyrekcja Eksploaticji Cystern Sp zoo (DEC)
ul Twarda 30, 00-831 Warsaw, Poland
Tel: (+48 22) 622 05 05; 697 91 94
Fax: (+48 22) 697 91 95

e-mail: marketing@decyst.com.pl
Web: http://www.decyst.com.pl

KVG Kesselwgen Vermietgesellschaft mbH
Herrengraben 74, D-20459 Hamburg, Germany
Tel: (+49 40) 36 80 40   Fax: (+49 40) 36 80 41 13
e-mail: info@kvg.mhs.compuserv.com
Executive Directors: Rainer Baumgarten,
   Gernot Schwayer
Managing Directors: Volker Grahl, Manfred Gürges

KVG Kesselwgen Vermietgesellschaft mbH
Kunigundbergstrasse 40, A-2380 Perchtoldsdorf,
Austria
Tel: (+43 1) 865 66 85   Fax: (+43 1) 86 56 68 59
Executive Directors: Gernot Schwayer,
   Rainer Baumgarten, Johannes Mansbart

Background

GATX Rail is a division of GATX Financial
Corporation, a wholly-owned subsidiary of GATX
Corporation. It acquired 100 per cent ownership of
KVG in December 2002.

Products

Rail wagon leasing, repair, maintenance and fleet
management services. GATX operates 65,000
wagons, 80 per cent of which are tank wagons.
Over half of the tank wagon fleet is employed
in chemicals traffic. The GATX tank wagon fleet
includes the TankTrain system, a series of tank
wagons interconnected with flexible hoses that
allow the entire string to be loaded and unloaded
from one connection.
   GATX offers tank wagons of every size for
handling any liquid commodity transported
by rail. The tank wagon fleet includes general
service, pressure, stainless steel, aluminium and
commodity-specific tank.
   The Airslide wagon is suitable for transporting
and unloading finely divided bulk chemical and
food products such as talc, flour, sugar, starch and
carbon black. For shippers who require pneumatic
unloading of their dry bulk commodities, GATX
now complements its Airslide wagon by offering
the Trinity-designed Power-Flo 15 lb/in^2, 5,125 ft^3
covered hopper wagon.
   Additions to the GATX fleet include 3,000 ft^3
covered hopper wagons for the transport of cement
and aggregates and jumbo covered hopper cars for
grain.

Vehicles

65,000.

Greenbrier Leasing Corporation

A subsidiary of the Greenbrier Companies Inc
One Centerpointe Drive, Suite 200, Lake Oswego,
Oregon 97035, USA
Tel: (+1 503) 684 70 00   Fax: (+1 503) 684 75 53
Web: http://www.gbrx.com

Products

Greenbrier Leasing Corporation specialises in
leasing freight wagons and in managing wagon
fleets for third parties. Greenbrier Leasing invests
heavily in research and development and since
1985 has sponsored the development of the
Gunderson range of double-stack container
wagons, centre-partition timber wagons and
high-capacity box cars. Since its acquisition of
TrentonWorks in 1995, Greenbrier Leasing has
also assisted in the development of covered
hoppers and box cars produced at the Nova Scotia
plant. The company also sponsored development
of Autostack, a system for transporting motor
vehicles in containers, and more recently Auto-
Max, a multilevel vehicle-carrying wagon.
   The Greenbrier Intermodal subsidiary claims to
be the leading provider of intermodal freight
wagons, for which it continually develops the market.

Vehicles

Greenbrier Leasing's fleet of owned, leased or
managed equipment totals more than 37,000
vehicles and includes a full range of types,
including: box cars; double-stack wagons;

covered and open-top hoppers; centre-partition timber wagons; mechanically and cryogenically refrigerated wagons; gondolas; wood-chip wagons; and the company's new Auto-Max vehicle-carrier.

HSBC Rail (UK) Ltd

PO Box 29499, London NW1 2ZF, UK
Tel: (+44 20) 73 80 50 40   Fax: (+44 20) 73 80 53 26

Key personnel
Head of HSBC Rail: Peter Aldridge
Head of Finance: David Mead
Heads of Commercial Services:
    Ellen Harwood, Bob Marrill
Head of Business Standards: Karin Kilbey
Head of Risk and Business Development:
    Paul Tweedale
Head of Engineering Services: Chris Moss

Political background
HSBC Rail (UK) Ltd is a member of the HSBC Group. It acquired the fleet as well as the leases with the Train Operating Companies created from the privatisation of British Rail.

Traction and rolling stock
HSBC Rail's portfolio of stock is predominantly electric, and covers some of the more modern types in Britain.
    The HSBC Rail has undertaken a wide range of refurbishment on its current rolling stock as well as being awarded contracts to supply both new passenger rolling stock and freight rolling stock in Europe and the UK. The company has recently entered the on-track plant equipment market.

HSBC Rail (UK) Ltd

Class	Train operating unit(s)	Number of vehicles in service Jan 2002
Locomotives		
86/2	Virgin Trains, Anglia Railways	53
91	GNER	31
66	Freightliner, TGOJ, GB Railfreight	41
Multiple-units		
168	Chiltern	9
170	ScotRail	27
306	First Great Eastern	3
310	Central Trains, C2C	179
313	Silverlink, West Anglia	192
315	WAGNm First Great Eastern	244
318	ScotRail	63
320	ScotRail	66
321	Silverlink, First Great Eastern	456
322	West Anglia Great Northern	20
334	ScotRail	120
365	WAGN, Connex South East	164
375	Connex South Eastern	294
421	Connex South Eastern, South West Trains	220
423	South West Trains, South Central	308
455	South Central	184
465/0	Connex South Eastern	388
483	Isle of Wight	12
Coaches		
Mark II	Virgin Trains, Anglia Railways, First Great Western, Scotrail	494
Mark IV	GNER	302

Invatra

Industrial de Vagones y Transportes SA
Poligono Industrial Alces, Alcazar de San Juan, Ciudad Real, Spain
Tel: (+34 926) 51 11 13

Vehicles
58 tank wagons.

KVG Kesselwagen Vermietgesellschaft mbH

Herrengraben 74, D-20459 Hamburg, Germany
Tel: (+49 40) 36 80 40   Fax: (+49 40) 36 80 41 13
e-mail: info@kvg.mhs.compuserv.com

Key personnel
Executive Directors: Rainer Baumgarten,
    Gernot Schwayer
Managing Directors: Volker Grahl, Manfred Gürges

Subsidiary
Jungenthal-Waggon GmbH
Am Hafen 29, D-30629 Hannover, Germany
Tel: (+49 511) 95 87 70   Fax: (+49 511) 958 77 15
Managing Directors: Volkmar Gassmann,
    Volker Grahl

Associated company
KVG Kesselwagen Vermietgesellschaft mbH
Kunigundbergstrasse 40, A-2380 Perchtoldsdorf, Austria
Tel: (+43 1) 865 66 85   Fax: (+43 1) 86 56 68 59
Executive Directors: Gernot Schwayer,
    Rainer Baumgarten, Johannes Mansbart

Background
KVG is a wholly-owned subsidiary of GATX Rail based in Illinois, USA.

Vehicles
KVG hires privately owned tank wagons and other specialised vehicles. The company owns a fleet of approximately 10,000 vehicles transporting light and heavy oil products, liquefied petroleum gases, acids, alkalis, solvents and other chemicals, powdered or granular products as well as standard goods wagons.

NACCO SA

40 rue La Boétie, F-75008 Paris, France
Tel: (+33 1) 45 61 56 20   Fax: (+33 1) 40 74 06 24
e-mail: info@naccorail.com
Web: //www.naccorail.com

Key personnel
President: David MacNaughton
Commercial Managers: François Guasp,
    Gregg MacNaughton, Christopher MacNaughton
Operations Manager: Yann Bonguardo

Subsidiary companies
NACCO GmbH
Lehmweg 17, D-20251 Hamburg, Germany
Tel: (+49 40) 32 08 58 30   Fax: (+49 40) 32 08 58 40
General Manager: Thomas von Berlepsch

NACCO UK Ltd
Trafford House, Chester Road, Stretford, Manchester, M32 0SJ, UK
Tel: (+44 161) 873 72 55   Fax: (+44 161) 877 28 08
General Manager: Bernard McDonnell

Representative offices in Prague, Budapest and Zarow.

Services
Founded in 1973, the company has a fleet of over 5,000 specialised wagons, the majority tank cars, covered hopper wagons and pressure discharge cars for the transport of petroleum products, chemicals, LPG and granular or powdered products.

UPDATED

NS Financial Services Company

Behan House, 10 Lower Mount Street, Dublin 2, Ireland
Tel: (+353 1) 638 13 80   Fax: (+353 1) 638 13 99
e-mail: hans.dejong@nsfinancialservices.ie
Web: http://www.nsfsc.com

Key personnel
Managing Director: B van Dijk
Manager, Marketing and Sales: J J M de Jong
Manager, Rolling Stock: F J van der Linden

Background
NS Financial Services Company (NSFSC) is an operational lease company active across Europe in financing and leasing of rolling stock. The company was established in Ireland and started trading in 1999. The primary focus is to provide finance and lease services to operators within the NS Group and other private operators in all liberated EU countries.

Vehicles
The company has fleets on lease with:
NS Reizigers: 125 dmus; 242 emus; 4 ICE-3M; 99 IC coaches; 14 electric locomotives.
Railion Benelux: 19 diesel locomotives; 23 electric locomotives; 531 wagons
Syntus: 22 dmus (LINT 41)
Connexxion: 2 dmus
NedTrain: 13 shunting locomotives

OEVA

Oesterreichische Eisenbahn-Verkehrs-Anstalt GmbH
Volksgartenstrasse 3, A-1010 Vienna, Austria
Tel: (+43 1) 52 33 62 10   Fax: (+43 1) 523 15 55

Key personnel
Managing Director: Gerhard W Schwertmann

Vehicles
900 owned and 1,350 managed.

On Rail

Gesellschaft für Vermietung und Verwaltung von Eisenbahnwaggons mbH
Steinesweg 10, D-40822 Mettmann, Germany
Tel: (+49 2104) 92 97-0; 92 97-50
Fax: (+49 2104) 252 54
e-mail: info@on-rail.com
Web: www.on-rail.com

Key personnel
Directors: Ulrich Swertz, Nathalie Tastevin

Vehicles
On Rail manages and leases a fleet of 3,200 private wagons, of which around 1,100 are tank wagons for the transport of light and heavy petroleum products, chemicals, pressurised gases and powders. On Rail also leases wagons for the transport of bulk goods and steel products.

UPDATED

Porterbrook Leasing Company Ltd

Burdett House, Becket Street, Derby DE1 1JP, UK
Tel: (+44 1332) 26 24 05   Fax: (+44 1332) 26 44 19
e-mail: enquiries@porterbrook.co.uk
Web: www.porterbrook.com

Key personnel
Managing Director: Paul Francis
Commercial Director: Keith Howard
Engineering Director: Tim Gilbert
Finance Director: Stefan McCormick
Operations Director: Alex White
Head of Legal and Compliance: Stephen McGurk
Head of Communications: Ian Pritchard

Political background
In April 1994 Porterbrook Leasing Company Ltd bought over one third of British Rail's rolling stock with leases in place with 17 train operators.
    Following a management employee buy-out in January 1996, Porterbrook was sold to Stagecoach in August of the same year and was subsequently bought by Abbey National Treasury Service plc (ANTS) in April 2000.
    Porterbrook's total investment to date in the UK rail market totals almost £1.6 billion in new trains and over £200 million on existing fleet refurbishment. The company now has leases with 19 out of the 25 passenger train operating companies as well as the four freight operating

companies. Porterbrook has won 60 per cent of the orders for new passenger trains since privatisation.

Traction and rolling stock

When British Rail was privatised, Porterbrook took over approximately one third of the total UK passenger rolling stock fleet. It currently leases rolling stock to 19 of the 25 UK train operating companies, including the freight operator and three out of the four UK freight operating companies.

The company's rolling stock fleet of over 5,000 vehicles includes electric and diesel locomotives, high-speed train, multiple units and freight wagons.

Since 1996, Porterbrook has invested in over 2,900 new vehicles with a total value of approximately £1.6 billion. As well as investing in new trains, Porterbrook finances the refurbishment and upgrading of existing stock to modern standards and has already invested £200 million in this process. The company is also able to purchase existing fleets, freeing up capital, and lease them back either in their current condition or refurbished.

Freight

Originally established to operate in the rail passenger market, Porterbrook expanded into rail freight. Soon after privatisation, Porterbrook undertook a purchase and leaseback arrangement on a freight fleet of 345 wagons and 70 locomotives. Since then, it has secured leases with the four freight operating companies.

UPDATED

Procor Limited

2001 Speers Road, Oakville, Ontario L6J 5E1, Canada
Tel: (+1 905) 827 41 11   Fax: (+1 905) 827 08 00
Web: http://www.procor.com

Key personnel

For a full list of personnel, see Procor Limited entry in *Freight vehicles and equipment* section.

Products

Leasing of tank and special-purpose freight wagons.

Vehicles

Over 22,000.

SGW

Société de Gerance de Wagons Grande Capacité
163 bis avenue de Clichy, F-75838 Paris Cedex 17, France
Tel: (+33 1) 40 25 37 00   Fax: (+33 1) 40 25 37 60

Key personnel

Chairman: Jacques Rolland
General Manager: Bernard Kail

Products

SGW caters exclusively for unit train movement of bulk freight suitable for open-wagon conveyance, such as coal, coke, ores, sand, stones and ballast, throughout Europe. It does not own wagons, but markets and manages the deployment of a pool of some 6,500 special purpose vehicles on behalf of wagon manufacturers, national and private industries, and private wagon leasing companies.

Siemens Dispolok GmbH

(Postal) Krauss-Maffei Strasse 2, D-80997 Munich, Germany
(Office) Georg-Reismüller-Strasse 32, D-80999 Munich, Germany
Tel: (+49 89) 88 99 20 15   Fax: (+49 89) 88 99 30 48
e-mail: ts.lmsp-dispolok@ts.siemens.de
Web: www.dispolok.com

Siemens' ES 64 U2 two-system Eurosprinter electric locomotive forming the part of Dispolok fleet
0525447

Siemens ES 64 U2 two-system Eurosprinter electric locomotive   **NEW**/0585281

Siemens Dispolok ER 20 Eurorunner diesel-electric locomotive   **NEW**/0585282

Key personnel

Chief Operating Officer: Dr Walter Breinl

Background

A wholly owned subsidiary of Siemens, Dispolok initially operated as a division of the company, having been established as a full service supplier in the locomotive leasing and rental market. It became an autonomous company in January 2001. Aiming at the upper end of the market, Dispolok offers modern, high-performance Siemens-locomotives, including traction for cross-border operations, and can also provide maintenance and repair services, as well as spare parts supply and driver training.

Vehicles

In 2005, Dispolok offered three locomotive types: the ES 64 F4/ES 64 U2, Siemens 6,400 kW second generation development of the EuroSprinter family of high-performance three-phase electric locomotives, the 'F4' a four-voltage equivalent similar to the DB Cargo Class 189, the 'U2' variant a dual-voltage (15 kV/25 kV AC) machine and the ER 20 2,000 kW EuroRunner diesel-electric passenger and freight locomotive; the ES 64 P 6,400 kW EuroSprinter prototype electric locomotive; By May 2005, Dispolok had procured 60 examples of the ES 64 U2, 30 examples of the ES 64 F4, 10 examples of the ER 20 and ordered a further 15 examples of the ES 64 F4.

19 operators using Dispolok locomotives and making service in five countries (Austria, Germany, Italy, Slovenia, Switzerland).

Dispolok also offers a full service option covering maintenance of the locomotives it leases.

UPDATED

Simotra SAS

A member of the VTG – Lehnkering Group
33 avenue du Maine, PO Box 50, F-75755 Paris Cedex 15, France
Tel: (+33 1) 40 47 33 00   Fax: (+33 1) 40 47 33 67

Key personnel

Managing Director: P Boucheteil
General Manager: P Charbonnier

Vehicles

8,351 rail wagons.

STVA

Société de Transports de Véhicules Automobiles
Immeuble Le Cardinet, PO Box 826, F-75828 Paris
Cedex 17, France
Tel: (+33 1) 44 85 56 78   Fax: (+33 1) 44 85 57 00
e-mail: stva@stva.com
Web: http://www.stva.com

Key personnel
Chairman of the Executive Board: J P Bernadet
Managing Director: J Elissèche
    Finance and Administration: J J Pronzae
    Deputy Managing Director Business
        Development: J Henry
    Advisor to the Chairman: S Charles

Vehicles
Automobile transporters; full service (predelivery
inspection) throughout Europe.

Touax

Groupe Touax
Tour Arago, 5 rue Bellini, F-92800 Puteaux La
Defense, France
Tel: (+33 1) 46 96 18 00   Fax: (+33 1) 46 96 18 18

Touax is an operator of rail vehicles in France and
USA and is active in the leasing market.

Transfesa

Transportes Ferroviarios Especiales SA
Musgo 1, Urb La Florida, E-28023 Madrid, Spain
Tel: (+34 91) 387 99 00   Fax: (+34 91) 372 90 59
e-mail: transfesa@transfesa.com
Web: www.transfesa.com

Key personnel
Chairman and Executive Director:
    Emilio Fernández Fernández
General Manager: Luis Del Campo Villaplana
Division Managers:
    Industrial Logistics Manager: Arturo Boix Faubell
Motor Vehicle Manager: Abraham Peralta Arroyo
Chemical and Bulk Products Manager:
    Juan Diego Pedrero Sancho
General Cargo Manager:
    Juan Diego Pedrero Sancho
Financial Manager: José González Rodríguez
Organisation and Resources Manager:
    Pablo Rodríguez Mosquera
Technical Manager: José L Sánchez Humanes
Controller: Carmen Romero de la Calle
Communications and Corporate Image Manager:
    Julián Gacimartín Quiñones
Road Transport Manager: Jaime González López

Services
The company is primarily engaged in activities
covering the management of transport, distribution
and warehousing of goods and logistic services.

Vehicles
The company owns a fleet of 8,000 wagons with
interchangeable axles, as well as 2,500 swapbodies.

UPDATED

Transitio AB

Engelbrektsgatan 31, SE-114 32 Stockholm, Sweden
Tel: (+46 8) 54 50 17 49   Fax: (+46 8) 54 50 17 40
e-mail: bf@transitio.se
Web: www.transitio.se

Key personnel
Managing Director: Bo Fredriksson
Technical Director: Johnny Karlsson

Administration Manager: Anna Gustavsson
Rolling Stock Manager: Björn Asplund
Administrative Manager: Anders Hakansson

Organisation
Transitio AB was established and is owned by
regional public transport authorities in Sweden to
procure, finance and manage rolling stock used for
deregulated rail services sponsored by the regions.
The company owns the vehicles it procures and
is responsible for their heavy overhaul and any
eventual upgrading or refurbishment. Trains
procured include Regina emus supplied by
Bombardier Transportation.

UPDATED

Trinity Chemical Industries Inc

PO Box 701436, Tulsa, Oklahoma 74170
Tel: (+1 918) 495 35 00 (+1 918) 485 35 61
e-mail: info@trinitychem.com
Web: www.trinitychem.com

Key personnel
President: Richard B Fenimore
Vice-President: Terry L Fisher
Chief Financial Officer: Danny Kittinger
Sales: Ryan Edwards

Services
Long- and short-term full service leases for tank
cars of various sizes and covered hoppers for
plastic pellet transport. Lined and unlined tank cars
are available and special equipment options include
stainless steel tanks, heating coils, insulation
and magnetic gauging devices. Trinity Chemical
Industries also undertakes vehicle cleaning and
fleet management services and undertakes sub-
leasing and/or storage for companies with excess
cars in their fleet.

NEW ENTRY

TTX Company

101 North Wacker Drive, Chicago, Illinois 60606, USA
Tel: (+1 312) 853 32 23
Fax: (+1 312) 984 37 90
Web: http://www.ttx.com

Key personnel
President and Chief Executive Officer: A F Reardon
Senior Vice-President, Fleet Management: T F Wells
Senior Vice-President, Equipment and
    Engineering: R S Hulick

Products
TTX owns, maintains and rents to North American
railways, including operators in Mexico, a fleet of
freight wagons, principally flat wagons, for the
movement of containers, road trailers and new
automobiles and boxcars.

Union Tank

Union Tank Car Co – A member of The Marmon
Group of companies
175 West Jackson Boulevard, Chicago, Illinois
60604, USA
Tel: (+1 312) 431 31 11
Fax: (+1 312) 431 50 03

Main works
151st and Railroad Avenue, East Chicago, Illinois
46312, USA

Key personnel
For a full list of personnel, see Union Tank entry in
Freight vehicles and equipment section.

Products
Steel, stainless steel and aluminium tank wagons
for carrying liquids and compressed gases.
Covered hopper wagons for bulk plastics.

Vehicles
49,000 for lease in the USA and Mexico.

VTG Aktiengesellschaft

Nagelsweg 34, D-20097 Hamburg, Germany
Tel: (+49 40) 235 40
Fax: (+49 40) 23 54 11 99

Key personnel
Executive Board: Dr Heiko Fischer (Chairman),
    Jürgen Hüllen (Technology and Operation),
    Dr Kai Kleeberg (Finance)

Background
Shareholdings include: Rail4chem Eisenbahnver-
kehrsgesellschft mbH, Essen (25.0%); Transpetrol
GmbH Internationale Eisenbahnspedition, Hamburg
(80.0%); VOTG Tanktainer GmbH, Hamburg (60.0%);
VTG France SAS, Paris (0%); VTG Rail España
SL, Madrid (100.0%); VTG Rail UK Ltd, Worcester
(100.0%); VTG Schweiz AG, Basel (100.0%); VTG
Austria Ges mbH, Vienna (100.0%); Waggon Holding
AG, Zug (50.0%).

Products
VTG is one of the major hiring companies of
rail tank cars, bulk freight cars and other types
of rail freight cars in Europe. VTG has a fleet of
approximately 37,000 rail car with about 300
types for almost all liquid or pourable products
(chemical, compressed gas and mineral oil rail
tank cars, as well as other rail freight cars). VTG
offers comprehensive rail logistics services
via the majority interest in Transpetrol GmbH
Internationale Eisenbahnspedition (Hamburg).
VTG's holding TRANSWAGGON Group (Hamburg
and Zug) provides rail logistics for the European
car and paper industry with its extensive fleet of
over 9,200 modern high-capacity freight and flat
cars. Through its holding in VOTG, the group is one
of the world's leading tank container operators with
approximately 5,800 tank containers.

Vehicles
22,000 tank and special purpose wagons; 5,000
automobile-carrying wagons; 7,000 general
purpose freight wagons.

UPDATED

Wagonmarket Spol sro

Ulica Rovná 594/5, PO Box 25, SK-05801 Poprad,
Slovakia
Tel: (+421 52) 716 42 01; 716 42 03; 716 42 05
Fax: (+421 52) 716 42 27; 716 42 18
e-mail: wagonmarket@trinityraileurope.com
Web: www.trinityraileurope.com

Key personnel
General Manager: František Štupák
Commercial Manager: Zuzana Straková
Technical Manager: Peter Streber

Background
Wagon Spol sro is part of Trinity Rail Group.

Products
Sale and purchase of freight wagons, bogies,
subassemblies, tank cars and relevant spare parts.
    Purchase, sale and rebuilding of used (second
hand) freight wagons and bogies.

UPDATED

CONSULTANCY SERVICES

Alphabetical listing

Accent Marketing and Research
Advanced Railway Research Centre
AEA Technology Rail
ALK Technologies Inc
ALSTOM Transport
Arcadis Infra BV
Ardanuy Ingeniera SA
AREP
Atkins Rail
Austria Rail Engineering
Balfour Beatty Rail Technologies Limited
Banverket Consulting
Bechtel Corporation
Best Impressions
Blue Print Rail Ltd
BMT Reliability Consultants Ltd
Booz, Allen & Hamilton Inc
Brown & Root Services
Brunel Railmotive GmbH
Cambridge Systematics Inc
CANAC Inc
Canarail Consultants Inc
Capita Symonds Group
Carillion Rail
Certifer
Charles River Associates Incorporated
CIE Consult
Colin Buchanan and Partners
Colston, Budd, Wardrop & Hunt
Corradine Group, The
COWI Consulting Engineers and Planners AS
CPCS Transcom
Creactive Design
Creadesign Oy
Crown Agents for Oversea Governments and Administrations Ltd
Currie & Brown
DCA Design International Ltd
DE-Consult
Delcan Corporation
Design and Projects Int Ltd
Design Research Unit
Design Triangle
DHA
DHV Group
Doxiadis Associates SA
Edwards and Kelcey Inc
Electrowatt Infra Ltd
Engage
ENOTRAC AG
EPV-GIV
Esveld Consulting Services BV
EurailTest
Eurostation SA
FaberMaunsell
First Engineering Ltd
Fleet Software
Frazer-Nash Consultancy Ltd
GHD-Transmark Pty Ltd

GIBBRail Ltd
HaCon Ingenieurgesellschaft mbH
Halcrow Group Ltd
Hatch Mott MacDonald Inc
HDR Engineering Inc
Heery Transportation Group (HTG)
High-Point Rendel Ltd
Hill International Inc
Hodgson and Hodgson Group Ltd
Holland Railconsult
HTM Consultancy BV
Hyder Consulting
ICF Consulting
Interfleet Technology Ltd
Intermetric GmbH
International Technical Services
Intraesa
iQR
Italcertifer
Italferr SpA
Jacobs Babtie
James Scott
Japan Railway Technical Service (JARTS)
Jones Garrard Move Ltd
Labbé Designers and Associates Inc
Lahmeyer International GmbH
Laramore, Douglass and Popham
LEK Consulting LLP
Lend Lease
Lester B Knight & Associates Inc
Lloyd's Register Rail Limited
Lockheed Martin Rail Systems
LogoMotive GmbH
LTK Engineering Services
Martyn Cornwall Design
MBD Design
McCormick Rankin Corporation
Mercer Management Consulting Inc
Mer Mec Spa
Metroconsult
Millbrook Proving Ground Limited
Modjeski & Masters Inc
Mott MacDonald Group
MTR Corporation Consultancy Services
MVA Asia Ltd
MVA Group
NEA Transport Research and Training
NedTrain Consulting
Nichols Group, The
Ødegaard & Danneskiold-Samsøe A/S
OMI Logistics Limited
Ove Arup & Partners
Owen Williams Group
Pacific Consultants International
Parsons Brinckerhoff Inc
Patrick Engineering Inc
Philips Projects BV
PricewaterhouseCoopers
Prima Services Group Ltd

Prose Ltd
Ptarmigan Transport Solutions Ltd
QinetiQ
QSS Group, The
Radermacher & Partner GmbH
Rail Air International Ltd (RAIL)
Rail Sciences Inc
Rail Services Australia
Railway Consultancy Ltd, The
Railway Technology Strategy Centre
RailWorks Corporation
Ramboll Sweden AB
Ranbury Management Group
RIQC – Rail Industry Quality Certification Ltd
RITES Ltd
R L Banks & Associates Inc
RMS Locotec
Roundel Design Group
RTA Rail Tec Arsenal Fahrzeugversuchsanlage GmbH
Schofield Lothian
SCI Verkehr GmbH
Scott Wilson Railways
Semaly SA
Seneca Group LLC, The
Serco Raildata
Serco Railtest Limited
Solvera Information Services Ltd
Southdowns Environmental Consultants Ltd
Steer Davies Gleave
Strukton Railinfra BV
STV Group
SwedeRail
Systra SA
TAMS Consultants Inc
TDI-AVE Design and Engineering Services
TERA
Thomas K Dyer Inc
Tilney Shane Ltd
TLC
TMG International Pty Ltd
Tractebel Development Engineering
Transportation Technology Center Inc
TransTec
Transurb Technirail SA
Transys Projects Ltd
Trauner
Trenitalia SpA
TRL Ltd (Transport Research Laboratory)
Trowers & Hamlins
TUC Rail
Tyréns Infraconsult AB
Urbitran
VannessGroup
Vossloh Information Technologies York Ltd
Wendell Cox Consultancy
Wilbur Smith Associates
Wilson, Ihrig & Associates
YTT International Inc

Company listing by country

AUSTRALIA
Colston, Budd, Wardrop & Hunt
GHD-Transmark Pty Ltd
iQR
Rail Services Australia
Ranbury Management Group
TMG International Pty Ltd

AUSTRIA
Austria Rail Engineering
RTA Rail Tec Arsenal Fahrzeugversuchsanlage
    GmbH

BELGIUM
Tractebel Development Engineering
Transurb Technirail SA

CANADA
CANAC Inc
Canarail Consultants Inc
CPCS Transcom
Delcan Corporation
International Technical Services
Labbé Designers & Associates Inc
McCormick Rankin Corporation

CHINA
MTR Corporation Consultancy Services
MVA Asia Ltd

DENMARK
COWI Consulting Engineers and Planners AS
Ødegaard & Danneskiold-Samsøe A/S

FINLAND
Creadesign Oy

FRANCE
ALSTOM Transport
AREP
Certifer
EurailTest
MBD Design
Semaly SA
Systra SA
TUC Rail

GERMANY
Brunel Railmotive GmbH
DE-Consult
EPV-GIV
HaCon Ingenieurgesellschaft mbH
Intermetric GmbH
Lahmeyer International GmbH
LogoMotive GmbH
Metroconsult
Radermacher & Partner GmbH
TLC
TransTec

GREECE
Doxiadis Associates SA

INDIA
RITES Ltd

IRELAND
CIE Consult

ITALY
Italcertifer
Italferr SpA

Mer Mec SpA
Trenitalia SpA

JAPAN
Japan Railway Technical Service (JARTS)
Pacific Consultants International
YTT International Inc

NETHERLANDS
Arcadis Infra BV
DHV Group
Esveld Consulting Services BV
Holland Railconsult
HTM Consultancy BV
NEA Transport Research and Training
NedTrain Consulting
Philips Projects BV
Strukton Railinfra BV

SPAIN
Ardanuy Ingenieria SA
Intraesa

SWEDEN
Banverket Consulting
Ramboll Sweden AB
SwedeRail
Tyréns Infraconsult AB

SWITZERLAND
Electrowatt Infra Ltd
ENOTRAC AG
Prose Ltd

UNITED KINGDOM
Accent Marketing and Research
Advanced Railway Research Centre
AEA Technology Rail
Atkins Rail
Balfour Beatty Rail Technologies Limited
Best Impressions
Blue Print Rail Ltd
BMT Reliability Consultants Ltd
Brown & Root Services
Capita Symonds Group
Carillion Rail
Colin Buchanan and Partners
Creactive Design
Crown Agents for Oversea Governments and
    Administrations Ltd
Currie & Brown
DCA Design International Ltd
Design and Projects Int Ltd
Design Research Unit
Design Triangle
Engage
FaberMaunsell
First Engineering Ltd
Fleet Software
Frazer-Nash Consultancy Ltd
GIBBRail Ltd
Halcrow Group Ltd
High-Point Rendel Ltd
Hodgson and Hodgson Group Ltd
Hyder Consulting
Interfleet Technology Ltd
Jacobs Babtie
James Scott
Jones Garrard Move Ltd
LEK Consulting LLP
Lend Lease
Lloyd's Register Rail Limited

Martyn Cornwall Design
Millbrook Proving Ground Limited
Mott MacDonald Group
MVA Group
Nichols Group, The
OMI Logistics Limited
Ove Arup & Partners
Owen Williams Group
PricewaterhouseCoopers
Prima Services Group Ltd
Ptarmigan Transport Solutions Ltd
QinetiQ
QSS Group, The
Rail Air International Ltd (RAIL)
Railway Consultancy Ltd, The
Railway Technology Strategy Centre
RIQC – Rail Industry Quality Certification Ltd
RMS Locotec
Roundel Design Group
Schofield Lothian
SCI Verkehr GmbH
Scott Wilson Railways
Serco Raildata
Serco Railtest Limited
Solvera Information Services Ltd
Southdowns Environmental Consultants Ltd
Steer Davies Gleave
TDI-AVE Design and Engineering Services
Tilney Shane Ltd
Transys Projects Ltd
TRL Ltd (Transport Research Laboratory)
Vossloh Information Technologies York Ltd

UNITED STATES OF AMERICA
ALK Technologies Inc
Bechtel Corporation
Booz, Allen & Hamilton Inc
Cambridge Systematics Inc
Charles River Associates Incorporated
Corradino Group, The
DHA
Edwards and Kelcey Inc
Hatch Mott MacDonald Inc
HDR Engineering Inc
Heery Transportation Group (HTG)
Hill International Inc
ICF Consulting
Laramore, Douglass and Popham
Lester B Knight & Associates Inc
Lockheed Martin Rail Systems
LTK Engineering Services
Mercer Management Consulting Inc
Modjeski & Masters Inc
Parsons Brinckerhoff Inc
Patrick Engineering Inc
Rail Sciences Inc
RailWorks Corporation
R L Banks & Associates Inc
Seneca Group LLC, The
STV Group
TAMS Consultants Inc
TERA
Thomas K Dyer Inc
Transportation Technology Center Inc
Trauner
Urbitran
Vanness Group
Wendell Cox Consultancy
Wilbur Smith Associates
Wilson, Ihrig & Associates

Accent Marketing and Research

Gable House, 14-16 Turnham Green Terrace, Chiswick, London W4 1QP, UK
Tel: (+44 20) 87 42 22 11   Fax: (+44 20) 87 42 19 91
e-mail: info@accent-mr.com
Web: http://www.accent-mr.com

Key personnel
Managing Director: Rob Sheldon
Directors: Kate Barber, Chris Heywood, Miranda Mayes

Capabilities
Accent is a full service research agency, with offices in London, Bristol, Edinburgh and Munich and the resources and equipment to undertake both qualitative and quantitative studies of significant size.

Accent is an expert in research using trade-off techniques and has been instrumental in the introduction and development of the technique in the UK, having conducted many studies using these methods for high-profile clients.

Accent's research in the rail industry includes: customer priorities, estimating demand, real-time information, strategy and policy, ticketing, value of time and vehicle design.

Contracts
Recent research studies have been commissioned by the following UK-based clients:
- Strategic Rail Authority: research (conducted with Mott MacDonald) to help inform SRA's Value of Rail programme. The SRA was interested in the users of lightly used parts of the rail system and the disadvantaged. Market research with rail users was undertaken in five areas of the UK.
- Waverly Railways Client Group: research (undertaken with Halcrow) to estimate the demand for a new railway line between Edinburgh and Galashiels. 600 CATI stated preference interviews were conducted with residents making commuting or leisure trips along the corridor.
- Strategic Rail Authority: following the launch of a public consultation on the fares policy for UK rail operators in July 2002, Accent (with Mott MacDonald) assisted in the framing of the questions in the consultation document and in analysing the responses from stakeholders and the general public.
- GNER: to undertake research to provide a hierarchy and willingness to pay value for existing and potential features of StandardPlus. Also a study into stakeholder attitudes towards GNER, including a programme of quantitative research of customer usage, awareness and satisfaction with the on-train catering facilities that GNER currently offers. The assessment of customer attitude to GNER following the Hatfield rail crash and the subsequent disruptions. Research to examine the price elasticity between Weekend First ticket prices and standards of service available. Research was also undertaken to help define GNER's smoking policy.
- London & Continental Railways (responsible for building the Channel tunnel Rail Link): commissioned research with local residents, businesses and key opinion formers to determine the most popular name for the new station being built at Ebbsfleet, UK.
- Department of Transport (UK): interviews conducted, relating to the development of Transport Direct, a new transport information service.
- Lancashire County Council (UK): a pre-feasibility study into demand for a station to be located in the town centre of Skelmersdale.
- SRA: as part of a panel led by Mott MacDonald, including Oxera, First Class Partnerships and Line by Line to provide economic and transport forecasting advice and development of solutions to specific issues including project evaluation, statistical advice, financial modelling and regulatory and competition issues.
- Ove Arup & Partners: investigation into the potential of building facilities along a freight line which runs near Liverpool Football Club's new stadium.
- ATOC: market research into the need for a new ticket designed to encourage the use of rail by groups of two people travelling together. Also the estimation of passengers' willingness to pay for new and refurbished rolling stock through stated preference and revealed preference analysis techniques.
- First Great Eastern: independent objective assessment of the cleanliness of every First Great Eastern train arriving at Liverpool Street station. Also research was carried out to determine customer priorities with respect to rolling stock.
- OPRAF: with the University of Newcastle, research was undertaken into a number of areas concerned with rail services and infrastructure.
- Northern Spirit, Scotrail and First North Western: the measurement of disbenefits of 'crowdedness' on trains operated by the three operators.
- BAA: research was conducted to establish the price that should be charged to passengers on the Heathrow Express service in First Class and Standard.
- Eurotunnel: a customer profile survey was conducted of Le Shuttle's users. Information was required regarding customers' activities at the terminals.

Advanced Railway Research Centre (ARRC)

The Innovation Centre, 217 Portobello, Sheffield S1 4DP, UK
Tel: (+44 114) 222 01 51   Fax: (+44 114) 222 01 55
e-mail: n.farquhar@sheffield.ac.uk
Web: http://www.arrc.shef.ac.uk

Key personnel
Director: Dr Mark Robinson
Chairman: Brian Clementson
City Freight Research Manager: Tom Zunder
Composite Research Manager: Dr Joe Carruthers
Rail Freight Research Manager: Phil Mortimer
Administrator: Nicki Farquhar

Capabilities
Enhancement of contact between industry and academia by an information service and a series of seminars; focal point in UK for European land transport projects by small businesses; funding of a programme of railway-related research; teaching and training modules for students and industry professionals.

Projects
Hycoprod (2000–2004): the major objective of Hycoprod is to design advanced composite production processes for the systematic manufacture of very large monocoque hybrid composite sandwich structure for the transportation sector.

Bestufs (2000–2004): the aim of Best Practice in Urban Freight Solutions is establishing and maintaining an open European network between urban freight transport experts, user group/association, on-going projects, interested cities, the relevant European Commission Directorates and representatives of national transport administrations in order to identify, describe and disseminate best practices, success criteria and bottlenecks with respect to the movement of goods in urban areas.

IN-HO-TRA (2000–2003): The Innovative Horizontal Transhipment (IN-HOT-RA) project will validate innovative horizontal transhipment technologies, their interoperability and the possibilities to integrate them into current intermodal transport operations, in order to make intermodal transport more effective, more competitive and to decrease the economic break even distance of intermodal transport.

Themis (2000–2004): the prime objective of Thematic Network Intermodal Services (THEMIS) is to co-ordinate on-going activities for research and development in the field of Intermodal Freight Transport (IFT) information systems in Europe, while providing at the same time a forum for dissemination and concertation activities among all parties involved.

Cargospeed (2001–2003): Cargospeed demonstrates new technology to prove intermodal transfer efficiency gains. The project itself includes the design and construction of a modified wagon equipped with movable well-floor, as well as new road/rail interchange technology and economic feasibility calculations.

Trainsafe (2002–2004): the Trainsafe Thematic Network has a remit which will co-ordinate, advise and support the activities necessary to implement the Safety Strand for European Rail Research. Trainsafe will create links to key actors and research emanating from other transport modes. The objective is to set up a network of experts, establish a programme of workshops chaired by leading industry personalities, identify centres of excellence, mobilise a network of excellence and run a website for the benefit of the European railway industry as a whole.

Composit (2001–2003): the focus on The Future Use of Composites In Transport (Composit) is a tool to develop and strengthen the use of composite across the transport modes by identifying the appropriate technologies and solutions, and this is justified by the fact that transport is a key generator of economic growth.

Moldova (2001–2002): the Advanced Railway Research Centre (ARRC) has been successful in securing European Commission TACIS funding to advise on the restructuring of the railway system in Moldova. Moldova achieved independence from the Soviet Union in the early 1990s. It is one of the poorest countries in economic terms in Europe. The railway system will be a vital component in the development of the national economy. The railway system operates on the 1,520 mm gauge, and mainly with equipment designed and built in the former Soviet Union. The railway operates on very orthodox operational grounds and has lacked investment in equipment, systems and management methods.

City Freight (2002–2005): the City Freight project proposes a comparison of a number of innovations in freight transport in different European Metropolitan Conurbations, taking into account the existing experiences and knowledge regarding urban and inter-urban transport and its effects. The field of research will cover technical aspects and other issues such as: new traffic and parking management methods aiming to dissuade certain categories of goods vehicles to enter the city centre; new urban planning principle aiming to influence positively the freight transport demand patterns.

VERIFIED

Advantage Technical Consulting

18 Lion and Lamb Yard, Farnham GU9 7LL, UK
Tel: (+44 1252) 73 85 00   Fax: (+44 1252) 82 37 99
e-mail: enquiries@advantage-business.co.uk
Web: www.advantage-business.co.uk

Key personnel
Rail Business Manager: David Angove
Tel: (+44 1252) 82 37 05, (+44 7767) 24 77 48
e-mail: david.angove@advantage-business.co.uk

Background
Advantage Technical Consulting is a division of Advantage Business Group, an independent technical, management and systems organisation based in the UK and working in the rail, air defence, nuclear, manufacturing, oil and gas and process industries.

Capabilities

Technical consultancy specialising in safety management system development, safety project planning and management, safety engineering, safety case development, reliability analysis, reliability growth, reliability performance improvement, project/programme assurance, requirements management and systems integration; systems modelling and simulation, asset management consulting and CMMS configuration/implementation, independent testing of high-integrity software, cost modelling, investment appraisal and business case development, technical and management training.

Projects

Provision of safety and approvals management for East London Line extension programme (SRA/Transport for London); safety management of the ERTMS programme for UK implementation (Arriva Trains); system design modelling for Southern Region Power Supply Upgrade programme (Network Rail); root cause failure analysis of points and track circuits (Network Rail); safety management, requirements management and systems integration management for the Piccadilly Line extension to Heathrow's Terminal 5 (metro lines); maintenance optimisation review for the Class 442 fleet (South West Trains); business modelling, support to franchise mechanisms, technical and safety risk assessments (London Docklands Light Railway), business case model for remote condition monitoring systems (Network Rail).

UPDATED

AEA Technology Rail

Jubilee House, 4 St Christopher's Way, Pride Park DE24 8KY, UK
Tel: (+44 870) 190 10 00   Fax: (+44 870) 190 10 08
e-mail: customer_service.rail@aeat.co.uk
Web: www.aeat.co.uk

Key personnel

Chief Executive: Eddie Morland

Subsidiaries

AEA Technology Rail BV
Concordiastraat 67, PO Box 8125, 3503 RC Utrecht, Netherlands
Tel: (+31 30) 300 51 00   Fax: (+31 30) 300 51 50
Web: www.nl.aeat.com

Key personnel

Managing Director: Dirk LeClercq

Services

Delivers technology services and solutions to the Dutch and wider European rail markets.
ERSA – European Rail Software Applications
5, rue Maruice Blin, ZI Metzgerhof, F-67500 Haguenau, France
Tel: (+33 3) 88 07 15 50   Fax: (+33 3) 88 07 15 51
e-mail: info@ersa-france.com
Web: www.ersa-france.com

Key personnel

Technical Director: Patrick Deutsch

Services

Develops software for European rail traffic management systems.
AEA Technology Global SA
C/Rosa de Lima 1 Bis, Edificio ALBA, E-28290 Las Matas, Madrid, Spain
Tel: (+34 916)30 22 03   Fax: (+34 916) 36 91 21

Key personnel

Managing Director: Robert Grant

Services

Supplies and installs technology on Spanish Railways, delivers technical consultancy and is closely associated with the Spanish High Speed Railway construction programme.

Background

AEA Technology Rail is a business division of the group, AEA Technology plc.

Capabilities

AEA Technology Rail specialises in consultancy and technology services for rail industries across the world. Combining hardware, software and specialised consultancy to improve decision-making across the entire railway system.

AEA Technology is committed to optimising long-term performance for its clients in terms of reliability, safety and profitability. With its specialist knowledge and broad rail experience it can provide simple and effective solutions to clients' needs.

AEA Technology Rail operates primarily in Western Europe and North America from offices in the UK, France, Holland, Spain and US.

UPDATED

ALK Technologies Inc

1000 Herrontown Road, Princeton, New Jersey 08540, USA
Tel: (+1 609) 683 02 20   Fax: (+1 609) 683 02 90
e-mail: consulting@alk.com
Web: http://www.alk.com

Key personnel

Senior Vice-President: Mark A Hornung

Services

ALK specialises in information technology products and services for the transportation industry. Capabilities include strategic planning, operations control systems, locomotive management, marshalling and scheduling, computer simulation of railway operations, and geographic information systems.

ALK's PC*Miler® | Rail is routing, mileage and mapping software for the North American rail network, for determining routes and mileages using city/state abbreviations or geographic codes, used for rate determination and negotiation, rolling stock management and mileage auditing, carrier selection and *ad valorem* tax reporting.

Projects

ALK has undertaken a number of strategic planning studies for major railways, especially involving mergers, consolidations, and network rationalisation. It has also developed locomotive management systems for Canadian National, Southern Pacific and Union Pacific (USA) and a pricing system for TFM (Mexico). ALK has acted as consultant to Norfolk Southern (USA) on an interline trip planning system, which preplans the marshalling sequence and train assignments for freight wagons before the beginning of their journey. It monitors the progress of each wagon and update connecting railways and the shipper of deviations from plan.

ALK, acting as a subcontractor to RailInc (USA), has been responsible for the collection and enhancement of the US Surface Transportation Board Waybill Sample, the key US railway traffic database, for over 20 years.

ALSTOM Transport

Alstom Transport SA
48 rue Albert Dhalenne, F-93482 Saint-Ouen Cedex, Paris, France
Tel: (+33 1) 41 66 90 00   Fax: (+33 1) 41 66 96 66
Web: www.transport.alstom.com

Key personnel

Senior Vice-President: Charles Carlier

Services

Caters to operator and maintenance provider needs for rolling stock, equipment, locomotives, track and infrastructure after-sales related services.

ALSTOM offers a range of services to support in-house capabilities or to out-source service and maintenance activities of operators, rolling stock owners or maintenance providers for all makes of transit or freight rolling stock, for track and other infrastructure.

It provides modernisation and upgrade solutions for older applications as well as renovation proficiency to extend product life, enhance performance and increase passenger comfort.

The company also offers parts and repair identification, servicing and distribution for all applications.

UPDATED

Arcadis Infra BV

PO Box 220, NL-3800 AE Amersfoort, Netherlands
Piet Mondriaanlaan 26, NL-3812 GV Amersfoort, Netherlands
Tel: (+31 33) 460 43 72   Fax: (+31 33) 477 20 00
e-mail: e.mak@arcadis.nl
Web: http://www.arcadis-global.com

Capabilities

Comprehensive consultancy services related to railway infrastructure projects, including: track design, signalling and safety systems, power supply, hardware and software development, and civil engineering structures such as bridges and tunnels. Specific services include: project preparation; surveys, including soil investigation, hydrological research and environmental impact surveys; studies, including feasibility, architectural and marketing studies; planning; design and engineering; computer-aided services, including CAD, CAE, CAM and CARE; project management; implementation, including procurement, supervision and contract management, testing, validation and commissioning; operation and maintenance; and institutional development, including corporate and organisation development, management consulting and training.

NEW ENTRY

Ardanuy Ingeneria SA

Avenida Europa 34, Edificio B, E-28023 Madrid, Spain
Tel: (+34 91) 799 45 00
Fax: (+34 91) 799 45 01
e-mail: ardanuy@ardanuy.com
Web: http://www.ardanuy.com

Key personnel

General Manager: Josep-Maria Ribes
Projects Manager: Carlos Alonso
Technical Assistance Manager: Felix Ardiaca

Other offices

C/Dr Ferran No 11, E-08034 Barcelona, Spain
Tel: (+34 93) 206 33 00   Fax: (+34 93) 206 33 01

Tas Rue de Treves 49, B-1040 Brussels, Belgium
Tel: (+32 2) 230 59 50   Fax: (+32 2) 230 70 35
e-mail: belgium@ardanuy.com

Capabilities

Studies, projects, works supervision and advice services for railways, metros and tramways (signalling, communications, overhead, power substations, track, rolling stock) and tunnels.

Projects

Analysis and bid evaluation for the new system of ATP in Bulgaria for BDŽ; project of legalisation and work management of the power and traction substations of the FC Madrid–Arganda route; safety facilities, telecommunications, catenary and track project for the Castellbisbal–Mollet section; specialised assistance for the development of basic engineering and technical assistance for the purchasing control of the Valparaiso–Viña del Mar traffic interconnection. Valparaiso Regional underground; Phare cross-border co-operation between Bulgaria and Greece. Dupnitza–Kulata Railways Project; NISA. Gas introduction in the border region; project of modernisation of signalling system on the East-West Corridor in Latvia; basic project for the metro system in Seville; Nudo de Trinidad–Montcada section of the high-speed line between Madrid and the French border.

VERIFIED

AREP

163 bis, avenue de Clichy, Impasse Chalabre, F-75017 Paris, France
Tel: (+33 1) 56 33 05 08
Fax: (+33 1) 53 42 28 29
e-mail: contact@arep.fr

Key personnel
President: Jean Marie Duthilleul
General Manager: Etienne Tricaud
Director, International Projects: Eric Dussiot

Background
AREP is part of the SNCF Group and a subsidiary of SNCF Participations.

Capabilities
AREP is a multidisciplinary engineering and consulting firm that designs and builds urban transport centres and exchange hubs and public spaces. It makes its competencies available to carriers, decision-makers and investors in different sectors through nine departments covering: urban planning and layout; design, engineering and site supervision; international projects; programming; interior layouts and design; structures; building engineers; design and installation of utility systems and economic viability studies.

VERIFIED

Atkins Rail

Euston Tower, 286 Euston Road, London NW1 3AT, UK
Tel: (+44 20) 71 21 20 00
Fax: (+44 20) 71 21 21 11
e-mail: rail@atkinsglobal.com
Web: http://www.atkinsglobal.com/rail

Key personnel
Managing Director: Tony Fletcher
Property Director: Dr Robert Davis
Control and Systems Director: Chris Thompson
Commercial and Marketing Director:
    Graham Clench
Managing Director, Rail Infrastructure:
    Charles Burch
Director, Rail Vehicles: Dave Saunders
Head of Communications: Dale Lawrence
Director of Civil Engineering: Richard Molloy
Director of Electrification: Bob Ducksbury

Subsidiary company
Atkins Danmark A/S
Arne Jacobsen Alle 17, 2300 Copenhagen, Denmark
Tel: (+45) 82 33 90 00   Fax: (+45) 82 33 90 01
e-mail: info-dk@atkinsglobal.com
Web: www.atkinsglobal.com
Managing Director: Preben Olsen

Regional offices
Atkins Rail UK regional office locations: Birmingham, Crewe, Croydon, Daventry, Derby, Glasgow, Manchester, Orpington, Swindon, Waterloo and York.

Overseas offices
Atkins office locations: Australia, Belgium, Brunei, Canada, China, Czech Republic, Denmark, France, Gibraltar, Greece, Hong Kong, Hungary, India, Ireland, Japan, Kuwait, Malaysia, Mexico, Oman, Poland, Portugal, Romania, Saudi Arabia, Singapore, Spain, Sweden, Thailand, UAE and US.

Capabilities
Atkins Rail can take a project through its complete life-cycle from feasibility studies, planning, conceptual design and detailed design to safety, reliability and risk assessments, project and contract management, life-cycle costing, design implementation and whole life asset management support. Experience includes work undertaken for main line, freight, suburban, metro and light rail systems in the UK and worldwide.

Projects
Atkins Rail is engaged on a number of significant alliances with players within the industry. Its partnership with Carillion and Network Rail together forms the North Staffs Alliance, which has enabled the completion of the £75 million modernisation of the West Coast Main Line (WCML). Atkins is also a member of the alliance with Network Rail, Balfour Beatty and Carillion to deliver the £600 million upgrade and renewal of the West Coast Main Line electrification system including 2,130 miles of re-wired overhead line equipment together with 22 new electrical substations.

The company is a partner in the equity consortium Metronet Rail BCV Limited, incorporating the London Underground Bakerloo, Central, Victoria and Waterloo and City lines, and Metronet Rail SSL Limited, incorporating the Metropolitan, District, Circle, Hammersmith and City and East London lines.

A strategic feasibility study has been completed for the SRA to establish whether there is a transport and business case for constructing a new high-speed railway line in the UK from London to the north.

The rail vehicle group is assisting Network Rail in its early measures to ensure compliance with regulations demanding the installation of the Train Protection and Early Warning System (TPWS). Atkins is the designer of the Skanska DMC Borough viaduct team, which is part of the Thameslink 2000 scheme.

The company's rail business continues to support major projects in Scandinavia and Far East Asia. Growth has been most notable in the Scandinavian markets, in December 2001 it opened an office in Malmö, complementing the two offices it has in Copenhagen and Aarhus, Denmark.

Atkins Denmark is currently working on the Bombardier interlocking support, a two-year project worth £1 million, the Copenhagen Circle Line upgrading worth £2 million and the Odense-Svendborg main line upgrading valued at £2 million.

UPDATED

Austria Rail Engineering

Österreichische Eisenbahn, Transport Planungs-und Beratungsgesellschaft mbH
PO Box 54, A-1072 Vienna, Austria
Tel: (+43 1) 526 93 31   Fax: (+43 1) 526 93 31 85
e-mail: are@aon.at

Key personnel
General Manager: Ing. Friedrich Pichler

Capabilities
Austria Rail Engineering (ARE), the lead company of the Austrian railway sector for international activities, is a broadly based transportation management and engineering organisation. Founded in 1979, ARE is transferring tried and tested Austrian railway technology and provides technical and advisory services to governments and private companies worldwide; planning, engineering, operating, marketing and maintenance expertise for rail-bound transportation systems is provided by ARE.

In 20 railway co-operations ARE is acting as a co-ordinator.

Projects
Current work includes projects in Southeast Asia and the Balkan countries.

UPDATED

Balfour Beatty Rail Technologies Limited

Midland House, Nelson Street, Derby DE1 2SA, UK
Tel: (+44 1332) 26 24 24   Fax: (+44 1332) 26 22 95
e-mail: hayley.green@bbrail.com
Web: www.bbrail.com

Key personnel
General Manager: Neil Andrew
Engineering and Development Director:
    Charles Penny
Proposals Manager: Martin Barnett
Technical Services Director: Andrew Curzon

Services
Balfour Beatty Rail Technologies Limited provides technical products and services to enhance railway infrastructure asset performance. Its main areas of expertise include specialist signalling services (condition and event monitoring, automated testing equipment and evaluation software such as Asset View, SwitchView and CATS), mechanised inspection (ultrasonics, track geometry, video) and related plant services such as grinding equipment, high output equipment and track geometry correction services. Balfour Beatty Rail Technologies Limited is involved in rail management and has developed RAMS, a software-based rail asset management system and ACFM, a rail surface defect measurement.

Balfour Beatty Rail Technologies Limited has also formed a partnership with XiTRACK Limited to deliver XiTRACK®, a geo-composite solution for track strengthening using a polymer compound. There is a strong research and development team within the company which develops step change solutions such as slabtrack; it also offers direct consultancy in value engineering, risk management and civil engineering sectors.

UPDATED

Banverket Consulting

Industridivisionen
SE-781 85 Borlänge, Sweden
Tel: (+46 243) 44 61 00   Fax: (+46 243) 44 61 10
e-mail: consulting@hk.banverket.se

Key personnel
Director: John-Olof Hermanson
Area Directors: Sture Åberg, Rolf Ericsson,
    Leif Malm, Lars Moberg, Lennart Eldh and
    Jan Nilsson

Background
Banverket Consulting is a consultancy unit within Banverket (the Swedish National Rail Administration). Banverket Consulting has run its operation since 1998 and is based in Borlänge, with area offices in six locations throughout Sweden.

Capabilities
Banverket Consulting offers a range of services and products: services in railway research, project planning, and project and construction management within the railway sector. Its operations are IT-intensive with technical planning tools integrated in common computer and CAD environments. Also expertise within the areas of marshalling technology, carrying capacity, power supply simulations and track geometry.

Bechtel Corporation

Corporate Headquarters
50 Beale Street, San Francisco, California 94105-1895, USA
Tel: (+1 415) 768 12 34   Fax: (+1 415) 768 90 38
Web: http://www.bechtel.com

Key personnel
Chairman and Chief Executive Officer: Riley Bechtel
President and Chief Operating Officer:
    Adiran Zaccaria
Executive Vice-President and Deputy Chief
    Operating Officer: Jude Laspa

Capabilities

Bechtel offers a broad spectrum of services including feasibility and environmental studies, architectural/engineering design, project management, engineering management, construction management, start-up and operations, and financial planning in addition to engineering, procurement and construction.

Bechtel's transportation experience includes over 20 urban rapid transit systems and more than 5,600 miles of railways. The company has been involved in most new transit projects in the USA (Washington metro; Boston rapid transit; San Diego light rail; Sacramento light rail; Atlanta MARTA; San Francisco BART; Baltimore rapid transit; MTA/LIRR East Side Access in New York; and the Los Angeles metro), in domestic main line projects, including the Alameda Corridor-East (ACE) freight corridor upgrade in California, and in key international transit and rail projects, such as the Caracas metro, the São Paulo metro, Taipei rapid transit, Attika Metro in Athens, South Korea high-speed rail, the Western Corridor Railway linking Kowloon (Hong Kong) with northwest New Territories, and the Channel Tunnel Rail Link, Thameslink 2000 and Jubilee Line Extension in the UK.

Bechtel was retained by London Underground Ltd to provide a fast-track push to the commissioning and completion of the Jubilee Line Extension in time to support the official opening of the Millennium Dome in Greenwich. The system was commissioned in three phases culminating in the provision of through passenger services from Stanmore to Stratford on 20 November 1999. The final passenger station, Westminster, opened on 22 December 1999.

Best Impressions

15 Starfield Road, Shepherds Bush, London W12 9SN
Tel: (+44 20) 87 40 64 43   Fax: (+44 20) 87 40 91 34
e-mail: talk2us@best-impressions.co.uk
Web: www.best-impressions.co.uk

Key personnel

Director: Ray Stenning

Capabilities

Leaflets, maps, brochures, liveries, branding and brand development, corporate identity, websites, marketing and vehicle styling.

Projects

Ongoing design and marketing work for Stagecoach UK Bus, South West Trains, Chiltern Railways, Trent Barton Buses, Oxford Bus Company, Wilts & Dorset, Solent Blueline, Surrey County Council, Buckinghamshire County Council.

UPDATED

Blue Print Rail Ltd

Strutt House, Bridge Foot, Belper, Derbyshire DE56 2UA, UK
Tel: (+44 1773) 82 83 59   Fax: (+44 1773) 82 83 49
e-mail: info@blue-print-rail.co.uk
Web: http://www.blue-print-rail.co.uk

Key personnel

Directors: S P Chadwick, R P Gibney

Capabilities

Blue Print Rail Ltd specialises in rolling stock design including freight wagons, passenger vehicles and locomotives. The company's client base includes a large number of the major railways in Europe and the southern hemisphere.

Blue Print Rail can make available a broad spectrum of specialised activities to support a client's existing design capability, or to take on the role of a complete design office including design scrutiny and vehicle acceptance in the UK.

To support the design and general consultancy services a number of specialist skills are available

in-house including: 3-D CAD modelling; structural analysis using FEA software; rail vehicle gauging including development of kinematic envelopes and clearance assessment; rail vehicle dynamics; bogie design; tender response; and preparation of vehicle specifications.

BMT Reliability Consultants Ltd

Trading as BMT Rail
12 Little Park Farm Road, Fareham PO15 5SU, UK
Tel: (+44 1489) 55 31 00   Fax: (+44 1489) 55 31 01
Web: www.bmtrcl.com

Key personnel

Managing Director: Jim Lambert
Rail Business Director: Jacque Reynolds
Commercial Director: Ashley Fookes
Business Development Director: Bob Smith

Capabilities

BMT's fields of expertise include specialist services, depot and infrastructure divisions. Engineering consultancy services to reduce risk and improve reliability, safety and through-life economics of railway assets and processes. The company develops and applies techniques which assist infrastructure suppliers and operators to assess and optimise rolling stock reliability, maintenance, safety, risk and cost.

Projects

Independent Safety Assessor for the dmus for Northern Ireland Railways (Translink); risk assessment programme management for the Class 465 modification programme; strategic maintenance review on both the Class 465 and Class 91 locomotive; GNER high-speed train (HST) impact minute reduction programme; project management of both the BT 41 bogie overhaul and the Silverlink Class 508 refurbishment; reliability and safety of the Heathrow Express train for Siemens and CAF; independent safety assessment of the Networker Classic for Bombardier Transportation (formerly Adtranz); maintenance optimisation of the Swanley Junction switched diamonds for Balfour Beatty Rail Maintenance; through-life cost model development for ALSTOM; risk analysis in support of the East London Line extension private finance initiative application for London Underground; corrosion management of Classes 313 and 321 for HSBC Rail (UK) Ltd.

UPDATED

Booz, Allen & Hamilton Inc

Transportation Consulting Division
101 California Street, Suite 3300, San Francisco, California 94111-5855, USA
Tel: (+1 415) 391 19 00   Fax: (+1 415) 627 42 83
Web: http://www.boozallen.com

Key personnel

Principal: William T Reed
Contacts:
Commercial Management Consulting:
   Michael Bulger (Tel: (+1 212) 551 67 24)
Technology and Government Sector:
   Marie Lerch (Tel: (+1 703) 902 55 59)
Australasia: Jodie Collins (+61 2) 93 21 19 31
Brazil: Luiz Verdi (Tel: (+55 11) 55 01 63 05)
France: Nathalie Lhuillery (+33 1) 44 34 30 41
Germany, Switzerland and Austria:
   Susanne Mathony (+49 89) 54 52 55 50

Capabilities

Booz, Allen & Hamilton conducts assignments for passenger and freight railways spanning a broad range of functional areas and issues: vehicle engineering; operations and productivity improvement; strategic planning and reliability, maintainability and safety systems.

Projects

Include Netherlands high-speed rail, Croydon Tramlink, Attiko Metro, Channel Tunnel Rail Link,

St Louis Metrolink, Bay Area Rapid Transit, Los Angeles Metro Rail, Hudson-Bergen Light Rail, San Francisco Municipal Railway, and State Rail Authority of New South Wales.

Brown & Root Services

Contacts
Americas Region
1550 Wilson Boulevard, Arlington, Virginia 22209, USA
e-mail: Transportation@halliburton.com

Asia Pacific Region
186 Greenhill Road, Parkside South, Australia 5063
e-mail: International@hallliburton.com

Europe/Africa Region
Hill Park Court, Springfield Drive, Leatherhead KT22 7NL, UK
Tel: (+44 1372) 86 35 72
Fax: (+44 1372) 86 33 58
e-mail: Consulting_uk@halliburton.com

Key personnel

President: Randy Harl
Director, Transportation, Europe & Africa:
   Danny Grand
Public Relations: Ken Beedle
Tel: (+44 1372) 86 66 22
e-mail: ken.beedle@halliburton.com
Parent company: Halliburton Company
Web: http://www.Halliburton.com

Capabilities

Project management, engineering, life cycle and programme management from conceptual studies through to construction and operations, maintenance and logistics for major railway programmes anywhere in the world. Brown & Root employs 20,000 people around the world. Recently, the company signed an agreement to collaborate with Holland Railconsult BV, of Utrecht, Netherlands, aiming to improve technical capabilities and project management skills in the railway industry.

Brunel Railmotive GmbH

Head office
Gross-Berliner-Damm 73D, D-12487 Berlin, Germany
Tel: (+49 30) 632 23 00   Fax: (+49 30) 63 22 30 99
Web: www.brunel.de

Key personnel

Managing Director: Carsten Siebeneich

Double-deck driving car on test stand for testing centre of gravity of the fully equipped vehicle body at Railmotive's Görlitz test facility 1034785

Capabilities

Engineering services in the development of rail vehicles including design of coach body/underframe, bogie, interior fittings and furnishings, ventilation/air conditioning, driver's cab, electrical fittings, circuit diagrams. Supply of simulation strength analysis and vehicle running dynamics, consulting in acoustical problems. Railmotive is accepted by EBA for vehicle testing including running dynamics tests, brake tests, strength tests and acoustical tests. The company supplies technical documentation on high standards as prescribed by DB AG. Railmotive meets standardised specifications such as UIC directives, international standard DB AG design directives.

At Görlitz, the company owns and operates test tracks and static vehicle testing facilities, for example – derailment testing.

UPDATED

Cambridge Systematics, Inc

Headquarters
100 Cambridge Park Drive, Suite 400, Cambridge, Massachusetts 02140, USA
Tel: (+1 617) 354 01 67   Fax: (+1 617) 354 15 42
Web: http://www.camsys.com

Key personnel

President: Dr Lance A Neumann
Chief Operating Officer: Robert E 'Chip' Taggart
Travel Forecasting/Market Research:
    Senior Vice-President: Marc R Cutler
    Senior Principal: Dr Moshe Ben-Akiva
    Principal: Maren L Outwater
    Principal: Thomas F Rossi
Transportation Planning/Policy Analysis:
    Senior Vice-President: Arlee T Reno
    Senior Vice-President: Steven M Pickrell
    Senior Associates: Louis H Lambert;
        Anita P Vandervalk-Ostrander
Commercial Vehicle/ITS:
    Vice-President: Dr Vassili Alexiadis
Economic Investment Planning:
    Principals: Laurie L Hussey; John G Kaliski;
        Christopher Wornum
Air Quality Conformity:
    Principal: John H Suhrbier
Intermodal Freight Planning:
    Senior Vice-President: Lance R Grenzeback
    Principals: John G Kaliski; Michael J Fisher;
        Gary E Maring
Public Transport Service and Policy Planning:
    Principals: Laurie L Hussey; Samuel T Lawton III;
        Robert G Stanley; Stephen D Decker
Information Technology:
    Principals: Michael J Markow;
        Dr Nicholas J Vlahos
    Senior Associate: Dr John C Sutton
    Vice-Presidents: Hyun A Park, Brad W Wright

Other offices

4445 Willard Avenue, Suite 300, Chevy Chase, Maryland 20815, USA
Tel: (+1 301) 347 01 00   Fax: (+1 301) 347 01 01

555 12th Street, Suite 1600, Oakland, California 94607
Tel: (+1 510) 873 87 00   Fax: (+1 510) 873 87 01

Civic Opera Building
20 North Wacker Drive, Suite 1475, Chicago, Illinois 60606, USA
Tel: (+1 312) 346 99 07   Fax: (+1 312) 346 99 08

1820 East Park Avenue, Suite 203, Tallahassee, Florida 32301, USA
Tel: (+1 850) 219 63 88   Fax: (+1 850) 219 63 89

Capabilities

In partnership with its clients, Cambridge Systematics' analytical techniques are applied in many areas including transportation planning and management; intelligent transportation systems; information technology; asset management; commercial vehicle operations, travel demand forecasting and modelling; and market research. More specialised work includes new technology assessments, congestion management/air quality planning, traffic and transit planning, multimodal planning, growth management, decision support, and Geographic Information Systems (GIS).

Projects

Work has been undertaken for many federal, state and local agencies in the United States, Europe, Asia and other countries, as well as for private clients.

Cambridge Systematics has been awarded two major contracts by the US General Services Administration (GSA) for Management, Organisational and Business Improvement Services (MOBIS). This is a 20-year contract for the provision of consulting services, survey services and programme integration and project management services to agencies that contract through GSA MOBIS.

For the Federal Highway Administration (FHWA), Cambridge Systematics is leading a team in developing a core of open behavioural algorithms in support of traffic simulation with a primary focus on microscopic modelling, with supporting documentation and validation data sets. The state of Connecticut awarded Cambridge Systematics the right to negotiate a contract for the Commercial Vehicle Information Systems and Networks (CVISN)/Performance and Registration Information Systems Management (PRISM) project. The CVISN/PRISM project will be conducted for the Departments of Information Technology, motor vehicles, public safety, revenue services and transportation.

The company is continuing to work on two major contracts awarded by the US Department of Transportation (DoT), Federal Highway Administration (FHWA), one of which is a task order contract with FHWA's office of operations. Cambridge Systematics supports the operations office's initiatives through technical and policy studies, demonstration projects, evaluations, technical assistance and technology transfer. The company also supports the FHWA's Office of Environment and Planning in the areas of funding, planning, infrastructure management and environmental provisions under a major, multitask order contract.

Developments

Development of ITS Deployment Analysis System (IDAS), an Intelligent Transport Systems (ITS) sketch planning tool, to assist public agencies and consultants with integrating the deployment of ITS into the transport planning process.

CANAC Inc

3950 Hickmore Street, St Laurent, Québec H4T 1K2, Canada
Tel: (+1 514) 734 47 10   Fax: (+1 514) 734 48 64
Web: www.canac.com

Key personnel

President and Chief Executive Officer:
    Allen Alexander
Executive Vice-President, Chief Financial Officer, Finance and Administration: Benson Lewis
Senior Vice-President Rail Services:
    Isaac Haboucha
Senior Vice-President Business Development Rail Operations Services: Kevin Haugh
Vice-President, Planning and Engineering:
    Andy Cebula
Vice-President, Training Services:
    Michael Cournoyer
Manager Finances: François Savard
Director, International Rail Services:
    Jean-François Leroux

Background

A subsidiary of Savage Companies, CANAC was established in 1971 and has completed over 800 major projects in all areas of railway operations and engineering.

Capabilities

CANAC Inc provides services to industrial rail users; freight, passenger and commuter railroads; and investor and government authorities.

Capabilities include planning and engineering using the latest technology, in-plant rail logistics where services range from yard design and optimisation studies to full contract switching and a rail technical training capability incorporating instructors/trainers and subject matter experts with a wide-ranging railroad operations course inventory.

UPDATED

Canarail Consultants Inc

1140 de Maisonneuve Boulevard West, Suite 1050, Montreal, Quebec H3A 1M8, Canada
Tel: (+1 514) 985 09 30   Fax: (+1 514) 985 09 29
e-mail: inbox@canarail.com
Web: www.canarail.com

Key personnel

President: Jim D Speilman
Executive Vice-President, Mechanical
    Engineering: Harry Aghjayan
Vice-President and Chief Engineer:
    Donald R Gillstrom
Vice-President, Business Development and Human
    Resources: Elizabeth Tadgell
Vice-President: Harry Aghjayan

Capabilities

Consulting services for the urban and railway transportation sectors. Providing expertise in civil and mechanical engineering, light rail transit systems, railway operations, signalling and telecommunications, financial, economic, marketing, training, institutional and environmental disciplines. Studies carried out by Canarail include feasibility studies, transportation planning, human resources, financial and economic services, and asset valuation. Training services include needs analysis, programme development, testing and certification of railway personnel and the delivery of technical training courses for the railway industry. Management services provide assistance to clients with the restructuring, commercialisation and the divestiture of government-owned railways to the private sector, including the preparation of enabling regulatory and legislative policies for private sector development.

Engineering services include conceptual and detailed design, environmental assessment, preparation of bid documents, procurement and tendering services, construction supervision and start-up assistance.

Projects

Côte d'Ivoire: Feasibility study for a new mining rail line and port facility.
Croatia: Croatian Railways modernisation and restructuring project, double-tracking and fibre-optic study.
Egypt: Consulting services for rehabilitation, modernisation and maintenance of locomotives and improvements.
El Salvador: Evaluation of the feasibility of rehabilitating the Salvadorian railway and preparation of an action plan either to re-invest in the railway or to divest from it.
Guinea: Mine haul road design.
Morocco: Preliminary design studies aimed at improving the capacity of freight cars on Morocco's phosphate lines.
Saudi Arabia: Preparation of detailed engineering and tender documents for the construction of a new railway line between Riyadh and the Port of Dammam.
Taiwan: On-going training, operations and management of Taiwan High Speed Rail Corporation Construction Railway.
Uganda/Kenya: Transaction advisor to the Government of Uganda with regard to the privatisation of the Uganda Railways. Joint concessioning with Kenya Railways has been included in this process.
Uruguay: Reorganisation and rehabilitation study for the government of Uruguay.
Uzbekistan: Assistance to Uzbekistan Railways with the procurement and technical aspects of

locomotive re-powering and foundry workshops modernisation.

Assistance with implementation of restructuring of Uzbekistan Railways.

UPDATED

Carillion Rail

Gloucester House, 65 Smallbrook Queensway, Birmingham B5 4HP, UK
Tel: (+44 121) 345 15 13   Fax: (+44 121) 345 15 18
e-mail: lesdeakin@gtrm.co.uk
Web: http://www.carillionrail.com

Key personnel
Managing Director: Paul Kirk
Commercial Director: Mike Hawe
Director of Projects: Bob Collard and Mark Cutler

Background
Carillion Rail is a trading name of GT Railway Maintenance Ltd and is wholly owned by Carillion plc. Carillion Rail was formally launched in April 2002 as a new rail business combining the resources of GTRM, Centrac and Carillion Infrastructure, and has three business groups: Projects; Infrastructure Maintenance; and Rail Services. GTRM Ltd was previously owned by a joint venture between Alstom and Carillion plc (formerly Tarmac Construction Ltd). Formerly known as Central Infrastructure Maintenance Co, GTRM Ltd was one of the British Rail Infrastructure Service units, sold off to Alstom and Carillion in 1996 as part of the British Rail privatisation.

Carillion Rail has taken over Swedish Rail Systems Entreprenad (SRSE), the Swedish-based rail infrastructure maintenance specialist which also has interests in Norway, Denmark and Finland. SRSE currently has major contracts in Sweden with Malmo City Tunnel and Inlandsbanen. It has new assets worth about SKr80 million.

Services
Carillion Rail offers a range of specialist services in the following areas: permanent way, signalling, electrification, structures, business development, planning, quality and safety and project management.

Certifer

Agence de Certification Ferroviaire
BP 45, 154 boulevard Harpignies, F-59300 Valenciennes, France
Tel: (+33 3) 27 28 35 00   Fax: (+33 3) 27 28 35 09
e-mail: certifer.valenciennes@wanadoo.fr
Web: www.certifer.asso.fr

Background
Certifer was founded in 1997 by French National Railways (SNCF), the Paris Public Transport Authority (RATP), the French Railway Industries Federation (FIF) and the National Research Institute for Transport and Transport Safety (INRETS). In 1998, the Union of Public Transport (UTP) and the French Railways Infrastructure Authority (RFF) became members.

Capabilities
Certifer assesses the compliance of railway products and services with statutes and regulations, technical specifications and standards. It also carries out audits, inspections, expert evaluations and evaluations for authorisation. Its competences cover rolling stock, signalling and control systems and infrastructure.

UPDATED

Charles River Associates Incorporated

John Hancock Tower, 200 Clarendon Street, Boston, Massachusetts 02116-5092, USA
Tel: (+1 617) 425 30 00   Fax: (+1 617) 425 31 32

Key personnel
President: James C Burrows
Director of Transportation Projects: Kevin Neels

Vice-Presidents: Kevin Neels, Daniel Brand, Michael Kemp, George Eads, Steve Grundman, Christopher Cavanaugh
Other Senior Staff: Jon Bottom, Larry Shughart, Harry Foster, Nils Von Hinton Reed, Mark Kiefer, Masroor Hasan, David Cuneo, Venkat Swaminathan

Background
Charles River Associates was established in 1965 and has a staff of 642.

Offices
London, UK; Melbourne and Brisbane, Australia; Brussels, Belgium; Dubai, UAE; Wellington and Auckland, New Zealand; Mexico City; Toronto, Canada; and in the USA: Boston, New York, Washington, Philadelphia, Chicago, Dallas, Houston, College Station, Salt Lake City, Oakland, Palo Alto, Pasadena.
Established: 1965
Staff: 280

Capabilities
All aspects of transport planning and evaluation, including ridership and revenue forecasting, capital investment planning and budgeting, major investment studies, financing and pricing, management and operations planning, mergers and acquisitions, and market research. Specialist work is undertaken in travel demand and revenue forecasting for new systems, project and programme evaluation including ITS, travel surveys, transport economics, urban planning and computer applications.

Projects
Include planning for the initial and subsequent capital programmes for upgrading the New York City metro, bus and commuter rail systems, and the Chicago RTA's long-range capital plan. Has also undertaken high-speed ground transport studies in several intercity corridors, both North American and overseas, and various highway toll studies.

CRA's policy analysis work has included developing analytical tools and studying the impact of federal policies on the transit industry. Comprehensive fare policy studies have been undertaken for several transit agencies.

For the US, DoT has provided information for formulation of federal policy concerning strategy studies, major capital investments, privatisation, and productivity improvements. Benefits and costs attributed to new starts, extensions, modernisation projects and vehicle purchase programmes have been examined.

Work has also been undertaken to evaluate existing and proposed ground access modes to a number of airports. Has also examined air quality regulations and impact of alternative fuels.

NEW ENTRY

CIE Consult

Heuston Station, Dublin 8, Ireland
Tel: (+353 1) 703 47 00
Fax: (+353 1) 703 47 25
e-mail: info@cieconsult.ie
Web: http://www.cieconsult.ie

Key personnel
General Manager: Barry Collins

Capabilities
CIE Consult draws on the resources and expertise of CIE Group of operating companies, Iarnród Éireann (the Irish state rail network), Bus Atha Cliath (the Dublin City bus operator) and Bus Éireann (operator of all other bus services) to provide transport-related consultancy service across the world.

Projects
Recent contracts include:
Botswana: Supervising engineer for a new railway signalling system; strategic concessioning advisor.

Bulgaria: Railway organisational restructuring and management development of the railway infrastructure company.
Lithuania: Support to Lithuanian Railways for restructuring and privatisation.
Macedonia: Assistance with restructuring study of transport investment needs in Macedonia.
Romania: FIDIC engineer for supervision and co-ordination of works for interlocking systems in four main railway stations.
Tanzania: Privatisation and concessioning of Tanzania Railways Corporation.
Zambia: Development of new legislative framework, post concessioning and strengthening the ministry.
Zimbabwe: Concessioning of Zimbabwe Railways.
Various: Support to the Phare PMU in Bulgaria and Slovakia; Central Asia Railways restructuring studies; Pakistan Railways corporatisation; rail privatisation study in Georgia; and Russian Railways MIS study.
Others:
Hungary: Study of investment needs of BKV, the Budapest urban transport operator, financed by the World Bank.
Kyrgyzstan: Study of urban passenger and urban roads for the three cities of Bishkek, Osh and Djalalabad for the World Bank.
Latvia: Public transport master plan study. EU Phare funded.
Mongolia: Urban transport components of the World Bank transport sector rehabilitation project. World Bank funded.
Poland: Gdansk Urban Transport Project financed by EBRD.
Romania: Bucharest urban transport study. EU phare funded.
Russian cities: Improving urban passenger transport in Russian cities, as a model for other former Soviet Union cities, with the World Bank.
Uzbekistan: World Bank-funded study of urban transport in five cities to identify World Bank investment opportunities.

Colin Buchanan and Partners

Newcombe House, 45 Notting Hill Gate, London W11 3PB, UK
Tel: (+44 20) 73 09 70 00
Fax: (+44 20) 73 09 09 06
e-mail: cbp@buchanan.co.uk
Web: www.cbuchanan.co.uk

Other offices
Belfast, Bristol, Dublin, Edinburgh, Galway, Limerick and Manchester.

Key personnel
Chairman: Malcolm Buchanan
Director of Development, Transport: Andreas Markides
Public Transport Director: Roland Niblett
Director of Demand Management, Traffic, TfL, Traffic Safety: Derek Turner
Economics Director: Paul Buchanan
Transport Planning Director: Atholl Noon
Planning, Regeneration and Urban Design Director: Mike Wrigley
Transport Modelling Director: Chris Pyatt
Director of Development and QMS: Mike Mogridge

Background
Colin Buchanan was founded in 1964 and employs a staff which includes transport planners, traffic engineers, planning and regeneration specialists, urban designers, traffic modellers, economists and market researchers.

Capabilities
Colin Buchanan is an established firm of transport, planning and economics consultants, who provide advice on rail planning, land use, urban development, highway design, parking strategy, computer software, market research, public transport provision, safety management, specialist surveys and economics. This includes heavy rail, light rail and intermediate capacity modes. Also

demand forecasting and service planning for bus services and the impact of mass car use on public services using the company's computer programme BUSMODEL.

UPDATED

Colston, Budd, Wardrop & Hunt

Suite 71, Chatswood Village, 47 Neridah Street, Chatswood, New South Wales 2067, Australia
Tel: (+61 2) 94 11 79 22   Fax: (+61 2) 94 11 28 31

Capabilities
Railway operating consultancy; modelling of system performance; proving of computerised schedules; traffic optimisation.

The Corradino Group

200 South Fifth Street, Suite 300 North First Trust Centre, Louisville, Kentucky 40202, USA
Tel: (+1 502) 587 72 21
Fax: (+1 502) 587 26 36
e-mail: louisville@corradino.com
Web: http://www.corradino.com

Key personnel
Chief Executive Officer: Joe C Corradino
President: Burt J Deutsch
Executive Vice President: Fred P Pool
Vice Presidents: Joe M Corradino, Steve Sullivan

Head Office
First Trust Centre – 300N, 59th Market Street, Louisville, Kentucky 40202, USA
Tel: (+1 502) 587 72 21
Fax: (+1 502) 587 26 36
e-mail: jcorradino@corradino.com
Web: http://www.corradino.com

Projects
Indianapolis Region Transit Plan; 1-73 Project feasibility study; 1-75 corridor study; Pontire, Michigan downtown plan; Owensboro Airport runway extension; Louisville Airport improvement programme.

Louisville Light Rail Alternatives Analysis, Transitional Study.

Under TARC's 2020 Plan, Corradino developed a long-range plan for transit development in TARC's service area, which helped TARC gain federal approval for a Major Investment Study for light rail transit. Corradino was also a consultant for the US$1.1 billion Miami rail rapid transit system and for the US$4.5 billion Los Angeles Metro rail transit system, for which it developed a computerised public/private development cash flow model.

COWI Consulting Engineers and Planners AS

Parallelvej 15, DK-2800 Lyngby, Denmark
Tel: (+45) 45 97 22 11   Fax: (+45) 45 97 22 12
e-mail: cowi@cowi.dk
Web: http://www.cowi.dk

Key personnel
Managing Director: Klaus H Ostenfeld
Director, Rail, Metro and Tunnel:
   Arne Steen Jacobsen

Other offices
Also in Norway (Oslo), Germany (Berlin), Belgium (Brussels), Spain (Madrid), Lithuania (Vilnius), Poland (Warsaw), Russia (Moscow), USA (San Francisco), Canada (Vancouver), Tanzania (Dar es Salaam), Nigeria (Lagos), Kenya (Nairobi), Uganda (Kampala), Ghana (Accra), Burkina Faso (Tenkodogo), South Africa (Johannesburg), Bahrain (Manama), United Arab Emirates (Dubai), Oman (Qurum), Qatar (Doha), Saudi Arabia (Dammam), Philippines (Manila), Thailand (Bangkok), China (Beijing) and Vietnam (Hanoi).

Corporate background
Since its foundation in 1930, the firm has been involved in more than 25,000 projects in 110 countries. The number of employees totals 2,100.

COWI is privately owned with the COWI Foundation as the majority shareholder.

Capabilities
COWI offers consulting services at all stages of multi-disciplinary large-scale railway projects from initial planning and engineering design to construction management and supervision and advice on operation and maintenance. COWI's services range from professional advice on a particular problem to total coverage of services required by public and private clients from idea to realisation of railway systems. Feasibility studies, development of tender solutions and contract documents, managing of tender procedures, contracting, authority approval management, contract administration, environmental management, risk management, system certification management and maintenance management, analysis for the maintenance and reinvestment of railway infrastructure and administration.

Recent contracts include: to develop a maintenance management system for the Danish and Norwegian railway agencies; project management and consultant on civil works design for the 21 km driverless metro system, worth £600 million, being developed in Copenhagen; consultant for the conceptual design and outline design of the 6.5 km Malmo City Railway tunnel linking Malmo Central Station to Copenhagen, worth £550 million; for EU PHARE railway upgrading projects in eastern Europe, and in Denmark, COWI is project manager for the S-train ring line running around the centre of Copenhagen and design of the Flintholm station, the largest in Denmark.

CPCS Transcom

72 Chamberlain Avenue, Ottawa, Ontario K1S 1V9, Canada
Tel: (+1 613) 237 25 00   Fax: (+1 613) 237 44 94
e-mail: ottawa@cpcstrans.com
Web: www.cpcstrans.com

Branch office
4 Lansing Square, Ontario M2J 1T1, Canada
Tel: (+1 416) 499 26 90   Fax: (+1 416) 499 29 29

Key personnel
Chairman and Chief Executive Officer: Greg Wood
President: D H Page

Capabilities
CPCS Transcom is a privately owned international consulting firm specialising in transportation, telecommunications and commercialisation/privatisation. It provides technical, advisory and training services to governments and the private sector in the planning, engineering, operating, marketing and maintenance of transportation and telecommunications systems. Since its establishment in 1969, CPCS Transcom has successfully completed over 700 projects in over 60 countries around the world.

UPDATED

Creactive Design

22 New Street, Leamington Spa CV31 1HP, UK
Tel: (+44 1926) 83 31 13   Fax: (+44 1926) 83 27 88
e-mail: info@creactive-design.co.uk
Web: http://www.creactive-design.co.uk

Key personnel
Directors Sales and Marketing:
   Neil Bates, Tony Hume

Other office
St John's Innovation Centre, Cowley Road, Cambridge CB4 0WS, UK
Tel: (+44 1223) 42 11 41   Fax: (+44 1223) 42 10 36
e-mail: hans@creactive-design.co.uk

Anglia Railways train livery   0087709

Standard class interior   0087710

Capabilities
Specialising in transport projects, Creactive Design provides a design resource for transport design with a team of designers, rolling stock engineers and ergonomists.

The company offers interior and exterior design for refurbished and new rolling stock, safety and emergency design. Environmental design including seating, street furniture, signage, telephone kiosks, lighting and station/stop design. Creactive designs and supplies mock-up models, rigs and prototypes.

Projects
Currently the company is designing catering trolleys for Intercity trains and a cab ventilation system for metro trains.

Recent contracts include: design for London Underground of new metro trains, KCRC West Rail, Hong Kong for IKK (Itochu, Kawasaki, Kinki); refurbishment and manufacture of D78 stock for London Underground; design of Warwick Parkway railway station for Chiltern Railways/Birse Rail, as well as for Anglia Railways, Bombardier Transportation and the Greater Nottingham Rapid Transit.

Creadesign Oy

Laivanvarustajankatu 5, FIN-00140 Helsinki, Finland
Tel: (+358 9) 251 21 00   Fax: (+358 9) 60 58 32
e-mail: info@creadesign.fi
Web: http://www.creadesign.fi

Key personnel
Managing Director: Hannu Kähönen

Capabilities
Industrial design, including rolling stock styling and interiors, from strategic concept design to product launch; corporate image; marketing communication.

Projects
Projects include interior and exterior design of the Variotram built by Bombardier Transportation and Talgo for HKL, Helsinki. Design work has also been undertaken for VR Ltd (Finnish Railways) and Talgo.

VERIFIED

Crown Agents for Oversea Governments and Administrations Ltd

St Nicholas House, St Nicholas Road, Sutton SM1 1EL, UK
Tel: (+44 181) 643 33 11   Fax: (+44 181) 643 82 32
e-mail: john.wrighton@crownagents.co.uk
Web: www.crownagents.com

Key personnel
Chief Executive: Jack Garvey
Director, Supply Chain Services: Keith White
Head of Railways: J A Wrighton

Capabilities
Crown Agents specialises in providing consultancy and procurement services associated with locomotives, rolling stock and track renewal and rehabilitation, and the provision of assistance with railway management reforms, restructuring and privatisation. Services are provided internationally with specialist knowledge of international procurement regulations and conditions.

Projects
Recent contracts include: feasibility studies for the manufacture of track-grinding trains in Russia (1997); cost verification and technical studies for track rehabilitation and upgrading in Russia (1999–2001); maintenance management of Class 36 locomotives, Tanzania Railways (1999–2002); contracts management for Railtrack recovery cranes overhaul and enhancement (1997–2000); and assistance with procurement, development and contracting for locomotive rehabilitation projects, Romania and Serbia Railways (2002–2003). Procurement of US$3 million timber sleepers for Pakistan Railways under Japanese financing (2002–2003).

UPDATED

Currie & Brown

Fortuna House, South Fifth Street, Milton Keynes MK9 2EU, UK
Tel: (+44 1908) 20 70 40   Fax: (+44 1908) 20 70 41
e-mail: colin.ashwood@curriebrown.com
Web: www.currieb.com

Key personnel
Chairman: A Angus McLean

Other offices
140 London Wall, London EC2Y 5DN, UK
Tel: (+44 20) 76 00 87 87   Fax: (+44 20) 77 26 23 98

Level 6, 67 Albert Avenue, Chatswood, Sydney New South Wales 2067, Australia
Tel: (+61 2) 94 15 16 00   Fax: (+61 2) 94 15 14 43

TM Hiroo Building, 7th Floor, 1-9-20 Hiroo, Shibuya-ku, Tokyo 150-0012, Japan
Tel: (+81 3) 34 42 66 42   Fax: (+81 3) 34 42 19 58

Suite 2047, 45 Rockefeller Plaza, New York, New York 10111-0100, US
Tel: (+1 212) 332 32 08   Fax: (+1 212) 332 32 09

Capabilities
Strategic procurement advice; procurement of works; commercial management; schedule and programme management; risk management; alliance and partnering advice; project management; project performance measurement; value management; and supply chain management.

Projects
Include: a range of services for signalling, telecommunications, electrification and track renewals for Network Rail in the UK; services connected with station reconstruction and refurbishment, health and safety measures, CCTV systems and maintenance works for London Underground Ltd; establishment of a Project Implementation Unit for Azerbaijan Railways; provision of contractor and third party management systems for Irish Rail; and services connected with signalling rehabilitation for Victoria Public Transport Corporation, Australia.

UPDATED

DCA Design International Ltd

19 Church Street, Warwick CV34 4AB, UK
Tel: (+44 1926) 49 94 61   Fax: (+44 1926) 40 11 34
e-mail: transport@dca-design.com
Web: www.dca-design.com/transport

Key personnel
Chairman and Managing Director: Rob Woolston
Directors: Rob Bassil, John Daly

Capabilities
Multidisciplinary design consultancy specialising in visual, ergonomic and component engineering aspects of transport design. Services include exterior styling, interior design, engineering and electronic design, corporate design, model-making, ergonomics/human factors, CAD, detailed drawing, computer visualising and animation, 'fast track' product development including 3D CAD. Large in-house workshop facilities enable construction of full-size mockups, prototypes and models. ISO 9001.

Projects
These have included designs for new rolling stock for London Underground Ltd's Central Line; visual and driver ergonomic aspects of the British Rail Class 90 and 91 locomotives; design of the Tangara double-deck commuter trainset for the State Rail Authority of New South Wales, Australia; Channel Tunnel shuttle wagons and locomotives for European and Canadian members of the ESCW and ESCL consortia; refurbishment of London Underground Ltd's Metropolitan Line trainsets; design of the British Rail Class 341 emu for the proposed CrossRail line; the Class 365 'Networker Express' trainset; Class 371 Thameslink 2000 trains; train and corporate design work for the new cross-border train services between Belfast and Dublin; new and refurbishment concepts for MTRC in Hong Kong; M6 double-deck cars for SNCB, Belgium; and mockups and detailed design of Virgin CrossCountry demus; TfL PPP ITT support; and Eurostar refurbishment seating design and supply and interior detail design and specification.

UPDATED

DE-Consult Deutsche Eisenbahn-Consulting GmbH

Bornitzstrasse 73-75, D-10365 Berlin, Germany
Tel: (+49 30) 63 43 11 15   Fax: (+49 30) 63 43 10 51
Web: http://www.de-consult.de

Business Department
International Transportation and Railway Service
Oskar-Sommer-Strasse 15, D-60596 Frankfurtam Main
Tel: (+49 69) 631 90   Fax: (+49 69) 631 92 95
e-mail: ITRS@de-consult.de
Web: http://www.de-consult.de

Key personnel
Supervisory Board Chairman: Martin Bay
Board of Managing Directors:
   Chair: Karl-Heinz Fleischmann
   Gerd Wiederwald
Head of Business Unit: Thomas Eckart

Background
DE-Consult is an independent transport consultancy firm founded in 1966. Its sole shareholder is DB ProjektBau GmbH, a 100 per cent subsidiary of Deutsche Bahn AG. It employs approximately 538 people worldwide.

Capabilities
Range of advisory services including planning and management of complex infrastructure projects involving long-distance and urban passenger and freight transport, rolling stock and workshops, management consultancy, operations planning, transport economics, finance and marketing, manpower development and training.

Projects
DE-Consult has been involved in over 1,000 projects internationally. High-speed projects in Germany, Korea, Taiwan, Spain and Italy; suburban transport systems projects in Thailand, Greece, Netherlands and Germany; rehabilitation projects in Eastern Europe, Africa, America and the Far East; freight transport projects in Germany, South America, Eastern Europe and Africa; training projects in South America, Eastern Europe, Africa and Asia.

Delcan Corporation

133 Wynford Drive, Toronto, Ontario M3C 1K1, Canada
Tel: (+1 416) 441 41 11   Fax: (+1 416) 441 41 31
e-mail: info@delcan.com
Web: www.delcan.com

Key personnel
President and Chief Operating Officer: Jim Kerr
President, National Engineering Technology: Jeanine Prince
Vice-President, Programme and Project Management: Charles Orolowitz

Overseas offices
Greece, Hong Kong, Israel, Taiwan, US (Atlanta, Chicago, Los Angeles, Salt Lake City and Washington, DC) and Venezuela.

Background
The DHV Group holds a 40 per cent interest in Delcan Corporation.

Capabilities
A full range of consulting services ranging from studies to the overall design and construction supervision of large railway infrastructures. Project management, feasibility studies, engineering, compliance auditing, implementation, operations and maintenance for urban transit, commuter, intercity passenger and freight railways.

UPDATED

Design and Projects Int Ltd

Wessex House, Upper Market Street, Eastleigh SO50 9FD, UK
Tel: (+44 23) 80 61 60 66   Fax: (+44 23) 80 61 60 68
e-mail: dpil@msn.com
Web: http://www.railwaymaintenance.com

Key personnel
Managing Director: Colin Brooks
Operations Director: Stuart Blyth

Capabilities
Design, supply and setting to work of equipment needed to overhaul, maintain, repair and clean rail vehicles and their components for metro, main line and suburban railway systems. This includes all depot/workshop, track and overhead catenary system maintenance equipment.
While the company normally executes contracts on a turnkey basis, projects can also be undertaken for equipment and supplier studies, and the design of one-off specialist equipment. The company can also provide other engineering and management support in relation to railway maintenance facilities, purpose equipment, design and supply of diagnostic test equipment for rail vehicles; supply of maintenance equipment for track work, signalling and all fixed systems.

Projects
Contracts include the supply of maintenance equipment for a people mover at Chep Lap Kok airport, Hong Kong, for Manila LRT3 and for the Arlanda Airport link.
Nottingham NET (Adtranz, March 2000), London Underground, Victoria Line depot upgrade (March 2000), Singapore Senbang/Punggol (MHI, September 1999) and Tashkent depot rehabilitation (September 1999).

Design Research Unit

The Old School, Exton Street, London SE1 8UE, UK
Tel: (+44 20) 76 33 97 11
Fax (+44 20) 72 61 03 33
e-mail: info@dru.co.uk
Web: http://www.dru.co.uk

Key personnel
Directors: Hugh Crawford, Maurice Green, Paul Cook, Peter Austin

Other offices

Design Research Unit International
2,103 Universal Trade Centre, 3 Arbuthnot Road,
Hong Kong
Tel: (+852) 23 77 47 37   Fax: (+852) 27 36 64 57
e-mail: general@designresearchunit.corn.hk

Design Research Unit Gulf (LLC)
PO Box 4233, Dubai, United Arab Emirates
Tel: (+971) 42 24 65 65   Fax: (+971) 42 28 09 69
e-mail: dru@emirates.net.ae

Capabilities

Station planning, architecture, interior design,
graphic design, wayfinding and urban planning.

Projects

Architectural and planning work has been undertaken
for many transport authorities. Current projects
include rail/light rail and metro systems in Bangkok,
Copenhagen, Hong Kong, London and Nottingham
(UK). Completed works include metro and rail
systems in Athens, Baghdad, Birmingham, Kuala
Lumpur, Singapore, Taipei and Toronto. Graphic
design and wayfinding for Network Rail, Docklands
Light Rail, South West Trains and Centro in UK.

Design Triangle

The Maltings, Burwell, Cambridge CB5 0HB, UK
Tel: (+44 1638) 74 30 70   Fax: (+44 1638) 74 34 93
e-mail: mail@designtriangle.co.uk
Web: http://www.designtriangle.com

Key personnel

Partners: Siep Wijsenbeek, Andrew Crawshaw,
   Andy Clark

Capabilities

Design Triangle specialises in the design of rail
vehicles: interiors, exteriors, cabs and components.
An integrated team of industrial designers and
engineers provides innovative concepts and develops
them through to manufacturing drawings and data.

Services include: research to improve understanding
of customer needs; innovative concepts for
competitive new products; 3-D CAD visualisation
and animation, allowing effective communication
of ideas; engineering development through design-
for-manufacture, analysis and the refinement of
forms; prototypes, allowing the testing and approval
of designs prior to manufacture; and consultancy.
Clients include service operators, major vehicle
manufacturers and component manufacturers.

Projects

Exterior and interior design of the Heathrow
Express train; exterior and interior design of
67 new carriages for Irish Rail Intercity for CAF as
part of an ongoing collaboration; tender designs
for Melbourne Hillside for ALSTOM; Spoornet
9E loco cab interior refurbishment for ALSTOM;
consultants to STIB, Brussels for the Tramway 2000;
exterior design and cab design of the Hong Kong
Airport Express train for the MTR Corporation,
including detrainment device; development of
exterior styling for the TKE train, also for the MTR
Corporation; consultancy for London Underground
planning standards; capacity and passenger flow
studies for Docklands Light Railway, UK; modular
seating prototypes for KAB Seating; design and
engineering for BAE Systems and Kawasaki for

*CAD design for Heathrow Express by Design
Triangle* • 0016444

Maryland double-deck cars; interior and exterior
design of RET Rotterdam metro; design of new
rolling stock for Metro de Madrid and many light
rail vehicles.
                                          VERIFIED

DHA

Delon Hampton & Associates
800 K Street NW, North Lobby, Suite 720,
Washington DC 20001, USA
Tel: (+1 202) 898 19 99   Fax: (+1 202) 371 20 73
e-mail: dhafbeach@aol.com

Key personnel

Chairman of the Board and Chief Executive
   Officer: Dr Delon Hampton (President of the
   American Society of Civil Engineers)
President and Chief Operating Officer: Elijah B Rogers
Executive Vice-President and Principal:
   Foster J Beach III, PE

Capabilities

Design, planning and inspection of rapid transit and
light rail systems and other transportation structures;
programme and construction management
services; planning, design and construction support
services and construction inspection services.

Projects

Projects include Program Management Oversight
consultant for the Federal Transit Authority
overseeing a five-year (1999–2004) project
for New Jersey Transit Corporation, Connecticut
Department of Transport, and Metro Northern
Railroad; construction management services
for the Los Angeles River Bridge for the
Almeda Corridor Railroad System completed 1999
and also for the Washington Boulevard/Santa Fe
grade separation; design and programme
management for the Memphis Area Transit Authority
during 1999; and assisting with evaluation of Tren
Urbano Transit System in San Juan Puerto Rico.

DHV Group

PO Box 219, NL-3800 AE Amersfoort, Netherlands
Tel: (+31 33) 468 37 00   Fax: (+31 33) 468 37 48
e-mail: info@dhv.nl
Web: http://www.dhv.com

Key personnel

Chairman of Executive Board: Renko G Campen
Executive Board: Bertrand M van Ee
Communications Manager: Jeannette van Enst

Background

DHV Group is an international consultancy and was
founded in 1917.

Capabilities

Management consultancy, advice, design and
engineering, project management, contract
management and operational management.

Projects

As a partner in the High-Speed Rail Link-South Project
Organisation, DHV developed noise barriers which
not only limit the noise nuisance for the surrounding
area but which can also be used for the sustainable
generation of electricity. The company has also been
closely involved in the planning and design of the five
tunnels required for the high-speed link.

DHV is also conducting research into techniques
of laying track beds on the soft terrain found in
delta areas.

Other recent projects include transport and
infrastructure in China.

Doxiadis Associates SA

13 Aegidon & Seneka str, GR-145 54 Nea Kifissia,
Athens, Greece
Tel: (+30 210) 624 63 00   Fax: (+30 210) 624 63 99

e-mail: doxiadis@doxiadis.com
Web: www.doxiadis.com

Key personnel

Managing Director: Yiannis Pasgianos
Vice-Chairman, Business Development:
   Anastasios C Antonopoulos

Other offices

Doxiadis Associates SA
PO Box 1574, Riyadh, 11441 Riyadh, Saudia Arabia
Tel: (+966 1) 476 28 00
e-mail: doxiadis@zajil.net

Doxiadis Associates SA
c/o Cato Manor Development Association, Intuthko
Junction, 750 Francois Road, Durban, South Africa
Tel: (+27 31) 261 66 40
e-mail: doxpk@iafrica.com

Background

Doxiadis Associates SA was established in 1951
and the majority shareholding was acquired by
Metrotech in 1999 through an increase of the
company's capital.

Capabilities

Transport planning and engineering design, traffic
management, analysis and design, urban planning,
project management, highway engineering,
construction supervision and maintenance. Also
participates in study teams for preparation of
comprehensive development, regional and urban
plans.

Since it was established the company has been
awarded major development projects in over
50 countries worldwide by clients that include
governments, private developers and international
organisations.
                                          UPDATED

Edwards and Kelcey Inc

299 Madison Avenue, PO Box 1936, Morristown,
New Jersey 07962-1936, USA
Tel: (+1 973) 267 05 55
Fax: (+1 973) 267 35 55
Web: http://www.ekorp.com

Key personnel

Chairman and Chief Executive Officer:
   Kevin J McMahon
Executive Vice-Presidents: Kenneth J Garrity;
   Richard M Hallahan; Mark G Pilla;
   Richard E Tangel

Offices

USA
Baltimore, Maryland; Boston North, Massachusetts;
Boston South, Massachusetts; Chicago, Illinois;
Cincinnati, Ohio; Edison, New Jersey; Fort
Lauderdale, Florida; Houston, Texas; Indianapolis,
Indiana; Jacksonville, Florida; Kittery, Maine;
Leesburg, Virginia; Londonderry, New Hampshire;
Manchester, New Hampshire; Miami, Florida;
Minneapolis, Minnesota; Morristown, New
Jersey; New York, New York; Palm Beach, Florida;
Philadelphia, Pennsylvania; Providence, Rhode
Island; Saratoga Springs, New York; Tampa, Florida;
Tarrytown, New York; Washington, DC; West
Chester, Pennsylvania.

International
Carolina, Puerto Rico.

Capabilities

Environmental, planning, design and construction
services for railways, mass transit, highways,
airports and ports. Urban transport services
include: terminals and stations, railways and
metros, tunnels, maintenance shops and yards,
track, catenary support structures, bridges, and
parking. Other services include alternatives analysis
and transport planning, patronage forecasting,
ridership surveys, urban freight movement, traffic
control systems, traffic impacts, circulation studies,
route and corridor selection, busways, cycleways
and pedestrianways.

Electrowatt Infra Ltd

Hardturmstrasse 161, PO Box CH-8037 Zurich, Switzerland
Tel: (+41 1) 355 55 55 Fax: (+41 1) 355 55 56
e-mail: martin.bachmann@ewi.ch
Web: www.ewi.ch

Key personnel
Head of BA Transportation: Johann Schmieder

Regional offices
Asia, eastern Europe, Germany, Latin America, Middle East, UK.

Background
Electrowatt Infra is a member of the Jaakko Pöyry Group.

Capabilities
Feasibility studies, environmental studies, modelling and data processing, economic assessment; safety/security consulting, planning and engineering of rail systems; specifications, tender documents, bid evaluations; supervision of manufacture and installation; project management; planning of timetables; consultancy for BOT projects, operation, maintenance and outsourcing.

Projects
Germany: construction supervision of sections of the high-speed railway line between Cologne and Frankfurt.
Poland: planning the upgrading of the railway route from Warsaw to Belarus.
Switzerland: design and construction supervision of the new Gotthard rail tunnel. Design of new railway Centre Line, Zurich.
Taiwan: Taiwan high speed rail, independent site and checking engineer.
Venezuela: Metro Valencia, overall project management.

UPDATED

Engage

A Shop, Derby Carriage Works, Litchurch Lane DE24 8AD, UK
Tel: (+44 1332) 29 99 88   Fax: (+44 1332) 25 17 64
e-mail: david.peel@engage-kgn.com
Web: http://www.engage-gkn.com

Key personnel
Business Development Manager: David Peel

Capabilities
Predictive analysis services for rail application including: structural analysis; joint analysis, thermal analysis, modal analysis and crashworthiness.

ENOTRAC AG

Postgaessli 23, CH-3661 Uetendorf, Switzerland
Tel: (+41 33) 345 62 22   Fax: (+41 33) 345 62 25
e-mail: info@enotrac.com
Web: http://www.enotrac.com

Key personnel
Executive: Heinz Voegeli

Subsidiary
ENOTRAC UK Ltd
Times House, Throwley Way, Sutton SM1 4AF, UK
Tel: (+44 20) 87 70 35 01   Fax: (+44 20) 87 70 35 02
e-mail: ziad.mouneimne@enotrac.com
Executive: Dr Ziad S Mouneimne

Capabilities
ENOTRAC provides consulting services covering systems engineering, feasibility studies, planning, technology evaluation, tender preparation and evaluation, asset replacement strategy, equipment specification, procurement support, software development, field tests, quality assurance, reliability and safety assessments, signalling compatibility studies and operational procedures.

For rolling stock, the services encompass performance evaluation, energy consumption, comparative assessment of traction equipment, rehabilitation and maintenance management.

For fleet management, ENOTRAC provides VIPSCARSIS, the software system for configuration, warranty, maintenance and modification management including LCC- and RAM-calculations, tailor-made for rolling stock and fixed installations. The services include process studies and consulting, workshops, training and full user support.

Power supply services include rating of equipment (substations, catenary), optimum substation spacing, reinforcement requirements, short-circuit calculations and protection, earthing, step and touch voltages, and energy, active and reactive power requirements and magnetic field computation. Optimised design is achieved by a powerful software suite developed in-house for multitrain simulation of complex AC and DC-supplied networks.

VERIFIED

EPV-GIV

Europrojekt Verkehr Gesellschaft für Ingenieurleistungen im Verkehrswesen mbH
Markgrafendamm 24, Haus 16, D-10245 Berlin, Germany
Tel: (+49 30) 29 38 06 20   Fax: (+49 30) 29 38 06 21
e-mail: geschaeftsleistung@epv-giv.de
Web: http://www.epv-giv.de

Background
EPV-GIV was created by the amalgamation in 1994 of EPV (Europrojekt Verkehr GmbH & Co) and GIV (Gesellschaft für Ingenieurleistungen im Verkehrswesen GmbH).

Capabilities
The company offers consultancy services in: project management; project financing; transport planning and technology; marketing; training and further education; and project co-ordination. These include the preparation and assessment of: analyses and studies for the development of traffic in Europe; strategic concepts for the development of transport; complex transport planning for countries and regions; studies regarding transport planning and analysis; and concepts, traffic streams, market and demand forecasting in the areas of passenger and freight traffic, in particular for railways.

In addition, EPV-GIV organises and co-ordinates: co-operation with partners at location; support during privatisation or company foundation and the formation of joint ventures; processing of offers for planning and construction of transportation routes; preparation of pre-qualification documents; co-operation with other consultancy firms; and participation in fairs and exhibitions.

Projects
EPV-GIV is active throughout Europe. Present activities are concentrated on Poland, the Czech Republic, the Baltic republics, Russia and other CIS states.

Esveld Consulting Services BV

PO Box 331, NL-5300 AH Zaltbommel, Netherlands
Tel: (+31 41) 801 63 69   Fax: (+31 41) 801 63 72

Key personnel
Director: Dr C Esveld

Capabilities
Consultancy in track technology.

ETC Transport Consultants GmbH

Am Karlsbad 11, D-10785 Berlin, Germany
PO Box 303150, D-10729 Berlin, Germany
Tel: (+49 30) 25 46 50   Fax: (+49 30) 25 46 51 01

e-mail: info@etc-consult.de
Web: http://www.etc-consult.de

Key personnel
Managing Directors: Rainer Obst
Key Consultants: Dr Christian Gleue, Michael Wagner, Hinrich Brümmer

Background
Established in 1967 as a division of Berliner Verkehrsbetreibe (BVG) (Berlin Transport Corporation) and founded as an independent company jointly by BVG and BC Berlin Consult in 1974, in 1995 the firm was renamed as ETC Transport Consultants GmbH, having merged with its subsidiary Ingenieurgesellschaft Verkehrs Berlin GmbH (IVB). ETC's shareholders since 2001 have been the Danish group COWI A/S and Deutsche Bahn AG (German Rail).

Capabilities
Feasibility studies and network planning; traffic and transport engineering; preliminary and detailed design (structural, mechanical, signalling, electrical and safety); management and financial studies, planning of operations, training and start-up operation, project management of turnkey projects, DP management, and management information systems (MIS).

Projects
Austria: Consulting services to Österreichische Bundesbahn (ÖBB, Austrian Federal Railways) regarding assessment of the existing and introduction of a new system of technical services in the field of maintenance of rolling stock.
Chile: Study regarding the development and optimisation of regional rail passenger transport in the Santiago de Chile region, pre-feasibility study on the introduction of an urban rail system, using an existing railway corridor in Santiago de Chile.
China: For Shanghai Metro Corporation (SMC), benchmarking study on large metro systems; organisation of a ten-year train overhaul and, consecutively, a five-year train overhaul for SMC as a training measures for their workshop staff; planning support to tge Quingdao municipality regarding the introduction of an MTR system; assistance to the Chinese government for the introduction of a technical standards system for metro and light rapid transport networks in Chinese cities; setting up a training centre for metro operational personnel including a simulator and training programmes; advisory services for the introduction of Metro and LRT systems; preliminary and detailed design of workshops in Shanghai, metro network studies in Chengdu, Quingdao and Shanghai.
Germany: Studies on and introduction of systems for the apportionment of revenue to transport companies belonging to large integrated transport systems; traffic forecasts and network planning; LRT and regional rail network studies, development of MIS and outline marketing strategies for the integrated transport systems in the Rhein-Main and Halle/Leipzig regions; studies on the development of tourism including outline marketing strategies; studies regarding simulators for staff training; studies on goods transport and logistics; for example, logistics for the Elbe bridge construction site near the city of Wittenberg; infrastructure and engineering projects for DB AG and S-Bahn Berlin; preliminary and final design for the upgrading of railway signalling systems in passenger transport services (PZB90, about 4,300 signals in about 1,100 stations) and in goods transport services (G_PZB90: about 1,900 signals in approximately 300 stations); project management for planning and construction works for the rehabilitation and upgrading of the Berlin–Rostock railway section (approximately 200 km); preliminary and detailed design of metro-, LRT- and urban workshops in Berlin.
Greece: Training programmes for operational personnel of Athens metro, preliminary and detailed design of workshops.
Mexico: Pre-feasibility studies on the introduction of urban rail-based passenger transport in Monterrey (using existing railway infrastructure) and Cancún (servicing tourist areas).

Mongolia: Project management and supervision of construction for the Zamyn Uud transhipment facility for oil products at the Mongolian–Chinese border, following the feasibility study and investment study on the improvement of the facility.

Switzerland: Development and improvement of the train path pricing system and train path sales strategy for Swiss Federal Railways (SBB) and some connecting private railways.

Turkey/Germany: Training of management personnel at Izmir metro.

Vietnam: Traffic forecast, financial and economic evaluation for rolling stock modernisation.

EurailTest

GIE EurailTest
Head office
75 avenue Parmentier, F-75544 Paris Cedex 11, France
Tel: (+33 1) 40 21 11 04   Fax: (+33 1) 40 21 24 21
e-mail: eurailtest@eurailtest.com
Web: www.eurailtest.com

Other offices
Paris
Tel: (+33 1) 40 21 11 04   Fax: (+33 1) 40 21 24 21

Lille
12, Place St Hubert, F-59000 Lille
Tel: (+33 3) 59 56 06 22   Fax: (+33 3) 59 56 06 42

Key personnel
Director: Nicolas Gravier
Deputy Director: Yves Thurin
Personal Assistant: Joelle Decourreges
Consulting Engineers:
  Rolling Stock: Catherine Dine, Christophe Delahaye
  Rolling Stock Components: Charles Cressan
  Infrastructure and Environment: François Viennot
  Urban Systems Transport: Cécile Moreau

Capabilities
Eurailtest provides a fully qualified comprehensive technical testing service for: rolling stock (characterisation of interaction wheel/track, pantograph/catenary and traction or brake performance); infrastructure (control/command systems (classical and ERTMS) and associated software, signalling equipment, track (electrical parameters and geometry); railway environment (acoustic, electromagnetic fields (EMF) and compatibility (EMC) with trackside installations).

A wide variety of services have been provided in Europe, both conventional rail and high-speed railways, freight and urban transport systems, from basic assessment to complete qualification testing for a global rail system.

Eurailtest is able to provide testing services for every stage of the life cycle of a product or complete system.

Projects
Major projects: all wagons: tank, containers, cars or trucks, combined rail and road transport. All locomotives: electric or diesel locomotives such as FRET SNCF, BR185 Railion, MaK 1000 by Vossloh Locomotives; emus or dmus: AGC (F), TER 2Nng (F), Silicio (P), Super Voyager (UK), automotive high speed: ICE3; TGV: AVE (Madrid Seville), KTX (Korea), TGV Duplex, TGB POS, ETR500. Urban transport systems: MF77, MF2000, Val de Rennes, M° Santiago and tramway including Bordeaux, Orléans, Caen, Nantes and Marseille. Tram-train: testing before commissioning (HSL in France, Eurotunnel, CTRL, Paris–Brussels, HSL in Korea. Acoustic and EMC for all types of rolling stock and infrastructures.

UPDATED

Eurostation SA

Rue Brogniezstraat 54, B-1070 Brussels, Belgium
Tel: (+32 2) 529 09 11
Fax: (+32 2) 522 23 79; 520 99 61
e-mail: eurostation@eurostation.be
Web: www.eurostation.be

Key personnel
President: Vincent Bourlard
Chief Executive Officer: Herwig Persoons

Capabilities
A subsidiary of Belgian National Railways, Eurostation combines architectural and civil engineering expertise with property acquisition, development and facilities management to optimise the commercial potential of stations.

Projects
Include: the development of properties at the high-speed train terminal at Brussels South (Midi); redevelopment of Antwerp station to create four operational levels, including facilities for the high-speed route to the Netherlands; and a feasibility study into a proposed major new station in the north of Brussels.

UPDATED

FaberMaunsell

Marlborough House, Upper Marlborough Road, St Albans, AL1 3UT, UK
Tel: (+44 20) 87 84 57 84
Fax: (+44 20) 87 57 00
Web: http://www.fabermaunsell.com

Key personnel
Chief Executive (International): Ken Dalton

Corporate background
FaberMaunsell is part of AECOM Technology Corporation and is the result of the merger of Oscar Faber, Maunsell Europe and Metcalf & Eddy UK.

UK offices
Aberdeen, Altrincham, Beckenham, Belfast, Birmingham, Bristol, Cardiff, Edinburgh, Exeter, Glasgow, Leeds, Leicester, London, Newcastle-upon-Tyne, Norwich, Redhill, St Albans, St Mellion, Warrington, Witham, York.

International offices
Athens, Brussels, Bucharest, Copenhagen, De Bilt, Dublin, Warsaw.

Capabilities
Services in the engineering and planning aspects of moving people and freight – estimation of travel demand, identification and design of infrastructure requirements, identification and development of operation requirements of transport facilities, management and institutional arrangements, economic and financial analysis, environmental impact analysis, traffic systems management, and design of traffic control systems including advanced transport telematics. Research into intelligent transport systems and road safety.

FaberMaunsell is currently developing new services covering issues such as Homeland Security relating to strategic transport networks and their vulnerability to attack. This work is being progressed and techniques developed in close co-operation with its parent company, AECOM.

UPDATED

First Engineering Ltd

Floor 8, Buchanan House, 58 Port Dundas Road, Glasgow G4 0HG, UK
Tel: (+44 141) 335 30 05   Fax: (+44 141) 335 30 06

Key personnel
Chief Executive Officer: Janette Anderson
Commercial Director: Robert Forbes

Background
Formerly known as Scotland Infrastructure Maintenance Co., First Engineering was one of the British Rail Infrastructure Services Units sold off as part of the British Rail privatisation. The company was sold in 1996 to a management buyout team known as TrackAction.

Capabilities
Railway infrastructure and civil engineering consultancy.

UPDATED

Fleet Software

3 Newton Business Centre, Thorncliffe Park Estate, Chapeltown, Sheffield S30 4PH, UK
Tel: (+44 114) 257 16 00   Fax: (+44 114) 257 16 09

Key personnel
Managing Director: John Rands

Projects
Developed TACT, software for rolling stock maintenance in the UK. It records individual parts performance and management of reliability issues. TACT software has been approved for use across the networks of National Express Group and Prism. TACT was first introduced to Supertram, Gatwick Express and in 1996 to Heathrow Express. It has now been extended to rolling stock manufacturers.

TACT technology was originally developed around the privatised bus industry when engineering costs became an issue.

Frazer-Nash Consultancy Ltd

Stonebridge House, Dorking Business Park, Dorking RH4 1HJ, UK
Tel: (+44 1306) 88 50 50   Fax: (+44 1306) 88 64 64
e-mail: info@fnc.co.uk
Web: http://www.fnc.co.uk

Key personnel
Managing Director: A G Milton
Engineering Director: WT Chester
Business and Information Systems Director:
  Dr C C H Guyott
Projects Director: P J Best
Financial and Commercial Director: R R Burge
Operations Director: Chris Edwards
Business Manager, Rail: Andy Lewis

Other offices
Bristol, Burton upon Trent (Midlands), Glasgow and Plymouth.

Capabilities
Frazer-Nash Consultancy provides a range of services to the rail industry in the UK and worldwide for the design and assessment of rolling stock structures, equipment and assemblies. Its principal areas of activity are project management services, bodyshell, bogie and underframe equipment design, crashworthiness, noise and vibration, safety reliability and maintainability, integrated logistics support. The company is active in other

For details of the latest updates to *Jane's World Railways* online and to discover the additional information available exclusively to online subscribers please visit
jwr.janes.com

sectors, enabling leading edge technologies and analytical techniques to be brought to rail projects.

Projects

Design and procurement support to development of rail points machinery; noise consultancy Heathrow Express; expert witness/technical support for litigation on emu brakes, cab, doors and bogies; design and installation of brakes for purpose built test equipment; design of engine and transmission underframe raft for dmu; predictive fire performance analysis for Pendolino vehicles; commercial assistance refurbishment project; gearbox dynamic analysis; brake rig instrumentation; transtop gauging accuracy; emu solebar burn through analysis; drivetrain analysis.

Support to HM Railways Inspectorate, UK into the effects of low-speed buffer-stop collision at Cannon Street station, London; crashworthiness design of European Nightstock for the former GEC Alsthom Metro Cammell Ltd; safety case development for luggage stowage for European Passenger Services; concept development and assessment to support the procurement specification for London CrossRail project; design support to Thorn Transit Systems International on the thermal performance of automated ticket machines in hostile environments; crashworthiness design and noise consultancy to CAF for the Heathrow Express emu; structural assessment of the refurbished KCRC rolling stock for the former GEC Alsthom Metro Cammell Ltd; infrastructure assessment for MTRC and escalator risk assessment work for London Underground; design of traction and auxiliary equipment rafts and dynamic gauging of train stops.

GHD-Transmark Pty Ltd

A joint venture company formed between Gutteridge Haskins & Davey Pty Ltd and Halcrow Transmark
39 Regent Street, Railway Square, New South Wales 2008, Australia
PO Box K839, Haymarket, New South Wales 1238, Australia
Tel: (+61 2) 96 90 70 37   Fax: (+61 2) 96 90 14 64
e-mail: transmarksyd@ghd.com.au

Projects

Illawarra electrification: State Rail Authority of New South Wales (value A$200 million).
East Hills-Cambelltown railway link: State Rail Authority of New South Wales (value A$80 million).
Implementation of train radio system: State Rail Authority of New South Wales (value A$80 million).
Automatic train fare collection: State Rail Authority of New South Wales.
VFT Sydney access studies: VFT Consortium.
Ultimo Pyrmont Light Rail Project (value A$100 million) Bondi Rail Link (ongoing, value A$100 million) New Southern Railway.

GIBBRail Ltd

Gibb House, London Road, Reading RG6 1BL, UK
Tel: (+44 118) 963 50 00   Fax: (+44 118) 949 10 54
e-mail: gibbrail@gibb.co.uk
Web: http://www.gibbltd.com

Key personnel

Managing Director, Transportation: Tony King
Director, Rail: Andy Collinson
Director, Transport Consulting: Nigel Ash
Technical Director, International: Brian Green
Technical Director, UK: Paul Dawkins

Offices

Reading (UK headquarters), Glasgow, York, Derby, Birmingham, London

Overseas offices

Belgium, Botswana, Bulgaria, Ethiopia, Ghana, Greece, Hong Kong, Hungary, Indonesia, Japan, Jordan, Kenya, Lebanon, Lesotho, Malawi, Mauritius, Oman, Poland, Portugal, Romania, Russian Federation, South Africa, Spain, Sudan,

Swaziland, Tanzania, Turkey, UAE, Georgia USA, Uganda, Zambia, Zimbabwe.

Background

GIBBRail is part of GIBB Ltd (formerly Sir Alexander Gibb & Partners Ltd) and is a specialist team within the LAWGIBB Group. The group includes professional railway system engineers and technical staff experienced in planning, operating engineering, development, procurement, construction and maintenance of high-speed, main line, metro and light rail projects for organisations in the UK and overseas. In-house capability includes project management; civil, structural, mechanical and electrical engineering; architecture; environmental and economic studies; topographical and geological surveys.

Projects

GIBBRail can undertake complex and routine maintenance inspections, multidisciplinary detailed designs typified by the remodelling of main line stations in Poland, Leeds station, UK, and the tram project in Croydon, UK. The company has carried out rehabilitation projects in Estonia and Latvia and detailed designs for 12 km of underground alignment, 20 km of tunnelling and 13 underground stations for Line 3 of the Rome metro. Other contracts include: independent engineer at Porto metro, Portugal; works supervision, advisers on EIB contract procedures at Latvia East-West Rail Corridor renewals; engineer for construction work on new rail connection to Ventspils Port Rail Terminal, Latvia; project management and delivery of track renewals at London Underground corporate track alliance project; full multidisciplinary design at Leeds Station remodelling, UK; consultancy services for tunnel under railway at Dublin Port Rail Tunnel, Ireland; engineer for supervision of contracting at St Petersburg Transportation and Commercial Centre, Russian Federation; detailed design at Croydon Tramlink, UK; government technical adviser for the Channel Tunnel Rail Link, UK; feasibility studies, detailed design on network development for Railtrack IOS, UK; civil engineering and plant engineering design work at Rome Metro Line C, Italy; strategic planning on the West Coast Mainline route modernisation, UK; project management of safety programme for Iarnród Eireann (Irish Railways), Ireland; system design on Thameslink 2000, UK; route modernisation feasibility study on East Coast Mainline, UK.

HaCon Ingenieurgesellschaft mbH

Head Office
Lister Strasse 15, D-30163 Hanover, Germany
Tel: (+49 511) 33 69 90   Fax: (+49 511) 336 99 99
e-mail: info@hacon.de
Web: http://www.hacon.de

Capabilities

HAFAS program system: timetable information on local intercity and air traffic routes and connections; UX-SIMU program system: interactive timetable planning and simulation; RASIM program system: simulation of marshalling operations and Radis program system: online information and scheduling management system.

Halcrow Group Ltd

Vineyard House, 44 Brook Green, Hammersmith, London W6 7BY, UK
Tel: (+44 20) 76 02 72 82   Fax: (+44 20) 76 03 00 95
e-mail: halcrow@halcrow.com
Web: www.halcrow.com

Key personnel

Executive Director – Rail: David Watters
Development Director – Urban and Light Rail: Glenn Gittoes
Development Director – Heavy Rail: Malcolm Cope
Development Director – Traction and Rolling Stock: Andrew Witkowski

Development Director – Rail Commercial: Michael Jamieson
Operations Director – Rail: John Irwin
Technical Director – Rail Engineering: Richard Lindop
Technical Director – Transportation Commercial: Richard Sumner
Communications Manager: Garry A Whitaker

Background

Formerly Sir William Halcrow and Partners, previous subsidiary companies now part of Halcrow Group included Transmark, Halcrow Fox, TME Torpy and Crouch Hogg and Waterman.

Capabilities

Halcrow is an independent provider of infrastructure-based business solutions. Specialising in the transport, water and property sectors, it offers professional consultancy resources for the planning, design and supervision of development on a global basis. Halcrow has a global network of 62 offices and is currently undertaking projects in over 70 countries.

Projects

Bulgaria: Track renewals PHARE.
Ireland: Dublin Metro LUAS.
Malaysia: Rawang – Ipoh double tracking, technical advisors. Light rail transit system 2, technical supervision and management of the detailed civil, architectural and systems design (joint venture).
Philippines: Manila Metro LRT Lines 1 and 2.
Poland: E20 and E30 railway rehabilitation.
South Korea: Seoul Metro, Line 4 technical advisors.
Thailand: SRT, track doubling, packages ST1, 2 and 4 signalling upgrade and railway rehabilitation. Blue Line depot design.
UK: Channel Tunnel Rail Link (as part of a joint venture). Docklands Light Railway – extension to City Airport and Woolwich. Edinburgh Airport Rail Link – design development.
US: 42nd Street tram.

UPDATED

Hatch Mott MacDonald Inc

Head office
27 Bleeker Street, Millburn, New Jersey 07041, USA
Tel: (+1 973) 379 34 00   Fax: (+1 973) 376 10 72
e-mail: railtransit@hatchmott.com; corporate@hatchmott.com
Web: http://www.hatchmott.com

Key personnel

President and Chief Executive Officer: Gordon A Smith
Director and COO: David P White
Directors: Jan J Feberwee, Peter Wickens, Ronald R Nolan, Timothy J Thirlwell
Secretary and Treasurer: Eric R Hartley

Other offices

3825 Hopyard Road, Suite 240 Pleasanton, California 94588, USA
Tel: (+1 925) 469 80 10   Fax: (+1 925) 469 80 11

161 North Beltline Highway, Mobile, Alabama 36608, USA
Tel: (+1 251) 343 43 66   Fax: (+1 251) 434 69 02

2727 Camino del Rio South, Suite 244, California 92108, USA
Tel: (+1 619) 858 15 95   Fax: (+1 619) 858 15 99

4100 South, Ferdon Boulevard, Suite -6, Crestview, Florida 32536, USA
Tel: (+1 850) 423 79 14   Fax: (+1 850) 423 79 20

11-C West 23rd Street, Panama City, Florida 32405, USA
Tel: (+1 850) 522 10 31

Canada office
2955 Speakman Drive, Sheridan Science & Technology Park, Mississauga, Ontario L5K 1B1
Tel: (+1 905) 855 20 10   Fax: (+1 905) 855 26 07

Capabilities

Engineering consulting services, project and construction management and planning and architectural services for rail and transit systems. Services include: planning, route selection and environmental assessment; civil engineering, including alignment, trackwork, structures, bridges and elevated guideways; tunnels in soft ground or rock, including planning, architecture and safety; building services; systems engineering including signalling, telecommunications, traction power and distribution, tunnel ventilation; programme and project management; and construction management.

Projects

Recent projects include: programme management services for Toronto Transit Commission's Rapid Transit Expansion Programme; design, project management and construction management for CN North America's St Clair River Tunnel between Sarnia, Canada and Port Huron, USA; construction management services for the construction of the Denver LRT system; application engineering services for the installation of an enhanced speed enforcement system at priority locations systemwide for New York City Transit Authority; detailed design of Ocean Parkway interlocking as part of the Brighton Beach Line resignalling programme for New York City Transit Authority; construction management services for the traction power system for the Montréal-Deux Montagnes route modernisation; and consulting and oversight services to Santa Clara County Transportation Agency on the design of trackwork, signals and telecommunications for the 12 mile Tasman Corridor LRT extension.

HDR Engineering Inc

1101 King Street, Suite 400, Alexandria, Virginia 22314, USA
Tel: (+1 703) 518 86 65   Fax: (+1 703) 518 85 78
e-mail: tsmithbe@hdrinc.com
Web: http://www.hdrinc.com

Key personnel

Senior Vice-President & National Director, Rail:
   Tom Smithberger

Capabilities

Railway structures design and inspection; trackwork design and inspection; construction administration services; metro (line and station) designs; terminal and transfer facilities design; light rail and peoplemover design; and environmental and security work.

Projects

Continuing projects with all North American Class I railroads, Amtrak and various commuter rail and transit agencies.

VERIFIED

Heery Transportation Group (HTG)

8201 Corporate Drive, Landover, Maryland 20785, USA
Tel: (+1 301) 306 01 18   Fax: (+1 301) 577 20 52

Key personnel

Vice-President: David Rankin

Capabilities

HTG offers planning, design and construction services for people mover, light rail, monorail, main line and high-speed railway systems. HTG focuses on the design and construction management of electrification systems, including traction power, operational simulations, catenary and third rail.

High-Point Rendel Ltd

61 Southwark Street, London SE1 1SA, UK
Tel: (+44 20) 76 54 04 00   Fax: (+44 20) 76 54 04 01

e-mail: london@highpointrendel.com
Web: http://www.hprendel.com

Key personnel

Group Chief Executive: Kelvin Hingley
Regional Chief Executive: John Bradley
Director, Capital Project Delivery, Major Projects:
   Richard Tappin
Director, Business and Management Services:
   David Canning
Sales and Marketing Director: Brian Roberts

Overseas offices

Bangladesh, Canada, Hong Kong, Indonesia, Libya, Malaysia, Singapore, Turkey, United Arab Emirates, USA

Capabilities

High-Point Rendel has experience of railway planning, design and maintenance of metros and LRT systems. Main areas of activity include: feasibility, economic and investment studies; survey, design and construction supervision for freight and urban railways; reconstruction, development and rehabilitation of railway infrastructure; and modernisation programmes for motive power, rolling stock and other equipment. Rendel Palmer & Tritton has also had a broad range of experience in the design and equipping of container and freight terminals; the design and development of loading and unloading installations for bulk transport; and in the development of railway containerisation and rail/port links.

Hill International Inc

1 Levitt Parkway, Willingboro, New Jersey 08046 USA
Tel: (+1 609) 871 58 00   Fax: (+1 609) 871 12 61
Web: http://www.hillintl.com

Key personnel

Chairman: Irvin E Richter

Capabilities

Engineering consultancy; project and construction management; project management supervision; construction claims analysis; expert witness testimony; claims prevention and dispute resolution.

Projects

Projects include: Istanbul LRT line, Turkey; Frankford Elevated reconstruction, Norristown Line reconstruction and RRD main line improvement programme, Philadelphia, USA; Los Angeles Metro Rail, USA; Tren Urbano rapid transit study, San Juan, Puerto Rico; Kearny rail connection, Kearny, USA; and Long Island Rail Road Richmond Hill improvements, Queens, USA.

Hodgson and Hodgson Group Ltd

Winnington Hall Mews, Northwich CW8 4DU, UK
Tel: (+44 1606) 765 93   Fax: (+44 1606) 743 15
e-mail: efitzpatrick.hodgsongroup.co.uk
Web: www.acoustic.co.uk

Key personnel

Chairman: G Balshaw-Jones
Managing Director: J Roberts
Technical Director: N Grundy
Sales Director: Margaret Narburgh
Commercial Director: P Rollinson
Export Sales Manager: E Fitzpatrick

Services

Acoustic consultancy services for bus and railway traction units, rolling stock and associated buildings.
   Recent projects have included Waterloo Eurostar Terminal (buildings), St Petersburg Rail Terminal (buildings), Barratt Housing Project (railside development), Eurotram (complete vehicle), Europa Transrapid (complete vehicle), MTRC Hong Kong (complete vehicle), Arlanda, Stockholm (complete

vehicle), Juniper, Turbostar and Electrostar and West Coast Main Line (rolling stock), Brush Engines (traction units), First Bus, Mellor Vancraft, Optare and Marshalls (engine/exhaust jacketing and moulded internal and external lining panels).

UPDATED

Holland Railconsult

Daalseplein 101, NL-3500 GW Utrecht, Netherlands
Tel: (+31 30) 265 42 20   Fax: (+31 30) 265 42 21
e-mail: information@hr.nl
Web: http://www.hr.nl

Key personnel

Executive Board Members: Gerrit Disberg;
   Wim Jol; Jan Moerkerk
Divisional Directors:
   Infrastructure: George Brouwer
   Urban Interchanges: Jan Garvelink
   Major Projects: Peter Otten
   International Consultancy: Rob Brugts

Overseas offices

Richard Wagnerstrasse, D-38106, Braunschweig, Germany
Tel: (+49 531) 380 23 95   Fax: (+49 531) 380 23 96

Praça Dom Pedro, IV, 74, 2D, P-1100-202, Cascais, Portugal
Tel: (+351 214) 86 98 16   Fax: (+351 214) 86 98 37

Railconsult Éireann & Partners
Arena House, Arena Road, Sandyford, Dublin 8, Ireland
Tel: (+353 1) 294 08 00   Fax: (+353 1) 294 08 20

Capabilities

The company is an engineering consultancy specialising in public transport and rail infrastructure, the design and engineering of guided transport systems including interchanges from feasibility studies, to planning, design and execution. The company employs 1,700 staff. Recently, the company signed an agreement to collaborate with Brown & Root Services (a Halliburton Company) aiming to improve their technical capabilities and project management skills in the railway industry.

Projects

Include European Railway Traffic Management Systems; Brussels 1997–2002; Xabregas Bridge in Portugal; pre-design of Trolihättan double track lift bridge in Sweden; maintenance and renewal assessments of Railtrack, UK; design of Ranstad light rail system including integration of the Hague Central Station in the Netherlands; environmental impact study of Betuwe high-capacity cargo railway link from Rotterdam harbour to the German border and Kijfhoek Marshalling Yard including 25kV power supply and design and contract mangement of catenary system, in the Netherlands.

HTM Consultancy BV

PO Box 16152, NL-2500 BD The Hague, Netherlands
Tel: (+31 70) 374 94 21   Fax: (+31 70) 374 94 22
e-mail: htmconsultancy@htm.net

Background

The company was formed in 1993 as HTA Transport Consultants by municipal transport operators in Amsterdam (GVBA), the Hague (HTM) and Rotterdam (RET). In 2001 HTM became the major shareholder and the company adopted its present name.
   HTM-Infra is a department within HTM Consultancy that designs, builds and maintains tram and light rail tracks.
   In 2000 a joint venture subsidiary, Mena Rail, was formed with an Egyptian engineering consultancy, Tecnico Construction and Trade. It is involved in upgrading the Alexandria light rail network and is also active in other Middle East countries.

Capabilities

Management of public transport organisations and processes; institutional relations; engineering and design of new rail infrastructure including design and project management, rehabilitation of tram, light rail, metro and bus vehicle fleets, workshops and depots; transport and feasibility studies; training; signalling and telecommunications; traffic management and vehicle control; automation and telematics; environmental aspects of urban public transport; marketing; traffic statistics; tariffs and ticketing; advice on equipment procurement, workshop practices and maintenance management; public transport security.

Projects

Recent projects include:

Benin: technical evaluation of the Benin Autobus bus project.

Egypt: implementation of electrified rapid mass transit systems; upgrading the Alexandria light rail system; a pilot project to upgrade the Ramleh light rail line, Alexandria.

Ireland: operations consultancy for the Luas light rail system, Dublin.

Israel: feasibility study into bus and light rail alternatives, Beer Sheva.

Netherlands: feasibility study into upgrading High Quality Bus (HQB) to light rail, Utrecht; transport studies and implementing a high quality public transport network, Utrecht.

Norway: light rail project track design, Bergen.

Poland: implementation of traffic and parking management controls, Krakow.

Russian Federation: feasibility into the local assembly and sale of Dutch buses.

Serbia and Montenegro: design of tram network infrastructure, Belgrade; urban transport upgrade project management and planning, Belgrade.

UK: review of light rail infrastructure design, Nottingham.

Europe-wide: study into airport public transport links.

Hyder Consulting

Head office
29 Bressenden Place, London SW1E 5DZ, UK
Tel: (+44 20) 73 16 60 00   Fax: (+44 20) 73 16 61 25
e-mail: corp.marketing@hyderconsulting.com
Web: www.hyderconsulting.com

Key personnel

Director, Transportation Planning: John Spiers
Director Rail: Peter Johnson
Director, Intelligent Transport Systems:
  Mike Hayward
Director, Urban Design and Transportation:
  John Birt

Associated companies

Hyder Consulting Middle East
PO Box 2774, Abu Dhabi, United Arab Emirates
Tel: (+971 2) 633 34 00   Fax: (+971 2) 633 07 46

Hyder Consulting (Australia) Pty Ltd
Level 5, 116 Miller Street, North Sydney, New South Wales 2060, Australia
Tel: (+61 2) 89 07 90 00   Fax: (+61 2) 89 07 90 01

Hyder Consulting (Hong Kong) Ltd
47th Floor, Hopewell Centre, 183 Queens Road East, Wanchai, Hong Kong
Tel: (+852) 29 11 22 33   Fax: (+852) 28 05 50 28

Capabilities

Management and advisory services including: economic and financial appraisal; operations and research management; tariff negotiation; public transportation planning, traffic modelling and forecasting; operations audits; technical audits; quality audits; maintenance management, environmental impact assessment; safety and reliability audits, training and certification, and value engineering.

Project planning and design including: design management; concept design and system selection; feasibility studies; outline designs and specification

for legislative approval; funding applications; operations and system planning; highway and traffic engineering; road safety engineering and audits; alignment and permanent way engineering; signalling and control systems; rolling stock; and freight terminal and depot design.

Project implementation including: preparation of contract documents; preparation of specifications; tender invitations and adjudication; project management and cost control; construction supervision; monitoring and quality control of procurement contracts; testing and commissioning; and operations and maintenance management.

Areas of activity include: main line, regional and suburban rail services; metros and urban transit systems including light rail and people movers; bus priority routes; freight logistics; bimodal and multimodal interchanges.

Projects

Recent projects include: the Bahrain traffic model and strategic plan; technical consultancy for Croydon Tramlink; Old Palace Yard, Westminster; Dubai parking control system; electronic road user charging projects in Hong Kong, New Zealand and UK; Western Sydney Bus Transitway; Melbourne City Link; London Bus Red Routes and London Bus Priority Network; London Congestion Charging, Project Evergreen (Chiltern Line, UK); technical consultancy for the London Underground PPP, and Transport for London framework contracts for consultancy services, project management and management consultancy.

Typical rail projects completed or in hand are:
Project Evergreen (Chiltern Line UK).
Technical consultancy for the London Underground PPP.
UK GNER, East Coast Main Line upgrade review.
Edinburgh Cross – rail feasibility.
Chiltern Railways, infrastructure upgrade feasibility.
Railtrack Midlands – structure design.
Railtrack North West, station improvements. Great Western – signalling design. West Coast Mainline – upgrade geotechnics study. Stratford Station Redevelopment (Western and Eastern Concourse) – structural and services design for new station concourse.
Istanbul metro, Turkey: electrical and mechanical systems design: supervision of implementation: commissioning and acceptance.
Thailand SRT Freight railway proposal: Feasibility study.
LAR Lai King Station, Hong Kong: detailed design of all E&M services and civil and structural aspects for the Lai King station, which is the interchange station of the Tsuen Wan line and Lantau Airport line.
Ankara LRT, Turkey: supervision of supply and installation of rolling stock, E&M subsystems, commissioning and acceptance.
Kuala Lumpur LRT system, Malaysia: checking engineer for the contracting consortium on Line 1, Stage 2.
New Southern Railway, Sydney, Australia: comprehensive review of technical aspects including contractual construction and environmental issues and tunnelling concepts.
Krakow, Poland: project appraisal of the financing and technology option.
West Rail, Hong Kong – Sham Shi Po section.
Guangzhou Metro China – civil, mechanical, electrical design and environmental services. Operations and maintenance advice.

UPDATED

ICF Consulting

9300 Lee Highway, Fairfax, Virginia 22031-1207, US
Tel: (+1 703) 934 36 03   Fax: (+1 703) 934 37 40
e-mail: info@icfconsulting.com
Web: www.icfconsulting.com

Other offices

Albany, New York; Charleston, South Carolina; Dayton, Ohio; Harrisburg, Philadelphia; Houston, Texas; Irvine, California; Lexington, Massachusetts; Los Angeles, California; Ogden, Utah;

Oklahomo City, Oklahoma; Providence, Rhode Island; Research Triangle Park, North Carolina; San Francisco, California; Washington, DC. London, UK; Moscow, Russian Federation; New Delhi, India; Rio de Janeiro, Brazil; Toronto, Ontario.

Key personnel

Chairman and Chief Executive Officer:
  Sudhakar Kesavan
Chief Financial Officer: Alan Stewart
Executive Vice-President Business Development:
  Peter Linquiti
Executive Vice-President Chief Operating Officer:
  John Wasson
Treasurer: Terrance C McGovern

Background

Since 1969 ICF Consulting has been serving major corporations, government at all levels and multinational institutions. More than 1,200 employees serve these clients from key business centres in the Americas, Asia and Europe.

Capabilities

ICF Consulting is a management, technology and policy consulting firm. The firm develops solutions to complex energy, environment, emergency management, homeland security, community development and transport issues. ICF Consulting's approach to these issues is strengthened by its expertise in information technology, organisational improvement, programme management and communications.

UPDATED

IndustrieHansa Consulting & Engineering

Klausenburger Strasse 4, D-81677 Munich, Germany
Tel: (+49 89) 93 08 00   Fax: (+49 89) 93 08 01 39
e-mail: info@industriehansa.de
Web: http://www.industriehansa.de

Background

IndustrieHansa Consulting & Engineering is a subsidiary of MCE VOEST.

Capabilities

Design and implementation of projects in the areas of head module, sanitary compartments and interior development. 3-D activities using CAD systems, Pro/Engineer, Unigraphics and Euclid. Data conversion and data transfer. FEM (Finite Elements Method) analysis for static and dynamic, linear and non-linear stability detection and also oscillation and frequency analyses.

INECO-TIFSA

Ingeniería y Economía del Transporte, SA
Tecnología e Investigación Ferroviaria, SA
P° de la Habana 138, E-28036 Madrid, Spain
Tel: (+34 91) 452 12 00   Fax: (+34 91) 452 13 00
e-mail: ineco@ineco.es
  tifsa@tifsa.es
Web: http://www.ineco.es
  http://www.tifsa@tifsa.es

Key personnel

Chairman: Antonio Fernández Gil
General Director, Advisor to the Chairman:
  Juan Barrón Benavente
General Directors: Juan Torrejón, Javier Cos, Marcos García Cruzado, Juan Batanero, José María Urgoiti

Capabilities

Railway engineering (track, electrification, signalling, telecommunications); consultancy (concessions, privatisations, demand analysis, feasibility studies); monitoring and supervision of railway works, technical assistance on site, integrated project management, railway planning and operations, rolling stock engineering (design, vehicle dynamics, quality control); railway maintenance (infrastructure, bridges, tunnels); intermodal transport, road and motorway engineering. Urban distribution and logistics and rail access to urban centres.

Projects

Overseas: Participation in SAMRAIL (Safety Management on Railways); development of a common safety management systems (SMS) for European railways.

Technical and economic feasibility study for a high-performance line in the Buenos Aires-Rosario corridor (Argentina).

SUR project - CPTM Brazil: INECO-TIFSA provides management and technical support to CPTM (Companhia Paulista de Trenes Metropolitanos).

Israeli railways: new technical assistance contract. Proposal of technical assistance for the development of the Trans-European Network of Goods Railways (TERFN).

National: High-speed railways programme in Spain: studies, infrastructure and superstructure, installation and telecommunications, railway systems, new stations, management of maintenance on existing lines, new technologies, environmental impact studies, control of assembly works, quality supervision and monitoring of track elements, prior to running.

Technical advice in the project and supervision during construction of 20 regional high-speed trainsets.

INECO-TIFSA has continued to work closely with different RENFE business units in many studies, projects and assistance assignments.

Consultancy and technical assistance for the control and supervision of the works in the Mediterranean Railway Corridor.

Master plan for the network railway access of Madrid; demand and mobility analysis studies for the suburban networks of Madrid, Bilbao, Seville and Cádiz.

Adaptation of Spanish railway legislation to EU directives.

Interfleet Technology Ltd

Interfleet House, Pride Parkway, Derby DE24 8HX, UK
Tel: (+44 1332) 22 33 30
Fax: (+44 1332) 22 33 31
e-mail: info@interfleet.co.uk
Web: www.interfleet-technology.com

Key personnel

Managing Director: David Rollin
Strategic Projects Director: Jonathan Wragg
Sales and Marketing Director: Peter Dudley
Operations Director: Neil Wilson
Development Director: David Curtis
Finance Director: Richard Tapping

Other offices

Exchange Tower, 1 Harbour Exchange Square, London E14 9GE, UK
Tel: (+44 20) 79 87 48 20   Fax: (+44 20) 79 87 48 30
e-mail: howarth.p@interfleet.co.uk
Director, London: Peter Howarth

Interfleet Technology Pty Ltd
Level 7, 333 George Street, Sydney, New South Wales 2000, Australia
Tel: (+61 2) 92 62 60 11   Fax: (+61 2) 92 62 60 77
e-mail: brianhastings@interfleet.aust.com
Regional Director: Brian Hastings

Suite 3a, Level 3, 320 Adelaide Street, Brisbane Queensland 4000, Australia
Tel: (+61 7) 32 29 69 00   Fax: (+61 7) 32 29 39 22
e-mail: markwishart@interfleet.aust.com
State Manager, Queensland: Mark Wishart

Level 15, 350 Collins Street, Melbourne, Victoria 3000, Australia
Tel: (+61 3) 86 05 48 28
Fax: (+61 3) 86 05 48 94
e-mail: petermetcalf@interfleet.aust.com
State Manager, Victoria: Peter Metcalf

Interfleet Technology AB
Box 35, SE-171 11 Solna, Sweden
Tel: (+46 8) 52 29 92 00 Fax (+46 8) 52 29 92 01
e-mail: sven.odeen@interfleet.se
Regional Director: Sven Ödeen

Interfleet Technology Inc
125 Strafford Avenue, Suite 130, Wayne, Pennsylvania 19087, US
Tel: (+1 610) 225 01 20   Fax: (+1 610) 225 01 21
e-mail: frawley.t@interfleetinc.com
Vice-President North America: Tom Frawley

Interfleet Technology GmbH
IM Media Park 4d, D-50670 Cologne, Germany
Tel: (+49 221) 57 73 34 00
Fax: (+49 221) 57 77 34 01
e-mail: sheldon.m@interfleet.co.uk
Regional Director, Commercial: Martin Sheldon

Background

Interfleet Technology works with worldwide clients on rail engineering and business solutions. Interfleet was created from the engineering arm of British Rail to become one of the world's first privatised rail technology consultancies. In 2003 Interfleet acquired the Swedish consultancy, TrainTech Engineering AB, renaming the company Interfleet Technology AB. In 2004 the company acquired RSE Management Ltd, a specialist in railway safety case auditing, from GB Railways Group plc.

In 2004 a German subsidiary, Interfleet GmbH, was established in Cologne.

Capabilities

Provision of engineering support for all vehicle types, from light to heavy rail, from commuter to high-speed and from passenger to freight, at every stage of traction and rolling stock life, for all types of projects. Total rail systems, encompassing all aspects of infrastructure and the physical railway environment. Interfleet is able to assist with infrastructure projects at both a strategic and a detail level. As business consultants and partners, Interfleet works with clients worldwide to implement new products, develop new processes and achieve maximum impact from investment programmes. Interfleet brings resources from all business disciplines, including contracting, finance, marketing, strategy and railway operations.

UPDATED

Intermetric GmbH

Industriestrasse 24, D-70595 Stuttgart, Germany
Tel: (+49 711) 780 03 92   Fax: (+49 711) 780 03 97
e-mail: geschaeftsfuehrung@intermetric.de
Web: http://www.intermetric.de

Capabilities

Track geometry design; surveying; plans; geotechnical measuring.

International Technical Services

400 Queens Avenue, London, Ontario N6B 1X9, Canada
Tel: (+1 519) 439 23 62
Fax: (+1 519) 675 18 68
e-mail: info@itsrail.com
Web: http://www.itsrail.com

Key personnel

President: Bill Graves
General Manager: Rick Girvin
Technical Manager: Warren Bjornson

Capabilities

Railway consulting services; project management and supervision; technical and management training; technical publications; fleet and facilities evaluation; operator/driver training; crane operator training; production planning; procurement planning; needs assessment/analysis; facilities management.

EMD locomotive specialists; EMD engine rebuild; locomotives and freight vehicle maintenance; air brake systems specialists; locomotive rebuilding and repowering; welding services (engine repair).

NEW ENTRY

Intraesa

Edificio Numancia 1, Calle Viriato 47, E-08014 Barcelona, Spain
Tel: (+34 93) 366 21 10
Fax: (+34 93) 439 17 69
e-mail: intraesa.bcn@comsa.com
Web: http://www.comsa.com

Key personnel

Managing Director: Félix Boronat

Offices

Alicante, Madrid, Tarragona

Background

Intraesa is a subsidiary of Comsa SA (see entry in *Permanent way components, equipment and services* section).

Capabilities

Feasibility, route and engineering studies, project and works management for high-speed main line, suburban and urban rail projects.

---

iQR

Railcentre 1, 305 Edward Street, Brisbane, Queensland, Australia 4000
GPO Box 1429, Brisbane, Queensland 4001
Tel: (+61 7) 32 32 33 90
Fax: (+61 7) 32 35 33 46
e-mail: sales@iqr.com.au
Web: www.iqr.com.au

Key personnel

General Manager: Michael Walsh
Marketing and Business Development Manager: Peter Harris
Sales Manager: Youfa Chen

Background

iQR was previously Queensland Rail Consultancy Services.

Capabilities

iQR is a government-owned corporation with rail solutions delivered to over 25 countries including Australia, China, Hong Kong, Malaysia, Taiwan, Thailand and the UK. iQR offers a variety of products and services including rollingstock, infrastructure, operations and specialist services: rollingstock products include locomotives, freight service and passenger vehicles with professional services such as asset management and driver training. Infrastructure hardware includes tracks, tunnels, bridges, overhead wiring, signalling, communications and yards with asset management and audit capabilities. Operations include train control, planning, scheduling and capacity management, with emergency incident response and environmental management. Specialist services include engineering studies, safety and environmental audits, level crossing safety audits, incident and derailment cause analysis, expert witness, rail transport system feasiblity, integrated rail solutions and training solutions to enable clients to become perpetually self-improving organisations.

iQR also provides a full range of professional services for the design, development and delivery of railway signalling systems from concept through to decommissioning, including: feasibility studies and estimates; detailed signalling design and independent design audit; definition of user requirements; preparation of contract specifications and systems integration; preparation of standards and procedures; signalling risk assessments and safety cases.

Permanent way construction on viaduct for KCRC West Rail in Hong Kong   **NEW**/1140310

25 kV AC traction power overhead wiring   **NEW**/1140309

iQR offers a wide range of services for the construction and maintenance of permanent way for suburban passenger lines, heavy haul freight lines, general freight lines and high-speed passenger lines. Capabilities include: design; project management; safety systems; environmental planning; site engineering and programmed maintenance. iQR is experienced in all types of track forms including ballasted, floating slab and low vibration track and can design, supply, install and maintain turnouts with conventional vees, full swing nose turnouts, heelless switches and scissor crossings.

iQR offers specialist electrical engineering advice and services in all phases of the railway electrification systems life-cycle. Services range from system analysis, planning and design through to construction, operation, maintenance, standards developments and audits in the following electric traction engineering fields: overhead traction distribution systems; traction system substations and power equipment; power supervisory control, protection and electrical sectioning; special power supply systems and services for railway facilities; electrical safety; electrical interference, power disturbances and harmonics; electrical energy management. iQR's capabilities include 25 kVAC

electrification in environments ranging from suburban and long-distance passenger services to heavy haul multiple-locomotive traffic.

iQR offers turnkey solutions for clients requiring design, project management, construction, testing, commissioning and maintenance of rail infrastructure.

See also entries for iQR in Freight vehicles and equipment, Passenger coach equipment and Vehicle maintenance sections.

Projects
iQR has delivered complete rail solutions to the Asian region for over 15 years to over 25 countries including Australia, Hong Kong, China, Malaysia, Singapore, Taiwan and the UK.

Kowloon Canton Railway Corporation, Hong Hong: joint venture partner and technical advisor to the CHCQ project, involving construction of the permanent way. This consists of ballasted, floating slab track and low vibration for the following extensions to the network: KCRC contract CC-1820 West Rail permanent way northern section from Kam Sheung Road to Teun Mun. KCRC contract CC-1860 Lok Ma Chau extension permanent way from Sheung Shui to Lok Ma Chau; KCRC contract CC-1850 East Rail extensions from Tai Wait to

Ma On Shan and from Hung Hom and East Tsim Sha (technical advisor only).

KTMB Malaysia, Rawang to Ipoh double-tracking and electrification. Project management and specialist technical advice on the double-tracking (1,000 mm gauge track) and electrification (25 kV AC).

UPDATED

Italcertifer

Italcertifer scpa
(Italian Institute for Railway Research and Certification)
Viale S Lavagnini 58, I-50123 Florence (I), Italy
Tel: (+39 055) 235 36 05; 35 36 53
Fax: (+39 055) 48 37 41
e-mail: commercial@italcertifer.com
Web: http://www.italcertifer.com

Key personnel
Managing Director: Enrico Mingozzi

Capabilities
Certification of railway systems and components, including: laboratory tests on rolling stock and components, on-track tests of rolling stock and components; tests on infrastructure components; tests on operational and functional characteristics of components and systems; research and technological development, tests and experimentation and training.

Italcertifer works in partnership with Rete Ferroviaria Italiana SpA, Trenitalia SpA, the Polytechnic University of Milan and the universities of Florence, Naples and Pisa.

VERIFIED

Italferr SpA

Via Marsala 53, I-00185 Rome, Italy
Tel: (+39 06) 497 51   Fax: (+39 06) 49 75 22 09
e-mail: italferr@italferr.it
Web: www.italferr.it

Key personnel
Chairman: Giuilio Burchi
Managing Director and Chief Executive:
  Riccardo Bonasso
Transactions and Trade: Francesco Ramacciotti
Production: Francesco Loffredo
International Activities and Operating Control:
  Claudio Collinvitti

Capabilities
Italferr SpA is the consulting engineering company of Italian Railways (FS) and was founded in 1984. Italian Railways is the sole shareholder with a share capital of EUR14.2 million. Italferr has a turnover of EUR194 million and has continuing projects worth around EUR45 billion.

The company's activities include traditional and high-speed rail as well as metro and non-rail transport systems. In Italy it carries out works on the high-speed systems, traditional lines and junctions, rolling stock maintenance and repair facilities, inter-regional links, alpine crossings, interconnections between main line and metropolitan areas, passenger stations and intermodal terminals; and technology development and upgrade for Italian Railways.

Services include transport development plans; feasibility studies; conceptual design and system definition; environmental impact appraisal; design of mono and multimodal transport systems; preliminary and detailed design; cost estimates, technical specifications and tender documents; maintenance manuals/construction site safety plans; preparation and evaluation of tenders; project management; supervision of works, testing, inspection, start-up assistance; technical assistance; procurement services; technical assistance and training; organisation and management studies; BOT and project financing assistance.

Projects

Italferr is responsible for the development of the EUR30 billion high-speed railway system covering three lines: Turin–Venice; Milan–Naples; and Milan–Genoa. The 1,200 km railway system will be an integral part of the European high-speed railway network. The project includes revamping urban junctions in Turin, Milan, Genoa, Verona, Venice, Bologna, Florence, Rome and Naples to enable them to cater for the new high-speed traffic.

Other contracts in Italy include new high-speed stations at Naples, Florence and Bologna; a computer interlocking system for Rome Termini; a peoplemover transport system for the city of Monza; preliminary and detailed design for a light rail system for Cagliari; preliminary and detailed design and works supervision for the Salerno metro; upgrading of infrastructure and technology over 6,400 km of the Italian railway network in order to meet traffic requirements imposed by the new high-speed system.

These projects have been carried out: works supervision of the Matsapha to Phuzumouya and Mpaka to the Mozambique border railway line; concession agreement and minimum safety works for the Djibouti-Ethiopia Border Railway (EC); feasibility study of the Asmara–Massawa railway line in Eritrea; study on competitiveness of railway transport on the Casablanca–Rabat–Meknes–Fes axis in Morocco conducted on behalf of the World Bank; study to improve railway interoperability between Italy and France; feasibility study for the Turin–Bussoleno railway line, an integral part of the new Turin–Lyon line which will connect the Italian high-speed network with its European counterpart through a long Alpine tunnel; Uzbekistan Railways modernisation project financed by the ADB; feasibility study on the development of railways and combined transport on Corridor IV (EC); project preparation support, modernisation of signalling and safety devices on the Divaca–Koper railway line, Slovenia (EC), ERTMS (European Railway Traffic Management System) pilot installation on the E-20 railway line (Kunowice–Warsaw Section), Poland (EC); modernisation of electric traction supply system of the E-20 railway line (Kunowice–Warsaw section, Poland (EC); technical assistance to SNCFR–Romanian Railways for management, technical and teacher training and for the reorganisation of existing training centres (EC); modernisation of Central Asian Railways' telecommunications for EC (TACIS); CEI Romanian Railways co-operation project for refurbishment of city station and rolling stock (EBRD); preliminary design and technical assistance for Lima Urban Electric Train Service in Peru; technical assistance for training to Ferrocar, Venezuela and to EFE, Chilean Railways; technical assistance to the concessionaire of the Pacific Railway Network, Colombia; and the project for the application of Directive 2001-16/EC on the interoperability of railways, Czech Railways (EC).

Current contracts include: technical assistance for the preparation of the detailed designs and tender documents for the improvement on the Mezotur–Gyoma line, Hungary (EC); review of railway rehabilitation in Central Asia (including links with other transport modes); upgrading of the Frejus–Modane line, France; technical study and railway operations scheme for the new Turin–Lyon railway line; consulting services for modernisation of track Bratislava Raca–Trnava, Section Bratislava Raca–Senkvice, Slovak Republic (EC); revision of the feasibility study for the rehabilitation of the railway line for the Hungarian border to Simeria, Romania (EC); Iraqi Transport Master Plan on behalf of the Coalition Provisional Authority; preliminary and detailed engineering of the New Puerto Cabello–La Encrucijada railway line, Venezuela; and in Syria, detailed design and tender documents of the Damascus–Daraa (Jordan Border) railway line; detailed design checking of Tabiyah–Al Bookamal railway line; and feasibility study and preliminary design of the Deir al Zoor Palmira railway line; and preliminary and detailed design, procurement assistance and works supervision of the Aleppo Coach Maintenance and Repair Centre.

UPDATED

Jacobs Babtie

95 Bothwell Street, Glasgow G2 7HX, UK
Tel: (+44 141) 204 25 11   Fax: (+44 141) 226 31 09
e-mail: marketing@jacobs.com
Web: www.jacobsbabtie.com

Key personnel

Vice-Presidents: Alan Craig, Mark Cubitt, Morris Murray, David Fawcett

International offices
(Associated companies)

Babtie Asia Ltd (Hong Kong)
15/F Cornwall House, 979 Kings Road, Quarry Bay, Hong Kong
Tel: (+852) 28 80 97 88   Fax: (+852) 25 65 55 61

Babtie China
121 Yanping Road, Sanhe Office Building, Room 9-D1, Shanghai, China 200042
Tel: (+86 21) 52 64 01 40
Fax: (+86 21) 52 64 01 41

Babtie Asia PTE Ltd (Singapore)
29 International Business Park, #03-01 Acer Building, Tower A, Singapore 609923
Tel: (+65) 68 90 19 60
Fax: (+65) 62 21 91 73

Babtie Consultants (India) Private Ltd
83 New York Tower 'A', Thaltej Cross Roads, Thaltej, Ahmedabad 380 054, Gujarat, India
Tel: (+91 79) 26 85 56 36
Fax: (+91 79) 26 84 35 27

PPU – Babtie spol s r o
Vyzlovska 2243/36, 100 00, Prague 1, Czech Republic
Tel: (+420 2) 74 81 25 37
Fax: (+420 2) 24 81 07 99

Babtie Pettit
Merrion House, Merrion Road, Dublin, Ireland D4
Tel: (+353 1) 269 56 66   Fax: (+353 1) 269 57 06

Babtie Consultants (India) Private Ltd
222, Solitaire Corporate Park, 151m M Vasanji Marg, Chakala, Andheri (E), Mumbai 4000 093, India
Tel: (+91 22) 56 97 00 65
Fax: (+91 22) 56 97 16 61

Babtie Middle East
PO Box 29000, National Bank of Abu Dhabi Building, 10th floor, Corniche Road (East), Abu Dhabi, UAE
Tel: (+71 26) 22 64 53

Babtie spol sro
Zlatnicka 10/1582, 110 00, Praha 1, Czech Republic
Tel: (+420 251) 01 92 31
Fax: (+420 251) 81 07 99

Background

Previously called Babtie Group, Jacobs Babtie is part of Jacobs Group. It is a technical and management consultancy operating in transport and development, environmental and utilities, property and structures, defence and energy, and partnerships and outsourcing markets both in the UK and internationally.

Services

Railways: include network appraisal, re-establishment of disused lines, extensions to the network, patronage forecasting, refurbishment of existing infrastructure, station developments, operational studies, assessment of freight transport, depot design and refurbishment, workshop plant and machinery, condition surveys, station design and refurbishment, geo-engineering advice, contaminated land, performance indicators, value engineering, asset delivery, procurement and specification, track alignment, freight operations and assessments, rolling stock, bridge inspection, assessment and design, signalling, tunnelling, geotechnical, electrical and mechanical, hydro-geology, environmental, surveying, quantity surveying.

Transportation: Jacobs Babtie's skills cover railways, highways, airports, public transport networks, ports and canals. Planning capability includes market research, strategic transport modelling, economic appraisals, feasibility and procurement strategies, pre-investment financial auditing, environmental impact assessments, training, value engineering and risk assessment, cost and contract consultancy, local transport plans.

Skills in construction methods, programming and value engineering include:
- tunnelling and ground engineering
- bridge design
- docks, ports and harbour design
- highway design, traffic control and telematics
- permanent way and railway depot design
- airports, pavement and ground lighting design
- busways and light rail infrastructure
- project management and cost consultancy.

Projects

Providing technical and procurement advice for Thameslink 2000, from scheme development to establishing the procurement method for the SRA. Dalton Park, Co Durham, provision of cost advice and value engineering services for Ballast plc.

UPDATED

James Scott

Division of AMEC Mechanical & Electrical Services Ltd
80-110 Finnieston Street, Glasgow G3 8LA, UK
Tel: (+44 141) 221 38 66   Fax: (+44 141) 226 30 68
e-mail: sales@isl.amec.co.uk
Web: http://www.amec.co.uk

Key personnel

Rail Business Manager: H MacEwan
Business Development: W Baird
Consulting Manager: M Scully
Branch Manager, Doncaster: C Brown
Business Development: D Atkinson

Other offices

Birmingham, London and Doncaster

Works address

Room 100, Dension House, PO Box 29 Doncaster DN1 1PD

Capabilities

James Scott offers a comprehensive electrical, mechanical and instrumentation service to the UK Railway infrastructure, from initial feasibility study through to installation and project management. Principal areas of expertise encompass:
Track infrastructure equipment including points heating, standby signalling supplies, approval of on-track plant and machinery.
Station facilities, including lifts and escalators, lighting design, air conditioning systems.
Infrastructure engineering support, including asset condition surveys, technical support and assistance with standards and legislative requirements.
Heavy workshops and train servicing facilities, including carriage washing facilities, fuel and oil delivery and recovery systems; depot facilities.

Projects

Include provision of improved stabling facilities on London Underground Northern Line, UK; transformer/rectifier renewals on Merseyrail, UK; various M & E design and installation packages on the Great Eastern resignalling scheme, UK; design and installation of points heating schemes, UK; swing bridge improvements, UK; refurbishment of funicular railway, UK; total renewal of electrical infrastructure on Blackpool Tramway, UK.

Current projects for Railtrack include a £15 million contract to design and install points heating, signalling generators, a control centre, plus signalling installation works and a £3 million design and survey lighting contract (1999–2003).

Japan Railway Technical Service (JARTS)

Taiyokan Building, 27-8 Hongo 2-chome, Bunkyo-ku, Tokyo 113-0033, Japan
Tel: (+81 3) 56 84 31 71; 31 79
Fax: (+81 3) 56 84 31 70; 31 80
e-mail: takahashik@jarts.or.jp

Key personnel
President: Hiroshi Komori
Senior Executive Vice-President:
  Osamu Matsumoto
Director Marketing: K Takahashi

Capabilities
Studies, surveys, design, planning specifications, preparation of contract documents, and project control and supervision of railway, high-speed rail, metro, monorail and advanced guided transit; construction of new lines; modernisation and improvement of track; electrification; modernisation of rolling stock; restructuring of railway management.

UPDATED

Jones Garrard Move Ltd

Jones Garrard Move Ltd, 31 Morland Avenue, Leicester LE2 2PF, UK
Tel: (+44 116) 270 11 18   Fax: (+44 116) 270 29 95
e-mail: michael-rodber@jonesgarrardmove.com
Web: www.jones-garrardmove.com

Key personnel
Director: Michael Rodber

Capabilities
Design: keeping transport brands on the move.

Projects
Eurostar, award winning design of train exterior. Vehicle interiors and specification, First Great Western. Brand new interior for high-speed train fleet. Marketing communication, Siemens/Network Rail, publication celebrating the Dorest Coast re-signalling project. Brand identity for Angel Trains. Internal brand development, Siemens. Promotion, FirstGroup, South West Trains. Environmental, Oeresund Link, design management for terminal area of Scandinavian road/rail link.

UPDATED

Jones Garrard has carried out projects for Eurostar   0021900

Labbé Designers & Associates Inc

5520 Chabot Bureau u 204, Montreal, Quebec, Canada H2H 2S7
Tel: (+1 514) 528 09 19   Fax: (+1 514) 528 60 83

Key personnel
President: Jean J Labbé

Background
Jean Labbé founded Labbé Designers & Associates in 1987.

Capabilities
Industrial designers in the rail and transportation market including rail vehicles, automated rapid transit systems, commuter cars, intercity cars, high-speed trains and locomotives.

Projects
Labbé designs have included the Bombardier Sea-Doo, BR400+, Mk VII monorail, the New York metro cars and the first high-speed train for Bombardier. Most recently Labbé has received a contract from Rotem to produce a state-of-the-art metro and it will produce the interior and exterior concept for the future development of high capacity metro cars by Rotem.

Lahmeyer International GmbH

Friedberger Strasse 173, D-61118 Bad Vilbel, Germany
Tel: (+49 6101) 55 0   Fax: (+49 6101) 55 22 22
e-mail: info@lahmeyer.de
Web: http://www.lahmeyer.de

Key personnel
Manager: Rainer Bothe
Transportation Division: Joachim Neumann;
  Heinz Saxer
  Offices in Berlin, Frankfurt, Munich, Stuttgart

Capabilities
The Lahmeyer International Group, with its 10 associated companies, is an independent engineering consultancy covering a spectrum of planning and consulting services in the fields of transportation, energy, hydropower and water resources, civil engineering and project management, technology and environmental sectors. The transportation division offers engineering services concerning transportation technology, railways and regional transport systems, tunnels and underground installations, bridges, roads, motorways, airports and specialised transportation facilities. The range of consultancy services encompasses studies, design and planning, tendering, project management, supervision and commissioning for high-speed railways, line-upgrading, stations, marshalling yards, intermodal terminals, depots and workshops, integrated transportation systems, suburban fast trains and feeder systems, underground and metro systems, tramways and light rail systems as well as maglev systems. Lahmeyer International possesses expertise and experience in specialised areas such as Life-Cycle Costing (LCC) in relation to rail traffic through observations and analyses in the RAMS sector (reliability, availability, maintainability and safety). For privately financed projects (BOT, PPP) it offers advisory and consultancy services to prospective owners or bidding consortia.

Projects
Include consultancy services for: the development of railway and combined transport links for the southern part of Corridor IX from Poland through Slovakia, Ukraine, Hungary, Romania, Bulgaria to Greece, comprising investigations of track infrastructure, electrification, bridges and tunnels, signalling/telecommunications, rolling stock review, operations and maintenance systems, as well as feasibility studies concerning proposals for improvements.
Design and construction supervision of a 20 km long underground metro system in Bangkok.
General consultant to German, Austrian and Italian Railway corporations concerning the Munich–Verona–Brenner Axis, providing technical assistance, risk analyses, economic and environmental impact assessments, marketing concepts, combined transport studies.
Preliminary, detailed and final design for two lines of the Athens metro.
Technical advisory services to the Financing Bank Syndicate of the Franco-British Channel Tunnel, comprising evaluation of construction contract, risk analyses, project monitoring.
General co-ordination, project management and site supervision for traffic facilities in the centre of Berlin. Detailed and approval design for the reconstruction and extension of the Munich Rapid Transit System. Detailed/approval design and site supervision for the connection of the Cologne-Rhine/Main new high-speed railway line to Cologne/Bonn airport.

Laramore, Douglass and Popham

332 South Michigan Ave, Suite 400, Chicago, Illinois 60604, USA
Tel: (+1 312) 427 84 86   Fax: (+1 312) 427 84 74
e-mail: postmaster@ldpgroup.com

Key personnel
President: Richard T Harvey
Senior Vice-President: H Saxena

Capabilities
Design and project management for electrified rapid transit and electric railway traction power supply and distribution systems.

Projects
Recent contracts include Red and Brown Line rehabilitation for Chicago Transit Authority; traction power upgrade for Chicago Transit Authority; traction power system consultant for Massachusetts Bay Transportation Authority; Fairgreen substation for Miami Valley Transit Authority.

LEK Consulting LLP

40 Grosvenor Place, London SW1X 7JL, UK
Tel: (+44 20) 73 89 72 00   Fax: (+44 20) 73 89 74 40
Web: www.lek.com

Other offices
Auckland, Bangkok, Beijing, Boston, Chicago, London, Los Angeles, Melbourne, Milan, Munich, Paris, San Francisco, Shanghai, Singapore, Sydney, Tokyo.

Key personnel
Directors: J I Goddard; J Simmons; A H Allum;
  P S Debenham; A J Scott

Capabilities
LEK is an international strategic consultancy and advises a broad range of transport-related clients worldwide. It has been active in public sector reform in the UK, Europe and Australia, and has advised passenger, freight, infrastructure and rolling stock organisations on key strategic and commercial issues. LEK has advised on a number of prominent European high-speed projects.
The company offers advice in strategic, economic, financial and operational areas, including privatisation and commercialisation advice, competition policy, traffic and financial forecasting, mergers and acquisitions and impact assessments of new rail technologies.

UPDATED

Lend Lease

142 Northolt Road, Harrow HA2 0EE, UK
Tel: (+44 20) 82 71 84 64   Fax: (+44 20) 82 71 80 26
Web: http://www.lendlease.com
  http://www.bovislendlease.com

Key personnel
Chief Executive Officer, Lend Lease: Ross Taylor
Chief Executive Officer, Lend Lease Europe:
  John Spanswick
Chief Executive Officer, Lend Lease Asia Pacific:
  Des Marks
Chief Executive Officer, Lend Lease Americas:
  Charles Bacon
Business Development: Mike Temple
Director, Lend Lease (Rail): Barry Taylor
Operations Director, Lend Lease (Rail):
  Nick Crossley

Capabilities

Lend Lease's real estate solutions business offers property-related services to clients involved in the creation, improvement or management of real estate assets. This includes expertise in development and capital raising, programme management, project and construction management, design and engineering, as well as facilities and asset management. The business includes Bovis Lend Lease. In the rail sector, Lend Lease works for government agencies, railway operators, developers and commercial entities on main line, suburban, metro and light railway projects in the UK, Europe, the Americas and Asia Pacific. Lend Lease is currently managing over £2,500 million-worth of railway projects. Services provided to the rail industry cover feasibility, project development, strategic planning and pre-construction phases through to development of procurement strategies, cost plans, construction planning and the tendering process – and to construction phase management, co-ordination and control, including quality, value, risk and safety management.

Projects

UK

SRA: Appointed to a panel of consultants providing project and commercial management services and currently involved in a range of enhancement programmes, all of which are vital to the SRA's strategy for transport. Current commissions include East Coast Main Line, Thameslink 2000, Freight Upgrade Programme and Strategic Regional Routes.

Network Rail (Railtrack): Involved in major route resignalling and modernisation projects, at feasibility, development and implementation stages on the West Coast, Great Western, Midlands, Southern and Channel Tunnel networks. Also involved in a number of major safety and compliance programmes.

Virgin Trains: Provision of project management support to Station Upgrade Programme.

Asia Pacific

Involved in track and station maintenance and upgrade and capital works programme management. Projects have included: Parramatta Rail Link, Sydney; Ultimo-Pyrmont Light Rail Transit Project, Sydney; Kuala Lumpur Central Station, Malaysia; and Auckland Rail Transit System, Auckland, New Zealand.

The Americas

Involved in track and maintenance and upgrade and capital works programme management. Projects have included: Long Island Railroad (on-call services), New York; rehabilitation of Times Square Station, New York; Grand Central Terminal modernisation, New York; New Jersey Transit (on-call services)/Rail Control Centre; and Newark International Airport, New Jersey.

Lester B Knight & Associates Inc

549 W Randolph Street, Chicago, Illinois 60606, USA
Tel: (+1 312) 346 21 00   Fax: (+1 312) 648 10 85

Key personnel
Vice-Presidents: Dominick J Gatto, Lee A Hoyt

Capabilities
Transport and environmental studies; railroads and rapid transit systems planning, design and construction, engineering and management; operations and maintenance.

Lloyd's Register Rail Limited

Dukesbridge Chambers, 1 Duke Street, Reading RG1 45A, UK
Tel: (+44 118) 955 61 00
Fax: (+44 118) 955 61 01/61 03
e-mail: enquiries@lrrail.com
Web: www.lrrail.com

Key personnel
Managing Director: P Thomas
Executive Directors: P Cheeseman, G Christmas, R Clutton, C H Porter, P Seller
Commercial Director: R Evans

Offices
Offices also in Birmingham, Belper, Bristol, Crewe, Derby, Glasgow, Leatherhead, London, Preston and York in the UK and also Hong Kong, Singapore, Sydney and Melbourne.

Background
Lloyd's Register Rail Limited is a wholly owned subsidiary of the Lloyd's Register Group and was previously MHA Systems Limited.

Capabilities
A multidisciplinary rail and transit consultancy, with expertise in human factors, systems engineering, safety critical systems, civil engineering and permanent way, signalling, control, communications and rolling stock. The company can carry out feasibility studies, project development and design, maintenance support and asset management surveys. Typical projects include technology strategies, new and post-design safety engineering, risk assessment, independent safety assessment and audit. In-depth knowledge of European interoperability requirements and international railway standards.

Contracts
Supply of project management, systems support and engineering services on major resignalling projects for Iarnród Éireann, and preparation of rail standards for Irish government. Safety cases for Virgin CrossCountry tilting rolling stock for Bombardier Transportation Group and ALSTOM Transport. Risk management framework for RailCorp, New South Wales.

Development and application of braking deficiency software for Network Rail; technical support and signalling design for Network Rail; technical support to develop new-generation control, train detection and interlocking systems; Notified Body for Network Rail WCRM and GWML.

Railway safety case audit for South West Trains.

UPDATED

Lockheed Martin Rail Systems

55 Charles Lindbergh Boulevard, Mitchel Field, New York 11553-3682, USA
Tel: (+1 516) 228 20 91   Fax: (+1 516) 228 18 97
e-mail: joe.tumbarello@lmco.com
Web: http://www.lockheedmartin.com

UK office
Lockheed Martin UK Integrated Systems
PO Box 41, North Harbour, Portsmouth PO6 3AU, UK
Tel: (+44 23) 92 56 54 00   Fax: (+44 23) 92 38 35 46
e-mail: bob.prothero@lmco.com
Web: http://www.lockheedmartin.co.uk

Capabilities
Systems integration provision for rail transport systems and projects.

Projects
In the USA, Lockheed Martin received a contract from the Illinois Department of Transportation to develop a Positive Train Control (PTC) system incorporating moving block technology to optimise track capacity and safety. The company was charged with managing the design, development, integration and test processes of the scheme by using integrated product teams to promulgate plans, manage and control processes and allocate resources during the project's life cycle.

In the UK, Lockheed Martin was designated Control Systems Integrator (CSI) by Railtrack for its West Coast Route Modernisation (WCRM) project, ensuring the scheme meets key agreed performance indicators and provides a foundation for future deployment of ERTMS technology. Responsibilities include: establishing baseline requirements; managing the development

of major system interfaces; defining and implementing a comprehensive human factors integration programme; support for Railtrack safety initiatives; systems level integration and test management; project management; and system level commissioning and integration. The Control Systems Integration of Phase 2 of the WCRM scheme will integrate the following elements into the overall WCRM programme: the Network Management Centre; train control system; conventional signalling; fixed bearer communications; and voice communications.

LogoMotive GmbH

Dr Carlo Schmid Strasse, 93a, D-90491 Nürnberg, Germany
Tel: (+49 911) 95 52 80   Fax: (+49 911) 955 28 19
e-mail: info@logomotive-nbg.com
Web: http://www.logomotive-nbg.com

Key personnel
Managing Director: Ulrich Hachmann

Capabilities
Services for the design, development and innovation of vehicles, systems and components for railways, automotive systems, special machines and plant construction. Specialising in acoustics, dynamics, structural mechanics, testing and overall systems. CAD design of complex structures such as carbodies, bogies and components. Concept design, detailed design, drawings, parts lists, including calculation (dynamics, acoustics, structural mechanics) to the relevant standards.

LTK Engineering Services

A member of the Klauder Group
100 West Butler Avenue, Ambler, Pennsylvania 19002, USA
Tel: (+1 215) 542 07 00   Fax: (+1 215) 542 76 76
Web: http://www.ltk.com

Key personnel
President: George N Dorshimer
Vice-Presidents: F H Landell, J S Gustafson, F W Frandsen, T B Furmaniak, C M Lawlor
Director, Business Development: David H Oglevee

Established: 1921
Staff: 230

Capabilities
Planning, engineering and design for urban transport, including design of passenger rail vehicles, communications and signalling systems, traction power systems, fare collection systems and rail vehicle maintenance facilities.

Projects
Portland Tri-Met: Systems Engineer for Banfield, Westside, Hillsboro and interstate extensions.
Amtrak: Rail vehicle engineering services for Northeast Corridor High Speed Rail Programme.
New York City Transit: Rail vehicle engineering services for R142, R142A, R143 and R160 programmes.
Seattle ST Link: Systems Engineer for the new light rail system.
Long Island Rail Road: Rail vehicle engineering services for M-7 emu programme.
New Jersey Transit: Engineering services for the Comet II overhaul and multi-level coaches.
Tren Urbano: Engineering services for new rapid transit car fleet.
Los Angeles County MTA: Rail vehicle engineering services for the LA Red, Blue, Green, Standard and Pasadena cars.
Sacramento RT: Engineering services for LRV1 and LRV3 programmes.
Dallas Area Rapid Transit: Engineering services for two LRV programmes.
Boston MBTA: Engineering services for the Green line routes 7 and 8, Red line No 3, Blue line No 4 and 5, and Orange line No 12 cars.

Philadelphia: Engineering services for SEPTA's M4 rapid transit car procurement.
Washington, DC: Engineering services for WMATA's original 766-car fleet.

Martyn Cornwall Design

Unit 15, Swan Court, 9 Tanner Street, London SE1 3LE, UK
Tel: (+44 20) 72 34 06 12   Fax: (+44 20) 74 03 38 68
e-mail: enq@mcdesign.co.uk

Key personnel
Director: Martyn Cornwall
Production Director: David French

Capabilities
Provision of comprehensive design services; strategic advice and planning of all aspects of design requirements; with a design team qualified to undertake corporate identity, graphic design, multimedia, web and interior design projects.

Projects
Include the development and implementation of new corporate identities for train operating companies in the UK, including train liveries and promotional material.

VERIFIED

MBD Design

11 rue Victor Hugo, F-93177 Bagnolet Cedex, France
Tel: (+33 1) 48 57 30 00   Fax: (+33 1) 48 57 41 31

Key personnel
Chairman: Yves Domergue
General Manager: Jean-Claude Marbach

Capabilities
MBD Design accepts commissions from network authorities and rolling stock manufacturers for both long- and short-term railway transportation projects.

Projects
Design studies have been undertaken for: BB 36000 and BB 27000 locomotives for SNCF; Singapore, Shanghai and Warsaw Metros for ALSTOM; TER (regional express train) TER 2N (double-deck regional express train) and AGC single-deck emu/dmu; MI 2N three-door double-deck car for the Paris RER; TRN (National rapid train) for SNCF.

McCormick Rankin Corporation

2655 North Sheridan Way, Mississauga, Ontario L5K 2P8, Canada
Tel: (+1 905) 823 85 00   Fax: (+1 905) 823 85 03
e-mail: mrc@mrc.ca
Web: www.mrc.ca

Key personnel
Chairman and Chief Executive Officer: Ian Williams
Senior Transit Personnel:
  Manager, Public Transit Services (Canada):
    Dale Turvey
  President, McCormick Rankin International:
    Ken Gosselin
  Rail Rapid Transit Planning, Design and
    Construction: Dennis R Callan
  Manager, McCormick Rankin Cagney Pty Ltd
    (Brisbane): Neil Cagney
  Transitway Planning/ITS Applications:
    Steve Schijns
  Operational Planning: Sean Rathwell

Other offices
Ottawa, Kingston, Kitchener, Halifax, Brisbane and Auckland.

Subsidiary companies
MR International (MRI)
McCormick Rankin Pty Ltd

McCormick Rankin Cagney Pty Ltd (Brisbane, Auckland)
Ecoplans Ltd (Environmental)

Capabilities
Expertise in planning, design and construction services for the full range of rapid transit technologies including busways, LRT, heavy rail, commuter rail and people mover systems; organisation and operation reviews of conventional transit systems serving urban populations from 15,000 to 500,000+; and development of stations, bus transit operations centres, transit ITS and specialised transit operations.

Projects
Completion of the Environmental Assessment Study (MIS) for the 31 km North-South LRT Line in Ottawa, Canada.
Complete design of the Bayview Station and ancillary facilities on the Toronto Transit Commission's new Sheppard Subway.
Environmental assessment for a 13 km Yonge–Spadina extension (loop) of the existing Toronto subway including ridership demand analysis, impact assessment and preliminary engineering. Planning and preliminary design of 30 km busway in Brisbane, Australia, including detailed design manuals and ITS applications for service monitoring, passenger information systems and the security network.
Development of a GO Transit interregional Bus Rapid Transit network, fully integrated with the existing Toronto TTC rapid transit and GO commuter rail networks.
Preparation of a GO Transit interregional Bus Rapid Transit network fully integrated with the existing Toronto TTC rapid transit and GO commuter rail networks.
Preparation of an LRT and busway implementation plan for the City of Ottawa, Canada.
Completion of a comprehensive review of the issues facing the Canadian transit industry, an inventory of international best practices specifically related to funding mechanisms and the development of a preferred plan for the involvement of Transport Canada.
Preparation of GO Transit's (Toronto) Georgetown Corridor Commuter Rail Upgrade study.
Introduction of BRT facilities for the City of Winnipeg within two corridors totalling 19 km in length.
Consulting services include preparation of design manual, the preliminary design of guideway, design of ITS application for passenger system and service monitoring and control, operation design of routes and schedules for the transitway and feeder services.

UPDATED

Mercer Management Consulting Inc

1166 Avenue of the Americas, New York, New York 10036, USA

Transportation Group
33 Hayden Avenue, Lexington, Massachusetts 02173, USA
Tel: (+1 617) 861 75 80   Fax: (+1 617) 862 39 35

Key personnel
Vice-President: Hugh Randall

UK office
1 Grosvenor Place, London SW1X 7HJ, UK
Tel: (+44 20) 72 35 54 44   Fax: (+44 20) 72 45 69 33

Key personnel
Vice-President: Matthew Vanderbroeck

Capabilities
Mercer's Transportation Group assists railways and other transport undertakings in the following areas: strategy development; privatisation and commercialisation planning and implementation; organisational restructuring and management development; process re-engineering; operations enhancement and cost reduction; marketing and market research; information management; acquisition and alliance planning; and litigation support.

Mer Mec SpA

Via Oberdan 70, I-70043 Monopoli (BA), Italy
Tel: (+39 080) 887 65 70   Fax: (+39 080) 887 40 28
e-mail: mermec@mermec.it
Web: www.mermec.it

Key personnel
Managing Director: Vito Pertosa
Marketing and Sales, Executive Director: Luca Ebreo
Marketing Manager: Pietro Stama
International Sales Manager: Carlo Evangelisti
Domestic Sales Manager: Mario Girolami
After Sales and Commissioning Manager:
  Guiseppe Aurisicchio
Quality Assurance Manager: Mauro Simone
Research and Development Programme Manager:
  Patrizia Sforza

Services
Professional services offered include: railway infrastructure monitoring and data management; vehicle commissioning, retrofit and maintenance.

UPDATED

Metroconsult

Helene-Weber-Allee 15, D-80637 Munich, Germany
Tel: (+49 89) 157 68 66   Fax: (+49 89) 157 24 73
e-mail: info@metroconsult.com
Web: www.metroconsult.com

Key personnel
Managing Director: Dr Ing Jürgen Rauch

Capabilities
Design of metro, railway stations and consultancy mainly to public transportation system owners; planners and industry; methods of increasing capacity of metros and commuter railway systems; design of innovative platform screen door systems for metro stations; planning of buildings for public infrastructure; integration of metro systems into the transportation system of a city, technically and by means of architectural design; design of innovative parking systems in cities and within limited space; design of commuter parking systems; development of materials of fire protection in confined spaces as there are tunnels and underground stations; technology development support to manufacturers; support for innovative concepts of railway vehicles for public transport.
Planning and consulting for building and operating stations and systems of public transport, development and management of innovations for the operation of public transport and stations including: passenger information systems; automatic passenger guidance; optimisation of operation; optimisation of weak points of operations; development of timetables/schedules, especially during peak hours for large commuter rail and metro systems; development of real time control methods for driverless passenger transport. Consulting for public transport operators, public authorities, supervising the transport market, consulting for the technical infrastructure.
Parking guidance systems; conception of modern intermodal station with park-and-ride; station operation technology, industrial development, new developments such as flexible platform doors. Planning for automated passenger counting for metros, train, buses; passenger guidance systems for metro; automated parking facilities; fire safe tunnel building and lining technology; planning camera surveillance – anti-terrorism measures in public transport and large public facilities; new card and fare collection technology modules and system modules for terrorism prevention to be combined with the transportation guidance system.

Projects
Passenger guidance system for the commuter railway network of Munich, Germany; several metro stations and parking systems in south Germany; vehicle parking related to public

transport; development of fire protection materials in confined spaces such as tunnels and underground stations for better orientation in daily operation and emergency situations; consultancy to several city authorities and to manufacturers in the public transport industry; development of platform door systems for metros; development of concepts for a light rail train with energy systems not yet used in railway technology.

Developments

In 2004 Metroconsult redeveloped the passenger guidance system into a transportation guidance system. Passenger counting, gained either by automatic passenger counting systems or by image processing software, combined with a camera surveillance system of metro trains and stations, automatically produces direct information to the dispatching system of a metro. This calculates real time the necessary number of trains, headway and train length needed for operation.

UPDATED

Millbrook Proving Ground Limited

Millbrook, Bedford MK45 2JQ, UK
Tel: (+44 1525) 40 84 08   Fax: (+44 1525) 40 84 68
Web: http://www.millbrook.co.uk

Key personnel

Head of Business Development: Damien Faysse

Products

Millbrook Technology Park provides test and development solutions for the automotive and rail industries. Millbrook has a unique combination of world class tracks and laboratories. Millbrook's areas of expertise include crashworthiness, systems durability, vehicle emission and fuel economy. In addition to the laboratories, the wide range of tracks available includes off road for passenger cars and military specifications, high-speed circuit, hill route and a full range of structural inputs.

VERIFIED

Modjeski & Masters Inc

PO Box 2345, Harrisburg, Pennsylvania 17105, USA
Tel: (+1 717) 790 95 65   Fax: (+1 717) 790 95 64
e-mail: harrisburg@modjeski.com
Web: http://www.modjeski.com

Key personnel

Principals: W B Conway; J M Kulicki; H E Waldner;
  D F Sorgenfrei; BT Martin; L K Huang; Z Prucz
Senior Associates: TY Soong; J E Prickett;
  L V Borden; M C Irwin; R A Little;
  J L O McKenney; R A Martino; W N Marianos, Jr   ·
Project Development: M F Britt

Other offices

New Orleans, Louisiana
Poughkeepsie, New York
Moorestown, New Jersey
Charleston, West Virginia
Edwardsville, Illinois
St Louis, Missouri

Capabilities

Rail structures; design and maintenance inspections of fixed and movable bridges.

Mott MacDonald Group

St Anne House, Wellesley Road, Croydon CR9 2UL, UK
Tel: (+44 20) 87 74 20 00   Fax: (+44 20) 86 81 57 06
e-mail: richard.williams@mottmac.com
  stuart.thompson@mottmac.com
Web: www.mottmac.com

Key personnel

Group Board Directors: M O Blackburn (Chairman),
  K J Howells (Managing), P M Chesworth,
  M S C Frame, K J Stovell, P J Wickens, F Turner
  (non-executive)
Transportation Managing Director: R E Williams
Transportation Directors: N Bristow, C Davis,
  A Finch, A Powderham, R Williams, M Wallwork,
  R F Davies
Key Directors: J D Corries, R Staniforth, A West,
  R Carter, M G Simpson, A R Walker, J Hughes,
  R N Dumolo, C Chalk, T O'Neill, S B Thompson,
  W Rankin, P Norgate, D A Hand

International offices

Australia, Bahamas, Bahrain, Bangladesh, Brazil, Bulgaria, Cambodia, Canada, China, Czech Republic, Egypt, Ethiopia, Hungary, Iceland, India, Indonesia, Ireland, Kazakhstan, Kuwait, Lesotho, Libya, Malaysia, New Zealand, Nigeria, Norway, Oman, Pakistan, Philippines, Poland, Portugal, Qatar, Romania, Russian Federation, Saudi Arabia, Serbia, Singapore, South Africa, Spain, Taiwan, Thailand, Turkey, Uganda, United Arab Emirates, US, Uzbekistan, Venezuela.

Capabilities

Mott MacDonald is an independent multidisciplinary engineering, management and development consultancy with a turnover approaching £500 million and 8,300 staff worldwide including chartered engineers, transportation planners, computer specialists, environmental scientists and support staff operating throughout 100 countries.

Capabilities encompass investigations, studies and technical feasibility reports, project definition, financial and environmental appraisal; safety assessment; preliminary and detailed design, contract preparation and rendering supervision, project scheduling, specification and procurement, quality control, cost and budget control, project implementation and construction management. The Mott MacDonald Group undertakes management and operational planning in the areas of traffic engineering, rail and transit operation and management related to different modes, supervision of inspection and testing of equipment during manufacture, testing and commissioning, investments planning, including development of transport models, demand forecasting, evaluation techniques on economics, financial, technical and environmental grounds and modal choice techniques.

The Group's consultancy services cover all forms of rail transport including heavy rail, metro, light rail and monorail. Design and implementation of urban public transport systems, comprehensive service in transport planning, civil and structural engineering and mechanical and electrical engineering; traffic engineering and highway planning; tunnel and station ventilation; train control, signalling and communications; studies of electromagnetic compatibility and safety from traction interference; rolling stock and traction power supply performance; rolling stock procurement advice.

Projects

Canada: Sheppard Subway Project: Programme managers for the Toronto Transit Commission's 6.4 km rapid transit subway project.
Czech Republic: PRaK, Czech Republic: feasibility study for rail link to Prague airport.
Denmark: Copenhagen Metro, Denmark: tunnel ventilation studies.
Holland: Support work on the Infraspeed Consortium's HSL-Zuid railway in Holland.
Hungary: Budapest Metro Line 4, Hungary: railway authorisation design.
India: Delhi Metro: Southern Section-civil architectural and M&E design for contract MC1B.
Indonesia: Jakarta Monorail: finalising a feasibility study into providing a mass transit monorail system.
Jakarta North-South integrated toll road and LRT project, Indonesia (transport planning and conceptual design)
Ireland: DART Ireland: power supply design for Greystones extension.
Luas: alignment and system design.

Malaysia: Putrajaya Light Rail, Malaysia: design and project management.
Kuala Lumpur Monorail, Malaysia: independent checking engineer.
Portugal: Linha do Norte upgrading, Portugal.
Porto Metro, Portugal: resident engineering services in joint venture with Geodata of Italy for Transmetro, 70 km of track and 66 stations.
Singapore: North East Line, Singapore: overhead catenary system design.
Marina Line, Singapore: concept design and performance specifications.
Sweden: Banverket, Sweden: signalling and ATP design services.
Taiwan: Taipei Department of Rapid Transit Systems: consultancy services, for mechanical and electrical commissioning.
Kaohsiung Metro, Taiwan: technical audit. Lead consultant to monitor and audit the design, construction and commissioning of the £10 billion BOT railway for Taiwan High Speed Rail Corporation (as part of the IREG consortium).
Thailand: Bangkok MRTA, Thailand: project management for 20 km underground metro system.
UK: Automatic Train Operation System (ATO), Glasgow, UK: Design and development of a replacement ATO system for the Glasgow underground.
Crossrail, UK: multi-disciplinary design services for line one for Crossrail, the 50/50 joint venture between Transport for London (TfL) and the Strategic Rail Authority (SRA).
Dorset Coast resignalling in partnership with Siemens on Railtrack's £20 million new signalling system.
Edinburgh Light Rail, a recently won project.
Heathrow Express: civil, mechanical and engineering design for 12 km of tunnel (plus open cut-and-cover works) with two underground stations; provision of an in-house developed geological and Geotechnical Data Management System (GDMS).
Heathrow Terminal 5: systems integrational assurance services for Heathrow Express and Piccadilly Line extensions into the new tunnel.
London Underground, UK: technical advice to London Underground Ltd's Chief Engineer's Directorate under a five-year framework agreement covering a diverse range of support including knowledge, risk and programme management.
Manchester Airport: Ground Transport Interchange, UK: lead consultant for new bus, rail interchange.
Manchester Light Rail.
London Underground Ltd, UK: asset and condition survey of earth structures.
LTS Rail: procurement of new rolling stock.
Merseyside Light Rail Transit, Liverpool, UK: proposals for progression of planned line 1, 2 and 3 of LRT network in Merseyside.
Nottingham Light Rapid Transit, Nottinghamshire, UK: review and refinement of existing LRT alignment options for extension of network.
OPRAF, UK: tilting train studies and specifications for West Coast Main Line.
Programme manager (in a joint venture with Fluor) for Railtrack's East Coast Main Line upgrade.
Railtrack, UK: safety of dual-voltage earthing arrangements on the North London line.
West Coast Main Line upgrade for Railtrack as consultant for feasibility and detailed design for whole route.
West Coast Route Modernisation, UK: multi-functional consultant.
US: LA Metro, USA (construction management for Red Line North Hollywood extension).
Long Island Rail Road East Side Access, New York: technical tunnel consultant to Bechtel/URS Greiner.

UPDATED

MTR Corporation Consultancy Services

MTR Tower, Telford Plaza, Kowloon Bay, Hong Kong
Tel: (+852) 29 93 23 17   Fax: (+852) 29 93 77 74
e-mail: jjdring@mtr.com.hk

Key personnel
General Manager, International Business:
Jonathan J Dring

Capabilities
MTR offers consultancy services to the railway industry in operating management, engineering management and maintenance fields. Based on its experience gained over 25 years in building and operating MTRC's urban rail system and Airport Express line, assistance can be given in: project preliminaries – financial proposals, feasibility studies, project definition and concept planning; project planning, design and construction – detailed planning, programming, design, system assurance, system integration, project management, construction management, inspections, testing and commissioning, operational readiness; railway operation and maintenance – station operation, train operation, central control, incident management, revenue services, documentation, infrastructure maintenance, railway system maintenance, asset management, management information systems; railway support services – safety management, quality management, environmental management, training, procurement, contract administration, inspection services, performance management; other related services – non-fare revenue, integrated property development, project financing, privatisation.

Projects
Recent projects include: operation, maintenance and system assurance for Kaohsiung metro; operations management for the automated people mover at Hong Kong International Airport; trackwork detailed design for both the Orange Line and the Red Line on the Kaohsiung Metro; Owner's Representative for Shanghai Metro Line R4; Owner's Advisor for Shanghai Metro Line M8; and operation and maintenance pre-operation support for Bangkok Metro and development of the national smart card system for the Netherlands.

MTR has also ventured into metro development in China. It has signed Memoranda of Understanding with local governments to develop Shenzhen Line 4 (on a BOT basis) and Beijing Line 4 (on a PPP basis).

UPDATED

MVA Asia Ltd

26/F, China Resources Building, 26 Harbour Road, Wanchai, Hong Kong
Tel: (+852) 25 29 70 37   Fax: (+852) 25 27 84 90
e-mail: mva@mva.com.hk
Web: www.mva-group.com

Key personnel
Director, MVA Asia Ltd: Chris Burley

Background
MVA Asia Ltd is part of the MVA Group, comprising 400 professional staff in traffic and transportation worldwide. MVA is a member of the international Systra Group and has a head office in Hong Kong and regional or local offices in most of the major countries in Asia.

Capabilities
MVA provides professional consultancy services in transport and traffic planning and management: transport infrastructure feasibility planning and appraisal; highway and road traffic engineering, planning and design; railway planning and operations; public transport systems, integration and intermodal facilities; integrated transport planning and design for development projects; patronage/traffic/revenue demand forecasting; financial and economic appraisal; policy and institutional assessment; social and market research.

Projects
MVA was commissioned by a private sector investor to carry out a preliminary analysis of the likely revenue and costs for a potential high-speed rail service linking Singapore, Malaysia and Thailand, with possible extension to China

via Laos. MVA has acted as transport adviser to investors in a number of urban rail projects. Services have included patronage and revenue studies, service planning, station/interchange planning, bus feeder systems.

Projects have included: studies on behalf of the LRT and monorail operators in Kuala Lumpur; LRT feasibility studies in Johor Bahru and Penang. Recent project experience in Thailand includes: transport studies for MRTA on the design of transfer facilities, bus restructuring and station area planning for the Blue Line metro system; Bangkok Transit System (Green Line – Skytrain) forecasts of future patronage and revenue. Taiwan four-stage model, used to provide patronage and revenue forecasts for the high-speed rail project; assessment of demand for new metro system in Vietnam; multimodal model of the Metro Manila area to assess a proposed suburban commuter rail system; the EDSA Metrostar Express light railway; Johor Bahru public transport model used to study bus route reorganisation and the role of light rail; review of pedestrian designs and transport interchange provisions for high-speed railway stations; review of pedestrian designs and provisions for high-speed railway stations, Taiwan; passenger origin-destination and route choice studies and surveys on BTS (Skytrain); planning of public square and pedestrian network for Guangzhou Liuhua and Suzhou railway station areas; study to improve passenger service and efficiency through rationalisation of the scheduled bus fleet, Madras.

NEW ENTRY

MVA Group

MVA Limited (UK, Middle East, North and South America)
MVA House, Victoria Way, Woking GU21 6DD, UK
Tel: (+44 1483) 72 80 51   Fax: (+44 1483) 75 52 07
e-mail: mail@mva.co.uk
Web: www.mva-group.com

Key personnel
Chief Executive Officer: Nigel Ash
UK contact: Andrew Jones
Hong Kong and China contact: Fred Brown
Asia Regions contact: Chris Burley

Associated companies
MVA Asia
26/F, China Resources Building, 26 Harbour Road, Wanchai, Hong Kong
Tel: (+852) 25 29 70 37   Fax: (+852) 25 27 84 90
e-mail: mva@mva.com.hk
Web: www.mva-group.com

MVA Singapore Pte Ltd
No 65 Club Street, Level 2, Singapore 069439
Tel: (+65) 62 27 32 52

MVA Thailand Ltd
37.F Unit F, Payatai Plaza, 128/405, 128 Phyatai Road, Thung-Phyatai, Rajthavee, Bangkok 10400, Thailand
Tel: (+662) 216 66 52
Fax: (+662) 216 66 51

MVA Shenzhen Ltd
Room 2229A, Wan Tong Building, 3002 East Sungang Road, Shenzhen, China
Tel: (+ 86) 755 25 86 78 60   Fax: (+86) 755 25 86 78 77

MVA Beijing
Room 505, Yue Tan Li Xiang Da Sha, A1.2 Lane, South Lishi Road, Xicheng District, Beijing 100045, China
Tel: (+86 10) 68 02 01 14
Fax: (+86 10) 68 02 01 14

Background
MVA was established in 1968 and has approximately 350 staff. MVA is a member of the Systra Group, international consulting engineers for rail and urban transport.

Capabilities
MVA provides sustainable transport solutions to clients across the transport sector. Clients include

governments, operators, agencies, developers and financiers. MVA delivers practical solutions in multi-modal planning, traffic engineering, public transport appraisal, demand and revenue forecasting, information management, intelligent transport systems, payment strategies, and social and market research.

Contracts
Recent contracts include advice on appraisal issues to Nottingham Express Transit, development of the business case for the Fastway guided bus system, public consultation and Transport and Works Act advice for Luton Translink, appraisal of options for reopening the Stirling–Alloa railway line, development of the North East Regional Planning Assessment for the Strategic Rail Authority, advice on smart card ticketing to the NoWcard Parnership of local authorities in the north west of the UK and customer satisfaction monitor for GNER.

Overseas contracts include demand forecasting for Hanoi tram, China cross boundary public transport study, Dubai light rail forecasting, advice to TransitLink on public transport integration, restructuring and fares issues in Singapore, Ho Chi Minh City Metro feasibility study, patronage and revenue studies for Bangkok Skytrain, advice on Taiwan high-speed railway, Beijing airport express line patronage and revenue study and preliminary design study for Shenzhen Metro Line 4.

UPDATED

NEA Transport Research and Training

Sir Winston Churchilllaan 297, PO Box 1969, NL-2280 DZ Rijswijk, Netherlands
Tel: (+31 70) 398 83 88   Fax: (+31 70) 395 41 86
e-mail: email@nea.nl
Web: www.nea.nl

Key personnel
Managing Director: Menno M Menist

Capabilities
NEA is an independent knowledge provider operating in the field of traffic, transport, infrastructure and logistics. NEA's activities encompass the economic and social aspects of both passenger and freight transport for all modes. In the field of research and consultancy, NEA has experience in modelling, forecasting and evaluation of international freight flows, including simulation, scenario building, economic impact analysis, socio-economic research, market research and mobility studies. Training experience includes development of course materials and case studies, training needs assessments, institutional strengthening, legal harmonisation and legal reform.

Projects
Research, training and consultancy assignments have been undertaken for governments, international agencies and organisations. Clients include the World Bank, Asian Development Bank, Commission of the European Communities, the Netherlands government, international branch organisations and the private sector.

UPDATED

NedTrain Consulting

PO Box 2016 – KTT 6, NL–3500 GA, Utrecht, Netherlands
Tel: (+31 30) 300 47 00   Fax: (+31 30) 300 48 00

Key personnel
Director: A J M Spaninks

Associated companies
NedTrain
PO Box 2016 – KTT 9, NL–3500 GA Utrecht, Netherlands
Tel: (+31 30) 300 46 60   Fax: (+31 30) 300 46 44
e-mail: info@nedtrain.nl
Web: www.nedtrain.nl

Background

NedTrain Consulting supports operators and owners of rolling stock during the entire life cycle of their fleet, supports manufacturers developing theur products and helps infra managers on the wheel-rail interface. NedTrain Consultancy is part of NedTrain, the maintenance partner of NS (the main Dutch Railway operator).

Capabilities

Consultancy geared towards support of rolling stock during all phases of the life cycle; strategic fleet management decisions, procurement support, tender evaluation, production monitoring, quality management, testing and commissioning, vehicle acceptance, reliability engineering and trouble shooting, design of vehicle modifications.

Projects

Recent projects include procurement support and project management for some major rolling stock projects (500 units), acceptance of light rail vehicles on the Dutch rail network, vehicle acceptance of BR189 with Siemens and DB, production monitoring for HSBC, RET and GVBA, safety cases for Virgin's Voyager, noise reduction projects for ProRail, NS Reizigers and various government departments and the business process redesign of the quality management department of a leading manufacturer.

Operation efficiency and cost reduction projects. Monitoring of rolling stock-infrastructure interference. Development and sales of rolling stock safety systems in co-operation with Bombardier.

UPDATED

The Nichols Group

2 Savile Row, London W1S 3PA, UK
Tel: (+44 20) 72 92 70 00   Fax: (+44 20) 72 92 52 00
e-mail: info@nichols.uk.com
Web: http://www.nicholsgroup.co.uk

Key personnel

Chairman and Chief Executive: Mike Nichols
Executive Team:
   Kathryn Nichols (Company Secretary and Corporate Services)
   Francis Nichols (Creative Director)
   Debra Rymer (Financial and Public Sector Services)
   Colin Britt (Transport Services)

Capabilities

The Nichols Group is an independent management consultancy specialising in advising on and managing large scale, complex and rapid changes covering both major capital investments and business change programmes. The group uses its expertise in project and programme management to initiate, develop and manage investments to maximise benefits for its clients. Since 1975 the Nichols Group has played a crucial role in many major transport schemes in the UK and southeast Asia.

Capabilities include strategic planning, programme management, project delivery and major project reviews, systems integration, risk management and training and development services.

Projects

Managed strategic planning for Railtrack (now Network Rail) – West Coast route modernisation planning; programme management for Strategic Rail Authority's Southern Region new trains programme; project delivery for Docklands Light Railway, various phases; systems development for Rail Safety and Standard Board by creating the industry plan of ERMTS; project management, sponsorship training and development for London Underground Ltd.

Ødegaard & Danneskiold-Samsøe A/S

Titangade 15, DK-2200 Copenhagen N, Denmark
Tel: (+45) 35 31 10 00   Fax: (+45) 35 31 10 01
e-mail: uds@oedan.dk
Web: http://www.odegaard.dk

Key personnel

Managing Director: John Ødegaard
Sales Manager, Chairman of the Board:
   Ulrik Danneskiold-Samsøe
Senior Consultants: Henrik W Thrane, Uffe Degn, Claus M Myllerup, Ulrik Møller Rasmussen, Morten Theill Jensen

Subsidiary

Ødegaard & Danneskiold-Samsøe Norge AS
Ørsnesalléen 17, N-3120 Tønsberg, Norway
Tel: (+47) 33 35 22 70   Fax: (+47) 33 35 22 71
Manager: Frank W Trulsen

Capabilities

Consulting engineers, specialising in noise and vibration control of trains and other modes of transport. The company's services apply to all phases of the life of a rail vehicle, from design through to operation. Services offered include: design advice, noise and vibration analyses and troubleshooting.

OMI Logistics Limited

2-10 Cawte Road Southampton SO15 3TD, UK
An OMI International plc Company
Tel: (+44 23) 80 90 82 00   Fax: (+44 23) 80 33 61 17
Web: http://www.omi-logistics.co.uk

Key personnel

Business Development Manager: John Churchman

Capabilities

Production of technical documentation for the major transportation industries.

Projects

OMI Logistics has recently delivered the first issue of a full suite of manuals at the end of an initial 21-month contract to the Siemens Transportation Systems consortium for Heathrow Express rolling stock.

Covering maintenance, operation, operator training and an illustrated parts catalogue, OMI compiled the hard copy manuals with supporting illustrations on CD-ROM from information provided by the pan-European group of manufacturers involved in the project, including CAF of Spain and Faively of France, as well as the German and UK arms of Siemens.

In addition, OMI worked on the signalling system for the Hong Kong Mass Transit System and provided documentation for new trains introduced on the London Underground Central Line.

Ove Arup & Partners

13 Fitzroy Street, London W1T 4BQ, UK
Tel: (+44 20) 76 36 15 31   Fax: (+44 20) 77 55 24 51

Key personnel

Chairman: Terry Hill
Transport Director: Ed Humphreys

Background

Founded in 1946, Arup Transportation comprises around 120 professional and technical staff operating through a network of UK offices including Birmingham, Bristol, Cardiff, Coventry, Edinburgh, London, Leeds, Manchester and Newcastle. Part of Arup Transport, it has over 6,000 staff based in 70 offices in more than 50 countries.

It has specialist staff in Australia, Hong Kong, Singapore, Johannesburg and New York and a network of offices throughout Europe, Asia, Australasia and North America.

Capabilities

Transport planning, environmental, economics, acoustics, geotechnical, civil, mechanical and electrical engineering, structural and building engineering services. Planning and design of urban and interurban transport systems; business case appraisal, specialists in rail, light rail and bus planning and appraisal, strategic transport planning studies, demand forecasting and modelling, information technology, bus priority and network planning, rail and LRT operations, traffic control systems, feasibility studies, design of infrastructure including bus and rail stations, station capacity, rail and road construction management; traffic engineering including parking control and strategy work; pedestrian movement and flow modelling.

Projects

Victoria Transport Interchange, Central London Congestion Charging, King's Cross Central, MEDA (European Mediterranean Transport Project), SYPTE framework projects, TERFFS extension; gauge corner cracking control strategy development; rail freight grant regime review and industry consultation; King's Cross rail and station planning; Channel Tunnel rail link; rail freight avoidable cost review, Manchester – crewe layout remodelling; level crossing signalling renewals for Railtrack LNE (Network Rail); Sunderland extension Metro-Railtrack signalling interface specification and implementation; Welwyn–Hitchin upgrade feasibility.

NEW ENTRY

Owen Williams Group

3 Duchess Place, Hagley Road, Birmingham B16 8NH, UK
Tel: (+44 121) 456 15 68   Fax: (+44 121) 456 17 57
Web: http://www.owenwilliams.co.uk

Key personnel

Chief Executive: R O M Williams
Marketing Manager: Donald Proud
Director – Highways: Dennis Hill
Director – Railways: Paul Marshall

Associated companies

Owen Wiliams Railways
Owen Williams Consultants
Primo FM Services

Capabilities

Railways and light rail; transport studies and economic assessments; highway and bridge engineering; highway maintenance management; bus priority; impact studies; park-and-ride studies; feasibility and privatisation studies; railway alignments and bridges; structure design; traffic control and communications; ground engineering.

VERIFIED

Pacific Consultants International

1-7-5 Sekido, Tama-shi, Tokyo 206-8550, Japan
Tel: (+81 42) 372 01 11   Fax: (+81 42) 372 63 64
e-mail: pci-info@pcitokyo.co.jp

Key personnel

President and Chief Executive Officer: Itaru Mae
Chief Operating Officer/EVP: Shota Morita
Managing Directors: Masahito Yamanaka;
   Kinichi Kato; Kimio Takeya; Jens Peter Henrichsen
Directors: Nobuo Endo; Katsuhide Nagayama;
   Masayoshi Taga

Capabilities

National and regional planning, demand forecast and master planning, feasibility and investment study, surveys and site selection, environmental studies, preliminary and detailed design, contract and tendering, construction supervision and administration, technical assistance and training project, quality control, budget control, contract and legal documentation, public private partnerships, operations and maintenance.

Contracts

Project management consulting services for: Delhi Mass Rapid Transport System (India); commuter rail upgrading and railway Bosphorus tube crossing (Turkey); Cairo Regional Area Transportation Study (Egypt); Port of Constantza – South Container Terminal Project (Romania); Kohat Tunnel and Access Road Project (Pakistan).

UPDATED

Parsons Brinckerhoff Inc

One Penn Plaza, New York, New York 10119, US
Tel: (+1 212) 465 50 00   Fax: (+1 212) 465 50 96
e-mail: pbinfo@pbworld.com
Web: www.pbworld.com

Key personnel
President and Chairman: Thomas J O'Neill
Controller: Richard A Schrader

Subsidiaries
Parsons Brinckerhoff Quade & Douglas Inc
One Penn Plaza, New York, New York 10119, US
Tel: (+1 212) 465 50 00   Fax: (+1 212) 465 50 96

PB Transit & Rail Systems Inc
Two Gateway Center, Newark, New Jersey 07102, US
Tel: (+1 973) 565 48 90   Fax: (+1 973) 648 08 88

Parsons Brinckerhoff Limited
Westbrook Mills, Godalming, Surrey GU7 2AZ, UK
Tel: (+44 1483) 52 84 00   Fax: (+44 1483) 52 85 38

Parsons Brinckerhoff (Asia) Limited
23rd floor, AIA Tower, 183 Electric Road, North Point, Hong Kong
Tel: (+852 2) 579 88 99   Fax: (+852 2) 856 99 03

Parsons Brinckerhoff Construction Services Inc
Spring Park Technology Center, 465 Spring Park Place, Herndon, Virginia 20170, US
Tel: (+1 703) 742 57 00   Fax: (+1 703) 742 58 00

Parsons Brinckerhoff Power Inc
Five Penn Plaza, 17th Floor, New York, New York 10001, US
Tel: (+1 212) 613 88 90   Fax: (+1 212) 613 88 88

PB Facilities Inc
One Penn Plaza, New York, New York 10119, US
Tel: (+1 212) 465 50 00   Fax: (+1 212) 465 50 96

PB Consult Inc
5 Penn Plaza, 17th Floor, New York, New York 1001, US
Tel: (+1 212) 613 88 00   Fax: (+1 212) 613 88 02

Major Operating Managers
President, Parsons Brinckerhoff Quade & Douglas, Inc: William D Smith
   Tel: (+1 212) 465 51 51
Chairman, PB Transit & Rail Systems Inc:
   Anthony Daniels
   Tel: (+1 415) 243 46 34
President: Tom Prendergast
   Tel: (+1 973) 565 48 10
Division Manager-Railroads: Bruce Pohlot
   Tel: (+1 973) 565 48 90
Managing Director, Parsons Brinckerhoff Limited:
   Tim Matthews
Business Development Manager (Rail): Alan Lee
Commercial Director: Greg Ayres
Railways Director: Michael Jenkins
   Tel: (+44 20) 77 98 24 00
Chief Operating Officer, Parson Brinckerhoff (Asia) Limited: Keith J Hawsworth
   Tel: (+852 2) 579 88 99
President, Parsons Brinckerhoff Construction Services, Inc: Christopher E Reseigh
   Tel: (+1 703) 742 57 01
President, PB Facilities Inc: William S Roman
   Tel: (+1 212) 465 50 23
President, PB Consult, Inc: Michael L Schneider
   Tel: (+1 212) 465 50 35

Corporate offices
Australia, Argentina, Belgium, China, Egypt, Hong Kong, India, Indonesia, Ireland, Israel, Japan, Kuwait, Lebanon, Macau, Malaysia, New Zealand, Philippines, Poland, Qatar, Saudi Arabia, Singapore, South Africa, South Korea, Spain, Switzerland, Taiwan, Thailand, Turkey, United Arab Emirates, UK and USA.

Capabilities
Parsons Brinckerhoff (PB) provides multidisciplinary planning, engineering, programme and construction management, and operations and maintenance services for all modes of rail transport – heavy rail and light rail, commuter and freight rail, automated people mover, and maglev and high-speed rail. PB is also equipped to respond to the industry's current critical needs in intermodal and multimodal transportation; track, structures and facility design and inspection; tunnels and subsurface structures; bridge design and inspection;ˑ signal and communications systems; electrification; equipment analysis; rolling stock; maintenance facilities and management of assets. It provides clients with a full range of support, including alternatives evaluation, systems design, fire/life/safety programmes, systems assurance, environmental health and safety, operations and maintenance plans, simulations, procurement support, and testing and start-up.

Projects
China: Shenzhen Metro Line 3: Line 3 is the latest addition to the new Shenzhen metro network. When completed in 2009 it will have a total length of 32.7 km (6.9 km underground section and 25.8 km above ground section). It will run in an east-west direction from Lo Wu to Longgang with 19 stations planned (six underground and 13 above ground) plus one depot at Wanggang and a stabling yard at Longtung. PB is the lead consultant and is working in association with STEDI, a local Chinese Design Institute (DI), to provide project management consultancy and design supervision services for the client – Shenzhen Metro No 3 Line Investment Co Ltd.

Egypt: Owner's representative for Egypt's National Authority for Tunnels, providing programme management services for the turnkey design, construction, and commissioning of Line 2 of the Greater Cairo Metro.

Hong Kong: Kowloon Canton Rail–West Rail Line: systemwide telecommunications, data and control systems front-end design, systemwide tunnel ventilation design, project management and design of Pat Heung Maintenance Centre, Kam Sheung Road Station, and the new West Rail Headquarters building. Design services for Tuen Mun Station and Siu Hong Station and vehicle inspection services. West Rail, opened in November 2003, was once Hong Kong's largest civil construction project. This strategically placed 30.5 km (19 mile) domestic passenger railway provides a much needed link for areas of increasing population in the North West New Territories and urban Kowloon.

MTRC Airport Express Line and Tung Chung Line: The 34 km (21.1 mile) Lantau and Airport Railway (LAR) provides service on two lines: the Airport Express and the Yung Chung Line, which are incorporated into Hong Kong's MTRC Airport Express system providing dedicated passenger rail service directly from Chek Lap Kok International Airport to Hong Kong Central. Mechanical and electrical design services for systemwide ventilation, power supply and trackside auxiliary systems. Station M&E design services for the Olympic, Kowloon and the Airport Ground Transportation Centre.

MTRC Tseung Kwan O Extension: A 9.5 km (5.9 mile) metro extension of the existing MTRC system in Hong Kong to the Tseung Kwan O new town in the eastern New Territories. Opened in August 2002, the new line is mostly underground to address environmental and aesthetic concerns and comprises five new stations, Yau Tong, Tiu Keng Leng, Tseung Kwan O, Hang Hau and Po Lam. PB conducted the feasibility study and later provided project management, full design services, and construction supervision of Po Lam and Hang Hau stations plus full mechanical and electrical design for Tseung Kan O and Tiu Keng Leng Stations.

Ma On Shan Extension: Opened on 21 December 2004 after a three year construction period, the Ma On Shan Rail line is the latest extension to be incorporated into the KCRC network. The line has a total of nine elevated stations and one maintenance centre and covers a distance of 11.4 km.

India: Delhi Metro: Project management and supervision of construction, general consultant services, contract administration, systems tender preparation, and procurement management to the Delhi Metro Rail Corporation for Phase 1, a 55 km (34 mile) combination subway and elevated system.

Ireland: Engineering feasibility and design of new Metro for Dublin, enhancement of main line suburban rail systems in Dublin Area (DASH), interconnector project for Irish Rail, and government advisor on the Light Rail in Dublin (LUAS).

Poland: Technical assistance to Polish Railways for project preparation for modernisation of the E-65 railway line.

Singapore: MRT North East Line: The line runs 20 km (12.4 miles) entirely underground through a variety of soft and hard ground conditions with 16 stations and one depot. PB is providing project management, detailed architectural, C/S and M/E rail design services for six stations, including two interchange stations, namely Kandang Kerbau, Farrer Park, Kovan, Hougang, Dhoby Ghaut and Outram Park Stations. This entails 6 km (3.7 miles) of tunnels, including 4.7 km (2.9 miles) of bored tunnel and 1.3 km (0.8 miles) of cut-and-cover tunnel. These tunnels range from 18 to 25 m (59 to 82 ft) deep with an inside diameter of 5.9 m (19.4 ft). The MRT NEL civil work was developed through a series of 12 design/build contracts, 11 for the stations and connecting line tunnels and one for the maintenance depot and storage yard.

MRT Circle Line (CCL): The CCL, which is set to be fully complete in 2010, will cut travelling time and allow commuters to bypass busy interchanges like City Hall and Raffles Place. CCL will be a fully underground orbital line linking all radial lines leading to the city. The line will interchange with the North-South Line, East-West Line and North-East Line. The CCL will be 33.3 km (20.7 miles) long with approx 29 stations. Starting from the Dhoby Ghaut Station, it will run through some of the busiest corridors in the city and end at HarbourFront Station on the North East Line. The project will be implemented in five stages. PB is providing the detailed architectural C/S and M/E engineer design service for stages 4 and 5. PB is also providing the M/E design services for the CCL Kim Chuan Depot and serving as independent checker for Contract 825 of Stage 1 CCL. Apart from providing consultancy services to the Land Transport Authority of Singapore, PB is also providing design services to three construction contracts for different contractors on CCL Stage 3.

Taiwan: Taiwan High-Speed Rail: feasibility study, programme management support, rail systems design, project controls support, system integration management, depot and station design, seismic evaluation, and construction management support services. The Taiwan High Speed Rail is a double-track, 2 345 km (214 mile) high-speed rail line that will include 40 km (30 miles) of tunnels, 260 km (160 miles) of viaducts and bridges, and 40 km (25 miles) of cut and fill.

Thailand: Bangkok Transit System Sky Train: This elevated automated line extends 23 km (14.3 miles) through the centre of Bangkok and has 23 stations. PB was appointed as the mechanical and electrical consultant with initial responsibility for the preparation and implementation of a tender evaluation plan for rolling stock, signalling system, operations, trackwork, power supply, communications and control system, depot design, maintenance philosophy and building services. PB's additional involvement covers system safety, quality assurance, environmental impact mitigation, reliability management, operating and maintenance costing, and contract negotiations.

Bangkok Blue Line: Construction supervision for depot and track work, and project management and systems engineering services for the concessionaire/operator. Blue Line Extensions and Orange Line: Project management and preliminary design for the proposed new route as part of the Bangkok underground transit system.

Bangkok Blue Line: Construction supervision for depot and track work, project management and systems engineering services for the concession/operator.

Blue Line Extensions and Orange Line: Project management and preliminary design for the proposed new route as part of the Bangkok underground transit system.

Turkey: Engineering, design, and supervision of new rail crossing of the Bospohorus River in Istanbul.

UK: Network Rail: Programme management support for West Coast Route Modernisation, train protection and warning system, and Southern Zone power supplies upgrade. Engineering feasibility and design of four packages on East Coast Main Line; Kings Cross Station improvements and additional platform; Virgin Cross-Country upgrade for high-speed tilting trains; feasibility study for upgrade of Midland main line; Thameslink resignalling, remodelling of track, signalling and station at Portsmouth Harbour; switch and crossing renewals, level crossing and signal box renewals and conversions; design of renewals of fixed telecom network (FTN); and telephone concentrator renewals. Detailed design of major new network management centre. Construction management of Strood and Higham Tunnel re-lining during a 12 month blockage.

Transport for London, London Underground Ltd and London Crossrail: Project controls and programme management support to Transport for London for implementation of public-private partnership for London Underground. Technical support to Chief Engineers Group on power supply, tunnel ventilation modelling, rolling stock specification, and study on air-conditioning of trains. Safety advice and planning supervision of new underground station at Kings Cross.

Strategic Rail Authority and London Crossrail: Advisor on Crossrail alternative routes, and development of electrical and mechanical systems, power supplies, and operational modelling. Engineering management of the East London Extension and National Telecoms strategic review.

Union Rail: Mechanical and electrical services design, tunnel ventilation design, safety assurance and RAMS for Contract 588 on the £2.5 billion Channel Tunnel Rail Link Project (CTRL).

Light Rail: Bank's advisor for Nottingham Express Transit (NET), engineering appraisal of extension to Midland Metro for Centro, evaluation of extensions to South Yorkshire Supertram, operational and safety advice on Croydon Tramlink, and tendor design for Leeds Supertram.

US: Metropolitan Atlanta Rapid Transit Authority (MARTA) transit system, Atlanta, GA: PB's association with MARTA began in 1966 when it established a joint venture to assist the newly created agency with the planning and development of a rapid transit system. The plans for the system have evolved since then to encompass 97 km (60 miles) of rapid transit lines, one express busway, 45 rail passenger stations, and park-and-ride facilities for more than 33,000 vehicles. Services provided by the joint venture for the heavy rail system include project management, engineering, development of design criteria, design of system-wide features and specific stations and line sections, quality control, and procurement and construction management.

Bay Area Rapid Transit (BART) Extension, San Francisco, California: PB provided General Engineering Consultancy (GEC) services for the expansion of the 75-mile heavy rail transit system whose planning, design and construction management were carried out in the 1950s–60s by a joint venture that included PB. In addition to its overall programme management and administrative responsibilites (25 section designers were managed), the joint venture provided preliminary and final design services for the trackwork, traction power, train control, communications, and fare collection systems, as well as construction management services that included design services during construction, DBE monitoring, constructibility reviews, claims avoidance, bid management, shop and as-built drawing management, quality assurance and materials testing, start-up and testing support and change-order management. Phase one of the project was divided into two parts that began in 1989 and were completed in 2004, during which time PB was a member of the Bay Area Transit Consultants (BATC) joint venture.

Dakota, Minnesota and Eastern Railroad, Powder River Basis, Coal Expansion Project, South Dakota and Wyoming: PB was responsible for the conceptual engineering in support of the proposed 450 route km (280 route miles) of new track in South Dakota and Wyoming, extending the railroad into the Powder River Basin, and the 965 route km (600 miles) of rehabilitated track and structures in South Dakota and Minnesota. The firm developed and evaluated alternative routes and conducted civil, geotechnical, structural and signalling studies as a prelude to developing capital and O&M cost estimates used in support of the railroad's applications to the Surface Transportation Board in February 1998. If built, the D M&E projects up to 100 million gross tons (MGT) of Powder River Basin low-sulphur coal would be carried on this upgraded line serving mid-west and upper Great Lakes electric utility companies.

California High-Speed Rail (Statewide-California): PB began assisting the California High-Speed Rail Authority in the preparation of a financial plan and an extensive outreach programme to secure funding and begin the implementation of a high-speed rail corridor network for California.

CSX North Bergen Yard Reconfiguration, North Bergen, New Jersey: PB was responsible for track, civil and structural design as well as environmental and permitting work.

METRA Circumferential Rail Alternative Study, Chicago, Illinois: Feasibility studies for providing service on two existing freight rail lines that travel around the Chicago metropolitan region.

Minnesota Intermodal Rail Terminal Study: PB undertook a three-part study to determine the rail operating parameters, develop an operating plan and conceptual designs of the proposed rail terminal and create a pro forma business plan for a new intermodal terminal facility to be shared by Burlington, Northern Santa Fe, Union Pacific, and the Canadian Pacific Railroads in the Minneapolis–St Paul Twin Cities area.

Railtran Commuter Rail and Intermodal Transportation Center, Fort Worth, Texas. Project management, project controls, systems engineering, and construction management services to the Fort Worth Transportation Authority (the T) for extension of service from South Irving to downtown Fort Worth.

Northeast Corridor (NEC) Strategic Plan: PB performed infrastructure evaluation and service planning as part of a 'first-of-a-kind' study to ensure the future success of Amtrak's Northeast Corridor strategic business unit (NEC SBU), which is responsible for one of Amtrak's core businesses.

MARC Commuter Rail Master Plan: PB provided railroad planning and operations services to address growth options for the 300 km (187 mile) system between northern portions of West Virginia and Maryland and the Baltimore and Washington, DC business districts.

UPDATED

Patrick Engineering Inc

4970 Varsity Drive, Lisle, Illinois 60532-4101, US
Tel: (+1 800) 799 70 50   Fax: (+1 630) 724 16 20
e-mail: sheath@patrickengineering.com
Web: www.patrickengineering.com

Key personnel

President: Daniel P Dietzler
Senior Vice-President, Transportation & Design
   Services: Ted W Lachus
Vice-President: Jeffrey C Schuh

Capabilities

Professional engineering, project management, and architectural services including civil, structural, mechanical, electrical and environmental engineering; architectural and surveying services including GPS; geotechnical investigations and drilling; design/build services for industrial systems and other railway projects. Patrick Engineering has provided services for Class I and short line railways in the US, Canada and in Central America.

Projects

Contracts completed recently include embankment stabilisation of triple main line embankment; major rail yard improvement; new intermodal terminal and emergency bridge replacement.

UPDATED

Philips Projects BV

Building TAM, PO Box 218, NL-5600 MD Eindhoven, Netherlands
Tel: (+31 40) 278 51 94   Fax: (+31 40) 278 69 14
e-mail: Rene.vanEijkelenburg@nl.ccmail.philips.com
Web: http://www.philips-projects.philips.com

Key personnel

Regional Director EMEA: René van Eijkelenburg

Offices

Austria, Belgium, Czech Repiblic, Denamrk, Finalnd, France, Germany, Greece, Italy, Netherlands, Poland, Portugal, Spain, Switzerland, and UK.

Capabilities

Railway stations systems overview, including: real time database, security management system; fire alarm system and public address management system, passenger tracking system; information display management system for timetables, baggage information, passengers information, station information and staff information; and media systems.

Projects

Video transmission system for CCTV on stations and platforms in the UK, including major London stations and Greater Manchester Metrolink; acoustics for PA in stations and platforms in Germany; system integration for PA systems for metros in France, Italy, Norway, and Singapore; public data and video for advertisements in the Netherlands. Philips is involved in station control systems for Italian railways; the new Dortmund railway station project and in the Telsul project in Portugal.

PricewaterhouseCoopers

1 Embankment Place, London, WC2N 6NN, UK
Tel: (+44 20) 72 13 47 43   Fax: (+44 20) 72 13 24 54
e-mail: chris.j.castles@uk.pwcglobal.com
Web: http://www.pcglobal.com

Key personnel

Partner, Transport Consultancy Services:
   Christopher Castles
Director: Hugh Ashton

Capabilities

Restructuring and privatisation of railways throughout the world including strategy advice on restructuring options, economic and financial advice and the evaluation and financing of large-scale systems development projects.

PwC provides a range of consulting and financial advisory services to railways and associated organisations covering policy, strategy, economics and financial advice, with extensive international experience in railway business strategy, restructuring, privatisation and regulation and providing detailed analytical support and advice. This covers all the major business issues affecting railways' performance including market analysis and forecasting, railway cost and profitability analysis, subsidy policy, track access charging, design of concessions and other forms of private sector participation, development of regulatory mechanisms, organisation restructuring and performance improvement.

Prima Services Group Ltd

Stirling House, 44 Richmond House, Kingston-upon-Thames KT2 5EE, UK
Tel: (+44 20) 85 49 37 20   Fax: (+44 20) 85 49 37 43

Key personnel
Managing Director: Don Clarke
Director: Graham Jones
Financial Director: Steve Clark

Associated company
Atlas Stirling Rail BV

Capabilities
Recruitment, training and contracting services to the heavy rail, mass transit and light rail industries. ISO 9002, Network Rail, SNCB, NS and London Underground approved. Permanent way and signal testing.

NEW ENTRY

Prose Ltd

Zürcherstrasse 41, CH-8400 Winterthur, Switzerland
Tel: (+41 52) 262 74 00   Fax: (+41 52) 262 74 01
e-mail: info@prose.ch
Web: http://www.prose.ch

Key personnel
General Manager: Berhard Huber
Management team:
 Stefan Bühler (Deputy General Manager, Manager of Test Laboratory), Christoph Gyr, Jochen Helmlinger, André Rohrbeck
Quality Manager: Peter Zwicky

Background
Founded in 1982, Prose acquired in 2001, the measuring and testing department of Bombardier Transportation (Switzerland), formerly SLM.

Capabilities
Consulting, project management, development and design of rolling stock, in particular, bogies. Measuring and testing of railway vehicles. Prose is an accredited test laboratory for railway rolling stock.

NEW ENTRY

Ptarmigan Transport Solutions Ltd

Tannoch House, Dunkeld Road, Bankfoot, Perth PH1 4AJ, UK
Tel: (+44 1738) 78 72 52
Web: www.ptarmigansolutions.co.uk

Key personnel
Managing Consultant: Stuart Newing-Davis
Engineering Consultant: John Kelly
Training Consultant: Brian Raven
Training Consultant: Cécile Dufossé
Head of Marketing: Milo Davies

Capabilities
Ptarmigan Transport Solutions provides full public transport and rail consultancy services. It offers commercial training, including ticket training in Shere, APTIS, SPORTIS, Avantix ticket formats, RJIS, Successful Presentation of Evidence (SPOE), penalty fares, protection, conflict avoidance, induction training, disability awareness, customer excellence, up-selling, rail geography, health and safety and barrier management. The company also provides temporary training staff to the railway industry.

NEW ENTRY

QinetiQ

Cody Technology Park, Ivey Road, Farnborough GU14 0LX, UK
Tel: (+44 8700) 10 09 42
Web: http://www.qinetiq.com

Key personnel
Chief Eexecutive Officer: Sir John Chisholm
Chief Financial Officer: Graham Love

Background
Formerly DERA, the UK's defence research agency, QinetiQ has 50 years of experience in transport research and development for military and government bodies.

Capabilities
System engineering, human sciences, asset management and project management. These cover the fields of: traffic control; rail safety; infrastructure management; management services; rolling stock engineering; and passenger services.

The QSS Group Ltd

QSS House, PO Box 464, London Road, Derby DE24 8ZL, UK
Tel: (+44 1332) 26 21 80   Fax: (+44 1332) 26 36 92
e-mail: enquiries@theqssgroup.co.uk
Web: www.theqssgroup.co.uk

Key personnel
Managing Director: Kenneth Mee
Director SQE Services: Peter Abbott
Director, Engineering Services: Tony Levy

Capabilities
The QSS Group Ltd offers customers practical assistance, consultancy and assessment services in management systems development (safety, quality, engineering and environmental management), risk assessment and performance improvement.

UPDATED

Radermacher & Partner GmbH

Elsenheimerstrasse 41, D-80687 Munich, Germany
Tel: (+49 89) 57 00 90   Fax: (+49 89) 57 00 91; 99
e-mail: mail@radermacher.de
Web: http://www.radermacher.de

Key personnel
Managing Directors: Hans-Josef Mayer,
 Erich F Pantele

Capabilities
Solution of strategic, operational and organisational tasks within engineering, supply chain management, production, quality. Clients are predominantly businesses with their own development and production. Radermacher has advised the German railway industry for a number of years.

Rail Air International Ltd (RAIL)

Suite 2, Blandel Bridge House, 56 Sloane Square, London SW1W 8AX, UK
Tel: (+44 20) 77 30 34 56   Fax: (+44 20) 78 23 69 69
e-mail: enquiries@railairintl.com
Web: http://www.railairintl.com

Key personnel
Director: Chris Cook
Consultant: Herb Pence

US office
315 Steinmetz Drive, Manchester, New Hampshire 03104, USA
Tel: (+1 603) 627 30 85

Capabilities
Provision of expertise in developing rail services to airports, ranging from specialised consultancy through to partnership in rail operations. RAIL also specialises in the in-town check-in and check-out, environmental, congestion and security issues connected with airport rail links.

VERIFIED

Rail Sciences Inc

411 North Clarendon Avenue, Scottdile, Georgia 30079, USA
Tel: (+1 404) 294 53 00   Fax: (+1 404) 294 54 23
e-mail: info@railsciences.com
Web: http://www.railsciences.com

Key personnel
President: Gary P Wolf
Vice-President: Warren B Egan

Capabilities
Railway consultancy specialising in the application of advanced analytical techniques to solve operational problems; accident and derailment analysis, rail line capacity simulation modelling, schedule feasibility, vehicle dynamics, operational planning and analysis, computer model development, dispatching control systems, driver training, testing and data acquisition, mechanical inspections; metalurgical and failure analysis; bearing failure analysis; vampire simulation modelling.

VERIFIED

Rail Services Australia

Level 13, Pacific Power Building, 201 Elizabeth Street, Sydney New South Wales 2000
Locked Bag A4090 South Sydney, New South Wales 1235
Tel: (+61 2) 92 24 37 02   Fax: (+61 2) 92 24 26 00
e-mail: railservicesaustralia.com.au

Key personnel
Chief Executive Officer: Terry Ogg
General Manager Workshops: Michael Peter
General Manager Corporate Services: Irina White
Chief Financial Officer: Frank Morrison
Manager Human Resources: Col Shrubb
General Manager, Business Development:
 Karl Mociak
General Manager Contracts: Geoff Baxter
General Manager Resources: Colin Andrews
Manager Legal Services: Susi Curtis

Capabilities
Asset management; design and investigation; project management; rolling stock; signalling, control and communications; electrical maintenance; bridge examination and maintenance; signalling rnaintenance; track monitoring and maintenance; railway turnouts and crossings; rail wagon maintenance, repair and modification; railway signalling, electrical and electronic equipment servicing; signal construction; track construction; civil construction; electrical construction. Rail Services Australia has particular skills in rail projects over difficult terrain or through high-density areas.

Projects
Contracts include the New South Wales New Southern Railway tunnel fitout; the Blackdown to Richmond (NSW) Infrastructure Works and Maintenance. Provider contract in association with Theiss Contractors. The Homebush Bay Olympic Loop and main line, interface works. Two significant projects for Australian Rail Track Corporation in South Australia and Western Australia; works are under way in Asia and the UK.

The Railway Consultancy Ltd

1st Floor South Tower, Crystal Palace Station, London SE19 2AZ, UK
Tel: (+44 20) 86 76 03 95   Fax: (+44 20) 87 78 74 39
e-mail: info@railcons.com
Web: www.railcons.com

Key personnel
Managing Director: Dr Nigel G Harris
Director: David R McIntosh
Senior Analyst: Luke Ripley

Capabilities

Planning, economics and management for metros and railway systems; demand estimation; train service simulations; timetable preparation; contingency planning; transport policy and management advice; business planning; specification and design of IT systems to assist the planning process; training courses on railway and transport planning issues.

Projects

Recent projects have included a performance audit of southern suburban railway operations at London Victoria; the economic assessment of different urban transport technologies for DETR, UK; database population for Network Rail; fares policy advice for MTRC Hong Kong; demand estimation for a proposed station at Catt Mill for Merseytravel; industrial research for equipment suppliers; a major study on the impacts of the privatisation of British Rail; an operational feasibility study into one of the proposed options for London's Crossrail project, for London Underground Ltd; business/demand study for service developments in the Stansted Airport/W Anglia corridor for the Strategic Rail Authority.

Key in-house tools available for projects include the GCOST™ model for estimating the passenger demand, revenue and time-saving impact of new stations, and the TRAKATTK train service simulation.

UPDATED

Railway Technology Strategy Centre

Centre for Transport Studies,
Department of Civil and Environmental Engineering,
Imperial College London, London SW7 2AZ, UK
Tel: (+44 20) 75 94 60 93   Fax: (+44 20) 75 94 61 07
e-mail: w.adeney@imperial.ac.uk
Web: http://www.rrsc.org.uk

Key personnel

Chairman: Professor T M Ridley
Director: Professor S Glaister
Senior Research Associates: Dr D J Graham,
    R J Anderson

Capabilities

The Railway Technology Strategy Centre (RTSC) was established in 1992 with funding from the former British Rail and now carries out projects on strategic, technology and economic issues for BR successors and railways elsewhere in the world.

Projects

The RTSC has recently completed a re-assessment of the economic case for CrossRail for the Corporation of London. It has also recently undertaken two interconnected studies on asset management on behalf of Lloyd's Register which covered urban mass transit along with a range of other industries.

The RTSC produced a cost model of signalling assets, and has assisted in the development of a 'trains' model, which considers the value of assets, and the systems interface between signalling, train and communication equipment.

In 1998, RTSC helped to update the Railtrack signalling strategy. This involved all relevant parts of the Railtrack organisation and the supply industry, as well as assessments of other railways and other industries.

The RTSC continues to serve as a benchmarking centre for a consortium of nine of the world's largest urban railways which is now known as CoMET (The Community of Metros). The participating metros are: Berlin, Hong Kong (MTRC), London, Mexico City, Moscow, Paris, New York, São Paulo and Tokyo (TRTA). The benchmarking study is dedicated to assisting metro railways in identifying and implementing best practice through the application of benchmarking comparisons and detailed case study evaluations.

Case study results have led to successful implementation of a number of projects. These include rolling stock reliability improvement by London Underground, advice on rolling stock procurement for STC (Mexico City), and on levels of fare regulation and service level for MTRC (Hong Kong). Following this, a second benchmarking group called Nova was formed for medium-sized metro systems. The ten current participants are Dublin, Glasgow, Hong Kong (KCRC), Madrid, Montreal, Naples, Newcastle, Oslo, Singapore and Taipei.

A mainline railway process benchmarking project was completed in Spring 2002 with the national railways of Germany (Deutsche Bahn), Italy (Trenitalia) and Spain (RENFE). The project includes case studies looking at revenue opportunities from the railways' telecom infrastructure and leverage of the customer base, organisation of, and revenue from, different sales channels and a comparison of rolling stock maintenance practices.

RailWorks Corporation

1104 Kenilworth Drive, Baltimore, Maryland 21204, USA

Key personnel

For a full list of personnel, see RailWorks Corporation entry in *Permanent way components, equipment and services* section.

Subsidiaries

Railworks Products and Services Group
Tel: (+1 412) 325 02 02
e-mail: wdonley@railworks.com
President: William R Donley
Vice-President of Business Development:
    George Caric

Railworks Track Systems Group
Tel: (+1 801) 366 93 39
President: Robert D Wolff

Railworks Transit Systems Group
Tel: (+1 914) 323 30 12
President: C William Moore

Pacific Northern Rail Contractors Inc
(a RailWorks Company)
Tel: (+1 604) 850 91 66
President: Henry Braun

Background

In September 2001, RailWorks Corporation and its operating subsidiaries in the US, voluntarily filed for protection under chapter 11 of the US Bankruptcy Code.

Capabilities

Integrated rail system services and products, active in new construction, rehabilitation, track repair and maintenance, signalling, communications, electrical and other track-related systems, and rail products manufacturing and supply.

Projects

New York City Transit Authority (NYCTA) White Plains – Phase II signal rehabilitation project, White Plains, New York.

New Jersey Transit (NJT) Morrisville Train Storage Yard, Falls Township Philadelphia: construct 12 new storage yard tracks and to install related crossovers, turnouts, electrical switch heaters and yard lighting. Modification of existing tracks, rehabilitation of two existing yard tracks, yard drainage, accesses roads and paving.

Minnesota Constructors Team, Hiawatha Light Rail Line, Minneapolis: extending from downtown Minneapolis, through Minneapolis, through the Minneapolis–St Paul International Airports to the Mall of America in Bloomington. The completed project will be approximately 11.4 miles long.

TransLink (Vancouver), Millennium Line Expansion of the Vancouver Skytrain, Vancouver, Canada: Linear induction motor and power rail installation, including the installation of 40,000 lineal metres of LIM (Linear Induction Motor) reaction rail and 80,000 lineal metres of a side mounted power rail including thermal expansion joints, power feeds, isolation joints, and coverboard. Also, core drilling and grouting of 120,00 support studs for the LIM rail and core drilling 60,000 holes to anchor the power rail insulator brackets.

New York City Transit Authority (NYCTA) Carnarsie Line, New York: complete installation of a new Communications Based Train Control System (CBTC) for 13 route miles of the Canarsie Line subway in association with Matra Transport International and Union Switch & Signal.

NYCTA, New York: Rehabilitate the train control signal system for 5.1 miles of track on the Flushing Line subway. Also to furnish and install eight escalators at five locations in the boroughs of Brooklyn and Manhattan.

Miami International Airport, Miami, Florida: Electrical and guideway construction services including automatic train control systems and 8,500 feet of concrete guideway in association with Sumitomo.

Panama Canal Railway, a venture between Kansas City Southern Industries and Lanco International: Reconstruction and expansion of track and facilities involving a 44-mile stretch of track, running parallel to the Canal from Colon to Balboa.

Vancouver Wharves, British Columbia, Canada: Rail yard construction services.

Long Island Rail Road, Long Island: Rehabilitation and improvements of the diesel yard.

IPSCO Steel, Mobile: Steel mill expansion project involving 30,000 crossties and additional switch ties.

US Lime and Minteral-Arkansas Lime, Batesville: The design/build project involves the removal and construction of 15,000 ft of track including three turnouts, grade crossing and bridgework.

Tuscola and Saginaw Bay Railroad, Michigan: Refurbishment project involving 20,000 crossties for the 12.5-mile branch line.

Ramboll Sweden AB

Box 4205, SE-102 65 Stockholm, Sweden
Tel: (+46 8) 615 60 00   Fax: (+46 8) 702 19 25
e-mail: info@ramboll.se
Web: www.ramboll.se

Key personnel

Executive: Per Lennart Karlsson

Background

Ramboll Sweden is a member of the Ramboll Group, which includes other companies in Denmark, Finland and Norway. The group is the largest engineering consultant in the Nordic Region. It employs over 4,000 multi-disciplined professionals and serves clients throughout the region and in many international countries.

Capabilities

Initial investigations, environmental studies, and feasibility studies; complete planning and design; preparation of tender documents; tender evaluation; procurement; construction supervision; and project management and cost control.

Scandiaconsult was recently bought out by Ramböll, a large Danish multi-disciplined engineering consultant, which has railway, geotechnical, structural, rail track and environmental specialists to complement the expertise of Scandiaconsult.

Projects

Projects completed include: the Arlanda rapid rail link between Arlanda Airport and Stockholm Central; the Öresund Link between Malmö, Sweden and Copenhagen, Denmark; the Stockholm Light Rapid Transit System; and the 450 km northern extension of the Bodö–Trondheim railway in northern Norway.

UPDATED

Ranbury Management Group

Level 15, 344 Queen Street, Brisbane, Queensland 4000, Australia
Tel: (+61 7) 32 11 23 00   Fax: (+61 7) 32 11 29 13
e-mail: ranburymgmt@ranbury.com.au
Web: http://www.ranbury.com.au

Key personnel

Directors: Nic Tilley, David Porter, Brad Johnson,
    Scott Kennelly

Capabilities

A multidisciplinary rail, property and infrastructure consultancy with management and engineering expertise in rolling stock design and manufacture; systems, signalling and communication; and the redevelopment of railway property. Ranbury's services also include business process management of assets including information systems, financial systems, accounting systems, compliance and strategic planning.

Projects

Projects include: infrastructure for Queensland Rail; heritage activities for Queensland Rail; alliance managers and project management services; design, manufacture and commission of the Cairns Tilt Train (Queensland Rail and EDI Rail); project management of the design, manufacture, commission and finance of the Sydney Millennium Train (EDI Rail); rail infrastructure construction interface Lang Park Redevelopment (Watpac/Multiplex); software project management, train operational systems (Queensland Rail); Port Botany Freight Project review (Rail Infrastructure Corporation); Cairns Transit Centre Development (Queensland Transport).

RIQC – Rail Industry Quality Certification Ltd

PO Box 464, London Road, Derby DE24 8ZL, UK
Tel: (+44 1332) 26 27 63   Fax: (+44 1332) 26 36 92
e-mail: the-qss-group@compuserve.com

Key personnel

Group Managing Director: Michael Winwood
Senior Executive: Laurie Fitch

Capabilities

Include supplier auditing and assessment; systems auditing, supervised and accredited by the United Kingdom Accreditation Service (UKAS) on behalf of the Department of Trade and Industry, to ISO 9000 Series standards.

RITES Ltd

Rites Bhawan, Plot No 1, Sector 29, Gurgaon, Haryana, 122 001, India
Tel: (+91 124) 257 16 66
Fax: (+91 124) 257 16 60
e-mail: info@rites.com
Web: www.rites.com

Key personnel

Chairman: P N Garg
Managing Director: V K Agarwal
Finance Director: B L Bagra
Technical Director: Anil Madan

Background

RITES, a Government of India Enterprise, was established in 1974 under the aegis of the Indian Railways. Over the last three decades, RITES has grown and diversified from being a rail consultant to a company of consultants, engineers and project managers. As a government enterprise under the Ministry of Railways, RITES continues to have the back up of the Indian Railways (IR), which carries over 12 million passengers and lifts 1.5 million tonnes of freight daily on the network which covers over 62,759 route km, 6,867 stations. RITES was previously called Rail India Technical and Economic Services Ltd.

Capabilities

Consultancy in the field of transportation, infrastructure and related technologies. It provides comprehensive consultancy services from concept to commissioning in the fields of railways, urban transport, urban development and urban engineering, roads and highways, airports, ropeway, inland waterways, ports and harbours, information technology and export packages of rolling stock and railway related equipment.

RITES diversified service packages include feasibility, design and detailed engineering,

multi-modal transport studies, project management and construction supervision, quality assurance and management, ISO-9000/ISO-14000, material engineering, economic and financial evaluation, financing plan and privatisation, property development, railway electrification, signalling and telecommunication, environmental impact assessment, training and human resource development. RITES employs nearly 2,000 staff including 1,200 specialists of high professional standing in the fields of engineering, management and planning.

Projects

Recent projects include:
Angola: Implementing railway rehabilitation projects in CFM Railway including supply of locomotives, coaches, pick-up vehicles and buses, rail-road vehicles, technical assistance and supply of equipment.
Bangladesh: Integrated export packages for broad-gauge (BG) diesel locomotives and maintenance of metre-gauge (MG) diesel locomotives, regional rail traffic enhancement study.
Colombia: Maintenance management services for rolling stock for Atlantic Railway.
Ethiopia: Detailed engineering, design review and advisory services for road projects.
Ghana: Design and construction supervision of sea wall for Defence Keta Sea Project.
India (domestic): Functioning as a major member of the international consortium commissioned as 'general consultants' for the Mass Rapid Transit System (MRTS) for Delhi. Major urban/regional planning assignments in Delhi and national capital region, detailed project report for MRTS, in Hyderabad, Delhi Phase II, Ahmedabad, feasibility study for high capacity bus system for Delhi and its suburbs. Design engineering services for rail bridges on the Quazikund–Baramulla line, across River Ganga, update study for Rohtang tunnel, designer's supervision of rail tunnel across Peer Panjal in Kashmir. Pre-feasibility study for a high-speed rail link between Ahmedabad and Mumbai. Development of crashworthy design of passenger coaches, frameless tank wagon for Indian railways. ISO 9000, ISO 1400, QS 9000 safety audit and consultancy in TS 16949-2002 to the public and private sectors. Operation and maintenance management for NTPC, Indraprastha Power Generation Corporation Ltd, Paradip Phosphate, West Bengal Power Development Corporation Ltd. Design engineering and construction supervision for National Highways and rural roads in various states under north-south and east-west corridors and Golden Quadrilateral under NHDP; independent consultants for construction of eight lane Delhi-Gurgaon section of NH-8 on BOT basis. Consultancy services for route re-alignment of transmission line in Katra–Quazikund new line, revamping of power distribution system in Uttaranchal State, project management of HT power transmission for TNVL. Consultancy cum project management for implementation of Advanced Traffic Management System (ATMS) and Electronic Toll Collection and Management for NHAI on NH2 and NH8 projects. EIA studies for Hydro-Electric projects in Chattisgarh and for State Electricity Board, Central Water Commission, National Frontier Railway, and various railway lines. Environmental and social impact assessment for Delhi Metro Corridor Phase II, Indo-Nepal ICD, Ahmedabad Metro. ADB funded EIA, SIA and RAP studies for Ministry of Railways.
Iran: Engineering supervision services for Karadj locomotive workshop.
Malaysia: Expert services for engineering design, construction and project management for railway track and electrification together with traffic and consultancy on route management system to Malaysian Railways.
Mozambique: One of the consortium members of the Beira Rail Corridor project and providing project management services.
Myanmar: Integrated export package for ten MG in-service locomotives with maintenance; techno-economic feasibility study of highway along river Kaladan from Indo-Myanmar border to Sittwe Port.
Nigeria: First assessment/pre-feasibility study of railway in Nigeria.

Saudi-Arabia: Project study for socio-economic analysis of new railway line for development of international freight corridor Riyadh–Jeddah, for UIC.
Senegal: Feasibility study for a new railway link between Zinginchor and Dhakar, Republic of Senegal. Supply of in-service diesel locomotives and MG day coaches to PTB, Senegal.
Sri Lanka: Study for restructuring and public-private partnership in Sri Lanka Railways, training of railway personnel and procurement consultancy services for North East emergency reconstruction programme.
Sudan: Supply of in-service and new locomotives, rehabilitation of locomotives, wagon spares and rehabilitation of traction motors.
Tanzania: Asset review and valuation study, leasing of 10 MG diesel electric locomotives, wagon rehabilitation project.
Uganda: Urban traffic improvements plan for the city of Kampala, supervision services for highway projects.
UK: Design support for overhead electric traction lines and upgrade of signalling for West Coast Main Line.
Uzbekistan: Consultancy services for computerisation of financial system of Uzbekistan Railways.

UPDATED

R L Banks & Associates Inc

1717 K Street NW, Washington DC 20006-5331, USA
Tel: (+1 202) 296 67 00   Fax: (+1 202) 296 37 00
e-mail: transport@rlbadc.com

Key personnel

Chief Executive: Robert L Banks
President: Charles H Banks
Vice-Presidents: Eslyn P Banks,
   Geroge K Withers Jr PE

Capabilities

Economic analyses, cost ascertainment, planning and policy development, concept engineering, privatisation and outsourcing of freight and passenger rail services, and negotiations with railways.

Contracts

Planning, evaluation, negotiation and implementation of rail passenger services in the 12 largest US metropolitan areas and many smaller cities/corridors.

Removal of hazardous waste by rail. Valuation of the Windsor and Hansport Railway in Nova Scotia. Identification of specific target waste disposal landfills with rail access on behalf of a planning committee for a municipal consortium in the Salt Lake Valley of Utah. Preparation of recommendations concerning removal and relocation of extensive railway trackage in the downtown Indianapolis area. Preparation and sponsorship of an extensive operations simulation as a foundation of a cost analysis supporting rate negotiations between an electric utility and Burlington Northern Santa Fe Railway.

NEW ENTRY

RMS Locotec

Rail Management Services
Vanguard Works, Bretton Street, Dewsbury WF12 9BJ, UK
Tel: (+44 1924) 46 50 50   Fax: (+44 1924) 46 54 22
e-mail: sales@rmslocotec.com
Web: www.rmslocotec.com

Key personnel

Managing Director: Lawrence Crossan
Commercial Manager: Derek Webb
Production Development Engineer: Peter Briddon

Capabilities

RMS was established to meet the demand for rail management services due to the impending changes in the UK rail industry, primarily on the freight side.

The services offered cover feasibility studies, project management, operations and commercial studies and specialist engineering services.

RMS can provide a rail operations and maintenance package, including provision of labour and resources, management of rail traffic, provision of an interface with rail operators and other suppliers, and management of fleets of locomotives, wagons or carriages. It can provide safety systems and offers staff training and certification for rail operations.

UPDATED

Roundel Ltd

7 Rosehart Mews, Westbourne Grove, London W11 3TY, UK
Tel: (+44 20) 72 21 19 51   Fax: (+44 20) 72 21 18 43
Web: http://www.roundel.com

Key personnel
Directors: Tony Howard, Michael Denny,
　Ian St John

Capabilities
Corporate identity design for transport systems and operators including service branding, livery design, signing and information design, branded environments and corporate communications.

Projects
Branding, livery design, environments and passenger information for Kowloon–Canton Railway; livery designs for Royal Train locomotives and Class 92 Channel Tunnel locomotives; branding and livery design for Railfreight, Great Western, Finnish State Railways and Southern; signing and information design for Docklands Light Railway and for all CTRL stations including St Pancras International; Light Rail system brands, environments and information design for Amey/ Bechtel; design guidelines for London Underground and corporate communication design for ABB.

VERIFIED

RTA Rail Tec Arsenal Fahrzeugversuchsanlage GmbH

Paukerwerkstrasse 3, A-1210 Vienna, Austria
Tel: (+43 1) 25 68 08 10   Fax: (+43 1) 25 68 08 16 00
e-mail: contact@rta.co.at
Web: www.rta.co.at

Key personnel
Managing Directors: Franz Hrachowitz,
　Wolfgang Palz
Technical/Scientific Director: Gabriel Haller

Background
Rail Tec Arsenal is an internationally active, neutral and independent research and testing institute for rail and road vehicles, new transport systems and technical facilities that are subject to extreme climatic conditions.

Rail Tec Arsenal operates two modern climatic wind tunnels designed to optimise thermal comfort in public transport vehicles (rail and road vehicles) and to investigate and improve the availability and safety of systems in sensitive industrial areas.

Rail Tec Arsenal was founded by Arsenal Research and the international rail vehicle industry in order to safeguard the facilities and specific expertise necessary for the performance of climatic tests on a long-term basis. The consortium comprises: Arsenal Research (26 per cent); ALSTOM Transport (14.8 per cent); Bombardier Transportation (29.6 per cent); Siemens Transportation Systems

Light rail vehicle undergoing climatic testing
0598245

Regional emu undergoing climatic testing
0598244

(14.8 per cent); Ansaldobreda and Firema Trasporti (together 14.8 per cent).

Capabilities
The Vienna Climatic Wind Tunnel, claimed by the company to be the world's largest, commenced operation on 1 January 2003. The testing facility consists of two separate climatic wind tunnels (CWT) for the testing of vehicles under extreme weather conditions. The large CWT is 100 m, long enough to accommodate a train consisting of a power car and two carriages. The highest air speed is close to 300 km/h. The dynamometer featuring one driven (power rating max 850 kW) and one non-driven axle, allows braking and traction tests to be carried out. At 31 m, the test section of the small CWT is long enough for a carriage, a trailer truck or a bus. The maximum air speed here is 120 km/h; the dynamometer (for road vehicles) has a maximum power rating of 250 kW. The air flow to the front of the test object can be shut off completely by means of flaps installed at the head of the test section, in order to simulate, for example, a stop and go cycle with doors opening. A soak room is directly attached to the smaller CWT. This can be used for temperature conditioning of vehicles (adaptation of material temperatures) but also for climatic cycling tests (for example,

thermal simulation of train passage through a tunnel in winter). Two separate preparation halls with controlled access provide room for setup, development and changeover.

UPDATED

Schofield Lothian

1 Swallow Court, Welwyn Garden City, AL7 1SB, UK
Tel: (+44 1707) 39 00 85   Fax: (+44 1707) 39 14 11
e-mail: info@schofields.com
Web: http://www.schofields.com

Other Office
Dublin, Ireland
Berlin, Germany
Hong Kong

Capabilities
Engineering management, interface management, clash checks, possession management, system integration, safety case and product acceptance, occupational health and safety, Her Majesty's Railways Inspectorate's approvals, Railway Act regulatory regime, train operations and logistics, risk management.

Projects
Connex South Eastern Limited: design and installation of CET facilities at Ashford and St Leonards Depots. London Underground's 'Infraco JNP' civils and asset maintenance providing project and construction expertise to senior programme managers. Railtrack, Balfour Beatty Rail, Westinghouse Signal Limited: West Coast Route modernisation, since 1996, supplying key support personnel to the Euston Alliance, delivering design and interface management, risk assessment and logging, handover and handback support, planning management, databases for the control of design and contract administration and more recently construction delivery. Railtrack Southern Zone: Shortlands Junction, providing advice for efficient co-ordination and management of environmental issues relating to the immediate area around Shortlands Junction and grade separation with management of the Transport and Works Order for promotion of this key project associated with the channel tunnel link. Thameslink: Luton Parkway Station, managing all the regulatory approvals necessary under the Access Conditions and the Railways Act.

SCI Verkehr GmbH

Schanzenstrasse 117, D-20357 Hamburg, Germany
Tel: (+49 40) 430 50 77   Fax: (+49 40) 430 60 77
e-mail: hamburg@sci.de
Web: www.sci.de

Main office
Hardefustrasse 11-13, D-50677 Cologne, Germany
Tel: (+49 221) 931 78 20   Fax: (+49 221) 931 78 78
e-mail: koeln@sci.de

Project office
Novalisstrasse 7, D-10115 Berlin, Germany
Tel: (+49 30) 283 45 40   Fax: (+49 30) 28 44 54 20
e-mail: berlin@sci.de

Background
SCI Verkehr has established itself in the railway market as an independent and consultant specialist having begun its activities as a business field in the company SCI Engineering Consulting. SCI Verkehr employs over 30 staff and in 2000 created a subsidiary, SCI Polska, based in Katowice, Poland.

For details of the latest updates to *Jane's World Railways* online and to discover the additional information available exclusively to online subscribers please visit
jwr.janes.com

Services

SCI Verkehr offers strategic management consulting service market analysis and feasibility studies for manufacturers of rail vehicles, transport equipment and transfer facilities, public and private transport companies, port authorities, interest groups and trade unions, authorities responsible for public transport as well as for traffic and economic departments of the federation, federal states and local authorities. SCI Multi Client Studies provide analyses and future prospects for the railway market.

NEW ENTRY

Scott Wilson Railways

Tricentre 3, Newbridge Square, Swindon SN1 1BY, UK
Tel: (+44 1793) 50 85 00   Fax: (+44 1793) 550 85 01
e-mail: rail.marketing@scottwilson.com
Web: www.scottwilson.com

Key personnel

Chief Executive: Hugh Blackwood
Director of Projects: Keith Wallace
Director of Engineering: David Coles
Director of Operations: Richard Jones
Commercial and Business Development Director:
  Robin Hawley
Human Resources: Anne Crosby
International Projects Director: Derek Holden
Technical Director, Permanent Way: John Perkin
Rail Operations and Planning Manager:
  Mark Roome
Technical Director, Signalling and
  Telecommunications: David Nye
Head of Railway Electrification Engineering:
  Rob Tidbury
Head of Electrical and Mechanical Engineering:
  Gareth Clarke
Finance Manager: Pat Forde
Commercial Manager: Martin Pickup
Systems Manager: Mick Fry
Marketing Manager: Lorraine Gee

Principal subsidiaries

Scott Wilson Kirkpatrick Ltd
Scott Wilson Pieshold
SWK Pavement Engineering Limited
Scott Neale and Partners

Background

Scott Wilson Railways was initially set up by the purchase of the two British Rail design offices in Swindon and Glasgow when BR was privatised in 1995. Since then it has continued to expand with its head office in Swindon and other locations in Birmingham, Crewe, Glasgow, London and York.

Scott Wilson Railways is one of the largest divisions within the Scott Wilson group.

Capabilities

Scott Wilson Railways is one of the largest divisions within the Scott Wilson Group and a market leader in the provision of multi-disciplinary railway services. Active in Europe, Africa and the Asia Pacific region, the company provides consultancy services that cover all aspects of rail infrastructure planning, design, project management, construction supervision and rail maintenance. The company is able to undertake major multidisciplinary railway projects worldwide. The company has over 500 railway specialists and is growing rapidly to meet client needs.

Scott Wilson Railways has the ability to deliver a range of infrastructure activities within the following areas: civil and structural engineering; construction management; electrical and mechanical engineering; environmental services; geotechnical services for railways; permanent way; project management; railway electrification; railway operations and planning; railway safety and training; railway systems assurance; signalling systems; telecommunications systems; trackbed investigations; tunnelling for railways. Within these areas the company also offers a range of services which include: management consultancy; advisory services; feasibility studies; detailed engineering design.

UPDATED

Semaly SA

25 Cours Emile Zola, F-69625 Villeurbanne, France
Tel: (+33 4) 72 69 60 00   Fax: (+33 4) 78 89 68 57
e-mail: semaly@semaly.com
Web: www.semaly.com

Other offices

SMM, Marseille, France
SIAS, Grenoble, France
GITRAM, Montpellier, France
GETAS, Strasbourg, France
OPC Marechaux, Paris, France
Semaly Ireland Ltd, Dublin, Ireland
Sisplan, Bologna, Italy
Semaly, Krakow, Poland
Semaly Lda, Lisbon, Portugal
EGIS Rail, Paris, France
EGIS Semaly Ltd, London, UK
EGIS SEMALY Inc, New York, USA

Key personnel

Chairman and Chief Executive Officer:
  Hervé Chaine
General Manager: Philippe Vuaillat
Deputy Managing Director: Olivier Bouvart
Technical Director: Christian Teillon
Business Development Director: Philippe Rouland
Marketing Manager: Sophie Perrillat-Charlaz

Background

Semaly's shareholders include Groupe EGIS 82 per cent, American Bechtel 16 per cent and SNCF 1 per cent.

Capabilities

From conceptual design to start up of services, Semaly, as an urban and railway transport engineering company, undertakes financial and economic studies, feasibility studies, preliminary and detailed design, construction management, testing and commissioning for public transport and rail systems.

Projects

Recent projects include:
Belgium: In association with Tractebel, operating and ridership studies for a Brussels area regional express network.
Central America: Semaly and partner ITI are in charge of the conceptual design studies, detailed design and assistance in the establishing of consultation file documents on the rail project expected to link Caguas with San Juan by 2008.
France: The LRT tramway project the Maréchaux Sud line, a first section of 8 km of a tramway line that will circle Paris, comprises changes to the Maréchaux Sud boulevards, as well as the major urban improvements spread out over the line; Paris Airport ADP project ownership assistance, planned network comprises: a main line (Line 1) approximately 3.3 km in length connecting airport stations 1 and 2 with: five stations, line sections on overpasses, in tunnels and overhead, one workshop garage housing the CCP, extensions to Line 1, a short LISA line linking terminal 2 S3 and 2E satellite terminals. Semaly, in conjunction with Setec, is responsible for the task of aiding the prime contractor in the supervision of the construction of an automatic passenger transport system for Roissy-Charles de Gaulle airport; high-speed line connection Rhine/Rhone Rivers: general project management, the project consists of the building of a high-speed train line between Dijon and Mulhouse; the territorial authorities for Corsica set up a complete renovation and modernisation plan for their rail network. Semaly was asked to produce the specifications manual for a future traffic management system and to oversee its implementation as far as the acceptance phase.
Ireland: Studies and technical assistance for Dublin's first light rail line in the final phase of the works, reception and tests.
Greece: Semaly and its partners have been entrusted by Attiko Metro with the studies of extension of the Athens Metro lines 2 and 3, partially designed to serve the new airport.
Le Reunion Island: The tram-train project will involve the crossing of nine towns, both through the town centres and the outskirts, as well as the boring of a 10 km long tunnel.

Portugal: Continuation of Porto's Metro Project, including design, construction and operation of four tram lines for the city of Porto. The contract also incudes the supply of 72 trains and two cars.
Poland: Advisory and technical assistance for Krakow 'Fast Tram' project.
Saudi Arabia: Semaly and its partner DAR Al-Handasah will handle the feasibility study and preliminary design of the Light Rail Transit project in Riyadh.
USA: In the framework of the 2nd Avenue project and Staten Island Metro project, Semaly is handling the implementation of the new signalling and remote transmissions metro systems.

UPDATED

The Seneca Group LLC

122 C Street NW, Suite 850, Washington DC 20001, USA
Tel: (+1 202) 783 58 61   Fax: (+1 202) 783 60 96
e-mail: mccarthy@senecagrp.com
Web: http://www.senecagrp.com

Key personnel

President: Chris McCarthy
Principals: David Soule, Phil Davila, John Pinto

Capabilities

The Seneca Group provides public and private sector clients expertise to manage projects, develop business opportunities, and secure funding and financing sources. In most instances the need for such assistance is short-term, and the client receives the benefit of essential expertise without the commitment to expand overhead costs.

Seneca offers project management, technical writing, financial analysis, product promotion, and public policy development.

Serco Raildata

Derwent House, rtc Business Park, London Road, Derby DE24 8UP, UK
Tel: (+44 1332) 26 35 84   Fax: (+44 1332) 26 24 38
e-mail: srds@serco.railtest.co.uk;
  kfretwell@serco.railtest.co.uk

Key personnel

General Manager: Kevin Fretwell
Application Manager: Mark Coley

Background

Serco Raildata is the specialist print and data management division of Serco Railtest Ltd.

Capabilities

Serco Raildata offers professional information management services, certified to BS EN ISO 9001: 2000 quality assurance standards. These include electronic delivery systems for rapid retrieval of structural text and graphics, database management services, parts cataloguing, document, drawing and parts management, data packaging, electronic delivery systems and archiving services.

UPDATED

Serco Railtest Ltd

Derwent House, rtc Business Park, London Road, Derby DE24 8UP, UK
Tel: (+44 1332) 26 26 26   Fax: (+44 1332) 26 20 84
e-mail: info@serco.railtest.co.uk
Web: www.sercorailtest.com

Key personnel

Contract Director: Richard Barraclough
General Manager: Richard Hobson
Engineering Innovations Manager: Bob Difusco

Subsidiary

Serco Raildata

Bombardier Transportation contracted Serco Railtest to undertake acceptance testing of the Virgin CrossCountry fleet of Voyager trains (Bombardier Transportation) 0129113

Network Rail's Network Management Train, based on a former HST high-speed diesel trainset, is operated by Serco Railtest to monitor track conditions 1036587

Background

Serco Railtest is part of the Serco Integrated Transport group; an operating company within Serco Group plc. Servicing a client base within the UK and international rail industry, Serco Railtest is an instrumentation/system integration and development company specialising in measurement in the railway environment.

Accredited to BS EN ISO 9001: 2000 the company is a licensed train operating company with a railway safety case, track access and nationwide depot access agreements.

Capabilities

Testing services, independent rail/vehicle acceptance testing in relation to investigation, analysis and solving vehicle and track/vehicle-related problems. Approved by Network Rail to type test new and modified traction, rolling stock and on-track plant. Accredited by Railway Safety as a Vehicle Acceptance Body (VAB). On-site engineering support for fault identification, reliability testing, field engineering and commissioning.

Infrastructure monitoring: The company operates the UK main line rail grinding service for Network Rail.

Materials evaluation: application of non-destructive testing techniques to the railway industry offering a global service supported by qualified engineers, laboratory, training and workshop facilities. Its training centre in Derby specialises in all methods of NDT inspection and trains, examines and assesses personnel in ultrasonic inspection of axles and magnetic particles to PCN standards.

UPDATED

Solvera Information Services Ltd

Chadsworth House, Wilmslow Road, Handforth, Wilmslow, Cheshire SK9 3HP, UK
Tel: (+44 1625) 25 60 00   Fax: (+44 1625) 53 62 46
Web: http://www.solvera.inform.com

Key personnel

Business Development Manager: John Churchman
Processing of technical information into operation and maintenance manuals, training manuals and promotional literature.

Supplier of documentation and training for the rail industry. Recent contracts include supply of training aids and manuals for New York City Transit Authority, Heathrow Express, and London Underground. Also supplier to ALSTOM, Hong Kong Mass Transit Corporation, Kowloon-Canton Rail Corporation and TGV Korea.

Southdowns Environmental Consultants Ltd

Suite A3, 16 Station Street, Lewes, East Sussex BN7 2DB, UK
Tel: (+44 1273) 48 81 86
Fax: (+44 1273) 48 81 87
e-mail: prw@southdowns.eu.com
Web: http://www.southdowns.eu.com

Key personnel

Directors: Patrick Williams, Dr R Hood, R H Method

Capabilities

Southdowns Environmental Consultants Ltd specialises in the measurement, calculation, evaluation and mitigation of environmental noise and vibration impacts from railways.

Projects

Washington Metro (1996); Channel Tunnel Rail Link (2003); West Coast Main Line (2003); London Tilbury Southend Line (1997); Dublin Light Rail (1999); Taiwan High Speed Rail (1998); Lantau Railway (1996); Docklands Light Rail (2003).

Steer Davies Gleave

28-32 Upper Ground, London SE1 9PD, UK
Tel: (+44 20) 79 19 85 00   Fax: (+44 20) 78 27 98 50
e-mail: sdginfo@sdgworld.net
Web: www.steerdaviesgleave.com

Key personnel

Managing Director: Peter Twelftree
Directors: Fred Beltrandi, Charles Russell, Tim Ryder, Tim Spencer, Luis Willumsen, Don Nutt, Andy Costin, Colin Rowland, Phil Bates
Chairman: Brian Martin

Other offices

West Riding House, 67 Albion Street, Leeds, LS1 5AA, UK
Tel: (+44 113) 242 99 55   Fax: (+44 113) 242 96 89
e-mail: leedsinfo@sdgworld.net

68-70 George Street, Edinburgh EH2 2LR, UK
Tel: (+44 131) 226 95 00   Fax: (+44 131) 226 95 01
e-mail: edinburghinfo@sdgworld.net

Suite 3, Bay Chambers, West Bute Street, Cardiff, CF10 5BB, UK
Tel: (+44 29) 20 49 79 92; 93   Fax: (+44 29) 20 49 83 02
e-mail: walesinfo@sdgworld.net

22-26 Vardon Avenue, East End, Adelaide, South Australia 5000, Australia
Tel: (+61 8) 82 23 16 77   Fax: (+61 8) 82 23 18 77
e-mail: australiainfo@sdgworld.net

Mariano Sánchez Fontecilla 310 Piso 16, Santiago, Chile
Tel: (+56 2) 473 69 00   Fax: (+56 2) 473 69 69
e-mail: chileinfo@sdgworld.net

Via Ugo Bassi 7, I-40121, Bologna, Italy
Tel: (+39 051) 656 93 81   Fax: (+39 051) 23 15 21
e-mail: italyinfo@sdgworld.net

1606 Ponce de Leon, Suite 500, San Juan 00909, Puerto Rico
Tel: (+1 787) 721 20 02
e-mail: puertoricoinfo@sdgworld.net

Plaza de España No 18, Torre de Madrid, Of 12-7, E-28008 Madrid, Spain
Tel: (+34 91) 541 86 96   Fax: (+34 91) 541 39 96
e-mail: spaininfo@sdgworld.net

Carrera 7, #71-21 Torrea officina 604, Edificio Avenida Chile A, Officina 604, Bogota, Colombia
Tel: (+57 1) 317 32 31   Fax: (+57 1) 317 31 50
e-mail: sdg601@cable.net.co

Capabilities

Policy; business development; strategic planning; advice on privatisation and open access regimes; economic regulation; passenger and freight demand and revenue forecasting; feasibility studies for new and/or reopened lines and stations; rolling stock procurement; franchise evaluation; performance indicators; project finance; public funding case development; operational analysis; marketing and market research; civil engineering, multi-modal freight interchanges, business strategy and policy, market research and insight, consultation and participation, modelling and forecasting, transport planning and regeneration, feasibility and design, environment, appraisal, project management, travel behaviour, tourism, visual communications, transport technology, freight and logistics.

Projects

Transmilénio, Colombia; Fenoco freight rail business case, Colombia; business case for passenger rail services, Santiago, Chile; Santiago–Valparaiso high-speed link, Chile; business case for Buenos Aires suburban railways, Argentina; concession bid support for urban rail services, Boston, USA; Arlanda airport rail link, Sweden; business case for Perpignan–Figueras high-speed line, Spain; rail feasibility study for Siracusa–Gela line, Sicily, Italy; light rail feasibility study for Alghero–Sassari line, Sardinia, Italy; high-speed rail business model,

Italy; rolling stock procurement and strategy for Trenitalia, Italy; audit of Gautrain high-speed link, Johannesburg, South Africa; franchise advice to Virgin Rail Group, First Group, Serco and Chiltern Railways, UK; Scottish Strategic Rail Study for Scottish Executive, UK; Mersey Electrics franchise evaluation, UK; track charging systems, EU; Sydney–Canberra high-speed rail, Australia; East–West Rail Link, Ebbw Vale rail strategy, Welsh Development Agency, UK; Thameslink 2000 transportation and business case for Network Rail (Railtrack at the time) and Strategic Rail Authority, UK; Crossrail business case for Strategic Rail Authority, UK; East Coast Main Line business case review; UK high-speed rail development study for the Commission for Integrated Transport; rail liberalisation impact study for European Commission.

UPDATED

Strukton Railinfra BV

Westkanaaldijk 2, NL-3542 DA Utrecht
PO Box 1025, NL-3600 BA Maarssen, Netherlands
Tel: (+31 302) 240 72 00   Fax: (+31 302) 48 66 01
Web: www.struktonrailinfra.com

Key personnel
Managing Directors: Ir A Schoots, Ir G J Vos
Public Relations: Drs I W J van Dam-Aaldijk
Marketing: J J Gozens (Strukton Railinfra Development & Technology)

Background
Strukton Railinfra is part of Strukton Groep NV.

Services
Innovation, engineering, maintenance management, information systems, monitoring systems, product development, project management, maintenance breakdown repair, renewal, new construction, power supply, underground infrastructure, cabling, telecommunications, rolling stock, traction, production and maintenance of high-output equipment, production welding, competence development and marketing monitoring.

UPDATED

STV Group

205 West Welsh Drive, Douglassville, Pennsylvania 19518, USA
Tel: (+1 610) 385 82 00   Fax: (+1 610) 385 85 01
e-mail: info@stvinc.com
Web: http://www.stvinc.com

STV Incorporated
225 Park Avenue South, New York, New York 10003, USA
Tel: (+1 212) 777 44 00   Fax: (+1 212) 529 52 37

Key personnel
Chairman and Chief Executive Officer: D M Servedio
President and Chief Operating Officer:
  M S Della Rocca
Key Rail Staff: W F Matts, M Gagliardi, T Spearing, B Clarke, D Borger, O Allen

Capabilities
Transport planning; system and facility design; rolling stock engineering; operations and maintenance analysis.

UPDATED

SwedeRail

PO Box 91780, SE-120 18 Stockholm, Sweden
Visiting address: Heliosgatan 3
Tel: (+46 8) 762 37 80   Fax: (+46 8) 10 62 43
e-mail: info@swederail.se
Web: www.swederail.se

Key personnel
Director: Gunnar Hallert
Project Directors: Bjorn Andersson;
  Per Ola Stromberg; Bo Marklund;
  Jan Gullbrandsson; Lars Moberg; Ulf Bernmar
Director: Ulf Halloff
Director: Bjorn Waldemarsson

Background
SwedeRail (Ltd) was established as an independent and international railway consultancy company in 1981 by the Swedish State Railways. During 2003 SwedeRail was acquired by the ÅF-group with more than 2,500 consultants. ÅF is one of the largest technical consultant companies in Sweden.

Capabilities
Transport consultancy including restructuring, market orientation, railway operation on deregulated market, management development and advice.

Projects
Capacity building project in transport sector, Serbia.

Research and development for the EU Frame Programme; management support in Kosovo, feasibility study in Guinea and Korea; technical assistance in Kenya; international training course on railway safety management; legal reforms in Kosovo, Bosnia-Herzegovina and Kazakhstan; various projects in Bangladesh, Bosnia, China, Croatia, Ethiopia, Portugal, Russian Federation and the Ukraine.

UPDATED

Capita Symonds Ltd

Railway and Transit Division, Capita Symonds House, Wood Street, East Grinstead RH19 1UU, UK
Tel: (+44 1342) 32 71 61   Fax: (+44 1342) 31 59 27
Web: www.capita.co.uk

Rail and Aviation Division
24-30 Holborn, London EC1N 2LX
Tel: (+44 20) 78 70 93 00   Fax: (+44 20) 78 70 93 99

Key personnel
Director, Railways & Transit: Roger D Sawyers
Associate Director, Engineering and Management:
  John M Mayne
Associate Director, Permanent Way: Steven Collins
Associate Director, Transport Planning and Economics: Peter Stanley
Division Director, Rail and Aviation: David Young

Capabilities
Multi-disciplinary consultants experienced in rail management and policy, and in the evaluation, design and commissioning of major infrastructure projects, cost and commercial management and engineering consultants; experience in passenger, freight and light rail; transport; transport management consultant to rail operators, banks and governments; infrastructure design.

Projects
Capita Symonds Ltd was responsible for the detailed design of the permanent way and structures for Contract 434 which provides the links from the Channel Tunnel Rail Link into Ashford International station (UK). The company is assisting various train operating companies to develop proposals for their passenger franchise bids in respect of both operational and infrastructure enhancements aspects. It continues to advise UK and international funders on a wide variety of rail projects including: Docklands Light Railway (UK), Lewisham (UK); Bankok Metro (Thailand); Chiltern Franchise (UK); and Gautrain (South Africa). The company is also appointed to the SRA's freight panel, providing assistance on freight-related issues. Other projects include project managing the development of new stations on the West London Line for the Strategic Rail Authority, the Ebbw Vale Rail re-opening and cost and commercial management to Network Rail and LUL/Infracos.

UPDATED

Systra SA

5 avenue du Coq, F-75009 Paris, France
Tel: (+33 1) 40 16 61 00   Fax: (+33 1) 40 16 61 04
e-mail: systra@systra.com
Web: www.systra.com

Key personnel
Chairman: Michel Cornil
President: Philippe Citroën
Vice-President, Finance and Legal Affairs:
  Jean-Claude Roynier
Vice-President, Human Resouces:
  Jean-Marie Champigny
Vice-President, Engineering: Yannic Bourbin
Vice-President, Europe: Jean-Pierre Orsi
Vice-President, Asia Pacific:
  Jean-Christophe Hugonnard
Vice-President, Americas, Africa and Middle-East:
  Alain Estève
Vice-President, France: Gérard Chaldoreille

Main Subsidiaries
MVA Ltd: UK
MVA: Hong-Kong
Systra Consulting: USA
CANARAIL: Canada
SOTEC Ingénierie: France

Other subsidiaries and participations:
Citilabs: UK/USA
Systra Ltd: UK
Mexistra: Mexico
Systracade: Chile
Eurometudes: Romania
Semto: Reunion Island
Systra: Philippines
Systra Shanghai: China

Background
Systra SA is an international engineering company and a subsidiary of SNCF and RATP Paris (36 per cent each).

Capabilities
Systra offers consulting and engineering services for all rail and urban transport infrastructure and systems (high-speed trains, conventional rail, mass transit, metro, light rail, automatic guided transit systems and buses). Capabilities include: transport planning and organisation, from master plan development to system feasibility analysis; project management; institutional organisation; design engineering; general building engineering; construction management; equipment manufacture supervision; operation and maintenance organisation; training; testing and commissioning; start-up and pre-revenue service operations; transport planning software.

Projects
High-speed rail projects
China: Beijing–Shanghai corridor project.
France: France/Italy: Lyon–Turin Ferroviaire.
Morocco: Economic study for a Casablanca–Marrakech–Agadir HSR project.
Republic of Korea Seou-Busan: KTX maintenance supervision.
Taiwan: Taipei–Kaoshiung: expertise for THSRC, checking and control on site.
US: Implementation plan for California HSR system.
UK: Channel Tunnel Rail Link: project design and work supervision; renovation of St Pancras Station, London.
Other projects
Middle-East and Africa: Euro-Mediterranean Transport Project (MEDA), Dubai preliminary LRT (UAE), North-South new rail line in Saudi Arabia; civil engineering for Algiers metro.
Europe: Optirails projects, OSE Business plan in Greece, railway restructuring studies in Bulgaria, Latvia. GSM-R network project in France. A number of LRT, BRT projects in France (Bordeaux, Toulouse, Mulhouse, Clermont-Ferrand, Toulon, Marseille, Rouen, Lyon, St Etienne, Nancy, Nantes and Nice)
Asia: Design for line 3 of Delhi metro (India), transport plans for Beijing and Suzhou, supervision of construction of Guangzhou, Shanghai and Shenzhen metros (China), assistance to Indian

railways in Mumbai, modernisation of the Hanoi-Vinh railway line (Vietnam), line 3 of the Manila LRT (Philippines), Kaohsiung metro in Taiwan. Feasibility study for Hanoi LRT.

South America: Extension of line 2 and studies for line 4 of Santiago metro, railway studies in Venezuela.

UPDATED

TAMS Consultants Inc

655 Third Avenue, New York, New York 10017, USA
Tel: (+1 212) 867 17 77   Fax: (+1 212) 697 63 54
e-mail: marketing@tamsconsultants.com
Web: http://www.tamsconsultants.com

Key personnel
President: Anthony R Dolcimascolo
Principals: Patrick J McAward Jr, Lyle H Hixenbaugh,
    Edward C Regan, G Barrie Heinzenknecht,
    Ronald H Axelrod, Frank A Baragona, Eric Cole,
    E Patrick Sorensen, Albert DiBernardo,
    Joseph Fiteni Jr, Kenneth F Standing,
    Charles F van Cook

Capabilities
TAMS offers international services in engineering, architecture, and planning. The firm has worked in more than 100 countries, providing comprehensive services for major ports, highways, railroads, bridges, airports, dams, agricultural and regional development, waste management, and urban planning projects.

TAMS has experience in the planning, design, and inspection of railroad facilities, ranging from the engineering of more than 4,000 km of railroads throughout the world to the design of major tunnels and stations. Projects include planning new lines through jungle or desert and rapid transit systems in US cities. Services provided by TAMS include location and alignment, trackwork, bridges, tunnels, and marshalling yards.

TDI-AVE Design and Engineering Services

A division of AVE Rail Products
High Technology Centre, Madeley Road, North Moons Moat, Redditch B98 9NB
Tel: (+44 1789) 49 03 70
e-mail: designstudio@tdi.uk.com
Web: www.tdi.uk.com

Key personnel
Design Director: Martin Permberton
Design Manager: Paul Salkeld
Sales Manager: Trevor Clews

Associate companies
AVE Rail Products
    Arrowvale Electronics
    Changzhou Evergreen AVE Ltd (China)
    Minitram Systems Ltd
    Transport Design International Pty Ltd (Australia)

Background
In 2005 Transport Design International UK (TDI), joined AVE Rail Products to form a new interior design division TDI-AVE Design and Engineering Services. AVE supplies complete rail vehicle interiors, components and electrical systems and has supplied nearly 2,000 vehicles including West Coast main line Pendilions, ALSTOM's Juniper vehicles, Coradia Class 180s fleet and Bombardier's Electrostars. TDI's portfolio, developed over the last 20 years, covers flagship projects in both light and heavy rail around the world. TDI employs specialist industrial designers, ergonomists and mechanical engineers to develop fully styled, fully engineered interior solutions and industrial design briefs. Typical commissions involve the design of saloons, vestibules, toilet modules, driving cabs and controls desks, detrainment systems, passenger seating, on-board catering equipment/galleys, IT and entertainment facilities.

Capabilities
Complete interior systems, including refurbishment kits which can be provided in piece or part, sub-assembled, or fully assembled modules, delivered straight to the line, complete with all equipment, electrical components and panel wiring.

TDI also provides a comprehensive service in the industrial design and branding of complete transport systems involving stations, street furniture, ticketing equipment, signage, vehicle exterior styling and liveries. Scale models and full-size mockups of complete vehicles and environments can be produced in-house and delivered on site.

In conjunction with strategic partners, TDI designs ultra light, low emission, low cost Minitrams. TDI's engineering team will produce complete bodyshell and rolling chassis solutions, including low emission drive lines, from specification through to supply of prototypes and production vehicles.

Projects
Current and recent projects include: production engineering of new interior systems across a number of new European vehicle platforms (Bombardier Transportation); design and supply of detrainment system (ALSTOM); tram concepts and system branding study for Birmingham Midland Metro Extensions (Centro); tram concepts and design of urban realm infrastructure for Phase 3 extension (Manchester Metrolink/Serco/SNC Lavalin); design of tube stock interiors and exteriors for Piccadilly Line (refurbishment); Northern Line and Jubilee Line Extensions (London Underground/Bombardier/ALSTOM); feasibility and concept design for Brisbane–Rockhampton–Cairns tilting train sets (Queensland Rail/Evans Deakin Industries); new designs for interurban Tangara emus, XPT seating, suburban stock refurbishment and flagship (4GT) Millennium double-deck train for Sydney (SRA, NSW, Australia/Goninan/Clyde Engineering); feasibility and concept design for new rolling stock (KCRC West Rail Hong Kong); feasibility and concept design for metro-car fleet refurbishment (MTRC Hong Kong).

NEW ENTRY

TERA International Group, Inc (TERA)

107 E. Holly Avenue, Suite 12, Sterling, Virginia 20164-5405, USA
Tel: (+1 703) 406 44 00   Fax: (+1 703) 406 15 50
Web: www.teraus.com

Key personnel
President: Asil Gezen
Vice-President: Lynn Harmon

Other offices
TERA Beijing
Sunshine 100, Suite C-3609, No. 2, Guanghua Road, Chaoyang District, Beijing 100026, China
Tel: (+86 10) 51 00 09 00   Fax: (+86 10) 51 00 09 01

TERA Sofia
Gotze Delchev Quarter, Block 108, En A F1, 3 Apt 7, Sofia 1404, Bulgaria
Tel: (+359 2) 958 20 40   Fax: (+359 2) 958 20 41

TERA Manila
41 Premium Street, GSIS Village Proj. 8, Quezon City, Metro Manila, Philippines
Tel: (+ 63 2) 929 59 67

TERA Hawaii
5324 Kalaniana'ole Hwy, Honolulu, Hawaii 96821, USA
Tel: (+1 808) 946 94 46   Fax: (+1 808) 946 94 47

Capabilities
TERA is a respected contract research and consulting firm founded in 1976.

Capabilities include: transportation economics and planning studies; technical, engineering and operational assessments; evaluation of the effectiveness and efficiency of transportation policies, institutions and programmes; economic and financial feasibility analysis; institutional and regulatory reviews; simulations and modelling of transportation systems; performance evaluation of existing infrastructure and identification of new infrastructure needs; preparation of business plans and information memoranda on infrastructure projects; preparation of technical, administrative and commercial specification for international and competitive tenders; traffic forecasting and analyses of tariffs based on competitive market conditions and customer needs; project presentations to lenders and investors to secure financing; environmental and customer needs; project presentations to lenders and investors to secure financing; environmental and social/poverty impact analyses; due diligence services for investors and lenders; and human resource utilisation and staff training.

Projects
TERA has successfully completed various projects in railway transportation in the USA, Central and Eastern Europe, Africa, the Middle East, Latin America and Central and Southern Asia. Project experience covers rail car ferry service operations, metro and light and heavy urban rail projects, computer simulation of freight and passenger traffic, locomotive workshop and other facility evaluations, financial and feasibility analyses including technical, environmental, poverty, institutional considerations; rail restructuring, commercialisation, concessioning and privatisation studies; and intermodal corridor transportation studies including double-stack, piggyback and trailer-truck on flat car operations.

TERA's railway project experience includes technical assessments of railway signalling and communications in Albania, Brazil, Bulgaria, China, Guatemala, Macedonia, Poland and Romania; track laying and maintenance machinery and equipment in Brazil, Egypt, Kazakhstan; locomotive and wagon workshops in Bulgaria, Malaysia, Thailand and 10 countries in Southern and Eastern Africa; market assessments for wheels and bogies, couplers, power components as well as rolling stock. Infrastructure evaluations, trade/traffic forecasting, operational and financial assessments and feasibility studies have been conducted in more than 50 countries worldwide.

Ultra-light railcar under development for Kalamata, Greece   *NEW*/1036999

Intermodal rail car ferry service for passenger and freight operations have been analysed around the world including the Caspian Sea, the Baltic Sea and Lake Michigan. TERA conducted regional rail corridor assessments for the Balkans and a feasibility study for a regional locomotive workshop and centralised part warehouse with unit exchange in Eastern and Southern Africa for the Union of African Railways. Metro/light rail system evaluations in India, Indonesia, Philippines, Romania, Singapore, Thailand, Turkey and Venezuela have been undertaken.

UPDATED

Thomas K Dyer Inc

HNTB Corporation
1762 Massachusetts Avenue, Lexington, Massachusetts 02420, USA
Tel: (+1 781) 862 20 75   Fax: (+1 781) 861 77 66
e-mail: dwoodbury@hntb.com
Web: http://www.hntb.com

Key personnel
Vice-Presidents: Douglas J Woodbury
Marketing: Kimberly Durkee

Other offices
Over 60 offices throughout the USA.

Background
Thomas K Dyer became part of HNTB Corporation in June 2001.

Capabilities
Architecture, engineering and planning, providing services throughout the USA and worldwide. With nearly 3,000 professional and technical employees, HNTB offers services within the disciplines of surface transportation, aviation, architecture, environmental engineering and construction services.

Tilney Shane Ltd

5 Heathmans Road, London SW6 4TJ, UK
Tel: (+44 20) 77 31 69 46
Fax: (+44 20) 77 36 33 56
e-mail: info@tilneyshane.co.uk
Web: http://www.tilneyshane.co.uk

Key personnel
Chief Executive: Kathy Tilney
Design Directors: Jonathan Wilson
Financial Director: Heather Shane

Capabilities
Design of interiors for transport systems and of environments suitable for high passenger densities; design of specialist rail vehicles including concepts, engineering development and textile design. Rail project design includes passenger flow analysis, CAD drawing production, mockup build supervision, technical and procurement specification.

TLC

Transport, Informatik, und Logistik-Consulting GmbH
Hallesches Ufer 30, D-10963 Berlin, Germany
Tel: (+49 611) 173 45 58
Fax: (+49 611) 173 41 13
e-mail: TLC.Sales@tlc.de

Key personnel
Managing Director: Dr Gerhard Pintag
International Sales: Dr Hubert Kreutzmann

Offices
TLC has branches in Wiesbaden, Frankfurt, Berlin, Vienna, Sofia and Bucharest.
Established 1988
Staff: 1,050

Background
TLC is 100 per cent owned by Deutsche Bahn AG of Germany.

Capabilities
Consultancy and logistics services for the transport industry, specialising particularly in business-process re-engineering. The company has been responsible for the design and implementation of computer software solutions, not only for the parent company (German Rail) but also for a large number of other customers in the transport industry. Increased emphasis being placed on contracts with customers outside Germany.

Projects
Include rail and local transport ticketing systems, train reservation systems, internet sales (including online ticketing fulfilment), freight management systems, infrastructure and property management systems. TLC is currently implementing the complete restructuring for DB's Cargo (freight) division.

TMG International Pty Ltd

13th Floor 39-41 York Street, Sydney, New South Wales 2000, Australia
Tel: (+61 2) 92 62 41 11   Fax: (+61 2) 92 62 41 10

Key personnel
Group Managing Director: Dale Coleman
Managing Director TMG Australia: Peter Thornton
Director Queensland: Keith Walker
Director Modelling Services: Alex Wardrop
Managing Director TMG International (Asia)
   Limited: Clive Yep

Other offices
98 South Road, Torrensville PO Box 253, Torrensville Plaza South Australia 5031, Australia
Tel: (+61 8) 84 43 41 33   Fax: (+61 8) 84 43 41 55
e-mail: tmgsa@senet.com.au

11/114 Albert Road, South Melbourne Victoria 3205, Australia
Tel: (+61 3) 96 96 01 06   Fax: (+61 3) 96 96 27 78

3rd Floor, 447 Upper Edward Street, Brisbane, Queensland 4000, Australia
Tel: (+61 7) 38 39 14 03   Fax: (+61 7) 38 31 67 11
e-mail: tmgbne@gil.com.au

TMG International Asia
Suite 2501, Golden Centre, 188 Des Voeux Road, Central Hong Kong
Tel: (+852 2) 815 65 99   Fax: (+852 2) 815 65 99
e-mail: tmgasia@nevigator.com

Capabilities
System modelling and simulation, timetable planning and development, operational and infrastructure planning, energy efficient train regulation, condition monitoring and dynamic analysis systems, maintenance analysis and planning, safety and risk analysis, operations and infrastructure benchmarking, asset management systems and support.

Tractebel Development Engineering

Avenue Ariane 7, B-1200 Brussels, Belgium
Tel: (+32 2) 773 75 11   Fax: (+32 2) 773 79 80
e-mail: development@tractebel.be
Web: http://www.tractebel.com

Key personnel
Rail and Road Department Manager: V Wilkin
Projects Manager: M Brismée

Capabilities
Tractebel's capabilities in railways, underground and light rail transit systems include: engineering infrastructures, superstructures, rolling stock, electricity supply, signalling, telecommunications, safety, and geotechnology; project management

and supervision of works; turnkey projects, and consultancy in town and country planning, freight/passenger transport systems, transport master plans and feasibility studies.

Transportation Technology Center Inc

A subsidiary of the Association of American Railroads
PO Box 11130, Pueblo, Colorado 81001, USA
Tel: (+1 719) 584 07 50   Fax: (+1 719) 584 06 72
e-mail: marketing@ttci.aar.com
Web: www.ttci.com

Subsidiary
TTCI (UK) Ltd
13 Fitzroy Street, London W1T 4BQ, UK
e-mail: firdausi_irani@ttci.aar.com
   richard_joy@ttci.aar.com

Key personnel
President: Roy A Allen
Managing Director: Firdausi Irani
Vice-President Technology: Keith L Hawthorne
Vice-President Business Development:
   Dr Albert J Reinschmidt
Vice-President and Chief Financial Officer:
   Dr Scott B Harvey
Senior Assistant, Vice-President Research and
   Development: Semih Kalay
Assistant Vice-President Strategic Planning
   Capabilities: Dr James R Lundgren
Assistant Vice-President Communications and
   Train Control Technologies: Alan L Polivka
Programme Manager: Richard Joy

Background
TTCI was formed on 1 January 1998 as a wholly owned subsidiary of the Association of American Railroads. About 275 engineers, technicians and support personnel make up the staff of TTCI.

TTCI (UK) Ltd was incorporated in June 2004 as a wholly owned subsidiary of the Transportation Technology Center, Inc. TTCI (UK) Ltd is supported by its parent organisation with extensive testing facilities and is a professional engineering consulting company that provides services to the European Union.

Capabilities
Full-scale vehicle on-track testing and specialised laboratory testing at Transportation Technology Center (TTC), a 52 sq-mile test facility located near Pueblo, Colorado, USA. The site provides over 81 km (50 miles) of specialised railroad test track, extensive track facilities for electric and dual-mode high-speed passenger, transit and commuter and freight testing. These tracks are used daily for track structure and vehicle performance testing, track and service worthiness, life-cycle and component reliability and ride comfort evaluation. Areas of expertise include: consulting services, test instrumentation, data collection and analysis, training, products including instrumented wheelsets, bogie curving performance detection systems, acoustic bearing defect detection system InteRRIS™ (Integrated Railway Industry Support Service).

As a subsidiary of the Association of American Railroads (AAR), TTCI provides for all the research and development needs of the North American rail industry.

Projects
Positive train control (Communication-based train control) related studies for fitment and track possession. UIC work project support in Rail Defect Management (JRP – 1) wheel/rail interface management (JRP – 2) and vehicle condition monitoring by means of wayside detection (JRP – 3). Investigation of rolling contact fatigue (Hatfield, UK incident). Track usage cost negotiations with the UK Office of Rail Regulation (ORR) and Railtrack (now Network Rail). Support to ORR – sponsored review of Network Rail's business plan. Application of advanced acoustic bearing technology to conduct a field survey of the condition of the wheel roller

Aerial view of the TTCI rail facility at Pueblo, Colorado, USA   ***NEW**/1114355*

bearings for a series of UK rolling stock. Metro De Caracas Bogies Reftrofit Programme. NYCt R142 crash testing. Acela high-speed train testing. RTRI gauge change train testing. Amtrak Express car performance testing. Wheel spalling investigation. Official AAR new car performance testing. General Motors Corporation EMD locomotive testing. Carajas Railway heavy axle load implementation testing.

UPDATED

TransTec Consult GmbH

Calenberger Esplanade 3, D-30169 Hanover, Germany
Tel: (+49 511) 929 75 70   Fax: (+49 511) 929 75 77
e-mail: info@transtec-consult.de
Web: http://www.transtec-consult.de

Key personnel
Managing Director: Rainer Johannsmeier,
  Stephen Kritzinger
Key Consultants: Hans Joachim Rönnau,
  Rainer Hesse, Dr Voker Stölting, Thomas Rieger,
  Gerhard Beckendorff

Background
TransTec Consult is an independent consulting firm for planning, design, implementation and operation of public transport systems.

Capabilities
Organisational and financing including: preparing operators and authorities for competitive market participation; development of institutional solutions; elaboration of procurement schemes and contract documents; life-cycle-cost estimation; cost-benefit analyses; funding and financing concepts; operating and investment cost evaluation. Planning and operation including: transport master plans, multimodal studies; patronage forecasting; planning of public transport supply; operational planning; system appraisals; marketing concepts; project management and institutional support. Systems and technology including: feasibility studies; operation and maintenance concepts; rolling stock, civil and technical infrastructure design; tender specification documents; support of transit system procurement and implementation; system acceptance and staff training and certification.

Projects
Germany: Hanover, support of authorities in preparing procurement of passenger rail services; Kiel, concept for further development of regional rail authority; Bremen, project management of regional LRT system; Neckar-Alb, operation concept for the introduction of a regional LRT (RegioStadtBahn) in the Neckar-Alb region; Brunswick, introduction of a regional LRT (RegioStadtBahn) in the Greater Brunswick area.
European Union: MARETOPE – development of tools for reorganisation of public transport.
Greece: Athens, design works for Athens tramway system.
Israel: Haifa, feasibility study on guided buses for upgrading public transport in Haifa.
Poland: Lower Silesia, public passenger rail services in the lower Silesia Voivodship.

Transurb Technirail SA

60 rue Raveinstein, Bte 18, B-1000 Brussels, Belgium
Tel: (+32 2) 548 53 40   Fax: (+32 2) 513 94 19
e-mail: a.willaert@transurb.com
Web: http://www.transurb.com

Key personnel
Chief Executive: Patrick Steyaert
Commercial Director: Etienne Deblon
Marketing Manager: Andy Willaert

Principal subsidiaries
TUC Rail, Belgium
Transurb, Argentina
Transurb, Gabon
Transurb, Malaysia
Transurb International Ltd, Russia
Technirail, Manila

Background
Transurb Technirail is a semi-public company and a subsidiary of SNCB/NMBS and also a daughter company of STIB (the urban transport company of Brussels) as well as of Belgian private transport-engineering consultants.

Capabilities
Transurb Technirail has been, since its creation in 1973, an expert company in railway and mass urban transit projects, for both transport of persons and freight.
  Transurb Technirail offers a diversified range of services to customers, worldwide including design, build, operation, maintenance, upgrade of project cycles relating to rolling stock.

Projects
The most significant projects include: the complete maintenance of Light Rail Transit System 1 in Manila, the maintenance, by way of a concession, of the Gabonese railway, the maintenance of bogies, brakes, wheels and air-conditioning systems on behalf of Eurotunnel, assistance to the operation of Ferrovias (Belgrano Norte Line) and Metrovías (Buenos Aires underground) franchise holders in Argentina, the design and manufacture of many railway-driving simulators, the rehabilitation of 650 wagons for the Romanian railway.

Transys Projects Ltd

2 Priestley Wharf, Holt Street, Aston Science Park, Birmingham B7 4BN, UK
Tel: (+44 121) 359 77 77   Fax: (+44 121) 359 18 11
e-mail: info@transysprojects.ltd.uk
Web: www.transysprojects.ltd.uk

Key personnel
Managing Director: Jeremy J Ashley
Sales and Marketing Director: Kevin Lane
Engineering Director: Karl J Barras
Financial Director: Dorothy J Lidster

Background
Transys Projects Ltd is an independent company.

Capabilities
Multidisciplined engineering consultancy and project management organisation, covering all aspects of passenger rail vehicles and their related support services. This covers mass transit vehicles, LRVs and tramcars as well as main line dmus, emus, and passenger coaches. Particularly specialising on UK, Railtrack Group Standards, Safety Cases and vehicle acceptance.
  Certified to BS EN ISO 9001 with 'Link-Up' accreditation in 26 relevant areas.
  Specific capabilities include:

Design Engineering
- design of complete rail vehicles or discrete areas
- concept to production drawings/schedules
- bodyshells, underframes, cabs, interiors
- assembly, component and system design
- mockups, modular design

Engineering Services
- optimise design to obtain requisite certification
- FEA, classical calculations, kinematics, dynamics
- specification and management of structural testing
- tender/tender response documents and purchase specifications

Electrical and Mechanical Systems, Engineering and Integration
- ability to project engineer a complete range of equipment, systems and services through design, testing and validation.
- vehicle mechanical equipment, systems, diesel drivelines and so on
- electrical/electronic systems including control and traction auxiliaries
- material choice, fire safety and testing
- vehicle performance, simulation on specified routes
- vehicle acceptance and safety case issues
- fault finding, trouble shooting

Project Support
- provide project management and planning
- select/monitor subcontractors and/or product performance
- vehicle maintenance, efficiency and reliability improvement
- developing technical literature and support information
- product support group undertakes modification and upgrade work.

Projects
Design engineered and project managed Class 323 and Malaysian emus plus Glasgow Underground trailer cars.
  Various recent tender support projects for major European vehicle manufacturers, covering emus, dmus and demus, predominantly for the UK but also for the Far East.
  Feasibility studies and engineering investigations for UK rail vehicle leasing companies (ROSCOs) as

well as turnkey projects. Various proposed design solutions for UK train operating companies (TOCs).

Complex underframe installation designs covering complete diesel drivelines, fuel tanks, brake frames, electric traction drives and auxiliary equipment cases. Interior finish design/engineering for several new multiple unit packages in the UK.

General engineering consultancy, detailed structural design, safety and maintenance support for Fiat Ferroviaria on their joint UK contract with Alstom Transport for Virgin's West Coast high-speed tilting trains.

Design support for Alusuisse Road and Rail. Various other contracts for engineering consultancy and maintenance support activities.

Turnkey provision including design, engineering, installation and commissioning of automatic sanding systems on more than 700 existing multiple-units in the UK. These cover various classes of dmus and emus to improve adhesion performance during braking and traction during the leaf-fall season.

UPDATED

Trauner

1617 JFK Boulevard, Suite 600, Philadelphia, Pennsylvania 19103, US
Tel: (+1 215) 814 64 00
Web: www.traunerconsulting.com

Key personnel
Vice President, Business Development:
  Tracy M Doyle

Capabilities
Most aspects of rail and road consultancy.

Projects
Trauner has been brought in to assure construction quality for the US$1.9 billion Alameda Corridor Transportation Authority (ACTA) rail project. The Corridor will be a freight track which runs north from the nation's largest seaports and is the first consolidated railroad link of its kind. The double-tracked Alameda Corridor will begin in the City of Compton before descending into a 10 mile long trench, finally emerging in Los Angeles and dispersing among the railroad mainlines. Construction on this project includes the digging of 33 ft deep trenches, new Los Angeles River crossings, and associated bridge structures. Trauner's professional engineers and consultants will be involved in the evaluation of the Design-Build Contractor's baseline schedules, design review and oversight, claims analysis and dispute resolution, and the preparation and delivery of training/workshops for ACTA personnel.

The Alameda Corridor will reduce traffic delays by eliminating conflicts at nearly 200 street-level railway crossings. The ACTA plan includes new overpasses to separate train and road traffic and the widening of Alameda Street from four lanes to six.

UPDATED

Trenitalia SpA

Italian State Railways
Rolling Stock Technology Department (UTMR)
Viala S Lavagnini 58, I-50129 Florence, Italy
Tel: (+39 055) 47 60 00   Fax: (+39 055) 48 19 05

Services
With more than 5,000 employees, UTMR is responsible for the maintenance of Italian State Railways (FS) traction and rolling stock. It also undertakes overhaul, refurbishment and conversion projects using the 13 facilities that form the company's Major Repairs Workshops (OGR) division. These resources and capabilities are offered to other operators.

TRL Limited (Transport Research Laboratory)

Crowthorne House, Nine Mile Ride, Wokingham, RG45 3GA, UK
Tel: (+44 1344) 77 00 07   Fax: (+44 1344) 77 08 80
e-mail: enquiries@trl.co.uk
Web: www.trl.co.uk

Key personnel
Head of Rail: Charles Oakley
Rail Marketing Campaign Manager:
  Katheryn Richardson

Capabilities
TRL's capabilities include: civil engineering: providing an independent and impartial service for research, consultancy and testing of major rail structures or components. Simulation: provided at TRL's advanced driving simulation centre which comprises a number of experts including modelling teams focusing on safety and vehicle engineering. TRL is at the forefront of developing safety strategies and advice is provided to governments, authorities and private sector businesses to manage risk. Crashworthiness: through the design of vehicle structures that manage to impact energy and allow the occupant compartments to remain intact, and the deceleration level to be tolerable for the driver and passengers. TRL provides transportation consultancy working on all aspects of rail based transport, from the pedestrian route access of railway stations to the application of new technology to improve real time information on the rail network. TRL also works to improve the environmental performance from informing policy through to influencing planning and delivery.

UPDATED

Trowers & Hamlins

Sceptre Court, 40 Tower Hill, London EC3N 4DX, UK
Tel: (+44 20) 74 23 80 00
Fax: (+44 20) 74 23 80 01
e-mail: enquiries@trowers.com
Web: http://www.trowers.com

Key personnel
Senior Partner: Jonathan Adlington
  Tel: (+44 20) 74 23 82 71
  e-mail: jadlington@trowers.com
Managing Partner: David Biggerstaff
  Tel: (+44 20) 74 23 85 92
  e-mail: dbiggerstaff@trowers.com
Head of Projects and Construction: David Mosey
  Tel: (+44 20) 74 23 83 70
  e-mail: dmosey@trowers.com
Transport Group Contact: Philip Wright
  Tel: (+44 20) 74 23 83 37
  e-mail: pwright@trowers.com

Other offices
Abu Dhabi, Bahrain, Cairo, Dubai, Exeter, London, Manchester.

Capabilities
Trowers & Hamlins is a full service international law firm with particular legal expertise in major rail projects, including both domestic and international projects. It is able to provide specialist advice on the planning, procurement, design, construction, operation, maintenance and renewal of rail infrastructure. The firm is also experienced in providing all legal advice arising from the structuring and restructuring of all types of railway systems including the privatisation of state-owned rail companies.

Trowers & Hamlins provides all aspects of legal advice as it impacts on the rail industry, and understands fully the regulatory, safety and operational requirements that are unique to the railway industry. The firm has extensive experience of the UK Private Finance Initiative and other forms of public-private funding and has been extensively involved in assignments for the Greek and Irish

governments advising on new PPP frameworks for the procurement of new rail infrastructure and the implementation of any new legislation required to facilitate such initiatives. The firm also has acted on several high value infrastructure projects in the Middle East using Islamic-compliant financing.

Contracts
UK: Channel Tunnel Rail Link (Railtrack plc, now Network Rail, and Union Railways South Limited). Advising on restructuring of high-speed rail link project. The CTRL is the largest infrastructure project undertaken in the UK. The project, which is the UK's first high-speed railway, is an integral part of the Trans-European High Speed Network. When completed it will halve journey times from central London to the Channel Tunnel as well as creating three new international stations.

The CTRL is being constructed under a private-public partnership contract between the UK government and London & Continental Railways Limited (LCR). Inter-Capital & Regional Rail, a consortium of National Express Group, British Airways and the French and Belgian national railways (SNCF and SNCB), have a management contract with LCR to run the Eurostar train service.
UK: Sunderland Metro Extension (Railtrack plc, now Network Rail). Advising Railtrack on a scheme proposed by the Tyne & Wear Passenger Executive to extend its metro system partly by using Railtrack's infrastructure stations and by constructing three new stations.
UK: Wembley Park station redevelopment (Infraco JNP Limited (now Tube Lines Limited)). Advising on the redevelopment of a major London Underground station and various congestion relief projects for selected Underground stations.
UK: London Underground Piccadilly, Northern and Jubilee Lines escalator works (Infraco JNP Limited, now Tube Lines Limited). Advising on a specialised long-term contract for the refurbishment of 196 escalators on the Jubilee, Northern and Piccadilly Lines. The contracts will encompass maintenance, renewal and refurbishment of the escalator lifts.
UK: London Underground PPP contract (Infraco JNP Limited, now Tube Lines Limited). Advising on the various issues (including risk analysis, PFI obligations and interfaces with third parties) on the public-private partnership for the London Underground. This entailed a number of connected contracts to be aligned with the PPP contract.
UK: West Coast Main Line modernisation project (Railtrack plc, now Network Rail). Advice in relation to the remodelling of Euston station, a major London rail terminal, and renewal of track and signalling.
UK: The Protection and Warning System contracts (Railtrack plc, now Network Rail). Advice and drafting on the contract for the national installation of the Train Protection and Warning System.
Ireland: Heuston–Kildare Route Capacity Upgrade (Córas Iompair Éireann). Advising on upgrade to 12 miles of track and three stations to increase capacity on a busy commuter link into Dublin's Heuston station.
Ireland: Department of Public Enterprise Policy Frameworks (Irish Department of Public Enterprise). Advising the Irish Republic's Department of Public Enterprise on the policy frameworks for Public Private Partnerships (PPPs) in the rail sector. The brief includes Dublin's proposed IR £1.4 billion (€1.8 billion) metro system and potentially some of the heavy rail networks also envisaged for the capital.
Ireland: Transport (Railway Infrastructure Act). (Irish Department of Public Enterprise). Advice to the Irish government on the provision of the Transport (Railway Infrastructure Act) 2002. This act pioneers the use of PPP structures (including that of BOT) for future rail projects in Ireland. The act established the Railway Procurement Agency which will procure all future rail projects in Ireland on a PPP basis. Also it ensures that legal measures have been put in place to facilitate and encourage the involvement of the private sector in such projects including the utilisation of private sector funding. Parallel to this, the act also prescribes a process for the Minister of Public Enterprise to grant railway orders expeditiously for the construction,

maintenance and operation of new railways and related infrastructure.
New Zealand: Railways restructuring. Providing strategic and legal advice in relation to the restructuring of the New Zealand railways.

TUC Rail

TUC Rail SA
Rue de France 91, B-1070 Brussels, Belgium
Tel: (+32 2) 529 78 20   Fax: (+32 2) 529 79 00
e-mail: info@tucrail.be
Web: http://www.tucrail.be

Capabilities

Preliminary studies for development of railway infrastructures, design of bridges, viaducts and tunnels, tracklaying, signalling, telecommunications, quality control, safety monitoring, validation and testing.

Tyréns Infraconsult AB

SE-118 86 Stockholm, Sweden
Tel: (+46 8) 429 00 00   Fax: (+46 8) 429 00 60
e-mail: info@tyrens.se
Web: http://www.tyrens.se

Capabilities

Include: environmental engineering and assessment; ground, water and sewerage planning; geotechnical engineering and rock mechanics; landscape and urban planning; cartography and surveying technology; and civil engineering, including railway, tunnel, street and highway design.

Projects

Projects on which Tyréns has provided services include: the Swedish National Rail Administration's Sveland Mälar valley, Arlanda and West Coast lines; Malmo City Tunnel; the Öresund Fixed Link; and for Norwegian State Railways, the reconstruction of Skien station, south of Oslo.

Urbitran

71 West 23rd Street, 11th Floor, New York, New York 10010, USA
Tel: (+1 212) 366 62 00   Fax: (+1 212) 366 62 14
e-mail: transit@urbitran.com
Web: http://www.urbitran.com

Key personnel

President: Michael Horodniceanu
Chief Operating Officer: Mac Ismail

Offices

New Haven, Connecticut; Holyoke, Massachusetts; Edison, New Jersey; Albany, New York.

Subsidiaries

Urbitran/Rosenbloom Architects (New York, New York)
Urbitran/Garmen (Montville, New Jersey)
UA Construction (New York, New York)

Capabilities

In the transit services sector: bus and rail operations studies, transit development plan preparation, origin and destination studies, paratransit planning, marketing and survey research, management performance reviews, welfare-to-work planning, public and community outreach and ridership forecasting.

Projects

Include: Connecticut DOT statewide bus system study; Baltimore regional transit study; Buffalo Intermodal Transportation Center ridership demand development; Rockingham County, Virginia, transportation development plan; Charlottesville, Virginia, transit development plan; Albany Corridor study; comprehensive review of transit services for Santa Clarita, California; North Carolina statewide technical assistance or transit planning services.

Vanness Group

830-132 A1A North, Suite 204, Ponte Vedra Beach, Florida 32082, USA
Tel: (+1 904) 280 18 98   Fax: (+1 904) 280 18 99
e-mail: vannessco@comcast.com
Web: www.transmatch.com/vanness-brackenbridge/

Key personnel

Principals: J Chris Rooney,
   Karl R Ziebarth, James Hanscom

Capabilities

Consultancy services for railways, governments and bilateral lending agencies concerning financial feasibility, restructuring and strategic planning issues. The Vanness Group has extensive experience in railway restructuring, strategic planning, financial planning, market analysis, organisational restructuring and policy development.

Projects

The Vanness Group has served clients in over 18 countries including Argentina, Australia, Bolivia, Brazil, Canada, Chile, Colombia, Ecuador, South Korea, Mexico, Myanmar (Burma), New Zealand, Spain, Sri Lanka, Thailand, Uganda and USA.
Projects include: US Department of Transportation: Advising railroad rehabilitation and infrastructure funding programme with respect to viability of proposed railways projects.
Amtrak, USA: As subcontractor, assisted Battelle Institute to vet the national passenger railway's five-year business plan on behalf of Inspector General of the US.
Asian Development Bank: Advised the ADB on the feasibility of financial assistance to upgrading the Royal Cambodian Railways.
Bolivian National Railways: Financial analysis of 'open access' policy and preparation of strategy for privatisation (capitalisation) of the railway.
CSX Transportation/Sea Land, USA: Adviser on value realisation strategies, network rationalisation, and acquisitions in the US and Latin America.
Canadian National Railways: Adviser in respect of strategic, marketplace, competitive and organisational issues.
Companhia Vale do Rio Dolce (CVRD): Managed a long-term engagement to migrate predominantly minerals systems to full service general freight and passenger operations.
Conrail USA: Supervised technical staff preparatory to privatisation.
Dakota, Minnesota and Eastern RR (US): Provided business and financial evaluation services relating to major way and works renovations.
Ferrocarriles Metropolitanos, Buenos Aires, Argentina: Formulated business strategy and assessed financial feasibility and budgets for suburban services in capital city.
Sri Lanka Railway (SLR): Advised Sri Lanka Ministry of Transport and World Bank on commercial strategies.
Texas Mexican Ry/TFM: Provided financial and business evaluation services relating to major way and works renovations.
Advising government of Uganda on privatisation/commercialisation of Uganda Railways Corporation.

UPDATED

Vossloh Information Technologies York Ltd

Jervaulx House, 6 St Mary's Court, Blossom Street, York YO24 1AH, UK
Tel: (+44 1904) 63 90 91   Fax: (+44 1904) 63 90 92
e-mail: info@vity.vossloh.com
Web: www.vit.vossloh.com

Key personnel

Managing Director: Roland F Albert
Head of International Business Development:
   Alastair Dick
UK Business Development Manager:
   David Clarke

Background

Vossloh Information Technologies York Ltd, formerly Comreco Rail Ltd, is a subsidiary of Vossloh Information Technologies GmbH, Kiel, Germany, which is part of the Vossloh Group.

Capabilities

RailPlan enables high-performance operations simulations including 'Monte Carlo' simulations of operational malfunctions. PowerPlan models the power supply system on the basis of the train movement data generated by RailPlan. TrainPlan creates timetables for all types of trains, with highly sophisticated conflict detection and resolution. Timetable Robustness Analyser identifies avoidable weak points of an existing timetable before operation. AccessPlan supports the online bid/offer process between infrastructure operator and train operating companies during the negotiation of track access. ResourcePlan creates crew and vehicle schedules using the timetable data generated in TrainPlan. ResourceManager manages rolling stock allocation in a real-time environment, using data from the ResourcePlan. All are available for licensed use and are used by Vossloh IT in consultancy work worldwide.

UPDATED

Wendell Cox Consultancy

PO Box 841, Belleville, Illinois 62222, USA
Tel: (+1 618) 632 85 07   Fax: (+1 618) 632 85 38
e-mail: policy@publicpwpose.com
Web: http://www.publicpwpose.com

Key personnel

Principal: Wendell Cox
Associate: Jean Love

Capabilities

Urban transport organisational design; competitive tendering; feasibility studies; strategic planning; legislation, planning and policy; privatisation; land use; smart growth.

Projects

Market analysis for US market entry, for UK and Italian companies (1997); review of Atlanta regional plans; review of Florida high-speed rail project; review of Austin regional plan; performance analysis of Texas public transport agencies; review of Charlotte light rail project; analysis of US urban rail analysis in relation to Auckland; New Zealand project; land use and smart growth presentations in the US, Europe, Australia and New Zealand.

Wilbur Smith Associates

1301 Gervais Street, 16th Floor, Columbia, South Carolina 29201-3356, USA
Tel: (+1 803) 758 45 00   Fax: (+1 803) 251 20 64
e-mail: intl@wilbursmith.com;
   mktg@wilbursmith.com
Web: www.wilbursmith.com

Postal address
PO Box 92, Columbia, South Carolina 29202-0092, USA

Key personnel

President: Hollis A Walker Jr
Regional Vice-President: D R Danforth
Senior Vice-President International Operations:
   Stephen W Schar
Vice-President Corporate Marketing:
   Gary A Schnelder
Director, Railways and Rail Transit: Glenn Michael
Public Relations Manager: Danielle M Gadow

Capabilities

Transportation consulting services covering rail, road, air and water systems, extending from planning, pre-feasibility and preliminary engineering through development of final design, contract documents, construction and maintenance services, training and technical assistance and field supervision. Management consultation services include planning, programming, budgeting and supervision of contractors and subcontractors. Services also include transit vehicle/station/facility design, design of architectural graphics and related visual communications systems, and interior space planning.

Projects

High Speed Rail: studies in California; Florida; Chicago, Illinois to Milwaukee, Wisconsin; Chicago, Illinois to St Louis, Missouri; the Pacific Northwest High Speed Rail Corridor; Oregon high speed rail capacity analysis; the Southeast high speed rail corridor improvement study.

Commuter Rail Studies: Ashtabula–Cleveland commuter rail study, Ohio; Virginia Railway Express capacity analysis; Contra Costa County commuter rail plan, California; Anchorage Metropolitan Area commuter rail study, Alaska; Florida Tri-County commuter rail extension study; Santa Clarita rail corridor study, California; Lincoln to Omaha rail passenger study, Nebraska.

Rail Freight Studies and Designs: State rail plans, grade crossing improvements, rail access to ports and industrial plants, track and structure evaluations, Intermodal Rail Yards; joint use evaluations, light density live evaluations; performance evaluations; merger studies.

Urban Rail Transit: feasibility studies; alternatives analysis; preliminary engineering and environmental studies; alternative financing plans; economic impact assessments.

Other Services/Experience: intermodal planning; geographic information systems; bridge inspection and rehabilitation; environmental services; underwater bridge inspections; highway/railway grade crossing surveys/closures; rail consolidation studies; and Public Private Partnerships.

UPDATED

Wilson, Ihrig & Associates

5776 Broadway, Oakland, California 94618, USA
Tel: (+1 510) 658 67 19
Fax: (+1 510) 652 44 41
Web: http://www.wiai.com

Key personnel
President: Steven L Wolfe

Capabilities
Acoustical design and vibration consulting with extensive experience in acoustical design of stations, line sections and facilities; vehicle noise, vibration and ride quality evaluation; assessment and prediction of ground-borne vibration; track fastener design, testing and specification; noise and vibration criteria development.

VERIFIED

YTT International Inc

2-33-4 Musashi-dai, Fuchu, Tokyo 183-0042, Japan
Tel: (+81 423) 28 15 15   Fax: (+81 423) 28 08 08
e-mail: niadanza@yttinc.co.jp

Key personnel
President: Yojiro Tawaragi
Vice-President: Shuhei Uchida
Directors: Iwao Yamamoto
General Manager, International Projects:
  Nicholas M Iadanza
Manager: Douglas W Martin
Administration: Sumiko Hanamura

Capabilities
Engineering, construction supervision and project management for rail vehicle projects; feasibility, planning, technical interface and co-ordination; rail vehicle specifications, inspection, testing, quality control and assurance, rail maintenance equipment, shop equipment, depot management and fare collection.

RAILWAY ASSOCIATIONS AND AGENCIES

RAILWAY ASSOCIATIONS AND AGENCIES

INTERNATIONAL

Arab Union of Railways (UACF)

PO Box 6599, Aleppo, Syria
Tel: (+963 21) 266 72 70; 56 11 Fax: (+963 21) 268 60 00
e-mail: uacf@scs-net.org

Key personnel
President of the Union: Eng Eid Abdel Kader Metwally, (Chairman of Egyptian Railways)
Vice-President of the Union: Eng Md Rabie Khlie (Director General des Ch de Fer Marocains)
Secretary General of the Union: Eng Mourhaf Sabouni (General Secretary of the Arab Union Railways)

Formed in 1979, the UACF stimulates co-operation between railways in Arab countries and co-ordinates their activities to ensure exchanges with each other and with international rail networks. Membership comprises railways in Algeria, Egypt, Iraq, Jordan (Aqaba and Hedjaz railways), Lebanon, Libya, Mauritania, Morocco, Saudi Arabia, Sudan, Syria (CFS and Syrian Hedjaz railways) and Tunisia, as well as a number of manufacturers, associations and railway-related organisations in several Arab countries.

The UACF also produces publications and stages events such as seminars and symposia, including a major biennial congress.

UPDATED

Association of Public Transport Operators

Union des Transports Publics (UTP)
5–7 rue d'Aumale, F-75009 Paris, France
Tel: (+33 1) 48 74 63 51   Fax: (+33 1) 40 16 11 72

Key personnel
President: Michel Cornil
Director General: T Soupault

The UTP represents the interests in France of domestic public transport operators.

VERIFIED

CLUB

Contactless Users Board
102 Esplanade de la Commune de Paris, F-93167 Noisy-Le-Grand Cedex, France
Tel: (+33 1) 43 03 93 59   Fax: (+33 1) 48 15 29 16
e-mail: helene.cron@contactless-club.assoc.fr
Web: http://www.contactless-club.com

Key personnel
Co-ordinator: Hélène Cron

Organisation
CLUB was established to enable transport operators to share experiences with the use of contactless smartcard techology for revenue collection. In 2003 membership included more than 160 operators.

NEW ENTRY

European Association for Railway Interoperability (AEIF)

221 avenue Louise, B-1050 Brussels, Belgium
Tel: (+32 2) 626 12 65 Fax: (+32 2) 626 12 61
Web: http://www.aeif.org

Technical and project office
66 boulevard de l'Impératrice, B-1000 Brussels, Belgium
Tel: (+32 2) 525 86 38   Fax: (+32 2) 525 96 39

Key personnel
Secretary General: Drewin Nieuwenhuis
Project Director: Bernard Alibert

Created in response to EU Directives, the AEIF is a 'joint representative body' drawn from European railway operating companies, infrastructure authorities and equipment suppliers to prepare technical specifications governing interoperability across the continent's networks. The UIC, UNIFE, UITP, UIP and UIRR are affiliate members. The organisation's work covers both high-speed and conventional networks.

European Company for the Financing of Railroad Rolling Stock (EUROFIMA)

Rittergasse 20, PO Box 1764, CH-4001 Basel, Switzerland
Tel: (+41 61) 287 33 40   Fax: (+41 61) 287 32 40
e-mail: info@eurofima.org
Web: www.eurofima.org

Key personnel
General Manager (Chief Executive Officer): Andre M Bovet
Senior Vice-Presidents: Bernard de Closset, Marco Termignone, Martin Fleischer

Shareholders
DB AG (23.70 per cent); SNCF (23.70 per cent); FS (13.50 per cent); SNCB (9.80 per cent); NS (5.80 per cent); RENFE (5.22 per cent); CFF (5.00 per cent); JŽ (Serbia and Montenegro) (2.30 per cent); SJ (2.00 per cent); CFL (2.00 per cent); ÖBB (2.00 per cent); CP (1.00 per cent); ČD (1.00 per cent); OSE (1.00 per cent); MÁV (0.50 per cent); HŽ (Croatia) (0.20 per cent); SŽ (Slovenia) (0.20 per cent); ŽBH (Bosnia and Herzegovina) (0.20 per cent); BDŽ (Bulgaria) (0.20 per cent); CFARYM (0.10 per cent); TCDD (0.04 per cent); DSB (0.02 per cent); NSB (0.02 per cent).

Activity
EUROFIMA finances railway equipment purchases for its shareholder national railways. Railway equipment is supplied to the national railways under equipment financing contracts. These contracts provide for periodical payments by the railways calculated to recover, over the life of a contract, repayments of principal and payments of interest on the funds borrowed, as well as EUROFIMA's expenses.

UPDATED

European Conference of Ministers of Transport (ECMT)

2 rue André Pascal, F-75755 Paris Cedex 16, France
Tel: (+33 1) 45 24 97 10   Fax: (+33 1) 45 24 97 42
e-mail: ecmt.contact@oecd.org
Web: http://www.oecd.org/cem

Key personnel
Secretary-General: Jack Short

Organisation
The European Conference of Ministers of Transport (ECMT) is an intergovernmental organisation established by a Protocol signed in Brussels on 17 October 1953. It is a forum in which ministers responsible for transport, and more specifically the inland transport sector, can co-operate on policy. Within this forum, ministers can openly discuss current problems and agree joint approaches for improving utilisation and rational development of European transport systems of international importance.

Member countries
Founding members (since 1953)
Austria, Belgium, Denmark, France, Germany, Greece, Italy, Luxembourg, Netherlands, Norway, Portugal, Spain, Sweden, Switzerland, Turkey, UK.

Other members
Albania, Armenia, Azerbaijan, Belarus, Bosnia Herzegovina, Bulgaria, Croatia, Czech Republic, Estonia, Finland, FYR Macedonia, Georgia, Hungary, Iceland, Ireland, Latvia, Lichtenstein, Lithuania, Malta, Moldova, Poland, Romania, Russian Federation, Serbia and Montenegro, Slovak Republic, Slovenia, Ukraine.

Associate members
Australia, Canada, Japan, Korea, New Zealand, USA.

Observer countries
Morocco.

VERIFIED

European Federation of Railway Trackwork Contractors (EFRTC)

2–4 route de Longwy, L-4830 Rodange, Luxembourg
Web: www.efrtc.org

Secretariat
ANIAF
Via Guattani 16, I-00161 Rome, Italy
Tel: (+39 06) 44 23 29 74   Fax: (+39 06) 84 56 75 55
e-mail: aniaf.99@flashnet.it

Key personnel
President: Jeremy Candfield
Secretary General: Roland Naggar

Organisation
An associate member of UNIFE, EFRTC promotes the common interests of specialist trackwork contractors in Europe. It has full members in Belgium, Estonia, France, Germany, Hungary, Italy, Luxembourg, the Netherlands, Portugal, Spain, Switzerland and the UK and an associate member in Bulgaria.

NEW ENTRY

European Rail Freight Association (ERFA)

Avenue de la Renaissance 1, B-1000 Brussels, Belgium
Tel: (+32 2) 733 55 63 Fax: (+32 2) 733 54 55
e-mail: monika.heiming@erfa.be
Web: www.erfa.be

Key personnel
President: Robert Spierings
Secretary-General: Monika Heiming

Organisation
Founded in 2003, ERFA is an association for European private and independent rail freight operators, representing the interests of companies wishing to take advantage of open access to the continent's networks.

NEW ENTRY

European Rail Infrastructure Managers Association (EIM)

21 rue de la Tourelle, B-1040 Brussels, Belgium
Tel: (+32 2) 234 37 70　Fax: (+32 2) 234 37 78
e-mail: info@eimrail.org
Web: http://www.eimrail.org

Key personnel
Secretary General: James Evans
Senior Technical Expert: Marc Falchi
External Communications: Susanne Kuschel
Information and Administration: Veerle Abeel

Formed in 2001, EIM represents the interest of rail infrastructure authorities in Europe at an international level.

Members
Banedanmark (Denmark)
Banverket (BV) (Sweden)
Gestor de Infraestructuras Ferroviarias (GIF) (Spain)
Jernbaneverket (JBV) (Norway)
Network Rail (UK)
ProRail (Netherlands)
Réseau Ferré de France (RFF) (France)
Ratahallintokeskus (RHK) (Finland)
Eurotunnel (France/UK)
REFER (Portugal)
AZP (Slovenia)

UPDATED

European Rail Research Advisory Council

66, Boulevard de l'Impératrice B-1000 Brussels, Belgium
Tel: (+32 2) 525 96 35/38　Fax: (+32 2) 525 96 39
Web: www.errac.org

Organisation
The European Rail Research Advisory Council (ERRAC) is an advisory body to the EU, representing Member States, the railway manufacturing and supply industry, rail operators and infrastructure managers, users, academia, environmental and urban planning organisation and the EU.

Its primary mission is to establish and carry forward a Strategic Rail Research Agenda that will influence all stakeholders in the planning of research programmes, particularly national and EU programmes. Its first meeting was held on 26 November 2001 in Cologne, Germany.

NEW ENTRY

European Rail Research Institute (ERRI)

Arthur van Schendelstraat 754, NL-3511 MK Utrecht, Netherlands
Tel: (+31 30) 232 42 52　Fax: (+31 30) 236 89 14
e-mail: rail_research@erri.nl
Web: http://www.erri.nl

Key personnel
Chairman of the Supervisory Board: Rod Muttram
Managing Director: Gunnar Gustafsson
General Manager: I Korpanec
Finance and Administration Director: R Hondelink
Technical Director: H Lagneau

Services
The European Rail Research Institute (ERRI) is a foundation under Dutch law within the International Union of Railways (UIC), carrying out research, studies and tests in fields of common interest. In 2000, ERRI had 32 European participants and 10 affiliates from 28 countries.

The Institute carries out collaborative research projects with various partners (railway operators, infrastructure owners and manufacturers, technical research centres, universities and so on) to boost the effectiveness of projects and allow

more flexibility and responsiveness to research requirements at a European level.

Research is targeted mainly at improved productivity, reduced operating costs, interoperability and environmental issues. Activities include the co-ordination of long-term research programmes.

Key skills include:
- Management and execution of studies, research programmes and test programes;
- Initiation and drafting of research and development programmes;
- Development and testing of railway equipment;
- Monitoring and application of emerging rail technologies.

ERRI has professional inhouse engineering and project management personnel, some directly employed and some seconded from participating railways under long-term contracts. Personnel for specific projects are also drawn from the engineering divisions of UIC railways, consultancies and universities. Individual specialists are sometimes contracted direct. ERRI uses the laboratories and testing facilities of the European railways for model and full-scale tests and demonstrations.

In 1998, ERRI launched a commercial branch, European Rail Services BV (ERS) (qv).

Subsidiaries
European Rail Services BV (ERS) is a wholly owned subsidiary of ERRI whose primary function is to provide ERRI products and services to non-UIC members.

ERSA (European Rail Software Applications) is a 60 per cent subsidiary of ERS. It specialises in the development and adaptation of simulation software for railway applications and also has the capability to carry out maintenance and support of other software packages.

European Rail Services BV (ERS)

Arthur van Schendelstraat 754, NL-3511 MK Utrecht, Netherlands
Tel: (+31 30) 232 48 23 Fax: (+31 30) 232 48 15
e-mail: ers@ers.erri.nl
Web: http://www.erri.nl

Key personnel
Managing Director: Bert Goote

European Rail Services BV handles all the commercial activities of the European Rail Research Institute's (ERRI), marketing its knowledge and experience in railway technology.

ERS products and services include reports and technical documents, drawings, software and consultancy. The client base is worldwide, encompassing mass transit and light rail authorities, academic institutions, manufacturers, infrastructure suppliers, freight and passenger operating companies, rolling stock leasing companies and consultancies. ERS also organises ERRI Interactive Conferences, promoting ERRI's R&D results to a wider audience.

In 1998, ERS set up a subsidiary, European Rail Software Applications (ERSA) (qv). ERSA's role is to develop software and to provide software services related to ERRI research projects.

Intergovernmental Organisation for International Carriage by Rail (OTIF)

Secretariat: Central Office for International Carriage by Rail (OCTI)
Gryphenhübeliweg 30, CH-3006 Berne, Switzerland
Tel: (+41 31) 359 10 10　Fax: (+41 31) 359 10 11
e-mail: info@otif.org

Key personnel
Chairman of Administrative Committee: M Aymeric

Director-General: S Schimming
Deputy to the Director-General: G E Mutz

UPDATED

International Air-Rail Organisation (IARO)

3rd Floor, 30 Eastbourne Terrace, London W2 6LE, UK
Tel: (+44 20) 87 50 66 32 Fax: (+44 20) 87 50 66 47
e-mail: int-airrail@baa.com
Web: http://www.airportrailwaysoftheworld.com
　　　http://www.iaro.com

Key personnel
Director-General: Andrew Sharp

The object of IARO is to spread world-class best practice and good practical ideas among people interested in rail links to airports. The organisation represents railways, airports, airlines and the supply industry. The first website address has been set up by IARO with the assistance of the International Air Transport Association (IATA) and includes all airports with a train service (and some with a bus-rail link).

Members include: Heathrow Express; Amsterdam Airport Schiphol; Manchester Airport; MTR Corporation Ltd, Hong Kong; Express Rail Link, Kuala Lumpur; Booz Allen Hamilton; the Chicago Transit Authority; and the Port Authority of New York and New Jersey. There is a regular newsletter,' Air Rail Express' and regular conferences and workshops.

UPDATED

International Association of Public Transport (UITP)

Rue Sainte Marie 6, B-1080 Brussels, Belgium
Tel: (+32 2) 673 61 00 Fax: (+32 2) 660 10 72
e-mail: communication@uitp.com
Web: www.uitp.com

Key personnel
President: Wolfgang Meyer
Vice-Presidents:
　Caetano Jannini Netto (São Paulo)
　Dmitry Gaev (Moscow)
　Johannes Sloth (Copenhagen)
　Dieter Ludwig (Karlsruhe)
　Jack C K So (Hong Kong)
　Enrico Mingardi (Rome)
　Tom Kaper (Den Haag)
　Ted Hesketh (Belfast)
　Richard J Simonetta (Atlanta)
Secretary-General: Hans Rat

UITP is the worldwide association of urban and regional passenger transport operators' authorities and suppliers. With over 2,000 members from nearly 80 countries, UITP promotes a better understanding of the potential of public transport and acts as the international network for all public transport professionals.

It acts as an international forum for the transport section to exchange information and ideas to further the position of public transport and is a platform for discussion between the industry, operators and authorities.

UPDATED

International Container Bureau (BIC)

Bureau International des Containers
167 rue de Courcelles, F-75017 Paris, France
Tel: (+33 1) 47 66 03 90　Fax: (+33 1) 47 66 08 91

Key personnel
General Secretary: J Rey

International Rail Transport Committee (CIT)

Weltpoststrasse 20, CH-3015 Bern 15, Switzerland
Tel: (+41 31) 350 01 90 Fax: (+41 31) 350 01 99
e-mail: info@cit-rail.org
Web: www.cit-rail.org

Key personnel
Secretary: Thomas Leimgruber

Founded in 1902, the International Rail Transport Committee (CIT) is a non- profit-making association under Swiss law. It has legal personality and has its headquarters in Bern. It is the railway organisation responsible for handling issues of international rail transport law.

Undertakings or associations of undertakings may become members if they provide transport in accordance with the Convention Concerning International Carriage by Rail (COTIF).

The CIT currently has as members about 102 railway organisations (carriers and infrastructure managers), shipping companies and road carriers from states that apply COTIF.

UPDATED

International Railway Congress Association (IRCA)

Association Internationale du Congrès des Chemins de Fer
Section 10, 85 Rue de France, B-1060 Brussels, Belgium
Tel: (+32 2) 520 78 31 Fax: (+32 2) 525 40 84
e-mail: secretariat@icref.org

Key personnel
Secretary General: A Martens (Former Director-General, Operations SNCB)

UPDATED

International Union of Railways (UIC)

Union Internationale des Chemins de Fer
16 rue Jean-Rey, F-75015 Paris, France
Tel: (+33 1) 44 49 20 20 Fax: (+33 1) 44 49 20 29
e-mail: (name)@uic.asso.fr
Web: /www.uic.asso.fr

Key personnel
Chairman: Benedikt Weibel (CFF/SBB)
Vice-Chairmen: Aad Veenman (NS)
Jean-Marie Bertrand (RFF) Ravindra Singh (Indian Railways)
Chief Executive: Philippe Roumeguère
Deputy Chief Executive: Werner Breitling
Communications Director: Paul Véron

The UIC is the world body for international railway o-operation, with more than 170 members on five continents. Its goals are to promote all forms of co-operation among railway ompanies, achieve technical harmonisation and interoperability and develop international rail transport at world level.

UPDATED

Latin American Railway Association (ALAF)

Asociación Latinoamericana De Ferrocarriles
Avda Belgrano 863, ler piso, 1092 Buenos Aires, Argentina
Tel: (+54 11) 43 42 72 71
e-mail: alaf@alaf.int.ar

Key personnel
Secretary General: Dr Lauro Ramírez López
Technical Secretary: Arnaldo Ercoli
Administration Secretary: Dr Jorge Gutracht

Background
Formed in 1964, ALAF is a non-governmental organisation representing railway undertakings and their supplies in Latin America. It is also the region's representative body on the UIC's world council.

UPDATED

Organisation for the Collaboration of Railways (OSJD)

UI Hoza 63/67, Warsaw PL-00 681, Poland
Tel: (+48 22) 657 36 54   Fax: (+48 22) 621 94 17
e-mail: osjd@osjd.org.pl
Web: www.osjd.org

Key personnel
Committee Chairman: Tadeusz Szozda
Secretary: Rastislav Chovan

UPDATED

Pan American Railway Congress Association (ACPF)

Asociación del Congreso Panamericano de Ferrocarriles
Casilla de Correo 129 Suc 1, 1332 Buenos Aires, Argentina
Tel: (+54 11) 49 81 06 25   Fax: (+54 11) 48 14 18 23
e-mail: acpt@nat.com.ar

Key personnel
President: Major General Eng Juan Carlos De Marchi
First Vice-President: Arq Eduardo Santos Castillo
General Secretary: Ing Alfredo Fernandez
Treasurer: Dr Ricardo S Tawil
Special Adviser: Dr Adalberto Rodríguez Giavarini (Director of International Organisation, Ministry of Foreign Affairs, Argentina)

The Community of European Railway and Infrastructure Companies (CER)

53 Avenue des Arts, B-1000 Brussels, Belgium
Tel: (+32 2) 213 08 70   Fax: (+32 2) 512 52 31
e-mail: contact@cer.be
Web: www.cer.be

Organisation
The Community of European Railway and Infrastructure Companies (CER) brings together 40 railway undertakings and infrastructure organisations from the European Union, the accession countries (Bulgaria, Croatia and Romania, as well as Bosnia-Herzegovina, Norway, Serbia and Montenegro and Switzerland).

All policy areas of significance to railway transport are dealt with by CER, which offers advice and recommendations to European policymakers. CER's interests span the whole spectrum of European transport policy: infrastructure planning, passenger and freight services, public service, the environment, research and development, and social dialogue.

NEW ENTRY

Union of African Railways (UAR)

Avenue Tombalbaye 869, PO Box 687, Kinshasa, Zaire
Tel: (+243 12) 238 61   Fax: (+243 12) 251 66

Key personnel
President: Hanson Sindowe
Vice-Presidents
    North Africa: Sudan Railway Corporation
    West Africa: Mali
    East Africa: Djibouti-Ethiopia Railways
    Central Africa: Congo
    Southern Africa: Tazara
General Secretariat (staff)
    Secretary-General: Robert G Nkana
    Administration and Finance: Canute Peter Shengena
    Translations: Nsanbu Seke

Association of European Railway Industries (UNIFE)

Avenue Louise 221 Bte 11, B-1050 Brussels, Belgium
Tel: (+32 2) 626 12 60 Fax: (+32 2) 626 12 61
e-mail: mail@unife.org
Web: www.unife.org

Key personnel
General Manager: Drewin Nieuwenhuis
Events and Information Officer: Corinne Dhainaut

The Association of European Railway Industries is an industrial organisation representing its members' interest towards the European institutions, rail operator associations and other business relations. It represents 80 of the largest and medium-sized companies of the railway supply industry and a further 800 suppliers of railway equipment are associated members through their national associations.

UPDATED

NATIONAL

Argentina

Chamber of Railway Industries

Cámara de Industriales Ferroviarios
Alsina 1607, Buenos Aires
Tel: (+54 1) 40 49 67; 55 71
Fax: (+54 1) 40 49 09 58

Key personnel
President: Eng E G Nottage
Secretary: E R Paduto

Members
Active members are Artimsa SA, Est Met A Longo SA, Ferromec SA, Petro Parts SA, Saft Nife Argentina SA, Servotron SA, Siderea SAIC yA and Siemens SA.

Australia

Australian Railway Industry Corporation Ltd (ARIC)

GPO Box 5301, Sydney, New South Wales 2001, Australia
Tel: (+61 2) 93 20 21 45   Fax: (+61 2) 93 90 21 05
e-mail: aric@aric.com.au
Web: www.aric.net.au

Key personnel

Chairman: Steven D Burraston
Executive Officer: Ian Robb

Organisation

ARIC is a non-profit, private sector organisation comprising members from the Australian railway supply industry as well as Australian rail operators, with some 'observer members' by virtue of their status in the industry. The organisation represents the interest of its members by promoting their products and services in world markets.

NEW ENTRY

Australasian Railway Association Inc

PO Box 4864, Kingston ACT 2604
Tel: (+61 2) 62 70 45 00 Fax: (+61 2) 62 73 55 81
e-mail: ara@ara.net.au
Web: www.ara.net.au

Office

Engineering House, Unit 17, Level 3, 11 National Circuit, Barton ACT 2600

Key personnel

Chairman: Stephen O'Donnell (Pacific National)
Chief Executive Officer: Bryan Nye
Government Relations: Phil Sochon
Code of Practice Manager: John Shalders
Manager Policy: Kathryn Rayner

The Australasian Railway Association Inc (ARA) is the leading industry body for the rail sector in Australia and New Zealand. It represents the interests of both private and government-owned rail operators in freight, passenger and tourist/heritage sectors, track owners, manufacturers of locomotives and rolling stock, suppliers of signalling and communications systems, maintenance and construction companies, freight forwarders, investment banks, legal firms, information technology and service providers, consultants and rail unions. Founded in 1994, the ARA currently has 166 members.

UPDATED

Rail Track Association of Australia

PO Box 346, Cherrybrook, New South Wales 21268, Australia
Tel: (+61 2) 671 65 55   Fax: (+61 2) 671 78 75

Key personnel

President: Bill Killinger
Vice-President and Secretary Treasurer: David Bull
Vice-President: Dave Hassall

Organisation

The objectives of the Rail Track Association Australia include the bringing together of specialists concerned with the design, construction, fabrication, operation and maintenance of railway tracks. To foster collaboration between railways, suppliers of track components and equipment and those providing services in the fields of design, consultancy or contracting. To seek participation of government and semi-government departments and bodies, universities, institutions and research facilities with a view to encouraging the optimisation of railway operations, especially in relation to railway track performance economics.

To initiate sponsor and or co-operate with others in research development and evaluation of railway track and its components. To sponsor the development of realistic standards in respect of track, its components and the practices employed in construction, operation and maintenance. To conduct meetings for the purpose of bringing together entities interested in the railway and its associated disciplines. To disseminate and seek out technical and other data from members and other interested entities. To promote the development and improvement of railways particularly in respect of railway track.

UPDATED

Railway Technical Society of Australasia

11 National Circuit, Barton, ACT 2600 Australia
Tel: (+61 62) 70 65 55   Fax: (+61 62) 73 14 88
e-mail: jarmstrong@eol.ieaust.org.au

Key personnel

National Chairman: Professor Philip Laird
Sydney Chapter Chair: Les McNaughton
Queensland Chapter Chair: Dr Luis Ferreira
Victoria Chapter Chair: D J Ferris
SA Chapter Chair: John Adams
Western Australian Chapter Chair: Shane Hinchliffe

The Society's mission is to provide a learned society and communal functions for individuals and groups in the railway industry and to provide practice-based opinion and advice for The Institution of Engineers, Australia.

VERIFIED

Austria

Federation of Cable Railways

Fachverband der Seilbahnen
PO Box 172, Wiedner Hauptstrasse 63, A-1045 Vienna
Tel: (+43 5) 909 00 31 66 Fax: (+43 5) 90 90 02 42
e-mail: seilbahnen@wko.at

Key personnel

Director: Dipl Ing Dr Ingo Karl
Manager: Dr Erik Wolf

UPDATED

Federation of Private Railways

Fachverband der Schienenbahnen
PO Box 172, Wiedner Hauptstrasse 63, A-1045 Vienna
Tel: (+43 5) 909 00 31 65 Fax: (+43 5) 90 90 02 42
e-mail: schienenbahnen@wko.at

Key personnel

General Director: Dr Csaba Szekely
Manager: Mag Robert Woppel

UPDATED

Brazil

National Public Transport Association (ANTP)

Associação Nacional de Transportes Públicos – ANTP
Alameda Santos 1000 – 7 andar, 01418-100 São Paulo
Tel: (+55 11) 33 71 22 99   Fax: (+55 11) 32 53 80 95
e-mail: antpsp@antp.org.br
Web: www.antp.org.br

Key personnel

President: Juarandir F R Fernandes
General Director: Nazareno Stanislau Affonso
Deputy Executive Directors: Cristina Maria Baddini Lucas, Eduardo Vasconcellos
Head of Finance: Francisco A Aquino

Organisation

Founded in 1977, ANTP's membership is drawn from Brazilian public and private sector urban transport operators, their suppliers, trade unions, other transport associations and universities and research institutions. Its aims include the development and diffusion of expertise and the proposal of policies and projects.

UPDATED

Canada

Association of Regional Railways of Canada

68 Robertson Road, Suite 105, Ottawa, Ontario K2H 8P5, Canada
Tel: (+1 613) 726 71 11   Fax: (+1 613) 726 71 39
Web: http://www.arrc.ca

Vancouver office
1080 Howe Street, Suite 700, Vancouver, British Columbia V6Z 2T1, Canada
Tel: (+1 604) 899 15 69   Fax: (+1 604) 488 14 89

Key personnel
President: Bob Ballantyne
Vice-Presidents: Cindy Hick, Forrest Hume

Organisation
The ARRC was formed in 2001 to represent the interests of regional railways and short lines in Canada.

UPDATED

Canadian Association of Railway Suppliers

PO Box 4459, Ottawa, Ontario K1S 5B4, Canada
Tel: (+1 613) 237 38 88   Fax: (+1 613) 237 48 88
e-mail: info@railwaysuppliers.ca
Web: www.railwaysuppliers.ca

Key personnel
President: Gord Patterson
Vice-President: Ric Duncan
Treasurer: Susan Barrie

Organisation
With over 100 members, the Canadian Association of Railway Suppliers represents the interests and enhances the visibility of companies supplying goods and services to Canadian railways.

UPDATED

Canadian Urban Transit Association

55 York Street, Suite 1401, Toronto, Ontario M5J 1R7, Canada
Tel: (+1 416) 365 98 00   Fax: (+1 416) 365 12 95
e-mail: transit@cutaactu.ca
Web: www.cutaactu.ca

Key personnel
President and Chief Executive Officer:
  Michael Roschlau

Ottowa office
1500 St Laurent Blvd, Ottawa, Ontario K1G 0Z8

Organisation
CUTA's mission is to establish public transit as the primary solution to urban mobility and to assist its members in the fulfilment of their mandates. A high number of urban transit systems in Canada from Victoria and Vancouver to Halifax and St John's are members. Membership includes 100 transit systems, 15 government agencies (federal, provincial and municipal), 250 business members (the firms or persons engaged in the manufacture or sale of transit equipment or services) including consultants, and 50 affiliates.

UPDATED

The Railway Association of Canada

99 Bank Street, Suite 1401, Ottawa, Ontario K1P 6B9
Tel: (+1 613) 567 85 91 Fax: (+1 613) 567 67 26
e-mail: rac@railcan.ca
Web: http://www.railcan.ca

Key personnel
Chairman: R J Ritchie
Vice-Chairman: P M Tellier
President and Chief Executive Officer: W A Rowat

Services
The Railway Association of Canada (RAC) represents 60 member freight, commuter, intercity, passenger and tourist railways, playing a major role in promoting the safety, viability and growth of the railway industry within Canada. The RAC's dedicated team of professionals co-ordinates the development of rules and recommended practices pertaining to operations and safety, which have made rail the safest mode of surface transport. RAC staff conduct the research, policy development and advocacy necessary to lobby all levels of government and transport-related businesses to promote rail's advantages and ensure a fair treatment among other modes.

The Association publishes Railway Trends, an annual statistical digest of the Canadian rail industry based on Canadian data, research reports, background and position papers, newsletters, an annual report and produces industry-related videos.

The Association's Institute for Railway Technology (IRT) works with 10 colleges throughout Canada, to establish pre-employment training programmes for the industry, which employs 40,000 increasingly skilled men and women in freight and passenger operations.

Member companies
Acadian Railway Trains LP
Agence métropolitaine de transport
Alberta Prairie Railway Excursions
Alberta RailNet Inc
Amtrak
Arnaud Railway Company
Athabasca Northern Railway Ltd
Barrie-Collingwood Railway
Burlington Northern (Manitoba) Ltd
Canadian National Railway
Canadian Pacific Railway
Cape Breton & Central Nova Scotia Railway
Capital Railway
Carlton Trail Railway Company
Cartier Railway Co
Central Manitoba Railway Inc
Central Western Railway
Charlevoix Railway Company Inc
Chemin de fer de la Matapédia et du Golfe Inc
CN
CSX Transportation Inc
E&N Railway Company (1998) Ltd
Essex Terminal Railway Company Ltd
Ferroequus Railway Company Ltd
Goderich-Exeter Railway Co Ltd
Great Canadian Railtour Company Ltd
Great Western Railway Ltd
Hudson Bay Railway
Huron Central Railway Inc
Kelowna Pacific Railway
Lakeland & Waterways Railway
Mackenzie Northern Railway
Montreal, Maine & Atlantic Railway, Ltd
New Brunswick East Coast Railway Inc
New Brunswick Southern Railway Company Ltd
Norfolk Southern Corporation
Okanagan Valley Railway
Ontario Northland Transportation Commission
Ontario Southland Railway Inc
Ottawa Central Railway Inc
Ottawa Valley Railway
Prairie Alliance for the Future
Québec Gatineau Railway Inc
Roberval and Saguenay Railway Company, The
South Simcoe Railway
Southern Manitoba Railway
Southern Ontario Railway
Southern Railway of British Columbia Ltd
St Lawrence & Atlantic Railroad (Québec) Inc
Sydney Coal Railway
Toronto Terminals Railway Company Ltd, The
Train Touristique L'Express de la Matapédia
Trillium Railway Company Ltd
VIA Rail Canada Inc
Wabush Lake Railway Company, Ltd
West Coast Express Ltd
White Pass & Yukon Route
Windsor & Hantsport Railway
Wisconsin Central Ltd

France

Fédération des Industries Ferroviaires (FIF)

12 rue Bixio, F-75007 Paris, France
Tel: (+33 1) 45 56 13 53 Fax: (+33 1) 47 05 29 17
Web: www.industrie-ferroviaire.com

Key personnel
President: Jean-Marie Bockel
General Delegate: J P Audoux

UPDATED

Public Transport Association (UTP)

Union des Transports Publics, 5-7 rue d'Aumale, F-75009 Paris, France
Tel: (+33 1) 48 74 63 51 Fax: (+33 1) 40 16 11 72
e-mail: lettre@utp.fr

Key personnel
President: Michel Cornil
Vice-President: Antoine Frérot; Philippe Segretain
Secretary General: Thierry Soupalt

UTP membership comprises rail and road public transport operators in France. The organisation protects the industry's interests and represents it domestically and at a European level.

SYCAFER

Groupement des Installations Ferroviaires Fixes (French Track Suppliers and Contractors Association)
12 rue Bixio, F-75007 Paris, France
Tel: (+33 1) 47 77 00 55 Fax: (+33 1) 47 05 52 49
e-mail: yvon.estelle@cofiger.fr

Key personnel
President: Jean-Louis Wagner
Vice-President: Gilles du Fou
General Secretary: Yvon Estellé
Deputy General Secretary: Dany Dupont-Weider

Germany

Association of German Transport Undertakings (VDV)

Kamekestrasse 37-39, D-50672 Cologne, Germany
Tel: (+49 221) 57 97 90   Fax: (+49 221) 51 42 72
e-mail: info@vdv.de
Web: http://www.vdv.de

Subsidiaries
VDV-Geschäftsstelle Berlin
Strasse des 17 Juni, D-10623 Berlin
Tel: (+49 30) 39 99 32-0   Fax: (+49 30) 39 99 32 15
e-mail: vdv-berlin.t-online.de
Manager: Dr Ing Martin Runkel
VDV-Geschäftsstelle Brüssel
c/o UITP, Brussels
Manager: Klaus J Meyer

Key personnel
President: Dr-Ing Eh Dieter Ludwig
Vice-Presidents:
   Dipl Kfm Günter Elste (Passenger Traffic)
   Dr Rolf Bender (Freight Traffic)
General Manager:
   Professor Dr-Ing Adolf Müller-Hellmann
Manager: Dr Thomas Muthesius (Passenger Traffic)
Manager: Dr Martin Henke (Freight Traffic)
Head of Marketing, Press and PR Department:
   Dipl-Volksw Friedhelm Bihn
   Dipl G Stephan Dnemüller (Freight Traffic)

Services
Trade association of the German public transport industry. Represents, among others, 51 'private', non-Deutsche Bahn, passenger railways and 147 freight, port and industrial railways. Very few railways still offer both passenger and freight services.   ***VERIFIED***

Association of Privately Owned Wagon Operators

Vereinigung der Privatgüterwagen Interessenten (VPI)
Schauenburger Strasse 52, D-20095 Hamburg, Germany
Tel: (+49 40) 450 50 86 Fax: (+49 40) 450 50 90
e-mail: mail@vpihamburg.de

Key personnel
President: Jürgen Hüllen
Vice-Presidents: Jürgen Bauer, Herbert Raab, Dieter Trapp
General Manager: Henning Traumann

UPDATED

Association of the Railway Supply Industry in Germany

Verband der Bahnindustrie in Deutschland (VDB) eV
Jägerstrasse 65, D-10117 Berlin
Tel: (+49 30) 20 62 89-0 Fax: (+49 30) 20 62 89 50
e-mail: info@bahnindustrie.info
Web: www.bahnindustrie.info

Key personnel
President: Dipl Ing Peter Witt
Directors: Dipl Ing Joachim Körber,
   Dipl Wi-Ing Norbert G Liebler

The organisation, which represents the railway supply industry in Germany, was formerly the German Railway Supply Industry Association.   ***UPDATED***

Rolled Steel Association Long Products Division/Railway Material

Walzstahl Vereinigung Abt Profilstahl
Sohnstrasse 65, D-40237, Düsseldorf
Tel: (+49 211) 670 71 88 Fax: (+49 211) 670 79 32

Key personnel
Secretary: H Bauer

Switch and Crossing Manufacturers Association

Fachverband Weichenbau
PO Box 1020, D-58010 Hagen
Tel: (+49 2331) 20 08; 29 Fax: (+49 2331) 20 08; 28

Key personnel
Chairman: Eckhard Bittel

Italy

ANIE

Italian Association of Electrotechnical and Electronics Industries
Via Algardi 2, I-20148 Milan, Italy
Tel: (+39 02) 326 41 Fax: (+39 02) 326 42 12

Key personnel
Chairman: Ing Gio Batta Clavarino
General Secretary: Ing Lorenzo Tringali-Casanuova

College of Italian Railway Engineers

Collegio Ingegneri Ferroviari Italiani
Via G Giolitti 34, I-00185 Rome
Tel: (+39 06) 488 21 29 Fax: (+39 06) 474 29 87
e-mail: mol1958@mclink.it

Key personnel
President: Dr Ing E Maestrini
Secretary: Dr Ing B Cirillo

Japan

Japan Overseas Rolling Stock Association (JORSA)

Tekko Building, 1-8-2 Marunouchi, Chiyoda-ku, Tokyo 100-0005
Tel: (+81 3) 32 01 31 45 Fax: (+81 3) 32 14 47 17
e-mail: infoweb@jorsa.or.jp

Key personnel
Senior Managing Director: S Suzuki
Director, Administration: Y Kurasawa

Established in 1953 the association is an umbrella organisation of 25 Japanese traders and manufacturers and promotes trading relations between members and railway operators and users.   ***UPDATED***

Japan Railway Engineers' Association

Tani Building, 1-28-6 Kameido, Kohtoh-ku, Tokyo 136-0071
Tel: (+81 3) 56 26 23 21 Fax: (+81 3) 56 26 23 25

Key personnel
Chairman: Koichi Sakata
Deputy Chairmen: Daisuke Takei (President, Railway Information Systems Co Ltd), Yukio Atimoti (Managing Director, East Japan Railway Co), Dr Eng Misao Sugawara (Tokyo University of Science)
Executive Director: Yoshio Ishiguro

Japan Society of Mechanical Engineers

Shinanomachi-Rengakan Building, Shinanomachi 35, Shinjuku-ku, Tokyo 160-0016
Tel: (+81 3) 53 60 35 00 Fax: (+81 3) 53 60 35 08
e-mail: library@jsme.or.jp
Web: www.jsme.or.jp

Key personnel
Secretary: K Fukuzawa

UPDATED

Railway Electrical Engineering Association of Japan

Kimigayo Building 4F, 3-20-15 Asakusabashi Taito-ku, Tokyo 111-0053
Tel: (+81 3) 38 61 86 78 Fax: (+81 3) 38 61 85 06
e-mail: info@rail-e.or.jp

Key personnel
President: Masanori Ozeki
Vice-Presidents: Nobuaki Maruyama, Tatsuyuki Enomoto, Shinichiro Otsuka
Senior Managing Director: Tatsumi Honda
Managing Directors: Hiroto Yasuhara, Sakuro Tsukuda

Netherlands

Holland Rail Industry

Boerhaavelaan 40, NL-2700 AD Zoetermeer
PO Box 190, NL-2700 AD Zoetermeer
Tel: (+31 79) 353 12 44
Fax: (+31 79) 353 13 65
e-mail: hri@fme.nl
Web: http://www.hollandrailindustry.nl

Key personnel
President: J A Pijnappels

Organisation
Holland Rail Industry is an association of manufacturers, engineers and consultants in the railway sector. It is an industrial branch association of FME-CWM (enterprises in the mechanical, metalworking, plastics, electronic and electrical engineering industries and allied sectors).

The objective of Holland Rail Industry is to strengthen the position of manufacturers of equipment for all sectors of railway activity, on the national and international market.

VERIFIED

South Africa

Railroad Association of South Africa (RRA)

PO Box 103, Maraisburg 1700, South Africa
Tel: (+27 11) 818 34 85
Fax: (+27 11) 474 35 82
e-mail: jit-rra@mweb.co.za
Web: http://www.rra.co.za

Key personnel
President: Ivor Evans
Immediate Past President: Gorman Zimba
Chief Executive Officer: John Thompson
Media and Research Officer: Allen Jorgensen

Organisation
The RRA, the membership of which includes both railway operators and their suppliers, acts as a voice for the industry nationally and provides it with a link to public authorities.

Spain

Association of Spanish Manufacturers of Railway Equipment

Asociación Nacional de Constructores Españoles de Material Ferroviario (Cemafe)
Príncipe de Vergara 74, 4a planta, E-28006 Madrid
Tel: (+34 91) 562 15 52
Fax: (+34 91) 562 19 22
e-mail: cemafe@cemafe.com
Web: http://www.cemafe.com

Organisation
Established in 1980, Cemafe represents and co-ordinates the interests of Spanish rolling stock and railway equipment manufacturing companies at national and international levels.

Institution of Spanish Railways

Fundación de los Ferrocarriles Españoles
Palacio de Fernán Núñez, C/ Santa Isabel 44, E-28012 Madrid, Spain
Tel: (+34 91) 527 61 72
Web: http://www.ffe.es

Key personnel
Director: Carlos Zapatero Ponte

The aims of the institution include: increasing the knowledge and use of rail; sponsoring studies into the rail transport mode; promotion of the socio-economic benefits and cultural importance of rail; and the conservation of the heritage of the railway in Spain. The institution's patrons include: RENFE; GIF; FEVE; FGC; and EuskoTren.

Spanish Private Wagon Owners Association

(Spanish Private Wagon Owners Association)
Asociación de Propietarios de Vagones de España
Juan Alvarez Mendizábal 30, 4° Centro, E-28008 Madrid
Tel: (+34 91) 547 82 86
Fax: (+34 91) 547 82 86

Key personnel
President: D Emilio Fernández Fernández
Vice-President: Bruno Torresano Guerreiro
Secretary: D Pablo Rodríguez Mosquera

Principal member companies
Cementos Alfa SA
LTF
Saltra
SEmat SA
Transfesa
Tudela Veguin SA
Member companies own 5,811 wagons, predominantly covered vehicles, tank wagons, car transporters and intermodal flat wagons. The association offers integrated transport solutions to meet logistics needs of the industry with interchangeable-axle rolling stock, swap bodies and trucks.

Sweden

Association of Swedish Train Operators

Branschföreningen Tågopertörerna
Box 16105, Blasieholmsgatan 5, SE-103 22 Stockholm, Sweden
Tel: (+46 8) 762 67 08
Fax: (+46 8) 762 67 09
e-mail: info@tagoperatorerna.se
Web: http://www.tagoperatorerna.se

Key personnel
Chairman: Sven-Olof Nehrer
Deputy Chairman: Lars Yngström
Chief Executive: Peder Wadman

Organisation
With a membership of some 20 companies, the association acts as a national representative body for train operators in Sweden. Its activities are carried out through its wholly owned subsidiary, Sveriges Tågoperatörer Service AB.

Switzerland

Swissrail Industry Association

Effingerstrasse 8, PO Box 7948, CH-3001 Bern
Tel: (+41 31) 398 50 50 Fax: (+41 31) 398 55 55
e-mail: admin@swissrail.com
Web: www.swissrail.com

Key personnel
General Manager: Walter Graeppi

Organisation
The Swissrail Industry Association, previously called Swissrail Export Association, was founded in 1977. It is a private law-based association with more than 70 member companies active abroad as well as in Switzerland, including leading consultants and industrial companies and a high number of innovative medium and small size enterprises in the field of public transport systems. Swiss Federal Railways, public transport authorities, including associated private railways and urban transport operators, the Federal Office of Transport and the Institute of Transportation, Traffic, Highway and Railway Engineering are supporting Swissrail activities.

Swissrail is the export promotion organisation for the Swiss railway industry.

UPDATED

United Kingdom

Association of Community Rail Partnerships (ACoRP)

Rail and River Centre, Slaithwaite Civic Hall, 15a New Street, Slaithwaite, Huddersfield, West Yorkshire HD7 5AB, UK
Tel: (+44 1484) 84 77 90 Fax: (+44 1484) 84 78 77
e-mail: info@acorp.uk.com
Web: www.acorp.uk.com

Key personnel
Chairman: Peter Roberts
General Manager: Paul Salveson

Organisation
ACoRP comprises some 40 rail partnerships created to co-ordinate local and community involvement with the running of rural and semi-rural railways in the UK. A not-for-profit association, it provides impartial advice and support for the development and provision of responsive and good quality services. ACoRP is funded by the Association of Train Operating Companies, the Countryside Agency, Esmée Fairbairn Foundation and the Strategic Rail Authority.

UPDATED

Centre for Rail Skills CFRS

PO Box 39685 London, W2 1XR, UK
Tel: (+44 0845) 345 27 00 Fax: (+44 0845) 345 27 18
Web: http://www.cfrs.org.uk

Key personnel
Chief Executive: Jackie Chappell
Business Development Manager: Angela Byrne

Capabilities
National occupational standards of competence, covering engineering and operations activities; national and Scottish vocational qualifications; advice on compliance with railway safety-critical work regulations; advice on implementation of competence management systems.

The CFRS membership list includes Iarnród Éirann, Ireland; in the UK members include Anglia Railways, Central Trains, Chiltern Trains, Direct Rail Services, Docklands Light Railway, Eurostar, EW&S, GNER, London Transport, Merseyrail Electrics, NEXUS, Tyne & Wear, Racal Telecommunications, Railtrack, ScotRail, Serco Metrolink, Stagecoach Supertram (Sheffield), South West Trains, Translink (Northern Ireland Railways), Virgin Trains, WAGN.

Engineering companies belonging to RITC include Amec, First Engineering, Centrac, Adtranz, Balfour Beatty Rail Renewals (Track Systems), Jarvis Training Management and Jarvis Fastline.

Associate members are Spoornet (South African Railways) and, in the UK, Heritage Railways, Creative Training Services, Four Counties Training and South Coast Training.

UPDATED

Derby and Derbyshire Rail Forum

Roman House, Friar Gate, Derby DE1 1XB, UK
Tel: (+44 1332) 71 63 76   Fax: (+44 1332) 71 63 77
e-mail: info@derbyrailforum.org.uk
Web: www.derbyrailforum.org.uk

Key personnel
Co-ordinator: Sarah Wolfe

Organisation
Derby and Derbyshire Rail Forum was established in 1993 to provide inward and outward trade missions, railway related trade exhibitions for Derby Rail Forum members, representation of Rail forum members as a whole and networking opportunities for the British rail industries. The Forum has a membership of 83 private and public sector organisations.

NEW ENTRY

The Chartered Institute of Logistics and Transport (UK)

Logistics and Transport Centre
Earlstrees Court, Earlstrees Road, Corby NN17 4AX, UK
Tel: (+44 1536) 7401 00 Fax: (+44 1536) 74 01 01
e-mail: enquiry@ciltuk.org.uk
Web: www.cilt.org.uk

London office
11/12 Buckingham Gate, London SW1E 6LB

Key personnel
Chief Executive: Geoff Newton
Director of Policy: Jonathan Bullock

Services
The Chartered Institute of Logistics and Transport (UK) (CILT) is the professional membership body for individuals and organisations involved in all disciplines, modes and aspects of logistics and transport. The Institute's 22,000 members have privileged access to a unique range of benefits and services, which support them personally and professionally throughout their careers and help connect them with worldwide expertise.

UPDATED

Institution of Civil Engineers

One Great George Street, London SW1P 3AA
Tel: (+44 20) 72 22 77 22 Fax: (+44 20) 72 22 75 00
e-mail: communications@ice.org.uk
Web: www.ice.org.uk

Key personnel
President: Colin J Clinton
Director General: Tom Foulkes
Marketing Manager: Andrew McMillan

UPDATED

Institution of Diesel and Gas Turbine Engineers

Bedford Heights, Manton Lane, Bedford MK41 7PH
Tel: (+44 1234) 24 13 40 Fax: (+44 1234) 35 54 93
e-mail: enquiries@idgte.org
Web: http://www.idgte.org

Key personnel
Director General: J H Blowes

VERIFIED

Institution of Electrical Engineers (IEE)

Savoy Place, London WC2R 0BL
Tel: (+44 20) 72 40 18 71 Fax: (+44 20) 72 40 77 35
e-mail: postmaster@iee.org.uk
Web: http://www.iee.org

Key personnel
Chief Executive: Dr A Roberts

VERIFIED

Institution of Mechanical Engineers Railway Division

1 Birdcage Walk, London SW1H 9JJ
Tel: (+44 20) 73 04 68 40 Fax: (+44 20) 79 73 01 82
e-mail: railway@imeche.org.uk
Web: www.imeche.org.uk/railway

Key personnel
Chairman: Andrew Lezala
Executive Officer: Sharon Dowds
Executive Assistant: Joanne Collett

The Railway Division of the Institution of Mechanical Engineers (IMechE) is one of eight engineering division of the institution and was formed in 1969 on the amalgamation of the former Institution of Locomotive Engineers with the IMechE.

The division consists of IMechE members of all grades professionally involved or interested in the science and practice of railway engineering. Its scope covers research, design, development, procurement, manufacture, operation, maintenance and disposal of traction, rolling stock, fixed equipment and their components, within rail, rapid transit and all forms of guided surface transport.

UPDATED

The Institution of Railway Operators

PO Box 128, Burgess Hill RH14 0UZ
Tel: (+44 1444) 24 89 31 Fax: (+44 1444) 24 63 92
e-mail: info@railwayoperators.org
Web: www.railwayoperators.org

Key personnel

Chairman: Richard Morris
Deputy Chairman: Chris Leah
Chief Executives: Chris Daughton

The objectives of The Institution of Railway Operators are to advance, for the benefit of the general public, the safe and reliable operation of the railways by improving the technical and general skills, knowledge and competence, and promoting the training of people engaged in the operation of railways in the UK.

UPDATED

Institution of Railway Signal Engineers (IRSE)

Savoy Hill House, Savoy Hill, London WC2R 0BS
Tel: (+44 20) 72 40 32 90
Fax: (+44 20) 72 40 32 81
Web: http://www.irse

Key personnel

Chief Executive: K W Burrage

The Institution of Railway Signal Engineers (IRSE) is the professional institution of all those engaged in, or associated with, railway signalling, telecommunication and allied professions. Founded in 1912, the Institution aims to advance, for the public benefit, the science and practice of signalling and telecommunications engineering within the industry and to maintain high standards of knowledge of the profession amongst the membership. The IRSE is active within Europe and worldwide as well as in the UK.

VERIFIED

The Locomotive & Carriage Institution

34 Camp Street, Derby DE1 3SD, UK
Tel: (+44 1332) 29 53 78   Fax: (+44 7952) 65 07 23
e-mail: alan.spencer@lococarriage.org.uk
Web: www.lococarriage.org.uk

Key personnel

President: Nick Agnew (Transport for London)
Vice-Presidents: Andrew Haines (MD UK Rail Division, First Group), Willi Frauenfelder (BLS (Bern Lotschbergbahn)
General Secretary: Alan Spencer (Fragonset Merlin Rail)
Membership Secretary: Peter Lindop (West Coast Traincare, ALSTOM)
Treasurer: Brian Ashfeld (Network Southeast, retired)

Organisation

The Institution holds monthly meetings/seminars (first Tuesday of each month, September – May)

covering many railway topics and arranges tours of railway and non-railway sites in the UK and elsewhere in Europe.

UPDATED

Permanent Way Institution

11 Caraway Place, Meir Park, Stoke-on-Trent, Staffordshire ST3 7FE, UK
Tel: (+44 1782) 39 78 80   Fax: (+44 1782) 39 75 46
e-mail: pwi.bjn@virgin.net
Web: www.permanentwayinstitution.com

Key personnel

President: Richard Spoors (2005–2006)
Secretary: Brian J Newman

UPDATED

Private Wagon Federation

Homelea, Westland Green, Little Hadham SG11 2AG
Tel: (+44 1279) 84 34 87   Fax: (+44 1279) 84 23 94

Key personnel

Chairman: John Jagger
Secretary-General: G Pratt

VERIFIED

Rail Civil Engineers' Association

One Great George Street, Westminster London SW1P 3AA, UK
Tel: (+44 20) 76 65 22 31   Fax: (+44 20) 77 99 13 25
e-mail: rcea@ice.org.uk

Membership is open to professional engineers who hold positions with responsibility for the development, design, construction or maintenance of infrastructure for railway operators.

VERIFIED

The Railway Forum

12 Grosvenor Place, London SW1X 7HH
Tel: (+44 20) 72 59 65 43   Fax: (+44 20) 72 59 65 44
e-mail: railinfo@railwayforum.com
Web: www.railwayforum.com

Key personnel

Chairman: Paul Kirk
Director-General: Adrian Lyons
External Communications and Events Organiser: Zoe Reeve

The Railway Forum is an industry-wide body sponsored by the majority of the train operating companies, including Eurostar, Network Rail, and the major rolling stock leasing companies, the Passenger Transport Executives, London

Underground, most of the infrastructure maintenance companies and many manufacturing and other businesses serving the UK rail industry. With over 60 members, the Railway Forum is complementary to the other rail associations, who are themselves members of the Forum: The Association of Train Operating Companies, The Railway Industry Association, and The Passenger Transport Executive Group. Its key role is to act as a think tank, information exchange and point of contact for those committed to and interested in the rail industry.

The mission of The Railway Forum is to promote the growth of a safe, efficient and affordable railway which meets its customers' needs; contribute actively to transport policy formulation, providing an industry-wide response to transport policy proposals from the government and the European Commission; sometimes raising issues which individual members may not wish to raise themselves; bring together the whole industry to discuss matters of common interest and strategies to address these; secure the most favourable environment for the development of the industry; promote and protect the interests and good name of the industry and those who work in it; promote the industry's achievements and innovations.

UPDATED

Railway Industry Association

22 Headfort Place, London SW1X 7RY
Tel: (+44 20) 72 01 07 77   Fax: (+44 20) 72 35 57 77
e-mail: ria@riagb.org.uk
Web: www.riagb.org.uk

Key personnel

Chairman: Haydn Abbott
Director General: Jeremy Candfield
Technical Director: Richard Gostling
Communications Director: Graham Coombs
Policy Director: Peter Loosley
Administration Manager: Barbara Williams

The trade association for UK-based manufacturers, maintainers, contractors, consultants, leasing companies and other providers of specialist services to the worldwide railway industry.

UPDATED

Wagon Building and Repairing Association

'Homelea', Westland Green, Little Hadham SG11 2AG, UK
Tel: (+44 1279) 84 34 87   Fax: (+44 1279) 84 23 94

Key personnel

Chairman: M Burge
Vice-Chairman: R Crutchley
Secretary: G Pratt

United States

American Association of Railroad Superintendents (AARS)

PO Box 456, Tinley Park, Illinois 60477-0456, USA
Tel: (+1 708) 342 02 10
Fax: (+1 708) 342 02 57

Key personnel

President: Robert A Kollmar
Administrative Manager: P A Weissmann
Treasurer: Donald A Orseno

UPDATED

American Association of State Highway and Transportation Officials (AASHTO)

444 North Capitol Street, NW, Suite 249, Washington DC 20001, USA
Tel: (+1 202) 624 58 13   Fax: (+1 202) 624 58 06
e-mail: penne@aashto.org
Web: http://www.transportation.org

Key personnel

Executive Director: John Horsley
Secretary, Standing Committee on Rail Transportation: Leo Penne

AASHTO represents States on safety; state-supported passenger services; high-speed programmes; and industry restructuring, covering both short line and major carriers.

American Public Transportation Association (APTA)

1666 K Street NW, Washington DC 20006, USA
Tel: (+1 202) 496 48 00
Fax: (+1 202) 496 43 24
Web: www.apta.com

Key personnel

President: William W Millar
Vice-Presidents:
  Chief of Staff: Karol J Popkin
  VP and Chief Counsel: Daniel Duff
  VP, Member Services: Anthony M Kouneski
  VP, Financial and Administration:
    C Samuel Kerns
  VP, Communications and Marketing:
    Rosemary Sheridan
  VP, Programme Management and Education
    Services: Pamela Boswell

The American Public Transportation Association (APTA) is a non-profit international association of more than 1,500 member organisations including public transportation systems; planning, design, construction and finance firms; product and service providers; academic institutions and state associations and departments of transportation. APTA members serve the public interest by providing safe, efficient and economical public transportation services and products.

UPDATED

American Railway Car Institute

50 F Street, NW (#7030), Washington DC 20001, USA
Tel: (+1 202) 347 46 64   Fax: (+1 202) 347 00 47
e-mail: rpi@rpi.org

Key personnel

Chairman: L Clark Wood (Greenbrier
  Manufacturing)
Executive Director: Thomas D Simpson

American Railway Engineering and Maintenance of Way Association (AREMA)

8201 Corporate Drive, Suite 1125, Landover, Maryland 20785
Tel: (+1 301) 459 32 00   Fax: (+1 301) 459 80 77
e-mail: bcaruso@arema.org
Web: http://www.arema.org

Key personnel

Executive Director/Chief Executive Officer:
  Dr Charles H Emely
Director of Administration: Elizabeth S Caruso

VERIFIED

American Short Line and Regional Railroad Association

50 F Street NW, 7th Floor, Washington DC 20001
Tel: (+1 202) 628 45 00   Fax: (+1 202) 628 64 30
Web: http://www.aslrra.org

Key personnel

President and Treasurer: Richard F Timmons
Executive Director, Administration:
  Stephen M Sullivan
Executive Director, Member Services:
  Kathleen M Cassidy
Executive Director, Federal and Industry
  Programmes: Matthew B Reilly
Executive Director, Traffic and E-Commerce:
  K Grant Ozburn

VERIFIED

Association of American Railroads

American Railroads Building, 50 F Street NW, Washington DC 20001
Tel: (+1 202) 639 21 00   Fax: (+1 202) 639 28 06
Web: http://www.aar.org

Key personnel

President and Chief Executive Officer:
  Edward R Hamberger
Senior Vice-President, Safety and Operations:
  Robert Vander Clute
Vice-President Administration and Finance:
  Jeff Marsh
Vice-President Communications: Peggy Wilhide
Vice-President Government Affairs:
  Obie O'Bannon
Vice-President Policy and Economics: Craig Rockey
Senior Vice-President, Law and General Counsel:
  Louis P Warchot

Subsidiaries

Transportation Technology Center Inc (TTCI)
PO Box 11130, Pueblo, Colorado 81001
Tel: (+1 719) 584 05 01   Fax: (+1 719) 584 07 11
President: Roy A Allen
See entry in *Consultancy services* section.

RAILINC
7001 Weston Parkway, Suite 200 Cary, North Carolina 27513
Tel: (+1 800) 544 72 45   Fax: (+1 919) 651 54 10
e-mail: csc@railinc.com
Web: http://www.railinc.com
President: James W Gardner

The AAR is the trade association for the rail industry. On behalf of the industry it engages in lobbying, standard-setting, communications and other public policy initiatives. AAR's subsidiarries, TTCI and RAILINC Corp engage in research and technology activities and information technology activities respectively.

UPDATED

The High Speed Ground Transportation Association (HSGTA)

1666 K Street, NW Suite 1100, Washington DC 20006-1215, USA
Tel: (+1 202) 261 60 20
Fax: (+1 202) 496 43 49
e-mail: info@hsgta.com
Web: http://www.hsgta.com

Key personnel

Chairman: Charles Quandel
Vice-Chairman: George Dorshimer
Assistant Treasurer: Leonard Parker
Secretary: Phyllis Wilkins

The HSGTA advocates the development and implementation of high-speed ground transport in North America. Members include suppliers, engineers, consultants, trade unions, public utilities, public officials and members of the public. The HSGTA produces a quarterly publication, *Speedlines*, and hosts an annual conference. Active in the pursuit of federal and state legislation on behalf of its members, with issues of advocacy including: full funding for Amtrak, the continuation of the Maglev Deployment Programme and the expansion of the Next Generation High Speed Rail Programme, particularly in the area of highway-railway grade crossings.

UPDATED

National Mediation Board

1301 K Street NW, Suite 250E, Washington DC 20572
Tel: (+1 202) 692 50 00   Fax: (+1 202) 692 50 80

Key personnel

Chairman: Ernest W Dubester
Member: Magdalena G Jacobsen
Chief of Staff: Stephen E Crable
Chief Financial Officer: June King
General Counsel: Ronald M Etters
Hearing Officers: Mary L Johnson,
  Benetta Mansfield, Sean Rogers

The National Railroad Construction and Maintenance Association Inc (NRC)

122 C Street NW, Suite 850, Washington DC 20001
Tel: (+1 202) 638 77 90   Fax: (+1 202) 638 10 45
e-mail: info@nrcma.org
Web: http://www.nrcma.org

Key personnel

President: Ray Chambers
Chairman of the Board: Rick Ebersold

NRC is a non-profit trade association of several hundred companies involved in the railroad construction industry. NRC members perform rail construction and maintenance work on every type of rail property: Class I, II, and III freight railroads, industry and military-owned track, transit, commuter and intercity passenger railroads. NRC companies also perform specialised work in areas such as monorails and port cargo crane rail systems.

National Railway Labor Conference

Suite 500, 1901 L Street NW, Washington DC 20036
Tel: (+1 202) 862 72 00

Key personnel

Chairman: Robert F Allen

National Transportation Safety Board (NTSB)

490 L'Enfant Plaza SW, Washington DC 20594, USA
Tel: (+1 202) 314 60 00   Fax: (+1 202) 314 64 97
Web: http://www.ntsb.gov

Key personnel

Board Members: John J Goglia, Jim Hall,
  John Hammerschmidt, George Black
Director, Office of Railroad Safety: Robert C Lauby

Regional offices

Chicago, Dallas, Fort Worth, Atlanta, Miami, Los Angeles, Denver, Seattle, Anchorage and Parsippany, New Jersey.
  NTSB is an independent federal accident investigation agency created in 1967. It ascertains probable cause, conducts special studies and assists federal agencies with rules and regulations.

Operation Lifesaver

1420 King Street, Suite 401, Alexandria, Virginia 22314
Tel: (+1 800) 537 62 24
e-mail: general@oli.org
Web: http://www.oli.org

Key personnel

President: Gerri Hall
Vice-President of Communication:
  Marmie Edwards

Operation Lifesaver is a national non-profit organisation that promotes public education and awareness, engineering and enforcement, to reduce collisions, injuries and fatalities at level crossings and to prevent railway trespass deaths and injuries on railway property.

VERIFIED

Port Terminal Railroad Association

8934 Manchester, Houston, Texas 77012, USA
Tel: (+1 713) 393 65 00   Fax: (+1 713) 393 66 73

Description

The Port Terminal Railroad Association is an association comprising any rail carrier that has trackage into Houston, Texas and wishes to be a member and the Port of Houston. Current members are the Burlington Northern Santa Fe, Texas Mexican Railway Company and the Union Pacific. Operating expenses are apportioned by wagons handled. The 51.6 km property includes 14.5 km equipped with CTC. The association leases 24 units of type MK1500D from MK Rail.

In 1998 the Association handled 517,945 wagonloads.

Railroad Human Resource Management Association (RHMA)

c/o Association of American Railroads
50 F Street NW, Room 3901, Washington DC 20001
Tel: (+1 202) 639 21 51   Fax: (+1 202) 639 28 06

Key personnel
Chairman: Paul N Austin
Vice-Chairman: Dennis J Cech
Secretary and Treasurer: Penny L Prue

Railroad Retirement Board

844 North Rush Street, Chicago, Illinois 60611-2092
Tel: (+1 312) 751 47 77   Fax: (+1 312) 751 71 54
Web: www.rrb.gov

Key personnel
Members: Michael S Schwartz (Chairman);
  V M Speakman, Jr (Labour member),
  Jerome F Kever (Carrier Member)

General Counsel: Steven A Bartholow
Director of Programs: Dorothy Isherwood
Senior Executive Officer and Director of
  Administration: Henry M Valiulis
Supervisor of Public Affairs: Anita J Rogers
Chief Financial Officer: Kenneth P Boehne
Chief Information Officer: Terri S Morgan
Chief Actuary: Frank J Buzzi

UPDATED

Railway Supply Institute

50 F Street, NW #7030, Washington, DC 20001, USA
Tel: (+1 202) 347 46 64
Fax: (+1 202) 347 00 47
e-mail: rpi@rpi.org
Web: www.rsiweb.org

29W 140 Butterfield Road, #103-A, Warrenville, Illinois 60555, USA
Tel: (+1 630) 393 01 06
Fax: (+1 630) 393 01 08
e-mail: rsupply@aol.com

Key personnel
Chairman:
  William P O'Donnell (Amsted Rail Group)
Vice-Chairman:
  Richard A Mathes (Standard Car Truck)
Executive Director – Washington:
  Thomas D Simpson
Executive Director – Chicago: Howard E Tonn

UPDATED

Railway Systems Suppliers, Inc

9304 New LaGrange Road, Suite 200, Louisville, Kentucky 40242, US
Tel: (+1 502) 327 77 74
Fax: (+1 502) 327 05 41
e-mail: rssi@rssi.org
Web: www.rssi.org

Key personnel
President: Jim Huntley (Erico, Inc)
Executive Vice-President:
  Jim Higginbottom (Okonite Co)
First Vice-President:
  George Kline (Safetran Systems Corp)
Second Vice-President:
  Franklin Brown (Dixie PreCast Inc)
Executive Director/Secretary-Treasurer:
  Donald F Remaley

Regional Railroads of America

122 C Street, NW, Suite 850, Washington DC 20001
Tel: (+1 202) 638 77 90
Fax: (+1 202) 638 10 45

Key personnel
Chairman: Peter Gilbertson, Anacostia & Pacific
Vice-Chairman: Mort Fuller, Genesee & Wyoming
Treasurer: Mike Barron, Ann Arbor RR

UPDATED

For details of the latest updates to *Jane's World Railways* online and to discover the additional information available exclusively to online subscribers please visit
jwr.janes.com

INDEX

Index

To help users of this title evaluate the published data, *Jane's Information Group* has divided entries into three categories.

N NEW ENTRY   Information on new equipment and/or systems appearing for the first time in the title.

V VERIFIED   The editor has made a detailed examination of the entry's content and checked its relevancy and accuracy for publication in the new edition to the best of his ability.

U UPDATED   During the verification process, significant changes to content have been made to reflect the latest position known to *Jane's* at the time of publication.

I

NOTES

NOTES

NOTES

NOTES

R
385 J33
05-06

Jane's world railways.
Central Business ADU REF
02/06